Immuno biology

SEVENTH EDITION

Janeway's immuno biology

SEVENTH EDITION

Kenneth Murphy
Washington University School of Medicine, St. Louis

Paul Travers
Anthony Nolan Research Institute, London

Mark Walport
The Wellcome Trust, London

With contributions by:

Michael Ehrenstein
University College London, Division of Medicine

Claudia Mauri
University College London, Division of Medicine

Allan Mowat
University of Glasgow

Andrey Shaw
Washington University School of Medicine, St. Louis

Garland Science
Taylor & Francis Group
NEW YORK AND LONDON

Vice President: Denise Schanck
Assistant Editor: Sigrid Masson
Text Editor: Eleanor Lawrence
Production Editor and Layout: Georgina Lucas
Copy Editor: Bruce Goatly
Illustration and Design: Matthew McClements, Blink Studio, Ltd.
Permissions Coordinator: Mary Dispenza
Indexer: Merrall-Ross International Ltd.

Immunobiology, Seventh Edition Interactive:
Storyboards by: Kenneth Murphy, Paul Travers, and Peter Walter
Narrated by: Julie Theriot
Animations, Interface Design, and Programming: Matthew McClements, Blink Studio, Ltd.
Senior Media Editor: Michael Morales

10-digit ISBN 0-8153-4123-7
0-8153-4290-X (International Student Edition)
13-digit ISBN 978-0-8153-4123-9
978-0-8153-4290-8 (International Student Edition)

Library of Congress Cataloging-in-Publication Data
Janeway, Charles.
Janeway's immunobiology. -- 7th ed. / Kenneth Murphy, Paul Travers, Mark Walport.
p. cm.
Includes index.
ISBN 0-8153-4123-7 (978-0-8153-4123-9)
I. Murphy, Kenneth. II. Travers, Paul. III. Walport, Mark. IV. Title.
QR181J37 2008
616.07'9--dc22
2007002499

Published by Garland Science, Taylor & Francis Group, LLC, an informa business
270 Madison Avenue, New York, NY 10016, US, and
2 Park Square, Milton Park, Abingdon, OX14 4RN, UK

Printed in the United States of America
15 14 13 12 11 10 9 8 7 6 5 4 3 2

Taylor & Francis Group, an informa business

Visit our web site at http://www.garlandscience.com

Preface

This book is intended as an introductory text for use in immunology courses for medical students, advanced undergraduate biology students, graduate students, and scientists in other fields who want to know more about the immune system. It presents the field of immunology from a consistent viewpoint, that of the host's interaction with an environment containing many species of potentially harmful microbes. The justification for this approach is that the absence of one or more components of the immune system is virtually always made clear by an increased susceptibility to one or more specific infections. Thus, first and foremost, the immune system exists to protect the host from infection, and its evolutionary history must have been shaped largely by this challenge. Other aspects of immunology, such as allergy, autoimmunity, graft rejection, and immunity to tumors, are treated as variations on this basic protective function. In these cases the nature of the antigen is the major variable.

In this seventh edition, all chapters have been updated to incorporate new observations that extend our knowledge and understanding of the immune system. Examples include new work on NK receptors, the deeper understanding of the role of activation-induced cytidine deaminase (AID) in the generation of antibody diversity, viral immunoevasins, cross-presentation of antigen to T cells, dendritic cell and T-cell subsets, and new innate receptors that recognize pathogens, to name but a few. Our chapter on evolution includes fascinating new insights into alternative forms of adaptive immunity in both invertebrates and higher organisms. The clinical chapters contain new sections on celiac disease and its mechanism, Crohn's disease, and immunological approaches to cancer treatment. New discussion questions have been provided at the end of each chapter. These questions can be used for review, or as the basis for discussion in class or in informal study groups. The accompanying CD-ROM, *Janeway's Immunobiology 7 Interactive,* has been expanded and enhanced.

Following a comprehensive overview of the immune system in Chapter 1, innate immunity is treated in Chapter 2 as an important protective system in its own right and a necessary precursor to the adaptive immune response. The coverage of Toll receptors and other pathogen-sensing mechanisms has been updated to reflect the rapid progress in this field in the past three years and the description of the different families of activating and inhibitory NK receptors has been revised to reflect the growing understanding of these. The material on pathogens at the beginning of Chapter 10 in previous editions has been moved to Chapter 2 to provide a more complete introduction to infection earlier in the book. After the discussion of innate immunity, the emphasis shifts to adaptive immunity, because we know far more about this subject, thanks to the efforts of the vast majority of immunologists. The central theme of the subsequent text is clonal selection of lymphocytes.

As in the sixth edition, we consider the two lymphoid lineages, B lymphocytes and T lymphocytes, together for most of the book, because it seems to us that essentially common mechanisms work in both cell types. An example is the rearrangement of gene segments to generate the receptors by which lymphocytes recognize antigen (Chapter 4). Chapter 5, on recognition of antigen, has been updated to include the cross-presentation of antigen by MHC class I molecules, and the interference by viral immunoevasins in antigen presentation. Chapter 6, on signaling, has been revised to describe the T-cell signaling pathways in more detail, with an extended and updated discussion of co-stimulatory signaling. We have substantially reorganized Chapter 7 so that the development of B cells and T cells are considered in separate sections.

Chapters 8 and 9 elaborate the effector functions of T cells and B cells separately, since different mechanisms are involved. We have updated and extended the treatment of dendritic cells and included recent research findings on the T_H17 and regulatory T-cell subsets (Chapter 8). We have taken the opportunity to refocus Chapter 10 on the dynamic nature of the immune response to infection, from innate immunity to the generation of immunological memory. We include recent advances in our understanding of the temporal changes in T-cell subsets during an immune response and of the nature of immunological memory. Because of its increasingly recognized role in immune defense, we have created a new chapter devoted entirely to mucosal immunity (Chapter 11).

The subsequent three chapters (Chapters 12-14) deal mainly with how diseases such as HIV/AIDS, autoimmunity, or allergy can be caused by inherited or acquired immunodeficiencies or by failure or malfunction of immunological mechanisms. As our understanding of the causes of disease grows, these chapters have been extended with descriptions of syndromes for which underlying gene defects have been newly identified. These chapters, which describe the failures of the immune system to maintain health, are followed by a chapter (Chapter 15) on how the immune response can be manipulated by vaccination and other means in attempts to combat not only infectious diseases, but also transplant rejections and cancer. These four chapters have been extensively revised and updated, especially in respect of the new generation of 'biological' therapeutics that are starting to enter medical practice.

The book ends with an updated chapter (Chapter 16) on the evolution of animal immune systems. The analysis of the genome sequences of both invertebrates and lower vertebrates has given us a new appreciation of the sophistication of invertebrate immune defenses and the discovery that our antibody and T-cell-based adaptive immune system is not the only way that adaptive immunity can be generated.

Diseases and immunological deficiencies are cross-referenced to the fifth edition of the clinical companion text *Case Studies in Immunology* by Raif Geha and Fred Rosen (ISBN: 9780815341451). This text presents major topics in immunology as background to a selection of real clinical cases that serve to reinforce and extend the basic science. There are five new cases in the fifth edition, bringing the total number to 47. A marginal icon in *Janeway's Immunobiology* provides the reader with a link to the relevant Case Study, where the science is applied in a clinical setting.

The seventh edition of *Janeway's Immunobiology* includes a CD-ROM with original immunological animations based on figures in the book and videos selected from visually compelling experiments. The animations have been revised and updated for this edition, and there are five new animations covering HIV Infection, DTH Response, Viral Evasins, Innate Recognition of Pathogens, and Pathogen Receptors. All the animations and videos are accompanied by a voice-over narration read by Julie Theriot, Stanford University School of Medicine. The CD-ROM also contains the figures and tables from the book pre-loaded into PowerPoint® presentations. There is also a JPEG archive of all these images. Instructors who adopt *Janeway's Immunobiology* for their course will have access to Garland Science Classwire™. The Classwire course management system allows instructors to build websites for their courses easily. It also serves as an online archive for instructors' resources. After registering for Classwire, instructors will be able to download all of the figures from *Immunobiology,* which are available in JPEG and PowerPoint formats, and all of the videos and animations on the CD. Instructors may also download resources from other Garland Science textbooks. Please visit the Garland Science website at www.garlandscience.com or e-mail science@garland.com for additional information on Classwire.

This edition has been retitled *Janeway's Immunobiology* in memory of Charles A. Janeway who originated this textbook and was its continual inspiration until his death in 2003. Andrey Shaw, Washington University School of Medicine, St. Louis, completely revised and updated Chapter 6 on signaling; Allan Mowat, University of Glasgow, did the same for Chapter 11 on mucosal immunity; and Claudia Mauri (Chapters 12 and 14) and Michael Ehrenstein (Chapters 13 and 15), University College London, revised and updated the clinical chapters. Appendix III, Cytokines and their Receptors, was updated and reorganized by Robert Schreiber, Washington University School of Medicine, St. Louis. Joost Oppenheim, National Cancer Institute, Washington, D.C., updated Appendix IV, Chemokines and their Receptors. We owe them all a great debt of gratitude for bringing their particular expertise to the book and for the care and effort they put into these revisions.

The editors, illustrators, and publishers are the essential glue that binds this book together. We have benefited from the editorial skills of Eleanor Lawrence, the book's developmental editor from its beginning, and from the creativity and artistic talent of Matt McClements, our illustrator since the second edition. Their 'institutional memory' sustains the overall coherence of this heavily updated edition. At Garland, Mike Morales has managed the creation of compelling animations that bring important concepts to life. None of these efforts would have borne fruit without the skillful (but patient) coordination of Sigrid Masson and the insightful suggestions and continuous support of our publisher, Denise Schanck. Kenneth Murphy would like to thank Theresa Murphy and Paul, Mike, Mark and Jason for their encouragement and support. Paul Travers would like to thank Rose Zamoyska for her infinite patience and support. Mark Walport would like to thank his wife, Julia, and children, Louise, Robert, Emily, and Fiona, for their unstinting support.

We would like to give great thanks to all those people who read parts or all of the chapters of the sixth edition, as well as drafts of the seventh edition, and advised us on how to improve them. They are listed by chapter on page ix. Every effort has been made to write a book that is error-free. Nonetheless, you may find them here and there, and it would be of great benefit if you took the trouble to communicate them to us.

Kenneth Murphy

Paul Travers

Mark Walport

Immunobiology Interactive

The *Immunobiology Interactive* CD-ROM contains the figures and tables from the book in PowerPoint® and JPEG formats. The videos and animations play in a self-contained interface and dynamically illustrate important concepts from the book. Videos are indicated with a (V).

Section I

2.1 Innate Recognition of Pathogens
2.2 Phagocytosis
2.3 Pathogen Recognition Receptors
2.4 Complement System
2.5 Rolling Adhesion
2.6 Leukocyte Rolling (V)
2.7 Leukocyte Extravasation

Section II

4.1 Somatic Recombination
5.1 MHC Class I Processing
5.2 Intracellular *Listeria* Infection (V)
5.3 MHC Class II Processing
5.4 Viral Evasins

Section III

6.1 Lipid Rafts
6.2 TCR Signaling
6.3 Activated T Cell (V)
6.4 MAP Kinase
6.5 Cytokine Signaling
6.6 Chemokine Signaling
6.7 Induction of Apoptosis
6.8 Apoptosis (V)
7.1 T-Cell Development
7.2 Lymph Node Development

Section IV

8.1 T-Cell Killing
8.2 Dendritic Cell Migration
8.3 TCR-APC Interaction
8.4 Immunological Synapse (V)
8.5 T-Cell Granule Release
10.1 The Immune Response
10.2 Chemotaxis (V)
10.3 Lymphocyte Trafficking
10.4 Lymphocyte Homing (V)
10.5 Germinal Centers

Section V

11.1 Antigenic Drift
11.2 Antigenic Shift
12.1 HIV Infection
13.1 DTH Response

Acknowledgments

We would like to thank the following experts who read parts or the whole of the sixth and/or seventh edition chapters indicated and provided us with invaluable advice in developing this new edition.

Chapter 1: Hans Acha-Orbea, Université de Lausanne; Leslie Berg, University of Massachusetts Medical Center; Michael Cancro, University of Pennsylvania; Elizabeth Godrick, Boston University; Michael Gold, University of British Columbia; Harris Goldstein, Albert Einstein College of Medicine; Kenneth Hunter, University of Nevada, Reno; Derek McKay, McMaster University; Eleanor Metcalf, Uniformed Services University of the Health Sciences, Maryland; Carol Reiss, New York University; Maria Marluce dos Santos Vilela, State University of Campinas Medical School, Brazil; Heather Zwickey, National College of Natural Medicine, Oregon.

Chapter 2: Alan Aderem, Institute for Systems Biology, Washington; John Atkinson, Washington University School of Medicine, St. Louis; Marco Colonna, Washington University School of Medicine, St. Louis; Jason Cyster, University of California, San Francisco; John Kearney, The University of Alabama, Birmingham; Lewis Lanier, University of California, San Francisco; Ruslan Medzhitov, Yale University School of Medicine; Alessandro Moretta, University of Genova, Italy; Gabriel Nunez, University of Michigan Medical School; Kenneth Reid, University of Oxford; Robert Schreiber, Washington University School of Medicine, St. Louis; Caetano Reis e Sousa, Cancer Research UK; Andrea Tenner, University of California, Irvine; Eric Vivier, Université de la Méditerranée Campus de Luminy; Wayne Yokoyama, Washington University School of Medicine, St. Louis.

Chapter 3: David Davies, NIDDK, National Institutes of Health, US; K. Christopher Garcia, Stanford University; David Fremont, Washington University School of Medicine; Bernard Malissen, Centre d'Immunologie Marseille-Luminy; Ellis Reinherz, Harvard Medical School; Roy Marriuzza, University of Maryland Biotechnology Institute; Robyn Stanfield, The Scripps Research Institute; Ian Wilson, The Scripps Research Institute.

Chapter 4: Fred Alt, Harvard Medical School; David Davies, NIDDK, National Institutes of Health, US; Amy Kenter, University of Illinois, Chicago; Michael Lieber, University of Southern California; John Manis, Harvard Medical School; Michael Neuberger, University of Cambridge; David Schatz, Yale University School of Medicine; Barry Sleckman, Washington University School of Medicine, St. Louis.

Chapter 5: Paul Allen, Washington University School of Medicine, St. Louis; Siamak Bahram, Centre de Recherche d'Immunologie et d'Hematologie; Michael Bevan, University of Washington; Peter Cresswell, Yale University School of Medicine; David Fremont, Washington University School of Medicine, St. Louis; K. Christopher Garcia, Stanford University; Ted Hansen, Washington University School of Medicine, St. Louis; Jim Kaufman, Institute for Animal Health, UK; Philippa Marrack, National Jewish Medical and Research Center, University of Colorado Health Sciences Center, Denver; Jim McCluskey, University of Melbourne, Victoria; Jacques Neefjes, The Netherlands Cancer Institute, Amsterdam; Chris Nelson, Washington University School of Medicine, St. Louis; Hans-Georg Rammensee, University of Tubingen, Germany; John Trowsdale, University of Cambridge; Emil Unanue, Washington University School of Medicine, St. Louis.

Chapter 6: Leslie Berg, University of Massachusetts Medical Center; John Cambier, University of Colorado Health Sciences Center; Doreen Cantrell, University of Dundee, UK; Andy Chan, Genentech, Inc.; Gary Koretzky, University of Pennsylvania School of Medicine; Gabriel Nunez, University of Michigan Medical School; Anton van der Merwe, University of Oxford; Andre Veillette, Institut de Recherches Cliniques de Montréal; Art Weiss, University of California, San Francisco.

Chapter 7: Avinash Bhandoola, University of Pennsylvania; B.J. Fowlkes, National Institutes of Health, US; Richard Hardy, Fox Chase Cancer Center, Philadelphia; Kris Hogquist, University of Minnesota; John Kearney, The University of Alabama, Birmingham; Dan Littman, New York University School of Medicine; John Monroe, University of Pennsylvania Medical School; David Raulet, University of California, Berkeley; Ellen Robey, University of California, Berkeley; Harinder Singh, University of Chicago; Barry Sleckman, Washington University School of Medicine, St. Louis; Brigitta Stockinger, National Institute for Medical Research, London; Paulo Vieira, Institut Pasteur, Paris; Harald von Boehmer, Harvard Medical School; Rose Zamoyska, National Institute for Medical Research, London.

Chapter 8: Rafi Ahmed, Emory University School of Medicine; Michael Bevan, University of Washington; Frank Carbone, University of Melbourne, Victoria; Bill Heath, University of Melbourne, Victoria; Tim Ley, Washington University School of Medicine, St. Louis; Anne O'Garra, The National Institute for Medical Research, London; Steve Reiner, University of Pennsylvania School of Medicine; Robert Schreiber, Washington University School of Medicine, St. Louis; Casey Weaver, The University of Alabama, Birmingham; Marco Colonna, Washington University School of Medicine, St. Louis.

Chapter 9: Michael Cancro, University of Pennsylvania; Robert H. Carter, The University of Alabama, Birmingham; John Kearney, The University of Alabama, Birmingham; Garnett Kelsoe, Duke University; Michael Neuberger, University of Cambridge.

Chapter 10-11: Rafi Ahmed, Emory University School of Medicine; Charles Bangham, Imperial College, London; Jason Cyster, University of California, San Francisco; David Gray, The University of Edinburgh; Dragana Jankovic, National Insitutes of Health; Michael Lamm, Case Western University; Antonio Lanzavecchia, Institute for Research in Biomedicine, Switzerland; Sara Marshall, Imperial College, London; Allan Mowat, University of Glasgow; Gabriel Nunez, University of Michigan Medical School; Michael Oldstone, The Scripps Research Insitute; Michael Russell, SUNY, Buffalo; Federica Sallusto, Institute for Research in Biomedicine, Switzerland; Philippe Sansonetti, Institut Pasteur, Paris; Alan Sher, National Institutes of Health, US.

Chapter 12: Mary Collins, University College, London; Alain Fischer, Groupe Hospitalier Necker-Enfants-Malades, Paris; Raif Geha, Harvard

Medical School; Paul Klenerman, University of Oxford; Dan Littman, New York University School of Medicine; Michael Malim, King's College; Sarah Rowland-Jones, University of Oxford; Adrian Thrasher, University College, London.

Chapter 13: Cezmi Akdis, Swiss Institute of Allergy and Asthma Research; Raif Geha, Harvard Medical School; Barry Kay, Imperial College, London; Gabriel Nunez, University of Michigan Medical School; Harald Renz, Philipps Universität Marburg, Germany; Alan Shaffer, Harvard Medical School.

Chapter 14: Antony Basten, The University of Sydney; Lucienne Chatenaud, Groupe Hospitalier Necker-Enfants-Malades, Paris; Maggie Dallman, Imperial College, London; Anne Davidson, Albert Einstein College of Medicine; Betty Diamond, Albert Einstein College of Medicine; Rikard Holmdahl, Lund University, Sweden; Laurence Turka, University of Pennsylvania School of Medicine; Kathryn Wood, University of Oxford.

Chapter 15: Filippo Belardinelli, Istituto Superiore di Sanita, Italy; Benny Chain, University College, London; Lucienne Chatenaud, Groupe Hospitalier Necker-Enfants-Malades, Paris; Robert Schreiber, Washington University School of Medicine, St. Louis; Ralph Steinman, The Rockefeller University; Richard Williams, Imperial College, London.

Chapter 16: Max Cooper, The University of Alabama, Birmingham; Jim Kaufman, Institute for Animal Health, UK; Gary Litman, University of South Florida; Ruslan Medzhitov, Yale University School of Medicine.

A special thanks to Matthew Vogt for careful reading of the first printing of the entire book.

Contents

Detailed Contents

PART I

AN INTRODUCTION TO IMMUNOBIOLOGY AND INNATE IMMUNITY

Basic Concepts in Immunology

1

Immunology is the study of the body's defense against infection. We live surrounded by microorganisms, many of which cause disease. Yet despite this continual exposure we become ill only rarely. How does the body defend itself? When infection does occur, how does the body eliminate the invader and cure itself? And why do we develop long-lasting immunity to many infectious diseases encountered once and overcome? These are the questions addressed by immunology, which we study to understand our body's defenses against infection at the cellular and molecular level.

Immunology is a relatively new science. Its origin is usually attributed to **Edward Jenner** (Fig. 1.1), who observed in the late 18th century that the relatively mild disease of cowpox, or vaccinia, seemed to confer protection against the often fatal disease of smallpox. In 1796, Jenner demonstrated that inoculation with cowpox could protect against smallpox. He called the procedure **vaccination**, and this term is still used to describe the inoculation of healthy individuals with weakened or attenuated strains of disease-causing agents to provide protection from disease. Although Jenner's bold experiment was successful, it took almost two centuries for smallpox vaccination to become universal, an advance that enabled the World Health Organization to announce in 1979 that smallpox had been eradicated (Fig. 1.2), arguably the greatest triumph of modern medicine.

Fig. 1.1 Edward Jenner. Portrait by John Raphael Smith. Reproduced courtesy of Yale University, Harvey Cushing/John Hay Whitney Medical Library.

When Jenner introduced vaccination he knew nothing of the infectious agents that cause disease: it was not until late in the 19th century that **Robert Koch** proved that infectious diseases are caused by microorganisms, each one responsible for a particular disease. We now recognize four broad categories of disease-causing microorganisms, or **pathogens**: viruses, bacteria, fungi, and the unicellular and multicellular eukaryotic organisms collectively termed parasites.

The discoveries of Koch and other great 19th-century microbiologists extended Jenner's strategy of vaccination to other diseases. In the 1880s, **Louis Pasteur** devised a vaccine against cholera in chickens, and developed a rabies vaccine that proved a spectacular success upon its first trial in a boy bitten by a rabid dog. These practical triumphs led to a search for the mechanism of protection and to the development of the science of immunology. In the early 1890s, **Emil von Behring** and **Shibasaburo Kitasato** discovered that

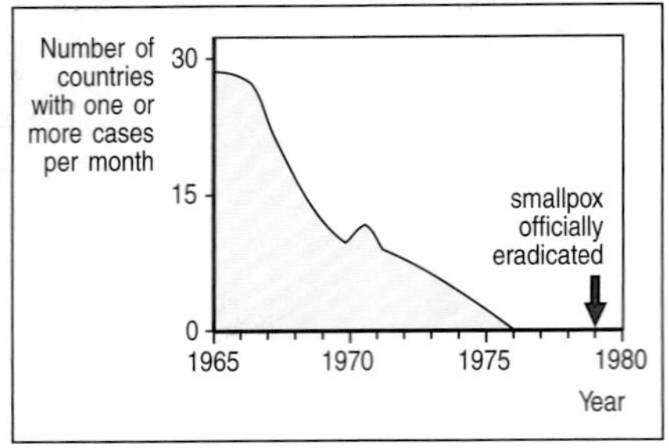

Fig. 1.2 The eradication of smallpox by vaccination. After a period of 3 years in which no cases of smallpox were recorded, the World Health Organization was able to announce in 1979 that smallpox had been eradicated, and vaccination stopped. A few laboratory stocks have been retained, however, and some fear that these are a source from which the virus might reemerge.

the serum of animals immune to diphtheria or tetanus contained a specific 'antitoxic activity' that could confer short-lived protection against the effects of diphtheria or tetanus toxins in people. This activity was due to the proteins we now call **antibodies**, which bound specifically to the toxins and neutralized their activity.

The responses we make against infection by potential pathogens are known as **immune responses**. A specific immune response, such as the production of antibodies against a particular pathogen or its products, is known as an **adaptive immune response** because it is developed during the lifetime of an individual as an adaptation to infection with that pathogen. In many cases, an adaptive immune response also results in the phenomenon known as immunological memory, which confers lifelong **protective immunity** to reinfection with the same pathogen. This is just one of the features that distinguish an adaptive immune response from the **innate immune response**, or innate immunity, which is always immediately available to combat a wide range of pathogens but does not lead to lasting immunity and is not specific for any individual pathogen. At the time that von Behring was developing serum therapy for diphtheria, innate immunity was known chiefly through the work of the great Russian immunologist **Elie Metchnikoff**, who discovered that many microorganisms could be engulfed and digested by phagocytic cells, which he called 'macrophages.' These cells are always present and ready to act, and are a frontline component of innate immune responses. In contrast, an adaptive immune response takes time to develop and is highly specific; antibodies against the influenza virus, for example, will not protect against poliovirus.

It quickly became clear that antibodies could be induced against a vast range of substances. Such substances were called **antigens** because they could stimulate *anti*body *gen*eration. Much later, it was discovered that antibody production is not the only function of adaptive immune responses, and the term antigen is now used to describe any substance that can be recognized and responded to by the adaptive immune system. The proteins, glycoproteins, and polysaccharides of pathogens are the antigens normally responded to by the immune system, but it can recognize and make a response to a much wider range of chemical structures—hence its ability to produce allergic immune responses to metals such as nickel, drugs such as penicillin, and organic chemicals in the leaves of poison ivy. Innate and adaptive immune responses together provide a remarkably effective defense system. Many infections are handled successfully by innate immunity and cause no disease; those that cannot be resolved trigger an adaptive immune response and, if overcome, are often followed by lasting immunological memory, which prevents disease if reinfection occurs.

The main focus of this book will be on the diverse mechanisms of adaptive immunity, by which the specialized white blood cells known as **lymphocytes** recognize and target pathogenic microorganisms or infected cells. We shall see, however, that the actions of the innate immune system are a prerequisite for the development of adaptive immunity and that cells involved in innate immune responses also participate in adaptive immune responses. Indeed, most of the ways in which the adaptive immune response destroys invading microorganisms depend upon linking antigen-specific recognition of the pathogen to the activation of the same destructive mechanisms that are used in innate immunity.

In this chapter we first introduce the principles of innate and adaptive immunity, the cells of the immune system, the tissues in which they develop, and the tissues through which they circulate. We then outline the specialized functions of the different types of cells and the mechanisms by which they eliminate infection.

Principles of innate and adaptive immunity.

The body is protected from infectious agents and the damage they cause, and from other harmful substances such as insect toxins, by a variety of effector cells and molecules that together make up the **immune system**. In this part of the chapter we discuss the main principles underlying immune responses and introduce the cells and tissues of the immune system on which an immune response depends.

1-1 Functions of the immune response.

To protect the individual effectively against disease, the immune system must fulfill four main tasks. The first is **immunological recognition**: the presence of an infection must be detected. This task is carried out both by the white blood cells of the innate immune system, which provide an immediate response, and by the lymphocytes of the adaptive immune system. The second task is to contain the infection and if possible eliminate it completely, which brings into play **immune effector functions** such as the complement system of blood proteins, antibodies, and the destructive capacities of lymphocytes and the other white blood cells. At the same time the immune response must be kept under control so that it does not itself do damage to the body. **Immune regulation**, or the ability of the immune system to self-regulate, is thus an important feature of immune responses, and failure of such regulation contributes to conditions such as allergy and autoimmune disease. The fourth task is to protect the individual against recurring disease due to the same pathogen. A unique feature of the adaptive immune system is that it is capable of generating **immunological memory**, so that having been exposed once to an infectious agent, a person will make an immediate and stronger response against any subsequent exposure to it; that is, they will have protective immunity against it. Finding ways of generating long-lasting immunity to pathogens that do not naturally provoke it is one of the greatest challenges facing immunologists today.

When an individual first encounters an infectious agent, the initial defenses against infection are physical and chemical barriers that prevent microbes from entering the body; these are not generally considered as part of the immune system proper and it is only when these barriers are overcome or evaded that the immune system comes into play. The first cells that respond are phagocytic white blood cells, such as macrophages, that form part of the innate immune system. These cells are able to ingest and kill microbes by producing a variety of toxic chemicals and powerful degradative enzymes. Innate immunity is of ancient origin—some form of innate defense against disease is found in all animals and plants. The macrophages of humans and other vertebrates, for example, are presumed to be the direct evolutionary descendants of the phagocytic cells present in simpler animals, such as those that Metchnikoff observed in the invertebrate sea stars.

Innate immune responses occur rapidly on exposure to an infectious organism. Overlapping with the innate immune response, but taking days rather than hours to develop, the adaptive immune system is capable of eliminating infections more efficiently than the innate immune response. It is present only in vertebrates and depends on the exquisitely specific recognition functions of lymphocytes, which have the ability to distinguish the particular pathogen and focus the immune response more strongly on it. These cells can recognize and respond to individual antigens by means of highly specialized **antigen receptors** on the lymphocyte surface. The billions of lymphocytes

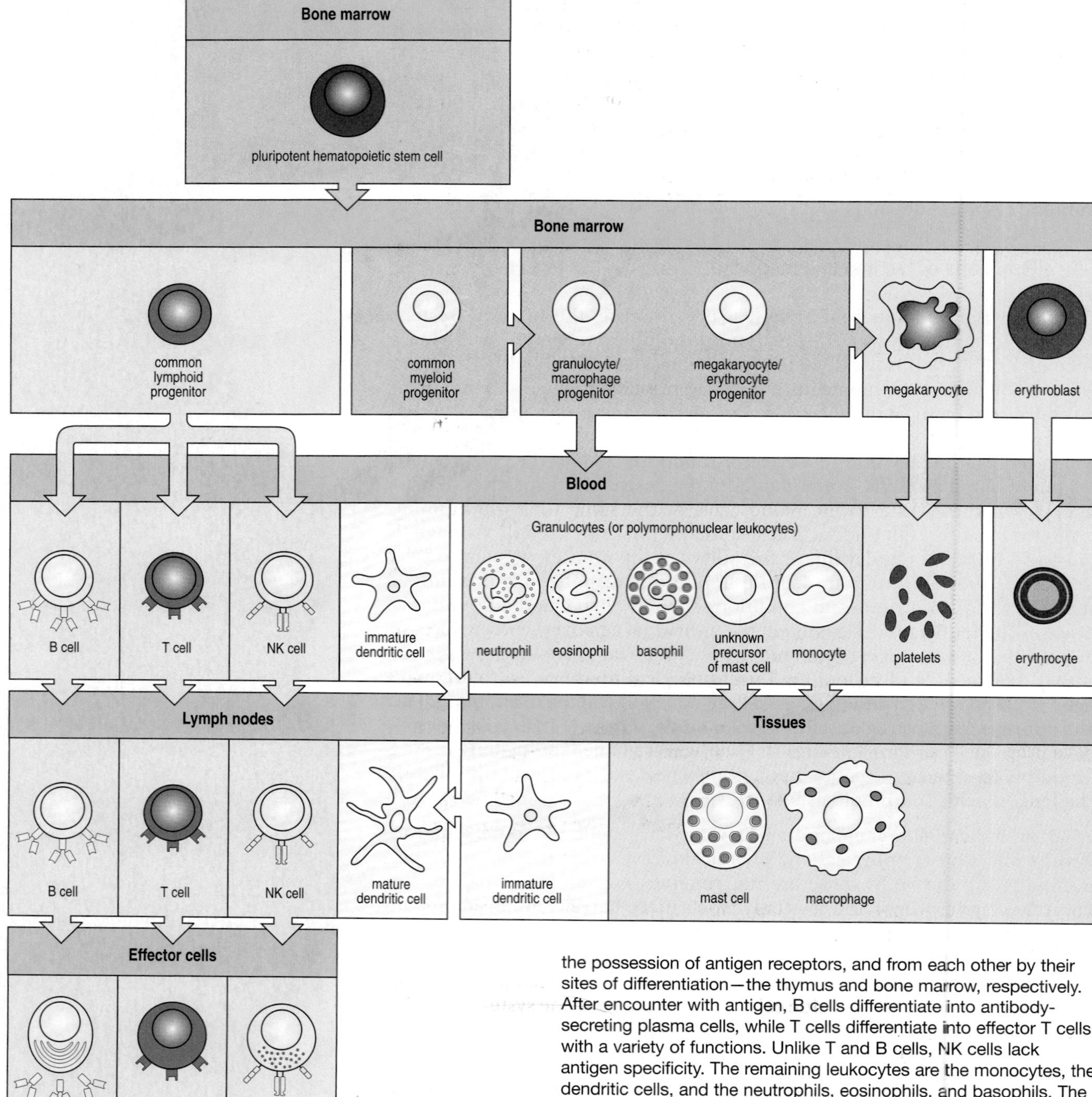

Fig. 1.3 All the cellular elements of the blood, including the cells of the immune system, arise from pluripotent hematopoietic stem cells in the bone marrow. These pluripotent cells divide to produce two types of stem cells. A common lymphoid progenitor gives rise to the lymphoid lineage (blue background) of white blood cells or leukocytes—the natural killer (NK) cells and the T and B lymphocytes. A common myeloid progenitor gives rise to the myeloid lineage (pink and yellow backgrounds), which comprises the rest of the leukocytes, the erythrocytes (red blood cells), and the megakaryocytes that produce platelets important in blood clotting. T and B lymphocytes are distinguished from the other leukocytes by the possession of antigen receptors, and from each other by their sites of differentiation—the thymus and bone marrow, respectively. After encounter with antigen, B cells differentiate into antibody-secreting plasma cells, while T cells differentiate into effector T cells with a variety of functions. Unlike T and B cells, NK cells lack antigen specificity. The remaining leukocytes are the monocytes, the dendritic cells, and the neutrophils, eosinophils, and basophils. The latter three circulate in the blood and are termed granulocytes, because of the cytoplasmic granules whose staining gives these cells a distinctive appearance in blood smears, or polymorphonuclear leukocytes, because of their irregularly shaped nuclei. Immature dendritic cells (yellow background) are phagocytic cells that enter the tissues; they mature after they have encountered a potential pathogen. The common lymphoid progenitor also gives rise to a minor subpopulation of dendritic cells, but for simplicity this developmental pathway has not been illustrated. However, as there are more common myeloid progenitor cells than there are common lymphoid progenitors, the majority of the dendritic cells in the body develop from common myeloid progenitors. Monocytes enter tissues, where they differentiate into phagocytic macrophages. The precursor cell that gives rise to mast cells is still unknown. Mast cells also enter tissues and complete their maturation there.

present in the body collectively possess a vast repertoire of antigen receptors, which enables the immune system to recognize and respond to virtually any antigen a person is likely to be exposed to. By making recognition and response specific for a particular pathogen, the adaptive immune response focuses the resources of the immune system on combating that pathogen, enabling the body to overcome pathogens that have evaded or overwhelmed innate immunity. Antibodies and activated lymphocytes produced in this phase of the response also persist after the original infection has been eliminated and prevent immediate reinfection. Lymphocytes are also responsible for the long-lasting immunity that is generated after a successful adaptive immune response to many pathogens, so that the response to a second exposure to the same microbe is both faster and greater in magnitude, even when it occurs many years later.

1-2 The cells of the immune system derive from precursors in the bone marrow.

Both innate and adaptive immune responses depend upon the activities of white blood cells or **leukocytes**. These cells all originate in the **bone marrow**, and many of them also develop and mature there. They then migrate to guard the peripheral tissues—some of them residing within tissues, others circulating in the bloodstream and in a specialized system of vessels called the **lymphatic system**, which drains extracellular fluid and free cells from tissues, transports them through the body as **lymph**, and eventually empties back into the blood system.

All the cellular elements of blood, including the red blood cells that transport oxygen, the platelets that trigger blood clotting in damaged tissues, and the white blood cells of the immune system, derive from the **hematopoietic stem cells** of the bone marrow. As these stem cells can give rise to all the different types of blood cells, they are often known as pluripotent hematopoietic stem cells. They give rise to stem cells of more limited developmental potential, which are the immediate progenitors of red blood cells, platelets, and the two main categories of white blood cells, the **lymphoid** and **myeloid** lineages. The different types of blood cells and their lineage relationships are summarized in Fig. 1.3.

1-3 The myeloid lineage comprises most of the cells of the innate immune system.

The **common myeloid progenitor** is the precursor of the macrophages, granulocytes, mast cells and dendritic cells of the innate immune system, and also of megakaryocytes and red blood cells, which we will not be concerned with here. The cells of the myeloid lineage are shown in Fig. 1.4.

Macrophages are resident in almost all tissues and are the mature form of **monocytes**, which circulate in the blood and continually migrate into tissues, where they differentiate. Together, monocytes and macrophages make up one of the three types of phagocytes in the immune system: the others are the granulocytes (the collective term for the white blood cells called neutrophils, eosinophils, and basophils) and the dendritic cells. Macrophages are relatively long-lived cells and perform several different functions throughout the innate immune response and the subsequent adaptive immune response. One is to engulf and kill invading microorganisms. In this phagocytic role they are an important first defense in innate immunity and also dispose of pathogens and infected cells targeted by an adaptive immune response. Both monocytes and macrophages are phagocytic, but most infections occur in the tissues, and so it is primarily macrophages that perform this important protective function. An additional and crucial role of macrophages is to

Fig. 1.4 Myeloid cells in innate and adaptive immunity. Cells of the myeloid lineage perform various important functions in the immune response. In the rest of the book these cells will be represented in the schematic form shown on the left. A photomicrograph of each cell type is shown in the center panels. Macrophages and neutrophils are primarily phagocytic cells that engulf pathogens and destroy them in intracellular vesicles, a function they perform in both innate and adaptive immune responses. Dendritic cells are phagocytic when they are immature and can take up pathogens; after maturing, they function as specialized cells that present pathogen antigens to T lymphocytes in a form they can recognize, thus activating T lymphocytes and initiating adaptive immune responses. Macrophages can also present antigens to T lymphocytes and can activate them. The other myeloid cells are primarily secretory cells that release the contents of their prominent granules upon activation via antibody during an adaptive immune response. Eosinophils are thought to be involved in attacking large antibody-coated parasites such as worms, whereas the function of basophils is less clear. Mast cells are tissue cells that trigger a local inflammatory response to antigen by releasing substances that act on local blood vessels; they are also important in allergic responses. Photographs courtesy of N. Rooney, R. Steinman, and D. Friend.

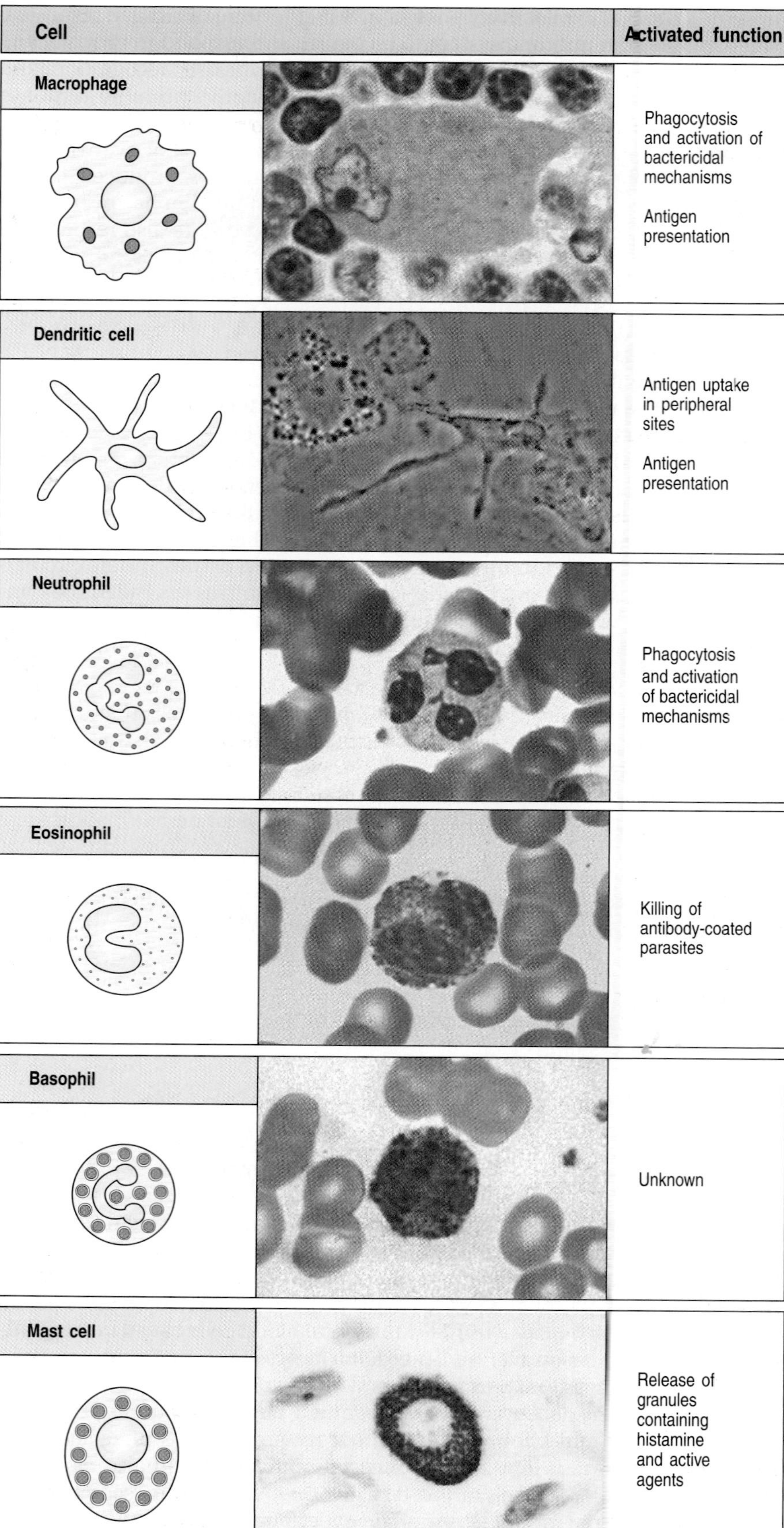

orchestrate immune responses: they help induce inflammation, which, as we shall see, is a prerequisite to a successful immune response, and they secrete signaling proteins that activate other immune-system cells and recruit them into an immune response. As well as their specialized role in the immune system, macrophages act as general scavenger cells in the body, clearing dead cells and cell debris.

The **granulocytes** are so called because they have densely staining granules in their cytoplasm; they are also called **polymorphonuclear leukocytes** because of their oddly shaped nuclei. There are three types of granulocytes—neutrophils, eosinophils, and basophils—which are distinguished by the different staining properties of the granules. In comparison with macrophages they are all relatively short-lived, surviving for only a few days, and are produced in increased numbers during immune responses, when they leave the blood to migrate to sites of infection or inflammation. The phagocytic **neutrophils** are the most numerous and most important cells in innate immune responses: they take up a variety of microorganisms by phagocytosis and efficiently destroy them in intracellular vesicles using degradative enzymes and other antimicrobial substances stored in their cytoplasmic granules. Their role is discussed in more detail in Chapter 2. Hereditary deficiencies in neutrophil function lead to overwhelming bacterial infection, which is fatal if untreated.

The protective functions of **eosinophils** and **basophils** are less well understood. Their granules contain a variety of enzymes and toxic proteins, which are released when the cell is activated. Eosinophils and basophils are thought to be important chiefly in defense against parasites, which are too large to be ingested by macrophages or neutrophils, but their main medical importance is their involvement in allergic inflammatory reactions, in which their effects are damaging rather than protective. We discuss the functions of these cells in Chapter 9 and their role in allergic inflammation in Chapter 13.

Mast cells, whose blood-borne precursors are not well defined, differentiate in the tissues. Although best known for their role in orchestrating allergic responses, which is discussed in Chapter 13, they are believed to play a part in protecting the internal surfaces of the body against pathogens and are involved in the response to parasitic worms. They have large granules in their cytoplasm that are released when the mast cell is activated; these help induce inflammation.

The **dendritic cells** are the third class of phagocytic cell of the immune system. They have long finger-like processes, like the dendrites of nerve cells, which gives them their name. Immature dendritic cells migrate through the bloodstream from the bone marrow to enter tissues. They both take up particulate matter by phagocytosis and continually ingest large amounts of the extracellular fluid and its contents by a process known as **macropinocytosis**. Like macrophages and neutrophils, they degrade the pathogens they take up, but their main role in the immune system is not the clearance of microorganisms. Instead, dendritic cells that have encountered invading microorganisms mature into cells capable of activating a particular class of lymphocytes—the T lymphocytes—which are described below. Dendritic cells do this by displaying pathogen antigens on their surface in such a way that they can be recognized and responded to by this type of lymphocyte. As we discuss later in the chapter, recognition of antigen alone is not sufficient to activate a T lymphocyte that has never encountered its antigen before. Mature dendritic cells, however, have additional properties that enable them to activate T lymphocytes. Cells that can present antigens to inactive T lymphocytes and activate them for the very first time are known as **antigen-presenting cells** (**APCs**), and such cells form a crucial link between the innate immune response and the adaptive immune response. Macrophages can also

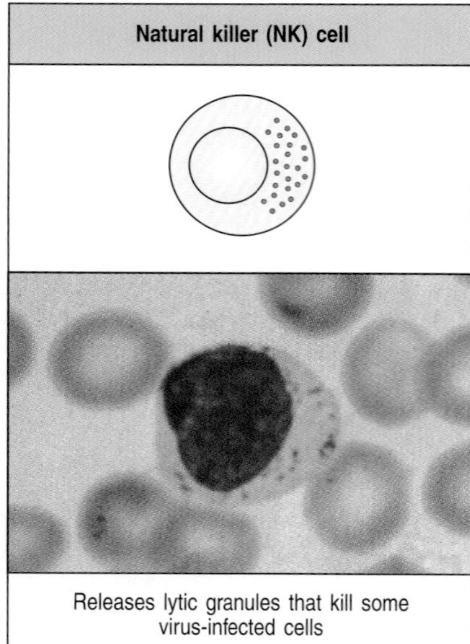

Natural killer (NK) cell
Releases lytic granules that kill some virus-infected cells

Fig. 1.5 Natural killer (NK) cells. These are large granular lymphoid-like cells with important functions in innate immunity, especially against intracellular infections, being able to kill other cells. Unlike lymphocytes, they lack antigen-specific receptors. Photograph courtesy of B. Smith.

act as antigen-presenting cells, and they are important in particular situations. Dendritic cells, however, are the cells that specialize in presenting antigens to lymphocytes and initiating adaptive immune responses.

1-4 The lymphoid lineage comprises the lymphocytes of the adaptive immune system and the natural killer cells of innate immunity.

The **common lymphoid progenitor** in the bone marrow gives rise to the antigen-specific lymphocytes of the adaptive immune system and also to a type of lymphocyte that responds to the presence of infection but is not specific for antigen, and is thus considered to be part of the innate immune system. This latter is a large cell with a distinctive granular cytoplasm and is called a **natural killer cell** (**NK cell**) (Fig. 1.5). These cells are able to recognize and kill some abnormal cells, for example some tumor cells and cells infected with herpes viruses. Their functions in innate immunity are described in Chapter 2.

We come finally to the antigen-specific lymphocytes, with which most of this book will be concerned: unless indicated otherwise, we shall use the term lymphocyte from now on to refer to the antigen-specific lymphocytes only. The immune system must be able to mount an immune response against any of the wide variety of different pathogens a person is likely to encounter during their lifetime. Lymphocytes collectively make this possible through the highly variable antigen receptors on their surface, by which they recognize and bind antigens. Each lymphocyte matures bearing a unique variant of a prototype antigen receptor, so that the population of lymphocytes expresses a huge repertoire of receptors that are highly diverse in their antigen-binding sites. Among the billion or so lymphocytes circulating in the body at any one time there will always be some that can recognize a given foreign antigen.

In the absence of an infection, most lymphocytes circulating in the body are small, featureless cells with few cytoplasmic organelles and much of the nuclear chromatin inactive, as shown by its condensed state (Fig. 1.6). This appearance is typical of inactive cells. It is hardly surprising that until the 1960s textbooks described these cells, now the central focus of immunology, as having no known function. Indeed, these small lymphocytes have no functional activity until they encounter their specific antigen. Lymphocytes that have not yet been activated by antigen are known as **naive lymphocytes**; those that have met their antigen, become activated, and have differentiated further into fully functional lymphocytes are known as **effector lymphocytes**.

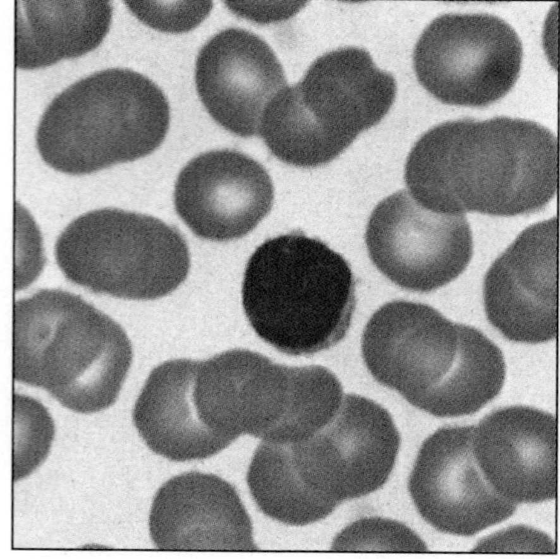

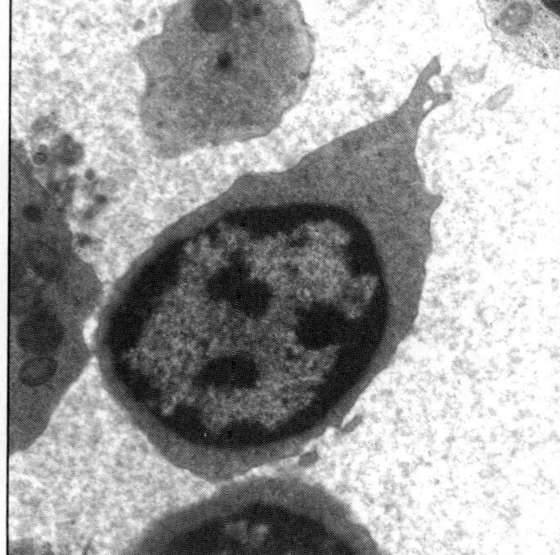

Fig. 1.6 Lymphocytes are mostly small and inactive cells. The left panel shows a light micrograph of a small lymphocyte in which the nucleus has been stained purple by the hematoxylin and eosin dye, surrounded by red blood cells (which have no nuclei). Note the darker purple patches of condensed chromatin of the lymphocyte nucleus, indicating little transcriptional activity, the relative absence of cytoplasm, and the small size. The right panel shows a transmission electron micrograph of a small lymphocyte. Again, note the evidence of functional inactivity: the condensed chromatin, the scanty cytoplasm, and the absence of rough endoplasmic reticulum. Photographs courtesy of N. Rooney.

There are two types of lymphocytes—**B lymphocytes** (**B cells**) and **T lymphocytes** (**T cells**)—each with quite different roles in the immune system and distinct types of antigen receptors. After antigen binds to a **B-cell antigen receptor**, or **B-cell receptor** (**BCR**), on the B-cell surface, the lymphocyte will proliferate and differentiate into **plasma cells**. These are the effector form of B lymphocytes and they produce antibodies, which are a secreted form of the B-cell receptor and have an identical antigen specificity. Thus the antigen that activates a given B cell becomes the target of the antibodies produced by that cell's progeny. Antibody molecules as a class are known as **immunoglobulins** (**Ig**), and so the antigen receptor of B lymphocytes is also known as **membrane immunoglobulin** (**mIg**) or **surface immunoglobulin** (**sIg**).

The **T-cell antigen receptor**, or **T-cell receptor** (**TCR**), is related to immunoglobulin but is quite distinct in its structure and recognition properties. After a T cell is activated by its first encounter with antigen it proliferates and differentiates into one of several different functional types of **effector T lymphocytes**. T-cell functions fall into three broad classes—killing, activation, and regulation. **Cytotoxic T cells** kill cells that are infected with viruses or other intracellular pathogens. **Helper T cells** provide essential additional signals that activate antigen-stimulated B cells to differentiate and produce antibody; some of these T cells can also activate macrophages to become more efficient at killing engulfed pathogens. We return to the functions of cytotoxic and helper T cells later in this chapter, and their actions are described in detail in Chapters 8 and 10. **Regulatory T cells** suppress the activity of other lymphocytes and help control immune responses; they are discussed in Chapters 8, 10, and 14.

During the course of an immune response, some of the B cells and T cells activated by antigen differentiate into **memory cells**, the lymphocytes that are responsible for the long-lasting immunity that can follow exposure to disease or vaccination. Memory cells will readily differentiate into effector cells on a second exposure to their specific antigen. Immunological memory is described in Chapter 10.

1-5 Lymphocytes mature in the bone marrow or the thymus and then congregate in lymphoid tissues throughout the body.

Lymphocytes circulate in the blood and the lymph and are also found in large numbers in **lymphoid tissues** or **lymphoid organs**, which are organized aggregates of lymphocytes in a framework of nonlymphoid cells. Lymphoid organs can be divided broadly into **central** or **primary lymphoid organs**, where lymphocytes are generated, and **peripheral** or **secondary lymphoid organs**, where mature naive lymphocytes are maintained and adaptive immune responses are initiated. The central lymphoid organs are the bone marrow and the **thymus**, an organ in the upper chest. The peripheral lymphoid organs comprise the **lymph nodes**, the **spleen** and the **mucosal lymphoid tissues** of the gut, the nasal and respiratory tract, the urogenital tract and other mucosa. The location of the main lymphoid tissues is shown schematically in Fig. 1.7, and we shall describe the individual peripheral lymphoid organs in more detail later in the chapter. Lymph nodes are interconnected by a system of lymphatic vessels, which drain extracellular fluid from tissues, through the lymph nodes, and back into the blood.

Both B and T lymphocytes originate in the bone marrow, but only the B lymphocytes mature there. The precursor T lymphocytes migrate to the thymus, from which they get their name, and mature there. The 'B' in B lymphocytes originally stood for the **bursa of Fabricius**, a lymphoid organ in young chicks in which lymphocytes mature; fortunately, it can stand equally well for bone marrow-derived. Once they have completed maturation, both types of

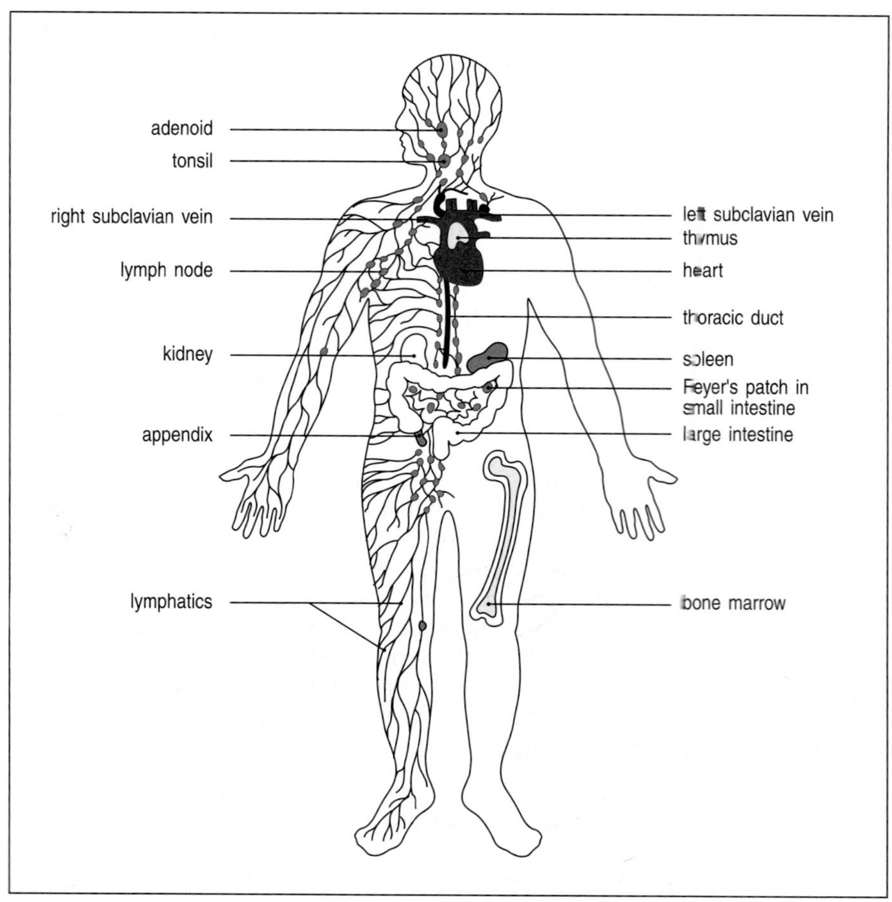

Fig. 1.7 The distribution of lymphoid tissues in the body. Lymphocytes arise from stem cells in bone marrow and differentiate in the central lymphoid organs (yellow)—B cells in the bone marrow and T cells in the thymus. They migrate from these tissues and are carried in the bloodstream to the peripheral lymphoid organs (blue). These include lymph nodes, spleen, and lymphoid tissues associated with mucosa, such as the gut-associated tonsils, Peyer's patches, and appendix. The peripheral lymphoid organs are the sites of lymphocyte activation by antigen, and lymphocytes recirculate between the blood and these organs until they encounter their specific antigen. Lymphatics drain extracellular fluid from the peripheral tissues, through the lymph nodes and into the thoracic duct, which empties into the left subclavian vein. This fluid, known as lymph, carries antigen taken up by dendritic cells and macrophages to the lymph nodes and recirculating lymphocytes from the lymph nodes back into the blood. Lymphoid tissue is also associated with other mucosa such as the bronchial linings (not shown).

lymphocytes enter the bloodstream as mature naive lymphocytes. They circulate through the peripheral lymphoid tissues, in which an adaptive immune response is initiated if a lymphocyte meets its corresponding antigen. Before this, however, an innate immune response to the infection has usually occurred, and we now look at how this alerts the rest of the immune system to the presence of a pathogen.

1-6 Most infectious agents activate the innate immune system and induce an inflammatory response.

The skin and the mucosal epithelia lining the airways and gut are the first defense against invading pathogens, forming a physical and chemical barrier against infection. Microorganisms that breach these defenses are met by cells and molecules that mount an immediate innate immune response. Macrophages resident in the tissues are the first line of defense against bacteria, for example, which they recognize by means of receptors that bind common constituents of many bacterial surfaces. Engagement of these receptors triggers the macrophage both to engulf the bacterium and degrade it internally, and to secrete proteins called cytokines and chemokines, as well as other biologically active molecules. Similar responses occur to viruses, fungi, and parasites. **Cytokine** is a general name for any protein that is secreted by cells and affects the behavior of nearby cells bearing appropriate receptors. **Chemokines** are secreted proteins that attract cells bearing chemokine receptors, such as neutrophils and monocytes, out of the bloodstream and into the infected tissue (Fig. 1.8). The cytokines and chemokines

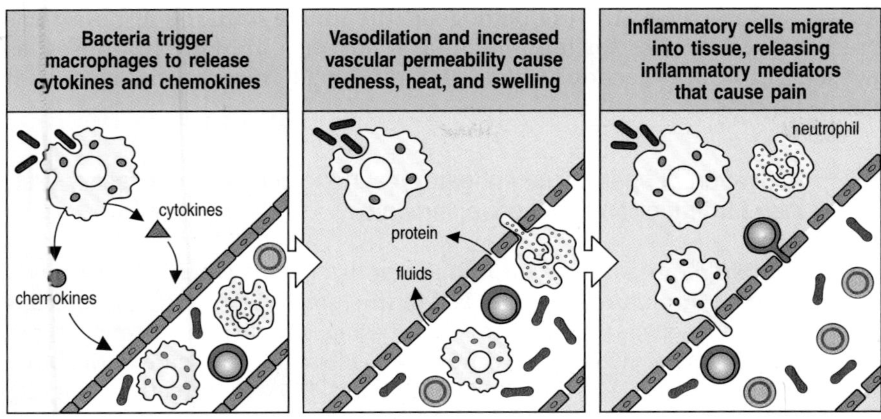
Bacteria trigger macrophages to release cytokines and chemokines
cytokines
chemokines
Vasodilation and increased vascular permeability cause redness, heat, and swelling
protein
fluids
Inflammatory cells migrate into tissue, releasing inflammatory mediators that cause pain
neutrophil

Fig. 1.8 Infection triggers an inflammatory response. Macrophages encountering bacteria or other types of microorganisms in tissues are triggered to release cytokines that increase the permeability of blood vessels, allowing fluid and proteins to pass into the tissues. They also produce chemokines, which direct the migration of neutrophils to the site of infection. The stickiness of the endothelial cells of the blood vessel wall is also changed, so that cells adhere to the wall and are able to crawl through it; first neutrophils and then monocytes are shown entering the tissue from a blood vessel. The accumulation of fluid and cells at the site of infection causes the redness, swelling, heat, and pain known collectively as inflammation. Neutrophils and macrophages are the principal inflammatory cells. Later in an immune response, activated lymphocytes can also contribute to inflammation.

released by activated macrophages initiate the process known as **inflammation**. Inflammation of an infected tissue has several beneficial effects in combating infection. It recruits cells and molecules of innate immunity out of the blood and into the tissue where they are needed to destroy the pathogen directly. In addition, it increases the flow of lymph bearing microbes and antigen-bearing cells to nearby lymphoid tissues, where they will activate lymphocytes and initiate the adaptive immune response. Finally, once the adaptive immune response has been triggered, inflammation also recruits the effectors of the adaptive immune system—antibody molecules and effector T cells—to the site of infection.

Local inflammation and the phagocytosis of invading bacteria may also be triggered as a result of the activation of a group of plasma proteins known collectively as **complement**. Activation of the complement system by bacterial surfaces leads to a cascade of proteolytic reactions that coats microbes, but not the body's own cells, with complement fragments. Complement-coated microbes are recognized and bound by specific **complement receptors** on macrophages, taken up by phagocytosis, and destroyed.

Inflammation is traditionally defined by the four Latin words *calor, dolor, rubor*, and *tumor*, meaning heat, pain, redness, and swelling, all of which reflect the effects of cytokines and other inflammatory mediators on the local blood vessels. Dilation and increased permeability of blood vessels during inflammation lead to increased local blood flow and the leakage of fluid into the tissues, and account for the heat, redness, and swelling. Cytokines and complement fragments have important effects on the endothelium lining blood vessels; endothelial cells themselves also produce cytokines in response to infection. The inflammatory cytokines produce changes in the adhesive properties of the endothelial cells, in turn causing circulating leukocytes to stick to the endothelial cells and migrate between them into the site of infection, to which they are attracted by chemokines. The migration of cells into the tissue and their local actions account for the pain.

The main cell types seen in the initial phase of an inflammatory response are macrophages and neutrophils, the latter being recruited into the inflamed, infected tissue in large numbers. Macrophages and neutrophils are thus also known as **inflammatory cells**. Like macrophages, neutrophils have surface receptors for common bacterial constituents and for complement, and they are the principal cells that engulf and destroy the invading microorganisms. The influx of neutrophils is followed a short time later by monocytes, which rapidly differentiate into macrophages, thus reinforcing and sustaining the innate immune response. More slowly, eosinophils also migrate into inflamed tissues and also contribute to the destruction of the invading microorganisms.

As well as directly destroying pathogens, the innate immune response has crucial consequences for the initiation of adaptive immune responses, as we see in the next section. It does this principally through the agency of dendritic cells.

1-7 Activation of specialized antigen-presenting cells is a necessary first step for induction of adaptive immunity.

The induction of an adaptive immune response begins when a pathogen is ingested by an immature dendritic cell in the infected tissue. These specialized phagocytic cells are resident in most tissues and, like macrophages, are long-lived compared with other white blood cells. They originate in the bone marrow (see Section 1-3), and while still not fully mature they migrate through the bloodstream to their peripheral stations, where they survey the local environment for pathogens.

Like macrophages and neutrophils, the immature dendritic cell carries receptors on its surface that recognize common features of many pathogens, such as **bacterial lipopolysaccharide**. Microbial components binding to these receptors stimulate the dendritic cell to engulf the pathogen and degrade it intracellularly. Immature dendritic cells are also continually taking up extracellular material, including virus particles and bacteria, by the receptor-independent mechanism of macropinocytosis, and are thus even able to internalize and degrade pathogens that their cell-surface receptors do not detect. The function of dendritic cells, however, is not primarily to destroy pathogens but to carry pathogen antigens to peripheral lymphoid organs and there present them to T lymphocytes. On taking up pathogens and their components, the dendritic cell migrates to peripheral lymphoid tissues, where it matures into a highly effective antigen-presenting cell. It displays fragments of pathogen antigens on its surface and also starts to produce cell-surface proteins known as **co-stimulatory molecules**, which, as their name suggests, provide signals that act together with antigen to stimulate the T lymphocyte to proliferate and differentiate into its final fully functional form (Fig. 1.9). Because B cells are not activated by most antigens without 'help' from activated helper T cells, the activation of naive T lymphocytes is an essential first stage in virtually all adaptive immune responses.

Activated dendritic cells also secrete cytokines that influence both innate and adaptive immune responses, making these cells essential gatekeepers that determine whether and how the immune system responds to the presence of infectious agents. We consider the maturation of dendritic cells and their central role in presenting antigens to naive T cells in Chapter 3.

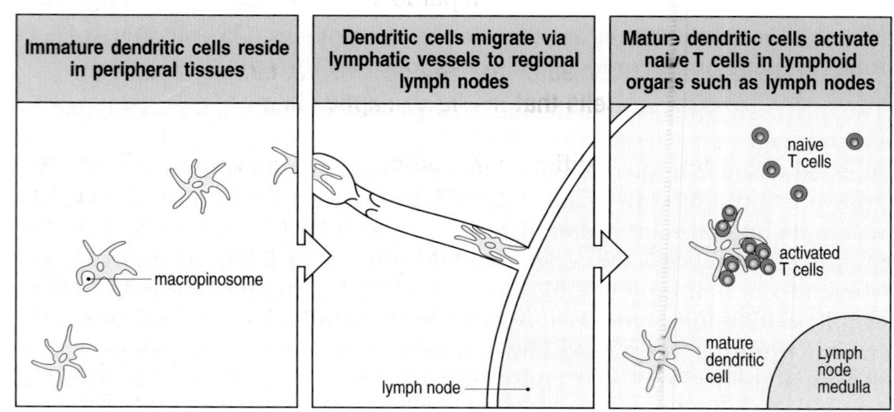
Immature dendritic cells reside in peripheral tissues
macropinosome
Dendritic cells migrate via lymphatic vessels to regional lymph nodes
lymph node
Mature dendritic cells activate naive T cells in lymphoid organs such as lymph nodes
naive T cells
activated T cells
mature dendritic cell
Lymph node medulla

Fig. 1.9 Dendritic cells initiate adaptive immune responses. Immature dendritic cells resident in a tissue take up pathogens and their antigens by macropinocytosis and by receptor-mediated endocytosis. They are stimulated by recognition of the presence of pathogens to migrate through the lymphatics to regional lymph nodes, where they arrive as fully mature non-phagocytic dendritic cells that express both antigen and the co-stimulatory molecules necessary to activate a naive T cell that recognizes the antigen, stimulating lymphocyte proliferation and differentiation.

1-8 The innate immune system provides an initial discrimination between self and nonself.

The defense systems of innate immunity are effective in combating many pathogens. They are constrained, however, by relying on a limited and invariant repertoire of receptors to recognize microorganisms. The pathogen-recognition receptors of macrophages, neutrophils, and dendritic cells recognize simple molecules and regular patterns of molecular structure known as **pathogen-associated molecular patterns (PAMPs)** that are present on many microorganisms but not on the body's own cells (Fig. 1.10). These receptors are known generally as **pattern recognition receptors (PRRs)**, and they recognize structures such as mannose-rich oligosaccharides, peptidoglycans and lipopolysaccharides in the bacterial cell wall, and unmethylated CpG DNA, which are common to many pathogens and have been conserved during evolution. The innate immune system is thus able broadly to distinguish self (the body) from nonself (pathogen) and to mount an attack against invaders. By becoming activated through their pattern recognition receptors, immature dendritic cells, which form part of the innate immune system, become able in their turn to activate naive T lymphocytes, as discussed in the previous section. Thus, the adaptive immune response is essentially initiated by an explicit recognition of nonself by the innate immune system.

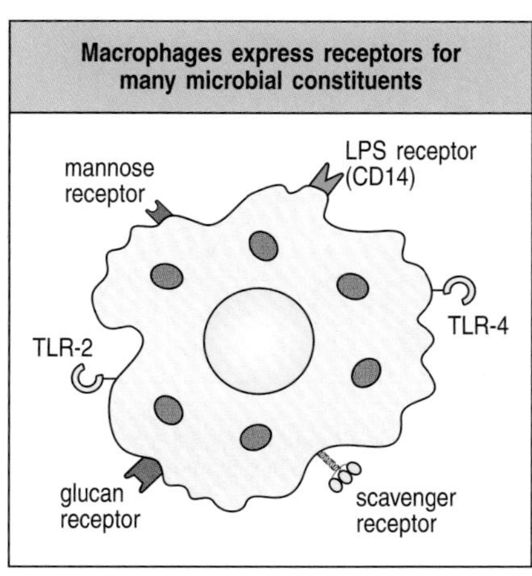

Fig. 1.10 Macrophages express a number of receptors that allow them to recognize different pathogens. Macrophages express a variety of receptors, each of which is able to recognize specific components of microbes. Some, like the mannose and glucan receptors and the scavenger receptor, bind cell-wall carbohydrates of bacteria, yeast, and fungi. The Toll-like receptors (TLRs) are an important family of pattern recognition receptors present on macrophages and other immune cells and are able to bind different microbial components; for example, TLR-2 binds cell-wall components of Gram-negative bacteria, whereas TLR-4 binds cell-wall components of Gram-positive bacteria. LPS: lipopolysaccharide.

The common pathogen constituents recognized by pattern recognition receptors are usually quite distinct from the pathogen-specific antigens that are recognized by lymphocytes. The requirement for microbial constituents other than the antigen to initiate an adaptive immune response was recognized experimentally long before the discovery of dendritic cells and their mode of activation. It was found that purified antigens such as proteins often did not evoke an immune response in an experimental immunization—that is, they were not **immunogenic**. To obtain adaptive immune responses to purified antigens, it was essential to add killed bacteria or bacterial extracts to the antigen. This additional material was termed an **adjuvant**, as it helped the response to the immunizing antigen (*adjuvare* is Latin for 'to help'). We know now that adjuvants are needed, at least in part, to activate dendritic cells to full antigen-presenting status in the absence of an infection. Finding suitable adjuvants is still an important part of vaccine preparation; we describe modern adjuvant formulations in Appendix I.

Microorganisms can evolve more rapidly than their hosts, and this may explain why the cells and molecules of the innate immune system recognize only molecular structures that have remained unchanged during evolution. As we see next, the recognition mechanism used by the lymphocytes of the adaptive immune response has evolved to overcome the constraints faced by the innate immune system. It enables the recognition of an almost infinite diversity of antigens, so that each different pathogen can be specifically targeted.

1-9 Lymphocytes activated by antigen give rise to clones of antigen-specific effector cells that mediate adaptive immunity.

Instead of bearing several different receptors, each recognizing a different feature shared by many pathogens, a naive lymphocyte bears antigen receptors specific for a single chemical structure. However, each lymphocyte emerging from the central lymphoid organs differs from the others in its receptor specificity. The diversity is generated by a unique genetic mechanism that operates during lymphocyte development in the bone marrow and the thymus to generate millions of different variants of the genes encoding the receptor molecules. This ensures that the lymphocytes in the body collectively carry millions of different antigen receptor specificities—the **lymphocyte receptor**

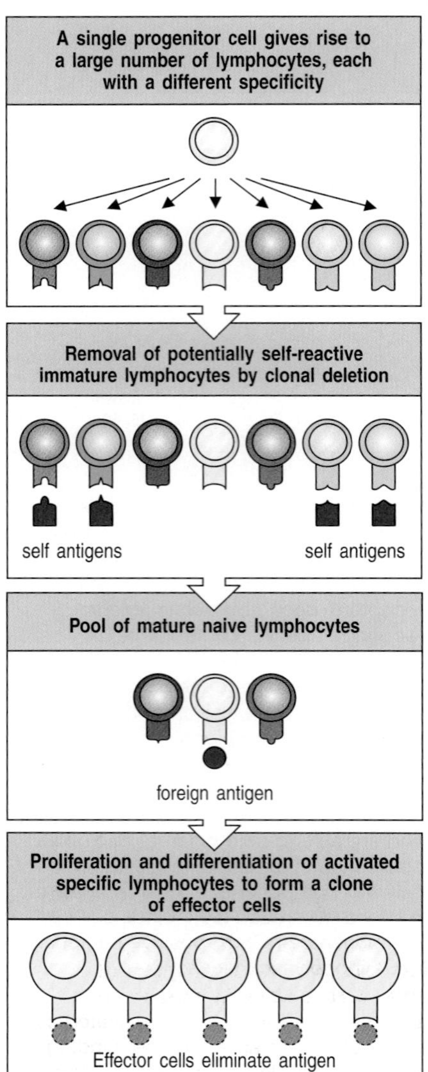

Fig. 1.11 Clonal selection. Each lymphoid progenitor gives rise to a large number of lymphocytes, each bearing a distinct antigen receptor. Lymphocytes with receptors that bind ubiquitous self antigens are eliminated before they become fully mature, ensuring tolerance to such self antigens. When a foreign antigen interacts with the receptor on a mature naive lymphocyte, that cell is activated and starts to divide. It gives rise to a clone of identical progeny, all of whose receptors bind the same antigen. Antigen specificity is thus maintained as the progeny proliferate and differentiate into effector cells. Once antigen has been eliminated by these effector cells, the immune response ceases, although some lymphocytes are retained to mediate immunological memory.

repertoire of the individual. These lymphocytes are continually undergoing a process akin to natural selection; only those lymphocytes that encounter an antigen to which their receptor binds will be activated to proliferate and differentiate into effector cells.

This selective mechanism was first proposed in the 1950s by **Macfarlane Burnet** to explain why a person produces antibodies against only those antigens to which he or she is exposed. Burnet postulated the preexistence in the body of many different potential antibody-producing cells, each having the ability to make antibody of a different specificity and displaying on its surface a membrane-bound version of the antibody: this serves as a receptor for the antigen. On binding antigen, the cell is activated to divide and to produce many identical progeny, a process known as **clonal expansion**; this **clone** of identical cells can now secrete **clonotypic** antibodies with a specificity identical to that of the surface receptor that first triggered activation and clonal expansion (Fig. 1.11). Burnet called this the **clonal selection theory** of antibody production.

1-10 Clonal selection of lymphocytes is the central principle of adaptive immunity.

Remarkably, at the time that Burnet formulated his theory, nothing was known of the antigen receptors of lymphocytes; indeed, the function of lymphocytes themselves was still obscure. Lymphocytes did not take center stage until the early 1960s, when **James Gowans** discovered that removal of the small lymphocytes from rats resulted in the loss of all known adaptive immune responses. These immune responses were restored when the small lymphocytes were replaced. This led to the realization that lymphocytes must be the units of clonal selection, and their biology became the focus of the new field of **cellular immunology**.

Clonal selection of lymphocytes with diverse receptors elegantly explained adaptive immunity, but it raised one significant conceptual problem. If the antigen receptors of lymphocytes are generated randomly during the lifetime of an individual, how are lymphocytes prevented from recognizing antigens on the tissues of the body and attacking them? **Ray Owen** had shown in the late 1940s that genetically different twin calves with a common placenta, and thus a shared placental blood circulation, were immunologically unresponsive, or **tolerant**, to one another's tissues: they did not make an immune

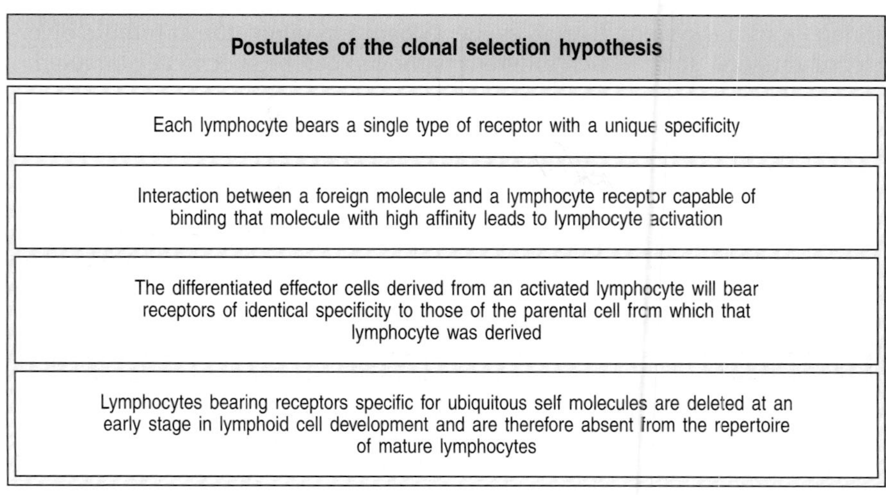

Postulates of the clonal selection hypothesis
Each lymphocyte bears a single type of receptor with a unique specificity
Interaction between a foreign molecule and a lymphocyte receptor capable of binding that molecule with high affinity leads to lymphocyte activation
The differentiated effector cells derived from an activated lymphocyte will bear receptors of identical specificity to those of the parental cell from which that lymphocyte was derived
Lymphocytes bearing receptors specific for ubiquitous self molecules are deleted at an early stage in lymphoid cell development and are therefore absent from the repertoire of mature lymphocytes

Fig. 1.12 The four basic principles of clonal selection.

response against each other. **Peter Medawar** then showed in 1953 that exposure to foreign tissues during embryonic development caused mice to become immunologically tolerant to these tissues. Burnet proposed that developing lymphocytes that are potentially self-reactive are removed before they can mature, a process known as **clonal deletion**. He has since been proved right in this too, although the mechanisms of **immunological tolerance** are still being worked out, as we shall see when we discuss the development of lymphocytes in Chapter 7.

Clonal selection of lymphocytes is the single most important principle in adaptive immunity. Its four basic postulates are listed in Fig. 1.12. The last of the problems posed by the clonal selection theory—how the diversity of lymphocyte antigen receptors is generated—was solved in the 1970s, when advances in molecular biology made it possible to clone the genes encoding antibody molecules.

1-11 The structure of the antibody molecule illustrates the central puzzle of adaptive immunity.

As discussed above, antibodies are the secreted form of the B cell's antigen receptor. Because they are produced in very large quantities in response to antigen, antibodies can be studied by traditional biochemical techniques; indeed, their structure was understood long before recombinant DNA technology made it possible to study the membrane-bound antigen receptors of B cells. The startling feature that emerged from the biochemical studies was that antibody molecules are composed of two distinct regions. One is a **constant region** that takes one of only four or five biochemically distinguishable forms; the other is a **variable region** that can be composed of an apparently infinite variety of different amino acid sequences, forming subtly different structures that allow antibodies to bind specifically to an equally vast variety of antigens. This division is illustrated in Fig. 1.13, where the antibody is depicted as a Y-shaped molecule. The variable region determines the antigen-binding specificity of the antibody. There are two identical variable regions in an antibody molecule, and it thus has two identical **antigen-binding sites**. The constant region determines the effector function of the antibody: that is, how the antibody disposes of the antigen once it is bound.

Each antibody molecule has a two-fold axis of symmetry and is composed of two identical **heavy chains** and two identical **light chains** (see Fig. 1.13,

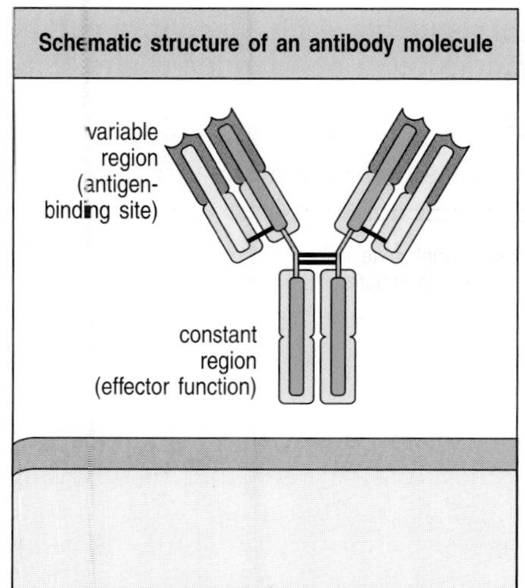

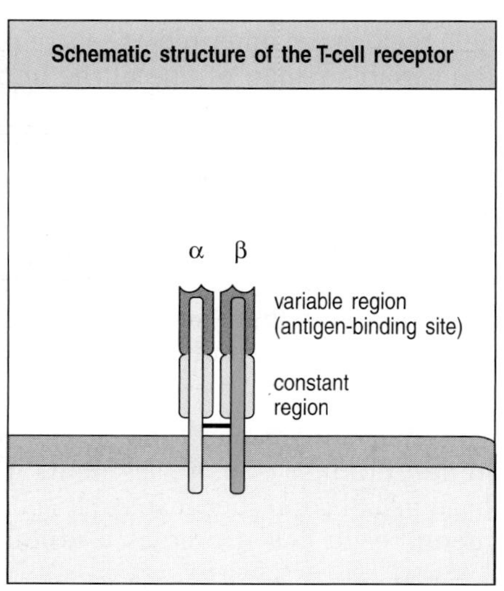

Fig. 1.13 Schematic structure of antigen receptors. Left panel: an antibody molecule, which is secreted by activated B cells as an antigen-binding effector molecule. A membrane-bound version of this molecule acts as the B-cell antigen receptor (not shown). An antibody is composed of two identical heavy chains (green) and two identical light chains (yellow). Each chain has a constant part (shaded blue) and a variable part (shaded red). Each arm of the antibody molecule is formed by a light chain and a heavy chain such that the variable parts of the two chains come together, creating a variable region that contains the antigen-binding site. The stem is formed from the constant parts of the heavy chains and takes a limited number of forms. This constant region is involved in the elimination of the bound antigen. Right panel: a T-cell antigen receptor. This is also composed of two chains, an α chain (yellow) and a β chain (green), which each have a variable and a constant part. As with the antibody molecule, the variable parts of the two chains create a variable region, which forms the antigen-binding site. The T-cell receptor is not produced in a secreted form.

where the heavy chains are in green and the light chains in yellow). Heavy and light chains both have variable and constant regions; the variable regions of a heavy and a light chain combine to form an antigen-binding site, so that both chains contribute to the antigen-binding specificity of the antibody molecule. The structure of antibody molecules is described in detail in Chapter 3, and the functional properties of antibodies conferred by their constant regions are considered in Chapters 4 and 9. For the time being, we are concerned only with the properties of antibody molecules as antigen receptors, and how the diversity of the variable regions is generated.

The T-cell receptor for antigen shows many similarities to the B-cell antigen receptor, and the two molecules are clearly related to each other evolutionarily; in fact, the T-cell receptor closely resembles a part of the antibody molecule. There are, however, important differences between the two molecules that, as we shall see, relate to their different roles within the immune system. The T-cell receptor, as shown in Fig. 1.13, is composed of two chains of roughly equal size, called the T-cell receptor α and β chains, each of which spans the T-cell membrane. Each chain has a variable region and a constant region, and the combination of the α- and β-chain variable regions creates a single site for binding antigen. This structure is described in detail in Chapter 3, and the way in which diversity is introduced into the variable regions is discussed in Chapter 4. As we will see, the organization of the genes for the antigen receptors and the way in which diversity is introduced to create a unique antigen-binding site is essentially the same for both the B-cell receptor and the T-cell receptor. There is, however, a crucial difference in the way in which the B-cell and T-cell receptors bind antigens; the T-cell receptor does not bind antigen molecules directly but instead recognizes fragments of antigens bound on the surface of other cells. The exact nature of the antigen recognized by T cells, and how the antigens are fragmented and carried to cell surfaces, is the subject of Chapter 5. A further difference from the antibody molecule is that there is no secreted form of the T-cell receptor; the function of the receptor is solely to signal to the T cell that it has bound its antigen, and the subsequent immunological effects depend on the actions of the T cells themselves, as we describe in Chapter 8.

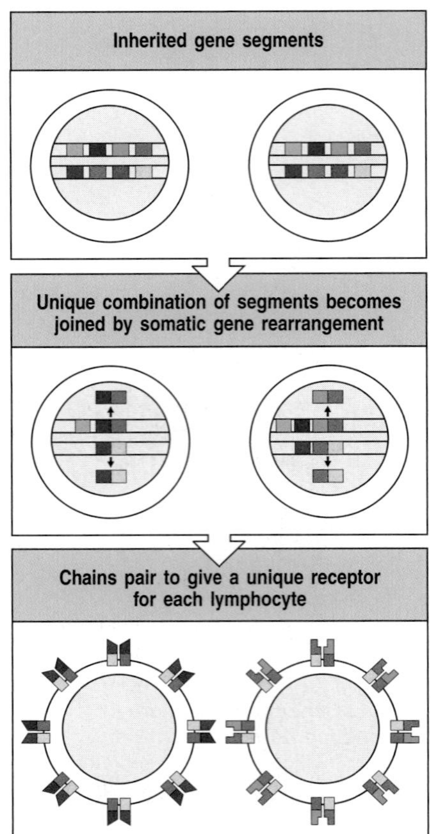
Inherited gene segments
Unique combination of segments becomes joined by somatic gene rearrangement
Chains pair to give a unique receptor for each lymphocyte

Fig. 1.14 The diversity of lymphocyte antigen receptors is generated by somatic gene-segment rearrangements. Different parts of the variable regions of antigen receptors are encoded by sets of gene segments. During a lymphocyte's development, one member of each set of gene segments is joined randomly to the others by an irreversible process of DNA recombination. The juxtaposed gene segments make up a complete gene that encodes the variable part of one chain of the receptor and is unique to that cell. This random rearrangement is repeated for the set of gene segments encoding the other chain. The rearranged genes are expressed to produce the two types of polypeptide chains. These come together to form a unique antigen receptor on the lymphocyte surface. Each lymphocyte bears many copies of its unique receptor.

1-12 Each developing lymphocyte generates a unique antigen receptor by rearranging its receptor gene segments.

How are antigen receptors with an almost infinite range of specificities encoded by a finite number of genes? This question was answered in 1976, when **Susumu Tonegawa** discovered that the genes for immunoglobulin variable regions are inherited as sets of **gene segments**, each encoding a part of the variable region of one of the immunoglobulin polypeptide chains (Fig. 1.14). During B-cell development in the bone marrow, these gene segments are irreversibly joined by DNA recombination to form a stretch of DNA encoding a complete variable region. Because there are many different gene segments in each set, and different gene segments are joined together in different cells, each cell generates unique genes for the variable regions of the heavy and light chains of the immunoglobulin molecule. Once these recombination events have succeeded in producing a functional receptor, further rearrangement is prohibited. Thus each lymphocyte expresses only one receptor specificity.

This mechanism has three important consequences. First, it enables a imited number of gene segments to generate a vast number of different proteins. Second, because each cell assembles a different set of gene segments, each cell expresses a unique receptor specificity. Third, because gene segment rearrangement involves an irreversible change in a cell's DNA, all the progeny of that cell will inherit genes encoding the same receptor specificity. This general scheme was later also confirmed for the genes encoding the antigen receptor on T cells.

The potential diversity of lymphocyte receptors generated in this way is enormous. Just a few hundred different gene segments can combine in different ways to generate thousands of different receptor chains. The diversity of lymphocyte receptors is further amplified by junctional diversity, created by adding or subtracting nucleotides in the process of joining the gene segments, and by the fact that each receptor is made by pairing two different variable chains, each encoded by distinct sets of gene segments. A thousand different chains of each type could thus generate 10^6 distinct antigen receptors through this combinatorial diversity. In this way, a small amount of genetic material can encode a truly staggering diversity of receptors. Only a subset of these randomly generated receptor specificities survive the selective processes that shape the peripheral lymphocyte repertoire; nevertheless, there are lymphocytes of at least 10^8 different specificities in an individual human at any one time. These provide the raw material on which clonal selection acts.

1-13 Immunoglobulins bind a wide variety of chemical structures, whereas the T-cell receptor is specialized to recognize foreign antigens as peptide fragments bound to proteins of the major histocompatibility complex.

In principle, almost any chemical structure can be recognized by the adaptive immune system as an antigen, but the usual antigens encountered in an infection are the proteins, glycoproteins, and polysaccharides of pathogens. An individual antigen receptor or antibody recognizes a small part of the molecular structure of an antigenic molecule, which is known as an **antigenic determinant** or **epitope** (Fig. 1.15). Macromolecular antigens such as proteins and glycoproteins usually have many different epitopes that can be recognized by different antigen receptors.

The antigen receptors of B cells and T cells are adapted to recognize antigen in two different ways, which reflects the roles their effector cells will eventually play in destroying pathogens. B cells are specialized to recognize the surface antigens on pathogens living outside cells and to differentiate into effector plasma cells that secrete antibodies to target these pathogens. B-cell receptors and their antibody counterparts are thus able to bind a wide variety of molecular structures.

Effector T cells, in contrast, have to deal with pathogens that have entered host cells, and also have to help activate B cells. To carry out these functions, the T-cell receptor is specialized to recognize antigens that have been generated inside cells and are being displayed on their surface. The recognition properties of the T-cell receptor reflect this: it recognizes only one type of antigen—peptides that have been produced in another host cell by the breakdown of proteins and are then displayed on the cell's surface. In addition, peptides are recognized only if they are bound to a particular type of cell-surface protein. These are the membrane glycoproteins known as **MHC molecules**, which are encoded in a cluster of genes called the **major histocompatibility complex**, which is abbreviated to **MHC**. The antigen recognized by T-cell receptors is thus a complex of a foreign peptide antigen and an MHC molecule (Fig. 1.16). We shall see how these compound antigens are recognized by T-cell receptors, and how they are generated, in Chapters 3 and 5, respectively.

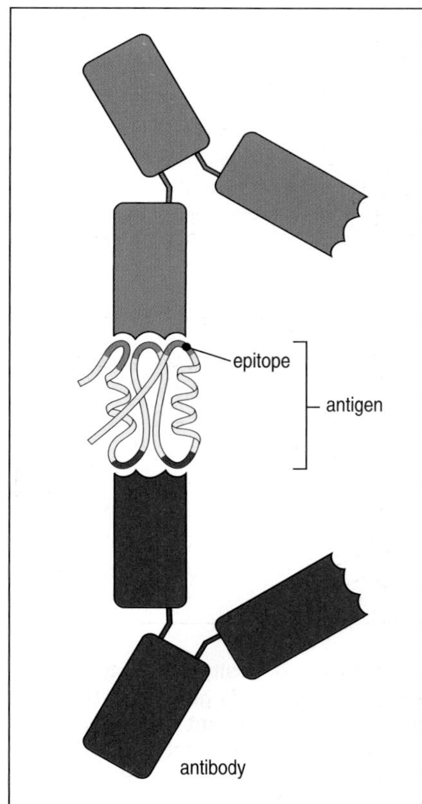

Fig. 1.15 Antigens are the molecules recognized by the immune response, while epitopes are sites within antigens to which antigen receptors bind. Antigens can be complex macromolecules such as proteins, as shown in yellow. Most antigens are larger than the sites on the antibody or antigen receptor to which they bind, and the actual portion of the antigen that is bound is known as the antigenic determinant, or epitope, for that receptor. Large antigens such as proteins can contain more than one epitope (indicated in red and blue), and thus may bind different antibodies.

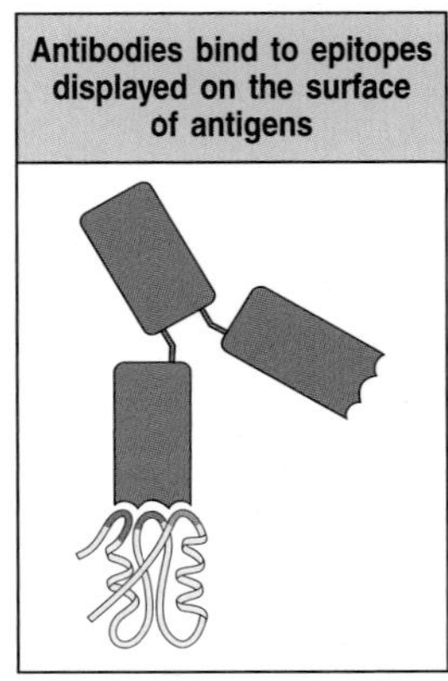

Antibodies bind to epitopes displayed on the surface of antigens

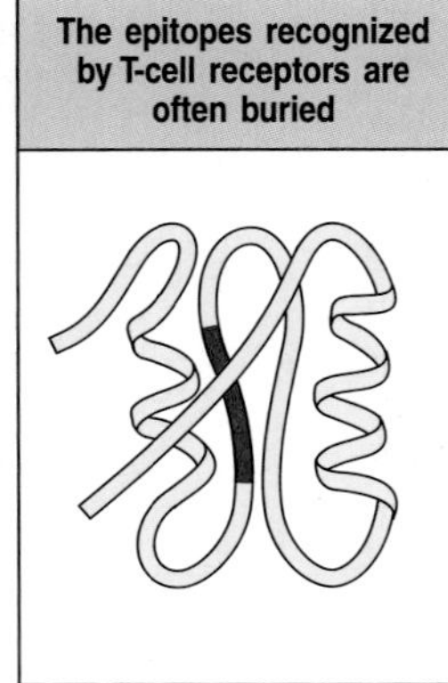

The epitopes recognized by T-cell receptors are often buried

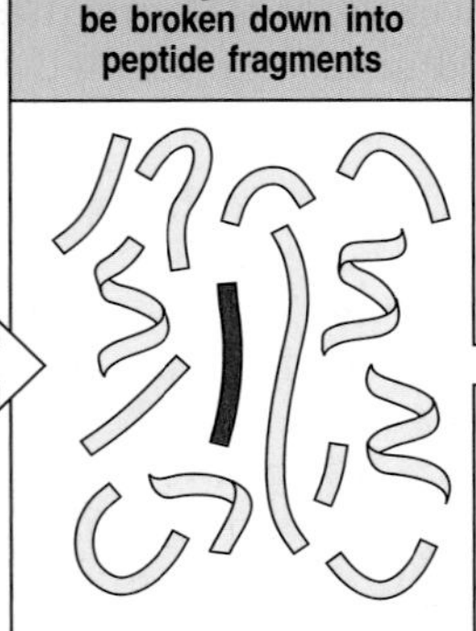

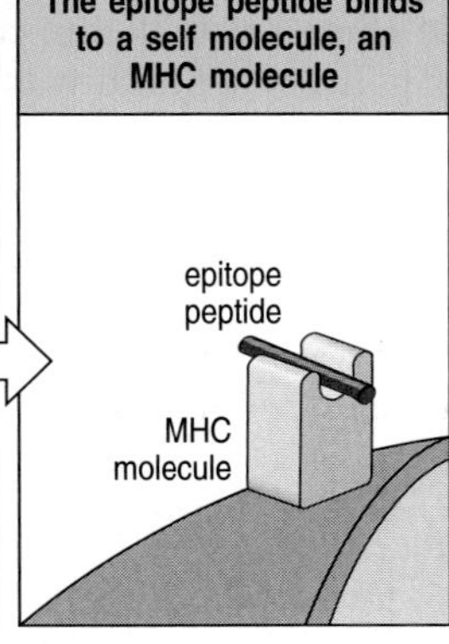

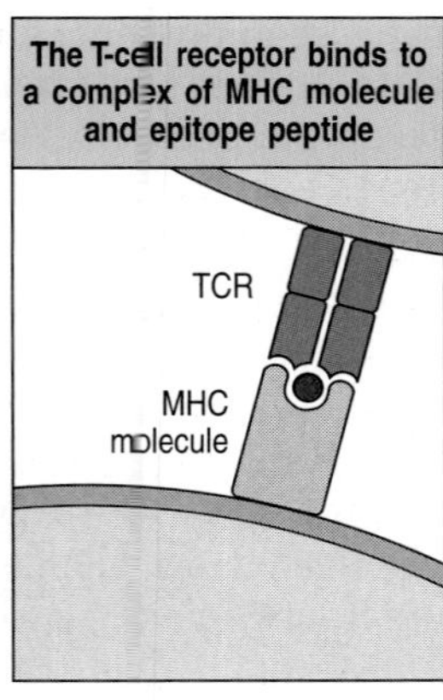

Fig. 1.16 An antibody binds an antigen directly, whereas a T-cell receptor binds a complex of antigen fragment and self molecule. Antibodies (first panel) bind directly to their antigens and recognize epitopes that form surface features of the antigen. In contrast, T-cell receptors can recognize epitopes that are buried within antigens and cannot be recognized directly (second panel). These antigens must first be degraded by proteinases (third panel), and the peptide epitope delivered to a self molecule, called an MHC molecule (fourth panel). It is in this form, as a complex of peptide and MHC molecule, that antigens are recognized by T-cell receptors (fifth panel).

1-14 The development and survival of lymphocytes is determined by signals received through their antigen receptors.

Equally amazing as the generation of millions of different antigen receptors is the shaping of this repertoire during lymphocyte development and the maintenance of an extensive repertoire in the periphery. How are potentially useful receptors maintained while those that could react against an individual's own antigens—their **self antigens**—are eliminated? How are the numbers of peripheral lymphocytes, and the percentages of B cells and T cells, kept relatively constant? The answer seems to be that throughout its lifetime, from its development in the central lymphoid organs onwards, the survival of a lymphocyte depends on signals received through its antigen receptor. If a lymphocyte fails to receive such survival signals, it dies by a form of cell suicide called **apoptosis** or **programmed cell death**. Lymphocytes that react strongly to self antigens are removed during development by clonal deletion, as predicted by Burnet's clonal selection theory, before they mature to a stage at which they could do damage. In contrast, a complete absence of signals from the antigen receptor during development can also lead to cell death. In addition, if a receptor is not used within a relatively short time of its entering the repertoire in the periphery, the cell bearing it dies, making way for new lymphocytes with different receptors. In this way, self-reactive receptors are eliminated, and receptors are tested to ensure they are potentially functional. The mechanisms that shape and maintain the lymphocyte receptor repertoire are examined in Chapter 7.

Apoptosis, derived from a Greek word meaning the falling of leaves from the trees, is a general means of regulating the number of cells in the body. It is responsible, for example, for the death and shedding of old skin and intestinal epithelial cells, and the turnover of liver cells. Each day the bone marrow produces millions of new neutrophils, monocytes, red blood cells, and lymphocytes, and this production must be balanced by an equal loss. Most white blood cells are relatively short lived and die by apoptosis. The dying cells are phagocytosed and degraded by specialized macrophages in the liver and spleen.

1-15 Lymphocytes encounter and respond to antigen in the peripheral lymphoid organs.

Pathogens can enter the body by many routes and set up an infection anywhere in the tissues, whereas lymphocytes are normally found only in the blood, lymph, and lymphoid organs. How, then, are they to meet? Antigen and lymphocytes eventually encounter each other in the peripheral lymphoid organs—the lymph nodes, spleen, and the mucosal lymphoid tissues (see Fig. 1.7). Mature naive lymphocytes are continually recirculating through these tissues, to which pathogen antigens are carried from sites of infection, primarily by dendritic cells. The peripheral lymphoid organs are specialized to trap antigen-bearing dendritic cells and to facilitate the initiation of adaptive immune responses.

Peripheral lymphoid tissues are composed of aggregations of lymphocytes in a framework of non-leukocyte stromal cells, which provide the basic structural organization of the tissue and provide survival signals to help sustain the life of the lymphocytes. Besides lymphocytes, peripheral lymphoid organs also contain resident macrophages and dendritic cells.

When an infection occurs in a tissue such as the skin, free antigen and antigen-bearing dendritic cells travel from the site of infection through the afferent lymphatic vessels into the **draining lymph nodes** (Fig. 1.17), peripheral lymphoid tissues where they activate antigen-specific lymphocytes. The activated lymphocytes then undergo a period of proliferation and differentiation, after which most leave the lymph nodes as effector cells via the efferent lymphatic vessel. This eventually returns them to the bloodstream (see Fig. 1.7), which then carries them to the tissues where they will act. This whole process takes about 4–6 days from the time that antigen is recognized, which means that an adaptive immune response to an antigen not encountered before does not become effective until about a week after infection. Naive lymphocytes that do not recognize their antigen also leave through the efferent lymphatic vessel and are returned to the blood, from which they continue to recirculate through lymphoid tissues until they recognize antigen or die.

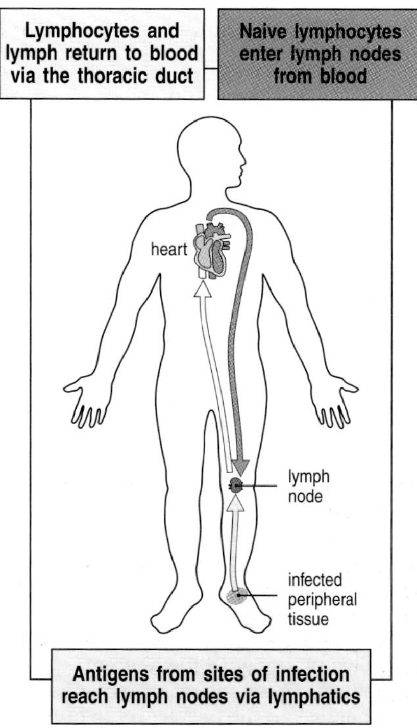
Lymphocytes and lymph return to blood via the thoracic duct
Naive lymphocytes enter lymph nodes from blood
heart
lymph node
infected peripheral tissue
Antigens from sites of infection reach lymph nodes via lymphatics

Fig. 1.17 Circulating lymphocytes encounter antigen in peripheral lymphoid organs. Naive lymphocytes recirculate constantly through peripheral lymphoid tissue, here illustrated as a popliteal lymph node—a lymph node situated behind the knee. In the case of an infection in the foot, this will be the draining lymph node, where lymphocytes may encounter their specific antigens and become activated. Both activated and non-activated lymphocytes are returned to the bloodstream via the lymphatic system.

The **lymph nodes** are highly organized lymphoid organs located at the points of convergence of vessels of the lymphatic system, which is the extensive system that collects extracellular fluid from the tissues and returns it to the blood (see Fig. 1.7). This extracellular fluid is produced continuously by filtration from the blood and is called lymph. Lymph flows away from the peripheral tissues under the pressure exerted by its continual production, and is carried by lymphatic vessels, or **lymphatics**. One-way valves in the lymphatic vessels prevent a reverse flow, and the movements of one part of the body in relation to another are important in driving the lymph along.

Afferent lymphatic vessels drain fluid from the tissues and carry pathogens and antigen-bearing cells from infected tissues to the lymph nodes (Fig. 1.18). Free antigens simply diffuse through the extracellular fluid to the lymph node, while the dendritic cells actively migrate into the lymph node under the influence of chemotactic chemokines. The same chemokines also attract lymphocytes from the blood, and these enter lymph nodes by squeezing through the walls of specialized blood vessels called **high endothelial venules** (**HEV**). In the lymph nodes, B lymphocytes are localized in **follicles**, which make up the outer **cortex** of the lymph node, with T cells more diffusely distributed in the surrounding **paracortical areas**, also referred to as the deep cortex or **T-cell zones** (see Fig. 1.18). Lymphocytes migrating from the blood into lymph nodes enter the paracortical areas first and, since they are attracted by the same chemokines, antigen-presenting dendritic cells and macrophages also become localized there. Free antigen diffusing through the lymph node can become trapped on these dendritic cells and macrophages. This juxtaposition of antigen, antigen-presenting cells, and naive T cells creates an ideal environment in the T-cell zone in which naive T cells can bind their antigen and thus become activated.

As noted earlier, activation of B cells usually requires not only antigen, which binds to the B-cell receptor, but also the cooperation of helper T cells, a type of effector T cell (see Section 1-4). The organization of the lymph node ensures that before entering the follicles, naive B cells pass through the T-cell zones, where they may encounter both their antigen and their cooperating helper T cells and become activated. Some of the B-cell follicles include **germinal centers**, where activated B cells are undergoing intense proliferation and differentiation into plasma cells.

In humans, the **spleen** is a fist-sized organ situated just behind the stomach (see Fig. 1.7). It has no direct connection with the lymphatic system; instead it collects antigen from the blood and is involved in immune responses to blood-borne pathogens. Lymphocytes enter and leave the spleen via blood vessels. The spleen also collects and disposes of senescent red blood cells. Its organization is shown schematically in Fig. 1.19. The bulk of the spleen is composed of **red pulp**, which is the site of red blood cell disposal. The lymphocytes surround the arterioles running through the spleen, forming isolated areas of **white pulp**. The sheath of lymphocytes around an arteriole is called the **periarteriolar lymphoid sheath** (**PALS**) and contains mainly T cells. Lymphoid follicles occur at intervals along it, and these contain mainly B cells. A so-called marginal zone surrounds the follicle; it has few T cells, is rich in macrophages and has a resident, non-circulating population of B cells known as **marginal zone B cells**, about which little is known; they are discussed in Chapter 7. Blood-borne microbes, soluble antigens and antigen:antibody complexes are filtered from the blood by macrophages and immature dendritic cells within the marginal zone. As with the migration of immature dendritic cells from peripheral tissues to the T-cell areas of lymph nodes, when dendritic cells in the splenic marginal zones take up antigens and become activated they migrate into the T-cell areas of the spleen, where they are able to present the antigens they carry to T cells.

Most pathogens enter the body through mucosal surfaces, and these are also exposed to a vast load of other potential antigens from the air, food, and the natural microbial flora of the body. Mucosal surfaces are protected by an extensive system of lymphoid tissues known generally as the **mucosal immune system** or **mucosa-associated tissues** (**MALT**). Collectively, the mucosal immune system is estimated to contain as many lymphocytes as all the rest of the body, and they form a specialized set of cells obeying somewhat different rules of recirculation from those in the other peripheral lymphoid organs. The **gut-associated lymphoid tissues** (**GALT**) include the **tonsils**, **adenoids**, **appendix**, and specialized structures called **Peyer's patches** in the

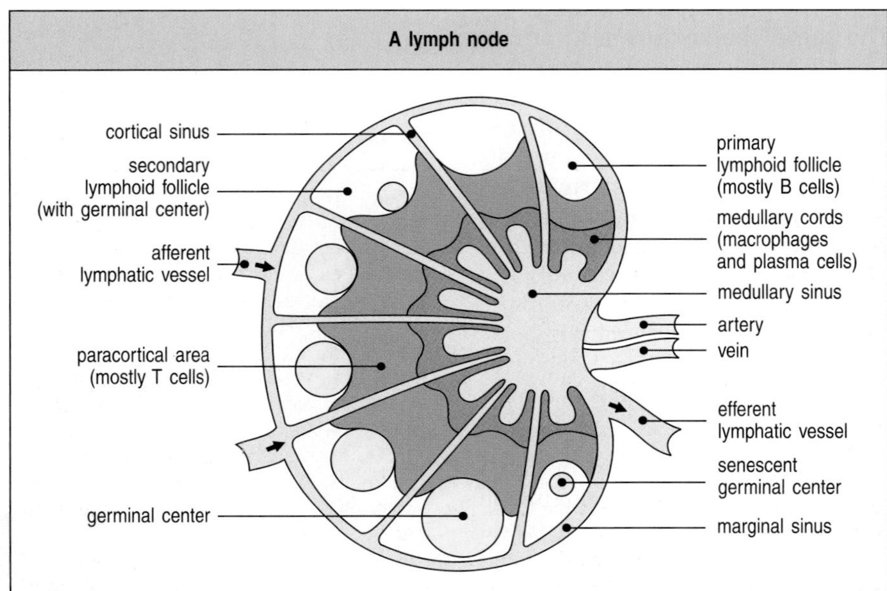

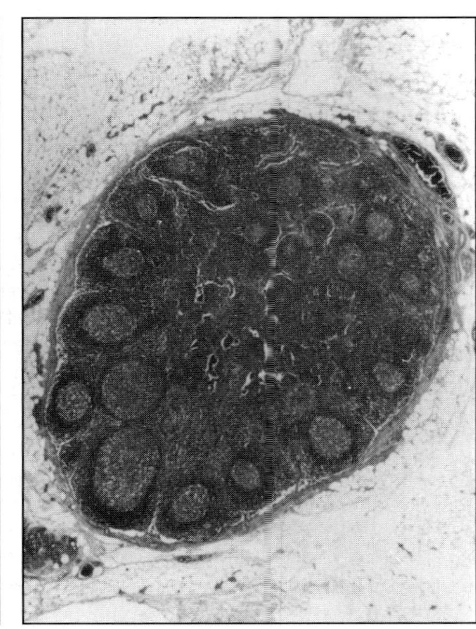

Fig. 1.18 Organization of a lymph node. As shown in the diagram on the left, which shows a lymph node in longitudinal section, a lymph node consists of an outermost cortex and an inner medulla. The cortex is composed of an outer cortex of B cells organized into lymphoid follicles, and deep, or paracortical, areas made up mainly of T cells and dendritic cells. When an immune response is under way, some of the follicles contain central areas of intense B-cell proliferation called germinal centers and are known as secondary lymphoid follicles. These reactions are very dramatic, but eventually die out as senescent germinal centers. Lymph draining from the extracellular spaces of the body carries antigens in phagocytic dendritic cells and phagocytic macrophages from the tissues to the lymph node via the afferent lymphatics. These migrate directly from the sinuses into the cellular parts of the node. Lymph leaves by the efferent lymphatics in the medulla. The medulla consists of strings of macrophages and antibody-secreting plasma cells known as the medullary cords. Naive lymphocytes enter the node from the bloodstream through specialized postcapillary venules (not shown) and leave with the lymph through the efferent lymphatic. The light micrograph shows a transverse section through a lymph node, with prominent follicles containing germinal centers. Magnification ×7. Photograph courtesy of N. Rooney.

small intestine, and they collect antigen from the epithelial surfaces of the gastrointestinal tract. In Peyer's patches, which are the most important and highly organized of these tissues, the antigen is collected by specialized epithelial cells called **microfold** or **M cells** (Fig. 1.20). The lymphocytes form

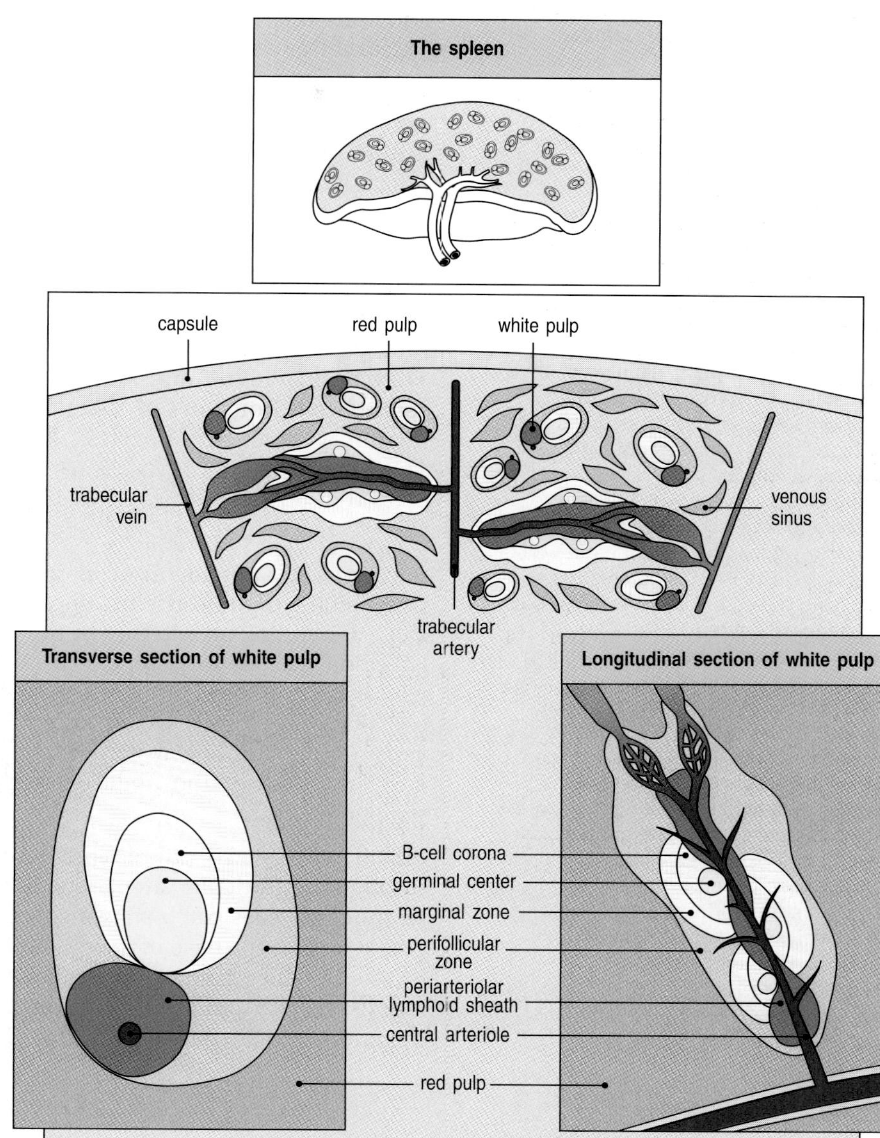

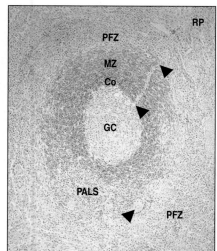

Fig. 1.19 Organization of the lymphoid tissues of the spleen. The schematic at top right shows that the spleen consists of red pulp (pink areas in the top panel), which is a site of red blood cell destruction, interspersed with the lymphoid white pulp. An enlargement of a small section of a human spleen (center) shows the arrangement of discrete areas of white pulp (yellow and blue) around central arterioles. Most of the white pulp is shown in transverse section, with two portions in longitudinal section. The bottom two schematics show enlargements of a transverse section (lower center) and longitudinal section (lower right) of white pulp. Surrounding the central arteriole is the periarteriolar lymphoid sheath (PALS), made up of T cells. Lymphocytes and antigen-loaded dendritic cells come together here. The follicles consist mainly of B cells; in secondary follicles a germinal center is surrounded by a B-cell corona. The follicles are surrounded by a so-called marginal zone of lymphocytes. In each area of white pulp, blood carrying both lymphocytes and antigen flows from a trabecular artery into a central arteriole. From this arteriole smaller blood vessels fan out, eventually terminating in a specialized zone in the human spleen called the perifollicular zone (PFZ), which surrounds each marginal zone. Cells and antigen then pass into the white pulp through open blood-filled spaces in the perifollicular zone. The light micrograph at bottom left shows a transverse section of white pulp of human spleen immunostained for mature B cells. Both follicle and PALS are surrounded by the perifollicular zone. The follicular arteriole emerges in the PALS (arrowhead at bottom) traverses the follicle, goes through the marginal zone and opens into the perifollicular zone (upper arrowheads). Co, follicular B-cell corona; GC, germinal center; MZ, marginal zone; RP, red pulp; arrowheads, central arteriole. Photograph courtesy of N.M. Milicevic.

a follicle consisting of a large central dome of B lymphocytes surrounded by smaller numbers of T lymphocytes. Dendritic cells resident within the Peyer's patch present the antigen to T lymphocytes. Lymphocytes enter Peyer's patches from the blood and leave through efferent lymphatics. Effector lymphocytes generated in Peyer's patches travel through the lymphatic system and into the bloodstream, from where they are disseminated back into mucosal tissues to carry out their effector actions.

Similar but more diffuse aggregates of lymphocytes are present in the respiratory tract and other mucosa: **nasal-associated lymphoid tissue** (**NALT**) and **bronchus-associated lymphoid tissue** (**BALT**) are present in the respiratory tract. Like the Peyer's patches, these mucosal lymphoid tissues are also overlaid by M cells, through which inhaled microbes and antigens that become trapped in the mucous covering of the respiratory tract can pass. The mucosal immune system is discussed in Chapter 11.

Although very different in appearance, the lymph nodes, spleen, and mucosa-associated lymphoid tissues all share the same basic architecture. They all operate on the same principle, trapping antigens and antigen-presenting cells from sites of infection and enabling them to present antigen to migratory small lymphocytes, thus inducing adaptive immune responses. The peripheral lymphoid tissues also provide sustaining signals to lymphocytes that do not encounter their specific antigen immediately, so that they survive and continue to recirculate.

Because they are involved in initiating adaptive immune responses, the peripheral lymphoid tissues are not static structures but vary quite dramatically, depending on whether or not infection is present. The diffuse mucosal lymphoid tissues may appear in response to infection and then disappear, whereas the architecture of the organized tissues changes in a more defined way during an infection. For example, the B-cell follicles of the lymph nodes expand as B lymphocytes proliferate to form germinal centers (see Fig. 1.18), and the entire lymph node enlarges, a phenomenon familiarly known as swollen glands.

Finally, specialized populations of lymphocytes can be found distributed throughout particular sites in the body rather than being found in organized lymphoid tissues. Such sites include the liver and the lamina propria of the gut, as well as the base of the epithelial lining of the gut, reproductive epithelia, and, in mice but not in humans, the epidermis. These lymphocyte populations seem to play an important role in protecting these tissues from infection, and are described further in Chapters 7 and 11.

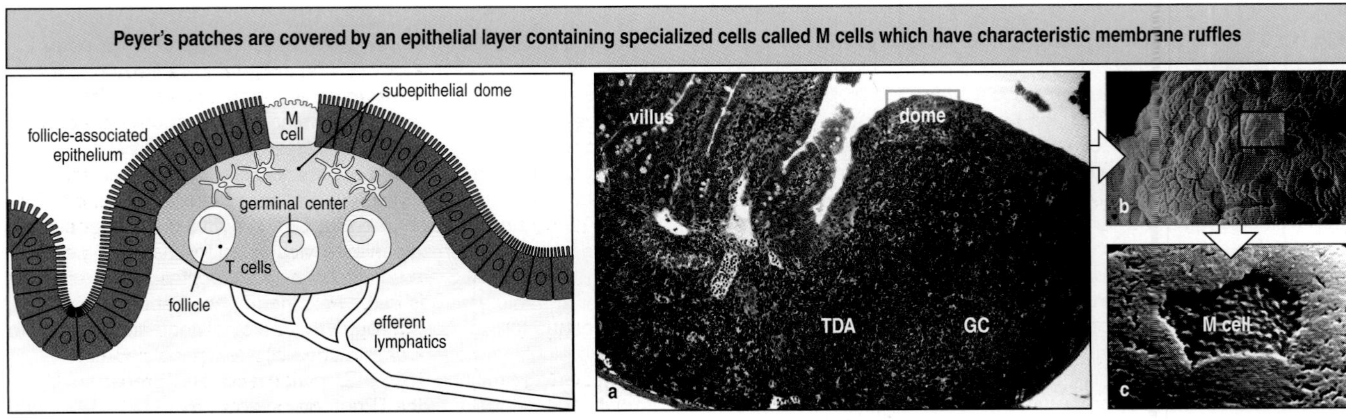

Fig. 1.20 Organization of a Peyer's patch in the gut mucosa. As the diagram on the left shows, a Peyer's patch contains numerous B-cell follicles with germinal centers. T cells occupy the areas between follicles, the T-cell-dependent areas. The layer between the surface epithelium and the follicles is known as the subepithelial dome, and is rich in dendritic cells, T cells, and B cells. Peyer's patches have no afferent lymphatics and the antigen enters directly from the gut across a specialized epithelium made up of so-called microfold (M) cells. Although this tissue looks very different from other lymphoid organs, the basic divisions are maintained. As in the lymph nodes, lymphocytes enter Peyer's patches from the blood across the walls of high endothelial venules (not shown), and leave via the efferent lymphatic. The light micrograph in panel a shows a section through a Peyer's patch in the gut wall of the mouse. The Peyer's patch can be seen lying beneath the epithelial tissues. GC, germinal center; TDA, T-Cell-dependent area. Panel b is a scanning electron micrograph of the follicle-associated epithelium boxed in (a), showing the M cells, which lack the microvilli and the mucous layer present on normal epithelial cells. Each M cell appears as a sunken area on the epithelial surface. Panel c is a higher-magnification view of the boxed area in (b), showing the characteristic ruffled surface of an M cell. M cells are the portal of entry for many pathogens and other particles. (a) Hematoxylin and eosin stain. Magnification ×100; (b) ×5000; (c) ×23,000. Source: Mowat, A., Viney, J.: *Immunol. Rev.* 1997, **156**:145–166.

1-16 Interaction with other cells as well as with antigen is necessary for lymphocyte activation.

As we have seen in Sections 1-3 and 1-6, peripheral lymphoid tissues are specialized not only to trap phagocytic cells that have ingested antigen but also to promote their interactions with lymphocytes that are needed to initiate an adaptive immune response.

All lymphocyte responses to antigen require not only the signal that results from antigen binding to lymphocyte receptors but also a second signal, which is delivered by another cell by means of cell-surface molecules known generally as co-stimulatory molecules (see Section 1-7). Naive T cells are usually stimulated by activated dendritic cells (Fig. 1.21, left panel), but for naive B cells (Fig. 1.21, right panel) the second signal is delivered by an activated helper T cell. Macrophages and B cells presenting foreign antigen on their surface can also be induced to express co-stimulatory molecules and thus can activate naive T cells. These three specialized antigen-presenting cells of the immune system are illustrated in Fig. 1.22. Dendritic cells are the most important of the three in this respect, with a central role in the initiation of adaptive immune responses.

The induction of co-stimulatory molecules is important in initiating an adaptive immune response because contact with antigen without accompanying co-stimulatory molecules inactivates naive lymphocytes rather than activating them, leading either to clonal deletion or an inactive state known as **anergy**. We shall return to this topic in Chapter 7. Thus, we need to add a final postulate to the clonal selection theory. A naive lymphocyte can only be activated by cells that bear not only specific antigen but also co-stimulatory molecules, whose expression is regulated by innate immunity.

1-17 Lymphocytes activated by antigen proliferate in the peripheral lymphoid organs, generating effector cells and immunological memory.

The great diversity of lymphocyte receptors means that there will usually be at least a few that can bind to a given foreign antigen. However, this number will be very small, certainly not enough to mount a response against a pathogen. To generate sufficient antigen-specific effector lymphocytes to fight an infection, a lymphocyte with an appropriate receptor specificity is

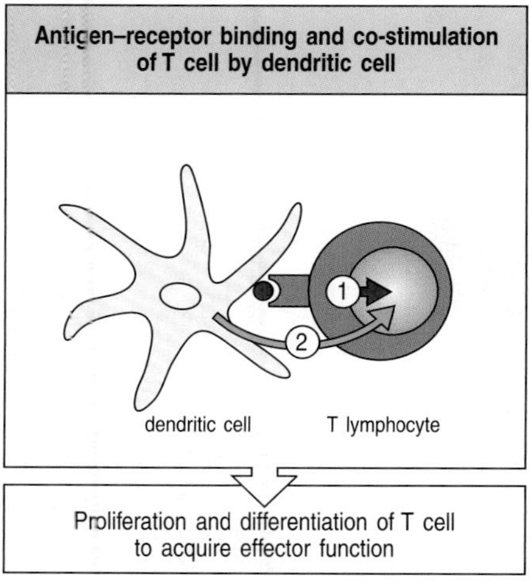

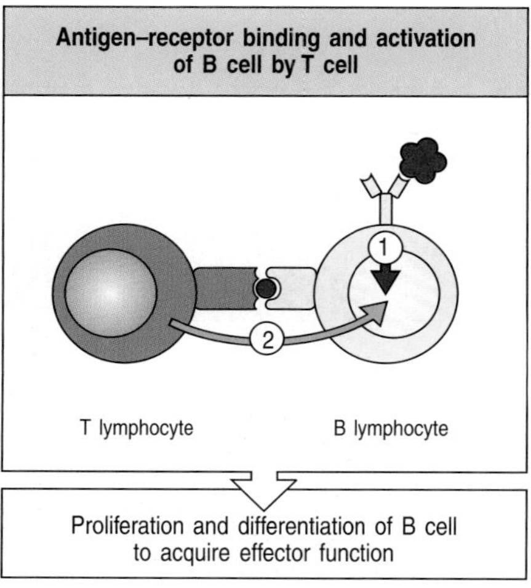

Fig. 1.21 Two signals are required for lymphocyte activation. As well as receiving a signal through their antigen receptor (signal 1), mature naive lymphocytes must also receive a second signal (signal 2) to become activated. For T cells (left panel), this second signal is delivered by an antigen-presenting cell such as the dendritic cell shown here. For B cells (right panel), the second signal is usually delivered by an activated T cell, which recognizes antigenic peptides taken up, processed, and presented by the B cell on its surface.

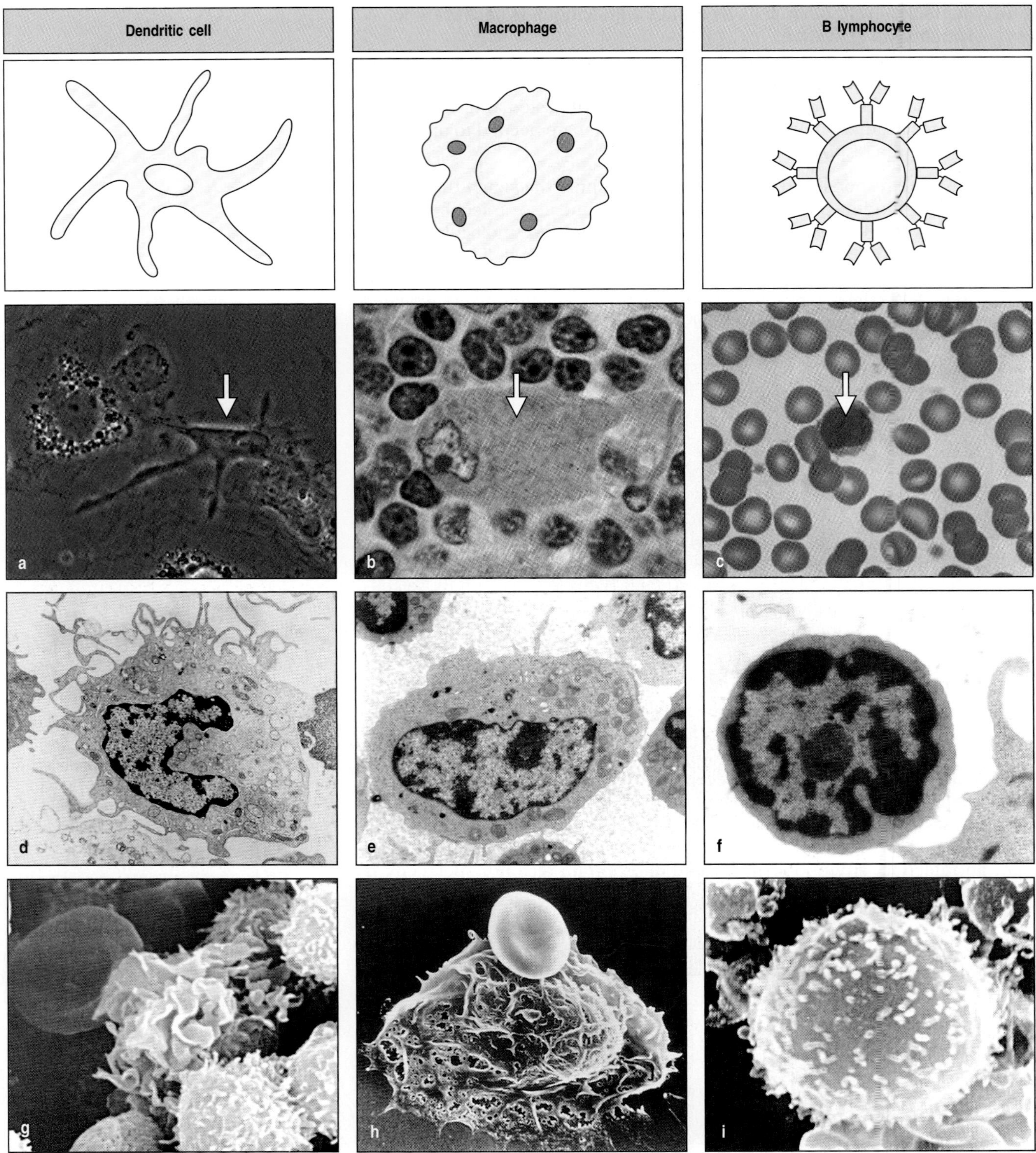

Fig. 1.22 The antigen-presenting cells. The three types of antigen-presenting cells are shown in the form in which they will be depicted throughout this book (top row), as they appear in the light microscope (second row; the relevant cell is indicated by an arrow), by transmission electron microscopy (third row) and by scanning electron microscopy (bottom row). Mature dendritic cells are found in lymphoid tissues and are derived from immature tissue dendritic cells that interact with many distinct types of pathogens. Macrophages are specialized to internalize extracellular pathogens, especially after they have been coated with antibody, and to present their antigens. B cells have antigen-specific receptors that enable them to internalize large amounts of specific antigen, process it, and present it. Photographs courtesy of R.M. Steinman (a); N. Rooney (b, c, e, f); S. Knight (d, g); and P.F. Heap (h, i).

activated first to proliferate. Only when a large clone of identical cells has been produced do these finally differentiate into effector cells. This clonal expansion is a feature common to all adaptive immune responses. On recognizing its specific antigen on an activated antigen-presenting cell, a naive lymphocyte stops migrating and enlarges. The chromatin in its nucleus becomes less dense, nucleoli appear, the volume of both the nucleus and the cytoplasm increases, and new mRNAs and new proteins are synthesized. Within a few hours, the cell looks completely different and is known as a **lymphoblast** (Fig. 1.23).

The lymphoblasts now begin to divide, normally duplicating themselves two to four times every 24 hours for 3–5 days, so that one naive lymphocyte gives rise to a clone of around 1000 daughter cells of identical specificity. These then differentiate into effector cells. In the case of B cells, the differentiated effector

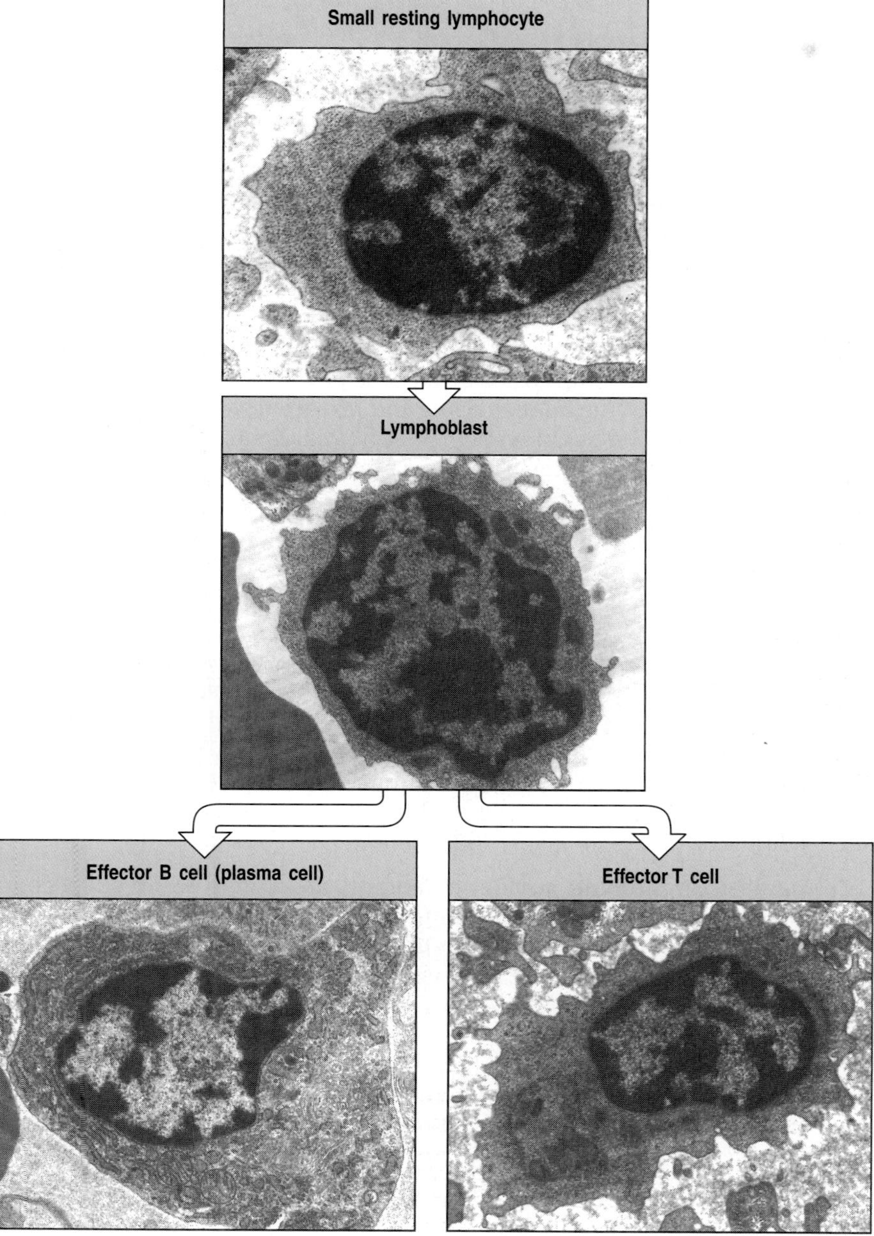

Fig. 1.23 Transmission electron micrographs of lymphocytes at various stages of activation to effector function. Small resting lymphocytes (top panel) have not yet encountered antigen. Note the scanty cytoplasm, the absence of rough endoplasmic reticulum, and the condensed chromatin, all indicative of an inactive cell. This could be either a T cell or a B cell. Small circulating lymphocytes are trapped in lymph nodes when their receptors encounter antigen on antigen-presenting cells. Stimulation by antigen induces the lymphocyte to become an active lymphoblast (center panel). Note the larger size, enlarged nucleus and more diffuse chromatin; again, T and B lymphoblasts are similar in appearance. This cell undergoes repeated division, which is followed by differentiation to effector function. The bottom panels show effector T and B lymphocytes. Note the large amount of cytoplasm, abundant mitochondria, and the presence of rough endoplasmic reticulum, all hallmarks of active cells. The rough endoplasmic reticulum is especially prominent in plasma cells (effector B cells), which are synthesizing and secreting very large amounts of protein in the form of antibody. Photographs courtesy of N. Rooney.

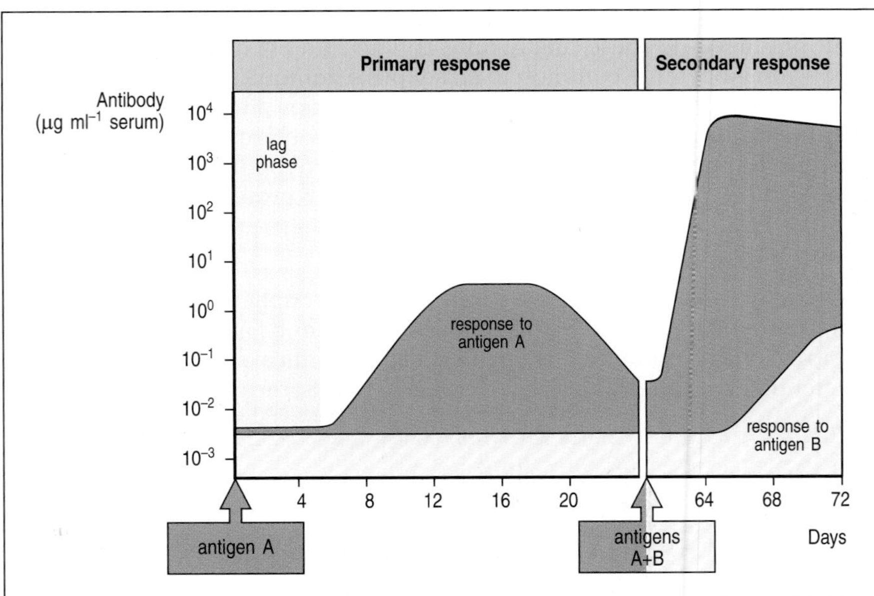

Fig. 1.24 The course of a typical antibody response. The first encounter with an antigen produces a primary response. Antigen A introduced at time zero encounters little specific antibody in the serum. After a lag phase (light blue), antibody against antigen A (dark blue) appears; its concentration rises to a plateau, and then gradually declines. This is typical of a primary response. When the serum is tested for antibody against another antigen, B (yellow), there is little present, demonstrating the specificity of the antibody response. When the animal is later challenged with a mixture of antigens A and B, a very rapid and intense secondary response to A occurs. This illustrates immunological memory, the ability of the immune system to make a second response to the same antigen more efficiently and effectively, providing the host with a specific defense against infection. This is the main reason for giving booster injections after an initial vaccination. Note that the response to B resembles the initial or primary response to A, as this is the first encounter of the host with antigen B.

cells are the plasma cells, which secrete antibody; in the case of T cells, the effector cells are cytotoxic T cells able to destroy infected cells or helper T cells that activate other cells of the immune system. Effector lymphocytes do not recirculate like naive lymphocytes. Some effector T cells detect sites of infection and migrate into them from the blood; others stay in the lymphoid tissues to activate B cells. Some antibody-secreting plasma cells remain in the peripheral lymphoid organs, but most plasma cells generated in the lymph nodes and spleen migrate to the bone marrow and take up residence there, pouring out antibodies into the blood system. Effector cells generated in the mucosal immune system generally stay within the mucosal tissues.

After a naive lymphocyte has been activated, it takes 4–5 days before clonal expansion is complete and the lymphocytes have differentiated into effector cells. The first adaptive immune response to a pathogen thus only occurs several days after the infection begins and has been detected by the innate immune system. Most of the lymphocytes generated by the clonal expansion in any given immune response eventually die. However, a significant number of activated antigen-specific B cells and T cells persist after antigen has been eliminated. These cells are known as **memory cells** and form the basis of immunological memory. They can be reactivated much more quickly than naive lymphocytes, which ensures a more rapid and effective response on a second encounter with a pathogen and thereby usually provides lasting protective immunity.

The characteristics of immunological memory are readily observed by comparing the antibody response of an individual to a first or **primary immunization** with the same response elicited in the same individual by a **secondary** or **booster immunization** with the same antigen. As shown in Fig. 1.24, the secondary antibody response occurs after a shorter lag phase, achieves a markedly higher level, and produces antibodies of higher affinity, or strength of binding, for the antigen. The increased affinity for antigen is called **affinity maturation** and is the result of events that select B-cell receptors, and thus antibodies, for progressively higher affinity for antigen during an immune response. Importantly, T-cell receptors do not undergo affinity maturation, and the lower threshold for activation of memory T cells compared with naive T cells results from a change in the responsiveness of the cell, not from a change in the receptor. We describe the mechanisms of these remarkable changes in Chapters 4, 9, and 10. The cellular basis of immunological memory is the clonal expansion and clonal differentiation of cells specific for the eliciting antigen, and it is therefore entirely antigen specific.

It is immunological memory that enables successful vaccination and prevents reinfection with pathogens that have been repelled successfully by an adaptive immune response. Immunological memory is the most important biological consequence of adaptive immunity, although its cellular and molecular basis is still not fully understood, as we shall see in Chapter 10.

Summary.

The early innate systems of defense, which depend on invariant receptors recognizing common features of pathogens, are crucially important, but they can be overcome by many pathogens and they do not lead to immunological memory. Recognizing a particular pathogen and providing enhanced protection against reinfection is unique to adaptive immunity. An adaptive immune response involves the selection and amplification of clones of lymphocytes bearing receptors that recognize the foreign antigen. This clonal selection provides the theoretical framework for understanding all the key features of an adaptive immune response. There are two major types of lymphocytes: B lymphocytes, which mature in the bone marrow and are the source of circulating antibodies, and T lymphocytes, which mature in the thymus and recognize peptides from pathogens presented by MHC molecules on infected cells or antigen-presenting cells. Each lymphocyte carries cell-surface receptors of a single antigen specificity. These receptors are generated by the random recombination of variable receptor gene segments and the pairing of distinct variable protein chains: heavy and light chains in immunoglobulins, or the two chains of T-cell receptors. This process produces a large collection of lymphocytes each bearing a distinct receptor, so that the total receptor repertoire can recognize virtually any antigen. If the receptor is specific for a ubiquitous self antigen, the lymphocyte is eliminated by encountering the antigen early in its development, while survival signals received through the antigen receptor select and maintain a repertoire of potentially useful lymphocytes. Adaptive immunity is initiated when an innate immune response fails to eliminate a new infection, and activated antigen-presenting cells bearing pathogen antigens are delivered to the draining lymphoid tissues. When a recirculating lymphocyte encounters its corresponding antigen in peripheral lymphoid tissues, it is induced to proliferate, and its progeny then differentiate into effector T and B lymphocytes that can eliminate the infectious agent. A subset of these proliferating lymphocytes differentiates into memory cells, ready to respond rapidly to the same pathogen if it is encountered again. The details of these processes of recognition, development, and differentiation form the main material of the central three parts of this book.

The effector mechanisms of adaptive immunity

We have seen in the first part of this chapter how naive lymphocytes are selected by antigen to differentiate into clones of activated effector lymphocytes. We now expand on the mechanisms by which activated effector lymphocytes target different pathogens for destruction in a successful adaptive immune response. The distinct lifestyles of different pathogens require different responses for both their recognition and their destruction (Fig. 1.25). B-cell receptors recognize antigens from the extracellular environment and

Fig. 1.25 The major types of pathogens confronting the immune system and some of the diseases they cause.

The immune system protects against four classes of pathogens		
Type of pathogen	**Examples**	**Diseases**
Extracellular bacteria, parasites, fungi	*Streptococcus pneumoniae* *Clostridium tetani* *Trypanosoma brucei* *Pneumocystis carinii*	Pneumonia Tetanus Sleeping sickness *Pneumocystis* pneumonia
Intracellular bacteria, parasites	*Mycobacterium leprae* *Leishmania donovani* *Plasmodium falciparum*	Leprosy Leishmaniasis Malaria
Viruses (intracellular)	Variola Influenza Varicella	Smallpox Flu Chickenpox
Parasitic worms (extracellular)	*Ascaris* *Schistosoma*	Ascariasis Schistosomiasis

differentiate into effector plasma cells that secrete antibody back into that environment. T-cell receptors are specialized to detect antigens that have been generated inside the body's cells, and this is reflected in the effector actions of T cells. Some effector T cells directly kill cells infected with intracellular pathogens such as viruses, while others participate in responses against extracellular pathogens by interacting with B cells to help them make antibody.

Most of the other effector mechanisms that dispose of pathogens targeted by an adaptive immune response are essentially identical to those of innate immunity and involve cells such as macrophages and neutrophils and proteins such as complement. Indeed, it seems likely that the vertebrate adaptive immune response evolved by the late addition of specific recognition by clonally distributed receptors to innate defense mechanisms already existing in invertebrates. This is discussed in Chapter 16. We begin by outlining the effector actions of antibodies, which depend almost entirely on recruiting cells and molecules of the innate immune system.

1-18 Antibodies deal with extracellular forms of pathogens and their toxic products.

Antibodies are found in the fluid component of blood, or plasma, and in extracellular fluids. Because body fluids were once known as humors, immunity mediated by antibodies is known as **humoral immunity**.

As we saw in Fig. 1.13, antibodies are Y-shaped molecules whose arms form two identical antigen-binding sites. These are highly variable from one molecule to another, providing the diversity required for specific antigen recognition. The stem of the Y is far less variable. There are only five major forms of this constant region of an antibody, and these are known as the antibody **classes** or **isotypes**. The constant region determines an antibody's functional properties—how it will engage with the effector mechanisms that dispose of antigen once it is recognized—and each class carries out its particular function by engaging a distinct set of effector mechanisms. We describe the antibody classes and their actions in Chapters 4 and 9.

The first and most direct way in which antibodies can protect against pathogens or their products is by binding to them and thereby blocking their

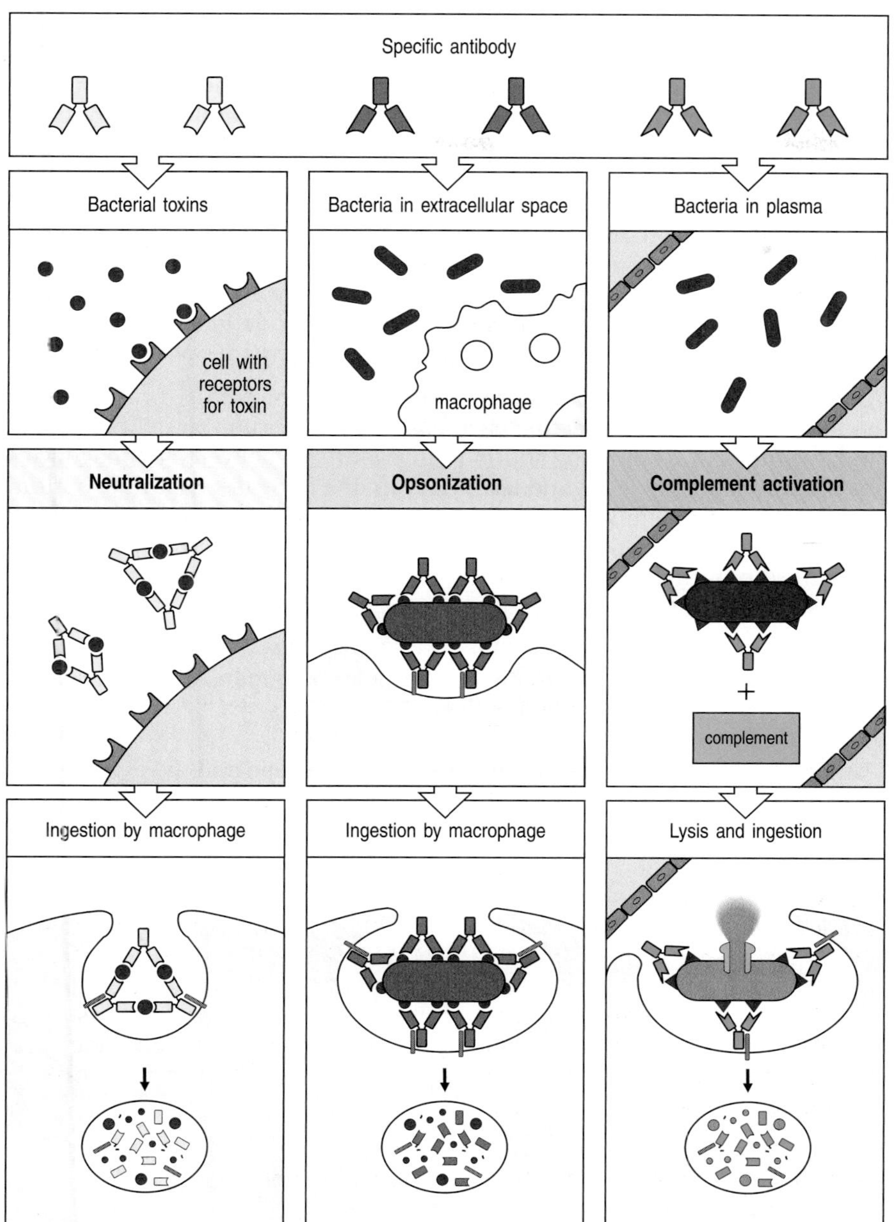

Fig. 1.26 Antibodies can participate in host defense in three main ways. The left panels show antibodies binding to and neutralizing a bacterial toxin, thus preventing it from interacting with host cells and causing pathology. Unbound toxin can react with receptors on the host cell, whereas the toxin:antibody complex cannot. Antibodies also neutralize complete virus particles and bacterial cells by binding and inactivating them. The antigen:antibody complex is eventually scavenged and degraded by macrophages. Antibodies coating an antigen render it recognizable as foreign by phagocytes (macrophages and neutrophils), which then ingest and destroy it; this is called opsonization. The middle panels show opsonization and phagocytosis of a bacterial cell. The right panels show activation of the complement system by antibodies coating a bacterial cell. Bound antibodies form a receptor for the first protein of the complement system, which eventually forms a protein complex on the surface of the bacterium that, in some cases, can kill the bacterium directly. More generally, complement coating favors the taking up and destroying of the bacterium by phagocytes. Thus, antibodies target pathogens and their toxic products for disposal by phagocytes.

access to cells that they might infect or destroy (Fig. 1.26, left panels). This is known as **neutralization** and is important for protection against viruses, which are prevented from entering cells and replicating, and against bacterial toxins.

For bacteria, however, binding by antibodies is not sufficient to stop their replication. In this case, the function of the antibody is to enable a phagocytic cell such as a macrophage or a neutrophil to ingest and destroy the bacterium. Many bacteria evade the innate immune system because they have an outer coat that is not recognized by the pattern recognition receptors of phagocytes. However, antigens in the coat can be recognized by antibodies, and phagocytes have receptors that bind the stems of the antibodies coating the bacterium, leading to phagocytosis (see Fig. 1.26, middle panels). The coating of pathogens and foreign particles in this way is known as **opsonization**.

The third function of antibodies is **complement activation**. Complement, which we discuss in detail in Chapter 2, is first activated in innate immunity by microbial surfaces, without the help of antibodies. But the constant regions of antibodies bound to bacterial surfaces form receptors for the first protein of the complement system, so that once antibodies are produced, complement activation increases. Complement components bound to the bacterial surface can directly destroy certain bacteria, and this is important in a few bacterial infections (see Fig. 1.26, right panels). The main function of complement, however, like that of antibodies, is to coat the pathogen surface and enable phagocytes to engulf and destroy bacteria that they would not otherwise recognize. Complement also enhances the bactericidal actions of phagocytes; indeed, it is so called because it 'complements' the activities of antibodies.

Antibodies of different classes are found in different compartments of the body and differ in the effector mechanisms they recruit, but all pathogens and free molecules bound by antibody are eventually delivered to phagocytes for ingestion, degradation, and removal from the body (see Fig. 1.26, bottom panels). The complement system and the phagocytes that antibodies recruit are not themselves antigen-specific; they depend upon antibody molecules to mark the particles as foreign. Producing antibodies is the sole effector function of B cells. T cells, by contrast, have a variety of effector actions.

1-19 T cells are needed to control intracellular pathogens and to activate B-cell responses to most antigens.

Pathogens are accessible to antibodies only in the blood and the extracellular spaces. However, some bacteria and parasites, and all viruses, replicate inside cells, where they cannot be detected by antibodies. The destruction of these

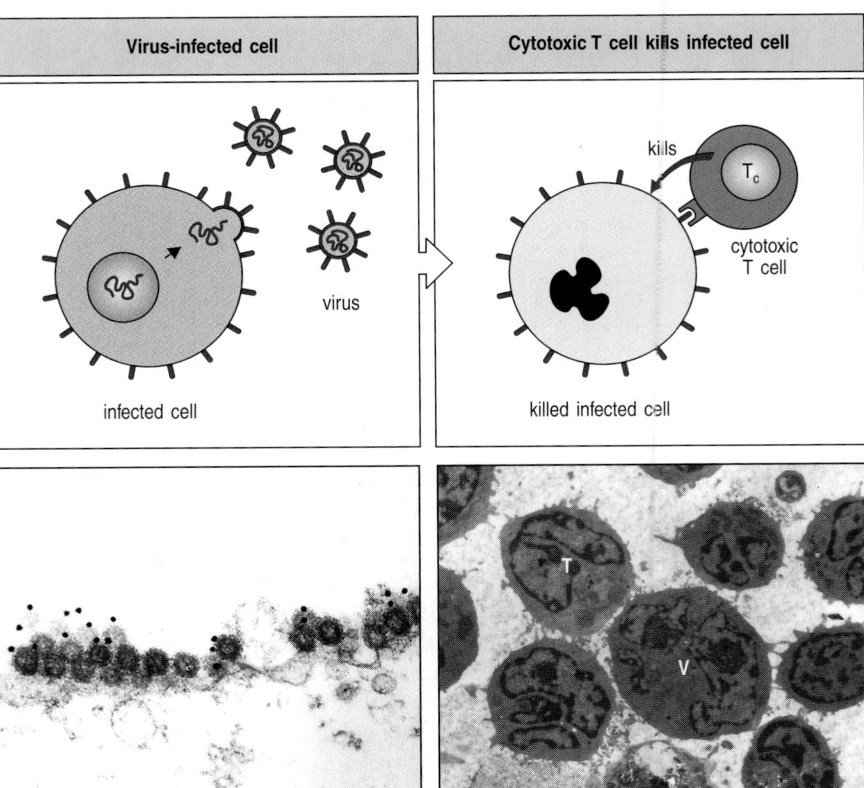

Fig. 1.27 Mechanism of host defense against intracellular infection by viruses. Cells infected by viruses are recognized by specialized T cells called cytotoxic T cells, which kill the infected cells directly. The killing mechanism involves the activation of enzymes known as caspases, which contain cysteine in their active site and cleave after aspartic acid. These in turn activate a cytosolic nuclease in the infected cell, which cleaves host and viral DNA. Panel a is a transmission electron micrograph showing the plasma membrane of a cultured CHO cell (the Chinese hamster ovary cell line) infected with influenza virus. Many virus particles can be seen budding from the cell surface. Some of these have been labeled with a monoclonal antibody that is specific for a viral protein and is coupled to gold particles, which appear as the solid black dots in the micrograph. Panel b is a transmission electron micrograph of a virus-infected cell (V) surrounded by cytotoxic T lymphocytes. Note the close apposition of the membranes of the virus-infected cell and the T cell (T) in the upper left corner of the micrograph, and the clustering of the cytoplasmic organelles in the T cell between its nucleus and the point of contact with the infected cell. Panel a courtesy of M. Bui and A. Helenius; panel b courtesy of N. Rooney.

invaders is the function of the T lymphocytes, which are responsible for the **cell-mediated immune responses** of adaptive immunity.

The action of cytotoxic T cells is the most direct. These effector T cells act against cells infected with viruses. Antigens derived from the virus multiplying inside the infected cell are displayed on the cell's surface, where they are recognized by the antigen receptors of cytotoxic T cells. These T cells can then control the infection by killing the infected cell before viral replication is complete and new viruses are released (Fig. 1.27).

From the end of their development in the thymus, T lymphocytes are composed of two main classes, one of which carries the cell-surface protein called **CD8** on its surface and the other bears a protein called **CD4**. These are not just random markers, but are important for a T cell's function, as they help to determine the interactions the T cell makes with other cells. Cytotoxic T cells carry CD8, while the class of T cells involved in activating the cells they recognize, rather than killing them, carry CD4.

CD8 T cells are destined to become cytotoxic T cells by the time they leave the thymus as naive lymphocytes. Naive CD4 T cells, in contrast, can differentiate into different types of effector T cells after their initial activation by antigen. The two major types of CD4 effector T cells are called **T_H1** and **T_H2** cells, although more have been described, as we shall see in Chapter 8. These two types are both involved in combating bacterial infections, but in quite different ways. T_H1 cells have a dual function. The first is to control certain intracellular bacterial infections. Some bacteria grow only in the intracellular membrane-bounded vesicles of macrophages; important examples are *Mycobacterium tuberculosis* and *M. leprae*, the pathogens that cause tuberculosis and leprosy, respectively. Bacteria phagocytosed by macrophages are usually destroyed in the lysosomes, which contain a variety of enzymes and antimicrobial substances. Mycobacteria and some other bacteria survive intracellularly because they prevent the vesicles they occupy from fusing with lysosomes (Fig. 1.28). These infections can be controlled by T_H1 cells that recognize bacterial antigens displayed on the

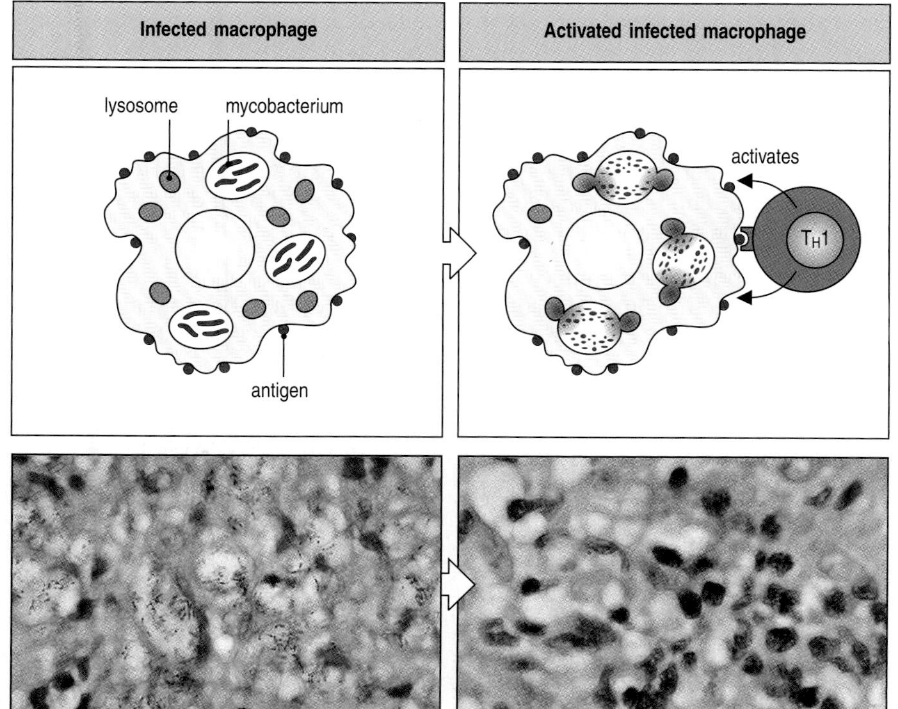

Fig. 1.28 Mechanism of host defense against intracellular infection by mycobacteria. Mycobacteria are engulfed by macrophages but resist being destroyed by preventing the intracellular vesicles in which they reside from fusing with lysosomes containing bactericidal agents. Thus the bacteria are protected from being killed. In resting macrophages, mycobacteria persist and replicate in these vesicles. When the phagocyte is recognized and activated by a T_H1 cell, however, the phagocytic vesicles fuse with lysosomes, and the bacteria can be killed. Macrophage activation is controlled by T_H1 cells, both to avoid tissue damage and to save energy. The light micrographs (bottom row) show resting (left) and activated (right) macrophages infected with mycobacteria. The cells have been stained with an acid-fast red dye to reveal mycobacteria. These are prominent as red-staining rods in the resting macrophages but have been eliminated from the activated macrophages. Photographs courtesy of G. Kaplan.

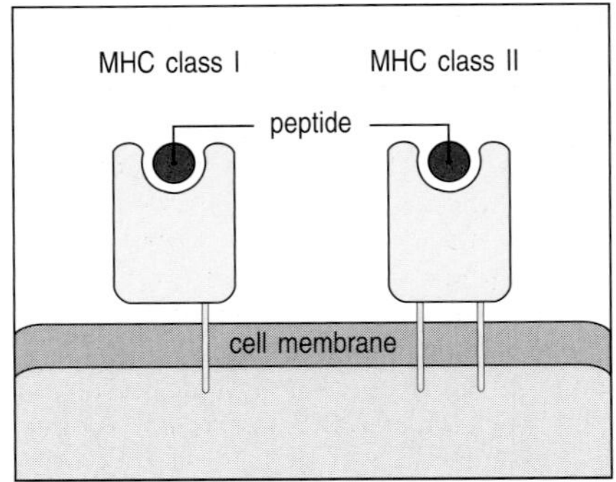

Fig. 1.29 MHC molecules on the cell surface display peptide fragments of antigens. MHC molecules are membrane proteins whose outer extracellular domains form a cleft in which a peptide fragment is bound. These fragments are derived from proteins degraded inside the cell, including both self and foreign protein antigens. The peptides are bound by the newly synthesized MHC molecule before it reaches the cell surface. There are two kinds of MHC molecules—MHC class I and MHC class II—with related but distinct structures and functions. Although not shown here for simplicity, both MHC class I and MHC class II molecules are trimers of two protein chains and the bound self or nonself peptide.

macrophage surface. T_H1 cells activate the infected macrophages, inducing the fusion of their lysosomes with the vesicles containing the bacteria and stimulating the macrophage's antibacterial mechanisms (Fig. 1.28). The second role of T_H1 cells is as helper T cells to stimulate the production of antibodies by producing co-stimulatory signals and interacting with B lymphocytes. We shall see in Chapter 9, when we discuss the humoral immune response in detail, that only a few antigens with special properties can activate naive B lymphocytes on their own; an accompanying co-stimulatory signal from T cells is usually required (see Fig. 1.21).

Whereas T_H1 cells have a dual role, helper T_H2 cells are entirely dedicated to the activation of naive B cells to produce antibody. The term 'helper T cell' is sometimes used by researchers to describe all CD4 T cells. It was originally coined, however, to describe T cells that 'help' B cells produce antibody, before the existence of two subtypes of CD4 T cells was recognized. When the macrophage-activating role of CD4 T cells was discovered, the designation 'helper' was extended to cover these as well (hence the H in T_H1). We consider this blanket usage confusing, and throughout this book we will only use the term 'helper T cell' in connection with the activation of B cells for antibody production, whether by T_H1 or T_H2 cells.

Naive T lymphocytes recognize their corresponding antigens on specialized antigen-presenting cells, which can also activate them. Effector T cells similarly recognize peptide antigens bound to MHC molecules, but in this case the T cell is already activated and thus does not need co-stimulatory signals.

1-20 CD4 and CD8 T cells recognize peptides bound to two different classes of MHC molecules.

The different types of effector T cells must be directed to act against the appropriate target cells. Antigen recognition is obviously crucial, but correct target recognition is also ensured by additional interactions between the CD8 and CD4 molecules on the T cells and the MHC molecules on the target cell.

As we saw in Section 1-13, T cells detect peptides derived from foreign antigens after antigens are degraded within cells, their peptide fragments are captured by MHC molecules, and this complex is displayed at the cell surface (see Fig. 1.16). There are two main types of MHC molecules, called **MHC class I** and **MHC class II**. They have slightly different structures but both have an elongated cleft in the extracellular surface of the molecule, in which a single peptide is trapped during the synthesis and assembly of the MHC molecule inside the cell. The MHC molecule bearing its cargo of peptide is transported to the cell surface, where it displays the peptide to T cells (Fig. 1.29).

The most important differences between the two classes of MHC molecule lie not in their structure but in the source of the peptides that they trap and carry to the cell surface. **MHC class I molecules** collect peptides derived from proteins

Fig. 1.30 MHC class I molecules present antigen derived from proteins in the cytosol. In cells infected with viruses, viral proteins are synthesized in the cytosol. Peptide fragments of viral proteins are transported into the endoplasmic reticulum (ER), where they are bound by MHC class I molecules, which then deliver the peptides to the cell surface.

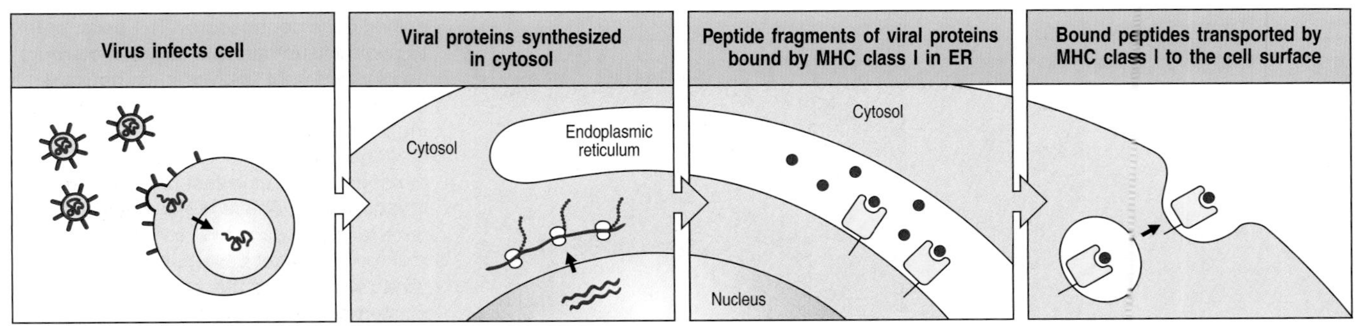

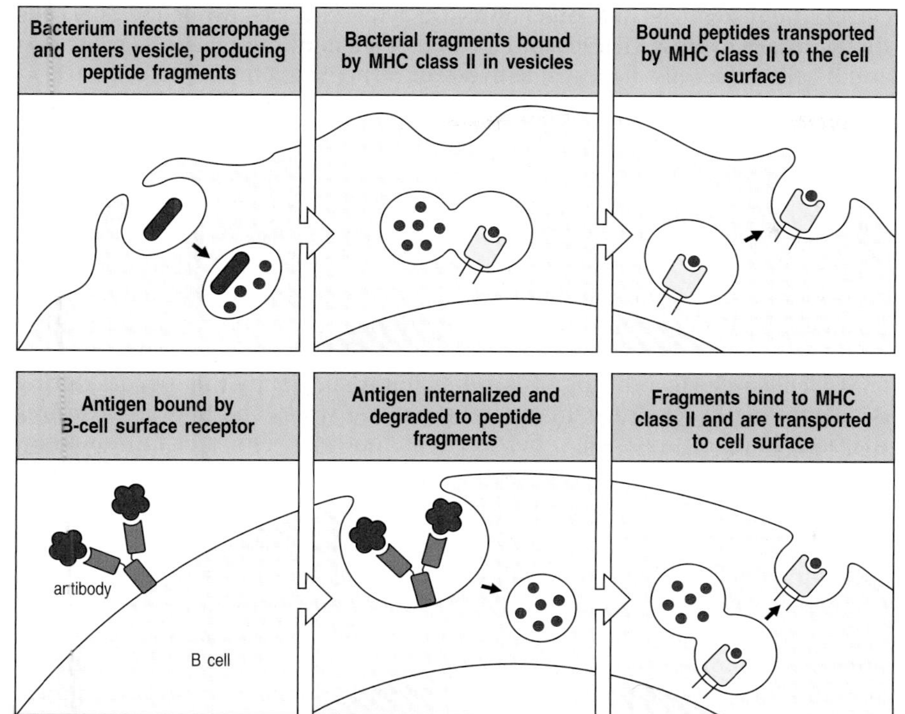
Bacterium infects macrophage and enters vesicle, producing peptide fragments
Bacterial fragments bound by MHC class II in vesicles
Bound peptides transported by MHC class II to the cell surface
Antigen bound by B-cell surface receptor
Antigen internalized and degraded to peptide fragments
Fragments bind to MHC class II and are transported to cell surface
antibody
B cell

Fig. 1.31 MHC class II molecules present antigen originating in intracellular vesicles. Some bacteria infect cells and grow in intracellular vesicles. Peptides derived from such bacteria are bound by MHC class II molecules and transported to the cell surface (top row). MHC class II molecules also bind and transport peptides derived from antigen that has been bound and internalized by B-cell antigen receptor-mediated uptake into intracellular vesicles (bottom row).

synthesized in the cytosol and are thus able to display fragments of viral proteins on the cell surface (Fig. 1.30). **MHC class II molecules** bind peptides derived from proteins in intracellular vesicles, and thus display peptides derived from pathogens living in macrophage vesicles or internalized by phagocytic cells and B cells (Fig. 1.31). We shall see in Chapter 5 exactly how peptides from these different sources are made available to the two types of MHC molecule.

Having reached the cell surface with their peptide cargo, the two classes of MHC molecule are recognized by different functional classes of T cell. This occurs because the CD8 molecule binds preferentially to MHC class I molecules, whereas CD4 binds preferentially to MCH class II molecules. Thus, MHC class I molecules bearing viral peptides are recognized by CD8-bearing cytotoxic T cells, which then kill the infected cell (Fig. 1.32); MHC class II molecules bearing peptides derived from pathogens taken up into vesicles are recognized by CD4-bearing T cells (Fig. 1.33). CD4 and CD8 are thus known as **co-receptors**, as they are inextricably involved in signaling to the T cell that the receptor has bound the correct antigen. Useful interactions are further ensured by the fact that all cells express MHC class I molecules, and thus any virus-infected

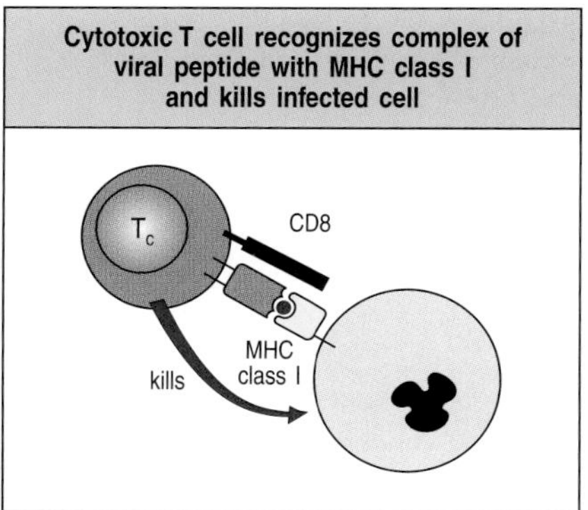
Cytotoxic T cell recognizes complex of viral peptide with MHC class I and kills infected cell
Tc
CD8
MHC class I
kills

Fig. 1.32 Cytotoxic CD8 T cells recognize antigen presented by MHC class I molecules and kill the cell. The peptide:MHC class I complex on virus-infected cells is detected by antigen-specific cytotoxic T cells. Cytotoxic T cells are preprogrammed to kill the cells they recognize.

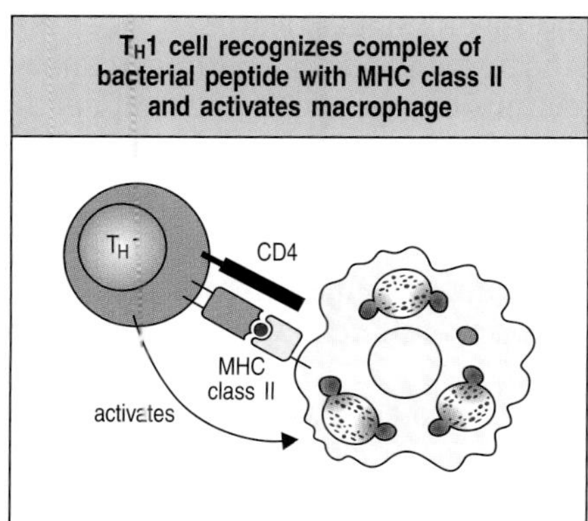
TH1 cell recognizes complex of bacterial peptide with MHC class II and activates macrophage
TH1
CD4
MHC class II
activates

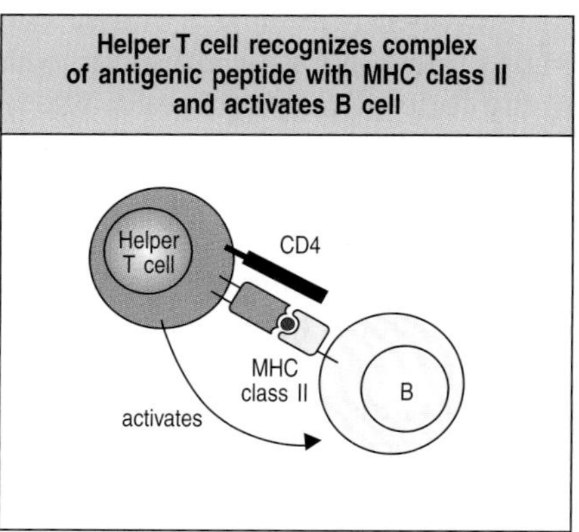
Helper T cell recognizes complex of antigenic peptide with MHC class II and activates B cell
Helper T cell
CD4
MHC class II
B
activates

Fig. 1.33 CD4 T cells recognize antigen presented by MHC class II molecules. On recognition of their specific antigen on infected macrophages, T_H1 cells activate the macrophage, leading to the destruction of the intracellular bacteria (left panel). When T_H2 or T_H1 helper T cells recognize antigen on B cells, they activate these cells to proliferate and differentiate into antibody-producing plasma cells (right panel).

cell can be recognized and killed by a CD8 cytotoxic T cell, while the only cells that normally express MHC class II molecules are dendritic cells, macrophages, and B cells—the cells that must activate, or be activated by, CD4 T cells.

Because the T-cell receptor is specific for a combination of peptide and MHC molecule, any given T-cell receptor will recognize either an MHC class I molecule or an MHC class II molecule. To be useful, T lymphocytes bearing antigen receptors that recognize MHC class I must also express CD8 co-receptors, whereas T lymphocytes bearing receptors specific for MHC class II must express CD4. The matching of a T-cell receptor with a co-receptor of the appropriate type occurs during lymphocyte development, and naive T cells emerge from the central lymphocyte organs bearing the correct combination of receptors and co-receptors. The maturation of T cells into either CD8 or CD4 T cells reflects a testing of T-cell receptor specificity that occurs during development. Exactly how this selective process works and how it maximizes the usefulness of the T-cell repertoire is a central question in immunology, and is a major topic of Chapter 7.

On recognizing their targets, the various types of effector T cells are stimulated to release different sets of effector molecules. These can directly affect their target cells or help to recruit other effector cells in ways we discuss in Chapter 8. These effector molecules include many cytokines, which have a crucial role in the clonal expansion of lymphocytes as well as in innate immune responses and in the effector actions of most cells of the immune system. Thus, understanding the actions of cytokines is central to understanding the various behaviors of the immune system. The actions of all the known cytokines are summarized in Appendix III, some are introduced in Chapter 2, and the T-cell-derived cytokines are discussed in Chapter 8.

1-21 Defects in the immune system result in increased susceptibility to infection.

We tend to take for granted the ability of our immune systems to free our bodies of infection and prevent its recurrence. In some people, however, parts of the immune system fail. In the most severe of these **immunodeficiency diseases**, adaptive immunity is completely absent, and death occurs in infancy from overwhelming infection unless heroic measures are taken. Other less catastrophic failures lead to recurrent infections with particular types of pathogen, depending on the particular deficiency. Much has been learned about the functions of the different components of the human immune system through the study of these immunodeficiencies, many of which are caused by inherited genetic defects.

More than 25 years ago, a devastating form of immunodeficiency appeared, the **acquired immune deficiency syndrome**, or **AIDS**, which is caused by an infectious agent, the human immunodeficiency viruses HIV-1 and HIV-2. This disease destroys T cells, dendritic cells, and macrophages bearing CD4, leading to infections caused by intracellular bacteria and other pathogens normally controlled by such cells. These infections are the major cause of death from this increasingly prevalent immunodeficiency disease, which is discussed fully in Chapter 12 together with the inherited immunodeficiencies.

1-22 Understanding adaptive immune responses is important for the control of allergies, autoimmune disease, and organ graft rejection.

The main function of our immune system is to protect the human host from infectious agents. However, many medically important diseases are associated with a normal immune response directed against an inappropriate antigen, often in the absence of infectious disease. Immune responses directed at non-infectious antigens occur in **allergy**, in which the antigen is an innocuous foreign substance, in **autoimmune disease**, in which the response is to a self antigen, and in **graft rejection**, in which the antigen is borne by a transplanted foreign cell. The major antigens provoking graft rejection are, in fact, the MHC molecules, as each of these is present in many different versions in the human population—that is, they are highly **polymorphic**—and most unrelated people differ in the set of MHC molecules they express. The MHC was originally recognized in mice as a gene locus, the **H2 locus**, that controlled the acceptance or rejection of transplanted tissues, while the human MHC molecules were first discovered after attempts to use skin grafts from donors to repair badly burned pilots and bomb victims during the Second World War. The patients rejected the grafts, which were recognized by their immune systems as being 'foreign'. What we call a successful immune response or a failure, and whether the response is considered harmful or beneficial to the host, depends not on the response itself but rather on the nature of the antigen and the circumstances in which the response occurs (Fig. 1.34).

Allergic diseases, which include asthma, are an increasingly common cause of disability in the developed world. Autoimmunity is also now recognized as the cause of many important diseases. An autoimmune response directed against pancreatic β cells is the leading cause of diabetes in the young. In allergies and autoimmune diseases, the powerful protective mechanisms of the adaptive immune response cause serious damage to the patient.

Immune responses to harmless antigens, to body tissues, or to organ grafts are, like all other immune responses, highly specific. At present, the usual way to treat these responses is with immunosuppressive drugs, which inhibit all immune responses, desirable and undesirable. If it were possible to suppress only those lymphocyte clones responsible for the unwanted response, the disease could be cured or the grafted organ protected without impeding protective immune responses. There is hope that this dream of antigen-specific immunoregulation to control unwanted immune responses could become a reality, as antigen-specific suppression of immune responses can be induced experimentally, although the molecular basis of this suppression is not fully understood. We shall discuss the present state of understanding of allergies, autoimmune disease, graft rejection, and immunosuppressive drugs in Chapters 13–15, and we shall see in Chapter 14 how the mechanisms of immune regulation are beginning to emerge from a better understanding of the functional subsets of lymphocytes and the cytokines that control them.

Antigen	Effect of response to antigen	
	Normal response	Deficient response
Infectious agent	Protective immunity	Recurrent infection
Innocuous substance	Allergy	No response
Grafted organ	Rejection	Acceptance
Self organ	Autoimmunity	Self tolerance
Tumor	Tumor immunity	Cancer

Fig. 1.34 Immune responses can be beneficial or harmful, depending on the nature of the antigen. Beneficial responses are shown in white, harmful responses in red shaded boxes. Where the response is beneficial, its absence is harmful.

1-23 Vaccination is the most effective means of controlling infectious diseases.

Although the specific suppression of immune responses must await advances in basic research on immune regulation and its application, the deliberate stimulation of an immune response by immunization, or vaccination, has achieved many successes in the two centuries since Jenner's pioneering experiment.

Mass immunization programs have led to the virtual eradication of several diseases that used to be associated with significant morbidity (illness) and mortality (Fig. 1.35). Immunization is considered so safe and so important that most states in the United States require children to be immunized against up to seven common childhood diseases. Impressive as these accomplishments are, there are still many diseases for which we lack effective vaccines. And even where vaccines for diseases such as measles or polio can be used effectively in developed countries, technical and economic problems can prevent their widespread use in developing countries, where mortality from these diseases is still high. The tools of modern immunology and molecular biology are being applied to develop new vaccines and improve old ones, and we discuss these advances in Chapter 15. The prospect of controlling these important diseases is tremendously exciting.

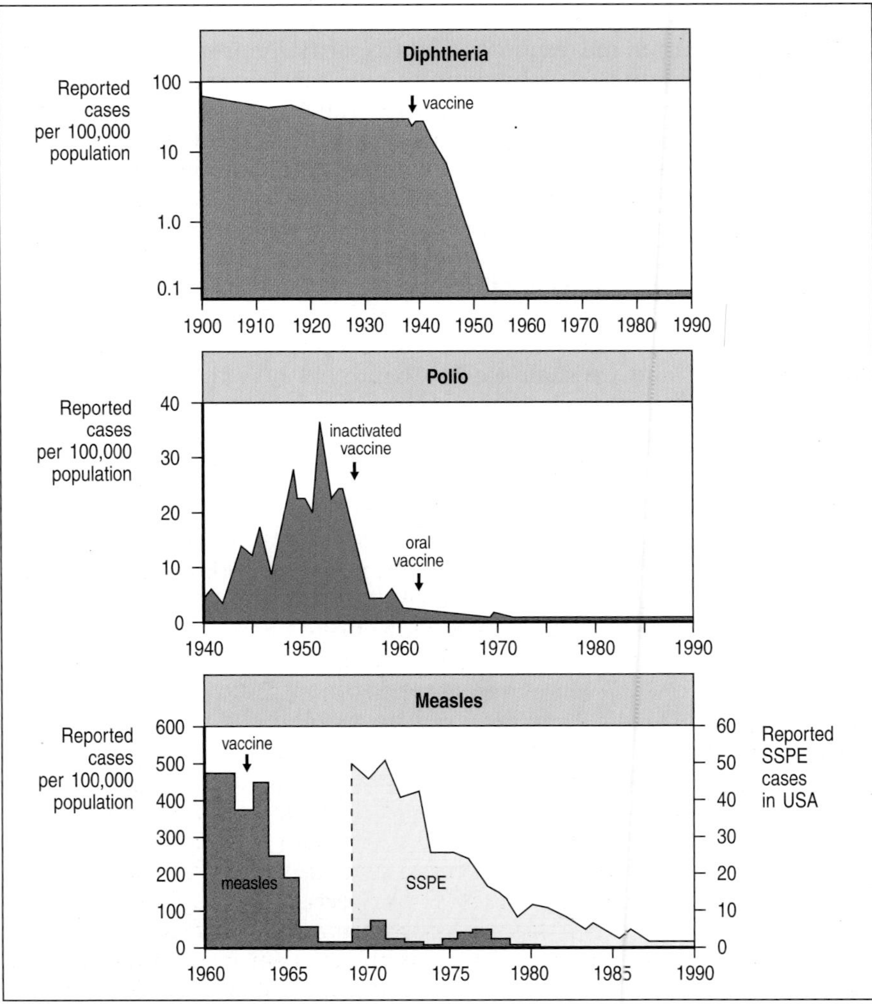

Fig. 1.35 Successful vaccination campaigns. Diphtheria, polio, and measles and its consequences have been virtually eliminated in the United States, as shown in these three graphs. SSPE stands for subacute sclerosing panencephalitis, a brain disease that is a late consequence of measles infection in a few patients. When measles was prevented, SSPE disappeared 15–20 years later. However, as these diseases have not been eradicated worldwide, immunization must be maintained in a very high percentage of the population to prevent their reappearance.

The guarantee of good health is a critical step toward population control and economic development. At a cost of pennies per person, great hardship and suffering can be alleviated.

Many serious pathogens have resisted efforts to develop vaccines against them, often because they can evade or subvert the protective mechanisms of an adaptive immune response. We examine some of the evasive strategies used by successful pathogens in Chapter 12. The conquest of many of the world's leading diseases, including malaria and diarrheal diseases (the leading killers of children) as well as the more recent threat from AIDS, depends on a better understanding of the pathogens that cause them and their interactions with the cells of the immune system.

Summary.

Lymphocytes have two distinct recognition systems specialized for the detection of extracellular and intracellular pathogens. B cells have cell-surface immunoglobulin molecules as receptors for antigen and, upon activation, secrete the immunoglobulin as soluble antibody that provides a defense against pathogens in the extracellular spaces of the body. T cells have receptors that recognize peptide fragments of intracellular pathogens transported to the cell surface by the glycoproteins of the MHC. Two classes of MHC molecules transport peptides from different intracellular compartments to the cell surface to present them to distinct types of effector T cells: cytotoxic CD8 T cells that kill infected target cells, and CD4 T cells that mainly activate macrophages and B cells. Thus, T cells are crucially important for both the humoral and cell-mediated responses of adaptive immunity. The adaptive immune response seems to have engrafted specific antigen recognition by highly diversified receptors onto innate defense systems, which have a central role in the effector actions of both B and T lymphocytes. The vital role of adaptive immunity in fighting infection is illustrated by the immunodeficiency diseases and the problems caused by pathogens that succeed in evading or subverting an adaptive immune response. The antigen-specific suppression of adaptive immune responses is the goal of treatment for important human diseases involving inappropriate activation of lymphocytes, whereas the specific stimulation of an adaptive immune response is the basis of successful vaccination.

Summary to Chapter 1.

The immune system defends the host against infection. Innate immunity serves as a first line of defense but lacks the ability to recognize certain pathogens and to provide the specific protective immunity that prevents reinfection. Adaptive immunity is based on clonal selection from a repertoire of lymphocytes bearing highly diverse antigen-specific receptors that enable the immune system to recognize any foreign antigen. In the adaptive immune response, antigen-specific lymphocytes proliferate and differentiate into clones of effector lymphocytes that eliminate the pathogen. Host defense requires different recognition systems and a wide variety of effector mechanisms to seek out and destroy the wide variety of pathogens in their various habitats within the body and at its external and internal surfaces. Not only can the adaptive immune response eliminate a pathogen but, in the process, it also generates increased numbers of differentiated memory lymphocytes through clonal selection, and this allows a more rapid and effective response upon reinfection. The regulation of immune responses, whether to suppress them when unwanted or to stimulate them in the prevention of infectious disease, is the major medical goal of research in immunology.

General references.

Historical background

Burnet, F.M.: *The Clonal Selection Theory of Acquired Immunity.* London, Cambridge University Press, 1959.

Gowans, J.L.: **The lymphocyte—a disgraceful gap in medical knowledge.** *Immunol. Today* 1996, **17**:288–291.

Landsteiner, K.: *The Specificity of Serological Reactions*, 3rd ed. Boston, Harvard University Press, 1964.

Metchnikoff, E.: *Immunity in the Infectious Diseases*, 1st ed. New York, Macmillan Press, 1905.

Silverstein, A.M.: *History of Immunology*, 1st ed. London, Academic Press, 1989.

Biological background

Alberts, B., Johnson, A., Lewis, J., Raff, M., Roberts, K. and Walter, P.: *Molecular Biology of the Cell*, 5th ed. New York, Garland Publishing, 2007.

Berg, J.M., Stryer, L. and Tymoczko, J.L.: *Biochemistry*, 5th ed. New York, W.H. Freeman, 2002.

Kaufmann, S.E., Sher, A. and Ahmed, R. (Eds): *Immunology of Infectious Diseases.* Washington, DC: ASM Press, 2001.

Mims, C., Nash, A. and Stephen, J.: *Mims' Pathogenesis of Infectious Disease*, 5th ed. London, Academic Press, 2001.

Geha, R.S. and Rosen, F.S.: *Case Studies in Immunology: A Clinical Companion*, 5th ed. New York, Garland Publishing, 2007.

Ryan, K.J. (ed): *Medical Microbiology*, 3rd ed. East Norwalk, CT, Appleton-Lange, 1994.

Lodish, H., Berk, A., Kaiser, C.A., Krieger, M., Scott, M.P., Bretscher, A., Ploegh, H., Matsudaira, P.: *Molecular Cell Biology*, 6th ed. New York, W.H. Freeman, 2008.

Primary journals devoted solely or primarily to immunology

Autoimmunity
Clinical and Experimental Immunology
Comparative and Developmental Immunology
European Journal of Immunology
Immunity
Immunogenetics
Immunology
Infection and Immunity
International Immunology
International Journal of Immunogenetics
Journal of Autoimmunity
Journal of Experimental Medicine
Journal of Immunology
Nature Immunology
Regional Immunology
Thymus

Primary journals with frequent papers in immunology

Cell
Current Biology
EMBO Journal
Journal of Biological Chemistry
Journal of Cell Biology
Journal of Clinical Investigation
Molecular Cell Biology
Nature
Nature Cell Biology
Nature Medicine
Proceedings of the National Academy of Sciences, USA
Science

Review journals in immunology

Advances in Immunology
Annual Reviews in Immunology
Contemporary Topics in Microbiology and Immunology
Current Opinion in Immunology
Immunogenetics Reviews
Immunological Reviews
Immunology Today
Nature Reviews Immunology
Research in Immunology
Seminars in Immunology
The Immunologist

Advanced textbooks in immunology, compendia, etc.

Lachmann, P.J., Peters, D.K., Rosen, F.S., and Walport, M.J. (eds): *Clinical Aspects of Immunology*, 5th ed. Oxford, Blackwell Scientific Publications, 1993.

Mak, T.W. and Simard, J.J.L.: *Handbook of Immune Response Genes.* New York, Plenum Press, 1998.

Paul, W.E. (ed): *Fundamental Immunology*, 5th ed. New York, Lippincott Williams & Wilkins, 2003.

Roitt, I.M. and Delves, P.J. (eds): *Encyclopedia of Immunology*, 2nd ed (4 vols). London/San Diego, Academic Press, 1998.

Innate Immunity 2

Throughout most of this book we will examine the ways in which the adaptive immune response protects the host from microorganisms that otherwise would cause disease. In this chapter, however, we will examine the role of the innate, nonadaptive defenses that form the earliest barriers to infection. The microorganisms that are encountered daily in the life of a healthy individual only occasionally cause perceptible disease. Most are detected and destroyed within minutes or hours by defense mechanisms that do not rely on the clonal expansion of antigen-specific lymphocytes (see Section 1-9) and thus do not require a prolonged period of induction: these are the mechanisms of **innate immunity**.

The time course and different phases of an encounter with a new pathogen are summarized in Fig. 2.1. Some innate immune mechanisms start acting immediately on encounter with infectious agents. Others are activated and amplified in the presence of infection and then return to baseline levels after the infection is finished. Innate immune mechanisms do not generate long-term protective immunological memory. Only if an infectious organism can breach these lines of defense will an adaptive immune response ensue, with the generation of antigen-specific effector cells that target the specific pathogen, and memory cells that provide long-lasting immunity against reinfection with the same microorganism. The power of adaptive immune responses is due to their antigen specificity, which we will be studying in the following chapters. However, they harness, and also depend upon, many of the effector mechanisms used by the innate immune system to remove pathogens, which we will describe in this chapter.

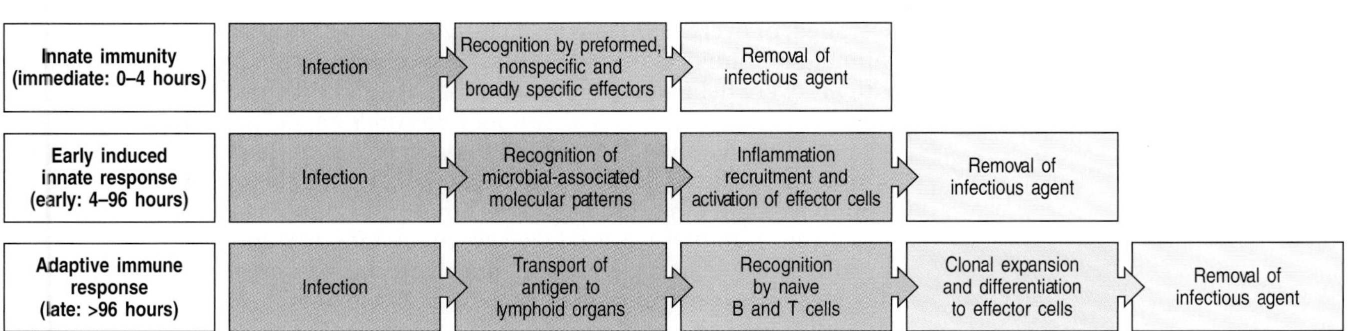

Fig. 2.1 The response to an initial infection occurs in three phases. These are the innate phase, the early induced innate response, and the adaptive immune response. The first two phases rely on the recognition of pathogens by germline-encoded receptors of the innate immune system, whereas adaptive immunity uses variable antigen-specific receptors that are produced as a result of gene segment rearrangements. Adaptive immunity occurs late, because the rare B and T cells specific for the invading pathogen must first undergo clonal expansion before they differentiate into effector cells that can clear the infection. The effector mechanisms that remove the infectious agent are similar or identical in each phase.

Whereas the adaptive immune system uses a large repertoire of receptors encoded by rearranging gene segments to recognize a huge variety of antigens (see Section 1-12), innate immunity depends upon germline-encoded receptors to recognize features that are common to many pathogens. In fact, the mechanisms of innate immunity discriminate very effectively between host cells and pathogens, and this ability to discriminate between self and nonself, and to recognize broad classes of pathogens, contributes to the induction of an appropriate adaptive immune response.

In the first part of the chapter we will consider the fixed defenses of the body: the epithelia that line the internal and external surfaces of the body, and the phagocytes that lie beneath all epithelial surfaces and that engulf and digest invading microorganisms. As well as killing microorganisms directly, these phagocytes induce the next phase of the innate immune response, inducing an inflammatory response that recruits new phagocytic cells and circulating effector molecules to the site of infection. Second, we will take a closer look at the ancient system of pattern-recognition receptors used by the phagocytic cells of the innate immune system to identify pathogens and distinguish them from self antigens. We shall see how, as well as directing the immediate destruction of pathogens, stimulation of some of these receptors on macrophages and dendritic cells leads to their becoming cells that can effectively present antigen to T lymphocytes, thus initiating an adaptive immune response. The third part of the chapter is devoted to a system of plasma proteins known as the complement system. This important element of so-called humoral innate immunity interacts with microorganisms to promote their removal by phagocytic cells. In the last part of the chapter we shall describe how the cytokines and chemokines produced by activated phagocytes and dendritic cells induce the later phases of the innate immune response, such as the so-called acute-phase response. We shall also meet another cell of the innate immune system, the natural killer cell (NK cell), which contributes to innate host defenses against viruses and other intracellular pathogens. In this phase, the first steps towards initiating an adaptive immune response take place, so that if the infection is not cleared by the innate responses, a full immune response will ensue.

The front line of host defense.

Microorganisms that cause disease in humans and animals enter the body at different sites and produce disease symptoms by a variety of mechanisms. Many different infectious agents can cause disease and damage to tissues, or pathology, and are referred to as **pathogenic microorganisms** or **pathogens**. In vertebrates, microbial invasion is initially countered by innate defenses that preexist in all individuals and begin to act within minutes of encounter with the infectious agent. Only when the innate host defenses are bypassed, evaded, or overwhelmed is an adaptive immune response required. Although innate immunity is obviously sufficient to prevent the body from being routinely overwhelmed by the vast number of microorganisms that live on and in it, pathogens, almost by definition, are microorganisms that have evolved ways of overcoming the body's innate defenses more effectively than other microorganisms. Once they have gained a hold, they require the concerted efforts of both innate and adaptive immune responses to clear them from the body. Even in these cases, however, the innate immune system usually performs a valuable delaying function, keeping pathogen numbers in check while the adaptive immune system gears up for action. In the first part of this chapter we will describe briefly the different types of pathogens and their invasive strategies, and then examine the innate defenses that, in most cases, prevent microorganisms from establishing an infection. We will examine the defense functions of the epithelial surfaces of the body, the role of antimicrobial peptides and proteins, and the defense of body tissues by phagocytic cells—the macrophages and neutrophils—which bind to and ingest invading microorganisms.

2-1 Infectious diseases are caused by diverse living agents that replicate in their hosts.

The agents that cause disease fall into five groups: viruses, bacteria, fungi, protozoa, and helminths (worms). Protozoa and worms are usually grouped together as parasites, and are the subject of the discipline of parasitology, whereas viruses, bacteria, and fungi are the subject of microbiology. In Fig. 2.2, the classes of microorganisms and parasites that cause disease are listed, with typical examples of each. The characteristic features of each pathogen are its mode of transmission, its mechanism of replication, its mechanism of **pathogenesis**—the means by which it causes disease—and the response it elicits from the host. The distinct habitats and life cycles of different pathogens mean that a range of different innate and adaptive immune mechanisms have to be deployed for their destruction.

Infectious agents can grow in all body compartments, as shown schematically in Fig. 2.3. We saw in Chapter 1 that two major compartments can be defined—extracellular and intracellular. Both innate and adaptive immune responses have different ways of dealing with pathogens found in these two compartments. Many bacterial pathogens live and replicate in extracellular spaces, either within tissues or on the surface of the epithelia that line body cavities. Extracellular bacteria are usually susceptible to killing by phagocytes, an important arm of the innate immune system, but some pathogens, such as *Staphylococcus* and *Streptococcus* species, are protected by a polysaccharide capsule that resists engulfment. This can be overcome to some extent by the help of another component of innate immunity—complement—which renders the bacteria more susceptible to phagocytosis. In the adaptive immune response, bacteria are rendered more susceptible to phagocytosis by a combination of antibodies and complement.

Obligate intracellular pathogens, such as all viruses, must invade host cells to replicate, whereas facultative intracellular pathogens, such as mycobacteria, can replicate either intracellularly or outside the cell. Intracellular pathogens must either be prevented from entering cells or be detected and eliminated once they have done so. They can be subdivided further into those that replicate free in the cell, such as viruses and certain bacteria (for example, *Chlamydia*, *Rickettsia*, and *Listeria*), and those, such as the mycobacteria, that replicate inside intracellular vesicles. Infectious agents that live intracellularly frequently cause disease by damaging or killing the cells that house them. The innate immune system has two general means of defense against this type of pathogen. Phagocytes may take up the pathogen before it enters cells, while NK cells can directly recognize and kill cells infected with some intracellular pathogens. NK cells are instrumental in keeping some viral infections in check until an adaptive immune response has been generated, after which cytotoxic T cells are able to take over the role of killing virus-infected cells. Pathogens that live inside macrophage vesicles may become susceptible to killing after activation of the macrophage as a result of NK cell or T cell actions (see Fig. 2.3).

Once pathogens have overcome the defenses of innate immunity, they grow and replicate in the body, causing markedly different diseases that reflect the diverse ways in which they damage tissues (Fig. 2.4). Many of the most

Fig. 2.2 A variety of microorganisms can cause disease. Pathogenic organisms are of five main types: viruses, bacteria, fungi, protozoa, and worms. Some well-known pathogens in each group are listed.

Some common causes of disease in humans			
Viruses	DNA viruses	Adenoviruses	Human adenoviruses (e.g., types 3, 4, and 7)
		Herpesviruses	Herpes simplex, varicella zoster, Epstein–Barr virus, cytomegalovirus, HHV-8
		Poxviruses	Variola, vaccinia virus
		Parvoviruses	Human parvovirus
		Papovaviruses	Papilloma virus
		Hepadnaviruses	Hepatitis B virus
	RNA viruses	Orthomyxoviruses	Influenza virus
		Paramyxoviruses	Mumps, measles, respiratory syncytial virus
		Coronaviruses	Cold viruses, SARS
		Picornaviruses	Polio, coxsackie, hepatitis A, rhinovirus
		Reoviruses	Rotavirus, reovirus
		Togaviruses	Rubella, arthropod-borne encephalitis
		Flaviviruses	Arthropod-borne viruses, (yellow fever, dengue fever)
		Arenaviruses	Lymphocytic choriomeningitis, Lassa fever
		Rhabdoviruses	Rabies
		Retroviruses	Human T-cell leukemia virus, HIV
Bacteria	Gram +ve cocci	Staphylococci	*Staphylococcus aureus*
		Streptococci	*Streptococcus pneumoniae, Strep. pyogenes*
	Gram –ve cocci	Neisseriae	*Neisseria gonorrhoeae, N. meningitidis*
	Gram +ve bacilli		*Corynebacterium diphtheriae, Bacillus anthracis, Listeria monocytogenes*
	Gram –ve bacilli		*Salmonella typhi, Shigella flexneri, Campylobacter jejuni, Vibrio cholerae, Yersinia pestis, Pseudomonas aeruginosa, Brucella melitensis, Haemophilus influenzae, Legionella pneumophilus, Bordetella pertussis*
	Firmicutes	Clostridia	*Clostridium tetani, C. botulinum, C. perfringens*
	Spirochaetes	Spirochetes	*Treponema pallidum, Borrelia burgdorferi, Leptospira interrogans*
	Actinobacteria	Mycobacteria	*Mycobacterium tuberculosis, M. leprae, M. avium*
	Protobacteria	Rickettsias	*Rickettsia prowazekii*
	Chlamydiae	Chlamydias	*Chlamydia trachomatis*
	Mollicutes	Mycoplasmas	*Mycoplasma pneumoniae*
Fungi	Ascomycetes		*Candida albicans, Cryptococcus neoformans, Aspergillus fumigatus, Histoplasma capsulatum, Coccidioides immitis, Pneumocystis carinii*
Protozoa			*Entamoeba histolytica, Giardia intestinalis, Leishmania donovani, Plasmodium falciparum, Trypanosoma brucei, Toxoplasma gondii, Cryptosporidium parvum*
Worms	Nematodes	Intestinal	*Trichuris trichura, Trichinella spiralis, Enterobius vermicularis, Ascaris lumbricoides, Ancylostoma duodenale, Strongyloides stercoralis*
		Tissues	*Onchocerca volvulus, Loa loa, Dracuncula medinensis*
	Flukes	Blood, liver	*Schistosoma mansoni, Clonorchis sinensis*

	Extracellular		Intracellular	
	Interstitial spaces, blood, lymph	**Epithelial surfaces**	**Cytoplasmic**	**Vesicular**
Site of infection		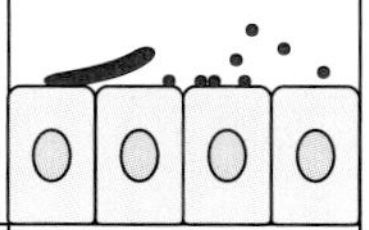	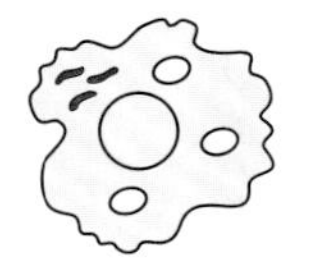	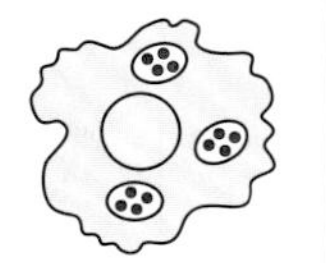
Organisms	Viruses Bacteria Protozoa Fungi Worms	*Neisseria gonorrhoeae* *Mycoplasma* spp. *Streptococcus pneumoniae* *Vibrio cholerae* *Escherichia coli* *Helicobacter pylori* *Candida albicans* Worms	Viruses *Chlamydia* spp. *Rickettsia* spp. *Listeria monocytogenes* Protozoa	*Mycobacterium* spp. *Salmonella typhimurium* *Yersinia pestis* *Listeria* spp. *Legionella pneumophila* *Cryptococcus neoformans* *Leishmania* spp. *Trypanosoma* spp. *Histoplasma*
Protective immunity	Complement Phagocytosis Antibodies	Antimicrobial peptides Antibodies, especially IgA	NK cells Cytotoxic T cells	T-cell and NK-cell dependent macrophage activation

Fig. 2.3 Pathogens can be found in various compartments of the body, where they must be combated by different host defense mechanisms. Virtually all pathogens have an extracellular phase in which they are vulnerable to the circulating molecules and cells of innate immunity and to the antibodies of the adaptive immune response. All these clear the microorganism primarily by promoting its phagocytosis by the phagocytes of the immune system. Intracellular phases of pathogens such as viruses are not accessible to these mechanisms; instead the infected cell is attacked by the NK cells of innate immunity or by the cytotoxic T cells of adaptive immunity. Activation of macrophages as a result of NK-cell or T-cell activity can induce the macrophage to kill pathogens that are living inside macrophage vesicles.

dangerous extracellular bacterial pathogens cause disease by releasing protein toxins, against which the innate immune system has little defense. The highly specific antibodies produced by the adaptive immune system are required to neutralize the action of such toxins (see Fig. 1.26). The damage caused by a particular infectious agent also depends on the site in which it grows; *Streptococcus pneumoniae* in the lung causes pneumonia, for example, whereas in the blood it causes a rapidly fatal systemic illness, pneumococcal sepsis.

As we will see in the following sections, for a microorganism to invade the body, it must first bind to or cross an epithelium. Intestinal pathogens such as *Salmonella typhi*, the causal agent of typhoid fever, or *Vibrio cholerae*, which causes cholera, are spread through fecally contaminated food and water, respectively. Immune responses to this type of pathogen occur in the specialized mucosal immune system once they have breached the epithelial barrier, as described in Chapter 11. The first defense against microorganisms invading through the gut consists of a healthy intestinal epithelium and intestinal flora, which competes with the pathogens for nutrients and epithelial attachment sites.

Most pathogenic microorganisms have evolved to be able to overcome innate immune responses and to continue to grow, making us ill. An adaptive immune response is required to eliminate them and to prevent subsequent reinfection. Other pathogens are never entirely eliminated by the immune system, and persist in the body for years. But most pathogens are not universally lethal. Those that have lived for thousands of years in the human population are highly evolved to exploit their human hosts; they cannot alter their pathogenicity without upsetting the compromise they have achieved with the human immune system. Rapidly killing every host it infects is no better for the long-term survival of a pathogen than being wiped out by the immune response before the microbe has had time to infect someone else. In short, we have adapted to live with our enemies, and they with us. Nevertheless, the

	Direct mechanisms of tissue damage by pathogens			Indirect mechanisms of tissue damage by pathogens		
	Exotoxin production	Endotoxin	Direct cytopathic effect	Immune complexes	Anti-host antibody	Cell-mediated immunity
Pathogenic mechanism						
Infectious agent	*Streptococcus pyogenes* *Staphylococcus aureus* *Corynebacterium diphtheriae* *Clostridium tetani* *Vibrio cholerae*	*Escherichia coli* *Haemophilus influenzae* *Salmonella typhi* *Shigella* *Pseudomonas aeruginosa* *Yersinia pestis*	Variola Varicella-zoster Hepatitis B virus Polio virus Measles virus Influenza virus Herpes simplex virus Human herpes virus 8 (HHV8)	Hepatitis B virus Malaria *Streptococcus pyogenes* *Treponema pallidum* Most acute infections	*Streptococcus pyogenes* *Mycoplasma pneumoniae*	*Mycobacterium tuberculosis* *Mycobacterium leprae* Lymphocytic choriomeningitis virus *Borrelia burgdorferi* *Schistosoma mansoni* Herpes simplex virus
Disease	Tonsilitis, scarlet fever Boils, toxic shock syndrome, food poisoning Diphtheria Tetanus Cholera	Gram-negative sepsis Meningitis, pneumonia Typhoid fever Bacillary dysentery Wound infection Plague	Smallpox Chickenpox, shingles Hepatitis Poliomyelitis Measles, subacute sclerosing panencephalitis Influenza Cold sores Kaposi's sarcoma	Kidney disease Vascular deposits Glomerulonephritis Kidney damage in secondary syphilis Transient renal deposits	Rheumatic fever Hemolytic anemia	Tuberculosis Tuberculoid leprosy Aseptic meningitis Lyme arthritis Schistosomiasis Herpes stromal keratitis

Fig. 2.4 Pathogens can damage tissues in a variety of different ways. The mechanisms of damage, representative infectious agents, and the common names of the diseases associated with each are shown. Exotoxins are released by microorganisms and act at the surface of host cells, for example by binding to receptors. Endotoxins, which are intrinsic components of microbial structure, trigger phagocytes to release cytokines that produce local or systemic symptoms. Many pathogens are cytopathic, directly damaging the cells they infect. Finally, an adaptive immune response to the pathogen can generate antigen:antibody complexes that activate neutrophils and macrophages, antibodies that can cross-react with host tissues, or T cells that kill infected cells. All of these have some potential to damage the host's tissues. In addition, neutrophils, the most abundant cells early in infection, release many proteins and small-molecule inflammatory mediators that both control infection and cause tissue damage (see Fig. 2.9).

recent concern about highly pathogenic strains of avian influenza, and the episode in 2002–2003 of SARS (severe acute respiratory syndrome), caused by a corona virus from bats that caused severe pneumonia in humans, remind us that new and deadly infections can transfer to humans from animal reservoirs. These are known as **zoonotic** infections—and we must be on the alert at all times for the emergence of new pathogens and new threats to health. The human immunodeficiency virus that causes AIDS (discussed in Chapter 12) serves as a warning that we remain constantly vulnerable.

2-2 Infectious agents must overcome innate host defenses to establish a focus of infection.

Our bodies are constantly exposed to microorganisms present in our environment, including infectious agents that have been shed from other infected individuals. Contact with these microorganisms may occur through external or internal epithelial surfaces: the respiratory tract mucosa provides a route of entry for airborne microorganisms, and the gastrointestinal mucosa does the same for microorganisms in food and water. Insect bites and wounds allow microorganisms to penetrate the skin, and direct contact between individuals offers opportunities for infection of the skin, the gut, and the mucosa of the reproductive tract (Fig. 2.5).

In spite of this exposure, infectious disease is fortunately quite infrequent. The epithelial surfaces of the body serve as an effective barrier against most microorganisms, and they are rapidly repaired if wounded. Furthermore, most of the microorganisms that do succeed in crossing an epithelial surface are efficiently removed by innate immune mechanisms that function in the underlying tissues. Thus, in most cases these defenses prevent infection from

Routes of infection for pathogens			
Route of entry	**Mode of transmission**	**Pathogen**	**Disease**
Mucosal surfaces			
Airway	Inhaled droplet	Influenza virus	Influenza
	Spores	*Neisseria meningitidis*	Meningococcal meningitis
		Bacillus anthracis	Inhalation anthrax
Gastrointestinal tract	Contaminated water or food	*Salmonella typhi*	Typhoid fever
		Rotavirus	Diarrhea
Reproductive tract	Physical contact	*Treponema pallidum*	Syphilis
		HIV	AIDS
External epithelia			
External surface	Physical contact	*Trichophyton*	Athlete's foot
Wounds and abrasions	Minor skin abrasions	*Bacillus anthracis*	Cutaneous anthrax
	Puncture wounds	*Clostridium tetani*	Tetanus
	Handling infected animals	*Francisella tularensis*	Tularemia
Insect bites	Mosquito bites (*Aedes aegypti*)	Flavivirus	Yellow fever
	Deer tick bites	*Borrelia burgdorferi*	Lyme disease
	Mosquito bites (*Anopheles*)	*Plasmodium* spp.	Malaria

Fig. 2.5 Pathogens infect the body through a variety of routes.

becoming established. It is difficult to know how many infections are repelled in this way, because they cause no symptoms and pass undetected. It is clear, however, that the microorganisms that a normal human being inhales or ingests, or that enter through minor wounds, are mostly held at bay or eliminated, because they seldom cause clinical disease.

Disease occurs when a microorganism succeeds in evading or overwhelming innate host defenses to establish a local site of infection, and then replicates there to allow its further transmission within our bodies. In some cases, such as the fungal disease athlete's foot, the initial infection remains local and does not cause significant pathology. In other cases, the infectious agent causes significant damage and serious illness as it spreads through the lymphatics or the bloodstream, invades and destroys tissues, or disrupts the body's workings with its toxins, as in the case of the agent of tetanus (*Clostridium tetani*), which secretes a powerful neurotoxin.

The spread of a pathogen is often initially countered by an inflammatory response that recruits more effector cells and molecules of the innate immune system from local blood vessels (Fig. 2.6.), while inducing clotting further downstream so that the microbe cannot spread through the blood. The induced responses of innate immunity act over several days, during which time an adaptive immune response gets under way in response to pathogen antigens delivered to local lymphoid tissue by dendritic cells (see Section 1-15). An adaptive immune response differs from innate immunity in its ability to target structures that are specific to particular strains and variants

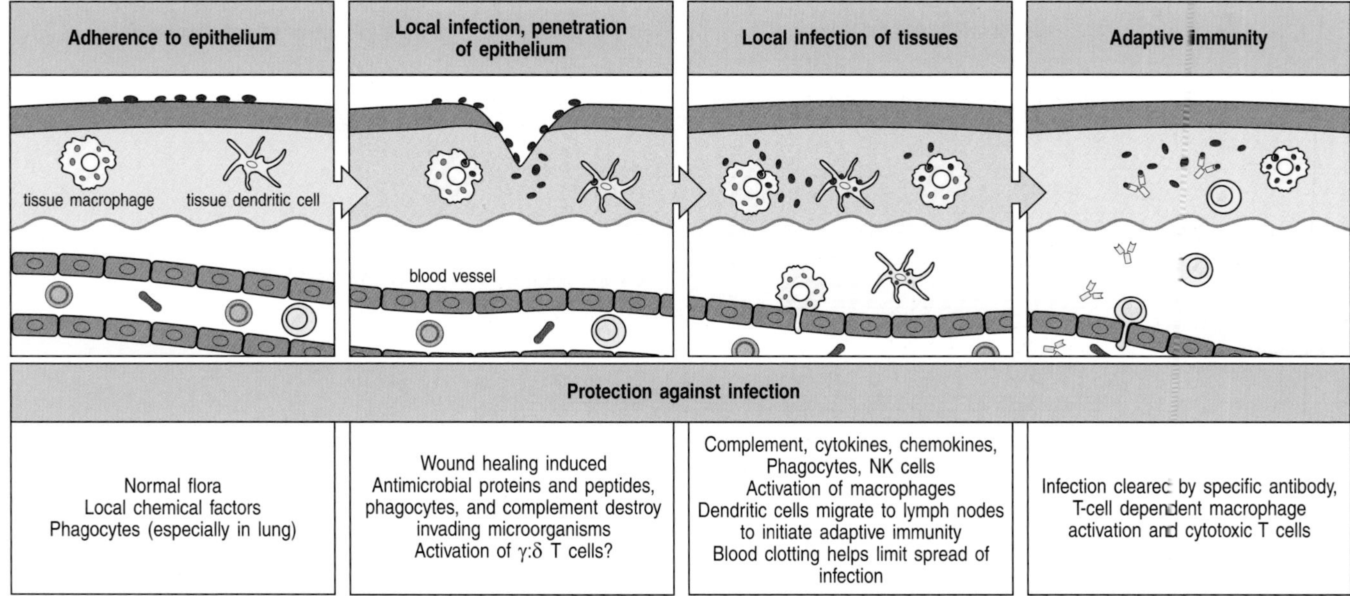

Fig. 2.6 An infection and the response to it can be divided into a series of stages. These are illustrated here for an infectious microorganism entering through a wound in the skin. The infectious agent must first adhere to the epithelial cells and then cross the epithelium. A local immune response may prevent the infection from becoming established. If not, it helps to contain the infection and also delivers the infectious agent, carried in lymph and inside dendritic cells, to local lymph nodes. This initiates the adaptive immune response and eventual clearance of the infection. The role of γ:δ T cells is uncertain, as we shall see in Section 2-34, and this is indicated by the question mark.

of pathogens. This response will usually clear the infection and protect the host against reinfection with the same pathogen, by producing effector cells and antibodies against the pathogen and by generating immunological memory of that pathogen.

2-3 The epithelial surfaces of the body make up the first lines of defense against infection.

Our body surfaces are defended by epithelia, which provide a physical barrier between the internal milieu and the external world that contains pathogens (Fig. 2.7). Epithelial cells are held together by tight junctions, which effectively form a seal against the external environment. Epithelia comprise the skin and the linings of the body's tubular structures—the gastrointestinal, respiratory, and urogenital tracts. Infections occur only when the pathogen can colonize or cross through these barriers, and because the dry, protective layers of the skin present a more formidable barrier, pathogen entry most often occurs through the internal epithelial surfaces, which make up the vast majority of epithelial surfaces of our bodies. The importance of epithelia in protection against infection is obvious when the barrier is breached, as in wounds, burns, and loss of the integrity of the body's internal epithelia, where

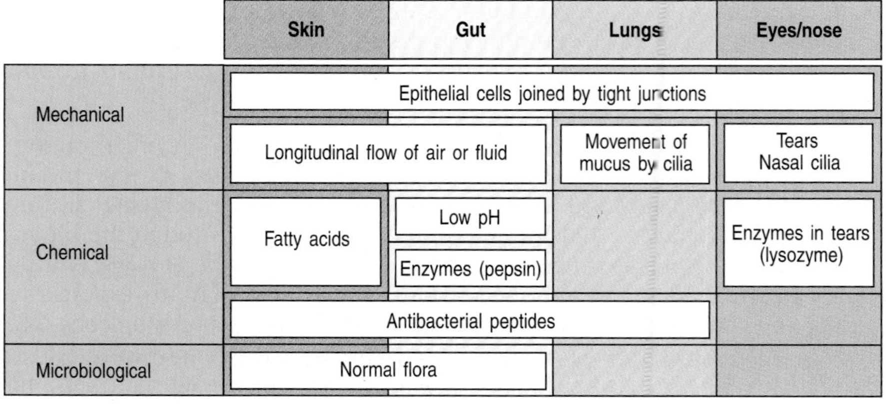

Skin
Gut
Lungs
Eyes/nose
Mechanical
Epithelial cells joined by tight junctions
Longitudinal flow of air or fluid
Movement of mucus by cilia
Tears
Nasal cilia
Chemical
Fatty acids
Low pH
Enzymes (pepsin)
Enzymes in tears (lysozyme)
Antibacterial peptides
Microbiological
Normal flora

Fig. 2.7 Many barriers prevent pathogens from crossing epithelia and colonizing tissues. Surface epithelia provide mechanical, chemical, and microbiological barriers to infection.

infection is a major cause of mortality and morbidity. In the absence of wounding or disruption, pathogens normally cross epithelial barriers by binding to molecules on the epithelial surfaces of our internal organs, or establish an infection by adhering to and colonizing these surfaces. This specific attachment allows the pathogen to infect the epithelial cell, to damage the epithelium so it can be crossed, or, in the case of colonizing pathogens, to avoid being dislodged by the flow of air or fluid across the epithelial surface.

The internal epithelia are known as **mucosal epithelia** because they secrete a viscous fluid called mucus, which contains many glycoproteins called mucins. Microorganisms coated in mucus may be prevented from adhering to the epithelium, and in mucosal epithelia such as that of the respiratory tract, microorganisms can be expelled in the flow of mucus driven by the beating of epithelial cilia. The efficacy of mucus flow in clearing infection is illustrated by people with defective mucus secretion or inhibition of ciliary movement, as occurs in the inherited disease cystic fibrosis. Such individuals frequently develop lung infections caused by bacteria that colonize the epithelial surface but do not cross it. In the gut, peristalsis is an important mechanism for keeping both food and infectious agents moving through the body. Failure of peristalsis is typically accompanied by the overgrowth of pathogenic bacteria within the lumen of the gut.

Our surface epithelia are more than mere physical barriers to infection; they also produce chemical substances that are microbicidal or that inhibit microbial growth. For example, the antibacterial enzymes lysozyme and phospholipase A are secreted in tears and saliva, and saliva contains several histatins—histidine-rich peptides with antimicrobial properties. The acid pH of the stomach and the digestive enzymes, bile salts, fatty acids, and lysolipids found in the upper gastrointestinal tract create a substantial chemical barrier to infection. Further down the intestinal tract, antibacterial and antifungal peptides called **cryptdins** or **α-defensins** are made by Paneth cells, which are resident in the base of the crypts in the small intestine beneath the epithelial stem cells. Related antimicrobial peptides, the **β-defensins**, are made by other epithelia, primarily in the respiratory and urogenital tracts, skin, and tongue. Antimicrobial peptides have a role in the immune defenses of many organisms, even in humans and other vertebrates that can make an adaptive immune response. Even more striking is resistance to infection in insects and other invertebrates, and even plants, in which innate immunity comprises the only system of host defense. In all these organisms, antimicrobial peptides are an important part of the defenses. Antimicrobial peptides such as the defensins are cationic peptides that are thought to kill bacteria by damaging the bacterial cell membrane.

Antimicrobial proteins that work by a different mechanism are secreted into the fluids that bathe the epithelial surfaces of the lung and gut. These proteins coat the surface of pathogens so that they are more easily phagocytosed by macrophages. They are members of a family of receptors able to recognize common features of microbial surfaces and are considered in detail later in this chapter.

In addition to these defenses, most epithelial surfaces are associated with a normal flora of nonpathogenic bacteria, known as **commensal** bacteria, that compete with pathogenic microorganisms for nutrients and for attachment sites on epithelial cells. This flora can also produce antimicrobial substances, such as the lactic acid produced by vaginal lactobacilli, some strains of which also produce antimicrobial peptides (bacteriocins). When the nonpathogenic bacteria are killed by antibiotic treatment, pathogenic microorganisms frequently replace them and cause disease. Under some circumstances commensal bacteria can cause disease. Their survival on our body surfaces is regulated by a balance between bacterial growth and elimination by the

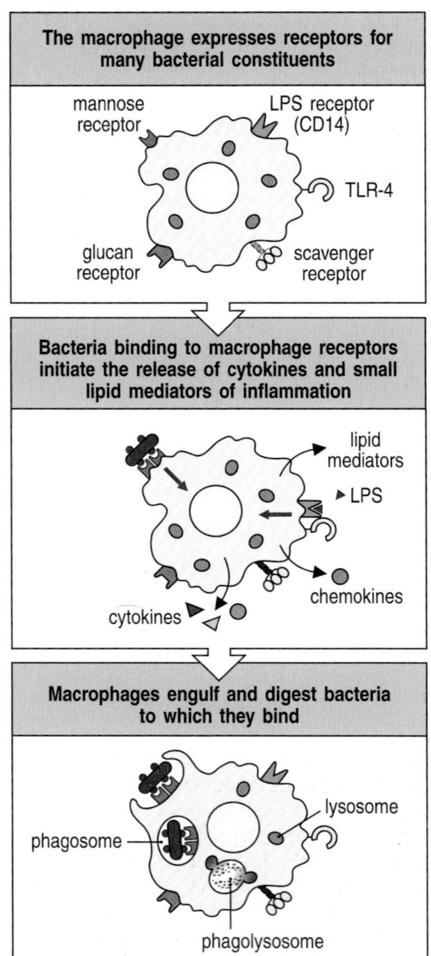
The macrophage expresses receptors for many bacterial constituents
mannose receptor
LPS receptor (CD14)
TLR-4
glucan receptor
scavenger receptor
Bacteria binding to macrophage receptors initiate the release of cytokines and small lipid mediators of inflammation
lipid mediators
LPS
chemokines
cytokines
Macrophages engulf and digest bacteria to which they bind
lysosome
phagosome
phagolysosome

Fig. 2.8 Macrophages are activated by pathogens and both engulf them and initiate inflammatory responses. Macrophages derive from circulating monocytes. They have many of the same characteristics but acquire new functions and new receptors when they become resting cells in connective tissues throughout the body. Macrophages express receptors for many bacterial components, including bacterial carbohydrates (mannose and glucan receptors), lipids (LPS receptor) and other pathogen-derived components (Toll-like receptors (TLRs) and scavenger receptor). Binding of bacteria to macrophage receptors stimulates the phagocytosis and uptake of pathogens into intracellular vesicles, where they are destroyed. Signaling through some receptors, such as the Toll receptors, in response to bacterial components causes the secretion of 'pro-inflammatory cytokines' such as interleukin-1β (IL-1β), IL-6, and tumor necrosis factor-α (TNF-α).

mechanisms of innate immunity; failures in this regulation, such as those caused by the inherited deficiencies of proteins of innate immunity that we discuss in Chapter 12, can allow normally nonpathogenic bacteria to grow excessively and cause disease.

2-4 After entering tissues, many pathogens are recognized, ingested, and killed by phagocytes.

If a microorganism crosses an epithelial barrier and begins to replicate in the tissues of the host, in most cases it is immediately recognized by the mononuclear phagocytes, or **macrophages**, that reside in these tissues. Macrophages mature continuously from monocytes that leave the circulation to migrate into tissues throughout the body. Macrophages in different tissues were historically given different names, for example, microglial cells in neural tissue and **Kupffer cells** in the liver; generically these cells are referred to as mononuclear phagocytes. They are found in especially large numbers in connective tissue, in the submucosal layer of the gastrointestinal tract, in the lung (where they are also found in both the interstitium and the alveoli), along certain blood vessels in the liver, and throughout the spleen, where they remove senescent blood cells. The second major family of phagocytes—the **neutrophils**, or **polymorphonuclear neutrophilic leukocytes** (**PMNs** or polys)—are short-lived cells that are abundant in the blood but are not present in normal, healthy tissues. Both of these phagocytic cells have a key role in innate immunity because they can recognize, ingest, and destroy many pathogens without the aid of an adaptive immune response.

As most microorganisms enter the body through the mucosa of the gut and respiratory system, the macrophages located in the submucosal tissues are the first cells to encounter most pathogens, but they are soon reinforced by the recruitment of large numbers of neutrophils to sites of infection. Macrophages and neutrophils recognize pathogens by means of cell-surface receptors that can discriminate between the surface molecules displayed by pathogens and those of the host. These receptors, which we will examine in more detail later in this chapter, include the macrophage mannose receptor, which is found on macrophages but not on monocytes or neutrophils; scavenger receptors, which bind many negatively charged ligands such as lipoteichoic acids, which are cell-wall components of Gram-positive bacteria; and CD14, found predominantly on monocytes and macrophages (Fig. 2.8). This binds the lipopolysaccharide present on the surface of Gram-negative bacteria and allows it to be recognized by other receptors called Toll-like receptors. In many cases, binding of a pathogen to these cell-surface receptors leads to **phagocytosis**, followed by the death of the pathogen inside the phagocyte. Phagocytosis is an active process, in which the bound pathogen is first surrounded by the phagocyte membrane and then internalized in a membrane-enclosed vesicle known as a **phagosome** or endocytic vacuole. The phagosome then becomes acidified, which kills most pathogens. In addition to being phagocytic, macrophages and neutrophils have membrane-enclosed granules, called **lysosomes**, that contain enzymes, proteins, and peptides that can attack the microbe. The phagosome fuses with one or more lysosomes to generate a **phagolysosome** in which the lysosomal contents are released to destroy the pathogen (see Fig. 2.8).

Upon phagocytosis, macrophages and neutrophils produce a variety of other toxic products that help kill the engulfed microorganism (Fig. 2.9). The most important of these are the antimicrobial peptides and nitric oxide (NO), the superoxide anion (O_2^-), and hydrogen peroxide (H_2O_2), which are directly toxic to bacteria. Nitric oxide is produced by a high-output form of nitric oxide synthase, iNOS2. Superoxide is generated by a multicomponent,

Class of mechanism	Specific products
Acidification	pH=~3.5–4.0, bacteriostatic or bactericidal
Toxic oxygen-derived products	Superoxide O_2^-, hydrogen peroxide H_2O_2, singlet oxygen $^1O_2^•$ hydroxyl radical $^•OH$, hypohalite OCl^-
Toxic nitrogen oxides	Nitric oxide NO
Antimicrobial peptides	Defensins and cationic proteins
Enzymes	Lysozyme—dissolves cell walls of some Gram-positive bacteria. Acid hydrolases—further digest bacteria
Competitors	Lactoferrin (binds Fe) and vitamin B_{12}-binding protein

Fig. 2.9 Bactericidal agents produced or released by phagocytes on the ingestion of microorganisms. Most of these agents are made by both macrophages and neutrophils. Some of them are toxic; others, such as lactoferrin, work by binding essential nutrients and preventing their uptake by the bacteria. The same substances can be released by phagocytes interacting with large antibody-coated surfaces such as parasitic worms or host tissues. As these agents are also toxic to host cells, phagocyte activation can cause extensive tissue damage during an infection.

membrane-associated NADPH oxidase in a process known as the **respiratory burst** because it is accompanied by a transient increase in oxygen consumption; the superoxide is converted by the enzyme superoxide dismutase into H_2O_2 (Fig. 2.10). Further chemical and enzymatic reactions produce a range of toxic chemicals from H_2O_2, including the hydroxyl radical (•OH) and hypochlorite (OCl^-) and hypobromite (OBr^-). Neutrophils are short-lived cells, dying soon after they have accomplished a round of phagocytosis. Dead and dying neutrophils are a major component of the **pus** that forms in some infections by extracellular bacteria, which are thus known as **pus-forming** or **pyogenic bacteria**. Macrophages, in contrast, are long-lived and continue to generate new lysosomes. Patients with a disease called chronic granulomatous disease have a genetic deficiency of NADPH oxidase, which means that their phagocytes do not produce the toxic oxygen derivatives characteristic of the respiratory burst and so are less able to kill ingested microorganisms and clear an infection. People with this defect are unusually susceptible to bacterial and fungal infections, especially in infancy.

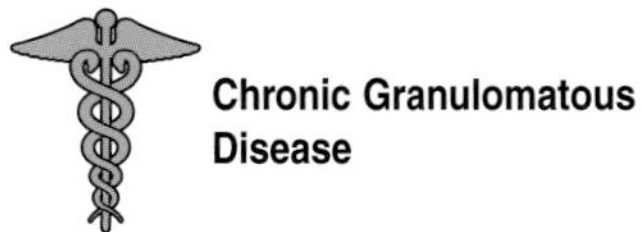
Chronic Granulomatous Disease

Macrophages can phagocytose pathogens and produce the respiratory burst immediately on encountering an infecting microorganism, and this can be sufficient to prevent an infection from becoming established. In the 19th century the cellular immunologist **Elie Metchnikoff** believed that the innate response of macrophages encompassed all host defense and, indeed, it is now clear that invertebrates, such as the sea star that he was studying, rely entirely

Fig. 2.10 The respiratory burst in macrophages and neutrophils is caused by a transient increase in oxygen consumption during the production of microbicidal oxygen metabolites. Ingestion of microorganisms activates the phagocyte to assemble the multisubunit enzyme NADPH oxidase from its components. The active enzyme converts molecular oxygen to the superoxide ion O_2^-, and other oxygen free radicals. Superoxide ion is then converted by the enzyme superoxide dismutase (SOD) to hydrogen peroxide (H_2O_2), which can kill microorganisms and is also converted by other enzymes and by chemical reactions with ferrous (Fe^{2+}) ions to microbicidal hypochlorite (OCl^-) and hydroxyl (•OH) radical.

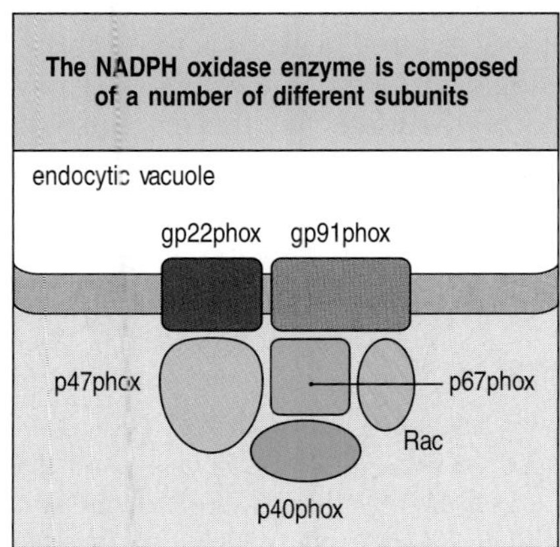
The NADPH oxidase enzyme is composed of a number of different subunits
endocytic vacuole
gp22phox
gp91phox
p47phox
p67phox
Rac
p40phox

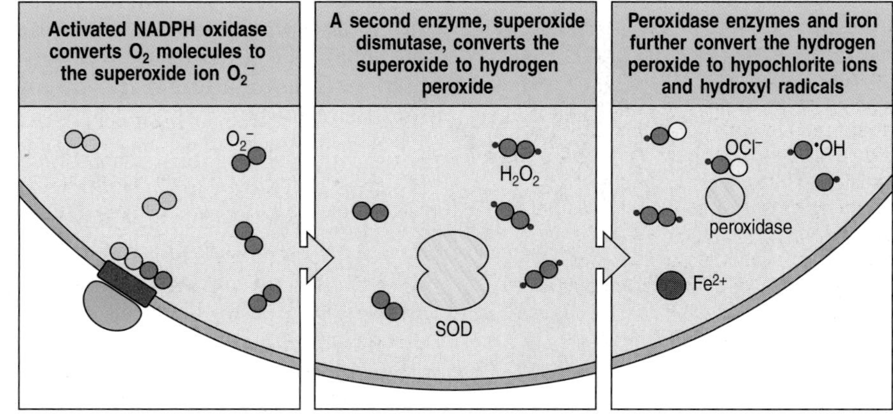
Activated NADPH oxidase converts O2 molecules to the superoxide ion O2−
O2−
A second enzyme, superoxide dismutase, converts the superoxide to hydrogen peroxide
H2O2
SOD
Peroxidase enzymes and iron further convert the hydrogen peroxide to hypochlorite ions and hydroxyl radicals
OCl−
•OH
peroxidase
Fe2+

on innate immunity to overcome infection. Although this is not the case in humans and other vertebrates, the innate response of macrophages still provides an important front line of defense that must be overcome if a microorganism is to establish an infection that can be passed on to a new host.

A key feature that distinguishes pathogenic from nonpathogenic microorganisms is their ability to overcome innate immune defenses. Pathogens have developed a variety of strategies to avoid immediate destruction by macrophages. As noted earlier, many extracellular pathogenic bacteria coat themselves with a thick polysaccharide capsule that is not recognized by any phagocyte receptor. Other pathogens, for example mycobacteria, have evolved ways to grow inside macrophage phagosomes by inhibiting their acidification and fusion with lysosomes. Without such devices, a microorganism must enter the body in sufficient numbers to simply overwhelm the immediate innate host defenses and to establish a focus of infection.

A second important effect of the interaction between pathogens and tissue macrophages is the activation of macrophages to release small proteins called cytokines and chemokines (chemoattractant cytokines), and other chemical mediators that set up a state of inflammation in the tissue and attract neutrophils and plasma proteins to the infection site. It is thought that the pathogen induces cytokine and chemokine secretion by signals delivered through some of the receptors to which it binds, and we will see later how this occurs in response to bacterial lipopolysaccharide. Receptors that signal the presence of pathogens and induce cytokines also have another important role. This is to induce the expression of so-called co-stimulatory molecules on macrophages and on **dendritic cells**, another type of phagocytic cell present in tissues, thus enabling these antigen-presenting cells to initiate an adaptive immune response (see Section 1-7).

The cytokines released by macrophages make an important contribution both to local inflammation and to other induced innate responses that occur in the first few days of a new infection. We will describe these induced innate responses and the role of individual cytokines in the last part of this chapter. However, because an inflammatory response is usually initiated within hours of infection or wounding, we will outline here how it occurs and how it contributes to host defense.

2-5 Pathogen recognition and tissue damage initiate an inflammatory response.

Inflammation has three essential roles in combating infection. The first is to deliver additional effector molecules and cells to sites of infection, to augment the killing of invading microorganisms by the front-line macrophages. The second is to induce local blood clotting, which provides a physical barrier to the spread of the infection in the bloodstream. The third is to promote the repair of injured tissue, a nonimmunological role that we will not discuss further. Inflammation at the site of infection is initiated by the response of macrophages to pathogens.

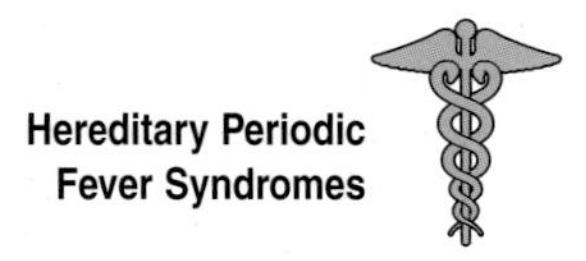

Inflammatory responses are operationally characterized by pain, redness, heat, and swelling at the site of an infection, reflecting four types of change in the local blood vessels, as shown in Fig. 2.11. The first is an increase in vascular diameter, leading to increased local blood flow—hence the heat and redness—and a reduction in the velocity of blood flow, especially along the inner walls of small blood vessels. The second change is that the endothelial cells lining the blood vessel are activated to express **cell-adhesion molecules** that promote the binding of circulating leukocytes. The combination of slowed blood flow and adhesion molecules allows leukocytes to attach to the

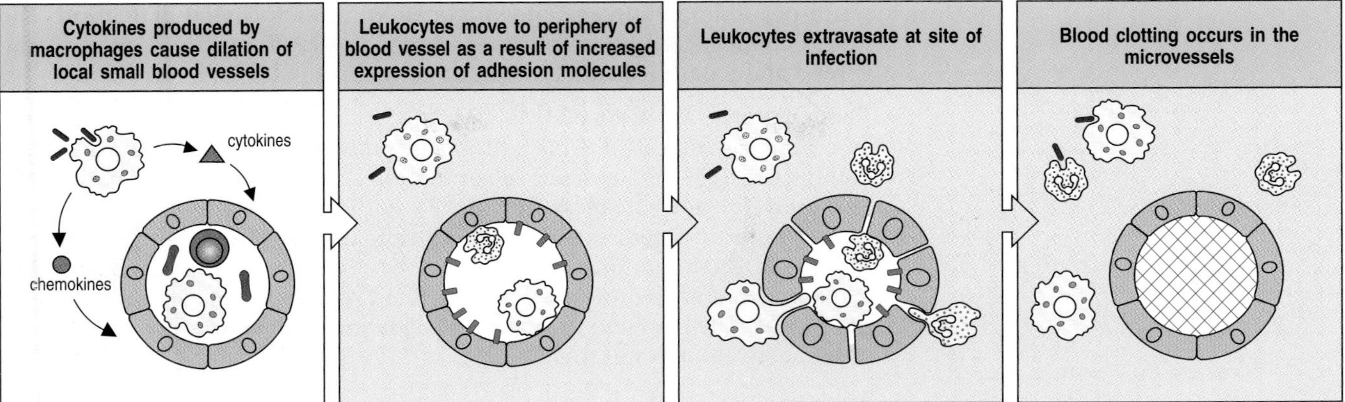

endothelium and migrate into the tissues, a process known as **extravasation**. All these changes are initiated by the cytokines and chemokines produced by activated macrophages.

Fig. 2.11 Infection stimulates macrophages to release cytokines and chemokines that initiate an inflammatory response. Cytokines produced by tissue macrophages at the site of infection cause the dilation of local small blood vessels and changes in the endothelial cells of their walls. These changes lead to the movement of leukocytes, such as neutrophils and monocytes, out of the blood vessel (extravasation) and into the infected tissue, guided by chemokines produced by the activated macrophages. The blood vessels also become more permeable, allowing plasma proteins and fluid to leak into the tissues. Together, these changes cause the characteristic inflammatory signs of heat, pain, redness, and swelling at the site of infection.

Once inflammation has begun, the first white blood cells attracted to the site are neutrophils. These are followed by monocytes, which differentiate into tissue macrophages (Fig. 2.12). Monocytes are also able to give rise to dendritic cells in the tissues, depending on the precise signals that they receive from their environment; for example, the cytokine granulocyte–macrophage colony-stimulating factor (GM-CSF), together with interleukin 4 (IL-4), will induce the monocyte to differentiate into a dendritic cell, whereas the cytokine macrophage colony-stimulating factor (M-CSF) induces differentiation into macrophages.

In the later stages of inflammation, other leukocytes such as eosinophils (see Section 1-3) and lymphocytes also enter the infected site. The third major change in local blood vessels is an increase in vascular permeability. Thus, instead of being tightly joined together, the endothelial cells lining the blood vessel walls become separated, leading to an exit of fluid and proteins from the blood and their local accumulation in the tissue. This accounts for the swelling, or **edema**, and pain—as well as the accumulation of plasma proteins that aid in host defense. The changes that occur in endothelium as a

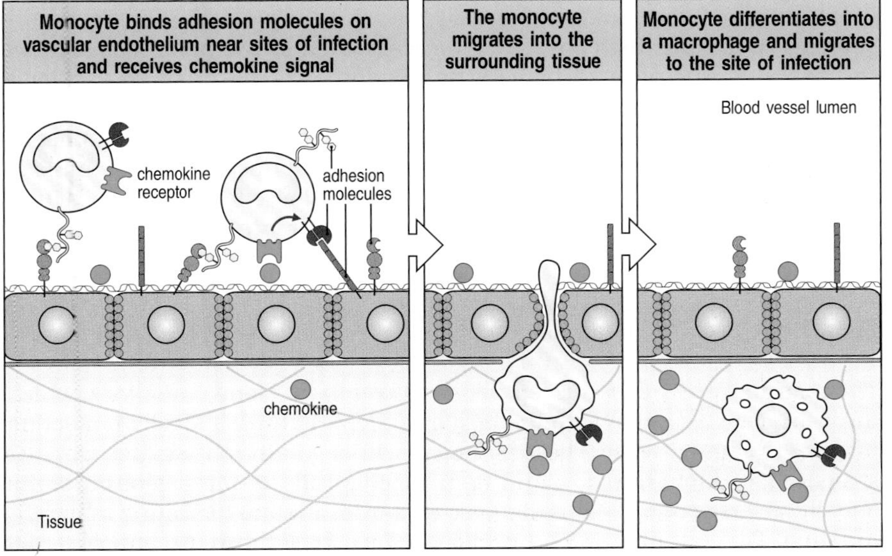

Monocyte binds adhesion molecules on vascular endothelium near sites of infection and receives chemokine signal
chemokine receptor
adhesion molecules
chemokine
Tissue
The monocyte migrates into the surrounding tissue
Monocyte differentiates into a macrophage and migrates to the site of infection
Blood vessel lumen

Fig. 2.12 Monocytes circulating in the blood leave the bloodstream to migrate toward sites of infection and inflammation. Adhesion molecules on the endothelial cells of the blood vessel wall first capture the monocyte and cause it to adhere to the vascular endothelium. Chemokines bound to the vascular endothelium then signal the monocyte to migrate across the endothelium into the underlying tissue. The monocyte, now differentiating into a macrophage, continues to migrate, under the influence of chemokines released during inflammatory responses, toward the site of infection. Monocytes leaving the blood in this way are also able to differentiate into dendritic cells (not shown), depending on the signals that they receive from their environment.

result of inflammation are known generally as **endothelial activation**. The fourth change, clotting in microvessels in the site of infection, prevents the spread of the pathogen via the blood.

These changes are induced by a variety of inflammatory mediators released as a consequence of the recognition of pathogens by macrophages. They include the lipid mediators of inflammation—**prostaglandins**, **leukotrienes**, and **platelet-activating factor** (**PAF**)—which are rapidly produced by macrophages through enzymatic pathways that degrade membrane phospholipids. Their actions are followed by those of the chemokines and cytokines that are synthesized and secreted by macrophages in response to pathogens. The cytokine **tumor necrosis factor-α** (**TNF-α**), for example, is a potent activator of endothelial cells.

As we will see in the third part of this chapter, another way in which pathogen recognition rapidly triggers an inflammatory response is through activation of complement. One of the cleavage products of the complement pathway is a peptide called C5a. C5a is a potent mediator of inflammation, with several different activities. In addition to increasing vascular permeability and inducing the expression of some adhesion molecules, it acts as a powerful chemoattractant for neutrophils and monocytes. C5a also activates phagocytes and local **mast cells** (see Section 1-3), which are in turn stimulated to release their granules containing the small inflammatory molecule histamine and the cytokine TNF-α.

If wounding has occurred, the injury to blood vessels immediately triggers two protective enzyme cascades. One is the **kinin system** of plasma proteases that is triggered by tissue damage to produce several inflammatory mediators, including the vasoactive peptide **bradykinin**. The kinin system is an example of a protease cascade, also known as a triggered-enzyme cascade, in which the enzymes are initially in an inactive, or **pro-enzyme** form. After the system is activated, one active protease cleaves and activates the next protease in the series, and so on. Bradykinin causes an increase in vascular permeability that promotes the influx of plasma proteins to the site of tissue injury. It also causes pain, which, although unpleasant to the victim, draws attention to the problem and leads to immobilization of the affected part of the body, which helps to limit the spread of the infection.

The **coagulation system** is another protease cascade that is triggered in the blood after damage to blood vessels. Its activation leads to the formation of a fibrin clot, whose normal role is to prevent blood loss. In regard to innate immunity, however, the clot physically bars the entry of infectious microorganisms to the bloodstream. The kinin cascade and the blood coagulation cascade are also triggered by activated endothelial cells, and so can have important roles in the inflammatory response to pathogens even if wounding or gross tissue injury has not occurred.Thus, within minutes of the penetration of tissues by a pathogen, the inflammatory response causes an influx of proteins and cells that may control the infection. It also forms a physical barrier in the form of blood clots to limit the spread of infection and makes the host fully aware of the local infection.

Summary.

The mammalian body is susceptible to infection by many pathogens, which must first make contact with the host and then establish a focus of infection in order to cause disease. These pathogens differ greatly in their lifestyles, the structures of their surfaces, and their mechanisms of pathogenesis, so an equally diverse set of defensive responses from the host immune system is required. The first phase of host defense consists of those mechanisms that are present and ready to resist an invader at any time. The epithelial surfaces of the body keep pathogens out and protect against colonization and against viruses and bacteria that enter the body through specialized cell-surface interactions. Their defense mechanisms include the prevention of pathogen adherence and the secretion of antimicrobial enzymes and peptides. Bacteria, viruses, and parasites that overcome these barriers are faced immediately by tissue macrophages equipped with surface receptors that can bind and phagocytose many different types of pathogen. This, in turn, leads to an inflammatory response, which causes the accumulation of phagocytic neutrophils and macrophages at the site of infection, which ingest and destroy the invading microorganisms.

Pattern recognition in the innate immune system.

Although the innate immune system lacks the fine specificity of adaptive immunity that is necessary to produce immunological memory, it can distinguish self from nonself. We have already seen how this occurs in the response of macrophages to pathogenic microbes. In this part of the chapter we will look more closely at the receptors that activate the innate immune response, including those that recognize pathogens directly and signal for a cellular innate immune response. Regular patterns of molecular structure are present on many microorganisms but not on the body's own cells. Proteins that recognize these features occur as receptors on macrophages, neutrophils, and dendritic cells, and as secreted molecules. Their general characteristics are contrasted with the antigen-specific receptors of adaptive immunity in Fig. 2.13. Unlike the antigen receptors described in Chapter 1, the receptors of the innate immune system are not clonally distributed; instead, a given set of receptors is present on all the cells of the same cell type. The binding of pathogen components by these receptors gives rise to very rapid responses, which are put into effect without the delay imposed by the need for activated

Receptor characteristic	Innate immunity	Adaptive immunity
Specificity inherited in the genome	Yes	No
Expressed by all cells of a particular type (e.g. macrophages)	Yes	No
Triggers immediate response	Yes	No
Recognizes broad classes of pathogens	Yes	No
Interacts with a range of molecular structures of a given type	Yes	No
Encoded in multiple gene segments	No	Yes
Requires gene rearrangement	No	Yes
Clonal distribution	No	Yes
Able to discriminate between even closely related molecular structures	No	Yes

Fig. 2.13 Comparison of the characteristics of recognition molecules of the innate and adaptive immune systems. The innate immune system uses receptors that are encoded by complete genes inherited through the germline. In contrast, the adaptive immune system uses antigen receptors encoded in gene segments that are assembled into complete T-cell and B-cell receptor genes during lymphocyte development, a process that leads to each individual cell expressing a receptor of unique specificity. Receptors of the innate immune system are deployed nonclonally (that is, by all the cells of a given cell type), whereas the antigen receptors of the adaptive immune system are clonally distributed on individual lymphocytes and their progeny.

lymphocytes to divide and differentiate during the development of an adaptive immune response.

The pattern recognition receptors of the innate immune system have several different functions. Many are phagocytic receptors that stimulate ingestion of the pathogens they recognize. Some are chemotactic receptors, which guide cells to sites of infection. A third function is to induce the production of effector molecules that contribute to the later, induced responses of innate immunity, and also to induce proteins that influence the initiation and nature of any subsequent adaptive immune response. In this part of the chapter we will first look at the recognition properties of some receptors that bind pathogens directly. We will then focus on an evolutionarily ancient pathogen recognition and signaling system, mediated by receptors called Toll-like receptors, that has a key role in defense against infection in plants, adult insects, and vertebrates, including mammals.

2-6 Receptors with specificity for pathogen molecules recognize patterns of repeating structural motifs.

Microorganisms typically bear repeating patterns of molecular structure on their surface. The cell walls of Gram-positive and Gram-negative bacteria, for example, are composed of a matrix of proteins, carbohydrates and lipids in a repetitive array, as shown in Figure 2.14. The lipoteichoic acids of Gram-positive bacterial cell walls and the lipopolysaccharide of the outer membrane of Gram-negative bacteria are, as we will see, important in the recognition of bacteria by the innate immune system. Other microbial components also have a repetitive structure. Bacterial flagella are made of repeated protein subunits, and bacterial DNA contains unmethylated repeats of the dinucleotide CpG. Viruses almost invariably express double-stranded RNA as part of their life cycles. These repetitive structures are known generally as **pathogen-associated molecular patterns** (**PAMPs**) and the receptors that recognize them as **pattern recognition receptors** (**PRRs**).

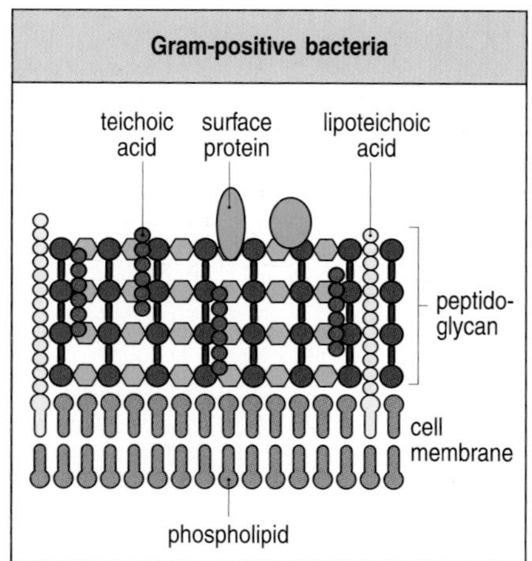

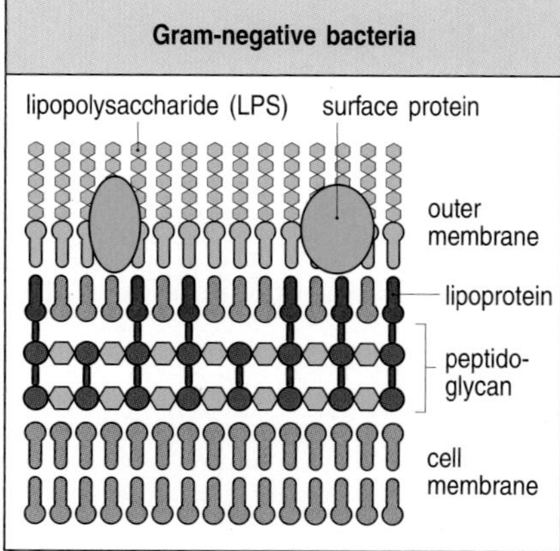

Fig. 2.14 The organization of the cell walls of Gram-positive and Gram-negative bacteria. Gram-positive bacteria (upper panel) have a cell wall composed of an outer layer of a repeating matrix of peptidoglycan molecules in which repeating *N*-acetylglucosamine (light blue hexagons) and *N*-acetylmuramic acid (purple circles) are cross-linked by peptide bridges into a dense three-dimensional network. Bacterial surface proteins and other molecules, such as teichoic acid, are embedded within this peptidoglycan layer, and lipoteichoic acids link the peptidoglycan layer to the bacterial cell membrane itself. The cell wall of Gram-negative bacteria (lower panel) is composed of an inner thin matrix of peptidoglycan and an outer lipid membrane, in which are embedded proteins and the lipopolysaccharide (LPS) characteristic of Gram-negative bacteria.

One such receptor is the **mannose-binding lectin** (**MBL**), which is present as a free protein in blood plasma. As we shall see in the next part of this chapter, it can initiate the lectin pathway of complement activation, but we shall discuss it briefly here as a good example of the recognition of molecular patterns. As illustrated in Fig. 2.15, pathogen recognition and discrimination from self by MBL is due to the recognition of a particular orientation of certain sugar residues, as well as their spacing, which is found only on microbes and not on host cells. Once formed, the MBL–pathogen complex is bound by phagocytes, either through interactions with MBL or through the phagocyte's receptors for complement, which also becomes bound to the pathogen. The outcome is the phagocytosis and killing of the pathogen (see Section 2-4) and the induction of other cellular responses, such as chemokine production. Coating of a particle with proteins that facilitate its phagocytosis is known as **opsonization**, and we will meet other examples of this defense strategy in this and later chapters.

MBL is a member of the collectin family of proteins, so called because they contain both collagen-like and lectin (sugar-binding) domains. Other members of this family are the **surfactant proteins A** and **D** (**SP-A** and **SP-D**), which are present in the fluid that bathes the epithelial surfaces of the lung. There they bind to and coat the surfaces of pathogens, making them more susceptible to phagocytosis by macrophages that have left the subepithelial tissues to enter the alveoli of the lung.

Phagocytes are also equipped with several cell-surface receptors that recognize pathogen surfaces directly. Among these is the **macrophage mannose**

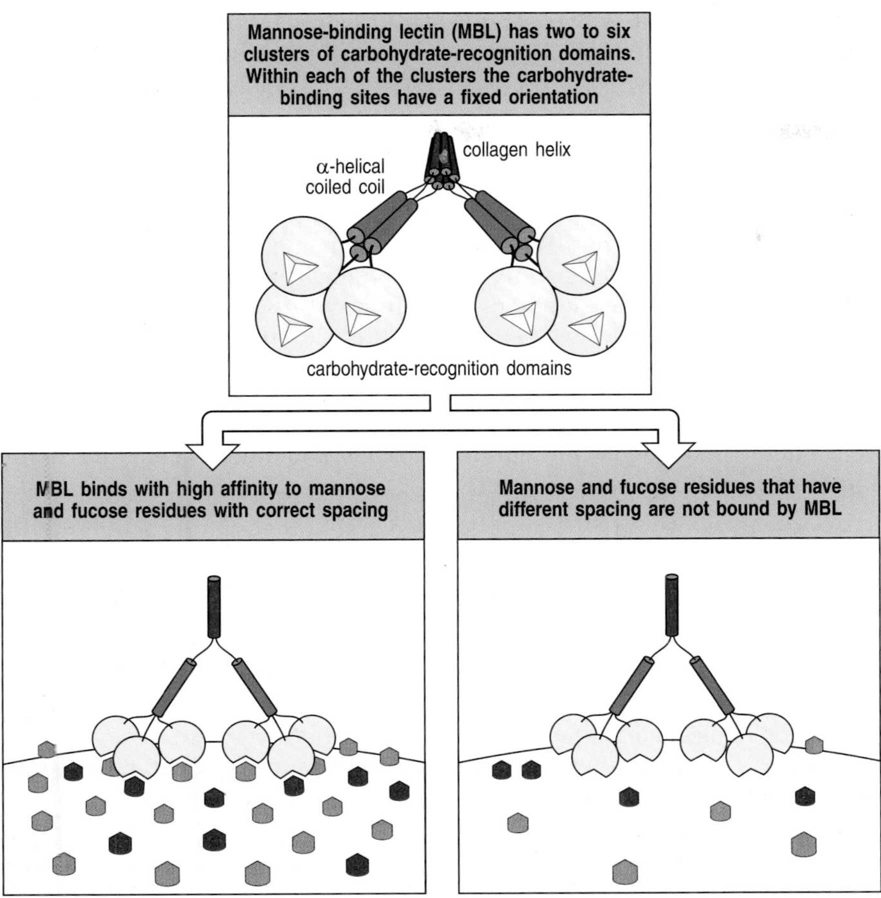
Mannose-binding lectin (MBL) has two to six clusters of carbohydrate-recognition domains. Within each of the clusters the carbohydrate-binding sites have a fixed orientation
α-helical coiled coil
collagen helix
carbohydrate-recognition domains
MBL binds with high affinity to mannose and fucose residues with correct spacing
Mannose and fucose residues that have different spacing are not bound by MBL

Fig. 2.15 M recogniz particu resid bin pa immun surfaces th arrangement o residues. The prese alone is not enough to the orientation of the bindin MBL is fixed, and only if the ma and fucose residues have the correc spacing will the MBL be able to bind. Once coated with MBL, bacteria are more susceptible to phagocytosis.

receptor (see Fig. 2.8). This receptor is a cell-bound C-type (calcium-dependent) lectin that binds certain sugars found on the surface of many bacteria and some viruses, including the human immunodeficiency virus (HIV). Its recognition properties are very similar to those of MBL (see Fig. 2.15) and, like that protein, it is a multipronged molecule with several carbohydrate-recognition domains. Because it is a transmembrane cell-surface receptor, however, it can function directly as a phagocytic receptor.

A second set of phagocytic receptors, called **scavenger receptors**, recognize various anionic polymers and acetylated low-density lipoproteins. These receptors are a structurally heterogeneous set of molecules, existing in at least six distinct molecular forms. Some scavenger receptors recognize structures that are shielded by sialic acid on normal host cells. There are other recognition targets, many of which still need to be characterized.

Not all receptors that recognize pathogen-specific molecules are phagocytic receptors. Bacterial polypeptides typically start with a formylated methionine residue, and the fMet-Leu-Phe (fMLP) receptor on macrophages and neutrophils binds these *N*-formylated peptides. This receptor is a chemotactic receptor and its ligation guides neutrophils to a site of infection. Binding of pathogens to some receptors on the macrophage surface not only stimulates phagocytosis but also sends signals to the cell that trigger the induced responses of innate immunity, as we will see later in this chapter. Stimulation of certain receptors by pathogen products also leads macrophages and dendritic cells to produce the cell-surface display of co-stimulatory molecules that enables them to act as antigen-presenting cells to T lymphocytes and initiate an adaptive immune response. The best-defined activation pathway of

te immune recognition by Toll-like receptors	
ke receptor	Ligand
R-1:TLR-2 heterodimer	Peptidoglycan Lipoproteins Lipoarabinomannan (mycobacteria) GPI (*T. cruzi*) Zymosan (yeast)
TLR-2:TLR-6 heterodimer	
TLR-3	dsRNA
TLR-4 dimer (plus MD-2 and CD14)	LPS (Gram-negative bacteria) Lipoteichoic acids (Gram-positive bacteria)
TLR-5	Flagellin
TLR-7	ssRNA
TLR-8	G-rich oligonucleotides
TLR-9	Unmethylated CpG DNA

Fig. 2.16 Innate immune recognition by Toll-like receptors. Each of the TLRs whose specificity is known recognizes one or more microbial molecular patterns, generally by direct interaction with molecules on the pathogen surface. Although some Toll-like receptor proteins form heterodimers (e.g. TLR-1:TLR-2), this is not the rule; TLR-4, for instance, can form only homodimers. GPI, glycosylphosphatidylinositol; *T. cruzi*, the protozoan parasite *Trypanosoma cruzi*; dsRNA, double-stranded RNA; ssRNA, single-stranded RNA.

this type is triggered through a family of evolutionarily conserved transmembrane receptors called the Toll-like receptors that seem to function exclusively as signaling receptors, and which we shall describe next.

2-7 The Toll-like receptors are signaling receptors that distinguish different types of pathogen and help direct an appropriate immune response.

The mammalian **Toll-like receptors** (**TLRs**) belong to an evolutionarily ancient recognition and signaling system, originally discovered as a result of its role in embryonic development in the fruitfly *Drosophila melanogaster*. It was subsequently found to have a role in the defense against bacterial and fungal infections in the adult insect and is now known to have a key role in response to infection in plants, adult insects, and vertebrates, including mammals. The receptor mediating these functions in *Drosophila* is known as Toll, and the homologous proteins in mammals and other animals are therefore known as the Toll-like receptors.

There are 10 expressed *TLR* genes in mice and humans, and each of the 10 proteins they produce is devoted to recognizing a distinct set of molecular patterns that are not found in normal vertebrates. These patterns are characteristic of components of pathogenic microorganisms at one or other stage of infection. Because there are only 10 functional *TLR* genes, the Toll-like receptors have limited specificity compared with the antigen receptors of the adaptive immune system and have evolved to recognize certain microbe-associated molecular patterns. But although the diversity of the Toll-like receptors is limited, they can recognize elements of most pathogenic microbes, as illustrated in Fig. 2.16.

Some mammalian TLRs act as cell-surface receptors, whereas others act intracellularly and are located in the membranes of endosomes, where they sense the presence of pathogens and pathogen components taken into the cell by endocytosis or macropinocytosis (Fig. 2.17). An important Toll-like receptor in the response to common bacterial infections is **TLR-4** on macrophages, which signals the presence of bacterial lipopolysaccharide by associating with CD14, the macrophage receptor for lipopolysaccharide, and an additional cellular protein, MD-2. TLR-4 is also involved in the immune response to at least one virus, the respiratory syncytial virus, although in this case the exact nature of the stimulating ligand is not yet known. Another mammalian Toll-like receptor, **TLR-2**, signals the presence of a different set of microbial constituents, which include the lipoteichoic acid (LTA) of Gram-positive bacteria and lipoproteins of Gram-negative bacteria, although how it recognizes these is not known. The responses made by cells to stimulation of the various TLRs are directed to dealing with the particular type of pathogen present. For example, stimulation of **TLR-3** by virus-derived double-stranded RNA leads to the production of an antiviral cytokine, interferon, as discussed in more detail later in the chapter. TLR-4 and TLR-2 induce similar but distinct signals, as shown by the

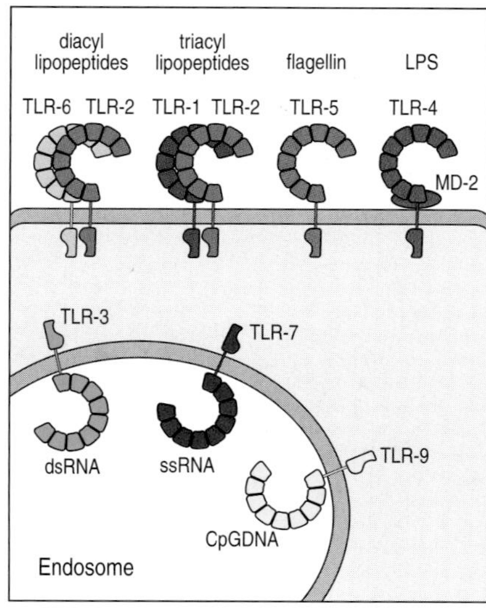

Fig. 2.17 The cellular locations of the mammalian Toll-like receptors. Some TLRs are located on the cell surface of dendritic cells and macrophages, where they are able to detect extracellular pathogen molecules. TLRs are thought to act as dimers; only those that form heterodimers are shown in dimeric form here. The rest act as homodimers. TLRs located intracellularly, in the wall of the endosome, can recognize microbial components, such as DNA, that are accessible only after the microbe has been broken down. The diacyl and triacyl peptides recognized by the heterodimeric receptors TLR-6:TLR-2 and TLR-1:TLR-2, respectively, are derived from the lipoteichoic acid of Gram-positive bacterial cell walls and the lipoproteins of the Gram-negative bacterial surface.

distinct responses resulting from lipopolysaccharide signaling through TLR-4 and LTA signaling through TLR-2; for example, while both lipopolysaccharide and LTA induce the production of TNF-α, lipopolysaccharide is also able to induce the production of interferon (IFN)-β.

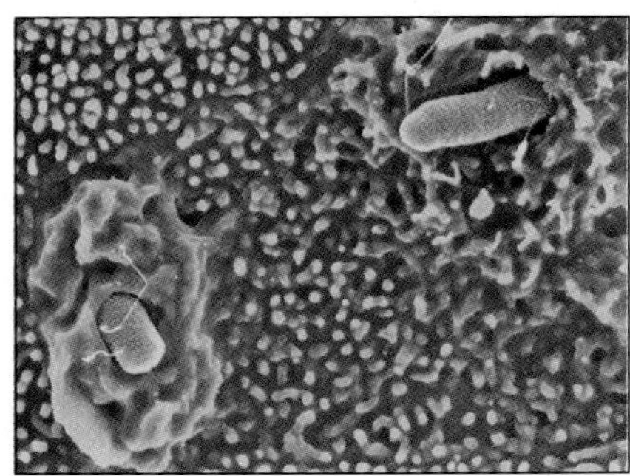

Fig. 2.18 Certain pathogens can invade directly through the intestinal or other internal epithelia. In the case of *Salmonella typhi*, which causes typhoid fever, shown here in the process of migrating through the intestinal epithelium, the flagellin proteins of the bacterial flagella are recognized by TLRs on macrophages and dendritic cells in the underlying tissues. This provokes an innate response that helps to control the infection. Photograph courtesy of J. Galan.

2-8 The effects of bacterial lipopolysaccharide on macrophages are mediated by CD14 binding to TLR-4.

Bacterial lipopolysaccharide (**LPS**) is a cell-wall component of Gram-negative bacteria such as *Salmonella* that has long been known to induce a reaction in the infected host. The systemic injection of LPS causes a collapse of the circulatory and respiratory systems, a condition known as **shock**. These dramatic effects of LPS are seen in humans as **septic shock**, which is caused by the overwhelming secretion of cytokines, particularly of TNF-α, as a result of an uncontrolled systemic bacterial infection, or **sepsis**. We will discuss the pathogenesis of septic shock later in this chapter and shall see that it is an undesirable consequence of the same effector actions of TNF-α that are important in containing local infections. LPS acts through TLR-4, and the benefits of TLR-4 signaling are clearly illustrated by mutant mice that lack TLR-4 function: although resistant to septic shock, they are highly sensitive to LPS-bearing pathogens such as *Salmonella typhimurium*, a natural pathogen of mice.

The same principles are seen in humans infected with *Salmonella typhi*, the cause of typhoid fever. This bacterium invades across the mucosal membranes (Fig. 2.18) but can be recognized by macrophages and other innate phagocytic cells because it expresses both LPS and flagellin, which allows it to activate two cell-surface TLRs—TLR-4 and **TLR-5** (see Fig. 2.16)—which leads to the production of TNF-α. Thus, systemic infection with *S. typhi* is capable of causing systemic shock in humans by the same mechanism provoked by *S. typhimurium* in mice, by inducing a systemic production of TNF-α.

However, *S. typhi*, in common with many pathogenic bacteria, has what is called a type III secretion system, a microhomolog of a syringe, which enables the bacterium to secrete molecules directly across the cell membrane of mammalian cells and into the cytosol. Using its type III secretion system, *S. typhi* is able to transfer into the macrophage cytosol protease molecules that inhibit the signaling pathway that leads to production of TNF-α; this may be a mechanism evolved by the bacterium to render the innate immune response less effective.

TLR-4 alone cannot recognize LPS; two other cell-surface proteins, CD14 and MD-2, are required. CD14 binds LPS and it is thought that the CD14:LPS complex is the actual ligand of TLR-4, although its direct binding to TLR-4 has yet to be demonstrated. MD-2 initially binds to TLR-4 within the cell and is necessary for the correct targeting of TLR-4 to the surface and its recognition of LPS. When the TLR-4:MD-2 complex interacts with LPS bound to CD14, this sends a signal to the cell nucleus that activates the transcription factor **NFκB** (Fig. 2.19). We will describe the signaling pathways used by TLRs in more detail in Chapter 6.

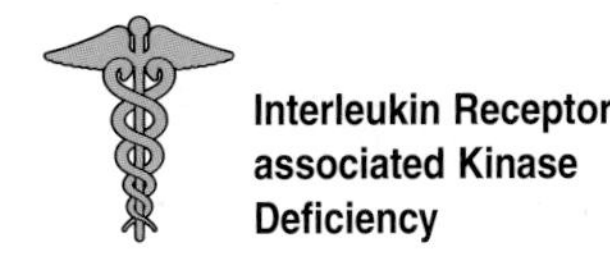

This NFκB signaling pathway was first discovered as the pathway used by the Toll receptor to determine dorso-ventral body pattern during embryogenesis in the fruitfly, and is often referred to as the Toll pathway. In the adult fly the same pathway leads to the formation of antimicrobial peptides in response to infection. Essentially the same signaling pathway is used by all the TLRs in their induction of innate immune responses in vertebrates, and a similar pathway is used by plants in their defense against viruses and other plant pathogens. Thus, the Toll pathway is an ancient signaling pathway that seems to be used in innate immunity in most, if not all, multicellular organisms.

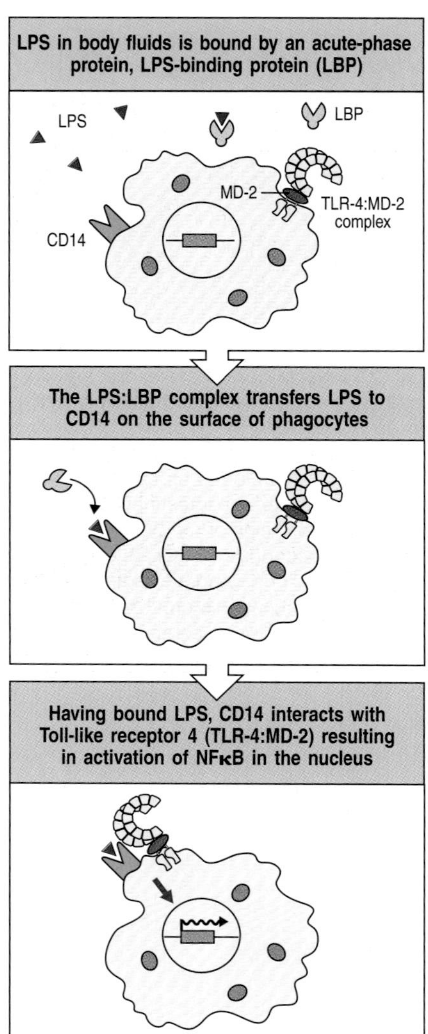

Fig. 2.19 Bacterial lipopolysaccharide signals through the Toll-like receptor TLR-4 to activate the transcription factor NFκB. In the plasma, LPS is bound by the soluble LPS-binding protein (LBP), which then loads its bound LPS onto the glycosylphosphatidylinositol (GPI)-tethered peripheral membrane protein CD14. This LPS:CD14 complex then triggers TLR-4, which is complexed with the protein MD-2, to signal to the nucleus, activating the transcription factor NFκB, which in turn activates genes encoding proteins involved in defense against infection.

2-9 The NOD proteins act as intracellular sensors of bacterial infection.

The TLRs are located in cellular membranes, either on the cell surface or in intracellular vesicles. Other proteins, sharing features of their ligand-binding domains with the TLRs, are present in the cytosol of the cell and are able to bind microbial products there and activate NFκB, thus initiating the same inflammatory processes as the TLRs (Fig. 2.20). These proteins are called **NOD1** and **NOD2**, because as well as a ligand-binding domain they contain a nucleotide-binding oligomerization (NOD) domain. They also contain protein domains that recruit caspases, a family of intracellular proteases, so the genes that encode the NOD proteins are referred to as members of the CARD family of genes—NOD1 is encoded by *CARD4* and NOD2 by *CARD15*.

The NOD proteins recognize fragments of bacterial cell-wall proteoglycans—NOD1 binds γ-glutamyl diaminopimelic acid (iE-DAP), a breakdown product of proteoglycans from Gram-negative bacteria, whereas NOD2 binds muramyl dipeptide, which is present in the proteoglycans of both Gram-positive and Gram-negative bacteria. Thus, NOD2 is able to act as a general sensor of bacterial infection, whereas NOD1 is more restricted to sensing the presence of Gram-negative bacteria. In keeping with this role, the NOD proteins are expressed in cells that are routinely exposed to bacteria—in epithelial cells that form the barrier that bacteria must cross to establish an infection in the body, and in the macrophages and dendritic cells that ingest bacteria that have succeeded in entering the body. As macrophages and dendritic cells also express TLRs that can recognize bacterial proteoglycans, in these cells signals from NOD1 and NOD2 act in addition to the signals from the TLRs. In epithelial cells, however, the expression of TLRs is weak or absent, and in these cells NOD1 is an important activator of the innate immune response. NOD2 seems to have a more specialized role, being strongly expressed in the Paneth cells of the gut, where it induces the expression of potent antimicrobial peptides, the α-defensins.

2-10 Activation of Toll-like receptors and NOD proteins triggers the production of pro-inflammatory cytokines and chemokines, and the expression of co-stimulatory molecules.

In humans and all other vertebrates investigated, activation of NFκB by the Toll and NOD pathways leads to the production of several important mediators of innate immunity, such as cytokines (Fig. 2.21) and chemokines (see Fig. 2.46). (Appendices III and IV provide a detailed listing of these important mediators.) The pathway also leads to the cell-surface expression of **co-stimulatory molecules** that are essential for the induction of adaptive immune responses. These proteins, called **B7.1** (**CD80**) and **B7.2** (**CD86**), are produced by both macrophages and tissue dendritic cells in response to LPS signaling through TLR-4 (Fig. 2.22). It is these proteins, along with the antigenic microbial peptides presented by MHC class II proteins on dendritic cells and macrophages (see Section 1-18), that activate naive CD4 T cells (Fig. 2.23). These cells, in turn, are needed to initiate most adaptive immune responses. To encounter a naive CD4 T cell, the antigen-presenting dendritic cell must migrate to a nearby lymph node through which circulating naive T cells pass, and this migration is stimulated by cytokines such as TNF-α, which are also induced by signaling through TLR-4. Thus, the activation of adaptive immunity depends on molecules induced as a consequence of the innate immune recognition of pathogens.

Substances such as LPS that induce co-stimulatory activity have been used for years in mixtures that are co-injected with protein antigens to enhance their immunogenicity. These substances are known as **adjuvants** (see

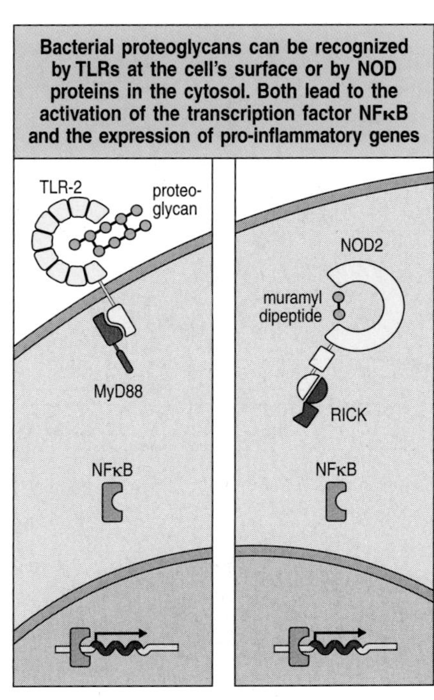
Bacterial proteoglycans can be recognized by TLRs at the cell's surface or by NOD proteins in the cytosol. Both lead to the activation of the transcription factor NFκB and the expression of pro-inflammatory genes
TLR-2
proteo-glycan
MyD88
NFκB
NOD2
muramyl dipeptide
RICK
NFκB

Fig. 2.20 Both membrane-bound and intracellular proteins act as bacterial sensors, recognizing bacterial proteoglycans and activating NFκB to induce the expression of pro-inflammatory genes. TLRs at the cell surface are able to bind microbial components—in the case of TLR-2, these are proteoglycans of the bacterial cell wall. Binding of bacterial proteoglycans to TLR-2 is signaled to the cell via the adaptor protein MyD88, leading to the activation of the transcription factor NFκB and its translocation into the nucleus to induce the expression of pro-inflammatory genes. Degradation of bacterial proteoglycans produces muramyl dipeptide, the ligand for the intracellular sensor of bacterial components, NOD2. By way of the adaptor protein kinase RICK (receptor-interacting serine–threonine kinase), NOD2 is able to activate NFκB, thus inducing the same pro-inflammatory genes as TLR-2.

Appendix I, Section A-4), and it was found empirically that the best adjuvants contained microbial components. A range of microbial components (see Fig. 2.16) can induce macrophages and tissue dendritic cells to express co-stimulatory molecules and cytokines. The exact profile of cytokines produced by the macrophage or dendritic cell varies according to the receptors stimulated and, as we shall see in Chapters 8 and 10, the cytokines secreted will in turn influence the functional character of the adaptive immune response that develops. In this way the ability of the innate immune system to discriminate between different types of pathogen is used to ensure an appropriate type of adaptive immune response.

Summary.

The innate immune system uses several different receptors to recognize and respond to pathogens. Those that recognize pathogen surfaces directly often bind to repeating patterns, for example of carbohydrate or lipid moieties, that are characteristic of microbial surfaces but are not found on host cells. Some of these receptors, such as the macrophage mannose receptor, directly

Cytokines secreted by macrophages and dendritic cells			
Cytokine	**Main producer**	**Acts upon**	**Effect**
IL-1	Macrophages Keratinocytes	Lymphocytes	Enhances responses
		Liver	Induces acute-phase protein secretion
IL-6	Macrophages Dendritic cells	Lymphocytes	Enhances responses
		Liver	Induces acute-phase protein secretion
CXCL8 (IL-8)	Macrophages Dendritic cells	Phagocytes	Chemoattractant for neutrophils
IL-12	Macrophages Dendritic cells	Naive T cells	Diverts immune response to T_H1, pro-inflammatory, cytokine secretion
TNF-α	Macrophages Dendritic cells	Vascular endothelium	Induces changes in vascular endothelium (expression of cell-adhesion molecules (E- and P-selectin), changes in cell–cell junctions with increased fluid loss, local blood clotting)

Fig. 2.21 Important cytokines secreted by macrophages in response to bacterial products include IL-1β, IL-6, CXCL8, IL-12, and TNF-α. TNF-α is an inducer of a local inflammatory response that helps to contain infections; it also has systemic effects, many of which are harmful (see Section 2-27). The chemokine CXCL8 is also involved in the local inflammatory response, helping to attract neutrophils into the site of infection. IL-1β, IL-6, and TNF-α have a crucial role in inducing the acute-phase response in the liver (see Section 2-28) and induce fever, which favors effective host defense in several ways. IL-12 activates the natural killer (NK) cells of the innate immune response, and favors the differentiation of CD4 T cells into the T_H1 subset during an adaptive immune response.

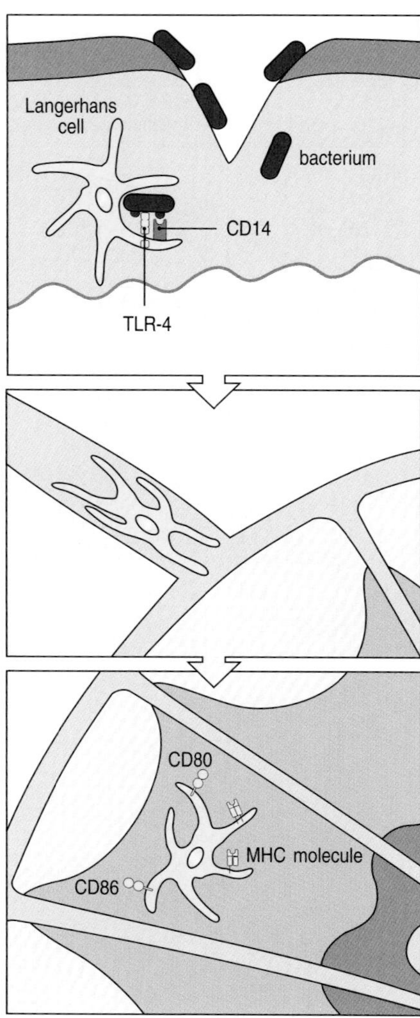

Fig. 2.22 Bacterial LPS induces changes in Langerhans cells, stimulating them to migrate and to initiate adaptive immunity by activating CD4 T cells. Langerhans cells are immature dendritic cells resident in the skin. During a bacterial infection they are activated by LPS through the TLR signaling pathway. This induces two types of changes in the cells. The first is a change in behavior and location. From being resting cells in the skin, the Langerhans cells become activated migrating cells in the afferent lymphatics, and eventually fully mature dendritic cells in the regional lymph nodes. The second is a drastic alteration in their cell-surface molecules. Resting Langerhans cells in the skin are highly phagocytic and macropinocytic but lack the ability to activate T lymphocytes. Mature dendritic cells in the lymph nodes have lost the ability to ingest antigen but have gained the ability to stimulate T cells. This is due to an increase in the number of MHC molecules on their surface and the expression of the co-stimulatory molecules CD80 (B7.1) and CD86 (B7.2).

stimulate phagocytosis, whereas others are produced as secreted molecules that promote the phagocytosis of pathogens by opsonization or by the activation of complement, as we shall see in the next part of this chapter. The innate immune system receptors that recognize pathogens also have an important role in signaling for the induced innate responses that are responsible for local inflammation, the recruitment of new effector cells, the containment of local infection, and the initiation of an adaptive immune response. Such signals can be transmitted through a family of signaling receptors, known as the Toll-like receptors (TLRs), that have been highly conserved across evolutionary time and serve to activate host defense through a signaling pathway that operates in most multicellular organisms. In vertebrates, TLRs also have a key role in enabling the initiation of adaptive immunity. TLR-4 detects the presence of Gram-negative bacteria through its association with the peripheral membrane protein CD14, which is a receptor for bacterial LPS. Other TLRs respond to other molecular patterns found on or in pathogens. TLRs activate the transcription factor NFκB, which then induces the transcription of a variety of genes, including those for the cytokines, chemokines, and co-stimulatory molecules that have essential roles in directing the course of the adaptive immune response later in infection. Whereas TLRs recognize the presence of bacteria and other microbes outside the cell, cytosolic proteins, the NOD proteins, detect similar bacterial products within the cytoplasm of the cell, activating the same NFκB pathway.

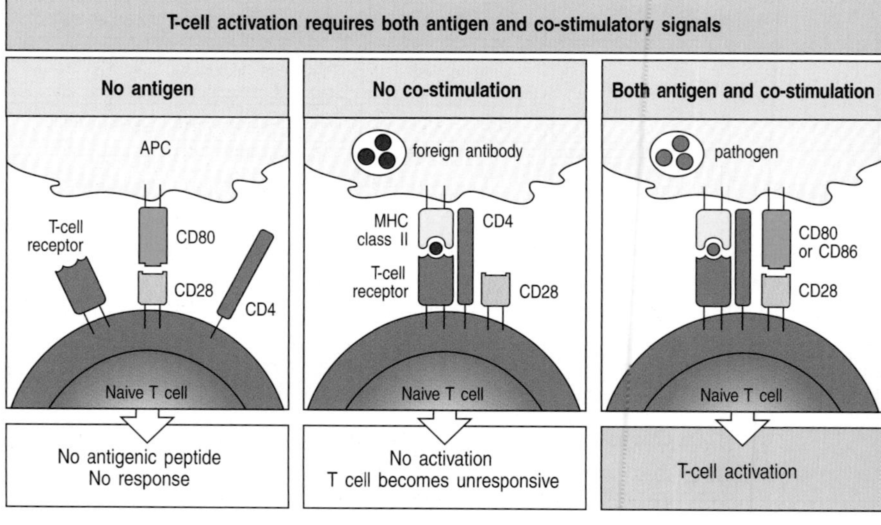

Fig. 2.23 For naive T cells to be activated by antigen, the antigen must be presented to them by an activated antigen-presenting cell that also expresses co-stimulatory molecules. Antigen is recognized by the T-cell receptor in the form of a peptide bound to an MHC molecule on an antigen-presenting cell (APC), such as a macrophage or dendritic cell. However, the T cell will be activated only if the antigen-presenting cell also expresses the co-stimulatory molecules CD80 or CD86.

The complement system and innate immunity.

Complement was discovered by **Jules Bordet** many years ago as a heat-labile component of normal plasma that augments the opsonization and killing of bacteria by antibodies. This activity was said to 'complement' the antibacterial activity of antibody; hence the name. Although first discovered as an effector arm of the antibody response, complement can also be activated early in infection in the absence of antibodies. Indeed, it now seems clear that complement first evolved as part of the innate immune system, where it still has an important role in coating pathogens and facilitating their destruction.

2-11 Complement is a system of plasma proteins that is activated by the presence of pathogens.

The **complement system** is made up of a large number of different plasma proteins that interact with one another both to opsonize pathogens and to induce a series of inflammatory responses that help to fight infection. A feature of the system is that several complement proteins are proteases that become activated only after cleavage, usually by another specific protease. In their inactive form, such enzymes are called pro-enzymes or **zymogens** and were first found in the gut. The digestive enzyme pepsin, for example, is stored inside cells and secreted as an inactive precursor, pepsinogen, which is only cleaved to pepsin in the acid environment of the stomach. The advantage to the host of not being autodigested is obvious.

The precursor zymogens of the complement system are widely distributed throughout body fluids and tissues. At sites of infection they are locally activated by the presence of the pathogen and trigger a series of powerful inflammatory events. The complement system becomes activated through a triggered-enzyme cascade in which an active complement protease generated by cleavage of its zymogen precursor then cleaves its substrate, another complement zymogen, to its active enzymatic form. This in turn cleaves and activates the next zymogen in the complement pathway. In this way, the activation of a small number of complement proteins at the start of the pathway is hugely amplified by each successive enzymatic reaction, resulting in the rapid generation of a disproportionately large complement response. The blood coagulation system mentioned earlier is another example of a triggered-enzyme cascade. In this case, a small injury to a blood vessel wall can lead to the development of a large blood clot.

A key site for the activation of the complement pathway is the surface of pathogens and there are three distinct pathways through which complement can be activated (Fig. 2.24). These pathways depend on different molecules for their initiation, but they converge to generate the same set of effector complement proteins. There are also three ways in which the complement system protects against infection (see Fig. 2.24). First, it generates large numbers of activated complement proteins that bind covalently to pathogens, opsonizing them for engulfment by phagocytes bearing receptors for complement. Second, the small fragments of some complement proteins act as chemoattractants to recruit more phagocytes to the site of complement activation, and also to activate these phagocytes. Third, the final components in the complement pathway damage certain bacteria by creating pores in the bacterial membrane.

As well as the direct effects of complement in eliminating infectious microorganisms, it also has an important role in activating the adaptive immune system. In part, this is a consequence of opsonization, because

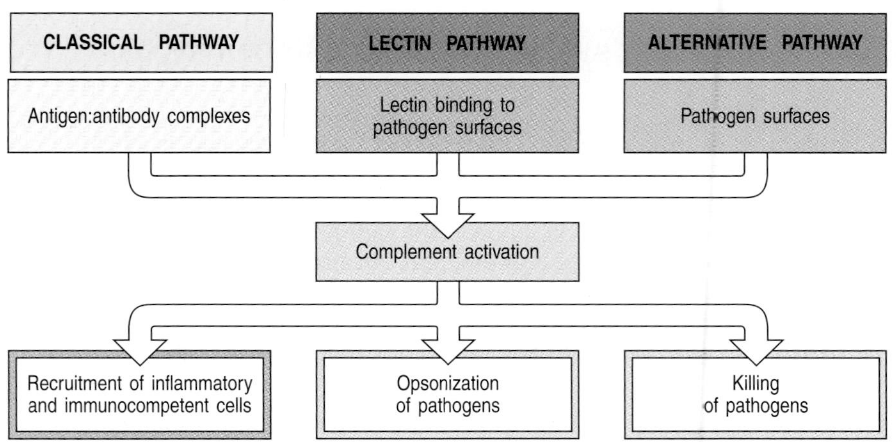

Fig. 2.24 Schematic overview of the complement cascade. There are three pathways of complement activation. One is the classical pathway, which is triggered by the binding of complement component C1q to antibody complexed with antigen by the direct binding of C1q to the pathogen surface, or by the binding of C1q to C-reactive protein bound to the pathogen. The second is the lectin pathway, which is triggered by mannose-binding lectin or the ficolin proteins, normal serum constituents that bind some encapsulated bacteria. The third is the alternative pathway, which is triggered directly on pathogen surfaces. All of these pathways generate a crucial enzymatic activity that, in turn, generates the effector molecules of complement. The three main consequences of complement activation are the opsonization of pathogens, the recruitment of inflammatory and immunocompetent cells, and the direct killing of pathogens.

antigen-presenting cells bear receptors for complement that enhance the uptake of complement-coated antigens and the presentation of these antigens to the adaptive immune system. In addition, as we will see in Chapter 9, B lymphocytes bear receptors for complement proteins that act as co-stimulators, enhancing the response of the B cell to complement-coated antigens.

Complement is not activated only by infectious organisms. Dying cells, such as those at sites of ischemic injury (injury to tissues caused by a lack of oxygen) can trigger complement activation. As complement-coated particles are more efficiently taken up by phagocytes, complement is important in the efficient disposal of dead, damaged, and apoptotic cells and, in doing so, protects against the development of autoimmunity—the attack by the immune system on the body's own antigens.

2-12 Complement interacts with pathogens to mark them for destruction by phagocytes.

In the early phases of an infection, the complement cascade can be activated on the surface of a pathogen through one or more of the three pathways shown in Fig. 2.25. The **classical pathway** is initiated by the binding of C1q, the first protein in the complement cascade, to the pathogen surface. C1q can bind to the surface of pathogens in one of three ways. It can bind directly to surface components of some bacteria, including certain proteins of bacterial cell walls and polyanionic surface structures such as lipoteichoic acid on Gram-positive bacteria. Second, C1q binds to C-reactive protein, an acute-phase protein of human plasma that binds to phosphocholine residues in bacterial polysaccharides, such as pneumococcal C polysaccharide; hence the name C-reactive protein. We shall discuss the acute-phase proteins of the early induced innate response later in the chapter. Third, C1q is a key link between the effector mechanisms of innate and adaptive immunity by binding to antibody:antigen complexes. The **lectin pathway** is initiated by the binding of carbohydrate-binding proteins to arrays of carbohydrates on the surface of pathogens. These carbohydrate-binding proteins include the lectin MBL, which binds to mannose-containing carbohydrates on bacteria or viruses, as described in Section 2-6, and the ficolins, which bind to the *N*-acetylglucosamine present on the surface of some pathogens. Finally, the **alternative pathway** of complement activation can be initiated by the binding of spontaneously activated complement component C3 in plasma to the surface of a pathogen.

In each pathway, a sequence of reactions generates a protease called a **C3 convertase**. These reactions are known as the 'early' events of complement

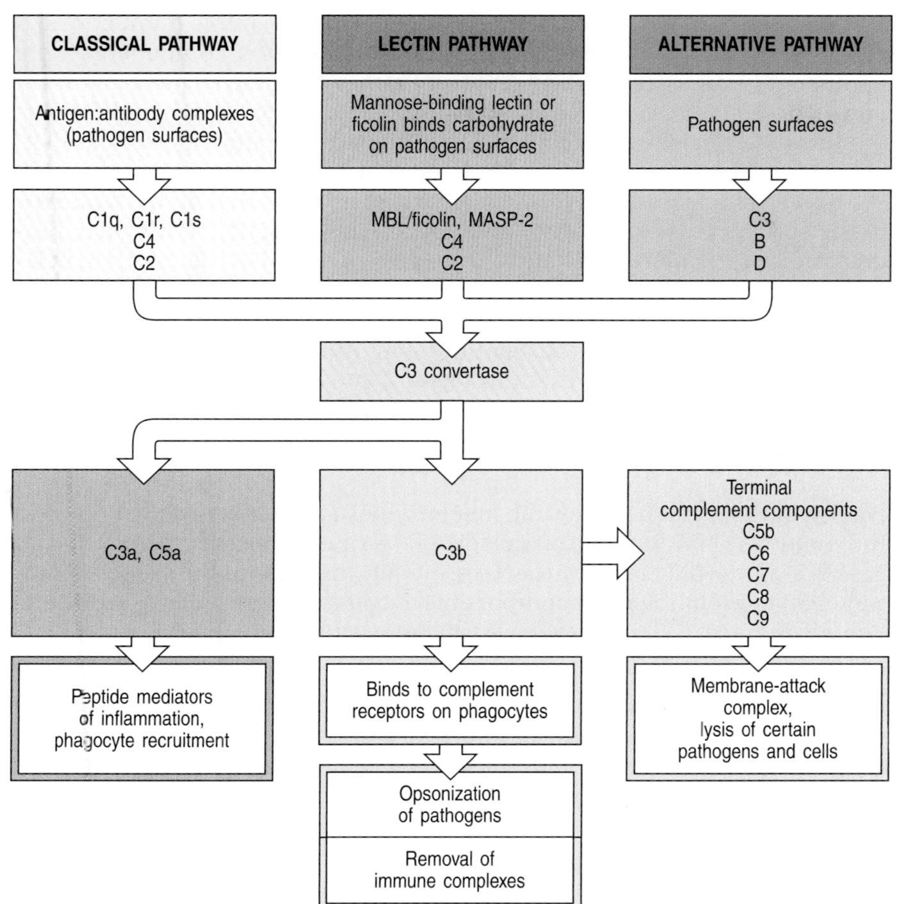

Fig. 2.25 Overview of the main components and effector actions of complement. Early on in all three pathways of complement activation a series of cleavage reactions culminates in the formation of an enzymatic activity called a C3 convertase, which cleaves complement component C3 into C3b and C3a. The production of the C3 convertase is the point at which the three pathways converge and the main effector functions of complement are generated. C3b binds covalently to the bacterial cell membrane and opsonizes the bacteria, enabling phagocytes to internalize them. C3a is a peptide mediator of local inflammation. C5a and C5b are generated by the cleavage of C5b by a C5 convertase formed by C3b bound to the C3 convertase (not shown in this simplified diagram). C5a is also a powerful peptide mediator of inflammation. C5b triggers the late events in which the terminal components of complement assemble into a membrane-attack complex that can damage the membrane of certain pathogens.

activation, and consist of triggered-enzyme cascades in which complement zymogens are successively cleaved to yield two fragments, the larger of which is an active serine protease. The active protease is retained at the pathogen surface; this ensures that the next complement zymogen in the pathway is also cleaved and activated at the pathogen surface. In contrast, the small peptide fragment is released from the site of the reaction and can act as a soluble mediator of inflammation.

The C3 convertases formed by these early events of complement activation are bound covalently to the pathogen surface. Here they cleave C3 to generate large amounts of **C3b**, the main effector molecule of the complement system, and C3a, a peptide mediator of inflammation. The C3b molecules act as opsonins; they bind covalently to the pathogen and thereby target it for destruction by phagocytes equipped with receptors for C3b. C3b also binds to the C3 convertase to form a **C5 convertase** that produces the most important and most potent inflammatory peptide, **C5a**, as well as a large active fragment, **C5b**, that initiates the 'late' events of complement activation. These comprise a sequence of polymerization reactions in which a set of complement proteins known as the terminal components interact to form a **membrane-attack complex**, which creates a pore in the cell membranes of certain pathogens that can lead to their death.

The nomenclature of complement proteins is an obstacle to understanding the system, and before discussing the complement cascade in more detail we will explain their naming. All components of the classical complement pathway and the membrane-attack complex are designated by the letter C

Functional protein classes in the complement system	
Binding to antigen:antibody complexes and pathogen surfaces	C1q
Binding to mannose on bacteria	MBL
Activating enzymes	C1r C1s C2a Bb D MASP-2
Membrane-binding proteins and opsonins	C4b C3b
Peptide mediators of inflammation	C5a C3a C4a
Membrane-attack proteins	C5b C6 C7 C8 C9
Complement receptors	CR1 CR2 CR3 CR4 C1qR
Complement-regulatory proteins	C1INH C4bp CR1 MCP DAF H I P CD59

Fig. 2.26 Functional protein classes in the complement system.

followed by a number. The native components have a simple number designation, for example, C1 and C2, but unfortunately the components were numbered in the order of their discovery rather than the sequence of reactions. The reaction sequence is C1, C4, C2, C3, C5, C6, C7, C8, and C9. The products of the cleavage reactions are designated by added lower-case letters, the larger fragment being designated b and the smaller a; thus, for example, C4 is cleaved to C4b, the large fragment of C4 that binds covalently to the surface of the pathogen, and C4a, a small fragment with weak pro-inflammatory properties. There is one exception to this rule of nomenclature. For C2, the larger fragment was originally named C2a, and it is this larger C2a component that contains the enzymatic activity. The components of the alternative pathway, instead of being numbered, are designated by different capital letters, for example factor B and factor D. As with the classical pathway, their cleavage products are designated by the addition of lower-case a and b: thus, the large fragment of B is called Bb and the small fragment Ba.

Finally, in the lectin pathway, the first enzymes to be activated are known as the *m*annose-binding-lectin-*a*ssociated *s*erine *p*roteases, **MASP-1** and **MASP-2**, after which the pathway is essentially the same as the classical pathway. Activated complement components are often designated by a horizontal line, for example, $\overline{\text{C2a}}$; however, we will not use this convention.

The formation of C3 convertase activity is pivotal in complement activation, leading to the production of the principal effector molecules, and initiating the late events. In the classical and lectin pathways the C3 convertase is formed from membrane-bound C4b complexed with C2a, designated C4b2a. In the alternative pathway, a homologous C3 convertase is formed from membrane-bound C3b complexed with Bb, C3bBb. The alternative pathway can act as an amplification loop for all three pathways, because it is initiated by the binding of C3b.

It is clear that a pathway leading to such potent inflammatory and destructive effects—and which, moreover, has a series of built-in amplification steps—is potentially dangerous and must be subject to tight regulation. One important safeguard is that key activated complement components are rapidly inactivated unless they bind to the pathogen surface on which their activation was initiated. There are also several points in the pathway at which regulatory proteins act on complement components to prevent the inadvertent activation of complement on host-cell surfaces, thereby protecting them from accidental damage. We will return to these regulatory mechanisms later.

We have now introduced all the relevant components of complement and are ready for a more detailed account of their functions. To help in distinguishing the different components according to their functions, we will use a color code in the tables in this part of the chapter. This is introduced in Fig. 2.26, where all the components of complement are grouped by function.

2-13 The classical pathway is initiated by activation of the C1 complex.

The classical pathway has a role in both innate and adaptive immunity. As we will see in Chapter 9, the first component of this pathway, C1q, links the adaptive humoral immune response to the complement system by binding to antibodies complexed with antigens. However, the classical pathway can be also be activated during innate immune responses. Antibodies called **natural antibodies** are produced by the immune system in the apparent absence of any infection. They have a broad specificity for self and microbial antigens, can react with many pathogens and, as in adaptive immunity, can activate complement through the binding of C1q. We will describe natural antibodies

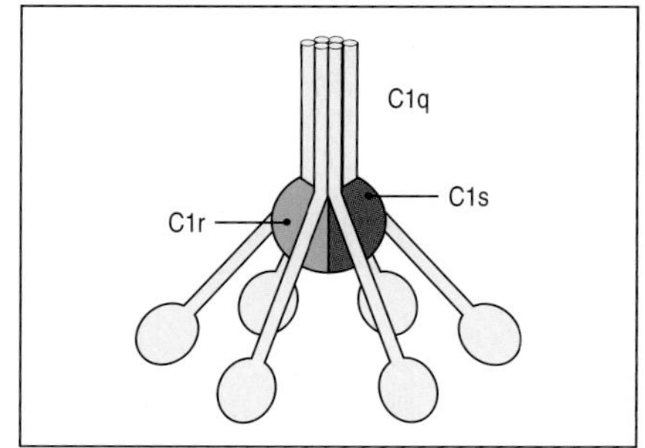

Fig. 2.27 The first protein in the classical pathway of complement activation is C1, which is a complex of C1q, C1r, and C1s. C1q is composed of six identical subunits with globular heads and long collagen-like tails, and has been described as looking like 'a bunch of tulips'. The tails combine to bind to two molecules each of C1r and C1s, forming the C1 complex C1q:C1r_2:C1s_2. The heads can bind to the constant regions of immunoglobulin molecules or directly to the pathogen surface, causing a conformational change in C1r, which then cleaves and activates the C1s zymogen. Photograph (× 500,000) courtesy of K.B.M. Reid.

and the specialized subset of lymphocytes that produce them in more detail in Section 2-34. Here it is only important to note that most of the natural antibody that is produced is of the class known as IgM and that this is the class that is most efficient at binding C1q; thus, natural antibodies provide an effective means by which complement activation can be directed to pathogen surfaces immediately upon infection.

An additional function of C1q in innate immunity is that it can bind directly to the surface of certain pathogens and thus trigger complement activation in the absence of antibody. It can, for example, bind to C-reactive protein bound to phosphocholine on bacteria. Therefore, the activation of C1 by natural antibody and directly by pathogen surfaces represents an important humoral arm of innate immunity.

C1q is part of the **C1 complex**, which comprises a single C1q molecule bound to two molecules each of the zymogens C1r and C1s. C1q itself is a hexamer, each subunit of which is in turn a trimer, forming a globular domain with a triple-helical collagen-like tail. In the C1q hexamer the six globular heads are linked together by their collagen-like tails, which surround the $(C1r{:}C1s)_2$ complex (Fig. 2.27). Binding of more than one of the C1q heads to a pathogen surface or to the constant region of antibodies, known as the Fc region, in an immune complex of antigen and antibody causes a conformational change in the $(C1r{:}C1s)_2$ complex, which leads to the activation of an autocatalytic enzymatic activity in C1r; the active form of C1r then cleaves its associated C1s to generate an active serine protease.

Once activated, the C1s enzyme acts on the next two components of the classical pathway, cleaving C4 and then C2 to generate two large fragments, C4b and C2a, which together form the C3 convertase of the classical pathway. In the first step, C1s cleaves C4 to produce C4b, which may bind covalently to the surface of the pathogen. The covalently attached C4b then binds one molecule of C2, making it susceptible, in turn, to cleavage by C1s. C1s cleaves C2 to produce the large fragment C2a, which is itself a serine protease. C4b2a, the complex of C4b with the active serine protease C2a, remains covalently linked to the surface of the pathogen as the C3 convertase of the classical pathway. Its most important activity is to cleave large numbers of C3 molecules to produce C3b molecules that coat the pathogen surface. At the same time, the other cleavage product, C3a, initiates a local inflammatory response. These reactions, which comprise the classical pathway of complement activation, are shown in schematic form in Fig. 2.28; the proteins involved, and their active forms, are listed in Fig. 2.29.

2-14 The lectin pathway is homologous to the classical pathway.

The lectin pathway uses proteins very similar to C1q to trigger the complement cascade. One of these proteins is the mannose-binding lectin (MBL) introduced earlier. It binds specifically to mannose residues, and to certain other sugars, which are present on many pathogen surfaces in a pattern that allows binding, as shown schematically in Fig. 2.15. On vertebrate cells, however,

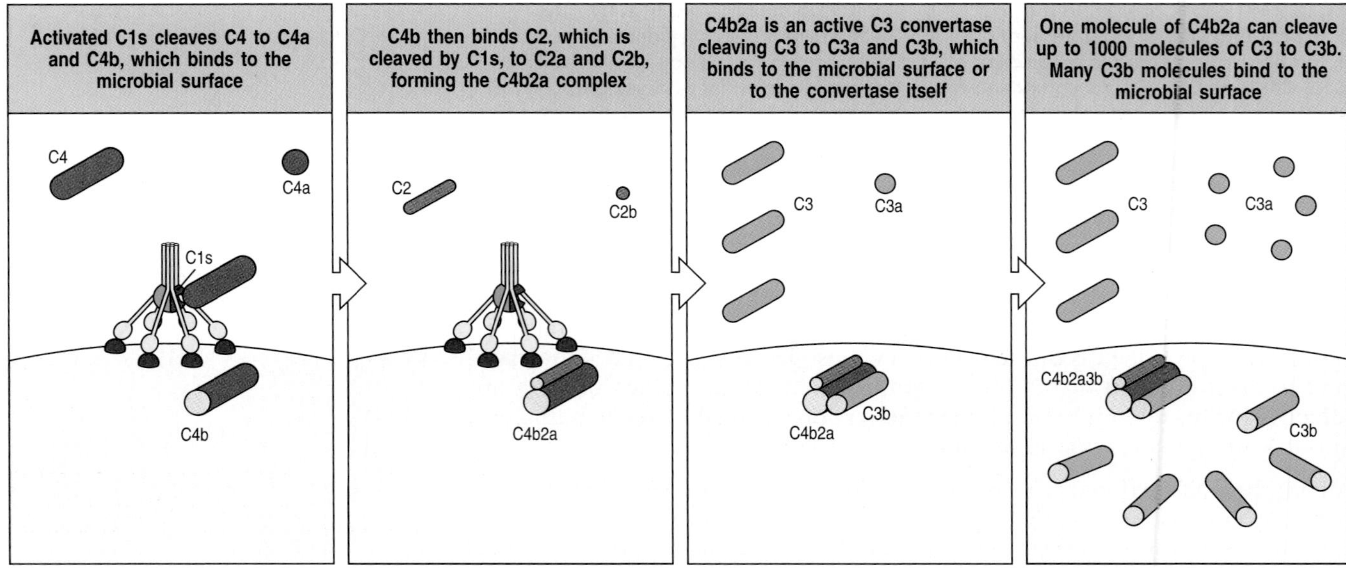

Activated C1s cleaves C4 to C4a and C4b, which binds to the microbial surface
C4
C4a
C1s
C4b
C4b then binds C2, which is cleaved by C1s, to C2a and C2b, forming the C4b2a complex
C2
C2b
C4b2a
C4b2a is an active C3 convertase cleaving C3 to C3a and C3b, which binds to the microbial surface or to the convertase itself
C3
C3a
C4b2a
C3b
One molecule of C4b2a can cleave up to 1000 molecules of C3 to C3b. Many C3b molecules bind to the microbial surface
C3
C3a
C4b2a3b
C3b

Fig. 2.28 The classical pathway of complement activation generates a C3 convertase that deposits large numbers of C3b molecules on the pathogen surface. The steps in the reaction are outlined here and detailed in the text. The cleavage of C4 by C1s exposes a reactive group on C4b that allows it to bind covalently to the pathogen surface. C4b then binds C2, making it susceptible to cleavage by C1s. The larger C2a fragment is the active protease component of the C3 convertase. It cleaves many molecules of C3 to produce C3b, which binds to the pathogen surface, and C3a, an inflammatory mediator.

mannose is covered by other sugar groups, especially by sialic acid. Thus, MBL is able to initiate complement activation by binding to pathogen surfaces while not becoming activated on host cells. It is present at low concentrations in the plasma of most individuals, and, as we will see in the last part of this chapter, its production by the liver is increased during the acute-phase reaction induced as part of the innate immune response.

MBL is a two- to six-headed molecule that, like C1q, forms a complex with two protease zymogens, which in the case of the MBL complex are the serine proteases MASP-1 and MASP-2 (Fig. 2.30). MASP-2 is closely related to C1r and C1s, and MASP-1 is a slightly more distantly related cousin; all four enzymes are likely to have evolved from the duplication of a gene for a common precursor. When the MBL complex binds to a pathogen surface, MASP-2 is activated to cleave C4 and C2. The role, if any, of MASP-1 in complement activation is uncertain; *in vitro* it is able to cleave C2 as efficiently as MASP-2, and so its role may be to enhance the activation of complement, even if it is unable to initiate it. Thus the lectin pathway initiates complement activation in a closely similar way to the classical pathway, forming a C3 convertase from C2a bound to C4b. People deficient in MBL or in MASP-2 experience a substantial increase in infections during early childhood, indicating the importance of the lectin pathway for host defense. The age window of susceptibility to infections associated with MBL deficiency illustrates the particular importance of innate host defense mechanisms in early childhood, which is a time before the child's adaptive immune responses are fully matured and after maternal antibodies transferred across the placenta and in colostrum have been lost.

Related in overall shape and function to MBL and Clq are the **ficolins**, which also bind carbohydrates on microbial surfaces and which, like the collectins, activate complement through the binding and activation of MASP-1 and MASP-2 (see Fig. 2.30). In humans, there are three ficolins: L-, M- and H-ficolin. Ficolins differ from the collectins in that instead of having a lectin domain attached to the collagen-like stalk, they have a fibrinogen-like domain; this fibrinogen-like domain binds carbohydrates and gives the ficolins their general specificity for oligosaccharides containing *N*-acetylglucosamine. In discussing the activation of complement by these innate activation molecules, we have used MBL as our prototype, but the ficolins

Fig. 2.29 The proteins of the classical pathway of complement activation.

Proteins of the classical pathway of complement activation

Native component	Active form	Function of the active form
C1 (C1q: $C1r_2$:$C1s_2$)	C1q	Binds directly to pathogen surfaces or indirectly to antibody bound to pathogens, thus allowing autoactivation of C1r
	C1r	Cleaves C1s to active protease
	C1s	Cleaves C4 and C2
C4	C4b	Covalently binds to pathogen and opsonizes it. Binds C2 for cleavage by C1s
	C4a	Peptide mediator of inflammation (weak activity)
C2	C2a	Active enzyme of classical pathway C3/C5 convertase: cleaves C3 and C5
	C2b	Precursor of vasoactive C2 kinin
C3	C3b	Many molecules of C3b bind to pathogen surface and act as opsonins. Initiates amplification via the alternative pathway. Binds C5 for cleavage by C2a
	C3a	Peptide mediator of inflammation (intermediate activity)

may be more important in practice, because their concentration in plasma is greater than that of MBL.

2-15 Complement activation is largely confined to the surface on which it is initiated.

We have seen that the classical and lectin pathways of complement activation are initiated by proteins that bind to pathogen surfaces. During the triggered-enzyme cascade that follows, it is important that activating events are confined to this same site, so that C3 activation also occurs on the surface of the pathogen and not in the plasma or on host-cell surfaces. This is achieved principally by the covalent binding of C4b to the pathogen surface. Cleavage

Mannose-binding lectin

MASP-1 MASP-1

MASP-2 MASP-2

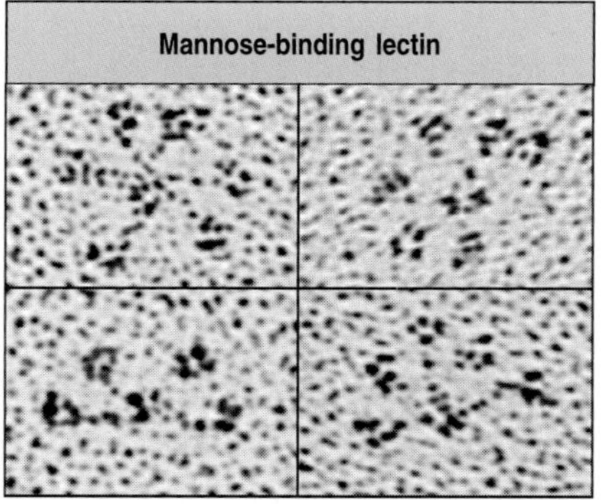

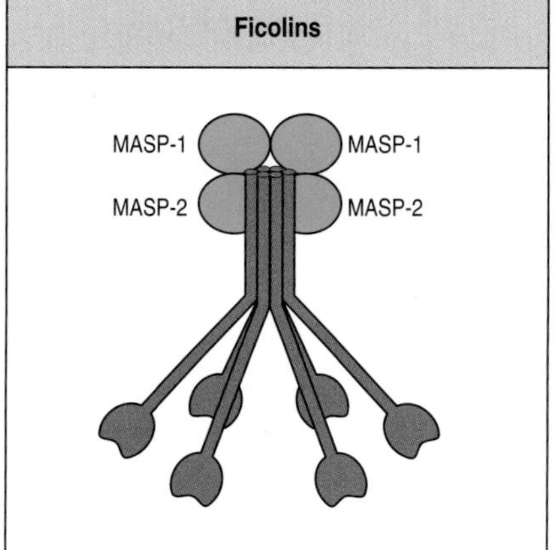

Fig. 2.30 The innate complement-activating molecules form a complex with serine proteases that resembles the complement C1 complex. Mannose-binding lectin (MBL) (top panel) forms clusters of two to six carbohydrate-binding heads around a central stalk formed from the collagen-like tails of the MBL monomers. This structure, easily discernible under the electron microscope (middle panels), closely resembles that of C1q. Associated with this complex are two serine proteases, MBL-associated serine protease 1 (MASP-1) and 2 (MASP-2). The structural disposition of the MASP proteins in the complex is not yet determined but it is likely that they interact with MBL in the same way in which C1r and C1s interact with C1q. On binding of MBL to bacterial surfaces, MASP-2 becomes activated and can then activate the complement system by cleaving and activating C4 and C2. The ficolins (bottom panel) resemble MBL in their overall structure, are associated with MASP-1 and MASP-2, and can activate C4 and C2 after binding to carbohydrate molecules present on microbial surfaces. The carbohydrate-binding domain of ficolins is a fibrinogen-like domain, rather than the lectin domain present in MBL. Photograph of MBL courtesy of K.B.M. Reid.

of C4 exposes a highly reactive thioester bond on the C4b molecule that allows it to bind covalently to molecules in the immediate vicinity of its site of activation. In innate immunity, C4 cleavage is catalyzed by a C1 or MBL complex bound to the pathogen surface, and C4b can bind adjacent proteins or carbohydrates on the pathogen surface. If C4b does not rapidly form this bond, the thioester bond is cleaved by reaction with water and this hydrolysis reaction irreversibly inactivates C4b (Fig. 2.31). This helps to prevent C4b from diffusing from its site of activation on the microbial surface and becoming coupled to host healthy cells.

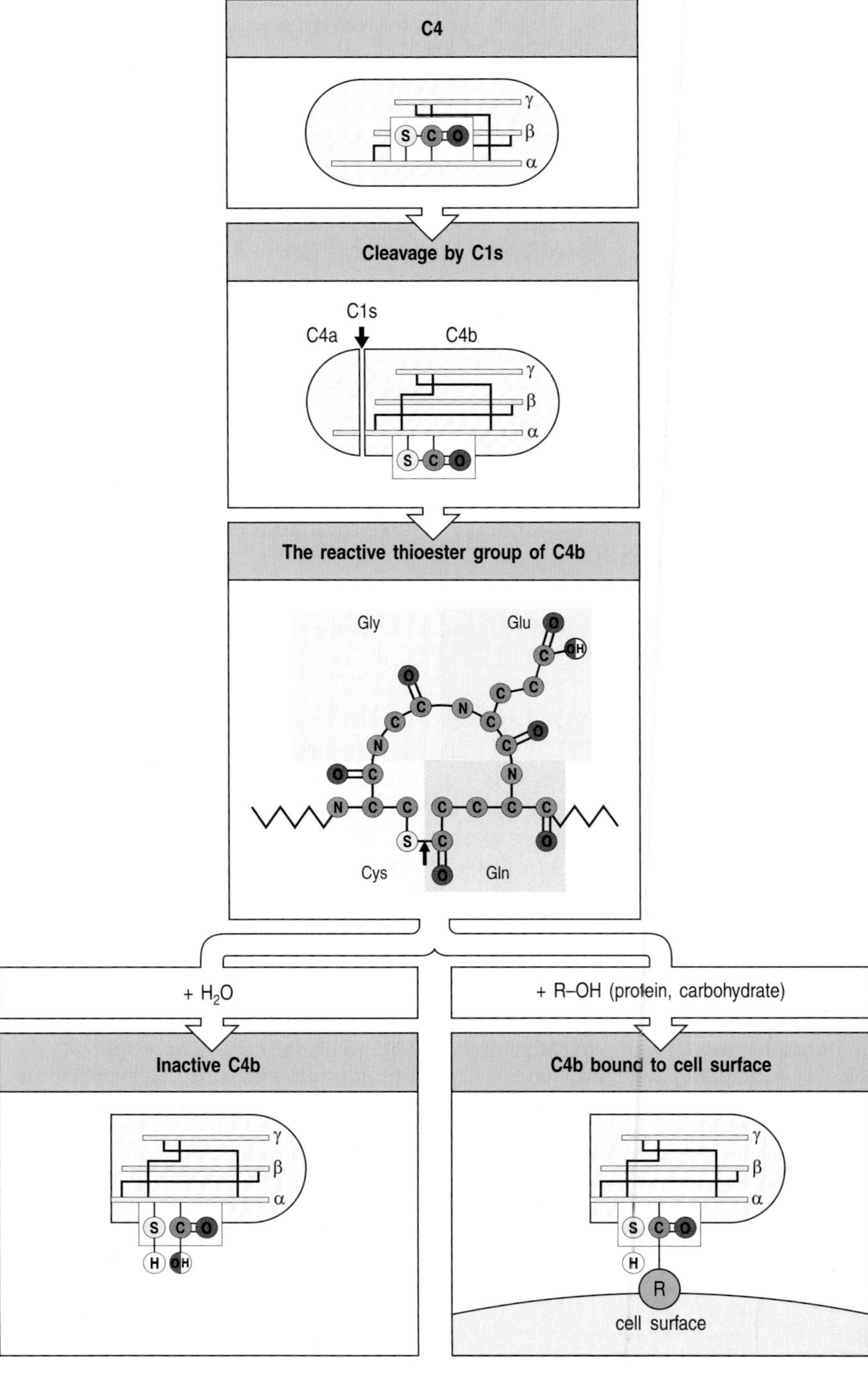

Fig. 2.31 Cleavage of C4 exposes an active thioester bond that causes the large fragment, C4b, to bind covalently to nearby molecules on the bacterial cell surface. Intact C4 consists of an α, a β, and a γ chain with a shielded thioester bond on the α chain. This is exposed when the α chain is cleaved by C1s to produce C4b. The thioester bond (marked by an arrow in the third panel) is rapidly hydrolyzed (that is, cleaved by water), inactivating C4b unless it reacts with hydroxyl or amino groups to form a covalent linkage with molecules on the pathogen surface. The homologous protein C3 has an identical reactive thioester bond that is also exposed on the C3b fragment when C3 is cleaved by C2a. The covalent attachment of C3b and C4b to the pathogen surface enables these molecules to act as opsonins and is important in confining complement activation to pathogen surfaces.

C2 becomes susceptible to cleavage by C1s only when it is bound by C4b, and the C2a serine protease is thereby also confined to the pathogen surface, where it remains associated with C4b, forming the C3 convertase C4b2a. The activation of C3 molecules thus also occurs at the surface of the pathogen. Furthermore, the C3b cleavage product is also rapidly inactivated unless it binds covalently by the same mechanism as C4b, and it therefore opsonizes only the surface on which complement activation has taken place. The opsonization of pathogens by C3b is more efficient when antibodies are bound to the pathogen surface, as phagocytes have receptors for both complement and antibody; these will be described in Chapter 9. Because the reactive forms of C3b and C4b are able to form a covalent bond with any adjacent protein or carbohydrate, when complement is activated by bound antibody a proportion of the reactive C3b or C4b will become linked to the antibody molecules themselves. This combination of antibody chemically cross-linked to complement is likely to be the most efficient trigger for phagocytosis.

2-16 Hydrolysis of C3 causes initiation of the alternative pathway of complement.

The alternative pathway is so named because it was discovered as a second, or 'alternative,' pathway for complement activation after the classical pathway had been defined. This pathway can proceed on many microbial surfaces in the absence of specific antibody, and it leads to the generation of a distinct C3 convertase designated C3bBb. In contrast to the classical and lectin pathways of complement activation, the alternative pathway does not depend on a pathogen-binding protein for its initiation; instead it is initiated through the spontaneous hydrolysis of C3, as shown in the top three panels of Fig. 2.32. The distinctive components of the pathway are listed in Fig. 2.33. Several mechanisms ensure that the activation pathway will proceed only on the surface of a pathogen or on damaged host cells, and not on normal host cells and tissues.

C3 is abundant in plasma, and C3b is produced at a significant rate by spontaneous cleavage (also known as 'tickover'). This occurs through the spontaneous hydrolysis of the thioester bond in C3 to form C3(H_2O), which has an altered conformation, allowing binding of the plasma protein **factor B**. The binding of B by C3(H_2O) then allows a plasma protease called **factor D** to cleave factor B to Ba and Bb, the latter remaining associated with C3(H_2O) to form the C3(H_2O)Bb complex. This complex is a fluid-phase C3 convertase, and although it is only formed in small amounts it can cleave many molecules of C3 to C3a and C3b. Much of this C3b is inactivated by hydrolysis, but some attaches covalently, through its reactive thioester group, to the surfaces of pathogens or of host cells. C3b bound in this way is able to bind factor B, allowing its cleavage by factor D to yield the small fragment Ba and the active protease Bb. This results in formation of the alternative pathway C3 convertase, C3bBb (Fig. 2.34).

2-17 Membrane and plasma proteins that regulate the formation and stability of C3 convertases determine the extent of complement activation under different circumstances.

The extent of complement activation is critically dependent on the stability of the C3bBb convertase. This stability is controlled by both positive and negative regulatory proteins. Normal host cells are protected from complement activation by several negative regulatory proteins, present in the plasma and in host-cell membranes, that protect normal host cells from the injurious effects of inappropriate complement activation. These proteins interact with

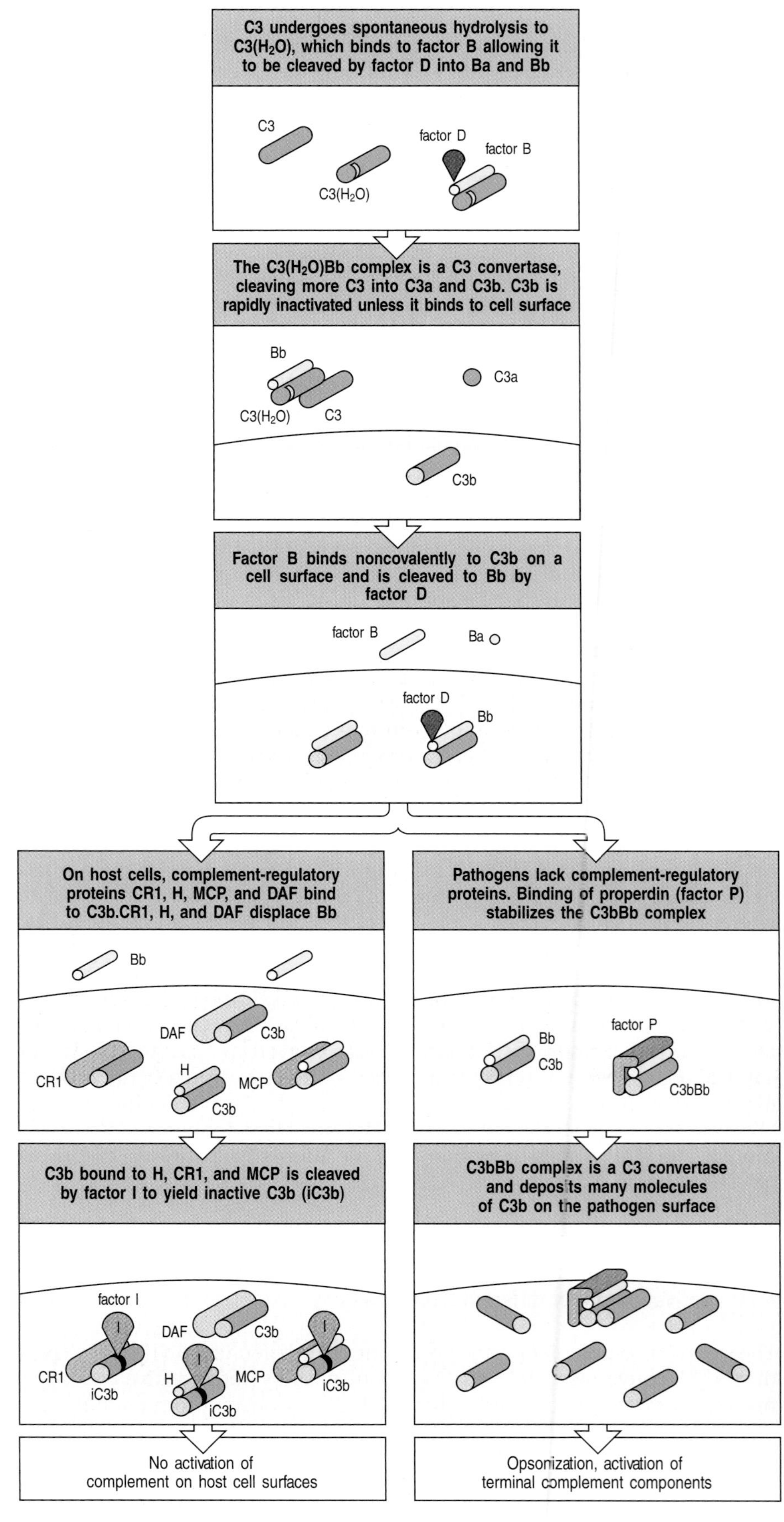

Fig. 2.32 Complement activated by the alternative pathway attacks pathogens while sparing host cells, which are protected by complement-regulatory proteins. The complement component C3 is cleaved spontaneously in plasma to give C3(H_2O), which binds factor B and enables the bound factor B to be cleaved by factor D (top panel). The resulting soluble C3 convertase cleaves C3 to give C3a and C3b, which can attach to host cells or pathogen surfaces (second panel). Covalently bound C3b binds factor B, which in turn is rapidly cleaved by factor D to Bb, which remains bound to C3b to form a C3 convertase, and Ba, which is released (third panel). If C3bBb forms on the surface of host cells (bottom left panels), it is rapidly inactivated by complement-regulatory proteins expressed by the host cell: complement receptor 1 (CR1), decay-accelerating factor (DAF), and membrane cofactor of proteolysis (MCP). Host cell surfaces also favor the binding of factor H from plasma. CR1, DAF, and factor H displace Bb from C3b, and CR1, MCP, and factor H catalyze the cleavage of bound C3b by the plasma protease factor I to produce inactive C3b (known as iC3b). Bacterial surfaces (bottom right panels) do not express complement-regulatory proteins and favor the binding of properdin (factor P), which stabilizes the C3bBb convertase activity. This convertase is the equivalent of C4b2a of the classical pathway (see Fig. 2.28).

Proteins of the alternative pathway of complement activation		
Native component	**Active fragments**	**Function**
C3	C3b	Binds to pathogen surface, binds B for cleavage by D, C3bBb is C3 convertase and $C3b_2Bb$ is C5 convertase
Factor B (B)	Ba	Small fragment of B, unknown function
	Bb	Bb is active enzyme of the C3 convertase C3bBb and C5 convertase $C3b_2Bb$
Factor D (D)	D	Plasma serine protease, cleaves B when it is bound to C3b to Ba and Bb
Properdin (P)	P	Plasma protein that stabilizes the C3bBb convertase on bacterial cells

Fig 2.33 The proteins of the alternative pathway of complement activation.

C3b and either prevent the convertase from forming or promote its rapid dissociation. Thus, a membrane-attached protein known as **decay-accelerating factor** (**DAF** or **CD55**) competes with factor B for binding to C3b on the cell surface, and can displace Bb from a convertase that has already formed. Convertase formation can also be prevented by cleaving C3b to its inactive derivative iC3b. This is achieved by a plasma protease, **factor I**, in conjunction with C3b-binding proteins that can act as cofactors, such as **membrane cofactor of proteolysis** (**MCP** or **CD46**), another host-cell membrane protein. Cell-surface complement receptor type 1 (CR1) has similar activities to DAF and MCP in the inhibition of C3 convertase formation and the promotion of the catabolism of C3b to inactive products, but has a more limited tissue distribution. **Factor H** is yet another complement-regulatory protein in plasma that binds C3b and, like CR1, it is able to compete with factor B to displace Bb from the convertase in addition to acting as a cofactor for factor I. Factor H binds preferentially to C3b bound to vertebrate cells because it has an affinity for the sialic acid residues present on the cell surface.

In contrast, when complement is activated on foreign surfaces, such as on bacterial surfaces, or indeed on host cells that have been damaged or modified by ischemia, viral infection, or antibody binding, C3 convertase enzymes are stabilized, which enables complement activation to proceed. On these cells a positive regulatory plasma protein, known as **properdin** or **factor P**, binds to the C3bBb convertase and enhances its stability, causing amplification of complement activation.

Fig. 2.34 The alternative pathway of complement activation can amplify the classical or the lectin pathway by forming an alternative C3 convertase and depositing more C3b molecules on the pathogen. C3b deposited by the classical or lectin pathway can bind factor B, making it susceptible to cleavage by factor D. The C3bBb complex is the C3 convertase of the alternative pathway of complement activation and its action, like that of C4b2a, results in the deposition of many molecules of C3b on the pathogen surface.

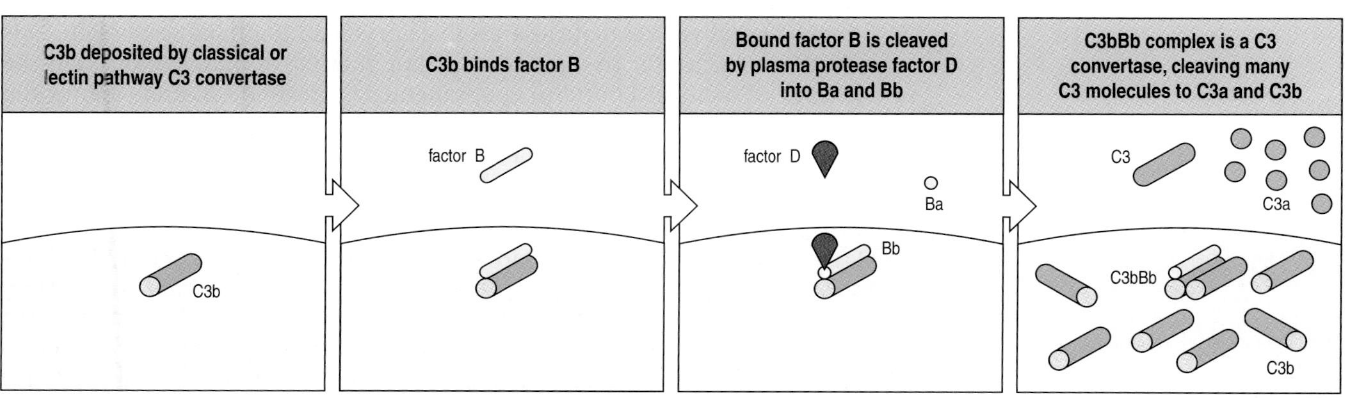

Step in pathway	Protein serving function in pathway			Relationship
	Alternative	Lectin	Classical	
Initiating serine protease	D	MASP	C1s	Homologous (C1s and MASP)
Covalent binding to cell surface	C3b	C4b		Homologous
C3/C5 convertase	Bb	C2b		Homologous
Control of activation	CR1 H	CR1 C4BP		Identical Homologous
Opsonization	C3b			Identical
Initiation of effector pathway	C5b			Identical
Local inflammation	C5a, C3a			Identical
Stabilization	P	None		Unique

Fig. 2.35 There is a close relationship between the factors of the alternative, lectin, and classical pathways of complement activation. Most of the factors are either identical or the products of genes that have duplicated and then diverged in sequence. The proteins C4 and C3 are homologous and contain the unstable thioester bond by which their large fragments, C4b and C3b, bind covalently to membranes. The genes encoding proteins C2 and B are adjacent in the class III region of the MHC and arose by gene duplication. Factor H, CR1, and C4BP regulatory proteins share a repeat sequence common to many complement-regulatory proteins. The greatest divergence between the pathways is in their initiation: in the classical pathway the C1 complex binds either to certain pathogens or to bound antibody and in the latter circumstance it serves to convert antibody binding into enzyme activity on a specific surface; in the lectin pathway, mannose-binding lectin (MBL) associates with a serine protease, activating MBL-associated serine protease (MASP), to serve the same function as C1r:C1s; in the alternative pathway this enzyme activity is provided by factor D.

Once formed on a surface that allows it to be stable, the C3bBb convertase rapidly cleaves yet more C3 to C3b, which can bind to the pathogen and either act as an opsonin or reinitiate the pathway to form another molecule of C3bBb convertase. Thus, the alternative pathway activates through an amplification loop that can proceed on the surface of a pathogen or on damaged host cells but not on normal host cells or tissues. This same amplification loop enables the alternative pathway to contribute to complement activation initially triggered through the classical or lectin pathways (see Fig. 2.25).

The C3 convertases resulting from activation of the classical and lectin pathways (C4b2b) and from the alternative pathway (C3bBb) are apparently distinct. However, understanding of the complement system is simplified somewhat by recognition of the close evolutionary relationships between the different complement proteins (Fig. 2.35). Thus the complement zymogens, factor B and C2, are closely related proteins encoded by homologous genes located in tandem within the major histocompatibility complex (MHC) on human chromosome 6. Furthermore, their respective binding partners, C3 and C4, both contain thioester bonds that provide the means of covalently attaching the C3 convertases to a pathogen surface. Only one component of the alternative pathway seems entirely unrelated to its functional equivalents in the classical and lectin pathways: the initiating serine protease, factor D. Factor D can also be singled out as the only activating protease of the complement system to circulate as an active enzyme rather than a zymogen. This is both necessary for the initiation of the alternative pathway through the cleavage of factor B bound to spontaneously activated C3, and safe for the host because factor D has no other substrate than factor B bound to C3b. This means that factor D finds its substrate only at a very low level in plasma, and at pathogen surfaces where the alternative pathway of complement activation is allowed to proceed.

Comparison of the different pathways of complement activation illustrates the general principle that most of the immune effector mechanisms that can be activated in a nonclonal fashion as part of the early nonadaptive response against infection have been harnessed during evolution to be used as effector mechanisms of adaptive immunity. It is almost certain that the adaptive response evolved by adding specific recognition to the original nonadaptive system. This is illustrated particularly clearly in the complement system, because here the components are defined and the functional homologs can be seen to be evolutionarily related.

2-18 Surface-bound C3 convertase deposits large numbers of C3b fragments on pathogen surfaces and generates C5 convertase activity.

The formation of C3 convertases is the point at which the three pathways of complement activation converge, because both the classical pathway and lectin pathway convertase C4b2a and the alternative pathway convertase C3bBb have the same activity, and they initiate the same subsequent events. They both cleave C3 to C3b and C3a. C3b binds covalently through its thioester bond to adjacent molecules on the pathogen surface; otherwise it is inactivated by hydrolysis. C3 is the most abundant complement protein in plasma, occurring at a concentration of 1.2 mg ml^{-1}, and up to 1000 molecules of C3b can bind in the vicinity of a single active C3 convertase (see Fig. 2.34). Thus, the main effect of complement activation is to deposit large quantities of C3b on the surface of the infecting pathogen, where it forms a covalently bonded coat that can signal the ultimate destruction of the pathogen by phagocytes.

The next step in the cascade is the generation of the C5 convertases. In the classical and the lectin pathways, a C5 convertase is formed by the binding of C3b to C4b2a to yield C4b2a3b. By the same token, the C5 convertase of the alternative pathway is formed by the binding of C3b to the C3 convertase to form $C3b_2Bb$. C5 is captured by these C5 convertase complexes through binding to an acceptor site on C3b, and is thus rendered susceptible to cleavage by the serine protease activity of C2a or Bb. This reaction, which generates C5b and C5a, is much more limited than cleavage of C3, because C5 can be cleaved only when it binds to C3b that is in turn bound to C4b2a or C3bBb to form the active C5 convertase complex. Thus, complement activated by all three pathways leads to the binding of large numbers of C3b molecules on the surface of the pathogen, the generation of a more limited number of C5b molecules, and the release of C3a and C5a (Fig. 2.36).

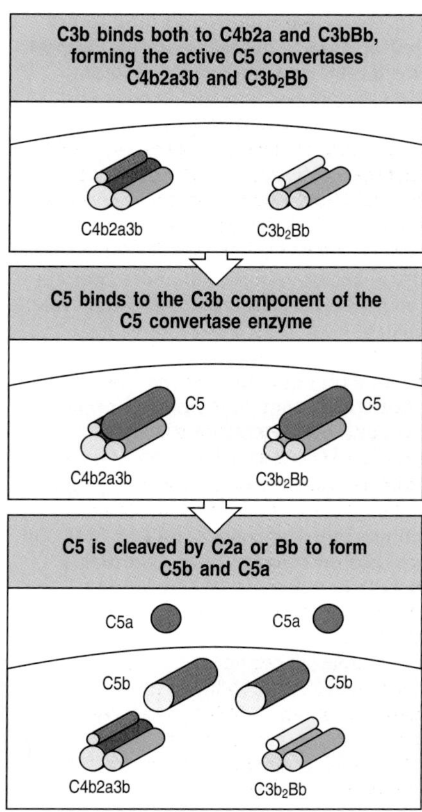

Fig. 2.36 Complement component C5 is cleaved when captured by a C3b molecule that is part of a C5 convertase complex. As shown in the top panel, C5 convertases are formed when C3b binds either the classical or lectin pathway C3 convertase C4b2a to form C4b2a3b, or the alternative pathway C3 convertase C3bBb to form C3b2Bb. C5 binds to C3b in these complexes (center panel). The bottom panel shows how C5 is cleaved by the active enzyme C2a or Bb to form C5b and the inflammatory mediator C5a. Unlike C3b and C4b, C5b is not covalently bound to the cell surface. The production of C5b initiates the assembly of the terminal complement components.

2-19 Ingestion of complement-tagged pathogens by phagocytes is mediated by receptors for the bound complement proteins.

The most important action of complement is to facilitate the uptake and destruction of pathogens by phagocytic cells. This occurs by the specific recognition of bound complement components by **complement receptors** (**CRs**) on phagocytes. These complement receptors bind pathogens opsonized with complement components: opsonization of pathogens is a major function of C3b and its proteolytic derivatives. C4b also acts as an opsonin but has a relatively minor role, largely because so much more C3b than C4b is generated.

The six known types of receptors for bound complement components are listed, with their functions and distributions, in Fig. 2.37. The best characterized is the C3b receptor **CR1** (**CD35**), which is expressed on both macrophages and neutrophils. Binding of C3b to CR1 cannot by itself stimulate phagocytosis, but it can lead to phagocytosis in the presence of other immune mediators that activate macrophages. For example, the small complement fragment C5a can activate macrophages to ingest bacteria bound to their CR1 receptors (Fig. 2.38). C5a binds to another receptor expressed by macrophages, the **C5a receptor**, which has seven membrane-spanning

Receptor	Specificity	Functions	Cell types
CR1 (CD35)	C3b, C4bi C3b	Promotes C3b and C4b decay Stimulates phagocytosis Erythrocyte transport of immune complexes	Erythrocytes, macrophages, monocytes, polymorphonuclear leukocytes, B cells, FDC
CR2 (CD21)	C3d, iC3b, C3dg Epstein–Barr virus	Part of B-cell co-receptor Epstein–Barr virus receptor	B cells, FDC
CR3 (Mac-1) (CD11b/CD18)	iC3b	Stimulates phagocytosis	Macrophages, monocytes, polymorphonuclear leukocytes, FDC
CR4 (gp150, 95) (CD11c/CD18)	iC3b	Stimulates phagocytosis	Macrophages, monocytes, polymorphonuclear leukocytes, dendritic cells
C5a receptor	C5a	Binding of C5a activates G protein	Endothelial cells, mast cells, phagocytes
C3a receptor	C3a	Binding of C3a activates G protein	Endothelial cells, mast cells, phagocytes

Fig. 2.37 Distribution and function of cell-surface receptors for complement proteins. There are several different receptors specific for different bound complement components and their fragments. CR1 and CR3 are especially important in inducing phagocytosis of bacteria with complement components bound to their surface. CR2 is found mainly on B cells, where it is also part of the B-cell co-receptor complex and the receptor by which the Epstein–Barr virus selectively infects B cells, causing infectious mononucleosis. CR1 and CR2 share structural features with the complement-regulatory proteins that bind C3b and C4b. CR3 and CR4 are integrins; CR3 is known to be important for leukocyte adhesion and migration, whereas CR4 is only known to function in phagocytic responses. The C5a and C3a receptors are seven-span G-protein-coupled receptors. FDC, follicular dendritic cells; these are not involved in innate immunity and are discussed in later chapters.

domains. Receptors of this type couple with intracellular guanine-nucleotide-binding proteins called G proteins, and the C5a receptor signals in this way. Proteins associated with the extracellular matrix, such as fibronectin, can also contribute to phagocyte activation; these are encountered when phagocytes are recruited to connective tissue and activated there.

Three other complement receptors—**CR2** (also known as **CD21**), **CR3** (**CD11b:CD18**), and **CR4** (**CD11c:CD18**)—bind to inactivated forms of C3b that remain attached to the pathogen surface. Like several other key components of complement, C3b is subject to regulatory mechanisms and can be cleaved into derivatives that cannot form an active convertase. One of the inactive derivatives of C3b, known as iC3b (see Section 2-17) acts as an opsonin in its own right when bound by the complement receptor CR3. Unlike the binding of iC3b to CR1, the binding of iC3b to CR3 is sufficient on its own to stimulate phagocytosis. A second breakdown product of C3b, called **C3dg**, binds only to CR2. CR2 is found on B cells as part of a co-receptor complex

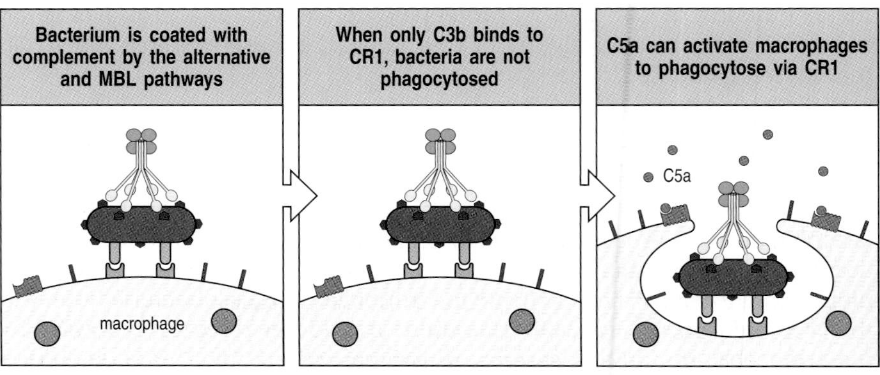

Bacterium is coated with complement by the alternative and MBL pathways
When only C3b binds to CR1, bacteria are not phagocytosed
C5a can activate macrophages to phagocytose via CR1
macrophage
C5a

Fig. 2.38 The anaphylatoxin C5a can enhance the phagocytosis of opsonized microorganisms. Activation of complement, either by the alternative or by the mannose-binding lectin pathway, leads to the deposition of C3b on the surface of the microorganism (left panel). C3b can be bound by the complement receptor CR1 on the surface of phagocytes, but this on its own is insufficient to induce phagocytosis (center panel). Phagocytes also express receptors for the anaphylatoxin C5a, and binding of C5a will now activate the cell to phagocytose microorganisms bound through CR1 (right panel).

that can augment the signal received through the antigen-specific immunoglobulin receptor. Thus, a B cell whose antigen receptor is specific for a given pathogen will receive a strongly augmented signal on binding this pathogen if it is also coated with C3dg. The activation of complement can therefore contribute to producing a strong antibody response (see Chapters 6 and 9). This example of how an innate humoral immune response can contribute to activating adaptive humoral immunity parallels the contribution made by the innate cellular response of macrophages and dendritic cells to the initiation of a T-cell response, which we will discuss later in this chapter.

The central role of opsonization by C3b and its inactive fragments in the destruction of extracellular pathogens can be seen in the effects of various complement deficiency diseases. Whereas individuals deficient in any of the late components of complement are relatively unaffected, showing only an increase in susceptibility to *Neisseria* infections, individuals deficient in C3 or in molecules that catalyze C3b deposition show increased susceptibility to infection by a wide range of extracellular bacteria, as we will see in Chapter 12.

2-20 Small fragments of some complement proteins can initiate a local inflammatory response.

The small complement fragments C3a, C4a, and C5a act on specific receptors (see Fig. 2.37) to produce local inflammatory responses. When produced in large amounts or injected systemically, they induce a generalized circulatory collapse, producing a shock-like syndrome similar to that seen in a systemic allergic reaction involving antibodies of the IgE class (which is discussed in Chapter 13). Such a reaction is termed **anaphylactic shock** and these small fragments of complement are therefore often referred to as **anaphylatoxins**. Of the three, C5a has the highest specific biological activity. All three induce smooth muscle contraction and increase vascular permeability, but C5a and C3a also act on the endothelial cells lining blood vessels to induce adhesion molecules. In addition, C3a and C5a can activate the mast cells that populate submucosal tissues to release mediators such as histamine and TNF-α that cause similar effects. The changes induced by C5a and C3a recruit antibody, complement, and phagocytic cells to the site of an infection (Fig. 2.39), and the increased fluid in the tissues hastens the movement of pathogen-bearing antigen-presenting cells to the local lymph nodes, contributing to the prompt initiation of the adaptive immune response.

C5a also acts directly on neutrophils and monocytes to increase their adherence to vessel walls, their migration toward sites of antigen deposition, and their ability to ingest particles. C5a also increases the expression of CR1 and CR3 on the surfaces of these cells. In this way, C5a, and to a smaller extent C3a and C4a, act in concert with other complement components to hasten the destruction of pathogens by phagocytes. C5a and C3a signal through transmembrane receptors that activate G proteins; thus, the action of C5a in attracting neutrophils and monocytes is analogous to that of chemokines, which also act via G proteins to control cell migration.

2-21 The terminal complement proteins polymerize to form pores in membranes that can kill certain pathogens.

One of the important effects of complement activation is the assembly of the terminal components of complement (Fig. 2.40) to form a membrane-attack complex. The reactions leading to the formation of this complex are shown schematically and microscopically in Fig. 2.41. The end result is a pore in the lipid bilayer membrane that destroys membrane integrity. This is thought to

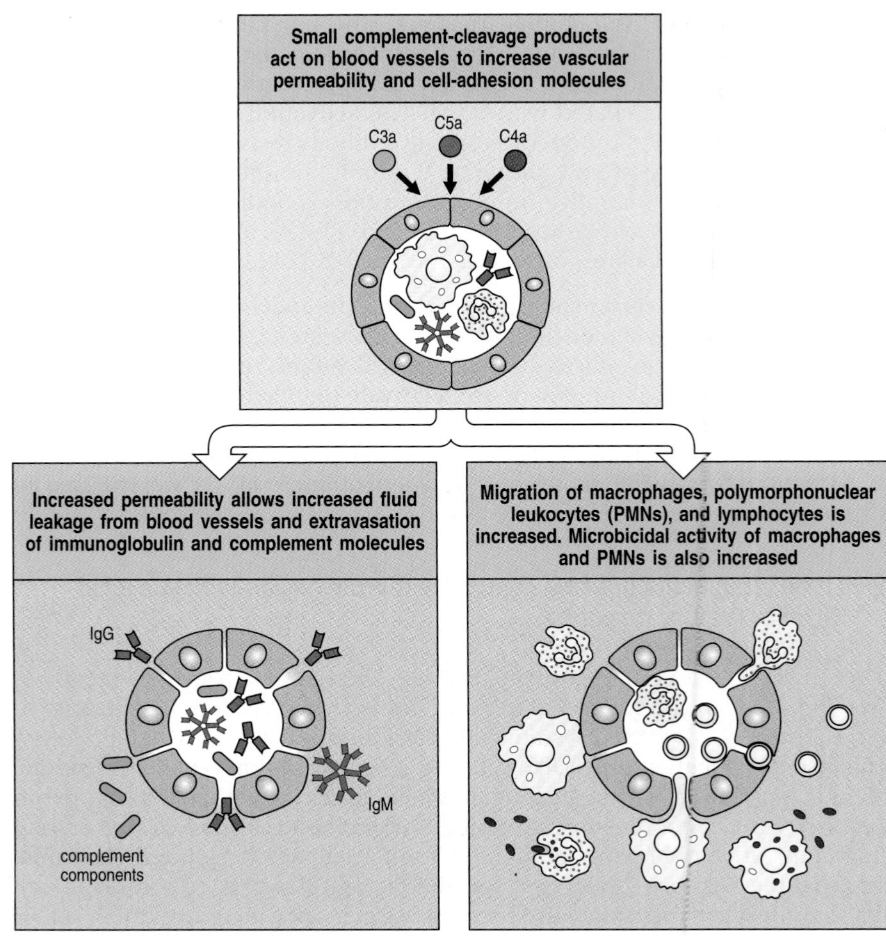

Fig. 2.39 Local inflammatory responses can be induced by small complement fragments, especially C5a. The small complement fragments are differentially active: C5a is more active than C3a, which is more active than C4a. They cause local inflammatory responses by acting directly on local blood vessels, stimulating an increase in blood flow, increased vascular permeability, and increased binding of phagocytes to endothelial cells. C5a also activates mast cells (not shown) to release mediators, such as histamine and TNF-α, that contribute to the inflammatory response. The increase in vessel diameter and permeability leads to the accumulation of fluid and protein. Fluid accumulation increases lymphatic drainage, bringing pathogens and their antigenic components to nearby lymph nodes. The antibodies, complement, and cells thus recruited participate in pathogen clearance by enhancing phagocytosis. The small complement fragments can also directly increase the activity of the phagocytes.

Fig. 2.40 The terminal complement components assemble to form the membrane-attack complex.

The terminal complement components that form the membrane-attack complex		
Native protein	Active component	Function
C5	C5a	Small peptide mediator of inflammation (high activity)
	C5b	Initiates assembly of the membrane-attack system
C6	C6	Binds C5b; forms acceptor for C7
C7	C7	Binds C5b6; amphiphilic complex inserts in lipid bilayer
C8	C8	Binds C5b67; initiates C9 polymerization
C9	$C9_n$	Polymerizes to C5b678 to form a membrane-spanning channel, lysing cell

kill the pathogen by destroying the proton gradient across the pathogen's cell membrane.

The first step in the formation of the membrane-attack complex is the cleavage of C5 by a C5 convertase to release C5b (see Fig. 2.36). In the next stages, shown in Fig. 2.41, C5b initiates the assembly of the later complement components and their insertion into the cell membrane. First, one molecule of C5b binds one molecule of C6, and the C5b6 complex then binds one molecule of C7. This reaction leads to a conformational change in the constituent molecules, with the exposure of a hydrophobic site on C7, which inserts into the lipid bilayer. Similar hydrophobic sites are exposed on the later components C8 and C9 when they are bound to the complex, allowing these proteins also to insert into the lipid bilayer. C8 is a complex of two proteins, C8β and C8α-γ. The C8β protein binds to C5b, and the binding of C8β to the membrane-associated C5b67 complex allows the hydrophobic domain of C8α-γ to insert into the lipid bilayer. Finally, C8α-γ induces the polymerization of 10–16 molecules of C9 into a pore-forming structure called the membrane-attack complex. The membrane-attack complex, shown schematically and by electron microscopy in Fig. 2.41, has a hydrophobic external face, allowing it to associate with the lipid bilayer, but a hydrophilic internal channel. The diameter of this channel is about 100 Å, allowing the free passage of solutes and water across the lipid bilayer. The disruption of the lipid bilayer leads to the loss of cellular homeostasis, the disruption of the proton gradient across the membrane, the penetration of enzymes such as lysozyme into the cell, and the eventual destruction of the pathogen.

Although the effect of the membrane-attack complex is very dramatic, particularly in experimental demonstrations in which antibodies against red blood cell membranes are used to trigger the complement cascade, the significance

Fig. 2.41 Assembly of the membrane-attack complex generates a pore in the lipid bilayer membrane. The sequence of steps and their approximate appearance are shown here in schematic form. C5b triggers the assembly of a complex of one molecule each of C6, C7, and C8, in that order. C7 and C8 undergo conformational changes, exposing hydrophobic domains that insert into the membrane. This complex causes moderate membrane damage in its own right, and also serves to induce the polymerization of C9, again with the exposure of a hydrophobic site. Up to 16 molecules of C9 are then added to the assembly to generate a channel 100 Å in diameter in the membrane. This channel disrupts the bacterial cell membrane, killing the bacterium. The electron micrographs show erythrocyte membranes with membrane-attack complexes in two orientations, end on and side on. Photographs courtesy of S. Bhakdi and J. Tranum-Jensen.

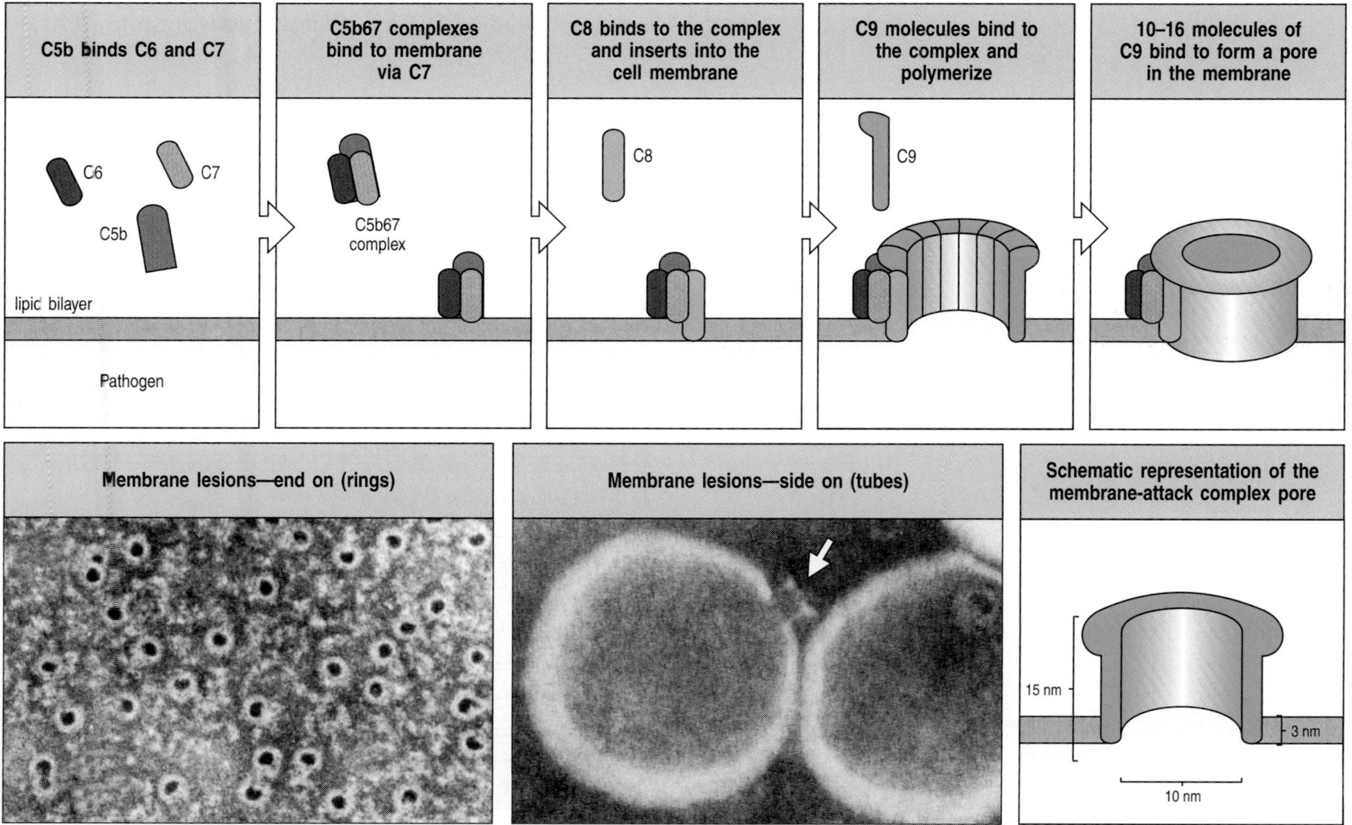

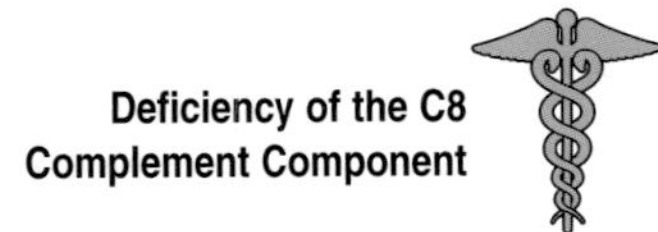

of these components in host defense seems to be quite limited. So far, deficiencies in complement components C5–C9 have been associated with susceptibility only to *Neisseria* species, the bacteria that cause the sexually transmitted disease gonorrhea and a common form of bacterial meningitis. Thus, the opsonizing and inflammatory actions of the earlier components of the complement cascade are clearly more important for host defense against infection. Formation of the membrane-attack complex seems to be important only for the killing of a few pathogens, although, as we will see in Chapter 14, it might well have a major role in immunopathology.

2-22 Complement control proteins regulate all three pathways of complement activation and protect the host from its destructive effects.

Given the destructive effects of complement, and the way in which its activation is rapidly amplified through a triggered-enzyme cascade, it is not surprising that there are several mechanisms to prevent its uncontrolled activation. As we have seen, the effector molecules of complement are generated through the sequential activation of zymogens, which are present in plasma in an inactive form. The activation of these zymogens usually occurs on a pathogen surface, and the activated complement fragments produced in the ensuing cascade of reactions usually bind nearby or are rapidly inactivated by hydrolysis. These two features of complement activation act as safeguards against uncontrolled activation. Even so, all complement components are activated spontaneously at a low rate in plasma, and activated complement components will sometimes bind proteins on host cells. The potentially damaging consequences are prevented by a series of complement control proteins, summarized in Fig. 2.42, which regulate the complement cascade at different points. As we saw in discussing the alternative pathway of complement activation (see Section 2-16), many of these control proteins specifically protect normal host cells while allowing complement activation to proceed

Fig. 2.42 The proteins that regulate the activity of complement

Regulatory proteins of the classical and alternative pathways	
Name (symbol)	**Role in the regulation of complement activation**
C1 inhibitor (C1INH)	Binds to activated C1r, C1s, removing them from C1q, and to activated MASP-2, removing it from MBL
C4-binding protein (C4BP)	Binds C4b, displacing C2a; cofactor for C4b cleavage by I
Complement receptor 1 (CR1)	Binds C4b, displacing C2a, or C3b displacing Bb; cofactor for I
Factor H (H)	Binds C3b, displacing Bb; cofactor for I
Factor I (I)	Serine protease that cleaves C3b and C4b; aided by H, MCP, C4BP, or CR1
Decay-accelerating factor (DAF)	Membrane protein that displaces Bb from C3b and C2a from C4b
Membrane cofactor protein (MCP)	Membrane protein that promotes C3b and C4b inactivation by I
CD59 (protectin)	Prevents formation of membrane-attack complex on autologous or allogeneic cells. Widely expressed on membranes

on pathogen surfaces. The complement control proteins therefore allow complement to distinguish self from nonself.

The reactions that regulate the complement cascade are shown in Fig. 2.43. The top two panels show how the activation of C1 is controlled by a plasma serine protease inhibitor or **serpin**, the C1 inhibitor (C1INH). C1INH binds the active enzymes C1r:C1s and causes them to dissociate from C1q, which remains bound to the pathogen. In this way, C1INH limits the time during which active C1s is able to cleave C4 and C2. By the same means, C1INH limits the spontaneous activation of C1 in plasma. Its importance can be seen in the C1INH deficiency disease **hereditary angioneurotic edema**, in which chronic spontaneous complement activation leads to the production of excess cleaved fragments of C4 and C2. The small fragment of C2, C2b, is further cleaved into a peptide, the C2 kinin, which causes extensive swelling—the most dangerous being local swelling in the trachea, which can lead to suffocation. Bradykinin, which has similar actions to C2 kinin, is also produced in an uncontrolled fashion in this disease, as a result of the lack of inhibition of another plasma protease, kallikrein, a component of the kinin system discussed in Section 2-5, which is activated by tissue damage and is also regulated by C1INH. This disease is fully corrected by replacing C1INH. The large activated fragments of C4 and C2, which normally combine to form the C3 convertase, do not damage host cells in such patients because C4b is rapidly inactivated in plasma (see Fig. 2.31) and the convertase does not form. Furthermore, any convertase that accidentally forms on a host cell is inactivated by the mechanisms described below.

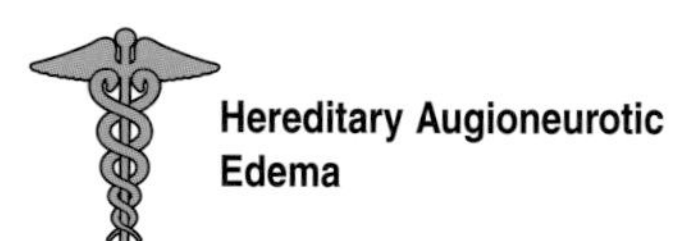

The thioester bond of activated C3 and C4 is extremely reactive and has no mechanism for distinguishing an acceptor hydroxyl or amine group on a host cell from a similar group on the surface of a pathogen. A series of protective mechanisms, mediated by other proteins, has evolved to ensure that the binding of a small number of C3 or C4 molecules to host cell membranes results in minimal formation of C3 convertase and little amplification of complement activation. We have already encountered most of these mechanisms in the description of the alternative pathway (see Fig. 2.32), but we will consider them again here because they are important regulators of the classical pathway convertase as well (see Fig. 2.43, second and third rows). The mechanisms can be divided into three categories. The first catalyze the cleavage of any C3b or C4b that does bind to host cells into inactive products. The complement-regulatory enzyme responsible is the plasma serine protease factor I; it circulates in active form but can cleave C3b and C4b only when they are bound to a membrane cofactor protein. In these circumstances, factor I cleaves C3b, first into iC3b and then further to C3dg, thus permanently inactivating it. C4b is similarly inactivated by cleavage into C4c and C4d. There are two cell-membrane proteins that bind C3b and C4b and possess cofactor activity for factor I; these are MCP and CR1 (see Section 2-17). Microbial cell walls lack these protective proteins and cannot promote the breakdown of C3b and C4b. Instead, these proteins act as binding sites for factor B and C2, promoting complement activation. The importance of factor I can be seen in people with genetically determined **factor I deficiency**. Because of uncontrolled complement activation, complement proteins rapidly become depleted and such people suffer repeated bacterial infections, especially with ubiquitous pyogenic bacteria.

There are also plasma proteins with cofactor activity for factor I. C4b is bound by a cofactor known as **C4b-binding protein** (**C4BP**), which acts mainly as a regulator of the classical pathway in the fluid phase. C3b is bound at cell membranes by cofactor proteins such as DAF and MCP. These regulatory molecules effectively compete with factor B for binding to C3b bound to cells. If factor B 'wins,' as typically happens on a pathogen surface, then more

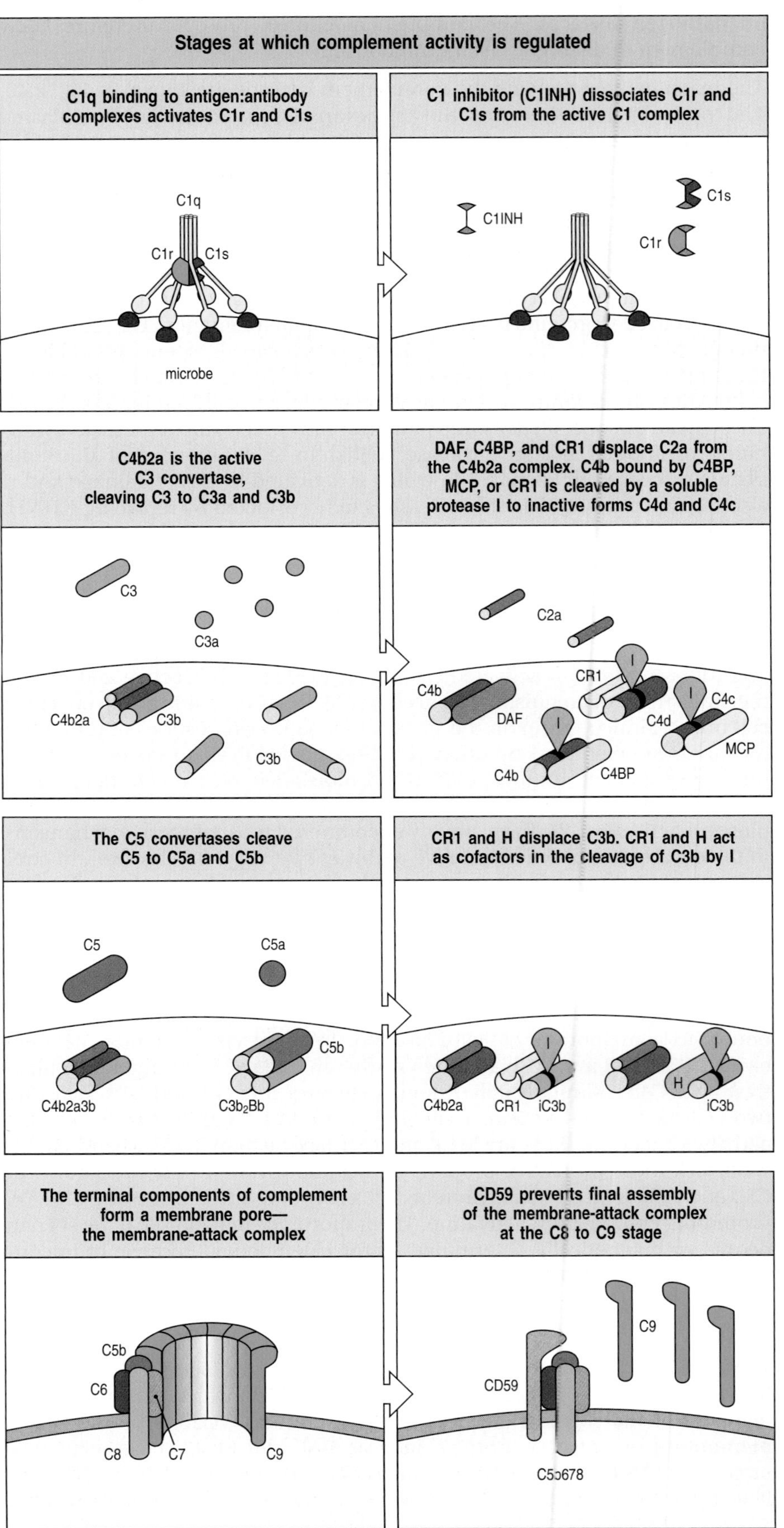

Fig. 2.43 Complement activation is regulated by a series of proteins that serve to protect host cells from accidental damage. These act on different stages of the complement cascade, dissociating complexes or catalyzing the enzymatic degradation of covalently bound complement proteins. Stages in the complement cascade are shown schematically down the left side of the figure, with the regulatory reactions on the right. The alternative pathway C3 convertase is similarly regulated by DAF, CR1, MCP, and factor H.

C3bBb C3 convertase forms and complement activation is amplified. If DAF and MCP win, as occurs on cells of the host, then the bound C3b is catabolized by factor I to iC3b and C3dg and complement activation is inhibited.

The critical balance between the inhibition and the activation of complement on cell surfaces is illustrated in individuals heterozygous for mutations in the regulatory proteins MCP, factor I, or factor H. In such individuals the concentration of functional regulatory proteins is reduced, and the tipping of the balance towards complement activation leads to a predisposition to **hemolytic uremic syndrome**, a condition characterized by damage to platelets and red blood cells and by kidney inflammation, as a result of ineffectively controlled complement activation.

The competition between DAF or MCP and factor B for binding to surface-bound C3b is an example of the second mechanism for inhibiting complement activation on host cells. Several proteins competitively inhibit the binding of C2 to cell-bound C4b and of factor B to cell-bound C3b, thereby inhibiting convertase formation. These proteins bind to C3b and C4b on the cell surface, and also mediate protection against complement through a third mechanism, which is to augment the dissociation of C4b2a and C3bBb convertases that have already formed. Host-cell membrane molecules that regulate complement through both these mechanisms include DAF and CR1, which promote the dissociation of convertase in addition to their cofactor activity. All the proteins that bind the homologous C4b and C3b molecules share one or more copies of a structural element called the short consensus repeat (SCR), complement control protein (CCP) repeat, or (especially in Japan) the sushi domain.

In addition to the mechanisms for preventing C3 convertase formation and C4 and C3 deposition on cell membranes, there are also inhibitory mechanisms that prevent the inappropriate insertion of the membrane-attack complex into membranes. We saw in Section 2-21 that the membrane-attack complex polymerizes onto C5b molecules created by the action of C5 convertase. This complex mainly inserts into cell membranes adjacent to the site of the C5 convertase; that is, close to the site of complement activation on a pathogen. However, some newly formed membrane-attack complexes may diffuse from the site of complement activation and insert into adjacent host-cell membranes. Several plasma proteins including, notably, vitronectin, also known as S-protein, bind to the C5b67 complex and thereby inhibit its random insertion into cell membranes. Host-cell membranes also contain an intrinsic protein, **CD59** or **protectin**, which inhibits the binding of C9 to the C5b678 complex (see Fig. 2.43, bottom row). CD59 and DAF are both linked to the cell surface by a glycosylphosphatidylinositol (GPI) tail, like many other peripheral membrane proteins. One of the enzymes involved in the synthesis of GPI tails is encoded on the X chromosome. In people with a somatic mutation in this gene in a clone of hematopoietic cells, both CD59 and DAF fail to function. This causes the disease **paroxysmal nocturnal hemoglobinuria**, which is characterized by episodes of intravascular red blood cell lysis by complement. Red blood cells that lack CD59 only are also susceptible to destruction as a result of spontaneous activation of the complement cascade.

Summary.

The complement system is one of the major mechanisms by which pathogen recognition is converted into an effective host defense against initial infection. Complement is a system of plasma proteins that can be activated directly by pathogens or indirectly by pathogen-bound antibody, leading to a cascade of reactions that occurs on the surface of pathogens and generates active components with various effector functions. There are three pathways of complement activation: the classical pathway, which is triggered directly by pathogen or indirectly by antibody binding to the pathogen surface; the lectin pathway; and the alternative pathway, which also provides an amplification loop for the other two pathways. All three pathways can be initiated independently of antibody as part of innate immunity. The early events in all pathways consist of a sequence of cleavage reactions in which the larger cleavage product binds covalently to the pathogen surface and contributes to the activation of the next component. The pathways converge with the formation of a C3 convertase enzyme, which cleaves C3 to produce the active complement component C3b. The binding of large numbers of C3b molecules to the pathogen is the central event in complement activation. Bound complement components, especially bound C3b and its inactive fragments, are recognized by specific complement receptors on phagocytic cells, which engulf pathogens opsonized by C3b and its inactive fragments. The small cleavage fragments of C3, C4, and especially C5 recruit phagocytes to sites of infection and activate them by binding to specific trimeric G-protein-coupled receptors. Together, these activities promote the uptake and destruction of pathogens by phagocytes. The molecules of C3b that bind the C3 convertase itself initiate the late events, binding C5 to make it susceptible to cleavage by C2a or Bb. The larger C5b fragment triggers the assembly of a membrane-attack complex, which can result in the lysis of certain pathogens. The activity of complement components is modulated by a system of regulatory proteins that prevent tissue damage as a result of inadvertent binding of activated complement components to host cells or of spontaneous activation of complement components in plasma.

Induced innate responses to infection.

In the final part of this chapter we will look at the induced responses of innate immunity. These depend on the cytokines and chemokines that are produced in response to pathogen recognition, and we will introduce these first. The chemokines are a large family of chemoattractant molecules with a central role in leukocyte migration. Adhesion molecules also have a central role in this process and we will also briefly consider them. We will then consider in more detail how macrophage-derived chemokines and cytokines promote the phagocytic response through the recruitment and production of fresh phagocytes, and the production of additional opsonizing molecules through the acute-phase reactions. We will also look at the role of the cytokines known as **interferons**, which are induced by viral infection, and at a class of lymphoid cells, known as **natural killer** (**NK**) **cells**, that are activated by interferons to contribute to innate host defense against viruses and other intracellular pathogens. We will also discuss the **innate-like lymphocytes** (**ILLs**), which contribute to rapid responses to infection by acting early but use a limited set of antigen-receptor gene segments (see Section 1-11) to make immunoglobulins and T-cell receptors. The induced innate responses can either succeed in clearing the infection or contain it while an adaptive response develops. Adaptive immunity harnesses many of the same effector mechanisms that are used by the innate immune system, but adaptive immunity is able to target them with much greater precision. Thus, antigen-specific T cells activate the microbicidal and cytokine-secreting properties of macrophages harboring pathogens, antibodies activate complement, act as direct opsonins for phagocytes, and stimulate NK cells to kill infected cells. In addition, the adaptive immune response uses cytokines and chemokines in a manner similar to that of innate immunity, to induce inflammatory responses that promote the influx of antibodies and effector lymphocytes to sites of infection. The effector mechanisms described here therefore serve as a primer for the later chapters that focus on adaptive immunity.

2-23 Activated macrophages secrete a range of cytokines that have a variety of local and distant effects.

Cytokines (see Appendix III) are small proteins (approximately 25 kDa) that are released by various cells in the body, usually in response to an activating stimulus, and they induce responses through binding to specific receptors. They can act in an autocrine manner, affecting the behavior of the cell that releases the cytokine, in a paracrine manner, affecting the behavior of adjacent cells, and some cytokines are stable enough to act in an endocrine manner, affecting the behavior of distant cells, although this depends on their ability to enter the circulation and on their half-life in the blood.

The cytokines secreted by macrophages in response to pathogens are a structurally diverse group of molecules that include interleukin-1β (IL-1β), IL-6, IL-12, TNF-α, and the chemokine CXCL8 (formerly known as IL-8). The name **interleukin (IL)** followed by a number (for example IL-1, IL-2, and so on) was coined in an attempt to develop a standardized nomenclature for molecules secreted by, and acting on, leukocytes. However, this became confusing when an ever-increasing number of cytokines with diverse origins, structures, and effects were discovered, and although the IL designation is still used, it is hoped that eventually a nomenclature based on cytokine structure will be developed. The cytokines are listed alphabetically, together with their receptors, in Appendix III. There are two major structural families of cytokines: the **hematopoietin family**, which includes growth hormones and also many interleukins with roles in both adaptive and innate immunity, and the **TNF family**, of which the prototype is TNF-α, which again functions in both innate and adaptive immunity and includes many members that are membrane-bound. Of the macrophage-derived interleukins, IL-6 belongs to the large family of hematopoietins, TNF-α is part of the TNF family, and IL-1 and IL-12 are structurally distinct. All have important local and systemic effects that contribute to both innate and adaptive immunity, and these are summarized in Fig. 2.44.

The recognition of different classes of pathogens by phagocytes and dendritic cells may involve signaling through different receptors, such as the different TLRs, and result in some variation in the cytokines induced. This is a way in which appropriate responses can be selectively activated as the released cytokines orchestrate the next phase of host defense. We will see how TNF-α, which is elicited by LPS-bearing pathogens, is particularly important in containing infection by these pathogens, and how the release of different chemokines can recruit and activate different types of effector cells.

2-24 Chemokines released by phagocytes and dendritic cells recruit cells to sites of infection.

Among the cytokines released by tissues in the earliest phases of infection are members of a family of chemoattractant cytokines known as **chemokines** (listed separately in Appendix IV). These small proteins induce directed chemotaxis in nearby responsive cells and were discovered only relatively recently. Because they were first detected in cytokine assays, they were initially named interleukins: interleukin-8 (now known as CXCL8) was the first chemokine to be cloned and characterized, and it remains typical of this family. All the chemokines are related in amino acid sequence, and their receptors

Activated macrophages secrete a range of cytokines

IL-1β | TNF-α | IL-6 | CXCL8 | IL-12

Local effects

IL-1β	TNF-α	IL-6	CXCL8	IL-12
Activates vascular endothelium Activates lymphocytes Local tissue destruction Increases access of effector cells	Activates vascular endothelium and increases vascular permeability, which leads to increased entry of IgG, complement, and cells to tissues and increased fluid drainage to lymph nodes	Lymphocyte activation Increased antibody production	Chemotactic factor recruits neutrophils, basophils, and T cells to site of infection	Activates NK cells Induces the differentiation of CD4 T cells into T_H1 cells

Systemic effects

IL-1β	TNF-α	IL-6
Fever Production of IL-6	Fever Mobilization of metabolites Shock	Fever Induces acute-phase protein production

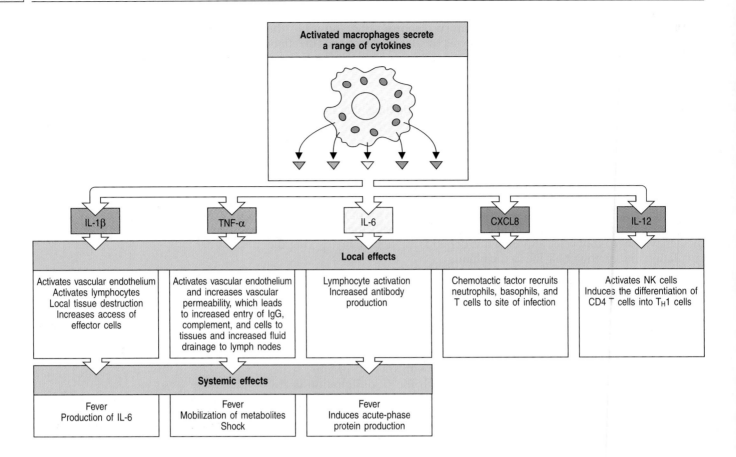

Fig. 2.44 Important cytokines secreted by macrophages in response to bacterial products include IL-1β, IL-6, CXCL8, IL-12, and TNF-α. TNF-α is an inducer of a local inflammatory response that helps to contain infections. It also has systemic effects, many of which are harmful (see Section 2-27). The chemokine CXCL8 is also involved in the local inflammatory response, helping to attract neutrophils to the site of infection. IL-1β, IL-6, and TNF-α have a critical role in inducing the acute-phase response in the liver (see Section 2-28) and induce fever, which favors effective host defense in various ways. IL-12 activates natural killer (NK) cells and favors the differentiation of CD4 T cells into the T_H1 subset in adaptive immunity.

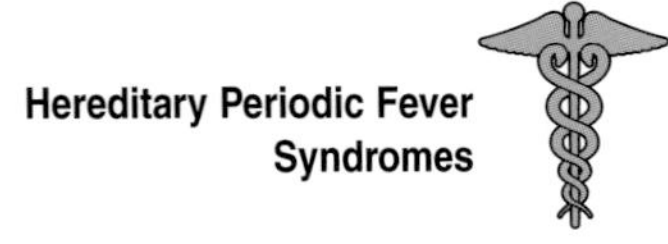

are all seven-span transmembrane proteins that signal through coupled G proteins. No atomic structure of a chemokine receptor has yet been determined, but they are similar to other seven-span G-protein-coupled receptors such as rhodopsin (Fig. 2.45) and the muscarinic acetylcholine receptor. Chemokines function mainly as chemoattractants for leukocytes, recruiting monocytes, neutrophils, and other effector cells from the blood to sites of infection. They can be released by many different types of cells and they serve to guide cells involved in innate immunity to sites of infection. They also guide the lymphocytes in adaptive immunity, as we will learn in Chapters 8–10. Some chemokines also function in lymphocyte development and migration, and in angiogenesis (the growth of new blood vessels). The properties of a variety of chemokines are listed in Fig. 2.46 (see also Appendix IV). It is striking that there are so many chemokines; this may reflect their importance in delivering cells to their correct location, as seems to be the case for lymphocytes.

Members of the chemokine family fall mostly into two broad groups—CC chemokines with two adjacent cysteines near the amino terminus, and CXC chemokines, in which the equivalent two cysteine residues are separated by a single amino acid. The two groups of chemokines act on different sets of receptors. CC chemokines bind to CC chemokine receptors, of which there are nine, designated CCR1–9. CXC chemokines bind to CXC receptors; there are six of these, CXCR1–6. These receptors are expressed on different cell types, which are consequently attracted by different chemokines. In general, CXC chemokines with a Glu-Leu-Arg tripeptide motif immediately before the first cysteine promote the migration of neutrophils. CXCL8 is an example of this type of chemokine. Other CXC chemokines that lack this motif, such as the B-lymphocyte chemokine (CXCL13), guide lymphocytes to their proper destination in the B-cell areas of the spleen, lymph nodes, and gut. The CC chemokines promote the migration of monocytes, lymphocytes, or other cell types. An example is macrophage chemoattractant protein-1 (CCL2). CXCL8 and CCL2 have similar, although complementary, functions: CXCL8 induces neutrophils to leave the bloodstream and migrate into the surrounding tissues; CCL2, by contrast, acts on monocytes, inducing their migration from the bloodstream to become tissue macrophages. Other CC chemokines such as CCL5 may promote the infiltration into tissues of a range of leukocytes, including effector T cells (see Section 10-6), with individual chemokines acting on different subsets of cells. The only known chemokine with only one cysteine (XCL1) was originally called lymphotactin; it is thought to attract T-cell precursors to the thymus by binding to XCR1. The chemokine fractalkine is unusual in several ways: it has three amino acid residues between the two cysteines, making it a CX_3CL chemokine. Most unusually, it exists in two forms, one tethered to the membrane of the endothelial and epithelial cells that express it, where it serves as an adhesion protein, and a soluble form released from the cell surface, which acts, like other chemokines, as a chemoattractant.

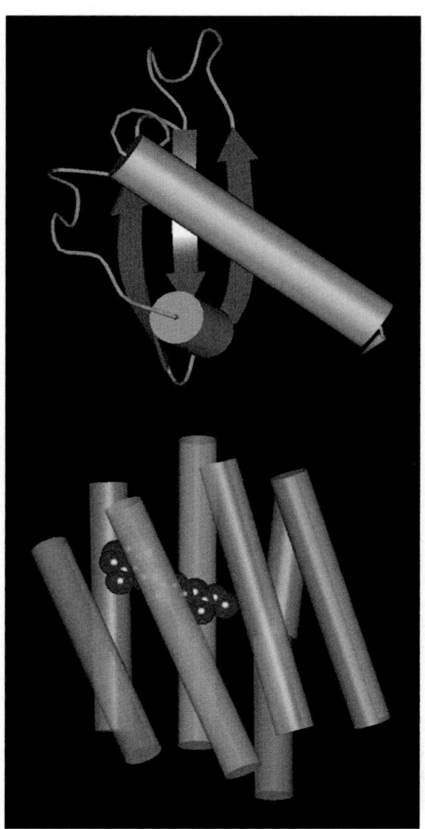

Fig. 2.45 Chemokines are a family of proteins of similar structures that bind to chemokine receptors, themselves part of a large family of G-protein-coupled receptors. The chemokines are represented here by CXCL8 (upper structure). The receptors for the chemokines are members of the family of seven-span receptors, which also includes the photoreceptor protein rhodopsin and many other receptors. They have seven transmembrane helices, and all the family interact with G proteins. The first solved structure of a seven-span membrane protein was of the bacterial protein bacteriorhodopsin; it is depicted here (lower structure), showing the orientation of the seven transmembrane helices (blue) with the bound ligand (in this case retinal) in red. Essentially all of this structure would be embedded within the cell membrane. Cylinders represent α helices and arrows β strands.

The role of chemokines such as CXCL8 and CCL2 in cell recruitment is twofold. First, they act on the leukocyte as it rolls along endothelial cells at sites of inflammation, converting this rolling into stable binding by triggering a change of conformation in the adhesion molecules known as leukocyte integrins. This allows the leukocyte to cross the blood vessel wall by squeezing between the endothelial cells, as we will see when we describe the process of extravasation. Second, the chemokine directs the migration of the leukocyte along a gradient of chemokine molecules bound to the extracellular matrix and the surfaces of endothelial cells. This gradient increases in concentration toward the site of infection.

Chemokines can be produced by a wide variety of cell types in response to bacterial products, viruses, and agents that cause physical damage, such as silica, alum, or the urate crystals that occur in gout. Thus, infection or physical damage to tissues sets in motion the production of chemokine gradients that can direct phagocytes to the sites where they are needed. In addition, peptides that act as chemoattractants for neutrophils are made by bacteria themselves. The fMLP peptide produced by bacteria is a potent chemotactic factor for inflammatory cells, especially neutrophils (see Section 2-6). The fMLP receptor is also a G-protein-coupled receptor, like the receptors for chemokines and for the complement fragments C5a, C3a, and C4a. Thus, there is a common mechanism for attracting neutrophils, whether by complement fragments, chemokines, or bacterial peptides. Neutrophils are the first cells to arrive in large numbers at a site of infection, with monocytes and immature dendritic cells being recruited later.

The complement peptide C5a and the chemokines CXCL8 and CCL2 also activate their respective target cells, so that not only are neutrophils and macrophages brought to potential sites of infection but, in the process, they are armed to deal with any pathogens they may encounter there. In particular, neutrophils exposed to CXCL8 and the cytokine TNF-α are activated to produce the respiratory burst that generates oxygen radicals and nitric oxide, and to release their stored lysosomal contents, thus contributing both to host defense and to the tissue destruction and pus formation seen in local sites, especially those infected with pyogenic bacteria.

Chemokines do not act alone in cell recruitment. They also require the action of vasoactive mediators to bring leukocytes close to the blood vessel endothelium (see Section 2-5) and cytokines such as TNF-α to induce the necessary adhesion molecules on the endothelial cells. We will return to the chemokines in later chapters, when they will be discussed in the context of the adaptive immune response. We will now turn to the molecules that mediate leukocyte adhesion to the endothelium, and then describe the process of leukocyte extravasation step-by-step as it has been shown for both neutrophils and monocytes.

Fig. 2.46 Properties of selected chemokines. Chemokines fall mainly into two related but distinct groups. The CXC chemokines, the genes for which are found mainly in a cluster on chromosome 17 in humans, have an amino acid residue (X) between two invariant cysteines (C) in the amino-terminal region. CC chemokines, which in humans are mostly encoded in one region of chromosome 4, have these two invariant cysteines adjacent. These classes can be divided further by the presence or absence of an amino acid triplet (ELR: glutamic acid–leucine–arginine) preceding the first of these invariant cysteines. All the chemokines that attract neutrophils have this motif, whereas most of the other CXC chemokines, including those that interact with receptors CXCR3, 4, and 5, lack it. A C chemokine with only one cysteine at this location, and fractalkine, a CX_3C chemokine, are encoded elsewhere in the genome. Each chemokine interacts with one or more receptors, and affects one or more types of cells. A comprehensive list of chemokines and their receptors is given in Appendix IV.

Class	Chemokine	Produced by	Receptors	Cells attracted	Major effects
CXC	CXCL8 (IL-8)	Monocytes Macrophages Fibroblasts Keratinocytes Endothelial cells	CXCR1 CXCR2	Neutrophils Naive T cells	Mobilizes, activates and degranulates neutrophils Angiogenesis
	CXCL7 (PBP, β-TG NAP-2)	Platelets	CXCR2	Neutrophils	Activates neutrophils Clot resorption Angiogenesis
	CXCL1 (GROα) CXCL2 (GROβ) CXCL3 (GROγ)	Monocytes Fibroblasts Endothelium	CXCR2	Neutrophils Naive T cells Fibroblasts	Activates neutrophils Fibroplasia Angiogenesis
	CXCL10 (IP-10)	Keratinocytes Monocytes T cells Fibroblasts Endothelium	CXCR3	Resting T cells NK cells Monocytes	Immunostimulant Antiangiogenic Promotes T_H1 immunity
	CXCL12 (SDF-1)	Stromal cells	CXCR4	Naive T cells Progenitor ($CD34^+$) B cells	B-cell development Lymphocyte homing Competes with HIV-1
	CXCL13 (BLC)	Stromal cells	CXCR5	B cells	Lymphocyte homing
CC	CCL3 (MIP-1α)	Monocytes T cells Mast cells Fibroblasts	CCR1, 3, 5	Monocytes NK and T cells Basophils Dendritic cells	Competes with HIV-1 Antiviral defense Promotes T_H1 immunity
	CCL4 (MIP-1β)	Monocytes Macrophages Neutrophils Endothelium	CCR1, 3, 5	Monocytes NK and T cells Dendritic cells	Competes with HIV-1
	CCL2 (MCP-1)	Monocytes Macrophages Fibroblasts Keratinocytes	CCR2B	Monocytes NK and T cells Basophils Dendritic cells	Activates macrophages Basophil histamine release Promotes T_H2 immunity
	CCL5 (RANTES)	T cells Endothelium Platelets	CCR1, 3, 5	Monocytes NK and T cells Basophils Eosinophils Dendritic cells	Degranulates basophils Activates T cells Chronic inflammation
	CCL11 (Eotaxin)	Endothelium Monocytes Epithelium T cells	CCR3	Eosinophils Monocytes T cells	Role in allergy
	CCL18 (DC-CK)	Dendritic cells	?	Naive T cells	Role in activating naive T cells
C	XCL1 (Lymphotactin)	CD8>CD4 T cells	CXCR1	Thymocytes Dendritic cells NK cells	Lymphocyte trafficking and development
CXXXC (CX_3C)	CX3CL1 (Fractalkine)	Monocytes Endothelium Microglial cells	CX_3CR1	Monocytes T cells	Leukocyte–endothelial adhesion Brain inflammation

2-25 Cell-adhesion molecules control interactions between leukocytes and endothelial cells during an inflammatory response.

The recruitment of activated phagocytes to sites of infection is one of the most important functions of innate immunity. Recruitment occurs as part of the inflammatory response and is mediated by cell-adhesion molecules that are induced on the surface of the local blood vessel endothelium. Before we consider the process of inflammatory cell recruitment, we will first describe some of the cell-adhesion molecules involved.

As with the complement components, a significant barrier to understanding cell-adhesion molecules is their nomenclature. Most adhesion molecules, especially those on leukocytes, which are relatively easy to analyze functionally, were named after the effects of specific monoclonal antibodies directed against them; only later were these characterized by gene cloning. Their names therefore bear no relation to their structure. For instance, the **leukocyte functional antigens LFA-1**, **LFA-2**, and **LFA-3** are actually members of two different protein families. In Fig. 2.47, the adhesion molecules are grouped according to their molecular structure, which is shown in schematic form, alongside their different names, sites of expression, and ligands. Three families of adhesion molecules are important for leukocyte recruitment. The **selectins** are membrane glycoproteins with a distal lectin-like domain that binds specific carbohydrate groups. Members of this family are induced on activated endothelium and initiate endothelium–leukocyte interactions by binding to fucosylated oligosaccharide ligands on passing leukocytes (see Fig. 2.47).

The next step in leukocyte recruitment depends on tighter adhesion, which is due to the binding of **intercellular adhesion molecules** (**ICAMs**) on the endothelium to heterodimeric proteins of the **integrin** family on leukocytes.

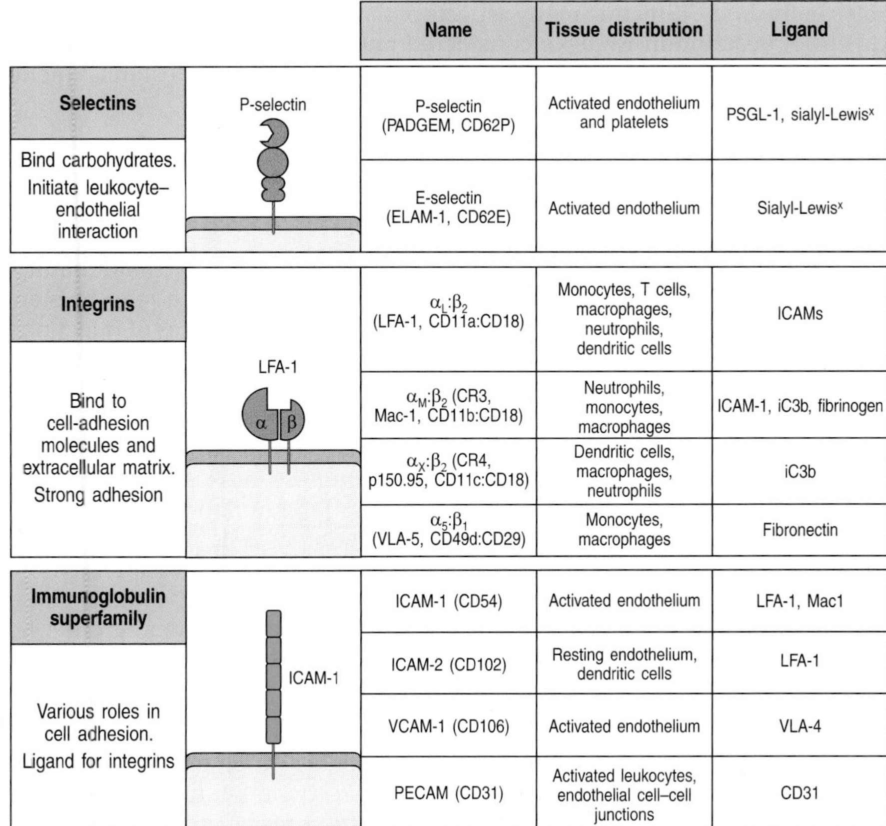

Family	Example	Name	Tissue distribution	Ligand
Selectins — Bind carbohydrates. Initiate leukocyte–endothelial interaction	P-selectin	P-selectin (PADGEM, CD62P)	Activated endothelium and platelets	PSGL-1, sialyl-Lewis^x
		E-selectin (ELAM-1, CD62E)	Activated endothelium	Sialyl-Lewis^x
Integrins — Bind to cell-adhesion molecules and extracellular matrix. Strong adhesion	LFA-1 (α, β)	α_L:β_2 (LFA-1, CD11a:CD18)	Monocytes, T cells, macrophages, neutrophils, dendritic cells	ICAMs
		α_M:β_2 (CR3, Mac-1, CD11b:CD18)	Neutrophils, monocytes, macrophages	ICAM-1, iC3b, fibrinogen
		α_X:β_2 (CR4, p150.95, CD11c:CD18)	Dendritic cells, macrophages, neutrophils	iC3b
		α_5:β_1 (VLA-5, CD49d:CD29)	Monocytes, macrophages	Fibronectin
Immunoglobulin superfamily — Various roles in cell adhesion. Ligand for integrins	ICAM-1	ICAM-1 (CD54)	Activated endothelium	LFA-1, Mac1
		ICAM-2 (CD102)	Resting endothelium, dendritic cells	LFA-1
		VCAM-1 (CD106)	Activated endothelium	VLA-4
		PECAM (CD31)	Activated leukocytes, endothelial cell–cell junctions	CD31

Fig. 2.47 Adhesion molecules involved in leukocyte interactions. Several structural families of adhesion molecules have a role in leukocyte migration, homing, and cell–cell interactions: the selectins, the integrins, and proteins of the immunoglobulin superfamily. The figure shows schematic representations of an example from each family, a list of other family members that participate in leukocyte interactions, their cellular distribution, and their ligand in adhesive interactions. The family members shown here are limited to those that participate in inflammation and other innate immune mechanisms. The same molecules and others participate in adaptive immunity and will be considered in Chapters 8 and 10. The nomenclature of the different molecules in these families is confusing because it often reflects the way in which the molecules were first identified rather than their related structural characteristics. Alternative names for each of the adhesion molecules are given in parentheses. Sulfated sialyl-Lewis^x, which is recognized by P- and E-selectin, is an oligosaccharide present on the cell-surface glycoproteins of circulating leukocytes. Sulfation can occur at either the sixth carbon atom of the galactose or the *N*-acetylglucosamine, but not both.

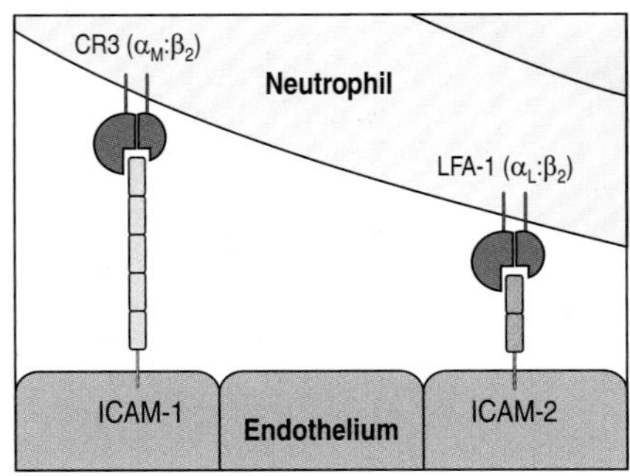

Fig. 2.48 Phagocyte adhesion to vascular endothelium is mediated by integrins. When vascular endothelium is activated by inflammatory mediators it expresses two adhesion molecules, namely ICAM-1 and ICAM-2. These are ligands for integrins expressed by phagocytes—α_M:β_2 (also called CR3, Mac-1, or CD11b:CD18) and $\alpha_L\beta_2$ (also called LFA-1 or CD11a:CD18).

The leukocyte integrins important for extravasation are **LFA-1** (α_L:β_2, also known as CD11a:CD18) and CR3 (α_M:β_2, complement receptor type 3, also known as CD11b:CD18 or Mac-1; we met CR3 in Section 2-19 as a receptor for iC3b, but that is just one of the ligands for this integrin) and they both bind to **ICAM-1** and to **ICAM-2** (Fig. 2.48). Strong adhesion between leukocytes and endothelial cells is promoted by the induction of ICAM-1 on inflamed endothelium and the activation of a conformational change in LFA-1 and CR3 that occurs in response to chemokine binding by the leukocyte. The importance of the leukocyte integrins in inflammatory cell recruitment is illustrated by the disease **leukocyte adhesion deficiency**, which stems from a defect in the β_2 chain common to both LFA-1 and CR3. People with this disease suffer from recurrent bacterial infections and impaired healing of wounds.

The activation of endothelium is driven by interactions with macrophage cytokines, particularly TNF-α, which induce the rapid externalization of granules called **Weibel–Palade bodies** in the endothelial cells. These granules contain preformed **P-selectin**, which is thus expressed within minutes on the surface of local endothelial cells after the production of TNF-α by macrophages. Shortly after the appearance of P-selectin on the cell surface, mRNA encoding **E-selectin** is synthesized, and within 2 hours the endothelial cells are expressing mainly E-selectin. Both these proteins interact with the sulfated sialyl-Lewis^x that is present on the surface of neutrophils.

Resting endothelium carries low levels of ICAM-2, apparently in all vascular beds. This may be used by circulating monocytes to navigate out of the vessels and into their tissue sites. This monocyte migration happens continuously and essentially ubiquitously. However, upon exposure to TNF-α, local expression of ICAM-1 is strongly induced on the endothelium of small vessels near or within the infectious focus. ICAM-1 in turn binds to LFA-1 or CR3 on circulating monocytes and polymorphonuclear leukocytes, in particular neutrophils, as shown in Fig. 2.48. Cell-adhesion molecules have many other roles in the body, directing many aspects of tissue and organ development. In this brief description we have considered only those that participate in the recruitment of inflammatory cells in the hours to days after the establishment of an infection.

2-26 Neutrophils make up the first wave of cells that cross the blood vessel wall to enter inflammatory sites.

The physical changes that accompany the initiation of the inflammatory response have been described in Section 2-5; here we give a step-by-step account of how the required effector cells are recruited into sites of infection. Under normal conditions, leukocytes flow in the center of small blood vessels, where blood flow is fastest. In inflammatory sites, the vessels are dilated and the slower blood flow allows the leukocytes to move out of the center of the blood vessel and interact with the vascular endothelium. Even in the absence of infection, monocytes migrate continuously into the tissues, where they differentiate into macrophages. During an inflammatory response, the induction of adhesion molecules on the endothelial cells by the infection focus, as well as induced changes in the adhesion molecules expressed on leukocytes, recruit large numbers of circulating leukocytes, initially neutrophils and later monocytes, into the site of an infection. The migration of leukocytes out of blood vessels, a process known as extravasation, is thought to occur in four steps. We will describe this process as it is known to occur for monocytes and neutrophils (Fig. 2.49).

The first step involves selectins. P-selectin appears on endothelial cell surfaces within a few minutes of exposure to leukotriene B4, the complement fragment C5a, or histamine, which is released from mast cells in response to

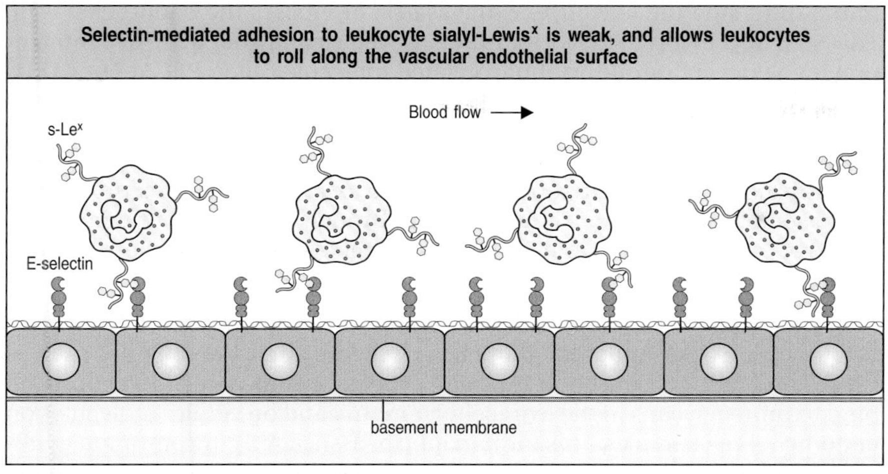

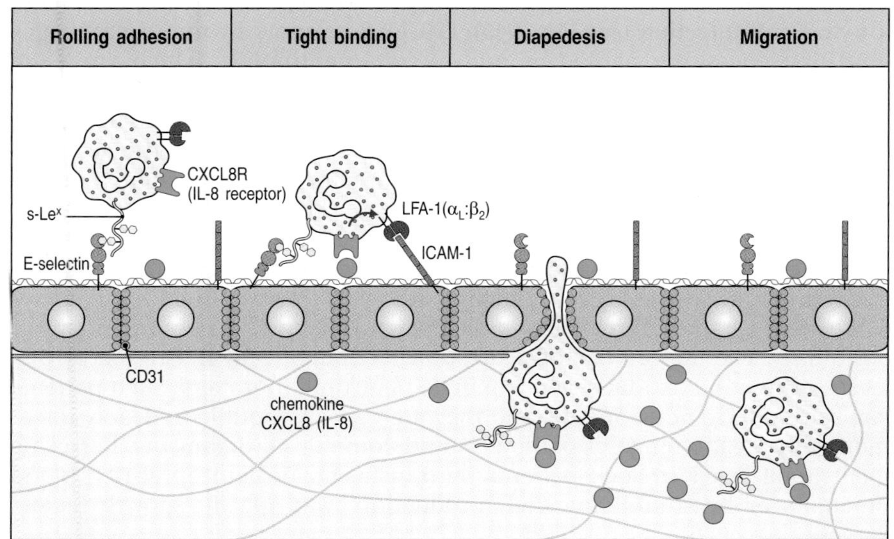

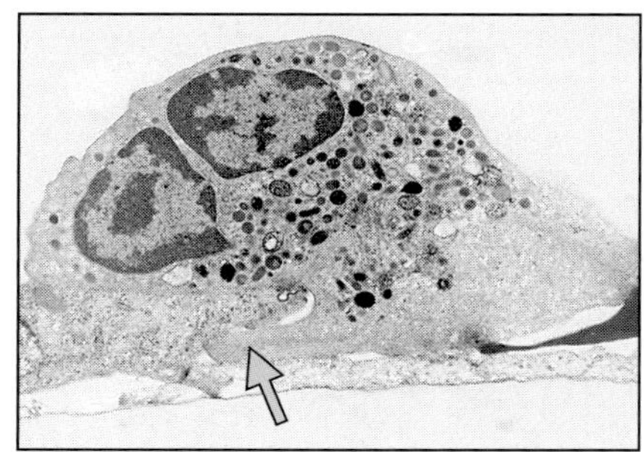

Fig. 2.49 Neutrophils leave the blood and migrate to sites of infection in a multi-step process involving adhesive interactions that are regulated by macrophage-derived cytokines and chemokines. The first step (top panel) involves the reversible binding of a neutrophil to vascular endothelium through interactions between selectins induced on the endothelium and their carbohydrate ligands on the neutrophil, shown here for E-selectin and its ligand the sialyl-Lewisx moiety (s-Le^x). This interaction cannot anchor the cells against the shearing force of the flow of blood, and they roll along the endothelium, continually making and breaking contact. The binding does, however, allow stronger interactions, which only result when binding of a chemokine such as CXCL8 to its specific receptor on the neutrophil triggers the activation of the integrins LFA-1 and CR3 (Mac-1) (not shown). Inflammatory cytokines such as TNF-α are also necessary to induce the expression of adhesion molecules such as ICAM-1 and ICAM-2, the ligands for these integrins, on the vascular endothelium. Tight binding between ICAM-1 and the integrins arrests the rolling and allows the neutrophil to squeeze between the endothelial cells forming the wall of the blood vessel (i.e., to extravasate). The leukocyte integrins LFA-1 and CR3 are required for extravasation, and for migration toward chemoattractants. Adhesion between molecules of CD31, expressed on both the neutrophil and the junction of the endothelial cells, is also thought to contribute to extravasation. The neutrophil also needs to traverse the basement membrane; it penetrates this with the aid of a matrix metalloproteinase enzyme that it expresses at the cell surface. Finally, the neutrophil migrates along a concentration gradient of chemokines (shown here as CXCL8) secreted by cells at the site of infection. The electron micrograph shows a neutrophil extravasating between endothelial cells. The blue arrow indicates the pseudopod that the neutrophil is inserting between the endothelial cells. Photograph (× 5500) courtesy of I. Bird and J. Spragg.

C5a. The appearance of P-selectin can also be induced by exposure to TNF-α or LPS, and both of these have the additional effect of inducing the synthesis of a second selectin, E-selectin, which appears on the endothelial cell surface a few hours later. These selectins recognize the sulfated sialyl-Lewis^X moiety of certain leukocyte glycoproteins that are exposed on the tips of leukocyte microvilli. The interaction of P-selectin and E-selectin with these glycoproteins allows monocytes and neutrophils to adhere reversibly to the vessel wall, so that circulating leukocytes can be seen to 'roll' along endothelium that has been treated with inflammatory cytokines (see Fig. 2.49, top panel). This adhesive interaction permits the stronger interactions of the next step in leukocyte migration.

This second step depends on interactions between the leukocyte integrins LFA-1 and CR3 with molecules on endothelium such as ICAM-1, which can also be induced on endothelial cells by TNF-α, and ICAM-2 (see Fig. 2.49, bottom panel). LFA-1 and CR3 normally adhere only weakly, but CXCL8 or other chemokines, bound to proteoglycans on the surface of endothelial cells, bind to specific chemokine receptors on the leukocyte and signal the cell to trigger a conformational change in LFA-1 and CR3 on the rolling leukocyte, which greatly increases the adhesive properties of the neutrophil. As a result of this greatly increased adhesion, the leukocyte attaches firmly to the endothelium and its rolling is arrested.

In the third step the leukocyte extravasates, or crosses the endothelial wall. This step also involves LFA-1 and CR3, as well as a further adhesive interaction involving an immunoglobulin-related molecule called **PECAM** or **CD31**, which is expressed both on the leukocyte and at the intercellular junctions of endothelial cells. These interactions enable the phagocyte to squeeze between the endothelial cells. It then penetrates the basement membrane with the aid of enzymes that break down the extracellular matrix proteins of the basement membrane. The movement through the basement membrane is known as **diapedesis**, and it enables phagocytes to enter the subendothelial tissues.

The fourth and final step in extravasation is the migration of leukocytes through the tissues under the influence of chemokines. Chemokines such as CXCL8 and CCL2 (see Section 2-24) are produced at the site of infection and bind to proteoglycans in the extracellular matrix and on similar molecules on endothelial cell surfaces. This forms a matrix-associated concentration gradient of chemokine on a solid surface along which the leukocyte can migrate to the focus of infection (see Fig. 2.49). CXCL8 is released by the macrophages that first encounter pathogens and recruits neutrophils, which enter the infected tissue in large numbers in the early part of the induced response. Their influx usually peaks within the first 6 hours of an inflammatory response, whereas monocytes can be recruited later, through the action of chemokines such as CCL2. Once in an inflammatory site, neutrophils are able to eliminate many pathogens by phagocytosis. They act as phagocytic effectors in an innate immune response through receptors that opsonize or capture infectious agents and their derived components through innate immune recognition as well as by recognizing pathogens directly. In addition, as we will see in Chapter 9, they act as phagocytic effectors in humoral adaptive immunity. The importance of neutrophils is dramatically illustrated by diseases or treatments that severely reduce neutrophil numbers. Such patients are said to have **neutropenia**, and they are very susceptible to deadly infection with a wide range of pathogens and commensal organisms. Restoring neutrophil levels in such patients by transfusion of neutrophil-rich blood fractions or by stimulating their production with specific growth factors largely corrects this susceptibility.

2-27 TNF-α is an important cytokine that triggers local containment of infection but induces shock when released systemically.

Inflammatory mediators also stimulate endothelial cells to express proteins that trigger blood clotting in the local small vessels, occluding them and thus cutting off blood flow. This can be important in preventing the pathogen from entering the bloodstream and spreading through the blood to organs all over the body. Instead, the fluid that has leaked into the tissue in the early phases of an infection carries the pathogen, usually enclosed in dendritic cells, via the lymph to the regional lymph nodes, where an adaptive immune response can be initiated. The importance of TNF-α in the containment of local infection is illustrated by experiments in which rabbits were infected locally with a bacterium. Normally, the infection would be contained at the site of the inoculation; if, however, an injection of anti-TNF-α antibody was also given to block the action of TNF-α, the infection spread via the blood to other organs.

Once an infection spreads to the bloodstream, however, the same mechanisms by which TNF-α so effectively contains local infection instead become catastrophic (Fig. 2.50). The presence of infection in the bloodstream, or sepsis, is accompanied by the release of TNF-α by macrophages in the liver, spleen, and other systemic sites. The systemic release of TNF-α causes vasodilation, which leads to a loss of blood pressure and increased vascular permeability, leading to a loss of plasma volume and eventually to shock. In septic shock, disseminated intravascular coagulation (blood clotting) is also triggered by TNF-α, leading to the generation of clots in many small vessels and the massive consumption of clotting proteins, thus causing the patient's ability to clot blood appropriately to be lost. This condition frequently leads to the failure of vital organs such as the kidneys, liver, heart, and lungs, which are quickly compromised by the failure of normal perfusion; consequently, septic shock has a very high mortality rate.

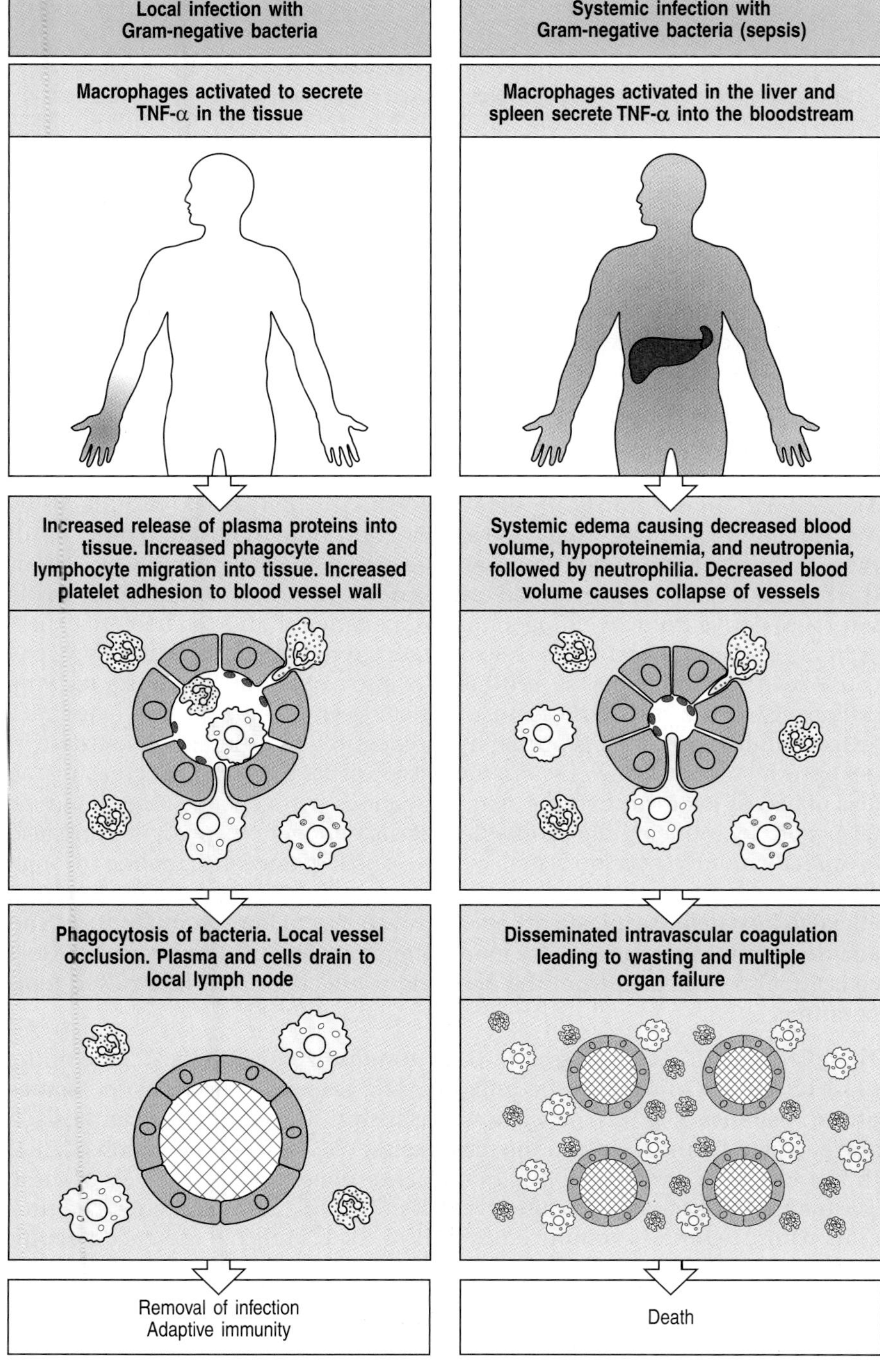
Local infection with Gram-negative bacteria
Systemic infection with Gram-negative bacteria (sepsis)
Macrophages activated to secrete TNF-α in the tissue
Macrophages activated in the liver and spleen secrete TNF-α into the bloodstream
Increased release of plasma proteins into tissue. Increased phagocyte and lymphocyte migration into tissue. Increased platelet adhesion to blood vessel wall
Systemic edema causing decreased blood volume, hypoproteinemia, and neutropenia, followed by neutrophilia. Decreased blood volume causes collapse of vessels
Phagocytosis of bacteria. Local vessel occlusion. Plasma and cells drain to local lymph node
Disseminated intravascular coagulation leading to wasting and multiple organ failure
Removal of infection
Adaptive immunity
Death

Fig. 2.50 The release of TNF-α by macrophages induces local protective effects, but TNF-α can be damaging when released systemically. The panels on the left show the causes and consequences of local release of TNF-α, and the panels on the right show the causes and consequences of systemic release. In both cases TNF-α acts on blood vessels, especially venules, to increase blood flow and vascular permeability to fluid, proteins, and cells, and to increase endothelial adhesiveness for leukocytes and platelets (center panels). Local release thus allows an influx of fluid, cells, and proteins into the infected tissue, where they participate in host defense. Later, blood clots form in the small vessels (lower left panel), preventing spread of infection via the blood, and the accumulated fluid and cells drain to regional lymph nodes, where an adaptive immune response is initiated. When there is a systemic infection, or sepsis, with bacteria that elicit TNF-α production, TNF-α is released into the blood by macrophages in the liver and spleen and acts in a similar way on all small blood vessels (lower right panel). The result is shock, disseminated intravascular coagulation with depletion of clotting factors, and consequent bleeding, multiple organ failure, and frequently death.

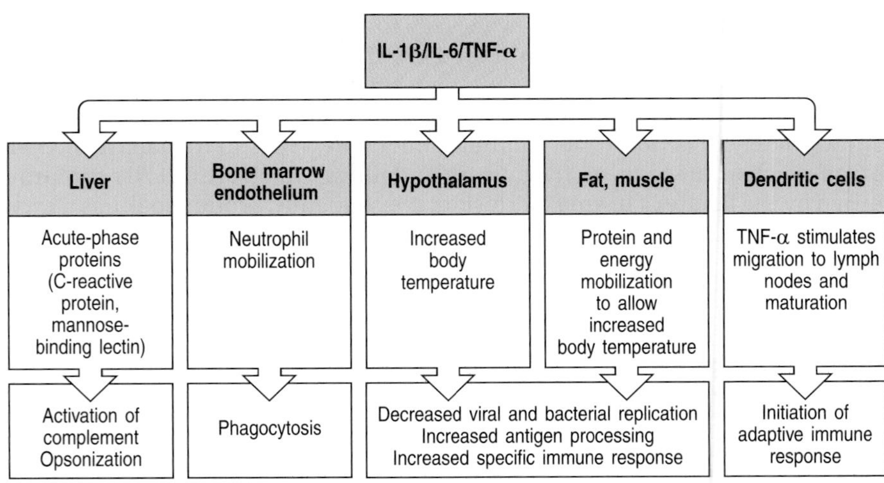

Fig. 2.51 The cytokines TNF-α, IL-1β, and IL-6 have a wide spectrum of biological activities that help to coordinate the body's responses to infection. IL-1β, IL-6, and TNF-α activate hepatocytes to synthesize acute-phase proteins, and bone marrow endothelium to release neutrophils. The acute-phase proteins act as opsonins, whereas the disposal of opsonized pathogens is augmented by the enhanced recruitment of neutrophils from the bone marrow. IL-1β, IL-6, and TNF-α are also endogenous pyrogens, raising body temperature, which is believed to help in eliminating infections. A major effect of these cytokines is to act on the hypothalamus, altering the body's temperature regulation, and on muscle and fat cells, altering energy mobilization to increase the body temperature. At higher temperatures, bacterial and viral replication is less efficient, whereas the adaptive immune response operates more efficiently.

Mice with a mutant TNF-α receptor gene are resistant to septic shock; however, such mice are also unable to control local infection. The features of TNF-α that make it so valuable in containing local infection are precisely those that give it a central role in the pathogenesis of septic shock. It is clear from the evolutionary conservation of TNF-α that its benefits in the former area far outweigh the devastating consequences of its systemic release.

2-28 Cytokines released by phagocytes activate the acute-phase response.

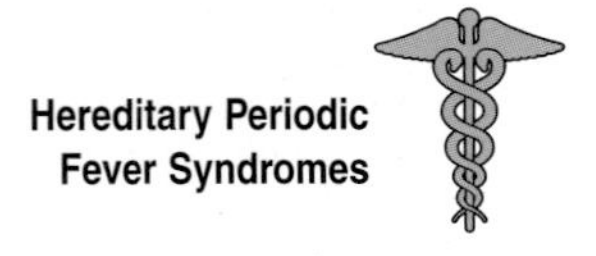

As well as their important local effects, the cytokines produced by macrophages have long-range effects that contribute to host defense. One of these is the elevation of body temperature, which is caused mainly by TNF-α, IL-1β, and IL-6. These are termed **endogenous pyrogens** because they cause fever and derive from an endogenous source rather than from bacterial components such as LPS, which is an **exogenous pyrogen**. Endogenous pyrogens cause fever by inducing the synthesis of prostaglandin E2 by the enzyme cyclooxygenase-2, the expression of which is induced by these cytokines. Prostaglandin E2 then acts on the hypothalamus, resulting in an increase in heat production by brown fat and increased vasoconstriction, decreasing the loss of excess heat through the skin. Exogenous pyrogens are able to induce fever both by inducing the production of the endogenous pyrogens and also by directly inducing cyclooxygenase-2 as a consequence of signaling through TLR-4, leading to the production of prostaglandin E2. Fever is generally beneficial to host defense; most pathogens grow better at lower temperatures and adaptive immune responses are more intense at elevated temperatures. Host cells are also protected from the deleterious effects of TNF-α at raised temperatures.

The effects of TNF-α, IL-1, and IL-6 are summarized in Fig. 2.51. One of the most important of these is the initiation of a response known as the **acute-phase response** (Fig. 2.52). This involves a shift in the proteins synthesized and secreted by the liver into the plasma and results from the action of IL-1, IL-6, and TNF-α on hepatocytes. In the acute-phase response, levels of some plasma proteins go down, while levels of others increase markedly. The proteins whose synthesis is induced by TNF-α, IL-1, and IL-6 are called **acute-phase proteins**. Several of these are of particular interest because they mimic the action of antibodies, but, unlike antibodies, these proteins have broad specificity for PAMPs and depend only on the presence of cytokines for their production.

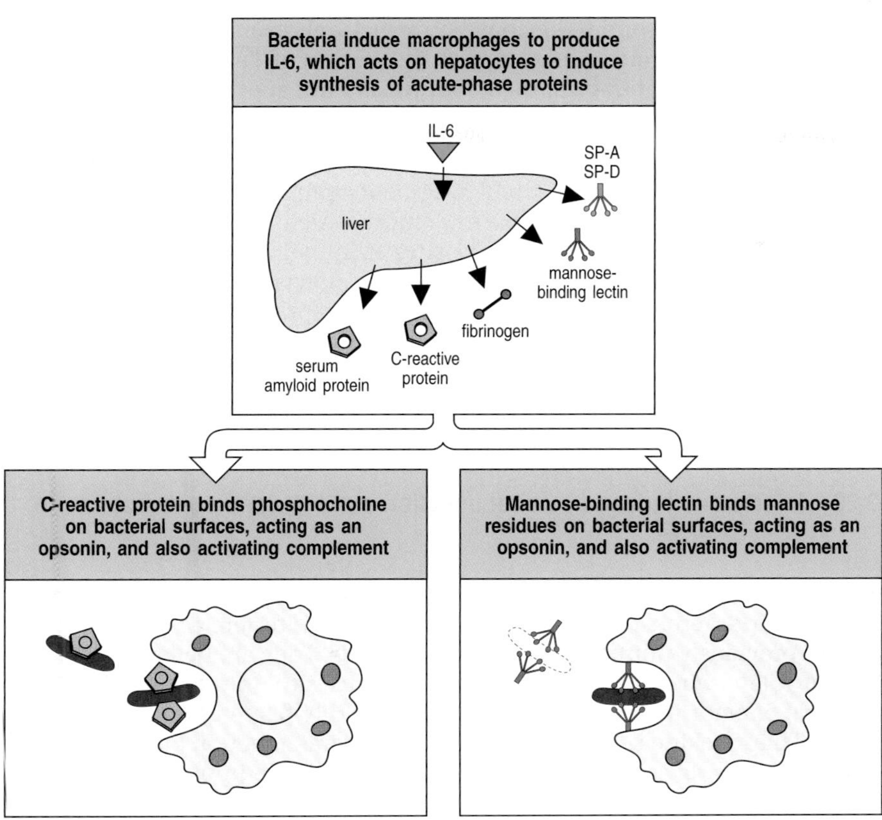

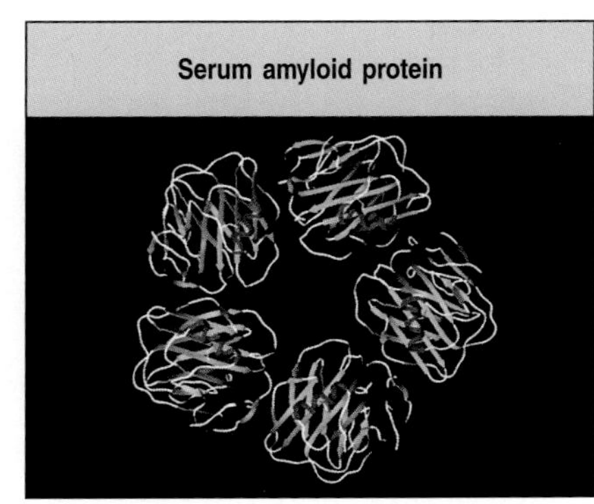

Fig. 2.52 The acute-phase response produces molecules that bind pathogens but not host cells. Acute-phase proteins are produced by liver cells in response to cytokines released by macrophages in the presence of bacteria. They include serum amyloid protein (SAP) (in mice but not humans), C-reactive protein (CRP), fibrinogen, and mannose-binding lectin (MBL). SAP and CRP are homologous in structure; both are pentraxins, forming five-membered discs, as shown for SAP (photograph on the right). CRP binds phosphocholine on certain bacterial and fungal surfaces but does not recognize it in the form in which it is found in host cell membranes. It both acts as an opsonin in its own right and activates the classical complement pathway by binding C1q to augment opsonization. MBL is a member of the collectin family, which includes the lung surfactant proteins SP-A and SP-D. It also resembles C1q in its structure. Like CRP, MBL can act as an opsonin in its own right, as can SP-A and SP-D. Model structure courtesy of J. Emsley.

One of these proteins, the **C-reactive protein**, is a member of the **pentraxin** protein family, so called because they are formed from five identical subunits. C-reactive protein is another example of a multipronged pathogen-recognition molecule, and binds to the phosphocholine portion of certain bacterial and fungal cell-wall lipopolysaccharides. Phosphocholine is also found in mammalian cell membrane phospholipids, but it cannot be bound by C-reactive protein. When C-reactive protein binds to a bacterium, it is not only able to opsonize it but can also activate the complement cascade by binding to C1q, the first component of the classical pathway of complement activation, as we learnt in Section 2-13. The interaction with C1q involves the collagen-like parts of C1q rather than the globular heads contacted by pathogen surfaces, but the same cascade of reactions is initiated.

The second acute-phase protein of interest is MBL, which we have already met as a pathogen-binding molecule (see Fig. 2.15) and as a trigger for the complement cascade (see Section 2-14). MBL is found in normal serum at low levels but is produced in increased amounts during the acute-phase response. It acts as an opsonin for monocytes, which, unlike tissue macrophages, do not express the macrophage mannose receptor. Two other proteins with opsonizing properties that are produced by the liver in increased quantities during the acute-phase response are the pulmonary surfactant proteins SP-A and SP-D (see Section 2-6). They are found along with macrophages in the alveolar fluid of the lung and are important in promoting the phagocytosis of respiratory pathogens such as *Pneumocystis carinii*, one of the main causes of pneumonia in patients with AIDS.

Thus, within a day or two, the acute-phase response provides the host with several proteins with the functional properties of antibodies but able to bind a broad range of pathogens. However, unlike the antibodies we shall learn

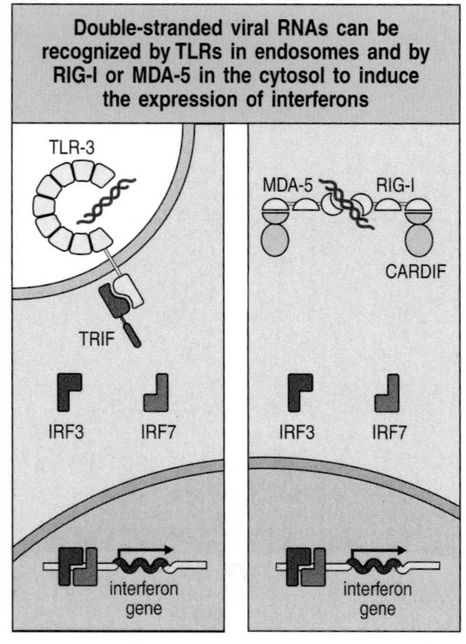
Double-stranded viral RNAs can be recognized by TLRs in endosomes and by RIG-I or MDA-5 in the cytosol to induce the expression of interferons
TLR-3
TRIF
IRF3
IRF7
interferon gene
MDA-5
RIG-I
CARDIF
IRF3
IRF7
interferon gene

Fig. 2.53 Double-stranded RNAs induce the expression of interferons by activating the interferon regulatory factors IRF3 and IRF7. Long double-stranded RNAs can be recognized by the Toll-like receptor TLR-3, which is present in endosomes (left panel). TLR-3 signals through an adaptor molecule, TRIF, to activate the transcription factors IRF3 and IRF7. Similarly, the intracellular receptors RIG-I and MDA-5 also bind long double-stranded RNAs (right panel) and activate IRF3 and IRF7, but in this case through the adaptor protein CARDIF (CARD adaptor inducing IFN-β). Activated IRF3 and IRF7 can form both homodimers (not shown) and IRF3:IRF7 heterodimers that enter the nucleus and activate the transcription of a number of genes, principally those for IFN-α and IFN-β.

about in Chapters 3 and 9 they have no structural diversity and are made in response to any stimulus that triggers the release of TNF-α, IL-1, and IL-6. Therefore, unlike antibodies, their synthesis is not specifically induced and targeted.

A final distant effect of the cytokines produced by phagocytes is to induce a **leukocytosis**, an increase in circulating neutrophils. The neutrophils come from two sources: the bone marrow, from which mature leukocytes are released in increased numbers; and sites in blood vessels where they are attached loosely to endothelial cells. Thus, the effects of these cytokines contribute to the control of infection while the adaptive immune response is being developed. As shown in Fig. 2.51, TNF-α also has a role in stimulating the migration of dendritic cells from their sites in peripheral tissues to the lymph node and in their maturation into nonphagocytic but highly co-stimulatory antigen-presenting cells.

2-29 Interferons induced by viral infection make several contributions to host defense.

Infection of cells with viruses induces the production of proteins that are known as interferons, because they were found to interfere with viral replication in previously uninfected tissue culture cells. They are believed to have a similar role *in vivo*, blocking the spread of viruses to uninfected cells. These antiviral effector molecules, called **IFN-α** and **IFN-β**, are quite distinct from **IFN-γ**. This cytokine is not directly induced by viral infection but is produced later and does have an important role in the adaptive immune response to intracellular pathogens, as we will see in later chapters. IFN-α, actually a family of several closely related proteins, and IFN-β, the product of a single gene, are synthesized by many cell types after their infection by diverse viruses. Interferon synthesis is thought to occur in response to the presence of double-stranded RNA, because synthetic double-stranded RNA is a potent inducer of IFN-α and IFN-β. Double-stranded RNA forms the genome of some viruses and might be made as part of the infectious cycle of all viruses. Although double-stranded RNA molecules are found in mammalian cells, they are present only as relatively short molecules, usually less than 100 nucleotides long, whereas double-stranded RNA genomes are thousands of nucleotides long. Therefore, long double-stranded RNA molecules might be the common element in interferon induction; such long molecules are recognized as a distinct molecular pattern by the Toll-like receptor TLR-3 (Fig. 2.53), which induces the synthesis of IFN-α and IFN-β.

Long double-stranded viral RNA can also induce the expression of interferons by activating the cytoplasmic proteins **RIG-I** and **MDA-5** (see Fig. 2.53). These proteins contain RNA helicase-like domains that bind double-stranded RNAs, and two CARD domains (see Section 2-9) that allow them to interact with adaptor proteins within the cell to deliver a signal that viral RNAs are present. For both RIG-I and MDA-5, the adaptors link the binding of double-stranded RNAs to the activation of the interferon-regulatory factors IRF3 and IRF7, transcription factors that induce the production of IFN-α and IFN-β.

Interferons make several contributions to defense against viral infection (Fig. 2.54). An obvious and important effect is the induction of a state of resistance to viral replication in all cells. IFN-α and IFN-β secreted by the infected cell bind to a common cell-surface receptor, known as the **interferon receptor**, on both the infected cell and nearby uninfected cells. The interferon receptor, like many other cytokine receptors, is coupled to a **Janus-family tyrosine kinase**, through which it signals. This signaling pathway, which we will describe in detail in Chapter 6, rapidly induces new gene transcription as the

Janus-family kinases directly phosphorylate signal-transducing activators of transcription known as **STAT proteins**. The phosphorylated STAT proteins enter the nucleus, where they activate the transcription of several different genes, including those encoding proteins that help to inhibit viral replication.

One such protein is the enzyme **oligoadenylate synthetase**, which polymerizes ATP into 2′–5′-linked oligomers (nucleotides in nucleic acids are normally linked 3′–5′). These activate an endoribonuclease that then degrades viral RNA. A second protein activated by IFN-α and IFN-β is a serine–threonine kinase called **PKR kinase**. This phosphorylates the eukaryotic protein synthesis initiation factor eIF-2, inhibiting translation and thus further contributing to the inhibition of viral replication. Another interferon-inducible protein, called **Mx**, is known to be required for cellular resistance to influenza virus replication. Mice that lack the gene for Mx are highly susceptible to infection with the influenza virus, whereas mice that can make Mx are not. Another way in which interferons act in innate immunity is to activate NK cells, which can kill virus-infected cells, as described in more detail in the next section.

Last, interferons have a more general role in the process by which recognition of pathogens by the innate immune system underlies and augments activation of the adaptive immune response. We have already discussed how recognition of double-stranded RNAs by TLR-3 can lead to the induction of IFN-α and IFN-β. Other TLRs, notably TLR-4, can also induce these interferons in response to the recognition of bacterial cell-wall components. Interferons in turn induce the expression of co-stimulatory molecules on macrophages and dendritic cells, which enables them to act as antigen-presenting cells that can fully activate T cells (see Section 1-7). Thus, a macrophage or dendritic cell that becomes activated when its Toll-like receptors bind pathogens is in turn able to signal to other macrophages and dendritic cells and recruit them to initiate an adaptive immune response. IFN-α and IFN-β also stimulate the increased expression of MHC class I molecules on all types of cells. The cytotoxic T lymphocytes of the adaptive immune system recognize virus-infected cells by the complexes of viral antigens and MHC class I molecules they display on their surface (see Fig. 1.30), and so, indirectly, interferons help promote the killing of virus-infected cells by CD8 cytotoxic T cells.

Almost all types of cells can produce IFN-α and IFN-β if necessary, but some cells seem to be specialized for the task. The **plasmacytoid dendritic cells** (also known as interferon-producing cells or natural interferon-producing cells) are circulating dendritic cells that accumulate in peripheral lymphoid tissues during an infection and can secrete up to 1000 times more interferon than other cell types. They are described in more detail in Chapter 8.

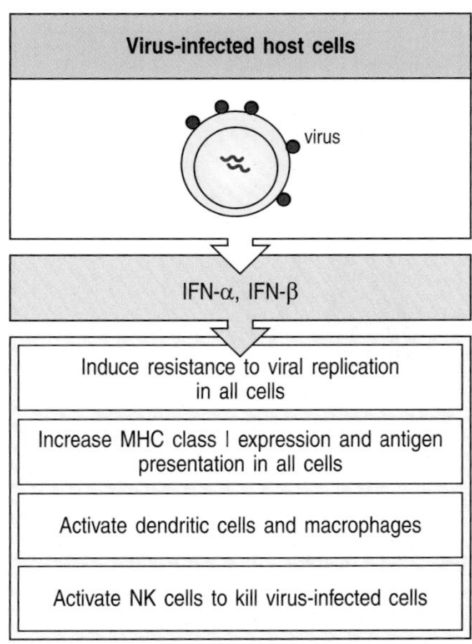

Fig. 2.54 Interferons are antiviral proteins produced by cells in response to viral infection. The interferons IFN-α and IFN-β have three major functions. First, they induce resistance to viral replication in uninfected cells by activating genes that cause the destruction of mRNA and inhibit the translation of viral and some host proteins. Second, they can induce MHC class I expression in most cell types in the body, thus enhancing their resistance to NK cells; they may also induce increased synthesis of MHC class I molecules in cells that are newly infected by virus, thus making them more susceptible to killing by CD8 cytotoxic T cells (see Chapter 8). Third, they activate NK cells, which then selectively kill virus-infected cells.

2-30 NK cells are activated by interferons and macrophage-derived cytokines to serve as an early defense against certain intracellular infections.

Natural killer cells (**NK cells**) develop in the bone marrow from the common lymphoid progenitor cell and circulate in the blood. They are larger than T and B lymphocytes, have distinctive cytoplasmic granules, and are functionally identified by their ability to kill certain lymphoid tumor cell lines *in vitro* without the need for prior immunization or activation. The mechanism of killing by NK cells is the same as that used by the cytotoxic T cells generated in an adaptive immune response; cytotoxic granules are released onto the surface of the bound target cell, and the effector proteins they contain penetrate the cell membrane and induce programmed cell death. However, NK cell killing is triggered by invariant receptors that recognize components of infected cell surfaces, and their known function in host defense is in the early

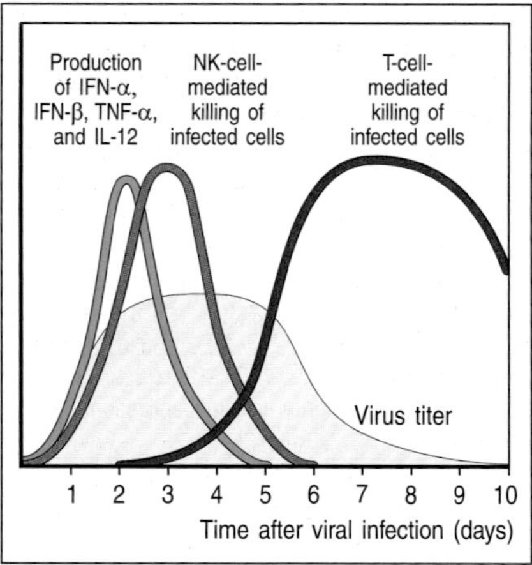

Fig. 2.55 Natural killer cells (NK cells) are an early component of the host response to virus infection. Experiments in mice have shown that IFN-α, IFN-β, and the cytokines TNF-α and IL-12 appear first, followed by a wave of NK cells, which together control virus replication but do not eliminate the virus. Virus elimination is accomplished when virus-specific CD8 T cells are produced. Without NK cells, the levels of some viruses are much higher in the early days of the infection, and can be lethal unless treated vigorously with antiviral compounds.

phases of infection with several intracellular pathogens, particularly herpes viruses and the protozoan parasite *Leishmania*. NK cells are classed as part of the innate immune system because of their invariant receptors.

NK cells are activated in response to interferons or macrophage-derived cytokines. Although NK cells that can kill sensitive targets can be isolated from uninfected individuals, this activity is increased 20–100-fold when NK cells are exposed to IFN-α and IFN-β or to the NK-cell-activating factor IL-12, which is one of the cytokines produced early in many infections. Activated NK cells serve to contain virus infections while the adaptive immune response is generating antigen-specific cytotoxic T cells that can clear the infection (Fig. 2.55). At present the only clue to the physiological function of NK cells in humans comes from rare patients deficient in these cells. Such patients prove to be highly susceptible to the early phases of herpes virus infection. A similar finding has recently been made in mice infected with murine cytomegalovirus, a herpes virus.

IL-12, acting in synergy with TNF-α, can also stimulate NK cells to secrete large amounts of the cytokine IFN-γ, and this is crucial in controlling some infections before the IFN-γ produced by activated CD8 cytotoxic T cells becomes available. This early production of IFN-γ by NK cells may also influence the CD4 T-cell response to infectious agents, inducing activated CD4 T cells to differentiate into inflammatory T_H1 cells able to activate macrophages (see Section 8-19).

2-31 NK cells possess receptors for self molecules that prevent their activation by uninfected cells.

If NK cells are to defend the body against infection with viruses and other pathogens, they must have some mechanism for distinguishing infected from uninfected cells. Exactly how this is achieved has not yet been worked out in every case, but it is thought that an NK cell is activated by a combination of direct recognition of changes in cell-surface glycoproteins induced by metabolic stress, such as malignant transformation or viral or bacterial infection, together with a recognition of 'altered self', which involves changes in the expression of MHC molecules. Altered expression of MHC class I molecules may be a common feature of cells infected by intracellular pathogens, because many of these pathogens have developed strategies to interfere with the ability of MHC class I molecules to capture and display peptides to T cells. Thus, one mechanism by which NK cells distinguish infected from uninfected cells is by recognizing alterations in MHC class I expression (Fig. 2.56).

NK cells are able to sense changes in the expression of MHC class I molecules by integrating the signals from two types of surface receptors, which together control their cytotoxic activity. One type is an activating receptor that triggers killing by the NK cell. Several classes of receptor provide this activation signal, including members of the immunoglobulin-like and the C-type lectin families of proteins. Stimulation of these receptors activates the NK cells, causing the release of cytokines such as IFN-γ and the directed killing of the stimulating cell through the release of cytotoxic granules containing granzymes and perforin. This killing mechanism is the same as that used by cytotoxic T cells and will be described in detail when we discuss the functions of that effector T-cell population in Chapter 8. NK cells also carry receptors for immunoglobulin, and binding of antibodies to these receptors activates NK cells to release their cytotoxic granules; this is known as antibody-dependent cellular cytotoxicity, or ADCC, and will be described in Chapter 9. A second set of receptors inhibits activation and prevents NK cells from killing normal host cells. These inhibitory receptors are specific for various MHC class I molecules, which helps to explain why NK cells selectively kill cells bearing low levels of

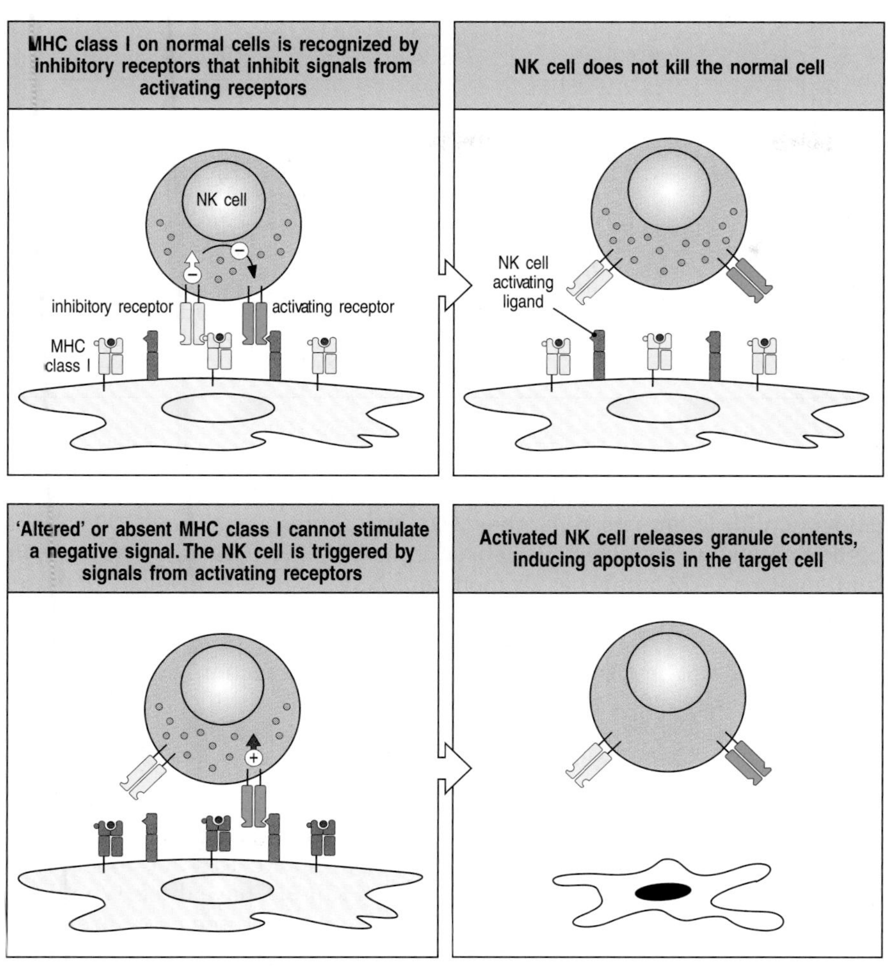

MHC class I on normal cells is recognized by inhibitory receptors that inhibit signals from activating receptors
NK cell
inhibitory receptor
activating receptor
MHC class I
NK cell does not kill the normal cell
NK cell activating ligand
'Altered' or absent MHC class I cannot stimulate a negative signal. The NK cell is triggered by signals from activating receptors
Activated NK cell releases granule contents, inducing apoptosis in the target cell

Fig. 2.56 NK cell killing depends on the balance between activating and inhibitory signals. NK cells have several different activating receptors that recognize common carbohydrate ligands on the surfaces of the cells of the body and signal the NK to kill the bound cell. However, NK cells are prevented from a wholesale attack by another set of receptors that recognize MHC class I molecules (which are present on almost all cell types) and inhibit killing by overruling the actions of the activating receptors. This inhibitory signal is lost when the target cells do not express MHC class I and perhaps also in cells infected with virus, many of which specifically inhibit MHC class I expression or alter its conformation so as to avoid recognition by CD8 T cells. Another possibility is that normal uninfected cells respond to IFN-α and IFN-β by increasing the expression of MHC class I molecules, making them resistant to killing by activated NK cells. In contrast, infected cells can fail to increase MHC class I expression, making them targets for activated NK cells.

MHC class I molecules but are prevented from killing cells with normal numbers. The higher the level of expression of MHC class I on a cell surface, the better protected it is against destruction by NK cells. This is why interferons, which induce the expression of MHC class I molecules, can protect uninfected host cells from NK cells. At the same time the interferon activates the NK cell to kill virus-infected cells.

The receptors that regulate the activity of NK cells fall into two large families (Fig. 2.57) that contain a number of other cell-surface receptors in addition to the NK receptors. One is composed of receptors homologous to C-type lectins; these are called **killer lectin-like receptors** (**KLRs**). The genes for KLRs are found within a gene cluster called the NK receptor complex, or NKC.

The other family of receptors is composed of receptors with immunoglobulin-like domains; hence their name **killer cell immunoglobulin-like receptors** (**KIRs**). The KIR genes form part of a larger cluster of immunoglobulin-like receptors known as the leukocyte receptor complex, or LRC. Both the NKC and LRC clusters are present in mice and in humans, but mice lack the KIR genes and therefore rely on the C-type lectin-like receptors of the NKC to control their NK-cell activity.

One complication in understanding the regulation of NK cell activity is that the same structural families of NK receptors contain both activating and inhibitory receptors. In humans and mice, NK cells express a heterodimer of two C-type lectins, called CD94 and NKG2, which interacts with nonpolymorphic MHC

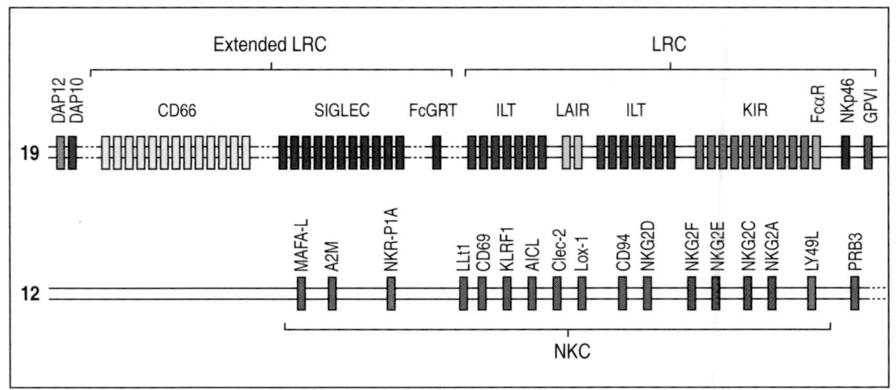

Fig. 2.57 The genes that encode NK receptors fall into two large families. The first, the leukocyte receptor complex (LRC), comprises a large cluster of genes encoding a family of proteins composed of immunoglobulin-like domains. These include the killer immunoglobulin-like receptors (KIRs) expressed by NK cells, the ILT (immunoglobulin-like transcript) class, and the leukocyte-associated immunoglobulin-like receptor (LAIR) gene families. The signaling lectins (SIGLECs) and members of the CD66 family are located nearby. In humans, this complex is located on chromosome 19. The second gene cluster is called the NK receptor complex (NKC), and encodes killer lectin-like receptors, a receptor family that includes the NKG2 proteins and CD94, with which an NKG2 molecule pairs to form a functional receptor. This complex is located on human chromosome 12. Some NK receptor genes are found outside of these two major gene clusters; for example, the genes for the natural cytotoxicity receptors NKp30 and NKp44 are located within the major histocompatibility complex on chromosome 6. Figure based on data courtesy of J. Trowsdale, University of Cambridge.

class I-like molecules, including HLA-E in humans and Qa-1 in mice, that bind leader peptide fragments from other MHC class I molecules. Thus, CD94:NKG2 could be sensitive to the presence of several different MHC class I variants. In humans, the NKG2 family has six members, NKG2A, B, C, D, E, and F. Of these, for example, NKG2A and B are inhibitory, whereas NKG2C is activating (Fig. 2.58). NKG2D is also activating but is distinct from the other NKG2 family members and will be discussed separately in the next section. In mice, Ly49H, a member of the Ly49 C-type lectin family, seems to be distinct, because binding of this molecule is an activating event that triggers a cytotoxic response, whereas binding to the other Ly49 members is inhibitory.

In the KIR receptor family, too, some members are activating whereas others are inhibitory. Different KIR genes also encode proteins with differing numbers of immunoglobulin domains; some, called KIR-2D, have two immunoglobulin domains while others, called KIR-3D, have three. Whether a KIR protein is activating or inhibitory depends on the presence or absence of specific signaling motifs in its cytoplasmic domain. Sequence motifs that are recognized by intracellular inhibitory adaptor proteins are found in KIR proteins that have long cytoplasmic tails; these proteins are designated KIR-2DL and KIR-3DL. KIR proteins with short cytoplasmic tails lack these inhibitory motifs, and instead associate with the activating adaptor protein DAP12 (also known as KARAP). Activating KIR receptors are therefore designated as KIR-2DS and KIR-3DS (see Fig. 2.58). Other inhibitory NK receptors specific for the products of the MHC class I loci are rapidly being defined, and all are members of either the immunoglobulin-like KIR family or the Ly49-like C-type lectins. Clearly, the regulation of NK-cell activity is complex, and whether any individual NK cell is activated by another cell will depend on the overall balance of activating and inhibitory receptors that the NK cell is expressing.

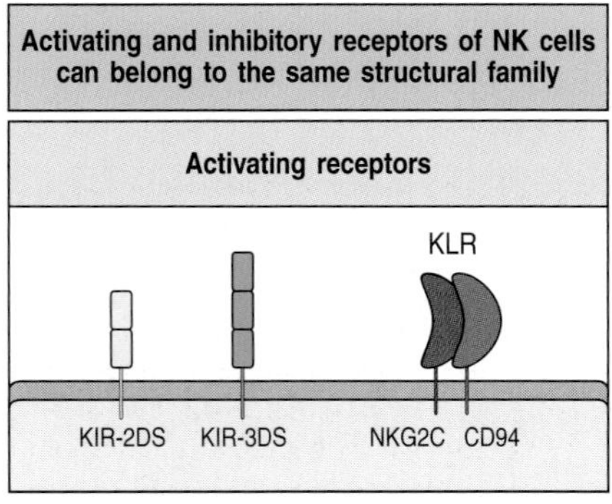

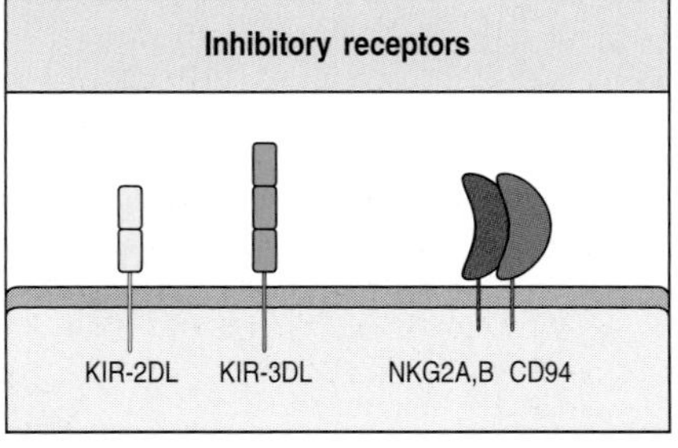

Fig. 2.58 The structural families of NK receptors encode both activating and inhibitory receptors. The families of killer immunoglobulin-like receptors (KIR) and killer lectin-like receptors (KLR) have members that send activating signals to the NK cell, as shown in the upper panel, and those that send inhibitory signals, as shown in the lower panel. KIR family members are designated according to the number of immunoglobulin-like domains they possess and by the length of their cytoplasmic tails. Activating KIR receptors have short cytoplasmic tails and bear the designation 'S'. These associate with a signaling adaptor protein, DAP12. The activating KLR receptor is a heterodimer of NKG2C with another C-type lectin family member, CD94. The inhibitory KIR receptors have longer cytoplasmic tails and are designated 'L'; these do not associate constitutively with adaptor proteins but contain signaling motifs that when phosphorylated are recognized by inhibitory phosphatases. Like the activating KLR, the inhibitory KLRs (NKG2A and NKG2B) form heterodimers with CD94.

The overall response of NK cells to differences in MHC expression is further complicated by the polymorphism of KIR genes; for example, for one of the KIR genes there are two alleles, one of which is activating and the other inhibitory. Moreover, the KIR gene cluster seems to be a very dynamic part of the human genome, because different numbers of activating and inhibitory KIR genes are found in different people. What advantage this diversity might have is not clear. As noted earlier, the KIR locus is not present in mice, and they use only the KLR molecules to regulate NK cell activity. So whatever the driving force is for the evolution of the KIR locus and its diversity, it may have arisen relatively recently in evolutionary terms.

Signaling by the inhibitory NK receptors suppresses the killing activity of NK cells. This means that NK cells will not kill healthy, genetically identical cells with normal expression of MHC class I molecules, such as the other cells of the body. Virus-infected cells, however, can become susceptible to being killed by NK cells by a variety of mechanisms. First, some viruses inhibit all protein synthesis in their host cells, so that synthesis of MHC class I proteins would be blocked in infected cells, even while their production in uninfected cells is being stimulated by the actions of interferon. The reduced level of MHC class I expression in infected cells would make them correspondingly less able to inhibit NK cells through their MHC-specific receptors, and they would therefore be more susceptible to being killed. Second, some viruses can selectively prevent the export of MHC class I molecules to the cell surface. This might allow the infected cell to evade recognition by cytotoxic T cells, but would make them susceptible to being killed by NK cells. Virus infection also alters the glycosylation of cellular proteins, perhaps allowing recognition by NK cell-activating receptors to dominate, or removing the normal ligand for the inhibitory receptors. This could allow infected cells to be detected even when the level of MHC class I expression has not changed.

Clearly, much remains to be learned about this innate mechanism of cytotoxic attack and its physiological relevance. The role of MHC class I molecules in allowing NK cells to detect intracellular infections is of particular interest because these same proteins govern the response of T cells to intracellular pathogens. It is possible that NK cells, which use a diverse set of nonclonal receptors to detect altered MHC, represent the modern remnants of the evolutionary forebears of T cells. Those T-cell ancestors went on to evolve rearranging genes that encode a vast repertoire of antigen-specific T-cell receptors geared to recognizing MHC molecules 'altered' by binding peptide antigens.

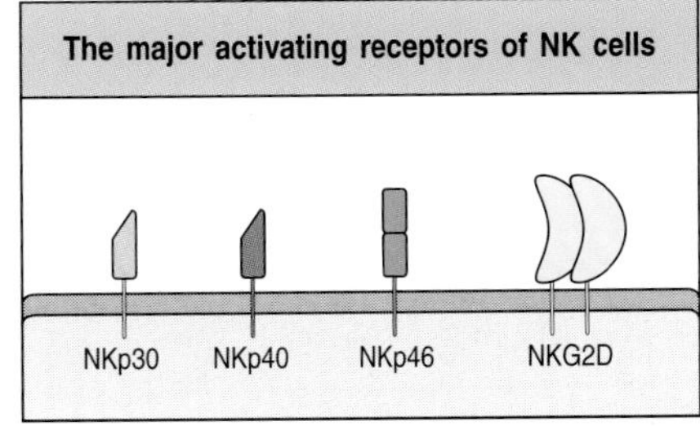

Fig. 2.59 The major activating receptors of NK cells are the natural cytotoxicity receptors and NKG2D. The natural cytotoxicity receptors are immunoglobulin-like proteins. NKp30 and NKp40 have an extracellular domain that resembles a single variable domain of an immunoglobulin molecule. They activate the NK cell through their association with homodimers of the CD3ζ chain or the Fc receptor γ chain (these are signaling proteins that also associate with other types of receptors and will be described in more detail in Chapter 6). NKp46 resembles the KIR-2D molecules in having two domains that resemble the constant domains of an immunoglobulin molecule. NKG2D is a member of the C-type lectin family and forms a homodimer.

2-32 NK cells bear receptors that activate their killer function in response to ligands expressed on infected cells or tumor cells.

In addition to the KIR and KLR receptors, which have a role in sensing the level of MHC class I expression on other cells, NK cells also express receptors that more directly sense the presence of infection or other perturbations in a cell. The most important activating receptors for the recognition of infected cells are the **natural cytotoxicity receptors** (**NCRs**) NKp30, NKp44, and NKp46, which are immunoglobulin-like receptors, and the C-type lectin family member NKG2D (Fig. 2.59). The ligands recognized by the natural cytotoxicity receptors are not well defined, although NKp46 is known to recognize heparan sulfate proteoglycans as well as some viral proteins.

NKG2D seems to have a specialized role in activating NK cells. Other NKG2 family members (NKG2A, C, and E) form heterodimers with CD94 and bind the MHC class I molecule HLA-E. NKG2D does neither; instead, the ligands for the NKG2D receptor are families of proteins that are distantly related to the MHC class I molecules but have a completely different function, being

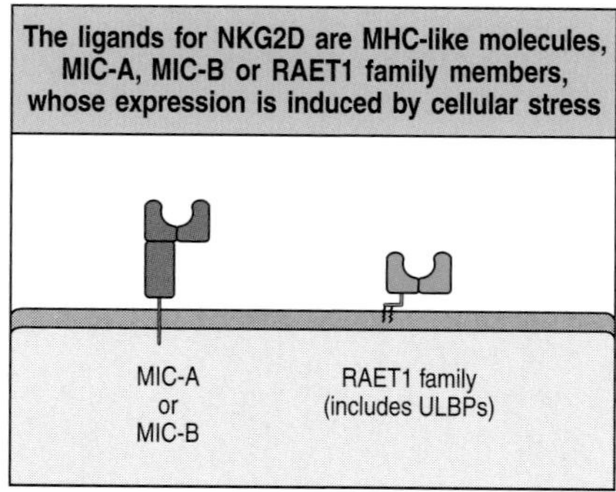
The ligands for NKG2D are MHC-like molecules, MIC-A, MIC-B or RAET1 family members, whose expression is induced by cellular stress
MIC-A
or
MIC-B
RAET1 family
(includes ULBPs)

Fig. 2.60 The ligands for the activating NK receptor NKG2D are proteins that are expressed in conditions of cellular stress. The MIC proteins MIC-A and MIC-B are MHC-like molecules induced on epithelial and other cells by stresses, such as heat shock, metabolic stress, or infection. RAET1 family members, including the subset designated as UL16-binding proteins (ULBPs), also resemble a portion of an MHC class I molecule, the α_1 and α_2 domains, and are attached to the cell via a glycophosphatidylinositol linkage.

produced in response to stress. The ligands in humans for NKG2D, as shown in Fig. 2.60, are the MHC class I-like MIC molecules, MIC-A and MIC-B, and the RAET1 protein family, which are homologous to the α_1 and α_2 domains of MHC class I molecules (which will be described when we discuss MHC molecule structure in Chapter 3). The RAET1 family has 10 members, 3 of which were initially characterized as ligands for the cytomegalovirus UL16 protein and are therefore also called UL16-binding proteins, or ULBPs. Mice do not express an equivalent of the MIC molecules, and the ligands for mouse NKG2D have a very similar structure to the RAET1 proteins, and are probably homologs of them. In fact, these ligands were first identified in mice as the retinoic acid early inducible 1 (Rae1) protein family.

The ligands for NKG2D are expressed in response to cellular or metabolic stress, and so are upregulated on cells that are infected with intracellular bacteria or with some viruses, such as cytomegalovirus, as well as on incipient tumor cells that have become malignantly transformed. Thus, recognition by NKG2D acts as a generalized 'danger' signal to the immune system. NKG2D is also expressed on activated macrophages and activated CD8 cytotoxic T cells, and recognition of NKG2D ligands by these cells provides a potent co-stimulatory signal that enhances their effector functions.

2-33 The NKG2D receptor activates a different signaling pathway from that of the other activating NK receptors.

As well as the ligands it recognizes, NKG2D also differs from other activating receptors of NK cells in the signaling pathway it engages within the cell. The other activating receptors, both the natural cytotoxicity receptors and activating KIRs, bind adaptor molecules such as the CD3ζ chain, the Fc receptor γ chain, and DAP12, which all contain specific signaling motifs called **immunoreceptor tyrosine-based activation motifs** (**ITAMs**). When the NK receptor binds its ligand, the ITAMS become phosphorylated, leading to the binding and activation of the intracellular tyrosine kinase Syk and further signaling events in the cell (see Section 6-17). In contrast, NKG2D binds a different adaptor protein, DAP10, which does not contain an ITAM sequence and instead activates the intracellular lipid kinase phosphatidylinositol-3-kinase (PI 3-kinase), initiating a different series of intracellular signaling events. Nevertheless, both signaling pathways result in the activation of the NK cell. In mice, the workings of NKG2D are even more complicated, because mouse NKG2D is produced in two alternatively spliced forms, one of which binds DAP12 while the other binds DAP10. Mouse NKG2D can thus activate both signaling pathways, whereas human NKG2D signals only through DAP10 to activate the PI 3-kinase pathway.

2-34 Several lymphocyte subpopulations behave as innate-like lymphocytes.

Receptor gene rearrangements are a defining characteristic of the lymphocytes of the adaptive immune system, and they allow the generation of an infinite variety of antigen receptors, each expressed by a different individual T cell or B cell (see Section 1-11). There are, however, several minor lymphocyte subsets that produce antigen receptors of this type but with only very limited diversity, encoded by a few common gene rearrangements. Because their receptors are relatively invariant and because they occur only in specific locations within the body, these lymphocytes do not need to undergo clonal expansion before responding effectively to the antigens they recognize; they are therefore known as **innate-like lymphocytes** (**ILLs**) (Fig. 2.61). To produce antigen receptors in these cells requires the recombinases RAG-1 and RAG-2;

Innate-like lymphocytes		
B-1 cells	Epithelial $\gamma:\delta$ cells	NK T cells
Make natural antibody, protect against infection with *Streptococcus pneumoniae*	Produce cytokines rapidly	Produce cytokines rapidly
Ligands not MHC associated	Ligands are MHC class IB associated	Ligands are lipids bound to CD1d
Cannot be boosted	Cannot be boosted	Cannot be boosted

Fig. 2.61 The three main classes of innate-like lymphocytes and their properties.

these proteins and their role in gene rearrangement in lymphocytes are described in Chapter 4. Because they express RAG-1 and RAG-2 and undergo the process of antigen receptor gene rearrangement, ILLs are, by definition, cells of the adaptive immune system. They behave, however, more like a part of the innate immune system, and so we will discuss them here.

One type of ILL is the subset of **γ:δ T cells** that resides within epithelia, such as the skin. γ:δ T cells are themselves a minor subset of the T cells introduced in Chapter 1. Their antigen receptors are composed of a γ chain and a δ chain, rather than the α and β chains that make up the antigen receptors on the majority subset of T cells involved in adaptive immunity. γ:δ T cells were discovered purely as a consequence of their having immunoglobulin-related receptors encoded by rearranged genes, and their function is still being clarified.

One of the most striking features of γ:δ T cells is their division into two highly distinct subsets. One is found in the lymphoid tissues of all vertebrates and, like B cells and the α:β T cells, has highly diversified T-cell receptors. In contrast, intraepithelial γ:δ T cells occur variably in different vertebrates, and commonly display receptors of very limited diversity, particularly in the skin and the female reproductive tract of mice, where the γ:δ T cells are essentially homogeneous in any one site. On the basis of this limited diversity and lack of recirculation, it has been proposed that intraepithelial γ:δ T cells may recognize ligands that are derived from the epithelium in which they reside, but which are expressed only when a cell has become infected. Candidate ligands are heat-shock proteins, MHC class Ib molecules (described in Chapter 5), and unorthodox nucleotides and phospholipids; there is evidence of recognition of all these ligands by γ:δ T cells.

Unlike α:β T cells, γ:δ T cells do not generally recognize antigen as peptides presented by MHC molecules; instead, they seem to recognize their target antigens directly, and thus could recognize and respond rapidly to molecules expressed by many different cell types. Recognition of molecules expressed as a consequence of infection, rather than recognition of pathogen-specific antigens themselves, would distinguish intraepithelial γ:δ T cells from other lymphocytes and would place them in the innate-like class.

Another subset of lymphocytes with antigen receptors of limited diversity is the **B-1** subset of B cells. B-1 cells are distinguished from other B cells by the cell-surface protein CD5 and have properties quite distinct from those of the conventional B cells that mediate adaptive humoral immunity. B-1 cells are in many ways analogous to intraepithelial γ:δ T cells: they arise early in embryonic development, they use a distinctive and limited set of gene rearrangements to make their receptors, they are self-renewing in tissues outside the central lymphoid organs, and they are the predominant lymphocyte in a distinctive

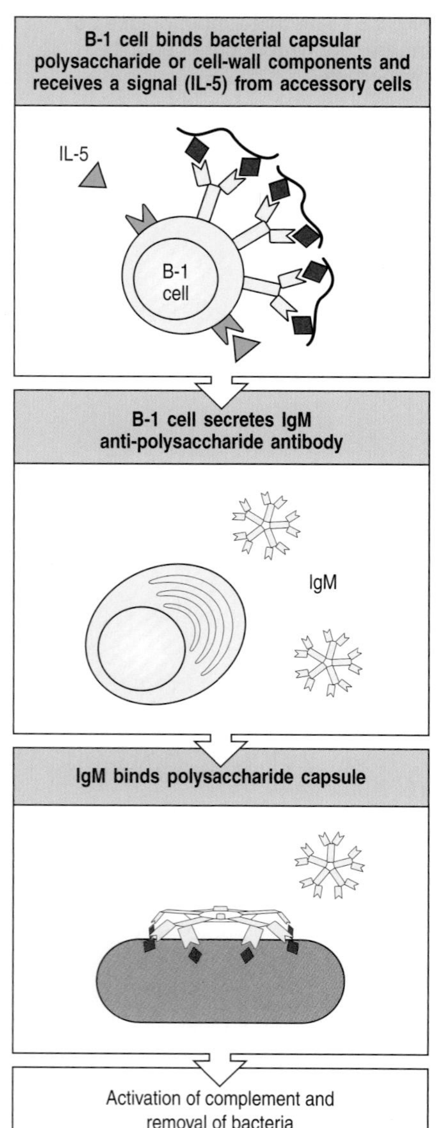
B-1 cell binds bacterial capsular polysaccharide or cell-wall components and receives a signal (IL-5) from accessory cells
IL-5
B-1 cell
B-1 cell secretes IgM anti-polysaccharide antibody
IgM
IgM binds polysaccharide capsule
Activation of complement and removal of bacteria

Fig. 2.62 B-1 cells might be important in the response to carbohydrate antigens such as bacterial polysaccharides. These responses occur rapidly, with antibody appearing within 48 hours after infection, presumably because there is a high frequency of precursors of the responding lymphocytes so that little clonal expansion is required. Unlike the responses to many other antigens, this response does not need the 'help' of T cells. In the absence of such help, only IgM is made (for reasons that will be explained in Chapter 9) and, in mice, these responses therefore clear bacteria mainly through the activation of complement, which is most efficient when the antibody is of the IgM isotype.

microenvironment—the peritoneal and pleural cavities. B-1 cells seem to make antibody responses mainly to polysaccharide antigens and can produce antibodies of the IgM class without needing help from T cells (Fig. 2.62). Although these responses can be augmented by T cells, they first appear within 48 hours of exposure to antigen, and so T cells cannot be involved. Thus, B-1 cells are not active in an antigen-specific adaptive immune response. The lack of an antigen-specific interaction with helper T cells might explain why immunological memory is not generated as a result of B-1 cell responses: repeated exposures to the same antigen elicit similar, or decreased, responses with each exposure. These responses, although generated by lymphocytes with rearranging receptors, therefore resemble innate rather than adaptive immune responses.

As with γ:δ T cells, the precise role of B-1 cells in host defense is uncertain. Mice that are deficient in B-1 cells are more susceptible to infection with *Streptococcus pneumoniae* because they fail to produce an anti-phosphocholine antibody that provides protection against this bacterium. A significant fraction of the B-1 cells can make antibodies of this specificity, and because no antigen-specific T-cell help is required, a potent response can be produced early in infection with this pathogen. Whether human B-1 cells have the same role is not certain.

A third subset of ILLs, known as **NK T cells**, exists both in the thymus and in peripheral lymphoid organs. These cells express an invariant T-cell receptor α chain, paired with one of three different β chains, and are able to recognize glycolipid antigens. The main response of NK T cells seems to be the rapid secretion of cytokines, including IL-4, IL-10, and IFN-γ, and it is thought that these cells may have a primarily regulatory function. We shall encounter these unusual cells again in Chapter 10.

From an evolutionary perspective, it is interesting to note that γ:δ T cells seem to defend the body surfaces, whereas B-1 cells defend the body cavity. Both cell types are relatively limited in their range of specificities and in the efficiency of their responses. It is possible that these two cell types represent a transitional phase in the evolution of the adaptive immune response, guarding the two main compartments of primitive organisms—the epithelial surfaces and the body cavity. It is not yet clear whether they are still critical to host defense or whether they represent an evolutionary relic. Nevertheless, because each cell type is prominent in certain sites in the body and contributes to responses against certain pathogens, they must be incorporated into our thinking about host defense.

A final component of the body's innate immune defenses are the IgM antibodies that are known as **natural antibodies**. This natural IgM is encoded by rearranged antibody genes that have not undergone further diversification by the process known as somatic hypermutation (which is described in Chapter 4). It makes up a considerable amount of the IgM circulating in humans and does not seem to be the consequence of an antigen-specific adaptive immune response to infection. It has a low affinity for many microbial pathogens and is highly cross-reactive, even binding to some self molecules. Furthermore, it is not known whether it is produced in response to the normal flora of the epithelial surfaces of the body or in response to self. However, it can have a role in host defense against *Streptococcus pneumoniae* by binding to phosphocholine in the bacterial cell envelope, leading to clearance of the bacteria before they become dangerous.

Summary.

Innate immunity uses a variety of induced effector mechanisms to clear an infection or, failing that, to hold it in check until the pathogen can be recognized by the adaptive immune system. These effector mechanisms are all regulated by germline-encoded receptor systems that are able to discriminate between normal self molecules on uninfected cells and infectious nonself ligands. Thus, the phagocyte's ability to discriminate between self and pathogen controls its release of pro-inflammatory chemokines and cytokines that act together to recruit more phagocytic cells to the site of infection. Especially prominent is the early recruitment of neutrophils that can recognize pathogens directly. Furthermore, cytokines released by tissue phagocytic cells induce fever, the production of acute-phase response proteins, including the pathogen-binding mannose-binding lectin and C-reactive protein, and the mobilization of antigen-presenting cells that induce the adaptive immune response. Viral pathogens are recognized by the cells in which they replicate, leading to the production of interferons that serve to inhibit viral replication and to activate NK cells, which in turn can distinguish infected from uninfected cells. As we will see later in this book, cytokines, chemokines, phagocytic cells, and NK cells are all effector mechanisms that are also employed in the adaptive immune response, which uses variable receptors to target specific pathogen antigens.

Summary to Chapter 2.

The innate system of host defense against infection is made up of several distinct components. The first of these are the barrier functions of the body's epithelia, which can simply prevent an infection from becoming established. Next, there are cells and molecules available to control or destroy the pathogen once it has breached the epithelial defenses. The most important of these are the tissue macrophages, which mediate cellular defense of the borders. The understanding of how the innate immune system recognizes pathogens is rapidly growing, and structural studies, such as those of mannose-binding lectin, have begun to reveal in detail how innate immune receptors can distinguish pathogen surfaces from those of host cells. Furthermore, the identification of the receptor for bacterial LPS and its link to human TLR-4 has opened a window onto innate immune recognition of microbe-associated molecular patterns. Recognition by the innate immune system leads to the elimination of invading pathogens through various effector mechanisms. Most of these have been known for a long time; indeed, the elimination of microorganisms by phagocytosis was the first immune response to be observed. However, more is being learned all the time; the chemokines, for example, have only been known of for about 15 years, and over 50 chemokine proteins have now been discovered. The complement system of proteins mediates humoral innate immunity of the tissue spaces and the blood. The induction of powerful effector mechanisms on the basis of innate immune recognition through germline-encoded receptors clearly has some dangers. Indeed, the double-edged sword embodied by the effects of the cytokine TNF-α—beneficial when it is released locally but disastrous when it is released systemically—illustrates the evolutionary knife-edge along which all innate mechanisms of host defense travel. The innate immune system can be viewed as a defense system that mainly frustrates the establishment of a focus of infection; however, even when it is inadequate to this function, it can still set the scene for the adaptive immune response, which forms an essential part of host defense in humans. Thus, having introduced the study of immunology with a consideration of innate immune function, we will now turn our attention to the adaptive immune response. This has been the focus of nearly all studies in immunology, because it is much easier to follow and experiment with by using reagents and responses that are specific for defined antigens.

Questions.

2.1 *The innate immune system uses two different strategies to identify pathogens: recognition of nonself and recognition of self. (a) Give examples of each and discuss how each example contributes to the ability of the organism to protect itself from infection. (b) What are the disadvantages of these different strategies?*

2.2 *The complement system gives rise to inflammatory signals, opsonins and molecules that lyse bacteria directly. (a) Describe the general properties of each class and discuss their utility in host defense. (b) Say which you think is most important in host defense, and why.*

2.3 *"The Toll receptors represent the most ancient pathways of host defense." Is this statement justified? Give an explanation for your answer.*

2.4 *Elie Metchnikoff discovered the protective role of macrophages by observing what happened in a starfish injured by a sea urchin spine. Describe the sequence of events that would follow were you to be speared by a sea urchin spine.*

2.5 *The complement system is a cascade of enzymes capable of producing powerful deleterious effects. (a) How is complement harnessed to protect us rather than creating harm? (b) What happens when things go wrong?*

2.6 *During their development and in order to perform their various functions efficiently, cells of the immune system must find their way to the correct part of the body. How do they manage to do this?*

General references.

Ezekowitz, R.A.B., and Hoffman, J.: **Innate immunity.** *Curr. Opin. Immunol.* 1998, **10**:9–53.

Fearon, D.T., and Locksley, R.M.: **The instructive role of innate immunity in the acquired immune response.** *Science* 1996, **272**:50–53.

Gallin, J.I., Goldstein, I.M., and Snyderman, R. (eds): *Inflammation—Basic Principles and Clinical Correlates*, 3rd ed. New York, Raven Press, 1999.

Janeway, C.A., Jr, and Medzhitov, R.: **Innate immune recognition.** *Annu. Rev. Immunol.* 2002, **20**:197–216.

Section references.

2-1 Infectious diseases are caused by diverse living agents that replicate in their hosts.

Kauffmann, S.H.E., Sher, A., and Ahmed, R.: *Immunology of Infectious Diseases.* Washington, DC, ASM Press, 2002.

Mandell, G.L., Bennett, J.E., and Dolin, R. (eds): *Principles and Practice of Infectious Diseases*, 4th ed. New York, Churchill Livingstone, 1995.

Salyers, A.A., and Whitt, D.D.: *Bacterial Pathogenesis: A Molecular Approach.* Washington, DC, ASM Press, 1994.

2-2 Infectious agents must overcome innate host defenses to establish a focus of infection.

Gibbons, R.J.: How microorganisms cause disease, in Gorbach, S.L., Bartlett, J.G., and Blacklow, N.R. (eds): *Infectious Diseases.* Philadelphia, W.B. Saunders Co., 1992.

Hornef, M.W., Wick, M.J., Rhen, M., and Normark, S : **Bacterial strategies for overcoming host innate and adaptive immune responses.** *Nat. Immunol.* 2002, **3**:1033–1040.

2-3 The epithelial surfaces of the body make up the first lines of defense against infection.

Gallo, R.L., Murakami, M., Ohtake, T., and Zaiou, M.: **Biology and clinical relevance of naturally occurring antimicrobial peptides.** *J. Allergy Clin. Immunol.* 2002, **110**:823–831.

Gudmundsson, G.H., and Agerberth, B.: **Neutrophil antibacterial peptides, multifunctional effector molecules in the mammalian immune system.** *J. Immunol. Methods* 1999, **232**:45–54.

Koczulla, A.R., and Bals, R.: **Antimicrobial peptides: current status and therapeutic potential.** *Drugs* 2003, **63**:389–406.

Risso, A.: **Leukocyte antimicrobial peptides: multifunctional effector molecules of innate immunity.** *J. Leukoc. Biol.* 2000, **68**:785–792.

Zaiou, M., and Gallo, R.L.: **Cathelicidins, essential gene-encoded mammalian antibiotics.** *J. Mol. Med.* 2002, **80**:549–561.

2-4 After entering tissues, many pathogens are recognized, ingested, and killed by phagocytes.

Aderem, A., and Underhill, D.M.: **Mechanisms of phagocytosis in macrophages.** *Annu. Rev. Immunol.* 1999, **17**:593–623.

Beutler, B., and Rietschel, E.T.: **Innate immune sensing and its roots: the story of endotoxin.** *Nat. Rev. Immunol.* 2003, **3**:169–176.

Bogdan, C., Rollinghoff, M., and Diefenbach, A.: **Reactive oxygen and reactive nitrogen intermediates in innate and specific immunity.** *Curr. Opin. Immunol.* 2000, **12**:64–76.

Dahlgren, C., and Karlsson, A.: **Respiratory burst in human neutrophils.** *J. Immunol. Methods* 1999, **232**:3–14.

Harrison, R.E., and Grinstein, S.: **Phagocytosis and the microtubule cytoskeleton.** *Biochem. Cell Biol.* 2002, **80**:509–515.

2-5 Pathogen recognition and tissue damage initiate an inflammatory response.

Chertov, O., Yang, D., Howard, O.M., and Oppenheim, J.J.: **Leukocyte granule proteins mobilize innate host defenses and adaptive immune responses.** *Immunol. Rev.* 2000, **177**:68–78.

Kohl, J.: **Anaphylatoxins and infectious and noninfectious inflammatory diseases.** *Mol. Immunol.* 2001, **38**:175–187.

Mekori, Y.A., and Metcalfe, D.D.: **Mast cells in innate immunity.** *Immunol. Rev.* 2000, **173**:131–140.

Svanborg, C., Godaly, G., and Hedlund, M.: **Cytokine responses during mucosal infections: role in disease pathogenesis and host defence.** *Curr. Opin. Microbiol.* 1999, **2**:99–105.

van der Poll, T.: **Coagulation and inflammation.** *J. Endotoxin Res.* 2001, **7**:301–304.

2-6 Receptors with specificity for pathogen molecules recognize patterns of repeating structural motifs.

Apostolopoulos, V., and McKenzie, I.F.: **Role of the mannose receptor in the immune response.** *Curr. Mol. Med.* 2001, **1**:469–474.

Feizi, T.: **Carbohydrate-mediated recognition systems in innate immunity.** *Immunol. Rev.* 2000, **173**:79–88.

Gough, P.J., and Gordon, S.: **The role of scavenger receptors in the innate immune system.** *Microbes Infect.* 2000, **2**:305–311.

Heine, H., and Lien, E.: **Toll-like receptors and their function in innate and adaptive immunity.** *Int. Arch. Allergy Immunol.* 2003, **130**:180–192.

Kaisho, T., and Akira, S.: **Critical roles of toll-like receptors in host defense.** *Crit. Rev. Immunol.* 2000, **20**:393–405.

Linehan, S.A., Martinez-Pomares, L., and Gordon, S.: **Macrophage lectins in host defence.** *Microbes Infect.* 2000, **2**:279–288.

Podrez, E.A., Poliakov, E., Shen, Z., Zhang, R., Deng, Y., Sun, M., Finton, P.J., Shan, L., Gugiu, B., Fox, P.L., *et al.*: **Identification of a novel family of oxidized phospholipids that serve as ligands for the macrophage scavenger receptor CD36.** *J Biol. Chem.* 2002, **277**:38503–38516.

Turner, M.W., and Hamvas, R.M.: **Mannose-binding lectin: structure, function, genetics and disease associations.** *Rev. Immunogenet.* 2000, **2**:305–322.

2-7 The Toll-like receptors are signaling receptors that distinguish different types of pathogen and help direct an appropriate immune response.

Barton, G.M., and Medzhitov, R.: **Toll-like receptors and their ligands.** *Curr. Top. Microbiol. Immunol.* 2002, **270**:81–92.

Lund, J.M., Alexopoulou, L., Sato, A., Karow, M., Adams, N.C., Gale, N.W., Iwasaki, A., and Flavell, R.A.: **Recognition of single-stranded RNA viruses by Toll-like receptor 7.** *Proc. Natl Acad. Sci. USA* 2004, **101**:5598–5603.

Lund, J., Sato, A., Akira, S., Medzhitov, R., and Iwasaki, A.: **Toll-like receptor 9-mediated recognition of Herpes simplex virus-2 by plasmacytoid dendritic cells.** *J. Exp. Med.* 2003, **198**:513-20.

Kawai, T., and Akira, S.: **Innate immune recognition of viral infection.** *Nat. Immunol.* 2006, **7**:131–137.

Medzhitov, R., and Janeway, C.A., Jr: **The toll receptor family and microbial recognition.** *Trends Microbiol.* 2000, **8**:452–456.

Peiser, L., De Winther, M.P., Makepeace, K., Hollinshead, M., Coull, P., Plested, J., Kodama, T., Moxon, E.R., and Gordon, S.: **The class A macrophage scavenger receptor is a major pattern recognition receptor for *Neisseria meningitidis* which is independent of lipopolysaccharide and not required for secretory responses.** *Infect. Immun.* 2002, **70**:5346–5354.

Salio, M., and Cerundolo, V.: **Viral immunity: cross-priming with the help of TLR3.** *Curr. Biol.* 2005, **15**:R336–R339.

2-8 The effects of bacterial lipopolysaccharide on macrophages are mediated by CD14 binding to TLR-4.

Beutler, B.: **Endotoxin, toll-like receptor 4, and the afferent limb of innate immunity.** *Curr. Opin. Microbiol.* 2000, **3**:23–28.

Beutler, B., and Rietschel, E.T.: **Innate immune sensing and its roots: the story of endotoxin.** *Nat. Rev. Immunol.* 2003, **3**:169–176.

2-9 The NOD proteins act as intracellular sensors of bacterial infection.

Abreu, M.T., Fukata, M., and Arditi, M.: **TLR signaling in the gut in health and disease.** *J. Immunol.* 2005, **174**:4453–4456.

Dziarski, R., and Gupta, D.: **Peptidoglycan recognition in innate immunity.** *J Endotoxin Res.* 2005, **11**:304–310.

Inohara, N., Chamaillard, M., McDonald C, Nunez G.: **NOD-LRR proteins: role in host-microbial interactions and inflammatory disease.** *Annu. Rev. Biochem.* 2005, **74**:355–383.

Strober, W., Murray, P.J., Kitani, A., and Watanabe, T.: **Signalling pathways and molecular interactions of NOD1 and NOD2.** *Nat. Rev. Immunol.* 2006, **6**:9–20.

2-10 Activation of Toll-like receptors and NOD proteins triggers the production of pro-inflammatory cytokines and chemokines, and the expression of co-stimulatory molecules.

Bowie, A., and O'Neill, L.A.: **The interleukin-1 receptor/Toll-like receptor superfamily: signal generators for pro-inflammatory interleukins and microbial products.** *J. Leukoc. Biol.* 2000, **67**:508–514.

Brightbill, H.D., Libraty, D.H., Krutzik, S.R., Yang, R.B., Belisle, J.T., Bleharski, J.R., Maitland, M., Norgard, M.V., Plevy, S.E., Smale, S.T., *et al.*: **Host defense mechanisms triggered by microbial lipoproteins through Toll-like receptors.** *Science* 1999, **285**:732–736.

Dalpke, A., and Heeg, K.: **Signal integration following Toll-like receptor triggering.** *Crit. Rev. Immunol.* 2002, **22**:217–250.

Heine, H., and Lien, E.: **Toll-like receptors and their function in innate and adaptive immunity.** *Int. Arch. Allergy Immunol.* 2003, **130**:180–192.

2-11 Complement is a system of plasma proteins that is activated by the presence of pathogens.

Tomlinson, S.: **Complement defense mechanisms.** *Curr. Opin. Immunol.* 1993, **5**:83–89.

2-12 Complement interacts with pathogens to mark them for destruction by phagocytes.

Frank, M.M.: **Complement deficiencies.** *Pediatr. Clin. North Am.* 2000, **47**:1339–1354.

Frank, M.M., and Fries, L.F.: **The role of complement in inflammation and phagocytosis.** *Immunol. Today* 1991, **12**:322–326.

2-13 The classical pathway is initiated by activation of the C1 complex.

Arlaud, G.J., Gaboriaud, C., Thielens, N.M., Budayova-Spano, M., Rossi, V., and Fontecilla-Camps, J.C.: **Structural biology of the C1 complex of complement unveils the mechanisms of its activation and proteolytic activity.** *Mol. Immunol.* 2002, **39**:383–394.

Cooper, N.R.: **The classical complement pathway. Activation and regulation of the first complement component.** *Adv. Immunol.* 1985, **37**:151–216.

2-14 The lectin pathway is homologous to the classical pathway.

Dodds, A.W.: **Which came first, the lectin/classical pathway or the alternative pathway of complement?** *Immunobiology* 2002, **205**:340–354.

Gal, P., and Ambrus, G.: **Structure and function of complement activating enzyme complexes: C1 and MBL-MASPs.** *Curr. Protein Pept. Sci.* 2001, **2**:43–59.

Jack, D.L., Klein, N.J., and Turner, M.W.: **Mannose-binding lectin: targeting the microbial world for complement attack and opsonophagocytosis.** *Immunol. Rev.* 2001, **180**:86–99.

Lu, J., Teh, C., Kishore, U., and Reid, K.B.: **Collectins and ficolins: sugar pattern recognition molecules of the mammalian innate immune system.** *Biochim. Biophys. Acta* 2002, **1572**:387–400.

Rabinovich, G.A., Rubinstein, N., and Toscano, M.A.: **Role of galectins in inflammatory and immunomodulatory processes.** *Biochim. Biophys. Acta* 2002, **1572**:274–284.

Schwaeble, W., Dahl, M.R., Thiel, S., Stover, C., and Jensenius, J.C.: **The mannan-binding lectin-associated serine proteases (MASPs) and MAp19: four components of the lectin pathway activation complex encoded by two genes.** *Immunobiology* 2002, **205**:455–466.

2-15 Complement activation is largely confined to the surface on which it is initiated.

Cicardi, M., Bergamaschini, L., Cugno, M., Beretta, A., Zingale, L.C., Colombo, M., and Agostoni, A.: **Pathogenetic and clinical aspects of C1 inhibitor deficiency.** *Immunobiology* 1998, **199**:366–376.

2-16 Hydrolysis of C3 causes initiation of the alternative pathway of complement.

Fijen, C.A., van den Bogaard, R., Schipper, M., Mannens, M., Schlesinger, M., Nordin, F.G., Dankert, J., Daha, M.R., Sjoholm, A.G., Truedsson, L., and Kuijper, E.J.: **Properdin deficiency: molecular basis and disease association.** *Mol. Immunol.* 1999, **36**:863–867.

Xu, Y., Narayana, S.V., and Volanakis, J.E.: **Structural biology of the alternative pathway convertase.** *Immunol. Rev.* 2001, **180**:123–135.

2-17 Membrane and plasma proteins that regulate the formation and stability of C3 convertases determine the extent of complement activation under different circumstances.

Fishelson, Z., Donin, N., Zell, S., Schultz, S., and Kirschfink, M.: **Obstacles to cancer immunotherapy: expression of membrane complement regulatory proteins (mCRPs) in tumors.** *Mol. Immunol.* 2003, **40**:109–123.

Golay, J., Zaffaroni, L., Vaccari, T., Lazzari, M., Borleri, G.M., Bernasconi, S., Tedesco, F., Rambaldi, A., Introna, M.: **Biologic response of B lymphoma cells to anti-CD20 monoclonal antibody rituximab *in vitro*: CD55 and CD59 regulate complement-mediated cell lysis.** *Blood* 2000, **95**:3900–3908.

Spiller, O.B., Criado-Garcia, O., Rodriguez De Cordoba, S., and Morgan, B.P.: **Cytokine-mediated up-regulation of CD55 and CD59 protects human hepatoma cells from complement attack.** *Clin. Exp Immunol.* 2000, **121**:234–241.

Varsano, S., Frolkis, I., Rashkovsky, L., Ophir, D., and Fishelson, Z.: **Protection of human nasal respiratory epithelium from complement-mediated lysis by cell-membrane regulators of complement activation.** *Am. J. Respir. Cell Mol. Biol.* 1996, **15**:731–737.

2-18 Surface-bound C3 convertase deposits large numbers of C3b fragments on pathogen surfaces and generates C5 convertase activity.

Rawal, N., and Pangburn, M.K.: **Structure/function of C5 convertases of complement.** *Int. Immunopharmacol.* 2001, **1**:415–422.

2-19 Ingestion of complement-tagged pathogens by phagocytes is mediated by receptors for the bound complement proteins.

Ehlers, M.R.: **CR3: a general purpose adhesion-recognition receptor essential for innate immunity.** *Microbes Infect.* 2000, **2**:289–294.

Fijen, C.A., Bredius, R.G., Kuijper, E.J., Out, T.A., De Haas, M., De Wit, A.P., Daha, M.R., and De Winkel, J.G.: **The role of Fcγ receptor polymorphisms and C3 in the immune defence against *Neisseria meningitidis* in complement-deficient individuals.** *Clin. Exp. Immunol.* 2000, **120**:338–345.

Linehan, S.A., Martinez-Pomares, L., and Gordon, S.: **Macrophage lectins in host defence.** *Microbes Infect.* 2000, **2**:279–288.

Ravetch, J.V.: **A full complement of receptors in immune complex diseases.** *J. Clin. Invest.* 2002, **110**:1759–1761.

Ross, G.D.: **Regulation of the adhesion versus cytotoxic functions of the Mac-1/CR3$\alpha_M\beta_2$-integrin glycoprotein.** *Crit. Rev. Immunol.* 2000, **20**:197–222.

2-20 Small fragments of some complement proteins can initiate a local inflammatory response.

Kildsgaard, J., Hollmann, T.J., Matthews, K.W., Bian, K. Murad, F., and Wetsel, R.A.: **Cutting edge: targeted disruption of the C3a receptor gene demonstrates a novel protective anti-inflammatory role for C3a in endotoxin-shock.** *J. Immunol.* 2000, **165**:5406–5409.

Kohl, J.: **Anaphylatoxins and infectious and noninfectious inflammatory diseases.** *Mol. Immunol.* 2001, **38**:175–187.

Monsinjon, T., Gasque, P., Ischenko, A., and Fontaine M.: **C3A binds to the seven transmembrane anaphylatoxin receptor expressed by epithelial cells and triggers the production of IL-8.** *FEBS Lett.* 2001, **487**:339–346.

Schraufstatter, I.U., Trieu, K., Sikora, L., Sriramarao, P., and DiScipio, R.: **Complement c3a and c5a induce different signal transduction cascades in endothelial cells.** *J. Immunol.* 2002, **169**:2102–2110.

2-21 The terminal complement proteins polymerize to form pores in membranes that can kill certain pathogens.

Bhakdi, S., and Tranum-Jensen, J.: **Complement lysis: a hole is a hole.** *Immunol. Today* 1991, **12**:318–320.

Parker, C.L., and Sodetz, J.M.: **Role of the human C8 subunits in complement-mediated bacterial killing: evidence that C8 γ is not essential.** *Mol. Immunol.* 2002, **39**:453–458.

Scibek, J.J., Plumb, M.E., and Sodetz, J.M.: **Binding of human complement C8 to C9: role of the N-terminal modules in the C8 α subunit.** *Biochemistry* 2002, **41**:14546–14551.

Wang, Y., Bjes, E.S., and Esser, A.F.: **Molecular aspects of complement-mediated bacterial killing. Periplasmic conversion of C9 from a protoxin to a toxin.** *J. Biol. Chem.* 2000, **275**:4687–4692.

2-22 Complement control proteins regulate all three pathways of complement activation and protect the host from its destructive effects.

Blom, A.M., Rytkonen, A., Vasquez, P., Lindahl, G., Dahlback, B., and Jonsson, A.B.: **A novel interaction between type IV pili of *Neisseria gonorrhoeae* and the human complement regulator C4B-binding protein.** *J. Immunol.* 2001, **166**:6764–6770.

Jiang, H., Wagner, E., Zhang, H., and Frank, M.M.: **Complement 1 inhibitor is a regulator of the alternative complement pathway.** *J. Exp. Med.* 2001, **194**:1609–1616.

Kirschfink, M.: **Controlling the complement system in inflammation.** *Immunopharmacology* 1997, **38**:51–62.

Kirschfink, M.: **C1-inhibitor and transplantation.** *Immunobiology* 2002, **205**:534–541.

Liszewski, M.K., Farries, T.C., Lublin, D.M., Rooney, I.A., and Atkinson, J.P.: **Control of the complement system.** *Adv. Immunol.* 1996, **61**:201–283.

Miwa, T., Zhou, L., Hilliard, B., Molina, H., and Song, W.C.: **Crry, but not CD59 and DAF, is indispensable for murine erythrocyte protection in vivo from spontaneous complement attack.** *Blood* 2002, **99**:3707–3716.

Pangburn, M.K.: **Host recognition and target differentiation by factor H, a regulator of the alternative pathway of complement.** *Immunopharmacology* 2000, **49**:149–157.

Singhrao, S.K., Neal, J.W., Rushmere, N.K., Morgan, B.P., and Gasque, P.: **Spontaneous classical pathway activation and deficiency of membrane regulators render human neurons susceptible to complement lysis.** *Am. J. Pathol.* 2000, **157**:905–918.

Smith, G.P., and Smith, R.A.: **Membrane-targeted complement inhibitors.** *Mol. Immunol.* 2001, **38**:249–255.

Suankratay, C., Mold, C., Zhang, Y., Lint, T.F., and Gewurz, H.: **Mechanism of complement-dependent haemolysis via the lectin pathway: role of the complement regulatory proteins.** *Clin. Exp. Immunol.* 1999, **117**:442–448.

Suankratay, C., Mold, C., Zhang, Y., Potempa, L.A., Lint, T.F., and Gewurz, H.: **Complement regulation in innate immunity and the acute-phase response: inhibition of mannan-binding lectin-initiated complement cytolysis by C-reactive protein (CRP).** *Clin. Exp. Immunol.* 1998, **113**:353–359.

Zipfel, P.F., Jokiranta, T.S., Hellwage, J., Koistinen, V., and Meri, S.: **The factor H protein family.** *Immunopharmacology* 1999, **42**:53–60.

2-23 Activated macrophages secrete a range of cytokines that have a variety of local and distant effects.

Larsson, B.M., Larsson, K., Malmberg, P., and Palmberg, L.: **Gram positive bacteria induce IL-6 and IL-8 production in human alveolar macrophages and epithelial cells.** *Inflammation* 1999, **23**:217–230.

Ozato, K., Tsujimura, H., and Tamura, T.: **Toll-like receptor signaling and regulation of cytokine gene expression in the immune system.** *Biotechniques* 2002, **Suppl**:66–69,70,72 C_3a, C5a.

Svanborg, C., Godaly, G., and Hedlund, M.: **Cytokine responses during mucosal infections: role in disease pathogenesis and host defence.** *Curr. Opin. Microbiol.* 1999, **2**:99–105.

2-24 Chemokines released by phagocytes and dendritic cells recruit cells to sites of infection.

Kunkel, E.J., and Butcher, E.C.: **Chemokines and the tissue-specific migration of lymphocytes.** *Immunity* 2002, **16**:1–4.

Luster, A.D.: **The role of chemokines in linking innate and adaptive immunity.** *Curr. Opin. Immunol.* 2002, **14**:129–135.

Matsukawa, A., Hogaboam, C.M., Lukacs, N.W., and Kunkel, S.L.: **Chemokines and innate immunity.** *Rev. Immunogenet.* 2000, **2**:339–358.

Ono, S.J., Nakamura, T., Miyazaki, D., Ohbayashi, M., Dawson, M., and Toda, M.: **Chemokines: roles in leukocyte development, trafficking, and effector function.** *J. Allergy Clin. Immunol.* 2003, **111**:1185–1199.

Scapini, P., Lapinet-Vera, J.A., Gasperini, S., Calzetti, F., Bazzoni, F., and Cassatella, M.A.: **The neutrophil as a cellular source of chemokines.** *Immunol. Rev.* 2000, **177**:195–203.

Yoshie, O.: **Role of chemokines in trafficking of lymphocytes and dendritic cells.** *Int. J. Hematol.* 2000, **72**:399–407.

2-25 Cell-adhesion molecules control interactions between leukocytes and endothelial cells during an inflammatory response.

Alon, R., and Feigelson, S.: **From rolling to arrest on blood vessels: leukocyte tap dancing on endothelial integrin ligands and chemokines at sub-second contacts.** *Semin. Immunol.* 2002, **14**:93–104.

Bunting, M., Harris, E.S., McIntyre, T.M., Prescott, S.M., and Zimmerman, G.A.: **Leukocyte adhesion deficiency syndromes: adhesion and tethering defects involving β 2 integrins and selectin ligands.** *Curr. Opin. Hematol.* 2002, **9**:30–35.

D'Ambrosio, D., Albanesi, C., Lang, R., Girolomoni, G., Sinigaglia, F., and Laudanna, C.: **Quantitative differences in chemokine receptor engagement generate diversity in integrin-dependent lymphocyte adhesion.** *J. Immunol.* 2002, **169**:2303–2312.

Johnston, B., and Butcher, E.C.: **Chemokines in rapid leukocyte adhesion triggering and migration.** *Semin. Immunol.* 2002, **14**:83–92.

Ley, K.: **Integration of inflammatory signals by rolling neutrophils.** *Immunol. Rev.* 2002, **186**:8–18.

Shahabuddin, S., Ponath, P., and Schleimer, R.P.: **Migration of eosinophils across endothelial cell monolayers: interactions among IL-5, endothelial-activating cytokines, and C-C chemokines.** *J. Immunol.* 2000, **164**:3847–3854.

Vestweber, D.: **Lymphocyte trafficking through blood and lymphatic vessels: more than just selectins, chemokines and integrins.** *Eur. J. Immunol.* 2003, **33**:1361–1364.

2-26 Neutrophils make up the first wave of cells that cross the blood vessel wall to enter inflammatory sites.

Bochenska-Marciniak, M., Kupczyk, M., Gorski, P., and Kuna, P.: **The effect of recombinant interleukin-8 on eosinophils' and neutrophils' migration in vivo and in vitro.** *Allergy* 2003, **58**:795–801.

Godaly, G., Bergsten, G., Hang, L., Fischer, H., Frendeus, B., Lundstedt, A.C., Samuelsson, M., Samuelsson, P., and Svanborg, C.: **Neutrophil recruitment, chemokine receptors, and resistance to mucosal infection.** *J. Leukoc. Biol.* 2001, **69**:899–906.

Gompertz, S., and Stockley, R.A.: **Inflammation—role of the neutrophil and the eosinophil.** *Semin. Respir. Infect.* 2000, **15**:14–23.

Lee, S.C., Brummet, M.E., Shahabuddin, S., Woodworth, T.G., Georas, S.N., Leiferman, K.M., Gilman, S.C., Stellato, C., Gladue, R.P., Schleimer, R.P., and Beck, L.A.: **Cutaneous injection of human subjects with macrophage inflammatory protein-1 α induces significant recruitment of neutrophils and monocytes.** *J. Immunol.* 2000, **164**:3392–3401.

Worthylake, R.A., and Burridge, K.: **Leukocyte transendothelial migration: orchestrating the underlying molecular machinery.** *Curr. Opin. Cell Biol.* 2001, **13**:569–577.

2-27 TNF-α is an important cytokine that triggers local containment of infection but induces shock when released systemically.

Cairns, C.B., Panacek, E.A., Harken, A.H., and Banerjee, A.: **Bench to bedside: tumor necrosis factor-α: from inflammation to resuscitation.** *Acad. Emerg. Med.* 2000, **7**:930–941.

Dellinger, R.P.: **Inflammation and coagulation: implications for the septic patient.** *Clin. Infect. Dis.* 2003, **36**:1259–1265.

Pfeffer, K.: **Biological functions of tumor necrosis factor cytokines and their receptors.** *Cytokine Growth Factor Rev.* 2003, **14**:185–191.

Sriskandan, S., and Cohen, J.: **Gram-positive sepsis. Mechanisms and differences from gram-negative sepsis.** *Infect. Dis. Clin. North Am.* 1999, **13**:397–412.

2-28 Cytokines released by phagocytes activate the acute-phase response.

Bopst, M., Haas, C., Car, B., and Eugster, H.P.: **The combined inactivation of tumor necrosis factor and interleukin-6 prevents induction of the major acute phase proteins by endotoxin.** *Eur. J. Immunol.* 1998, **28**:4130–4137.

Ceciliani, F., Giordano, A., and Spagnolo, V.: **The systemic reaction during inflammation: the acute-phase proteins.** *Protein Pept. Lett.* 2002, **9**:211–223.

He, R., Sang, H., and Ye, R.D.: **Serum amyloid A induces IL-8 secretion through a G protein-coupled receptor, FPRL1/LXA4R.** *Blood* 2003, **101**:1572–1581.

Horn, F., Henze, C., and Heidrich, K.: **Interleukin-6 signal transduction and lymphocyte function.** *Immunobiology* 2000, **202**:151–167.

Mold, C., Rodriguez, W., Rodic-Polic, B., and Du Clos, T.W.: **C-reactive protein mediates protection from lipopolysaccharide through interactions with Fc γ R.** *J. Immunol.* 2002, **169**:7019–7025.

Sheth, K., and Bankey, P.: **The liver as an immune organ.** *Curr. Opin. Crit. Care* 2001, **7**:99–104.

Volanakis, J.E.: **Human C-reactive protein: expression, structure, and function.** *Mol. Immunol.* 2001, **38**:189–197.

2-29 Interferons induced by viral infection make several contributions to host defense.

Kawai, T., and Akira, S.: **Innate immune recognition of viral infection.** *Nat. Immunol.* 2006, **7**:131–137.

Meylan, E., and Tschopp, J.: **Toll-like receptors and RNA helicases: two parallel ways to trigger antiviral responses.** *Mol. Cell* 2006, **22**:561–569.

Pietras, E.M., Saha, S.K., and Cheng, G.: **The interferon response to bacterial and viral infections.** *J. Endotoxin Res.* 2006, **12**:246–250.

2-30 NK cells are activated by interferons and macrophage-derived cytokines to serve as an early defense against certain intracellular infections.

Biron, C.A., Nguyen, K.B., Pien, G.C., Cousens, L.P., and Salazar-Mather, T.P.: **Natural killer cells in antiviral defense: function and regulation by innate cytokines.** *Annu. Rev. Immunol.* 1999, **17**:189–220.

Carnaud, C., Lee, D., Donnars, O., Park, S.H., Beavis, A., Koezuka, Y., and Bendelac, A.: **Cutting edge. Cross-talk between cells of the innate immune system: NKT cells rapidly activate NK cells.** *J. Immunol.* 1999, **163**:4647–4650.

Dascher, C.C., and Brenner, M.B.: **CD1 antigen presentation and infectious disease.** *Contrib. Microbiol.* 2003, **10**:164–182.

Godshall, C.J., Scott, M.J., Burch, P.T., Peyton, J.C., and Cheadle, W.G.: **Natural killer cells participate in bacterial clearance during septic peritonitis through interactions with macrophages.** *Shock* 2003, **19**:144–149.

Orange, J.S., Fassett, M.S., Koopman, L.A., Boyson, J.E., and Strominger, J.L.: **Viral evasion of natural killer cells.** *Nat. Immunol.* 2002, **3**:1006–1012.

Salazar-Mather, T.P., Hamilton, T.A., and Biron, C.A.: **A chemokine-to-cytokine-to-chemokine cascade critical in antiviral defense.** *J. Clin. Invest.* 2000, **105**:985–993.

Seki, S., Habu, Y., Kawamura, T., Takeda, K., Dobashi, H., Ohkawa, T., and Hiraide, H.: **The liver as a crucial organ in the first line of host defense: the roles of Kupffer cells, natural killer (NK) cells and NK1.1 Ag$^+$ T cells in T helper 1 immune responses.** *Immunol. Rev.* 2000, **174**:35–46.

2-31 NK cells possess receptors for self molecules that prevent their activation by uninfected cells.

Borrego, F., Kabat, J., Kim, D.K., Lieto, L., Maasho, K., Pena, J., Solana, R., and Coligan, J.E.: **Structure and function of major histocompatibility complex (MHC) class I specific receptors expressed on human natural killer (NK) cells.** *Mol. Immunol.* 2002, **38**:637–660.

Boyington, J.C., and Sun, P.D.: **A structural perspective on MHC class I recognition by killer cell immunoglobulin-like receptors.** *Mol. Immunol.* 2002, **38**:1007–1021.

Brown, M.G., Dokun, A.O., Heusel, J.W., Smith, H.R., Beckman, D.L., Blattenberger, E.A., Dubbelde, C.E., Stone, L.R., Scalzo, A.A., and Yokoyama, W.M.: **Vital involvement of a natural killer cell activation receptor in resistance to viral infection.** *Science* 2001, **292**:934–937.

Robbins, S.H., and Brossay, L.: **NK cell receptors: emerging roles in host defense against infectious agents.** *Microbes Infect.* 2002, **4**:1523–1530.

Trowsdale, J.: **Genetic and functional relationships between MHC and NK receptor genes.** *Immunity* 2001, **15**:363–374.

Vilches, C., and Parham, P.: **KIR: diverse, rapidly evolving receptors of innate and adaptive immunity.** *Annu. Rev. Immunol.* 2002, **20**:217–251.

2-32 NK cells bear receptors that activate their killer function in response to ligands expressed on infected cells or tumor cells.

Gasser, S., Orsulic, S., Brown, E.J., and Raulet, D.H.: **The DNA damage pathway regulates innate immune system ligands of the NKG2D receptor.** *Nature* 2005, **436**:1186–1190.

Moretta, L., Bottino, C., Pende, D., Castriconi, R., Mingari, M.C., Moretta, A.: **Surface NK receptors and their ligands on tumor cells.** *Semin. Immunol.* 2006, **18**:151–158.

Parham, P.: **MHC class I molecules and KIRs in human history, health and survival.** *Nat. Rev. Immunol.* 2005, **5**:201–214.

Stewart, C.A., Vivier, E., and Colonna, M.: **Strategies of natural killer cell recognition and signaling.** *Curr. Top. Microbiol. Immunol.* 2006, **298**:1–21.

2-33 The NKG2D receptor activates a distinct signaling pathway from that of the other activating NK receptors.

Gonzalez, S., Groh, V., and Spies, T.: **Immunobiology of human NKG2D and its ligands.** *Curr. Top. Microbiol. Immunol.* 2006, **298**:121–138.

Upshaw, J.L., and Leibson, P.J.: **NKG2D-mediated activation of cytotoxic lymphocytes: unique signaling pathways and distinct functional outcomes.** *Semin. Immunol.* 2006, **18**:167–175.

Vivier, E., Nunes, J.A., and Vely, F.: **Natural killer cell signaling pathways.** *Science* 2004, **306**:1517–1519.

2-34 Several lymphocyte subpopulations behave as innate-like lymphocytes.

Bos, N.A., Cebra, J.J., and Kroese, F.G.: **B-1 cells and the intestinal microflora.** *Curr. Top. Microbiol. Immunol.* 2000, **252**:211–220.

Chan, W.L., Pejnovic, N., Liew, T.V., Lee, C.A., Groves, R., and Hamilton, H.: **NKT cell subsets in infection and inflammation.** *Immunol. Lett.* 2003, **85**:159–163.

Chatenoud, L.: **Do NKT cells control autoimmunity?** *J. Clin. Invest.* 2002, **110**:747–748.

Galli, G., Nuti, S., Tavarini, S., Galli-Stampino, L., De Lalla, C., Casorati, G., Dellabona, P., and Abrignani, S.: **CD1d-restricted help to B cells by human invariant natural killer T lymphocytes.** *J. Exp. Med. 2003*, **197**:1051–1057.

Kronenberg, M., and Gapin, L.: **The unconventional lifestyle of NKT cells.** *Nat. Rev. Immunol.* 2002, **2**:557–568.

Reid, R.R., Woodcock, S., Prodeus, A.P., Austen, J., Kobzik, L., Hechtman, H., Moore, F.D., Jr, and Carroll, M.C.: **The role of complement receptors CD21/CD35 in positive selection of B-1 cells.** *Curr. Top. Microbiol. Immunol.* 2000, **252**:57–65.

Sharif, S., Arreaza, G.A., Zucker, P., Mi, Q.S., and Delovitch, T.L.: **Regulation of autoimmune disease by natural killer T cells.** *J. Mol. Med.* 2002, **80**:290–300.

Stober, D., Jomantaite, I., Schirmbeck, R., and Reimann, J.: **NKT cells provide help for dendritic cell-dependent priming of MHC class I-restricted CD8$^+$ T cells *in vivo*.** *J. Immunol.* 2003, **170**:2540–2548.

Zinkernagel, R.M.: **A primitive T cell-independent mechanism of intestinal mucosal IgA responses to commensal bacteria.** *Science* 2000, **288**:2222–2226.

PART II

THE RECOGNITION OF ANTIGEN

Antigen Recognition by B-cell and T-cell Receptors

3

Innate immune responses initially defend the body against infection, but these only work to control pathogens that have certain molecular patterns or that induce interferons and other secreted yet nonspecific defenses, as we have learned in Chapter 2. To effectively fight the wide range of pathogens an individual will encounter, the lymphocytes of the adaptive immune system have evolved to recognize a great variety of different antigens from bacteria, viruses, and other disease-causing organisms. The antigen-recognition molecules of B cells are the **immunoglobulins** (**Ig**). These proteins are produced by B cells in a vast range of antigen specificities, each B cell producing immunoglobulin of a single specificity (see Sections 1-11 and 1-12). Membrane-bound immunoglobulin on the B-cell surface serves as the cell's receptor for antigen, and is known as the **B-cell receptor** (**BCR**). Immuno-globulin of the same antigen specificity is secreted as **antibody** by terminally differentiated B cells—the plasma cells. The secretion of antibodies, which bind pathogens or their toxic products in the extracellular spaces of the body, is the main effector function of B cells in adaptive immunity.

Antibodies were the first molecules involved in specific immune recognition to be characterized, and they are still the best understood. The antibody molecule has two separate functions: one is to bind specifically to molecules from the pathogen that elicited the immune response; the other is to recruit other cells and molecules to destroy the pathogen once the antibody is bound to it. For example, binding by antibody neutralizes viruses and marks pathogens for destruction by phagocytes and complement, as described in Section 1-18. Recognition and effector functions are structurally separated in the antibody molecule, one part of which specifically binds to the antigen whereas the other engages the various elimination mechanisms. The antigen-binding region varies extensively between antibody molecules and is thus known as the **variable region** or **V region**. The variability of antibody molecules allows each antibody to bind a different specific antigen, and the total repertoire of antibodies made by a single individual is large enough to ensure that virtually any structure can be recognized. The region of the antibody molecule that engages the effector functions of the immune system does not vary in the same way and is thus known as the **constant region** or **C region**. It comes in five main forms, each of which is specialized for activating different effector mechanisms. The membrane-bound B-cell receptor does not have these effector functions, because the C region remains inserted in the membrane of the B cell. Its function is to recognize and bind antigen by the V regions exposed on the surface of the cell, thus transmitting a signal that causes B-cell activation, leading to clonal expansion and specific antibody production.

The antigen-recognition molecules of T cells are made solely as membrane-bound proteins and function only to signal T cells for activation. These **T-cell receptors** (**TCRs**) are related to immunoglobulins both in their protein

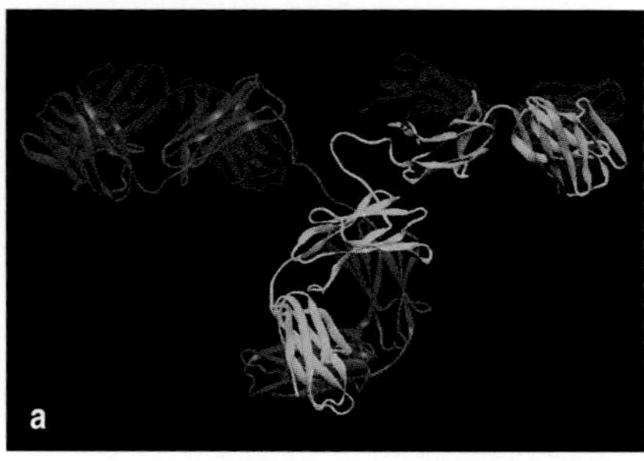

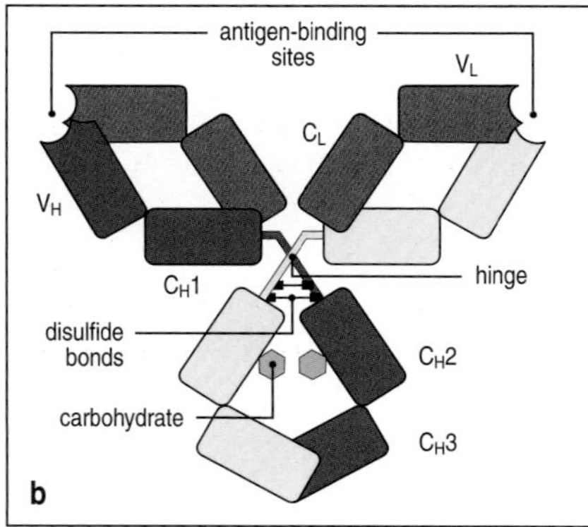

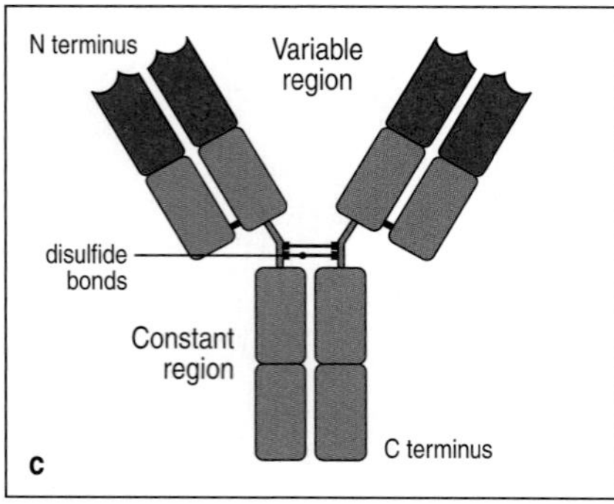

Fig. 3.1 Structure of an antibody molecule. Panel a illustrates a ribbon diagram based on the X-ray crystallographic structure of an IgG antibody, showing the course of the backbones of the polypeptide chains. Three globular regions form a Y shape. The two antigen-binding sites are at the tips of the arms, which are tethered to the trunk of the Y by a flexible hinge region. A schematic representation of the structure in a is given in panel b, illustrating the four-chain composition and the separate domains comprising each chain. Panel c shows a simplified schematic representation of an antibody molecule that will be used throughout this book. C terminus, carboxy terminus; N terminus, amino terminus. Panel a courtesy of A. McPherson and L. Harris.

structure—having both V and C regions—and in the genetic mechanism that produces their great variability, which will be discussed in Chapter 4. The T-cell receptor differs from the B-cell receptor in an important way, however: it does not recognize and bind antigen directly, but instead recognizes short peptide fragments of protein antigens, which are bound to proteins known as **MHC molecules** on the surfaces of cells.

The MHC molecules are glycoproteins encoded in the large cluster of genes known as the **major histocompatibility complex** (**MHC**). Their most striking structural feature is a cleft running across their outermost surface, in which a variety of peptides can be bound. MHC molecules are highly **polymorphic**: that is, each type of MHC molecule occurs in many different versions within the population. Most people are therefore heterozygous for the MHC molecules: they express two different forms of each type of MHC molecule, which increases the range of pathogen-derived peptides that can be bound. T-cell receptors recognize features of both the peptide antigen and the MHC molecule to which it is bound. This introduces an extra dimension to antigen recognition by T cells, known as **MHC restriction**, because any given T-cell receptor is specific not simply for a foreign peptide antigen but for a unique combination of a peptide and a particular MHC molecule. We shall discuss MHC polymorphism and its consequences for T-cell antigen recognition and T-cell development in Chapters 5 and 7, respectively.

In this chapter we focus on the structure and antigen-binding properties of immunoglobulins and T-cell receptors. Both receptors are also associated with intracellular signaling complexes, which transmit the antigen-binding signal onwards into the cell; these are described in Chapter 6. Although B cells and T cells recognize foreign molecules in distinct fashions, the receptor molecules they use for this task are very similar in structure. We will see how this basic structure can accommodate great variability in antigen specificity, and how it enables immunoglobulins and T-cell receptors to perform their functions as the antigen-recognition molecules of the adaptive immune response.

The structure of a typical antibody molecule.

Antibodies are the secreted form of the B-cell receptor. As they are soluble and secreted in large quantities, antibodies are easily obtainable and easily studied. For this reason, most of what we know about the B-cell receptor comes from the study of antibodies.

Antibody molecules are roughly Y-shaped molecules consisting of three equal-sized portions, connected by a flexible tether. Three schematic representations of antibody structure, which has been determined by X-ray crystallography, are shown in Fig. 3.1. The aim of this part of the chapter is to explain how this structure is formed and how it allows antibody molecules to perform their dual tasks—binding on the one hand to a wide variety of antigens, and on the other hand to a limited number of effector molecules and cells. As we will see, each of these tasks is performed by separable parts of the molecule. The two arms of the Y end in regions that vary between different antibody molecules—the V regions. These are involved in antigen binding, whereas the stem of the Y, or the C region, is far less variable and is the part that interacts with effector cells and molecules.

All antibodies are constructed in the same way from paired heavy and light polypeptide chains, and the generic term immunoglobulin is used for all such proteins. Within this general category, however, five different **classes** of immunoglobulins—**IgM**, **IgD**, **IgG**, **IgA**, and **IgE**—can be distinguished by their C region. More subtle differences confined to the V region account for the specificity of antigen binding. We will use the IgG antibody molecule as an example to describe the general structural features of immunoglobulins.

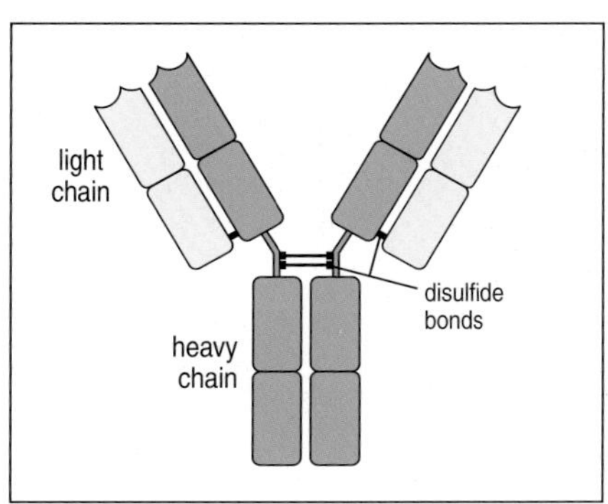

Fig. 3.2 Immunoglobulin molecules are composed of two types of protein chains: heavy chains and light chains. Each immunoglobulin molecule is made up of two heavy chains (green) and two light chains (yellow) joined by disulfide bonds so that each heavy chain is linked to a light chain and the two heavy chains are linked together.

3-1 IgG antibodies consist of four polypeptide chains.

IgG antibodies are large molecules with a molecular weight of approximately 150 kDa and are composed of two different kinds of polypeptide chain. One, of approximately 50 kDa, is termed the **heavy** or **H chain**, and the other, of 25 kDa, is termed the **light** or **L chain** (Fig. 3.2). Each IgG molecule consists of two heavy chains and two light chains. The two heavy chains are linked to each other by disulfide bonds and each heavy chain is linked to a light chain by a disulfide bond. In any given immunoglobulin molecule, the two heavy chains and the two light chains are identical, giving an antibody molecule two identical antigen-binding sites (see Fig. 3.1) and thus the ability to bind simultaneously to two identical structures.

Two types of light chain, termed **lambda** (**λ**) and **kappa** (**κ**), are found in antibodies. A given immunoglobulin has either κ chains or λ chains, never one of each. No functional difference has been found between antibodies having λ or κ light chains, and either type of light chain can be found in antibodies of any of the five major classes. The ratio of the two types of light chains varies from species to species. In mice, the average κ to λ ratio is 20:1, whereas in humans it is 2:1 and in cattle it is 1:20. The reason for this variation is unknown. Distortions of this ratio can sometimes be used to detect the abnormal proliferation of a clone of B cells. These would all express the identical light chain, and thus an excess of λ light chains in a person might indicate the presence of a B-cell tumor producing λ chains.

The class, and thus the effector function, of an antibody is defined by the structure of its heavy chain. There are five main heavy-chain classes or **isotypes**, some of which have several subtypes, and these determine the functional activity of an antibody molecule. The five major classes of immunoglobulin are **immunoglobulin M** (**IgM**), **immunoglobulin D** (**IgD**), **immunoglobulin G** (**IgG**), **immunoglobulin A** (**IgA**), and **immunoglobulin E** (**IgE**). Their heavy chains are denoted by the corresponding lower-case Greek letter (μ, δ, γ, α, and ε, respectively). IgG is by far the most abundant immunoglobulin and has several subclasses (IgG1, 2, 3, and 4 in humans). Their distinctive functional properties are conferred by the carboxy-terminal part of the heavy chain, where it is not associated with the light chain. We will describe the structure and functions of the different heavy-chain isotypes in Chapter 4. The general structural features of all the isotypes are similar, and we will consider IgG, the most abundant isotype in plasma, as a typical antibody molecule.

The structure of the B-cell receptor structure is identical to that of its corresponding antibody except for a small portion of the carboxy terminus of the heavy-chain C region. In the B-cell receptor the carboxy terminus is a hydrophobic sequence that anchors the molecule in the membrane, and in the antibody it is a hydrophilic sequence that allows secretion.

3-2 Immunoglobulin heavy and light chains are composed of constant and variable regions.

The amino acid sequences of many immunoglobulin heavy and light chains have been determined and reveal two important features of antibody molecules. First, each chain consists of a series of similar, although not identical, sequences, each about 110 amino acids long. Each of these repeats corresponds to a discrete, compactly folded region of protein structure known as a protein domain. The light chain is made up of two such **immunoglobulin domains**, whereas the heavy chain of the IgG antibody contains four (see Fig. 3.1a). This suggests that the immunoglobulin chains have evolved by repeated duplication of an ancestral gene corresponding to a single domain.

The second important feature revealed by comparisons of amino acid sequences is that the amino-terminal sequences of both the heavy and light chains vary greatly between different antibodies. The variability in sequence is limited to approximately the first 110 amino acids, corresponding to the first domain, whereas the remaining domains are constant between immunoglobulin chains of the same isotype. The amino-terminal variable domains (V domains) of the heavy and light chains (V_H and V_L, respectively) together make up the V region of the antibody and confer on it the ability to bind specific antigen, whereas the constant domains (C domains) of the heavy and light chains (C_H and C_L, respectively) make up the C region (see Fig. 3.1b and c). The multiple heavy-chain C domains are numbered from the amino-terminal end to the carboxy terminus, for example C_H1, C_H2, and so on.

3-3 The antibody molecule can readily be cleaved into functionally distinct fragments.

The protein domains described above associate to form larger globular domains. Thus, when fully folded and assembled, an antibody molecule comprises three equal-sized globular portions joined by a flexible stretch of polypeptide chain known as the **hinge region** (see Fig. 3.1b). Each arm of this Y-shaped structure is formed by the association of a light chain with the amino-terminal half of a heavy chain, whereas the trunk of the Y is formed by the pairing of the carboxy-terminal halves of the two heavy chains. The association of the heavy and light chains is such that the V_H and V_L domains are paired, as are the C_H1 and C_L domains. The C_H3 domains pair with each other but the C_H2 domains do not interact; carbohydrate side chains attached to the C_H2 domains lie between the two heavy chains. The two antigen-binding sites are formed by the paired V_H and V_L domains at the ends of the two arms of the Y (see Fig. 3.1b).

Proteolytic enzymes (proteases) that cleave polypeptide sequences have been used to dissect the structure of antibody molecules and to determine which parts of the molecule are responsible for its various functions. Limited digestion with the protease papain cleaves antibody molecules into three fragments (Fig. 3.3). Two fragments are identical and contain the antigen-binding activity. These are termed the **Fab fragments**, for **F**ragment **a**ntigen **b**inding. The Fab fragments correspond to the two identical arms of the antibody molecule, each of which consists of a complete light chain paired with the V_H and C_H1 domains of a heavy chain. The other fragment contains no antigen-binding activity but was originally observed to crystallize readily, and for this reason it was named the **Fc fragment**, for **F**ragment **c**rystallizable. This fragment corresponds to the paired C_H2 and C_H3 domains and is the part of the antibody molecule that interacts with effector molecules and cells. The functional differences between heavy-chain isotypes lie mainly in the Fc fragment.

The protein fragments obtained after proteolysis are determined by where the protease cuts the antibody molecule in relation to the disulfide bonds that link the two heavy chains. These lie in the hinge region between the C_H1 and C_H2 domains and, as illustrated in Fig. 3.3, papain cleaves the antibody molecule on the amino-terminal side of the disulfide bonds. This releases the two arms of the antibody as separate Fab fragments, whereas in the Fc fragment the carboxy-terminal halves of the heavy chains remain linked.

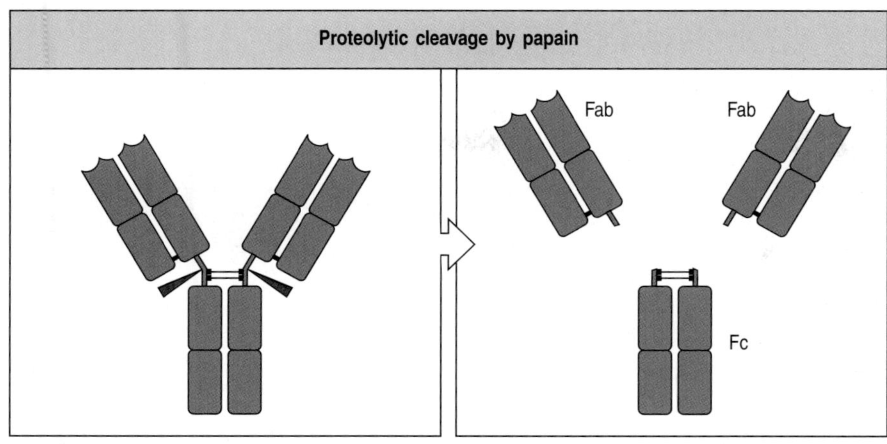

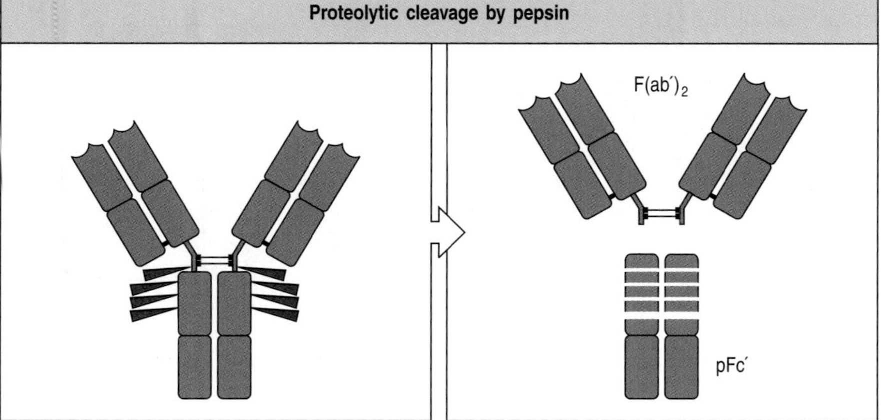

Fig. 3.3 The Y-shaped immunoglobulin molecule can be dissected by partial digestion with proteases. Upper panels: papain cleaves the immunoglobulin molecule into three pieces, two Fab fragments and one Fc fragment. The Fab fragment contains the V regions and binds antigen. The Fc fragment is crystallizable and contains C regions. Lower panels: pepsin cleaves immunoglobulin to yield one $F(ab')_2$ fragment and many small pieces of the Fc fragment, the largest of which is called the pFc′ fragment. $F(ab')_2$ is written with a prime because it contains a few more amino acids than Fab, including the cysteines that form the disulfide bonds.

Another protease, pepsin, cuts in the same general region of the antibody molecule as papain but on the carboxy-terminal side of the disulfide bonds (see Fig. 3.3). This produces a fragment, the **$F(ab')_2$ fragment**, in which the two antigen-binding arms of the antibody molecule remain linked. In this case the remaining part of the heavy chain is cut into several small fragments. The $F(ab')_2$ fragment has exactly the same antigen-binding characteristics as the original antibody but is unable to interact with any effector molecule. It is thus of potential value in the therapeutic applications of antibodies as well as in research into the functional role of the Fc portion.

Genetic engineering techniques also now permit the construction of many different antibody-related molecules. One important type is a truncated Fab comprising the V domain of a heavy chain linked by a stretch of synthetic peptide to a V domain of a light chain. This is called **single-chain Fv**, named from **F**ragment **v**ariable. Fv molecules may become valuable therapeutic agents because of their small size, which allows them to penetrate tissues readily. For example, Fv molecules specific for tumor antigens and coupled to protein toxins have potential applications in tumor therapy, as discussed in Chapter 15.

3-4 The immunoglobulin molecule is flexible, especially at the hinge region.

The region that links the Fc and Fab portions of the antibody molecule is in reality not a rigid hinge but a flexible tether, allowing independent movement

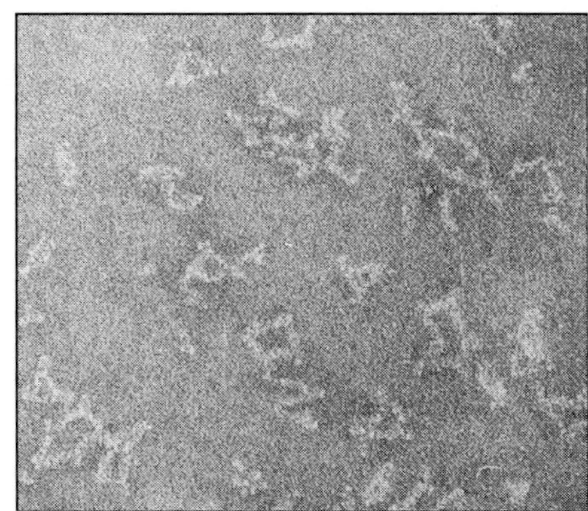

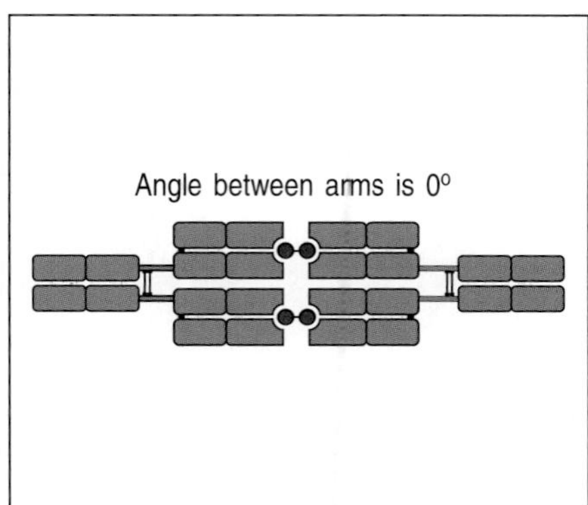
Angle between arms is 0°

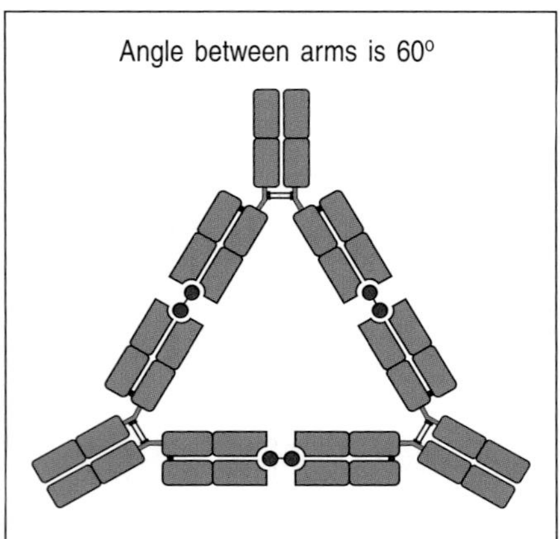
Angle between arms is 60°

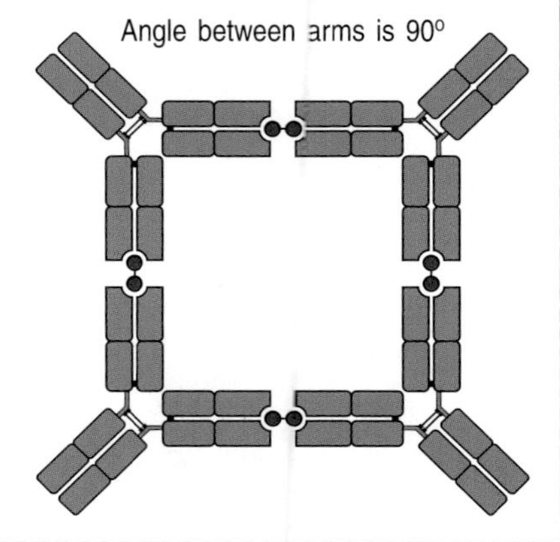
Angle between arms is 90°

Fig. 3.4 Antibody arms are joined by a flexible hinge. An antigen consisting of two hapten molecules (red balls in diagrams) that can cross-link two antigen-binding sites is used to create antigen:antibody complexes, which can be seen in the electron micrograph. Linear, triangular, and square forms are seen, with short projections or spikes. Limited pepsin digestion removes these spikes (not shown in the figure), which therefore correspond to the Fc portion of the antibody; the F(ab′)$_2$ pieces remain cross-linked by antigen. The interpretation of the complexes is shown in the diagrams. The angle between the arms of the antibody molecules varies, from 0° in the antibody dimers, through 60° in the triangular forms, to 90° in the square forms, showing that the connections between the arms are flexible. Photograph (× 300,000) courtesy of N.M. Green.

of the two Fab arms. This flexibility is revealed by studies of antibodies bound to small antigens known as **haptens**. These are molecules of various types that are typically about the size of a tyrosine side chain. Although haptens can be specifically recognized by antibody, they can only stimulate the production of anti-hapten antibodies when linked to a protein (see Appendix I, Section A-1). Two identical hapten molecules joined by a short flexible region can link two or more anti-hapten antibodies, forming dimers, trimers, tetramers, and so on, which can be seen by electron microscopy (Fig. 3.4). The shapes formed by these complexes show that antibody molecules are flexible at the hinge region. Some flexibility is also found at the junction between the V and C domains, allowing bending and rotation of the V domain relative to the C domain. For example, in the antibody molecule shown in Fig. 3.1a, not only are the two hinge regions clearly bent differently, but the angle between the V and C domains in each of the two Fab arms is also different. This range of motion has led to the junction between the V and C domains being referred to as a 'molecular ball-and-socket joint.' Flexibility at both the hinge and the V–C junction enables the two arms of an antibody molecule to bind to sites some distance apart, such as the repeating sites on bacterial cell-wall polysaccharides. Flexibility at the hinge also enables antibodies to interact with the antibody-binding proteins that mediate immune effector mechanisms.

3-5 The domains of an immunoglobulin molecule have similar structures.

As we saw in Section 3-2, immunoglobulin heavy and light chains are composed of a series of discrete protein domains, all of which have a similar folded structure. Within this basic three-dimensional structure there are distinct differences between V and C domains. The structural similarities and

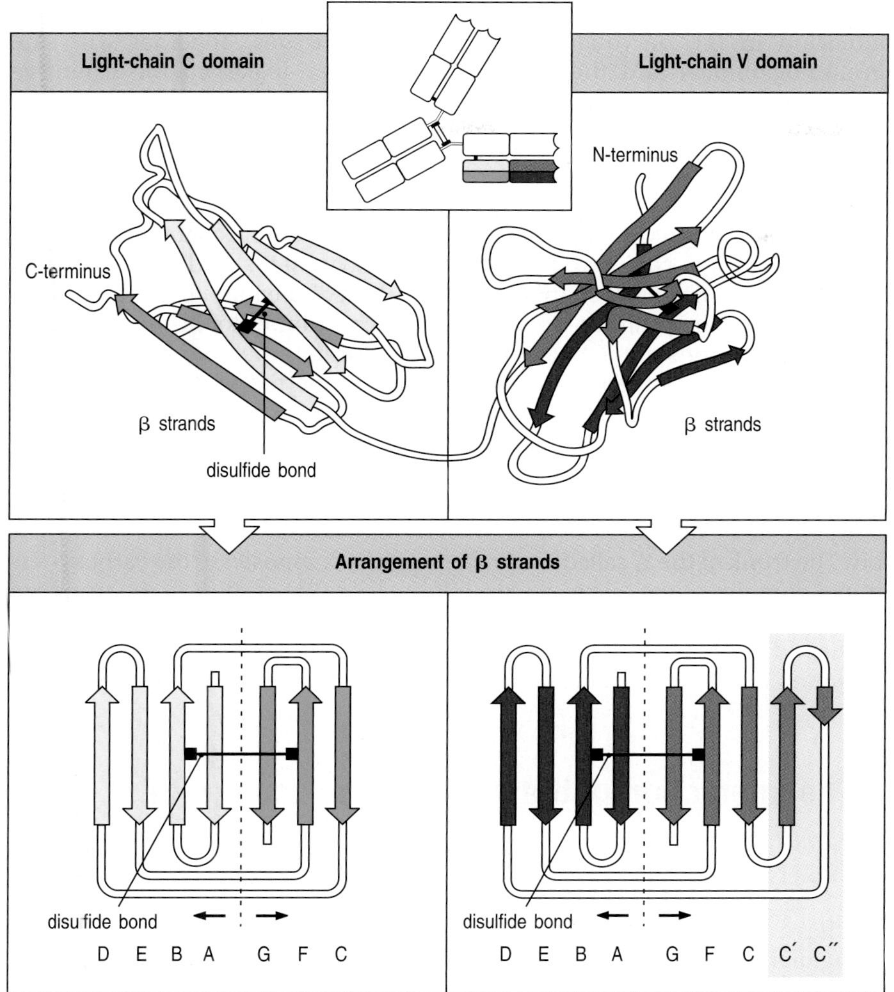

Fig. 3.5 The structure of immunoglobulin constant and variable domains. The upper panels show schematically the folding pattern of the constant (C) and variable (V) domains of an immunoglobulin light chain. Each domain is a barrel-shaped structure in which strands of polypeptide chain (β strands) running in opposite directions (antiparallel) pack together to form two β sheets (shown in yellow and green for the C domain and red and blue for the V domain), which are held together by a disulfide bond. The way the polypeptide chain folds to give the final structure can be seen more clearly when the sheets are opened out, as shown in the lower panels. The β strands are lettered sequentially with respect to the order of their occurrence in the amino acid sequence of the domains; the order in each β sheet is characteristic of immunoglobulin domains. The β strands C′ and C″ that are found in the V domains but not in the C domains are indicated by a blue shaded background. The characteristic four-strand plus three-strand (C-region type domain) or four-strand plus five-strand (V-region type domain) arrangements are typical immunoglobulin superfamily domain building blocks, found in a whole range of other proteins as well as antibodies and T-cell receptors.

differences can be seen in the diagram of a light chain in Fig. 3.5. Each domain is constructed from two **β sheets**, which are elements of protein structure made up of strands of the polypeptide chain (**β strands**) packed together; the sheets are linked by a disulfide bridge and together form a roughly barrel-shaped structure, known as a **β barrel**. The distinctive folded structure of the immunoglobulin protein domain is known as the **immunoglobulin fold**.

Both the essential similarity of V and C domains and the critical difference between them are most clearly seen in the bottom panels of Fig. 3.5, where the cylindrical domains are opened out to reveal how the polypeptide chain folds to create each of the β sheets and how it forms flexible loops as it changes direction. The main difference between the V and C domains is that the V domain is larger, with an extra loop. The flexible loops of the V domains form the antigen-binding site of the immunoglobulin molecule.

Many of the amino acids that are common to C and V domains of immunoglobulin chains lie in the core of the immunoglobulin fold and are essential for its stability. For that reason, other proteins with sequences similar to those of immunoglobulins are believed to form domains of similar structure, and in many cases this has been demonstrated by crystallography. These **immunoglobulin-like domains** are present in many other proteins of the immune system, and in proteins involved in cell–cell recognition and adhesion in the nervous system and other tissues. Together with the immunoglobulins and the T-cell receptors, they make up the extensive **immunoglobulin superfamily**.

Summary.

The IgG antibody molecule is made up of four polypeptide chains, comprising two identical light chains and two identical heavy chains, and can be thought of as forming a flexible Y-shaped structure. Each of the four chains has a variable (V) region at its amino terminus, which contributes to the antigen-binding site, and a constant (C) region, which determines the isotype. The isotype of the heavy chain determines the functional properties of the antibody. The light chains are bound to the heavy chains by many noncovalent interactions and by disulfide bonds, and the V regions of the heavy and light chains pair in each arm of the Y to generate two identical antigen-binding sites, which lie at the tips of the arms of the Y. The possession of two antigen-binding sites allows antibody molecules to cross-link antigens and to bind them much more stably. The trunk of the Y, called the Fc fragment, is composed of the carboxy-terminal domains of the heavy chains. Joining the arms of the Y to the trunk are the flexible hinge regions. The Fc fragment and hinge regions differ in antibodies of different isotypes, thus determining their functional properties. However, the overall organization of the domains is similar in all isotypes.

The interaction of the antibody molecule with specific antigen.

We have described the structure of the antibody molecule and how the V regions of the heavy and light chains fold and pair to form the antigen-binding site. In this part of the chapter we look at the antigen-binding site in more detail. We discuss the different ways in which antigens can bind to antibody and address the question of how variation in the sequences of the antibody V domains determines the specificity for antigen.

3-6 Localized regions of hypervariable sequence form the antigen-binding site.

The V regions of any given antibody molecule differ from those of every other. Sequence variability is not, however, distributed evenly throughout the V region but is concentrated in certain segments, as can be clearly seen in what is termed a **variability plot** (Fig. 3.6), in which the amino acid sequences of many different antibody V regions are compared. Three particularly variable segments can be identified in both the V_H and V_L domains. They are designated **hypervariable regions** and are denoted HV1, HV2, and HV3. In the heavy chains they run roughly from residues 30 to 36, 49 to 65, and 95 to 103, respectively, while in the light chains they are located roughly at residues 28 to 35, 49 to 59, and 92 to 103, respectively. The most variable part of the domain is in the HV3 region. The regions between the hypervariable regions, which comprise the rest of the V domain, show less variability and are termed the **framework regions**. There are four such regions in each V domain, designated FR1, FR2, FR3, and FR4.

The framework regions form the β sheets that provide the structural framework of the domain, whereas the hypervariable sequences correspond to

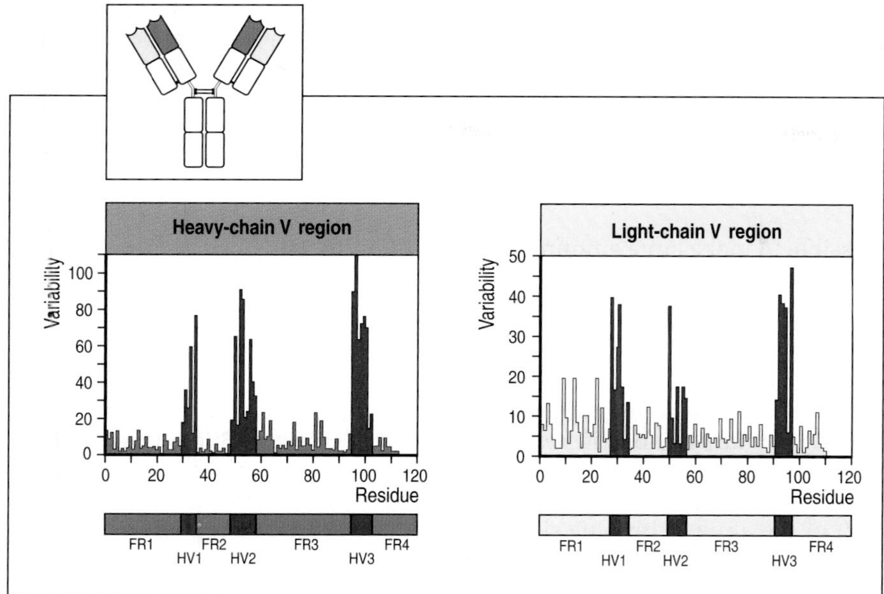

Fig. 3.6 There are discrete regions of hypervariability in V domains. A variability plot derived from comparison of the amino acid sequences of several dozen heavy-chain and light-chain V domains is shown. At each amino acid position the degree of variability is the ratio of the number of different amino acids seen in all of the sequences together to the frequency of the most common amino acid. Three hypervariable regions (HV1, HV2, and HV3) are indicated in red and are also known as the complementarity-determining regions, CDR1, CDR2, and CDR3. They are flanked by less variable framework regions (FR1, FR2, FR3, and FR4, shown in blue or yellow).

three loops at the outer edge of the β barrel, which are juxtaposed in the folded domain (Fig. 3.7). Thus, not only is diversity concentrated in particular parts of the V domain sequence, but it is localized to a particular region on the surface of the molecule. When the V_H and V_L domains are paired in the antibody molecule, the hypervariable loops from each domain are brought together, creating a single hypervariable site at the tip of each arm of the molecule. This is the binding site for antigen, the **antigen-binding site** or **antibody combining site**. The six hypervariable loops determine antigen specificity by forming a surface complementary to the antigen, and are more commonly termed the **complementarity-determining regions**, or **CDRs** (there are three CDRs from each of the heavy and light chains, namely CDR1, CDR2, and CDR3). Because CDRs from both V_H and V_L domains contribute to the antigen-binding site, it is the combination of the heavy and the light chain, and not either alone, that determines the final antigen specificity. Thus, one way in which the immune system is able to generate antibodies of different specificities is by generating different combinations of heavy- and light-chain V regions. This means of producing variability is known as **combinatorial diversity**; we will encounter a second form of combinatorial diversity when we consider in Chapter 4 how the genes encoding the heavy- and light-chain V regions are created from smaller segments of DNA.

3-7 Antibodies bind antigens via contacts with amino acids in CDRs, but the details of binding depend upon the size and shape of the antigen.

In early investigations of antigen binding to antibodies, the only available sources of large quantities of a single type of antibody molecule were tumors of antibody-secreting cells. The antigen specificities of these antibodies were unknown, so many compounds had to be screened to identify ligands that could be used to study antigen binding. In general, the substances found to bind to these antibodies were haptens (see Section 3-4) such as phosphocholine or vitamin K_1. Structural analysis of complexes of antibodies with their hapten ligands provided the first direct evidence that the hypervariable regions form the antigen-binding site, and demonstrated the structural basis of specificity for the hapten. Subsequently, with the discovery of methods of generating **monoclonal antibodies** (see Appendix I, Section A-12), it became

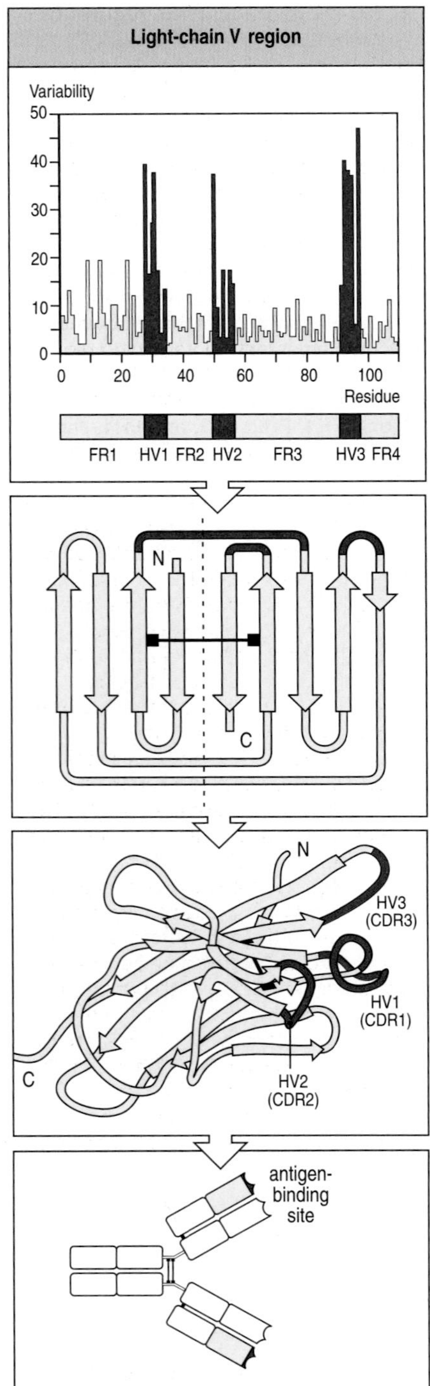

Fig. 3.7 The hypervariable regions lie in discrete loops of the folded structure. When the hypervariable regions (CDRs) are positioned on the structure of a V domain it can be seen that they lie in loops that are brought together in the folded structure. In the antibody molecule, the pairing of a heavy and a light chain brings together the hypervariable loops from each chain to create a single hypervariable surface, which forms the antigen-binding site at the tip of each arm. C, carboxy terminus; N, amino terminus.

possible to make large amounts of pure antibody specific for a given antigen. This has provided a more general picture of how antibodies interact with their antigens, confirming and extending the view of antibody–antigen interactions derived from the study of haptens.

The surface of the antibody molecule formed by the juxtaposition of the CDRs of the heavy and light chains is the site to which an antigen binds. Clearly, as the amino acid sequences of the CDRs are different in different antibodies, so are the shapes of the surfaces created by these CDRs. As a general principle, antibodies bind ligands whose surfaces are complementary to that of the antigen-binding site. A small antigen, such as a hapten or a short peptide, generally binds in a pocket or groove lying between the heavy- and light-chain V domains (Fig. 3.8a and b). Some antigens, such as proteins, can be the same size as, or larger than, the antibody itself. In these cases, the interface between antigen and antibody is often an extended surface that involves all the CDRs and, in some cases, part of the framework region as well (Fig. 3.8c). This surface need not be concave but can be flat, undulating, or even convex. In some cases, antibody molecules with elongated CDR3 loops can protrude a 'finger' into recesses in the surface of the antigen, as shown in Fig. 3.8d, where an antibody binding to the HIV gp120 antigen projects a long loop against its target.

3-8 Antibodies bind to conformational shapes on the surfaces of antigens.

The biological function of antibodies is to bind to pathogens and their products, and to facilitate their removal from the body. An antibody generally recognizes only a small region on the surface of a large molecule such as a polysaccharide or protein. The structure recognized by an antibody is called an **antigenic determinant** or **epitope**. Some of the most important pathogens have polysaccharide coats, and antibodies that recognize epitopes formed by the sugar subunits of these molecules are essential in providing immune protection from such pathogens. In many cases, however, the antigens that provoke an immune response are proteins. For example, protective antibodies against viruses recognize viral coat proteins. In all such cases, the structures recognized by the antibody are located on the surface of the protein. Such sites are likely to be composed of amino acids from different parts of the polypeptide chain that have been brought together by protein folding. Antigenic determinants of this kind are known as **conformational** or **discontinuous epitopes** because the structure recognized is composed of segments of the protein that are discontinuous in the amino acid sequence of the antigen but are brought together in the three-dimensional structure. In contrast, an epitope composed of a single segment of polypeptide chain is termed a **continuous** or **linear epitope**. Although most antibodies raised against intact, fully folded proteins recognize discontinuous epitopes, some will bind peptide fragments of the protein. Conversely, antibodies raised against peptides of a protein or against synthetic peptides corresponding to part of its sequence are occasionally found to bind to the natural folded protein. This makes it possible, in some cases, to use synthetic peptides in vaccines that aim at raising antibodies against a pathogen protein.

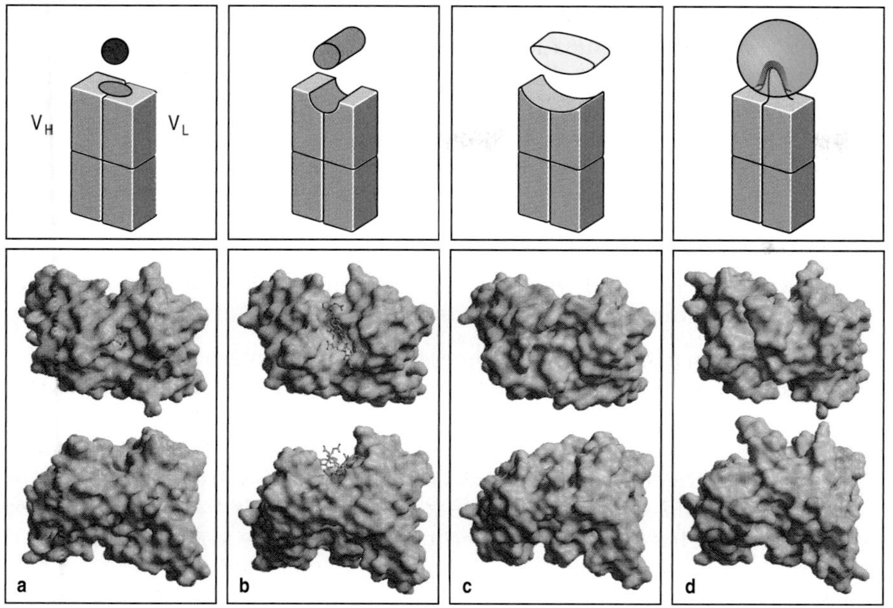

Fig. 3.8 Antigens can bind in pockets, or grooves, or on extended surfaces in the binding sites of antibodies. The panels in the top row show schematic representations of the different types of binding sites in a Fab fragment of an antibody: first panel, pocket; second panel, groove; third panel, extended surface; and fourth panel, protruding surface. Below are examples of each type. Panel a: the top image shows the molecular surface of the interaction of a small hapten with the complementarity-determining regions (CDRs) of a Fab fragment as viewed looking into the antigen-binding site. The ferrocene hapten, shown in green, is bound into the antigen-binding pocket (yellow). In the bottom image (and in those of panels b, c, and d) the molecule has been rotated by about 90° to give a side-on view of the binding site. Panel b: in a complex of an antibody with a peptide from the human immunodeficiency virus (HIV), the peptide (green) binds along a groove (yellow) formed between the heavy- and light-chain V domains. Panel c: complex between hen egg-white lysozyme and the Fab fragment of its corresponding antibody (HyHel5). The surface on the antibody that comes into contact with the lysozyme is colored yellow. All six CDRs of the antigen-binding site are involved in the binding. Panel d: an antibody molecule against the HIV gp120 antigen has elongated CDR3 loops that protrude into a recess in the surface of the antigen. The structure of the complex between this antibody and gp120 has not been solved, and so the yellow area on the images in the lower panels represents the extent of the CDR regions rather than the actual region of contact between antibody and antigen. Photographs courtesy of I.A. Wilson and R.L. Stanfield.

3-9 Antigen–antibody interactions involve a variety of forces.

The interaction between an antibody and its antigen can be disrupted by high salt concentrations, by extremes of pH, by detergents, and sometimes by competition with high concentrations of the pure epitope itself. The binding is therefore a reversible noncovalent interaction. The forces, or bonds, involved in these noncovalent interactions are outlined in Fig. 3.9.

Electrostatic interactions occur between charged amino acid side chains, as in salt bridges. Interactions also occur between electric dipoles, as in hydrogen bonds, or can involve short-range van der Waals forces. High salt concentrations and extremes of pH disrupt antigen–antibody binding by

Noncovalent forces	Origin	
Electrostatic forces	Attraction between opposite charges	$-\overset{\oplus}{N}H_3 \quad \overset{\ominus}{O}OC-$
Hydrogen bonds	Hydrogen shared between electronegative atoms (N,O)	$>\underset{\delta^-}{N}-\underset{\delta^+}{H}--\underset{\delta^-}{O}=C<$
Van der Waals forces	Fluctuations in electron clouds around molecules oppositely polarize neighboring atoms	$\delta^+ \; \delta^-$ ⇌ $\delta^- \; \delta^+$
Hydrophobic forces	Hydrophobic groups interact unfavorably with water and tend to pack together to exclude water molecules. The attraction also involves van der Waals forces	H_2O δ^+ δ^- δ^- δ^+ H_2O H_2O H_2O

Fig. 3.9 The noncovalent forces that hold together the antigen:antibody complex. Partial charges found in electric dipoles are shown as δ^+ or δ^-. Electrostatic forces diminish as the inverse square of the distance separating the charges, whereas van der Waals forces, which are more numerous in most antigen–antibody contacts, fall off as the sixth power of the separation and therefore operate only over very short ranges. Covalent bonds never occur between antigens and naturally produced antibodies.

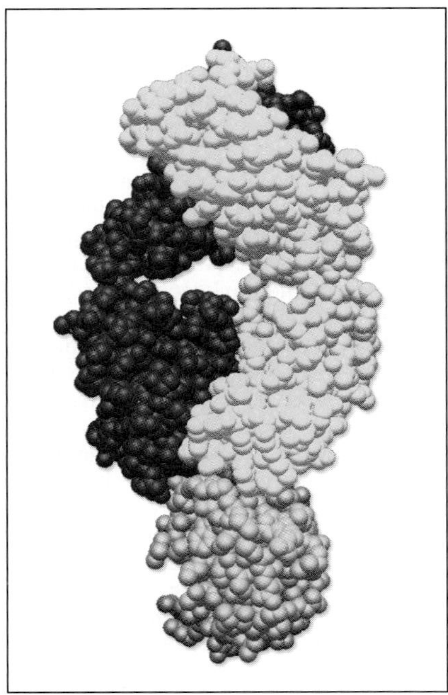

Fig. 3.10 The complex of lysozyme with the antibody D1.3. The interaction of the Fab fragment of D1.3 with hen egg-white lysozyme is shown, with the lysozyme in blue, the heavy chain in purple and the light chain in yellow. A glutamine residue of lysozyme, shown in red, protrudes between the two V domains of the antigen-binding site and makes hydrogen bonds that are important to the antigen–antibody binding. Courtesy of R.J. Poljak.

weakening electrostatic interactions and/or hydrogen bonds. This principle is employed in the purification of antigens by using affinity columns of immobilized antibodies, and vice versa for antibody purification (see Appendix I, Section A-5). Hydrophobic interactions occur when two hydrophobic surfaces come together to exclude water. The strength of a hydrophobic interaction is proportional to the surface area that is hidden from water. For some antigens, hydrophobic interactions probably account for most of the binding energy. In some cases, water molecules are trapped in pockets in the interface between antigen and antibody. These trapped water molecules, especially those between polar amino acid residues, may also contribute to binding and hence to the specificity of the antibody.

The contribution of each of these forces to the overall interaction depends on the particular antibody and antigen involved. A striking difference between antibody interactions with protein antigens and most other natural protein–protein interactions is that antibodies often have many aromatic amino acids in their antigen-binding sites. These amino acids participate mainly in van der Waals and hydrophobic interactions, and sometimes in hydrogen bonds. Tyrosine, for example, can participate in both hydrogen bonding and in hydrophobic interactions; it is therefore particularly suitable for providing diversity in antigen recognition and is over-represented in antigen-binding sites. In general, the hydrophobic and van der Waals forces operate over very short ranges and serve to pull together two surfaces that are complementary in shape: hills on one surface must fit into valleys on the other for good binding to occur. In contrast, electrostatic interactions between charged side chains, and hydrogen bonds bridging oxygen and/or nitrogen atoms, accommodate specific features or reactive groups while strengthening the interaction overall. Amino acids that possess charged side chains, such as arginine, are also over-represented at antigen-binding sites.

An example of a reaction involving a specific amino acid in the antigen can be seen in the complex of hen egg-white lysozyme with the antibody D1.3 (Fig. 3.10), where strong hydrogen bonds are formed between the antibody and a particular glutamine in the lysozyme molecule that protrudes between the V_H and V_L domains. Lysozymes from partridge and turkey have another amino acid in place of the glutamine and do not bind to this antibody. In the high-affinity complex of hen egg-white lysozyme with another antibody, HyHel5 (see Fig. 3.8c), two salt bridges between two basic arginines on the surface of the lysozyme interact with two glutamic acids, one each from the V_H CDR1 and CDR2 loops. Lysozymes that lack one of the two arginine residues show a 1000-fold decrease in affinity for HyHel5. Overall surface complementarity must have an important role in antigen–antibody interactions, but in most antibodies that have been studied at this level of detail only a few residues make a major contribution to the binding energy and hence to the final specificity of the antibody. Although many antibodies naturally bind their ligands with high affinity, genetic engineering by site-directed mutagenesis can tailor an antibody to bind even more strongly to its epitope.

Summary.

X-ray crystallographic analysis of antigen:antibody complexes has shown that the hypervariable loops (complementarity-determining regions, CDRs) of immunoglobulin V regions determine the binding specificity of an antibody. With protein antigens, an antibody molecule contacts the antigen over a broad area of its surface that is complementary to the surface recognized on the antigen. Electrostatic interactions, hydrogen bonds, van der Waals forces, and hydrophobic interactions can all contribute to binding. Depending on the size of the antigen, amino acid side chains in most or all of the CDRs make contact with antigen and determine both the specificity and the affinity of the interaction. Other parts of the V region play little part in the direct contact with the antigen but provide a stable structural framework for the CDRs and help to determine their position and conformation. Antibodies raised against intact proteins usually bind to the surface of the protein and make contact with residues that are discontinuous in the primary structure of the molecule; they may, however, occasionally bind peptide fragments of the protein, and antibodies raised against peptides derived from a protein can sometimes be used to detect the native protein molecule. Peptides binding to antibodies usually bind in the cleft between the V regions of the heavy and light chains, where they make specific contact with some, but not necessarily all, of the CDRs. This is also the usual mode of binding for carbohydrate antigens and small molecules such as haptens.

Antigen recognition by T cells.

In contrast to the immunoglobulins, which interact with pathogens and their toxic products in the extracellular spaces of the body, T cells only recognize foreign antigens that are displayed on the surfaces of the body's own cells. These antigens can derive from pathogens such as viruses or intracellular bacteria, which replicate within cells, or from pathogens or their products that cells have internalized by endocytosis from the extracellular fluid.

T cells detect the presence of an intracellular pathogen because the infected cells display on their surface peptide fragments of the pathogen's proteins. These foreign peptides are delivered to the cell surface by specialized host-cell glycoproteins—the MHC molecules. These are encoded in a large cluster of genes that were first identified by their powerful effects on the immune response to transplanted tissues. For that reason, the gene complex was called the major histocompatibility complex (MHC) and the peptide-binding glycoproteins are known as MHC molecules. The recognition of antigen as a small peptide fragment bound to an MHC molecule and displayed at the cell surface is one of the most distinctive features of T cells, and will be the focus of this part of the chapter. How the peptide fragments of antigen are generated and become associated with MHC molecules will be described in Chapter 5.

We describe here the structure and properties of the T-cell receptor (TCR). As might be expected from their function as highly variable antigen-recognition structures, the genes for T-cell receptors are closely related to those for immunoglobulins. There are, however, important differences between T-cell receptors and immunoglobulins that reflect the special features of antigen recognition by T cells.

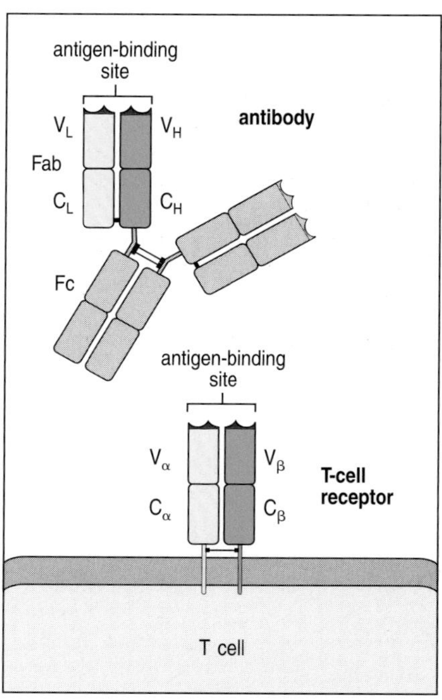

Fig. 3.11 The T-cell receptor resembles a membrane-bound Fab fragment. The Fab fragment of an antibody molecule is a disulfide-linked heterodimer, each chain of which contains one immunoglobulin C domain and one V domain; the juxtaposition of the V domains forms the antigen-binding site (see Section 3-6). The T-cell receptor is also a disulfide-linked heterodimer, with each chain containing an immunoglobulin C-like domain and an immunoglobulin V-like domain. As in the Fab fragment, the juxtaposition of the V domains forms the site for antigen recognition.

3-10 The T-cell receptor is very similar to a Fab fragment of immunoglobulin.

T-cell receptors were first identified by using monoclonal antibodies that bound to a single cloned T-cell line: such antibodies either specifically inhibit antigen recognition by the clone or specifically activate it by mimicking the antigen (see Appendix I, Section A-19). These **clonotypic** antibodies were then used to show that each T cell bears about 30,000 identical antigen receptors on its surface, each receptor consisting of two different polypeptide chains, termed the **T-cell receptor α (TCRα)** and **β (TCRβ)** chains, linked by a disulfide bond. The **α:β heterodimers** are very similar in structure to the Fab fragment of an immunoglobulin molecule (Fig. 3.11), and they account for antigen recognition by most T cells. A minority of T cells bear an alternative,

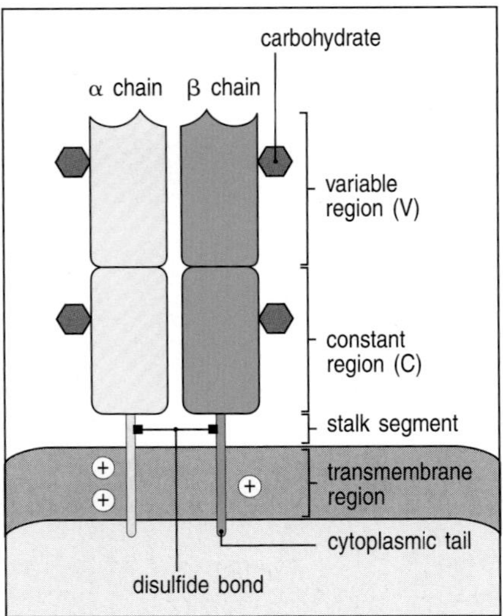

Fig. 3.12 Structure of the T-cell receptor. The T-cell receptor heterodimer is composed of two transmembrane glycoprotein chains, α and β. The extracellular portion of each chain consists of two domains, resembling immunoglobulin V and C domains, respectively. Both chains have carbohydrate side chains attached to each domain. A short stalk segment, analogous to an immunoglobulin hinge region, connects the immunoglobulin-like domains to the membrane and contains the cysteine residue that forms the interchain disulfide bond. The transmembrane helices of both chains are unusual in containing positively charged (basic) residues within the hydrophobic transmembrane segment. The α chain carries two such residues; the β chain has one.

but structurally similar, receptor made up of a different pair of polypeptide chains designated **γ** and **δ**. The **γ:δ T-cell receptors** seem to have different antigen-recognition properties from the **α:β T-cell receptors**, and the function of γ:δ T cells in immune responses is not yet entirely clear (see Section 2-34). In the rest of this chapter we shall use the term T-cell receptor to mean the α:β receptor, except where specified otherwise. Both types of T-cell receptor differ from the membrane-bound immunoglobulin that serves as the B-cell receptor in two main ways. A T-cell receptor has only one antigen-binding site, whereas a B-cell receptor has two, and T-cell receptors are never secreted, whereas immunoglobulin can be secreted as antibody.

The first insights into the structure and function of the α:β T-cell receptor came from studies of cloned cDNA encoding the receptor chains. The amino acid sequences predicted from the cDNA showed clearly that both chains of the T-cell receptor have an amino-terminal variable (V) region with homology to an immunoglobulin V domain, a constant (C) region with homology to an immunoglobulin C domain, and a short stalk segment containing a cysteine residue that forms the interchain disulfide bond (Fig. 3.12). Each chain spans the lipid bilayer by a hydrophobic transmembrane domain, and ends in a short cytoplasmic tail. These close similarities of T-cell receptor chains to the heavy and light immunoglobulin chains first enabled prediction of the structural resemblance of the T-cell receptor heterodimer to a Fab fragment of immunoglobulin.

The three-dimensional structure of the T-cell receptor has since been determined by X-ray crystallography and the structure is indeed similar to that of a Fab fragment. The T-cell receptor chains fold in much the same way as those of a Fab fragment (Fig. 3.13a), although the final structure appears a little shorter and wider. There are, however, some distinct structural differences between T-cell receptors and Fab fragments. The most striking is in the C_α domain, where the fold is unlike that of any other immunoglobulin-like domain. The half of the domain that is juxtaposed with the C_β domain forms a β sheet similar to that found in other immunoglobulin-like domains, but the other half of the domain is formed of loosely packed strands and a short segment of α helix (Fig. 3.13b). In a C_α domain the intramolecular disulfide bond, which in immunoglobulin-like domains normally joins two β strands, joins a β strand to this segment of α helix.

There are also differences in the way in which the domains interact. The interface between the V and C domains of both T-cell receptor chains is more extensive than in antibodies. The interaction between the C_α and C_β domains is distinctive as it might be assisted by carbohydrate, with a sugar group from the C_α domain making a number of hydrogen bonds to the C_β domain (see Fig. 3.13b). Finally, a comparison of the variable binding sites shows that, although the CDR loops align fairly closely with those of antibody molecules, there is some relative displacement (see Fig. 3.13c). This is particularly marked in the V_α CDR2 loop, which is oriented at roughly right angles to the equivalent loop in antibody V domains, as a result of a shift in the β strand that anchors one end of the loop from one face of the domain to the other. A strand displacement also causes a change in the orientation of the V_β CDR2

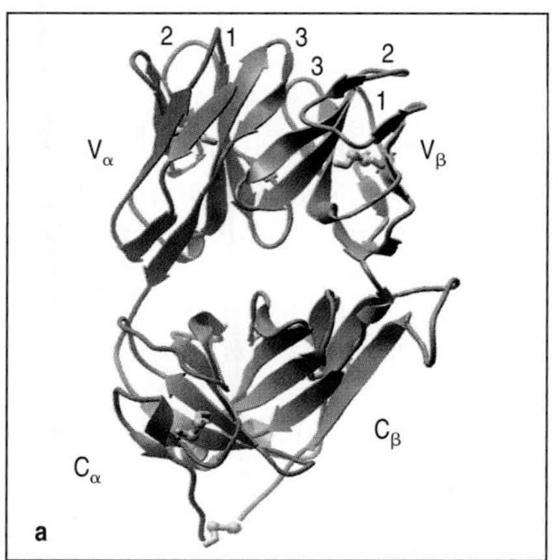

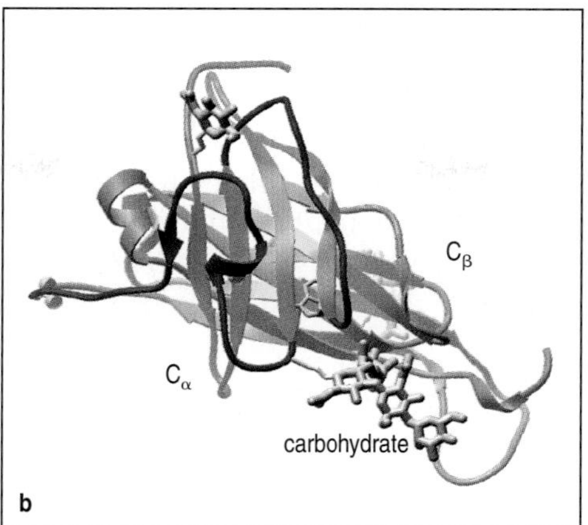

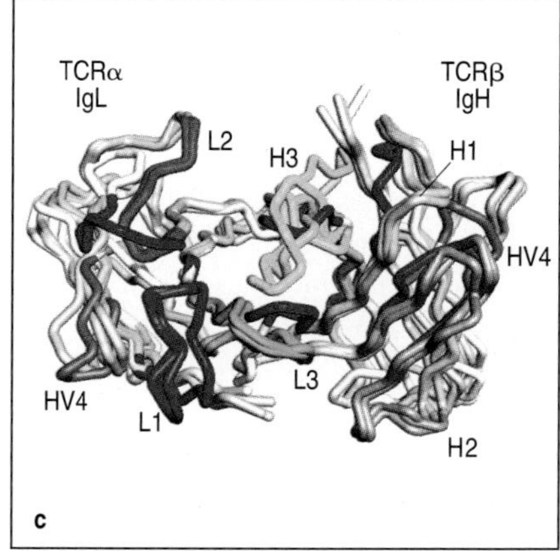

Fig. 3.13 The crystal structure of an α:β T-cell receptor resolved at 2.5 Å. In panels a and b the α chain is shown in pink and the β chain in blue. Disulfide bonds are shown in green. In panel a, the T-cell receptor is viewed from the side as it would sit on a cell surface, with the CDR loops that form the antigen-binding site (labeled 1, 2, and 3) arrayed across its relatively flat top. In panel b, the $C_\alpha$and C_β domains are shown. The C_α domain does not fold into a typical immunoglobulin-like domain; the face of the domain away from the C_β domain is mainly composed of irregular strands of polypeptide rather than β sheet. The intramolecular disulfide bond joins a β strand to this segment of α helix. The interaction between the C_α and C_β domains is assisted by carbohydrate (colored gray and labeled on the figure), with a sugar group from the C_α domain making hydrogen bonds to the C_β domain. In panel c, the T-cell receptor is shown aligned with the antigen-binding sites from three different antibodies. This view is looking down into the binding site. The V_α domain of the T-cell receptor is aligned with the V_L domains of the antigen-binding sites of the antibodies, and the V_β domain is aligned with the V_H domains. The CDRs of the T-cell receptor and immunoglobulin molecules are colored, with CDRs 1, 2, and 3 of the TCR shown in red and the HV4 loop in orange. For the immunoglobulin V domains, the CDR1 loops of the heavy chain (H1) and light chain (L1) are shown in light and dark blue, respectively, and the CDR2 loops (H2, L2) in light and dark purple, respectively. The heavy-chain CDR3 loops (H3) are in yellow; the light-chain CDR3s (L3) are in bright green. The HV4 loops of the TCR (orange) have no hypervariable counterparts in immunoglobulins. Model structures courtesy of I.A. Wilson.

loop in some V_β domains whose structures are known. As relatively few crystallographic structures have been solved to this level of resolution, it remains to be seen to what degree all T-cell receptors share these features, and whether there are more differences to be discovered.

3-11 A T-cell receptor recognizes antigen in the form of a complex of a foreign peptide bound to an MHC molecule.

Antigen recognition by T-cell receptors clearly differs from recognition by B-cell receptors and antibodies. Antigen recognition by B cells involves the direct binding of immunoglobulin to the intact antigen and, as discussed in Section 3-8, antibodies typically bind to the surface of protein antigens, contacting amino acids that are discontinuous in the primary structure but are brought together in the folded protein. T cells, in contrast, respond to short

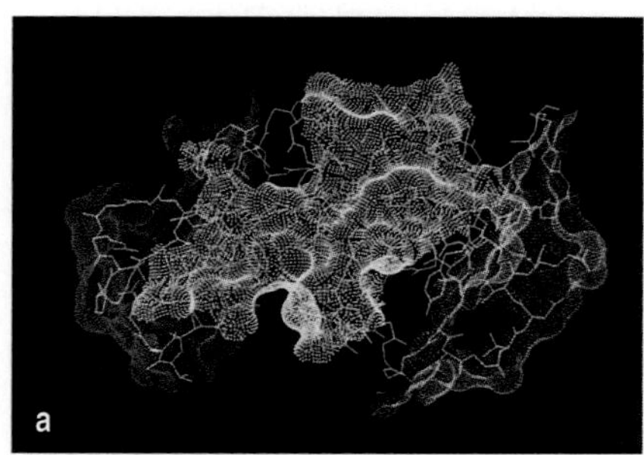

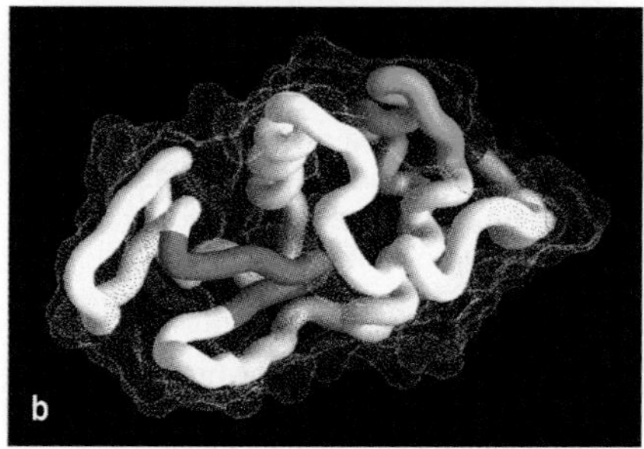

Fig. 3.14 Differences in the recognition of hen egg-white lysozyme by immunoglobulins and T-cell receptors. Antibodies can be shown by X-ray crystallography to bind epitopes on the surface of proteins, as shown in panel a, where the epitopes for three antibodies are shown in different colors on the surface of hen egg-white lysozyme (see also Fig. 3.10). In contrast, the epitopes recognized by T-cell receptors need not lie on the surface of the molecule, because the T-cell receptor recognizes not the antigenic protein itself but a peptide fragment of the protein. The peptides corresponding to two T-cell epitopes of lysozyme are shown in panel b. One epitope, shown in blue, lies on the surface of the protein but a second, shown in red, lies mostly within the core and is inaccessible in the folded protein. For this residue to be accessible to the T-cell receptor, the protein must be unfolded and processed. Panel a courtesy of S. Sheriff.

contiguous amino acid sequences in proteins. These sequences are often buried within the native structure of the protein and thus cannot be recognized directly by T-cell receptors unless the protein is unfolded and processed into peptide fragments (Fig. 3.14). We shall see in Chapter 5 how this occurs.

The nature of the antigen recognized by T cells became clear with the realization that the peptides that stimulate T cells are recognized only when bound to an MHC molecule. The ligand recognized by the T cell is thus a complex of peptide and MHC molecule. The evidence for involvement of the MHC in T-cell recognition of antigen was at first indirect, but it has recently been proved conclusively by stimulating T cells with purified peptide:MHC complexes. The T-cell receptor interacts with this ligand by making contacts with both the MHC molecule and the antigen peptide.

3-12 There are two classes of MHC molecules with distinct subunit composition but similar three-dimensional structures.

There are two classes of MHC molecules—**MHC class I** and **MHC class II**—which differ in both their structure and expression pattern on the tissues of the body. As shown in Figs 3.15 and 3.16, MHC class I and MHC class II molecules are closely related in overall structure but differ in their subunit composition.

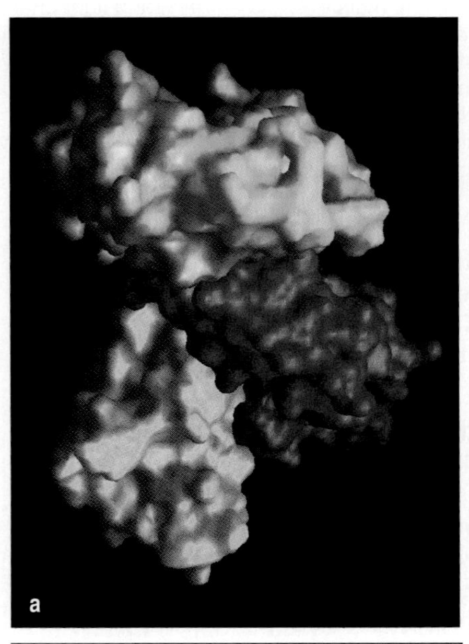

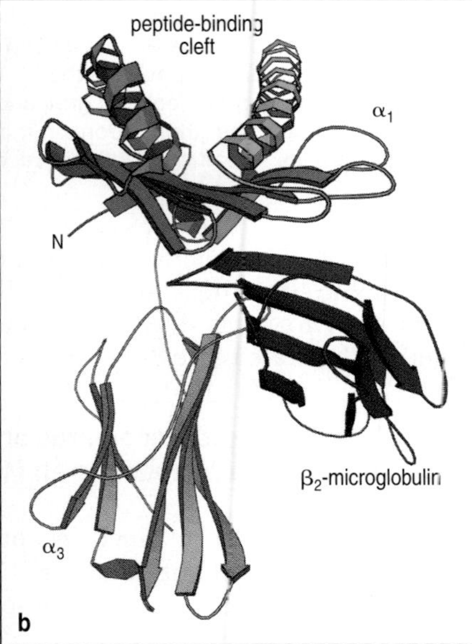

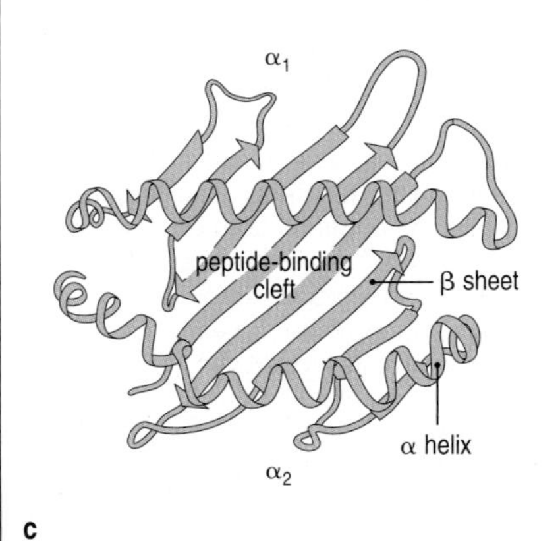

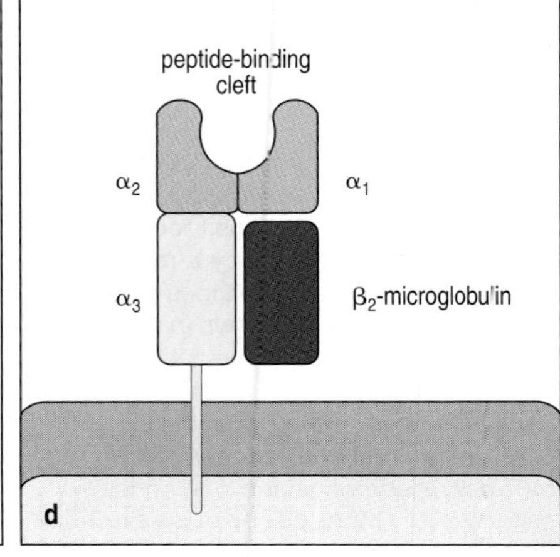

Fig. 3.15 The structure of an MHC class I molecule determined by X-ray crystallography. Panel a shows a computer graphic representation of a human MHC class I molecule, HLA-A2, which has been cleaved from the cell surface by the enzyme papain. The surface of the molecule is shown, colored according to the domains shown in panels b–d and described below. Panels b and c show a ribbon diagram of that structure. Shown schematically in panel d, the MHC class I molecule is a heterodimer of a membrane-spanning α chain (molecular weight 43 kDa) bound noncovalently to β_2-microglobulin (12 kDa), which does not span the membrane. The α chain folds into three domains: α_1, α_2, and α_3. The α_3 domain and β_2-microglobulin show similarities in amino acid sequence to immunoglobulin C domains and have similar folded structures, whereas the α_1 and α_2 domains fold together into a single structure consisting of two separated α helices lying on a sheet of eight antiparallel β strands. The folding of the α_1 and α_2 domains creates a long cleft or groove, which is the site at which peptide antigens bind to the MHC molecules. The transmembrane region and the short stretch of peptide that connects the external domains to the cell surface are not seen in panels a and b because they have been removed by the digestion with papain. As can be seen in panel c, looking down on the molecule from above, the sides of the cleft are formed from the inner faces of the two α helices; the β-pleated sheet formed by the pairing of the α_1 and α_2 domains creates the floor of the cleft.

In both classes, the two paired protein domains nearest to the membrane resemble immunoglobulin domains, whereas the two domains furthest away from the membrane fold together to create a long cleft, or groove, which is the site at which a peptide binds. Purified peptide:MHC class I and peptide:MHC class II complexes have been characterized structurally, allowing us to describe in detail both the MHC molecules themselves and the way in which they bind peptides.

MHC class I molecules (see Fig. 3.15) consist of two polypeptide chains. One chain, the α chain, is encoded in the MHC (on chromosome 6 in humans) and is noncovalently associated with a smaller chain, **β_2-microglobulin**, which is not polymorphic and is encoded on a different chromosome—chromosome 15 in humans. Only the class I α chain spans the membrane. The complete molecule has four domains, three formed from the MHC-encoded α chain, and one contributed by β_2-microglobulin. The α_3 domain and β_2-microglobulin closely resemble immunoglobulin-like domains in their folded structure. The folded α_1 and α_2 domains form the walls of a cleft on the surface of the molecule; this is where the peptide binds and is known as the **peptide-binding cleft** or **peptide-binding groove**. The MHC molecules are highly polymorphic and the major differences between the different forms are located in the peptide-binding cleft, influencing which peptides will bind and thus the specificity of the dual antigen presented to T cells.

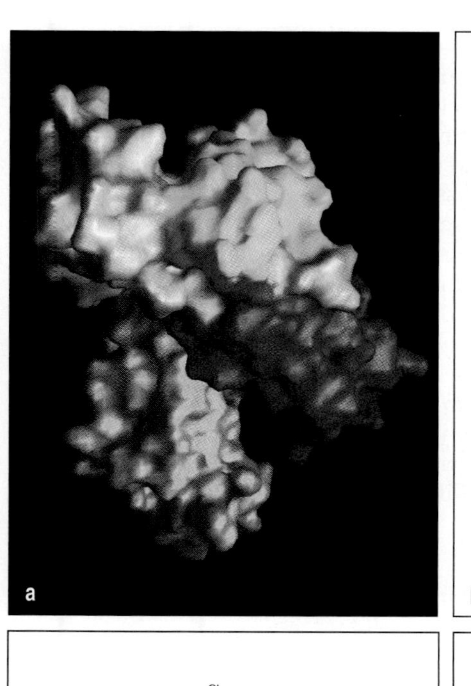

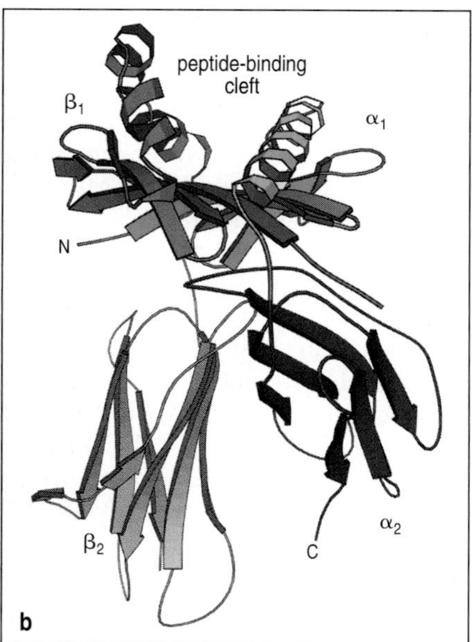

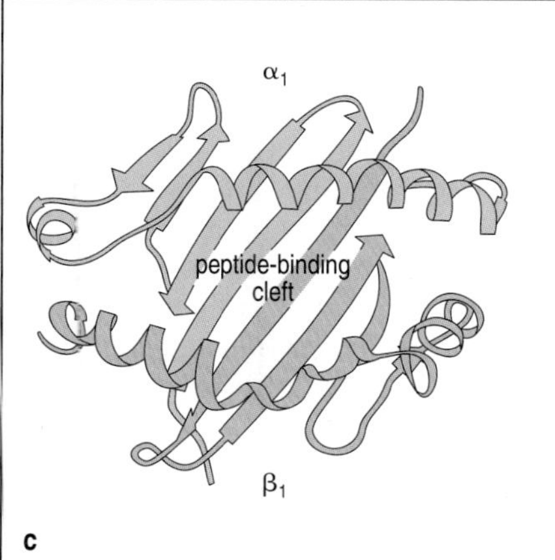

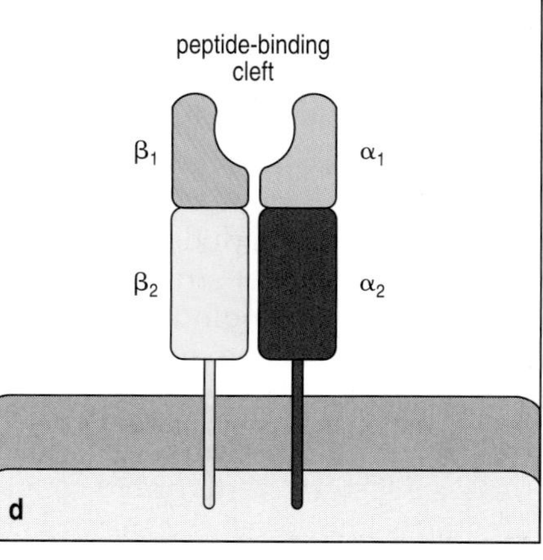

Fig. 3.16 MHC class II molecules resemble MHC class I molecules in overall structure. The MHC class II molecule is composed of two transmembrane glycoprotein chains, α (34 kDa) and β (29 kDa), as shown schematically in panel d. Each chain has two domains, and the two chains together form a compact four-domain structure similar to that of the MHC class I molecule (compare with panel d of Fig. 3.15). Panel a shows a computer graphic representation of the surface of the MHC class II molecule, in this case the human protein HLA-DR1, and panel b shows the equivalent ribbon diagram. The α_2 and β_2 domains, like the α_3 and β_2-microglobulin domains of the MHC class I molecule, have amino acid sequence and structural similarities to immunoglobulin C domains; in the MHC class II molecule the two domains forming the peptide-binding cleft are contributed by different chains and are therefore not joined by a covalent bond (see panels c and d). Another important difference, not apparent in this diagram, is that the peptide-binding groove of the MHC class II molecule is open at both ends.

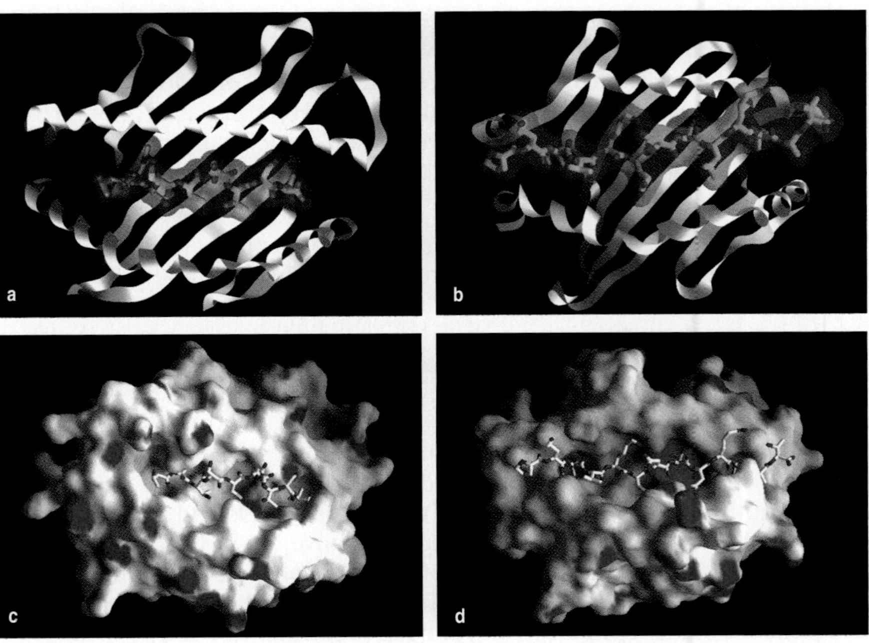

Fig. 3.17 MHC molecules bind peptides tightly within the cleft. When MHC molecules are crystallized with a single synthetic peptide antigen, the details of peptide binding are revealed. In MHC class I molecules (panels a and c) the peptide is bound in an elongated conformation with both ends tightly bound at either end of the cleft. In MHC class II molecules (panels b and d) the peptide is also bound in an elongated conformation but the ends of the peptide are not tightly bound and the peptide extends beyond the cleft. The upper surface of the peptide:MHC complex is recognized by T cells, and is composed of residues of the MHC molecule and the peptide. In representations c and d, the electrostatic potential of the MHC molecule surface is shown, with blue areas indicating a positive potential and red a negative potential.

An MHC class II molecule consists of a noncovalent complex of two chains, α and β, both of which span the membrane (see Fig. 3.16). The MHC class II α chain is a different protein from the class I α chain. The MHC class II α and β chains are both encoded within the MHC. The crystallographic structure of the MHC class II molecule shows that it is folded very much like the MHC class I molecule, but in MHC class II molecules the peptide-binding cleft is formed by two domains from different chains, the α_1 and β_1 domains. The major differences lie at the ends of the peptide-binding cleft, which are more open in MHC class II molecules than in MHC class I molecules. Consequently, the ends of a peptide bound to an MHC class I molecule are substantially buried within the molecule, whereas the ends of peptides bound to MHC class II molecules are not. In both MHC class I and class II molecules, bound peptides are sandwiched between the two α-helical segments of the MHC molecule (Fig. 3.17). The T-cell receptor interacts with this compound ligand, making contacts both with the MHC molecule and with the peptide antigen. The sites of major polymorphism in MHC class II molecules are again located in the peptide-binding cleft.

3-13 Peptides are stably bound to MHC molecules, and also serve to stabilize the MHC molecule on the cell surface.

An individual can be infected by a wide variety of pathogens, the proteins of which will not generally have peptide sequences in common. If T cells are to be alerted to all possible infections, the MHC molecules on each cell (both class I and class II) must be able to bind stably to many different peptides. This behavior is quite distinct from that of other peptide-binding receptors, such as those for peptide hormones, which usually bind only a single type of peptide. The crystal structures of peptide:MHC complexes have helped to show how a single binding site can bind peptides with high affinity while retaining the ability to bind a wide variety of different peptides.

An important feature of the binding of peptides to MHC molecules is that the peptide is bound as an integral part of the MHC molecule's structure, and MHC molecules are unstable when peptides are not bound. Stable peptide binding is important, because otherwise peptide exchanges occurring at the cell surface would prevent peptide:MHC complexes from being reliable indicators of infection or of uptake of a specific antigen. When MHC molecules are purified from cells, their stably bound peptides co-purify with them, and this has enabled the peptides bound by particular MHC molecules to be analyzed. Peptides are released from the MHC molecules by denaturing the complex in acid; they can then be purified and sequenced. Pure synthetic peptides can also be incorporated into previously empty MHC molecules and the structure of the complex determined, revealing details of the contacts between the MHC molecule and the peptide. From such studies a detailed picture of the binding interactions has been built up. We first discuss the peptide-binding properties of MHC class I molecules.

3-14 MHC class I molecules bind short peptides of 8–10 amino acids by both ends.

Binding of a peptide to an MHC class I molecule is stabilized at both ends of the peptide-binding cleft by contacts between atoms in the free amino and carboxy termini of the peptide and invariant sites that are found at each end of the cleft in all MHC class I molecules (Fig. 3.18). These are thought to be the main stabilizing contacts for peptide:MHC class I complexes because synthetic peptide analogs lacking terminal amino and carboxyl groups fail to bind stably to MHC class I molecules. Other residues in the peptide serve as additional anchors. Peptides that bind to MHC class I molecules are usually 8–10 amino acids long. Longer peptides are thought to be able to bind, particularly if they can bind at their carboxy terminus, but are subsequently cleaved by exopeptidases present in the endoplasmic reticulum, which is where MHC class I molecules bind peptides. The peptide lies in an elongated conformation along the cleft; variations in peptide length seem to be accommodated, in most cases, by a kinking in the peptide backbone. However, two examples of MHC class I molecules in which the peptide is able to extend out of the cleft at the carboxy terminus suggest that some length variation can also be accommodated in this way.

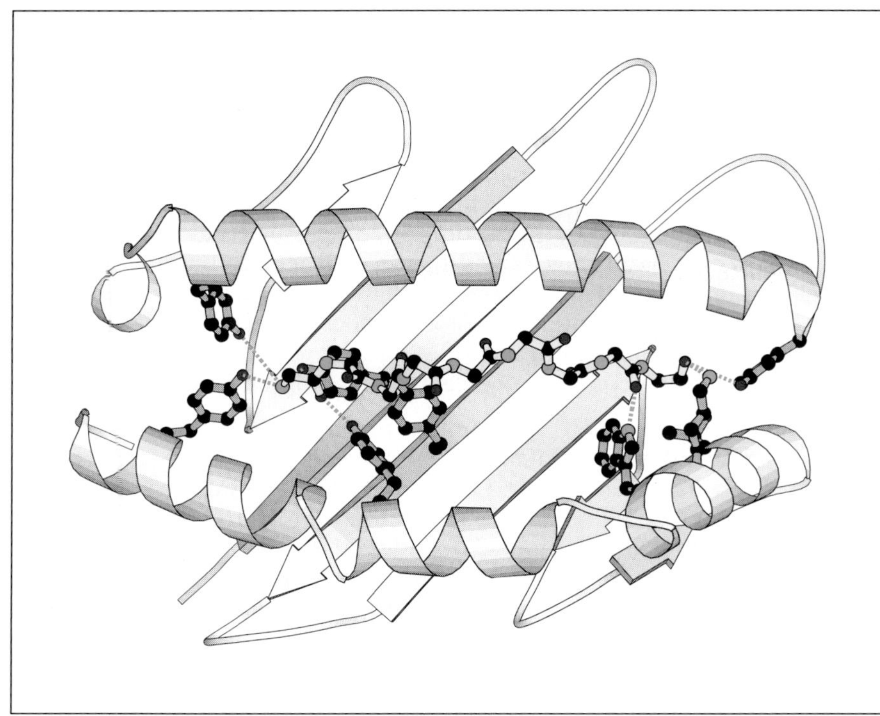

Fig. 3.18 Peptides are bound to MHC class I molecules by their ends. MHC class I molecules interact with the backbone of a bound peptide (shown in yellow) through a series of hydrogen bonds and ionic interactions (shown as dotted blue lines) at each end of the peptide. The amino terminus of the peptide is to the left, the carboxy terminus to the right. Black circles are carbon atoms; red are oxygen; blue are nitrogen. The amino acid residues in the MHC molecule that form these bonds are common to all MHC class I molecules, and their side chains are shown in full (in gray) on a ribbon diagram of the MHC class I groove. A cluster of tyrosine residues common to all MHC class I molecules forms hydrogen bonds to the amino terminus of the bound peptide, while a second cluster of residues forms hydrogen bonds and ionic interactions with the peptide backbone at the carboxy terminus and with the carboxy terminus itself.

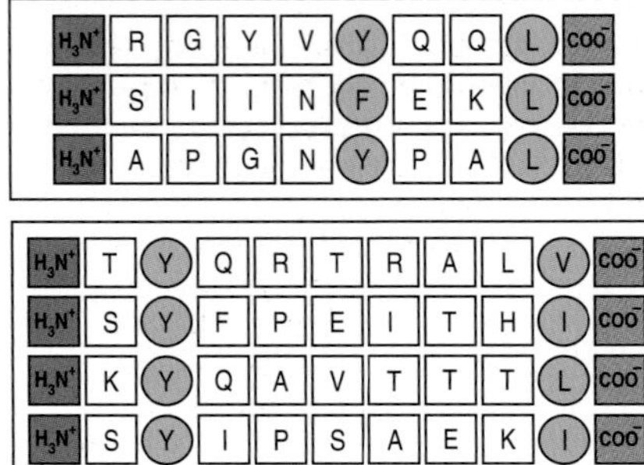

Fig. 3.19 Peptides bind to MHC molecules through structurally related anchor residues. Peptides eluted from two different MHC class I molecules are shown in the upper and lower panels, respectively. The anchor residues (green) differ for peptides that bind different alleles of MHC class I molecules but are similar for all peptides that bind to the same MHC molecule. The anchor residues that bind a particular MHC molecule need not be identical, but are always related (for example, phenylalanine (F) and tyrosine (Y) are both aromatic amino acids, whereas valine (V), leucine (L), and isoleucine (I) are all large hydrophobic amino acids). Peptides also bind to MHC class I molecules through their amino (blue) and carboxy (red) termini.

These interactions give all MHC class I molecules their broad peptide-binding specificity. In addition, MHC molecules are highly polymorphic. There are hundreds of different versions, or **alleles**, of the MHC class I genes in the human population as a whole, and each individual carries only a small selection. The main differences between the allelic MHC variants are found at certain sites in the peptide-binding cleft, resulting in different amino acids in key peptide-interaction sites in the different MHC variants. The consequence of this is that the different MHC variants preferentially bind different peptides. The peptides that can bind to a given MHC variant have the same or very similar amino acid residues at two or three particular positions along the peptide sequence. The amino acid side chains at these positions insert into pockets in the MHC molecule that are lined by the polymorphic amino acids. Because the binding of these side chains anchors the peptide to the MHC molecule, the peptide residues involved have been called **anchor residues**. Both the position and identity of these anchor residues can vary, depending on the particular MHC class I variant that is binding the peptide. However, most peptides that bind to MHC class I molecules have a hydrophobic (or sometimes basic) anchor residue at the carboxy terminus (Fig. 3.19). Whereas changing an anchor residue will in most cases prevent the peptide from binding, not every synthetic peptide of suitable length that contains these anchor residues will bind the appropriate MHC class I molecule, and so the overall binding must also depend on the nature of the amino acids at other positions in the peptide. In some cases, particular amino acids are preferred in certain positions, whereas in others the presence of particular amino acids prevents binding. These additional amino acid positions are called 'secondary anchors'. These features of peptide binding enable an individual MHC class I molecule to bind a wide variety of different peptides, yet allow different MHC class I allelic variants to bind different sets of peptides.

3-15 The length of the peptides bound by MHC class II molecules is not constrained.

Peptide binding to MHC class II molecules has also been analyzed by elution of bound peptides and by X-ray crystallography, and differs in several ways from peptide binding to MHC class I molecules. Peptides that bind to MHC class II molecules are at least 13 amino acids long and can be much longer. The clusters of conserved residues that bind the two ends of a peptide in MHC class I molecules are not found in MHC class II molecules, and the ends of the peptide are not bound. Instead, the peptide lies in an extended conformation along the MHC class II peptide-binding cleft. It is held there both by peptide side chains that protrude into shallow and deep pockets lined by polymorphic residues and by interactions between the peptide backbone and side chains of conserved amino acids that line the peptide-binding cleft in all MHC class II molecules (Fig. 3.20). Although there are fewer crystal structures of MHC class II-bound peptides than of peptides bound to MHC class I, the available data show that amino acid side chains at residues 1, 4, 6, and 9 of an MHC class II-bound peptide can be held in these binding pockets.

The binding pockets of MHC class II molecules accommodate a greater variety of side chains than those of MHC class I molecules, making it more difficult to define anchor residues and to predict which peptides will be able to bind a particular MHC class II molecule (Fig. 3.21). Nevertheless, by comparing the sequences of known binding peptides it is usually possible to detect a pattern of permissive amino acids for each of the different alleles of MHC class II molecules, and to model how the amino acids of this peptide sequence motif will interact with the amino acids that make up the peptide-binding cleft. Because the peptide is bound by its backbone and allowed to emerge from both ends of the binding groove, there is, in principle, no upper

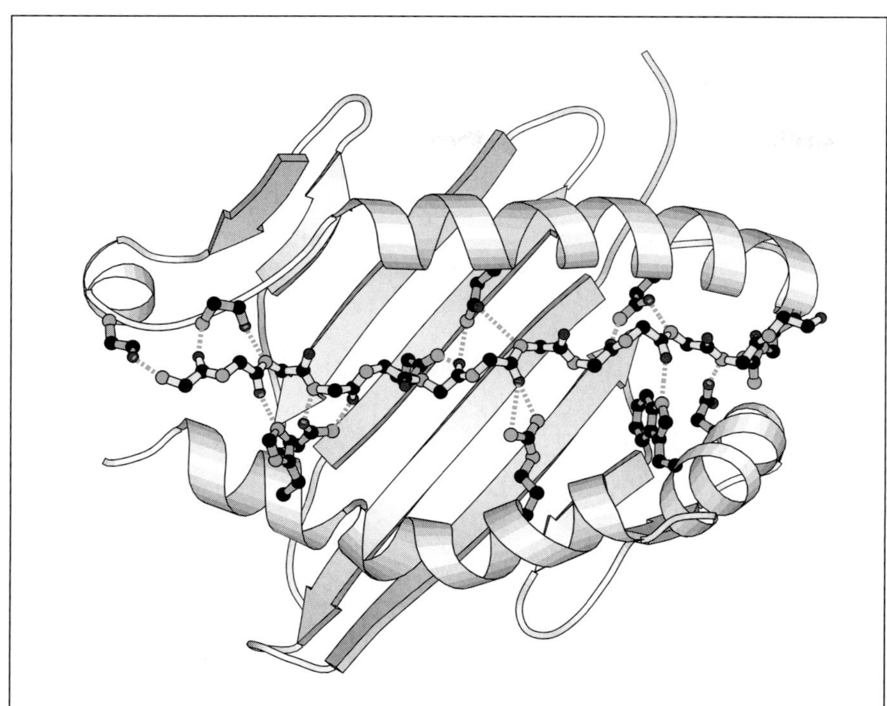

Fig. 3.20 Peptides bind to MHC class II molecules by interactions along the length of the binding groove. A peptide (yellow; shown as the peptide backbone only, with the amino terminus to the left and the carboxy terminus to the right) is bound by an MHC class II molecule through a series of hydrogen bonds (dotted blue lines) that are distributed along the length of the peptide. The hydrogen bonds toward the amino terminus of the peptide are made with the backbone of the MHC class II polypeptide chain, whereas throughout the peptide's length bonds are made with residues that are highly conserved in MHC class II molecules. The side chains of these residues are shown in gray on the ribbon diagram of the MHC class II groove.

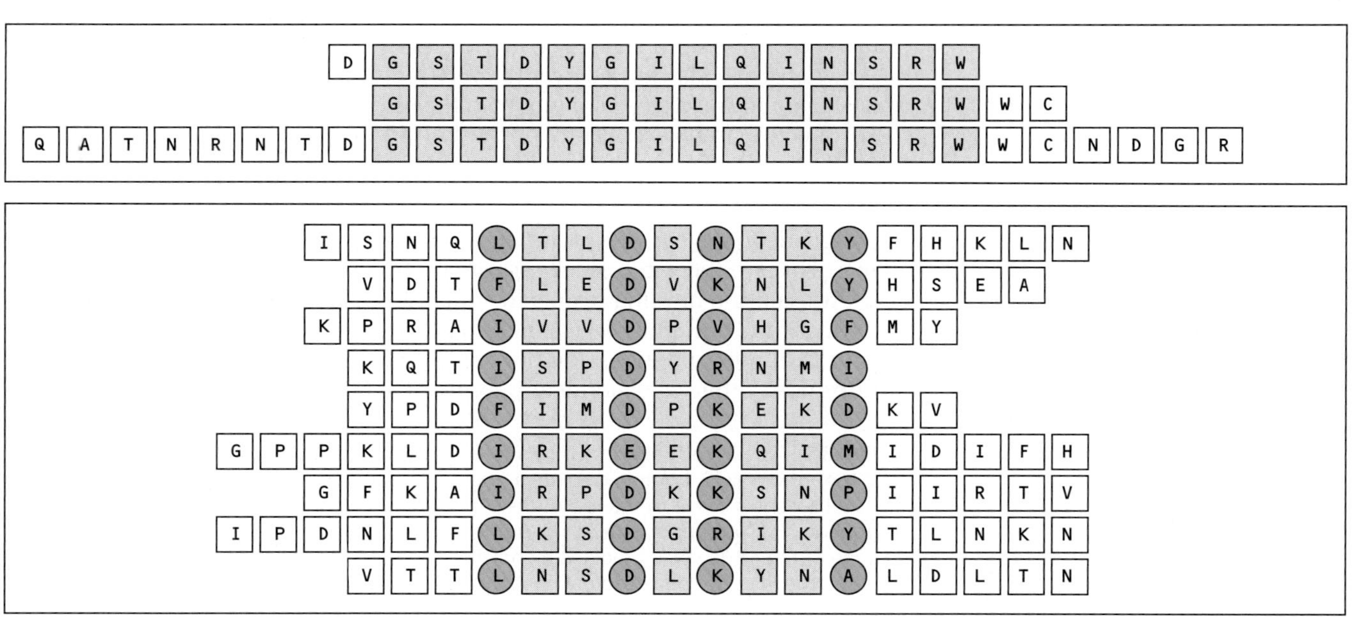

Fig. 3.21 Peptides that bind MHC class II molecules are variable in length and their anchor residues lie at various distances from the ends of the peptide. The sequences of a set of peptides that bind to the mouse MHC class II A^k allele are shown in the upper panel. All contain the same core sequence (shaded) but differ in length. In the lower panel, different peptides binding to the human MHC class II allele HLA-DR3 are shown. Anchor residues are shown as green circles. The lengths of these peptides can vary, and so by convention the first anchor residue is denoted as residue 1. Note that all of the peptides share a hydrophobic residue in position 1, a negatively charged residue (aspartic acid (D) or glutamic acid (E)) in position 4, and a tendency to have a basic residue (lysine (K), arginine (R), histidine (H), glutamine (Q), or asparagine (N)) in position 6 and a hydrophobic residue (for example, tyrosine (Y), leucine (L), phenylalanine (F)) in position 9

limit to the length of peptides that could bind to MHC class II molecules. However, it seems that longer peptides bound to MHC class II molecules are trimmed by peptidases to a length of 13–17 amino acids in most cases. Like MHC class I molecules, MHC class II molecules that lack bound peptide are unstable, but the critical stabilizing interactions that the peptide makes with the MHC class II molecule are not yet known.

3-16 The crystal structures of several peptide:MHC:T-cell receptor complexes show a similar T-cell receptor orientation over the peptide:MHC complex.

At the time that the first X-ray crystallographic structure of a T-cell receptor was published, a structure of the same T-cell receptor bound to a peptide:MHC class I ligand was also produced. This structure (Fig. 3.22), which had been forecast by site-directed mutagenesis of the MHC class I molecule, showed the T-cell receptor aligned diagonally over the peptide and the peptide-binding cleft, with the TCRα chain lying over the α_2 domain and the amino-terminal end of the bound peptide, the TCRβ chain lying over the α_1 domain and the carboxy-terminal end of the peptide, and the CDR3 loops of both TCRα and TCRβ chains meeting over the central amino acids of the peptide. The T-cell receptor is threaded through a valley between the two high points on the two surrounding α helices that form the walls of the peptide-binding cleft.

Analysis of other peptide:MHC class I:T-cell receptor complexes and of peptide:MHC class II:T-cell receptor complexes (Fig. 3.23) shows that all have a very similar orientation, particularly for the V_α domain, although some variability does occur in the location and orientation of the V_β domain. In this orientation, the V_α domain makes contact primarily with the amino terminus of the bound peptide, whereas the V_β domain contacts primarily the carboxy terminus of the bound peptide. Both chains also interact with the α helices of the MHC class I molecule (see Fig. 3.22). The T-cell receptor contacts are not symmetrically distributed over the MHC molecule: whereas the V_α CDR1 and CDR2 loops are in close contact with the helices of the peptide:MHC complex around the amino terminus of the bound peptide, the β-chain CDR1 and CDR2 loops, which interact with the complex at the carboxy terminus of the bound peptide, have variable contributions to the binding.

Comparison of the three-dimensional structure of an unliganded T-cell receptor and the same T-cell receptor complexed to its peptide:MHC ligand shows that the binding results in some degree of conformational change, or 'induced fit,' particularly within the V_α CDR3 loop. It has also been shown that subtly different peptides can have strikingly different effects on the recognition of an otherwise identical peptide:MHC ligand by the same T cell. The

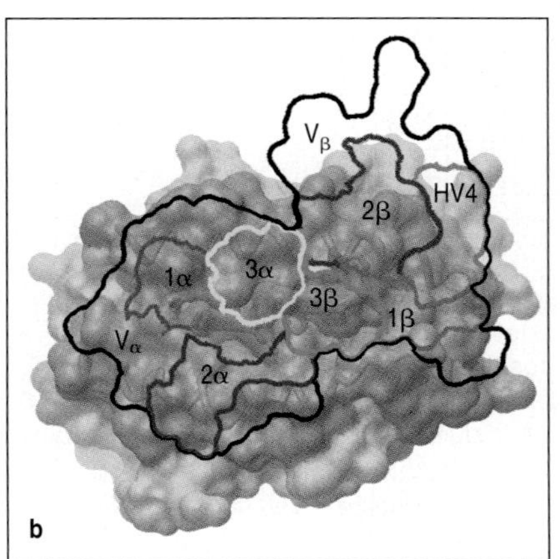

Fig. 3.22 The T-cell receptor binds to the peptide:MHC complex. Panel a: the T-cell receptor binds to the top of the peptide:MHC complex, straddling, in the case of the MHC class I molecule shown here, both the α_1 and α_2 domain helices. The CDRs of the T-cell receptor are indicated in color: the CDR1 and CDR2 loops of the β chain in light and dark blue, respectively; and the CDR1 and CDR2 loops of the α chain in light and dark purple, respectively. The α chain CDR3 loop is in yellow, and the β chain CDR3 loop is in green. The β chain HV4 loop is in red. The thick yellow line P1–P8 is the bound peptide. Panel b: the outline of the T-cell receptor antigen-binding site (thick black line) is superimposed on the top surface of the peptide:MHC complex (the peptide is shaded dull yellow). The T-cell receptor lies diagonally across the peptide:MHC complex, with the α and β CDR3 loops of the T-cell receptor (3α, 3β, yellow and green, respectively) contacting the center of the peptide. The α-chain CDR1 and CDR2 loops (1α, 2α, light and dark purple, respectively) contact the MHC helices at the amino terminus of the bound peptide, whereas the β-chain CDR1 and CDR2 loops (1β, 2β, light and dark blue, respectively) make contact with the helices at the carboxy terminus of the bound peptide. Courtesy of I.A. Wilson.

flexibility in the CDR3 loop demonstrated by these two structures helps to explain how the T-cell receptor can adopt conformations that can recognize related, but different, ligands.

From an examination of the available structures it is hard to predict whether the main binding energy is contributed by T-cell receptor contacts with the bound peptide, or by T-cell receptor contacts with the MHC molecule. Measurements of the kinetics of T-cell receptor binding to peptide:MHC ligands suggest that the interactions between the T-cell receptor and the MHC molecule might predominate at the start of the contact, guiding the receptor into the correct position where a second, more detailed, interaction with the peptide as well as the MHC molecule dictates the final outcome of the interaction—binding or dissociation. As with antibody–antigen interactions, only a few amino acids at the interface may provide the essential contacts that determine the specificity and strength of binding. It is known that alterations as simple as changing a leucine to isoleucine in the peptide are sufficient to alter a T-cell response from strong killing to absolutely no response at all. Studies show that mutations of single residues in the presenting MHC molecules can have the same effect. Thus, the specificity of T-cell recognition involves both the peptide and its presenting MHC molecule. This dual specificity underlies the MHC restriction of T-cell responses, a phenomenon that was observed long before the peptide-binding properties of MHC molecules were known. We will recount the story of how MHC restriction was discovered when we return to the issue of how MHC polymorphism affects antigen recognition by T cells in Chapter 5. Another consequence of this dual specificity is a need for T-cell receptors to be able to interact appropriately with the antigen-presenting surface of MHC molecules. It seems that some inherent specificity for MHC molecules is encoded in the T-cell receptor genes, and there is selection during T-cell development for a repertoire of receptors able to interact appropriately with the particular MHC molecules present in that individual. We discuss the evidence for this in Chapter 7.

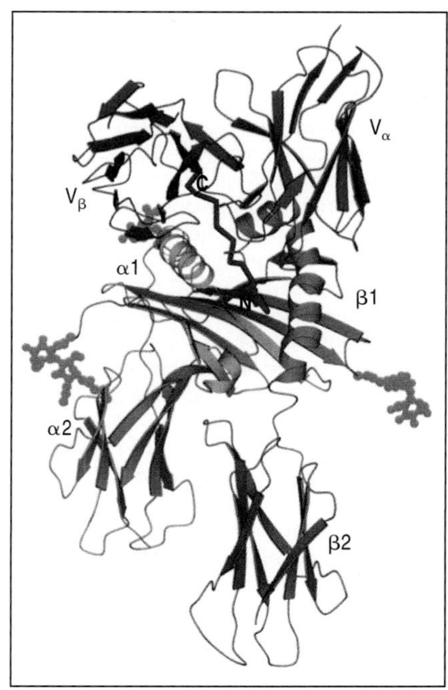

Fig. 3.23 The T-cell receptor interacts with MHC class I and MHC class II molecules in a similar fashion. The structure of a T-cell receptor binding to an MHC class II molecule has been determined, and shows the T-cell receptor binding to an equivalent site, and in an equivalent orientation, to the way in which T-cell receptors bind to MHC class I molecules (see Fig. 3.22). Only the V_α and V_β domains of the T-cell receptor are shown, colored in blue. The peptide is colored red, and carbohydrate residues are indicated in gray. The T-cell receptor sits in a shallow saddle formed between the α-helical regions of the MHC class II α (yellow-green) and β chain (orange), at roughly 90° to the long axis of the MHC class II molecule and the bound peptide. Courtesy of E.L. Reinherz and J-H. Wang.

3-17 The CD4 and CD8 cell-surface proteins of T cells are required to make an effective response to antigen.

As well as engaging a peptide:MHC ligand with its antigen receptor, T cells make additional interactions with the MHC molecule that stabilize the interaction and are required for the T cell to respond effectively to antigen. T cells fall into two major classes that have different effector functions and are distinguished by the expression of the cell-surface proteins **CD4** and **CD8**. CD8 is carried by cytotoxic T cells, while CD4 is carried by T cells whose function it is to activate other cells (see Section 1-19). CD4 and CD8 were known as markers for these different functional sets of T cells for some time before it became clear that this distinction was based on their ability to recognize the different classes of MHC molecules: CD8 recognizes MHC class I molecules and CD4 recognizes MHC class II molecules. During antigen recognition, CD4 or CD8 molecules (depending on the type of T cell) associate on the T-cell surface with the T-cell receptor and bind to invariant sites on the MHC portion of the composite peptide:MHC ligand, away from the peptide-binding site. This binding is required for the T cell to make an effective response, and so CD4 and CD8 are called **co-receptors**.

CD4 is a single-chain molecule composed of four immunoglobulin-like domains (Fig. 3.24). The first two domains (D1 and D2) of the CD4 molecule are packed tightly together to form a rigid rod about 60 Å long, which is joined by a flexible hinge to a similar rod formed by the third and fourth domains (D3 and D4). There is some evidence that CD4 forms homodimers on the T cell suface that are functional in recognition of MHC class II, although the structural basis of their formation is still uncertain. CD4 binds MHC class II mole-

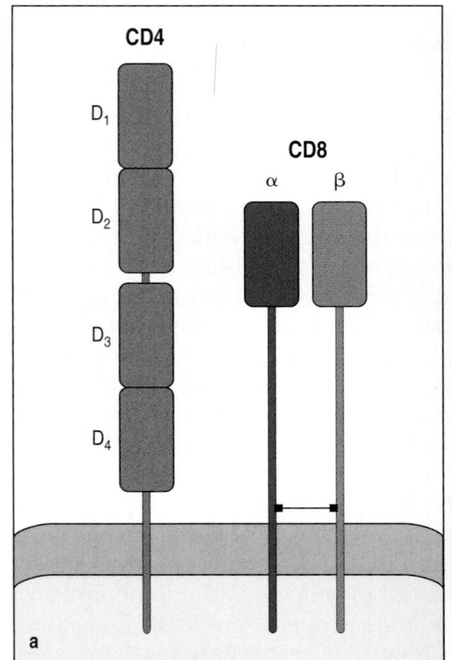

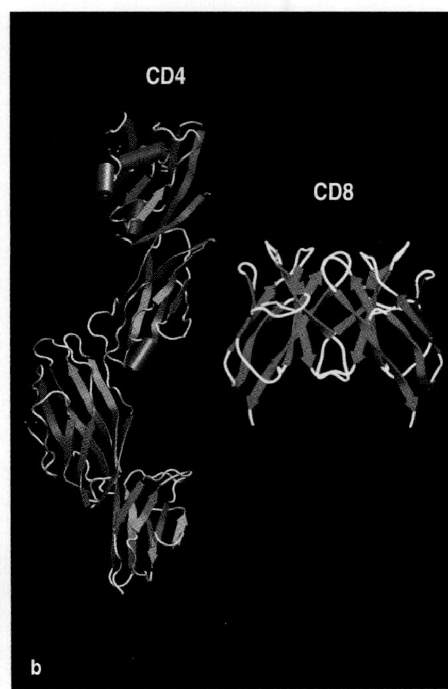

Fig. 3.24 The structures of the CD4 and CD8 co-receptor molecules. The CD4 molecule contains four immunoglobulin-like domains, shown in schematic form in panel a, and as a ribbon diagram of the crystal structure in panel b. The amino-terminal domain, D_1, is similar in structure to an immunoglobulin V domain. The second domain, D_2, although clearly related to an immunoglobulin domain, is different from both V and C domains and has been termed a C2 domain. The first two domains of CD4 form a rigid rodlike structure that is linked to the two carboxy-terminal domains by a flexible link. The binding site for MHC class II molecules is thought to involve mainly the D_1 domain. The CD8 molecule is a heterodimer of an α and a β chain covalently linked by a disulfide bond; an alternative form of CD8 exists as a homodimer of α chains. The heterodimer is depicted in panel a, whereas the ribbon diagram in panel b is of the homodimer. CD8α and CD8β chains have very similar structures, each having a single domain resembling an immunoglobulin V domain and a stretch of polypeptide chain, believed to be in a relatively extended conformation, that anchors the V-like domain to the cell membrane.

cules through a region that is located mainly on a lateral face of the first domain, D1. CD4 binds to a hydrophobic crevice formed at the junction of the α_2 and β_2 domains of the MHC class II molecule. This site is well away from the site where the T-cell receptor binds (Fig. 3.25a), and so the CD4 molecule and the T-cell receptor can bind simultaneously to the same peptide:MHC class II complex. The intracellular portion of CD4 interacts strongly with a cytoplasmic tyrosine kinase called Lck, and brings Lck close to the intracellular signaling components associated with the T-cell receptor. This results in enhancement of the signal that is generated when the T-cell receptor binds its ligand, as we discuss in Chapter 6. When CD4 and the T-cell receptor simultaneously bind to the same MHC class II:peptide complex, the T cell is about a hundredfold more sensitive to the antigen than if CD4 were absent.

The structure of CD8 is quite different. It is a disulfide-linked dimer of two different chains, called α and β, each containing a single immunoglobulin-like domain linked to the membrane by a segment of extended polypeptide (see Fig. 3.24). This segment is extensively glycosylated, which is thought to maintain it in an extended conformation and protect it from cleavage by proteases. CD8α chains can form homodimers, although these are not found when the CD8β chains are present. The CD8α homodimer may have a specific function in recognizing a specialized subset of nonclassical MHC class I molecules that we describe in Chapter 5.

CD8 binds weakly to an invariant site in the α_3 domain of an MHC class I molecule (see Fig. 3.25b). Although only the interaction of the CD8α homodimer with MHC class I is known in detail, this shows that the MHC class I binding site of the CD8 α:β heterodimer is formed by the interaction of the CD8α and β chains. In addition, CD8 (most probably through the α chain) interacts with residues in the base of the α_2 domain of the MHC class I molecule. The strength of binding of CD8 to the MHC class I molecule is influenced by the glycosylation of the CD8 molecule; increased numbers of sialic acid residues added to the CD8 carbohydrate structures decrease the strength of the interaction. The pattern of sialylation of CD8 changes during the maturation of T cells and also on activation, and this is likely to have a role in modulating antigen recognition.

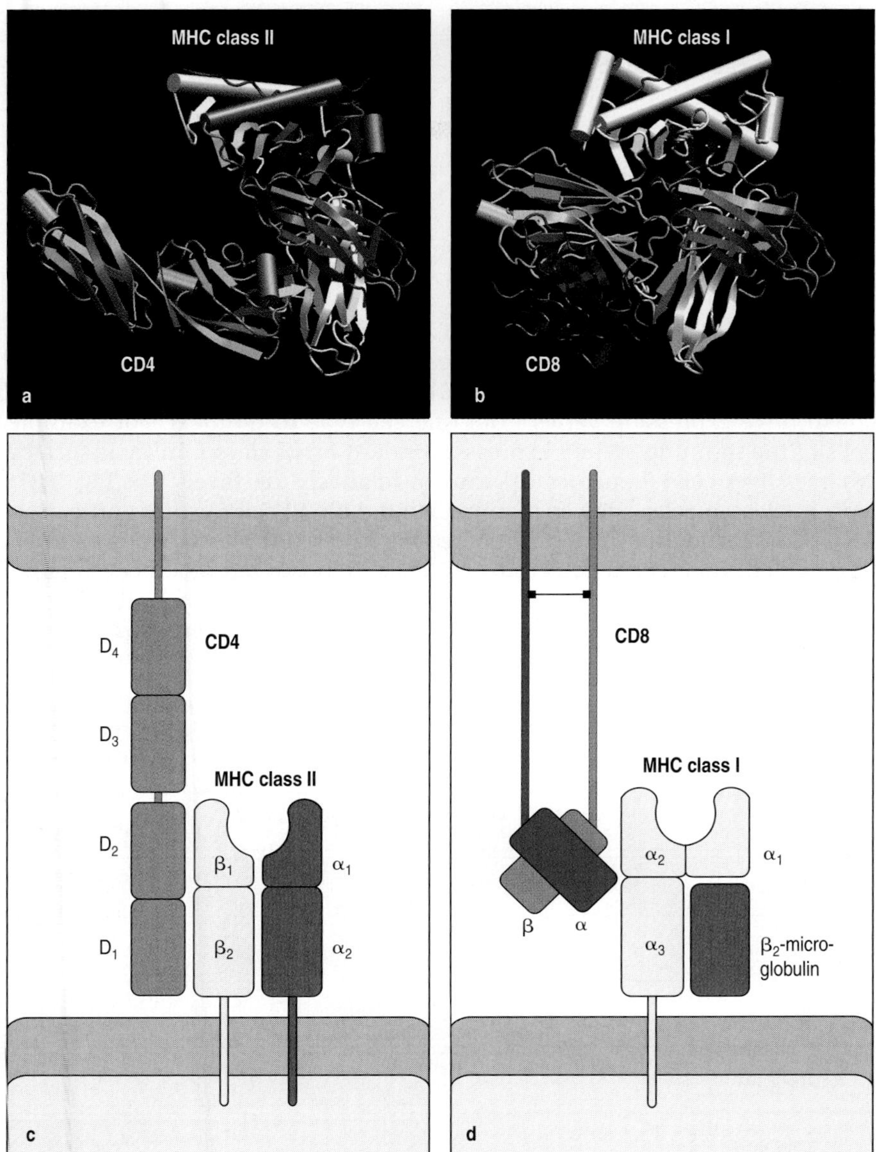

Fig. 3.25 The binding sites for CD4 and CD8 on MHC class II and class I molecules lie in the immunoglobulin-like domains. The binding sites for CD4 and CD8 on the MHC class II and class I molecules, respectively, lie in the immunoglobulin-like domains nearest to the membrane and distant from the peptide-binding cleft. The binding of CD4 to an MHC class II molecule is shown as a structure graphic in panel a and schematically in panel c. The α chain of the MHC class II molecule is shown in pink, and the β chain in white, while CD4 is in gold. Only the D_1 and D_2 domains of the CD4 molecule are shown in panel a. The binding site for CD4 lies at the base of the β_2 domain of an MHC class II molecule, in the hydrophobic crevice between the β_2 and α_2 domains. The binding of CD8 to an MHC class I molecule is shown in panel b and schematically in panel d. The class I heavy chain and β_2-microglobulin are shown in white and pink, respectively, and the two chains of the CD8 dimer are shown in light and dark purple. The structure is actually of the binding of the CD8α homodimer, but the CD8α:β heterodimer is believed to bind in the same way. The binding site for CD8 on the MHC class I molecule lies in a similar position to that of CD4 in the MHC class II molecule, but CD8 binding also involves the base of the α_1 and α_2 domains, and thus the binding of CD8 to MHC class I is not completely equivalent to the binding of CD4 to MHC class II.

By binding to the membrane-proximal domains of the MHC class I and class II molecules, the co-receptors leave the upper surface of the MHC molecule exposed and free to interact with a T-cell receptor, as shown for CD8 in Fig. 3.26. Both CD4 and CD8 bind Lck—in the case of the CD8α:β heterodimer through the cytoplasmic tail of the α chain—and bring it into close proximity to the T-cell receptor. And as with CD4, the presence of CD8 increases the sensitivity of T cells to antigen presented by MHC class I molecules about a hundredfold. Thus, CD4 and CD8 have similar functions and bind to the same approximate location in MHC class I and MHC class II molecules, even though the structures of the two co-receptor proteins are only distantly related.

3-18 The two classes of MHC molecules are expressed differentially on cells.

MHC class I and MHC class II molecules have distinct distributions among cells that reflect the different effector functions of the T cells that recognize them (Fig. 3.27). MHC class I molecules present peptides from pathogens,

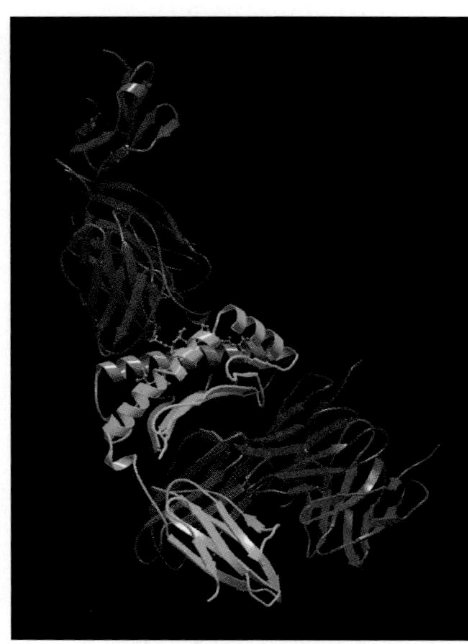

Fig. 3.26 CD8 binds to a site on MHC class I molecules distant from that to which the T-cell receptor binds. The relative positions of the T-cell receptor and CD8 molecules bound to the same MHC class I molecule can be seen in this hypothetical reconstruction of the interaction of an MHC class I molecule (the α chain is shown in green; β_2-microglobulin (dull yellow) can be seen faintly in the background) with a T-cell receptor and CD8. The α and β chains of the T-cell receptor are shown in pink and purple, respectively. The CD8 structure is that of a CD8α homodimer, but is colored to represent the likely orientation of the subunits in the heterodimer, with the CD8α subunit in red and the CD8β subunit in blue. Courtesy of G. Gao.

commonly viruses, to CD8 cytotoxic T cells, which are specialized to kill any cell that they specifically recognize. Because viruses can infect any nucleated cell, almost all such cells express MHC class I molecules, although the level of constitutive expression varies from one cell type to the next. For example, cells of the immune system express abundant MHC class I on their surface, whereas liver cells (hepatocytes) express relatively low levels (see Fig. 3.27). Non-nucleated cells, such as mammalian red blood cells, express little or no MHC class I, and thus the interior of red blood cells is a site in which an infection can go undetected by cytotoxic T cells. Because red blood cells cannot support viral replication, this is of no great consequence for viral infection, but it might be the absence of MHC class I that allows the *Plasmodium* parasites that cause malaria to live in this privileged site.

In contrast, the main function of the CD4 T cells that recognize MHC class II molecules is to activate other effector cells of the immune system. Thus MHC

Fig. 3.27 The expression of MHC molecules differs between tissues. MHC class I molecules are expressed on all nucleated cells, although they are most highly expressed in hematopoietic cells. MHC class II molecules are normally expressed only by a subset of hematopoietic cells and by thymic stromal cells, although they may be expressed by other cell types on exposure to the inflammatory cytokine IFN-γ.
*In humans, activated T cells express MHC class II molecules, whereas in mice all T cells are MHC class II-negative.
†In the brain, most cell types are MHC class II-negative, but microglia, which are related to macrophages, are MHC class II-positive.

Tissue	MHC class I	MHC class II
Lymphoid tissues		
T cells	+++	+*
B cells	+++	+++
Macrophages	+++	++
Dendritic cells	+++	+++
Epithelial cells of the thymus	+	+++
Other nucleated cells		
Neutrophils	+++	–
Hepatocytes	+	–
Kidney	+	–
Brain	+	–†
Non-nucleated cells		
Red blood cells	–	–

class II molecules are normally found on B lymphocytes, dendritic cells, and macrophages—cells that participate in immune responses—but not on other tissue cells (see Fig. 3.27). When CD4 T cells recognize peptides bound to MHC class II molecules on B cells, they stimulate the B cells to produce antibody. Similarly, CD4 T cells recognizing peptides bound to MHC class II molecules on macrophages activate these cells to destroy the pathogens in their vesicles. We shall see in Chapter 8 that MHC class II molecules are also expressed on specialized antigen-presenting cells, the dendritic cells, in lymphoid tissues where naive T cells encounter antigen and are first activated. The expression of both MHC class I and MHC class II molecules is regulated by cytokines, in particular interferons, released in the course of immune responses. Interferon-γ (IFN-γ), for example, increases the expression of MHC class I and MHC class II molecules, and can induce the expression of MHC class II molecules on certain cell types that do not normally express them. Interferons also enhance the antigen-presenting function of MHC class I molecules by inducing the expression of key components of the intracellular machinery that enables peptides to be loaded onto the MHC molecules.

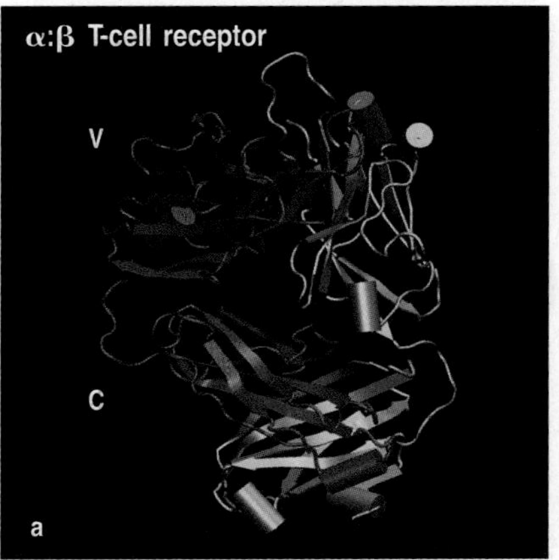

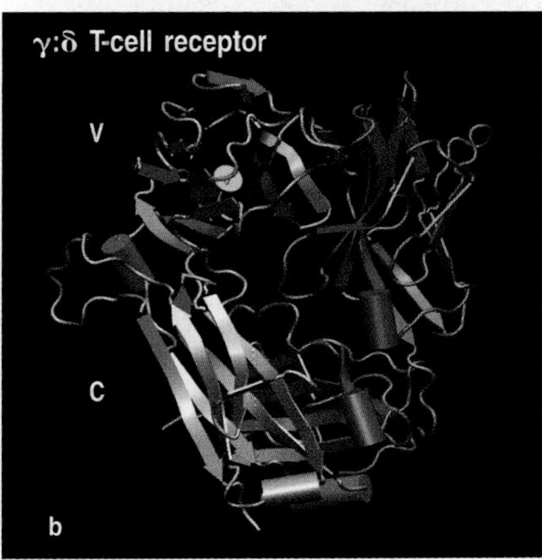

Fig. 3.28 Structures of α:β and γ:δ T-cell receptors. The structures of the α:β and the γ:δ T-cell receptors have both been determined by X-ray crystallography. The α:β T-cell receptor is shown in panel a, with the α chain colored red and the β chain blue. Panel b shows the γ:δ receptor, with the γ chain colored purple and the δ chain pink. Both receptors have very similar structures, somewhat resembling that of a Fab fragment of an immunoglobulin molecule. The C_δ domain is more like an immunoglobulin domain than is the corresponding C_α domain of the α:β T-cell receptor.

3-19 A distinct subset of T cells bears an alternative receptor made up of γ and δ chains.

During the search for the gene for the TCRα chain, another T-cell receptor-like gene was unexpectedly discovered. This gene was named TCRγ, and its discovery led to a search for further T-cell receptor genes. Another receptor chain was identified by using antibody against the predicted sequence of the γ chain and was called the δ chain. It was soon discovered that a minority population of T cells bore a distinct type of T-cell receptor made up of γ:δ heterodimers rather than α:β heterodimers. The development of these cells is described in Sections 7-11 and 7-12.

The crystallographic structure of a γ:δ T-cell receptor reveals that, as expected, it is similar in shape to α:β T-cell receptors (Fig. 3.28). γ:δ T-cell receptors may be specialized to bind certain kinds of ligands, including heat-shock proteins and nonpeptide ligands such as phosphorylated ligands or mycobacterial lipid antigens. It seems likely that γ:δ T-cell receptors are not restricted by the 'classical' MHC class I and class II molecules. They may bind the free antigen, much as immunoglobulins do, and/or they may bind to peptides or other antigens presented by nonclassical MHC-like molecules. These are proteins that resemble MHC class I molecules but are relatively nonpolymorphic and are described in Chapter 5. We still know little about how γ:δ T-cell receptors bind antigen and thus how these cells function, and what their role is in immune responses. The structure and rearrangement of the genes for γ:δ T-cell receptors is covered in Sections 4-11 and 7-12.

Summary.

The receptor for antigen on most T cells, the α:β T-cell receptor, is composed of two protein chains, TCRα and TCRβ, and resembles in many respects a single Fab fragment of immunoglobulin. T-cell receptors are always membrane-bound. α:β T-cell receptors do not recognize antigen in its native state, as do the immunoglobulin receptors of B cells, but recognize a composite ligand of a peptide antigen bound to an MHC molecule. MHC molecules are highly polymorphic glycoproteins encoded by genes in the major histocompatibility complex (MHC). Each MHC molecule binds a wide variety of different peptides, but the different variants each preferentially recognize sets of peptides with particular sequence and physical features. The peptide antigen is generated intracellularly, and bound stably in a peptide-binding cleft on the surface of the MHC molecule. There are two classes of MHC molecule, and these are bound in their nonpolymorphic domains by CD8 and CD4 molecules that distinguish two different functional classes of α:β T cells. CD8 binds MHC class I molecules and can bind simultaneously to the same class I peptide:MHC complex being recognized by a T-cell receptor, thus acting as a co-receptor and enhancing the T-cell response; CD4 binds MHC class II molecules and acts as a co-receptor for T-cell receptors that recognize class II peptide:MHC ligands. A T-cell receptor interacts directly both with the antigenic peptide and with polymorphic features of the MHC molecule that displays it, and this dual specificity underlies the MHC restriction of T-cell responses. A second type of T-cell receptor, composed of a γ and a δ chain, is structurally similar to the α:β T-cell receptor but seems to bind different ligands, including nonpeptide ligands. It is thought not to be MHC restricted. It is found on a minority population of T cells, the γ:δ T cells.

Summary to Chapter 3.

B cells and T cells use different, but structurally similar, molecules to recognize antigen. The antigen-recognition molecules of B cells are immunoglobulins, and are made both as a membrane-bound receptor for antigen, the B-cell receptor, and as secreted antibodies that bind antigens and elicit humoral effector functions. The antigen-recognition molecules of T cells, in contrast, are made only as cell-surface receptors. Immunoglobulins and T-cell receptors are highly variable molecules, with the variability concentrated in that part of the molecule, the variable (V) region, that binds to antigen. Immunoglobulins bind a wide variety of chemically different antigens, whereas the major α:β type of T-cell receptor predominantly recognizes peptide fragments of foreign proteins bound to the MHC molecules that are ubiquitous on cell surfaces.

Binding of antigen by immunoglobulins has chiefly been studied with antibodies. The binding of antibody to its corresponding antigen is highly specific, and this specificity is determined by the shape and physicochemical properties of the antigen-binding site. The part of the antibody that elicits effector functions, once the variable part has bound an antigen, is located at the other end of the molecule from the antigen-binding sites, and is termed the constant region. There are five main functional classes of antibodies, each encoded by a different type of constant region. As we will see in Chapter 9, these interact with different components of the immune system to incite an inflammatory response and eliminate the antigen.

T-cell receptors differ in several respects from the B-cell immunoglobulins. One is the absence of a secreted form of the receptor. This reflects the functional differences between T cells and B cells. B cells deal with pathogens and their protein products circulating within the body; secretion of a soluble antigen-recognition molecule by the activated B cell after antigen has been encountered enables them to mop up antigen effectively throughout the extracellular spaces of the body. T cells, in contrast, are specialized for cell–cell interactions. They either kill cells that are infected with intracellular pathogens and that bear foreign antigenic peptides on their surface, or interact with cells of the immune system that have taken up foreign antigen and are displaying it on the cell surface. They thus have no requirement for a soluble, secreted receptor.

The second distinctive feature of the T-cell receptor is that it recognizes a composite ligand made up of the foreign peptide bound to a self-MHC molecule.

This means that T cells can interact only with a body cell displaying the antigen, not with the intact pathogen or protein. Each T-cell receptor is specific for a particular combination of peptide and a self-MHC molecule.

MHC molecules are encoded by a family of highly polymorphic genes; although each individual expresses several of these genes, this represents only a small selection of all possible variants. During T-cell development, the T-cell receptor repertoire is selected so that the T cells of each individual recognize antigen only in conjunction with their own MHC molecules. Expression of multiple variant MHC molecules, each with a different peptide-binding repertoire, helps to ensure that T cells from an individual will be able to recognize at least some peptides generated from nearly every pathogen.

Questions.

3.1 *The immunoglobulin superfamily is one of the most abundant families of protein domain structures. (a) What are the characteristics of an immunoglobulin domain and how do the various subtypes of these domains differ? (b) What regions of the V-type immunoglobulin domain contributes to its complementarity determining regions (CDRs) and how do the V-type and C-type immunoglobulin domains differ in those regions.*

3.2 *How do antibodies, which all have the same basic shape, recognize antigens of a wide variety of different shapes?*

3.3 *T cells must deliver effector functions appropriate for the cellular localization of pathogens, whereas B cells are not so constrained. (a) How does this account for the different recognition properties of B- and T-cell antigen receptors? (b) Describe the similarities and differences between B- and T-cell antigen receptors. (c) Given these differences, what is the essential difference in the function of B cells and T cells.*

3.4 *There are two kinds of MHC molecules: class I and class II. (a) What role do MHC molecules have in the activation of antigen-specific T cells? (b) Explain how the peptide-binding region of MHC class I and class II molecules can be so similar, even though one is encoded by a single gene and the other is encoded by two different genes. (c) If MHC class I and class II peptide-binding regions are so similar, how can T cells functionally distinguish between antigens presented by MHC class I and class II molecules?*

General references.

Ager, A., Callard, R., Ezine, S., Gerard, C., and Lopez-Botet, M.: **Immune receptor Supplement.** *Immunol. Today* 1996, 17.

Davies, D.R., and Chacko, S.: **Antibody structure.** *Acc. Chem. Res.* 1993, **26**:421–427.

Frazer, K., and Capra, J.D.: **Immunoglobulins: structure and function**, in Paul W.E. (ed): *Fundamental Immunology*, 4th ed. New York, Raven Press, 1998.

Garcia, K.C., Teyton, L., and Wilson, I.A.: **Structural basis of T cell recognition.** *Annu. Rev. Immunol.* 1999, **17**:369–397.

Germain, R.N.: **MHC-dependent antigen processing and peptide presentation: providing ligands for T lymphocyte activation.** *Cell* 1994, **76**:287–299.

Honjo, T., and Alt, F.W. (eds): *Immunoglobulin Genes,* 2nd ed. London, Academic Press, 1996.

Moller, G. (ed): **Origin of major histocompatibility complex diversity.** *Immunol. Rev.* 1995, **143**:5–292.

Poljak, R.J.: **Structure of antibodies and their complexes with antigens.** *Mol. Immunol.* 1991, **28**:1341–1345.

Section references.

3-1 IgG antibodies consist of four polypeptide chains.

Edelman, G.M.: **Antibody structure and molecular immunology.** *Scand. J. Immunol.* 1991, **34**:4–22.

Faber, C., Shan, L., Fan, Z., Guddat, L.W., Furebring, C., Ohlin, M., Borrebaeck, C.A.K., and Edmundson, A.B.: **Three-dimensional structure of a human Fab with high affinity for tetanus toxoid.** *Immunotechnology* 1998, **3**:253–270.

Harris, L.J., Larson, S.B., Hasel, K.W., Day, J., Greenwood, A., and McPherson, A.: **The three-dimensional structure of an intact monoclonal antibody for canine lymphoma.** *Nature* 1992, **360**:369–372.

3-2 Immunoglobulin heavy and light chains are composed of constant and variable regions.

Han, W.H., Mou, J.X., Sheng, J., Yang, J., and Shao, Z.F.: **Cryo-atomic force microscopy— new approach for biological imaging at high resolution.** *Biochemistry* 1995, **34**:8215–8220.

3-3 The antibody molecule can readily be cleaved into functionally distinct fragments.

Porter, R.R.: **Structural studies of immunoglobulins.** *Scand. J. Immunol.* 1991, **34**:382–389.

Yamaguchi, Y., Kim, H., Kato, K., Masuda, K., Shimada, I., and Arata, Y.: **Proteolytic fragmentation with high specificity of mouse IgG—mapping of proteolytic cleavage sites in the hinge region.** *J. Immunol. Meth.* 1995, **181**:259–267.

3-4 The immunoglobulin molecule is flexible, especially at the hinge region.

Gerstein, M., Lesk, A.M., and Chothia, C.: **Structural mechanisms for domain movements in proteins.** *Biochemistry* 1994, **33**:6739–6749.

Jimenez, R., Salazar, G., Baldridge, K.K., and Romesberg, F.E.: **Flexibility and molecular recognition in the immune system.** *Proc. Natl Acad. Sci. USA* 2003, **100**:92–97.

Saphire, E.O., Stanfield, R.L., Crispin, M.D., Parren, P.W., Rudd, P.M., Dwek, R.A., Burton, D.R., and Wilson, I.A.: **Contrasting IgG structures reveal extreme asymmetry and flexibility.** *J. Mol. Biol.* 2002, **319**:9–18.

3-5 The domains of an immunoglobulin molecule have similar structures.

Barclay, A.N., Brown, M.H., Law, S.K., McKnight, A.J., Tomlinson, M.G., and van der Merwe, P.A. (eds): *The Leukocyte Antigen Factsbook*, 2nd ed. London, Academic Press, 1997.

Brummendorf, T., and Lemmon, V.: **Immunoglobulin superfamily receptors: *cis*-interactions, intracellular adapters and alternative splicing regulate adhesion.** *Curr. Opin. Cell Biol.* 2001, **13**:611–618.

Marchalonis, J.J., Jensen, I., and Schluter, S.F.: **Structural, antigenic and evolutionary analyses of immunoglobulins and T cell receptors.** *J. Mol. Recog.* 2002, **15**:260–271.

Ramsland, P.A., and Farrugia, W.: **Crystal structures of human antibodies: a detailed and unfinished tapestry of immunoglobulin gene products.** *J. Mol. Recog.* 2002, **15**:248–259.

3-6 Localized regions of hypervariable sequence form the antigen-binding site.

Chitarra, V., Alzari, P.M., Bentley, G.A., Bhat, T.N., Eisele, J.L., Houdusse, A., Lescar, J., Souchon, H., and Poljak, R.J.: **Three-dimensional structure of a heteroclitic antigen-antibody cross-reaction complex.** *Proc. Natl Acad. Sci. USA* 1993, **90**:7711–7715.

Decanniere, K., Muyldermans, S., and Wyns, L.: **Canonical antigen-binding loop structures in immunoglobulins: more structures, more canonical classes?** *J. Mol. Biol.* 2000, **300**:83–91.

Gilliland, L.K., Norris, N.A., Marquardt, H., Tsu, T.T., Hayden, M.S., Neubauer, M.G., Yelton, D.E., Mittler, R.S., and Ledbetter, J.A.: **Rapid and reliable cloning of antibody variable regions and generation of recombinant single-chain antibody fragments.** *Tissue Antigens* 1996, **47**:1–20.

Johnson, G., and Wu, T.T.: **Kabat Database and its applications: 30 years after the first variability plot.** *Nucleic Acids Res.* 2000, **28**:214–218.

Wu, T.T., and Kabat, E.A.: **An analysis of the sequences of the variable regions of Bence Jones proteins and myeloma light chains and their implications for antibody complementarity.** *J. Exp. Med.* 1970, **132**:211–250.

Xu, J., Deng, Q., Chen, J., Houk, K.N., Bartek, J., Hilvert, D., and Wilson, I.A.: **Evolution of shape complementarity and catalytic efficiency from a primordial antibody template.** *Science* 1999, **286**:2345–2348.

3-7 Antibodies bind antigens via contacts with amino acids in CDRs, but the details of binding depend upon the size and shape of the antigen.

&

3-8 Antibodies bind to conformational shapes on the surfaces of antigens.

Ban, N., Day, J., Wang, X., Ferrone, S., and McPherson, A.: **Crystal structure of an anti-anti-idiotype shows it to be self-complementary.** *J. Mol. Biol.* 1996, **255**:617–627.

Davies, D.R., and Cohen, G.H.: **Interactions of protein antigens with antibodies.** *Proc. Natl Acad. Sci. USA* 1996, **93**:7–12.

Decanniere, K., Desmyter, A., Lauwereys, M., Ghahroudi, M.A., Muyldermans, S., and Wyns, L.: **A single-domain antibody fragment in complex with RNase A: non-canonical loop structures and nanomolar affinity using two CDR loops.** *Structure Fold. Des.* 1999, **7**:361–370.

Padlan, E.A.: **Anatomy of the antibody molecule.** *Mol. Immunol.* 1994, **31**:169–217.

Saphire, E.O., Parren, P.W., Pantophlet, R., Zwick, M.B., Morris, G.M., Rudd, P.M., Dwek, R.A., Stanfield, R.L., Burton, D.R., and Wilson, I.A.: **Crystal structure of a neutralizing human IGG against HIV-1: a template for vaccine design.** *Science* 2001, **293**:1155–1159.

Stanfield, R.L., and Wilson, I.A.: **Protein–peptide interactions.** *Curr. Opin. Struct. Biol.* 1995, **5**:103–113.

Tanner, J.J., Komissarov, A.A., and Deutscher, S.L.: **Crystal structure of an antigen-binding fragment bound to single-stranded DNA.** *J. Mol. Biol.* 2001, **314**:807–822.

Wilson, I.A., and Stanfield, R.L.: **Antibody–antigen interactions: new structures and new conformational changes.** *Curr. Opin. Struct. Biol.* 1994, **4**:857-867.

3-9 Antigen–antibody interactions involve a variety of forces.

Braden, B.C., and Poljak, R.J.: **Structural features of the reactions between antibodies and protein antigens.** *FASEB J.* 1995, **9**:9–16.

Braden, B.C., Goldman, E.R., Mariuzza, R.A., and Poljak, R.J.: **Anatomy of an antibody molecule: structure, kinetics, thermodynamics and mutational studies of the antilysozyme antibody D1.3.** *Immunol. Rev.* 1998, **163**:45–57.

Ros, R., Schwesinger, F., Anselmetti, D., Kubon, M., Schäfer, R., Plückthun, A., and Tiefenauer, L.: **Antigen binding forces of individually addressed single-chain Fv antibody molecules.** *Proc. Natl Acad. Sci. USA* 1998, **95**:7402–7405.

3-10 The T-cell receptor is very similar to a Fab fragment of immunoglobulin.

Al-Lazikani, B., Lesk, A.M., and Chothia, C.: **Canonical structures for the hypervariable regions of T cell $\alpha\beta$ receptors.** *J. Mol. Biol.* 2000, **295**:979–995.

Kjer-Nielsen, L., Clements, C.S., Brooks, A.G., Purcell, A.W., McCluskey, J., and Rossjohn, J.: **The 1.5 Å crystal structure of a highly selected antiviral T cell receptor provides evidence for a structural basis of immunodominance.** *Structure (Camb.)* 2002, **10**:1521–1532.

Machius, M., Cianga, P., Deisenhofer, J., and Ward, E.S.: **Crystal structure of a T cell receptor Vα11 (AV11S5) domain: new canonical forms for the first and second complementarity determining regions.** *J. Mol. Biol.* 2001, **310**:689–698.

3-11 A T-cell receptor recognizes antigen in the form of a complex of a foreign peptide bound to an MHC molecule.

Baker, B.M., Gagnon, S.J., Biddison, W.E., and Wiley, D.C.: **Conversion of a T cell antagonist into an agonist by repairing a defect in the TCR/peptide/MHC interface: implications for TCR signaling.** *Immunity* 2000, **13**:475–484.

Davis, M.M., Boniface, J.J., Reich, Z., Lyons, D., Hampl, J., Arden, B., and Chien, Y.: **Ligand recognition by $\alpha\beta$ T cell receptors.** *Annu. Rev. Immunol.* 1998, **16**:523–544.

Hennecke, J., and Wiley, D.C.: **Structure of a complex of the human $\alpha\beta$ T cell receptor (TCR) HA1.7, influenza hemagglutinin peptide, and major histocompatibility complex class II molecule, HLA-DR4 (DRA*0101 and DRB1*0401): insight into TCR cross-restriction and alloreactivity.** *J. Exp. Med.* 2002, **195**:571–581.

Hennecke, J., Carfi, A., and Wiley, D.C.: **Structure of a covalently stabilized complex of a human $\alpha\beta$ T-cell receptor, influenza HA peptide and MHC class II molecule, HLA-DR1.** *EMBO J.* 2000, **19**:5611–5624.

Luz, J.G., Huang, M., Garcia, K.C., Rudolph, M.G., Apostolopoulos, V., Teyton, L., and Wilson, I.A.: **Structural comparison of allogeneic and syngeneic T cell receptor–peptide–major histocompatibility complex complexes: a buried alloreactive mutation subtly alters peptide presentation substantially increasing V_β interactions.** *J. Exp. Med.* 2002, **195**:1175–1186.

3-12 There are two classes of MHC molecules with distinct subunit composition but similar three-dimensional structures.

&

3-13 Peptides are stably bound to MHC molecules, and also serve to stabilize the MHC molecule on the cell surface.

Bouvier, M.: **Accessory proteins and the assembly of human class I MHC molecules: a molecular and structural perspective.** *Mol. Immunol.* 2003, **39**:697–706

Dessen, A., Lawrence, C.M., Cupo, S., Zaller, D.M., and Wiley, D.C.: **X-ray crystal structure of HLA-DR4 (DRA*0101, DRB1*0401) complexed with a peptide from human collagen II.** *Immunity* 1997, **7**:473–481.

Fremont, D.H., Hendrickson, W.A., Marrack, P., and Kappler, J.: **Structures of an MHC class II molecule with covalently bound single peptides.** *Science* 1996, **272**:1001–1004.

Fremont, D.H., Matsumura, M., Stura, E.A., Peterson, P.A. and Wilson, I.A.: **Crystal structures of two viral peptides in complex with murine MHC class 1 H-2K^b.** *Science* 1992, **257**:919–927.

Fremont, D.H., Monnaie, D., Nelson, C.A., Hendrickson, W.A., and Unanue, E.R.: **Crystal structure of I-Ak in complex with a dominant epitope of lysozyme.** *Immunity* 1998, **8**:305–317.

Macdonald, W.A., Purcell, A.W., Mifsud, N.A., Ely, L.K., Williams, D.S., Chang, L., Gorman, J.J., Clements, C.S., Kjer-Nielsen, L., Koelle, D.M., Burrows, S.R., Tait, B.D., Holdsworth, R., Brooks, A.G., Lovrecz, G.O., Lu, L., Rossjohn, J., and McCluskey, J.: **A naturally selected dimorphism within the HLA-B44 supertype alters class I structure, peptide repertoire, and T cell recognition.** *J. Exp. Med.* 2003, **198**:679–691.

Zhu, Y., Rudensky, A.Y., Corper, A.L., Teyton, L., and Wilson, I.A.: **Crystal structure of MHC class II I-Ab in complex with a human CLIP peptide: prediction of an I-Ab peptide-binding motif.** *J. Mol. Biol.* 2003, **326**:1157–1174.

3-14 MHC class I molecules bind short peptides of 8–10 amino acids by both ends.

Bouvier, M., and Wiley, D.C.: **Importance of peptide amino and carboxyl termini to the stability of MHC class I molecules.** *Science* 1994, **265**:398–402.

Govindarajan, K.R., Kangueane, P., Tan, T.W., and Ranganathan, S.: **MPID: MHC-Peptide Interaction Database for sequence–structure–function information on peptides binding to MHC molecules.** *Bioinformatics* 2003, **19**:309–310.

Saveanu, L., Fruci, D., and van Endert, P.: **Beyond the proteasome: trimming, degradation and generation of MHC class I ligands by auxiliary proteases.** *Mol. Immunol.* 2002, **39**:203–215.

Weiss, G.A., Collins, E.J., Garboczi, D.N., Wiley, D.C., and Schreiber, S.L.: **A tricyclic ring system replaces the variable regions of peptides presented by three alleles of human MHC class I molecules.** *Chem. Biol.* 1995, **2**:401–407.

3-15 The length of the peptides bound by MHC class II molecules is not constrained.

Conant, S.B., and Swanborg, R.H.: **MHC class II peptide flanking residues of exogenous antigens influence recognition by autoreactive T cells.** *Autoimmun. Rev.* 2003, **2**:8–12.

Guan, P., Doytchinova, I.A., Zygouri, C., and Flower, D.R.: **MHCPred: a server for quantitative prediction of peptide-MHC binding.** *Nucleic Acids Res.* 2003, **31**:3621–3624.

Lippolis, J.D., White, F.M., Marto, J.A., Luckey, C.J., Bullock, T.N., Shabanowitz, J., Hunt, D.F., and Engelhard, V.H.: **Analysis of MHC class II antigen processing by quantitation of peptides that constitute nested sets.** *J. Immunol.* 2002, **169**:5089–5097.

Park, J.H., Lee, Y.J., Kim, K.L., and Cho, E.W.: **Selective isolation and identification of HLA-DR-associated naturally processed and presented epitope peptides.** *Immunol. Invest.* 2003, **32**:155–169.

Rammensee, H.G.: **Chemistry of peptides associated with MHC class I and class II molecules.** *Curr. Opin. Immunol.* 1995, **7**:85–96.

Rudensky, A.Y., Preston-Hurlburt, P., Hong, S.C., Barlow, A., and Janeway, C.A., Jr.: **Sequence analysis of peptides bound to MHC class II molecules.** *Nature* 1991, **353**:622.

Sercarz, E.E., and Maverakis, E.: **MHC-guided processing: binding of large antigen fragments.** *Nat. Rev. Immunol.* 2003, **3**:621–629.

Sinnathamby, G., and Eisenlohr, L.C.: **Presentation by recycling MHC class II molecules of an influenza hemagglutinin-derived epitope that is revealed in the early endosome by acidification.** *J. Immunol.* 2003, **170**:3504–3513.

3-16 The crystal structures of several peptide:MHC:T-cell receptor complexes show a similar T-cell receptor orientation over the peptide:MHC complex.

Buslepp, J., Wang, H., Biddison, W.E., Appella, E., and Collins, E.J.: **A correlation between TCR Vα docking on MHC and CD8 dependence: implications for T cell selection.** *Immunity* 2003, **19**:595–606.

Ding, Y.H., Smith, K.J., Garboczi, D.N., Utz, U., Biddison, W.E., and Wiley, D.C.: **Two human T cell receptors bind in a similar diagonal mode to the HLA-A2/Tax peptide complex using different TCR amino acids.** *Immunity* 1998, **8**:403–411.

Kjer-Nielsen, L., Clements, C.S., Purcell, A.W., Brooks, A.G., Whisstock, J.C., Burrows, S.R., McCluskey, J., and Rossjohn, J.: **A structural basis for the selection of dominant αβ T cell receptors in antiviral immunity.** *Immunity* 2003, 18:53–64.

Garcia, K.C., Degano, M., Pease, L.R., Huang, M., Peterson, P.A., Leyton, L., and Wilson, I.A.: **Structural basis of plasticity in T cell receptor recognition of a self peptide-MHC antigen.** *Science* 1998, **279**:1166–1172.

Sant'Angelo, D.B., Waterbury, G., Preston-Hurlburt, P., Yoon, S.T., Medzhitov, R., Hong, S.C., and Janeway, C.A., Jr.: **The specificity and orientation of a TCR to its peptide-MHC class II ligands.** *Immunity* 1996, **4**:367–376.

Reiser, J.B., Darnault, C., Gregoire, C., Mosser, T., Mazza, G., Kearney, A., van der Merwe, P.A., Fontecilla-Camps, J.C., Housset, D., and Malissen, B.: **CDR3 loop flexibility contributes to the degeneracy of TCR recognition.** *Nat. Immunol.* 2003, **4**:241–247.

Teng, M.K., Smolyar, A., Tse, A.G.D., Liu, J.H., Liu, J., Hussey, R.E., Nathenson, S.G., Chang, H.C., Reinherz, E.L., and Wang, J.H.: **Identification of a common docking topology with substantial variation among different TCR–MHC–peptide complexes.** *Curr. Biol.* 1998, **8**:409–412.

3-17 The CD4 and CD8 cell-surface proteins of T cells are required to make an effective response to antigen.

Gao, G.F., Tormo, J., Gerth, U.C., Wyer, J.R., McMichael, A.J., Stuart, D.I., Bell, J.I., Jones, E.Y., and Jakobsen, B.Y.: **Crystal structure of the complex between human CD8αα and HLA-A2.** *Nature* 1997, **387**:630–634.

Gaspar, R., Jr., Bagossi, P., Bene, L., Matko, J., Szollosi, J., Tozser, J., Fesus, L., Waldmann, T.A., and Damjanovich, S.: **Clustering of class I HLA oligomers with CD8 and TCR: three-dimensional models based on fluorescence resonance energy transfer and crystallographic data.** *J. Immunol.* 2001, **166**:5078–5086.

Kim, P.W., Sun, Z.Y., Blacklow, S.C., Wagner, G., and Eck, M.J.: **A zinc clasp structure tethers Lck to T cell coreceptors CD4 and CD8.** *Science* 2003, **301**:1725–1728.

Moldovan, M.C., Yachou, A., Levesque, K., Wu, H., Hendrickson, W.A., Cohen, E.A., and Sekaly, R.P.: **CD4 dimers constitute the functional component required for T-cell activation.** *J. Immunol.* 2002, **169**:6261–6268.

Wang, J.H., and Reinherz, E.L.: **Structural basis of T cell recognition of peptides bound to MHC molecules.** *Mol. Immunol.* 2002, **38**:1039–1049.

Wu, H., Kwong, P.D., and Hendrickson, W.A.: **Dimeric association and segmental variability in the structure of human CD4.** *Nature* 1997, **387**:527–530.

Zamoyska, R.: **CD4 and CD8: modulators of T cell receptor recognition of antigen and of immune responses?** *Curr. Opin. Immunol.* 1998, **10**:82–86.

3-18 The two classes of MHC molecules are expressed differentially on cells.

Steimle, V., Siegrist, C.A., Mottet, A., Lisowska-Grospierre, B., and Mach, B.: **Regulation of MHC class II expression by interferon-γ mediated by the transactivator gene CIITA.** *Science* 1994, **265**:106–109.

3-19 A distinct subset of T cells bears an alternative receptor made up of γ and δ chains.

Allison, T.J., and Garboczi, D.N.: **Structure of γδ T cell receptors and their recognition of non-peptide antigens.** *Mol. Immunol.* 2002, **38**:1051–1061.

Allison, T.J., Winter, C.C., Fournie, J.J., Bonneville, M., and Garboczi, D.N.: **Structure of a human γδ T-cell antigen receptor.** *Nature* 2001, **411**:820–824.

Carding, S.R., and Egan, P.J.: **γδ T cells: functional plasticity and heterogeneity.** *Nat. Rev. Immunol.* 2002, **2**:336–345.

Das, H., Wang, L., Kamath, A., and Bukowski, J.F.: **$V_{\gamma}2V_{\delta}2$ T-cell receptor-mediated recognition of aminobisphosphonates.** *Blood* 2001, **98**:1616–1618.

Wilson, I.A., and Stanfield, R.L.: **Unraveling the mysteries of γδ T cell recognition.** *Nat. Immunol.* 2001, **2**:579–581.

Wu, J., Groh, V., and Spies, T.: **T cell antigen receptor engagement and specificity in the recognition of stress-inducible MHC class I-related chains by human epithelial γ δ T cells.** *J. Immunol.* 2002, **169**:1236–1240.

The Generation of Lymphocyte Antigen Receptors

4

Lymphocyte antigen receptors, in the form of immunoglobulins on B cells and T-cell receptors on T cells, are the means by which lymphocytes sense the presence of antigens in their environment. Individual lymphocytes bear numerous copies of a single antigen receptor with a unique antigen-binding site, which determines the antigens that the lymphocyte can bind. Because each person possesses billions of lymphocytes, these cells collectively enable a response to a great variety of antigens. The wide range of antigen specificities in the antigen-receptor repertoire is due to variation in the amino acid sequence at the antigen-binding site, which is made up from the variable (V) regions of the receptor protein chains. In each chain the V region is linked to an invariant constant (C) region, which provides effector or signaling functions.

Given the importance of a diverse repertoire of lymphocyte receptors in the defense against infection, it is not surprising that a complex and elegant genetic mechanism has evolved for generating these highly variable proteins. Each receptor-chain variant cannot be encoded in full in the genome, as this would require more genes for antigen receptors than there are genes in the entire genome. Instead, we will see that the V regions of the receptor chains are encoded in several pieces—so-called gene segments. These are assembled in the developing lymphocyte by somatic DNA recombination to form a complete V-region sequence, a mechanism known generally as **gene rearrangement**. A fully assembled V-region sequence is made up of two or three types of gene segment, each of which is present in multiple copies in the germline genome. The selection of a gene segment of each type during gene rearrangement occurs at random, and the large number of possible combinations accounts for much of the diversity of the receptor repertoire.

In the first part of this chapter we describe the intrachromosomal gene rearrangements that generate the primary repertoire of V regions of immunoglobulin and T-cell receptor genes. The mechanism of gene rearrangement is common to both B cells and T cells, and its evolution was probably critical to the evolution of the vertebrate adaptive immune system. The antigen receptors expressed after these primary gene rearrangements provide the repertoire of diverse antigen specificities of naive B cells and T cells.

Immunoglobulins can be synthesized as either transmembrane receptors or secreted antibodies, unlike T-cell receptors, which only exist as transmembrane receptors. In the second part of the chapter we shall see how the transition from the production of transmembrane immunoglobulins by activated B cells to the production of secreted antibodies by plasma cells is achieved. The C regions of antibodies have important effector functions in an immune response, and we also briefly consider here the different types of antibody C regions and their properties, a topic that we return to in more detail in Chapter 9.

In the last part of the chapter we consider three kinds of secondary modifications that can take place in rearranged immunoglobulin genes in B cells but do not occur in T cells. These all provide further diversity in the antibody repertoire that helps make the antibody response more effective over time. One is a process known as somatic hypermutation, which introduces point mutations into the V regions of rearranged immunoglobulin genes in activated B cells, producing some variants that bind more strongly to the antigen. The second is a modification called gene conversion, which in some species has a more significant role than combinatorial diversity in diversifying rearranged V regions during the development of immature B cells. The third modification is the limited, but functionally important, sequential expression of different immunoglobulin C regions in activated B cells by a process called class switching, which enables antibodies with the same antigen specificity but different functional properties to be produced.

Primary immunoglobulin gene rearrangement

Virtually any substance can be the target of an antibody response, and the response to even a single epitope comprises many different antibody molecules, each with a subtly different specificity for the epitope and a unique **affinity**, or binding strength. The total number of antibody specificities available to an individual is known as the **antibody repertoire** or **immunoglobulin repertoire**, and in humans is at least 10^{11}. The number of antibody specificities present at any one time is, however, limited by the total number of B cells in an individual, as well as by each individual's previous encounters with antigens.

Before it was possible to examine the immunoglobulin genes directly, there were two main hypotheses for the origin of this diversity. The **germline theory** held that there is a separate gene for each different immunoglobulin chain and that the antibody repertoire is largely inherited. In contrast, **somatic diversification theories** proposed that the observed repertoire is generated from a limited number of inherited V-region sequences that undergo alteration within B cells during the individual's lifetime. Cloning of the immunoglobulin genes revealed that elements of both theories were correct and that the DNA sequence encoding each V region is generated by rearrangements of a relatively small group of inherited gene segments. Diversity is further enhanced by the process of somatic hypermutation in mature activated B cells. Thus the somatic diversification theory was essentially correct, although the concept of multiple germline genes embodied in the germline theory also proved true.

4-1 Immunoglobulin genes are rearranged in antibody-producing cells.

In nonlymphoid cells, the gene segments encoding the greater part of the V region of an immunoglobulin chain are a considerable distance away from the sequence encoding the C region. In mature B lymphocytes, however, the assembled V-region sequence lies much nearer the C region, as a consequence of gene rearrangement. Rearrangement within the immunoglobulin genes was originally discovered about 30 years ago, when the techniques of restriction enzyme analysis first made it possible to study the organization of the immunoglobulin genes in both B cells and nonlymphoid cells. In these procedures, chromosomal DNA is first cut with a restriction enzyme, and the DNA fragments containing particular V- and C-region sequences are identified by hybridization with radiolabeled DNA probes specific for the relevant DNA sequences. In germline DNA from nonlymphoid cells the V- and

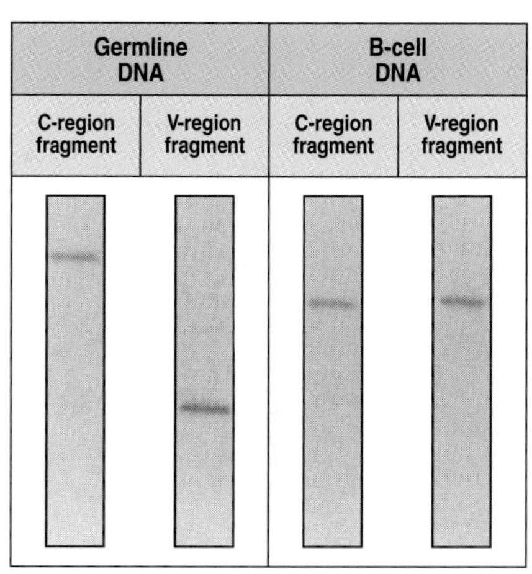

Fig. 4.1 Immunoglobulin genes are rearranged in B cells. In the original experiments by Hozumi and Tonegawa, DNA fragment sizes were determined by measuring the hybridization of radiolabeled probes to restriction fragments isolated from gel slices after electrophoresis. Later, the Southern blot, in which the electrophoresed fragments are transferred to a nitrocellulose membrane, became the technique of choice. The two photographs on the left (germline DNA) show a Southern blot of a restriction enzyme digest of DNA from nonlymphoid cells from a normal person. The locations of immunoglobulin DNA sequences are identified by hybridization with V- and C-region probes. The V and C regions are found in distinct DNA fragments in the nonlymphoid DNA. The two photographs on the right (B-cell DNA) are of the same restriction digest of DNA from peripheral blood lymphocytes from a patient with chronic lymphocytic leukemia (see Chapter 7), in which a single clone of B cells is greatly expanded. The malignant B cells express the V region from which the V-region probe was obtained and, owing to their predominance in the cell population, this unique rearrangement can be detected. In this DNA, the V and C regions are found in the same fragment, which is a different size from either the C- or the V-region germline fragments. Although not shown in this figure, a population of normal B lymphocytes has many different rearranged genes, so they yield a smear of DNA fragment sizes, which are not visible as a crisp band. Photographs courtesy of S. Wagner and L. Luzzatto.

C-region sequences are on separate DNA fragments, but in DNA from an antibody-producing B cell they are on the same DNA fragment, showing that a rearrangement of the DNA has occurred. A typical experiment is shown in Fig. 4.1.

This simple experiment showed that segments of genomic DNA within the immunoglobulin genes are rearranged in cells of the B-lymphocyte lineage, but not in other cells. This process of rearrangement is known as **somatic recombination**, to distinguish it from the meiotic recombination that takes place during the production of gametes.

4-2 Complete genes that encode a variable region are generated by the somatic recombination of separate gene segments.

The V region, or V domain, of an immunoglobulin heavy or light chain is encoded by more than one gene segment. For the light chain, the V domain is encoded by two separate DNA segments. The first encodes the first 95–101 amino acids, the greater part of the domain, and is called a **variable** or **V gene segment**. The second encodes the remainder of the domain (up to 13 amino acids) and is called a **joining** or **J gene segment**.

The rearrangements that produce a complete immunoglobulin light-chain gene are shown in Fig. 4.2 (center panel). The joining of a V and a J gene segment creates an exon that encodes the whole light-chain V region. In the unrearranged DNA, the V gene segments are located relatively far away from the C region. The J gene segments are located close to the C region, however, and the joining of a V gene segment to a J gene segment also brings the V gene segment close to a C-region sequence. The J gene segment of the rearranged V region is separated from a C-region sequence only by a short intron. In the experiment shown in Fig. 4.1, the germline DNA fragment identified by the V-region probe contains the V gene segment, and that identified by the C-region probe actually contains both the J gene segment and the C-region sequence. To make a complete immunoglobulin light-chain messenger RNA, the V-region exon is joined to the C-region sequence by RNA splicing after transcription (see Fig. 4.2).

A heavy-chain V region is encoded in three gene segments. In addition to the V and J gene segments (denoted V_H and J_H to distinguish them from the light-chain V_L and J_L), there is a third gene segment called the **diversity** or **D_H gene**

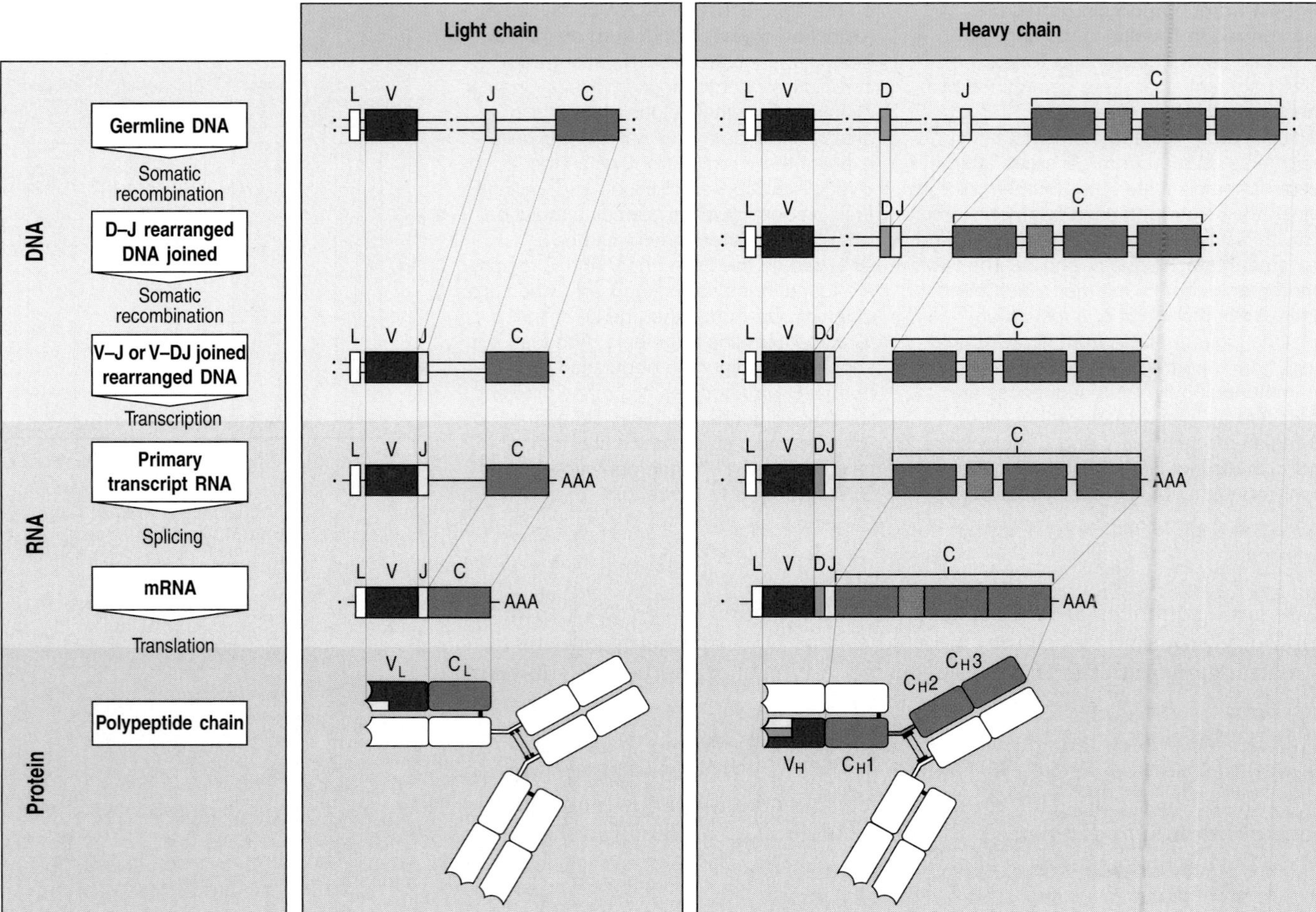

Fig. 4.2 V-region genes are constructed from gene segments. Light-chain V-region genes are constructed from two segments (center panel). A variable (V) and a joining (J) gene segment in the genomic DNA are joined to form a complete light-chain V-region exon. Immunoglobulin chains are extracellular proteins and the V gene segment is preceded by an exon encoding a leader peptide (L), which directs the protein into the cell's secretory pathways and is then cleaved. The light-chain C region is encoded in a separate exon and is joined to the V-region exon by splicing of the light-chain RNA to remove the L-to-V and the J-to-C introns. Heavy-chain V regions are constructed from three gene segments (right panel). First, the diversity (D) and J gene segments join, then the V gene segment joins to the combined DJ sequence, forming a complete V_H exon. A heavy-chain C-region gene is encoded by several exons. The C-region exons, together with the leader sequence, are spliced to the V-domain sequence during processing of the heavy-chain RNA transcript. The leader sequence is removed after translation and the disulfide bonds that link the polypeptide chains are formed. The hinge region is shown in purple.

segment, which lies between the V_H and J_H gene segments. The recombination process that generates a complete heavy-chain V region is shown in Fig. 4.2 (right panel), and occurs in two separate stages. In the first, a D_H gene segment is joined to a J_H gene segment; then a V_H gene segment rearranges to DJ_H to make a complete V_H-region exon. As with the light-chain genes, RNA splicing joins the assembled V-region sequence to the neighboring C-region gene.

4-3 Multiple contiguous V gene segments are present at each immunoglobulin locus.

For simplicity we have discussed the formation of a complete V-region sequence as though there were only a single copy of each gene segment. In fact, there are multiple copies of all the gene segments in germline DNA. It is the random selection of just one gene segment of each type that makes possible the great diversity of V regions among immunoglobulins. The numbers of functional gene segments of each type in the human genome, as determined by gene cloning and sequencing, are shown in Fig. 4.3. Not all the gene segments discovered are functional, as a proportion have accumulated mutations that prevent them from encoding a functional protein. These are termed 'pseudogenes.' Because there are many V, D, and J gene segments in germline DNA, no single one is essential. This reduces the evolutionary pressure on each gene segment to remain intact, and has resulted in a relatively large number of pseudogenes. Since some of these can undergo rearrangement just like a normal gene segment, a significant proportion of rearrangements incorporate a pseudogene and will thus be nonfunctional.

We saw in Section 3-1 that there are three sets of immunoglobulin chains, the heavy chains and two equivalent types of light chain, the κ and λ chains. The immunoglobulin gene segments forming each of these chains are organized into three clusters or **genetic loci**—the κ, λ, and heavy-chain loci—each of which can assemble a complete V-region sequence. Each locus is on a different chromosome and is organized slightly differently, as shown for the human loci in Fig. 4.4. At the λ light-chain locus, located on chromosome 22, a cluster of V_λ gene segments is followed by four sets of J_λ gene segments each linked to a single C_λ gene. In the κ light-chain locus, on chromosome 2, the cluster of V_κ gene segments is followed by a cluster of J_κ gene segments, and then by a single C_κ gene. The organization of the heavy-chain locus, on chromosome 14, resembles that of the κ locus, with separate clusters of V_H, D_H, and J_H gene segments and of C_H genes. The heavy-chain locus differs in one important way: instead of a single C region, it contains a series of C regions arrayed one after the other, each of which corresponds to a different isotype. B cells initially express the heavy-chain isotypes μ and δ (see Section 3-1), which is accomplished by alternative mRNA splicing and leads to expression of immunoglobulins IgM and IgD, as we shall see in Section 4-14. The expression of other isotypes, such as γ (giving IgG), occurs at a later stage through DNA rearrangements referred to as class switching, which is described in Section 4-20.

Number of functional gene segments in human immunoglobulin loci

Segment	Light chains		Heavy chain
	κ	λ	H
Variable (V)	40	30	40
Diversity (D)	0	0	25
Joining (J)	5	4	6

Fig. 4.3 The numbers of functional gene segments for the V regions of human heavy and light chains. These numbers are derived from exhaustive cloning and sequencing of DNA from one individual and exclude all pseudogenes (mutated and nonfunctional versions of a gene sequence). As a result of genetic polymorphism, the numbers will not be the same for all people.

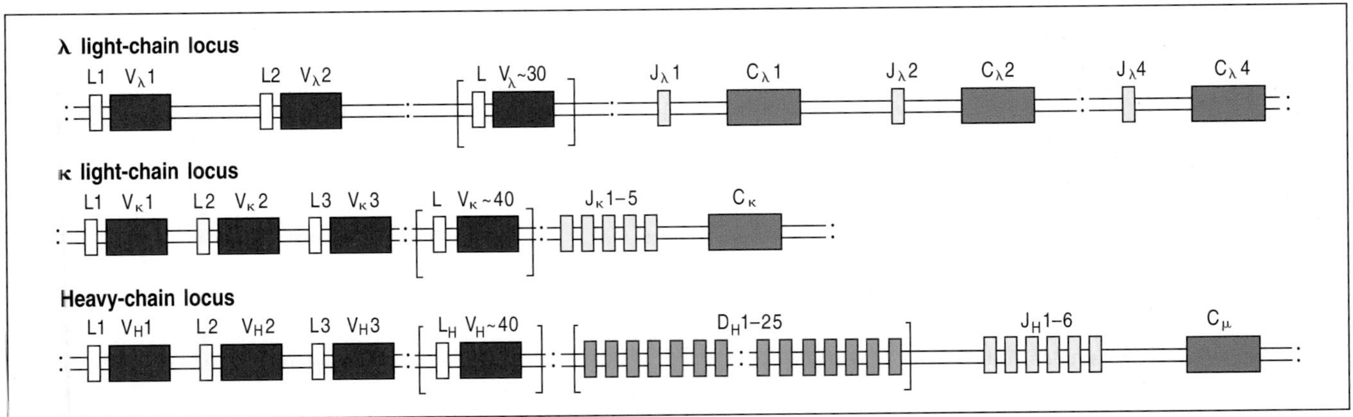

Fig. 4.4 The germline organization of the immunoglobulin heavy- and light-chain loci in the human genome. The genetic locus for the λ light chain (chromosome 22) has about 30 functional V_λ gene segments and four pairs of functional J_λ gene segments and C_λ genes. The κ locus (chromosome 2) is organized in a similar way, with about 40 functional V_κ gene segments accompanied by a cluster of five J_κ gene segments but with a single C_κ gene. In approximately 50% of individuals, the entire cluster of V_κ gene segments has undergone an increase by duplication (not shown, for simplicity). The heavy-chain locus (chromosome 14) has about 40 functional V_H gene segments and a cluster of around 25 D_H segments lying between these V_H gene segments and six J_H gene segments. The heavy-chain locus also contains a large cluster of C_H genes that are shown in Fig. 4.17. For simplicity we have shown only a single C_H gene in this diagram without illustrating its separate exons, have omitted pseudogenes, and have shown all V gene segments in the same orientation. L, leader sequence. This diagram is not to scale: the total length of the heavy-chain locus is over 2 megabases (2 million bases), whereas some of the D gene segments are only six bases long.

The human V gene segments can be grouped into families in which each member shares at least 80% DNA sequence identity with all others in the family. Both the heavy-chain and κ-chain V gene segments can be subdivided into seven families, and there are eight families of V_λ gene segments. The families can be grouped into clans, made up of families that are more similar to each other than to families in other clans. Human V_H gene segments fall into three clans. All the V_H gene segments identified from amphibians, reptiles, and mammals also fall into the same three clans, suggesting that these clans existed in a common ancestor of these modern animal groups. Thus, the V gene segments that we see today have arisen by a series of gene duplications and diversification through evolutionary time.

4-4 Rearrangement of V, D, and J gene segments is guided by flanking DNA sequences.

For a complete immunoglobulin or T-cell receptor chain to be expressed, DNA rearrangements must take place at the correct locations relative to the V, D, or J gene segment coding regions. In addition, joins must be regulated such that a V gene segment joins to a D or J and not to another V. DNA rearrangements are guided by conserved noncoding DNA sequences that are found adjacent to the points at which recombination takes place and are called **recombination signal sequences** (**RSSs**). A recombination signal sequence consists of a conserved block of seven nucleotides—the **heptamer** 5′CACAGTG3′—which is always contiguous with the coding sequence, followed by a nonconserved region known as the **spacer**, which is either 12 or 23 base pairs (bp) long, followed by a second conserved block of nine nucleotides—the **nonamer** 5′ACAAAAACC3′ (Fig. 4.5). The spacers vary in sequence but their conserved lengths correspond to one turn (12 bp) or two turns (23 bp) of the DNA double helix. This brings the heptamer and nonamer sequences to the same side of the DNA helix, where they can be bound by the complex of proteins that catalyzes recombination. The heptamer–spacer–nonamer sequence motif—the RSS—is always found directly adjacent to the coding sequence of V, D, or J gene segments. Recombination normally occurs between gene segments located on the same chromosome. A gene segment flanked by an RSS with a 12-bp spacer typically can be joined only to one flanked by a 23-bp spacer RSS. This is known as the **12/23 rule**. Thus, for the heavy chain, a D_H gene segment can be joined to a J_H gene segment and a V_H gene segment to a D_H gene segment, but V_H gene segments cannot be joined to J_H gene segments directly, as both V_H and J_H gene segments are flanked by 23-bp spacers and the D_H gene segments have 12-bp spacers on both sides (see Fig. 4.5).

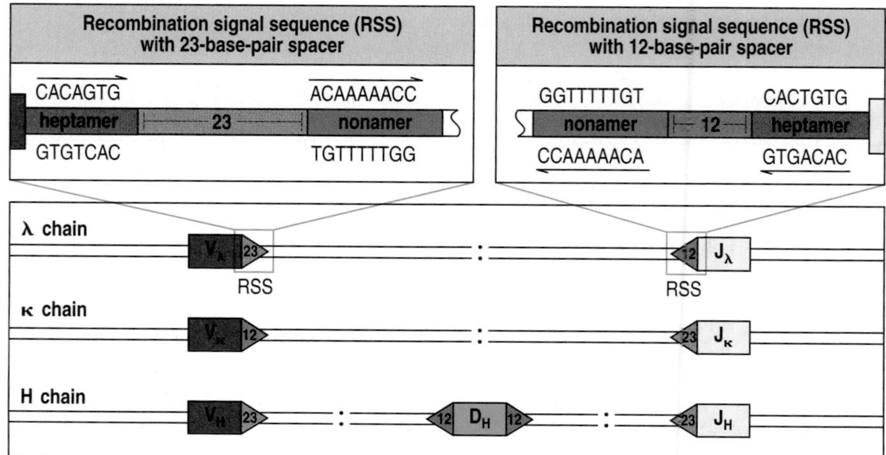

Fig. 4.5 Recombination signal sequences are conserved heptamer and nonamer sequences that flank the gene segments encoding the V, D, and J regions of immunoglobulins. Recombination signal sequences (RSSs) are composed of heptamer (CACAGTG) and nonamer (ACAAAAACC) sequences that are separated either by 12 bp or approximately 23 bp nucleotides. The heptamer–12-bp spacer–nonamer motif is depicted here as an orange arrowhead; the motif that includes the 23-bp spacer is depicted as a purple arrowhead. Joining of gene segments almost always involves a 12-bp and a 23-bp RSS—the 12/23 rule. The arrangement of RSSs in the V (red), D (green), and J (yellow) gene segments of heavy (H) and light (λ and κ) chains of immunoglobulin is shown here. Note that according to the 12/23 rule, the arrangement of RSSs in the immunoglobulin heavy-chain gene segments precludes direct V-to-J joining.

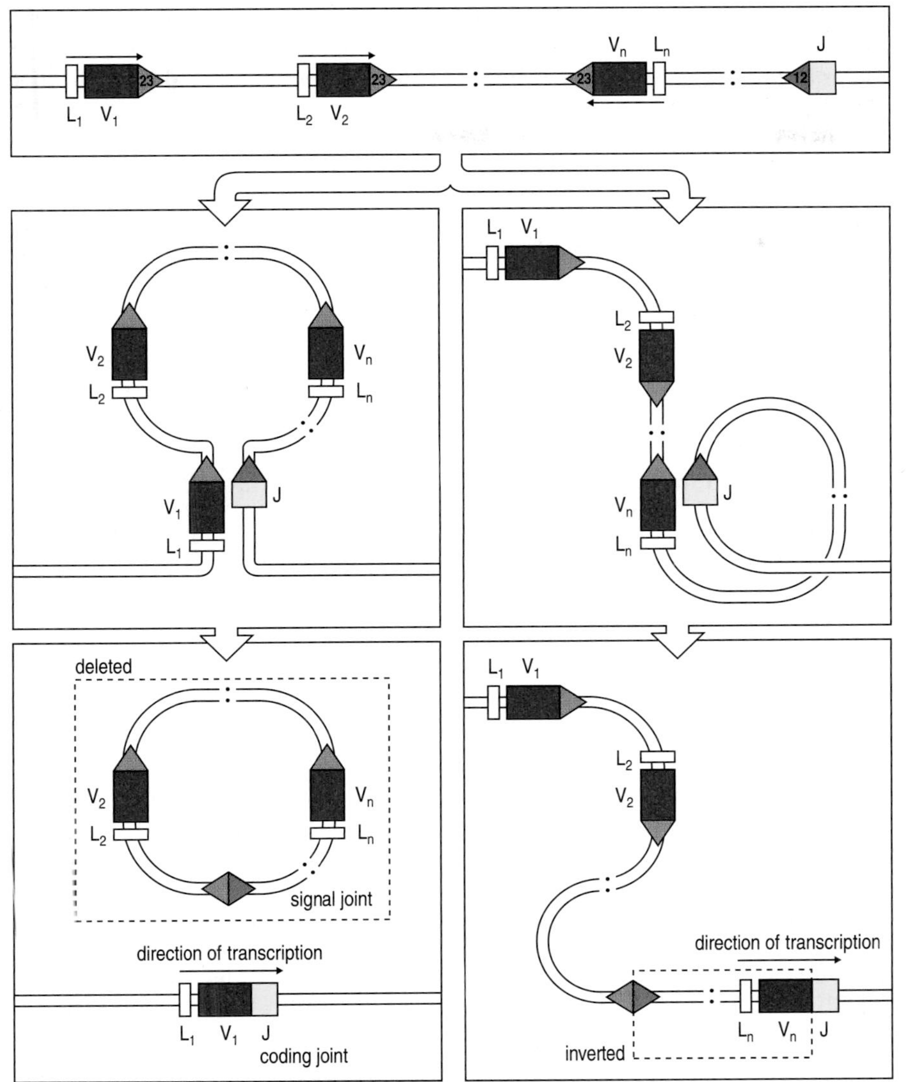

Fig. 4.6 V-region gene segments are joined by recombination. In every V-region recombination event, the recombination signal sequences (RSSs) flanking the gene segments are brought together to allow recombination to take place. The 12-bp-spaced RSSs are shown in orange, the 23-bp-spaced RSSs in purple. For simplicity, the recombination of a light-chain gene is illustrated; for a heavy-chain gene, two separate recombination events are required to generate a functional V region. In most cases, the two segments undergoing rearrangement (the V and J gene segments in this example) are arranged in the same transcriptional orientation in the chromosome (left panels), and juxtaposition of the RSSs results in the looping out of the intervening DNA. Recombination occurs at the ends of the heptamer sequences in the RSSs, creating the so-called signal joint and releasing the intervening DNA in the form of a closed circle. Subsequently, the joining of the V and J gene segments creates the coding joint in the chromosomal DNA. In other cases, illustrated in the right panels, the V and J gene segments are initially oriented in opposite transcriptional directions. Bringing together the RSSs in this case requires a more complex looping of the DNA. Joining the ends of the two heptamer sequences now results in the inversion and integration of the intervening DNA into a new position on the chromosome. Again, the joining of the V and J segments creates a functional V-region exon.

Recall from Section 3-6 that the antigen-binding region of an immunoglobulin is formed by three hypervariable regions. The first two hypervariable regions, CDR1 and CDR2, are encoded in the V gene segment itself. The third hypervariable region, CDR3, is encoded by the additional DNA sequence that is created by the joining of the V and J gene segments for the light chain, and the V, D and J gene segments for the heavy chain. Additional diversity in the antibody repertoire can result from the generation of CDR3 regions that seem to result from the joining of one D gene segment to another D gene segment. Although infrequent, such D–D joining would seem to violate the 12/23 rule, and it is not clear how these rare rearrangements are generated. In humans, D–D joining is found in approximately 5% of antibodies and is the major mechanism accounting for the unusually long CDR3 loops found in some heavy chains.

The mechanism of DNA rearrangement is similar for the heavy- and light-chain loci, although only one joining event is needed to generate a light-chain gene but two for a heavy-chain gene. When two gene segments are in the same orientation in the DNA, rearrangement involves the looping-out and deletion of the DNA between them (Fig. 4.6, left panels), but if the gene segments have opposite transcriptional orientations the intervening DNA meets a different fate (Fig. 4.6, right panels). In the latter case, the intervening DNA is retained in the chromosome in an inverted orientation. This mode of recombination is less common, but accounts for about half of all V_κ to J_κ joins in humans because the orientation of half the V_κ gene segments is opposite to that of the J_κ gene segments.

4-5 The reaction that recombines V, D, and J gene segments involves both lymphocyte-specific and ubiquitous DNA-modifying enzymes.

The molecular mechanism of V-region rearrangement, or **V(D)J recombination**, is illustrated in Fig. 4.7. The two RSSs are brought together by interactions between proteins that specifically recognize the length of spacer and thus enforce the 12/23 rule for recombination. The DNA molecule is then broken in two places and rejoined in a different configuration. The ends of the heptamer sequences are joined precisely in a head-to-head fashion to form a **signal joint**; when the joining segments are in direct orientation, the signal joint is in a circular piece of extrachromosomal DNA (see Fig. 4.6, left panels), which is lost from the genome when the cell divides. The V and J gene segments, which remain on the chromosome, join to form what is called the **coding joint**. In the case of rearrangement by inversion (see Fig 4.6, right panels), the signal joint is also retained within the chromosome, and the region of DNA between the V gene segment and the RSS of the J gene segment is inverted to form the coding joint. As we shall see later, the coding joint junction is imprecise, and consequently generates much additional variability in the V-region sequence.

The complex of enzymes that act in concert to carry out somatic V(D)J recombination is termed the **V(D)J recombinase**. The lymphoid-specific components of the recombinase are called **RAG-1** and **RAG-2**, and are encoded by two recombination-activating genes, *RAG-1* and *RAG-2*. This pair of genes is expressed in developing lymphocytes only while they are engaged in assembling their antigen receptors, as described in more detail in Chapter 7, and they are essential for V(D)J recombination. Indeed, the *RAG* genes expressed together can confer on nonlymphoid cells such as fibroblasts the capacity to rearrange exogenous segments of DNA containing the appropriate RSSs; this is how RAG-1 and RAG-2 were initially discovered.

The other proteins in the recombinase complex are mainly ubiquitous DNA-modifying proteins that are involved in the repair of DNA double-strand breaks and the modification of the ends of broken DNA strands. One is **Ku**, which is a heterodimer (Ku70:Ku80); this forms a ring around the DNA and associates tightly with a protein kinase catalytic subunit, DNA-PKcs, to form the **DNA-dependent protein kinase** (**DNA-PK**). Another is the protein **Artemis**, which has nuclease activity. The DNA ends are finally joined together by the enzyme **DNA ligase IV**, which forms a complex with the DNA repair protein XRCC4.

V(D)J recombination is a multistep enzymatic process in which the first reaction is an endonucleolytic cleavage requiring the coordinated activity of both RAG proteins. Initially, two RAG protein complexes, each containing RAG-1, RAG-2, and high-mobility group chromatin proteins, recognize and align the two RSSs that target the cleavage reaction (see Fig. 4.7). RAG-1 is thought to specifically recognize the nonamer of the RSS. At this stage, the 12/23 rule is established through mechanisms that are still poorly understood. The endonuclease activity of the RAG protein complexes, which is thought to reside in RAG-1, then makes two single-strand DNA breaks at sites just 5′ of each bound RSS, leaving a free 3′-OH group at the end of each coding segment. The 3′-OH group then attacks the phosphodiester bond on the other strand, creating a DNA 'hairpin' at the end of the gene segment coding region

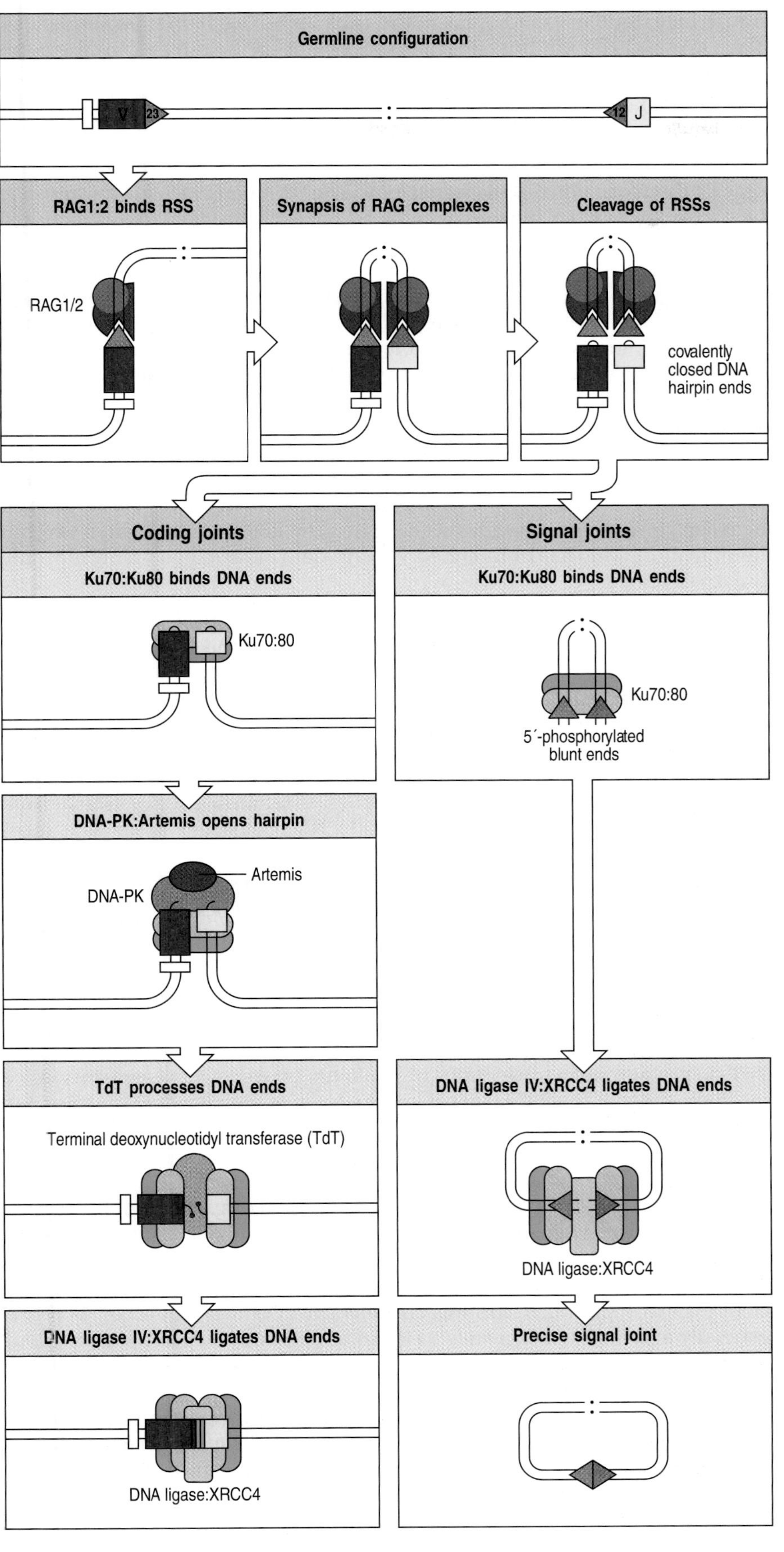

Fig. 4.7 Enzymatic steps in RAG-dependent V(D)J rearrangement. Gene segments containing recombination signal sequences (RSSs) (triangles) undergo rearrangement beginning with the binding of RAG-1, RAG-2 (blue and purple), and high-mobility group (HMG) proteins (not shown) to an RSS flanking the coding sequences to be joined (second row). After binding of two RAG complexes to two RSSs, a presumed synapsis occurs in which the two complexes are brought together. In the cleavage step, the endonuclease activity of the RAG complex initially cuts a phosphodiester bond of the DNA backbone to create a 3′-hydroxyl group precisely between the coding segment and its RSS. This newly created 3′-OH then reacts with a phosphodiester bond on the opposite DNA strand to generate a blunt 5′-phosphorylated DNA end at the heptamer sequence of the RSS and a hairpin on the coding end. Subsequently, these two DNA ends are resolved in slightly different ways. At the coding ends (left panels) essential proteins such as Ku70:Ku80 (green) bind to the hairpin. The DNA-PK:Artemis complex (purple) then joins the complex and its endonuclease activity opens the DNA hairpin at a random site to yield either a flush or a single-strand extended DNA end (depending on the exact site of hairpin cleavage). This DNA end is then modified by TdT (pink) and exonuclease activities, which randomly create diverse and imprecise ends (this process is shown in more detail in Fig. 4.8). The ends are finally ligated by DNA ligase IV (turquoise) in association with XRCC4 (green). At the signal ends (right panels), the two 5′ phosphorylated blunt ends at the heptamer sequences are bound by Ku70:Ku80 but are not further modified. Instead, a complex of DNA ligase IV:XRCC4 joins the two signal ends precisely to form the signal joint.

and a flush double-strand break at the ends of the two heptamer sequences. The DNA ends do not float apart, however, but are held tightly in the complex until the join is completed. The 5′ DNA ends are held together by Ku and are precisely joined by a complex of DNA ligase IV and XRCC4 to form the signal joint.

Formation of the coding joint is more complex but seems to occur more quickly than formation of the signal joint. The DNA ends with the hairpins are held together by Ku, which recruits the DNA-PKcs subunit. Artemis recruited to the complex is activated by phosphorylation by DNA-PK and then opens the DNA hairpins by making a single-strand nick in the DNA. This nicking can happen at various points along the hairpin, which leads to sequence variability in the final joint. The DNA repair enzymes in the complex modify the opened hairpins by removing nucleotides, while at the same time the lymphoid-specific enzyme **terminal deoxynucleotidyl transferase** (**TdT**), which is also part of the recombinase complex, adds nucleotides randomly to the single-strand ends. Addition and deletion of nucleotides can occur in any order and one does not necessarily precede the other. Finally, DNA ligase IV joins the processed ends together, thus reconstituting a chromosome that includes the rearranged gene. This repair process creates diversity in the joint between gene segments while ensuring that the RSS ends are ligated without modification and that unintended genetic damage such as a chromosome break is avoided.

The recombination mechanism controlled by the RAG proteins shares many interesting features with the mechanism by which the integrases of retroviruses catalyze the insertion of retroviral DNA into the genome, and also with the transposition mechanism used by transposons—mobile genetic elements that encode their own transposase, allowing them to excise from one site in the genome and reinsert themselves elsewhere. Even the structure of the *RAG* genes themselves, which lie close together in the chromosome and lack the usual introns of mammalian genes, is reminiscent of a transposon. Indeed, it has recently been shown that the RAG complex can act as a transposase *in vitro*. These features have provoked speculation that the RAG complex originated as a transposase whose function was adapted by evolution in vertebrates to allow V gene segment recombination, thus leading to the advent of the vertebrate adaptive immune system; this is discussed in Chapter 16. Consistent with this idea, no genes homologous with the *RAG* genes have been found in nonvertebrates.

Omenn Syndrome

The *in vivo* roles of the enzymes involved in V(D)J recombination have been established through natural or artificially induced mutations. Mice lacking TdT do not add extra nucleotides to the joints between gene segments. Mice in which either of the *RAG* genes is knocked out, or which lack DNA-PKcs, Ku, or Artemis, suffer a complete block in lymphocyte development at the gene-rearrangement stage, or make only trivial numbers of B and T cells. They are said to suffer from **s**evere **c**ombined **i**mmune **d**eficiency (**SCID**). The original ***scid*** mutation was discovered some time before the components of the recombination pathway were identified and was subsequently identified as a mutation in DNA-PKcs. Consistent with its proposed role, mice lacking DNA-PK function have defects in coding joint, but not signal joint formation. Mice deficient in DNA-PKcs, Ku, or Artemis are defective in double-strand break repair generally and are hypersensitive to ionizing radiation (which produces double-strand breaks). In humans, mutations in *RAG1* or *RAG2* that result in partial V(D)J recombinase activity are responsible for an inherited disorder called **Omenn syndrome**, which is characterized by an absence of circulating B cells and an infiltration of skin by activated oligoclonal T lymphocytes. Defects in Artemis in humans produce a combined immunodeficiency of B and T cells that is associated with increased radiosensitivity and is called RS-SCID.

4-6 The diversity of the immunoglobulin repertoire is generated by four main processes.

The gene rearrangement that combines gene segments to form a complete V-region exon generates diversity in two ways. First, there are multiple different copies of each type of gene segment, and different combinations of gene segments can be used in different rearrangement events. This **combinatorial diversity** is responsible for a substantial part of the diversity of V regions. Second, **junctional diversity** is introduced at the joints between the different gene segments as a result of the addition and subtraction of nucleotides by the recombination process. A third source of diversity is also combinatorial, arising from the many possible different combinations of heavy- and light-chain V regions that pair to form the antigen-binding site in the immunoglobulin molecule. The two means of generating combinatorial diversity alone could give rise, in theory, to approximately 1.9×10^6 different antibody molecules (see Section 4-7). Coupled with junctional diversity, it is estimated that as many as 10^{11} different receptors could make up the repertoire of receptors expressed by naive B cells. Finally, **somatic hypermutation**, which we discuss later in this chapter, introduces point mutations into the rearranged V-region genes of activated B cells, creating further diversity that can be selected for enhanced binding to antigen.

4-7 The multiple inherited gene segments are used in different combinations.

There are multiple copies of the V, D, and J gene segments, each of which can contribute to an immunoglobulin V region. Many different V regions can therefore be made by selecting different combinations of these segments. For human κ light chains, there are approximately 40 functional V_κ gene segments and 5 J_κ gene segments, and thus potentially 200 different V_κ regions. For λ light chains there are approximately 30 functional V_λ gene segments and 4 J_λ gene segments, yielding 120 possible V_λ regions. So, in all, 320 different light chains can be made as a result of combining different light-chain gene segments. For the heavy chains of humans, there are 40 functional V_H gene segments, approximately 25 D_H gene segments, and 6 J_H gene segments, and thus around 6000 different possible V_H regions ($40 \times 25 \times 6 = 6000$). During B-cell development, rearrangement at the heavy-chain gene locus to produce a heavy chain is followed by several rounds of cell division before light-chain gene rearrangement takes place, resulting in the same heavy chain being paired with different light chains in different cells. Because both the heavy- and the light-chain V regions contribute to antibody specificity, each of the 320 different light chains could be combined with each of the approximately 6000 heavy chains to give around 1.9×10^6 different antibody specificities.

This theoretical estimate of combinatorial diversity is based on the number of germline V gene segments contributing to functional antibodies (see Fig. 4.3); the total number of V gene segments is larger, but the additional gene segments are pseudogenes and do not appear in expressed immunoglobulin molecules. In practice, combinatorial diversity is likely to be less than one might expect from the calculations above. One reason is that not all V gene segments are used at the same frequency; some are common in antibodies, while others are found only rarely. Also, not every heavy chain can pair with every light chain: certain combinations of V_H and V_L regions will not form a stable molecule. Cells in which heavy and light chains fail to pair may undergo further light-chain gene rearrangement until a suitable chain is produced or they will be eliminated. Nevertheless, it is thought that most heavy and light chains can pair with each other, and that this type of combinatorial diversity has a major role in forming an immunoglobulin repertoire with a wide range of specificities.

4-8 Variable addition and subtraction of nucleotides at the junctions between gene segments contributes to the diversity of the third hypervariable region.

As noted earlier, of the three hypervariable loops in an immunoglobulin chain, CDR1 and CDR2 are encoded within the V gene segment. CDR3, however, falls at the joint between the V gene segment and the J gene segment, and in the heavy chain it is partly encoded by the D gene segment. In both heavy and light chains, the diversity of CDR3 is significantly increased by the addition and deletion of nucleotides at two steps in the formation of the junctions between gene segments. The added nucleotides are known as P-nucleotides and N-nucleotides and their addition is illustrated in Fig. 4.8.

P-nucleotides are so called because they make up palindromic sequences added to the ends of the gene segments. As described in Section 4-5, the RAG proteins generate DNA hairpins at the coding ends of the V, D, or J segments, after which Artemis catalyzes a single-stranded cleavage at a random point within the coding sequence but near the original point at which the hairpin was first formed. When this cleavage occurs at a different point from the initial break induced by the RAG1/2 complex, a single-stranded tail is formed from a few nucleotides of the coding sequence plus the complementary nucleotides from the other DNA strand (see Fig. 4.8). In most light-chain gene rearrangements, DNA repair enzymes then fill in complementary nucleotides on the single-stranded tails, which would leave short palindromic sequences (the P-nucleotides) at the joint if the ends were rejoined without any further exonuclease activity.

In heavy-chain gene rearrangements and in a proportion of human light-chain gene rearrangements, however, **N-nucleotides** are added by a quite different mechanism before the ends are rejoined. N-nucleotides are so called because they are non-template-encoded. They are added by the enzyme TdT to the single-stranded ends of the coding DNA after hairpin cleavage. After the addition of up to 20 nucleotides, the two single-stranded stretches form

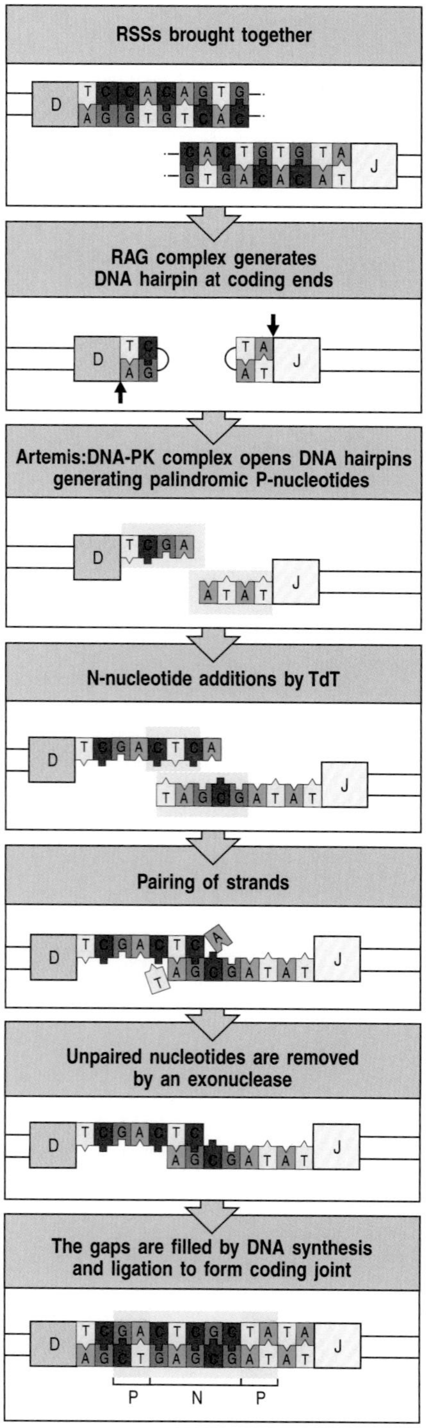

Fig. 4.8 The introduction of P- and N-nucleotides diversifies the joints between gene segments during immunoglobulin gene rearrangement. The process is illustrated for a D_H to J_H rearrangement (first panel); however, the same steps occur in V_H to D_H and in V_L to J_L rearrangements. After formation of the DNA hairpins (second panel), the two heptamer sequences are ligated to form the signal joint (not shown here), while Artemis:DNA-PK cleave the DNA hairpin at a random site (indicated by the arrows) to yield a single-stranded DNA end (third panel). Depending on the site of cleavage, this single-stranded DNA may contain nucleotides that were originally complementary in the double-stranded DNA and which therefore form short DNA palindromes, such as TCGA and ATAT, as indicated by the blue-shaded box. Such stretches of nucleotides that originate from the complementary strand are known as P-nucleotides. For example, the sequence GA at the end of the D segment shown is complementary to the preceding sequence TC. Where the enzyme terminal deoxynucleotidyl transferase (TdT) is present, nucleotides are added at random to the ends of the single-stranded segments (fourth panel), indicated by the shaded box surrounding these non-templated, or N, nucleotides. The two single-stranded ends then pair (fifth panel). Exonuclease trimming of unpaired nucleotides (sixth panel) and repair of the coding joint by DNA synthesis and ligation (bottom panel) leaves both the P- and N-nucleotides present in the final coding joint (indicated by light-blue shading). The randomness of insertion of P- and N-nucleotides makes an individual P–N region virtually unique and a valuable marker for following an individual B-cell clone as it develops, for instance in studies of somatic hypermutation (see Fig. 4.25).

complementary base pairs. Repair enzymes then trim off non-matching nucleotides, synthesize complementary DNA to fill in the remaining single-stranded gaps, and ligate the new DNA to the palindromic region (see Fig. 4.8). TdT is maximally expressed during the period in B-cell development when the heavy-chain gene is being assembled, and so N-nucleotides are common in their V–D and D–J junctions. N-nucleotides are less common in light-chain genes, which undergo rearrangement after heavy-chain genes (see Chapter 7).

Nucleotides can also be deleted at gene segment junctions. This is accomplished by as-yet-unidentified exonucleases. Thus, a heavy-chain CDR3 can be shorter than even the smallest D segment. In some instances it is difficult, if not impossible, to recognize the D segment that contributed to CDR3 formation because of the excision of most of its nucleotides. Deletions may also erase the traces of P-nucleotide palindromes introduced at the time of hairpin opening. For this reason, many completed V(D)J joins do not show obvious evidence of P-nucleotides. As the total number of nucleotides added by these processes is random, the added nucleotides often disrupt the reading frame of the coding sequence beyond the joint. Such frameshifts will lead to a nonfunctional protein, and DNA rearrangements leading to such disruptions are known as **nonproductive rearrangements**. As roughly two in every three rearrangements will be nonproductive, many B-cell progenitors never succeed in producing functional immunoglobulin and therefore never become mature B cells. Thus, junctional diversity is achieved only at the expense of considerable cell wastage. We discuss this further in Chapter 7.

Summary.

The extraordinary diversity of the immunoglobulin repertoire is achieved in several ways. Perhaps the most important factor enabling this diversity is that V regions are encoded by separate gene segments (V, D, and J gene segments), which are brought together by a somatic recombination process (V(D)J recombination) to produce a complete V-region exon. Many different gene segments are present in the genome of an individual, thus providing a heritable source of diversity that this combinatorial mechanism can use. Unique lymphocyte-specific recombinases, the RAG proteins, are absolutely required to catalyze this rearrangement, and the evolution of RAG proteins coincided with the appearance of the modern vertebrate adaptive immune system. Another substantial fraction of the functional diversity of immunoglobulins comes from the joining process itself. Variability at the joints between gene segments is generated by the insertion of random numbers of P- and N-nucleotides and by the variable deletion of nucleotides at the ends of some segments. The association of different light- and heavy-chain V regions to form the antigen-binding site of an immunoglobulin molecule contributes further diversity. The combination of all these sources of diversity generates a vast primary repertoire for antibody specificities. Additional changes in the rearranged V regions, introduced by somatic hypermutation (discussed later in the chapter), add even greater diversity to this primary repertoire.

T-cell receptor gene rearrangement.

The mechanism by which B-cell antigen receptors are generated is such a powerful means of creating diversity that it is not surprising that the antigen receptors of T cells bear structural resemblances to immunoglobulins and are generated by the same mechanism. In this part of the chapter we describe the

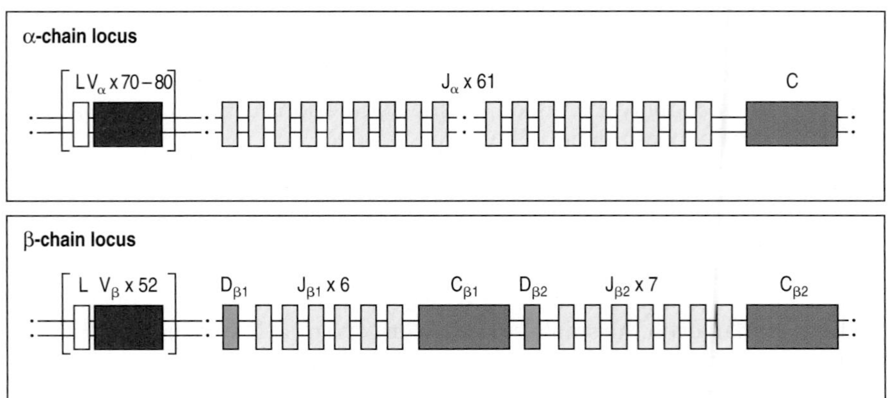

Fig. 4.9 The germline organization of the human T-cell receptor α and β loci. The arrangement of the gene segments resembles that at the immunoglobulin loci, with separate variable (V), diversity (D), joining (J) gene segments, and constant (C) genes. The TCRα locus (chromosome 14) consists of 70–80 V_α gene segments, each preceded by an exon encoding the leader sequence (L). How many of these V_α gene segments are functional is not known exactly. A cluster of 61 J_α gene segments is located a considerable distance from the V_α gene segments. The J_α gene segments are followed by a single C gene, which contains separate exons for the constant and hinge domains and a single exon encoding the transmembrane and cytoplasmic regions (not shown). The TCRβ locus (chromosome 7) has a different organization, with a cluster of 52 functional V_β gene segments located distantly from two separate clusters each containing a single D gene segment together with six or seven J gene segments and a single C gene. Each TCRβ C gene has separate exons encoding the constant domain, the hinge, the transmembrane region, and the cytoplasmic region (not shown). The TCRα locus is interrupted between the J and V gene segments by another T-cell receptor locus—the TCRδ locus (not shown here; see Fig. 4.14).

organization of the T-cell receptor loci and the generation of the genes for the individual T-cell receptor chains.

4-9 The T-cell receptor gene segments are arranged in a similar pattern to immunoglobulin gene segments and are rearranged by the same enzymes.

Like immunoglobulin light and heavy chains, T-cell receptor α and β chains each consist of a variable (V) amino-terminal region and a constant (C) region (see Section 3-10). The organization of the TCRα and TCRβ loci is shown in Fig. 4.9. The organization of the gene segments is broadly homologous to that of the immunoglobulin gene segments (see Sections 4-2 and 4-3). The TCRα locus, like those for the immunoglobulin light chains, contains V and J gene segments (V_α and J_α). The TCRβ locus, like that for the immunoglobulin heavy chain, contains D gene segments in addition to V_β and J_β gene segments. The T-cell receptor gene segments rearrange during T-cell development to form complete V-domain exons (Fig. 4.10). T-cell receptor gene rearrangement takes place in the thymus; the order and regulation of the rearrangements are dealt with in detail in Chapter 7. Essentially, however, the mechanics of gene rearrangement are similar for B and T cells. The T-cell receptor gene segments are flanked by 12-bp and 23-bp spacer recombination signal sequences (RSSs) that are homologous to those flanking immunoglobulin gene segments (Fig. 4.11, and see Section 4-4) and are recognized by the same enzymes. The DNA circles resulting from gene rearrangement (see Fig. 4.6) are known as T-cell receptor excision circles (TRECs) and are used as markers for T cells that have recently emigrated from the thymus. All known defects in genes that control V(D)J recombination affect T cells and B cells equally, and animals with these genetic defects lack functional lymphocytes altogether (see Section 4-5). A further shared feature of immunoglobulin and T-cell receptor gene rearrangement is the presence of P- and N-nucleotides in the junctions between the V, D, and J gene segments of the rearranged TCRβ gene. In T cells, P- and N-nucleotides are also added between the V and J gene segments of all rearranged TCRα genes, whereas only about half the V–J joints in immunoglobulin light-chain genes are modified by N-nucleotide addition, and these are often left without any P-nucleotides as well (Fig. 4.12, and see Section 4-8).

The main differences between the immunoglobulin genes and those encoding T-cell receptors reflect the fact that all the effector functions of B cells depend upon secreted antibodies whose different heavy-chain C-region isotypes trigger distinct effector mechanisms. The effector functions of T cells, in

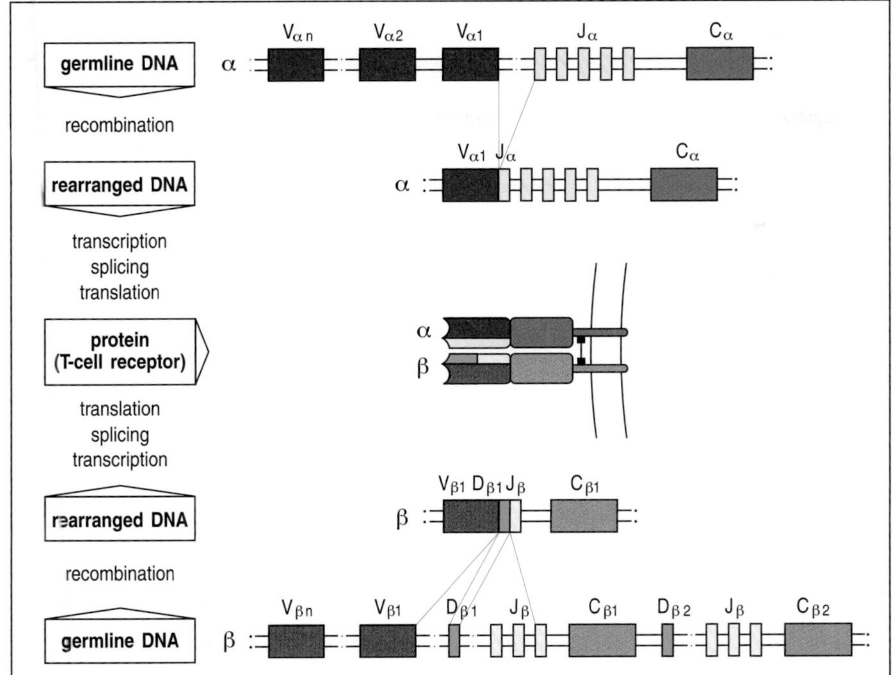

Fig. 4.10 T-cell receptor α- and β-chain gene rearrangement and expression. The TCRα- and β-chain genes are composed of discrete segments that are joined by somatic recombination during development of the T cell. Functional α- and β-chain genes are generated in the same way that complete immunoglobulin genes are created. For the α chain (upper part of figure), a V_α gene segment rearranges to a J_α gene segment to create a functional V-region exon. Transcription and splicing of the VJ_α exon to C_α generates the mRNA that is translated to yield the T-cell receptor α-chain protein. For the β chain (lower part of figure), like the immunoglobulin heavy chain, the variable domain is encoded in three gene segments, V_β, D_β, and J_β. Rearrangement of these gene segments generates a functional VDJ_β V-region exon that is transcribed and spliced to join to C_β; the resulting mRNA is translated to yield the T-cell receptor β chain. The α and β chains pair soon after their synthesis to yield the α:β T-cell receptor heterodimer. Not all J gene segments are shown, and the leader sequences preceding each V gene segment are omitted for simplicity.

contrast, depend upon cell–cell contact and are not mediated directly by the T-cell receptor, which serves only for antigen recognition. Thus, the C regions of the TCRα and TCRβ loci are much simpler than those of the immunoglobulin heavy-chain locus. There is only one Cα gene and, although there are two Cβ genes, they are very closely homologous and there is no known functional distinction between their products. The T-cell receptor C-region genes encode only transmembrane polypeptides.

4-10 T-cell receptors concentrate diversity in the third hypervariable region.

The three-dimensional structure of the antigen-recognition site of a T-cell receptor looks much like that of an antibody molecule (see Sections 3-11 and 3-7, respectively). In an antibody, the center of the antigen-binding site is formed by the CDR3s of the heavy and light chains. The structurally equivalent third hypervariable loops (CDR3s) of the T-cell receptor α and β chains, to which the D and J gene segments contribute, also form the center of the antigen-binding site of a T-cell receptor; the periphery of the site consists of

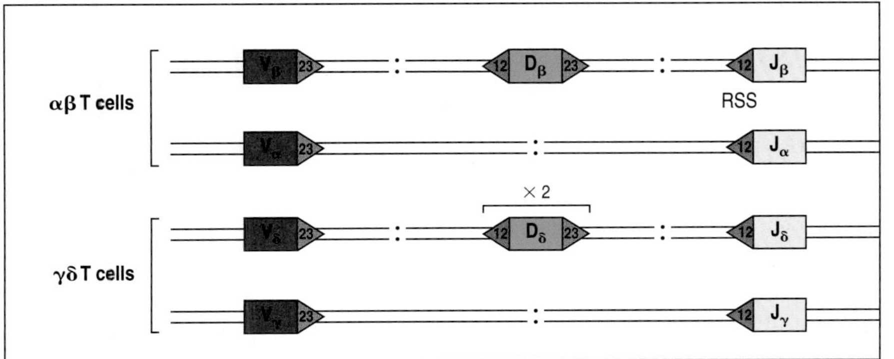

Fig. 4.11 Recombination signal sequences flank T-cell receptor gene segments. As in the immunoglobulin gene loci (see Fig. 4.5), the individual gene segments at the TCRα and TCRβ loci are flanked by heptamer–spacer–nonamer recombination signal sequences (RSSs). RSS motifs containing 12-bp spacers are depicted here as orange arrowheads, and those containing 23-bp spacers are shown in purple. Joining of gene segments almost always follows the 12/23 rule. Because of the disposition of heptamer and nonamer RSSs in the TCRβ and TCRδ loci, direct V_β to J_β joining is in principle allowed by the 12/23 rule (unlike in the immunoglobulin heavy-chain gene), although this occurs very rarely owing to other types of regulation that occur.

Element	Immunoglobulin		α:β T-cell receptors	
	H	κ+λ	β	α
Variable segments (V)	40	70	52	~70
Diversity segments (D)	25	0	2	0
D segments read in three frames	rarely	–	often	–
Joining segments (J)	6	5(κ) 4(λ)	13	61
Joints with N- and P-nucleotides	2	50% of joints	2	1
Number of V gene pairs	1.9×10^6		5.8×10^6	
Junctional diversity	$\sim 3 \times 10^7$		$\sim 2 \times 10^{11}$	
Total diversity	$\sim 5 \times 10^{13}$		$\sim 10^{18}$	

Fig. 4.12 The numbers of human T-cell receptor gene segments and the sources of T-cell receptor diversity compared with those of immunoglobulins. Note that only about half of human κ chains contain N-nucleotides. Somatic hypermutation as a source of diversity in immunoglobulins is not included in this figure because it does not occur in T cells.

the CDR1 and CDR2 loops, which are encoded within the germline V gene segments for the α and β chains. The extent and pattern of variability in T-cell receptors and immunoglobulins reflect the distinct nature of their ligands. Whereas the antigen-binding sites of immunoglobulins must conform to the surfaces of an almost infinite variety of different antigens, and thus come in a wide variety of shapes and chemical properties, the ligand for the major class of human T-cell receptor (α:β) is always a peptide bound to an MHC molecule. As a group, the antigen-recognition sites of T-cell receptors would therefore be predicted to have a less variable shape, with most of the variability focused on the bound antigenic peptide occupying the center of the surface in contact with the receptor. Indeed, the less variable CDR1 and CDR2 loops of a T-cell receptor will mainly contact the relatively less variable MHC component of the ligand, whereas the highly variable CDR3 regions will mainly contact the unique peptide component (Fig. 4.13).

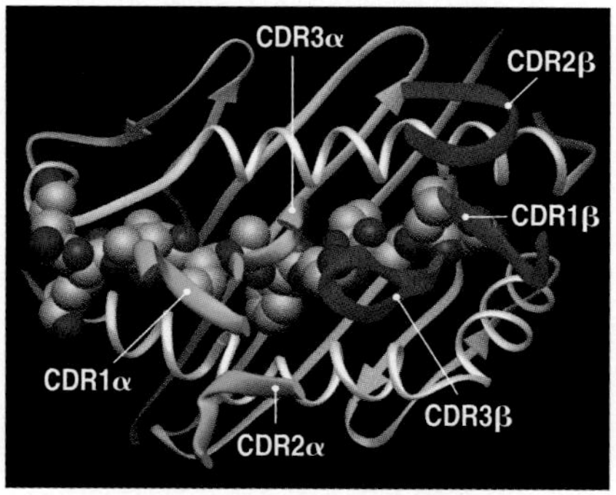

Fig. 4.13 The most variable parts of the T-cell receptor interact with the peptide bound to an MHC molecule. The positions of the CDR loops of a T-cell receptor are shown as colored tubes, which in this figure are superimposed onto the peptide:MHC complex (MHC, gray; peptide, yellow-green with O atoms in red and N atoms in blue). The CDR loops of the α chain are in green, while those of the β chain are in magenta. The CDR3 loops lie in the center of the interface between the TCR and the peptide:MHC complex, and make direct contacts with the antigenic peptide.

The structural diversity of T-cell receptors is attributable mainly to combinatorial and junctional diversity generated during the process of gene rearrangement. It can be seen from Fig. 4.12 that most of the variability in T-cell receptor chains is in the junctional regions, which are encoded by V, D, and J gene segments and modified by P- and N-nucleotides. The TCRα locus contains many more J gene segments than either of the immunoglobulin light-chain loci: in humans, 61 J_α gene segments are distributed over about 80 kb of DNA, whereas immunoglobulin light-chain loci have only 5 J gene segments at most (see Fig. 4.12). Because the TCRα locus has so many J gene segments, the variability generated in this region is even greater for T-cell receptors than for immunoglobulins. Thus, most of the diversity resides in the CDR3 loops that contain the junctional region and form the center of the antigen-binding site.

4-11 γ:δ T-cell receptors are also generated by gene rearrangement.

A minority of T cells bear T-cell receptors composed of γ and δ chains (see Section 3-19). The organization of the TCRγ and TCRδ loci (Fig. 4.14) resembles

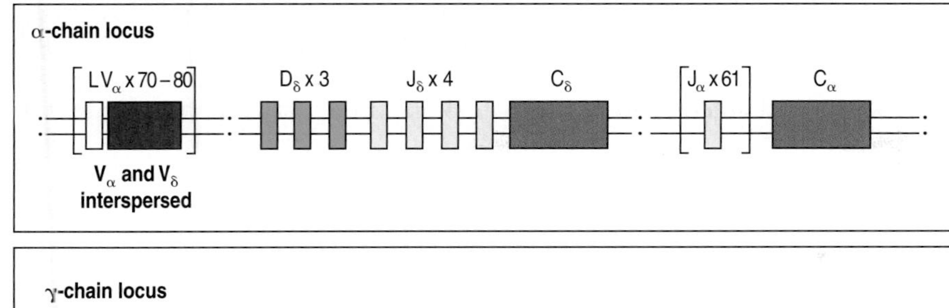

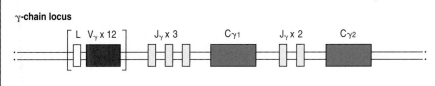

Fig. 4.14 The organization of the T-cell receptor γ- and δ-chain loci in humans. The TCRγ and TCRδ loci, like the TCRα and TCRβ loci, have discrete V, D, and J gene segments, and C genes. Uniquely, the locus encoding the δ chain is located entirely within the α-chain locus. The three D_δ gene segments, four J_δ gene segments, and the single δ C gene lie between the cluster of V_α gene segments and the cluster of J_α gene segments. There are two V_δ gene segments located near the δ C gene, one just upstream of the D regions and one in inverted orientation just downstream of the C gene (not shown). In addition, there are six V_δ gene segments interspersed among the V_α gene segments. Five are shared with V_α and can be used by either locus and one is unique to the δ locus. The human TCRγ locus resembles the TCRβ locus in having two C genes each with its own set of J gene segments. The mouse γ locus (not shown) has a more complex organization and there are three functional clusters of γ gene segments, each containing V and J gene segments and a C gene. Rearrangement at the γ and δ loci proceeds as for the other T-cell receptor loci, with the exception that during TCRδ rearrangement two D segments can be used in the same gene. The use of two D segments greatly increases the variability of the δ chain, mainly because extra N-region nucleotides can be added at the junction between the two D gene segments as well as at the V–D and D–J junctions.

that of the TCRα and TCRβ loci, although there are important differences. The cluster of gene segments encoding the δ chain is found entirely within the TCRα locus, between the V_α and the J_α gene segments. V_δ genes are interspersed with the V_α genes but are located primarily in the 3′ region of the locus. Because all V_α gene segments are oriented such that rearrangement will delete the intervening DNA, any rearrangement at the α locus results in the loss of the δ locus (Fig. 4.15). There are substantially fewer V gene segments at the TCRγ and TCRδ loci than at either the TCRα or TCRβ loci or any of the immunoglobulin loci. Increased junctional variability in the δ chains may compensate for the small number of V gene segments and has the effect of focusing almost all the variability in the γ:δ receptor in the junctional region. As we have seen for the α:β T-cell receptors, the amino acids encoded by the junctional regions lie at the center of the T-cell receptor binding site.

T cells bearing γ:δ receptors are a distinct lineage of T cells whose functions are at present unclear. The ligands for these receptors are also largely unknown. Some γ:δ T-cell receptors seem able to recognize antigen directly, much as antibodies do, without presentation by an MHC molecule or antigen processing. Detailed analysis of the rearranged V regions of γ:δ T-cell receptors shows that they resemble the V regions of antibody molecules more than those of α:β T-cell receptors.

Summary.

T-cell receptors are structurally similar to immunoglobulins and are encoded by homologous genes. T-cell receptor genes are assembled by somatic recombination from sets of gene segments in the same way that the immunoglobulin genes are. Diversity is, however, distributed differently in immunoglobulins and T-cell receptors: the T-cell receptor loci have roughly the same number of V gene segments but more J gene segments, and there is greater diversification of the junctions between gene segments during gene rearrangement. Moreover, functional T-cell receptors are not known to diversify their V genes after rearrangement through somatic hypermutation. This leads to a T-cell receptor in which the highest diversity is in the central part of the receptor,

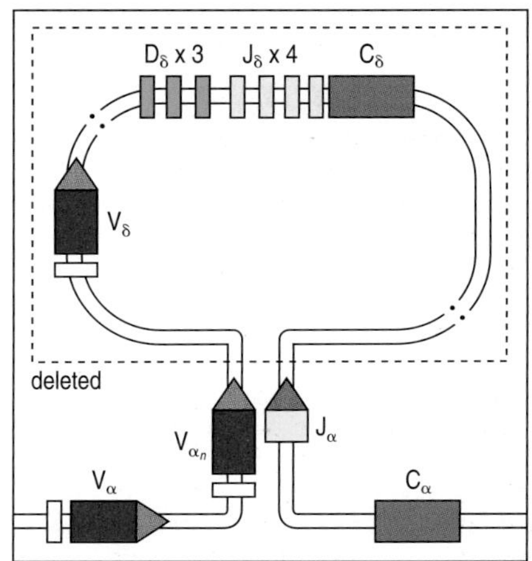

Fig. 4.15 Deletion of the TCRδ locus is induced by rearrangement of a V_α to J_α gene segment. The TCRδ locus is entirely contained within the chromosomal region containing the TCRα locus. When any V region in the V_α/V_δ region rearranges to any one of the J_α segments, the intervening region, and the entire V_δ locus, is deleted. Thus, V_α rearrangement prevents any continued expression of a V_δ gene and precludes lineage development down the γ:δ pathway.

which in the case of α:β T-cell receptors contacts the bound peptide fragment of the ligand. Most of the diversity among γ:δ T-cell receptors is also in CDR3, but how this affects ligand binding is less clear because γ:δ T cells directly recognize poorly characterized ligands which in some cases are independent of MHC molecules.

Structural variation in immunoglobulin constant regions.

So far in this chapter we have focused on the structural variation in antigen receptors that results from the assembly of the V regions. We now turn to the C regions. The C regions of T-cell receptors have no functional purpose beyond supporting the V regions and anchoring the molecule in the membrane, and we shall not discuss them further here. Immunoglobulins, in contrast, can be made as both a transmembrane receptor and a secreted antibody, and the C domains of antibodies are crucial to their diverse effector functions.

Immunoglobulins are made in several different classes, which are distinguished by their heavy chains. Different heavy chains are produced in a given clone of B cells by linking different heavy-chain C regions (C_H) to the rearranged V_H gene. All classes of immunoglobulin produced by a clone of B cells thus have the same V region. In the heavy-chain locus, the different C regions are encoded in separate genes located downstream of the V-region segments. Initially, naive B cells use only the first two of these, the C_μ and C_δ genes, which are expressed along with the associated assembled V-region sequence to produce transmembrane IgM and IgD on the surface of the naive B cell. During the course of an antibody response, activated B cells can switch to the expression of C_H genes other than C_μ and C_δ by a type of somatic recombination known as class switching. Along with other mechanisms that further diversify immunoglobulins, class switching will be discussed in the last part of this chapter. In contrast to heavy-chain C regions, light-chain C regions (C_L) do not provide specific effector function other than structural attachment for V regions, do not undergo class switching, and there appear to be no functional differences between λ and κ light chains.

In this part of the chapter we consider the structural features that distinguish the C_H regions of antibodies of the five major classes, and discuss some of their special properties. The functions of the different antibody classes are considered in more detail in Chapter 9. We also explain how the same antibody gene can generate both membrane-bound immunoglobulin and secreted immunoglobulin through alternative mRNA splicing.

4-12 Different classes of immunoglobulins are distinguished by the structure of their heavy-chain constant regions.

The five main classes of immunoglobulin are IgM, IgD, IgG, IgE, and IgA, all of which can occur as transmembrane antigen receptors or secreted antibodies. In humans, IgG can be further subdivided into four subclasses (IgG1, IgG2, IgG3, and IgG4), and IgA antibodies are found as two subclasses (IgA1 and IgA2). The IgG subclasses in humans are named in order of the abundance of the antibodies in serum, with IgG1 being the most abundant. The different heavy chains that define these classes are known as isotypes and are designated by the lower-case Greek letters μ, δ, γ, ε, and α, as shown in Fig. 4.16, which lists the major physical and functional properties of the different human antibody classes. IgM forms pentamers in serum, which accounts for its high molecular weight. Secreted IgA can occur as either a monomer or a dimer. Sequence differences between immunoglobulin heavy chains cause the various isotypes to differ in several characteristic respects. These include the number and location of interchain disulfide bonds, the number of attached oligosaccharide moieties, the number of C domains, and the length of the hinge region (Fig. 4.17). IgM and IgE heavy chains contain an extra C domain that replaces the hinge region found in γ, δ, and α chains. The absence of the hinge region does not imply that IgM and IgE molecules lack flexibility; electron micrographs of IgM molecules binding to ligands show that the Fab arms can bend relative to the Fc portion. However, such a difference in structure may have functional consequences that are not yet characterized. Different isotypes and subtypes also differ in their ability to engage various effector functions, as described later. The distinct properties of the different C regions are encoded by different immunoglobulin C_H genes that are present in a cluster located 3′ of the J_H segments. We describe the rearrangement process by which the V region becomes associated with a different C_H gene in Section 4-20.

4-13 The constant region confers functional specialization on the antibody.

Antibodies can protect the body in a variety of ways. In some cases it is enough for the antibody simply to bind antigen. For instance, by binding tightly to a toxin or virus, an antibody can prevent it from recognizing its receptor on a host cell (see Fig. 1.24). The antibody V regions are sufficient for

	Immunoglobulin								
	IgG1	IgG2	IgG3	IgG4	IgM	IgA1	IgA2	IgD	IgE
Heavy chain	γ_1	γ_2	γ_3	γ_4	μ	α_1	α_2	δ	ε
Molecular weight (kDa)	146	146	165	146	970	160	160	184	188
Serum level (mean adult mg ml^{-1})	9	3	1	0.5	1.5	3.0	0.5	0.03	5×10^{-5}
Half-life in serum (days)	21	20	7	21	10	6	6	3	2
Classical pathway of complement activation	++	+	+++	−	++++	−	−	−	−
Alternative pathway of complement activation	−	−	−	−	−	+	−	−	−
Placental transfer	+++	+	++	−/+	−	−	−	−	−
Binding to macrophage and phagocyte Fc receptors	+	−	+	−/+	−	+	+	−	+
High-affinity binding to mast cells and basophils	−	−	−	−	−	−	−	−	+++
Reactivity with staphylococcal Protein A	+	+	−/+	+	−	−	−	−	−

Fig. 4.16 The physical properties of the human immunoglobulin isotypes. IgM is so called because of its size: although monomeric IgM is only 190 kDa, it normally forms pentamers, known as macroglobulin (hence the M), of very large molecular weight (see Fig. 4.20). IgA dimerizes to give a molecular weight of around 390 kDa in secretions. IgE antibody is associated with immediate-type hypersensitivity. When fixed to tissue mast cells, IgE has a much longer half-life than its half-life in plasma shown here.

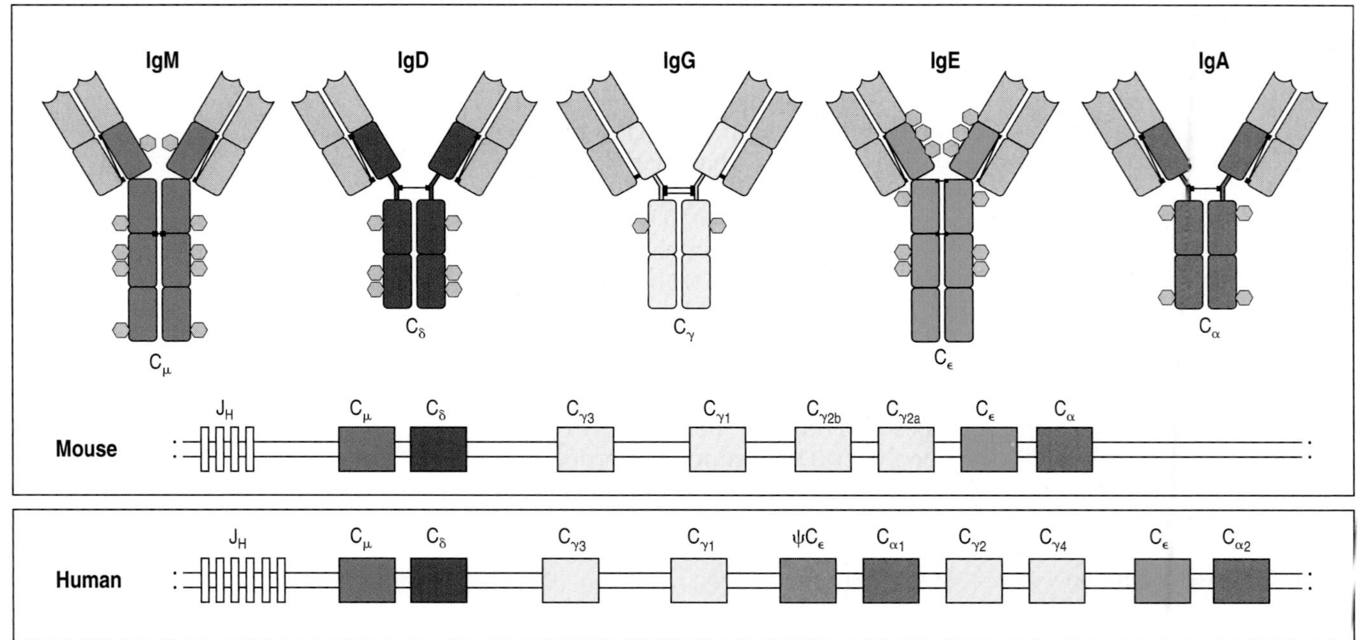

Fig. 4.17 The immunoglobulin isotypes are encoded by a cluster of immunoglobulin heavy-chain C-region genes. The general structure of the main immunoglobulin isotypes (upper panel) is indicated, with the immunoglobulin domain indicated as a rectangle. These are encoded by separate heavy-chain C-region genes arranged in a cluster in both mouse and human (lower panel). The constant region of the heavy chain for each istoype is indicated by the same color as the C-region gene segment that encodes it. IgM and IgE lack a hinge region but each contains an extra heavy-chain domain. Note the differences in the numbers and locations of the disulfide bonds (black lines) linking the chains. The isotypes also differ in the distribution of N-linked carbohydrate groups, shown as hexagons. In humans, the cluster shows evidence of evolutionary duplication of a unit consisting of two γ genes, an ε gene, and an α gene. One of the ε genes is a pseudogene (ψ); hence only one subtype of IgE is expressed. For simplicity, other pseudogenes are not illustrated, and the exon details within each C gene are not shown. The classes of immunoglobulins found in mice are called IgM, IgD, IgG1, IgG2a, IgG2b, IgG3, IgA, and IgE.

this activity. The C region is essential, however, for recruiting the help of other cells and molecules to destroy and dispose of pathogens to which the antibody has bound.

The C regions (Fc portions) of antibodies have three main effector functions. First, the Fc portions of certain isotypes are recognized by specialized **Fc receptors** expressed by immune effector cells. Fcγ receptors present on the surface of phagocytic cells such as macrophages and neutrophils bind the Fc portions of IgG1 and IgG3 antibodies, thus facilitating the phagocytosis of pathogens coated with these antibodies. The Fc portion of IgE binds to a high-affinity Fcε receptor on mast cells, basophils, and activated eosinophils, enabling these cells to respond to the binding of specific antigen by releasing inflammatory mediators. Second, the Fc portions of antigen:antibody complexes can bind to complement (see Fig. 1.24) and initiate the complement cascade, which helps to recruit and activate phagocytes, aids the engulfment of microbes by phagocytes, and can also directly destroy pathogens. Third, the Fc portion can deliver antibodies to places they would not reach without active transport. These include mucous secretions, tears, and milk (IgA), and the fetal blood circulation by transfer from the pregnant mother (IgG). In both cases, the Fc portion engages a specific receptor that actively transports the immunoglobulin through cells to reach different body compartments.

The role of the Fc portion in these effector functions has been demonstrated by studying immunoglobulins that have had one or other Fc domain cleaved off enzymatically (see Section 3-3) or, more recently, by genetic engineering, which permits detailed mapping of the amino acid residues within the Fc that are needed for particular functions. Many microorganisms have responded to the destructive potential of the Fc portion by evolving proteins that either bind or cleave it, and so prevent it from working; examples are Protein A and Protein G of *Staphylococcus* and Protein D of *Haemophilus*. Researchers have exploited these proteins to help map the Fc and as immunological reagents (see Appendix I, Section A-10). Not all immunoglobulin classes have the same capacity to engage each of the effector functions. The different functional properties of each heavy-chain isotype are summarized in Fig. 4.16. For example, IgG1 and IgG3 have a higher affinity than IgG2 for the most common type of Fc receptor.

4-14 Mature naive B cells express both IgM and IgD at their surface.

The immunoglobulin C_H genes form a large cluster spanning about 200 kb to the 3′ side of the J_H gene segments (see Fig. 4.17). Each C_H gene is split into several exons (not shown in the figure), each corresponding to an individual immunoglobulin domain in the folded C region. The gene encoding the μ C region lies closest to the J_H gene segments, and therefore closest to the assembled V_H-region exon (VDJ exon) after DNA rearrangement. Once rearrangement is completed, a complete μ heavy-chain transcript is produced. Any J_H gene segments remaining between the assembled V gene and the C_μ gene are removed during RNA processing to generate the mature mRNA. μ heavy chains are therefore the first to be expressed and IgM is the first immunoglobulin to be produced during B-cell development.

Immediately 3′ to the μ gene lies the δ gene, which encodes the C region of the IgD heavy chain (see Fig 4.17). IgD is coexpressed with IgM on the surface of almost all mature B cells, although this isotype is secreted in only small amounts by plasma cells and its function is unknown. Indeed, mice lacking the C_δ exons seem to have essentially normal immune systems. B cells expressing IgM and IgD have not undergone class switching, which, as we will see, entails an irreversible change in the DNA. Instead, these cells produce a long primary mRNA transcript that is differentially cleaved and spliced to yield one of two distinct mRNA molecules. In one, the VDJ exon is linked to the C_μ exons to encode a μ heavy chain, and in the other the VDJ exon is linked to the C_δ exons to encode a δ heavy chain (Fig. 4.18). The processing of the long mRNA transcript is developmentally regulated, with immature B cells making mostly the μ transcript and mature B cells making mostly δ along with some μ. When a B cell is activated it ceases to coexpress IgD with IgM, either because μ and δ sequences have been removed as a consequence of a class switch or, in IgM-secreting plasma cells, because transcription from the V_H promoter no longer extends through the C_δ exons.

Fig. 4.18 Coexpression of IgD and IgM is regulated by RNA processing. In mature B cells, transcription initiated at the V_H promoter extends through both C_μ and C_δ exons. This long primary transcript is then processed by cleavage and polyadenylation (AAA), and by splicing. Cleavage and polyadenylation at the μ site (pA1) and splicing between C_μ exons yields an mRNA encoding the μ heavy chain (left panel). Cleavage and polyadenylation at the δ site (pA2) and a different pattern of splicing that removes the C_μ exons yields mRNA encoding the δ heavy chain (right panel). For simplicity we have not shown all the individual C-region exons.

4-15 Transmembrane and secreted forms of immunoglobulin are generated from alternative heavy-chain transcripts.

Immunoglobulins of all classes can be produced either as a membrane-bound receptor or as secreted antibodies. All B cells initially express the

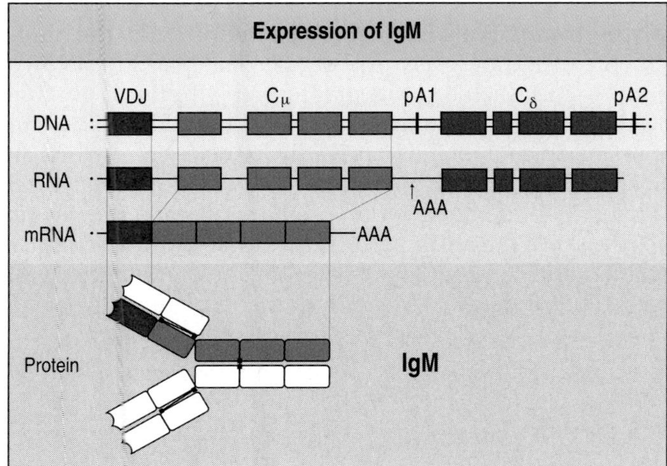

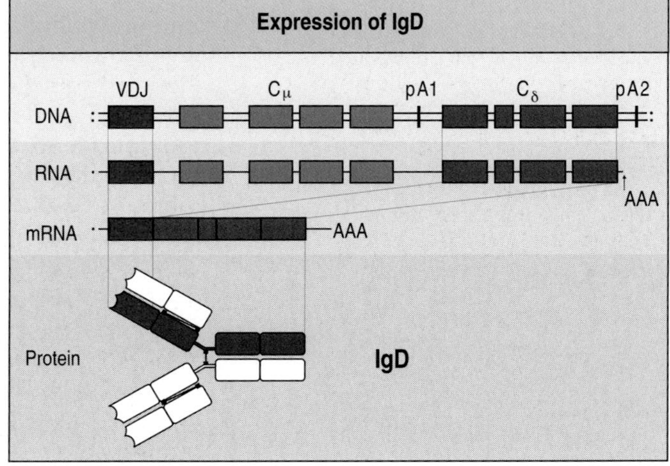

transmembrane form of IgM; after stimulation by antigen, some of their progeny differentiate into plasma cells producing IgM antibodies, whereas others undergo class switching to express transmembrane immunoglobulins of a different class followed by the production of secreted antibody of the new class. The membrane forms of all immunoglobulin classes are monomers comprising two heavy and two light chains: IgM and IgA polymerize only when they have been secreted. In its membrane-bound form, the immunoglobulin heavy chain has a hydrophobic transmembrane domain of about 25 amino acid residues at the carboxy terminus, which anchors it to the surface of the B lymphocyte. This domain is absent from the secreted form, whose carboxy terminus is a hydrophilic secretory tail. The carboxy termini of the transmembrane and secreted forms of immunoglobulin heavy chains are encoded in two different exons, and production of the two forms is achieved by alternative RNA processing (Fig. 4.19). The last two exons of each C_H gene contain the sequences encoding the secreted and the transmembrane regions respectively; if the primary transcript is cleaved and polyadenylated at a site downstream of these exons, the sequence encoding the carboxy terminus of the secreted form is removed by splicing, and the cell-surface form of immunoglobulin is produced. Alternatively, if the primary transcript is cleaved at the polyadenylation site located before the last two exons, only the secreted molecule can be produced. This differential RNA processing is illustrated for C_μ in Fig. 4.19, but it occurs in the same way for all isotypes. In activated B cells that differentiate to become antibody-secreting plasma cells, most of the transcripts are spliced to the secreted rather than the transmembrane form of whichever heavy-chain isotype the B cell happens to be expressing.

4-16 IgM and IgA can form polymers.

Although all immunoglobulin molecules are constructed from a basic unit of two heavy and two light chains, both IgM and IgA can form multimers of these basic units (Fig. 4.20). IgM and IgA C regions include a 'tailpiece' of 18 amino acids that contains a cysteine residue essential for polymerization. A separate 15 kDa polypeptide chain called the J chain promotes polymerization by linking to the cysteines of this tailpiece, which is found only in the secreted forms of the μ and α chains. (This J chain should not be confused with the immunoglobulin J region encoded by a J gene segment; see Section 4-2.) In the case of IgA, polymerization is required for transport through epithelia, as we discuss in Chapter 9. IgM molecules are found as pentamers, and occasionally hexamers (without J chain), in plasma, whereas IgA is found mainly as a dimer in mucous secretions, but as a monomer in plasma.

Immunoglobulin polymerization is also thought to be important in the binding of antibody to repetitive epitopes. An antibody molecule has at least two identical antigen-binding sites, each of which has a given affinity, or binding strength, for antigen (see Appendix I, Section A-9). If the antibody attaches to multiple identical epitopes on a target antigen, it will dissociate only when all binding sites dissociate. The dissociation rate of the whole antibody will therefore be much slower than the dissociation rate for a single binding site; multiple binding sites thus give the antibody a greater total binding strength, or **avidity**. This consideration is particularly relevant for pentameric IgM, which has 10 antigen-binding sites. IgM antibodies frequently recognize

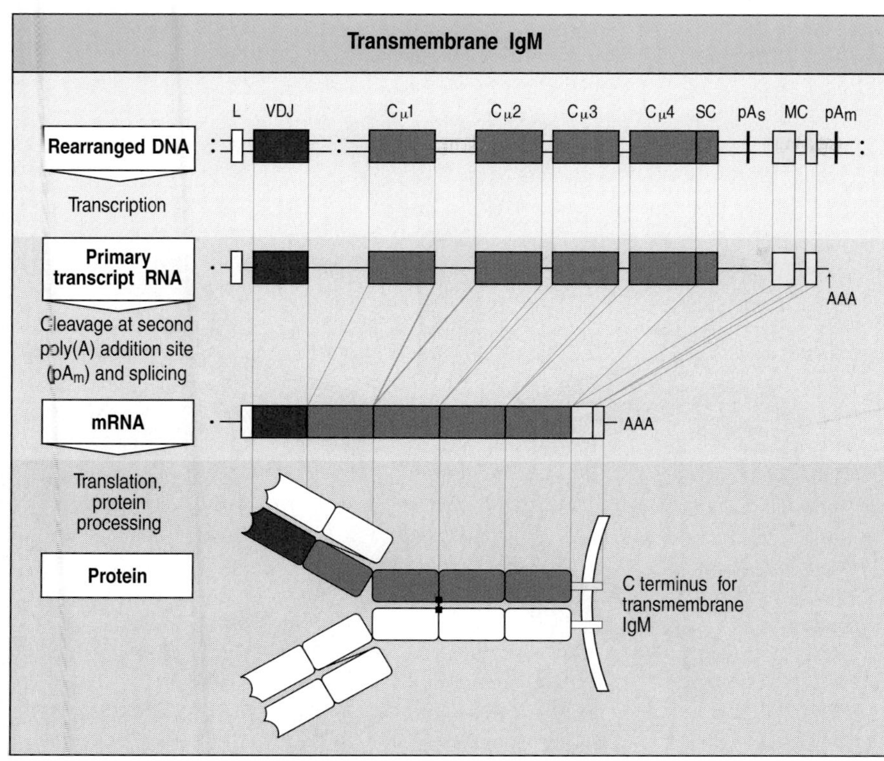

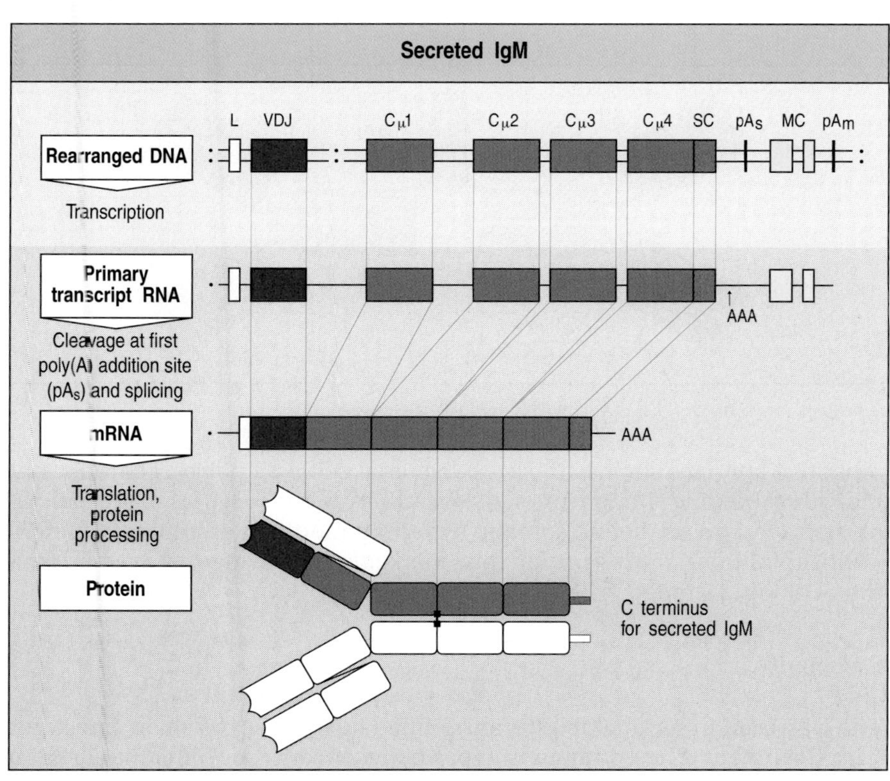

Fig. 4.19 Transmembrane and secreted forms of immunoglobulins are derived from the same heavy-chain sequence by alternative RNA processing. Each heavy-chain C gene has two exons (membrane-coding (MC), yellow) that encode the transmembrane region and cytoplasmic tail of the transmembrane form, and a secretion-coding (SC) sequence (orange) that encodes the carboxy terminus of the secreted form. In the case of IgD, the SC sequence is present on a separate exon, but for the other isotypes, including IgM as shown here, the SC sequence is contiguous with the last C-domain exon. The events that dictate whether a heavy-chain RNA will result in a secreted or transmembrane immunoglobulin occur during processing of the initial transcript. Each heavy-chain C gene has two potential polyadenylation sites (shown as pA_s and pA_m). In the upper panel, the transcript is cleaved and polyadenylated (AAA) at the second site (pA_m). Splicing between a site located between the $C_\mu 4$ exon and the SC sequence, and a second site at the 5′ end of the MC exons, results in removal of the SC sequence and joining of the MC exons to the $C_\mu 4$ exon. This generates the transmembrane form of the heavy chain.

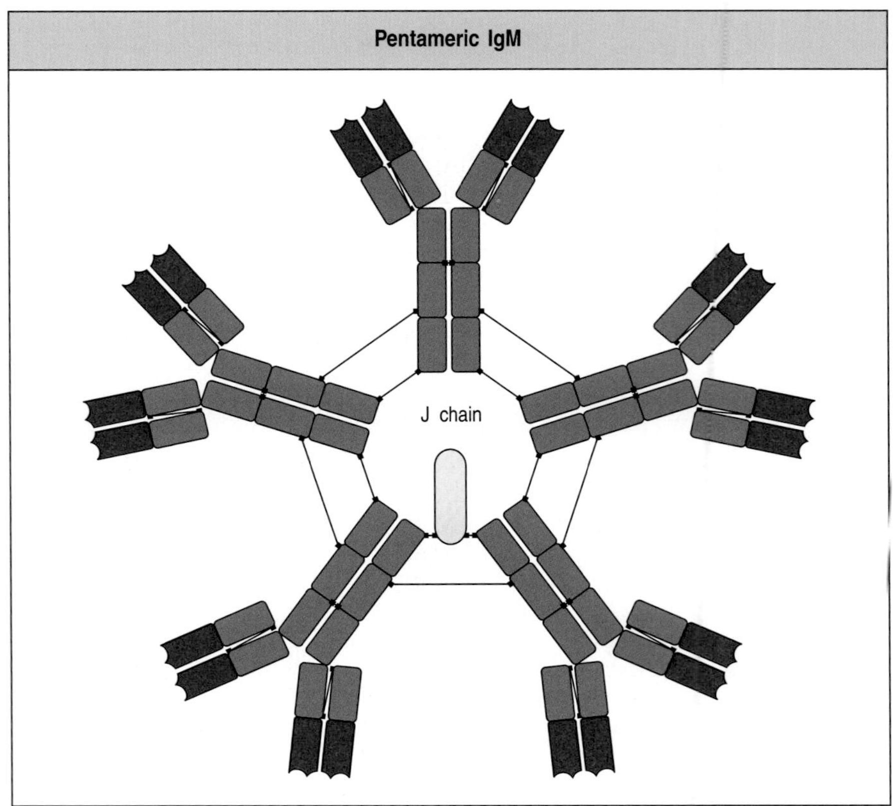

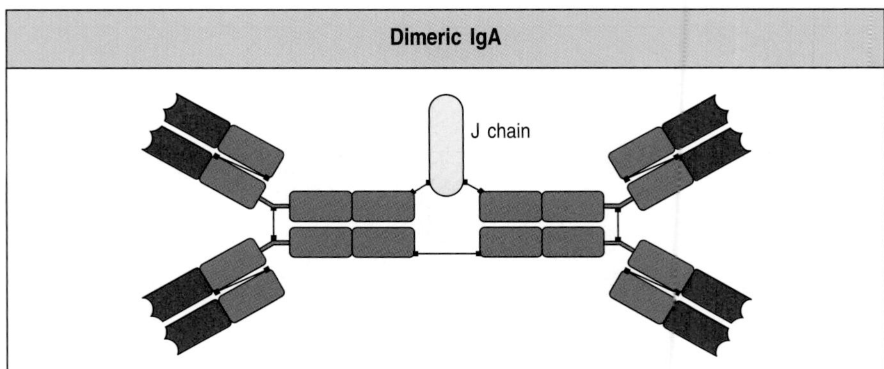

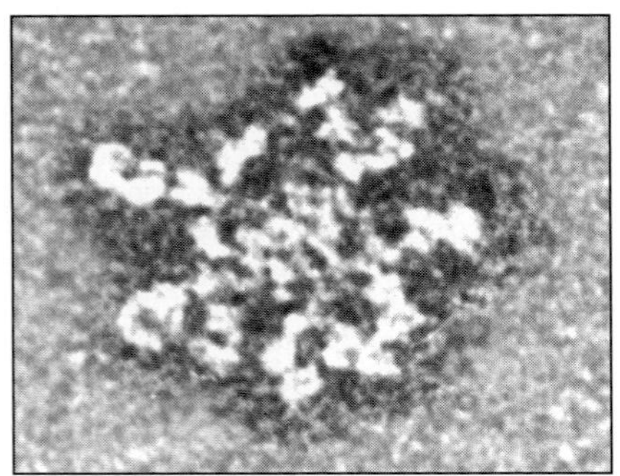

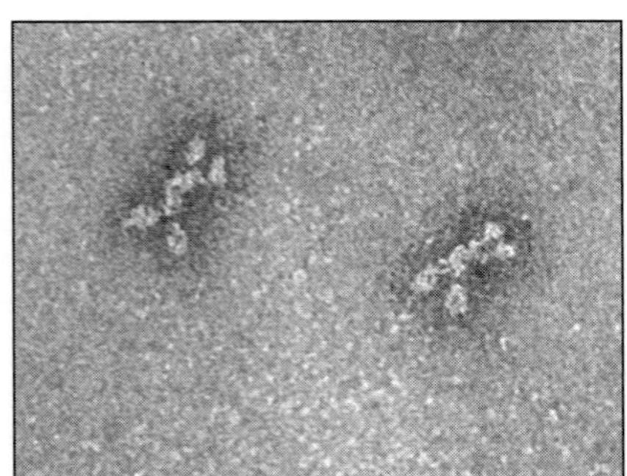

Fig. 4.20 The IgM and IgA molecules can form multimers. IgM and IgA are usually synthesized as multimers in association with an additional polypeptide chain, the J chain. In pentameric IgM, the monomers are cross-linked by disulfide bonds to each other and to the J chain. The top left panel shows an electron micrograph of an IgM pentamer, showing the arrangement of the monomers in a flat disc. IgM can also form hexamers that lack a J chain. In dimeric IgA, the monomers have disulfide bonds to the J chain as well as to each other. The bottom left panel shows an electron micrograph of dimeric IgA. Photographs (× 900,000) courtesy of K.H. Roux and J.M. Schiff.

repetitive epitopes such as those on bacterial cell-wall polysaccharides, but individual binding sites are often of low affinity because IgM is made early in immune responses, before somatic hypermutation and affinity maturation. Multisite binding makes up for this, markedly improving the overall functional binding strength.

Summary.

The classes of immunoglobulins are defined by their heavy-chain C regions, with the different heavy-chain isotypes being encoded by different C-region genes. The heavy-chain C-region genes lie in a cluster 3′ to the V and J gene segments. A productively rearranged V-region exon is initially expressed in association with μ and δ C_H genes, which are coexpressed in naive B cells by alternative splicing of an mRNA transcript that contains both the μ and δ C_H exons. In addition, B cells can express any class of immunoglobulin as a membrane-bound antigen receptor or as secreted antibody. This is achieved by differential splicing of mRNA to include exons that encode either a hydrophobic membrane anchor or a secretable tailpiece. The antibody that a B cell secretes upon activation thus recognizes the antigen that initially activated the B cell via its antigen receptor. The same V-region exon can subsequently be associated with any one of the other isotypes to direct the production of antibodies of different classes. This process of class switching is described in the next part of the chapter.

Secondary diversification of the antibody repertoire.

The RAG-mediated V(D)J recombination described in the first part of the chapter is responsible for the initial antibody repertoire of B cells developing in the bone marrow. These somatic mutations—in the form of gene rearrangements—assemble the genes that produce the primary repertoire of immunoglobulins, and this takes place without interaction of B cells with antigen. Although this primary repertoire is large, further diversification can occur that enhances both the ability of immunoglobulins to recognize and bind to foreign antigens and the effector capacities of the expressed antibodies. This secondary phase of diversification occurs in activated B cells and is largely driven by antigen. Diversification is achieved through three mechanisms—**somatic hypermutation**, **gene conversion**, and **class switching** or **class switch recombination**—which alter the sequence of the secreted immunoglobulin in distinct ways (Fig. 4.21). Class switch recombination involves the C region only: it replaces the original C_{μ} heavy-chain C region with an alternative C

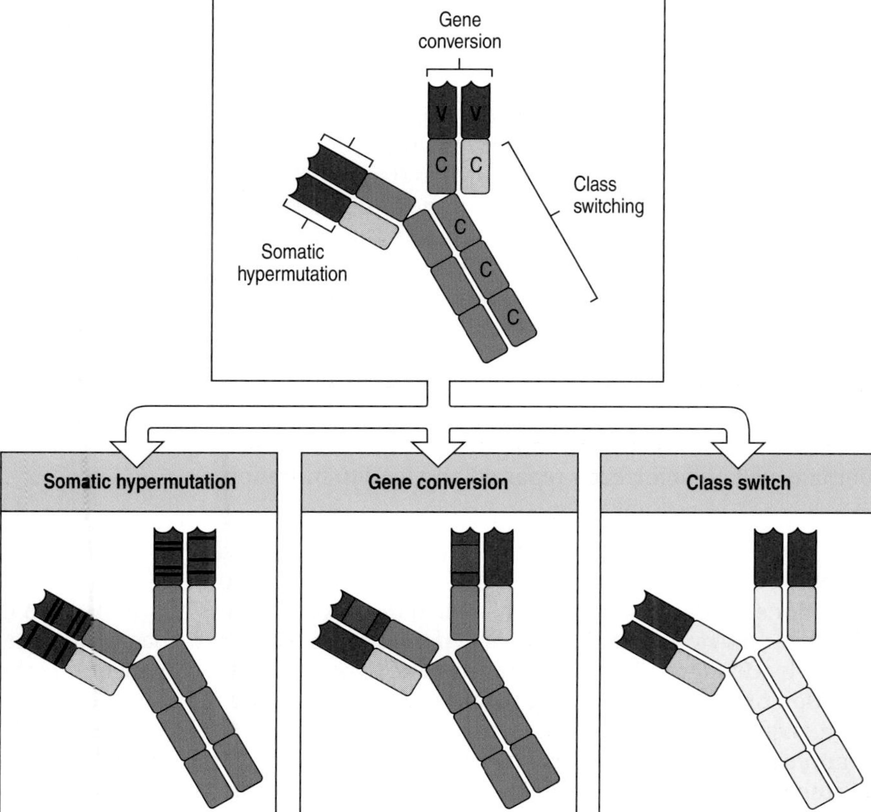

Fig. 4.21 The primary antibody repertoire is diversified by three processes that modify the rearranged immunoglobulin gene. The primary antibody repertoire is initially composed of IgM containing variable regions produced by V(D)J recombination. This broad range of reactivity can further modified by somatic hypermutation, gene conversion, and class switch recombination at the immunoglobulin loci. Somatic hypermutation results in mutations (shown as blue lines) being introduced into the heavy-chain and light-chain V regions (red), altering the affinity of the antibody for its antigen. In gene conversion, the rearranged V region is modified by the introduction of sequences derived from V gene segment pseudogenes, creating additional antibody specificities. In class switch recombination, the initial μ heavy-chain C-regions (blue) are replaced by heavy-chain regions of another isotype (shown as yellow), modifying the effector activity of the antibody but not its antigen specificity.

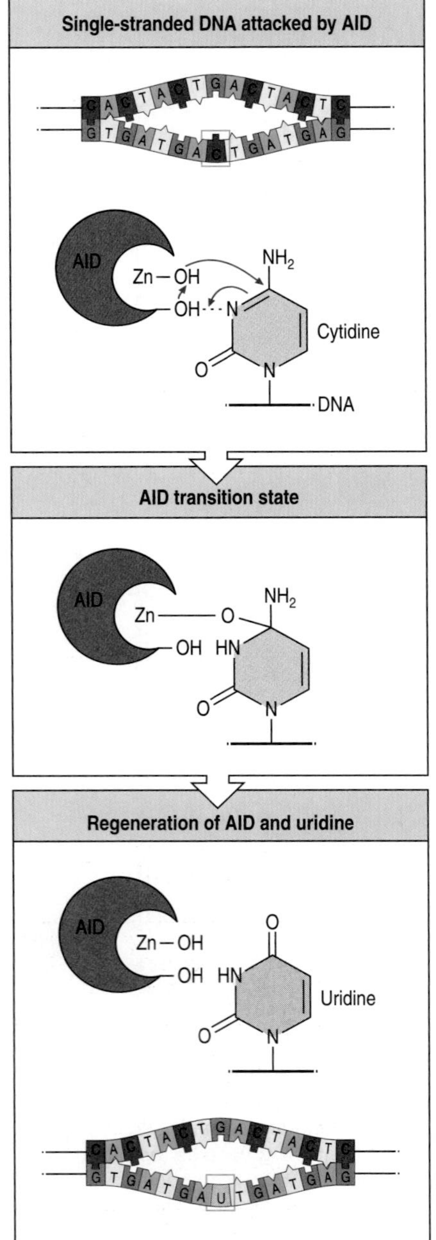

Fig 4.22 Activation-induced cytidine deaminase (AID) is the initiator of mutations in somatic hypermutation, gene conversion, and class switching. The activity of AID, which is expressed only in B cells, requires access to the cytidine side chain of a single-stranded DNA molecule (first panel), which is normally prevented by the hydrogen bonding in double-stranded DNA. AID initiates a nucleophilic attack on the cytosine ring (second panel), which is resolved by the deamination of the cytidine to form uridine (third panel).

region, thereby increasing the functional diversity of the immunoglobulin repertoire. Somatic hypermutation and gene conversion affect the V region. Somatic hypermutation diversifies the antibody repertoire by introducing point mutations into the V regions of both chains, which alters the affinity of the antibody for antigen. Gene conversion diversifies the primary antibody repertoire in some animals, replacing blocks of sequence in the V regions with sequences derived from the V regions of pseudogenes. Like RAG-mediated V(D)J recombination, these processes involve somatic mutation of the immunoglobulin genes, but unlike V(D)J recombination all are initiated by an enzyme called **activation-induced cytidine deaminase** (**AID**), which is expressed specifically in B cells; they do not occur in T-cell receptor genes. The initiation mechanism underlying all these processes is similar and so we will start with a general description of the enzymes involved.

4-17 Activation-induced cytidine deaminase introduces mutations in genes transcribed in B cells.

The enzyme AID was initially identified as a gene that is expressed specifically upon activation of B cells. Its importance for antibody diversification was revealed by analysis of mice engineered to lack AID expression, which showed a lack of somatic hypermutation and class switch recombination. Humans with mutations in AID have also been identified that lack class switching and somatic hypermutation. The sequence of AID is related to that of a protein known for short as APOBEC1 (apolipoprotein B mRNA editing catalytic polypeptide 1), which converts cytosine in apolipoprotein B mRNA to uracil by deamination, and so initially AID was thought to act as an mRNA cytidine deaminase. Although this is still a possibility, current evidence suggests that AID can also act as a DNA cytidine deaminase, directly deaminating cytidine residues in the immunoglobulin genes to uridine. AID can bind to and deaminate single-stranded DNA but not double-stranded DNA. Thus, the DNA double helix must be temporarily unwound locally for AID to act, and this seems to happen as a result of transcription of nearby sequences. By analogy with other cytidine deaminases, it is thought that AID initiates a nucleophilic attack on the pyrimidine ring of the exposed cytidine (Fig. 4.22). Additional ubiquitous DNA repair enzymes cooperate with AID to alter the single-stranded DNA sequence further (Fig. 4.23). The uracil residue produced by AID can be the substrate for the base-excision repair enzyme uracil-DNA glycosylase (UNG), which removes the pyrimidine base to form an abasic site in the DNA. The apurinic/apyrimidinic endonuclease 1 (APE1) can excise the rest of the residue, thereby introducing a single-strand nick in the DNA at the site of the original cytosine. UNG and APE1 act in all cells to efficiently repair the frequent cytosine-to-uracil conversions and abasic sites that occur as a result of spontaneous damage to DNA. AID is active only in activated B cells, and by substantially increasing the amount of DNA damage present in the immunoglobulin genes it greatly increases the chance that this damage will be incorrectly repaired and lead to mutation.

The three types of alterations can lead to quite distinct types of mutation in the immunoglobulin gene, with the extent of the initial change in the DNA roughly corresponding to the nature of the final mutation (Fig. 4.24). These mutations are described in more detail in the next three sections. If the DNA is acted on only by AID, only somatic hypermutation will occur. Abasic sites generated by UNG can also give rise to somatic hypermutation by nucleotide substitution on replication. Single-strand nicks generated by APE1 are thought to be a required signal to initiate the process of templated replication using homologous sequences that occurs in gene conversion. Finally, a high density of single-strand nicks in specific regions flanking C-region genes is thought to generate the staggered double-strand breaks required for class switching.

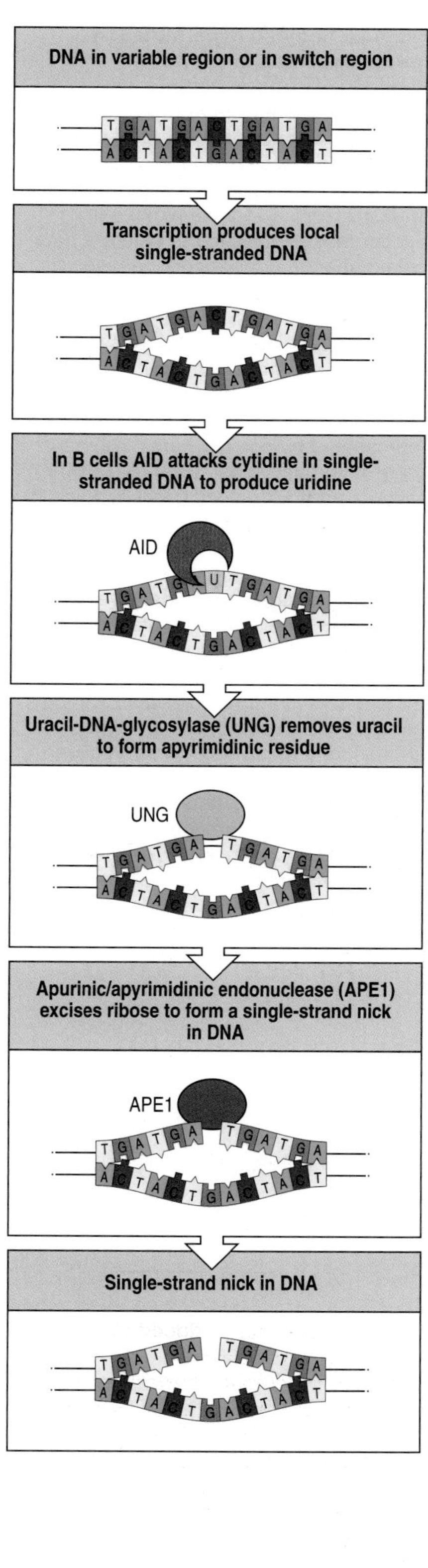

Fig 4.23 Generation of single-strand nicks in DNA by sequential action of AID, uracil-DNA-glycosylase (UNG) and apurinic/apyrimidinic endonuclease 1 (APE1). Double- stranded DNA (first panel) is made accessible to AID by localized transcription that unwinds the DNA helix (second panel). AID, which is specifically expressed in activated B cells, acts to convert cytidine residues to uradines (third panel). The ubiquitous DNA repair enzymes UNG and APE1 can then act on uradine first to remove the uracil ring forming an abasic site (fourth panel), and then to excise the abasic ribose residue from the DNA strand (fifth panel), leading to the formation of a single stranded nick in the DNA (sixth panel).

4-18 Rearranged V-region genes are further diversified by somatic hypermutation.

Somatic hypermutation operates on B cells in peripheral lymphoid organs after functional immunoglobulin genes have been assembled. It introduces point mutations throughout the rearranged V region exon at a very high rate, giving rise to mutant B-cell receptors on the surface of the B cells (Fig. 4.25). In mice and humans, somatic hypermutation occurs in germinal centers only after mature B cells have been activated by their corresponding antigen and it also requires signals from activated T cells. Somatic hypermutation is preferentially targeted to rearranged V regions, which are being actively transcribed in B cells, and does not occur in inactive loci, because AID requires a single-stranded DNA substrate. Other transcribed genes expressed in the

Fig. 4.24 AID initiates processes that lead to somatic hypermutation, gene conversion and class switch recombination. Somatic hypermutation will occur by transition mutations (C to T, or G to A) when the uracil produced by AID action is recognized as a T by DNA polymerases. If UNG has generated an abasic site, non-templated replication across the site can generate either transition or transversion mutations. Gene conversion seems to be triggered by the presence of single-strand nicks, followed by DNA replication that uses homologous pseudogenes as a template for repair. If single-strand nicks are simultaneously converted to staggered double-strand breaks in two different regions flanking C-region genes (switch regions), the cell's machinery for repairing double-strand breaks can lead to rejoining of the breaks in a way that leads to class switching.

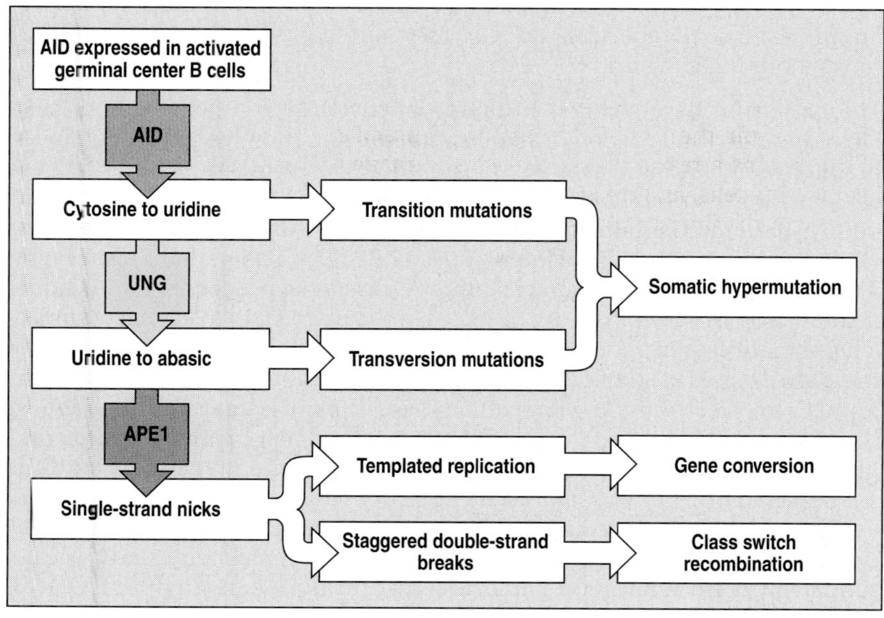

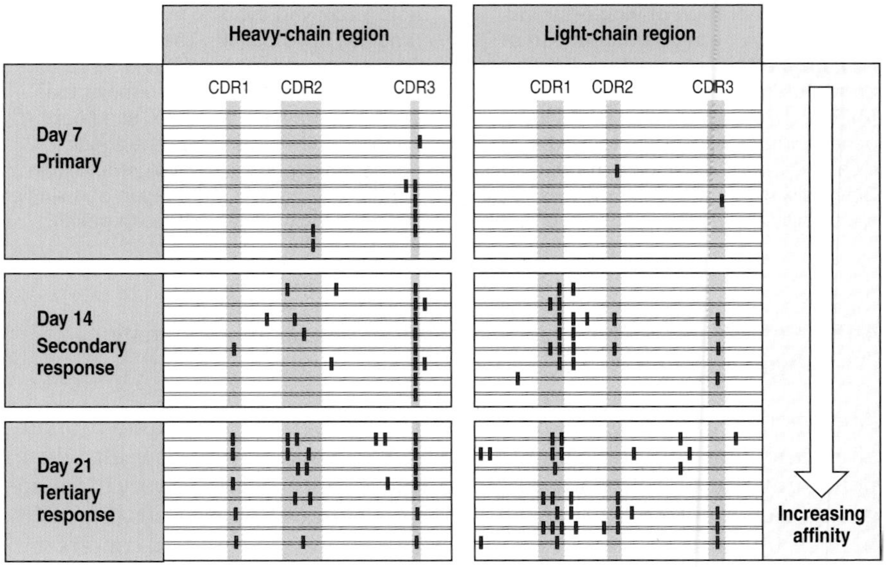

Fig. 4.25 Somatic hypermutation introduces mutations into the rearranged immunoglobulin variable regions that improve antigen binding. In some circumstances it is possible to follow the process of somatic hypermutation by sequencing immunoglobulin variable regions from hybridomas established at different time points after immunization. The result of one such experiment is depicted here. Each variable region is represented by a horizontal line, on which the positions of the complementarity-determining regions, CDR1, CDR2 and CDR3, are shown as shaded regions. Mutations are represented by colored bars. Within a few days of immunization, the V regions within a particular clone of responding B cells begin to acquire mutations, and over the course of the next week more mutations accumulate (top panels). B cells whose variable regions have accumulated deleterious mutations and can no longer bind antigen die. B cells whose variable regions have acquired mutations that improve antigen binding are able to compete more effectively for antigen, and receive signals that drive their proliferation and expansion. This process of mutation and selection can continue in the lymph node germinal center through multiple cycles in secondary and tertiary immune responses (center and bottom panels). In this way, over time, the antigen-binding efficiency of the antibody response is improved.

B cell, such as the C regions, are not as affected, whereas rearranged V_H and V_L genes are mutated even if they are nonproductive rearrangements and are being transcribed but not expressed as a protein.

Somatic hypermutation in functional V-region genes has various consequences. Mutations that alter amino acid sequences in the conserved framework regions will tend to disrupt basic antibody structure and are selected against because the process takes place in the germinal center, where clones of B cells compete with each other for interaction with antigen. Clones with the highest affinity for antigen are favored for survival. Some of the mutant immunoglobulin molecules bind antigen better than the original B-cell receptors, and B cells expressing them are preferentially selected to mature into antibody-secreting cells. This gives rise to a phenomenon called **affinity maturation** of the antibody population, which we discuss in more detail in Chapters 9 and 10. The result of selection for enhanced binding to antigen is that base changes that alter amino acid sequences, and thus protein structure, tend to be clustered in the CDRs, whereas silent mutations that preserve amino acid sequence and do not alter protein structure are scattered throughout the V region.

The pattern of base changes in nonproductive V-region genes, in contrast, illustrates the result of somatic hypermutation without selection for enhanced binding to antigen and may reveal the underlying process better. The base changes are distributed throughout the V region, but not completely randomly: there are certain mutation 'hotspots' that indicate a preference for characteristic short motifs of four or five nucleotides, and perhaps also certain ill-defined secondary structural features. As discussed in Section 4-17, cytidine deamination by the enzyme AID is thought to be the main mechanism underlying somatic hypermutation. Cytidine deamination to uracil explains some of the known biases of somatic hypermutation such as transition-type mutations from C to T or G to A. More difficult to explain is how deamination at C residues could give rise to mutations at A–T base pairs, which are also common in somatic hypermutation. It is possible that when repair mechanisms are triggered by a U–G mismatch, nicks in DNA are created and there is more extensive and error-prone repair through DNA replication, leading to mutation at adjacent A–T base pairs. When a single-strand nick is created by APE1, similarly relaxed replication could also lead to non-templated transversion

Activation-induced Cytidine Deaminase (AID) Deficiency

mutations. The relationship between these mutational mechanisms and the repair of double-strand breaks in DNA, which are also associated with the mutation of V regions, is unknown.

In contrast to B cells, all the diversity in T-cell receptors is generated during gene rearrangement, and somatic hypermutation of rearranged V regions does not occur in T cells. This means that variability in the CDR1 and CDR2 regions is limited to that of the germline V gene segments, and that most diversity is focused on the CDR3 regions. The strongest argument for why T cells lack somatic hypermutation is that hypermutation is simply an adaptive specialization for B cells to make very high-affinity secreted antibodies that will efficiently carry out their effector functions. Because T cells do not need this capacity, and because deleterious changes in receptor-binding specificities in mature T cells are potentially more damaging to the immune response than are those in B cells, somatic hypermutation in T cells has never evolved.

Certain issues surrounding somatic hypermutation are still unresolved. For example, it is not clear why mutations seem to be selectively targeted to immunoglobulin genes, although it is suspected that the immunoglobulin enhancers and promoters are involved. Specific sequences within these regions that target a gene for mutation still remain to be defined, however. In addition, the immunoglobulin promoters may recruit highly error-prone repair polymerases that can replicate through damaged DNA regions.

4-19 In some species, most immunoglobulin gene diversification occurs after gene rearrangement.

In birds, rabbits, cows, pigs, sheep, and horses there is little or no germline diversity in the V, D, and J gene segments that are rearranged to form the genes for the initial B-cell receptors, and the rearranged V-region sequences are identical or similar in most immature B cells. These B cells then migrate to specialized microenvironments, the best known of which is the bursa of Fabricius in chickens. Here, B cells proliferate rapidly, and their rearranged immunoglobulin genes undergo further diversification. In birds and rabbits this occurs mainly by **gene conversion**, a process by which short sequences in the expressed rearranged V-region gene are replaced with sequences from an upstream V gene segment pseudogene (Fig. 4.26). It seems that gene conversion is related to somatic hypermutation in its mechanism, because gene conversion in a chicken B-cell line has been shown to require AID. The single-strand nicks generated by APE1 subsequent to cytosine deamination are thought to be the signal that initiates a homology-directed repair process in which a homologous V gene segment is used as the template for the DNA replication that repairs the V-region gene.

In sheep and cows, immunoglobulin diversification is the result of somatic hypermutation, which occurs in an organ known as the ileal Peyer's patch. Somatic hypermutation, independent of T cells and a particular driving antigen, also contributes to immunoglobulin diversification in birds, sheep, and rabbits.

4-20 Class switching enables the same assembled V_H exon to be associated with different C_H genes in the course of an immune response.

The V_H-region exons expressed by any given B cell are determined during its early differentiation in the bone marrow and, although they may subsequently be modified by somatic hypermutation, no further V(D)J recombination occurs. All the progeny of that B cell will therefore express the same V_H

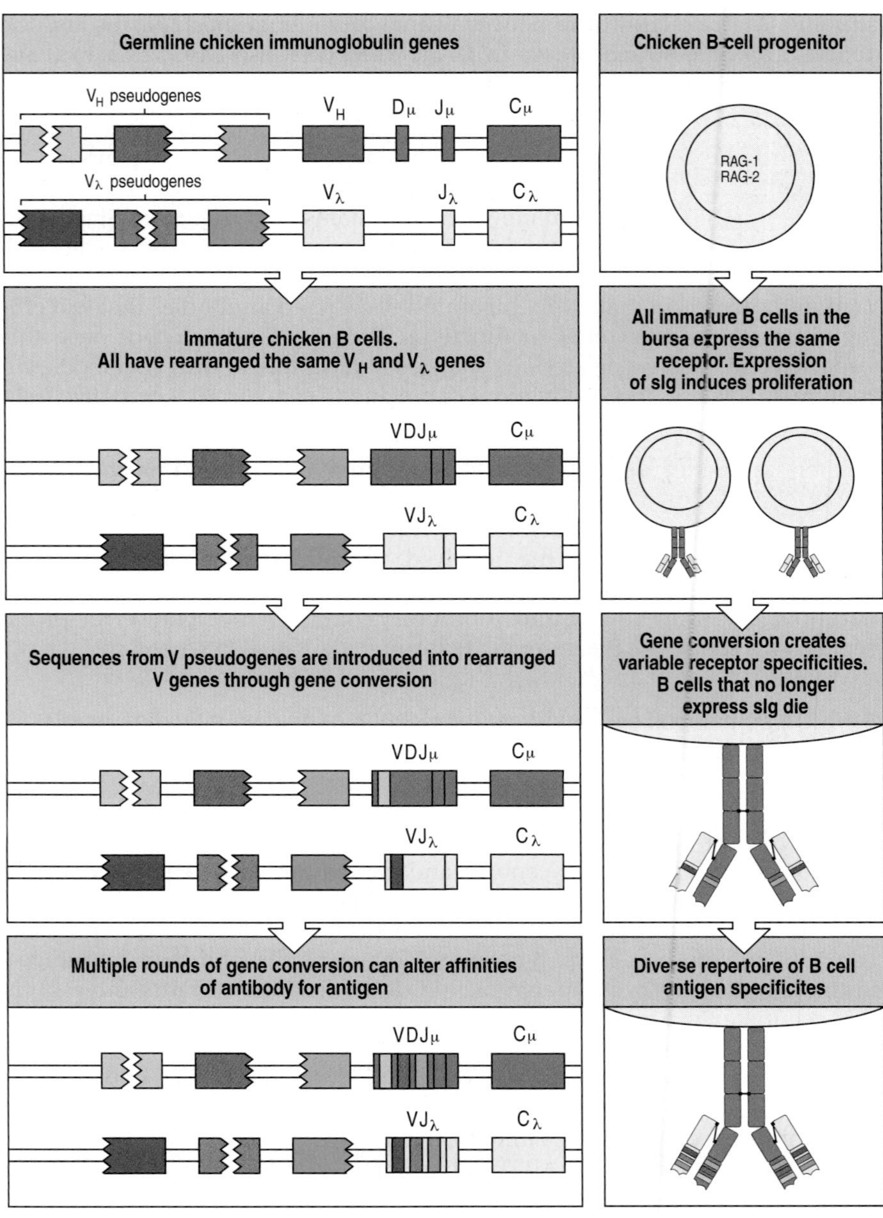

Fig. 4.26 The diversification of chicken immunoglobulins occurs through gene conversion. In chickens, the immunoglobulin diversity that can be created by V(D)J recombination is extremely limited. Initially, there is only one active V and one J gene segment and 15 D gene segments for the chicken heavy-chain gene and one active V and J gene segment at the single light-chain locus (top left panel). Primary gene rearrangement can thus produce only a very limited number of receptor specificities (second panels). Immature B cells expressing this receptor migrate to the bursa of Fabricius, where the cross-linking of surface immunoglobulin (sIg) induces cell proliferation (second panels). Gene conversion events introduce sequences from adjacent V gene segment pseudogenes into the expressed gene, creating diversity in the receptors (third panel). Some of these gene conversions will inactivate the previously expressed gene (not shown). If a B cell can no longer express sIg after such a gene conversion, it is eliminated. Repeated gene conversion events can continue to diversify the repertoire (bottom panels).

gene. In contrast, several different C-region isotypes can be expressed in the B cell's progeny as the cells mature and proliferate in the course of an immune response. The first antigen receptors expressed by B cells are IgM and IgD, and the first antibody produced in an immune response is always IgM. Later in the immune response, the same assembled V region may be expressed in IgG, IgA, or IgE antibodies. This change is known as class switching (or **isotype switching**), and, unlike the expression of IgD, it involves irreversible DNA recombination. It is stimulated in the course of an immune response by external signals such as cytokines released by T cells or mitogenic signals delivered by pathogens, as we discuss further in Chapter 9. Here we are concerned with the molecular basis of the class switch.

Switching from IgM to the other immunoglobulin classes occurs only after B cells have been stimulated by antigen. It is achieved through class switch recombination, which is a type of nonhomologous DNA recombination that is guided by stretches of repetitive DNA known as **switch regions**. Switch regions lie in the intron between the J_H gene segments and the C_μ gene, and at equivalent sites upstream of the genes for each of the other heavy-chain isotypes, with the exception of the δ gene, the expression of which is not dependent on DNA rearrangement (Fig. 4.27, first panel). When a B cell switches from the coexpression of IgM and IgD to the expression of another subtype, DNA recombination occurs between S_μ and the S region immediately upstream of the gene for that isotype. In such a recombination event the

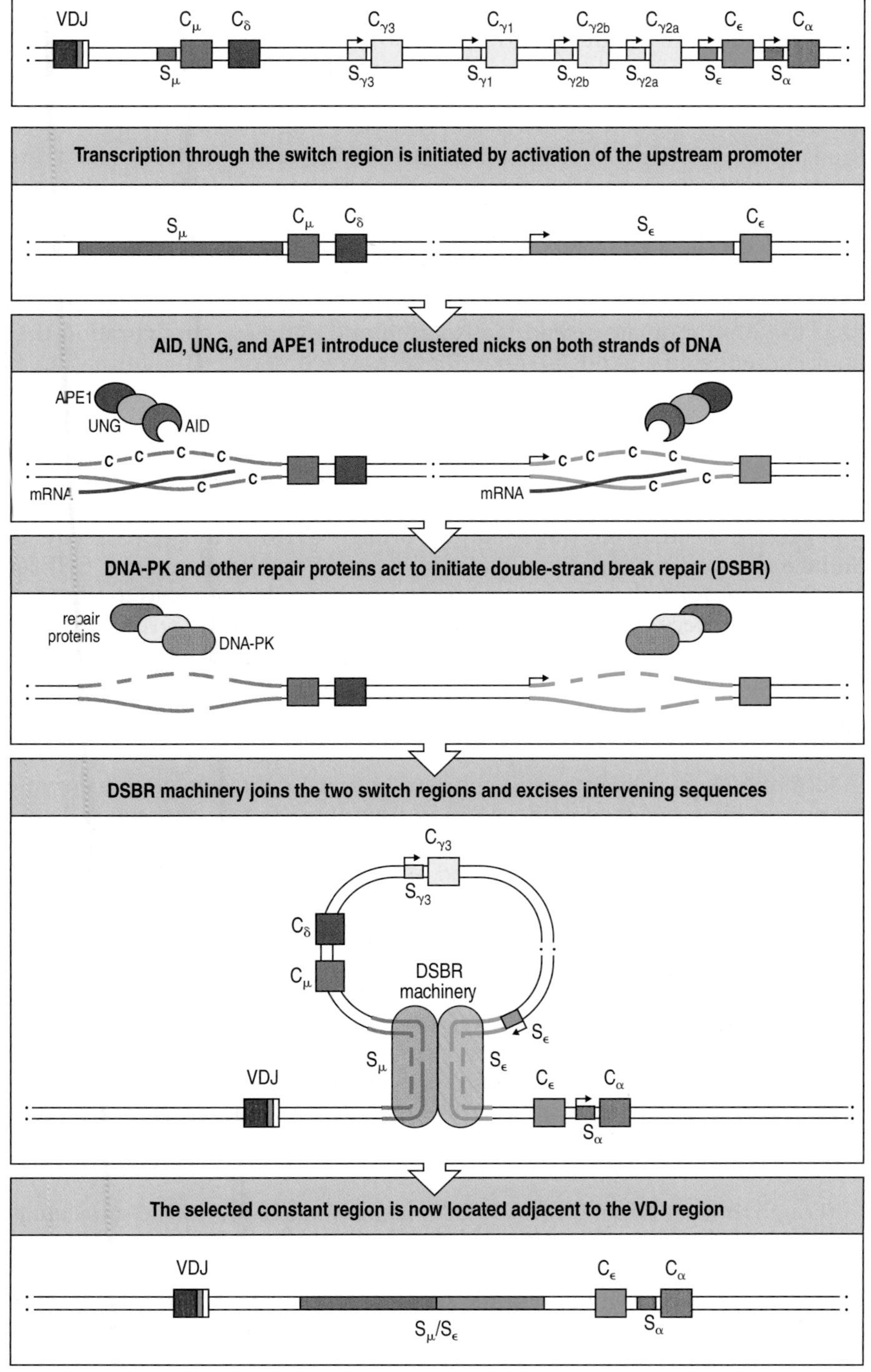

Fig. 4.27 Class switching involves recombination between specific switch signals. Switching between the μ and ε isotypes in the mouse heavy-chain locus is illustrated in this figure. Repetitive DNA sequences, switch regions (S), that guide class switching are found upstream of each of the immunoglobulin C-region genes, with the exception of the δ gene. Switching is guided by the initiation of transcription through these regions from promoters (arrows) located upstream of each S. Because of the nature of the repetitive sequences, transcription through S regions generates R-loops (extended regions of single-stranded DNA formed by the non-template strand), which serve as substrates for AID, and subsequently for UNG and APE1. These activities introduce a high density of single-strand nicks in the non-template DNA strand, and presumably a smaller number of nicks into the template strand as well. Staggered nicks are converted to double-strand breaks by a mechanism that is not yet understood. These breaks are putatively recognized by the cell's double-strand break repair machinery in a process that involves the participation of DNA-PKcs and other repair proteins. The two switch regions, in this case S_μ and S_ε, are brought together by this machinery, and class switching is completed by excision of the intervening region of DNA (including C_μ, C_δ) and ligation of the S_μ and S_ε regions.

C_δ coding regions and all of the intervening DNA between it and the S region undergoing rearrangement are deleted. Fig. 4.27 illustrates switching from C_μ to C_ε in the mouse. All switch recombination events produce genes that can encode a functional protein because the switch sequences lie in introns and therefore cannot cause frameshift mutations.

As noted in Section 4-17, AID can act only on single-stranded DNA. It is known that transcription through the switch regions is required for efficient class switching, and this transcription is presumably necessary to open up the DNA and allow AID access to cytidine residues in the switch regions. The sequences of the switch regions have characteristics that may promote accessibility of the unwound DNA to AID when they are being transcribed. First, the non-template strand is G-rich. S_μ consists of about 150 repeats of the sequence $(GAGCT)_n(GGGGGT)$, where n is usually 3 but can be as many as 7. The sequences of the other switch regions (S_γ, S_α, and S_ε) differ in detail, but all contain repeats of the GAGCT and GGGGGT sequences. It is thought that transcription produces bubble-like structures, called **R-loops**, that are formed when the transcribed RNA displaces the non-template strand of the DNA double helix (see Fig. 4.27). It has been suggested that the RNA–DNA hybrid that is formed when transcribing switch regions particularly favors R-loop formation, although there are other theoretical structures that the template strand might adopt that promote switching. Whatever the case, it seems that the non-template strand is displaced and adopts a configuration that makes the region a good substrate for AID, which initiates the formation of single-strand nicks at the sites of C residues. In addition, particular sequences, such as AGCT, may be particularly good substrates for AID, and because they are palindromic they may allow AID to act on the cytidine residues of both strands concurrently, introducing multiple single-strand nicks on both strands that eventually lead to a double-strand break. Whatever the precise mechanism, transcription through the switch regions seems to induce the generation of double-strand breaks in these regions. Cellular mechanisms for repairing double-strand breaks could then lead to the non-homologous recombination between switch regions that results in class switching, with the ends to be joined being brought together by alignment of repetitive sequences common to the different switch regions. Rejoining of the DNA ends would then lead to excision of all DNA between the two switch regions and formation of a chimeric region at the junction.

Activation-induced Cytidine Deaminase (AID) Deficiency

A lack of AID completely blocks class switching. A deficiency in this enzyme in humans is associated with a form of immunodeficiency known as hyper IgM type 2 syndrome, which is characterized by an absence of immunoglobulins other than IgM and is discussed in Chapter 12. A deficiency of UNG in both mouse and humans also severely impairs class switching, which is further evidence for the sequential actions of AID and UNG as described in Section 4-17. The involvement of double-strand break repair is shown by the fact that switching is markedly reduced in mice lacking Ku proteins. As Ku proteins are also essential for the rejoining of DNA during V(D)J recombination (see Section 4-5), the experiment to show its involvement in class switching was carried out in mice with pre-rearranged heavy and light-chain transgenes. Deficiencies in other DNA-repair proteins such as DNA-PKcs also impair class switching, most probably because they are required for the DNA pairing and end-joining processes.

Although they both involve DNA rearrangement and some of the same enzymatic machinery, class switch recombination is unlike V(D)J recombination in several ways. First, all class switch recombination is productive; second, it uses different recombination signal sequences and does not require the RAG enzymes; third, it happens after antigen stimulation and not during B-cell development in the bone marrow; and fourth, the switching process is not random but is directed by external signals such as those provided by T cells, as discussed in Chapter 9.

Summary

Immunoglobulin genes rearranged by V(D)J recombination can be further diversified by somatic hypermutation, gene conversion, and class switching, which all rely on DNA repair and recombination processes initiated by the enzyme activation-induced cytidine deaminase (AID). Unlike V(D)J recombination, this secondary diversification occurs only in B cells and, in the case of somatic hypermutation and class switching, only after B-cell activation by antigen. Somatic hypermutation diversifies the V region by the introduction of point mutations. Where this results in a greater affinity for the antigen, activated B cells producing the mutated immunoglobulin are selected for survival, which in turn results in an increase in the affinity of antibodies for the antigen as the immune response proceeds. Class switching does not affect the V region but increases the functional diversity of immunoglobulins by replacing the C_μ region in the immunoglobulin gene first expressed with another heavy-chain C region to produce IgG, IgA, or IgE antibodies. Class switching provides antibodies with the same antigen specificity but distinct effector capacities. Gene conversion is the main mechanism used to provide a diverse immunoglobulin repertoire in animals in which only limited diversity can be generated from the germline genes by V(D)J recombination. It involves the replacement of segments of the rearranged V region by sequences derived from pseudogenes.

Summary to Chapter 4.

Lymphocyte receptors are remarkably diverse, and developing B cells and T cells use the same basic mechanism to achieve this diversity. In each cell, functional genes for the immunoglobulin and T-cell receptor chains are assembled by somatic recombination from sets of separate gene segments that together encode the V region. The substrates for the joining process are arrays of V, D, and J gene segments, which are similar in all the antigen-receptor gene loci, although there are some important differences in the details of their arrangement. The lymphoid-specific proteins RAG-1 and RAG-2 direct the V(D)J recombination process in both T and B cells. These proteins function in concert with ubiquitous DNA-modifying enzymes and at least one other lymphoid-specific enzyme, TdT, to complete the gene rearrangements. As each type of gene segment is present in multiple, slightly different, versions, the random selection of one gene segment from each set for assembly is the source of substantial potential diversity. During the process of assembly, a high degree of functionally important diversity is introduced at the gene segment junctions through imprecise joining mechanisms. This diversity is concentrated in the DNA encoding the CDR3 loops of the receptors, which lie at the center of the antigen-binding site. The independent association of the two chains of immunoglobulins or T-cell receptors to form a complete antigen receptor multiplies the overall diversity available. In addition, mature B cells that are activated by antigen initiate a process of somatic point mutation of the V-region DNA, which creates numerous variants of the original assembled V regions. An important difference between immunoglobulins and T-cell receptors is that immunoglobulins exist in both membrane-bound forms (B-cell receptors) and secreted forms (antibodies). The ability to express both a secreted and a membrane-bound form of the same molecule is due to differential splicing of the heavy-chain mRNA to include exons that encode different forms of the carboxy terminus. Heavy-chain C regions contain three or four immunoglobulin domains, whereas the

Event	Process	Nature of change	Process occurs in:	
			B cells	T cells
V-region assembly	Somatic recombination of DNA	Irreversible	Yes	Yes
Junctional diversity	Imprecise joining, N-sequence insertion in DNA	Irreversible	Yes	Yes
Transcriptional activation	Activation of promoter by proximity to the enhancer	Irreversible but regulated	Yes	Yes
Switch recombination	Somatic recombination of DNA	Irreversible	Yes	No
Somatic hypermutation	DNA point mutation	Irreversible	Yes	No
IgM, IgD expression on surface	Differential splicing of RNA	Reversible, regulated	Yes	No
Membrane vs secreted form	Differential splicing of RNA	Reversible, regulated	Yes	No

Fig. 4.28 Changes in immunoglobulin and T-cell receptor genes that occur during B-cell and T-cell development and differentiation. Those changes that establish immunological diversity are all irreversible, as they involve changes in B-cell or T-cell DNA. Certain changes in the organization of DNA or in its transcription are unique to B cells. Somatic hypermutation has not been observed in functional T-cell receptors. The B-cell-specific processes, such as switch recombination, allow the same V region to be attached to several functionally distinct heavy-chain C regions, and thereby create functional diversity in an irreversible manner. By contrast, the expression of IgM versus IgD, and of membrane-bound versus secreted forms of all immunoglobulin types, can in principle be reversibly regulated.

T-cell receptor chains have only one. Finally, B cells are able to increase the diversity of immunoglobulins by three mechanisms that involve AID-dependent somatic mutation of the primary repertoire—somatic hypermutation, gene conversion and class switching. Somatic hypermutation and gene conversion increase diversity by changes to the V regions of immunoglobulin genes. Antibodies also have a variety of effector functions that are mediated by their C regions. Class switching allows the use of several alternative heavy-chain C regions with the same V region, thus producing antibodies with the same specificity but different effector functions. In this way, the progeny of a single B cell can express several different antibody classes, thus maximizing the possible effector functions of a given antigen-specific antibody. The changes in immunoglobulin and T-cell receptor genes that occur during B-cell and T-cell development are summarized in Fig. 4.28.

Questions.

4.1 *(a) What are the two kinds of somatic rearrangements of DNA that occur in the immunoglobulin gene locus? (b) Compare and contrast the mechanisms that generate these types of rearrangements. (c) Which one of these kinds of rearrangements also occurs in the loci that encode the T cell receptor? (d) What would be the consequences of AID activity occurring in T cells?*

4.2 *(a) What are the critical lymphocyte-specific genes involved in V(D)J recombination? (b) What are their main enzymatic activities? (c) Which of these activities are preferentially used in the formation of rearranged heavy-chain genes in comparison with light-chain genes? (d) Which if any of these activities is used only in the processing of coding joins? Of signal joins? (e) How does this explain precise signal joins compared with imprecise coding joins?*

4.3 *The complete V(D)J recombination process uses both tissue-specific (B cells and T cells) as well as tissue nonspecific (that is, ubiquitously expressed) enzymatic activities. (a) Discuss two nonspecific enzymatic activities that are required for the completion of V(D)J joining. (b) Why do these activities not result in inappropriate V(D)J DNA rearrangements in other tissues?*

4.4 *(a) Discuss the four main processes that generate diversity of the lymphocyte repertoire. (b) Which of these processes is not shared by both B and T cells? (c) How does this difference relate to the kinds of DNA rearrangements that occur in B cells and T cells? (d) What other processes occur in B cells that do not occur in T cells, and why?*

4.5 *(a) What are the physiologic functions of antibody gene class switching? (b) How is class switching regulated by the environment or by interactions with pathogens?*

General references.

Casali, P., and Silberstein, L.E.S. (eds): **Immunoglobulin gene expression in development and disease.** *Ann. N.Y. Acad. Sci.* 1995, **764**.

Fugmann, S.D., Lee, A.I., Shockett, P.E., Villey, I.J., and Schatz, D.G.: **The RAG proteins and V(D)J recombination: complexes, ends, and transposition.** *Annu. Rev. Immunol.* 2000, **18**:495–527.

Papavasiliou, F.N. and Schatz, D.G.: **Somatic hypermutation of immunoglobulin genes: merging mechanisms for genetic diversity.** *Cell* 2002, **109 Suppl**:S35–S44.

Section references.

4-1 Immunoglobulin genes are rearranged in antibody-producing cells.

Hozumi, N., and Tonegawa, S.: **Evidence for somatic rearrangement of immunoglobulin genes coding for variable and constant regions.** *Proc. Natl Acad. Sci. USA* 1976, **73**:3628–3632.

Tonegawa, S., Brack, C., Hozumi, N., and Pirrotta, V.: **Organization of immunoglobulin genes.** *Cold Spring Harbor Symp. Quant. Biol.* 1978, **42**:921–931.

Waldmann, T.A.: **The arrangement of immunoglobulin and T-cell receptor genes in human lymphoproliferative disorders.** *Adv. Immunol.* 1987, **40**:247–321.

4-2 Complete genes that encode a variable region are generated by the somatic recombination of separate gene segments.

Early, P., Huang, H., Davis, M., Calame, K., and Hood, L.: **An immunoglobulin heavy chain variable region gene is generated from three segments of DNA: V_H, D and J_H.** *Cell* 1980, **19**:981–992.

Tonegawa, S., Maxam, A.M., Tizard, R., Bernard, O., and Gilbert, W.: **Sequence of a mouse germ-line gene for a variable region of an immunoglobulin light chain.** *Proc. Natl Acad. Sci. USA* 1978, **75**:1485–1489.

4-3 Multiple contiguous V gene segments are present at each immunoglobulin locus.

Cook, G.P., and Tomlinson, I.M.: **The human immunoglobulin V-H repertoire.** *Immunol. Today* 1995, **16**:237–242.

Kofler, R., Geley, S., Kofler, H., and Helmberg, A.: **Mouse variable-region gene families—complexity, polymorphism, and use in nonautoimmune responses.** *Immunol. Rev.* 1992, **128**:5–21.

Maki, R., Traunecker, A., Sakano, H., Roeder, W., and Tonegawa, S.: **Exon shuffling generates an immunoglobulin heavy chain gene.** *Proc. Natl Acad. Sci. USA* 1980, **77**:2138–2142.

Matsuda, F., and Honjo, T.: **Organization of the human immunoglobulin heavy-chain locus.** *Adv. Immunol.* 1996, **62**:1–29.

Thiebe, R., Schable, K.F., Bensch, A., Brensing-Kuppers, J., Heim, V., Kirschbaum, T., Mitlohner, H., Ohnrich, M., Pourrajabi, S., Roschenthaler, F., Schwendinger, J., Wichelhaus, D., Zocher, I., and Zachau, H.G.: **The variable genes and gene families of the mouse immunoglobulin kappa locus.** *Eur. J. Immunol.* 1999, **29**:2072–2081.

4-4 Rearrangement of V, D, and J gene segments is guided by flanking DNA sequences.

Grawunder, U., West, R.B., and Lieber, M.R.: **Antigen receptor gene rearrangement.** *Curr. Opin. Immunol.* 1998, **10**:172–180.

Max, E.E., Seidman, J.G., and Leder, P.: **Sequences of five potential recombination sites encoded close to an immunoglobulin kappa constant region gene.** *Proc. Natl Acad. Sci. USA* 1979, **76**:3450–3454.

Sakano, H., Huppi, K., Heinrich, G., and Tonegawa, S.: **Sequences at the somatic recombination sites of immunoglobulin light-chain genes.** *Nature* 1979, **280**:288–294.

4-5 The reaction that recombines V, D, and J gene segments involves both lymphocyte-specific and ubiquitous DNA-modifying enzymes.

Agrawal, A., and Schatz, D.G.: **RAG1 and RAG2 form a stable postcleavage synaptic complex with DNA containing signal ends in V(D)J recombination.** *Cell* 1997, **89**:43–53.

Blunt, T., Finnie, N.J., Taccioli, G.E., Smith, G.C.M., Demengeot, J., Gottlieb, T.M., Mizuta, R., Varghese, A.J., Alt, F.W., Jeggo, P.A., and Jackson, S.P.: **Defective DNA-dependent protein kinase activity is linked to V(D)J recombination and DNA-repair defects associated with the murine–scid mutation.** *Cell* 1995, **80**:813–823.

Gu, Z., Jin, S., Gao, Y., Weaver, D.T., and Alt, F.W.: **Ku70-deficient embryonic stem cells have increased ionizing radiosensitivity, defective DNA end-binding activity, and inability to support V(D)J recombination.** *Proc. Natl Acad. Sci. USA* 1997, **94**:8076–8081.

Jung, D., Giallourakis, C., Mostoslavsky, R., and Alt, F.W.: **Mechanism and control of V(D)J recombination at the immunoglobulin heavy chain locus.** *Annu. Rev. Immunol.* 2006, **24**:541–570.

Li, Z.Y., Otevrel, T., Gao, Y.J., Cheng, H.L., Seed, B., Stamato, T.D., Taccioli, G.E., and Alt, F.W.: **The XRCC4 gene encodes a novel protein involved in DNA double-strand break repair and V(D)J recombination.** *Cell* 1995, **83**:1079–1089.

Moshous, D., Callebaut, I., de Chasseval, R., Corneo, B., Cavazzana-Calvo, M., Le Deist, F., Tezcan, I., Sanal, O., Bertrand, Y., Philippe, N., Fischer, A., and de Villartay, J.P.: **Artemis, a novel DNA double-strand break repair/V(D)J recombination protein, is mutated in human severe combined immune deficiency.** *Cell* 2001, **105**:177–186.

Oettinger, M.A., Schatz, D.G., Gorka, C., and Baltimore, D.: **RAG-1 and RAG-2, adjacent genes that synergistically activate V(D)J recombination.** *Science* 1990, **248**:1517–1523.

Villa, A., Santagata, S., Bozzi, F., Giliani, S., Frattini, A., Imberti, L., Gatta, L.B., Ochs, H.D., Schwarz, K., Notarangelo, L.D., Vezzoni, P., and Spanopoulou, E.: **Partial V(D)J recombination activity leads to Omenn syndrome.** *Cell* 1998, **93**:885–896.

4-6 The diversity of the immunoglobulin repertoire is generated by four main processes.

Fanning, L.J., Connor, A.M., and Wu, G.E.: **Development of the immunoglobulin repertoire.** *Clin. Immunol. Immunopathol.* 1996, **79**:1–14.

Weigert, M., Perry, R., Kelley, D., Hunkapiller, T., Schilling, J., and Hood, L.: **The joining of V and J gene segments creates antibody diversity.** *Nature* 1980, **283**:497–499.

4-7 The multiple inherited gene segments are used in different combinations.

Lee, A., Desravines, S., and Hsu, E.: **IgH diversity in an individual with only one million B lymphocytes.** *Dev. Immunol.* 1993, **3**:211–222.

4-8 Variable addition and subtraction of nucleotides at the junctions between gene segments contributes to the diversity in the third hypervariable region.

Gauss, G.H., and Lieber, M.R.: **Mechanistic constraints on diversity in human V(D)J recombination.** *Mol. Cell. Biol.* 1996, **16**:258–269.

Komori, T., Okada, A., Stewart, V., and Alt, F.W.: **Lack of N regions in antigen receptor variable region genes of TdT-deficient lymphocytes.** *Science* 1993, **261**:1171–1175.

Weigert, M., Gatmaitan, L., Loh, E., Schilling, J., and Hood, L.: **Rearrangement of genetic information may produce immunoglobulin diversity.** *Nature* 1978, **276**:785–790.

4-9 The T-cell receptor gene segments are arranged in a similar pattern to immunoglobulin gene segments and are rearranged by the same enzymes.

Rowen, L., Koop, B.F., and Hood, L.: **The complete 685-kilobase DNA sequence of the human β T cell receptor locus.** *Science* 1996, **272**:1755–1762.

Shinkai, Y., Rathbun, G., Lam, K.P., Oltz, E.M., Stewart, V., Mendelsohn, M., Charron, J., Datta, M., Young, F., Stall, A.M., and Alt, F.W.: **RAG-2 deficient mice lack mature lymphocytes owing to inability to initiate V(D)J rearrangement.** *Cell* 1992, **68**:855–867.

4-10 T-cell receptors concentrate diversity in the third hypervariable region.

Davis, M.M., and Bjorkman, P.J.: **T-cell antigen receptor genes and T-cell recognition.** *Nature* 1988, **334**:395–402.

Garboczi, D.N., Ghosh, P., Utz, U., Fan, Q.R., Biddison, W.E., and Wiley, D.C.: **Structure of the complex between human T-cell receptor, viral peptide and HLA-A2.** *Nature* 1996, **384**:134–141.

Hennecke, J., and Wiley, D.C.: **T cell receptor-MHC interactions up close.** *Cell* 2001, **104**:1–4.

Hennecke, J., Carfi, A., and Wiley, D.C.: **Structure of a covalently stabilized complex of a human alphabeta T-cell receptor, influenza HA peptide and MHC class II molecule, HLA-DR1.** *EMBO J.* 2000, **19**:5611–5624.

Jorgensen, J.L., Esser, U., Fazekas de St. Groth, B., Reay, P.A., and Davis, M.M.: **Mapping T-cell receptor–peptide contacts by variant peptide immunization of single-chain transgenics.** *Nature* 1992, **355**:224–230.

4-11 γ:δ T-cell receptors are also generated by gene rearrangement.

Chien, Y.H., Iwashima, M., Kaplan, K.B., Elliott, J.F., and Davis, M.M.: **A new T-cell receptor gene located within the alpha locus and expressed early in T-cell differentiation.** *Nature* 1987, **327**:677–682.

Hayday, A.C., Saito, H., Gillies, S.D., Kranz, D.M., Tanigawa, G., Eisen, H.N., and Tonegawa, S.: **Structure, organization, and somatic rearrangement of T cell gamma genes.** *Cell* 1985, **40**:259–269.

Lafaille, J.J., DeCloux, A., Bonneville, M., Takagaki, Y., and Tonegawa, S.: **Junctional sequences of T cell receptor gamma delta genes: implications for gamma delta T cell lineages and for a novel intermediate of V-(D)-J joining.** *Cell* 1989, **59**:859–870.

Tonegawa, S., Berns, A., Bonneville, M., Farr, A.G., Ishida, I., Ito, K., Itohara, S., Janeway, C.A., Jr., Kanagawa, O., Kubo, R., *et al.*: **Diversity, development, ligands, and probable functions of gamma delta T cells.** *Adv. Exp. Med. Biol.* 1991, **292**:53–61.

4-12 Different classes of immunoglobulins are distinguished by the structure of their heavy-chain constant regions.

Davies, D.R., and Metzger, H.: **Structural basis of antibody function.** *Annu. Rev. Immunol.* 1983, **1**:87–117.

4-13 The constant region confers functional specialization on the antibody.

Helm, B.A., Sayers, I., Higginbottom, A., Machado, D.C., Ling, Y., Ahmad, K., Padlan, E.A., and Wilson, A.P.M.: **Identification of the high affinity receptor binding region in human IgE.** *J. Biol. Chem.* 1996, **271**:7494–7500.

Jefferis, R., Lund, J., and Goodall, M.: **Recognition sites on human IgG for Fcγ receptors—the role of glycosylation.** *Immunol. Lett.* 1995, **44**:111–117.

Sensel, M.G., Kane, L.M., and Morrison, S.L.: **Amino acid differences in the N-terminus of C_H2 influence the relative abilities of IgG2 and IgG3 to activate complement.** *Mol. Immunol.* **34**:1019–1029.

4-14 Mature naive B cells express both IgM and IgD at their surface.

Abney, E.R., Cooper, M.D., Kearney, J.F., Lawton, A.R., and Parkhouse, R.M.: **Sequential expression of immunoglobulin on developing mouse B lymphocytes: a systematic survey that suggests a model for the generation of immunoglobulin isotype diversity.** *J. Immunol.* 1978, **120**:2041–2049.

Blattner, F.R. and Tucker, P.W.: **The molecular biology of immunoglobulin D.** *Nature* 1984, **307**:417–422.

Goding, J.W., Scott, D.W., and Layton, J.E.: **Genetics, cellular expression and function of IgD and IgM receptors.** *Immunol. Rev.* 1977, **37**:152–186.

4-15 Transmembrane and secreted forms of immunoglobulin are generated from alternative heavy-chain transcripts.

Early, P., Rogers, J., Davis, M., Calame, K., Bond, M., Wall, R., and Hood, L.: **Two mRNAs can be produced from a single immunoglobulin μ gene by alternative RNA processing pathways.** *Cell* 1980, **20**:313–319.

Peterson, M.L., Gimmi, E.R., and Perry, R.P.: **The developmentally regulated shift from membrane to secreted μ mRNA production is accompanied by an increase in cleavage-polyadenylation efficiency but no measurable change in splicing efficiency.** *Mol. Cell. Biol.* 1991, **11**:2324–2327.

Rogers, J., Early, P., Carter, C., Calame, K., Bond, M., Hood, L., and Wall, R.: **Two mRNAs with different 3′ ends encode membrane-bound and secreted forms of immunoglobulin mu chain.** *Cell* 1980, **20**:303–312.

4-16 IgM and IgA can form polymers.

Hendrickson, B.A., Conner, D.A., Ladd, D.J., Kendall, D., Casanova, J.E., Corthesy, B., Max, E.E., Neutra, M.R., Seidman, C.E., and Seidman, J.G.: **Altered hepatic transport of IgA in mice lacking the J chain.** *J. Exp. Med.* 1995, **182**:1905–1911.

Niles, M.J., Matsuuchi, L., and Koshland, M.E.: **Polymer IgM assembly and secretion in lymphoid and nonlymphoid cell-lines—evidence that J chain is required for pentamer IgM synthesis.** *Proc. Natl Acad. Sci. USA* 1995, **92**:2884–2888.

4-17 Activation-induced cytidine deaminase introduces mutations in genes transcribed in B cells.

Muramatsu, M., Kinoshita, K., Fagarasan, S., Yamada, S., Shinkai, Y., and Honjo, T.: **Class switch recombination and hypermutation require activation-induced cytidine deaminase (AID), a potential RNA editing enzyme.** *Cell* 2000, **102**:553–563.

Petersen-Mahrt, S.K., Harris, R.S., and Neuberger, M.S.: **AID mutates *E. coli* suggesting a DNA deamination mechanism for antibody diversification.** *Nature* 2002, **418**:99–103.

Yu, K., Huang, F.T., and Lieber, M.R.: **DNA substrate length and surrounding sequence affect the activation-induced deaminase activity at cytidine.** *J. Biol. Chem* 2004, **279**:6496–6500.

4-18 Rearranged V genes are further diversified by somatic hypermutation.

Basu, U., Chaudhuri, J., Alpert, C., Dutt, S., Ranganath, S., Li, G., Schrum, J.P., Manis, J.P., and Alt, F.W.: **The AID antibody diversification enzyme is regulated by protein kinase A phosphorylation.** *Nature* 2005, **438**:508–511.

Betz, A.G., Rada, C., Pannell, R., Milstein, C., and Neuberger, M.S.: **Passenger transgenes reveal intrinsic specificity of the antibody hypermutation mechanism: clustering, polarity, and specific hot spots.** *Proc. Natl Acad. Sci. USA* 1993, **90**:2385–2388.

Chaudhuri, J., Khuong, C., and Alt, F.W.: **Replication protein A interacts with AID to promote deamination of somatic hypermutation targets.** *Nature* 2004, **430**:992–998.

Di Noia, J. and Neuberger, M.S.: **Altering the pathway of immunoglobulin hypermutation by inhibiting uracil-DNA glycosylase.** *Nature* 2002, **419**:43–48.

McKean, D., Huppi, K., Bell, M., Straudt, L., Gerhard, W., and Weigert, M.: **Generation of antibody diversity in the immune response of BALB/c mice to influenza virus hemagglutinin.** *Proc. Natl Acad. Sci. USA* 1984, **81**:3180–3184.

Weigert, M.G., Cesari, I.M., Yonkovich, S.J., and Cohn, M.: **Variability in the lambda light chain sequences of mouse antibody.** *Nature* 1970, **228**:1045–1047.

4-19 In some species, most immunoglobulin gene diversification occurs after gene rearrangement.

Harris, R.S., Sale, J.E., Petersen-Mahrt, S.K., and Neuberger, M.S.: **AID is essential for immunoglobulin V gene conversion in a cultured B cell line.** *Curr. Biol.* 2002, **12**:435–438.

Knight, K.L., and Crane, M.A.: **Generating the antibody repertoire in rabbit.** *Adv. Immunol.* 1994, **56**:179–218.

Reynaud, C.A., Bertocci, B., Dahan, A., and Weill, J.C.: **Formation of the chicken B-cell repertoire—ontogeny, regulation of Ig gene rearrangement, and diversification by gene conversion.** *Adv. Immunol.* 1994, **57**:353–378.

Reynaud, C.A., Garcia, C., Hein, W.R., and Weill, J.C.: **Hypermutation generating the sheep immunoglobulin repertoire is an antigen independent process.** *Cell* 1995, **80**:115–125.

Vajdy, M., Sethupathi, P., and Knight, K.L.: **Dependence of antibody somatic diversification on gut-associated lymphoid tissue in rabbits.** *J. Immunol.* 1998, **160**:2725–2729.

4-20 Class switching enables the same assembled V_H exon to be associated with different C_H genes in the course of an immune response.

Chaudhuri, J., and Alt, F.W.: **Class-switch recombination: interplay of transcription, DNA deamination and DNA repair.** *Nat. Rev. Immunol.* 2004, **4**:541–552.

Jung, S., Rajewsky, K., and Radbruch, A.: **Shutdown of class switch recombination by deletion of a switch region control element.** *Science* 1993, **259**:984.

Revy, P., Muto, T., Levy, Y., Geissmann, F., Plebani, A., Sanal, O., Catalan, N., Forveille, M., Dufourcq-Lagelouse, R., Gennery, A., *et al.*: **Activation-induced cytidine deaminase (AID) deficiency causes the autosomal recessive form of the hyper-IgM syndrome (HIGM2).** *Cell* 2000, **102**:565–575.

Sakano, H., Maki, R., Kurosawa, Y., Roeder, W., and Tonegawa, S.: **Two types of somatic recombination are necessary for the generation of complete immunoglobulin heavy-chain genes.** *Nature* 1980, **286**:676–683.

Shinkura, R., Tian, M., Smith, M., Chua, K., Fujiwara, Y., and Alt, F.W.: **The influence of transcriptional orientation on endogenous switch region function.** *Nat. Immunol.* 2003, **4**:435–441.

Antigen Presentation to T Lymphocytes

5

In an adaptive immune response, antigen is recognized by two distinct sets of highly variable receptor molecules—the immunoglobulins that serve as antigen receptors on B cells and the antigen-specific receptors of T cells. As we saw in Chapter 3, T cells only recognize antigens that are displayed on cell surfaces. These antigens may be derived from pathogens that replicate within cells, such as viruses or intracellular bacteria, or from pathogens or their products that cells take up by endocytosis from the extracellular fluid. Infected cells display on their surface peptide fragments derived from the pathogens' proteins and can thus be detected by T cells. These foreign peptides are delivered to the cell surface by specialized host-cell glycoproteins, the MHC molecules, which were also described in Chapter 3. The MHC molecules are encoded in a large cluster of genes that were first identified by their potent effects on the immune response to transplanted tissues. For that reason, the gene complex is called the **major histocompatibility complex** (**MHC**).

We begin by discussing the mechanisms by which protein antigens are degraded into peptides inside cells and the peptides are then carried to the cell surface bound to MHC molecules. We will see that the two different classes of MHC molecules, known as MHC class I and MHC class II, obtain peptides at different cellular locations. Peptides from the cytosol are bound to MHC class I molecules and are recognized by CD8 T cells, whereas peptides generated in intracellular vesicles are bound to MHC class II molecules and recognized by CD4 T cells. The two functional subsets of T cells are thereby activated to initiate the destruction of pathogens resident in these two different cellular compartments. Some CD4 T cells activate naive B cells that have internalized specific antigen, and thus also stimulate the production of antibodies against extracellular pathogens and their products.

The second part of this chapter focuses on the MHC class I and II genes and their remarkable genetic variability. There are several different MHC molecules in each class and each of their genes is highly polymorphic, with many variants in the population. MHC polymorphism has a profound effect on antigen recognition by T cells, and the combination of multiple genes and polymorphism greatly extends the range of peptides that can be presented to T cells by each individual and each population at risk from an infectious pathogen. We shall also see that the MHC contains genes other than those for the MHC molecules, and that the products of many of these genes are involved in the production of peptide:MHC complexes.

We shall also consider a group of proteins, encoded both within and outside the MHC, that are similar to MHC class I molecules but have limited polymorphism. They have a variety of functions, one of which is the presentation of microbial lipid antigens to T cells and NK cells.

The generation of T-cell receptor ligands.

The protective function of T cells depends on their ability to recognize cells that are harboring pathogens or that have internalized pathogens or their products. T cells do this by recognizing peptide fragments of pathogen-derived proteins in the form of complexes of peptides and MHC molecules on the cell surface. The generation of peptides from an intact antigen involves modification of the native protein and is commonly referred to as **antigen processing**. The display of the peptide at the cell surface by the MHC molecule is referred to as **antigen presentation**. We have already described the structure of MHC molecules and seen how they bind peptide antigens in a cleft on their outer surface (see Sections 3-13 to 3-16). In this chapter we will look at how peptides are generated from pathogens present in the cytosol or in the vesicular compartment and are loaded onto MHC class I or MHC class II molecules, respectively.

5-1 The MHC class I and class II molecules deliver peptides to the cell surface from two intracellular compartments.

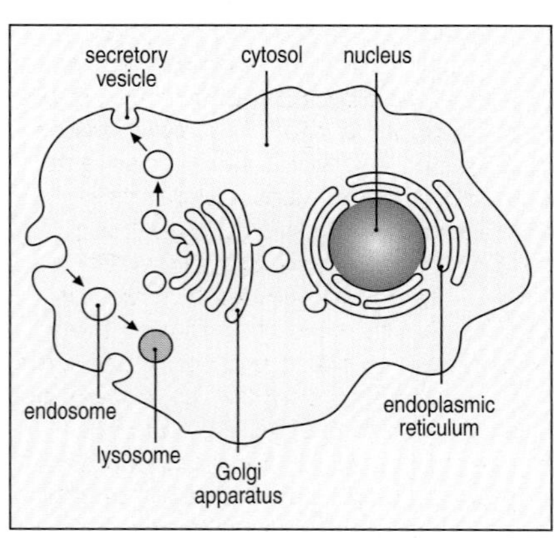

Fig. 5.1 There are two major intracellular compartments, separated by membranes. The first is the cytosol, which also communicates with the nucleus via the nuclear pores in the nuclear membrane. The second is the vesicular system, which comprises the endoplasmic reticulum, Golgi apparatus, endosomes, lysosomes, and other intracellular vesicles. The vesicular system can be thought of as continuous with the extracellular fluid. Secretory vesicles bud off from the endoplasmic reticulum and are transported via fusion with Golgi membranes to move vesicular contents out of the cell, whereas extracellular material is taken up by endocytosis into endosomes.

Infectious agents can replicate in either of two distinct intracellular compartments (Fig. 5.1). Viruses and certain bacteria replicate in the cytosol or in the contiguous nuclear compartment (Fig. 5.2, first panel), whereas many pathogenic bacteria and some eukaryotic parasites replicate in the endosomes and lysosomes that form part of the vesicular system (Fig. 5.2, third panel). Exogenous antigens derived from extracellular pathogens or other pathogen-infected cells can also enter the cytosol of specialized antigen-presenting cells (Fig. 5.2, second panel), as we will describe in more detail later. The immune system has different strategies for eliminating pathogens from the cytosol and the endosomal system. Cells infected with viruses or with cytosolic bacteria are eliminated by **cytotoxic T cells**; these T cells are distinguished by the co-receptor molecule CD8 (see Section 3-17). The function of CD8 T cells is to kill infected cells; this is an important means of eliminating sources of new viral particles and obligate cytosolic bacteria, thus freeing the host from infection.

Pathogens and their products in the vesicular compartments of cells are detected by a different class of T cell, distinguished by the co-receptor molecule CD4 (see Section 3-17). **CD4 T cells** have several distinct activities, which are displayed by different effector CD4 subsets. The first subsets to be recognized were T_H1 cells, which activate macrophages to kill the intravesicular pathogens they harbor and also provide help to B cells to make antibody, and T_H2 cells, which function in response to parasites and help in antibody production. A recently identified CD4 T-cell subset called T_H17 is named for its production of the pro-inflammatory cytokine interleukin 17. In certain situations CD4 T cells have cytotoxic activity similar to that of CD8 T cells. For example, virus-specific human CD4 T cells can kill B lymphocytes infected with Epstein–Barr virus (EBV). Other subsets include at least two types of regulatory CD4 T cells: one is derived during development in the thymus, and the others are generated in the periphery during an immune response.

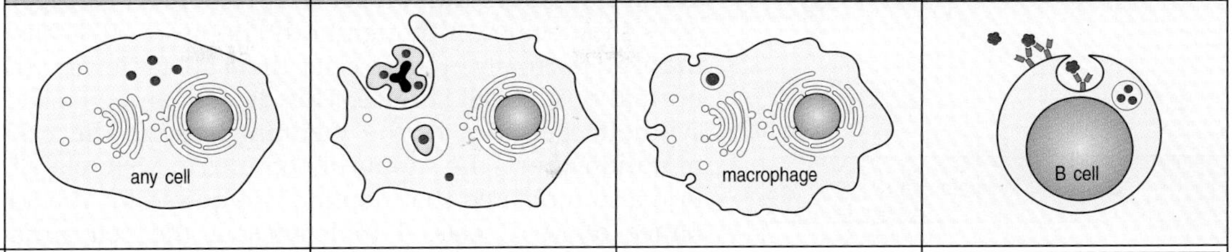

	Cytosolic pathogens	Cross-presentation of exogenous antigens	Intravesicular pathogens	Extracellular pathogens and toxins
	any cell		macrophage	B cell
Degraded in	Cytosol	Cytosol (by retrotranslocation)	Endocytic vesicles (low pH)	Endocytic vesicles (low pH)
Peptides bind to	MHC class I	MHC class I	MHC class II	MHC class II
Presented to	Effector CD8 T cells	Naive CD8 T cells	Effector CD4 T cells	Effector CD4 T cells
Effect on presenting cell	Cell death	The presenting cell, usually a dendritic cell, activates the CD8 T cell	Activation to kill intravesicular bacteria and parasites	Activation of B cells to secrete Ig to eliminate extracellular bacteria/toxins

Fig. 5.2 Pathogens and their products can be found in either the cytosolic or the vesicular compartment of cells. First panel: all viruses and some bacteria replicate in the cytosolic compartment. Their antigens are presented by MHC class I molecules to CD8 T cells. Second panel: exogenous antigens from a dying virus-infected cell that is phagocytosed by a dendritic cell can be retrotransported into the cytosol, where they can be degraded and loaded onto MHC class I molecules. Such cross-presentation is important in enabling dendritic cells to activate naive CD8 T cells specific for viruses that do not infect dendritic cells themselves. Third panel: other bacteria and some parasites are taken up into endosomes, usually by specialized phagocytic cells such as macrophages. Here they are killed and degraded, or in some cases are able to survive and proliferate within the vesicle. Their antigens are presented by MHC class II molecules to CD4 T cells. Fourth panel: proteins derived from extracellular pathogens may enter the intracellular vesicular system by binding to cell-surface receptors followed by endocytosis. This is illustrated here for proteins bound by the surface immunoglobulin of B cells (the endoplasmic reticulum and Golgi apparatus have been omitted for simplicity). The B cells present these antigens to CD4 helper T cells, which can then stimulate the B cells to produce antibody. Other types of cells that bear receptors for the Fc regions of antibody molecules can also internalize antigens in this way and are able to activate T cells.

Microbial antigens may enter the vesicular compartment in either of two ways. Some bacteria, including the mycobacteria that cause tuberculosis and leprosy, invade macrophages and flourish in intracellular vesicles. Other bacteria proliferate outside cells, where they cause tissue damage by secreting toxins and other proteins. These bacteria and their toxic products can be internalized by phagocytosis, endocytosis, or macropinocytosis into the intracellular vesicles of cells that then present the pathogen's antigens to T cells. Antigen-presenting cells include the dendritic cells that specialize in initiating T-cell responses (see Section 1-7), macrophages that specialize in taking up particulate material (see Section 2-4), and B cells that efficiently internalize specific antigen by receptor-mediated endocytosis of antigen bound to their surface immunoglobulin (Fig. 5.2, fourth panel).

MHC class I molecules deliver peptides originating in the cytosol to the cell surface, where they are recognized by CD8 T cells, whereas MHC class II molecules deliver peptides originating in the vesicular system to the cell surface, where they are recognized by CD4 T cells. As we saw in Section 3-17, the specificity of this reaction is due to the fact that CD8 and CD4 bind MHC class I and MHC class II molecules, respectively. The different activities of CD8 and CD4 T cells can largely be viewed as adapted to deal with pathogens found in different cellular compartments but, as we shall see, there is significant cross-talk between these two pathways.

5-2 Peptides that bind to MHC class I molecules are actively transported from the cytosol to the endoplasmic reticulum.

The polypeptide chains of proteins destined for the cell surface, including the chains of MHC molecules, are translocated during synthesis into the lumen of the endoplasmic reticulum. Here, the two chains of each MHC molecule fold correctly and assemble with each other. This means that the peptide-binding site of the MHC class I molecule is formed in the lumen of the endoplasmic reticulum and is never exposed to the cytosol. The antigen fragments that bind to MHC class I molecules, however, are typically derived from viral proteins made in the cytosol. This raised the question: How are peptides derived from proteins in the cytosol able to bind to MHC class I molecules for delivery to the cell surface?

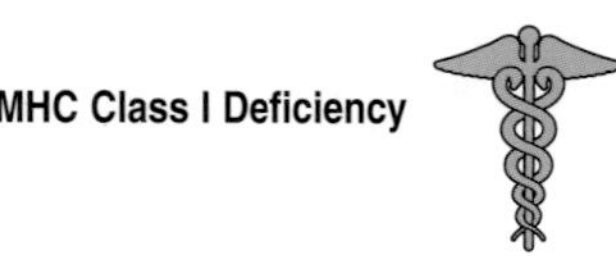

The answer is that peptides are transported from the cytosol by proteins in the endoplasmic reticulum membrane. The first clues to this delivery mechanism came from mutant cells with a defect in antigen presentation by MHC class I molecules. Although both chains of MHC class I molecules are synthesized normally in these cells, there are far fewer MHC class I proteins than normal on the cell surface. This defect can be corrected by the addition of synthetic peptides to the medium bathing the cells, suggesting that the mutation affects the supply of peptides to MHC class I molecules. This mutation also indicated that peptide is required for the appearance and maintenance of MHC class I molecules at the cell surface, and was the first indication that MHC molecules are unstable in the absence of bound peptide. Analysis of the DNA of the mutant cells showed that two genes encoding members of the ATP-binding cassette (ABC) family of proteins are mutant or absent in these cells. ABC proteins mediate the ATP-dependent transport of ions, sugars, amino acids, and peptides across membranes in many types of cells, including bacteria. The two ABC proteins missing from the mutant cells are normally associated with the endoplasmic reticulum membrane. Transfection of the mutant cells with both genes restores presentation of peptides by the cell's MHC class I molecules. These proteins are now called ***t*ransporters associated with *a*ntigen *p*rocessing-1** and **-2** (**TAP1** and **TAP2**). The two TAP proteins form a heterodimer (Fig. 5.3), and mutations in either TAP gene can prevent antigen presentation by MHC class I molecules. Viral infection of the cell increases the delivery of cytosolic peptides into the endoplasmic reticulum. The genes *TAP1* and *TAP2* map within the MHC (see Section 5-11) and are inducible by interferons, which are produced in response to viral infection.

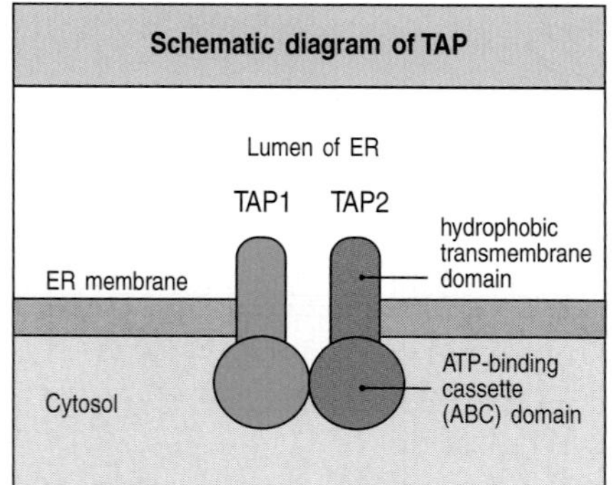

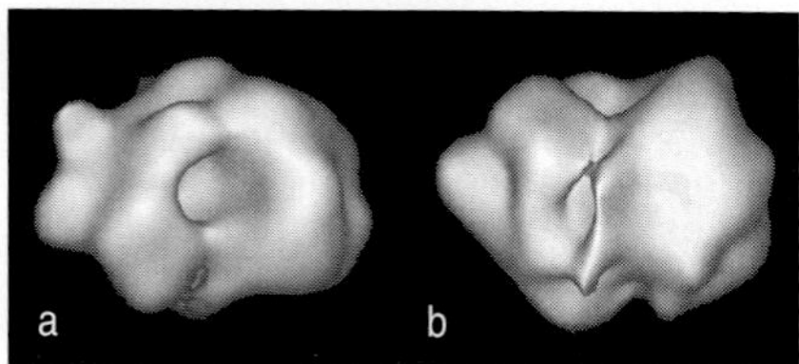

Fig. 5.3 TAP1 and TAP2 form a peptide transporter in the endoplasmic reticulum membrane. All transporters of the ATP-binding cassette (ABC) family are composed of four domains, as shown in the upper panel of this figure: two hydrophobic transmembrane domains that have multiple transmembrane regions, and two ATP-binding domains. TAP1 and TAP2 each encode a polypeptide chain with one hydrophobic and one ATP-binding domain; the two chains assemble into a heterodimer to form a four-domain transporter. From similarities between the TAP molecules and other members of the ABC transporter family, it is believed that the ATP-binding domains lie within the cytosol, whereas the hydrophobic domains project through the membrane into the lumen of the endoplasmic reticulum (ER) to form a channel through which peptides can pass, as shown in the lower panel in an electron microscopic reconstruction of the structure of the TAP1:TAP2 heterodimer. View a shows the ER luminal surface of the TAP transporter, looking down onto the top of the transmembrane domains, while view b shows the molecule in the plane of the membrane. The ATP-binding domains form two lobes beneath the transmembrane domains and are not visible in this view. TAP structures courtesy of G. Velarde.

In assays *in vitro* using normal cell fractions, microsomal vesicles that mimic the endoplasmic reticulum will internalize peptides, which then bind to MHC class I molecules already present in the microsome lumen. Vesicles from TAP1- or TAP2-deficient cells do not transport peptides. Peptide transport into the normal microsomes requires ATP hydrolysis, proving that the TAP1:TAP2 complex is an ATP-dependent peptide transporter. Similar experiments show that the TAP complex has some specificity for the peptides it will transport. It prefers peptides of between 8 and 16 amino acids, with hydrophobic or basic residues at the carboxy terminus—the exact features of peptides that bind MHC class I molecules (see Section 3-14)—and has a bias against proline in the first three amino-terminal residues. Discovery of TAP explained how viral peptides gain access to the lumen of the endoplasmic reticulum and are bound by MHC class I molecules, but left open the question of how these peptides are generated.

5-3 Peptides for transport into the endoplasmic reticulum are generated in the cytosol.

Proteins in cells are continually being degraded and replaced with newly synthesized proteins. Much cytosolic protein degradation is carried out by a large, multicatalytic protease complex called the **proteasome**. This is a large cylindrical complex of some 28 subunits, arranged in four stacked rings of seven subunits each. It has a hollow core lined by the active sites of the proteolytic subunits. Proteins to be degraded are introduced into the core of the proteasome and are there broken down into short peptides, which are then released.

Various lines of evidence implicate the proteasome in the production of peptide ligands for MHC class I molecules. The proteasome is part of the ubiquitin-dependent degradation pathway for cytosolic proteins, and experimentally tagging proteins with ubiquitin results in more efficient presentation of their peptides by MHC class I molecules. Inhibitors of the proteolytic activity of the proteasome also inhibit antigen presentation by MHC class I molecules. Whether the proteasome is the only cytosolic protease capable of generating peptides for transport into the endoplasmic reticulum is not known.

Two subunits of the proteasome, called LMP2 (or b1i) and LMP7 (or b5i), are encoded within the MHC near *TAP1* and *TAP2*. Along with MHC class I and TAP proteins, their expression is induced by interferons, which are produced in response to viral infections. LMP2 and LMP7 substitute for two constitutively expressed proteasome subunits. A third subunit, MECL-1 (also known as b2i), which is not encoded within the MHC, is also induced by interferons and also displaces a constitutive proteasome subunit. The proteasome can therefore exist in two forms—the constitutive proteasome, present in all cells, and the **immunoproteasome**, found in cells stimulated with interferons. The three inducible subunits of the immunoproteasome and their constitutive counterparts are thought to be the active proteases. The replacement of constitutive components by their interferon-inducible counterparts seems to change the enzymatic specificity of the proteasome such that there is increased cleavage of polypeptides after hydrophobic residues, and reduced cleavage after acidic residues. This produces peptides with carboxy-terminal residues that are preferred anchor residues for binding to most MHC class I molecules and are also the preferred structures for transport by TAP.

The production of antigenic peptides of the right length is enhanced by a further modification of the proteasome induced by interferon-γ (IFN-γ). This is the binding to the proteasome of a protein complex called the PA28 proteasome-activator complex. PA28 is a six- or seven-membered ring composed of

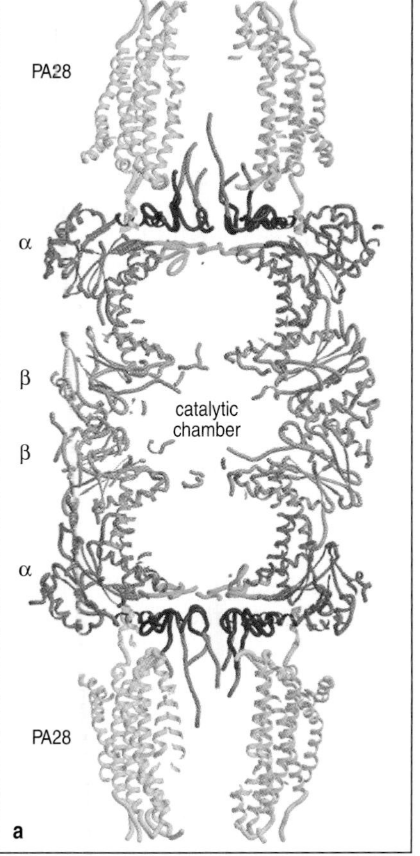

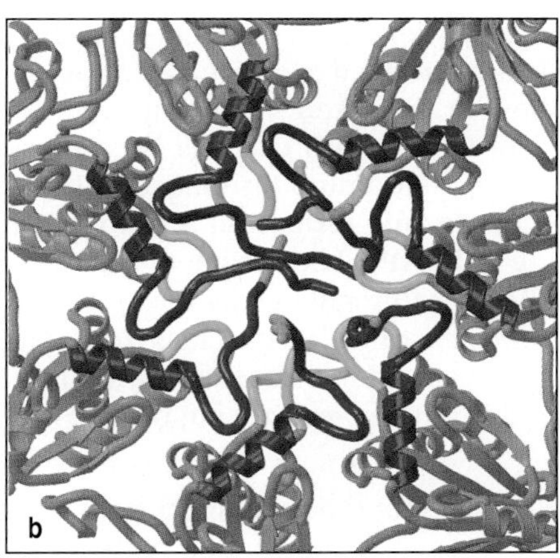

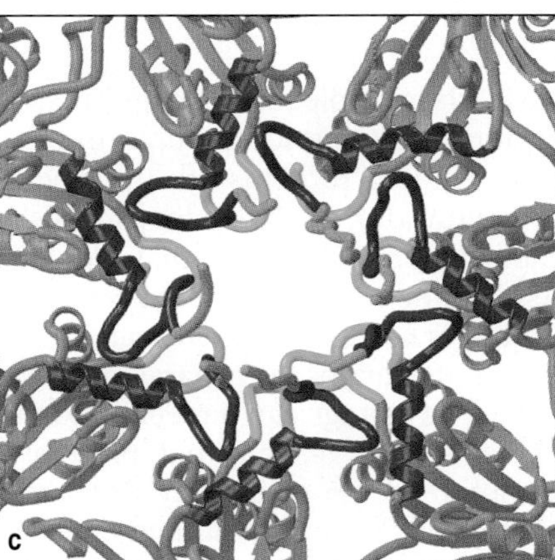

Fig. 5.4 The PA28 proteasome activator binds to either end of the proteasome. Panel a: the heptamer rings of the PA28 proteasome activator (yellow) interact with the α subunits (pink) at either end of the core proteasome (the β subunits that make up the catalytic cavity of the core are in blue). Within this region is the α-annulus (green), a narrow ring-like opening that is normally blocked by other parts of the α subunits (shown in red). Panel b: close-up view of the α-annulus. Panel c: binding of PA28 (not shown, for simplicity) to the proteasome changes the conformation of the α subunits, moving those parts of the molecule that block the α-annulus and thus opening the end of the cylinder. Courtesy of F. Whitby.

two proteins, PA28α and PA28β, both of which are induced by IFN-γ. The PA28 rings bind to either or both ends of the proteasome cylinder and, by opening up the ends, increase the rate at which peptides are released from the proteasome (Fig. 5.4). As well as simply providing more peptides, this increased rate of flow will allow potentially antigenic peptides to escape additional processing that might destroy their antigenicity.

Translation of self or pathogen-derived mRNAs in the cytoplasm not only generates properly folded proteins, but also a significant amount—possibly up to 30%—of peptides and proteins that are known as ***d*****efective *ri*****bosomal *p*****roducts** (**DRiPs**). These include peptides translated from introns in improperly spliced mRNAs, translations of frameshifts, and improperly folded proteins. DRiPs are recognized and tagged by ubiquitin for rapid degradation by the proteasome. This seemingly wasteful process ensures that both self proteins and proteins derived from pathogens generate abundant peptides for delivery to the proteasome for eventual presentation by MHC class I proteins. The proteasome can also increase the pool of peptides through an excision–splicing mechanism, in which an internal segment of a protein is removed and the surrounding noncontiguous polypeptide segments are joined and used as the peptide presented by MHC class I. Although it is not yet clear how frequently excision–splicing occurs, there are several examples of tumor-specific CD8 T cells that recognize peptide antigens formed in this way.

The proteasome produces peptides that are ready for delivery into the endoplasmic reticulum. At this stage, cellular chaperones, such as the TCP-1 ring complex (TRiC), a group II chaperone, protect these peptides from complete degradation in the cytoplasm. Many of these peptides are, however, longer than can be bound by MHC class I molecules. Thus, cleavage in the proteasome may not be the only processing of antigens for MHC class I molecules. There is very good evidence that the carboxy-terminal ends of peptide antigens are indeed produced by cleavage in the proteasome. The amino termini may be produced by another mechanism. Peptides too long to bind MHC class I molecules can still be transported into the endoplasmic reticulum, where their amino termini can be trimmed by an aminopeptidase called the ***e*****ndoplasmic *r*****eticulum *a*****minopeptidase associated with *a*****ntigen *p*****rocessing** (**ERAAP**). Like other components of the antigen-processing pathway, ERAAP is upregulated by IFN-γ. Mice lacking ERAAP have defective loading of peptides onto MHC class I molecules and have impaired CD8 T-cell responses, indicating that ERAAP is an essential and unique aminopeptidase in this antigen-processing pathway.

5-4 Retrograde transport from the endoplasmic reticulum to the cytosol enables exogenous proteins to be processed for cross-presentation by MHC class I molecules.

MHC class I molecules can also present peptides derived from membrane and secreted proteins, for example, the glycoproteins of viral envelopes. Membrane and secreted proteins are translocated into the lumen of the endoplasmic reticulum during their biosynthesis, yet the peptides bound by MHC class I molecules bear evidence that such proteins have been degraded in the cytosol. Asparagine-linked carbohydrate moieties commonly present on membrane-bound or secreted proteins can be removed in the cytosol by an enzyme reaction that changes the asparagine residue into aspartic acid, and this diagnostic sequence change can be seen in some peptides presented by MHC class I molecules. It now seems that proteins in the endoplasmic reticulum can be returned to the cytosol by the same translocation system that transported them into the endoplasmic reticulum in the first place. This newly discovered mechanism, known as **retrograde translocation** (**retrotranslocation**), may be the normal mechanism by which proteins in the endoplasmic reticulum are turned over and misfolded proteins removed. Once back in the cytosol, the polypeptides are degraded by the proteasome. The resulting peptides are then transported back into the lumen of the endoplasmic reticulum via TAP and loaded onto MHC class I molecules.

Because of this retrotranslocation mechanism, MHC class I molecules can also present peptides derived from proteins from other cells that have been taken up into the vesicular system from the extracellular environment. These can include, for example, proteins from virus-infected cells or from a tissue transplant. The presentation of exogenous antigens by MHC class I molecules to CD8 T cells is called **cross-presentation** (see Fig. 5.2) and was first recognized in the mid-1970s, long before the mechanism was understood. In an early experiment documenting cross-presentation, spleen cells from a mouse of one MHC type, H-2^b, were injected into a recipient H-$2^{b \times d}$ mouse (which carries both the b and d MHC types). The mice also differed in their genetic background outside the MHC. Surprisingly, some CD8 T cells responded to 'foreign' antigens expressed by the immunizing cells, although one would have expected only a CD4 T-cell response to these exogenous antigens. These responses were restricted by the recipient's H-2^d MHC class I molecules. This result was interpreted to mean that CD8 T cells could react to peptides that were derived from the immunizing cells but were presented by a host MHC class I molecule.

Recognition of cross-presentation preceded the recognition that retrotranslocation was involved, and how the exogenously derived proteins were delivered to the cytosol of the host cell was at first a puzzle. The precise biochemical machinery involved in retrotranslocation is the subject of current research, but once exogenous proteins have reached the cytosol they can be degraded by the proteasome and their peptides transported back into the endoplasmic reticulum and loaded onto MHC class I molecules. Cross-presentation occurs not only for antigens on tissue or cell grafts, as in the original experiment described above, but also in response to viral, bacterial, and tumor antigens. Cross-presentation occurs particularly well in a subset of dendritic cells that express CD8 on their surface; they are particularly efficient at taking up exogenous antigens into the endosomal system by phagocytosis and translocating them from there into the cytosol for processing and subsequent presentation by MHC class I molecules. This pathway is important in activating naive CD8 T cells against viruses that do not infect antigen-presenting cells such as dendritic cells.

5-5 Newly synthesized MHC class I molecules are retained in the endoplasmic reticulum until they bind a peptide.

Binding a peptide is an important step in the assembly of a stable MHC class I molecule. When the supply of peptides into the endoplasmic reticulum is disrupted, as in *TAP* mutant cells, newly synthesized MHC class I molecules are held in the endoplasmic reticulum in a partly folded state. This explains why cells with mutations in *TAP1* or *TAP2* fail to express MHC class I molecules at the cell surface. The folding and assembly of a complete MHC class I molecule (see Fig. 3.20) depends on the association of the MHC class I α chain first with β_2-microglobulin and then with peptide, and this process involves a number of accessory proteins with chaperone-like functions. Only after peptide has bound is the MHC class I molecule released from the endoplasmic reticulum and allowed to travel to the cell surface.

In humans, newly synthesized MHC class I α chains that enter the endoplasmic reticulum membranes bind to the chaperone protein **calnexin**, which retains the MHC class I molecule in a partly folded state in the endoplasmic reticulum (Fig. 5.5). Calnexin also associates with partly folded T-cell receptors, immunoglobulins, and MHC class II molecules, and so has a central role in the assembly of many immunological proteins. When β_2-microglobulin binds to the α chain, the partly folded α:β_2-microglobulin heterodimer dissociates from calnexin and now binds to a complex of proteins called the MHC class I loading complex. One component of this complex—**calreticulin**—is similar to calnexin and probably carries out a similar chaperone function. A second component of the complex is the TAP-associated protein **tapasin**, encoded by a gene within the MHC. Tapasin forms a bridge between MHC class I molecules and TAP, allowing the partly folded α:β_2-microglobulin heterodimer to await the transport of a suitable peptide from the cytosol. A third component of this complex is the chaperone **Erp57**, a thiol oxidoreductase that may have a role in breaking and reforming the disulfide bond in the MHC class I α_2 domain during peptide loading. Calnexin, Erp57, and calreticulin bind to a number of glycoproteins during their assembly in the endoplasmic reticulum and seem to be part of the cell's general quality-control mechanism.

The last component of the MHC class I loading complex is the TAP molecule itself, whose role as a transporter is also the best understood. The other components seem to be essential to maintain the MHC class I molecule in a state receptive to peptide and also to carry out a peptide-editing function, allowing the exchange of low-affinity peptides bound to the MHC class I molecule for peptides of higher affinity. Certainly, cells defective in either calreticulin or tapasin show defects in the assembly of MHC class I molecules and express class I complexes at the cell surface that contain suboptimal, low-affinity peptides.

The binding of a peptide to the partly folded heterodimer finally releases it from the MHC class I loading complex. The fully folded MHC class I molecule and its bound peptide can now leave the endoplasmic reticulum and be transported to the cell surface. It is not yet clear whether the complex directly

Fig. 5.5 MHC class I molecules do not leave the endoplasmic reticulum unless they bind peptides. Newly synthesized MHC class I α chains assemble in the endoplasmic reticulum with a membrane-bound protein, calnexin. When this complex binds β_2-microglobulin (β_2m), the MHC class I α:β_2m dimer dissociates from calnexin, and the partly folded MHC class I molecule then binds to the peptide transporter TAP by interacting with one molecule of the TAP-associated protein tapasin. The chaperone molecules calreticulin and Erp57 also bind to form part of this complex. The MHC class I molecule is retained within the endoplasmic reticulum until released by the binding of a peptide, which completes the folding of the MHC molecule. Even in the absence of infection, there is a continual flow of peptides from the cytosol into the ER. Defective ribosomal products (DRiPs) and old proteins marked for destruction are degraded in the cytoplasm by the proteasome to generate peptides that are transported into the lumen of the endoplasmic reticulum by TAP, as shown here, and some will bind to MHC class I molecules. Once a peptide has bound to the MHC molecule, the peptide:MHC complex leaves the endoplasmic reticulum and is transported through the Golgi apparatus and finally to the cell surface.

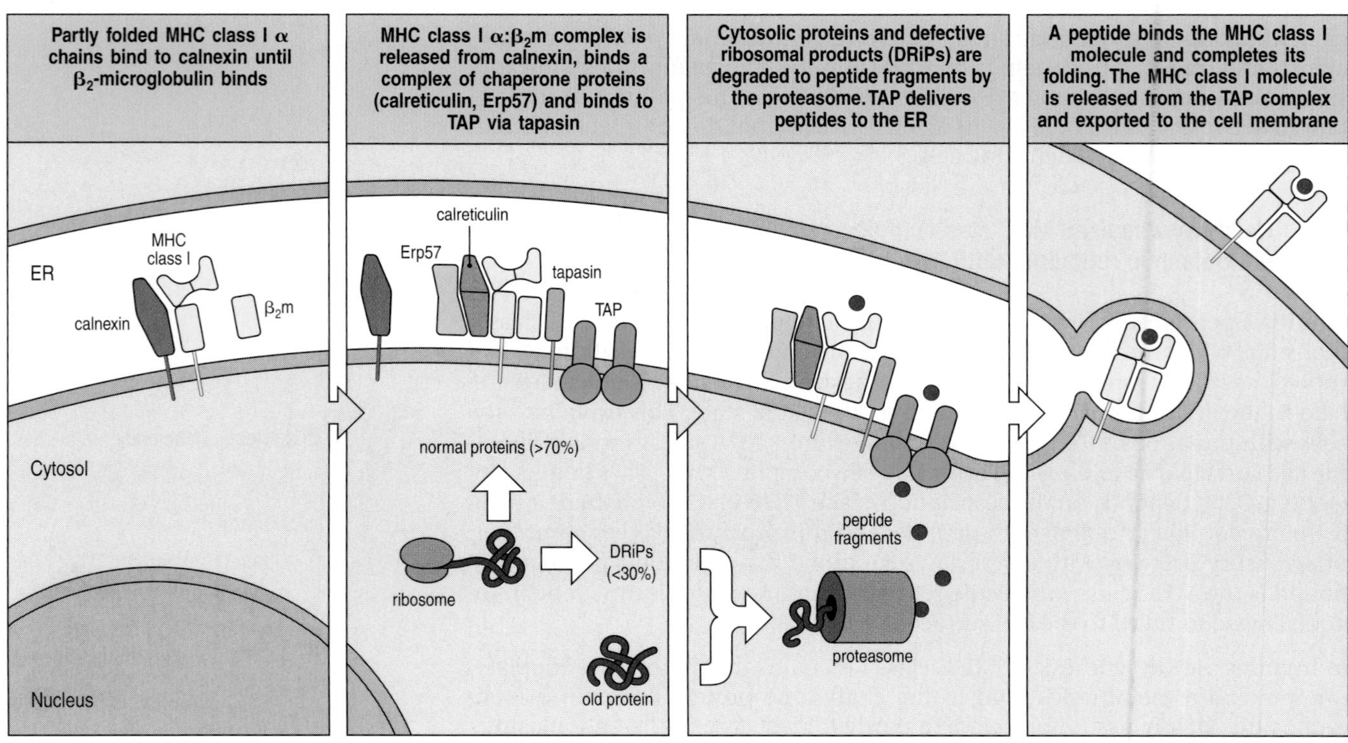

loads peptides onto MHC class I molecules or whether binding to it simply allows the MHC class I molecule to scan the peptides transported by TAP before they diffuse throughout the lumen of the endoplasmic reticulum or are transported back into the cytosol. Most of the peptides transported by TAP will not bind to the MHC molecules in that cell and are rapidly cleared out of the endoplasmic reticulum; there is evidence that they are transported back into the cytosol by an ATP-dependent transport mechanism distinct from that of TAP, known as the Sec61 complex.

In cells with mutant *TAP* genes, the MHC class I molecules in the endoplasmic reticulum are unstable and are eventually translocated back into the cytosol, where they are degraded. Thus, the MHC class I molecule must bind a peptide to complete its folding and be transported onward. In uninfected cells, peptides derived from self proteins fill the peptide-binding cleft of the mature MHC class I molecules and are carried to the cell surface. In normal cells, MHC class I molecules are retained in the endoplasmic reticulum for some time, which suggests that they are present in excess of peptide. This is very important for the immunological function of MHC class I molecules because they must be immediately available to transport viral peptides to the cell surface if the cell becomes infected.

5-6 Many viruses produce immunoevasins that interfere with antigen presentation by MHC class I molecules.

The presentation of viral peptides by MHC class I molecules at a cell surface signals CD8 T cells to kill the infected cell. Some viruses produce proteins, called **immunoevasins**, that enable the virus to evade immune recognition by preventing the appearance of peptide:MHC class I complexes on the infected cell (Fig. 5.6). Some viral immunoevasins block peptide entry into the endoplasmic reticulum by targeting the TAP transporter (Fig. 5.7, upper

Virus	Protein	Category	Mechanism
Herpes simplex virus 1	ICP47	Blocks peptide entry to endoplasmic reticulum	Blocks peptide binding to TAP
Human cytomegalovirus (HCMV)	US6		Inhibits TAP ATPase activity
Bovine herpes virus	UL49.5		Inhibits TAP peptide transport
Adenovirus	E19	Retention of MHC class I in endoplasmic reticulum	Competitive inhibitor of tapasin
HCMV	US3		Blocks tapasin function
Murine cytomegalovirus (CMV)	M152		Unknown
HCMV	US2	Degradation of MHC class I (dislocation)	Transports some newly synthesized MHC class I molecules into cytosol
Murine gamma herpes virus 68	mK3		E3-ubiquitin ligase activity
Murine CMV	m4	Binds MHC class I at cell surface	Interferes with recognition by cytotoxic lymphocytes by an unknown mechanism

Fig. 5.6 Immunoevasins produced by viruses interfere with the processing of antigens that bind to MHC class I molecules.

panel). The herpes simplex virus, for example, produces a protein, ICP47, that binds to the cytosolic surface of TAP and prevents peptides from entering the transporter. The US6 protein from human cytomegalovirus prevents peptide transport by inhibiting TAP ATPase activity, and the UL49.5 protein from bovine herpes virus inhibits TAP peptide transport. Viruses can also prevent peptide:MHC complexes from reaching the cell surface by retaining MHC class I molecules in the endoplasmic reticulum (Fig. 5.7, middle panel). The adenovirus E19 protein interacts with certain MHC class I proteins and contains a motif that retains the protein complex in the endoplasmic reticulum. E19 also prevents the tapasin–TAP interaction required for peptide loading onto the MHC class I molecule. Several viral proteins can catalyze the degradation of newly synthesized MHC class I molecules by a process known as **dislocation**, which initiates the pathway that is normally used to degrade misfolded endoplasmic reticulum proteins by directing them back into the cytosol. For example, the US11 protein of human cytomegalovirus binds nascent MHC class I molecules and in conjunction with a ubiquitous host endoplasmic reticulum membrane protein, derlin-1, delivers them into the cytosol, where they are degraded (Fig. 5.7, lower panel). Most viral immunoevasins come from DNA viruses such as the herpes virus family, which have large genomes and whose strategy of replication in the host involves a period of latency or quiescence.

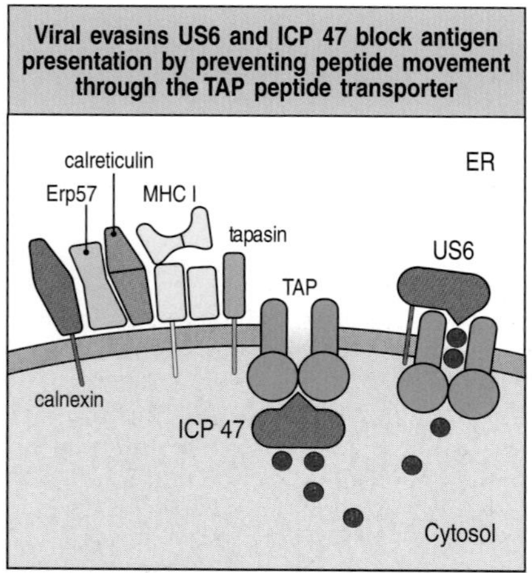
Viral evasins US6 and ICP 47 block antigen presentation by preventing peptide movement through the TAP peptide transporter
calreticulin
ER
Erp57
MHC I
tapasin
US6
TAP
calnexin
ICP 47
Cytosol

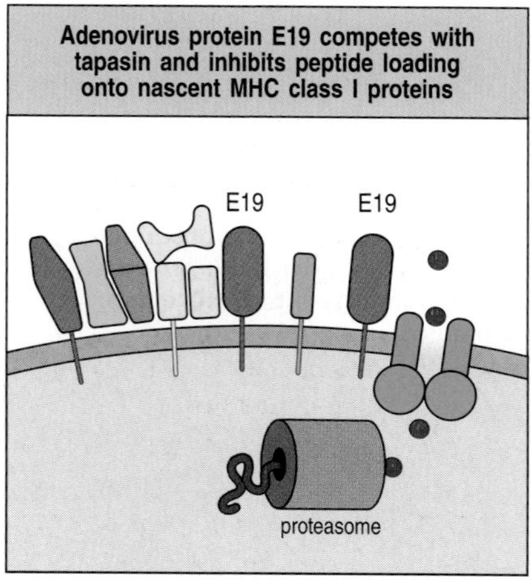
Adenovirus protein E19 competes with tapasin and inhibits peptide loading onto nascent MHC class I proteins
E19
E19
proteasome

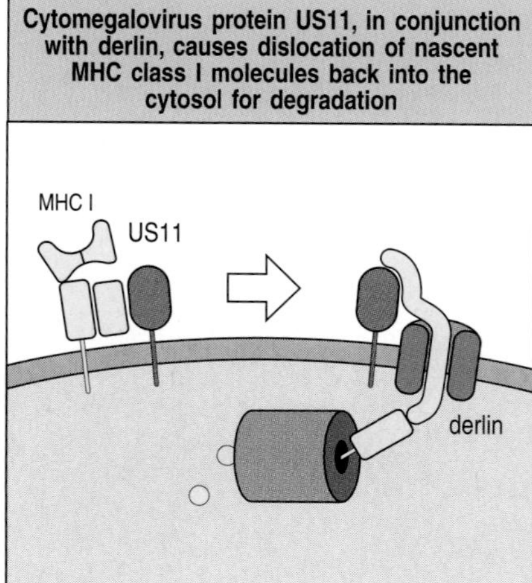
Cytomegalovirus protein US11, in conjunction with derlin, causes dislocation of nascent MHC class I molecules back into the cytosol for degradation
MHC I
US11
derlin

5-7 Peptides presented by MHC class II molecules are generated in acidified endocytic vesicles.

Several classes of pathogens, including the protozoan parasite *Leishmania* and the mycobacteria that cause leprosy and tuberculosis, replicate inside intracellular vesicles in macrophages. Because they reside in membrane-bounded vesicles, the proteins of these pathogens are not usually accessible to proteasomes in the cytosol. Instead, after activation of the macrophage, proteins in vesicles, which are often globular proteins stabilized by intramolecular disulfide bonds, are reduced and degraded by proteases within the vesicles into peptide fragments that bind to MHC class II molecules. In this way they are delivered to the cell surface, where they can be recognized by CD4 T cells. Extracellular pathogens and proteins that are internalized into endocytic vesicles are also processed by this pathway and their peptides are presented to CD4 T cells (Fig. 5.8).

Most of what we know about protein processing in the endocytic pathway has come from experiments in which simple proteins are fed to macrophages and are taken up by endocytosis; in this way the processing of added antigen can be quantified. Proteins that bind to surface immunoglobulin on B cells and are internalized by receptor-mediated endocytosis are processed by the same pathway. Proteins that enter cells through endocytosis are delivered to

Fig. 5.7 The peptide-loading complex in the endoplasmic reticulum is targeted by viral immunoevasins. The top panel shows blockade of peptide entry to the endoplasmic reticulum (ER). The cytosolic ICP47 protein from HSV-1 prevents peptides from binding to TAP in the cytosol, whereas the US6 protein from human CMV interferes with the ATP-dependent transfer of peptides through TAP. The center panel shows the retention of MHC class I molecules in the ER by the adenovirus E19 protein. This binds certain MHC molecules and retains them in the ER through an ER-retention motif, at the same time competing with tapasin to prevent association with TAP and peptide loading. The bottom panel shows how the human CMV US11 protein associates with newly synthesized MHC class I molecules and directs them back into the cytosol through an ER membrane channel, derlin-1. Once in the cytosol the MHC protein is marked for degradation in the proteasome.

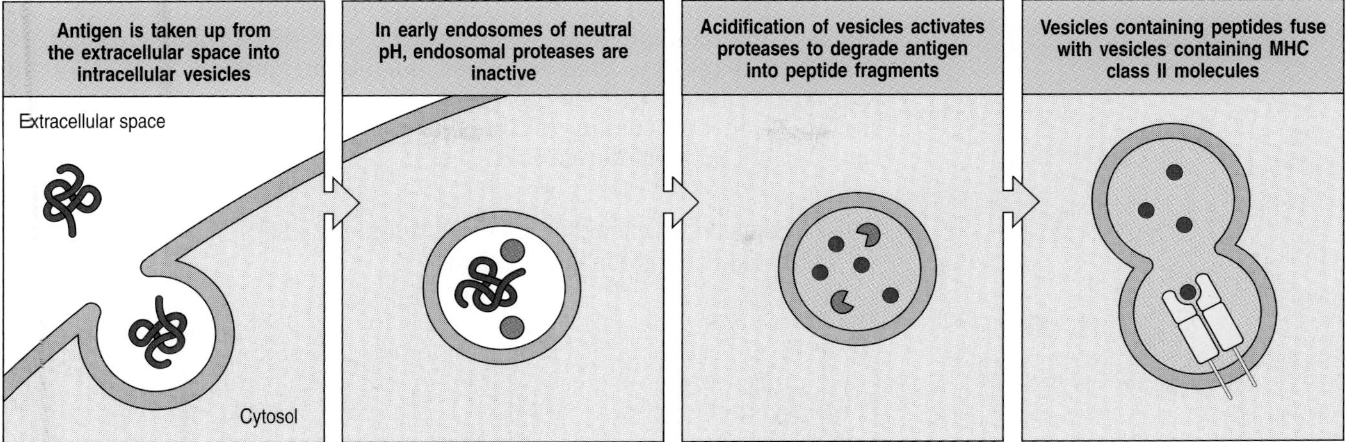

Antigen is taken up from the extracellular space into intracellular vesicles
In early endosomes of neutral pH, endosomal proteases are inactive
Acidification of vesicles activates proteases to degrade antigen into peptide fragments
Vesicles containing peptides fuse with vesicles containing MHC class II molecules
Extracellular space
Cytosol

Fig. 5.8 Peptides that bind to MHC class II molecules are generated in acidified endocytic vesicles. In the case illustrated here, extracellular foreign antigens, such as bacteria or bacterial antigens, have been taken up by an antigen-presenting cell such as a macrophage or immature dendritic cell. In other cases, the source of the peptide antigen may be bacteria or parasites that have invaded the cell to replicate in intracellular vesicles. In both cases the antigen-processing pathway is the same. The pH of the endosomes containing the engulfed pathogens decreases progressively, activating proteases within the vesicles to degrade the engulfed material. At some point on their pathway to the cell surface, newly synthesized MHC class II molecules pass through such acidified vesicles and bind peptide fragments of the antigen, transporting the peptides to the cell surface.

endosomes, which become increasingly acidic as they progress into the interior of the cell, eventually fusing with lysosomes. The endosomes and lysosomes contain proteases, known as acid proteases, that are activated at low pH and eventually degrade the protein antigens contained in the vesicles. Larger particulate material internalized by phagocytosis or macropinocytosis can also be handled by this pathway of antigen processing.

Drugs such as chloroquine that raise the pH of endosomes, making them less acidic, inhibit the presentation of antigens that enter the cell in this way, suggesting that acid proteases are responsible for processing internalized antigen. Among these acid proteases are the cysteine proteases cathepsins B, D, S, and L, the last of which is the most active enzyme in this family. Antigen processing can be mimicked to some extent by the digestion of proteins with these enzymes *in vitro* at acid pH. Cathepsins S and L may be the predominant proteases involved in the processing of vesicular antigens; mice that lack cathepsin B or cathepsin D show normal antigen processing, whereas mice with no cathepsin S show some deficiencies in antigen processing. It is likely that the overall repertoire of peptides produced within the endosomal pathway reflects the activities of the many proteases that are found in the endosomal and lysosomal compartments.

Disulfide bonds, particularly intramolecular disulfide bonds, may need to be reduced before proteins that contain them can be digested in endosomes. An IFN-γ-induced thiol reductase present in the endosomal compartment—**IFN-γ-induced lysosomal thiol reductase (GILT)**—carries out this role in the antigen-processing pathway.

MHC class II molecules primarily present peptides from proteins from the vesicular pathway, and MHC class I molecules present peptides derived from intracellular proteins. As described in Section 5-4, however, there is cross-talk between these pathways, allowing the cross-presentation of exogenous proteins by MHC class I molecules. Conversely, it should not be surprising that a significant number of peptides bound to MHC class II molecules arise from proteins, such as actin and ubiquitin, that are cytostolic in location. The most likely mechanism by which cytosolic proteins are processed for MHC class II presentation is the normal process of protein turnover known as **autophagy**, in which cytosolic proteins and organelles are delivered to lysosomes for degradation. Autophagy is constitutive but can be enhanced by cellular stresses such as starvation, when the cell must catabolize intracellular proteins to provide energy. In the process of **microautophagy**, cytosol is continuously internalized into the vesicular system by lysosomal invaginations, whereas in **macroautophagy**, which is induced by starvation, a double-membraned autophagosome engulfs cytosol and fuses with lysosomes. A third autophagy pathway uses the heat-shock cognate protein 70 (Hsc70) and the lysosome-associated membrane protein-2 (LAMP-2) to transport cytosolic proteins to lysosomes. Autophagy has been demonstrated in the processing of the Epstein–Barr virus nuclear antigen 1 (EBNA-1) for presentation to CD4 T cells.

5-8 The invariant chain directs newly synthesized MHC class II molecules to acidified intracellular vesicles.

The function of MHC class II molecules is to bind peptides generated in the intracellular vesicles of macrophages, immature dendritic cells, B cells, and other antigen-presenting cells and to present these peptides to CD4 T cells. The biosynthetic pathway for MHC class II molecules, like that of other cell-surface glycoproteins, starts with their translocation into the endoplasmic reticulum, and they must therefore be prevented from binding prematurely to peptides transported into the endoplasmic reticulum lumen or to the cell's own newly synthesized polypeptides. Because the endoplasmic reticulum is richly endowed with unfolded and partly folded polypeptide chains, a general mechanism is needed to prevent their binding in the open-ended MHC class II molecule peptide-binding groove.

Binding is prevented by the assembly of newly synthesized MHC class II molecules with a protein known as the MHC class II-associated **invariant chain (Ii)**. The invariant chain forms trimers, with each subunit binding noncovalently to an MHC class II α:β heterodimer (Fig. 5.9). An Ii chain binds to the MHC class II molecule with part of its polypeptide chain lying within the peptide-binding groove, thus blocking the groove and preventing the binding of either peptides or partly folded proteins. While this complex is being assembled in the endoplasmic reticulum, its component parts are associated with calnexin. Only when assembly is completed to produce a nine-chain complex is the complex released from calnexin for transport out of the endoplasmic reticulum. When it is part of the nine-chain complex, the MHC class II molecule cannot bind peptides or unfolded proteins, so that peptides present in the endoplasmic reticulum are not usually presented by MHC class II molecules. There is evidence that in the absence of invariant chains many MHC class II molecules are retained in the endoplasmic reticulum as complexes with misfolded proteins.

The invariant chain has a second function, which is to target delivery of the MHC class II molecules to a low-pH endosomal compartment where peptide loading can occur. The complex of MHC class II α:β heterodimers with Ii is retained for 2–4 hours in this compartment. During this time, Ii is cleaved by

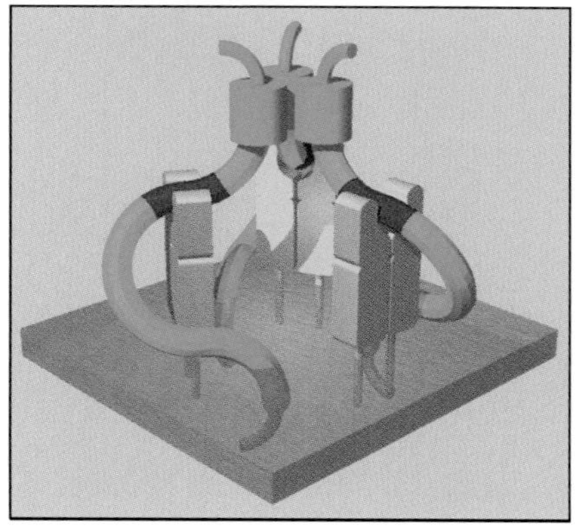

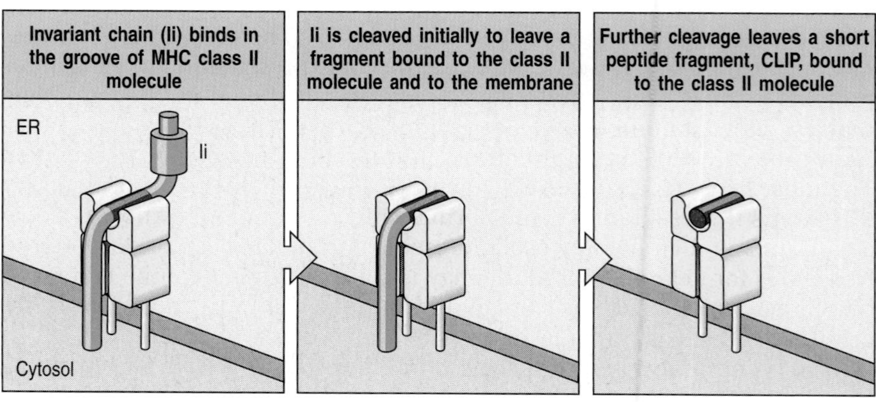
Invariant chain (Ii) binds in the groove of MHC class II molecule
Ii is cleaved initially to leave a fragment bound to the class II molecule and to the membrane
Further cleavage leaves a short peptide fragment, CLIP, bound to the class II molecule
ER
Ii
Cytosol

Fig. 5.9 The invariant chain is cleaved to leave a peptide fragment, CLIP, bound to the MHC class II molecule. A model of the trimeric invariant chain bound to MHC class II α:β heterodimers is shown on the left. The CLIP portion is shown in red, the rest of the invariant chain in green, and the MHC class II molecules in yellow. In the endoplasmic reticulum, the invariant chain (Ii) binds to MHC class II molecules with the CLIP section of its polypeptide chain lying along the peptide-binding groove (model and left of three panels). After transport into an acidified vesicle, Ii is cleaved, initially just at one side of the MHC class II molecule (center panel). The remaining portion of Ii (known as the leupeptin-induced peptide or LIP fragment) retains the transmembrane and cytoplasmic segments that contain the signals that target Ii:MHC class II complexes to the endosomal pathway. Subsequent cleavage (right panel) of LIP leaves only a short peptide still bound by the class II molecule; this peptide is the CLIP fragment. Model structure courtesy of P. Cresswell.

acid proteases such as cathepsin S in several steps, as shown in Fig. 5.9. The initial cleavages generate a truncated form of Ii that remains bound to the MHC class II molecule and retains it within the proteolytic compartment. A subsequent cleavage releases the MHC class II molecule from the membrane-associated fragment of Ii, leaving a short fragment of Ii called **CLIP** (for ***cl**ass II-**a**ssociated **i**nvariant-chain **p**eptide*) still bound to the MHC molecule. MHC class II molecules associated with CLIP cannot bind other peptides. CLIP must either dissociate or be displaced to allow a peptide to bind to the MHC molecule and enable the complex to be delivered to the cell surface. Cathepsin S cleaves Ii in most MHC class II-positive cells, including antigen-presenting cells, whereas cathepsin L seems to substitute for cathepsin S in thymic cortical epithelial cells.

The endosomal compartment in which Ii is cleaved and MHC class II molecules encounter peptide is not yet clearly defined. Most newly synthesized MHC class II molecules are brought toward the cell surface in vesicles, which at some point fuse with incoming endosomes. However, there is also evidence that some MHC class II:Ii complexes are first transported to the cell surface and then re-internalized into endosomes. In either case, MHC class II:Ii complexes enter the endosomal pathway and there encounter and bind peptides derived from self proteins. Immunoelectron-microscopy using antibodies tagged with gold particles to localize Ii and MHC class II molecules within cells suggests that Ii is cleaved and peptides bind to MHC class II molecules in a particular endosomal compartment called the **MIIC** (***M**HC class **II** **c**ompartment*), late in the endosomal pathway (Fig. 5.10).

As with MHC class I molecules, MHC class II molecules in uninfected cells bind peptides derived from self proteins. MHC class II molecules that do not bind peptide after dissociation from the invariant chain are unstable in the acidic pH of the endosome, and are rapidly degraded.

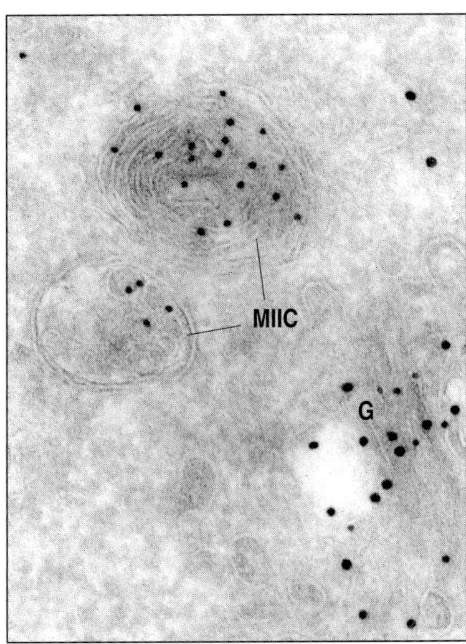

Fig. 5.10 MHC class II molecules are loaded with peptide in a specialized intracellular compartment. MHC class II molecules are transported from the Golgi apparatus (labeled G in this electron micrograph of an ultrathin section of a B cell) to the cell surface via specialized intracellular vesicles called the MHC class II compartment (MIIC). These have a complex morphology, showing internal vesicles and sheets of membrane. Antibodies labeled with different-sized gold particles identify the presence of both MHC class II molecules (small gold particles) and the invariant chain (large gold particles) in the Golgi, whereas only MHC class II molecules are detectable in the MIIC. This compartment is therefore thought to be the one in which the invariant chain is cleaved and peptide loading occurs. Photograph (× 135,000) courtesy of H.J. Geuze.

5-9 A specialized MHC class II-like molecule catalyzes loading of MHC class II molecules with peptides.

Another component of the vesicular antigen-processing pathway was revealed by analysis of mutant human B-cell lines with a defect in antigen presentation. MHC class II molecules in these cell lines assemble correctly with the invariant chain and seem to follow the normal vesicular route. However, they fail to bind peptides derived from internalized proteins and often arrive at the cell surface with the CLIP peptide still bound.

The defect in these cells lies in an MHC class II-like molecule called **HLA-DM** in humans (H-2M in mice). The HLA-DM genes are found near the TAP and LMP (now also known as PSMB) genes in the MHC class II region (see Fig. 5.12); they encode an α chain and a β chain that closely resemble those of other MHC class II molecules. The HLA-DM molecule is not expressed at the cell surface, however, but is found predominantly in the MIIC compartment. HLA-DM binds to and stabilizes empty MHC class II molecules that would otherwise aggregate; in addition, it catalyzes both the release of the CLIP fragment from MHC class II:CLIP complexes and the binding of other peptides to the empty MHC class II molecule (Fig. 5.11). However, the HLA-DM molecule itself does not bind peptides, and the open groove found in other MHC class II molecules is closed in HLA-DM.

HLA-DM also catalyzes the release of unstably bound peptides from MHC class II molecules. In the presence of a mixture of peptides capable of binding to MHC class II molecules, as occurs in the MIIC, HLA-DM continuously binds and rebinds to peptide:MHC class II complexes, removing weakly bound peptides and allowing other peptides to replace them. Antigens presented by MHC

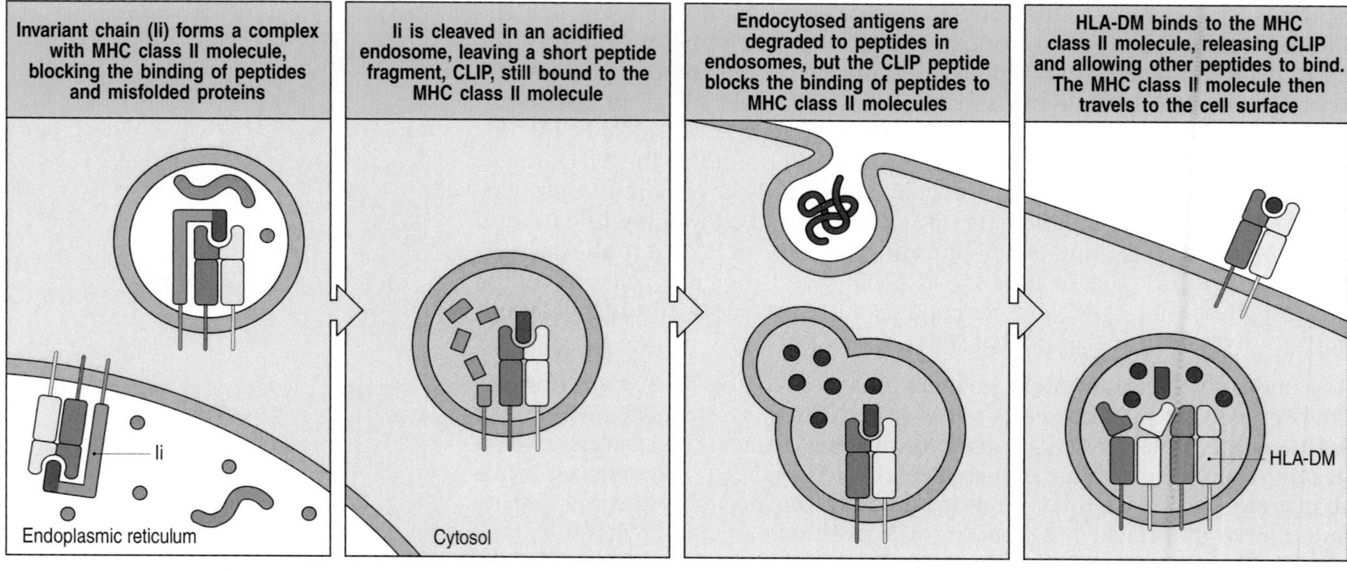

Invariant chain (li) forms a complex with MHC class II molecule, blocking the binding of peptides and misfolded proteins
li
Endoplasmic reticulum
li is cleaved in an acidified endosome, leaving a short peptide fragment, CLIP, still bound to the MHC class II molecule
Cytosol
Endocytosed antigens are degraded to peptides in endosomes, but the CLIP peptide blocks the binding of peptides to MHC class II molecules
HLA-DM binds to the MHC class II molecule, releasing CLIP and allowing other peptides to bind. The MHC class II molecule then travels to the cell surface
HLA-DM

Fig. 5.11 HLA-DM facilitates the loading of antigenic peptides onto MHC class II molecules. The invariant chain (li) binds to newly synthesized MHC class II molecules and blocks the binding of peptides and unfolded proteins in the endoplasmic reticulum and during the transport of the MHC class II molecule into acidified endocytic vesicles (first panel). In such vesicles, proteases cleave the invariant chain, leaving the CLIP peptide bound to the MHC class II molecule (second panel). Pathogens and their proteins are broken down into peptides within acidified endosomes, but these peptides cannot bind to MHC class II molecules that are occupied by CLIP (third panel). The class II-like molecule, HLA-DM, binds to MHC class II:CLIP complexes, catalyzing the release of CLIP and the binding of antigenic peptides (fourth panel).

class II molecules may have to persist on the surface of antigen-presenting cells for some days before encountering T cells able to recognize them. The ability of HLA-DM to remove unstably bound peptides, sometimes called **peptide editing**, ensures that the peptide:MHC class II complexes displayed on the surface of the antigen-presenting cell survive long enough to stimulate the appropriate CD4 cells.

A second atypical MHC class II molecule, called **HLA-DO** (H-2O in mice), is produced in thymic epithelial cells and B cells. This molecule is a heterodimer of the HLA-DOα chain and the HLA-DOβ chain (see Fig. 5.12). HLA-DO is not present at the cell surface, being found only in intracellular vesicles, and it does not seem to bind peptides. Instead, it acts as a negative regulator of HLA-DM, binding to it and inhibiting both the HLA-DM-catalyzed release of CLIP from, and the binding of other peptides to, MHC class II molecules. Expression of the HLA-DOβ chain is not increased by IFN-γ, whereas the expression of HLA-DM is. Thus, during inflammatory responses, in which IFN-γ is produced by T cells and NK cells, the increased expression of HLA-DM is able to overcome the inhibitory effects of HLA-DO. Why the antigen-presenting ability of thymic epithelial cells and B cells should be regulated in this way is not known; in thymic epithelial cells the purpose may be to select developing CD4 T cells by using a repertoire of self peptides different from those to which they will be exposed as mature T cells. The role of HLA-DM in facilitating the binding of peptides to MHC class II molecules parallels that of TAP in facilitating peptide binding to MHC class I molecules. Thus it seems likely that specialized mechanisms of delivering peptides have coevolved with the MHC molecules themselves. It is also likely that pathogens have evolved strategies to inhibit the loading of peptides onto MHC class II molecules, much as viruses have found ways of subverting antigen processing and presentation through the MHC class I molecules.

5-10 Stable binding of peptides by MHC molecules provides effective antigen presentation at the cell surface.

For MHC molecules to perform their essential function of signaling intracellular infection, the peptide:MHC complex must be stable at the cell surface. If the complex were to dissociate too readily, the pathogen in the infected cell could escape detection. In addition, MHC molecules on uninfected cells could pick up peptides released by MHC molecules on infected cells and falsely signal to cytotoxic T cells that a healthy cell is infected, triggering its unwarranted destruction. The tight binding of peptides by MHC molecules makes both of these undesirable outcomes unlikely.

The persistence of a peptide:MHC complex on a cell can be measured by its ability to stimulate T cells, while the fate of the MHC molecules themselves can be followed directly by specific staining. In this way it can be shown that specific peptide:MHC complexes on living cells are lost from the surface and re-internalized as part of natural protein turnover at the same rate as the MHC molecules themselves, indicating that peptide binding is essentially irreversible. This stable binding enables even rare peptides to be transported efficiently to the cell surface by MHC molecules, and allows long-term display of these complexes on the surface of the infected cell. This satisfies the first of the requirements for effective antigen presentation.

The second requirement is that if a peptide should dissociate from a cell-surface MHC molecule, peptides from the surrounding extracellular fluid would not be able to bind to the empty peptide-binding site. In fact, removal of the peptide from a purified MHC class I molecule has been shown to require denaturation of the protein. When peptide dissociates from an MHC class I molecule at the surface of a living cell, the molecule changes conformation, the β_2-microglobulin dissociates, and the α chain is internalized and rapidly degraded. Thus, most empty MHC class I molecules are quickly lost from the cell surface.

At neutral pH, empty MHC class II molecules are more stable than empty MHC class I molecules, yet empty MHC class II molecules are also removed from the cell surface. They aggregate readily, and internalization of such aggregates may account for their loss from the surface. Moreover, peptide loss from MHC class II molecules is most likely when the molecules transit through acidified endosomes as part of the normal process of cell-membrane recycling. At acidic pH, MHC class II molecules are able to bind peptides that are present in the vesicles, but those that fail to do so are rapidly degraded.

Thus, both MHC class I and class II molecules are effectively prevented from acquiring peptides directly from the surrounding extracellular fluid. This ensures that T cells act selectively on infected cells or on cells specialized for pathogen ingestion and display, while sparing surrounding healthy cells.

Summary.

The most distinctive feature of antigen recognition by T cells is the form of the ligand recognized by the T-cell receptor. This comprises a peptide derived from a foreign antigen and bound to an MHC molecule. MHC molecules are cell-surface glycoproteins with a peptide-binding groove that can bind a wide variety of different peptides. The MHC molecule binds the peptide in an intracellular location and delivers it to the cell surface, where the combined ligand can be recognized by a T cell. There are two classes of MHC molecule, MHC class I and MHC class II, which acquire peptides at different intracellular sites, and which activate CD8 and CD4 T cells, respectively. MHC class I molecules are synthesized in the endoplasmic reticulum and acquire their peptides at this location. The peptides loaded onto MHC class I are derived from proteins degraded in the cytosol by a multicatalytic protease, the proteasome. Peptides generated by proteasomes are transported into the endoplasmic reticulum by a heterodimeric ATP-binding protein called TAP and are then available for binding by partly folded MHC class I molecules that are held tethered to TAP. Peptide binding is an integral part of

MHC class I molecule assembly, and must occur before the MHC class I molecule can complete its folding and leave the endoplasmic reticulum for the cell surface. The proteasome degrades normal cytosolic proteins, allowing the detection and elimination of cytosolic pathogens, such as viruses, by CD8 T cells specialized to kill any cells displaying foreign peptides. The proteasome can also degrade proteins that have been transported into the cytosol from the vesicular system by retrograde transportation. This can occur, for example, when a dendritic cell has engulfed dead cells killed by a virus. Transporting exogenous viral antigens into the cytosol allows uninfected dendritic cells to process and present these antigens to naive CD8 T cells in the phenomenon of cross-presentation, which is important for the generation of effective immune responses.

In contrast to peptide loading of MHC class I, MHC class II molecules do not acquire their peptide ligands in the endoplasmic reticulum because of their early association with the invariant chain (Ii), which binds to and blocks their peptide-binding groove. They are targeted by Ii to an acidic endosomal compartment where—in the presence of active proteases, in particular cathepsin S, and with the help of HLA-DM, a specialized MHC class II-like molecule that catalyzes peptide loading—Ii is released and other peptides are bound. MHC class II molecules thus bind peptides from proteins that are degraded in endosomes. There they can capture peptides from pathogens that have entered the vesicular system of macrophages, or from antigens internalized by immature dendritic cells or the immunoglobulin receptors of B cells. The process of autophagy can deliver cytosolic proteins to the vesicular system for presentation by MHC class II. The CD4 T cells that recognize peptide:MHC class II complexes have a variety of specialized effector activities. Subsets of CD4 T cells activate macrophages to kill the intravesicular pathogens they harbor, help B cells to secrete immunoglobulins against foreign molecules, and regulate immune responses.

The major histocompatibility complex and its functions.

The function of MHC molecules is to bind peptide fragments derived from pathogens and display them on the cell surface for recognition by the appropriate T cells. The consequences are almost always deleterious to the pathogen—virus-infected cells are killed, macrophages are activated to kill bacteria living in their intracellular vesicles, and B cells are activated to produce antibodies that eliminate or neutralize extracellular pathogens. Thus, there is strong selective pressure in favor of any pathogen that has mutated in such a way that it escapes presentation by an MHC molecule.

Two separate properties of the MHC make it difficult for pathogens to evade immune responses in this way. First, the MHC is **polygenic**: it contains several different MHC class I and MHC class II genes, so that every individual possesses a set of MHC molecules with different ranges of peptide-binding specificities. Second, the MHC is highly **polymorphic**; that is, there are multiple variants of each gene within the population as a whole. The MHC genes are, in fact, the most polymorphic genes known. In this section we describe the organization of the genes in the MHC and discuss how the variation in MHC molecules arises. We will also see how the effect of polygeny and polymorphism on the range of peptides that can be bound contributes to the ability of the immune system to respond to the multitude of different and rapidly evolving pathogens.

5-11 Many proteins involved in antigen processing and presentation are encoded by genes within the major histocompatibility complex.

The major histocompatibility complex is located on chromosome 6 in humans and chromosome 17 in the mouse and extends over at least 4×10^6 base pairs. In humans it contains more than 200 genes. As work continues to define the genes within and around the MHC, it becomes difficult to establish precise boundaries for this locus, which it is now thought could span as much as 7×10^6 base pairs. The genes encoding the α chains of MHC class I molecules and the α and β chains of MHC class II molecules are linked within the complex; the genes for β_2-microglobulin and the invariant chain are on different chromosomes (chromosomes 15 and 5, respectively, in humans, and chromosomes 2 and 18 in the mouse). Figure 5.12 shows the general organization of the MHC class I and II genes in human and mouse. In humans these genes are called ***h**uman **l**eukocyte **a**ntigen* or **HLA** genes, because they were first discovered through antigenic differences between white blood cells from different individuals; in the mouse they are known as the **H-2** genes. The murine MHC class II genes were in fact first identified as genes that controlled whether an immune response was made to a given antigen and were originally called **Ir** (**Immune response**) **genes**. Because of this, the murine MHC class II *A* and *E* genes are frequently referred to as *I-A* and *I-E*; but this terminology should not be confused with MHC class I genes.

There are three class I α-chain genes in humans, called *HLA-A*, *-B*, and *-C*. There are also three pairs of MHC class II α- and β-chain genes, called *HLA-DR*, *-DP*, and *-DQ*. In many people, however, the HLA-DR cluster contains an extra β-chain gene whose product can pair with the DRα chain. This means that the three sets of genes can give rise to four types of MHC class II molecules. All the MHC class I and class II molecules can present peptides to T cells, but each protein binds a different range of peptides (see Sections 3-14 and 3-15). Thus, the presence of several different genes for each MHC class means that any one individual is equipped to present a much broader range of peptides than if only one MHC molecule of each class were expressed at the cell surface.

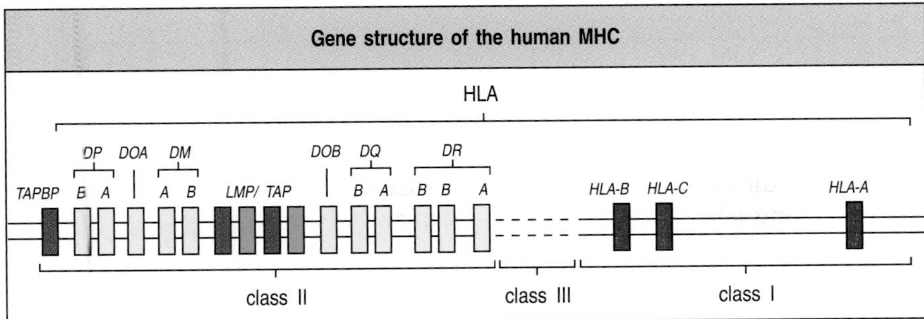

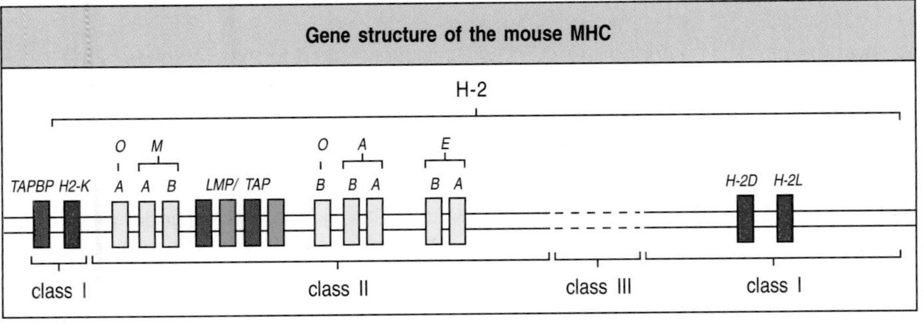

Fig. 5.12 The genetic organization of the major histocompatibility complex (MHC) in human and mouse. The organization of the principal MHC genes is shown for both humans (where the MHC is called HLA and is on chromosome 6) and mice (in which the MHC is called H-2 and is on chromosome 17). The organization of the MHC genes is similar in both species. There are separate clusters of MHC class I genes (red) and MHC class II genes (yellow), although in the mouse an MHC class I gene (*H-2K*) seems to have translocated relative to the human MHC so that the class I region in mice is split in two. In both species there are three main class I genes, which are called *HLA-A*, *HLA-B*, and *HLA-C* in humans, and *H2-K*, *H2-D*, and *H2-L* in the mouse. Each of these encodes the α chain of the respective MHC class I protein (HLA-A, HLA-B, etc.). The other subunit of an MHC class I molecule, β_2-microglobulin, is encoded by a gene located on a different chromosome—chromosome 15 in humans and chromosome 2 in the mouse. The class II region includes the genes for the α and β chains (designated *A* and *B*, respectively, in the gene names) of the MHC class II molecules HLA-DR, -DP, and -DQ (H-2A and -E in the mouse). In addition, the genes for the TAP1:TAP2 peptide transporter, the *LMP* genes that encode proteasome subunits, the genes encoding the DMα and DMβ chains (*DMA* and *DMB*), the genes encoding the α and β chains of the DO molecule (*DOA* and *DOB*, respectively), and the gene encoding tapasin (*TAPBP*) are also in the MHC class II region. The so-called class III genes encode various other proteins with functions in immunity (see Fig. 5.13).

Figure 5.13 shows a more detailed map of the human MHC locus. An inspection of this map shows that many genes within this locus participate in antigen processing or antigen presentation, or have other functions related to the innate or adaptive immune response. The two *TAP* genes lie in the MHC class II region, in close association with the *LMP* genes, whereas the gene encoding tapasin (*TAPBP*), which binds to both TAP and empty MHC class I molecules, lies at the edge of the MHC nearest the centromere (see Fig. 5.13). The genetic linkage of the MHC class I genes (whose products deliver cytosolic peptides to the cell surface) with the *TAP*, tapasin, and proteasome (*LMP*) genes (whose products deliver cytosolic peptides into the endoplasmic reticulum) suggests that the entire MHC has been selected during evolution for antigen processing and presentation.

When cells are treated with the interferons IFN-α, -β, or -γ, there is a marked increase in the transcription of MHC class I α-chain and β_2-microglobulin

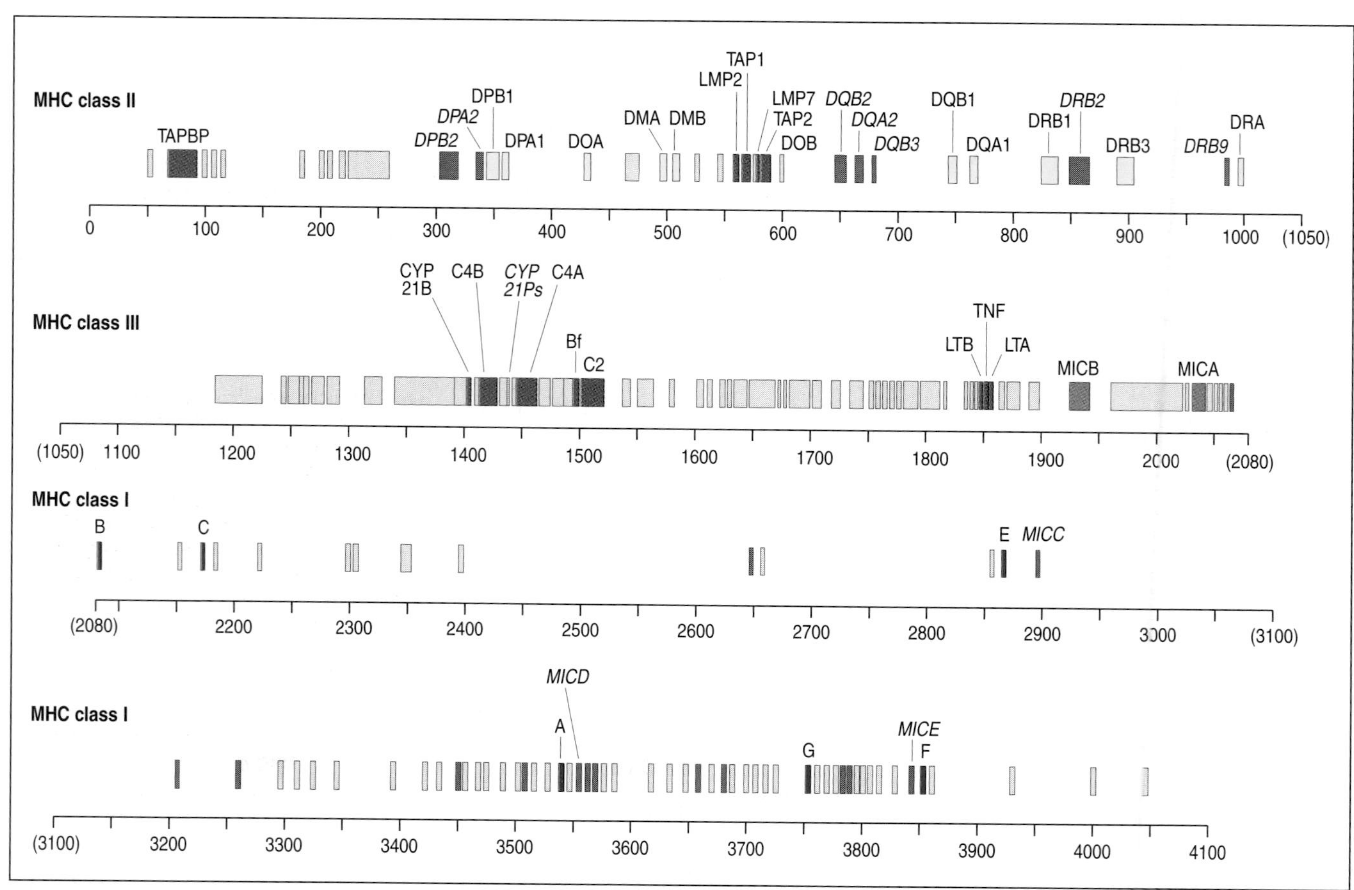

Fig. 5.13 Detailed map of the human MHC. The organization of the class I, class II, and class III regions of the human MHC is shown, with approximate genetic distances given in thousands of base pairs (kbp). Most of the genes in the class I and class II regions are mentioned in the text. The additional genes indicated in the class I region (for example, E, F, and G) are class I-like genes, encoding class Ib molecules; the additional class II genes are pseudogenes. The genes shown in the class III region encode the complement proteins C4 (two genes, shown as C4A and C4B), C2, and factor B (shown as Bf) as well as genes that encode the cytokines tumor necrosis factor-α (TNF) and lymphotoxin (LTA, LTB). Closely linked to the C4 genes is the gene encoding 21-hydroxylase (shown as CYP 21B), an enzyme involved in steroid biosynthesis. In this figure, genes shown in dark gray and named in italic are pseudogenes. The genes are color coded, with the MHC class I genes being shown in red, except for the MIC genes, which are shown in blue; these are distinct from the other class I-like genes and are under different transcriptional control. The MHC class II genes are shown in yellow. Genes in the MHC region that have immune functions but are not related to the MHC class I and class II genes are shown in purple.

genes and of the proteasome, tapasin, and *TAP* genes. Interferons are produced early in viral infections as part of the innate immune response, as described in Chapter 2, so this effect increases the ability of cells to process viral proteins and present the resulting virus-derived peptides at the cell surface. This helps to activate the appropriate T cells and initiate the adaptive immune response in response to the virus. The coordinated regulation of the genes encoding these components may be facilitated by the linkage of many of them in the MHC.

The *HLA-DM* genes, which encode the DM molecule, whose function is to catalyze peptide binding to MHC class II molecules (see Section 5-9), are clearly related to the MHC class II genes. The *DOA* and *DOB* genes, which encode the DOα and DOβ subunits of the DO molecule, a negative regulator of DM, are also clearly related to the MHC class II genes. The classical MHC class II genes, along with the invariant-chain gene and the genes encoding DMα, DMβ, and DOα, but not DOβ, are coordinately regulated. This distinct regulation of MHC class II genes by IFN-γ, which is made by activated T cells of the T_H1 type as well as by activated CD8 and NK cells, allows T cells responding to bacterial infections to increase the expression of those molecules concerned with the processing and presentation of intravesicular antigens. Expression of all these molecules is induced by IFN-γ (but not by IFN-α or -β), via the production of a transcriptional activator known as **MHC class II transactivator** (**CIITA**). An absence of CIITA causes severe immunodeficiency due to nonproduction of MHC class II molecules. Finally, the MHC locus contains many so-called 'nonclassical' MHC genes, which resemble MHC class I genes in structure. We will return to these genes, often called MHC class Ib genes, in Section 5-18, after we complete our discussion of the classical MHC genes.

5-12 The protein products of MHC class I and class II genes are highly polymorphic.

Because of the polygeny of the MHC, every person expresses at least three different antigen-presenting MHC class I molecules and three (or sometimes four) MHC class II molecules on his or her cells (see Section 5-11). In fact, the number of different MHC molecules expressed on the cells of most people is greater because of the extreme polymorphism of the MHC (Fig. 5.14) and the codominant expression of MHC gene products.

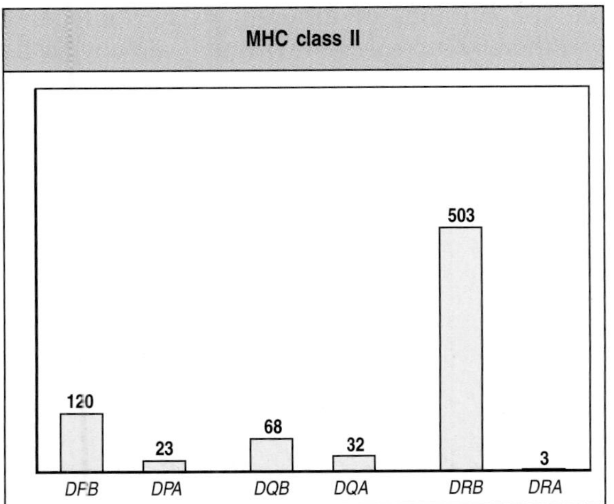

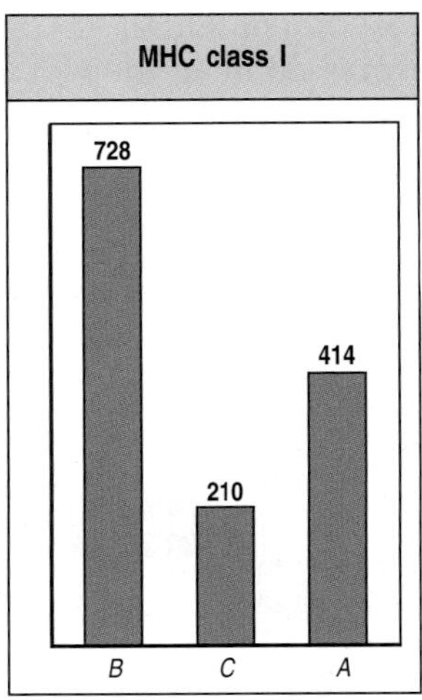

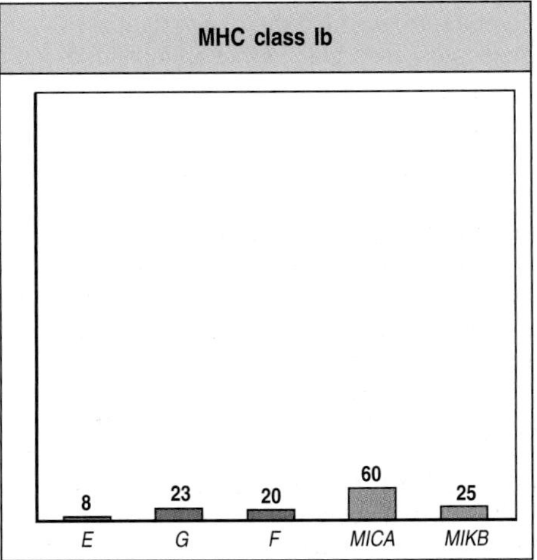

Fig. 5.14 Human MHC genes are highly polymorphic. With the notable exception of the DRα locus, which is functionally monomorphic, each locus has many alleles. Shown in this figure as the heights of the bars are the number of different HLA alleles currently assigned by the WHO Nomenclature Committee for Factors of the HLA System as of January 2006.

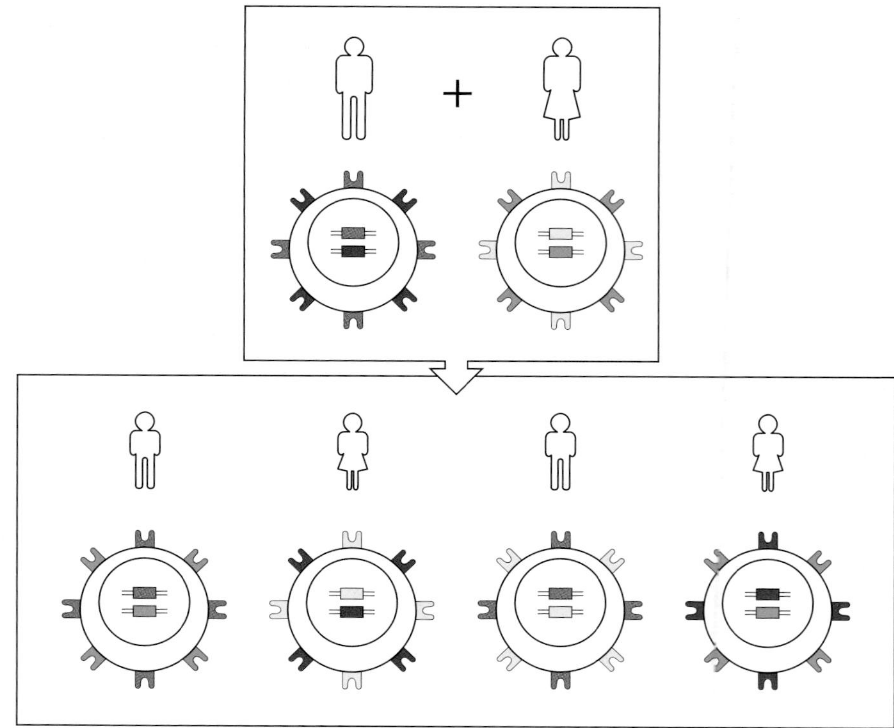

Fig. 5.15 Expression of MHC alleles is codominant. The MHC is so polymorphic that most individuals are likely to be heterozygous at each locus. Alleles are expressed from both MHC haplotypes in any one individual, and the products of all alleles are found on all expressing cells. In any mating, four possible combinations of haplotypes can be found in the offspring; thus siblings are also likely to differ in the MHC alleles they express, there being one chance in four that an individual will share both haplotypes with a sibling. One consequence of this is the difficulty of finding suitable donors for tissue transplantation.

The term polymorphism comes from the Greek *poly*, meaning many, and *morphe*, meaning shape or structure. As used here, it means within-species variation at a gene locus, and thus in its protein product; the variant genes that can occupy the locus are termed alleles. There are more than 400 alleles of some human MHC class I and class II genes, far more than the number of alleles of other genes found within the MHC locus (Fig. 5.14). Each MHC class I and class II allele is relatively frequent in the population, so there is only a small chance that the corresponding MHC loci on both homologous chromosomes of an individual will have the same allele; most individuals will be **heterozygous** at MHC loci. The particular combination of MHC alleles found on a single chromosome is known as an **MHC haplotype**. Expression of MHC alleles is codominant, with the protein products of both the alleles at a locus being expressed in the cell, and both gene products being able to present antigens to T cells (Fig. 5.15). The extensive polymorphism at each locus thus has the potential to double the number of different MHC molecules expressed in an individual and thereby increases the diversity already available through polygeny (Fig. 5.16).

Fig. 5.16 Polymorphism and polygeny both contribute to the diversity of MHC molecules expressed by an individual. The high polymorphism of the classical MHC loci ensures a diversity in MHC gene expression in the population as a whole. However, no matter how polymorphic a gene, no individual can express more than two alleles at a single gene locus. Polygeny, the presence of several different related genes with similar functions, ensures that each individual produces a number of different MHC molecules. Polymorphism and polygeny combine to produce the diversity of MHC molecules seen both within an individual and in the population at large.

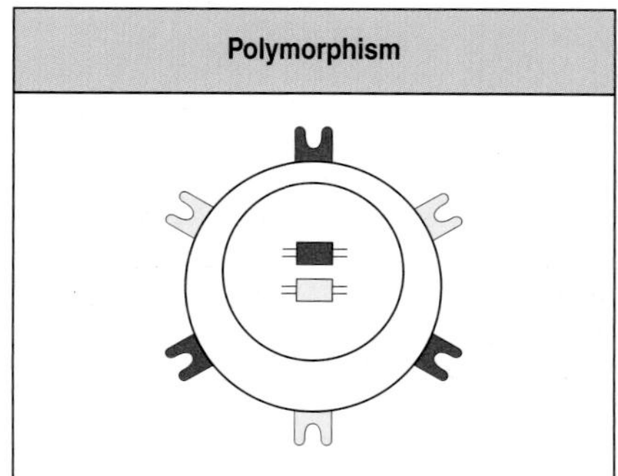

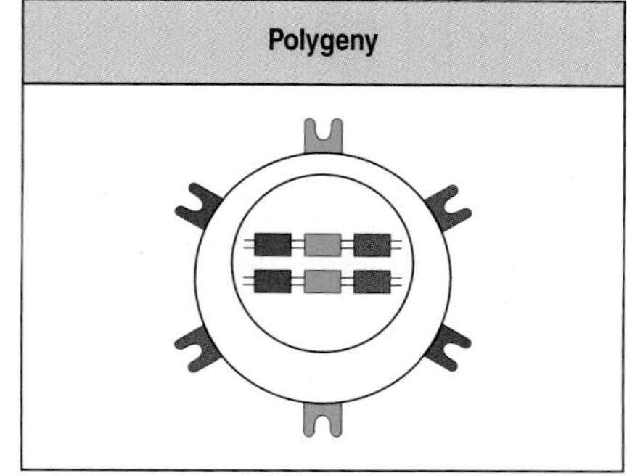

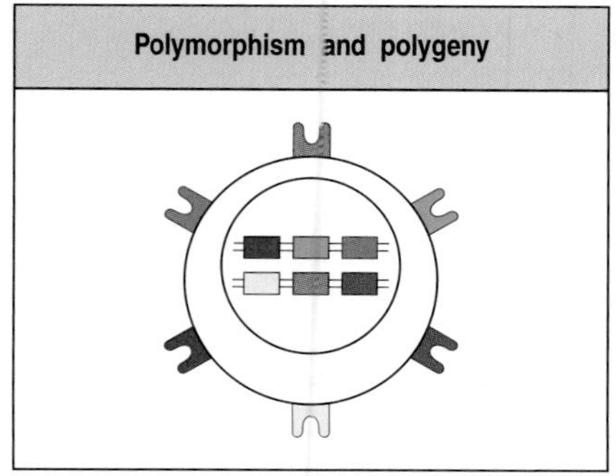

Thus, with three MHC class I genes and a possible four sets of MHC class II genes on each chromosome 6, a person typically expresses six different MHC class I molecules and eight different MHC class II molecules on his or her cells. For the MHC class II genes, the number of different MHC molecules may be increased still further by the combination of α and β chains encoded by different chromosomes (so that two α chains and two β chains can give rise to four different proteins, for example). In mice it has been shown that not all combinations of α and β chains can form stable dimers and so, in practice, the exact number of different MHC class II molecules expressed depends on which alleles are present on each chromosome.

All MHC class I and II proteins are polymorphic to a greater or lesser extent, with the exception of the DRα chain and its mouse homolog Eα. These chains do not vary in sequence between different individuals and are said to be **monomorphic**. This might indicate a functional constraint that prevents variation in the DRα and Eα proteins, but no such special function has been found. Many mice, both domestic and wild, have a mutation in the Eα gene that prevents synthesis of the Eα protein. They thus lack cell-surface H-2E molecules, so if H2-E molecules do have a special function it is unlikely to be essential.

Evidence suggests that the MHC arose after the evolutionary divergence of the agnathans (jawless vertebrates) by multiple gene duplications of an unknown ancestral gene to generate class I and class II genes, which have undergone further genetic divergence. MHC polymorphisms at individual MHC genes seem to have been also strongly selected by evolutionary pressures. Several genetic mechanisms contribute to the generation of new alleles. Some new alleles arise by point mutations and others by **gene conversion**, a process in which a sequence in one gene is replaced, in part, by sequences from a different gene (Fig. 5.17).

The effects of selective pressure in favor of polymorphism can be seen clearly in the pattern of point mutations in the MHC genes. Point mutations can be classified as replacement substitutions, which change an amino acid, or silent substitutions, which simply change the codon but leave the amino acid the same. Replacement substitutions occur within the MHC at a higher frequency relative to silent substitutions than would be expected, providing evidence that polymorphism has been actively selected for in the evolution of the MHC. The next few sections describe how MHC polymorphisms benefit the immune response and how pathogen-driven selection can account for the high large number of MHC alleles.

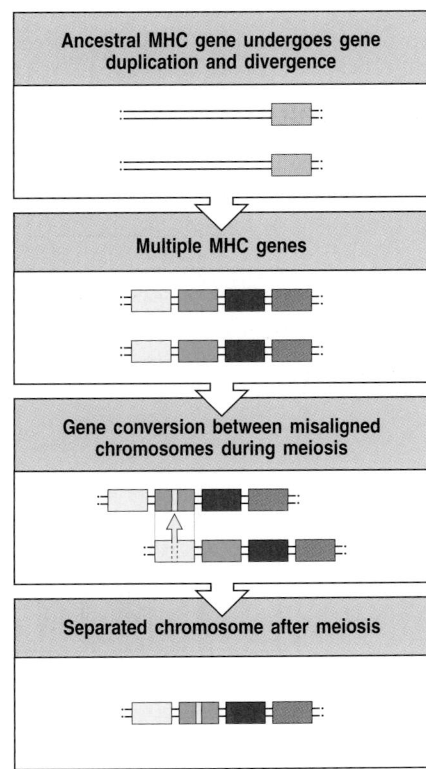

Fig. 5.17 Gene conversion can create new alleles by copying sequences from one MHC gene to another. Multiple MHC genes of generally similar structure were derived over evolutionary time by duplication of an unknown ancestral MHC gene (gray) followed by genetic divergence. Further interchange between these genes can occur by a process known as gene conversion, in which sequences can be transferred from part of one gene to a similar gene. For this to happen, the two genes must become apposed during meiosis. This can occur as a consequence of the misalignment of the two paired homologous chromosomes when there are many copies of similar genes arrayed in tandem—somewhat like buttoning in the wrong buttonhole. During the process of crossing-over and DNA recombination, a DNA sequence from one chromosome is sometimes copied to the other, replacing the original sequence. In this way, several nucleotide changes can be inserted all at once into a gene and can cause several simultaneous amino acid changes between the new gene sequence and the original gene. Because of the similarity of the MHC genes to each other and their close linkage, gene conversion has occurred many times in the evolution of MHC alleles.

5-13 MHC polymorphism affects antigen recognition by T cells by influencing both peptide binding and the contacts between T-cell receptor and MHC molecule.

The products of individual MHC alleles can differ from one another by up to 20 amino acids, making each variant protein quite distinct. Most of the differences are localized to exposed surfaces of the extracellular domain furthest from the membrane, and to the peptide-binding groove in particular (Fig. 5.18). We have seen that peptides bind to MHC class I and class II molecules through the interaction of specific anchor residues with peptide-binding pockets in the peptide-binding groove (see Sections 3-14 and 3-15). Many of the polymorphisms in MHC molecules alter the amino acids that line these pockets and thus change the pockets' binding specificities. This in turn changes the anchor residues of peptides that can bind to each MHC allele. The set of anchor residues that allows binding to given allele of an MHC class I or class II molecule is called a **sequence motif**, and this can be used to predict peptides within a protein that might bind that allele (Fig. 5.19). Such predictions may be very important in designing peptide vaccines.

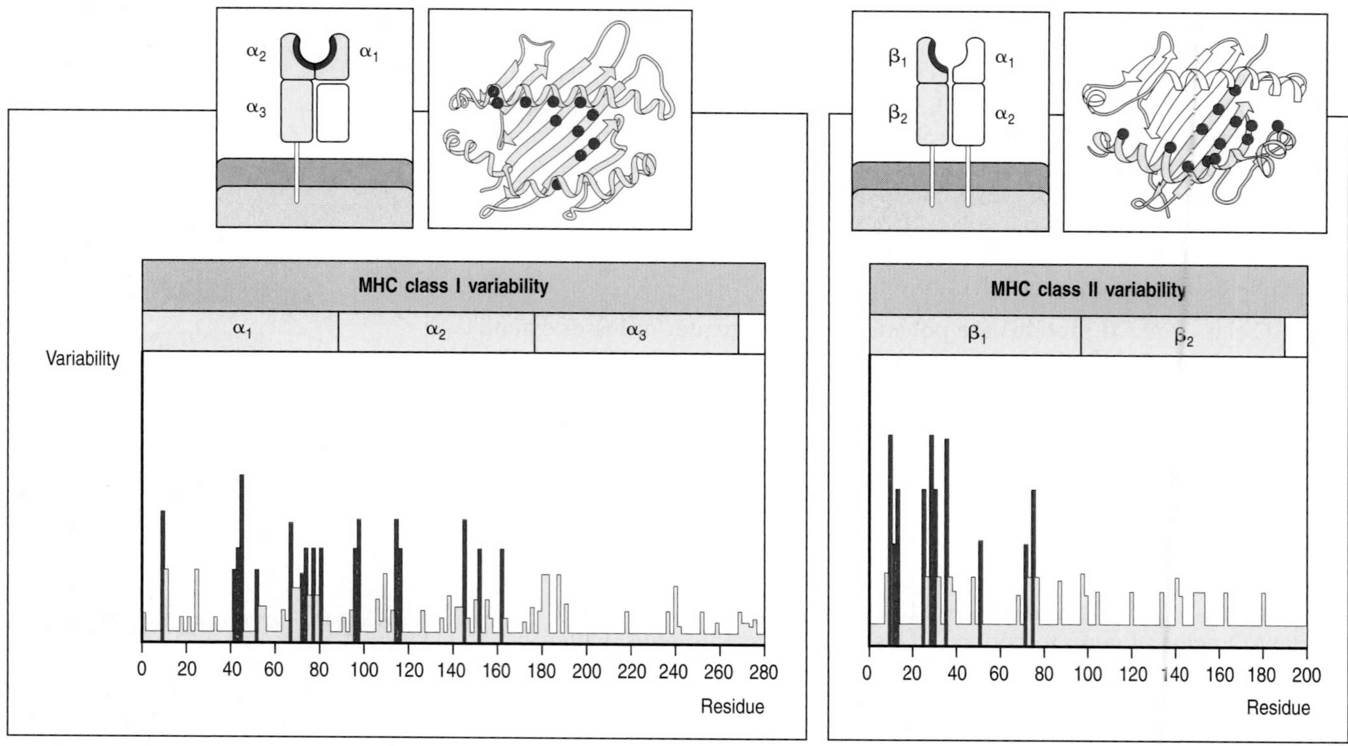

Fig. 5.18 Allelic variation occurs at specific sites within MHC molecules. Variability plots of the amino acid sequences of MHC molecules show that the variation arising from genetic polymorphism is restricted to the amino-terminal domains (α_1 and α_2 domains of MHC class I molecules, and α_1 and β_1 domains of MHC class II molecules). These are the domains that form the peptide-binding cleft. Moreover, allelic variability is clustered in specific sites within the amino-terminal domains, lying in positions that line the peptide-binding cleft, either on the floor of the groove or directed inward from the walls. For the MHC class II molecule, the variability of the HLA-DR alleles is shown. For HLA-DR, and its homologs in other species, the α chain is essentially invariant and only the β chain shows significant polymorphism.

In rare cases, processing of a protein will not generate any peptides with a suitable sequence motif for binding to any of the MHC molecules expressed by an individual. When this happens, the individual fails to respond to the antigen. Such failures in responsiveness to simple antigens were first reported in inbred animals, where they were called **immune response (Ir) gene** defects. These defects were mapped to genes within the MHC long before the structure or function of MHC molecules was understood, and they were the first clue to the antigen-presenting function of MHC molecules. We now understand that Ir gene defects are common in inbred strains of mice because they are homozygous at all their MHC loci, which limits the range of peptides they can present to T cells. Ordinarily, MHC polymorphism guarantees a sufficient number of different MHC molecules in a single individual to make this type of nonresponsiveness unlikely, even to relatively simple antigens such as small toxins. This has obvious importance for host defense.

Initially, the only evidence linking Ir gene defects to the MHC was genetic—mice of one MHC genotype could make antibody in response to a particular antigen, whereas mice of a different MHC genotype, but otherwise genetically identical, could not. The MHC genotype was somehow controlling the ability of the immune system to detect or respond to specific antigens, but it was not clear at the time that direct recognition of MHC molecules was involved.

Later experiments showed that the antigen specificity of T-cell recognition was controlled by MHC molecules. The immune responses affected by the Ir genes were known to depend on T cells, and this led to a series of experiments in mice to ascertain how MHC polymorphism might control T-cell responses. The earliest of these experiments showed that T cells could be activated only by macrophages or B cells that shared MHC alleles with the mouse in which the T cells originated. This was the first evidence that antigen recognition by

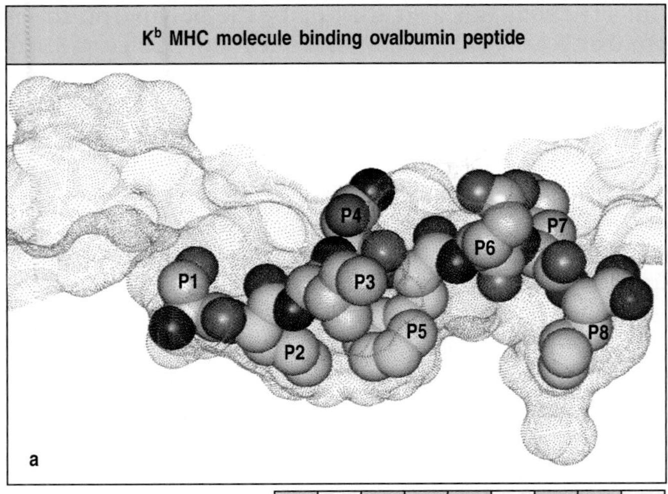

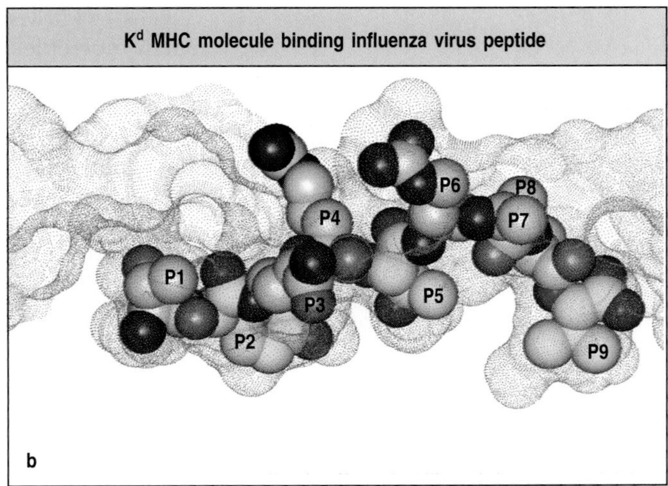

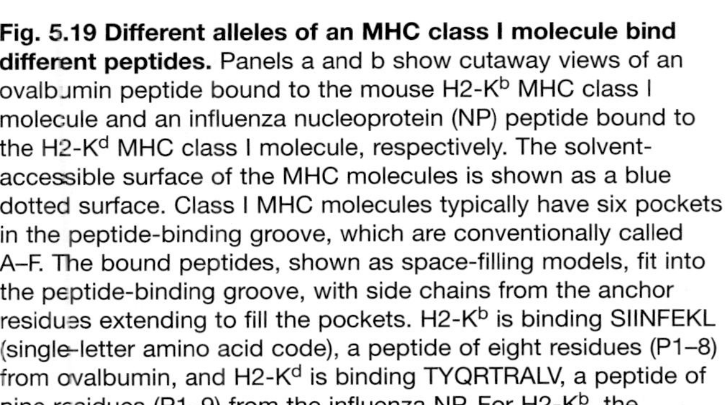

	P1	P2	P3	P4	—	P5	P6	P7	P8
Ovalbumin (257–264)	S	I	I	N		F	E	K	L
HBV SA (208–215)	I	L	S	P		F	L	P	L
Influenza NS2 (114–121)	R	T	F	S		F	Q	L	I
LCMV NP (205–212)	Y	T	V	K		Y	P	N	L
VSV NP (52–59)	R	G	Y	V		Y	Q	G	L
Sendai virus NP (324–332)	F	A	P	G	N	Y	P	A	L

	P1	P2	P3	P4	P5	P6	P7	P8	P9
Influenza NP (147–155)	T	Y	Q	R	T	R	A	L	V
tERK2 kinase (136–144)	Q	Y	I	H	S	A	N	V	L
P198 (14–22)	K	Y	Q	A	V	T	T	T	L
P. yoelii CS (280–288)	S	Y	V	P	S	A	E	Q	I
P. berghei CS (25)	G	Y	I	P	S	A	E	K	I
JAK1 kinase (367–375)	S	Y	F	P	E	I	T	H	I

Fig. 5.19 Different alleles of an MHC class I molecule bind different peptides. Panels a and b show cutaway views of an ovalbumin peptide bound to the mouse H2-K^b MHC class I molecule and an influenza nucleoprotein (NP) peptide bound to the H2-K^d MHC class I molecule, respectively. The solvent-accessible surface of the MHC molecules is shown as a blue dotted surface. Class I MHC molecules typically have six pockets in the peptide-binding groove, which are conventionally called A–F. The bound peptides, shown as space-filling models, fit into the peptide-binding groove, with side chains from the anchor residues extending to fill the pockets. H2-K^b is binding SIINFEKL (single-letter amino acid code), a peptide of eight residues (P1–8) from ovalbumin, and H2-K^d is binding TYQRTRALV, a peptide of nine residues (P1–9) from the influenza NP. For H2-K^b, the sequence motif is determined by the C pocket, which binds the P5 side chain of the peptide (a tyrosine (Y) or a phenylalanine (F)), and the F pocket, which binds the P8 residue (a non-aromatic hydrophobic side chain from leucine (L), isoleucine (I), methionine (M), or valine (V)). For H2-K^d, the sequence motif is primarily determined by the B and F pockets, which bind the P2 and P9 side chains of the peptide, respectively. The B pocket accommodates a tyrosine side chain. The F pocket binds either leucine, isoleucine, or valine. Beneath the structures are shown known sequence motifs from peptides that bind to each MHC molecule, respectively. An extensive collection of motifs can be found at http://www.syfpeithi.de. Structures courtesy of V.E. Mitaksov and D. Fremont.

T cells depends on the presence of specific MHC molecules in the antigen-presenting cell—the phenomenon we now know as MHC restriction, as discussed in Chapter 3.

The clearest example of this feature of T-cell recognition came, however, from studies of virus-specific cytotoxic T cells, for which Peter Doherty and Rolf Zinkernagel were awarded the Nobel Prize in 1996. When mice are infected with a virus, they generate cytotoxic T cells that kill cells infected with the virus, while sparing uninfected cells or cells infected with unrelated viruses. The cytotoxic T cells are thus virus-specific. The additional and striking outcome of their experiments was the demonstration that the ability of cytotoxic

T cells to kill virus-infected cells was also affected by the polymorphism of MHC molecules. Cytotoxic T cells induced by viral infection in mice of MHC genotype a (MHC^a) would kill any MHC^a cell infected with that virus but would not kill cells of MHC genotype b, or c, and so on, even if they were infected with the same virus. In other words, cytotoxic T cells kill cells infected by virus only if those cells express self MHC. Because the MHC genotype 'restricts' the antigen specificity of the T cells, this effect was called MHC restriction. Together with the earlier studies on both B cells and macrophages, this work showed that MHC restriction is a critical feature of antigen recognition by all functional classes of T cells.

We now know that MHC restriction is due to the fact that the binding specificity of an individual T-cell receptor is not for its peptide antigen alone but for the complex of peptide and MHC molecule (see Chapter 3). MHC restriction is explained in part by the fact that different MHC molecules bind different peptides. In addition, some of the polymorphic amino acids in MHC molecules are located in the α helices that flank the peptide-binding groove, but have side chains oriented toward the exposed surface of the peptide:MHC complex that can directly contact the T-cell receptor (see Figs 5.18 and 3.22). It is therefore not surprising that T cells can readily distinguish between a peptide bound to MHC^a and the same peptide bound to MHC^b. This restricted recognition may sometimes be caused both by differences in the conformation of the bound peptide imposed by the different MHC molecules and by direct recognition of polymorphic amino acids in the MHC molecule itself. Thus, the specificity of a T-cell receptor is defined both by the peptide it recognizes and by the MHC molecule bound to it (Fig. 5.20).

5-14 Alloreactive T cells recognizing nonself MHC molecules are very abundant.

A Kidney Graft for Complications of Autoimmune Insulin-Dependent Diabetes Mellitus

The discovery of MHC restriction also helped to explain the otherwise puzzling phenomenon of recognition of nonself MHC in the rejection of organs and tissues transplanted between members of the same species. Transplanted organs from donors bearing MHC molecules that differ from

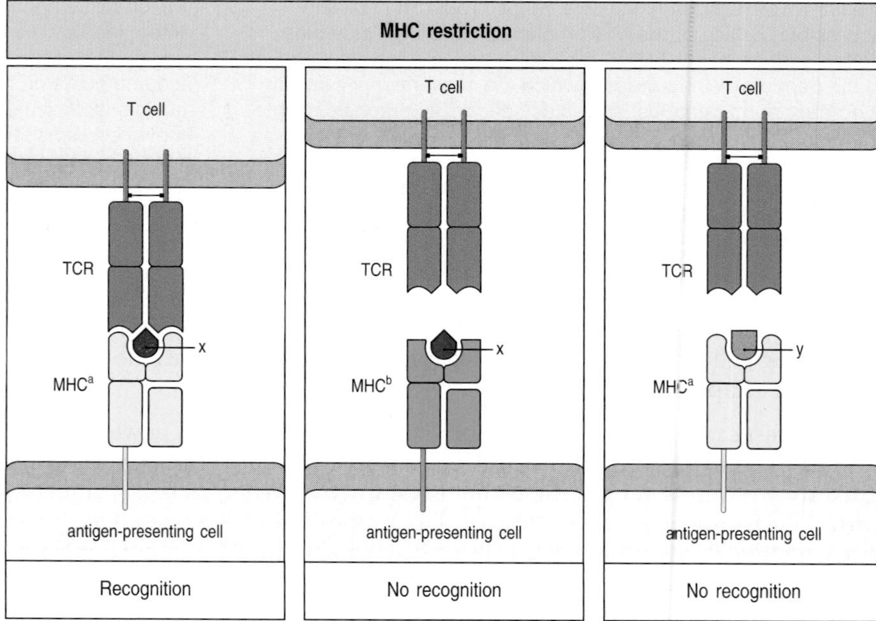

Fig. 5.20 T-cell recognition of antigens is MHC restricted. The antigen-specific T-cell receptor (TCR) recognizes a complex of an antigenic peptide and a self MHC molecule. One consequence of this is that a T cell specific for peptide x and a particular MHC allele, MHC^a (left panel), will not recognize the complex of peptide x with a different MHC allele, MHC^b (center panel), or the complex of peptide y with MHC^a (right panel). The co-recognition of a foreign peptide and MHC molecule is known as MHC restriction because the MHC molecule is said to restrict the ability of the T cell to recognize antigen. This restriction may either result from direct contact between the MHC molecule and T-cell receptor or be an indirect effect of MHC polymorphism on the peptides that bind or on their bound conformation.

those of the recipient—even by as little as one amino acid—are rapidly rejected owing to the presence in any individual of large numbers of T cells that react to nonself, or **allogeneic**, MHC molecules. Early studies on T-cell responses to allogeneic MHC molecules used the **mixed lymphocyte reaction**, in which T cells from one individual are mixed with lymphocytes from a second individual. If the T cells of this individual recognize the other individual's MHC molecules as 'foreign,' the T cells will divide and proliferate. (The lymphocytes from the second individual are usually prevented from dividing by irradiation or treatment with the cytostatic drug mitomycin C.) Such studies have shown that roughly 1–10% of all T cells in an individual will respond to stimulation by cells from another, unrelated, member of the same species. This type of T-cell response is called an **alloreaction** or **alloreactivity** because it represents the recognition of allelic polymorphism in allogeneic MHC molecules.

Before the role of the MHC molecules in antigen presentation was understood, it was a mystery why so many T cells should recognize nonself MHC molecules, as there is no reason that the immune system should have evolved a defense against tissue transplants. Once it was realized that T-cell receptors have evolved to recognize foreign peptides in combination with polymorphic MHC molecules, alloreactivity became easier to explain. We now know of at least two processes that can contribute to the high frequency of alloreactive T cells (Fig. 5.21). First, T cells developing in the thymus have gone through a stringent positive selection process that favors the survival of cells whose T-cell receptors interact weakly with the self-MHC molecules that are expressed in the thymus (this is described in detail in Chapter 7). It is thought that selecting T-cell receptors for their interaction with one type of MHC molecule increases the likelihood that they will cross-react with other (nonself) MHC variants. Second, it seems that the T-cell receptor genes encode an inherent ability to recognize MHC molecules. This was shown by experiments that observed frequent alloreactivity in T cells that are artificially driven to mature in animals lacking MHC class I and class II, and in which positive selection in the thymus cannot occur.

Alloreactivity thus represents the cross-reactivity of T-cell receptors for nonself-peptide:nonself MHC complexes (see Fig. 5.21). The interaction is, however, influenced by the bound peptide as well as by the MHC molecule. At one

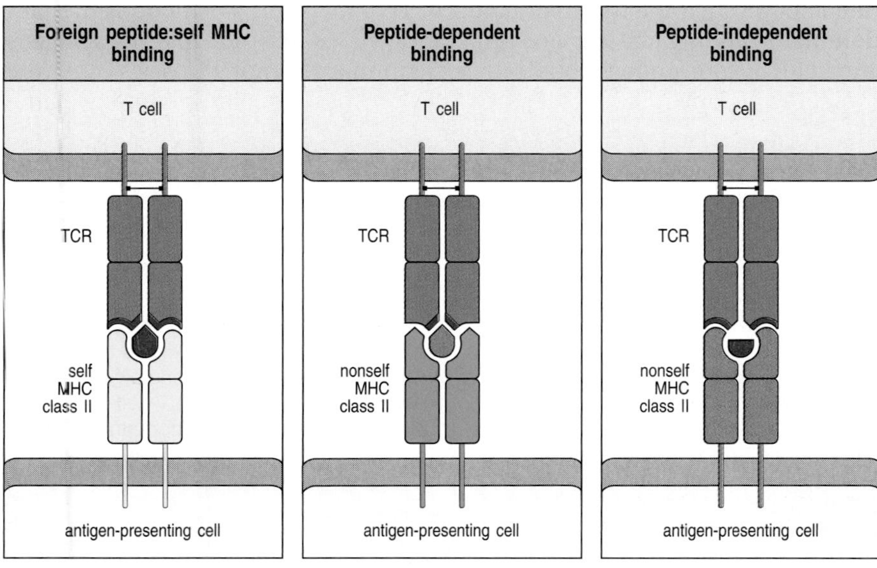

Fig. 5.21 Two modes of cross-reactive recognition that may explain alloreactivity. A T cell that is specific for a nonself or foreign peptide:self MHC combination (left panel) may cross-react with peptides presented by other, nonself (allogeneic), MHC molecules. This may come about in either of two ways. Most commonly, the peptides bound to the allogeneic MHC molecule fit well to the T-cell receptor (TCR), allowing binding even if there is not a good fit with the MHC molecule (center panel). Alternatively, but less often, the allogeneic MHC molecule may provide a better fit to the T-cell receptor, giving a tight binding that is thus less dependent on the peptide that is bound to the MHC molecule (right panel).

extreme, there are alloreactive T cells that interact strongly with one specific peptide:MHC complex, but not with the same nonself MHC molecule bound to different peptides (see Fig. 5.21, center panel). Peptide-dependent alloreactive T cells might be triggered in response to peptides bound by nonself MHC molecules on the transplanted tissues that differ from the peptides bound by the host's own MHC. At the other extreme, there are peptide-independent alloreactive T cells that interact with nonself MHC molecules without a strict peptide requirement (see Fig. 5.21, right panel). In practice, alloreactive responses against a transplanted organ are likely to represent the activity of many alloreactive T cells of each type, and it is not possible to determine the fraction of the reaction represented by each type.

5-15 Many T cells respond to superantigens.

Superantigens are a distinct class of antigens that stimulate a primary T-cell response similar in magnitude to a response to allogeneic MHC molecules. Such responses were first observed in mixed lymphocyte reactions using lymphocytes from strains of mice that were MHC identical but otherwise genetically distinct. The antigens provoking this reaction were originally designated **minor lymphocyte stimulating (Mls)** antigens, and it seemed reasonable to suppose that they might be functionally similar to the MHC molecules themselves. We now know that this is not true, however. The Mls antigens found in these mouse strains are encoded by retroviruses, such as the mouse mammary tumor virus, that have become stably integrated at various sites in the mouse chromosomes. Mls proteins act as superantigens because they have a distinctive mode of binding to both MHC and T-cell receptor molecules that enables them to stimulate very large numbers of T cells. Superantigens are produced by many different pathogens, including bacteria, mycoplasmas, and viruses, and the responses they provoke are helpful to the pathogen rather than the host.

Superantigens are unlike other protein antigens, in that they are recognized by T cells without being processed into peptides that are captured by MHC molecules. Indeed, fragmentation of a superantigen destroys its biological activity, which depends on binding as an intact protein to the outside surface of an MHC class II molecule that has already bound peptide. In addition to binding MHC class II molecules, superantigens are able to bind the V_β region of many T-cell receptors (Fig. 5.22). Bacterial superantigens bind mainly to the V_β CDR2 loop, and to a smaller extent to the V_β CDR1 loop and an additional loop called the hypervariable 4 or HV4 loop. The HV4 loop is the predominant binding site for viral superantigens, at least for the Mls antigens encoded by the endogenous mouse mammary tumor viruses. Thus, the α-chain V region and the CDR3 of the β chain of the T-cell receptor have little

Bacterial superantigen e.g. SE,TSST-1
antigen-presenting cell
MHC class II
β α
TCR
α β
T cell

Viral superantigen
antigen-presenting cell
α β
α β
T cell

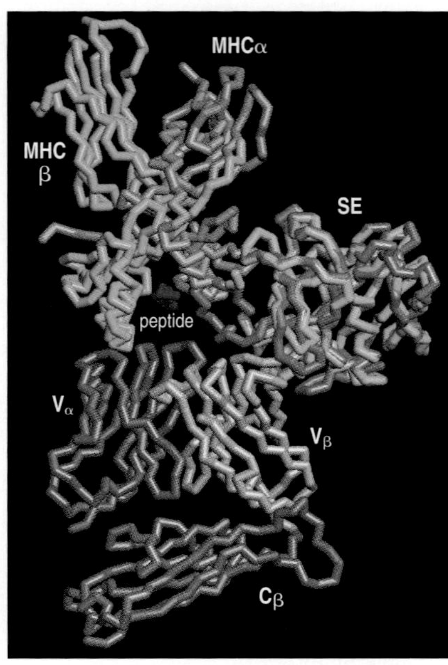

Fig. 5.22 Superantigens bind directly to T-cell receptors and to MHC molecules. Superantigens can bind independently to MHC class II molecules and to T-cell receptors, binding to the V_β domain of the T-cell receptor (TCR), away from the complementarity-determining regions, and to the outer faces of the MHC class II molecule, outside the peptide-binding site (top panels). The bottom panel shows a reconstruction of the interaction between a T-cell receptor, an MHC class II molecule and a staphylococcal enterotoxin (SE) superantigen, produced by superimposing separate structures of an enterotoxin:MHC class II complex onto an enterotoxin:T-cell receptor complex. The two enterotoxin molecules (actually SEC3 and SEB) are shown in turquoise and blue, binding to the α chain of the class II molecule (yellow) and to the β chain of the T-cell receptor (colored gray for the V_β domain and pink for the C_β domain). Molecular model courtesy of H.M. Li, B.A. Fields, and R.A. Mariuzza.

effect on superantigen recognition, which is determined largely by the germline-encoded V gene segments that encode the expressed V_β chain. Each superantigen is specific for one or a few of the different V_β gene products, of which there are 20–50 in mice and humans; a superantigen can thus stimulate 2–20% of all T cells.

This mode of stimulation does not prime an adaptive immune response specific for the pathogen. Instead, it causes a massive production of cytokines by CD4 T cells, the predominant responding population of T cells. These cytokines have two effects on the host: systemic toxicity and suppression of the adaptive immune response. Both these effects contribute to microbial pathogenicity. Among the bacterial superantigens are the **staphylococcal enterotoxins** (**SEs**), which cause food poisoning, and the **toxic shock syndrome toxin-1** (**TSST-1**), the etiologic principle in toxic shock syndrome.

The role of viral superantigens in human disease is less clear. The best-characterized viral superantigens remain the mouse mammary tumor virus superantigens, which are common as endogenous antigens in mice.

5-16 MHC polymorphism extends the range of antigens to which the immune system can respond.

Most polymorphic genes encode proteins that vary by only one or a few amino acids, whereas the different allelic variants of MHC proteins differ by up to 20 amino acids. The extensive polymorphism of the MHC proteins has almost certainly evolved to outflank the evasive strategies of pathogens. Pathogens can avoid an immune response either by evading detection or by suppressing the ensuing response. The requirement that pathogen antigens must be presented by an MHC molecule provides two possible ways of evading detection. One is through mutations that eliminate from the pathogen's proteins all peptides able to bind MHC molecules. The Epstein–Barr virus provides an example. In regions of south-east China and in Papua New Guinea there are small isolated populations in which about 60% of the people carry the HLA-A11 allele. Many isolates of the Epstein–Barr virus obtained from these populations have mutations in a dominant peptide epitope normally presented by HLA-A11; the mutant peptides no longer bind to HLA-A11 and cannot be recognized by HLA-A11-restricted T cells. This strategy is plainly much more difficult to follow if there are many different MHC molecules, and the presence of different loci encoding functionally related proteins may have been an evolutionary adaptation by hosts to this strategy by pathogens.

In large outbred populations, polymorphism at each locus can potentially double the number of different MHC molecules expressed by an individual, as most individuals will be heterozygotes. Polymorphism has the additional advantage that individuals in a population will differ in the combinations of MHC molecules that they express and will therefore present different sets of peptides from each pathogen. This makes it unlikely that all individuals in a population will be equally susceptible to a given pathogen, and its spread will therefore be limited. The fact that exposure to pathogens over an evolutionary timescale can select for the expression of particular MHC alleles is indicated by the strong association of the HLA-B53 allele with recovery from a potentially lethal form of malaria. This allele is very common in people from West Africa, where malaria is endemic, and rare elsewhere, where lethal malaria is uncommon.

Similar arguments apply to a second strategy for evading recognition. Pathogens that can block the presentation of their peptides by MHC molecules can avoid the adaptive immune response. Adenoviruses encode a protein that binds to MHC class I molecules in the endoplasmic reticulum and prevents their transport to the cell surface, thus preventing the recognition of viral peptides by CD8 cytotoxic T cells. This MHC-binding protein must interact with a polymorphic region of the MHC class I molecule, because some allelic variants are retained in the endoplasmic reticulum by the adenoviral protein whereas others are not. Increasing the variety of MHC molecules expressed reduces the likelihood that a pathogen will be able to block presentation by all of them and completely evade an immune response.

These arguments raise a question: If having three MHC class I loci is better than having one, why are there not far more? The probable explanation is that each time a distinct MHC molecule is added to the repertoire, all T cells that can respond to the self peptide bound to that MHC molecule must be removed to maintain self tolerance. It seems that the number of MHC loci present in humans and mice is about optimal to balance out the advantages of presenting an increased range of foreign peptides and the disadvantages of a loss of T cells from the repertoire.

5-17 A variety of genes with specialized functions in immunity are also encoded in the MHC.

In addition to the highly polymorphic 'classical' MHC class I and class II genes, there are many 'nonclassical' MHC genes in the MHC locus (see Fig. 5.13) encoding MHC class I-type molecules that show comparatively little polymorphism (see Fig. 5.14); many of these have yet to be assigned a function. They are linked to the class I region of the MHC, and their exact number varies greatly between species and even between members of the same species. These genes have been termed **MHC class Ib** genes; like MHC class I genes, many, but not all, associate with β_2-microglobulin when expressed on the cell surface. Their expression on cells is variable, both in the amount present at the cell surface and in tissue distribution. The characteristics of several MHC class Ib genes are shown in Fig. 5.23.

One mouse MHC class Ib molecule, H2-M3, can present peptides with *N*-formylated amino termini, which is of interest because all bacteria initiate protein synthesis with *N*-formylmethionine. Cells infected with cytosolic bacteria can be killed by CD8 T cells that recognize *N*-formylated bacterial peptides bound to H2-M3. Whether an equivalent MHC class Ib molecule exists in humans is not known.

Two other closely related murine MHC class Ib genes, *T22* and *T10*, are expressed by activated lymphocytes and are recognized by a subset of γ:δ T cells. The exact function of the T22 and T10 proteins is unclear, but it has been proposed that this interaction allows the regulation of activated lymphocytes by γ:δ T cells.

The other genes that map within the MHC include some that encode complement components (for example, C2, C4, and factor B) and some that encode cytokines—for example, tumor necrosis factor-α (TNF-α) and lymphotoxin (TNF-β)—all of which have important functions in immunity. These genes lie in the so-called 'MHC class III' region (see Fig. 5.13).

Many studies have established associations between susceptibility to certain diseases and particular alleles of MHC genes (see Chapter 14), and we now have considerable insight into how polymorphism in the classical MHC class I and class II genes can affect resistance or susceptibility. Most MHC-influenced traits or diseases are known or suspected to have an immunological cause, but this is not so for all of them: many genes lying within the MHC have no known or suspected immunological function. For example, the class Ib gene *M10* encodes a protein that is recognized by pheromone receptors in the vomeronasal organ. *M10* could thus influence mating preference, a trait that has been linked to the MHC locus in rodents.

The gene for HLA-H, which has been renamed *HFE* (see Fig. 5.23), lies some 3×10^6 base pairs from HLA-A. Its protein product is expressed on cells in the intestinal tract and has a function in iron metabolism, regulating the uptake of dietary iron into the body, most probably through interactions with the transferrin receptor that reduce its affinity for iron-loaded transferrin. Individuals defective for this gene have an iron-storage disease, hereditary hemochromatosis, in which an abnormally high level of iron is retained in the liver and other organs. Mice lacking β_2-microglobulin, and hence defective in the expression of all class I molecules, have a similar iron overload. Another MHC gene with a nonimmune function encodes the enzyme 21-hydroxylase, which, when deficient, causes congenital adrenal hyperplasia and, in severe cases, salt-wasting syndrome. Even where a disease-related gene is clearly homologous to immune-system genes, as is the case with *HFE*, the disease mechanism may not be immune related. Disease associations mapping to the MHC must therefore be interpreted with caution, in the light of a detailed understanding of its genetic structure and the functions of its individual genes. Much remains to be learned about the latter and about the significance of all the genetic variation localized within the MHC. For instance, in humans the complement component C4 comes in two versions, C4A and C4B, and different individuals have variable numbers of the gene for each type in their genomes, but the adaptive significance of this genetic variability is not well understood.

5-18 Specialized MHC class I molecules act as ligands for the activation and inhibition of NK cells.

Some MHC class Ib genes, for example the members of the **MIC** gene family, are under a different regulatory control from the classical MHC class I genes and are induced in response to cellular stress (such as heat shock). There are five MIC genes, but only two—*MICA* and *MICB*—are expressed and produce protein products (see Fig. 5.23). They are expressed in fibroblasts and epithelial cells, particularly in intestinal epithelial cells, and have a role in innate immunity or in the induction of immune responses in circumstances in which interferons are not produced. The MIC-A and MIC-B proteins are recognized by the NKG2D receptor expressed by NK cells, γ:δ T cells, and some CD8 T cells, and can activate these cells to kill MIC-expressing targets. NKG2D is an 'activating' member of the NKG2 family of NK-cell receptors; its cytoplasmic domain lacks the inhibitory sequence motif found in other members of this family, which act as inhibitory receptors (see Sections 2-31 and 2-32). NKG2D is coupled to the adaptor protein DAP10, which transmits the signal into the interior of the cell by interacting with and activating intracellular phosphatidylinositol 3-kinase.

Even more distantly related to MHC class I genes are a small family of proteins known in humans as the UL16-binding proteins (ULBPs) or the RAET1 proteins (see Fig. 5.23); the homologous proteins in mice are known as Rae1 (retinoic acid early inducible 1) and H60. These proteins also bind the NKG2D receptor, as described in Section 2-32. They seem to be expressed under conditions of cellular stress, such as when cells are infected with pathogens or have undergone transformation. By expressing ULBPs, stressed or infected cells can activate NKG2D on NK cells, γ:δ T cells and CD8 cytotoxic α:β T cells, and so be recognized and eliminated.

The human MHC class Ib molecule HLA-E and its murine counterpart Qa-1 (see Fig. 5.23) have a specialized role in cell recognition by NK cells. HLA-E and Qa-1 bind a very restricted subset of nonpolymorphic peptides, called

	Class 1b molecule						Receptors or interacting proteins			
	Human	Mouse	Expression pattern	Associates with β_2m	Poly-morphism	Ligand	T-cell receptor	NK receptor	Other	Biological function
MHC encoded	HLA-C (class 1a)		Ubiquitous	Yes	High	Peptide	TCR	KIRs		Activate T cells Inhibit NK cells
MHC encoded		H2-M3	Limited	Yes	Low	fMet peptide	TCR			Activate CTLs with bacterial peptides
MHC encoded		T22 T10	Splenocytes	Yes	Low	None	γ:δ TCR			Regulation of activated splenocytes
MHC encoded	HLA-E	Qa-1	Ubiquitous	Yes	Low	MHC leader peptides (Qdm)		NKG2A NKG2C		NK cell inhibition
MHC encoded	HLA-F		Widely expressed	Yes	Low	Peptide?		LILRB1 LILRB2		Unknown
MHC encoded	HLA-G		Maternal/fetal interface	Yes	Low	Peptide	TCR	LILRB1		Modulate maternal/fetal interaction
MHC encoded	MIC-A MIC-B		GI tract, widely expressed	No	Moderate	None		NKG2D		Stress-induced activation of NK and CD8 cells
MHC encoded		TL	Small intestine epithelium	Yes	Low	None	CD8α:α			Potential modulation of T-cell activation
MHC encoded		M10	Vomeronasal neurons	Yes	Low	Unknown			Vomeronasal receptor V2R	Pheromone detection
Non-MHC encoded	ULBPs	MULT1 H60, Rae1	Limited	No	Low	None		NKG2D		Induced NK-cell-activating ligand
Non-MHC encoded	MR1	MR1	Ubiquitous	Yes	None	Unknown		LILRB2		Control of inflammatory response
Non-MHC encoded	CD1a–CD1e	CD1d	Limited	Yes	None	Lipids glycolipids	α:β TCR			Activate T cells against bacterial lipids
Non-MHC encoded		Mill1 Mill2	Ubiquitous	Yes?	Low	Unknown	Unknown			Unknown
Non-MHC encoded	HFE	HFE	Liver and gut	Yes	Low	None			Transferrin receptor	Iron homeostasis
Non-MHC encoded	FcRn	FcRn	Maternal/fetal interface	Yes	Low	None			Fc (IgG)	Shuttle maternal IgG to fetus (passive immunity)
Non-MHC encoded	ZAG	ZAG	Bodily fluid	No	None	Fatty acid				Lipid homeostasis

Fig. 5.23 MHC class Ib proteins and their functions. MHC class Ib proteins are encoded both within the MHC locus and on other chromosomes. The functions of some MHC class Ib proteins are unrelated to the immune response, but many have a role in innate immunity by interacting with receptors on NK cells (see text).

Qa-1 determinant modifier (Qdm), that are derived from the leader peptides of other HLA class I molecules. These peptide:HLA-E complexes can bind to the receptor NKG2A, which is present on NK cells in a complex with the cell-surface molecule CD94 (see Section 2-32). NKG2A is an inhibitory member of the NKG2 family, and upon HLA-E engagement it inhibits the cytotoxic activity of the NK cell. Thus, a cell that expresses either HLA-E or Qa-1 is not killed by NK cells.

Two other MHC class Ib molecules, HLA-F and HLA-G (see Fig. 5.23), may also inhibit cell killing by NK cells. HLA-G is expressed on fetus-derived placental cells that migrate into the uterine wall. These cells express no classical MHC class I molecules and cannot be recognized by CD8 T cells but, unlike other cells lacking such proteins, they are not killed by NK cells. This seems to be because HLA-G is recognized by another inhibitory receptor on NK cells, the leukocyte immunoglobulin-like receptor subfamily B member 1 (LILRB1), also called ILT-2 or LIR-1, which prevents the NK cell from killing the placental cells. HLA-F is expressed in a variety of tissues, although it is usually not detected at the cell surface except, for example, on some monocyte cell lines or on virus-transformed lymphoid cells. HLA-F is also thought to interact with LILRB1.

5-19 The CD1 family of MHC class I-like molecules is encoded outside the MHC and presents microbial lipids to CD1-restricted T cells.

Some MHC class I-like genes map outside the MHC region. One small family of such genes is called **CD1** and is expressed on dendritic cells, monocytes, and some thymocytes. Humans have five CD1 genes, CD1a through e, whereas mice express only two highly homologous versions of CD1d, namely CD1d1 and CD1d2. CD1 proteins can present antigens to T cells, but they have two features that distinguish them from classical MHC class I molecules. The first is that CD1, although similar to an MHC class I molecule in its subunit organization and association with β_2-microglobulin, behaves like an MHC class II molecule. It is not retained within the endoplasmic reticulum by association with the TAP complex but is targeted to vesicles, where it binds its ligand. The second unusual feature is that, unlike MHC class I, CD1 molecules have a hydrophobic channel that is specialized for binding hydrocarbon alkyl chains. This confers on CD1 molecules an ability to bind and present glycolipids.

CD1 molecules are classified into group 1, comprising CD1a, CD1b, and CD1c, and group 2, containing only CD1d; CD1e is considered intermediate. Group 1 molecules bind microbially derived glycolipids, phospholipids, and lipopeptide antigens, such as the mycobacterial membrane components mycolic acid, glucose monomycolate, phosphoinositol mannosides, and lipoarabinomannan. Group 2 CD1 molecules are thought to bind mainly self lipid antigens such as sphingolipids and diacylglycerols. Structural studies show that the CD1 molecule has a deep binding groove into which the glycolipid antigens bind. Group 1 molecules bind their antigens by anchoring the alkyl chains in the hydrophobic groove, which orients the variable carbohydrate headgroups or other hydrophilic parts of these molecules protruding from the end of the binding groove, allowing recognition by the T-cell receptors on CD1-restricted T cells.

Whereas T cells that recognize antigen presented by MHC class I and class II molecules express CD8 and CD4, respectively, the T cells that recognize lipids presented by CD1 molecules express neither CD4 nor CD8. Most of the T cells recognizing lipids presented by group 1 CD1 molecules have a diverse repertoire of α:β receptors; in contrast, CD1d-restricted T cells are less diverse, many using the same TCRα chain ($V_\alpha 24$–$J_\alpha 18$ in humans).

It seems that the CD1 proteins have evolved as a separate lineage of antigen-presenting molecules able to present microbial lipids and glycolipids to T cells. Just as peptides are loaded onto classical MHC proteins at various cellular locations, the various CD1 proteins are transported differently through the endoplasmic reticulum and endocytic compartments, providing access to different lipid antigens. Transport is regulated by an amino acid sequence motif at the terminus of the cytoplasmic domain of the CD1 protein, which controls interaction with adaptor-protein (AP) complexes. CD1a lacks this binding motif and moves to the cell surface, where it is transported only through the early endocytic compartment. CD1c and CD1d have motifs that interact with the adaptor AP-2 and are transported through early and late endosomes; CD1d is also targeted to lysosomes. CD1b and murine CD1d bind AP-2 and AP-3 and can be transported through late endosomes, lysosomes, and the MIIC. CD1 proteins can thus bind lipids delivered into and processed within the endocytic pathway, such as by the internalization of mycobacteria or the ingestion of lipoarabinomannans mediated by the mannose receptor (see Section 2-6).

Summary.

The major histocompatibility complex (MHC) of genes consists of a linked set of genetic loci encoding many of the proteins involved in antigen presentation to T cells, most notably the MHC class I and class II glycoproteins (the MHC molecules) that present peptides to the T-cell receptor. The outstanding feature of the MHC molecules is their extensive polymorphism. This polymorphism is of critical importance in antigen recognition by T cells. A T cell recognizes antigen as a peptide bound by a particular allelic variant of an MHC molecule, and will not recognize the same peptide bound to other MHC molecules. This behavior of T cells is called MHC restriction. Most MHC alleles differ from one another by multiple amino acid substitutions, and these differences are focused on the peptide-binding site and surface-exposed regions that make direct contact with the T-cell receptor. At least three properties of MHC molecules are affected by MHC polymorphism: the range of peptides bound; the conformation of the bound peptide; and the direct interaction of the MHC molecule with the T-cell receptor. Thus, the highly polymorphic nature of the MHC has functional consequences, and the evolutionary selection for this polymorphism suggests that it is critical to the role of the MHC molecules in the immune response. Powerful genetic mechanisms generate the variation that is seen among MHC alleles, and a compelling argument can be made that selective pressure to maintain a wide variety of MHC molecules in the population comes from infectious agents.

Within the MHC locus there are also a large number of genes whose structure is closely related to the MHC class I molecules—the so-called nonclassical, or class Ib, MHC. Although some of these genes serve purposes that are unrelated to the immune system, many are involved in recognition by activating and inhibitory receptors expressed by NK cells, $\gamma{:}\delta$ T cells as well as $\alpha{:}\beta$ T cells. MHC class Ib proteins called CD1 molecules are encoded outside the MHC locus and bind lipids and glycolipid antigens for presentation to T cells.

Summary to Chapter 5.

Normally, the antigen receptors on T cells recognize different self peptides that are bound by self-MHC molecules. During an infection, the antigen receptors on T cells recognize complexes of pathogen-derived peptides bound to MHC molecules on a target cell surface. There are two classes of MHC molecules—MHC class I molecules, which bind stably to peptides derived from proteins synthesized and degraded in the cytosol, and MHC class II molecules, which bind stably to peptides derived from proteins degraded in endocytic vesicles. In addition to being bound by the T-cell receptor, the two classes of MHC molecules are differentially recognized by the two co-receptor molecules, CD8 and CD4, which characterize the two major subsets of T cells. CD8 T cells recognize peptide:MHC class I complexes and are activated to kill cells displaying foreign peptides derived from cytosolic pathogens, such as viruses. Exogenous antigens, such as those obtained during phagocytosis of viral antigens by dendritic cells, can be delivered from the vesicular system to the cytosol, a process known as cross-presentation, for loading and presentation by MHC class I molecules. This pathway is important in the initial activation of CD8 T cells by dendritic cells. CD4 T cells recognize peptide:MHC class II complexes and are specialized to activate other immune effector cells, for example B cells or macrophages, to act against the foreign antigens or pathogens that they have taken up. Thus, the two classes of MHC molecule deliver peptides from different intracellular compartments to the cell surface, where they are recognized by different types of T cells that perform the appropriate effector function.

There are several genes for each class of MHC molecule, arranged in clusters within a larger region known as the major histocompatibility complex (MHC). Within the MHC, the genes for the MHC molecules are closely linked to genes involved in the degradation of proteins into peptides, the formation of the complex of peptide and MHC molecule, and the transport of these complexes to the cell surface. Because the several different genes for the MHC class I and class II molecules are highly polymorphic and are expressed in a codominant fashion, each individual expresses a number of different MHC class I and class II molecules. Each different MHC molecule can bind stably to a range of different peptides, and thus the MHC repertoire of each individual can recognize and bind many different peptide antigens. Because the T-cell receptor binds a combined peptide:MHC ligand, T cells show MHC-restricted antigen recognition, such that a given T cell is specific for a particular peptide bound to a particular MHC molecule. The MHC locus contains many nonclassical MHC genes, many of which participate in immune responses by interacting with other receptors besides T-cell receptor, such as the NKG2D receptor expressed by NK cells. These MHC class Ib molecules can provide both activating and inhibitory signals and participate in innate immunity and immunoregulation.

Questions.

5.1 MHC class I and class II molecules are structurally and functionally homologous, yet have dissimilar pathways of assembly and delivery to the cell surface. (a) Describe how these differences in assembly and delivery are integrated with the different functions of class I and class II molecules. (b) How do these functions relate to source from which class I or class II MHC receive peptides? (c) Given that the processes of cross-presentation and autophagy can redirect antigens from various sources for processing by alternative pathways, how do these processes alter your answer to (b)?

5.2 Viral pathogens have acquired diverse mechanisms to evade the immune response. (a) Describe the steps at which viruses can prevent recognition of viral antigens by CD8 T cells, and provide a specific example for each. (b) Of the examples of viral evasion presented in this chapter, most were concerned with antigens presented by class I MHC. Why might there be more examples of viral inhibition of antigen presentation by class I MHC than by class II MHC? (c) Suggest a reason that large DNA viruses might use this mechanisms more than small RNA viruses.

5.3 "The MHC is an antigen presentation operon." To what extent is the statement an accurate description of the MHC, and what factors may have been responsible for this organization?

5.4 Many of the proteins encoded within the MHC exist in the population in multiple forms, or allelic variants. (a) What genetic events give rise to this variation and what are its functional consequences? (b) In some cases, particular combinations of alleles of the different MHC genes are found to be present at a far higher frequency than chance would predict. What are the possible mechanisms that might explain this finding?

5.5 Rejection of transplanted tissues can result from alloreactivity of the T-cell repertoire against the MHC of the transplant. (a) Describe the processes that generate alloreactivity. (b) Discuss the relationship between alloreactivity and MHC restriction of the T-cell repertoire. (c) How was the phenomenon of MHC restriction discovered? (d) What is the role of peptides in alloreactivity?

5.6 Many genes outside the MHC locus encode proteins that are structurally and functionally related to class I MHC proteins. (a) Discuss the cell types that recognize various 'non-classical' MHC proteins and what their functions are. (b) Discuss the kinds of ligand(s), if any, that are presented by various members of these proteins.

General references.

Bodmer, J.G., Marsh, S.G.E., Albert, E.D., Bodmer, W.F., DuPont, B., Erlich, H.A., Mach, B., Mayr, W.R., Parham, P., Saszuki, T., *et al.*: **Nomenclature for factors of the HLA system, 1991.** *Tissue Antigens* 2000, **56**:289–290.

Germain, R.N.: **MHC-dependent antigen processing and peptide presentation: providing ligands for T lymphocyte activation.** *Cell* 1994, **76**:287–299.

Klein, J.: *Natural History of the Major Histocompatibility Complex.* New York, J. Wiley & Sons, 1986.

Moller, G. (ed.): **Origin of major histocompatibility complex diversity.** *Immunol. Rev.* 1995, **143**:5–292.

Section references.

5-1 The MHC class I and class II molecules deliver peptides to the cell surface from two intracellular compartments.

Brocke, P., Garbi, N., Momburg, F., and Hammerling, G.J.: **HLA-DM, HLA-DO and tapasin: functional similarities and differences.** *Curr. Opin. Immunol.* 2002, **14**:22–29.

Gromme, M., and Neefjes, J.: **Antigen degradation or presentation by MHC class I molecules via classical and non-classical pathways.** *Mol. Immunol.* 2002, **39**:181–202.

Villadangos, J.A.: **Presentation of antigens by MHC class II molecules: getting the most out of them.** *Mol. Immunol.* 2001, **38**:329–346.

Williams, A., Peh, C.A., and Elliott, T.: **The cell biology of MHC class I antigen presentation.** *Tissue Antigens* 2002, **59**:3–17.

5-2 Peptides that bind to MHC class I molecules are actively transported from the cytosol to the endoplasmic reticulum.

Gorbulev, S., Abele, R., and Tampe, R.: **Allosteric crosstalk between peptide-binding, transport, and ATP hydrolysis of the ABC transporter TAP.** *Proc. Natl Acad. Sci. USA* 2001, **98**:3732–3737.

Lankat-Buttgereit, B., and Tampe, R.: **The transporter associated with antigen processing: function and implications in human diseases.** *Physiol. Rev.* 2002, **82**:187–204.

Townsend, A., Ohlen, C., Foster, L., Bastin, J., Lunggren, H.G., and Karre, K.: **A mutant cell in which association of class I heavy and light chains is induced by viral peptides.** *Cold Spring Harbor Symp. Quant. Biol.* 1989, **54**:299–308.

Uebel, S., and Tampe, R.: **Specificity of the proteasome and the TAP transporter.** *Curr. Opin. Immunol.* 1999, **11**:203–208.

5-3 Peptides for transport into the endoplasmic reticulum are generated in the cytosol.

Goldberg, A.L., Cascio, P., Saric, T., and Rock, K.L.: **The importance of the proteasome and subsequent proteolytic steps in the generation of antigenic peptides.** *Mol. Immunol.* 2002, **39**:147–164.

Hammer, G.E., Gonzalez, F., Champsaur, M., Cado, D., and Shastri, N.: **The aminopeptidase ERAAP shapes the peptide repertoire displayed by major histocompatibility complex class I molecules.** *Nat. Immunol.* 2006, **7**:103–112.

Rock, K.L., York, I.A., Saric, T., and Goldberg, A.L.: **Protein degradation and the generation of MHC class I-presented peptides.** *Adv. Immunol.* 2002, **80**:1–70.

Schubert, U., Anton, L.C., Gibbs, J., Norbury, C.C., Yewdell, J.W., and Bennink, J.R.: **Rapid degradation of a large fraction of newly synthesized proteins by proteasomes.** *Nature* 2000, **404**:770–774.

Serwold, T., Gonzalez, F., Kim, J., Jacob, R., and Shastri, N.: **ERAAP customizes peptides for MHC class I molecules in the endoplasmic reticulum.** *Nature* 2002, **419**:480–483.

Shastri, N., Schwab, S., and Serwold, T.: **Producing nature's gene-chips: the generation of peptides for display by MHC class I molecules.** *Annu. Rev. Immunol.* 2002, **20**:463–493.

Sijts, A., Sun, Y., Janek, K., Kral, S., Paschen, A., Schadendorf, D., and Kloetzel, P.M.: **The role of the proteasome activator PA28 in MHC class I antigen processing.** *Mol. Immunol.* 2002, **39**:165–169.

Vigneron, N., Stroobant, V., Chapiro, J., Ooms, A., Degiovanni, G., Morel, S., van der Bruggen, P., Boon, T., and Van den Eynde, B.J.: **An antigenic peptide produced by peptide splicing in the proteasome.** *Science* 2004, **304**:587–590.

5-4 Retrograde transport from the endoplasmic reticulum to the cytosol enables exogenous proteins to be processed for cross-presentation by MHC class I molecules.

Ackerman, A.L., and Cresswell, P.: **Cellular mechanisms governing cross-presentation of exogenous antigens.** *Nat. Immunol.* 2004, **5**:678–684.

Bevan, M.J.: **Minor H antigens introduced on H-2 different stimulating cells cross-react at the cytotoxic T cell level during in vivo priming.** *J. Immunol.* 1976, **117**:2233–2238.

Bevan, M.J.: **Helping the CD8$^+$ T cell response.** *Nat. Rev. Immunol.* 2004, **4**:595–602.

Groothius, T.A.M., and Neefjes, J.: **The many roads to cross-presentation.** *J. Exp. Med.* 2005, **202**:1313–1318.

5-5 Newly synthesized MHC class I molecules are retained in the endoplasmic reticulum until they bind a peptide.

Bouvier, M.: **Accessory proteins and the assembly of human class I MHC molecules: a molecular and structural perspective.** *Mol. Immunol.* 2003, **39**:697–706.

Gao, B., Adhikari, R., Howarth, M., Nakamura, K., Gold, M.C., Hill, A.B., Knee, R., Michalak, M., and Elliott, T.: **Assembly and antigen-presenting function of MHC class I molecules in cells lacking the ER chaperone calreticulin.** *Immunity* 2002, **16**:99–109.

Grandea, A.G. III, and Van Kaer, L.: **Tapasin: an ER chaperone that controls MHC class I assembly with peptide.** *Trends Immunol.* 2001, **22**:194–199.

Pilon, M., Schekman, R., and Romisch, K.: **Sec61p mediates export of a misfolded secretory protein from the endoplasmic reticulum to the cytosol for degradation.** *EMBO J.* 1997, **16**:4540-4548.

Van Kaer, L.: **Accessory proteins that control the assembly of MHC molecules with peptides.** *Immunol. Res.* 2001, **23**:205–214.

Williams, A., Peh, C.A., and Elliott, T.: **The cell biology of MHC class I antigen presentation.** *Tissue Antigens* 2002, **59**:3–17.

Williams, A.P., Peh, C.A., Purcell, A.W., McCluskey, J., and Elliott, T.: **Optimization of the MHC class I peptide cargo is dependent on tapasin.** *Immunity* 2002, **16**:509–520.

5-6 Many viruses produce immunoevasins that interfere with antigen presentation by MHC class I molecules.

Lilley, B.N., and Ploegh, H.L.: **A membrane protein required for dislocation of misfolded proteins from the ER.** *Nature* 2004, **429**:834–840.

Lilley, B.N., and Ploegh, H.L.: **Viral modulation of antigen presentation: manipulation of cellular targets in the ER and beyond.** *Immunol. Rev.* 2005, **207**:126–144.

Lybarger, L., Wang, X., Harris, M., and Hansen, T.H.: **Viral immune evasion molecules attack the ER peptide-loading complex and exploit ER-associated degradation pathways.** *Curr. Opin. Immunol.* 2005, **17**:79–87.

5-7 Peptides presented by MHC class II molecules are generated in acidified endocytic vesicles.

Godkin, A.J., Smith, K.J., Willis, A., Tejada-Simon, M.V., Zhang, J., Elliott, T., and Hill, A.V.: **Naturally processed HLA class II peptides reveal highly conserved immunogenic flanking region sequence preferences that reflect antigen processing rather than peptide–MHC interactions.** *J. Immunol.* 2001, 166:6720–6727.

Hiltbold, E.M., and Roche, P.A.: **Trafficking of MHC class II molecules in the late secretory pathway.** *Curr. Opin. Immunol.* 2002, 14:30–35.

Hsieh, C.S., deRoos, P., Honey, K., Beers, C., and Rudensky, A.Y.: **A role for cathepsin L and cathepsin S in peptide generation for MHC class II presentation.** *J. Immunol.* 2002, 168:2618–2625.

Lennon-Dumenil, A.M., Bakker, A.H., Wolf-Bryant, P., Ploegh, H.L., and Lagaudriere-Gesbert, C.: **A closer look at proteolysis and MHC-class-II-restricted antigen presentation.** *Curr. Opin. Immunol.* 2002, 14:15–21.

Maric, M., Arunachalam, B., Phan, U.T., Dong, C., Garrett, W.S., Cannon, K.S., Alfonso, C., Karlsson, L., Flavell, R.A., and Cresswell, P.: **Defective antigen processing in GILT-free mice.** *Science* 2001, **294**:1361–1365.

Pluger, E.B., Boes, M., Alfonso, C., Schroter, C.J., Kalbacher, H., Ploegh, H.L., and Driessen, C.: **Specific role for cathepsin S in the generation of antigenic peptides in vivo.** *Eur. J. Immunol.* 2002, **32**:467–476.

Schwarz, G., Brandenburg, J., Reich, M., Burster, T., Driessen, C., and Kalbacher, H.: **Characterization of legumain.** *Biol. Chem.* 2002, **383**:1813–1816.

5-8 The invariant chain directs newly synthesized MHC class II molecules to acidified intracellular vesicles.

Gregers, T.F., Nordeng, T.W., Birkeland, H.C., Sandlie, I., and Bakke, O.: **The cytoplasmic tail of invariant chain modulates antigen processing and presentation.** *Eur. J. Immunol.* 2003, **33**:277–286.

Hiltbold, E.M., and Roche, P.A.: **Trafficking of MHC class II molecules in the late secretory pathway.** *Curr. Opin. Immunol.* 2002, **14**:30–35.

Kleijmeer, M., Ramm, G., Schuurhuis, D., Griffith, J., Rescigno, M., Ricciardi-Castagnoli, P., Rudensky, A.Y., Ossendorp, F., Melief, C.J., Stoorvogel, W., and Geuze, H.J.: **Reorganization of multivesicular bodies regulates MHC class II antigen presentation by dendritic cells.** *J. Cell Biol.* 2001, **155**:53–63.

van Lith, M., van Ham, M., Griekspoor, A., Tjin, E., Verwoerd, D., Calafat, J., Janssen, H., Reits, E., Pastoors, L., and Neefjes, J.: **Regulation of MHC class II antigen presentation by sorting of recycling HLA-DM/DO and class II within the multivesicular body.** *J. Immunol.* 2001, **167**:884–892.

5-9 A specialized MHC class II-like molecule catalyzes loading of MHC class II molecules with peptides.

Pathak, S.S., Lich, J.D., and Blum, J.S.: **Cutting edge: editing of recycling class II:peptide complexes by HLA-DM.** *J. Immunol.* 2001, **167**:632–635.

Qi, L., and Ostrand-Rosenberg, S.: **H2-O inhibits presentation of bacterial superantigens, but not endogenous self antigens.** *J. Immunol.* 2001, **167**:1371–1378.

Van Kaer, L.: **Accessory proteins that control the assembly of MHC molecules with peptides.** *Immunol. Res.* 2001, **23**:205–214.

Zarutskie, J.A., Busch, R., Zavala-Ruiz, Z., Rushe, M., Mellins, E.D., and Stern, L.J.: **The kinetic basis of peptide exchange catalysis by HLA-DM.** *Proc. Natl Acad. Sci. USA* 2001, 98:12450–12455.

5-10 Stable binding of peptides by MHC molecules provides effective antigen presentation at the cell surface.

Apostolopoulos, V., McKenzie, I.F., and Wilson, I.A.: **Getting into the groove: unusual features of peptide binding to MHC class I molecules and implications in vaccine design.** *Front. Biosci.* 2001, **6**:D1311–D1320.

Buslepp, J., Zhao, R., Donnini, D., Loftus, D., Saad, M., Appella, E., and Collins, E.J.: **T cell activity correlates with oligomeric peptide-major histocompatibility complex binding on T cell surface.** *J. Biol. Chem.* 2001, **276**:47320–47328.

Hill, J.A., Wang, D., Jevnikar, A.M., Cairns, E., and Bell, D.A.: **The relationship between predicted peptide-MHC class II affinity and T-cell activation in a HLA-DRβ1*0401 transgenic mouse model.** *Arthritis Res. Ther.* 2003, **5**:R40–R48.

Su, R.C., and Miller, R.G.: **Stability of surface H-2K^b, H-2D^b, and peptide-receptive H-2K^b on splenocytes.** *J. Immunol.* 2001, **167**:4869–4877.

5-11 Many proteins involved in antigen processing and presentation are encoded by genes within the major histocompatibility complex.

Aguado, B., Bahram, S., Beck, S., Campbell, R.D., Forbes, S.A., Geraghty, D., Guillaudeux, T., Hood, L., Horton, R., Inoko, H., *et al.* (The MHC Sequencing Consortium): **Complete sequence and gene map of a human major histocompatibility complex.** *Nature* 1999, **401**:921–923.

Chang, C.H., Gourley, T.S., and Sisk, T.J.: **Function and regulation of class II transactivator in the immune system.** *Immunol. Res.* 2002, **25**:131–142.

Kumnovics, A., Takada, T., and Lindahl, K.F.: **Genomic organization of the mammalian MHC.** *Annu. Rev. Immunol.* 2003, **21**:629–657.

Lefranc, M.P.: **IMGT, the international ImMunoGeneTics database.** *Nucleic Acids Res.* 2003, **31**:307–310.

5-12 The protein products of MHC class I and class II genes are highly polymorphic.

Gaur, L.K., and Nepom, G.T.: **Ancestral major histocompatibility complex DRB genes beget conserved patterns of localized polymorphisms.** *Proc. Natl Acad. Sci. USA* 1996, **93**:5380–5383.

Marsh, S.G.: **Nomenclature for factors of the HLA system, update December 2002.** *Eur. J. Immunogenet.* 2003, **30**:167–169.

Robinson, J., and Marsh, S.G.: **HLA informatics. Accessing HLA sequences from sequence databases.** *Methods Mol. Biol.* 2003, **210**:3–21.

Robinson, J., Waller, M.J., Parham, P., de Groot, N., Bontrop, R., Kennedy, L.J., Stoehr, P., and Marsh, S.G.: **IMGT/HLA and IMGT/MHC: sequence databases for the study of the major histocompatibility complex.** *Nucleic Acids Res.* 2003, **31**:311–314.

5-13 MHC polymorphism affects antigen recognition by T cells by influencing both peptide binding and the contacts between T-cell receptor and MHC molecule.

Falk, K., Rotzschke, O., Stevanovic, S., Jung, G., and Rammensee, H.G.: **Allele-specific motifs revealed by sequencing of self-peptides eluted from MHC molecules.** *Nature* 1991, **351**:290–296.

Garcia, K.C., Degano, M., Speir, J.A., and Wilson, I.A.: **Emerging principles for T cell receptor recognition of antigen in cellular immunity.** *Rev. Immunogenet.* 1999, **1**:75–90.

Hillig, R.C., Coulie, P.G., Stroobant, V., Saenger, W., Ziegler, A., and Hulsmeyer, M.: **High-resolution structure of HLA-A*0201 in complex with a tumour-specific antigenic peptide encoded by the MAGE-A4 gene.** *J. Mol. Biol.* 2001, **310**:1167–1176.

Katz, D.H., Hamaoka, T., Dorf, M.E., Maurer, P.H., and Benacerraf, B.: **Cell interactions between histoincompatible T and B lymphocytes. IV. Involvement of immune response (Ir) gene control of lymphocyte interaction controlled by the gene.** *J. Exp. Med.* 1973, **138**:734–739.

Kjer-Nielsen, L., Clements, C.S., Brooks, A.G., Purcell, A.W., Fontes, M.R., McCluskey, J., and Rossjohn, J.: **The structure of HLA-B8 complexed to an immunodominant viral determinant: peptide-induced conformational changes and a mode of MHC class I dimerization.** *J. Immunol.* 2002, **169**:5153–5160.

Rosenthal, A.S., and Shevach, E.M.: **Function of macrophages in antigen recognition by guinea pig T lymphocytes. I. Requirement for histocompatible macrophages and lymphocytes.** *J. Exp. Med.* 1973, **138**:1194–1212.

Wang, J.H., and Reinherz, E.L.: **Structural basis of T cell recognition of peptides bound to MHC molecules.** *Mol. Immunol.* 2002, **38**:1039–1049.

Zinkernagel, R.M., and Doherty, P.C.: **Restriction of *in vivo* T-cell mediated cytotoxicity in lymphocytic choriomeningitis within a syngeneic or semiallogeneic system.** *Nature* 1974, **248**:701–702.

5-14 Alloreactive T cells recognizing nonself MHC molecules are very abundant.

Hennecke, J., and Wiley, D.C.: **Structure of a complex of the human alpha/beta T cell receptor (TCR) HA1.7, influenza hemagglutinin peptide, and major histocompatibility complex class II molecule, HLA-DR4 (DRA*0101 and DRB1*0401): insight into TCR cross-restriction and alloreactivity.** *J. Exp. Med.* 2002, **195**:571–581.

Jankovic, V., Remus, K., Molano, A., and Nikolich-Zugich, J.: **T cell recognition of an engineered MHC class I molecule: implications for peptide-independent alloreactivity.** *J. Immunol.* 2002, **169**:1887–1892.

Merkenschlager, M., Graf, D., Lovatt, M., Bommhardt, U., Zamoyska, R., and Fisher, A.G.: **How many thymocytes audition for selection?** *J. Exp. Med.* 1997, **186**:1149–1158.

Nesic, D., Maric, M., Santori, F.R., and Vukmanovic, S.: **Factors influencing the patterns of T lymphocyte allorecognition.** *Transplantation* 2002, **73**:797–803.

Reiser, J.B., Darnault, C., Guimezanes, A., Gregoire, C., Mosser, T., Schmitt-Verhulst, A.M., Fontecilla-Camps, J.C., Malissen, B., Housset, D., and Mazza, G.: **Crystal structure of a T cell receptor bound to an allogeneic MHC molecule.** *Nat. Immunol.* 2000, **1**:291–297.

Speir, J.A., Garcia, K.C., Brunmark, A., Degano, M., Peterson, P.A., Teyton, L., and Wilson, I.A.: **Structural basis of 2C TCR allorecognition of H-2Ld peptide complexes.** *Immunity* 1998, **8**:553–562.

Zerrahn, J., Held, W., and Raulet, D.H.: **The MHC reactivity of the T cell repertoire prior to positive and negative selection.** Cell 1997, **88**:627–636.

5-15 Many T cells respond to superantigens.

Acha-Orbea, H., Finke, D., Attinger, A., Schmid, S., Wehrli, N., Vacheron, S., Xenarios, I., Scarpellino, L., Toellner, K.M., MacLennan, I.C., and Luther, S.A.: **Interplays between mouse mammary tumor virus and the cellular and humoral immune response.** *Immunol. Rev.* 1999, **168**:287–303.

Alouf, J.E., and Muller-Alouf, H.: **Staphylococcal and streptococcal superantigens: molecular, biological and clinical aspects.** *Int. J. Med. Microbiol.* 2003, **292**:429–440.

Macphail, S.: **Superantigens: mechanisms by which they may induce, exacerbate and control autoimmune diseases.** *Int. Rev. Immunol.* 1999, **18**:141–180.

Kappler, J.W., Staerz, U., White, J., and Marrack, P.: **T cell receptor Vb elements which recognize Mls-modified products of the major histocompatibility complex.** *Nature* 1988, **332**:35–40.

Sundberg, E.J., Li, H., Llera, A.S., McCormick, J.K., Tormo, J., Schlievert, P.M., Karjalainen, K., and Mariuzza, R.A.: **Structures of two streptococcal superantigens bound to TCR beta chains reveal diversity in the architecture of T cell signaling complexes.** *Structure* 2002, **10**:687–699.

Torres, B.A., Perrin, G.Q., Mujtaba, M.G., Subramaniam, P.S., Anderson, A.K., and Johnson, H.M.: **Superantigen enhancement of specific immunity: antibody production and signaling pathways.** *J. Immunol.* 2002, **169**:2907–2914.

White, J., Herman, A., Pullen, A.M., Kubo, R., Kappler, J.W., and Marrack, P.: **The Vb-specific super antigen staphylococcal enterotoxin B: stimulation of mature T cells and clonal deletion in neonatal mice.** *Cell* 1989, **56**:27–35.

5-16 MHC polymorphism extends the range of antigens to which the immune system can respond.

Hill, A.V., Elvin, J., Willis, A.C., Aidoo, M., Allsopp, C.E.M., Gotch, F.M., Gao, X.M., Takiguchi, M., Greenwood, B.M., Townsend, A.R.M., *et al.*: **Molecular analysis of the association of B53 and resistance to severe malaria.** *Nature* 1992, **360**:435–440.

Martin, M.P., and Carrington, M.: **Immunogenetics of viral infections.** *Curr. Opin. Immunol.* 2005, **17**:510–516.

Messaoudi, I., Guevara Patino, J.A., Dyall, R., LeMaoult, J., and Nikolich-Zugich, J.: **Direct link between *mhc* polymorphism, T cell avidity, and diversity in immune defense.** *Science* 2002, **298**:1797–1800.

Potts, W.K., and Slev, P.R.: **Pathogen-based models favouring MHC genetic diversity.** *Immunol. Rev.* 1995, **143**:181–197.

5-17 A variety of genes with specialized functions in immunity are also encoded in the MHC.

Alfonso, C., and Karlsson, L.: **Nonclassical MHC class II molecules.** *Annu. Rev. Immunol.* 2000, **18**:113–142.

Allan, D.S., Lepin, E.J., Braud, V.M., O'Callaghan, C.A., and McMichael, A.J.: **Tetrameric complexes of HLA-E, HLA-F, and HLA-G.** *J. Immunol. Methods* 2002, **268**:43–50.

Gao, G.F., Willcox, B.E., Wyer, J.R., Boulter, J.M., O'Callaghan, C.A., Maenaka, K., Stuart, D.I., Jones, E.Y., Van Der Merwe, P.A., Bell, J.I., and Jakobsen, B.K.: **Classical and nonclassical class I major histocompatibility complex molecules exhibit subtle conformational differences that affect binding to CD8αα.** *J. Biol. Chem.* 2000, **275**:15232–15238.

Powell, L.W., Subramaniam, V.N., and Yapp, T.R.: **Haemochromatosis in the new millennium.** *J. Hepatol.* 2000, **32**:48–62.

5-18 Specialized MHC class I molecules act as ligands for the activation and inhibition of NK cells.

Borrego, F., Kabat, J., Kim, D.K., Lieto, L., Maasho, K., Pena, J., Solana, R., and Coligan, J.E.: **Structure and function of major histocompatibility complex (MHC) class I specific receptors expressed on human natural killer (NK) cells.** *Mol. Immunol.* 2002, **38**:637–660.

Boyington, J.C., Riaz, A.N., Patamawenu, A., Coligan, J.E., Brooks, A.G., and Sun, P.D.: **Structure of CD94 reveals a novel C-type lectin fold: implications for the NK cell-associated CD94/NKG2 receptors.** *Immunity* 1999, **10**:75–82.

Braud, V.M., and McMichael, A.J.: **Regulation of NK cell functions through interaction of the CD94/NKG2 receptors with the nonclassical class I molecule HLA-E.** *Curr. Top. Microbiol. Immunol.* 1999, **244**:85–95.

Lanier, L.L.: **NK cell recognition.** *Annu. Rev. Immunol.* 2005, **23**:225–274.

Lopez-Botet, M., and Bellon, T.: **Natural killer cell activation and inhibition by receptors for MHC class I.** *Curr. Opin. Immunol.* 1999, **11**:301–307.

Lopez-Botet, M., Bellon, T., Llano, M., Navarro, F., Garcia, P., and de Miguel, M.: **Paired inhibitory and triggering NK cell receptors for HLA class I molecules.** *Hum. Immunol.* 2000, **61**:7–17.

Lopez-Botet, M., Llano, M., Navarro, F., and Bellon, T.: **NK cell recognition of non-classical HLA class I molecules.** *Semin. Immunol.* 2000, 12:109–119.

Rodgers, J.R., and Cook, R.G.: **MHC class Ib molecules bridge innate and acquired immunity.** *Nat. Rev. Immunol.* 2005, **5**:459–471.

Vales-Gomez, M., Reyburn, H., and Strominger, J.: **Molecular analyses of the interactions between human NK receptors and their HLA ligands.** *Hum. Immunol.* 2000, **61**:28–38.

5-19 The CD1 family of MHC class I-like molecules is encoded outside the MHC and presents microbial lipids to CD1-restricted T cells.

Gadola, S.D., Zaccai, N.R., Harlos, K., Shepherd, D., Castro-Palomino, J.C., Ritter, G., Schmidt, R.R., Jones, E.Y., and Cerundolo, V.: **Structure of human CD1b with bound ligands at 2.3 Å, a maze for alkyl chains.** *Nat. Immunol.* 2002, **3**:721–726.

Hava, D.L., Brigl, M., van den Elzen, P., Zajonc, D.M., Wilson, I.A., Brenner, M.B.: **CD1 assembly and the formation of CD1-antigen complexes.** *Curr. Opin. Immunol.* 2005, **17**:88–94.

Jayawardena-Wolf, J., and Bendelac, A.: **CD1 and lipid antigens: intracellular pathways for antigen presentation.** *Curr. Opin. Immunol.* 2001, **13**:109–113.

Moody, D.B., and Besra, G.S.: **Glycolipid targets of CD1-mediated T-cell responses.** *Immunology* 2001, **104**:243–251.

Moody, D.B., and Porcelli, S.A.: **CD1 trafficking: invariant chain gives a new twist to the tale.** *Immunity* 2001, **15**:861–865.

Moody, D.B., and Porcelli, S.A.: **Intracellular pathways of CD1 antigen presentation.** *Nat. Rev. Immunol.* 2003, **3**:11–22.

PART III

THE DEVELOPMENT OF MATURE LYMPHOCYTE RECEPTOR REPERTOIRES

Signaling Through Immune System Receptors

6

Specific changes in the extracellular environment are sensed by the immune system and result in the activation of immune-system cells. Cells communicate with their environment through a variety of cell-surface receptors that recognize and bind molecules present in the extracellular environment. Although the lymphocyte antigen receptors were historically the best studied, the operation of a wide variety of other receptors on lymphocytes and other cells of the immune system is also now well understood. The intracellular signals generated by these receptors and how they alter the cells' behavior is the main topic of this chapter.

The challenge that faces all cells that respond to external stimuli is how the recognition of a stimulus is able to effect changes within the cell. All the extracellular signals we shall consider in this chapter are received at the outer surface of the cell and transmitted across the plasma membrane by transmembrane receptor proteins, which are instrumental in converting information outside the cell into an intracellular biochemical event. Once inside the cell, the signal is transmitted along **intracellular signaling pathways** composed of proteins that interact with each other in a variety of ways. The signal is converted into different biochemical forms—the process known as **signal transduction**—distributed to different sites in the cell, and sustained and amplified as it proceeds toward its destinations. In the signaling pathways we shall consider in this chapter, the final destination of most signals is the nucleus, and the primary cellular response is the alteration of gene expression: this leads in turn to the synthesis of new proteins such as cytokines, chemokines, cell-adhesion molecules, and other cell-surface proteins, and to cellular events such as cell division, cell differentiation, and in some cases, cell death.

We start this chapter by discussing some general principles of intracellular signaling. We then outline the pathways that are involved in the activation of a naive lymphocyte when it encounters its specific antigen. As well as the signals a lymphocyte receives through its antigen receptors and co-receptors, we discuss briefly the co-stimulatory signaling that is necessary to activate naive T cells and, in most cases, naive B cells. In the last part of the chapter we look at a selection of other signaling pathways used by immune-system cells, including those from the cytokine receptors, Toll-like receptors, and the death receptors that stimulate apoptosis.

General principles of signal transduction.

In this part of the chapter we review briefly some general principles of receptor action and signal transduction that are common to many of the pathways discussed here. We start with the cell-surface receptors through which cells receive extracellular signals.

6-1 Transmembrane receptors convert extracellular signals into intracellular biochemical events.

All cell-surface receptors that have a signaling function are either transmembrane proteins themselves or form parts of protein complexes that link the exterior and interior of the cell. Different classes of receptors transduce extracellular signals in a variety of ways: a common theme in the receptors covered in this chapter is that ligand binding results in the activation of an enzymatic activity. The enzymes most commonly associated with receptor activation are the **protein kinases**. This large group of enzymes catalyzes the covalent attachment of a phosphate group to a protein in the reversible process known as **protein phosphorylation**. The receptor-associated protein kinases are normally inactive, but when a ligand binds to the extracellular part of the receptor they become active and transmit the signal onward by phosphorylating and activating other signaling molecules inside the cell.

In animals, protein kinases phosphorylate proteins on three amino acids—tyrosine, serine, or threonine. Most of the enzyme-linked receptors we discuss in detail in this chapter activate **tyrosine protein kinases**. Tyrosine kinases are specific for tyrosine residues, while serine/threonine kinases phosphorylate serine and threonine residues. In general, protein tyrosine phosphorylation is a much rarer modification than serine/threonine phosphorylation, and is found mainly in signaling pathways.

In one large group of receptors, the kinase activity is an intrinsic part of the cytoplasmic portion of the receptor (Fig. 6.1, top panel). Receptor tyrosine kinases of this type include those for many growth factors; lymphocyte receptors of this type include Kit and FLT3, which are expressed on developing lymphocytes and are discussed in Chapter 7. The receptor for transforming growth factor-β (TGF-β), a cytokine produced by activated T_H2 cells, is a receptor serine/threonine kinase.

Even more important to the function of mature lymphocytes are a class of receptors that have no intrinsic enzymatic activity themselves but whose cytoplasmic tails are noncovalently associated with a cytoplasmic tyrosine kinase. Ligand binding to the extracellular domain of these receptors causes activation of the associated enzyme, which carries out the signal transduction function of the receptor (Fig. 6.1, bottom panel). The antigen receptors and many cytokine receptors are of this type.

Both these classes of receptors are activated when ligand binding causes the dimerization or clustering of individual receptor molecules, thus bringing the associated kinases together. Clustering activates the enzymes, which then phosphorylate the receptor tails or other proteins associated with the receptor. This phosphorylation event is the initial intracellular signal generated by ligand binding.

The role of protein kinases in cell signaling is not confined to receptor activation, and they are found at many different stages in intracellular signaling pathways. For example, they often act at the final step in the pathway to activate the cell's response machinery. Protein kinases figure largely in cell

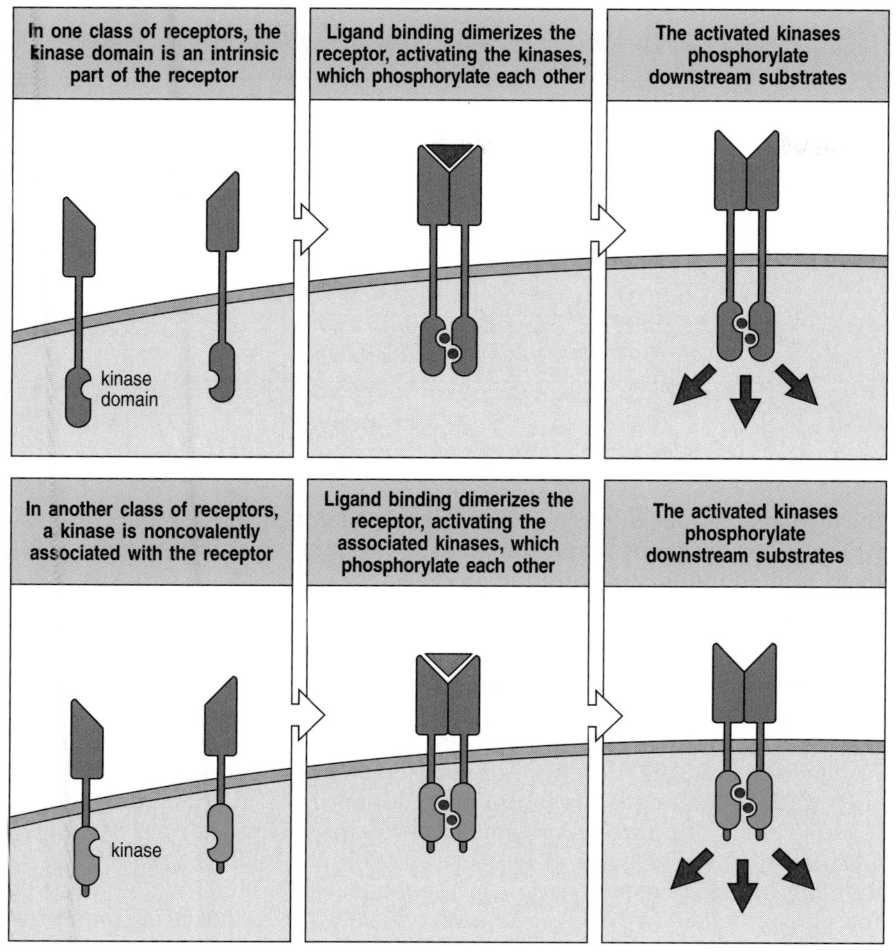

In one class of receptors, the kinase domain is an intrinsic part of the receptor
Ligand binding dimerizes the receptor, activating the kinases, which phosphorylate each other
The activated kinases phosphorylate downstream substrates
kinase domain
In another class of receptors, a kinase is noncovalently associated with the receptor
Ligand binding dimerizes the receptor, activating the associated kinases, which phosphorylate each other
The activated kinases phosphorylate downstream substrates
kinase

Fig. 6.1. Two types of receptors that signal through protein kinases are used in the immune system. In both these receptors, the information that a ligand has bound to the extracellular portion is converted into the activation of protein kinase activity on the cytoplasmic side of the membrane. In one receptor class (top panels), the kinase activity is part of the receptor itself. Ligand binding results in clustering of the receptor, activation of the catalytic activity, and the consequent phosphorylation of the receptor tails and other substrates, transmitting the signal onward. In the second class of receptors (bottom panels), the receptor itself does not have any enzymatic activity. Instead, enzymes in the cytoplasm are either constitutively associated with the cytoplasmic portion of the receptor or are induced to associate with it after ligand has bound to the extracellular part of the receptor. Receptor dimerization or clustering then activates the associated enzyme. In all the receptors of these types that we will encounter in this chapter, the enzyme is tyrosine kinase.

signaling because phosphorylation and dephosphorylation—the removal of a phosphate group—are the means of regulating the activity of many enzymes, transcription factors, and other proteins. Equally important to the working of the signaling pathway is that phosphorylation generates sites on proteins to which other signaling proteins can bind.

Phosphate groups are removed from proteins by a large class of enzymes called **protein phosphatases** (see Fig. 6.8). Different classes of protein phosphatases remove phosphatases from phosphotyrosine or from phosphoserine/phosphothreonine. Specific dephosphorylation by phosphatases is one important means of regulating signaling pathways by resetting a protein back to its original state and thus switching signaling off.

6-2 Intracellular signal transduction often takes place in large multiprotein signaling complexes.

Signal transduction by transmembrane receptors informs the cell interior that the receptor has encountered its ligand. This is just the first step in a multistep process. A cascade of intracellular signaling is initiated that will direct the various different biochemical responses that characterize a specific cellular response. The intracellular signaling pathways that lead away from receptors are composed of series of proteins that interact with each other to pass on the signal. The collection of specific enzyme activities assembled into a multiprotein complex determines the specific character of

Fig. 6.2 Signaling proteins interact with each other and with lipid signaling molecules via modular protein domains. A few of the most common protein domains used by immune-system signaling proteins are listed, together with some proteins containing the domain that are mentioned in this chapter or elsewhere in the book, and the general class of ligand that they bind. The right-hand column lists specific examples of a protein motif bound (in single-letter amino-acid code) or, for the phosphoinositide-binding domains, the particular phosphoinositide they bind. PI3K, PI 3-kinase. All these domains are used in many other non-immune signaling pathways as well.

Protein domain	Found in	Ligand class	Example of ligand
SH2	Lck, ZAP-70, Fyn, Src, Grb2, PLC-γ, STAT, Cbl, Btk, Itk, SHIP, Vav, SAP, PI3K	phosphotyrosine	pYXXZ
SH3	Lck, Fyn, Src, Grb2, Btk, Itk, Tec, Fyb, Nck, GADS	proline	PXXP
PH	Tec, PLC-γ, Akt, Btk, Itk, SOS	phosphoinositides	PIP_3
PX	$P40^{phox}$, $P47^{phox}$, PLD	phosphoinositides	PIP_2
PDZ	CARMA1	C termini of proteins	IESDV, VETDV

the response. Some pathways include some, but not all, of the same enzymes, which allows different signal transduction systems to be built from a relatively limited number of common modules.

The assembly of large signaling complexes involves specific interactions involving a variety of **protein interaction domains** (Fig. 6.2). In the pathways we shall consider in this chapter, the most important mechanism underlying the formation of signaling complexes is the specific phosphorylation of protein tyrosine residues. Phosphotyrosines are binding sites for a number of protein domains, the most important of which in the pathways we will consider is the **SH2 domain** (**Src homology 2 domain**). SH2 domains are found in a wide variety of intracellular signaling proteins, where they are associated with many different types of enzymatic or other functional domains. SH2 domains bind to phosphotyrosine in a sequence-specific fashion, recognizing the phosphorylated tyrosine (pY) and, typically, the amino acid three positions away (pYXXZ, where X is any amino acid and Z is a specific amino acid).

In pathways leading from tyrosine kinase-associated receptors, **scaffold proteins** and **adaptor proteins** are used to assemble multiprotein signaling complexes. Scaffolds and adaptors do not have enzymatic activity; their function is to recruit other proteins to a signaling complex so that they can interact with each other. Scaffolds are larger proteins that can, for example, become tyrosine phosphorylated on multiple sites and can thus recruit many different proteins (Fig. 6.3, top panel). By determining which proteins are recruited to a pathway, scaffolds can define the character of a particular signaling response. This function of tyrosine phosphorylation in generating binding sites may explain why it is so commonly used in signaling pathways.

Adaptors are smaller proteins with usually no more than two or three domains whose function is to link two proteins together. The adaptor protein Grb2, for example, binds to a phosphotyrosine residue on a receptor or scaffold via an SH2 domain and to another signaling protein, SOS, which contains proline-rich motifs, via its SH3 binding domains (Fig. 6.3, bottom panel). Thus, Grb2 functions as an adaptor to link tyrosine phosphorylation of a receptor to the next stage of signaling.

6-3 The activation of some receptors generates small-molecule second messengers.

After an initial intracellular signal has been generated, the information is then transmitted to the intracellular targets that will carry out the appropriate cellular response. In many cases, the signaling pathway involves the activation of

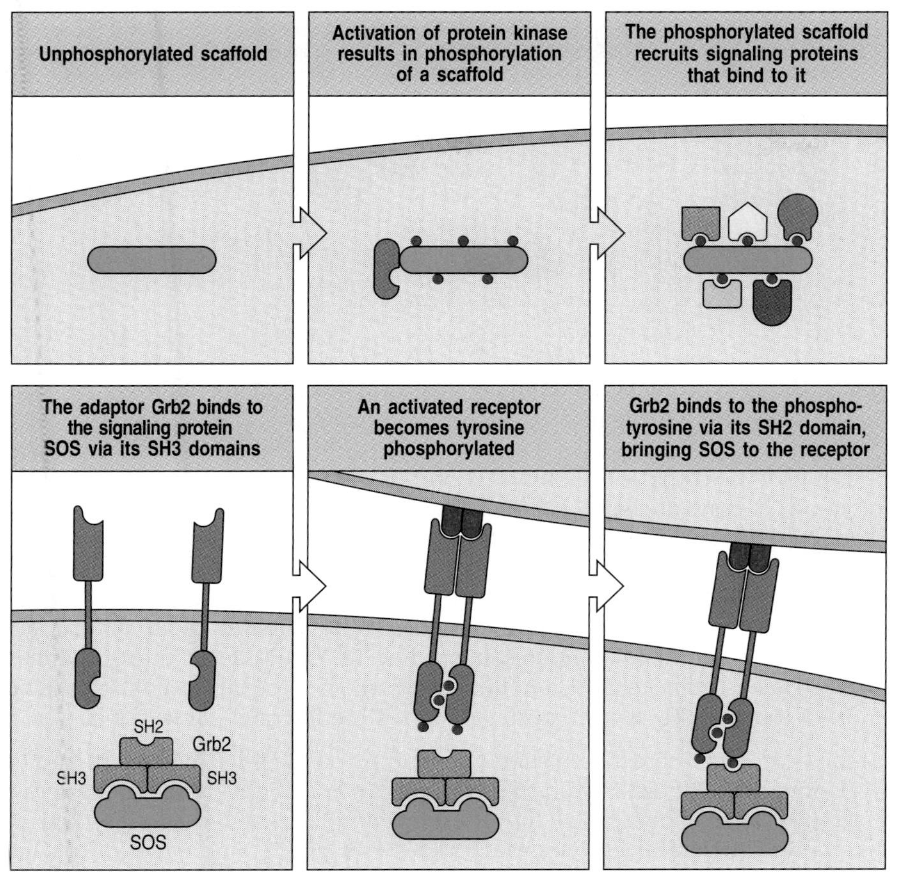

Fig. 6.3 Assembly of signaling complexes is mediated by scaffold and adaptor proteins. Assembly of signaling complexes is an important aspect of signal transduction. This is often achieved through scaffold and adaptor proteins. Scaffolds function to bring many different signaling proteins together (top panel). They generally have numerous potential sites of tyrosine phosphorylation which, after phosphorylation, can recruit many different proteins that contain SH2 domains. Which set of proteins is recruited determines the character of the signaling response. An adaptor protein functions to bring two different proteins together (bottom panel). Grb2, the adaptor protein shown here, contains two SH3 domains and an SH2 domain. With the SH3 domains it can, for example, bind proline-rich sites on the signaling molecule SOS (which we shall meet again later in the chapter). Activation and tyrosine phosphorylation of a receptor generates a binding site for the SH2 domain of Grb2, resulting in the recruitment of SOS to the activated receptor.

enzymes that produce small-molecule biochemical mediators known as **second messengers** (Fig. 6.4). These mediators can diffuse throughout the cell, enabling the signal to activate a variety of target proteins. They are also a means of amplifying the initial signal, as one activated enzyme molecule can produce hundreds of second messenger molecules. The second messengers generated by receptors that signal via tyrosine kinases include calcium ions (Ca^{2+}) and a variety of membrane lipids. Although the latter are confined to membranes, they can move within them. A second messenger binding to its target protein typically induces a conformational change that allows the protein to be activated.

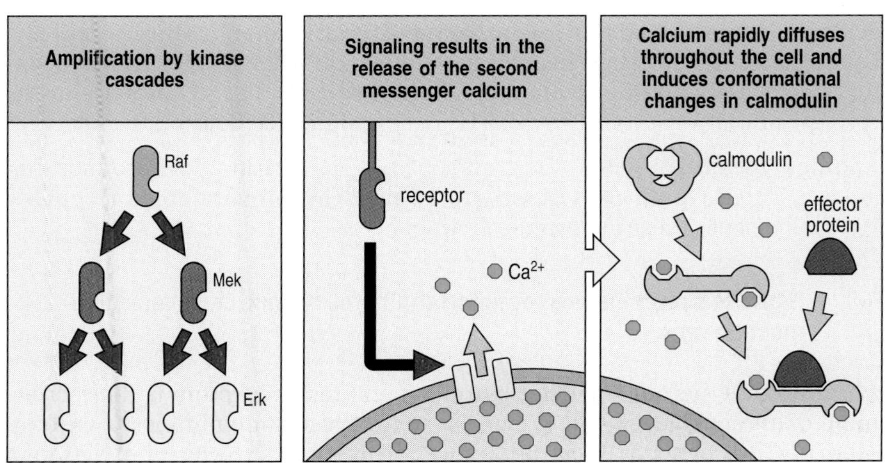

Fig. 6.4 Signaling pathways amplify the initial signal. Amplification of the initial signal is an important element of most signal transduction pathways. One means of amplification is a kinase cascade (left-hand panel), in which protein kinases successively phosphorylate and activate each other. In this example, taken from a commonly used kinase cascade, activation of the kinase Raf results in the phosphorylation and activation of a second kinase, Mek, that phosphorylates yet another kinase, Erk. As each kinase can phosphorylate many different substrate molecules, the signal is amplified at each step, resulting in a huge amplification of the initial signal. Another method of signal amplification is the generation of second messengers (right-hand panels). In the example illustrated here, signaling results in the release of the second messenger calcium (Ca^{2+}) from intracellular stores or its influx from the extracellular environment. The large number of Ca^{2+} ions can potentially activate many downstream signaling molecules, such as the calcium-binding protein calmodulin. Calcium binding induces a conformational change in calmodulin, which allows it to bind to and regulate a variety of effector proteins.

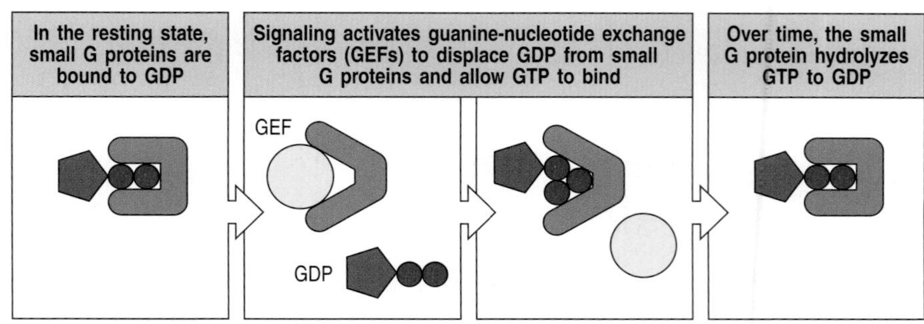

In the resting state, small G proteins are bound to GDP
Signaling activates guanine-nucleotide exchange factors (GEFs) to displace GDP from small G proteins and allow GTP to bind
Over time, the small G protein hydrolyzes GTP to GDP
GEF
GDP

Fig. 6.5 Small G proteins are switched from inactive to active states by guanine-nucleotide exchange factors and the binding of GTP. Ras is a small GTP-binding protein with intrinsic GTPase activity. In its resting state, Ras is bound to GDP. Receptor signaling activates guanine-nucleotide exchange factors (GEFs), which can bind to small G proteins such as Ras and displace GDP, allowing GTP to bind in its place (center panels). The GTP-bound form of Ras can then bind to a large number of effectors, recruiting them to the membrane. Over time, the intrinsic GTPase activity of Ras will result in the hydrolysis of GTP to GDP. GTPase-activating proteins (GAPs) can accelerate the hydrolysis of GTP to GDP, thus shutting off the signal more rapidly.

6-4 Small G proteins act as molecular switches in many different signaling pathways.

A family of monomeric GTP-binding proteins known as the **small G proteins** or **small GTPases** are key components of a number of signaling pathways that lead from tyrosine kinase-associated receptors. The most important small G proteins in lymphocyte signaling are the **Ras family**, which includes Ras, Rac, Rho, and Cdc42. Ras is involved in many different pathways leading to cell proliferation, and mutations that lock Ras in the active state are among the commonest mutations found in cancers. Rac, Rho and Cdc42 control changes in a cell's actin cytoskeleton, and this aspect of T-cell receptor signaling will be looked at in Chapter 8, as it is crucial to the function of effector T cells.

Small G proteins exist in two states, depending on whether they are bound to GTP or to GDP. The GDP-bound form is inactive but is converted into the active form by exchange of the GDP for GTP, a reaction mediated by proteins known as **guanine nucleotide exchange factors** (**GEFs**; Fig. 6.5). Binding of GTP results in a conformational change in the G protein that allows it to bind to a wide variety of different targets. Thus, GTP binding functions like an on/off switch.

The GTP-bound form does not remain permanently active but is rapidly converted into the inactive GDP-bound form by the intrinsic GTPase activity in the G protein, which removes a phosphate group from the bound GTP. This reaction is accelerated by regulatory cofactors known as **GTPase-activating proteins** (**GAPs**). G proteins are thus usually present in the inactive GDP-bound state and are activated only transiently in response to a signal from an activated receptor.

The GEFs are the key to G-protein activation and are recruited to the site of receptor activation at the cell membrane by binding to adaptor proteins. Once recruited, they are able to activate Ras or other small G proteins, which are themselves localized to the inner surface of the plasma membrane via fatty acids that are attached to the G protein posttranslationally. Thus, G proteins act as molecular switches, becoming switched on when a cell-surface receptor is activated and then switching off automatically. Each G protein has its own specific GEFs and GAPs, which helps to confer specificity on the pathway.

Another type of G protein is the group of larger heterotrimeric G proteins that are associated with a class of receptors called G-protein-coupled receptors. We shall discuss these later in the chapter.

6-5 Signaling proteins are recruited to the membrane by a variety of mechanisms.

An important step in signaling by transmembrane receptors is the recruitment of intracellular signaling proteins to the plasma membrane. As we have seen, one mechanism of recruitment can be the tyrosine phosphorylation of

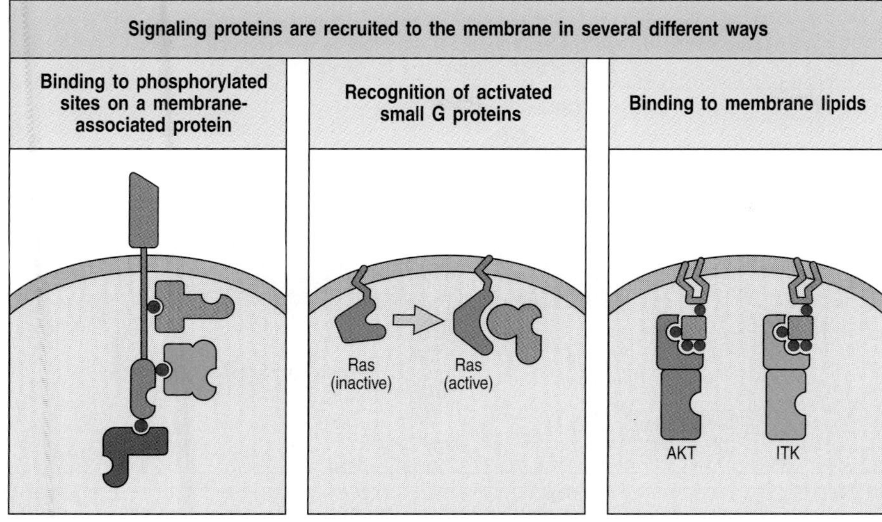
Signaling proteins are recruited to the membrane in several different ways
Binding to phosphorylated sites on a membrane-associated protein
Recognition of activated small G proteins
Binding to membrane lipids
Ras (inactive)
Ras (active)
AKT
ITK

Fig. 6.6 Signaling proteins can be recruited to the membrane in a variety of ways. As the activated receptor is usually located at the plasma membrane, an important aspect of intracellular signaling is the recruitment of signaling proteins to the membrane. Tyrosine phosphorylation of proteins associated with the membrane, such as the receptor itself, will recruit phosphotyrosine-binding proteins (left panel). Small G proteins such as Ras can be membrane-associated via lipid linkages, and when activated are able to bind a wide variety of signaling proteins (center panel). Signaling proteins are also recruited to the membrane via binding to lipid signaling molecules that are generated in the membrane as a result of receptor activation. In this example, the activation of the lipid-modifying enzyme PI 3-kinase (PI3K) at the membrane results in the localized production of the membrane lipid PIP_3 by phosphorylation of PIP_2. Signaling proteins, such as the kinase Akt or the kinase ITK, have PH or PX domains (see Fig. 6.2) that bind to PIP_3. Thus, the production of lipids such as PIP_3 recruits signaling molecules to the membrane.

the receptor itself (or an associated scaffold) and the subsequent recruitment of SH2-domain-containing signaling proteins to the receptor (Fig. 6.6). Another mechanism is the activation of membrane-associated small G proteins, which can then recruit signaling molecules to the membrane.

A third means of recruitment is the local production of modified membrane lipids as a result of receptor activation. These lipids are produced by phosphorylation of the membrane phospholipid phosphatidylinositol by enzymes known as **phosphatidylinositol kinases**, which are activated as a result of receptor signaling. The inositol head group of phosphatidylinositol is a sugar ring that can be phosphorylated singly or multiply in several positions to generate a wide variety of derivatives. The ones that will mainly concern us here are phosphatidylinositol 3,4-bisphosphate (PIP_2) and phosphatidylinositol 3,4,5-trisphosphate (PIP_3), which is generated from PIP_2 by the enzyme **phosphatidylinositol 3-kinase** (**PI 3-kinase**) (see Fig. 6.6). PI 3-kinase is recruited to the membrane by the interaction of its SH2 domain with a tyrosine-phosphorylated receptor tail. The membrane phosphoinositides are produced rapidly after receptor activation and are short-lived, which makes them ideal signaling molecules. PIP_3 is recognized specifically by proteins containing a pleckstrin homology (PH) domain or a PX domain (see Fig. 6.2), and one of its functions is to recruit such proteins to the membrane.

6-6 Signal transduction proteins are organized in the plasma membrane in structures called lipid rafts.

Recent evidence suggests that recruitment of signaling proteins to the plasma membrane may also be regulated by its lipid composition. In eukaryotic cells, different types of lipids segregate in the membrane to form structures known as glycolipid-enriched microdomains (GEMs), detergent-insoluble glycolipid-rich domains (DIG), or more simply, **lipid rafts** (Fig. 6.7). Lipid rafts are small, cholesterol-rich areas in the cell membrane that were originally discovered because they are resistant to solubilization by mild detergents. They are enriched in particular lipids, notably sphingolipids and cholesterol, which suggests that their segregation is based on the differences in the biophysical properties of lipids, analogous to a phase separation. In most cells, lipid rafts can constitute about 25–50% of the total plasma membrane. They are thought to be dynamic structures that can change in size and whose protein composition is constantly changing.

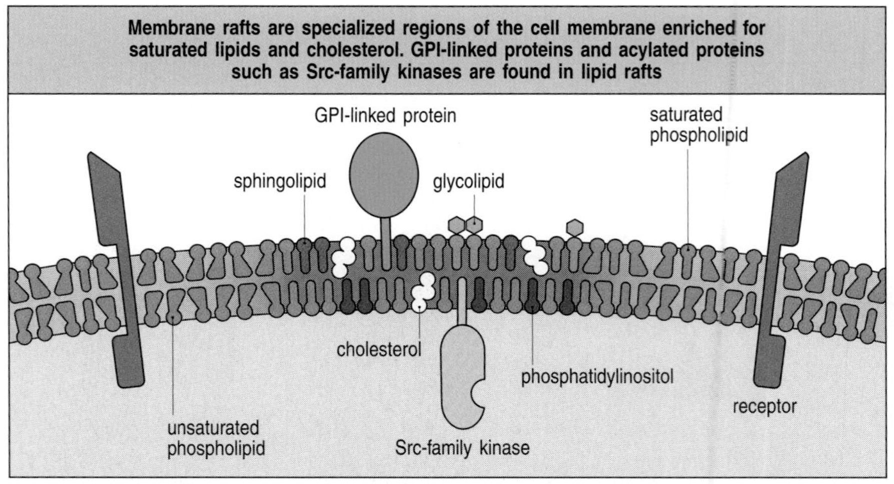
Membrane rafts are specialized regions of the cell membrane enriched for saturated lipids and cholesterol. GPI-linked proteins and acylated proteins such as Src-family kinases are found in lipid rafts
GPI-linked protein
saturated phospholipid
sphingolipid
glycolipid
cholesterol
phosphatidylinositol
receptor
unsaturated phospholipid
Src-family kinase

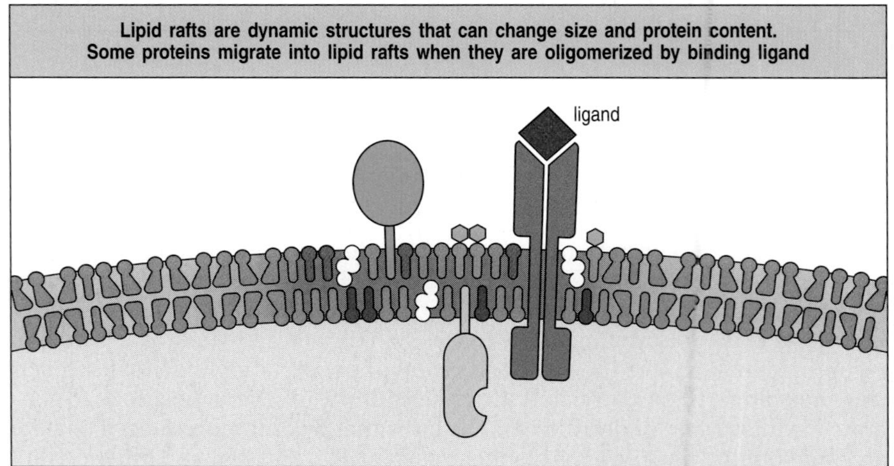
Lipid rafts are dynamic structures that can change size and protein content. Some proteins migrate into lipid rafts when they are oligomerized by binding ligand
ligand

Fig. 6.7 Signaling molecules are associated with specialized regions of the membrane called membrane rafts. Cell membranes contain a mixture of different phospholipids containing saturated and unsaturated fatty-acid chains (top panel). Intrinsic physical differences between lipids and proteins that associate preferentially with different lipids cause the generation of specialized membrane domains. Because saturated phospholipids can pack more closely together, regions of the membrane enriched for saturated phospholipids are more rigid than those that have more unsaturated phospholipids. Such regions also contain a higher proportion of cholesterol than the rest of the membrane, which also increases membrane rigidity. These specialized membrane microdomains are called 'membrane rafts' or 'lipid rafts,' on account of their specialized composition. Lipid rafts are enriched for other saturated lipids such as sphingolipids and glycolipids; these are restricted to the external face of the membrane. The phospholipid phosphatidylinositol is enriched in the inner leaflet of the bilayer of these lipid rafts. Various proteins are associated with lipid rafts, such as GPI-linked proteins and those intracellular proteins that have certain acyl modifications such as the palmitoyl-linked Src-family kinases. Other proteins can migrate into the rafts. Receptors that are found outside lipid rafts can migrate into rafts once the receptor has been oligomerized by binding ligand (bottom panel).

Interest in lipid rafts was initially stimulated by the finding that they are enriched in certain signaling proteins, suggesting that they might be the sites in the membrane where most signaling occurs. One possibility is that receptors move into lipid rafts to facilitate their interaction with important signaling proteins. Many lipid-raft proteins have lipid attachments, suggesting that their enrichment in lipid rafts is due to the association of these lipids with particular membrane lipids. Proteins such as Thy-1, which is linked to the plasma membrane via glycosylphosphatidylinositol (GPI), are preferentially found in lipid rafts, as are proteins modified with fatty acids such as palmitate. However, none of these proteins is exclusively associated with rafts, as they are also found in other regions of the membrane.

6-7 Protein degradation has an important role in terminating signaling responses.

Just as important as mechanisms for initiating signaling are those that turn it off. Signaling is most commonly terminated by targeted protein degradation or by the dephosphorylation of signaling proteins by protein phosphatases (Fig. 6.8). Proteins are most commonly targeted for destruction by the covalent attachment of one or more molecules of the small protein **ubiquitin**. Ubiquitin is attached to lysine residues on target proteins by enzymes known as ubiquitin ligases, which also determine the substrate specificity of the reaction. An important ubiquitin ligase in immunology is **Cbl**, which selects its targets via its SH2 domain. Cbl can thus bind to specific tyrosine-phosphorylated

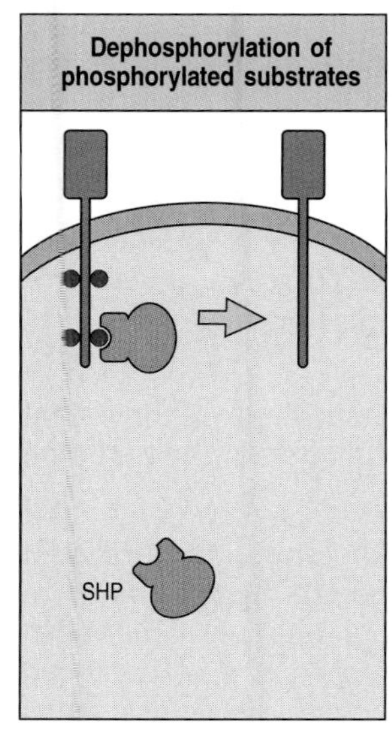

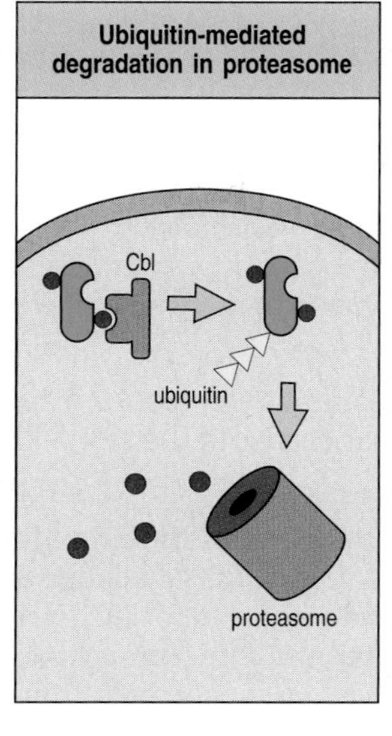

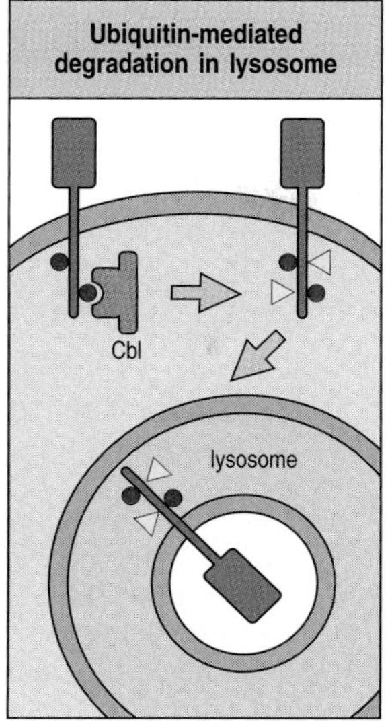

Fig. 6.8. Signaling must be turned off as well as turned on. The inability to terminate a signaling pathway can result in serious diseases such as autoimmunity or cancer. As a significant proportion of signaling events depend on protein phosphorylation, protein phosphatases, such as SHP, have an important part in shutting down signaling pathways (left panel). Another common mechanism for terminating signaling is regulated protein degradation (center and right panels). Phosphorylated proteins recruit ubiquitin ligases, such as Cbl, that add the small protein ubiquitin to proteins, thus targeting them for degradation. Cytoplasmic proteins are targeted by ubiquitination for destruction in the proteasome (center panel). Membrane receptors that become ubiquitinated are internalized and transported to the lysosome for destruction (right panel).

targets, causing them to become ubiquitinated. Proteins that recognize ubiquitin then target the ubiquitinated proteins to degradative pathways. Ubiquitin-tagged membrane proteins, such as receptors, are degraded in lysosomes. Ubiquitin tagging of cytosolic proteins targets them to the proteasome (see Fig. 6.8).

Summary.

Cell-surface receptors serve as the front line of a cell's interaction with its environment, sensing extracellular events and converting them into biochemical signals for the cell. As most receptors sit in the plasma membrane, a critical step in the transduction of extracellular signals to the interior of the cell is recruitment of intracellular proteins to the membrane and changes in the composition of the membrane surrounding the receptor. Once inside the cell, the signal is transmitted onward by intracellular proteins, which often form large multiprotein complexes, the specific composition of the complex determining the character of the signaling response. Formation of signaling complexes is mediated by the wide variety of interaction domains found in proteins. In many cases, the signal is amplified inside the cell by the enzymatic production of small-molecule signaling intermediates called second messengers. Termination of signaling involves protein dephosphorylation and regulated protein degradation.

Antigen receptor signaling and lymphocyte activation.

The ability of T cells and B cells to recognize and respond to their specific antigen is central to adaptive immunity. As described in Chapters 3 and 4, the B-cell antigen receptor and the T-cell antigen receptor are made up of antigen-binding chains—the heavy and light immunoglobulin chains in the B-cell receptor, and the TCRα and TCRβ chains in the T-cell receptor. These variable antigen-binding chains have exquisite specificity for antigen but no intrinsic signaling capacity. In the fully functional antigen–receptor complex

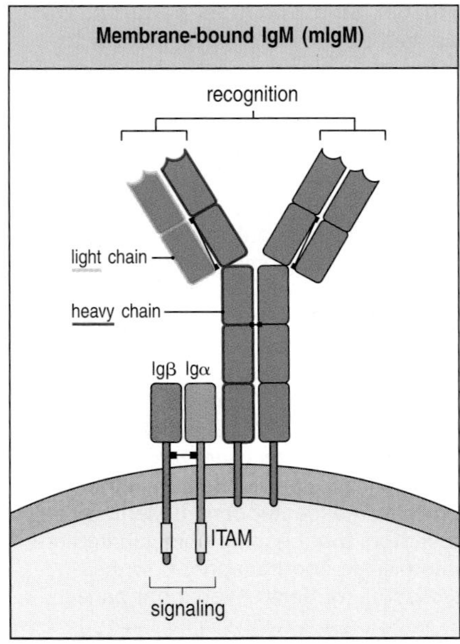

Fig. 6.9 The B-cell receptor complex is made up of cell-surface immunoglobulin with one each of the invariant proteins Igα and Igβ. The immunoglobulin recognizes and binds antigen but cannot itself generate a signal. It is associated with antigen-nonspecific signaling molecules—Igα and Igβ. These each have a single immunoreceptor tyrosine-based activation motif (ITAM), shown as a yellow segment, in their cytosolic tails that enables them to signal when the B-cell receptor is ligated with antigen. Igα and Igβ form a disulfide-linked heterodimer that is associated with the heavy chains, but it is not known which binds to the heavy chain.

they are associated with invariant accessory proteins that initiate signaling when the receptors bind extracellular antigen. Assembly with these accessory proteins is also essential for transport of the receptor to the cell surface. In this part of the chapter we describe the structure of the antigen–receptor complexes on B cells and T cells and the signaling pathways that lead from them.

Binding of antigen to a naive lymphocyte is not on its own sufficient for activation. So we shall also look at the signaling from co-receptors and co-stimulatory receptors that helps to activate a naive lymphocyte.

6-8 The variable chains of antigen receptors are associated with invariant accessory chains that carry out the signaling function of the receptor.

The antigen-binding portion of the B-cell receptor has no signaling function itself. On the cell surface, the antigen-binding immunglobulin is associated with invariant accessory protein chains, called **Igα** and **Igβ**, that are required both for its transport to the surface and for the signaling function of the B-cell receptor. The fully functional protein complex is often known as the **B-cell receptor complex**. Igα and Igβ associate with immunoglobulin heavy chains destined for the cell membrane and enable their transport to the cell surface, thus ensuring that only fully assembled B-cell receptor complexes are present on the cell. Igα and Igβ are single-chain proteins composed of an amino-terminal immunoglobulin-like domain connected via a transmembrane domain to a cytoplasmic tail. They form a disulfide-linked heterodimer that is noncovalently associated with each surface immunoglobulin molecule. The complete B-cell receptor is thought to be a complex of six chains—two identical light chains, two identical heavy chains, one Igα and one Igβ (Fig. 6.9).

One copy of a conserved sequence motif called an **immunoreceptor tyrosine-based activation motif** (**ITAM**) is present in each Igα and Igβ chain and is essential for the receptor's signaling ability. This motif is also present in the signaling chains of the T-cell receptor and in the signaling chains of the NK-cell receptors described in Chapter 2, as well as in the receptors for immunoglobulin (Fc receptors) present on mast cells, macrophages, monocytes, neutrophils, and NK cells. ITAMs contain tyrosine residues that are phosphorylated by associated kinases when the receptor binds its ligand, providing sites for the recruitment of signaling proteins as described earlier in the chapter. They are composed of two YXXL/I motifs separated by about six to nine amino acids, where Y is tyrosine, L is leucine, I is isoleucine, and X represents any amino acid. The canonical ITAM sequence is ...YXX[L/I]X_{6-9}YXX[L/I]....

In T cells, the highly variable TCRα:β heterodimer (see Chapter 4) is also not sufficient on its own to make up a complete cell-surface receptor. When cells were transfected with cDNAs encoding the TCRα and TCRβ chains, the heterodimers formed were degraded and did not appear on the cell surface. This implied that other molecules are required for the T-cell receptor to be expressed on the cell surface. These are the CD3γ, CD3δ, and CD3ε protein chains, which together form the **CD3 complex**, and the ζ chain, which is present as a disulfide-linked homodimer. The CD3 proteins have an extracellular immunoglobulin-like domain, whereas the ζ chain is distinct in having only a short extracellular domain.

Although the exact stoichiometry of the **T-cell receptor complex** is not definitively established, it is thought that the receptor α chain interacts with a CD3δ:CD3ε dimer and the ζ dimer while the receptor β chain interacts with a CD3γ:CD3ε dimer (Fig. 6.10). These interactions are mediated by two positive charges in the TCRα transmembrane region and one in the TCRβ transmembrane domain. Negative charges in the CD3 and ζ transmembrane domains interact with the positive charges in α and β. Assembly of CD3 with the α:β

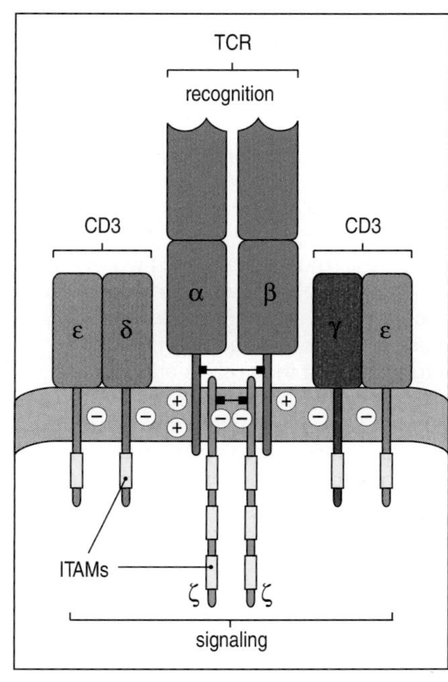

Fig. 6.10 The T-cell receptor complex is made up of antigen-recognition proteins and invariant signaling proteins. The T-cell receptor α:β heterodimer (TCR) recognizes and binds its peptide:MHC ligand, but cannot signal to the cell that antigen has bound. In the functional receptor complex, α:β heterodimers are associated with a complex of four other signaling chains (two ε, one δ, one γ) collectively called CD3, which are required for the cell-surface expression of the antigen-binding chains and for signaling. The cell-surface receptor complex is also associated with a homodimer of ζ chains, which also contain sequences that can signal to the interior of the cell upon antigen binding. All of the chains contain a similar signaling motif called an ITAM. Each CD3 chain has one ITAM (yellow segment), whereas each ζ chain has three. The transmembrane regions of each chain have either positive or negative charges as shown. It is now thought that one of the positive charges of the α chain interacts with the two negative charges of the CD3 δ:ε dimer, while the other positive charge interacts with the ζ homodimer. The positive charge of the β chain interacts with the negative charges in the CD3 γ:ε dimer.

heterodimer stabilizes the dimer and allows the complex to be transported to the plasma membrane. This ensures that all T-cell receptors present on the plasma membrane are properly assembled. Recent evidence suggests that the composition of the T-cell receptor complex is dynamic and can change after stimulation of the receptor by its ligand.

Signaling from the T-cell receptor complex is due to the presence in CD3ε, γ, δ and ζ of an ITAM motif like those present in Igα and Igβ. CD3γ, δ, and ε each have a single ITAM, while each of the two ζ chains has three copies. This gives the T-cell receptor complex a total of 10 ITAMs.

6-9 Lymphocytes are extremely sensitive to their specific antigens.

To make an effective immune response, T cells and B cells must be able to respond to their specific antigen even when it is present at extremely low levels. This is especially important for T cells, as the antigen-presenting cell will display many different peptides from both self and foreign proteins on its surface and so the number of peptide:MHC complexes specific for a particular T-cell receptor is likely to be very low. A naive CD4 T cell can become activated when only about 10–50 antigenic peptide:MHC complexes are displayed on the surface of the antigen-presenting cell. An effector CD8 cytotoxic T cell is even more sensitive: it can apparently be stimulated to kill by between 1 and 3 peptide:MHC complexes on its target cell. B cells become activated when about 20 B-cell receptors are engaged.

The lymphocyte antigen receptors are tyrosine kinase-associated receptors and, as explained in Section 6-1, most receptors of this type become activated when two or more receptor proteins cluster as a result of ligand binding. In the case of the B-cell receptor, binding of a monovalent antigen to a single receptor complex will not produce a signal. Signaling is initiated only when two or more receptors are linked together, or **cross-linked**, by a multivalent antigen. This was first shown by experiments using specific antibody and antibody fragments as ligands for the receptor (Fig. 6.11). The clustering of B-cell receptors caused by cross-linking promotes the activation of their associated tyrosine kinases and the generation of an intracellular signal.

How antigen binding stimulates T-cell activation is less clear, and a number of mechanisms have been proposed. None of these has been ruled out by experiment, and some aspects of all of them may be involved. Antibodies that bind to T-cell receptors and cross-link them can activate T cells *in vitro*, which suggests that receptor clustering could be a mechanism for activating T cells. However, because antigenic peptides are vastly outnumbered by other peptides displayed on the surface of the antigen-binding cell, cross-linking of

Fig. 6.11 Activation of B cells occurs via cross-linking of the B-cell receptor. As shown in the left panel, Fab fragments of an anti-immunoglobulin can bind to the receptors but cannot cross-link them; they also fail to activate B cells. $F(ab')_2$ fragments of the same anti-immunoglobulin, which have two binding sites, can bridge two receptors (center panel), and thus signal, albeit weakly, to the B cell. The most effective activation occurs when receptors are extensively cross-linked by first adding the $F(ab')_2$ fragments and then rabbit antibody molecules that bind and cross-link the bound $F(ab')_2$ fragments (right panel). In a natural situation, multivalent antigens can lead to extensive receptor cross-linking.

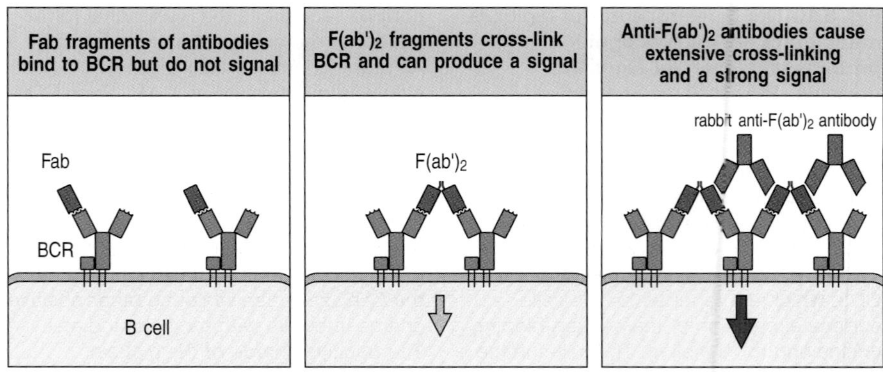

receptors via the dimerization of ligand is unikely. One suggestion is that receptor clustering may not be required; instead, antigen binding induces changes in the conformation of the T-cell receptor or changes in the composition of the signaling complex and it is this that generates the signal (Fig. 6.12).

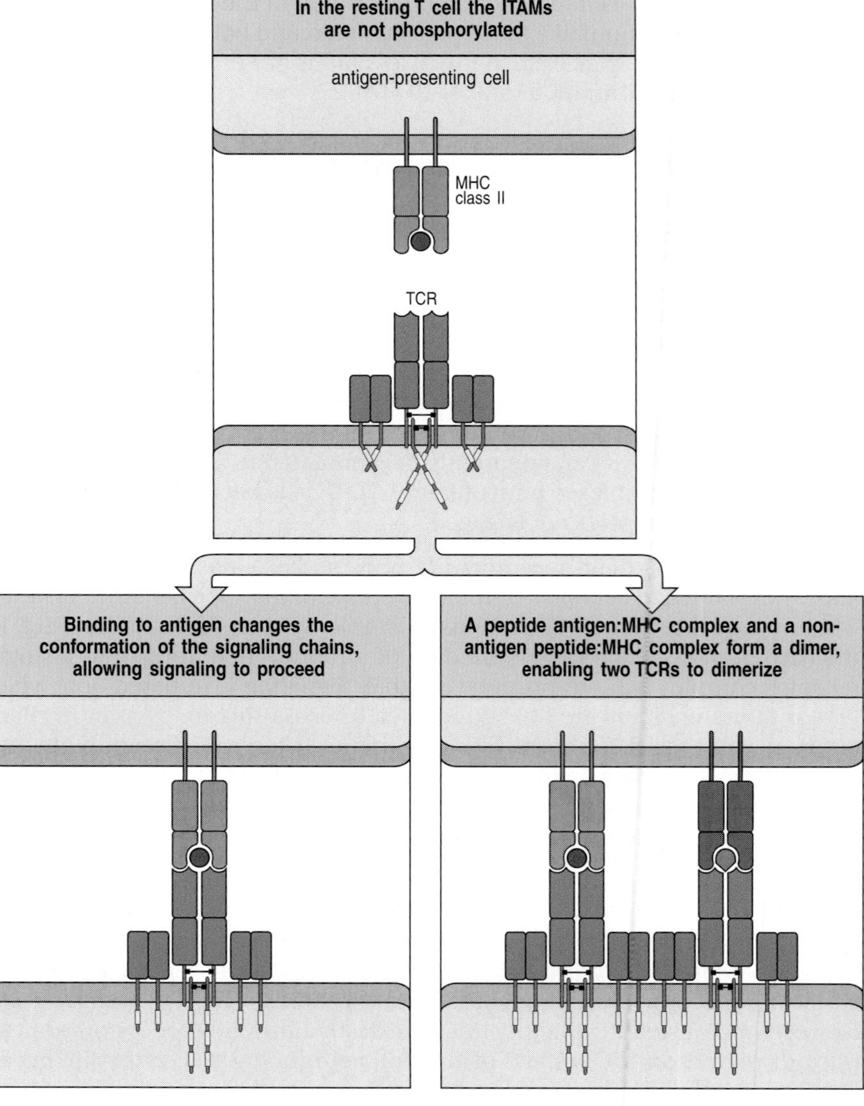

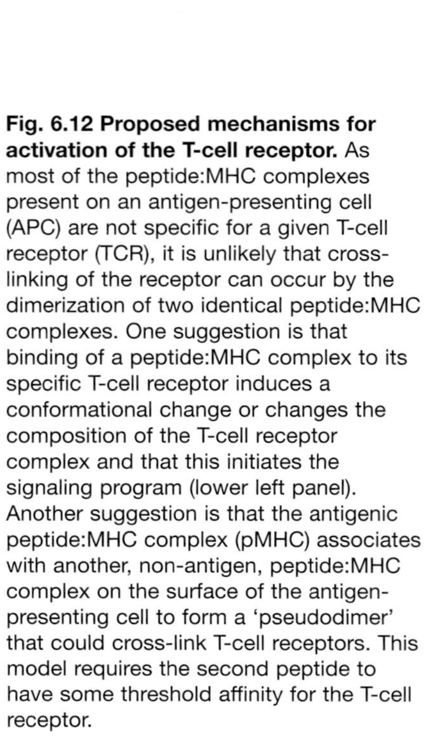

Fig. 6.12 Proposed mechanisms for activation of the T-cell receptor. As most of the peptide:MHC complexes present on an antigen-presenting cell (APC) are not specific for a given T-cell receptor (TCR), it is unlikely that cross-linking of the receptor can occur by the dimerization of two identical peptide:MHC complexes. One suggestion is that binding of a peptide:MHC complex to its specific T-cell receptor induces a conformational change or changes the composition of the T-cell receptor complex and that this initiates the signaling program (lower left panel). Another suggestion is that the antigenic peptide:MHC complex (pMHC) associates with another, non-antigen, peptide:MHC complex on the surface of the antigen-presenting cell to form a 'pseudodimer' that could cross-link T-cell receptors. This model requires the second peptide to have some threshold affinity for the T-cell receptor.

Other proposals do involve clustering. For example, a second hypothesis is that signaling is initiated by T-cell receptor dimerization through the recognition of 'pseudo-dimeric' peptide:MHC complexes containing one antigen peptide:MHC complex and one self-peptide:MHC complex on the surface of the antigen-presenting cell (see Fig. 6.12).

A third suggestion is that receptor activation is promoted by the formation of the **immunological synapse**. This structure forms around the site of contact between a T cell and its antigen-presenting cell as a consequence of a reorganization of T-cell membrane proteins (Fig. 6.13). T-cell receptors and associated co-receptor and signaling proteins are concentrated at the contact site, while proteins that inhibit signaling, such as tyrosine phosphatases, are excluded. In some cases, the contact surface organizes into two zones: a central zone known as the central supramolecular activation complex (**c-SMAC**) and an outer zone known as the peripheral supramolecular activation complex (**p-SMAC**). The c-SMAC contains most of the signaling proteins known to be important in T-cell activation. The p-SMAC is notable mainly for the presence of the integrin LFA-1 and the cytoskeletal protein talin. The function of the immunological synapse is currently the focus of much research, but it is thought to have an important role in regulating signaling. As we shall see in Chapter 8, it is also involved in the directed secretion of cytokines and cytotoxins by effector T cells in contact with their target cells.

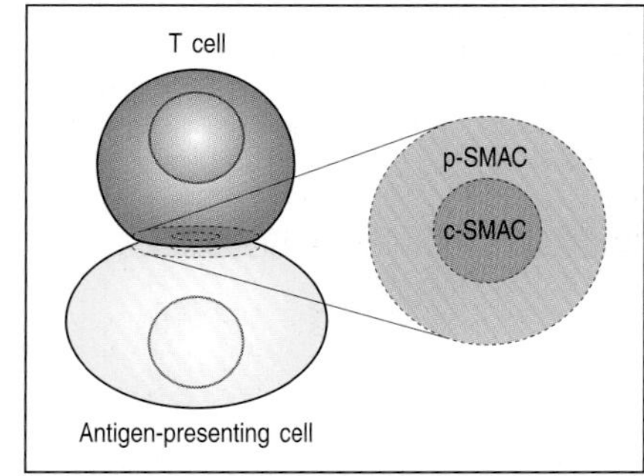

c-SMAC	p-SMAC
TCR CD2 CD4 CD8 CD28 PKC-θ	LFA-1 ICAM-1 talin

Fig. 6.13 Proteins in the contact area between the T cell and the antigen-presenting cell form a structure called an immunological synapse. The center of the contact area is enriched in T-cell receptors, the co-receptors CD4 and CD8, the co-stimulatory receptor CD28, the adhesion molecule CD2, and the signaling protein kinase PKC-θ (see Section 6-16). This zone is called the central supramolecular activation complex (c-SMAC). Outside the c-SMAC is a zone that is enriched in the integrin LFA-1, the cell-adhesion molecule ICAM-1, and the cytoskeletal protein talin and is called the peripheral supramolecular activation complex (p-SMAC).

6-10 Antigen binding leads to phosphorylation of the ITAM sequences associated with the antigen receptors.

Phosphorylation of both tyrosines in the ITAMs serves as the first intracellular signal that the lymphocyte has detected its specific antigen. As the signaling pathways are very similar, we will focus first on the signals transduced by the T-cell receptor and follow this signaling pathway into the nucleus. We will then return to the B-cell receptor.

In T cells, two protein tyrosine kinases of the Src family – Lck and Fyn – are thought to be responsible for phosphorylation of the ITAMs in the T-cell receptor (Fig. 6.14). **Lck** is mostly constitutively associated with the cytoplasmic domain of the co-receptor molecules CD4 and CD8 (see Section 3-17), and **Fyn** associates weakly with the cytoplasmic domains of the ζ and CD3 chains. It is still not clear how antigen recognition actually stimulates the ability of Fyn and Lck to phosphorylate the ITAMs, but it is likely to involve some type of receptor clustering event (see Section 6-9).

Optimal signaling through the T-cell receptor complex occurs when it is associated with the co-receptors CD4 or CD8. CD4 binds to MHC class II molecules and thus clusters with T-cell receptors that recognize peptide:MHC class II ligands (see Section 3-17). Similarly, CD8 binds to MHC class I molecules and thus clusters with T-cell receptors restricted to MHC class I. Association of the T-cell receptor with the appropriate co-receptor helps to stimulate signal transduction by bringing the Lck tyrosine kinase associated with the co-receptor together with the ITAMs and other targets associated with the cytoplasmic domains of the T-cell receptor complex (see Fig. 6.14). Co-receptors are also thought to stabilize the low-affinity interaction between the T-cell receptor and an MHC molecule.

The activation of the Src-family kinases is the first step in a signaling pathway that passes the signal on to many different molecules. Like many other signaling proteins, Src-family kinases are associated with the inner leaflet of the plasma membrane, which facilitates their association with receptors. Src kinases are targeted to the membrane by posttranslational attachment of myristate; some Src kinases are additionally modified with palmitate, which targets them to lipid rafts (see Section 6-6).

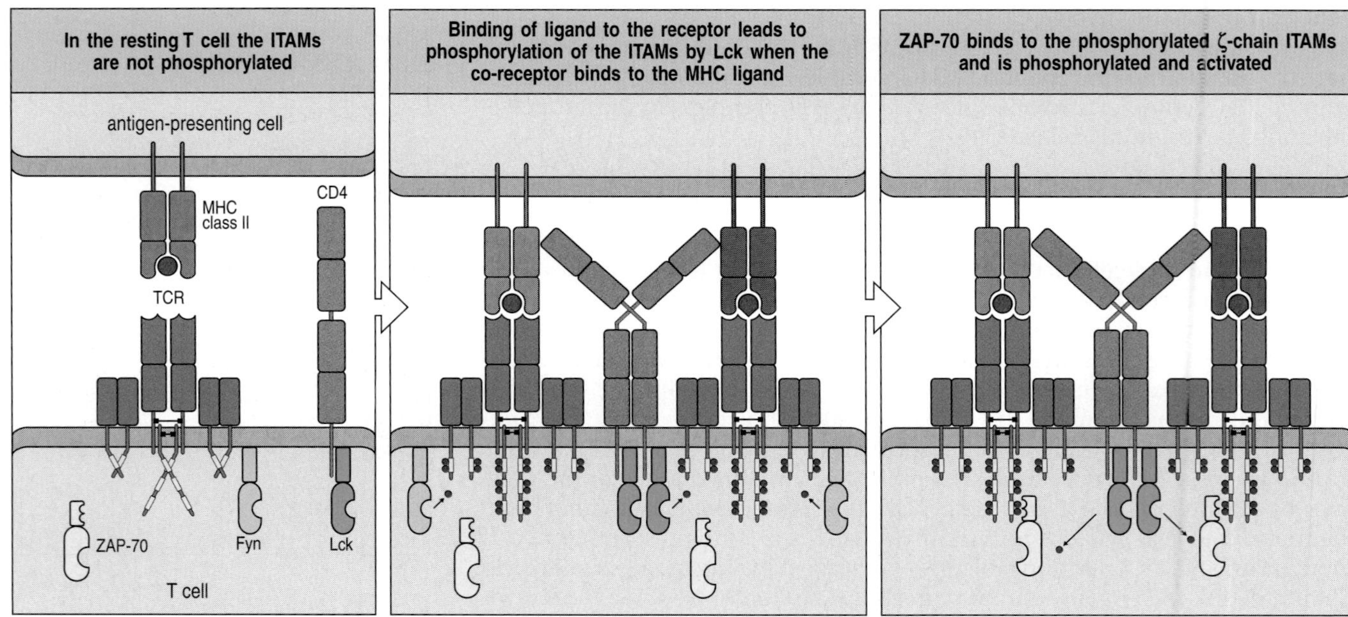
In the resting T cell the ITAMs are not phosphorylated
antigen-presenting cell
MHC class II
CD4
TCR
ZAP-70
Fyn
Lck
T cell
Binding of ligand to the receptor leads to phosphorylation of the ITAMs by Lck when the co-receptor binds to the MHC ligand
ZAP-70 binds to the phosphorylated ζ-chain ITAMs and is phosphorylated and activated

Fig. 6.14 Clustering of co-receptors with the TCR can enhance phosphorylation of the T-cell receptor. When T-cell receptors and co-receptors are brought together by binding peptide:MHC complexes on the surface of an antigen-presenting cell, recruitment of the co-receptor-associated kinase Lck and activation of receptor-associated kinases such as Fyn lead to phosphorylation of the CD3γ, δ, and ε ITAMs as well as those on the ζ chain (first and second panels). The tyrosine kinase ZAP-70 binds to the phosphorylated ITAMs of the ζ chain and is subsequently phosphorylated and activated by Lck (third panel). The crystal structure of CD4 suggests that when a single CD4 binds to a peptide:MHC complex, the Lck associated with the cytoplasmic domain is too far away to phosphorylate the T-cell receptor bound to the same MHC molecule (note that CD4 bends to contact an MHC molecule). This supports the idea that clustering of T-cell receptors and CD4 molecules is required to enable Lck to phosphorylate a neighboring T-cell receptor in the cluster.

Src-family kinases have an SH3 domain and an SH2 domain preceding the kinase domain, and are kept inactive by intramolecular interactions between these domains and the rest of the protein, which depend on phosphorylation of an inhibitory tyrosine at the carboxy terminus of the protein and interaction of the SH3 domains with a linker domain between the SH2 and kinase domains (Fig. 6.15). A protein tyrosine kinase called **C-terminal Src kinase** (**Csk**) phosphorylates the inhibitory tyrosine. Dephosphorylation of the carboxy-terminal tyrosine or engagement of the SH2 or SH3 domains with binding ligands releases the kinase from its inactive conformation. Activation is further stimulated by phosphorylation of the kinase on a tyrosine in the catalytic domain. In

Fig. 6.15 General scheme for the activation of Src kinases. Src kinases contain SH3 (blue) and SH2 (red) domains preceding the kinase domain (green). In the inactive state, the kinase domain is tethered by interactions with both the SH2 and SH3 domains, which constrains the mobility of the two lobes of the kinase domain. The SH2 domain interacts with a phosphorylated tyrosine at the carboxyl terminus of the kinase domain. The SH3 domain interacts with a proline sequence (P) contained in a linker sequence between the SH2 domain and the kinase domain (colored line). This tethers the SH3 domain against the upper lobe of the kinase domain. Release of either the SH2 or SH3 domain can activate the kinase activity. Dephosphorylation of the carboxyl terminal tyrosine by the phosphatase CD45 results in release of the SH2 domain and kinase activation. Binding of a ligand to the SH3 would cause the release of the SH3 domain from the kinase, resulting in kinase activation. Rephosphorylation of the carboxyl terminal tyrosine by the C-terminal Src kinase (CSK) or loss of the SH3 ligand returns the kinase to the inactive state.

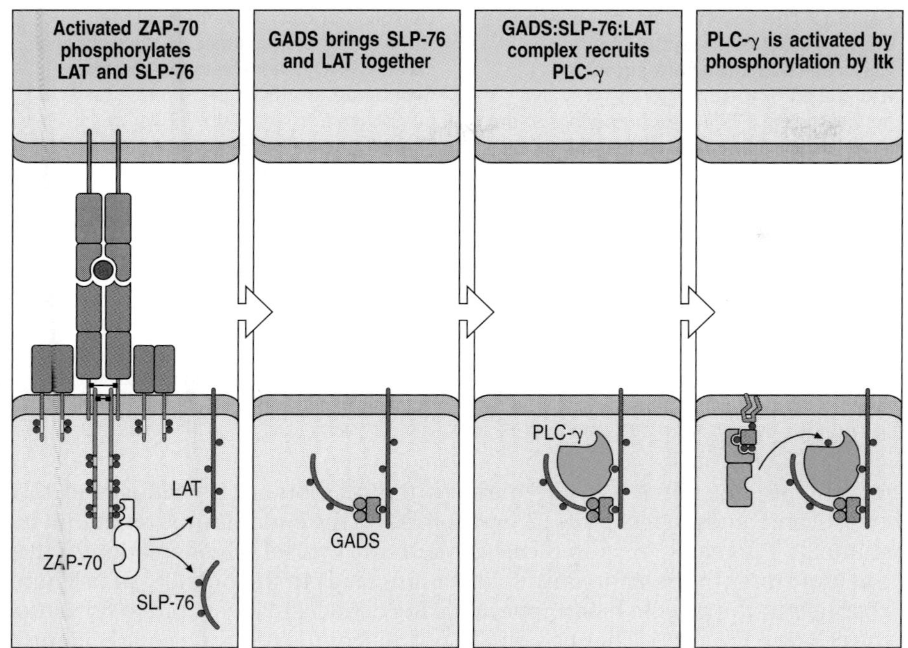
Activated ZAP-70 phosphorylates LAT and SLP-76
GADS brings SLP-76 and LAT together
GADS:SLP-76:LAT complex recruits PLC-γ
PLC-γ is activated by phosphorylation by Itk
LAT
ZAP-70
SLP-76
GADS
PLC-γ

Fig. 6.16 The recruitment and activation of phospholipase C-γ by LAT and SLP-76 is a crucial step in T-cell activation. ZAP-70 phosphorylates and recruits the scaffold proteins LAT and SLP-76 to the activated receptor complex. An adaptor, GADS, holds tyrosine-phosphorylated LAT and SLP-76 together. Phospholipase C-γ (PLC-γ) binds to phosphorylated sites in both LAT and SLP-76. Activation of PLC-γ requires phosphorylation by one of the Tec family kinases, Itk, which is recruited to the membrane by the production of PIP_3, a product of activated PI 3-kinase, and by interactions of Itk with phosphorylated SLP-76. Once phosphorylated by Itk, phospholipase C-γ is active.

lymphocytes, the tyrosine phosphatase CD45, which can dephosphorylate both tyrosine phosphorylation sites, has an important role in maintaining Src kinases in a partially active, dephosphorylated state.

6-11 In T cells, fully phosphorylated ITAMs bind the kinase ZAP-70 and enable it to be activated.

The phosphorylated YXXL/I motif is a binding site for an SH2 domain (see Fig. 6.2) and the precise spacing of the two motifs in an ITAM suggests that this is a binding site for a signaling protein with two SH2 domains. In T cells, this is the tyrosine kinase **ZAP-70** (**ζ-chain-associated protein**), which is responsible for further signaling. ZAP-70 has two tandem SH2 domains that can be simultaneously engaged by both phosphorylated tyrosines in the ITAM. The affinity of the phosphorylated YXXL sequence for a single SH2 domain is low; binding of both SH2 domains to the doubly phosphorylated ITAM is significantly stronger and confers specificity on ZAP-70 binding. Once recruited to the phosphorylated receptor ZAP-70 is phosphorylated and activated by the co-receptor-associated Src kinase Lck (see Fig. 6.14).

6-12 Activated ZAP-70 phosphorylates scaffold proteins that mediate many of the downstream effects of antigen receptor signaling.

Once activated, ZAP-70 phosphorylates the scaffold proteins **LAT** (**linker of activated T cells**) and **SLP-76**. LAT and SLP-76 seem to function together, as they can be linked by the adaptor protein GADS. This seems to be important for their function, as mice lacking GADS have defects in T-cell activation. LAT is a transmembrane protein, which facilitates its interaction with ZAP-70, and it is posttranslationally modified by palmitate, which promotes its interaction with lipid rafts (see Section 6-6).

Phospholipase C-γ (**PLC-γ**) is one of the key signaling molecules recruited by the phosphorylation of LAT and SLP-76 (Fig. 6.16). PLC-γ catalyzes the breakdown of the membrane lipid PIP_2 (see Section 6-5) to generate two breakdown products, the second messenger **inositol 1,4,5-trisphosphate** (**IP_3**) and the membrane lipid **diacylglycerol** (**DAG**) (Fig. 6.17). DAG stays confined to the

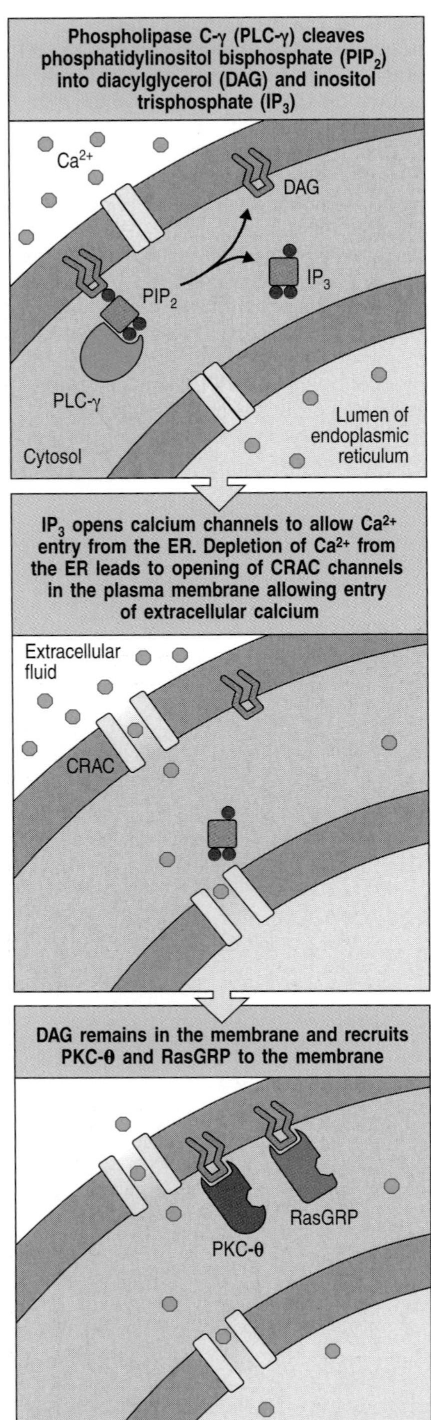

Phospholipase C-γ (PLC-γ) cleaves phosphatidylinositol bisphosphate (PIP2) into diacylglycerol (DAG) and inositol trisphosphate (IP3)
Ca2+
DAG
PIP2
IP3
PLC-γ
Lumen of endoplasmic reticulum
Cytosol
IP3 opens calcium channels to allow Ca2+ entry from the ER. Depletion of Ca2+ from the ER leads to opening of CRAC channels in the plasma membrane allowing entry of extracellular calcium
Extracellular fluid
CRAC
DAG remains in the membrane and recruits PKC-θ and RasGRP to the membrane
RasGRP
PKC-θ

Fig. 6.17 The enzyme phospholipase C-γ cleaves inositol phospholipids to generate two important signaling molecules. Phosphatidylinositol bisphosphate (PIP_2) is a component of the inner leaflet of the plasma membrane. When PLC-γ is activated by phosphorylation, it cleaves PIP_2 into two parts, inositol trisphosphate (IP_3), which diffuses away from the membrane into the cytosol, and diacylglycerol (DAG), which stays in the membrane. Both these molecules are important in signaling. IP_3 binds to a receptor in the endoplasmic reticulum (ER) membrane, opening calcium channels and allowing calcium ions (Ca^{2+}) to enter the cytosol from stores in the ER. Depletion of calcium in the ER then stimulates the opening of calcium channels, called CRAC channels, in the plasma membrane, enabling calcium to enter the cytoplasm from the extracellular space. Thus, there are two phases of calcium release, an early phase from intracellular stores and a later phase from outside the cell. DAG binds and recruits signaling proteins to the membrane, most importantly the Ras-GEF called RasGRP and a serine/threonine kinase called protein kinase C-θ (PKC-θ). Recruitment of RasGRP to the plasma membrane activates Ras, and PKC-θ activation results in the activation of the transcription factor NFκB.

membrane, but diffuses in the plane of the membrane. IP_3 diffuses in the cytosol and binds to receptors (IP_3 receptors) on the endoplasmic reticulum to stimulate the release of stored calcium into the cytosol. The depletion of the calcium stores in the endoplasmic reticulum leads to the opening of calcium channels in the plasma membrane, allowing extracellular calcium to flow into the cell (see Fig. 6.17). These channels, not yet completely molecularly identified, are known as **CRAC channels** (***c**alcium **r**elease-**a**ctivated **c**alcium channels*). The gene product of *ORAI1*, which is mutated in some cases of severe combined immunodeficiency, was recently shown to form at least part of the CRAC channel.

The activation of PLC-γ marks an important step, because after this point the antigen signaling pathway splits into three distinct branches, each of which ends in the activation of a different transcription factor. These signaling pathways are not exclusive to lymphocytes but are versions of pathways used in many different cell types. The signaling pathways from the T-cell receptor are summarized in Fig. 6.18. The combined actions of calcium and DAG activate these three signaling pathways. The importance of their actions is shown by the observation that treatment of T cells with both phorbol myristate acetate (an analog of DAG) and ionomycin (a pore-forming drug that allows extracellular calcium to flow into the cell) can largely reconstitute the effects of T-cell activation. Not surprisingly for such a central step in the antigen signaling pathway, the activation of PLC-γ is under a complex set of controls and we shall consider these first before returning to the final stages of the pathways.

6-13 PLC-γ is activated by Tec tyrosine kinases.

PLC-γ is recruited to the membrane by binding to the phosphorylated scaffolds LAT and SLP-76 (see Fig. 6.16), but this does not activate its catalytic activity. Activation requires phosphorylation by a member of the **Tec family** of cytoplasmic tyrosine kinases. Three Tec kinases are expressed in lymphoid cells: Tec, Itk and Bruton's tyrosine kinase (Btk). Itk is the family member expressed mainly in T lymphocytes. Itk is recruited to the receptor-based signaling complex, where it is phosphorylated and activated by Lck. Tec kinases contain PH, SH2, and SH3 domains and are recruited to the plasma membrane by their PH domain, which interacts with PIP_3 on the inner face of the cell membrane (see Fig. 6.16). PIP_3 is generated by activation of PI 3-kinase, and although it is not exactly clear how the T-cell receptor activates PI 3-kinase, an important PI 3-kinase activator in this context is the co-stimulatory receptor CD28, which we discuss later. Itk is also recruited to the phosphorylated scaffolds by its SH2 and SH3 domains. Thus, the coordinated activation of PI 3-kinase and tyrosine phosphorylation of the scaffold is required to recruit Itk to the plasma membrane, where it can be

X-linked Severe Combined Immunodeficiency

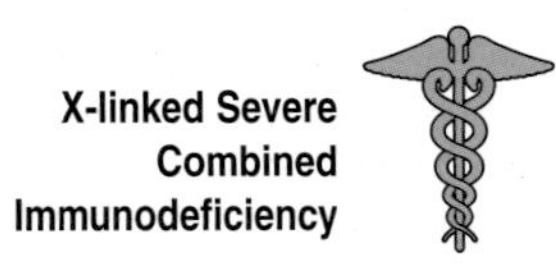

phosphorylated by Lck. Once activated, Tec kinases phosphorylate and then activate PLC-γ.

6-14 Activation of the small G protein Ras activates a MAP kinase cascade, resulting in the production of the transcription factor AP-1.

The DAG generated by PLC-γ diffuses in the plasma membrane, where it activates a variety of proteins that can bind to DAG. The most important of

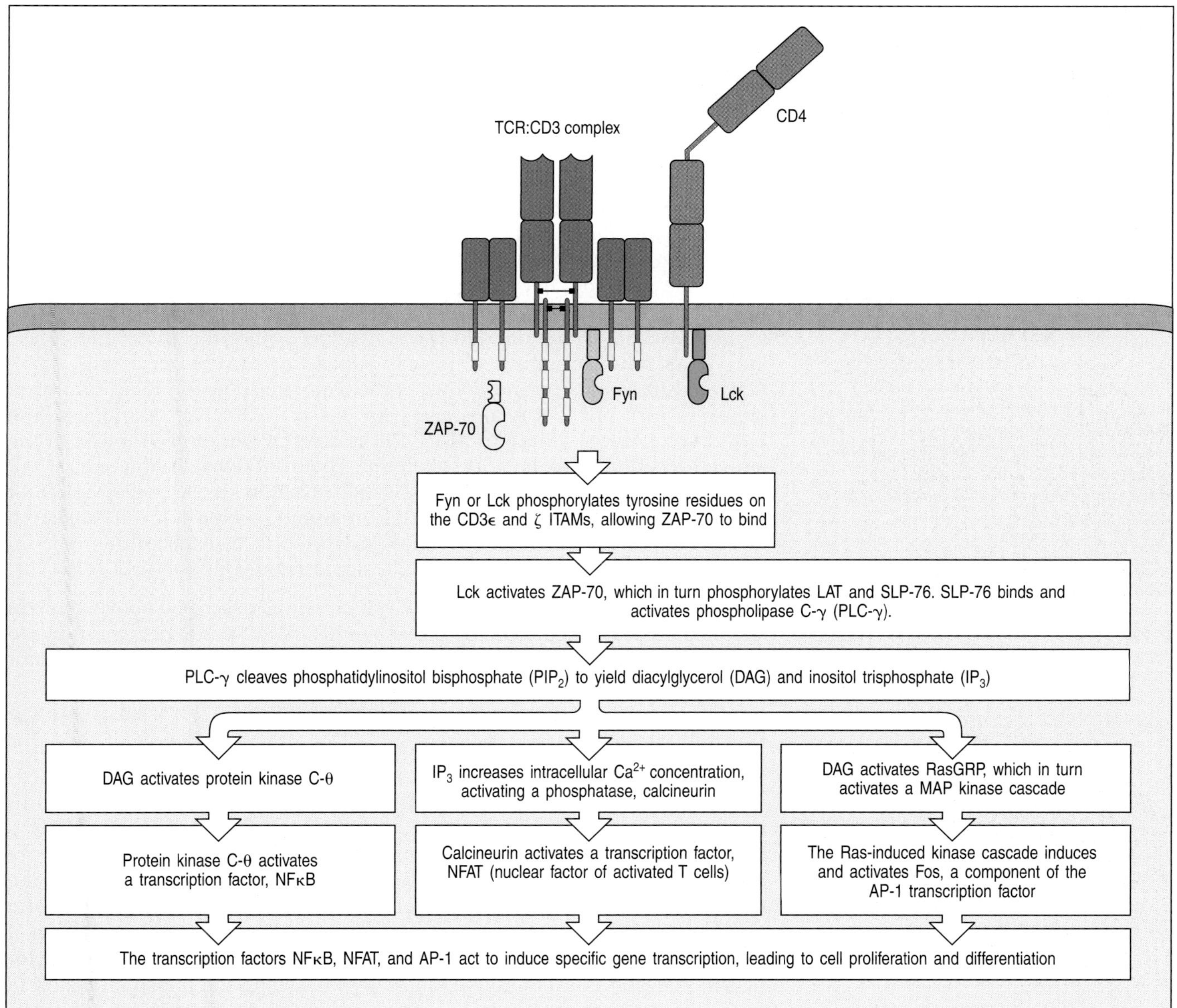

CD4
TCR:CD3 complex
Fyn
Lck
ZAP-70
Fyn or Lck phosphorylates tyrosine residues on the CD3ε and ζ ITAMs, allowing ZAP-70 to bind
Lck activates ZAP-70, which in turn phosphorylates LAT and SLP-76. SLP-76 binds and activates phospholipase C-γ (PLC-γ).
PLC-γ cleaves phosphatidylinositol bisphosphate (PIP_2) to yield diacylglycerol (DAG) and inositol trisphosphate (IP_3)
DAG activates protein kinase C-θ
IP_3 increases intracellular Ca^{2+} concentration, activating a phosphatase, calcineurin
DAG activates RasGRP, which in turn activates a MAP kinase cascade
Protein kinase C-θ activates a transcription factor, NFκB
Calcineurin activates a transcription factor, NFAT (nuclear factor of activated T cells)
The Ras-induced kinase cascade induces and activates Fos, a component of the AP-1 transcription factor
The transcription factors NFκB, NFAT, and AP-1 act to induce specific gene transcription, leading to cell proliferation and differentiation

Fig. 6.18 Simplified outline of the intracellular signaling pathways initiated by the T-cell receptor complex and its co-receptor. The T-cell receptor complex and co-receptor (in this example the CD4 molecule) are associated with Src-family protein kinases Fyn and Lck, respectively. It is thought that binding of a peptide:MHC ligand to the T-cell receptor and co-receptor and clustering of T-cell receptors and CD4 molecules brings together CD4 with the T-cell receptor complex. Phosphorylation of the ITAMs in CD3ε, γ, and δ and the ζ chain enables them to bind the cytosolic tyrosine kinase ZAP-70. ZAP-70 recruited to the T-cell receptor complex is phosphorylated and activated by Lck. Activated ZAP-70 phosphorylates the adaptor proteins LAT and SLP-76, which in turn leads to membrane recruitment of PLC-γ and its phosphorylation and activation by Tec kinases. Activated PLC-γ initiates three important signaling pathways that culminate in the activation of transcription factors in the nucleus. Together, NFκB, NFAT, and AP-1 act in the nucleus to initiate gene transcription that results in the differentiation, proliferation, and effector actions of T cells. This diagram is a highly simplified version of the pathways, showing the main events only.

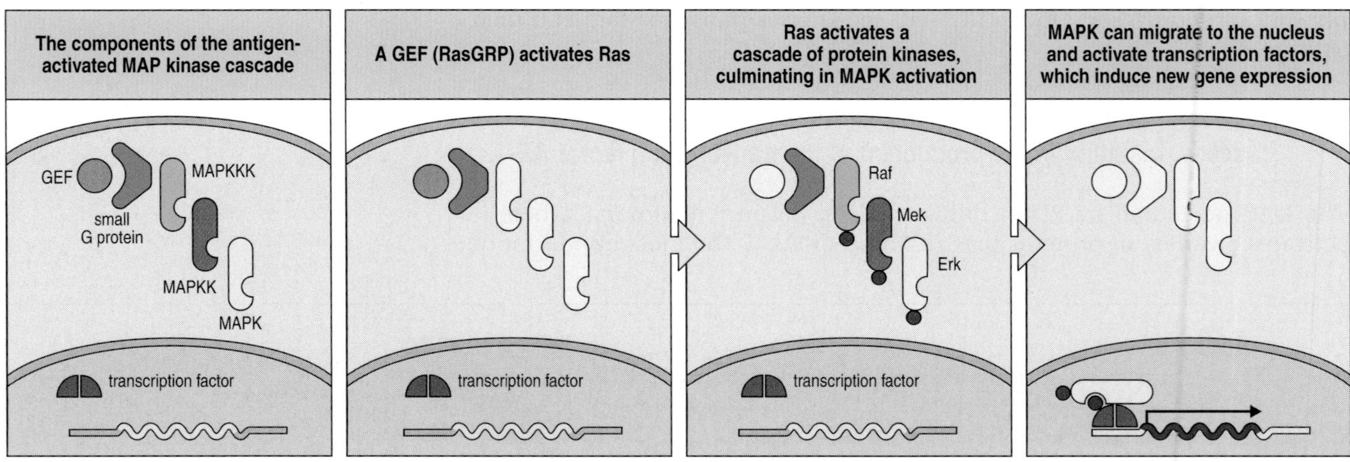

Fig. 6.19 MAP kinase cascades activate transcription factors. All MAP kinase cascades share the same general features. They are initiated by a small G protein, which is switched from an inactive state to an active state by a guanine-nucleotide exchange factor (GEF). The small G protein activates the first enzyme of the cascade, a protein kinase called a MAP kinase kinase kinase (MAPKKK), which phosphorylates a second kinase called a MAP kinase kinase (MAPKK), which in turn phosphorylates and activates a MAP kinase (MAPK) (first panel). In the example shown in the three right-hand panels, the GEF RasGRP activates Ras, leading to the sequential activation of the kinases Raf, Mek, and Erk. Phosphorylation and activation of Erk releases it from the complex so that it can diffuse within the cell and enter the nucleus. Phosphorylation of transcription factors by Erk results in new gene transcription.

these in regard to antigen signaling are the serine/threonine kinase **protein kinase C** and the protein **RasGRP**, which is a GTP-exchange factor that specifically activates the small G protein Ras (see Section 6-4). We will look first at the pathway that starts with the activation of RasGRP. This activates Ras, which then triggers a three-kinase relay system often called the **MAP kinase cascade**, which ends in the activation of a serine/threonine kinase known as mitogen-activated protein kinase or **MAP kinase** (Fig. 6.19). Activated Ras binds to and activates the first kinase in the relay, and each kinase in turn phosphorylates and activates the next. The first kinase (the MAP kinase kinase kinase or MAPKKK) is a serine/threonine kinase; in the antigen receptor pathway it is called Raf. The next kinase in the relay (MAP kinase kinase or MAPKK) is a dual-specificity protein kinase called MEK, that phosphorylates both a tyrosine and a threonine residue on MAP kinase to activate it. The particular MAP kinase activated as a result of this relay in B cells and T cells is called extracellular signal-related kinase (Erk).

As well as being activated via the PLC-γ pathway as described above, Ras can also be activated via another GTP-exchange factor, SOS. SOS is recruited to the signaling complex around the activated antigen receptor by the adaptor protein Grb2, which binds to the phosphorylated scaffold formed by LAT/SLP-76 in T cells or by the functionally analogous B-cell linker protein (BLNK) in B cells.

One of the most important functions of Ras–MAP kinase activation is the activation of transcription factors and new gene expression. Erk activation promotes the formation of the transcriptional regulator **AP-1**, which is a heterodimer composed of one monomer each from the Fos and Jun families of transcription factors (Fig. 6.20). Active Erk stimulates Fos transcription via the phosphorylation of the transcription factor Elk-1, which cooperates with another transcription factor, serum response factor, to initiate transcription of the *fos* gene. The Jun transcription factor is constitutively present in the cytoplasm. Activation of the protein kinase JNK results in the phosphorylation of Jun and its translocation into the nucleus, where it combines with Fos to form AP-1. Details of how JNK is activated by T-cell signaling are not currently known.

6-15 The transcription factor NFAT is indirectly activated by Ca^{2+}.

We shall now look at the signaling pathways initiated by the rise in the concentration of free Ca^{2+} in the cytosol (see Section 6-12). Ca^{2+} indirectly activates a transcription factor called **NFAT** (**nuclear factor of activated T cells**). This is something of a misnomer, because NFAT transcription factors are

Fig. 6.20 The transcription factor AP-1 is formed as a result of the Ras/MAP kinase signaling pathway. Phosphorylation of the MAP kinase Erk activated as a result of the Ras–MAP kinase cascade allows Erk to enter the nucleus, where it phosphorylates the transcription factor Elk-1, which binds to the serum response element (SRE) in the promoter of the gene for the transcription factor c-Fos, stimulating its transcription. At the same time, phosphorylation of another MAP kinase, Jun kinase (JNK), enables it to phosphorylate the transcription factor c-Jun, which is constitutively present in the cytoplasm. Phosphorylated c-Jun then enters the nucleus, where it dimerizes with c-Fos to make AP-1.

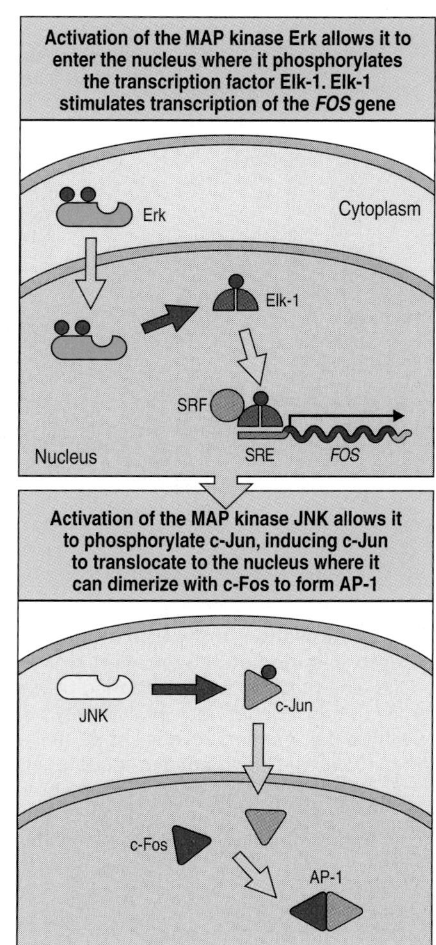

expressed ubiquitously. NFAT is present in the cytoplasm of resting cells, and in the absence of signals it is kept there by phosphorylation by serine/threonine kinases, including glycogen synthase kinase 3 (GSK3) and casein kinase 2 (CK2). The phosphorylation blocks recognition of the nuclear localization sequence of NFAT, thus preventing its entry into the nucleus (Fig. 6.21).

NFAT is released from the cytosol by the action of the enzyme **calcineurin**, a serine/threonine protein phosphatase that is activated by the increase in intracellular free Ca^{2+} that accompanies lymphocyte activation. The binding of Ca^{2+} to a protein called **calmodulin** causes a change in conformation that allows calmodulin to bind to and activate a wide variety of enzymes (see Fig. 6.21). One of these is calcineurin. Dephosphorylation of NFAT by calcineurin allows the nuclear localization sequence to be recognized, and NFAT enters the nucleus (see Fig 6.18).

The importance of NFAT in T-cell activation is illustrated by the effects of selective inhibitors of calcineurin called cyclosporin A and FK506 (tacrolimus). By inhibiting calcineurin these drugs prevent the formation of active NFAT. T cells express low levels of calcineurin, so they are more sensitive to inhibition of this pathway than are many other cell types. Both cyclosporin A and FK506 thus act as effective immunosuppressants with only limited side-effects. These drugs are widely used to prevent the rejection of organ transplants, and are discussed further in Chapter 14.

6-16 The transcription factor NFκB is activated by the actions of protein kinase C.

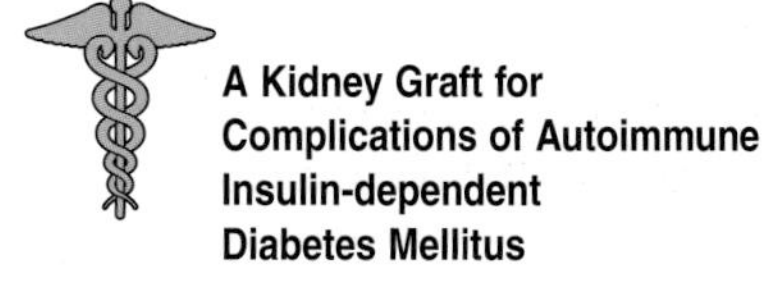

A Kidney Graft for Complications of Autoimmune Insulin-dependent Diabetes Mellitus

The third downstream signaling pathway leading from PLC-γ results in the activation of a specific isoform of protein kinase C, PKC-θ, by the combined

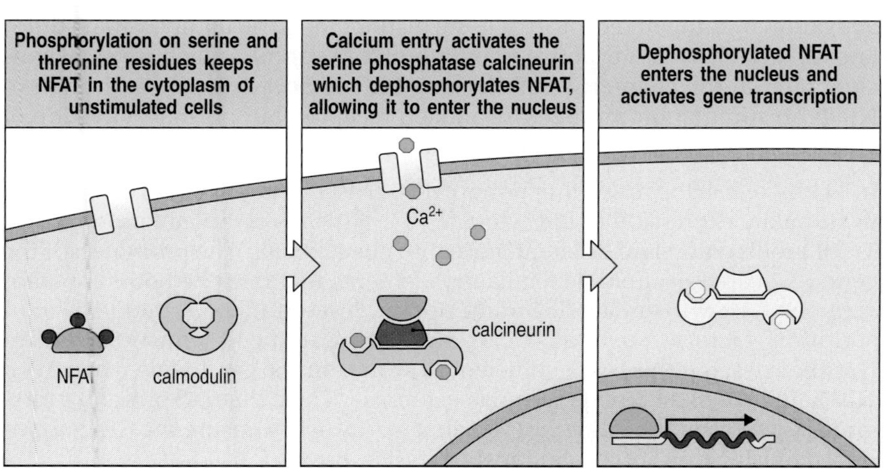

Phosphorylation on serine and threonine residues keeps NFAT in the cytoplasm of unstimulated cells
NFAT
calmodulin
Calcium entry activates the serine phosphatase calcineurin which dephosphorylates NFAT, allowing it to enter the nucleus
Ca^{2+}
calcineurin
Dephosphorylated NFAT enters the nucleus and activates gene transcription

Fig. 6.21 The transcription factor NFAT is regulated by calcium signaling. NFAT is maintained in the cytoplasm by phosphorylation on serine and threonine. Calcium entering the cell binds to calmodulin, and the Ca^{2+}:calmodulin complex binds to the serine/threonine phosphatase calcineurin, activating it. Calcineurin then dephosphorylates NFAT, allowing NFAT to translocate into the nucleus. In the nucleus, NFAT binds to promoter elements and activates the transcription of various genes.

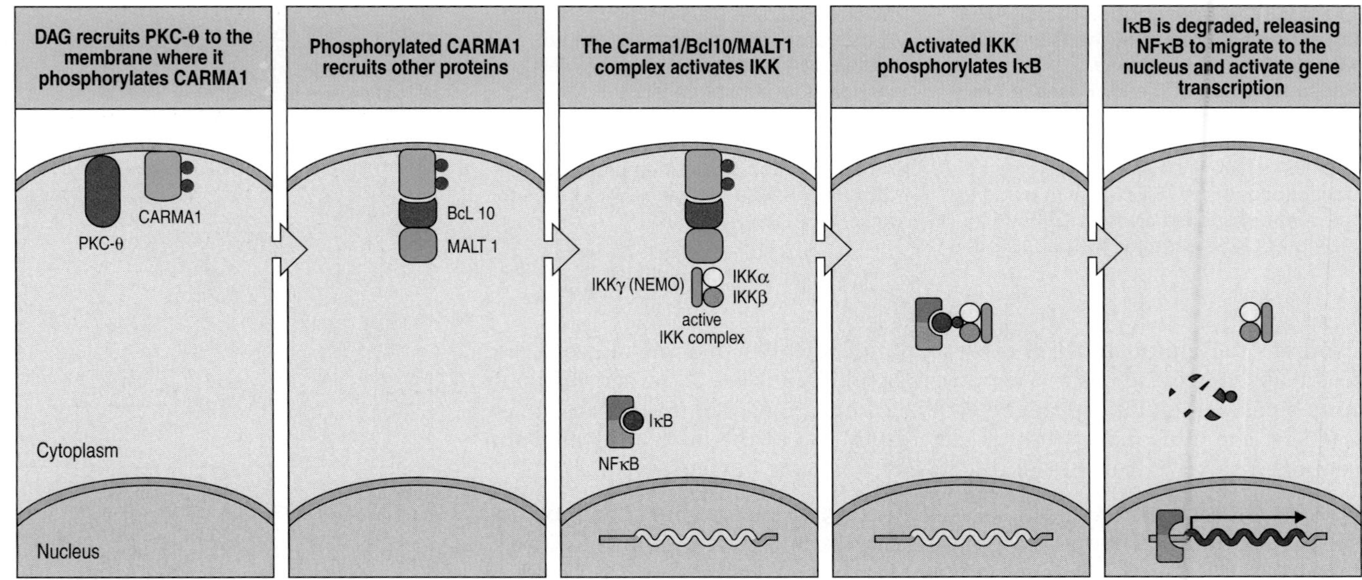

DAG recruits PKC-θ to the membrane where it phosphorylates CARMA1
Phosphorylated CARMA1 recruits other proteins
The Carma1/Bcl10/MALT1 complex activates IKK
Activated IKK phosphorylates IκB
IκB is degraded, releasing NFκB to migrate to the nucleus and activate gene transcription
CARMA1
PKC-θ
BcL 10
MALT 1
IKKγ (NEMO)
IKKα
IKKβ
active
IKK complex
IκB
NFκB
Cytoplasm
Nucleus

Fig 6.22 Activation of the transcription factor NFκB by antigen receptors is mediated by protein kinase C. NFκB exists in an unstimulated cell as a dimer formed of two members of the Rel family of transcription factors, typically p65Rel and p50Rel, bound to a third component, inhibitor of κB (IκB), which keeps NFκB in the cytoplasm. During antigen-receptor signaling, the production of diacylglycerol (DAG) results in membrane recruitment and activation of protein kinase C (PKC-θ). This phosphorylates a scaffold protein called CARMA1, which binds other proteins (Bcl10, MALT1) to form a membrane-associated complex that recruits and activates the serine/threonine kinase IκB kinase (IKK) complex (IKKα:IKKβ:IKKγ (NEMO)). This phosphorylates IκB, stimulating its ubiquitination, which targets IκB for degradation in the proteasome. Released from IκB, NFκB is now able to translocate to the nucleus to stimulate transcription of its target genes. A defect in NEMO that prevents NFκB activation causes immunodeficiency, among other symptoms.

X-linked Hypohydrotic Ectodermal Dysplasia and Immunodeficiency

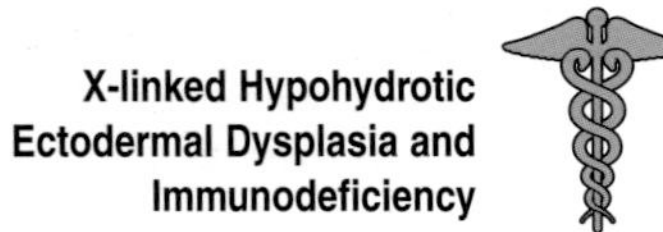

actions of DAG and Ca^{2+}. This in turn results in the transcription factor NFκB being released from its inhibitor in the cytoplasm and entering the nucleus. NFκB is the general name for a member of a family of homo- and heterodimeric transcription factors made up of the Rel family of proteins. The most common NFκB activated in lymphocytes is a heterodimer of p50:p65Rel. The dimer is held in an inactive state in the cytoplasm by binding to an inhibitory protein called inhibitor of κB (IκB) (Fig. 6.22). Activation of a complex of serine kinases, IκB kinase (IKK), results in the phosphorylation, ubiquitination, and subsequent degradation of IκB, with the consequent release of NFκB. This can then enter the nucleus. Note that the pathway of activation by antigen receptors is quite distinct from the pathway that stimulates NFκB release in response to inflammatory stimuli, which we shall consider later in the chapter: T cells lacking PKC-θ show defective activation of NFκB on stimulation through the antigen receptor, but normal activation of NFκB in response to inflammatory stimuli.

In T cells, one of the major functions of AP-1, NFAT, and NFκB is to act together to stimulate expression of the cytokine IL-2, which is essential for promoting T-cell proliferation and differentiation into effector cells. The promoter for the gene *IL-2* contains multiple regulatory elements that must be bound by transcription factors to initiate *IL-2* transcription. Some are already bound by transcription factors, such as Oct1, that are produced constitutively in lymphocytes, but this is not sufficient to switch on the gene. Only when AP-1, NFAT, and NFκB all bind is the gene expressed. Thus, the *IL-2* promoter integrates signals from the different signaling pathways to ensure that IL-2 is produced only in appropriate circumstances (Fig. 6.23).

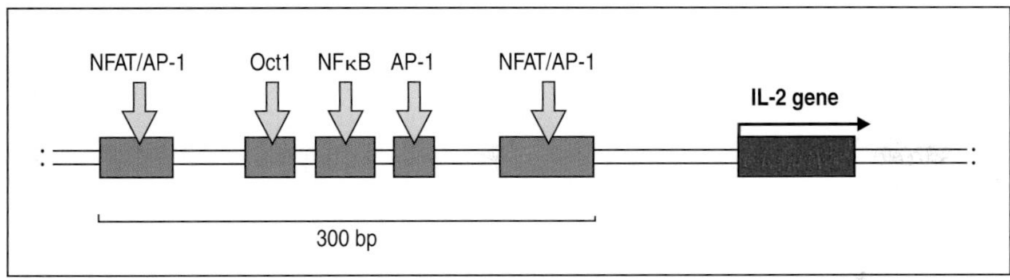
NFAT/AP-1
Oct1
NFκB
AP-1
NFAT/AP-1
IL-2 gene
300 bp

Fig. 6.23 Multiple signaling pathways converge on the IL-2 promoter. AP-1, NFAT and NFκB binding to the IL-2 promoter together integrate multiple signaling pathways into a single output, the production of IL-2. MAP kinase activates AP-1; calcium activates NFAT; protein kinase C activates NFκB. All three pathways are required to stimulate IL-2 transcription. Both NFAT and AP-1 must bind to one type of promoter element. Oct1 is a transcription factor that is required for *IL-2* transcription. Unlike the other transcription factors, it is constitutively bound to the promoter and is therefore not regulated by TCR signaling.

6-17 The logic of B-cell receptor signaling is similar to that of T-cell receptor signaling but some of the signaling components are specific to B cells.

There are many similarities between signaling from T-cell receptors and B-cell receptors. As with the T-cell receptor, the antigen-specific chains of the B-cell receptor are associated with ITAM-containing signaling chains, in this case Igα and Igβ (see Fig. 6.9). In B cells, three protein tyrosine kinases of the Src family—Fyn, Blk, and Lyn—are thought to be responsible for phosphorylation of the ITAMs (Fig. 6.24). These kinases associate with resting receptors via a low-affinity interaction with the unphosphorylated ITAMs in Igα and Igβ. After the receptors bind a multivalent antigen, which cross-links them, the receptor-associated kinases are activated and phosphorylate the tyrosine residues in the ITAMs. B cells do not express ZAP-70; instead, a closely related tyrosine kinase, Syk, containing two SH2 domains, is recruited to the phosphorylated ITAM. In contrast to ZAP-70, which requires additional Lck phosphorylation for activation, Syk is activated simply by its binding to the phosphorylated site.

The B-cell equivalent to the co-receptors CD4 and CD8 is a complex of cell-surface proteins—CD19, CD21, and CD81—which is known as the **B-cell co-receptor** (Fig. 6.25). As with T cells, antigen-dependent signaling from the B-cell receptor is enhanced if the B-cell co-receptor is simultaneously bound by its ligand and clusters with the antigen receptor. CD21 (also known as complement receptor 2, CR2) is a receptor for the C3d fragment of complement. This means that antigens such as bacterial pathogens on which C3d is bound (see Chapter 2) can cross-link the B-cell receptor with the CD21:CD19:CD81 complex. This induces phosphorylation of the cytoplasmic tail of CD19 by B-cell receptor-associated tyrosine kinases, which in turn leads to the binding of Src-family kinases, the augmentation of signaling through the B-cell receptor itself, and the recruitment of PI 3-kinase (see Section 6-5). PI 3-kinase initiates an additional signaling pathway to that leading from the B-cell receptor (see Fig. 6.25). Thus, the B-cell co-receptor serves to strengthen the signal resulting from antigen recognition. The role of the third component of the B-cell receptor complex, CD81 (TAPA-1), is as yet unknown.

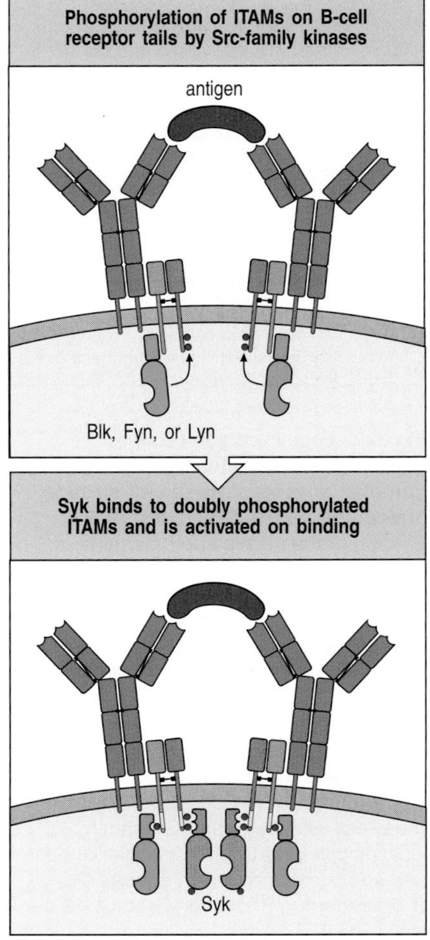
Phosphorylation of ITAMs on B-cell receptor tails by Src-family kinases
antigen
Blk, Fyn, or Lyn
Syk binds to doubly phosphorylated ITAMs and is activated on binding
Syk

Fig. 6.24 Src-family kinases are associated with antigen receptors and phosphorylate the tyrosines in ITAMs to create binding sites for Syk and Syk activation via transphosphorylation. The membrane-bound Src-family kinases Fyn, Blk, and Lyn associate with the B-cell antigen receptor by binding to ITAMs, either (as shown in the figure) through their amino-terminal domains or by binding a single phosphorylated tyrosine through their SH2 domains. After ligand binding and receptor clustering, they phosphorylate tyrosines in the ITAMs on the cytoplasmic tails of Igα and Igβ. Subsequently, Syk binds to the phosphorylated ITAMs of the Igβ chain. Because there are at least two receptor complexes in each cluster, Syk molecules become bound in close proximity and can activate each other by transphosphorylation, thus initiating further signaling.

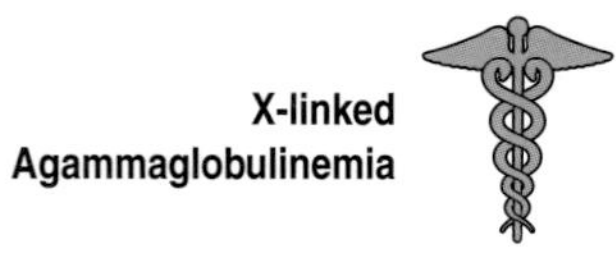
X-linked
Agammaglobulinemia

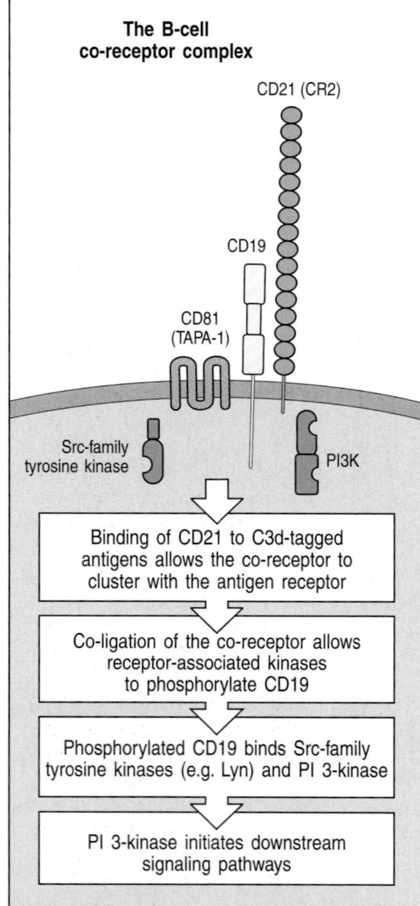
The B-cell
co-receptor complex
CD21 (CR2)
CD19
CD81
(TAPA-1)
Src-family
tyrosine kinase
PI3K
Binding of CD21 to C3d-tagged antigens allows the co-receptor to cluster with the antigen receptor
Co-ligation of the co-receptor allows receptor-associated kinases to phosphorylate CD19
Phosphorylated CD19 binds Src-family tyrosine kinases (e.g. Lyn) and PI 3-kinase
PI 3-kinase initiates downstream signaling pathways

Fig. 6.25 B-cell antigen receptor signaling is modulated by a co-receptor complex of at least three cell-surface molecules, CD19, CD21, and CD81. Binding of the cleaved complement fragment C3d to antigen allows the tagged antigen to bind to both the B-cell receptor and the cell-surface protein CD21 (complement receptor 2, CR2), a component of the B-cell co-receptor complex. Cross-linking and clustering of the co-receptor with the antigen receptor results in phosphorylation of tyrosine residues in the cytoplasmic domain of CD19 by protein kinases associated with the B-cell receptor; other Src-family kinases can bind to phosphorylated CD19 and so augment signaling through the B-cell receptor. Phosphorylated CD19 can also bind PI 3-kinase.

Once activated, Syk phosphorylates the scaffold protein **BLNK** (also known as SLP-65). Like LAT in T cells, BLNK has multiple sites for tyrosine phosphorylation and recruits a variety of SH2-containing proteins, including enzymes and adaptor proteins, to form several distinct multiprotein signaling complexes that can act in concert. As in T cells, a key signaling protein is the enzyme phospholipase C-γ, which is activated with the aid of the B-cell specific Tec kinase Btk and hydrolyzes PIP_2 to form DAG and IP_3. As discussed above for the T-cell receptor, signaling by calcium and DAG leads to the activation of downstream transcription factors. The B-cell receptor signaling pathway is summarized in Fig. 6.26. A deficiency in Btk (which is encoded by a gene on the X chromosome) prevents the development and functioning of B cells, resulting in the disease X-linked agammaglobulinemia.

6-18 ITAMs are also found in other receptors on leukocytes that signal for cell activation.

Other immune-system receptors also use ITAM-containing accessory chains to transduce activating signals (Fig. 6.27). One example is FcγRIII (CD16); this is a receptor for IgG that triggers antibody-dependent cell-mediated cytotoxicity (ADCC) by NK cells, which we will consider in Chapter 9; CD16 is also found on macrophages and neutrophils, where it facilitates the uptake and destruction of antibody-bound pathogens. To signal, FcγRIII must associate with either the ζ chain found also in the T-cell receptor complex, or with a second member of the same protein family known as the Fcγ chain. The Fcγ chain is also the signaling component of another receptor—the Fcε receptor I (FcεRI) on mast cells. As we discuss in Chapter 12, this receptor binds IgE antibodies and on cross-linking by allergens it triggers the degranulation of mast cells. Lastly, many activating receptors on NK cells are associated with DAP12, another ITAM-containing protein.

Several viral pathogens seem to have acquired ITAM-containing receptors from their hosts. These include the Epstein–Barr virus (EBV), whose *LMP2A* gene encodes a membrane protein with a cytoplasmic tail containing an ITAM. This enables EBV to trigger B-cell proliferation by using the downstream signaling pathways discussed in Section 6-17 and the preceding sections. Another virus that expresses an ITAM-containing protein is the Kaposi sarcoma herpes virus (KSHV or HHV8), which also causes malignant transformation and proliferation of the cells it infects.

6-19 The cell-surface protein CD28 is a co-stimulatory receptor for naive T cells.

The signaling through the T-cell receptor complex described in the previous sections is not by itself sufficient to activate a naive T cell. As noted in Chapter 1, antigen-presenting cells that can activate naive T cells bear cell-surface proteins known as **co-stimulatory molecules** or co-stimulatory ligands. These interact with cell-surface receptors, known as **co-stimulatory receptors**, on the naive T cell to transmit a signal that is required, along with antigen stimulation, for T-cell activation—this signal is often known as 'signal 2.' We discuss the immunological consequences of this requirement in detail in Chapter 8. The best understood of these co-stimulatory receptors is **CD28**. Despite the fact that there are many known effects of CD28 signaling, the precise nature of the co-stimulatory signal and why it is required for T-cell activation have not yet been determined.

CD28 is present on the surface of all naive T cells and binds the co-stimulatory ligands **B7.1** (CD80) and **B7.2** (CD86), which are expressed mainly on specialized antigen-presenting cells such as dendritic cells (Fig. 6.28). To become activated, the naive lymphocyte must engage both antigen and a co-stimulatory ligand on the same antigen-presenting cell. The requirement for CD28 signaling thus means that naive T cells can be activated only by professional antigen-presenting cells and not by other bystander cells that might happen to carry the antigen on their surface. Because co-stimulatory ligands are induced on antigen-presenting cells by infection (see Chapter 2), this also helps ensure that T cells are activated only in response to infection. It is thought that CD28 signaling aids antigen-dependent T-cell activation mainly by promoting T-cell proliferation, cytokine production, and cell survival. All these effects are mediated by signaling motifs present in the cytoplasmic domain of CD28.

After engagement by B7 molecules, CD28 becomes tyrosine phosphorylated on a non-ITAM motif, YXXM, which enables it to recruit and activate PI 3-kinase (see Fig. 6.28, left panel). This results in the production of PIP_3, which recruits the serine/threonine kinase **Akt** (also known as protein kinase B) to the membrane via Akt's PH domain (Fig. 6.6). Akt becomes activated and

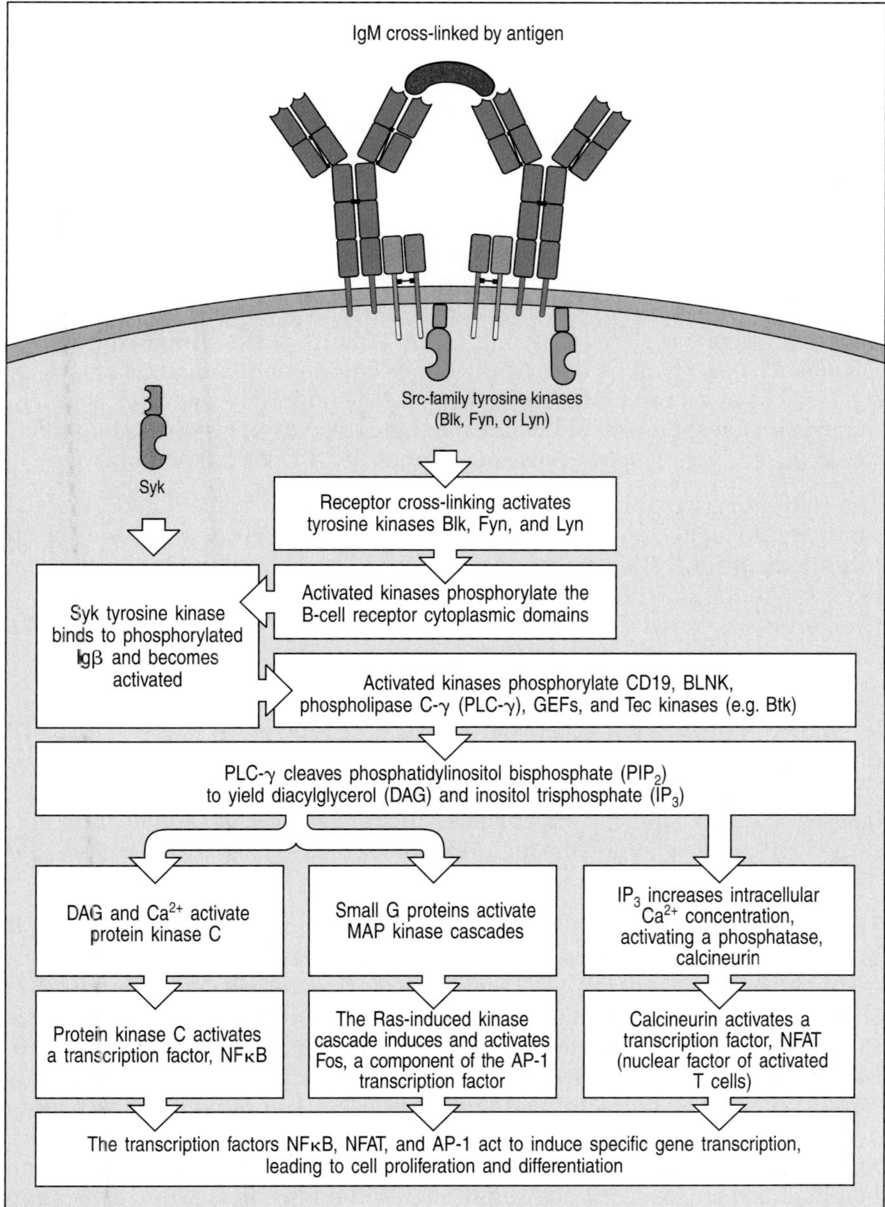
IgM cross-linked by antigen
Src-family tyrosine kinases (Blk, Fyn, or Lyn)
Syk
Receptor cross-linking activates tyrosine kinases Blk, Fyn, and Lyn
Activated kinases phosphorylate the B-cell receptor cytoplasmic domains
Syk tyrosine kinase binds to phosphorylated Igβ and becomes activated
Activated kinases phosphorylate CD19, BLNK, phospholipase C-γ (PLC-γ), GEFs, and Tec kinases (e.g. Btk)
PLC-γ cleaves phosphatidylinositol bisphosphate (PIP2) to yield diacylglycerol (DAG) and inositol trisphosphate (IP3)
DAG and Ca2+ activate protein kinase C
Small G proteins activate MAP kinase cascades
IP3 increases intracellular Ca2+ concentration, activating a phosphatase, calcineurin
Protein kinase C activates a transcription factor, NFκB
The Ras-induced kinase cascade induces and activates Fos, a component of the AP-1 transcription factor
Calcineurin activates a transcription factor, NFAT (nuclear factor of activated T cells)
The transcription factors NFκB, NFAT, and AP-1 act to induce specific gene transcription, leading to cell proliferation and differentiation

Fig. 6.26 Simplified outline of the intracellular signaling pathways initiated by cross-linking of B-cell receptors by antigen. Cross-linking of surface immunoglobulin molecules activates the receptor-associated Src-family protein tyrosine kinases Blk, Fyn, and Lyn. The receptor-associated kinases phosphorylate the ITAMs in the receptor complex, which bind and activate the cytosolic protein kinase Syk, whose activation has been described in Fig. 6.24. Syk then phosphorylates other targets, including the adaptor protein BLNK, which helps to recruit Tec kinases that in turn phosphorylate and activate the enzyme phospholipase C-γ. PLC-γ cleaves the membrane phospholipid PIP_2 into IP_3 and DAG, thus initiating two of the three main signaling pathways to the nucleus. IP_3 releases Ca^{2+} from intracellular and extracellular sources, and Ca^{2+}-dependent enzymes are activated, whereas DAG activates protein kinase C with the help of Ca^{2+}. The third main signaling pathway is initiated by guanine-nucleotide exchange factors (GEFs) that become associated with the receptor and activate small GTP-binding proteins such as Ras. These in turn trigger protein kinase cascades (MAP kinase cascades) that lead to the activation of MAP kinases that move into the nucleus and phosphorylate proteins that regulate gene transcription. This scheme is a simplification of the events that actually occur during signaling, showing only the main events and pathways.

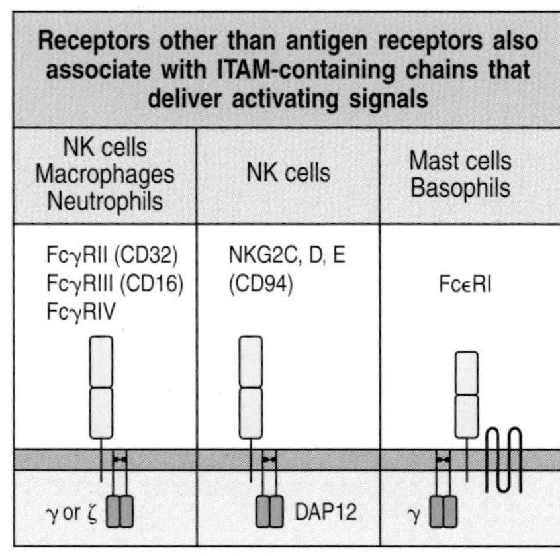

Fig. 6.27 Other receptors that pair with ITAM-containing chains can deliver activating signals. Cells other than B and T cells have receptors that pair with accessory chains containing ITAMs, which are phosphorylated when the receptor is cross-linked. These receptors deliver activating signals. The Fcγ receptor III (CD16) is found on NK cells, macrophages, and neutrophils. Binding of IgG to this receptor activates the killing function of the NK cell, leading to the process known as antibody-dependent cell-mediated cytotoxicity (ADCC). Activating receptors on NK cells, such as NKG2C, NKG2D, and NKG2E, also associate with ITAM-containing signaling chains. The Fcε receptor (FcεRI) is found on mast cells and basophils. It binds to IgE antibodies with very high affinity. When antigen subsequently binds to the IgE, the mast cell is triggered to release granules containing inflammatory mediators. The γ chain associated with the Fc receptors, and the DAP12 chain that associates with the NK killer-activating receptors, also contain one ITAM per chain and are present as homodimers.

can then phosphorylate a variety of downstream proteins. One of its effects is to promote cell survival by inhibiting the cell-death pathway that we discuss later in this chapter; another is to stimulate the cell's metabolism by increasing the utilization of glucose.

Activated CD28 also enhances the T-cell receptor signal directly. It becomes phosphorylated on another motif (YXN), which recruits the adaptor protein Grb2 (see Fig. 6.28, middle panel). This means that CD28 can potentially activate the Ras–MAP kinase signaling pathway through recruitment of the GTP-exchange factor SOS (see Section 6-2), leading to activation of the MAP kinase Erk. The cytoplasmic tail of CD28 also carries a proline-rich motif (PXXP) that binds the SH3 domains of the Src-family kinase Lck and the Tec kinase Itk (see Fig. 6.28, right panel). Engagement of the SH3 domain of these tyrosine kinases removes the inhibitory influence of the domain on their catalytic activity (see Section 6-10). CD28 can therefore enhance T-cell receptor signaling by promoting the enzymatic activity of Lck and Itk, and thus eventually stimulating the production of IL-2.

6-20 Inhibitory receptors on lymphocytes help regulate immune responses.

CD28 is only one of a family of receptors that are expressed by lymphocytes and bind to B7-family ligands. Some, such as the receptor ICOS, which is discussed in Chapter 8, act as activating receptors, but others inhibit signaling by the antigen receptors and are important in regulating the immune response. Inhibitory CD28-related receptors expressed by T cells include **CTLA-4** (CD152) and **PD-1** (**programmed death-1**), while the **B and T lymphocyte attenuator** (**BTLA**) is expressed by both T cells and B cells. Of these, CTLA-4 is arguably the most important. It is induced on activated T cells and has a crucial role in regulating T-cell signaling. CTLA-4 binds to the same co-stimulatory ligands (B7.1 and B7.2) as does CD28, but its engagement inhibits signaling by the T-cell receptor rather than enhancing it. The importance of CTLA-4 in regulating T-cell responses is shown by the phenotype of CTLA-4-deficient mice, which die at a young age because of uncontrolled T-cell proliferation.

The inhibitory signaling pathway induced by CTLA-4 is mediated by a distinct amino acid sequence called an **immunoreceptor tyrosine-based inhibitory motif** (**ITIM**) in the cytoplasmic tail of the protein. In this motif, a large hydrophobic residue such as isoleucine (I) or valine (V) occurs two residues upstream of a tyrosine (Y) that is followed by two amino acids and a leucine (L) to give the amino acid sequence ...[I/V]XYXX[L/I]... (Fig. 6.29).

When the tyrosine is phosphorylated, an ITIM can recruit either of two inhibitory phosphatases, called **SHP** (SH2-containing phosphatase) and **SHIP** (SH2-containing inositol phosphatase), via their SH2 domains. SHP is a protein tyrosine phosphatase and removes phosphate groups added by tyrosine kinases. SHIP is an inositol phosphatase and removes the phosphate from PIP_3 to give PIP_2, thus reversing the recruitment of proteins such as Tec kinases and Akt to the cell membrane.

PD-1 is induced transiently on activated T cells, B cells and myeloid cells. It binds to two ligands, both members of the B7 family called **PD-L1** (programmed death ligand-1, B7-H1) and **PD-L2** (programmed death ligand-2, B7-DC). PD-L1 is constitutively expressed on a wide variety of cells, whereas PD-L2 expression is induced on antigen-presenting cells during inflammation. As PD-L1 itself is constitutively expressed, regulation of PD-1 expression may have a critical role in controlling T-cell responses. For example, inflammatory cytokine signaling can repress PD-1 expression, enhancing T-cell responses. Mice lacking PD-1 gradually develop autoimmunity, presumably because of an inability to regulate T-cell activation. In chronic infections, PD-1 expression reduces the effector activity of T cells; this helps to limit potential damage to bystander cells, but at the expense of pathogen clearance. PD-1 has two cytoplasmic ITIMs that become phosphorylated after its engagement by ligand, and can recruit both SHP and SHIP. BTLA is expressed on activated T cells and B cells. Like PD-1 and CTLA-4, BTLA signals through ITIMs that recruit SHP. Unlike other CD28-family members, however, BTLA does not interact with B7 ligands but binds a member of the tumor necrosis factor (TNF) receptor family called the herpes virus entry molecule (HVEM), which is highly expressed on resting T cells and immature dendritic cells.

Other structural types of receptors on B cells and T cells also contain ITIMs and can inhibit cell activation when ligated along with the antigen receptors. One example is the receptor **FcγRIIB-1** on B cells, which binds the Fc region of IgG. It has long been known that the activation of naive B cells in response to antigen can be inhibited by soluble IgG antibodies that recognize the same

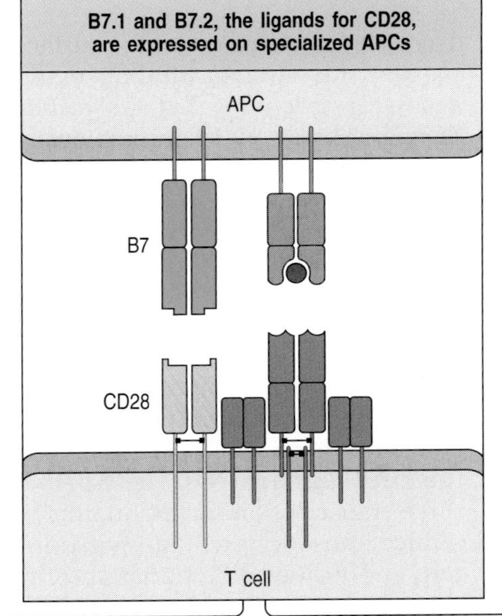

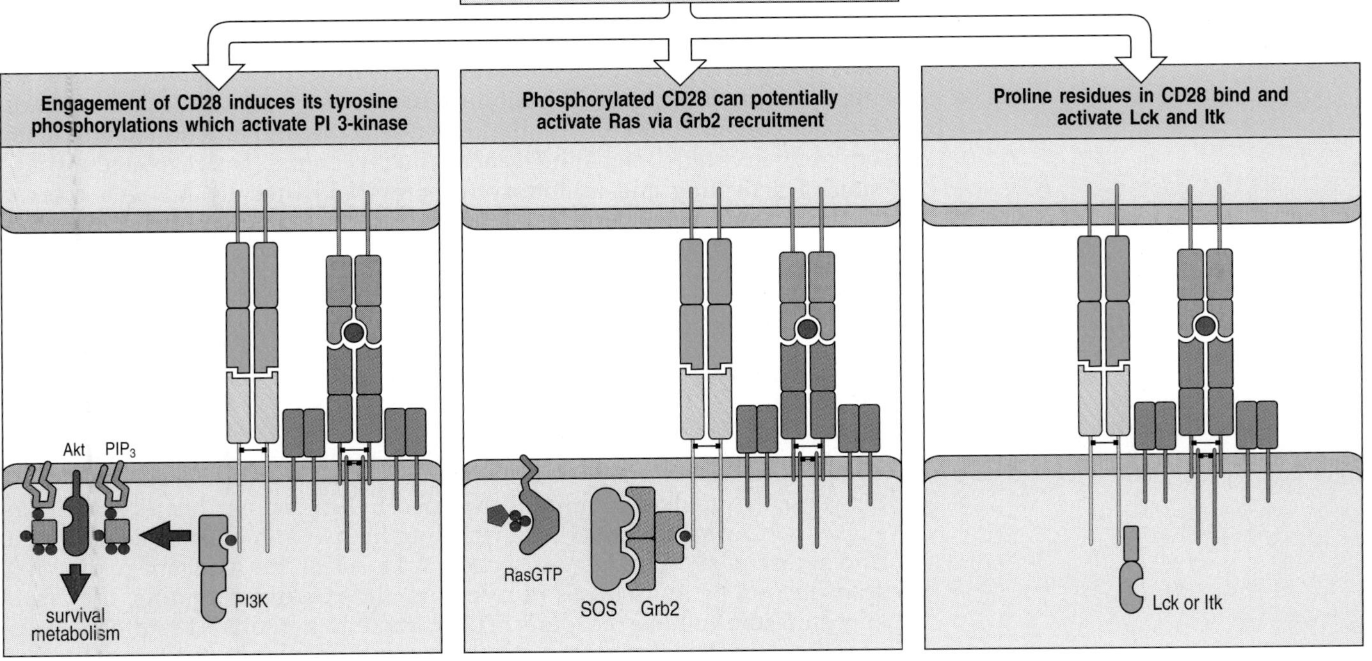

Fig. 6.28 The co-stimulatory protein, CD28, transduces a variety of different signals. The ligands for CD28, namely B7.1 and B7.2, are expressed only on specialized antigen-presenting cells (APCs) such as dendritic cells. Engagement of CD28 induces its tyrosine phosphorylation which activates PI 3-kinase (PI3K), with subsequent production of PIP_3 and activation of the protein kinase Akt. Activated Akt enhances cell survival and upregulates cell metabolism. The phosphorylated tyrosine can also potentially recruit the adaptor Grb2. Grb2 is bound to SOS, stimulating Ras activation or another molecule called Vav, an activator of the actin cytoskeleton. Lastly, proline motifs in the cytoplasmic domain can bind to and stimulate the tyrosine kinase activity of Lck and Itk.

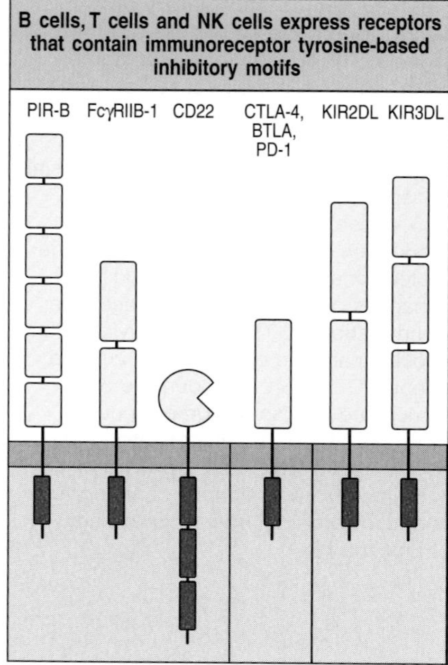
B cells, T cells and NK cells express receptors that contain immunoreceptor tyrosine-based inhibitory motifs
PIR-B
FcγRIIB-1
CD22
CTLA-4, BTLA, PD-1
KIR2DL
KIR3DL

Fig. 6.29 Some lymphocyte cell-surface receptors contain motifs involved in downregulating activation. Several receptors that transduce signals that inhibit lymphocyte or NK-cell activation contain a motif called an ITIM (immunoreceptor tyrosine-based inhibitory motif) in their cytoplasmic tails. ITIMs bind to various phosphatases that, when activated, inhibit signals derived from ITAM-containing receptors.

antigen and therefore co-ligate the B-cell receptor with this Fc receptor. The FcγRIIB-1 ITIM recruits SHIP into a complex with the B-cell receptor to block the actions of PI 3-kinase. Another inhibitory receptor on B cells is the trans-membrane protein **CD22**, which contains an ITIM that interacts with SHP.

The ITIM motif is also an important motif in signaling by receptors on NK cells that inhibit the killer activity of these cells (see Section 2-30). These inhibitory receptors recognize MHC class I molecules and transmit signals that inhibit the release of cytotoxic granules when NK cells recognize healthy uninfected cells (see Section 2-31).

Thus, signaling through receptors containing ITAMs and ITIMs can precisely control the intensity and nature of the final signal received by the cell. In some cases, an ITIM-containing receptor can completely block signaling from activating receptors.

Summary.

The antigen receptors on the surface of lymphocytes are multiprotein complexes with antigen-binding extracellular components interacting with accessory receptors that are responsible for signaling from the receptor. Signaling by many immunologically important receptors is mediated by a tyrosine-containing signaling motif known as the ITAM. Activation of the receptors by antigen results in phosphorylation of the ITAM by Src-family kinases. The phosphorylated ITAM then recruits another tyrosine kinase known as ZAP-70 in T cells and Syk in B cells. Activation of ZAP-70 and Syk result in the phosphorylation of scaffolds called LAT and SLP-76 in T cells and BLNK in B cells. The most important of the signaling proteins recruited and activated by these phosphorylated scaffolds is phospholipase C-γ, which when activated, generates inositol trisphosphate (IP_3) and diacylglycerol (DAG). IP_3 has an important role in inducing changes in intracellular calcium concentrations, while DAG is involved in activating protein kinase C-θ and the small G protein Ras. These pathways ultimately result in the activation of three transcription factors, namely AP-1, NFAT, and NFκB, which together induce transcription of the cytokine IL-2, which is essential for the proliferation and further differentiation of the activated lymphocyte. An important secondary signaling system is provided by the CD28 family of co-stimulatory proteins, which bind members of the B7 family of proteins. Activating members of the CD28 family are important in ensuring the activation of T cells by the appropriate target cell. Inhibitory members of this and other receptor families contain inhibitory motifs known as ITIMs and function to attenuate or completely block signaling by activating receptors. The regulated expression of activating and inhibitory receptors and their ligands generates a sophisticated level of control of immune responses that is only beginning to be understood.

Other receptors and signaling pathways.

Lymphocytes are normally studied in terms of their responsiveness to antigen. However, they and other immune-system cells bear many other receptors that make them aware of events occurring both in their immediate neighborhood and at distant sites. In the next section, we focus on the mechanism of signal transduction by four classes of receptors: the cytokine receptors, the death receptors, the Toll-like receptors (TLRs), and the chemokine receptors.

6-21 Cytokines typically activate fast signaling pathways that end in the nucleus.

One of the main ways in which the cells of the immune system communicate with each other and with the other cells of the body is through a class of small secreted proteins known as cytokines, some of which were introduced in Chapter 2. They are usually secreted in response to an extracellular stimulus, and they may act on the cells that produce them, on other cells in the immediate vicinity, or on cells at a distance after being carried in blood or tissue fluids. Cytokines affect cell behavior in a variety of ways and, as we will see in subsequent chapters, they have key roles in controlling the growth, development, functional differentiation and activation of lymphocytes and other leukocytes. The cytokines secreted by activated and effector T cells are critical to these cells' functions in the immune system. Cytokines produce immediate responses in the cells they affect, and their signaling properties reflect this. Their receptors activate particularly direct signaling pathways to effect rapid changes in gene expression in the nucleus.

6-22 Cytokine receptors form dimers or trimers on ligand binding.

One large structurally related class of cytokine receptors, the hemopoietin family of receptors, are tyrosine kinase-associated receptors that form dimers when their cytokine ligand binds. As in antigen receptor clustering, this dimerization initiates intracellular signaling from the tyrosine kinases associated with the cytoplasmic domains of the receptor. In some types of cytokine receptor, the dimer is composed of two identical subunits, in others it has two different subunits. An important feature of cytokine signaling is the large variety of different receptor combinations that occur. The large diversity of receptors used in cytokine signaling is described in more detail in Chapter 8 (see Fig. 8.35).

The second class of cytokine receptors contains those for cytokines of the TNF family. These are structurally unrelated to the receptors described above but also have to cluster to become activated. Cytokines of this family, such as TNF-α and lymphotoxin, act as trimers, and binding of ligand induces the clustering of three identical receptor subunits. Some cytokines of the TNF family are not secreted but are either transmembrane proteins or proteins that remain associated with the cell surface.

6-23 Cytokine receptors are associated with the JAK family of tyrosine kinases which activate STAT transcription factors.

The signaling chains of hemopoietin family of cytokine receptors are noncovalently associated with protein tyrosine kinases of the **Janus kinase (JAK) family**—so called because they have two tandem kinase-like domains and thus resemble the two-headed mythical Roman god Janus. There are four members of the JAK family, Jak1, Jak2, Jak3 and Tyk2. As mice deficient for individual JAK family members show different phenotypes, each kinase must have a distinct function. Presumably, the use of different combinations of JAKs by different cytokine receptors enables a diversity of signaling responses.

The dimerization or clustering of the signaling chains allows the JAKs to cross-phosphorylate each other, thus stimulating their kinase activity. The activated JAKs then phosphorylate their associated receptors on specific tyrosine residues to generate binding sites for proteins with SH2 domains (Fig. 6.30). Some of the tyrosine-phosphorylated sites recruit SH2-containing latent transcription factors known as **signal transducers and activators of transcription** (**STATs**).

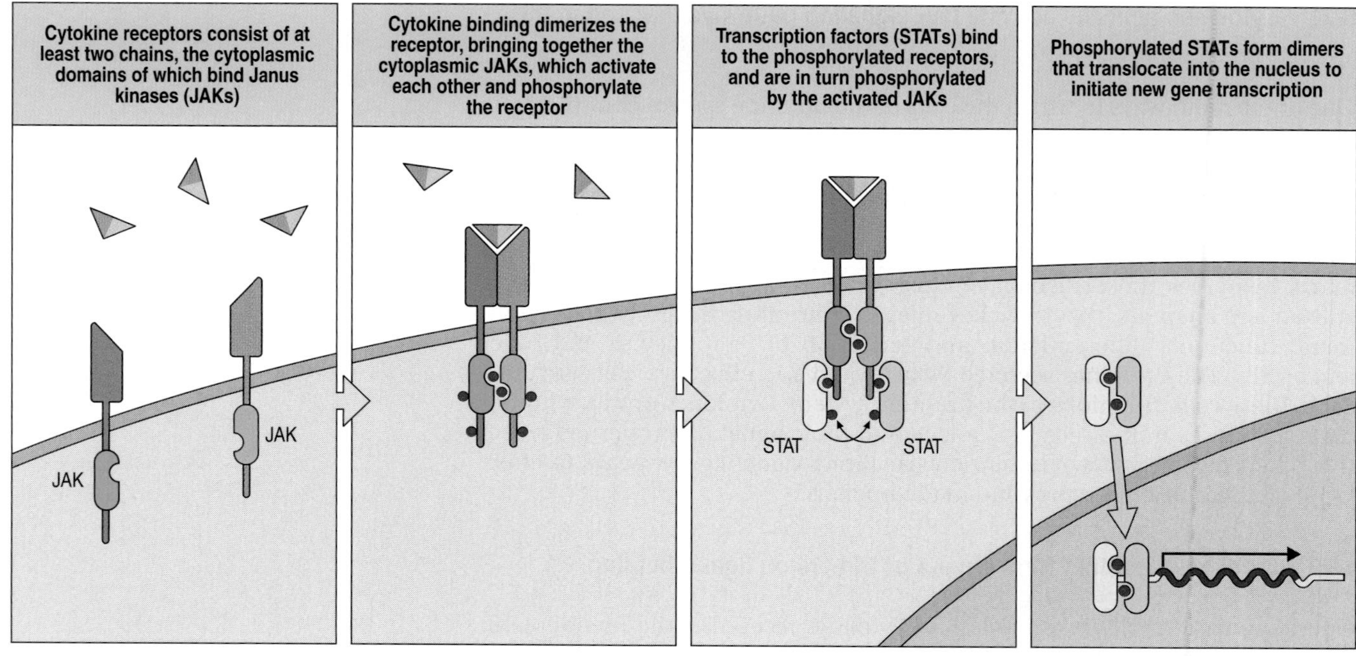

Cytokine receptors consist of at least two chains, the cytoplasmic domains of which bind Janus kinases (JAKs)
JAK
JAK
Cytokine binding dimerizes the receptor, bringing together the cytoplasmic JAKs, which activate each other and phosphorylate the receptor
Transcription factors (STATs) bind to the phosphorylated receptors, and are in turn phosphorylated by the activated JAKs
STAT
STAT
Phosphorylated STATs form dimers that translocate into the nucleus to initiate new gene transcription

Fig. 6.30 Cytokine receptors signal using a rapid pathway called the JAK–STAT pathway. Many cytokines act via receptors that are associated with cytoplasmic Janus kinases (JAKs). The receptor consists of at least two chains, each associated with a specific JAK (first panel). Binding of dimeric ligand results in dimerization of the receptor chains, bringing together the JAKs, which can phosphorylate each other, thus activating themselves. The activated JAKs then phosphorylate tyrosines in the receptor tails (second panel). Members of the STAT (signal transducers and activators of transcription) family of proteins (which contain SH2 domains) bind to the tyrosine phosphorylated receptors and are themselves phosphorylated by the JAKs (third panel). After phosphorylation, STAT proteins dimerize by binding phosphotyrosine residues in SH2 domains and translocate to the nucleus (last panel), where they bind to and activate the transcription of a variety of genes important for adaptive immunity.

There are seven STATs (1–5, and 6a and 6b). The specificity of a particular STAT for a particular receptor is determined by the recognition of the distinctive phosphotyrosine sequence on the activated receptor by the SH2 domain of the STAT. Recruitment of a STAT to the activated receptor brings the STAT close to an activated JAK, which can then phosphorylate it. This leads to a conformational change in the STAT that allows it to bind to another STAT to form a dimer. STATs can form homodimers or heterodimers. The phosphorylated STAT dimers now dissociate from the receptors and enter the nucleus, where they act as transcription factors to initiate the expression of selected genes. These STAT-regulated genes include genes that contribute to the growth and differentiation of particular subsets of lymphocytes. An example of the specificity of STAT-mediated transcription is that STAT4 is essential for T_H1 cell development, whereas STAT6 is required for T_H2 cell development.

STATs are activated not only by cytokine receptors but also by some other types of receptors expressed in immune cells. In addition, STAT-mediated transcription is not the only pathway that can be initiated by cytokine receptors. Cytokine receptors can, for example, activate the Ras–MAP kinase pathway and the phosphatidylinositol pathway. Relatively little is known about how cytokine receptors activate these pathways, but it is possible that the ability of closely related cytokines to induce distinct biological responses may result from the selective activation of different combinations of multiple possible signaling pathways.

6-24 Cytokine signaling is terminated by a negative feedback mechanism.

As cytokines have so many and such powerful effects, the activation of cytokine signaling pathways must be tightly controlled; breakdown in control can lead to significant pathological effects. A variety of cytokine-specific inhibitory mechanisms ensure that cytokine signaling pathways can be efficiently terminated. As cytokine receptor signaling depends on tyrosine phosphorylation, dephosphorylation of the receptor complex by tyrosine phosphatases is one important means of termination. A variety of tyrosine phosphatases have been implicated in the dephosphorylation of cytokine receptors, JAKs, and STATs; these include SHP, CD45, and the T-cell phosphatase (TCPTP).

Cytokine signaling can also be terminated by a negative feedback process involving specific inhibitors that are induced by cytokine activation. One class of inhibitors contains the SOCS proteins, which terminate signaling in a variety of ways, including promoting the ubiquitination and subsequent degradation of receptors, JAKs, and STATs. Another class of inhibitory proteins consists of the protein inhibitors of activated STAT proteins (PIAS proteins), which also seem to be involved in promoting the degradation of receptors and pathway components.

6-25 The receptors that induce apoptosis activate specialized intracellular proteases called caspases.

Programmed cell death or **apoptosis** (see Section 1-14) is a normal process that is crucial to the proper development and function of the immune system. In particular, it has an important role in the termination of immune responses by getting rid of cells that are no longer needed after an infection has been cleared. It also has a key role in lymphocyte development in removing developing lymphocytes that fail to generate functional antigen receptors (see Chapter 4) or that have produced potentially autoreactive receptors, as discussed in Chapter 7. Apoptosis is a regulated process that is induced by specific extracellular signals (or in some cases by the lack of signals required for survival) and proceeds by a series of cellular events that include plasma membrane blebbing, changes in the distribution of membrane lipids, and enzymatic fragmentation of chromosomal DNA.

Two general pathways are involved in signaling cell death. One, called the **extrinsic pathway of apoptosis**, is mediated by the activation of so-called **death receptors** by extracellular ligands. Engagement of the ligand stimulates apoptosis in the receptor-bearing cell. The other pathway is known as the **intrinsic** or **mitochondrial pathway of apoptosis** and mediates apoptosis in response to noxious stimuli including ultraviolet irradiation, chemotherapeutic drugs, starvation, or lack of the growth factors required for survival. Common to both pathways is the activation of specialized proteases called aspartic-acid-specific cysteine proteases or **caspases**.

Like many other proteases, caspases are synthesized as inactive pro-caspases, in which the catalytic domain is inhibited by an adjacent pro-domain. Pro-caspases are activated by other caspases that cleave the protein to release the inhibitory pro-domain. There are two classes of caspases involved in the apoptotic pathway: **initiator caspases** promote apoptosis by cleaving and activating other caspases, and the **effector caspases** are the ones that initiate the cellular changes associated with apoptosis. The extrinsic pathway uses the initiator caspase, caspase 8, while the intrinsic pathway uses caspase 9. Both pathways use caspases 3, 6, and 7 as effector caspases. The effector caspases cleave a variety of proteins that are critical for cellular integrity and also activate enzymes that promote the death of the cell. For example, they cleave and degrade nuclear proteins, such as lamin B, that are required for the structural integrity of the nucleus, and activate the endonucleases that fragment the chromosomal DNA.

We shall first consider the apoptosis pathway leading from death receptors, as these are involved in many immune-system functions. The activation of caspase 8 is the critical step in the apoptosis pathway and starts with the recruitment of this initiator pro-caspase to the activated death receptor.

The death receptors are members of the large TNF receptor family but are distinguished from other receptors in this family by a conserved domain known as the **death domain** (**DD**) in the cytoplasmic part of the receptor. Of those death receptors that are expressed in immune-system cells, **Fas** (CD95) and **TNFR-I** are the best understood. Fas and its ligand **FasL** are expressed widely, not only in the immune system. Fas-mediated cell death occurs in numerous contexts, including the protection of immunologically privileged sites (see Chapter 11) and the regulation and termination of immune responses (see Chapter 8). The signaling pathway resulting from the stimulation of Fas by FasL is shown in Fig. 6.31.

The first step in Fas-mediated apoptosis is the binding of FasL, which results in its clustering. Death domains specifically bind to other death domains, and, on clustering, the Fas death domains recruit adaptor proteins that contain both a death domain and an additional domain that can bind a pro-caspase (see Fig. 6.31). Each type of receptor recruits a specific adaptor; with Fas it is called FADD (Fas-associated via death domain). In addition to the death domain, FADD contains a domain known as a death effector domain (DED), which allows FADD to recruit the initiator, caspase, pro-caspase 8, directly via interactions with a similar domain on the enzyme. The high local concentration of caspase 8 around the receptors allows it to cleave itself, resulting in its self-activation. Once activated, the initiator caspase, caspase 8, is released from the receptor complex and can activate the downstream effector caspases.

A similar, but distinct, pathway is used by TNFR-I when stimulated by its ligand TNF-α. In some cells, TNFR-I signaling induces apoptosis; in other cells, TNFR-I signaling induces the induction of pro-inflammatory response genes. What determines whether apoptosis or new gene transcription is activated is not known. The current hypothesis suggests that the two different responses are regulated by two different signaling complexes that can be assembled by TNFR-I. In both cases, the receptor first recruits an adaptor called TRADD, and then the pathways diverge. When TRADD binds FADD, the pathway proceeds to apoptosis as in Fig. 6.32. In other conditions, however, TRADD recruits a serine/threonine kinase called RIP (receptor-interacting protein) and an adaptor called TRAF-2 (TNF receptor associated factor-2). Using a pathway that is not yet known, RIP mediates the activation of NFκB through the activation of IKK. TRAF-2 stimulates a MAP kinase signaling pathway that results in the activation of JNK and the transcription factor Jun, a component of the AP-1 complex (see Fig. 6.20).

Fig. 6.31 Binding of Fas ligand to Fas initiates the extrinsic pathway of apoptosis. The cell-surface receptor Fas contains a so-called death domain in its cytoplasmic tail. When Fas ligand (FasL) binds Fas, this trimerizes the receptor (first panel). The adaptor protein FADD (also known as MORT-1) also contains a death domain and can bind to the clustered death domains of Fas (second panel). FADD also contains a domain called a death effector domain (DED) that allows it to recruit the pro-caspase 8 (which also contains a DED domain) (third panel). Clustered pro-caspase 8 activates itself to release an active caspase into the cytoplasm (not shown).

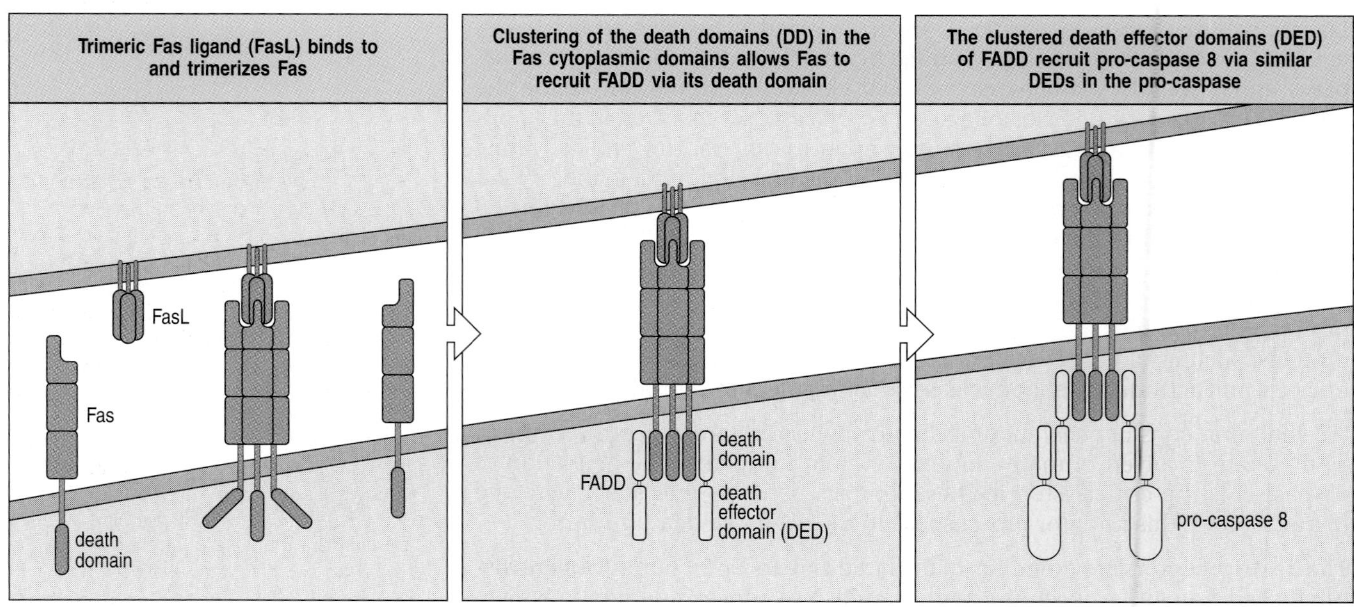
Trimeric Fas ligand (FasL) binds to and trimerizes Fas
FasL
Fas
death domain
Clustering of the death domains (DD) in the Fas cytoplasmic domains allows Fas to recruit FADD via its death domain
death domain
FADD
death effector domain (DED)
The clustered death effector domains (DED) of FADD recruit pro-caspase 8 via similar DEDs in the pro-caspase
pro-caspase 8

6-26 The intrinsic pathway of apoptosis is mediated by release of cytochrome *c* from mitochondria.

Apoptosis by the intrinsic pathway is triggered when the cell is stressed by exposure to noxious stimuli or does not receive extracellular signals that are required for cell survival. The critical step is the release of cytochrome *c* from mitochondria, which triggers the activation of caspases. Once in the cytoplasm, cytochrome *c* binds to a protein called Apaf-1 (apoptotic protease activating factor-1) stimulating its oligomerization. The Apaf-1 oligomer then recruits an initiator caspase, pro-caspase 9. Aggregation of caspase 9 permits its self-cleavage, and frees it to stimulate the activation of effector caspases as in the death receptor pathways (Fig. 6.33).

The release of cytochrome *c* is controlled by interactions between members of the Bcl-2 family of proteins. The **Bcl-2 family** of proteins are defined by the presence of one or more Bcl-2 homology (BH) domains and can be divided into two general groups: members that promote apoptosis and members that inhibit apoptosis (Fig. 6.34). Some pro-apoptotic Bcl-2 family members, such as Bax, Bak, and Bok (referred to as executioners), bind to mitochondrial membranes and can directly cause cytochrome *c* release. How they do this is still not known, but they may form pores in the membranes.

The anti-apoptotic Bcl-2 family members are induced by stimuli that promote cell survival. The best known of the anti-apoptotic proteins is Bcl-2 itself. The *Bcl-2* gene was first identified as an oncogene in a B-cell lymphoma, and its overexpression in tumors makes the cells more resistant to apoptotic stimuli and thus more likely to progress to an invasive cancer and difficult to kill. Other members of the inhibitory family include Bcl-X_L and Bcl-W. Anti-apoptotic proteins function by binding to the mitochondrial membrane to block the release of cytochrome *c*. The precise mechanism of inhibition is not clear, but they may function by directly blocking the function of the pro-apoptotic family members.

A second family of pro-apoptotic Bcl-2 family members are sentinels and are activated by apoptotic stimuli. Once activated, these proteins, which include Bad, Bid, and PUMA, can either act to block the activity of the anti-apoptotic proteins or act directly to stimulate the activity of the executioner pro-apoptotic proteins.

6-27 Microbes and their products act via Toll-like receptors to activate NFκB.

The 10 Toll-like receptors (TLRs) in humans (11 in mice) are a class of pattern recognition receptors that act in innate immunity. The ligands they bind and their roles in innate immunity are discussed in detail in Chapter 2. Structurally, the TLRs are single-pass transmembrane proteins characterized by multiple copies of a leucine-rich motif in their extracellular domain and a shared motif called TIR (for Toll–IL-1 receptor) in their cytoplasmic domain. This motif is also present in the receptor for the cytokine IL-1, which suggests that the TLRs and the IL-1 receptor use similar signaling pathways.

TLR signaling induces a diverse range of responses that regulate the production of inflammatory cytokines, chemotactic factors, and antimicrobial products (see Chapter 2). Many different signaling proteins are induced by TLR activation, including various MAP kinases and PI 3-kinase. The most important signaling pathway leading from TLRs, however, is the activation of NFκB, and this pathway is initiated by the TIR domain. This pathway is highly conserved in multicellular organisms and thus represents a very ancient pathway involved in defense against infection.

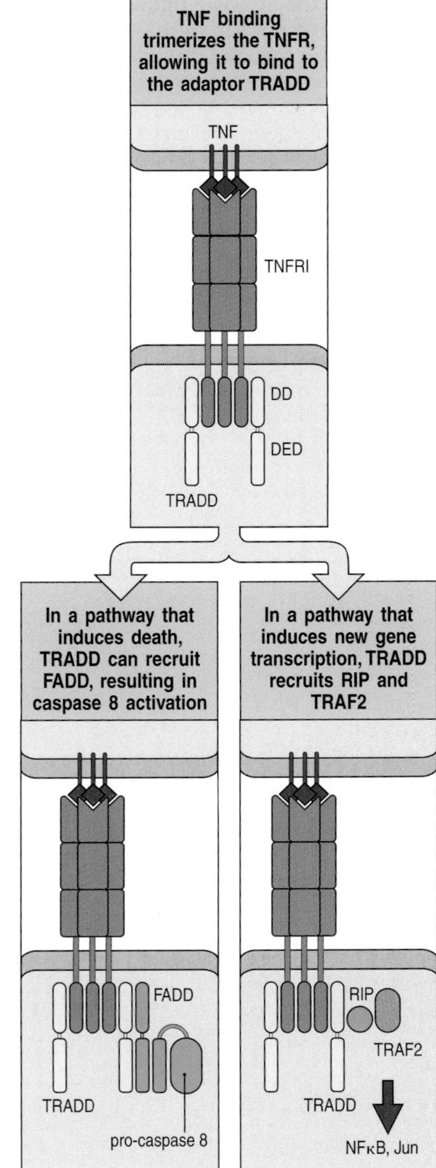

Fig. 6.32 Signaling by the TNF receptor TNFR-I. Like Fas, TNFR-I contains a cytoplasmic death domain (DD), which recruits the adaptor TRADD, which also contains a death domain. TRADD can assemble two different signaling complexes. Through a DD–DD interaction, TRADD can recruit FADD, resulting in caspase 8 activation and apoptosis (bottom left panel; see also Fig. 6.31). In a second pathway, TRADD can also recruit a serine/threonine kinase called RIP and an adaptor protein called TRAF-2. RIP activates IKK, resulting in the activation of NFκB. TRAF-2 stimulates the JNK signaling pathway, resulting in the phosphorylation of Jun. It is not known how one pathway is chosen over another.

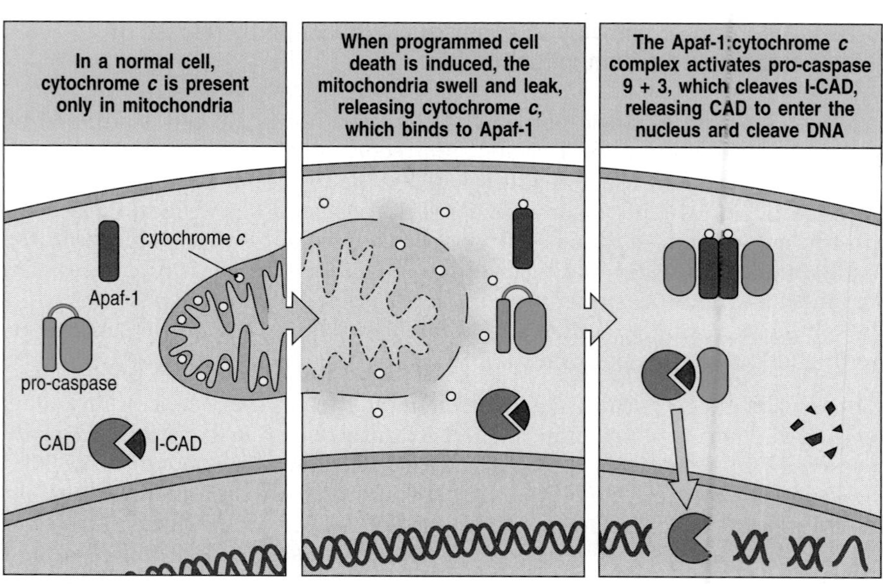

In a normal cell, cytochrome *c* is present only in mitochondria
When programmed cell death is induced, the mitochondria swell and leak, releasing cytochrome *c*, which binds to Apaf-1
The Apaf-1:cytochrome *c* complex activates pro-caspase 9 + 3, which cleaves I-CAD, releasing CAD to enter the nucleus and cleave DNA
cytochrome *c*
Apaf-1
pro-caspase
CAD
I-CAD

Fig. 6.33 In the intrinsic pathway, cytochrome *c* release from mitochondria induces programmed cell death. In normal cells, cytochrome *c* is confined to the mitochondria (first panel). However, during stimulation of the intrinsic pathway, the mitochondria swell, allowing the cytochrome *c* to leak out into the cytosol (second panel). There it interacts with the protein Apaf-1, forming a cytochrome *c*:Apaf-1 complex which recruits pro-caspase 9. Clustering of pro-caspase 9 activates it, allowing it to cleave downstream caspases, such as caspase 3, resulting in the activation of enzymes such as I-CAD, which can cleave DNA (third panel).

Like the death domains, TIR domains bind other TIR domains. Ligand binding to a TLR receptor induces a conformational change that allows its intracellular TIR domain to bind an adaptor protein that also contains a TIR domain. There are five identified TIR adaptors, the best known of which is MyD88 (myeloid differentiation factor 88). Many of the differences in signaling between different TLRs can be attributed to the use of different adaptors.

We will illustrate here the signaling pathway used by TLR-4, which is the receptor for bacterial lipopolysaccharide (LPS) on macrophages, neutrophils, and dendritic cells. LPS binds first to a circulating LPS-binding protein (LBP), which enables it to bind to the cell-surface protein CD14 (Fig. 6.35). The ligand-bound CD14 then interacts with TLR-4. TLR-4 mediates two signaling pathways known, respectively, as the MyD88-dependent and the MyD88-independent pathways. In the MyD88-dependent pathway, TLR-4 directly recruits MyD88 to the cytoplasmic tail. This adaptor has a TIR domain at one end, by which it binds the receptor, and a death domain at the other. After binding to the receptor, the MyD88 death domain recruits and activates a serine/threonine protein kinase known as IL1-receptor associated kinase (IRAK), which also possesses a death domain. The activated IRAK then binds the adaptor TRAF-6. TRAF-6 activates a MAPKKK called TAK1, and TAK1 phosphorylates and activates the IKK complex. As discussed in Section 6-16, IKK liberates NFκB from its inhibitor IκB so that it can translocate to the nucleus. In addition, TAK1 also stimulates the activation of JNK and another class of MAP kinases called the p38 family.

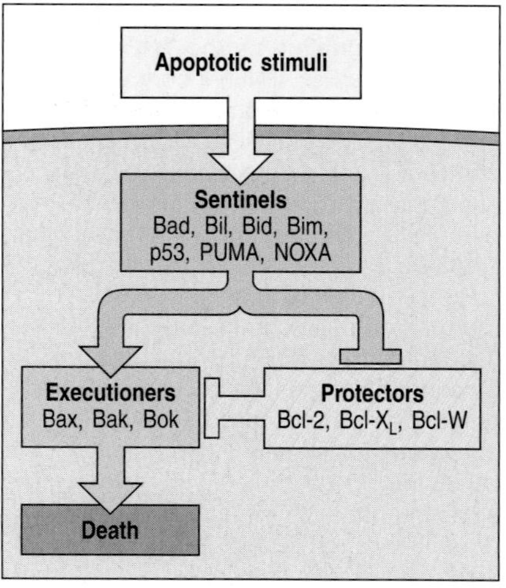

Fig. 6.34 General scheme of intrinsic pathway regulation by Bcl-2-family proteins. Extracellular apoptotic stimuli activate a group of pro-apoptotic (sentinel) proteins. Sentinel proteins can function either to block the protection provided by pro-survival, protector proteins or to directly activate pro-apoptotic, executioner proteins. In mammalian cells, apoptosis is mediated by the executioner proteins Bax, Bak, and Bok. In normal cells, these proteins are prevented from acting by the protector proteins (Bcl-2, Bcl-X_L, and Bcl-W). The release of activated executioner proteins cause the release of cytochrome *c* and subsequent cell death, as shown in Fig. 6.33.

TLR-4 can also signal via a MyD88-independent pathway to stimulate the production of the antiviral protein interferon (IFN)-β (see Section 2-29). As illustrated in Fig. 6.36, TLR4 can recruit another TIR-domain-containing adaptor called TRIF. Like MyD88, activated TRIF can bind to TRAF-6 to activate NFκB. Unlike MyD88, TRIF can also bind to unusual kinases called IκKε and TBK1. These kinases activate transcription factors called IRFs (interferon regulatory factors) that are involved in stimulating interferon (IFN)-β transcription. Thus, the adaptor TRIF enables TLR-4 signaling to induce production of IFN-β in addition to activation of NFκB.

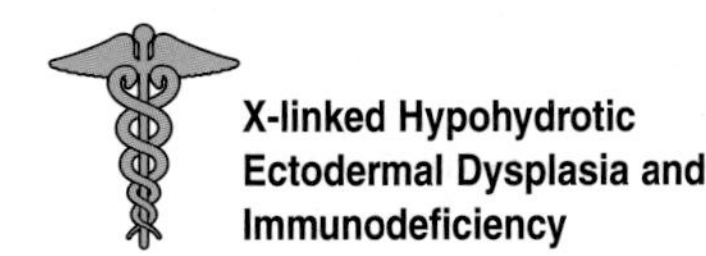

6-28 Bacterial peptides, mediators of inflammatory responses, and chemokines signal through members of the G-protein-coupled receptor family.

Another way in which cells in the innate immune system are able to detect the presence of infection is by binding bacterial peptides containing *N*-formylmethionine, or fMet, a modified amino acid that is uniquely present in prokaryotes. The receptor that recognizes these peptides is known as the fMet-Leu-Phe (fMLP) receptor, after a tripeptide for which it has a high affinity, although it is not restricted to binding just this tripeptide. The fMLP receptor belongs to an ancient and widely distributed family of receptors that have seven membrane-spanning segments; the best-characterized members of this family are the photoreceptors rhodopsin and bacteriorhodopsin. In the immune system, members of this family of receptors have several essential roles; the receptors for the anaphylatoxins (see Section 2-20) and for chemokines (see Section 2-24) belong to this family.

All receptors of this family use the same mechanism of signaling; ligand binding activates a member of a class of GTP-binding proteins called **G proteins**. These are sometimes called 'heterotrimeric G proteins'—to distinguish them from the family of 'small' GTPases typified by Ras—because each is made up of three subunits: Gα, Gβ, and Gγ. The Gα subunit is similar to the single subunit of the small GTPases and works in the same way, being active when GTP

Fig 6.35 Toll-like receptors activate NFκB. Toll-like receptors (TLRs) activate NFκB by a pathway that is different in its early stages from that leading from the antigen receptors or TNF receptors. TLRs signal via a domain in their cytoplasmic tails called a TIR domain, which recruits a family of adaptor proteins that also contain a TIR domain. The best studied of these adaptors is MyD88. In addition to its TIR domain, MyD88 contains a death domain (DD), by which it activates and recruits the serine/threonine kinase IRAK. Activated IRAK recruits the adaptor TRAF-6, which stimulates the activation of TAK1, a MAPKKK. TAK1 stimulates the activation of IKK, resulting in the destruction of IκB and the activation of NFκB. TAK1 also stimulates the activation of the MAP kinases JNK and p38.

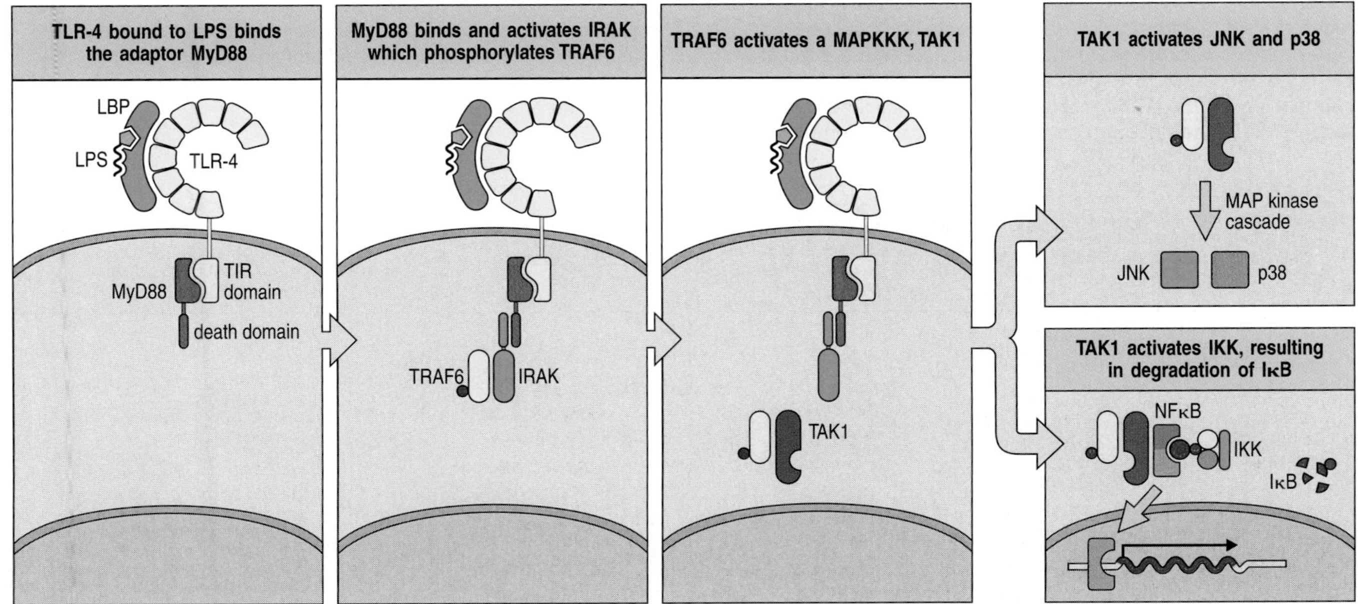

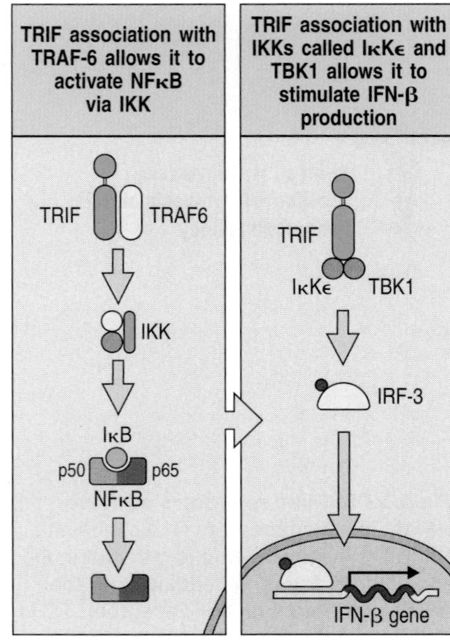

Fig. 6.36 MyD88-independent signaling from TLRs is mediated by TRIF. TLR-4 also signals through what is known as the MyD88-independent signaling pathway. In this pathway, a TIR-domain-containing adaptor protein called TRIF is recruited to the receptor instead of MyD88. TRIF can bind directly to TRAF-6 and can therefore stimulate activation of NFκB. TRIF can also activate two serine/threonine kinases called IκKε and TBK1. The activation of these kinases stimulates interferon-regulatory factor (IRF), a transcription factor that stimulates transcription of the gene for interferon (IFN)-β.

is bound and inactive when GDP is bound. Roughly 20 different heterotrimeric G proteins are known, each interacting with different cell-surface receptors and transmitting signals to different intracellular pathways. In the resting state, the G protein is inactive, not associated with the receptor, and has a molecule of GDP bound to the α subunit. When the receptor binds its ligand, a conformational change in the receptor allows it to bind the G protein, which results in the displacement of the GDP from the G protein and its replacement with GTP. The G protein now dissociates into two components, the α subunit and a complex of β and γ subunits; each of these components can interact with other cellular components to transmit and amplify the signal. The intrinsic GTPase activity of the α subunit results in the hydrolysis of GTP to GDP, and thus allowing the α and βγ subunits to reassociate (Fig. 6.37). Since the intrinsic rate of GTP hydrolysis by α subunits is relatively slow, the activity of heterotrimeric G protein signaling in vivo is regulated by a family of GTPase activating proteins known as RGS proteins, which accelerate the rate of GTP hydrolysis.

Important enzyme targets for the active G-protein subunits are adenylate cyclase, which produces the second messenger cyclic AMP, phospholipase C, whose activation gives rise to IP_3 and Ca^{2+}, tyrosine kinases like BTK and regulators of Ras family G proteins. These second messengers in turn activate a variety of intracellular pathways that affect cell metabolism, motility, gene

Fig. 6.37 Seven-transmembrane-domain receptors signal by coupling with heterotrimeric G proteins. Seven transmembrane-domain receptors such as the chemokine receptors signal through GTP-binding proteins known as heterotrimeric G proteins. In the inactive state, the α subunit of the G protein is bound to GDP and associated with two other subunits called the β and γ subunits. When the ligand binds to the receptor, the receptor interacts with the G protein complex resulting in replacement of GDP with GTP. This triggers the disassociation of the complex into two parts, the α subunit and the β/γ subunit, both of which can activate other proteins at the inner surface of the cell membrane. The activated response ceases when the intrinsic GTPase activity of the α subunit cleaves the GTP to GDP, allowing the α and β/γ subunits to reassociate.

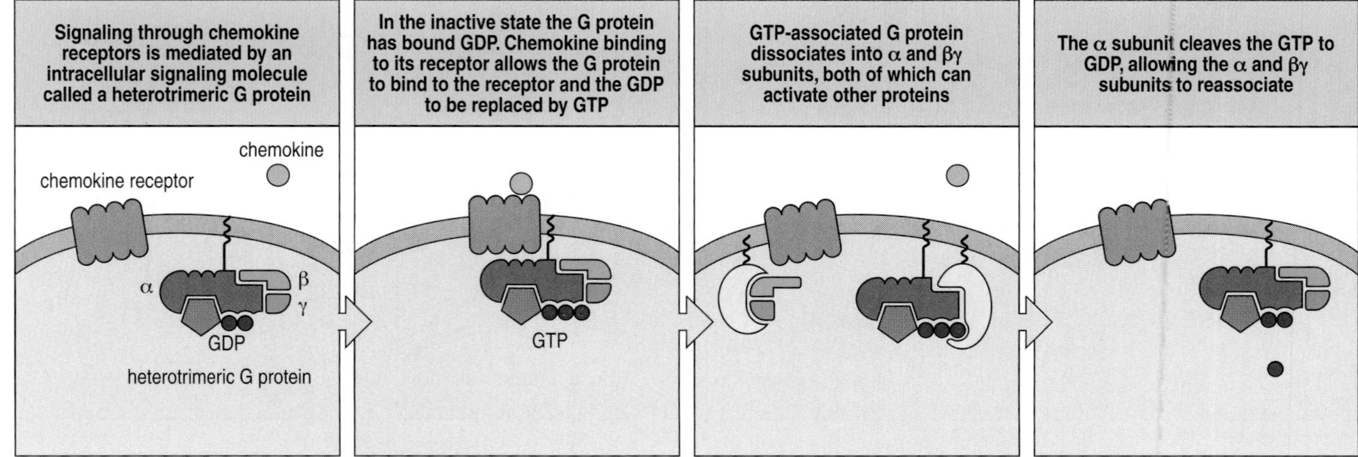

expression, and cell division. Thus, activation of G-protein-coupled receptors can have a wide variety of effects depending on the precise nature of the receptor and the G proteins that it interacts with, as well as the different downstream pathways that are activated in different cell types.

Summary.

Many different signals govern lymphocyte behavior, only some of which are delivered via the antigen receptor. Lymphocyte development, activation, and longevity are clearly influenced by the antigen receptor, but these processes are also regulated by other extracellular signals. Other signals are delivered in a variety of ways. An ancient signaling pathway with a role in host defense leads rapidly from the IL-1 receptor or the Toll-like receptors to initiate the degradation of the inhibitory protein IκB and the release of the transcription factor NFκB, which can then enter the nucleus and activate the transcription of specific genes, many involved in innate immunity. Most cytokines signal through an express pathway that links receptor-associated JAK kinases to pre-formed STAT transcription proteins, which after phosphorylation dimerize through their SH2 domains and head for the nucleus. Activated lymphocytes are programmed to die when the Fas receptor that they express binds the Fas ligand. This transmits a death signal, which activates a protease cascade that triggers apoptosis. Lymphocyte apoptosis is inhibited by some members of the intracellular Bcl-2 family and promoted by others. Working out the complete picture of the signals processed by lymphocytes as they develop, circulate, respond to antigen, and die is an immensely exciting prospect.

Summary to Chapter 6.

Signaling by cell-surface receptors of many different sorts is crucial to the ability of the immune system to respond appropriately to foreign pathogens. The importance of these signaling pathways is demonstrated by the many diseases that are due to aberrant signaling, which include both immunodeficiency diseases and autoimmune diseases. Common features of many signaling pathways are the generation of second messengers such as calcium and phosphoinositides and the activation of both serine/threonine and tyrosine kinases. An important concept in the initiation of signaling pathways by receptor proteins is the recruitment of signaling proteins to the plasma membrane and the assembly of multiprotein signaling complexes. In many cases, signal transduction leads to the activation of transcription factors that lead directly or indirectly to the proliferation, differentiation, and effector function of activated lymphocytes. Other roles of signal transduction are to mediate changes in the cytoskeleton important for cell functions like migration and shape changes.

While we are beginning to understand the basic circuitry of signal transduction pathways, it is important to keep in mind that we do not yet understand why these pathways are so complex. The complexity of signaling pathways could have roles in properties like amplification, robustness, diversity and efficiency of signaling responses. An important goal for the future will be to understand how the design principles of each signaling pathway contributes to the particular quality and sensitivity needed for specific signaling responses.

Questions.

6.1 Discuss the role of phosphotyrosine in signal transduction.

6.2 Describe different mechanisms used to recruit signaling molecules to the plasma membrane.

6.3 What are some of the advantages of using complexes of many signaling proteins for signal transduction?

6.4 How are G proteins regulated?

6.5 Describe how phospholipase C-γ is activated by T-cell receptor signaling.

6.6 Describe three different pathways used by immune-system cells to activate NFκB.

6.7 Name at least three differences between T-cell and B-cell receptor signaling.

6.8 Speculate why CD28 family members are both positive and negative regulators of T-cell activation.

6.9 Compare and contrast the intrinsic versus the extrinsic pathway of apoptosis.

6.10 Suggest some reasons why signaling pathways are so complicated.

General references.

Alberts, B., Johnson, A., Lewis, J., Raff, M., Roberts, K. and Walter, P.: *Molecular Biology of the Cell*, 5th ed. New York: Garland Science, 2008.

Gomperts, B., Kramer, I., Tatham, P.: *Signal Transduction.* San Diego: Elsevier, 2002.

Section references.

6-1 Transmembrane receptors convert extracellular signals into intracellular biochemical events.

Lin, J., and Weiss, A.: **T cell receptor signalling.** *J. Cell Sci.* 2001, **114**:243–244.

Weiss, A., and Littman, D.R.: **Signal transduction by lymphocyte antigen receptors.** *Cell* 1994, **76**:263–274.

6-2 Intracellular signal transduction often takes place in large multiprotein signaling complexes.

Pawson, T.: **Specificity in signal transduction: from phosphotyrosine-SH2 domain interactions to complex cellular systems.** *Cell* 2004, **116**:191–203.

Pawson, T., and Nash, P.: **Assembly of cell regulatory systems through protein interaction domains.** *Science* 2003, **300**:445–452.

Pawson, T., and Scott, J.D.: **Signaling through scaffold, anchoring, and adaptor proteins.** *Science* 1997, **278**:2075–2080.

6-3 The activation of some receptors generates small-molecule second messengers.

Kresge, N., Simoni, R.D., and Hill, R.L.: **Earl W. Sutherland's discovery of cyclic adenine monophosphate and the second messenger system.** *J. Biol Chem.* 2005, **280**:39–40.

Rall, T.W., and Sutherland, E.W.: **Formation of a cyclic adenine ribonucleotide by tissue particles.** *J. Biol Chem.*1958, **232**:1065–1076.

6-4 Small G proteins act as molecular switches in many different signaling pathways.

Cantrell, D.A.: **GTPases and T-cell activation.** *Immunol. Rev.* 2003, **192**:122–130.

Etienne-Manneville, S., and Hall, A.: **Rho GTPases in cell biology.** *Nature* 2002, **420**:629–635.

Mitin, N., Rossman, K.L., and Der, C.J.: **Signaling interplay in Ras superfamily function.** *Curr. Biol.* 2005, **15**:R563–R574.

6-5 Signaling proteins are recruited to the membrane by a variety of mechanisms.

Buday, L.: **Membrane-targeting of signaling molecules by SH2/SH3 domain-containing adaptor proteins.** *Biochim. Biophys. Acta* 1999, **1422**:187–204.

Kanai, F., Liu, H., Field, S.J., Akbary, H., Matsuo, T., Brown, G.E., Cantley, L.C., and Yaffe, M.B.: **The PX domains of p47phox and p40phox bind to lipid products of PI(3)K.** *Nat. Cell Biol.* 2001, **3**:675–678.

Kholodenko, B.N., Hoek, J.B., and Westerhoff, H.V.: **Why cytoplasmic signaling** proteins should be recruited to cell membranes. *Trends Cell Biol.* 2000, **10**:173–178.

Lemmon, M.A.: **Phosphoinositide recognition domains.** *Traffic* 2003, **4**:201–213.

6-6 Signal transduction proteins are organized in the plasma membrane in structures called lipid rafts.

Hancock, J.F.: **Lipid rafts: contentious only from simplistic standpoints.** *Nat. Rev. Mol. Cell Biol.* 2006, **7**:456–462.

Harder, T.: **Lipid raft domains and protein networks in T-cell receptor signal transduction.** *Curr. Opin. Immunol.* 2004, **16**:353–359.

Horejsi, V.: **Lipid rafts and their roles in T-cell activation.** *Microbes Infect.* 2005, **7**:310–316.

Shaw, A.S.: **Lipids rafts, now you see them, now you don't.** *Nat. Immunol.* 2006, **7**:1139-1142

6-7 Protein degradation has an important role in terminating signaling responses.

Ciechanover, A.: **Proteolysis: from the lysosome to ubiquitin and the proteasome.** *Nat. Rev. Mol. Cell Biol.* 2005, **6**:79–87.

Katzmann, D.J., Odorizzi, G., and Emr, S.D.: **Receptor downregulation and multivesicular-body sorting.** *Nat. Rev. Mol. Cell Biol.* 2002, **3**:893–905.

Liu, Y.C., Penninger, J., and Karin, M.: **Immunity by ubiquitylation: a reversible process of modification.** *Nat. Rev. Immunol.* 2005, **5**:941–952.

6-8 The variable chains of the T cell receptor are associated with invariant accessory chains that perform the signaling function of the receptor.

Call, M.E., Pyrdol, J., Wiedmann, M., and Wucherpfennig, K.W.: **The organizing principle in the formation of the T cell receptor-CD3 complex.** *Cell* 2002, **111**:967–979.

Exley, M., Terhorst, C., and Wileman, T.: **Structure, assembly and intracellular transport of the T cell receptor for antigen.** *Semin. Immunol.* 1991, **3**:283–297.

6-9 Lymphocytes are extremely sensitive to their specific antigens.

Gil, D., Schamel, W.W., Montoya, M., Sanchez-Madrid, F., and Alarcon, B.: **Recruitment of Nck by CD3 epsilon reveals a ligand-induced conformational change essential for T cell receptor signaling and synapse formation.** *Cell* 2002, **109**:901–912.

Harding, C.V., and Unanue, E.R.: **Quantitation of antigen-presenting cell MHC class II/peptide complexes necessary for T-cell stimulation.** *Nature* 1990, **346**:574–576.

Irvine, D.J., Purbhoo, M.A., Krogsgaard, M., and Davis, M.M.: **Direct observation of ligand recognition by T cells.** *Nature* 2002, **419**:845–849.

Krogsgaard, M., Li, Q.J., Sumen, C., Huppa, J.B., Huse, M., and Davis, M.M.: **Agonist/endogenous peptide-MHC heterodimers drive T cell activation and sensitivity.** *Nature* 2005, **434**:238–243.

Li, Q.J., Dinner, A.R., Qi, S., Irvine, D.J., Huppa, J.B., Davis, M.M., and Chakraborty, A.K.: **CD4 enhances T cell sensitivity to antigen by coodinating Lck accumulation at the immunological synapse.** *Nat. Immunol.* 2004, **5**:791–799.

6-10 Antigen binding leads to phosphorylation of the ITAM sequences associated with the antigen receptors.

Irving, B.A., and Weiss, A.: **The cytoplasmic domain of the T cell receptor zeta chain is sufficient to couple to receptor-associated signal transduction pathways.** *Cell* 1991, **64**:891–901.

Letourneur, F., and Klausner, R.D.: **Activation of T cells by a tyrosine kinase activation domain in the cytoplasmic tail of CD3 epsilon.** *Science* 1992, **255**:79–82.

Romeo, C. and Seed, B.: **Cellular immunity to HIV activated by CD4 fused to T cell or Fc receptor polypeptides.** *Cell* 1991, **64**:1037–1046.

6-11 In T cells, fully phosphorylated ITAMs bind the kinase ZAP-70 and enable it to be activated.

Chan, A.C., Dalton, M., Johnson, R., Kong, G.H., Wang, T., Thoma, R., and Kurosaki, T.: **Activation of ZAP-70 kinase activity by phosphorylation of tyrosine 493 is required for lymphocyte antigen receptor function.** *EMBO J.* 1995, **14**:2499–2508.

Chan, A.C., Iwashima, M., Turck, C.W., and Weiss, A.: **ZAP-70: a 70 kd protein-tyrosine kinase that associates with the TCR zeta chain.** *Cell* 1992, **71**:649–662.

Gauen, L.K., Zhu, Y., Letourneur, F., Hu, Q., Bolen, J.B., Matis, L.A., Klausner, R.D., and Shaw, A.S.: **Interactions of p59fyn and ZAP-70 with T-cell receptor activation motifs: defining the nature of a signalling motif.** *Mol. Cell Biol.* 1994, **14**:3729–3741.

Iwashima, M., Irving, B.A., van Oers, N.S., Chan, A.C., and Weiss, A.: **Sequential interactions of the TCR with two distinct cytoplasmic tyrosine kinases.** *Science* 1994, **263**:1136–1139.

6-12 Activated Syk and ZAP-70 phosphorylate scaffold proteins that mediate many of the downstream effects of antigen receptor signaling

Janssen, E., and Zhang, W.: **Adaptor proteins in lymphocyte activation.** *Curr. Opin. Immunol.* 2003, **15**:269–276.

Jordan, M.S., Singer, A.L., and Koretzky, G.A.: **Adaptors as central mediators of signal transduction in immune cells.** *Nat. Immunol.* 2003, **4**:110–116.

Samelson, L.E.: **Signal transduction mediated by the T cell antigen receptor: the role of adapter proteins.** *Annu. Rev. Immunol.* 2002, **20**:371–394.

6-13 PLC-γ is activated by Tec tyrosine kinases.

Berg, L.J., Finkelstein, L.D., Lucas, J.A., and Schwartzberg, P.L.: **Tec family kinases in T lymphocyte development and function.** *Annu. Rev. Immunol.* 2005, **23**:549–600.

Lewis, C.M., Broussard, C., Czar, M.J., and Schwartzberg, P.L.: **Tec kinases: modulators of lymphocyte signaling and development.** *Curr. Opin. Immunol.* 2001, **13**:317–325.

6-14 Activation of the small G protein Ras activates a MAP kinase cascade, resulting in the production of the transcription factor AP-1.

Downward, J., Graves, J.D., Warne, P.H., Rayter, S., and Cantrell, D.A.: **Stimulation of p21ras upon T-cell activation.** *Nature* 1990, **346**:719–723.

Leevers, S.J., and Marshall, C.J.: **Activation of extracellular signal-regulated kinase, ERK2, by p21ras oncoprotein.** *EMBO J.* 1992, **11**:569–574.

Thomas, G.: **MAP kinase by any other name smells just as sweet.** *Cell* 1992, **68**:3–6.

6-15 The transcription factor NFAT is indirectly activated by Ca^{2+}.

Hogan, P.G., Chen, L., Nardone, J., and Rao, A.: **Transcriptional regulation by calcium, calcineurin, and NFAT.** *Genes Dev.* 2003, **17**:2205–2232.

Macian, F., Lopez-Rodriguez, C., and Rao, A.: **Partners in transcription: NFAT and AP-1.** *Oncogene* 2001, **20**:2476–2489.

6-16 The transcription factor NFκB is activated by the actions of protein kinase C.

Matsumoto, R., Wang, D., Blonska, M., Li, H., Kobayashi, M., Pappu, B., Chen, Y., Wang, D., and Lin, X.: **Phosphorylation of CARMA1 plays a critical role in T cell receptor-mediated NF-κB activation.** *Immunity* 2005, **23**:575–585.

Rueda, D., and Thome, M.: **Phosphorylation of CARMA1: the link(er) to NF-κB activation.** *Immunity* 2005, **23**:551–553.

Sommer, K., Guo, B., Pomerantz, J.L., Bandaranayake, A.D., Moreno-Garcia, M.E., Ovechkina, Y.L., and Rawlings, D.J.: **Phosphorylation of the CARMA1 linker controls NF-κB activation.** *Immunity* 2005, **23**:561–574.

6-17 The logic of B-cell receptor signaling is similar to that of T-cell receptor signaling but some of the signaling components are specific to B cells.

Cambier, J.C., Pleiman, C.M., and Clark, M.R.: **Signal transduction by the B cell antigen receptor and its coreceptors.** *Annu. Rev. Immunol.* 1994, **12**:457–486.

DeFranco, A.L., Richards, J.D., Blum, J.H., Stevens, T.L., Law, D.A., Chan, V.W., Datta, S.K., Foy, S.P., Hourihane, S.L., Gold, M.R., *et al.*: **Signal transduction by the B-cell antigen receptor.** *Ann. NY Acad. Sci.* 1995, **766**:195–201.

Kurosaki, T.: **Functional dissection of BCR signaling pathways.** *Curr. Opin. Immunol.* 2000, **12**:276–281.

6-18 ITAMs are also found in other receptors on leukocytes that signal for cell activation.

Daeron, M.: **Fc receptor biology.** *Annu. Rev. Immunol.* 1997, **15**:203–234.

Lanier, L.L., and Bakker, A.B.: **The ITAM-bearing transmembrane adaptor DAP12 in lymphoid and myeloid cell function.** *Immunol. Today* 2000, **21**:611–614.

6-19 The cell-surface protein CD28 is a co-stimulatory receptor for naive T cells.

Acuto, O., and Michel, F.: **CD28-mediated co-stimulation: a quantitative support for TCR signalling.** *Nat. Rev. Immunol.* 2003, **3**:939–951.

Frauwirth, K.A., Riley, J.L., Harris, M.H., Parry, R.V., Rathmell, J.C., Plas, D.R., Elstrom, R.L., June, C.H., and Thompson, C.B.: **The CD28 signaling pathway regulates glucose metabolism.** *Immunity* 2002, **16**:769–777.

6-20 Inhibitory receptors on lymphocytes help regulate immune responses.

Chen, L.: **Co-inhibitory molecules of the B7-CD28 family in the control of T-cell immunity.** *Nat. Rev. Immunol.* 2004, **4**:336–347.

Lanier, L.L.: **NK cell receptors.** *Annu. Rev. Immunol.* 1998, **16**:359–393.

McVicar, D.W., and Burshtyn, D.N.: **Intracellular signaling by the killer immunoglobulin-like receptors and Ly49.** *Sci. STKE* **2001**:re1. doi:10.1126/stke.2001.75.re1.

Moretta, A., Biassoni, R., Bottino, C., and Moretta, L.: **Surface receptors delivering opposite signals regulate the function of human NK cells.** *Semin. Immunol.* 2000, **12**:129–138.

Riley, J.L., and June, C.H.: **The CD28 family: a T-cell rheostat for therapeutic control of T-cell activation.** *Blood* 2005, **105**:13–21.

Rudd, C.E., and Schneider, H.: **Unifying concepts in CD28, ICOS and CTLA4 co-receptor signalling.** *Nat. Rev. Immunol.* 2003, **3**:544–556.

Sharpe, A.H., and Freeman, G.J.: **The B7-CD28 superfamily.** *Nat. Rev. Immunol.* 2002, **2**:116–126.

Tomasello, E., Blery, M., Vely, F., and Vivier, E.: **Signaling pathways engaged by NK cell receptors: double concerto for activating receptors, inhibitory receptors and NK cells.** *Semin. Immunol.* 2000, **12**:139–147.

6-21 Cytokines typically activate fast signaling pathways that end in the nucleus.

Fu, X.Y.: **A transcription factor with SH2 and SH3 domains is directly activated by an interferon α-induced cytoplasmic protein tyrosine kinase(s).** *Cell* 1992, **70**:323–335.

Schindler, C., Shuai, K., Prezioso, V.R., and Darnell, J.E., Jr.: **Interferon-dependent tyrosine phosphorylation of a latent cytoplasmic transcription factor.** *Science* 1992, **257**:809–813.

6-22 Cytokine receptors form dimers or trimers on ligand binding.

de Vos, A.M., Ultsch, M., and Kossiakoff, A.A.: **Human growth hormone and extracellular domain of its receptor: crystal structure of the complex.** *Science* 1992, **255**:306–312.

Ihle, J.N.: **Cytokine receptor signalling.** *Nature* 1995, **377**:591–594.

6-23 Cytokine receptors are associated with the JAK family of tyrosine kinases which activate STAT transcription factors.

Leonard, W.J., and O'Shea, J.J.: **Jaks and STATs: biological implications.** *Annu. Rev. Immunol.* 1998, **16**:293–322.

Levy, D.E., and Darnell, J.E., Jr.: **Stats: transcriptional control and biological impact.** *Nat. Rev. Mol. Cell Biol.* 2002, **3**:651–662.

6-24 Cytokine signaling is terminated by a negative feedback mechanism.

Shuai, K., and Liu, B.: **Regulation of JAK-STAT signalling in the immune system.** *Nat. Rev. Immunol.* 2003, **3**:900–911.

Yasukawa, H., Sasaki, A., and Yoshimura, A.: **Negative regulation of cytokine signaling pathways.** *Annu. Rev. Immunol.* 2000, **18**:143–164.

6-25 The receptors that induce apoptosis activate specialized intracellular proteases called caspases.

Aggarwal, B.B.: **Signalling pathways of the TNF superfamily: a double-edged sword.** *Nat. Rev. Immunol.* 2003, **3**:745–756.

Bishop, G.A.: **The multifaceted roles of TRAFs in the regulation of B-cell function.** *Nat. Rev. Immunol.* 2004, **4**:775–786.

Siegel, R.M.: **Caspases at the crossroads of immune-cell life and death.** *Nat. Rev. Immunol.* 2006, **6**:308–317.

6-26 The intrinsic pathway of apoptosis is mediated by release of cytochrome *c* from mitochondria.

Borner, C.: **The Bcl-2 protein family: sensors and checkpoints for life-or-death decisions.** *Mol. Immunol.* 2003, **39**:615–647.

Hildeman, D.A., Zhu, Y., Mitchell, T.C., Kappler, J., and Marrack, P.: **Molecular mechanisms of activated T cell death *in vivo*.** *Curr. Opin. Immunol.* 2002, **14**:354–359.

Strasser, A.: **The role of BH3-only proteins in the immune system.** *Nat. Rev. Immunol.* 2005, **5**:189–200.

6-27 Microbes and their products act via Toll-like receptors to activate NFκB.

Akira, S., and Takeda, K.: **Toll-like receptor signalling.** *Nat. Rev. Immunol.* 2004, **4**:499–511.

Barton, G.M., and Medzhitov, R.: **Toll-like receptor signaling pathways.** *Science* 2003, **300**:1524–1525.

Beutler, B.: **Inferences, questions and possibilities in Toll-like receptor signalling.** *Nature* 2004, **430**:257–263.

Beutler, B., Hoebe, K., Du, X., and Ulevitch, R.J.: **How we detect microbes and respond to them: the Toll-like receptors and their transducers.** *J. Leukoc. Biol.* 2003, **74**:479–485.

6-28 Bacterial peptides, mediators of inflammatory responses, and chemokines signal through members of the G-protein-coupled receptor family.

Gerber, B.O., Meng, E.C., Dotsch, V., Baranski, T.J., and Bourne, H.R.: **An activation switch in the ligand binding pocket of the C5a receptor.** *J. Biol. Chem.* 2001, **276**:3394–3400.

Pierce, K.L., Premont, R.T., and Lefkowitz, R.J.: **Seven-transmembrane receptors.** *Nat. Rev. Mol. Cell. Biol.* 2002, **3**:639-650.

Proudfoot, A.E.: **Chemokine receptors: multifaceted therapeutic targets.** *Nat. Rev. Immunol.* 2002, **2**:106–115.

The Development and Survival of Lymphocytes

7

As described in Chapters 3 and 4, the antigen receptors carried by B and T lymphocytes are immensely variable in their antigen specificity, enabling an individual to make immune responses against the wide range of pathogens encountered during a lifetime. This diverse repertoire of B-cell receptors and T-cell receptors is generated during the development of B cells and T cells from their uncommitted precursors. The production of new lymphocytes, or **lymphopoiesis**, takes place in specialized lymphoid tissues—the **central lymphoid tissues**—which are the bone marrow for most B cells and the thymus for most T cells. Lymphocyte precursors originate in the bone marrow, but whereas B cells complete most of their development there, the precursors of most T cells migrate to the thymus, where they develop into mature T cells. B cells also originate and develop in the fetal liver and the neonatal spleen. Some T cells that form specialized populations within the gut epithelium may migrate as immature precursors from the bone marrow to develop in sites called 'cryptopatches' just under the intestinal epithelial crypts. We will mainly focus here on the development of bone marrow derived B cells and thymus-derived T cells.

In the fetus and the juvenile, the central lymphoid tissues are the sources of large numbers of new lymphocytes, which migrate to populate the **peripheral lymphoid tissues** such as lymph nodes, spleen and mucosal lymphoid tissue. In mature individuals, the development of new T cells in the thymus slows down, and T-cell numbers are maintained through long-lived individual T cells together with the division of mature T cells outside the central lymphoid organs. New B cells, in contrast, are continually produced from the bone marrow, even in adults.

Chapter 4 described the structure of the antigen receptor genes expressed by B and T cells, introduced the mechanisms controlling the DNA rearrangements needed to assemble a complete antigen receptor, and explained how these processes can generate an antigen receptor repertoire of high diversity. This chapter builds on that foundation to explain how B and T lymphocytes themselves develop from a common progenitor through a series of stages, and how each of these stages tests for the proper assembly of antigen receptors.

Once an antigen receptor has been formed, rigorous testing is required to select lymphocytes that carry useful antigen receptors—that is, antigen receptors that are capable of recognizing pathogens and yet will not react against an individual's own cells. Given the incredible diversity of receptors that the rearrangement process can generate, it is important that those lymphocytes that mature are likely to be useful in recognizing and responding to foreign antigens, especially as an individual can express only a small fraction of the total possible receptor repertoire in his or her lifetime. We describe how the specificity and affinity of the receptor for self ligands are tested to make the determination of whether the immature lymphocyte will survive into the mature repertoire, or die. In general, it seems that developing lymphocytes whose receptors interact weakly with self antigens, or bind self antigens in a particular way, receive a signal that enables them to survive; this type of selection is known as **positive selection**. Positive selection is particularly critical in the development of α:β T cells, which recognize composite antigens consisting of peptides bound to MHC molecules, because it ensures that an individual's T cells will be able to respond to peptides bound to his or her MHC molecules.

In contrast, lymphocytes with strongly self-reactive receptors must be eliminated to prevent autoimmune reactions; this process of **negative selection** is one of the ways in which the immune system is made self-tolerant. The default fate of developing lymphocytes, in the absence of any signal being received from the receptor, is death, and as we will see, the vast majority of developing lymphocytes die before emerging from the central lymphoid organs or before completing maturation in the peripheral lymphoid organs.

In this chapter we describe the different stages of the development of B cells and T cells in mice and humans, from the uncommitted stem cell up to the mature, functionally specialized lymphocyte with its unique antigen receptor ready to respond to a foreign antigen. The final stages in the life history of a mature lymphocyte, in which an encounter with a foreign antigen activates it to become an effector or memory lymphocyte, are discussed in Chapters 8–10. The chapter is divided into five parts. The first two describe B-cell and T-cell development, respectively. Although there are similarities in these two processes, we present B-cell and T-cell development separately insofar as they take place in separate central lymphoid compartments. We then look at the processes of positive and negative selection of T cells in the thymus. Next, we describe the fate of the newly generated lymphocytes as they leave the central lymphoid organs and migrate to the peripheral lymphoid tissues, where further maturation occurs. Mature lymphocytes continually recirculate between the blood and peripheral lymphoid tissues (see Chapter 1) and, in the absence of infection, their numbers remain relatively constant, despite the continual production of new lymphocytes. We look at the factors that govern the survival of naive lymphocytes in the peripheral lymphoid organs, and the maintenance of lymphocyte homeostasis. Finally, we describe some lymphoid tumors; these are cells that have escaped from the normal controls on cell proliferation and are also of interest because they capture features of the different developmental stages of B cells and T cells.

Development of B lymphocytes

The main phases of a B-cell's life history are shown in Fig. 7.1. The stages in B-lymphocyte development are defined mainly by the various steps in the assembly and expression of functional antigen-receptor genes, and by the appearance of features that distinguish the different functional types of B and T cells. At each step of lymphocyte development, the progress of gene rearrangement is monitored and the major recurring theme in this phase of development is that successful gene rearrangement leads to the production of a protein chain that serves as a signal for the cell to progress to the next stage. We will see that although a developing B cell is presented with opportunities for multiple rearrangements that increase the likelihood of expressing a functional antigen receptor, there are also specific checkpoints that reinforce the requirement that each B cell expresses just one receptor specificity. We shall start by looking at how the earliest recognizable cells of the B-cell lineage develop from the pluripotent hematopoietic stem cell in the bone marrow, and at what point the B-cell and T-cell lineages divide.

7-1 Lymphocytes derive from hematopoietic stem cells in the bone marrow.

The cells of the lymphoid lineage—B cells, T cells, and NK cells—are all derived from common lymphoid progenitors, which themselves derive from

Fig. 7.1 B cells develop in the bone marrow and migrate to peripheral lymphoid organs, where they can be activated by antigens. In the first phase of development, progenitor B cells in the bone marrow rearrange their immunoglobulin genes. This phase is independent of antigen but is dependent on interactions with bone marrow stromal cells (first panels). It ends in an immature B cell that carries an antigen receptor in the form of cell-surface IgM and can now interact with antigens in its environment. Immature B cells that are strongly stimulated by antigen at this stage either die or are inactivated in a process of negative selection, thus removing many self-reactive B cells from the repertoire (second panels). In the third phase of development, the surviving immature B cells emerge into the periphery and mature to express IgD as well as IgM. They can now be activated by encounter with their specific foreign antigen in a secondary lymphoid organ (third panels). Activated B cells proliferate, and differentiate into antibody-secreting plasma cells and long-lived memory cells (fourth panels).

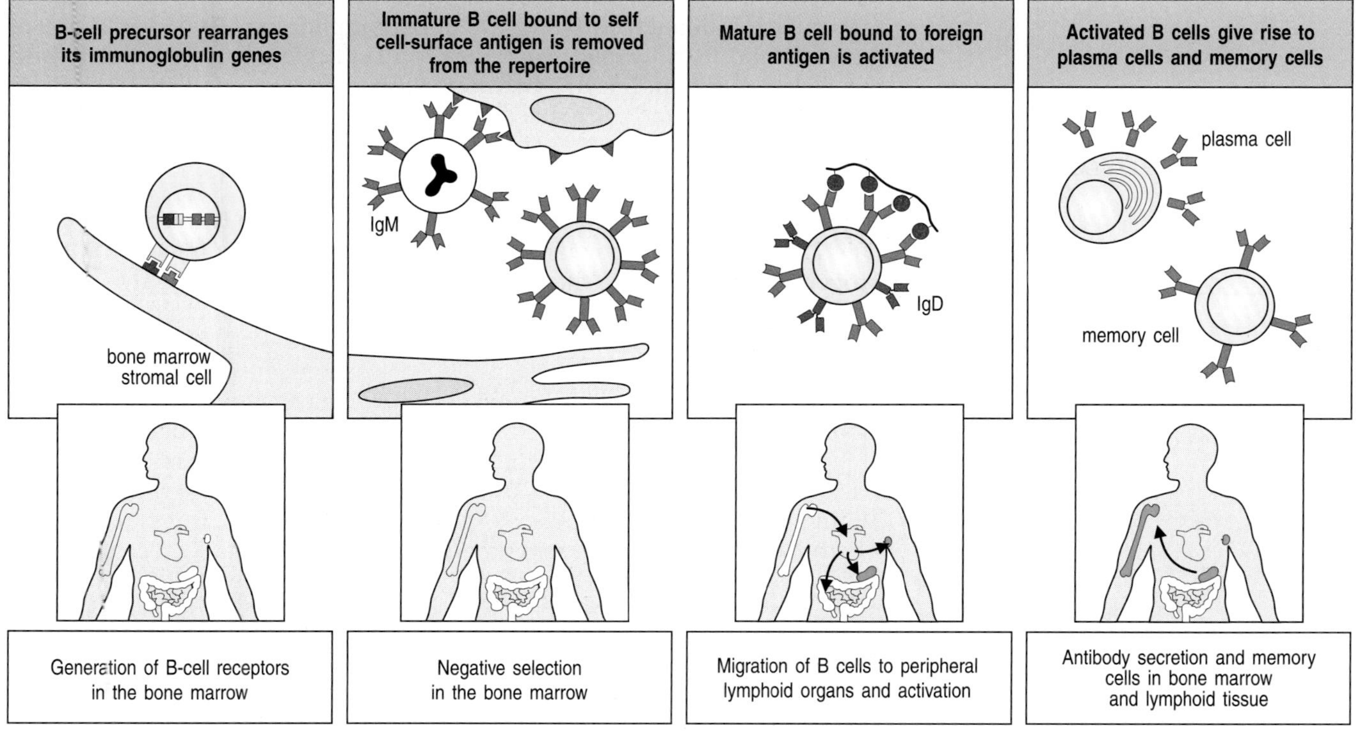
B-cell precursor rearranges its immunoglobulin genes
Immature B cell bound to self cell-surface antigen is removed from the repertoire
Mature B cell bound to foreign antigen is activated
Activated B cells give rise to plasma cells and memory cells
bone marrow stromal cell
IgM
IgD
plasma cell
memory cell
Generation of B-cell receptors in the bone marrow
Negative selection in the bone marrow
Migration of B cells to peripheral lymphoid organs and activation
Antibody secretion and memory cells in bone marrow and lymphoid tissue

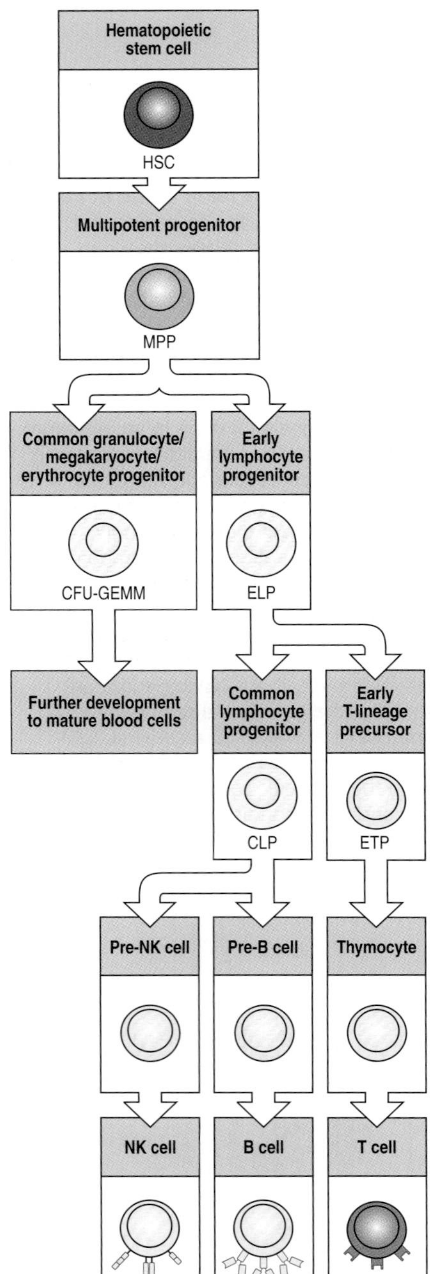

Fig. 7.2 A pluripotent hematopoietic stem cell generates all the cells of the immune system. In the bone marrow or other hematopoietic sites, the pluripotent stem cell gives rise to cells with progressively more limited potential. The multipotent progenitor (MPP), for example, has lost its stem-cell properties. The first branch leads to cells with myeloid and erythroid potential on the one hand (CFU-GEMM) and on the other to the early lymphoid progenitor (ELP), with lymphoid potential. The former give rise to all nonlymphoid cellular blood elements, including circulating monocytes and granulocytes and the macrophages and dendritic cells that reside in tissues and secondary lymphoid organs (not shown). The ELP can give rise to NK cells, T cells or B cells through successive stages of differentiation in either the bone marrow or thymus. The common lymphoid progenitor (CLP) is so called because it was thought to be the stage that gave rise to both the B-cell and T-cell lineages, but although it can give rise to T cells and B cells in culture, it is not clear whether it does so *in vivo*. There may be considerable plasticity in these pathways, in that in certain circumstances progenitor cells may switch their commitment. For example, a progenitor cell may give rise to either B cells or macrophages; however, for simplicity these alternative pathways are not shown. Some dendritic cells are also thought to be derived from the lymphoid progenitor.

the pluripotent **hematopoietic stem cells** that give rise to all blood cells (see Fig. 1.3). Development from the precursor stem cell into cells that are committed to becoming B cells or T cells follows certain basic principles of cell differentiation. Properties that are essential for the function of the mature cell are gradually acquired, along with the loss of properties that are more characteristic of the immature cell. In the case of lymphocyte development, cells become committed first to the lymphoid, as opposed to the myeloid, lineage, and then to either the B-cell or the T-cell lineages (Fig. 7.2).

The specialized microenvironment of the bone marrow provides signals both for the development of lymphocyte progenitors from hematopoietic stem cells and for the subsequent differentiation of B cells. Such signals act on the developing lymphocytes to switch on key genes that direct the developmental program. In the bone marrow, the external signals are produced by the network of specialized non-lymphoid connective-tissue **stromal cells** that interact intimately with the developing lymphocytes. The contribution of the stromal cells is twofold. First, they form specific adhesive contacts with the developing lymphocytes by interactions between cell-adhesion molecules and their ligands. Second, they provide soluble and membrane-bound cytokines and chemokines that control lymphocyte differentiation and proliferation.

Numerous factors secreted by bone marrow have roles in B-cell development (Fig. 7.3). The hematopoietic stem cell first differentiates into **multipotent progenitor cells** (**MPPs**), which can produce both lymphoid and myeloid cells but are no longer self-renewing stem cells. Multipotent progenitors express a cell-surface receptor tyrosine kinase known as FLT3 (originally called stem-cell kinase 1 (STK1) in humans and Flt3/Flk2 in mice) that binds the membrane-bound FLT3 ligand on stromal cells. Signaling through FLT3 is needed for differentiation to the next stage, the **common lymphoid progenitor** (**CLP**). This stage is called the common lymphoid progenitor because in the past it was thought to be the stage that gave rise to both the B-cell and T-cell lineages. Although it can give rise to T cells and B cells in culture, it is not yet clear whether the common lymphoid progenitor does so *in vivo*. A preceding stage has been identified, called the **early lymphoid progenitor (ELP)**, that gives rise to the T-cell precursors that migrate from the bone marrow to the thymus and also to the common lymphoid progenitor (see Fig. 7.2).

Lymphocyte differentiation is accompanied by expression of the receptor for interleukin-7 (IL-7), which is induced by FLT3 signaling together with the

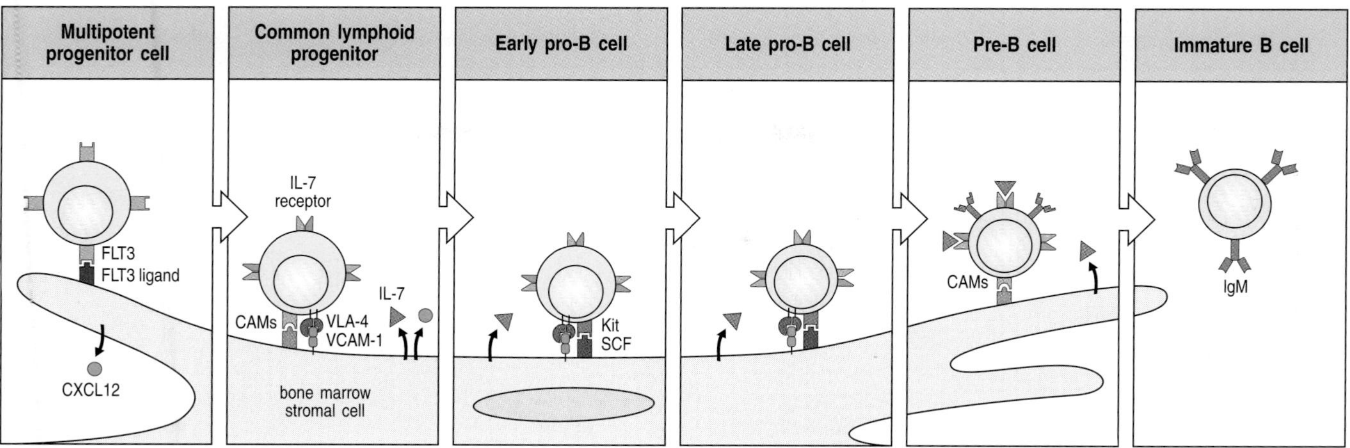

Fig. 7.3 The early stages of B-cell development are dependent on bone marrow stromal cells. Interaction of B-cell progenitors with bone marrow stromal cells is required for development to the immature B-cell stage. The designations pro-B cell and pre-B cell refer to defined phases of B-cell development as described in Fig. 7.6. Multipotent progenitor cells express the receptor tyrosine kinase FLT3, which binds to its ligand on stromal cells. Signaling through FLT3 is required for differentiation to the next stage, the common lymphoid progenitor. The receptor for interleukin-7 (IL-7) is present from this stage, and IL-7 produced by stromal cells is required for the development of B-lineage cells. The chemokine CXCL12 (SDF-1) acts to retain stem cells and lymphoid progenitors to appropriate stromal cells in the bone marrow. Progenitor cells bind to the adhesion molecule VCAM-1 on stromal cells through the integrin VLA-4 and also interact through other cell-adhesion molecules (CAMs). The adhesive interactions promote the binding of the receptor tyrosine kinase Kit (CD117) on the surface of the pro-B cell to stem-cell factor (SCF) on the stromal cell, which activates the kinase and induces the proliferation of B-cell progenitors.

activity of the transcription factor PU.1. The cytokine IL-7, secreted by stromal cells, is essential for the growth and survival of developing B cells in mice (but possibly not in humans) and of T cells in both mice and humans. Another essential factor is stem-cell factor (SCF), a membrane-bound cytokine present on stromal cells that stimulates the growth of hematopoietic stem cells and earliest B-lineage progenitors. SCF interacts with the receptor tyrosine kinase Kit on the precursor cells (see Fig. 7.3). The chemokine CXCL12 (stromal cell-derived factor 1, SDF-1) is also essential for the early stages of B-cell development. It is produced constitutively by the stromal cells, and one of its roles may be to retain developing B-cell precursors in the marrow microenvironment. Thymic stroma-derived lymphopoietin (**TSLP**) resembles IL-7 and binds a receptor sharing the common γ chain of the IL-7 receptor. TSLP may promote B-cell development in the embryonic liver and, in the perinatal period at least, in the mouse bone marrow.

The common lymphoid progenitor gives rise to the earliest B-lineage cell, the **pro-B cell** (see Fig. 7.3), in which immunoglobulin gene rearrangement begins. A definitive B-cell fate is specified by induction of the B-lineage-specific transcription factor E2A, which is present as two alternatively spliced forms called E12 and E47, and the early B-cell factor (EBF) (Fig. 7.4). It is thought that IL-7 signaling promotes the expression of E2A, which cooperates with the transcription factor PU.1 to induce the expression of EBF. Together, E2A and EBF act to drive the expression of proteins that determine the pro-B cell state.

As B-lineage cells mature, they migrate within the marrow, remaining in contact with the stromal cells. The earliest stem cells lie in a region called the **endosteum**, which is adjacent to the inner surface of the bone. Developing B-lineage cells make contact with reticular stromal cells in the trabecular spaces, and as they mature they move toward the central sinus of the marrow cavity. The final stages of development of immature B cells into mature B cells occur in peripheral lymphoid organs such as the spleen.

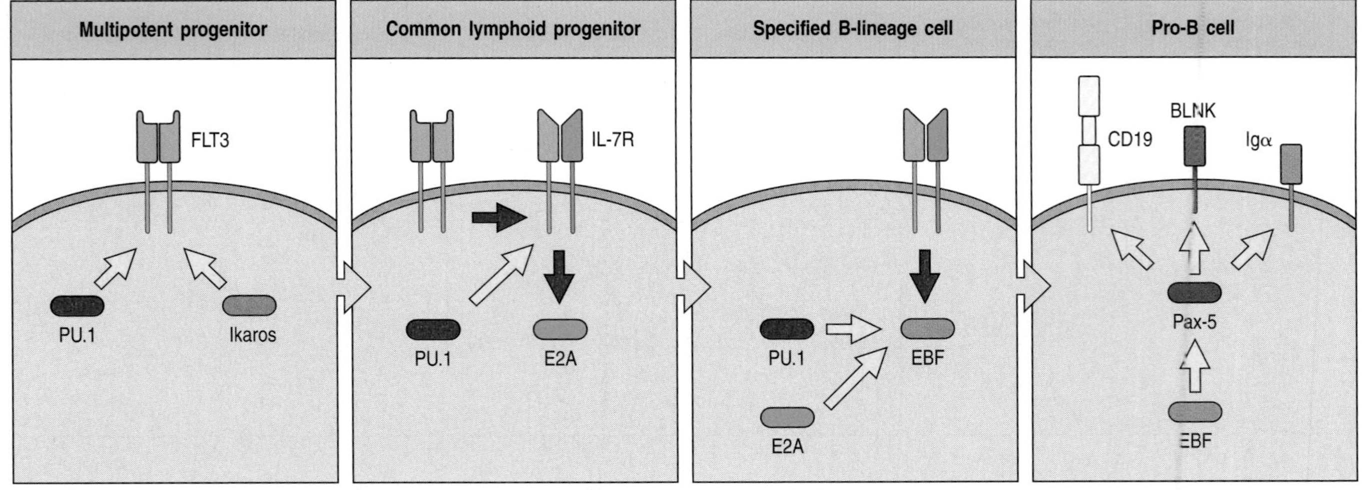

Fig 7.4 Early stages of B-cell development in the mouse are orchestrated by gene regulatory networks of transcription factors and growth factor receptors. The transcription factors PU.1 and Ikaros expressed in the multipotent progenitor cell promote expression of FLT3, which interacts with a ligand expressed on bone marrow stromal cells (see Fig. 7.3). FLT3 signaling acts in concert with PU.1 to induce the expression of the IL-7 receptor. IL-7, secreted by stromal cells, is required for growth and survival of developing B cells in mice, and acts to induce E2A in the common lymphoid progenitor. Together with PU.1 and E2A, IL-7 subsequently induces expression of EBF, which marks a specified B-lineage cell, and then Pax-5, which directs the expression of B-cell-specific proteins such as the B-cell co-receptor component CD19, the signaling protein Igα, and the scaffold protein BLNK (see Chapter 6) by pro-B cells.

7-2 B-cell development begins by rearrangement of the heavy-chain locus.

The stages of B-cell development are, in the order that they occur: **early pro-B cell**, **late pro-B cell**, **large pre-B cell**, **small pre-B cell**, and **mature B cell** (Fig. 7.5). Only one gene locus is rearranged at a time, in a fixed sequence. Both B cells and T cells rearrange the locus that contains D gene segments first: for B cells this is the immunoglobulin heavy-chain (IgH) locus. As shown in Fig. 7.5, expression of a functional heavy chain allows the formation of the **pre-B-cell receptor**, which is the signal to the cell to proceed to the next stage of development, the rearrangement of a light-chain gene. The transcription factors E2A and EBF in the early pro-B cell induce the expression of several key proteins that enable gene rearrangement to occur, including the RAG-1 and RAG-2 components of the V(D)J recombinase (see Chapter 4). Thus, E2A and EBF allow the initiation of V(D)J recombination at the heavy-chain locus and the expression of a heavy chain. In the absence of E2A or EBF, even the earliest identifiable stage in B-cell development, D to J_H joining, fails to occur.

Another key protein induced by E2A and EBF is the transcription factor Pax-5, one isoform of which is known as the B-cell activator protein (BASP). Among the targets of Pax-5 are the gene for the B-cell co-receptor component CD19 and the gene for Igα, a signaling component of both the pre-B-cell receptor and the B-cell receptor (see Section 6-8). In the absence of Pax-5, pro-B cells fail to develop further down the B-cell pathway but can be induced to give rise to T cells and myeloid cell types, indicating that Pax-5 is required for commitment of the pro-B cell to the B-cell lineage. Pax-5 also induces the expression of the B-cell linker protein (BLNK), a signaling molecule that is required for further development of the pro-B cell and for signaling from the mature B-cell antigen receptor (see Section 6-17). The temporal expression of some of the surface proteins, receptors and transcription factors required for B-cell development are listed in Fig. 7.6.

Although the V(D)J recombinase system operates in both B- and T-lineage cells and uses the same core enzymes, rearrangements of T-cell receptor genes do not occur in B-lineage cells, nor do complete rearrangements of immunoglobulin genes occur in T cells. The ordered rearrangement events that do occur are associated with lineage-specific low-level transcription of the gene segments about to be joined.

Rearrangement of the immunoglobulin heavy-chain locus begins in early pro-B cells with D to J_H joining (Fig. 7.7). This typically occurs at both alleles of the heavy-chain locus, at which point the cell becomes a late pro-B cell.

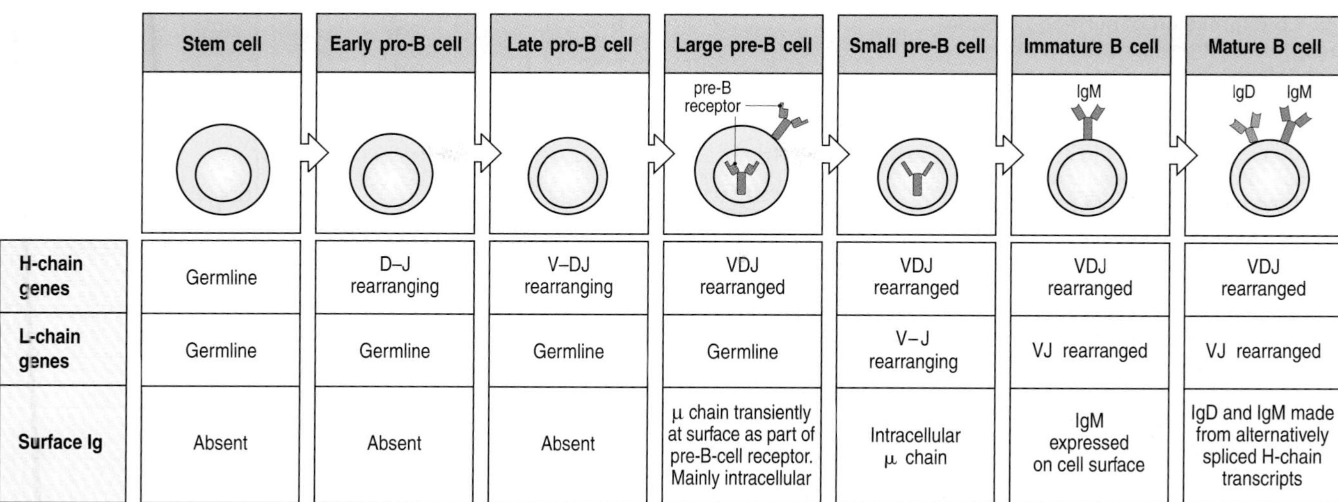

Stem cell
Early pro-B cell
Late pro-B cell
Large pre-B cell
Small pre-B cell
Immature B cell
Mature B cell
pre-B receptor
IgM
IgD IgM
H-chain genes
Germline
D–J rearranging
V–DJ rearranging
VDJ rearranged
VDJ rearranged
VDJ rearranged
VDJ rearranged
L-chain genes
Germline
Germline
Germline
Germline
V–J rearranging
VJ rearranged
VJ rearranged
Surface Ig
Absent
Absent
Absent
μ chain transiently at surface as part of pre-B-cell receptor. Mainly intracellular
Intracellular μ chain
IgM expressed on cell surface
IgD and IgM made from alternatively spliced H-chain transcripts

Fig. 7.5 The development of a B-lineage cell proceeds through several stages marked by the rearrangement and expression of the immunoglobulin genes. The stem cell has not yet begun to rearrange its immunoglobulin (Ig) gene segments; they are in the germline configuration as found in all nonlymphoid cells. The heavy-chain (H-chain) locus rearranges first. Rearrangement of a D gene segment to a J_H gene segment occurs in early pro-B cells, generating late pro-B cells in which V_H to DJ_H rearrangement occurs. A successful VDJ_H rearrangement leads to the expression of a complete immunoglobulin heavy chain as part of the pre-B-cell receptor, which is found mainly in the cytoplasm and to some degree on the surface of the cell. Once this occurs, the cell is stimulated to become a large pre-B cell, which proliferates. Large pre-B cells then cease dividing and become small resting pre-B cells, at which point they cease expression of the surrogate light chains and express the μ heavy chain alone in the cytoplasm. When the cells are again small, they re-express the RAG proteins and start to rearrange the light-chain (L-chain) genes. Upon successfully assembling a light-chain gene, a cell becomes an immature B cell that expresses a complete IgM molecule at the cell surface. Mature B cells produce a δ heavy chain as well as a μ heavy chain, by a mechanism of alternative mRNA splicing, and are marked by the additional appearance of IgD on the cell surface.

Most D to J_H joins in humans are potentially useful, because most human D gene segments can be translated in all three reading frames without encountering a stop codon. Thus, there is no need of a special mechanism for distinguishing successful D to J_H joins, and at this early stage there is also no need to ensure that only one allele undergoes rearrangement. Indeed, given the likely rate of failure at later stages, starting off with two successfully rearranged D–J sequences is an advantage.

To produce a complete immunoglobulin heavy chain, the late pro-B cell now proceeds with a second rearrangement of a V_H gene segment to a DJ_H sequence. In contrast to D to J_H rearrangement, V_H to DJ_H rearrangement occurs first on only one chromosome. A successful rearrangement leads to the production of intact μ heavy chains, after which V_H to DJ_H rearrangement ceases and the cell becomes a pre-B cell. Pro-B cells that do not produce a μ chain are eliminated, and at least 45% of pro-B cells are lost at this stage. In at least two out of three cases the first V_H to DJ_H rearrangement is nonproductive, and rearrangement then occurs on the other chromosome, again with a theoretical two in three chance of failure. A rough estimate of the chance of generating a pre-B cell is thus 55% $(1/3 + (2/3 \times 1/3) = 0.55)$. The actual frequency is somewhat lower, because the V gene segment repertoire contains pseudogenes that can rearrange yet have major defects that prevent the expression of a functional protein. An initial nonproductive rearrangement need not signal the immediate failure of development of the pro-B cell, because it is possible for most loci to undergo successive rearrangements on the same chromosome, and where that fails, the locus on the other chromosome will rearrange.

The diversity of the B-cell antigen-receptor repertoire is enhanced at this stage by the enzyme terminal deoxynucleotidyl transferase (TdT). TdT is expressed by the pro-B cell and adds nontemplated nucleotides (N-nucleotides) at the joints between rearranged gene segments (see Section 4-8). In adult humans it is expressed in pro-B cells during heavy-chain gene rearrangement, but its expression declines at the pre-B-cell stage during light-chain gene rearrangement. This explains why N-nucleotides are found in the V–D and D–J joints of nearly all heavy-chain genes but only in about a quarter of human light-chain joints. N-nucleotides are rarely found in mouse light-chain V–J joints, showing that TdT is switched off slightly earlier in the development of mouse B cells. In fetal development, when the peripheral immune system is first being supplied with T and B lymphocytes, TdT is expressed at low levels, if at all.

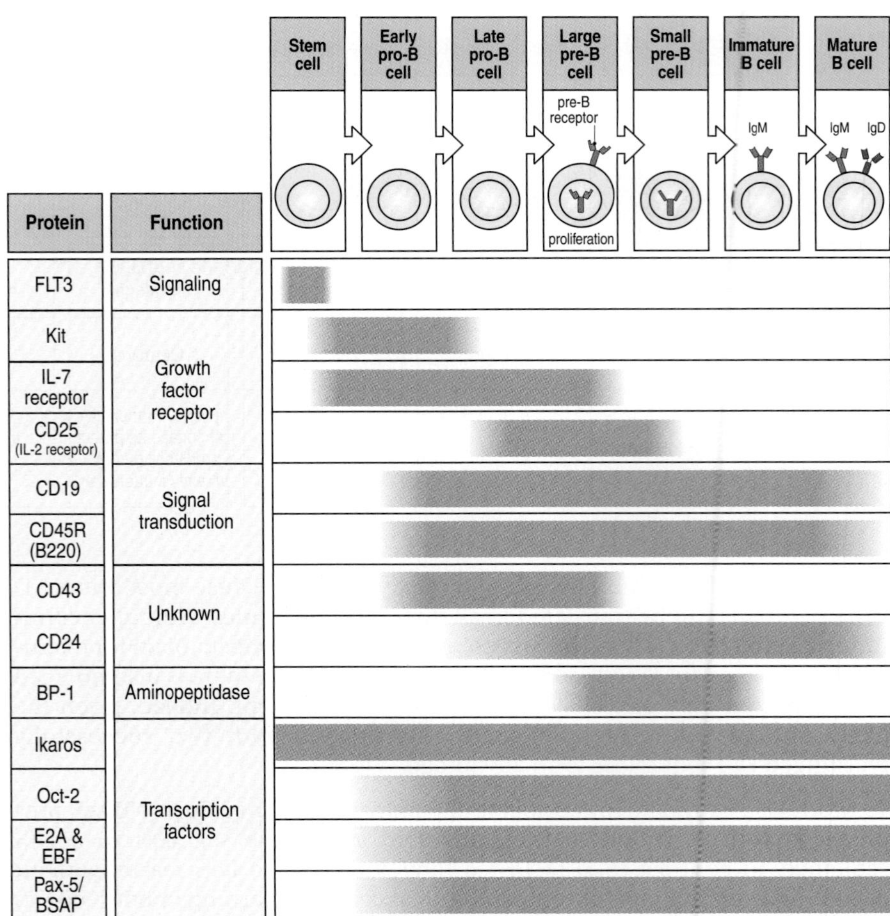

Fig. 7.6 Expression of surface proteins, receptors and transcription factors in B-cell development. The stages of B-cell development corresponding to those shown in Fig. 7.5 are shown at the top of the figure. The receptor FLT3 is expressed on hematopoietic stem cells and the common lymphoid progenitor. The earliest B-lineage surface markers are CD19 and CD45R (B220 in the mouse), which are expressed throughout B-cell development. A pro-B cell is also distinguished by the expression of CD43 (a marker of unknown function), Kit (CD117), and the IL-7 receptor. A late pro-B cell starts to express CD24 (a marker of unknown function) and the IL-2 receptor CD25. A pre-B cell is phenotypically distinguished by the expression of the enzyme BP-1, whereas Kit and the IL-7 receptor are no longer expressed. The actions of the listed transcription factors in B-cell development are discussed in the text, with the exception of the octamer transcription factor, Oct-2, which binds the octamer ATGCAAAT found in the heavy-chain promoter and elsewhere.

7-3 The pre-B-cell receptor tests for successful production of a complete heavy chain and signals for proliferation of pro-B cells.

The imprecise nature of V(D)J recombination is a double-edged sword. Although it produces increased diversity in the antibody repertoire, it also results in unsuccessful rearrangements. Pro-B cells therefore need a way of testing that a potentially functional heavy chain has been produced. They do this by incorporating the heavy chain into a receptor that can signal its successful production. This test, however, takes place in the absence of light chains, which have not yet rearranged. Instead, pro-B cells produce two invariant 'surrogate' proteins that have a structural resemblance to the light chain and together can pair with the μ chain to form the pre-B-cell receptor (pre-BCR) (see Fig. 7.7). The pre-B-cell receptor signals to the pro-B cell that a productive rearrangement has been made.

The surrogate chains are encoded by non-rearranging genes separate from the antigen-receptor loci and their expression is induced by the transcription factors E2A and EBF. One is called **λ5** because of its close resemblance to the C domain of the λ light chain; the other, called **VpreB**, resembles a light-chain V domain but has an extra region at the amino-terminal end. Other proteins expressed by the pre-B cell are also required for the formation of a functional receptor complex and are essential for B-cell development. The invariant proteins Igα (CD79α) and Igβ (CD79β) are components of both the pre-B-cell receptor and the B-cell receptor complexes on the cell surface. Igα and Igβ transduce signals from these receptors by interacting with intracellular tyrosine kinases through their cytoplasmic tails (see Section 6-8). Igα and Igβ are

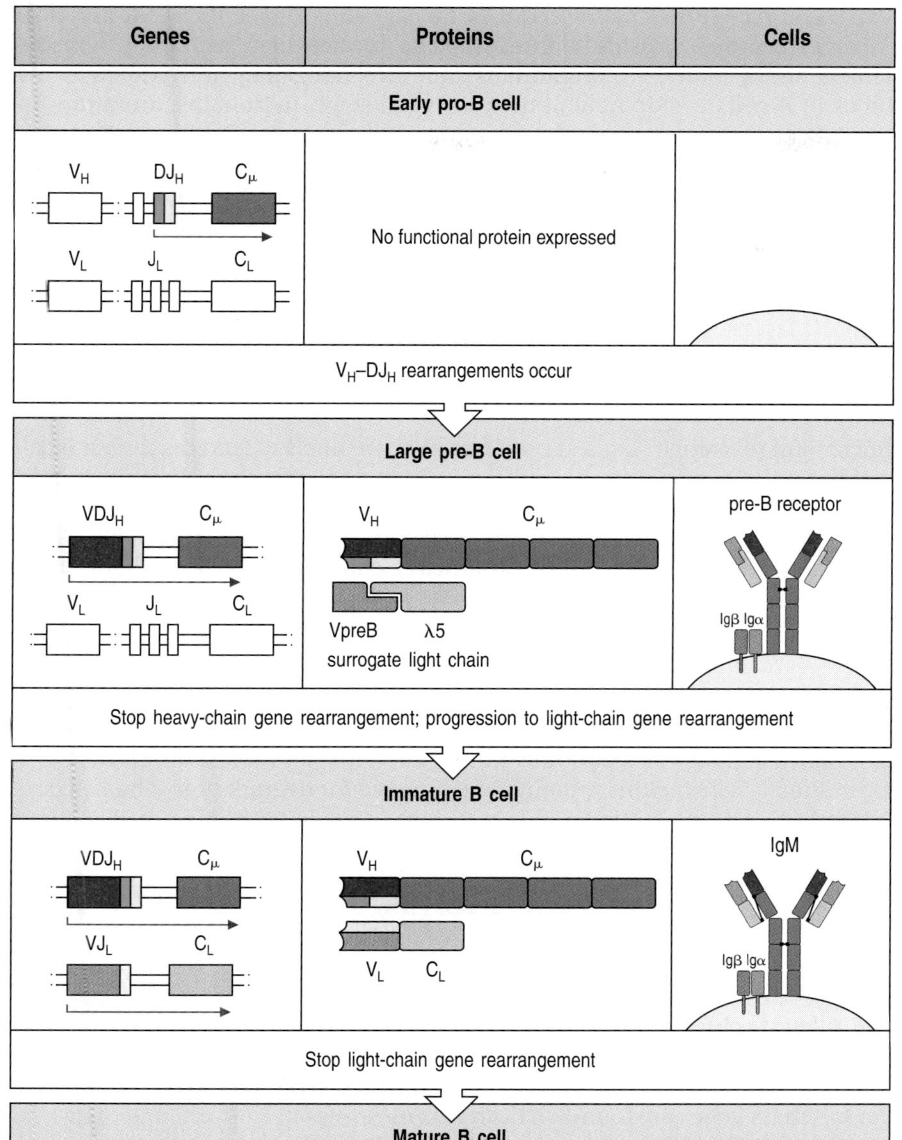

Fig. 7.7 A productively rearranged immunoglobulin gene is immediately expressed as a protein by the developing B cell. In early pro-B cells, heavy-chain gene rearrangement is not yet complete and no functional μ protein is expressed, although transcription occurs (red arrow), as shown in the top panel. As soon as a productive heavy-chain gene rearrangement has taken place, μ chains are expressed by the cell in a complex with two other chains, λ5 and VpreB, which together make up a surrogate light chain. The whole immunoglobulin-like complex is known as the pre-B-cell receptor (second panel). It is also associated with two other protein chains, Igα (CD79α) and Igβ (CD79β), in the cell. These associated chains signal the B cell to halt heavy-chain gene rearrangement, and drive the transition to the large pre-B cell stage by inducing proliferation. The progeny of large pre-B cells stop dividing and become small pre-B cells, in which light-chain gene rearrangements commence. Successful light-chain gene rearrangement results in the production of a light chain that binds the μ chain to form a complete IgM molecule, which is expressed together with Igα and Igβ at the cell surface, as shown in the third panel. Signaling via these surface IgM molecules is thought to trigger the cessation of light-chain gene rearrangement.

expressed from the pro-B-cell stage until the death of the cell or its terminal differentiation into an antibody-secreting plasma cell.

Formation of the pre-B-cell receptor is an important checkpoint in B-cell development that mediates the transition between the pro-B cell and the pre-B cell. In mice that either lack λ5 or have mutant heavy-chain genes that cannot produce the transmembrane domain, the pre-B-cell receptor cannot be formed and B-cell development is blocked after heavy-chain gene rearrangement. The pre-B-cell receptor complex is expressed transiently, perhaps because the production of λ5 mRNA stops as soon as pre-B-cell receptors begin to be formed. The pre-B-cell receptor is expressed at low levels on the surface of pre-B cells, but it is not clear whether it interacts with an external ligand. Whatever the precise mechanism of activating pre-B-cell receptor signaling may be, the expression of the receptor halts rearrangement of the heavy-chain locus and induces proliferation of the pro-B cell, initiating transition to the large pre-B cell, which will begin rearrangement of the light-chain locus.

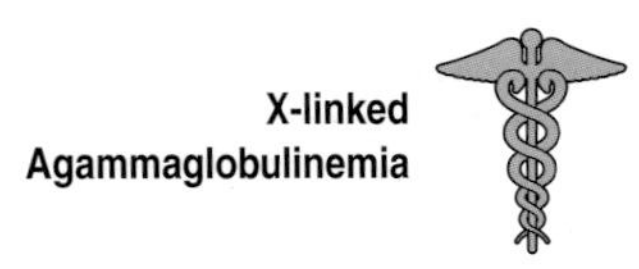

Pre-B-cell receptor signaling requires the signaling molecule BLNK and also involves Bruton's tyrosine kinase (Btk), an intracellular Tec-family tyrosine kinase (see Section 6-13). In humans and mice, deficiency of BLNK leads to a block in B-cell development at the pro-B-cell stage. In humans, mutations in the *Btk* gene cause a profound B-lineage-specific immune deficiency, **Bruton's X-linked agammaglobulinemia** (**XLA**), in which no mature B cells are produced. In humans, the block in B-cell development caused by mutations at the *XLA* locus is almost total, interrupting the transition from pre-B cell to immature B cell. A similar, though less severe, defect called **X-linked immunodeficiency** or **xid** arises from mutations in the corresponding gene in mice.

7-4 Pre-B-cell receptor signaling inhibits further heavy-chain locus rearrangement and enforces allelic exclusion.

Successful rearrangements at both heavy-chain alleles could result in a B cell producing two receptors of different antigen specificities. To prevent this, signaling by the pre-B-cell receptor enforces **allelic exclusion**, the state in which only one of the two alleles of a gene is expressed in a diploid cell. Allelic exclusion, which occurs at both the heavy-chain locus and the light-chain loci, was discovered more than 30 years ago and provided one of the original pieces of experimental support for the theory that one lymphocyte expresses one type of antigen receptor (Fig. 7.8).

Signaling from the pre-B-cell receptor promotes heavy-chain allelic exclusion in three ways. First, it reduces V(D)J recombinase activity by directly reducing the expression of *RAG-1* and *RAG-2*. Second, it further reduces levels of RAG-2 by indirectly causing this protein to be targeted for degradation, which occurs when RAG-2 is phosphorylated in response to the entry of the pro-B cell into S phase (the DNA synthesis phase) of the cell cycle. Finally, pre-B-cell receptor signaling reduces access of the heavy-chain locus to the recombinase machinery, although the precise details are not clear. At a later stage of B-cell development, RAG proteins will again be expressed in order to carry out light-chain locus rearrangement, but at that point the heavy-chain locus does not undergo further rearrangement. In the absence of pre-B-cell receptor signaling, allelic exclusion of the heavy-chain locus does not occur. For example, in λ5 knockout mice, in which the pre-B-cell receptor is not formed and the signal for V_H to DJ_H rearrangement to stop is not given, rearrangements of the heavy-chain genes are found on both chromosomes in all B-cell precursors, so that about 10% of the cells have two productive VDJ_H rearrangements.

7-5 Pre-B cells rearrange the light-chain locus and express cell-surface immunoglobulin.

The transition from the pro-B cell to the large pre-B-cell stage is accompanied by several rounds of cell division, expanding the population of cells

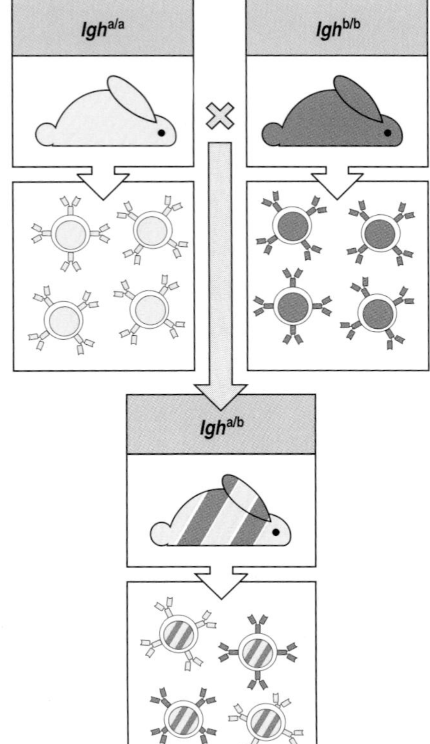

Fig. 7.8 Allelic exclusion in individual B cells. Most species have genetic polymorphisms of the constant regions of their immunoglobulin heavy-chain and light-chain genes; these are known as allotypes (see Appendix I, Section A-10). In rabbits, for example, all of the B cells in an individual homozygous for the *a* allele of the immunoglobulin heavy-chain locus ($Igh^{a/a}$) will express immunoglobulin of allotype a, whereas in an individual homozygous for the *b* allele ($Igh^{b/b}$) all the B cells make immunoglobulin of allotype b. In a heterozygous animal ($Igh^{a/b}$), which carries the *a* allele at one of the Igh loci and the *b* allele on the other, individual B cells can be shown to express surface immunoglobulin either of the a-allotype or b-allotype, but not both. This allelic exclusion reflects the productive rearrangement of only one of the two parental *Igh* alleles, because the production of a successfully rearranged immunoglobulin heavy chain forms a pre-B-cell receptor, which signals the cessation of further heavy-chain gene rearrangement.

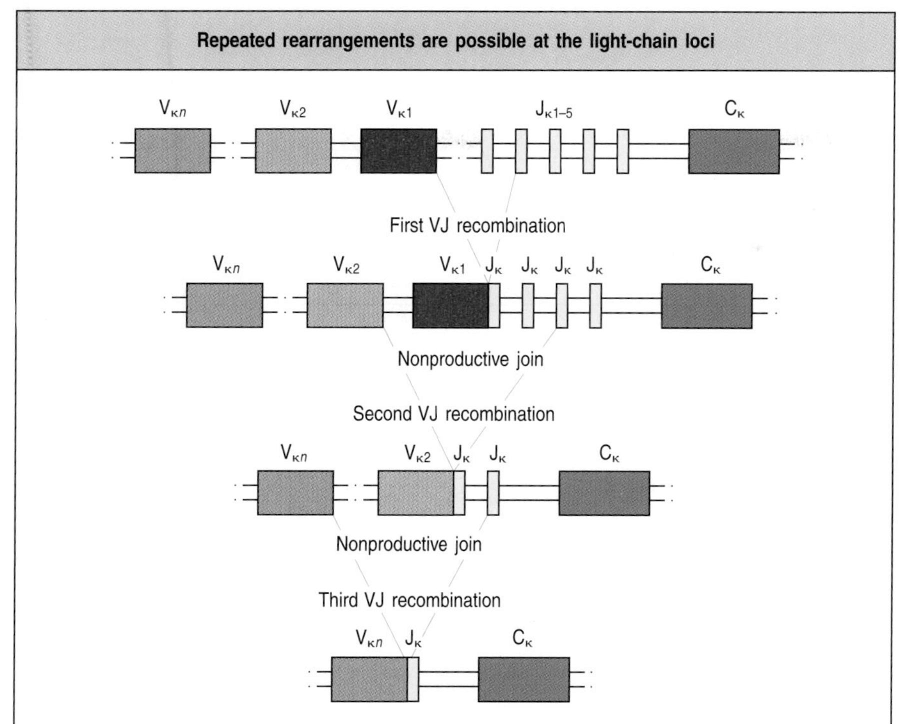

Fig. 7.9 Nonproductive light-chain gene rearrangements can be rescued by further gene rearrangement. The organization of the light-chain loci in mice and humans offers many opportunities for the rescue of pre-B cells that initially make an out-of-frame rearrangement. Light-chain rescue is illustrated for the human κ locus. If the first rearrangement is nonproductive, a 5′ V_κ gene segment can recombine with a 3′ J_κ gene segment to remove the out-of-frame join located between them and to replace it with a new rearrangement. In principle, this can happen up to five times on each chromosome, because there are five functional J_κ gene segments in humans. If all rearrangements of κ-chain genes fail to yield a productive light-chain join, λ-chain gene rearrangement may succeed (not shown; see Fig. 7.11).

with successful in-frame joins by about 30–60-fold before they become resting small pre-B cells. A large pre-B cell with a particular rearranged heavy-chain gene therefore gives rise to numerous small pre-B cells. RAG proteins are produced again in the small pre-B cells, and rearrangement of the light-chain locus begins. Each of these cells can make a different rearranged light-chain gene and so cells with many different antigen specificities are generated from a single pre-B cell, which makes an important contribution to overall B-cell receptor diversity.

Light-chain rearrangement also exhibits allelic exclusion. Rearrangements at the light-chain locus generally take place at only one allele at a time. The light-chain loci lack D segments and rearrangement occurs by V to J joining, and if a particular VJ rearrangement fails to produce a functional light chain, repeated rearrangements of unused V and J gene segments at the same allele can occur (Fig. 7.9). Several attempts at productive rearrangement of a light-chain gene can therefore be made on one chromosome before initiating any rearrangements on the second chromosome. This greatly increases the chances of eventually generating an intact light chain, especially as there are two different light-chain loci. As a result, many cells that reach the pre-B-cell stage succeed in generating progeny that bear intact IgM molecules and can be classified as **immature B cells**. Figure 7.10 lists some of the proteins involved in V(D)J recombination and shows how their expression is regulated throughout B-cell development. Figure 7.11 summarizes the stages of B-cell development up to the point of assembly of a complete surface immunoglobulin, indicating the points at which developing B cells can be lost as a result of failure to produce a productive join.

As well as allelic exclusion, light chains also display **isotypic exclusion**; that is, the expression of only one type of light chain—κ or λ—by an individual B cell. In mice and humans, the κ light-chain locus tends to rearrange before the λ locus. This was first deduced from the observation that myeloma cells secreting λ light chains generally have both their κ and λ light-chain genes rearranged, whereas in myelomas secreting κ light chains, generally only the

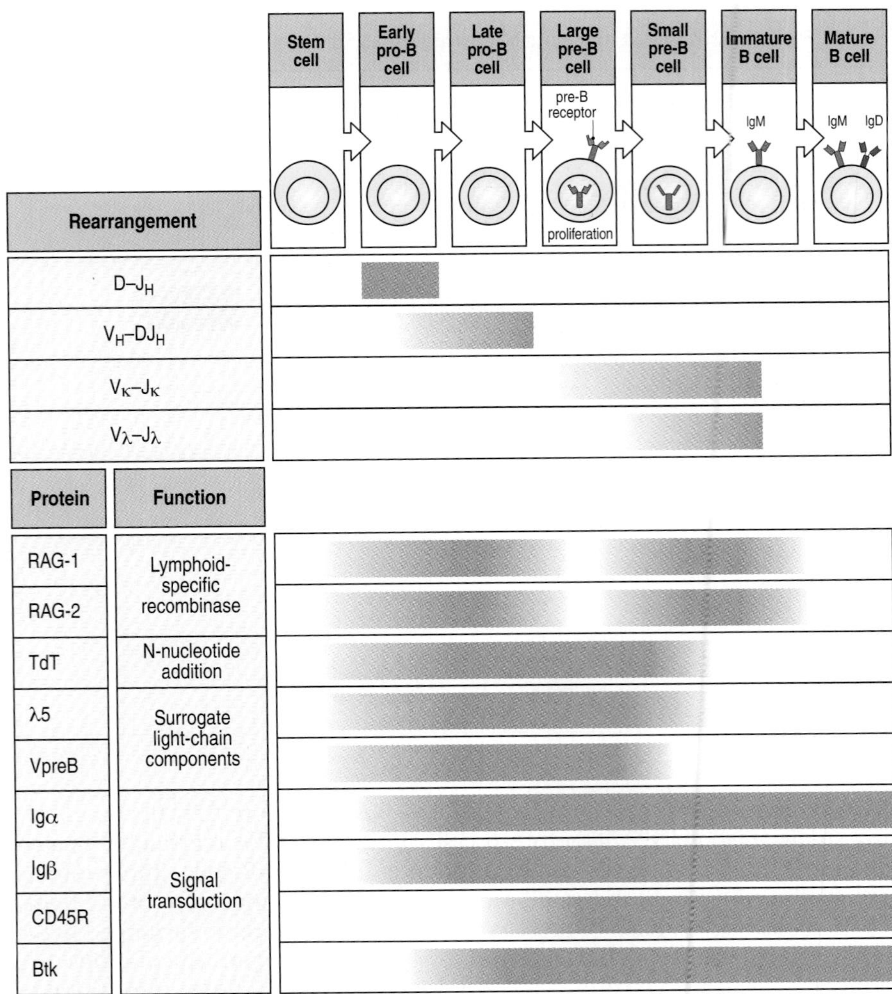

Fig 7.10 Expression of proteins involved in gene rearrangement and the production of pre-B-cell and B-cell receptors. The proteins listed here have been included because of their proven importance in the developmental sequence, largely on the basis of studies in mice. Also shown is the temporal sequence of gene rearrangement. Their individual contributions to B-cell development are discussed in the text. Signaling proteins and transcription factors involved in early B-lineage development are listed in Fig. 7.6.

κ genes are rearranged. This order is occasionally reversed, however, and λ gene rearrangement does not absolutely require the previous rearrangement of the κ genes. The ratios of κ-expressing versus λ-expressing mature B cells vary from one extreme to the other in different species. In mice and rats it is 95% κ to 5% λ, in humans it is typically 65%:35%, and in cats it is 5%:95%, the opposite of that in mice. These ratios correlate most strongly with the number of functional V_κ and V_λ gene segments in the genome of the species. They also reflect the kinetics and efficiency of gene segment rearrangements. The κ:λ ratio in the mature lymphocyte population is useful in clinical diagnostics, because an aberrant κ:λ ratio indicates the dominance of one clone and the presence of a lymphoproliferative disorder, which may be malignant.

7-6 Immature B cells are tested for autoreactivity before they leave the bone marrow.

Once a rearranged light chain pairs with a μ chain, IgM can be expressed on the cell surface (sIgM) and the pre-B cell becomes an **immature B cell**. At this stage, the antigen receptor is first tested for tolerance to self antigens. The tolerance produced in the B-cell repertoire at this stage of development is known as **central tolerance** because it arises in a central lymphoid organ, the bone marrow. As we shall see later in the chapter and in Chapter 14, self-reactive B cells that escape this test and go on to mature may still be removed

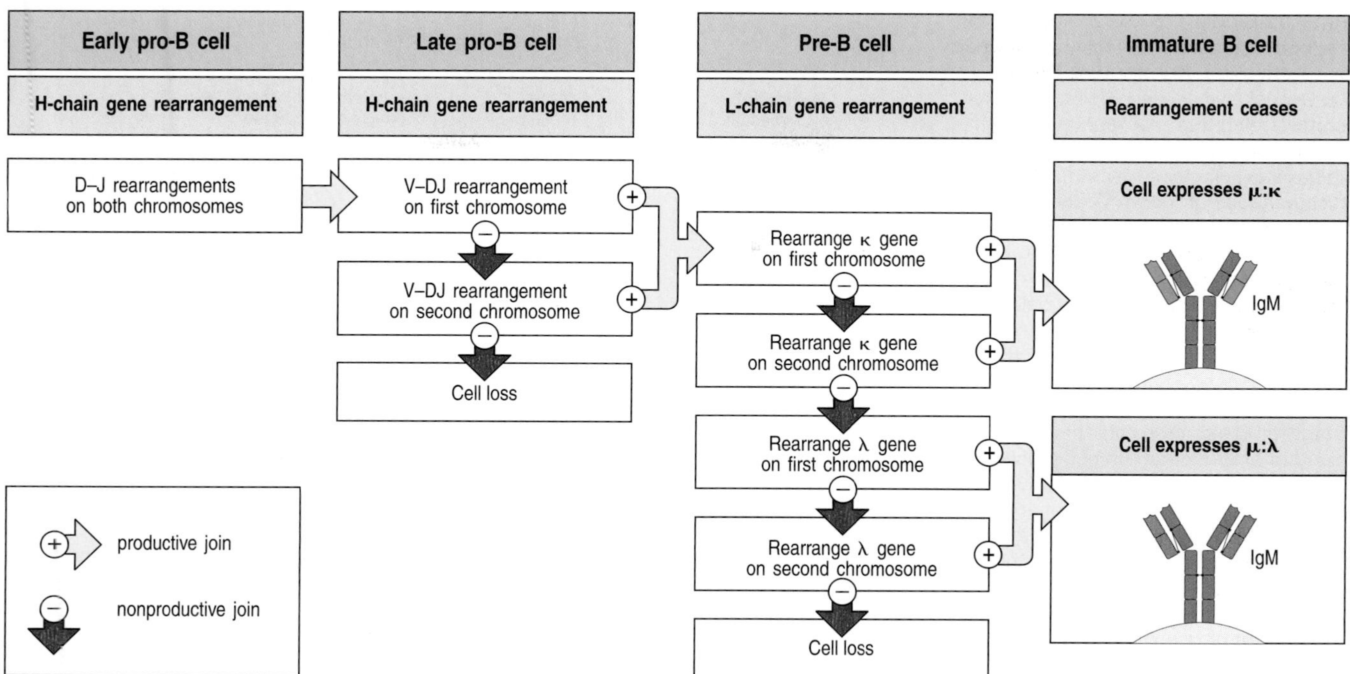

Fig. 7.11 The steps in immunoglobulin gene rearrangement at which developing B cells can be lost. The developmental program rearranges the heavy-chain (H-chain) locus first and then the light-chain (L-chain) loci. Cells are allowed to progress to the next stage when a productive rearrangement has been achieved. Each rearrangement has about a one in three chance of being successful, but if the first attempt is nonproductive, development is suspended and there is a chance for one or more further attempts, so by simple mathematics four in nine rearrangements give rise to a heavy chain. The scope for repeated rearrangements is greater at the light-chain loci (see Fig. 7.9), so that fewer cells are lost between the pre-B and immature B-cell stages than in the pro-B to pre-B transition.

from the repertoire after they have left the bone marrow. This process is known as **peripheral tolerance**.

In the bone marrow, the fate of the immature B cell depends on signals delivered from sIgM on interaction with its environment. sIgM associates with Igα and Igβ to form a functional B-cell receptor complex (see Section 6-8). Igα signaling is particularly important in dictating the emigration of B cells from the bone marrow and/or their survival in the periphery, because mice that express Igα with a truncated cytoplasmic domain that cannot signal to the interior of the cell show a fourfold reduction in the number of immature B cells in the marrow and a hundredfold reduction in the number of peripheral B cells.

Immature B cells that have no strong reactivity to self antigens are allowed to mature. They leave the marrow via sinusoids that enter the central sinus and are carried by the venous blood supply to the spleen. If, however, the newly expressed receptor encounters a strongly cross-linking antigen in the bone marrow—that is, if the B cell is strongly self-reactive—development is arrested and the cell will not mature. This was first demonstrated by experiments in which antigen receptors on immature B cells were experimentally stimulated *in vivo* using anti-μ chain antibodies (see Appendix I, Section A-10); the outcome was elimination of the immature B cells.

More recent experiments using mice expressing transgenes that enforce the expression of self-reactive B-cell receptors have confirmed these earlier findings but have also shown that immediate elimination is not the only possible outcome of binding to a self antigen. Instead, there are four possible fates for self-reactive immature B cells, depending on the nature of the ligand they recognize (Fig. 7.12). These fates are: cell death by apoptosis or clonal deletion; the production of a new receptor by a process known as receptor editing; the induction of a permanent state of unresponsiveness, or anergy, to antigen; and immunological ignorance. An immunologically ignorant cell is defined as one that has affinity for a self antigen but does not sense the self antigen because the antigen is sequestered, is in low concentration, or does not activate the B-cell receptor. Because ignorant cells can be (and in fact are)

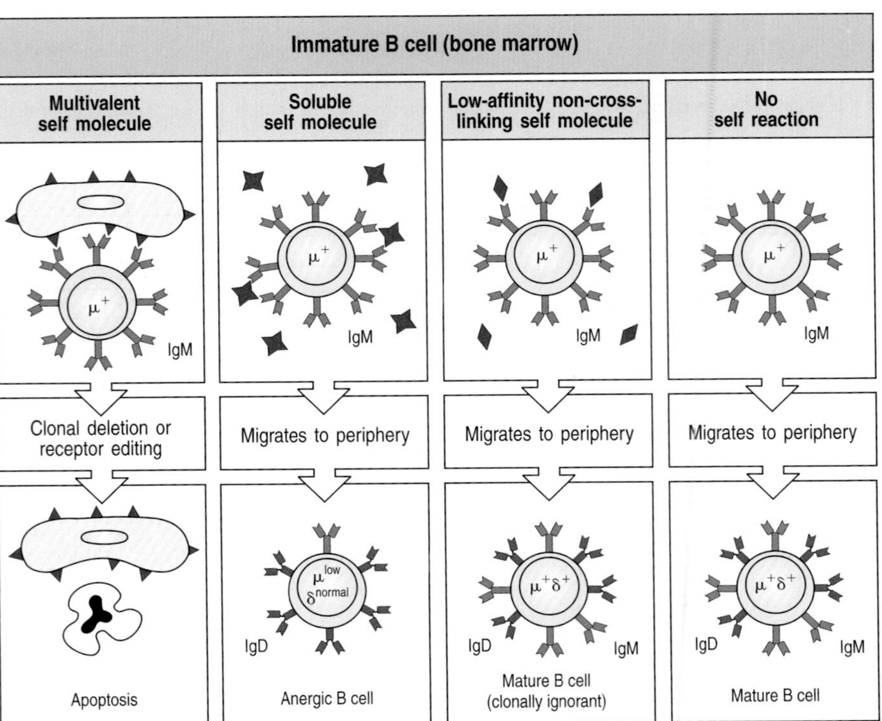

Fig. 7.12 Binding to self molecules in the bone marrow can lead to the death or inactivation of immature B cells. First panels: when developing B cells express receptors that recognize multivalent ligands, for example ubiquitous self cell-surface molecules such as those of the MHC, these receptors are deleted from the repertoire. The B cells either undergo receptor editing (see Fig. 7.13), so that the self-reactive receptor specificity is deleted, or the cells themselves undergo programmed cell death or apoptosis (clonal deletion). Second panels: immature B cells that bind soluble self antigens able to cross-link the B-cell receptor are rendered unresponsive to the antigen (anergic) and bear little surface IgM. They migrate to the periphery where they express IgD but remain anergic; if in competition with other B cells in the periphery, they are rapidly lost. Third panels: immature B cells that bind monovalent antigens or soluble self antigens with low affinity do not receive any signal as a result of this interaction and mature normally to express both IgM and IgD at the cell surface. Such cells are potentially self reactive and are said to be clonally ignorant because their ligand is present but is unable to activate them. Fourth panels: immature B cells that do not encounter antigen mature normally; they migrate from the bone marrow to the peripheral lymphoid tissues, where they may become mature recirculating B cells bearing both IgM and IgD on their surface.

activated under certain conditions such as inflammation or when self antigens reach unusually high concentrations, they should not be considered inert, and they are fundamentally different from nonreactive cells that could never be activated by self antigens.

Clonal deletion, or the removal of cells of a particular antigen specificity from the repertoire, seems to predominate when the interacting self antigen is multivalent. The effect of an encounter with a multivalent antigen was tested in mice transgenic for both chains of an immunoglobulin specific for H-2K^b MHC class I molecules. In such mice, nearly all the B cells that develop bear the anti-MHC immunoglobulin as sIgM. If the transgenic mouse does not express H-2K^b, normal numbers of B cells develop, all bearing transgene-encoded anti-H-2K^b receptors. However, in mice expressing both H-2K^b and the immunoglobulin transgenes, B-cell development is blocked. Normal numbers of pre-B cells and immature B cells are found, but B cells expressing the anti-H-2K^b immunoglobulin as sIgM never mature to populate the spleen and lymph nodes; instead, most of these immature B cells die in the bone marrow by apoptosis.

Clonal deletion is, however, not the only outcome for lymphocytes with autoreactive receptors. There is an interval before cell death during which the autoreactive B cell can be rescued by further gene rearrangements that replace the autoreactive receptor with a new receptor that is not self reactive. This mechanism is termed **receptor editing** (Fig. 7.13). When an immature B cell first produces sIgM, RAG protein is still being made. If the receptor is not self-reactive, the absence of sIgM cross-linking allows gene rearrangement to cease; B-cell development continues, with RAG proteins eventually disappearing as B cells reach full maturity in the spleen. For an autoreactive receptor, however, an encounter with the self antigen results in strong cross-linking of sIgM, further development is halted, and RAG expression continues. Light-chain gene rearrangement therefore continues, as described in Fig. 7.9. These secondary rearrangements can rescue immature self-reactive

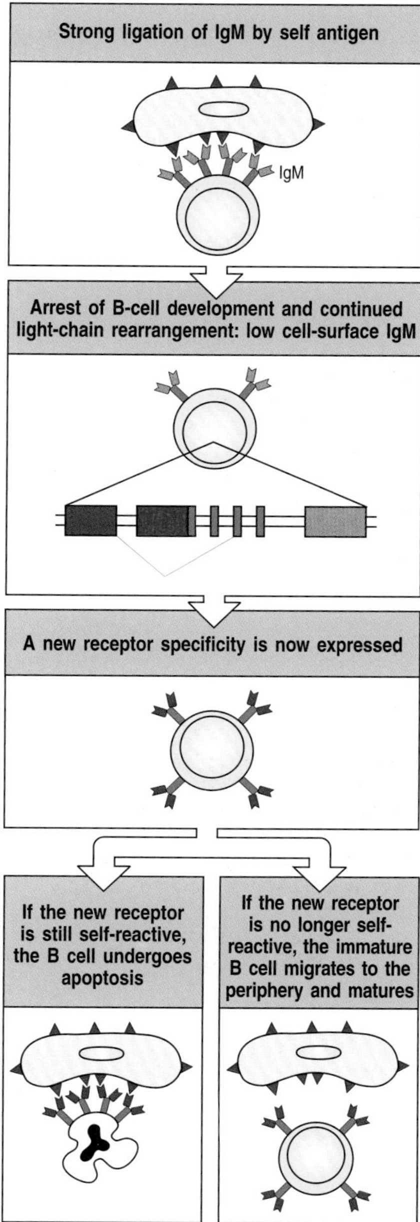

Fig. 7.13 Replacement of light chains by receptor editing can rescue some self-reactive B cells by changing their antigen specificity. When a developing B cell expresses antigen receptors that are strongly cross-linked by multivalent self antigens such as MHC molecules on cell surfaces (top panel), its development is arrested. The cell decreases surface expression of IgM and does not turn off the RAG genes (second panel). Continued synthesis of RAG proteins allows the cell to continue light-chain gene rearrangement. This usually leads to a new productive rearrangement and the expression of a new light chain, which combines with the previous heavy chain to form a new receptor (receptor editing; third panel). If this new receptor is not self-reactive, the cell is 'rescued' and continues normal development much like a cell that had never reacted with self (bottom right panel). If the cell remains self-reactive it may be rescued by another cycle of rearrangement, but if it continues to react strongly with self it will undergo programmed cell death or apoptosis and be deleted from the repertoire (clonal deletion; bottom left panel).

B cells by deleting the self-reactive light chain gene and replacing it with another sequence. If the light chain expressed by this new rearrangement is not autoreactive, the B cell continues normal development. If the receptor remains autoreactive, rearrangement continues until a non-autoreactive receptor is produced or V and J gene segments are exhausted. Cells that remain autoreactive then undergo apoptosis.

Receptor editing has been shown definitively in mice bearing transgenes for autoantibody heavy and light chains that have been placed within the immunoglobulin loci by homologous recombination (see Appendix I, Section A-47 for details of this method). The transgene imitates a primary gene rearrangement and is surrounded by unused endogenous gene segments. In mice that express the antigen recognized by the transgene-encoded receptor, the mature B cells that emerge into the periphery have used these surrounding gene segments for rearrangements that replace the autoreactive light-chain transgene with a non-autoreactive gene.

It is not clear whether receptor editing occurs at the heavy-chain locus. There are no available D segments at a rearranged heavy-chain locus, so a new rearrangement cannot simply occur by the normal mechanism and remove the pre-existing one. Instead, a process of V_H replacement may use embedded recombination signal sequences in a recombination event that displaces the V gene segment from the self-reactive rearrangement and replaces it with a new V gene segment. This has been observed in some B-cell tumors, but whether it occurs during normal B-cell development in response to signals from autoreactive B-cell receptors is not certain.

It was originally thought that the successful production of a heavy chain and a light chain caused the almost instantaneous shutdown of light-chain locus rearrangement and that this ensured both allelic and isotypic exclusion. The unexpected ability of self-reactive B cells to continue to rearrange their light-chain genes, even after having made a productive rearrangement, has raised questions about this supposed mechanism of allelic exclusion.

The decline in the level of RAG protein that follows a successful non-autoreactive rearrangement is undoubtedly crucial to maintaining allelic exclusion, because it reduces the chance of a subsequent rearrangement. Furthermore, an additional productive rearrangement would not necessarily breach allelic exclusion. If it occurred on the same chromosome it would eliminate the existing productive rearrangement, and if it occurred on the other chromosome it would be nonproductive in two out of three cases. Thus, the fall in the level of RAG protein could be the principal, if not the sole, mechanism behind allelic exclusion at the light-chain locus. Consistent with this, it seems that allelic exclusion is not absolute, because there are rare B cells that express two light chains.

We have so far discussed the fate of newly formed B cells that undergo multivalent cross-linking of their sIgM. Immature B cells that encounter more weakly cross-linking self antigens of low valence, such as small soluble proteins, respond differently. In this situation, self-reactive B cells tend to be inactivated and enter a state of permanent unresponsiveness, or **anergy**, but do not immediately die (see Fig. 7.12). Anergic B cells cannot be activated by their specific antigen even with help from antigen-specific T cells. Again, this phenomenon was elucidated using transgenic mice. Hen egg-white lysozyme (HEL) was expressed in soluble form from a transgene in mice that were also transgenic for high-affinity anti-HEL immunoglobulin. The HEL-specific B cells matured but could not respond to antigen. Anergic cells retain their IgM within the cell and transport little to the surface. In addition, they develop a partial block in signal transduction so that, despite normal levels of HEL-binding sIgD, the cells cannot be stimulated by cross-linking this receptor. It seems that signal transduction is blocked before the phosphorylation of the Igα and Igβ chains, although the exact step is not yet known. The signaling defect may involve the inability of B-cell receptors on anergic B cells to enter regions of the cell membrane in which other important signaling molecules normally segregate; they cannot then transmit a complete signal after antigen binding. Cells that have received an anergizing signal may also increase the expression of molecules that inhibit signaling.

The migration of anergic B cells within peripheral lymphoid organs is also impaired, and their lifespan and ability to compete with immunocompetent B cells are compromised. In normal circumstances, in which B cells binding a soluble self antigen are rare, the self-reactive anergic B cells are detained in the T-cell areas of peripheral lymphoid tissues and are excluded from lymphoid follicles. Anergic B cells cannot be activated by T cells, because all the T cells will be tolerant to soluble antigens. Instead they die relatively quickly, presumably because they fail to get survival signals from T cells. This ensures that the long-lived pool of peripheral B cells is purged of potentially self-reactive cells.

The fourth potential fate of self-reactive immature B cells is that nothing happens to them; they remain in a state of **immunological ignorance** of their self antigen (see Fig. 7.12). It is clear that some B cells with a weak but definite affinity for a self antigen mature as though they were not self-reactive at all. Such B cells do not respond to their self antigen because it interacts so weakly with the receptor that little, if any, intracellular signal is generated. Alternatively, some self-reactive B cells may not encounter their antigen at this stage because it is not accessible to B cells developing in the bone marrow and spleen. The fact that such B cells are allowed to mature reflects the balance that the immune system strikes between purging all self-reactivity and maintaining the ability to respond to pathogens. If the elimination of self-reactive cells were too efficient, the receptor repertoire might become too limited and thus unable to recognize a wide variety of pathogens. Some autoimmune disease might be the price of this balance, because it is very likely that these low-affinity self-reactive lymphocytes can be activated and cause disease under certain circumstances. Thus, these cells can be thought of as the seeds of autoimmune disease. Normally, however, ignorant B cells will be held in check by a lack of T-cell help, the continued inaccessibility of the self antigen, or the tolerance that can be induced in mature B cells, which is described later in this chapter and in Chapter 14, in the context of autoimmune disease.

Summary.

Up to this point we have followed B-cell development from the earliest progenitors in the bone marrow to the immature B cell that is ready to emerge into the peripheral lymphoid tissue. The heavy-chain locus is rearranged first and, if this is successful, a μ heavy chain is produced that combines with surrogate light chains to form the pre-B-cell receptor; this is the first checkpoint in B-cell development. Production of the pre-B-cell receptor signals successful heavy-chain gene rearrangement, and causes cessation of this rearrangement, thus enforcing allelic exclusion. It also initiates pre-B-cell proliferation, generating numerous progeny in which subsequent light-chain rearrangement can be attempted. If the initial light-chain gene rearrangement is productive, a complete immunoglobulin B-cell receptor is formed, gene rearrangement again ceases, and the B cell continues its development. If the first light-chain gene rearrangement is unsuccessful, rearrangement continues until either a productive rearrangement is made or all available J regions are used up. If no productive rearrangement is made, the developing B cell dies. In the next section, we turn to T-cell development in the thymus; after this we return to consider B and T cells together as they populate the peripheral lymphoid tissues.

T-cell development in the thymus.

T cells develop from progenitors that are derived from the pluripotent hematopoietic stem cells in the bone marrow and migrate through the blood to the thymus, where they mature (Fig. 7.14); for this reason they are called thymus-dependent (T) lymphocytes or T cells. The development of T cells parallels B-cell development in many ways, including the orderly and stepwise rearrangement of antigen-receptor genes, the sequential testing for successful gene rearrangement and the eventual formation of a complete heterodimeric antigen receptor. In addition, T-cell development in the thymus includes some

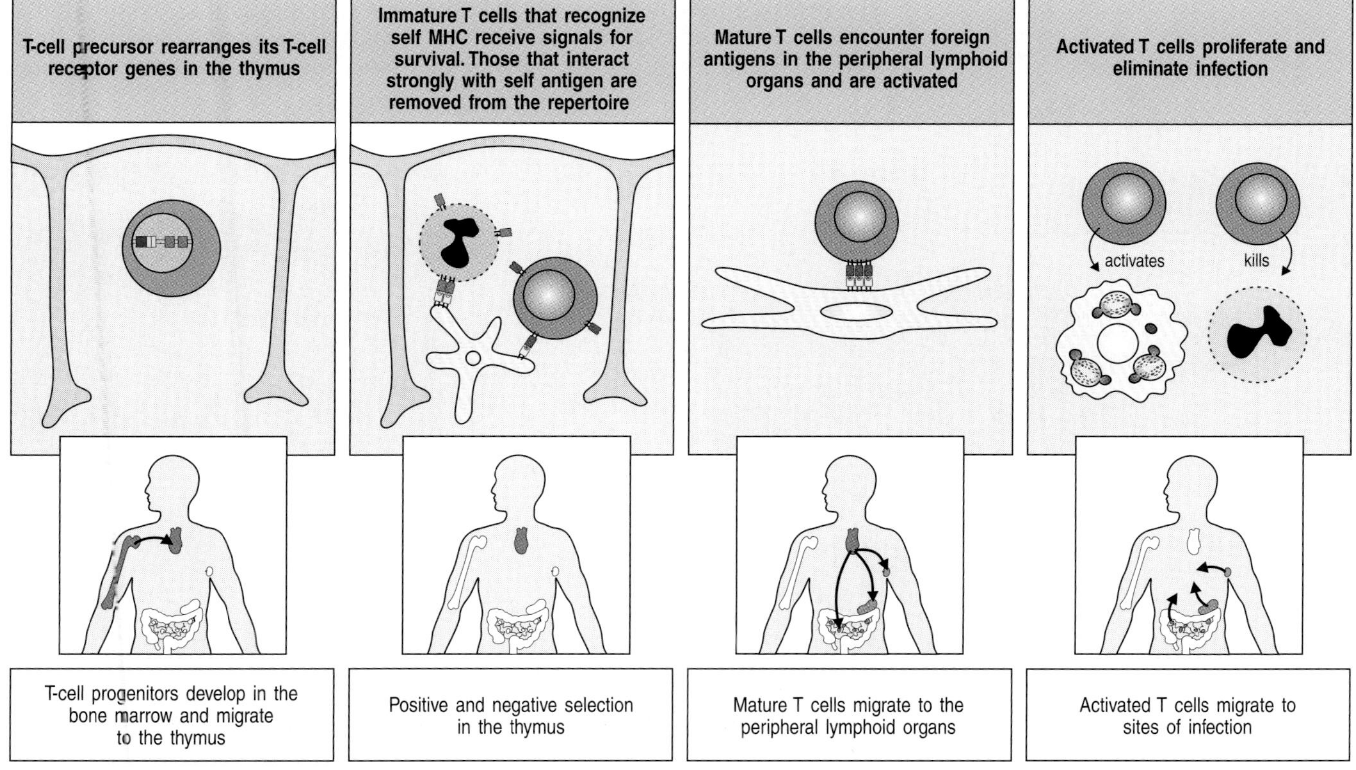

T-cell precursor rearranges its T-cell receptor genes in the thymus
Immature T cells that recognize self MHC receive signals for survival. Those that interact strongly with self antigen are removed from the repertoire
Mature T cells encounter foreign antigens in the peripheral lymphoid organs and are activated
Activated T cells proliferate and eliminate infection
activates
kills
T-cell progenitors develop in the bone marrow and migrate to the thymus
Positive and negative selection in the thymus
Mature T cells migrate to the peripheral lymphoid organs
Activated T cells migrate to sites of infection

Fig. 7.14 T cells undergo development in the thymus and migrate to the peripheral lymphoid organs, where they are activated by foreign antigens. T-cell precursors migrate from the bone marrow to the thymus, where the T-cell receptor genes are rearranged (first panels); α:β T-cell receptors that are compatible with self-MHC molecules transmit a survival signal on interacting with thymic epithelium, leading to positive selection of the cells that bear them. Self-reactive receptors transmit a signal that leads to cell death and are thus removed from the repertoire in a process of negative selection (second panels). T cells that survive selection mature and leave the thymus to circulate in the periphery; they repeatedly leave the blood to migrate through the peripheral lymphoid organs, where they may encounter their specific foreign antigen and become activated (third panels). Activation leads to clonal expansion and differentiation into effector T cells. These are attracted to sites of infection, where they can kill infected cells or activate macrophages (fourth panels); others are attracted into B-cell areas, where they help to activate an antibody response (not shown).

processes that do not happen for B cells, such as the generation of two distinct lineages of T cells, the γ:δ lineage and the α:β lineage, which express distinct antigen-receptor genes. Developing T cells also undergo an extensive selection process that depends on interactions with thymic cells and that shapes the mature repertoire of T cells to ensure self-MHC restriction as well as self tolerance. We begin with a general overview of the stages of thymocyte development and its relationship to thymic anatomy before considering gene rearrangement and the mechanisms of selection.

7-7 T-cell progenitors originate in the bone marrow, but all the important events in their development occur in the thymus.

The thymus is situated in the upper anterior thorax, just above the heart. It consists of numerous lobules, each clearly differentiated into an outer cortical region—the **thymic cortex**—and an inner **medulla** (Fig. 7.15). In young individuals, the thymus contains large numbers of developing T-cell precursors embedded in a network of epithelia known as the **thymic stroma**, which provides a unique microenvironment for T-cell development analogous to that provided for B cells by the stromal cells of the bone marrow.

T lymphocytes develop from a lymphoid progenitor in the bone marrow that also gives rise to B lymphocytes. Some of these progenitors leave the bone marrow and migrate to the thymus (see Fig. 7.14). In the thymus, the progenitor cell receives a signal, most probably from stromal cells, that is transduced through a receptor called Notch1 to switch on specific genes. Notch signaling is widely used in animal development to specify tissue differentiation; in lymphocyte development, the Notch signal instructs the precursor to commit to the T-cell lineage rather than the B-cell lineage. Although the details are still incomplete, Notch signaling is required throughout T-cell development and is also thought to help regulate other T-cell lineage choices, including the α:β versus γ:δ choice and the CD4 versus CD8 decision.

The thymic epithelia arise early in embryonic development from endoderm-derived structures known as the third pharyngeal pouch and the third branchial cleft. Together these epithelial tissues form a rudimentary thymus,

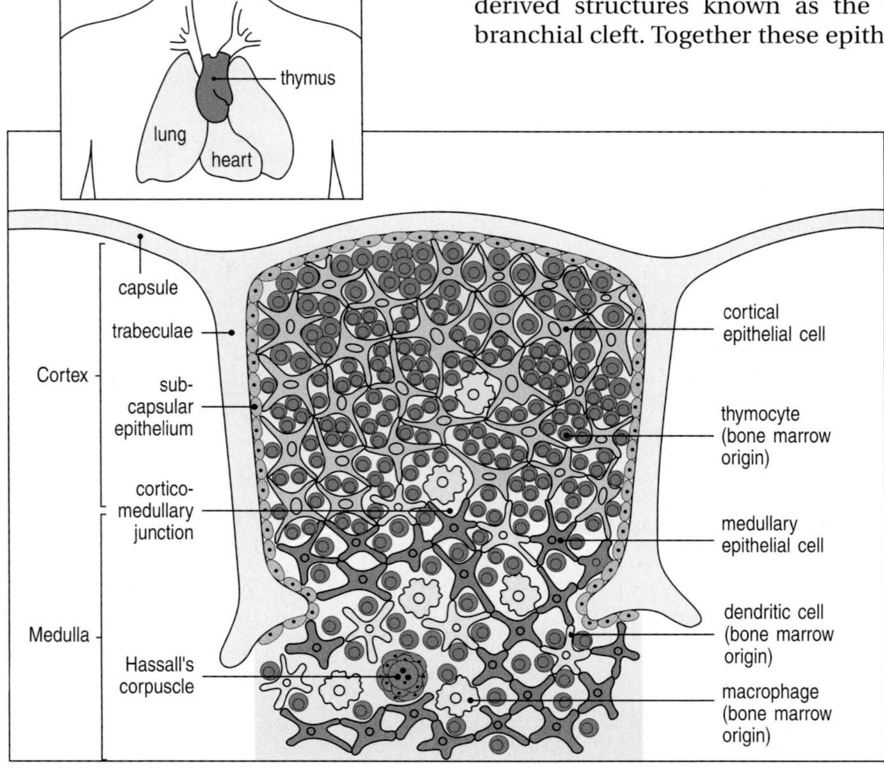

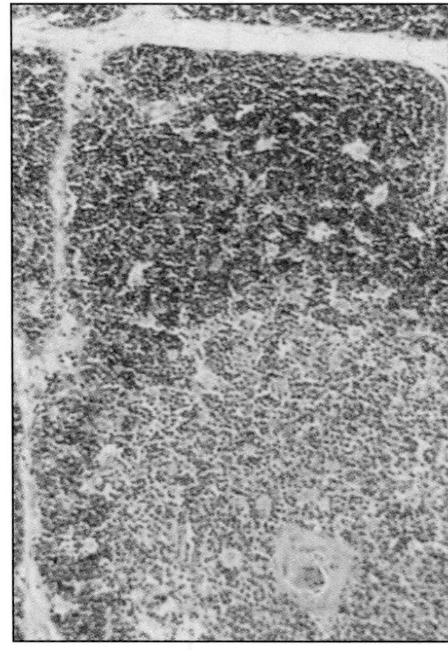

Fig. 7.15 The cellular organization of the human thymus. The thymus, which lies in the midline of the body, above the heart, is made up of several lobules, each of which contains discrete cortical (outer) and medullary (central) regions. As shown in the diagram on the left, the cortex consists of immature thymocytes (dark blue), branched cortical epithelial cells (pale blue), with which the immature cortical thymocytes are closely associated, and scattered macrophages (yellow), which are involved in clearing apoptotic thymocytes. The medulla consists of mature thymocytes (dark blue) and medullary epithelial cells (orange), along with macrophages (yellow) and dendritic cells (yellow) of bone marrow origin. Hassall's corpuscles are probably also sites of cell destruction. The thymocytes in the outer cortical cell layer are proliferating immature cells, whereas the deeper cortical thymocytes are mainly immature T cells undergoing thymic selection. The photograph shows the equivalent section of a human thymus, stained with hematoxylin and eosin. The cortex is darkly staining, whereas the medulla is lightly stained. The large body in the medulla is a Hassall's corpuscle. Photograph courtesy of C.J. Howe.

or **thymic anlage**. This is then colonized by cells of hematopoietic origin that give rise to large numbers of **thymocytes**, which are committed to the T-cell lineage, and to the **intrathymic dendritic cells**. The thymocytes are not simply passengers within the thymus: they influence the arrangement of the thymic epithelial cells on which they depend for survival, inducing the formation of a reticular epithelial structure that surrounds the developing thymocytes (Fig. 7.16). The thymus is independently colonized by numerous macrophages, also of bone marrow origin.

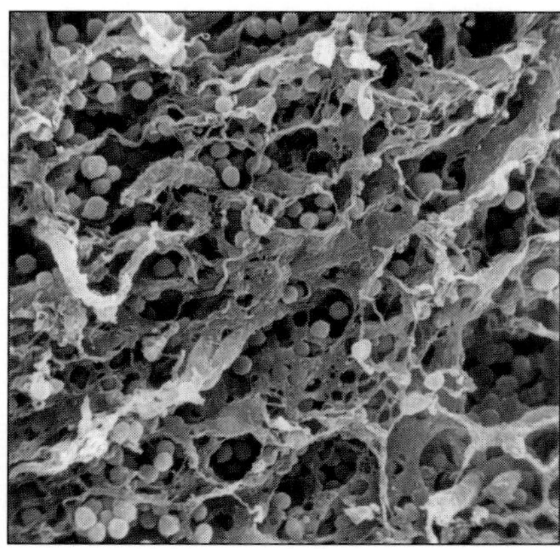

Fig. 7.16 The epithelial cells of the thymus form a network surrounding developing thymocytes. In this scanning electron micrograph of the thymus, the developing thymocytes (the spherical cells) occupy the interstices of an extensive network of epithelial cells. Photograph courtesy of W. van Ewijk.

The cellular architecture of the human thymus is illustrated in Fig. 7.15. Bone marrow derived cells are differentially distributed between the thymic cortex and medulla. The cortex contains only immature thymocytes and scattered macrophages, whereas more mature thymocytes, along with dendritic cells and macrophages, are found in the medulla. This reflects the different developmental events that occur in these two compartments.

The importance of the thymus in immunity was first discovered through experiments on mice; indeed, most of our knowledge of T-cell development in the thymus comes from the mouse. It was found that surgical removal of the thymus (**thymectomy**) at birth resulted in immunodeficient mice, focusing interest on this organ at a time when the difference between T and B lymphocytes in mammals had not been defined. Much evidence, including observations on immunodeficient children, has since confirmed the importance of the thymus in T-cell development. In **DiGeorge's syndrome** in humans and in mice with the ***nude*** mutation, the thymus does not form and the affected individual produces B lymphocytes but few T lymphocytes. DiGeorge's syndrome is a complex combination of cardiac, facial, endocrine, and immune defects associated with deletions of chromosome 22q11, while the *nude* mutation is due to a defect in the gene for Whn, a transcription factor required for terminal epithelial cell differentiation; the name *nude* was given to the mutation because it also causes hairlessness.

The crucial role of the thymic stroma in inducing the differentiation of bone marrow derived precursor cells can be demonstrated by tissue grafts between two mutant mice, each lacking mature T cells for a different reason. In *nude* mice the thymic epithelium fails to differentiate, whereas in *scid* mice B and T lymphocytes fail to develop because of a defect in antigen-receptor gene rearrangement (see Section 4-5). Reciprocal grafts of thymus and bone marrow between these immunodeficient strains show that *nude* bone marrow precursors develop normally in a *scid* thymus (Fig. 7.17). Thus, the defect in *nude* mice is in the thymic stromal cells. Transplanting a *scid* thymus into *nude* mice leads to T-cell development. However, *scid* bone marrow cannot develop T cells, even in a wild-type recipient.

In mice, the thymus continues to develop for 3–4 weeks after birth, whereas in humans it is fully developed at birth. The rate of T-cell production by the thymus is greatest before puberty. After puberty, the thymus begins to shrink and the production of new T cells in adults is lower, although it does continue throughout life. In both mice and humans, removal of the thymus after puberty is not accompanied by any notable loss of T-cell function or numbers. Thus, it seems that once the T-cell repertoire is established, immunity can be sustained without the production of significant numbers of new T cells; the pool of peripheral T cells is instead maintained by long-lived T cells and also by some division of mature T cells.

7-8 T-cell precursors proliferate extensively in the thymus but most die there.

T-cell precursors arriving in the thymus from the bone marrow spend up to a week differentiating there before they enter a phase of intense proliferation.

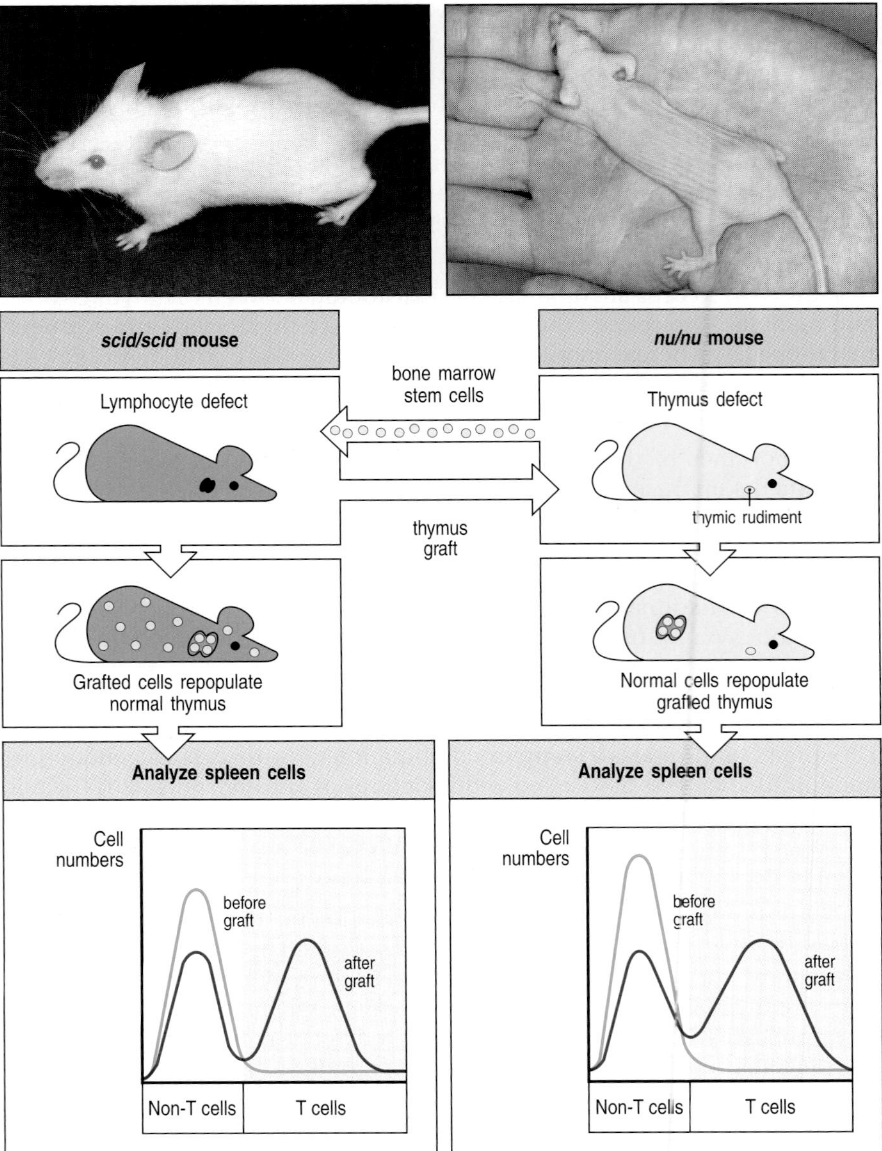

Fig. 7.17 The thymus is critical for the maturation of bone marrow derived cells into T cells. Mice with the *scid* mutation (upper left photograph) have a defect that prevents lymphocyte maturation, whereas mice with the *nude* mutation (upper right photograph) have a defect that affects the development of the cortical epithelium of the thymus. T cells do not develop in either strain of mouse: this can be demonstrated by staining spleen cells with antibodies specific for mature T cells and analyzing them in a flow cytometer (see Appendix I, Section A-22), as represented by the blue line in the graphs in the bottom panels. Bone marrow cells from *nude* mice can restore T cells to *scid* mice (red line in graph on left), showing that, in the correct environment, the nude bone marrow cells are intrinsically normal and capable of producing T cells. Thymic epithelial cells from *scid* mice can induce the maturation of T cells in *nude* mice (red line in graph on right), demonstrating that the thymus provides the essential microenvironment for T-cell development.

In a young adult mouse the thymus contains about 10^8 to 2×10^8 thymocytes. About 5×10^7 new cells are generated each day; however, only about 10^6 to 2×10^6 (roughly 2–4%) of these leave the thymus each day as mature T cells. Despite the disparity between the numbers of T cells generated daily in the thymus and the number leaving, the thymus does not continue to grow in size or cell numbers. This is because about 98% of the thymocytes that develop in the thymus also die in the thymus. No widespread damage is seen, indicating that death is by apoptosis rather than by necrosis (see Section 1-14).

Changes in the plasma membrane of cells undergoing apoptosis lead to their rapid phagocytosis, and apoptotic bodies, which are the residual condensed chromatin of apoptotic cells, are seen inside macrophages throughout the thymic cortex (Fig. 7.18). This apparently profligate waste of thymocytes is a crucial part of T-cell development because it reflects the intensive screening that each thymocyte undergoes for the ability to recognize self-peptide:self-MHC complexes and for self tolerance.

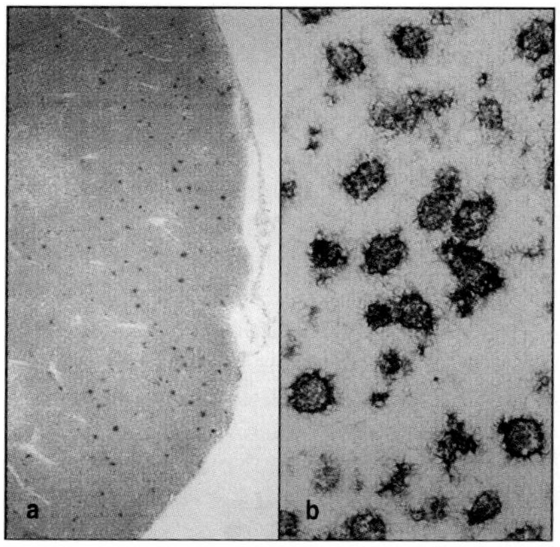

Fig. 7.18 Developing T cells that undergo apoptosis are ingested by macrophages in the thymic cortex. Panel a shows a section through the thymic cortex and part of the medulla in which cells have been stained for apoptosis with a red dye. Thymic cortex is to the right of the photograph. Apoptotic cells are scattered throughout the cortex but are rare in the medulla. Panel b shows a section of thymic cortex at higher magnification that has been stained red for apoptotic cells and blue for macrophages. The apoptotic cells can be seen within macrophages. Magnifications: panel a, × 45; panel b, × 164. Photographs courtesy of J. Sprent and C. Surh.

7-9 Successive stages in the development of thymocytes are marked by changes in cell-surface molecules.

Like developing B cells, developing thymocytes pass through a series of distinct phases. These are marked by changes in the status of T-cell receptor genes and in the expression of the T-cell receptor, and by changes in the expression of cell-surface proteins such as the CD3 complex (see Section 6-8) and the co-receptor proteins CD4 and CD8 (see Section 3-17). These surface changes reflect the state of functional maturation of the cell, and particular combinations of cell-surface proteins are used as markers for T cells at different stages of differentiation. The principal stages are summarized in Fig. 7.19. Two distinct lineages of T cells—α:β and γ:δ, which have different types of T-cell receptor—are produced early in T-cell development. Later, α:β T cells develop into two distinct functional subsets, CD4 and CD8 T cells.

When progenitor cells first enter the thymus from the bone marrow, they lack most of the surface molecules characteristic of mature T cells, and their receptor genes are unrearranged. These cells give rise to the major population of α:β T cells and the minor population of γ:δ T cells. If injected into the peripheral circulation, these lymphoid progenitors can even give rise to B cells and NK cells. Interactions with the thymic stroma trigger an initial phase of differentiation along the T-cell lineage pathway followed by cell proliferation and the expression of the first cell-surface molecules specific for T cells, for example CD2 and (in mice) Thy-1. At the end of this phase, which can last about a week, the thymocytes bear distinctive markers of the T-cell lineage but do not express any of the three cell-surface markers that define mature T cells. These are the CD3:T-cell receptor complex and the co-receptors CD4 or CD8. Because of the absence of CD4 and CD8 such cells are called **double-negative thymocytes** (see Fig. 7.19).

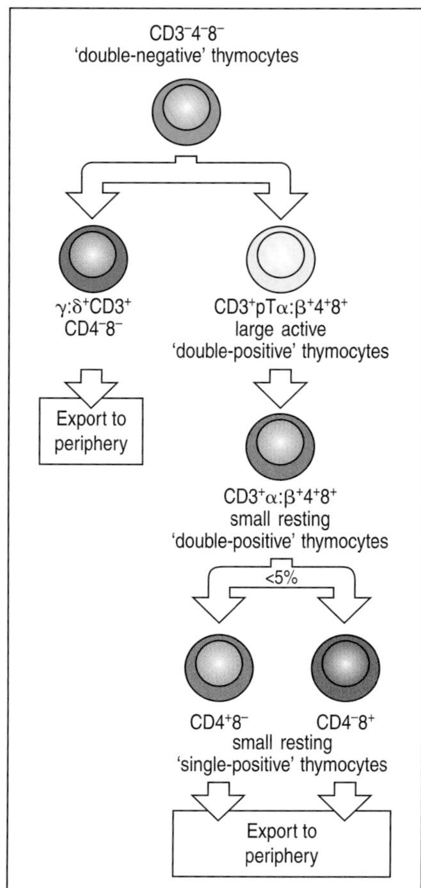

Fig. 7.19 Two distinct lineages of thymocytes are produced in the thymus. CD4, CD8, and T-cell receptor complex molecules (CD3, and the T-cell receptor α and β chains) are important cell-surface molecules for identifying thymocyte subpopulations. The earliest cell population in the thymus does not express any of these proteins, and because they do not express CD4 or CD8 they are called 'double-negative' thymocytes. These cells include precursors that give rise to two T-cell lineages: the minority population of γ:δ T cells (which lack CD4 or CD8 even when mature), and the majority α:β T-cell lineage. The development of prospective α:β T cells proceeds through stages in which both CD4 and CD8 are expressed by the same cell; these are known as 'double-positive' thymocytes. These cells enlarge and divide. Later, they become small resting double-positive cells that express low levels of the T-cell receptor. Most thymocytes die within the thymus after becoming small double-positive cells, but those cells whose receptors can interact with self-peptide:self-MHC molecular complexes lose expression of either CD4 or CD8 and increase the level of expression of the T-cell receptor. The outcome of this process is the 'single-positive' thymocytes, which, after maturation, are exported from the thymus as mature single-positive CD4 or CD8 T cells.

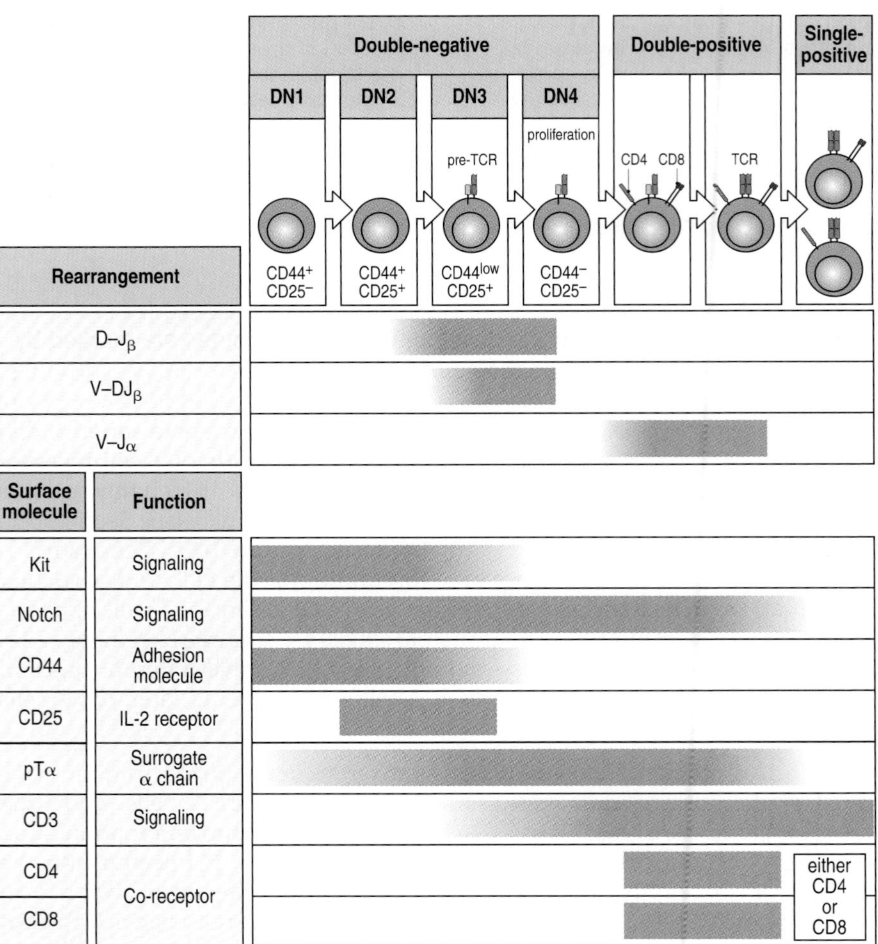

Fig. 7.20 The correlation of stages of α:β T-cell development in the mouse thymus with the program of gene rearrangement and the expression of cell-surface proteins. Lymphoid precursors are triggered to proliferate and become thymocytes committed to the T-cell lineage through interactions with the thymic stroma. These double-negative (DN1) cells express CD44 and Kit, and at a later stage (DN2) express CD25, the α chain of the IL-2 receptor. After this, the DN2 ($CD44^+$ $CD25^+$) cells begin to rearrange the β-chain locus, becoming $CD44^{low}$ and Kit^{low} as this occurs, and become DN3 cells. The DN3 cells are arrested in the $CD44^{low}$ $CD25^+$ stage until they productively rearrange the β-chain locus; the in-frame β chain then pairs with a surrogate chain called pTα to form the pre-T-cell receptor (pre-TCR) and is expressed on the cell surface, which triggers entry into the cell cycle. Expression of small amounts of pTα:β on the cell surface in association with CD3 signals the cessation of β-chain gene rearrangement and triggers rapid cell proliferation, which causes the loss of CD25. The cells are then known as DN4 cells. Eventually, the DN4 cells cease to proliferate and CD4 and CD8 are expressed. The small $CD4^+$ $CD8^+$ double-positive cells begin efficient rearrangement at the α-chain locus. The cells then express low levels of an α:β T-cell receptor and the associated CD3 complex and are ready for selection. Most cells die by failing to be positively selected or as a consequence of negative selection, but some are selected to mature into CD4 or CD8 single-positive cells and eventually to leave the thymus. Expression of some other cell-surface proteins is depicted with respect to the stages of thymocyte development. The proteins listed here are a selection of those known to be associated with early T-lineage development and have been included because of their proven importance in the developmental sequence, largely on the basis of studies in mice. Their individual contributions to T-cell development are discussed in the text.

In the fully developed thymus, the immature double-negative T cells constitute about 60% of the thymocytes that lack both CD4 and CD8. This pool (about 5% of all thymocytes) also includes two populations of more mature T cells that belong to minority lineages. One of these, representing about 20% of the double-negative cells in the thymus, comprises cells that have rearranged and are expressing the genes encoding the γ:δ T-cell receptor; we will return to these cells in Section 7-12. The second, representing another 20% of all double-negative thymocytes, includes cells bearing α:β T-cell receptors of very limited diversity. These cells also express the NK1.1 receptor commonly found on NK cells; they are therefore known as **NK T cells**. NK T cells are activated as part of the early response to many infections; they differ from the major lineage of α:β T cells in recognizing CD1 molecules rather than MHC class I or MHC class II molecules (see Section 5-18) and they are not shown in Fig. 7.19. In this and subsequent discussions, we reserve the term 'double-negative thymocytes' for the immature thymocytes that do not yet express a complete T-cell receptor molecule. These cells give rise to both γ:δ and α:β T cells (see Fig. 7.19), although most of them develop along the α:β pathway.

The α:β pathway is shown in more detail in Fig. 7.20. The double-negative stage can be further subdivided into four stages based on expression of the adhesion molecule CD44, CD25 (the α chain of the IL-2 receptor), and Kit, the receptor for SCF (see Section 7-1). At first, double-negative thymocytes express Kit and CD44 but not CD25 and are called **DN1** cells; in these cells, the genes encoding both chains of the T-cell receptor are in the germline configuration. As thymocytes mature, they begin to express CD25 on their surface and are called **DN2** cells; later, expression of CD44 and Kit is reduced, and they are called **DN3** cells.

Rearrangement of the T-cell receptor β-chain locus begins in DN2 cells with some D_β to J_β rearrangements and continues in DN3 cells with V_β to DJ_β rearrangement. Cells that fail to make a successful rearrangement of the β-chain locus remain at the DN3 ($CD44^{low}$ $CD25^+$) stage and soon die, whereas cells that make productive β-chain gene rearrangements and express the β chain lose expression of CD25 once again and progress to the **DN4** stage, in which they proliferate. The functional significance of the transient expression of CD25 is unclear: T cells develop normally in mice in which the IL-2 gene has been deleted by gene knockout (see Appendix I, Section A-47). By contrast, Kit is quite important for the development of the earliest double-negative thymocytes, in that mice lacking Kit have a much smaller number of double-negative T cells. In addition, the IL-7 receptor is also essential for early T-cell development, because there is a severe block to development when it is defective. Finally, continuous Notch signaling is important for progression through each of these stages of T-cell development.

In DN3 thymocytes, the expressed β chains pair with a surrogate pre-T-cell receptor α chain called **pTα** (pre-T-cell α), which allows the assembly of a complete **pre-T-cell receptor** that is analogous in structure and function to the pre-B-cell receptor. The pre-T-cell receptor is expressed on the cell surface in a complex with the CD3 molecules that provide the signaling components of T-cell receptors (see Section 6-8). The assembly of the CD3:pre-T-cell receptor complex leads to cell proliferation, the arrest of further β-chain gene rearrangement and the expression of both CD8 and CD4. These **double-positive thymocytes** make up the vast majority of thymocytes. Once the large double-positive thymocytes cease to proliferate and become small double-positive cells, the α-chain locus begins to rearrange. As we will see later in this chapter, the structure of the α locus (see Section 4-9) allows multiple successive attempts at rearrangement, so that it is successfully rearranged in most developing thymocytes. Thus, most double-positive cells produce an α:β T-cell receptor during their relatively short life span.

Small double-positive thymocytes initially express low levels of the T-cell receptor. Most of these receptors cannot recognize self-peptide:self-MHC molecular complexes; they will fail positive selection and the cells will die. In contrast, those double-positive cells that recognize self-peptide:self-MHC complexes and can therefore be positively selected, go on to mature, and express high levels of the T-cell receptor. Concurrently, they cease to express one or other of the two co-receptor molecules, becoming either CD4 or CD8 **single-positive thymocytes**. Thymocytes also undergo negative selection during and after the double-positive stage, which eliminates those cells capable of responding to self antigens. About 2% of the double-positive thymocytes survive this dual screening and mature as single-positive T cells that are gradually exported from the thymus to form the peripheral T-cell repertoire. The time between the entry of a T-cell progenitor into the thymus and the export of its mature progeny is estimated to be about 3 weeks in the mouse.

7-10 Thymocytes at different developmental stages are found in distinct parts of the thymus.

The thymus is divided into two main regions, a peripheral cortex and a central medulla (see Fig. 7.15). Most T-cell development takes place in the cortex; only mature single-positive thymocytes are seen in the medulla. Initially, progenitors from the bone marrow enter at the cortico-medullary junction and migrate to the outer cortex (Fig. 7.21). At the outer edge of the cortex, in the subcapsular region of the thymus, large immature double-negative

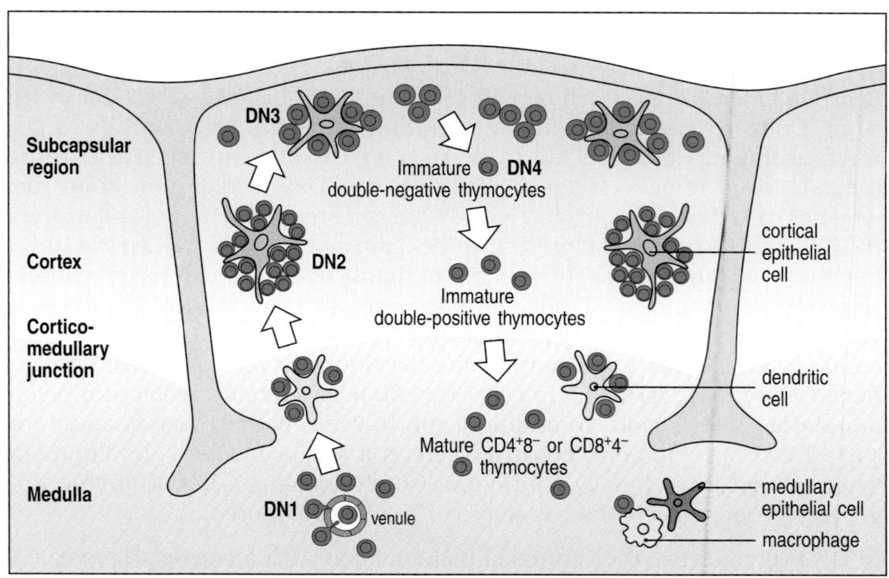

Fig. 7.21 Thymocytes at different developmental stages are found in distinct parts of the thymus. The earliest precursor thymocytes enter the thymus from the bloodstream via venules near the cortico-medullary junction. Ligands that interact with the receptor Notch1 are expressed in the thymus and act on the immigrant cells to commit them to the T-cell lineage. As these cells differentiate through the early $CD4^-$ $CD8^-$ double-negative (DN) stages described in the text, they migrate through the cortico-medullary junction and to the outer cortex. DN3 cells reside near the subcapsular region. As the progenitor matures further to the $CD4^+$ $CD8^+$ double-positive stage, it migrates back through the cortex. Finally, the medulla contains only mature single-positive T cells, which eventually leave the thymus.

thymocytes proliferate vigorously; these cells are thought to represent committed thymocyte progenitors and their immediate progeny and will give rise to all subsequent thymocyte populations. Deeper in the cortex, most of the thymocytes are small double-positive cells. The cortical stroma is composed of epithelial cells with long branching processes that express both MHC class II and MHC class I molecules on their surface. The thymic cortex is densely packed with thymocytes, and the branching processes of the thymic cortical epithelial cells make contact with almost all cortical thymocytes (see Fig. 7.16). Contact between the MHC molecules on thymic cortical epithelial cells and the receptors of developing T cells has a crucial role in positive selection, as we will see later in this chapter.

After positive selection, developing T cells migrate from the cortex to the medulla. The medulla contains fewer lymphocytes, and those present are predominantly the newly matured single-positive T cells that will eventually leave the thymus. The medulla plays a role in negative selection. The antigen-presenting cells in this environment include dendritic cells that express co-stimulatory molecules, which are generally absent from the cortex. In addition, specialized medullary epithelial cells present peripheral antigens for the induction of self tolerance. Cortical and medullary epithelial cells develop from a common progenitor, which expresses the surface antigen MTS24. The differentiation of the two types of epithelia is presumably critical to the proper function of the thymus.

7-11 T cells with $\alpha:\beta$ or $\gamma:\delta$ receptors arise from a common progenitor.

T cells bearing $\gamma:\delta$ receptors differ from $\alpha:\beta$ T cells in the types of antigen they recognize, in the pattern of expression of the CD4 and CD8 co-receptors, and in their anatomical distribution in the periphery. The two types of T cells also differ in function, although relatively little is known about the functions of $\gamma:\delta$ T cells and the ligands they recognize (see Sections 2-34 and 3-19). Different genetic loci are used to make these two types of T-cell receptors, as described in Section 4-11. The T-cell developmental program must control which lineage a precursor commits to and must also ensure that a mature T cell expresses receptor components of only one lineage. The gene rearrangements found in thymocytes and in mature $\gamma:\delta$ and $\alpha:\beta$ T cells suggest that these two cell lineages diverge from a common precursor after certain gene

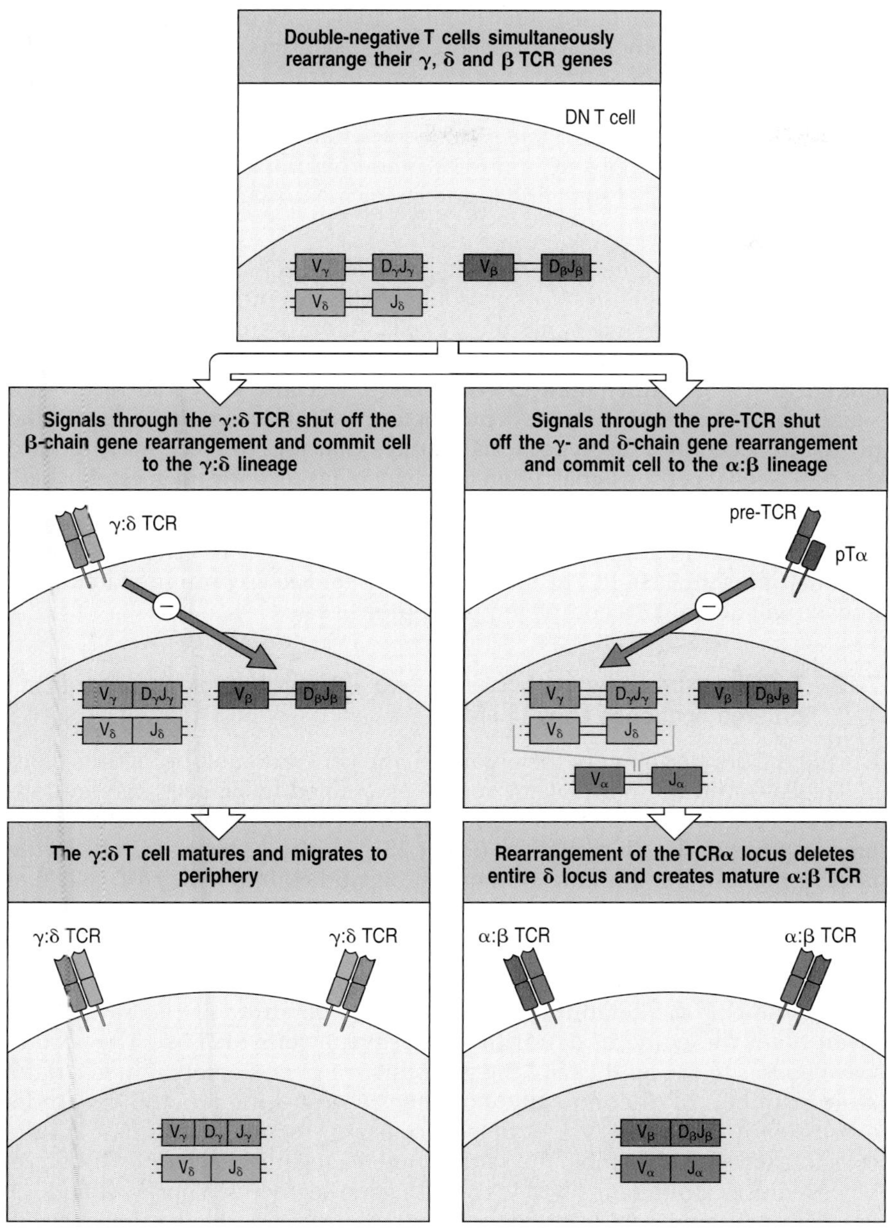

Fig. 7.22 Signals through the γ:δ T-cell receptor and the pre-T-cell receptor compete to determine the fate of thymocytes. During the development of T cells in the thymus, double-negative (DN) thymocytes begin to rearrange the γ, δ, and β T-cell receptor loci simultaneously (top panel). If a complete γ:δ T-cell receptor is formed before a successful β-chain gene rearrangement has led to the production of the pre-T-cell receptor (left panels), the thymocyte receives signals through the γ:δ receptor, which shuts off further rearrangement of the β-chain gene and commits the cell to the γ:δ lineage. This cell then matures into a γ:δ T cell and migrates out of the thymus into the peripheral circulation (bottom left panel). If a functional β chain is formed before a complete γ:δ receptor, it pairs with the pTα to generate a pre-T-cell receptor (right panels). In this case, the developing thymocyte receives a signal through the pre-T-cell receptor, shuts off rearrangements of the γ and δ loci, and commits to the α:β lineage. The thymocyte passes from the DN3 stage through the proliferating DN4 stage into the double-positive stage, at which the TCRα-chain locus rearranges and a mature α:β T-cell receptor is produced (bottom right panel). α-chain locus rearrangement deletes the δ genes, thus precluding the production of a γ:δ receptor on the same cell.

rearrangements have already occurred (Fig. 7.22). Mature γ:δ T cells can contain rearranged β-chain genes, although 80% of these are nonproductive, and mature α:β T cells often contain rearranged, but mostly out-of-frame, γ-chain genes.

The β, γ, and δ loci undergo rearrangement almost simultaneously in developing thymocytes. The decision of a precursor to commit to the γ:δ or the α:β lineage is thought to depend on whether a functional γ chain and a functional δ chain, and thus a functional γ:δ receptor, are produced before a functional β chain which can pair with pTα to create the pre-T-cell receptor (β:pTα) (see Section 7-9). It is thought that the γ:δ T-cell receptor delivers a stronger signal to the T-cell precursor than is delivered by the pre-T-cell receptor and that this stronger signal leads to γ:δ commitment, whereas the weaker signaling by the pre-T-cell receptor leads to α:β commitment. Some evidence suggests that the strength of Notch signaling may also contribute to the choice of cell fate.

In most precursors there is a successful β-chain gene rearrangement before successful rearrangement of both γ and δ has occurred. The production of a pre-T-cell receptor then arrests further gene rearrangement and signals the thymocyte to proliferate, to express its co-receptor genes, and eventually to start rearranging the α-chain genes. It is known that the β:pTα receptor signals constitutively via the tyrosine kinase Lck and does not seem to need a ligand on thymic stroma. This signaling is crucial for the further development of an α:β T cell.

It seems likely that signals through the pre-T-cell receptor commit the cell to the α:β lineage (see Fig. 7.22). One problem with this model, however, is how to explain the occurrence of mature γ:δ cells that carry productive rearrangements at the β-chain locus. One way to reconcile this is if these cells had committed to the γ:δ rather than the α:β lineage because they had received a signal from an assembled γ:δ receptor before having assembled a functional pre-T-cell receptor. This hypothesis requires that the γ:δ T-cell receptor and the pre-T-cell receptor signal differently, which has recently been established.

Once the α-chain locus starts rearranging after a pre-T-cell receptor signal, the δ-chain gene segments located within the α-chain locus are deleted as an extrachromosomal circle. This further ensures that cells committed to the α:β lineage will not make a complete γ:δ receptor.

7-12 T cells expressing particular γ- and δ-chain V regions arise in an ordered sequence early in life.

During the development of the organism, the generation of the various types of T cells—even the particular V region assembled in γ:δ cells—is developmentally controlled. The first T cells to appear during embryonic development carry γ:δ T-cell receptors (Fig 7.23). In the mouse, in which the development of the immune system can be studied in detail, γ:δ T cells first appear in discrete waves or bursts, with the T cells in each wave populating distinct sites in the adult animal.

The first wave of γ:δ T cells populates the epidermis; the T cells become wedged among the keratinocytes and adopt a dendritic-like form that has given them the name of **dendritic epidermal T cells** (**dETCs**). The second wave homes to the epithelia of the reproductive tract. Remarkably, given the large number of theoretically possible rearrangements, the receptors expressed by these early waves of γ:δ T cells are essentially invariant. All the cells in each wave assemble the same V_γ and V_δ regions. Each different wave, however, uses a different set of V, D, and J gene segments. Thus, certain V, D, and J gene segments are selected for rearrangement at particular times during embryonic development; the reasons for this limitation are poorly understood. There are no N-nucleotides contributing additional diversity at the junctions between V, D, and J gene segments, reflecting the absence of the enzyme TdT from these fetal T cells.

After these initial waves, T cells are produced continuously rather than in bursts, and α:β T cells predominate, making up more than 95% of thymocytes. The γ:δ T cells produced at this stage are different from those of the early waves. They have considerably more diverse receptors, for which several different V gene segments have been used, and the receptor sequences have abundant N-nucleotide additions. Most of these γ:δ T cells, like α:β T cells, are found in peripheral lymphoid tissues rather than in epithelial sites.

The developmental changes in V gene segment usage and N-nucleotide addition in murine γ:δ T cells parallel changes in B-cell populations during fetal development, which are discussed later. Their functional significance is unclear, however, and not all of these changes in the pattern of receptors expressed by γ:δ T cells occur in humans. The dETCs, for example, do not seem to have exact human counterparts, although there are γ:δ T cells in the human reproductive and gastrointestinal tracts. The mouse dETCs may serve as sentinel cells that are activated upon local tissue damage or as cells that regulate inflammatory processes.

7-13 Successful synthesis of a rearranged β chain allows the production of a pre-T-cell receptor that triggers cell proliferation and blocks further β-chain gene rearrangement.

We shall now return to the development of α:β T cells. The rearrangement of the β- and α-chain loci follows a sequence that closely parallels the rearrangement of immunoglobulin heavy-chain and light-chain loci during B-cell development (see Sections 7-2 and 7-5). As shown in Fig. 7.24, the β-chain gene segments rearrange first, with the $D_β$ gene segments rearranging to $J_β$ gene segments, and this is followed by $V_β$ gene segments to $DJ_β$ gene rearrangement. If no functional β chain can be synthesized from these rearrangements, the cell will not be able to produce a pre-T-cell receptor and will die unless it makes successful rearrangements at both the γ and δ loci (see

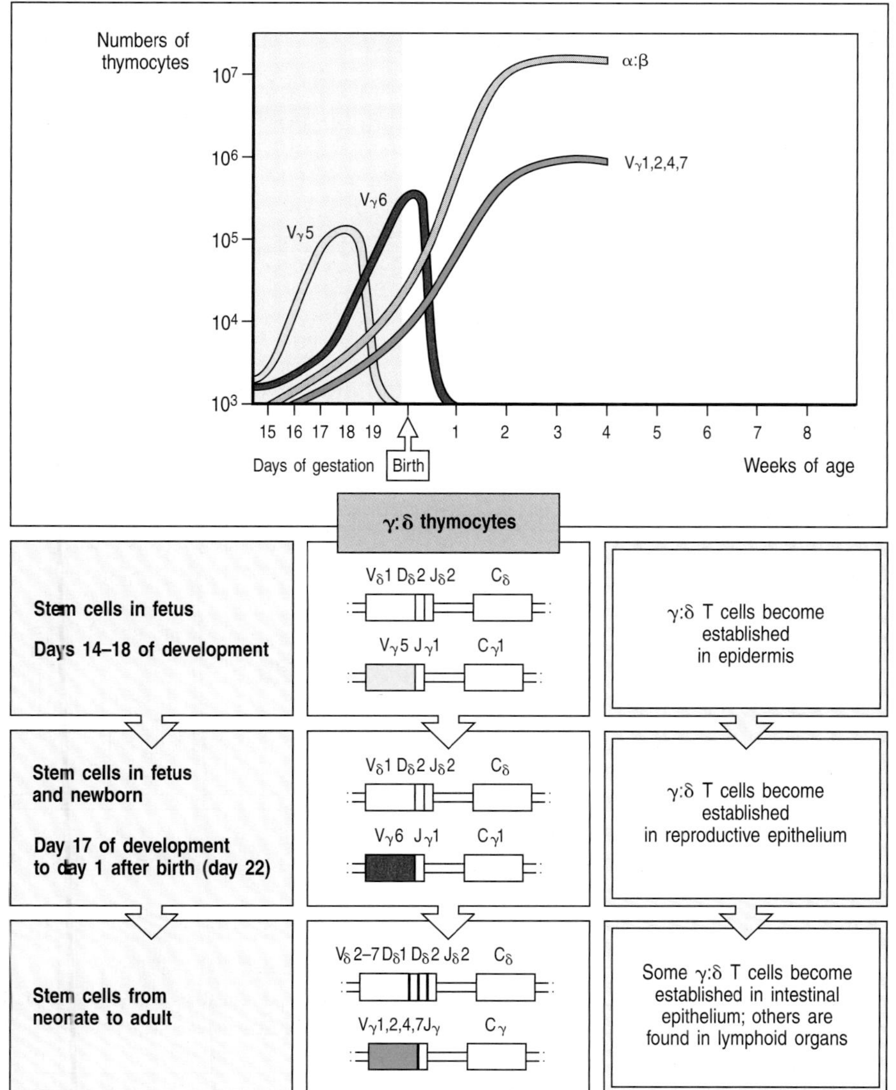

Fig. 7.23 The rearrangement of T-cell receptor γ and δ genes in the mouse proceeds in waves of cells expressing different $V_γ$ and $V_δ$ gene segments. At about 2 weeks of gestation, the $C_{γ1}$ locus is expressed with its closest V gene ($V_{γ5}$). After a few days, $V_{γ5}$-bearing cells decline (upper panel) and are replaced by cells expressing the next most proximal gene, $V_{γ6}$. Both these rearranged γ chains are expressed with the same rearranged δ-chain gene, as shown in the lower panels, and there is little junctional diversity in either the $V_γ$ or the $V_δ$ chain. As a consequence, most of the γ:δ T cells produced in each of these early waves have the same specificity, although the antigen recognized in each case is not known. The $V_{γ5}$-bearing cells become established selectively in the epidermis, whereas the $V_{γ6}$-bearing cells become established in the epithelium of the reproductive tract. After birth, the α:β T-cell lineage becomes dominant and, although γ:δ T cells are still produced, they are a much more heterogeneous population, bearing receptors with a great deal of junctional diversity. Note that $V_γ$ gene segments are described using the system proposed by Tonegawa.

Section 7-12). However, unlike B cells with nonproductive immunoglobulin heavy-chain gene rearrangements, thymocytes with nonproductive β-chain VDJ rearrangements can be rescued by further rearrangement, which is possible because of the two clusters of D_β and J_β gene segments upstream of two C_β genes (see Fig. 4.9). For this reason, the likelihood of a productive VDJ join at the β locus is somewhat higher than the 55% chance of a productive immunoglobulin heavy-chain gene arrangement.

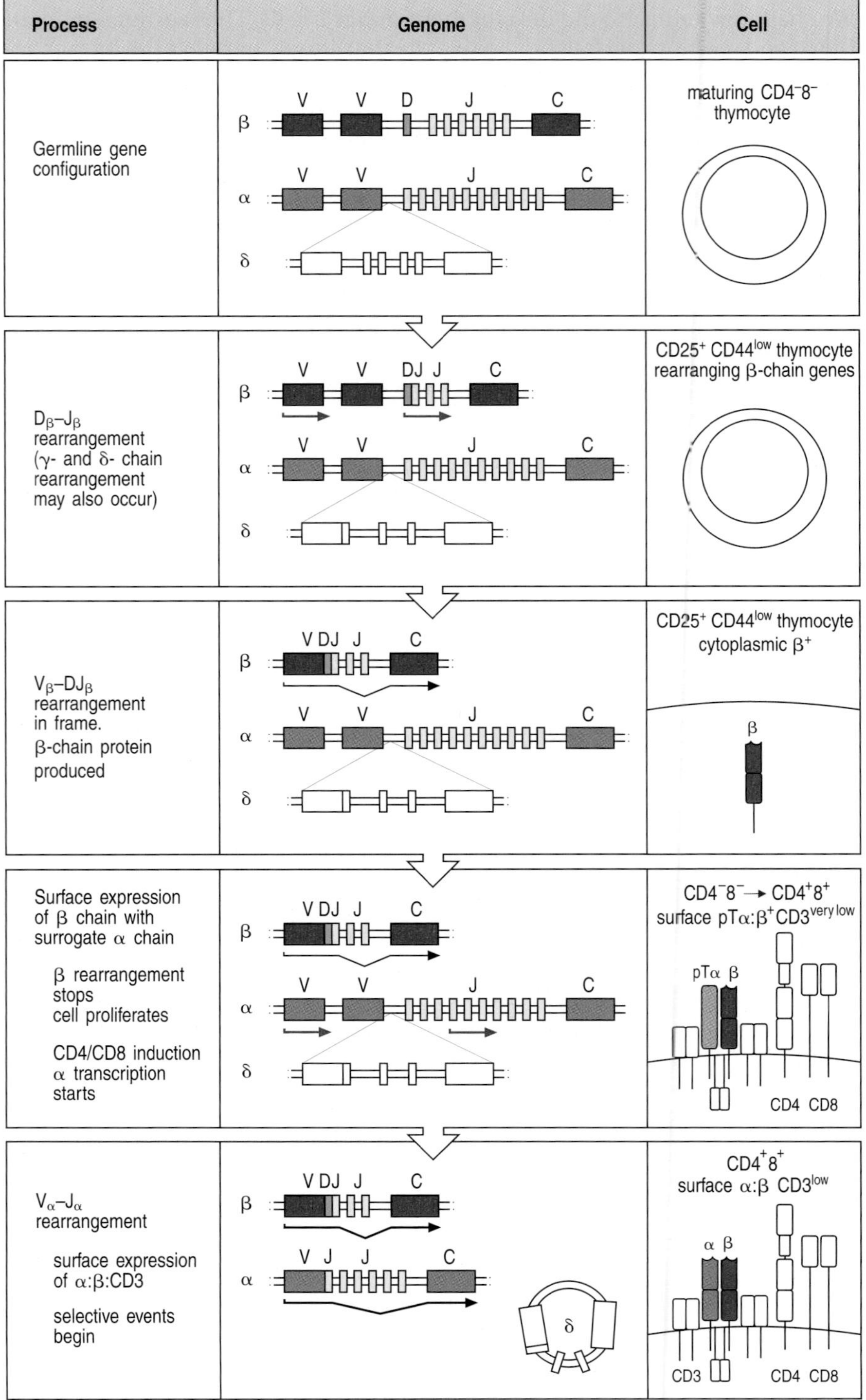

Fig. 7.24 The stages of gene rearrangement in α:β T cells. The sequence of gene rearrangements is shown, together with an indication of the stage at which the events take place and the nature of the cell-surface receptor molecules expressed at each stage. The β-chain locus rearranges first, in CD4⁻ CD8⁻ double-negative thymocytes expressing CD25 and low levels of CD44. As with immunoglobulin heavy-chain genes, D to J gene segments rearrange before V gene segments rearrange to DJ (second and third panels). It is possible to make up to four attempts to generate a productive rearrangement at the β-chain locus, because there are four D gene segments with two sets of J gene segments associated with each TCR β chain locus (not shown). The productively rearranged gene is expressed initially within the cell and then at low levels on the cell surface. It associates with pTα, a surrogate 33 kDa α chain that is equivalent to λ5 in B-cell development, and this pTα:β heterodimer forms a complex with the CD3 chains (fourth panel). The expression of the pre-T-cell receptor signals the developing thymocytes to halt β-chain gene rearrangement and to undergo multiple cycles of division. At the end of this proliferative burst, the CD4 and CD8 molecules are expressed, the cell ceases cycling, and the α chain is now able to undergo rearrangement. The first α-chain gene rearrangement deletes all δ D, J, and C gene segments on that chromosome, although these are retained as a circular DNA, proving that these are nondividing cells (bottom panel). This permanently inactivates the δ-chain gene. Rearrangements at the α-chain locus can proceed through several cycles, because of the large number of V_α and J_α gene segments, so that productive rearrangements almost always occur. When a functional α chain is produced that pairs efficiently with the β chain, the $CD3^{low}$ $CD4^+$ $CD8^+$ thymocyte is ready to undergo selection for its ability to recognize self peptides in association with self-MHC molecules.

Once a productive β-chain gene rearrangement has occurred, the β chain is expressed together with the invariant partner chain pTα and the CD3 molecules (see Fig. 7.24) and is transported to the cell surface. The β:pTα complex is a functional pre-T-cell receptor analogous to the μ:VpreB:λ5 pre-B-cell receptor complex in B-cell development (see Section 7-3). Expression of the pre-T-cell receptor at the DN3 stage of thymocyte development triggers the phosphorylation and degradation of RAG-2, halting β-chain gene rearrangement and thus ensuring allelic exclusion at the β locus. This signal induces the DN4 stage in which rapid cell proliferation occurs, and eventually the co-receptor proteins CD4 and CD8 are expressed. The pre-T-cell receptor signals constitutively via the cytoplasmic protein kinase Lck, a Src-family tyrosine kinase (see Fig. 6.14), and does not seem to require a ligand on the thymic epithelium. Lck subsequently associates with the co-receptor proteins. In mice genetically deficient in Lck, T-cell development is arrested before the CD4 CD8 stage and no α-chain gene rearrangements can be made.

The role of the expressed β chain in suppressing further β-locus rearrangement can be demonstrated in mice containing a rearranged TCRβ transgene: these mice express the transgenic β chain on virtually 100% of their T cells, showing that rearrangement of the endogenous β-chain genes is strongly suppressed. The importance of pTα has been shown in mice deficient in pTα, in which there is a hundredfold decrease in α:β T cells and an absence of allelic exclusion at the β locus.

During the proliferation of DN4 cells triggered by expression of the pre-T-cell receptor, the *RAG-1* and *RAG-2* genes are repressed. Hence, no rearrangement of the α-chain locus occurs until the proliferative phase ends, when *RAG-1* and *RAG-2* are transcribed again, and the functional RAG-1:RAG-2 complex accumulates. This ensures that each cell in which a β-chain gene has been successfully rearranged gives rise to many CD4 CD8 thymocytes. Once the cells stop dividing, each of them can independently rearrange its α-chain genes, so that a single functional β chain can be associated with many different α chains in the progeny cells. During the period of α-chain gene rearrangement, α:β T-cell receptors are first expressed and selection by self-peptide:self-MHC complexes in the thymus can begin.

As T cells progress from the double-negative to the double-positive and finally the single-positive stage, there is a distinct pattern of expression of proteins involved in rearrangement and signaling, and also of transcription factors that most probably control the expression of important T-cell genes such as those for the T-cell receptor itself (Fig. 7.25). TdT, the enzyme responsible for the insertion of N-nucleotides, is expressed throughout T-cell receptor gene rearrangement; N-nucleotides are found at the junctions of all rearranged α and β genes. Lck and another tyrosine kinase, ZAP-70, are both expressed from an early stage in thymocyte development. As well as its key role in signaling from the pre-T-cell receptor, Lck is also important for γ:δ T-cell development. In contrast, gene knockout studies (see Appendix I, Section A-47) show that ZAP-70, although expressed from the double-negative stage onward, has a role later: it promotes the development of single-positive thymocytes from double-positive thymocytes. Fyn, a Src-family kinase similar to Lck, is expressed at increasing levels from the double-positive stage onward. It is not essential for the development of α:β thymocytes as long as Lck is present, but is required for the development of NK T cells.

Finally, several transcription factors have been identified that guide the development of thymocytes from one stage to the next. Ikaros and GATA-3 are expressed in early T-cell progenitors; in the absence of either, T-cell development is generally disrupted. Moreover, these molecules also have roles in the normal functioning of mature T cells. In contrast, Ets-1, though also expressed in early progenitors, is not essential for T-cell development,

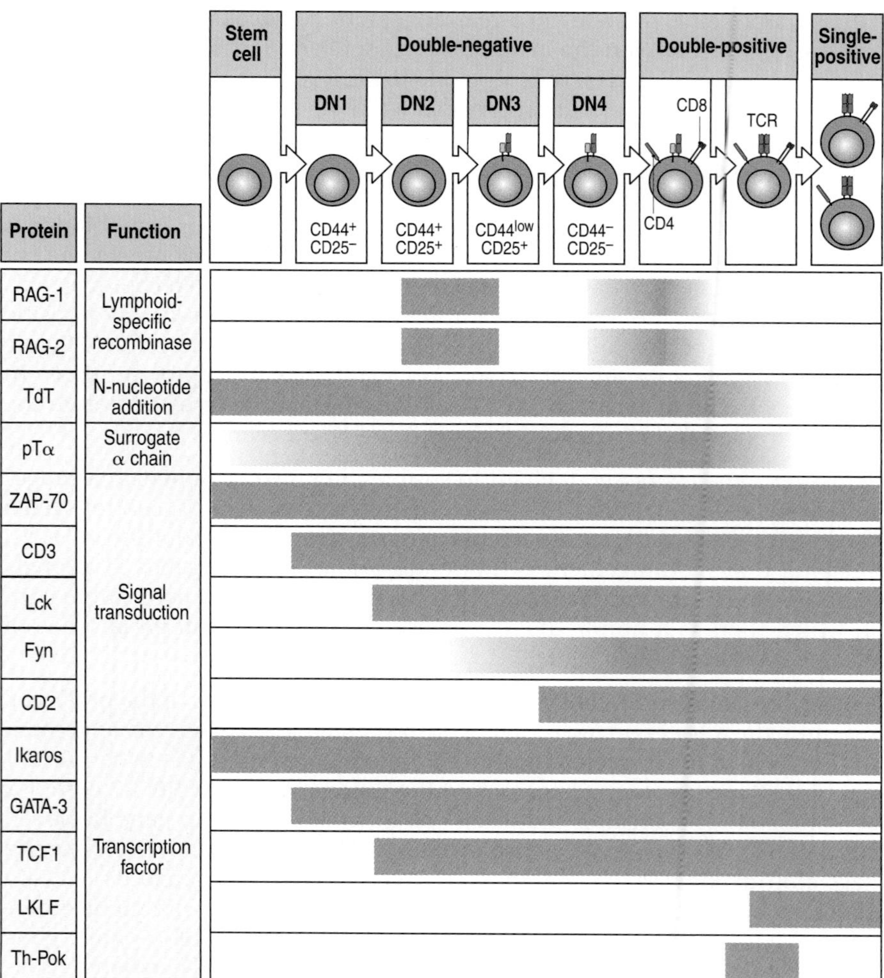

Fig. 7.25 The temporal pattern of expression of some cellular proteins important in early T-cell development. Expression is depicted with regard to the stages of thymocyte development as determined by cell-surface marker expression. The proteins listed are a selection of those known to be associated with early T-lineage development and have been included because of their proven importance in the developmental sequence, largely on the basis of studies in mice. Some of these proteins are involved in gene rearrangement and signaling through receptors, and their individual contributions are discussed in the text. Several transcription factors have been identified that guide the development of thymocytes from one stage to the next by regulating gene expression. Ikaros and GATA-3 are expressed in early T-cell progenitors; in the absence of either, T-cell development is generally disrupted. These proteins also have roles in mature T cells. In the absence of TCF1 (T-cell factor-1), double-negative T cells that make productive β-chain gene rearrangements do not proliferate in response to the pre-T-cell receptor signal, thus preventing the efficient production of double-positive thymocytes. LKLF (lung Kruppel-like factor) is first expressed at the single-positive stage; if absent, thymocytes exhibit a defect in emigration to populate peripheral lymphoid tissues, due in part to their failure to express receptors involved in trafficking, such as the sphingo-1-phosphate (S1P) receptor, $S1P_1$ (see Chapter 8). The transcription factor Ets-1 (not shown on this figure) is not essential for T-cell development, but mice lacking this factor do not make NK cells.

although mice lacking this factor do not make NK cells. TCF1 (T-cell factor-1) is first expressed during the double-negative stage. In its absence, double-negative T cells that make productive β-chain gene rearrangements do not proliferate as usually seen in response to the pre-T-cell receptor signal, preventing the efficient production of double-positive thymocytes. Thus, transcription factors expressed at various developmental stages control normal thymocyte development by controlling the expression of appropriate genes.

7-14 T-cell α-chain genes undergo successive rearrangements until positive selection or cell death intervenes.

The T-cell receptor α-chain genes are comparable to the immunoglobulin κ and λ light-chain genes in that they do not have D gene segments and are rearranged only after their partner receptor chain has been expressed. As with the immunoglobulin light-chain genes, repeated attempts at α-chain gene rearrangement are possible, as illustrated in Fig. 7.26. The presence of multiple V_α gene segments, and about 60 J_α gene segments spread over some 80 kb of DNA, allows many successive V_α to J_α rearrangements to take place at both α-chain alleles. This means that T cells with an initial nonproductive α-gene rearrangement are much more likely to be rescued by a subsequent rearrangement than are B cells with a nonproductive light-chain gene rearrangement.

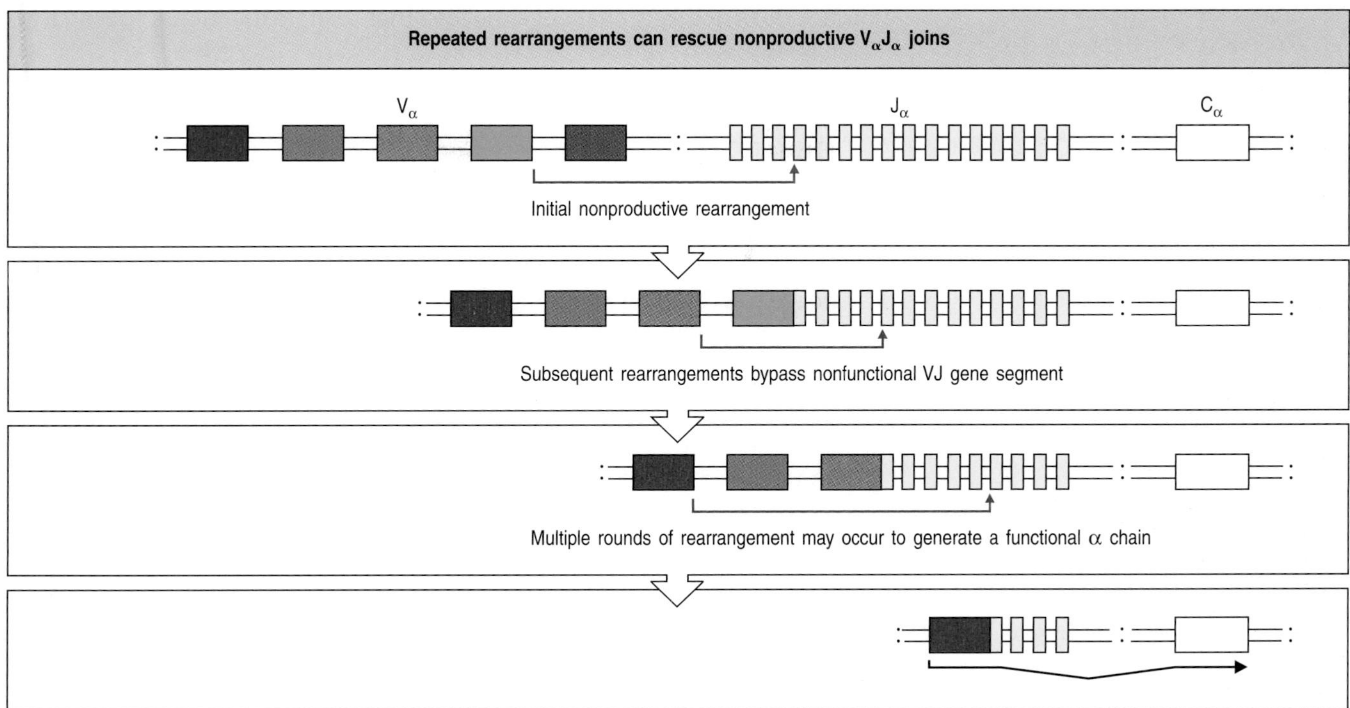
Repeated rearrangements can rescue nonproductive $V_\alpha J_\alpha$ joins
V_α
J_α
C_α
Initial nonproductive rearrangement
Subsequent rearrangements bypass nonfunctional VJ gene segment
Multiple rounds of rearrangement may occur to generate a functional α chain

Fig. 7.26 Multiple successive rearrangement events can rescue nonproductive T-cell receptor α-chain gene rearrangements. The multiplicity of V and J gene segments at the α-chain locus allows successive rearrangement events to 'leapfrog' over previously rearranged VJ segments, deleting any intervening gene segments. The α-chain rescue pathway resembles that of the immunoglobulin κ light-chain genes (see Section 7-5), but the number of possible successive rearrangements is greater. α-chain gene rearrangement continues until either a productive rearrangement leads to positive selection or the cell dies.

One key difference between B and T cells is that the final assembly of an immunoglobulin leads to the cessation of gene rearrangement and initiates the further differentiation of the B cell, whereas in T cells rearrangement of the V_α gene segments continues unless there is signaling by a self-peptide:self-MHC complex that positively selects the receptor. This means that many T cells have in-frame rearrangements on both chromosomes and thus can produce two types of α chains. This is possible because expression of the T-cell receptor is not in itself sufficient to shut off gene rearrangement. Continued rearrangements on both chromosomes can allow several different α chains to be produced successively as well as simultaneously in each developing T cell and to be tested for self-peptide:self-MHC recognition in partnership with the same β chain. This phase of gene rearrangement lasts for 3 or 4 days in the mouse and ceases only when positive selection occurs as a consequence of receptor engagement, or when the cell dies. One can predict that if the frequency of positive selection is sufficiently low, roughly one in three mature T cells will express two productively rearranged α chains at the cell surface. This was confirmed recently for both human and mouse T cells. Thus, in the strict sense, T-cell receptor α-chain genes are not subject to allelic exclusion. However, as we will see in the next part of this chapter, only T-cell receptors that are positively selected for self-peptide:self-MHC recognition can function in self-MHC-restricted responses. The regulation of α-chain gene rearrangement by positive selection therefore ensures that each T cell has only a single functional specificity, even if two different α chains are expressed.

T cells with dual specificity might be expected to give rise to inappropriate immune responses if the cell is activated through one receptor yet can act upon target cells recognized by the second receptor. However, only one of the two receptors is likely to be able to recognize peptide presented by a self-MHC molecule. This is because once the cell has been positively selected, α-chain gene rearrangement ceases. Thus, the existence of cells with two α-chain genes productively rearranged and two α chains expressed at the cell surface does not truly challenge the idea that a single functional specificity is expressed by each cell.

Summary.

The thymus provides a specialized and architecturally organized microenvironment for the development of mature T cells. Precursors of T cells migrate from the bone marrow to the thymus, where they interact with environmental cues such as ligands for the Notch receptor that drives commitment to the T lineage. Developing thymocytes decide between three alternative T-cell lineages: γ:δ T cells, NK T cells, and α:β T cells. The α:β T cells pass through a series of stages distinguished by the differential expression of CD44 and CD25, CD3:T-cell receptor complex proteins, and the co-receptors CD4 and CD8. T-cell development is accompanied by extensive cell death, reflecting the intensive selection of T cells and the elimination of those with inappropriate receptor specificities. Most steps in T-cell development take place in the thymic cortex, whereas the medulla contains mainly mature T cells. In differentiating T cells, receptor genes rearrange according to a defined program similar to that of B cells, but with the added complexity that T-cell progenitors must choose between more than a single lineage, developing either into T cells bearing γ:δ T-cell receptors or α:β T-cell receptors. Early in ontogeny, the production of γ:δ T cells predominates over α:β T cells, and these cells populate several peripheral tissues, including the skin, reproductive epithelium and intestine. Later, more than 90% of thymocytes express α:β T-cell receptors. In developing thymocytes, the γ, δ, and β genes rearrange virtually simultaneously; signaling by a functional γ:δ receptor commits the precursor toward the γ:δ lineage and these cells halt further gene rearrangement and do not express CD4 and CD8 co-receptors. Production of a functionally rearranged β-chain gene and signaling by the pre-T-cell receptor commits the precursor to the α:β lineage.

Up to this point, thymocyte development has been independent of antigen. From this point onward, developmental decisions depend on the interaction of the α:β T-cell receptor with its peptide:MHC ligands. Clearly, the engagement of any particular T-cell receptor with a self-peptide:self-MHC ligand will depend on the receptor's specificity. Thus, the next phase of α-chain gene rearrangement marks an important change in the forces shaping the destiny of the T cell.

Positive and negative selection of T cells.

For T-cell precursors that are committed to the α:β lineage at the DN3 stage, a phase of vigorous proliferation follows in the DN4 stage of development in the subcapsular region. Subsequently, these cells differentiate first into immature CD8-single positive (ISP) cells and then into double-positive (DP) cells that express low levels of the T-cell receptor and both the CD4 and CD8 co-receptors and move into the deeper regions of the thymic cortex. These double-positive cells have a life-span of only about 3 to 4 days unless they are rescued by engagement of their T-cell receptor. The rescue of double-positive cells from programmed cell death and their maturation into CD4 or CD8 single-positive cells is the process known as positive selection. Only about 10–30% of the T-cell receptors generated by gene rearrangement will be able to recognize self-peptide:self-MHC complexes and thus function in self-MHC-restricted responses to foreign antigens (see Chapter 4); those that

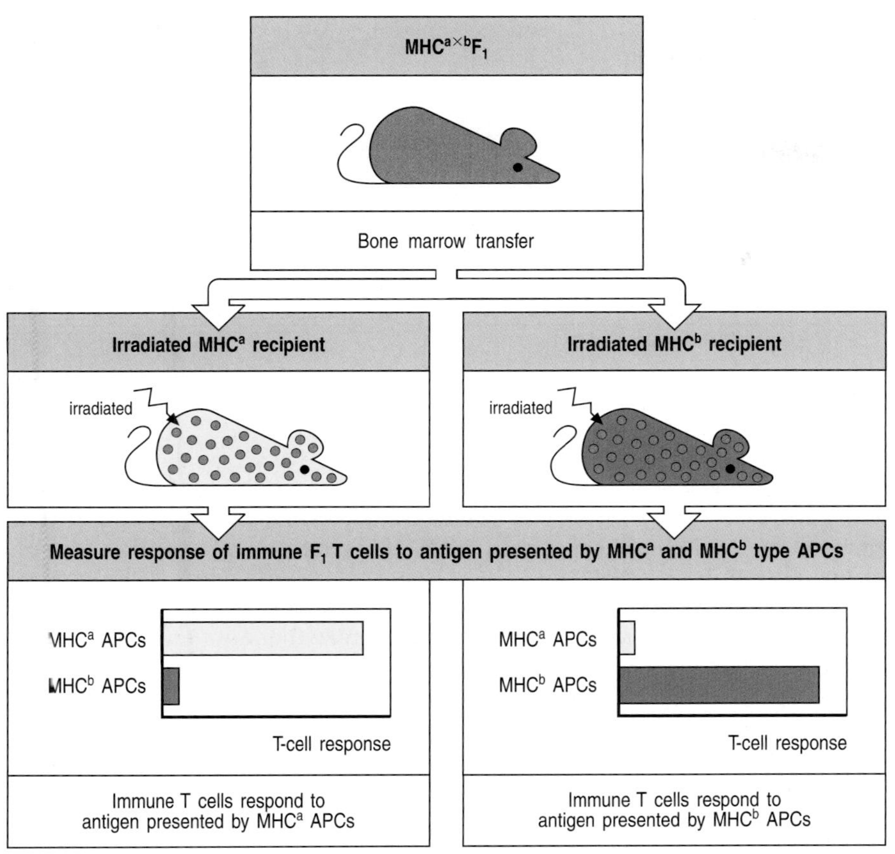

Fig. 7.27 Positive selection is revealed by bone marrow chimeric mice. As shown in the top two sets of panels, bone marrow from an MHC$^{a\times b}$ F$_1$ hybrid mouse is transferred to a lethally irradiated recipient mouse of either parental MHC type (MHC^a or MHC^b). When these chimeric mice are immunized with antigen, the antigen can be presented by the bone marrow derived MHC$^{a\times b}$ antigen-presenting cells (APCs) in association with both MHC^a and MHC^b molecules. The T cells from an MHC$^{a\times b}$ F$_1$ mouse include cells that respond to antigen presented by APCs from MHC^a mice and cells that respond to APCs from MHC^b mice (not shown). But when immunized T cells from the chimeric animals are tested *in vitro* with APCs bearing MHC^a or MHC^b only, they respond far better to antigen presented by the MHC molecules of the recipient MHC type, as shown in the bottom panels. This shows that the T cells have undergone positive selection for MHC restriction in the recipient thymus.

have this capability are selected for survival in the thymus. Double-positive cells also undergo negative selection: T cells whose receptors recognize self-peptide:self-MHC complexes too strongly undergo apoptosis, thus eliminating potentially self-reactive cells. In this section we examine the interactions between developing double-positive thymocytes and different thymic components and examine the mechanisms by which these interactions shape the mature T-cell repertoire.

7-15 The MHC type of the thymic stroma selects a repertoire of mature T cells that can recognize foreign antigens presented by the same MHC type.

Positive selection was first demonstrated in classic experiments using mice whose bone marrow had been completely replaced by bone marrow from a mouse of different MHC genotype but otherwise genetically identical. These mice are known as **bone marrow chimeras** (see Appendix I, Section A-43). The recipient mouse is first irradiated to destroy all its own lymphocytes and bone marrow progenitor cells; after bone marrow transplantation, all bone marrow derived cells will be of the donor genotype. These will include all lymphocytes, as well as the antigen-presenting cells they interact with. The rest of the animal's tissues, including the nonlymphoid stromal cells of the thymus, will be of the recipient MHC genotype.

In the experiments that demonstrated positive selection (Fig. 7.27), the donor mice were F$_1$ hybrids derived from MHC^a and MHC^b parents and thus were of the MHC$^{a\times b}$ genotype. The irradiated recipients were one of the parental

Bone marrow donor	Recipient	Mice contain APC of type:	Secondary T-cell responses to antigen presented *in vitro* by APC of type:	
			MHC^a APC	MHC^b APC
MHC$^{a \times b}$	MHC^a	MHC$^{a \times b}$	Yes	No
MHC$^{a \times b}$	MHC^b	MHC$^{a \times b}$	No	Yes
MHC^a	MHC^b	MHC^a	No	No
MHC^a	MHC^b + MHC^b APC	MHC^a + MHC^b	No	Yes

Fig. 7.28 Summary of T-cell responses to immunization in bone marrow chimeric mice. A set of bone marrow chimeric mice with different combinations of donor and recipient MHC types were made. These mice were then immunized and their T cells were isolated. These were then tested *in vitro* for a secondary immune reaction by using MHC^a or MHC^b antigen-presenting cells (APCs). The results are indicated in the last two columns. T cells can make antigen-specific immune responses far better if the APCs present in the host at the time of priming share at least one MHC molecule with the thymus in which the T cells developed.

strains, either MHC^a or MHC^b. Because of MHC restriction, individual T cells recognize either MHC^a or MHC^b, but not both. Normally, roughly equal numbers of the MHC$^{a \times b}$ T cells from MHC$^{a \times b}$ F_1 hybrid mice will recognize antigen presented by MHC^a or MHC^b. However, in bone marrow chimeras in which T cells of MHC$^{a \times b}$ genotype develop in an MHC^a thymus, T cells immunized to a particular antigen turn out to recognize that antigen mainly, if not exclusively, when it is presented by MHC^a molecules, even though the antigen-presenting cells display antigen bound to both MHC^a and MHC^b. These experiments demonstrated that the MHC molecules present in the environment in which T cells develop determine the MHC restriction of the mature T-cell receptor repertoire.

A similar kind of experiment, using grafts of thymic tissue, showed that the radioresistant cells of the thymic stroma are responsible for positive selection. In these experiments, the recipient animals were athymic *nude* or thymectomized mice of MHC$^{a \times b}$ genotype that were given thymic stromal grafts of MHC^a genotype. Thus, all their cells except the thymic stroma carried both MHC^a and MHC^b. The MHC$^{a \times b}$ bone marrow cells of these mice matured into T cells that recognized antigens presented by MHC^a but not by MHC^b. This result showed that it is the MHC molecules expressed by the thymic stromal cells that determine what mature T cells consider to be self MHC. These results also argued that the MHC-restriction phenomenon in the immunized bone marrow chimeras could be mediated in the thymus, presumably by selecting T cells as they develop.

The chimeric mice used to demonstrate positive selection produce normal T-cell responses to foreign antigens. In constrast, chimeras made by injecting MHC^a bone marrow cells into MHC^b animals cannot make normal T-cell responses. This is because most of the T cells in these animals have been selected to recognize peptides when they are presented by MHC^b, but most of the antigen-presenting cells that they encounter as mature T cells in the periphery are bone marrow derived MHC^a cells. T cells will therefore fail to recognize antigen presented by antigen-presenting cells of their own MHC type, and T cells can be activated in these animals only if antigen-presenting cells of the MHC^b type are injected together with the antigen. Thus, for a bone marrow graft to reconstitute T-cell immunity, there must be at least one MHC molecule in common between donor and recipient (Fig. 7.28).

7-16 Only thymocytes whose receptors interact with self-peptide:self-MHC complexes can survive and mature.

Bone marrow chimeras and thymic grafting provided evidence that MHC molecules in the thymus influence the MHC-restricted T-cell repertoire.

However, mice transgenic for rearranged T-cell receptor genes provided the first conclusive evidence that the interaction of the T cell with self-peptide:self-MHC complexes is necessary for the survival of immature T cells and their maturation into naive CD4 or CD8 T cells. For these experiments, the rearranged α- and β-chain genes were cloned from a T-cell clone (see Appendix I, Section A-24) whose origin, antigen specificity, and MHC restriction were known. When such genes are introduced into the mouse genome, these transgenes are expressed early during thymocyte development and the rearrangement of endogenous T-cell receptor genes is inhibited; endogenous β-chain gene rearrangement is inhibited completely but that of α-chain genes is inhibited only incompletely. The result is that most of the developing thymocytes express the T-cell receptor encoded by the transgenes.

By introducing a T-cell receptor transgene specific for a known MHC genotype, the effect of MHC molecules on the maturation of thymocytes with receptors of known specificity can be studied directly without the need for immunization and analysis of effector function. These studies showed that thymocytes bearing a particular T-cell receptor could develop to the double-positive stage in thymuses that expressed different MHC molecules from those in which the cell bearing the T-cell receptor originally developed. However, these transgenic thymocytes developed beyond the double-positive stage and became mature T cells only if the thymus expressed the same self-MHC molecule as that on which the original T-cell clone was selected (Fig. 7.29). Such experiments have also established the fate of T cells that fail positive selection. Rearranged receptor genes from a mature T cell specific for a peptide presented by a particular MHC molecule were introduced into a recipient mouse lacking that MHC molecule, and the fate of the thymocytes was investigated by staining with antibodies specific for the transgenic receptor. Antibodies against other molecules such as CD4 and CD8 were used at the same time to mark the stages of T-cell development. It was found that cells that fail to recognize the MHC molecules present on the thymic epithelium never progress further than the double-positive stage and die in the thymus within 3 or 4 days of their last division.

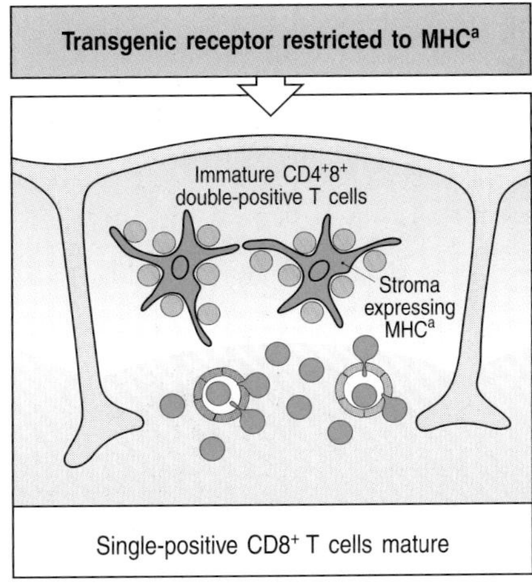

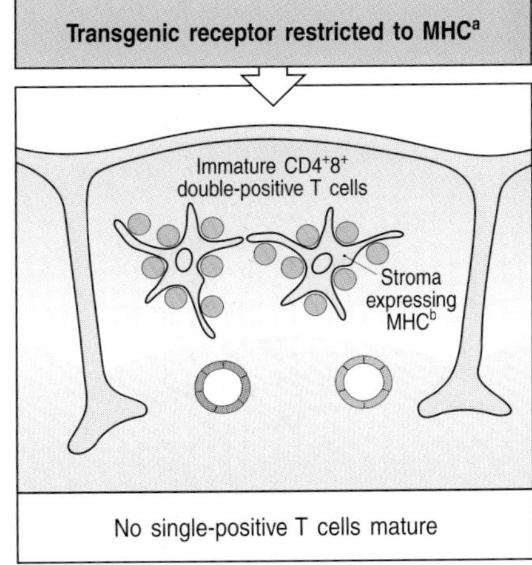

Fig. 7.29 Positive selection is demonstrated by the development of T cells expressing rearranged T-cell receptor transgenes. In mice transgenic for rearranged α:β T-cell receptor genes, the maturation of T cells depends on the MHC haplotype expressed in the thymus. If the transgenic mice express the same MHC haplotype in their thymic stromal cells as the mouse from which the rearranged TCRα-chain and TCRβ-chain genes were cloned (both MHC^a, top panels), then the T cells expressing the transgenic T-cell receptor will develop from the double-positive stage (pale green) into mature T cells (dark green), in this case mature CD8$^+$ single-positive cells. If the MHC^a-restricted TCR transgenes are crossed into a different MHC background (MHC^b, yellow) (bottom panel), then developing T cells expressing the transgenic receptor will progress to the double-positive stage but will fail to mature further. This failure is due to the absence of an interaction between the transgenic T-cell receptor with MHC molecules on the thymic cortex, and thus no signal for positive selection is delivered.

7-17 Positive selection acts on a repertoire of T-cell receptors with inherent specificity for MHC molecules.

Positive selection acts on a repertoire of receptors whose specificity is determined by a combination of germline gene segments and junctional regions whose diversity is randomly created as the genes rearrange (see Section 4-8). It seems, however, that T-cell receptors exhibit a bias toward recognition of MHC molecules even before positive selection. If the binding specificity of the unselected repertoire were completely random, only a very small proportion of thymocytes would be expected to recognize any MHC molecule. However, it seems that the variable CDR1 and CDR2 loops of both chains of the T-cell receptor, which are encoded within the germline V gene segments (see Section 4-10), give the T-cell receptor an intrinsic specificity for MHC molecules. This is evident from the way in which these two regions contact MHC molecules in crystal structures (see Section 3-16). An inherent specificity for MHC molecules has also been shown by examining mature T cells that represent an unselected repertoire of receptors. Such T cells can be generated in fetal thymic organ cultures, using thymuses that do not express either MHC class I or MHC class II molecules, by substituting the binding of anti-β-chain antibodies and anti-CD4 antibodies for the receptor engagement responsible for normal positive selection. When the reactivity of these antibody-selected CD4 T cells is tested, roughly 5% can respond to any one MHC class II genotype and, because they developed without selection by MHC molecules, this must reflect specificity inherent in the germline V gene segments. This germline-encoded specificity for MHC molecules should

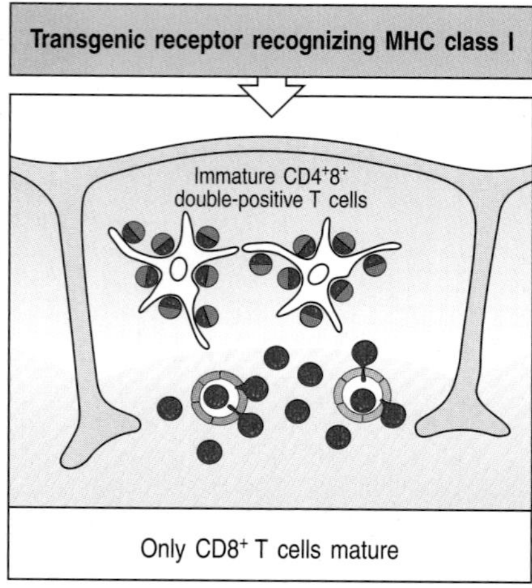

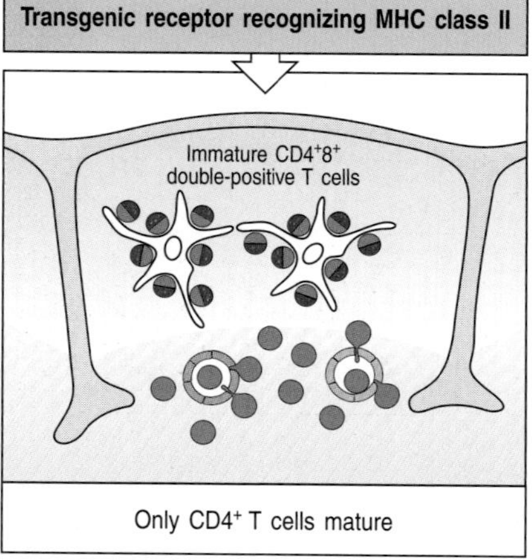

Fig. 7.30 The MHC molecules that induce positive selection determine co-receptor specificity. In mice transgenic for T-cell receptors restricted by an MHC class I molecule (top panel), the mature T cells that develop all have the CD8 (red) phenotype. In mice transgenic for receptors restricted by an MHC class II molecule (bottom panel), all mature T cells have the CD4 (blue) phenotype. In both cases, normal numbers of immature, double-positive thymocytes (half blue, half red) are found. The specificity of the T-cell receptor determines the outcome of the developmental pathway, ensuring that the only T cells that mature are those equipped with a co-receptor that is able to bind the same self-MHC molecule as the T-cell receptor.

significantly increase the proportion of receptors that can be positively selected in any individual.

7-18 Positive selection coordinates the expression of CD4 or CD8 with the specificity of the T-cell receptor and the potential effector functions of the T cell.

At the time of positive selection, the thymocyte expresses both CD4 and CD8 co-receptor molecules. By the end of thymic selection, mature thymocytes ready for export to the periphery will cease to express one of these co-receptors and will belong to one of the following three categories: conventional CD4 or CD8 T cells, or a subset of regulatory T cells expressing CD4 and high levels of CD25. Moreover, almost all mature T cells that express CD4 have receptors that recognize peptides bound to self-MHC class II molecules and are programmed to become cytokine-secreting cells. In contrast, most of the cells that express CD8 have receptors that recognize peptides bound to self-MHC class I molecules and are programmed to become cytotoxic effector cells. Thus, positive selection also determines the cell-surface phenotype and functional potential of the mature T cell, selecting the appropriate co-receptor for efficient antigen recognition and the appropriate program for the T cell's eventual functional differentiation in an immune response.

Experiments with mice transgenic for rearranged T-cell receptor genes show clearly that the specificity of the T-cell receptor for self-peptide:self-MHC molecule complexes determines which co-receptor a mature T cell will express. If the transgenes encode a T-cell receptor specific for antigen presented by self-MHC class I molecules, mature T cells that express the transgenic receptor are CD8 T cells. Similarly, in mice made transgenic for a receptor that recognizes antigen with self-MHC class II molecules, mature T cells that express the transgenic receptor are CD4 T cells (Fig. 7.30).

The importance of MHC molecules in this selection is illustrated by the class of human immunodeficiency diseases known as **bare lymphocyte syndromes**, which are caused by mutations that lead to an absence of MHC molecules on lymphocytes and thymic epithelial cells. People who lack MHC class II molecules have CD8 T cells but only a few, highly abnormal, CD4 T cells; a similar result has been obtained in mice in which MHC class II expression has been eliminated by targeted gene disruption (see Appendix I, Section A-47). Likewise, mice and humans that lack MHC class I molecules lack CD8 T cells. Thus, MHC class II molecules are absolutely required for CD4 T-cell development, whereas MHC class I molecules are similarly required for CD8 T-cell development.

In mature T cells, the co-receptor functions of CD8 and CD4 depend on their respective abilities to bind invariant sites on MHC class I and MHC class II molecules (see Section 3-17). Co-receptor binding to an MHC molecule is also required for normal positive selection, as shown for CD4 in the experiment discussed in the next section. Thus, positive selection depends on engagement of both the antigen receptor and co-receptor with an MHC molecule, and determines the survival of single-positive cells that express only the appropriate co-receptor. The exact mechanism whereby lineage commitment is coordinated with receptor specificity remains to be established, however. It seems that the developing thymocyte integrates signals from both the antigen receptor and the co-receptor to determine its fate. Co-receptor-associated Lck signals are most effectively delivered when CD4 rather than CD8 is engaged as a co-receptor, and these Lck signals play a large part in the decision to become a mature CD4 cell. It seems that when a T cell receives a signal inducing positive selection through the T-cell receptor, it first down-regulates both CD4 and CD8, after which it re-expresses CD4, regardless of

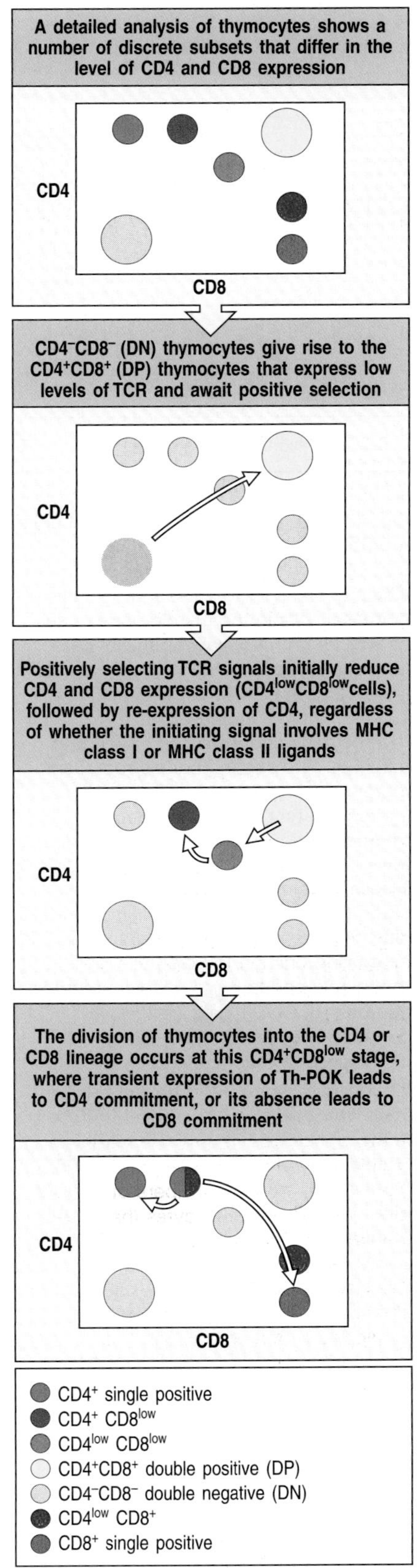

Fig. 7.31 Stages in the positive selection of α:β T cells as identified by FACS analysis. The diagram represents a summary of the results of FACS analysis (see Appendix I, Fig. A.25) of thymic populations of thymocytes at various stages with reference to the co-receptor molecules CD4 and CD8. Each colored circle represents a subset of thymocytes at a different stage of development. Double-negative (DN) cells that have successfully rearranged a β chain and are expressing a pre-T-cell receptor (pre-TCR) undergo proliferation, followed by expression of the co-receptors CD8 and CD4. Rearrangement of the α-chain locus occurs in these cells, with expression of a T-cell receptor on the cell surface first at low and then at intermediate levels. In these cells, signaling is co-receptor dependent. If the expressed T-cell receptor successfully interacts with MHC molecules on thymic stroma to induce positive selection, the cell initially reduces expression of both CD8 and CD4, followed by a subsequent increased expression of CD4, to generate the $CD4^+CD8^{low}$ population. If selection was provided by an MHC class II molecule, signaling in the $CD4^+CD8^{low}$ T cell is of a longer duration and commitment to CD4 occurs, with maintenance of CD4 and loss of CD8 expression. If the selection was provided by an MHC class I molecule, signaling in the $CD4^+CD8^{low}$ T cell will be of shorter duration, and this leads to commitment to the CD8 lineage, with re-expression of CD8 and loss of CD4. Th-POK, T-helper-inducing POZ/Kruppel-like transcription factor.

whether the T-cell receptor has been engaged by MHC class I or MHC class II molecules (Fig. 7.31). One model proposes that the strength or duration of signaling upon CD4 re-expression determines the lineage choice. If the cell is being selected by MHC class II, the re-expression of CD4 provides a stronger or more sustained signal, mediated in part by Lck, and this is responsible for further differentiation along the CD4 pathway, with the complete loss of CD8. If the cell is being selected by MHC class I, re-expression of CD4 will not lead to further signaling via Lck; this weaker signal in turn determines CD8 commitment, with a subsequent loss of CD4 expression and the re-expression later of CD8.

It is a general principle of lineage commitment that different signals must be created to activate lineage-specific factors and generate a divergence of developmental programming. For example, the transcription factor Th-POK (T-helper-inducing POZ/Kruppel-like) (see Fig. 7.31) is essential for development of the CD4 lineage from double-positive thymocytes, as shown by the fact that a naturally occurring loss-of-function mutation in Th-POK causes the redirection of MHC class II-restricted thymocytes toward the CD8 lineage. Although much remains to be discovered about this process in developing α:β thymocytes, it is clear that the different signals that are created result in a divergence of functional programming, so that the ability to express genes involved in the killing of target cells, for example, develops in CD8 T cells but not in most CD4 T cells, whereas the potential to express various cytokine genes develops in CD4 T cells and, to a lesser extent, in CD8 T cells.

The majority of double-positive thymocytes that undergo positive selection develop into either CD4 or CD8 single-positive T cells. However, the thymus also generates a minority population of T cells that express CD4 but not CD8 and that seem to represent a distinct lineage of T cells that regulate the actions of other T cells. These cells also express high levels of the surface proteins CD25 and CTLA-4 (see Section 6-20) and the Forkhead transcription factor FoxP3 and are known as **natural regulatory T cells (T_{reg} cells)**. The basis for the selection and development of this T-cell subset is currently not known.

7-19 Thymic cortical epithelial cells mediate positive selection of developing thymocytes.

The thymus transplantation studies described in Section 7-15 suggested that stromal cells were important for positive selection. These cells form a web of

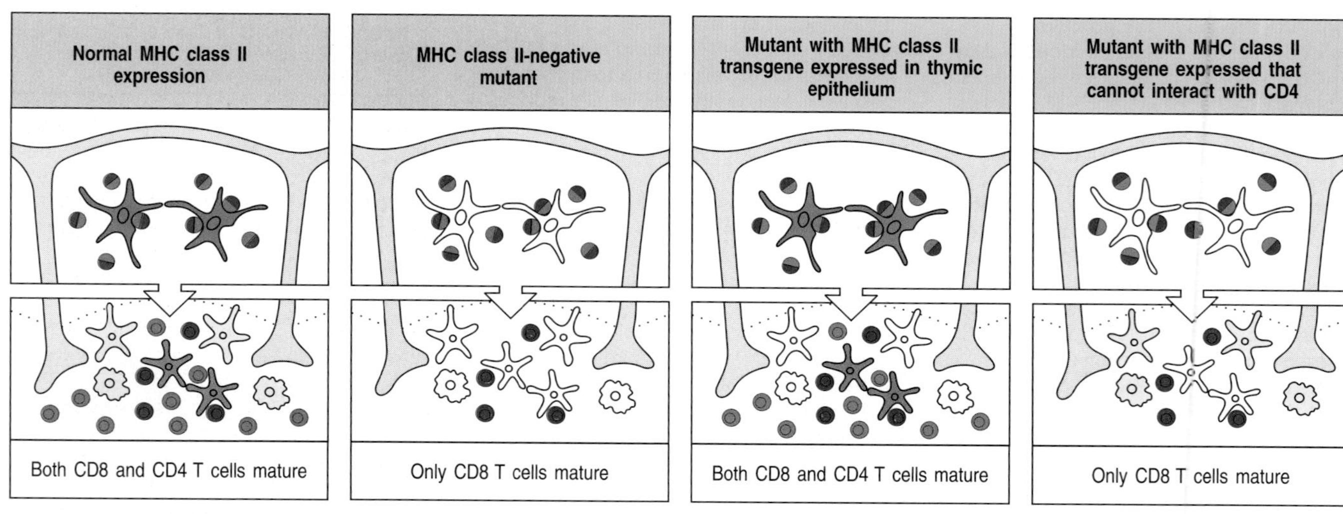
Normal MHC class II expression
MHC class II-negative mutant
Mutant with MHC class II transgene expressed in thymic epithelium
Mutant with MHC class II transgene expressed that cannot interact with CD4
Both CD8 and CD4 T cells mature
Only CD8 T cells mature
Both CD8 and CD4 T cells mature
Only CD8 T cells mature

Fig. 7.32 Thymic cortical epithelial cells mediate positive selection. In the thymus of normal mice (first panels), which express MHC class II molecules on epithelial cells in the thymic cortex (blue) as well as on medullary epithelial cells (orange) and bone marrow derived cells (yellow), both CD4 (blue) and CD8 (red) T cells mature. Double-positive thymocytes are shown as half red and half blue. The second panels represent mutant mice in which MHC class II expression has been eliminated by targeted gene disruption; in these mice, few CD4 T cells develop, although CD8 T cells develop normally. In MHC class II-negative mice containing an MHC class II transgene engineered so that it is expressed only on the epithelial cells of the thymic cortex (third panels), normal numbers of CD4 T cells mature. In contrast, if a mutant MHC class II molecule with a defective CD4-binding site is expressed (fourth panel), positive selection of CD4 T cells does not take place. This indicates that the cortical epithelial cells are the critical cell type mediating positive selection and that the MHC class II molecule needs to be able to interact with the CD4 protein.

cell processes that make close contacts with the double-positive T cells undergoing positive selection (see Fig. 7.16), and T-cell receptors can be seen clustering with MHC molecules at the sites of contact. Direct evidence that thymic cortical epithelial cells mediate positive selection comes from an ingenious manipulation of mice whose MHC class II genes have been eliminated by targeted gene disruption (Fig. 7.32). Mutant mice that lack MHC class II molecules do not normally produce CD4 T cells. To test the role of the thymic epithelium in positive selection, an MHC class II gene was placed under the control of a promoter that restricted its expression to thymic cortical epithelial cells. This was then introduced as a transgene into these mutant mice, and CD4 T-cell development was restored. A further variant of this experiment shows that, to promote the development of CD4 T cells, the MHC class II molecule on the thymic cortical epithelium must be able to interact effectively with CD4. Thus, when the MHC class II transgene expressed in the thymus contains a mutation that prevents its binding to CD4, very few CD4 T cells develop. Equivalent studies of CD8 interaction with MHC class I molecules show that co-receptor binding is also necessary for normal positive selection of CD8 cells.

The critical role of the thymic cortical epithelium in positive selection raises the question of whether there is anything distinctive about the antigen-presenting properties of these cells. This is not clear at present; however, thymic epithelium may differ from other tissues in the proteases used to degrade the invariant chain (Ii) during the passage of MHC class II molecules to the cell surface (see Section 5-8). The protease cathepsin L dominates in thymic cortical epithelium, whereas cathepsin S seems to be most important in peripheral tissues. Consequently, CD4 T-cell development is severely impaired in cathepsin L knockout mice. Thymic epithelial cells do seem to have a relatively high density of MHC class II molecules on their surface that retain the invariant chain-associated peptide (CLIP) (see Fig. 5.9). Another reason that the thymic stromal cells are critical may simply be that these are the cells that are in anatomical proximity to the developing thymocytes during the period allowed for positive selection, and there are very few macrophages and dendritic cells in the cortex.

7-20 T cells that react strongly with ubiquitous self antigens are deleted in the thymus.

When the T-cell receptor of a mature naive T cell is ligated by a peptide:MHC complex displayed on a specialized antigen-presenting cell in a peripheral

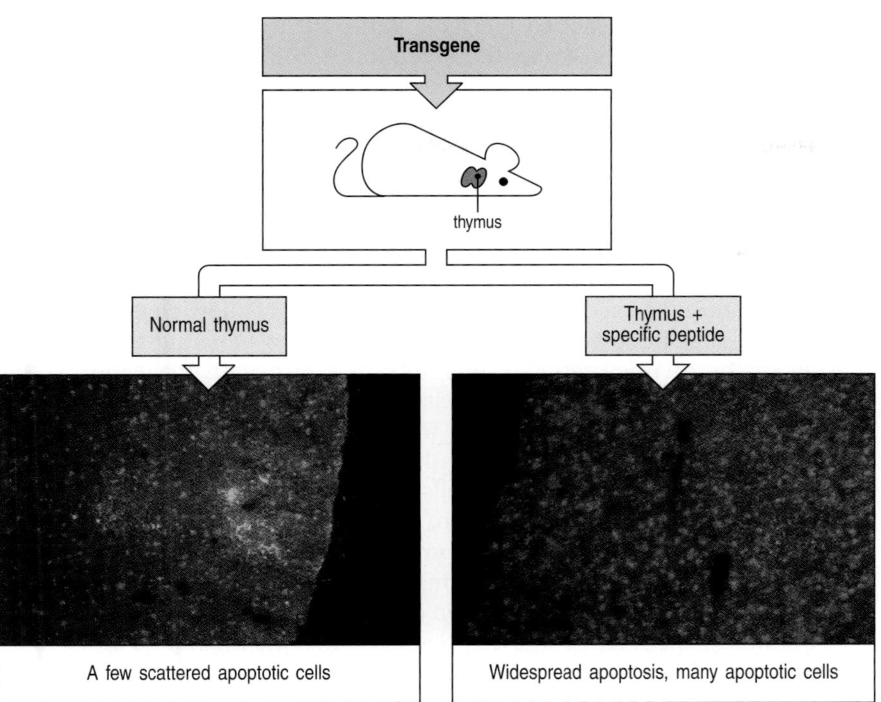

Fig. 7.33 T cells specific for self antigens are deleted in the thymus. In mice transgenic for a T-cell receptor that recognizes a known peptide antigen complexed with self MHC, all the T cells have the same specificity. In the absence of the peptide, most thymocytes mature and emigrate to the periphery. This can be seen in the bottom left panel, where a normal thymus is stained with antibody to identify the medulla (in green), and by the TUNEL technique (see Appendix I, Section A-32) to identify apoptotic cells (in red). If the mice are injected with the peptide that is recognized by the transgenic T-cell receptor, massive cell death occurs in the thymus, as shown by the increased numbers of apoptotic cells in the right-hand bottom panel. Photographs courtesy of A. Wack and D. Kioussis.

lymphoid organ, the T cell is activated to proliferate and produce effector T cells. In contrast, when the T-cell receptor of a developing thymocyte is similarly ligated by antigen on stromal or bone marrow derived cells in the thymus, it dies by apoptosis. This response of immature T cells to stimulation by antigen is the basis of negative selection. Elimination of these T cells in the thymus prevents their potentially harmful activation later on if they should encounter the same peptides when they are mature T cells.

Negative selection has been demonstrated by the use of artificial and naturally occurring self peptides. Negative selection of thymocytes reactive to an artificial self peptide was demonstrated with TCR-transgenic mice in which the majority of thymocytes express a T-cell receptor specific for a peptide of ovalbumin bound to an MHC class II molecule. When these mice were injected with the ovalbumin peptide, most of the CD4 CD8 double-positive thymocytes in the thymic cortex die by apoptosis (Fig. 7.33). Negative selection to a natural self peptide was observed with TCR-transgenic mice expressing T-cell receptors specific for self peptides expressed only in male mice. Thymocytes bearing these receptors disappear from the developing T-cell population in male mice at the CD4 CD8 double-positive stage of development, and no single-positive cells bearing the transgenic receptors mature. By contrast, in female mice, which lack the male-specific peptide, the transgenic T cells mature normally. Negative selection to male-specific peptides has also been demonstrated in normal mice and also occurs by deletion of T cells.

The stage of development at which negative selection occurs can differ depending on the particular experimental system and the particular self antigen. For example, TCR-transgenic mice can express functional T-cell receptors earlier than normal mice during development and have a very high frequency of cells in the thymus reactive to any particular peptide. These features may cause negative selection to occur earlier in TCR-transgenic mice than in normal mice. A slightly more physiological system for evaluating negative selection involves the transgenic expression of only a β chain of the T-cell receptor reactive to a peptide derived from moth cytochrome *c*. In such transgenic mice, the β chain pairs with endogenous α chains, but the

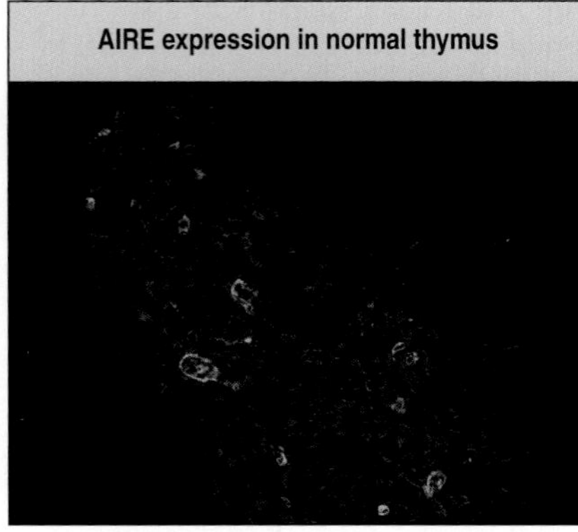

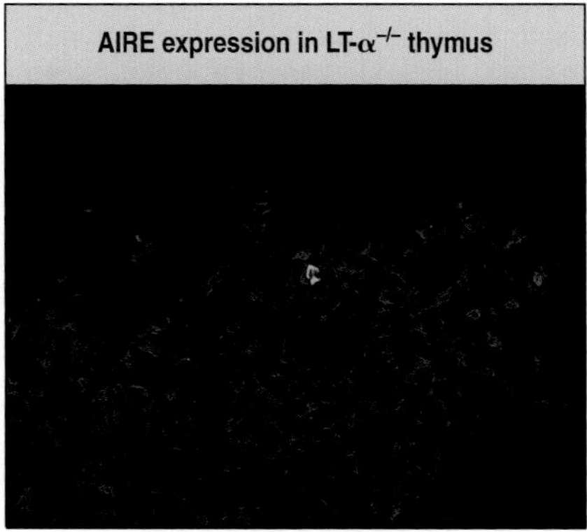

Fig. 7.34 AIRE is expressed in the medulla of the thymus and promotes the expression of proteins normally expressed in peripheral tissues. Expression of AIRE by thymic medullary cells is regulated by lymphotoxin (LT)-α, which signals through the LT-β receptor. Top panel: AIRE expression (green) is shown in the wild-type thymic medulla by immunofluorescence; expression of the thymic medullary epithelial marker MTS10 is shown in red. Bottom panel: Expression of AIRE by thymic medullary cells is reduced in LT-$\alpha^{-/-}$ mice. Photographs courtesy of R.K. Chin and Y.-X. Fu.

frequency of peptide-reactive T cells is sufficient for detection using peptide:MHC tetramers (see Appendix I, Section A-28). These studies indicate that negative selection can occur throughout all stages of development, and that positive and negative selection may not necessarily be sequential processes.

These experiments illustrate the principle that self-peptide:self-MHC complexes encountered in the thymus purge the mature T-cell repertoire of T cells bearing self-reactive receptors. One obvious problem with this scheme is that many tissue-specific proteins, such as pancreatic insulin, would not be expected to be expressed in the thymus. However, it is now clear that many such 'tissue-specific' proteins actually are expressed by some stromal cells present in the thymic medulla; thus, intrathymic negative selection could apply even to proteins that are otherwise restricted to tissues outside the thymus. The expression of such proteins in the thymic medulla is controlled by a gene called ***AIRE* (autoimmune regulator)** by an as-yet unknown mechanism. *AIRE* is expressed in stromal cells located in the thymic medulla (Fig. 7.34). Mutations in *AIRE* give rise to an autoimmune disease known as **autoimmune polyglandular syndrome type I** or **autoimmune polyendocrinopathy–candidiasis–ectodermal dystrophy** (**APECED**), highlighting the important role of that intrathymic expression of tissue-specific proteins in maintaining tolerance to self. *AIRE* expression in the medulla is induced by lymphotoxin (LT) signaling; in mice deficient in LT-α and its receptor, expression of *AIRE* is reduced (see Fig. 7.34). In these mice, expression of insulin in the thymic medulla is decreased compared with normal mice, and peripheral tolerance to insulin is impaired. Thus, negative selection of developing T cells involves interactions with ubiquitous self antigens and tissue-restricted self antigens, and can take place in both the thymic cortex and the thymic medulla.

It is not clear that *AIRE* accounts for the expression of all self proteins in the thymus. Thus, negative selection in the thymus may not remove all T cells reactive to self antigens that appear exclusively in other tissues or are expressed at different stages in development. There are, however, several mechanisms operating in the periphery that can prevent mature T cells from responding to tissue-specific antigens; these will be discussed in Chapter 13, when we consider the problem of autoimmune responses and their avoidance.

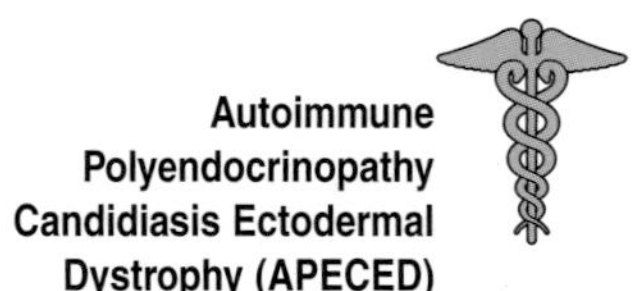

7-21 Negative selection is driven most efficiently by bone marrow derived antigen-presenting cells.

As discussed above, negative selection occurs throughout thymocyte development, both in the thymic cortex and the medulla, and so seems to be mediated by several different cell types. However, there appears to be a hierarchy in the effectiveness of cells in mediating negative selection. The most important seem to be bone marrow derived dendritic cells and macrophages. These are antigen-presenting cells that also activate mature T cells in peripheral lymphoid tissues, as we shall see in Chapter 8. The self antigens presented by these cells are therefore the most important source of potential autoimmune responses, and T cells responding to such self peptides must be eliminated in the thymus.

Experiments using bone marrow chimeric mice have clearly shown the role of thymic macrophages and dendritic cells in negative selection. In these experiments, MHC$^{a \times b}$ F_1 bone marrow is grafted into one of the parental strains (MHC^a in Fig. 7.35). The MHC$^{a \times b}$ T cells developing in the grafted animals are thus exposed to MHC^a thymic epithelium. Bone marrow derived dendritic cells and macrophages will, however, express both MHC^a and MHC^b. The bone marrow chimeras will tolerate skin grafts from either MHC^a or MHC^b animals (see Fig. 7.35), and from the acceptance of both grafts we can infer that the developing T cells are not self-reactive for either of the two MHC antigens. The only cells that could present self-peptide:MHC^b complexes to thymocytes, and thus induce tolerance to MHC^b, are the bone marrow derived cells. The dendritic cells and macrophages are therefore assumed to have a crucial role in negative selection.

In addition, both the thymocytes themselves and the thymic epithelial cells can cause the deletion of self-reactive cells. Such reactions may normally be of secondary significance compared with the dominant role of bone marrow derived cells. In patients undergoing bone marrow transplantation from an unrelated donor, however, where all the thymic macrophages and dendritic cells are of donor type, negative selection mediated by thymic epithelial cells can assume a special importance in maintaining tolerance to the recipient's own antigens.

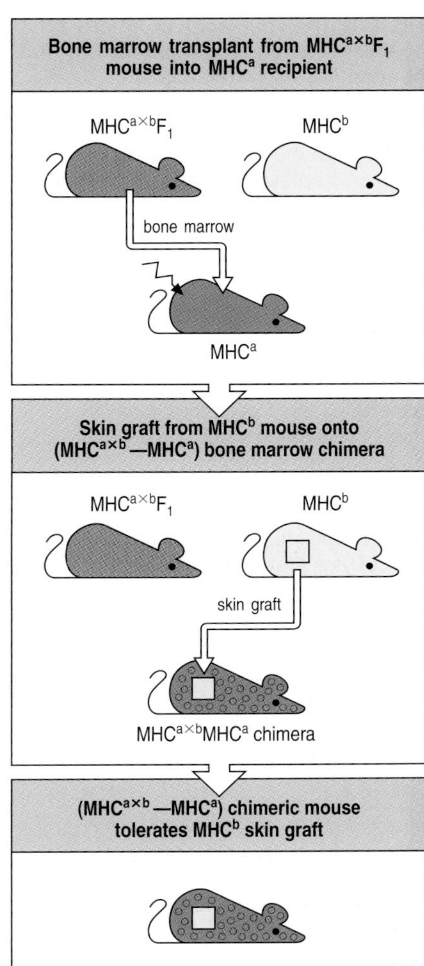

Fig. 7.35 Bone marrow derived cells mediate negative selection in the thymus. When MHCa$^{a\times b}$ F$_1$ bone marrow is injected into an irradiated MHC^a mouse, the T cells mature on thymic epithelium expressing only MHC^a molecules. Nevertheless, the chimeric mice are tolerant to skin grafts expressing MHC^b molecules (provided that these grafts do not present skin-specific peptides that differ between strains a and b). This implies that the T cells whose receptors recognize self antigens presented by MHC^b have been eliminated in the thymus. Because the transplanted MHC$^{a\times b}$ F$_1$ bone marrow cells are the only source of MHC^b molecules in the thymus, bone marrow derived cells must be able to induce negative selection.

7-22 The specificity and/or the strength of signals for negative and positive selection must differ.

We have explained that T cells undergo both positive selection for self-MHC restriction and negative selection for self tolerance by interacting with self-peptide:self-MHC complexes expressed on stromal cells in the thymus. An unresolved issue is how the interaction of the T-cell receptor with self-peptide:self-MHC complexes distinguishes between these opposite outcomes. First, more receptor specificities must be positively selected than are negatively selected. Otherwise, all the cells that were positively selected in the thymic cortex would be eliminated by negative selection, and no T cells would ever be produced (Fig. 7.36). Second, the consequences of the interactions that lead to positive and negative selection must differ: cells that recognize self-peptide:self-MHC complexes on cortical epithelial cells are induced to mature, whereas those whose receptors might confer strong and potentially damaging autoreactivity are induced to die.

One hypothesis to account for the differences between positive and negative selection states that the outcome of peptide:MHC binding by thymocyte T-cell receptors depends on the strength of signal delivered by the receptor and co-receptor on binding, and that this will, in turn, depend upon both the affinity of the T-cell receptor for the peptide:MHC complex and the density of the complex on a thymic cortical epithelial cell. Weak signaling is proposed to rescue thymocytes from apoptosis, leading to positive selection; strong signaling would induce apoptosis and thus negative selection. Because more complexes are likely to bind weakly than strongly, this will result in the positive selection of a larger repertoire of cells than are negatively selected. A second hypothesis proposes that the quality of signal delivered by the receptor, and not just the number of receptors engaged, distinguishes positive from negative selection. According to the strength of signal model, a specific peptide:MHC complex could either drive positive or negative selection for a particular T-cell receptor, depending on its density on the cell surface. In contrast, according to the quality of signal model, changes in peptide:MHC density would not affect the quality of signaling hypothesis. Experiments have not yet unambiguously distinguished between these two ideas. However, differences in the activation of downstream signaling pathways distinguish positive and negative selection, and differential activation of the MAP kinase pathway by the T-cell receptor (see Chapter 6) has been proposed to mediate the opposing outcomes of positive and negative selection. Evidence suggests that positive selection is a result of low or sustained levels

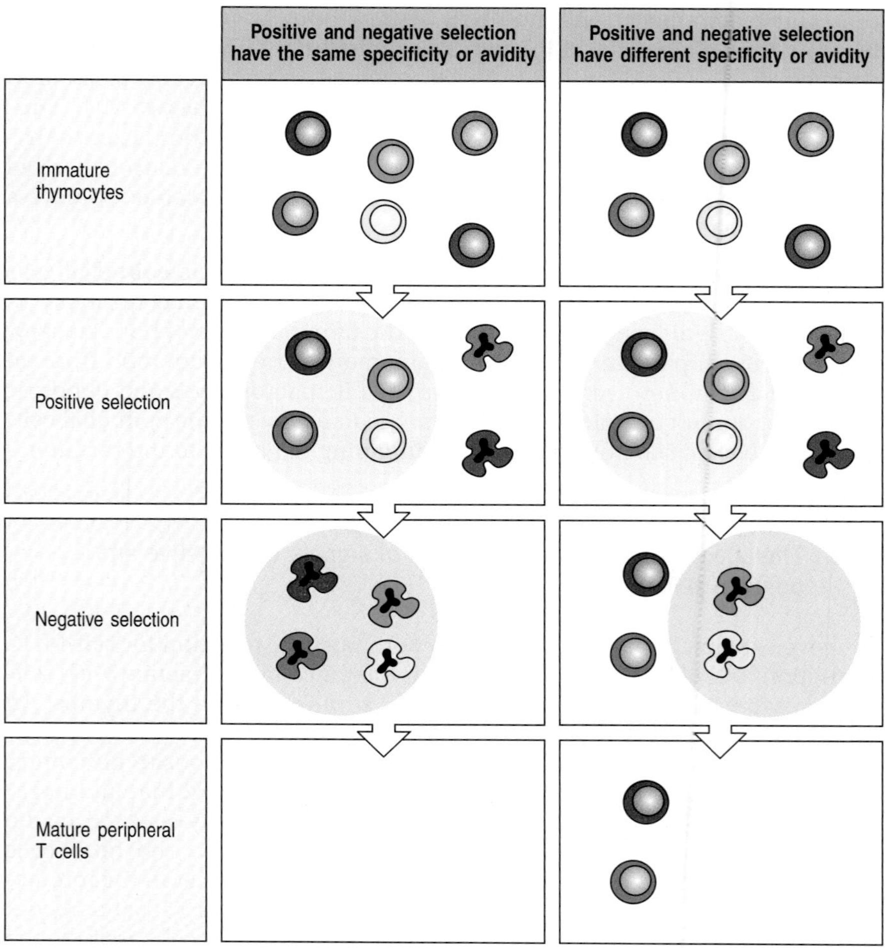

Fig. 7.36 The specificity or affinity of positive selection must differ from that of negative selection. Immature T cells are positively selected in such a way that only those thymocytes whose receptors can engage the peptide:MHC complexes on thymic epithelium mature, giving rise to a population of thymocytes restricted for self MHC. Negative selection removes those thymocytes whose receptors can be activated by self peptides complexed with self-MHC molecules, giving a self-tolerant population of thymocytes. If the specificity and avidity of positive and negative selection were the same (left panels), all the T cells that survive positive selection would be deleted during negative selection. Only if the specificity and avidity of negative selection are different from those of positive selection (right panels) can thymocytes mature into T cells.

of activation of the protein kinase ERK, and that negative selection occurs with higher levels of ERK activation along with activation of the related protein kinases JNK and p38 (see Section 6-14).

Summary.

The stages of thymocyte development up to the expression of the pre-T-cell receptor—including the decision between commitment to either the α:β or the γ:δ lineage—are all independent of peptide:MHC interactions. With the successful rearrangement of α chain genes and expression of the α:β T-cell receptor, thymocytes undergo further development that is determined by the nature of their particular TCR with self peptides presented by the MHC molecules on the thymic stroma. CD4 CD8 double-positive thymocytes whose receptors interact with self-peptide:self-MHC complexes expressed on thymic cortical epithelial cells are positively selected, and become mature CD4 or CD8 single-positive cells. T cells that react too strongly with self antigens are deleted in the thymus, a process driven most efficiently by bone marrow derived antigen-presenting cells. The outcome of positive and negative selection is the generation of a mature T-cell repertoire that is both MHC-restricted and self-tolerant. The paradox that recognition of self-peptide:self-MHC ligands by the T-cell receptor can lead to two opposing effects, namely positive and negative selection, remains unsolved. Its solution will come from a full understanding of the ligand–receptor interactions, the signal transduction mechanisms, and the physiology of each step of the process.

Survival and maturation of lymphocytes in peripheral lymphoid tissues.

Once B and T lymphocytes complete their development in the central lymphoid tissues, they are carried in the blood to the peripheral lymphoid tissues. These tissues have a highly organized architecture, with distinct areas in which B cells and T cells reside, which is determined by interactions between the lymphocytes and the other cell types that make up the lymphoid tissues. The survival and maturation of T lymphocytes reaching the peripheral lymphoid tissue depend on further interactions with their self ligands as well as with neighboring cells. Before considering the factors governing the survival and maturation of newly formed lymphocytes in the periphery, we will briefly look at the organization and development of these tissues and the signals that guide lymphocytes to their correct locations within them. Normally, a lymphocyte will leave peripheral lymphoid tissue and recirculate via lymph and blood (see Section 1-15), continually reentering lymphoid tissues until antigen is encountered or the lymphocyte dies. When it meets its antigen, the lymphocyte stops recirculating, proliferates, and differentiates, as described in Chapters 8–10. When a lymphocyte dies, its place is taken by a newly formed lymphocyte; this enables a turnover of the receptor repertoire and ensures that lymphocyte numbers remain constant.

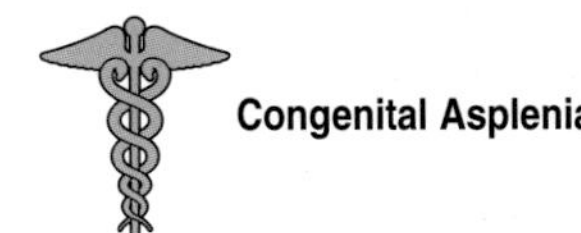

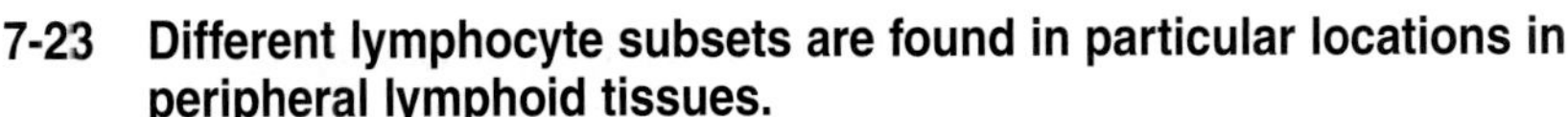

7-23 Different lymphocyte subsets are found in particular locations in peripheral lymphoid tissues.

As we saw in Chapter 1, the various peripheral lymphoid organs are organized roughly along the same lines, with distinct areas of B cells and T cells, and also contain macrophages, dendritic cells, and nonleukocyte stromal cells. The lymphoid tissue of the spleen is the white pulp, whose overall design is illustrated in Fig. 1.19. Each area of white pulp is demarcated by a **marginal sinus**, a vascular network that branches from the central arteriole. The **marginal zone** of the white pulp, the outer border of which is the edge of the marginal sinus, is a highly organized region whose function is poorly understood. It has few T cells but is rich in macrophages and contains a unique population of B cells, the **marginal zone B cells**, which do not recirculate. Pathogens reaching the bloodstream are efficiently trapped in the marginal zone by the macrophages, and it could be that marginal zone B cells are uniquely adapted to provide the first responses to such pathogens.

The white pulp contains clearly separated areas of T cells and B cells. T cells are clustered around the central arteriole, and the globular B-cell areas or follicles are located farther out. Some follicles may contain **germinal centers**, in which B cells involved in an adaptive immune response are proliferating and undergoing somatic hypermutation (see Section 4-18). In follicles with germinal centers, the resting B cells that are not part of the immune response are pushed outward to make up the **mantle zone** around the proliferating lymphocytes. The antigen-driven production of germinal centers will be described in detail when we consider B-cell responses in Chapter 9.

Other types of cells are found within the B-cell and T-cell areas. The B-cell zone contains a network of **follicular dendritic cells (FDCs)**, which are concentrated mainly in the area of the follicle most distant from the central arteriole. Follicular dendritic cells have long processes, from which they get their name, and these are in contact with B cells. Follicular dendritic cells are a distinct type of cell from the dendritic cells we have encountered previously (see Section 1-3) in that they are not leukocytes and are not derived from bone marrow precursors; in addition, they are not phagocytic and do not express MHC class II proteins. Follicular dendritic cells seem to be specialized to capture antigen in the form of immune complexes—complexes of antigen, antibody, and complement. The immune complexes are not internalized but remain intact on the surface of the follicular dendritic cell, where the antigen can be recognized by B cells. Follicular dendritic cells are also important in the development of B-cell follicles.

T-cell zones contain a network of bone marrow derived dendritic cells, sometimes known as **interdigitating dendritic cells** from the way in which their processes interweave among the T cells. There are two subtypes of these dendritic cells, distinguished by characteristic cell-surface proteins; one expresses the α chain of CD8, whereas the other is CD8 negative but expresses CD11b:CD18, an integrin that is also expressed by macrophages.

As in the spleen, the T cells and B cells in lymph nodes are organized into discrete T-cell and B-cell areas (see Fig. 1.18). B-cell follicles have a similar structure and composition to those in the spleen and are located just under the outer capsule of the lymph node. T-cell zones surround the follicles in the paracortical areas. Unlike the spleen, lymph nodes have connections to both the blood system and the lymphatic system. Lymph enters into the subcapsular space, which is also known as the marginal sinus, and brings in antigen and antigen-bearing dendritic cells from the tissues.

The muscosa-associated lymphoid tissues (MALT) are associated with the body's epithelial surfaces that provide physical barriers against infection. Peyer's patches are part of the MALT and are lymph node-like structures interspersed at intervals just beneath the gut epithelium. They have B-cell follicles and T-cell zones (see Fig. 1.20), and the gut epithelial cells overlying them lack the typical brush border. Instead, these M cells are adapted to channel antigens and pathogens from the gut lumen to the Peyer's patch (see Section 1-15). Peyer's patches and similar tissue present in the tonsils provide specialized sites where B cells can become committed to synthesizing IgA. The stromal cells of the MALT secrete the cytokine TGF-β, which has been shown to induce IgA secretion by B cells in culture. In addition, as discussed in Section 7-12, during fetal development waves of γ:δ T cells with specific γ- and δ-gene rearrangements leave the thymus and migrate to these epithelial barriers. The mucosal immune system is discussed in more detail in Chapter 11.

7-24 The development and organization of peripheral lymphoid tissues are controlled by proteins of the tumor necrosis factor family.

Once lymphocytes enter the spleen or lymph node, how do they find their way to their respective zones? As the next section describes, they are directed there mainly by responses to chemokines; B and T cells have different sets of receptors that respond to chemokines differentially secreted in the T and B zones. But this raises the question of how these zones develop in the first place and how they come to secrete specific chemokines.

Surprisingly, members of the tumor necrosis factor (TNF)/TNF receptor (TNFR) family, which were originally thought to be involved in inflammation and cell death, have a critical role in the development and maintenance of normal lymphoid architecture. This has been best demonstrated in a series of knockout mice in which either the ligand or its receptor has been inactivated (Fig. 7.37). These knockouts have complicated phenotypes, which is partly due to the fact that individual TNF-family proteins can bind to multiple receptors and, conversely, many receptors can bind more than one protein. In addition, it seems clear that there is some overlapping function or cooperation between TNF-family proteins. Nonetheless, some general conclusions can be drawn.

Lymph node development depends on the expression in the developing tissue of a subset of TNF-family proteins known as the lymphotoxins (LTs), and different types of lymph nodes depend on signals from different LTs. LT-α_3, a soluble homotrimer of the LT-α chain, supports the development of cervical and mesenteric lymph nodes, and possibly lumbar and sacral lymph nodes. All these lymph nodes drain mucosal sites. LT-α_3 probably exerts its effects by binding to TNFR-I and possibly also to another TNFR-family member called HVEM. The membrane-bound heterotrimer comprising LT-α and the distinct protein chain LT-β (LT-α_2:β_1) binds only to the LT-β receptor and supports the development of all the other lymph nodes. In addition, Peyer's patches do not form in the absence of the LT-α_2:β_1 heterotrimer. These effects are not reversible in adult animals and there are certain critical developmental periods during which the absence or inhibition of these LT-family proteins will irrevocably prevent the development of lymph nodes and Peyer's patches.

Although the spleen develops in mice deficient in any of the known TNF or TNFR family members, its architecture is abnormal in many of these mutants (see Fig. 7.37). LT (most probably the membrane-bound heterotrimer) is required for the normal segregation of T-cell and B-cell zones. TNF-α, binding to TNFR-I, also contributes to the organization of the white pulp: when TNF-α signals are disrupted, B cells surround T-cell zones in a ring rather than in discrete follicles. In addition, the marginal zones are not well defined when TNF-α or its receptor is absent. Perhaps most importantly, follicular dendritic cells are not found in mice that lack TNF-α or TNFR-I. These mice do have lymph nodes and Peyer's patches, because they express LTs, but these structures lack follicular dendritic cells. Similarly, mice that cannot form membrane-bound LT-α_2:β_1 or signal through it also lack normal follicular dendritic cells in the spleen and any residual lymph nodes. Unlike the disruption of lymph node development, which is irreversible, the disorganized lymphoid architecture in the spleen is reversible when the missing TNF-family member is restored. B cells are the likely source for the membrane-bound LT because normal B cells can restore follicular dendritic cells and follicles when transferred to RAG-deficient recipients (which lack lymphocytes). A similar role for B cells in the development of the M cells that lie over Peyer's patches was recently discovered. In this case it seems that signals independent of LT-α are required, because B cells deficient in LT-α will still restore the development of M cells in Peyer's patches.

Fig. 7.37 Normal architecture of the secondary lymphoid organs requires TNF family members and their receptors. The role of TNF family members in the development of peripheral lymphoid organs has been deduced mainly from the study of knockout mice deficient in one or more TNF family ligands or receptors. Some receptors bind more than one ligand, and some ligands bind more than one receptor, complicating the effects of their deletion. (Note that receptors are named for the first ligand known to bind them.) The defects are organized here with respect to the two main receptors, TNFR-I and the LT-β receptor, along with a relatively newly recognized receptor, the herpes virus entry mediator (HVEM), which may also be involved in lymphoid organization. In some cases, the loss of ligands that bind the same receptor leads to different phenotypes. This is due to the ability of the ligand to bind another receptor, and is indicated in the figure. In addition, the LT-α protein chain contributes to two distinct ligands, the trimer LT-α_3 and the heterodimer LT-α_2:β_1, each of which has a distinct receptor. In general, signaling through the LT-β receptor is required for lymph node and follicular dendritic cell development and normal splenic architecture, whereas signaling through TNFR-I is also required for follicular dendritic cells and normal splenic architecture but not for lymph node development.

		Effects seen in knockout (KO) mice				
Receptor	Ligands	Spleen	Peripheral lymph node	Mesenteric lymph node	Peyer's patch	Follicular dendritic cells
TNFR-I	TNF-α LT-α_3	Distorted architecture	Present in TNF-α KO Absent in LT-α KO owing to lack of LT-β signals	Present	Reduced	Absent
LT-β receptor	TNF-α LT-α_2/β_1 LIGHT	Distorted No marginal zones	Absent	Present in LT-β KO Absent in LT-β receptor KO	Absent	Absent
HVEM	LT-α_3 LIGHT	Although both LT-α and LIGHT can bind HVEM, there is no known role for HVEM signaling in organogenesis				

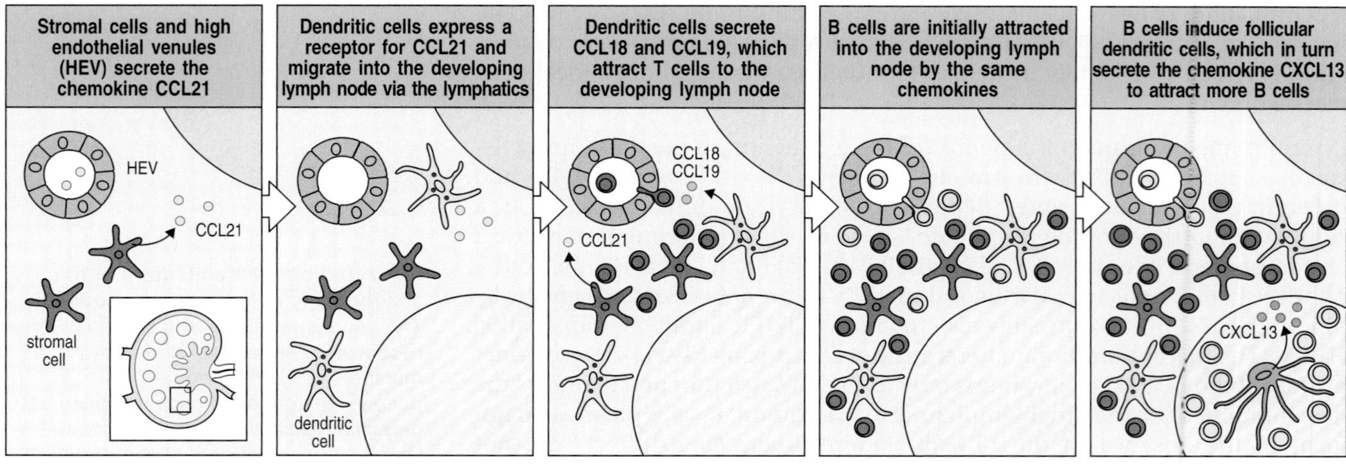

Stromal cells and high endothelial venules (HEV) secrete the chemokine CCL21
Dendritic cells express a receptor for CCL21 and migrate into the developing lymph node via the lymphatics
Dendritic cells secrete CCL18 and CCL19, which attract T cells to the developing lymph node
B cells are initially attracted into the developing lymph node by the same chemokines
B cells induce follicular dendritic cells, which in turn secrete the chemokine CXCL13 to attract more B cells
HEV
CCL21
stromal cell
dendritic cell
CCL18
CCL19
CCL21
CXCL13

Fig. 7.38 The organization of a lymphoid organ is orchestrated by chemokines. The cellular organization of lymphoid organs is initiated by stromal cells and vascular endothelial cells, which secrete the chemokine CCL21 (first panel). Dendritic cells with a receptor for CCL21, CCR7, are attracted to the site of the developing lymph node by CCL21 (second panel); it is not known whether at the earliest stages of lymph node development immature dendritic cells enter from the bloodstream or via the lymphatics, as they do later in life. Once in the lymph node, the dendritic cells express the chemokines CCL18 (also called DC-CK1) and CCL19, for which T cells express receptors. Together, the chemokines secreted by stromal cells and dendritic cells attract T cells to the developing lymph node (third panel). The same combination of chemokines also attracts B cells into the developing lymph node (fourth panel). The B cells are able to either induce the differentiation of the non-leukocyte follicular dendritic cells (which are a distinct lineage from the bone marrow derived dendritic cells) or direct their recruitment into the lymph node. Once present, the follicular dendritic cells secrete a chemokine, CXCL13, which is a chemoattractant for B cells. The production of CXCL13 drives the organization of B cells into discrete B-cell areas (follicles) around the follicular dendritic cells and contributes to the further recruitment of B cells from the circulation into the lymph node (fifth panel).

7-25 The homing of lymphocytes to specific regions of peripheral lymphoid tissues is mediated by chemokines.

Newly formed lymphocytes enter the spleen via the blood, exiting first in the marginal sinus, from which they migrate to the appropriate areas of the white pulp. Lymphocytes that survive their passage through the spleen most probably leave via venous sinuses in the red pulp. In lymph nodes, lymphocytes enter from the blood through the walls of specialized blood vessels, the high endothelial venules (HEVs), which are located within the T-cell zones. Naive B cells migrate through the HEVs in the T-cell area and come to rest in the follicle where, unless they encounter their specific antigen and become activated, they remain for about a day. B cells and T cells leave in the lymph via the efferent lymphatic, which returns them eventually to the blood. The precise location of B cells, T cells, macrophages, and dendritic cells in peripheral lymphoid tissue is controlled by chemokines, which are produced by both stromal cells and bone marrow derived cells (Fig. 7.38).

B cells constitutively express the chemokine receptor CXCR5 and are attracted to the follicles by the ligand for this receptor, the chemokine CXCL13 (B-lymphocyte chemokine, BLC). The most likely source of CXCL13 is the follicular dendritic cell, possibly along with other follicular stromal cells. B cells are, in turn, the source of the LT that is required for the development of follicular dendritic cells. This reciprocal dependence of B cells and follicular dendritic cells illustrates the complex web of interactions that organizes peripheral lymphoid tissues. T cells can also express CXCR5, although at a lower level, and this may explain how T cells are able to enter B-cell follicles, which they do on activation, to participate in the formation of the germinal center.

T-cell localization to the T zones involves two chemokines, CCL19 (MIP-3β) and CCL21 (secondary lymphoid chemokine, SLC). Both of these bind the receptor CCR7, which is expressed by T cells; mice that lack CCR7 do not form normal T zones and have impaired primary immune responses. CCL21 is produced by stromal cells of the T zone in spleen, and by the endothelial cells of HEVs in lymph nodes and Peyer's patches. Another source of CCL19 and CCL21 is the interdigitating dendritic cells, which are also prominent in the T zones. Indeed, dendritic cells themselves express CCR7 and will localize to T zones even in RAG-deficient mice. Thus, in lymph node development, the T zone might be organized first through the attraction of dendritic cells and T cells by CCL21 produced by stromal cells. This organization would then be reinforced by CCL21 and CCL19 secreted by resident mature dendritic cells, which in turn attract more T cells and immature dendritic cells.

B cells—particularly activated ones—also express CCR7, but at lower levels than do T cells or dendritic cells. This may account for their characteristic migration pattern, which is first through the T zone (where they may linger if activated) and then to the B-cell follicle. Although the cellular organization of T-cell and B-cell areas in lymph nodes and Peyer's patches has been less well studied, it seems likely that it is controlled by similar chemokines and receptors.

7-26 Lymphocytes that encounter sufficient quantities of self antigens for the first time in the periphery are eliminated or inactivated.

Autoreactive lymphocytes have been purged from the population of new lymphocytes in the central lymphoid organs; however, this is effective only for autoantigens that are expressed in or could reach these organs. Not all potential self antigens are expressed in central lymphoid organs. Some, like the thyroid product thyroglobulin, are tissue specific and/or are compartmentalized so that little if any is available in the circulation. Therefore, newly emigrated self-reactive lymphocytes that encounter autoantigens for the first time in the periphery must be eliminated or inactivated. This is the tolerance mechanism known as peripheral tolerance. Lymphocytes that encounter self antigens *de novo* in the periphery can have three fates, much like those that recognize such antigens in central lymphoid organs: deletion, anergy, or survival (also known as ignorance).

Mature B cells that encounter a strongly cross-linking antigen in the periphery will undergo clonal deletion. This was elegantly shown in studies of B cells expressing immunoglobulin specific for H-2K^b MHC class I molecules. These cells are deleted even when, in transgenic animals, the expression of the H-2K^b molecule is restricted to the liver by the use of a liver-specific gene promoter. B cells that encounter strongly cross-linking antigens in the periphery undergo apoptosis directly, unlike their counterparts in the bone marrow, which attempt further receptor rearrangements. The different outcomes may be due to the fact that the B cells in the periphery are more mature and can no longer rearrange their light-chain loci.

As with immature B cells, mature B cells that encounter and bind an abundant soluble antigen become anergized. This was demonstrated in mice by placing the *HEL* transgene under the control of an inducible promoter that can be regulated by changes in the diet. It is thus possible to induce the production of lysozyme at any time and thereby study its effects on HEL-specific B cells at different stages of maturation. These experiments have shown that both mature and immature B cells are inactivated when they are chronically exposed to soluble antigen.

The situation is similar for T cells. Again, our understanding of the fates of autoreactive T cells in the periphery comes mainly from the study of T-cell receptor transgenic mice. In some cases, T cells reacting to self antigens in the periphery are eliminated, though this may follow a brief period of activation and cell division known as **activation-induced cell death**. In other cases, these cells may be rendered anergic. When studied *in vitro*, these anergic T cells prove refractory to signals given through the T-cell receptor.

If the encounter of mature lymphocytes with self antigens leads to cell death or anergy, why does this not happen in a mature lymphocyte that recognizes a pathogen-derived antigen? The answer is that infection sets up inflammation, which induces inflammatory cytokines and the expression of co-stimulatory molecules on the antigen-presenting cells. In the absence of these signals, however, the interaction of a mature lymphocyte with an antigen seems to result in a tolerance-inducing or **tolerogenic** signal from the antigen receptor. This was recently demonstrated *in vivo* for T cells. In the

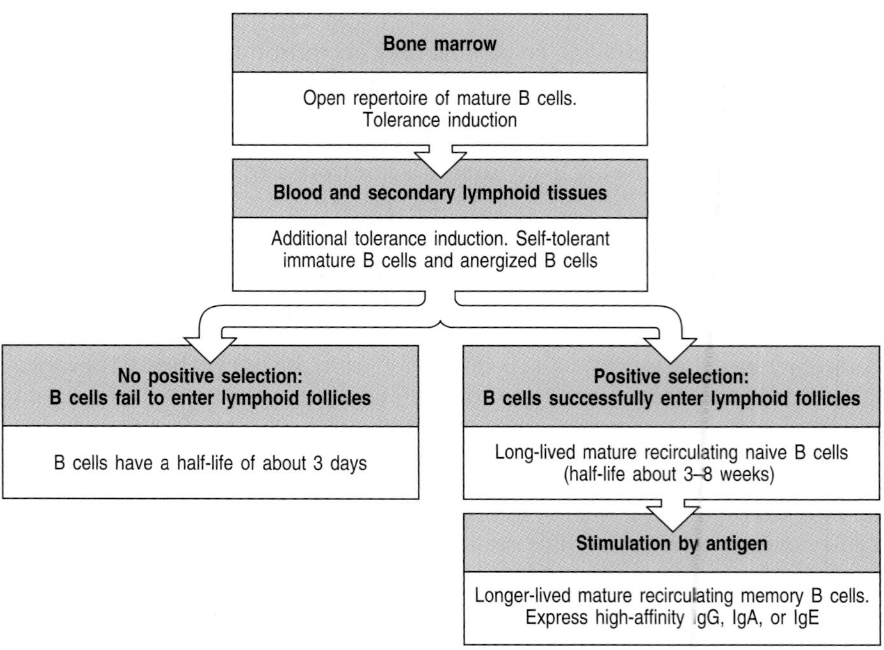

Fig. 7.39 Proposed population dynamics of conventional B cells. B cells are produced as receptor-positive immature B cells in the bone marrow. The most avidly self-reactive B cells are removed at this stage. B cells then migrate to the periphery, where they enter the secondary lymphoid tissues. It is estimated that $10–20 \times 10^6$ B cells are produced by the bone marrow and exported each day in a mouse, and an equal number are lost from the periphery. There seem to be two classes of peripheral B cells: long-lived B cells and short-lived B cells. The short-lived B cells are, by definition, recently formed B cells. Most of the turnover of short-lived B cells might result from B cells that fail to enter lymphoid follicles. In some cases this is a consequence of being rendered anergic by binding to soluble self antigen; for the remaining immature B cells, entry into lymphoid follicles is thought to entail some form of positive selection. Thus, the remainder of the short-lived B cells fail to join the long-lived pool because they are not positively selected. About 90% of all peripheral B cells are relatively long-lived mature B cells that seem to have undergone positive selection in the periphery. These mature naive B cells recirculate through peripheral lymphoid tissues and have a half-life of 6–8 weeks in mice. Memory B cells, which have been activated previously by antigen and T cells, are thought to have a longer life.

absence of infection and inflammation, quiescent dendritic cells can still present self antigens to T cells, but the consequences of a naive T cell recognizing self antigen in these circumstances are either activation-induced cell death or anergy. Thus, when the innate immune system is not activated, antigens presented by dendritic cells may lead to T-cell tolerance rather than T-cell activation.

7-27 Most immature B cells arriving in the spleen are short-lived and require cytokines and positive signals through the B-cell receptor for maturation and survival.

When B cells emerge from bone marrow into the periphery, they are still functionally immature, expressing high levels of sIgM but little sIgD. Most of these immature cells will not survive to become fully mature B cells bearing low levels of sIgM and high levels of sIgD. Fig. 7.39 shows the possible fates of newly produced B cells that enter the periphery. The daily output of new B cells from the bone marrow is roughly 5–10% of the total B-lymphocyte population in the steady-state peripheral pool. The size of this pool seems to remain constant in unimmunized animals, and so the stream of new B cells needs to be balanced by the removal of an equal number of peripheral B cells. However, the majority of peripheral B cells are long-lived and only 1–2% of these die each day. Most of the B cells that die are in the short-lived immature peripheral B-cell population, of which more than 50% die every 3 days. The failure of most newly formed B cells to survive for more than a few days in the periphery seems to be due to competition between peripheral B cells for access to the follicles in the peripheral lymphoid tissues. If newly produced immature B cells do not enter a follicle, their passage through the periphery is halted and they eventually die. The limited number of lymphoid follicles cannot accommodate all of the B cells generated each day and so there is continual competition for entry.

The follicle seems to provide signals necessary for B-cell survival, particularly the TNF family member BAFF (for B-cell activating factor belonging to the TNF

Property	B-1 cells	Conventional B-2 cells	Marginal zone B cells
When first produced	Fetus	After birth	After birth
N-regions in VDJ junctions	Few	Extensive	Yes
V-region repertoire	Restricted	Diverse	Partly restricted
Primary location	Body cavities (peritoneal, pleural)	Secondary lymphoid organs	Spleen
Mode of renewal	Self-renewing	Replaced from bone marrow	Long-lived
Spontaneous production of immunoglobulin	High	Low	Low
Isotypes secreted	IgM >> IgG	IgG > IgM	IgM > IgG
Response to carbohydrate antigen	Yes	Maybe	Yes
Response to protein antigen	Maybe	Yes	Yes
Requirement for T-cell help	No	Yes	Sometimes
Somatic hypermutation	Low to none	High	?
Memory development	Little or none	Yes	?

Fig. 7.40 A comparison of the properties of B-1 cells, conventional B cells (B-2 cells), and marginal zone B cells. B-1 cells can develop in unusual sites in the fetus, such as the omentum, in addition to the liver. B-1 cells predominate in the young animal, although they probably can be produced throughout life. Being produced mainly during fetal and neonatal life, their rearranged variable-region sequences contain few N-nucleotides. In contrast, marginal zone B cells accumulate after birth and do not reach peak levels in the mouse until 8 weeks of age. B-1 cells are best thought of as a partly activated self-renewing pool of lymphocytes that are selected by ubiquitous self and foreign antigens. Because of this selection, and possibly because the cells are produced early in life, the B-1 cells have a restricted repertoire of variable regions and antigen-binding specificities. Marginal zone B cells also have a restricted repertoire that may be selected by a set of antigens similar to those that select B-1 cells. B-1 cells seem to be the major population of B cells in certain body cavities, most probably because of exposure at these sites to antigens that drive B-1 cell proliferation. Marginal zone B cells remain in the marginal zone of the spleen and are not thought to recirculate. Partial activation of B-1 cells leads to the secretion of mainly IgM antibody; B-1 cells contribute much of the IgM that circulates in the blood. The limited diversity of both the B-1 and marginal zone B-cell repertoire and the propensity of these cells to react with common bacterial carbohydrate antigens suggest that they carry out a more primitive, less adaptive, immune response than conventional B cells (B-2 cells). In this regard they are comparable to γ:δ T cells.

family), which is secreted by several cell types, and its receptor BAFF-R, which is expressed by B cells. The BAFF/BAFF-R pair has been shown to have an important role in follicular B-cell survival, because mutants lacking BAFF-R have mainly immature B cells and few long-lived peripheral B cells.

Peripheral B cells also include memory B cells, which differentiate from mature B cells after their first encounter with antigen; we will return to B-cell memory in Chapter 10. Competition for follicular entry favors mature B cells that are already established in the relatively long-lived and stable peripheral B-cell pool. Mature B cells have undergone phenotypic changes that might make their access to the follicles easier; for example, they express the receptor CXCR5 for the chemoattractant CXCL13, which is expressed by follicular dendritic cells (see Fig. 7.37). They also have increased expression of the B-cell co-receptor component CR2 (CD21), which affects the signaling capacity of the B cell.

Continuous signaling through the B-cell receptor also has a positive role in the maturation and continued recirculation of peripheral B cells. A clever method of inactivating the B-cell receptor in mature B cells by conditional gene deletion has demonstrated that continuous expression of the B-cell receptor is required for B-cell survival. Mice that lack the tyrosine kinase Syk, which is involved in signaling from the B-cell receptor (see Section 6-12), fail to develop mature B cells although they do have immature B cells. Thus, a Syk-transduced signal may be required for final B-cell maturation or for the survival of mature B cells. Although each B-cell receptor has a unique specificity, such signaling need not depend on antigen-specific interactions; the receptor could, for example, be responsible for 'tonic' signaling, in which a weak but important signal is generated by the assembly of the receptor complex and infrequently triggers some or all of the downstream signaling events.

7-28 B-1 cells and marginal zone B cells are distinct B-cell subtypes with unique antigen receptor specificity.

The receptor specificity is important in shaping the peripheral B-cell pools that derive from immature B cells that reach the spleen. This is most clearly shown in the role of the B-cell receptor and antigen in the selection of two subsets of B cells that do not reside in B-cell follicles: the so-called **B-1 cells** or **CD5$^+$ B cells** and the marginal zone B cells.

B-1 cells are a unique subset of B cells comprising about 5% of all B cells in mice and humans, and are the major population in rabbits. B-1 cells express the cell-surface protein CD5, have high levels of sIgM but little sIgD, and are found primarily in the peritoneal and pleural cavity fluid. These cells appear first during fetal development (Fig. 7.40) and are called B-1 cells because their development precedes that of the conventional B cells whose development has been discussed up to now—and which are called **B-2 cells**. It is clear that antigen specificity affects the fate of B-1 cells and/or their precursors, in that certain autoantigens and environmental antigens encountered in the periphery drive the expansion and maintenance of B-1 cells. Some of these antigens, such as phosphocholine, are encountered on the surface of bacteria that colonize the gut.

There is some debate about the origin of B-1 cells. It is not yet clear whether they arise as a distinct lineage from a unique precursor cell or differentiate to the B-1 phenotype from a precursor cell that could also give rise to B-2 cells. In the mouse, fetal liver produces mainly B-1 cells, whereas adult bone marrow generates predominantly B-2 cells, and this has been interpreted as support for the unique precursor hypothesis. However, the weight of evidence favors the idea that commitment to the B-1 or B-2 subset is due to a selection step, rather than being a distinct lineage difference such as that between $\gamma{:}\delta$ and $\alpha{:}\beta$ T cells.

Marginal zone B cells, so called because they reside in the marginal sinus of the white pulp in the spleen, are another unique subset of B cells. They seem to be resting mature B cells, yet they have a different set of surface proteins from the major follicular population of B cells. For example, they express lower levels of CD23, a C-type lectin, and high levels of both the MHC class I-like molecule CD1 (see Section 5-19) and two receptors for the C3 fragment of complement, CR1 (CD35) and CR2 (CD21). Marginal zone B cells have restricted antigen specificities, biased towards common environmental and self antigens, and may be adapted to provide a quick response if such antigens enter the bloodstream. They may not require T-cell help to become activated. Functionally and phenotypically, marginal zone B cells resemble B-1 cells; recent experiments suggest they are positively selected by certain self antigens, much as B-1 cells are. They are, however, distinct both in location and in surface protein expression; for example, marginal zone B cells do not have high levels of CD5.

The functions of B-1 cells and marginal zone B cells are being clarified. Their locations suggest a role for B-1 cells in defending the body cavities, and for marginal zone B cells in defense against pathogens that penetrate the bloodstream. The restricted repertoire of receptors in both cell types seems to equip them for a function in the early, nonadaptive phase of an immune response (see Section 2-34). Indeed, the V gene segments that are used to encode the receptors of B-1 and marginal zone B cells might have evolved by natural selection to recognize common bacterial antigens, thus allowing them to contribute to the very early phases of the adaptive immune response. In practice, it is found that B-1 cells make little contribution to adaptive immune responses to most protein antigens, but contribute strongly to some antibody responses against carbohydrate antigens. Moreover, a large proportion of the IgM that normally circulates in the blood of unimmunized mice derives from B-1 cells. The existence of these so-called **natural antibodies**, which are highly cross-reactive and bind with low affinity to both microbial and self antigens, supports the view that B-1 cells are partly activated because they are selected for self-renewal by ubiquitous self and environmental antigens.

7-29 T-cell homeostasis in the periphery is regulated by cytokines and self-MHC interactions.

When T cells have expressed their receptors and co-receptors, and matured within the thymus for a further week or so, they emigrate to the periphery. Unlike B cells emigrating from bone marrow, only relatively small numbers of T cells are exported from the thymus, roughly $1–2 \times 10^6$ per day in the mouse. In the absence of infection, the size and composition of the peripheral pool of naive T cells is regulated by mechanisms that keep it at a roughly constant size and composed of diverse but potentially functional T-cell receptors. Such regulatory processes are known as **homeostasis**. These homeostatic mechanisms involve both cytokines and signals received through the T-cell receptor in response to its interaction with self-MHC molecules.

The requirement for the cytokine IL-7 and interactions with self-peptide:self-MHC complexes for T-cell survival in the periphery has been shown experimentally. If T cells are transferred from their normal environment to recipients lacking MHC molecules, or lacking the 'correct' MHC molecules that originally selected the T cells, they do not survive long. In contrast, if T cells are transferred into recipients that have the correct MHC molecules, they survive. Contact with the appropriate self-peptide:self-MHC complex as they circulate through peripheral lymphoid organs leads mature naive T cells to undergo infrequent cell division. This slow increase in T-cell numbers must be balanced by a slow loss of T cells, such that the number of T cells remains roughly constant. Most probably, this loss occurs among the daughters of the dividing naive cells.

Where do the mature naive CD4 and CD8 T cells encounter their positively selecting ligands? Current evidence favors self-MHC molecules on dendritic cells resident in the T-cell zones of peripheral lymphoid tissues. These cells are similar to the dendritic cells that migrate to the lymph nodes from other tissues but lack sufficient co-stimulatory potential to induce full T-cell activation. The study of peripheral positive selection is in its infancy, however, and a clear picture has yet to emerge. Memory T cells are also part of the peripheral T-cell pool, and we return to their regulation in Chapter 10.

Summary.

The organization of the peripheral lymphoid tissues is controlled by proteins of the TNF family and their receptors (TNFRs). The interaction between B cells expressing lymphotoxin and follicular dendritic cells expressing the receptor TNFR-I generates signals necessary for establishing the normal architecture of the spleen and lymph nodes. The homing of B and T cells to distinct areas of lymphoid tissue involves attraction by specific chemokines. B and T lymphocytes that survive selection in the bone marrow and thymus are exported to the peripheral lymphoid organs. Most of the newly formed B cells that emigrate from the bone marrow die soon after their arrival in the periphery, thus keeping the number of circulating B cells fairly constant. A small number mature and become longer-lived naive B cells. T cells leave the thymus as fully mature cells and are produced in smaller numbers than B cells. The fate of mature lymphocytes in the periphery is still controlled by their antigen receptors. In the absence of an encounter with their specific foreign antigen, naive lymphocytes require some tonic signaling through their antigen receptors for long-term survival.

T cells are generally long-lived and are thought to be slowly self-renewing in the peripheral lymphoid tissues, being maintained by repeated contacts with self-peptide:self-MHC complexes that can be recognized by the T-cell receptor but do not cause T-cell activation, in combination with signals derived from IL-7. The evidence for receptor-mediated survival signals is clearest for T cells, but they also seem to be needed for B-1 cells and marginal zone B cells, in which case they may promote differentiation, expansion, and survival, and most probably also for B-2 cells, in which case they promote survival without expansion. The lymphoid follicle, through which B cells must circulate to survive, seems to provide signals for their maturation and survival. A few ligands that select B-1 and marginal zone B cells are known, but in general the ligands involved in B-cell selection are unknown. The distinct minority subpopulations of lymphocytes, such as the B-1 cells, marginal zone B cells, γ:δ T cells, and the double-negative T cells with α:β receptors of very limited diversity, have different developmental histories and functional properties from conventional B-2 cells and α:β T cells and are likely to be regulated independently of these majority B-cell and T-cell populations.

Lymphoid tumors.

Individual B cells or T cells can undergo neoplastic transformation and can then give rise to either blood-borne leukemias or tissue-resident lymphomas. The characteristics of the different lymphoid tumors reflect the developmental stage of the cell from which the tumor derives. All lymphoid tumors except those derived from very early uncommitted cells have characteristic gene rearrangements that allow their placement in the B or T lineage. These rearrangements are frequently accompanied by chromosomal translocations, often between a locus involved in generating the antigen receptor and a cellular proto-oncogene. These next three sections briefly introduce these tumors and describe some of their basic properties.

7-30 B-cell tumors often occupy the same site as their normal counterparts.

Tumors may retain many characteristics of the cell type from which they arose. This is clearly illustrated by B-cell tumors. Tumors corresponding to essentially all stages of B-cell development have been found in humans, from the earliest stages to the myelomas that represent malignant outgrowths of plasma cells (Fig. 7.41). Furthermore, each type of tumor may retain its characteristic homing properties. Thus, a tumor that resembles mature, germinal center, or memory cells homes to follicles in lymph nodes and spleen, giving rise to a **follicular center cell lymphoma**, whereas a tumor of plasma cells usually disperses to many different sites in the bone marrow much as normal plasma cells do, from which comes the clinical name of **multiple myeloma** (tumor of bone marrow). These similarities mean that it is often possible to use tumor cells, which are available in large

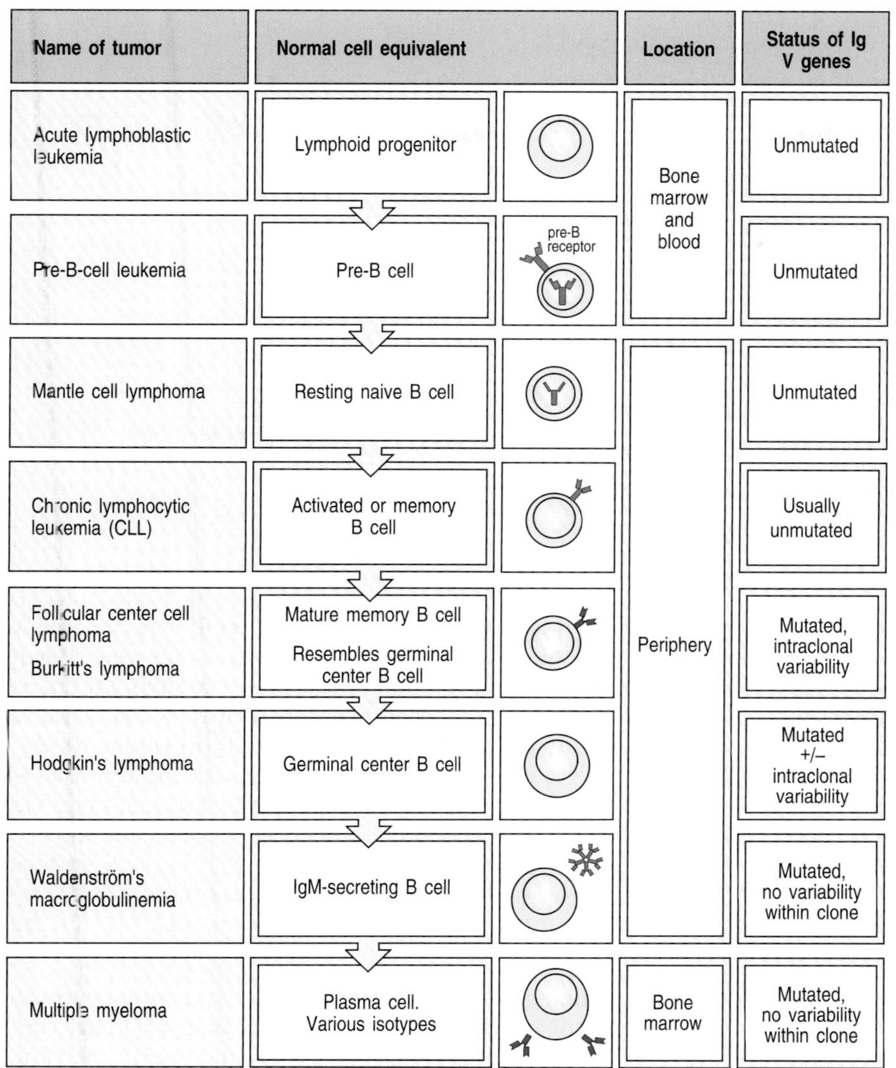

Name of tumor
Normal cell equivalent
Location
Status of Ig V genes
Acute lymphoblastic leukemia
Lymphoid progenitor
Unmutated
Bone marrow and blood
Pre-B-cell leukemia
Pre-B cell
pre-B receptor
Unmutated
Mantle cell lymphoma
Resting naive B cell
Unmutated
Chronic lymphocytic leukemia (CLL)
Activated or memory B cell
Usually unmutated
Follicular center cell lymphoma
Burkitt's lymphoma
Mature memory B cell
Resembles germinal center B cell
Periphery
Mutated, intraclonal variability
Hodgkin's lymphoma
Germinal center B cell
Mutated +/– intraclonal variability
Waldenström's macroglobulinemia
IgM-secreting B cell
Mutated, no variability within clone
Multiple myeloma
Plasma cell. Various isotypes
Bone marrow
Mutated, no variability within clone

Fig. 7.41 B-cell tumors represent clonal outgrowths of B cells at various stages of development. Each type of tumor cell has a normal B-cell equivalent, homes to similar sites, and has behavior similar to that cell. The tumor called multiple myeloma is made up of cells that appear much like the plasma cells from which they derive; they secrete immunoglobulin and are found predominantly in the bone marrow. The most enigmatic of B-cell tumors is Hodgkin's disease, which consists of two cell phenotypes: a lymphoid cell and a large, odd-looking cell known as a Reed–Sternberg (RS) cell. The RS cell seems to derive from a germinal center B cell that has a decreased expression of surface immunoglobulin, possibly owing to somatic mutation. Chronic lymphocytic leukemia (CLL) was previously thought to derive from the B-1 lineage because it expresses CD5, but recent gene expression profile studies of CLL suggest that it resembles an activated or memory B cell. Many lymphomas and myelomas can go through a preliminary, less aggressive, lymphoproliferative phase, and some mild lymphoproliferations seem to be benign.

quantities, to study the cell-surface molecules and signaling pathways responsible for lymphocyte homing and other cellular functions.

The clonal nature of B-lineage tumors is clearly illustrated by the identical immunoglobulin gene rearrangements found in different cells from a particular patient's lymphoma. This is useful for clinical diagnosis, because tumor cells can be detected by sensitive assays for these homogeneous rearrangements (Fig. 7.42). Indeed, the presence of rearrangements at the B-cell receptor loci is highly indicative of a tumor's B-cell origin, just as rearrangements at the T-cell receptor loci indicate a T-cell origin. This approach has proved useful in typing acute lymphoblastic leukemia, a common malignancy of childhood. Most of these have rearrangements of the heavy-chain loci but not the light-chain loci, indicating their origin from a pre-B cell and consistent with their relatively undifferentiated phenotype. Some of these have rearranged light chains as well and may have arisen from a slightly more developed precursor. A few lymphoblastic leukemias have rearrangements at the T-cell receptor loci and thus are not of B-cell origin.

Similarly, gene rearrangement status helped to identify the origin of a class of tumors known as **Hodgkin's disease**. The bizarre-looking cell that is characteristic

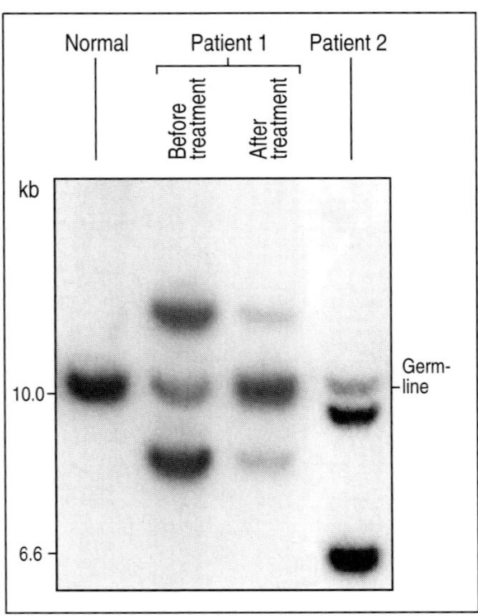

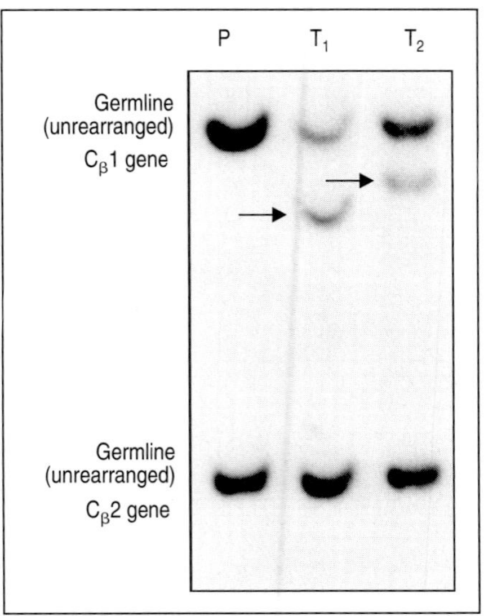

Fig. 7.42 Clonal analysis of B-cell and T-cell tumors. DNA analysis of tumor cells by Southern blotting techniques can detect and monitor lymphoid malignancy. Left panel: B-cell tumor analysis. In a sample from a healthy person (left lane), immunoglobulin genes are in the germline configuration in non-B cells, so a digest of their DNA with a suitable restriction endonuclease yields a single germline DNA fragment when probed with an immunoglobulin heavy-chain J-region probe (J_H). Normal B cells present in this sample make many different rearrangements to J_H, producing a spectrum of 'bands' each so faint that it is undetectable. By contrast, in samples from patients with B-cell malignancies (patient 1 and patient 2), in which a single cell has given rise to all the tumor cells in the sample, two extra bands are seen with the J_H probe. These bands are characteristic of each patient's tumor and result from the rearrangement of both alleles of the J_H gene in the original tumor cells. The intensity of the bands compared with that of the germline band gives an indication of the abundance of tumor cells in the sample. After antitumor treatment (see patient 1), the intensity of the tumor-specific bands can be seen to diminish. kb, kilobases. Right panel: the unique rearrangement events in each T cell can be used similarly to identify tumors of T cells by Southern blotting. The probe used in this case was for the T-cell receptor β-chain constant regions ($C_{\beta}1$ and $C_{\beta}2$). DNA from a placenta (P), a tissue in which the T-cell receptor genes are not rearranged, shows one prominent band for each region. DNA from peripheral blood lymphocytes from two patients suffering from T-cell tumors (T_1 and T_2) gives additional bands that correspond to specific rearrangements (arrowed) that are present in a large number of cells (the tumor). As with B cells, no bands deriving from rearranged genes in normal T cells also present in the patients' samples can be seen, because no one rearranged band is present at sufficient concentration to be detected in this assay. Left panel: photograph courtesy of T.J. Vulliamy and L. Luzzatto. Right panel: photograph courtesy of T. Diss.

of Hodgkin's disease, known as a **Reed–Sternberg (RS) cell**, was previously thought to be of T-cell or dendritic cell origin. DNA analysis has now shown that these cells have rearranged immunoglobulin genes, classifying them as outgrowths of a single B cell. How the originally transformed B cell changes morphology to become an RS cell is not known. Curiously, in Hodgkin's disease, RS cells are sometimes a minority population; the more numerous surrounding cells are usually polyclonal T and B cells that may be reacting to the RS cells or to a soluble factor that they secrete. One of the reasons that the origin of RS cells was unclear is that in nearly all cases they lack surface immunoglobulin. We now know that in many cases the reason for loss of surface immunoglobulin is a somatic mutation that inactivates one of the immunoglobulin V-region genes.

Whether the immunoglobulin genes in a B-cell tumor contain somatic mutations also provides important information on its origin. Mutated V genes suggest that the cell of origin had passed through a germinal center reaction. Pre-B-cell leukemias and most **chronic lymphocytic leukemias (CLLs)** have no mutations. By contrast, cells of follicular lymphoma or **Burkitt's lymphoma**, arising from germinal center B cells, express mutated V genes. If the V genes from several different lymphoma lines of these types from the same patient are sequenced, minor variations (intraclonal variations) are seen because somatic hypermutation can continue in tumor cells. Later-stage B-cell tumors such as multiple myelomas contain mutated genes but do not display intraclonal variation, because later in B-cell development somatic hypermutation has ceased. Some caution is needed in generalizing from the somatic mutation status of immunoglobulin genes because it is not entirely clear that mutation is restricted to germinal centers, and memory cells may pass through a germinal center reaction and not acquire any somatic mutations.

DNA microarray-based gene-expression analysis has allowed the comprehensive description and comparison of the genes that are expressed in tumor cells and in normal cells (see Appendix I, Section A-35). This approach has provided insights into how tumors relate to normal tissues, and permits more precise classification and insight into the biology of tumor cells. This work has confirmed previous classifications based on homing patterns and allows further subdivision of tumor types. For example, diffuse non-Hodgkin's lymphoma can be subdivided into groups that resemble either activated B cells or germinal center B cells. This may have prognostic significance, as tumors that resemble germinal center cells respond much better to therapy. The

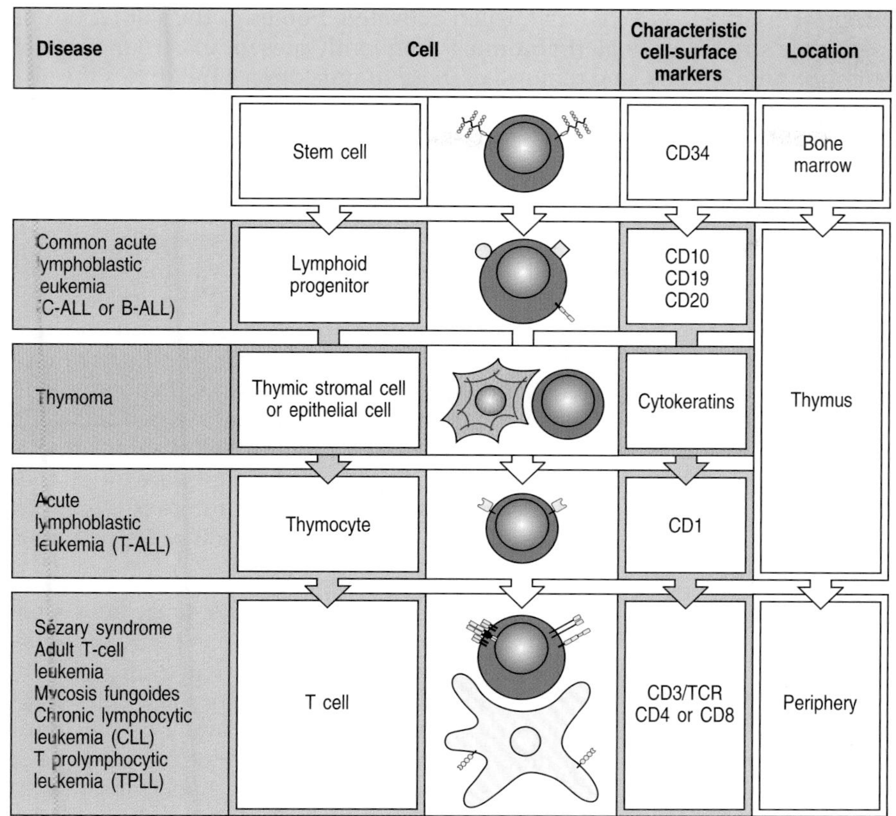

Disease
Cell
Characteristic cell-surface markers
Location
Stem cell
CD34
Bone marrow
Common acute lymphoblastic leukemia (C-ALL or B-ALL)
Lymphoid progenitor
CD10
CD19
CD20
Thymoma
Thymic stromal cell or epithelial cell
Cytokeratins
Thymus
Acute lymphoblastic leukemia (T-ALL)
Thymocyte
CD1
Sézary syndrome
Adult T-cell leukemia
Mycosis fungoides
Chronic lymphocytic leukemia (CLL)
T prolymphocytic leukemia (TPLL)
T cell
CD3/TCR
CD4 or CD8
Periphery

Fig. 7.43 T-cell tumors represent monoclonal outgrowths of normal cell populations. Each distinct T-cell tumor has a normal equivalent and retains many of the properties of the cell from which it develops. However, tumors of T cells lack the intermediates in the T-cell developmental pathway. Some of these tumors represent a massive outgrowth of a rare cell type. For example, acute lymphoblastic leukemia is derived from the lymphoid progenitor cell. One T-cell-related tumor is also included. Thymomas derive from thymic stromal or epithelial cells. Some characteristic cell-surface markers for each stage are also shown. For example, CD10 (common acute lymphoblastic leukemia antigen or CALLA) is used as a marker for common acute lymphoblastic leukemia. Note that T-cell chronic lymphocytic leukemia (CLL) cells express CD8, whereas the other T-cell tumors listed express CD4. Adult T-cell leukemia is caused by the retrovirus HTLV-1.

analysis of CLLs by gene-expression profiling is particularly revealing. Because these tumors express CD5 and typically lack somatic mutations, they were thought for many years to arise from a B-1 cell precursor (see Section 7-28). Gene-expression analysis revealed little resemblance to normal CD5 B cells, however, and instead suggested a relationship to a resting B cell, possibly a memory-type B cell, which agrees with the fact that some CLLs do have somatic mutations. The mutated and unmutated CLLs express nearly all the same genes, with the exception of a unique subset of genes expressed by the mutated CLLs, which is probably responsible for their benign prognosis.

7-31 T-cell tumors correspond to a small number of stages of T-cell development.

Tumors of T-lineage cells have been identified but, unlike B-cell tumors, few that correspond to intermediate stages in T-cell development have been identified in humans. Instead, the tumors resemble either mature T cells or, as in **acute lymphoblastic leukemia**, the earliest type of lymphoid progenitor (Fig. 7.43). One possible reason for the rarity of tumors corresponding to intermediate stages is that immature T cells are programmed to die unless rescued within a very narrow time window by positive selection (see Section 7-14). Thymocytes might simply not remain long enough in the intermediate stages of their development to provide an opportunity for malignant transformation. Thus, only T cells that are transformed at earlier stages, or that are transformed after the T cell has matured, are seen frequently as tumors.

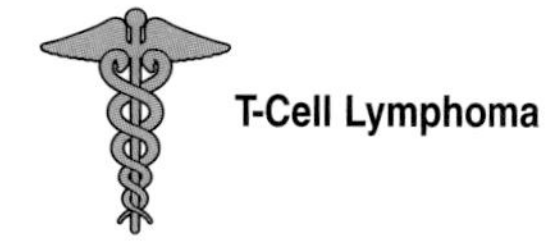

As with B cells, the behavior of mature T-cell tumors has provided insight into different aspects of T-cell biology, and vice versa. For example, **cutaneous T-cell lymphomas**, which home to the skin and proliferate slowly, are clonal

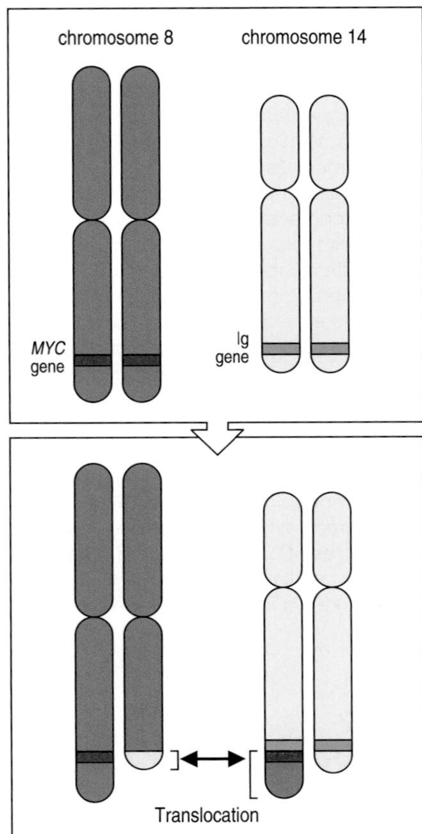

Fig. 7.44 Specific chromosomal rearrangements are found in some lymphoid tumors. Chromosomal rearrangements that join one of the immunoglobulin genes to a cellular oncogene can cause aberrant expression of the oncogene due to proximity to the immunoglobulin regulatory sequences. Such rearrangements are frequently associated with B-cell tumors. In the example shown, characteristic of Burkitt's lymphoma, the translocation of the oncogene *MYC* from chromosome 8 (top panel) to the immunoglobulin heavy-chain locus on chromosome 14 (bottom panel) results in the deregulated expression of *MYC* and the unregulated growth of the B cell. The immunoglobulin gene on the normal chromosome 14 is usually productively rearranged, and the tumors that result from such translocations generally have a mature B-cell phenotype and express immunoglobulin.

outgrowths of a CD4 T cell that, when activated, homes to the skin. A tumor of thymic stroma, called a **thymoma**, is frequently present in certain types of autoimmune disease, and removal of these tumors often ameliorates the disease. The reasons for this are not yet known.

7-32 B-cell lymphomas frequently carry chromosomal translocations that join immunoglobulin loci to genes that regulate cell growth.

The unregulated accumulation of cells of a single clone, which is the most striking characteristic of tumors, is caused by mutations that release the founder cell from the normal restraints on its growth or prevent its normal programmed death. In B-cell tumors the disruption of normal cellular homeostatic controls is often associated with an aberrant immunoglobulin gene rearrangement, in which one of the immunoglobulin loci is joined to a gene on another chromosome. This genetic fusion with another chromosome is known as a **translocation**, and in B-cell tumors translocations are found that disrupt the expression and function of genes important for controlling cell growth. Cellular genes that cause cancer when their function or expression is disrupted are termed **oncogenes**.

Translocations give rise to chromosomal abnormalities that are visible microscopically in metaphase. Characteristic translocations are seen in different B-cell tumors and reflect the involvement of a particular oncogene in each tumor type. Characteristic translocations that involve the T-cell receptor loci are also seen in T-cell tumors. Immunoglobulin and T-cell receptor loci are sites at which double-strand DNA breaks occur during gene rearrangement, and during class switching and somatic hypermutation in B cells, so it is not surprising that they are especially likely to be sites of chromosomal translocation.

The analysis of chromosomal abnormalities has revealed much about the regulation of B-cell growth and the disruption of growth control in tumor cells. In Burkitt's lymphoma cells, the *MYC* oncogene on chromosome 8 is recombined with an immunoglobulin locus by translocations that involve either chromosome 14 (heavy chain) (Fig. 7.44), chromosome 2 (κ light chain), or chromosome 22 (λ light chain). The Myc protein is known to be involved in the control of the cell cycle in normal cells. The translocation deregulates expression of the Myc protein, which leads to an increased proliferation of B cells, although other mutations elsewhere in the genome are also needed before a B-cell tumor results.

Other B-cell tumors, particularly follicular lymphomas, carry a chromosomal translocation of immunoglobulin genes to the oncogene *bcl-2*, increasing the production of Bcl-2 protein. The Bcl-2 protein prevents apoptosis in B-lineage cells (see Section 6-26), so its abnormal expression allows some B cells to survive and accumulate beyond their normal life-span. During this time, further genetic changes can occur that lead to malignant transformation. Proof that Bcl-2 rearrangement and consequent overexpression can promote lymphoma comes from mice carrying a constitutively overexpressed *bcl-2* transgene. These mice tend to develop B-cell lymphomas in later life. Similarly, the gene *bcl-6* is commonly rearranged in diffuse large B-cell lymphomas and is thought to have a causative role in the transformation of these cells.

Summary.

Very rarely, an individual B cell or T cell undergoes mutation and gives rise to a leukemia or lymphoma. Different lymphoid tumors can exhibit properties that reflect the stage of the cell from which the tumor derives, such as the growth pattern and location. Most lymphoid tumors, except those derived from very early uncommitted cells, exhibit characteristic gene rearrangements that indicate their derivation from either a B- or T-lineage precursor. These rearrangements are frequently accompanied by chromosomal translocations, often between a locus involved in generating the antigen receptor and a cellular proto-oncogene, such as the immunoglobulin locus and the *MYC* oncogene. Detailed gene-expression analysis of these tumors is revealing their origins as well as the key genes involved in malignant transformation. Such studies are already aiding diagnosis and are likely to lead to specific therapies.

Summary to Chapter 7.

In this chapter we have learned about the formation of the B-cell and T-cell lineages from a primitive lymphoid progenitor. The somatic gene rearrangements that generate the highly diverse repertoire of antigen receptors—immunoglobulin for B cells and the T-cell receptor for T cells—occur in the early stages of the development of T cells and B cells from a common bone marrow derived lymphoid progenitor. Mammalian B-cell development takes place in fetal liver and, after birth, in the bone marrow; T cells also originate in the bone marrow but undergo most of their development in the thymus. Much of the somatic recombination machinery, including the RAG proteins that are an essential part of the V(D)J recombinase, is common to both. Also common to B and T cells is the fact that gene rearrangement proceeds successively at each gene locus, beginning with loci that contain D genes. The first step in B-cell development is the rearrangement of the locus for the immunoglobulin heavy chain, and for T cells the β chain. In each case, the developing cell is allowed to proceed to the next stage of development only if the rearrangement has produced an in-frame sequence that can be translated into a protein expressed on the cell surface: either the pre-B-cell receptor or the pre-T cell receptor. Cells that do not generate successful rearrangements for both receptor chains die by apoptosis. The course of conventional B-cell development is summarized in Fig. 7.45, and that of α:β T cells in Fig. 7.46.

Once a functional antigen receptor has appeared on the cell surface, the lymphocyte is tested in two ways. Positive selection tests for the potential usefulness of the antigen receptor, whereas negative selection removes self-reactive cells from the lymphocyte repertoire, rendering it tolerant to the antigens of the body. Positive selection is particularly important for T cells, because it ensures that only cells bearing T-cell receptors that can recognize antigen in combination with self-MHC molecules will continue to mature. Positive selection also coordinates the choice of co-receptor expression. CD4 becomes expressed by T cells harboring MHC class II restricted receptors, and CD8 by cells harboring MHC class I restricted receptors. This ensures the optimal use of these receptors in responses to pathogens. For B cells, positive selection seems to occur at the final transition from immature to mature B cells, which occurs in peripheral lymphoid tissues. Tolerance is enforced at different stages throughout the development of both B and T cells, and positive selection likewise seems to represent a continuous process.

B and T cells surviving development in the central lymphoid organs emigrate to the periphery, where they home to occupy specific sites. The organization of the peripheral lymphoid organs, such as spleen and lymph nodes, involves interactions between cells expressing TNF and TNFR family proteins. The homing of B and T cells to different parts of these peripheral tissues involves their expression of distinct chemokine receptors and the secretion of specific chemokines by various stromal elements. Maturation and survival of B and T lymphocytes in these peripheral tissues involves other specific factors. B cells receive survival signals in the follicle through interaction with BAFF. Naive T cells require the cytokines IL-7 and IL-15 for survival and homeostatic proliferation, along with signals received through the T-cell receptor interacting

Fig. 7.45 A summary of the development of human conventional B-lineage cells. The state of the immunoglobulin genes, the expression of some essential intracellular proteins, and the expression of some cell-surface molecules are shown for successive stages of B-2-cell development. The immunoglobulin genes undergo further changes during antigen-driven B-cell differentiation, such as class switching and somatic hypermutation (see Chapter 4), which are evident in the immunoglobulins produced by memory cells and plasma cells. These antigen-dependent stages are described in more detail in Chapter 9.

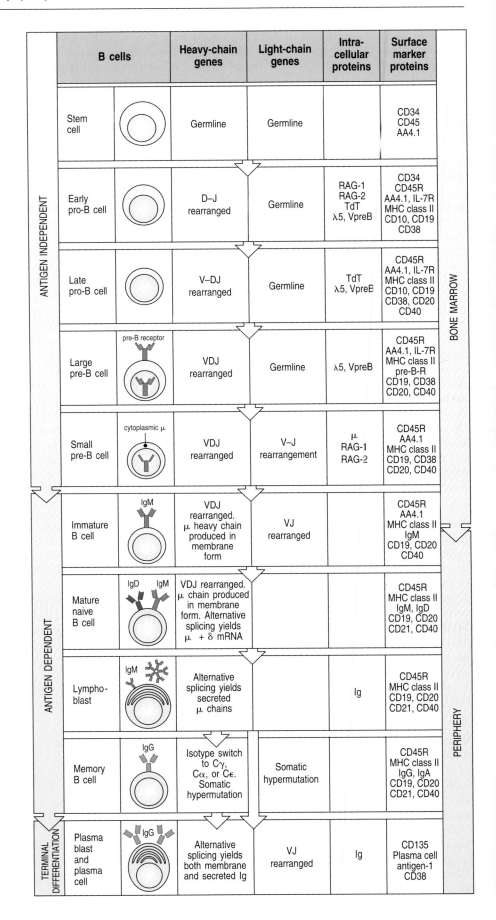

Phase	B cells		Heavy-chain genes	Light-chain genes	Intra-cellular proteins	Surface marker proteins	Location
ANTIGEN INDEPENDENT	Stem cell		Germline	Germline		CD34 CD45 AA4.1	BONE MARROW
ANTIGEN INDEPENDENT	Early pro-B cell		D–J rearranged	Germline	RAG-1 RAG-2 TdT λ5, VpreB	CD34 CD45R AA4.1, IL-7R MHC class II CD10, CD19 CD38	BONE MARROW
ANTIGEN INDEPENDENT	Late pro-B cell		V–DJ rearranged	Germline	TdT λ5, VpreB	CD45R AA4.1, IL-7R MHC class II CD10, CD19 CD38, CD20 CD40	BONE MARROW
ANTIGEN INDEPENDENT	Large pre-B cell	pre-B receptor	VDJ rearranged	Germline	λ5, VpreB	CD45R AA4.1, IL-7R MHC class II pre-B-R CD19, CD38 CD20, CD40	BONE MARROW
ANTIGEN INDEPENDENT	Small pre-B cell	cytoplasmic μ	VDJ rearranged	V–J rearrangement	μ RAG-1 RAG-2	CD45R AA4.1 MHC class II CD19, CD38 CD20, CD40	BONE MARROW
ANTIGEN DEPENDENT	Immature B cell	IgM	VDJ rearranged. μ heavy chain produced in membrane form	VJ rearranged		CD45R AA4.1 MHC class II IgM CD19, CD20 CD40	BONE MARROW
ANTIGEN DEPENDENT	Mature naive B cell	IgD IgM	VDJ rearranged. μ chain produced in membrane form. Alternative splicing yields μ + δ mRNA			CD45R MHC class II IgM, IgD CD19, CD20 CD21, CD40	PERIPHERY
ANTIGEN DEPENDENT	Lympho-blast	IgM	Alternative splicing yields secreted μ chains		Ig	CD45R MHC class II CD19, CD20 CD21, CD40	PERIPHERY
ANTIGEN DEPENDENT	Memory B cell	IgG	Isotype switch to Cγ, Cα, or Cε. Somatic hypermutation	Somatic hypermutation		CD45R MHC class II IgG, IgA CD19, CD20 CD21, CD40	PERIPHERY
TERMINAL DIFFERENTIATION	Plasma blast and plasma cell	IgG	Alternative splicing yields both membrane and secreted Ig	VJ rearranged	Ig	CD135 Plasma cell antigen-1 CD38	PERIPHERY

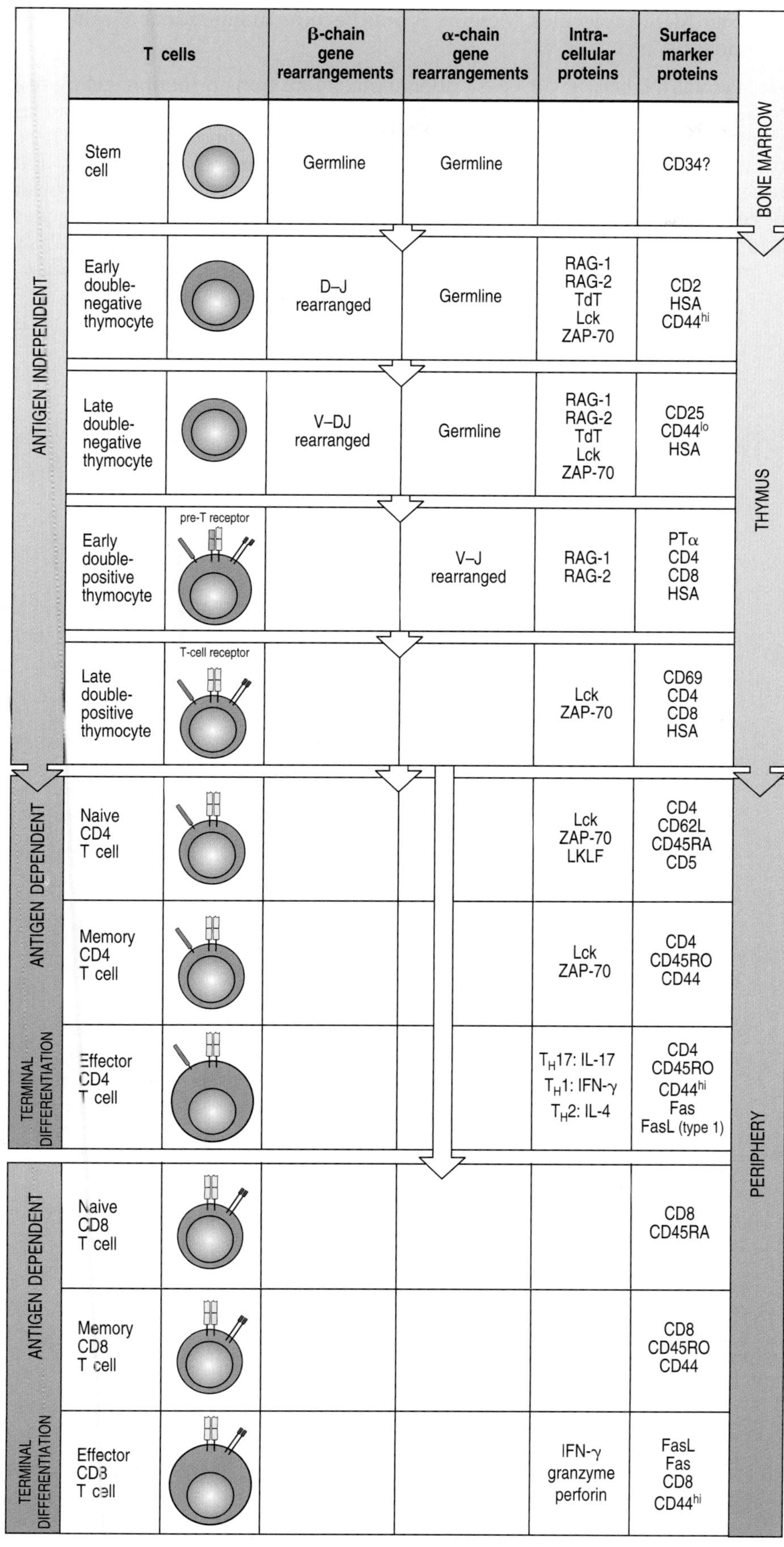

Phase	T cells	β-chain gene rearrangements	α-chain gene rearrangements	Intracellular proteins	Surface marker proteins	Site
ANTIGEN INDEPENDENT	Stem cell	Germline	Germline		CD34?	BONE MARROW
ANTIGEN INDEPENDENT	Early double-negative thymocyte	D–J rearranged	Germline	RAG-1 RAG-2 TdT Lck ZAP-70	CD2 HSA $CD44^{hi}$	THYMUS
ANTIGEN INDEPENDENT	Late double-negative thymocyte	V–DJ rearranged	Germline	RAG-1 RAG-2 TdT Lck ZAP-70	CD25 $CD44^{lo}$ HSA	THYMUS
ANTIGEN INDEPENDENT	Early double-positive thymocyte (pre-T receptor)		V–J rearranged	RAG-1 RAG-2	PTα CD4 CD8 HSA	THYMUS
ANTIGEN INDEPENDENT	Late double-positive thymocyte (T-cell receptor)			Lck ZAP-70	CD69 CD4 CD8 HSA	THYMUS
ANTIGEN DEPENDENT	Naive CD4 T cell			Lck ZAP-70 LKLF	CD4 CD62L CD45RA CD5	PERIPHERY
ANTIGEN DEPENDENT	Memory CD4 T cell			Lck ZAP-70	CD4 CD45RO CD44	PERIPHERY
TERMINAL DIFFERENTIATION	Effector CD4 T cell			T_H17: IL-17 T_H1: IFN-γ T_H2: IL-4	CD4 CD45RO $CD44^{hi}$ Fas FasL (type 1)	PERIPHERY
ANTIGEN DEPENDENT	Naive CD8 T cell				CD8 CD45RA	PERIPHERY
ANTIGEN DEPENDENT	Memory CD8 T cell				CD8 CD45RO CD44	PERIPHERY
TERMINAL DIFFERENTIATION	Effector CD8 T cell			IFN-γ granzyme perforin	FasL Fas CD8 $CD44^{hi}$	PERIPHERY

Fig. 7.46 A summary of the development of human α:β T cells. The state of the T-cell receptor genes, the expression of some essential intracellular proteins, and the expression of some cell-surface molecules are shown for successive stages of α:β T-cell development. Note that because the T-cell receptor genes do not undergo further changes during antigen-driven development, only the phases during which they are actively undergoing rearrangement in the thymus are indicated. The antigen-dependent phases of CD4 and CD8 cells are depicted separately, and are detailed in Chapter 8.

with self-MHC molecules. Memory B cells become independent of self-MHC interactions.

Occasionally, B cells and T cells undergo malignant transformation, giving rise to tumors that have escaped normal growth controls while retaining most features of the parent cell, including its characteristic homing pattern. These tumors frequently carry translocations involving the antigen-receptor loci and other genes that are intimately involved in the regulation of lymphocyte growth or cell death; thus, these translocations have been informative about the genes and proteins that regulate lymphocyte homeostasis. Gene-expression analysis is providing powerful and comprehensive insights into the origins of lymphocyte tumors as well as the origins of many tumors of nonlymphoid origin.

Questions.

7.1 *B-cell development in the bone marrow shares many features with T-cell development in the thymus. (a) What are the two major goals of lymphocyte development? (b) Discuss the ordered steps of receptor rearrangement in B and T cells, drawing the parallels between the two cell types. (c) What is the function of the pre-B-cell receptor and the pre-T-cell receptor? (d) Why do T cells develop in the thymus and B cells develop in the bone marrow?*

7.2 *Lymphocyte development is notable for huge cell losses at several steps. (a) What are the major reasons that lymphocytes die without progressing beyond the pre-T-cell or pre-B-cell stage? (b) What is the major reason that lymphocytes die after reaching the immature stage of expressing a complete TCR or BCR?*

7.3 *Discuss the process of positive selection of T cells in the thymus. (a) Where does it take place? (b) What are the ligands? (c) When (at what stage) during T-cell development does positive selection occur? (d) Describe how the choice between expression of the co-receptor—CD4 or CD8—occurs, and identify any known regulators of this process.*

7.4 *Peripheral lymphoid tissues become organized through communication between several kinds of cells and several kinds of receptor interactions. (a) What families of molecules are critical for the proper organization of peripheral lymphoid tissues? (b) Which are important for organizing the B-cell zones? (c) Which are important for organizing the T-cell zone?*

7.5 *There are three main subsets of B cells: follicular, marginal zone, and B-1. Compare and contrast their development and functions, covering at least five different categories.*

7.6 *What does the presence or absence of somatic hypermutations in immunoglobulin V regions of B-lineage tumors tell us about the origin of the neoplastic cells?*

General references.

Casali, P., and Silberstein, L.E.S. (eds): *Immunoglobulin Gene Expression in Development and Disease.* New York, New York Academy of Sciences, 1995.

Loffert, D., Schaal, S., Ehlich, A., Hardy, R.R., Zou, Y.R., Muller, W., and Rajewsky, K.: **Early B-cell development in the mouse—insights from mutations introduced by gene targeting.** *Immunol. Rev.* 1994, **137**:135–153.

Melchers, F., ten Boekel, E., Seidl, T., Kong, X.C., Yamagami, T., Onishi, K., Shimizu, T., Rolink, A.G., and Andersson, J.: **Repertoire selection by pre-B-cell receptors and B-cell receptors, and genetic control of B-cell development from immature to mature B cells.** *Immunol. Rev.* 2000, **175**:33–46.

Starr, T.K., Jameson, S.C., and Hogquist, K.A.: **Positive and negative selection of T cells.** *Annu. Rev. Immunol.* 2003, **21**:139–176.

von Boehmer, H.: **The developmental biology of T lymphocytes.** *Annu. Rev. Immunol.* 1993, **6**:309–326.

Section references.

7-1 Lymphocytes derive from hematopoietic stem cells in the bone marrow.

Akashi, K., Kondo, M., Cheshier, S., Shizuru, J., Gandy, K., Domen, J., Mebius, R., Traver, D., and Weissman, I.L.: **Lymphoid development from stem cells and the common lymphocyte progenitors.** *Cold Spring Harbor Symp. Quant. Biol.* 1999, **64**:1–12.

Bhandoola, A., and Sambandam, A.: **From stem cell to T cell: one route or many?** *Nat. Rev. Immunol.* 2006, **6**:117–126.

Funk, P.E., Kincade, P.W., and Witte, P.L.: **Native associations of early hematopoietic stem-cells and stromal cells isolated in bone-marrow cell aggregates.** *Blood* 1994, **83**:361–369.

Jacobsen, K., Kravitz, J., Kincade, P.W., and Osmond, D.G.: **Adhesion receptors on bone-marrow stromal cells—in vivo expression of vascular cell adhesion molecule-1 by reticular cells and sinusoidal endothelium in normal and γ-irradiated mice.** *Blood* 1996, **87**:73–82.

Nagasawa, T., Hirota, S., Tachibana, V., Takakura, N., Nishikawa, S., Kitamura, Y., Yoshida, V., Kikutani, H., and Kishimoto, T.: **Defects of B-cell lymphopoiesis and bone-marrow myelopoiesis in mice lacking the CXC chemokine PBSF/SDF-1.** *Nature* 1996, **382**:635–638.

Singh, H., Medina, K.L., and Pongubala, J.M.: **Contingent gene regulatory networks and B cell fate specification.** *Proc. Natl Acad. Sci. USA* 2005, **102**:4949–4953.

7-2 B-cell development begins by rearrangement of the heavy-chain locus.

Allman, D., Li, J., and Hardy, R.R.: **Commitment to the B lymphoid lineage occurs before DH-JH recombination.** *J. Exp. Med.* 1999, **189**:735–740.

Allman, D., Lindsley, R.C., DeMuth, W., Rudd, K., Shinton, S.A., and Hardy, R.R.: **Resolution of three nonproliferative immature splenic B cell subsets reveals multiple selection points during peripheral B cell maturation.** *J. Immunol.* 2001, **167**:6834–6840.

Ehrlich, A., and Kuppers, R.: **Analysis of immunoglobulin gene rearrangements in single B cells.** *Curr. Opin. Immunol.* 1995, **7**:281–284.

Hardy, R.R., Carmack, C.E., Shinton, S.A., Kemp, J.D., and Hayakawa, K.: **Resolution and characterization of pro-B and pre-pro-B cell stages in normal mouse bone marrow.** *J. Exp. Med.* 1991, **173**:1213–1225.

Osmond, D.G., Rolink, A., and Melchers, F.: **Murine B lymphopoiesis: towards a unified model.** *Immunol. Today* 1998, **19**:65–68.

ten Boekel, E., Melchers, F., and Rolink, A.: **The status of Ig loci rearrangements in single cells from different stages of B-cell development.** *Int. Immunol.* 1995, **7**:1013–1019.

7-3 The pre-B-cell receptor tests for successful production of a complete heavy chain and signals for proliferation of pro-B cells.

Grawunder, U., Leu, T.M.J., Schatz, D.G., Werner, A., Rolink, A.G., Melchers, F., and Winkler, T.H.: **Down-regulation of Rag1 and Rag2 gene expression in pre-B cells after functional immunoglobulin heavy-chain rearrangement.** *Immunity* 1995, **3**:601–608.

Monroe, J.G.: **ITAM-mediated tonic signalling through pre-BCR and BCR complexes.** *Nat. Rev. Immunol.* 2006 **6**:283–294.

7-4 Pre-B-cell receptor signaling inhibits further heavy-chain locus rearrangement and enforces allelic exclusion.

Loffert, D., Ehlich, A., Muller, W., and Rajewsky, K.: **Surrogate light-chain expression is required to establish immunoglobulin heavy-chain allelic exclusion during early B-cell development.** *Immunity* 1996, **4**:133–144.

Melchers, F., ten Boekel, E., Yamagami, T., Andersson, J., and Rolink, A.: **The roles of preB and B cell receptors in the stepwise allelic exclusion of mouse IgH and L chain gene loci.** *Semin. Immunol.* 1999, **11**:307–317.

7-5 Pre-B cells rearrange the light-chain locus and express cell-surface immunoglobulin.

Arakawa, H., Shimizu, T., and Takeda, S.: **Reevaluation of the probabilities for productive rearrangements on the κ-loci and λ-loci.** *Int. Immunol.* 1996, **8**:91–99.

Gorman, J.R., van der Stoep, N., Monroe, R., Cogne, M., Davidson, L., and Alt, F.W.: **The Igκ 3′ enhancer influences the ratio of Igκ versus Igλ B lymphocytes.** *Immunity* 1996, **5**:241–252.

Hesslein, D.G., and Schatz, D.G.: **Factors and forces controlling V(D)J recombination.** *Adv. Immunol.* 2001, **78**:169–232.

Kee, B.L., and Murre, C.: **Transcription factor regulation of B lineage commitment.** *Curr. Opin. Immunol.* 2001, **13**:180–185.

Sleckman, B.P., Gorman, J.R., and Alt, F.W.: **Accessibility control of antigen receptor variable region gene assembly—role of cis-acting elements.** *Annu. Rev. Immunol.* 1996, **14**:459–481.

Takeda, S., Sonoda, E., and Arakawa, H.: **The κ–λ ratio of immature B cells.** *Immunol. Today* 1996, **17**:200–201.

7-6 Immature B cells are tested for autoreactivity before they leave the bone marrow.

Casellas, R., Shih, T.A., Kleinewietfeld, M., Rakonjac, J., Nemazee, D., Rajewsky, K., and Nussenzweig, M.C.: **Contribution of receptor editing to the antibody repertoire.** *Science* 2001, **291**:1541–1544.

Chen, C., Nagy, Z., Radic, M.Z., Hardy, R.R., Huszar, D., Camper, S.A., and Weigert, M.: **The site and stage of anti-DNA B-cell deletion.** *Nature* 1995, **373**:252–255.

Cornall, R.J., Goodnow, C.C., and Cyster, J.G.: **The regulation of self-reactive B cells.** *Curr. Opin. Immunol.* 1995, **7**:804–811.

Melamed, D., Benschop, R.J., Cambier, J.C., and Nemazee, D.: **Developmental regulation of B lymphocyte immune tolerance compartmentalizes clonal selection from receptor selection.** *Cell* 1998, **92**:173–182.

Nemazee, D.: **Receptor editing in lymphocyte development and central tolerance.** *Nat. Rev. Immunol.* 2006, **6**:728–740.

Prak, E.L., and Weigert, M.: **Light-chain replacement—a new model for antibody gene rearrangement.** *J. Exp. Med.* 1995, **182**:541–548.

Tiegs, S.L., Russell, D.M., and Nemazee, D.: **Receptor editing in self-reactive bone marrow B cells.** *J. Exp. Med.* 1993, **177**:1009–1020.

7-7 T-cell progenitors migrate from the bone marrow, but all the important events in their development occur in the thymus.

Anderson, G., Moore, N.C., Owen, J.J.T., and Jenkinson, E.J.: **Cellular interactions in thymocyte development.** *Annu. Rev. Immunol.* 1996, **14**:73–99.

Carlyle, J.R., and Zúñiga-Pflücker, J.C.: **Requirement for the thymus in αβ T lymphocyte lineage commitment.** *Immunity* 1998, **9**:187–197.

Ciofani, M., Knowles, G., Wiest, D., von Boehmer, H., and Zúñiga-Pflücker, J.: **Stage-specific and differential Notch dependency at the αβ and γδ T lineage bifurcation.** *Immunity* 2006, **25**:105–116.

Cordier, A.C., and Haumont, S.M.: **Development of thymus, parathyroids, and ultimobranchial bodies in NMRI and nude mice.** *Am. J. Anat.* 1980, **157**:227–263.

Gordon J., Wilson, V.A., Blair, N.F., Sheridan, J., Farley, A., Wilson, L., Manley, N.R., and Blackburn, CC. **Functional evidence for a single endodermal origin for the thymic epithelium**. *Nat. Immunol.* 2004, **5**:546–553.

Nehls, M., Kyewski, B., Messerle, M., Waldschütz, R., Schüddekopf, K., Smith, A.J.H., and Boehm, T.: **Two genetically separable steps in the differentiation of thymic epithelium.** *Science* 1996, **272**:886–889.

van Ewijk, W., Hollander, G., Terhorst, C., and Wang, B.: **Stepwise development of thymic microenvironments *in vivo* is regulated by thymocyte subsets.** *Development* 2000, **127**:1583–1591.

Zúñiga-Pflücker, J.C., and Lenardo, M.J.: **Regulation of thymocyte development from immature progenitors.** *Curr. Opin. Immunol.* 1996, **8**:215–224.

7-8 T-cell precursors proliferate extensively in the thymus but most die there.

Shortman, K., Egerton, M., Spangrude, G.J., and Scollay, R.: **The generation and fate of thymocytes.** *Semin. Immunol.* 1990, **2**:3–12.

Surh, C.D., and Sprent, J.: **T-cell apoptosis detected *in situ* during positive and negative selection in the thymus.** *Nature* 1994, **372**:100–103.

7-9 Successive stages in the development of thymocytes are marked by changes in cell-surface molecules.

Borowski, C., Martin, C., Gounari, F., Haughn, L., Aifantis, I., Grassi, F., and von Boehmer, H.: **On the brink of becoming a T cell.** *Curr. Opin. Immunol.* 2002, **14**:200–206.

Saint-Ruf, C., Ungewiss, K., Groetrrup, M., Bruno, L., Fehling, H.J., and von Boehmer, H.: **Analysis and expression of a cloned pre-T-cell receptor gene.** *Science* 1994, **266**:1208–1212.

Shortman, K., and Wu, L.: **Early T lymphocyte progenitors.** *Annu. Rev. Immunol.* 1996, **14**:29–47.

7-10 Thymocytes at different developmental stages are found in distinct parts of the thymus.

Benz, C., Heinzel, K., and Bleul, C.C.: **Homing of immature thymocytes to the subcapsular microenvironment within the thymus is not an absolute requirement for T cell development.** *Eur. J. Immunol.* 2004, **34**:3652–3663.

Bleul, C.C., and Boehm, T.: **Chemokines define distinct microenvironments in the developing thymus.** *Eur. J. Immunol.* 2000, **30**:3371–3379.

Picker, L.J., and Siegelman, M.H.: **Lymphoid tissues and organs**, in Paul, W.E. (ed): *Fundamental Immunology*, 3rd ed. New York, Raven Press, 1993.

Ueno, T., Saito F., Gray, D.H.D., Kuse, S., Hieshima, K., Nakano, H., Kakiuchi, T., Lipp, M., Boyd, R.L., and Takahama, Y.: **CCR7 signals are essential for cortex–medulla migration of developing thymocytes.** *J. Exp. Med.* 2004, **200**:493–505.

7-11 T cells with α:β or γ:δ receptors arise from a common progenitor.

Fehling, H.J., Gilfillan, S., and Ceredig, R.: **Alpha β/γ δ lineage commitment in the thymus of normal and genetically manipulated mice.** *Adv. Immunol.* 1999, **71**:1–76.

Hayday, A.C., Barber, D.F., Douglas, N., and Hoffman, E.S.: **Signals involved in γ/δ T cell versus α/β T cell lineage commitment.** *Semin. Immunol.* 1999, **11**:239–249.

Hayes, S.M., and Love, P.E.: **Distinct structure and signaling potential of the γ/δ TCR complex.** *Immunity* 2002, **16**:827–838.

Kang, J., and Raulet, D.H.: **Events that regulate differentiation of α β TCR+ and γ δ TCR+ T cells from a common precursor.** *Semin. Immunol.* 1997, **9**:171–179.

Kang, J., Coles, M., Cado, D., and Raulet, D.H.: **The developmental fate of T cells is critically influenced by TCRγδ expression.** *Immunity* 1998, **8**:427–438.

Lauzurica, P., and Krangel, M.S.: **Temporal and lineage-specific control of T-cell receptor α/δ gene rearrangement by T-cell receptor α and δ enhancers.** *J. Exp. Med.* 1994, **179**:1913–1921.

Livak, F., Petrie, H.T., Crispe, I.N., and Schatz, D.G.: **In-frame TCR δ gene rearrangements play a critical role in the αβ/γδ T cell lineage decision.** *Immunity* 1995, **2**:617–627.

Sleckman, B.P., Bassing, C.H., Bardon, C.G., Okada, A., Khor, B., Bories, J.C., Monroe, R., and Alt, F.W.: **Accessibility control of variable region gene assembly during T-cell development.** *Immunol. Rev.* 1998, **165**:121–130.

7-12 T cells expressing particular γ- and δ-chain V regions arise in an ordered sequence early in life.

Ciofani, M., Knowles, G.C., Wiest, D.L., von Boehmer, H., and Zuniga-Pflucker, J.C.: **Stage-specific and differential notch dependency at the αβ and γδ T lineage bifurcation.** *Immunity* 2006, **25**:105–116.

Dunon, D., Courtois, D., Vainio, O., Six, A., Chen, C.H., Cooper, M.D., Dangy, J.P., and Imhof, B.A.: **Ontogeny of the immune system: γδ and αβ T cells migrate from thymus to the periphery in alternating waves.** *J. Exp. Med.* 1997, **186**:977–988.

Havran, W.L., and Boismenu, R.: **Activation and function of γδ T cells.** *Curr. Opin. Immunol.* 1994, **6**:442–446.

7-13 Successful synthesis of a rearranged β chain allows the production of a pre-T-cell receptor that triggers cell proliferation and blocks further β-chain gene rearrangement.

Borowski, C, Li, X, Aifantis, I, Gounari, F, and von Boehmer, H.: **Pre-TCRα and TCRα are not interchangeable partners of TCRβ during T lymphocyte development.** *J. Exp. Med.* 2004, **199**:607–615.

Dudley, E.C., Petrie, H.T., Shah, L.M., Owen, M.J., and Hayday, A.C.: **T-cell receptor β chain gene rearrangement and selection during thymocyte development in adult mice.** *Immunity* 1994, **1**:83–93.

Philpott, K.I., Viney, J.L., Kay, G., Rastan, S., Gardiner, E.M., Chae, S., Hayday, A.C., and Owen, M.J.: **Lymphoid development in mice congenitally lacking T cell receptor α β-expressing cells.** *Science* 1992, **256**:1448–1453.

von Boehmer, H., Aifantis, I., Azogui, O., Feinberg, J., Saint-Ruf, C., Zober, C., Garcia, C., and Buer, J.: **Crucial function of the pre-T-cell receptor (TCR) in TCR β selection, TCR β allelic exclusion and αβ versus γδ lineage commitment.** *Immunol. Rev.* 1998, **165**:111–119.

7-14 T-cell α-chain genes undergo successive rearrangements until positive selection or cell death intervenes.

Buch, T., Rieux-Laucat, F., Förster, I., and Rajewsky, K.: **Failure of HY-specific thymocytes to escape negative selection by receptor editing.** *Immunity* 2002, **16**:707–718.

Hardardottir, F., Baron, J.L., and Janeway, C.A., Jr.: **T cells with two functional antigen-specific receptors.** *Proc. Natl Acad. Sci. USA* 1995, **92**:354–358.

Huang, C.-Y., Sleckman, B.P., and Kanagawa, O.: **Revision of T cell receptor α chain genes is required for normal T lymphocyte development.** *Proc. Natl Acad. Sci. USA* 2005, **102**:14356–14361.

Marrack, P., and Kappler, J.: **Positive selection of thymocytes bearing αβ T cell receptors.** *Curr. Opin. Immunol.* 1997, **9**:250–255.

Padovan, E., Casorati, G., Dellabona, P., Meyer, S., Brockhaus, M., and Lanzavecchia, A.: **Expression of two T-cell receptor α chains: dual receptor T cells.** *Science* 1993, **262**:422–424.

Petrie, H.T., Livak, F., Schatz, D.G., Strasser, A., Crispe, I.N., and Shortman, K.: **Multiple rearrangements in T-cell receptor α-chain genes maximize the production of useful thymocytes.** *J. Exp. Med.* 1993, **178**:615–622.

7-15 The MHC type of the thymic stroma selects a repertoire of mature T cells that can recognize foreign antigens presented by the same MHC type.

Fink, P.J., and Bevan, M.J.: **H-2 antigens of the thymus determine lymphocyte specificity.** *J. Exp. Med.* 1978, **148**:766–775.

Zinkernagel, R.M., Callahan, G.N., Klein, J., and Dennert, G.: **Cytotoxic T cells learn specificity for self H-2 during differentiation in the thymus.** *Nature* 1978, **271**:251–253.

7-16 Only thymocytes whose receptors interact with self-peptide:self-MHC complexes can survive and mature.

Hogquist, K.A., Tomlinson, A.J., Kieper, W.C., McGargill, M.A., Hart, M.C., Naylor, S., and Jameson, S.C.: **Identification of a naturally occurring ligand for thymic positive selection.** *Immunity* 1997, **6**:389–399.

Huessman, M., Scott, B., Kisielow, P., and von Boehmer, H.: **Kinetics and efficacy of positive selection in the thymus of normal and T-cell receptor transgenic mice.** *Cell* 1991, **66**:533–562.

Stefanski, H.E., Mayerova, D., Jameson, S.C., and Hogquist, K.A.: **A low affinity TCR ligand restores positive selection of CD8⁺ T cells *in vivo*.** *J. Immunol.* 2001, **166**:6602–6607.

7-17 Positive selection acts on a repertoire of T-cell receptors with inherent specificity for MHC molecules.

Merkenschlager, M., Graf, D., Lovatt, M., Bommhardt, U., Zamoyska, R., and Fisher, A.G.: **How many thymocytes audition for selection?** *J. Exp. Med.* 1997, **186**:1149–1158.

Zerrahn, J., Held, W., and Raulet, D.H.: **The MHC reactivity of the T cell repertoire prior to positive and negative selection.** *Cell* 1997, **88**:627–636.

7-18 Positive selection coordinates the expression of CD4 or CD8 with the specificity of the T-cell receptor and the potential effector functions of the T cell.

Basson, M.A., Bommhardt, U., Cole, M.S., Tso, J.Y., and Zamoyska, R.: **CD3 ligation on immature thymocytes generates antagonist-like signals appropriate for CD8 lineage commitment, independently of T cell receptor specificity.** *J. Exp. Med.* 1998, **187**:1249–1260.

Bommhardt, U., Cole, M.S., Tso, J.Y., and Zamoyska, R.: **Signals through CD8 or CD4 can induce commitment to the CD4 lineage in the thymus.** *Eur. J. Immunol.* 1997, **27**:1152–1163.

Germain, R.N.: **T-cell development and the CD4–CD8 lineage decision.** *Nat. Rev. Immunol.* 2002, **2**:309–322.

Lundberg, K., Heath, W., Kontgen, F., Carbone, F.R., and Shortman, K.: **Intermediate steps in positive selection: differentiation of $CD4^{+}8^{int}$ TCR^{int} thymocytes into $CD4^{-}8^{+}TCR^{hi}$ thymocytes.** *J. Exp. Med.* 1995, **181**:1643–1651.

Singer, A., Bosselut, R., and Bhandoola, A.: **Signals involved in CD4/CD8 lineage commitment: current concepts and potential mechanisms.** *Semin. Immunol.* 1999, **11**:273–281.

von Boehmer, H., Kisielow, P., Lishi, H., Scott, B., Borgulya, P., and Teh, H.S.: **The expression of CD4 and CD8 accessory molecules on mature T cells is not random but correlates with the specificity of the $\alpha\beta$ receptor for antigen.** *Immunol. Rev.* 1989, **109**:143–151.

7-19 Thymic cortical epithelial cells mediate positive selection of developing thymocytes.

Cosgrove, D., Chan, S.H., Waltzinger, C., Benoist, C., and Mathis, D.: **The thymic compartment responsible for positive selection of CD4⁺ T cells.** *Int. Immunol.* 1992, **4**:707–710.

Ernst, B.B., Surh, C.D., and Sprent, J.: **Bone marrow-derived cells fail to induce positive selection in thymus reaggregation cultures.** *J. Exp. Med.* 1996, **183**:1235–1240.

Fowlkes, B.J., and Schweighoffer, E.: **Positive selection of T cells.** *Curr. Opin. Immunol.* 1995, **7**:188–195.

7-20 T cells that react strongly with ubiquitous self antigens are deleted in the thymus.

Kishimoto, H., and Sprent, J.: **Negative selection in the thymus includes semimature T cells.** *J. Exp. Med.* 1997, **185**:263–271.

Zal, T., Volkmann, A., and Stockinger, B.: **Mechanisms of tolerance induction in major histocompatibility complex class II-restricted T cell specific for a blood-borne self antigen.** *J. Exp. Med.* 1994, **180**:2089–2099.

7-21 Negative selection is driven most efficiently by bone marrow derived antigen-presenting cells.

Matzinger, P., and Guerder, S.: **Does T cell tolerance require a dedicated antigen-presenting cell?** *Nature* 1989, **338**:74–76.

Sprent, J., and Webb, S.R.: **Intrathymic and extrathymic clonal deletion of T cells.** *Curr. Opin. Immunol.* 1995, **7**:196–205.

Webb, S.R., and Sprent, J.: **Tolerogenicity of thymic epithelium.** *Eur. J. Immunol.* 1990, **20**:2525–2528.

7-22 The specificity and/or the strength of signals for negative and positive selection must differ.

Alberola-Ila, J., Hogquist, K.A., Swan, K.A., Bevan, M.J., and Perlmutter, R.M.: **Positive and negative selection invoke distinct signaling pathways.** *J. Exp. Med.* 1996, **184**:9–18.

Ashton-Rickardt, P.G., Bandeira, A., Delaney, J.R., Van Kaer, L., Pircher, H.P., Zinkernagel, R.M., and Tonegawa, S.: **Evidence for a differential avidity model of T-cell selection in the thymus.** *Cell* 1994, **76**:651–663.

Bommhardt, U., Basson, M.A., Krummrei, U., and Zamoyska, R.: **Activation of the extracellular signal-related kinase/mitogen-activated protein kinase pathway discriminates CD4 versus CD8 lineage commitment in the thymus.** *J. Immunol.* 1999, **163**:715–722.

Bommhardt, U., Scheuring, Y., Bickel, C., Zamoyska, R., and Hunig, T.: **MEK activity regulates negative selection of immature $CD4^{+}CD8^{+}$ thymocytes.** *J. Immunol.* 2000, **164**:2326–2337.

Hogquist, K.A., Jameson, S.C., Heath, W.R., Howard, J.L., Bevan, M.J., and Carbone, F.R.: **T-cell receptor antagonist peptides induce positive selection.** *Cell* 1994, **76**:17–27.

7-23 Different lymphocyte subsets are found in particular locations in peripheral lymphoid tissues.

Liu, Y.J.: **Sites of B lymphocyte selection, activation, and tolerance in spleen.** *J. Exp. Med.* 1997, **186**:625–629.

Loder, F., Mutschler, B., Ray, R.J., Paige, C.J., Sideras, P., Torres, R., Lamers, M.C., and Carsetti, R.: **B cell development in the spleen takes place in discrete steps and is determined by the quality of B cell receptor-derived signals.** *J. Exp. Med.* 1999, **190**:75–89.

Mebius, R.E.: **Organogenesis of lymphoid tissues.** *Nat. Rev. Immunol.* 2003, **3**:292–303.

7-24 The development and organization of peripheral lymphoid tissues are controlled by proteins of the tumor necrosis factor family.

Douni, E., Akassoglou, K., Alexopoulou, L., Georgopoulos, S., Haralambous, S., Hill, S., Kassiotis, G., Kontoyiannis, D., Pasparakis, M., Plows, D., Probert, L., and Kollias, G.: **Transgenic and knockout analysis of the role of TNF in immune regulation and disease pathogenesis.** *J. Inflamm.* 1996, **47**:27–38.

Fu, Y.X., and Chaplin, D.D.: **Development and maturation of secondary lymphoid tissues.** *Annu. Rev. Immunol.* 1999, **17**:399–433.

Mariathasan, S., Matsumoto, M., Baranyay, F., Nahm, M.H., Kanagawa, O., and

Chaplin, D.D.: **Absence of lymph nodes in lymphotoxin-α (LTα)-deficient mice is due to abnormal organ development, not defective lymphocyte migration.** *J. Inflamm.* 1995, **45**:72–78.

7-25 The homing of lymphocytes to specific regions of peripheral lymphoid tissues is mediated by chemokines.

Ansel, K.M., and Cyster, J.G.: **Chemokines in lymphopoiesis and lymphoid organ development.** *Curr. Opin. Immunol.* 2001, **13**:172–179.

Cyster, J.G.: **Chemokines and cell migration in secondary lymphoid organs.** *Science* 1999, **286**:2098–2102.

Cyster, J.G.: **Leukocyte migration: scent of the T zone.** *Curr. Biol.* 2000, **10**:R30–R33.

Cyster, J.G., Ansel, K.M., Reif, K., Ekland, E.H., Hyman, P.L., Tang, H.L., Luther, S.A., and Ngo, V.N.: **Follicular stromal cells and lymphocyte homing to follicles.** *Immunol. Rev.* 2000, **176**:181–193.

7-26 Lymphocytes that encounter sufficient quantities of self antigens for the first time in the periphery are eliminated or inactivated.

Arnold, B.: **Levels of peripheral T cell tolerance.** *Transpl. Immunol.* 2002, **10**:109–114.

Cyster, J.G., Hartley, S.B., and Goodnow, C.C.: **Competition for follicular niches excludes self-reactive cells from the recirculating B-cell repertoire.** *Nature* 1994, **371**:389–395.

Goodnow, C.C., Crosbie, J., Jorgensen, H., Brink, R.A., and Basten, A.: **Induction of self-tolerance in mature peripheral B lymphocytes.** *Nature* 1989, **342**:385–391.

Lam, K.P., Kuhn, R., and Rajewsky, K.: **In vivo ablation of surface immunoglobulin on mature B cells by inducible gene targeting results in rapid cell death.** *Cell* 1997, **90**:1073–1083.

Russell, D.M., Dembic, Z., Morahan, G., Miller, J.F.A.P., Burki, K., and Nemazee, D.: **Peripheral deletion of self-reactive B cells.** *Nature* 1991, **354**:308–311.

Steinman, R.M., and Nussenzweig, M.C.: **Avoiding horror autotoxicus: the importance of dendritic cells in peripheral T cell tolerance.** *Proc. Natl Acad. Sci. USA* 2002, **99**:351–358.

7-27 Most immature B cells arriving in the spleen are short-lived and require cytokines and positive signals through the B-cell receptor for maturation and survival.

Allman, D.M., Ferguson, S.E., Lentz, V.M., and Cancro, M.P.: **Peripheral B cell maturation. II. Heat-stable antigen^{hi} splenic B cells are an immature developmental intermediate in the production of long-lived marrow-derived B cells.** *J. Immunol.* 1993, **151**:4431–4444.

Harless, S.M., Lentz, V.M., Sah, A.P., Hsu, B.L., Clise-Dwyer, K., Hilbert, D.M., Hayes, C.E., and Cancro, M.P.: **Competition for BLyS-mediated signaling through Bcmd/BR3 regulates peripheral B lymphocyte numbers.** *Curr. Biol.* 2001, **11**:1986–1989.

Levine, M.H., Haberman, A.M., Sant'Angelo, D.B., Hannum, L.G., Cancro, M.P., Janeway, C.A., Jr., and Shlomchik, M.J.: **A B-cell receptor-specific selection step governs immature to mature B cell differentiation.** *Proc. Natl Acad. Sci. USA* 2000, **97**:2743–2748.

Rolink, A.G., Tschopp, J., Schneider, P., and Melchers, F.: **BAFF is a survival and maturation factor for mouse B cells.** *Eur. J. Immunol.* 2002, **32**:2004–2010.

7-28 B-1 cells and marginal zone B cells are distinct B-cell subtypes with unique antigen receptor specificity.

Clarke, S.H., and Arnold, L.W.: **B-1 cell development: evidence for an uncommitted immunoglobulin (Ig)M$^+$ B cell precursor in B-1 cell differentiation.** *J. Exp. Med.* 1998, **187**:1325–1334.

Hardy, R.R., and Hayakawa, K.: **A developmental switch in B lymphopoiesis.** *Proc. Natl Acad. Sci. USA* 1991, **88**:11550–11554.

Hayakawa, K., Asano, M., Shinton, S.A., Gui, M., Allman, D., Stewart, C.L., Silver, J., and Hardy, R.R.: **Positive selection of natural autoreactive B cells.** *Science* 1999, **285**:113–116.

Martin, F., and Kearney, J.F.: **Marginal-zone B cells.** *Nat. Rev. Immunol.* 2002, **2**:323–335.

7-29 T-cell homeostasis in the periphery is regulated by cytokines and self-MHC interactions.

Freitas, A.A., and Rocha, B.: **Peripheral T cell survival.** *Curr. Opin. Immunol.* 1999, **11**:152–156.

Judge, A.D., Zhang, X., Fujii, H., Surh, C.D., and Sprent, J.: **Interleukin 15 controls both proliferation and survival of a subset of memory-phenotype CD8$^+$ T cells.** *J. Exp. Med.* 2002, **196**:935–946.

Kassiotis, G., Garcia, S., Simpson, E., and Stockinger, B.: **Impairment of immunological memory in the absence of MHC despite survival of memory T cells.** *Nat. Immunol.* 2002, **3**:244–250.

Ku, C.C., Murakami, M., Sakamoto, A., Kappler, J., and Marrack, P.: **Control of homeostasis of CD8$^+$ memory T cells by opposing cytokines.** *Science* 2000, **288**:675–678.

Murali-Krishna, K., Lau, L.L., Sambhara, S., Lemonnier, F., Altman, J., and Ahmed, R.: **Persistence of memory CD8 T cells in MHC class I-deficient mice.** *Science* 1999, **286**:1377–1381.

Seddon, B., Tomlinson, P., and Zamoyska, R.: **IL-7 and T cell receptor signals regulate homeostasis of CD4 memory cells.** *Nat. Immunol.* 2003, **4**:680–686.

7-30 B-cell tumors often occupy the same site as their normal counterparts.

Alizadeh, A.A., and Staudt, L.M.: **Genomic-scale gene expression profiling of normal and malignant immune cells.** *Curr. Opin. Immunol.* 2000, **12**:219–225.

Cotran, R.S., Kumar, V., and Robbins, S.L.: **Diseases of white cells, lymph nodes, and spleen**. Robbins, S.L. (ed): *Pathologic Basis of Disease*, 5th ed. Philadelphia, W.B. Saunders, 1994.

7-31 T-cell tumors correspond to a small number of stages of T-cell development.

Hwang, L.Y., and Baer, R.J.: **The role of chromosome translocations in T cell acute leukemia.** *Curr. Opin. Immunol.* 1995, **7**:659–664.

Rabbitts, T.H.: **Chromosomal translocations in human cancer.** *Nature* 1994, **372**:143–149.

7-32 B-cell lymphomas frequently carry chromosomal translocations that join immunoglobulin loci to genes that regulate cell growth.

Cory, S.: **Regulation of lymphocyte survival by the Bcl-2 gene family.** *Annu. Rev. Immunol.* 1995, **13**:513–543.

Rabbitts, T.H.: **Chromosomal translocations in human cancer.** *Nature* 1994, **372**:143–149.

Yang, E., and Korsmeyer, S.J.: **Molecular thanatopsis—a discourse on the Bcl-2 family and cell death.** *Blood* 1996, **88**:386–401.

PART IV

THE ADAPTIVE IMMUNE RESPONSE

T Cell-Mediated Immunity

8

An adaptive immune response is induced when an infection overwhelms innate defense mechanisms. The pathogen continues to replicate and antigen accumulates. Together with the changed cellular environment produced by innate immunity, this triggers the adaptive immune response. Some infections may be dealt with solely by innate immunity, as discussed in Chapter 2; they will be eliminated early and produce few symptoms and little damage. But most pathogens, almost by definition, can overcome the innate immune system, and adaptive immunity is essential for defense against them. This is shown by the immunodeficiency syndromes that are associated with failure of specific parts of the adaptive immune response; these will be discussed in Chapter 12. In the next three chapters, we will learn how the adaptive immune response involving the antigen-specific T cells and B cells is initiated and deployed. We will consider T cell-mediated immune responses first, in this chapter, and humoral immunity—the antibody response produced by B cells—in Chapter 9. In Chapter 10 we will combine what we have learned in Chapters 8 and 9 to present a dynamic view of adaptive immune responses to pathogens, including a discussion of one of its most important features—immunological memory.

Once T cells have completed their development in the thymus, they enter the bloodstream. On reaching a peripheral lymphoid organ they leave the blood to migrate through the lymphoid tissue, returning via the lymphatics to the bloodstream to recirculate between blood and peripheral lymphoid tissues. Mature recirculating T cells that have not yet encountered their specific antigens are known as **naive T cells**. To participate in an adaptive immune response, a naive T cell must meet its specific antigen, presented to it as a peptide:MHC complex on the surface of an antigen-presenting cell, and be induced to proliferate and differentiate into cells that have acquired new activities that contribute to removing the antigen. These cells are called **effector T cells** and act very rapidly when they encounter their specific antigen on other cells. Because of their requirement to recognize peptide antigens presented by MHC molecules, all effector T cells act on other host cells, not on the pathogen itself. The cells on which effector T cells act will be referred to as their **target cells**.

	CD8 cytotoxic T cells	CD4 T_H1 cells	CD4 T_H2 cells	CD4 T_H17 cells	CD4 regulatory T cells (various types)
Types of effector T cell	CTL	T_H1	T_H2	T_H17	T_{reg}
Main functions in adaptive immune response	Kill virus-infected cells	Activate infected macrophages Provide help to B cells for antibody production	Provide help to B cells for antibody production, especially switching to IgE	Enhance neutrophil response	Suppress T-cell responses
Pathogens targeted	Viruses (e.g. influenza, rabies, vaccinia) Some intracellular bacteria	Microbes that persist in macrophage vesicles (e.g. mycobacteria, *Listeria*, *Leishmania donovani*, *Pneumocystis carinii*) Extracellular bacteria	Helminth parasites	Extracellular bacteria (e.g. *Salmonella enterica*)	

Fig. 8.1 The roles of effector T cells in cell-mediated and humoral immune responses. Cell-mediated immune responses are directed principally at intracellular pathogens. They involve the destruction of infected cells by cytotoxic CD8 T cells, or the destruction of intracellular pathogens in macrophages activated by CD4 T_H1 cells. CD4 T_H2 cells and T_H1 cells contribute to humoral immunity by stimulating the production of antibodies by B cells and inducing class switching. All classes of antibody contribute to humoral immunity, which is directed principally at extracellular pathogens. Both cell-mediated and humoral immunity are involved in many infections. CD4 T_H17 cells help to recruit neutrophils to sites of infection early in the adaptive immune response, which is also a response aimed mainly at extracellular pathogens. Regulatory T cells tend to suppress the adaptive immune response and are important in preventing immune responses from becoming uncontrolled and in preventing autoimmunity.

On antigen recognition, naive T cells differentiate into several functional classes of effector T cells that are specialized for different activities. CD8 T cells recognize pathogen peptides presented by MHC class I molecules, and naive CD8 T cells all differentiate into cytotoxic effector T cells that recognize and kill infected cells. CD4 T cells have a more flexible repertoire of effector activities. After recognition of pathogen peptides presented by MHC class II molecules, naive CD4 T cells can differentiate down distinct pathways that generate effector subsets with different immunological functions. The main CD4 effector subsets currently distinguished are T_H1, T_H2, and T_H17, which activate their target cells, and several regulatory T cell subsets that have inhibitory activity that limits the extent of immune activation (Fig. 8.1).

The activation of naive T cells in response to antigen, and their subsequent proliferation and differentiation into effector cells, constitutes a **primary cell-mediated immune response**. Effector T cells differ in many ways from their naive precursors, and these changes equip them to respond quickly and efficiently when they encounter specific antigen on target cells. In this chapter we will describe the specialized mechanisms of T cell-mediated cytotoxicity and of macrophage activation by effector T cells, which make up the major components of **cell-mediated immunity**. The other main function of effector T cells is to provide help to B cells to trigger antibody production. We will only touch on this in this chapter and will discuss it in detail in Chapter 9. At the same time as providing effector T cells, the primary T-cell response also generates **memory T cells**, long-lived cells that give an accelerated response to antigen, which yields protection from subsequent challenge by the same pathogen. We will discuss T-cell and B-cell immunological memory together in Chapter 10.

In this chapter, we will see how naive T cells are activated to proliferate and produce effector T cells the first time they encounter their specific antigen. The activation and clonal expansion of a naive T cell on its initial encounter with antigen is often called **priming**, to distinguish it from the responses of effector T cells to antigen on their target cells and the responses of primed memory T cells. The initiation of adaptive immunity is one of the most compelling narratives in immunology. As we will learn, the activation of naive T cells is controlled by a variety of signals: in the nomenclature developed by the late Charles Janeway and used in this book these are called signal 1, signal 2, and signal 3. A naive T cell recognizes antigen in the form of a peptide:MHC complex on the surface of a specialized antigen-presenting cell, as discussed in Chapter 5. Antigen-specific activation of the T-cell receptor delivers signal 1; interaction of **co-stimulatory molecules** on antigen-presenting cells with ligands on T cells delivers signal 2; and cytokines that control differentiation into different types of effector cells deliver signal 3. All these events are set in motion by much earlier signals that arise from the initial detection of the pathogens by the innate immune system. These signals, which were predicted and later identified by Janeway, are delivered to cells of the innate immune system by receptors such as the Toll-like receptors (TLRs), which recognize pathogen-associated molecular patterns that signify the presence of nonself (see Chapter 2). As we will see in this chapter, those signals are essential to activate antigen-presenting cells so that they are able, in turn, to activate naive T cells.

By far the most important antigen-presenting cells in the activation of naive T cells are the highly specialized **dendritic cells**, whose major function is to ingest and present antigen. Tissue dendritic cells take up antigen at sites of infection and are activated as part of the innate immune response. This induces their migration to local lymphoid tissue and their maturation into cells that are highly effective at presenting antigen to recirculating naive T cells. In the first part of this chapter we shall see how naive T cells and dendritic cells meet in the peripheral lymphoid organs, and how dendritic cells become activated to full antigen-presenting cell status.

Entry of naive T cells and antigen-presenting cells into peripheral lymphoid organs.

Adaptive immune responses are initiated in the peripheral lymphoid organs—lymph nodes, spleen, and the mucosa-associated lymphoid tissues such as the Peyer's patches in the gut. This means that for a T-cell immune response to be induced in response to an infection, the rare naive T cells specific for the appropriate antigens must meet dendritic cells presenting those antigens in a peripheral lymphoid organ. An infection can originate in virtually any site in the body, however, and so the pathogen's antigens must be brought to peripheral lymphoid organs. We shall see in this part of the chapter how dendritic cells pick up antigen at sites of infection and travel to local lymphoid organs, where they mature into cells that can both present antigen to T cells and activate them. Free antigens, such as bacteria and virus particles, also travel through lymphatics and in the blood directly to lymphoid organs, where they can be taken up and presented by antigen-presenting cells. As we learned in Chapter 1, naive T cells are continuously recirculating through the peripheral lymphoid tissues, surveying the antigen-presenting cells for foreign antigens. We shall look first at how this cellular traffic is orchestrated by chemotactic cytokines (chemokines) and adhesion molecules, which direct naive T cells out of the blood and into the lymphoid organs.

8-1 Naive T cells migrate through peripheral lymphoid tissues, sampling the peptide:MHC complexes on dendritic cell surfaces.

Naive T cells circulate from the bloodstream into lymph nodes, spleen and mucosa-associated lymphoid tissues and back to the blood (see Fig. 1.17 for the overall circulation in respect of a lymph node). This enables them to make contact with thousands of dendritic cells in the lymphoid tissues every

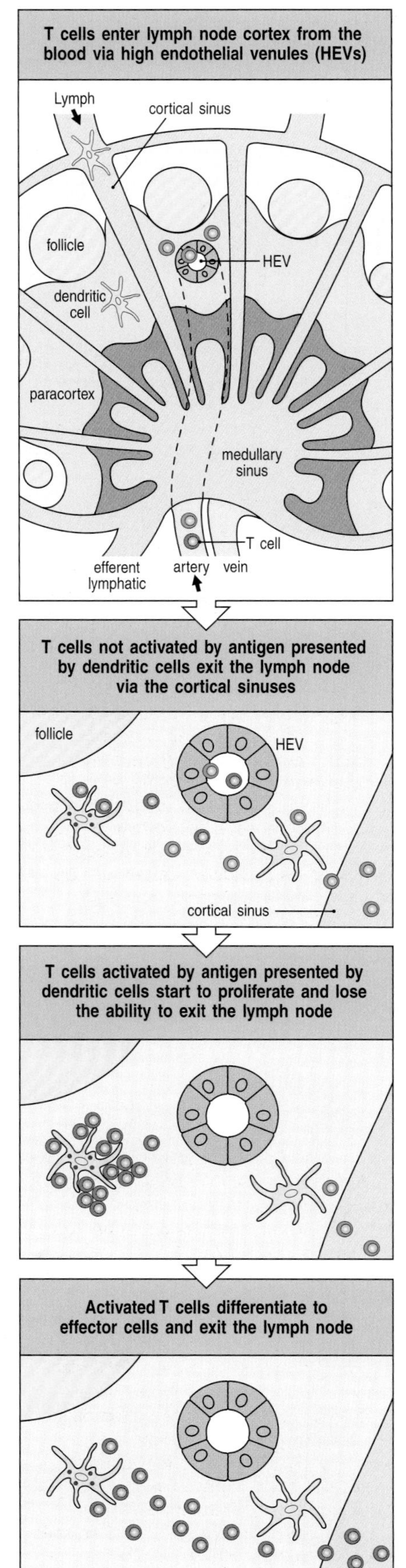
T cells enter lymph node cortex from the blood via high endothelial venules (HEVs)
Lymph
cortical sinus
follicle
HEV
dendritic cell
paracortex
medullary sinus
T cell
efferent lymphatic
artery
vein
T cells not activated by antigen presented by dendritic cells exit the lymph node via the cortical sinuses
follicle
HEV
cortical sinus
T cells activated by antigen presented by dendritic cells start to proliferate and lose the ability to exit the lymph node
Activated T cells differentiate to effector cells and exit the lymph node

day and sample the peptide:MHC complexes on the surfaces of the dendritic cells. Each T cell thus has a high probability of encountering antigens derived from any pathogens that have set up an infection in whatever location in the body (Fig. 8.2). Naive T cells that do not encounter their specific antigen exit from the lymphoid tissue via the efferent lymphatics, eventually reenter the bloodstream, and continue recirculating. When a naive T cell recognizes its specific antigen on the surface of a mature dendritic cell, however, it ceases to migrate. It proliferates for several days, undergoing **clonal expansion** and differentiation, and gives rise to a clone of effector T cells of identical antigen specificity. At the end of this period, the effector T cells become able to exit into the efferent lymphatics and reenter the bloodstream, through which they migrate to the sites of infection. The exception to this type of recirculation is the spleen, which has no connection with the lymphatic system; all cells enter the spleen from the blood and exit directly back into it.

The efficiency with which T cells screen each antigen-presenting cell in lymph nodes is very high, as can be seen by the rapid trapping of antigen-specific T cells in a single lymph node containing antigen: all of the antigen-specific T cells in a sheep were trapped in one lymph node within 48 hours of antigen deposition (Fig. 8.3). Such efficiency is crucial for the initiation of an adaptive immune response, as only one naive T cell in 10^4–10^6 is likely to be specific for a particular antigen, and adaptive immunity depends on the activation and expansion of these rare cells.

8-2 Lymphocyte entry into lymphoid tissues depends on chemokines and adhesion molecules.

The migration of naive T cells into peripheral lymphoid tissues depends on their binding to high endothelial venules (HEVs) through interactions that are not antigen-specific. These nonspecific cell–cell interactions are governed by cell-adhesion molecules, some of which we encountered in Chapter 2 when we considered the recruitment of neutrophils and monocytes to sites of infection in an innate immune response (see Fig. 2.12). In this chapter we shall mainly use lymph nodes and spleen as our examples. The circulation of lymphocytes through mucosal tissues and their activation there follow similar principles, but differ in some details; these will be described in Section 11-6.

The main classes of adhesion molecules involved in lymphocyte interactions are the selectins, the integrins, members of the immunoglobulin superfamily, and some mucin-like molecules. Entry of lymphocytes into lymph nodes occurs in distinct stages that include the initial rolling of lymphocytes along the endothelial surface, activation of integrins, firm adhesion, and transmigration or **diapedesis** across the endothelial layer into the paracortical areas,

Fig. 8.2 Naive T cells encounter antigen during their recirculation through peripheral lymphoid organs. Naive T cells recirculate through peripheral lymphoid organs, such as the lymph node shown here, entering from the arterial blood via the specialized vascular endothelium of high endothelial venules (HEVs). Entry into the lymph node is regulated by chemokines that direct the T cells' migration through the HEV wall and into the paracortical areas, where they encounter mature dendritic cells (top panel). Those T cells shown in green do not encounter their specific antigen; they receive a survival signal through their interaction with self-peptide:self-MHC complexes and IL-7, and leave the lymph node through the lymphatics to return to the circulation (second panel). T cells shown in blue encounter their specific antigen on the surface of mature dendritic cells, lose their ability to exit from the node and become activated to proliferate and to differentiate into effector T cells (third panel). After several days, these antigen-specific effector T cells regain the expression of receptors needed to exit the node, leave via the efferent lymphatics, and enter the circulation in greatly increased numbers (bottom panel).

the T-cell zones (Fig. 8.4). These stages are regulated by a coordinated interplay of adhesion molecules and chemokines. Most adhesion molecules have fairly broad roles in immune responses, being involved not only in lymphocyte migration but also in interactions between naive T cells and antigen-presenting cells, interactions between effector T cells and their targets, interactions of other types of leukocytes with endothelium (such as the entry of monocytes and neutrophils into infected tissue), and T-cell–B-cell interactions.

The selectins (Fig. 8.5) are important for specifically guiding leukocytes to particular tissues, a phenomenon known as leukocyte **homing**. **L-selectin** (CD62L) is expressed on leukocytes, whereas P-selectin (CD62P) and E-selectin (CD62E) are expressed on vascular endothelium (see Section 2-25). L-selectin on naive T cells guides their exit from the blood into peripheral lymphoid tissues by initiating a light attachment to the wall of the HEV that results in the T cells rolling along the endothelium surface (see Fig. 8.4). P-selectin and E-selectin are expressed on the vascular endothelium at sites of infection and serve to recruit effector cells into the infected tissue. Selectins are cell-surface molecules with a common core structure, distinguished from each other by the presence of different lectin-like domains in their extracellular portion (see Fig. 2.48). The lectin domains bind to particular sugar groups, and each selectin binds to a cell-surface carbohydrate. L-selectin binds to the carbohydrate moiety—sulfated sialyl-Lewis^{x}—of mucin-like molecules called **vascular addressins**, which are expressed on the surface of vascular endothelial cells. Two of these addressins, **CD34** and **GlyCAM-1** (see Fig. 8.5), are expressed on high endothelial venules in lymph nodes. A third, **MAdCAM-1** (see Fig. 8.5), is expressed on endothelium in mucosa, and guides lymphocyte entry into mucosal lymphoid tissue such as the Peyer's patches in the gut.

The interaction between L-selectin and the vascular addressins is responsible for the specific homing of naive T cells to lymphoid organs. On its own, however, it does not enable the cell to cross the endothelial barrier into the lymphoid tissue. This requires a concerted interaction between integrins and chemokines.

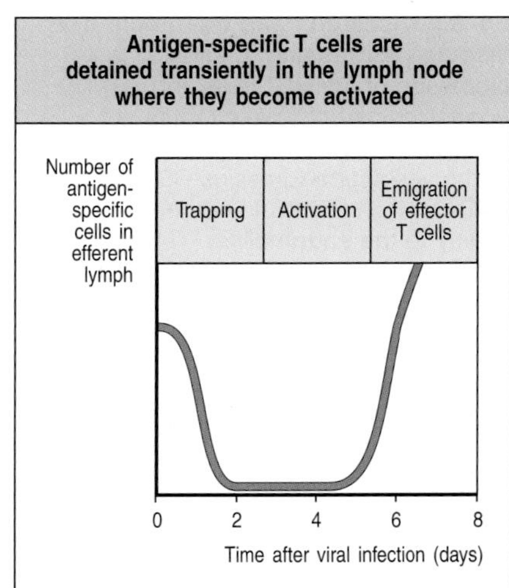

Fig. 8.3 Trapping and activation of antigen-specific naive T cells in lymphoid tissue. Naive T cells entering the lymph node from the blood encounter antigen-presenting dendritic cells in the lymph node cortex. T cells that recognize their specific antigen bind stably to the dendritic cells and are activated through their T-cell receptors, resulting in the production of effector T cells. By 5 days after the arrival of antigen, activated effector T cells are leaving the lymph node in large numbers via the efferent lymphatics. Lymphocyte recirculation and recognition are so effective that all the naive T cells in the peripheral circulation specific for a particular antigen can be trapped by that antigen in one node within 2 days.

8-3 Activation of integrins by chemokines is responsible for the entry of naive T cells into lymph nodes.

Entry of naive T cells into lymph nodes and other peripheral lymphoid organs requires the actions of two other protein families—the integrins and the

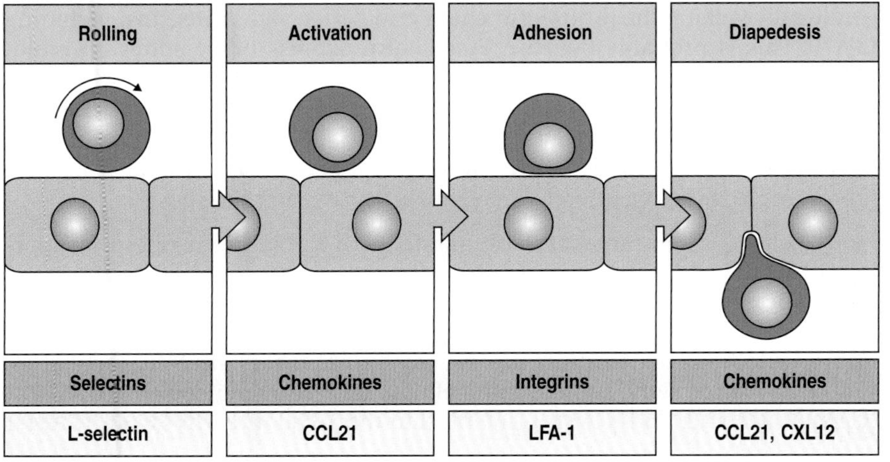

Fig. 8.4 Lymphocyte entry into a lymph node from the blood occurs in distinct stages involving the activity of adhesion molecules, chemokines and chemokine receptors. Naive T cells are induced to roll along the surface of a high endothelial venule (HEV) by the interactions of selectins expressed by the T cells with vascular addressins on the endothelial cell membranes. Chemokines present at the HEV surface activate receptors on the T cell, and chemokine signaling induces an increase in the affinity of T-cell integrins for adhesion molecules on the HEV. This induces strong adhesion. After adhesion, the T cells follow gradients of chemokines to pass through the HEV wall into the paracortical region of the lymph node.

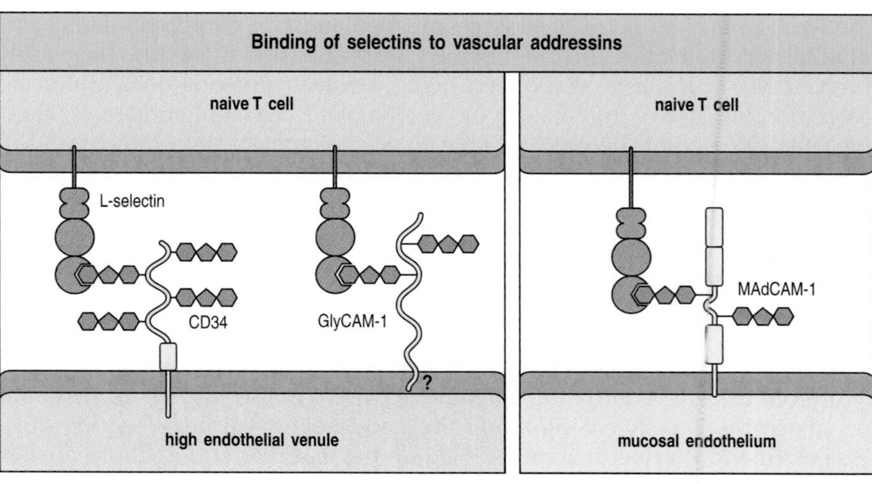

Fig. 8.5 L-selectin and the mucin-like vascular addressins. L-selectin is expressed on naive T cells and recognizes carbohydrate motifs. Its binding to sulfated sialyl-Lewis^x moieties on the vascular addressins CD34 and GlyCAM-1 on HEVs binds the lymphocyte weakly to the endothelium. The relative importance of CD34 and GlyCAM-1 in this interaction is unclear. GlyCAM-1 is expressed exclusively on HEVs but has no transmembrane region and it is unclear how it is attached to the membrane; CD34 has a transmembrane anchor and is expressed in appropriately glycosylated form only on HEV cells, although it is found in other forms on other endothelial cells. The addressin MAdCAM-1 is expressed on mucosal endothelium and guides lymphocytes to mucosal lymphoid tissue. The icon shown represents mouse MAdCAM-1, which contains an IgA-like domain closest to the cell membrane; human MAdCAM-1 has an elongated mucin-like domain and lacks the IgA-like domain.

immunoglobulin superfamily. These proteins also have a crucial role in the subsequent interactions of lymphocytes with antigen-presenting cells and later with their target cells. The integrins are a large family of cell-surface proteins involved in adhesion between cells, and between cells and the extracellular matrix. Integrins bind tightly to their ligands after receiving signals that induce a change in their conformation. For example, as we saw in Chapter 2, signaling by chemokines activates integrins on leukocytes to bind tightly to the vascular surface in preparation for their migration into sites of inflammation. Similarly, chemokines that are present at the luminal surface of the HEV activate integrins expressed on the naive T cell during their migration into lymphoid organs (see Fig. 8.4).

The integrins were introduced in Chapter 2, so we will just review their most important characteristics here. An integrin molecule consists of a large α chain that pairs noncovalently with a smaller β chain. There are several subfamilies of integrins, broadly defined by their common β chains. We will be concerned chiefly with the **leukocyte integrins**, which have a common β_2 chain with distinct α chains (Fig. 8.6). All T cells express the β_2 integrin α_L:β_2 (CD11a:CD18), better known as leukocyte functional antigen-1 (LFA-1). This leukocyte integrin is also present on macrophages and neutrophils, and is involved in their recruitment to sites of infection (see Section 2-25). LFA-1 has a similar role in both naive and effector T cells in enabling their migration out of the blood.

LFA-1 is also important in the adhesion of both naive and effector T cells to their target cells. Nevertheless, T-cell responses can be normal in individuals genetically lacking the β_2 integrin chain and hence all β_2 integrins, including LFA-1. This is probably because T cells also express other adhesion molecules, including the immunoglobulin superfamily member CD2 and β_1 integrins, that may be able to compensate for the absence of LFA-1. Expression of the β_1 integrins increases significantly at a late stage in T-cell activation, and they are thus often called **VLAs**, for **very late activation antigens**; they are important in directing effector T cells to inflamed target tissues.

Many cell-adhesion molecules are members of the immunoglobulin superfamily, which also includes the antigen receptors of T and B cells, the T-cell co-receptors CD4 and CD8, the B-cell co-receptor component CD19, and the invariant domains of MHC molecules. At least five adhesion molecules of the immunoglobulin superfamily are especially important in T-cell activation (Fig. 8.7). Three very similar intercellular adhesion molecules (ICAMs)—ICAM-1, ICAM-2, and ICAM-3—all bind to the T-cell integrin LFA-1. ICAM-1

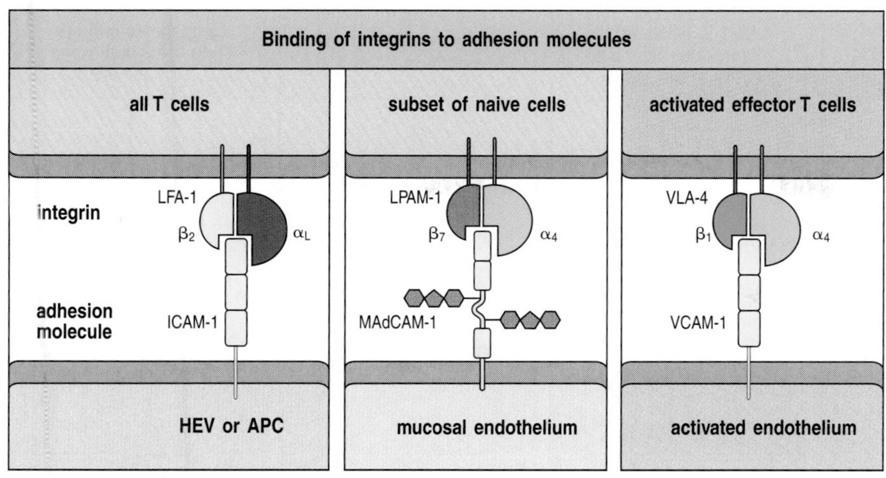

Fig. 8.6 Integrins are important in T-lymphocyte adhesion. Integrins are heterodimeric proteins containing a β chain, which defines the class of integrin, and an α chain, which defines the different integrins within a class. The α chain is larger than the β chain and contains binding sites for divalent cations that may be important in signaling. LFA-1 (integrin α_L:β_2) is expressed on all leukocytes. It binds ICAMs and is important in cell migration and in the interactions of T cells with antigen-presenting cells (APCs) or target cells; it is expressed at higher levels on effector T cells than on naive T cells. Lymphocyte Peyer's patch adhesion molecule (LPAM-1 or integrin α_4:β_7) is expressed by a subset of naive T cells and contributes to lymphocyte entry into mucosal lymphoid tissues by supporting adhesive interactions with vascular addressin MAdCAM-1. VLA-4 (integrin α_4:β_1) is expressed strongly after T-cell activation. It binds to VCAM-1 on activated endothelium and is important for recruiting effector T cells into sites of infection.

and ICAM-2 are expressed on endothelium as well as on antigen-presenting cells; binding to these molecules enables lymphocytes to migrate through blood vessel walls. ICAM-3 is expressed only on naive T cells and is thought to have an important role in the adhesion of T cells to antigen-presenting cells, particularly dendritic cells. In addition to binding LFA-1, ICAM-3 binds with high affinity to a lectin called **DC-SIGN** that is found only on dendritic cells. The two remaining immunoglobulin superfamily adhesion molecules, **CD58** (formerly known as LFA-3) on the antigen-presenting cell and CD2 on the T cell, bind to each other; this interaction synergizes with that of ICAM-1 or ICAM-2 with LFA-1.

As in phagocyte migration, naive T cells are specifically attracted into the lymph node by chemokines secreted by cells in the lymph node. The chemokines bind to proteoglycans in the extracellular matrix and high endothelial venule wall, forming a chemical gradient, and are recognized by receptors on the naive T cell. The extravasation of naive T cells is prompted by the chemokine **CCL21** (secondary lymphoid tissue chemokine, SLC). This is expressed by vascular high endothelial cells and the stromal cells of lymphoid tissues, and binds to the chemokine receptor **CCR7** on naive T cells, stimulating activation of the intracellular receptor-associated G-protein subunit $G\alpha_i$ (see Section 6-28). The resulting intracellular signaling rapidly increases the affinity of integrin binding by a mechanism that is not yet fully worked out.

The entry of a naive T cell into a lymph node is shown in detail in Fig. 8.8. Initial rolling along the high endothelial venule surface is mediated by L-selectin. Contact of naive T cells with the CCL21 in the high endothelial

Immunoglobulin superfamily

Various roles in cell adhesion. Ligands for integrins

CD2

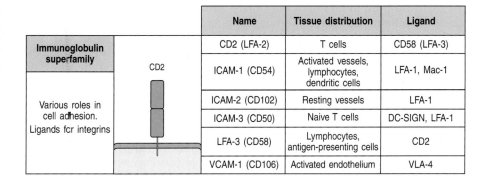

Name	Tissue distribution	Ligand
CD2 (LFA-2)	T cells	CD58 (LFA-3)
ICAM-1 (CD54)	Activated vessels, lymphocytes, dendritic cells	LFA-1, Mac-1
ICAM-2 (CD102)	Resting vessels	LFA-1
ICAM-3 (CD50)	Naive T cells	DC-SIGN, LFA-1
LFA-3 (CD58)	Lymphocytes, antigen-presenting cells	CD2
VCAM-1 (CD106)	Activated endothelium	VLA-4

Fig. 8.7 Immunoglobulin superfamily adhesion molecules involved in leukocyte interactions. Adhesion molecules of the immunoglobulin superfamily bind to adhesion molecules of various types, including integrins (LFA-1 and VLA-4), other immunoglobulin superfamily members (the CD2–CD58 (LFA-3) interaction) and lectins (DC-SIGN). These interactions have a role in lymphocyte migration, homing, and cell–cell interactions; most of the molecules listed here have been introduced in Fig. 2.47.

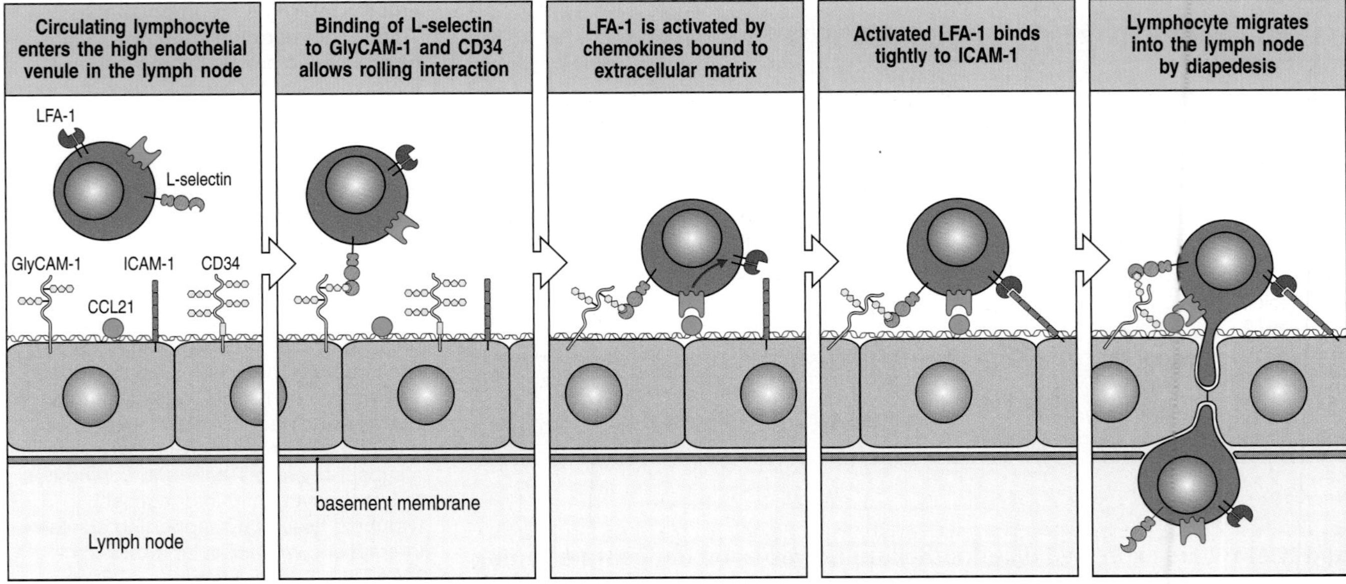
Circulating lymphocyte enters the high endothelial venule in the lymph node
Binding of L-selectin to GlyCAM-1 and CD34 allows rolling interaction
LFA-1 is activated by chemokines bound to extracellular matrix
Activated LFA-1 binds tightly to ICAM-1
Lymphocyte migrates into the lymph node by diapedesis
LFA-1
L-selectin
GlyCAM-1
ICAM-1
CD34
CCL21
basement membrane
Lymph node

Fig. 8.8 Lymphocytes in the blood enter lymphoid tissue by crossing the walls of high endothelial venules. The first step is the binding of L-selectin on the lymphocyte to sulfated carbohydrates (sulfated sialyl-Lewis^x) of GlyCAM-1 and CD34 on the HEV. Local chemokines such as CCL21 bound to a proteoglycan matrix on the HEV surface stimulate chemokine receptors on the T cell, leading to the activation of LFA-1. This causes the T cell to bind tightly to ICAM-1 on the endothelial cell, allowing migration across the endothelium. As in the case of neutrophil migration (see Fig. 2.49), matrix metalloproteinases on the lymphocyte surface (not shown) enable it to penetrate the basement membrane.

venule causes the integrin LFA-1 on the naive T cell to become activated, increasing its affinity for ICAM-2 and ICAM-1. ICAM-2 is expressed constitutively on all endothelial cells, whereas in the absence of inflammation, ICAM-1 is expressed only on the high endothelial cells of peripheral lymphoid tissues. The mobility of integrin in the T-cell membrane is also increased by chemokine stimulation, so that integrin molecules migrate into the area of cell–cell contact. This produces stronger binding, which arrests the T cell on the endothelial surface and thus enables it to enter the lymphoid tissue.

The interplay of chemokines and cell-adhesion molecules, together with the architecture of the peripheral lymphoid organs (see Figs 1.18–1.20), virtually guarantees the contact of foreign antigen with the T-cell receptors specific for it. Once naive T cells have arrived in the T-cell zone via the high endothelial venules, CCR7 directs their migration toward sources of CCL21, made by stromal cells in the T-cell zones, and **CCL19**, a second chemokine ligand for CCR7, which is made by T-cell zone stromal cells and to a lesser degree by dendritic cells, which are also most concentrated in the areas traversed by T cells. Mature dendritic cells also produce the chemokine **CCL18** (**DC-CK**), which attracts naive T cells. Once in the T-cell zone, the naive T cells scan the surfaces of dendritic cells for specific peptide:MHC complexes. If they find their antigen and bind to it, they are trapped in the lymph node. If they are not activated by antigen, naive T cells soon leave the lymph node (see Fig. 8.2).

T cells exit from a lymph node via the cortical sinuses, which lead into the medullary sinus and hence into the efferent lymphatic vessel. The egress of T cells from peripheral lymphoid organs involves the lipid molecule **sphingosine 1-phosphate** (**S1P**). This has chemotactic activity and signaling properties similar to those of chemokines, in that the receptors for S1P are G-protein-coupled receptors; S1P signaling activates $G\alpha_i$. S1P is produced by phosphorylation of the cell lipid sphingosine, and can be degraded by S1P lyases or by S1P phosphatases. There seems to be a S1P concentration gradient between the lymphoid tissues and lymph or blood, such that naive T cells expressing a S1P receptor are drawn away from the lymphoid tissues and back into circulation.

T cells activated by antigen in the lymphoid organs downregulate the expression of S1P receptors for several days, and thus cannot respond to the S1P gradient; they therefore do not exit during this period. After several days of proliferation, the effector T cells re-express S1P receptors and are once again able to migrate in response to the S1P gradient. The regulation of the exit of both naive and effector lymphocytes from peripheral lymphoid organs by S1P is the basis for a new kind of immunosuppressive drug, FTY720. FTY720 inhibits immune responses in animal models of transplantation and autoimmunity by preventing lymphocytes from returning to the circulation, causing rapid lymphopenia. *In vivo*, FTY720 becomes phosphorylated, and mimics S1P as an agonist at S1P receptors. Phosphorylated FTY720 may inhibit lymphocyte exit by effects on endothelial cells that increase tight junction formation and close exit portals or by chronic activation of S1P receptors, leading to inactivation and downregulation of the receptor.

8-4 T-cell responses are initiated in peripheral lymphoid organs by activated dendritic cells.

Peripheral lymphoid organs were first shown to be important in the initiation of adaptive immune responses by ingenious experiments in which a flap of skin was isolated from the body wall so that it had a blood circulation but no lymphatic drainage. Antigen placed in the flap did not elicit a T-cell response, showing that T cells do not become sensitized in the infected tissue itself. Pathogens and their products must therefore be transported to lymphoid tissues. Antigens introduced directly into the bloodstream are picked up by antigen-presenting cells in the spleen. Pathogens infecting other sites, such as a wound in the skin, are transported in lymph and trapped in the lymph nodes nearest to the site of infection (see Section 1-15). Pathogens infecting mucosal surfaces are transported directly across the mucosa into lymphoid tissues such as the tonsils or the Peyer's patches of the gut.

The delivery of antigen from a site of infection to the nearest lymphoid tissue is actively aided by the innate immune response. One response of innate immunity is an inflammatory reaction at the site of infection that increases the rate of entry of blood plasma into the infected tissues and thus increases the drainage of extracellular fluid into the lymph, taking with it free antigen that is then carried to lymphoid tissues. Even more important for the initiation of the adaptive response is the induced maturation of tissue dendritic cells that have taken up particulate and soluble antigens at the site of infection (Fig. 8.9). Immature dendritic cells that reside in the tissues can be activated via their TLRs, which signal the presence of pathogens (see Fig. 2.16), or by tissue damage, or by cytokines produced during the inflammatory response. Dendritic cells respond to these signals by migrating to the lymph node and expressing the co-stimulatory molecules that are required, in addition to antigen, for the activation of naive T cells. In the lymphoid tissues these mature dendritic cells present antigen to naive T lymphocytes and activate any antigen-specific T cells to divide and mature into effector cells that reenter the circulation.

Macrophages, which are found in most tissues including lymphoid tissue, and B cells, which are located primarily in lymphoid tissue, can be similarly induced through the same antigen-nonspecific receptors to express co-stimulatory molecules and act as antigen-presenting cells. The distribution of dendritic cells, macrophages, and B cells in a lymph node is shown schematically in Fig. 8.10. Only these three cell types express the specialized co-stimulatory molecules required to activate naive T cells; furthermore, they all express these molecules only when activated in the context of infection. Dendritic cells can take up, process, and present antigens from all types of sources, are present mainly in the T-cell areas, and overwhelmingly drive the initial clonal expansion and differentiation of naive T cells into effector

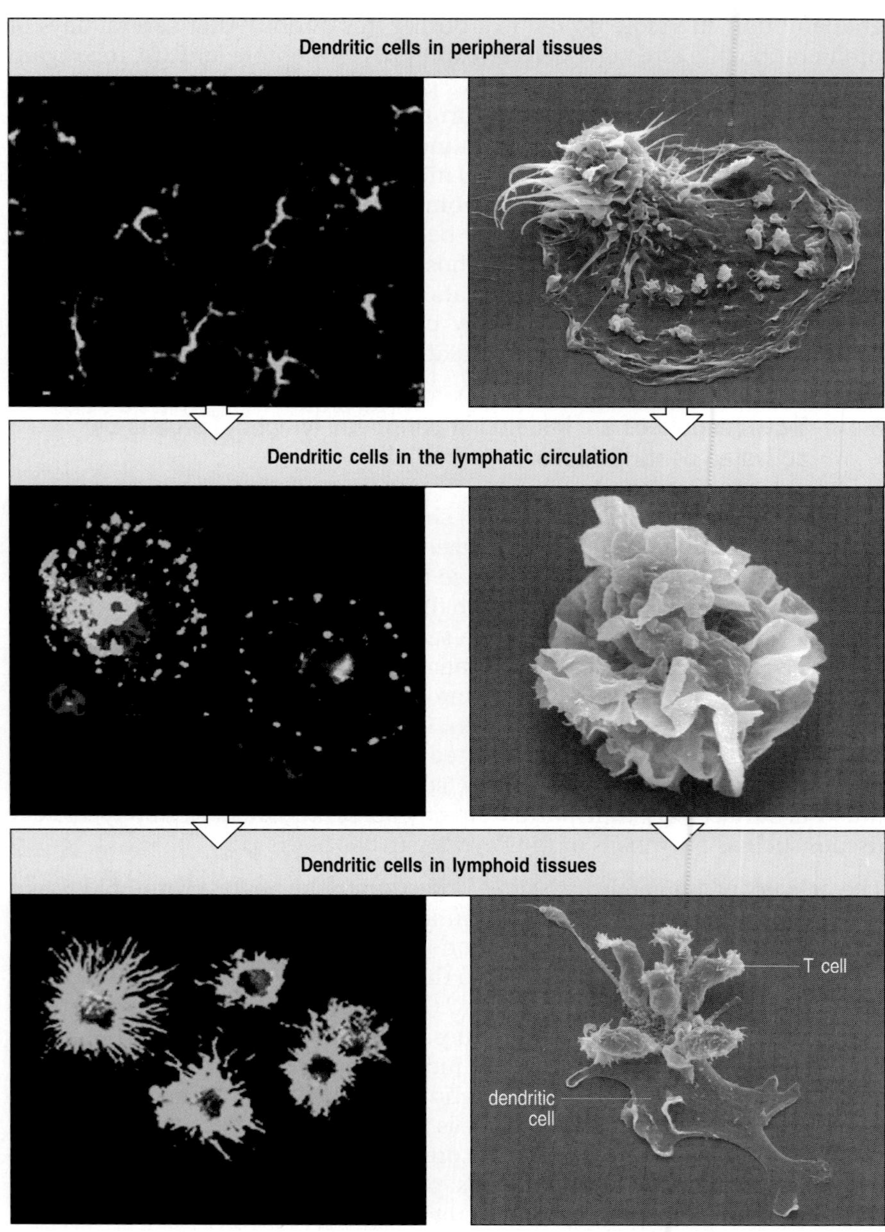

Fig. 8.9 Dendritic cells in different stages of maturation. The left panels show fluorescence micrographs of dendritic cells stained for MHC class II molecules in green and for a lysosomal protein in red. The right panels show scanning electron micrographs of single dendritic cells. Immature dendritic cells (top panels) have many long processes, or dendrites, from which the cells get their name. The cell bodies are difficult to distinguish in the left panel, but the cells contain many endocytic vesicles that stain both for MHC class II molecules and for the lysosomal protein; when these two colors overlap they give rise to a yellow fluorescence. The immature cells are activated and leave the tissues to migrate through the lymphatics to secondary lymphoid tissues. During this migration their morphology changes. The dendritic cells stop phagocytosing antigen, and the staining for lysosomal proteins is beginning to be distinct from that for MHC class II molecules (center left panel). The dendritic cell now has many folds of membrane (right panel), which gave these cells their original name of 'veil' cells. Finally, in the lymph nodes, they become mature dendritic cells that express high levels of peptide:MHC complexes and co-stimulatory molecules, and are very good at stimulating naive CD4 and naive CD8 T cells. These cells do not phagocytose and the red staining of lysosomal proteins is quite distinct from the green-stained MHC class II molecules displayed at high density on many dendritic processes (bottom left panel). The typical morphology of a mature dendritic cell is shown on the right, as it interacts with a T cell. Fluorescent micrographs courtesy of I. Mellman, P. Pierre, and S. Turley. Scanning electron micrographs courtesy of K. Dittmar.

T cells. Macrophages and B cells specialize in processing and presenting antigens from ingested pathogens and soluble antigens, respectively, and mainly interact with already primed effector CD4 T cells.

8-5 There are two different functional classes of dendritic cells.

Dendritic cells arise from both myeloid and lymphoid progenitors within the bone marrow; they emerge from the bone marrow to migrate via the blood to tissues throughout the body, and also directly to peripheral lymphoid organs. At least two broad classes of dendritic cells are recognized: the so-called **conventional dendritic cells** (**cDC**), by which we mean those dendritic cells that seem to participate most directly in antigen presentation and activation of naive T cells; and **plasmacytoid dendritic cells** (**pDC**), a distinct lineage that

Fig. 8.10 Antigen-presenting cells are distributed differentially in the lymph node. Dendritic cells are found throughout the cortex of the lymph node in the T-cell areas. Macrophages are distributed throughout but are found mainly in the marginal sinus, where the afferent lymph collects before percolating through the lymphoid tissue, and also in the medullary cords, where the efferent lymph collects before passing via the efferent lymphatics into the blood. B cells are found mainly in the follicles. The three types of antigen-presenting cell are thought to be adapted to present different types of pathogen or pathogen products, but mature dendritic cells are by far the strongest activators of naive T cells.

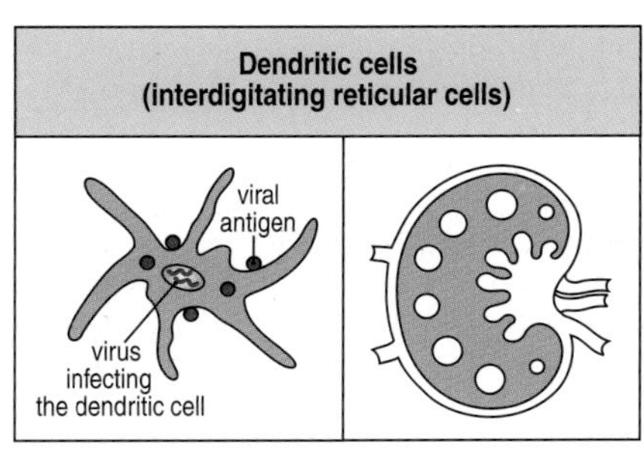

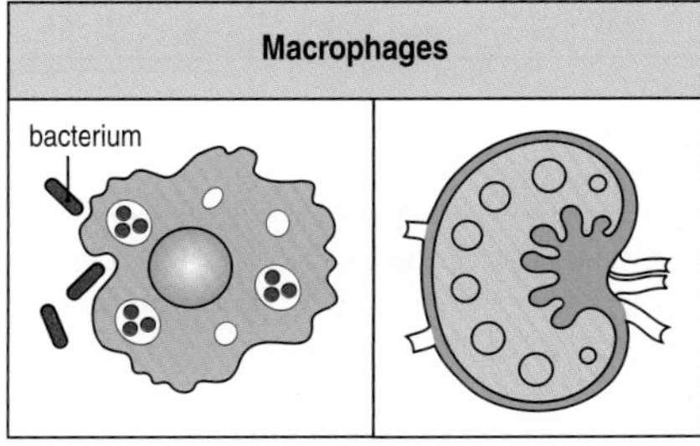

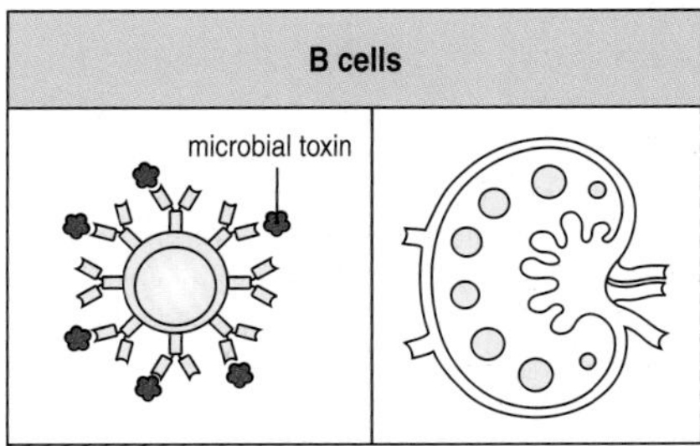

generate large amounts of interferons, particularly in response to viral infections, but do not seem to be as important for activating naive T cells (Fig. 8.11). Throughout this book, all mentions of dendritic cells refer to conventional dendritic cell function unless noted otherwise.

Dendritic cells can be identified by their expression of specific surface molecules. Dendritic cells, macrophages, and monocytes express different integrin α chains and thus display distinct β_2 integrins on their surface. The predominant leukocyte integrin on conventional dendritic cells is α_X:β_2, also known as **CD11c:CD18** or complement receptor 4 (CR4). This integrin is a receptor for the complement C3 cleavage product iC3b, fibrinogen, and ICAM-1. In the mouse, CD11c-positive dendritic cells can be further divided into three subsets expressing CD4, or the CD8α homodimer, or neither. It is not yet clear whether the differential expression of these markers is functionally significant, but these 'CD11c-bright' dendritic-cell subsets may differ in the production of cytokines such as IL-12, which could have effects on the subsequent adaptive immune response, as we shall see. In contrast, monocytes and macrophages express low levels of CD11c, and predominantly express the integrin α_M:β_2, also called **CD11b:CD18** or **Mac-1**. Plasmacytoid dendritic cells also do not express high levels of CD11c, and have been identified by the expression of specific markers, such as blood dendritic cell antigen 2 (BDCA-2, a C-type lectin) in humans, or the sialic-acid binding immunoglobulin-like lectin H (Siglec-H) in mice, both of which may function in pathogen recognition.

Dendritic cells are found under most surface epithelia, and in most solid organs such as the heart and kidneys. There they have an immature phenotype that is associated with low levels of MHC proteins and the B7 co-stimulatory molecules (see Section 2-10), and so are not yet equipped to stimulate naive T cells. Immature dendritic cells also share with their close relatives, the macrophages, the ability to recognize and ingest pathogens through receptors that recognize pathogen-associated molecular patterns, and they are very active in taking up antigens by phagocytosis by means of receptors

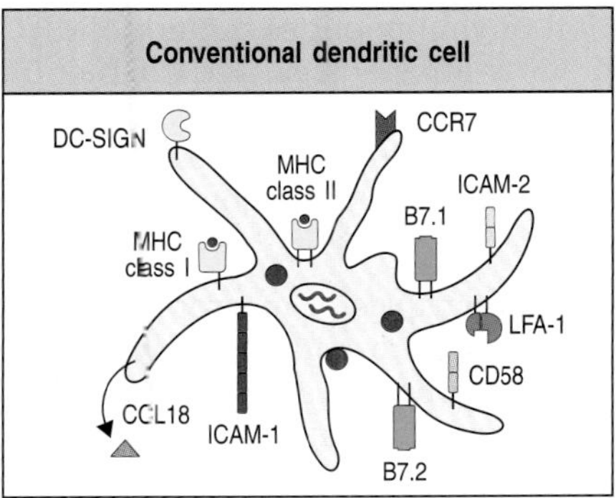

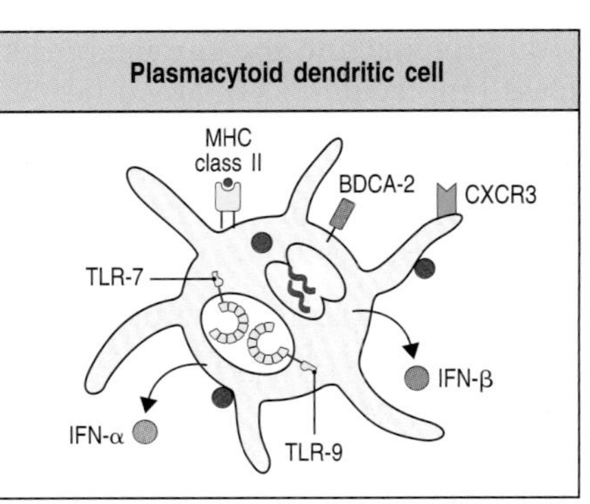

Fig. 8.11 Conventional and plasmacytoid dendritic cells have different roles in the immune response. Mature conventional dendritic cells (left panel) are primarily concerned with the activation of naive T cells. There are several subsets of conventional dendritic cells, but these all process antigen efficiently, and when they are mature they express MHC proteins and co-stimulatory molecules for priming naive T cells. The cell-surface proteins expressed by the mature dendritic cell are described in the text. Immature dendritic cells do not bear many of the cell-surface molecules shown here but have numerous surface receptors that recognize pathogen molecules, including most of the Toll-like receptors (TLRs). Plasmacytoid dendritic cells (right panel) are sentinels primarily for viral infections and secrete large amounts of class I interferons. This category of dendritic cell is less efficient in priming naive T cells but they express the intracellular receptors TLR-7 and TLR-9 for sensing viral infections.

Routes of antigen processing and presentation by dendritic cells					
	Receptor-mediated phagocytosis	Macropinocytosis	Viral infection	Cross-presentation after phagocytic or macropinocytic uptake	Transfer from incoming dendritic cell to resident dendritic cell
Type of pathogen presented	Extracellular bacteria	Extracellular bacteria, soluble antigens, virus particles	Viruses	Viruses	Viruses
MHC molecules loaded	MHC class II	MHC class II	MHC class I	MHC class I	MHC class I
Type of naive T cell activated	CD4 T cells	CD4 T cells	CD8 T cells	CD8 T cells	CD8 T cells

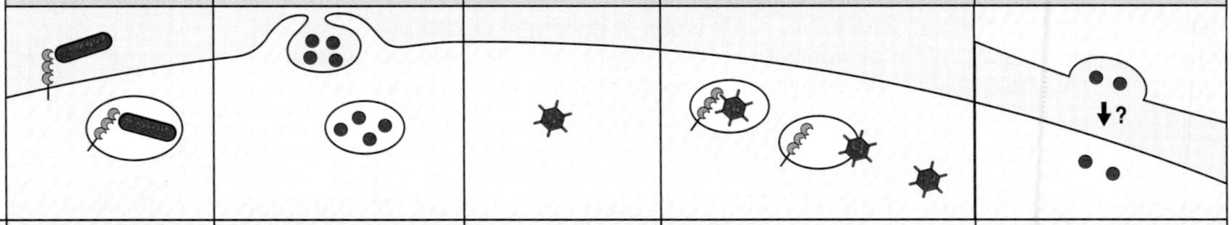

Fig. 8.12 The different routes by which dendritic cells can take up, process, and present protein antigens. Uptake of antigens into the endocytic system, either by receptor-mediated phagocytosis or by macropinocytosis, is considered to be the major route for delivering peptides to MHC class II molecules for presentation to CD4 T cells (first two panels). Production of antigens in the cytosol, for example as a result of viral infection, is thought to be the major route for delivering peptides to MHC class I molecules for presentation to CD8 T cells (third panel). It is possible, however, for exogenous antigens taken into the endocytic pathway to be delivered into the cytosol for eventual delivery to MHC class I molecules for presentation to CD8 T cells, a process called cross-presentation (fourth panel). Finally, antigens seem to be transmitted from one dendritic cell to another for presentation to CD8 T cells, although the details of this route are still unclear (fifth panel).

such as the lectin DEC 205. Other extracellular antigens are taken up non-specifically by the process of **macropinocytosis**, in which large volumes of surrounding fluid are engulfed.

8-6 Dendritic cells process antigens from a wide array of pathogens.

Their various mechanisms for taking up extracellular material enable dendritic cells to present antigens from virtually any type of pathogen (Fig. 8.12). The first route is via phagocytic receptors such as the mannose receptor and DEC 205. These receptors recognize a wide variety of bacteria and viruses. Antigens taken up in this way enter the endocytic pathway, where they can be processed and presented on MHC class II molecules (see Chapter 5) for recognition by CD4 T cells. Some microbes have evolved to escape recognition by phagocytic receptors (see Chapter 2), but these pathogens can be taken up by tissue dendritic cells by macropinocytosis and enter the endocytic pathway that way (see Fig. 8.12).

A second route is entry directly into the cytosol, for example through virus infection. Dendritic cells are particularly important in stimulating T-cell responses to viruses, which fail to induce co-stimulatory activity in other types of antigen-presenting cells. Dendritic cells are susceptible to infection by quite a large number of viruses, which enter the cell by binding to cell-surface proteins that act as entry receptors for the virus. Such viruses enter the dendritic cell's cytoplasm and synthesize their proteins using the dendritic cell's protein-synthesis machinery, leading to the proteasomal processing of viral proteins and the surface presentation of viral peptides bound to MHC class I molecules, as in any other type of virus-infected cell (see Chapter 5). This enables dendritic cells to present antigen to and activate naive CD8 T cells, whose T-cell receptors recognize antigen presented on MHC class I molecules. Effector CD8 T cells are cytotoxic cells that can recognize and kill virus-infected cells.

Uptake of extracellular virus particles by phagocytosis or macropinocytosis into the endocytic pathway can also result in the presentation of viral peptides on MHC class I molecules, a phenomenon known as cross-presentation. This is due to antigen processing via an alternative pathway to the usual endocytic pathway, as described in Section 5-4. By this pathway, viruses that

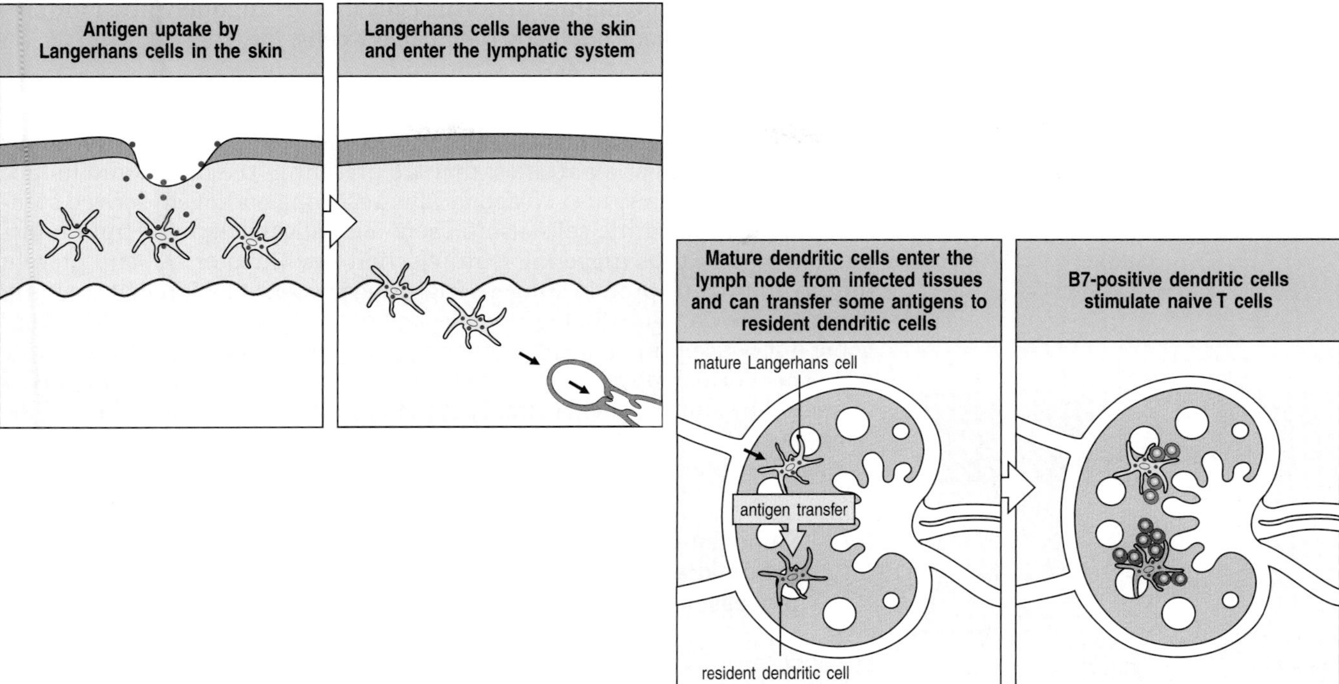

Antigen uptake by Langerhans cells in the skin
Langerhans cells leave the skin and enter the lymphatic system
Mature dendritic cells enter the lymph node from infected tissues and can transfer some antigens to resident dendritic cells
mature Langerhans cell
antigen transfer
resident dendritic cell
B7-positive dendritic cells stimulate naive T cells

Fig. 8.13 Langerhans cells take up antigen in the skin, migrate to the peripheral lymphoid organs, and present foreign antigens to T cells. Langerhans cells (yellow) are immature dendritic cells. They ingest antigen in various ways but have no co-stimulatory activity (first panel). In the presence of infection, they take up antigen locally and then migrate to the lymph nodes (second panel). There they differentiate into dendritic cells that can no longer ingest antigen but have co-stimulatory activity. Now they can prime both naive CD8 or CD4 T cells. In the case of some viral infections, such as herpes simplex virus, some dendritic cells arriving from the site of infection seem able to transfer antigen to resident dendritic cells (orange) in the lymph nodes (third panel) for presentation of class I MHC-restricted antigens to naive CD8 T cells (fourth panel).

are not able to infect dendritic cells can still stimulate effective antiviral responses by CD8 T cells. Thus, any virus infection can lead to the generation of cytotoxic CD8 effector T cells. In addition, viral peptides presented on the dendritic cell's MHC class II molecules will activate naive CD4 T cells, which will lead to the production of CD4 effector T cells that stimulate the production of antiviral antibodies by B cells and produce cytokines that enhance the immune response.

In some cases, such as infections with herpes simplex or influenza viruses, the dendritic cells that migrate to the lymph nodes from peripheral tissues may not be the same cells that finally present antigen to naive T cells. In herpes simplex infection, for example, immature dendritic cells resident in the skin, called Langerhans cells, capture antigen in the skin and transport it to the draining lymph nodes (Fig. 8.13). There, some antigen is transferred to a CD8-positive subset of dendritic cells resident in the lymph node, which seem to be the dominant dendritic cells responsible for priming naive CD8 T cells for development into antiviral cytotoxic T cells in this disease. This means that antigens from viruses that infect but rapidly kill dendritic cells can still be presented by uninfected dendritic cells that take up antigen by cross-presentation and have been activated via their TLRs and by chemokines.

Langerhans cells are typical of immature conventional dendritic cells. They are actively phagocytic and contain large granules known as Birbeck granules. These are an endosomal recycling compartment that is formed where langerin, a transmembrane lectin with mannose-binding specificity, accumulates. In the presence of an infection in the skin, the Langerhans cells will pick up pathogen antigens by any of the routes outlined above. Encounter with pathogens also triggers their migration to the regional lymph nodes (see Fig. 8.13). Here they rapidly lose the ability to take up antigen but briefly increase the synthesis of MHC molecules. On arriving in the lymph node they also express co-stimulatory B7 molecules and large numbers of adhesion molecules, which enable them to interact with antigen-specific T cells. In this way the Langerhans cells capture antigens from invading pathogens and differentiate into mature dendritic cells that are uniquely fitted for presenting these antigens and activating naive T cells.

Dendritic cells are believed to present antigens from fungi and parasites as well as from viruses and bacteria. For example, immature dendritic cells resident in the spleen are ideally suited to sample antigens from infectious agents, such as malaria parasites, that are present in the blood, and induce strong T-cell immunity to such agents after receiving pathogen-derived stimuli to mature. Dendritic cells also present alloantigens deriving from transplanted organs, thus triggering graft rejection (see Chapter 14), and present the environmental protein antigens triggering the sensitization that results in allergies (see Chapter 13). In principle, any nonself antigen will be immunogenic if it is taken up and subsequently presented by an activated dendritic cell. The normal physiology of dendritic cells is to migrate, and this is increased by stimuli, such as transplantation, that activate the linings of the lymphatics; this is why dendritic cells are so potent at stimulating reactions against transplanted tissues.

8-7 Pathogen-induced TLR signaling in immature dendritic cells induces their migration to lymphoid organs and enhances antigen processing.

We shall now look at the steps in dendritic cell maturation in more detail. Working together in ways that are not yet completely understood, TLR signaling and signals received from chemokines convert the immature dendritic cell residing in peripheral tissues into mature dendritic cells arriving in the lymphoid tissues. When an infection occurs, dendritic cells recognize pathogen molecules such as bacterial lipopolysaccharide (LPS) or mannose residues by means of receptors such as TLRs and DEC 205, and this triggers their activation (Fig. 8.14, top panel). These signals are of key importance in determining whether an adaptive immune response will be initiated. Multiple members of the TLR family are expressed on tissue dendritic cells and are thought to be involved in detecting and signaling the presence of the various classes of pathogens (see Fig. 2.16). In humans, conventional dendritic cells express all

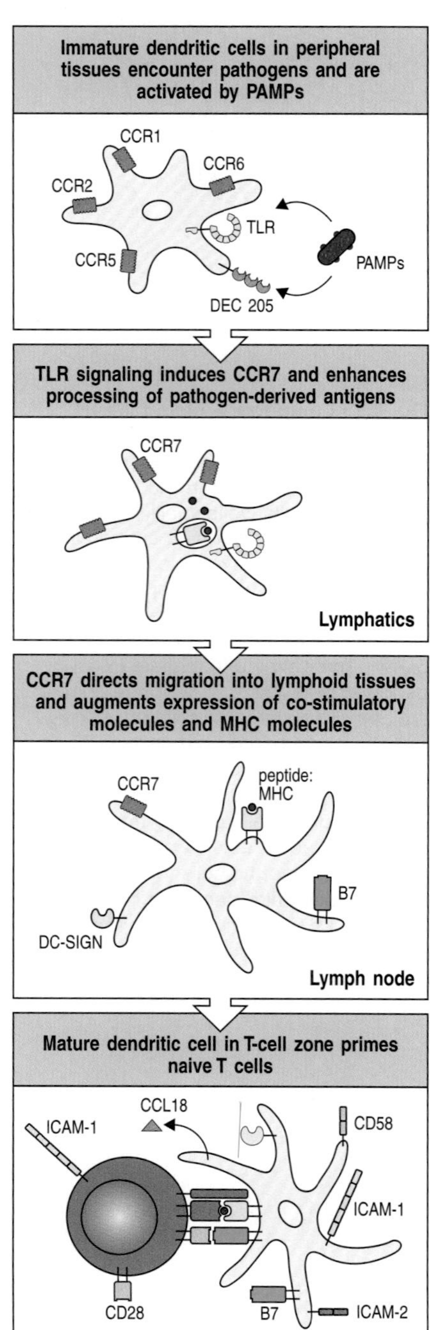

Fig. 8.14 Conventional dendritic cells mature through at least two definable stages to become potent antigen-presenting cells in peripheral lymphoid tissue. Immature dendritic cells originate from bone marrow progenitors and migrate via the blood, from which they enter and populate most tissues, including some direct entry into peripheral lymphoid tissues. Entry to particular tissues is based on the particular chemokine receptors they express: CCR1, CCR2, CCR5, CCR6, CXCR1, and CXCR2 (not all shown here, for simplicity). Immature dendritic cells in tissues are highly phagocytic via receptors such as DEC 205 and are actively macropinocytic, but they do not express co-stimulatory molecules. They carry most of the different types of Toll-like receptors (TLRs) (see the text). At sites of infection, immature dendritic cells are exposed to pathogens, leading to activation of their TLRs (top panel). TLR signaling causes the dendritic cells to become licensed and begin to undergo maturation, which involves induction of the chemokine receptor CCR7. TLR signaling also increases the processing of antigens taken up into phagosomes (second panel). Dendritic cells expressing CCR7 are sensitive to CCL19 and CCL21, which directs them to the draining lymphoid tissues. CCL19 and CCL21 provide further maturation signals, which result in higher levels of co-stimulatory B7 molecules and MHC molecules. They also express high levels of the dendritic-cell-specific adhesion molecule DC-SIGN (third panel). In the draining lymph node, mature conventional dendritic cells have become powerful activators of naive T cells but are no longer phagocytic. They express B7.1, B7.2, and high levels of MHC class I and class II molecules, as well as high levels of the adhesion molecules ICAM-1, ICAM-2, LFA-1, DC-SIGN, and CD58 (bottom panel).

known TLRs except for TLR-9, which is, however, expressed by plasmacytoid dendritic cells along with TLR-1 and TLR-7, and other TLRs to a lesser degree. Other receptors that can bind pathogens, such as receptors for complement, or phagocytic receptors such as the mannose receptor, may contribute to dendritic cell activation as well as to phagocytosis.

TLR signaling results in a significant alteration in the chemokine receptors expressed by dendritic cells, which facilitates their entry into peripheral lymphoid tissues (Fig 8.14, second panel). This change in dendritic cell behavior is often called **licensing**, as the cells are now embarked on the program of differentiation that will enable them to activate T cells. TLR signaling induces expression of the receptor CCR7, which makes the activated dendritic cells sensitive to the chemokine CCL21 produced by lymphoid tissue and induces their migration through the lymphatics and into the local lymphoid tissues. Whereas T cells have to cross the high endothelial venule wall to leave the blood and reach the T-cell zones, dendritic cells entering via the afferent lymphatics can migrate directly into the T-cell zones from the marginal sinus.

Pathogen-derived proteins that enter the immature dendritic cell via phagocytosis are processed in the endocytic compartment for presentation by MHC class II molecules (see Fig 8.14, second panel). It has recently been recognized that the efficiency of antigen processing by this endocytic compartment is greatly increased by signals delivered by TLRs. This was shown by experiments in which the generation of peptide:MHC complexes could be traced to phagocytosed particles containing specific antigenic proteins and/or TLR ligands. Phagosomes in which the antigenic proteins were co-localized on a particle with TLR ligands, such as bacterial LPS, efficiently generated specific peptide:MHC complexes, whereas in the absence of co-localized TLR ligands there were fewer or no peptide:MHC complexes. This seems to be a mechanism for linking TLR signaling within a phagosome to antigen processing and peptide:MHC loading within the same phagosome, and it enables dendritic cells to classify the various antigen sources into those that represent self and those that represent nonself. This mechanism preferentially delivers pathogen-derived peptides into the pool of peptide:MHC complexes that are transported to the dendritic cell surface, where they can be presented to naive T cells in the context of co-stimulation.

As well as inducing migration into lymphoid tissue, CCL21 signaling through CCR7 is thought to contribute to the further maturational changes that take place in activated dendritic cells, so that by the time they arrive in the T-cell zone of the lymphoid organs they have a completely different phenotype (Fig 8.14, third panel). As mature dendritic cells within lymphoid tissues, they are no longer able to engulf antigens by phagocytosis or macropinocytosis. They now express very high levels of long-lived MHC class I and MHC class II molecules, however, which enables them to stably present peptides from pathogens already taken up and processed. Equally importantly, by this time they also have high levels of co-stimulatory B7 molecules on their surface. These are two structurally related transmembrane glycoproteins called B7.1 (CD80) and B7.2 (CD86), which deliver co-stimulatory signals by interacting with receptors on naive T cells. Mature dendritic cells also express very high levels of adhesion molecules, including DC-SIGN, and they secrete the chemokine CCL18, which specifically attracts naive T cells. Together, these properties enable the dendritic cell to stimulate strong responses in naive T cells (Fig 8.14, bottom panel).

Despite their enhanced presentation of pathogen-derived antigens, mature dendritic cells also present some self peptides, which could present a problem for the maintenance of self-tolerance. The T-cell receptor repertoire has, however, been purged in the thymus of receptors that recognize self peptides presented by dendritic cells (see Chapter 7), and thus T-cell responses against ubiquitous self antigens are avoided. In addition, tissue dendritic cells reaching the end of their life span in the tissues without having been activated by infection also travel via the lymphatics to local lymphoid tissue. They will bear self-peptide:MHC complexes on their surface, derived from the breakdown of their own proteins and tissue proteins present in the extracellular fluid. But because these cells do not express the appropriate co-stimulatory molecules, they do not have the same capacity to activate naive T cells as activated, mature dendritic cells. Although the details are still unclear, it is thought that presentation of self peptides by such immature, or unlicensed, dendritic cells instead induces a state of unresponsiveness in naive T cells to these antigens.

Intracellular degradation of pathogens is thought to reveal pathogen components, other than peptides, that trigger dendritic cell activation. For example, bacterial or viral DNA containing unmethylated CpG dinucleotide motifs induces the rapid activation of plasmacytoid dendritic cells, probably as a consequence of the recognition of the DNA by TLR-9, which is present in intracellular vesicles (see Fig. 2.17). Exposure to bacterial DNA activates NFκB and mitogen-activated protein kinase (MAP kinase) signaling pathways (see Fig. 6.35), leading to the production of cytokines such as IL-6, IL-12, IL-18, and interferon (IFN)-α and IFN-γ by the dendritic cell. In turn, these cytokines act on the dendritic cells themselves to augment the expression of co-stimulatory molecules. Heat-shock proteins are another internal bacterial constituent that can activate the antigen-presenting function of dendritic cells. Some viruses are thought to be recognized by TLRs inside the dendritic cell as a consequence of the production of double-stranded RNA in the course of their replication. As discussed in Section 2-29, viral infection also induces the production of IFN-α and IFN-β by all types of infected cells; both these interferons can further activate dendritic cells to increase the expression of co-stimulatory molecules.

The induction of co-stimulatory activity in antigen-presenting cells by common microbial constituents is believed to allow the immune system to distinguish antigens borne by infectious agents from antigens associated with innocuous proteins, including self proteins. Indeed, many foreign proteins do not induce an immune response when injected on their own, presumably because they fail to induce co-stimulatory activity in antigen-presenting cells. When such protein antigens are mixed with bacteria, however, they become immunogenic, because the bacteria induce the essential co-stimulatory activity in cells that ingest the protein. Bacteria used in this way are known as adjuvants (see Appendix I, Section A-4). We will see in Chapter 14 how self proteins mixed with bacterial adjuvants can induce autoimmune diseases, illustrating the crucial importance of the regulation of co-stimulatory activity in the discrimination of self from nonself.

8-8 Plasmacytoid dendritic cells detect viral infections and generate abundant type I interferons and pro-inflammatory cytokines.

The conventional dendritic cells discussed in the preceding sections are primarily concerned with activating naive T cells. The plasmacytoid dendritic cell lineage has an important adjunct role in modifying the immune response, particularly to viruses. These dendritic cells express CXCR3, a receptor for the chemokines CXCL9, CXCL10, and CXCL11, which are induced in lymphoid tissue by the cytokine IFN-γ. Plasmacytoid dendritic cells thus migrate from the blood into lymph nodes in which there is an ongoing inflammatory response to a pathogen. Human plasmacytoid dendritic cells were initially recognized as a rare population of peripheral blood cells that make abundant type I interferons (IFN-α and IFN-β) in response to viruses. Such cells, also called **interferon-producing cells (IPCs)**, lacked surface marker proteins that would identify them as T cells, B cells, monocytes, or NK cells, but they did express MHC class II molecules, suggesting a lymphoid derivation. Eventually, specific markers were identified, such as BDCA-2 and Siglec-H (see Section 8-5), which respectively distinguish human and mouse plasmacytoid dendritic cells from other leukocyte populations.

Plasmacytoid dendritic cells express a subset of TLRs, particularly TLR-7 and TLR-9. These TLRs are located in the endosomal compartment and provide sensitivity to single-stranded RNA viruses and to the non-methylated CpG residues present in the genomes of many DNA viruses. The requirement for TLR-9 in sensing infections caused by DNA viruses has been demonstrated, for example, by the inability of TLR-9-deficient plasmacytoid dendritic cells to generate type I interferons in response to herpes simplex virus. It is thought that some of the markers specific for these cells, such as Siglec-H, play a part in the capture and delivery of viruses or other pathogens to the intracellular TLRs. In addition, human and mouse plasmacytoid dendritic cells can generate the pro-inflammatory cytokine IL-12, although the amount may be less than that produced by conventional dendritic cells. As we saw in Section 2-29, type I interferons stimulate a rapid antiviral response in uninfected somatic cells; these interferons also have the effect of promoting dendritic cell development and maturation from blood monocytes. Plasmacytoid dendritic cells express fewer MHC class II and co-stimulatory molecules on their surface and process antigens less efficiently than conventional dendritic cells. For these reasons, plasmacytoid dendritic cells are not as effective in supporting the proliferation of naive antigen-specific T cells and so are not thought to be important for directly initiating T-cell immune responses.

They may, however, act as helper cells for antigen presentation by conventional dendritic cells. An interplay between conventional and plasmacytoid dendritic cells has been revealed by studies in mice infected with the intracellular bacterium *Listeria monocytogenes*. Normally, stimulation with bacteria, or with a synthetic TLR-9 ligand containing CpG, induces conventional dendritic cells to produce a rapid pulse of the cytokine IL-15, followed by sustained generation of IL-12. The IL-12 produced by conventional dendritic cells is important in driving the particular type of CD4 T-cell response that is effective against these bacteria, as we shall see later. When either IL-15 or plasmacytoid dendritic cells were experimentally eliminated, the production of IL-12 by conventional dendritic cells decreased, and the mice become susceptible to *Listeria*. It seems that the IL-15 produced as a result of TLR stimulation acts in an autocrine loop to induce the expression of the transmembrane protein **CD40** on the conventional dendritic cells. At the same time, signaling via TLR-9 induces the expression of the transmembrane protein **CD40 ligand** (CD40L or CD154; so called because it binds to CD40) in the plasmacytoid dendritic cell. This enables the plasmacytoid dendritic cells to trigger CD40 signaling in the conventional dendritic cells, which has the effect of sustaining their production of IL-12.

8-9 Macrophages are scavenger cells that can be induced by pathogens to present foreign antigens to naive T cells.

The two other cell types that can act as antigen-presenting cells to naive T cells are macrophages and B cells. As we learned in Chapter 2, many of the microorganisms that enter the body are engulfed and destroyed by phagocytes, which provide an innate, antigen-nonspecific first line of defense against infection. However, pathogens have developed many mechanisms to avoid elimination by innate immunity. One such is to resist the killing properties of phagocytes. Macrophages that have bound and ingested microorganisms, but have failed to destroy them, contribute to the adaptive immune response by acting as antigen-presenting cells. As we will see later in this chapter, the adaptive immune response is in turn able to enhance the microbicidal and phagocytic capacities of these cells so that they can kill the pathogen.

As well as being resident in tissues, macrophages are found in lymphoid organs (see Fig. 8.10). They are present in many areas of the lymph node, especially in the marginal sinus, where the afferent lymph enters the lymphoid tissue, and in the medullary cords, where the efferent lymph collects before flowing into the blood (see Fig. 1.18). Their main role is to ingest microbes and particulate antigens and so prevent them from entering the blood. Although macrophages process ingested microbes and antigens and display peptide antigens on their surface in conjunction with co-stimulatory molecules, it is thought that their main function in lymphoid tissues is as scavengers of pathogens and of apoptotic lymphocytes.

Resting macrophages have few or no MHC class II molecules on their surface and do not express B7. The expression of both MHC class II molecules and B7 is induced by the ingestion of microorganisms and recognition of their foreign molecular patterns. Macrophages, like tissue dendritic cells, have a variety of receptors that recognize microbial surface components, including the mannose receptor, the scavenger receptor, complement receptors, and several TLRs (see Chapter 2). These receptors are involved in the ingestion of microorganisms by phagocytosis and in signaling for the secretion of pro-inflammatory cytokines, which recruit and activate more phagocytes. The phagocytic receptors function similarly to those on tissue dendritic cells and thus allow the macrophage to function as an antigen-presenting cell. Once bound, microorganisms are engulfed and degraded in the phagosomes and phagolysosomes, generating peptides that can be presented by MHC class II molecules. At the same time, the receptors recognizing these microorganisms transmit a signal that leads to the expression of MHC class II molecules and B7.

Macrophages continuously scavenge dead or dying cells, which are rich sources of self antigens, so it is particularly important that they should not activate T cells in the absence of microbial infection. The Kupffer cells of the liver sinusoids and the macrophages of the splenic red pulp, in particular, remove large numbers of dying cells from the blood daily. Kupffer cells express little MHC class II and no TLR-4, the receptor that signals the presence of bacterial LPS. Thus, although they generate large amounts of self peptides in their endosomes, these macrophages are not likely to elicit an autoimmune response.

At present, there is very little evidence that macrophages ever initiate T-cell immunity, so it is likely that their expression of co-stimulatory molecules is more important for expanding primary or secondary responses already initiated by dendritic cells. This might be envisaged to be important for effector or memory T cells that enter the site of infection.

8-10 B cells are highly efficient at presenting antigens that bind to their surface immunoglobulin.

Macrophages cannot take up soluble antigens efficiently. B cells, by contrast, are uniquely adapted to bind specific soluble molecules through their cell-surface immunoglobulin and will internalize the bound molecules by receptor-mediated endocytosis. If the antigen contains a protein component, the B cell will process the internalized protein to peptide fragments and then display

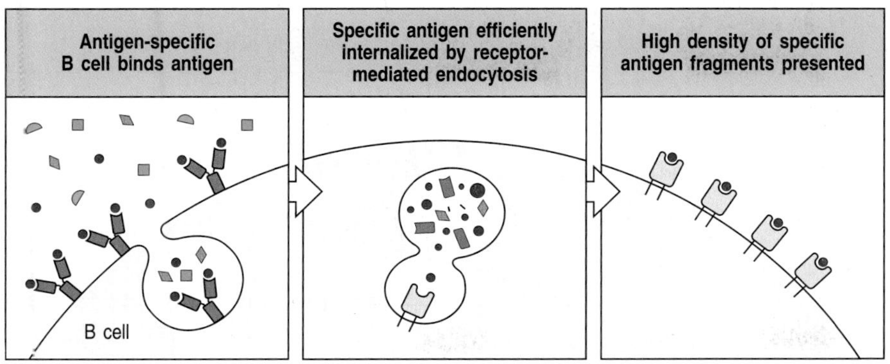
Antigen-specific B cell binds antigen
Specific antigen efficiently internalized by receptor-mediated endocytosis
High density of specific antigen fragments presented
B cell

Fig. 8.15 B cells can use their surface immunoglobulin to present specific antigen very efficiently to T cells. Surface immunoglobulin allows B cells to bind and internalize specific antigen very efficiently, especially if the antigen is present as a soluble protein, as most toxins are. The internalized antigen is processed in intracellular vesicles where it binds to MHC class II molecules. The vesicles are transported to the cell surface where the foreign-peptide:MHC class II complexes can be recognized by T cells. When the protein antigen is not specific for the B-cell receptor, its internalization is inefficient and only a few fragments of such proteins are subsequently presented at the B-cell surface (not shown).

peptide fragments as peptide:MHC class II complexes. This mechanism of antigen uptake is extremely efficient, concentrating the specific antigen in the endocytic pathway. B cells also constitutively express high levels of MHC class II molecules, and so high levels of specific peptide:MHC class II complexes appear on the B-cell surface (Fig. 8.15). This pathway of antigen presentation allows B cells to be targeted by antigen-specific CD4 T cells, which drive their differentiation, as we will see in Chapter 9.

B cells do not constitutively express co-stimulatory activity but, as with dendritic cells and macrophages, they can be induced by various microbial constituents to express B7 molecules. In fact, B7.1 was first identified as a protein on B cells activated by LPS, and B7.2 is predominantly expressed by B cells *in vivo*. These observations help to explain why it is essential to co-inject bacterial adjuvants to produce an immune response to soluble proteins such as ovalbumin, hen egg-white lysozyme, and cytochrome *c*, which may require B cells as antigen-presenting cells. The requirement for induced co-stimulatory activity also helps to explain why, although B cells present soluble proteins efficiently, they are unlikely to initiate an immune response to soluble self proteins in the absence of infection.

Although much of what we know about the immune system in general, and about T-cell responses in particular, has been learned from the study of immune responses to soluble protein immunogens presented by B cells, it is not clear how important B cells are in priming naive T cells in natural immune responses. Soluble protein antigens are not abundant during infections; most natural antigens, such as bacteria and viruses, are particulate, and soluble bacterial toxins act by binding to cell surfaces and so are present only at low concentrations in solution. Some natural immunogens enter the body as soluble molecules: examples are insect toxins, anticoagulants injected by blood-sucking insects, snake venoms, and many allergens. However, tissue dendritic cells could also be responsible for activating naive T cells that recognize these antigens, as they can take them up by macropinocytosis. Although tissue dendritic cells could not concentrate these antigens in the same way as antigen-specific B cells do, they may be more likely to encounter a naive T cell with the appropriate antigen specificity than would the limited number of antigen-specific B cells. The chances of a B cell encountering a T cell that can recognize the peptide antigens it displays are greatly increased once a naive T cell has been detained in lymphoid tissue by finding its antigen on the surface of a dendritic cell.

The three types of antigen-presenting cell are compared in Fig. 8.16. In each of these cell types the expression of co-stimulatory activity is controlled so as to provoke responses against pathogens while avoiding immunization against self.

Fig. 8.16 The properties of the various antigen-presenting cells. Dendritic cells, macrophages, and B cells are the main cell types involved in the presentation of foreign antigens to naive T cells. These cells vary in their means of antigen uptake, MHC class II expression, co-stimulator expression, the type of antigen they present effectively, their locations in the body, and their surface adhesion molecules (not shown).

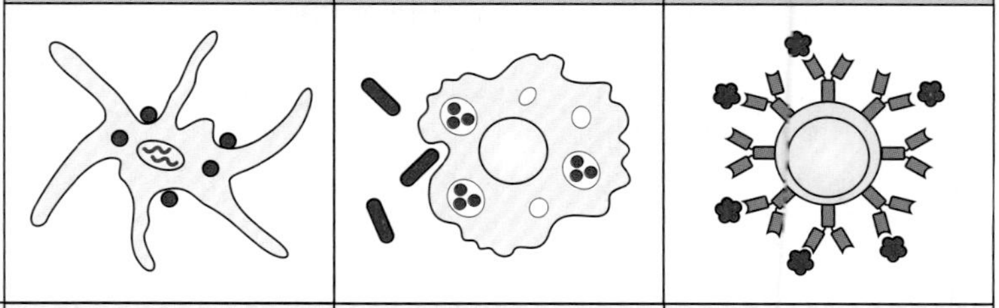

	Dendritic cells	Macrophages	B cells
Antigen uptake	+++ Macropinocytosis and phagocytosis by tissue dendritic cells Viral infection	Phagocytosis +++	Antigen-specific receptor (Ig) ++++
MHC expression	Low on tissue dendritic cells High on dendritic cells in lymphoid tissues	Inducible by bacteria and cytokines – to +++	Constitutive Increases on activation +++ to ++++
Co-stimulator delivery	Constitutive by mature, nonphagocytic lymphoid dendritic cells ++++	Inducible – to +++	Inducible – to +++
Antigen presented	Peptides Viral antigens Allergens	Particulate antigens Intracellular and extracellular pathogens	Soluble antigens Toxins Viruses
Location	Ubiquitous throughout the body	Lymphoid tissue Connective tissue Body cavities	Lymphoid tissue Peripheral blood

Summary.

An adaptive immune response is generated when naive T cells contact mature, activated antigen-presenting cells in the peripheral lymphoid organs. To ensure that rare antigen-specific T cells survey the body effectively for equally rare pathogen-bearing antigen-presenting cells, T cells continuously recirculate through the lymphoid organs and thus can sample antigens brought in by antigen-presenting cells from many different sites of infection. The migration of naive T cells into lymphoid organs is guided by the chemokine receptor CCR7, which binds the chemokine CCL21 produced by stromal cells in the T-cell zones of peripheral lymphoid organs. L-selectin expressed by naive T cells initiates their rolling along the specialized surfaces of high endothelial venules, and contact with CCL21 there induces a switch in the integrin LFA-1 expressed by T cells to a configuration with affinity for the ICAM-1 expressed on the venule endothelium. This initiates strong adhesion, diapedesis, and migration of the T cells into the T-cell zone. There, naive T cells meet antigen-bearing dendritic cells. There are two main populations of dendritic cells, CD11c-positive conventional dendritic cells and plasmacytoid dendritic cells. Conventional dendritic cells continuously survey peripheral tissues for invading pathogens and are the dendritic cells responsible for activating naive lymphocytes. Contact with pathogens delivers signals to dendritic cells via TLRs and other receptors that accelerate antigen processing and the production of foreign peptide:self-MHC complexes. TLR signaling also induces expression of CCR7 by the dendritic cells, which directs their migration to T-cell zones of peripheral lymphoid organs where they encounter naive T cells and activate them.

Some other cell types are able to act as antigen-presenting cells to naive T cells, although dendritic cells are the most potent activators of naive T cells and are thought to initiate most T-cell responses to pathogenic microorganisms. Macrophages efficiently ingest particulate antigens such as bacteria and are induced by infectious agents to express MHC class II molecules and co-stimulatory activity. The unique ability of B cells to bind and internalize soluble protein antigens via their receptors, and then display processed peptides as peptide:MHC complexes, may be important in activating T cells to provide antigen-specific help to B cells. In all three types of antigen-presenting cells, the expression of co-stimulatory molecules is activated in response to signals from receptors that also function in innate immunity to signal the presence of infectious agents.

Priming of naive T cells by pathogen-activated dendritic cells.

T-cell responses are initiated when a mature naive CD4 or CD8 T cell encounters a properly activated antigen-presenting cell displaying the appropriate peptide:MHC ligand. We have described the trafficking of naive T cells and dendritic cells to specific zones of peripheral lymphoid organs where they can encounter each other in the T-cell zones. We will now describe the generation of effector T cells from naive T cells. The activation and differentiation of naive T cells, often called priming, is distinct from the later responses of effector T cells to antigen on their target cells, and from the responses of primed memory T cells to subsequent encounters with the same antigen. Priming of naive CD8 T cells generates cytotoxic T cells capable of directly killing pathogen-infected cells. CD4 cells develop into a diverse array of effector types depending on the nature of the signals they receive during their priming. CD4 effector activity can include cytotoxicity, but more frequently it involves the secretion of a set of cytokines that directs the target cell to make a particular response.

8-11 Cell-adhesion molecules mediate the initial interaction of naive T cells with antigen-presenting cells.

As they migrate through the cortical region of the lymph node, naive T cells bind transiently to each antigen-presenting cell that they encounter. Mature dendritic cells bind naive T cells very efficiently through interactions between LFA-1, ICAM-3, and CD2 on the T cell, and ICAM-1, ICAM-2, DC-SIGN, and CD58 on the dendritic cell (Fig. 8.17). The binding of ICAM-3 to DC-SIGN is unique to the interaction between dendritic cells and T cells, whereas the other adhesion molecules synergize in the binding of lymphocytes to all three types of antigen-presenting cells. Perhaps because of this synergy, the precise role of each adhesion molecule has been difficult to distinguish. People lacking LFA-1 can have normal T-cell responses, and this also seems to be true for genetically engineered mice lacking CD2. It would not be surprising if there were enough redundancy in the molecules mediating T-cell adhesive interactions to enable immune responses to occur in the absence of any one of them; such molecular redundancy has been observed in other complex biological processes.

The transient binding of naive T cells to antigen-presenting cells is crucial in providing time for T cells to sample large numbers of MHC molecules on each antigen-presenting cell for the presence of a specific peptide. In those rare cases in which a naive T cell recognizes a peptide:MHC ligand, signaling through the T-cell receptor induces a conformational change in LFA-1 that greatly increases its affinity for ICAM-1 and ICAM-2. This conformational change is the same as that induced by signaling through chemokine receptors

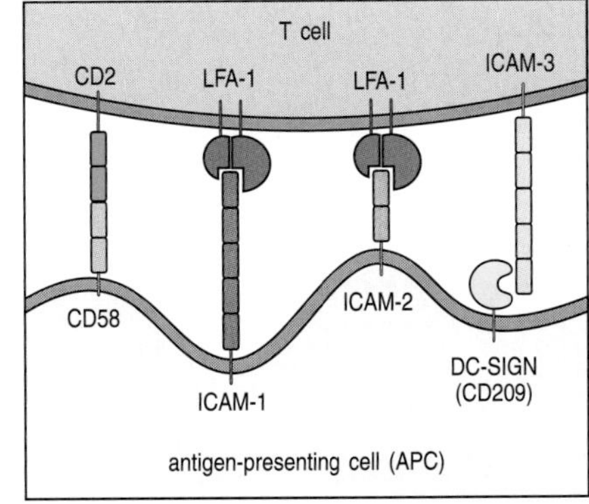

Fig. 8.17 Cell-surface molecules of the immunoglobulin superfamily are important in the interactions of lymphocytes with antigen-presenting cells. In the initial encounter of T cells with antigen-presenting cells, CD2 binding to CD58 on the antigen-presenting cell synergizes with LFA-1 binding to ICAM-1 and ICAM-2. One interaction that seems to be exclusive to the interaction of naive T cells with dendritic cells is that between ICAM-3 on the naive T cell and DC-SIGN (CD209), a C-type lectin that is specific to dendritic cells and binds ICAM-3 with high affinity. LFA-1 is the α_L:β_2 integrin heterodimer CD11a:CD18. ICAM-1, -2, and -3 are also known as CD54, CD102, and CD50, respectively.

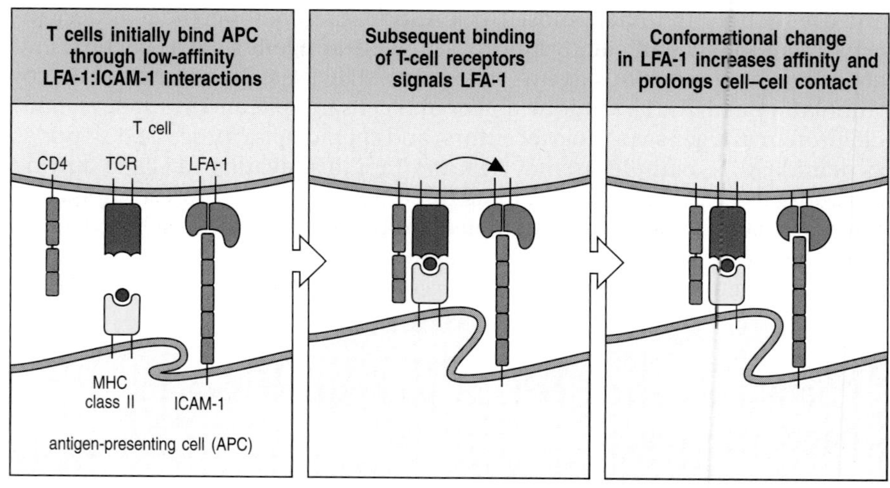

Fig. 8.18 Transient adhesive interactions between T cells and antigen-presenting cells are stabilized by specific antigen recognition. When a T cell binds to its specific ligand on an antigen-presenting cell, intracellular signaling through the T-cell receptor (TCR) induces a conformational change in LFA-1 that causes it to bind with higher affinity to ICAMs on the antigen-presenting cell. The T cell shown here is a CD4 T cell.

during the migration of naive T cells into a peripheral lymphoid organ (see Section 8-2). The change in LFA-1 stabilizes the association between the antigen-specific T cell and the antigen-presenting cell (Fig. 8.18). The association can persist for several days, during which time the naive T cell proliferates and its progeny, which also adhere to the antigen-presenting cell, differentiate into effector T cells.

Most encounters of T cells with antigen-presenting cells do not, however, result in the recognition of an antigen. In this case, the T cell must be able to separate efficiently from the antigen-presenting cell so that it can continue to migrate through the lymph node, eventually leaving via the efferent lymphatic vessels to reenter the blood and continue circulating. Dissociation, like stable binding, may also involve signaling between the T cell and the antigen-presenting cells, but little is known of its mechanism.

8-12 Antigen-presenting cells deliver three kinds of signals for clonal expansion and differentiation of naive T cells.

Priming of naive T cells is controlled by several signals. As explained in the introduction to this chapter, we will adopt a terminology that divides these signals into three types: signal 1, signal 2, and signal 3. Signal 1 comprises those antigen-specific signals derived from the interaction of a specific peptide:MHC complex with the T-cell receptor. Engagement of the T-cell receptor with its peptide antigen is essential for activating a naive T cell, but even if the co-receptor—CD4 or CD8—is also ligated, this does not on its own stimulate the T cell to proliferate and differentiate into effector T cells. The antigen-specific clonal expansion of a naive T cell involves at least two other kinds of signals, which are generally delivered by the same antigen-presenting cell. These additional signals have been divided into the co-stimulatory signals that are primarily involved in promoting (or inhibiting) the survival and expansion of the T cells (signal 2), and those that are primarily involved in directing T-cell differentiation into the different subsets of effector T cells (signal 3) (Fig. 8.19).

The best-characterized co-stimulatory molecules that deliver signal 2 are the B7 molecules (see Section 8-6). These homodimeric members of the immunoglobulin superfamily are found exclusively on the surfaces of cells, such as dendritic cells, that stimulate naive T-cell proliferation. Their role in co-stimulation has been demonstrated by transfecting fibroblasts that express a T-cell ligand with genes encoding B7 molecules and showing that

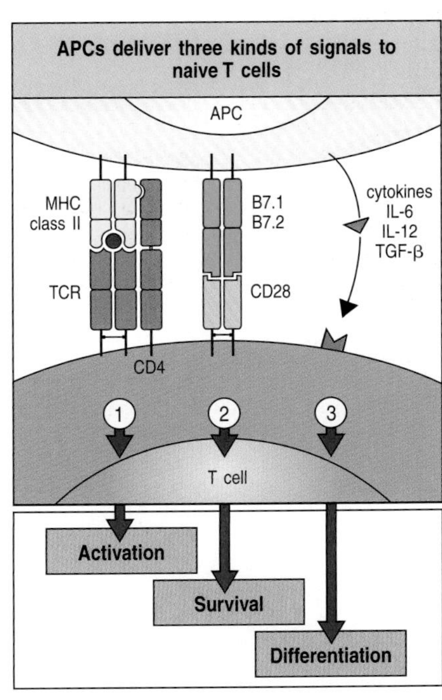

Fig. 8.19 Three kinds of signals are involved in activation of naive T cells by antigen-presenting cells. Binding of the foreign-peptide:self-MHC complex by the T-cell receptor and, in this example, a CD4 co-receptor, transmits a signal (arrow 1) to the T cell that antigen has been encountered. Effective activation of naive T cells requires a second signal (arrow 2), the co-stimulatory signal, to be delivered by the same antigen-presenting cell (APC). In this example, CD28 on the T cell encountering B7 molecules on the antigen-presenting cell delivers signal 2, whose net effect is the increased survival and proliferation of the T cell that has received signal 1. ICOS and members of the TNF receptor family may also provide co-stimulatory signals. For CD4 T cells in particular, different pathways of differentiation produce subsets of effector T cells that carry out different effector responses, depending on the nature of a third signal (arrow 3) delivered by the antigen-presenting cell. Cytokines are commonly, but not exclusively, involved in directing this differentiation.

the fibroblasts could then stimulate the clonal expansion of naive T cells. The receptor for B7 molecules on the T cell is **CD28**, yet another member of the immunoglobulin superfamily. Ligation of CD28 by B7 molecules or by anti-CD28 antibodies is necessary for the optimal clonal expansion of naive T cells, whereas anti-B7 antibodies, which inhibit the binding of B7 molecules to CD28, have been shown experimentally to inhibit T-cell responses. Although other molecules have been reported to co-stimulate naive T cells, so far only the B7 molecules have been shown definitively to provide co-stimulatory signals for naive T cells in normal immune responses.

8-13 CD28-dependent co-stimulation of activated T cells induces expression of the T-cell growth factor interleukin-2 and the high-affinity IL-2 receptor.

Naive T cells can live for many years, dividing infrequently and equally infrequently undergoing apoptosis. They are found as small resting cells with condensed chromatin and scanty cytoplasm and synthesize little RNA or protein. On activation, they reenter the cell cycle and divide rapidly to produce the large numbers of progeny that will differentiate into effector T cells. Their proliferation and differentiation are driven by the cytokine **interleukin-2** (**IL-2**), which is produced by the activated T cell itself.

The initial encounter with specific antigen in the presence of a co-stimulatory signal triggers entry of the T cell into the G1 phase of the cell cycle; at the same time, it also induces the synthesis of IL-2 along with the α chain of the IL-2 receptor (also known as CD25). The IL-2 receptor has three chains: α, β, and γ (Fig. 8.20). Resting T cells express a form of this receptor composed of β and γ chains that binds IL-2 with moderate affinity, allowing resting T cells to respond to very high concentrations of IL-2. Association of the α chain with the β and γ heterodimer creates a receptor with a much higher affinity for IL-2, allowing the cell to respond to very low concentrations of IL-2. Binding of IL-2 to the high-affinity receptor then triggers progression through the rest of the cell cycle (Fig. 8.21). T cells activated in this way can divide two or three times a day for several days, allowing one cell to give rise to a clone of thousands of cells that all bear the same receptor for antigen. IL-2 is a survival factor for these cells, and the removal of IL-2 from activated T cells results in their death; it also promotes the differentiation of these activated cells into effector T cells.

Antigen recognition by the T-cell receptor induces the synthesis or activation of the transcription factors NFAT, AP-1, and NFκB (see Chapter 6), which bind to the promoter region of the IL-2 gene and are essential to activate its transcription. Co-stimulation through CD28 contributes to the production of IL-2 in at least two ways. First, signals from CD28 bound by B7 molecules increase the production of AP-1 and NFκB, and this increases the initiation of transcription

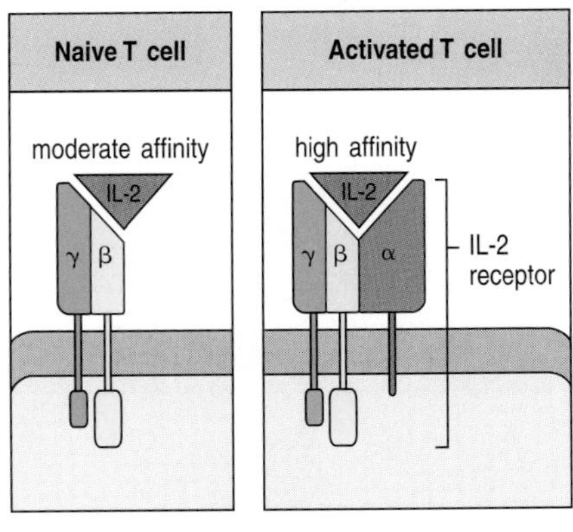

Fig. 8.20 High-affinity IL-2 receptors are three-chain structures that are present only on activated T cells. On resting T cells, the β and γ chains are expressed constitutively. They bind IL-2 with moderate affinity. Activation of T cells induces the synthesis of the α chain and the formation of the high-affinity heterotrimeric receptor. The β and γ chains show similarities in amino-acid sequence to cell-surface receptors for growth hormone and prolactin, each of which also regulates cell growth and differentiation.

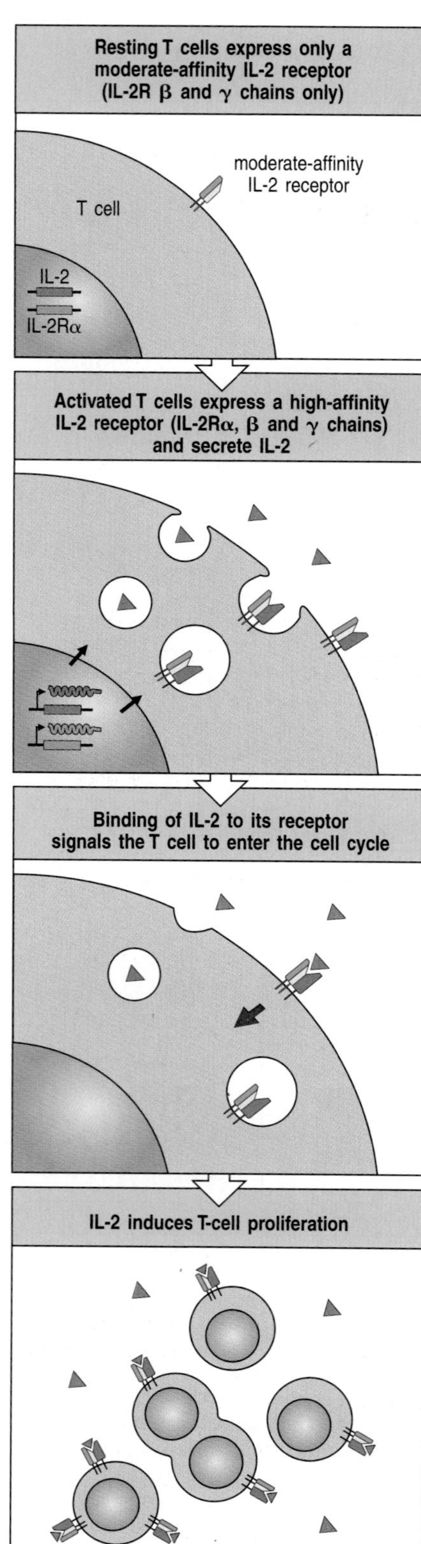

Fig. 8.21 Activated T cells secrete and respond to IL-2. Activation of naive T cells in the presence of co-stimulation through CD28 signaling induces the expression and secretion of IL-2 and the expression of high-affinity IL-2 receptors. IL-2 binds to the high-affinity IL-2 receptors to promote T-cell growth in an autocrine fashion.

of IL-2 mRNA approximately threefold. The second effect of signaling through CD28 is thought to be the stabilization of IL-2 mRNA, which increases the production of IL-2 protein 20–30-fold. These two effects together increase IL-2 protein production as much as a hundredfold. Cytokine mRNAs are very short-lived because of an 'instability' sequence in their 3′ untranslated region. The unstable RNA prevents sustained cytokine production and release, thus enabling cytokine activity to be tightly regulated. When a T cell recognizes specific antigen in the absence of co-stimulation through its CD28 molecule, little IL-2 is produced and the T cell does not proliferate. Thus, the most important function of the co-stimulatory signal is to promote the synthesis of IL-2.

The central importance of IL-2 in initiating adaptive immune responses is exploited by drugs commonly used to suppress undesirable immune responses such as transplant rejection. The immunosuppressive drugs cyclosporin A and FK506 (tacrolimus or fujimycin) inhibit IL-2 production by disrupting signaling through the T-cell receptor, whereas rapamycin (sirolimus) inhibits signaling through the IL-2 receptor. Cyclosporin A and rapamycin act synergistically to inhibit immune responses by preventing the IL-2-driven clonal expansion of T cells. The mode of action of these drugs will be considered in detail in Chapter 15.

8-14 Signal 2 can be modified by additional co-stimulatory pathways.

Once a naive T cell is activated it expresses a number of proteins in addition to CD28 that contribute to sustaining or modifying the co-stimulatory signal that drives clonal expansion and differentiation. These other co-stimulatory proteins generally belong to either the CD28 family of receptors or the tumor necrosis factor (TNF)/TNF receptor families.

CD28-related proteins are induced on activated T cells and modify the co-stimulatory signal as the T-cell response develops. One is called inducible co-stimulator, or **ICOS**, and binds a ligand known as **LICOS** (the ligand of ICOS, or B7h), a relative of B7.1 and B7.2. LICOS is produced on activated dendritic cells, monocytes, and B cells, but its contribution to immune responses has not yet been clearly defined. Although ICOS resembles CD28 in driving T-cell proliferation, it does not induce IL-2 but seems to regulate the expression of other cytokines made by the CD4 T-cell subsets.

Another CD28-related protein is **CTLA-4** (CD152), an additional receptor for B7 molecules. CTLA-4 is very similar to CD28 in sequence, and the two proteins are encoded by closely linked genes. However, CTLA-4 binds B7 molecules about 20 times more avidly than does CD28 and delivers an inhibitory signal to the activated T cell (Fig. 8.22). This makes the activated progeny of a naive T cell less sensitive to stimulation by the antigen-presenting cell and limits production of the cytokine interleukin-2 (IL-2), the main cytokine that drives T-cell proliferation. Thus, binding of CTLA-4 to B7 molecules is essential for limiting the proliferative response of activated T cells to antigen and B7. This was confirmed by producing mice with a disrupted CTLA-4 gene; such mice develop a fatal disorder characterized by a massive overgrowth of lymphocytes.

TNF family molecules can also deliver co-stimulatory signals. **CD27** is a TNF receptor family protein constitutively expressed on naive T cells that binds **CD70** on dendritic cells and delivers a potent co-stimulatory signal to T cells early in the activation process. The TNF receptor family molecule CD40 on dendritic cells (see Section 8-8) binds to CD40 ligand expressed on T cells and initiates a two-way signaling that transmits activating signals to the T cell and also induces the antigen-presenting cell to express B7 molecules, thus stimulating further T-cell proliferation. The role of the CD40–CD40 ligand pair in sustaining the development of a T-cell response is demonstrated in mice lacking CD40 ligand; when these mice are immunized, the clonal expansion of responding T cells is curtailed at an early stage. The T-cell molecule **4-1BB** (CD137) and its ligand **4-1BBL**, which is expressed on activated dendritic cells, macrophages, and B cells, are another pair of TNF family co-stimulators. As with CD40 ligand and CD40, the effects of this interaction are bidirectional, with both the T cell and the antigen-presenting cell receiving activating signals; this type of interaction is sometimes referred to as the T-cell:antigen-presenting cell dialog.

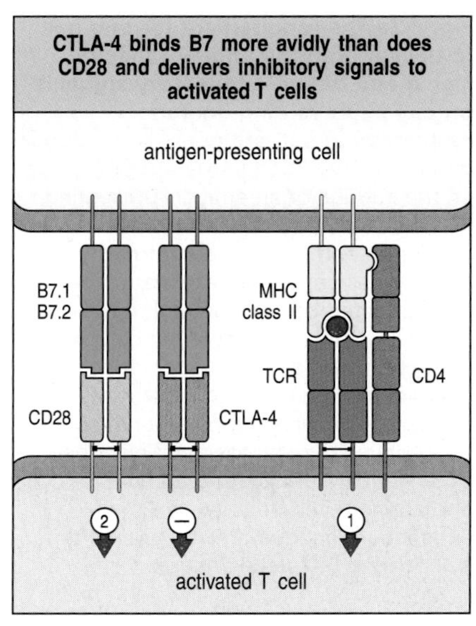

Fig. 8.22 CTLA-4 is an inhibitory receptor for B7 molecules. Naive T cells express CD28, which delivers a co-stimulatory signal on binding B7 molecules (see Fig. 8.19), thereby driving the survival and expansion of T cells that encounter specific antigen presented by a B7-positive antigen-presenting cell. Once activated, T cells express increased levels of CTLA-4 (CD152). CTLA-4 has a higher affinity than CD28 for B7 molecules and thus binds most or all of the B7 molecules, serving to regulate the proliferative phase of the response.

8-15 Antigen recognition in the absence of co-stimulation leads to functional inactivation or clonal deletion of peripheral T cells.

As we saw in Section 7-20, ubiquitous self proteins will be presented by antigen-presenting cells in the thymus and will induce clonal deletion of the T cells reactive to them. However, many proteins have specialized functions and are made only by the cells of certain tissues. Thus, peptides from some tissue-specific proteins may not be displayed on the MHC molecules of thymic cells, and T cells specific for them are unlikely to be deleted in the thymus. An important factor in avoiding autoimmune responses to these tissue-specific proteins is the absence of co-stimulatory activity on tissue cells. Because the IL-2 gene is regulated by signals derived from both the T-cell receptor pathway and the CD28 pathway, effective activation of naive T cells requires the simultaneous delivery of antigen-specific and co-stimulatory signals. Naive T cells recognizing self peptides on tissue cells lacking co-stimulatory molecules are not activated; instead, they are thought to enter a state of anergy (Fig. 8.23). An anergic T cell is refractory to activation by specific antigen even when the antigen is subsequently presented to it by an antigen-presenting cell expressing co-stimulatory molecules, and this provides a means of sustained self-tolerance. The requirement that the same cell present both the specific antigen and the co-stimulatory signal is therefore important in preventing destructive immune responses to self tissues (Fig. 8.24). In the absence of this requirement, self-tolerance could be

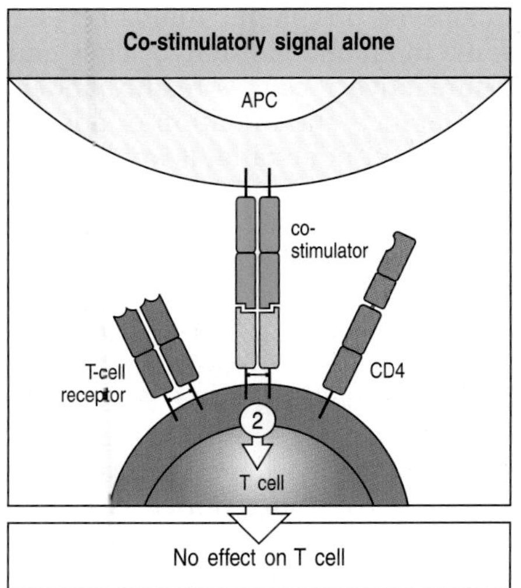

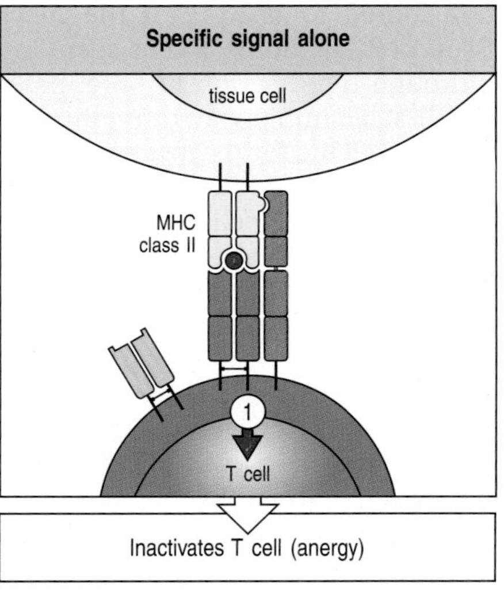

Fig. 8.23 T-cell tolerance to antigens expressed on tissue cells results from antigen recognition in the absence of co-stimulation. An antigen-presenting cell (APC) will neither activate nor inactivate a T cell if the appropriate antigen is not present on the APC surface, even if it expresses a co-stimulatory molecule and can deliver signal 2 (left panel). However, when a T cell recognizes antigen in the absence of co-stimulatory molecules, it receives signal 1 alone and is inactivated (right panel). This allows self antigens expressed on tissue cells to induce tolerance in the peripheral T-cell population.

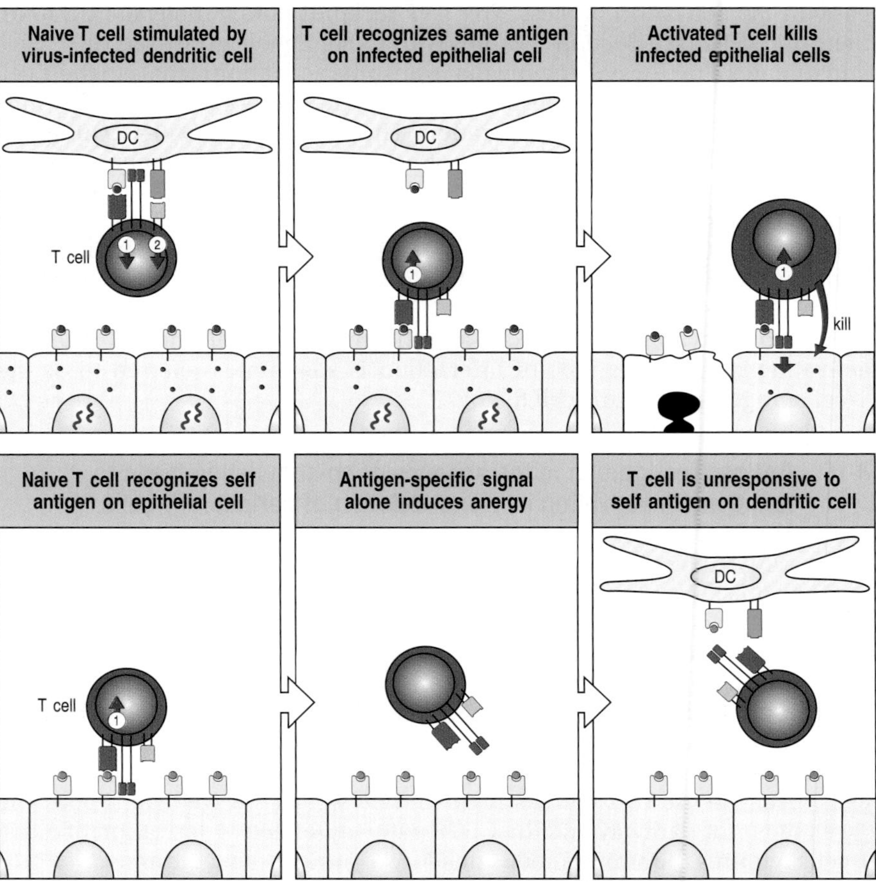

Fig. 8.24 The requirement for one cell to deliver both the antigen-specific signal and the co-stimulatory signal is crucial in preventing immune responses to self antigens. In the upper panels, a T cell recognizes a viral peptide on the surface of an antigen-presenting dendritic cell (DC) and is activated to proliferate and differentiate into an effector cell capable of eliminating any virus-infected cell. In contrast, naive T cells recognizing antigen on cells that cannot provide co-stimulation become anergic, as when a T cell recognizes a self antigen expressed by an uninfected epithelial cell (lower panels). This T cell does not differentiate into an effector cell and cannot even be activated by a subsequent encounter with a dendritic cell bearing the same antigen.

broken if naive autoreactive T cells recognized self antigens on tissue cells and could subsequently be co-stimulated separately by interaction with an antigen-presenting cell, either locally or at a distant site.

The molecular mechanism of anergy in T cells is not fully understood. The most important change is that anergic T cells do not produce IL-2, and so do not proliferate and differentiate into effector cells when they encounter antigen. Anergy has only been formally demonstrated *in vitro*, although there is *in vivo* evidence of anergy to various antigens, and it is generally considered to be one of the mechanisms of peripheral tolerance (see Section 7-26). Some T cells seem to persist in an anergic state *in vivo*, and although the deletion of potentially autoreactive T cells is readily understood as a simple way to maintain self-tolerance, the retention of anergic T cells specific for tissue antigens is less easy to understand. It would seem more efficient to eliminate such cells; indeed, binding of the T-cell receptor on peripheral T cells in the absence of co-stimulators can lead to programmed cell death instead of anergy. One possible explanation for the retention of anergic cells is that they play a part in preventing responses by naive, non-anergic T cells to foreign antigens that mimic self-peptide:self-MHC complexes. Anergic T cells could recognize and bind to such complexes on antigen-presenting cells without responding, and thus could compete with the potentially autoreactive naive T cells of the same specificity. In this way, anergic T cells could prevent the accidental activation of autoreactive T cells by infectious agents, thus actively contributing to tolerance. Another possible explanation is that the anergic cells are in fact regulatory T cells, as there are similarities in the phenotype. They both fail to

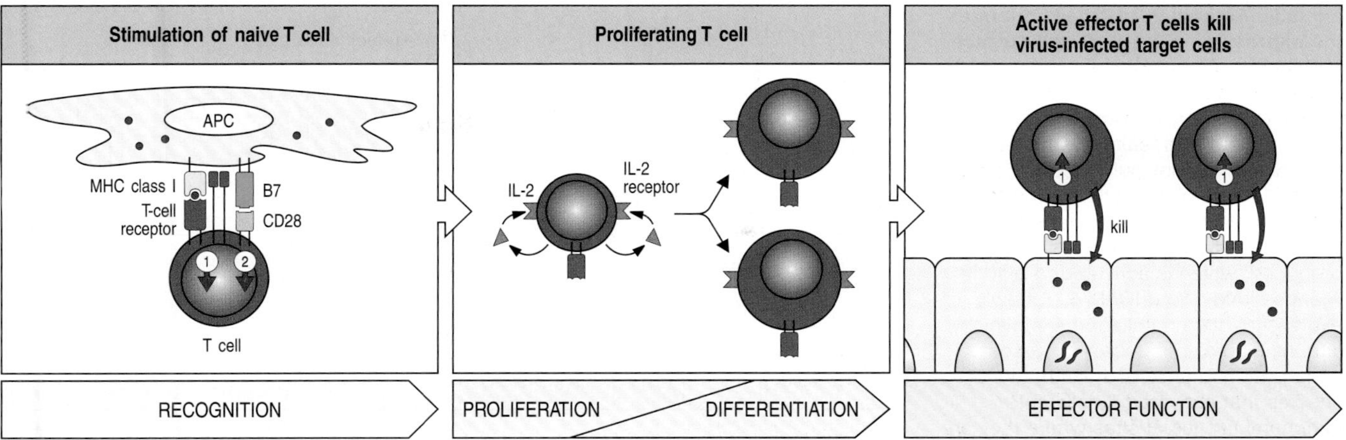

Fig. 8.25 Effector T cells can respond to their target cells without co-stimulation. A naive T cell that recognizes antigen on the surface of an antigen-presenting cell and receives the required two signals (arrows 1 and 2, left panel) becomes activated, and both secretes and responds to IL-2. IL-2-driven clonal expansion (center panel) is followed by the differentiation of the T cells to effector cell status. Once the cells have differentiated into effector T cells, any encounter with specific antigen triggers their effector actions without the need for co-stimulation. Thus, as illustrated here, a cytotoxic T cell can kill virus-infected target cells that express only the peptide:MHC ligand and not co-stimulatory signals (right panel).

proliferate or produce IL-2 *in vitro* in response to stimulation by their specific antigen. If it turns out that the anergic and regulatory T-cell populations overlap *in vivo*, then anergy might be a means of actively maintaining tolerance to self antigens.

8-16 Proliferating T cells differentiate into effector T cells that do not require co-stimulation to act.

Late in the proliferative phase of the T-cell response induced by IL-2, after 4–5 days of rapid growth, activated T cells differentiate into effector T cells that can synthesize all the effector molecules required for their specialized functions as helper or cytotoxic T cells. In addition, all classes of effector T cells have undergone changes that distinguish them from naive T cells. One of the most important is in their activation requirements: once a T cell has differentiated into an effector cell, encounter with its specific antigen results in immune attack without the need for co-stimulation (Fig. 8.25).

This applies to all classes of effector T cells. Its importance is particularly easy to understand for CD8 cytotoxic T cells, which must be able to act on any cell infected with a virus, whether or not the infected cell can express co-stimulatory molecules. However, it is also important for the effector function of CD4 cells, as effector CD4 T cells must be able to activate B cells and macrophages that have taken up antigen even if these cells are not expressing co-stimulatory molecules.

Changes are also seen in the cell-adhesion molecules and receptors expressed by effector T cells. They express higher levels of LFA-1 and CD2 than naive T cells, but lose cell-surface L-selectin and therefore cease to recirculate through lymph nodes. Instead, they express the integrin VLA-4, which allows them to bind to vascular endothelium bearing the adhesion molecule VCAM-1, which is expressed at sites of inflammation. This allows the effector T cells to enter sites of infection and put their armory of effector proteins to good use. These changes in the T-cell surface are summarized in Fig. 8.26.

8-17 T cells differentiate into several subsets of functionally different effector cells.

Before we explain the ways in which T cells become differentiated, we will briefly introduce the different subsets of effector T cells, and their general functions in immune responses. Naive T cells fall into two large groups, one

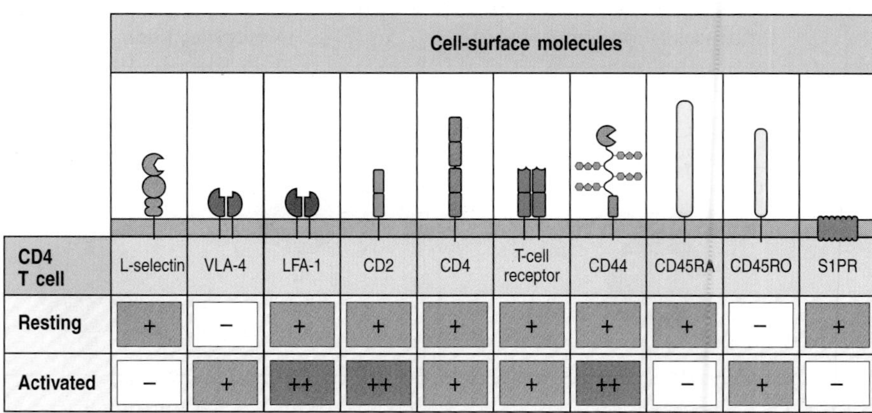

CD4 T cell	Cell-surface molecules									
	L-selectin	VLA-4	LFA-1	CD2	CD4	T-cell receptor	CD44	CD45RA	CD45RO	S1PR
Resting	+	–	+	+	+	+	+	+	–	+
Activated	–	+	++	++	+	+	++	–	+	–

Fig. 8.26 Activation of T cells changes the expression of several cell-surface molecules. The example here is a CD4 T cell. Resting naive T cells express L-selectin, through which they home to lymph nodes, with relatively low levels of other adhesion molecules such as CD2 and LFA-1. Upon activation, expression of L-selectin ceases and, instead, increased amounts of the integrin LFA-1 are produced, which is activated to bind its ligands ICAM-1 and ICAM-2. A newly expressed integrin called VLA-4, which acts as a homing receptor for vascular endothelium at sites of inflammation, ensures that activated T cells enter peripheral tissues at sites where they are likely to encounter infection. Activated T cells also have a higher density of the adhesion molecule CD2 on their surface, increasing the avidity of the interaction with potential target cells, and a higher density of the adhesion molecule CD44. The isoform of CD45 expressed changes, by alternative splicing of the RNA transcript of the CD45 gene, so that activated T cells express the CD45RO isoform, which associates with the T-cell receptor and CD4. This change makes the T cell more sensitive to stimulation by lower concentrations of peptide:MHC complexes. Finally, the sphingosine 1-phosphate receptor (S1PR) is expressed by resting naive T cells, allowing the egress of cells that do not become activated. Downregulation of S1PR for several days after activation prevents T-cell egress during the period of proliferation and differentiation. After several days, it is expressed again, allowing effector cells to exit from the lymphoid tissues.

of which carries the co-receptor CD8 on its surface and the other bears the co-receptor CD4. CD8 T cells all differentiate into **CD8 cytotoxic T cells** (also sometimes called **cytotoxic lymphocytes** or **CTLs**), which kill their target cells (Fig. 8.27). They are important in the defense against intracellular pathogens, especially viruses. Virus-infected cells display fragments of viral proteins as peptide:MHC class I complexes on their surface and these are recognized by CD8 cytotoxic T cells.

CD4 T cells, in contrast, differentiate into a number of different effector T cells with a variety of functions. The main functional subsets of CD4 effector T cells now recognized are **T_H1**, **T_H2**, **T_H17** and the **regulatory T cells**. These subsets, particularly T_H1, T_H2, and T_H17, are defined on the basis of the different cytokines that they secrete. The first of these subsets to be distinguished were the T_H1 and T_H2 cells (see Fig. 8.27). T_H1 cells have a dual function. One is to control bacteria that can set up intravesicular infections in macrophages, such as the mycobacteria that cause tuberculosis and leprosy. These bacteria are taken up by macrophages in the usual way but can evade the killing mechanisms described in Chapter 2. If a T_H1 cell recognizes bacterial antigens displayed on the surface of an infected macrophage, it will interact with the infected cell to activate it further, stimulating the macrophage's microbicidal activity to enable it to kill the intracellular bacteria. The second role of T_H1 cells is to stimulate the production of antibodies against extracellular pathogens by producing co-stimulatory signals for antigen-activated naive B lymphocytes. T_H1 cells also induce class switching of activated B cells to produce particular antibody isotypes.

T_H2 cells carry out a similar function of activating naive B cells and inducing class switching. T_H2 cells are required in particular for the switching of B cells to produce the IgE class of antibody, whose primary role is to fight parasite infections, as we shall see in Chapter 9. IgE is also the antibody responsible for allergies, and thus T_H2 differentiation is of additional medical interest, as discussed in Chapter 13. In their roles of providing help for antibody production, both T_H1 and T_H2 cells are often called **helper T cells** (see Section 1-4). We shall describe the macrophage-activating functions of T_H1 cells later in this chapter; the helper functions of T_H1 and T_H2 cells in antibody production will be described in Chapter 9. A much more recently described subset of CD4 effector T cells are the T_H17 cells. They are induced early in the adaptive immune response to extracellular bacteria and seem to be involved in stimulating the neutrophil response that helps to clear such bacteria (see Fig. 8.27).

All the effector T cells described above are involved in activating their target cells to make responses that help clear the pathogen from the body. The other CD4 T cells found in the periphery have a different function. These are the regulatory T cells, whose function is to suppress T-cell responses rather than activate them. They are involved in limiting the immune response and preventing autoimmune responses. Two main groups of regulatory T cells are currently recognized. One subset becomes committed to a regulatory fate while still in the thymus; they are known as the **natural regulatory T cells** (**T_{reg}**). Other subsets of CD4 regulatory T cells with different phenotypes have been recognized more recently and are thought to differentiate from naive CD4 T cells in the periphery under the influence of particular environmental conditions. This group is known as **adaptive regulatory T cells**.

Fig. 8.27 CD8 cytotoxic T cells and T_H1, T_H2, and T_H17 CD4 effector T cells are specialized to deal with different classes of pathogens. CD8 cytotoxic cells (left panels) kill target cells that display peptide fragments of cytosolic pathogens, most notably viruses, bound to MHC class I molecules at the cell surface. T_H1 cells (second panels) and T_H2 cells (third panels) both express the CD4 co-receptor and recognize fragments of antigens degraded within intracellular vesicles, displayed at the cell surface by MHC class II molecules. T_H1 cells produce cytokines that activate macrophages, enabling them to destroy intracellular microorganisms more efficiently. They can also activate B cells to produce strongly opsonizing antibodies belonging to certain IgG subclasses (IgG1 and IgG3 in humans, and their homologs IgG2a and IgG2b in the mouse). T_H2 cells, in contrast, produce cytokines that drive B cells to differentiate and produce immunoglobulins of other types, especially IgE, and are responsible for initiating B-cell responses by activating naive B cells to proliferate and secrete IgM. The various types of immunoglobulins together make up the effector molecules of the humoral immune response. T_H17 cells (fourth panels) are a recently recognized subset of CD4 effector T cells. They induce local epithelial and stromal cells to produce chemokines that recruit neutrophils to sites of infection early in the adaptive immune response. The remaining subset of effector T cells are the regulatory T cells (right panels), a heterogeneous class of cells that suppress T-cell activity and help prevent the development of autoimmunity during immune responses.

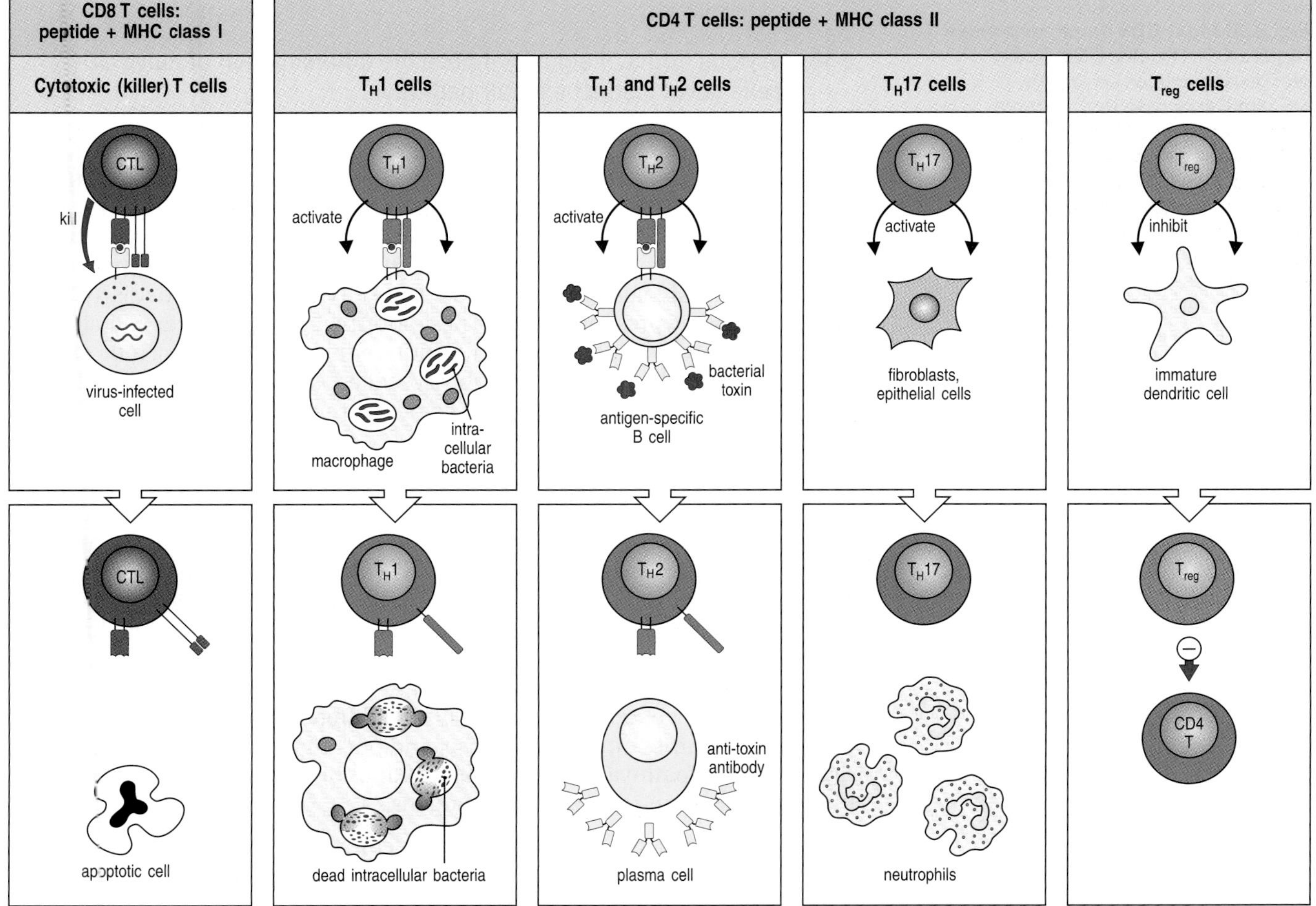

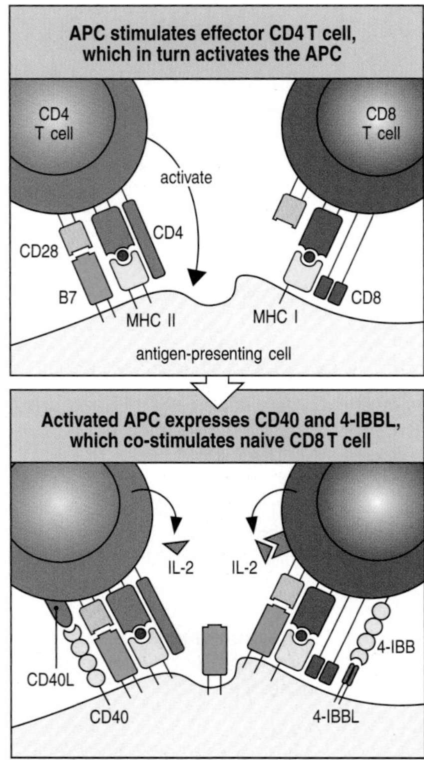

Fig. 8.28 Most CD8 T-cell responses require CD4 T cells. CD8 T cells recognizing antigen on weakly co-stimulatory cells may become activated only in the presence of CD4 T cells bound to the same antigen-presenting cell (APC). This happens mainly by an effector CD4 T cell recognizing antigen on the antigen-presenting cell and being triggered to induce increased levels of co-stimulatory activity on the antigen-presenting cell. The CD4 T cells also produce abundant IL-2 and thus help drive CD8 T-cell proliferation. This may in turn activate the CD8 T cell to make its own IL-2.

8-18 CD8 T cells can be activated in different ways to become cytotoxic effector cells.

Having had a brief overview of effector T cells and their functions, we will now see how they are derived from naive T cells. Naive CD8 T cells differentiate into cytotoxic cells, and perhaps because the effector actions of these cells are so destructive, naive CD8 T cells require more co-stimulatory activity to drive them to become activated effector cells than do naive CD4 T cells. This requirement can be met in two ways. The simplest is activation by mature dendritic cells, which have high intrinsic co-stimulatory activity. These cells can directly stimulate CD8 T cells to synthesize the IL-2 that drives their own proliferation and differentiation, and this property has been exploited to generate cytotoxic T-cell responses against tumors, as we will see in Chapter 15.

Such direct priming of CD8 cells by virus-infected antigen-presenting cells may occur in some settings, but in the majority of viral infections it seems that CD8 T-cell activation requires extra help. This is provided by CD4 effector T cells that recognize related antigens on the surface of the same antigen-presenting cell (Fig. 8.28). It is thought that the actions of the CD4 T cell are needed to compensate for inadequate co-stimulation of naive CD8 T cells by the virus-infected antigen-presenting cell; the recruitment of an effector CD4 T cell activates the antigen-presenting cell to express higher levels of co-stimulatory activity. Dendritic cells carry CD40 on their cell surface (see Section 8-8), and binding of this by CD40 ligand on the CD4 T cell induces B7 expression in the dendritic cell and enables it to co-stimulate the naive CD8 T cell directly. CD4 cells may also contribute by producing IL-2, which acts to promote CD8 T-cell differentiation.

8-19 Various forms of signal 3 induce the differentiation of naive CD4 T cells down distinct effector pathways.

The differentiation of CD4 T cells is more varied than that of CD8 T cells. Whereas CD8 effector T cells seem to take on a uniform cytotoxic phenotype, CD4 T cells can differentiate into several different kinds of effector subsets that act on other cells to promote distinct outcomes. The fate of the progeny of a naive CD4 T cell is largely decided during the initial priming period and is regulated by signals provided by the local environment, particularly by the priming antigen-presenting cell. These are the signals we will call signal 3. Naive CD4 T cells are now known to differentiate into at least four effector subtypes—T_H1, T_H2, T_H17, and the so-called adaptive regulatory T cells, which may be a heterogeneous subset, acting through the secretion of a variety of inhibitory cytokines (Fig. 8.29).

The differentiation of the T_H1 and T_H2 subsets is the best understood and we shall deal with them first. These subsets are distinguished principally by their production of specific cytokines, such as IFN-γ and IL-2 by T_H1 cells, and IL-4 and IL-5 by T_H2 cells. One or other of the T_H1 and T_H2 subsets are often found to predominate in immune responses that become chronic, such as in autoimmunity or in allergies. In most acute responses to infection, however, it is likely that both T_H1 and T_H2 cells are involved in making an effective response. A great deal has been learned about the mechanism by which these two subsets are generated. The decision to differentiate into T_H1 or T_H2 cells occurs early in the immune response, and one important determinant of the differentiation pathway is the mix of cytokines produced by cells of the innate immune system in response to pathogens.

In the case of T_H1 development, signal 3 comprises the cytokines IFN-γ and IL-12, which favor the differentiation of CD4 T cells to T_H1 when present at an early stage in T-cell activation. As described in Section 6-23, many key

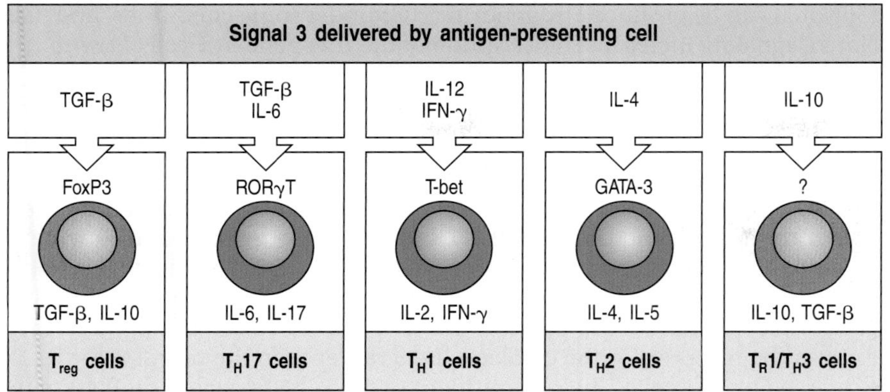

Fig. 8.29 Variation in signal 3 causes naive CD4 T cells to acquire several distinct types of effector functions. Naive CD4 T cells respond to specific peptide:MHC class II complexes and to co-stimulatory molecules by making IL-2 and proliferating. Antigen-presenting cells, principally dendritic cells, generate various cytokines or express surface proteins that act as signal 3 to induce the development of CD4 T cells into distinct types of effector cells. The specific form of signal 3 depends on the environmental conditions, such as the exposure to various pathogens. When pathogens are absent, a relative abundance of TGF-β and the lack of IL-6, IFN-γ, and IL-12 favors the development of FoxP3-expressing adaptive T_{reg} cells. Early in infection, IL-6 produced by dendritic cells acts with TGF-β to induce T_H17 cells expressing the transcription factor ROR-γT, which are amplified by IL-23. Later, dendritic cells and other antigen-presenting cells produce cytokines that promote either T_H1 (IFN-γ and IL-12) or T_H2 (IL-4 and Notch ligands) and suppress T_H17 development. T_H1 and T_H2 cells express the T-bet and GATA-3 transcription factors, respectively. Other adaptive regulatory subsets (T_R1 and T_H3) require IL-10 signaling during the differentiation of CD4 T cells. Characteristic cytokines produced by each effector subset are shown.

cytokines, including IFN-γ and IL-12, stimulate the JAK–STAT intracellular signaling pathway, resulting in the activation of specific genes. JAKs (Janus tyrosine kinases) and STATs (signal-transducing activators of transcription) are present as families of proteins, and different members can be activated to achieve different effects. Differentiation into the T_H1 subset is promoted by the activation of STAT1 in antigen-stimulated naive T cells by IFN-γ. In the context of infection, the required IFN-γ will initially be produced by cells of the innate immune system, such as NK cells, dendritic cells and macrophages, because the IFN-γ gene in resting naive CD4 T cells is switched off.

STAT1 in turn induces the expression of another transcription factor, T-bet, which switches on the IFN-γ gene in the T cell and also induces expression of the signaling subunit of the IL-12 receptor. These T cells are now committed to becoming T_H1 cells. The cytokine IL-12, again produced by innate immune cells such as dendritic cells, can then act through this receptor, via a signaling pathway that activates STAT4, to promote further expansion and differentiation of the committed T_H1 cells. These effector T_H1 cells will generate copious IFN-γ when they recognize antigen on a target cell, thus reinforcing the signal for the differentiation of more T_H1 cells. Thus, recognition of a particular type of pathogen by the innate immune system initiates a chain reaction that links the innate response to the adaptive immune response. For example, bacterial infections induce dendritic cells and macrophages to produce IL-12, favoring the emergence of T_H1 effector cells. These promote effector functions such as macrophage activation, which is required to clear infections caused by mycobacteria and *Listeria* for example, and the provision of help for antibody production against extracellular bacteria.

T_H2 development is favored by a different signal 3, in this case IL-4 (see Fig. 8.29). This cytokine is the most powerful trigger for inducing T_H2 development from naive CD4 T cells. If encountered while the naive T cells are being activated by antigen, the IL-4 signal activates STAT6, which promotes expression of the transcription factor GATA-3 in the T cell. GATA-3 is a powerful activator of the genes for several cytokines characteristically produced by T_H2 cells, which include IL-4, and also induces its own expression. In this way, GATA-3 both induces and maintains T_H2 differentiation. It is still unclear whether there is a single distinct source for the IL-4 that initially triggers the T_H2 response. Recent evidence has suggested that certain proteins secreted by activated dendritic cells can lead to the activation of the IL-4 and GATA-3 genes in T cells, thus starting a cascade of positive feedback for differentiation as T_H2 cells as a result of continued IL-4 secretion. It is thought that these dendritic cell signals may be ligands for the Notch receptor on T cells (which we discussed in Chapter 7 in relation to its role in the development of T cells in the thymus). Although the details are still incomplete, it seems that ligands for Notch can be produced by dendritic cells under some conditions, and that Notch signaling increases transcription of the IL-4 gene in T cells *in vitro*.

T_H1 and T_H2 cells have been analyzed in detail because conditions have been found that allow them to be generated in abundance and maintained *in vitro*. Other functional subsets of CD4 cells have been recognized more recently, and their properties and the conditions under which they differentiate are, as yet, less well characterized. The CD4 T_H17 cells are distinguished by their ability to produce the cytokine IL-17 but not IFN-γ or IL-4, and they have recently been recognized as a distinct effector lineage (see Fig. 8.29). Naive CD4 T cells commit to the T_H17 lineage when both IL-6 and transforming growth factor (TGF)-β are present, but IL-4 and IL-12 are absent, and they express the receptor for the cytokine IL-23 rather than the receptor for IL-12 expressed by T_H1 cells. The commitment to the T_H17 lineage is thought to be under the control of the transcription factor RORγT, which is induced in these conditions and which drives expression of the receptor for IL-23. The expansion and further development of T_H17 effector activity seem to require IL-23, in a similar way to the requirement for IL-12 for effective T_H1 responses.

The other effector T-cell subsets that may differentiate from naive CD4 T cells are the adaptive regulatory T cells (see Fig. 8.29). They produce the cytokines IL-10 and TGF-β, which are cytokines that inhibit rather than activate T-cell responses. These inhibitory cytokines provide these cells with regulatory activity, which is important in maintaining self-tolerance during strong immune responses to pathogens.

The consequences of inducing the development of these various CD4 subsets are profound: the selective production of T_H1 cells leads to cell-mediated immunity and the production of opsonizing antibody classes (predominantly IgG), whereas the production of predominantly T_H2 cells provides humoral immunity, especially IgM, IgA, and IgE. T_H17 cells seem to be important in the recruitment of neutrophils to control the early stages of an infection, and the regulatory T-cell subsets restrain inflammation and maintain tolerance.

A striking example of the difference that different T-cell subsets can make to the outcome of infection is seen in leprosy, a disease caused by infection with *Mycobacterium leprae*. *M. leprae*, like *M. tuberculosis*, grows in macrophage vesicles, and effective host defense requires macrophage activation by T_H1 cells. In patients with **tuberculoid leprosy**, in which T_H1 cells are preferentially induced, few live bacteria are found, little antibody is produced, and, although skin and peripheral nerves are damaged by the inflammatory responses associated with macrophage activation, the disease progresses slowly and the patient usually survives. However, when T_H2 cells are preferentially induced, the main response is humoral, the antibodies produced cannot reach the intracellular bacteria, and the patients develop **lepromatous leprosy**, in which *M. leprae* grows abundantly in macrophages, causing gross tissue destruction that is eventually fatal.

8-20 Regulatory CD4 T cells are involved in controlling adaptive immune responses.

The regulatory T cells found in the periphery are a heterogeneous group of cells with different developmental origins. One subset of regulatory T cells becomes committed to a regulatory fate during their development in the thymus (see Section 7-18). These are the natural T regulatory cells (natural T_{reg}). They are CD4-positive cells that also express the α chain of the IL-2 receptor (CD25), and high levels of the L-selectin receptor CD62L, and represent about 10–15% of the CD4 T cells in the human circulation. Natural T_{reg} express the transcription factor FoxP3, which interferes with the interaction between

AP-1 and NFAT at the IL-2 promoter, preventing transcriptional activation of the IL-2 gene (see Section 8-13). Natural T_{reg} are potentially self-reactive T cells that express conventional α:β T-cell receptors and seem to be selected in the thymus by high-affinity binding to MHC molecules containing self peptides. It is currently unknown whether they are activated to express their regulatory function in the periphery by the same self ligands that selected them in the thymus or by other self or nonself antigens. Once activated, they may mediate their effects in a contact-dependent fashion, although some evidence suggests they may also function by secreting IL-10 and TGF-β, cytokines that can inhibit T-cell proliferation (see Fig. 8.29). IL-10 can also affect the differentiation of dendritic cells, inhibiting their secretion of IL-12 and thus impairing their ability to promote T-cell activation and T_H1 differentiation. Failure of natural T_{reg} function is known to be involved in several autoimmune syndromes and is described in more detail in Chapter 14. In addition to their ability to prevent autoimmune disease *in vivo*, natural T_{reg} have been shown to suppress antigen-specific T-cell proliferation and T-cell proliferation in response to allogeneic cells *in vitro*.

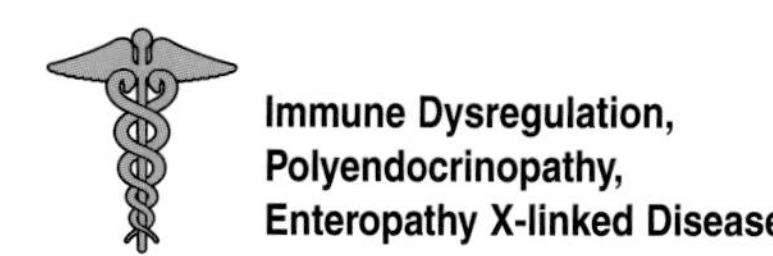

Adaptive regulatory T cells, in contrast, develop in the periphery from apparently uncommitted naive CD4 T cells (see Fig. 8.29). They are a heterogeneous group that include several subsets of T cells with different phenotypes, different properties and different conditions favoring their differentiation. One subset of these adaptive regulatory T cells, called $\mathbf{T_H3}$, is found in the mucosal immune system (see Section 11-13). T_H3 cells produce IL-4, IL-10, and TGF-β; they are distinguished from T_H2 cells by their production of TGF-β. T_H3 cells may be predominantly of mucosal origin and be activated by the mucosal presentation of antigen. They seem to function to suppress or control immune responses in the mucosae, which form barriers to the microbe-laden world. Lack of these cells is linked to autoimmune disease in the gut and to inflammatory bowel disease. Giving animals large amounts of self antigen orally, which induces so-called oral tolerance (see Section 11-13), can sometimes lead to unresponsiveness to these antigens when given by other routes and can prevent autoimmune disease. Induction of oral tolerance leads to the generation or expansion of T_H3 cells, which could have a role in this mechanism.

Another subset of adaptive regulatory T cells is called $\mathbf{T_R1}$. T_R1 cells have been produced *in vitro* and are also likely to be present *in vivo*. T_R1 cells can be cultured *in vitro* in the presence of high concentrations of IL-10, and their development is also favored by IFN-α. They secrete the inhibitory cytokine TGF-β but not IL-4, which helps distinguish them from T_H3 cells. The natural origin of T_R1 cells is not clear. Immature dendritic cells presenting antigens in the absence of inflammatory stimuli could be the origin of the IFN-α and IL-10 that initiate their development.

More recently, another population of adaptive regulatory T cells has been described in which FoxP3 expression is induced in naive CD4 T cells in the periphery under conditions in which there is a predominance of TGF-β, rather than IFN-γ, IL-12, or IL-4, in the environment. These adaptive regulatory CD4 T cells can produce TGF-β and may exert direct suppression through other mechanisms as well. The relationship between these cells and T_H3 and T_R1 cells is currently unclear.

IL-10 suppresses T-cell responses directly by reducing the production of IL-2, TNF-α, and IL-5 by T cells, and indirectly by inhibiting antigen presentation by reducing the expression of MHC and co-stimulatory molecules by antigen-presenting cells. TGF-β similarly blocks T-cell cytokine production, cell division, and killing ability. Not all the effects of IL-10 and TGF-β are immunosuppressive, however: IL-10 can enhance B-cell survival and maturation into plasma cells and increase the activity of CD8 T cells. Nevertheless, the dominant effects *in vivo* of both IL-10 and TGF-β are immunosuppressive, as shown by the fact that mice lacking either cytokine are prone to autoimmune disease.

Summary.

The crucial first step in adaptive immunity is the activation or priming of naive antigen-specific T cells by antigen-presenting cells within the lymphoid tissues and organs through which they are constantly passing. The most distinctive feature of antigen-presenting cells is the expression of cell-surface co-stimulatory molecules, of which the B7 molecules are the most important in natural responses to infection. Naive T cells will respond to antigen only when the antigen-presenting cell presents both a specific antigen to the T-cell receptor and a B7 molecule to CD28, the receptor for B7 on the T cell.

The activation of naive T cells leads to their proliferation and the differentiation of their progeny into effector T cells. Proliferation and differentiation depend on the production of cytokines, in particular IL-2, which binds to a high-affinity receptor on the activated T cell. T cells whose antigen receptors are ligated in the absence of co-stimulatory signals fail to make IL-2 and instead become anergic or die. This dual requirement for both receptor ligation and co-stimulation by the same antigen-presenting cell helps to prevent naive T cells from responding to self antigens on tissue cells, which lack co-stimulatory activity.

Antigen-stimulated proliferating T cells develop into effector T cells, the critical event in most adaptive immune responses. Various forms of signal 3 contribute to the type of effector T cell that develops in response to an infection. The nature of signal 3 is in turn influenced by the response made by the innate immune system when it first recognizes the pathogen. Once an expanded clone of T cells achieves effector function, its progeny can act on any target cell that displays antigen on its surface. Effector T cells have a variety of functions. CD8 cytotoxic T cells recognize virus-infected cells and kill them. T_H1 effector cells promote the activation of macrophages, and together they make up cell-mediated immunity. Both T_H2 and T_H1 cells coordinate the activation of B cells to produce different classes of antibody, thus driving the humoral immune response. T_H17 cells enhance the acute inflammatory response to infection by recruiting neutrophils to sites of infection. Regulatory CD4 T-cell subsets restrain the immune response by producing inhibitory cytokines, sparing surrounding tissues from collateral damage.

General properties of effector T cells and their cytokines.

All T-cell effector functions involve the interaction of an effector T cell with a target cell displaying specific antigen. The effector proteins released by the T cells are focused on the target by mechanisms that are activated by antigen recognition. The focusing mechanism is common to all types of effector T cells, whereas their effector actions depend on the array of membrane and secreted proteins that they express or release upon ligation of their antigen receptors. The different types of effector T cells are specialized to deal with different types of pathogens, and the effector molecules that they are programmed to produce cause distinct and appropriate effects on the target cell.

8-21 Effector T-cell interactions with target cells are initiated by antigen-nonspecific cell-adhesion molecules.

Once an effector T cell has completed its differentiation in the lymphoid tissue, it must find target cells that are displaying the peptide:MHC complex that it recognizes. Some T_H2 cells encounter their B-cell targets without leaving the lymphoid tissue, as we discuss further in Chapter 9. However, most of the effector T cells emigrate from their site of activation in lymphoid tissues and enter the blood via the thoracic duct. Because of the cell-surface changes that have occurred during differentiation, they can now migrate into tissues, particularly at sites of infection. They are guided to these sites by changes in the adhesion molecules expressed on the endothelium of the local blood vessels as a result of infection, and by local chemotactic factors.

The initial binding of an effector T cell to its target, like that of a naive T cell to an antigen-presenting cell, is an antigen-nonspecific interaction mediated by LFA-1 and CD2. The levels of LFA-1 and of CD2 are two to fourfold higher on effector T cells than on naive T cells, and so effector T cells can bind efficiently to target cells that have less ICAM and CD58 on their surface than do antigen-presenting cells. This interaction is transient unless recognition of antigen on the target cell by the T-cell receptor triggers an increase in the affinity of the T-cell's LFA-1 for its ligands. The T cell then binds more tightly to its target and remains bound for long enough to release its effector molecules. CD4 effector T cells, which activate macrophages or induce B cells to secrete antibody, must maintain contact with their targets for relatively long periods. Cytotoxic T cells, by contrast, can be observed under the microscope attaching to and dissociating from successive targets relatively rapidly as they kill them (Fig. 8.30). Killing of the target, or some local change in the T cell, allows the effector T cell to detach and address new targets. How CD4 effector T cells disengage from their antigen-negative targets is not known, although evidence suggests that CD4 binding to MHC class II molecules without engagement of the T-cell receptor provides a signal for the cell to detach.

8-22 Binding of the T-cell receptor complex directs the release of effector molecules and focuses them on the target cell.

When binding to their specific antigenic peptide:self-MHC complexes or to self-peptide:self-MHC complexes, the T-cell receptors and their associated co-receptors cluster at the site of cell–cell contact, forming what is called the **supramolecular adhesion complex** (**SMAC**) or the **immunological synapse**. Other cell-surface molecules also cluster here. For example, the tight binding of LFA-1 to ICAM-1 induced by ligation of the T-cell receptor creates a molecular seal that surrounds the T-cell receptor and its co-receptor (Fig. 8.31).

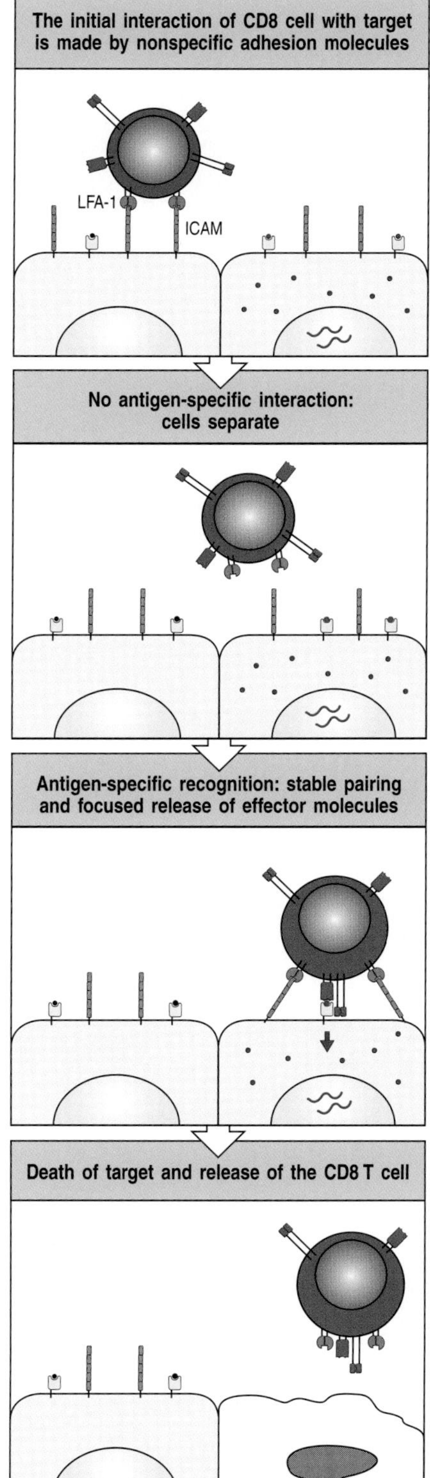
The initial interaction of CD8 cell with target is made by nonspecific adhesion molecules
LFA-1
ICAM
No antigen-specific interaction: cells separate
Antigen-specific recognition: stable pairing and focused release of effector molecules
Death of target and release of the CD8 T cell

Fig. 8.30 Interactions of T cells with their targets initially involve nonspecific adhesion molecules. The major initial interaction is between LFA-1 on the T cell, illustrated here as a cytotoxic CD8 T cell, and ICAM-1 or ICAM-2 on the target cell (top panel). This binding allows the T cell to remain in contact with the target cell and to scan its surface for the presence of specific peptide:MHC complexes. If the target cell does not carry the specific antigen, the T cell disengages (second panel) and can scan other potential targets until it finds the specific antigen (third panel). Signaling through the T-cell receptor increases the strength of the adhesive interactions, prolonging the contact between the two cells and stimulating the T cell to deliver its effector molecules. The T cell then disengages (bottom panel).

Outer ring (red) pSMAC	Inner circle (green) cSMAC
LFA-1:ICAM-1	TCR, CD4, CD28 MHC:peptide

Fig. 8.31 The area of contact between an effector T cell and its contact forms an immunological synapse. A confocal fluorescence micrograph of the area of contact between a CD4 T cell and a B cell (as viewed through one of the cells) is shown. Proteins in the contact area between the T cell and the antigen-presenting cell form a structure called the immunological synapse, also known as the supramolecular adhesion complex (SMAC), which is organized into two distinct regions: the outer, or peripheral SMAC (pSMAC), indicated by the red ring; and the inner, or central SMAC (cSMAC), indicated in bright green. The cSMAC is enriched in the T-cell receptor (TCR), CD4, CD8, CD28, and CD2. The pSMAC is enriched for the integrin LFA-1 and the cytoskeletal protein talin. Photograph courtesy of A. Kupfer.

Clustering of the T-cell receptors signals a reorientation of the cytoskeleton that polarizes the effector cell so as to focus the release of effector molecules at the site of contact with the target cell. This is illustrated for a cytotoxic T cell in Fig. 8.32. An important intermediary in the effect of T-cell signaling on the cytoskeleton is the Wiskott-Aldrich sydrome protein (WASP), defects in which result in the inability of T cells to become polarized, among other effects, and cause an immune deficiency syndrome for which the protein is named (see Section 12-15). WASP is activated via T-cell receptor signaling through several pathways, for example, by an adaptor protein called Nck or by the small GTP-binding proteins Cdc42 and Rac1, which are activated by the adaptor protein Vav. Polarization starts with the local reorganization of the cortical actin cytoskeleton at the site of contact; this in turn leads to the reorientation of the microtubule-organizing center (MTOC), the center from which the microtubule cytoskeleton is produced, and of the Golgi apparatus (GA), through which most proteins destined for secretion travel. In the cytotoxic T cell, the cytoskeletal reorientation focuses exocytosis of the preformed cytotoxic granules at the site of contact with its target cell. The polarization of a T cell also focuses the secretion of soluble effector molecules whose synthesis is induced *de novo* by ligation of the T-cell receptor. For example, the secreted cytokine IL-4, which is the principal effector molecule of T_H2 cells, is confined and concentrated at the site of contact with the target cell (see Fig. 9.6).

Thus, the antigen-specific T-cell receptor controls the delivery of effector signals in three ways: it induces the tight binding of effector cells to their target cells to create a narrow space in which effector molecules can be concentrated; it focuses their delivery at the site of contact by inducing a reorientation of the secretory apparatus of the effector cell; and it triggers their synthesis and/or release. All these mechanisms contribute to targeting the action of effector molecules onto the cell bearing a specific antigen. Effector T-cell activity is thus highly selective for the appropriate target cells, even though the effector molecules themselves are not antigen-specific.

8-23 The effector functions of T cells are determined by the array of effector molecules that they produce.

The effector molecules produced by effector T cells fall into two broad classes: **cytotoxins**, which are stored in specialized cytotoxic granules and released by CD8 cytotoxic T cells (see Fig. 8.32), and cytokines and related membrane-associated proteins, which are synthesized *de novo* by all effector T cells. Cytotoxins are the principal effector molecules of cytotoxic T cells and will be discussed in Section 8-28. Their release in particular must be tightly regulated because they are not specific: they can penetrate the lipid bilayer and trigger apoptosis in any cell. By contrast CD4 effector T cells act mainly through the production of cytokines and membrane-associated proteins, and

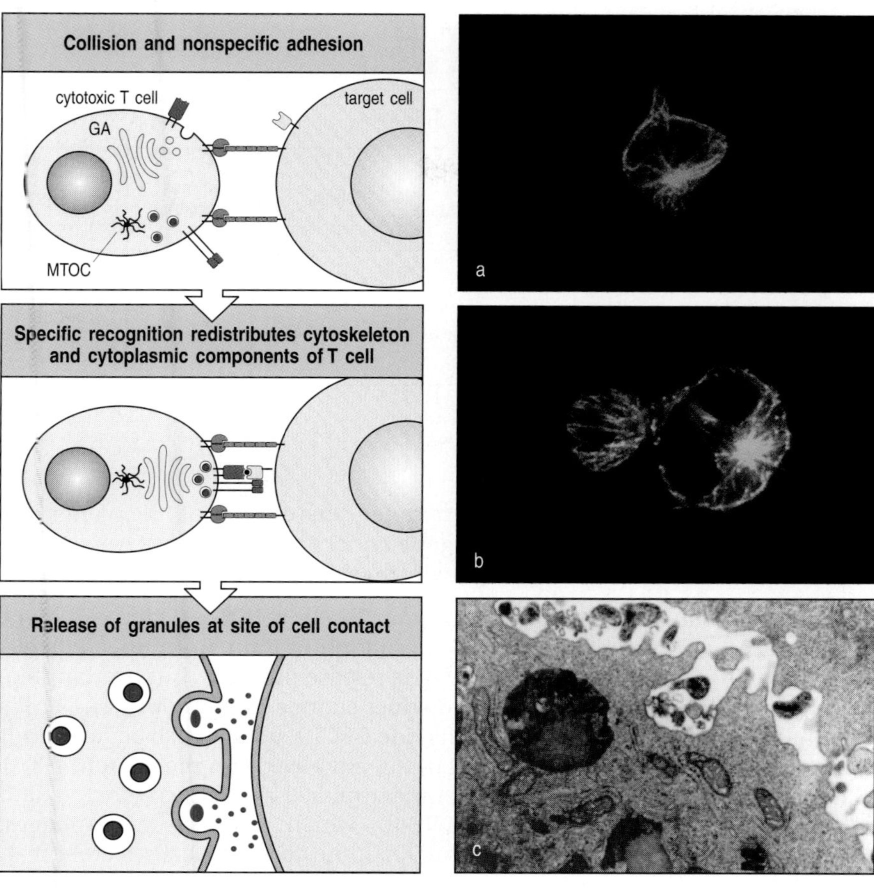
Collision and nonspecific adhesion
cytotoxic T cell
target cell
GA
MTOC
a
Specific recognition redistributes cytoskeleton and cytoplasmic components of T cell
b
Release of granules at site of cell contact
c

Fig. 8.32 The cellular polarization of T cells during specific antigen recognition allows effector molecules to be focused on the antigen-bearing target cell. The example illustrated here is a CD8 cytotoxic T cell. Cytotoxic T cells contain specialized lysosomes called cytotoxic granules (shown in red in the left panels), which contain cytotoxic proteins. Initial binding to a target cell through adhesion molecules does not have any effect on the location of the cytotoxic granules. Binding of the T-cell receptor causes the T cell to become polarized: reorganization within the cortical actin cytoskeleton at the site of contact aligns the microtubule-organizing center (MTOC), which in turn aligns the secretory apparatus, including the Golgi apparatus (GA), towards the target cell. Proteins stored in cytotoxic granules derived from the Golgi are then directed specifically onto the target cell. The photomicrograph in panel a shows an unbound, isolated cytotoxic T cell. The microtubule cytoskeleton is stained in green and the cytotoxic granules in red. Note how the granules are dispersed throughout the T cell. Panel b depicts a cytotoxic T cell bound to a (larger) target cell. The granules are now clustered at the site of cell–cell contact in the bound T cell. The electron micrograph in panel c shows the release of granules from a cytotoxic T cell. Panels a and b courtesy of G. Griffiths. Panel c courtesy of E. Podack.

their actions are restricted to cells bearing MHC class II molecules and expressing receptors for these proteins.

The main effector molecules of T cells are summarized in Fig. 8.33. The cytokines are a diverse group of proteins and we will review them briefly before discussing the T-cell cytokines and their actions. Soluble cytokines and membrane-associated molecules often act in combination to mediate these effects.

8-24 Cytokines can act locally or at a distance.

Cytokines are small soluble proteins secreted by one cell that can alter the behavior or properties of the cell itself or of another cell. They are produced by many cells in addition to those of the immune system. We have already discussed the cytokines released by phagocytic cells in Chapter 2, where we dealt with the inflammatory reactions that are important in innate immunity. Here we are concerned mainly with the cytokines that mediate the effector functions of T cells. Cytokines produced by lymphocytes are often called **lympho-kines**, but this nomenclature can be confusing because some lymphokines are also secreted by nonlymphoid cells; we will therefore use the generic term 'cytokine' for all of them. Most cytokines produced by T cells are given the name **interleukin** (**IL**) followed by a number: we have encountered several interleukins already in this chapter. The cytokines produced by T cells are shown in Fig. 8.34, and a more comprehensive list of cytokines of immunological interest is in Appendix III. Most cytokines have a multitude of different biological effects when tested at high concentration

CD8 T cells: peptide + MHC class I		CD4 T cells: peptide + MHC class II							
Cytotoxic (killer) T cells		T_H1 cells		T_H2 cells		T_H17 cells		T_{reg} cells	
Cytotoxic effector molecules	Others	Macrophage-activating effector molecules	Others	B-cell-activating effector molecules	Others	Neutrophil recruitment	Others	Suppressive cytokines	Others
Perforin Granzymes Granulysin Fas ligand	IFN-γ LT-α TNF-α	IFN-γ GM-CSF TNF-α CD40 ligand Fas ligand	IL-3 LT-α CXCL2 (GROβ)	IL-4 IL-5 IL-13 CD40 ligand	IL-3 GM-CSF IL-10 TGF-β CCL11 (eotaxin) CCL17 (TARC)	IL-17A IL-17F IL-6	TNF CXCL1 (GROα)	IL-10 TGF-β	GM-CSF

Fig. 8.33 The different types of effector T cell subsets produce different effector molecules. CD8 T cells are predominantly killer T cells that recognize peptide:MHC class I complexes. They release perforin (which helps deliver granzymes into the target cell), granzymes (which are pro-proteases that are activated intracellularly to trigger apoptosis in the target cell), and often also produce the cytokine IFN-γ. They also carry the membrane-bound effector molecule Fas ligand (CD178). When this binds to Fas (CD95) on a target cell it activates apoptosis in the Fas-bearing cell. The various functional subsets of CD4 T cells recognize peptide:MHC class II complexes. T_H1 cells are specialized to activate macrophages that are infected by or have ingested pathogens; they secrete IFN-γ to activate the infected cell, as well as other effector molecules. They can express membrane-bound CD40 ligand and/or Fas ligand. CD40 ligand triggers activation of the target cell, whereas Fas ligand triggers the death of Fas-bearing targets, and so which molecule is expressed strongly influences T_H1 function. T_H2 cells are specialized for promoting immune responses to parasites and also promote allergic responses. They provide help in B-cell activation and secrete the B-cell growth factors IL-4, IL-5, IL-9, and IL-13. The principal membrane-bound effector molecule expressed by T_H2 cells is CD40 ligand, which binds to CD40 on B cells and induces B-cell proliferation and isotype switching (see Chapter 9). T_H17 cells produce members of the IL-17 family and IL-6, and promote acute inflammation by helping to recruit neutrophils to sites of infection. T_{reg} cells, of which there are several types, produce inhibitory cytokines such as IL-10 and TGF-β and exert inhibitory actions through unknown mechanisms that are dependent on cell contact

in biological assays *in vitro*, but targeted disruption of the genes for cytokines and cytokine receptors by gene knockout in mice (see Appendix I, Section A-47) has helped to clarify their physiological roles.

The main cytokine released by CD8 effector T cells is IFN-γ, which can block viral replication or even lead to the elimination of virus from infected cells without killing them. CD4 effector subsets release different, but overlapping, sets of cytokines, which define their distinct actions in immunity. T_H17 cells secrete IL-17, IL-6, TNF, and the chemokine CXCL1, all of which act to recruit neutrophils to sites of infection early in the adaptive immune response. T_H1 cells secrete IFN-γ, which is the main macrophage-activating cytokine, and LT-α (also called lymphotoxin or TNF-β), which activates macrophages, inhibits B cells, and is directly cytotoxic for some cells. T_H2 cells secrete IL-4, IL-5, IL-9, and IL-13, as well as having CD40 ligand on their surface, all of which activate B cells, and IL-10, which inhibits macrophage activation. During the earliest stages of activation, provided that co-stimulatory signals are present, CD4 T cells produce IL-2, and only very small amounts of IL-4 and IFN-γ.

We have already discussed in Section 8-22 how binding of the T-cell receptor orchestrates the polarized release of these cytokines so that they are concentrated at the site of contact with the target cell. Furthermore, most of the soluble cytokines have local actions that synergize with those of the membrane-bound effector molecules. The effect of all these molecules is therefore combinatorial, and, because the membrane-bound effectors can bind only to receptors on an interacting cell, this is another mechanism by which selective effects of cytokines are focused on the target cell. The effects of some cytokines are further confined to target cells by tight regulation of their synthesis: the synthesis of IL-2, IL-4, and IFN-γ is controlled by mRNA instability (see Section 8-13), so that their secretion by T cells does not continue after the interaction with a target cell ends.

Some cytokines have more distant effects. IL-3 and GM-CSF (see Fig. 8.34) are released by T_H1 and T_H2 cells and act on bone marrow cells to stimulate the production of macrophages and granulocytes, both of which are important nonspecific effector cells in both humoral and cell-mediated immunity. IL-3 and GM-CSF also stimulate the production of dendritic cells from bone marrow precursors. The predominant T cells activated in allergic reactions are T_H2 cells, and the IL-5 they produce can increase the production of eosinophils, which contribute to the late phase of an allergic reaction (see Chapter 13). Whether a given cytokine effect is local or more distant is likely to reflect the amounts released, the degree to which this release is focused on the target cell, and the stability of the cytokine *in vivo*.

Cytokine	T-cell source	Effects on					Effect of gene knockout
		B cells	T cells	Macrophages	Hematopoietic cells	Other somatic cells	
Interleukin-2 (IL-2)	Naive, T_H1, some CD8	Stimulates growth and J-chain synthesis	Growth	–	Stimulates NK cell growth	–	↓T-cell responses IBD
Interferon-γ (IFN-γ)	T_H1, CTL	Differentiation IgG2a synthesis (mouse)	Inhibits T_H2 cell growth	Activation, ↑MHC class I and class II	Activates NK cells	Antiviral ↑ MHC class I and class II	Susceptible to mycobacteria, some viruses
Lymphotoxin (LT, TNF-β)	T_H1, some CTL	Inhibits	Kills	Activates, induces NO production	Activates neutrophils	Kills fibroblasts and tumor cells	Absence of lymph nodes Disorganized spleen
Interleukin-4 (IL-4)	T_H2	Activation, growth IgG1, IgE ↑MHC class II induction	Growth, survival	Inhibits macrophage activation	↑Growth of mast cells	–	No T_H2
Interleukin-5 (IL-5)	T_H2	Mouse: Differentiation IgA synthesis	–	–	↑Eosinophil growth and differentiation	–	Reduced eosinophilia
Interleukin-10 (IL-10)	T_H2 (human: some T_H1), T_{reg}	↑MHC class II	Inhibits T_H1	Inhibits cytokine release	Co-stimulates mast cell growth	–	IBD
Interleukin-3 (IL-3)	T_H1, T_H2 some CTL	–	–	–	Growth factor for progenitor hematopoietic cells (multi-CSF)	–	–
Tumor necrosis factor-α (TNF-α)	T_H1, some T_H2 some CTL	–	–	Activates, induces NO production	–	Activates microvascular endothelium	Susceptibility to Gram –ve sepsis
Granulocyte-macrophage colony-stimulating factor (GM-CSF)	T_H1, some T_H2 some CTL	Differentiation	Inhibits growth?	Activation Differentiation to dendritic cells	↑Production of granulocytes and macrophages (myelopoiesis) and dendritic cells	–	–
Transforming growth factor-β (TGF-β)	CD4 T cells (T_{reg})	Inhibits growth IgA switch factor	Inhibits growth, promotes survival	Inhibits activation	Activates neutrophils	Inhibits/ stimulates cell growth	Death at ~10 weeks
Interleukin-17 (IL-17)	CD4 T cells (T_H17) macrophages	–	–	–	Stimulates neutrophil recruitment	Stimulates fibroblasts and epithelial cells to secrete chemokines	–

Fig. 8.34 The nomenclature and functions of well-defined T-cell cytokines. Each cytokine has multiple activities on different cell types. Major activities of effector cytokines are highlighted in red. The mixture of cytokines secreted by a given cell type produces many effects through what is called a 'cytokine network.' ↑, increase; ↓, decrease; CTL, cytotoxic lymphocyte; NK, natural killer; CSF, colony-stimulating factor; IBD, inflammatory bowel disease; NO, nitric oxide.

8-25 Cytokines and their receptors fall into distinct families of structurally related proteins.

Cytokines can be grouped by structure into different families and their receptors can likewise be grouped (Fig. 8.35). We have encountered members of some of these families in Chapter 2 and have given an overview of the chemokines there (see Section 2-24). We will focus here on the hematopoietins, the TNF family, and IFN-γ because of their important roles in T-cell effector function. Members of the TNF family act as trimers, most of which are membrane-bound and so are quite distinct in their properties from the other cytokines. Nevertheless, they share some important properties with the

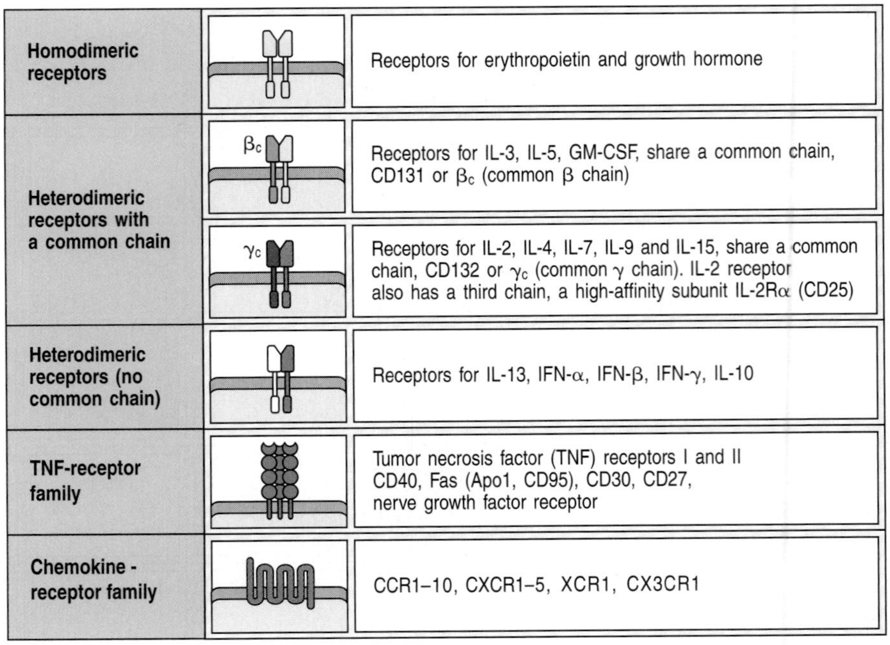

Receptor family		Examples
Homodimeric receptors		Receptors for erythropoietin and growth hormone
Heterodimeric receptors with a common chain	β_c	Receptors for IL-3, IL-5, GM-CSF, share a common chain, CD131 or β_c (common β chain)
	γ_c	Receptors for IL-2, IL-4, IL-7, IL-9 and IL-15, share a common chain, CD132 or γ_c (common γ chain). IL-2 receptor also has a third chain, a high-affinity subunit IL-2Rα (CD25)
Heterodimeric receptors (no common chain)		Receptors for IL-13, IFN-α, IFN-β, IFN-γ, IL-10
TNF-receptor family		Tumor necrosis factor (TNF) receptors I and II CD40, Fas (Apo1, CD95), CD30, CD27, nerve growth factor receptor
Chemokine - receptor family		CCR1–10, CXCR1–5, XCR1, CX3CR1

Fig. 8.35 Cytokine receptors belong to families of receptor proteins, each with a distinctive subunit structure. A large number of cytokines signal through members of the hematopoietin receptor superfamily, named after the first member of this family to be described, the receptor for erythropoietin. The hematopoietin receptor superfamily includes homodimeric and heterodimeric receptors, which are subdivided into families on the basis of protein sequence and subunit structure. Examples of this superfamily are given in the first three rows. In these receptors, one chain often defines the ligand specificity of the receptor, whereas the β or γ chain confers the intracellular signaling function. The receptor for the macrophage cytokine IL-6 (see Chapter 2) is also a member of the hematopoietin superfamily, but signals through a different accessory chain from β_c or γ_c. Receptors for interferons and interferon-like cytokines are heterodimeric receptors that make up another, smaller, cytokine receptor family. Other families of cytokine receptors are the tumor necrosis factor receptor (TNFR) family and the chemokine receptor family, the latter belonging to the very large superfamily of G-protein-coupled receptors. Each family member is a variant with a distinct specificity, performing a particular function on the cell that expresses it. For the TNFR family, the ligands act as trimers and may be associated with the cell membrane rather than being secreted. The diagrams indicate the representations of these receptors that you will encounter throughout this book. Some of these receptors have been mentioned already in this book, some will occur in later chapters, and some are important examples from other biological systems

soluble T-cell cytokines, because they are also synthesized *de novo* upon antigen recognition by T cells and affect the behavior of the target cell.

Many of the soluble cytokines made by effector T cells are members of the hematopoietin family. These cytokines and their receptors can be further divided into subfamilies characterized by functional similarities and genetic linkage. For instance, IL-3, IL-4, IL-5, IL-13, and GM-CSF are related structurally, their genes are closely linked in the genome, and all are major cytokines produced by T_H2 cells. In addition, they bind to closely related receptors, which form a family of cytokine receptors. The IL-3, IL-5, and GM-CSF receptors share a common β chain. Another subgroup of cytokine receptors is defined by their use of the γ chain of the IL-2 receptor. This chain is shared by receptors for the cytokines IL-2, IL-4, IL-7, IL-9, and IL-15 and is now called the common γ chain (γ_c). More distantly related, the receptor for IFN-γ is a member of a small family of cytokine receptors with some similarities to the hematopoietin receptor family. This family also includes the receptor for IFN-α and IFN-β, and the IL-10 and IL-13 receptors.

Overall, the structural, functional, and genetic relations between the cytokines and their receptors suggest that they may have diversified in parallel during the evolution of increasingly specialized effector functions. These specific functional effects depend on intracellular signaling events that are triggered by the cytokines binding to their specific receptors. The hematopoietin and interferon receptors all signal through the JAK–STAT pathway and activate different combinations of STATs with different effects, as described in Section 8-19.

8-26 The TNF family of cytokines are trimeric proteins that are usually associated with the cell surface.

TNF-α is made by T cells in soluble and membrane-associated forms, both of which are made up of three identical protein chains (a homotrimer; see Fig. 2.44). LT-α (formerly known as TNF-β) now more commonly called lymphotoxin (LT)-α, can be produced as a secreted homotrimer but is usually linked to the cell surface by forming heterotrimers with a third, membrane-associated, member of this family called LT-β. The receptors for these molecules, TNFR-I and TNFR-II, form homotrimers when bound to either TNF-α or LT. The trimeric structure is characteristic of all members of the TNF family, and the ligand-induced trimerization of their receptors seems to be the critical event in initiating signaling.

Most effector T cells express members of the TNF protein family as cell-surface molecules. The most important in T-cell effector function are TNF-α, LT-α, Fas ligand (CD178), and CD40 ligand, the latter two always being cell-surface associated. These proteins all bind receptors that are members of the TNFR family; TNFR-I and TNFR-II can each interact with either TNF-α or LT-α, whereas Fas ligand and CD40 ligand bind respectively to the transmembrane proteins Fas (CD95) and CD40 on target cells. Fas contains a 'death' domain in its cytoplasmic tail and binding of Fas by Fas ligand induces death by apoptosis in the Fas-bearing cell (see Fig. 6.29). Other TNFR-family members, including TNFR-I, are also associated with death domains and can also induce apoptosis. Thus, TNF-α and LT-α can induce apoptosis by binding to TNFR-I.

CD40 ligand is particularly important for CD4 T-cell effector function; it is induced on T_H1 and T_H2 cells, and delivers activating signals to B cells and macrophages through CD40. The cytoplasmic tail of CD40 lacks a death domain; instead, it seems to be linked downstream to proteins called TRAFs (TNF-receptor-associated factors). CD40 is involved in the activation of B cells and macrophages; the ligation of CD40 on B cells promotes growth and isotype switching, whereas CD40 ligation on macrophages induces them to secrete TNF-α and become receptive to much lower concentrations of IFN-γ. Deficiency in CD40 ligand expression is associated with immunodeficiency, as we will learn in Chapters 9 and 14.

Summary.

Interactions between effector T cells and their targets are initiated by transient antigen-nonspecific adhesion between the cells. T-cell effector functions are elicited only when peptide:MHC complexes on the surface of the target cell are recognized by the receptor on an effector T cell. This recognition event triggers the effector T cell to adhere more strongly to the antigen-bearing target cell and to release its effector molecules directly at the target cell, leading to the activation or death of the target. The immunological consequences of antigen recognition by an effector T cell are determined largely by the set of effector molecules that it produces on binding a specific target cell. CD8 cytotoxic T cells store preformed cytotoxins in specialized cytotoxic granules whose release is tightly focused at the site of contact with the infected target cell, thus killing it without killing any uninfected cells nearby. Cytokines and members of the TNF family of membrane-associated effector proteins are synthesized *de novo* by most effector T cells. T_H1 cells express effector proteins that activate macrophages, and cytokines that induce class switching to certain antibody classes. T_H2 cells express B-cell-activating effector proteins and secrete cytokines that promote class switching to antibodies involved in anti-parasitic and allergic type responses. T_H17 cells secrete IL-17, which recruits acute inflammatory cells such as neutrophils to the site of infection. Membrane-associated effector molecules can deliver signals only to an interacting cell bearing the appropriate receptor, whereas soluble cytokines can act on cytokine receptors expressed locally on the target cell, or on hematopoietic cells at a distance. The actions of cytokines and membrane-associated effector molecules through their specific receptors, together with the effects of the cytotoxins released by CD8 cells, account for most of the effector functions of T cells.

T cell-mediated cytotoxicity.

All viruses, and some bacteria, multiply in the cytoplasm of infected cells; indeed, a virus is a highly sophisticated parasite that has no biosynthetic or metabolic apparatus of its own and, in consequence, can replicate only inside cells. Although susceptible to antibody before they enter cells, once they have done so these pathogens are not accessible to antibodies and can be eliminated only by the destruction or modification of the infected cells on which they depend. This role in host defense is largely filled by CD8 cytotoxic T cells, although CD4 T cells may also acquire cytotoxic capacities. The crucial role of cytotoxic T cells in limiting such infections is seen in the increased susceptibility of animals artificially depleted of these T cells, or of mice or humans that lack the MHC class I molecules that present antigen to CD8 T cells. The elimination of infected cells without the destruction of healthy tissue requires the cytotoxic mechanisms of CD8 T cells to be both powerful and accurately targeted.

8-27 Cytotoxic T cells can induce target cells to undergo programmed cell death.

Cells can die in various ways. Physical or chemical injury, such as the deprivation of oxygen that occurs in heart muscle during a heart attack or membrane damage with antibody and complement, leads to cell disintegration or necrosis. The dead or necrotic tissue is taken up and degraded by phagocytic cells, which eventually clear the damaged tissue and heal the wound. The other form of cell death is known as programmed cell death, which can be by apoptosis or by autophagic cell death. Apoptosis is a normal cellular response that is crucial in the tissue remodeling that occurs during development and metamorphosis in all multicellular animals. As we saw in Chapter 7, most

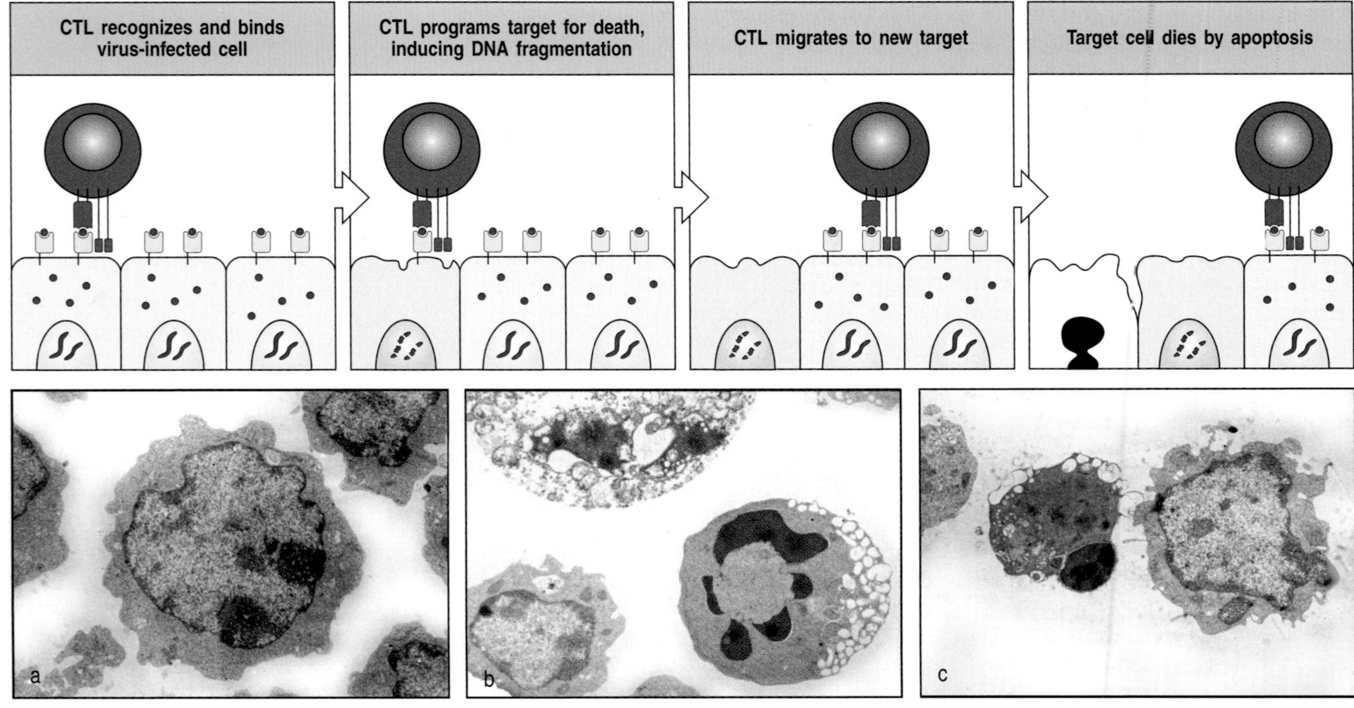

Fig. 8.36 Cytotoxic CD8 T cells can induce apoptosis in target cells. Specific recognition of peptide:MHC complexes on a target cell (top panels) by a cytotoxic CD8 T cell (CTL) leads to the death of the target cell by apoptosis. Cytotoxic T cells can recycle to kill multiple targets. Each killing requires the same series of steps, including receptor binding and the directed release of cytotoxic proteins stored in granules. The process of apoptosis is shown in the micrographs (bottom panels), where panel a shows a healthy cell with a normal nucleus. Early in apoptosis (panel b) the chromatin becomes condensed (red) and, although the cell sheds membrane vesicles, the integrity of the cell membrane is retained, in contrast to the necrotic cell in the upper part of the same field. In late stages of apoptosis (panel c), the cell nucleus (middle cell) is very condensed, no mitochondria are visible, and the cell has lost much of its cytoplasm and membrane through the shedding of vesicles. Photographs (× 3500) courtesy of R. Windsor and E. Hirst.

thymocytes die by apoptosis when they fail positive selection. Early changes seen in apoptosis are nuclear blebbing, alteration in cell morphology, and eventually, fragmentation of the DNA. The cell then destroys itself from within, shrinking by shedding membrane-bound vesicles, and degrading itself until little is left. A hallmark of apoptosis is the fragmentation of nuclear DNA into pieces of 200 base pairs through the activation of nucleases that cleave the DNA between nucleosomes. As we described in Chapter 5, autophagy is the process of degrading senescent or abnormal proteins and organelles. In autophagic programmed cell death, large vacuoles degrade cellular organelles before the condensation and destruction of the nucleus that is characteristic of apoptosis.

Cytotoxic T cells kill their targets by inducing them to undergo apoptosis (Fig. 8.36). When cytotoxic T cells are mixed with target cells and rapidly brought into contact by centrifugation, they can induce antigen-specific target cells to die within 5 minutes, although death can take hours to become fully evident. The rapidity of this response reflects the release of preformed effector molecules, which activate an apoptotic pathway within the target cell.

A mechanism for inducing apoptosis that does not depend on cytotoxic granules involves members of the TNF family, particularly Fas and Fas ligand. In contrast to the killing of infected tissue cells, this mechanism is used mainly to regulate lymphocyte numbers. Activated lymphocytes express both Fas and Fas ligand, and thus activated cytotoxic T cells can kill other lymphocytes through the activation of caspases, which induces apoptosis in the target lymphocyte. Thus, Fas–Fas ligand interactions are important in terminating lymphocyte proliferation after the pathogen initiating an immune response has been cleared. As well as cytotoxic T cells, T_H1 cells, and some T_H2 cells have been shown to be able to kill cells by this pathway. The importance of Fas in maintaining lymphocyte homeostasis can be seen from the effects of mutations in the genes encoding Fas and Fas ligand. Mice and humans with a mutant form of Fas develop a lymphoproliferative disease associated with severe autoimmunity, which is described in more detail in Section 14-19. A mutation in the gene encoding Fas ligand in another mouse strain creates a nearly identical phenotype. These mutant phenotypes represent the best-characterized examples of generalized autoimmunity caused by single-gene defects.

As well as killing the host cell, the apoptotic mechanism may also act directly on cytosolic pathogens. For example, the nucleases that are activated in apoptosis to destroy cellular DNA can also degrade viral DNA. This prevents the assembly of virions and thus the release of infectious virus, which could otherwise infect nearby cells. Other enzymes activated in the course of apoptosis may destroy nonviral cytosolic pathogens. Apoptosis is therefore preferable to necrosis as a means of killing infected cells; in cells dying by necrosis, intact pathogens are released from the dead cell and these can continue to infect healthy cells or can parasitize the macrophages that ingest them.

8-28 Cytotoxic effector proteins that trigger apoptosis are contained in the granules of CD8 cytotoxic T cells.

The principal mechanism of cytotoxic T-cell action is the calcium-dependent release of specialized **cytotoxic granules** upon recognition of antigen on the surface of a target cell. Cytotoxic granules are modified lysosomes that contain at least three distinct classes of cytotoxic effector proteins that are expressed specifically in cytotoxic T cells (Fig. 8.37). Such proteins are stored in the cytotoxic granules in an active form, but conditions within the granules prevent them from functioning until after their release. One of these cytotoxic proteins, known as **perforin**, acts in the delivery of the contents of cytotoxic granules to target-cell membranes. The importance of perforin in

Protein in granules of cytotoxic T cells	Actions on target cells
Perforin	Aids in delivering contents of granules into the cytoplasm of target cell
Granzymes	Serine proteases, which activate apoptosis once in the cytoplasm of the target cell
Granulysin	Has antimicrobial actions and can induce apoptosis

Fig. 8.37 Cytotoxic effector proteins released by cytotoxic T cells.

cytotoxicity is well illustrated in mice that have had their perforin gene knocked out. They are severely defective in their ability to mount a cytotoxic T-cell response to many, but not all, viruses. Another class of cytotoxic proteins comprises a family of serine proteases, called **granzymes**, of which there are 5 in humans and 10 in the mouse. The third cytotoxic protein, **granulysin**, which is expressed in humans but not mice, has antimicrobial activity and at high concentrations is also able to induce apoptosis in target cells. Granules that store perforin, granzymes, and granulysin can be seen in CD8 cytotoxic effector cells in infected tissue.

Both perforin and granzymes are required for effective cell killing. Their separate roles have been investigated in experiments that rely on similarities between the cytotoxic granules of CD8 T cells and the granules of mast cells, which are more easily studied. The release of mast-cell granules occurs on cross-linking of a cell-surface receptor for IgE, just as the release of cytotoxic granules from T cells occurs after the aggregation of T-cell receptors at the immunological synapse. The mechanism of signaling for granule release is thought to be the same or similar in both cases, because both the IgE receptor and the T-cell receptor have ITAM motifs in their cytoplasmic domains, and their cross-linking leads to tyrosine phosphorylation of the ITAMs (see Chapter 6). When a mast-cell line is transfected with the genes for perforin or a granzyme, the gene products are stored in mast-cell granules, and when the cell is activated these granules are released. When transfected with the gene encoding perforin alone, mast cells can kill other cells, but large numbers of transfected cells are needed because the killing is very inefficient. In contrast, mast cells transfected with the gene encoding granzyme B alone are unable to kill other cells. However, when perforin-transfected mast cells are also transfected with the gene encoding granzyme B, the cells or their purified granules become as effective at killing targets as granules from cytotoxic cells. It had been thought that perforin acted by causing a pore to form in the target cell plasma membrane, through which granzymes would enter. However, it seems that perforin and granzymes form multimeric complexes with the proteoglycan **serglycin**, which is the primary proteoglycan of cytotoxic granules and acts as a scaffold (Fig 8.38). Granzyme B does not simply diffuse from the extracellular space through a perforin pore as once thought; rather, it is delivered in the form of multimeric complexes into the cytosol without the apparent formation of a pore in the plasma membrane, a mechanism more similar to that of virus entry. Although the exact mechanism is not yet defined, perforin seems to act as the translocator of these complexes and to mediate the release of the bound granzyme into the cytosol.

The granzymes trigger apoptosis in the target cell by activating caspases. Granzyme B cleaves and activates caspase-3, which is a cysteine protease that cuts after aspartic acid residues (hence the name caspase). Caspase-3

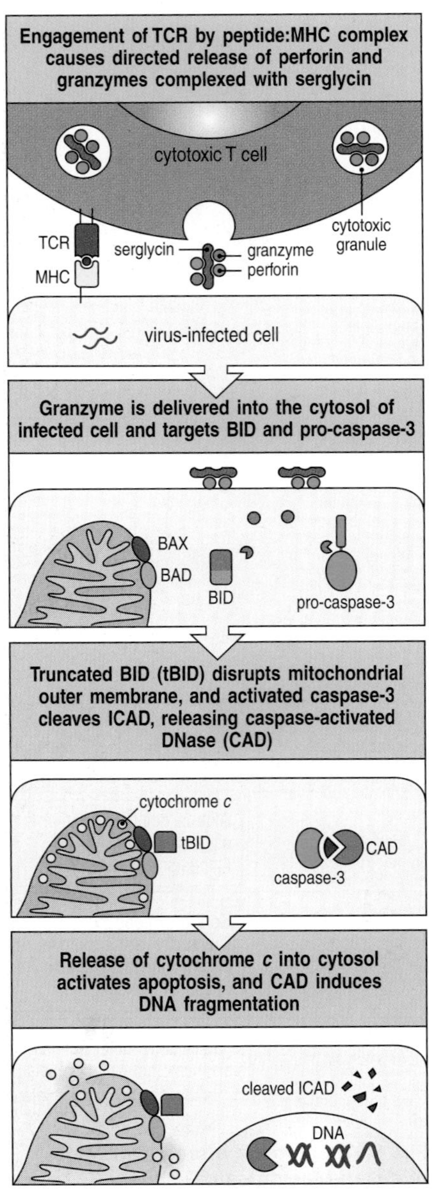

Engagement of TCR by peptide:MHC complex causes directed release of perforin and granzymes complexed with serglycin
cytotoxic T cell
cytotoxic granule
TCR
MHC
serglycin
granzyme
perforin
virus-infected cell
Granzyme is delivered into the cytosol of infected cell and targets BID and pro-caspase-3
BAX
BAD
BID
pro-caspase-3
Truncated BID (tBID) disrupts mitochondrial outer membrane, and activated caspase-3 cleaves ICAD, releasing caspase-activated DNase (CAD)
cytochrome c
tBID
CAD
caspase-3
Release of cytochrome c into cytosol activates apoptosis, and CAD induces DNA fragmentation
cleaved ICAD
DNA

Fig. 8.38 Perforin, granzymes and serglycin are released from cytotoxic granules and deliver granzymes into the cytosol of target cells to induce apoptosis. Recognition of its antigen on a virus-infected cell by a cytotoxic CD8 T cell induces the release of the contents of its cytotoxic granules in a directed fashion. Perforin and granzymes, complexed with the proteoglycan serglycin, are delivered as a complex to the membrane of the target cell (top panel). By an unknown mechanism, perforin directs the entry of the granule contents into the cytosol of the target cell without apparent pore formation, and the introduced granzymes then act on specific intracellular targets such as the proteins BID and pro-caspase-3. Either directly or indirectly, the granzymes cause the cleavage of BID into truncated BID (tBID) and the cleavage of pro-caspase-3 to an active caspase (second panel). tBID acts on mitochondria to release cytochrome c into the cytosol, and activated caspase-3 targets ICAD to release caspase-activated DNase (CAD) (third panel). Cytochrome c in the cytosol promotes apoptosis, and CAD fragments the DNA (bottom panel).

activates a caspase proteolytic cascade, which eventually activates the caspase-activated deoxyribonuclease (CAD) by cleaving an inhibitory protein (ICAD) that binds to and inactivates CAD. This nuclease is believed to be the enzyme that degrades the DNA (see Fig. 8.38). Granzyme B also activates other pathways of cell death. One important target of is the protein BID (for BH3-interacting domain death agonist protein). When BID is cleaved, either directly by granzyme B or indirectly by activated caspase-3, the mitochondrial outer membrane becomes disrupted, causing the release from the mitochondrial intermembrane space of pro-apoptotic molecules such as cytochrome *c*. Other granzymes are thought to promote apoptosis by targeting different cellular components.

Cells undergoing programmed cell death are rapidly ingested by phagocytic cells, which recognize a change in the cell membrane: phosphatidylserine, which is normally found only in the inner leaflet of the membrane, replaces phosphatidylcholine as the predominant phospholipid in the outer leaflet. The ingested cell is broken down and completely digested by the phagocyte without the induction of co-stimulatory proteins. Thus, apoptosis is normally an immunologically 'quiet' process; that is, apoptotic cells do not normally contribute to or stimulate immune responses.

8-29 Cytotoxic T cells are selective and serial killers of targets expressing a specific antigen.

When cytotoxic T cells are offered a mixture of equal amounts of two target cells, one bearing a specific antigen and the other not, they kill only the target cell bearing the specific antigen. The 'innocent bystander' cells and the cytotoxic T cells themselves are not killed. The cytotoxic T cells are probably not killed because release of the cytotoxic effector molecules is highly polarized. As we saw in Fig. 8.32, cytotoxic T cells orient their Golgi apparatus and microtubule-organizing center to focus secretion on the point of contact with a target cell. Granule movement toward the point of contact is shown in Fig. 8.39. Cytotoxic T cells attached to several different target cells reorient their secretory apparatus toward each cell in turn and kill them one by one, strongly suggesting that the mechanism whereby cytotoxic mediators are released allows attack at only one point of contact at any one time. The narrowly focused action of CD8 cytotoxic T cells allows them to kill single infected cells in a tissue without creating widespread tissue damage (Fig. 8.40) and is of crucial importance in tissues where cell regeneration does not occur, as with the neurons of the central nervous system, or is very limited, as in the pancreatic islets.

Cytotoxic T cells can kill their targets rapidly because they store preformed cytotoxic proteins in forms that are inactive in the environment of the cytotoxic

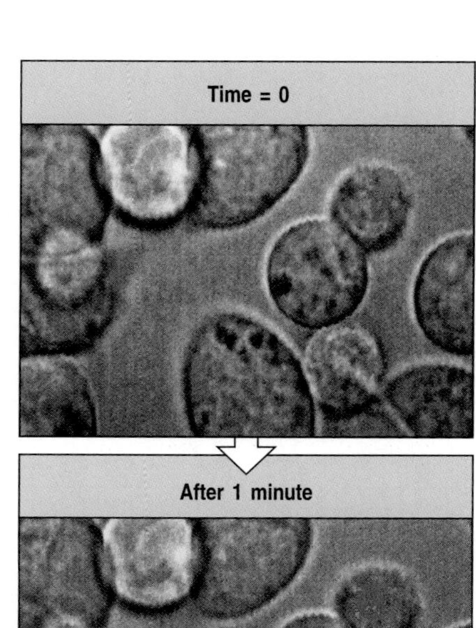

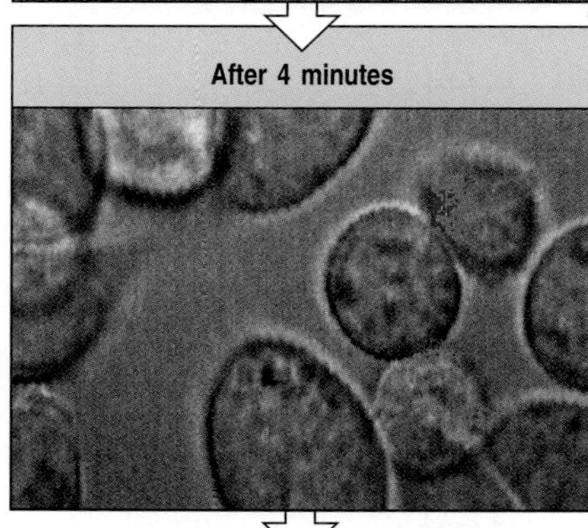

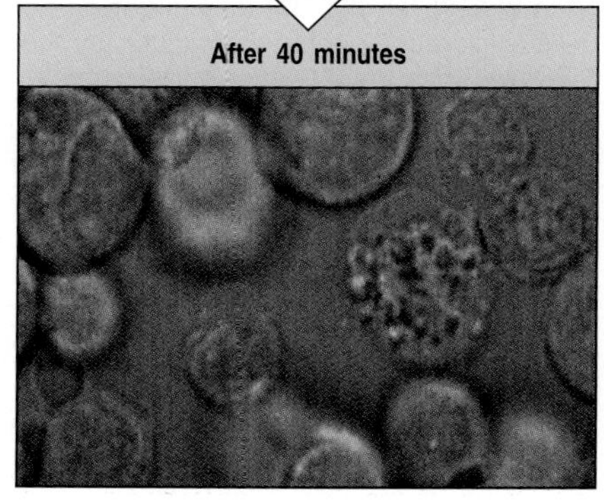

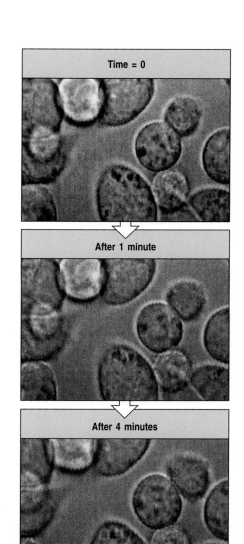

Fig. 8.39 Effector molecules are released from T-cell granules in a highly polar fashion. The granules of cytotoxic T cells can be labeled with fluorescent dyes, allowing them to be seen under the microscope, and their movements can be followed by time-lapse photography. Here we show a series of pictures taken during the interaction of a cytotoxic T cell with a target cell, which is eventually killed. In the top panel, at time 0, the T cell (upper right) has just made contact with a target cell (diagonally below). At this time, the granules of the T cell, labeled with a red fluorescent dye, are distant from the point of contact. In the second panel, after 1 minute has elapsed, the granules have begun to move towards the target cell, a move that has essentially been completed in the third panel, after 4 minutes. After 40 minutes, in the last panel, the granule contents have been released into the space between the T cell and the target, which has begun to undergo apoptosis (note the fragmented nucleus). The T cell will now disengage from the target cell and can recognize and kill other targets. Photographs courtesy of G. Griffiths

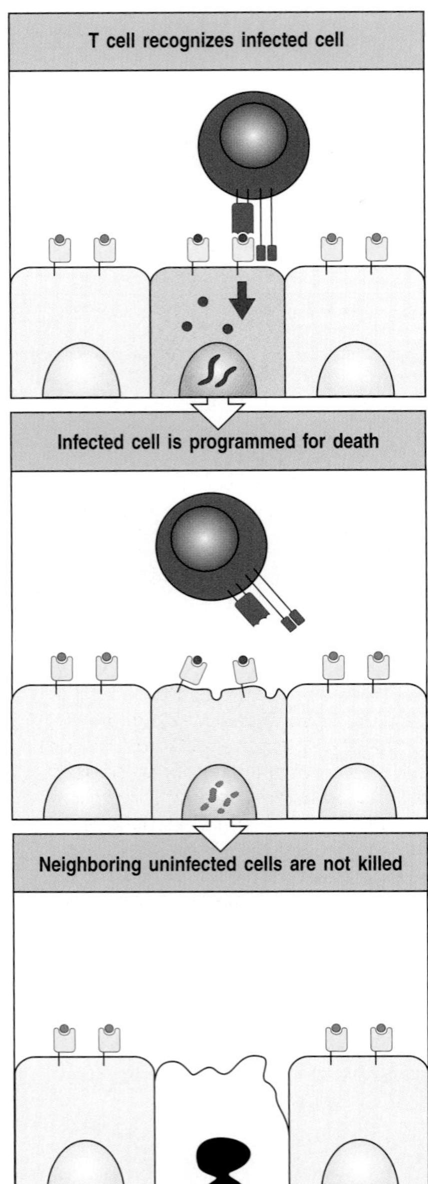

T cell recognizes infected cell
Infected cell is programmed for death
Neighboring uninfected cells are not killed

Fig. 8.40 Cytotoxic T cells kill target cells bearing specific antigen while sparing neighboring uninfected cells. All the cells in a tissue are susceptible to killing by the cytotoxic proteins of armed effector CD8 T cells, but only infected cells are killed. Specific recognition by the T-cell receptor identifies which target cell to kill, and the polarized release of the cytotoxic granules (not shown) ensures that neighboring cells are spared.

granule. Cytotoxic proteins are synthesized and loaded into the granules during the first encounter of a naive cytotoxic precursor T cell with its specific antigen. Ligation of the T-cell receptor similarly induces *de novo* synthesis of perforin and granzymes in effector CD8 T cells, so that the supply of cytotoxic granules is replenished. This makes it possible for a single CD8 T cell to kill a series of targets in succession.

8-30 Cytotoxic T cells also act by releasing cytokines.

Inducing apoptosis in target cells is the main way in which CD8 cytotoxic T cells eliminate infection. However, most CD8 cytotoxic T cells also release the cytokines IFN-γ, TNF-α, and LT-α, which contribute to host defense in several ways. IFN-γ inhibits viral replication directly, and induces the increased expression of MHC class I molecules and of other proteins that are involved in peptide loading of these newly synthesized MHC class I molecules in infected cells. This increases the chance that infected cells will be recognized as target cells for cytotoxic attack. IFN-γ also activates macrophages, recruiting them to sites of infection both as effector cells and as antigen-presenting cells. TNF-α and LT-α can synergize with IFN-γ in macrophage activation and in killing some target cells through their interaction with TNFR-I, which induces apoptosis (see Section 8-26). Thus, effector CD8 cytotoxic T cells act in a variety of ways to limit the spread of cytosolic pathogens. The relative importance of each of these mechanisms is being rapidly determined through gene knockouts in mice.

Summary.

Effector CD8 cytotoxic T cells are essential in host defense against pathogens that live in the cytosol: most commonly these will be viruses. These cytotoxic T cells can kill any cell harboring such pathogens by recognizing foreign peptides that are transported to the cell surface bound to MHC class I molecules. CD8 cytotoxic T cells perform their killing function by releasing three types of preformed cytotoxic proteins: the granzymes, which seem able to induce apoptosis in any type of target cell, perforin, which acts in the delivery of granzymes into the target cell, and granulysin. These properties allow the cytotoxic T cell to attack and destroy virtually any cell infected with a cytosolic pathogen. The membrane-bound Fas ligand, expressed by CD8 and some CD4 T cells, may also induce apoptosis by binding to Fas on some target cells, but this pathway is probably more important in removing Fas-bearing activated lymphocytes after an infection has been cleared and in maintaining lymphocyte homoeostasis. CD8 cytotoxic T cells also produce IFN-γ, which inhibits viral replication and is an important inducer of MHC class I molecule expression and macrophage activation. Cytotoxic T cells kill infected targets with great precision, sparing adjacent normal cells. This precision is crucial in minimizing tissue damage while allowing the eradication of infected cells.

Macrophage activation by T_H1 cells.

Some microorganisms, such as mycobacteria, are intracellular pathogens that grow primarily in the phagosomes of macrophages, shielded from the effects of both antibodies and cytotoxic T cells. These microbes maintain themselves in the usually hostile environment of the phagocyte by inhibiting the fusion of lysosomes to the phagosomes in which they grow, or by preventing the acidification of these vesicles that is required to activate lysosomal proteases. Such microorganisms can be eliminated when the macrophage is activated by a T_H1 cell. T_H1 cells act by synthesizing membrane-associated proteins and a range of soluble cytokines whose local and distant actions coordinate the immune response to these intracellular pathogens. T_H1 effector cells also activate macrophages to kill recently ingested pathogens, and can activate B cells to secrete a limited but highly effective set of immunoglobulin isotypes, as described in Chapter 9.

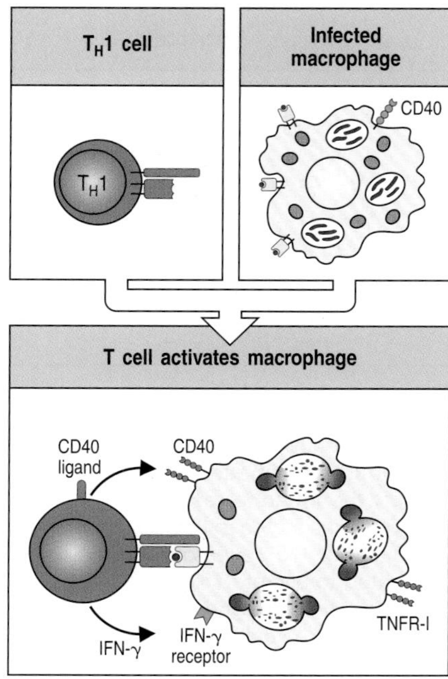

Fig. 8.41 T_H1 cells activate macrophages to become highly microbicidal. When an effector T_H1 cell specific for a bacterial peptide contacts an infected macrophage, the T cell is induced to secrete the macrophage-activating factor IFN-γ and to express CD40 ligand. Together these newly synthesized T_H1 proteins activate the macrophage.

8-31 T_H1 cells have a central role in macrophage activation.

Several important pathogens live within macrophages, whereas many others are ingested by macrophages from the extracellular fluid. In many cases the macrophage is able to destroy such pathogens without the need for T-cell activation, as we saw in Chapter 2. In several clinically important infections, however, the pathogens chronically infect the macrophage and incapacitate it, and CD4 T cells are needed to provide additional activating signals for the macrophage to be able to destroy its pathogen load. The boost to antimicrobial mechanisms in macrophages is known as **macrophage activation** and is the principal effector action of T_H1 cells. Among the extracellular pathogens that are killed when macrophages are activated is *Pneumocystis carinii*; this fungal opportunist is a common cause of death in people with AIDS because of their deficiency of CD4 T cells. Macrophage activation can be measured by the ability to damage a broad spectrum of microbes as well as certain tumor cells. Macrophage effects on extracellular targets extends to healthy tissue cells, which means that macrophages must normally be maintained in an unactivated state.

Macrophages require two signals for activation. One of these is provided by IFN-γ; the other can be provided in a variety of ways and sensitizes the macrophage to respond to IFN-γ. Effector T_H1 cells can deliver both signals. IFN-γ is the most characteristic cytokine produced by T_H1 cells on interacting with their specific target cells, and the CD40 ligand expressed by the T_H1 cell delivers the sensitizing signal by contacting CD40 on the macrophage (Fig. 8.41). CD8 T cells are also an important source of IFN-γ and can activate macrophages presenting antigens derived from cytosolic proteins; mice lacking MHC class I molecules, and thus having no CD8 T cells, show increased susceptibility to some parasite infections. Macrophages can also be made more sensitive to IFN-γ by very small amounts of bacterial LPS, and this latter pathway may be particularly important when CD8 T cells are the primary source of the IFN-γ. It is also possible that membrane-associated TNF-α or LT-α can substitute for CD40 ligand in macrophage activation. These cell-associated molecules apparently stimulate the macrophage to secrete TNF-α, and antibody against TNF-α can inhibit macrophage activation. T_H2 cells are inefficient macrophage activators because they produce IL-10, a cytokine that can deactivate macrophages, and they also do not produce IFN-γ. They do express CD40 ligand, however, and can deliver the contact-dependent signal required to activate macrophages to respond to IFN-γ.

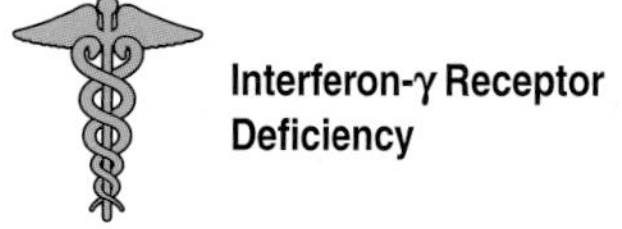

Within minutes of the recognition of specific antigen by CD8 cytotoxic T cells, directed exocytosis of preformed perforins and granzymes induces target cells to die by apoptosis. In contrast, when T_H1 cells encounter specific antigen, they must induce *de novo* transcription of the effector cytokines and cell-surface molecules through which they act. This induction starts within an hour of contact and requires hours to complete, rather than minutes, so T_H1 cells must adhere to their target cells for far longer than cytotoxic T cells. The newly synthesized cytokines are then delivered directly through microvesicles of the constitutive secretory pathway to the site of contact between the T-cell membrane and the macrophage. It is thought that the

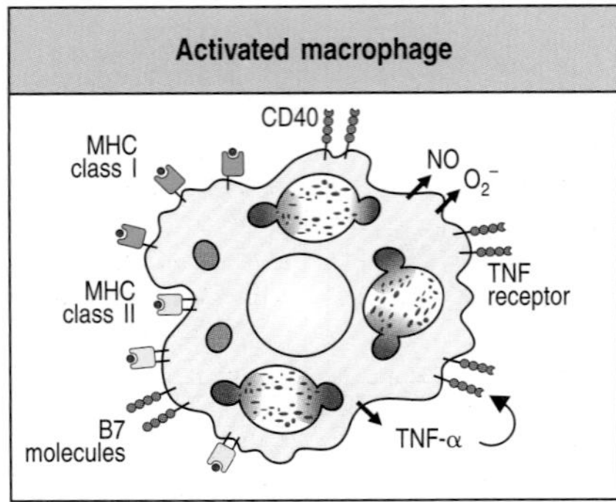

Fig. 8.42 Activated macrophages undergo changes that greatly increase their antimicrobial effectiveness and amplify the immune response. Activated macrophages increase their expression of CD40 and of TNF receptors, and are stimulated to secrete TNF-α. This autocrine stimulus synergizes with IFN-γ secreted by T_H1 cells to increase the antimicrobial action of the macrophage, in particular by inducing the production of nitric oxide (NO) and superoxide (O_2^-). The macrophage also upregulates its B7 molecules in response to binding to CD40 ligand on the T cell, and increases its expression of MHC class II molecules, thus allowing further activation of resting CD4 T cells.

newly synthesized cell-surface CD40 ligand is also expressed in this polarized fashion. This means that, although all macrophages have receptors for IFN-γ, the macrophage actually displaying antigen to the T_H1 cell is far more likely to become activated by it than are neighboring uninfected macrophages.

8-32 Activation of macrophages by T_H1 cells promotes microbial killing and must be tightly regulated to avoid tissue damage.

T_H1 cells activate infected macrophages through cell contact and the focal secretion of IFN-γ. This generates a series of biochemical responses that convert the macrophage into a potent antimicrobial effector cell (Fig. 8.42). Activated macrophages fuse their lysosomes more efficiently to phagosomes, exposing intracellular or recently ingested microbes to a variety of microbicidal lysosomal enzymes. Activated macrophages also make oxygen radicals and nitric oxide (NO), both of which have potent antimicrobial activity, as well as synthesizing antimicrobial peptides and proteases that can be released to attack extracellular parasites.

Additional changes in the activated macrophage help to amplify the immune response. The number of B7 molecules, CD40, MHC class II molecules, and TNF receptors on the macrophage surface increases, making the cell more effective at presenting antigen to fresh T cells, and more responsive to CD40 ligand and TNF-α. TNF-α produced by the activated macrophage can synergize with IFN-γ produced by T_H1 cells in macrophage activation, particularly in the induction of the reactive nitrogen metabolite NO, which has broad antimicrobial activity. The NO is produced by the enzyme inducible NO synthase (iNOS), and mice that have had the gene for iNOS knocked out are highly susceptible to infection by several intracellular pathogens. Activated macrophages also secrete IL-12, which directs the differentiation of activated naive CD4 T cells into T_H1 effector cells (see Section 8-19). These and many other surface and secreted molecules of activated macrophages are instrumental in the effector actions of macrophages in cell-mediated responses; cytokines secreted by macrophages are also important in humoral immune responses and in recruiting other immune cells to sites of infection.

Because activated macrophages are extremely effective in destroying pathogens, one may ask why macrophages are not simply maintained in a state of constant activation. Besides the fact that macrophages consume huge quantities of energy to maintain the activated state, macrophage activation *in vivo* is usually associated with localized tissue destruction that apparently results from the release of oxygen radicals, NO, and proteases, which are toxic to host cells as well as to pathogens. The release of toxic mediators by activated macrophages is important in host defense because it enables them to attack large extracellular pathogens that they cannot ingest, such as parasitic worms. However, this is achieved only at the expense of tissue damage. Tight regulation of the activity of macrophages by T_H1 cells thus allows the specific and effective deployment of this powerful means of host defense while minimizing local tissue damage and energy consumption.

Activated effector T cells are the main source of the IFN-γ that activates macrophages, and thus control of macrophage activation is tightly linked to control of IFN-γ synthesis in the T cell. This seems to be achieved by regulating the half-life of the mRNA encoding IFN-γ. IFN-γ mRNA, like that encoding some other cytokines, such as IL-2, contains an instability sequence $(AUUUA)_n$ in its 3′ untranslated region that greatly reduces its half-life and limits the period of cytokine production. Activation of the T cell seems to induce the synthesis of a new protein that promotes cytokine mRNA degradation: treatment of activated effector T cells with the protein synthesis inhibitor cycloheximide greatly increases the level of cytokine mRNA. The rapid destruction of cytokine mRNA, together with the focal delivery of IFN-γ at the point of contact between the activated T_H1 cell and its macrophage target, thus limits the action of the effector T cell to the infected macrophage. In addition, macrophage activation is markedly inhibited by cytokines such as TGF-β and IL-10. Several of these inhibitory cytokines are produced by CD4 T_H2 cells, and so the induction of T_H2 cells is important for limiting macrophage activation.

8-33 T_H1 cells coordinate the host response to intracellular pathogens.

The activation of macrophages by T_H1 cells expressing CD40 ligand and secreting IFN-γ is central to the host response to pathogens that proliferate in macrophage vesicles. In mice whose IFN-γ gene or CD40 ligand gene has been destroyed by targeted gene disruption, the production of antimicrobial agents by macrophages is impaired, and the animals succumb to sublethal doses of *Mycobacterium* and *Leishmania* species. Macrophage activation is also crucial in controlling vaccinia virus. Mice lacking TNF receptors are more susceptible to these pathogens. However, although IFN-γ and CD40 ligand are probably the most important effector molecules synthesized by T_H1 cells, the immune response to pathogens that proliferate in macrophage vesicles is complex, and other cytokines secreted by T_H1 cells may also be crucial in coordinating these responses (Fig. 8.43). For example, macrophages that are chronically infected with intracellular bacteria may lose the ability to become activated, and such cells could provide a reservoir of infection shielded from immune attack. Activated T_H1 cells can also express Fas ligand and thus kill a limited range of target cells that express Fas, including macrophages, thereby destroying these infected cells.

Some intravesicular bacteria, such as some mycobacteria and *Listeria monocytogenes*, can escape from cell vesicles and enter the cytoplasm, where they are not susceptible to macrophage activation. Their presence can, however, be detected by CD8 cytotoxic T cells, which can release them by killing the cell. The pathogens released when macrophages are killed either by T_H1 cells or by CD8 cytotoxic T cells can be taken up by freshly recruited macrophages still capable of activation to antimicrobial activity.

Fig. 8.43 The immune response to intracellular bacteria is coordinated by activated T_H1 cells. The activation of T_H1 cells by infected macrophages results in the synthesis of cytokines that both activate the macrophage and coordinate the immune response to intracellular pathogens. IFN-γ and CD40 ligand synergize in activating the macrophage, which allows it to kill engulfed pathogens. Chronically infected macrophages lose the ability to kill intracellular bacteria, and Fas ligand or LT-α produced by the T_H1 cell can kill these macrophages, releasing the engulfed bacteria, which are taken up and killed by fresh macrophages. In this way, IFN-γ and LT-β synergize in the removal of intracellular bacteria. IL-2 produced by T_H1 cells induces T-cell proliferation and potentiates the release of other cytokines. IL-3 and GM-CSF stimulate the production of new macrophages by acting on hematopoietic stem cells in the bone marrow. New macrophages are recruited to the site of infection by the actions of TNF-α, LT-β, and other cytokines on vascular endothelium, which signal macrophages to leave the bloodstream and enter the tissues. A chemokine with macrophage chemotactic activity (CXCL2) signals macrophages to migrate into sites of infection and accumulate there. Thus, the T_H1 cell coordinates a macrophage response that is highly effective in destroying intracellular infectious agents.

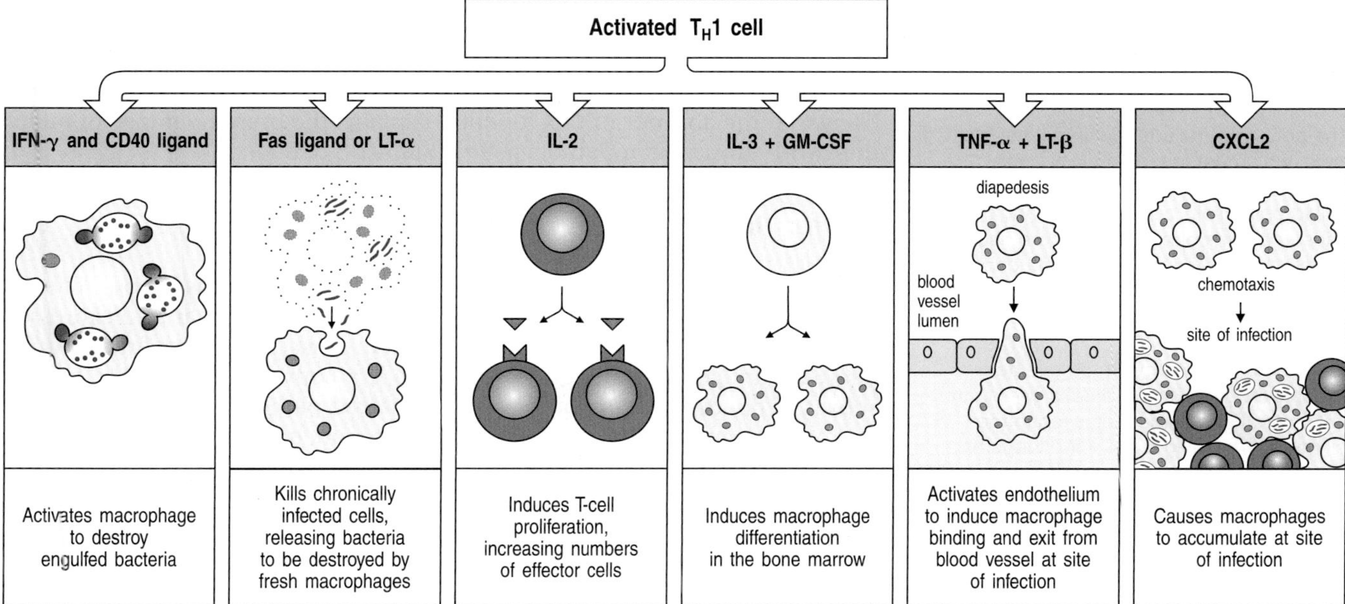

Activated T_H1 cell
IFN-γ and CD40 ligand
Fas ligand or LT-α
IL-2
IL-3 + GM-CSF
TNF-α + LT-β
CXCL2
diapedesis
blood vessel lumen
chemotaxis
site of infection
Activates macrophage to destroy engulfed bacteria
Kills chronically infected cells, releasing bacteria to be destroyed by fresh macrophages
Induces T-cell proliferation, increasing numbers of effector cells
Induces macrophage differentiation in the bone marrow
Activates endothelium to induce macrophage binding and exit from blood vessel at site of infection
Causes macrophages to accumulate at site of infection

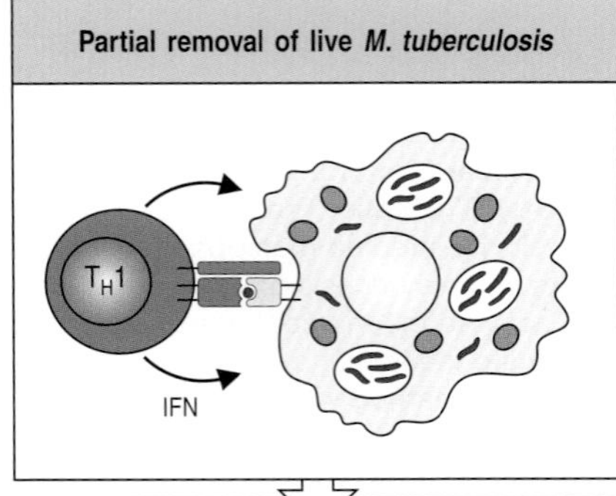

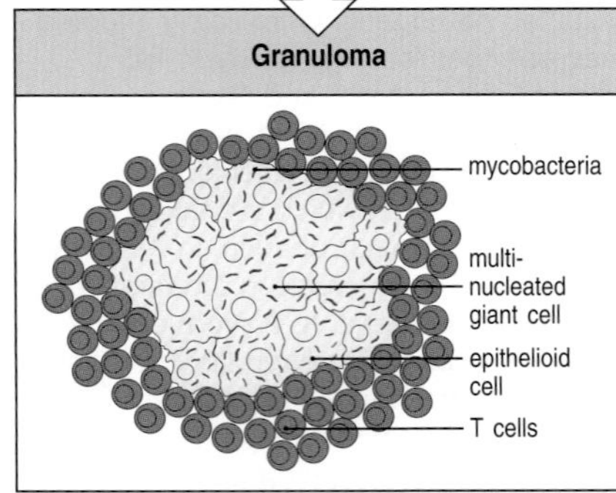

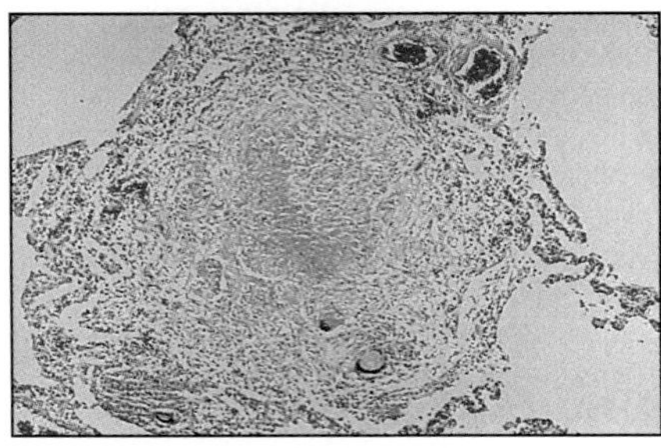

Fig. 8.44 Granulomas form when an intracellular pathogen or its constituents cannot be totally eliminated. When mycobacteria (red) resist the effects of macrophage activation, a characteristic localized inflammatory response called a granuloma develops. This consists of a central core of infected macrophages. The core may include multinucleate giant cells, which are fused macrophages, surrounded by large macrophages often called epithelioid cells, but in granulomas caused by mycobacteria the core usually becomes necrotic. Mycobacteria can persist in the cells of the granuloma. The central core is surrounded by T cells, many of which are CD4-positive. The exact mechanisms by which this balance is achieved, and how it breaks down, are unknown. Granulomas, as seen in the bottom panel, also form in the lungs and elsewhere in a disease known as sarcoidosis, which may be caused by inapparent mycobacterial infection. Photograph courtesy of J. Orrell.

Another very important function of T_H1 cells is the recruitment of phagocytic cells to sites of infection. T_H1 cells recruit macrophages by two mechanisms. First, they make the hematopoietic growth factors IL-3 and GM-CSF, which stimulate the production of new phagocytic cells in the bone marrow. Second, TNF-α and LT-α, which are secreted by T_H1 cells at sites of infection, change the surface properties of endothelial cells so that phagocytes adhere to them. Chemokines such as CXCL2, which is produced by T_H1 cells in the inflammatory response, serve to direct the migration of monocytes through the vascular endothelium and into the infected tissue (see Section 2-24).

When microbes effectively resist the microbicidal effects of activated macrophages, chronic infection with inflammation can develop. Often, this has a characteristic pattern, consisting of a central area of macrophages surrounded by activated lymphocytes. This pathological pattern is called a granuloma (Fig. 8.44). Giant cells consisting of fused macrophages can form in the center of these granulomas. A granuloma serves to 'wall off' pathogens that resist destruction. T_H2 cells seem to participate in granulomas along with T_H1 cells, perhaps by regulating their activity and preventing widespread tissue damage. In tuberculosis, the center of the large granulomas can become isolated and the cells there die, probably from a combination of lack of oxygen and the cytotoxic effects of activated macrophages. As the dead tissue in the center resembles cheese, this process is called caseation necrosis. Thus, the activation of T_H1 cells can cause significant pathology. Their non-activation, however, leads to the more serious consequence of death from disseminated infection, which is now seen frequently in patients with AIDS and concomitant mycobacterial infection.

Summary.

CD4 T cells that can activate macrophages have a critical role in host defense against those intracellular and extracellular pathogens that resist killing after being engulfed by macrophages. Macrophages are activated by membrane-bound signals delivered by activated T_H1 cells and by the potent macrophage-activating cytokine IFN-γ, which is secreted by activated T cells. Once activated, the macrophage can kill intracellular and ingested bacteria. Activated macrophages can also cause local tissue damage, and this explains why their activity is strictly regulated by antigen-specific T cells. T_H1 cells produce a range of cytokines, chemokines, and surface molecules that not only activate infected macrophages but also kill chronically infected senescent macrophages, stimulate the production of new macrophages in bone marrow, and recruit fresh macrophages to sites of infection. Thus, T_H1 cells control and coordinate host defense against certain intracellular pathogens. It is likely that the absence of this function explains the preponderance of infections with intracellular pathogens in adult AIDS patients.

Summary to Chapter 8.

An adaptive immune response is initiated when naive T cells encounter specific antigen on the surface of an antigen-presenting cell that also expresses the co-stimulatory molecules B7.1 and B7.2. In most cases, these first encounters with antigen are thought to occur with a conventional (CD11c-bright) subset of dendritic cells that have encountered pathogens in the periphery and become activated through innate recognition, have taken up antigen at a site of infection, and have migrated to local lymphoid tissue. The dendritic cell may mature to become a potent direct activator of naive T cells, or it may transfer antigen to dendritic cells resident in peripheral lymphoid organs for cross-presentation to naive CD8 T cells. Plasmacytoid dendritic cells contribute to rapid responses against viruses by the production of type I interferons. Once activated by encounter with an antigen-presenting dendritic cell,

T cells produce IL-2, which drives them to proliferate and differentiate into several types of effector T cells. All T-cell effector functions involve cell–cell interactions. When effector T cells recognize specific antigen on target cells, they release mediators that act directly on the target cell, altering its behavior. The triggering of effector T cells by peptide:MHC complexes is independent of co-stimulation, so that any infected target cell can be activated or destroyed by an effector T cell. CD8 cytotoxic T cells kill target cells infected with cytosolic pathogens, removing sites of pathogen replication. CD4 T cells can become specialized effectors that promote inflammatory (T_H1), humoral or allergic (T_H2), or acute (T_H17) responses to pathogens. CD4 T_H1 cells activate macrophages to kill intracellular parasites. CD4 T cells are also essential in the activation of B cells to secrete the antibodies that mediate humoral immune responses directed against extracellular pathogens. T_H17 cells help enhance the neutrophil response to extracellular pathogens. Thus, effector T cells control virtually all known effector mechanisms of the adaptive immune response. In addition, subsets of CD4 regulatory T cells are produced that help control and limit immune responses by suppressing T-cell activity.

Questions.

8.1 *Dendritic cells migrate through tissues, providing a surveillance mechanism for infection by pathogens. (a) What lineage of cells are dendritic cells, and what types are there? (b) Describe how dendritic cells identify the presence of infection in peripheral tissues and initiate an immune response to it in the lymph nodes or secondary lymphoid tissues. (c) What mechanisms prevent dendritic cells from initiating immune responses to self antigens?*

8.2 *Activation of a naive T cell requires interaction with an antigen-presenting cell, such as a dendritic cell. (a) Which molecules on T cells are involved in this process, and what do they interact with on the antigen-presenting cell? (b) What consequences would you expect if these molecules were deficient in an individual? (c) What scope do these molecules offer for the design of anti-inflammatory or immunosuppressive drugs?*

8.3 *In some particle-physics experiments, coincidence detection—the simultaneous measurement of the same event by two separate detectors—is used to discriminate real events from spurious fluctuations in the detector systems. How do the requirements for T-cell activation follow the same principle in (a) the recognition of pathogens, or (b) the prevention of autoimmune reactions?*

8.4 *Consider the claim "T-cell effector functions are primarily mediated by secreted products." (a) To what extent is this statement true for CD4 cells and for CD8 T cells? (b) Describe the roles of T-cell membrane-bound effector molecules in the immune response.*

8.5 *CD4 T cells acquire several distinct phenotypes, which have been called separate lineages. (a) Describe the known CD4 subsets and correlate their immunological functions with their specific effector mechanisms. (b) What properties of these subsets agree or disagree with the notion that they represent distinct lineages of cells? (c) Describe the role of antigen-presenting cells and pathogens in generating each subset. (d) Discuss how antigen-presenting cells and CD4 T-cell subsets are related to the maintenance of tolerance.*

General references.

Dustin, M.L.: **Coordination of T-cell activation and migration through formation of the immunological synapse.** *Ann. N.Y. Acad. Sci.* 2003, **987**:51–59.

Ihle, J.N.: **Cytokine receptor signaling.** *Nature* 1995, **377**:591–594.

Janeway, C.A., and Bottomly, K.: **Signals and signs for lymphocyte responses.** *Cell* 1994, **76**:275–285.

Mosmann, T.R., Li, L., Hengartner, H., Kagi, D., Fu, W., and Sad, S.: **Differentiation and functions of T cell subsets.** *Ciba Found. Symp.* 1997, **204**:148–154; discussion 154–158.

Snyder, J.E., and Mosmann, T.R.: **How to 'spot' a real killer.** *Trends Immunol.* 2003, **24**:231–232.

Springer, T.A.: **Traffic signals for lymphocyte recirculation and leukocyte emigration: the multistep paradigm.** *Cell* 1994, **76**:301–314.

Tseng, S.Y., and Dustin, M.L.: **T-cell activation: a multidimensional signaling network.** *Curr. Opin. Cell Biol.* 2002, **14**:575–580.

Section references.

8-1 Naive T cells migrate through peripheral lymphoid tissues, sampling the peptide:MHC complexes on dendritic cell surfaces.

Caux, C., Ait-Yahia, S., Chemin, K., de Bouteiller, O., Dieu-Nosjean, M.C., Homey, B., Massacrier, C., Vanbervliet, B., Zlotnik, A., and Vicari, A.: **Dendritic cell biology and regulation of dendritic cell trafficking by chemokines.** *Springer Semin. Immunopathol.* 2000, **22**:345–369.

Dupuis, M., Denis-Mize, K., LaBarbara, A., Peters, W., Charo, I.F., McDonald, D.M., and Ott, G.: **Immunization with the adjuvant MF59 induces macrophage trafficking and apoptosis.** *Eur. J. Immunol.* 2001, **31**:2910–2918.

Itano, A.A., and Jenkins, M.K.: **Antigen presentation to naive CD4 T cells in the lymph node.** *Nat. Immunol.* 2003, **4**:733–739.

Picker, L.J., and Butcher, E.C.: **Physiological and molecular mechanisms of lymphocyte homing.** *Annu. Rev. Immunol.* 1993, **10**:561–591.

Steptoe, R.J., Li, W., Fu, F., O'Connell, P.J., and Thomson, A.W.: **Trafficking of APC from liver allografts of Flt3L-treated donors: augmentation of potent allostimulatory cells in recipient lymphoid tissue is associated with a switch from tolerance to rejection.** *Transpl. Immunol.* 1999, **7**:51–57.

Yoshino, M., Yamazaki, H., Nakano, H., Kakiuchi, T., Ryoke, K., Kunisada, T., and Hayashi, S.: **Distinct antigen trafficking from skin in the steady and active states.** *Int. Immunol.* 2003, **15**:773–779.

8-2 Lymphocyte entry into lymphoid tissues depends on chemokines and adhesion molecules.

Hogg, N., Henderson, R., Leitinger, B., McDowall, A., Porter, J., and Stanley, P.: **Mechanisms contributing to the activity of integrins on leukocytes.** *Immunol. Rev.* 2002, **186**:164–171.

Kunkel, E.J., Campbell, D.J., and Butcher, E.C.: **Chemokines in lymphocyte trafficking and intestinal immunity.** *Microcirculation* 2003, **10**:313–323.

Madri, J.A., and Graesser, D.: **Cell migration in the immune system: the evolving interrelated roles of adhesion molecules and proteinases.** *Dev. Immunol.* 2000, **7**:103–116.

Rosen, S.D.: **Ligands for L-selectin: homing, inflammation, and beyond.** *Annu. Rev. Immunol.* 2004, **22**:129–156.

von Andrian, U.H., and Mempel, T.R.: **Homing and cellular traffic in lymph nodes.** *Nat. Rev. Immunol.* 2003, **3**:867–878.

8-3 Activation of integrins by chemokines is responsible for entry of naive T cells into lymph nodes.

Cyster, J.G.: **Chemokines, sphingosine-1-phosphate, and cell migration in secondary lymphoid organs.** *Annu. Rev. Immunol.* 2005, **23**:127–159.

Iwata, S., Kobayashi, H., Miyake-Nishijima, R., Sasaki, T., Souta-Kuribara, A., Nori, M., Hosono, O., Kawasaki, H., Tanaka, H., and Morimoto, C.: **Distinctive signaling pathways through CD82 and β1 integrins in human T cells.** *Eur. J. Immunol.* 2002, **32**:1328–1337.

Laudanna, C., Kim, J.Y., Constantin, G., and Butcher, E.: **Rapid leukocyte integrin activation by chemokines.** *Immunol. Rev.* 2002, **186**:37–46.

Lo, C.G., Lu, T.T., and Cyster, J.G.: **Integrin-dependence of lymphocyte entry into the splenic white pulp.** *J. Exp. Med.* 2003, **197**:353–361.

Rosen, H., and Goetzl E.J.: **Sphingosine 1-phosphate and its receptors: an autocrine and paracrine network.** *Nat. Rev. Immunol.* 2005, **5**:560–570.

Takagi, J., and Springer, T.A.: **Integrin activation and structural rearrangement.** *Immunol. Rev.* 2002, **186**:141–163.

8-4 T-cell responses are initiated in peripheral lymphoid organs by activated dendritic cells.

Miller, M.J., Wei, S.H., Cahalan, M.D., and Parker, I.: **Autonomous T cell trafficking examined *in vivo* with intravital two-photon microscopy.** *Proc. Natl Acad. Sci. USA* 2003, **100**:2604–2609.

Miller, M.J., Wei, S.H., Parker, I., and Cahalan, M.D.: **Two-photon imaging of lymphocyte motility and antigen response in intact lymph node.** *Science* 2002, **296**:1869–1873.

Schlienger, K., Craighead, N., Lee, K.P., Levine, B.L., and June, C.H.: **Efficient priming of protein antigen-specific human CD4⁺ T cells by monocyte-derived dendritic cells.** *Blood* 2000, **96**:3490–3498.

Thery, C., and Amigorena, S.: **The cell biology of antigen presentation in dendritic cells.** *Curr. Opin. Immunol.* 2001, **13**:45–51.

8-5 There are two different functional classes of dendritic cells.

Ardavin, C.: **Origin, precursors and differentiation of mouse dendritic cells.** *Nat. Rev. Immunol.* 2003, **3**:582–590.

Belz, G.T., Carbone, F.R., and Heath, W.R.: **Cross-presentation of antigens by dendritic cells.** *Crit. Rev. Immunol.* 2002, **22**:439–448.

Gatti, E., and Pierre, P.: **Understanding the cell biology of antigen presentation: the dendritic cell contribution.** *Curr. Opin. Cell Biol.* 2003, **15**:468–473.

Guermonprez, P., Valladeau, J., Zitvogel, L., Thery, C., and Amigorena, S.: **Antigen presentation and T cell stimulation by dendritic cells.** *Annu. Rev. Immunol.* 2002, **20**:621–667.

Shortman, K., and Liu, Y.L.: **Mouse and human dendritic cell subtypes.** *Nat. Rev. Immunol.* 2002, **21**:151–161.

8-6 Dendritic cells process antigens from a wide array of pathogens.
&

8-7 Pathogen-induced TLR signaling in immature dendritic cell induces their migration to lymphoid organs and enhances antigen processing.

Allan, R.S., Waithman, J., Bedoui, S., Jones, C.M., Villadangos, J.A., Zhan, Y., Lew, A.M., Shortman, K., Heath, W.R., and Carbone, F.R.: **Migratory dendritic cells transfer antigen to a lymph node-resident dendritic cell population for efficient CTL priming.** *Immunity* 2006, **25**:153–162.

Bachman, M.F., Kopf, M., and Marsland, B.J.: **Chemokines: more than just road signs.** *Nat. Rev. Immunol.* 2006, **6**:159–164.

Blander, J.M., and Medzhitov, R.: **Toll-dependent selection of microbial antigens for presentation by dendritic cells.** *Nature* 2006, **440**:808–812.

Reis e Sousa, C.: **Toll-like receptors and dendritic cells: for whom the bug tolls.** *Semin. Immunol.* 2004, **16**:27–34.

8-8 Plasmacytoid dendritic cells detect viral infections and generate abundant type I interferons and pro-inflammatory cytokines.

Asselin-Paturel, C., and Trinchieri, G.: **Production of type I interferons: plasmacytoid dendritic cells and beyond.** *J. Exp. Med.* 2005, **202**:461–465.

Blasius, A.L., and Colonna, M.: **Sampling and signaling in plasmacytoid**

dendritic cells: the potential roles of Siglec-H. *Trends Immunol.* 2006, **27**:255–260.

Colonna, M., Trinchieri, G., and Liu, Y.J.: **Plasmacytoid dendritic cells in immunity.** *Nat. Immunol.* 2004, **5**:1219–1226.

Krug, A., Veeraswamy, R., Pekosz, A., Kanagawa, O., Unanue, E.R., Colonna, M., and Cella M.: **Interferon-producing cells fail to induce proliferation of naive T cells but can promote expansion and T helper 1 differentiation of antigen-experienced unpolarized T cells.** *J. Exp. Med.* 2003, **197**:899–906.

Kuwajima, S., Sato, T., Ishida, K., Tada, H., Tezuka, H., and Ohteki, T.: **Interleukin 15–dependent crosstalk between conventional and plasmacytoid dendritic cells is essential for CpG-induced immune activation.** *Nat. Immunol.* 2006, **7**:740–746.

8-9 Macrophages are scavenger cells that can be induced by pathogens to present foreign antigens to naive T cells.

Barker, R.N., Erwig, L.P., Hill, K.S., Devine, A., Pearce, W.P., and Rees, A.J.: **Antigen presentation by macrophages is enhanced by the uptake of necrotic, but not apoptotic, cells.** *Clin. Exp. Immunol.* 2002, **127**:220–225.

Underhill, D.M., Bassetti, M., Rudensky, A., and Aderem, A.: **Dynamic interactions of macrophages with T cells during antigen presentation.** *J. Exp. Med.* 1999, **190**:1909–1914.

Zhu, F.G., Reich, C.F., and Pisetsky, D.S.: **The role of the macrophage scavenger receptor in immune stimulation by bacterial DNA and synthetic oligonucleotides.** *Immunology* 2001, **103**:226–234.

8-10 B cells are highly efficient at presenting antigens that bind to their surface immunoglobulin.

Guermonprez, P., England, P., Bedouelle, H., and Leclerc, C.: **The rate of dissociation between antibody and antigen determines the efficiency of antibody-mediated antigen presentation to T cells.** *J. Immunol.* 1998, **161**:4542–4548.

Shirota, H., Sano, K., Hirasawa, N., Terui, T., Ohuchi, K., Hattori, T., and Tamura, G.: **B cells capturing antigen conjugated with CpG oligodeoxynucleotides induce Th1 cells by elaborating IL-12.** *J. Immunol.* 2002, **169**:787–794.

Zaliauskiene, L., Kang, S., Sparks, K., Zinn, K.R., Schwiebert, L.M., Weaver, C.T., and Collawn, J.F.: **Enhancement of MHC class II-restricted responses by receptor-mediated uptake of peptide antigens.** *J. Immunol.* 2002, **169**:2337–2345.

8-11 Cell-adhesion molecules mediate the initial interaction of naive T cells with antigen-presenting cells.

Bromley, S.K., Burack, W.R., Johnson, K.G., Somersalo, K., Sims, T.N., Sumen, C., Davis, M.M., Shaw, A.S., Allen, P.M., and Dustin, M.L.: **The immunological synapse.** *Annu. Rev. Immunol.* 2001, **19**:375–396.

Friedl, P., and Brocker, E.B.: **TCR triggering on the move: diversity of T-cell interactions with antigen-presenting cells.** *Immunol. Rev.* 2002, **186**:83–89.

Gunzer, M., Schafer, A., Borgmann, S., Grabbe, S., Zanker, K.S., Brocker, E.B., Kampgen, E., and Friedl, P.: **Antigen presentation in extracellular matrix: interactions of T cells with dendritic cells are dynamic, short lived, and sequential.** *Immunity* 2000, **13**:323–332.

Montoya, M.C., Sancho, D., Vicente-Manzanares, M., and Sanchez-Madrid, F.: **Cell adhesion and polarity during immune interactions.** *Immunol. Rev.* 2002, **186**:68–82.

Wang, J., and Eck, M.J.: **Assembling atomic resolution views of the immunological synapse.** *Curr. Opin. Immunol.* 2003, **15**:286–293.

8-12 Antigen-presenting cells deliver three kinds of signals for clonal expansion and differentiation of naive T cells.

Bour-Jordan, H., and Bluestone, J.A.: **CD28 function: a balance of costimulatory and regulatory signals.** *J. Clin. Immunol.* 2002, **22**:1–7.

Gonzalo, J.A., Delaney, T., Corcoran, J., Goodearl, A., Gutierrez-Ramos, J.C., and Coyle, A.J.: **Cutting edge: the related molecules CD28 and inducible costimulator deliver both unique and complementary signals required for optimal T-cell activation.** *J. Immunol.* 2001, **166**:1–5.

Kapsenberg, M.L.: **Dendritic-cell control of pathogen-driven T-cell polarization.** *Nat. Rev. Immunol.* 2003, **3**:984–993.

Wang, S., Zhu, G., Chapoval, A.I., Dong, H., Tamada, K., Ni, J., and Chen, L.: **Costimulation of T cells by B7-H2, a B7-like molecule that binds ICOS.** *Blood* 2000, **96**:2808–2813.

8-13 CD28–dependent costimulation of activated T cells induces expression of the T-cell growth factor interleukin-2 and high affinity IL-2 receptor.

Appleman, L.J., Berezovskaya, A., Grass, I., and Boussiotis, V.A.: **CD28 costimulation mediates T cell expansion via IL-2-independent and IL-2-dependent regulation of cell cycle progression.** *J. Immunol.* 2000, **164**:144–151.

Chang, J.T., Segal, B.M., and Shevach, E.M.: **Role of costimulation in the induction of the IL-12/IL-12 receptor pathway and the development of autoimmunity.** *J. Immunol.* 2000, **164**:100–106.

Gaffen, S.L.: **Signaling domains of the interleukin 2 receptor.** *Cytokine* 2001, **14**:63–77.

Michel, F., Attal-Bonnefoy, G., Mangino, G., Mise-Omata, S., and Acuto, O.: **CD28 as a molecular amplifier extending TCR ligation and signaling capabilities.** *Immunity* 2001, **15**:935–945.

Zhou, X.Y., Yashiro-Ohtani, Y., Nakahira, M., Park, W.R., Abe, R., Hamaoka, T., Naramura, M., Gu, H., and Fujiwara, H.: **Molecular mechanisms underlying differential contribution of CD28 versus non-CD28 costimulatory molecules to IL-2 promoter activation.** *J. Immunol.* 2002, **168**:3847–3854.

8-14 Signal 2 can be modified by additional co-stimulatory pathways.

Greenwald, R.J., Freeman, G.J., and Sharpe, A.H.: **The B7 family revisited.** *Annu. Rev. Immunol.* 2005, **23**:515–548.

Watts, T.H.: **TNF/TNFR family members in costimulation of T cell responses.** *Annu. Rev. Immunol.* 2005, **23**:23–68.

8-15 Antigen recognition in the absence of co-stimulation leads to functional inactivation or clonal deletion of peripheral T cells.

Schwartz, R.H.: **T cell anergy.** *Annu. Rev. Immunol.* 2003, **21**:305–334.

Vanhove, B., Laflamme, G., Coulon, F., Mougin, M., Vusio, P., Haspot, F., Tiollier, J., and Soulillou, J.P.: **Selective blockade of CD28 and not CTLA-4 with a single-chain Fv-α1-antitrypsin fusion antibody.** *Blood* 2003, **102**:564–570.

Wekerle, T., Blaha, P., Langer, F., Schmid, M., and Muehlbacher, F.: **Tolerance through bone marrow transplantation with costimulation blockade.** *Transpl. Immunol.* 2002, **9**:125–133.

8-16 Proliferating T cells differentiate into effector T cells that do not require co-stimulation to act.

Gudmundsdottir, H., Wells, A.D., and Turka, L.A.: **Dynamics and requirements of T cell clonal expansion *in vivo* at the single-cell level: effector function is linked to proliferative capacity.** *J. Immunol.* 1999, **162**:5212–5223.

London, C.A., Lodge, M.P., and Abbas, A.K.: **Functional responses and costimulator dependence of memory CD4$^+$ T cells.** *J. Immunol.* 2000, **164**:265–272.

Schweitzer, A.N., and Sharpe, A.H.: **Studies using antigen-presenting cells lacking expression of both B7-1 (CD80) and B7-2 (CD86) show distinct requirements for B7 molecules during priming versus restimulation of Th2 but not Th1 cytokine production.** *J. Immunol.* 1998, **161**:2762–2771.

8-17 T cells differentiate into several subsets of functionally different effector cells.

Abbas, A.K., Murphy, K.M., and Sher, A.: **Functional diversity of helper T lymphocytes.** *Nature* 1996, **383**:787–793.

Glimcher, L.H., and Murphy, K.M.: **Lineage commitment in the immune system: the T helper lymphocyte grows up.** *Genes Dev.* 2000 **14**:1693–1711.

Sakaguchi, S., Ono, M., Setoguchi, R., Yagi, H., Hori, S., Fehervari, Z., Shimizu, J.,

Takahashi, T., and Nomura, T.: **Foxp3+ CD25+ CD4+ natural regulatory T cells in dominant self-tolerance and autoimmune disease.** *Immunol. Rev.* 2006, **212**:8–27.

8-18 CD8 T cells can be activated in different ways to become cytotoxic effector cells.

Andreasen, S.O., Christensen, J.E., Marker, O., and Thomsen, A.R.: **Role of CD40 ligand and CD28 in induction and maintenance of antiviral CD8⁺ effector T cell responses.** *J. Immunol.* 2000, **164**:3689–3697.

Blazevic, V., Trubey, C.M., and Shearer, G.M.: **Analysis of the costimulatory requirements for generating human virus-specific *in vitro* T helper and effector responses.** *J. Clin. Immunol.* 2001, **21**:293–302.

Croft, M.: **Co-stimulatory members of the TNFR family: keys to effective T-cell immunity?** *Nat. Rev. Immunol.* 2003, **3**:609–620.

Liang, L., and Sha, W.C.: **The right place at the right time: novel B7 family members regulate effector T cell responses.** *Curr. Opin. Immunol.* 2002, **14**:384–390.

Seder, R.A., and Ahmed, R.: **Similarities and differences in CD4⁺ and CD8⁺ effector and memory T cell generation.** *Nat. Immunol.* 2003, **4**:835–842.

Weninger, W., Manjunath, N., and von Andrian, U.H.: **Migration and differentiation of CD8⁺ T cells.** *Immunol. Rev.* 2002, **186**:221–233.

8-19 Various forms of signal 3 induce naive CD4 T-cell differentiation down distinct effector pathways.

Ansel, K.M., Lee, D.U., and Rao, A.: **An epigenetic view of helper T cell differentiation.** *Nat. Immunol.* 2003, **4**:616–623.

Murphy, K.M., and Reiner, S.L.: **The lineage decisions of helper T cells.** *Nat. Rev. Immunol.* 2002, **2**:933–944.

Nath, I., Vemuri, N., Reddi, A.L., Jain, S., Brooks, P., Colston, M.J., Misra, R.S., and Ramesh, V.: **The effect of antigen presenting cells on the cytokine profiles of stable and reactional lepromatous leprosy patients.** *Immunol. Lett.* 2000, **75**:69–76.

Stockinger, B., Bourgeois, C., and Kassiotis, G.: **CD4⁺ memory T cells: functional differentiation and homeostasis.** *Immunol. Rev.* 2006, **211**:39–48.

Szabo, S.J., Sullivan, B.M., Peng, S.L., and Glimcher, L.H.: **Molecular mechanisms regulating Th1 immune responses.** *Annu. Rev. Immunol.* 2003, **21**:713–758.

Veldhoen, M., Hocking, R.J., Atkins, C.J., Locksley, R.M., and Stockinger, B.: **TGFβ in the context of an inflammatory cytokine milieu supports de novo differentiation of IL-17-producing T cells.** *Immunity* 2006, **24**:179–189.

Weaver, C.T., Harrington, L.E., Mangan, P.R., Gavrieli, M., and Murphy, K.M.: **Th17: an effector CD4 lineage with regulatory T cell ties.** *Immunity*, 2006, **24**:677–688.

8-20 Regulatory CD4 T cells are involved in controlling adaptive immune responses.

Fantini, M.C., Becker, C., Monteleone, G., Pallone, F., Galle, P.R., and Neurath, M.F.: **TGF-β induces a regulatory phenotype in CD4⁺CD25⁻ T cells through Foxp3 induction and down-regulation of Smad7.** *J. Immunol.* 2004, **172**:5149–5153.

Fontenot, J.D., and Rudensky, A.Y.: **A well adapted regulatory contrivance: regulatory T cell development and the forkhead family transcription factor Foxp3.** *Nat. Immunol.* 2005, **6**:331–337.

Roncarolo, M.G., Bacchetta, R., Bordignon, C., Narula, S., and Levings, M.K.: **Type 1 T regulatory cells.** *Immunol. Rev.* 2001, **182**:68–79.

Sakaguchi, S.: **Naturally arising Foxp3-expressing CD25⁺CD4⁺ regulatory T cells in immunological tolerance to self and non-self.** *Nat. Immunol.* 2005, **6**:345–352.

8-21 Effector T-cell interactions with target cells are initiated by antigen-nonspecific cell-adhesion molecules.

Dustin, M.L.: **Role of adhesion molecules in activation signaling in T lymphocytes.** *J. Clin. Immunol.* 2001, **21**:258–263.

van der Merwe, P.A., and Davis, S.J.: **Molecular interactions mediating T cell antigen recognition.** *Annu. Rev. Immunol.* 2003, **21**:659–684.

8-22 Binding of the T-cell receptor complex directs the release of effector molecules and focuses them on the target cell.

Bossi, G., Trambas, C., Booth, S., Clark, R., Stinchcombe, J., and Griffiths, G.M.: **The secretory synapse: the secrets of a serial killer.** *Immunol. Rev.* 2002, **189**:152–160.

Dustin, M.L.: **Coordination of T-cell activation and migration through formation of the immunological synapse.** *Ann. N.Y. Acad. Sci.* 2003, **987**:51–59.

Montoya, M.C., Sancho, D., Vicente-Manzanares, M., and Sanchez-Madrid, F.: **Cell adhesion and polarity during immune interactions.** *Immunol. Rev.* 2002, **186**:68–82.

Trambas, C.M., and Griffiths, G.M.: **Delivering the kiss of death.** *Nat. Immunol.* 2003, **4**:399–403.

8-23 The effector functions of T cells are determined by the array of effector molecules that they produce.

&

8-24 Cytokines can act locally or at a distance.

Guidotti, L.G., and Chisari, F.V.: **Cytokine-mediated control of viral infections.** *Virology* 2000, **273**:221–227.

Harty, J.T., Tvinnereim, A.R., and White, D.W.: **CD8⁺ T cell effector mechanisms in resistance to infection.** *Annu. Rev. Immunol.* 2000, **18**:275–308.

Santana, M.A., and Rosenstein, Y.: **What it takes to become an effector T cell: the process, the cells involved, and the mechanisms.** *J. Cell Physiol.* 2003, **195**:392–401.

8-25 Cytokines and their receptors fall into distinct families of structurally related proteins.

Basler, C.F., and Garcia-Sastre, A.: **Viruses and the type I interferon antiviral system: induction and evasion.** *Int. Rev. Immunol.* 2002, **21**:305–337.

Boulay, J.L., O'Shea, J.J., and Paul, W.E.: **Molecular phylogeny within type I cytokines and their cognate receptors.** *Immunity* 2003, **19**:159–163.

Collette, Y., Gilles, A., Pontarotti, P., and Olive, D.: **A co-evolution perspective of the TNFSF and TNFRSF families in the immune system.** *Trends Immunol.* 2003, **24**:387–394.

Proudfoot, A.E.: **Chemokine receptors: multifaceted therapeutic targets.** *Nat. Rev. Immunol.* 2002, **2**:106–115.

Taniguchi, T., and Takaoka, A.: **The interferon-α/β system in antiviral responses: a multimodal machinery of gene regulation by the IRF family of transcription factors.** *Curr. Opin. Immunol.* 2002, **14**:111–116.

Wong, M.M., and Fish, E.N.: **Chemokines: attractive mediators of the immune response.** *Semin. Immunol.* 2003, **15**:5–14.

8-26 The TNF family of cytokines are trimeric proteins that are usually associated with the cell surface.

Hehlgans, T., and Mannel, D.N.: **The TNF–TNF receptor system.** *Biol. Chem.* 2002, **383**:1581–1585.

Locksley, R.M., Killeen, N., and Lenardo, M.J.: **The TNF and TNF receptor superfamilies: integrating mammalian biology.** *Cell* 2001, **104**:487–501.

Screaton, G., and Xu, X.N.: **T cell life and death signalling via TNF-receptor family members.** *Curr. Opin. Immunol.* 2000, **12**:316–322.

Theill, L.E., Boyle, W.J., and Penninger, J.M.: **RANK-L and RANK: T cells, bone loss, and mammalian evolution.** *Annu. Rev. Immunol.* 2002, **20**:795–823.

Zhou, T., Mountz, J.D., and Kimberly, R.P.: **Immunobiology of tumor necrosis factor receptor superfamily.** *Immunol. Res.* 2002, **26**:323–336.

8-27 Cytotoxic T cells can induce target cells to undergo programmed cell death.

Barry, M., and Bleackley, R.C.: **Cytotoxic T lymphocytes: all roads lead to death.** *Nat. Rev. Immunol.* 2002, **2**:401–409.

Green, D.R., Droin, N., and Pinkoski, M.: **Activation-induced cell death in T cells.** *Immunol. Rev.* 2003, **193**:70–81.

Greil, R., Anether, G., Johrer, K., and Tinhofer, I.: **Tracking death dealing by Fas** and TRAIL in lymphatic neoplastic disorders: pathways, targets, and therapeutic tools. *J. Leukoc. Biol.* 2003, **74**:311–330.

Medana, I.M., Gallimore, A., Oxenius, A., Martinic, M.M., Wekerle, H., and Neumann, H.: **MHC class I-restricted killing of neurons by virus-specific CD8+ T lymphocytes is effected through the Fas/FasL, but not the perforin pathway.** *Eur. J. Immunol.* 2000, **30**:3623–3633.

Russell, J.H., and Ley, T.J.: **Lymphocyte-mediated cytotoxicity.** *Annu. Rev. Immunol.* 2002, **20**:323–370.

Wallin, R.P., Screpanti, V., Michaelsson, J., Grandien, A., and Ljunggren, H.G.: **Regulation of perforin-independent NK cell-mediated cytotoxicity.** *Eur. J. Immunol.* 2003, **33**:2727–2735.

Zimmermann, K.C., and Green, D.R.: **How cells die: apoptosis pathways.** *J. Allergy Clin. Immunol.* 2001, **108**:S99–S103.

8-28 Cytotoxic effector proteins that trigger apoptosis are contained in the granules of CD8 cytotoxic T cells.

Barry, M., Heibein, J.A., Pinkoski, M.J., Lee, S.F., Moyer, R.W., Green, D.R., and Bleackley, R.C.: **Granzyme B short-circuits the need for caspase 8 activity during granule-mediated cytotoxic T-lymphocyte killing by directly cleaving Bid.** *Mol. Cell Biol.* 2000, **20**:3781–3794.

Grossman, W.J., Revell, P.A., Lu, Z.H., Johnson, H., Bredemeyer, A.J., and Ley, T.J.: **The orphan granzymes of humans and mice.** *Curr. Opin. Immunol.* 2003, **15**:544–552.

Lieberman, J.: **The ABCs of granule-mediated cytotoxicity: new weapons in the arsenal.** *Nat. Rev. Immunol.* 2003, **3**:361–370.

Metkar, S.S., Wang B., Aguilar-Santelises, M., Raja, S.M., Uhlin-Hansen, L., Podack, E., Trapani, J.A., and Froelich, C.J.: **Cytotoxic cell granule-mediated apoptosis: perforin delivers granzyme B–serglycin complexes into target cells without plasma membrane pore formation.** *Immunity* 2002, **16**:417–428.

Smyth, M.J., Kelly, J.M., Sutton, V.R., Davis, J.E., Browne, K.A., Sayers, T.J., and Trapani, J.A.: **Unlocking the secrets of cytotoxic granule proteins.** *J. Leukoc. Biol.* 2001, **70**:18–29.

Yasukawa, M., Ohminami, H., Arai, J., Kasahara, Y., Ishida, Y., Fujita, S.: **Granule exocytosis, and not the fas/fas ligand system, is the main pathway of cytotoxicity mediated by alloantigen-specific CD4+ as well as CD8+ cytotoxic T lymphocytes in humans.** *Blood* 2000, **95**:2352–2355.

8-29 Cytotoxic T cells are selective and serial killers of targets expressing a specific antigen.

Bossi, G., Trambas, C., Booth, S., Clark, R., Stinchcombe, J., and Griffiths, G.M.: **The secretory synapse: the secrets of a serial killer.** *Immunol. Rev.* 2002, **189**:152–160.

Stinchcombe, J.C., Bossi, G., Booth, S., and Griffiths, G.M.: **The immunological synapse of CTL contains a secretory domain and membrane bridges.** *Immunity* 2001, **15**:751–761.

Trambas, C.M., and Griffiths, G.M.: **Delivering the kiss of death.** *Nat. Immunol.* 2003, **4**:399–403.

8-30 Cytotoxic T cells also act by releasing cytokines.

Amel-Kashipaz, M.R., Huggins, M.L., Lanyon, P., Robins, A., Todd, I., and Powell, R.J.: **Quantitative and qualitative analysis of the balance between type 1 and type 2 cytokine-producing CD8− and CD8+ T cells in systemic lupus erythematosus.** *J. Autoimmun.* 2001, **17**:155–163.

Kemp, R.A., and Ronchese, F.: **Tumor-specific Tc1, but not Tc2, cells deliver protective antitumor immunity.** *J. Immunol.* 2001, **167**:6497–6502.

Prezzi, C., Casciaro, M.A., Francavilla, V., Schiaffella, E., Finocchi, L., Chircu, L.V., Bruno, G., Sette, A., Abrignani, S., and Barnaba, V.: **Virus-specific CD8+ T cells with type 1 or type 2 cytokine profile are related to different disease activity in chronic hepatitis C virus infection.** *Eur. J. Immunol.* 2001, **31**:894–906.

Woodland, D.L., and Dutton, R.W.: **Heterogeneity of CD4+ and CD8+ T cells.** *Curr. Opin. Immunol.* 2003, **15**:336–342.

8-31 T_H1 cells have a central role in macrophage activation.

Monney, L., Sabatos, C.A., Gaglia, J.L., Ryu, A., Waldner, H., Chernova, T., Manning, S., Greenfield, E.A., Coyle, A.J., Sobel, R.A., Freeman, G.J., and Kuchroo, V.K.: **Th1-specific cell surface protein Tim-3 regulates macrophage activation and severity of an autoimmune disease.** *Nature* 2002, **415**:536–541.

Munoz Fernandez, M.A., Fernandez, M.A., and Fresno, M.: **Synergism between tumor necrosis factor-α and interferon-γ on macrophage activation for the killing of intracellular *Trypanosoma cruzi* through a nitric oxide-dependent mechanism.** *Eur. J. Immunol.* 1992, **22**:301–307.

Stout, R., and Bottomly, K.: **Antigen-specific activation of effector macrophages by interferon-γ producing (T_H1) T-cell clones: failure of IL-4 producing (T_H2) T-cell clones to activate effector functions in macrophages.** *J. Immunol.* 1989, **142**:760–765.

Shaw, G., and Kamen, R.: **A conserved AU sequence from the 3′ untranslated region of GM-CSF mRNA mediates selective mRNA degradation.** *Cell* 1986, **46**:659–667.

8-32 Activation of macrophages by T_H1 cells promotes microbial killing and must be tightly regulated to avoid tissue damage.

Duffield, J.S.: **The inflammatory macrophage: a story of Jekyll and Hyde.** *Clin. Sci. (Lond.)* 2003, **104**:27–38.

James, D.G.: **A clinicopathological classification of granulomatous disorders.** *Postgrad. Med. J.* 2000, **76**:457–465.

Labow, R.S., Meek, E., and Santerre, J.P.: **Model systems to assess the destructive potential of human neutrophils and monocyte-derived macrophages during the acute and chronic phases of inflammation.** *J. Biomed. Mater. Res.* 2001, **54**:189–197.

Wigginton, J.E., and Kirschner, D.: **A model to predict cell-mediated immune regulatory mechanisms during human infection with *Mycobacterium tuberculosis*.** *J. Immunol.* 2001, **166**:1951–1967.

8-33 T_H1 cells coordinate the host response to intracellular pathogens.

Alexander, J., Satoskar, A.R., and Russell, D.G.: ***Leishmania* species: models of intracellular parasitism.** *J. Cell. Sci.* 1999, **112**:2993–3002.

Berberich, C., Ramirez-Pineda, J.R., Hambrecht, C., Alber, G., Skeiky, Y.A., and Moll, H.: **Dendritic cell (DC)-based protection against an intracellular pathogen is dependent upon DC-derived IL-12 and can be induced by molecularly defined antigens.** *J. Immunol.* 2003, **170**:3171–3179.

Biedermann, T., Zimmermann, S., Himmelrich, H., Gumy, A., Egeter, O., Sakrauski, A.K., Seegmuller, I., Voigt, H., Launois, P., Levine, A.D., Wagner, H., Heeg, K., Louis, J.A., and Rocken, M.: **IL-4 instructs T_H1 responses and resistance to *Leishmania major* in susceptible BALB/c mice.** *Nat. Immunol.* 2001, **2**:1054–1060.

Koguchi, Y., and Kawakami, K.: **Cryptococcal infection and Th1-Th2 cytokine balance.** *Int. Rev. Immunol.* 2002, **21**:423–438.

Neighbors, M., Xu, X., Barrat, F.J., Ruuls, S.R., Churakova, T., Debets, R., Bazan, J.F., Kastelein, R.A., Abrams, J.S., and O'Garra, A.: **A critical role for interleukin 18 in primary and memory effector responses to *Listeria monocytogenes* that extends beyond its effects on interferon gamma production.** *J. Exp. Med.* 2001, **194**:343–354.

The Humoral Immune Response

9

Many of the bacteria that cause infectious disease in humans multiply in the extracellular spaces of the body, and most intracellular pathogens spread by moving from cell to cell through the extracellular fluids. The extracellular spaces are protected by the **humoral immune response**, in which antibodies produced by B cells cause the destruction of extracellular microorganisms and prevent the spread of intracellular infections. Activation of naive B cells is triggered by antigen and usually requires helper T cells; the activated B cells then differentiate into antibody-secreting **plasma cells** (Fig. 9.1) and memory B cells. In this chapter we use the general term helper T cell to mean any effector CD4 T cell, either T_H1 or T_H2, that can activate a B cell (see Chapter 8).

Antibodies contribute to immunity in three main ways (see Fig. 9.1). The first is known as **neutralization**. To enter cells, viruses and intracellular bacteria bind to specific molecules on the target cell surface. Antibodies that bind to the pathogen can prevent this and are said to neutralize the pathogen. Neutralization by antibodies is also important in preventing bacterial toxins from entering cells. Second, antibodies protect against bacteria that multiply outside cells mainly by facilitating uptake of the pathogen by phagocytes. Coating the surface of a pathogen to enhance phagocytosis is called opsonization. Antibodies bound to the pathogen are recognized by phagocytic cells by means of receptors called Fc receptors that bind to the antibody constant region (C region).Third, antibodies coating a pathogen can activate the proteins of the complement system by the classical pathway, as described in Chapter 2. Complement proteins bound to the pathogen surface opsonize the pathogen by binding complement receptors on phagocytes. Other complement components recruit phagocytic cells to the site of infection, and the terminal components of complement can lyse certain microorganisms directly by forming pores in their membranes. Which effector mechanisms are engaged in a particular response is determined by the heavy-chain isotype of the antibodies produced, which determines their class (see Chapter 4).

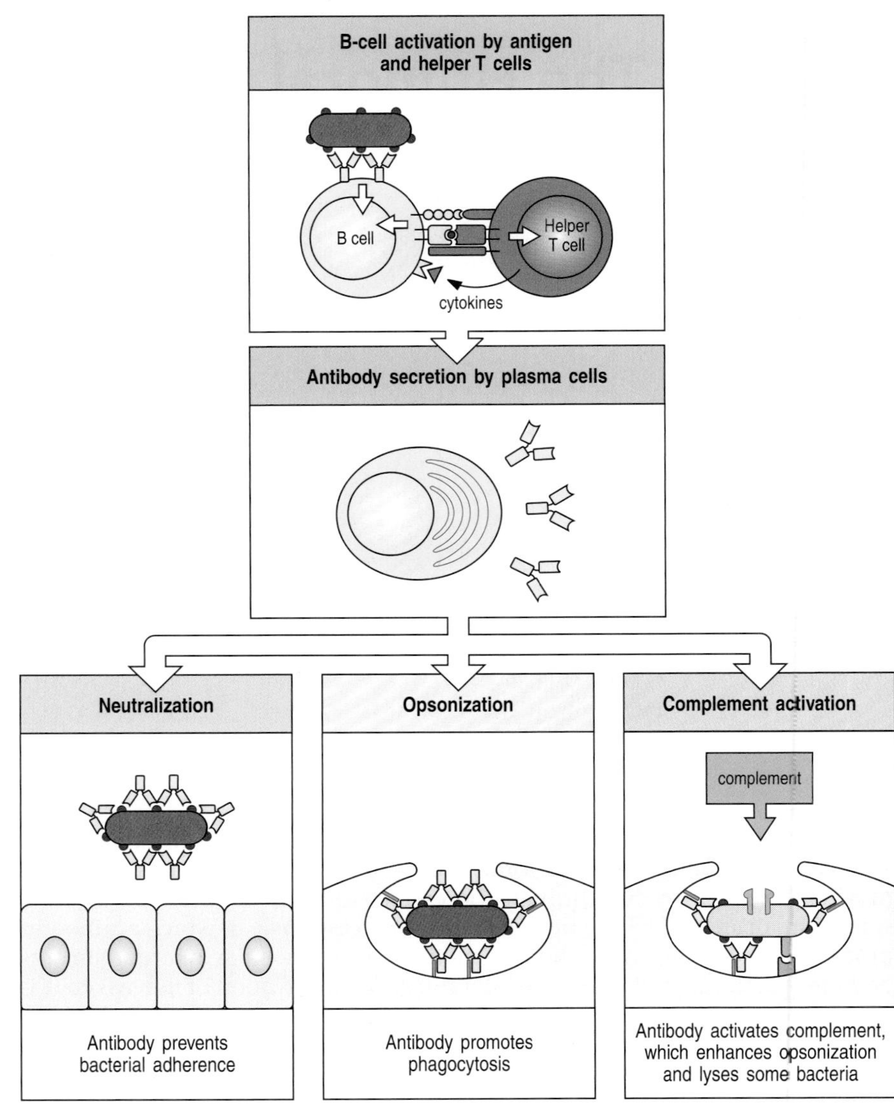

Fig. 9.1 The humoral immune response is mediated by antibody molecules secreted by plasma cells. Antigen that binds to the B-cell antigen receptor signals B cells and is, at the same time, internalized and processed into peptides that activate armed helper T cells. Signals from the bound antigen and from the helper T cell induce the B cell to proliferate and differentiate into plasma cells secreting specific antibody (top two panels). These antibodies protect the host from infection in three main ways. First, they can inhibit the toxic effects or infectivity of pathogens by binding to them: this is termed neutralization (bottom left panel). Second, by coating the pathogens they can enable accessory cells that recognize the Fc portions of arrays of antibodies to ingest and kill the pathogen, a process called opsonization (bottom center panel). Third, antibodies can trigger the activation of the complement system. Complement proteins can strongly enhance opsonization, and can directly kill certain bacterial cells (bottom right panel).

In the first part of this chapter we describe the interactions of naive B cells with antigen and with helper T cells that lead to the activation of B cells and antibody production. Some important microbial antigens can provoke antibody production without the help of T cells, and we shall also consider these responses here. Most antibody responses undergo a process called affinity maturation, in which antibodies of greater affinity for their target antigen are produced by the somatic hypermutation of antibody variable-region (V-region) genes. The molecular mechanism of somatic hypermutation was described in Chapter 4 and here we look at its immunological consequences. We also revisit class switching (see Chapter 4), which produces antibodies of different functional classes and confers functional diversity on the antibody response. Both affinity maturation and class switching occur only in B cells and require T-cell help. In the rest of the chapter we look in detail at the various effector mechanisms by which antibodies contain and eliminate infections. Like the T-cell response, the humoral immune response produces immunological memory, and this will be discussed in Chapter 10.

B-cell activation and antibody production.

The surface immunoglobulin that serves as the **B-cell antigen receptor** (**BCR**) can bind a vast variety of chemical structures. In the context of natural infections it binds native proteins, glycoproteins, and polysaccharides, as well as whole virus particles and bacterial cells, by recognizing epitopes on their surfaces. It has two roles in B-cell activation. First, like the antigen receptor on T cells, it signals to the cell's interior when antigen is bound (see Chapter 6). Second, the B-cell antigen receptor delivers the bound antigen to intracellular sites, where it can be degraded to give peptides that are returned to the B-cell surface bound to MHC class II molecules (see Chapter 5). These peptide:MHC class II complexes are recognized by antigen-specific helper T cells that have already differentiated in response to the same pathogen, as described in Chapter 8. The effector T cells make cytokines that cause the B cell to proliferate and its progeny to differentiate into antibody-secreting cells and into memory B cells. Some microbial antigens can activate B cells directly in the absence of T-cell help, and the ability of B cells to respond directly to these antigens provides a rapid response to many important pathogens. However, the fine tuning of antibody responses to increase the affinity of the antibody for the antigen and the switching to most immunoglobulin classes other than IgM depend on the interaction of antigen-stimulated B cells with helper T cells and other cells in the peripheral lymphoid organs. Thus, antibodies induced by microbial antigens alone tend to have lower affinity and to be less functionally versatile than those induced with T-cell help.

9-1 The humoral immune response is initiated when B cells that bind antigen are signaled by helper T cells or by certain microbial antigens alone.

It is a general rule in adaptive immunity that naive antigen-specific lymphocytes are difficult to activate by antigen alone. As we saw in Chapter 8, priming of naive T cells requires a co-stimulatory signal from professional antigen-presenting cells; naive B cells also require accessory signals that can come either from a helper T cell or, in some cases, directly from microbial constituents.

Antibody responses to protein antigens require antigen-specific T-cell help. These antigens are unable to induce antibody responses in animals or humans who lack T cells, and they are therefore known as **thymus-dependent** or **TD antigens**. To receive T-cell help, the B cell must be displaying antigen on its surface in a form a T cell can recognize. This occurs when antigen bound by surface immunoglobulin on the B cell is internalized and returned to the cell surface as peptides bound to MHC class II molecules. Helper T cells that recognize the peptide:MHC complex then deliver activating signals to the B cell (Fig. 9.2, top two panels). Thus, protein antigens binding to B cells

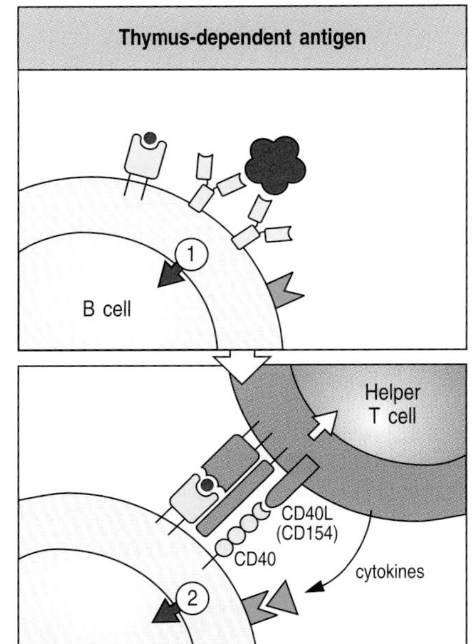

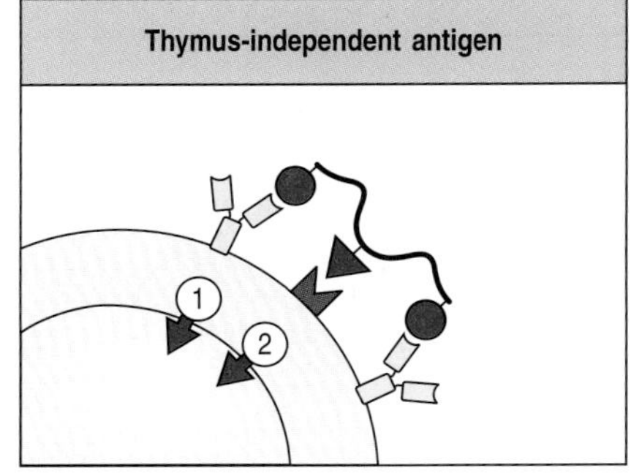

Fig. 9.2 A second signal is required for B-cell activation by either thymus-dependent or thymus-independent antigens. The first signal (indicated as 1 in the figure) required for B-cell activation is delivered through its antigen receptor (top panel). For thymus-dependent antigens, the second signal (indicated as 2) is delivered by a helper T cell that recognizes degraded fragments of the antigen as peptides bound to MHC class II molecules on the B-cell surface (center panel); the interaction between CD40 ligand (CD40L, also called CD154) on the T cell and CD40 on the B cell contributes an essential part of this second signal. For thymus-independent antigens, the second signal can be delivered by the antigen itself (bottom panel), either through direct binding of a part of the antigen to a receptor of the innate immune system (purple), or simply by extensive cross-linking of the membrane IgM by a polymeric antigen (not shown).

both provide a specific signal to the B cell by cross-linking its antigen receptors and allow the B cell to attract antigen-specific T-cell help. When an activated helper T cell recognizes and binds a peptide:MHC class II complex on the B-cell surface it induces the B cell to proliferate and differentiate into antibody-producing plasma cells (Fig. 9.3). The requirement for T-cell help means that before a B cell can be induced to make antibody against the proteins of an infecting pathogen, CD4 T cells specific for peptides from this pathogen must be activated to produce helper T cells. This occurs when the naive T cells interact with dendritic cells presenting the appropriate peptides, as described in Chapter 8.

Although peptide-specific helper T cells are required for B-cell responses to protein antigens, many microbial constituents, such as bacterial polysaccharides, can induce antibody production in the absence of helper T cells. These microbial antigens are known as **thymus-independent** or **TI antigens** because they can induce antibody responses in individuals who have no T lymphocytes. The second signal required to activate antibody production against TI antigens is either provided directly by recognition of a common microbial constituent (see Fig. 9.2, bottom panel) or by massive cross-linking of B-cell receptors, which would occur when a B cell binds repeating epitopes on the bacterial cell. Thymus-independent antibody responses provide some protection against extracellular bacteria, and we will return to them later.

9-2 B-cell responses to antigen are enhanced by co-ligation of the B-cell co-receptor.

B-cell responsiveness to antigen is greatly enhanced by signaling through the cell-surface **B-cell co-receptor complex**, as described in Section 6-17. The co-receptor complex is composed of three proteins: CD19, CD21, and CD81. CD21 (also known as complement receptor 2, CR2) is a receptor for the complement fragments C3d and C3dg (see Section 2-19). When complement is activated, either by the innate pathways or by antibody bound to an antigen such as a bacterial cell, activated complement components are deposited on the antigen itself. When the B-cell receptor binds to the antigen in such a complex, CD21 can bind the complement, bringing together the B-cell receptor

Fig. 9.3 Armed helper T cells stimulate the proliferation and then the differentiation of antigen-binding B cells. The specific interaction of an antigen-binding B cell with an armed helper T cell leads to the expression of the B-cell stimulatory molecule CD40 ligand (CD154) on the helper T-cell surface and to the secretion of the B-cell stimulatory cytokines IL-4, IL-5, and IL-6, which drive the proliferation and differentiation of the B cell into antibody-secreting plasma cells. Alternatively, an activated B cell can become a memory cell.

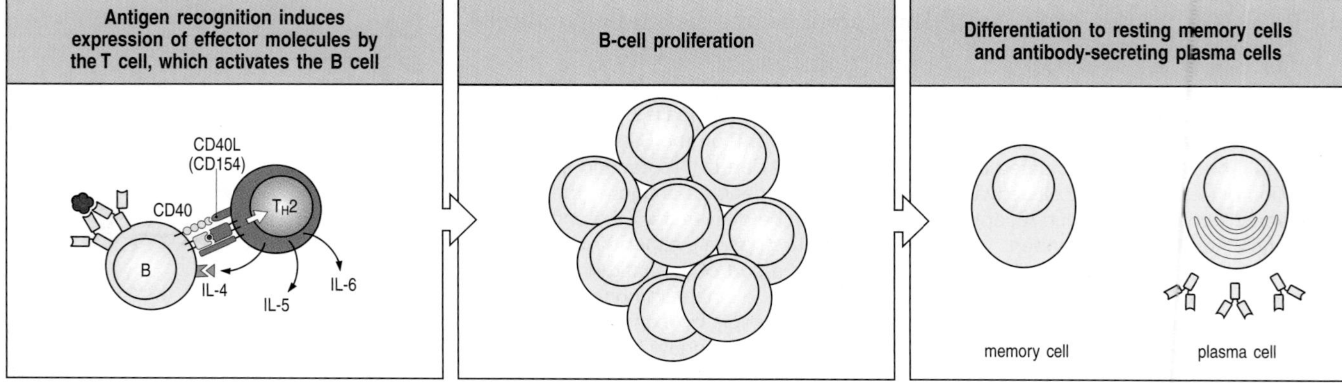
Antigen recognition induces expression of effector molecules by the T cell, which activates the B cell
CD40L (CD154)
CD40
T_H2
B
IL-4
IL-5
IL-6
B-cell proliferation
Differentiation to resting memory cells and antibody-secreting plasma cells
memory cell
plasma cell

and the co-receptor and generating signals through CD19 that activate a PI 3-kinase signaling pathway and co-stimulate the B-cell response (see Section 6-17). It is thought that the signaling pathways activated by CD21 augment the intracellular signal that leads directly to differentiation and antibody production, induce co-stimulatory molecules on the B cell, thus making it more effective at eliciting T-cell help, and increase the receptor-mediated uptake of antigen. Which of these effects has most influence on enhancing B-cell responsiveness is not yet known.

The presence of the B-cell co-receptor powerfully amplifies antibody responses, as the complexes that the antibody makes with antigen and C3dg produce a more potent antigen, leading to more efficient B-cell activation and antibody production. The effect of co-ligation of the B-cell receptor and co-receptor is dramatically demonstrated when mice are immunized with hen egg-white lysozyme coupled to three linked molecules of C3dg. In this case the dose of modified lysozyme needed to induce antibody in the absence of added adjuvant is as little as 1/10,000 of that needed with the unmodified lysozyme.

9-3 Helper T cells activate B cells that recognize the same antigen.

A given B cell can only be activated by helper T cells that respond to the same antigen; this is called **linked recognition**. Although the epitope recognized by the helper T cell must be linked to that recognized by the B cell, the two cells need not recognize identical epitopes. Indeed, we saw in Chapter 5 that T cells can recognize peptides derived from the core region of proteins that are quite distinct from the surface epitopes on the same protein recognized by B cells. For more complex natural antigens, such as viruses and bacteria, which are composed of multiple proteins and carry both protein and carbohydrate epitopes, the T cell and the B cell might not even recognize the same molecule. It is, however, crucial that the peptide recognized by the T cell is physically associated with the antigen recognized by the B cell, so that the B cell will produce an appropriate peptide after internalization of the antigen bound to its B-cell receptors.

For example, by recognizing an epitope on a viral protein coat, a B cell can bind and internalize a complete virus particle. After internalization, the virus particle is degraded and peptides from internal viral proteins as well as coat proteins can be displayed by MHC class II molecules on the B-cell surface. Helper T cells that have been primed earlier in the infection by dendritic cells presenting these internal peptides can then activate the B cell to make antibodies that recognize the coat protein (Fig. 9.4).

The specific activation of the B cell by its **cognate** T cell—that is, a helper T cell primed by the same antigen—depends on the ability of the antigen-specific B cell to concentrate the appropriate peptide on its surface MHC class II molecules. B cells that bind a particular antigen are up to 10,000 times more efficient at displaying peptide fragments of that antigen on their MHC class II molecules than are B cells that do not bind the antigen. A helper T cell

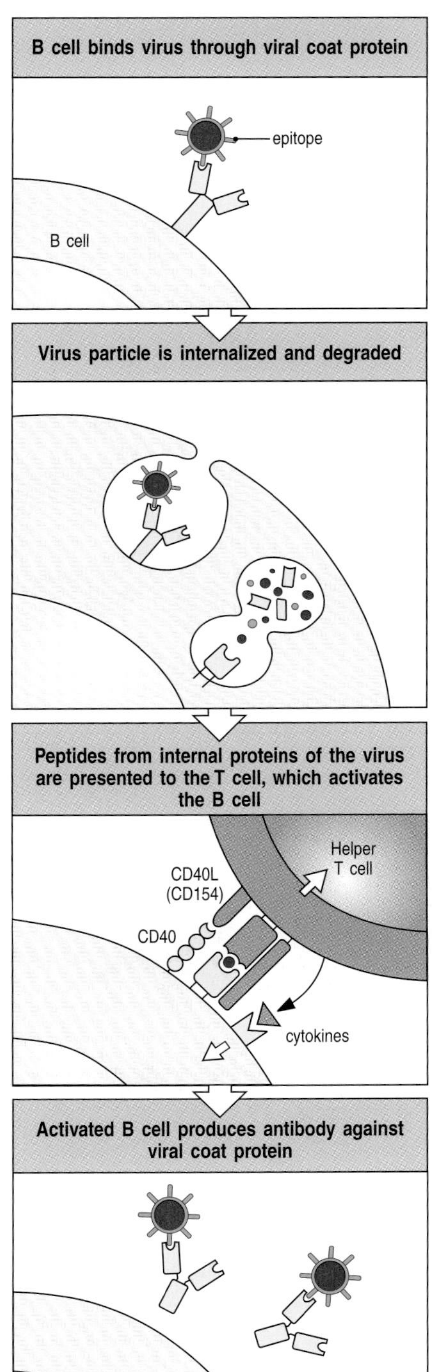
B cell binds virus through viral coat protein
epitope
B cell
Virus particle is internalized and degraded
Peptides from internal proteins of the virus are presented to the T cell, which activates the B cell
Helper T cell
CD40L (CD154)
CD40
cytokines
Activated B cell produces antibody against viral coat protein

Fig. 9.4 B cells and helper T cells must recognize epitopes of the same molecular complex in order to interact. An epitope on a viral coat protein is recognized by the surface immunoglobulin on a B cell, and the virus is internalized and degraded. Peptides derived from viral proteins, including internal proteins, are returned to the B-cell surface bound to MHC class II molecules (see Chapter 5). Here, these complexes are recognized by helper T cells, which help to activate the B cells to produce antibody against the coat protein.

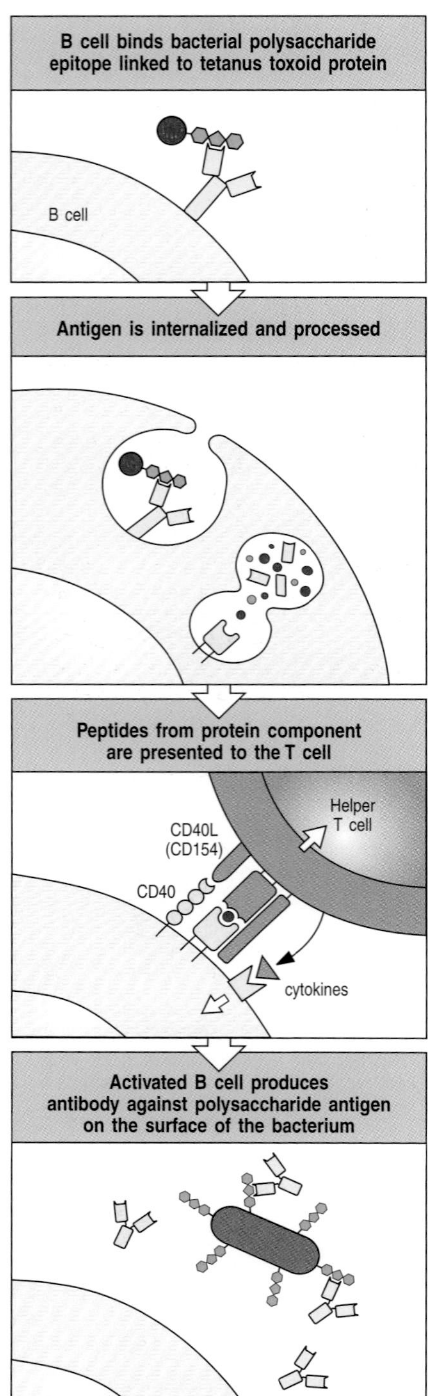
B cell binds bacterial polysaccharide epitope linked to tetanus toxoid protein
B cell
Antigen is internalized and processed
Peptides from protein component are presented to the T cell
Helper T cell
CD40L (CD154)
CD40
cytokines
Activated B cell produces antibody against polysaccharide antigen on the surface of the bacterium

Fig. 9.5 Protein antigens attached to polysaccharide antigens allow T cells to help polysaccharide-specific B cells. The vaccine against *Haemophilus influenzae* type b is a conjugate of bacterial polysaccharide and the tetanus toxoid protein. The B cell recognizes and binds the polysaccharide, internalizes and degrades the whole conjugate and then displays toxoid-derived peptides on surface MHC class II molecules. Helper T cells generated in response to earlier vaccination against the toxoid recognize the complex on the B-cell surface and activate the B cell to produce anti-polysaccharide antibody. This antibody can then protect against infection with *H. influenzae* type b.

will therefore help only those B cells whose receptors bind an antigen containing the peptide recognized by the T cell.

The requirement for linked recognition has important consequences for the regulation and manipulation of the humoral immune response. One is that linked recognition helps to ensure self tolerance, because it means that an autoimmune response will occur only if both a self-reactive T cell and a self-reactive B cell are present at the same time. This is discussed further in Chapter 14. Another important application of linked recognition is in the design of vaccines, such as that used to immunize infants against *Haemophilus influenzae* type b. This bacterial pathogen can infect the covering of the brain, called the meninges, causing meningitis. Protective immunity to this pathogen is mediated by antibodies against its capsular polysaccharide. Although adults make very effective thymus-independent responses to these polysaccharide antigens, such responses are weak in the immature immune system of the infant. To make an effective vaccine for use in infants, therefore, the polysaccharide is linked chemically to tetanus toxoid, a foreign protein against which infants are routinely and successfully vaccinated (see Chapter 15). B cells that bind the polysaccharide component of the vaccine can be activated by helper T cells specific for peptides of the linked toxoid (Fig. 9.5).

Linked recognition was originally discovered through studies of the production of antibodies against haptens (see Appendix I, Section A-1). Haptens are small chemical groups that cannot elicit antibody responses on their own because they cannot cross-link B-cell receptors and they cannot recruit T-cell help. When coupled to a carrier protein, however, they become immunogenic, because the protein will carry multiple hapten groups that can now cross-link B-cell receptors. In addition, T-cell dependent responses are possible because T cells can be primed to respond to peptides derived from the protein. Accidental coupling of a hapten to a protein is responsible for the allergic responses shown by many people to the antibiotic penicillin, which reacts with host proteins to form a coupled hapten that can stimulate an antibody response, as we will learn in Chapter 13.

9-4 Antigenic peptides bound to self-MHC class II molecules on B cells trigger helper T cells to make membrane-bound and secreted molecules that can activate a B cell.

Recognition of peptide:MHC class II complexes on B cells triggers helper T cells to synthesize both cell-bound and secreted effector molecules that synergize in activating the B cell. One particularly important T-cell effector molecule is the TNF family member CD40 ligand, which binds to CD40 on B cells. CD40 is a member of the TNF-receptor family of cytokine receptors (see Section 8-26) and is involved in activating important phases of the B-cell response, such as B-cell proliferation, immunoglobulin class switching, and somatic hypermutation. Binding of CD40 by CD40 ligand helps to drive the

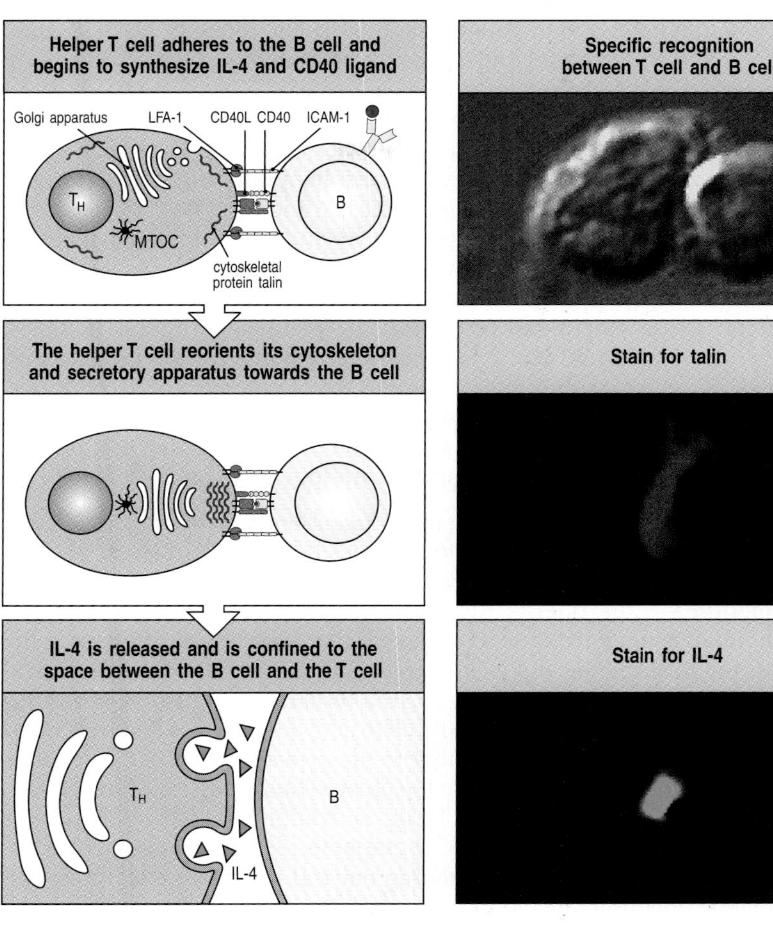
Helper T cell adheres to the B cell and begins to synthesize IL-4 and CD40 ligand
Golgi apparatus
LFA-1
CD40L
CD40
ICAM-1
T_H
B
MTOC
cytoskeletal protein talin
The helper T cell reorients its cytoskeleton and secretory apparatus towards the B cell
IL-4 is released and is confined to the space between the B cell and the T cell
T_H
B
IL-4
Specific recognition between T cell and B cell
Stain for talin
Stain for IL-4

Fig. 9.6 When a helper T cell encounters an antigen-binding B cell, it becomes polarized and secretes IL-4 and other cytokines as well as the cell-associated TNF family member CD40 ligand at the point of cell–cell contact. On binding antigen on the B cell through its T-cell receptor, the helper T cell is induced to express CD40 ligand, which binds to CD40 on the B cell. As shown in the top left panel, the tight junction formed between the cells after antigen-specific binding seems to be sealed by a ring of adhesion molecules, with LFA-1 on the T cell interacting with ICAM-1 on the B cell. The cytoskeleton becomes polarized, as revealed by the relocation of the cytoskeletal protein talin (stained red in the right center panel) to the point of cell–cell contact, and the secretory apparatus (the Golgi apparatus) is reoriented by the cytoskeleton toward the point of contact with the B cell. As shown in the bottom panels, cytokines are released at the point of contact. The bottom right panel shows IL-4 (stained green) confined to the space between the B cell and the helper T cell. MTOC, microtubule-organizing center. Photographs courtesy of A. Kupfer.

resting B cell into the cell cycle and is essential for B-cell responses to thymus-dependent antigens. It also causes the B cell to increase its expression of co-stimulatory molecules, especially those of the B7 family. These provide important signals that sustain T-cell growth and differentiation, thus enhancing the mutual T cell–B cell interaction.

When exposed to a mixture of artificially synthesized CD40 ligand and the cytokine interleukin-4 (IL-4) *in vitro*, B cells are stimulated to proliferate. IL-4 is made by T_H2 cells when they recognize their specific ligand on the B-cell surface, and IL-4 and CD40 ligand are thought to synergize in driving the clonal expansion of B cells that precedes antibody production *in vivo*. IL-4 is secreted in a polar fashion by the T_H2 cell and is focused at the site of contact with the B cell (Fig. 9.6) so that it acts selectively on the antigen-specific target B cell. B-cell proliferation is therefore the result of a combination of B-cell receptor and CD40 ligation, along with IL-4 and other signals derived from direct T-cell contact. Some of these additional contact signals have recently been elucidated. They involve other TNF/TNF-receptor family members, including **CD30** and **CD30 ligand** (now also called CD153), and 4-1BB (CD137) on T cells with 4-1BB ligand on the B cell, as well as homologs of B7 and CD28, including **B7-RP** and ICOS, respectively. The soluble TNF-family cytokine BAFF (see Section 7-27) is secreted by dendritic cells and macrophages and acts as a survival factor for differentiating B cells. After several rounds of proliferation, B cells can differentiate into antibody-secreting plasma cells. Two additional cytokines, IL-5 and IL-6, both secreted by helper T cells, contribute to these later stages of B-cell activation.

9-5 B cells that have bound antigen via their B-cell receptor are trapped in the T-cell zones of secondary lymphoid tissues.

One of the most puzzling features of the antibody response is how a B cell manages to encounter a helper T cell with an appropriate antigen specificity. This question arises because the frequency of naive lymphocytes specific for any given antigen is estimated to be between 1 in 10,000 and 1 in 1,000,000. Thus, the chance of an encounter between a T lymphocyte and a B lymphocyte that recognize the same antigen should be between 1 in 10^8 and 1 in 10^{12}. An additional difficulty is that T cells and B cells mostly occupy quite distinct zones in peripheral lymphoid tissue—the **T-cell areas** and the **primary lymphoid follicles**, respectively (see Figs 1.18–1.20). When circulating naive B cells migrate into these tissues through high endothelial venules they first enter the T-cell zones, and usually move quickly through the zone into the primary follicle. As with naive T-cell activation, the answer to the question posed at the beginning of the paragraph seems to lie in the antigen-specific trapping of circulating B cells.

We saw in Chapter 8 how recirculating naive T cells are trapped very efficiently in the T-cell zone of secondary lymphoid tissues by recognizing their peptide antigen presented on dendritic cells, and are activated to helper T cell status there. Ingenious experiments using mice transgenic for rearranged immunoglobulin genes have shown that B cells that have bound antigens in the blood or in the intracellular fluids are trapped at the border of the T-cell and B-cell zones of peripheral lymphoid tissue by a similar mechanism (Fig. 9.7). An encounter with antigen signals a naive B cell to activate the adhesion molecules it bears on its surface in a similar manner to the activation that occurs when a naive T cell encounters its antigen (see Fig. 8.18). Thus, once they have bound antigen, migrating B cells are arrested by the activation of adhesion molecules such as LFA-1 and the engagement of chemokine receptors such as CCR7, a receptor for CCL19 and CCL21. Circulating naive B cells may encounter and bind pathogen antigens in the blood stream or as free antigens brought into the lymphoid tissues by the lymph. Dendritic cells can also present antigens to B cells. Dendritic cells can passively bind some antigens directly and others in the form of antigen:antibody complexes. In this capacity, they act as filters that sit in lymphoid tissues and concentrate antigens coming in from a site of infection, so that the chances of a B cell meeting its cognate antigen are increased.

Trapping of antigen-bearing B cells at the borders of T-cell zones provides an elegant solution to the problem of getting B cells together with their appropriate helper T cells. B cells that are already in a lymphoid follicle when they

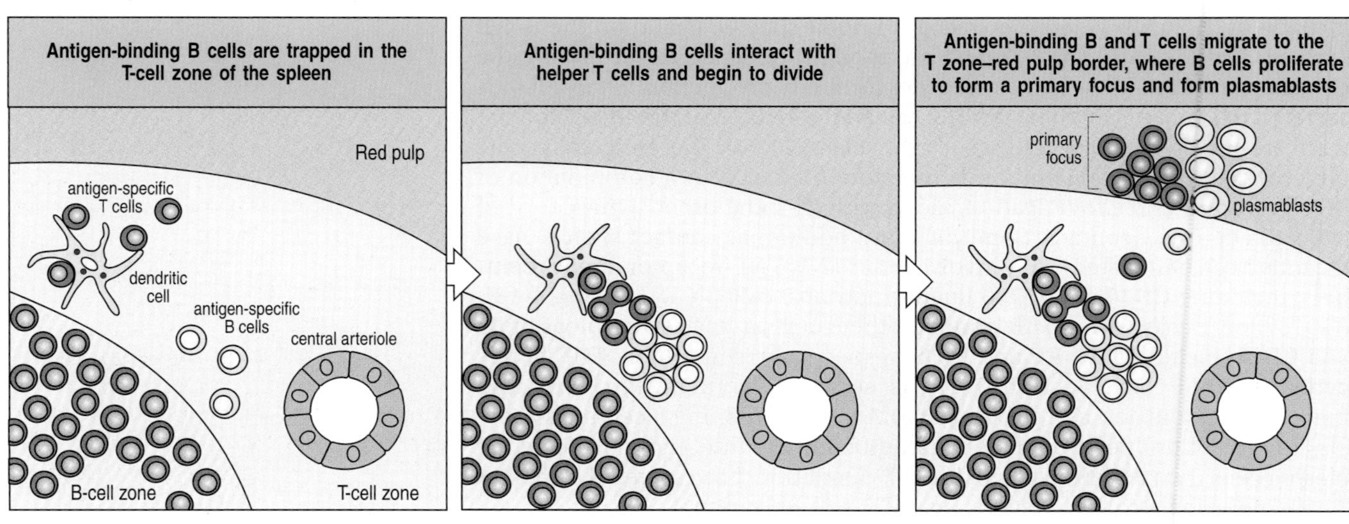

Antigen-binding B cells are trapped in the T-cell zone of the spleen
Red pulp
antigen-specific T cells
dendritic cell
antigen-specific B cells
central arteriole
B-cell zone
T-cell zone
Antigen-binding B cells interact with helper T cells and begin to divide
Antigen-binding B and T cells migrate to the T zone–red pulp border, where B cells proliferate to form a primary focus and form plasmablasts
primary focus
plasmablasts

Fig. 9.7 Antigen-binding B cells meet T cells at the border between the T-cell and B-cell zones in secondary lymphoid tissue. Activation of B cells in the spleen is shown here. Upon entry into the spleen from the blood through the marginal sinus (not shown), naive T cells and B cells home to different regions, as described in Chapter 7. If T cells encounter their antigen on the surface of an antigen-presenting cell, such as a dendritic cell, in the T-cell zone, they become activated, some differentiating into helper T cells (left panel). If B cells specific for the same antigen encounter it, either in the blood or tissue fluids or localized on the surface of dendritic cells in the lymphoid tissues, they are arrested in the T-cell zone, near the T zone–B zone border, where they may encounter activated helper T cells specific for the same antigen. This interaction gives rise to an initial proliferation of B cells (center panel). In spleen, the activated lymphocytes then migrate to the border of the T-cell zone and the red pulp, where they continue to proliferate and where the B cells differentiate into plasmablasts, forming a so-called primary focus (right panel). In lymph nodes, the primary focus arises in the medullary cords (see Fig. 9.9).

encounter antigen most probably also migrate to the border between T zone and B zone. Thus, B cells that have bound antigen become selectively trapped in precisely the correct location to maximize their chance of encountering a helper T cell that can activate them. Antigen-stimulated B cells that fail to interact with T cells that recognize the same antigen die within 24 hours.

After their initial encounter, the B cells and their cognate T cells migrate from the T zone–B zone border to continue their proliferation and differentiation. In the spleen they move to the border of the T zone and the red pulp. Here they establish a **primary focus** of clonal expansion (see Fig. 9.7). In lymph nodes the primary focus is located in the medullary cords, where lymph drains out of the node. Primary foci appear about 5 days after an infection or immunization with an antigen not previously encountered, which correlates with the time needed for helper T cells to differentiate.

9-6 Antibody-secreting plasma cells differentiate from activated B cells.

Both T cells and B cells proliferate in the primary focus for several days, and this constitutes the first phase of the primary humoral immune response. Some of these proliferating B cells differentiate into antibody-synthesizing **plasmablasts** in the primary focus. Others may migrate into the lymphoid follicle and differentiate further there before becoming plasma cells, as we will describe later. Plasmablasts are cells that have begun to secrete antibody, yet are still dividing and still express many of the characteristics of activated B cells that allow their interaction with T cells. After a few more days the plasmablasts stop dividing and either die or differentiate further into plasma cells. The differentiation of a B cell into a plasma cell is accompanied by many morphological changes that reflect its commitment to the production of large amounts of secreted antibody. Some of the plasma cells remain in the lymphoid organs, where they are short lived, while the majority migrate to the bone marrow and continue antibody production there.

The properties of resting B cells, plasmablasts, and plasma cells are compared in Fig. 9.8. Plasmablasts and plasma cells have abundant cytoplasm dominated by multiple layers of rough endoplasmic reticulum (see Fig. 1.23). The nucleus shows a characteristic pattern of peripheral chromatin condensation, a prominent perinuclear Golgi apparatus is visible, and the cisternae

	Property					
	Intrinsic			Inducible		
B-lineage cell	Surface Ig	Surface MHC class II	High-rate Ig secretion	Growth	Somatic hyper-mutation	Class switch
Resting B cell	High	Yes	No	Yes	Yes	Yes
Plasmablast	High	Yes	Yes	Yes	Unknown	Yes
Plasma cell	Low	No	Yes	No	No	No

Fig. 9.8 Plasma cells secrete antibody at a high rate but can no longer respond to antigen or helper T cells. Resting naive B cells have membrane-bound immunoglobulin (usually IgM and IgD) and MHC class II molecules on their surface. Their V genes do not carry somatic mutations. They can take up antigen and present it to helper T cells, which then induce the B cells to proliferate, switch the isotype of the immunoglobulin they are producing (class or isotype switching), and undergo somatic hypermutation; however, B cells do not secrete significant amounts of antibody. Plasmablasts have an intermediate phenotype. They do secrete antibody, but they retain substantial surface immunoglobulin and MHC class II molecules and so can continue to take up and present antigen to T cells. Plasma cells are terminally differentiated B cells that secrete antibodies. They can no longer interact with helper T cells because they have very low levels of surface immunoglobulin and lack MHC class II molecules, although they have usually already undergone class switching and somatic hypermutation. Plasma cells have also lost the ability to change the class of their antibody or to undergo further somatic hypermutation.

of the endoplasmic reticulum are rich in immunoglobulin, which in a plasma cell makes up 10–20% of all the protein synthesized. Although plasmablasts still express B7 co-stimulatory molecules and MHC class II molecules, plasma cells do not. Thus, plasma cells can no longer present antigen to helper T cells, although these T cells may still provide important signals for plasma-cell differentiation and survival, such as IL-6 and CD40 ligand. Plasmablasts express surface immunoglobulin, which is expressed on plasma cells only at low levels. Nonetheless, these low levels of surface immunoglobulin may be physiologically important, because recent evidence suggests that the survival of plasma cells may be determined in part by their ability to continue to bind antigen. Plasma cells have a range of life spans. Some survive for only days to a few weeks after their final differentiation, whereas others are very long lived and account for the persistence of antibody responses.

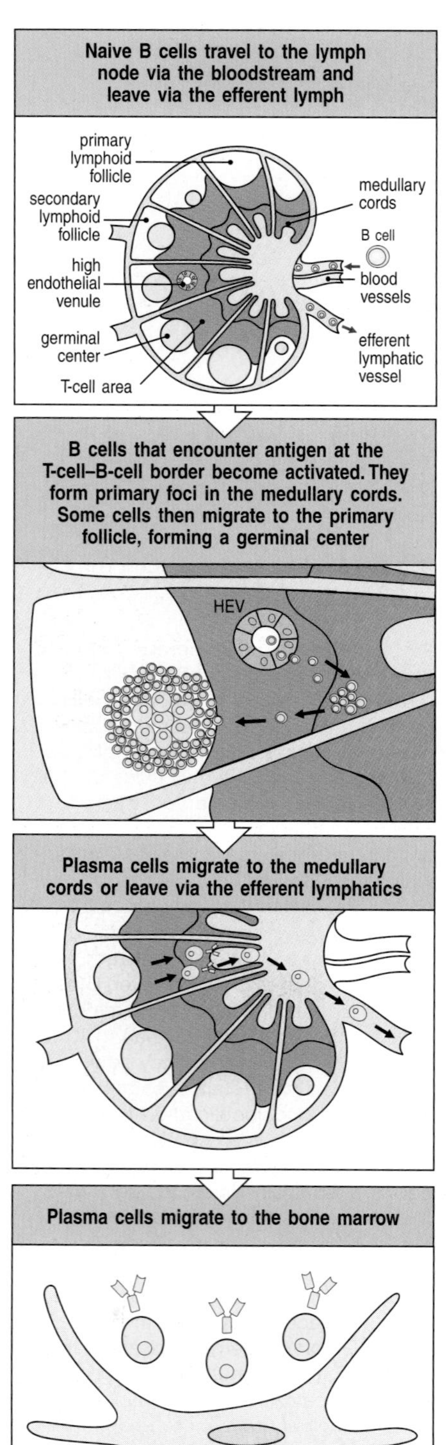

9-7 The second phase of a primary B-cell immune response occurs when activated B cells migrate to follicles and proliferate to form germinal centers.

Some of the B cells that proliferate early in the immune response take a more circuitous route before they become plasma cells. Together with their associated T cells, they migrate into a primary lymphoid follicle (Fig. 9.9), where they continue to proliferate and ultimately form a **germinal center** (Fig. 9.10). Primary follicles are present in unstimulated lymph nodes in the absence of infection and contain resting B cells clustered around a dense network of processes extending from a specialized cell type, the **follicular dendritic cell** (**FDC**). Follicular dendritic cells attract both naive and activated B cells into the follicles by secreting the chemokine CXCL13, which is recognized by the receptor CXCR5 on B cells (see Section 7-25).

It is unclear whether the cells that seed a germinal center come from cells initially activated at the T zone–B zone border or from cells that arise later in primary foci, or from both sources. Germinal centers are composed mainly of proliferating B cells, but antigen-specific T cells make up about 10% of germinal center lymphocytes and provide indispensable help to the B cells. The germinal center is essentially an island of cell division that sets up amid a sea of resting B cells in the primary follicles. Proliferating germinal center B cells displace the resting B cells toward the periphery of the follicle, forming a **mantle zone** of resting cells around the center. A follicle containing a germinal center is known as a **secondary follicle** (see Fig. 9-9). The germinal center grows in size as the immune response proceeds, and then shrinks and finally disappears when the infection is cleared. Germinal centers are present for about 3–4 weeks after initial antigen exposure.

Fig. 9.9 Activated B cells form germinal centers in lymphoid follicles. Activation of B cells in a lymph node is shown here. Top panel: naive circulating B cells enter lymph nodes from the blood via high endothelial venules; if they do not encounter antigen, they leave via the efferent lymphatic vessel. Second panel: if antigen-specific B cells encounter both their antigen and activated helper T cells specific for the same antigen, they become activated. Some B cells activated at the T cell–B cell border form a primary focus in the medullary cords, while others migrate to form a germinal center within a primary follicle. Germinal centers are sites of rapid B-cell proliferation and differentiation. Follicles in which germinal centers have formed are known as secondary follicles. Within the germinal center, B cells begin their differentiation into either antibody-secreting plasma cells or memory B cells. Third and fourth panels: plasma cells leave the germinal center and migrate to the medullary cords or leave the lymph node altogether via the efferent lymphatics and migrate to the bone marrow.

The early events in the primary focus lead to the prompt secretion of specific antibody that immediately protects the infected individual. In contrast, the germinal center reaction provides for a more effective later response, should the pathogen establish a chronic infection or the host become reinfected. To this end, B cells undergo several important modifications in the germinal center. These include somatic hypermutation, which alters the V regions of immunoglobulin genes and enables a process called affinity maturation, which selects for the survival of mutated B cells with high affinity for the antigen, and class switching, which allows these selected B cells to express a variety of effector functions in the form of antibodies of different classes. The selected B cells will differentiate either into memory B cells, whose function is described in Chapter 10, or into plasma cells, which will begin to secrete higher-affinity and class-switched antibody during the latter part of the primary immune response.

The germinal center is a site of intense cell proliferation, with B cells dividing every 6–8 hours. Initially, these rapidly proliferating B cells markedly reduce their expression of surface immunoglobulin, particularly of IgD. These B cells are termed **centroblasts**. As time goes on, some B cells reduce their rate of division and begin to express higher levels of surface immunoglobulin. These are termed **centrocytes** and they presumably arise from centroblasts. The centroblasts at first proliferate in the dark zone of the germinal center (see Fig. 9.10), so called because the proliferating cells are densely packed. With further development, B cells begin to fill the light zone of the germinal center, an area of the follicle that is more richly supplied with follicular dendritic cells and less densely packed with cells. It was thought originally that only the centroblasts in the dark zone proliferated, whereas centrocytes in the light zone did not divide. Indeed, this may be true in chronic germinal centers found in inflamed tonsils that have been surgically removed. However, in newly forming germinal centers in mice, proliferation can occur in both light and dark zones, and proliferative cells in the dark zone can express moderate amounts of immunoglobulin on their surface. Follicular dendritic cells, which originally were most prominent in the light zone, seem to react to germinal center formation, and their dendritic processes become more evident

Fig. 9.10 Germinal centers are formed when activated B cells enter lymphoid follicles. The germinal center is a specialized microenvironment in which B-cell proliferation, somatic hypermutation, and selection for strength of antigen binding all occur. Closely packed centroblasts form the so-called 'dark zone' of the germinal center, as can be seen in the lower part of the photomicrograph in the center, which shows a high-power view of a section through a human tonsillar germinal center. The photomicrograph on the right shows a lower-power view of a tonsillar germinal center; B cells are found in the dark zone, light zone, and mantle zone. Proliferating cells are stained green for Ki67, an antigen expressed in nuclei of dividing cells, revealing the centroblasts in the dark zone. The dense network of follicular dendritic cells, stained red, mainly occupies the light zone. Cells in the light zone are also proliferating, though to a lesser degree in most germinal centers. Small recirculating B cells occupy the mantle zone at the edge of the B-cell follicle. Large masses of CD4 T cells, stained blue, can be seen in the T-cell zones, which separate the follicles. There are also significant numbers of T cells in the light zone of the germinal center; CD4 staining in the dark zone is associated mainly with CD4-positive phagocytes. Photographs courtesy of I. MacLennan.

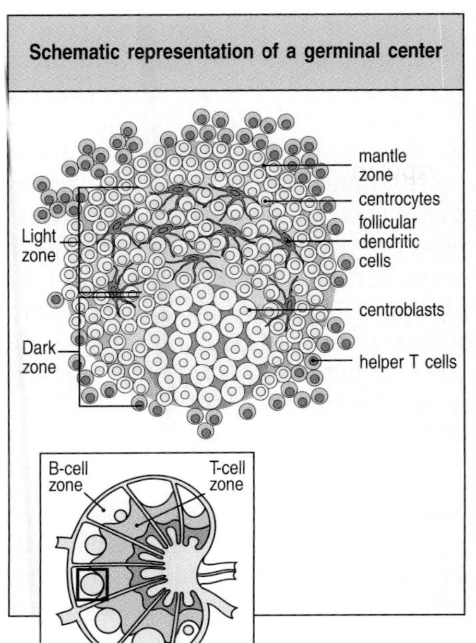

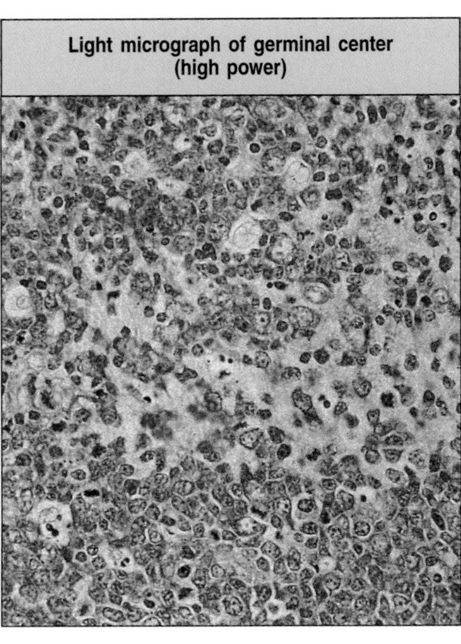

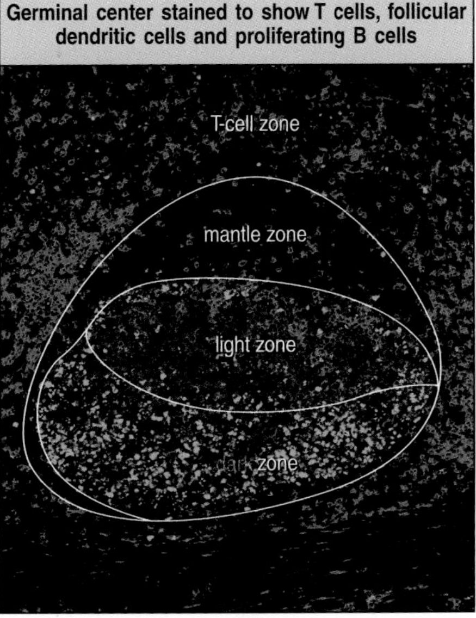

throughout the germinal center as it develops. The result is that a mature germinal center at day 15 after immunization more closely resembles a light zone, with few of the classic dark-zone characteristics. This view of germinal center evolution may help to explain how B cells with high affinity for the stimulating antigen are selected, as we now discuss.

9-8 Germinal center B cells undergo V-region somatic hypermutation, and cells with mutations that improve affinity for antigen are selected.

What is known about the molecular mechanism of somatic hypermutation was described in Sections 4-17 and 4-18 as one of the secondary mechanisms for generating more diversity among antibodies. Here we describe the signals that initiate hypermutation and the biological consequences of mutation in activated B cells. Somatic hypermutation is normally restricted to B cells that are proliferating in germinal centers. Studies *in vitro* have shown, however, that B cells can be induced to undergo hypermutation outside germinal centers when their B-cell receptors are cross-linked and they receive help, including cytokines and stimulation by CD40 ligand, from activated T cells.

Unlike the primary mechanisms of immunoglobulin diversification (see Sections 4-1 to 4-6), which generate B cells with radically differing B-cell receptors, somatic hypermutation has the potential to create a series of related clones of B cells that differ subtly in their specificity and affinity for antigen. This is because somatic hypermutation generally involves individual point mutations that change only a single amino acid. Immunoglobulin V-region genes accumulate mutations at a rate of about one base pair change per 10^3 base pairs per cell division. The mutation rate in the rest of the DNA is much lower: around one base pair change per 10^{10} base pairs per cell division. These mutations also affect some DNA flanking the rearranged V gene, but they generally do not extend into the C-region exons. Thus, random point mutations are somehow targeted to the rearranged V genes in a B cell. Because each of the expressed heavy-chain and light-chain V-region genes is encoded by about 360 base pairs, and about three out of every four base changes results in an altered amino acid, every second B cell will acquire a mutation in its receptor at each division.

The point mutations accumulate in a stepwise manner as the descendants of an individual B cell (B-cell clones) proliferate in the germinal center. Mutations can affect the ability of a B cell to bind antigen and thus will affect the fate of the B cell in the germinal center (Fig. 9.11). Most mutations have a negative impact on the ability of the B-cell receptor to bind the original antigen, either preventing the production of a correctly folded immunoglobulin molecule or changing the complementarity-determining regions in such a way that antigen binding is reduced or abolished. Such mutations are disastrous for the cells that harbor them; these cells are eliminated by apoptosis either because they can no longer make a functional B-cell receptor or because they cannot compete with sibling cells that bind antigen more strongly. Deleterious mutation is a frequent event, and germinal centers are filled with apoptotic B cells that are quickly engulfed by macrophages, giving rise to the characteristic **tingible body macrophages**, which contain dark-staining nuclear debris in their cytoplasm and are a long-recognized histologic feature of germinal centers.

Less frequently, mutations improve the affinity of a B-cell receptor for antigen. Cells that harbor these mutations are efficiently selected and expanded. Whether expansion is due to the prevention of cell death and/or an enhancement of cell division is still unclear. In either case it is clear that selection is incremental. After each round of mutation, a B cell begins to express the new receptor, and it determines the cell's fate, whether favorable or unfavorable.

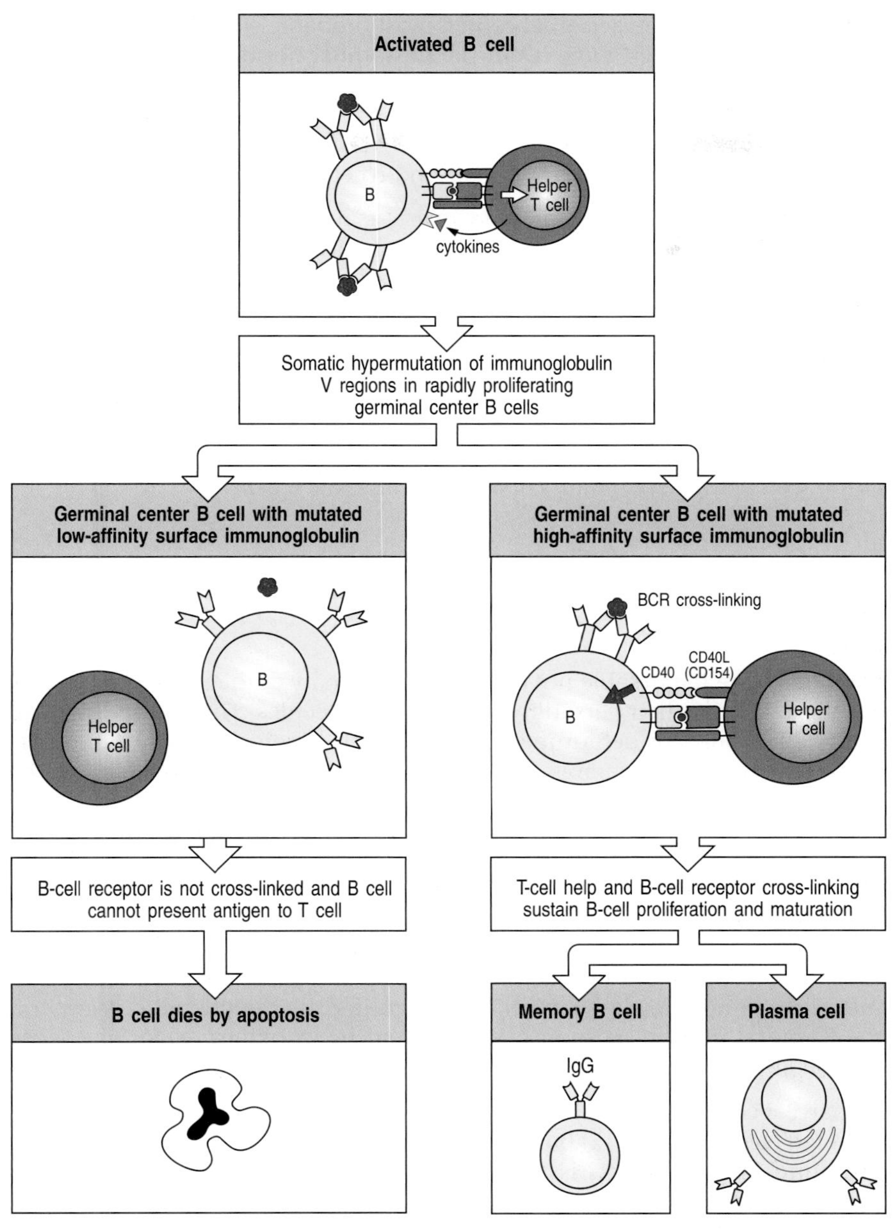

Fig. 9.11 Activated B cells undergo rounds of mutation and selection for higher-affinity mutants in the germinal center, resulting in high-affinity antibody-secreting plasma cells and high-affinity memory B cells. B cells are first activated outside follicles by the combination of antigen and T cells (top panel). They migrate to germinal centers (not shown), where the remaining events occur. Somatic hypermutation can result in amino acid replacements in immunoglobulin V regions that affect the fate of the B cell. Mutations that result in a B-cell receptor (BCR) of lower affinity for the antigen (left panels) will prevent the B cell from being activated as efficiently, because both B-cell receptor cross-linking and the ability of the B cell to present peptide antigen to T cells are reduced. This results in the B cell dying by apoptosis. In this way, low-affinity cells are purged from the germinal center. Most mutations are either negative or neutral (not shown) and thus the germinal center is a site of massive B-cell death as well as of proliferation. Some mutations, however, will improve the ability of the B-cell receptor to bind antigen. This increases the B cell's chance of interacting with T cells, and thus of proliferating and surviving (right panels). Surviving cells undergo repeated cycles of mutation and selection during which some of the progeny B cells undergo differentiation to either memory B cells or plasma cells (bottom right panels) and leave the germinal center. The signals that control these differentiation decisions are unknown.

If favorable, the cell undergoes another round of division and mutation, and the expression and selection process is repeated. In this way the affinity and specificity of positively selected B cells are continually refined during the germinal center response, the process known as **affinity maturation**. The fact that both centroblasts and centrocytes proliferate and can express immunoglobulin explains how mutation and positive selection can take place simultaneously throughout the germinal center without the need for migration back and forth between the dark and light zones.

Evidence of positive and negative selection is seen in the pattern of somatic hypermutations in V regions of B cells that have survived passage through the germinal center (see Section 4-18). The existence of negative selection is shown by the relative scarcity of amino acid replacements in the framework regions, reflecting the loss of cells that had mutated any one of the many residues that are critical for immunoglobulin V-region folding. Negative selection is an important force in the germinal center, most probably eliminating more than one in every two cells. Were it not for substantial negative selection, B cells dividing three to four times per day in a single germinal center would quickly create enough progeny to overwhelm the entire organism; more than a billion cells could be created in 10 days in a single germinal center. Instead, a germinal center actually contains a few thousand B cells at its peak.

The mark of positive selection, in contrast, is an accumulation of numerous amino acid replacements in the complementarity-determining regions (see Fig. 4.25). The consequence of these cycles of proliferation, mutation, and selection, all of which happen within the germinal center, is that the average affinity of the population of responding B cells for its antigen increases over time, largely explaining the observed phenomenon of affinity maturation of the antibody response. The selection process can be quite stringent: although 50–100 B cells may seed the germinal center, most of them leave no progeny, and by the time the germinal center reaches maximum size, it is typically composed of the descendants of only one or a few B cells.

9-9 Class switching in thymus-dependent antibody responses requires expression of CD40 ligand by the helper T cell and is directed by cytokines.

Antibodies are remarkable not only for the diversity of their antigen-binding sites but also for their versatility as effector molecules. The specificity of an antibody response is determined by the antigen-binding site, which consists of the two variable V domains, V_H and V_L. In contrast, the effector action of the antibody is determined by the isotype of its heavy-chain C region (see Section 3-1). A given heavy-chain V domain can become associated with the C region of any isotype through the process of class switching (see Section 4-20), which takes place after B cells become activated in the T-cell zones of lymphoid organs and can continue in the primary foci and in a proportion of the cells in the germinal center. We will see later in this chapter how antibodies of each class contribute to the elimination of pathogens. The DNA rearrangements that underlie class switching and confer this functional diversity on the humoral immune response are directed by cytokines, especially those released by effector CD4 T cells.

All naive B cells express cell-surface IgM and IgD, and IgM is the first antibody secreted (see Section 4-15) but makes up less than 10% of the immunoglobulin found in plasma; IgG is the most abundant. Much of the antibody in plasma has therefore been produced by B cells that have undergone class switching. Little IgD antibody is produced at any time, so the early stages of the antibody response are dominated by IgM antibodies. Later, IgG and IgA are the predominant antibody classes, with IgE contributing a small but biologically important part of the response. The overall predominance of IgG is also due in part to its longer lifetime in the plasma (see Fig. 4.16).

Productive interactions between B cells and helper T cells are essential for class switching to occur. This is demonstrated by people who have a genetic deficiency of CD40 ligand, which is required for these interactions. Class switching is greatly reduced in such individuals and they have abnormally high levels of IgM in their plasma. This condition is thus known as **hyper IgM syndrome**. Despite the absence of CD40 ligand these people do make IgM antibodies in response to thymus-dependent antigens, which indicates that in the B-cell response, CD40L–CD40 interactions are most important in enabling a sustained immune response that includes class switching. Other defects that interfere with class switching, such as a deficiency of CD40, or of the enzyme activation-induced cytidine deaminase (AID), which is essential for the class switch recombination process, also result in forms of hyper IgM

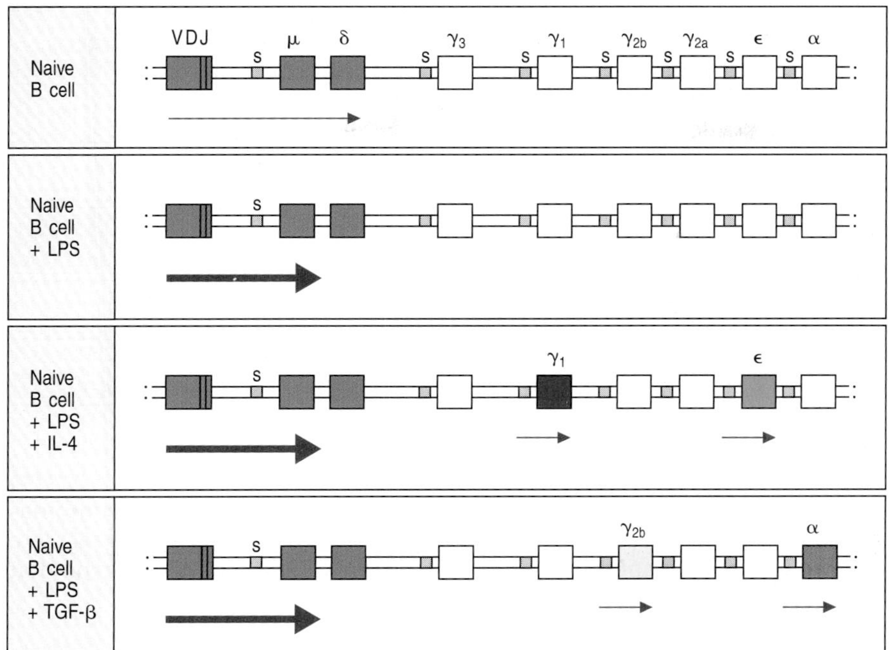

Fig. 9.12 Class switching is preceded by transcriptional activation of heavy-chain C-region genes. Resting naive B cells transcribe the genes for the heavy-chain isotypes μ and δ at a low rate, giving rise to surface IgM and IgD. Bacterial lipopolysaccharide (LPS), which can activate B cells independently of antigen, induces IgM secretion. In the presence of IL-4, however, $C_{\gamma}1$ and C_{ε} are transcribed at a low rate, presaging switches to IgG1 and IgE production. The transcripts originate before the 5′ end of the region to which switching occurs, and do not code for protein. Similarly, TGF-β gives rise to $C_{\gamma 2b}$ and C_{α} transcripts and drives switching to IgG2b and IgA. It is not known what determines which of the two transcriptionally activated heavy-chain C genes undergoes switching. Red arrows indicate transcription. The figure shows class switching in the mouse.

syndome and are discussed in Chapter 12. Much of the IgM in hyper IgM syndromes may be induced by thymus-independent antigens on the pathogens that chronically infect these patients, who suffer from severe humoral immunodeficiency.

The mechanism of class switching, and the switch regions between which recombination occurs to translocate the rearranged V region in front of different C regions, are discussed in detail in Section 4-20. The selection of a C region as the target for the recombination process is not random, however, but is regulated by cytokines that are produced by helper T cells and other cells during the immune response. Most of what is known about the regulation of class switching by helper T cells has come from experiments *in vitro* in which mouse B cells are exposed to various nonspecific stimuli, such as bacterial lipopolysaccharide (LPS), along with purified cytokines (Fig. 9.12). These experiments show that different cytokines preferentially induce switching to different isotypes. In the mouse, IL-4 preferentially induces switching to IgG1 ($C\gamma_1$) and IgE ($C\varepsilon$), whereas transforming growth factor (TGF)-β induces switching to IgG2b ($C\gamma_{2b}$) and IgA ($C\alpha$). T_H2 cells make both of these cytokines, and also make IL-5, which promotes IgA secretion by cells that have already undergone switching. Although T_H1 cells are relatively poor initiators of antibody responses, they participate in class switching by releasing interferon (IFN)-γ, which preferentially induces switching to IgG2a and IgG3. The role of cytokines in directing B cells to make the different antibody isotypes is summarized in Fig. 9.13. Such a directed mechanism is supported by the observation that individual B cells frequently undergo switching to the same C gene on both chromosomes, even though the antibody heavy chain is being expressed from only one of the chromosomes.

Cytokines induce class switching in part by stimulating the production of RNA transcripts from the switch recombination sites that lie 5′ to each heavy-chain C gene (see Fig. 9.12). When activated B cells are exposed to IL-4, for example, transcription from a site upstream of the switch regions of $C\gamma_1$ and $C\varepsilon$ can be detected a day or two before switching occurs. Interestingly, each of the cytokines that induce switching seems to induce transcription from the

Fig. 9.13 Different cytokines induce switching to different antibody classes. The individual cytokines induce (violet) or inhibit (red) the production of certain antibody classes. Much of the inhibitory effect is probably the result of directed switching to a different class. These data are drawn from experiments with mouse cells.

Role of cytokines in regulating expression of antibody classes							
Cytokines	IgM	IgG3	IgG1	IgG2b	IgG2a	IgE	IgA
IL-4	Inhibits	Inhibits	Induces		Inhibits	Induces	
IL-5							Augments production
IFN-γ	Inhibits	Induces	Inhibits		Induces	Inhibits	
TGF-β	Inhibits	Inhibits		Induces			Induces

switch regions of two different heavy-chain C genes, but specific recombination occurs in one or other of these genes only. Thus, helper T cells regulate the production of antibody by B cells and also the heavy-chain isotype that determines the effector function of the antibody.

9-10 Ligation of the B-cell receptor and CD40, together with direct contact with T cells, are all required to sustain germinal center B cells.

Germinal center B cells are inherently prone to dying, and to survive they must receive specific signals. It was originally discovered *in vitro* that germinal center B cells could be kept alive by simultaneously cross-linking their B-cell receptors and ligating their cell-surface CD40. *In vivo* these signals are delivered by antigen and T cells, respectively. Additional signals are also required for survival, which are delivered by direct contact with T cells. The nature of these signals is still obscure but they might include ICOS and B7-RP (see Section 9-4) and other members of the TNF/TNF-receptor family.

Fig. 9.14 Immune complexes bind to the surface of follicular dendritic cells. Radiolabeled antigen localizes to, and persists in, lymphoid follicles of draining lymph nodes (see the light micrograph and the schematic representation below, showing a germinal center in a lymph node). Radiolabeled antigen has been injected 3 days previously and its localization in the germinal center is shown by the intense dark staining. The antigen is in the form of antigen:antibody:complement complexes bound to Fc and complement receptors on the surface of the follicular dendritic cell, as depicted schematically for immune complexes bound to both Fc and CR3 receptors in the right panel and inset. These complexes are not internalized. Antigen can persist in this form for long periods. Photograph courtesy of J. Tew.

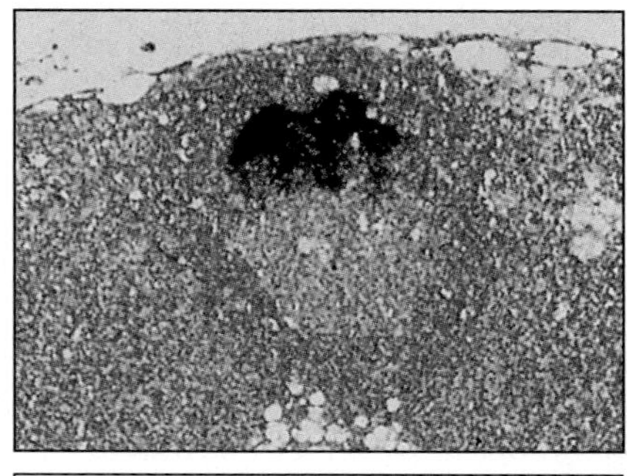

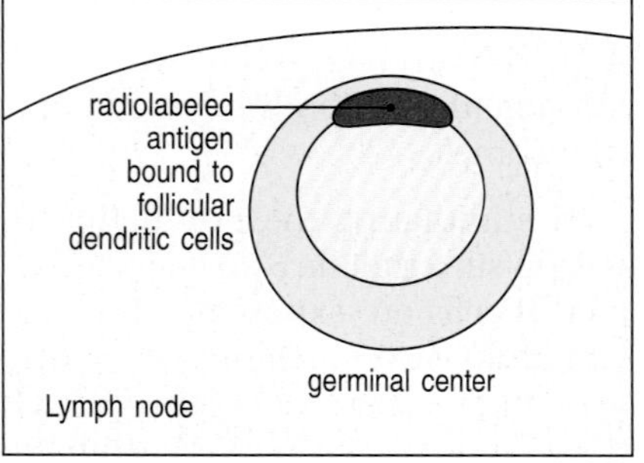

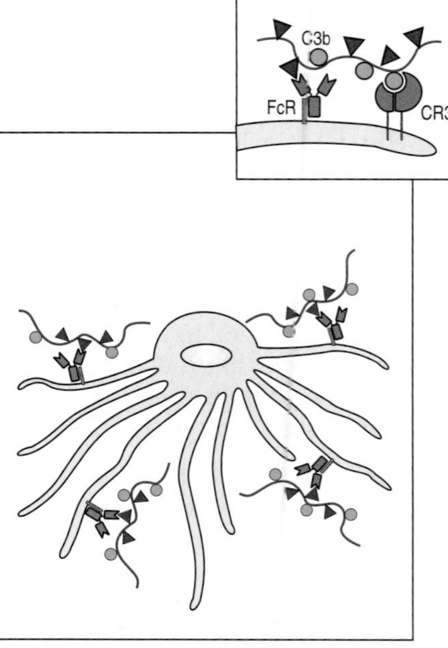

The source of antigen in the germinal center has been a matter of some controversy. Antigen can be trapped and stored for long periods in the form of immune complexes on follicular dendritic cells (Figs 9.14 and 9.15), and it was therefore assumed that this was the antigen that sustained germinal center B-cell proliferation. Although this may be true under certain circumstances, there is now evidence that antigen on follicular dendritic cells is not required to sustain a normal germinal center response. Indeed, the role of the antigen depot on these cells is unknown, although one role could be to maintain long-lived plasma cells. Where, then, does the antigen that sustains the germinal center come from? Under normal circumstances it is most likely that live pathogens will continue to provide antigens until they are eliminated by the immune response, after which the germinal center decays. Immunizations with protein antigens are usually given in a form that slowly releases the antigen over time, which mimics the situation with live pathogens. Indeed, it is difficult to stimulate germinal center formation by immunization without either a live replicating pathogen or a sustained release of antigen in adjuvant (see Appendix I, Section A-4).

How the various signals that maintain the germinal center exert their effects on B cells is not completely understood. The combined signals from the B-cell receptor and CD40 seem to increase the expression of a protein called Bcl-X_L, a relative of Bcl-2, which promotes B-cell survival by inhibiting apoptosis (see Section 6-26). There may be many other signals yet to be discovered that promote B-cell differentiation.

9-11 Surviving germinal center B cells differentiate into either plasma cells or memory cells.

The purpose of the germinal center reaction is to enhance the later part of the primary immune response. Germinal center B cells differentiate first into plasmablasts, at which stage they undergo somatic hypermutation and some also may undergo class switching. Some then differentiate into plasma cells under the control of a regulatory protein, **BLIMP-1** (B-lymphocyte-induced maturation protein 1). This is a transcriptional repressor in B cells that switches off genes required for B-cell proliferation in the germinal center, and for class switching and affinity maturation. B cells in which BLIMP-1 is induced become plasma cells, and they cease proliferating, increase the synthesis and secretion of immunoglobulins, and change their cell-surface properties. This involves downregulating the chemokine receptor CXCR5, which recognizes CXCL13 (see Section 9-7), and upregulating

Fig. 9.15 Immune complexes bound to follicular dendritic cells form iccosomes, which are released and can be taken up by B cells in the germinal center. Follicular dendritic cells have a prominent cell body and many dendritic processes. Immune complexes, bound to complement and Fc receptors on the follicular dendritic cell surface, become clustered, forming prominent 'beads' along the dendrites. An intermediate form of follicular dendritic cell is shown (left panel) with both straight filiform dendrites and others that are becoming beaded. These beads are shed from the cell as iccosomes (immune complex-coated bodies), which can bind (center panel) and be taken up by B cells in the germinal center (right panel). In the center and right panels, the iccosome has been formed with immune complexes containing horseradish peroxidase, which is electron-dense and therefore appears dark in the transmission electron micrographs. Photographs courtesy of A.K. Szakal.

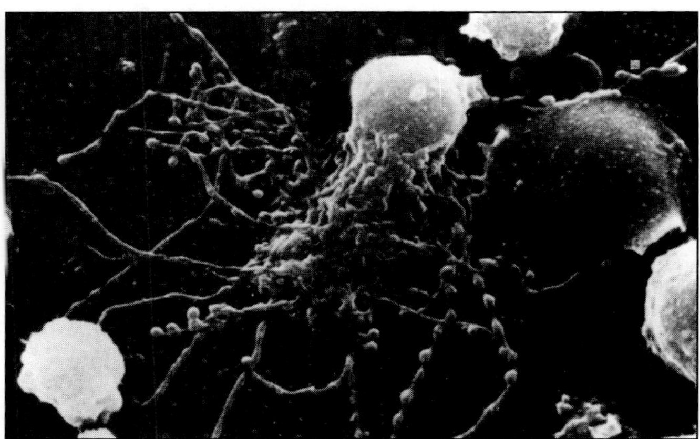
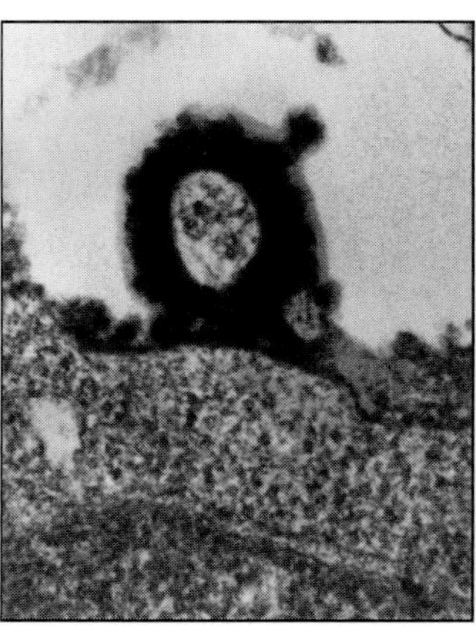
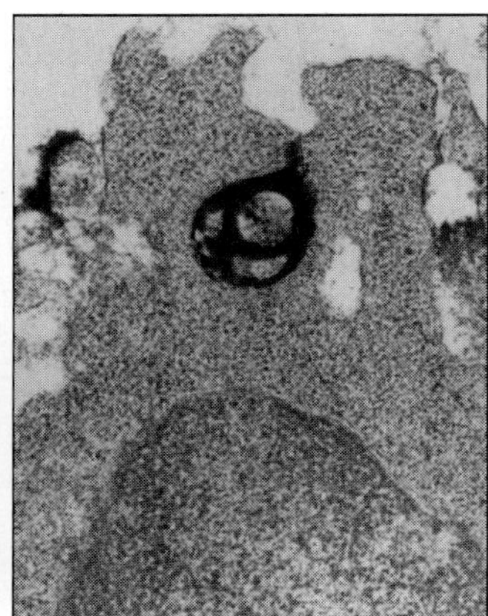

CXCR4 and $\alpha_4:\beta_1$ integrins, so that the plasma cells can now leave the germinal centers and home to peripheral tissues. Some plasma cells from germinal centers in lymph node or spleen migrate to the bone marrow, where a subset live for a long period, while others migrate to the splenic red pulp. B cells that have been activated in germinal centers in mucosal tissues, and which are predominantly switched to IgA production, stay within the mucosal system. Plasma cells obtain signals from stromal cells that are essential for their survival and can be very long lived. These long-lived plasma cells are a source of long-lasting high-affinity antibody.

Other germinal center B cells differentiate into **memory B cells**. Memory B cells are long-lived descendants of cells that were once stimulated by antigen and had proliferated in the germinal center. They divide very slowly if at all; they express surface immunoglobulin but secrete no antibody, or do so only at a low rate. Because the precursors of memory B cells once participated in a germinal center reaction, memory B cells inherit the genetic changes that occurred in germinal center cells, including somatic hypermutation and the gene rearrangements that result in a class switch. The signals that control which differentiation path a B cell takes, and even whether at any given point the B cell continues to divide instead of differentiating, are still being investigated. We discuss memory B cells in Chapter 10.

9-12 B-cell responses to bacterial antigens with intrinsic ability to activate B cells do not require T-cell help.

Although antibody responses to most protein antigens are dependent on helper T cells, humans and mice with T-cell deficiencies nevertheless make antibodies against many bacterial antigens. This is because the special properties of some bacterial polysaccharides, polymeric proteins, and lipopolysaccharides enable them to stimulate naive B cells in the absence of T-cell help. These antigens are known as **thymus-independent antigens** (**TI antigens**) because they stimulate strong antibody responses in athymic individuals. The nonprotein bacterial products cannot elicit classical T-cell responses, yet they induce antibody responses in normal individuals. Although responses to TI antigens can occur in mice that lack all T cells and natural killer (NK) cells, if such cells are activated during the course of a physiological immune response (for example by other protein antigens or via the innate immune system) they can affect the TI immune response. In particular, cytokines secreted by T cells, NK T cells, or NK cells can affect the isotype of the antibody secreted. NK T cells (see Section 7-9) are particularly intriguing as cells that might influence the TI response to nonprotein antigens because the T-cell receptors on these cells recognize certain polysaccharides bound to unconventional MHC class I or class I-like molecules such as CD1 (see Section 5-19).

Thymus-independent antigens fall into two classes, which activate B cells by two different mechanisms. **TI-1 antigens** possess an intrinsic activity that can directly induce B-cell division. At high concentration these molecules cause the proliferation and differentiation of most B cells regardless of their antigen specificity; this is known as **polyclonal activation** (Fig. 9.16, top two panels). TI-1 antigens are thus often called **B-cell mitogens**, a mitogen being a substance that induces cells to undergo mitosis. An example of a B-cell mitogen and TI-1 antigen is LPS, which binds to LPS-binding protein and CD14 (see Chapter 2), which then associates with the activating receptor TLR-4 on B cells. LPS activates B cells only at doses at least 100 times those needed to activate dendritic cells. Thus, when B cells are exposed to concentrations of TI-1 antigens that are 10^3–10^5 times lower than those used for polyclonal activation, only those B cells whose B-cell receptors also specifically bind the TI-1 molecules become activated. At these low antigen concentrations, sufficient amounts of TI-1 for B-cell activation can only be concentrated on the B-cell surface with the aid of this specific binding (Fig. 9.16, bottom two panels).

It is likely that, as with any pathogen antigen, concentrations of TI-1 antigens are low during the early stages of infections *in vivo*; thus, only antigen-specific B cells are likely to be activated and these will produce antibodies specific for the TI-1 antigen that may in turn neutralize the toxic effects of these molecules. Such responses have an important role in defense against several extracellular pathogens, as they arise earlier than thymus-dependent responses because they do not require previous priming and clonal expansion of helper T cells. However, TI-1 antigens are inefficient inducers of affinity maturation and memory B cells, both of which require antigen-specific T-cell help.

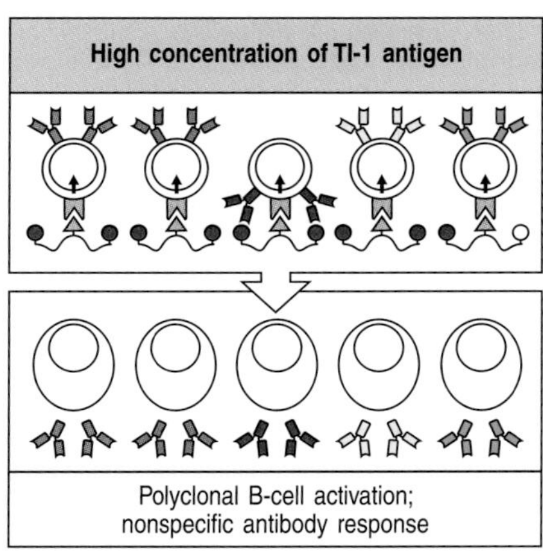

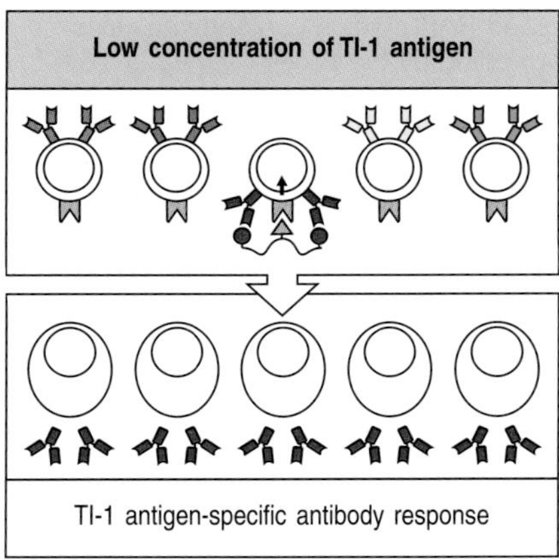

Fig. 9.16 Thymus-independent type 1 antigens (TI-1 antigens) are polyclonal B-cell activators at high concentrations, whereas at low concentrations they induce an antigen-specific antibody response. At high concentration, the signal delivered by the B cell-activating moiety of TI-1 antigens is sufficient to induce proliferation and antibody secretion by B cells in the absence of specific antigen binding to surface immunoglobulin. Thus, all B cells respond (top panels). At low concentration, only B cells specific for the TI-1 antigen bind enough of it to focus its B-cell activating properties onto the B cell; this gives a specific antibody response to epitopes on the TI-1 antigen (lower panels).

9-13 B-cell responses to bacterial polysaccharides do not require peptide-specific T-cell help.

The second class of thymus-independent antigens consists of molecules such as bacterial capsular polysaccharides that have highly repetitive structures. These thymus-independent antigens, called **TI-2 antigens**, contain no intrinsic B cell-stimulating activity. Whereas TI-1 antigens can activate both immature and mature B cells, TI-2 antigens can activate only mature B cells; immature B cells, as we saw in Section 7-6, are inactivated by repetitive epitopes. Infants do not make antibodies against polysaccharide antigens efficiently, and this might be because most of their B cells are immature. Responses to several TI-2 antigens are made prominently by B-1 cells (also known as CD5 B cells), which comprise an autonomously replicating subpopulation of B cells, and by marginal zone B cells, another unique subset of nonrecirculating B cells that line the border of the splenic white pulp (see Section 7-28). Although B-1 cells arise early in development, young children do not make a fully effective response to carbohydrate antigens until about 5 years of age. In contrast, marginal zone B cells are rare at birth and accumulate with age; they might therefore be responsible for most physiological TI-2 responses, which also increase with age.

TI-2 antigens most probably act by simultaneously cross-linking a critical number of B-cell receptors of mature B cells specific for the antigen (Fig. 9.17, left panels). There is also evidence that dendritic cells and macrophages provide co-stimulatory signals for the initial activation of B cells by TI-2 antigens, signals that are necessary for the survival of the antigen-specific B cell and its differentiation into a plasmablast secreting IgM. One of these co-stimulatory signals is the TNF-family cytokine BAFF, which is secreted by the dendritic cell and interacts with the receptor TACI on the B cell.

Excessive cross-linking of B-cell receptors renders mature B cells unresponsive or anergic, just as it does immature B cells. Thus, the density of TI-2 antigen epitopes presented to the B cell is critical: if it is too low, receptor cross-linking is insufficient to activate the cell; if too high, the B cell becomes anergic.

B-cell responses to TI-2 antigens provide a prompt and specific response to an important class of pathogen. Many common extracellular bacterial pathogens are surrounded by a polysaccharide capsule that enables them to resist ingestion by phagocytes. The bacteria not only escape direct destruction by phagocytes but also avoid stimulating T-cell responses through the presentation of bacterial peptides by macrophages. Antibody that is produced rapidly in response to this polysaccharide capsule without the help

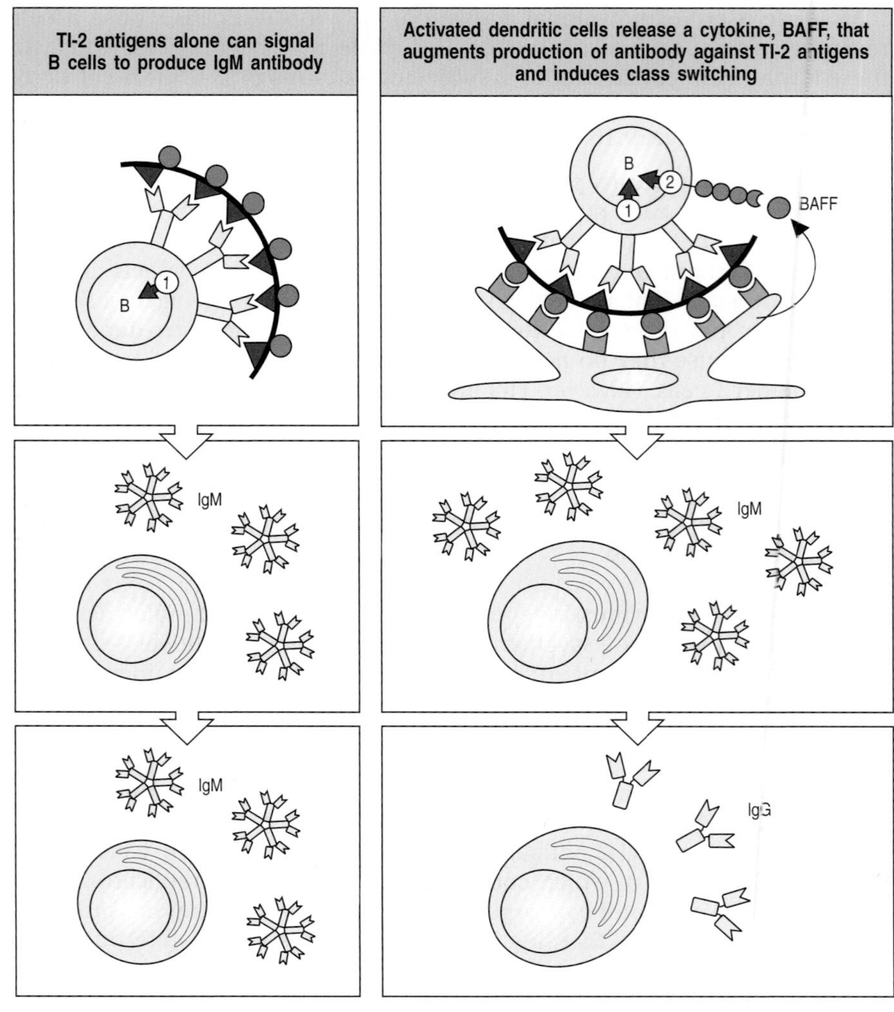

Fig. 9.17 B-cell activation by thymus-independent type 2 antigens (TI-2 antigens) requires, or is greatly enhanced by, cytokines. Multiple cross-linking of the B-cell receptor by TI-2 antigens can lead to IgM antibody production (left panels), but there is evidence that in addition, cytokines greatly augment these responses and lead to isotype switching as well (right panels). It is not clear where such cytokines are produced but one possibility is that dendritic cells, which may be able to bind the antigen through innate immune system receptors on their surface and so present it to the B cells, secrete a soluble TNF-family cytokine called BAFF, which can activate class switching by the B cell.

of peptide-specific T cells can coat these bacteria, promoting their ingestion and destruction by phagocytes.

As well as producing IgM, thymus-independent responses can include switching to certain other antibody classes, such as IgG3 in the mouse. This is probably the result of help from dendritic cells (Fig. 9.17, right panels), which provide secreted cytokines such as BAFF, and membrane-bound signals to nearby proliferating plasmablasts as they respond to TI antigens.

The IgM and IgG antibodies induced by TI-2 antigens are likely to be an important part of the humoral immune response in many bacterial infections. We mentioned earlier the importance of antibodies against the capsular polysaccharide of *Haemophilus influenzae* type b, a TI-2 antigen, in protective immunity to this bacterium. A further example of the importance of TI-2 responses can be seen in patients with an immunodeficiency disease known as Wiskott–Aldrich syndrome, which is described in more detail in Section 12-15. These patients can respond, although poorly, to protein antigens but fail to make antibody against polysaccharide antigens and are highly susceptible to infection with encapsulated bacteria. Thus, the TI responses are important components of the humoral immune response to nonprotein antigens that do not engage peptide-specific T-cell help. The distinguishing features of thymus-dependent, TI-1, and TI-2 antibody responses are summarized in Fig. 9.18.

	TD antigen	TI-1 antigen	TI-2 antigen
Antibody response in infants	Yes	Yes	No
Antibody production in congenitally athymic individual	No	Yes	Yes
Antibody response in absence of all T cells	No	Yes	No
Primes T cells	Yes	No	No
Polyclonal B-cell activation	No	Yes	No
Requires repeating epitopes	No	No	Yes
Examples of antigen	Diphtheria toxin Viral hemagglutinin Purified protein derivative (PPD) of *Mycobacterium tuberculosis*	Bacterial lipopoly-saccharide *Brucella abortus*	Pneumococcal polysaccharide *Salmonella* polymerized flagellin Dextran Hapten-conjugated Ficoll (polysucrose)

Fig. 9.18 Properties of different classes of antigen that elicit antibody responses.

Summary.

B-cell activation by many antigens requires both binding of the antigen by the B-cell surface immunoglobulin—the B-cell receptor—and interaction of the B cell with antigen-specific helper T cells. Helper T cells recognize peptide fragments derived from the antigen internalized by the B cell and displayed by the B cells as peptide:MHC class II complexes. Helper T cells stimulate the B cell through the binding of CD40 ligand on the T cell to CD40 on the B cell, through the interaction of other TNF–TNF-receptor family ligand pairs, and by the directed release of cytokines. Activated B cells also provide signals to T cells, for example via B7-family molecules, that promote the T cells' continued activation. The initial interaction occurs at the border of the T-cell and B-cell areas of secondary lymphoid tissue, where both antigen-specific helper T cells and antigen-specific B cells are trapped as a consequence of binding antigen. Further interactions between T cells and B cells continue after migration into the B-cell zone or follicle and the formation of a germinal center.

Helper T cells induce a phase of vigorous B-cell proliferation and direct the differentiation of the clonally expanded progeny of naive B cells into either antibody-secreting plasma cells or memory B cells. During the differentiation of activated B cells, the antibody class can change in response to cytokines released by helper T cells, and the antigen-binding properties of the antibody can change by somatic hypermutation of V-region genes. Somatic hypermutation and selection for high-affinity binding occur in the germinal centers. Helper T cells control these processes by selectively activating cells that have retained their specificity for the antigen and by inducing proliferation and differentiation into plasma cells and memory B cells. Some nonprotein antigens stimulate B cells in the absence of linked recognition by peptide-specific helper T cells. Responses to these thymus-independent antigens are accompanied by only limited class switching and do not induce memory B cells. However, such responses have a crucial role in host defense against pathogens whose surface antigens cannot elicit peptide-specific T-cell responses.

The distribution and functions of immunoglobulin classes.

Extracellular pathogens can find their way to most sites in the body, and antibodies must be equally widely distributed to combat them. Most classes of antibodies are distributed by diffusion from their site of synthesis, but specialized transport mechanisms are required to deliver antibodies to epithelial surfaces lining the lumina of organs such as the lungs and intestine. The distribution of antibodies is determined by their heavy-chain isotype, which can limit their diffusion or enable them to engage specific transporters that deliver them across epithelia. In this part of the chapter we describe the mechanisms by which antibodies of different isotypes are directed to the compartments of the body in which their particular effector functions are appropriate, and discuss the protective functions of antibodies that result solely from their binding to pathogens. In the last part of the chapter we discuss the effector cells and molecules that are specifically engaged by different isotypes.

9-14 Antibodies of different classes operate in distinct places and have distinct effector functions.

Pathogens most commonly enter the body across the epithelial barriers of the mucosa lining the respiratory, digestive, and urogenital tracts, or through damaged skin. Less often, insects, wounds, or hypodermic needles introduce microorganisms directly into the blood. Antibodies protect all the body's mucosal surfaces, tissues, and blood from such infections; these antibodies serve to neutralize the pathogen or promote its elimination before it can establish a significant infection. Antibodies of different isotypes are adapted to function in different compartments of the body. Because a given V region can become associated with any C region through class switching (see Section 4-20), the progeny of a single B cell can produce antibodies that share the same specificity yet provide all of the protective functions appropriate for each body compartment.

The first antibodies to be produced in a humoral immune response are always IgM, because IgM can be expressed without class switching (see Fig. 4.18). These early IgM antibodies are produced before B cells have undergone somatic hypermutation and therefore tend to be of low affinity. However, IgM molecules form pentamers whose 10 antigen-binding sites can bind simultaneously to multivalent antigens such as bacterial capsular polysaccharides. This compensates for the relatively low affinity of the IgM monomers by multipoint binding that confers high overall avidity. As a result of the large size of the pentamers, IgM is mainly found in the blood and, to a smaller extent, the lymph. The pentameric structure of IgM makes it especially effective in activating the complement system, as we will see in the last part of this chapter. Infection of the bloodstream has serious consequences unless it is controlled quickly, and the rapid production of IgM and its efficient activation of the complement system are important in controlling such infections. Some IgM is produced in secondary and subsequent responses, and also after somatic hypermutation, although other classes dominate the later phases of the antibody response. IgM is also produced by B-1 cells that reside in the peritoneal cavity and the pleural spaces. These cells are naturally activated and secrete antibodies against environmental pathogens, thus providing in these body cavities a preformed repertoire of IgM antibodies that can recognize invading pathogens (see Sections 2-34 and 7-28).

Antibodies of the other classes—IgG, IgA, and IgE—are smaller, and diffuse easily out of the blood into the tissues. IgA can form dimers, as we saw in Chapter 4, but IgG and IgE are always monomeric. The affinity of the individual antigen-binding sites for their antigen is therefore critical for the effectiveness of these antibodies, and most of the B cells expressing these classes have been selected for increased affinity of antigen-binding in germinal centers. IgG is the principal class in the blood and extracellular fluid, whereas IgA is the principal class in secretions, the most important being those of the epithelium lining the intestinal and respiratory tracts. IgG efficiently opsonizes pathogens for engulfment by phagocytes and activates the complement system, but IgA is a less potent opsonin and a weak activator of complement. This difference is not surprising, because IgG operates mainly in the body tissues, where accessory cells and molecules are available, whereas IgA operates mainly on epithelial surfaces where complement and phagocytes are not normally present; therefore IgA functions chiefly as a neutralizing antibody. IgA is also produced by plasma cells that differentiate from class-switched B cells in lymph nodes and spleen, and acts as a neutralizing antibody in extracellular spaces and in the blood. This IgA is monomeric and is predominantly of the subclass IgA1; the ratio of IgA1 to IgA2 in the blood is 10:1. The IgA antibodies produced by plasma cells in the gut are dimeric and predominantly of subclass IgA2; the ratio of IgA2 to IgA1 in the gut is 3:2.

Finally, IgE antibody is present only at very low levels in blood or extracellular fluid but is bound avidly by receptors on mast cells that are found just beneath the skin and mucosa and along blood vessels in connective tissue. Antigen binding to this cell-associated IgE triggers mast cells to release powerful chemical mediators that induce reactions, such as coughing, sneezing, and vomiting, that in turn can expel infectious agents, as discussed later in this chapter when we describe the receptors that bind immunoglobulin C regions and engage effector functions. The distribution and main functions of antibodies of the different classes are summarized in Fig. 9.19.

Functional activity	IgM	IgD	IgG1	IgG2	IgG3	IgG4	IgA	IgE
Neutralization	+	–	++	++	++	++	++	–
Opsonization	+	–	+++	*	++	+	+	–
Sensitization for killing by NK cells	–	–	++	–	++	–	–	–
Sensitization of mast cells	–	–	+	–	+	–	–	+++
Activates complement system	+++	–	++	+	+++	–	+	–

Distribution	IgM	IgD	IgG1	IgG2	IgG3	IgG4	IgA	IgE
Transport across epithelium	+	–	–	–	–	–	+++ (dimer)	–
Transport across placenta	–	–	+++	+	++	+/–	–	–
Diffusion into extravascular sites	+/–	–	+++	+++	+++	+++	++ (monomer)	+
Mean serum level (mg ml^{-1})	1.5	0.04	9	3	1	0.5	2.1	3×10^{-5}

Fig. 9.19 Each human immunoglobulin class has specialized functions and a unique distribution. The major effector functions of each class (+++) are shaded in dark red, whereas lesser functions (++) are shown in dark pink, and very minor functions (+) in pale pink. The distributions are marked similarly, with actual average levels in serum being shown in the bottom row. IgA has two subclasses, IgA1 and IgA2. The IgA column refers to both. *IgG2 can act as an opsonin in the presence of an Fc receptor of the appropriate allotype, found in about 50% of white people.

9-15 Transport proteins that bind to the Fc regions of antibodies carry particular isotypes across epithelial barriers.

In the mucosal immune system, IgA-secreting plasma cells are found predominantly in the lamina propria, which lies immediately below the basement membrane of many surface epithelia. From there the IgA antibodies can be transported across the epithelium to its external surface, for example to the lumen of the gut or the bronchi (Fig. 9.20). IgA antibody synthesized in the lamina propria is secreted as a dimeric IgA molecule associated with a single J chain (see Fig. 4.20). This polymeric form of IgA binds specifically to a receptor called the poly-Ig receptor, which is present on the basolateral surfaces of the overlying epithelial cells. When the poly-Ig receptor has bound a molecule of dimeric IgA, the complex is internalized and carried through the cytoplasm of the epithelial cell in a transport vesicle to its luminal surface. This process is called transcytosis. IgM also binds to the poly-Ig receptor and can be secreted into the gut by the same mechanism. Upon reaching the luminal surface of the enterocyte, the antibody is released into the secretions by proteolytic cleavage of the extracellular domain of the polymeric Ig receptor. The cleaved extracellular domain of the polymeric Ig receptor is known as secretory component (frequently abbreviated to SC) and remains associated with the antibody (this is shown in more detail in Fig. 11.13). Secretory component is bound to the part of the Fc region of IgA that contains the binding site for the Fcα receptor I, which is why secretory IgA does not bind to this receptor. Secretory component serves several physiological roles. It binds to mucins in mucus, acting as 'glue' to bind secreted IgA to the mucous layer on the luminal surface of the gut epithelium, where the antibody binds and neutralizes gut pathogens and their toxins (see Fig. 9.20). Secretory component also protects the antibodies against cleavage by gut enzymes.

Some molecules of dimeric IgA diffuse from the lamina propria into the extracellular spaces of the tissues, draining into the bloodstream before being excreted into the gut via the bile (this pathway is described in more detail in Section 11-8). It is therefore not surprising that patients with obstructive jaundice, a condition in which bile is not excreted, show a marked increase in dimeric IgA in the plasma.

The principal sites of IgA synthesis and secretion are the gut, the respiratory epithelium, the lactating breast, and various other exocrine glands such as the salivary and tear glands. It is believed that the primary functional role of

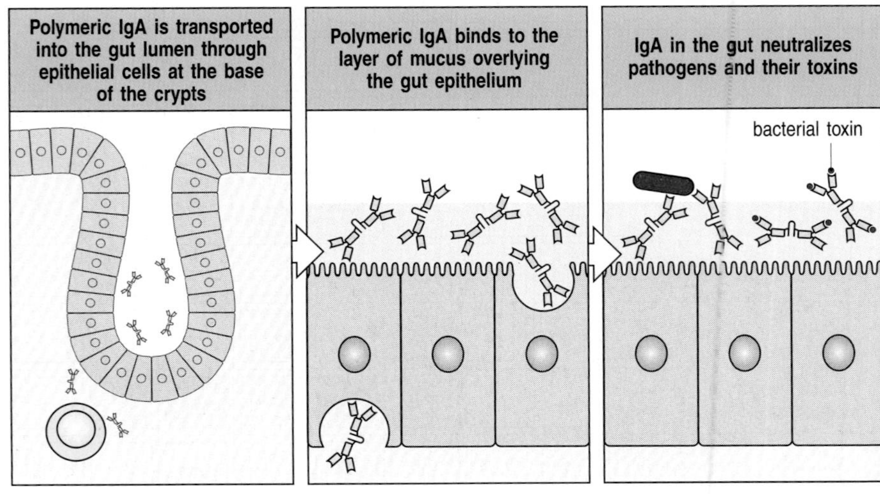
Polymeric IgA is transported into the gut lumen through epithelial cells at the base of the crypts
Polymeric IgA binds to the layer of mucus overlying the gut epithelium
IgA in the gut neutralizes pathogens and their toxins
bacterial toxin

Fig. 9.20 The major class of antibody present in the lumen of the gut is secretory dimeric IgA. This is synthesized by plasma cells in the lamina propria and transported into the lumen of the gut through epithelial cells at the base of the crypts. Dimeric IgA binds to the layer of mucus overlying the gut epithelium and acts as an antigen-specific barrier to pathogens and toxins in the gut lumen.

IgA antibodies is to protect epithelial surfaces from infectious agents, just as IgG antibodies protect the extracellular spaces of the internal tissues. IgA antibodies prevent the attachment of bacteria or toxins to epithelial cells and the absorption of foreign substances, and provide the first line of defense against a wide variety of pathogens. IgA is also thought to have an additional role in the gut, that of regulating the gut microflora.

Newborn infants are especially vulnerable to infection, having had no previous exposure to the microbes in the environment they enter at birth. IgA antibodies are secreted in breast milk and are thereby transferred to the gut of the newborn infant, where they provide protection from newly encountered bacteria until the infant can synthesize its own protective antibody. IgA is not the only protective antibody that a mother passes on to her baby. Maternal IgG is transported across the placenta directly into the bloodstream of the fetus during intrauterine life; human babies at birth have as high a level of plasma IgG as their mothers, and with the same range of antigen specificities. The selective transport of IgG from mother to fetus is due to an IgG transport protein in the placenta, FcRn, which is closely related in structure to MHC class I molecules. Despite this similarity, FcRn binds IgG quite differently from the binding of peptide to MHC class I molecules, because its peptide-binding groove is occluded. It binds to the Fc portion of IgG molecules (Fig. 9.21). Two molecules of FcRn bind one molecule of IgG, bearing it across the placenta. In some rodents, FcRn also delivers IgG to the circulation of the neonate from the gut lumen. Maternal IgG is also ingested by the newborn animal in its mother's milk and colostrum, the protein-rich fluid secreted by the early postnatal mammary gland. In this case, FcRn transports the IgG from the lumen of the neonatal gut into the blood and tissues. Interestingly, FcRn is also found in adults in the gut and liver and on endothelial cells. Its function in adults is to maintain the levels of IgG in plasma, which it does by binding antibody, endocytosing it, and recycling it to the blood, thus preventing its excretion.

By means of these specialized transport systems, mammals are supplied from birth with antibodies against pathogens common in their environments. As

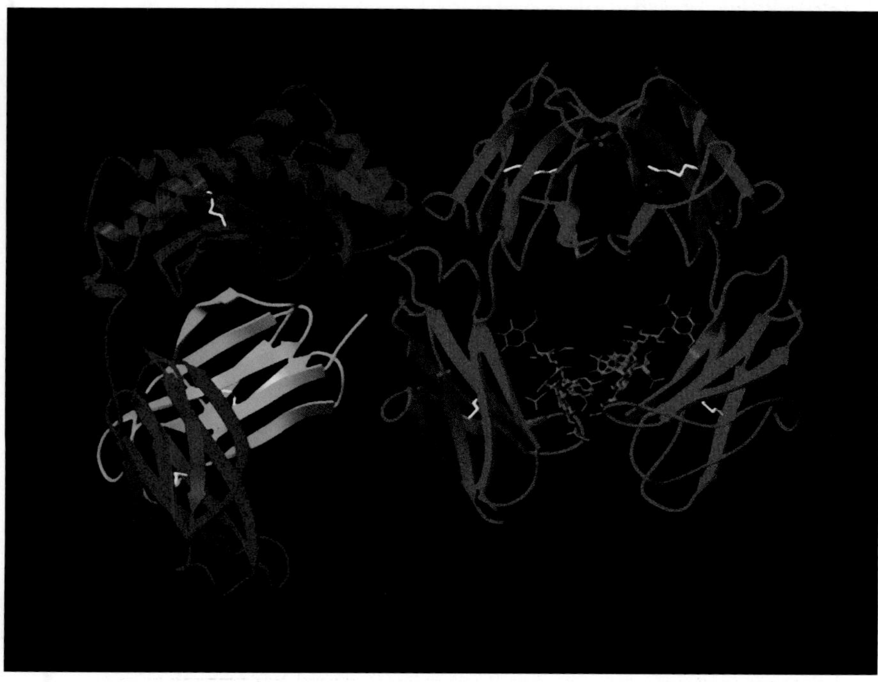

Fig. 9.21 FcRn binds to the Fc portion of IgG. The structure of a molecule of FcRn (blue and green) is shown bound to one chain of the Fc portion of IgG (red), at the interface of the Cγ2 and Cγ3 domains, with the Cγ2 region at the top. The β_2-microglobulin component of the FcRn is green. The dark-blue structure attached to the Fc portion of IgG is a carbohydrate chain, reflecting glycosylation. FcRn transports IgG molecules across the placenta in humans and also across the gut in rats and mice. It also has a role in maintaining the levels of IgG in adults. Although only one molecule of FcRn is shown binding to the Fc portion, it is thought that it takes two molecules of FcRn to capture one molecule of IgG. Courtesy of P. Björkman.

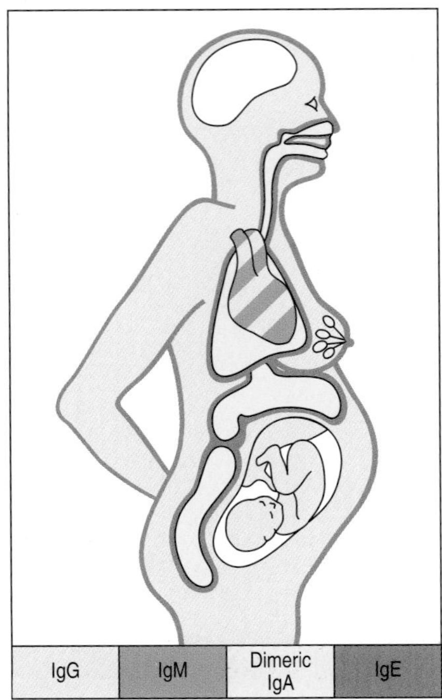

Fig. 9.22 Immunoglobulin classes are selectively distributed in the body. IgG and IgM predominate in plasma, whereas IgG and monomeric IgA are the major antibodies in extracellular fluid within the body. Dimeric IgA predominates in secretions across epithelia, including breast milk. The fetus receives IgG from the mother by transplacental transport. IgE is found mainly associated with mast cells just beneath epithelial surfaces (especially of the respiratory tract, gastrointestinal tract, and skin). The brain is normally devoid of immunoglobulin.

they mature and make their own antibodies of all isotypes, these are distributed selectively to different sites in the body (Fig. 9.22). Thus, throughout life, class switching and the distribution of antibody classes through the body provide effective protection against infection in extracellular spaces.

9-16 High-affinity IgG and IgA antibodies can neutralize bacterial toxins.

Many bacteria cause disease by secreting proteins called toxins, which damage or disrupt the function of the host's cells (Fig. 9.23). To have an effect, a toxin must interact specifically with a molecule that serves as a receptor on the surface of the target cell. In many toxins the receptor-binding domain is on one polypeptide chain but the toxic function is carried by a second chain. Antibodies that bind to the receptor-binding site on the toxin molecule can prevent the toxin from binding to the cell and thus protect the cell from attack (Fig. 9.24). Antibodies that act in this way to neutralize toxins are referred to as neutralizing antibodies.

Most toxins are active at nanomolar concentrations: a single molecule of diphtheria toxin can kill a cell. To neutralize toxins, therefore, antibodies must be able to diffuse into the tissues and bind the toxin rapidly and with high affinity. The ability of IgG antibodies to diffuse easily throughout the

Disease	Organism	Toxin	Effects *in vivo*
Tetanus	*Clostridium tetani*	Tetanus toxin	Blocks inhibitory neuron action, leading to chronic muscle contraction
Diphtheria	*Corynebacterium diphtheriae*	Diphtheria toxin	Inhibits protein synthesis, leading to epithelial cell damage and myocarditis
Gas gangrene	*Clostridium perfringens*	Clostridial toxin	Phospholipase activation, leading to cell death
Cholera	*Vibrio cholerae*	Cholera toxin	Activates adenylate cyclase, elevates cAMP in cells, leading to changes in intestinal epithelial cells that cause loss of water and electrolytes
Anthrax	*Bacillus anthracis*	Anthrax toxic complex	Increases vascular permeability, leading to edema, hemorrhage, and circulatory collapse
Botulism	*Clostridium botulinum*	Botulinum toxin	Blocks release of acetylcholine, leading to paralysis
Whooping cough	*Bordetella pertussis*	Pertussis toxin	ADP-ribosylation of G proteins, leading to lymphoproliferation
		Tracheal cytotoxin	Inhibits cilia and causes epithelial cell loss
Scarlet fever	*Streptococcus pyogenes*	Erythrogenic toxin	Vasodilation, leading to scarlet fever rash
		Leukocidin Streptolysins	Kill phagocytes, allowing bacterial survival
Food poisoning	*Staphylococcus aureus*	Staphylococcal enterotoxin	Acts on intestinal neurons to induce vomiting. Also a potent T-cell mitogen (SE superantigen)
Toxic-shock syndrome	*Staphylococcus aureus*	Toxic-shock syndrome toxin	Causes hypotension and skin loss. Also a potent T-cell mitogen (TSST-1 superantigen)

Fig. 9.23 Many common diseases are caused by bacterial toxins. These toxins are all exotoxins—proteins secreted by the bacteria. High-affinity IgG and IgA antibodies protect against these toxins. Bacteria also have nonsecreted endotoxins, such as lipopolysaccharide, which are released when the bacterium dies. The endotoxins are also important in the pathogenesis of disease, but there the host response is more complex because the innate immune system has receptors for some of these (see Chapter 2).

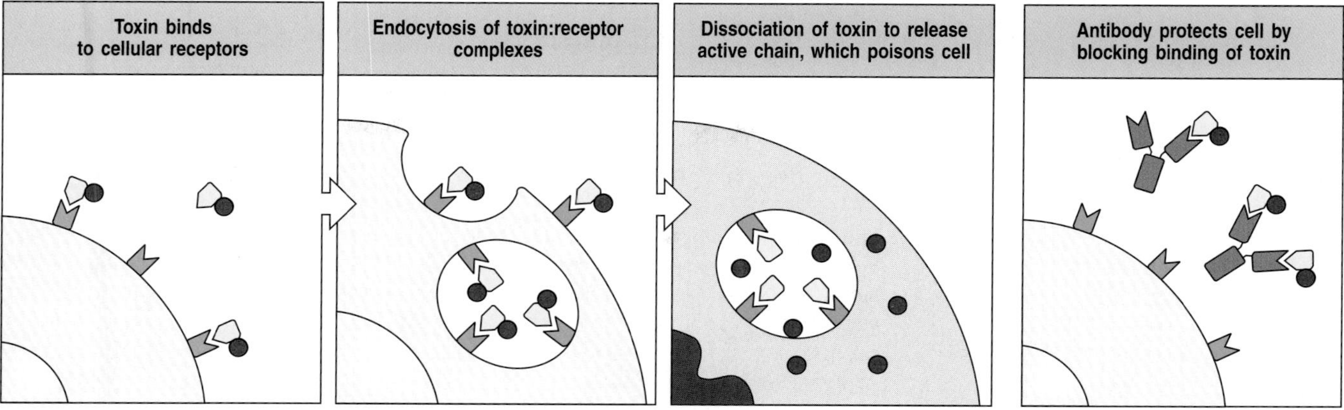

Toxin binds to cellular receptors
Endocytosis of toxin:receptor complexes
Dissociation of toxin to release active chain, which poisons cell
Antibody protects cell by blocking binding of toxin

Fig. 9.24 Neutralization of toxins by IgG antibodies protects cells from their damaging action. Many bacteria (as well as venomous insects and snakes) cause their damaging effects by elaborating toxic proteins (see Fig. 9.23). These toxins are usually composed of several distinct moieties. One part of the toxin molecule binds a cellular receptor, which enables the molecule to be internalized. Another part of the toxin molecule then enters the cytoplasm and poisons the cell. Antibodies that inhibit toxin binding can prevent, or neutralize, these effects.

extracellular fluid and their high affinity make these the principal neutralizing antibodies for toxins found in tissues. IgA antibodies similarly neutralize toxins at the mucosal surfaces of the body.

Diphtheria and tetanus toxins are two bacterial toxins in which the toxic and receptor-binding functions are on separate protein chains. It is therefore possible to immunize individuals, usually as infants, with modified toxin molecules in which the toxic chain has been denatured. These modified toxins, called toxoids, lack toxic activity but retain the receptor-binding site. Thus, immunization with the toxoid induces neutralizing antibodies that protect against the native toxin.

Some insect or animal venoms are so toxic that a single exposure can cause severe tissue damage or death, and for these the adaptive immune response is too slow to be protective. Exposure to these venoms is a rare event, and protective vaccines have not been developed for use in humans. Instead, neutralizing antibodies are generated by immunizing other species, such as horses, with insect and snake venoms to produce anti-venom antibodies (antivenins). These antivenins are injected into exposed individuals to protect them against the toxic effects of the venom. Transfer of antibodies in this way is known as passive immunization (see Appendix I, Section A-37).

9-17 High-affinity IgG and IgA antibodies can inhibit the infectivity of viruses.

Animal viruses infect cells by binding to a particular cell-surface receptor, often a cell-type-specific protein that determines which cells they can infect. The hemagglutinin of influenza virus, for example, binds to terminal sialic acid residues on the carbohydrates of glycoproteins present on epithelial cells of the respiratory tract. It is known as hemagglutinin because it recognizes and binds to similar sialic acid residues on chicken red blood cells and agglutinates these red blood cells. Antibodies against the hemagglutinin can prevent infection by the influenza virus. Such antibodies are called virus-neutralizing antibodies and, as with the neutralization of toxins, high-affinity IgA and IgG antibodies are particularly important.

Many antibodies that neutralize viruses do so by directly blocking viral binding to surface receptors (Fig. 9.25). However, viruses are sometimes successfully neutralized when only a single molecule of antibody is bound to a virus particle that has many receptor-binding proteins on its surface. In these cases, the

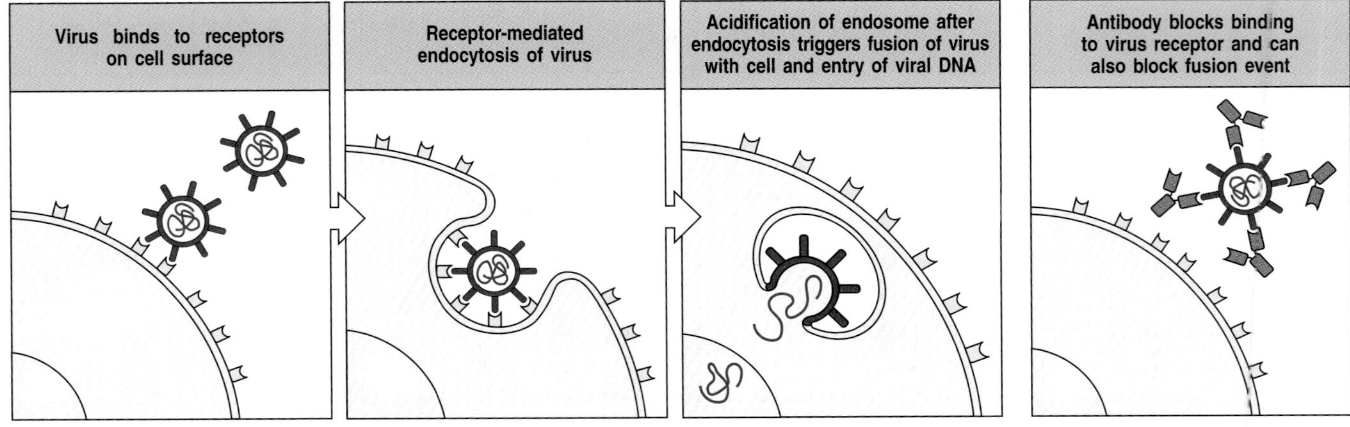

Virus binds to receptors on cell surface
Receptor-mediated endocytosis of virus
Acidification of endosome after endocytosis triggers fusion of virus with cell and entry of viral DNA
Antibody blocks binding to virus receptor and can also block fusion event

Fig. 9.25 Viral infection of cells can be blocked by neutralizing antibodies. For a virus to multiply within a cell, it must introduce its genes into the cell. The first step in entry is usually the binding of the virus to a receptor on the cell surface. For enveloped viruses, as shown in the figure, entry into the cytoplasm requires fusion of the viral envelope and the cell membrane. For some viruses this fusion event takes place on the cell surface (not shown); for others it can occur only within the more acidic environment of endosomes, as shown here. Nonenveloped viruses must also bind to receptors on cell surfaces but they enter the cytoplasm by disrupting endosomes. Antibodies bound to viral surface proteins neutralize the virus, inhibiting either its initial binding to the cell or its subsequent entry.

antibody must cause some change in the virus that disrupts its structure and either prevents it from interacting with its receptors or interferes with the fusion of the virus membrane with the cell surface after the virus has engaged its surface receptor.

9-18 Antibodies can block the adherence of bacteria to host cells.

Many bacteria have cell-surface molecules called adhesins that enable them to bind to the surface of host cells. This adherence is critical to the ability of these bacteria to cause disease, whether they subsequently enter the cell, as do *Salmonella* species, or remain attached to the cell surface as extracellular pathogens (Fig. 9.26). *Neisseria gonorrhoeae,* the causative agent of the sexually transmitted disease gonorrhea, has a cell-surface protein known as pilin, which enables the bacterium to adhere to the epithelial cells of the urinary and reproductive tracts and is essential to its infectivity. Antibodies against pilin can inhibit this adhesive reaction and prevent infection.

IgA antibodies secreted onto the mucosal surfaces of the intestinal, respiratory, and reproductive tracts are particularly important in preventing infection by inhibiting the adhesion of bacteria, viruses, or other pathogens to the epithelial cells lining these surfaces. The adhesion of bacteria to cells within tissues can also contribute to pathogenesis, and IgG antibodies against adhesins protect tissues from damage much as IgA antibodies protect at mucosal surfaces.

9-19 Antibody:antigen complexes activate the classical pathway of complement by binding to C1q.

Another way in which antibodies can protect against infection is by activation of the cascade of complement proteins. We described these proteins in Chapter 2 because they can also be activated on pathogen surfaces in the absence of antibody, as part of the innate immune response. Complement activation proceeds via a series of proteolytic cleavage reactions, in which inactive components present in plasma are cleaved to form proteolytic enzymes that attach covalently to the pathogen surface. All known pathways of complement activation converge to generate the same set of effector actions: the pathogen surface or immune complex is coated with covalently attached fragments (principally C3b) that act as opsonins to promote uptake and removal by phagocytes. At the same time, small peptides with inflammatory and chemotactic activity are released (principally C5a) so that phagocytes are recruited to the site. In addition, the terminal complement components can form a membrane-attack complex that damages some bacteria.

Antibodies initiate complement activation by a pathway known as the classical pathway because it was the first pathway of complement activation to be discovered. The full details of this pathway, and of the other two known pathways of complement activation, are given in Chapter 2, but we describe here how antibody is able to initiate the classical pathway after binding to pathogen, or after forming immune complexes.

The first component of the classical pathway of complement activation is C1, which is a complex of three proteins called C1q, C1r, and C1s. Two molecules each of C1r and C1s are bound to each molecule of C1q (see Fig. 2.27). Complement activation is initiated when antibodies attached to the surface of a pathogen bind C1q. C1q can be bound by either IgM or IgG antibodies but, because of the structural requirements of binding to C1q, neither of these antibody classes can activate complement in solution; the cascade is initiated only when the antibodies are bound to multiple sites on a cell surface, normally that of a pathogen.

The C1q molecule has six globular heads joined to a common stem by long, filamentous domains that resemble collagen molecules; the whole C1q complex has been likened to a bunch of six tulips held together by the stems. Each globular head can bind to one Fc domain, and binding of two or more globular heads activates the C1q molecule. In plasma the pentameric IgM molecule has a planar conformation that does not bind C1q (Fig. 9.27, left panel); however, binding to the surface of a pathogen deforms the IgM pentamer so that it looks like a staple (Fig. 9.27, right panel), and this distortion exposes binding sites for the C1q heads. Although C1q binds with low affinity to some subclasses of IgG in solution, the binding energy required for C1q activation is achieved only when a single molecule of C1q can bind two or more IgG molecules that are held within 30–40 nm of each other as a result of binding antigen. This requires that multiple molecules of IgG be bound to a single pathogen. For this reason, IgM is much more efficient than IgG in activating complement. The binding of C1q to a single bound IgM molecule, or to two or more bound IgG molecules (Fig. 9.28), leads to the activation of an enzymatic activity in C1r, triggering the complement cascade. This translates antibody binding into the activation of the complement cascade, which, as we learned in Chapter 2, can also be triggered by the direct binding of C1q to the pathogen surface.

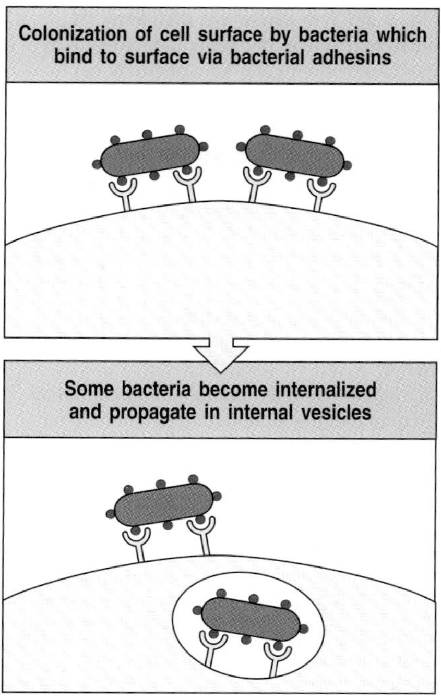

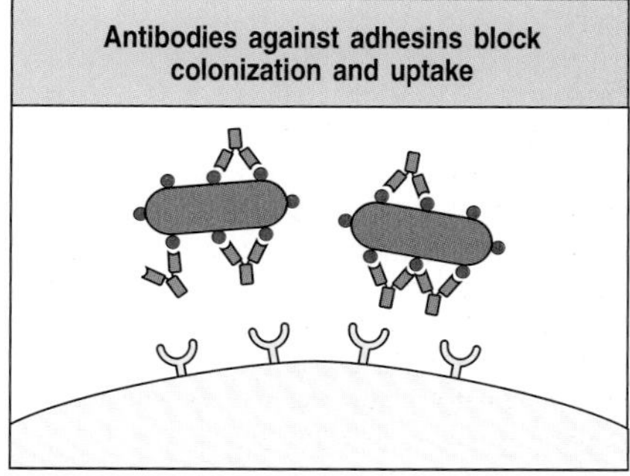

Fig. 9.26 Antibodies can prevent the attachment of bacteria to cell surfaces. Many bacterial infections require an interaction between the bacterium and a cell-surface receptor. This is particularly true for infections of mucosal surfaces. The attachment process involves very specific molecular interactions between bacterial adhesins and their receptors on host cells; antibodies against bacterial adhesins can block such infections.

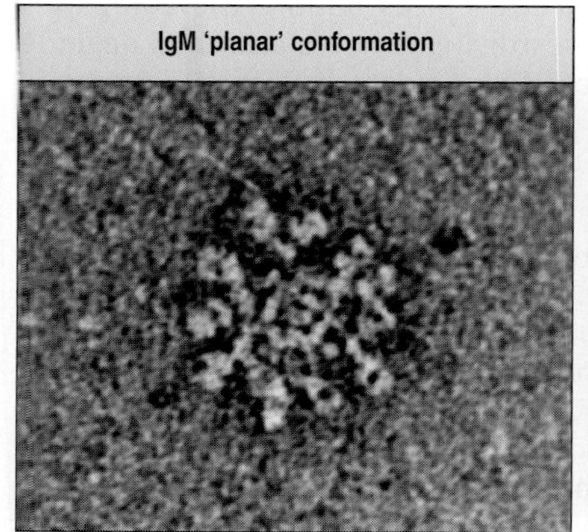

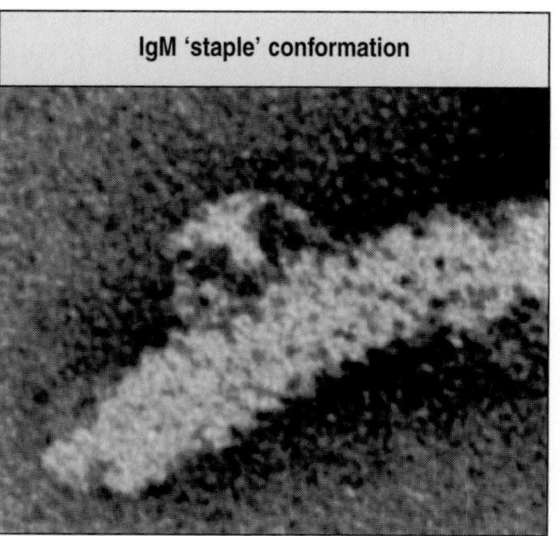

Fig. 9.27 The two conformations of IgM. The left panel shows the planar conformation of soluble IgM; the right panel shows the 'staple' conformation of IgM bound to a bacterial flagellum. Photographs (× 760,000) courtesy of K.H. Roux.

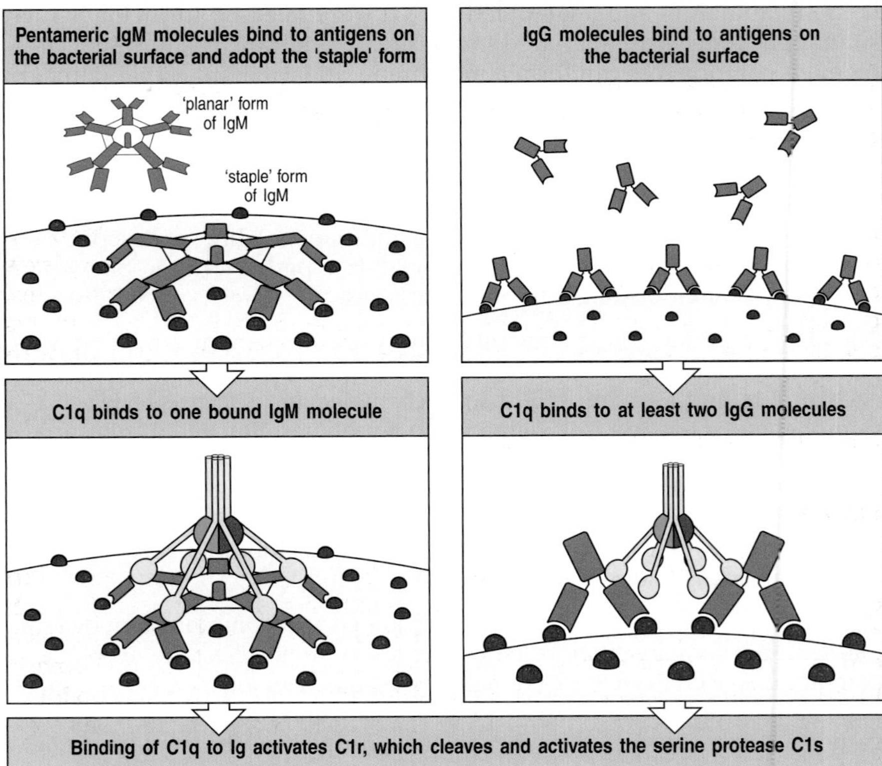

Pentameric IgM molecules bind to antigens on the bacterial surface and adopt the 'staple' form
'planar' form of IgM
'staple' form of IgM
IgG molecules bind to antigens on the bacterial surface
C1q binds to one bound IgM molecule
C1q binds to at least two IgG molecules
Binding of C1q to Ig activates C1r, which cleaves and activates the serine protease C1s

Fig. 9.28 The classical pathway of complement activation is initiated by the binding of C1q to antibody on a surface such as a bacterial surface. In the left panels one molecule of IgM, bent into the 'staple' conformation by binding several identical epitopes on a pathogen surface, allows the globular heads of C1q to bind to its Fc pieces on the surface of the pathogen. In the right panels, multiple molecules of IgG bound on the surface of a pathogen allow the binding of a single molecule of C1q to two or more Fc pieces. In both cases, the binding of C1q activates the associated C1r, which becomes an active enzyme that cleaves the pro-enzyme C1s, generating a serine protease that initiates the classical complement cascade (see Chapter 2).

9-20 Complement receptors are important in the removal of immune complexes from the circulation.

Many small soluble antigens form antibody:antigen complexes known as immune complexes containing too few molecules of IgG to be readily bound to the Fcγ receptors that we discuss in the next part of the chapter. These antigens include toxins bound by neutralizing antibodies and debris from dead microorganisms. Such immune complexes are found after most infections and are removed from the circulation through the action of complement. The soluble immune complexes trigger their own removal by activating complement, again through the binding of C1q, leading to the covalent binding of the activated components C4b and C3b to the complex, which is then cleared from the circulation by the binding of C4b and C3b to complement receptor 1 (CR1) on the surface of erythrocytes. The erythrocytes transport the bound complexes of antigen, antibody, and complement to the liver and spleen. Here, macrophages bearing CR1 and Fc receptors remove the complexes from the erythrocyte surface without destroying the cell, and then degrade them (Fig. 9.29). Even larger aggregates of particulate antigen and antibody can be made soluble after activation of the classical complement pathway and the consequent binding of C3b to the aggregates; these can then be removed by binding to complement receptors.

Systemic Lupus Erythematosus

Immune complexes that are not removed tend to deposit in the basement membranes of small blood vessels, most notably those of the renal glomerulus, where the blood is filtered to form urine. Immune complexes that pass through the basement membrane of the glomerulus bind to the complement receptor CR1 on the renal podocytes, cells that lie beneath the basement membrane. The functional significance of these receptors in the kidney is unknown; however, they have an important role in the pathology of some autoimmune diseases.

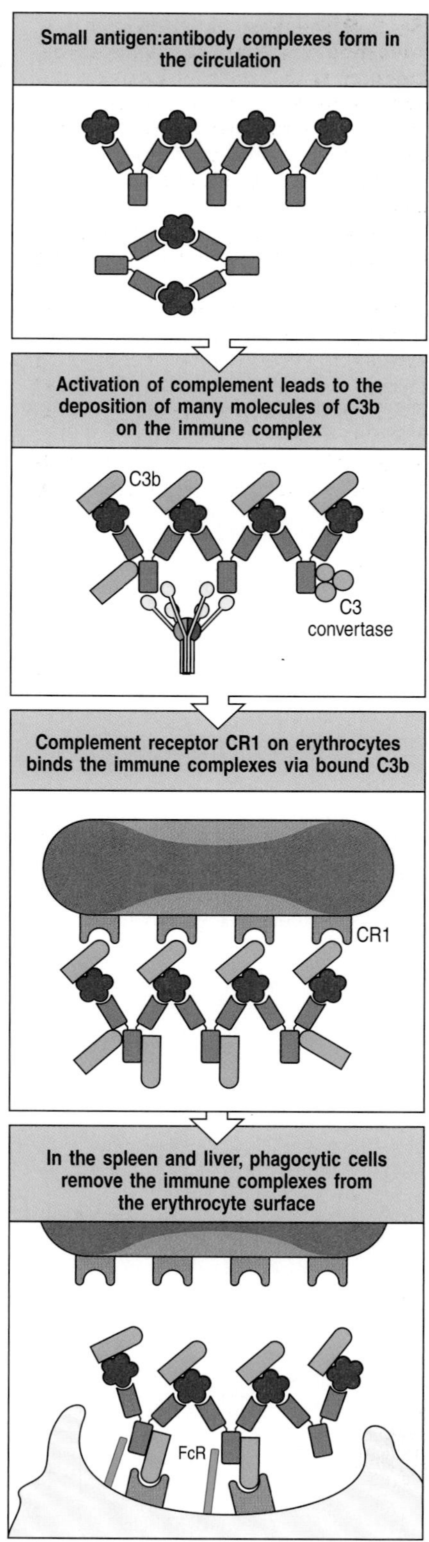

Small antigen:antibody complexes form in the circulation
Activation of complement leads to the deposition of many molecules of C3b on the immune complex
C3b
C3 convertase
Complement receptor CR1 on erythrocytes binds the immune complexes via bound C3b
CR1
In the spleen and liver, phagocytic cells remove the immune complexes from the erythrocyte surface
FcR

Fig. 9.29 Erythrocyte CR1 helps to clear immune complexes from the circulation. CR1 on the erythrocyte surface has an important role in the clearance of immune complexes from the circulation. Immune complexes bind to CR1 on erythrocytes, which transport them to the liver and spleen, where they are removed by macrophages expressing receptors for both Fc and bound complement components.

In the autoimmune disease systemic lupus erythematosus, which we describe in Chapter 14, excessive levels of circulating immune complexes cause huge deposits of antigen, antibody, and complement on the podocytes, damaging the glomerulus; kidney failure is the principal danger in this disease. Immune complexes can also be a cause of pathology in patients with deficiencies in the early components of complement. Such patients do not clear immune complexes effectively and they also suffer tissue damage, especially in the kidneys, in a similar way.

Summary.

The T-cell dependent antibody response begins with IgM secretion but quickly progresses to the production of additional antibody classes. Each class is specialized both in its localization in the body and in the functions it can perform. IgM antibodies are found mainly in blood; they are pentameric in structure. IgM is specialized to activate complement efficiently upon binding antigen and to compensate for the low affinity of a typical IgM antigen-binding site. IgG antibodies are usually of higher affinity and are found in blood and in extracellular fluid, where they can neutralize toxins, viruses, and bacteria, opsonize them for phagocytosis, and activate the complement system. IgA antibodies are synthesized as monomers, which enter blood and extracellular fluids, or as dimeric molecules by plasma cells in the lamina propria of various mucosal tissues. IgA dimers are selectively transported across the epithelial layer into sites such as the lumen of the gut, where they neutralize toxins and viruses and block the entry of bacteria across the intestinal epithelium. Most IgE antibody is bound to the surface of mast cells that reside mainly just below body surfaces; antigen binding to this IgE triggers local defense reactions. Antibodies can defend the body against extracellular pathogens and their toxic products in several ways. The simplest is by direct interactions with pathogens or their products, for example by binding to the active sites of toxins and neutralizing them or by blocking their ability to bind to host cells through specific receptors. When antibodies of the appropriate isotype bind to antigens, they can activate the classical pathway of complement, which leads to the elimination of the pathogen by the various mechanisms described in Chapter 2. Soluble immune complexes of antigen and antibody also fix complement and are cleared from the circulation via complement receptors on red blood cells.

The destruction of antibody-coated pathogens via Fc receptors.

The ability of high-affinity antibodies to neutralize toxins, viruses, or bacteria can protect against infection but does not, on its own, solve the problem of how to remove the pathogens and their products from the body. Moreover, many pathogens cannot be neutralized by antibody and must be destroyed by other means. Many pathogen-specific antibodies do not bind to neutralizing targets on pathogen surfaces and thus need to be linked to other effector mechanisms to play their part in host defense. We have already seen how the binding of antibody to antigen can activate complement. Another important defense mechanism is the activation of a variety of **accessory effector cells** bearing receptors called **Fc receptors** because they are specific for the Fc portion of antibodies. These receptors facilitate the phagocytosis of neutralized microorganisms and resistant extracellular pathogens by macrophages, dendritic cells, and neutrophils. Other nonphagocytic cells—NK cells, eosinophils, basophils, and mast cells (see Fig. 1.4)—are triggered to secrete stored mediators when their Fc receptors are engaged. These mechanisms maximize the effectiveness of all antibodies regardless of where they bind. Fc receptor-bearing cells are activated when their Fc receptors are aggregated by binding to the multiple Fc regions of antibody molecules coating a pathogen. They can also be activated by soluble mediators; these include products of the complement cascade, which can itself be activated by antibody, as we have seen.

9-21 The Fc receptors of accessory cells are signaling receptors specific for immunoglobulins of different classes.

The Fc receptors are a family of cell-surface molecules that bind the Fc portion of immunoglobulins. Each member of the family recognizes immunoglobulin of one or a few closely related heavy-chain isotypes through a recognition domain on the α chain of the Fc receptor. Most Fc receptors are themselves members of the immunoglobulin gene superfamily. Different cell types bear different sets of Fc receptors, and the isotype of the antibody thus determines which types of cells will be engaged in a given response. The different Fc receptors, the cells that express them, and their specificities for different antibody classes are shown in Fig. 9.30.

Most Fc receptors function as part of a multisubunit complex. Only the α chain is required for specific recognition; the other chains are required for

Fig. 9.30 Distinct receptors for the Fc region of the different immunoglobulin classes are expressed on different accessory cells. The subunit structure and binding properties of these receptors and the cell types expressing them are shown. The exact chain composition of any receptor can vary from one cell type to another. For example, FcγRIII in neutrophils is expressed as a molecule with a glycosylphosphatidylinositol membrane anchor, without γ chains, whereas in NK cells it is a transmembrane molecule associated with γ chains. The FcγRII-B1 differs from the FcγRII-B2 by the presence of an additional exon in the intracellular region. This exon prevents the FcγRII-B1 from being internalized after cross-linking. The binding affinities are taken from data on human receptors. *Only some allotypes of FcγRII-A bind IgG2. †In these cases Fc receptor expression is inducible rather than constitutive. ‡In eosinophils, the molecular weight of the CD89α chain is 70–100 kDa.

Receptor	FcγRI (CD64)	FcγRII-A (CD32)	FcγRII-B2 (CD32)	FcγRII-B1 (CD32)	FcγRIII (CD16)	FcεRI	FcαRI (CD89)	Fcα/μR
Structure 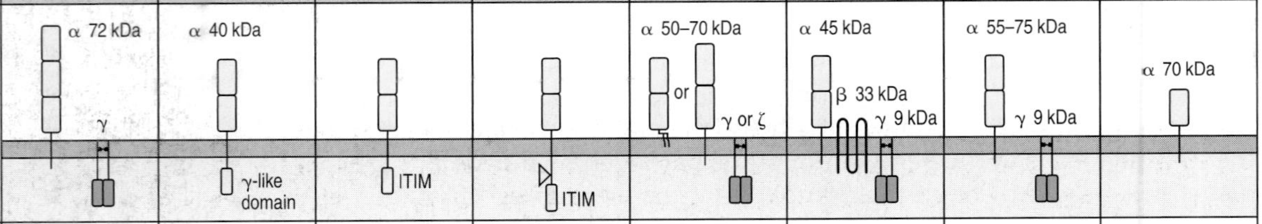	α 72 kDa γ	α 40 kDa γ-like domain	ITIM	ITIM	α 50–70 kDa or γ or ζ	α 45 kDa β 33 kDa γ 9 kDa	α 55–75 kDa γ 9 kDa	α 70 kDa
Binding	IgG1 $10^8 M^{-1}$	IgG1 $2 \times 10^6 M^{-1}$	IgG1 $2 \times 10^6 M^{-1}$	IgG1 $2 \times 10^6 M^{-1}$	IgG1 $5 \times 10^5 M^{-1}$	IgE $10^{10} M^{-1}$	IgA1, IgA2 $10^7 M^{-1}$	IgA, IgM $3 \times 10^9 M^{-1}$
Order of affinity	1) IgG1=IgG3 2) IgG4 3) IgG2	1) IgG1 2) IgG3=IgG2* 3) IgG4	1) IgG1=IgG3 2) IgG4 3) IgG2	1) IgG1=IgG3 2) IgG4 3) IgG2	IgG1=IgG3		IgA1=IgA2	1) IgM 2) IgA
Cell type	Macrophages Neutrophils† Eosinophils† Dendritic cells	Macrophages Neutrophils Eosinophils Platelets Langerhans cells	Macrophages Neutrophils Eosinophils	B cells Mast cells	NK cells Eosinophils Macrophages Neutrophils Mast cells	Mast cells Eosinophils† Basophils	Macrophages Eosinophils‡ Neutrophils	Macrophages B cells
Effect of ligation	Uptake Stimulation Activation of respiratory burst Induction of killing	Uptake Granule release (eosinophils)	Uptake Inhibition of stimulation	No uptake Inhibition of stimulation	Induction of killing (NK cells)	Secretion of granules	Uptake Induction of killing	Uptake

ort to the cell surface and for signal transduction when an Fc region nd. Some Fcγ receptors, the Fcα receptor I, and the high-affinity or for IgE use a γ chain for signaling; the γ chain, which is closely l to the ζ chain of the T-cell receptor complex, associates noncova- with the Fc-binding α chain. The human FcγRII-A is a single-chain tor in which the cytoplasmic domain of the α chain replaces the func- f the γ chain. FcγRII-B1 and FcγRII-B2 are also single-chain receptors unction as inhibitory receptors because they contain an ITIM that ges the inositol 5′-phosphatase SHIP (see Section 6-20). Although the prominent function of Fc receptors is the activation of accessory cells tack pathogens, they can also contribute in other ways to immune onses. For example, the FcγRII-B receptors negatively regulate B cells, cells, macrophages, and neutrophils by adjusting the threshold at h immune complexes will activate these cells. Fc receptors expressed endritic cells enable them to ingest antigen:antibody complexes and to ent antigenic peptides to T cells.

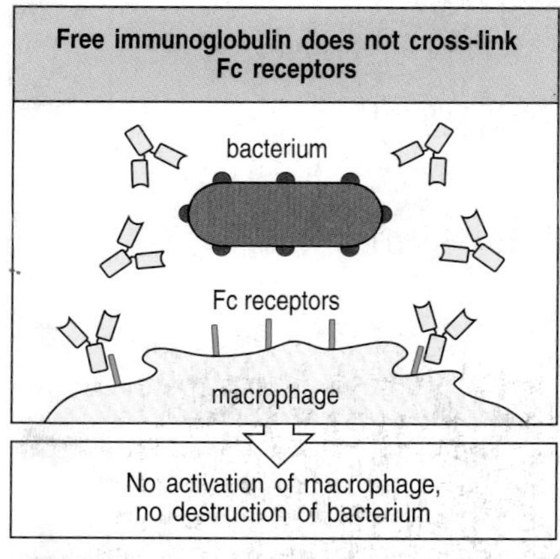

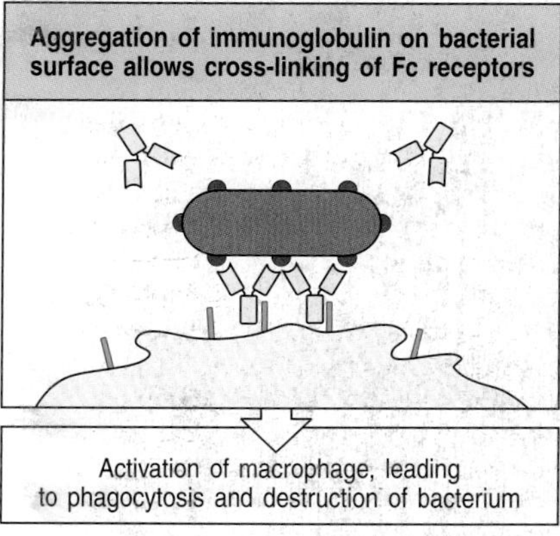

Fig. 9.31 Bound antibody is distinguishable from free immunoglobulin by its state of aggregation. Free immunoglobulin molecules bind most Fc receptors with very low affinity and cannot cross-link Fc receptors. Antigen-bound immunoglobulin, however, binds to Fc receptors with high avidity because several antibody molecules that are bound to the same surface bind to multiple Fc receptors on the surface of the accessory cell. This Fc receptor cross-linking sends a signal to activate the cell bearing it. With Fc receptors that have ITIMs, the result is inhibition.

Fc receptors on phagocytes are activated by antibodies bound to the surface of pathogens and enable the phagocytes to ingest and destroy pathogens.

gocytes are activated by IgG antibodies, especially IgG1 and IgG3, that d to specific Fcγ receptors on the phagocyte surface (see Fig. 9.30). ause phagocyte activation can initiate an inflammatory response and se tissue damage, it is essential that the Fc receptors on phagocytes be e to distinguish antibody molecules bound to a pathogen from the much ger number of free antibody molecules that are not bound to anything. is distinction is made possible by the aggregation or multimerization of tibodies that occurs when they bind to multimeric antigens or to multiva- t antigenic particles such as viruses and bacteria. Fc receptors on the rface of a cell bind antibody-coated particles with higher avidity than munoglobulin monomers, and this is the principal mechanism by which ound antibodies are distinguished from free immunoglobulin (Fig. 9.31). he result is that Fc receptors enable cells to detect pathogens through ound antibody molecules. Thus, specific antibody together with Fc recep- rs gives to phagocytic cells that lack intrinsic specificity the ability to iden- and remove pathogens and their products from the extracellular spaces.

e most important Fc-bearing cells in humoral immune responses are the phagocytic cells of the monocytic and myelocytic lineages, particularly macrophages and neutrophils (see Chapter 2). Many bacteria are directly recognized, ingested, and destroyed by phagocytes, and these bacteria are not pathogenic in normal individuals. Bacterial pathogens, however, often have polysaccharide capsules that allow them to resist direct engulfment by phagocytes. Such bacteria become susceptible to phagocytosis when they are coated with antibody and complement that engages the Fcγ or Fcα receptors and CR1 on phagocytic cells, triggering bacterial uptake (Fig. 9.32). The stimulation of phagocytosis by complement-coated antigens binding complement receptors is particularly important early in the immune response, before isotype-switched antibodies have been made. Capsular polysaccharides belong to the TI-2 class of thymus-independent antigens (see Section 9-11) and can therefore stimulate the early production of IgM antibodies, which are very effective at activating the complement system. IgM binding to encapsulated bacteria thus triggers the opsonization of these bacteria by complement and their prompt ingestion and destruction by phagocytes bearing complement receptors. Recently, an Fc receptor for IgM has been discovered, suggesting that IgM may promote phagocytosis directly *in vivo*.

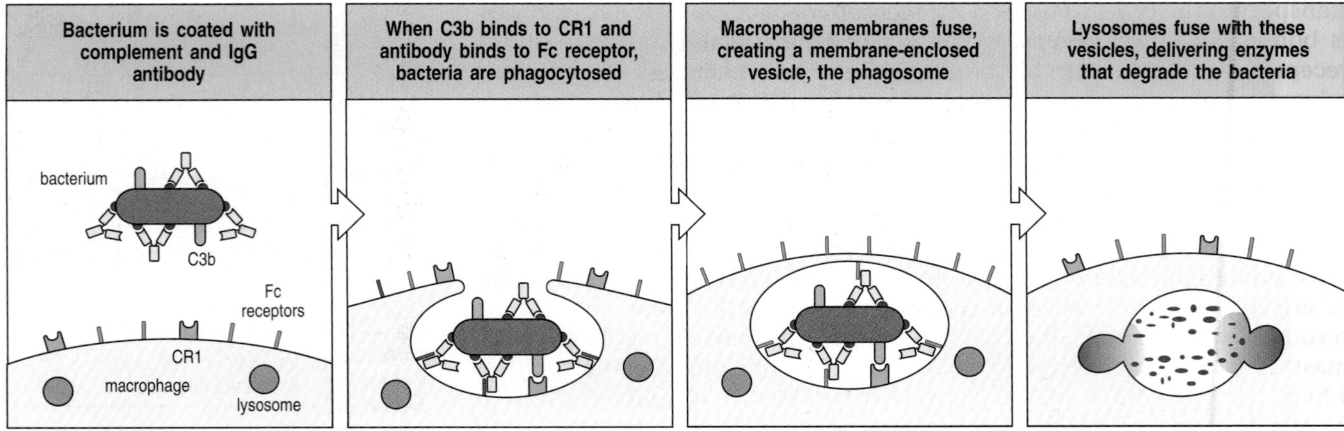

Bacterium is coated with complement and IgG antibody
When C3b binds to CR1 and antibody binds to Fc receptor, bacteria are phagocytosed
Macrophage membranes fuse, creating a membrane-enclosed vesicle, the phagosome
Lysosomes fuse with these vesicles, delivering enzymes that degrade the bacteria
bacterium
C3b
Fc receptors
CR1
macrophage
lysosome

Fig. 9.32 Fc and complement receptors on phagocytes trigger the uptake and degradation of antibody-coated bacteria. Many bacteria resist phagocytosis by macrophages and neutrophils. Antibodies bound to these bacteria, however, enable them to be ingested and degraded through the interaction of the multiple Fc domains arrayed on the bacterial surface with Fc receptors on the phagocyte surface. Antibody coating also induces activation of the complement system and the binding of complement components to the bacterial surface. These can interact with complement receptors (for example CR1) on the phagocyte. Fc receptors and complement receptors synergize in inducing phagocytosis. Bacteria coated with IgG antibody and complement are therefore more readily ingested than those coated with IgG alone. Binding of Fc and complement receptors signals the phagocyte to increase the rate of phagocytosis, to fuse lysosomes with phagosomes, and to increase its bactericidal activity.

The process of internalization and destruction of microorganisms is greatly enhanced by interactions between the molecules coating an opsonized microorganism and their receptors on the phagocyte surface. When an antibody-coated pathogen binds to Fcγ receptors on the surface of a phagocyte, for example, the cell surface extends around the surface of the particle through successive binding of Fcγ receptors to the antibody Fc regions bound to the pathogen surface. This is an active process triggered by the stimulation of Fcγ receptors. Endocytosis leads to enclosure of the particle in an acidified cytoplasmic vesicle called a phagosome. The phagosome then fuses with one or more lysosomes to generate a phagolysosome, releasing the lysosomal enzymes into the phagosome interior, where they destroy the bacterium (see Fig. 9.32). The process of bacterial destruction in the phagolysosome was described in detail in Section 2-4.

Some particles are too large for a phagocyte to ingest; parasitic worms are one example. In this case the phagocyte attaches to the surface of the antibody-coated parasite via its Fcγ, Fcα, or Fcε receptors, and the lysosomes fuse with the attached surface membrane. This reaction discharges the contents of the lysosome onto the surface of the parasite, damaging it directly in the extracellular space. Thus, Fcγ and Fcα receptors can trigger the internalization of external particles by phagocytosis or the externalization of internal vesicles by exocytosis. The principal leukocytes involved in the destruction of bacteria are macrophages and neutrophils, but large parasites such as helminths are more usually attacked by eosinophils (Fig. 9.33). Cross-linking of IgE bound to the high-affinity FcεRI usually results in exocytosis. We will see in the next three sections that NK cells and mast cells also release mediators stored in their vesicles when their Fc receptors are aggregated.

9-23 Fc receptors activate NK cells to destroy antibody-coated targets.

Infected cells are usually destroyed by T cells that have been activated by foreign peptides bound to cell-surface MHC molecules. However, virus-infected cells can also signal the presence of intracellular infection by expressing on their surfaces viral proteins that can be recognized by antibodies. Cells bound by such antibodies can then be killed by a specialized non-T, non-B lymphoid cell called a natural killer cell (NK cell), which we met in Chapter 2. NK cells are large lymphoid cells with prominent intracellular granules; they make up a small fraction of peripheral blood lymphoid cells. They bear no known antigen-specific receptors but are able to recognize and kill a limited range of abnormal cells. They were first discovered because of their ability to kill some tumor cells, but are now known to have an important role in innate immunity.

The destruction of antibody-coated target cells by NK cells is called **antibody-dependent cell-mediated cytotoxicity** (**ADCC**) and is triggered when antibody bound to the surface of a cell interacts with Fc receptors on the NK cell (Fig. 9.34). NK cells express the receptor FcγRIII (CD16), which recognizes the IgG1 and IgG3 subclasses. The mechanism of attack is exactly analogous to that of cytotoxic T cells, involving the release of cytoplasmic granules containing perforin and granzymes (see Section 8-28). The importance of ADCC in defense against infection with bacteria or viruses has not yet been fully established. However, ADCC represents yet another mechanism by which, through engaging an Fc receptor, antibodies can direct an antigen-specific attack by an effector cell that itself lacks specificity for antigen.

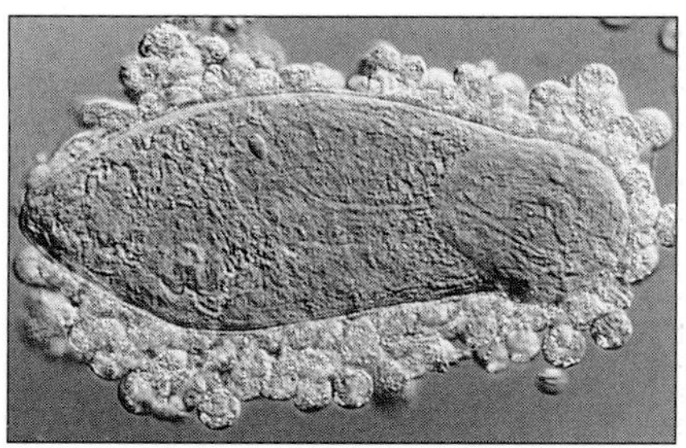

Fig. 9.33 Eosinophils attacking a schistosome larva in the presence of serum from an infected patient. Large parasites, such as worms, cannot be ingested by phagocytes; however, when the worm is coated with antibody, especially IgE, eosinophils can attack it through their binding to the high-affinity FcεRI. Similar attacks can be mounted by other Fc receptor-bearing cells on various large targets. These cells release the toxic contents of their granules directly onto the target, a process known as exocytosis. Photograph courtesy of A. Butterworth.

9-24 Mast cells, basophils, and activated eosinophils bind IgE antibody via the high-affinity Fcε receptor.

When pathogens cross epithelial barriers and establish a local focus of infection, the host must mobilize its defenses and direct them to the site of pathogen growth. One mechanism by which this is achieved is to activate a specialized cell type known as a **mast cell**. Mast cells are large cells containing distinctive cytoplasmic granules that contain a mixture of chemical mediators, including histamine, that act rapidly to make local blood vessels more permeable. Mast cells have a distinctive appearance after staining with the dye toluidine blue that makes them readily identifiable in tissues (see Fig. 1.4). They are found in particularly high concentrations in vascularized connective tissues just beneath epithelial surfaces, including the submucosal tissues of the gastrointestinal and respiratory tracts and the dermis that lies just below the surface of the skin.

Mast cells have Fc receptors specific for IgE (FcεRI) and IgG (FcγRIII) and can be activated to release their granules, and to secrete lipid inflammatory mediators and cytokines, via antibody bound to these receptors. We saw earlier that most Fc receptors bind stably to the Fc regions of antibodies only when the antibodies have themselves bound antigen. In contrast, FcεRI binds monomeric IgE antibodies with a very high affinity, measured at approximately $10^{10}\ M^{-1}$. Thus, even at the low levels of IgE found circulating in normal individuals, a substantial portion of the total IgE is bound to the FcεRI on mast cells in tissues and on circulating basophils. Eosinophils can also express Fc receptors, but they express FcεRI only when activated and recruited to an inflammatory site.

Although mast cells are usually stably associated with bound IgE, they are not activated simply by the binding of monomeric antigens to this IgE. Mast-cell

Fig. 9.34 Antibody-coated target cells can be killed by NK cells in antibody-dependent cell-mediated cytotoxicity (ADCC). NK cells (see Chapter 2) are large granular non-T, non-B lymphoid cells that have FcγRIII (CD16) on their surface. When these cells encounter cells coated with IgG antibody, they rapidly kill the target cell. The importance of ADCC in host defense or tissue damage is still controversial.

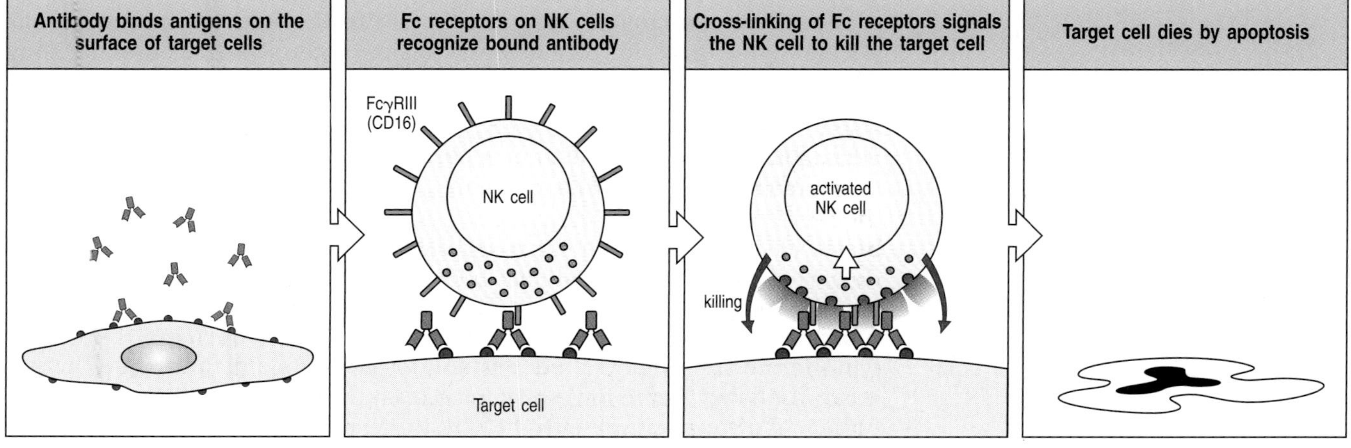
Antibody binds antigens on the surface of target cells
Fc receptors on NK cells recognize bound antibody
Cross-linking of Fc receptors signals the NK cell to kill the target cell
Target cell dies by apoptosis
FcγRIII (CD16)
NK cell
Target cell
activated NK cell
killing

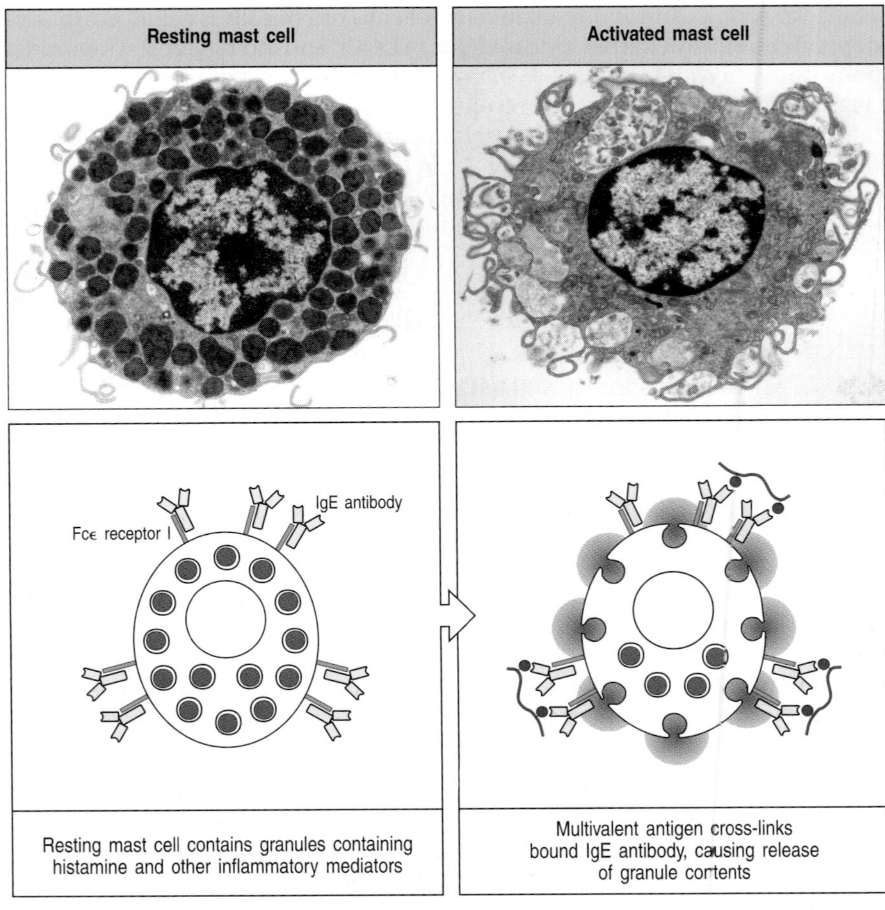

Resting mast cell
Activated mast cell
IgE antibody
Fcε receptor I
Resting mast cell contains granules containing histamine and other inflammatory mediators
Multivalent antigen cross-links bound IgE antibody, causing release of granule contents

Fig. 9.35 IgE antibody cross-linking on mast-cell surfaces leads to a rapid release of inflammatory mediators. Mast cells are large cells found in connective tissue that can be distinguished by secretory granules containing many inflammatory mediators. They bind stably to monomeric IgE antibodies through the very high-affinity Fcε receptor I. Antigen cross-linking of the bound IgE antibody molecules triggers rapid degranulation, releasing inflammatory mediators into the surrounding tissue. These mediators trigger local inflammation, which recruits cells and proteins required for host defense to sites of infection. These cells are also triggered during allergic reactions when allergens bind to IgE on mast cells. Photographs courtesy of A.M. Dvorak.

activation occurs only when the bound IgE is cross-linked by multivalent antigen. This signal activates the mast cell to release the contents of its granules, which occurs in seconds (Fig. 9.35), to synthesize and release lipid mediators such as prostaglandin D_2 and leukotriene C4, and to secrete cytokines such as TNF-α, thereby initiating a local inflammatory response. Degranulation releases the stored histamine, causing a local increase in blood flow and vascular permeability that quickly leads to an accumulation of fluid and blood proteins, including antibodies, in the surrounding tissue. Shortly afterwards there is an influx of blood-borne cells such as polymorphonuclear leukocytes and, later, macrophages, eosinophils, and effector lymphocytes. This influx can last from a few minutes to a few hours and produces an inflammatory response at the site of infection. Thus, mast cells are part of the front-line host defenses against pathogens that enter the body across epithelial barriers.

9-25 IgE-mediated activation of accessory cells has an important role in resistance to parasite infection.

Mast cells are thought to serve at least three important functions in host defense. First, their location near body surfaces allows them to recruit both specific and nonspecific effector elements to sites where infectious agents are most likely to enter the internal milieu. Second, they increase the flow of lymph from sites of antigen deposition to the regional lymph nodes, where naive lymphocytes are first activated. Third, their ability to trigger muscular contraction can contribute to the physical expulsion of pathogens from the lungs or the gut. Mast cells respond rapidly to the binding of antigen to surface-bound IgE antibodies, and their activation leads to the recruitment and activation of basophils and eosinophils, which contribute further to the IgE-mediated response. There is increasing evidence that such IgE-mediated responses are crucial to defense against parasite infestation.

A role for mast cells in the clearance of parasites is suggested by the accumulation of mast cells in the intestine, known as **mastocytosis**, that accompanies helminth infection, and by observations in W/W^V mutant mice, which have a profound mast-cell deficiency caused by mutation of the gene c-*kit*. These mutant mice show impaired clearance of the intestinal nematodes *Trichinella spiralis* and *Strongyloides* species. Clearance of *Strongyloides* is even more impaired in W/W^V mice that lack IL-3 and so also fail to produce basophils. Thus, both mast cells and basophils seem to contribute to defense against these helminth parasites. Other evidence also points to the importance of IgE antibodies and eosinophils in defense against parasites. Infections by certain classes of parasites, particularly helminths, are strongly associated with the production of IgE antibodies and the presence of an abnormally large number of eosinophils (eosinophilia) in blood and tissues. Furthermore, experiments in mice show that depletion of eosinophils by polyclonal anti-eosinophil antisera increases the severity of infection with the parasitic helminth *Schistosoma mansoni*. Eosinophils seem to be directly responsible for helminth destruction; examination of infected tissues shows degranulated eosinophils adhering to helminths, and experiments *in vitro* have shown that eosinophils can kill *S. mansoni* in the presence of specific IgE, IgG, or IgA anti-schistosome antibodies (see Fig. 9.31).

The role of IgE, mast cells, basophils, and eosinophils can also be seen in resistance to the feeding of blood-sucking ixodid ticks. Skin at the site of a tick bite shows degranulated mast cells and an accumulation of basophils and eosinophils that are degranulated, an indicator of recent activation. Resistance to subsequent feeding by these ticks develops after the first exposure, suggesting a specific immunological mechanism. Mice deficient in mast cells show no such acquired resistance to ticks, and in guinea pigs the depletion of either basophils or eosinophils by specific polyclonal antibodies also reduces resistance to tick feeding. Finally, recent experiments have shown that resistance to ticks in mice is mediated by specific IgE antibody.

Thus, many clinical studies and experiments support a role for this system of IgE binding to the high-affinity FcεRI in host resistance to pathogens that enter across epithelia. We will see in Chapter 13 that the same system accounts for many of the symptoms in allergic diseases such as asthma and hay fever, and in the life-threatening response known as systemic anaphylaxis.

Summary.

Antibody-coated pathogens are recognized by effector cells through Fc receptors that bind to an array of constant regions (Fc portions) provided by the pathogen-bound antibodies. Binding activates the cell and triggers destruction of the pathogen, through either phagocytosis or granule release, or through both. Fc receptors comprise a family of proteins, each of which recognizes immunoglobulins of particular isotypes. Fc receptors on macrophages and neutrophils recognize the constant regions of IgG or IgA antibodies bound to a pathogen and trigger the engulfment and destruction of bacteria coated with IgG or IgA. Binding to the Fc receptor also induces the production of microbicidal agents in the intracellular vesicles of the phagocyte. Eosinophils are important in the elimination of parasites too large to be engulfed: they bear Fc receptors specific for the constant region of IgG, as well as high-affinity receptors for IgE; aggregation of these receptors triggers the release of toxic substances onto the surface of the parasite. NK cells, tissue mast cells, and blood basophils also release their granule contents when their Fc receptors are engaged. The high-affinity receptor for IgE is expressed constitutively by mast cells and basophils, and is induced in activated eosinophils. It differs from other Fc receptors in that it can bind free monomeric antibody, thus enabling an immediate response to pathogens at their site of first entry into the tissues. When IgE bound to the surface of a mast cell is aggregated by binding to antigen, it triggers the release of histamine and many other mediators that increase the blood flow to sites of infection; it thereby recruits antibodies and effector cells to these sites. Mast cells are found principally below epithelial surfaces of the skin and beneath the basement membrane of the digestive and respiratory tracts. Their activation by innocuous substances is responsible for many of the symptoms of acute allergic reactions, as will be described in Chapter 13.

Summary to Chapter 9.

The humoral immune response to infection involves the production of antibody by plasma cells derived from B lymphocytes, the binding of this antibody to the pathogen, and the elimination of the pathogen by phagocytic cells and molecules of the humoral immune system. The production of antibody usually requires the action of helper T cells specific for a peptide fragment of the antigen recognized by the B cell. The B cell then proliferates and differentiates, first at the T zone–B zone boundary in secondary lymphoid tissues, then at the T zone–red pulp border, and finally in the germinal center, where somatic hypermutation diversifies the B-cell receptors expressed by a clone of B cells. The B cells that bind antigen most avidly are selected for further differentiation by the continual requirement for contact with antigen and the requirement to present antigen-derived peptides to helper T cells in the germinal center. These events allow the affinity of antibodies to increase over the course of an immune response, especially in repeated responses to the same antigen. Helper T cells also direct class switching, leading to the production of antibody of various classes that can be distributed to various body compartments.

IgM is produced naturally by B-1 cells, as well as early in the response of conventional or B-2 cells. IgM has a major role in protecting against infection in the bloodstream, whereas isotypes secreted later, such as IgG, diffuse into the tissues. Certain pathogens that have highly repeating antigenic determinants and express mitogens that intrinsically stimulate B cells can elicit IgM and some IgG independently of T-cell help. Such antigens are called TI antigens, and the antibody elicited by these antigens can provide an early protective immune response. Multimeric IgA is produced in the lamina propria and transported across epithelial surfaces, whereas IgE is made in small amounts and binds avidly to the surface of mast cells. Antibodies that bind with high affinity to critical sites on toxins, viruses, and bacteria can neutralize them. However, pathogens and their products are destroyed and removed from the body largely through uptake into phagocytes and degradation inside these cells. Antibodies that coat pathogens bind to Fc receptors on phagocytes, which are thereby triggered to engulf and destroy the pathogen. Binding of antibody C regions to Fc receptors on other cells leads to the exocytosis of stored mediators; this is particularly important in parasite infections, in which Fcε-expressing mast cells and activated eosinophils are triggered by the binding of antigen to IgE antibody to release inflammatory mediators directly onto parasite surfaces. Antibodies can also initiate the destruction of pathogens by activating the complement system. Complement components can opsonize pathogens for uptake by phagocytes, recruit phagocytes to sites of infection, and directly destroy pathogens by creating pores in their cell membrane. Receptors for complement components and Fc receptors often synergize in activating the uptake and destruction of pathogens and immune complexes. Thus, the humoral immune response is targeted to the infecting pathogen through the production of specific antibody; however, the effector actions of that antibody are determined by its heavy-chain isotype, which determines its class, and are the same for all pathogens bound by antibody of a particular class.

Questions.

9.1 *Describe the requirements for the activation of naive B cells by a thymus-dependent antigen.*

9.2 *Compare and contrast mature B cells, plasmablasts and plasma cells in terms of their proliferation, antibody secretion, life span and location in the body.*

9.3 *Compare and contrast the properties and functions of antibodies of the IgM and IgG classes.*

9.4 *Compare and contrast the B-cell response to the two types of thymus-independent antigen.*

9.5 *Which of the antibody classes mainly activates mast cells? How does it do so and what are the results. Which type of pathogen is this class of antibody mainly directed against? What unwanted reaction is this antibody also responsible for?*

9.6 *Describe two different ways in which antibodies other than IgM could be produced against a polysaccharide antigen.*

9.7 *Describe the process responsible for the phenomenon of affinity maturation of the antibody response. Where does affinity maturation mainly take place?*

9.8 *How do antibodies interact with the complement system to rid the body of pathogens?*

9.9 *Which classes of maternal antibodies would you expect to find in a breast-fed newborn infant and how have they got there?*

General references.

Liu, Y J., Zhang, J., Lane, P.J., Chan, E.Y., and MacLennan, I.C.: **Sites of specific B cell activation in primary and secondary responses to T cell-dependent and T cell-independent antigens.** *Eur. J. Immunol.* 1991, **21:**2951–2962.

Metzger, H. (ed): *Fc Receptors and the Action of Antibodies*, 1st edn. Washington, DC, American Society for Microbiology, 1990.

Rajewsky, K.: **Clonal selection and learning in the antibody system.** *Nature* 1996, **381:**751–758.

Section references.

9-1 The humoral immune response is initiated when B cells that bind antigen are signaled by helper T cells or by certain microbial antigens alone.

Gulbranson-Judge, A., and MacLennan, I.: **Sequential antigen-specific growth of T cells in the T zones and follicles in response to pigeon cytochrome c.** *Eur. J. Immunol.* 1996, **26:**1830–1837.

9-2 B-cell responses to antigen are enhanced by co-ligation of the B-cell co-receptor.

Barrington, R.A., Zhang, M., Zhong, X., Jonsson, H., Holodick, N., Cherukuri, A., Pierce, S.K., Rothstein, T.L., and Carroll, M.C.: **CD21/CD19 coreceptor signaling promotes B cell survival during primary immune responses.** *J. Immunol.* 2005, **175:**2859–2867.

Fearon, D.T., and Carroll, M.C.: **Regulation of B lymphocyte responses to foreign and self-antigens by the CD19/CD21 complex.** *Annu. Rev. Immunol.* 2000, **18:**393–422.

O'Rourke, L., Tooze, R., and Fearon, D.T.: **Co-receptors of B lymphocytes.** *Curr. Opin. Immunol.* 1997, **9:**324–329.

Rickert, R.C.: **Regulation of B lymphocyte activation by complement C3 and the B cell coreceptor complex.** *Curr. Opin. Immunol.* 2005, **17:**237–243.

9-3 Helper T cells activate B cells that recognize the same antigen.

Eskola, J., Peltola, H., Takala, A.K., Kayhty, H., Hakulinen, M., Karanko, V., Kela, E., Rekola, P., Ronnberg, P.R., Samuelson, J.S., et al. **Efficacy of *Haemophilus influenzae* type b polysaccharide-diphtheria toxoid conjugate vaccine in infancy.** *N. Engl. J. Med.* 1987, **317**:717–722.

Lanzavecchia, A.: **Receptor-mediated antigen uptake and its effect on antigen presentation to class II-restricted T lymphocytes.** *Annu. Rev. Immunol.* 1990, **8**:773–793.

MacLennan, I.C.M., Gulbranson-Judge, A., Toellner, K.M., Casamayor-Palleja, M., Chan, E., Sze, D.M.Y., Luther, S.A., and Orbea, H.A.: **The changing preference of T and B cells for partners as T-dependent antibody responses develop.** *Immunol. Rev.* 1997, **156**:53–66.

McHeyzer-Williams, L.J., Malherbe, L.P., and McHeyzer-Williams, M.G.: **Helper T cell-regulated B cell immunity.** *Curr. Top. Microbiol. Immunol.* 2006, **311**:59–83.

Parker, D.C.: **T cell-dependent B-cell activation.** *Annu. Rev. Immunol.* 1993, **11**:331–340.

9-4 Antigenic peptides bound to self-MHC class II molecules trigger helper T cells to make membrane-bound and secreted molecules that can activate a B cell.

Jaiswal, A.I., and Croft, M.: **CD40 ligand induction on T cell subsets by peptide-presenting B cells.** *J. Immunol.* 1997, **159**:2282–2291.

Kalled, S.L.: **Impact of the BAFF/BR3 axis on B cell survival, germinal center maintenance and antibody production.** *Semin. Immunol.* 2006, **18**:290–296.

Lane, P., Traunecker, A., Hubele, S., Inui, S., Lanzavecchia, A., and Gray, D.: **Activated human T cells express a ligand for the human B cell-associated antigen CD40 which participates in T cell-dependent activation of B lymphocytes.** *Eur. J. Immunol.* 1992, **22**:2573–2578.

Mackay, F., and Browning, J.L.: **BAFF: a fundamental survival factor for B cells.** *Nat. Rev. Immunol.* 2002, **2**:465–475.

Noelle, R.J., Roy, M., Shepherd, D.M., Stamenkovic, I., Ledbetter, J.A., and Aruffo, A.: **A 39-kDa protein on activated helper T cells binds CD40 and transduces the signal for cognate activation of B cells.** *Proc. Natl Acad. Sci. USA* 1992, **89**:6550–6554.

Shanebeck, K.D., Maliszewski, C.R., Kennedy, M.K., Picha, K.S., Smith, C.A., Goodwin, R.G., and Grabstein, K.H.: **Regulation of murine B cell growth and differentiation by CD30 ligand.** *Eur. J. Immunol.* 1995, **25**:2147–2153.

Sharpe, A.H., and Freeman, G.J.: **The B7-CD28 superfamily.** *Nat. Rev. Immunol.* 2002, **2**:116–126.

Valle, A., Zuber, C.E., Defrance, T., Djossou, O., De, R.M., and Banchereau, J.: **Activation of human B lymphocytes through CD40 and interleukin 4.** *Eur. J. Immunol.* 1989, **19**:1463–1467.

Yoshinaga, S.K., Whoriskey, J.S., Khare, S.D., Sarmiento, U., Guo, J., Horan, T., Shih, G., Zhang, M., Coccia, M.A., Kohno, T. *et al.*: **T-cell co-stimulation through B7RP-1 and ICOS.** *Nature* 1999, **402**:827–832.

9-5 B cells that have bound antigen via their B-cell receptor are trapped in the T-cell zone of secondary lymphoid tissues.

Cahalan, M.D., and Parker, I.: **Close encounters of the first and second kind: T-DC and T-B interactions in the lymph node.** *Semin. Immunol.* 2005, **17**:442–451.

Garside, P., Ingulli, E., Merica, R.R., Johnson, J.G., Noelle, R.J., and Jenkins, M.K.: **Visualization of specific B and T lymphocyte interactions in the lymph node.** *Science* 1998, **281**:96–99.

Jacob, J., Kassir, R., and Kelsoe, G.: **In situ studies of the primary immune response to (4-hydroxy-3-nitrophenyl)acetyl. I. The architecture and dynamics of responding cell population.** *J. Exp. Med.* 1991, **173**:1165–1175.

Okada, T., and Cyster, J.G.: **B cell migration and interactions in the early phase of antibody responses.** *Curr. Opin. Immunol.* 2006, **18**:278–285.

Pape, K.A., Kouskoff, V., Nemazee, D., Tang, H.L., Cyster, J.G., Tze, L.E., Hippen, K.L., Behrens, T.W., and Jenkins, M.K.: **Visualization of the genesis and fate of isotype-switched B cells during a primary immune response.** *J. Exp. Med.* 2003, **197**:1677–1687.

9-6 Antibody-secreting plasma cells differentiate from activated B cells.

Moser, K., Tokoyoda, K., Radbruch, A., MacLennan, I., Manz, R.A.: **Stromal niches, plasma cell differentiation and survival.** *Curr. Opin. Immunol.* 2006, **18**:265–270.

Radbruch, A., Muehlinghaus, G., Luger, E.O., Inamine, A., Smith, K.G., Dorner, T., Hiepe, F.: **Competence and competition: the challenge of becoming a long-lived plasma cell.** *Nat. Rev. Immunol.* 2006, **6**:741–750.

Sciammas, R., and Davis, M.M.: **Blimp-1; immunoglobulin secretion and the switch to plasma cells.** *Curr. Top. Microbiol. Immunol.* 2005, **290**:201–224.

Shapiro-Shelef M, Calame K.: **Regulation of plasma-cell development.** *Nat. Rev. Immunol.* 2005, **5**:230–242.

9-7 The second phase of the primary B-cell immune response occurs when activated B cells migrate to follicles and proliferate to form germinal centers.

Brachtel, E.F., Washiyama, M., Johnson, G.D., Tenner-Racz, K., Racz, P., and MacLennan, I.C.: **Differences in the germinal centres of palatine tonsils and lymph nodes.** *Scand. J. Immunol.* 1996, **43**:239–247.

Camacho, S.A., Kosco-Vilbois, M.H., and Berek, C.: **The dynamic structure of the germinal center.** *Immunol. Today* 1998, **19**:511–514.

Cozine, C.L., Wolniak, K.L., and Waldschmidt, T.J..: **The primary germinal center response in mice.** *Curr. Opin. Immunol.* 2005, **17**:298–302.

Jacob, J., and Kelsoe, G.: **In situ studies of the primary immune response to (4-hydroxy-3-nitrophenyl)acetyl. II. A common clonal origin for periarteriolar lymphoid sheath-associated foci and germinal centers.** *J. Exp. Med.* 1992, **176**:679–687.

Jacob, J., Przylepa, J., Miller, C., and Kelsoe, G.: **In situ studies of the primary immune response to (4-hydroxy-3-nitrophenyl)acetyl. III. The kinetics of V region mutation and selection in germinal center B cells.** *J. Exp. Med.* 1993, **178**:1293–1307.

Kelsoe, G.: **The germinal center: a crucible for lymphocyte selection.** *Semin. Immunol.* 1996, **8**:179–184.

MacLennan, I.C.: **Germinal centers still hold secrets.** *Immunity* 2005, **22**:656–657.

MacLennan, I.C.M.: **Germinal centers.** *Annu. Rev. Immunol.* 1994, **12**:117–139.

9-8 Germinal center B cells undergo V-region somatic hypermutation, and cells with mutations that improve affinity for antigen are selected.

Clarke, S.H., Huppi, K., Ruezinsky, D., Staudt, L., Gerhard, W., and Weigert, M.: **Inter- and intraclonal diversity in the antibody response to influenza hemagglutinin.** *J. Exp. Med.* 1985, **161**:687–704.

Jacob, J., Kelsoe, G., Rajewsky, K., and Weiss, U.: **Intraclonal generation of antibody mutants in germinal centres.** *Nature* 1991, **354**:389–392.

Li, Z., Woo, C.J., Iglesias-Ussel, M.D., Ronai, D., and Scharff, M.D.: **The generation of antibody diversity through somatic hypermutation and class switch recombination.** *Genes Dev.* 2004, **18**:1–11.

Odegard, V.H., and Schatz, D.G.: **Targeting of somatic hypermutation.** *Nat. Rev. Immunol.* 2006, **6**:573–583.

Shlomchik, M.J., Litwin, S., and Weigert, M.: **The influence of somatic mutation on clonal expansion.** *Prog. Immunol. Proc. 7th Int. Cong. Immunol.* 1990, **7**:415–423.

Ziegner, M., Steinhauser, G., and Berek, C.: **Development of antibody diversity in single germinal centers: selective expansion of high-affinity variants.** *Eur. J. Immunol.* 1994, **24**:2393–2400.

9-9 Class switching in thymus-dependent antibody responses requires expression of CD40 ligand by the helper T cell and is directed by cytokines.

Francke, U., and Ochs, H.D.: **The CD40 ligand, gp39, is defective in activated T cells from patients with X-linked hyper-IgM syndrome.** *Cell* 1993, **72**:291–300.

Jumper, M., Splawski, J., Lipsky, P., and Meek, K.: **Ligation of CD40 induces sterile transcripts of multiple Ig H chain isotypes in human B cells.** *J. Immunol.* 1994, **152**:438–445.

Litinskiy, M.B., Nardelli, B., Hilbert, D.M., He, B., Schaffer, A., Casali, P., and Cerutti, A.: **DCs induce CD40-independent immunoglobulin class switching through BLyS and APRIL.** *Nat. Immunol.* 2002, **3**:822–829.

MacLennan, I.C., Toellner, K.M., Cunningham, A.F., Serre, K., Sze, D.M., Zuniga, E., Cook, M.C., and Vinuesa, C.G.: **Extrafollicular antibody responses.** *Immunol. Rev.* 2003, **194**:8–18.

Snapper, C.M., Kehry, M.R., Castle, B.E., and Mond, J.J.: **Multivalent, but not divalent, antigen receptor cross-linkers synergize with CD40 ligand for induction of Ig synthesis and class switching in normal murine B cells.** *J. Immunol.* 1995, **154:**1177–1187.

Stavnezer, J.: **Immunoglobulin class switching.** *Curr. Opin. Immunol.* 1996, **8:**199–205.

9-10 Ligation of the B-cell receptor and CD40, together with direct contact with T cells, are all required to sustain germinal center B cells.

Han, S., Hathcock, K., Zheng, B., Kepler, T.B., Hodes, R., and Kelsoe, G.: **Cellular interaction in germinal centers. Roles of CD40 ligand and B7-2 in established germinal centers.** *J. Immunol.* 1995, **155:**556–567.

Hannum, L.G., Haberman, A.M., Anderson, S.M., and Shlomchik, M.J.: **Germinal center initiation, variable gene region hypermutation, and mutant B cell selection without detectable immune complexes on follicular dendritic cells.** *J. Exp. Med.* 2000, **192:**931–942.

Humphrey, J.H., Grennan, D., and Sundaram, V.: **The origin of follicular dendritic cells in the mouse and the mechanism of trapping of immune complexes on them.** *Eur. J. Immunol.* 1984, **14:**859–864.

Liu, Y.J., Joshua, D.E., Williams, G.T., Smith, C.A., Gordon, J., and MacLennan, I.C.M.: **Mechanism of antigen-driven selection in germinal centres.** *Nature* 1989, **342:**929–931.

Wang, Z., Karras, J.G., Howard, R.G., and Rothstein, T.L.: **Induction of bcl-x by CD40 engagement rescues sIg-induced apoptosis in murine B cells.** *J. Immunol.* 1995, **155:**3722–3725.

9-11 Surviving germinal center B cells differentiate into either plasma cells or memory cells.

Coico, R.F., Bhogal, B.S., and Thorbecke, G.J.: **Relationship of germinal centers in lymphoid tissue to immunologic memory. IV. Transfer of B cell memory with lymph node cells fractionated according to their receptors for peanut agglutinin.** *J. Immunol.* 1983, **131:**2254–2257.

Koni, P.A., Sacca, R., Lawton, P., Browning, J.L., Ruddle, N.H., and Flavell, R.A.: **Distinct roles in lymphoid organogenesis for lymphotoxins α and β revealed in lymphotoxin β-deficient mice.** *Immunity* 1997, **6:**491–500.

Matsumoto, M., Lo, S.F., Carruthers, C.J.L., Min, J., Mariathasan, S., Huang, G., Plas, D.R., Martin, S.M., Geha, R.S., Nahm, M.H., and Chaplin, D.D.: **Affinity maturation without germinal centres in lymphotoxin-α-deficient mice.** *Nature* 1996, **382:**462–466.

Omori, S.A., Cato, M.H., Anzelon-Mills, A., Puri, K.D., Shapiro-Shelef, M., Calame, K., and Rickert, R.C.: **Regulation of class-switch recombination and plasma cell differentiation by phosphatidylinositol 3-kinase signaling.** *Immunity*, 2006, **25:**545–557.

Radbruch, A., Muehlinghaus, G., Luger, E.O., Inamine, A., Smith, K.G., Dorner, T., and and Hiepe, F.: **Competence and competition: the challenge of becoming a long-lived plasma cell.** *Nat. Rev. Immunol.* 2006, **6**:741–750.

Schebesta, M., Heavey, B., and Busslinger, M.: **Transcriptional control of B-cell development.** *Curr. Opin. Immunol.* 2002, **14:**216–223.

Tew, J.G., DiLosa, R.M., Burton, G.F., Kosco, M.H., Kupp, L.I., Masuda, A., and Szakal, A.K.: **Germinal centers and antibody production in bone marrow.** *Immunol. Rev.* 1992, **126:**99–112.

9-12 B-cell responses to bacterial antigens with intrinsic ability to activate B cells do not require T-cell help.

Anderson J., Coutinho, A., Lernhardt, W., and Melchers, F.: **Clonal growth and maturation to immunoglobulin secretion in vitro of every growth-inducible B lymphocyte.** *Cell* 1977, **10:**27–34.

Dubois, B., Vanbervliet, B., Fayette, J., Massacrier, C., Kooten, C.V., Briere, F., Banchereau, J., and Caux, C.: **Dendritic cells enhance growth and differentiation of CD40-activated B lymphocytes.** *J. Exp. Med.* 1997, **185:**941–952.

Garcia De Vinuesa, C., Gulbranson-Judge, A., Khan, M., O'Leary, P., Cascalho, M., Wabl, M., Klaus, G.G., Owen, M.J., and MacLennan, I.C.: **Dendritic cells associated with plasmablast survival.** *Eur. J. Immunol.* 1999, **29:**3712–3721.

9-13 B-cell responses to bacterial polysaccharides do not require peptide-specific T-cell help.

Balazs, M., Martin, F., Zhou, T., and Kearney, J.: **Blood dendritic cells interact with splenic marginal zone B cells to initiate T-independent immune responses.** *Immunity* 2002, **17:**341–352.

Craxton, A., Magaletti, D., Ryan, E.J., and Clark, E.A.: **Macrophage- and dendritic cell-dependent regulation of human B-cell proliferation requires the TNF family ligand BAFF.** *Blood* 2003, **101:**4464–4471.

Fagarasan, S., and Honjo, T.: **T-independent immune response: new aspects of B cell biology.** *Science* 2000, **290:**89–92.

MacLennan, I., and Vinuesa, C.: **Dendritic cells, BAFF, and APRIL: innate players in adaptive antibody responses.** *Immunity* 2002, **17:**341–352.

Mond, J.J., Lees, A., and Snapper, C.M.: **T cell-independent antigens type 2.** *Annu. Rev. Immunol.* 1995, **13:**655–692.

Snapper, C.M.: **Differential regulation of protein- and polysaccharide-specific Ig isotype production in vivo in response to intact Streptococcus pneumoniae.** *Curr. Protein Pept. Sci.* 2006, **7**:295–305.

Snapper, C.M., Shen, Y., Khan, A.Q,. Colino, J., Zelazowski, P., Mond, J.J., Gause, W.C., and Wu, Z.Q.: **Distinct types of T-cell help for the induction of a humoral immune response to Streptococcus pneumoniae.** *Trends Immunol.* 2001, **22**:308–311.

9-14 Antibodies of different isotypes operate in distinct places and have distinct effector functions.

Cebra, J.J.: **Influences of microbiota on intestinal immune system development.** *Am. J. Clin. Nutr.* 1999, **69:**1046S-1051S.

Clark, M.R.: **IgG effector mechanisms.** *Chem. Immunol.* 1997, **65:**88–110.

Herrod, H.G.: **IgG subclass deficiency.** *Allergy Proc.* 1992, **13:**299–302.

Janeway, C.A., Rosen, F.S., Merler, E., and Alper, C.A.: *The Gamma Globulins*, 2nd edn. Boston, Little Brown and Co., 1967.

Suzuki, K., Meek, B., Doi, Y., Muramatsu, M., Chiba, T., Honjo, T., and Fagarasan, S.: **Aberrant expansion of segmented filamentous bacteria in IgA-deficient gut.** *Proc. Natl Acad. Sci. USA* 2004, **101:**1981–1986.

Ward, E.S., and Ghetie, V.: **The effector functions of immunoglobulins: implications for therapy.** *Ther. Immunol.* 1995, **2:**77–94.

9-15 Transport proteins that bind to the Fc regions of antibodies carry particular isotypes across epithelial barriers.

Burmeister, W.P., Gastinel, L.N., Simister, N.E., Blum, M.L., and Bjorkman, P.J.: **Crystal structure at 2.2 Å resolution of the MHC-related neonatal Fc receptor.** *Nature* 1994, **372:**336–343.

Corthesy, B., and Kraehenbuhl, J.P.: **Antibody-mediated protection of mucosal surfaces.** *Curr. Top. Microbiol. Immunol.* 1999, **236:**93–111.

Ghetie, V., and Ward, E.S.: **Multiple roles for the major histocompatibility complex class I-related receptor FcRn.** *Annu. Rev. Immunol.* 2000, **18:**739–766.

Lamm, M.E.: **Current concepts in mucosal immunity. IV. How epithelial transport of IgA antibodies relates to host defense.** *Am. J. Physiol.* 1998, **274:**G614–G617.

Mostov, K.E.: **Transepithelial transport of immunoglobulins.** *Annu. Rev. Immunol.* 1994, **12:**63–84.

Simister, N.E., and Mostov, K.E.: **An Fc receptor structurally related to MHC class I antigens.** *Nature* 1989, **337:**184–187.

9-16 High-affinity IgG and IgA antibodies can neutralize bacterial toxins.
&
9-17 High-affinity IgG and IgA antibodies can inhibit the infectivity of viruses.

Brandtzaeg, P.: **Role of secretory antibodies in the defence against infections.** *Int. J. Med. Microbiol.* 2003, **293**:3–15.

Mandel, B.: **Neutralization of polio virus: a hypothesis to explain the mechanism and the one hit character of the neutralization reaction.** *Virology* 1976, **69:**500–510.

Possee, R.D., Schild, G.C., and Dimmock, N.J.: **Studies on the mechanism of neutralization of influenza virus by antibody: evidence that neutralizing antibody (anti-haemagglutinin) inactivates influenza virus in vivo by inhibiting virion transcriptase activity.** *J. Gen. Virol.* 1982, **58:**373–386.

Roost, H.P., Bachmann, M.F., Haag, A., Kalinke, U., Pliska, V., Hengartner, H., and Zinkernagel, R.M.: **Early high-affinity neutralizing anti-viral IgG responses without further overall improvements of affinity.** *Proc. Natl Acad. Sci. USA* 1995, **92:**1257–1261.

Sougioultzis, S., Kyne, L., Drudy, D., Keates, S., Maroo, S., Pothoulakis, C., Giannasca, P.J., Lee, C.K., Warny, M., Monath, T.P., and Kelly, C.P.: **Clostridium difficile toxoid vaccine in recurrent C. difficile-associated diarrhea.** *Gastroenterology* 2005, **128**:764–770.

9-18 Antibodies can block the adherence of bacteria to host cells.

Fischetti, V.A., and Bessen, D.: **Effect of mucosal antibodies to M protein in colonization by group A streptococci,** in Switalski, L., Hook, M., and Beachery, E. (eds): *Molecular Mechanisms of Microbial Adhesion.* New York, Springer, 1989.

Wizemann, T.M., Adamou, J.E., and Langermann, S.: **Adhesins as targets for vaccine development.** *Emerg. Infect. Dis.* 1999, **5**:395–403.

9-19 Antibody:antigen complexes activate the classical pathway of complement by binding to C1q.

Cooper, N.R.: **The classical complement pathway. Activation and regulation of the first complement component.** *Adv. Immunol.* 1985, **37:**151–216.

Perkins, S.J., and Nealis, A.S.: **The quaternary structure in solution of human complement subcomponent $C1r_2C1s_2$.** *Biochem. J.* 1989, **263:**463–469.

9-20 Complement receptors are important in the removal of immune complexes from the circulation.

Nash, J.T., Taylor, P.R., Botto, M., Norsworthy, P.J., Davies, K.A., and Walport, M.J.: **Immune complex processing in C1q-deficient mice.** *Clin. Exp. Immunol.* 2001, **123:**196–202.

Nash, J.T., Taylor, P.R., Botto, M., Norsworthy, P.J., Davies, K.A., Walport, M.J., Schifferli, J.A., and Taylor, J.P.: **Physiologic and pathologic aspects of circulating immune complexes.** *Kidney Int.* 1989, **35:**993–1003.

Schifferli, J.A., Ng, Y.C., and Peters, D.K.: **The role of complement and its receptor in the elimination of immune complexes.** *N. Engl. J. Med.* 1986, **315:**488–495.

Walport, M.J., Davies, K.A., and Botto, M.: **C1q and systemic lupus erythematosus.** *Immunobiology* 1998, **199:**265–285.

9-21 The Fc receptors of accessory cells are signaling receptors specific for immunoglobulins of different isotypes.

Kinet, J.P., and Launay, P.: **Fc α/microR: single member or first born in the family?** *Nat. Immunol.* 2000, **1:**371–372.

Ravetch, J.V., and Bolland, S.: **IgG Fc receptors.** *Annu. Rev. Immunol.* 2001, **19:**275–290.

Ravetch, J.V., and Clynes, R.A.: **Divergent roles for Fc receptors and complement *in vivo.*** *Annu. Rev. Immunol.* 1998, **16:**421–432.

Shibuya, A., Sakamoto, N., Shimizu, Y., Shibuya, K., Osawa, M., Hiroyama, T., Eyre, H.J., Sutherland, G.R., Endo, Y., Fujita, T. *et al*: **Fc α/μ receptor mediates endocytosis of IgM-coated microbes.** *Nat. Immunol.* 2000, **1:**441–446.

Stefanescu, R.N., Olferiev M., Liu, Y., and Pricop, L.: **Inhibitory Fc gamma receptors: from gene to disease.** *J. Clin. Immunol.* 2004, **24**:315–326.

9-22 Fc receptors on phagocytes are activated by antibodies bound to the surface of pathogens and enable the phagocytes to ingest and destroy pathogens.

Gounni, A.S., Lamkhioued, B., Ochiai, K., Tanaka, Y., Delaporte, E., Capron, A., Kinet, J.P., and Capron, M.: **High-affinity IgE receptor on eosinophils is involved in defence against parasites.** *Nature* 1994, **367:**183–186.

Karakawa, W.W., Sutton, A., Schneerson, R., Karpas, A., and Vann, W.F.: **Capsular antibodies induce type-specific phagocytosis of capsulated *Staphylococcus aureus* by human polymorphonuclear leukocytes.** *Infect. Immun.* 1986, **56:**1090–1095.

9-23 Fc receptors activate NK cells to destroy antibody-coated targets.

Lanier, L.L., and Phillips, J.H.: **Evidence for three types of human cytotoxic lymphocyte.** *Immunol. Today* 1986, **7:**132.

Leibson, P.J.: **Signal transduction during natural killer cell activation: inside the mind of a killer.** *Immunity* 1997, **6:**655–661.

Sulica, A., Morel, P., Metes, D., and Herberman, R.B.: **Ig-binding receptors on human NK cells as effector and regulatory surface molecules.** *Int. Rev. Immunol.* 2001, **20**:371–414.

Takai, T.: **Multiple loss of effector cell functions in FcR γ-deficient mice.** *Int. Rev. Immunol.* 1996, **13:**369–381.

9-24 Mast cells, basophils, and activated eosinophils bind IgE antibody via the high-affinity Fcε receptor.

Beaven, M.A., and Metzger, H.: **Signal transduction by Fc receptors: the FcεRI case.** *Immunol. Today* 1993, **14:**222–226.

Kalesnikoff, J., Huber, M., Lam, V., Damen, J.E., Zhang, J., Siraganian, R.P., and Krystal, G.: **Monomeric IgE stimulates signaling pathways in mast cells that lead to cytokine production and cell survival.** *Immunity* 2001, **14:**801–811.

Sutton, B.J., and Gould, H.J.: **The human IgE network.** *Nature* 1993, **366:**421–428.

9-25 IgE-mediated activation of accessory cells has an important role in resistance to parasite infection.

Capron, A., and Dessaint, J.P.: **Immunologic aspects of schistosomiasis.** *Annu. Rev. Med.* 1992, **43:**209–218.

Capron, A., Riveau, G., Capron, M., and Trottein, F.: **Schistosomes: the road from host-parasite interactions to vaccines in clinical trials.** *Trends Parasitol.* 2005, **21**:143–149.

Grencis, R.K.: **Th2-mediated host protective immunity to intestinal nematode infections.** *Philos. Trans. R. Soc. Lond. B Biol. Sci.* 1997, **352**:1377–1384.

Grencis, R.K., Else, K.J., Huntley, J.F., and Nishikawa, S.I.: **The in vivo role of stem cell factor (c-kit ligand) on mastocytosis and host protective immunity to the intestinal nematode *Trichinella spiralis* in mice.** *Parasite Immunol.* 1993, **15:**55–59.

Kasugai, T., Tei, H., Okada, M., Hirota, S., Morimoto, M., Yamada, M., Nakama, A., Arizono, N., and Kitamura, Y.: **Infection with *Nippostrongylus brasiliensis* induces invasion of mast cell precursors from peripheral blood to small intestine.** *Blood* 1995, **85:**1334–1340.

Ushio, H., Watanabe, N., Kiso, Y., Higuchi, S., and Matsuda, H.: **Protective immunity and mast cell and eosinophil responses in mice infested with larval *Haemaphysalis longicornis* ticks.** *Parasite Immunol.* 1993, **15:**209–214.

Dynamics of Adaptive Immunity

10

Throughout this book we have examined the separate ways in which the innate and the adaptive immune responses protect the individual from invading microorganisms. In this chapter we consider how the cells and molecules of the immune system work as an integrated defense system to eliminate or control an infectious agent and how the adaptive immune system provides long-lasting protective immunity. This is the first of several chapters that consider how the immune system functions as a whole in health and disease. The next chapter describes the role and specializations of the mucosal immune system, which forms the front-line defense against most pathogens. Subsequent chapters examine how immune defenses can fail or unwanted immune responses occur, and how the immune response can be manipulated to benefit the individual.

In Chapter 2, we saw how innate immunity is brought into play in the earliest phases of an infection. Pathogens, however, have developed strategies that allow them on occasion to elude or overcome innate immune defenses and to establish a focus of infection from which they can spread. In these circumstances, the innate immune response sets the scene for the induction of an adaptive immune response. In the **primary immune response**, which occurs to a pathogen encountered for the first time, several days are required for the clonal expansion and differentiation of naive lymphocytes into effector T cells and antibody-secreting B cells, as described in Chapters 8 and 9. In most cases, these cells and antibodies will effectively target the pathogen for elimination (Fig. 10.1).

During this period, specific **immunological memory** is also established. This ensures a rapid reinduction of antigen-specific antibody and effector T cells on subsequent encounters with the same pathogen, thus providing long-lasting and often lifelong protection against it. Immunological memory is discussed in the last part of the chapter. Memory responses differ in several ways from primary responses. We discuss the reasons for this, and what is known about how immunological memory is maintained.

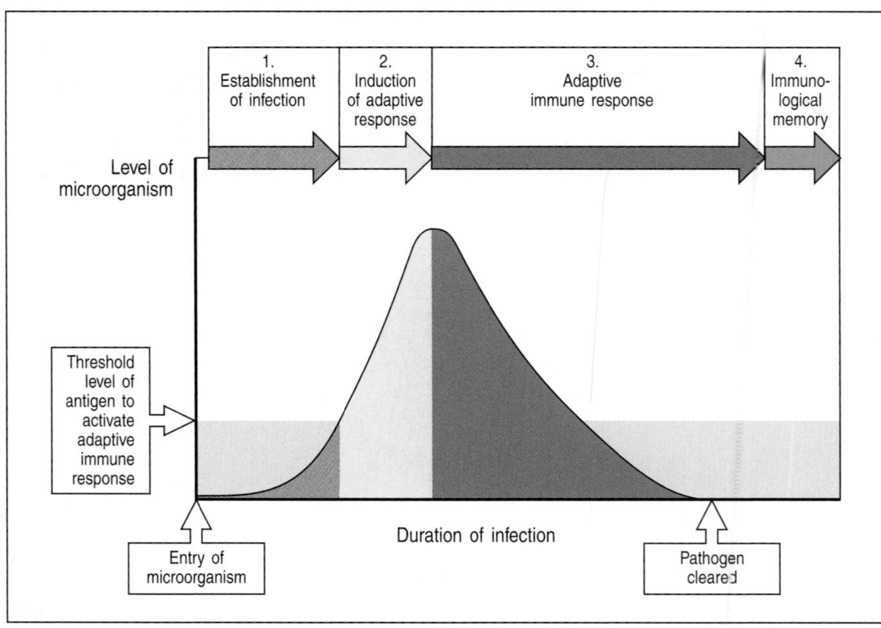

Fig. 10.1 The course of a typical acute infection that is cleared by an adaptive immune reaction. 1. The level of infectious agent increases as the pathogen replicates. 2. When numbers of the pathogen exceed the threshold dose of antigen required for an adaptive response, the response is initiated; the pathogen continues to grow, retarded only by responses of the innate immune system. At this stage, immunological memory also starts to be induced. 3. After 4–7 days, effector cells and molecules of the adaptive response start to clear the infection. 4. When the infection is cleared and the dose of antigen falls below the response threshold, the response ceases, but antibody, residual effector cells, and immunological memory provide lasting protection against reinfection in most cases.

The course of the immune response to infection.

The immune response is a dynamic process, and both its nature and intensity change over time. It starts with the relatively nonspecific responses of innate immunity and becomes both more focused on the pathogen and more powerful as the adaptive immune response is initiated and rapidly develops. In this part of the chapter we discuss how the different phases of an immune response are orchestrated in space and time, how the response develops in both strength and precision, how changes in specialized cell-surface molecules and chemokines guide effector lymphocytes to the appropriate site of action, and how these cells are regulated during the different stages.

An innate immune response is an essential prerequisite to a primary adaptive immune response, because the co-stimulatory molecules induced on cells of the innate immune system during their interaction with microorganisms are essential for the activation of the antigen-specific lymphocytes (see Chapter 8). Cells of the innate immune system hand on other important signals in the form of secreted cytokines that influence the characteristics of the adaptive response and tailor it to the type of pathogen encountered. For this to happen, cells from different locations must engage to coordinate the specific activation of naive T cells and B cells, and the migration of cells to precise locations within lymphoid tissues is thus critical for the coordination of an adaptive response.

10-1 The course of an infection can be divided into several distinct phases.

An infection can be broken down into various stages (see Fig. 2.6), but in Chapter 2 we considered in detail only the responses of innate immunity. Here, we return to the various stages of an infection but will now integrate the adaptive immune response into the picture.

In the first stage of infection with a pathogen, a new host is exposed to infectious particles either shed by an infected individual or present in the environment. The numbers, route, mode of transmission, and stability of an infectious agent outside the host determine its infectivity. Some pathogens, such as the anthrax bacterium, are spread by spores that are highly resistant to heat and drying, whereas others, such as the human immunodeficiency virus (HIV), are spread only by the exchange of bodily fluids or tissues because they are unable to survive as infectious agents outside the body.

The first contact with a new host occurs through an epithelial surface, such as the skin or the mucosal surfaces of the respiratory, gastrointestinal, or urogenital tracts. As most pathogens gain entry to the body through mucosal surfaces, the immune responses that occur in this specialized compartment of the immune system are of great importance and are considered in detail in Chapter 11. After making contact, an infectious agent must establish a focus of infection. It must either adhere to the epithelial surface and colonize it, or penetrate it to replicate in the tissues (Fig. 10.2, first two panels). Wounds and insect and tick bites that breach the epidermal barrier help some pathogens to get through the skin. Many microorganisms are repelled or kept in check at this stage by innate immunity through a variety of germline-encoded receptors that discriminate between foreign microbial and self host-cell surfaces, or between infected and normal cells (see Chapter 2). These responses are not as effective as adaptive immune responses, which can afford to be more powerful because they are antigen specific and thus target the pathogen precisely. However, they can prevent an infection from being established or, failing that, can contain it and prevent the spread of the pathogen into the bloodstream while an adaptive immune response develops.

Only when a microorganism has successfully established a focus of infection in the host does disease occur. With the possible exception of lung infections, in which the primary infection can cause life-threatening disease, little damage will be caused unless the agent spreads from the original focus or secretes toxins that can spread to other parts of the body. Extracellular

Fig. 10.2 Infections and the responses to them can be divided into a series of stages. These are illustrated here for an infectious microorganism (red) entering across a wound in an epithelium. The microorganism must first adhere to epithelial cells and then invade beyond the epithelium. A local nonadaptive response helps to contain the infection and delivers antigen to local lymph nodes, leading to adaptive immunity and clearance of the infection. The role of γ:δ T cells is uncertain, as indicated by the question mark.

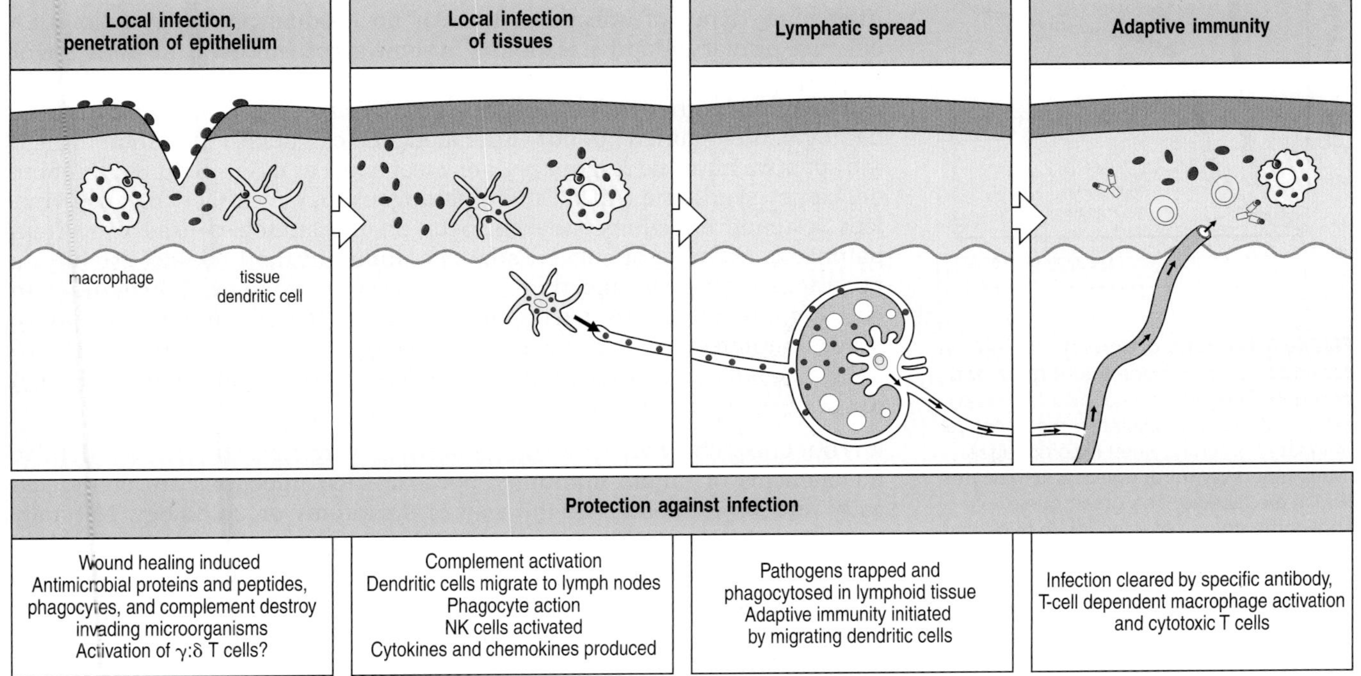
Local infection, penetration of epithelium
Local infection of tissues
Lymphatic spread
Adaptive immunity
macrophage
tissue dendritic cell
Protection against infection
Wound healing induced
Antimicrobial proteins and peptides, phagocytes, and complement destroy invading microorganisms
Activation of γ:δ T cells?
Complement activation
Dendritic cells migrate to lymph nodes
Phagocyte action
NK cells activated
Cytokines and chemokines produced
Pathogens trapped and phagocytosed in lymphoid tissue
Adaptive immunity initiated by migrating dendritic cells
Infection cleared by specific antibody, T-cell dependent macrophage activation and cytotoxic T cells

pathogens spread by direct extension of the infection through the lymphatics or the bloodstream. Usually, spread into the bloodstream occurs only after the lymphatic system has been overwhelmed. Obligate intracellular pathogens spread from cell to cell; they do so either by direct transmission from one cell to the next or by release into the extracellular fluid and reinfection of both adjacent and distant cells. In contrast, some of the bacteria that cause gastroenteritis exert their effects without spreading into the tissues. They establish a site of infection on the luminal surface of the epithelium lining the gut and cause no direct pathology themselves, but secrete toxins that cause damage either *in situ* or after crossing the epithelial barrier and entering the circulation.

Most infectious agents show a significant degree of host specificity, causing disease in only one or a few related species. What determines host specificity for every agent is not known, but the requirement for attachment to a particular cell-surface molecule is one critical factor. As other interactions with host cells are also commonly needed to support replication, most pathogens have a limited host range. The molecular mechanisms of host specificity comprise an area of research known as molecular pathogenesis, which is outside the scope of this book.

Adaptive immunity is triggered when an infection eludes or overwhelms the innate defense mechanisms and generates a threshold level of antigen (see Fig. 10.1). Adaptive immune responses are then initiated in the local lymphoid tissue, in response to antigens presented by dendritic cells activated during the course of the innate immune response (Fig. 10.2, second and third panels). Antigen-specific effector T cells and antibody-secreting B cells are generated by clonal expansion and differentiation over several days, as described in greater detail in Chapters 8 and 9. During this time, the induced responses of innate immunity, such as the acute-phase responses and interferon production (see Sections 2-28 and 2-29), continue to function. Eventually, antigen-specific T cells and then antibodies are released into the blood and recruited to the site of infection (Fig. 10.2, fourth panel). A cure involves the clearance of extracellular infectious particles by antibodies and the clearance of intracellular residues of infection through the actions of effector T cells.

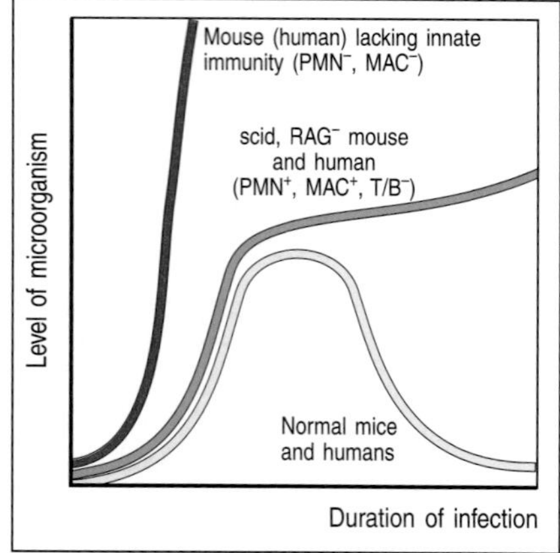

Fig. 10.3 The time course of infection in normal and immunodeficient mice and humans. The red curve shows the rapid growth of microorganisms in the absence of innate immunity, when macrophages (MAC) and polymorphonuclear leukocytes (PMN) are lacking. The green curve shows the course of infection in mice and humans that have innate immunity but have no T or B lymphocytes and so lack adaptive immunity. The yellow curve shows the normal course of an infection in immunocompetent mice or humans.

After many types of infections, little or no residual pathology follows an effective primary adaptive response. In some cases, however, the infection or the response to it causes significant tissue damage. In yet other cases, such as infection with cytomegalovirus or *Mycobacterium tuberculosis*, the pathogen is contained but not eliminated, and can persist in a latent form. If the adaptive immune response is later weakened, as it is in acquired immune deficiency syndrome (AIDS), these pathogens may resurface to cause virulent systemic infections. We will focus on the strategies used by certain pathogens to evade or subvert adaptive immunity, and thereby establish a persistent, or chronic, infection, in the first part of Chapter 12. In addition to clearing the infectious agent, an effective adaptive immune response prevents reinfection. For some infectious agents, this protection is essentially absolute, whereas for others infection is reduced or attenuated upon reexposure to the pathogen.

It is not known how many infections are dealt with solely by the nonadaptive mechanisms of innate immunity, because such infections are eliminated early and produce little in the way of symptoms or pathology. Naturally occurring deficiencies in nonadaptive defenses are rare, so it has seldom been possible to study their consequences. Innate immunity does, however, seem to be essential for effective host defense, as shown by the progression of infection in mice that lack components of innate immunity but have an intact adaptive immune system (Fig. 10.3). Adaptive immunity is also essential, as shown by the immunodeficiency syndromes associated with defects in various components of the adaptive immune response.

10-2 The nonspecific responses of innate immunity are necessary for an adaptive immune response to be initiated.

The establishment of a focus of infection in tissues and the response of the innate immune system produce changes in the immediate environment. Many of these changes have been described in earlier chapters, but we review them briefly here to provide a cohesive framework for the induction of adaptive immunity.

In a bacterial infection, the first thing that usually happens is that the infected tissue becomes inflamed. This is initially the result of the activation of the resident macrophages by bacterial components such as lipopolysaccharide (LPS) acting through Toll-like receptors (TLRs) on the macrophage. The cytokines and chemokines secreted by the activated macrophages, especially the cytokine tumor necrosis factor-α (TNF-α), induce numerous changes in the endothelial cells of nearby blood capillaries, a process known as endothelial cell activation. Inflammation also results from the activation of complement, resulting in the production of the anaphylatoxins C3a and C5a, which are able to activate vascular endothelium. In a primary infection, complement is activated mainly via the alternative and MBL pathways (see Fig. 2.25).

Activation of the vascular endothelium causes the release of Weibel–Palade bodies from within the endothelial cells, which deliver the cell-adhesion molecule P-selectin to the endothelial cell surface (see Section 2-25). Activation also induces the transcription and translation of RNA encoding E-selectin, which then appears on the endothelial cell surface. These two selectins cause leukocytes, including neutrophils and monocytes, to adhere to and roll on the endothelial surface. The cytokines also induce the production of the adhesion molecule ICAM-1 on endothelial cells. By binding to adhesion molecules, such as LFA-1, on neutrophils and monocytes, ICAM-1 strengthens their interaction with endothelial cells and aids their entry in large numbers into the infected tissue to form an inflammatory focus (see Fig. 2.49). As monocytes mature into tissue macrophages and become activated in their turn, additional inflammatory cells are attracted into the infected tissue and the inflammatory response is maintained and reinforced. The inflammatory response can be thought of as putting up a flag on the endothelial cells to signal the presence of infection, but as yet the response is entirely nonspecific for the pathogen's antigens.

A second crucial effect of infection is the activation of specialized antigen-presenting cells, the dendritic cells residing in most tissues, as described in Sections 8-4 to 8-6. Dendritic cells take up antigen in the infected tissues and, like macrophages, they are activated through innate immune receptors that respond to common constituents of pathogens, such as TLRs (Section 2-7) and NOD proteins (Section 2-9). Activated dendritic cells increase their synthesis of MHC class II molecules and, most importantly, begin to express the co-stimulatory molecules B7.1 and B7.2 on their surface. As described in Chapter 8, these antigen-presenting cells migrate away from the infected tissue through the lymphatics, along with their antigen cargo, to enter peripheral lymphoid tissues, where they initiate the adaptive immune response. They arrive in large numbers at the draining lymph nodes, or other nearby lymphoid tissue, attracted by the chemokines CCL19, CCL20, and CCL21, which are produced by the lymph-node stroma and high endothelial venules.

When dendritic cells arrive in the lymphoid tissues, they seem to have reached their final destination. They eventually die in these tissues, but before this their role is to activate antigen-specific naive T lymphocytes. Naive lymphocytes are continually passing through the lymph nodes, which they enter from the blood across the walls of high endothelial venules (see Fig. 8.8). Those naive T cells that recognize antigen on the surface of dendritic cells are activated, and divide and mature into effector cells that reenter the circulation. Where there is a local infection, the changes induced by inflammation in the walls of nearby venules induce these effector T cells to leave the blood vessel and migrate into the site of infection.

Thus, the local release of cytokines and chemokines at the site of infection has far-reaching consequences. As well as recruiting neutrophils and macrophages, which are not specific for antigen, the changes induced in the blood vessel walls also enable newly activated effector T lymphocytes to enter infected tissue.

10-3 Cytokines made in the earliest phase of an infection influence differentiation of CD4 T cells toward the T_H17 subset.

The differentiation of naive CD4 T cells into distinct classes of CD4 effector T cells—T_H17, T_H1, or T_H2, or regulatory subsets (see Chapter 8)—occurs during the progression of an infection, and depends on the effects of the infection on the antigen-presenting cells. The conditions created by dendritic cells during the initial contact of T cells with their antigen has an impact on the outcome of an adaptive immune response, determining the relative amounts of the different types of T cells produced. In turn, the T-cell subsets generated influence the extent of macrophage activation, the extent of neutrophil or eosinophil recruitment to the site of infection, and which classes of antibody will predominate.

The cellular and transcriptional mechanisms that control this decision-making step in CD4 T-cell differentiation have become better defined in recent years and were introduced in Chapter 8. It is clear that cytokines present during the initial phase of CD4 T-cell activation greatly influence their subsequent differentiation.

The first subset of effector T cells to be generated in response to infection is often T_H17. Upon encountering a pathogen, the earliest response of dendritic cells is to synthesize IL-6 along with TGF-β. In the absence of IL-4, IFN-γ, or IL-12, these two cytokines induce naive CD4 T cells to differentiate into T_H17 cells, and not T_H1 or T_H2 cells (Fig. 10.4, right panels). The T_H17 cells leave the lymph node and migrate to distant sites of infection. There, they encounter pathogen antigens and are stimulated to synthesize and release cytokines, which include various members of the IL-17 family such as IL-17A and IL-17E (also known as IL-25). The receptor for IL-17 is expressed ubiquitously on cells such as fibroblasts, epithelial cells, and keratinocytes. IL-17 induces these cells to secrete various cytokines, including IL-6, the chemokines CXCL8 and CXCL2, and the hematopoietic factors granulocyte colony-stimulating factor (G-CSF) and granulocyte-macrophage colony-stimulating factor (GM-CSF). The chemokines can act directly to recruit neutrophils, while the actions of G-CSF and GM-CSF feed back to the bone marrow to augment the production of neutrophils and macrophages. These cytokines may also alter the local differentiation of local monocytes into macrophages, but this has not been confirmed experimentally in the context of T_H17 cells.

Thus, one important action of IL-17 at sites of infection is to induce local cells to secrete cytokines and chemokines that attract neutrophils. T_H17 cells also produce IL-22, a cytokine related to IL-10. IL-22 acts cooperatively with IL-17 to induce the expression of antimicrobial peptides, such as β-defensins, by the keratinocytes of the epidermis. In this way, the presence of pathogen-specific

T_H17 cells serves as an efficient amplifier of an acute inflammatory response by the innate immune system at sites of early infection. CD4 T cells that acquire the T_H17 phenotype are not the only cells that can produce IL-17 in response to infections. CD8 T cells have also been shown to produce abundant IL-17.

The cytokine environment is also influential in preventing the immune system from making inappropriate responses to self antigens or those of commensal organisms, the microorganisms that normally inhabit the body. Even in the absence of infection, dendritic cells take up self and environmental antigens and eventually carry them to secondary lymphoid tissues, where they may meet antigen-specific naive T cells. Regulatory mechanisms prevent the immune system from mounting a damaging adaptive immune response in such circumstances. The usual pro-inflammatory signals are not present and dendritic cells are not activated; instead they seem to actively generate tolerance to the antigens the T cells are encountering (Fig. 10.4, left panels). These dendritic cells produce the cytokine TGF-β, but not the other cytokines that can affect CD4 T-cell differentiation. TGF-β on its own inhibits the proliferation and differentiation of T_H17, T_H1, and T_H2 cells, and when a naive CD4 T cell encounters its cognate peptide:MHC ligand in the presence of TGF-β, it acquires the phenotype of a regulatory T cell, in that it can inhibit the activation of other T cells. Regulatory T cells induced in this way outside the central lymphoid organs are called adaptive regulatory cells and some express the transcription factor FoxP3 (see Section 8-20). The regulatory cells, in theory, should not be specific for pathogen antigens—which they have not yet encountered—but should rather be specific for either self antigens or peptides from commensal organisms. Other FoxP3-expressing regulatory CD4 T cells seem to acquire their regulatory phenotype in the thymus, and these are often called natural regulatory T cells (see Section 7-18).

The reciprocal pathways for the development of T_H17 cells and regulatory T cells seem to be based on an evolutionarily ancient system of activation and inactivation, because proteins similar to TGF-β and IL-17 are present in invertebrates that possess primitive intestinal immune systems. This might suggest that the dichotomy between T_H17 cells and regulatory T cells is largely concerned with maintaining the lymphocyte balance in tissues exposed to large numbers of potential pathogens, such as the mucosae of the gut and the lungs, where a rapid response to infection is critical. For example, IL-17-producing T cells play an important role in mice in resistance to infections of the lung by Gram-negative bacteria such as *Klebsiella pneumoniae.* Mice lacking the receptor for IL-17 are significantly more susceptible to lung infection by this pathogen than are normal mice, and they show decreased production of G-CSF and CXCL2 and poorer recruitment of neutrophils to infected lungs. T_H17 cells also promote resistance to the gut nematode *Nippostrongylus brasiliensis*. This effect seems to be due to the induction or recruitment by IL-17E of a population of non-T non-B leukocytes, perhaps similar to basophils, that secrete the T_H2 cytokines IL-4, IL-5, and IL-13. These cytokines, particularly IL-13, promote resistance to *N. brasiliensis* by, for example, inducing its expulsion from the gut by augmenting the production of mucus (see Chapter 11).

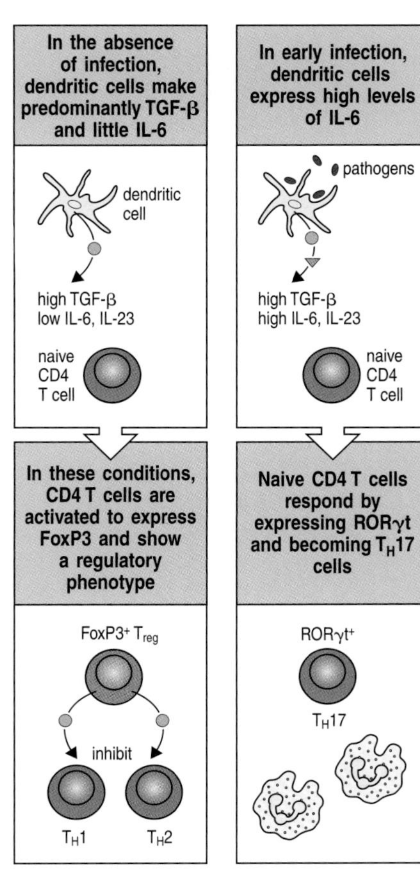

Fig. 10.4 In early infection, differentiation of naive CD4 T cells shifts from a regulatory to a T_H17 program. The balance between TGF-β and IL-6 production acts to induce either the transcription factor FoxP3, which is characteristic of regulatory T cells, or RORγt (an 'orphan' member of the nuclear receptor family), which is characteristic of T_H17 cells. In the absence of infection, TGF-β production by dendritic cells dominates, and IL-6 production is low. In these conditions, T cells that do encounter their cognate antigen will be induced to express FoxP3 and predominantly acquire a regulatory phenotype, while those that do not encounter antigen remain naive. During early infection, dendritic cells rapidly start to produce IL-6, before the production of other cytokines such as IL-12; under these conditions, naive T cells start to express RORγt and become T_H17 cells. The cytokines produced by this T-cell subset, IL-17 and IL-17F, induce cells such as epithelium to secrete chemokines that attract inflammatory cells such as neutrophils.

10-4 Cytokines made in the later stages of an infection influence differentiation of CD4 T cells toward T_H1 or T_H2 cells.

T_H1 and T_H2 cells were the first of the CD4 effector subsets to be identified and analyzed; however, as we have just seen, they are not the first to be generated in response to pathogens. Highly polarized T_H1 or T_H2 responses typically arise during prolonged or chronic infections, when the complete removal of the pathogen requires the specialized effector activities of these T-cell subsets. As the immune response progresses, the production of TGF-β and IL-6 by dendritic cells seems to decline and cytokines that induce naive T cells to commit to either T_H1 or T_H2 predominate.

T_H1 responses tend to be induced by viruses and by bacterial and protozoan pathogens that can survive inside macrophage intracellular vesicles. In the case of viruses, the T_H1 response is generally involved in helping to activate the CD8 cytotoxic T cells that will recognize virus-infected cells and destroy them (see Chapter 8). T_H1 cells can also induce the production of some subclasses of IgG antibodies, which can neutralize virus particles in the blood and extracellular fluid. In the case of mycobacteria, and of protozoa such as *Leishmania* and *Toxoplasma*, the role of T_H1 cells is to activate macrophages to a degree that will destroy the invaders. Experiments *in vitro* have shown that naive CD4 T cells initially stimulated in the presence of IL-12 and IFN-γ tend to develop into T_H1 cells (Fig. 10.5, left panels). In part, this is because these cytokines induce or activate the transcription factors leading to T_H1 development, and in part because IFN-γ inhibits the proliferation of T_H2 cells, as described in Chapter 8. NK cells and CD8 cells are also both activated in response to infections with viruses and some other intracellular pathogens, as discussed in Chapters 2 and 8, and both produce abundant IFN-γ. Dendritic cells and macrophages produce IL-12. Thus, CD4 T-cell responses in these infections tend eventually to be dominated by T_H1 cells.

Signals that stimulate dendritic cells to release IL-12 include the chemokines CCL3, CCL4, and CCL5. These are produced by many activated cell types, including macrophages, dendritic cells themselves, and endothelial cells. These chemokines bind to the receptors CCR5 and CCR1 on dendritic cells, promoting the production of IL-12, and attract T_H1 cells, which also bear these receptors. Production of IL-12 by dendritic cells is also stimulated by

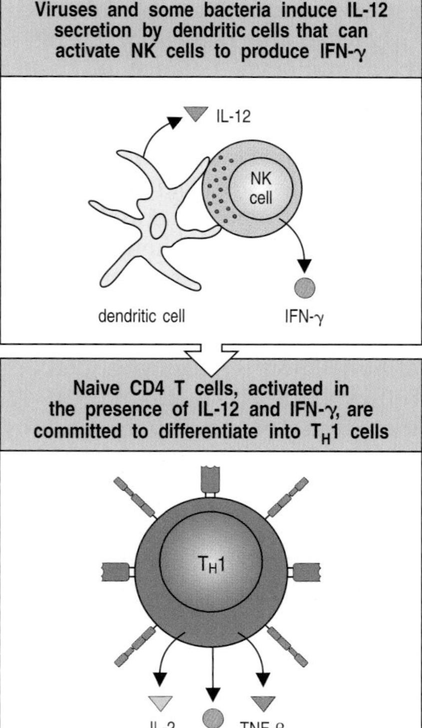

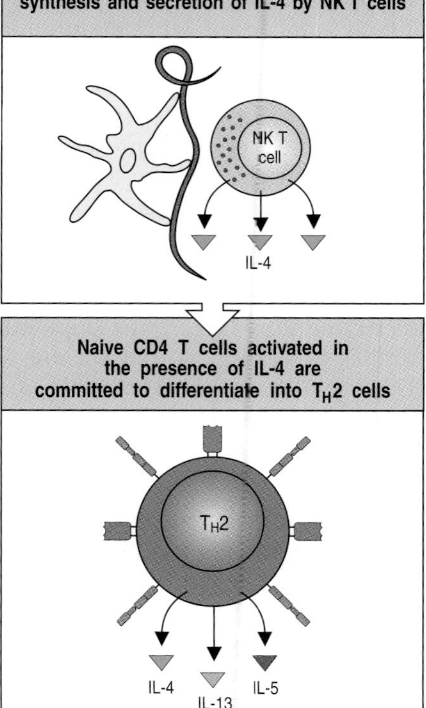

Fig. 10.5 The differentiation of naive CD4 T cells into other subclasses of effector T cells is influenced by cytokines elicited by the pathogen. Left panels: many pathogens, especially intracellular bacteria and viruses, activate dendritic cells to produce IL-12 and NK cells to produce IFN-γ. These cytokines cause proliferating CD4 T cells to differentiate into T_H1 cells. NK cells can be induced by certain stimuli and adjuvants to migrate into lymph nodes, where they could promote T_H1 responses. Right panels: IL-4, which can be produced by various cells, is made in response to parasitic worms and some other pathogens and acts on proliferating CD4 T cells to cause them to become T_H2 cells. An NK T cell is shown as a source of IL-4 here, but these cells are not the only source of IL-4 that can promote T_H2 responses (see text). The mechanisms by which these cytokines induce the selective differentiation of CD4 T cells is discussed in Section 8-19 and Fig. 8.29. Selective induction of transcription factors induced by cytokine binding to cytokine receptors leads to the activation of these two different fates.

IFN-γ and prostaglandin E2 produced in sites of inflammation, or by the binding of TLRs on the dendritic cell surface by bacterial ligands such as LPS.

The importance of TLRs in driving dendritic cells to produce IL-12 has been shown in mice lacking the adaptor protein MyD88, a component of an intracellular signaling pathway activated by the stimulation of some TLRs (see Section 6-27). Mice deficient in MyD88 do not survive a challenge with *T. gondii*, which normally elicits a strong T_H1 response. Dendritic and other cells from mice lacking MyD88 failed to produce IL-12 in response to parasite antigens, and the animals failed to mount a T_H1 response (Fig. 10.6).

Pathogen-driven T_H2 development is less well understood, and this is an active area of research. Much of what we know about how innate immunity controls the adaptive immune response is based on pathogens that drive T_H1 responses. These pathogens activate the production of cytokines such as IFN-γ and IL-12 through TLR signaling pathways. For T_H2 responses, the mechanisms linking innate immunity to regulation of the adaptive T_H2 response are less clear and somewhat controversial. Naive CD4 T cells activated in the presence of IL-4, especially when IL-6 is also present, tend to differentiate into T_H2 cells (see Fig. 10.5, right panels). Some pathogens, such as helminths and other extracellular parasites, consistently induce development of T_H2 responses, and do so in a manner that requires IL-4 signaling and the pathways of T_H2 development described in Chapter 8. It is still not clear, however, how these pathogens are initially sensed by the immune system and stimulate the delivery of T_H2-inducing signals. One possibility is that because such pathogens do not actively drive the production of IL-12 and IFN-γ, the small amounts of IL-4 produced by some cells can become the dominant factor in the environment.

The source of the IL-4 that initiates the primary T_H2 response is also not yet entirely clear. Once differentiated, effector T_H2 cells themselves are a source of IL-4, which reinforces the development of more T_H2 cells (see Section 8-19), but this cytokine can be produced by other cells besides conventional T cells, such as NKT cells (see Section 2-34), and such sources could contribute to the initial T_H2 development (see Fig. 10.5). Mast cells are also potent producers of IL-4 after stimulation and can migrate to peripheral lymphoid organs, making them a possible candidate for an early source of IL-4. Other evidence indicates that some ligands for TLRs might deliver signals to dendritic cells that cause them to produce cytokines favoring T_H2 development rather than T_H1. Dendritic cells make more IL-10 and less IL-12 when stimulated by some ligands for TLR-2, including bacterial lipoproteins, peptidoglycans, and zymosan, a carbohydrate derivative of yeast cell walls, compared with other TLR ligands. These ligands could therefore favor T_H2 development. Finally, recent evidence suggests that dendritic cells can produce ligands for the receptor protein Notch on T cells, and that Notch signaling augments the production of IL-4 by naive T cells, again favouring T_H2 development.

Fig. 10.6 Infection may trigger T_H1 polarization through Toll-like receptor signaling pathways. The adaptor protein MyD88 is a key component of Toll-like receptor signaling. Wild-type mice and mice deficient in MyD88 were infected intraperitoneally with the protozoan parasite *Toxoplasma gondii* (left panel). Five days after infection, mice lacking MyD88 showed a severe reduction in the amount of IL-12 in plasma compared with wild-type mice (middle panel), and dendritic cells from the spleens of these animals failed to produce IL-12 on stimulation with antigens from *T. gondii*. The mice lacking MyD88 also failed to produce a large IFN-γ response (middle panel) and a T_H1 response to the infection, and died at about 2 weeks after infection (right panel). In contrast, wild-type mice produced a strong IL-12, IFN-γ, and T_H1 response, controlled the infection and survived.

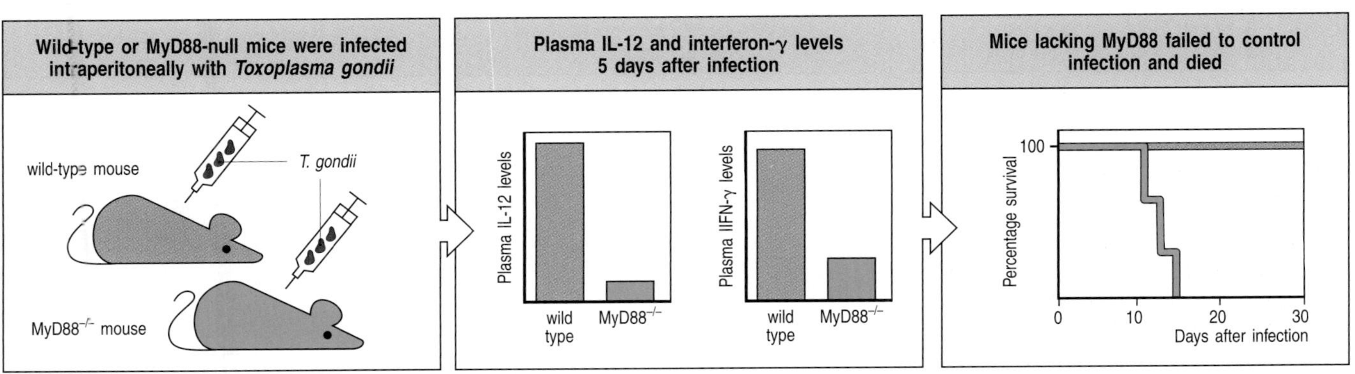

10-5 The distinct subsets of CD4 T cells can regulate each other's differentiation.

The various subsets of CD4 T cells—T_{reg}, T_H17, T_H1 and T_H2—each have very different functions. T_{reg} cells maintain tolerance and limit immunopathology, while T_H17 cells amplify acute inflammation at sites of early infection. T_H1 cells are crucial for cell-mediated immunity due to phagocytes, and also provide help for antibody production. T_H2 cells are associated with a response producing high levels of neutralizing antibodies (IgG and IgA), or IgE and mast-cell activation. This latter response promotes barrier immunity against many parasites by augmenting the production of mucus at epithelial surfaces, thus creating a barrier to their colonization, and also promotes their expulsion from the body.

We have already seen how T_H17 cells are induced by the presence of both IL-6 and TGF-β early in an infection (see Section 10-3). However, when IFN-γ (typically produced by T_H1 cells) or IL-4 (typically produced by T_H2 cells) are also present, TGF-β and IL-6 do not efficiently generate T_H17 cells, as it seems that the signals delivered by IFN-γ and IL-4 can dominate over those delivered by TGF-β and IL-6, and drive either T_H1 or T_H2 development. Thus, as T_H1 or T_H2 cells emerge and begin to produce their cytokines, the early T_H17 response becomes inhibited (Fig. 10.7).

There is also cross-regulation between T_H1 and T_H2 cells. IL-10, which is a product of T_H2 cells, can inhibit the development of T_H1 cells by suppressing the production of IL-12 by dendritic cells, whereas IFN-γ, a product of T_H1 cells, can prevent the production of T_H2 cells (see Fig. 10.7). If a particular CD4 T-cell subset is activated first or preferentially in a response, it can thus suppress the development of the other subset. The overall effect is that certain responses, especially chronic responses, are eventually dominated by either a T_H2 or a T_H1 response, and once one subset becomes dominant it is often hard to shift the response to the other subset. In many infections, however, there is a mixed T_H1 and T_H2 response.

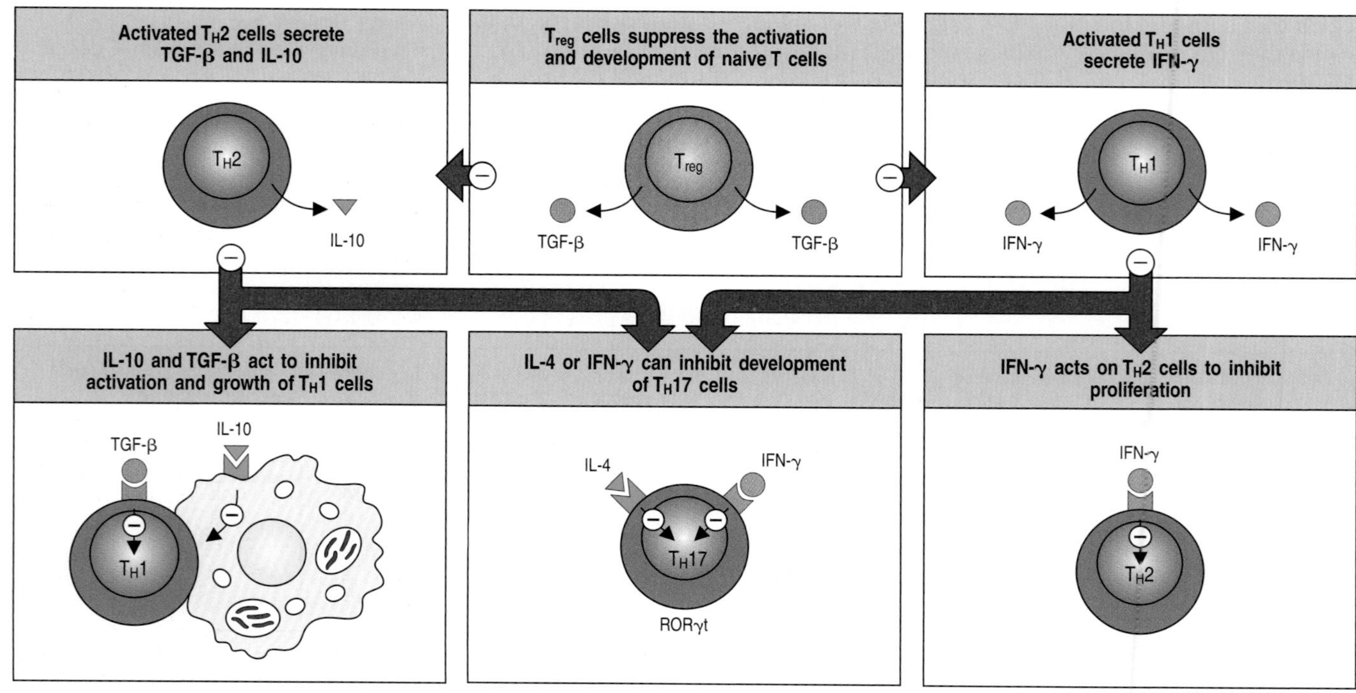

Fig. 10.7 The subsets of CD4 T cells each produce cytokines that can negatively regulate the development or effector activity of other subsets. In the absence of infection, in homeostatic conditions, the TGF-β produced by T_{reg} cells can inhibit the activation of naive T cells, thus preventing the development of a T_H17, T_H1 or T_H2 response (upper panels). During an infection, T_H17 cells are the first to emerge, in response to the IL-6 now produced by dendritic cells. As the T_H17 response develops, regulatory T cells become downregulated and the amount of TGF-β in the environment decreases. Once T_H1 or T_H2 cells emerge, their cytokines inhibit T_H17 development (lower center panel) and also cross-inhibit each other's activity. T_H2 cells make IL-10, which acts on macrophages to inhibit T_H1 activation, perhaps by blocking macrophage IL-12 synthesis, and TGF-β, which acts directly on the T_H1 cells to inhibit their growth (left panels). T_H1 cells make IFN-γ, which blocks the growth of T_H2 cells (right panels). These effects allow either subset to dominate a response by suppressing the outgrowth of cells of the other subset.

NK T cells, a class of innate-like lymphocytes, might also regulate T_H1 versus T_H2 development in the direction of T_H2 (see Fig. 10.5). Many NK T cells express CD4, but some lack both CD4 and CD8 (see Section 7-9). These cells express the cell-surface marker NK1.1 normally associated with NK cells, but have an α:β T-cell receptor, which uses a restricted, nearly invariant α chain composed of the gene segments $V_α14–J_α28$ in mice and the equivalent gene segments, $V_α24–J_α18$, in humans (see Section 5-19). The development of NK T cells in the thymus depends not on the expression of MHC class I or class II molecules but instead on the MHC class IB molecules known as CD1 proteins (see Section 5-19), which are expressed in the thymus and bind self lipids.

Expression of CD1 proteins in tissues outside the thymus can be induced by infection, and they can present microbial lipids to T cells. At least some NK T cells recognize specific glycolipid antigens presented by CD1d. On activation, NK T cells secrete very large amounts of IL-4 and IFN-γ, and may provide an initial source of cytokines that polarize a T-cell response, particularly in the direction of T_H2 cells. NK T cells are not the only T cells that recognize antigens presented by CD1 molecules. CD1b presents the bacterial lipid mycolic acid to α:β T cells, and other CD1 molecules are recognized by γ:δ T cells.

The killer lymphocytes of innate immunity, the NK cells, may contribute to T_H1 development (see Fig. 10.5). NK cells are not normally found within lymph nodes, but injection of mice with certain adjuvants, or with mature dendritic cells, can induce their recruitment to lymph nodes via expression of the chemokine receptor CXCR3 by the NK cell. As NK cells produce abundant IFN-γ, but little IL-4, they may act in lymph nodes during infections to direct the development of T_H1 cells.

The interplay of cytokines in CD4 T-cell differentiation, and indeed throughout the whole immune response, is important in human disease, as indicated by studies showing that the pattern of cytokines present can differ in different diseases and between individuals with a given disease and infected but asymptomatic people. The effects of cytokines on CD4 T-cell differentiation *in vivo* are difficult to study in humans, however, and so the links between cytokine action and disease have been explored mainly in mouse models, where polarized responses are easier to study.

For example, BALB/c mice are genetically susceptible to infection with the protozoan parasite *Leishmania major*, which requires a T_H1 response to clear it. When BALB/c mice are infected experimentally, their CD4 T cells fail to differentiate into T_H1 cells; instead, they make T_H2 cells, which are unable to activate macrophages to inhibit leishmanial growth. In contrast, C57BL/6 mice respond by producing T_H1 cells that protect the animal by activating infected macrophages to kill *L. major*. This genetic difference in the immune response seems to result from a population of memory cells that are specific for gut-derived antigens but cross-react with an antigen, LACK (*Leishmania* analog of the receptors of activated C kinase), expressed by the *Leishmania* parasite. For unknown reasons, these memory cells produce IL-4 in BALB/c mice but not in C57BL/6 mice. In BALB/c mice, the IL-4 secreted by these cells during *Leishmania* infection drives new *Leishmania*-specific CD4 T cells to become T_H2 cells, which leads eventually to the failure to eliminate the pathogen, and death. The preferential development of T_H2 rather than T_H1 cells in BALB/c mice can be reversed if IL-4 is blocked in the first days of infection by injecting anti-IL-4 antibody, but this treatment is ineffective after a week or so of infection, demonstrating the crucial importance of the early exposure to cytokines for decisions by naive T cells (Fig. 10.8).

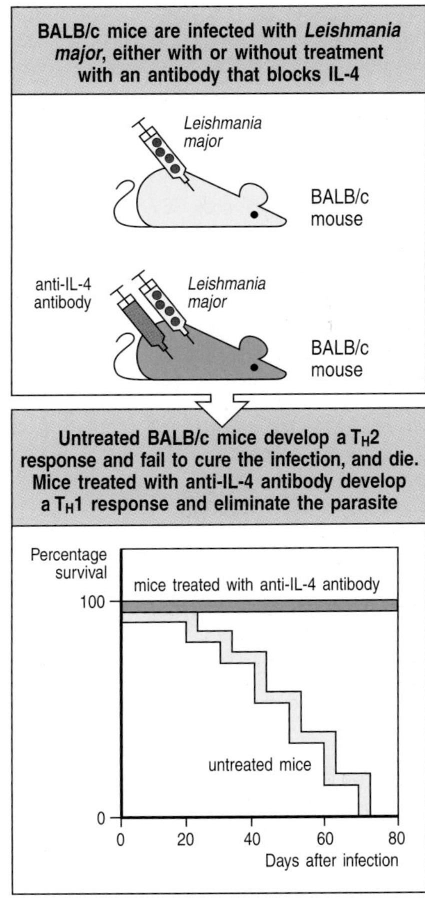

Fig. 10.8 The development of CD4 subsets can be manipulated by altering the cytokines acting during the early stages of infection. Elimination of infection with the intracellular protozoan parasite *Leishmania major* requires a T_H1 response, because IFN-γ is needed to activate the macrophages that provide protection. BALB/c mice are normally susceptible to *L. major* because they generate a T_H2 response to the pathogen. This is because they produce IL-4 early during infection and this induces naive T cells into the T_H2 lineage (see the text). Treatment of BALB/c mice with neutralizing anti-IL-4 antibodies at the beginning of infection inhibits this IL-4 and prevents the diversion of naive T cells toward the T_H2 lineage, and these mice develop a protective T_H1 response.

It is sometimes possible to shift the balance between T_H1 and T_H2 cells by administering appropriate cytokines. IL-2 and IFN-γ have been used to stimulate cell-mediated immunity in diseases such as lepromatous leprosy and

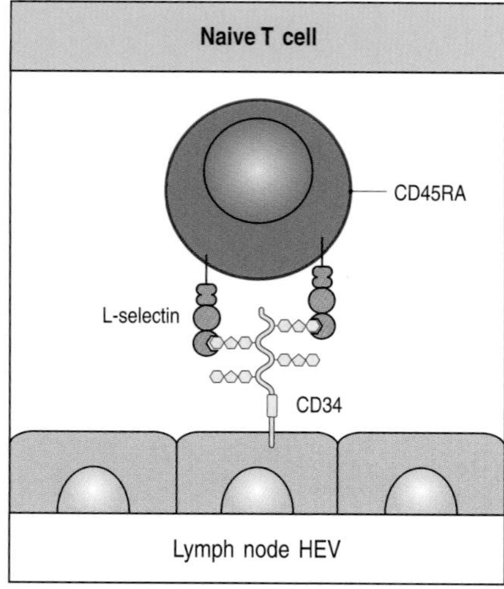

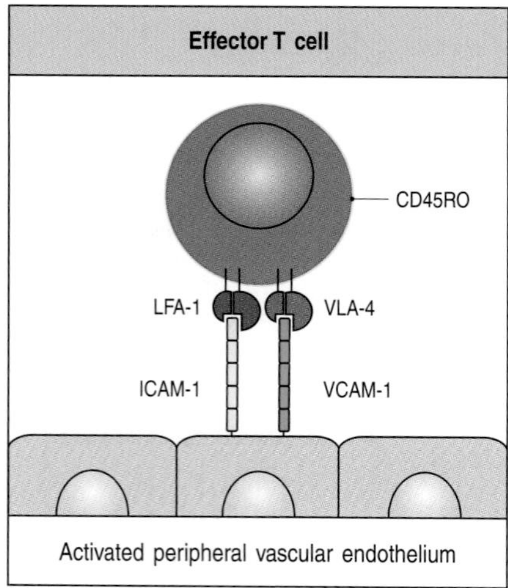

Fig. 10.9 Effector T cells change their surface molecules, allowing them to home to sites of infection. Naive T cells home to lymph nodes through the binding of L-selectin to sulfated carbohydrates displayed by various proteins, such as CD34 and GlyCAM-1, on the high endothelial venule (HEV, top panel). After encounter with antigen, may of the differentiated effector T cells lose expression of L-selectin, leave the lymph node about 4–5 days later, and now express the integrin VLA-4 and increased levels of LFA-1. These bind to VCAM-1 and ICAM-1, respectively, on peripheral vascular endothelium at sites of inflammation (bottom panel). On differentiating into effector cells, T cells also alter their splicing of the mRNA encoding the cell-surface protein CD45. The CD45RO isoform expressed by effector T cells lacks one or more exons that encode extracellular domains present in the CD45RA isoform expressed by naive T cells, and somehow makes effector T cells more sensitive to stimulation by specific antigen.

can cause both a local resolution of the lesion and a systemic change in T-cell responses.

CD8 T cells are also able to regulate the immune response by producing cytokines. Effector CD8 T cells can, in addition to their familiar cytotoxic function, also respond to antigen by secreting cytokines typical of either T_H1 or T_H2 cells. Such CD8 T cells, called T_C1 or T_C2 by analogy with the T_H subsets, seem to be responsible for the development of leprosy in its lepromatous rather than its tuberculoid form. As we saw in Chapter 8, lepromatous leprosy is due to the predominance of a T_H2 cell response, which does not clear the bacteria. Patients with the less destructive tuberculoid leprosy make T_C1 cells whose cytokines induce T_H1 cells, which can activate macrophages to rid the body of its burden of leprosy bacilli. Patients with lepromatous leprosy have CD8 T cells that suppress the T_H1 response by making IL-10 and TGF-β. The expression of these cytokines could explain the suppression of CD4 T cells by CD8 T cells that has been observed in various situations.

Another factor that influences the differentiation of CD4 T cells into distinct effector subsets is the amount and exact sequence of the antigenic peptide that initiates the response. Large amounts of peptide, which are presented at a high density on the surface of antigen-presenting cells, or peptides that interact strongly with the T-cell receptor, tend to stimulate T_H1 responses, whereas a low density of peptide or peptides that bind weakly tend to elicit T_H2 responses. These effects do not seem to be due to differences in signaling through the T-cell receptor but may involve changes in the overall balance of different cytokines produced by the cells involved in activating naive T cells.

Such differences could be important in certain circumstances. For instance, allergy is caused by the production of IgE antibody, which requires high levels of IL-4 but does not occur in the presence of IFN-γ, a powerful inhibitor of IL-4-driven class switching to IgE. Antigens that elicit IgE-mediated allergy are generally delivered in minute doses and elicit T_H2 cells that make IL-4 and no IFN-γ. It is also relevant that allergens do not elicit any of the known innate immune responses, which in general produce cytokines that tend to bias CD4 T-cell differentiation toward T_H1 cells. Finally, allergens are delivered to humans in minute doses across a thin mucosa, such as that of the lung. Something about this route of sensitization allows even potent generators of T_H1 responses such as *L. major* to induce T_H2 responses.

10-6 Effector T cells are guided to sites of infection by chemokines and newly expressed adhesion molecules.

The full activation of naive T cells takes 4–5 days and is accompanied by marked changes in their homing behavior. Effector cytotoxic CD8 T cells must travel from the peripheral lymphoid tissue in which they have been activated to attack and destroy infected cells. Effector CD4 T_H1 cells must also leave the lymphoid tissues to activate macrophages at the site of infection. Most effector T cells cease production of L-selectin, which mediates homing to the lymph nodes, whereas the expression of other adhesion molecules is increased (Fig. 10.9). One important change is a marked increase in synthesis of the integrin $\alpha_4{:}\beta_1$, also known as VLA-4. This binds to the adhesion molecule VCAM-1, a member of the immunoglobulin superfamily that is induced on activated endothelial cell surfaces and initiates the extravasation of the effector T cells. Thus, if the innate immune response has already activated the endothelium at the site of infection, as described in Section 10-2, effector T cells will be rapidly recruited.

In the early stage of an immune response, only a few of the effector T cells that enter the infected tissue will be expected to be specific for pathogen, because any effector T cell specific for any antigen will also be able to enter. Specificity of the reaction is maintained, however, because only those effector T cells that recognize pathogen antigens will carry out their function, destroying infected cells or specifically activating pathogen-loaded macrophages. By the peak of an adaptive immune response, after several days of clonal expansion and differentiation, most of the recruited T cells will be specific for the infecting pathogen.

Not all infections trigger innate immune responses that activate local endothelial cells, and it is not so clear how effector T cells are guided to the sites of infection in these cases. However, activated T cells seem to enter all tissues in very small numbers, perhaps via adhesive interactions such as the binding of P-selectin on the endothelial cells to its ligand **P-selectin glycoprotein ligand-1** (**PSGL-1**), which is expressed by activated T cells, and could thus encounter their antigens even in the absence of a previous inflammatory response.

One or a few specific effector T cells encountering antigen in a tissue can thus initiate or augment a potent local inflammatory response that recruits many more effector lymphocytes and nonspecific inflammatory cells to that site. Effector T cells that recognize pathogen antigens in the tissues produce cytokines such as TNF-α, which activates endothelial cells to express E-selectin, VCAM-1, and ICAM-1, and chemokines such as CCL5 (see Fig. 2.46), which acts on effector T cells to activate their adhesion molecules. VCAM-1 and ICAM-1 on endothelial cells bind VLA-4 and LFA-1, respectively, on effector T cells, recruiting more of these cells into tissues that contain the antigen. At the same time, monocytes and polymorphonuclear leukocytes are recruited by adhesion to E-selectin. TNF-α and IFN-γ released by activated T cells also act synergistically to change the shape of endothelial cells, allowing increased blood flow, increased vascular permeability, and increased emigration of leukocytes, fluid, and protein into the infected tissue. Thus, at the later stages of an infection, the protective effects of macrophages secreting TNF-α and other pro-inflammatory cytokines at the infected site (see Section 2-24) are reinforced by the actions of effector T cells.

By contrast, effector T cells that enter tissues but do not recognize their antigen are rapidly lost. They either enter the lymph from the tissues and eventually return to the bloodstream, or undergo apoptosis. Most of the T cells in the afferent lymph that drains tissues are memory or effector T cells, which characteristically express the CD45RO isoform of the cell-surface molecule CD45 and lack L-selectin (see Fig. 10.9). Effector T cells and memory T cells have similar phenotypes, as we discuss later, and both seem to be committed to migration through potential sites of infection. As well as allowing effector T cells to clear all sites of infection, this pattern of migration allows them to contribute, along with memory cells, to protecting the host against reinfection with the same pathogen.

Expression of particular adhesion molecules can direct different subsets of effector T cells to specific sites. As we shall see in Chapter 11, the peripheral immune system is compartmentalized, such that different populations of lymphocytes migrate through different lymphoid compartments and—after activation—through the different tissues they serve. This is achieved by the selective expression of adhesion molecules that bind selectively to tissue-specific addressins. In this context, the adhesion molecules are often known as **homing receptors**. For example, some activated T cells specifically populate the skin. They are induced during activation to express an adhesion molecule named **cutaneous lymphocyte antigen (CLA)** (Fig. 10.10). This is a glycosylated isoform of PSGL-1 that binds to E-selectin on cutaneous vascular endothelium. CLA-expressing T lymphocytes also produce the chemokine receptor CCR4. This binds the chemokine CCL17 (TARC), which is present at

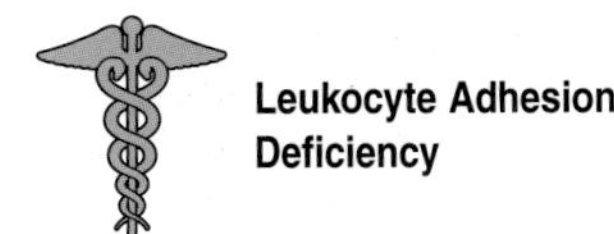

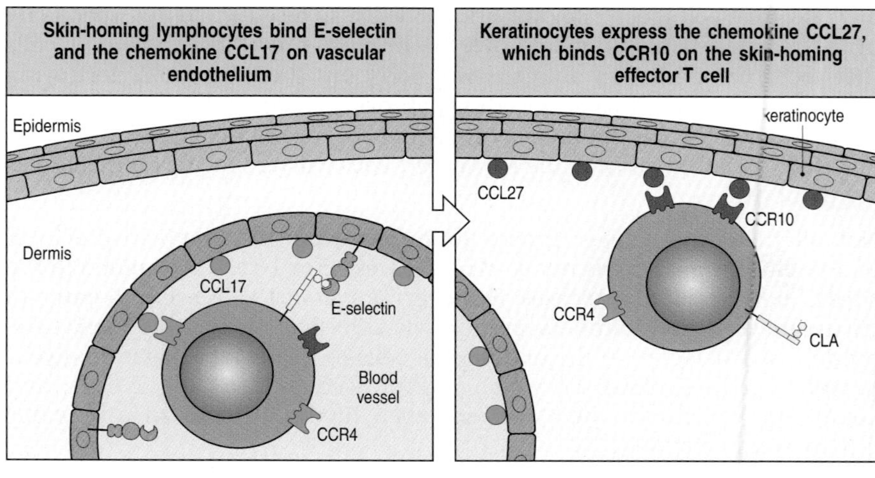

Fig. 10.10 Skin-homing T cells use specific combinations of integrins and chemokines to migrate specifically to the skin. Left panel: a skin-homing lymphocyte binds to the endothelium lining a cutaneous blood vessel by interactions between cutaneous lymphocyte antigen (CLA) and constitutively expressed E-selectin on the endothelial cells. The adhesion is strengthened by an interaction between lymphocyte chemokine receptor CCR4 and the endothelial chemokine CCL17. Right panel: once through the endothelium, keratinocytes of the epidermis anchor the effector T lymphocyte by the chemokine CCL27 that they produce, which binds to the receptor CCR10 on lymphocytes.

high levels on the endothelium of cutaneous blood vessels. Interaction of CLA with E-selectin causes the T cell to roll against the wall of the vessel, and the signal delivered by endothelial CCL17 is thought to cause the arrest of lymphocytes and to bring about their adhesion to the wall, probably by inducing tight binding of integrin, as described for the action of CCL21 on naive T cells (see Section 8-3). In addition to CCR4, skin-homing T cells carry the chemokine receptor CCR10 (GPR-2), which binds the chemokine CCL27 (CTACK) expressed by keratinocytes, the epithelial cells of the skin.

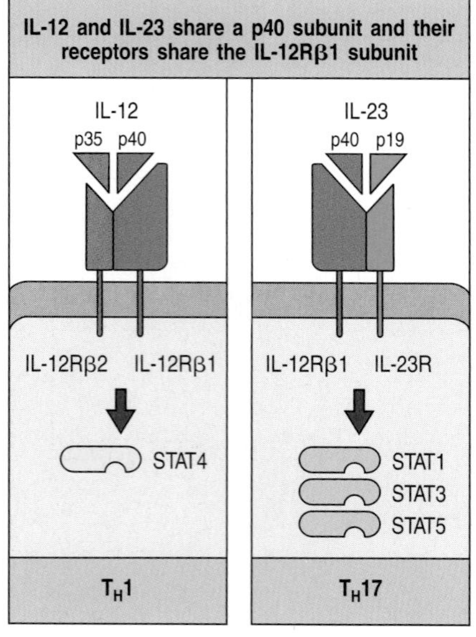

Fig. 10.11 IL-12 and IL-23 share subunits and receptor components. The dimeric cytokines IL-12 and IL-23 share the p40 subunit, and the receptors for IL-12 and IL-23 have the IL-12Rβ1 subunit in common. IL-12 signaling activates the transcriptional activators STAT1, STAT3, and STAT4, but its action in increasing IFN-γ production is due to STAT4. IL-23 activates several other STATs, but activates STAT4 only weakly. Both cytokines augment the activity and proliferation of the CD4 subsets that express receptors for them; T_H1 cells express IL-12R, and T_H17 cells express IL-23R. Mice deficient in p40 lack expression of both of these cytokines, and manifest immune defects as a result of deficient T_H1 and T_H17 activities.

10-7 Differentiated effector T cells are not a static population but continue to respond to signals as they carry out their effector functions.

The commitment of CD4 T cells to become distinct lineages of effector cells begins in peripheral lymphoid tissues, such as lymph nodes, as described in Sections 10-3 and 10-4. However, the effector activities of these cells once they enter sites of infection are not defined simply by the signals received in the lymphoid tissues. Instead, evidence suggests that there is continuous regulation of the expansion and the effector activities of differentiated CD4 cells, in particular of T_H17 and T_H1 cells.

As noted earlier, commitment of naive T cells to become T_H17 cells is triggered by exposure to TGF-β and IL-6; commitment to T_H1 cells is initially triggered by IFN-γ. These initial conditions are not, however, sufficient to generate complete or effective T_H17 or T_H1 responses. In addition, each T-cell subset also requires stimulation by another cytokine—IL-23 in the case of T_H17 cells and IL-12 in the case of T_H1 cells. IL-23 and IL-12 are closely related in structure; each is a heterodimer and they share a subunit. IL-23 is composed of one p40 and one p19 subunit, whereas IL-12 has the p40 subunit and a unique p35 subunit. Committed T_H17 cells express a receptor for IL-23, whereas T_H1 cells express a receptor for IL-12. The receptors for IL-12 and IL-23 are also related, sharing a common subunit (Fig. 10.11).

IL-23 and IL-12 amplify the activities of T_H17 and T_H1 cells, respectively. Like many other cytokines, they both act through the JAK–STAT intracellular signaling pathway (see Fig. 6.30). IL-23 signaling activates the intracellular transcriptional activators STAT1, STAT3, and STAT5, but activates STAT4 very weakly. In contrast, IL-12 activates STAT1 and STAT3 and also strongly activates STAT4. IL-23 does not initiate the commitment of naive CD4 T cells to T_H17 cells, but it does stimulate their expansion. Many *in vivo* responses that depend on IL-17 are diminished in the absence of IL-23. For example, mice lacking the IL-23-specific subunit p19 show decreased production of IL-17 and IL-17F in the lung after infection by *Klebsiella pneumoniae*.

Mice lacking the p40 subunit, which is shared by IL-12 and IL-23, are deficient in both IL-23 and IL-12. This fact caused some confusion before the separate role of IL-23 in T_H17 activity was appreciated. For example, it had been thought that the brain inflammation occurring in experimental autoimmune encephalomyelitis (EAE) in mice was due to IFN-γ and T_H1 cells. This interpretation was initially based on an analysis of p40-deficient mice, which do not show brain inflammation in EAE. However, p35-deficient mice, lacking IL-12 but maintaining IL-23, are susceptible to EAE. It turns out that the brain inflammation in EAE is largely a result of the activity of IL-17 and T_H17 cells.

IL-12 regulates the effector activity of committed T_H1 cells at sites of infection, but other cytokines such as IL-18 may also be involved. Studies of two different pathogens have shown that the initial differentiation of T_H1 cells is not sufficient for protection, and that continuous signals are required. Mice deficient in p40 can resist initial infection by *T. gondii* as long as IL-12 is administered continuously. If IL-12 is administered during the first 2 weeks of infection, p40-deficient mice survive the initial infection and establish a latent chronic infection characterized by cysts containing the pathogen. When IL-12 administration is stopped, however, these mice gradually reactivate the latent cysts and the animals eventually die of toxoplasmic encephalitis. IFN-γ production by pathogen-specific T cells decreases in the absence of IL-12 but could be restored by IL-12 administration. Similarly, the adoptive transfer of differentiated T_H1 cells from mice cured of *L. major* protects RAG-deficient mice infected by *L. major*, but cannot protect p40-deficient mice (Fig. 10.12). Together, these experiments suggest that T_H1 cells continue to respond to signals during an infection, and that continuous IL-12 is needed to sustain the effectiveness of differentiated T_H1 cells against at least some pathogens.

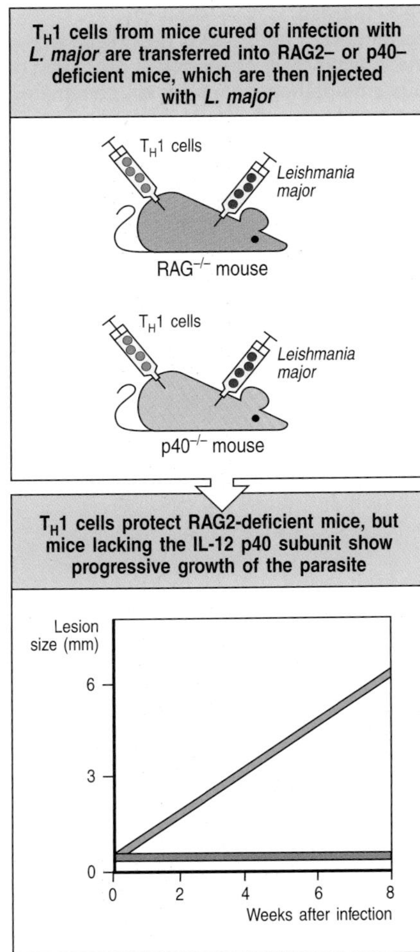

T_H1 cells from mice cured of infection with *L. major* are transferred into RAG2– or p40–deficient mice, which are then injected with *L. major*
T_H1 cells
Leishmania major
$RAG^{-/-}$ mouse
T_H1 cells
Leishmania major
$p40^{-/-}$ mouse
T_H1 cells protect RAG2-deficient mice, but mice lacking the IL-12 p40 subunit show progressive growth of the parasite
Lesion size (mm)
6
3
0
0
2
4
6
8
Weeks after infection

Fig. 10.12 Continuous IL-12 is required for resistance to pathogens requiring T_H1 responses. Mice that have eliminated an infection with *Leishmania major* and have generated T_H1 cells specific to the pathogen were used as a source of T cells that were adoptively transferred into either RAG2-deficient mice, which lack T cells and B cells and cannot control *L. major* infection but can produce IL-12, or into mice lacking p40, which cannot produce IL-12. On subsequent infection of the RAG2-deficient mice, lesions did not enlarge because the transferred T_H1 cells conferred immunity. But despite the fact that the transferred cells were already differentiated T_H1 cells, they did not confer resistance to IL-12 p40-deficient mice in which a continuous source of IL-12 was not present.

10-8 Primary CD8 T-cell responses to pathogens can occur in the absence of CD4 help.

Many CD8 T-cell responses require help from CD4 T cells (see Section 8-18). This is typically the case where the antigen recognized by the CD8 T cells is derived from an agent that does not cause inflammation on initial infection. In such circumstances, CD4 T-cell help is required to activate dendritic cells to become able to stimulate a complete CD8 T-cell response, an activity that has been described as licensing of the antigen-presenting cell (see Section 8-7). Licensing involves the induction of co-stimulatory molecules such as B7, CD40 and 4-1BBL on the dendritic cell, which can then deliver signals that fully activate naive CD8 T cells (see Fig. 8.28). Licensing enforces a requirement for dual recognition of an antigen by the immune system by both CD4 and CD8 T cells, which provides a useful safeguard against autoimmunity. Dual recognition is also seen in the cooperation between T cells and B cells for antibody generation (see Chapter 9). However, not all CD8 T-cell responses require such help.

Some infectious agents, such as the intracellular Gram-positive bacterium *Listeria monocytogenes* and the Gram-negative bacterium *Burkholderia pseudomallei*, provide the inflammatory environment required to license dendritic cells and thus can induce primary CD8 T-cell responses without help from CD4 T cells. These pathogens carry a number of immunostimulatory signals, such as ligands for TLRs, and so can directly activate antigen-presenting cells to express the co-stimulatory molecules B7 and CD40. Thus, fully activated dendritic cells presenting *Listeria* or *Burkholderia* antigens can activate naive antigen-specific CD8 T cells without the help of CD4 T cells, and can induce them to undergo clonal expansion (Fig. 10.13). The activated

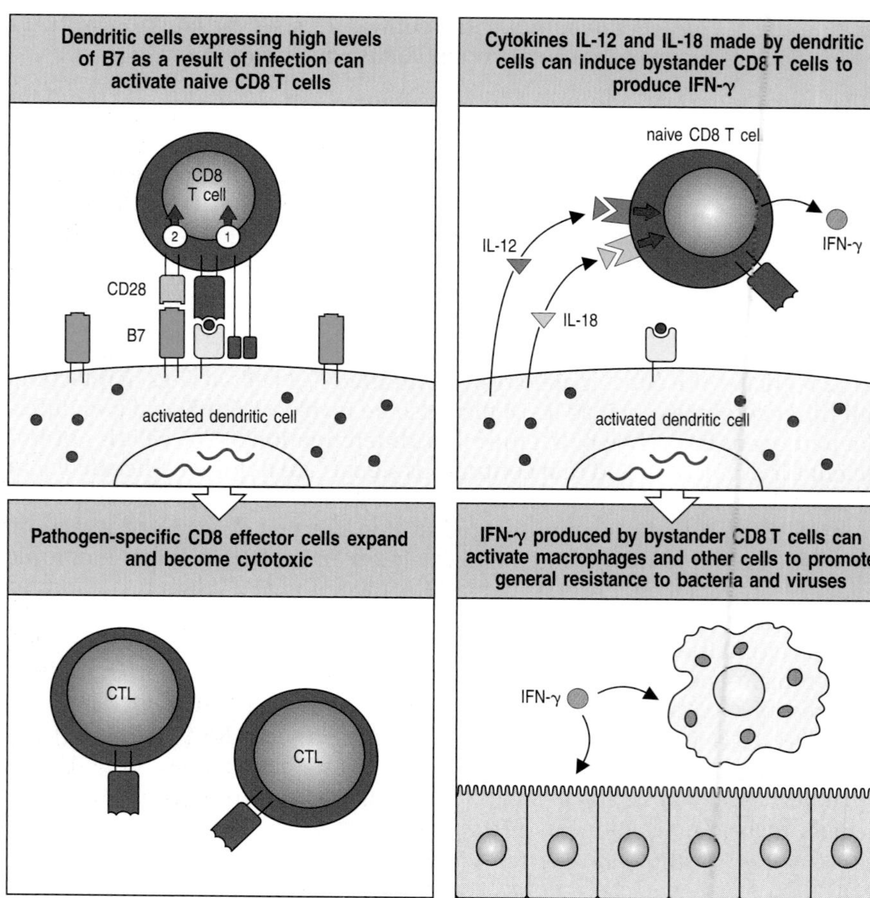

Fig. 10.13 Naive CD8 T cells can be activated directly by potent antigen-presenting cells through their T-cell receptor or through the action of cytokines. Left panels: naive CD8 T cells that encounter peptide:MHC class I complexes on the surface of dendritic cells expressing high levels of co-stimulatory molecules as a result of the inflammatory environment produced by some pathogens (upper left panel) are activated to proliferate in response, eventually differentiating into cytotoxic CD8 T cells (lower left panel). Right panels: activated dendritic cells also produce the cytokines IL-12 and IL-18, whose combined effect on CD8 T cells rapidly induces production of IFN-γ (upper right panel). This activates macrophages for the destruction of intracellular bacteria and can promote antiviral responses in other cells (lower right panel).

dendritic cell also secretes cytokines such as IL-12 and IL-18, which act on naive CD8 T cells in a so-called 'bystander' effect to induce them to produce IFN-γ, which in turn induces other protective effects (see Fig. 10.13).

Primary CD8 T-cell responses to *L. monocytogenes* were examined in mice that were genetically deficient in MHC class II molecules and thus lacked CD4 T cells (see Section 7-18). The numbers of CD8 T cells specific for a particular antigen expressed by the pathogen were measured by using MHC tetramers (see Appendix I, Section A-28). On day 7 after infection, wild-type mice and mice lacking CD4 T cells show equivalent expansion, and equivalent cytotoxic capacity, of pathogen-specific CD8 T cells. Mice lacking CD4 T cells cleared the initial infection by *L. monocytogenes* as effectively as wild-type mice. These experiments clearly show that protective responses can be generated by pathogen-specific CD8 T cells without CD4 T-cell help. However, as we will see later, the nature of the CD8 memory response is different and is diminished in the absence of CD4 T-cell help.

A second pathway of CD8 T-cell activation independent of T-cell help is also independent of antigen. Naive antigen-nonspecific CD8 T cells can be activated very early in infection by IL-12 and IL-18 in a 'bystander effect', and produce cytokines such as IFN-γ that help to progress the protective immune response (see Fig. 10.13). Mice infected with *L. monocytogenes* or *B. pseudomallei* rapidly produce a strong IFN-γ response, which is essential for their survival. The source of this IFN-γ seems to be both the NK cells of innate immunity and naive CD8 T cells, which begin to secrete it within the first few hours after infection. This is believed to be too soon for any significant expansion of pathogen-specific CD8 T cells, which would initially be too rare to contribute in an antigen-specific manner, and too soon for the differentiation of T_H1 cells that could help activate the CD8 T cells. The production of IFN-γ by both NK and CD8 T cells at this early time can be blocked experimentally by antibodies against IL-12 and IL-18, suggesting that these cytokines are responsible. The source of the IL-12 and IL-18 was not identified in this experiment, but they are produced by macrophages and dendritic cells in response to activation via TLRs. These experiments indicate that naive CD8 T cells can contribute nonspecifically in a kind of innate defense, not requiring CD4 T cells, in response to early signals of infection.

10-9 Antibody responses develop in lymphoid tissues under the direction of CD4 helper T cells.

Migration out of lymphoid tissues is clearly important for the effector actions of antigen-specific CD8 cytotoxic T cells, T_H17 and T_H1 cells. However, another important effector function of CD4 T cells, both T_H1 and T_H2 cells, depends on their interactions with B cells, and these interactions occur in the lymphoid tissues themselves. B cells specific for a protein antigen cannot be activated to proliferate, form germinal centers, or differentiate into plasma cells until they encounter a helper T cell that is specific for one of the peptides derived from that antigen. Humoral immune responses to protein antigens thus cannot occur until after antigen-specific helper T cells have been generated.

One of the most interesting questions in immunology is how two antigen-specific lymphocytes—the naive antigen-binding B cell and the helper T cell—find one another to initiate a T-cell dependent antibody response. As we learned in Chapter 9, the likely answer lies in the migratory path of B cells through the lymphoid tissues and the presence of helper T cells on that path (Fig. 10.14). If B cells binding their specific antigen in the T-cell zone of peripheral lymphoid organs receive specific signals from helper T cells, they proliferate in the T-cell areas (see Fig. 10.14, second panel). In the absence of T-cell signals, these antigen-stimulated B cells die within 24 hours of arriving in the T-cell zone. B cells that do not contact their antigen enter the lymphoid follicles and eventually continue to recirculate between lymph, blood, and peripheral lymphoid tissues.

About 5 days after primary immunization, primary foci of proliferating B cells appear in the T-cell areas, which correlates with the time needed for helper T cells to differentiate. Some of the B cells activated in the primary focus may migrate to the medullary cords of the lymph node, or to those parts of the red pulp that are next to the T-cell zones of the spleen, where they become plasma cells and secrete specific antibody for a few days (see Fig. 10.14, third panel). Others migrate to the follicle (see Fig. 10.14, fourth panel), where they proliferate further, forming a germinal center in which they undergo somatic hypermutation and affinity maturation—the production of B cells with receptors of greater affinity for the antigen (see Sections 4-18 and 9-8).

Antigen is retained for very long periods in lymphoid follicles in the form of antigen:antibody complexes on the surface of the local follicular dendritic cells. The antigen:antibody complexes, which become coated with fragments of C3, are held on the cell by receptors for the complement fragments (CR1, CR2, and CR3) as well as by a nonphagocytic Fc receptor (see Fig. 9.14). The function of this antigen is unclear, as there is evidence that it is not absolutely required for the stimulation of B cells in the germinal center (see Section 9-10), but it may regulate the long-term antibody response.

The proliferation, somatic hypermutation, and selection of higher-affinity B cells in the germinal centers during a primary antibody response have been

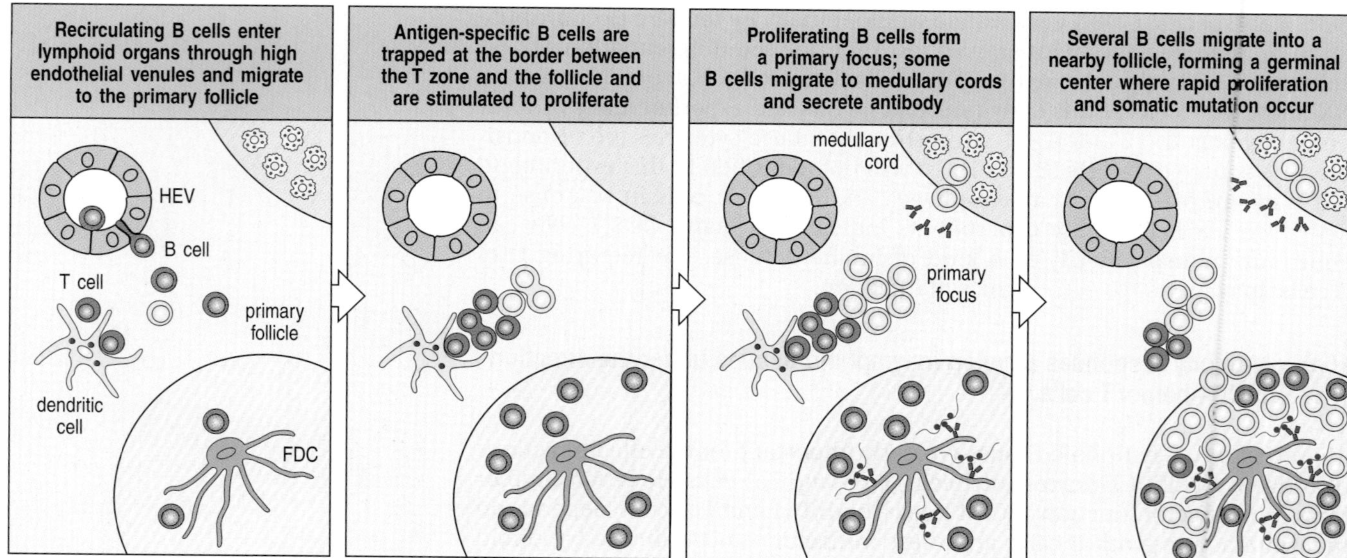

Fig. 10.14 Peripheral lymphoid tissues provide an environment where antigen-specific naive B cells interact with helper T cells specific for the same antigen. First panel: T cells specific for a foreign protein (blue cells) become activated to helper cell status in the T-cell zone by antigen-presenting dendritic cells. A few of the naive B cells entering through the HEV will express receptors specific for the same foreign protein (yellow cells), but most will not (brown cells). Second panel: B cells that do not contact their antigen in the T-cell zone pass straight through it and enter the lymphoid follicles, from which they will continue their recirculation through the peripheral lymphoid tissues. The rare antigen-specific naive B cells take up the foreign protein via their B-cell antigen receptors, and present its peptides on MHC proteins to antigen-specific T cells. Thus, B cells and T cells specific for the same foreign protein are able to interact as the B cells migrate through the T-cell zone. Third panel: the interaction with T cells stimulates the antigen-specific B cells to proliferate and form a primary focus, and results in some isotype switching. Some of the activated B cells migrate to the medullary cords, where they divide, differentiate into plasma cells, and secrete antibody for a few days. Fourth panel: other activated B cells migrate into primary lymphoid follicles, where they proliferate rapidly to form a germinal center with the help of antigen-specific helper T cells (blue). The germinal center is the site of somatic hypermutation and selection of high-affinity B cells (affinity maturation) (see Chapter 9). Antigen (red) that is trapped in the form of immune complexes (antigen:antibody:complement complexes) on the surface of follicular dendritic cells (FDC) may be involved in B-cell stimulation during affinity maturation.

described in Chapter 9. The adhesion molecules and chemokines that govern the migratory behavior of B cells are likely to be very important to these processes but, as yet, little is known of their nature. The chemokine/receptor pair CXCL13/CXCR5, which controls B-cell migration into the follicle, may be important, particularly for B cells homing to the germinal center. Another chemokine receptor, CCR7, which is expressed strongly on T cells and weakly on B cells, may be involved in temporarily directing B cells to the interface with the T-cell zone. The ligands for CCR7 are CCL19 and CCL21, which are abundant in the T-cell zone and could attract B cells that have increased expression of CCR7.

10-10 Antibody responses are sustained in medullary cords and bone marrow.

The B cells activated in primary foci migrate either to adjacent follicles or to local extrafollicular sites of proliferation. B cells grow exponentially in these sites for 2–3 days and undergo six or seven cell divisions before the progeny come out of the cell cycle and form antibody-producing plasma cells *in situ* (Fig. 10.15, upper panel). Most of these plasma cells have a life span of 2–4 days, after which they undergo apoptosis. About 10% of plasma cells in these extrafollicular sites live longer; their origin and ultimate fate are unknown. The B cells that migrate to the primary follicles to form germinal centers undergo class switching and affinity maturation before either becoming memory cells or leaving the germinal center to become relatively long-lived antibody-producing cells (see Sections 9-7 to 9-9).

These B cells leave germinal centers as plasmablasts (pre-plasma cells). Plasmablasts originating in the follicles of Peyer's patches and mesenteric lymph nodes migrate via lymph to the blood and then enter the lamina propria of the gut and other epithelial surfaces. Those originating in peripheral lymph node or splenic follicles migrate to the bone marrow (see Fig. 10.15, lower panel). In these distant sites of antibody production, the plasmablasts differentiate into plasma cells, which have a life span of months to years. These are thought to provide the antibody that can last in the blood for years after an initial immune response. Whether this supply of plasma cells is replenished by the continual but occasional differentiation of memory cells is not yet known. Studies of responses to nonreplicating antigens show that germinal centers are present for only 3–4 weeks after initial antigen exposure. However, small numbers of B cells continue to proliferate in the follicles for months. These may be the precursors of antigen-specific plasma cells in the mucosa and bone marrow throughout the subsequent months and years.

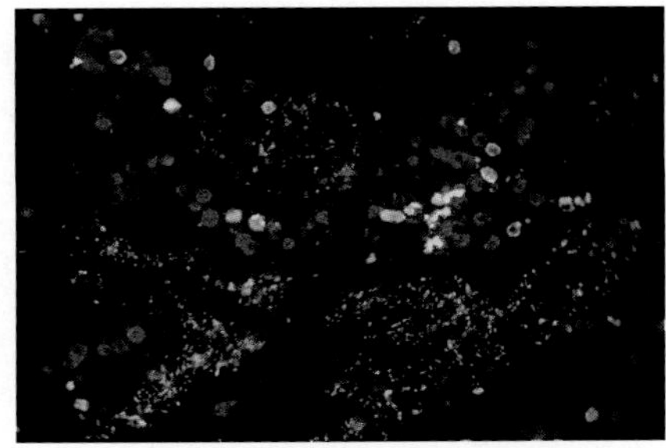

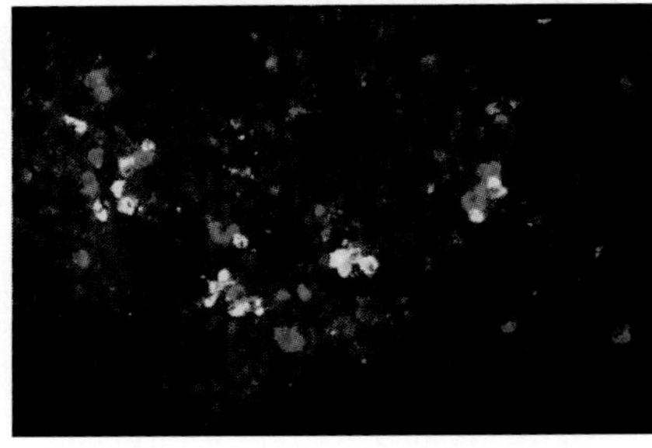

Fig. 10.15 Plasma cells are dispersed in medullary cords and bone marrow. In these sites they secrete antibody at high rates directly into the blood for distribution to the rest of the body. In the upper micrograph, plasma cells in lymph node medullary cords are stained green (with fluorescein anti-IgA) if they are secreting IgA, and red (with rhodamine anti-IgG) if they are secreting IgG. The plasma cells in these local extrafollicular sites are short lived (2–4 days). The lymphatic sinuses are outlined by green granular staining selective for IgA. In the lower micrograph, longer-lived plasma cells (3 weeks to 3 months or more) in the bone marrow are revealed with antibodies specific for light chains (fluorescein anti-λ and rhodamine anti-κ stain). Plasma cells secreting immunoglobulins containing λ light chains show on this micrograph as yellow. Those secreting immunoglobulins containing κ light chains stain red. Photographs courtesy of P. Brandtzaeg.

10-11 The effector mechanisms used to clear an infection depend on the infectious agent.

Most infections engage both the cell-mediated and the humoral aspects of immunity, and in many cases both are helpful in clearing or containing the pathogen and setting up protective immunity, as shown in Fig. 10.16, although the relative importance of the different effector mechanisms, and the effective classes of antibody involved, vary with different pathogens. As we learned in Chapter 8, cytotoxic T cells are important in destroying virus-infected cells, and in some viral diseases they are the predominant class of lymphocytes present in the blood during a primary infection. Nevertheless, the role of antibodies in clearing viruses from the body and preventing them from getting a hold should not be forgotten. Ebola virus causes a hemorrhagic fever and is one of the most lethal viruses known, but some patients do survive and some people even become infected but remain asymptomatic. In both cases, a strong antiviral IgG response early in infection seems to be essential for survival. The antibody response seems to clear the virus from the bloodstream and gives the patient time to activate cytotoxic T cells. In contrast, this antibody response did not occur in infections that proved fatal, the virus continued to replicate, and even though there was some activation of T cells, the disease progressed.

Cytotoxic T cells are also required for the destruction of cells infected with some intracellular bacterial pathogens, such as *Rickettsia*, the causative agent of typhus. In contrast, mycobacteria, which live inside macrophage vesicles, are mainly kept in check by CD4 T_H1 cells, which activate infected macrophage to kill the bacteria. Antibodies are the principal immune reactants that clear primary infections with common extracellular bacteria such as *Staphylococcus aureus* and *Streptococcus pneumoniae*. The IgM and IgG antibodies produced against components of the bacterial surface coat opsonize the bacteria and make them more susceptible to phagocytosis.

Figure 10.16 also indicates the mechanisms involved in immunity to reinfection, or protective immunity, against the pathogens listed. Inducing protective immunity is the goal of vaccine development, and to achieve this it is necessary to induce an adaptive immune response that has both the antigen specificity and the appropriate functional elements to combat the particular pathogen concerned. Pathogens carry multiple epitopes for both B cells and T cells, and thus generate diverse antibody and T-cell responses, but not all of these will be equally effective in clearing the disease. Protective immunity consists of two components—immune reactants, such as antibody or effector T cells generated in the initial infection or by vaccination, and long-lived immunological memory (Fig. 10.17), which we consider in the last part of this chapter.

The type of antibody or effector T cell that offers protection depends on the infectious strategy and lifestyle of the pathogen. Thus, when opsonizing antibodies such as IgG1 are present (see Section 9-14), opsonization and phagocytosis of extracellular pathogens will be more efficient. If specific IgE is present, then pathogens will also be able to activate mast cells, rapidly initiating an inflammatory response through the release of histamine and leukotrienes. In many cases the most efficient protective immunity is mediated by neutralizing antibody that can prevent pathogens from establishing an infection, and most of the established vaccines against acute childhood viral infections work primarily by inducing protective antibodies. Effective immunity against polio virus, for example, requires preexisting antibody (see Fig. 10.16), because the virus rapidly infects motor neurons and destroys them unless it is immediately neutralized by antibody and prevented from spreading within the body. In polio, specific IgA on mucosal epithelial surfaces also neutralizes the virus before it enters the tissues. Thus, protective immunity can involve effector mechanisms (IgA in this case) that do not operate in the elimination of the primary infection.

When a primary adaptive immune response is successful in halting an infection, it will often clear the primary infection from the body by the effector mechanisms discussed in Chapters 8 and 9. However, as we discuss in Chapter 12, many pathogens evade complete clearance and persist for the life of the

Fig. 10.16 Different effector mechanisms are used to clear primary infections with different classes of pathogens and to protect against reinfection. The defense mechanisms used to clear a primary infection are identified by red shaded boxes. Yellow shading indicates a role in protective immunity. Paler shades indicate less well established mechanisms. It is clear that classes of pathogens elicit similar protective immune responses, reflecting similarities in their lifestyles. The CD4 responses indicated on this diagram refer only to those involved in macrophage activation. In addition, in virtually all diseases, helper CD4 T-cell responses will be involved in stimulating antibody production, class switching, and the production of memory cells.

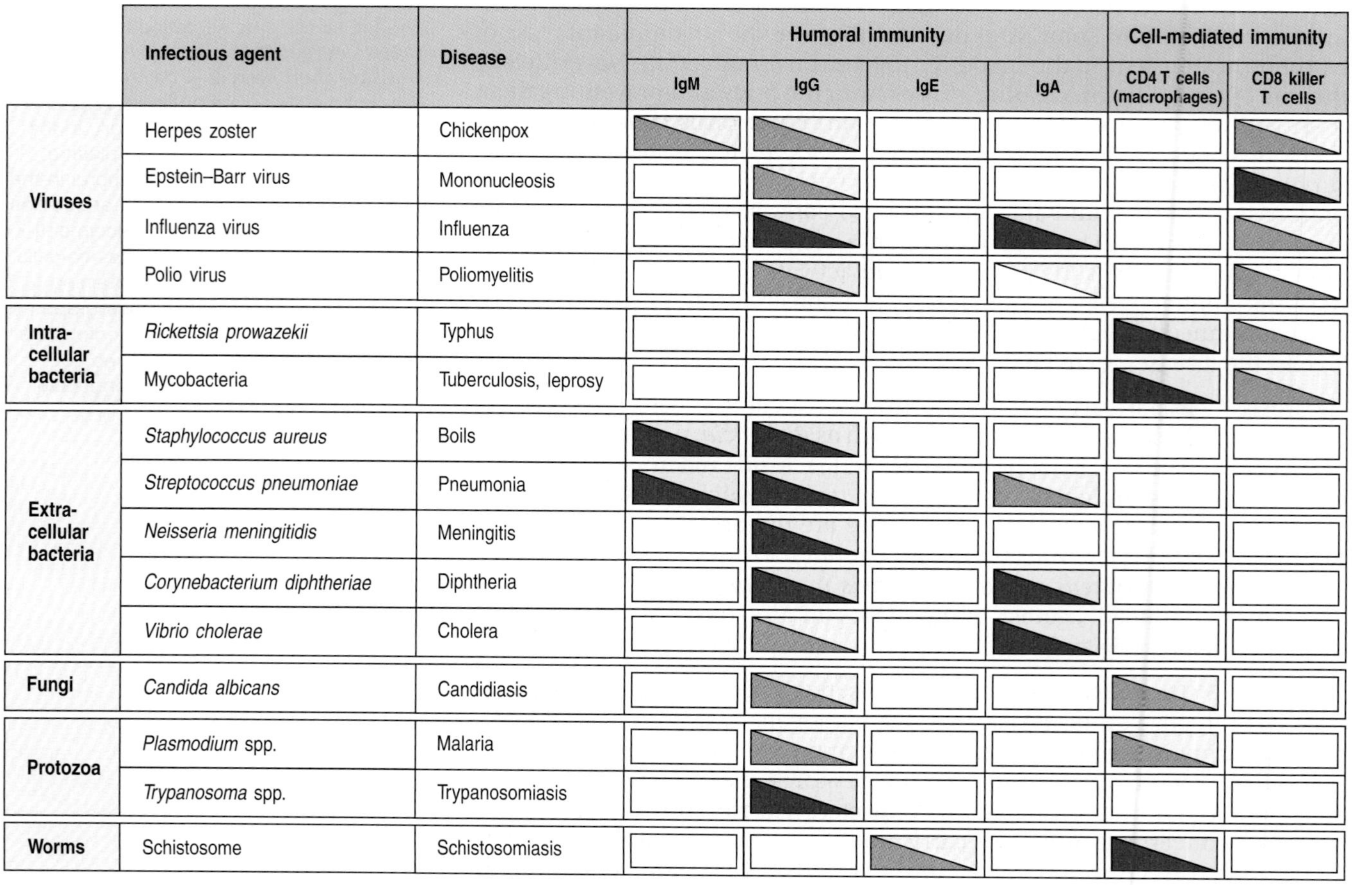

	Infectious agent	Disease	Humoral immunity				Cell-mediated immunity	
			IgM	IgG	IgE	IgA	CD4 T cells (macrophages)	CD8 killer T cells
Viruses	Herpes zoster	Chickenpox						
	Epstein–Barr virus	Mononucleosis						
	Influenza virus	Influenza						
	Polio virus	Poliomyelitis						
Intra-cellular bacteria	*Rickettsia prowazekii*	Typhus						
	Mycobacteria	Tuberculosis, leprosy						
Extra-cellular bacteria	*Staphylococcus aureus*	Boils						
	Streptococcus pneumoniae	Pneumonia						
	Neisseria meningitidis	Meningitis						
	Corynebacterium diphtheriae	Diphtheria						
	Vibrio cholerae	Cholera						
Fungi	*Candida albicans*	Candidiasis						
Protozoa	*Plasmodium* spp.	Malaria						
	Trypanosoma spp.	Trypanosomiasis						
Worms	Schistosome	Schistosomiasis						

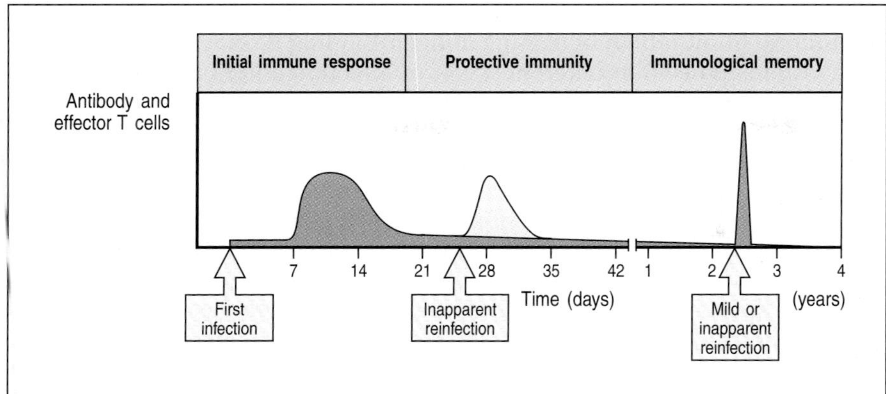

Fig. 10.17 Protective immunity consists of preformed immune reactants and immunological memory. The first time a particular pathogen is encountered, pathogen-specific antibody and effector T cells are produced. After the infection has been cleared, their levels gradually decline. An early reinfection with the same pathogen is rapidly cleared by these preformed immune reactants. There are few symptoms, but levels of immune reactants are found to increase temporarily (light blue peak). Reinfection years afterwards leads to an immediate and rapid increase in pathogen-specific antibody and effector T cells as a result of immunological memory, and disease symptoms are mild or inapparent.

host. The virus herpes zoster, which causes chickenpox on primary infection, then lies latent in the body for years without causing disease, but can later in life, or if the body is stressed, become reactivated and cause shingles.

10-12 Resolution of an infection is accompanied by the death of most of the effector cells and the generation of memory cells.

When an infection is effectively repelled by the adaptive immune system, two things occur. The actions of effector cells remove the specific stimulus that originally recruited them. In the absence of this stimulus, the cells then undergo 'death by neglect,' removing themselves by apoptosis. The dying cells are rapidly cleared by phagocytes and other cells, which recognize the membrane lipid phosphatidylserine. This lipid is normally found only on the inner surface of the plasma membrane, but in apoptotic cells it rapidly redistributes to the outer surface, where it can be recognized by specific receptors on many cells. Thus, not only does the ending of infection lead to the removal of the pathogen, it also leads to the loss of most of the pathogen-specific effector cells.

However, some of the effector cells are retained, and these provide the raw material for memory T-cell and B-cell responses. These are crucially important to the operation of the adaptive immune system. Memory T cells, in particular, are retained virtually forever. The mechanisms underlying the decision to induce apoptosis in most effector cells and to retain only a few are only now being discovered and are still not well understood. It seems likely that the answers will lie in the cytokines produced by the environment and by the T cells themselves, and in the affinity of the T-cell receptors for their antigens.

Summary.

The adaptive immune response is required for effective protection of the host against pathogenic microorganisms. The response of the innate immune system to pathogens helps to initiate the adaptive immune response. Interactions with pathogens lead to the activation of dendritic cells to full antigen-presenting cell status and to the production of cytokines that direct the quality of the CD4 T-cell response. Pathogen antigens are transported to local lymphoid organs by the migrating dendritic cells and are presented to antigen-specific naive T cells that continuously recirculate through the lymphoid organs. T-cell priming and the differentiation of effector T cells occur here on the surface of antigen-loaded dendritic cells, and the effector T cells either leave the lymphoid organ to effect cell-mediated immunity in sites of infection in the tissues or remain in the lymphoid organ to participate in humoral immunity by activating antigen-binding B cells. Distinct types of CD4 responses occur at different stages of infection and to different types of pathogens. During initial stages of infection, cytokines made by activated dendritic cells drive T_H17 responses, which are potent inducers of acute inflammation at sites of infection. In more chronic infections, other cytokines begin to drive either T_H1 or T_H2 responses, and cytokines from these cells begin to shut down T_H17 differentiation. CD8 T cells have an important role in protective immunity, especially in protecting the host against infection by viruses and intracellular infections by *Listeria* and other microbial pathogens that have special means for entering the host cell's cytoplasm. Primary CD8 T-cell responses to pathogens usually require CD4 T-cell help, but can occur in response to some pathogens without such help. CD4-independent responses can lead either to the generation and expansion of antigen-specific cytotoxic T cells or to the nonspecific activation of naive CD8 T cells to secrete IFN-γ, which in turn contributes to host protection. Ideally, the adaptive immune response eliminates the infectious agent and provides the host with a state of protective immunity against reinfection with the same pathogen.

Immunological memory

Having considered how an appropriate primary immune response is mounted, we now turn to how long-lasting protective immunity is generated. The establishment of immunological memory is perhaps the most important consequence of an adaptive immune response, as it enables the immune system to respond more rapidly and effectively to pathogens that have been encountered previously, and prevents them from causing disease. Memory responses, which are called **secondary immune responses**, **tertiary immune responses**, and so on, depending on the number of exposures to antigen, also differ qualitatively from primary responses. This is particularly clear in the antibody response, in which the characteristics of antibodies produced in secondary and subsequent responses are distinct from those produced in the primary response to the same antigen. Memory T-cell responses can also be distinguished qualitatively from the responses of naive or effector T cells. The principal focus of this part of the chapter is the altered character of memory responses, although we also discuss emerging explanations of how immunological memory persists after exposure to antigen.

10-13 Immunological memory is long-lived after infection or vaccination.

Most children in developed countries are now vaccinated against measles virus; before vaccination was widespread, most were naturally exposed to this virus and suffered from an acute, unpleasant, and potentially dangerous illness. Whether through vaccination or infection, children exposed to the virus acquire long-term protection from measles, lasting for most people for the whole of their life. The same is true of many other acute infectious diseases: this state of protection is a consequence of immunological memory.

The basis of immunological memory has been hard to explore experimentally. Although the phenomenon was first recorded by the ancient Greeks and has been exploited routinely in vaccination programs for over 200 years, it is just now becoming clear that memory reflects a small population of specialized **memory cells** formed during the adaptive immune response that can persist in the absence of the antigen that originally induced them. This mechanism of maintaining memory is consistent with the finding that only individuals who were themselves previously exposed to a given infectious agent are immune, and that memory is not dependent on repeated exposure to infection as a result of contacts with other infected individuals. This was established by observations made of populations on remote islands, where a virus such as measles can cause an epidemic, infecting all people living on the island at that time, after which the virus disappears for many years. On reintroduction from outside the island, the virus does not affect the original population but causes disease in those people born since the first epidemic.

A recent study attempted to determine the duration of immunological memory by evaluating responses in people who received vaccinia, the virus used to immunize against smallpox. As smallpox was eradicated in 1978, it is presumed that their responses represent true immunological memory, and are not due to restimulation from time to time by the smallpox virus. The study found strong vaccinia-specific CD4 and CD8 T-cell memory responses as long as 75 years after the original immunization, and from the strength of these responses estimated that the memory response had an approximate half-life of between 8 and 15 years. Half-life represents the time a response takes to reduce to 50% of its original strength. Titers of antivirus antibody remained stable, without measurable decline.

These findings show that immunological memory need not be maintained by repeated exposure to infectious virus. Instead, it is most likely that memory is sustained by long-lived antigen-specific lymphocytes that were induced by the original exposure and that persist until a second encounter with the pathogen. Although most of the memory cells are in a resting state, careful studies have shown that a small percentage are dividing at any one time. What stimulates this infrequent cell division is unclear, but it is likely that cytokines produced either constitutively or during antigen-specific immune responses directed at other, non-cross-reactive, antigens are responsible. The number of memory cells for a given antigen is highly regulated, remaining practically constant during the memory phase, which reflects a control mechanism that maintains a balance between cell proliferation and cell death.

Immunological memory can be measured experimentally in various ways. Adoptive transfer assays (see Appendix I, Section A-42) of lymphocytes from animals immunized with simple, nonliving antigens have been favored for such studies, because the antigen cannot proliferate. In these experiments, the existence of memory cells is measured purely in terms of the transfer of specific responsiveness from an immunized, or 'primed,' animal to a nonimmunized recipient, as tested by a subsequent immunization with the antigen. Animals that received memory cells have a faster and more robust response to antigen challenge than do controls that did not receive cells, or that received cells from nonimmune donors.

Experiments like these have shown that when an animal is first immunized with a protein antigen, functional helper T-cell memory against that antigen appears abruptly and reaches a maximum after 5 days or so. Functional antigen-specific B-cell memory appears some days later, because B-cell activation cannot begin until helper T cells are available, and B cells must then enter a phase of proliferation and selection in lymphoid tissue. By 1 month after immunization, memory B cells are present at their maximum level. These levels of memory cells are then maintained with little alteration for the lifetime of the animal. It is important to recognize that the functional memory elicited in these experiments can be due to the precursors of memory cells as well as the memory cells themselves. These precursors are probably activated T and B cells, some of whose progeny will later differentiate into memory cells. Thus, precursors to memory can appear very shortly after immunization, even though resting memory-type lymphocytes may not yet have developed.

In the following sections we look in more detail at the changes that occur in lymphocytes after antigen priming that lead to the development of resting memory lymphocytes, and discuss the mechanisms that might account for these changes.

10-14 Memory B-cell responses differ in several ways from those of naive B cells.

Immunological memory in B cells can be examined quite conveniently *in vitro* by isolating B cells from immunized mice and restimulating them with antigen in the presence of helper T cells specific for the same antigen. The response observed will be due to **memory B cells**. When compared with a primary B-cell response seen when B cells are isolated from unimmunized mice and stimulated with the same antigen, it is clear that the response generated by memory B cells differs both quantitatively and qualitatively from that generated from naive B cells (Fig. 10.18). B cells that can respond to the antigen increase in frequency by up to 100-fold after their initial priming in the primary immune response. They also produce antibody of higher average affinity than unprimed B lymphocytes as a result of the process of affinity maturation. Thus both clonal expansion and clonal differentiation contribute to B-cell memory.

A primary antibody response is characterized by the initial rapid production of IgM, accompanied by an IgG response, due to class switching, which lags slightly behind it (Fig. 10.19). The secondary antibody response is characterized in its first few days by the production of relatively small amounts of IgM antibody and much larger amounts of IgG antibody, with some IgA and IgE. At the beginning of the secondary response, the source of these antibodies is memory B cells generated in the primary response that are already switched from IgM to these more mature isotypes and express IgG, IgA, or IgE on their surface, as well as a somewhat higher level of MHC class II molecules and B7.1 than is typical of naive B cells.

The average affinity of IgG antibodies increases throughout the primary response and continues to increase during the ongoing secondary and subsequent responses (see Fig. 10.19). The higher affinity of memory B cells for antigen and their higher levels of cell-surface MHC class II molecules facilitate antigen uptake and presentation, which together with an increased expression of co-stimulatory molecules allows memory B cells to initiate their crucial interactions with helper T cells at lower doses of antigen than do naive B cells. This means that B-cell differentiation and antibody production start earlier after antigen stimulation than in the primary response. The secondary

	Source of B cells	
	Unimmunized donor Primary response	**Immunized donor Secondary response**
Frequency of antigen-specific B cells	$1{:}10^4 - 1{:}10^5$	$1{:}10^2 - 1{:}10^3$
Isotype of antibody produced	IgM > IgG	IgG, IgA
Affinity of antibody	Low	High
Somatic hypermutation	Low	High

Fig. 10.18 The generation of secondary antibody responses from memory B cells is distinct from the generation of the primary antibody response. These responses can be studied and compared by isolating B cells from immunized and unimmunized donor mice, and stimulating them in culture in the presence of antigen-specific effector T cells. The primary response usually consists of antibody molecules made by plasma cells derived from a quite diverse population of precursor B cells specific for different epitopes of the antigen and with receptors with a range of affinities for the antigen. The antibodies are of relatively low affinity overall, with few somatic mutations. The secondary response derives from a far more limited population of high-affinity B cells, which have, however, undergone significant clonal expansion. Their receptors and antibodies are of high affinity for the antigen and show extensive somatic mutation. The overall effect is that although there is usually only a 10–100-fold increase in the frequency of activatable B cells after priming, the quality of the antibody response is radically altered, in that these precursors induce a far more intense and effective response.

response is characterized by a more vigorous and earlier generation of plasma cells than in the primary response, thus accounting for the almost immediate abundant production of IgG (see Fig. 10.19).

The distinction between primary and secondary antibody responses is most clearly seen in those cases in which the primary response is dominated by antibodies that are closely related to each other and show little, if any, somatic hypermutation. This occurs in inbred mouse strains in response to certain haptens that are recognized by a limited set of naive B cells. The antibodies produced are encoded by the same V_H and V_L genes in all animals of the strain, suggesting that these variable regions have been selected during evolution for the recognition of determinants on pathogens that happen to cross-react with some haptens. As a result of the uniformity of the primary response, changes in the antibody molecules produced in secondary responses to the same antigens are easy to observe. These differences include not only numerous somatic hypermutations in antibodies containing the dominant V-regions but also the addition of antibodies containing V_H and V_L gene segments not detected in the primary response. These are thought to derive from B cells that were activated at low frequency during the primary response, and thus were not detected, and which differentiated into memory B cells.

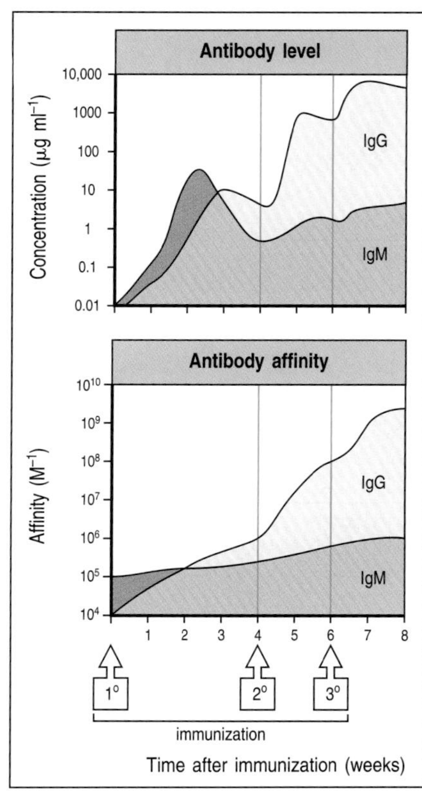

Fig. 10.19 Both the affinity and the amount of antibody increase with repeated immunization. The upper panel shows the increase in antibody concentration with time after a primary (1°), followed by a secondary (2°) and a tertiary (3°), immunization; the lower panel shows the increase in the affinity of the antibodies (affinity maturation). Affinity maturation is seen largely in IgG antibody (as well as in IgA and IgE, which are not shown) coming from mature B cells that have undergone isotype switching and somatic hypermutation to yield higher-affinity antibodies. The blue shading represents IgM on its own; the yellow shading IgG, and the green shading the presence of both IgG and IgM. Although some affinity maturation occurs in the primary antibody response, most arises in later responses to repeated antigen injections. Note that these graphs are on a logarithmic scale; it would otherwise be impossible to represent the overall increase of around a million-fold in the concentration of specific IgG antibody from its initial level.

10-15 Repeated immunization leads to increasing affinity of antibody due to somatic hypermutation and selection by antigen in germinal centers.

In secondary and subsequent immune responses, any antibodies persisting from previous responses are immediately available to bind the newly introduced pathogen. These antibodies divert the antigen to phagocytes for degradation and disposal (see Section 9-22) and if there is sufficient antibody to clear or inactivate the pathogen completely, it is possible that no secondary immune response will ensue. If antigen persists, a secondary B-cell response will be initiated in the peripheral lymphoid organs. Antibodies remaining from the primary response, and those produced early in the secondary response, are important in driving the considerable increase in antibody affinity that occurs during the secondary response (see Fig. 10.19). This is because only memory B cells whose receptors bind the antigen with sufficient avidity to compete with the preexisting antibody will take up free antigen, process it and present it on their surface—and thus be able to get help from T cells.

Like a primary immune response, a secondary B-cell response begins with the proliferation of B cells and T cells at the interface between the T-cell and B-cell zones. Memory T cells can enter nonlymphoid tissues as a result of changes in cell-surface molecules that affect migration and homing (see Section 10-6), but it is thought that memory B cells continue to recirculate through the same secondary lymphoid compartments as naive B cells, principally the follicles of the spleen, lymph nodes, and the Peyer's patches of the gut mucosa. Some memory B cells can also be found in the marginal zones of the spleen (see Fig. 1.19), although it is not clear whether these represent a distinct subset of memory B cells.

Memory B cells that have picked up antigen present peptide:MHC class II complexes to their cognate effector helper T cells surrounding and infiltrating the germinal centers. Contact between the antigen-presenting B cells and helper T cells leads to an exchange of activating signals and the rapid proliferation of both activated antigen-specific B cells and helper T cells. As the higher-affinity memory B cells compete most effectively for antigen, only these B cells are efficiently stimulated in the secondary immune response. Reactivated B cells that have not yet undergone differentiation into plasma

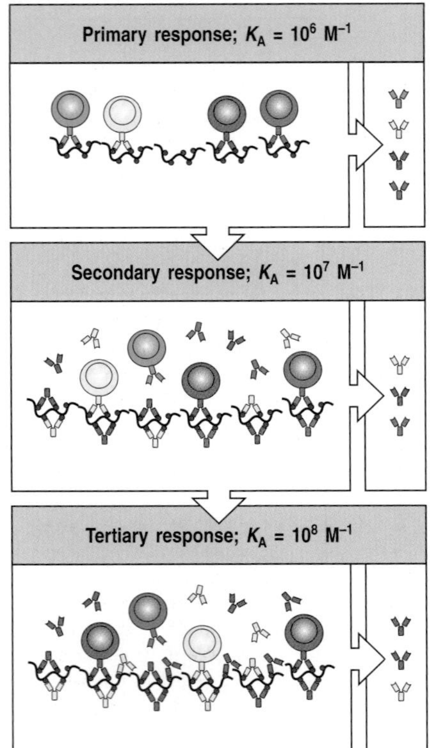

Fig. 10.20 The mechanism of affinity maturation in an antibody response. At the beginning of a primary response, B cells with receptors of a wide variety of affinities (K_A), most of which will bind antigen with low affinity, take up antigen, present it to helper T cells, and become activated to produce antibody of varying and relatively low affinity (top panel). These antibodies then bind and clear antigen, so that only those B cells with receptors of the highest affinity can continue to capture antigen and interact effectively with helper T cells. Such B cells will therefore be selected to undergo further expansion and clonal differentiation, and the antibodies they produce will dominate a secondary response (middle panel). These higher-affinity antibodies will in turn compete for antigen and select for the activation of B cells bearing receptors of still higher affinity in the tertiary response (bottom panel).

cells migrate into the follicle and become germinal center B cells. There, they enter a second round of proliferation, during which the DNA encoding their immunoglobulin V domains undergoes somatic hypermutation, before differentiating into antibody-secreting plasma cells (see Section 9-8). The affinity of the antibodies produced rises progressively and rapidly, because B cells with the highest-affinity antigen receptors produced by somatic hypermutation bind antigen most efficiently, and will be selected to proliferate by their interactions with antigen-specific helper T cells in the germinal center (Fig. 10.20).

10-16 Memory T cells are increased in frequency compared with naive T cells specific for the same antigen and have distinct activation requirements and cell-surface proteins that distinguish them from effector T cells.

Because the T-cell receptor does not undergo class switching or somatic hypermutation, it is not as easy to identify a memory T cell unequivocally as it is to identify a memory B cell. After immunization, the number of T cells reactive to a given antigen increases markedly as effector T cells are produced, and then falls back to persist at a level 100–1000-fold above the initial frequency for the rest of the life of the animal or person (Fig. 10.21). These persisting cells are designated **memory T cells**. They are long-lived cells with a particular set of cell-surface proteins, responses to stimuli, and expression of genes that control cell survival. Overall, their cell-surface proteins are similar to those of effector cells, but there are some distinctive differences (Fig. 10.22). In B cells, there is an obvious distinction between effector and memory cells, because effector B cells are terminally differentiated plasma cells that have already been activated to secrete antibody until they die.

A major problem in experiments aimed at establishing the existence of memory T cells is that many assays for T-cell effector function take several days, during which the putative memory T cells are reinduced to effector cell status. Thus, assays requiring several days do not distinguish preexisting effector cells from memory T cells, because memory cells can acquire effector activity during the period of the assay. This problem does not apply to cytotoxic T cells, however, because cytotoxic effector T cells can program a target cell for lysis in 5 minutes, whereas memory CD8 T cells need more time

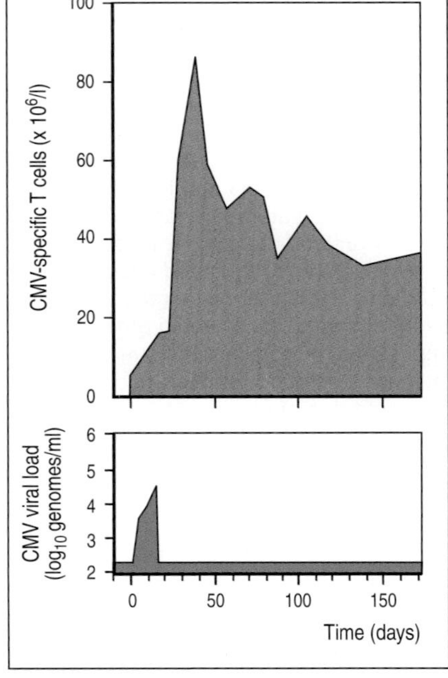

Fig. 10.21 Generation of memory T cells after a virus infection. After an infection, in this case a reactivation of latent cytomegalovirus (CMV), the number of T cells specific for viral antigen increases dramatically and then falls back to give a sustained low level of memory T cells. The top panel shows the numbers of T cells (orange); the bottom panel shows the course of the virus infection (blue), as estimated by the amount of viral DNA in the blood. Data courtesy of G. Aubert.

Protein	Naive	Effector	Memory	Comments
CD44	+	+++	+++	Cell-adhesion molecule
CD45RO	+	+++	+++	Modulates T-cell receptor signaling
CD45RA	+++	+	+++	Modulates T-cell receptor signaling
CD62L	+++	–	Some +++	Receptor for homing to lymph node
CCR7	+++	+/–	Some +++	Chemokine receptor for homing to lymph node
CD69	–	+++	–	Early activation antigen
Bcl-2	++	+/–	+++	Promotes cell survival
Interferon-γ	–	+++	+++	Effector cytokine; mRNA present and protein made on activation
Granzyme B	–	+++	+/–	Effector molecule in cell killing
FasL	–	+++	+	Effector molecule in cell killing
CD122	+/–	++	++	Part of receptor for IL-15 and IL-2
CD25	–	++	–	Part of receptor for IL-2
CD127	++	–	+++	Part of receptor for IL-7
Ly6C	+	+++	+++	GPI-linked protein
CXCR4	+	+	++	Receptor for chemokine CXCL12; controls tissue migration
CCR5	+/–	++	Some +++	Receptor for chemokines CCL3 and CCL4; tissue migration

Fig. 10.22 Expression of many proteins alters when naive T cells become memory T cells. Proteins that are differently expressed in naive T cells, effector T cells and memory T cells include adhesion molecules, which govern interactions with antigen-presenting cells and endothelial cells; chemokine receptors, which affect migration to lymphoid tissues and sites of inflammation; proteins and receptors that promote the survival of memory cells; and proteins that are involved in effector functions, such as granzyme B. Some changes also increase the sensitivity of the memory T cell to antigen stimulation. Many of the changes that occur in memory T cells are also seen in effector cells, but some, such as expression of the cell-surface proteins CD25 and CD69, are specific to effector T cells; others, such as expression of the survival factor Bcl-2, are limited to long-lived memory T cells. This list represents a general picture that applies to both CD4 and CD8 T cells in mice and humans, but some details that may differ between these sets of cells have been omitted for simplicity.

than this to be reactivated to become cytotoxic. Thus, their cytotoxic actions will appear later than those of any preexisting effector cells, even though they can become activated without undergoing DNA synthesis, as shown by studies carried out in the presence of mitotic inhibitors.

Recently it has become possible to track particular clones of antigen-specific CD8 T cells by staining them with tetrameric peptide:MHC complexes (see Appendix I, Section A-28). It has been found that the number of antigen-specific CD8 T cells increases markedly during an infection, and then decreases by up to 100-fold; nevertheless, this final level is distinctly higher than before priming. These cells continue to express some markers characteristic of activated cells, such as CD44, but stop expressing other activation markers, such as CD69. In addition they express more Bcl-2, a protein that promotes cell survival and may be responsible for the long half-life of memory CD8 cells.

The α subunit of the IL-7 receptor (IL-7Rα or CD127) may be a good marker for activated T cells that will become long-lived memory cells (see Fig. 10.22). Naive T cells express IL-7Rα, but it is rapidly lost upon activation and is not expressed by most effector T cells. For example, during the peak of the effector response against lymphocytic choriomeningitis virus (LCMV) in mice, around day 7 of the infection, a small population of approximately 5% of CD8 effector T cells expressed high levels of IL-7Rα. Adoptive transfer of these cells, but not the effector T cells expressing low levels of IL-7Rα, could provide functional CD8 T-cell memory to uninfected mice (Fig. 10.23). This experiment suggests that the early maintenance, or the reexpression, of IL-7Rα identifies effector CD8 T cells that generate memory T cells, although it is still not known whether, and how, this process is regulated. Memory T cells are more sensitive to restimulation by antigen than are naive T cells, and more quickly and more vigorously produce cytokines such as IFN-γ in response to such stimulation.

The issue of memory has been more difficult to address directly for CD4 T-cell responses, in part because their responses are smaller than those of CD8 T cells and also because, until recently, there were no peptide:MHC class II reagents similar to the peptide:MHC class I tetramers. Nevertheless, the transfer and priming of naive T cells carrying T-cell receptor transgenes that give the T cells a known peptide:MHC specificity made it possible to visualize memory CD4 T cells. They appear as a long-lived population of cells that share some surface characteristics of activated effector T cells but are distinct from effector T cells in that they require additional restimulation before acting on target cells. Changes in three cell-surface proteins—L-selectin, CD44, and CD45—that occur on the putative memory CD4 T cells after exposure to antigen are particularly significant. L-selectin is lost on most memory CD4 T cells, whereas CD44 levels are increased on all memory T cells; these changes contribute to directing the migration of memory T cells from the blood into the tissues rather than directly into lymphoid tissue. The isoform of CD45 changes because of alternative splicing of exons that encode the extracellular domain of CD45, leading to isoforms, such as CD45RO, that are smaller and more readily associated with the T-cell receptor and facilitate antigen recognition (see Fig. 10.22). These changes are characteristic of cells that have been activated to become effector T cells, yet some of the cells on which these changes have occurred have many characteristics of resting CD4 T cells, suggesting that they represent memory CD4 T cells. Only after reexposure to antigen on an antigen-presenting cell do they achieve effector T-cell status and acquire all the characteristics of T_H2 or T_H1 cells, secreting IL-4 and IL-5, or IFN-γ, respectively.

It therefore seems reasonable to designate these cells as memory CD4 T cells and to surmise that naive CD4 T cells can differentiate either into effector T cells or into memory T cells that can later be activated to effector status. As with memory CD8 T cells, the field is currently being revolutionized by the direct staining of CD4 T cells with peptide:MHC class II tetramers (see Appendix I, Section A-28). This technique allows one not only to identify

Fig. 10.23 Expression of the IL-7 receptor (IL-7R) indicates which CD8 effector T cells can generate robust memory responses. Mice expressing a T-cell receptor (TCR) transgene specific for a viral antigen from lymphocytic choriomeningitis virus (LCMV) were infected and effector cells were collected on day 11. Effector CD8 T cells expressing high levels of IL-7R (IL-7R^{hi}, blue) were separated and transferred into one group of naive mice, and effector CD8 T cells expressing low IL-7R (IL7R^{lo}, green) were transferred into another group. Three weeks after transfer, the mice were challenged with a bacterium engineered to express the original viral antigen, and the numbers of responding transferred T cells (detected by their expression of the transgenic TCR) were measured at various times after challenge. Only the transferred IL-7R^{hi} effector cells could generate a robust expansion of CD8 T cells after the secondary challenge.

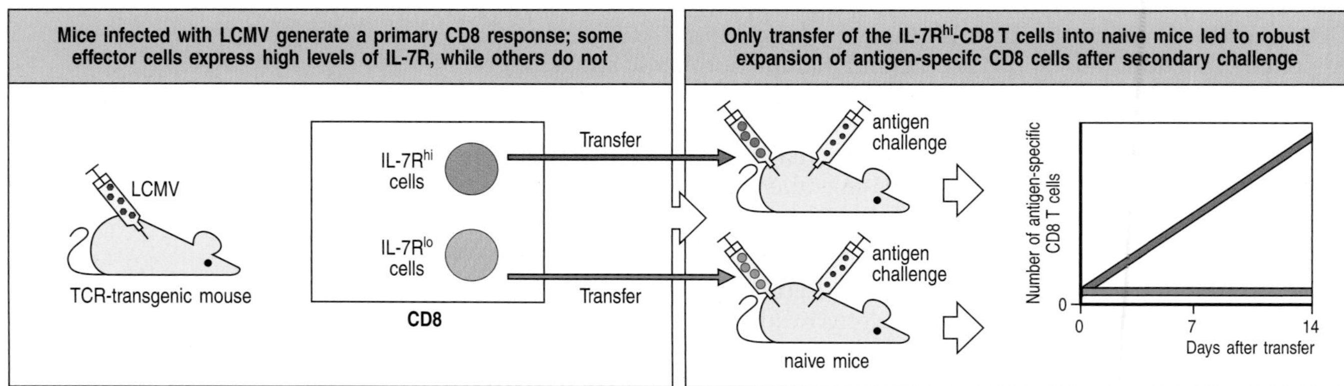

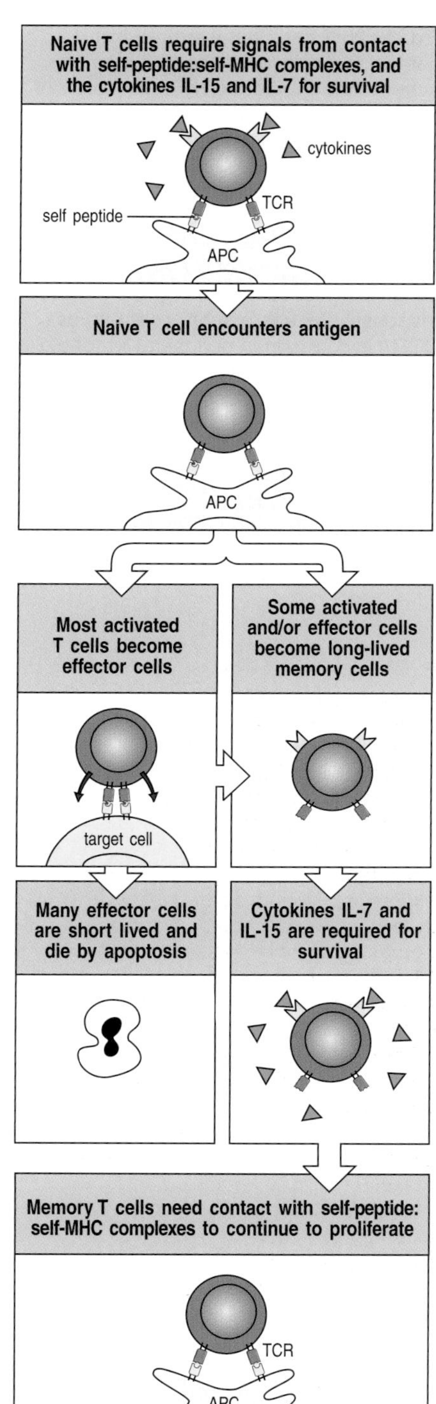

Fig. 10.24 Naive T cells and memory T cells have different requirements for survival. For their survival in the periphery, naive T cells require periodic stimulation with the cytokines IL-7 and IL-15 and with self antigens presented by MHC molecules. On priming with its specific antigen, a naive T cell divides and differentiates. Most of the progeny differentiate into relatively short-lived effector cells, but some become long-lived memory T cells, which need to be sustained by cytokines but do not require contact with self-peptide:self-MHC complexes purely for survival. However, contact with self antigens does seem necessary for memory T cells to continue to proliferate and thus keep up their numbers in the memory pool.

antigen-specific CD4 T cells but also, by using intracellular cytokine staining (see Appendix I, Section A-27), to determine whether they are T_H1 or T_H2 cells. These improvements in the identification and phenotyping of CD4 T cells will rapidly increase our knowledge of these hitherto mysterious cells and could contribute valuable comparative information on naive, memory, and effector CD4 T cells.

The homeostatic mechanisms governing the survival of memory T cells differ from those for naive T cells. Memory T cells divide more frequently than do naive T cells, and their expansion is controlled by a balance between proliferation and cell death. As with naive cells, survival of memory T cells requires stimulation by the cytokines IL-7 and IL-15. IL-7 is required for the survival of both CD4 and CD8 memory T cells, but in addition, IL-15 is critical for the long-term survival and proliferation of CD8 memory T cells under normal conditions. For memory CD4 T cells, the role of IL-15 is still controversial.

As well as cytokine stimulation, naive T cells also require contact with self-peptide:self-MHC complexes for their long-term survival in the periphery (see Section 7-29), but it seems that memory T cells do not have this requirement. It has been found, however, that memory T cells surviving after transfer to MHC-deficient hosts have some defects in typical T-cell memory functions, indicating that stimulation by self-peptide:self-MHC complexes may be required for their continued proliferation and optimal function (Fig. 10.24).

10-17 Memory T cells are heterogeneous and include central memory and effector memory subsets.

It has recently been discovered that both CD4 and CD8 T cells can differentiate into two types of memory cells with distinct activation characteristics (Fig. 10.25). One type is called an **effector memory cell** because it can rapidly mature into an effector T cell and secrete large amounts of IFN-γ, IL-4, and IL-5 early after restimulation. These cells lack the chemokine receptor CCR7 but express high levels of β_1 and β_2 integrins, as well as receptors for inflammatory chemokines. This profile suggests that these effector memory cells are specialized for rapidly entering inflamed tissues. The other type is called a **central memory cell**. It expresses CCR7 and would therefore be expected to recirculate more easily to the T-cell zones of peripheral lymphoid tissues, as do naive T cells. Central memory cells are very sensitive to cross-linking of their T-cell receptors and rapidly express CD40 ligand in response; however, they take longer than effector memory cells to differentiate into effector T cells and thus do not secrete such large amounts of cytokines early after restimulation.

The distinction between central memory cells and effector memory cells has been made both in humans and in the mouse. However, this general distinction does not imply that each subset is a uniform population. Within the CCR7-expressing central memory subset there are extensive differences in the expression of other markers, particularly receptors for other chemokines. For example, within the CCR7-positive central memory cells there is a subset

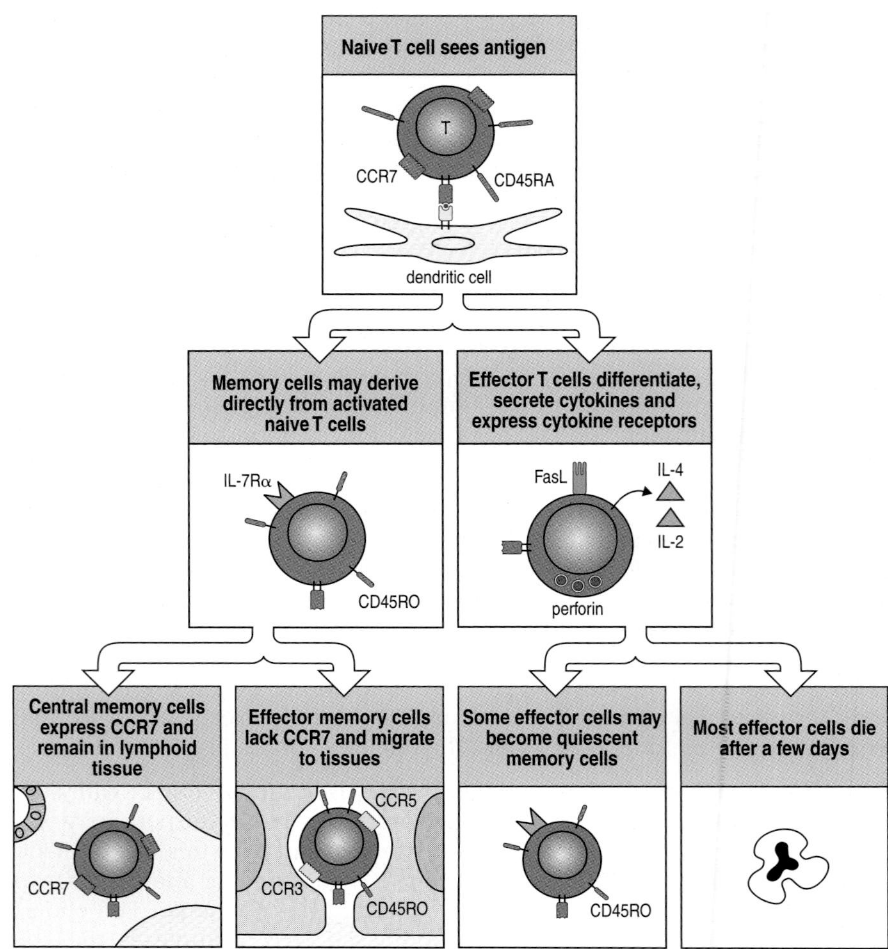

Fig. 10.25 T cells differentiate into central memory and effector memory subsets distinguished by expression of the chemokine receptor CCR7. Quiescent memory cells bearing the characteristic CD45RO surface protein can arise from activated effector cells (right half of diagram) or directly from activated naive T cells (left half of diagram). Two types of quiescent memory T cells can derive from the primary T-cell response. Central memory cells express CCR7 and remain in peripheral lymphoid tissues after restimulation. The other type of memory cells, effector memory cells, mature rapidly into effector T cells after restimulation and secrete large amounts of IFN-γ, IL-4, and IL-5. They do not express the receptor CCR7, but express receptors (CCR3 and CCR5) for inflammatory chemokines.

of cells that express CXCR5, a receptor for CXCL13, a chemokine made in B-cell follicles. These CXCR5-positive central memory T cells have been called **follicular helper cells**; they produce IL-2 and provide help for B cells.

On stimulation by antigen, central memory cells rapidly lose expression of CCR7 and differentiate into effector memory cells. Effector memory cells are also heterogeneous in the chemokine receptors they express, and have been classified according to chemokine receptors typical of T_H1, such as CCR5, and of T_H2, such as CCR4. Central memory cells are not yet committed to particular effector lineages and even effector memory cells are not fully committed to the T_H1 or the T_H2 lineage, although there is some correlation between their eventual output of T_H1 or T_H2 cells and the chemokine receptors expressed. Further stimulation with antigen seems to drive the differentiation of effector memory cells gradually into the distinct effector T-cell lineages.

10-18 CD4 T-cell help is required for CD8 T-cell memory and involves CD40 and IL-2 signaling.

We described earlier how primary CD8 T-cell responses to *Listeria monocytogenes* can occur in mice that lack CD4 T cells. After 7 days of infection, wild-type mice and mice lacking CD4 T cells show equivalent expansion and activity of pathogen-specific CD8 effector T cells (see Section 10-8). They are not, however, equally able to generate memory CD8 T cells. Mice that lack

CD4 T cells due to a deficiency in MHC class II were found to generate much weaker secondary responses, characterized by many fewer expanding memory CD8 T cells specific for the pathogen. In this experiment, the *Listeria* carried a gene for the protein ovalbumin, and it was the response to this protein that was measured as a marker for CD8 T-cell memory (Fig. 10.26). CD4 T cells in these mice are lacking both during the primary response and in any secondary challenge, and so the requirement for CD4 T cells could be either in the initial programming of CD8 T cells during their primary activation to enable memory development, or, alternatively, in providing help only during the secondary memory response.

This question was resolved by observations that memory CD8 T cells that developed in the absence of CD4 help showed a greatly reduced ability to proliferate even after they were transferred into wild-type mice. This indicates that it is their programming to become memory cells that is deficient and not simply a lack of CD4 T-cell help at the time of secondary responses. The requirement for CD4 help in CD8 memory generation has also been demonstrated by experiments in which CD4 T cells were depleted by antibody treatment or in which mice were deficient in the CD4 gene. These experiments indicate that CD4 T-cell help is necessary for programming naive CD8 T cells to be able to generate memory cells that are capable of robust expansion in a secondary immune response.

The mechanism underlying this effect of CD4 T cells is not completely understood, but it is likely to involve at least two types of signals to the CD8 T cell—those received through CD40 and those received through the IL-2 receptor. CD8 T cells that do not express CD40 are unable to generate memory T cells. Although many cells could potentially express the CD40 ligand needed to stimulate CD40, it is most likely that CD4 T cells are the source of this signal.

The requirement for IL-2 signaling in programming CD8 memory was discovered by using CD8 T cells with a genetic deficiency in the IL-2Rα subunit, which were therefore unable to respond to IL-2. Because IL-2Rα signaling is required for the development of T_{reg} cells, mice lacking IL-2Rα develop a lymphoproliferative disorder. However, this disorder does not develop in mice that are mixed bone marrow chimeras harboring both wild-type and IL-2Rα-deficient cells, and these chimeras can be used to study the behavior of IL-2Rα-deficient cells. When these chimeric mice were infected with lymphocytic choriomeningitis virus (LCMV) and their responses were tested, memory CD8 responses were found to be defective specifically in the T cells lacking IL-2Rα.

CD4 T cells also appear to provide help in maintaining the numbers of CD8 memory T cells, and this seems to be distinct from their effect in programming naive CD8 T cells to become memory cells. When CD8 memory T cells are

Fig. 10.26 CD4 T cells are required for the development of functional CD8 memory T cells. Mice that do not express MHC class II molecules (MHC II$^{-/-}$) fail to develop CD4 T cells. Wild-type and MHC II$^{-/-}$ mice were infected with *Listeria monocytogenes* expressing the model antigen ovalbumin (LM-OVA). After 7 days, the numbers of OVA-specific CD8 T cells can be measured using specific MHC tetramers that contain an OVA peptide, and therefore bind to T-cell receptors that react with this antigen. After 7 days of infection, mice lacking CD4 T cells have the same number of OVA-specific CD8 T cells as wild-type mice. However, when mice are allowed to recover for 60 days, during which time memory T cells develop, and are then rechallenged with LM-OVA, the mice lacking CD4 T cells fail to expand CD8 memory cells specific to OVA, whereas there is a strong CD8 memory response in the wild-type mice.

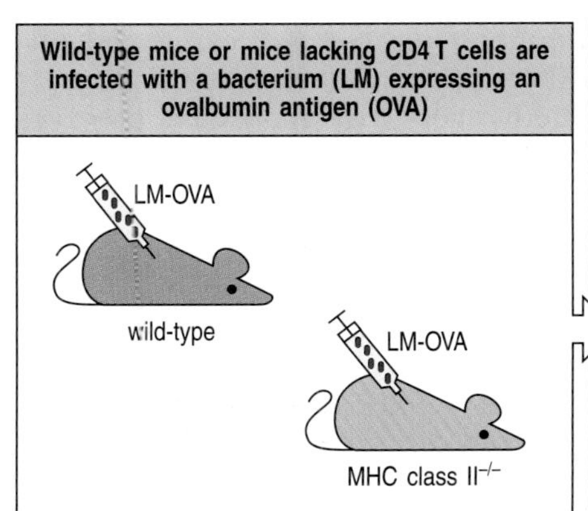
Wild-type mice or mice lacking CD4 T cells are infected with a bacterium (LM) expressing an ovalbumin antigen (OVA)
LM-OVA
wild-type
LM-OVA
MHC class II$^{-/-}$

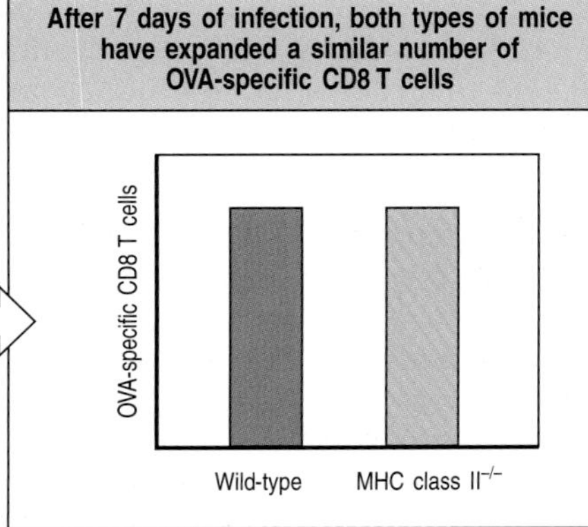
After 7 days of infection, both types of mice have expanded a similar number of OVA-specific CD8 T cells
OVA-specific CD8 T cells
Wild-type
MHC class II$^{-/-}$

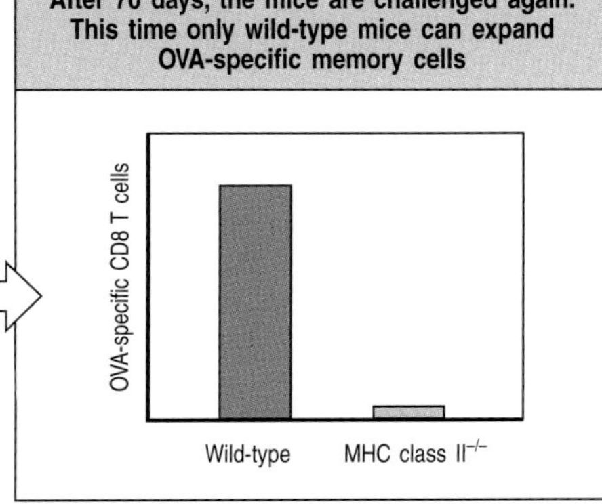
After 70 days, the mice are challenged again. This time only wild-type mice can expand OVA-specific memory cells
OVA-specific CD8 T cells
Wild-type
MHC class II$^{-/-}$

transferred into immunologically naive mice, the presence or absence of CD4 T cells in the recipient influences the maintenance of the CD8 memory cells. Transfer of CD8 memory cells into mice lacking CD4 T cells is followed by a gradual decrease in the number of memory cells in comparison with a similar transfer into wild-type mice. Also, CD8 effector cells transferred into mice lacking CD4 T cells had a relative impairment of CD8 effector functions. These experiments show that the CD4 T cells activated during an immune response have a significant impact on the quantity and quality of the CD8 T-cell response, even when they are not needed for the initial CD8 T-cell activation. CD4 T cells help to program naive CD8 T cells to be able to generate memory T cells, help to promote efficient effector activity, and help to maintain memory T-cell numbers.

10-19 In immune individuals, secondary and subsequent responses are mainly attributable to memory lymphocytes.

In the normal course of an infection, a pathogen proliferates to a level sufficient to elicit an adaptive immune response and then stimulates the production of antibodies and effector T cells that eliminate the pathogen from the body. Most of the effector T cells then die, and antibody levels gradually decline, because the antigens that elicited the response are no longer present at the level needed to sustain it. We can think of this as feedback inhibition of the response. Memory T and B cells remain, however, and maintain a heightened ability to mount a response to a recurrence of infection with the same pathogen.

The antibody and memory lymphocytes remaining in an immunized individual also largely prevent the activation of naive B and T cells on a subsequent encounter with the same antigen. This can be shown by passively transferring antibody or memory T cells to naive recipients; when the recipient is then immunized with the same antigen, naive lymphocytes do not respond. Responses to other antigens are, however, unaffected.

This phenomenon has been put to practical use to prevent Rh^- mothers from making an immune response to a Rh^+ fetus, which can result in hemolytic disease of the newborn (see Appendix I, Section A-11). If anti-Rh antibody is given to the mother before she is first exposed to her child's Rh^+ red blood cells, her response will be inhibited. The mechanism of this suppression is likely to involve the antibody-mediated clearance and destruction of red blood cells from the fetus that have entered the mother, thus preventing naive B cells and T cells from mounting an immune response. Memory B-cell responses are, however, not inhibited by antibody, so the Rh^- mothers at risk must be identified and treated before a primary response has occurred. Because of their high affinity for antigen and alterations in their B-cell receptor signaling requirements, memory B cells are much more sensitive to the small amounts of antigen that cannot be efficiently cleared by the passive anti-Rh antibody. The ability of memory B cells to be activated to produce antibody, even when exposed to preexisting antibody, also allows secondary antibody responses to occur in individuals who are already immune.

The presence of antigen-specific memory T cells also prevents the activation of naive T cells to the same antigen, as shown by the suppression of naive T-cell activation after the adoptive transfer of immune T cells to naive syngeneic mice. This effect has been shown most clearly for cytotoxic T cells. It is possible that, once reactivated, the memory CD8 T cells regain cytotoxic activity sufficiently rapidly to kill professional antigen-presenting cells, such as dendritic cells, before these can activate naive CD8 T cells.

These suppressive mechanisms might also explain the phenomenon known as **original antigenic sin**. This term was coined to describe the tendency of people to make antibodies only against epitopes expressed on the first influenza virus variant to which they are exposed, even in subsequent infections with variants that bear additional, highly immunogenic, epitopes (Fig. 10.27). Antibodies against the original virus will tend to suppress responses of naive B cells specific for the new epitopes. This might benefit the host by using only those B cells that can respond most rapidly and effectively to the virus. This pattern is broken only if the person is exposed to an influenza virus that lacks all epitopes seen in the original infection, because now no preexisting antibodies bind the virus and naive B cells are able to respond.

Summary.

Protective immunity against reinfection is one of the most important consequences of adaptive immunity. Protective immunity depends not only on preformed antibody and effector T cells but most importantly on the establishment of a population of lymphocytes that mediate long-lived immunological memory. The capacity of these cells to respond rapidly to restimulation with the same antigen can be transferred to naive recipients by primed B and T cells. The precise changes that distinguish naive, effector, and memory lymphocytes are now being characterized, and include the regulation of expression of receptors for cytokines, such as IL-7, that help to maintain these cells, and the regulation of chemokine receptors, such as CCR7, that distinguish between functional subsets of memory cells. The advent of receptor-specific reagents—MHC tetramers—has allowed an analysis of the relative contributions of clonal expansion and differentiation to the memory phenotype. Memory B cells can be distinguished by changes in their immunoglobulin genes because of isotype switching and somatic hypermutation, and secondary and subsequent immune responses are characterized by antibodies with increasing affinity for the antigen. The complex interplay between CD4 and CD8 T cells in regulating memory is being revealed. Although CD8 T cells can generate effective primary responses in the absence of help from CD4 T cells, it is becoming clear that CD4 T cells play an integral role in regulating CD8 T-cell memory. These issues will be

Fig. 10.27 When individuals who have been infected with one variant of influenza virus are infected with a second or third variant, they make antibodies only against epitopes that were present on the initial virus. A child infected for the first time with an influenza virus at 2 years of age makes a response to all epitopes (left panel). At age 5 years, the same child exposed to a different influenza virus responds preferentially to those epitopes shared with the original virus, and makes a smaller than normal response to new epitopes on the virus (middle panel). Even at age 20 years, this commitment to respond to epitopes shared with the original virus, and the subnormal response to new epitopes, is retained (right panel). This phenomenon is called 'original antigenic sin.'

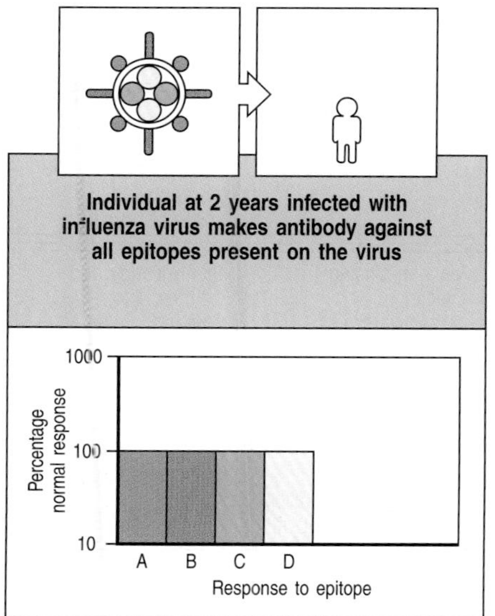
Individual at 2 years infected with influenza virus makes antibody against all epitopes present on the virus
1000
100
10
Percentage normal response
A B C D
Response to epitope

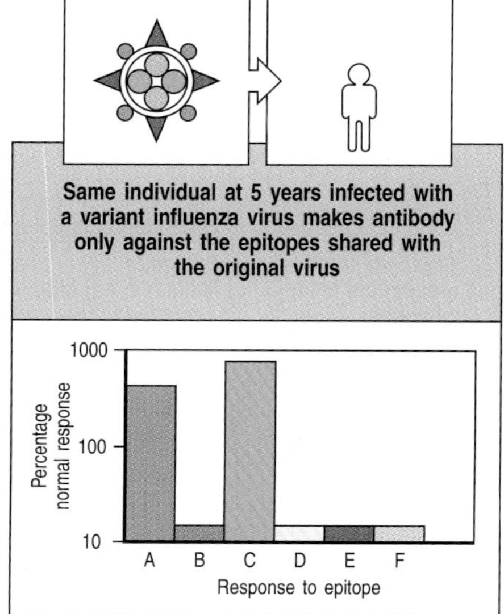
Same individual at 5 years infected with a variant influenza virus makes antibody only against the epitopes shared with the original virus
1000
100
10
Percentage normal response
A B C D E F
Response to epitope

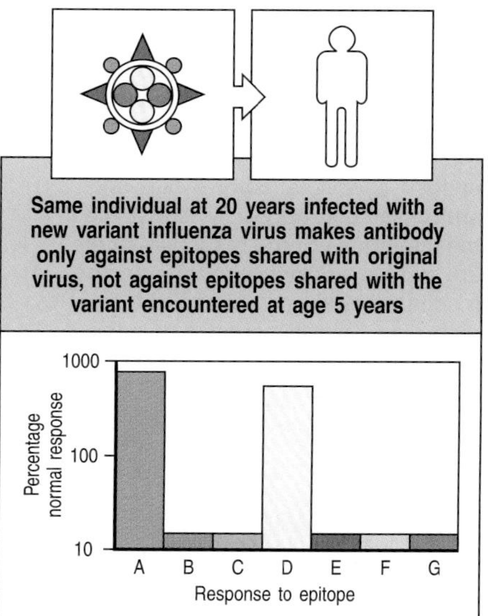
Same individual at 20 years infected with a new variant influenza virus makes antibody only against epitopes shared with original virus, not against epitopes shared with the variant encountered at age 5 years
1000
100
10
Percentage normal response
A B C D E F G
Response to epitope

critical in understanding, for example, how to design effective vaccines for diseases such as HIV/AIDS.

Summary to Chapter 10.

Vertebrates resist infection by pathogenic microorganisms in several ways. The innate defenses can act immediately and may succeed in repelling the infection, but if not they are followed by a series of induced early responses that help to contain the infection as adaptive immunity develops. These first two phases of the immune response rely on recognizing the presence of infection by using the nonclonotypic receptors of the innate immune system. They are summarized in Fig. 10.28 and covered in detail in Chapter 2. Specialized subsets of T cells and B cells, which can be viewed as intermediates between innate and adaptive immunity, include NK T cells, which can help to bias the CD4 T-cell response toward a T_H1 or a T_H2 phenotype, and NK cells, which can be recruited to lymph nodes and secrete IFN-γ, and thus promote a T_H1 response. The third phase of an immune response is the adaptive immune response (see Fig. 10.28), which is mounted in the peripheral lymphoid tissue that serves the particular site of infection and takes several days to develop, because T and B lymphocytes must encounter their specific antigen, proliferate, and differentiate into effector cells. T-cell dependent B-cell responses cannot be initiated until antigen-specific T cells have had a chance to proliferate and differentiate. Once an adaptive immune response has occurred, the antibodies and effector T cells are dispersed via the circulation and recruited into the infected tissues; the infection is usually controlled and the pathogen is contained or eliminated. The final effector mechanisms used to clear an infection depend on the type of infectious agent, and in most cases they are the same as those employed in the early phases of immune defense; only the recognition mechanism changes and is more selective (see Fig. 10.28).

	Phases of the immune response		
	Immediate (0–4 hours)	Early (4–96 hours)	Late (96–100 hours)
	Nonspecific Innate No memory No specific T cells	Nonspecific + specific Inducible No memory No specific T cells	Specific Inducible Memory Specific T cells
Barrier functions	Skin, epithelia	Local inflammation (C5a) Local TNF-α	IgA antibody in luminal spaces IgE antibody on mast cells Local inflammation
Response to extracellular pathogens	Phagocytes Alternative and MBL complement pathway	Mannan-binding lectin C-reactive protein T-independent B-cell antibody Complement	IgG antibody and Fc receptor-bearing cells IgG, IgM antibody + classical complement pathway
Response to intracellular bacteria	Macrophages	Activated NK-dependent macrophage activation IL-1, IL-6,TNF-α, IL-12	T-cell activation of macrophages by IFN-γ
Response to virus-infected cells	Natural killer (NK) cells	IFN-α and IFN-β IL-12-activated NK cells	Cytotoxic T cells IFN-γ

Fig. 10.28 The components of the three phases of the immune response against different classes of microorganisms. The mechanisms of innate immunity that operate in the first two phases of the immune response are described in Chapter 2, and thymus-independent (T-independent) B-cell responses are covered in Chapter 9. The early phases contribute to the initiation of adaptive immunity, and they influence the functional character of the antigen-specific effector T cells and antibodies that appear on the scene in the late phase of the response. There are striking similarities in the effector mechanisms at each phase of the response; the main change is in the recognition structures used.

An effective adaptive immune response leads to a state of protective immunity. This state consists of the presence of effector cells and molecules produced in the initial response, and of immunological memory. Immunological memory is manifested as a heightened ability to respond to pathogens that have previously been encountered and successfully eliminated. Memory T and B lymphocytes have the property of being able to transfer immune memory to naive recipients. The precise mechanism of the maintenance of immunological memory, which is arguably the crucial feature of adaptive immunity, seems to be due to the presence of certain cytokines and to homeostatic interactions with self-MHC:self-peptide complexes. The artificial induction of protective immunity, which includes immunological memory, by vaccination is the most outstanding accomplishment of immunology in the field of medicine. The understanding of how this is accomplished is now catching up with its practical success. However, as we will see in Chapter 12, many pathogens do not induce protective immunity that completely eliminates the pathogen, so we will need to learn what prevents this before we can prepare effective vaccines against these pathogens.

Questions.

10.1 *Communication is critical in any large enterprise. (a) How is the body alerted to an invasion by microbes, and (b) how does it ensure that its responses reach the site of infection?*

10.2 *The immune system responds to particular classes of pathogens in different ways. What properties of viruses and bacteria are used to activate T_H1 responses to them, and what host cells provide the information about the type of pathogen present?*

10.3 *Differentiated T cells require continued signals to maintain their function. (a) What signals do T_H1 cells need? (b) What advantages might the requirement for continued signals have? What disadvantages?*

10.4 *The different subsets of effector T cells regulate each other's development. What advantage might derive from the fact that cytokines produced by T_H1 or T_H2 cells inhibit the differentiation of T_H17 cells?*

10.5 *One could question the need for immunological memory. Invertebrates get by pretty well without it. After all, if you survive an infection once, you should be able to survive it again. If you fail to survive the first infection, memory is of no help. (a) What are the advantages of immunological memory that counter this argument? What features of pathogens might have driven the evolution of immunological memory? (b) Innate immune responses do not evoke memory. What properties of the adaptive immune response make the immunological memory it develops of greater value? In what way could these properties be a disadvantage?*

10.6 *Memory responses differ from primary immune responses in a number of important properties. Name three ways in which they differ, and describe the underlying mechanism(s) involved in each case.*

10.7 *(a) Discuss the relative roles of cytokine signals and signals received through the T-cell receptor in the survival and function of memory T cells. (b) Compare and contrast their requirements and responses to such signals with those of naive T cells.*

Section references.

10-1 The course of an infection can be divided into several distinct phases.

Mandell, G., Bennett, J., and Dolin, R. (eds): *Principles and Practice of Infectious Diseases*, 5th ed. New York, Churchill Livingstone, 2000.

10-2 The nonspecific responses of innate immunity are necessary for an adaptive immune response to be initiated.

Fearon, D.T., and Carroll, M.C.: **Regulation of B lymphocyte responses to foreign and self-antigens by the CD19/CD21 complex.** *Annu. Rev. Immunol.* 2000, **18**:393–422.

Fearon, D.T., and Locksley, R.M.: **The instructive role of innate immunity in the acquired immune response.** *Science* 1996, **272**:50–53.

Janeway, C.A., Jr.: **The immune system evolved to discriminate infectious nonself from noninfectious self.** *Immunol. Today* 1992, **13**:11–16.

10-3 Cytokines made in the earliest phase of an infection influence differentiation of CD4 T cells toward the T_H17 subset.

Dillon, S., Agrawal, A., Van Dyke, T., Landreth, G., McCauley, L., Koh, A., Maliszewski, C., Akira, S., and Pulendran, B.: **A Toll-like receptor 2 ligand stimulates Th2 responses in vivo, via induction of extracellular signal-regulated kinase mitogen-activated protein kinase and c-Fos in dendritic cells.** *J. Immunol.* 2004, **172**:4733–4743.

Fallon, P.G., Ballantyne, S.J., Mangan, N.E., Barlow, J.L., Dasvarma, A., Hewett, D.R., McIlgorm, A., Jolin, H.E., and McKenzie, A.N.J. : **Identification of an interleukin (IL)-25-dependent cell population that provides IL-4, IL-5, and IL-13 at the onset of helminth expulsion.** *J. Exp. Med.* 2006, **203**:1105–1116.

Fossiez, F., Djossou, O., Chomarat, P., Flores-Romo, L., Ait-Yahia, S., Maat, C., Pin, J.J., Garrone, P., Garcia, E., Saeland, S., et al.: **T cell interleukin-17 induces stromal cells to produce proinflammatory and hematopoietic cytokines.** *J. Exp. Med.* 1996, **183**:2593–2603.

Happel, K.I., Zheng, M., Young, E., Quinton, L.J., Lockhart, E., Ramsay, A.J., Shellito, J.E., Schurr, J.R., Bagby, G.J., Nelson, S., et al.: **Cutting edge: roles of Toll-like receptor 4 and IL-23 in IL-17 expression in response to *Klebsiella pneumoniae* infection.** *J. Immunol.* 2003, **170**:4432–4436.

Tato, C.M., and O'Shea, J.J.: **What does it mean to be just 17?** *Nature* 2006, **441**:166–168.

Ye, P., Rodriguez, F.H., Kanaly, S., Stocking, K.L., Schurr, J., Schwarzenberger, P., Oliver, P., Huang, W., Zhang, P., Zhang, J., et al.: **Requirement of interleukin 17 receptor signaling for lung CXC chemokine and granulocyte colony-stimulating factor expression, neutrophil recruitment, and host defense.** *J. Exp. Med.* 2001, **194**:519–527.

10-4 Cytokines made in the later stages of an infection influence differentiation of CD4 T cells toward T_H1 or T_H2 cells.

Amsen, D., Blander, J.M., Lee, G.R., Tanigaki, K., Honjo, T., and Flavell, R.A.: **Instruction of distinct CD4 T helper cell fates by different Notch ligands on antigen-presenting cells.** *Cell* 2004, **117**:515–526.

Bendelac, A., Rivera, M.N., Park, S.H., and Roark, J.H.: **Mouse CD1-specific NK1 T cells: development, specificity, and function.** *Annu. Rev. Immunol.* 1997, **15**:535–562.

Finkelman, F.D., Shea-Donohue, T., Goldhill, J., Sullivan, C.A., Morris, S.C., Madden, K.B., Gauser, W.C., and Urban, J.F., Jr.: **Cytokine regulation of host defense against parasitic intestinal nematodes.** *Annu. Rev. Immunol.* 1997, **15**:505–533.

Hsieh, C.S., Macatonia, S.E., Tripp, C.S., Wolf, S.F., O'Garra, A., and Murphy, K.M.: **Development of T_H1 CD4$^+$ T cells through IL-12 produced by *Listeria*-induced macrophages.** *Science* 1993, **260**:547–549.

Jankovic, D., Sher, A., and Yap, G.: **Th1/Th2 effector choice in parasitic infection: decision making by committee.** *Curr. Opin. Immunol.* 2001, **13**:403–409.

Moser, M., and Murphy, K.M.: **Dendritic cell regulation of T_H1-T_H2 development.** *Nat. Immunol.* 2000, **1**:199–205.

Pulendran, B., and Ahmed, R.: **Translating innate immunity into immunological memory: implications for vaccine development.** *Cell* 2006, **124**:849–863.

10-5 The distinct subsets of CD4 T cells can regulate each other's differentiation.

Constant, S.L., and Bottomly, K.: **Induction of Th1 and Th2 CD4$^+$ T cell responses: the alternative approaches.** *Annu. Rev. Immunol.* 1997, **15**:297–322.

Croft, M., Carter, L., Swain, S.L., and Dutton, R.W.: **Generation of polarized antigen-specific CD8 effector populations: reciprocal action of interleukin-4 and IL-12 in promoting type 2 versus type 1 cytokine profiles.** *J. Exp. Med.* 1994, **180**:1715–1728.

Grakoui, A., Donermeyer, D.L., Kanagawa, O., Murphy, K.M. and Allen, P.M.: **TCR-independent pathways mediate the effects of antigen dose and altered peptide ligands on Th cell polarization.** *J. Immunol.* 1999, **162**:1923–1930.

Harrington, L.E., Hatton, R.D., Mangan, P.R., Turner, H., Murphy, T.L., Murphy, K.M., and Weaver, C.T.: **Interleukin 17-producing CD4$^+$ effector T cells develop via a lineage distinct from the T helper type 1 and 2 lineages.** *Nat. Immunol.* 2005, **6**:1123–1132.

Julia, V., McSorley, S.S., Malherbe, L., Breittmayer, J.P., Girard-Pipau, F., Beck, A., and Glaichenhaus, N.: **Priming by microbial antigens from the intestinal flora determines the ability of CD4$^+$ T cells to rapidly secrete IL-4 in BALB/c mice infected with *Leishmania major*.** *J. Immunol.* 2000, **165**:5637–5645.

Martin-Fontecha, A., Thomsen, L.L., Brett, S., Gerard, C., Lipp, M., Lanzavecchia, A., and Sallusto, F.: **Induced recruitment of NK cells to lymph nodes provides IFN-γ for T_H1 priming.** *Nat. Immunol.* 2004, **5**:1260–1265.

Nakamura, T., Kamogawa, Y., Bottomly, K., and Flavell, R.A.: **Polarization of IL-4- and IFN-γ-producing CD4$^+$ T cells following activation of naive CD4+ T cells.** *J. Immunol.* 1997, **158**:1085–1094.

Seder, R.A., and Paul, W.E.: **Acquisition of lymphokine producing phenotype by CD4$^+$ T cells.** *Annu. Rev. Immunol.* 1994, **12**:635–673.

Wang, L.F., Lin, J.Y., Hsieh, K.H., and Lin, R.H.: **Epicutaneous exposure of protein antigen induces a predominant T_H2-like response with high IgE production in mice.** *J. Immunol.* 1996, **156**:4079–4082.

10-6 Effector T cells are guided to sites of infection by chemokines and newly expressed adhesion molecules.

MacKay, C.R., Marston, W., and Dudler, L.: **Altered patterns of T-cell migration through lymph nodes and skin following antigen challenge.** *Eur. J. Immunol.* 1992, **22**:2205–2210.

Romanic, A.M., Graesser, D., Baron, J.L., Visintin, I., Janeway, C.A., Jr., and Madri, J.A.: **T cell adhesion to endothelial cells and extracellular matrix is modulated upon transendothelial cell migration.** *Lab. Invest.* 1997, **76**:11–23.

Sallusto, F., Kremmer, E., Palermo, B., Hoy, A., Ponath, P., Qin, S., Forster, R., Lipp, M., and Lanzavecchia, A.: **Switch in chemokine receptor expression upon TCR stimulation reveals novel homing potential for recently activated T cells.** *Eur. J. Immunol.* 1999, **29**:2037–2045.

10-7 Differentiated effector T cells are not a static population but continue to respond to signals as they carry out their effector functions.

Cua, D.J., Sherlock, J., Chen, Y., Murphy, C.A., Joyce, B., Seymour, B., Lucian, L., To, W., Kwan, S., Churakova, T., et al.: **Interleukin-23 rather than interleukin-12 is the critical cytokine for autoimmune inflammation of the brain.** *Nature* 2003, **421**:744–748.

Ghilardi, N., Kljavin, N., Chen, Q., Lucas, S., Gurney, A.L., and De Sauvage, F.J.: **Compromised humoral and delayed-type hypersensitivity responses in IL-23-deficient mice.** *J. Immunol.* 2004, **172**:2827–2833.

Gran, B., Zhang, G.X., Yu, S., Li, J., Chen, X.H., Ventura, E.S., Kamoun, M., and Rostami, A.: **IL-12p35-deficient mice are susceptible to experimental autoimmune encephalomyelitis: evidence for redundancy in the IL-12 system in the** induction of central nervous system autoimmune demyelination. *J. Immunol.* 2002, **169**:7104–7110.

Parham, C., Chirica, M., Timans, J., Vaisberg, E., Travis, M., Cheung, J., Pflanz, S., Zhang, R., Singh, K.P., Vega, F., et al.: **A receptor for the heterodimeric cytokine IL-23 is composed of IL-12Rβ1 and a novel cytokine receptor subunit, IL-23R.** *J. Immunol.* 2002, **168**:5699–5708.

Park, A.Y., Hondowics, B.D., and Scott, P.: **IL-12 is required to maintain a Th1 response during *Leishmania major* infection.** *J. Immunol.* 2000, **165**:896–902.

Stobie, L., Gurunathan, S., Prussin, C., Sacks, D.L., Glaichenhaus, N., Wu, C.Y., and Seder, R.A.: **The role of antigen and IL-12 in sustaining Th1 memory cells in vivo: IL-12 is required to maintain memory/effector Th1 cells sufficient to mediate protection to an infectious parasite challenge.** *Proc. Natl Acad. Sci. USA* 2000, **97**:8427–8432.

Yap, G., Pesin, M., and Sher, A.: **Cutting edge: IL-12 is required for the maintenance of IFN-γ production in T cells mediating chronic resistance to the intracellular pathogen *Toxoplasma gondii*.** *J. Immunol.* 2000, **165**:628–631.

10-8 Primary CD8 T-cell responses to pathogens can occur in the absence of CD4 help.

Lertmemongkolchai, G., Cai, G., Hunter, C.A., and Bancroft, G.J.: **Bystander activation of CD8 T cells contributes to the rapid production of IFN-γ in response to bacterial pathogens.** *J. Immunol.* 2001, **166**:1097–1105.

Rahemtulla, A., Fung-Leung, W.P., Schilham, M.W., Kundig, T.M., Sambhara, S.R., Narendran, A., Arabian, A., Wakeham, A., Paige, C.J., Zinkernagel, R.M., et al.: **Normal development and function of CD8+ cells but markedly decreased helper cell activity in mice lacking CD4.** *Nature* 1991, **353**:180–184.

Schoenberger, S.P., Toes, R.E., van der Voort, E.I., Offringa, R., and Melief, C.J.: **T-cell help for cytotoxic T lymphocytes is mediated by CD40–CD40L interactions.** *Nature* 1998, **393**:480–483.

Sun, J.C., and Bevan, M.J.: **Defective CD8 T cell memory following acute infection without CD4 T-cell help.** *Science* 2003, **300**:339–349.

10-9 Antibody responses develop in lymphoid tissues under the direction of CD4 helper T cells.

Jacob, J., Kassir, R., and Kelsoe, G.: **In situ studies of the primary immune response to (4-hydroxy-3-nitrophenyl)acetyl. I. The architecture and dynamics of responding cell populations.** *J. Exp. Med.* 1991, **173**:1165–1175.

Kelsoe, G., and Zheng, B.: **Sites of B-cell activation *in vivo*.** *Curr. Opin. Immunol.* 1993, **5**:418–422.

Liu, Y.J., Zhang, J., Lane, P.J., Chan, E.Y., and MacLennan, I.C.: **Sites of specific B cell activation in primary and secondary responses to T cell-dependent and T cell-independent antigens.** *Eur. J. Immunol.* 1991, **21**:2951–2962.

MacLennan, I.C.M.: **Germinal centres.** *Annu. Rev. Immunol.* 1994, **12**:117–139.

10-10 Antibody responses are sustained in medullary cords and bone marrow.

Benner, R., Hijmans, W., and Haaijman, J.J.: **The bone marrow: the major source of serum immunoglobulins, but still a neglected site of antibody formation.** *Clin. Exp. Immunol.* 1981, **46**:1–8.

Manz, R.A., Thiel, A., and Radbruch, A.: **Lifetime of plasma cells in the bone marrow.** *Nature* 1997, **388**:133–134.

Slifka, M.K., Antia, R., Whitmire, J.K., and Ahmed, R.: **Humoral immunity due to long-lived plasma cells.** *Immunity* 1998, **8**:363–372.

Takahashi, Y., Dutta, P.R., Cerasoli, D.M., and Kelsoe, G.: **In situ studies of the primary immune response to (4-hydroxy-3-nitrophenyl)acetyl. V. Affinity maturation develops in two stages of clonal selection.** *J. Exp. Med.* 1998, **187**:885–895.

10-11 The effector mechanisms used to clear an infection depend on the infectious agent.

Baize, S., Leroy, E.M., Georges-Courbot, M.C., Capron, M., Lansoud-Soukate, J., Debre, P., Fisher-Hoch, S.P., McCormick, J.B., and Georges, A.J.: **Defective humoral responses and extensive intravascular apoptosis are associated with fatal outcome in Ebola virus-infected patients.** *Nat. Med.* 1999, **5**:423–426.

Kaufmann, S.H.E., Sher, A., and Ahmed, R. (eds): *Immunology of Infectious Diseases.* Washington DC, ASM Press, 2002.

Mims, C.A.: *The Pathogenesis of Infectious Disease*, 5th ed. London, Academic Press, 2000.

10-12 Resolution of an infection is accompanied by the death of most of the effector cells and the generation of memory cells.

Murali-Krishna, K., Altman, J.D., Suresh, M., Sourdive, D.J., Zajac, A.J., Miller, J.D., Slansky, J., and Ahmed, R.: **Counting antigen-specific CD8 T cells: a reevaluation of bystander activation during viral infection.** *Immunity* 1998, **8**:177–187.

Webb, S., Hutchinson, J., Hayden, K., and Sprent, J.: **Expansion/deletion of mature T cells exposed to endogenous superantigens *in vivo*.** *J. Immunol.* 1994, **152**:586–597.

10-13 Immunological memory is long-lived after infection or vaccination.

Black, F.L., and Rosen, L.: **Patterns of measles antibodies in residents of Tahiti and their stability in the absence of re-exposure.** *J. Immunol.* 1962, **88**:725–731.

Hammarlund, E., Lewis, M.W., Hansen, S.G., Strelow, L.I., Nelson, J.A., Sexton, G.J., Hanifin, J.M., and Slifka, M.K.: **Duration of antiviral immunity after smallpox vaccination.** *Nat. Med.* 2003, **9**:1131–1137.

Kassiotis, G., Garcia, S., Simpson, E., and Stockinger, B.: **Impairment of immunological memory in the absence of MHC despite survival of memory T cells.** *Nat. Immunol.* 2002, **3**:244–250.

Ku, C.C., Murakami, M., Sakamoto, A., Kappler, J., and Marrack, P.: **Control of homeostasis of CD8+ memory T cells by opposing cytokines.** *Science* 2000, **288**:675–678.

Murali-Krishna, K., Lau, L.L., Sambhara, S., Lemonnier, F., Altman, J., and Ahmed, R.: **Persistence of memory CD8 T cells in MHC class I-deficient mice.** *Science* 1999, **286**:1377–1381.

Seddon, B., Tomlinson, P., and Zamoyska, R.: **Interleukin 7 and T cell receptor signals regulate homeostasis of CD4 memory cells.** *Nat. Immunol.* 2003, **4**:680–686.

10-14 Memory B-cell responses differ in several ways from those of naive B cells.

Berek, C., and Milstein, C.: **Mutation drift and repertoire shift in the maturation of the immune response.** *Immunol. Rev.* 1987, **96**:23–41.

Cumano, A., and Rajewsky, K.: **Clonal recruitment and somatic mutation in the generation of immunological memory to the hapten NP.** *EMBO J.* 1986, **5**:2459–2468.

Klein, U., Tu, Y., Stolovitzky, G.A., Keller, J.L., Haddad, J., Jr., Miljkovic, V., Cattoretti, G., Califano, A., and Dalla-Favera, R.: **Transcriptional analysis of the B cell germinal center reaction.** *Proc. Natl Acad. Sci. USA* 2003, **100**:2639–2644.

Tarlinton, D.: **Germinal centers: form and function.** *Curr. Opin. Immunol.* 1998, **10**:245–251.

10-15 Repeated immunization leads to increasing affinity of antibody due to somatic hypermutation and selection by antigen in germinal centers.

Berek, C., Jarvis, J.M., and Milstein, C.: **Activation of memory and virgin B cell clones in hyperimmune animals.** *Eur. J. Immunol.* 1987, **17**:1121–1129.

Liu, Y.J., Zhang, J., Lane, P.J., Chan, E.Y., and MacLennan, I.C.: **Sites of specific B cell activation in primary and secondary responses to T cell-dependent and T cell-independent antigens.** *Eur. J. Immunol.* 1991, **21**:2951–2962.

Siskind, G.W., Dunn, P., and Walker, J.G.: **Studies on the control of antibody synthesis: II. Effect of antigen dose and of suppression by passive antibody on the affinity of antibody synthesized.** *J. Exp. Med.* 1968, **127**:55–66.

10-16 Memory T cells are increased in frequency compared with naive T cells specific for the same antigen and have distinct activation requirements and cell-surface proteins that distinguish them from effector T cells.

Bradley, L.M., Atkins, G.G., and Swain, S.L.: **Long-term CD4+ memory T cells from the spleen lack MEL-14, the lymph node homing receptor.** *J. Immunol.* 1992, **148**:324–331.

Hataye, J., Moon, J.J., Khoruts, A., Reilly, C., and Jenkins, M.K.: **Naive and memory CD4+ T cell survival controlled by clonal abundance.** *Science* 2006, **312**:114–116.

Kaech, S.M., Hemby, S., Kersh, E., and Ahmed, R.: **Molecular and functional profiling of memory CD8 T cell differentiation.** *Cell* 2002, **111**:837–851.

Kaech, S.M., Tan, J.T., Wherry, E.J., Konieczny, B.T., Surh, C.D., and Ahmed, R.: **Selective expression of the interleukin 7 receptor identifies effector CD8 T cells that give rise to long-lived memory cells.** *Nat. Immunol.* 2003, **4**:1191–1198.

Rogers, P.R., Dubey, C., and Swain, S.L.: **Qualitative changes accompany memory T cell generation: faster, more effective responses at lower doses of antigen.** *J. Immunol.* 2000, **164**:2338–2346.

Wherry, E.J., Teichgraber, V., Becker, T.C., Masopust, D., Kaech, S.M., Antia, R., von Andrian, U.H., and Ahmed, R.: **Lineage relationship and protective immunity of memory CD8 T cell subsets.** *Nat. Immunol.* 2003, **4**:225–234.

10-17 Memory T cells are heterogeneous and include central memory and effector memory subsets.

Lanzavecchia, A., and Sallusto, F.: **Understanding the generation and function of memory T cell subsets.** *Curr. Opin. Immunol.* 2005, **17**:326–332.

Sallusto, F., Geginat, J., and Lanzavecchia, A.: **Central memory and effector memory T cell subsets: function, generation, and maintenance.** *Annu. Rev. Immunol.* 2004, **22**:745–763.

Sallusto, F., Lenig, D., Forster, R., Lipp, M., and Lanzavecchia, A.: **Two subsets of memory T lymphocytes with distinct homing potentials and effector functions.** *Nature* 1999, **401**:708–712.

10-18 CD4 T-cell help is required for CD8 T-cell memory and involves CD40 and IL-2 signaling.

Bourgeois, C., and Tanchot, C.: **CD4 T cells are required for CD8 T cell memory generation.** *Eur. J. Immunol.* 2003, **33**:3225–3231.

Bourgeois, C., Rocha, B., and Tanchot, C.: **A role for CD40 expression on CD8 T cells in the generation of CD8 T cell memory.** *Science* 2002, **297**:2060–2063.

Janssen, E.M., Lemmens, E.E., Wolfe, T., Christen, U., von Herrath, M.G., and Schoenberger, S.P.: **CD4 T cells are required for secondary expansion and memory in CD8 T lymphocytes.** *Nature* 2003, **421**:852–856.

Shedlock, D.J., and Shen, H.: **Requirement for CD4 T cell help in generating functional CD8 T cell memory.** *Science* 2003, **300**:337–339.

Tanchot, C., and Rocha, B.: **CD8 and B cell memory: same strategy, same signals.** *Nat. Immunol.* 2003, **4**:431–432.

Sun, J.C., Williams, M.A., and Bevan, M.J.: **CD4 T cells are required for the maintenance, not programming, of memory CD8 T cells after acute infection.** *Nat. Immunol.* 2004, **9**:927–933.

Williams, M.A., Tyznik, A.J., and Bevan, M.J.: **Interleukin-2 signals during priming are required for secondary expansion of CD8 memory T cells.** *Nature* 2006, **441**:890–893.

10-19 In immune individuals, secondary and subsequent responses are mainly attributable to memory lymphocytes.

Fazekas de St Groth, B., and Webster, R.G.: **Disquisitions on original antigenic sin. I. Evidence in man.** *J. Exp. Med.* 1966, **140**:2893–2898.

Fridman, W.H.: **Regulation of B cell activation and antigen presentation by Fc receptors.** *Curr. Opin. Immunol.* 1993, **5**:355–360.

Pollack, W., Gorman, J.G., Freda, V.J., Ascari, W.Q., Allen, A.E., and Baker, W.J.: **Results of clinical trials of RhoGAm in women.** *Transfusion* 1968, **8**:151–153.

The Mucosal Immune System

11

A series of anatomically distinct compartments can be distinguished within the immune system, each of which is specially adapted to generate a response to antigens encountered in a particular set of tissues. In previous chapters we have mainly discussed the adaptive immune responses that are initiated in peripheral lymph nodes and spleen, the compartments that respond to antigens that have entered the tissues or spread into the blood. These are the immune responses most studied by immunologists, as they are the responses evoked when antigens are administered by injection. There is, however, an additional compartment of the adaptive immune system, of even greater size, located near the surfaces where most pathogens actually invade. This is the **mucosal immune system**, which is the subject of this chapter.

The organization of the mucosal immune system.

The epithelial surfaces of the body are exposed to large amounts of antigen from which they are separated by only a thin layer of cells, the epithelium. These tissues are essential for life and so require continual and effective protection against invasion. This is partly maintained by the epithelium itself acting as a physical barrier; however, this can be breached relatively easily, meaning that the more sophisticated mechanisms of the innate and adaptive immune systems also play crucial roles. These are the functions of the mucosal immune system. The innate defenses of mucosal tissues were described in Chapter 2; in this chapter we focus on the adaptive mucosal immune system.

11-1 The mucosal immune system protects the internal surfaces of the body.

The mucosal immune system comprises the gastrointestinal tract, the upper and lower respiratory tracts, and the urogenital tract. It also includes the exocrine glands associated with these organs, such as the pancreas, the conjunctivae and lachrymal glands of the eye, the salivary glands, and the lactating breast (Fig. 11.1). The mucosal surfaces represent an enormous area to be protected. The human small intestine, for instance, has a surface area of almost 400 m^2, which is 200 times that of the skin. Because of their physiological functions in gas exchange (the lungs), food absorption (the gut), sensory activities (eyes, nose, mouth, and throat), and reproduction (uterus and vagina), the mucosal surfaces are thin and permeable barriers to the interior of the body. The importance of these tissues to life means that it is critical to have effective defense mechanisms in place to protect them from invasion. Equally, their fragility and permeability create obvious vulnerability to infection, and it is not surprising that the vast majority of infectious agents invade

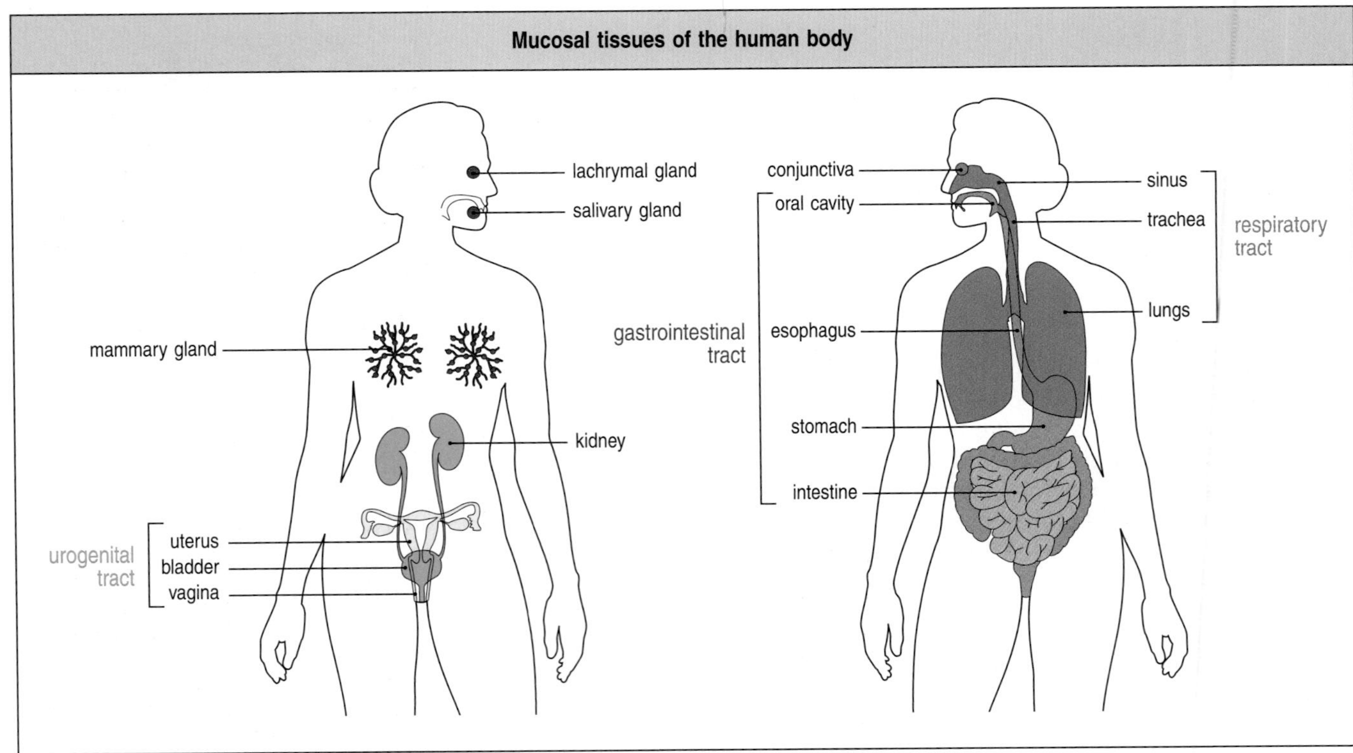

Fig. 11.1 The mucosal immune system. The tissues of the mucosal immune system are the lymphoid organs associated with the intestine, respiratory tract and urogenital tract, as well as the oral cavity and pharynx and the glands associated with these tissues, such as the salivary glands and lachrymal glands. The lactating breast is also part of the mucosal immune system.

the human body by these routes (Fig. 11.2). Diarrheal diseases, acute respiratory infections, pulmonary tuberculosis, measles, whooping cough and worm infestations continue to be the major causes of death throughout the world, especially in infants in developing countries. To this must be added the human immunodeficiency virus (HIV), a pathogen whose natural route of entry via a mucosal surface is often overlooked.

A second important point to bear in mind when considering the immunobiology of mucosal surfaces is that they are also portals of entry for a vast array of foreign antigens that are not pathogenic. This is best seen in the gut, which is exposed to enormous quantities of food proteins—an estimated 10–15 kg per year per person. At the same time, the healthy large intestine is colonized by at least 1000 species of microorganisms that live in symbiosis with their host and are known as **commensal** microorganisms. These are mostly bacteria and are present at levels of at least 10^{12} organisms per milliliter in the colon contents, making them the most numerous cells in the body. In normal circumstances they do no harm and are beneficial to their host in many ways.

As food proteins and commensal bacteria contain many foreign antigens, they are fully capable of being recognized by the adaptive immune system. Generating protective immune responses against these harmless agents would, however, be inappropriate and wasteful. Indeed, aberrant immune responses of this kind are now believed to be the cause of some relatively common diseases, including celiac disease (caused by a response to the wheat protein gluten) and inflammatory bowel diseases such as Crohn's disease (a response to commensal bacteria). As we shall see, the intestinal mucosal immune system has evolved the means of distinguishing harmful pathogens from antigens in food and the natural gut flora. Similar issues are faced at other mucosal surfaces such as the respiratory tract. Protective immunity against pathogens is essential, but, as in the gut, many of the antigens entering the respiratory tract are derived from commensal organisms, pollen, and other innocuous environmental material.

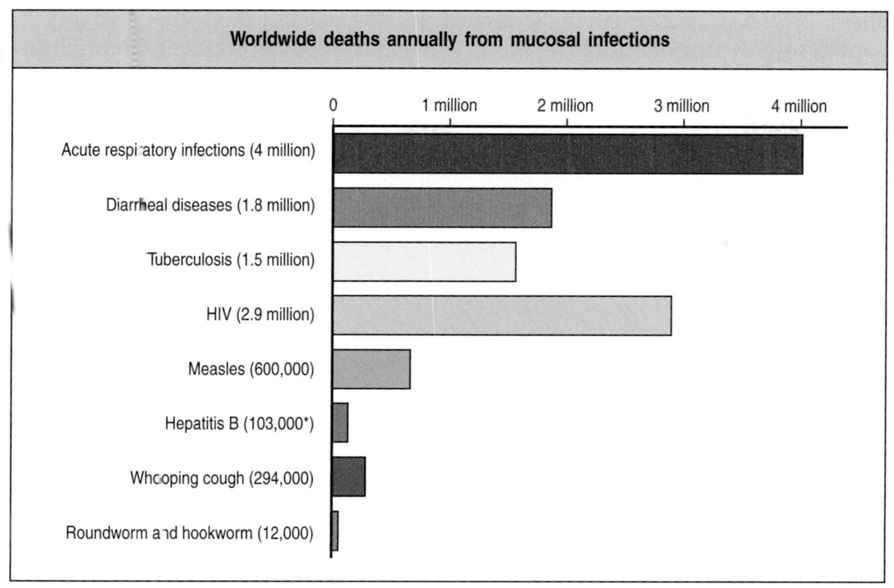

Fig. 11.2 Mucosal infections are one of the biggest health problems worldwide. Most of the pathogens that cause the deaths of large numbers of people are those of mucosal surfaces or enter the body through these routes. Respiratory infections are caused by numerous bacteria (such as *Streptococcus pneumoniae*, *Haemophilus influenzae*, which cause pneumonia, and *Bordetella pertussis*, the cause of whooping cough) and viruses (influenza and respiratory syncytial virus). Diarrheal diseases are caused by both bacteria (such as the cholera bacterium *Cholera vibrio*) and viruses (such as rotaviruses). The bacterium *Mycobacterium tuberculosis*, which causes tuberculosis, also enters through the respiratory tract. Measles manifests itself as a systemic disease, but it originally enters via the oral/respiratory route. The human immunodeficiency virus (HIV) that causes AIDS enters through the mucosa of the urogenital tract or is secreted into breast milk and is passed from mother to child in this way. Hepatitis B is also a sexually transmitted virus. Finally, parasitic worms inhabiting the intestine cause chronic debilitating disease and premature death. Most of these deaths, especially those from acute respiratory and diarrheal diseases, occur in children under 5 years old in the developing world and there are still no effective vaccines against many of these pathogens. *Does not include deaths from liver cancer of cirrhosis resulting from chronic infection. Estimated mortality data (for 2002) from *World Health Report 2004* (World Health Organization).

11-2 The mucosal immune system may be the original vertebrate immune system.

From the point of view of traditional immunology, the mucosal immune system has been considered to be an unusual and relatively minor subcompartment of the immune system. In terms of size and function, this is an inaccurate description. As a result of its physiologically critical role and extent of exposure to antigens, the mucosal immune system forms the largest part of the body's immune tissues, containing approximately three-quarters of all lymphocytes and producing the majority of immunoglobulin in healthy individuals. When compared with lymph nodes and spleen (which in this chapter we will call the **systemic immune system**), the mucosal immune system has many unique and unusual features. The main distinctive features are listed in Fig. 11.3.

The mucosal immune system may well have been the first part of the vertebrate adaptive immune system to evolve. The gut was the first differentiated organ in animals to require defense against invasion, and organized lymphoid tissues are

Distinctive features of the mucosal immune system	
Anatomical features	Intimate interactions between mucosal epithelia and lymphoid tissues
	Discrete compartments of diffuse lymphoid tissue and more organized structures such as Peyer's patches, isolated lymphoid follicles, and tonsils
	Specialized antigen-uptake mechanisms, e.g. M cells in Peyer's patches, adenoids, and tonsils
Effector mechanisms	Activated/memory T cells predominate even in the absence of infection
	Nonspecifically activated 'natural' effector/regulatory T cells present
Immunoregulatory environment	Active downregulation of immune responses (e.g. to food and other innocuous antigens) predominates
	Inhibitory macrophages and tolerance-inducing dendritic cells

Fig. 11.3 Distinctive features of the mucosal immune system. The mucosal immune system is bigger, encounters a wider range of antigens, and encounters them much more frequently, than the rest of the immune system—what we call in this chapter the systemic immune system. This is reflected in distinctive anatomical features, specialized mechanisms for uptake of antigen, and unusual effector and regulatory responses that are designed to prevent unwanted immune responses to food and other innocuous antigens.

first found in vertebrates in the guts of primitive cartilaginous fishes. Two important central lymphoid organs—the thymus and the avian bursa of Fabricius—derive from the embryonic intestine. For all these reasons, it has been proposed that the mucosal immune system represents the original vertebrate immune system, and that the spleen and lymph nodes are later specializations.

11-3 Mucosa-associated lymphoid tissue is located in anatomically defined compartments in the gut.

Many of the anatomical and immunological principles underlying the mucosal immune system apply to all its constituent tissues; in this chapter we use the intestine as our example. Lymphocytes and other immune-system cells such as macrophages and dendritic cells are found throughout the intestinal tract, both in organized tissues and scattered throughout the surface epithelium of the mucosa and an underlying layer of connective tissue called the **lamina propria**. The organized lymphoid tissues in the gut are known as the **gut-associated lymphoid tissues** (**GALT**) (Fig. 11.4). They have the anatomically compartmentalized structure typical of peripheral lymphoid organs, and are sites at which immune responses are initiated. The cells scattered throughout the epithelium and the lamina propria comprise the effector cells of the local immune response.

The organized lymphoid tissues of the GALT include the Peyer's patches and solitary lymphoid follicles of the intestine, the appendix, the tonsils and adenoids in the throat, and the mesenteric lymph nodes. The **palatine tonsils**, **adenoids** and **lingual tonsils** consist of large aggregates of secondary lymphoid tissue covered by a layer of squamous epithelium and form a ring, known as Waldeyer's ring, at the back of the mouth at the entrance of the gut and airways (Fig. 11.5). They often become extremely enlarged in childhood

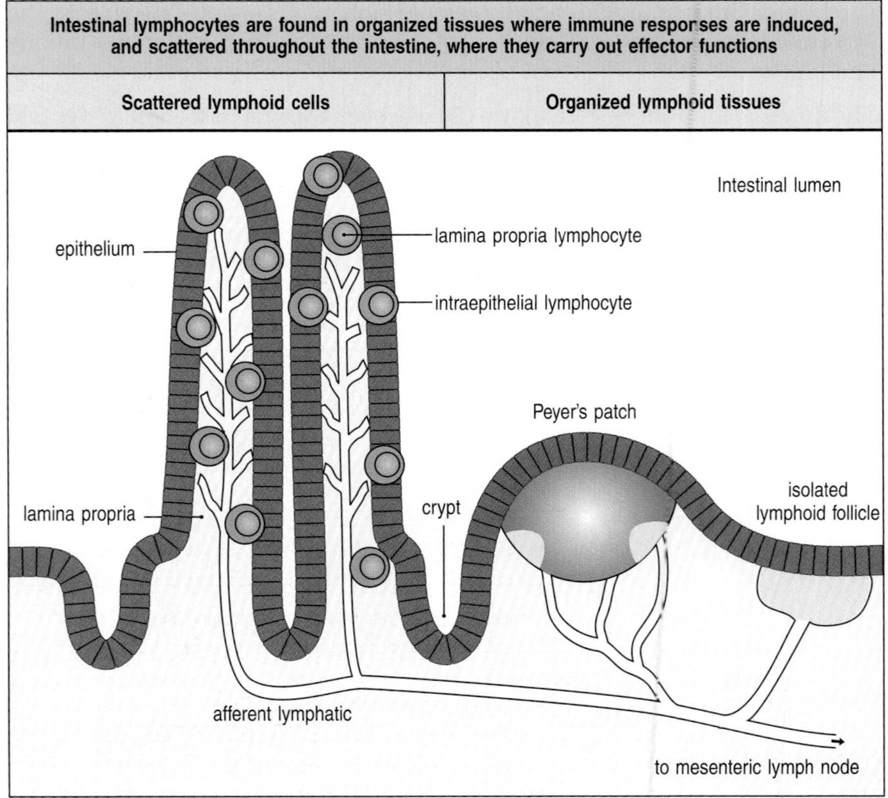

Fig. 11.4 Gut-associated lymphoid tissues and lymphocyte populations. The intestinal mucosa of the small intestine is made up of finger-like processes (villi) covered by a thin layer of epithelial cells (red) that are responsible for digestion of food and absorption of nutrients. These epithelial cells are replaced continually by new cells that derive from stem cells in the crypts. The tissue layer underlying the epithelium is called the lamina propria, and will be colored pale brown throughout this chapter. Lymphocytes are found in a number of discrete compartments in the intestine, with the organized lymphoid tissues such as Peyer's patches and isolated lymphoid follicles forming what is known as the gut-associated lymphoid tissues (GALT). These tissues lie in the wall of the intestine itself, separated from the contents of the intestinal lumen by the single layer of epithelium. The draining lymph nodes for the gut are the mesenteric lymph nodes (see Fig. 11.11), which are connected to Peyers' patches and the intestinal mucosa by afferent lymphatic vessels and are the largest lymph nodes in the body. Together, these organized tissues are the sites of antigen presentation to T cells and B cells and are responsible for the induction phase of immune responses. Peyer's patches and mesenteric lymph nodes contain discrete T-cell areas (blue) and B-cell follicles (yellow) while the isolated follicles comprise mainly B cells. Many lymphocytes are found scattered throughout the mucosa outside the organized lymphoid tissues and these are effector cells—effector T cells and antibody-secreting plasma cells. Effector lymphocytes are found both in the epithelium and in the lamina propria. Lymphatics also drain from the lamina propria to the mesenteric lymph nodes.

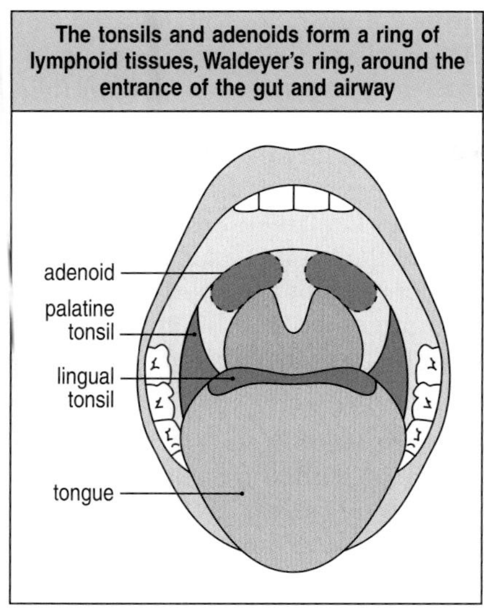
The tonsils and adenoids form a ring of lymphoid tissues, Waldeyer's ring, around the entrance of the gut and airway
adenoid
palatine tonsil
lingual tonsil
tongue

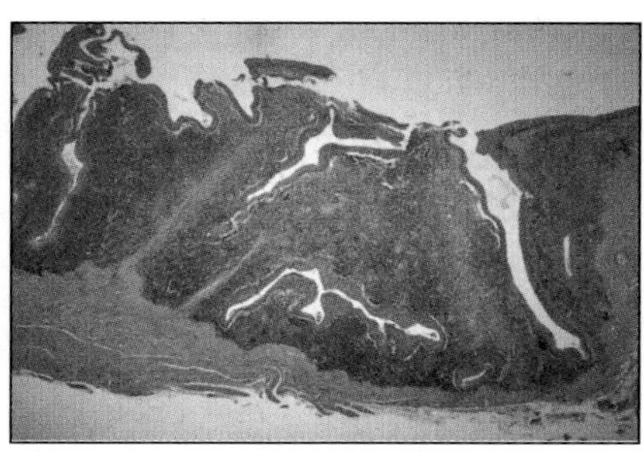

Fig. 11.5 A ring of lymphoid organs called Waldeyer's ring surrounds the entrance to the intestine and respiratory tract. The adenoids lie at either side of the base of the nose, while the palatine tonsils lie at either side of the back of the oral cavity. The lingual tonsils are discrete lymphoid organs on the base of the tongue. The micrograph shows a section through an inflamed human tonsil. In the absence of inflammation, the tonsils and adenoids normally comprise areas of organized tissue with both B-cell and T-cell areas, covered by a layer of squamous epithelium (at top of photo). The surface contains deep crevices (crypts) that increase the surface area but can easily become sites of infection. Hematoxylin and eosin staining. Magnification ×100.

because of recurrent infections, and in the past were victims of a vogue for surgical removal. A reduced IgA response to oral polio vaccination has been seen in individuals who have had their tonsils and adenoids removed.

Some secondary lymphoid organs of the GALT are found within the wall of the intestine; these are the **Peyer's patches** of the small intestine, the **appendix** (which is another frequent victim of the surgeon's knife), and the **isolated lymphoid follicles** of the large intestine. Peyer's patches are extremely important sites for the initiation of immune responses in the gut and have a distinctive appearance, forming dome-like aggregates of lymphoid cells that project into the intestinal lumen (Fig. 11.6). Each Peyer's patch consists of a large number of B-cell follicles with germinal centers, together with smaller

Fig. 11.6 A Peyer's patch and its specialized surface epithelium. Panel a: Peyer's patches are organized lymphoid tissues lying in the submucosal layer of the intestinal wall. Each comprises numerous, highly active B-cell follicles with germinal centres (GC), as well as intervening T-cell dependent areas (TDA) and a layer between the surface epithelium and the follicles known as the subepithelial dome, which is rich in dendritic cells, T cells, and B cells (see Fig. 1.20 for a schematic view of a Peyer's patch). The surface epithelium is known as the follicle-associated epithelium and is a single layer of columnar epithelial cells. Panel b: scanning electron micrograph of the follicle-associated epithelium of the mouse Peyer's patch shown boxed in (a) reveals microfold (M) cells, which lack the microvilli and the layer of mucus that is present on normal epithelial cells. Each M cell appears as a sunken area on the epithelial surface. Panel c: a higher-magnification view of the boxed area in (b) shows the characteristic ruffled surface of an M cell. M cells are the portal of entry for many pathogens and other particles. (a) Hematoxylin and eosin staining. Magnification ×100. (b) ×5000. (c) ×23,000.

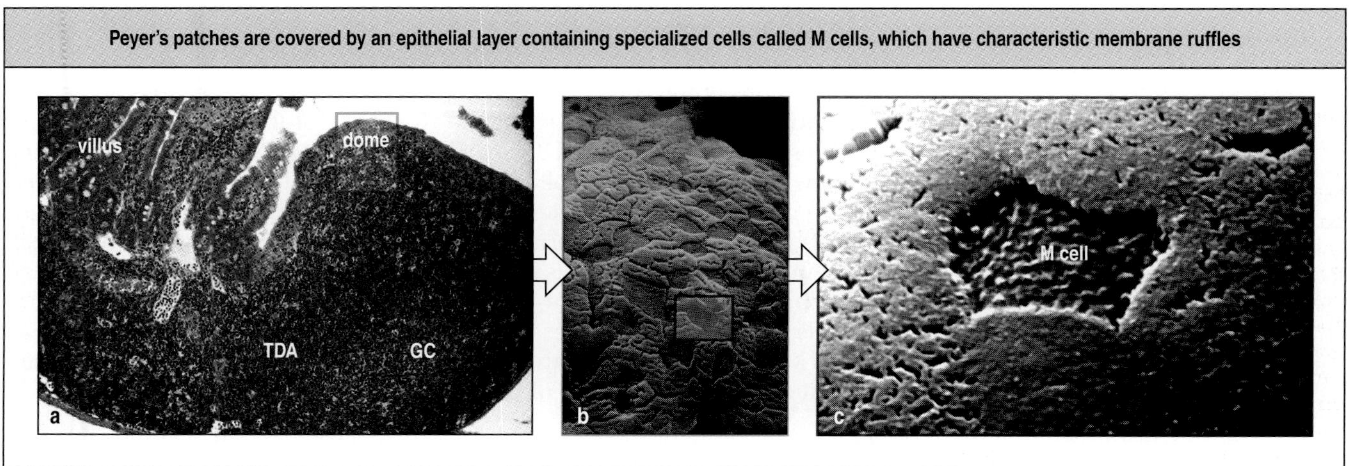
Peyer's patches are covered by an epithelial layer containing specialized cells called M cells, which have characteristic membrane ruffles
villus
dome
TDA
GC
a
b
M cell
c

T-cell areas that are found between and immediately below the follicles. The subepithelial dome is rich in dendritic cells, T cells and B cells. Overlying the lymphoid tissues and separating them from the gut lumen is a layer of follicle-associated epithelium. This contains conventional intestinal epithelial cells known as enterocytes and a smaller number of specialized epithelial cells called **microfold cells** (**M cells**), which have a folded luminal surface instead of the microvilli present on enterocytes. Unlike enterocytes, M cells do not secrete digestive enzymes or mucus and lack a thick surface glycocalyx. They are therefore readily accessible to organisms and particles within the gut lumen and are the route by which antigen enters the Peyer's patch from the lumen. The follicle-associated epithelium also contains lymphocytes and dendritic cells.

In addition to the Peyer's patches, which are visible to the naked eye, numerous **isolated lymphoid follicles** can be identified microscopically in the small and large intestines. Like Peyer's patches, these are composed of an epithelium containing M cells overlying organized lymphoid tissue, but they contain mainly B cells and develop only after birth, whereas Peyer's patches are present in the fetal gut. Similar isolated follicles are found in the wall of the upper respiratory tract, where they are known as the **bronchus-associated lymphoid tissues** (**BALT**), and in the lining of the nose, where they are called **nasal-associated lymphoid tissue** (**NALT**). The term **mucosa-associated lymphoid tissues** (**MALT**) is sometimes used to refer collectively to these similar tissues found in mucosal organs. Peyer's patches and isolated lymphoid follicles are connected by lymphatics to the draining **mesenteric lymph nodes**, located in the connective tissue that tethers the intestine to the rear wall of the abdomen. These are the largest lymph nodes in the body and play a crucial role in initiating and shaping immune responses to intestinal antigens.

The immune responses that are generated when antigen is recognized in one of the tissues of the GALT are quite distinct from those stimulated in lymph nodes or spleen when antigen is given into a tissue such as the skin, muscle or bloodstream. This is because the microenvironment of the GALT has its own characteristic content of lymphoid cells, hormones, and other immunomodulatory factors. The mesenteric lymph nodes and Peyer's patches differentiate independently of the systemic immune system during fetal development, with the involvement of specific chemokines and receptors of the tumor necrosis factor (TNF) family (Fig. 11.7; see also Section 7-24). The differences between the GALT and the systemic lymphoid organs are thus imprinted early in life and are independent of exposure to antigen.

11-4 The intestine has distinctive routes and mechanisms of antigen uptake.

Antigens present at mucosal surfaces must be transported across an epithelial barrier before they can stimulate the mucosal immune system. Peyer's patches are highly adapted for the uptake of antigen from the intestinal lumen. The M cells in the follicle-associated epithelium are continually taking up molecules and particles from the gut lumen by endocytosis or phagocytosis (Fig. 11.8). This material is transported through the interior of the cell in membrane-bound vesicles to the basal cell membrane, where it is released into the extracellular space—a process known as **transcytosis**. Because M cells are much more accessible than enterocytes, a number of pathogens target M cells to gain access to the subepithelial space, even though they then find themselves in the heart of the intestinal adaptive immune system.

The basal cell membrane of an M cell is extensively folded, forming a pocket that encloses lymphocytes and dendritic cells. The dendritic cells take up the

Control of development of the GALT compared with systemic lymphoid tissues										
	Protein required for tissue development									
Tissue	TNFR1	LT-α	LT-β	LTβR	TRANCE	IL-7R	β7	L-sel	CXCR5	NKκB2
Peyer's patch	+	+	+	+	–	+	+/–	–	+/–	+
Mesenteric lymph node	–	+	–	+	+	–	–	+/–	–	–
Systemic lymph node	+/–	+	+/–	+	+	–	–	+	–	+/–

Fig. 11.7 The fetal development of intestinal lymphoid tissues is controlled by a specific set of cytokines. Experiments in knockout mice show that the mesenteric lymph nodes and Peyer's patches differ from each other, and from lymph nodes in other parts of the body, in the signals that are required for their development in fetal and early neonatal life. The development of all these lymphoid tissues requires an interchange of signals between lymphoid-tissue inducer cells, and local stromal cells. Signals from the stromal cells induce the lymphoid-tissue inducer cells to express lymphotoxin (LT)-α and -β subunits. These can form homotrimers (LT-α_3) or heterotrimers (LT-α_1:β_2); LT-α_1:β_2 acts on local stromal cells via the LT-β receptor, and this receptor is required for the development of all the lymphoid tissues considered here, as is the production of the LT-α subunit. Stimulation of stromal cells via the LT-β receptor leads to the expression of adhesion molecules such as VCAM-1 and the production of chemokines such as CCL19, CCL21, and CXCL13, all of which recruit lymphocytes into the developing organ, as well as more lymphoid-tissue inducer cells. Mesenteric lymph nodes are the first lymphoid tissues to develop in the fetus. Lymphoid-tissue inducer cells in these sites produce LT-α_1:β_2 in response to the TNF-family cytokine TRANCE produced by the stromal cells, but knockout experiments in mice show that the LT-β subunit is not essential for mesenteric lymph node development and that it can be replaced by another TNF-family molecule, LIGHT, which can also bind the LT-β receptor. The development of Peyer's patches is absolutely dependent on the presence of both LT-α and LT-β subunits, which are produced by lymphoid-tissue inducer cells in response to IL-7 produced by stromal cells. Lymphoid-tissue inducer cells are also uniquely recruited to Peyer's patches via their CXCR5 receptors, and the TNF receptor TNFR-I is also involved in the development of Peyer's patches but not of the other tissues shown here. In respect of LT signals, the requirements of the peripheral lymph nodes are more similar to those of the mesenteric lymph node. The differences in the requirements for LT subunits and receptors probably reflect subtle differences in the signaling pathways used in the different sites. Adhesion molecules are also involved in lymphoid tissue development. Peyer's patches develop normally in the absence of L-selectin but are partly dependent on the integrin α_4:β_7 and are entirely absent if both these proteins are lacking. Mesenteric lymph nodes also require either L-selectin or α_4:β_7 integrin, but develop normally in the absence of either. Systemic lymph nodes require only L-selectin for their development.

transported material released from the M cells and process it for presentation to T lymphocytes. These dendritic cells are in a particularly favorable position to acquire gut antigens, and they are recruited to this region in response to chemokines that are released constitutively by the epithelial cells. The chemokines include CCL20 (MIP-3α) and CCL9 (MIP-1γ), which bind to the receptors CCR6 and CCR1, respectively, on the dendritic cell (see Appendix IV for a listing of chemokines and their receptors). The antigen-loaded dendritic cells then migrate from the dome region to the T-cell areas of the Peyer's patch, where they meet naive, antigen-specific T cells; they may also migrate via the draining lymphatics to the mesenteric lymph nodes, where they will also encounter naive T cells. The dendritic cells in Peyer's patches have a preferential ability to imprint the T cells they activate with gut-homing properties, a process we shall discuss later.

Dendritic cells are also abundant in the wall of the intestine, mainly in the lamina propria (Fig. 11.9). Some of these cells can make their way into the epithelium, or send processes through the epithelial layer without disturbing its integrity. Dendritic cell motility is increased in response to local bacterial infection, and these cells can be seen to acquire bacteria from the lumen before returning with them to the lamina propria. This behavior allows mucosal dendritic cells to acquire antigens across an intact epithelial barrier without the need for M cells. After taking up antigens from the intestinal lumen, lamina propria dendritic cells transport them to the T-cell areas of

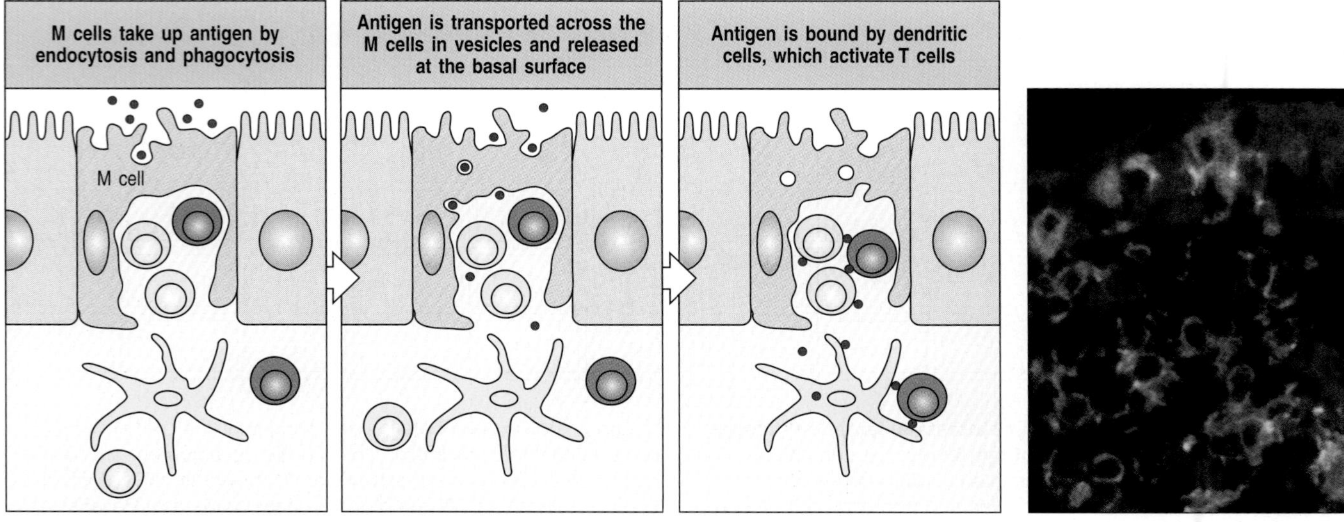

M cells take up antigen by endocytosis and phagocytosis
M cell
Antigen is transported across the M cells in vesicles and released at the basal surface
Antigen is bound by dendritic cells, which activate T cells

Fig. 11.8 Uptake and transport of antigens by M cells. As shown in the first three panels, M cells in the follicle-associated epithelium of Peyer's patches have convoluted basal membranes that form 'pockets' within the epithelial layer, allowing close contact with lymphocytes and other cells. This favors the local transport of antigens that have been taken up from the intestine by the M cells and their delivery to dendritic cells for antigen presentation. The micrograph of part of a Peyer's patch on the right shows epithelial cells (dark blue) some of which are M cells that form pockets where T cells (red) and B cells (green) accumulate. The cells have been stained with fluorescently labeled antibodies specific for individual cell types.

mesenteric lymph nodes via the afferent lymphatics that drain the intestinal wall. Similar populations of dendritic cells that take up local antigens and migrate to the draining lymph nodes are found in the lung and other mucosal surfaces.

11-5 The mucosal immune system contains large numbers of effector lymphocytes even in the absence of disease.

In addition to the organized lymphoid organs, a mucosal surface contains enormous numbers of lymphocytes and other leukocytes scattered throughout the tissue. The majority of the scattered lymphocytes have the appearance of cells that have been activated by antigen, and they comprise the effector T cells and plasma cells of the mucosal immune system. In the intestine, effector cells are found in two main compartments: the epithelium and the lamina propria (Fig. 11.10). These tissues are quite distinct in immunological terms, despite being separated only by a thin layer of basement membrane. The epithelium contains mainly lymphocytes, the vast majority of which are CD8 T cells. The lamina propria is much more heterogeneous, with large numbers of CD4 and CD8 T cells, as well as plasma cells, macrophages,

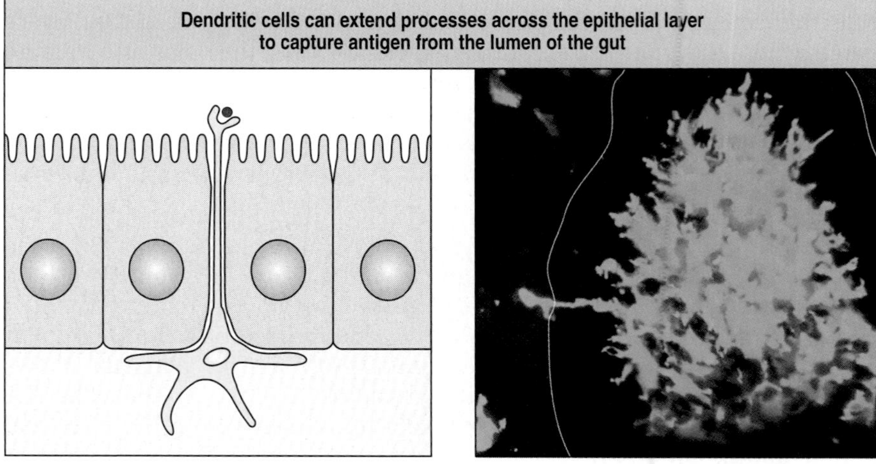

Dendritic cells can extend processes across the epithelial layer to capture antigen from the lumen of the gut

Fig. 11.9 Capture of antigens from intestine by dendritic cells in the lamina propria. Dendritic cells can extend processes between the cells of the epithelium without disturbing its integrity. These cell processes can pick up antigen, such as bacteria, from the gut lumen. The micrograph shows dendritic cells (stained green with a fluorescent tag on the CD11c molecule) in the lamina propria of a villus of mouse small intestine. The epithelium is not stained and appears black, but its luminal (outer) surface is shown by the white line. The process from the dendritic cell has squeezed between two epithelial cells and its tip is present in the lumen of the intestine. Magnification ×200. Micrograph from Niess, J.H., *et al.*: *Science*. 2005, **307**: 254-258.

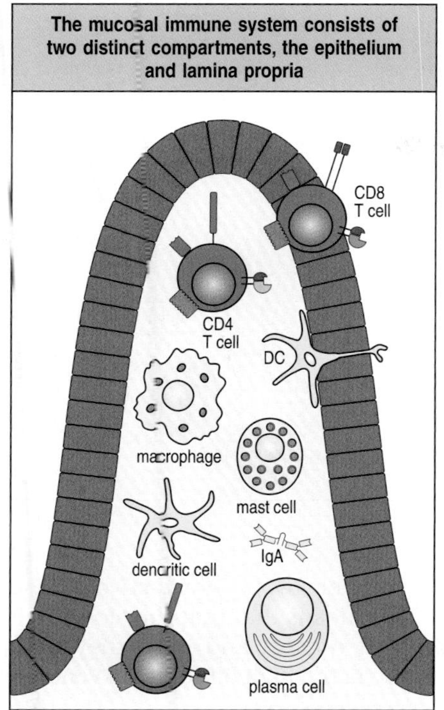

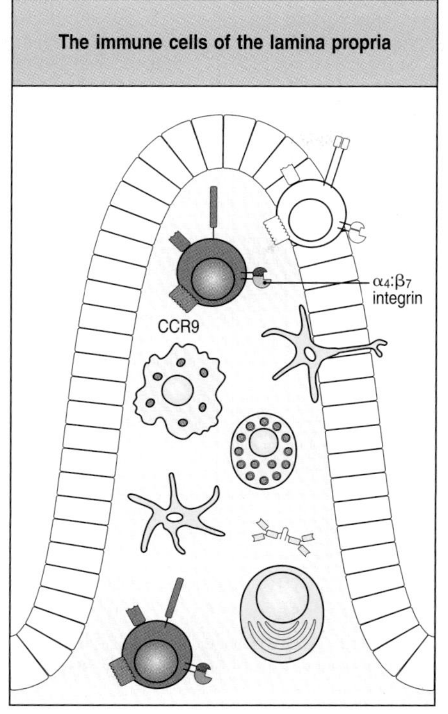

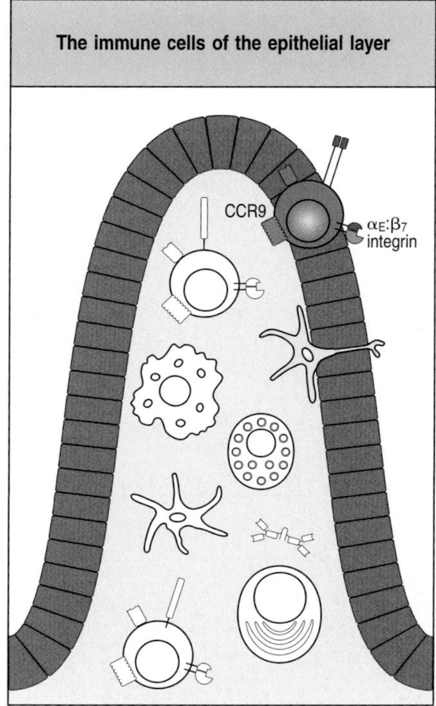

Fig. 11.10 The lamina propria and epithelium of the intestinal mucosa are discrete lymphoid compartments. The lamina propria contains a heterogeneous mixture of IgA-producing plasma cells, lymphocytes with a 'memory' phenotype (see Chapter 10), conventional CD4 and CD8 effector T cells, dendritic cells, macrophages, and mast cells. T cells in the lamina propria of the small intestine express the integrin α_4:β_7 and the chemokine receptor CCR9, which attracts them into the tissue from the bloodstream. Intraepithelial lymphocytes express CCR9 and the integrin α_E:β_7, which binds to E-cadherin on epithelial cells. They are mostly CD8 T cells, some of which express the conventional α:β form of CD8 and others the CD8α:α homodimer. CD4 T cells predominate in the lamina propria, whereas CD8 T cells predominate in the epithelium.

dendritic cells and occasional eosinophils and mast cells. Neutrophils are rare in the healthy intestine, although their numbers increase rapidly during inflammatory disease or infection. The total number of lymphocytes in the epithelium and lamina propria probably exceeds that of most other parts of the body.

The healthy intestinal mucosa therefore displays many characteristics of a chronic inflammatory response—namely, the presence of numerous effector lymphocytes and other leukocytes in the tissues. This is the result of the local responses that are continually being made to the myriad of innocuous antigens that are bombarding the mucosal surfaces. Overt disease is rare, however, indicating that there are powerful regulatory mechanisms that prevent these local responses from getting out of hand.

11-6 The circulation of lymphocytes within the mucosal immune system is controlled by tissue-specific adhesion molecules and chemokine receptors.

The arrival of effector lymphocytes in the mucosal surface layer is the outcome of a series of events in which the homing characteristics of lymphocytes change as they become activated. The life history of mucosal lymphocytes starts with the emergence of naive T cells and B cells from the thymus and bone marrow, respectively. At this point, the naive lymphocytes circulating in the bloodstream are not predetermined as to which compartment of the immune system they will end up in. Naive lymphocytes arriving at Peyer's patches and mesenteric lymph nodes enter them through high endothelial venules (Fig. 11.11). As in the other peripheral lymphoid organs, entry is controlled by the chemokines CCL21 and CCL19, which are released from peripheral lymphoid tissues and bind the receptor CCR7 on naive lymphocytes. If the naive lymphocytes do not see their antigen, they exit from the lymphoid organ via the efferent lymphatics and return to the bloodstream. If

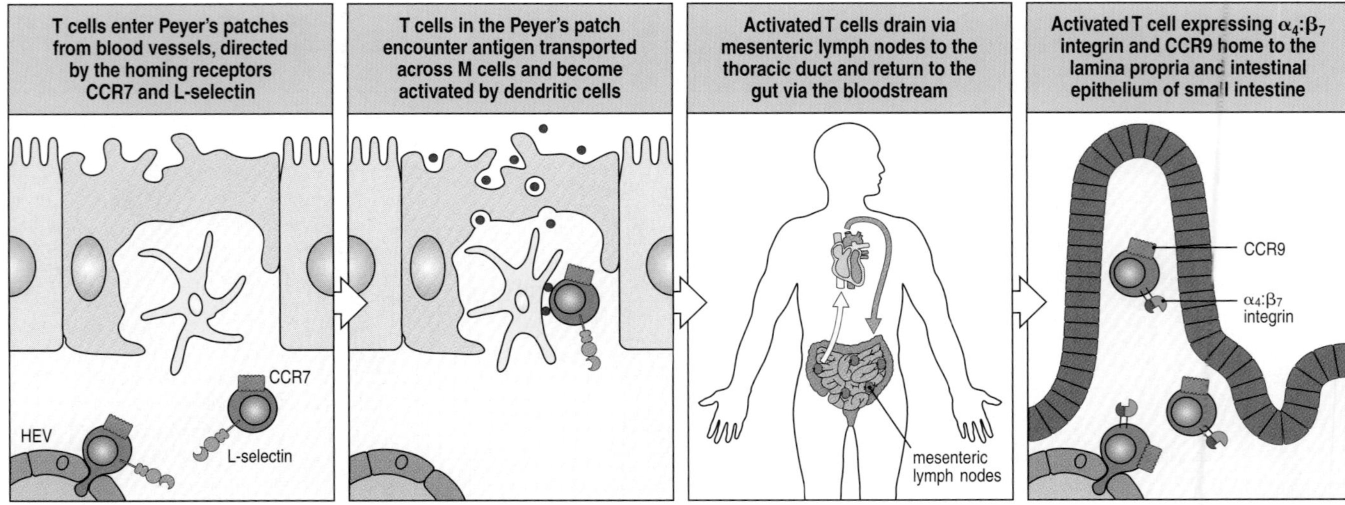

T cells enter Peyer's patches from blood vessels, directed by the homing receptors CCR7 and L-selectin
T cells in the Peyer's patch encounter antigen transported across M cells and become activated by dendritic cells
Activated T cells drain via mesenteric lymph nodes to the thoracic duct and return to the gut via the bloodstream
Activated T cell expressing α4:β7 integrin and CCR9 home to the lamina propria and intestinal epithelium of small intestine
CCR7
HEV
L-selectin
mesenteric lymph nodes
CCR9
α4:β7 integrin

Fig. 11.11 Priming of naive T cells and the redistribution of effector T cells in the intestinal immune system. Naive T cells carry the chemokine receptor CCR7 and L-selectin, which direct their entry into Peyer's patches via high endothelial venules (HEV). In the T-cell area they encounter antigen that has been transported into the lymphoid tissue by M cells and is presented by local dendritic cells. During activation, and under the selective control of gut-derived dendritic cells, the T cells lose L-selectin and acquire the chemokine receptor CCR9 and the integrin α_4:β_7. After activation, but before full differentiation, the primed T cells exit from the Peyer's patch via the draining lymphatics, passing through the mesenteric lymph node to enter the thoracic duct. The thoracic duct empties into the bloodstream, delivering the activated T cells back to the wall of the intestine. Here T cells bearing CCR9 and α_4:β_7 are attracted specifically to leave the bloodstream and enter the lamina propria of the villus.

they encounter antigen in the GALT, the lymphocytes become activated and lose expression of CCR7 and L-selectin. This means that that they lose their homing preference for peripheral lymphoid organs, and once they have exited from them they are unable to reenter via the high endothelial venules.

Effector mucosal lymphocytes leave the mucosal lymphoid organs in which they were activated and travel back to the mucosa. Lymphocytes activated in Peyer's patches leave via the lymphatics, pass through mesenteric lymph nodes, and eventually end up in the thoracic duct. From there they circulate in the bloodstream throughout the body (see Fig. 11.11) and selectively reenter the mucosal tissues via the small blood vessels in the lamina propria. Antigen-specific B cells are primed as IgM-producing B cells in the Peyer's patch, undergo switching to IgA production there and enter the lamina propria as IgA-producing plasma cells.

Gut-specific homing is partly determined by the expression of α_4:β_7 integrin on the lymphocytes. This binds to the mucosal vascular addressin **MAdCAM-1**, which is found mainly on the endothelial cells that line the blood vessels within the gut wall (Fig. 11.12). Lymphocytes originally primed in the gut are also lured back as a result of tissue-specific expression of chemokines by the gut epithelium. CCL25 (TECK) is expressed by the epithelium of the small intestine and is a ligand for the chemokine receptor CCR9 expressed on gut-homing T cells and B cells. Even within the intestine there seems to be regional specialization of chemokine expression. The colon and salivary glands express CCL28 (MEC, mucosal epithelial chemokine), which is a ligand for the receptor CCR10 on gut-homing lymphocytes and attracts IgA-producing B lymphoblasts.

Only lymphocytes that first encounter antigen in a gut-associated secondary lymphoid organ are induced to express gut-specific homing receptors and integrins. This induction is a specific property of GALT dendritic cells and is mediated in part by retinoic acid, which is a derivative of vitamin A produced by the action of the enzyme retinal dehydrogenase expressed in intestinal dendritic cells. These dendritic cells selectively imprint the expression of integrin α_4:β_7 and CCR9 when presenting antigen and activating naive T cells, whereas dendritic cells from non-mucosal lymphoid tissues induce activated T cells to express α_4:β_1 integrin, cutaneous lymphocyte antigen (CLA) and the chemokine receptor CCR4, for example, which direct them to tissues such as the skin (see Section 10-6). These tissue-specific consequences of lymphocyte priming in the GALT explain why vaccination against intestinal infections

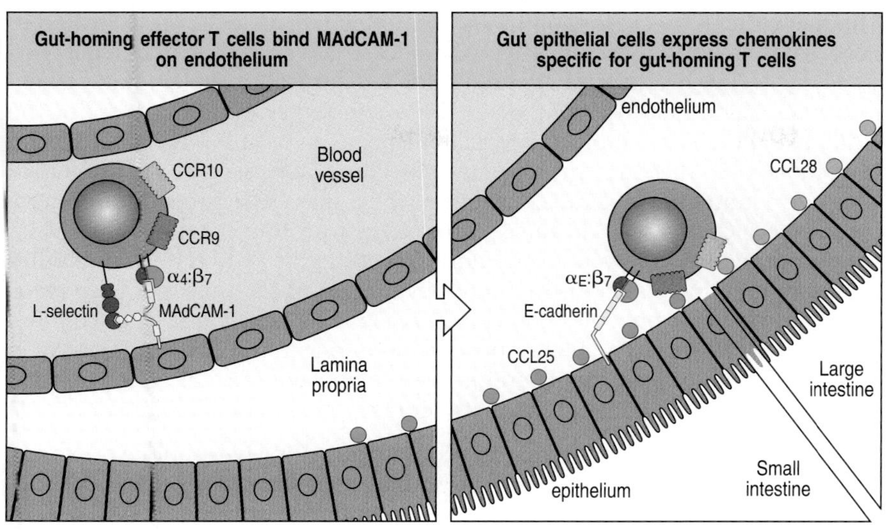

Fig. 11.12 Molecular control of intestine-specific homing of lymphocytes. Left panel: T and B lymphocytes primed by antigen in the gut-associated lymphoid tissues arrive as effector lymphocytes in the bloodstream supplying the intestinal wall (see Fig. 11.11). The lymphocytes express the integrin α_4:β_7, which binds specifically to MAdCAM-1 expressed selectively on the endothelium of blood vessels in mucosal tissues. This provides the adhesion signal needed for the emigration of cells into the lamina propria. Right panel: if primed in the small intestine, the effector lymphocytes also express the chemokine receptor CCR9, which allows them to respond to CCL25 (green circles) produced by epithelial cells of the small intestine; this enhances selective recruitment. Effector lymphocytes that have been primed in the large intestine do not express CCR9 but do express CCR10. This may respond to CCL28 (blue circles) produced by colon epithelial cells to fulfill a similar function. Lymphocytes that will enter the epithelial layer stop expressing the α_4:β_7 integrin and instead express the α_E:β_7 integrin. The receptor for this is E-cadherin on the epithelial cells. These interactions may help keep lymphocytes in the epithelium once they have entered it.

requires immunization by a mucosal route, because other routes, such as subcutaneous or intramuscular immunization, do not involve dendritic cells with the correct imprinting properties.

11-7 Priming of lymphocytes in one mucosal tissue can induce protective immunity at other mucosal surfaces.

MAdCAM-1 is not restricted entirely to the blood vessels of the intestine, but is also found on the vasculature in the other mucosal surfaces. As a result, lymphocytes that have been primed in the GALT, for example, can recirculate as effector cells to the respiratory tract, urogenital tract and lactating breast. Thus, the mucosal immune system forms a unified recirculation compartment, referred to as the **common mucosal immune system**, which is distinct from other parts of the immune system. This has several important implications for vaccine development, as it allows immunization by one mucosal route to be used to protect against infection at another mucosal surface. This has been illustrated in many experimental models, the most interesting being the ability of nasal immunization to prime immune responses in the urogenital tract against HIV. In addition, the induction of IgA antibody production in the lactating breast by natural infection or vaccination at mucosal surfaces elsewhere, such as the intestine, is an important means of generating protective immunity that is transmitted to infants by the passive transfer of antibodies in milk.

11-8 Secretory IgA is the class of antibody associated with the mucosal immune system.

The dominant class of antibody in the mucosal immune system is IgA, which is produced locally by plasma cells present in the mucosal wall. This class of antibody is found in humans in two isotypic forms, IgA1 and IgA2. The nature of IgA differs between the two main compartments in which it is found, the blood and mucosal secretions. In the blood, IgA is found mainly as a monomer and is produced in the bone marrow by plasma cells derived from B cells that have been activated in lymph nodes; the ratio of IgA1 to IgA2 in blood is about 10:1. In the mucosal tissues, IgA is produced almost exclusively as a dimer linked by a J chain, and the ratio of IgA1 to IgA2 in mucosa is about 3:2.

The naive B-cell precursors of the IgA-secreting plasma cells are activated in Peyer's patches and mesenteric lymph nodes. Class switching of naive B lymphocytes to IgA production occurs under the control of the cytokine transforming growth factor-β (TGF-β) in the organized lymphoid tissues of the GALT using the same molecular mechanisms as in lymph nodes and spleen (the molecular mechanisms of class switching are discussed in detail in Chapter 4 and the general consequences of class switching for immune responses in Chapter 9). Around 5 grams of IgA is produced in the mucosal tissues of humans each day, considerably exceeding the production of all other immunoglobulin classes in the body. Several common intestinal pathogens possess proteolytic enzymes that can cleave IgA1, whereas IgA2 is much more resistant to cleavage. The higher proportion of plasma cells secreting IgA2 in the gut lamina propria may therefore be the consequence of selective pressure by pathogens against individuals with low IgA2 levels in the gut.

After B-cell activation and differentiation, the resulting lymphoblasts express the mucosal homing integrin $\alpha_4{:}\beta_7$, as well as the chemokine receptors CCR9 and CCR10. The localization of IgA-secreting plasma cells to mucosal tissues is achieved by the mechanisms that we considered in Section 11-6. Once in the lamina propria, plasma cells synthesize and secrete intact J-chain-linked IgA dimers into the subepithelial space (Fig. 11.13). To reach its target antigens in the gut lumen, the IgA then has to be transported across the epithelium. This is done by immature epithelial cells located at the base of the intestinal crypts, which express the **polymeric immunoglobulin receptor** (**poly-Ig receptor**) on their basolateral surfaces. This receptor has a high affinity for J-chain-linked polymeric immunoglobulins such as dimeric IgA, and transports the antibody by transcytosis to the luminal surface of the epithelium, where it is released by proteolytic cleavage of the extracellular domain of the poly-Ig receptor. Part of the cleaved receptor remains associated with the IgA and is known as **secretory component** (frequently abbreviated to SC). The resulting antibody is now referred to as **secretory IgA**.

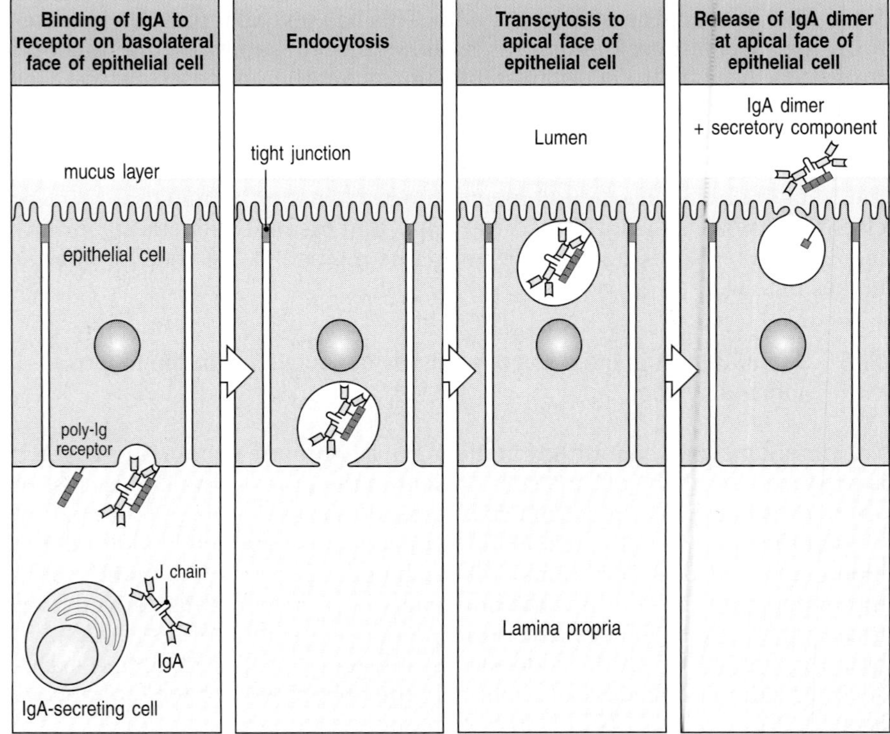

Fig. 11.13 Transcytosis of IgA antibody across epithelia is mediated by the poly-Ig receptor, a specialized transport protein. Most IgA antibody is synthesized in plasma cells lying just beneath epithelial basement membranes of the gut, the respiratory epithelia, the tear and salivary glands, and the lactating mammary gland. The IgA dimer bound to a J chain diffuses across the basement membrane and is bound by the poly-Ig receptor on the basolateral surface of the epithelial cell. The bound complex undergoes transcytosis by which it is transported in a vesicle across the cell to the apical surface, where the poly-Ig receptor is cleaved to leave the extracellular IgA-binding component bound to the IgA molecule as the so-called secretory component. Carbohydrate on the secretory component binds to mucins in mucus and holds the IgA at the epithelial surface. The residual piece of the poly-Ig receptor is nonfunctional and is degraded. IgA is transported across epithelia in this way into the lumina of several organs that are in contact with the external environment.

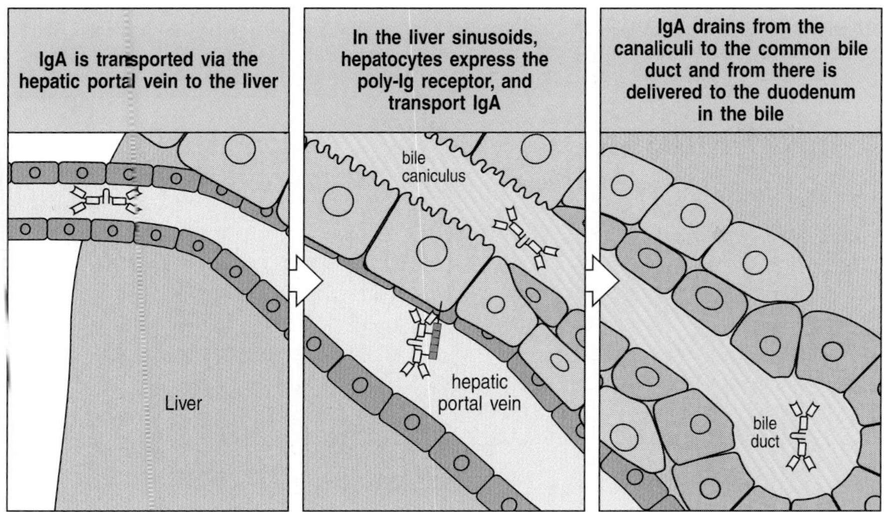
IgA is transported via the hepatic portal vein to the liver
In the liver sinusoids, hepatocytes express the poly-Ig receptor, and transport IgA
IgA drains from the canaliculi to the common bile duct and from there is delivered to the duodenum in the bile
Liver
bile caniculus
hepatic portal vein
bile duct

Fig. 11.14 The hepatobiliary route of IgA secretion. In some species, the direct transport of dimeric IgA across the intestinal epithelium is complemented by secretion via the liver. Excess dimeric IgA produced in the intestinal wall is taken up into the portal veins, which drain from the lamina propria to the liver. In the liver, these blood vessels (sinusoids) are lined by cells expressing the poly-Ig receptor, which transport the dimeric IgA across the vessel walls into adjacent vessels carrying bile (canaliculi). These canaliculi drain into the common bile duct, which empties into the upper small intestine, delivering its cargo of secretory IgA.

In some animals there is a second route of IgA secretion into the intestine—the **hepatobiliary route** (Fig. 11.14). In this case, dimeric IgA antibodies that do not bind the poly-Ig receptor on epithelial cells are taken up into the portal veins in the lamina propria, which drain intestinal blood to the liver. In the liver these small veins (sinusoids) are lined by hepatocytes that express the poly-Ig receptor on their basal surface, allowing the uptake and transcytosis of IgA into the adjoining bile ducts. In this way, secretory IgA antibodies can be delivered directly into the upper small intestine via the common bile duct. In addition, IgA antibodies that have bound antigens in the lumen may be taken back into the gut wall via epithelial cells and eliminated from the body by the hepatobiliary route. Although highly efficient in rats, rabbits, and chickens, this route does not seem to be of great significance in humans, in whom hepatocytes do not express the poly-Ig receptor.

IgA secreted into the gut lumen binds to the layer of mucus coating the epithelial surface via carbohydrate determinants in secretory component. Its retention near the epithelial surface means that it can prevent the adherence of microorganisms, as well as neutralizing their toxins or enzymes (Fig. 11.15). In addition to this activity in the lumen of the intestine, IgA inside epithelial cells has been found to neutralize bacterial lipopolysaccharide that has penetrated the epithelial cells. Secretory IgA has little capacity to activate the classical pathway of complement or to act as an opsonin, and so cannot induce inflammation. Its main function is to limit the access of pathogens to the mucosal surfaces, without risking inflammatory damage to these fragile tissues. Intestinal IgA also has an important role in the symbiotic relationship between an individual and their gut commensal bacteria, helping to restrict these organisms to the gut lumen. The IgA repertoire in the gut includes antibodies specific for antigens expressed by commensal bacteria; these antibody specificities are not found in serum except in pathological circumstances when commensal bacteria invade the bloodstream.

In mice, a significant proportion of intestinal IgA antibody is produced by lymphocytes of the B-1 subset (see Section 7-28). B-1 cells arise from precursor B cells in the peritoneal cavity, display a restricted immunoglobulin repertoire, and produce antibody against certain antigens without T-cell help. As yet, there is little evidence for this source of IgA in humans, in whom all secretory IgA responses involve somatic hypermutation and seem to be T-cell dependent. Nevertheless, its occurrence in mice may offer a glimpse into the evolutionary history of specific antibody responses.

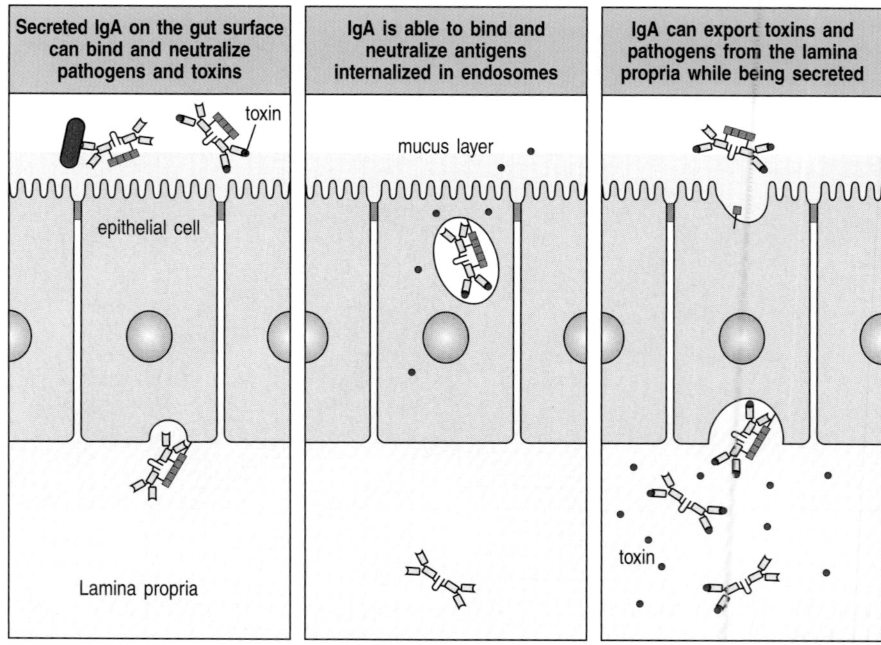
Secreted IgA on the gut surface can bind and neutralize pathogens and toxins
toxin
epithelial cell
Lamina propria
IgA is able to bind and neutralize antigens internalized in endosomes
mucus layer
IgA can export toxins and pathogens from the lamina propria while being secreted
toxin

Fig. 11.15 Secretory IgA has several functions in epithelial surfaces. First panel: IgA adsorbs to the layer of mucus covering the epithelium, where it can neutralize pathogens and their toxins, preventing their access to tissues and inhibiting their functions. Second panel: antigen internalized by the epithelial cell can meet and be neutralized by IgA in endosomes. Third panel: toxins or pathogens that have reached the lamina propria encounter pathogen-specific dimeric IgA in the lamina propria, and the resulting complexes are excreted into the lumen across the epithelial cell as the IgA is secreted by means of the poly-Ig receptor.

11-9 IgA deficiency is common in humans but may be overcome by secretory IgM.

Selective deficiency of IgA production is the commonest primary immune deficiency in humans, occurring in around 1 in 500 to 1 in 700 individuals in populations of Caucasian origin, although it is somewhat rarer in other ethnic groups. A higher incidence of atopic and autoimmune disease has been reported in people with IgA deficiency, but most individuals are normal, and mucosal infections are not more prevalent than usual unless there is also a deficiency in IgG2 production. This probably reflects the ability of IgM to replace IgA as the predominant antibody in secretions, and increased numbers of IgM-producing plasma cells are indeed found in the intestinal mucosa of IgA-deficient patients. Because IgM is a J-chain-linked polymer, it is bound efficiently by the poly-Ig receptor and is transported across epithelial cells into the lumen as secretory IgM. The importance of this back-up mechanism is shown in knockout mice, in which animals lacking IgA alone have a normal phenotype, but those lacking the poly-Ig receptor are susceptible to mucosal infections.

11-10 The mucosal immune system contains unusual T lymphocytes.

T lymphocytes are abundant in the mucosal tissues, not only in the organized tissues of the MALT but also scattered throughout the mucosa. In the intestine, scattered T cells are found in two distinct locations, the lamina propria and the epithelium (see Fig. 11.4). The T-cell population of the lamina propria has a ratio of CD4:CD8 T cells of 3:1 or more, similar to that in systemic lymphoid tissues. Most of these cells have markers associated with antigen-experienced effector or memory T cells, such as CD45RO in humans (see Section 10-16). They also express the gut-homing markers CCR9 and α_4:β_7 integrin, as well as receptors for pro-inflammatory chemokines such as CCL5 (RANTES). Lamina propria T cells proliferate poorly when stimulated by mitogens or antigen, but they secrete large amounts of cytokines such as interferon (IFN)-γ, interleukin (IL)-5, and IL-10, even in the normal intestine and in the

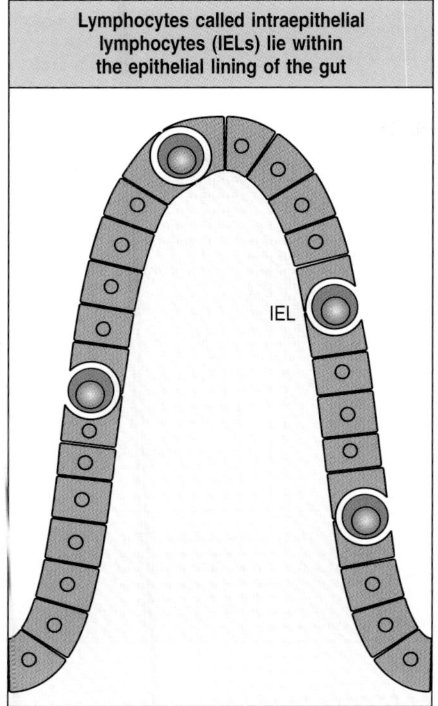

Lymphocytes called intraepithelial lymphocytes (IELs) lie within the epithelial lining of the gut
IEL

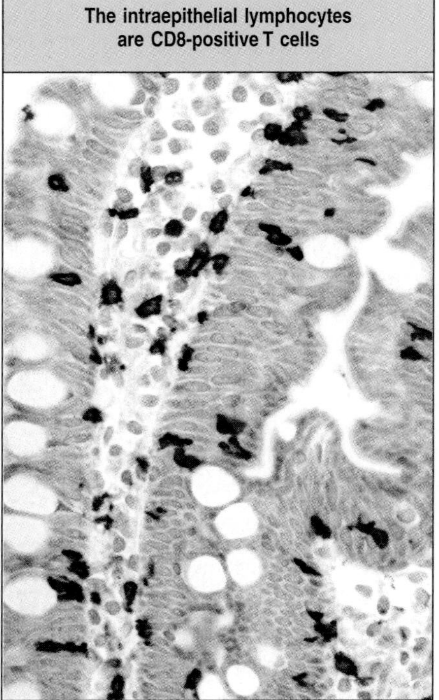

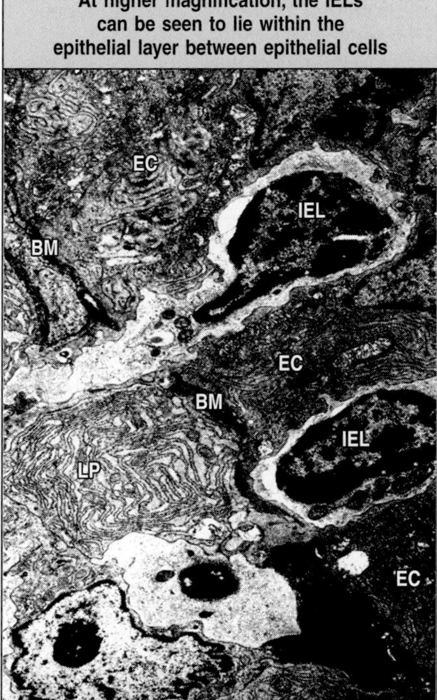

Fig. 11.16 Intraepithelial lymphocytes. The epithelium of the small intestine contains a large population of lymphocytes known as intraepithelial lymphocytes (IELs) (left panel). The micrograph in the center is of a section through human small intestine in which CD8 T cells have been stained brown with a peroxidase-labeled monoclonal antibody. Most of the lymphocytes in the epithelium are CD8 T cells. Magnification ×400. The electron micrograph on the right shows that the IELs lie between epithelial cells (EC) on the basement membrane (BM) separating the lamina propria (LP) from the epithelium. One IEL can be seen having crossed the basement membrane into the epithelium, leaving a trail of cytoplasm in its wake. Magnification ×8000.

absence of inflammation. In conditions such as celiac disease and inflammatory bowel diseases, lamina propria CD4 T cells are clearly the principal effector T cells responsible for causing the local tissue damage, but their function in the healthy gut is uncertain. They may assist IgA production by local plasma cells, or they may be regulatory T cells that are involved in preventing hypersensitivity reactions to food proteins and commensal bacteria, as described later in this chapter. Activated CD8 T cells are also present in the lamina propria and are capable of both cytokine production and cytotoxic activity during a protective immune response to pathogens and in inflammation.

The lymphocytes found in the epithelium—the **intraepithelial lymphocytes (IEL)**—are quite distinct in character (Fig. 11.16). There are 10–15 lymphocytes for every 100 epithelial cells in the normal small intestine, meaning that this is one of the largest populations of lymphocytes in the body. More than 90% of intraepithelial lymphocytes are T cells, and around 80% of these carry CD8, in complete contrast to the lymphocytes in the lamina propria. As in the lamina propria, however, most intraepithelial lymphocytes have an activated appearance, as well as having intracellular granules containing perforin and granzymes, as seen in effector cytotoxic T cells. The T-cell receptors of most of this lymphocyte population show evidence of restricted use of V(D)J gene segments, indicating that they may expand locally in response to a relatively small number of antigens. The intraepithelial lymphocytes of the small intestine express the chemokine receptor CCR9, but have the $\alpha_E{:}\beta_7$ integrin on their surface, instead of the $\alpha_4{:}\beta_7$ integrin found on other gut-homing T cells. The receptor for $\alpha_E{:}\beta_7$ integrin is E-cadherin on the surface of epithelial cells, and this interaction may assist these lymphocytes to remain in the epithelium (see Fig. 11.12).

The origin and functions of the intraepithelial lymphocytes are controversial. In young animals and in the adults of some species, there are unusually large numbers of $\gamma{:}\delta$ T cells in the gut epithelium. In normal adult mice and humans, however, $\gamma{:}\delta$ T cells are found in similar numbers in the epithelium

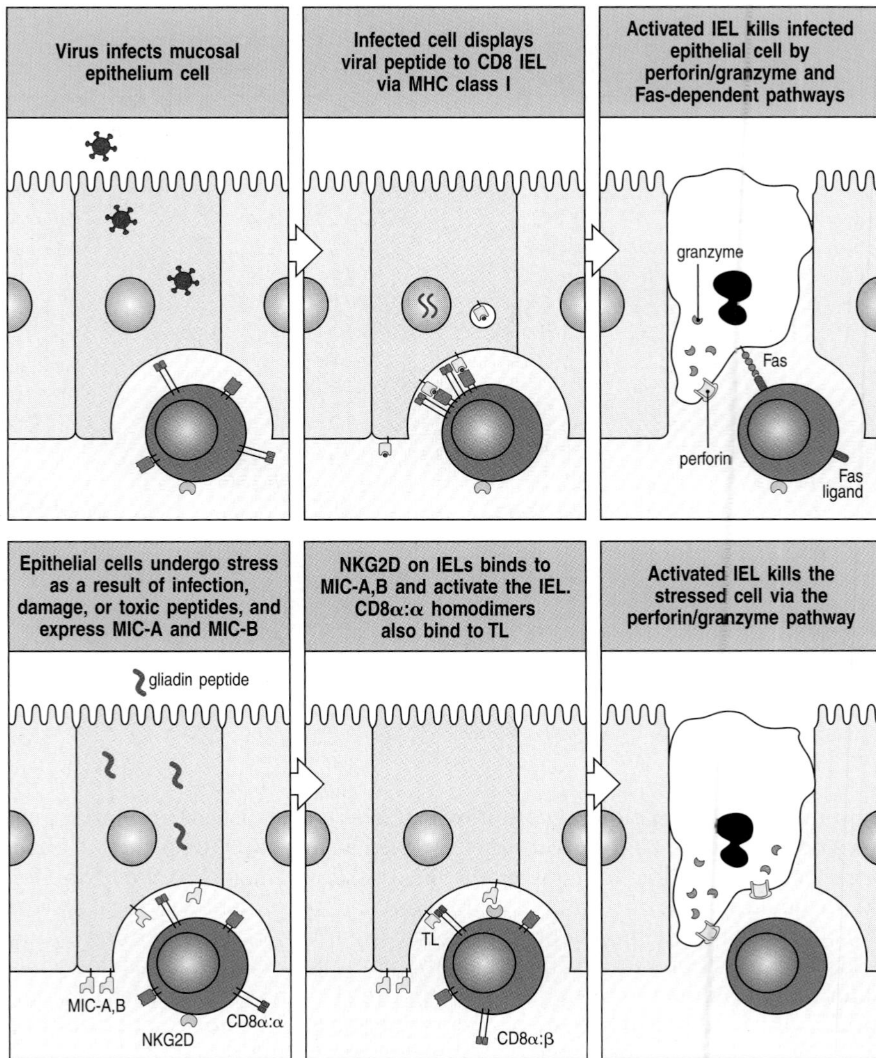

Fig. 11.17 Functions of intraepithelial lymphocytes. There are two main types of intraepithelial lymphocytes (IELs). As shown in the top panels, one type (type a IELs) are conventional CD8 cytotoxic T cells that recognize peptides derived from viruses or other intracellular pathogens bound to classical MHC class I molecules on infected epithelial cells. The activated IEL recognizes specific peptide:MHC complexes by using its α:β T-cell receptor, with the CD8α:β heterodimer as co-receptor. The IEL releases perforin and granzyme, which kill the infected cell. Apoptosis of epithelial cells can also be induced by the binding of Fas ligand on the T cell to Fas on the epithelial cell. In the bottom panels, epithelial cells that have been stressed by infection or altered cell growth, or by a toxic peptide from the protein α-gliadin (a component of gluten), upregulate expression of the non-classical MHC class I molecules MIC-A and MIC-B and produce IL-15. Neighboring IELs are activated by IL-15 and recognize MIC-A and MIC-B by using the receptor NKG2D (see Section 2-32); these are called type b IELs. They also kill the epithelial cells by releasing perforin and granzyme. These IELs carry the CD8α:α homodimer, and this protein may also contribute to their recognition of infected cells by binding directly to the non-classical MHC class I molecule TL, encoded in the T region of the MHC, which is present on epithelial cells.

and bloodstream. In mice, around 50% of intraepithelial lymphocytes express the unusual homodimeric α:α form of CD8, and they are divided into two groups on the basis of which form of CD8 is expressed. One type, designated type a, are conventional T cells bearing α:β T-cell receptors and the CD8α:β heterodimer. They are derived from naive CD8 T cells activated in the Peyer's patches as described earlier, and function as conventional class I MHC-restricted cytotoxic T cells, killing virus-infected cells, for example (Fig. 11.17, top panels). They also secrete effector cytokines such as IFN-γ.

The second class of intraepithelial lymphocytes, designated type b, comprises T cells expressing the CD8 α homodimer (CD8 α:α). These have either an α:β or a γ:δ T-cell receptor. The receptors of the α:β T cells in this group do not, however, bind conventionalpeptide:MHC ligands, but instead bind a number of other ligands, including MHC class Ib molecules (see Sections 5-17 and 5-18). Unlike type a intraepithelial T cells, many of the type b T cells do not undergo conventional positive and negative selection in the thymus (see Chapter 7) and express apparently autoreactive T-cell receptors. The absence of the CD8 α:β protein, however, means that these T cells have low affinity for conventional peptide:MHC complexes and thus cannot act as self-reactive effector cells.

Until recently, it was believed that type b intraepithelial lymphocytes were derived by extrathymic T-cell differentiation that occurred entirely in the gut itself, perhaps in the lymphoid aggregates known as **cryptopatches** that are found in the intestinal wall. However, subsequent work suggests that cryptopatches may simply be the sites where lymphoid-tissue inducer cells (see Section 7-24) accumulate. In response to postnatal antigen stimulation, these give rise to the small B-cell-rich isolated lymphoid follicles (see Section 11-3). It now seems that all intraepithelial lymphocytes, including those of type b, require the thymus for differentiation, although those expressing the CD8α homodimer may escape conventional negative selection by self antigens as a result of their low affinity for self MHC molecules. Instead, expression of the CD8α homodimer may enable a process of so-called **agonist selection**, in which late double-negative/early double-positive T cells are positively selected in the thymus by relatively high-affinity ligands, not unlike the process that is thought to drive selection of CD4 CD25 T_{reg} and NKT cells (see Chapter 7). The intraepithelial lymphocyte precursors then exit from the thymus before they are fully differentiated and undergo further maturation in the intestine, which may involve additional positive selection on non-classical MHC molecules expressed on the epithelium. In some mouse strains, one of the selecting molecules in the gut is the thymus leukemia antigen (TL), which is a non-classical MHC class I molecule that does not present antigenic peptides. TL is expressed by intestinal epithelial cells and binds the CD8α homodimer directly and with high affinity.

In addition to agonist selection, type b intraepithelial lymphocytes share several other properties of cells of the innate immune system, including the constitutive expression of cytotoxic activity and of pro-inflammatory cytokines and chemokines, as well as of receptors for these molecules. All intraepithelial lymphocytes express high levels of the activating C-type lectin NK receptor NKG2D (see Sections 2-31 and 2-32). This binds to two MHC-like molecules—MIC-A and MIC-B—that are expressed on intestinal epithelial cells in response to cellular injury and stress. The injured cells can then be recognized and killed by the intraepithelial lymphocytes. These lymphocytes may thus be considered in evolutionary terms as being at the interface between innate and adaptive immunity. Their role in the gut may be the rapid recognition and elimination of epithelial cells that express an abnormal phenotype as a result of stress or infection (Fig. 11.17, bottom panels). There is also evidence that intraepithelial lymphocytes are important for controlling the subsequent repair of the mucosa, a function particularly associated with the γ:δ subset of these T cells, which have a similar role in skin repair. These functions of intraepithelial lymphocytes may also be involved in causing disease. For example, the MIC-A-dependent cytotoxic activity of these T cells is enhanced in celiac disease, which is associated with epithelial damage and increased numbers of intraepithelial lymphocytes. This activation is mediated by IL-15, which is released by epithelial cells in response to certain components of gluten.

Summary.

The mucosal tissues of the body such as the intestine and respiratory tract are exposed continuously to enormous amounts of different antigens, which can be either pathogenic invaders or harmless materials such as foods and commensal organisms. Potential immune responses to this antigen load are controlled by a distinct compartment of the immune system, the mucosal immune system, which is the largest in the body and possesses many unique features. These include distinctive routes and processes for the uptake and presentation of antigens, exploiting M cells to transport antigens over the epithelium of Peyer's patches, and unusual populations of dendritic cells that imprint the T cells they activate with gut-homing properties. Lymphocytes primed in the mucosa-associated lymphoid tissues acquire specific homing receptors, allowing them to redistribute preferentially back to mucosal surfaces as effector cells. Exposure to antigen outside the mucosal immune system cannot reproduce these effects. The mucosa-associated lymphoid tissues also generate different effector responses from those in other parts of the body, including unique forms of innate immunity. The adaptive immune response in mucosal tissues is characterized by the production of secretory dimeric IgA, and by the presence of distinctive populations of effector T cells whose functional and phenotypic properties are highly influenced by their anatomical location.

The mucosal response to infection and regulation of mucosal immune responses.

The major role of the mucosal immune response is defense against infectious agents, which can include all forms of microorganisms from viruses to multicellular parasites. This means that the host must be able to generate a wide spectrum of immune responses tailored to meet the challenge of individual pathogens, and it is equally unsurprising that many microbes have evolved means of adapting to and subverting the host response. To be able to ensure adequate responses to pathogens, the mucosal immune system needs to recognize harmless antigens but must not produce equivalent effector responses to them. A major role of this compartment of the immune system is to balance these competing demands, and these mechanisms form the major focus of the following sections.

11-11 Enteric pathogens cause a local inflammatory response and the development of protective immunity.

Despite the array of innate immune mechanisms in the gut, and stiff competition from the indigenous flora, the gut is the most frequent site of infection by pathogenic microorganisms. These include many viruses, enteric bacteria such as *Salmonella* and *Shigella* species, protozoa such as *Entamoeba histolytica*, and helminth parasites such as tapeworms and pinworms (Fig. 11.18). These pathogens cause disease in many ways, but certain common features of infection are crucial to understanding how they stimulate a productive immune response by the host. The key to this in the gut, as elsewhere in the body, is the activation of the innate immune system.

Innate mechanisms eliminate most intestinal infections rapidly and without significant spread beyond the intestine. Activation of local inflammatory cells via pattern-recognition receptors such as the Toll-like receptors (TLRs) is important in this process, but intestinal epithelial cells themselves also contribute significantly and are not simply passive victims of infection. Epithelial cells do not express TLRs or CD14 (an essential part of the TLR-4 complex that detects bacterial lipopolysaccharide) on their apical surface and so are probably unable to sense bacteria that are in the intestinal lumen. They do bear TLR-5 on their basal surfaces, allowing them to recognize flagellin (the protein of which bacterial flagella are made) on bacteria that have managed to cross the epithelial barrier. Mutant mice that lack this receptor show increased susceptibility to infection by *Salmonella*, for example. They also carry TLRs in intracellular vascuoles that can detect pathogens and their products that have been internalized by endocytosis (Fig. 11.19).

Epithelial cells also have intracellular sensors that can react to microorganisms or their products that enter the cytoplasm (see Fig. 11.19). These sensors include the nucleotide-binding oligomerization domain proteins NOD1 and NOD2, which are related to the TLRs (see Section 2-9). These proteins are also known as CARD4 and CARD15, respectively, because they contain a caspase-recruitment domain. NOD1 recognizes a muramyl tripeptide containing diaminopimelic acid that is found only in the cell walls of Gram-negative bacteria; NOD2 recognizes a muramyl dipeptide found in the peptidoglycans of most bacteria, and epithelial cells defective in NOD2 are less resistant to infection by intracellular bacteria. Oligomerization of NOD1 or NOD2 as a result of ligand binding enables them to bind and activate the protein kinase

Intestinal pathogens and human disease	
Bacteria	
Salmonella typhi	Typhoid fever
Salmonella paratyphi	Enteric fever (paratyphoid)
Salmonella enteritidis	Food poisoning
Vibrio cholera	Cholera
Shigella dysenteriae, flexneri, sonnei	Dysentery
Enteropathogenic *E. coli* (EPEC)	Gastroenteritis, systemic infection
Enterohemolytic *E. coli* (EHEC)	Gastroenteritis, systemic infection
Enterotoxigenic *E. coli* (ETEC)	Gastroenteritis, 'travelers diarrhea'
Enteroaggregative *E. coli* (EAEC)	Gastroenteritis, systemic infection
Yersinia enterocolitica	Gastroenteritis, systemic infection
Clostridium difficile	Necrotizing enterocolitis
Campylobacter jejuni	Gastroenteritis
Staphylococcus aureus	Gastroenteritis
Bacillus cereus	Gastroenteritis
Clostridium perfringens	Gastroenteritis
Helicobacter pylori	Gastritis, peptic ulcer, gastric cancer
Mycobacterium tuberculosis	Intestinal TB
Listeria monocytogenes	Foodborne infection
Viruses	
Rotaviruses	Gastroenteritis
Norwalk-like viruses	'Winter vomiting' disease
Astroviruses	'Winter vomiting' disease
Adenoviruses	'Winter vomiting' disease
Parasites	
Protozoa	
Giardia lamblia	Gastroenteritis
Blastocystis hominis	Gastroenteritis (esp. in immunocompromised hosts)
Toxoplasma gondii	Gastroenteritis, systemic disease (esp. in immunocompromised hosts)
Cryptosporidium parvum	Gastroenteritis (esp. in immunocompromised hosts)
Entamoeba histolytica	Amebic dysentery + liver abscesses
Microsporidium species	Diarrheal disease
Helminths	
Ascaris lumbricoides	Roundworm infection of small intestine
Necator americanus	Hookworm infection of small intestine
Strongyloides species	Roundworm infection of small intestine
Enterobius species	Pinworm infection of large intestine
Trichinella spiralis	Trichinosis
Trichuris trichiura	Whipworm infection of large intestine
Taenia species	Tapeworm infections
Schistosoma species	Schistosomiasis: enteritis, mesenteric vein infection

Fig. 11.18 Intestinal pathogens and infectious disease in humans. Many species of bacteria, viruses, and parasites can cause disease in the human intestine.

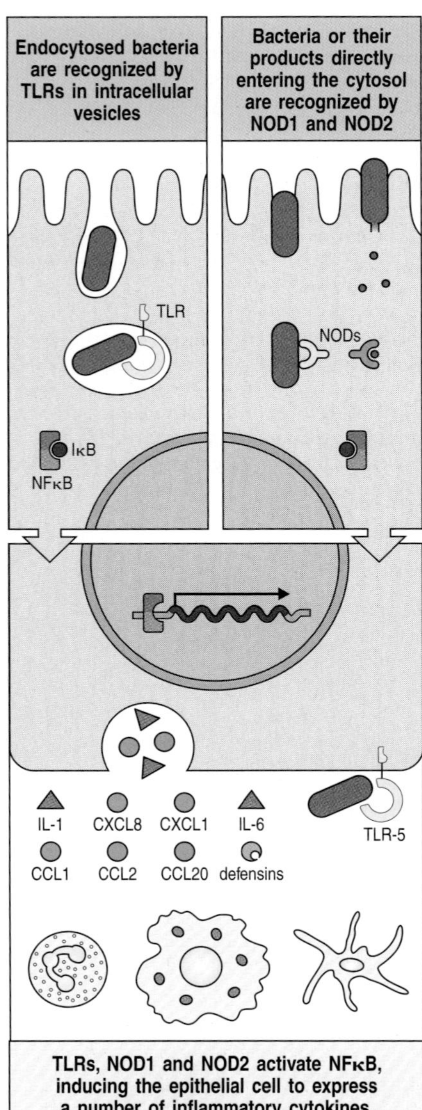

Endocytosed bacteria are recognized by TLRs in intracellular vesicles
Bacteria or their products directly entering the cytosol are recognized by NOD1 and NOD2
TLR
NODs
IκB
NFκB
IL-1
CXCL8
CXCL1
IL-6
TLR-5
CCL1
CCL2
CCL20
defensins
TLRs, NOD1 and NOD2 activate NFκB, inducing the epithelial cell to express a number of inflammatory cytokines, chemokines, and other mediators. These in turn activate neutrophils, macrophages, and dendritic cells

Fig. 11.19 Epithelial cells play a critical role in innate defense against pathogens. Toll-like receptors (TLRs) are present in intracellular vesicles or on the basolateral surface of epithelial cells, where they recognize different components of invading bacteria. NOD1 and NOD2 pattern-recognition receptors are found in the cytoplasm and recognize cell-wall peptides from bacteria. Both TLRs and NODs activate the NFκB pathway, leading to the generation of pro-inflammatory responses by epithelial cells. These include the production of chemokines such as CXCL8, CXCL1 (GROα), CCL1, and CCL2, which attract neutrophils and macrophages, and CCL20 and β-defensin, which attract immature dendritic cells in addition to possessing antimicrobial properties. The cytokines IL-1 and IL-6 are also produced and activate macrophages and other components of the acute inflammatory response. The epithelial cells also express MIC-A and MIC-B and other stress-related non-classical MHC molecules, which can be recognized by cells of the innate immune system. IκB, inhibitor of NFκB.

RICK (also known as Rip2 and CARDIAK) through that protein's caspase-recruitment domain. This results in the activation of the NFκB pathway in the epithelial cells, which leads to the release of cytokines, chemokines, and the antimicrobial defensins (see Section 2-3). The NFκB pathway is shown in detail in Fig. 6.21. Other epithelial cell products include the chemokine CXCL8 (IL-8), which is a potent neutrophil chemoattractant, and the chemokines CCL2, CCL3, CCL4, and CCL5, which are chemoattractants for monocytes, eosinophils, and T cells. Infected epithelial cells also increase their production of CCL20, which attracts immature dendritic cells via the receptor CCR6. In this way, the onset of infection triggers an influx of inflammatory cells and lymphocytes into the mucosa from the bloodstream, which assists the induction of a specific immune response to the antigens of the infectious agent.

Injury and stress to the enterocytes lining the gut stimulates the expression of non-classical MHC molecules, such as MIC-A and MIC-B (see Fig. 11.17). These proteins can be recognized by the receptor NKG2D on local cytotoxic lymphocytes, which are then activated to kill the infected epithelial cells, thereby promoting repair and recovery of the injured mucosa.

11-12 The outcome of infection by intestinal pathogens is determined by a complex interplay between the microorganism and the host immune response.

Many enteric pathogens need to exploit host mechanisms of antigen uptake via M cells and inflammation as part of their invasive strategy. Poliovirus, reoviruses, and some retroviruses are transported through M cells by transcytosis, and initiate infection in tissues distant from the intestine after delivery into the subepithelial space. HIV may use a similar route into the lymphoid tissue of the rectal mucosa, where it first encounters and infects dendritic cells. Many of the most important enteric bacterial pathogens also gain entry to the body through M cells. These include *Salmonella typhi*, the causative agent of typhoid, *Salmonella typhimurium*, a major cause of bacterial food poisoning, *Shigella* species that cause dysentery, and *Yersinia pestis*, which causes plague. After entry into the M cell, these bacteria produce factors that reorganize the M-cell cytoskeleton in a manner that encourages their transcytosis.

M cells are not the only port of entry into the mucosa. Some intestinal bacteria such as *Clostridium difficile* or *Vibrio cholerae* produce high levels of secreted protein toxins, allowing them to cause disease without the need to invade the epithelium. Other bacteria, such as enteropathogenic and enterohemolytic *E. coli*, have specialized means of attaching to and invading epithelial cells, enabling them to cause intestinal damage and produce harmful

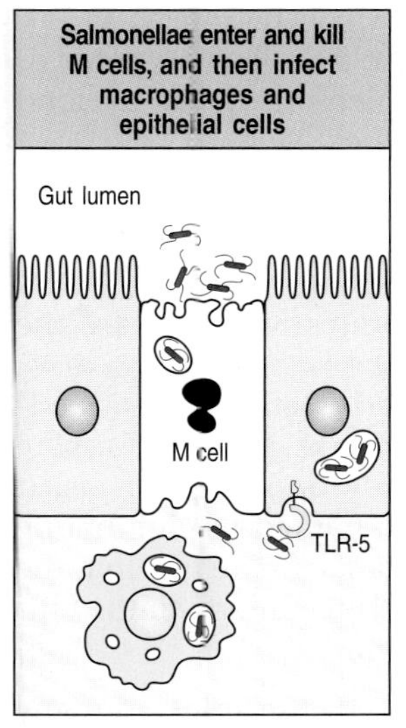

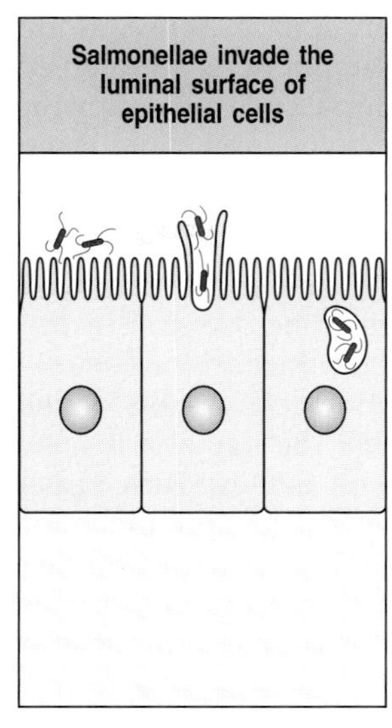

Salmonellae invade the luminal surface of epithelial cells

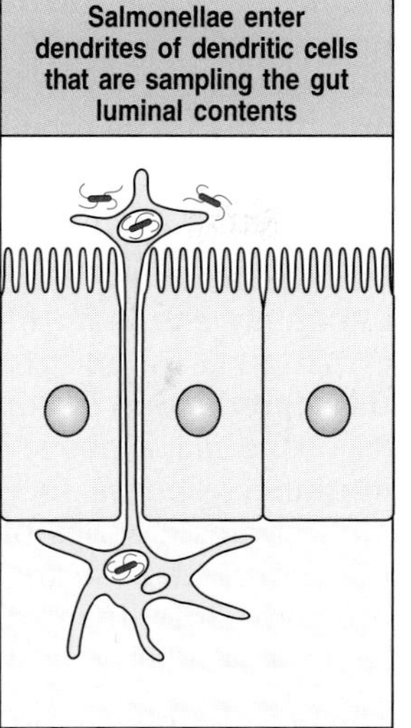

Salmonellae enter dendrites of dendritic cells that are sampling the gut luminal contents

Fig. 11.20 *Salmonella typhimurium*, an important cause of food poisoning, can penetrate the gut epithelial layer by three routes. In the first route (left panel), *S. typhimurium* adheres to and enters M cells, which it then kills by causing apoptosis. Having penetrated the epithelium, it infects macrophages and gut epithelial cells. Epithelial cells express TLR-5 on their basal membrane; this binds flagellin on the salmonella flagella, activating an inflammatory response via the NFκB pathway. Salmonellae can also invade gut epithelial cells directly by adherence of their fimbriae (fine threadlike protrusions) to the luminal epithelial surface (middle panel). In the third route of entry, dendritic cells sampling the gut lumen extrude dendrites between epithelial cells. These effectively breach the epithelial layer and can be infected by salmonellae in the lumen (right panel).

toxins from an intracellular location. Viruses such as rotaviruses also invade enterocytes directly. Some of the entry mechanisms used by salmonellae are shown in Fig. 11.20 and those of shigellae in Fig. 11.21.

Once delivered into the subepithelial space, pathogenic bacteria and viruses are able to cause more widespread infection in a variety of ways. Paradoxically, the host inflammatory response is an additional and often essential part of this invasive process. Bacteria transcytosed through M cells are free to interact with TLRs on inflammatory cells such as macrophages and on the basal surfaces of adjacent epithelial cells. In addition, after being ingested by phagocytes, many of these microbes induce caspase-dependent apoptosis of the phagocyte. All this stimulates the production of a cascade of inflammatory

Fig. 11.21 *Shigella flexneri*, a cause of bacterial dysentery, infects intestinal epithelial cells, triggering activation of the NFκB pathway. *Shigella flexneri* binds to M cells and is translocated beneath the gut epithelium (first panel). The bacteria infect intestinal epithelial cells from their basal surface and are released into the cytoplasm (second panel). The lipopolysaccharide (LPS) on the shigellae binds to and oligomerizes the protein NOD1; oligomerized NOD1 binds the protein kinase RICK, which triggers activation of the NFκB pathway, leading to transcription of genes for chemokines and cytokines (third panel). Activated epithelial cells release the chemokine CXCL8 (IL-8), which acts as a neutrophil chemoattractant (fourth panel). IκK, IκB kinase; IκB, inhibitor of NFκB.

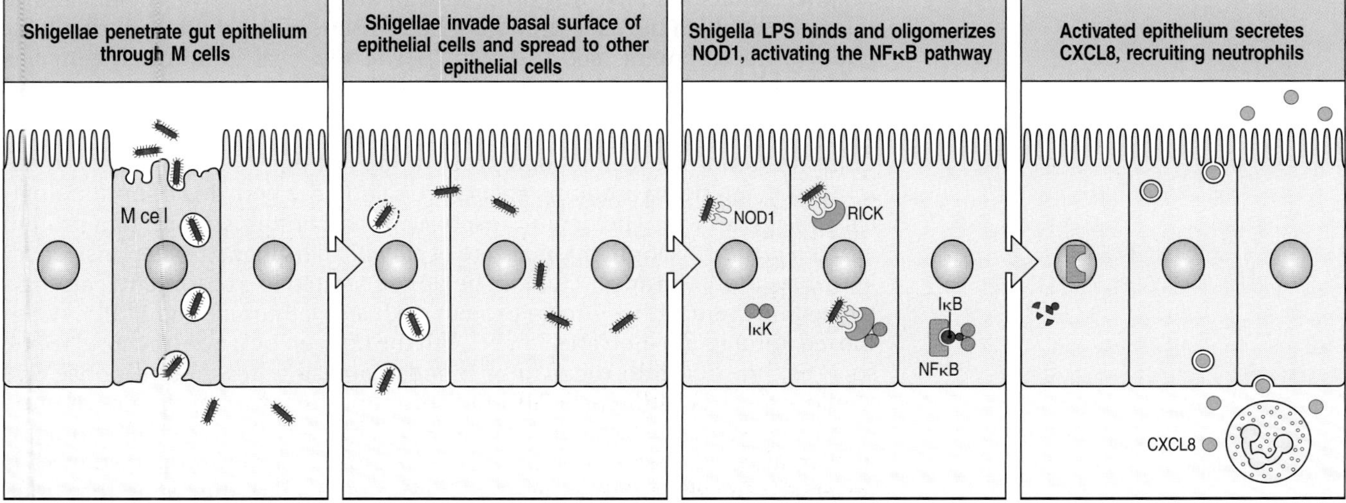

mediators of the innate immune response, among which IL-1β and TNF-α dramatically loosen the tight junctions between epithelial cells. This removes the normal barrier to bacterial invasion, allowing microorganisms to flood into the intestinal tissue from the lumen and extend the infection.

Despite their apparent benefit to the invader, it is important to remember that the principal role of the mediators and cells induced by the innate immune response is to help initiate the adaptive immune response that will ultimately eliminate the microbe. Central to this protective effect are the cytokines IL-12 and IL-18 produced by infected macrophages. These drive IFN-γ production by antigen-specific T cells, which in turn enhances the ability of the macrophage to kill the bacteria it has ingested. Thus, the innate immune response to enteric bacteria has apparently opposing effects. It orchestrates a series of potent effector mechanisms aimed at eliminating infection, but these are exploited by the invading organism. The fact that the protective immune response wins out in most cases is testimony to the efficiency and adaptability of the mucosal immune system.

The host–pathogen interaction is further complicated by the ability of many enteric microbes to modulate the host inflammatory response. For instance, *Yersinia* species produce Yop proteins, which can both inhibit the inflammatory response and block phagocytosis and intracellular killing of microbes by phagocytes. *Salmonella typhi* creates its own safe haven within phagosomes by modifying the phagosome membrane and preventing the recruitment of killing mechanisms. *Shigella*, in contrast, resides in the cytoplasm of epithelial cells, where it remodels the actin cytoskeleton, creating a molecular machinery that allows its direct cell-to-cell spread without exposure to the immune system. All these microorganisms also induce apoptosis in phagocytic cells, thus disabling an important arm of the inflammatory response as well as enhancing their spread. The immunomodulatory molecules produced by these bacteria are frequently essential to their ability to cause disease, which emphasizes their vital role in the bacterial life cycle.

11-13 The mucosal immune system must maintain a balance between protective immunity and homeostasis to a large number of different foreign antigens.

The majority of antigens encountered by the normal intestinal immune system are not derived from pathogens but come from food and commensal bacteria. These are not only harmless but are in fact highly beneficial to the host. Antigens of this kind normally do not induce an immune response, despite the fact that, like any other foreign antigens, there will be no central tolerance to them because they were not present in the thymus during lymphocyte development (see Chapter 7). The mucosal immune system has developed sophisticated means of discriminating between pathogens and innocuous antigens.

Contrary to popular belief, food proteins are not digested completely in the intestine; significant amounts are absorbed into the body in an immunologically relevant form. The default response to oral administration of a protein antigen is the development of a state of specific peripheral unresponsiveness known as **oral tolerance**. This can be demonstrated in experimental animals by feeding them with a foreign protein such as ovalbumin (Fig. 11.22). When the fed animals are then challenged with the antigen by a non-mucosal route, such as injection into the skin or bloodstream, the immune response one would expect is blunted or absent. This suppression of systemic immune responses is long-lasting and is antigen-specific: responses to other antigens are not affected. A similar suppression of subsequent immune responses is observed after the administration of inert proteins into the respiratory tract,

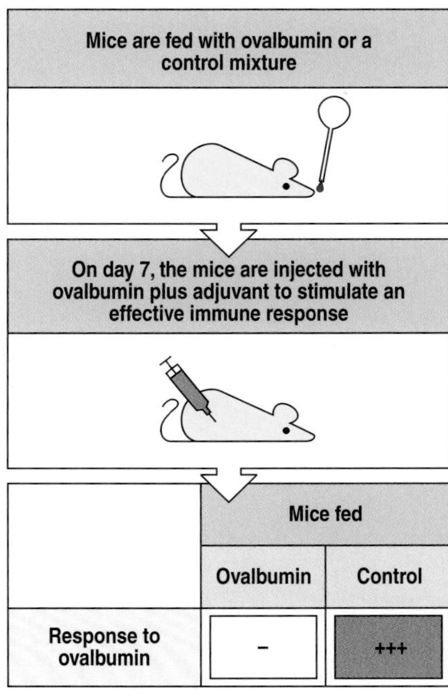

Fig. 11.22 Immune priming and oral tolerance are different outcomes of intestinal exposure to antigen. Top panel: the intestinal immune system generates protective immunity against antigens that are a threat to the host, such as pathogenic organisms and their products. IgA antibodies are produced locally, serum IgG and IgA are made, and the appropriate effector T cells are activated in the intestine and elsewhere. When the antigen is encountered again, there is effective memory, ensuring rapid protection. Harmless antigens such as food proteins or antigens from commensal bacteria induce a phenomenon known as oral tolerance. They lack the danger signals needed to activate local antigen-presenting cells, or do not invade sufficiently to cause inflammation. In the case of food proteins, there is no local IgA antibody production and no primary systemic antibody response, nor are effector T cells activated. As shown in the lower panels, oral tolerance can be induced by feeding a protein such as ovalbumin to a normal mouse. First, mice are fed with either ovalbumin or a different protein as a control. Seven days later, the mice are immunized subcutaneously with ovalbumin and an adjuvant; 2 weeks later, systemic immune responses such as serum antibodies and T-cell function are measured. Mice fed with ovalbumin have a lower ovalbumin-specific systemic immune response than those fed with the control protein.

giving rise to the concept of **mucosal tolerance** as the usual response to such antigens delivered via a mucosal surface.

All aspects of the peripheral immune response can be affected by oral tolerance, although T-cell dependent effector responses and IgE production tend to be more inhibited than serum IgG antibody responses. Thus, the systemic immune responses most susceptible to oral tolerance are those that are usually associated with tissue inflammation. Mucosal immune responses to the antigen are also prevented, meaning that the phenomenon extends to both peripheral and local tissues. A breakdown in oral tolerance is believed to occur in celiac disease. In this condition, genetically susceptible individuals generate IFN-γ-producing, CD4 T-cell responses against the protein gluten found in wheat, and the resulting inflammation destroys the upper small intestine (see Section 13-15).

The mechanisms of oral tolerance to protein antigens are only partly understood but are likely to include anergy or deletion of antigen-specific T cells and the generation of regulatory T cells of different types. These can be found in Peyer's patches and mesenteric lymph nodes and can migrate back to the lamina propria, as well as influencing responses elsewhere in the body. As explained in Chapter 8, regulatory T cells can act in a variety of ways, but regulatory CD4 T cells producing transforming growth factor-β (TGF-β) are particularly associated with oral tolerance. They are sometimes referred to as **T_H3 cells** (see Section 8-20). TGF-β has many immunosuppressive properties and also stimulates B-cell switching to IgA. Together, these properties could help prevent active immunity to food proteins by favoring tolerance of effector T cells specific for these antigens and the production of non-inflammatory IgA antibodies. IL-10 produced by regulatory T cells may also be involved in oral tolerance; it plays an important role in the equivalent tolerance that occurs to certain potential antigens introduced via the respiratory route.

In addition to its physiological role in preventing inappropriate food-related immune responses, mucosal tolerance has proved useful as a means of preventing inflammatory disease in experimental animal models. Oral or intranasal administration of appropriate antigens has been found to be extremely effective in preventing or even treating type 1 diabetes mellitus, experimental arthritis, encephalomyelitis and other autoimmune diseases in animals. Up to now, clinical trials using mucosal tolerance to treat the equivalent diseases in humans have been less successful, but it remains a potentially attractive means of inducing antigen-specific tolerance in clinical situations.

Celiac Disease 

Commensal bacteria also do not provoke a systemic primary immune response, but there is no active tolerance to these antigens in the systemic lymphoid system—instead they seem to be ignored. They do, however, stimulate local IgA antibody production in the intestine and there is active suppression of local effector T-cell responses. When effector T cell responses do occur against food proteins or commensal bacteria, diseases such as celiac disease and Crohn's disease (see Sections 13-15 and 13-21) may develop.

11-14 The healthy intestine contains large quantities of bacteria but does not generate productive immunity against them.

We each harbor more than 1000 species of commensal bacteria in our intestine, and they are present in the largest numbers in the colon and lower ileum. Despite the fact that these bacteria collectively weigh about 1 kg, for most of the time we cohabit with our intestinal bacterial flora in a happy symbiotic relationship. Nevertheless, they do represent a potential threat, as is shown when the integrity of the intestinal epithelium is damaged, allowing large numbers of commensal bacteria to enter the mucosa. This can occur when the blood flow to the gut is compromised by trauma, infection, or blood vessel disease, for example, or toxic shock syndrome (see Fig. 9.23). In these circumstances, normally innocuous gut bacteria, such as non-pathogenic *E. coli*, can cross the mucosa, invade the bloodstream, and cause fatal systemic infection.

The normal gut flora has an essential role in maintaining health. Its members assist in the metabolism of dietary constituents such as cellulose, as well as degrading toxins and producing essential cofactors such as vitamin K_1 and short-chain fatty acids. By having direct effects on epithelial cells, the commensal bacteria are also essential for maintaining the normal barrier function of the epithelium. Another important property of commensal organisms is that they interfere with the ability of pathogenic bacteria to colonize and invade the gut. Commensals do this partly by competing for space and nutrients, but they can also directly inhibit the pro-inflammatory signaling pathways that pathogens stimulate in epithelial cells and that are needed for invasion. The protective role of the commensal flora is dramatically illustrated by the adverse effects of broad-spectrum antibiotics. These antibiotics can kill large numbers of commensal gut bacteria, thereby creating an ecological niche for bacteria that would not otherwise be able to compete successfully with the normal flora. One example of a bacterium that grows in the antibiotic-treated gut and can cause a severe infection is *Clostridium difficile*; this produces two toxins, which can cause severe bloody diarrhoea associated with mucosal injury (Fig. 11.23). Triggering of TLRs by commensal bacteria is also important in protecting against inflammation in the intestine, because mice lacking TLR-2, TLR-9, or the TLR signaling adaptor protein MyD88 are much more susceptible to the induction of experimental inflammatory bowel diseases. This protective effect of TLR seems to involve the epithelial cells being made more resistant to inflammation-induced damage.

Commensal bacteria and their products are recognized by the adaptive immune system. The scale of this phenomenon is illustrated by the study of **germ-free** (or **gnotobiotic**) animals, in which there is no colonization of the gut by microorganisms. These animals have marked reductions in the size of all peripheral lymphoid organs, low serum immunoglobulin levels and reduced immune responses of all types. The intestinal secretions of normal animals contain high levels of secretory IgA directed at commensal bacteria. In addition, normal individuals contain T cells that can recognize commensal bacteria, although as with food proteins, effector T-cell responses are not usually generated against these antigens. Commensal bacteria may induce a state of systemic immune unresponsiveness similar to the oral tolerance

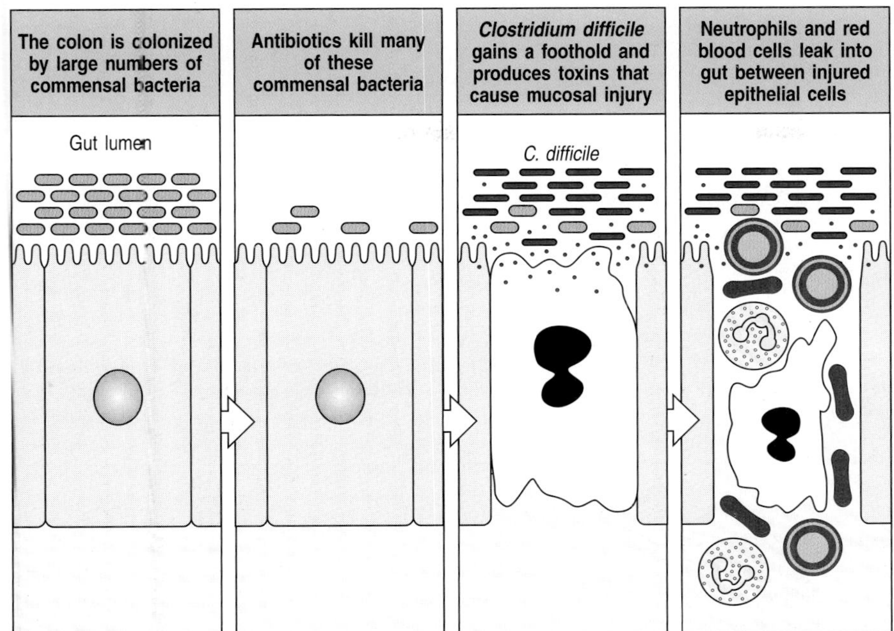

The colon is colonized by large numbers of commensal bacteria
Antibiotics kill many of these commensal bacteria
Clostridium difficile gains a foothold and produces toxins that cause mucosal injury
Neutrophils and red blood cells leak into gut between injured epithelial cells
Gut lumen
C. difficile

Fig. 11.23 Infection by *Clostridium difficile*. Treatment with antibiotics causes massive death of the commensal bacteria that normally colonize the colon. This allows pathogenic bacteria to proliferate and to occupy an ecological niche that is normally occupied by harmless commensal bacteria. *Clostridium difficile* is an example of a pathogen producing toxins that can cause severe bloody diarrhea in patients treated with antibiotics.

found with protein antigens, but this is uncertain. Unlike pathogenic bacteria, commensals do not possess the virulence factors necessary for penetrating the epithelium and cannot disseminate throughout the body. Therefore, the systemic immune system seems to remain ignorant of their presence, despite the fact that they are clearly recognizable by lymphocytes in the GALT.

This compartmentalization seems to occur because the only route of entry into the body for gut commensal bacteria is by capture by M cells in Peyer's patches, with subsequent transfer to local dendritic cells that migrate no further than to a mesenteric lymph node. Dendritic cells loaded with commensal bacteria can directly activate naive B cells to become IgA-expressing B lymphocytes, which will then redistribute to the lamina propria as IgA-secreting plasma cells. In the presence of commensal bacteria, however, there is constitutive production by gut epithelial cells and mesenchymal cells of TGF-β, thymic stromal lymphopoietin (TSLP), and prostaglandin E_2 (PGE_2), all of which tend to maintain local dendritic cells in a quiescent state with low levels of co-stimulatory molecules. When such cells present antigens to naive CD4 T cells in the mesenteric lymph node, this results in the differentiation of the naive T cells into anti-inflammatory or regulatory T cells (T_{reg}), rather than into the effector T_H1 and T_H2 cells induced by pathogen invasion (Fig. 11.24). The combined effects of the presence of commensal bacteria are therefore the production of local IgA antibodies that inhibit the adherence to and penetration of the epithelium by commensal bacteria, together with the inhibition of effector T cells that could cause inflammation. Thus, localized uptake of commensal bacteria by dendritic cells in the GALT results in responses that are anatomically compartmentalized and that avoid the activation of inflammatory effector cells.

In addition to the processes that actively regulate local immune responses to commensal bacteria in an antigen-specific manner, nonspecific factors also contribute to maintaining the local symbiotic relationship (see Fig. 11.15). The inability of commensal bacteria to penetrate the intact epithelium, together with the lack of TLRs and CD14 on the luminal surface of epithelial cells, means that these bacteria cannot induce the inflammation that loosens the epithelial barrier in the way that pathogens do.

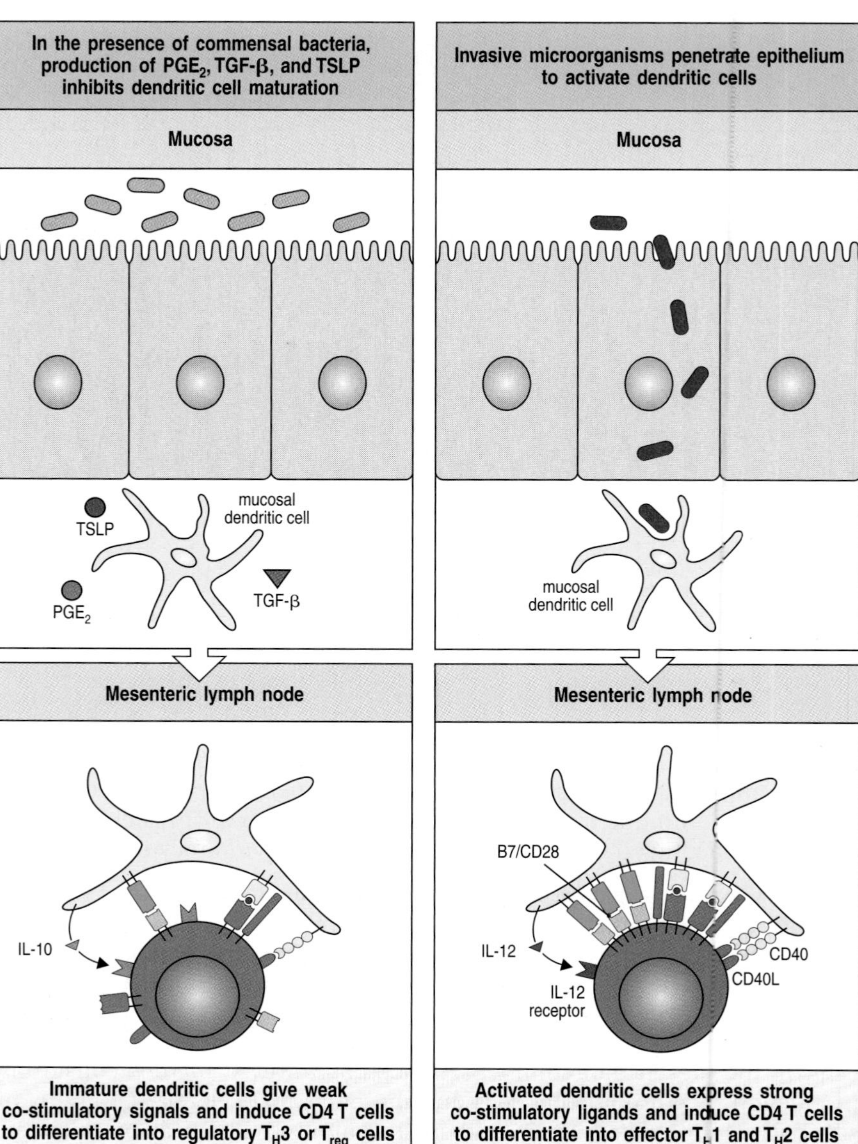

Fig. 11.24 Mucosal dendritic cells regulate the induction of tolerance and immunity in the intestine. Under normal conditions (left panels), dendritic cells are present in the mucosa underlying the epithelium and can acquire antigens from foods or commensal organisms. They take these antigens to the draining mesenteric lymph node, where they present them to naive CD4 T cells. There is, however, constitutive production by epithelial cells and mesenchymal cells of molecules such as TGF-β, thymic stromal lymphopoietin (TSLP), and prostaglandin E_2 (PGE_2) that maintain the local dendritic cells in a quiescent state with low levels of co-stimulatory molecules, so that when they present antigen to naive CD4 T cells, anti-inflammatory or regulatory T cells are generated. These recirculate back to the intestinal wall and maintain tolerance to the harmless antigens. Invasion by pathogens or a massive influx of commensal bacteria (right panels) overcomes these homeostatic mechanisms, resulting in full activation of local dendritic cells and their expression of co-stimulatory molecules and pro-inflammatory cytokines such as IL-12. Presentation of antigen to naive CD4 T cells in the mesenteric lymph node by these dendritic cells causes differentiation into effector T_H1 and T_H2 cells, leading to a full immune response.

Commensal bacteria also actively inhibit pro-inflammatory, NFκB-mediated signaling responses induced in epithelial cells by pathogenic bacteria. This inhibition may involve preventing the activation of NFκB by inhibiting the degradation of the inhibitor protein IκB (the protein that keeps NFκB in a complex in the cytoplasm), or by promoting the export of NFκB from the nucleus via the peroxisome proliferator activated receptor-γ (PPARγ) (Fig. 11.25).

If commensal bacteria do cross the epithelium in small numbers, their lack of virulence factors means that they cannot resist uptake and killing by phagocytic cells in the way that pathogens can, and they are rapidly destroyed. As a result, commensal organisms can remain associated with the mucosal surface without invading it or provoking inflammation and a consequent adaptive immune response. In parallel, the lack of tolerance to these bacteria in the systemic immune system means that it will be able to generate protective immunity to them if they do manage to enter the body through a damaged intestinal barrier.

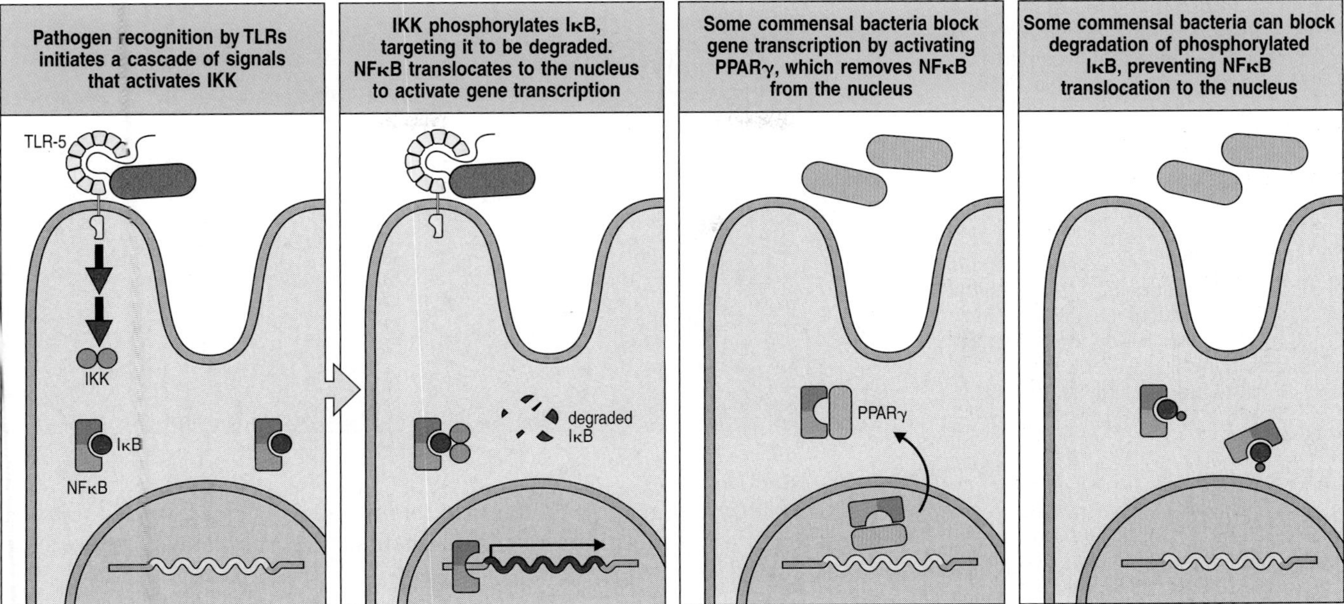

Fig. 11.25 Commensal bacteria can prevent inflammatory responses in the intestine. The pro-inflammatory transcription factor NFκB pathway is activated in epithelial cells via the ligation of TLRs by pathogens (first two panels). Commensal bacteria have been found to inhibit this pathway and thus prevent inflammation. One way is by activation of the nuclear receptor PPARγ, leading to the export of NFκB from the nucleus (third panel). Another is by blocking the degradation of the inhibitor IκB and thus retaining NFκB in the cytoplasm (fourth panel).

11-15 Full immune responses to commensal bacteria provoke intestinal disease.

It is now generally accepted that potentially aggressive T cells that respond to commensal bacteria are always present in normal animals but are usually kept in check by active regulation. If these regulatory mechanisms fail, unrestricted immune responses to commensals lead to inflammatory bowel diseases such as Crohn's disease (see Section 13-21). This is shown by animal models that have defects in immunoregulatory mechanisms involving IL-10 and TGF-β, or in which the epithelial barrier has been disrupted, allowing commensal bacteria to penetrate in large numbers. Under these conditions, systemic immune responses are generated against commensal bacterial antigens such as flagellin. Strong inflammatory T-cell responses are also generated in the mucosa, leading to severe intestinal damage. These are typically T_H1-dependent responses, which involve the production of IFN-γ and TNF-α and are driven by IL-12 or IL-23 (see Fig. 11.24, right panels). In all cases, these disorders are entirely dependent on the presence of commensal bacteria, because they can be prevented by treatment with antibiotics or in germ-free animals. It is not known whether all commensal species can provoke the inflammation, or whether only certain species do this.

Around 30% of patients with Crohn's disease carry a non-functional mutation in the *NOD2* gene, underlining the likely role of abnormal responsiveness to commensal bacteria in the disease.

11-16 Intestinal helminths provoke strong T_H2-mediated immune responses.

The intestines of virtually all animals and humans, except those living in the developed world, are colonized by large numbers of helminth parasites (Fig. 11.26). Although many of these infections may be cleared rapidly by the generation of an effective response, they are also important causes of debilitating and chronic disease in humans and animals. In these circumstances, the parasite persists for long periods apparently undisturbed by the host's attempts to expel it and causes disease by competing with the host for

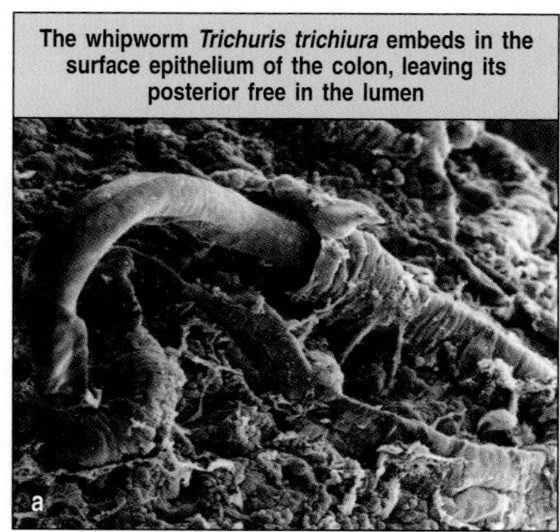

The whipworm *Trichuris trichiura* embeds in the surface epithelium of the colon, leaving its posterior free in the lumen
a

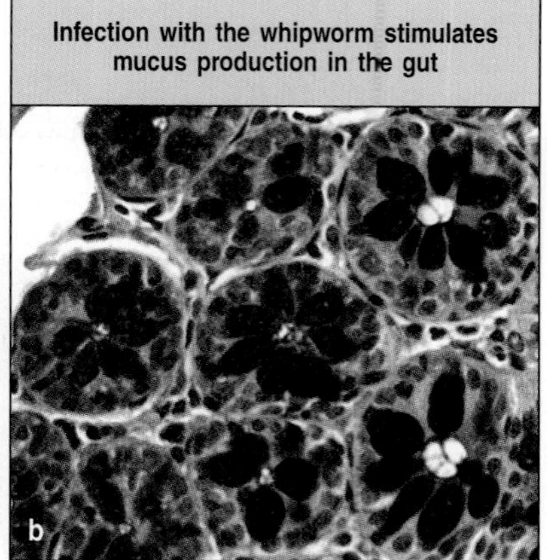

Infection with the whipworm stimulates mucus production in the gut
b

Fig. 11.26 Intestinal helminth infection. Panel a: the whipworm *Trichuris trichiura* is a helminth parasite that lives partly embedded in intestinal epithelial cells. This scanning electron micrograph of mouse colon shows the head of the parasite buried in an epithelial cell and its posterior lying free in the lumen. Panel b: a cross-section of crypts from the colon of a mouse infected with *T. trichiura* shows the markedly increased production of mucus by goblet cells in the intestinal epithelium. The mucus is seen as large droplets in vesicles inside the goblet cells and stains dark blue with periodic acid–Schiff reagent. Magnification ×400.

nutrients, or by causing local damage to epithelial cells or blood vessels. In addition, the host immune response against these parasites can produce many harmful effects.

The exact nature of the host–pathogen interaction in helminth infections depends very much on the particular type of parasite involved. Some remain within the lumen, whereas others invade and colonize epithelial cells; others invade beyond the intestine and spend part of their life cycle in other tissues such as the liver, lung, or muscle; some are found only the small intestine and others inhabit the large intestine. In virtually all cases, the dominant protective immune response is generated by CD4 T_H2 cells, whereas a T_H1 response does not clear the pathogen and tends to produce an inflammatory reaction that damages the mucosa (Fig. 11.27). A T_H2 response is polarized by worm products acting on the dendritic cells presenting worm antigens. This may drive T_H2 responses directly (by unknown mechanisms) and/or prevent IL-12 production and the generation of T_H1 cells. Although the exact role of each component of the response varies depending on the parasite, production of the cytokines IL-3, IL-4, IL-5, IL-9, and IL-13 by the T_H2 cells results in high levels of IgE antibody and recruits mast cells and eosinophils to the wall of the intestine. IL-4 and IL-13 stimulate the switching of B cells to IgE production, and IL-13 also has direct effects in enhancing the production of mucus by goblet cells, increasing contractility of the smooth muscle cells of the intestine and increasing the migration and turnover of epithelial cells. IL-5 recruits and activates eosinophils, which can have direct toxic effects on pathogens by releasing cytotoxic molecules such as major basic protein (MBP). Eosinophils bear Fc receptors for IgG and can display antibody-dependent cell-mediated cytotoxicity (ADCC) against IgG-coated parasites (see Fig. 9.33).

IL-3 and IL-9 recruit and activate a specialized population of mast cells, known as **mucosal mast cells**, which are armed by the IgE produced by the class-switched B cells (see Section 9-24). These mast cells differ from their counterparts in other tissues by having only small numbers of IgE receptors and producing very little histamine. When antigen binds to the receptor-bound IgE, the mucosal mast cells produce large amounts of other preformed inflammatory mediators, such as prostaglandins, leukotrienes, and several proteases including the mucosal mast cell protease (MMCP-1). This can remodel the intestinal mucosal tissues by digesting the basement membrane between the epithelium and the lamina propria, and may also have direct effects on parasites. Together, the mast cell-derived mediators increase vascular permeability, induce leukocyte recruitment, increase intestinal motility and stimulate the production of mucus by goblet cells, all of which helps to create a hostile microenvironment for the parasite. Mast cells also produce large quantities of TNF-α, which may assist in killing parasites and infected epithelial cells. TNF-α is, however, also an important cause of the inflammation and intestinal damage that occurs in such infections.

An additional important component of the host response to parasitic worms is an accelerated turnover of epithelial cells (see Fig. 11.27, first panel). This helps to eliminate parasites that have attached to the epithelium and reduces the surface area available for colonization. It occurs partly because epithelial cells in the crypt sense the loss of damaged cells from the surface layer and divide more rapidly in an attempt to repair the damage. Increased epithelial cell turnover is also a direct and specific effect of the IL-13 produced by T cells, NK cells, and NK T cells in the presence of infection. Although it makes life difficult for the parasite, the enhanced epithelial turnover also compromises intestinal function, because the newly produced epithelial cells are immature and are defective in absorptive and digestive activity. The host immune response in intestinal helminth infections has to tread an extremely fine line,

Fig. 11.27 Protective and pathological responses to intestinal helminths. Most intestinal helminths can induce both protective and pathological immune responses by CD4 T cells. T_H2 responses tend to create an unfriendly environment for the parasite (see the text for details), leading to expulsion and protective immunity. If the antigen-presenting cells produce IL-12 on contact with pathogen antigens, however, the CD4 T-cell response is polarized to predominantly T_H1 effector T cells, which do not clear the pathogen. The stimuli that induce IL-12 production in these circumstances are still unknown. If not balanced by a protective T_H2 response, the T_H1 response leads to persistent infection and chronic intestinal pathology. It is likely that both responses are present in most situations and that there is a continuous spectrum between them.

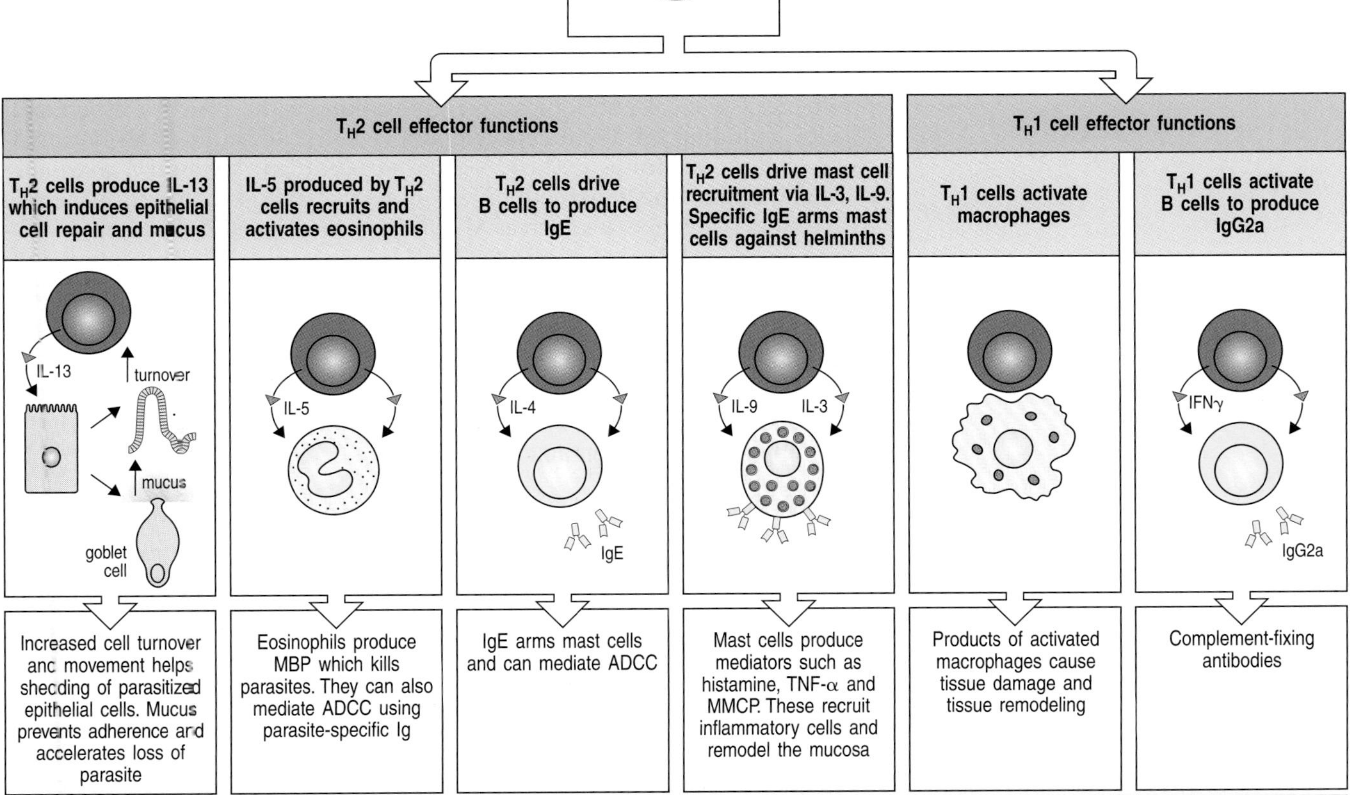

because the most efficient aspects of the protective immune response are also likely to produce deleterious effects in the local environment.

Some intestinal helminths are the ultimately adapted chronic infectious agents, having evolved sophisticated means of persisting for long periods in the host in the face of an ongoing immune response. They modulate the host immune response in several ways. These include the production of mediators that dampen the innate inflammatory response, and the expression of decoy receptors for inflammatory cytokines and chemokines. In addition, several molecules secreted by helminths modify the differentiation of T cells, often encouraging the generation of IL-10-producing regulatory T cells at the expense of effector cells. This may involve the downregulation of IL-12 production by dendritic cells via interference with TLR signaling, or stimulation of the production of inhibitory cytokines such as IL-10 and TGF-β. The overall effect of these processes is to balance the production and inflammatory potential of cytokines such as IFN-γ and TNF-α. Regulatory T cells will attempt to modulate both T_H1 and T_H2 responses, producing a state of persistent infection in the absence of severe damage to the host.

These opposing immunological processes operate simultaneously in many parasitic infections, rather as we saw in the response to commensal bacteria but to a more exaggerated extent. This can result in an intestine that appears badly inflamed but that can retain some physiological function despite harboring large numbers of live, multicellular parasites.

11-17 Other eukaryotic parasites provoke protective immunity and pathology in the gut.

The intestinal immune system has to contend with a variety of unicellular eukaryotic parasites, mainly protozoa such as *Giardia lamblia*, *Cryptosporidium parvum*, and *Toxoplasma gondii*. *Giardia lamblia* is a widespread non-invasive waterborne microorganism that is an important cause of intestinal inflammation. Protective immunity to *G. lamblia* is associated with the production of local antibodies and infiltration of the mucosa by effector T cells, including intraepithelial lymphocytes, but immunity can be inefficient, leading to chronic disease. *Cryptosporidium parvum* and *T. gondii* are normally opportunistic infections, being most commonly found in people with immune deficiencies such as AIDS. They are intracellular pathogens that require both CD4 T_H1 cells and CD8 T cells to clear them. Chronic infection is associated with marked pathology caused by overproduction of IFN-γ and TNF-α by T cells and macrophages, respectively.

11-18 Dendritic cells at mucosal surfaces favor the induction of tolerance under physiological conditions and maintain the presence of physiological inflammation.

We have seen in previous sections how the immune system in the normal intestine and other mucosal surfaces is biased to avoid making active immune responses against the majority of antigens encountered. But the antigens are still recognized, and potent protective immune responses must be, and are, generated against pathogens when required. How can these apparently opposing needs be met without compromising the health of the host? The answer seems to be in the interactions between local dendritic cells and factors in the mucosal microenvironment (see Fig. 11.24). Dendritic cells are constantly patrolling the mucosal surface, sampling antigen and carrying it to the T-cell areas of the GALT. This high throughput of dendritic cells in and out of the mucosa is constitutive, and does not depend on the presence of pathogens or other inflammatory stimuli.

Recent experiments show that dendritic cells in the Peyer's patches and lamina propria produce IL-10 rather than pro-inflammatory cytokines such as IL-12, and under normal conditions the usual outcome of presentation of antigens to T cells by these dendritic cells is the induction of tolerance and/or local IgA responses. As we discussed, this quiescent behavior of dendritic cells is not simply a default response to the lack of pro-inflammatory signals but seems to be actively maintained by factors in the local environment. These include TSLP and TGF-β released by epithelial cells, as well as mediators such as PGE_2 released by stromal cells. As a result, dendritic cells that have acquired antigen from the intestinal lumen can still migrate to the draining mesenteric lymph node but lack the co-stimulatory molecules necessary for activating naive T cells when they arrive (see Fig. 11.24). Intestinal dendritic cells of this kind may produce mediators such as IL-10 that directly favor the development of regulatory T cells. In addition, they retain the ability to induce gut-homing molecules on T cells, ensuring that any functional consequences will be restricted to the mucosa.

Fortunately for health, this predominantly inhibitory microenvironment can be changed by the presence of invasive pathogens or of adjuvants, allowing dendritic cells to be fully activated and productive immunity to be induced when required (see Fig. 11.24). The ability of mucosal dendritic cells to change their behavior rapidly and with high sensitivity probably reflects the fact that even in the absence of overt infection, both the inflammatory and regulatory components of the immune response are likely to be operating simultaneously in the mucosa. The term **physiological inflammation** is used to describe the appearance of the normal intestine, which contains large numbers of lymphocytes and other cells that are normally associated with chronic inflammation and are generally not present in other organs in the absence of disease. This 'inflammation' is driven mainly by the presence of commensal bacteria and to a lesser extent by food antigens, and it is essential for the normal function of both the intestine and the mucosal immune system. It probably also ensures that dendritic cells are always in a state of high readiness to respond appropriately to changes in their local environment.

As well as combating infection, these regulatory interactions may have had a wider influence on the evolution of the gut and the immune system, being one of the factors underlying the hygiene hypothesis (see Section 13-4). According to this idea, the human immune system has evolved in the face of continued exposure to intestinal helminths, whose immunomodulatory products have helped to condition the polarization of responses to other foreign antigens. With the increasing cleanliness of the human environment, our immune system is no longer exposed to this influence during the critical early period of life, allowing hypersensitivity reactions of all kinds to develop unchecked against autoantigens and harmless environmental materials.

Summary.

The immune system in the mucosa has to distinguish between potential pathogens and harmless antigens, generating strong effector responses to pathogens but remaining unresponsive to foods and commensals. Pathogenic microorganisms such as enteric bacteria use a number of strategies to invade, often exploiting the host's antigen-uptake and inflammatory mechanisms as well as modulating different components of the immune response. The strong immune reaction they provoke normally results in elimination of the infection. In contrast, food proteins induce an active form of immunological tolerance, which may be mediated by regulatory T cells producing IL-10 and/or TGF-β. Immune recognition of commensal bacteria is restricted entirely to the mucosal immune system, as they are presented to T cells by dendritic cells that migrate from the intestinal wall and lodge in the draining mesenteric lymph node. This ensures systemic ignorance but active mucosal tolerance and the production of local IgA antibodies that restrict colonization by the microorganisms. Because commensal bacteria have many beneficial effects for the host, these immunoregulatory processes are important in allowing the bacteria to live in peaceful coexistence with the immune system.

Another source of intestinal antigen is intestinal helminths, which frequently produce chronic infections, partly because they produce several factors that can modulate the host immune system. The dominant protective response against helminths is T_H2 mediated, with the involvement of mast cells and eosinophils and the production of TNF-α. Such a response can also damage the intestine, and the immune system maintains a balance between protective immunity and immunopathology. The absence of helminth-derived immunomodulatory factors may contribute to the increased incidence of allergic and inflammatory diseases in developed countries.

The key factor deciding between the generation of protective immunity and immune tolerance in the gut mucosa is the activation status of the local dendritic cells. The default is quiescent dendritic cells that lack the full expression of co-stimulatory molecules but can present antigen to T cells and thus polarize the T-cell response towards the differentiation of gut-seeking regulatory T cells. Nevertheless, the dendritic cells can still respond fully to invading organisms and inflammatory signals when required, allowing the priming of T cells to effector status. When the normal regulatory processes break down, inflammatory diseases can occur. As a consequence of these competing, but interacting, needs of the immune response, the intestine normally has the appearance of physiological inflammation, which helps maintain normal function of the gut and immune system.

Summary to Chapter 11.

The mucosal immune system is a large and complex apparatus that plays a crucial role in health, not just by protecting physiologically vital organs but also by helping to regulate the tone of the entire immune system and prevent disease. The peripheral lymphoid organs focused on by most immunologists may be a recent specialization of an original template that evolved in mucosal tissues. The mucosal surfaces of the body are highly vulnerable to infection and possess a complex array of innate and adaptive mechanisms of immunity. The adaptive immune system of the mucosa-associated lymphoid tissues differs from that of the rest of the peripheral lymphoid system in several respects: the immediate juxtaposition of mucosal epithelium and lymphoid tissue; diffuse lymphoid tissue as well as more organized lymphoid organs; specialized antigen uptake mechanisms; the predominance of activated/memory lymphocytes even in the absence of infection; the production of polymeric secretory IgA as the predominant antibody; and the downregulation of immune responses to innocuous antigens such as food antigens and commensal microorganisms. No systemic immune response can normally be detected to these antigens. In contrast, pathogenic microorganisms induce strong protective responses. The key factor in the decision between tolerance and the development of powerful adaptive immune responses is the context in which antigen is presented to T lymphocytes in the mucosal immune system. When there is no inflammation, presentation of antigen to T cells by antigen-presenting cells occurs in the absence of full co-stimulation, tending to induce the differentiation of regulatory T cells. By contrast, pathogenic microorganisms crossing the mucosa induce an inflammatory response in the tissues, which stimulates the maturation of antigen-presenting cells and their expression of co-stimulatory molecules, thus favoring a protective T-cell response.

Questions.

11.1. *Describe the processes that allow a specific CD4 T cell to be primed against antigen in the intestine and discuss how the resulting effector T cells can return to the intestinal surface.*

11.2. *Discuss how IgA antibodies gain access to the intestinal lumen and outline how these antibodies might contribute to defense against infection.*

11.3. *What populations of T lymphocytes are found in the intestinal mucosa and what roles do they play in host defense?*

11.4. *Compare and contrast the host response to commensal and invasive bacteria in the intestine, indicating the immunological consequences of these different effects.*

11.5. *We are exposed to foreign antigens in large quantities in the food that we eat. (a) Why do we not mount effective immune responses against these food antigens? (b) How does the immune system distinguish between food antigens and antigens that are potentially harmful?*

11.6. *Describe how different aspects of the host immune response may produce either protective immunity or tissue damage during infection by an intestinal worm.*

General references.

Brandtzaeg, P., Farstad, I.N., Johansen, F.E., Morton, H.C., Norderhaug, I.N., and Yamanaka, T.: **The B-cell system of human mucosae and exocrine glands.** *Immunol. Rev.* 1999, **171**:45–87.

MacDonald, T.T.: **The mucosal immune system.** *Parasite Immunol.* 2003, **25**:235–246.

Mowat, A.M.: **Anatomical basis of tolerance and immunity to intestinal antigens.** *Nat. Rev. Immunol.* 2003, **3**:331–341.

Section references.

11-1 The mucosal immune system protects the internal surfaces of the body.

Bienenstock, J., and McDermott, M.R.: **Bronchus- and nasal-associated lymphoid tissues.** *Immunol. Rev.* 2005, **206**:22–31.

Hooper, L.V., and Gordon, J.I.: **Commensal host-bacterial relationships in the gut.** *Science* 2001, **292**:1115–1118.

Kiyono, H., and Fukuyama, S.: **NALT-versus Peyer's-patch-mediated mucosal immunity.** *Nat. Rev. Immunol.* 2004, **4**:699–710.

The World Health Report: World Health Organization, Geneva, 2004.

Wira, C.R., Fahey, J.V., Sentman, C.L., Pioli, P.A., and Shen, L.: **Innate and adaptive immunity in female genital tract: cellular responses and interactions.** *Immunol. Rev.* 2005, **206**:306–335.

11-2 The mucosal immune system may be the original vertebrate immune system.

Cheroutre, H.: **Starting at the beginning: new perspectives on the biology of mucosal T cells.** *Annu. Rev. Immunol.* 2004, **22**:217–246.

Fagarasan, S.: **Intestinal IgA synthesis: a primitive form of adaptive immunity that regulates microbial communities in the gut.** *Curr. Top. Microbiol. Immunol.* 2006, **308**:137–153.

Matsunaga, T., and Rahman, A.: **In search of the origin of the thymus: the thymus and GALT may be evolutionarily related.** *Scand. J. Immunol.* 2001, **53**:1–6.

11-3 Mucosa-associated lymphoid tissue is located in anatomically defined compartments in the gut.

Brandtzaeg, P., and Pabst, R.: **Let's go mucosal: communication on slippery ground.** *Trends Immunol.* 2004, **25**:570–577.

Fagarasan, S., and Honjo, T.: **Regulation of IgA synthesis at mucosal surfaces.** *Curr. Opin. Immunol.* 2004, **16**:277–283.

Finke, D., and Meier, D.: **Molecular networks orchestrating GALT development.** *Curr. Top. Microbiol. Immunol.* 2006, **308**:19–57.

Kraal, G., Samsom, J.N., and Mebius, R.E.: **The importance of regional lymph nodes for mucosal tolerance.** *Immunol. Rev.* 2006, **213**:119–130.

Mowat, A.M., and Viney, J.L.: **The anatomical basis of intestinal immunity.** *Immunol. Rev.* 1997, **156**:145–166.

Newberry, R.D., and Lorenz, R.G.: **Organizing a mucosal defense.** *Immunol. Rev.* 2005, **206**:6–21.

Pabst, O., Herbrand, H., Worbs, T., Friedrichsen, M., Yan, S., Hoffmann, M.W., Korner, H., Bernhardt, G., Pabst, R., and Forster, R.: **Cryptopatches and isolated lymphoid follicles: dynamic lymphoid tissues dispensable for the generation of intraepithelial lymphocytes.** *Eur. J. Immunol.* 2005, **35**:98–107.

11-4 The intestine has distinctive routes and mechanisms of antigen uptake.

Chieppa, M., Rescigno, M., Huang, A.Y., and Germain, R.N.: **Dynamic imaging of dendritic cell extension into the small bowel lumen in response to epithelial cell TLR engagement.** *J. Exp. Med.* 2006, **203**:2841–2852.

Chirdo, F.G., Millington, O.R., Beacock-Sharp, H., and Mowat, A.M.: **Immunomodulatory dendritic cells in intestinal lamina propria.** *Eur. J. Immunol.* 2005, **35**:1831–1840.

Jang, M.H., Kweon, M.N., Iwatani, K., Yamamoto, M., Terahara, K., Sasakawa, C., Suzuki, T., Nochi, T., Yokota, Y., Rennert, P.D., *et al.*: **Intestinal villous M cells: an antigen entry site in the mucosal epithelium.** *Proc. Natl Acad. Sci. USA* 2004, **101**:6110–6115.

Jang, M.H., Sougawa, N., Tanaka, T., Hirata, T., Hiroi, T., Tohya, K., Guo, Z., Umemoto, E., Ebisuno, Y., Yang, B.G., *et al.*: **CCR7 is critically important for migration of dendritic cells in intestinal lamina propria to mesenteric lymph nodes.** *J. Immunol.* 2006, **176**:803–810.

Mach, J., Hshieh, T., Hsieh, D., Grubbs, N., and Chervonsky A.: **Development of intestinal M cells.** *Immunol. Rev.* 2005, **206**:177–189.

Neutra, M.R., Mantis, N.J., and Kraehenbuhl, J.P.: **Collaboration of epithelial cells with organized mucosal lymphoid tissues.** *Nat. Immunol.* 2001, **2**:1004–1009.

Niess, J.H., Brand, S., Gu, X., Landsman, L., Jung, S., McCormick, B.A., Vyas, J.M., Boes, M., Ploegh, H.L., Fox, J.G., *et al.*: **CX3CR1-mediated dendritic cell access to the intestinal lumen and bacterial clearance.** *Science* 2005, **307**:254–258.

Rescigno, M., Urbano, M., Valzasina, B., Francolini, M., Rotta, G., Bonasio, R., Granucci, F., Kraehenbuhl, J.P., and Ricciardi-Castagnoli, P.: **Dendritic cells express tight junction proteins and penetrate gut epithelial monolayers to sample bacteria.** *Nat. Immunol.* 2001, **2**:361–367.

Salazar-Gonzalez, R.M., Niess, J.H., Zammit, D.J., Ravindran, R., Srinivasan, A., Maxwell, J.R., Stoklasek, T., Yadav, R., Williams, I.R., Gu, X., *et al.*: **CCR6-mediated dendritic cell activation of pathogen-specific T cells in Peyer's patches.** *Immunity* 2006, **24**:623–632.

Shreedhar, V.K., Kelsall, B.L., and Neutra, M.R.: **Cholera toxin induces migration of dendritic cells from the subepithelial dome region to T- and B-cell areas of Peyer's patches.** *Infect. Immun.* 2003, **71**:504–509.

Zhao, X., Sato, A., Dela Cruz, C.S., Linehan, M., Luegering, A., Kucharzik, T., Shirakawa, A.K., Marquez, G., Farber, J.M., Williams, I., *et al.*: **CCL9 is secreted by the follicle-associated epithelium and recruits dome region Peyer's patch CD11b+ dendritic cells.** *J. Immunol.* 2003, **171**:2797–2803.

11-5 The mucosal immune system contains large numbers of effector lymphocytes even in the absence of disease.

Agace, W.W., Roberts, A.I., Wu, L., Greineder, C., Ebert, E.C., and Parker, C.M.: **Human intestinal lamina propria and intraepithelial lymphocytes express receptors specific for chemokines induced by inflammation.** *Eur. J. Immunol.* 2000, **30**:819–826.

Brandtzaeg, P., and Johansen, F.E.: **Mucosal B cells: phenotypic characteristics, transcriptional regulation, and homing properties.** *Immunol. Rev.* 2005, **206**:32–63.

11-6 The circulation of lymphocytes within the mucosal immune system is controlled by tissue-specific adhesion molecules and chemokine receptors.

Iwata, M., Hirakiyama, A., Eshima, Y., Kagechika, H., Kato, C., and Song, S.Y.: **Retinoic acid imprints gut-homing specificity on T cells.** *Immunity* 2004, **21**:527–538.

Johansen, F.E., Baekkevold, E.S., Carlsen, H.S., Farstad, I.N., Soler, D., and Brandtzaeg, P.: **Regional induction of adhesion molecules and chemokine receptors explains disparate homing of human B cells to systemic and mucosal effector sites: dispersion from tonsils.** *Blood* 2005, **106**:593–600.

Johansson-Lindbom, B., and Agace, W.W.: **Generation of gut-homing T cells and their localization to the small intestinal mucosa.** *Immunol. Rev.* 2007, **215**:226–242.

Kunkel, E.J., and Butcher, E.C.: **Plasma-cell homing.** *Nat. Rev. Immunol.* 2003, **3**:822–829.

Mora, J.R., Bono, M.R., Manjunath, N., Weninger, W., Cavanagh, L.L., Rosemblatt, M., and Von Andrian, U.H.: **Selective imprinting of gut-homing T cells by Peyer's patch dendritic cells.** *Nature* 2003, **424**:88–93.

Mora, J.R., Iwata, M., Eksteen, B., Song, S.Y., Junt, T., Senman, B., Otipoby, K.L., Yokota, A., Takeuchi, H., Ricciardi-Castagnoli, P., *et al.*: **Generation of gut-homing IgA-secreting B cells by intestinal dendritic cells.** Science 2006, **314**:1157–1160.

Salmi, M., and Jalkanen, S.: **Lymphocyte homing to the gut: attraction, adhesion, and commitment.** *Immunol. Rev.* 2005, **206**:100–113.

11-7 Priming of lymphocytes in one mucosal tissue can induce protective immunity at other mucosal surfaces.

Holmgren, J., and Czerkinsky, C.: **Mucosal immunity and vaccines.** *Nat. Med.* 2005, **11**:S45–S53.

Johansen, F.E., Baekkevold, E.S., Carlsen, H.S., Farstad, I.N., Soler, D., and Brandtzaeg, P.: **Regional induction of adhesion molecules and chemokine receptors explains disparate homing of human B cells to systemic and mucosal effector sites: dispersion from tonsils.** *Blood* 2005, **106**:593–600.

11-8 Secretory IgA is the class of antibody associated with the mucosal immune system.

Corthesy, B.: **Roundtrip ticket for secretory IgA: role in mucosal homeostasis?** *J. Immunol.* 2007, **178**: 27–32.

Fagarasan, S.: **Intestinal IgA synthesis: a primitive form of adaptive immunity that regulates microbial communities in the gut.** *Curr. Top. Microbiol. Immunol.* 2006, **308**:137–153.

Fagarasan, S., and Honjo, T.: **Regulation of IgA synthesis at mucosal surfaces.** *Curr. Opin Immunol.* 2004, **16**:277–283.

Favre, L., Spertini, F., and Corthesy, B.: **Secretory IgA possesses intrinsic modulatory properties stimulating mucosal and systemic immune responses.** *J. Immunol.* 2005, **175**:2793–2800.

Johansen, F.E., and Brandtzaeg, P.: **Transcriptional regulation of the mucosal IgA system.** *Trends Immunol.* 2004, **25**:150–157.

Macpherson, A.J., Gatto, D., Sainsbury, E., Harriman, G.R., Hengartner, H., and Zinkernagel, R.M.: **A primitive T cell-independent mechanism of intestinal mucosal IgA responses to commensal bacteria.** *Science* 2000, **288**:2222–2226.

Mora, J.R., Iwata, M., Eksteen, B., Song, S.Y., Junt, T., Senman, B., Otipoby, K.L., Yokota, A., Takeuchi, H., Ricciardi-Castagnoli, P., *et al.* **Generation of gut-homing IgA-secreting B cells by intestinal dendritic cells.** *Science* 2006, **314**:1157–1160.

11-9 IgA deficiency is common in humans but may be overcome by secretory IgM.

Cunningham-Rundles, C.: **Physiology of IgA and IgA deficiency.** *J. Clin. Immunol.* 2001, **21**:303–309.

Johansen, F.E., Pekna, M., Norderhaug, I.N., Haneberg, B., Hietala, M.A., Krajci, P., Betsholtz, C., and Brandtzaeg, P.: **Absence of epithelial immunoglobulin A transport, with increased mucosal leakiness, in polymeric immunoglobulin receptor/secretory component-deficient mice.** *J. Exp. Med.* 1999, **190**:915–922.

11-10 The mucosal immune system contains unusual T lymphocytes.

Agace, W.W., Roberts, A.I., Wu, L., Greineder, C., Ebert, E.C., and Parker, C.M.: **Human intestinal lamina propria and intraepithelial lymphocytes express receptors specific for chemokines induced by inflammation.** *Eur. J. Immunol.* 2000, **30**:819–826.

Bendelac, A., Bonneville, M., and Kearney, J.F. **Autoreactivity by design: innate B and T lymphocytes.** *Nat. Rev. Immunol.* 2001, **1**:177–186.

Cheroutre, H.: **IELs: enforcing law and order in the court of the intestinal epithelium.** *Immunol. Rev.* 2005, **206**:114–131.

Eberl, G., and Littman, D.R.: **Thymic origin of intestinal $\alpha\beta$ T cells revealed by fate mapping of RORγt$^+$ cells.** *Science* 2004, **305**:248–251.

Guy-Grand, D., Azogui, O., Celli, S., Darche, S., Nussenzweig, M.C., Kourilsky, P., and Vassalli, P.: **Extrathymic T cell lymphopoiesis: ontogeny and contribution to gut intraepithelial lymphocytes in athymic and euthymic mice.** *J. Exp. Med.* 2003, **197**:333–341.

Lefrancois, L., and Puddington, L.: **Intestinal and pulmonary mucosal T cells: local heroes fight to maintain the status quo.** *Annu. Rev. Immunol.* 2006, **24**:681–704.

Leishman, A.J., Gapin, L., Capone, M., Palmer, E., MacDonald, H.R., Kronenberg, M., and Cheroutre, H.: **Precursors of functional MHC class I- or class II-restricted CD8$\alpha\alpha^+$ T cells are positively selected in the thymus by agonist self-peptides.** *Immunity* 2002, **16**:355–364.

Makita, S., Kanai, T., Oshima, S., Uraushihara, K., Totsuka, T., Sawada, T., Nakamura, T., Koganei, K., Fukushima, T., and Watanabe, M.: **CD4$^+$CD25^{bright} T cells in human intestinal lamina propria as regulatory cells.** *J. Immunol.* 2004, **173**:3119–3130.

Pabst, O., Herbrand, H., Worbs, T., Friedrichsen, M., Yan, S., Hoffmann, M.W., Korner, H., Bernhardt, G., Pabst, R., and Forster, R.: **Cryptopatches and isolated lymphoid follicles: dynamic lymphoid tissues dispensable for the generation of intraepithelial lymphocytes.** *Eur. J. Immunol.* 2005, **35**:98–107.

Staton, T.L., Habtezion, A., Winslow, M.M., Sato, T., Love, P.E., and Butcher, E.C.: **CD8$^+$ recent thymic emigrants home to and efficiently repopulate the small intestine epithelium.** *Nat. Immunol.* 2006, **7**:482–488.

11-11 Enteric pathogens cause a local inflammatory response and the development of protective immunity.

Cario, E.: **Bacterial interactions with cells of the intestinal mucosa: Toll-like receptors and NOD2.** *Gut* 2005, **54**:1182–1193.

Fritz, J.H., Ferrero, R.L., Philpott, D.J., and Girardin, S.E.: **Nod-like proteins in immunity, inflammation and disease.** *Nat. Immunol.* 2006, **7**:1250–1257.

Gewirtz, A.T., Navas, T.A., Lyons, S., Godowski, P.J., and Madara, J.L.: **Cutting edge: bacterial flagellin activates basolaterally expressed TLR5 to induce epithelial proinflammatory gene expression.** *J. Immunol.* 2001, **167**:1882–1885.

Girardin, S.E., Travassos, L.H., Herve, M., Blanot, D., Boneca, I.G., Philpott, D.J., Sansonetti, P.J., and Mengin-Lecreulx, D.: **Peptidoglycan molecular requirements allowing detection by Nod1 and Nod2.** *J. Biol. Chem.* 2003, **278**:41702–41708.

Holmes, K.V., Tresnan, D.B., and Zelus, B.D.: **Virus–receptor interactions in the enteric tract. Virus–receptor interactions.** *Adv. Exp. Med. Biol.* 1997, **412**:125–133.

Masumoto, J., Yang, K., Varambally, S., Hasegawa, M., Tomlins, S.A., Qiu, S., Fujimoto, Y., Kawasaki, A., Foster, S.J., Horie, Y., *et al.*: **Nod1 acts as an intracellular receptor to stimulate chemokine production and neutrophil recruitment in vivo.** *J. Exp. Med.* 2006, **203**:203–213.

Mumy, K.L., and McCormick, B.A.: **Events at the host–microbial interface of the gastrointestinal tract. II. Role of the intestinal epithelium in pathogen-induced inflammation.** *Am. J. Physiol. Gastrointest. Liver Physiol.* 2005, **288**:G854–G859.

Pothoulakis, C., and LaMont, J.T.: **Microbes and microbial toxins: paradigms for microbial-mucosal interactions. II. The integrated response of the intestine to *Clostridium difficile* toxins.** *Am. J. Physiol. Gastrointest. Liver Physiol.* 2001, **280**:G178–G183.

Salazar-Gonzalez, R.M., Niess, J.H., Zammit, D.J., Ravindran, R., Srinivasan, A., Maxwell, J.R., Stoklasek, T., Yadav, R., Williams, I.R., Gu, X., *et al.*: **CCR6-mediated dendritic cell activation of pathogen-specific T cells in Peyer's patches.** *Immunity* 2006, **24**:623–632.

Sansonetti, P.J.: **War and peace at mucosal surfaces.** *Nat. Rev. Immunol.* 2004, **4**:953–964.

Selsted, M.E., and Ouellette, A.J.: **Mammalian defensins in the antimicrobial immune response.** *Nat. Immunol.* 2005, **6**:551–657.

Uematsu, S., Jang, M.H., Chevrier, N., Guo, Z., Kumagai, Y., Yamamoto, M., Kato, H., Sougawa, N., Matsui, H., Kuwata, H., *et al.*: **Detection of pathogenic intestinal bacteria by Toll-like receptor 5 on intestinal CD11c+ lamina propria cells.** *Nat. Immunol.* 2006, **7**:868–874.

11-12 The outcome of infection by intestinal pathogens is determined by a complex interplay between the microorganism and the host immune response.

Cornelis, G.R.: **The *Yersinia* Ysc-Yop 'type III' weaponry.** *Nat. Rev. Mol. Cell Biol.* 2002, **3**:742–752.

Cossart, P., and Sansonetti, P.J.: **Bacterial invasion: the paradigms of enteroinvasive pathogens.** *Science* 2004, **304**:242–248.

Owen, R.: **Uptake and transport of intestinal macromolecules and microorganisms by M cells in Peyer's patches—a personal and historical perspective.** *Semin. Immunol.* 1999, **11**:1–7.

Sansonetti, P.J.: **War and peace at mucosal surfaces.** *Nat. Rev. Immunol.* 2004, **4**:953–964.

Sansonetti, P.J., and Di Santo, J.P.: **Debugging how bacteria manipulate the immune response.** *Immunity* 2007, **26**:149–161.

11-13 The mucosal immune system must maintain a balance between protective immunity and homeostasis to a large number of different foreign antigens.

Iweala, O.I., and Nagler, C.R.: **Immune privilege in the gut: the establishment

and maintenance of non-responsiveness to dietary antigens and commensal flora. *Immunol. Rev.* 2006, **213**:82–100.

Kraal, G., Samsom, J.N., and Mebius, R.E.: **The importance of regional lymph nodes for mucosal tolerance.** *Immunol. Rev.* 2006, **213**:119–130.

Macdonald, T.T., and Monteleone, G.: **Immunity, inflammation, and allergy in the gut.** *Science* 2005, **307**:1920–1925.

Strobel, S., and Mowat, A.M.: **Oral tolerance and allergic responses to food proteins.** *Curr. Opin. Allergy Clin. Immunol.* 2006, **6**:207–213.

Sun, J.B., Raghavan, S., Sjoling, A., Lundin, S., and Holmgren, J.: **Oral tolerance induction with antigen conjugated to cholera toxin B subunit generates both Foxp3+CD25+ and Foxp3–CD25– CD4+ regulatory T cells.** *J. Immunol.* 2006, **177**:7634–7644.

Worbs, T., Bode, U., Yan, S., Hoffmann, M.W., Hintzen, G., Bernhardt, G., Forster, R., and Pabst, O.: **Oral tolerance originates in the intestinal immune system and relies on antigen carriage by dendritic cells.** *J. Exp. Med.* 2006, **203**:519–527.

11-14 The healthy intestine contains large quantities of bacteria but does not generate productive immunity against them.

Araki, A., Kanai, T., Ishikura, T., Makita, S., Uraushihara, K., Iiyama, R., Totsuka, T., Takeda, K., Akira, S., and Watanabe, M.: **MyD88-deficient mice develop severe intestinal inflammation in dextran sodium sulfate colitis.** *J. Gastroenterol.* 2005, **40**:16–23.

Backhed, F., Ley, R.E., Sonnenburg, J.L., Peterson, D.A., and Gordon, J.I.: **Host–bacterial mutualism in the human intestine.** *Science* 2005, **307**:1915–1920.

Gad, M., Pedersen, A.E., Kristensen, N.N., and Claesson, M.H.: **Demonstration of strong enterobacterial reactivity of CD4+CD25–T cells from conventional and germ-free mice which is counter-regulated by CD4+CD25+ T cells.** *Eur. J. Immunol.* 2004, **34**:695–704.

Hooper, L.V.: **Bacterial contributions to mammalian gut development.** *Trends Microbiol.* 2004, **12**:129–134.

Kelly, D., Campbell, J.I., King, T.P., Grant, G., Jansson, E.A., Coutts, A.G., Pettersson, S., and Conway, S.: **Commensal anaerobic gut bacteria attenuate inflammation by regulating nuclear-cytoplasmic shuttling of PPAR-gamma and RelA.** *Nat. Immunol.* 2004, **5**:104–112.

Lee, J., Mo, J.H., Katakura, K., Alkalay, I., Rucker, A.N., Liu, Y.T., Lee, H.K., Shen, C., Cojocaru, G., Shenouda, S., *et al.*: **Maintenance of colonic homeostasis by distinctive apical TLR9 signalling in intestinal epithelial cells.** *Nat. Cell Biol.* 2006, **8**:1327–1336.

Lotz, M., Gutle, D., Walther, S., Menard, S., Bogdan, C., and Hornef, M.W.: **Postnatal acquisition of endotoxin tolerance in intestinal epithelial cells.** *J. Exp. Med.* 2006, **203**:973–984.

Macpherson, A.J., and Uhr, T.: **Induction of protective IgA by intestinal dendritic cells carrying commensal bacteria.** *Science* 2004, **303**:1662–1665.

Mueller, C., and Macpherson, A.J.: **Layers of mutualism with commensal bacteria protect us from intestinal inflammation.** *Gut* 2006, **55**:276–284.

Neish, A.S., Gewirtz, A.T., Zeng, H., Young, A.N., Hobert, M.E., Karmali, V., Rao, A.S., and Madara, J.L.: **Prokaryotic regulation of epithelial responses by inhibition of IκB-α ubiquitination.** *Science* 2000, **289**:1560–1563.

Rakoff-Nahoum, S., Hao, L., and Medzhitov, R.: **Role of toll-like receptors in spontaneous commensal-dependent colitis.** *Immunity* 2006, **25**:319–329.

Sansonetti, P.J.: **War and peace at mucosal surfaces.** *Nat. Rev. Immunol.* 2004, **4**:953–964.

Tien, M.T., Girardin, S.E., Regnault, B., Le Bourhis, L., Dillies, M.A., Coppee, J.Y., Bourdet-Sicard, R., Sansonetti, P.J., and Pedron, T.: **Anti-inflammatory effect of *Lactobacillus casei* on *Shigella*-infected human intestinal epithelial cells.** *J. Immunol.* 2006, **176**:1228–1237.

Wang, Q., McLoughlin, R.M., Cobb, B.A., Charrel-Dennis, M., Zaleski, K.J., Golenbock, D., Tzianabos, A.O., and Kasper, D.L.: **A bacterial carbohydrate links innate and adaptive responses through Toll-like receptor 2.** *J. Exp. Med.* 2006, **203**:2853–2863.

11-15 Full immune responses to commensal bacteria provoke intestinal disease.

Elson, C.O., Cong, Y., McCracken, V.J., Dimmitt, R.A., Lorenz, R.G., and Weaver, C.T.: **Experimental models of inflammatory bowel disease reveal innate, adaptive, and regulatory mechanisms of host dialogue with the microbiota.** *Immunol. Rev.* 2005, **206**:260–276.

Kullberg, M.C., Jankovic, D., Feng, C.G., Hue, S., Gorelick, P.L., McKenzie, B.S., Cua, D.J., Powrie, F., Cheever, A.W., Maloy, K.J., *et al.*: **IL-23 plays a key role in *Helicobacter hepaticus*-induced T cell-dependent colitis.** *J. Exp. Med.* 2006, **203**:2485–2494.

Lodes, M.J., Cong, Y., Elson, C.O., Mohamath, R., Landers, C.J., Targan, S.R., Fort, M., and Hershberg, R.M.: **Bacterial flagellin is a dominant antigen in Crohn's disease.** *J. Clin. Invest.* 2004, **113**:1296–1306.

Macdonald, T.T., and Monteleone, G.: **Immunity, inflammation, and allergy in the gut.** *Science* 2005, **307**:1920–1925.

Rescigno, M., and Nieuwenhuis, E.E.: **The role of altered microbial signaling via mutant NODs in intestinal inflammation.** *Curr. Opin. Gastroenterol.* 2007, **23**:21–26.

11-16 Intestinal helminths provoke strong T_H2-mediated immune responses.

Cliffe, L.J., and Grencis, R.K.: **The *Trichuris muris* system: a paradigm of resistance and susceptibility to intestinal nematode infection.** *Adv. Parasitol.* 2004, **57**:255–307.

Cliffe, L.J., Humphreys, N.E., Lane, T.E., Potten, C.S., Booth, C., and Grencis, R.K.: **Accelerated intestinal epithelial cell turnover: a new mechanism of parasite expulsion.** *Science* 2005, **308**:1463–1465.

Dixon, H., Blanchard, C., Deschoolmeester, M.L., Yuill, N.C., Christie, J.W., Rothenberg, M.E., and Else, K.J.: **The role of Th2 cytokines, chemokines and parasite products in eosinophil recruitment to the gastrointestinal mucosa during helminth infection.** *Eur. J. Immunol.* 2006, **36**:1753–1763.

Lawrence, C.E., Paterson, Y.Y., Wright, S.H., Knight, P.A., and Miller, H.R.: **Mouse mast cell protease-1 is required for the enteropathy induced by gastrointestinal helminth infection in the mouse.** *Gastroenterology* 2004, **127**:155–165.

Maizels, R.M., and Yazdanbakhsh, M.: **Immune regulation by helminth parasites: cellular and molecular mechanisms.** *Nat. Rev. Immunol.* 2003, **3**:733–744.

Specht, S., Saeftel, M., Arndt, M., Endl, E., Dubben, B., Lee, N.A., Lee, J.J., and Hoerauf, A.: **Lack of eosinophil peroxidase or major basic protein impairs defense against murine filarial infection.** *Infect. Immun.* 2006, **74**:5236–5243.

Vliagoftis, H., and Befus, A.D.: **Rapidly changing perspectives about mast cells at mucosal surfaces.** *Immunol. Rev.* 2005, **206**:190–203.

Voehringer, D., Shinkai, K., and Locksley, R.M.: **Type 2 immunity reflects orchestrated recruitment of cells committed to IL-4 production.** *Immunity* 2004, **20**:267–277.

Zaiss, D.M., Yang, L., Shah, P.R., Kobie, J.J., Urban, J.F., and Mosmann, T.R.: **Amphiregulin, a T_H2 cytokine enhancing resistance to nematodes.** *Science* 2006, **314**:1746.

11-17 Other eukaryotic parasites provoke protective immunity and pathology in the gut.

Buzoni-Gatel, D., Schulthess, J., Menard, L.C., and Kasper, L.H.: **Mucosal defences against orally acquired protozoan parasites, emphasis on *Toxoplasma gondii* infections.** *Cell. Microbiol.* 2006, **8**:535–544.

Dalton, J.E., Cruickshank, S.M., Egan, C.E., Mears, R., Newton, D.J., Andrew, E.M., Lawrence, B., Howell, G., Else, K.J., Gubbels, M.J., *et al.*: **Intraepithelial $\gamma\delta^+$ lymphocytes maintain the integrity of intestinal epithelial tight junctions in response to infection.** *Gastroenterology* 2006, **131**:818–829.

Eckmann, L.: **Mucosal defences against *Giardia*.** *Parasite Immunol.* 2003, **25**:259–270.

11-18 Dendritic cells at mucosal surfaces favor the induction of tolerance under physiological conditions and maintain the presence of physiological inflammation.

Annacker, O., Coombes, J.L., Malmstrom, V., Uhlig, H.H., Bourne, T., Johansson-Lindbom, B., Agace W.W., Parker, C.M., and Powrie, F.: **Essential role for CD103 in the T cell-mediated regulation of experimental colitis.** *J. Exp. Med.* 2005, **202**:1051–1061.

Chirdo, F.G., Millington, O.R., Beacock-Sharp, H., and Mowat, A.M.: **Immunomodulatory dendritic cells in intestinal lamina propria.** *Eur. J. Immunol.* 2005, **35**:1831–1840.

Dunne, D.W., and Cooke, A.: **A worm's eye view of the immune system: consequences for evolution of human autoimmune disease.** *Nat. Rev. Immunol.* 2005, **5**:420–426.

Kelsall, B.L., and Leon, F.: **Involvement of intestinal dendritic cells in oral tolerance, immunity to pathogens, and inflammatory bowel disease.** *Immunol. Rev.* 2005, **206**:132–148.

Maizels, R.M., and Yazdanbakhsh, M.: **Immune regulation by helminth parasites: cellular and molecular mechanisms.** *Nat. Rev. Immunol.* 2003, **3**:733–744.

Milling, S.W., Yrlid, U., Jenkins, C., Richards, C.M., Williams, N.A., and MacPherson, G.: **Regulation of intestinal immunity: effects of the oral adjuvant *Escherichia coli* heat-labile enterotoxin on migrating dendritic cells.** *Eur. J. Immunol.* 2007, **37**:87–99.

Rescigno, M.: **CCR6$^+$ dendritic cells: the gut tactical-response unit.** *Immunity* 2006, **24**:508–510.

Rimoldi, M., Chieppa, M., Salucci, V., Avogadri, F., Sonzogni, A., Sampietro, G.M., Nespoli, A., Viale, G., Allavena, P., and Rescigno, M.: **Intestinal immune homeostasis is regulated by the crosstalk between epithelial cells and dendritic cells.** *Nat. Immunol.* 2005, **6**:507–514.

Sato, A., Hashiguchi, M., Toda, E., Iwasaki, A., Hachimura, S., and Kaminogawa, S.: **CD11b+ Peyer's patch dendritic cells secrete IL-6 and induce IgA secretion from naive B cells.** *J. Immunol.* 2003, **171**:3684–3690.

Zaph, C., Troy, A.E., Taylor, B.C., Berman-Booty, L.D., Guild, K.J., Du, Y., Yost, E.A., Gruber, A.D., May, M.J., Greten, F.R., *et al.*: **Epithelial-cell-intrinsic IKK-β expression regulates intestinal immune homeostasis.** *Nature* 2007, **446**, 552–556.

PART V

THE IMMUNE SYSTEM IN HEALTH AND DISEASE

Failures of Host Defense Mechanisms

12

In the normal course of an infection, the infectious agent first triggers an innate immune response that causes symptoms. The foreign antigens of the infectious agent, enhanced by signals from the innate immune response, then induce an adaptive immune response that clears the infection and establishes a state of protective immunity. This does not always happen, however, and in this chapter we examine three circumstances in which there are failures of host defense against pathogens: the avoidance or subversion of a normal immune response by the pathogen; inherited failures of immune defenses because of gene defects; and the acquired immune deficiency syndrome (AIDS), a generalized susceptibility to infection that is itself due to the failure of the host to control and eliminate the human immunodeficiency virus (HIV).

To propagate itself, a pathogen must replicate in the infected host and spread to new hosts. Common pathogens must therefore grow without activating too vigorous an immune response, but they must not kill the host too quickly. The most successful pathogens persist either because they do not elicit an immune response or because they evade the response once it has occurred. Over millions of years of coevolution with their hosts, pathogens have developed various strategies for avoiding destruction by the immune system, and these are examined in the first part of this chapter.

In the second part of the chapter we turn to the **immunodeficiency diseases**, in which host defense fails. In most of these diseases, a defective gene results in the elimination of one or more components of the immune system, leading to heightened susceptibility to infection with particular classes of pathogens. Immunodeficiency diseases caused by defects in T- or B-lymphocyte development, phagocyte function, and complement components have all been discovered. In the last part of the chapter, we consider how the persistent infection of immune system cells by HIV leads to AIDS, example of an acquired deficiency. The study of all these immunodeficiencies has already contributed greatly to our understanding of host defense mechanisms and, in the longer term, might help to provide new methods of controlling or preventing infectious diseases, including AIDS.

Evasion and subversion of immune defenses.

Just as vertebrates have developed many different defenses against pathogens, so pathogens have evolved numerous ways of overcoming them. These range from resisting phagocytosis to avoiding recognition by the adaptive immune system and even actively suppressing immune responses. We start by looking at how some pathogens keep one step ahead of the adaptive immune response.

12-1 Antigenic variation allows pathogens to escape from immunity.

One way in which an infectious agent can evade surveillance by the immune system is by altering its antigens; this is known as **antigenic variation** and is particularly important for extracellular pathogens, which are generally eliminated by antibodies against their surface structures (see Chapter 9). There are three main forms of antigenic variation. First, many infectious agents exist in a wide variety of antigenic types. There are, for example, 84 known types of *Streptococcus pneumoniae*, an important cause of bacterial pneumonia, which differ in the structure of their polysaccharide capsules. The different types are distinguished by using specific antibodies as reagents in serological tests and so are often known as **serotypes**. Infection with one serotype can lead to type-specific immunity, which protects against reinfection with that type but not with a different serotype. Thus, from the point of view of the adaptive immune system, each serotype of *S. pneumoniae* represents a distinct organism, with the result that essentially the same pathogen can cause disease many times in the same individual (Fig. 12.1).

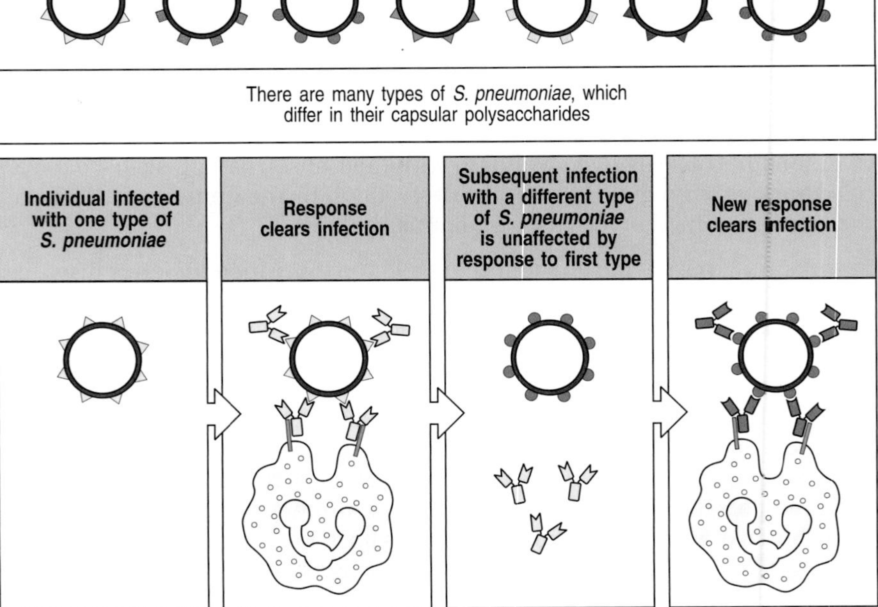
Streptococcus pneumoniae
There are many types of S. pneumoniae, which differ in their capsular polysaccharides
Individual infected with one type of S. pneumoniae
Response clears infection
Subsequent infection with a different type of S. pneumoniae is unaffected by response to first type
New response clears infection

Fig. 12.1 Host defense against *Streptococcus pneumoniae* is type specific. The different strains of *S. pneumoniae* have antigenically distinct capsular polysaccharides. The capsule prevents effective phagocytosis until the bacterium is opsonized by specific antibody and complement, allowing phagocytes to destroy it. Antibody against one type of *S. pneumoniae* does not cross-react with the other types, so an individual immune to one type has no protective immunity to a subsequent infection with a different type. An individual must generate a new adaptive immune response each time he or she is infected with a different type of *S. pneumoniae*.

A second, more dynamic, mechanism of antigenic variation is an important feature of the influenza virus. At any one time, a single virus type is responsible for most cases of influenza throughout the world. The human population gradually develops protective immunity to this type, chiefly through neutralizing antibody directed against the viral hemagglutinin, the main surface protein of the influenza virus. Because the virus is rapidly cleared from immune individuals, it might be in danger of running out of potential hosts if it had not evolved two distinct ways of changing its antigenic type (Fig. 12.2).

The first of these is called **antigenic drift** and is caused by point mutations in the genes encoding hemagglutinin and a second surface protein, neuraminidase. Every 2–3 years a variant flu virus arises with mutations that allow it to evade neutralization by the antibodies present in the population. Other mutations affect epitopes in the proteins that are recognized by T cells, in particular CD8 cytotoxic T cells, and so cells infected with the mutant virus also escape destruction. People immune to the old flu virus are thus susceptible to the new variant, but because the changes in the viral proteins are relatively minor, there is still some cross-reaction with antibodies and memory T cells produced against the previous variant, and most of the population still has some level of immunity (see Fig. 10.27). An epidemic resulting from antigenic drift is usually relatively mild.

The other type of antigenic change in influenza virus is known as **antigenic shift** and is due to major changes in the hemagglutinin of the virus. Antigenic shifts cause global pandemics of severe disease, often with substantial mortality, as the new hemagglutinin is recognized poorly, if at all, by antibodies

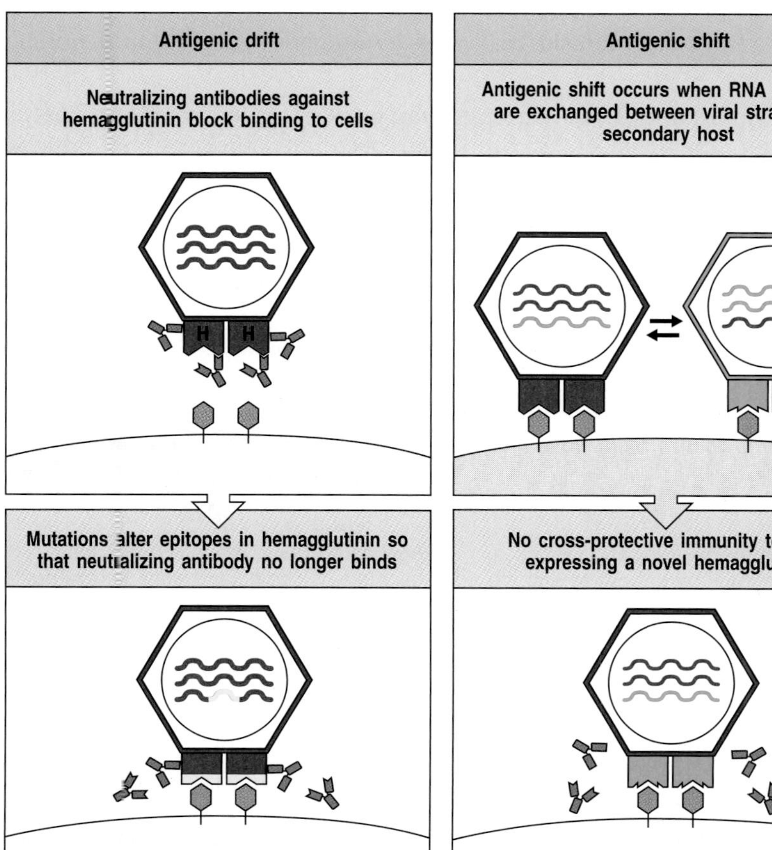
Antigenic drift
Neutralizing antibodies against hemagglutinin block binding to cells
H
H
Mutations alter epitopes in hemagglutinin so that neutralizing antibody no longer binds
Antigenic shift
Antigenic shift occurs when RNA segments are exchanged between viral strains in a secondary host
No cross-protective immunity to virus expressing a novel hemagglutinin

Fig. 12.2 Two types of variation allow repeated infection with type A influenza virus. Neutralizing antibody that mediates protective immunity is directed at the viral surface protein hemagglutinin (H), which is responsible for viral binding to and entry into cells. Antigenic drift (left panels) involves the emergence of point mutants with altered binding sites for protective antibodies on the hemagglutinin. The new virus can grow in a host that is immune to the previous strain of virus, but as T cells and some antibodies can still recognize epitopes that have not been altered, the new variants cause only mild disease in previously infected individuals. Antigenic shift (right panels) is a rare event involving the reassortment of the segmented RNA viral genomes of two different influenza viruses, probably in a bird or a pig. These antigen-shifted viruses have large changes in their hemagglutinin, and therefore T cells and antibodies produced in earlier infections are not protective. These shifted strains cause severe infection that spreads widely, causing the influenza pandemics that occur every 10–50 years. There are eight RNA molecules in each viral genome, but for simplicity only three are shown.

and T cells directed against the previous variant. Antigenic shift is due to reassortment of the segmented RNA genome of the human influenza virus and animal influenza viruses in an animal host, in which the hemagglutinin gene from the animal virus replaces the hemagglutinin gene in the human virus.

The third mechanism of antigenic variation in pathogens involves programmed gene rearrangements. The most striking example occurs in African trypanosomes, where changes in the major surface antigen occur repeatedly within the same infected host. African trypanosomes are insect-borne protozoan parasites that replicate in the extracellular spaces of tissues and cause the disease known as trypanosomiasis or sleeping sickness. The trypanosome is coated with a single type of glycoprotein, the variant-specific glycoprotein (VSG), which elicits a potent protective antibody response that rapidly clears most of the parasites. The trypanosome genome, however, contains about 1000 VSG genes, each encoding a protein with distinct antigenic properties. A VSG gene is expressed by being placed into the active expression site in the parasite genome. Only one VSG is expressed at a time, and it can be changed by a gene rearrangement that places a new VSG gene into the expression site (Fig. 12.3). So, by using gene rearrangement to change the VSG protein produced, trypanosomes keep one step ahead of an immune system capable of generating many distinct antibodies by gene rearrangement. The few trypanosomes with changed surface glycoproteins are not affected by the antibodies previously made by the host, and these variants multiply and cause a recurrence of disease (see Fig. 12.3, bottom panel). Antibodies are then made against the new VSG, and the whole cycle is repeated. The chronic cycles of antigen clearance lead to immune-complex damage and inflammation, and eventually to neurological damage, resulting finally in the coma that gives sleeping sickness its name. The cycles of evasive action make trypanosome infections very difficult for the immune system to defeat, and they are a major health problem in Africa. Malaria is another serious and widespread disease caused by a protozoan parasite that varies its antigens to avoid elimination by the immune system.

Antigenic variation by DNA rearrangement also occurs in bacteria and helps to account for the success of two important bacterial pathogens—*Salmonella typhimurium*, a common cause of salmonella food poisoning, and *Neisseria gonorrhoeae*, which causes gonorrhea, a major sexually transmitted disease and an increasing public health problem in the United States. *S. typhimurium* regularly alternates two versions of its surface flagellin protein. Inversion of a segment of DNA containing the promoter for one flagellin gene turns off expression of the gene and allows the expression of a second flagellin gene that encodes an antigenically distinct protein. *N. gonorrhoeae* has several variable antigens, the most important of which is the pilin protein, which is responsible for adherence of the bacterium to a mucosal surface. Like the VSGs of the African trypanosome, there is more than one pilin gene variant,

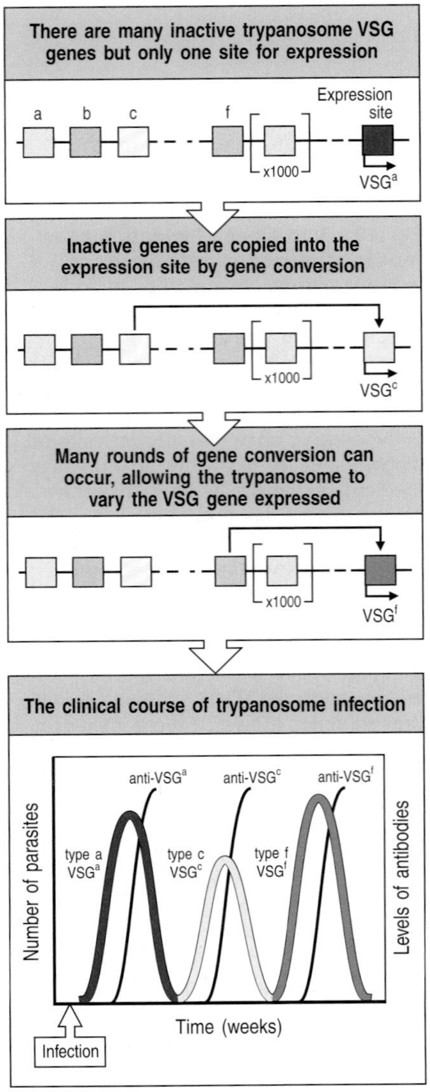

Fig. 12.3 Antigenic variation in trypanosomes allows them to escape immune surveillance. The surface of a trypanosome is covered with a variant-specific glycoprotein (VSG). Each trypanosome has about 1000 genes encoding different VSGs, but only the gene in a specific expression site within the telomere at one end of the chromosome is active. Although several genetic mechanisms have been observed for changing the VSG gene expressed, the usual mechanism is gene conversion. An inactive gene, which is not at the telomere, is copied and transposed into the telomeric expression site, where it becomes active. When an individual is first infected, antibodies are raised against the VSG initially expressed by the trypanosome population. A small number of trypanosomes spontaneously switch their VSG gene to a new type, and although the host antibody eliminates the initial variant, the new variant is unaffected. As the new variant grows, the whole sequence of events is repeated.

only one of which is active at any given time. From time to time, a different pilin gene replaces the active gene under the control of the pilin promoter. All these mechanisms of antigenic variation help the pathogen to evade an otherwise specific and effective immune response.

12-2 Some viruses persist *in vivo* by ceasing to replicate until immunity wanes.

Once they have entered cells, viruses usually betray their presence to the immune system by directing the synthesis of viral proteins, fragments of which are displayed on the MHC molecules of the infected cell where they can be detected by T lymphocytes. To replicate, a virus must make viral proteins, and rapidly replicating viruses that produce acute illnesses are readily detected by T cells, which normally control them. Some viruses, however, can enter a state known as **latency**, in which the virus is not being replicated. In the latent state, the virus does not cause disease; however, because there are no viral peptides to signal its presence it cannot be eliminated. Latent infections can be reactivated, and this results in recurrent illness.

An example is herpes simplex virus (HSV), the cause of cold sores, which infects epithelial cells and spreads to sensory neurons serving the infected area. An effective immune response controls the epithelial infection, but the virus persists in a latent state in the sensory neurons. Factors such as sunlight, bacterial infection, or hormonal changes reactivate the virus, which then travels down the axons of the sensory neuron and reinfects the epithelial tissues (Fig. 12.4). At this point, the immune response again becomes active and controls the local infection by killing the epithelial cells, producing a new sore. This cycle can be repeated many times.

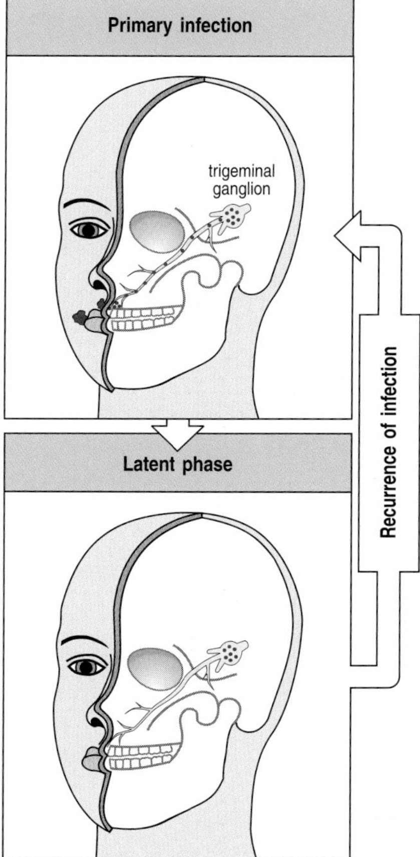

Fig. 12.4 Persistence and reactivation of herpes simplex virus infection. The initial infection in the skin is cleared by an effective immune response, but residual infection persists in sensory neurons such as those of the trigeminal ganglion, whose axons innervate the lips. When the virus is reactivated, usually by some environmental stress and/or alteration in immune status, the skin in the area served by the nerve is reinfected from virus in the ganglion and a new cold sore results. This process can be repeated many times.

There are two reasons why the sensory neuron remains infected: first, the virus is quiescent and generates few virus-derived peptides to present on MHC class I molecules; second, neurons carry very low levels of MHC class I molecules, which makes it harder for CD8 cytotoxic T cells to recognize infected neurons and attack them. The low level of MHC class I expression might be beneficial, because it reduces the risk that neurons, which do not regenerate or do so only very slowly, will be attacked inappropriately by cytotoxic T cells. It does, however, make neurons unusually vulnerable to persistent infections. Herpes viruses often enter latency: herpes zoster (varicella zoster), the virus that causes chickenpox, remains latent in one or a few dorsal root ganglia after the acute illness is over, and can be reactivated by stress or immunosuppression. It then spreads down the nerve and reinfects the skin to cause the disease **shingles**, which is marked by the reappearance of the classic varicella rash in the area of skin served by the infected dorsal root. Unlike herpes simplex, in which reactivation occurs frequently, herpes zoster usually reactivates only once in a lifetime in an immunocompetent host.

Yet another herpes virus, the Epstein–Barr virus (EBV), establishes a persistent infection in most individuals. EBV enters latency in B cells after a primary infection that often passes without being diagnosed. In a minority of infected individuals, usually those who contract the virus in adulthood, the initial acute infection of B cells is more severe, causing the disease known as **infectious mononucleosis** or glandular fever. EBV infects B cells by binding to CR2 (CD21), a component of the B-cell co-receptor complex, and to MHC class II molecules. In the primary infection most of the infected cells proliferate and produce virus, leading in turn to the proliferation of antigen-specific T cells and the excess of mononuclear white cells in the blood that gives the disease its name. Virus is released from the B cells, destroying them in the process, and virus can be recovered from saliva. The infection is eventually controlled by virus-specific CD8 cytotoxic T cells, which kill the infected proliferating

B cells. A fraction of memory B lymphocytes become latently infected, however, and EBV remains quiescent in these cells.

These two forms of infection are accompanied by quite different patterns of expression of viral genes. EBV has a large DNA genome encoding more than 70 proteins. Many of these are required for viral replication and are expressed by the replicating virus, providing a source of viral peptides by which infected cells can be recognized. In a latent infection, in contrast, the virus survives within the host B cells without replicating, and a very limited set of viral proteins is expressed. One of these is the Epstein–Barr nuclear antigen 1 (EBNA-1), which is needed to maintain the viral genome. EBNA-1 interacts with the proteasome (see Section 5-3) to prevent its own degradation into peptides that would otherwise elicit a T-cell response.

Latently infected B cells can be isolated by culturing B cells from individuals who have apparently cleared their EBV infection: in the absence of T cells, latently infected cells retaining the EBV genome become transformed into so-called immortal cell lines, the equivalent of tumorigenesis *in vitro*. Infected B cells occasionally undergo malignant transformation *in vivo*, giving rise to a B-cell lymphoma called Burkitt's lymphoma (see Section 7-30). In this lymphoma, expression of the peptide transporters TAP-1 and TAP-2 is downregulated (see Section 5-6), and so cells are unable to process endogenous antigens for presentation on HLA class I molecules (the human MHC class I). This deficiency provides one explanation for how these tumors escape attack by CD8 cytotoxic T cells. Patients with acquired and inherited immunodeficiencies of T-cell function have an increased risk of developing EBV-associated lymphomas, presumably as a result of a failure of immune surveillance.

12-3 Some pathogens resist destruction by host defense mechanisms or exploit them for their own purposes.

Some pathogens induce a normal immune response but have evolved specialized mechanisms for resisting its effects. For instance, some bacteria that are engulfed by macrophages have evolved means of avoiding destruction by these phagocytes and instead use macrophages as their primary host cell. *Mycobacterium tuberculosis*, for example, is taken up by macrophages but prevents the fusion of the phagosome with the lysosome, protecting itself from the bactericidal actions of the lysosomal contents.

Other microorganisms, such as *Listeria monocytogenes*, escape from the phagosome into the cytoplasm of the macrophage, where they multiply. They then spread to adjacent cells in the tissue without emerging into the extracellular environment. They do this by hijacking the cytoskeletal protein actin, which assembles into filaments at the rear of the bacterium. The actin filaments drive the bacteria forward into vacuolar projections to adjacent cells; the vacuoles are then lysed by the *Listeria*, releasing the bacteria into the cytoplasm of the adjacent cell. In this way *Listeria* avoids attack by antibodies, but the infected cells are still susceptible to killing by cytotoxic T cells. The protozoan parasite *Toxoplasma gondii* generates its own vesicle, which does not fuse with any cellular vesicle and thus isolates the parasite from the rest of the cell. This might render peptides derived from *T. gondii* less available for loading onto MHC molecules.

The spirochete bacterium *Treponema pallidum*, the cause of syphilis, can avoid elimination by antibodies and establish a persistent and extremely damaging infection in tissues. *T. pallidum* is believed to avoid recognition by antibodies by coating its surface with host proteins until it has invaded tissues such as the central nervous system, where it is less accessible to antibodies. Another spirochete, the tick-borne *Borrelia burgdorferi*, is the cause

Viral strategy	Specific mechanism	Result	Virus examples
Inhibition of humoral immunity	Virally encoded Fc receptor	Blocks effector functions of antibodies bound to infected cells	Herpes simplex Cytomegalovirus
	Virally encoded complement receptor	Blocks complement-mediated effector pathways	Herpes simplex
	Virally encoded complement control protein	Inhibits complement activation by infected cell	Vaccinia
Inhibition of inflammatory response	Virally encoded chemokine receptor homolog, e.g., β-chemokine receptor	Sensitizes infected cells to effects of β-chemokine; advantage to virus unknown	Cytomegalovirus
	Virally encoded soluble cytokine receptor, e.g., IL-1 receptor homolog, TNF receptor homolog, interferon-γ receptor homolog	Blocks effects of cytokines by inhibiting their interaction with host receptors	Vaccinia Rabbit myxoma virus
	Viral inhibition of adhesion molecule expression, e.g., LFA-3 ICAM-1	Blocks adhesion of lymphocytes to infected cells	Epstein–Barr virus
	Protection from NFκB activation by short sequences that mimic TLRs	Blocks inflammatory responses elicited by IL-1 or bacterial pathogens	Vaccinia
Blocking of antigen processing and presentation	Inhibition of MHC class I expression	Impairs recognition of infected cells by cytotoxic T cells	Herpes simplex Cytomegalovirus
	Inhibition of peptide transport by TAP	Blocks peptide association with MHC class I	Herpes simplex
Immunosuppression of host	Virally encoded cytokine homolog of IL-10	Inhibits T_H1 lymphocytes Reduces interferon-γ production	Epstein–Barr virus

Fig. 12.5 Mechanisms used by viruses of the herpes and pox families to subvert the host immune system.

of Lyme disease, which occurs as a result of chronic infection by the bacterium. Some strains of *B. burgdorferi* may avoid lysis by complement by coating themselves in the complement inhibitory protein factor H made by the host (see Section 2-17), which binds to receptor proteins in the bacterium's outer membrane.

Finally, many viruses subvert particular parts of the immune system. The mechanisms used include the capture of cellular genes for cytokines or cytokine receptors, the synthesis of complement-regulatory molecules, the inhibition of MHC class I molecule synthesis or assembly (as observed in EBV infections), and the production of decoy proteins that mimic the TIR domains that are part of the TLR/IL-1 receptor signaling pathway (see Fig. 6.34). The human cytomegalovirus produces a protein called UL18, which is homologous to an HLA class I molecule. By the interaction of UL18 with the receptor protein LIR-1, an inhibitory receptor on NK cells, the virus is thought to provide an inhibitory signal to the innate immune response (see Section 2-31).

Subversion of immune responses is one of the most rapidly expanding areas of research into host–pathogen relationships. Examples of how members of the herpes and poxvirus families subvert host responses are shown in Fig. 12.5.

12-4 Immunosuppression or inappropriate immune responses can contribute to persistent disease.

Many pathogens suppress immune responses in general. For example, staphylococci produce toxins, such as the **staphylococcal enterotoxins** and **toxic shock syndrome toxin-1**, that act as superantigens. Superantigens are proteins that bind the antigen receptors of very large numbers of T cells (see Section 5-15), stimulating them to produce cytokines that cause a severe inflammatory illness—**toxic shock**. The stimulated T cells proliferate and then rapidly undergo apoptosis, leaving a generalized immunosuppression together with the deletion of certain families of peripheral T cells.

Bacillus anthracis, the cause of anthrax, also suppresses immune responses through the release of a toxin. Anthrax is contracted by inhalation of, contact with, or ingestion of *B. anthracis* endospores and is often fatal if the endospores become disseminated throughout the body. *B. anthracis* produces a toxin called anthrax lethal toxin, which is a complex of two proteins—lethal factor and protective antigen. The main role of the protective antigen is to route the lethal factor into the host-cell cytosol. Lethal factor is a metalloproteinase with a unique specificity for MAP kinase kinases, components of many intracellular signaling pathways, and induces apoptosis of infected macrophages and abnormal maturation of dendritic cells. This results in the disruption of the immunological effector pathways that might otherwise delay bacterial growth.

Many other pathogens cause a mild or transient immunosuppression during acute infection. These forms of suppressed immunity are poorly understood but important, as they often make the host susceptible to secondary infections by common environmental microorganisms. Another clinically important immune suppression is caused by major trauma, burns, and occasionally surgery. Generalized infection is a common cause of death in severely burned patients. The reasons for this immunosuppression are not fully understood.

The measles virus can cause a relatively long-lasting immunosuppression after an infection, which is a particular problem in malnourished or undernourished children. In spite of the widespread availability of an effective vaccine, measles still accounts for 10% of the global mortality of children under 5 years old and is the eighth leading cause of death worldwide. Malnourished children are the main victims, and the cause of death is usually a secondary bacterial infection, particularly pneumonia, caused by measles-induced immunosuppression. This immunosuppression can last for several months after the disease is over and is associated with reduced T- and B-cell function. An important factor in measles-induced immunosuppression is the infection of dendritic cells by the measles virus. The infected dendritic cells render T lymphocytes generally unresponsive to antigen by mechanisms that are not yet understood, and it seems likely that this is the immediate cause of the immunosuppression.

The RNA virus hepatitis C (HCV) infects the liver and causes acute and chronic hepatitis, liver cirrhosis, and in some cases hepatocellular carcinoma. Immune responses probably have an important role in the clearance of HCV infection, but in more than 70% of cases HCV sets up a chronic infection. Although HCV mainly infects the liver during the early stage of a primary infection, the virus subverts the adaptive immune response by interfering with dendritic cell activation and maturation. This leads to inadequate activation of CD4 T cells and a consequent lack of T_H1 cell differentiation, which is thought to be responsible for the infection becoming chronic, most probably because of the lack of CD4 T-cell help to activate naive CD8 cytotoxic T cells. There is evidence that the decrease in levels of viral antigen seen after antiviral treatment improves CD4 T-cell help and allows the restoration of cytotoxic CD8 T-cell function and memory CD8 T-cell function. The delay in dendritic cell maturation caused by HCV is thought to synergize with another property of the virus that helps it to evade an immune response. The RNA polymerase that the virus uses to replicate its genome lacks proofreading capacity. This contributes to a high viral mutation rate and thus a change in its antigenicity, which allows it to evade adaptive immunity.

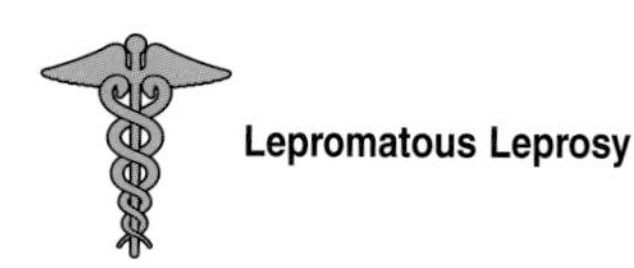

Leprosy, which we discussed in Section 8-19, is a more complex example of immunosuppression by an infection. In lepromatous leprosy, cell-mediated immunity is profoundly depressed, *M. leprae*-infected cells are present in great profusion, and cellular immune responses to many other antigens are suppressed (Fig. 12.6). This leads to a state called anergy, which in this context

Infection with *Mycobacterium leprae* can result in different clinical forms of leprosy

There are two polar forms, tuberculoid and lepromatous leprosy, but several intermediate forms also exist

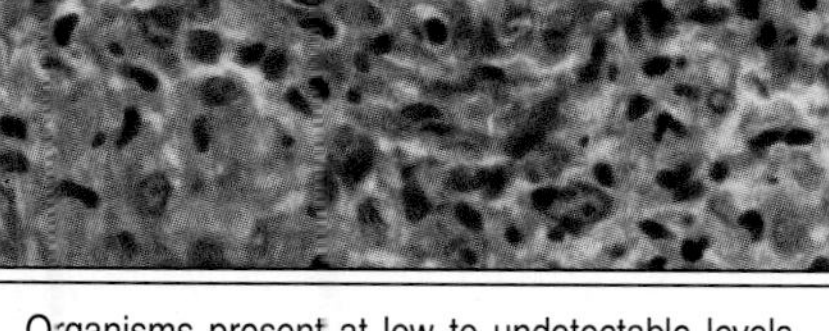

Tuberculoid leprosy	Lepromatous leprosy
Organisms present at low to undetectable levels	Organisms show florid growth in macrophages
Low infectivity	High infectivity
Granulomas and local inflammation. Peripheral nerve damage	Disseminated infection. Bone, cartilage, and diffuse nerve damage
Normal serum immunoglobulin levels	Hypergammaglobulinemia
Normal T-cell responsiveness. Specific response to *M. leprae* antigens	Low or absent T-cell responsiveness. No response to *M. leprae* antigens

Cytokine patterns in leprosy lesions

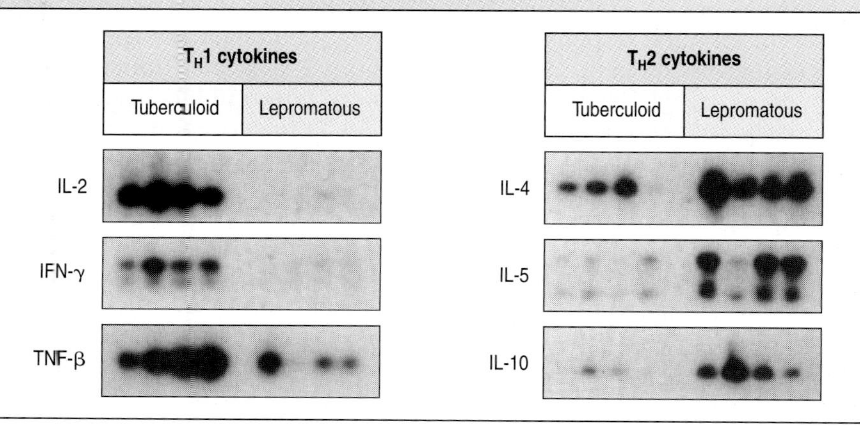

Fig. 12.6 T-cell and macrophage responses to *Mycobacterium leprae* are sharply different in the two polar forms of leprosy. Infection with *M. leprae*, whose cells stain as small dark red dots in the photographs, can lead to two very different forms of disease (top panels). In tuberculoid leprosy (left), growth of the organism is well controlled by T_H1-like cells that activate infected macrophages. The tuberculoid lesion contains granulomas and is inflamed, but the inflammation is local and causes only local effects, such as peripheral nerve damage. In lepromatous leprosy (right), infection is widely disseminated and the bacilli grow uncontrolled in macrophages; in the late stages of disease there is major damage to connective tissues and to the peripheral nervous system. There are several intermediate stages between these two polar forms. The lower panel shows Northern blots demonstrating that the cytokine patterns in the two polar forms of the disease are sharply different, as shown by the analysis of RNA isolated from lesions of four patients with lepromatous leprosy and four patients with tuberculoid leprosy. Cytokines typically produced by T_H2 cells (IL-4, IL-5, and IL-10) dominate in the lepromatous form, whereas cytokines produced by T_H1 cells (IL-2, IFN-γ, and TNF-β) dominate in the tuberculoid form. It therefore seems that T_H1-like cells predominate in tuberculoid leprosy, and T_H2-like cells in lepromatous leprosy. IFN-γ would be expected to activate macrophages, enhancing the killing of *M. leprae*, whereas IL-4 can actually inhibit the induction of bactericidal activity in macrophages. Photographs courtesy of G. Kaplan; cytokine patterns courtesy of R.L. Modlin.

specifically means the absence of delayed-type hypersensitivity to a wide range of antigens unrelated to *M. leprae* (see Section 7-6 for the more general definition of anergy in use in other contexts). In tuberculoid leprosy, in contrast, there is strong cell-mediated immunity with macrophage activation that controls, but does not eradicate, infection. Most of the pathology in tuberculoid leprosy is caused by the ongoing localized inflammatory response to the mycobacteria that persist.

12-5 Immune responses can contribute directly to pathogenesis.

Tuberculoid leprosy is just one example of an infection in which the pathology is caused largely by the immune response, the phenomenon known as **immunopathology**. This is true to some degree in most infections; for example, the fever that accompanies a bacterial infection is caused by the release of cytokines by macrophages. One medically important example of immunopathology is the wheezy bronchiolitis caused by **respiratory syncytial virus** (**RSV**) infection. Bronchiolitis caused by RSV is the major cause of admission of young children to hospital in the Western world, with as many as 90,000 admissions and 4500 deaths each year in the United States alone. The first indication that the immune response to the virus might have a role in the pathogenesis of this disease came from the observation that infants vaccinated with an alum-precipitated killed virus preparation had a more severe illness than children who did not receive the vaccine. This occurred because the vaccine failed to induce neutralizing antibodies but succeeded in producing effector T_H2 cells. When the vaccinated children encountered the virus, the T_H2 cells released interleukins IL-3, IL-4, and IL-5, which induced bronchospasm, increased the secretion of mucus, and increased tissue eosinophilia. Mice can be infected with RSV and develop a disease similar to that seen in humans.

Another example of a pathogenic immune response is the response to schistosome eggs. Schistosomes lay eggs in the hepatic portal vein. Some reach the intestine and are shed in the feces, spreading the infection; other eggs lodge in the portal circulation of the liver, where they elicit a potent immune response leading to chronic inflammation, hepatic fibrosis, and eventually liver failure. This process reflects the excessive activation of T_H1 cells and can be modulated by T_H2 cells, IL-4, or CD8 T cells, which can also produce IL-4.

12-6 Regulatory T cells can affect the outcome of infectious disease.

Some pathogens avoid an immune response by interacting with regulatory T cells, which have been discussed in Section 8-19. Natural CD4 CD25 regulatory T cells (T_{reg}) arise in the thymus and migrate to the periphery where they help to maintain tolerance, as we discuss in Chapter 14, and are thought to control immune responses by suppressing proliferation of lymphocytes recognizing autoantigens. Other CD4 regulatory T cells arise from the differentiation of naive CD4 T cells in the periphery. Interaction between regulatory T cells and pathogens can generate either a protective response in favor of the host or, if it leads to the suppression of immune responses, can act as a mechanism of immune evasion for the pathogen. Examples of the latter include chronic persistent infections such as HCV, and perhaps HIV. Patients infected with HCV have higher numbers of recirculating natural T_{reg} cells than healthy individuals, and *in vitro* depletion of T_{reg} cells enhances cytotoxic lymphocyte responses against the virus. During infections with the protozoan parasite *Leishmania major*, T_{reg} cells accumulate in the dermis, where they impair the ability of effector T cells to eliminate pathogens from this site.

In contrast, studies in both humans and mice have shown that the inflammation occurring during ocular infections with HSV is limited by the presence of T_{reg} cells. If these cells are depleted from mice before HSV infection, a more severe disease results, even when smaller doses of virus are used to cause infection. T_{reg} cells also restrain inflammation in the pulmonary disease that occurs in immunodeficient mice infected with the opportunistic fungal pathogen *Pneumocystis carinii*, which is a common pathogen in immunodeficient humans.

Summary.

Infectious agents can cause recurrent or persistent disease by avoiding normal host defense mechanisms or by subverting them to promote their own replication. There are many different ways of evading or subverting the immune response. Antigenic variation, latency, resistance to immune effector mechanisms, and suppression of the immune response all contribute to persistent and medically important infections. In some cases the immune response is part of the problem: some pathogens use immune activation to spread infection, and others would not cause disease if it were not for the immune response. Each of these mechanisms teaches us something about the nature of the immune response and its weaknesses, and each requires a different medical approach to prevent or to treat infection.

Immunodeficiency diseases.

Immunodeficiencies occur when one or more components of the immune system are defective. The immunodeficiencies are classified as primary or secondary. Primary immunodeficiencies are caused by mutations in any of a large number of genes that are involved in or control immune responses. Clinical manifestations of primary immunodeficiencies are highly variable; they commonly include recurrent and often overwhelming infections in very young children, but allergy, abnormal proliferation of lymphocytes, and autoimmunity can also occur. In contrast, secondary immunodeficiencies are acquired as a consequence of other diseases, or are secondary to environmental factors such as starvation, or are an adverse consequence of medical intervention.

By examining which infectious diseases accompany a particular inherited or acquired immunodeficiency, we can see which components of the immune system are important in the response to particular agents. The inherited immunodeficiencies also reveal how interactions between different cell types contribute to the immune response and to the development of T and B lymphocytes. Finally, these inherited diseases can lead us to the defective gene, often revealing new information about the molecular basis of immune processes and providing the necessary information for diagnosis, for genetic counseling, and eventually the possibility of gene therapy.

12-7 A history of repeated infections suggests a diagnosis of immunodeficiency.

Patients with immune deficiency are usually detected clinically by a history of recurrent infection with the same or similar pathogens. The type of infection is a guide to which part of the immune system is deficient. Recurrent infection by pyogenic, or pus-forming, bacteria suggests a defect in antibody, complement, or phagocyte function, reflecting the role of these parts of the immune system in defense against such infections. In contrast, a history of persistent fungal skin infection, such as cutaneous candidiasis, or recurrent viral infections is more suggestive of a defect in host defense mediated by T lymphocytes.

12-8 Inherited immunodeficiency diseases are caused by recessive gene defects.

Before the advent of antibiotics, it is likely that most individuals with inherited immune defects died in infancy or early childhood because of their susceptibility to particular classes of pathogens. Such cases would not have been easy to identify, because many normal infants also died of infection. Most of the gene defects that cause inherited immunodeficiencies are recessive, and many are caused by mutations in genes on the X chromosome. As males have only one X chromosome, all males who inherit an X chromosome carrying a defective gene will show the disease. In contrast, female carriers with one defective X chromosome are usually healthy because their immune system develops from stem cells that are naturally selected for those in which X-inactivation has shut down the X chromosome bearing the mutated gene. Immunodeficiency diseases that affect various steps in B- and T-lymphocyte development have been described, as have defects in surface molecules that are important for T- or B-cell function. Defects in phagocytic cells, in complement, in cytokines, in cytokine receptors, and in molecules that mediate effector responses also occur. Thus, immunodeficiency can be caused by defects in either the adaptive or the innate immune system. Examples of immunodeficiency diseases are listed in Fig. 12.7. None are very common (a selective deficiency in IgA being the most frequently reported), and some are extremely rare. Some of these individual diseases are described in later sections.

The use of gene knockout techniques in mice (see Appendix I, Section A-47) has allowed the creation of many immunodeficient states that are adding rapidly to our knowledge of the contribution of individual proteins to normal immune function. Nevertheless, human immunodeficiency diseases are still the best source of insight into the normal pathways of defense against infectious diseases in humans. For example, a deficiency of antibody, of complement, or of phagocytic function each increases the risk of infection by certain pyogenic bacteria. This shows that the normal pathway of host defense against such bacteria is the binding of antibody followed by the fixation of complement, which allows the opsonized bacteria to be taken up by phagocytic cells and killed. Breaking any of the links in this chain of events causes a similar immunodeficient state.

Immunodeficiencies also teach us about the redundancy of defense mechanisms against infectious disease. The first two people to be discovered with a hereditary deficiency of complement were healthy immunologists. This teaches us two lessons. The first is that there are multiple protective immune mechanisms against infection; for example, although there is abundant evidence that complement deficiency increases susceptibility to pyogenic infection, not every human with complement deficiency suffers from recurrent infections. The second lesson concerns the phenomenon of **ascertainment artifact**. When an unusual observation is made in a patient with disease, there is a temptation to seek a causal link between that observation and the disease; however, no one would suggest that complement deficiency causes a genetic predisposition to becoming an immunologist. Complement deficiency was discovered in immunologists because they used their own blood in their experiments. If a particular measurement is made only in a highly

Name of deficiency syndrome	Specific abnormality	Immune defect	Susceptibility
Severe combined immune deficiency	See Fig. 12.14		General
DiGeorge's syndrome	Thymic aplasia	Variable numbers of T and B cells	General
MHC class I deficiency	TAP mutations	No CD8 T cells	Chronic lung and skin inflammation
MHC class II deficiency	Lack of expression of MHC class II	No CD4 T cells	General
Wiskott–Aldrich syndrome	X-linked; defective WASP gene	Defective anti-polysaccharide antibody, impaired T-cell activation responses, and T_{reg} dysfunction	Encapsulated extracellular bacteria
X-linked agammaglobulinemia	Loss of Btk tyrosine kinase	No B cells	Extracellular bacteria, viruses
Hyper IgM syndrome	AID deficiency CD40 ligand deficiency CD40 deficiency NEMO (IKK) deficiency	No isotype switching and/or no somatic hypermutation	Extracellular bacteria *Pneumocystis carinii* *Cryptosporidium parvum*
Common variable immunodeficiency	ICOS deficiency, other unknown	Defective IgA and IgG production	Extracellular bacteria
Selective IgA	Unknown; MHC-linked	No IgA synthesis	Respiratory infections
Phagocyte deficiencies	Many different	Loss of phagocyte function	Extracellular bacteria and fungi
Complement deficiencies	Many different	Loss of specific complement components	Extracellular bacteria especially *Neisseria* spp.
X-linked lymphoproliferative syndrome	SAP (SH2D1A) mutant	Inability to control B-cell growth	EBV-driven B-cell tumors
Ataxia telangiectasia	Mutation of kinase domain of ATM	T cells reduced	Respiratory infections
Bloom's syndrome	Defective DNA helicase	T cells reduced Reduced antibody levels	Respiratory infections

Fig. 12.7 Human immunodeficiency syndromes. The specific gene defect, the consequence for the immune system, and the resulting disease susceptibilities are listed for some common and some rare human immunodeficiency syndromes. Syndromes that lead to severe combined immunodeficiency are listed separately in Fig. 12.14. AID, activation-induced cytidine deaminase; ATM, ataxia telangiectasia-mutated protein; EBV, Epstein–Barr virus; IKKγ, γ subunit of the kinase IKK; TAP, transporters associated with antigen processing; WASP, Wiskott–Aldrich syndrome protein.

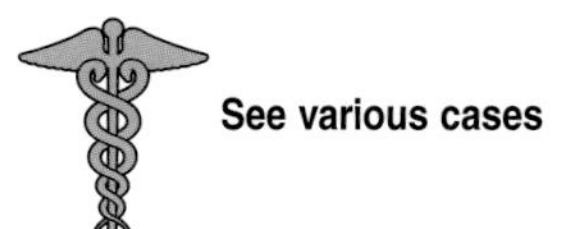

See various cases

selected group of patients with a particular disease, it is inevitable that the only abnormal results will be discovered in patients with that disease. This is the source of ascertainment artifacts and emphasizes the importance of studying appropriate controls.

12-9 The main effect of low levels of antibody is an inability to clear extracellular bacteria.

Pyogenic bacteria have polysaccharide capsules that are not directly recognized by the receptors on macrophages and neutrophils that stimulate phagocytosis. The bacteria escape immediate elimination by the innate immune response and are successful extracellular pathogens. Normal individuals can clear infections by pyogenic bacteria because antibody and complement

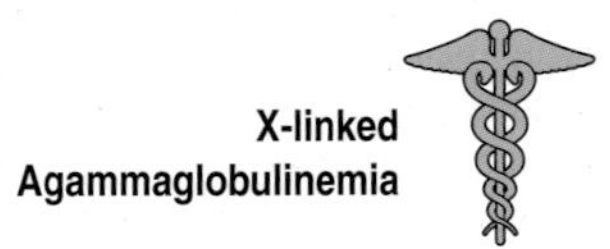
X-linked
Agammaglobulinemia

opsonize the bacteria, enabling phagocytes to ingest and destroy them. The principal effect of deficiencies in antibody production is therefore a failure to control infections by this type of bacterium. Susceptibility to some viral infections, most notably those caused by enteroviruses, is also increased in cases of antibody deficiency because of the importance of antibodies in neutralizing viruses that enter the body through the gut.

The first description of an immunodeficiency disease was Ogden C. Bruton's account, in 1952, of the failure of a male child to produce antibody. Because this defect is inherited in an X-linked fashion and is characterized by the absence of immunoglobulin in the serum, it was called **Bruton's X-linked agammaglobulinemia** (**XLA**). The absence of antibody can be detected by immunoelectrophoresis (Fig. 12.8). Since then, many more defects of antibody production have been described, most of them the consequence of failures in the development or activation of B lymphocytes. Infants with these diseases are usually identified as a result of recurrent infections with pyogenic bacteria such as *Streptococcus pneumoniae* and the occurrence of chronic infections with viruses such as hepatitis B and C, poliovirus, and ECHO virus.

The defective gene in XLA encodes a protein tyrosine kinase called Btk (Bruton's tyrosine kinase), which is a member of the Tec family of kinases (see Section 6-13). Btk is expressed in neutrophils as well as in B cells, although only B cells are defective in XLA patients, in whom B-cell maturation is largely arrested at the pre-B-cell stage. It is likely that Btk is required to couple the pre-B-cell receptor to nuclear events that lead to pre-B-cell growth and differentiation (see Section 7-9). In patients with Btk deficiencies, some B cells mature despite the defect, suggesting that signals transmitted by Tec family kinases are not absolutely required.

Because the gene responsible for XLA is on the X chromosome, it is possible to identify female carriers by analyzing X-chromosome inactivation in their B cells. During embryonic development, female mammals randomly inactivate one of their two X chromosomes. Because Btk is required for B-lymphocyte development, only cells in which the normal allele of *btk* is active can develop into mature B cells. Thus, in female carriers of a mutant *btk* gene, all B cells have the normal X chromosome as the active X. In contrast, the active X chromosomes in the T cells and macrophages of carriers are an equal mixture of normal and *btk* mutant X chromosomes. This fact allowed female carriers of XLA

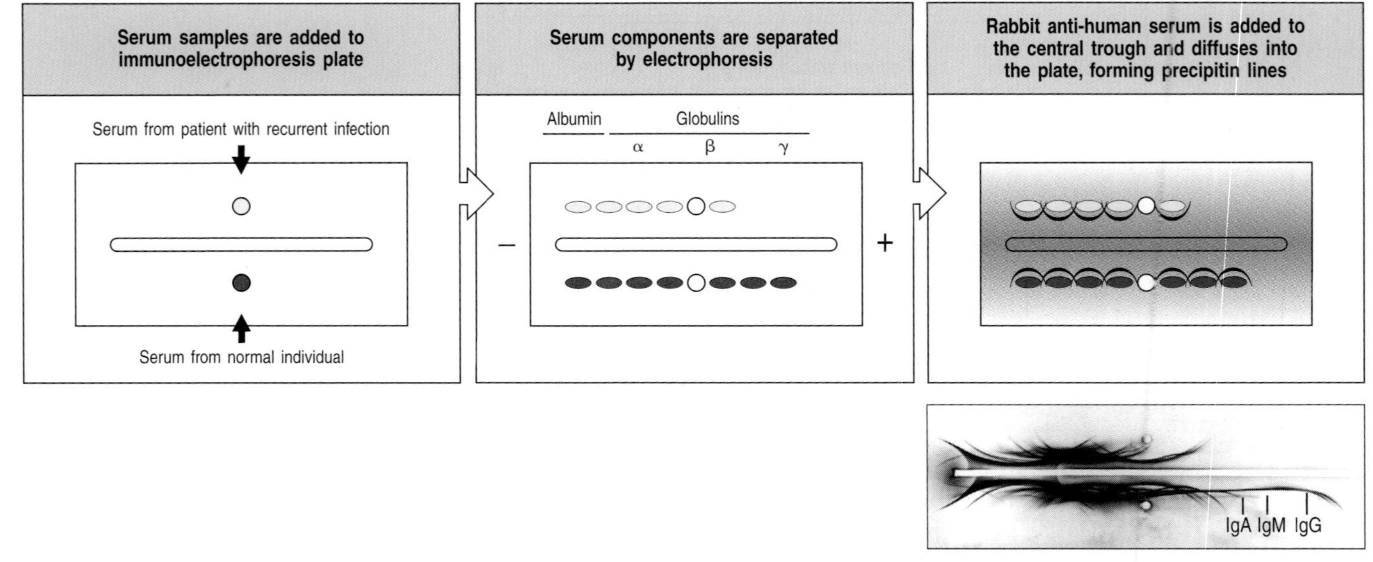
Serum samples are added to immunoelectrophoresis plate
Serum from patient with recurrent infection
Serum from normal individual
Serum components are separated by electrophoresis
Albumin
Globulins
α
β
γ
–
+
Rabbit anti-human serum is added to the central trough and diffuses into the plate, forming precipitin lines
IgA IgM IgG

Fig. 12.8 Immunoelectrophoresis reveals the absence of several distinct immunoglobulin isotypes in serum from a patient with X-linked agammaglobulinemia (XLA). Serum samples from a normal control and from a patient with recurrent bacterial infection caused by the absence of antibody production, as reflected in an absence of gamma globulins, are separated by electrophoresis on an agar-coated slide. Antiserum raised against whole normal human serum and containing antibodies against many of its different proteins is put in a trough down the middle; each antibody forms an arc of precipitation with the protein it recognizes. The position of each arc is determined by the electrophoretic mobility of the serum protein; immunoglobulins migrate to the gamma-globulin region of the gel. The absence of immunoglobulins in a patient who has X-linked agammaglobulinemia is shown in the panel at the bottom, where several arcs are missing from the patient's serum (upper set). These are IgA, IgM, and several subclasses of IgG, each recognized in normal serum (lower set) by antibodies in the antiserum against human serum proteins. Bottom panel from the collection of the late C.A. Janeway Sr.

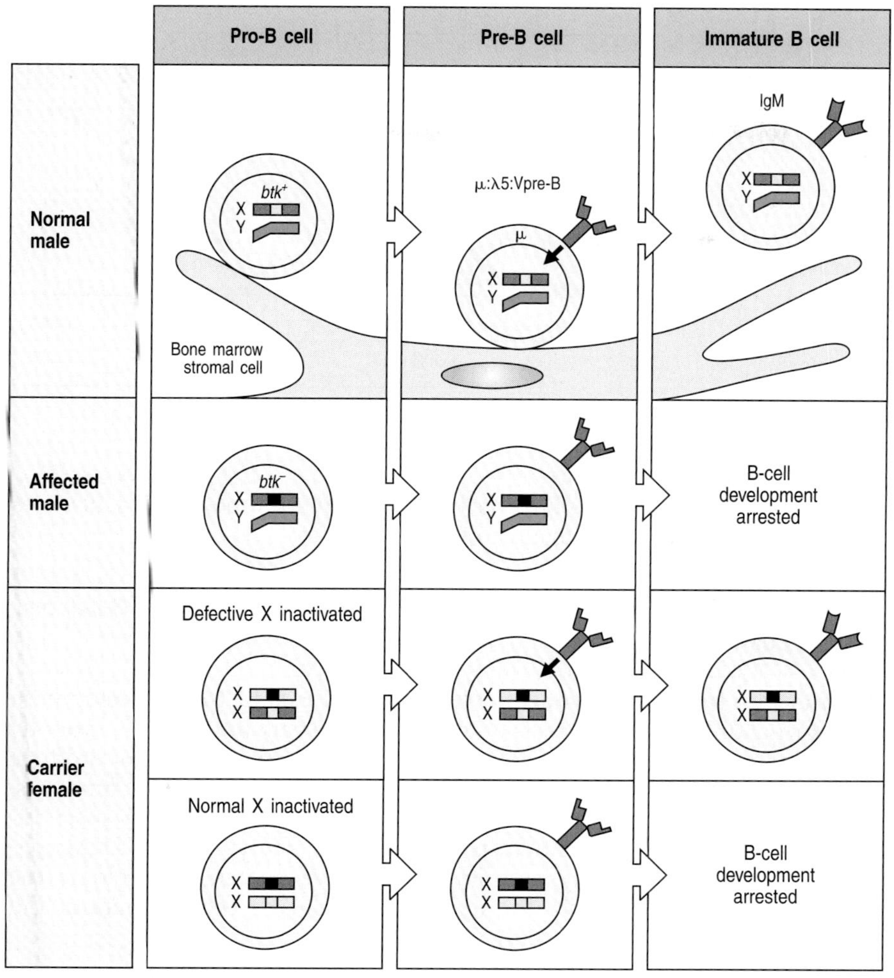

Fig. 12.9 The product of the *BTK* gene is important for B-cell development. In X-linked agammaglobulinemia (XLA), a protein tyrosine kinase of the Tec family called Btk, which is encoded on the X chromosome, is defective. In normal individuals, B-cell development proceeds through a stage in which the pre-B-cell receptor, consisting of μ:λ5:V pre-B (see Section 7-3), transduces a signal via Btk, triggering further B-cell development. In males with XLA, no signal can be transduced and, although the pre-B-cell receptor is expressed, the B cells develop no further. In female mammals, including humans, one of the two X chromosomes in each cell is permanently inactivated early in development. Because the choice of which chromosome to inactivate is random, half of the pre-B cells in a carrier female will have inactivated the chromosome with the wild-type *btk* gene, meaning that they can express only the defective *btk* gene and cannot develop further. Therefore, in the carrier, mature B cells always have the nondefective X chromosome active. This is in sharp contrast to all other cell types, which have the nondefective X chromosome active in only half of their cells. Nonrandom X-chromosome inactivation in a particular cell lineage is a clear indication that the product of the X-linked gene is required for the development of cells of that lineage. It is also sometimes possible to identify the stage at which the gene product is required, by detecting the point in development at which X-chromosome inactivation develops bias. Using this kind of analysis, one can identify carriers of X-linked traits such as XLA without needing to know the nature of the mutant gene.

to be identified even before the nature of the Btk protein was known. Nonrandom X inactivation only in B cells also shows conclusively that the *btk* gene is required the development of B cells but not for that of other cell types, and that Btk must act within the B cells themselves rather than in stromal cells or other cells required for B-cell development (Fig. 12.9).

The commonest humoral immune defect is in fact a transient deficiency in immunoglobulin production that occurs in the first 6–12 months of life. A newborn infant has antibody levels comparable to those of the mother because of the transplacental transport of maternal IgG (see Section 9-15). As the IgG is catabolized, antibody levels gradually decrease until the infant begins to produce useful amounts of its own IgG at about 6 months of age (Fig. 12.10). Thus, IgG levels are quite low between the ages of 3 months and 1 year, when the infant's own IgG antibody responses are poor, and this can lead to a period of heightened susceptibility to infection. This is especially true for premature babies, who begin with lower levels of maternal IgG and also reach immune competence later after birth.

People with pure B-cell defects resist many pathogens successfully. However, effective defense against a subset of extracellular pyogenic bacteria, including staphylococci and streptococci, requires the opsonization of these bacteria with specific antibody. In people with B-cell defects, these infections can be suppressed with antibiotics and with monthly infusions of human immunoglobulin collected from a large pool of donors. Because there are

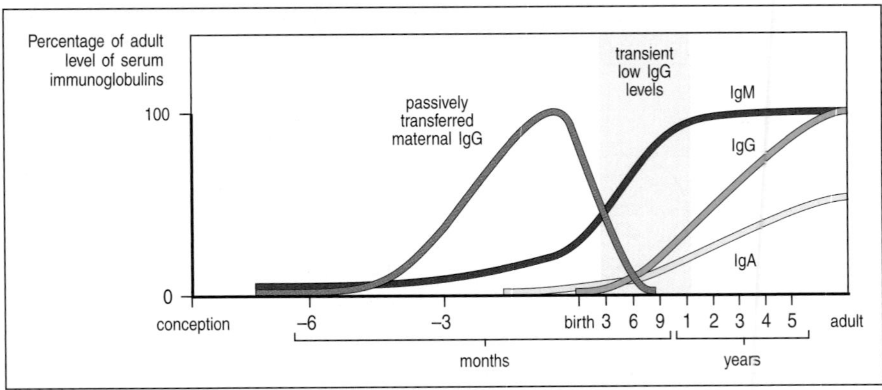

Fig. 12.10 Immunoglobulin levels in newborn infants fall to low levels at about 6 months of age. Babies are born with high levels of IgG, which is actively transported across the placenta from the mother during gestation. After birth, the production of IgM starts almost immediately; the production of IgG, however, does not begin for about 6 months, during which time the total level of IgG falls as the maternally acquired IgG is catabolized. Thus, IgG levels are low from about the age of 3 months to 1 year, which can lead to susceptibility to disease.

antibodies against many pathogens in this pooled immunoglobulin, it serves as a fairly successful shield against infection.

12-10 Some antibody deficiencies can be due to defects in either B-cell or T-cell function.

Patients with **hyper IgM syndrome** have normal B- and T-cell development and normal or high serum levels of IgM, but make very limited IgM antibody responses against antigens that require T-cell help. They also produce immunoglobulin isotypes other than IgM and IgD only in trace amounts, indicating that class switching is impaired. This makes them highly susceptible to infection with extracellular pathogens. Five causes of hyper IgM syndrome have been distinguished, and these have helped to elucidate the pathways that are essential for normal class-switch recombination and somatic hypermutation in B cells.

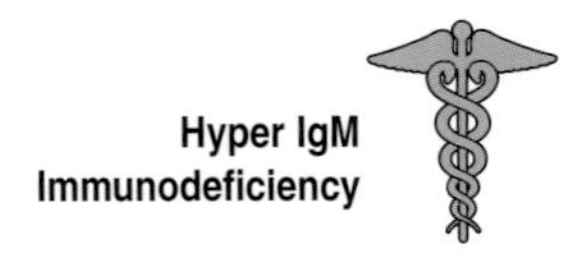

The commonest form of hyper IgM syndrome is **X-linked hyper IgM syndrome**, which is caused by mutations in the gene encoding CD40 ligand (CD154), which is on the X chromosome. CD40 ligand is normally expressed on activated T cells, enabling them to engage the CD40 protein on B cells and activate them (see Section 9-4). In cases of CD40 ligand deficiency, B cells do not have their CD40 engaged although they themselves are normal. As we saw in Chapter 4, CD40 ligand interaction with CD40 is also essential for the induction of isotype switching and the formation of germinal centers (Fig. 12.11).

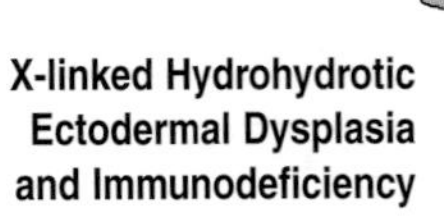

A very similar syndrome has been identified in patients with mutations in two other genes. Not unexpectedly, one is the gene encoding CD40 on chromosome 20, mutations in which have been found in several patients with a recessive variant of hyper IgM syndrome. A different mutated gene was found in patients with a rare developmental disorder called **hypohydrotic ectodermal dysplasia with immunodeficiency**, in which patients lack sweat glands, have abnormal hair and tooth development, and also show hyper IgM syndrome. In this disease, also known as **NEMO deficiency**, mutations were found in the gene encoding the protein NEMO (also known as IKKγ, a subunit of the kinase IKK), which is an essential component of the intracellular signaling pathway leading to activation of the transcription factor NFκB (see Fig. 6.22).

This group of hyper IgM syndromes show that mutations at these different places in the pathway triggered by CD40 ligand on T cells binding to CD40 on B cells all result in a similar immunodeficiency syndrome. These patients show reduced protection against various microorganisms, mostly pyogenic bacteria and mycobacteria.

Patients with X-linked hyper IgM syndrome also have defects in cell-mediated immunity. They are susceptible to infection with *P. carinii*, which is normally killed by activated macrophages. This susceptibility is thought to be due, at least in part, to the inability of their T cells to deliver an activating signal to infected macrophages by engaging the CD40 on these cells. A defect in T-cell activation could also contribute to the profound immunodeficiency in these patients, because studies on mice that lack CD40 ligand have revealed a failure of antigen-specific T cells to expand in response to primary immunization with antigen.

Another type of hyper IgM syndrome is an intrinsic defect in B cells caused by mutations in the gene encoding the enzyme activation-induced cytidine deaminase (AID; see Section 4-17). It is associated with a milder form of immunodeficiency than the other forms of the syndrome. Patients with **AID deficiency** are more susceptible than normal to severe bacterial infections but not to opportunistic infections such as *P. carinii*. The B cells in these patients fail to switch antibody isotype and also have much reduced somatic hypermutation. The consequence is that immature B cells accumulate in abnormal germinal centers, causing enlargement of the lymph nodes and spleen. AID is expressed only in B cells that have been triggered to undergo class switching or hypermutation, thus demonstrating its unique role in these two processes. The milder degree of immunodeficiency associated with hyper IgM syndrome due to AID deficiency in comparison with that associated with CD40 ligand, CD40, or NEMO deficiency is because AID deficiency only causes the failure of antibody responses, whereas deficiency of the other proteins is associated with defects in both B- and T-cell function. Yet another cause of hyper IgM syndrome was identified recently in a small number of patients who have normal somatic hypermutation and normal AID function but defective isotype switching. This genetic abnormality has yet to be defined.

A fourth example of predominantly humoral immunodeficiency is **common variable immunodeficiency** (**CVID**). In this syndrome there is usually a deficiency in IgM, IgG, and IgA together. At least some cases of CVID are familial. CVID is a disorder in which the function of both B and T cells is impaired; it displays a range of different symptoms. Patients are susceptible to recurrent infections and have decreased serum immunoglobulin and abnormal antibody responses. Autoimmune conditions and gastrointestinal diseases have also been reported in some patients with CVID. Children with CVID are more susceptible than normal to infections of the middle ear (otitis media), and may develop infections in the joints, bones, skin, and parotid glands.

The condition is not as severe as some of the other immunodeficiencies, and most patients are not generally diagnosed until adulthood. A significant percentage of cases of CVID, and a smaller proportion of cases of single IgA deficiency, are associated with a genetic defect in a transmembrane protein called the TNF-like receptor transmembrane activator and CAML interactor (TACI). This is the receptor for the cytokine BAFF, which is secreted by dendritic cells and provides co-stimulatory and survival signals for B-cell activation and class switching (see Section 9-13).

Another genetic defect that has been linked to a small percentage of people with CVID is a deficiency of the co-stimulatory molecule ICOS. As described in Section 8-14, ICOS is an inducible co-stimulatory molecule that is upregulated on T cells when they are activated. The effects of ICOS deficiency have confirmed its essential role in T-cell help for the later stages of B-cell differentiation, including class switching and the formation of memory cells.

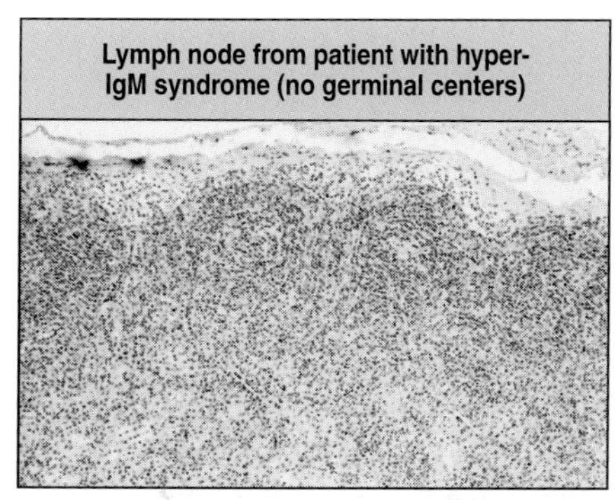

Lymph node from patient with hyper-IgM syndrome (no germinal centers)

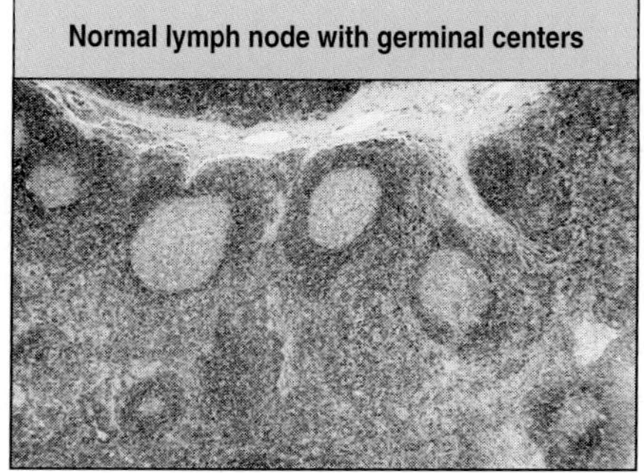

Fig. 12.11 Patients with X-linked hyper IgM syndrome are unable to activate their B cells fully. Lymphoid tissues in patients with hyper IgM syndrome are devoid of germinal centers (top panel), unlike a normal lymph node (bottom panel). B-cell activation by T cells is required both for isotype switching and for the formation of germinal centers, where extensive B-cell proliferation takes place. Photographs courtesy of R. Geha and A. Perez-Atayde.

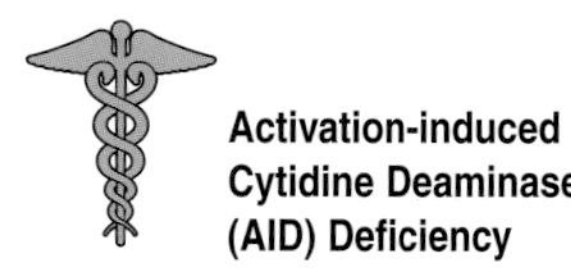

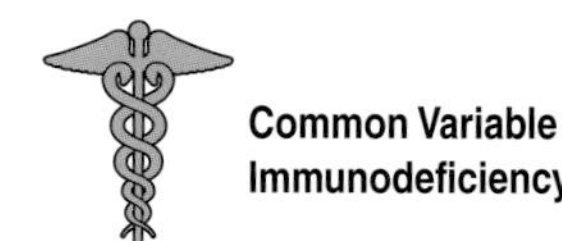

12-11 Defects in complement components cause defective humoral immune function.

Not surprisingly, the spectrum of infections associated with complement deficiencies overlaps substantially with that seen in patients with deficiencies in antibody production (Fig. 12.12). Defects in the activation of C3, and in C3 itself, are associated with a wide range of pyogenic infections, including *S. pneumoniae*, emphasizing the important role of C3 as an opsonin that promotes the phagocytosis of bacteria. In contrast, defects in the membrane-attack components of complement (C5–C9) have more limited effects, and result exclusively in susceptibility to *Neisseria* species. This indicates that defense against these bacteria, which can survive intracellularly, is mediated by extracellular lysis by the membrane-attack complex. Data from large population studies in Japan, where endemic *N. meningitidis* infection is rare, show that the risk each year of infection with this organism is approximately 1 in 2,000,000 to a normal person. This compares with a risk of 1 in 200 in the same population to a person with an inherited deficiency of one of the membrane-attack complex proteins—a 10,000-fold increase in risk. The early components of the classical complement pathway are particularly important for the elimination of immune complexes and apoptotic cells, which can cause significant pathology in autoimmune diseases such as systemic lupus erythematosus. This aspect of inherited complement deficiency is discussed in Chapter 14.

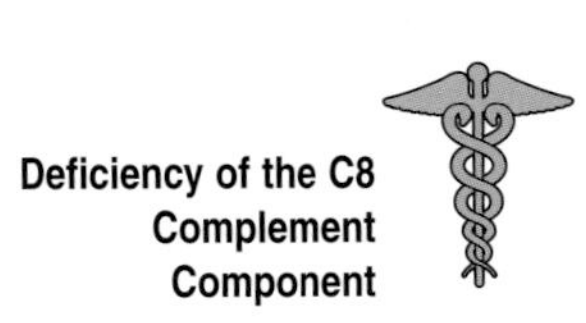

Another set of diseases is caused by defects in complement control proteins. People lacking decay-accelerating factor (DAF) and CD59, which protect the body's cell surfaces from complement activation, destroy their own red blood cells. This results in the disease paroxysmal nocturnal hemoglobinuria, as discussed in Section 2-22. A more striking consequence of the loss of a complement regulatory protein is seen in patients with C1-inhibitor defects, which causes the syndrome known as **hereditary angioneurotic edema**. As well as inhibiting the serine proteases C1r and C1s, and thus regulating the initiation of the classical pathway of complement activation, C1 inhibitor inhibits two serine proteases that participate in the contact activation system of blood clotting—factor XIIa (activated Hageman factor) and kallikrein.

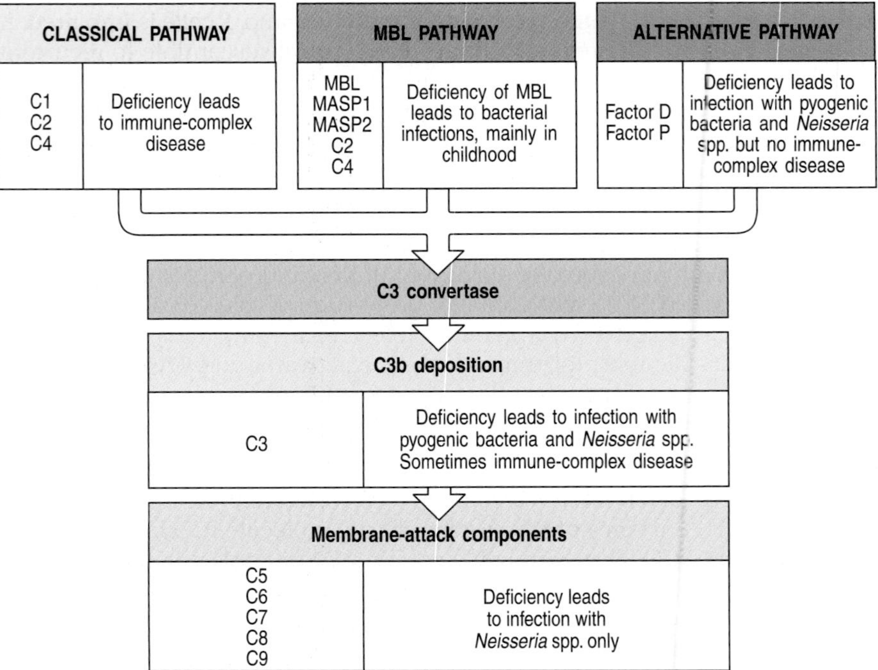

Fig. 12.12 Defects in complement components are associated with susceptibility to certain infections and accumulation of immune complexes. Defects in the early components of the alternative pathway and in C3 lead to susceptibility to extracellular pathogens, particularly pyogenic bacteria. Defects in the early components of the classical pathway predominantly affect the processing of immune complexes and the clearance of apoptotic cells, leading to immune-complex disease. Deficiency of mannose-binding lectin (MBL), the recognition molecule of the mannose-binding lectin pathway, is associated with bacterial infections, mainly in early childhood. Defects in the membrane-attack components are associated only with susceptibility to strains of *Neisseria* species, the causative agents of meningitis and gonorrhea, implying that the effector pathway is important chiefly in defense against these organisms.

Deficiency of C1 inhibitor leads to a failure to regulate both the blood clotting and complement activation pathways, leading to excessive production of vasoactive mediators that cause fluid accumulation in the tissues (edema) and epiglottal swelling that can result in suffocation. These mediators are bradykinin, produced by the cleavage of high molecular weight kininogen by kallikrein, and the C2 kinin, produced by the activity of C1s on C2b.

Deficiences in mannose-binding lectin (MBL), which initiates complement activation in innate immunity (see Section 2-12), are relatively common (5% of the population). MBL deficiency may be associated with a mild immunodeficiency with an excess of bacterial infection in early childhood.

12-12 Defects in phagocytic cells permit widespread bacterial infections.

Deficiencies in phagocyte numbers or function can be associated with severe immunodeficiency; indeed, a total absence of neutrophils is incompatible with survival in a normal environment. There are three types of phagocyte immunodeficiencies, caused by genes encoding proteins that control phagocyte production, phagocyte interaction and phagocyte killing of microrganisms, respectively. We consider each in turn. Inherited deficiencies of neutrophil production (**neutropenias**) are classified either as **severe congenital neutropenias** or as **cyclic neutropenias**. In severe congenital neutropenia, which can be inherited as a dominant or recessive trait, the neutrophil count is persistently extremely low, at less than 0.2×10^9 per liter of blood (normal numbers are $3–5.5 \times 10^9$/liter), and patients depend on a successful bone marrow transplant for survival. Cyclic neutropenia is a dominantly inherited disease in which neutrophil numbers fluctuate from near normal to very low or none, with an approximate cycle time of 21 days. Other bone marrow derived cells—monocytes, platelets, lymphocytes, and reticulocytes—undergo smaller fluctuations in numbers with the same periodicity.

Surprisingly, mutations in human neutrophil elastase (*ELA2*) cause cyclic neutropenia and also cause a significant fraction of dominant severe congenital neutropenia. The mutations lead to the production of dysfunctional elastases and this in turn leads to the production of a toxic intracellular protein that blocks neutrophil maturation. Heterozygous mutations in the oncogene *GFI1*, which encodes a transcriptional repressor, have been detected in three patients with neutropenias. This finding arose from the unexpected observation that mice lacking the protein Gfi1 are neutropenic. Closer analysis revealed that mutation in mouse *Gfi1* affects the expression of *Ela2*, providing a link between these two genes in a common pathway of myeloid cell differentiation. How the mutant elastase causes a 21-day cycle in neutropenia and the effects on other bone marrow cell types is still a mystery.

Intermittent neutropenia is also characteristic of patients with **Shwachman–Diamond syndrome**, another rare example of an autosomal recessive immunodeficiency. This syndrome is characterized by skeletal abnormalities, exocrine pancreatic insufficiency and bone marrow dysfunction. A mutation in a gene called *SBDS* has been identified in 89% of unrelated individuals with Shwachman–Diamond syndrome. *SBDS* is a member of a gene family that includes genes involved in RNA processing, suggesting that the syndrome may be due to a dysfunction in RNA metabolism essential for hematopoiesis, chondrogenesis (formation of cartilage), and the development of the exocrine pancreas.

Defects in the migration of phagocytic cells to extravascular sites of infection can cause serious immunodeficiency. Leukocytes reach such sites by emigrating from blood vessels in a tightly regulated process (see Fig. 2.49). The first stage is the rolling adherence of leukocytes to endothelial cells, through the binding of a fucosylated tetrasaccharide ligand known as sialyl-Lewis^x on the leukocyte to E-selectin and P-selectin on endothelium. The second stage is the tight adherence of the leukocytes to the endothelium through the binding of leukocyte β_2 integrins such as CD11b:CD18 (Mac-1:CR3) to counter-receptors on endothelial cells. The third and final stage is the transmigration of leukocytes through the endothelium along gradients of chemokines originating from the site of tissue injury.

Deficiencies in the molecules involved in each of these stages can prevent neutrophils and macrophages from reaching sites of infection to ingest and destroy bacteria. Reduced rolling adhesion has been described in patients with a lack of sialyl-Lewis^x antigen caused by a deficiency in a putative GDP-fucose transporter that participates in fucosylation during sialyl-Lewis^x biosynthesis. Deficiencies in the leukocyte integrin common β_2 subunit CD18 prevent the migration of leukocytes to sites of infection by abolishing the leukocyte's ability to adhere tightly to the endothelium, and are the cause of **leukocyte adhesion deficiency**. A third genetic defect that prevents neutrophil migration was identified in the *Rac2* gene. The Rac2 protein is a member of the Rho family of GTPases that regulate neutrophil activation and cytoskeletal function. All these deficiencies lead to infections that are resistant to antibiotic treatment and that persist despite an apparently effective cellular and humoral adaptive immune response. Acquired neutropenia associated with chemotherapy, malignancy, or aplastic anemia is also associated with a similar spectrum of severe pyogenic bacterial infections.

Leukocyte Adhesion Deficiency

Warts, hypogammaglobulinemia, infections and myelokathexis syndrome (**WHIM**) is a rare neutropenia that has recently been linked to a heterozygous mutation in the gene encoding the chemokine receptor CXCR4. Although the surface expression of CXCR4 is normal, the mutation seems to affect the cytoplasmic tail domain. CXCR4 is the receptor for CXCL12 and is expressed by myeloid cells, B cells, and naive T cells, as well as by neurons. There are fewer B cells in circulation than normal and the condition is associated with hypogammaglobulinemia, although patients are able to mount an almost normal antibody response upon immunization. Patients affected by this condition are predisposed to infections with chronic papilloma virus, as evidenced by large numbers of skin and cervical warts.

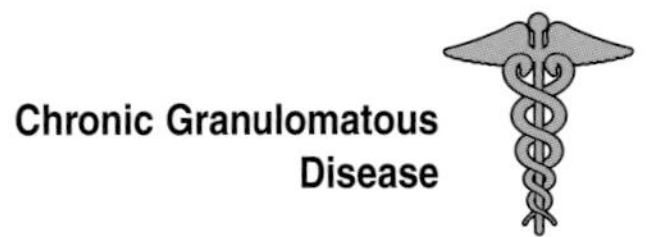

Most of the other known defects in phagocytic cells affect their ability to kill intracellular bacteria or ingest extracellular bacteria (Fig. 12.13). Patients with **chronic granulomatous disease** are highly susceptible to bacterial and fungal infections and form granulomas as a result of an inability to kill bacteria ingested by phagocytes (see Fig. 8.44). The defect in this case is in the production of reactive oxygen species (ROS) such as the superoxide anion (see Section 2-4). Discovering the molecular defect in this disease gave weight to

Fig. 12.13 Defects in phagocytic cells are associated with persistence of bacterial infection. Defects in the leukocyte integrins with a common β_2 subunit (CD18) or defects in the selectin ligand, sialyl-Lewis^x, prevent phagocytic cell adhesion and migration to sites of infection (leukocyte adhesion deficiency). The respiratory burst is defective in chronic granulomatous disease, glucose-6-phosphate dehydrogenase (G6PD) deficiency, and myeloperoxidase deficiency. In chronic granulomatous disease, infections persist because macrophage activation is defective, leading to chronic stimulation of CD4 T cells and hence to granulomas. Vesicle fusion in phagocytes is defective in Chediak–Higashi syndrome. These diseases illustrate the critical role of phagocytes in removing and killing pathogenic bacteria.

Type of defect/name of syndrome	Associated infections or other diseases
Leukocyte adhesion deficiency	Widespread pyogenic bacterial infections
Chronic granulomatous disease	Intracellular and extracellular infection, granulomas
G6PD deficiency	Defective respiratory burst, chronic infection
Myeloperoxidase deficiency	Defective intracellular killing, chronic infection
Chediak–Higashi syndrome	Intracellular and extracellular infection, granulomas

the idea that these killed bacteria directly; this notion has since been challenged by the finding that the generation of ROS is not itself sufficient to kill target microorganisms. It is now thought that ROS causes an influx of K^+ ions into the phagocytic vacuole, increasing the pH to the optimal level for the action of microbicidal peptides and proteins, which are the key agents in killing the invading microorganism.

Several different genetic defects, affecting any one of the four constituent proteins of the NADPH oxidase expressed in neutrophils and monocytes (see Section 2-4), can cause chronic granulomatous disease. Patients with the disease have chronic bacterial infections, which in some cases lead to the formation of granulomas. Deficiencies in glucose-6-phosphate dehydrogenase and myeloperoxidase also impair intracellular bacterial killing and lead to a similar, although less severe, phenotype. Finally, in **Chediak–Higashi syndrome**, a complex syndrome characterized by partial albinism, abnormal platelet function, and severe immunodeficiency, a defect in a protein called CSH1, which is involved in intracellular vesicle formation and trafficking, causes a failure of lysosomes and phagosomes to fuse properly; the phagocytes in these patients have enlarged granules and impaired intracellular killing ability. This defect also impairs the general secretory pathway; the consequences of this are described in Section 12-19.

12-13 Defects in T-cell differentiation can result in severe combined immunodeficiencies.

Although patients with B-cell defects can deal with many pathogens adequately, patients with defects in T-cell development are highly susceptible to a broad range of infectious agents. This demonstrates the central role of T-cell differentiation and maturation in adaptive immune responses to virtually all antigens. Because such patients make neither specific T-cell dependent antibody responses nor cell-mediated immune responses, and thus cannot develop immunological memory, they are said to suffer from **severe combined immunodeficiency** (**SCID**).

Several genetic defects can lead to the SCID phenotype. A common denominator in all children with SCID is an intrinsic arrest of T-cell differentiation, often associated with defective differentiation of B cells and in some genetic types deficiencies in NK cells as well. The affected children suffer from severe opportunistic infections such as adenoviruses, Epstein–Barr virus, *Candida albicans*, and *P. carinii*, and they normally die in their first year of life unless given antibody and a bone marrow transplant. Figure 12.14 lists the major causes of SCID.

X-linked SCID (**XSCID**) is the most frequent form of SCID and is sometimes known as the 'bubble boy disease' after a boy with XSCID who lived in a protective bubble for more than a decade before he died after an unsuccessful bone marrow transplant. Patients with XSCID have a mutation in the gene for the interleukin-2 receptor (IL-2R) common gamma chain (γ_c). Several cytokine receptors, including those for IL-2, IL-4, IL-7, IL-9, IL-15, and IL-21, share γ_c, and thus all are defective in this type of SCID. As a result of this genetic defect, T cells and NK cells fail to develop normally, whereas B-cell numbers, but not function, are normal. A clinically and immunologically indistinguishable type of SCID is associated with an inactivating mutation in one of the proteins in the signaling pathway from the γ_c and other cytokine receptors, the kinase Jak3 (see Section 6-23). This mutation causes the development of abnormal T and NK cells, but the development of B cells is unaffected.

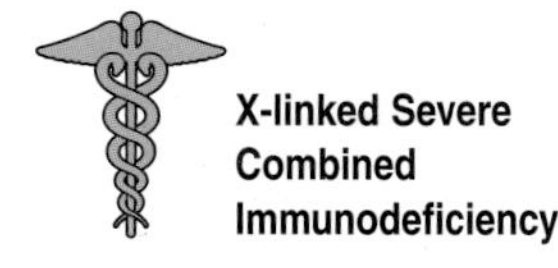

Other immunodeficiencies in humans and mice have unraveled some of the roles of individual cytokines and their receptors in T-cell and NK-cell development. For example, a child has been reported with SCID who lacked NK cells

Fig. 12.14 Severe combined immunodeficiency syndromes. The known causes of severe combined immunodeficiency (SCID) in humans and mice are listed. The defective gene, the cellular process affected, and the T-cell, B-cell, and NK-cell phenotype are given. ADA, adenosine deaminase.

Disease	Gene defect	Mechanism affected	Phenotype	
			Human	Mouse
XSCID	IL-2 receptor γ chain	Cytokine signaling	$T^-B^+NK^-$	$T^-B^-NK^-$
	JAK3	Cytokine signaling	$T^-B^+NK^-$	$T^-B^-NK^-$
	IL-7 receptor	Cytokine signaling	$T^-B^+NK^+$	$T^-B^-NK^+$
RAG deficiency Omenn syndrome	*RAG1*	Antigen receptor recombination	$T^-B^-NK^+$	$T^-B^-NK^+$
	RAG2	Antigen receptor recombination	$T^-B^-NK^+$	$T^-B^-NK^+$
	Artemis	Antigen receptor recombination	$T^-B^-NK^+$	$T^-B^-NK^+$
ADA deficiency	*ADA*	Metabolism	$T^-B^-NK^-$	$T^-B^-NK^-$

and T cells but had normal genes for γ_c and Jak3 kinase. It transpired that he had a deficiency of the common β chain, β_c, shared by the IL-2 and IL-15 receptors. This single child and mice with targeted mutations in the β-chain gene define a key role for IL-15 as a growth factor for the development of NK cells, as well as a role for IL-15 in T-cell maturation and trafficking. Mice with targeted mutations in IL-15 or the α chain of its receptor have no NK cells and relatively normal T-cell development, but they show reduced T-cell homing to peripheral lymphoid tissues and a reduction in the number of CD8-positive T cells.

Humans with a deficiency of the IL-7 receptor α chain have no T cells but normal levels of NK cells, illustrating that IL-7 signaling is essential for the development of T cells but not for that of NK cells. In humans and mice whose T cells show defective production of IL-2 after receptor stimulation, T-cell development itself is normal. The more limited effects of individual cytokine defects are in contrast to the global defects in T- and NK-cell development in patients with XSCID.

As in all serious T-cell deficiencies, patients with XSCID do not make effective antibody responses to most antigens, although their B cells seem normal. Because the gene defect is on the X chromosome, one can determine whether the lack of B-cell function is solely a consequence of the lack of T-cell help by examining X-chromosome inactivation in B cells of unaffected carriers (see Section 12-9). Most, but not all, naive IgM-positive B cells from female carriers of XSCID have inactivated the defective X chromosome rather than the normal one, showing that B-cell development is affected by, but is not wholly dependent on, the γ_c chain. Mature memory B cells that have undergone class switching have inactivated the defective X chromosome almost without exception. This might reflect the fact that the γ_c chain is also part of the receptors for IL-4 and IL-21. Thus, B cells that lack this chain will have defective IL-4 and IL-21 receptors and will not proliferate in T-cell dependent antibody responses (see Section 9-4).

A second type of autosomally inherited SCID is due to **adenosine deaminase (ADA) deficiency** and **purine nucleotide phosphorylase (PNP) deficiency**. These enzyme defects affect purine degradation, and both result in an accumulation of nucleotide metabolites that are particularly toxic to developing

T cells. B cells are also severely compromised in these patients, more so in ADA than in PNP deficiency.

12-14 Defects in antigen receptor gene rearrangement result in SCID.

A third set of defects leading to SCID are those that cause failures of DNA rearrangement in developing lymphocytes. For example, defects in either the *RAG-1* or *RAG-2* gene result in the arrest of lymphocyte development because of a failure to rearrange the antigen receptor genes. Thus there is a complete lack of T and B cells in mice with genetically engineered defects in the *RAG* genes, and in patients with autosomally inherited forms of SCID who lack a functional RAG protein. There are other children with mutations in either *RAG-1* or *RAG-2* who can nonetheless make a small amount of functional RAG protein, allowing a small amount of V(D)J recombination. This latter group suffers from a distinctive and severe disease called **Omenn's syndrome**, which, in addition to increased susceptibility to multiple opportunistic infections, has clinical features very similar to graft-versus-host disease (see Section 14-35), with rashes, eosinophilia, diarrhea, and enlargement of the lymph nodes. Normal or increased numbers of T cells, all of which are activated, are found in these unfortunate children. A possible explanation for this phenotype is that very low levels of *RAG* activity allow some limited T-cell receptor gene recombination. No B cells are found, however, and it may be that B cells have more stringent requirements for *RAG* activity. The T cells that are produced in patients with Omenn's syndrome have an abnormal and highly restricted receptor repertoire, both in the thymus and in the periphery, where they have undergone activation and clonal expansion. The clinical features strongly suggest that these peripheral T cells are autoreactive and are responsible for the graft-versus-host phenotype.

Another group of patients with autosomal SCID has a phenotype very similar to that of a mutant mouse strain called *scid*; *scid* mice have an abnormal sensitivity to ionizing radiation as well as having SCID. They produce very few mature B and T cells because there is a failure of DNA rearrangement in their developing lymphocytes; only rare VJ or VDJ joins are seen, and most of these are abnormal. The underlying defect in *scid* mice is in the enzyme DNA-dependent protein kinase (DNA-PK_{CS}), which is involved in antigen receptor gene rearrangement (see Section 4-5). A different mutation found in some people with autosomal SCID is in the protein Artemis, which acts in the same pathway as DNA-PK_{CS}. Artemis is an exonuclease that forms a complex with, and is activated by, DNA-PK_{CS}. The normal role of the Artemis:DNA-PK_{CS} complex is to open the hairpin structures to allow the formation of the VDJ joins, to complete the process of VDJ recombination.

Other defects in enzymes involved in DNA repair and recombination are associated with combinations of immunodeficiency, increased sensitivity to the damaging effects of ionizing radiation, and the development of cancer. One example is **Bloom's syndrome**, a disease caused by mutations in a DNA helicase, an enzyme that unwinds double-stranded DNA. Another is **ataxia telangiectasia** (**AT**), in which the underlying defect is in a protein called ATM (ataxia telangiectasia-mutated), which contains a kinase domain thought to be involved in intracellular signaling in response to DNA damage. The enzyme DNA ligase IV, which joins the DNA together in V(D)J recombination (see Section 4-5), is lacking in a small group of patients with a syndrome similar to ataxia telangiectasia, in which both V(D)J recombination and class switching are impaired. DNA ligase IV is a component of the general nonhomologous end-joining pathway of DNA repair and joins broken DNA in a variety of repair processes. Defective repair of DNA breaks also causes an increased susceptibility to cancer, in both lymphoid and other tissues.

12-15 Defects in signaling from T-cell antigen receptors can cause severe immunodeficiency.

Several gene defects have been described that interfere with signaling through T-cell receptors, and thus with the activation of T cells necessary for an adaptive immune response. Patients who lack CD3γ chains have low levels of T-cell receptors at the cell surface and have defective T-cell responses. Patients making low levels of mutant CD3ε chains are also deficient in T-cell activation. Although not strictly classifiable as SCID, patients who make a defective form of the cytosolic protein tyrosine kinase ZAP-70, which transmits signals from the T-cell receptor (see Section 6-11), have been described. CD4 T cells emerge from the thymus in normal numbers, whereas CD8 T cells are absent. However, the CD4 T cells that mature fail to respond to stimuli that normally activate the cells through the T-cell receptor, and patients with defective ZAP-70 are severely immunodeficient. Another lymphocyte signaling defect that leads to severe immunodeficiency is caused by mutations in the tyrosine phosphatase CD45. Humans and mice with CD45 deficiency show a marked reduction in peripheral T-cell numbers and abnormal B-cell maturation.

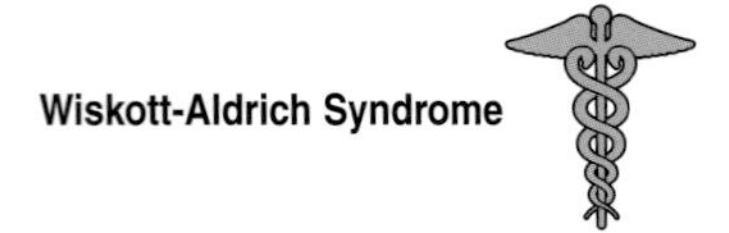

Wiskott–Aldrich syndrome (**WAS**) has shed new light on the molecular basis of signaling and immune synapse formation between a variety of cells in the immune system. The disease also affects platelets and was first described as a blood-clotting disorder, but it is also associated with immunodeficiency due to impaired lymphocyte function, leading to reduced T-cell numbers, defective NK-cell cytotoxicity, and a failure of antibody responses to encapsulated bacteria. WAS is caused by a defective gene on the X chromosome, encoding a protein called WAS protein (WASP). WASP is expressed in all hematopoietic cell lineages and is likely to be a key regulator of lymphocyte and platelet development and function through its effects on the actin cytoskeleton, which is critical for immune synapse formation and the polarization of effector T cells (see Section 8-22). It has also recently been suggested that WASP is required for the suppressive function of natural T_{reg} cells. WASP has a key role in transducing signals to the cytoskeletal framework of cells because it activates the Arp2/3 complex, which is essential for initiating actin polymerization. In patients with WAS, and in mice whose WASP gene has been knocked out, T cells fail to respond normally to mitogens or to the cross-linking of their T-cell receptors. Several signaling pathways leading from the T-cell receptor are known to activate WASP. One pathway involves the scaffold protein SLP-76, which serves as the binding site for an adaptor protein, Nck, which in turn binds WASP. WASP can also be activated by small GTP-binding proteins, notable Cdc42 and Rac1, which can themselves be activated via the T-cell receptor signaling through the adaptor protein Vav.

12-16 Genetic defects in thymic function that block T-cell development result in severe immunodeficiencies.

A disorder of thymic development, associated with SCID and a lack of body hair, has been known for many years in mice; the mutant strain is descriptively named ***nude*** (see Section 7-7). A small number of children have been described with the same phenotype. In both mice and humans this syndrome is caused by mutations in a gene called *FOXN1* (also known as *WHN*), which is on chromosome 17 in humans and encodes a transcription factor selectively expressed in skin and thymus. FOXN1 is necessary for the differentiation of thymic epithelium and the formation of a functional thymus. In patients with a mutation in *FOXN1,* the lack of thymic function prevents thymic-dependent T-cell development. In many cases, B-cell development is normal in individuals with the mutation, yet the response to nearly all pathogens is profoundly impaired because of the lack of T cells.

DiGeorge's syndrome is another disorder in which the thymic epithelium fails to develop normally, resulting in SCID. The genetic abnormality underlying this complex developmental disorder is a deletion within one copy of chromosome 22. The deletion varies between 1.5 and 5 megabases in extent, with the smallest deletion that causes the syndrome containing approximately 24 genes. The relevant gene within this interval is *TBX1*, which encodes a transcription factor, T-box 1. DiGeorge's syndrome is caused by the deletion of a single copy of this gene. Without the proper inductive thymic environment T cells cannot mature, and both T-cell dependent antibody production and cell-mediated immunity are impaired. Patients with this syndrome have normal levels of serum immunoglobulin but an absence of or incomplete development of the thymus and parathyroids, with varying degrees of T-cell immunodeficiency.

Defects in the expression of MHC molecules can lead to severe immunodeficiency as a result of effects on the positive selection of T cells in the thymus. Individuals with **bare lymphocyte syndrome** lack expression of all MHC class II molecules on their cells. Because the thymus lacks MHC class II molecules, CD4 T cells cannot be positively selected and few develop. The antigen-presenting cells in these individuals also lack MHC class II molecules and so the few CD4 T cells that do develop cannot be stimulated by antigen. MHC class I expression is normal, and CD8 T cells develop normally. However, such people suffer from severe immunodeficiency, illustrating the central importance of CD4 T cells in adaptive immunity to most pathogens.

MHC class II deficiency is caused not by mutations in the MHC genes themselves but by mutations in one of several genes encoding gene-regulatory proteins that are required for the transcriptional activation of MHC class II promoters. Four complementing gene defects (known as groups A, B, C, and D) have been defined in patients who fail to express MHC class II molecules, which suggests that the products of at least four different genes are required for the normal expression of these proteins. Genes corresponding to each complementation group have been identified: the *MHC class II transactivator*, or *CIITA*, is mutated in group A, and the genes *RFXANK*, *RFX5*, and *RFXAP* are mutated in groups B, C, and D, respectively. These last three encode proteins that are components of a multimeric complex, RFX, which is involved in controlling transcription. RFX binds a DNA sequence named an X-box, which is present in the promoter region of all MHC class II genes.

A more limited immunodeficiency, associated with chronic respiratory bacterial infections and skin ulceration with vasculitis, has been observed in a small number of patients who have almost no cell-surface MHC class I molecules—a condition known as **MHC class I deficiency**. Affected individuals have normal levels of mRNA encoding MHC class I molecules and normal production of MHC class I proteins, but very few of the proteins reach the cell surface. This defect is similar to that in the *TAP* mutant cells mentioned in Section 5-2, and mutations in either *TAP1* or *TAP2*, which encode the subunits of the peptide transporter, have been found in patients with MHC class I deficiency. By analogy with MHC class II deficiency, the absence of MHC class I molecules on the surfaces of thymic epithelial cells results in a lack of CD8 T cells expressing the $\alpha:\beta$ T-cell receptor, but patients do have $\gamma:\delta$ CD8 T cells, whose development is independent of the thymus. People with MHC class I deficiency are not abnormally susceptible to viral infections, which is surprising given the key role of MHC class I presentation and of cytotoxic CD8 $\alpha:\beta$ T cells in combating viral infections. There is, however, evidence for TAP-independent pathways for the presentation of certain peptides by MHC class I molecules, and the clinical phenotype of TAP1- and TAP2-deficient patients indicates that these pathways may be sufficient to allow viruses to be controlled.

12-17 The normal pathways for host defense against intracellular bacteria are pinpointed by genetic deficiencies of IFN-γ and IL-12 and their receptors.

A small number of families have been identified that contain several individuals who suffer from persistent and eventually fatal attacks by intracellular pathogens, especially mycobacteria and salmonellae. These people typically suffer from the ubiquitous, environmental, nontuberculous strains of mycobacteria, such as *Mycobacterium avium*. They may also develop disseminated infection after vaccination with *Mycobacterium bovis* bacillus Calmette–Guérin (BCG), the strain of *M. bovis* that is used as a live vaccine against *M. tuberculosis*. Susceptibility to these infections is conferred by a variety of mutations that abolish the function of any of the following: the cytokine IL-12; the IL-12 receptor; or the receptor for interferon (IFN)-γ and its signalling pathway. Mutations have been found in the p40 subunit of IL-12, the IL-12 receptor β_1 chain, and the two subunits (R1 and R2) of the IFN-γ receptor. p40 is shared by IL-12 and IL-23, and so p40 deficiency can cause both IL-12 and IL-23 deficiency. A mutation in STAT1, a protein in the signaling pathway activated after ligation of the IFN-γ receptor, is also associated in humans with increased susceptibility to mycobacterial infections. Similar susceptibility to intracellular bacterial infection is seen in mice with induced mutations in these same genes, and also in mice lacking tumor necrosis factor (TNF)-α or the TNF p55 receptor gene. Why tuberculosis itself is not seen more often in patients with these defects, especially since *M. tuberculosis* is more virulent than *M. avium* and *M. bovis*, remains unclear.

Mycobacteria and salmonellae enter dendritic cells and macrophages, where they can reproduce and multiply. At the same time they provoke an immune response that occurs in several stages and eventually controls the infection with the help of CD4 T cells. First, lipoproteins and peptidoglycans on the surface of the bacteria ligate receptors on macrophages and dendritic cells as they enter the cells. These receptors include the Toll-like receptors (TLRs) (see Section 2-7), particularly TLR-2, and the mannose receptor, and their ligation stimulates nitric oxide (NO) production within the cells, which is toxic to bacteria. Signaling by the TLRs stimulates the release of IL-12, which in turn drives NK cells to produce IFN-γ in the early phase of the immune response. IL-12 also stimulates antigen-specific CD4 T cells to release IFN-γ and TNF-α. These cytokines activate and recruit more macrophages to the site of infection, resulting in the formation of granulomas (see Section 8-33).

The key role of IFN-γ in activating macrophages to kill intracellular bacteria is dramatically illustrated by the failure to control infection in patients who are genetically deficient in either of the two subunits of its receptor. In the total absence of IFN-γ receptor expression, granuloma formation is much reduced, showing a role for this receptor in granuloma development. In contrast, if the underlying mutation is associated with the presence of low levels of functional receptor, granulomas form but the macrophages within them are not sufficiently activated to be able to control the division and spread of the mycobacteria. It is important to remember that this cascade of cytokine reactions is occurring in the context of cognate interactions between antigen-specific CD4 T cells and the macrophages and dendritic cells harboring the intracellular bacteria. T-cell receptor ligation and co-stimulation of the phagocyte by, for example, the interaction between CD40 and CD40 ligand, sends signals that help activate infected phagocytes to kill the intracellular bacteria.

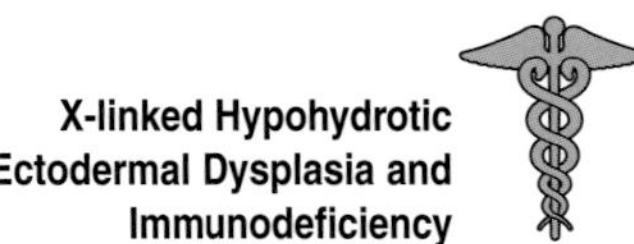

Unusual mycobacterial infections have been reported in several patients with NEMO deficiency (see Section 12-10). This leads to impaired NFκB activation and thus affects many cellular responses, including those to TLR ligands and TNF-α, which stimulate this signaling pathway. The conclusion to be drawn from these diseases is that pathways controlled by TLRs and NFκB seem to be important in immune responses against a collection of unrelated pathogens, whereas the IL-12/IL-23/IFN-γ pathway is especially important for immunity to mycobacteria and salmonellae but not to other pathogens.

12-18 X-linked lymphoproliferative syndrome is associated with fatal infection by Epstein–Barr virus and with the development of lymphomas.

The Epstein–Barr virus we encountered earlier in the chapter (see Section 12-2) can transform B lymphocytes and is used to immortalize clones of B cells in the laboratory. Transformation does not normally happen *in vivo* in humans because EBV infection is actively controlled and maintained in a latent state by cytotoxic T cells with specificity for B cells expressing EBV antigens. In the presence of T-cell immunodeficiency, however, this control mechanism can break down and a potentially lethal B-cell lymphoma may develop. One of the situations in which this occurs is the rare immunodeficiency **X-linked lymphoproliferative syndrome**, which results from mutations in a gene named SH2-domain containing gene 1A (*SH2D1A*). This encodes a protein named SAP (an acronym for SLAM-associated protein; SLAM is itself an acronym for signaling lymphocyte activation molecule). Boys with SAP deficiency typically develop overwhelming EBV infection, and sometimes lymphomas, during childhood. EBV infection in this condition is usually fatal and is associated with uncontrolled inflammation and necrosis of the liver. Thus, SAP must have a vital, nonredundant role in the normal control of EBV infection.

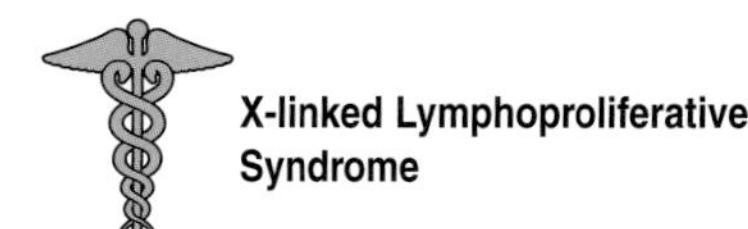

The function of SAP is partly understood. The SH2 domain of the protein interacts with the cytoplasmic tails of the transmembrane receptors SLAM and 2B4 (which are structurally homologous to each other), and to the T-cell adhesion molecule CD2. SLAM is expressed on activated T cells, whereas 2B4 is found on T cells, B cells, and NK cells. One function of SAP is to allow the recruitment of the tyrosine kinase FynT to these receptors. This activates an intracellular signaling cascade that inhibits the production of IFN-γ after T-cell receptor ligation while not affecting IL-2 production. In the absence of SAP, T cells produce increased amounts of IFN-γ, and this could lead to a skewing of immune responses toward the T_H1 cell subset. Boys with X-linked lymphoproliferative syndrome produce markedly more IFN-γ than normal in response to primary infection by EBV.

There are two hypotheses to explain the pathogenesis of the fatal EBV infection seen in children with defects in SAP. The first is that the failure of T cells to kill B cells expressing antigens from multiplying EBV allows uncontrolled infection. The second is that the abnormal cytokine response of T cells presented with EBV peptides by infected B cells causes severe inflammatory injury by mechanisms that we discuss in the next section. Some cases of lymphoma in young boys have now been found associated with mutations in the *SH2D1A* gene in the absence of any evidence of EBV infection. This raises the possibility that *SH2D1A* might be a tumor suppressor gene in its own right, in addition to controlling a virus that can contribute to tumor formation. In view of the fact that EBV persists in memory B cells (see Section 12-2), B-cell depletion has been successfully used to treat patients that have overwhelming EBV infection.

12-19 Genetic abnormalities in the secretory cytotoxic pathway of lymphocytes cause uncontrolled lymphoproliferation and inflammatory responses to viral infections.

A small group of inherited immunodeficiency diseases also affect skin pigmentation, causing albinism. The link between these two apparently unrelated phenotypes is a defect in the regulated secretion of lysosomes.

Many cell types derived from the bone marrow, including lymphocytes, granulocytes, and mast cells, have in common the property of regulated secretion of lysosomes. In response to specific stimuli they exocytose secretory lysosomes that contain specialized collections of proteins. Prominent among other cell types that show this regulated secretion of lysosomes are melanocytes, the pigment cells of the skin. The contents of the secretory lysosomes differ depending on cell type. In melanocytes, melanin is the major component, whereas in cytotoxic T cells, secretory lysosomes contain the cytolytic proteins perforin, granulysin, and granzymes (see Section 8-28). Although the contents of the granules differ between cell types, the fundamental mechanisms for their secretion do not, and this explains how inherited disorders affecting the regulated secretion of lysosomes can cause the combination of albinism and immunodeficiency.

We learned in Section 12-18 that X-linked lymphoproliferative syndrome is associated with uncontrolled inflammation in response to EBV infection. In that respect it is very similar to a group of diseases that cause a syndrome known as the **hemophagocytic syndrome** in which there is a dysregulated expansion of CD8 cytotoxic lymphocytes that is associated with macrophage activation. The clinical manifestations of the disease are due to an inflammatory response caused by an increased release of pro-inflammatory cytokines such as IFN-γ, TNF, IL-6, IL-10, and macrophage colony-stimulating factor (M-CSF). These mediators are secreted by activated T lymphocytes and macrophages that infiltrate all tissues, causing tissue necrosis and organ failure. The activated macrophages phagocytose blood cells, including erythrocytes and leukocytes, which gives the syndrome its name. Some of these hemophagocytic syndromes are inherited, and these can be classified into two types according to the nature of the gene defect. In the first type the effects of the mutation are confined to lymphocytes or other cells of the immune system because the mutated protein is located in the granules of NK and cytotoxic lymphocytes. In the second type, the genetic abnormality is located in the regulated secretory pathway of lysosomes and affects all cell types that use this pathway; in these cases albinism may also result.

An unpleasant disease named **familial hemophagocytic lymphohistiocytosis** (**FHL**) is caused by an inherited deficiency of the cytotoxic protein perforin. This is a lymphocyte-specific disorder, in which polyclonal CD8-positive T cells accumulate in lymphoid tissue and other organs, in association with activated hemophagocytic macrophages. The progressive inflammation is lethal unless checked by immunosuppressive therapy. In mice that lack perforin, no immediate defect is observed, but when the mice are infected with lymphocytic choriomeningitis virus (LCV) or other viruses, a disease resembling human FHL develops, driven by an uncontrolled virus-specific T-cell response. This rare syndrome beautifully demonstrates a role for CD8-positive lymphocytes in limiting T-cell immune responses, for example in response to viral infection, by perforin-dependent cytotoxic mechanisms. When this mechanism fails, uncontrolled activated T cells kill their host. Perforin is also critical for NK-cell cytotoxicity, which is impaired in FHL.

Examples of inherited diseases that affect the regulated secretion of lysosomes are Chediak–Higashi syndrome, caused by mutations in a protein, CHS1, that regulates lysosomal trafficking, and **Griscelli syndrome**, caused by mutations in a small GTPase, Rab27a, that controls the movement of vesicles within cells. Two other types of Griscelli syndrome have been identified, in which patients have pigmentary changes only and no immunological deficiency. In Chediak–Higashi syndrome, abnormal giant lysosomes accumulate in melanocytes, neutrophils, lymphocytes, eosinophils, and platelets. The hair is typically a metallic silver color, vision is poor because of abnormalities in retinal pigment cells, and platelet dysfunction causes increased bleeding. Children with the syndrome suffer from recurrent severe infections because of a failure of T-cell, neutrophil, and NK-cell function. After a few years, hemophagocytic lymphohistiocytosis develops, with fatal consequences if untreated. Antibiotics are needed to treat and prevent infections, and immunosuppression to deal with the uncontrolled inflammation; only bone marrow transplantation offers any real hope to patients with Chediak–Higashi disease.

12-20 Bone marrow transplantation or gene therapy can be useful to correct genetic defects.

It is frequently possible to correct the defects in lymphocyte development that lead to SCID and some other immunodeficiency phenotypes by replacing the defective component, generally by bone marrow transplantation. The major difficulties in these therapies result from MHC polymorphism. To be useful, the graft must share some MHC alleles with the host. As we learned in Section 7-15, the MHC alleles expressed by the thymic epithelium determine which T cells can be positively selected. When bone marrow cells are used to restore immune function to individuals with a normal thymic stroma, both the T cells and the antigen-presenting cells are derived from the graft. Therefore, unless the graft shares at least some MHC alleles with the recipient, the T cells that are selected on host thymic epithelium cannot be activated by graft-derived antigen-presenting cells (Fig. 12.15). There is also a danger that mature, post-thymic T cells in donor bone marrow might recognize the host as foreign and attack it, causing **graft-versus-host disease** (**GVHD**) (Fig. 12.16, top panel). This can be overcome by depleting the donor bone marrow of mature T cells. Bone marrow recipients are usually treated with irradiation that kills their own lymphocytes, thus making space for the grafted bone marrow cells and minimizing the threat of **host-versus-graft disease** (**HVGD**) (Fig. 12.16, third panel). In patients with the SCID phenotype, however, there is little problem with the host response to the transplanted bone marrow because the patient is immunodeficient.

Now that specific gene defects are being identified, a different approach to correcting these inherited immune deficiencies can be attempted. The strategy involves extracting a sample of the patient's own bone marrow cells, inserting a normal copy of the defective gene with the use of a retrovirus-derived construct, and returning them to the patient by transfusion. This approach, called **somatic gene therapy**, should correct the gene defect. Moreover, in immunodeficient patients, it might be possible to reinfuse the bone marrow into the patient without the usual irradiation used to suppress the recipient's bone marrow function. This approach has been used successfully to treat X-linked SCID and ADA deficiency. However, because most lymphocytes divide regularly, thus diluting out the new gene, the treatment had to be repeated regularly.

Sadly, this success has been followed by a major setback: two of the nine children whose immunodeficiency was corrected by this gene therapy developed

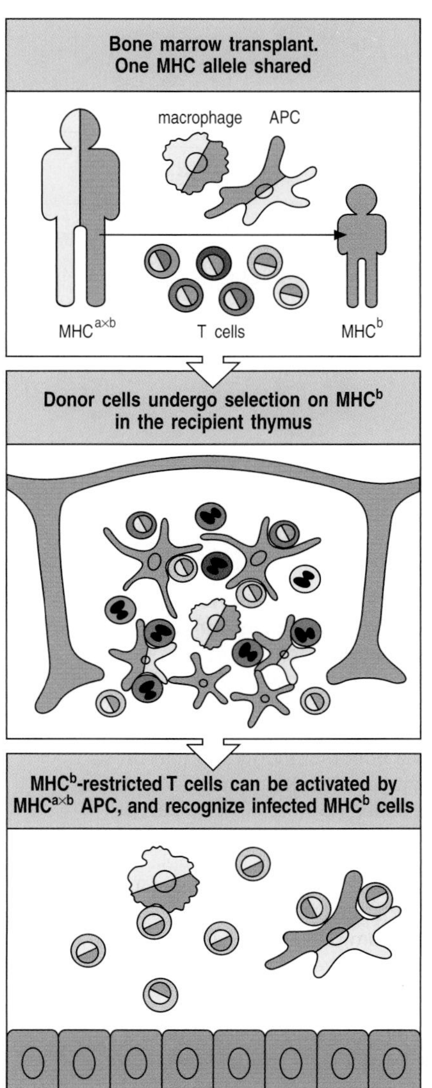

Fig. 12.15 Bone marrow donor and recipient must share at least some MHC molecules to restore immune function. A bone marrow transplant from a genetically different donor is illustrated in which the donor marrow cells share some MHC molecules with the recipient. The shared MHC type is designated 'b' and illustrated in blue; the MHC type of the donor marrow that is not shared is designated 'a' and shown in yellow. In the recipient, developing donor lymphocytes are positively selected on MHC^b on thymic epithelial cells and negatively selected by the recipient's stromal epithelial cells and at the cortico-medullary junction by encounter with dendritic cells derived from both the donor bone marrow and residual recipient dendritic cells. The negatively selected cells are shown as apoptotic cells. The donor-derived antigen-presenting cells (APCs) in the periphery can activate T cells that recognize MHC^b molecules; the activated T cells can then recognize the recipient's infected MHC^b-bearing cells.

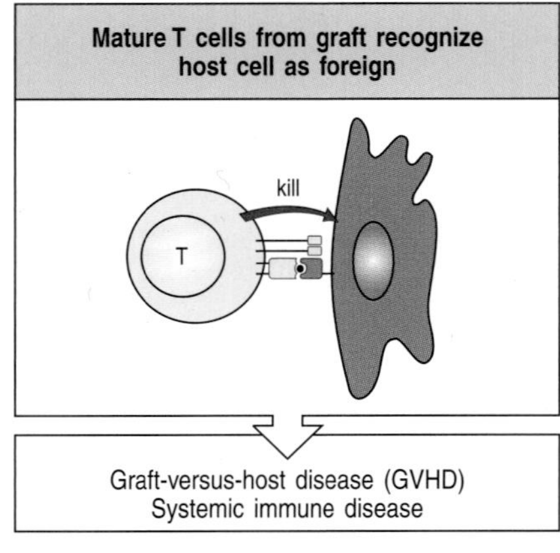
Mature T cells from graft recognize host cell as foreign
kill
T
Graft-versus-host disease (GVHD)
Systemic immune disease

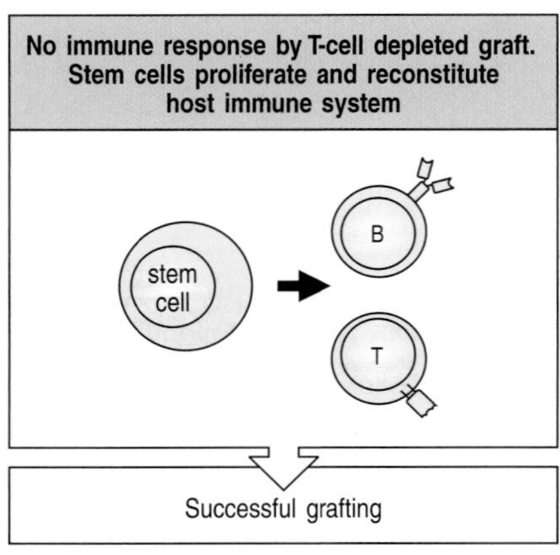
No immune response by T-cell depleted graft. Stem cells proliferate and reconstitute host immune system
stem cell
B
T
Successful grafting

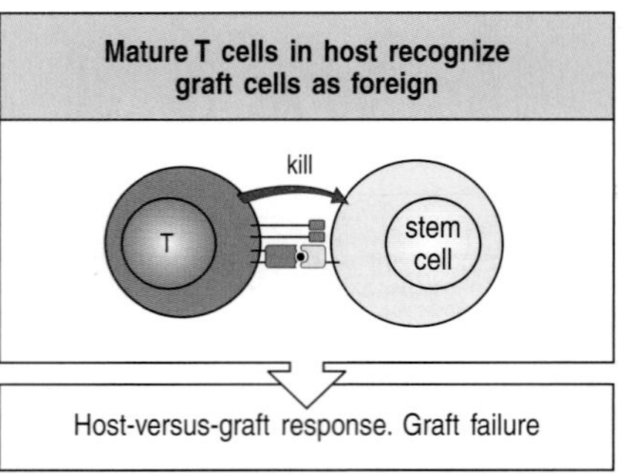
Mature T cells in host recognize graft cells as foreign
kill
T
stem cell
Host-versus-graft response. Graft failure

Fig. 12.16 Bone marrow grafting can be used to correct immunodeficiencies caused by defects in lymphocyte maturation, but two problems can arise. First, if there are mature T cells in the bone marrow, they can attack cells of the host by recognizing their MHC antigens, causing graft-versus-host disease (top panel). This can be prevented by T-cell depletion of the donor bone marrow (center panel). Second, if the recipient has competent T cells, these can attack the bone marrow stem cells (bottom panel). This causes failure of the graft by the usual mechanism of transplant rejection (see Chapter 13).

a T-cell tumor. This was caused by integration of the retroviral vector used for the gene therapy close to the promoter for a proto-oncogene called *LMO2*, a gene that regulates hematopoiesis.

12-21 Secondary immunodeficiencies are major predisposing causes of infection and death.

The primary immunodeficiencies have taught us much about the biology of specific proteins of the immune system. Fortunately, these conditions are rare. In contrast, secondary immunodeficiency is extremely common and important in everyday medical practice. Malnutrition devastates many populations around the world, and a major feature of malnutrition is secondary immunodeficiency. This particularly affects cell-mediated immunity, and death in famines is frequently caused by infection. Measles, which itself causes immunosuppression (see Section 12-4), is an important cause of death in malnourished children. In the developed world, measles is an unpleasant illness but major complications are uncommon. In contrast, measles in the malnourished has a high mortality. Tuberculosis is another important infection in the malnourished. In mice, protein starvation causes immunodeficiency by affecting antigen-presenting cell function, but in humans it is not clear how malnourishment specifically affects immune responses. Links between the endocrine and immune systems may provide part of the answer. Adipocytes (fat cells) produce the hormone leptin, and leptin levels are related directly to the amount of fat present in the body; leptin levels fall in starvation. Both mice and humans with genetic leptin deficiency have reduced T-cell responses, and in mice the thymus atrophies. In both starved mice and those with inherited leptin deficiency these abnormalities can be reversed by the administration of leptin.

Secondary immunodeficiency states are also associated with hematopoietic tumors such as leukemia and lymphomas. Depending on the specific type, leukemia may be associated with an excess of neutrophils (neutrophilia) or a deficiency (neutropenia). In either case, neutrophil dysfunction increases susceptibility to bacterial and fungal infections, as described in Section 12-12. Destruction or invasion of peripheral lymphoid tissue by lymphomas or metastases from other cancers can promote opportunistic infections. Surgical removal of the spleen, or destruction of spleen function by certain diseases, is associated with a lifelong predisposition to overwhelming infection by *S. pneumoniae*, graphically illustrating the role of mononuclear phagocytic cells within the spleen in the clearance of this organism from blood. Patients who have lost spleen function should be vaccinated against pneumococcal infection and are often recommended to take prophylactic antibiotics throughout their life.

Unfortunately, a major complication of the cytotoxic drugs used to treat cancer is immunosuppression and increased susceptibility to infection. These drugs kill all dividing cells, and cells of the bone marrow and lymphoid systems are the main unwanted targets of these agents. Infection is thus one of the major side effects of cytotoxic drug therapy. This is also the case when these and similar drugs are used therapeutically as immunosuppressants. Another undesirable side-effect of medical intervention is the increased risk of infection around implanted medical devices such as catheters, artificial heart valves, and artificial joints. These act as privileged sites for the development of infections that resist easy elimination by antibiotics. These implanted materials lack the innate defensive mechanisms of normal body tissues and act as a 'protected' matrix for the growth of bacteria and fungi. Catheters used for peritoneal dialysis or for the infusion of drugs or fluids into the circulation can also act as a conduit for bacteria to bypass the normal defensive barrier of the skin.

Summary.

Genetic defects can occur in almost any molecule involved in the immune response. These defects give rise to characteristic deficiency diseases, which, although rare, provide much information about the development and functioning of the immune system in normal humans. Inherited immunodeficiencies illustrate the vital role of the adaptive immune response and T cells in particular, without which both cell-mediated and humoral immunity fails. They have provided information about the separate roles of B lymphocytes in humoral immunity and of T lymphocytes in cell-mediated immunity, the importance of phagocytes and complement in humoral and innate immunity, and the specific functions of several cell-surface or signaling molecules in the adaptive immune response. There are also some inherited immune disorders whose causes we still do not understand. The study of these diseases will undoubtedly teach us more about the normal immune response and its control. Acquired defects in the immune system, the secondary immunodeficiencies, are much commoner than the primary inherited immunodeficiencies, and starvation is a major cause of immunodeficiency and death worldwide. In the next section we consider the pandemic of acquired immune deficiency syndrome caused by infection with the HIV virus.

Acquired immune deficiency syndrome.

The most extreme case of immune suppression caused by a pathogen is the **acquired immune deficiency syndrome** (**AIDS**) caused by infection with the **human immunodeficiency virus** (**HIV**). HIV infection leads to a gradual loss of immune competence, allowing infection with organisms that are not normally pathogenic. The earliest documented case of HIV infection in humans was reported in a sample of serum from Kinshasa (Democratic Republic of Congo) that was stored in 1959. It was not until 1981, however, that the first cases of AIDS were officially reported. The disease is characterized by a susceptibility to infection with opportunistic pathogens or by the occurrence of an aggressive form of Kaposi's sarcoma or B-cell lymphoma, accompanied by a profound decrease in the number of CD4 T cells.

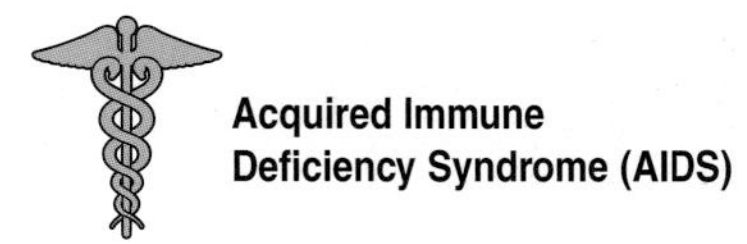

Because the disease seemed to be spread by contact with body fluids, the cause was suspected to be a new virus, and by 1983 the virus responsible, HIV, was isolated and identified. It is now clear that there are at least two types of HIV—HIV-1 and HIV-2—which are closely related. HIV-2 is endemic in West Africa and is now spreading in India. Most AIDS worldwide is, however, caused by the more virulent HIV-1. Both viruses seem to have spread to humans from other primate species, and the best evidence from nucleotide sequence relationships suggests that HIV-1 has passed to humans on at least three independent occasions from the chimpanzee, *Pan troglodytes*, whereas HIV-2 originated in the sooty mangabey, *Cercocebus atys*.

HIV-1 shows marked genetic variability and is classified on the basis of nucleotide sequence into three major groups named M (main), O (outlier), and N (non-M, non-O). These are only distantly related to each other and are thought to have passed into humans by independent transmissions from chimpanzees. The M group is the major cause of AIDS worldwide and is genetically diversified into subtypes, sometimes known as clades, that are designated by letters A to K; in different parts of the world different subtypes predominate. The ancestry of the M group of HIV-1 to a common parent has been traced by phylogenetic analysis and it is most likely that the subtypes of the M virus have evolved within humans after the original transmission of the virus from a chimpanzee, which harbors the related simian immunodeficiency virus (SIV_{cpz}). It has been estimated that the common ancestor of the M group may date back as far as 1915–1941; if correct, this means that HIV-1 has been infecting humans in central Africa for far longer than had been thought.

HIV infection does not immediately cause AIDS, and the mechanisms of how it does, and whether all HIV-infected patients will progress to overt disease, are incompletely understood. Nevertheless, accumulating evidence clearly implicates the growth of the virus in CD4 T cells, and the immune response to it, as the central keys to the puzzle of AIDS. HIV is now a worldwide pandemic, and although great strides are being made in understanding the pathogenesis and epidemiology of the disease, the number of infected people around the world continues to grow at an alarming rate, presaging the death of many people from AIDS for many years to come. Estimates from the World Health Organization are that more than 25 million people have died from AIDS since the beginning of the epidemic, and that there are currently around 44 million people alive with HIV infection (Fig. 12.17). Most of these are living in sub-Saharan Africa, where approximately 7.4% of young adults are infected. In some countries within this region, such as Zimbabwe and Botswana, over 25% of adults are infected. There are growing epidemics of HIV infection and AIDS in China and in India, where surveys in several states have shown a 1–2% prevalence of HIV infection in pregnant women. The incidence of HIV infection is rising faster in Eastern Europe and Central Asia than in the rest of the world. About one-third of those infected with HIV are aged between 15 and 24 years, and the majority are unaware that they carry the virus.

12-22 Most individuals infected with HIV progress over time to AIDS.

Many viruses cause an acute but limited infection that induces lasting protective immunity. Others, such as herpes viruses, set up a latent infection that is not eliminated but is adequately controlled by an adaptive immune response (see Section 12-2). Infection with HIV, however, seems rarely to lead to an immune response that can prevent ongoing replication of the virus. Although the initial acute infection does seem to be controlled by the immune system, HIV continues to replicate and infect new cells.

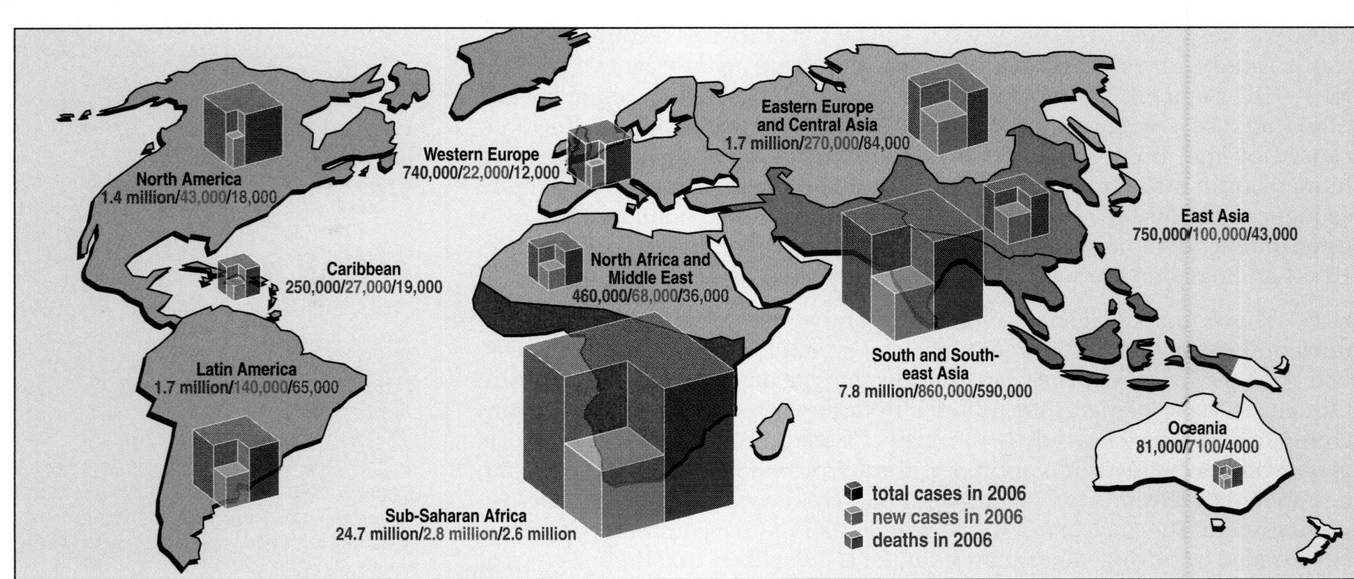

Fig. 12.17 HIV infection is spreading on all continents. The number of HIV-infected individuals is large and is increasing. Worldwide in 2006 there were around 40 million individuals infected with HIV, including some 5 million new cases, and more than 3 million deaths from AIDS. Data are estimated numbers of adults and children living with HIV/AIDS at the end of 2006 (*AIDS Epidemic Update* UNAIDS/World Health Organization; 2006).

Infection with HIV generally occurs after the transfer of body fluids from an infected person to an uninfected one. HIV infection is most commonly spread by sexual intercourse, contaminated needles used for intravenous drug delivery, and the therapeutic use of infected blood or blood products, although this last route of transmission has largely been eliminated in the developed world, where blood products are screened routinely for the presence of HIV. An important route of virus transmission is from an infected mother to her baby at birth or through breast milk. Rates of transmission from an infected mother to a child range from as low as 11% to as high as 60% depending on the severity of the infection (as judged by the viral load) and the frequency of breastfeeding. Antiviral drugs such as zidovudine (AZT) or nevirapine administered during pregnancy significantly reduce the amount of virus passed to the newborn infant, thus reducing the frequency of transmission.

The virus is carried mainly in infected cells that express CD4, which acts as the receptor for the virus, along with a co-receptor, usually the chemokine receptors CCR5 or CXCR4, or as a free virus in blood, semen, vaginal fluid, or mother's milk. The gastrointestinal and genital mucosae are the dominant sites of primary infection. The virus actively multiplies and spreads in the lymphoid compartment of mucosal tissues; this is followed by systemic infection of other peripheral lymphoid organs.

The **acute phase** is clinically characterized by an influenza-like illness in up to 80% of cases, with an abundance of virus (viremia) in the peripheral blood and a marked decrease in the numbers of circulating CD4 T cells. The diagnosis at this stage is usually missed unless there is a high index of suspicion. This acute viremia is associated in virtually all patients with the activation of CD8 T cells, which kill HIV-infected cells, and subsequently with antibody production, or **seroconversion**. The cytotoxic T-cell response is thought to be important in controlling virus levels, which peak and then decline, as the CD4 T-cell counts rebound to around 800 cells μl^{-1} (the normal value is around 1200 cells μl^{-1}).

By 3–4 months after infection, the symptoms of acute viremia have usually passed. The level of virus that persists in the blood plasma at this stage of infection is usually the best indicator of future disease progression. Almost all people who are infected with HIV will eventually develop AIDS, after a period of apparent quiescence of the disease known as clinical latency or the **asymptomatic period** (Fig. 12.18). This period is not silent, however, for there is persistent replication of the virus, and a gradual decline in the function and numbers of CD4 T cells until eventually such patients have few CD4 T cells left. At this point, which can occur anywhere between 6 months and 20 years

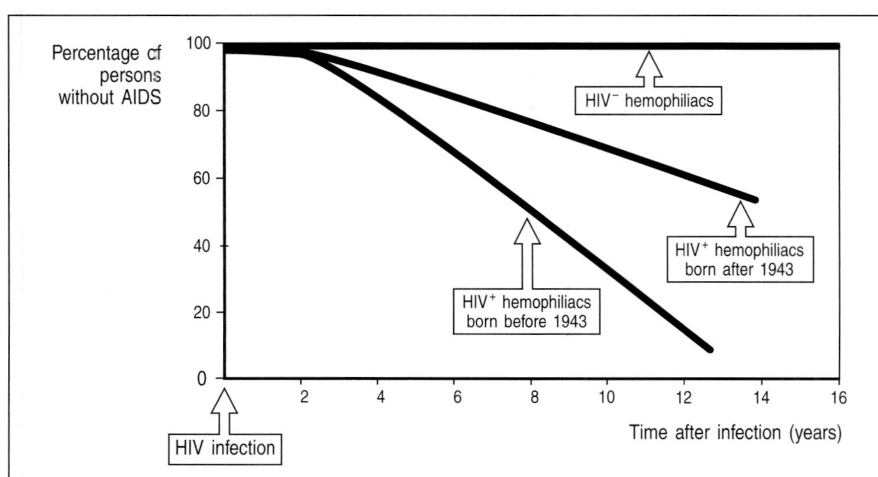

Fig. 12.18 Most HIV-infected individuals progress to AIDS over a period of years. The incidence of AIDS increases progressively with time after infection. Men who have sex with men (MSM) and hemophiliacs are two of the groups at highest risk in the West—MSM from sexually transmitted virus, and hemophiliacs from infected human blood used to replace clotting factor VIII. In Africa, spread is mainly by heterosexual intercourse. Hemophiliacs are now protected by the screening of blood products and the use of recombinant factor VIII. Neither MSM nor hemophiliacs who have not been infected with HIV show any evidence of AIDS. The majority of hemophiliacs contracted HIV by contaminated blood. Disease progression to AIDS is depicted here. The age of the individual seems to have a significant role in the rate of progression of the development of HIV. More than 80% of those aged more than 40 at the time of infection progress to AIDS over 13 years, in comparison with approximately 50% of those aged less than 40 over a comparable time. There are a few individuals who, although infected with HIV, seem not to progress to develop AIDS.

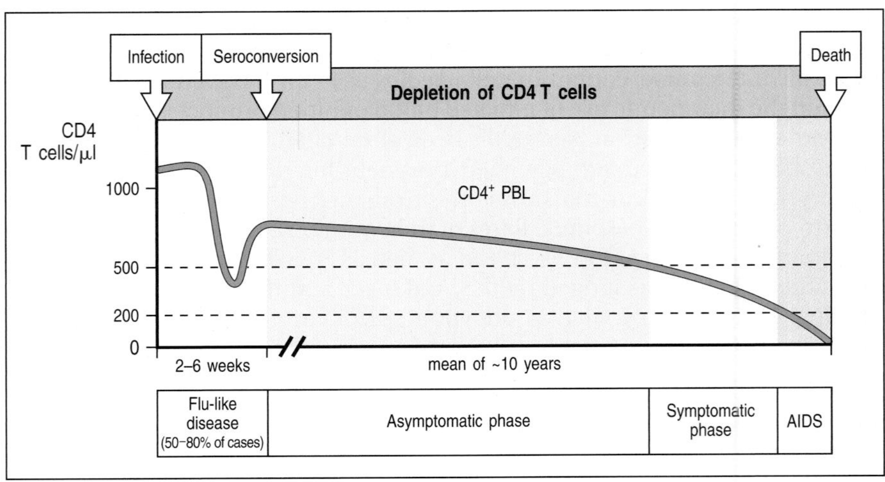

Fig. 12.19 The typical course of untreated infection with HIV. The first few weeks are typified by an acute influenza-like viral illness, sometimes called seroconversion disease, with high titers of virus in the blood. An adaptive immune response follows, which controls the acute illness and largely restores levels of CD4 T cells ($CD4^+$ PBL) but does not eradicate the virus. Opportunistic infections and other symptoms become more frequent as the CD4 T-cell count falls, starting at about 500 cells/μl^{-1}. The disease then enters the symptomatic phase. When CD4 T-cell counts fall below 200 cells/μl^{-1}, the patient is said to have AIDS. Note that CD4 T-cell counts are measured for clinical purposes in cells per microliter (cells/μl^{-1}), rather than cells per milliliter (cells/ml^{-1}), the unit used elsewhere in this book.

or more after the primary infection, the period of clinical latency ends and opportunistic infections begin to appear.

There are at least three dominant mechanisms for the loss of CD4 T cells in HIV infection. First, there is evidence for direct viral killing of infected cells; second, there is increased susceptibility to the induction of apoptosis in infected cells; and third, there is killing of infected CD4 T cells by CD8 cytotoxic lymphocytes that recognize viral peptides. In addition, regeneration of new T cells is also defective in infected people, suggesting infection and destruction of the progenitors of CD4 T cells within the thymus. This may also provide an explanation for the rapid progression of disease in infants.

The typical course of an infection with HIV is illustrated in Fig. 12.19. However, it has become increasingly clear that the course of the disease can vary widely. Thus, although most people infected with HIV go on to develop AIDS and ultimately die of opportunistic infections or cancer, this is not true of all individuals. A small percentage of people seroconvert, making antibodies against many HIV proteins, but do not seem to have progressive disease, in that their CD4 T cell counts and other measures of immune competence are maintained. These long-term non-progressors have unusually low levels of circulating virus and are being studied intensively to discover how they are able to control their HIV infection. A second group consists of seronegative people who have been highly exposed to HIV yet remain disease-free and virus-negative. Some of these people have specific cytotoxic lymphocytes and T_H1 lymphocytes directed against infected cells, which confirms that they have been exposed to HIV or possibly to noninfectious HIV antigens. It is not clear whether this immune response accounts for clearing the infection, but it is of considerable interest for the development and design of vaccines.

12-23 HIV is a retrovirus that infects CD4 T cells, dendritic cells, and macrophages.

HIV is an enveloped retrovirus whose structure is shown in Fig. 12.20. Each virus particle, or virion, contains two copies of an RNA genome and numerous copies of essential enzymes that are required for the initial steps of infection and genome replication, before new viral proteins are produced. The viral genome is reverse transcribed into DNA in the infected cell by the viral **reverse transcriptase**, and the DNA is integrated into the host-cell chromosomes with the aid of the viral **integrase**. RNA transcripts are produced from the integrated viral DNA and serve both as mRNAs to direct the synthesis of viral proteins and later as the RNA genomes of new viral particles. These escape from the cell by budding from the plasma membrane, each enclosed in a membrane envelope. HIV belongs to a group of retroviruses called the **lentiviruses**, from the Latin *lentus*, meaning slow, because of the gradual course of the diseases that they cause. These viruses persist and continue to replicate for many years before causing overt signs of disease.

The ability of HIV to enter particular cell types, known as its cellular **tropism**, is determined by the expression of specific receptors for the virus on the surfaces of those cells. HIV enters cells by means of a complex of two non-covalently associated viral glycoproteins, gp120 and gp41, in the viral envelope. The gp120 portion of the glycoprotein complex binds with high affinity to the cell-surface molecule CD4. The virus thus binds to CD4 T cells and to dendritic cells and macrophages, which also express some CD4. Before fusion and entry of the virus, gp120 must also bind to a co-receptor in the membrane of the host cell. Several chemokine receptors can serve as co-receptors for HIV entry. The major co-receptors are CCR5, which is predominantly expressed on dendritic cells, macrophages, and CD4 T cells, and CXCR4, which is expressed on activated T cells. After binding of gp120 to the receptor and co-receptor, gp41 then causes fusion of the viral envelope with the cell's plasma membrane, allowing the viral genome and associated viral proteins to enter the cytoplasm. This fusion process has provided a target for drug therapy. Peptide analogs of the carboxy-terminal peptide of gp41 inhibit fusion of the viral envelope and plasma membrane, and administration of one such peptide, called T-20, to patients with HIV infection caused an approximately 20-fold decline in HIV RNA levels in plasma.

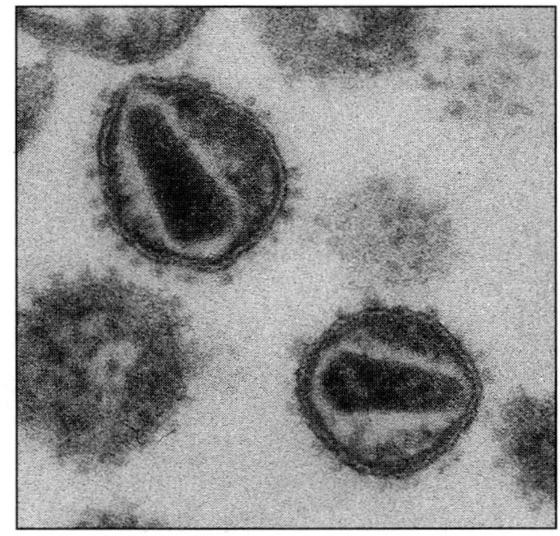

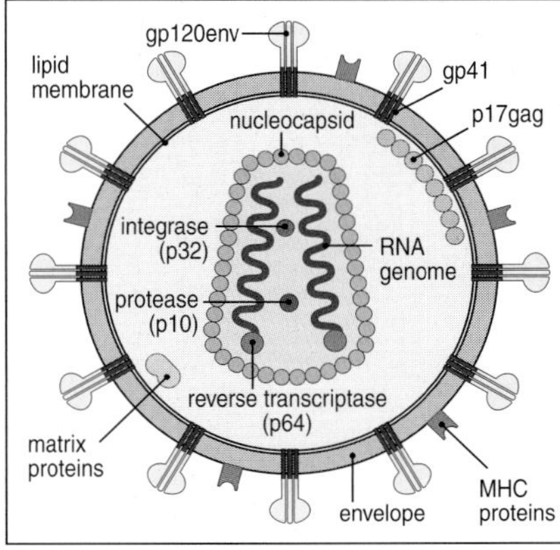

Fig. 12.20 The virion of human immunodeficiency virus (HIV). The virus illustrated is HIV-1, the leading cause of AIDS. The reverse transcriptase, integrase, and viral protease enzymes are packaged in the virion and are shown schematically in the viral capsid. In reality, many molecules of these enzymes are contained in each virion. Photograph courtesy of H. Gelderblom.

HIV mutates rapidly within the course of its replication in the body. This gives rise to many different variants in a single infection and also within the population as a whole. Different variants infect different cell types, and their tropism is determined to a large degree by which chemokine receptor the virus uses as co-receptor. The co-receptor used by variants of HIV that are associated with primary infections is CCR5, which binds the CC chemokines CCL3, CCL4, and CCL5, and these variants require only a low level of CD4 on the cells they infect. The variants of HIV that use CCR5 infect dendritic cells, macrophages, and T cells *in vivo*, and are now usually designated as 'R5' viruses, reflecting their chemokine receptor usage. In contrast, 'X4' viruses preferentially infect CD4 T cells and use CXCR4 (the receptor for chemokine CXCL12) as a co-receptor.

It seems that R5 isolates of HIV are preferentially transmitted by sexual contact, because they are the dominant viral phenotype found in newly infected individuals. Virus is disseminated from an initial reservoir of infected dendritic cells and macrophages, and there is evidence for an important role for mucosal lymphoid tissue in this process. Mucosal epithelia, which are constantly exposed to foreign antigens, provide a milieu of immune system activity in which HIV can readily replicate. Infection occurs across two types of epithelia. The mucosa of the vagina, penis, cervix, and anus is covered by stratified squamous epithelium, which is an epithelium composed of several layers of cells. A second type of epithelium, composed of a single layer of cells, is present in the rectum and in the endocervix.

A complex ferrying mechanism seems to transfer HIV picked up by dendritic cells in squamous epithelia to CD4 T cells in lymphoid tissue. *In vitro* studies have shown that HIV attaches to monocyte-derived dendritic cells by the binding of viral gp120 to C-type lectin receptors such as langerin (CD207), the mannose receptor (CD206), and DC-SIGN. A portion of the bound virus is rapidly taken up into vacuoles, where it remains for days in an infectious state. In this way the virus is protected during the passage of dendritic cells to lymphoid tissue and remains stable until it encounters a susceptible CD4 T cell (Fig. 12.21). The existence of this transport mechanism confirms the

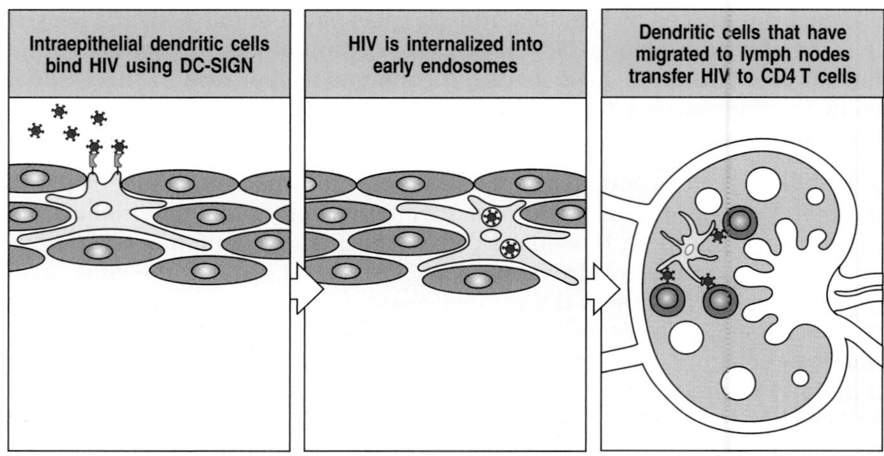

Fig. 12.21 Dendritic cells initiate infection by transporting HIV from mucosal surfaces to lymphoid tissue. HIV adheres to the surface of intraepithelial dendritic cells by the binding of viral gp120 to DC-SIGN (left panel). It gains access to dendritic cells at sites of mucosal injury or possibly to dendritic cells that have protruded between epithelial cells to sample the external world. Dendritic cells internalize HIV into mildly acidic early endosomes and migrate to lymphoid tissue (center panel). The HIV is translocated back to the cell surface, and when the dendritic cell encounters CD4 T cells in a secondary lymphoid tissue, the HIV is transferred to the T cell (right panel).

suggestion that HIV can either infect CD4 cells directly or via the immunological synapse formed between dendritic cells and CD4 T cells.

Epithelial cells from the single-layer epithelium covering rectum and endocervix express CCR5 and another HIV-binding molecule, glycosphingolipid galactosyl ceramide, and have been shown to selectively translocate R5 HIV variants, but not X4, through the epithelial monolayer, thus allowing HIV to bind to and infect submucosal CD4 T cells and dendritic cells. Infection of CD4 T cells via CCR5 occurs early in the course of infection and continues, with activated CD4 T cells accounting for the major production of HIV throughout infection. Late in infection, in approximately 50% of cases, the viral phenotype switches to the X4 type that infects T cells via CXCR4 co-receptors, and this is followed by a rapid decline in CD4 T-cell count and progression to AIDS.

12-24 Genetic variation in the host can alter the rate of progression of disease.

The rate of progression of HIV infection toward AIDS can be modified by the genetic makeup of the infected person. Genetic variation in HLA type is one factor: alleles HLA-B57 and HLA-B27 are associated with a better prognosis, and HLA-B35 with more rapid disease progression. Homozygosity of HLA class I (HLA-A, HLA-B and HLA-C) is associated with more rapid progression, presumably because the T-cell response to infection is less diverse. Certain polymorphisms of the killer cell immunoglobulin-like receptors (KIRs) present on NK cells (see Section 2-31), in particular the receptor KIR-3DS1 in combination with certain alleles of HLA-B, delay the progression to AIDS.

The clearest case of host genetic variation affecting HIV infection is a mutant allele of CCR5 that when homozygous effectively blocks HIV-1 infection, and when heterozygous slows AIDS progression. This is discussed in more detail in the next section. Mutations that affect the production of cytokines such as IL-10 and IFN-γ have also been implicated in the restriction of HIV progression. Genes that influence the progression to AIDS are listed in Fig. 12.22.

12-25 A genetic deficiency of the co-receptor CCR5 confers resistance to HIV infection *in vivo*.

Evidence for the importance of chemokine receptors in HIV infection has come from studies in a small group of individuals with a high risk of exposure to HIV-1 but who remain seronegative. Cultures of lymphocytes and macrophages from these people were relatively resistant to HIV infection and were found to secrete high levels of the chemokines CCL3, CCL4, and CCL5 in response to inoculation with HIV. The resistance of these rare individuals to HIV infection has now been explained by the discovery that they are homozygous for an allelic, nonfunctional variant of CCR5 called Δ32, which is caused by a 32-base-pair deletion from the coding region that leads

Fig. 12.22 Genes that influence progression to AIDS in humans. E, an effect that acts early in progression to AIDS; L, acts late in AIDS progression; ?, plausible mechanism of action with no direct support. Reprinted with permission from Macmillan Publishers Ltd: *Nat. Genet.* S.J O'Brien, G.W. Nelson, **36**: 565–574, © 2004.

Genes that influence progression to AIDS				
Gene	**Allele**	**Mode**	**Effect**	**Mechanism of action**
HIV entry				
CCR5	Δ32	Recessive	Prevents infection	Knockout of CCR5 expression
		Dominant	Prevents lymphoma (L)	Decreases available CCR5
			Delays AIDS	
	P1	Recessive	Accelerates AIDS (E)	Increases CCR5 expression
CCR2	I64	Dominant	Delays AIDS	Interacts with and reduces CXCR4
CCL5	In1.1c	Dominant	Accelerates AIDS	Decreases CCL5 expression
CXCL12	3´A	Recessive	Delays AIDS (L)	Impedes CCR5–CXCR4 transition (?)
CXCR6	E3K	Dominant	Accelerates *P. carinii* pneumonia (L)	Alters T-cell activations (?)
CCL2-CCL7-CCL11	H7	Dominant	Enhances infection	Stimulates immune resoponse (?)
Cytokine anti-HIV				
IL10	5´A	Dominant	Limits infection	Decreases IL-10 expression
			Accelerates AIDS	
IFN-G	–179T	Dominant	Accelerates AIDS (E)	
Acquired immunity, cell mediated				
HLA	A, B, C	Homozygous	Accelerates AIDS	Decreases breadth of HLA class I epitope recognition
	*B*27*	Codominant	Delays AIDS	Delays HIV-1 escape
	*B*57*			
	*B*35-Px*		Accelerates AIDS	Deflects CD8-mediated T-cell clearance of HIV-1
Acquired immunity, innate				
KIR3DS1	3DS1	Epistatic with HLA-Bw4	Delays AIDS	Clears HIV^+, HLA^- cells (?)

to a frameshift mutation and a truncated protein. The gene frequency of this mutant allele in caucasoid populations is quite high, at 0.09 (that is, about 10% of the population are heterozygous carriers of the allele and about 1% are homozygous). The mutant allele has not been found in Japanese or in black Africans from Western or Central Africa. Heterozygous deficiency of CCR5 might provide some protection against sexual transmission of HIV infection and a modest reduction in the rate of progression of the disease. In addition to the structural polymorphism of the gene, variation in the promoter region of the CCR5 gene has been found in both Caucasian and African Americans. Different promoter variants were associated with different rates of progression of disease.

These results provide dramatic confirmation that CCR5 is the major macrophage and T-lymphocyte co-receptor used by HIV to establish primary infection *in vivo*, and offers the possibility that primary infection might be blocked by antagonists of the CCR5 receptor. Indeed, there is preliminary evidence that low molecular weight inhibitors of this receptor can block infection of macrophages by HIV *in vitro*. Such inhibitors might be the precursors of useful drugs that could be taken by mouth to prevent infection. Such drugs are very unlikely to provide complete protection, however, because a very small number of individuals who are homozygous for the nonfunctional variant of CCR5 are infected with HIV. These individuals seem to have suffered from primary infection by X4 strains of the virus.

12-26 HIV RNA is transcribed by viral reverse transcriptase into DNA that integrates into the host-cell genome.

Once the virus has entered cells, it replicates in the same way as other retroviruses. One of the proteins carried in the virus particle is the viral reverse transcriptase, which transcribes the viral RNA into a complementary DNA (cDNA) copy. The viral cDNA is then integrated into the host-cell genome by the viral integrase, which also enters the cell with the viral RNA. The integrated cDNA copy is known as the **provirus**. The entire infectious cycle is shown in Fig. 12.23. In activated CD4 T cells, virus replication is initiated by transcription of the provirus, as we discuss in the next section. However, HIV can, like other retroviruses, establish a latent infection in which the provirus remains quiescent. This seems to occur in memory CD4 T cells and in dormant macrophages, and these cells are thought to be an important reservoir of infection.

The HIV genome consists of nine genes flanked by long terminal repeat sequences (LTR). The latter are required for the integration of the provirus into the host-cell DNA and contain binding sites for gene regulatory proteins that control the expression of the viral genes. Like other retroviruses, HIV has three major genes—*gag*, *pol*, and *env* (Fig. 12.24). The *gag* gene encodes the structural proteins of the viral core, *pol* encodes the enzymes involved in viral replication and integration, and *env* encodes the viral envelope glycoproteins. The *gag* and *pol* mRNAs are translated to give polyproteins—long polypeptide chains that are then cleaved by the **viral protease** (also encoded by *pol*) into individual functional proteins. The product of the *env* gene, gp160, has to be cleaved by a host-cell protease into gp120 and gp41, which are then assembled as trimers into the viral envelope. As shown in Fig. 12.24, HIV has six other, smaller, genes encoding proteins that affect viral replication and infectivity in various ways. Two of these, Tat and Rev, perform regulatory functions that are essential for viral replication. The remaining four—Nef, Vif, Vpr, and Vpu—are essential for efficient virus production *in vivo*.

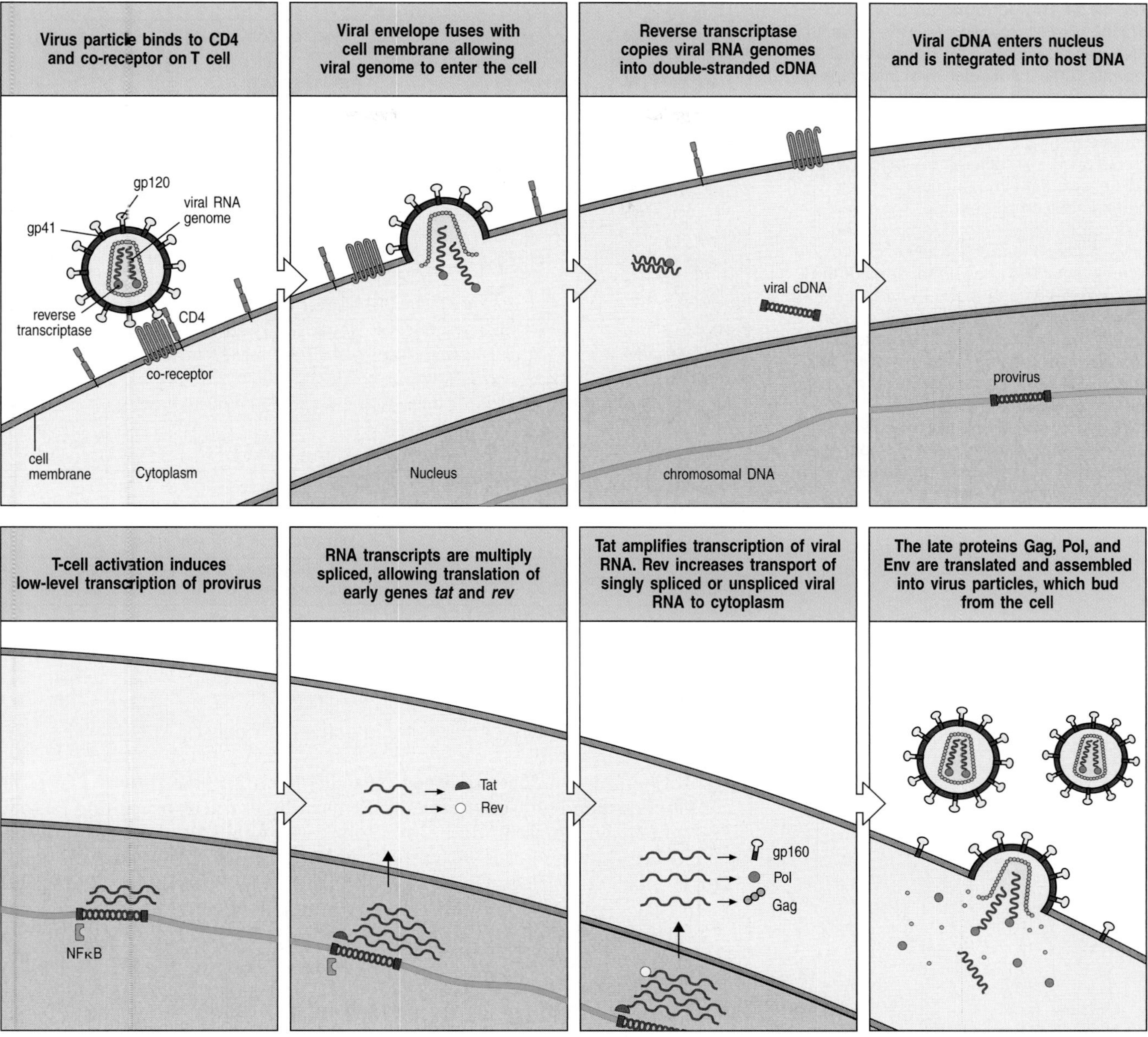

Fig. 12.23 The life cycle of HIV. Top row: the virus binds to CD4 using gp120, which is altered by CD4 binding so that it now also binds a chemokine receptor that acts as a co-receptor for viral entry. This binding releases gp41, which causes fusion of the viral envelope with the cell membrane and release of the viral core into the cytoplasm. Once in the cytoplasm, the viral core releases the RNA genome, which is reverse-transcribed into double-stranded cDNA using the viral reverse transcriptase. The double-stranded cDNA migrates to the nucleus in association with the viral integrase and the Vpr protein and is integrated into the cell genome, becoming a provirus. Bottom row: activation of CD4 T cells induces the expression of the transcription factors NFκB and NFAT, which bind to the proviral LTR and initiate transcription of the HIV genome. The first viral transcripts are extensively processed, producing spliced mRNAs encoding several regulatory proteins, including Tat and Rev. Tat both enhances transcription from the provirus and binds to the RNA transcripts, stabilizing them in a form that can be translated. Rev binds the RNA transcripts and transports them to the cytosol. As levels of Rev increase, less extensively spliced and unspliced viral transcripts are transported out of the nucleus. The singly spliced and unspliced transcripts encode the structural proteins of the virus, and unspliced transcripts, which are also the new viral genomes, are packaged with these proteins to form many new virus particles.

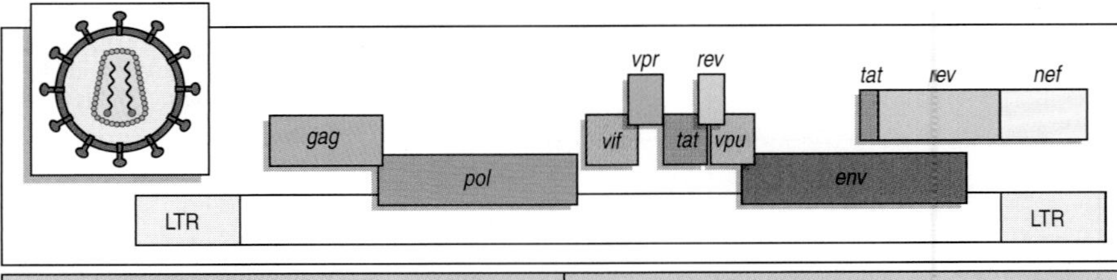

Gene		Gene product/function
gag	Group-specific antigen	Core proteins and matrix proteins
pol	Polymerase	Reverse transcriptase, protease, and integrase enzymes
env	Envelope	Transmembrane glycoproteins. gp120 binds CD4 and CCR5; gp41 is required for virus fusion and internalization
tat	Transactivator	Positive regulator of transcription
rev	Regulator of viral expression	Allows export of unspliced and partially spliced transcripts from nucleus
vif	Viral infectivity	Affects particle infectivity
vpr	Viral protein R	Transport of DNA to nucleus. Augments virion production. Cell-cycle arrest
vpu	Viral protein U	Promotes intracellular degradation of CD4 and enhances release of virus from cell membrane
nef	Negative-regulation factor	Augments viral replication *in vivo* and *in vitro.* Decreases CD4, MHC class I and II expression

Fig. 12.24 The genomic organization of HIV. Like all retroviruses, HIV-1 has an RNA genome flanked by long terminal repeats (LTR) involved in viral integration and in regulation of transcription of the viral genome. The genome can be read in three frames and several of the viral genes overlap in different reading frames. This allows the virus to encode many proteins in a small genome. The three main protein products—Gag, Pol, and Env—are synthesized by all infectious retroviruses. The known functions of the different genes and their products are listed. The products of *gag*, *pol*, and *env* are known to be present in the mature viral particle, together with the viral RNA. The mRNAs for Tat, Rev, and Nef proteins are produced by splicing of viral transcripts, so their genes are split in the viral genome. In the case of Nef, only one exon, shown in yellow, is translated.

12-27 Replication of HIV occurs only in activated T cells.

The production of infectious virus particles from an HIV provirus in CD4 T cells is stimulated by T-cell activation. This induces the transcription factors NFκB and NFAT, which bind to promoters in the viral LTR, thereby initiating the transcription of viral RNA by the cellular RNA polymerase II. This transcript is spliced in various ways to produce mRNAs for the viral proteins. The Gag and Gag–Pol proteins are translated from unspliced mRNA; Vif, Vpr, Vpu, and Env are translated from singly spliced viral mRNA; Tat, Rev, and Nef are translated from multiply spliced mRNA. Tat greatly enhances the transcription of viral RNA from the provirus by the RNA polymerase II complex. It binds to the transcriptional activation region (TAR) in the provirus 5′LTR in a complex with the cellular cyclin T1 and its partner, cyclin-dependent kinase 9 (CDK9), to form a complex that phosphorylates RNA polymerase and stimulates its RNA elongation activity. The expression of the cyclin T1–CDK9 complex is greatly increased in activated T cells compared with quiescent ones. Together with the increased expression of NFκB and NFAT in activated T cells, this may explain the ability of HIV to lie dormant in resting T cells and replicate in activated T cells (Fig. 12.25).

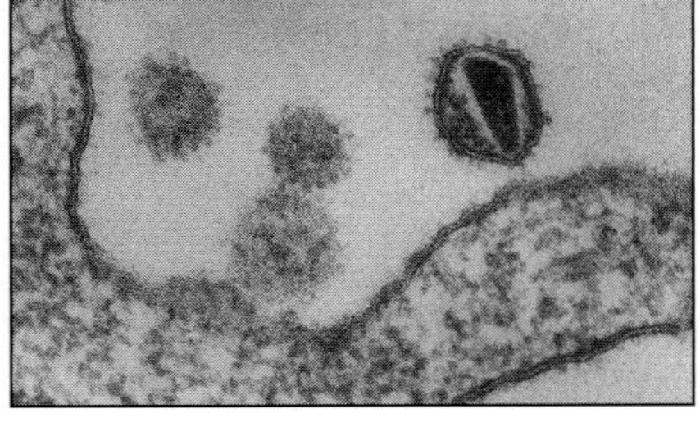
Fig. 12.25 HIV replicates in activated CD4 T cells. A complete virion can be clearly seen on the right. Photograph courtesy of H. Gelderblom.

Eukaryotic cells have mechanisms to prevent the export from the cell nucleus of incompletely spliced mRNA transcripts. This could pose a problem for a retrovirus that is dependent on the export of unspliced, singly spliced, and multiply spliced mRNA species to translate the full complement of viral proteins. The Rev protein is the viral solution to this problem. Export from the nucleus and translation of the three HIV proteins encoded by the fully spliced mRNA transcripts—Tat, Nef, and Rev—occurs early after viral infection by means of the normal host cellular mechanisms of mRNA export. The expressed Rev protein then enters the nucleus and binds to a specific viral RNA sequence, the Rev response element (RRE). In the presence of *rev*, RNA is exported from the nucleus before it can be spliced, so that the structural proteins and RNA genome can be produced. Rev also binds to a cellular transport protein named Crm1, which engages a host pathway for exporting mRNA species through nuclear pores into the cytoplasm.

When the provirus is first activated, Rev levels are low, the transcripts are translocated slowly from the nucleus, and thus multiple splicing events can occur. In this way, more Tat and Rev are produced, and Tat in turn ensures that more viral transcripts are made. Later, when Rev levels have increased, the transcripts are translocated rapidly from the nucleus unspliced or only singly spliced. These unspliced or singly spliced transcripts are translated to produce the structural components of the viral core and envelope, together with the reverse transcriptase, the integrase, and the viral protease, all of which are needed to make new viral particles. The complete, unspliced transcripts that are exported from the nucleus late in the infectious cycle are required for the translation of *gag* and *pol* and are also destined to be packaged with the proteins as the RNA genomes of the new virus particles.

The success of virus replication also depends on the proteins Nef, Vif, Vpr, and Vpu. Vif (viral infectivity factor) is an RNA-binding protein that accumulates in the cytoplasm and on the plasma membrane of infected cells. Vif acts to overcome a natural cellular defense against retroviruses. Cells express a cytidine deaminase called APOBEC, which can be incorporated into virions. This enzyme, which belongs to the same protein family as the activation-induced cytidine deaminase (AID) (see Section 12-10), catalyzes the conversion of deoxycytidine to deoxyuridine in the first strand of reverse-transcribed viral cDNA, thus destroying its ability to encode viral proteins. Vif induces the transport of APOBEC to proteasomes, where it is degraded. The expression of Nef (negative regulation factor) early in the viral life cycle induces T-cell activation and the establishment of a persistent state of HIV infection. Nef inhibits the expression of MHC class I molecules on the infected cells, thus making them less likely to be killed by cytotoxic T cells. It also inhibits the MHC class II-restricted presentation of peptides to CD4 T cells, thus inhibiting the generation of an antiviral immune response. The function of Vpr (viral protein R) is not fully understood, but it has various activities that enhance viral production and release. Vpu (viral protein U) is unique to HIV-1 and variants of SIV, and is required for the maturation of progeny virions and their efficient release.

12-28 Lymphoid tissue is the major reservoir of HIV infection.

Although HIV load and turnover are usually measured in terms of the RNA present in virions in the blood, a major reservoir of HIV infection is lymphoid tissue, in which infected CD4 T cells, monocytes, macrophages, and dendritic cells are found. In addition, HIV is trapped in the form of immune complexes on the surface of follicular dendritic cells in the germinal center. These cells are not themselves infected but may act as a store of infective virions. Several other potential reservoirs for HIV-1 that may contribute to its long-term persistence are infected cells in the central nervous system, the gastrointestinal system, and the male urogenital tract.

From studies of patients receiving drug treatment, it is estimated that more than 95% of the virus that can be detected in the plasma is derived from productively infected CD4 T cells that have a very short half-life of about 2 days (see Fig. 12.27). Virus-producing CD4 T cells are found in the T-cell areas of lymphoid tissue, and these are thought to succumb to infection while being activated in an immune response. Latently infected CD4 memory T cells that become reactivated by antigen also produce virus that can spread to other activated CD4 T cells. Unfortunately, latently infected CD4 memory T cells have an extremely long mean half-life of around 44 months. This means that drug therapy may never be able to eliminate an HIV infection and therefore needs to be administered throughout life. In addition to cells that are productively or latently infected, a further large population of cells are infected by defective proviruses, which do not produce infectious virus.

Macrophages and dendritic cells seem to be able to harbor replicating HIV without necessarily being killed by it, and they are believed to be an important reservoir of infection; they also serve as a means of spreading virus to other tissues, such as the brain. Although the function of macrophages as antigen-presenting cells does not seem to be compromised by HIV infection, it is thought that the virus causes abnormal patterns of cytokine secretion that could account for the wasting that commonly occurs in AIDS patients late in their disease.

12-29 An immune response controls but does not eliminate HIV.

Infection with HIV generates an immune response that contains the virus but only very rarely, if ever, eliminates it. The time course of various elements in the adaptive immune response to HIV is shown, together with the levels of infectious virus in plasma, in Fig. 12.26. The initial acute phase that occurs as the adaptive immune response develops is followed by the chronic, semi-stable phase that eventually culminates in AIDS. Current thinking indicates that virus-mediated cytopathicity is very important during early infection and that this results in a substantial depletion of CD4 T cells, in particular in the mucosal tissues. After the acute phase there is a good initial recovery, but cytotoxic lymphocytes directed against HIV-infected cells, immune activation (direct and indirect), cytopathicity, and insufficient T-cell regeneration combine to establish the chronic state, during which immunodeficiency develops. In this section we consider in turn the roles of CD8 cytotoxic T cells, CD4 T cells, antibody and soluble factors in the immune response to HIV infection that ultimately fails to contain the infection.

Studies of peripheral blood cells from infected individuals reveal cytotoxic T cells specific for viral peptides that can kill infected cells *in vitro*. *In vivo*, cytotoxic T cells can be seen to invade sites of HIV replication and they could, in theory, be responsible for killing many productively infected cells before any infectious virus is released, thereby containing viral load at the quasi-stable levels that are characteristic of the asymptomatic period. Evidence for the clinical importance of the control of HIV-infected cells by CD8 cytotoxic T cells comes from studies relating the numbers and activity of CD8 T cells to viral load. An inverse correlation was found between the number of CD8

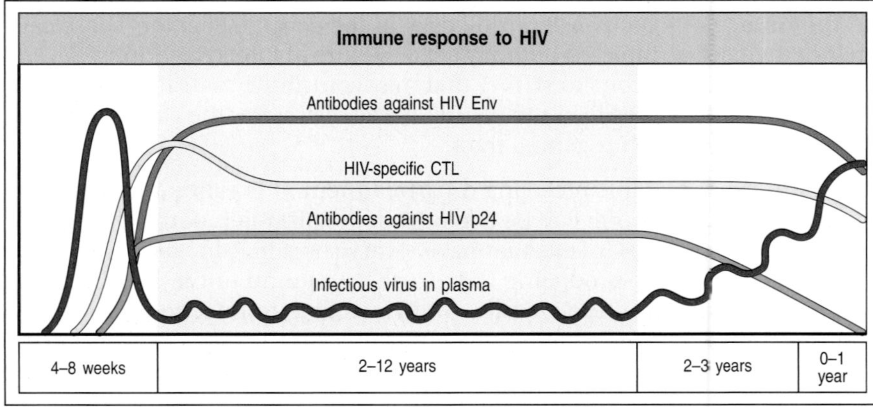

Fig. 12.26 The immune response to HIV. Infectious virus is present at relatively low levels in the peripheral blood of infected individuals during a prolonged asymptomatic phase, during which the virus is replicated persistently in lymphoid tissues. During this period, CD4 T-cell counts gradually decline, although antibodies and CD8 cytotoxic T cells directed against the virus remain at high levels. Two different antibody responses are shown in the figure, one to the envelope protein (Env) of HIV, and one to the core protein p24. Eventually, the levels of antibody and HIV-specific cytotoxic T lymphocytes (CTLs) also decline, and there is a progressive increase in infectious HIV in the peripheral blood.

T cells carrying a receptor specific for an HLA-A2-restricted HIV peptide and the amount of viral RNA in the plasma. Similarly, patients with high levels of HIV-specific CD8 T cells showed slower progression of disease than those with low levels. There is also direct evidence from experiments in macaques infected with SIV that CD8 cytotoxic T cells control retrovirus-infected cells *in vivo*. Treatment of infected animals with monoclonal antibodies that remove CD8 T cells was followed by a large increase in viral load.

A variety of factors produced by CD4, CD8, and NK cells are important in antiviral immunity. Evidence for a non-cytotoxic suppressor activity of CD8 cells on HIV-1 came from the observation that peripheral blood mononuclear cells (PMBCs) from seropositive asymptomatic individuals failed to replicate HIV-1 *in vitro*, but that depletion of CD8 T cells, but not other cells (for example, NK cells) from this PMBC fraction led to an increase in viral replication. The inhibition is now known to be mediated by secreted proteins. Chemokines such as CCL5, CCL3, and CCL4 are released at the site of infection and inhibit virus spread (without killing the cell) by competing with R5 strains of HIV-1 for the engagement of co-receptor CCR5, whereas factors still unknown compete with R4 strains for binding to CXCR4. Cytokines such as IFN-α and IFN-γ may also be involved in controlling virus spread, but a mechanism for this is not clear.

In addition to being a major target for HIV infection, three pieces of evidence show that CD4 T cells also have an important role in the host response to HIV-infected cells. First, an inverse correlation is found between the strength of CD4 T-cell proliferative responses to HIV antigen and viral load. Second, some patients who did not progress to AIDS long after infection by HIV showed strong CD4 T-cell proliferative responses. Third, early treatment of acutely infected individuals with antiretroviral drugs was associated with a recovery in CD4 proliferative responses to HIV antigens. If this antiretroviral therapy was stopped, the CD4 responses persisted in some of these people and were associated with reduced levels of viremia. However, the infection persisted in all patients and it is likely that immunological control of the infection will ultimately fail. If CD4 T-cell responses are essential for the control of HIV infection, then the fact that HIV is tropic for these cells and kills them may be the explanation for the long-term inability of the host immune response to control the infection.

Antibodies against gp120 and gp41 envelope viral antigens are produced in response to infection but, as with T cells, are unable to clear the infection. The antibodies react well with purified antigens *in vitro* and with viral debris, but bind poorly to intact enveloped virions or to infected cells. This suggests that the native conformation of these antigens, which are heavily glycosylated, is not accessible to naturally produced antibodies. The evidence is strong that antibodies cannot significantly modify established disease. Nevertheless the passive administration of antibodies against HIV can protect experimental animals from mucosal infection by HIV, and this offers hope that an effective vaccine might be developed that could prevent new infections.

The mutations that occur as HIV replicates can allow the resulting virus variants to escape recognition by neutralizing antibody or cytotoxic T cells and contribute to the long-term failure of the immune system to contain the infection. An immune response is often dominated by T cells specific for particular epitopes—the **immunodominant** epitopes—and mutations in immunodominant HIV peptides presented by MHC class I molecules have been found. Mutant peptides have been found to inhibit T cells responsive to the wild-type epitope, thus allowing both the mutant and wild-type viruses to survive. Inhibitory mutant peptides have also been reported in hepatitis B virus infections, and similar mutant immunodominant peptides might contribute to the persistence of other viral infections.

Infections	
Parasites	*Toxoplasma* spp. *Cryptosporidium* spp. *Leishmania* spp. *Microsporidium* spp.
Intracellular bacteria	*Mycobacterium tuberculosis* *Mycobacterium avium intracellulare* *Salmonella* spp.
Fungi	*Pneumocystis carinii* *Cryptococcus neoformans* *Candida* spp. *Histoplasma capsulatum* *Coccidioides immitis*
Viruses	Herpes simplex Cytomegalovirus Herpes zoster
Malignancies	
Kaposi's sarcoma – (HHV8) Non-Hodgkin's lymphoma, including EBV-positive Burkitt's lymphoma Primary lymphoma of the brain	

Fig. 12.27 A variety of opportunistic pathogens and cancers can kill AIDS patients. Infections are the major cause of death in AIDS, with respiratory infection with *P. carinii* and mycobacteria being the most prominent. Most of these pathogens require effective macrophage activation by CD4 T cells or effective cytotoxic T cells for host defense. Opportunistic pathogens are present in the normal environment but cause severe disease primarily in immunocompromised hosts, such as AIDS patients and cancer patients. AIDS patients are also susceptible to several rare cancers, such as Kaposi's sarcoma (associated with human herpes virus 8 (HHV8)) and various lymphomas, suggesting that immune surveillance of the causative herpes viruses by T cells can normally prevent such tumors (see Chapter 15).

An exciting development in the study of HIV immunity is the identification of a number of cellular proteins that can target HIV replication. The enzyme APOBEC (see Section 12-27) causes extensive mutation of newly formed HIV cDNA, thus destroying its coding and replicative capacity. APOBEC is active in resting CD4 T cells but is degraded in infected CD4 T cells, providing yet another reason that resting CD4 T cells are resistant to infection. The powerful antiretroviral action of APOBEC has provoked considerable interest in finding small molecules that interfere with its virus-induced degradation. Another cytoplasmic protein, TRIM 5α, limits HIV-1 infections in rhesus monkeys, probably by targeting the viral capsid and preventing the uncoating and release of viral RNA.

12-30 The destruction of immune function as a result of HIV infection leads to increased susceptibility to opportunistic infection and eventually to death.

When CD4 T-cell numbers decline below a critical level, cell-mediated immunity is lost, and infections with a variety of opportunistic microbes appear (Fig. 12.27). Typically, resistance is lost early to oral *Candida* species and to *M. tuberculosis*, which is manifested as an increased prevalence of thrush (oral candidiasis) and tuberculosis, respectively. Later, patients suffer from shingles, caused by the activation of latent herpes zoster, from EBV-induced B-cell lymphomas, and from Kaposi's sarcoma, a tumor of endothelial cells that probably represents a response both to cytokines produced in the infection and to a herpes virus called HHV-8 that was identified in these lesions. Pneumonia caused by the fungus *P. carinii* is common and was often fatal before effective antifungal therapy was introduced. Co-infection by hepatitis C virus is common and associated with more rapid progression of hepatitis. In the final stages of AIDS, infection with cytomegalovirus or *M. avium* complex is more prominent. It is important to note that not all patients with AIDS get all these infections or tumors, and there are other tumors and infections that are less prominent but still significant. Fig. 12.27 lists the commonest opportunistic infections and tumors, most of which are normally controlled by robust CD4 T cell-mediated immunity that wanes as the CD4 T-cell counts drop toward zero (see Fig. 12.19).

12-31 Drugs that block HIV replication lead to a rapid decrease in titer of infectious virus and an increase in CD4 T cells.

Studies with powerful drugs that completely block the cycle of HIV replication indicate that the virus is replicating rapidly at all phases of infection, including the asymptomatic phase. Two viral proteins in particular have been the target of drugs aimed at arresting viral replication. These are the viral reverse transcriptase, which is required for synthesis of the provirus, and the viral protease, which cleaves the viral polyproteins to produce the virion proteins and viral enzymes. The reverse transcriptase is inhibited by nucleoside analogs such as zidovudine (AZT), which was the first anti-HIV drug to be licensed in the United States. Inhibitors of the reverse transcriptase and the protease prevent the establishment of further infection in uninfected cells. Cells that are already infected can continue to produce virions because, once the provirus is established, reverse transcriptase is not needed to make new virus particles, while the viral protease acts at a very late maturation step of the virus, and inhibition of the protease does not prevent virus from being released. However, in both cases, the released virions are not infectious and further cycles of infection and replication are prevented.

The introduction of combination therapy with a cocktail of viral protease inhibitors and nucleoside analogs, also known as **highly active antiretroviral**

therapy (**HAART**), dramatically reduced mortality and morbidity in patients with advanced HIV infection in the United States between 1995 and 1997 (Fig. 12.28). Many patients treated with HAART show a rapid and dramatic reduction in viremia, eventually maintaining levels of HIV RNA close to the limit of detection (50 copies/ml of plasma) for a long period (Fig. 12.29).

HAART therapy is also accompanied by a slow but steady increase in CD4 T cells, despite the fact that many other compartments of the immune system remain compromised. Although HAART is effective at treating HIV infection, the maximum effect of this therapy is prevented by the viral reservoirs established early in infection. Cessation of HAART leads to a rapid rebound of virus multiplication, implying that patients will require treatment indefinitely. Finally, because of the serious side-effects and the cost, HAART is not affordable for most of the world.

It is unclear how the virus particles are removed so rapidly from the circulation after the initiation of HAART therapy. It seems most likely that they are opsonized by specific antibody and complement and removed by phagocytic cells of the mononuclear phagocyte system. Opsonized HIV particles can also be trapped on the surface of follicular dendritic cells in lymphoid follicles, which are known to capture antigen:antibody complexes and retain them for prolonged periods.

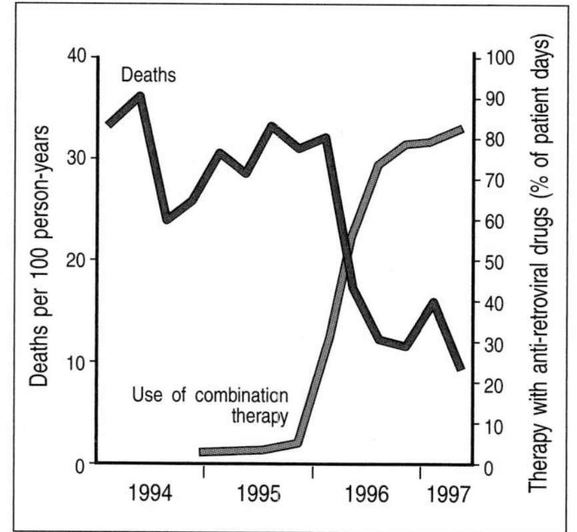

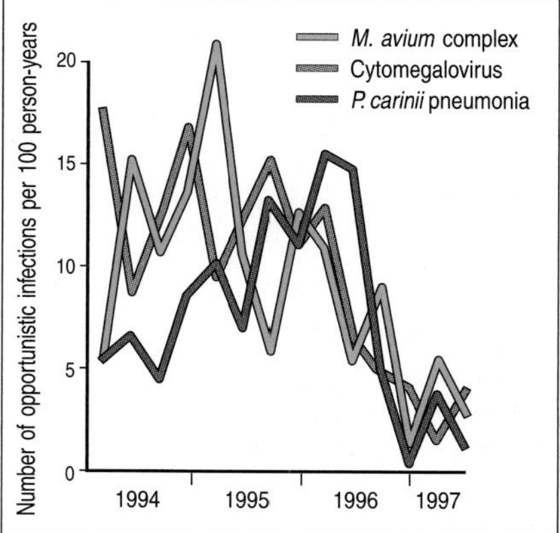

Fig. 12.28 The morbidity and mortality in advanced HIV infection fell in the United States in parallel with the introduction of combination anti-retroviral drug therapy. The upper graph shows the number of deaths, expressed each calendar quarter as the deaths per 100 person-years. The lower graph shows the decline in opportunistic infections caused by cytomegalovirus, *Pneumocystis carinii*, and *Mycobacterium avium*, during the same period of time. Figure based on data from F. Palella.

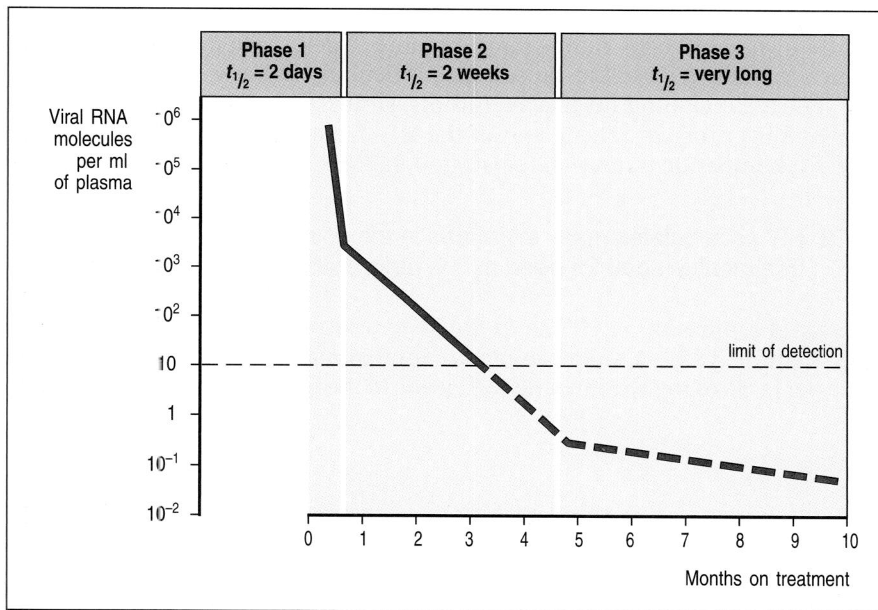

Fig. 12.29 The time-course of the reduction of HIV circulating in the blood on drug treatment. The production of new HIV particles can be arrested for prolonged periods by combinations of protease inhibitors and viral reverse-transcriptase inhibitors. After the initiation of such treatment, virus production is curtailed as these cells die and no new cells are infected. The half-life of virus decay occurs in three phases. The first phase has a half-life of about 2 days, reflecting the half-life of productively infected CD4 T cells, and lasts for about 2 weeks, during which time viral production declines as the lymphocytes that were productively infected at the onset of treatment die. Released virus is rapidly cleared from the circulation, where it has a half-life ($t_{1/2}$) of 6 hours, and there is a decrease in virus levels in plasma of more than 95% during this first phase. The second phase lasts for about 6 months and has a half-life of about 2 weeks. During this phase, virus is released from infected macrophages and from resting, latently infected CD4 T cells stimulated to divide and develop productive infection. It is thought that there is then a third phase of unknown length that results from the reactivation of integrated provirus in memory T cells and other long-lived reservoirs of infection. This reservoir of latently infected cells might remain present for many years. Measurement of this phase of viral decay is impossible at present because viral levels in plasma are below detectable levels (dotted line). Data courtesy of G.M. Shaw.

The other issue raised by studies of drug treatment is the effect of HIV replication on the population dynamics of CD4 T cells. The decline in plasma viremia is accompanied by a steady increase in CD4 T-cell counts in peripheral blood: what is the source of the new CD4 T cells that appear once treatment is started? Three complementary mechanisms have been established for the recovery in CD4 T-cell numbers. The first of these is a redistribution of CD4 T memory cells from lymphoid tissues into the circulation as viral replication is controlled; this occurs within weeks of starting treatment. The second is the reduction in the abnormal levels of immune activation as the HIV infection is controlled, associated with reduced cytotoxic T lymphocyte killing of infected CD4 T cells. The third is much slower and is caused by the emergence of new naive T cells from the thymus. Although the thymus involutes with aging, evidence that these later-arriving cells are indeed of thymic origin is provided by the observation that they contain T-cell receptor excision circles (TRECs) (see Section 4-9).

Given that the latent reservoirs of infection represent the main cause of failure to eradicate the virus by drug treatment, ways of flushing out this reservoir have been explored. One strategy includes the administration of cytokines such as IL-2, IL-6, and TNF-α that favor viral transcription and replication in cells latently harboring the virus, thus facilitating the actions of HAART. IL-2 is one of the few T cell-activating cytokines that has been trialed in the treatment of AIDS to boost the depleted immune system. Despite its lack of effect on the clearance of HIV-1 RNA, IL-2 treatment induces an approximate sixfold increase in the CD4 T-cell count when administered in combination with antiretroviral therapy, with the increase being predominantly in naive T cells rather than memory T cells. Whether IL-2 will have clinical benefit remains to be tested, particularly in view of the associated side-effects, including flu-like symptoms, sinus congestion, low blood pressure and liver toxicity. The stages of the HIV life cycle that are being considered as therapeutic targets are illustrated in Fig. 12.30.

12-32 HIV accumulates many mutations in the course of infection, and drug treatment is soon followed by the outgrowth of drug-resistant variants.

The rapid replication of HIV, with the generation of 10^9 to 10^{10} virions every day, is coupled with a mutation rate of approximately 3×10^{-5} per nucleotide base per cycle of replication, and thus leads to the generation of many variants

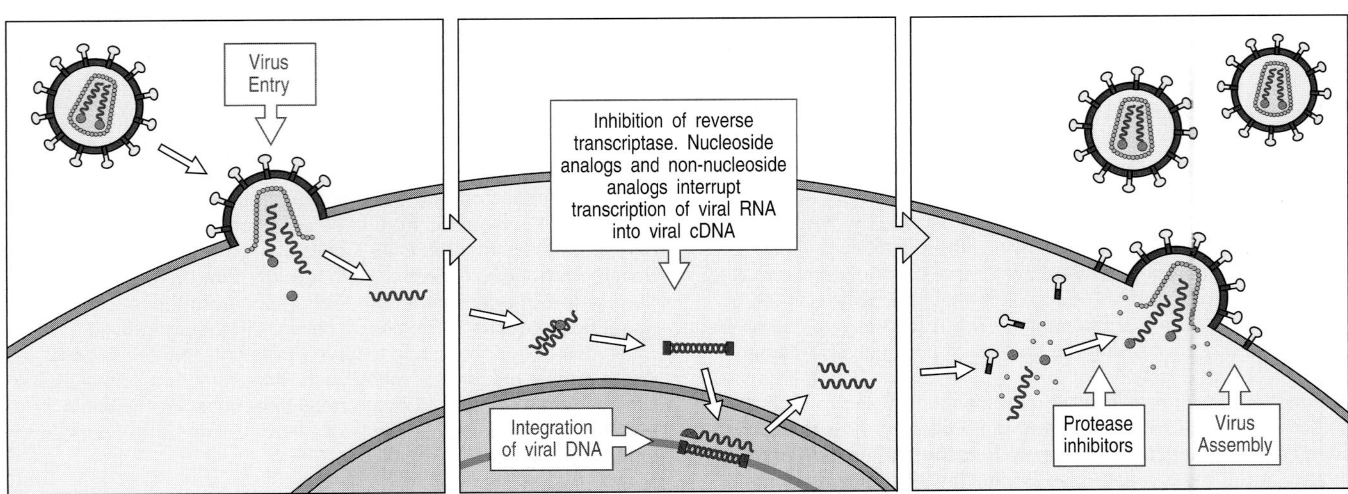

Fig. 12.30 Possible targets for interference with the HIV life cycle. In principle, HIV could be attacked by therapeutic drugs at multiple points in its life cycle: virus entry, inhibition of reverse transcriptase, insertion of viral cDNA into cellular DNA by the viral integrase, cleavage of viral polyproteins by the viral protease, and assembly and budding of infectious virions. As yet, only drugs that inhibit reverse transcriptase and protease action have been developed. There are eight nucleoside analog inhibitors and three non-nucleoside inhibitors of reverse transcriptase available, and seven protease inhibitors. Combination therapy using different kinds of drugs is more effective than using a single drug.

of HIV in a single infected patient in the course of a day. This high mutation rate arises from the error-prone nature of retroviral replication. Reverse transcriptase lacks the proofreading mechanisms associated with cellular DNA polymerases, and the RNA genomes of retroviruses are copied into DNA with relatively low fidelity. The transcription of the proviral DNA into RNA by RNA polymerase II is also a low-fidelity process. A rapidly replicating persistent virus that is going through these two steps repeatedly in the course of an infection can accumulate many mutations, and numerous variants of HIV, sometimes called **quasi-species**, are found within a single infected individual. This phenomenon was first recognized in HIV and has since proved to be common to all lentiviruses.

As a consequence of its high variability, HIV rapidly develops resistance to antiviral drugs. When a drug is administered, variants of the virus with mutations conferring resistance to the drug emerge and multiply until the previous levels of virus are regained. Resistance to some of the protease inhibitors requires only a single mutation and appears after only a few days (Fig. 12.31); resistance to some of the inhibitors of reverse transcriptase develops in a similarly short time. In contrast, resistance to the nucleoside analog zidovudine takes months to develop, as it requires three or four mutations to occur in the viral reverse transcriptase. As a result of the relatively rapid appearance of resistance to all known anti-HIV drugs, successful drug treatment depends on combination therapy (see Section 12-31). It might also be important to begin treatment early in the course of an infection, thereby reducing the chance that a variant virus will have accumulated all the mutations necessary to resist the entire cocktail.

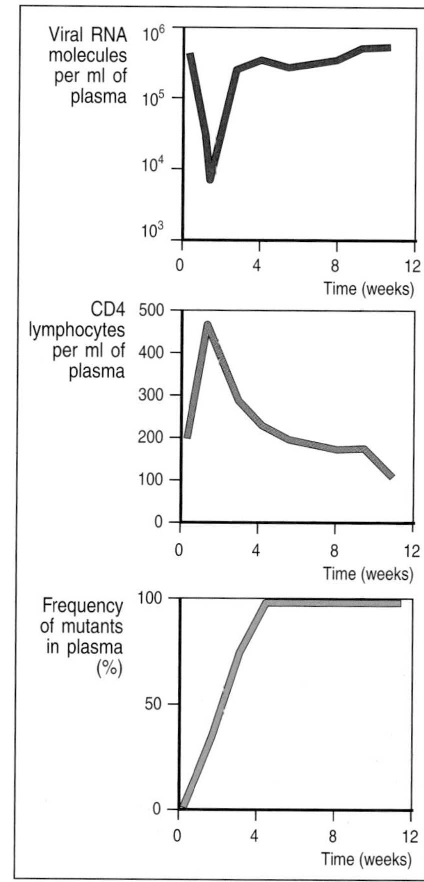

Fig. 12.31 Resistance of HIV to protease inhibitors develops rapidly. After the administration of a single protease inhibitor to a patient with HIV there is a precipitous fall in viral RNA levels in plasma with a half-life of about 2 days (top panel). This is accompanied by an initial rise in the number of CD4 T cells in peripheral blood (center panel). Within days of starting the drug, mutant drug-resistant variants can be detected in plasma (bottom panel) and in peripheral blood lymphocytes. After only 4 weeks of treatment, viral RNA levels and CD4 lymphocyte levels have returned to their original pre-drug levels, and 100% of plasma HIV is present as the drug-resistant mutant.

12-33 Vaccination against HIV is an attractive solution but poses many difficulties.

A safe and effective vaccine for the prevention of HIV infection and AIDS is the ultimate goal, but its attainment is fraught with difficulties that have not been faced in developing vaccines against other diseases. The main problem is the nature of the infection itself, featuring a virus that proliferates extremely rapidly and causes sustained infection in the face of strong cytotoxic T-cell and antibody responses. The development of vaccines that could be administered to patients already infected, to boost immune responses and prevent progression to AIDS, has been considered, as well as prophylactic vaccines that would be given to prevent initial infection. The development of therapeutic vaccination in those already infected would be extremely difficult. As discussed in the previous section, HIV evolves in individual patients by the selective proliferation of mutant viruses that escape recognition by antibodies and cytotoxic T cells. The ability of the virus to persist in latent form as a transcriptionally silent provirus invisible to the immune system might also prevent even an immunized person from clearing an infection once it has been established.

There is more hope for prophylactic vaccination to prevent new infection. But even here, the lack of effect of the normal immune response and the sheer scale of sequence diversity among HIV strains in the population as a whole is a significant challenge. Patients infected with one strain of virus do not seem to be resistant to closely related strains, ruling out a universal vaccine. For example, a patient infected with HIV-1 clade AE was successfully treated for 28 months, but 3 months after ceasing treatment he contracted an infection with a clade B HIV-1 as a result of sexual encounters in Brazil, where this clade is endemic. Cases of superinfection, where two strains simultaneously infect the same cell, have also been described. Key among the difficulties is our uncertainty over what form protective immunity to HIV might take. It is not known whether antibodies, responses by CD4 T cells or by CD8 cytotoxic T cells, or all three, are necessary to achieve protective immunity, and which epitopes might provide the targets for protective immunity.

However, against this pessimistic background, there are grounds for hope that successful vaccines can be developed. Of particular interest are rare groups of people who have been exposed often enough to HIV to make it virtually certain that they should have become infected but who have not developed the disease. In some cases this is due to an inherited deficiency in the chemokine receptor used as co-receptor for HIV entry (see Section 12-25). However, this mutant chemokine receptor does not occur in Africa, where one such group has been identified. A small group of Gambian and Kenyan sex workers who are estimated to have been exposed to many HIV-infected male partners each month for up to 5 years were found to lack antibody responses but to have cytotoxic T-cell responses to a variety of peptide epitopes from HIV. These women seem to have been naturally immunized against HIV. Follow-up of a number of them showed that about 10% subsequently acquired HIV infection. Paradoxically, HIV infection was found more frequently in women who had reduced their sex work and thereby their regular exposure to the virus. A possible explanation is that the absence of repeated exposure to HIV antigens led to a loss of the cytotoxic T-cell response, thus rendering the women susceptible to infection.

Various strategies are being tried in an attempt to develop vaccines against HIV. Many successful vaccines against other viral diseases contain a live attenuated strain of the virus, which raises an immune response but does not cause disease (see Section 15-23). There are substantial difficulties in the development of live attenuated vaccines against HIV, not least the worry of recombination between vaccine strains and wild-type viruses, which would lead to reversion to virulence. An alternative approach is the use of DNA vaccination, a technique that we discuss in Section 15-27. DNA vaccination against HIV followed by the administration of a recombinant modified vaccinia boost, containing HIV antigens, has been piloted in primate experiments and was successful in preventing infection from an intrarectal challenge given 7 months after the booster vaccination. For every success in the route to HIV vaccination there is a setback, however. Rhesus monkeys were vaccinated with a DNA vaccine against SIV together with an IL-2 fusion protein and then challenged with a pathogenic SIV–HIV hybrid. Six months after the challenge, one of the monkeys developed an AIDS-like illness that was associated with the emergence of a virus carrying a point mutation in an immunodominant Gag epitope that is recognized by cytotoxic T cells. This is a beautiful but depressing example of the ability of HIV to escape immune control under the pressure of a cytotoxic T-cell response.

Subunit vaccines, which induce a response against only some proteins in the virus, have also been made. One such vaccine made from the envelope protein gp120 has been tested on chimpanzees. This vaccine proved to be specific to the precise strain of virus used to make it and was therefore useless in protection against natural infection. Subunit vaccines are also less efficient at inducing prolonged cytotoxic T-cell responses. In spite of the results in chimpanzees, a recombinant gp120 protein vaccine has been trialed in uninfected human volunteers. A small number of the volunteers subsequently contracted HIV-1 infection, the course of which was not modified by the previous vaccination.

In addition to the biological obstacles to developing effective HIV vaccines, there are difficult ethical issues. It would be unethical to conduct a vaccine trial without trying at the same time to minimize the exposure of a vaccinated population to the virus itself. The effectiveness of a vaccine can, however, only be assessed in a population in which the exposure rate to the virus is high enough to assess whether vaccination protects against infection. This means that initial vaccine trials might have to be conducted in countries where the incidence of infection is very high and public health measures have not yet succeeded in reducing the spread of HIV.

12-34 Prevention and education are one way in which the spread of HIV and AIDS can be controlled.

The one way in which we know we can protect against infection with HIV is by avoiding contact with body fluids, such as semen, blood, blood products, or milk, from people who are infected. Indeed, it has been repeatedly demonstrated that this precaution, simple enough in the developed world, is sufficient to prevent infection, because healthcare workers can take care of AIDS patients for long periods without seroconversion or signs of infection.

For this strategy to work, however, one must be able to periodically test people at risk of infection with HIV, so that they can take the steps necessary to avoid passing the virus to others. This, in turn, requires strict confidentiality and mutual trust. A barrier to the control of HIV is the reluctance of individuals to find out whether or not they are infected, especially as one of the consequences of a positive HIV test is stigmatization by society. As a result, infected individuals can unwittingly infect many others. Balanced against this is the success of combination drug therapy (see Section 12-31), which provides an incentive for potentially infected people to identify the presence of infection and gain the benefits of treatment. Responsibility is at the heart of AIDS prevention, and a law guaranteeing the rights of people infected with HIV might go a long way toward encouraging responsible behavior. The rights of HIV-infected people are protected in a few countries. The problem in the less-developed nations, where elementary health precautions are extremely difficult to establish, is more profound.

Summary.

Infection with the human immunodeficiency virus (HIV) is the cause of acquired immune deficiency syndrome (AIDS). This worldwide epidemic is now spreading at an alarming rate, especially through heterosexual contact in less-developed countries. HIV is an enveloped retrovirus that replicates in cells of the immune system. Viral entry requires the presence of CD4 and a particular chemokine receptor, and the viral cycle is dependent on transcription factors found in activated T cells. Infection with HIV causes a loss of CD4 T cells and an acute viremia that rapidly subsides as cytotoxic T-cell responses develop, but HIV infection is not eliminated by this immune response. Uninfected cells become activated and also subsequently die, which is a key feature distinguishing HIV from non-pathogenic natural infections of African primates with various SIVs. HIV establishes a state of persistent infection in which the virus is continually replicating in newly infected cells. Current treatment consists of combinations of viral protease inhibitors together with nucleoside analogs that inhibit the reverse transcriptase, causing a rapid decrease in virus levels and a slower increase in CD4 T-cell counts. The main effect of HIV infection is the destruction of CD4 T cells, which occurs through the direct cytopathic effects of HIV infection and through killing by CD8 cytotoxic T cells. As the CD4 T-cell counts wane, the body becomes progressively more susceptible to opportunistic infection. Eventually, most HIV-infected individuals develop AIDS and die; however, a small minority (3–7%) remain healthy for many years with no apparent ill effects of infection. We hope to be able to learn from these people how infection with HIV can be controlled. The existence of these people, and of others who seem to have been naturally immunized against infection, gives hope that it will be possible to develop effective vaccines against HIV.

Summary to Chapter 12.

Whereas most infections elicit protective immunity, most successful pathogens have developed some means of at least partly resisting the immune response, and they can result in serious, persistent disease. Some individuals have inherited deficiencies in different components of the immune system, making them highly susceptible to certain classes of infectious agents. Persistent infection and the inherited immunodeficiency diseases illustrate the importance of innate and adaptive immunity in an effective host defense against infection, and present huge challenges for future immunological research. The human immunodeficiency virus (HIV), which leads to acquired autoimmune disease syndrome (AIDS), combines the characteristics of a persistent infectious agent with the ability to create immunodeficiency in its human host, a combination that is usually slowly lethal to the patient. The key to fighting new pathogens such as HIV is to develop our understanding of the basic properties of the immune system and its role in combating infection more fully.

Questions.

12.1 *List the different ways in which viruses can evade the immune system. Which of these strategies lead to chronic infection, and why?*

12.2 *Discuss the factors that allow herpes viruses to maintain latent infections in the host, and how reactivation occurs so the virus can spread from one host to another.*

12.3 *From what you have learned about* Leishmania *infection in other chapters (for example, Chapters 8 and 10), discuss how the accumulation of T_{reg} cells in the dermis is likely to impair the elimination of the pathogen from this site.*

12.4 *Hepatitis C virus is thought to interfere with dendritic cell activation and maturation. (a) How does this enable the virus to establish a chronic infection? (b) How else might HCV evade the immune response?*

12.5 *Discuss the general importance of a balanced, rather than a polarized, CD4 T-cell and cytokine response to infection. Illustrate your answer using a named pathogen. In what disease is a polarized response more beneficial, and why?*

12.6 *List the causes of immunodeficiencies affecting T lymphocytes. Why do these generally affect immune responses more severely than deficiencies involving B cells only?*

12.7 *What do people with inherited and acquired immunodeficiencies teach us about the normal mechanism of host protection against tuberculosis?*

12.8 *How does infection by HIV cause AIDS?*

12.9 *Why is it difficult to make an HIV vaccine?*

12.10 *Why can HIV infection not be cured by drug therapy?*

General references.

Chapel, H., Geha, R., and Rosen, F.: **Primary immunodeficiency diseases: an update.** *Clin. Exp. Immunol.* 2003, **132**:9–15.

Cohen, O.J., Kinter, A., and Fauci, A.S.: **Host factors in the pathogenesis of HIV disease.** *Immunol. Rev.* 1997, **159**:31–48.

De Cock, K.M., Mbori-Ngacha, D., and Marum, E.: **Shadow on the continent: public health and HIV/AIDS in Africa in the 21st century.** *Lancet* 2002, **360**:67–72.

De Cock, K.M.: **Epidemiology and the emergence of human immunodeficiency virus and acquired immune deficiency syndrome.** *Phil. Trans. R. Soc. Lond B* 2001, **356**:795–798.

Fischer, A., Cavazzana-Calvo, M., De-Saint-Basile, G., DeVillartay, J.P., Di-Santo, J.P., Hivroz, C., Rieux-Laucat, F., and Le-Deist, F.: **Naturally occurring primary deficiencies of the immune system.** *Annu. Rev. Immunol.* 1997, **15**:93–124.

Hill, A.V.: **The immunogenetics of human infectious diseases.** *Annu. Rev. Immunol.* 1998, **16**:593–617.

Korber, B., Muldoon, M., Theiler, J., Gao, F., Gupta, R., Lapedes, A., Hahn, B.H., Wolinsky, S., and Bhattacharya, T.: **Timing the ancestor of the HIV-1 pandemic strains.** *Science* 2000, **288**:1789–1796.

Lederberg, J.: **Infectious history.** *Science* 2000, **288**:287–293.

McNicholl, J.M., Downer, M.V., Udhayakumar, V., Alper, C.A., and Swerdlow, D.L.: **Host–pathogen interactions in emerging and re-emerging infectious diseases: a genomic perspective of tuberculosis, malaria, human immunodeficiency virus infection, hepatitis B, and cholera.** *Annu. Rev. Public Health* 2000, **21**:15–46.

Royce, R.A., Sena, A., Cates, W., Jr., and Cohen, M.S.: **Sexual transmission of HIV.** *N. Engl. J. Med.* 1997, **336**:1072–1078.

Tortorella, D., Gewurz, B.E., Furman, M.H., Schust, D.J., and Ploegh, H.L.: **Viral subversion of the immune system.** *Annu. Rev. Immunol.* 2000, **18**:861–926.

Xu, X.N., Screaton, G.R., and McMichael, A.J.: **Virus infections: escape, resistance, and counterattack.** *Immunity* 2001, **15**:867–870.

Zinkernagel, R.M.: **Immunology taught by viruses.** *Science* 1996, **271**:173–178.

Section references.

12-1 Antigenic variation allows pathogens to escape from immunity.

Clegg, S., Hancox, L.S., and Yeh, K.S.: ***Salmonella typhimurium* fimbrial phase variation and FimA expression.** *J. Bacteriol.* 1996, **178**:542–545.

Cossart, P.: **Host/pathogen interactions. Subversion of the mammalian cell cytoskeleton by invasive bacteria.** *J. Clin. Invest.* 1997, **99**:2307–2311.

Donelson, J.E., Hill, K.L., and El-Sayed, N.M.: **Multiple mechanisms of immune evasion by African trypanosomes.** *Mol. Biochem. Parasitol.* 1998, **91**:51–66.

Gibbs, M.J., Armstrong, J.S., and Gibbs, A.J.: **Recombination in the hemagglutinin gene of the 1918 'Spanish flu'.** *Science* 2001, **293**:1842–1845.

Hatta, M., Gao, P., Halfmann, P., and Kawaoka, Y.: **Molecular basis for high virulence of Hong Kong H5N1 influenza A viruses.** *Science* 2001, **293**:1840–1842.

Kuppers, R.: **B cells under the influence: transformation of B cells by Epstein-Barr virus.** *Nat. Rev. Immunol.* 2003, **3**:801–812.

Laver, G., and Garman, E.: **Virology. The origin and control of pandemic influenza.** *Science* 2001, **293**:1776–1777.

Ressing, M.E., Keating, S.E., van Leeuwen, D., Koppers-Lalic, D., Pappworth, I.Y., Wiertz, E.J., and Rowe, M.: **Impaired transporter associated with antigen processing-dependent peptide transport during productive EBV infection.** *J. Immunol.* 2005, **174**:6829–6838.

Rudenko, G., Cross, M., and Borst, P.: **Changing the end: antigenic variation orchestrated at the telomeres of African trypanosomes.** *Trends Microbiol.* 1998, **6**:113–116.

Seifert, H.S., Wright, C.J., Jerse, A.E., Cohen, M.S., and Cannon, J.G.: **Multiple gonococcal pilin antigenic variants are produced during experimental human infections.** *J. Clin. Invest.* 1994, **93**:2744–2749.

Webster, R.G.: **Virology. A molecular whodunit.** *Science* 2001, **293**:1773–1775.

12-2 Some viruses persist *in vivo* by ceasing to replicate until immunity wanes.

Cohen, J.I.: **Epstein–Barr virus infection.** *N. Engl. J. Med.* 2000, **343**:481–492.

Ehrlich, R.: **Selective mechanisms utilized by persistent and oncogenic viruses to interfere with antigen processing and presentation.** *Immunol. Res.* 1995, **14**:77–97.

Garcia Blanco, M.A., and Cullen, B.R.: **Molecular basis of latency in pathogenic human viruses.** *Science* 1991, **254**:815–820.

Hahn, G., Jores, R., and Mocarski, E.S.: **Cytomegalovirus remains latent in a common precursor of dendritic and myeloid cells.** *Proc. Natl Acad. Sci. USA* 1998, **95**:3937–3942.

Ho, D.Y.: **Herpes simplex virus latency: molecular aspects.** *Prog. Med. Virol.* 1992, **39**:76–115.

Longnecker, R., and Miller, C.L.: **Regulation of Epstein–Barr virus latency by latent membrane protein 2.** *Trends Microbiol.* 1996, **4**:38–42.

Macsween, K.F., and Crawford, D.H.: **Epstein–Barr virus—recent advances.** *Lancet Infect. Dis.* 2003, **3**:131–140.

Mitchell, B.M., Bloom, D.C., Cohrs, R.J., Gilden, D.H., and Kennedy, P.G.: **Herpes simplex virus-1 and varicella-zoster virus latency in ganglia.** *J. Neurovirol.* 2003, **9**:194–204.

Nash, A.A.: **T cells and the regulation of herpes simplex virus latency and reactivation.** *J. Exp. Med.* 2000, **191**:1455–1458.

Wensing, B., and Farrell, P.J.: **Regulation of cell growth and death by Epstein–Barr virus.** *Microbes Infect.* 2000, **2**:77–84.

Yewdell, J.W., and Hill, A.B.: **Viral interference with antigen presentation.** *Nat. Immunol.* 2002, **2**:1019-1025.

12-3 Some pathogens resist destruction by host defense mechanisms or exploit them for their own purposes.

Alcami, A., and Koszinowski, U.H.: **Viral mechanisms of immune evasion.** *Trends Microbiol.* 2000, **8**:410–418.

Arvin, A.M.: **Varicella-zoster virus: molecular virology and virus–host interactions.** *Curr. Opin. Microbiol.* 2001, **4**:442–449.

Brander, C., and Walker, B.D.: **Modulation of host immune responses by clinically relevant human DNA and RNA viruses.** *Curr. Opin. Microbiol.* 2000, **3**:379–386.

Cooper, S.S., Glenn, J., and Greenberg, H.B.: **Lessons in defense: hepatitis C, a case study.** *Curr. Opin. Microbiol.* 2000, **3**:363–365.

Connolly, S.E., and Benach, J.L.: **The versatile roles of antibodies in *Borrelia* infections.** *Nat. Rev. Microbiol.* 2005, **3**:411–420.

Cosman, D., Fanger, N., Borges, L., Kubin, M., Chin, W., Peterson, L., and Hsu, M.L.: **A novel immunoglobulin superfamily receptor for cellular and viral MHC class I molecules.** *Immunity* 1997, **7**:273–282.

Gewurz, B.E., Gaudet, R., Tortorella, D., Wang, E.W., and Ploegh, H.L.: **Virus subversion of immunity: a structural perspective.** *Curr. Opin. Immunol.* 2001, **13**:442–450.

Hadler, J.L.: **Learning from the 2001 anthrax attacks: immunological characteristics.** *J. Infect. Dis.* 2007, **195**:163–164.

Lauer, G.M., and Walker, B.D.: **Hepatitis C virus infection.** *N. Engl. J. Med.* 2001, **345**:41–52.

McFadden, G., and Murphy, P.M.: **Host-related immunomodulators encoded by poxviruses and herpesviruses.** *Curr. Opin. Microbiol.* 2000, **3**:371–378.

Miller, J.C., and Stevenson, B.: ***Borrelia burgdorferi* erp genes are expressed at different levels within tissues of chronically infected mammalian hosts.** *Int. J. Med. Microbiol.* 2006, **296** Suppl 40:185–194.

Park, J.M., Greten, F.R., Li, Z.W., and Karin, M.: **Macrophage apoptosis by anthrax lethal factor through p38 MAP kinase inhibition.** *Science* 2002, **297**:2048–2051.

Radolf, J.D.: **Role of outer membrane architecture in immune evasion by *Treponema pallidum* and *Borrelia burgdorferi*.** *Trends Microbiol.* 1994, **2**:307–311.

Sinai, A.P., and Joiner, K.A.: **Safe haven: the cell biology of nonfusogenic pathogen vacuoles.** *Annu. Rev. Microbiol.* 1997, **51**:415–462.

12-4 Immunosuppression or inappropriate immune responses can contribute to persistent disease.

Auffermann-Gretzinger, S., Keeffe, E.B., and Levy, S.: **Impaired dendritic cell maturation in patients with chronic, but not resolved, hepatitis C virus infection.** *Blood* 2001, **97**:3171–3176.

Bhardwaj, N.: **Interactions of viruses with dendritic cells: a double-edged sword.** *J. Exp. Med.* 1997, **186**:795–799.

Bloom, B.R., Modlin, R.L., and Salgame, P.: **Stigma variations: observations on suppressor T cells and leprosy.** *Annu. Rev. Immunol.* 1992, **10**:453–488.

Fleischer, B.: **Superantigens.** *APMIS* 1994, **102**:3–12.

Kanto, T., Hayashi, N., Takehara, T., Tatsumi, T., Kuzushita, T., Ito, A., Sasaki, Y., Kasahara, A., and Hori, M.: **Impaired allostimulatory capacity of peripheral blood dendritic cells recovered from hepatitis C virus-infected individuals.** *J. Immunol.* 1999, **162**:5584–5591.

Lerat, H., Rumin, S., Habersetzer, F., Berby, F., Trabaud, M.A., Trepo, C., and Inchauspe, G.: ***In vivo* tropism of hepatitis C virus genomic sequences in hematopoietic cells: influence of viral load, viral genotype, and cell phenotype.** *Blood* 1998, **91**:3841–3849.

Salgame, P., Abrams, J.S., Clayberger, C., Goldstein, H., Convit, J., Modlin, R.L., and Bloom, B.R.: **Differing lymphokine profiles of functional subsets of human CD4 and CD8 T cell clones.** *Science* 1991, **254**:279–282.

Swartz, M.N.: **Recognition and management of anthrax—an update.** *N. Engl. J. Med.* 2001, **345**:1621–1626.

12-5 Immune responses can contribute directly to pathogenesis.

Cheever, A.W., and Yap, G.S.: **Immunologic basis of disease and disease regulation in schistosomiasis.** *Chem. Immunol.* 1997, **66**:159–176.

Doherty, P.C., Topham, D.J., Tripp, R.A., Cardin, R.D., Brooks, J.W., and Stevenson, P.G.: **Effector CD4⁺ and CD8⁺ T-cell mechanisms in the control of respiratory virus infections.** *Immunol. Rev.* 1997, **159**:105–117.

Openshaw, P.J.: **Immunopathological mechanisms in respiratory syncytial virus disease.** *Springer Semin. Immunopathol.* 1995, **17**:187–201.

Varga, S.M., Wang, X., Welsh, R.M., and Braciale, T.J.: **Immunopathology in RSV infection is mediated by a discrete oligoclonal subset of antigen-specific CD4⁺ T cells.** *Immunity* 2001, **15**:637–646.

12-6 Regulatory T cells can affect the outcome of infectious disease.

Rouse, B.T., Sarangi, P.P., and Suvas, S.: **Regulatory T cells in virus infections.** *Immunol. Rev.* 2006, **212**:272–286.

Waldmann, H., Adams, E., Fairchild, P., and Cobbold, S.: **Infectious tolerance and the long-term acceptance of transplanted tissue.** *Immunol. Rev.* 2006,**212**:301–313.

12-7 A history of repeated infections suggests a diagnosis of immunodeficiency.

Carneiro-Sampaio, M., and Coutinho, A.: **Immunity to microbes: lessons from primary immunodeficiencies.** *Infect. Immun.* 2007, **75**:1545–1555.

Cunningham-Rundles, C., and Ponda, P.P.: **Molecular defects in T- and B-cell primary immunodeficiency diseases.** *Nat. Rev. Immunol.* 2005, **5**:880–892.

Rosen, F.S., Cooper, M.D., and Wedgwood, R.J.: **The primary immunodeficiencies.** *N. Engl. J. Med.* 1995, **333**:431–440.

12-8 Inherited immunodeficiency diseases are caused by recessive gene defects.

Fischer, A.: **Inherited disorders of lymphocyte development and function.** *Curr. Opin. Immunol.* 1996, **8**:445–447.

Kokron, C.M., Bonilla, F.A., Oettgen, H.C., Ramesh, N., Geha, R.S., and Pandolfi, F.: **Searching for genes involved in the pathogenesis of primary immunodeficiency diseases: lessons from mouse knockouts.** *J. Clin. Immunol.* 1997, **17**:109–126.

Smart, B.A., and Ochs, H.D.: **The molecular basis and treatment of primary immunodeficiency disorders.** *Curr. Opin. Pediatr.* 1997, **9**:570–576.

Smith, C.I., and Notarangelo, L.D.: **Molecular basis for X-linked immunodeficiencies.** *Adv. Genet.* 1997, **35**:57–115.

12-9 The main effect of low levels of antibody is an inability to clear extracellular bacteria.

Bruton, O.C.: **Agammaglobulinemia.** *Pediatrics* 1952, **9**:722–728.

Burrows, P.D., and Cooper, M.D.: **IgA deficiency.** *Adv. Immunol.* 1997, **65**:245–276.

Desiderio, S.: **Role of Btk in B cell development and signaling.** *Curr. Opin. Immunol.* 1997, **9**:534–540.

Fuleihan, R., Ramesh, N., and Geha, R.S.: **X-linked agammaglobulinemia and immunoglobulin deficiency with normal or elevated IgM: immunodeficiencies of B cell development and differentiation.** *Adv. Immunol.* 1995, **60**:37–56.

Lee, M.L., Gale, R.P., and Yap, P.L.: **Use of intravenous immunoglobulin to prevent or treat infections in persons with immune deficiency.** *Annu. Rev. Med.* 1997, **48**:93–102.

Notarangelo, L.D.: **Immunodeficiencies caused by genetic defects in protein kinases.** *Curr. Opin. Immunol.* 1996, **8**:448–453.

Ochs, H.D., and Wedgwood, R.J.: **IgG subclass deficiencies.** *Annu. Rev. Med.* 1987, **38**:325–340.

Preud'homme, J.L., and Hanson, L.A.: **IgG subclass deficiency.** *Immunodefic. Rev.* 1990, **2**:129–149.

12-10 Some antibody deficiencies can be due to defects in either B-cell or T-cell function.

Doffinger, R., Smahi, A., Bessia, C., Geissmann, F., Feinberg, J., Durandy, A., Bodemer, C., Kenwrick, S., Dupuis-Girod, S., Blanche, S., *et al.*: **X-linked anhidrotic ectodermal dysplasia with immunodeficiency is caused by impaired NF-κB signaling.** *Nat. Genet.* 2001, **27**:277–285.

Durandy, A., and Honjo, T.: **Human genetic defects in class-switch recombination (hyper-IgM syndromes).** *Curr. Opin. Immunol.* 2001, **13**:543–548.

Ferrari, S., Giliani, S., Insalaco, A., Al Ghonaium, A., Soresina, A.R., Loubser, M., Avanzini, M.A., Marconi, M., Badolato, R., Ugazio, A.G., *et al.*: **Mutations of CD40 gene cause an autosomal recessive form of immunodeficiency with hyper IgM.** *Proc. Natl Acad. Sci. USA* 2001, **98**:12614–12619.

Grimbacher, B., Hutloff, A., Schlesier, M., Glocker, E., Warnatz, K., Drager, R., Eibel, H., Fischer, B., Schaffer, A.A., Mages, H.W., *et al.*: **Homozygous loss of ICOS is associated with adult-onset common variable immunodeficiency.** *Nat. Immunol.* 2003, **4**:261–268.

Harris, R.S., Sheehy, A.M., Craig, H.M., Malim, M.H., and Neuberger, M.S.: **DNA deamination: not just a trigger for antibody diversification but also a mechanism for defense against retroviruses.** *Nat. Immunol.* 2003, **4**:641–643.

12-11 Defects in complement components cause defective humoral immune function.

Botto, M., Dell'Agnola, C., Bygrave, A.E., Thompson, E.M., Cook, H.T., Petry, F., Loos, M., Pandolfi, P.P., and Walport, M.J.: **Homozygous C1q deficiency causes glomerulonephritis associated with multiple apoptotic bodies.** *Nat. Genet.* 1998, **19**:56–59.

Colten, H.R., and Rosen, F.S.: **Complement deficiencies.** *Annu. Rev. Immunol.* 1992, **10**:809–834.

Dahl, M., Tybjaerg-Hansen, A., Schnohr, P., and Nordestgaard, B.G.: **A population-based study of morbidity and mortality in mannose-binding lectin deficiency.** *J. Exp. Med.* 2004, **199**:1391–1399.

Ochsenbein, A.F., and Zinkernagel, R.M.: **Natural antibodies and complement**

link innate and acquired immunity. *Immunol. Today* 2000, **21**:624–630.

Walport, M.J.: **Complement. First of two parts.** *N. Engl. J. Med.* 2001, **344**:1058–1066.

Walport, M.J.: **Complement. Second of two parts.** *N. Engl. J. Med.* 2001, **344**:1140–1144.

12-12 Defects in phagocytic cells permit widespread bacterial infections.

Ambruso, D.R., Knall, C., Abell, A.N., Panepinto, J., Kurkchubasche, A., Thurman, G., Gonzalez-Aller, C., Hiester, A., deBoer, M., Harbeck, R.J., *et al.*: **Human neutrophil immunodeficiency syndrome is associated with an inhibitory Rac2 mutation.** *Proc. Natl Acad. Sci. USA* 2000, **97**:4654–4659.

Andrews, T., and Sullivan, K.E.: **Infections in patients with inherited defects in phagocytic function.** *Clin. Microbiol. Rev.* 2003, **16**:597–621.

Aprikyan, A.A., and Dale, D.C.: **Mutations in the neutrophil elastase gene in cyclic and congenital neutropenia.** *Curr. Opin. Immunol.* 2001, **13**:535–538.

Ellson, C.D., Davidson, K., Ferguson, G.J., O'Connor, R., Stephens, L.R., and Hawkins, P.T.: **Neutrophils from *p40phox$^{-/-}$* mice exhibit severe defects in NADPH oxidase regulation and oxidant-dependent bacterial killing.** *J. Exp. Med.* 2006, **203**:1927–1937.

Fischer, A., Lisowska Grospierre, B., Anderson, D.C., and Springer, T.A.: **Leukocyte adhesion deficiency: molecular basis and functional consequences.** *Immunodefic. Rev.* 1988, **1**:39–54.

Goldblatt, D., and Thrasher, A.J.: **Chronic granulomatous disease.** *Clin. Exp. Immunol.* 2000, **122**:1–9.

Luhn, K., Wild, M.K., Eckhardt, M., Gerardy-Schahn, R., and Vestweber, D.: **The gene defective in leukocyte adhesion deficiency II encodes a putative GDP-fucose transporter.** *Nat. Genet.* 2001, **28**:69–72.

Malech, H.L., and Nauseef, W.M.: **Primary inherited defects in neutrophil function: etiology and treatment.** *Semin. Hematol.* 1997, **34**:279–290.

Rotrosen, D., and Gallin, J.I.: **Disorders of phagocyte function.** *Annu. Rev. Immunol.* 1987, **5**:127–150.

Spritz, R.A.: **Genetic defects in Chediak–Higashi syndrome and the beige mouse.** *J. Clin. Immunol.* 1998, **18**:97–105.

12-13 Defects in T-cell differentiation can result in severe combined immunodeficiencies.

Leonard, W.J.: **The molecular basis of X linked severe combined immunodeficiency.** *Annu. Rev. Med.* 1996, **47**:229–239.

Buckley, R.H., Schiff, R.I., Schiff, S.E., Markert, M.L., Williams, L.W., Harville, T.O., Roberts, J.L., and Puck, J.M.: **Human severe combined immunodeficiency: genetic, phenotypic, and functional diversity in one hundred eight infants.** *J. Pediatr.* 1997, **130**:378–387.

Stephan, J.L., Vlekova, V., Le Deist, F., Blanche, S., Donadieu, J., De Saint-Basile, G., Durandy, A., Griscelli, C., and Fischer, A.: **Severe combined immunodeficiency: a retrospective single-center study of clinical presentation and outcome in 117 patients.** *J. Pediatr.* 1993, **123**:564–572.

Hirschhorn, R.: **Adenosine deaminase deficiency: molecular basis and recent developments.** *Clin. Immunol. Immunopathol.* 1995, **76**:S219–S227.

12-14 Defects in antigen receptor gene rearrangement result in SCID.

Bosma, M.J., and Carroll, A.M.: **The SCID mouse mutant: definition, characterization, and potential uses.** *Annu. Rev. Immunol.* 1991, **9**:323–350.

Fugmann, S.D.: **DNA repair: breaking the seal.** *Nature* 2002, **416**:691–694.

Gennery, A.R., Cant, A.J., and Jeggo, P.A.: **Immunodeficiency associated with DNA repair defects.** *Clin. Exp. Immunol.* 2000, **121**:1–7.

Lavin, M.F., and Shiloh, Y.: **The genetic defect in ataxia-telangiectasia.** *Annu. Rev. Immunol.* 1997, **15**:177–202.

Moshous, D., Callebaut, I., de Chasseval, R., Corneo, B., Cavazzana-Calvo, M., Le Deist, F., Tezcan, I., Sanal, O., Bertrand, Y., Philippe, N., *et al.*: **Artemis, a novel DNA double-strand break repair/V(D)J recombination protein, is mutated in human severe combined immune deficiency.** *Cell* 2001, **105**:177–186.

12-15 Defects in signaling from T-cell antigen receptors can cause severe immunodeficiency.

Arnaiz Villena, A., Timon, M., Corell, A., Perez Aciego, P., Martin Villa, J.M., and Regueiro, J.R.: **Brief report: primary immunodeficiency caused by mutations in the gene encoding the CD3-gamma subunit of the T-lymphocyte receptor.** *N. Engl. J. Med.* 1992, **327**:529–533.

Castigli, E., Pahwa, R., Good, R.A., Geha, R.S., and Chatila, T.A.: **Molecular basis of a multiple lymphokine deficiency in a patient with severe combined immunodeficiency.** *Proc. Natl Acad. Sci. USA* 1993, **90**:4728–4732.

DiSanto, J.P., Keever, C.A., Small, T.N., Nicols, G.L., O'Reilly, R.J., and Flomenberg, N.: **Absence of interleukin 2 production in a severe combined immunodeficiency disease syndrome with T cells.** *J. Exp. Med.* 1990, **171**:1697–1704.

DiSanto, J.P., Rieux Laucat, F., Dautry Varsat, A., Fischer, A., and de Saint Basile, G.: **Defective human interleukin 2 receptor gamma chain in an atypical X chromosome-linked severe combined immunodeficiency with peripheral T cells.** *Proc. Natl Acad. Sci. USA* 1994, **91**:9466–9470.

Gilmour, K.C., Fujii, H., Cranston, T., Davies, E.G., Kinnon, C., and Gaspar, H.B.: **Defective expression of the interleukin-2/interleukin-15 receptor beta subunit leads to a natural killer cell-deficient form of severe combined immunodeficiency.** *Blood* 2001, **98**:877–879.

Humblet-Baron, S., Sather, B., Anover, S., Becker-Herman, S., Kasprowicz, D.J., Khim, S., Nguyen, T., Hudkins-Loya, K., Alpers, C.E., Ziegler, S.F., *et al.*: **Wiskott-Aldrich syndrome protein is required for regulatory T cell homeostasis.** *J Clin Invest.* 2007, **117**: 407–418.

Kung, C., Pingel, J.T., Heikinheimo, M., Klemola, T., Varkila, K., Yoo, L.I., Vuopala, K., Poyhonen, M., Uhari, M., Rogers, M., *et al.*: **Mutations in the tyrosine phosphatase CD45 gene in a child with severe combined immunodeficiency disease.** *Nat. Med.* 2000, **6**:343–345.

Ochs, H.D.: **The Wiskott–Aldrich syndrome.** *Springer Semin. Immunopathol.* 1998, **9**:435–458.

Roifman, C.M., Zhang, J., Chitayat, D., and Sharfe, N.: **A partial deficiency of interleukin-7R alpha is sufficient to abrogate T-cell development and cause severe combined immunodeficiency.** *Blood* 2000, **96**:2803–2807.

Snapper, S.B., and Rosen, F.S.: The Wiskott–Aldrich syndrome protein (WASP): roles in signaling and cytoskeletal organization. *Annu. Rev. Immunol.* 1999, **17**:905–929.

12-16 Genetic defects in thymic function that block T-cell development result in severe immunodeficiencies.

Masternak, K., Barras, E., Zufferey, M., Conrad, B., Corthals, G., Aebersold, R., Sanchez, J.C., Hochstrasser, D.F., Mach, B., and Reith, W.: **A gene encoding a novel RFX-associated transactivator is mutated in the majority of MHC class II deficiency patients.** *Nat. Genet.* 1998, **20**:273–277.

Adriani, M., Martinez-Mir, A., Fusco, F., Busiello, R., Frank, J., Telese, S., Matrecano, E., Ursini, M.V., Christiano, A.M., and Pignata, C.: **Ancestral founder mutation of the nude (FOXN1) gene in congenital severe combined immunodeficiency associated with alopecia in Southern Italy population.** *Ann. Hum. Genet.* 2004, **68**:265–268.

Coffer, P.J., and Burgering, B.M.: **Forkhead-box transcription factors and their role in the immune system.** *Nat. Rev. Immunol.* 2004, **4**:889–899.

Gadola, S.D., Moins-Teisserenc, H.T., Trowsdale, J., Gross, W.L., and Cerundolo, V.: **TAP deficiency syndrome.** *Clin. Exp. Immunol.* 2000, **121**:173–178.

Grusby, M.J., and Glimcher, L.H.: **Immune responses in MHC class II-deficient mice.** *Annu. Rev. Immunol.* 1995, **13**:417–435.

Pignata, C., Gaetaniello, L., Masci, A.M., Frank, J., Christiano, A., Matrecano, E., and Racioppi, L.: **Human equivalent of the mouse Nude/SCID phenotype: long-term evaluation of immunologic reconstitution after bone marrow transplantation.** *Blood* 2001, **97**:880–885.

Schinke, M., and Izumo, S.: **Deconstructing DiGeorge syndrome.** *Nat.Genet.* 2001, **27**:238–240.

Steimle, V., Reith, W., and Mach, B.: **Major histocompatibility complex class II deficiency: a disease of gene regulation.** *Adv. Immunol.* 1996, **61**:327–340.

12-17 The normal pathways for host defense against intracellular bacteria are pinpointed by genetic deficiencies of IFN-γ and IL-12 and their receptors.

Casanova, J.L., and Abel, L.: **Genetic dissection of immunity to mycobacteria: the human model.** *Annu. Rev. Immunol.* 2002, **20**:581–620.

Dupuis, S., Dargemont, C., Fieschi, C., Thomassin, N., Rosenzweig, S., Harris, J., Holland, S.M., Schreiber, R.D., and Casanova, J.L.: **Impairment of mycobacterial but not viral immunity by a germline human STAT1 mutation.** *Science* 2001, **293**:300–303.

Keane, J., Gershon, S., Wise, R.P., Mirabile-Levens, E., Kasznica, J., Schwieterman, W.D., Siegel, J.N., and Braun, M.M.: **Tuberculosis associated with infliximab, a tumor necrosis factor α-neutralizing agent.** *N. Engl. J. Med.* 2001, **345**:1098–1104.

Lammas, D.A., Casanova, J.L., and Kumararatne, D.S.: **Clinical consequences of defects in the IL-12-dependent interferon-γ (IFN-γ) pathway.** *Clin. Exp. Immunol.* 2000, **121**:417–425.

Newport, M.J., Huxley, C.M., Huston, S., Hawrylowicz, C.M., Oostra, B.A., Williamson, R., and Levin, M.: **A mutation in the interferon-γ-receptor gene and susceptibility to mycobacterial infection.** *N. Engl. J. Med.* 1996, **335**:1941–1949.

Shtrichman, R., and Samuel, C.E.: **The role of γ interferon in antimicrobial immunity.** *Curr. Opin. Microbiol.* 2001, **4**:251–259.

Van de Vosse, E., Hoeve, M.A., and Ottenhoff, T.H.: **Human genetics of intracellular infectious diseases: molecular and cellular immunity against mycobacteria and salmonellae.** *Lancet Infect Dis* 2004, **4**:739–749.

12-18 X-linked lymphoproliferative syndrome is associated with fatal infection by Epstein–Barr virus and with the development of lymphomas.

Latour, S., Gish, G., Helgason, C.D., Humphries, R.K., Pawson, T., and Veillette, A.: **Regulation of SLAM-mediated signal transduction by SAP, the X-linked lymphoproliferative gene product.** *Nat. Immunol.* 2001, **2**:681–690.

Milone, M.C., Tsai, D.E., Hodinka, R.L., Silverman, L.B., Malbran, A., Wasik, M.A., and Nichols, K.E.: **Treatment of primary Epstein-Barr virus infection in patients with X-linked lymphoproliferative disease using B-cell-directed therapy.** *Blood* 2005, **105**:994–996.

Morra, M., Howie, D., Grande, M.S., Sayos, J., Wang, N., Wu, C., Engel, P., and Terhorst, C.: **X-linked lymphoproliferative disease: a progressive immunodeficiency.** *Annu. Rev. Immunol.* 2001, **19**:657–682.

Nichols, K.E., Koretzky, G.A., and June, C.H.: **SAP: natural inhibitor or grand SLAM of T-cell activation?** *Nat. Immunol.* 2001, **2**:665–666.

Satterthwaite, A.B., Rawlings, D.J., and Witte, O.N.: **DSHP: a 'power bar' for sustained immune responses?** *Proc. Natl Acad. Sci. USA* 1998, **95**:13355–13357.

12-19 Genetic abnormalities in the secretory cytotoxic pathway of lymphocytes cause uncontrolled lymphoproliferation and inflammatory responses to viral infections.

de Saint, B.G., and Fischer, A.: **The role of cytotoxicity in lymphocyte homeostasis.** *Curr. Opin. Immunol.* 2001, **13**:549–554.

Dell'Angelica, E.C., Mullins, C., Caplan, S., and Bonifacino, J.S.: **Lysosome-related organelles.** *FASEB J.* 2000, **14**:1265–1278.

Huizing, M., Anikster, Y., and Gahl, W.A.: **Hermansky–Pudlak syndrome and Chediak–Higashi syndrome: disorders of vesicle formation and trafficking.** *Thromb. Haemost.* 2001, **86**:233–245.

Menasche, G., Pastural, E., Feldmann, J., Certain, S., Ersoy, F., Dupuis, S., Wulffraat, N., Bianchi, D., Fischer, A., Le Deist, F., and de Saint, B.G.: **Mutations in RAB27A cause Griscelli syndrome associated with haemophagocytic syndrome.** *Nat. Genet.* 2000, **25**:173–176.

Stinchcombe, J.C., and Griffiths, G.M.: **Normal and abnormal secretion by haemopoietic cells.** *Immunology* 2001, **103**:10–16.

12-20 Bone marrow transplantation or gene therapy can be useful to correct genetic defects.

Fischer, A., Le Deist, F., Hacein-Bey-Abina, S., Andre-Schmutz, I., de Saint, B.G., de Villartay, J.P., and Cavazzana-Calvo, M.: **Severe combined immunodeficiency. A model disease for molecular immunology and therapy.** *Immunol. Rev.* 2005, **203**:98–109.

Pesu, M., Candotti, F., Husa, M., Hofmann, S.R., Notarangelo, L.D., and O'Shea, J.J.: **Jak3, severe combined immunodeficiency, and a new class of immunosuppressive drugs.** *Immunol Rev.* 2005, **203**:127–142.

Anderson, W.F.: **Human gene therapy.** *Nature* 1998, **392**:25–30.

Candotti, F., and Blaese, R.M.: **Gene therapy of primary immunodeficiencies.** *Springer Semin. Immunopathol.* 1998, **19**:493–508.

Fischer, A., Hacein-Bey, S., and Cavazzana-Calvo, M.: **Gene therapy of severe combined immunodeficiencies.** *Nat. Rev. Immunol.* 2002, **2**:615–621.

Fischer, A., Haddad, E., Jabado, N., Casanova, J.L., Blanche, S., Le Deist, F., and Cavazzana-Calvo, M.: **Stem cell transplantation for immunodeficiency.** *Springer Semin. Immunopathol.* 1998, **19**:479–492.

Hacein-Bey-Abina, S., Le Deist, F., Carlier, F., Bouneaud, C., Hue, C., De Villartay, J.P., Thrasher, A.J., Wulffraat, N., Sorensen, R., Dupuis-Girod, S., *et al.*: **Sustained correction of X-linked severe combined immunodeficiency by *ex vivo* gene therapy.** *N. Engl. J. Med.* 2002, **346**:1185–1193.

Hacein-Bey-Abina, S., Von Kalle, C., Schmidt, M., McCormack, M.P., Wulffraat, N., Leboulch, P., Lim, A., Osborne, C.S., Pawliuk, R., Morillon, E., *et al.*: **LMO2-associated clonal T cell proliferation in two patients after gene therapy for SCID-X1.** *Science* 2003, **302**:415–419.

Kohn, D.B., Hershfield, M.S., Carbonaro, D., Shigeoka, A., Brooks, J., Smogorzewska, E.M., Barsky, L.W., Chan, R., Burotto, F., Annett, G., *et al.*: **T lymphocytes with a normal ADA gene accumulate after transplantation of transduced autologous umbilical cord blood $CD34^+$ cells in ADA-deficient SCID neonates.** *Nat. Med.* 1998, **4**:775–780.

Onodera, M., Ariga, T., Kawamura, N., Kobayashi, I., Ohtsu, M., Yamada, M., Tame, A., Furuta, H., Okano, M., Matsumoto, S., *et al.*: **Successful peripheral T-lymphocyte-directed gene transfer for a patient with severe combined immune deficiency caused by adenosine deaminase deficiency.** *Blood* 1998, **91**:30–36.

Rosen, F.S.: **Successful gene therapy for severe combined immunodeficiency.** *N. Engl. J. Med.* 2002, **346**:1241–1243.

12-21 Secondary immunodeficiencies are major predisposing causes of infection and death.

Chandra, R.K.: **Nutrition, immunity and infection: from basic knowledge of dietary manipulation of immune responses to practical application of ameliorating suffering and improving survival.** *Proc. Natl Acad. Sci. USA* 1996, **93**:14304–14307.

Lord, G.M., Matarese, G., Howard, J.K., Baker, R.J., Bloom, S.R., and Lechler, R.I.: **Leptin modulates the T-cell immune response and reverses starvation-induced immunosuppression.** *Nature* 1998, **394**:897–901.

12-22 Most individuals infected with HIV progress over time to AIDS.

Gao, F., Bailes, E., Robertson, D.L., Chen, Y., Rodenburg, C.M., Michael, S.F., Cummins, L.B., Arthur, L.O., Peeters, M., Shaw, G.M., *et al.*: **Origin of HIV-1 in the chimpanzee *Pan troglodytes troglodytes*.** *Nature* 1999, **397**:436–441.

Heeney, J.L., Dalgleish, A.G., and Weiss, R.A.: **Origins of HIV and the evolution of resistance to AIDS.** *Science* 2006, **313**:462–466.

Baltimore, D.: **Lessons from people with nonprogressive HIV infection.** *N. Engl. J. Med.* 1995, **332**:259–260.

Barre-Sinoussi, F.: **HIV as the cause of AIDS.** *Lancet* 1996, **348**:31–35.

Kirchhoff, F., Greenough, T.C., Brettler, D.B., Sullivan, J.L., and Desrosiers, R.C.: **Brief report: absence of intact nef sequences in a long-term survivor with non-progressive HIV-1 infection.** *N. Engl. J. Med.* 1995, **332**:228–232.

Pantaleo, G., Menzo, S., Vaccarezza, M., Graziosi, C., Cohen, O.J., Demarest, J.F., Montefiori, D., Orenstein, J.M., Fox, C., Schrager, L.K., *et al.*: **Studies in subjects with long-term nonprogressive human immunodeficiency virus infection.** *N. Engl. J. Med.* 1995, **332**:209–216.

Peckham, C., and Gibb, D.: **Mother-to-child transmission of the human immunodeficiency virus.** *N. Engl. J. Med.* 1995, **333**:298–302.

Rosenberg, P.S., and Goedert, J.J.: **Estimating the cumulative incidence of HIV infection among persons with haemophilia in the United States of America.** *Stat. Med.* 1998, **17**:155–168.

Volberding, P.A.: **Age as a predictor of progression in HIV infection.** *Lancet* 1996, **347**:1569–1570.

Wang, W.K., Essex, M., McLane, M.F., Mayer, K.H., Hsieh, C.C., Brumblay, H.G., Seage, G., and Lee, T.H.R.: **Pattern of gp120 sequence divergence linked to a lack of clinical progression in human immunodeficiency virus type 1 infection.** *Proc. Natl Acad. Sci. USA* 1996, **93**:6693–6697.

12-23 HIV is a retrovirus that infects CD4 T cells, dendritic cells, and macrophages.

Bomsel, M., and David, V.: **Mucosal gatekeepers: selecting HIV viruses for early infection.** *Nat. Med.* 2002, **8**:114–116.

Cammack, N.: **The potential for HIV fusion inhibition.** *Curr. Opin. Infect. Dis.* 2001, **14**:13–16.

Chan, D.C., and Kim, P.S.: **HIV entry and its inhibition.** Cell 1998, **93**:681–684.

Connor, R.I., Sheridan, K.E., Ceradini, D., Choe, S., and Landau, N.R.: **Change in coreceptor use correlates with disease progression in HIV-1—infected individuals.** *J. Exp. Med.* 1997, **185**:621–628.

Farber, J.M., and Berger, E.A.: **HIV's response to a CCR5 inhibitor: I'd rather tighten than switch!** *Proc. Natl Acad. Sci. USA* 2002, **99**:1749–1751.

Grouard, G., and Clark, E.A.: **Role of dendritic and follicular dendritic cells in HIV infection and pathogenesis.** *Curr. Opin. Immunol.* 1997, **9**:563–567.

Kilby, J.M., Hopkins, S., Venetta, T.M., DiMassimo, B., Cloud, G.A., Lee, J.Y., Alldredge, L., Hunter, E., Lambert, D., Bolognesi, D., *et al.*: **Potent suppression of HIV-1 replication in humans by T-20, a peptide inhibitor of gp41-mediated virus entry.** *Nat.Med.* 1998, **4**:1302–1307.

Kwon, D.S., Gregorio, G., Bitton, N., Hendrickson, W.A., and Littman, D.R.: **DC-SIGN-mediated internalization of HIV is required for trans-enhancement of T cell infection.** *Immunity* 2002, **16**:135–144.

Moore, J.P., Trkola, A., and Dragic, T.: **Co-receptors for HIV-1 entry.** *Curr. Opin. Immunol.* 1997, **9**:551–562.

Pohlmann, S., Baribaud, F., and Doms, R.W.: **DC-SIGN and DC-SIGNR: helping hands for HIV.** *Trends Immunol.* 2001, **22**:643–646.

Root, M.J., Kay, M.S., and Kim, P.S.: **Protein design of an HIV-1 entry inhibitor.** *Science* 2001, **291**:884–888.

Sol-Foulon, N., Moris, A., Nobile, C., Boccaccio, C., Engering, A., Abastado, J.P., Heard, J.M., van Kooyk, Y., and Schwartz, O.: **HIV-1 Nef-induced upregulation of DC-SIGN in dendritic cells promotes lymphocyte clustering and viral spread.** *Immunity* 2002, **16**:145–155.

Unutmaz, D., and Littman, D.R.: **Expression pattern of HIV-1 coreceptors on T cells: implications for viral transmission and lymphocyte homing.** *Proc. Natl Acad. Sci. USA* 1997, **94**:1615–1618.

Wyatt, R., and Sodroski, J.: **The HIV-1 envelope glycoproteins: fusogens, antigens, and immunogens.** *Science* 1998, **280**:1884–1888.

12-24 Genetic variation in the host can alter the rate of progression of disease.

Bream, J.H., Ping, A., Zhang, X., Winkler, C., and Young, H.A.: **A single nucleotide polymorphism in the proximal IFN-gamma promoter alters control of gene transcription.** *Genes Immun.* 2002, **3**:165–169.

Martin, M.P., Gao, X., Lee, J.H., Nelson, G.W., Detels, R., Goedert, J.J., Buchbinder, S., Hoots, K., Vlahov, D., Trowsdale, J., *et al.*: **Epistatic interaction between KIR3DS1 and HLA-B delays the progression to AIDS.** *Nat. Genet.* 2002, **31**:429–434.

Shin, H.D., Winkler, C., Stephens, J.C., Bream, J., Young, H., Goedert, J.J., O'Brien, T.R., Vlahov, D., Buchbinder, S., Giorgi, J., *et al.*: **Genetic restriction of HIV-1 pathogenesis to AIDS by promoter alleles of IL10.** *Proc. Natl Acad. Sci. USA* 2000, **97**:14467–14472.

12-25 A genetic deficiency of the co-receptor CCR5 confers resistance to HIV infection in vivo.

Berger, E.A., Murphy, P.M., and Farber, J.M.: **Chemokine receptors as HIV-1 coreceptors: roles in viral entry, tropism, and disease.** *Annu. Rev. Immunol.* 1999, **17**:657–700.

Gonzalez, E., Kulkarni, H., Bolivar, H., Mangano, A., Sanchez, R., Catano, G., Nibbs, R.J., Freedman, B.I., Quinones, M.P., Bamshad, M.J., *et al.*: **The influence of CCL3L1 gene-containing segmental duplications on HIV-1/AIDS susceptibility.** *Science* 2005, **307**:1434–1440.

Lehner, T.: **The role of CCR5 chemokine ligands and antibodies to CCR5 coreceptors in preventing HIV infection.** *Trends Immunol.* 2002, **23**:347–351.

Littman, D.R.: **Chemokine receptors: keys to AIDS pathogenesis?** *Cell* 1998, **93**:677–680.

Liu, R., Paxton, W.A., Choe, S., Ceradini, D., Martin, S.R., Horuk, R., Macdonald, M.E., Stuhlmann, H., Koup, R.A., and Landau, N.R.: **Homozygous defect in HIV-1 coreceptor accounts for resistance of some multiply exposed individuals to HIV 1 infection.** *Cell* 1996, **86**:367–377.

Murakami, T., Nakajima, T., Koyanagi, Y., Tachibana, K., Fujii, N., Tamamura, H., Yoshida, N., Waki, M., Matsumoto, A., Yoshie, O., *et al.*: **A small molecule CXCR4 inhibitor that blocks T cell line-tropic HIV-1 infection.** *J. Exp. Med.* 1997, **186**:1389–1393.

Samson, M., Libert, F., Doranz, B.J., Rucker, J., Liesnard, C., Farber, C.M., Saragosti, S., Lapoumeroulie, C., Cognaux, J., Forceille, C., *et al.*: **Resistance to HIV-1 infection in Caucasian individuals bearing mutant alleles of the CCR 5 chemokine receptor gene.** *Nature* 1996, **382**:722–725.

Yang, A.G., Bai, X., Huang, X.F., Yao, C., and Chen, S.: **Phenotypic knockout of HIV type 1 chemokine coreceptor CCR-5 by intrakines as potential therapeutic approach for HIV-1 infection.** *Proc. Natl Acad. Sci. USA* 1997, **94**:11567–11572.

12-26 HIV RNA is transcribed by viral reverse transcriptase into DNA that integrates into the host-cell genome.

Andrake, M.D., and Skalka, A.M.R.: **Retroviral integrase, putting the pieces together.** *J. Biol. Chem.* 1995, **271**:19633–19636.

Baltimore, D.: **The enigma of HIV infection.** *Cell* 1995, **82**:175–176.

McCune, J.M.: **Viral latency in HIV disease.** *Cell* 1995, **82**:183–188.

Wei, P., Garber, M.E., Fang, S.M., Fischer, W.H., and Jones, K.A.: **A novel CDK9-associated C-type cyclin interacts directly with HIV-1 Tat and mediates its high-affinity, loop-specific binding to TAR RNA.** *Cell* 1998, **92**:451–462.

12-27 Replication of HIV occurs only in activated T cells.

Cullen, B.R.: **Connections between the processing and nuclear export of mRNA: evidence for an export license?** *Proc. Natl Acad. Sci. USA* 2000, **97**:4–6.

Cullen, B.R.: **HIV-1 auxiliary proteins: making connections in a dying cell.** *Cell* 1998, **93**:685–692.

Emerman, M., and Malim, M.H.: **HIV-1 regulatory/accessory genes: keys to unraveling viral and host cell biology.** *Science* 1998, **280**:1880–1884.

Fujinaga, K., Taube, R., Wimmer, J., Cujec, T.P., and Peterlin, B.M.: **Interactions between human cyclin T, Tat, and the transactivation response element (TAR) are disrupted by a cysteine to tyrosine substitution found in mouse cyclin T.** *Proc. Natl Acad. Sci. USA* 1999, **96**:1285–1290.

Kinoshita, S., Su, L., Amano, M., Timmerman, L.A., Kaneshima, H., and Nolan, G.P.: **The T-cell activation factor NF-ATc positively regulates HIV-1 replication** and gene expression in T cells. *Immunity* 1997, **6**:235–244.

Pollard, V.W., and Malim, M.H.: **The HIV-1 Rev protein.** Annu. Rev. Microbiol. 1998, **52**:491–532.

Subbramanian, R.A., and Cohen, E.A.: **Molecular biology of the human immuno-deficiency virus accessory proteins.** *J. Virol.* 1994, **68**:6831–6835.

Trono, D.: **HIV accessory proteins: leading roles for the supporting cast.** *Cell* 1995, **82**:189–192.

12-28 Lymphoid tissue is the major reservoir of HIV infection.

Burton, G.F., Masuda, A., Heath, S.L., Smith, B.A., Tew, J.G., and Szakal, A.K.: **Follicular dendritic cells (FDC) in retroviral infection: host/pathogen perspectives.** *Immunol. Rev.* 1997, **156**:185–197.

Chun, T.W., Carruth, L., Finzi, D., Shen, X., DiGiuseppe, J.A., Taylor, H., Hermankova, M., Chadwick, K., Margolick, J., Quinn, T.C., *et al.*: **Quantification of latent tissue reservoirs and total body viral load in HIV-1 infection.** *Nature* 1997, **387**:183–188.

Clark, E.A.: **HIV: dendritic cells as embers for the infectious fire.** *Curr. Biol.* 1996, **6**:655–657.

Finzi, D., Blankson, J., Siliciano, J.D., Margolick, J.B., Chadwick, K., Pierson, T., Smith, K., Lisziewicz, J., Lori, F., Flexner, C., *et al.*: **Latent infection of CD4$^+$ T cells provides a mechanism for lifelong persistence of HIV-1, even in patients on effective combination therapy.** *Nat. Med.* 1999, **5**:512–517.

Haase, A.T.: **Population biology of HIV-1 infection: viral and CD4$^+$ T cell demographics and dynamics in lymphatic tissues.** *Annu. Rev. Immunol.* 1999, **17**:625–656.

Orenstein, J.M., Fox, C., and Wahl, S.M.: **Macrophages as a source of HIV during opportunistic infections.** *Science* 1997, **276**:1857–1861.

Palella, F.J., Jr., Delaney, K.M., Moorman, A.C., Loveless, M.O., Fuhrer, J., Satten, G.A., Aschman, D.J., and Holmberg, S.D.: **Declining morbidity and mortality among patients with advanced human immunodeficiency virus infection. HIV Outpatient Study Investigators.** *N. Engl. J. Med.* 1998, **338**:853–860.

Pierson, T., McArthur, J., and Siliciano, R.F.: **Reservoirs for HIV-1: mechanisms for viral persistence in the presence of antiviral immune responses and antiretroviral therapy.** *Annu. Rev. Immunol.* 2000, **18**:665–708.

Wong, J.K., Hezareh, M., Gunthard, H.F., Havlir, D.V., Ignacio, C.C., Spina, C.A., and Richman, D.D.: **Recovery of replication-competent HIV despite prolonged suppression of plasma viremia.** *Science* 1997, **278**:1291–1295.

12-29 An immune response controls but does not eliminate HIV.

Barouch, D.H., and Letvin, N.L.: **CD8$^+$ cytotoxic T lymphocyte responses to lentiviruses and herpesviruses.** *Curr. Opin. Immunol.* 2001, **13**:479–482.

Chiu, Y.L., Soros, V.B., Kreisberg, J.F., Stopak, K., Yonemoto, W., and Greene, W.C.: **Cellular APOBEC3G restricts HIV-1 infection in resting CD4$^+$ T cells.** *Nature* 2005, **435**:108–114.

Evans, D.T., O'Connor, D.H., Jing, P., Dzuris, J.L., Sidney, J., da Silva, J., Allen, T.M., Horton, H., Venham, J.E., Rudersdorf, R.A., *et al.*: **Virus-specific cytotoxic T-lymphocyte responses select for amino-acid variation in simian immunodeficiency virus Env and Nef.** *Nat. Med.* 1999, **5**:1270–1276.

Goulder, P.J., Sewell, A.K., Lalloo, D.G., Price, D.A., Whelan, J.A., Evans, J., Taylor, G.P., Luzzi, G., Giangrande, P., Phillips, R.E., *et al.*: **Patterns of immunodominance in HIV-1-specific cytotoxic T lymphocyte responses in two human histocompatibility leukocyte antigens (HLA)-identical siblings with HLA-A*0201 are influenced by epitope mutation.** *J. Exp. Med.* 1997, **185**:1423–1433.

Johnson, W.E., and Desrosiers, R.C.: **Viral persistance: HIV's strategies of immune system evasion.** *Annu. Rev. Med.* 2002, **53**:499–518.

Poignard, P., Sabbe, R., Picchio, G.R., Wang, M., Gulizia, R.J., Katinger, H., Parren, P.W., Mosier, D.E., and Burton, D.R.: **Neutralizing antibodies have limited effects on the control of established HIV-1 infection *in vivo*.** *Immunity* 1999, **10**:431–438.

Price, D.A., Goulder, P.J., Klenerman, P., Sewell, A.K., Easterbrook, P.J., Troop, M., Bangham, C.R., and Phillips, R.E.: **Positive selection of HIV-1 cytotoxic T lymphocyte escape variants during primary infection.** *Proc. Natl Acad. Sci. USA* 1997, **94**:1890–1895.

Schmitz, J.E., Kuroda, M.J., Santra, S., Sasseville, V.G., Simon, M.A., Lifton, M.A., Racz, P., Tenner-Racz, K., Dalesandro, M., Scallon, B.J., *et al.*: **Control of viremia in simian immunodeficiency virus infection by CD8$^+$ lymphocytes.** *Science* 1999, **283**:857–860.

Stremlau, M., Owens, C.M., Perron, M.J., Kiessling, M., Autissier, P., and Sodroski, J.: **The cytoplasmic body component TRIM5alpha restricts HIV-1 infection in Old World monkeys.** *Nature* 2004, **427**:848–583.

12-30 The destruction of immune function as a result of HIV infection leads to increased susceptibility to opportunistic infection and eventually to death.

Badley, A.D., Dockrell, D., Simpson, M., Schut, R., Lynch, D.H., Leibson, P., and Paya, C.V.: **Macrophage-dependent apoptosis of CD4$^+$ T lymphocytes from HIV-infected individuals is mediated by FasL and tumor necrosis factor.** *J. Exp. Med.* 1997, **185**:55–64.

Ho, D.D., Neumann, A.U., Perelson, A.S., Chen, W., Leonard, J.M., and Markowitz, M.: **Rapid turnover of plasma virions and CD4 lymphocytes in HIV-1 infection.** *Nature* 1995, **373**:123–126.

Kedes, D.H., Operskalski, E., Busch, M., Kohn, R., Flood, J., and Ganem, D.R.: **The seroepidemiology of human herpesvirus 8 (Kaposi's sarcoma associated herpesvirus): distribution of infection in KS risk groups and evidence for sexual transmission.** *Nat. Med.* 1996, **2**:918–924.

Kolesnitchenko, V., Wahl, L.M., Tian, H., Sunila, I., Tani, Y., Hartmann, D.P., Cossman, J., Raffeld, M., Orenstein, J., Samelson, L.E., and Cohen, D.I.: **Human immunodeficiency virus 1 envelope-initiated G2-phase programmed cell death.** *Proc. Natl Acad. Sci. USA* 1995, **92**:11889–11893.

Lauer, G.M., and Walker, B.D.: **Hepatitis C virus infection.** *N. Engl. J. Med.* 2001, **345**:41–52.

Miller, R.: **HIV-associated respiratory diseases.** *Lancet* 1996, **348**:307–312.

Pantaleo, G., and Fauci, A.S.: **Apoptosis in HIV infection.** *Nat. Med.* 1995, **1**:118–120.

Zhong, W.D., Wang, H., Herndier, B., and Ganem, D.R.: **Restricted expression of Kaposi sarcoma associated herpesvirus (human herpesvirus 8) genes in Kaposi sarcoma.** *Proc. Natl Acad. Sci. USA* 1996, **93**:6641–6646.

12-31 Drugs that block HIV replication lead to a rapid decrease in titer of infectious virus and an increase in CD4 T cells.

Boyd, M., and Reiss, P.: **The long-term consequences of antiretroviral therapy: a review.** *J. HIV Ther.* 2006, **11**:26–35.

Carcelain, G., Debre, P., and Autran, B.: **Reconstitution of CD4$^+$ T lymphocytes in HIV-infected individuals following antiretroviral therapy.** *Curr. Opin. Immunol.* 2001, **13**:483–488.

Chun, T.W., and Fauci, A.S.: **Latent reservoirs of HIV: obstacles to the eradication of virus.** *Proc Natl Acad Sci USA* 1999, **96**:10958–10961.

Ho, D.D.: **Perspectives series: host/pathogen interactions. Dynamics of HIV-1 replication *in vivo*.** *J. Clin. Invest.* 1997, **99**:2565–2567.

Lempicki, R.A., Kovacs, J.A., Baseler, M.W., Adelsberger, J.W., Dewar, R.L., Natarajan, V., Bosche, M.C., Metcalf, J.A., Stevens, R.A., Lambert, L.A., *et al.*: **Impact of HIV-1 infection and highly active antiretroviral therapy on the kinetics of CD4$^+$ and CD8$^+$ T cell turnover in HIV-infected patients.** *Proc. Natl Acad. Sci. USA* 2000, **97**:13778–13783.

Lipsky, J.J.: **Antiretroviral drugs for AIDS.** *Lancet* 1996, **348**:800–803.

Lundgren, J.D., and Mocroft, A.: **The impact of antiretroviral therapy on AIDS and survival.** *J. HIV Ther.* 2006, **11**:36–38.

Palella, F.J., Jr., Delaney, K.M., Moorman, A.C., Loveless, M.O., Fuhrer, J., Satten, G.A., Aschman, D.J., and Holmberg, S.D.: **Declining morbidity and mortality among patients with advanced human immunodeficiency virus infection. HIV Outpatient Study Investigators.** *N. Engl. J. Med.* 1998, **338**:853–860.

Perelson, A.S., Essunger, P., Cao, Y.Z., Vesanen, M., Hurley, A., Saksela, K., Markowitz, M., and Ho, D.D.: **Decay characteristics of HIV-1-infected compartments during combination therapy.** *Nature* 1997, **387**:188–191.

Pau, A.K., and Tavel, J.A.: **Therapeutic use of interleukin-2 in HIV-infected** patients. *Curr. Opin. Pharmacol.* 2002, **2**:433–439.

Smith, D.: **The long-term consequences of antiretroviral therapy.** *J. HIV Ther.* 2006, **11**:24–25.

Smith, K.A.: **To cure chronic HIV infection, a new therapeutic strategy is needed.** *Curr. Opin. Immunol.* 2001, **13**:617–624.

Wei, X., Ghosh, S.K., Taylor, M.E., Johnson, V.A., Emini, E.A., Deutsch, P., Lifson, J.D., Bonhoeffer, S., Nowak, M.A., Hahn, B.H., *et al.*: **Viral dynamics in human immunodeficiency virus type 1 infection.** *Nature* 1995, **373**:117–122.

12-32 HIV accumulates many mutations in the course of infection, and drug treatment is soon followed by the outgrowth of drug-resistant variants.

Bonhoeffer, S., May, R.M., Shaw, G.M., and Nowak, M.A.: **Virus dynamics and drug therapy.** *Proc. Natl Acad. Sci. USA* 1997, **94**:6971–6976.

Condra, J.H., Schleif, W.A., Blahy, O.M., Gabryelski, L.J., Graham, D.J., Quintero, J.C., Rhodes, A., Robbins, H.L., Roth, E., Shivaprakash, M., *et al.*: ***In vivo* emergence of HIV-1 variants resistant to multiple protease inhibitors.** *Nature* 1995, **374**:569–571.

Finzi, D., and Siliciano, R.F.: **Viral dynamics in HIV-1 infection.** *Cell* 1998, **93**:665–671.

Katzenstein, D.: **Combination therapies for HIV infection and genomic drug resistance.** *Lancet* 1997, **350**:970–971.

Moutouh, L., Corbeil, J., and Richman, D.D.: **Recombination leads to the rapid emergence of HIV 1 dually resistant mutants under selective drug pressure.** *Proc. Natl Acad. Sci. USA* 1996, **93**:6106–6111.

12-33 Vaccination against HIV is an attractive solution but poses many difficulties.

Amara, R.R., Villinger, F., Altman, J.D., Lydy, S.L., O'Neil, S.P., Staprans, S.I., Montefiori, D.C., Xu, Y., Herndon, J.G., Wyatt, L.S., *et al.*: **Control of a mucosal challenge and prevention of AIDS by a multiprotein DNA/MVA vaccine.** *Science* 2001, **292**:69–74.

Baba, T.W., Liska, V., Hofmann-Lehmann, R., Vlasak, J., Xu, W., Ayehunie, S., Cavacini, L.A., Posner, M.R., Katinger, H., Stiegler, G., *et al.*: **Human neutralizing monoclonal antibodies of the IgG1 subtype protect against mucosal simian–human immunodeficiency virus infection.** *Nat. Med.* 2000, **6**:200–206.

Barouch, D.H., Kunstman, J., Kuroda, M.J., Schmitz, J.E., Santra, S., Peyerl, F.W., Krivulka, G.R., Beaudry, K., Lifton, M.A., Gorgone, D.A., *et al.*: **Eventual AIDS vaccine failure in a rhesus monkey by viral escape from cytotoxic T lymphocytes.** *Nature* 2002, **415**:335–339.

Burton, D.R.: **A vaccine for HIV type 1: the antibody perspective.** *Proc. Natl Acad. Sci. USA* 1997, **94**:10018–10023.

Kaul, R., Rowland-Jones, S.L., Kimani, J., Dong, T., Yang, H.B., Kiama, P., Rostron, T., Njagi, E., Bwayo, J.J., MacDonald, K.S., *et al.*: **Late seroconversion in HIV-resistant Nairobi prostitutes despite pre-existing HIV-specific CD8$^+$ responses.** *J. Clin. Invest.* 2001, **107**:341–349.

Letvin, N.L.: **Strategies for an HIV vaccine.** *J. Clin. Invest.* 2002, **110**:15–20.

Letvin, N.L., Barouch, D.H., and Montefiori, D.C.: **Prospects for vaccine protection against HIV-1 infection and AIDS.** *Annu. Rev. Immunol.* 2002, **20**:73–99.

Letvin, N.L., and Walker, B.D.: **HIV versus the immune system: another apparent victory for the virus.** *J. Clin. Invest.* 2001, **107**:273–275.

MacQueen, K.M., Buchbinder, S., Douglas, J.M., Judson, F.N., McKirnan, D.J., and Bartholow, B.: **The decision to enroll in HIV vaccine efficacy trials: concerns elicited from gay men at increased risk for HIV infection.** *AIDS Res. Hum. Retroviruses* 1994, **10 Suppl 2**:S261–S264.

Mascola, J.R., and Nabel, G.J.: **Vaccines for the prevention of HIV-1 disease.** *Curr. Opin. Immunol.* 2001, **13**:489–495.

Mascola, J.R., Stiegler, G., VanCott, T.C., Katinger, H., Carpenter, C.B., Hanson, C.E., Beary, H., Hayes, D., Frankel, S.S., Birx, D.L., and Lewis, M.G.: **Protection of macaques against vaginal transmission of a pathogenic HIV-1/SIV chimeric virus by passive infusion of neutralizing antibodies.** *Nat. Med.* 2000, **6**:207–210.

Robert-Guroff, M.: **IgG surfaces as an important component in mucosal protection.** *Nat. Med.* 2000, **6**:129–130.

Shiver, J.W., Fu, T.M., Chen, L., Casimiro, D.R., Davies, M.E., Evans, R.K., Zhang, Z.Q., Simon, A.J., Trigona, W.L., Dubey, S.A., *et al.*: **Replication-incompetent adenoviral vaccine vector elicits effective anti-immunodeficiency-virus immunity.** *Nature* 2002, **415**:331–335.

12-34 Prevention and education are one way in which the spread of HIV and AIDS can be controlled.

Coates, T.J., Aggleton, P., Gutzwiller, F., Des-Jarlais, D., Kihara, M., Kippax, S., Schechter, M., and van-den-Hoek, J.A.: **HIV prevention in developed countries.** *Lancet* 1996, **348**:1143–1148.

Decosas, J., Kane, F., Anarfi, J.K., Sodji, K.D., and Wagner, H.U.: **Migration and AIDS.** *Lancet* 1995, **346**:826–828.

Dowsett, G.W.: **Sustaining safe sex: sexual practices, HIV and social context.** *AIDS* 1993, **7** Suppl. 1:S257–S262.

Kimball, A.M., Berkley, S., Ngugi, E., and Gayle, H.: **International aspects of the AIDS/HIV epidemic.** *Annu. Rev. Public. Health* 1995, **16**:253–282.

Kirby, M.: **Human rights and the HIV paradox.** *Lancet* 1996, **348**:1217–1218.

Nelson, K.E., Celentano, D.D., Eiumtrakol, S., Hoover, D.R., Beyrer, C., Suprasert, S., Kuntolbutra, S., and Khamboonruang, C.: **Changes in sexual behavior and a decline in HIV infection among young men in Thailand.** *N. Engl. J. Med.* 1996, **335**:297–303.

Weniger, B.G., and Brown, T.: **The march of AIDS through Asia.** *N. Engl. J. Med.* 1996, **335**:343–345.

Allergy and Hypersensitivity

13

The adaptive immune response is a critical component of host defense against infection and is essential for normal health. Unfortunately, adaptive immune responses are also sometimes elicited by antigens not associated with infectious agents, and this can cause serious disease. One circumstance in which this occurs is when harmful immune reactions known generally as **hypersensitivity reactions** are made in response to inherently harmless 'environmental' antigens such as pollen, food, and drugs.

Hypersensitivity reactions were classified into four types by Coombs and Gell (Fig. 13.1). **Allergy**, the commonest type of hypersensitivity, is often equated with **type I hypersensitivity reactions**, which are immediate-type hypersensitivity reactions mediated by IgE antibodies, but many of the allergic diseases discussed below also have features characteristic of other types of hypersensitivity, particularly of T cell-mediated type IV hypersensitivity reactions. In the majority of allergies, such as those to food, pollen, and house dust, reactions occur because of the individual has become **sensitized** to an innocuous antigen—the **allergen**—by producing IgE antibodies against it. Subsequent exposure to the allergen triggers the activation of IgE-binding cells, including mast cells and basophils, in the exposed tissue, leading to a series of responses that are characteristic of allergy and are known as **allergic reactions**. Allergic reactions can, however, be independent of IgE; T lymphocytes have the predominant role in allergic contact dermatitis.

	Type I	Type II		Type III	Type IV		
Immune reactant	IgE	IgG		IgG	T_H1 cells	T_H2 cells	CTL
Antigen	Soluble antigen	Cell- or matrix-associated antigen	Cell-surface receptor	Soluble antigen	Soluble antigen	Soluble antigen	Cell-associated antigen
Effector mechanism	Mast-cell activation	Complement, FcR^+ cells (phagocytes, NK cells)	Antibody alters signaling	Complement, phagocytes	Macrophage activation	IgE production, eosinophil activation, mastocytosis	Cytotoxicity
	Ag	platelets + complement		immune complex, blood vessel, + complement	IFN-γ, T_H1, chemokines, cytokines, cytotoxins	T_H2, IL-4, IL-5, eotaxin, cytotoxins, inflammatory mediators	CTL
Example of hypersensitivity reaction	Allergic rhinitis, asthma, systemic anaphylaxis	Some drug allergies (e.g. penicillin)	Chronic urticaria (antibody against $Fc\epsilon RI\alpha$)	Serum sickness, Arthus reaction	Contact dermatitis, tuberculin reaction	Chronic asthma, chronic allergic rhinitis	Graft rejection

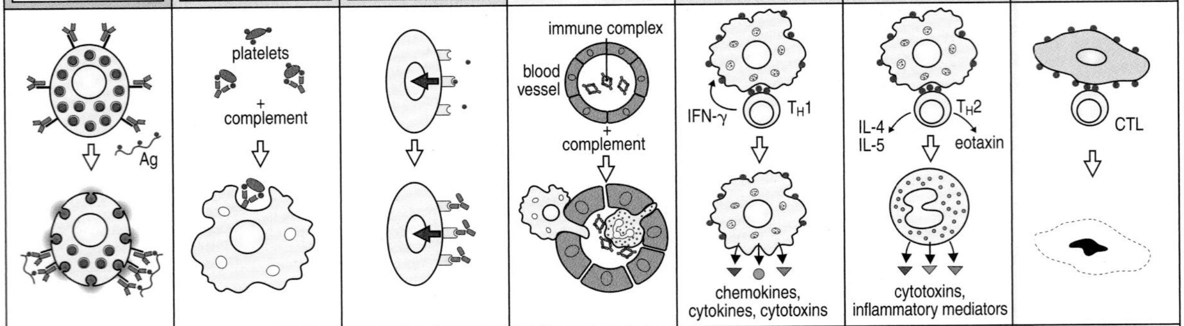

Fig. 13.1 Hypersensitivity reactions are mediated by immunological mechanisms that cause tissue damage. Four types of hypersensitivity reactions are generally recognized. Types I–III are antibody-mediated and are distinguished by the different types of antigens recognized and the different classes of antibody involved. Type I responses are mediated by IgE, which induces mast-cell activation, whereas types II and III are mediated by IgG, which can engage complement-mediated and phagocytic effector mechanisms to varying degrees, depending on the subclass of IgG and the nature of the antigen involved. Type II responses are directed against cell-surface or matrix antigens, whereas type III responses are directed against soluble antigens, and the tissue damage involved is caused by responses triggered by immune complexes. A special category of type II responses involves IgG antibodies against cell-surface receptors that disrupt the normal functions of the receptor, either by causing uncontrollable activation or by blocking receptor function. Type IV hypersensitivity reactions are T-cell mediated and can be subdivided into three groups. In the first group, tissue damage is caused by the activation of macrophages by T_H1 cells, which results in an inflammatory response. In the second, damage is caused by the activation by T_H2 cells of inflammatory responses in which eosinophils predominate; in the third, damage is caused directly by cytotoxic T cells (CTL).

The biological role of IgE is in protective immunity, especially in response to parasitic worms, which are prevalent in less developed countries. In the industrialized countries, allergic IgE responses to innocuous antigens predominate and are an important cause of disease (Fig. 13.2). Almost half the population of North America and Europe have allergies to one or more common environmental antigens and, although rarely life-threatening, these cause much distress and lost time from school and work. Much more is known about the pathophysiology of IgE-mediated responses than about the normal physiological role of IgE, probably because the prevalence of allergy in industrialized societies has doubled in the past 10–15 years.

In this chapter we first consider the mechanisms that favor the sensitization of an individual to an allergen through the production of IgE. We then describe the allergic reaction itself—the pathological consequences of the interaction between allergen and the IgE bound to the high-affinity Fcε receptor on mast cells and basophils. Finally, we consider the causes and consequences of other types of immunological hypersensitivity reactions.

IgE-mediated allergic reactions			
Syndrome	**Common allergens**	**Route of entry**	**Response**
Systemic anaphylaxis	Drugs Serum Venoms Food, e.g. peanuts	Intravenous (either directly or following oral absorption into the blood)	Edema Increased vascular permeability Laryngeal edema Circulatory collapse Death
Acute urticaria (wheal-and-flare)	Animal hair Insect bites Allergy testing	Through skin Systemic	Local increase in blood flow and vascular permeability
Seasonal rhinoconjunctivitis (hay fever)	Pollens (ragweed, trees, grasses) Dust-mite feces	Inhalation	Edema of nasal mucosa Sneezing
Asthma	Danders (cat) Pollens Dust-mite feces	Inhalation	Bronchial constriction Increased mucus production Airway inflammation
Food allergy	Tree nuts Shellfish Peanuts Milk Eggs Fish Soy Wheat	Oral	Vomiting Diarrhea Pruritis (itching) Urticaria (hives) Anaphylaxis (rarely)

Fig. 13.2 IgE-mediated reactions to extrinsic antigens. All IgE-mediated responses involve mast-cell degranulation, but the symptoms experienced by the patient can be very different depending on whether the allergen is injected, inhaled, or eaten, and depending also on the dose of the allergen.

Sensitization and the production of IgE.

IgE is produced both by plasma cells in lymph nodes draining the site of antigen entry and by plasma cells at the site of the allergic reaction, where germinal centers develop within the inflamed tissue. IgE differs from other antibody isotypes in being predominantly localized in tissues, where it is tightly bound to mast-cell surfaces through the high-affinity IgE receptor FcεRI (see Section 9-22). Binding of antigen to IgE cross-links these receptors, causing the release of chemical mediators from the mast cells that can lead to a type I hypersensitivity reaction. Basophils also express FcεRI and so can display surface-bound IgE and take part in type I hypersensitivity reactions. How an initial antibody response comes to be dominated by IgE is still being worked out. In this part of the chapter we describe the current understanding of the factors that contribute to this process.

13-1 Allergens are often delivered transmucosally at low dose, a route that favors IgE production.

Certain antigens and routes of antigen presentation to the immune system favor the production of IgE, which is driven by CD4 T_H2 cells (see Section 9-9). Much human allergy is caused by a limited number of small inhaled proteins that reproducibly elicit IgE production in susceptible individuals. We inhale many different proteins that do not induce IgE production; this raises the question of what is unusual about those proteins that are common allergens. Although we do not yet have a complete answer, some general properties have emerged (Fig. 13.3). Most allergens are relatively small,

Features of inhaled allergens that may promote the priming of T_H2 cells that drive IgE responses	
Protein, often with carbohydrate side chains	Only proteins induce T-cell responses
Enzymatically active	Allergens are often proteases
Low dose	Favors activation of IL-4-producing CD4 T cells
Low molecular weight	Allergen can diffuse out of particle into mucus
Highly soluble	Allergen can be readily eluted from particle
Stable	Allergen can survive in desiccated particle
Contains peptides that bind host MHC class II	Required for T-cell priming

Fig. 13.3 Properties of inhaled allergens. The typical characteristics of inhaled allergens are described in this table.

highly soluble proteins that are carried on dry particles such as pollen grains or mite feces. On contact with the mucosa of the airways, for example, the soluble allergen elutes from the particle and diffuses into the mucosa. Allergens are typically presented to the immune system at very low doses. It has been estimated that the maximum exposure of a person to the common pollen allergens in ragweed (*Ambrosia* species) does not exceed 1 μg per year. Yet many people develop irritating and even life-threatening T_H2-driven IgE antibody responses to these minute doses of allergen. It should be emphasized, however, that only some people who are exposed to these substances make IgE antibodies against them.

It seems likely that presenting an antigen across a mucosal epithelium and at very low doses is a particularly efficient way of inducing T_H2-driven IgE responses. IgE antibody production requires help from T_H2 cells that produce interleukin-4 (IL-4) and IL-13, and it can be inhibited by T_H1 cells that produce interferon-γ (IFN-γ) (see Fig. 9.13). Low doses of antigen can favor the activation of T_H2 cells over T_H1 cells (see Section 10-5), and many common allergens are delivered to the respiratory mucosa by the inhalation of a low dose. In the respiratory mucosa these allergens encounter dendritic cells that take up and process protein antigens very efficiently and thus become activated. In some circumstances, mast cells and eosinophils can also present antigen to T cells and promote the differentiation of T_H2 cells.

13-2 Enzymes are frequent triggers of allergy.

Several lines of evidence suggest that the natural role of IgE is in the defense against parasitic worms (see Section 11-16). Many of these invade their hosts by secreting proteolytic enzymes that break down connective tissue and allow the parasite access to internal tissues, and it has been proposed that these enzymes are particularly active at promoting T_H2 responses. This idea receives some support from the many examples of allergens that are enzymes. The major allergen in the feces of the house dust mite (*Dermatophagoides pteronyssimus*), which is responsible for allergy in about 20% of the North American population, is a cysteine protease known as Der p 1. This enzyme has been found to cleave occludin, a protein component of intercellular tight junctions. This reveals one possible reason for the allergenicity of certain enzymes. By destroying the integrity of the tight junctions between epithelial cells, Der p 1 may gain abnormal access to subepithelial antigen-presenting cells, resident mast cells, and eosinophils (Fig. 13.4).

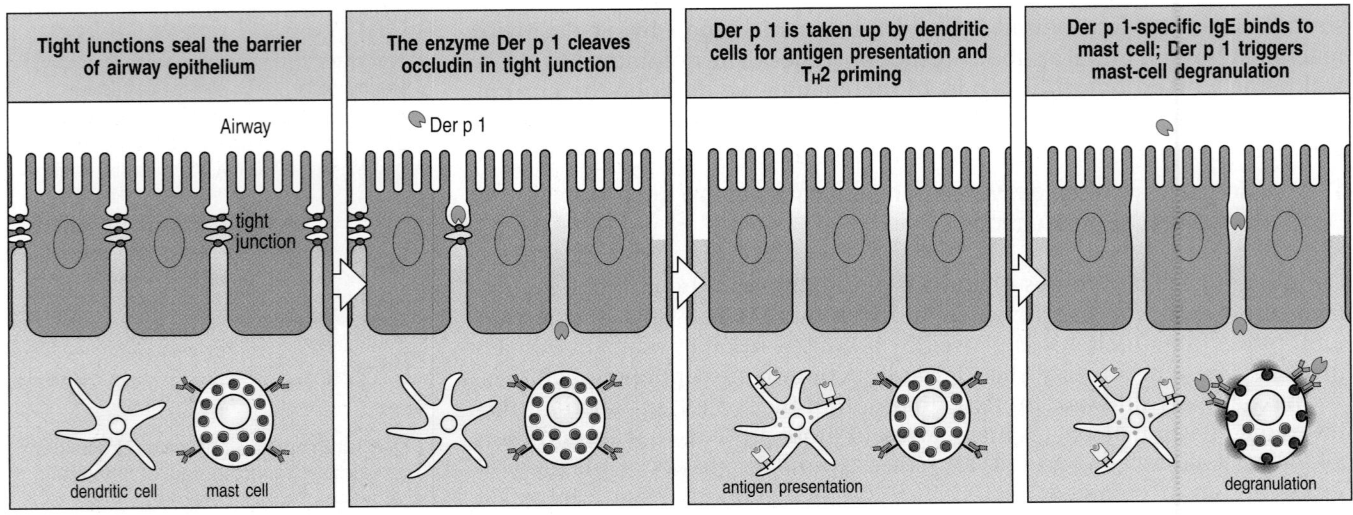

Fig. 13.4 The enzymatic activity of some allergens enables the penetration of epithelial barriers. The epithelial barrier of the airways is formed by tight junctions between the epithelial cells. Fecal pellets from the house dust mite, *Dermatophagoides pteronyssimus*, contain a proteolytic enzyme, Der p 1, that acts as an allergen. It cleaves occludin, a protein that helps to maintain the tight junctions, and thus destroys the barrier function of the epithelium. Mite fecal antigens can then pass through and be taken up by dendritic cells in subepithelial tissue. Der p 1 is taken up by dendritic cells, which are activated and move to lymph nodes (not shown), where they act as antigen-presenting cells, inducing the production of T_H2 cells specific for Der p 1 and the production of Der p 1-specific IgE. Der p 1 can then bind directly to specific IgE on the resident mast cells, triggering mast-cell activation.

The tendency of proteases to induce IgE production is highlighted by individuals with Netherton's disease (Fig. 13.5), which is characterized by high levels of IgE and multiple allergies. The defect in this disease is the lack of a protease inhibitor called SPINK5, which is thought to inhibit the proteases released by bacteria such as *Staphylococcus aureus*, thus raising the possibility that protease inhibitors might be novel therapeutic targets in some allergic disorders. The cysteine protease papain, derived from the papaya fruit, is used as a meat tenderizer and causes allergy in workers preparing the enzyme; such allergies are called **occupational allergies**. Not all allergens are enzymes, however; two allergens identified from filarial worms are enzyme inhibitors, for example. Many protein allergens derived from plants have been identified and sequenced, but their biochemical functions are currently obscure. Thus, there seems to be no systematic association between enzymatic activity and allergenicity.

Knowledge of the identity of allergenic proteins can be important to public health and can have economic significance, as illustrated by the following cautionary tale. Some years ago, the gene for a protein from brazil nuts that encodes a protein rich in methionine and cysteine was transferred by genetic engineering into soy beans intended for animal feed. This was done to improve the nutritional value of soy beans, which are intrinsically poor in these sulfur-containing amino acids. This experiment led to the discovery that the protein, 2S albumin, was the major brazil nut allergen. Injection of extracts of the genetically modified soy beans into the epidermis triggered an allergic skin response in people with an allergy to brazil nuts. As there could be no guarantee that the modified soy beans could be kept out of the human food chain if they were produced on a large scale, development of this genetically modified food was abandoned.

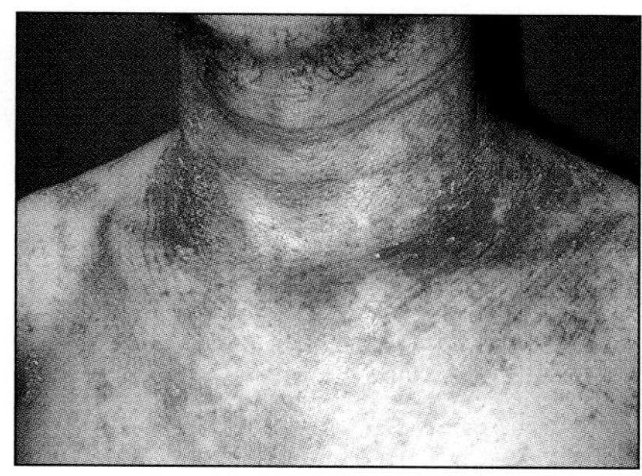

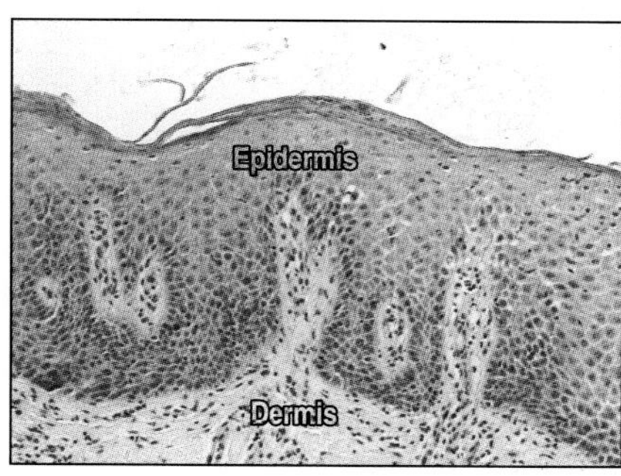

Fig. 13.5 Netherton's syndrome illustrates the association of proteases with the development of high levels of IgE and allergy. This 26-year-old man with Netherton's syndome, caused by a deficiency in the protease inhibitor SPINK5, had persistent erythroderma, recurrent infections of the skin and elsewhere, and multiple food allergies associated with high serum IgE levels. In the top photograph, large erythematous plaques covered with scales and erosions are visible over the upper trunk. The lower panel shows a section through the skin of the same patient. Note the psoriasis-like hyperplasia of the epidermis. Neutrophils are also present in the epidermis. In the dermis, a perivascular infiltrate containing both mononuclear cells and neutrophils is evident. Source: Sprecher, E., *et al.*: *Clin. Exp. Dermatol.* 2004, **29**:513–517.

13-3 Class switching to IgE in B lymphocytes is favored by specific signals.

The immune response leading to IgE production is driven by two main groups of signals. The first consists of the signals that favor the differentiation of naive T cells to a T_H2 phenotype. The second comprises the action of cytokines and co-stimulatory signals from T_H2 cells that stimulate B cells to switch to the production of IgE antibodies.

The fate of a naive CD4 T cell responding to a peptide presented by a dendritic cell is determined by the cytokines it is exposed to before and during this response, and by the intrinsic properties of the antigen, the antigen dose, and the route of presentation. Exposure to IL-4, IL-5, IL-9, and IL-13 favors the development of T_H2 cells, whereas IFN-γ and IL-12 (and its relatives IL-23 and IL-27) favor T_H1-cell development (see Section 8-19). Immune defenses against multicellular parasites are found mainly at the sites of parasite entry—under the skin and in the mucosal-associated lymphoid tissues of the airways and the gut. Cells of the innate and adaptive immune systems at these sites are specialized to secrete cytokines that promote a T_H2-cell response. Dendritic cells taking up antigen in these tissues migrate to regional lymph nodes, where they tend to drive antigen-specific naive CD4 T cells to become effector T_H2 cells; T_H2 cells themselves secrete IL-4, IL-5, IL-9, and IL-13, thus maintaining an environment in which further differentiation of T_H2 cells is favored.

There is evidence that the mix of cytokines and chemokines in the environment polarizes both dendritic cells and T cells in respect of T_H2 differentiation. The chemokines CCL2, CCL7, and CCL13, for example, act on activated monocytes to suppress their production of IL-12, and thereby promote T_H2 responses. In general, however, it seems that an interaction between antigen-presenting dendritic cells and naive T cells in the absence of inflammatory stimuli induced by bacterial or viral infection tends to polarize T-cell differentiation toward T_H2 cells. In contrast, if antigen is encountered by dendritic cells in the context of pro-inflammatory signals, then the dendritic cells are stimulated to produce T_H1-polarizing cytokines such as IL-12, IL-23, and IL-27.

The cytokines and chemokines produced by T_H2 cells both amplify the T_H2 response and stimulate the class switching of B cells to IgE production. As we saw in Chapter 9, IL-4 or IL-13 provide the first signal that switches B cells to IgE production. Cytokines IL-4 and IL-13 activate the Janus-family tyrosine kinases Jak1 and Jak3 (see Section 6-23), which ultimately leads to phosphorylation of the transcriptional regulator STAT6 in T and B lymphocytes. Mice lacking functional IL-4, IL-13, or STAT6 have impaired T_H2 responses and impaired IgE switching, demonstrating the key importance of these cytokines and their signaling pathways. The second signal is a co-stimulatory interaction between CD40 ligand on the T-cell surface and CD40 on the B-cell surface. This interaction is essential for all antibody class switching; patients with the X-linked hyper IgM syndrome have a deficiency of CD40 ligand and produce no IgG, IgA, or IgE (see Section 12-10).

The IgE response, once initiated, can be amplified by mast cells and basophils, which also drive IgE production (Fig. 13.6). These cells express FcεRI, and when they are activated by antigen cross-linking their FcεRI-bound IgE, they express cell-surface CD40 ligand and secrete IL-4. Like T_H2 cells, therefore, they can drive class switching and IgE production by B cells. The interaction between these specialized granulocytes and B cells can occur at the site of the allergic reaction, because B cells are observed to form germinal centers at inflammatory foci. One goal of therapy for allergies is to block this amplification process and thus prevent allergic reactions from becoming self-sustaining.

13-4 Both genetic and environmental factors contribute to the development of IgE-mediated allergy.

Studies have found that as many as 40% of people in the populations of Western industrialized countries show an exaggerated tendency to mount IgE responses to a wide variety of common environmental allergens. This state is called **atopy**, has a strong familial basis, and is influenced by several genetic loci. Atopic individuals have higher total levels of IgE in the circulation and higher levels of eosinophils than their normal counterparts and are more susceptible to allergic diseases such as hay fever and asthma. Environment and genetic variation each account for about 50% of the risk of allergic diseases such as asthma. Genome-wide linkage scans have uncovered a number of distinct susceptibility genes for the allergic diseases atopic dermatitis and

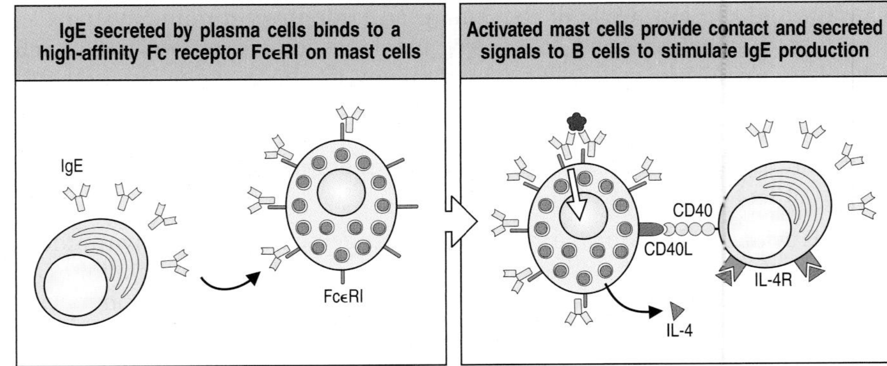

IgE secreted by plasma cells binds to a high-affinity Fc receptor FcεRI on mast cells
IgE
FcεRI
Activated mast cells provide contact and secreted signals to B cells to stimulate IgE production
CD40
CD40L
IL-4R
IL-4

Fig. 13.6 Antigen binding to IgE on mast cells leads to amplification of IgE production. Left panel: IgE secreted by plasma cells binds to the high-affinity IgE receptor on mast cells (illustrated here) and basophils. Right panel: when the surface-bound IgE is cross-linked by antigen, these cells express CD40 ligand (CD40L) and secrete IL-4, which in turn binds to IL-4 receptors (IL-4R) on the activated B cell, stimulating class switching by B cells and the production of more IgE. These interactions can occur *in vivo* at the site of allergen-triggered inflammation, for example in bronchus-associated lymphoid tissue.

asthma, although there is little overlap between the two, suggesting that the genetic predisposition differs somewhat (Fig. 13.7). In addition, there are many ethnic differences in the susceptibility genes for the same disease. Several of the chromosome regions associated with allergy or asthma are also associated with the inflammatory disease psoriasis and autoimmune diseases, suggesting the presence of genes that are involved in exacerbating inflammation (see Fig. 13.7).

One candidate susceptibility gene for asthma and atopic dermatitis, at chromosome 11q12–13, encodes the β subunit of the high-affinity IgE receptor (FcεRI). Another region of the genome associated with disease, 5q31–33, contains at least four types of candidate gene that might be responsible for increased susceptibility. First, there is a cluster of tightly linked genes for cytokines that promote T_H2 responses by enhancing IgE class switching, eosinophil survival, and mast-cell proliferation. This cluster includes the genes for IL-3, IL-4, IL-5, IL-9, IL-13, and granulocyte–macrophage colony-stimulating factor (GM-CSF). In particular, genetic variation in the promoter region of the IL-4 gene has been associated with raised IgE levels in atopic individuals. The variant promoter directs increased expression of a reporter gene in experimental systems and thus might produce increased IL-4 *in vivo*. Atopy has also been associated with a gain-of-function mutation of the α subunit of the IL-4 receptor that causes increased signaling after ligation of the receptor.

A second set of genes in this region of chromosome 5 is the TIM family (for *T* cell, *i*mmunoglobulin domain and *m*ucin domain), which encode T-cell

Fig. 13.7 Susceptibility loci identified by genome screens for asthma, atopic dermatitis, and other immune disorders. Only loci with significant linkages are indicated. Clustering of disease-susceptibility genes is found for the MHC on chromosome 6p21, and also in several other genomic regions. There is in fact little overlap between susceptibility genes for asthma and atopic dermatitis, suggesting that specific genetic factors are involved in both. There is also some overlap between susceptibility genes for asthma and those for autoimmune diseases, and between those for the inflammatory skin disease psoriasis and atopic dermatitis. Adapted from Cookson, W.: *Nat. Rev. Immunol.* 2004, **4**:978–988.

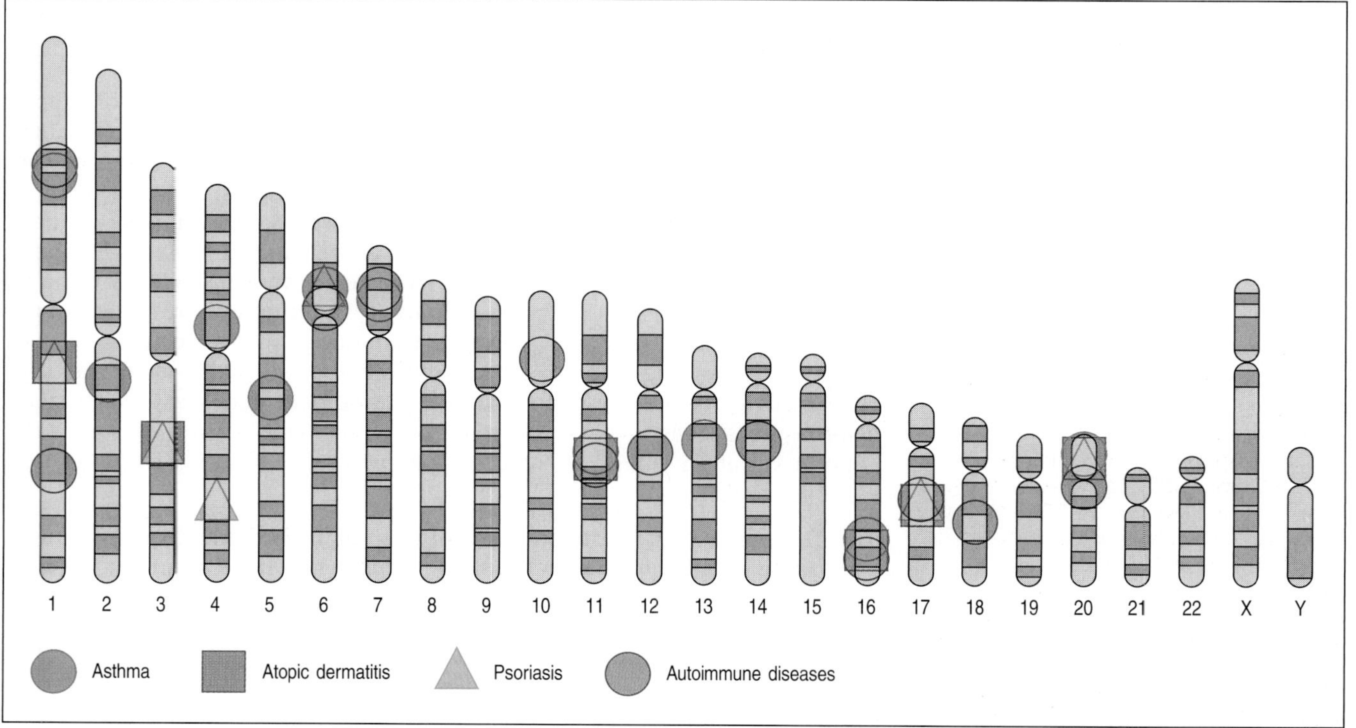

surface proteins. In mice, Tim-3 protein is specifically expressed on T_H1 cells and negatively regulates T_H1 responses, whereas Tim-2 (and to a lesser extent Tim-1) is preferentially expressed in, and negatively regulates, T_H2 cells. Mouse strains that carry different variants of the TIM genes differ both in their susceptibility to allergic inflammation of the airways and in the production of IL-4 and IL-13 by their T cells. Inherited variation in the TIM genes in humans has been correlated with levels of airway hyperreactivity, the condition in which a nonspecific irritant causes contraction of bronchial smooth muscle similar to that seen in asthma. The third candidate susceptibility gene in this part of the genome encodes p40, one of the two subunits of IL-12. This cytokine promotes T_H1 responses, and genetic variation in p40 expression that could cause reduced production of IL-12 was found to be associated with more severe asthma. A fourth candidate susceptibility gene, that encoding the β-adrenergic receptor, is encoded in this region. Variation in this receptor might be associated with alteration in smooth muscle responsiveness to endogenous and pharmacological ligands.

This complexity illustrates a common challenge in identifying the genetic basis of complex disease traits. Relatively small regions of the genome, identified as containing genes for altered disease susceptibility, may contain many good candidates, judging by their known physiological activities. Identifying the correct gene, or genes, may require studies of several very large populations of patients and controls. For chromosome 5q31–33, for example, it is still too early to know how important each of the different polymorphisms is in the complex genetics of atopy.

A second type of inherited variation in IgE responses is linked to the HLA class II region (the human MHC class II region) and affects responses to specific allergens, rather than a general susceptibility to atopy. IgE production in response to particular allergens is associated with certain HLA class II alleles, implying that particular peptide:MHC combinations might favor a strong T_H2 response; for example, IgE responses to several ragweed pollen allergens are associated with haplotypes containing the HLA class II allele *DRB1*1501*. Many people are therefore generally predisposed to make T_H2 responses and are specifically predisposed to respond to some allergens more than others. However, allergies to drugs such as penicillin show no association with HLA class II or the presence or absence of atopy.

There are also likely to be genes that affect only particular aspects of allergic disease. In asthma, for example, there is evidence that different genes affect at least three aspects of the disease—IgE production, the inflammatory response, and clinical responses to particular treatments. Polymorphism of the gene on chromosome 20 encoding ADAM33, a metalloproteinase expressed by bronchial smooth muscle cells and lung fibroblasts, has been associated with asthma and bronchial hyperreactivity. This is likely to be an example of genetic variation in the pulmonary inflammatory response and in the pathological anatomical changes that occur in the airways (airway remodeling), leading to increased susceptibility to asthma. Some of the best-characterized genetic polymorphisms of candidate genes associated with asthma are shown in Fig. 13.8, together with possible ways in which the genetic variation might affect the type of disease that develops and its response to drugs.

The prevalence of atopic allergy, and of asthma in particular, is increasing in economically advanced regions of the world, an observation that is best explained by environmental factors. The four main candidate environmental factors are changes in exposure to infectious diseases in early childhood, environmental pollution, allergen levels, and dietary changes. Pollution has been blamed for an increase in the incidence of non-allergic cardiopulmonary diseases such as chronic bronchitis, but an association with allergy

Gene	Nature of polymorphism	Possible mechanism of association
IL-4	Promoter variant	Variation in expression of IL-4
IL-4 receptor α chain	Structural variant	Increased signaling in response to IL-4
High-affinity IgE receptor β chain	Structural variant	Variation in consequences of IgE ligation by antigen
MHC class II genes	Structural variants	Enhanced presentation of particular allergen-derived peptides
T-cell receptor α locus	Microsatellite markers	Enhanced T-cell recognition of certain allergen-derived peptides
ADAM33	Structural variants	Variation in airway remodeling
β_2-Adrenergic receptor	Structural variants	Increased bronchial hyperreactivity*
5-Lipoxygenase	Promoter variant	Variation in leukotriene production†
TIM gene family	Promoter and structural variants	Regulation of the T_H1/T_H2 balance

Fig. 13.8 Candidate susceptibility genes for asthma. *May also affect response to bronchodilator therapy with β_2-adrenergic agonists. †Patients with alleles associated with reduced enzyme production failed to show a beneficial response to a drug inhibitor of 5-lipoxygenase. This is an example of a pharmacogenetic effect, in which genetic variation affects the response to medication.

has been less easy to demonstrate. There is, however, increasing evidence for an interaction between allergens and pollution, particularly in genetically sensitive individuals. Diesel exhaust particles are the best-studied pollutant in this context; they increase IgE production 20–50-fold when combined with allergen, with an accompanying shift to T_H2 cytokine production. Reactive oxidant chemicals seem to be generated and individuals less able to deal with this onslaught may be at increased risk of allergic disease. Genes that might be governing this susceptibility are *GSTP1* and *GSTM*, the members of the glutathione-*S*-transferase superfamily, as people with variant alleles of these genes showed airway hyperreactivity when exposed to allergen. Indeed, genetic factors may explain why the epidemiological evidence for an association between pollution and allergy remains moderate at best, as it may only apply to genetically sensitive individuals.

A decrease in exposure to microbial pathogens as a possible cause of the increase in allergy has also received much attention since the idea was first mooted in 1989. This is known as the 'hygiene hypothesis' (Fig. 13.9). The proposition is that less hygienic environments, specifically environments that predispose to infections early in childhood, help to protect against atopy and asthma. This implies that T_H2 responses predominate over T_H1 responses by default in early childhood, and that the immune system is reprogrammed toward more T_H1-dominated responses by the cytokine response to early infections.

There is much evidence in support of this hypothesis, but also some observations that are difficult to reconcile with it. In favor, there is evidence for a bias toward T_H2 responses in newborn infants, in whom dendritic cells produce less IL-12 and T cells produce less IFN-γ than in older children and adults. There is also evidence that exposure to childhood infections, with the important exception of some respiratory infections that we consider below, helps to protect against the development of atopic allergic disease. Younger

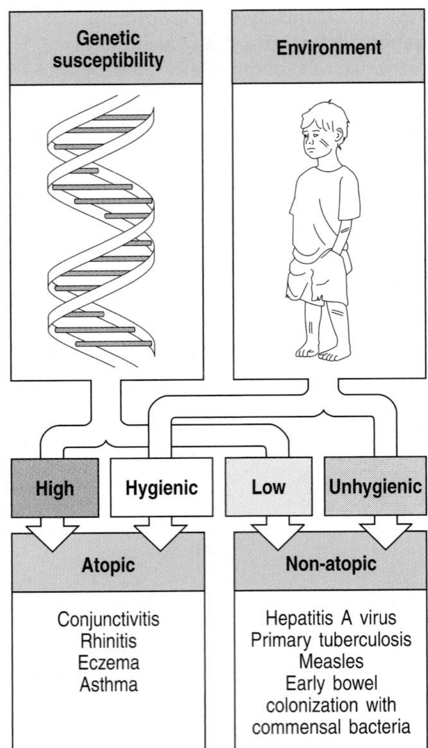

Fig. 13.9 Genes, the environment, and atopic allergic diseases. Both inherited and environmental factors are important determinants of the likelihood of developing atopic allergic disease. Some known genes that influence the development of asthma are shown in Fig. 13.8. The postulate of the 'hygiene hypothesis' is that exposure to some infectious agents in childhood drives the immune system toward a general state of T_H1 responsiveness and non-atopy. In contrast, children with genetic susceptibility to atopy and who live in an environment with low exposure to infectious disease tend to mount T_H2 responses, which naturally predominate in the neonatal period of life. It is these children who are thought to be most susceptible to the development of atopic allergic disease.

children from families with three or more older siblings, and children aged less than 6 months who are exposed to other children in daycare facilities—situations linked to a greater exposure to infections—are somewhat protected against atopy and asthma. Furthermore, early colonization of the gut by commensal bacteria such as lactobacilli and bifidobacteria, or infection by gut pathogens such as *Toxoplasma gondii* (which stimulates a T_H1 response) or *Helicobacter pylori* are associated with a reduced prevalence of allergic disease.

A history of infection with measles or hepatitis A virus, or a positive tuberculin skin test (suggesting previous exposure and an immune response to *Mycobacterium tuberculosis*), also seem to have a negative association with atopy. The human counterpart of the murine Tim-1 protein, which might be important in determining airway hyperreactivity and the production of IL-4 and IL-13 by T cells, is the cellular receptor for hepatitis A virus. The infection of T cells by hepatitis A virus could thus directly influence their differentiation and cytokine production, limiting the development of T_H2 responses.

In contrast to these negative associations between childhood infection and the development of atopy and asthma, there is evidence that children who have had attacks of bronchiolitis associated with respiratory syncytial virus (RSV) infection are more prone to developing asthma later on. This effect of RSV may depend on the age at first infection. Infection of neonatal mice with RSV was followed by a decreased IFN-γ response compared with mice challenged at 4 or 8 weeks of age. When these mice were rechallenged at 12 weeks of age with RSV infection, animals that had been primarily infected as neonates suffered from more severe lung inflammation than animals infected at 4 or 8 weeks of age (Fig. 13.10). Similarly, children hospitalized with RSV infection have a skewed ratio of cytokine production away from IFN-γ toward IL-4, the cytokine that induces T_H2 responses. All these findings suggest that an infection that evokes a T_H1 immune response early in life might reduce the likelihood of T_H2 responses later in life, and vice versa.

The biggest 'fly in the ointment' for the hygiene theory, however, is the strong negative correlation between infection by helminths (such as hookworm and schistosomes) and the development of allergy. A study in Venezuela showed that children treated for a prolonged period with antihelminthic agents had a higher prevalence of atopy compared with untreated and heavily parasitized children. As we have seen, however, helminths are strong drivers of T_H2 responses, and it is difficult to reconcile this with the idea that the polarization of T-cell responses toward T_H1 is a general mechanism by which infection protects against atopy.

These observations have led to a modification of the hygiene hypothesis known as the **counter-regulation hypothesis**. This proposes that all types of infection might protect against the development of atopy by driving the production of cytokines such as IL-10 and transforming growth factor (TGF)-β, which downregulate both T_H1 and T_H2 responses (see Section 8-19). In hygienic environments, children suffer fewer infections, resulting in a reduced production of these cytokines. Neither the molecular pathways induced by microbial exposure nor the tolerance-inducing responses in the host have yet been identified, but there are a variety of microbial products with immunoregulatory potential. For instance, exposure of dendritic cells to various Toll-like receptor (TLR) ligands, such as bacterial lipopolysaccharide (the ligand for TLR-4), CpG DNA (the ligand for TLR-9), or pro-inflammatory mediators such as IFN-γ can stimulate the production of indoleamine 2,3-dioxygenase (IDO), an enzyme that degrades the essential amino acid tryptophan. Dendritic cells expressing IDO can suppress T_H2-driven inflammation and promote the differentiation of regulatory T cells, providing both immediate and long-term protective effects against allergy. Genetic factors

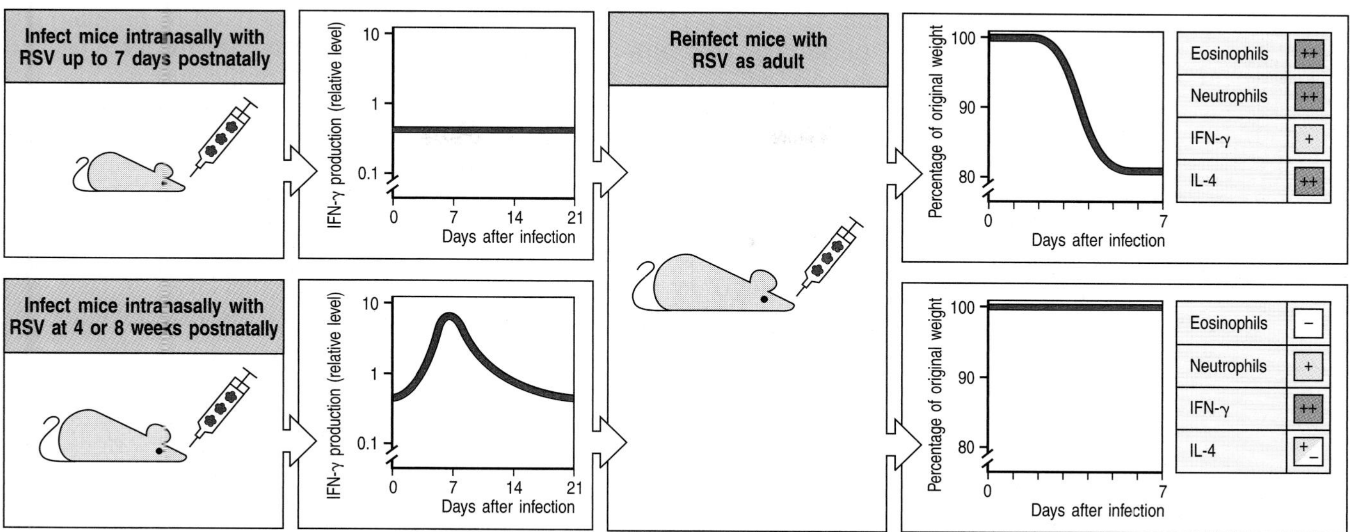

Fig. 13.10 Response to infection and rechallenge of mice with respiratory syncytial virus (RSV) according to the age of primary infection. Mice respond to infection with RSV in different ways according to the age of primary infection. The graphs on the left show the IFN-γ response after either neonatal infection (upper panel) or infection at 4 or 8 weeks of age (lower panel). Mice infected neonatally fail to produce IFN-γ. The right-hand panels show the consequences of reinfection with RSV of the two cohorts of mice when adult. The mice that had the primary infection as neonates show weight loss and a severe inflammatory response to reinfection, with infiltration of the lungs by eosinophils and neutrophils, accompanied by production of the T_H2 cytokine IL-4. In contrast, mice that had the primary infection at 4 or 8 weeks of age show no weight loss, mild neutrophil infiltration, and production of the T_H1 cytokine IFN-γ.

may also have a bearing on this type of regulation because newborn infants with a genetic predisposition to allergy have been found to have impaired regulatory T-cell function.

13-5 Regulatory T cells can control allergic responses.

Peripheral blood mononuclear cells (PBMCs) from atopic individuals have a tendency to secrete T_H2 cytokines after nonspecific stimulation via the T-cell receptor, whereas those from non-atopic individuals do not. This has led to the suggestion that regulatory mechanisms have an important role in preventing IgE responses to allergens. Regulatory T cells, in particular, are receiving considerable attention with regard to all types of immunologically mediated disease. The different types of regulatory T cells (see Section 8-17) may all have a role in modulating allergy. Natural regulatory T cells (CD4 CD25 T_{reg} cells) from atopic individuals are defective in suppressing T_H2 cytokine production compared with those from non-atopic individuals, and this defect is even more pronounced during the pollen season. More evidence comes from mice deficient in the transcription factor FoxP3, the master switch for producing CD4 CD25 T_{reg} cells, which develop manifestations of allergy including eosinophilia, hyper IgE, and allergic airway inflammation, suggesting that these result from the absence of regulatory T cells. This syndrome could be partially reversed by a concomitant deficiency in STAT6, which independently prevents the development of the T_H2 response (see Section 13-3).

Regulatory T cells might also be induced by IDO secreted by a variety of cell types (see Section 13-4). Dendritic cells secrete IDO on activation through stimulation of the receptor TLR-9 by ligands containing unmethylated CpG DNA. IDO secretion from resident lung cells stimulated in this way has been shown to ameliorate experimental asthma in mice.

Summary.

Allergic reactions are the result of the production of specific IgE antibody against common, innocuous antigens. Allergens are small antigens that commonly provoke an IgE antibody response. Such antigens normally enter the body at very low doses by diffusion across mucosal surfaces and therefore trigger a T_H2 response. The differentiation of naive allergen-specific T cells into T_H2 cells is also favored by cytokines such as IL-4 and IL-13. Allergen-specific T_H2 cells producing IL-4 and IL-13 drive allergen-specific B cells to produce IgE. The specific IgE produced in response to the allergen binds to the high-affinity receptor for IgE on mast cells, basophils, and activated eosinophils. IgE production can be amplified by these cells because, upon activation, they produce IL-4 and CD40 ligand. The tendency to IgE overproduction is influenced by genetic and environmental factors. Once IgE is produced in response to an allergen, reexposure to the allergen triggers an allergic response. Immunoregulation is critical in the control of allergic disease through a variety of mechanisms, including regulatory T cells. We describe the mechanism and pathology of the allergic responses themselves in the next part of the chapter.

Effector mechanisms in allergic reactions.

Allergic reactions are triggered when allergens cross-link preformed IgE bound to the high-affinity receptor FcεRI on mast cells. Mast cells line the body surfaces and serve to alert the immune system to local infection. Once activated, they induce inflammatory reactions by secreting chemical mediators stored in preformed granules and by synthesizing prostaglandins, leukotrienes, and cytokines after activation occurs. In allergy, they provoke very unpleasant reactions to innocuous antigens that are not associated with invading pathogens that need to be expelled. The consequences of IgE-mediated mast-cell activation depend on the dose of antigen and its route of entry; symptoms range from the irritating sniffles of hay fever when pollen is inhaled, to the life-threatening circulatory collapse that occurs in systemic anaphylaxis (Fig. 13.11). The immediate allergic reaction caused by mast-cell degranulation is followed by a more sustained inflammation, known as the late-phase response. This late response involves the recruitment of other

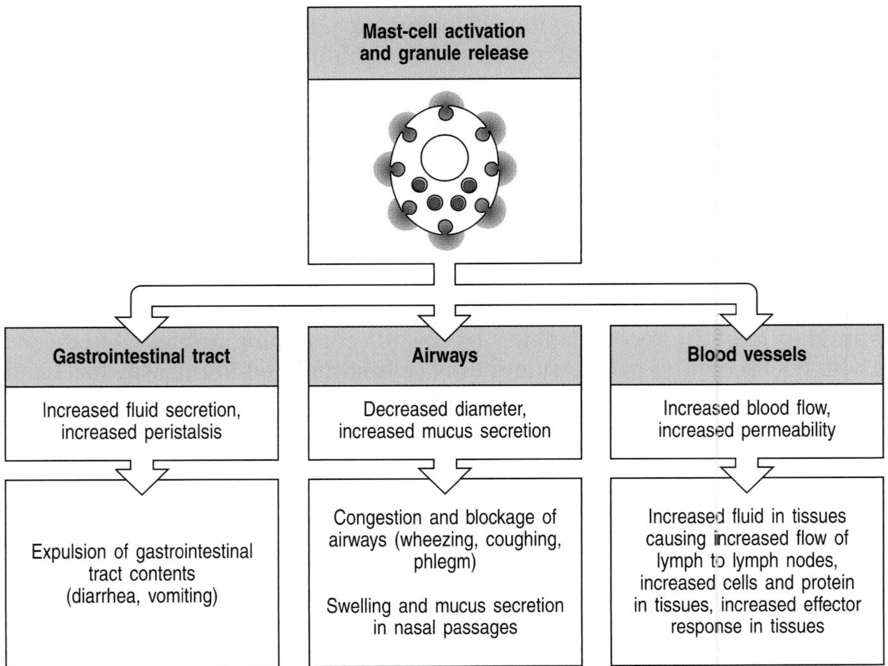

Fig. 13.11 Mast-cell activation has different effects on different tissues.

effector cells, notably T_H2 lymphocytes, eosinophils, and basophils, which contribute significantly to the immunopathology of an allergic response.

13-6 Most IgE is cell-bound and engages effector mechanisms of the immune system by different pathways from other antibody isotypes.

Antibodies engage effector cells such as mast cells by binding to receptors specific for the Fc constant regions. Most antibodies engage Fc receptors only after the antibody variable region has bound specific antigen, forming an immune complex of antigen and antibody. IgE is an exception, because it is captured by the high-affinity Fcε receptor in the absence of bound antigen. This means that, unlike other antibodies, which are found mainly in body fluids, IgE is mostly found fixed in the tissues on mast cells that bear this receptor as well as on circulating basophils and activated eosinophils. The ligation of the cell-bound IgE antibody by specific antigen triggers the activation of these cells at the sites of antigen entry into the tissues. The release of inflammatory lipid mediators, cytokines, and chemokines at sites of IgE-triggered reactions recruits eosinophils and basophils to augment the type I hypersensitivity response. It also recruits other effector cells, including T lymphocytes, that can mediate a local type IV hypersensitivity response.

There are two types of IgE-binding Fc receptor. The first, FcεRI, is a high-affinity receptor of the immunoglobulin superfamily that binds IgE on mast cells, basophils, and activated eosinophils (see Section 9-24). When the cell-bound IgE is cross-linked by specific antigen, FcεRI transduces an activating signal. High levels of IgE, such as those that exist in people with allergic diseases or parasite infections, can result in a marked increase in FcεRI on the surface of mast cells, an enhanced sensitivity of such cells to activation by low concentrations of specific antigen, and a markedly increased IgE-dependent release of chemical mediators and cytokines.

The second IgE receptor, FcεRII, usually known as **CD23**, is a C-type lectin and is structurally unrelated to FcεRI; it binds IgE with low affinity. CD23 is present on many cell types, including B cells, activated T cells, monocytes, eosinophils, platelets, follicular dendritic cells, and some thymic epithelial cells. This receptor was thought to be crucial for the regulation of IgE levels, but mouse strains in which the CD23 gene has been inactivated still develop relatively normal polyclonal IgE responses. Nevertheless, CD23 does seem to be involved in enhancing IgE antibody levels in some situations. Responses against a specific antigen are known to be increased in the presence of the same antigen complexed with IgE, but such enhancement fails to occur in mice that lack the CD23 gene. This has been interpreted to indicate that CD23 on antigen-presenting cells has a role in the capture of antigen complexed with IgG.

13-7 Mast cells reside in tissues and orchestrate allergic reactions.

Mast cells were described by Ehrlich in the mesentery of rabbits and named *Mastzellen* ('fattened cells'). Like basophils, mast cells contain granules rich in acidic proteoglycans that take up basic dyes. Mast cells are derived from hematopoietic stem cells but mature locally, often residing near surfaces exposed to pathogens and allergens. The major factors for mast-cell growth and development include stem-cell factor (the ligand for the receptor tyrosine kinase Kit), IL-3, and T_H2-associated cytokines such as IL-4 and IL-9. Mice with defective Kit lack differentiated mast cells, and although they produce IgE they cannot make IgE-mediated inflammatory responses. This shows that such responses depend almost exclusively on mast cells. Mast-cell activation depends on the activation of phosphatidylinositol 3-kinase (PI 3-kinase) in mast cells by Kit, and pharmacological inactivation of the p110δ isoform of PI 3-kinase protects mice against allergic responses. p110δ is thus a potential target for therapy in allergy and other mast-cell related pathologies.

Mast cells express FcεRI constitutively on their surface and are activated when antigens cross-link IgE bound to these receptors (see Fig. 9.35). Different levels of stimulation result in varied responses; for instance, low levels of allergen resulting in low receptor occupancy provide a strong signal leading to allergic inflammation. Conversely, higher levels of antigen occupancy can induce the production of immunoregulatory cytokines such as IL-10. Thus, mast cells display various responses depending upon the signals they receive.

Mast cell degranulation begins within seconds, releasing an array of preformed and newly generated inflammatory mediators (Fig. 13.12). Among these are **histamine**—a short-lived vasoactive amine that causes an immediate increase in local blood flow and vessel permeability—and enzymes such as mast-cell chymase, tryptase, and serine esterases. These enzymes can in turn activate matrix metalloproteinases, which break down tissue matrix proteins, causing tissue destruction. Large amounts of the cytokine tumor necrosis factor (TNF)-α are also released by mast cells after activation. Some comes from stores in mast-cell granules; some is newly synthesized by the activated mast cells. TNF-α activates endothelial cells, causing an increased expression of adhesion molecules, which promotes the influx of inflammatory leukocytes and lymphocytes into tissues (see Chapter 2).

On activation, mast cells also synthesize and release chemokines, lipid mediators such as prostaglandins, leukotrienes, and platelet-activating factor (PAF), and cytokines such as IL-4 and IL-13, which perpetuate the T_H2 response. These mediators contribute to both the acute and the chronic inflammatory responses. The lipid mediators, in particular, act rapidly to cause smooth muscle contraction, increased vascular permeability, and the

Fig. 13.12 Molecules released by mast cells on activation. Mast cells produce a wide variety of biologically active proteins and other chemical mediators. The enzymes and toxic mediators listed in the first two rows are released from the preformed granules. The cytokines, chemokines, and lipid mediators are synthesized after activation.

Class of product	Examples	Biological effects
Enzyme	Tryptase, chymase, cathepsin G, carboxypeptidase	Remodel connective tissue matrix
Toxic mediator	Histamine, heparin	Toxic to parasites Increase vascular permeability Cause smooth muscle contraction
Cytokine	IL-4, IL-13	Stimulate and amplify T_H2-cell response
	IL-3, IL-5, GM-CSF	Promote eosinophil production and activation
	TNF-α (some stored preformed in granules)	Promotes inflammation, stimulates cytokine production by many cell types, activates endothelium
Chemokine	CCL3	Attracts monocytes, macrophages, and neutrophils
Lipid mediator	Prostaglandins D_2, E_2 Leukotrienes B4, C4	Cause smooth muscle contraction Increase vascular permeability Stimulate mucus secretion
	Platelet-activating factor	Attracts leukocytes Amplifies production of lipid mediators Activates neutrophils, eosinophils, and platelets

secretion of mucus, and also induce the influx and activation of leukocytes, which contribute to the late phase of the allergic response. The lipid mediators derive from membrane phospholipids, which are cleaved to release the precursor molecule arachidonic acid. This molecule can be modified via two pathways to give rise to prostaglandins, thromboxanes, and leukotrienes. Prostaglandin D_2 is the major prostaglandin produced by mast cells and recruits T_H2 cells, eosinophils, and basophils, all of which express its receptor protein (PTGDR). Prostaglandin D_2 is critical to the development of allergic diseases such as asthma, and polymorphisms in PTGDR have been linked to an increased risk of developing asthma. The leukotrienes, especially B4 and C4, are also important in sustaining inflammatory responses in tissues. Many anti-inflammatory drugs are inhibitors of arachidonic acid metabolism. Aspirin, for example, is an inhibitor of the enzyme cyclooxygenase and blocks the production of prostaglandins.

IgE-mediated activation of mast cells thus orchestrates an important inflammatory cascade that is amplified by the recruitment of several cell types including eosinophils, basophils, T_H2 lymphocytes, B cells, and dendritic cells. The physiological importance of this reaction is as a defense against parasite infection (see Section 9-25). In allergy, however, the acute and chronic inflammatory reactions triggered by mast-cell activation have important pathophysiological consequences, as seen in the diseases associated with allergic responses to environmental antigens. Increasingly, mast cells are also considered to have a role in immunoregulation as well as being drivers of pro-inflammatory reactions. High concentrations of allergen, leading to high occupancy of the receptor IgE, produce immunoregulatory rather than inflammatory consequences. Mast cells can also participate in autoimmune reactions.

13-8 Eosinophils are normally under tight control to prevent inappropriate toxic responses.

Eosinophils are granulocytic leukocytes that originate in bone marrow. They are so called because their granules, which contain arginine-rich basic proteins, are colored bright orange by the acidic stain eosin. Only very small numbers of these cells are normally present in the circulation; most eosinophils are found in tissues, especially in the connective tissue immediately underneath respiratory, gut, and urogenital epithelium, implying a likely role for these cells in defense against invading organisms. Eosinophils have two kinds of effector functions. First, on activation they release highly toxic granule proteins and free radicals, which can kill microorganisms and parasites but also cause significant tissue damage in allergic reactions. Second, activation induces the synthesis of chemical mediators such as prostaglandins, leukotrienes, and cytokines. These amplify the inflammatory response by activating epithelial cells and recruiting and activating more eosinophils and leukocytes (Fig. 13.13). Eosinophils also secrete a number of proteins involved in airway tissue remodeling.

The activation and degranulation of eosinophils is strictly regulated, as their inappropriate activation would be harmful to the host. The first level of control acts on the production of eosinophils by the bone marrow. Few eosinophils are produced in the absence of infection or other immune stimulation. But when T_H2 cells are activated, cytokines such as IL-5 are released that increase the production of eosinophils in the bone marrow and their release into the circulation. However, transgenic animals overexpressing IL-5 have increased numbers of eosinophils (**eosinophilia**) in the circulation but not in their tissues, indicating that the migration of eosinophils from the circulation into tissues is regulated separately, by a second set of controls. The key molecules in this case are CC chemokines. Most of these cause chemotaxis

Fig. 13.13 Eosinophils secrete a range of highly toxic granule proteins and other inflammatory mediators.

Class of product	Examples	Biological effects
Enzyme	Eosinophil peroxidase	Toxic to targets by catalyzing halogenation Triggers histamine release from mast cells
	Eosinophil collagenase	Remodels connective tissue matrix
	Matrix metalloproteinase-9	Matrix protein degradation
Toxic protein	Major basic protein	Toxic to parasites and mammalian cells Triggers histamine release from mast cells
	Eosinophil cationic protein	Toxic to parasites Neurotoxin
	Eosinophil-derived neurotoxin	Neurotoxin
Cytokine	IL-3, IL-5, GM-CSF	Amplify eosinophil production by bone marrow Cause eosinophil activation
	TGF-α, TGF-β	Epithelial proliferation, myofibroblast formation
Chemokine	CXCL8 (IL-8)	Promotes influx of leukocytes
Lipid mediator	Leukotrienes C4, D4, E4	Cause smooth muscle contraction Increase vascular permeability Increase mucus secretion
	Platelet-activating factor	Attracts leukocytes Amplifies production of lipid mediators Activates neutrophils, eosinophils, and platelets

of several types of leukocytes, but three are particularly important in attracting and activating eosinophils, and have been named the **eotaxins**: CCL11 (eotaxin 1), CCL24 (eotaxin 2), and CCL26 (eotaxin 3).

The eotaxin receptor on eosinophils, CCR3, is quite promiscuous and binds other CC chemokines, including CCL7, CCL13, and CCL5, which also induce eosinophil chemotaxis and activation. Identical or similar chemokines stimulate mast cells and basophils. For example, eotaxins attract basophils and cause their degranulation, and CCL2, which binds to CCR2, similarly activates mast cells in both the presence and the absence of antigen. CCL2 can also promote the differentiation of naive T cells to T_H2 cells; T_H2 cells also carry the receptor CCR3 and migrate toward eotaxins. It is striking that these interactions between different chemokines and their receptors show a high degree of overlap and redundancy; we do not understand the significance of this complexity. However, these findings show that families of chemokines, as well as cytokines, can coordinate certain kinds of immune response.

A third set of controls regulates the state of eosinophil activation. In their nonactivated state, eosinophils do not express high-affinity IgE receptors and have a high threshold for release of their granule contents. After activation by cytokines and chemokines this threshold drops, FcεRI is expressed, and the number of Fcγ receptors and complement receptors on the cell surface also increases. The eosinophil is now primed to carry out its effector activity—degranulation in response to antigen that cross-links specific IgE bound to FcεRI on the eosinophil surface.

13-9 Eosinophils and basophils cause inflammation and tissue damage in allergic reactions.

What were later to be defined as eosinophils were observed in the 19th century in the first pathological description of fatal *status asthmaticus*, but the precise role of these cells in allergic disease is still unclear. In a local allergic reaction, mast-cell degranulation and T_H2 activation cause eosinophils to accumulate in large numbers and to become activated. Eosinophils can also present antigens to T cells and secrete T_H2 cytokines. Eosinophils seem to promote the apoptosis of T_H1 cells, and their promotion of T_H2-cell expansion may be partly due to a relative reduction in T_H1-cell numbers. Their continued presence is characteristic of chronic allergic inflammation and they are thought to be major contributors to tissue damage.

Basophils are also present at the site of an inflammatory reaction and growth factors for basophils are very similar to those for eosinophils; they include IL-3, IL-5, and GM-CSF. There is evidence for reciprocal control of the maturation of the stem-cell population into basophils or eosinophils. For example, TGF-β in the presence of IL-3 suppresses eosinophil differentiation and enhances that of basophils. Basophils are normally present in very low numbers in the circulation and seem to have a similar role to that of eosinophils in defense against pathogens. Like eosinophils, they are recruited to the sites of allergic reactions. Basophils express FcεRI on the cell surface and, on activation by cytokines or antigen, they release histamine from the basophilic granules after which they are named; they also produce IL-4 and IL-13.

Eosinophils, mast cells, and basophils can interact with each other. Eosinophil degranulation releases **major basic protein**, which in turn causes degranulation of mast cells and basophils. This effect is augmented by any of the cytokines that affect eosinophil and basophil growth, differentiation, and activation, such as IL-3, IL-5, and GM-CSF.

13-10 Allergic reactions can be divided into immediate and late-phase responses.

The inflammatory response after IgE-mediated mast-cell activation occurs as an immediate reaction, starting within seconds, and a late reaction, which takes up to 8–12 hours to develop. These reactions can be distinguished clinically (Fig. 13.14). The **immediate reaction** is due to the activity of histamine,

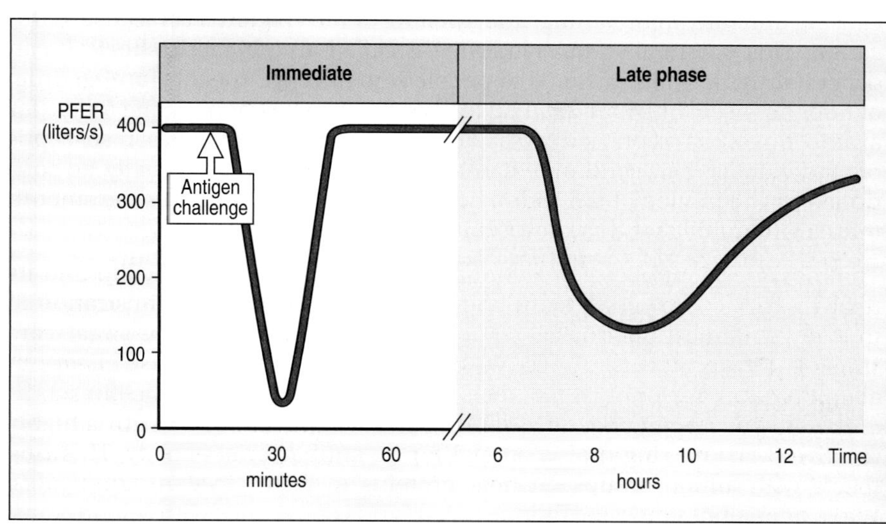

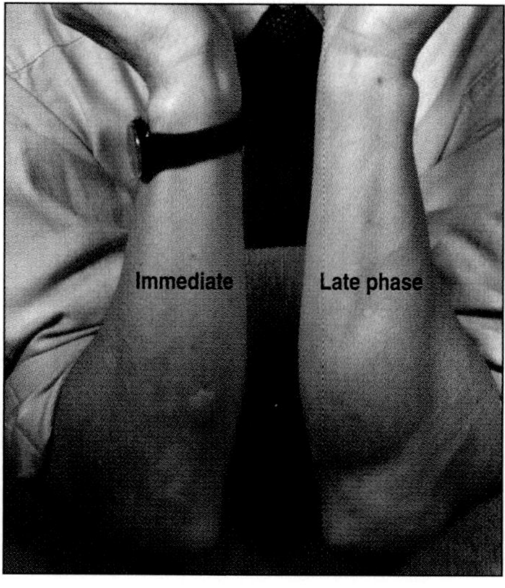

Fig. 13.14 Allergic reactions can be divided into an immediate response and a late-phase response. Left panel: the response to an inhaled antigen can be divided into early and late responses. An asthmatic response in the lungs with narrowing of the airways caused by the constriction of bronchial smooth muscle can be measured as a fall in the peak expiratory flow rate (PEFR). The immediate response peaks within minutes after antigen inhalation and then subsides. Six to eight hours after antigen challenge, there is a late-phase response that also results in a fall in the PEFR. The immediate response is caused by the direct effects on blood vessels and smooth muscle of rapidly metabolized mediators such as histamine and lipid mediators released by mast cells. The late-phase response is caused by the effects of an influx of inflammatory leukocytes attracted by chemokines and other mediators released by mast cells during and after the immediate response. Right panel: a wheal-and-flare allergic reaction develops within a minute or two of superficial injection of antigen into the epidermis and lasts for up to 30 minutes. The more widespread edematous response characteristic of the late phase develops approximately 6 hours later and can persist for some hours. The photograph shows an intradermal skin challenge with allergen showing a 15-minute wheal and flare (early-phase) reaction (left) and a 6-hour late-phase reaction (right). The allergen was grass pollen extract. Photograph courtesy of S.R. Durham.

prostaglandins, and other preformed or rapidly synthesized mediators that cause a rapid increase in vascular permeability and the contraction of smooth muscle. The **late-phase reaction**, which occurs in about 50% of patients with an early-phase response, is caused by the induced synthesis and release of mediators including prostaglandins, leukotrienes, chemokines, and cytokines such as IL-5 and IL-13 from the activated mast cells and basophils (see Fig. 13.12). These recruit other leukocytes, including eosinophils and T_H2 lymphocytes, to the site of inflammation. Late-phase reactions are associated with a second phase of smooth muscle contraction mediated by T cells, with sustained edema, and with tissue remodeling such as smooth muscle hypertrophy (an increase in size due to cell growth) and hyperplasia (an increase in the number of cells).

Allergic Asthma

The late-phase reaction and its long-term sequel, **chronic allergic inflammation**, which is in essence a type IV hypersensitivity reaction (see Fig. 13.1), contribute to much serious long-term illness, such as chronic asthma. The chronic phase of asthma is characterized by the presence of both T_H1 cytokines (such as IFN-γ) and T_H2 cytokines, though the latter seem to predominate.

13-11 The clinical effects of allergic reactions vary according to the site of mast-cell activation.

When reexposure to allergen triggers an allergic reaction, the effects are focused on the site at which mast-cell degranulation occurs. In the immediate response, the preformed mediators released are short-lived, and their potent effects on blood vessels and smooth muscles are therefore confined to the vicinity of the activated mast cell. The more sustained effects of the late-phase response are also focused on the site of initial allergen-triggered activation, and the particular anatomy of this site may determine how readily the inflammation can be resolved. Thus, the clinical syndrome produced by an allergic reaction depends critically on three variables: the amount of allergen-specific IgE present; the route by which the allergen is introduced; and the dose of allergen (Fig. 13.15).

Acute Systematic Anaphylaxis

If an allergen is introduced directly into the bloodstream or is rapidly absorbed from the gut, the connective tissue mast cells associated with all blood vessels can become activated. This activation causes a very dangerous syndrome called **systemic anaphylaxis**. Disseminated mast-cell activation has a variety of potentially fatal effects: the widespread increase in vascular permeability leads to a catastrophic loss of blood pressure; airways constrict, causing difficulty in breathing; and swelling of the epiglottis can cause suffocation. This potentially fatal syndrome is called **anaphylactic shock**. It can occur if drugs are administered to people who have IgE specific for that drug, or after an insect bite in individuals allergic to insect venom. Some foods, for example peanuts or brazil nuts, can cause systemic anaphylaxis in susceptible individuals. This syndrome can be rapidly fatal but can usually be controlled by the immediate injection of epinephrine, which relaxes the smooth muscle and inhibits the cardiovascular effects of anaphylaxis.

The most frequent allergic reactions to drugs occur with penicillin and its relatives. In people with IgE antibodies against penicillin, administration of the drug by injection can cause anaphylaxis and even death. Great care should be taken to avoid giving a drug to patients with a past history of allergy to that drug or one that is closely related structurally. Penicillin acts as a hapten (see Appendix I, Section A-1); it is a small molecule with a highly reactive β-lactam ring that is crucial for its antibacterial activity. This ring reacts with amino groups on host proteins to form covalent conjugates. When penicillin is ingested or injected, it forms conjugates with self proteins, and the penicillin-modified self peptides provoke a T_H2 response in some individuals. These T_H2 cells then activate penicillin-binding B cells to produce IgE antibody against the penicillin hapten. Thus, penicillin acts both as the B-cell antigen and, by modifying self peptides, as the T-cell antigen. When penicillin is injected intravenously into an allergic individual, the penicillin-modified proteins can cross-link IgE molecules on tissue mast cells and circulating basophils and thus cause anaphylaxis.

Fig. 13.15 The dose and route of allergen administration determine the type of IgE-mediated allergic reaction that results. There are two main anatomical distributions of mast cells: those associated with vascularized connective tissues, called connective tissue mast cells, and those found in submucosal layers of the gut and respiratory tract, called mucosal mast cells. In an allergic individual, all of these are loaded with IgE directed against specific allergens. The overall response to an allergen then depends on which mast cells are activated. Allergen in the bloodstream activates connective tissue mast cells throughout the body, resulting in the systemic release of histamine and other mediators. Subcutaneous administration of allergen activates only local connective tissue mast cells, leading to a local inflammatory reaction. Inhaled allergen, penetrating across epithelia, activates mainly mucosal mast cells, causing smooth muscle contraction in the lower airways; this leads to bronchoconstriction and difficulty in expelling inhaled air. Mucosal mast-cell activation also increases the local secretion of mucus by epithelial cells and causes irritation. Similarly, ingested allergen penetrates across gut epithelia, causing vomiting due to intestinal smooth muscle contraction and diarrhea due to outflow of fluid across the gut epithelium. Food allergens can also be disseminated in the bloodstream, causing urticaria (hives) when the allergen reaches the skin.

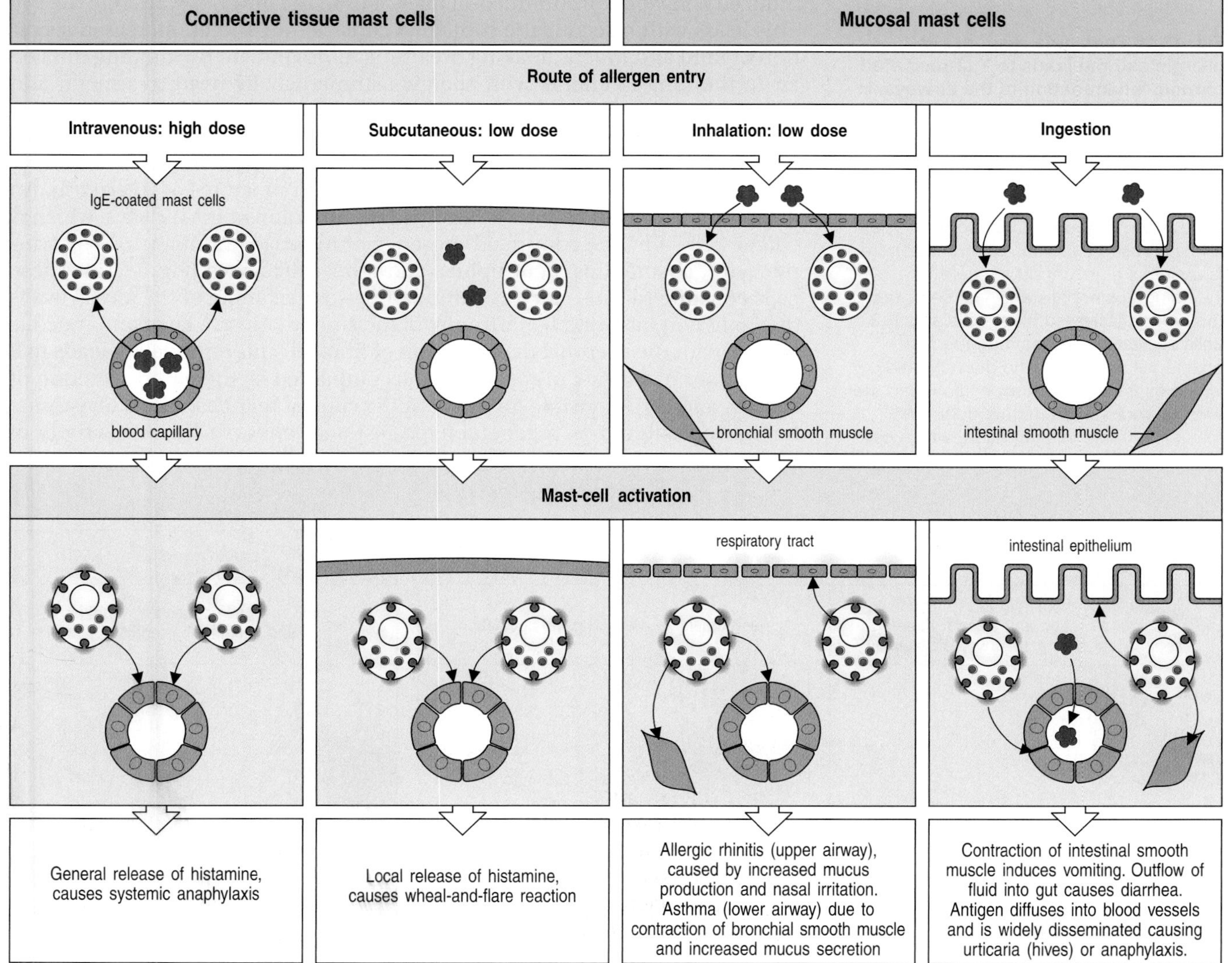

13-12 Allergen inhalation is associated with the development of rhinitis and asthma.

Inhalation is the most common route of allergen entry. Many people have mild allergies to inhaled antigens, manifesting as sneezing and a runny nose. This is called **allergic rhinitis** and results from the activation of mucosal mast cells beneath the nasal epithelium by allergens such as pollens that release their protein contents, which can then diffuse across the mucous membranes of the nasal passages. Allergic rhinitis is characterized by intense itching and sneezing, local edema leading to blocked nasal passages, a nasal discharge, which is typically rich in eosinophils, and irritation of the nose as a result of histamine release. A similar reaction to airborne allergens deposited on the conjunctiva of the eye is called **allergic conjunctivitis**. Allergic rhinitis and conjunctivitis are commonly caused by environmental allergens that are present only during certain seasons of the year. For example, hay fever (known clinically as seasonal rhinoconjunctivitis) is caused by a variety of allergens, including certain grass and tree pollens. Summer and autumnal symptoms may be caused by weed pollen, such as that of ragweed or the spores of fungi such as *Alternaria*. Perennial allergens such as cat dander and house dust mites can be a cause of year-round misery.

A more serious syndrome is **allergic asthma**, which is triggered by allergen-induced activation of submucosal mast cells in the lower airways (Fig. 13.16). This leads within seconds to bronchial constriction and an increased secretion of fluid and mucus, making breathing more difficult by trapping inhaled air in the lungs. Patients with allergic asthma usually need treatment, and asthmatic attacks can be life threatening. The same allergens that cause allergic rhinitis and conjunctivitis commonly cause asthma attacks. For example, respiratory arrest caused by severe attacks of asthma in the summer or autumn has been associated with the inhalation of spores of *Alternaria*. An important feature of asthma is chronic inflammation of the airways, which is characterized by the continued presence of increased numbers of T_H2 lymphocytes, eosinophils, neutrophils, and other leukocytes (Fig. 13.17). These cells conspire to cause **airway remodeling**—a thickening of the airway walls due to hyperplasia and hypertrophy of the smooth muscle layer and mucous glands, with the eventual development of fibrosis. This remodeling leads to a permanent narrowing of the airways accompanied by increased secretion of mucus, and is responsible for many of the clinical manifestations of asthma. In chronic asthmatics, a general hyperresponsiveness or **hyperreactivity** of the airways to non-immunological stimuli also often develops.

Fig. 13.16 The acute response in allergic asthma leads to T_H2-mediated chronic inflammation of the airways. In sensitized individuals, cross-linking of specific IgE on the surface of mast cells by an inhaled allergen triggers them to secrete inflammatory mediators, causing increased vascular permeability, contraction of bronchial smooth muscle, and increased secretion of mucus. There is an influx of inflammatory cells, including eosinophils and T_H2 cells, from the blood. Activated mast cells and T_H2 cells secrete cytokines that augment eosinophil activation and degranulation, which causes further tissue injury and the entry of more inflammatory cells. The result is chronic inflammation, which can cause irreversible damage to the airways.

Acute responses

Inflammatory mediators cause increased mucus secretion and smooth muscle contraction leading to airway obstruction

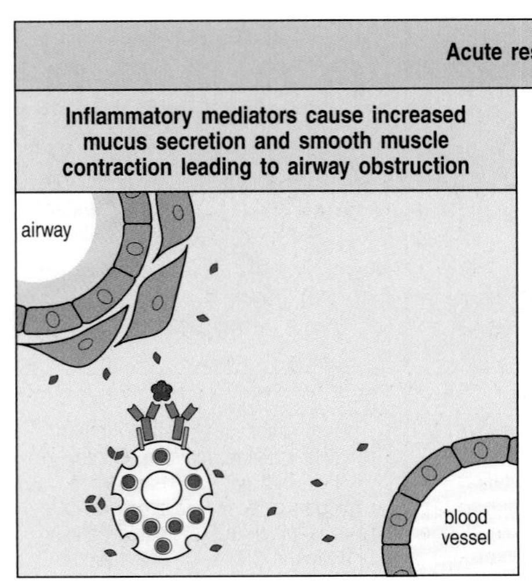

Recruitment of cells from the circulation

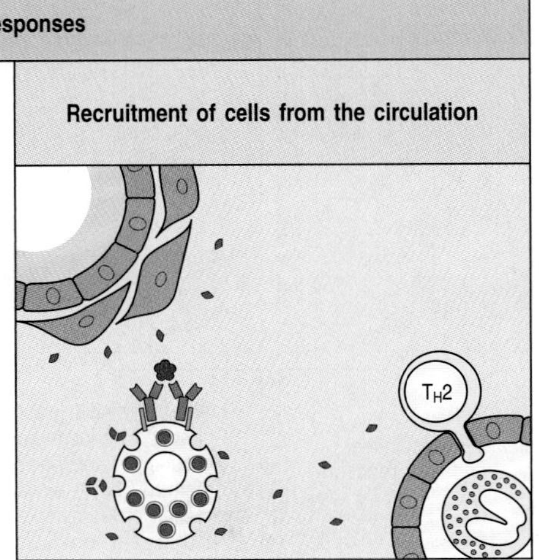

Chronic response

Chronic response caused by cytokines and eosinophil products

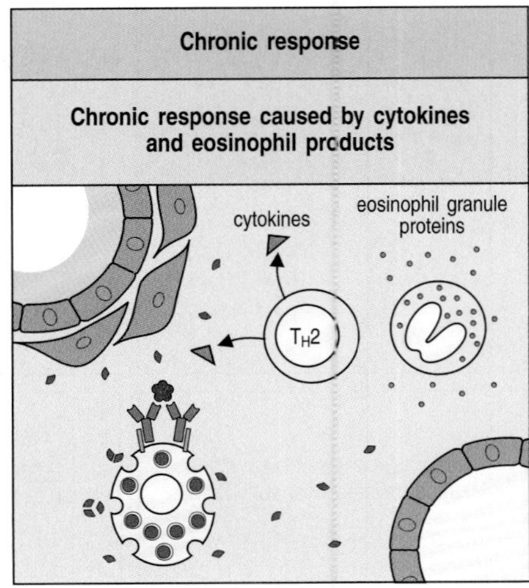

The direct action of T_H2 cytokines such as IL-9 and IL-13 on airway epithelial cells may have a dominant role in one of the major features of the disease, the induction of goblet-cell metaplasia, which is the increased differentiation of epithelial cells as goblet cells, and the consequent increase in secretion of mucus. Lung epithelial cells can also produce the chemokine receptor CCR3 and at least two of the ligands for this receptor—CCL5 and CCL11. These chemokines enhance the T_H2 response by attracting more T_H2 cells and eosinophils to the damaged lungs. The direct effects of T_H2 cytokines and chemokines on airway smooth muscle cells and lung fibroblasts cause the apoptosis of epithelial cells and airway remodeling, induced in part by the production of TGF-β, which has numerous effects on the epithelium, ranging from inducing apoptosis to stimulating cell proliferation.

A disease resembling human asthma develops in mice that lack the transcription factor T-bet, which is required for T_H1 differentiation (see Section 8-19), and in which T-cell responses are thought to be skewed to T_H2. These mice have increased levels of the T_H2 cytokines IL-4, IL-5, and IL-13 and develop airway inflammation involving lymphocytes and eosinophils (Fig. 13.18). They also develop nonspecific airway hyperreactivity to non-immunological stimuli, similar to that seen in human asthma. These changes occur in the absence of any exogenous inflammatory stimulus and show that, in extreme circumstances, a genetic imbalance toward T_H2 responses can cause allergic disease. The involvement of eosinophils in asthma seems somewhat different in humans and in mice. In human asthma patients, the number of eosinophils is directly associated with the severity of asthma. In mice deficient in eosinophils, however, the only consistent finding relevant to asthma pathophysiology is a reduction in airway remodeling without a reduction in airway hyperreactivity.

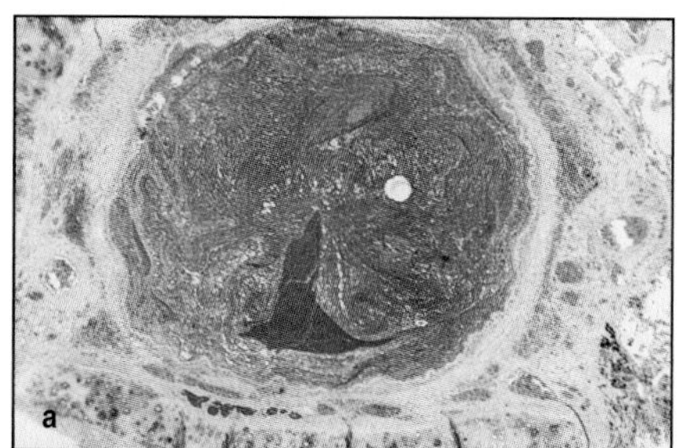

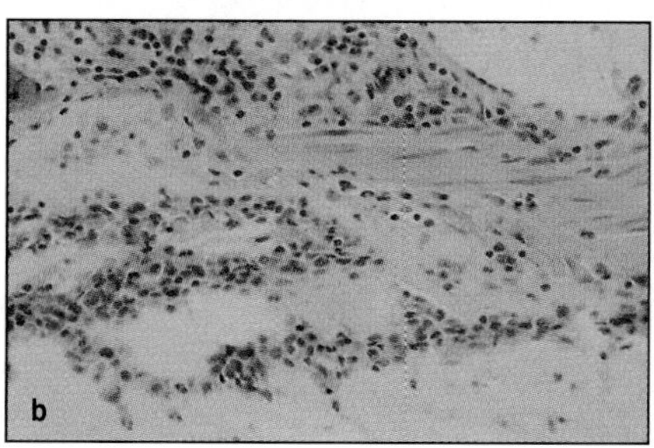

Fig. 13.17 Morphological evidence of chronic inflammation in the airways of an asthmatic patient. Panel a shows a section through a bronchus of a patient who died of asthma; there is almost total occlusion of the airway by a mucous plug. In panel b, a close-up view of the bronchial wall shows injury to the epithelium lining the bronchus, accompanied by a dense inflammatory infiltrate that includes eosinophils, neutrophils, and lymphocytes. Photographs courtesy of T. Krausz.

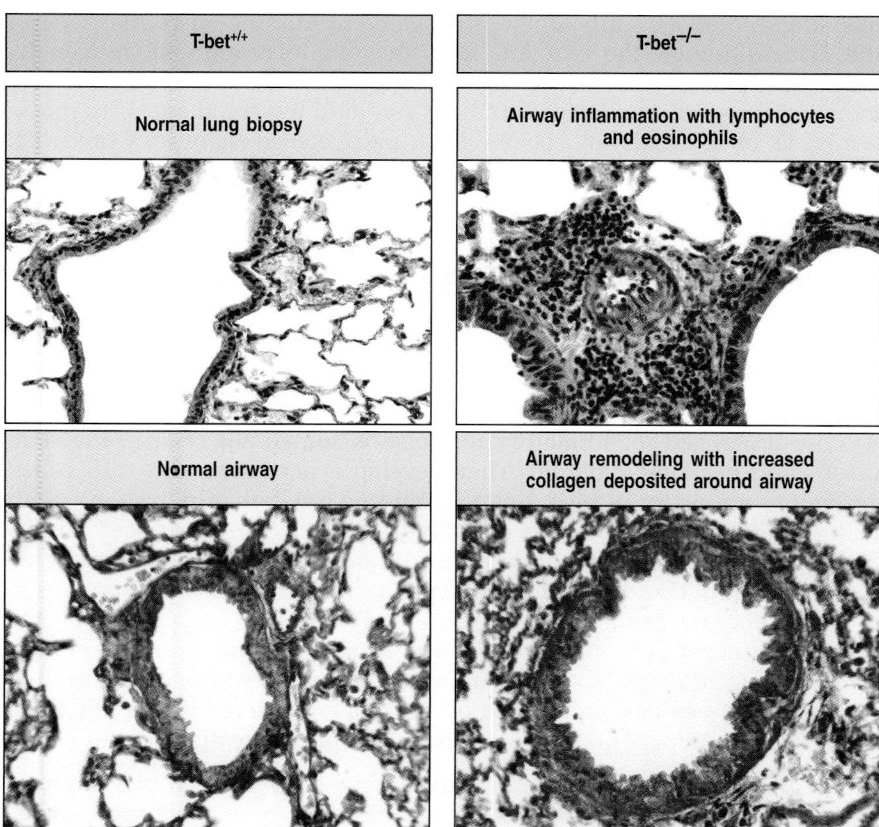

Fig. 13.18 Mice lacking the transcription factor T-bet develop asthma and T-cell responses polarized toward T_H2. T-bet binds to the promoter of the gene encoding IL-2 and is present in T_H1 but not T_H2 cells. Mice with a gene-targeted deletion of T-bet (T-bet–/–) developed a spontaneous asthma-like phenotype in the lungs. Left-hand panels: lung and airways in normal mice. Right-hand panels: T-bet-deficient mice showed lung inflammation, with lymphocytes and eosinophils around the airway and blood vessels (top) and airway remodeling with increased collagen around the airway (bottom). Photographs courtesy of L. Glimcher.

Although allergic asthma is initially driven by a response to a specific allergen, the subsequent chronic inflammation seems to be perpetuated even in the apparent absence of exposure to allergen. The airways become characteristically hyperreactive, and factors other than antigen can trigger asthma attacks. Asthmatics characteristically show hyperresponsiveness to environmental chemical irritants such as cigarette smoke and sulfur dioxide. Viral or, to a smaller extent, bacterial respiratory tract infections can also exacerbate the disease by inducing a T_H2-dominated local response.

13-13 Skin allergy is manifested as urticaria or chronic eczema.

The same dichotomy between immediate and delayed responses is seen in cutaneous allergic responses. The skin forms an effective barrier to the entry of most allergens but it can be breached by the local injection of small amounts of allergen, for example by a stinging insect. The entry of allergen into the epidermis or dermis causes a localized allergic reaction. Local mast-cell activation in the skin leads immediately to a local increase in vascular permeability, which causes extravasation of fluid and swelling. Mast-cell activation also stimulates the release of chemicals from local nerve endings by a nerve axon reflex, causing the vasodilation of surrounding cutaneous blood vessels, which causes redness of the surrounding skin. The resulting skin lesion is called a **wheal-and-flare reaction**. About 8 hours later, a more widespread and sustained edematous response appears in some individuals as a consequence of the late-phase response (see Fig. 13.14). A disseminated form of the wheal-and-flare reaction, known as **urticaria** or hives, sometimes appears when ingested allergens enter the bloodstream and reach the skin. Histamine released by mast cells activated by allergen in the skin causes large, itchy, red swellings of the skin.

Allergists take advantage of the immediate response to test for allergy by injecting minute amounts of potential allergens into the epidermal layer of the skin. Although the reaction after the administration of antigen by intraepidermal injection is usually very localized, there is a small risk of inducing systemic anaphylaxis. Another standard test for allergy is to measure levels of IgE antibody specific for a particular allergen in a sandwich ELISA (see Appendix I, Section A-6).

Although acute urticaria is commonly caused by allergens, the causes of chronic urticaria, in which the urticarial rash can recur over long periods, are less well understood. It seems likely that up to one-third of cases of chronic urticaria are caused by autoantibodies against the α chain of FcεRI, and are thus due to autoimmunity. This is an example of a type II hypersensitivity reaction (see Fig. 13.1) in which an autoantibody against a cellular receptor triggers cellular activation, in this case causing mast-cell degranulation with resulting urticaria.

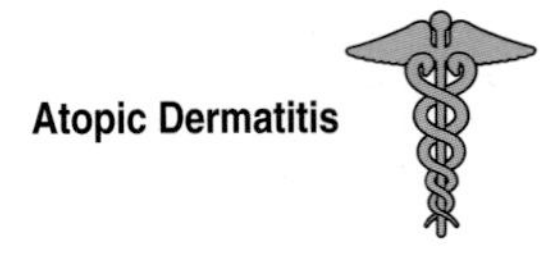

A more prolonged inflammatory response is sometimes seen in the skin, most often in atopic children. They develop a persistent skin rash called **eczema** or **atopic dermatitis**, due to a chronic inflammatory response with features of tissue remodeling and fibrosis similar to those seen in the bronchial walls of patients with asthma. Although allergy is often considered solely in the context of a T_H2 phenotype, in human disease (as opposed to murine models) both T_H1 and T_H2 cytokines can contribute to the immunopathogenesis. Atopic dermatitis is an excellent example of this. About one-third of patients show minimal, if any, elevation of IgE in their sera, and T_H1-cell development is preferentially observed in the lesions of atopic dermatitis patients with a persistent history of the disease.

Innate immune responses due to activation of TLRs by microbial products can exacerbate atopic dermatitis. Activation of these receptors usually initiates a T_H1-cell response by stimulating the production of IL-12 and IL-18.

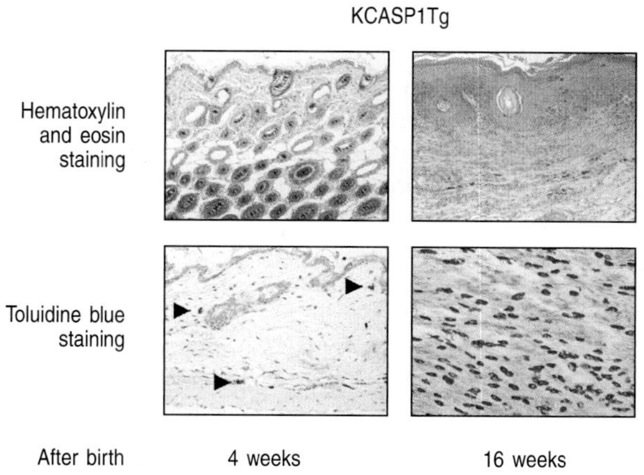

KCASP1Tg

Hematoxylin and eosin staining

Toluidine blue staining

After birth 4 weeks 16 weeks

Genotypes	Dermatitis	IgE	Mast cells
Wild type	—	control	control
KCASP1Tg	+++	+++	+++
$Stat6^{-/-}KCASP1Tg$	+++	ND	+
$IL\text{-}18^{-/-}KCASP1Tg$		+	control
KIL-18Tg	+++*	+++	+++

*Onset is delayed

Fig. 13.19 A deficiency of IL-18 prevents the development of atopic dermatitis in susceptible mice. KCASP1Tg mice overexpress the enzyme caspase-1 in their keratinocytes and develop a condition similar to human atopic dermatitis. Left panel: in skin sections stained by hematoxylin and eosin (HE) (top row), the lesions are characterized by hyperkeratosis and dense infiltration with leukocytes and lymphocytes. When stained by toluidine blue (bottom row), a dense accumulation of mast cells can be seen. The dark-purple stained cells are mast cells. Far greater numbers of mast cells (indicated by arrows) are present in the lesion at 16 weeks compared with 4 weeks. Right panel: KCASP1Tg mice deficient in STAT6 have serum IgE concentrations below detectable levels, but still suffer from similar changes in the skin, whereas KCASP1Tg mice deficient in IL-18 are free from dermatitis. This suggests that T_H2 cytokines are not important in this model. KIL-18Tg mice, which overexpress mature IL-18 in their keratinocytes, show the same symptoms as KCASP1Tg mice except for the delay in disease onset. KCASP1Tg, keratinocyte-specific caspase-1-transgenic mice; KIL-18Tg, keratinocyte-specific mature IL-18-transgenic mice; ND, not detectable. Photographs courtesy of Tsutsui, H., *et al.*: *Immunol. Rev.* 2004, 202: 115–138.

An experimental situation in which these cytokines are overproduced is in mutant mice that overexpress the enzyme caspase-1 specifically in their keratinocytes (KCASP1Tg mice). These mice are born healthy but develop cutaneous changes similar to human atopic dermatitis and start frequent scratching at around 8–10 weeks after birth. Serum IgE and IgG levels also begin to rise at that time. The overexpression of caspase-1 leads to increased apoptosis of keratinocytes, but also to increased levels of IL-1 and IL-18, because caspase-1 is required to activate these cytokines. As the mice grow, the skin lesions expand and the disease becomes more severe. The mice are, however, completely protected from developing the condition when they are made deficient in IL-18, and thus do not develop a strong T_H1 response. They are not protected when made deficient in STAT6, which leads to a lack of a T_H2-cell response (Fig. 13.19). This type of allergy has been classified as **innate-type allergy**, in contrast to T_H2-dependent classical allergy.

T_H2 responses are, however, important in natural atopic dermatitis and may lead indirectly to exacerbation of the disease by making the individual more susceptible to certain infections. For example, individuals with atopic dermatitis are more susceptible to cutaneous inflammation after vaccination with vaccinia virus. The increased susceptibility results from the spread of the vaccinia virus due to the actions of the T_H2 cytokines IL-4 and IL-13. The T_H2 response also inhibits the production of the antimicrobial peptide cathelicidin, which is normally induced as a result of stimulation of TLR-3. Thus, one could envisage a vicious circle of infection triggering atopic dermatitis resulting in increased susceptibility to further infection.

13-14 Allergy to foods causes systemic reactions as well as symptoms limited to the gut.

Genuine food allergy affects about 1–4% of American and European populations, although food intolerances and dislikes are ubiquitous and often misnamed 'allergy' by the sufferer. About one-quarter of true food allergy in the United States and Europe is accounted for by allergy to peanuts, which is increasing in incidence—tripling in the past 5 years. Food allergy causes approximately 30,000 anaphylactic reactions each year in the United States, including 200 deaths. This is a significant public-health problem, especially in school, where children may be unwittingly exposed to peanuts, which are present in many different foods. Fig. 13.20 illustrates the risk factors for the development of food allergy.

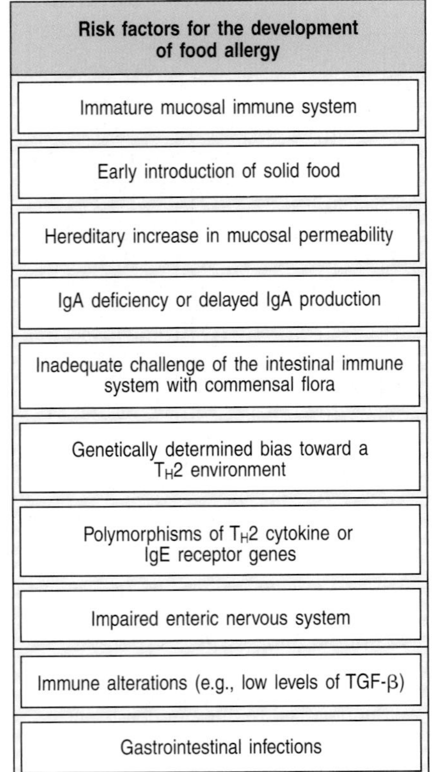

Risk factors for the development of food allergy
Immature mucosal immune system
Early introduction of solid food
Hereditary increase in mucosal permeability
IgA deficiency or delayed IgA production
Inadequate challenge of the intestinal immune system with commensal flora
Genetically determined bias toward a T_H2 environment
Polymorphisms of T_H2 cytokine or IgE receptor genes
Impaired enteric nervous system
Immune alterations (e.g., low levels of TGF-β)
Gastrointestinal infections

Fig. 13.20 Risk factors for the development of food allergy.

One of the characteristic features of food allergens is a high degree of resistance to digestion by pepsin in the stomach. This allows them to reach the mucosal surface of the small intestine as intact allergens. When an allergen is eaten, two types of allergic responses are seen. Activation of mucosal mast cells associated with the gastrointestinal tract leads to transepithelial fluid loss and smooth muscle contraction, causing diarrhea and vomiting. For reasons that are not fully understood, connective tissue mast cells in the dermis and subcutaneous tissues can also be activated after ingestion of allergen, presumably by allergen that has been absorbed into the bloodstream, and this results in urticaria. Urticaria is a common reaction when penicillin is given orally to a patient who already has penicillin-specific IgE antibodies. Ingestion of food allergens can also lead to the development of asthma and of generalized anaphylaxis, accompanied by cardiovascular collapse. Certain foods, most importantly peanuts, tree nuts, and shellfish, are particularly associated with this type of life-threatening response. Food allergy can be either IgE mediated, as in asthma or systemic anaphylaxis, or non-IgE mediated. An important example of the latter is celiac disease.

13-15 Celiac disease is a model of antigen-specific immunopathology.

Celiac disease is a chronic condition of the upper small intestine caused by an immune response directed at gluten, a complex of proteins present in wheat, oats, and barley. Elimination of gluten from the diet restores normal gut function, but must be continued throughout life. The pathology of celiac disease is characterized by the loss of the slender, finger-like villi formed by the intestinal epithelium (a condition termed villous atrophy), together with an increase in the size of the sites in which epithelial cells are renewed (crypt hyperplasia) (Fig. 13.21). These pathological changes result in the loss of the mature epithelial cells that cover the villi and which normally absorb and digest food, and is accompanied by severe inflammation of the intestinal wall, with increased numbers of T cells, macrophages, and plasma cells in the lamina propria, as well as increased numbers of lymphocytes in the epithelial layer. Gluten seems to be the only food protein that provokes intestinal inflammation in this way, a property that reflects gluten's ability to stimulate both specific and innate immunity in genetically susceptible individuals.

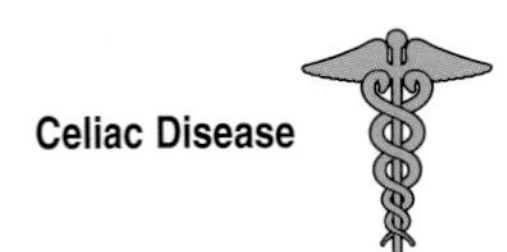

Celiac disease shows an extremely strong genetic predisposition, with more than 95% of patients expressing the HLA-DQ2 class II MHC allele, and there is an 80% concordance in monozygotic twins (that is, if one twin develops it, there is an 80% probability that the other will), but only a 10% concordance in dizygotic twins. Nevertheless, most individuals expressing HLA-DQ2 do not develop celiac disease despite the almost universal presence of gluten in the Western diet. Thus, other genetic factors must make important contributions to susceptibility.

Most evidence indicates that celiac disease requires the aberrant priming of IFN-γ-producing CD4 T cells by antigenic peptides present in α-gliadin, one of the major proteins in gluten. It is generally accepted that only a limited number of peptides can provoke an immune response leading to celiac disease. This is likely to be due to the unusual structure of the peptide-binding groove of the HLA-DQ2 molecule. The key step in the immune recognition of α-gliadin is the deamidation of its peptides by the enzyme tissue transglutaminase (tTG), which converts selected glutamine residues to negatively charged glutamic acid. Only peptides containing negatively charged residues in certain positions bind strongly to HLA-DQ2, and thus the transamination reaction promotes the formation of peptide:HLA-DQ2 complexes, which can activate antigen-specific CD4 T cells (Fig. 13.22). Multiple peptide epitopes can be generated from gliadin. Activated gliadin-specific

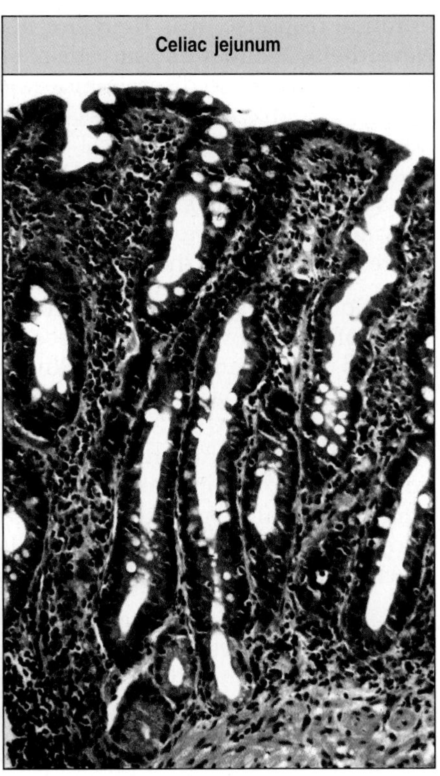

Figure 13.21 The pathological features of celiac disease. Left: the surface of the normal small intestine is folded into finger-like villi, which provide an extensive surface for nutrient absorption. Right: The local immune response against the food protein α-gliadin provokes destruction of the villi. In parallel, there is lengthening and increased mitotic activity in the underlying crypts where new epithelial cells are produced. There is also a marked inflammatory infiltrate in the intestinal mucosa, with increased numbers of lymphocytes in the epithelial layer and accumulation of CD4 T cells, plasma cells, and macrophages in the deeper layer, the lamina propria. Because the villi contain all the mature epithelial cells that digest and absorb foodstuffs, their loss results in life-threatening malabsorption and diarrhea. Photographs courtesy of Allan Mowat.

CD4 T cells accumulate in the lamina propria, producing IFN-γ, a cytokine that leads to intestinal inflammation.

Celiac disease is entirely dependent on the presence of a foreign antigen (gluten) and is not associated with a specific immune response against antigens in the tissue—the intestinal epithelium—that is damaged during the

Fig. 13.22 Molecular basis of immune recognition of gluten in celiac disease. After the digestion of gluten by gut digestive enzymes, deamidation of epitopes by tissue transglutaminase leads to their binding to HLA-DQ molecules and priming of the immune system.

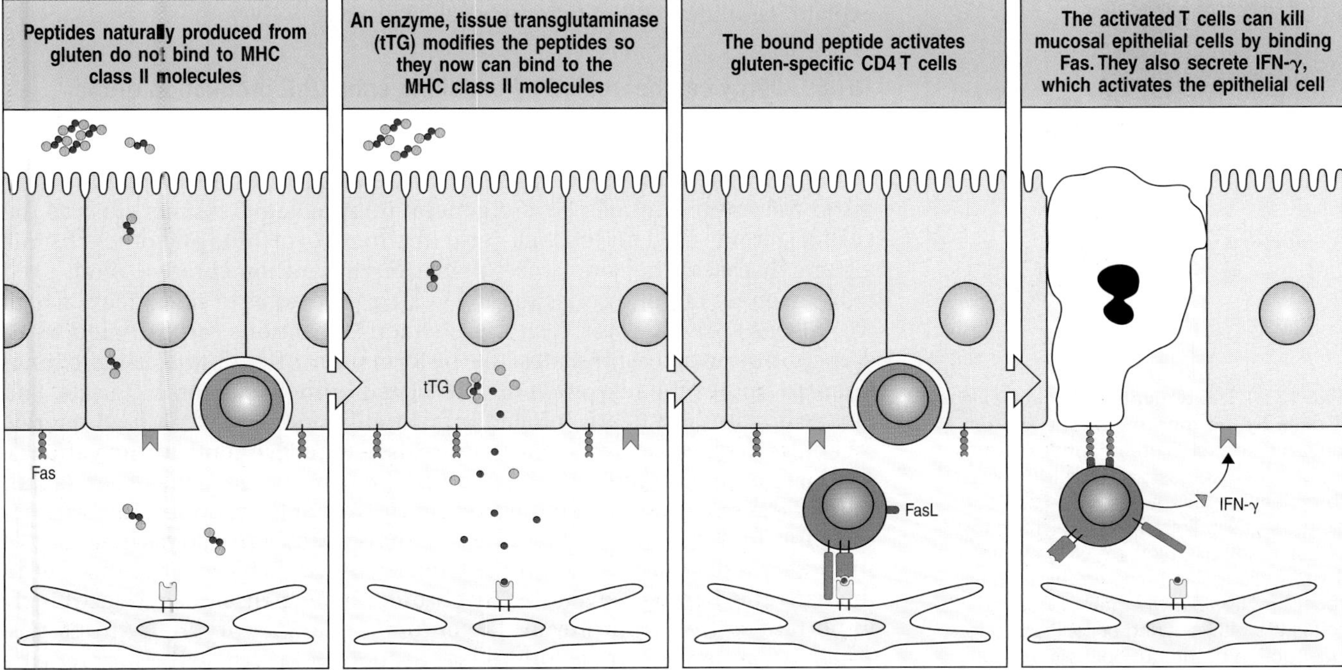

Peptides naturally produced from gluten do not bind to MHC class II molecules
An enzyme, tissue transglutaminase (tTG) modifies the peptides so they now can bind to the MHC class II molecules
The bound peptide activates gluten-specific CD4 T cells
The activated T cells can kill mucosal epithelial cells by binding Fas. They also secrete IFN-γ, which activates the epithelial cell
Fas
tTG
FasL
IFN-γ

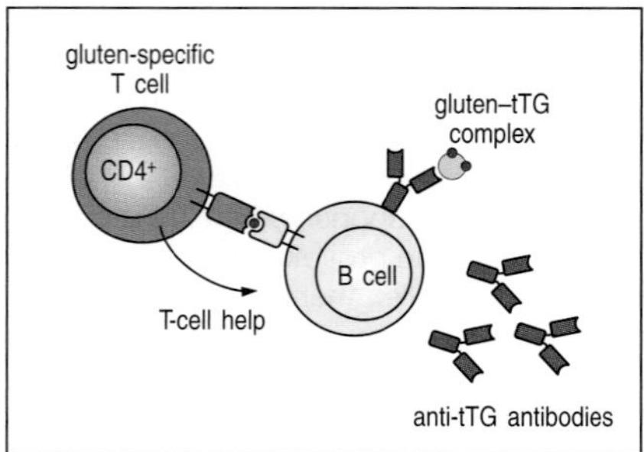

Fig. 13.23 A hypothesis to explain antibody production against tissue transglutaminase (tTG) in the absence of T cells specific for tTG in celiac patients. tTG-reactive B cells endocytose gluten–tTG complexes and present gluten peptides to the gluten-specific T cells. The stimulated T cells can now provide help to these B cells, which produce autoantibodies against tTG.

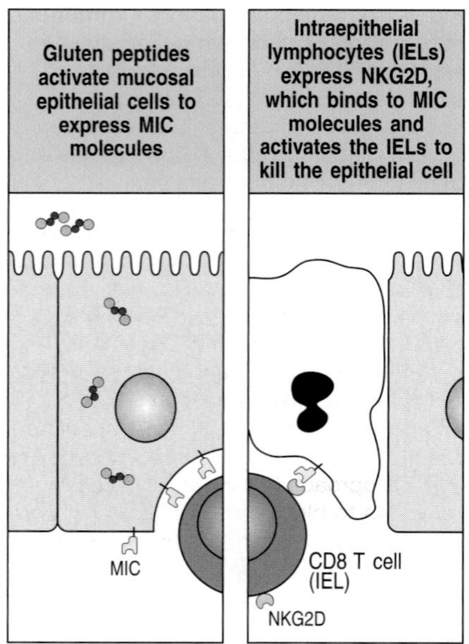

Fig. 13.24 The activation of cytotoxic T cells by the innate immune system in celiac disease. Gluten peptides can induce the expression of the MHC class Ib molecules MIC-A and MIC-B on gut epithelial cells. Intraepithelial lymphocytes (IELs), many of which are CD8 cytotoxic T cells, recognize these proteins via the receptor NKG2D, which activates the IELs to kill the MIC-bearing cells, leading to destruction of the gut epithelium.

immune response. It is therefore not considered an autoimmune disease. Nevertheless, autoantibodies against tissue transglutaminase are found in all patients with celiac disease; indeed, the presence of serum IgA antibodies against this enzyme is used as a sensitive and specific test for the disease. Interestingly, no tTG-specific T cells have been found and it has been proposed that gluten-reactive T cells provide help to B cells reactive to tissue transglutaminase. In support of this hypothesis, gluten can complex with the enzyme and therefore could be taken up by tTG-reactive B cells (Fig. 13.23). There is no evidence that these autoantibodies contribute to tissue damage.

Chronic T-cell responses against food proteins are normally prevented by the development of oral tolerance (see Section 11-13). Why this breaks down in patients with celiac disease is unknown. The properties of the HLA-DQ2 molecule provide a partial explanation, but there must be additional factors because most HLA-DQ2-positive individuals do not develop celiac disease and the high concordance rates in monozygotic twins indicate a role for additional genetic factors. Polymorphisms in the gene for CTLA-4 or in other immunoregulatory genes may be associated with susceptibility. There could also be differences in how individuals digest gliadin in the intestine, so that differing amounts survive for deamidation and presentation to T cells.

The gluten protein also seems to have several properties that contribute to pathogenesis. As well as its relative resistance to digestion, there is mounting evidence that some gliadin-derived peptides stimulate the innate immune system by inducing the release of IL-15 by intestinal epithelial cells. This process is antigen-nonspecific and involves peptides that cannot be bound by HLA-DQ2 molecules or recognized by CD4 T cells. IL-15 release leads to the activation of dendritic cells in the lamina propria, as well as the upregulation of MIC-A expression by epithelial cells. CD8 T cells in the mucosal epithelium can be activated via their NKG2D receptors, which recognize MIC-A, and they can kill MIC-A-expressing epithelial cells via these same NKG2D receptors (Fig. 13.24). Triggering of these innate immune responses by α-gliadin may create some intestinal damage on its own and also induce some of the co-stimulatory events necessary for initiating an antigen-specific CD4 T-cell response to other parts of the α-gliadin molecule. The ability of gluten to stimulate both innate and adaptive immune responses may thus explain its unique ability to induce celiac disease.

13-16 Allergy can be treated by inhibiting either IgE production or the effector pathways activated by the cross-linking of cell-surface IgE.

Current drug treatments for allergic disease either treat the symptoms only, as do the antihistamines, or are general immunosuppressants such as the corticosteroids used for the long-term treatment of asthma and other chronic allergic diseases. They are largely palliative, rather than curative, often need to be taken for life, and consequently incur a wide range of side effects, which we discuss in Chapter 15. Anaphylactic reactions are treated with epinephrine, which stimulates the re-formation of endothelial tight junctions, promotes the relaxation of constricted bronchial smooth muscle, and also stimulates the heart. Inhaled bronchodilators that act on β-adrenergic receptors to relax constricted muscle are used to relieve acute asthma attacks. Antihistamines that block the histamine H_1 receptor reduce the urticaria that follows the release of histamine from mast cells and basophils. Relevant H_1 receptors include those on blood vessels that cause increased permeability of the vessel wall, and those on unmyelinated nerve fibers that are thought to mediate the itching sensation. In chronic allergic disease it is extremely important to treat and prevent the chronic inflammatory injury to tissues. Topical or systemic corticosteroids (see Section 15-1) are used to suppress the chronic inflammatory changes seen in asthma, rhinitis, and eczema. What is really needed, however, is a means of regulating the T-cell response to the allergenic peptide antigen in an antigen-specific manner.

Some of the newer approaches to the treatment and prevention of allergy that attempt to do this are set out in Fig. 13.25. Two treatments are commonly used in clinical practice—one is **desensitization** or **specific allergen immunotherapy** and the other is blockade of the effector pathways. There are also several approaches still in the experimental stage. In desensitization the aim is to restore tolerance to the allergen by reducing its tendency to induce IgE production. The key to this therapy seems to be the induction of regulatory T cells secreting IL-10 and/or TGF-β, which skew the response away from IgE (see Section 13-3). Beekeepers exposed to multiple stings are often naturally protected from severe allergic reactions such as anaphylaxis through a mechanism involving IL-10-secreting T cells. Similarly, specific allergen immunotherapy for sensitivity to insect venom and airborne allergens induces the increased production of IL-10 and in some cases TGF-β, as well as the induction of IgG isotypes, particularly IgG4, an isotype selectively promoted by IL-10. Patients are desensitized by injection with escalating doses of allergen, starting with tiny amounts, an injection schedule that gradually decreases the IgE-dominated response. Allergen injection immunotherapy in fact seems to downregulate both T_H1- and T_H2-driven hypersensitivity disease, in line with its presumed induction of T_{reg} cells. Recent evidence shows that desensitization is also associated with a reduction in the numbers of late-phase inflammatory cells at the site of the allergic reaction. A potential complication of the desensitization approach is the risk of inducing IgE-mediated allergic responses. This strategy is not always successful, for example in treating severe reactions against food allergens such as peanut allergy.

An alternative, and still experimental, approach to desensitization is vaccination with peptides derived from common allergens. This procedure induces

Target step	Mechanism of treatment	Specific approach
T_H2 activation	Induce regulatory T cells	Injection of specific antigen or peptides Administration of cytokines, e.g., IFN-γ, IL-10, IL-12, TGF-β Use of adjuvants such as CpG oligodeoxynucleotides to stimulate T_H1 response
Activation of B cell to produce IgE	Block co-stimulation Inhibit T_H2 cytokines	Inhibit CD40L Inhibit IL-4 or IL-13
Mast-cell activation	Inhibit effects of IgE binding to mast cell	Blockade of IgE receptor
Mediator action	Inhibit effects of mediators on specific receptors Inhibit synthesis of specific mediators	Antihistamine drugs Lipooxygenase inhibitors
Eosinophil-dependent inflammation	Block cytokine and chemokine receptors that mediate eosinophil recruitment and activation	Inhibit IL-5 Block CCR3

Fig. 13.25 Approaches to the treatment of allergy. Possible methods of inhibiting allergic reactions are shown. Two approaches are in regular clinical use. The first is the injection of specific antigen in desensitization regimes, which is believed to restore tolerance to the allergen—perhaps through the production of regulatory T cells. The second clinically useful approach is the use of specific inhibitors to block the synthesis or effects of inflammatory mediators produced by mast cells.

T-cell anergy (see Section 8-15), which is associated with multiple changes in T-cell phenotype, including the production of IL-10 and upregulation of the cell-surface protein CD5. IgE-mediated responses are not induced by the peptides because IgE, in contrast to T cells, can recognize only the intact antigen. A difficulty with this approach is that an individual's responses to peptides are restricted by their MHC class II alleles; patients with different MHC class II molecules therefore respond to different allergen-derived peptides. One possible solution is the use of peptides that contain short sequences with multiple overlapping MHC-binding motifs that would provide coverage for the majority of the population.

Another vaccination strategy that shows promise in experimental models of allergy is the use of oligodeoxynucleotides rich in unmethylated CpG as adjuvants for desensitization regimes. These oligonucleotides mimic the CpG motifs in bacterial DNA and strongly promote T_H1 responses, probably through the stimulation of TLR-9 in dendritic cells (see Section 8-7). The mechanism of action of adjuvants is discussed in Appendix I, Section A-4.

The signaling pathways that enhance the IgE response in allergic disease are also potential targets for therapy. Inhibitors of IL-4, IL-5, and IL-13 would be predicted to reduce IgE responses, but redundancy between some of the activities of these cytokines might make this approach difficult to implement in practice. A second approach to manipulating the response is to give cytokines that promote T_H1-type responses. IFN-γ, IFN-α, and IL-12 have each been shown to reduce IL-4-stimulated IgE synthesis *in vitro*, and IFN-γ and IFN-α have been shown to reduce IgE synthesis *in vivo*. Administration of IL-12 to patients with mild allergic asthma caused a decrease in the number of eosinophils in blood and sputum but had no effect on immediate or late-phase responses to inhaled allergen. The treatment with IL-12 was accompanied by quite severe flu-like symptoms in most patients, which is likely to limit its possible therapeutic value.

Another target for therapeutic intervention might be the high-affinity IgE receptor. An effective competitor for IgE at this receptor could prevent the binding of IgE to the surfaces of mast cells, basophils, and eosinophils. Clinical trials have taken place of a humanized mouse anti-IgE monoclonal antibody named omalizumab, which binds to the portion of IgE that ligates the high-affinity IgE receptor. Because IgE is present in plasma at low levels it was possible to give a large molar excess of omalizumab that caused a decrease in IgE levels of more than 95%. This was accompanied by down-regulation of the numbers of high-affinity IgE receptors on basophils and mast cells. This antibody blocked both the immediate and late-phase responses to experimentally inhaled allergen. Patients with asthma and allergic rhinitis receiving omalizumab in clinical trials had fewer exacerbations than patients on the placebo, and were able to reduce their use of corticosteroids. The efficacy of this agent, which has led to its being licensed for use to treat patients with asthma, provides a clear-cut demonstration of the importance of IgE in the atopic allergic diseases. Targeting the inhibitory receptor FcγRIIb is a potential new therapy for allergy to cat's dander. A chimeric fusion protein consisting of human Fcγ and the cat allergen Fel d 1 blocked the skin reaction in a mouse model of cat allergy and inhibited the release of inflammatory mediators from basophils. This inhibition is specific to the allergen.

A further approach to treatment would be to block the recruitment of eosinophils to sites of allergic inflammation. The eotaxin receptor CCR3 is a potential target in this context. The production of eosinophils in bone marrow and their exit into the circulation might also be reduced by a blockade of IL-5 action. Studies using anti-IL-5 treatment have not been encouraging, however: anti-IL-5 did reduce the numbers of eosinophils in blood and sputum but did not alter immediate and late-phase responses to inhaled allergen or airway hyperreactivity to histamine.

Summary.

The allergic response to innocuous antigens reflects the pathophysiological aspects of a defensive immune response whose physiological role is to protect against helminth parasites. It is triggered by the binding of antigen to IgE antibodies bound to the high-affinity IgE receptor FcεRI on mast cells. Mast cells are strategically distributed beneath the mucosal surfaces of the body and in connective tissue. Antigen cross-linking the IgE on their surface causes them to release large amounts of inflammatory mediators. The resulting inflammation can be divided into early events, characterized by short-lived mediators such as histamine, and later events that involve leukotrienes, cytokines, and chemokines, which recruit and activate eosinophils and basophils. The late phase of this response can evolve into chronic inflammation, characterized by the presence of effector T cells and eosinophils, which is most clearly seen in chronic allergic asthma.

Hypersensitivity diseases.

In this part of the chapter we focus on immunological responses involving IgG antibodies or specific T cells that cause adverse hypersensitivity reactions. Although these effector arms of the immune response normally participate in protective immunity to infection, they occasionally react with noninfectious antigens to produce acute or chronic hypersensitivity reactions. Although the mechanisms initiating the various forms of hypersensitivity are different, much of the pathology is due to the same immunological effector mechanisms. We also consider here a newly characterized category of hypersensitivity disease, in which certain variants of the genes regulating inflammatory responses cause the inappropriate triggering of inflammation, leading to severe disease.

13-17 Innocuous antigens can cause type II hypersensitivity reactions in susceptible individuals by binding to the surfaces of circulating blood cells.

Antibody-mediated destruction of red blood cells (hemolytic anemia) or platelets (thrombocytopenia) can be caused by some drugs, including the antibiotics penicillin and cephalosporin. These are examples of **type II hypersensitivity reactions** in which the drug binds to the cell surface and serves as a target for anti-drug IgG antibodies that cause destruction of the cell (see Fig. 13.1). The anti-drug antibodies are made in only a minority of people and it is not clear why these individuals make them. The cell-bound antibody triggers the clearance of the cell from the circulation, predominantly by tissue macrophages in the spleen, which bear Fcγ receptors.

13-18 Systemic disease caused by immune-complex formation can follow the administration of large quantities of poorly catabolized antigens.

Type III hypersensitivity reactions can arise with soluble antigens (see Fig. 13.1). The pathology is caused by the deposition of antigen:antibody

Drug-Induced
Serum Sickness

aggregates, or **immune complexes**, in particular tissues and sites. Immune complexes are generated in all antibody responses, but their pathogenic potential is determined, in part, by their size and by the amount, affinity, and isotype of the responding antibody. Larger aggregates fix complement and are readily cleared from the circulation by the mononuclear phagocyte system. However, the small complexes that form when antigen is in excess tend to be deposited in blood vessel walls. There they can ligate Fc receptors on leukocytes, leading to leukocyte activation and tissue injury.

A local type III hypersensitivity reaction called an **Arthus reaction** (Fig. 13.26) can be triggered in the skin of sensitized individuals who possess IgG antibodies against the sensitizing antigen. When antigen is injected into the skin, circulating IgG antibody that has diffused into the skin forms immune complexes locally. The immune complexes bind Fc receptors such as FcγRIII on mast cells and other leukocytes, generating a local inflammatory response and increased vascular permeability. Fluid and cells, especially polymorphonuclear leukocytes, then enter the site of inflammation from local blood vessels. The immune complexes also activate complement, leading to the production of the complement fragment C5a. This is a key participant in the inflammatory reaction because it interacts with C5a receptors on leukocytes to activate these cells and attract them to the site of inflammation (see Section 2-5). Both C5a and FcγRIII have been shown to be required for the experimental induction of an Arthus reaction in the lung by macrophages in the walls of the alveoli, and they are probably required for the same reaction induced by mast cells in the skin and the linings of joints (synovia).

A systemic type III hypersensitivity reaction, known as **serum sickness**, can result from the injection of large quantities of a poorly catabolized foreign antigen. This illness was so named because it frequently followed the administration of therapeutic horse antiserum. In the pre-antibiotic era, antiserum made by immunizing horses was often used to treat pneumococcal pneumonia; the specific anti-pneumococcal antibodies in the horse serum would help the patient to clear the infection. In much the same way, **antivenin** (serum from horses immunized with snake venoms) is still used today as a source of neutralizing antibodies to treat people suffering from the bites of poisonous snakes. The increasing use of monoclonal antibodies in the treatment of disease (for example, anti-TNF-α in rheumatoid arthritis) has led to the development of serum sickness in a small minority of patients.

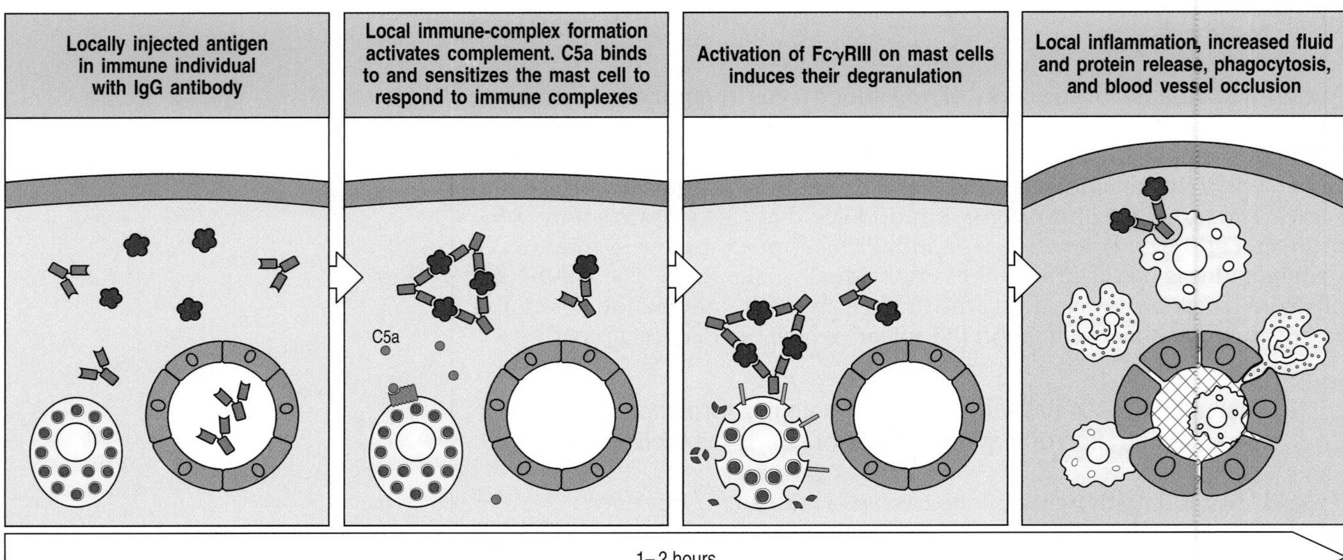
Locally injected antigen in immune individual with IgG antibody
Local immune-complex formation activates complement. C5a binds to and sensitizes the mast cell to respond to immune complexes
C5a
Activation of FcγRIII on mast cells induces their degranulation
Local inflammation, increased fluid and protein release, phagocytosis, and blood vessel occlusion
1–2 hours

Fig. 13.26 The deposition of immune complexes in tissues causes a local inflammatory response known as an Arthus reaction (type III hypersensitivity reaction). In individuals who have already made IgG antibody against an antigen, the same antigen injected into the skin forms immune complexes with IgG antibody that has diffused out of the capillaries. Because the dose of antigen is low, the immune complexes are only formed close to the site of injection, where they activate mast cells bearing Fcγ receptors (FcγRIII). The complement component C5a seems important in sensitizing the mast cell to respond to immune complexes. As a result of mast-cell activation, inflammatory cells invade the site, and blood vessel permeability and blood flow are increased. Platelets also accumulate inside the vessel at the site, ultimately leading to vessel occlusion.

Serum sickness occurs 7–10 days after the injection of horse serum, an interval that corresponds to the time required to mount an IgG-switched primary immune response against the foreign antigens. The clinical features of serum sickness are chills, fever, rash, arthritis, and sometimes glomerulonephritis (inflammation of the glomeruli of the kidneys). Urticaria is a prominent feature of the rash, implying a role for histamine derived from mast-cell degranulation. In this case, the mast-cell degranulation is triggered by the ligation of cell-surface FcγRIII by IgG-containing immune complexes.

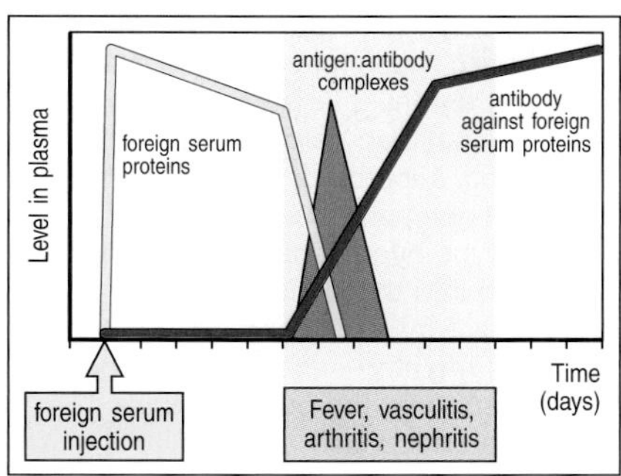

Fig. 13.27 Serum sickness is a classic example of a transient immune complex-mediated syndrome. An injection of a foreign protein or proteins leads to an antibody response. These antibodies form immune complexes with the circulating foreign proteins. The complexes are deposited in small vessels and activate complement and phagocytes, inducing fever and the symptoms of vasculitis, nephritis, and arthritis. All these effects are transient and resolve when the foreign protein is cleared.

The course of serum sickness is illustrated in Fig. 13.27. The onset of disease coincides with the development of antibodies against the abundant soluble proteins in the foreign serum; these antibodies form immune complexes with their antigens throughout the body. These immune complexes fix complement and can bind to and activate leukocytes bearing Fc and complement receptors; these in turn cause widespread tissue damage. The formation of immune complexes causes clearance of the foreign antigen, so serum sickness is usually a self-limiting disease. Serum sickness after a second dose of antigen follows the kinetics of a secondary antibody response (see Section 10-14), with symptoms typically appearing within a day or two.

Pathological immune-complex deposition is seen in other situations in which antigen persists. One is when an adaptive antibody response fails to clear the infecting pathogen, as occurs in subacute bacterial endocarditis or chronic viral hepatitis. In these situations, the replicating pathogen is continuously generating new antigen in the presence of a persistent antibody response, with the consequent formation of abundant immune complexes. These are deposited within small blood vessels, with consequent injury in many tissues and organs, including the skin, kidneys, and nerves.

Immune-complex disease also occurs when inhaled allergens provoke IgG rather than IgE antibody responses, perhaps because they are present at relatively high levels in the air. When a person is reexposed to high doses of such allergens, immune complexes form in the walls of alveoli in the lung. This leads to the accumulation of fluid, protein, and cells in the alveolar wall, slowing blood–gas interchange and compromising lung function. This type of reaction is more likely to occur in occupations such as farming, where there is repeated exposure to hay dust or mold spores, and the resulting disease is known as **farmer's lung**. If exposure to antigen is sustained, the lining of the lungs can be permanently damaged.

13-19 Delayed-type hypersensitivity reactions are mediated by T_H1 cells and CD8 cytotoxic T cells.

Unlike the immediate hypersensitivity reactions described so far, which are mediated by antibodies, **delayed-type hypersensitivity** or **type IV hypersensitivity reactions** are mediated by antigen-specific effector T cells. These function in essentially the same way as they do in a response to a pathogen, as described in Chapter 8. The causes and consequences of some syndromes in which type IV hypersensitivity responses predominate are listed in Fig. 13.28. These responses can be transferred between experimental animals by purified T cells or cloned T-cell lines. Much of the inflammation seen in some of the allergic diseases described in the earlier parts of this chapter is in fact due to delayed-type hypersensitivity.

The prototypic delayed-type hypersensitivity reaction is an artifact of modern medicine—the tuberculin test (see Appendix I, Section A-38). This is used to determine whether an individual has previously been infected with *M. tuberculosis*. Small amounts of tuberculin—a complex mixture of peptides and carbohydrates derived from *M. tuberculosis*—are injected intradermally. In people who have been exposed to the bacterium, either by infection or by

Type IV hypersensitivity reactions are mediated by antigen-specific effector T cells		
Syndrome	**Antigen**	**Consequence**
Delayed-type hypersensitivity	Proteins: Insect venom Mycobacterial proteins (tuberculin, lepromin)	Local skin swelling: Erythema Induration Cellular infiltrate Dermatitis
Contact hypersensitivity	Haptens: Pentadecacatechol (poison ivy) DNFB Small metal ions: Nickel Chromate	Local epidermal reaction: Erythema Cellular infiltrate Vesicles Intraepidermal abscesses
Gluten-sensitive enteropathy (celiac disease)	Gliadin	Villous atrophy in small bowel Malabsorption

Fig. 13.28 Type IV hypersensitivity responses. These reactions are mediated by T cells and all take some time to develop. They can be grouped into three syndromes, according to the route by which antigen passes into the body. In delayed-type hypersensitivity the antigen is injected into the skin; in contact hypersensitivity it is absorbed into the skin; and in gluten-sensitive enteropathy it is absorbed by the gut. DNFB, dinitrofluorobenzene.

immunization with the BCG vaccine (an attenuated form of *M. tuberculosis*), a local T cell-mediated inflammatory reaction evolves over 24–72 hours. The response is caused by T_H1 cells, which enter the site of antigen injection, recognize complexes of peptide:MHC class II molecules on antigen-presenting cells, and release inflammatory cytokines such as IFN-γ and TNF-β. These stimulate the expression of adhesion molecules on endothelium and increase local blood vessel permeability, allowing plasma and accessory cells to enter the site, thus causing a visible swelling (Fig. 13.29). Each of these phases takes several hours and so the fully developed response only appears 24–48 hours after challenge. The cytokines produced by the activated T_H1 cells and their actions are shown in Fig. 13.30.

Very similar reactions are observed in several cutaneous hypersensitivity responses. These can be elicited by either CD4 or CD8 T cells, depending on the pathway by which the antigen is processed. Typical antigens that cause cutaneous hypersensitivity responses are highly reactive small molecules that can easily penetrate intact skin, especially if they cause itching that leads to scratching. These chemicals then react with self proteins, creating

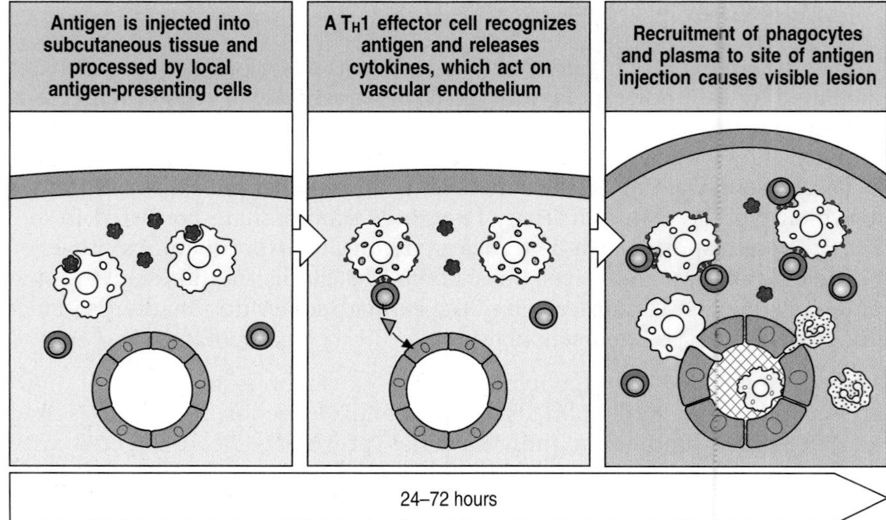
Antigen is injected into subcutaneous tissue and processed by local antigen-presenting cells
A T_H1 effector cell recognizes antigen and releases cytokines, which act on vascular endothelium
Recruitment of phagocytes and plasma to site of antigen injection causes visible lesion
24–72 hours

Fig. 13.29 The stages of a delayed-type hypersensitivity reaction. The first phase involves uptake, processing, and presentation of the antigen by local antigen-presenting cells. In the second phase, T_H1 cells that were primed by a previous exposure to the antigen migrate into the site of injection and become activated. Because these specific cells are rare, and because there is little inflammation to attract cells into the site, it can take several hours for a T cell of the correct specificity to arrive. These cells release mediators that activate local endothelial cells, recruiting an inflammatory cell infiltrate dominated by macrophages and causing the accumulation of fluid and protein. At this point the lesion becomes apparent.

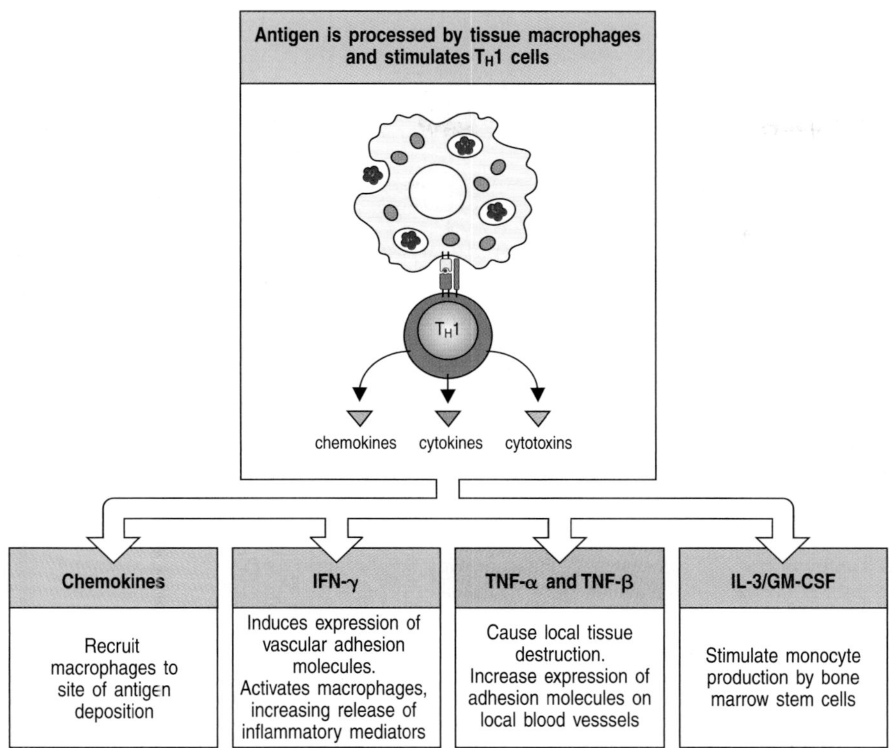
Antigen is processed by tissue macrophages and stimulates T_H1 cells
T_H1
chemokines cytokines cytotoxins
Chemokines
Recruit macrophages to site of antigen deposition
IFN-γ
Induces expression of vascular adhesion molecules. Activates macrophages, increasing release of inflammatory mediators
TNF-α and TNF-β
Cause local tissue destruction. Increase expression of adhesion molecules on local blood vesssels
IL-3/GM-CSF
Stimulate monocyte production by bone marrow stem cells

Fig. 13.30 The delayed-type (type IV) hypersensitivity response is directed by chemokines and cytokines released by antigen-stimulated T_H1 cells. Antigen in the local tissues is processed by antigen-presenting cells and presented on MHC class II molecules. Antigen-specific T_H1 cells that recognize the antigen locally at the site of injection release chemokines and cytokines that recruit macrophages to the site of antigen deposition. Antigen presentation by the newly recruited macrophages then amplifies the response. T cells can also affect local blood vessels through the release of TNF-α and TNF-β, and stimulate the production of macrophages through the release of IL-3 and GM-CSF. T_H1 cells activate macrophages through the release of IFN-γ and TNF-α, and kill macrophages and other sensitive cells through the cell-surface expression of the Fas ligand.

hapten:protein complexes that can be processed to hapten:peptide complexes capable of being presented by MHC molecules and recognized by T cells as foreign antigens. There are two phases to a cutaneous hypersensitivity response: sensitization and elicitation. During the sensitization phase, cutaneous Langerhans cells take up and process antigen, and migrate to regional lymph nodes, where they activate T cells (see Fig. 8.13) with the consequent production of memory T cells, which end up in the dermis. In the elicitation phase, a subsequent exposure to the sensitizing chemical leads to antigen presentation to memory T cells in the dermis, with the release of T-cell cytokines such as IFN-γ and IL-17. This stimulates the keratinocytes of the epidermis to release IL-1, IL-6, TNF-α, GM-CSF, the chemokine CXCL8, and the interferon-inducible chemokines CXCL11 (IP-9), CXCL10 (IP-10), and CXCL9 (Mig; monokine induced by IFN-γ). The cytokines and chemokines enhance the inflammatory response by inducing the migration of monocytes into the lesion and their maturation into macrophages, and by attracting more T cells (Fig. 13.31).

The rash produced by contact with poison ivy (Fig. 13.32) is caused by a CD8 T-cell response to a chemical in the poison ivy leaf called pentadecacatechol. This compound is lipid-soluble and can therefore cross the cell membrane and modify intracellular proteins. The modified proteins generate modified peptides within the cytosol, and these are translocated into the endoplasmic reticulum and delivered to the cell surface bound to MHC class I molecules. CD8 T cells recognizing the peptides cause damage either by killing the eliciting cell or by secreting cytokines such as IFN-γ. The well-studied chemical picryl chloride produces a CD4 T-cell hypersensitivity reaction. It modifies extracellular self proteins, which are then processed by the exogenous pathway (see Section 5-5) into modified self peptides that bind to self-MHC class II molecules and are recognized by T_H1 cells. When sensitized T_H1 cells recognize these complexes, they produce extensive inflammation by activating macrophages (see Fig. 13.31). Because the chemicals in these examples

Contact Sensitivity to Poison Ivy

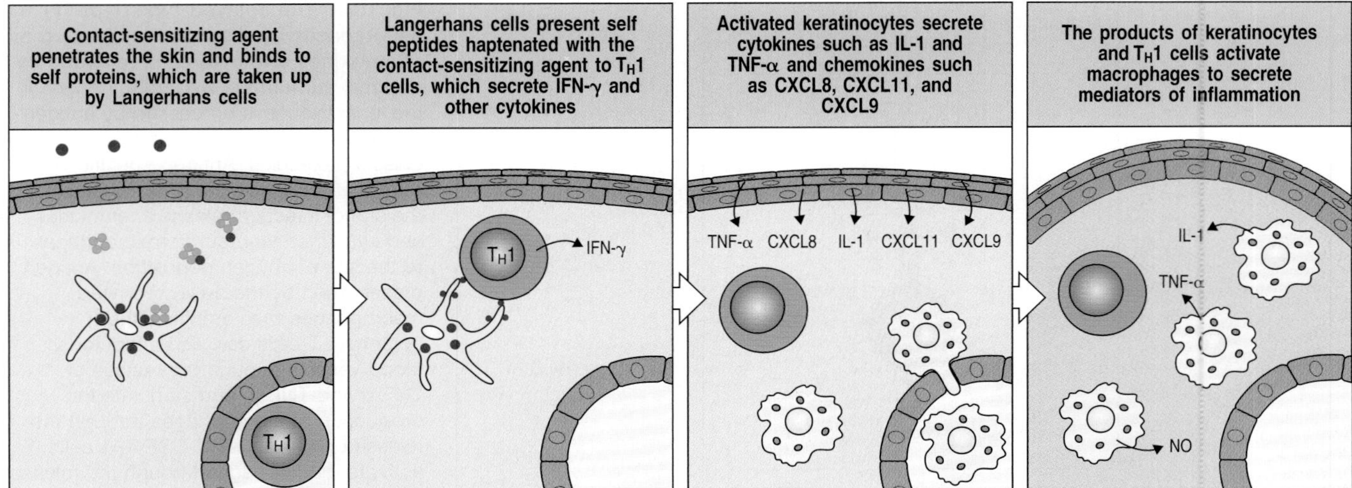

Fig. 13.31 Elicitation of a delayed-type hypersensitivity response to a contact-sensitizing agent. A contact-sensitizing agent is a small highly reactive molecule that can easily penetrate intact skin. It binds covalently as a hapten to a variety of endogenous proteins, which are taken up and processed by Langerhans cells, the major antigen-presenting cells of skin. These present haptenated peptides to effector T_H1 cells (which must have been previously primed in lymph nodes and then have traveled back to the skin). These then secrete cytokines such as IFN-γ that stimulate keratinocytes to secrete further cytokines and chemokines. These in turn attract monocytes and induce their maturation into activated tissue macrophages, which contribute to the inflammatory lesions depicted in Fig. 13.32. NO, nitric oxide.

are delivered by contact with the skin, the rash that follows is called a **contact hypersensitivity reaction**.

Some insect proteins also elicit a delayed-type hypersensitivity response. However, the early phases of the host reaction to an insect bite are often IgE-mediated or result from the direct effects of insect venoms. Important delayed-type hypersensitivity responses to divalent cations such as nickel have also been observed. These divalent cations can alter the conformation or the peptide binding of MHC class II molecules, and thus provoke a T-cell response. Finally, although this section has focused on the role of T cells in inducing delayed-type hypersensitivity reactions, there is evidence that antibody and complement might also have a role. Mice deficient in B cells, antibody, or complement show impaired contact hypersensitivity reactions. In particular, IgM antibodies (produced in part by B1 cells), which activate the complement cascade, facilitate the initiation of these reactions.

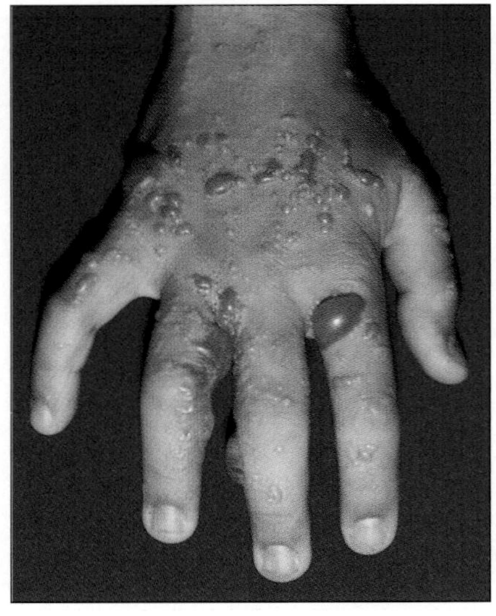

Fig. 13.32 Blistering skin lesions on hand of patient with poison ivy contact dermatitis. Photograph courtesy of R. Geha.

13-20 Mutation in the molecular regulators of inflammation can cause hypersensitive inflammatory responses resulting in 'autoinflammatory disease.'

We have seen throughout this book that host defense against infection depends on the engagement by the immune system of effector mechanisms that limit the spread of infection and kill the infectious agent. In this chapter we have seen how inappropriate responses to noninfectious immunological stimuli may cause diseases as diverse as asthma and hypersensitivity to nickel. There is a very fine balance between host underresponsiveness to infectious stimuli, allowing the uncontrolled spread of infection, and overresponsiveness, killing not only the infection but potentially also the host. There are a small number of diseases in which mutations in genes that control the life, death, and activities of inflammatory cells are associated with severe inflammatory disease. These conditions represent a failure to limit damage during inflammation and immune responses to infection and are known as **autoinflammatory diseases** (Fig. 13.33).

The name **familial Mediterranean fever** (**FMF**) describes the key features of one such severe inflammatory illness, inherited as an autosomal recessive disorder. The pathogenesis of FMF was a total mystery until its cause was discovered to be mutations in the gene encoding the protein pyrin, named to reflect its association with fever. This gene was also discovered by a second group of researchers at much the same time and named marenostrin, after the Latin name *mare nostrum* for the Mediterranean sea. The name pyrin has stuck and has been extended to describe a domain in this protein that is the prototype of 'pyrin domains' found in some proteins involved in apoptosis.

A disease with similar clinical manifestations is **familial Hibernian fever** (**FHF**) (also known as **TNF-receptor associated periodic syndrome** (**TRAPS**)). Although inherited as an autosomal dominant disease, it was romantically thought to have been a variant of FMF brought to Ireland by sailors of the Spanish Armada, until genetic analysis showed it to be caused by mutations in a completely different gene, that encoding the TNFR-I receptor (a receptor for TNF-α). Patients have reduced levels of TNFR-1, which leads to increased levels of TNF-α in the circulation because it is not mopped up by receptors. The disease responds to therapeutic blockade with anti-TNF agents such as etanercept, a soluble TNF receptor fortuitously developed to treat patients with rheumatoid arthritis (see Section 15-8). Both FMF and FHF are characterized by episodic attacks of severe inflammation associated with fever, an acute-phase response, severe malaise and, in FMF, attacks of pleural or peritoneal inflammation known as pleurisy and peritonitis, respectively. Mutations in the gene encoding CD2-binding protein 1 (CD2BP1), a pyrin-interacting protein, are associated with another dominantly inherited autoinflammatory syndrome—**pyogenic arthritis, pyoderma gangrenosum, and acne** (**PAPA**). These mutations accentuate the interaction between pyrin and CD2BP1.

Fig. 13.33 The autoinflammatory diseases.

Disease (common abbreviation)	Clinical features	Inheritance	Mutated gene	Protein (alternative name)
Familial Mediterranean fever (FMF)	Periodic fever, serositis (inflammation of the pleural and/or peritoneal cavity), arthritis, acute-phase response	Autosomal recessive	*MEFV*	Pyrin (marenostrin)
TNF-receptor associated periodic syndrome (TRAPS) (also known as familial Hibernian fever)	Periodic fever, myalgia, rash, acute-phase response	Autosomal dominant	*TNFRSF1A*	TNF-α 55 kDa receptor (TNFR-I)
Pyogenic arthritis, pyoderma gangrenosum and acne (PAPA)		Autosomal dominant	*PTSTPIP*	CD2-binding protein 1
Muckle–Wells syndrome	Periodic fever, urticarial rash, joint pains, conjunctivitis, progressive deafness	Autosomal dominant	*CIAS1*	Cryopyrin
Familial cold autoinflammatory syndrome (FCAS) (familial cold urticaria)	Cold-induced periodic fever, urticarial rash, joint pains, conjunctivitis			
Chronic infantile neurologic, cutanean articular syndrome (CINCA)	Neonatal onset recurrent fever, urticarial rash, chronic arthropathy, facial dysmorphia, neurologic involvement		*CIAS1*	Cryopyrin
Hyper-IgD syndrome (HIDS)	Periodic fever, elevated IgD levels, lymphadenopathy	Autosomal recessive	*MVK*	Mevalonate synthase
Blau syndrome	Granulomatous inflammation of skin, eye, and joints	Autosomal dominant	*NOD2 (CARD15)*	NOD2 (CARD15)
Crohn's disease	Granulomatous inflammatory bowel disease, sometimes eye, skin, and joint granulomata	Complex trait		

How mutations in pyrin cause FMF is not known, but the pyrin domain is found in proteins that participate in pathways leading to the activation of caspases involved in the proteolytic processing and activation of the inflammatory cytokines pro-1β and pro-IL-18, and in apoptosis. It is not difficult to envisage how unregulated cytokine activity and defective apoptosis could result in a failure to control inflammation. In mice, an absence of pyrin causes an increased sensitivity to lipopolysaccharide and a defect in macrophage apoptosis. A related protein, named cryopyrin, encoded by the gene *CIAS1*, is mutated in the episodic inflammatory diseases **Muckle–Wells syndrome** and **familial cold autoinflammatory syndrome** (**FCAS**). These dominantly inherited syndromes present with episodes of fever—which is induced by exposure to cold in the case of FCAS—as well as urticarial rash, joint pains, and conjunctivitis. Mutations in *CIAS1* are also associated with the autoinflammatory disorder **chronic infantile neurologic cutaneous and articular syndrome** (**CINCA**), in which short recurrent fever episodes are common, although severe arthropathic, neurologic, and dermatologic symptoms predominate. Both pyrin and cryopyrin are predominantly expressed in leukocytes and in cells that act as barriers to pathogens, such as intestinal epithelial cells. The stimuli that modulate pyrin and related molecules include inflammatory cytokines and lipopolysaccharide. The mechanism underlying these diseases is not fully understood but is thought to be a failure to regulate NFκB and IL-1 production. Indeed, Muckle–Wells syndrome responds dramatically to the drug anakinra, an antagonist of the receptor for IL-1.

Not all autoinflammatory diseases are caused by mutations in genes involved in the regulation of apoptosis. **Hyper IgD syndrome** (**HIDS**), which is associated with attacks of fever starting in infancy, high levels of IgD in serum, and lymphadenopathy, is caused by mutations that result in a partial deficiency of mevalonate kinase, an enzyme in the pathway for the synthesis of isoprenoids and cholesterol. It is not yet clear how this enzyme deficiency causes the autoinflammatory disease.

13-21 Crohn's disease is a relatively common inflammatory disease with a complex etiology.

The heritable autoinflammatory diseases just described are fortunately rare, although they illustrate well the importance of the precise regulation of inflammatory responses. A much commoner inflammatory disease is **Crohn's disease**, an intestinal disorder of the type known generally as inflammatory bowel disease. The other main inflammatory bowel disease is ulcerative colitis. Crohn's disease is thought to result from an abnormal overresponsiveness to the normal commensal gut flora; unlike the autoinflammatory diseases discussed previously, it has multiple genetic risk factors. Patients have episodes of severe inflammation that commonly affect the terminal ileum—hence the alternative name of regional ileitis for this disease—but any part of the gastrointestinal tract can be involved. The disease is characterized by a chronic inflammation of the mucosa and submucosa of the intestine that includes the prominent development of granulomatous lesions (Fig. 13.34) similar to those seen in the type IV hypersensitivity responses discussed in Section 13-19. Genetic analysis of patients with Crohn's disease and their families has identified a disease-susceptibility gene named *NOD2* (also known as *CARD15*) that is expressed predominantly in monocytes, dendritic cells, and the Paneth cells of the small intestine. Mutations and uncommon polymorphic variants of the NOD2 protein are strongly associated with the presence of Crohn's disease, with around 30% of patients carrying a loss-of-function mutation in *NOD2*. Mutations in the same gene are also the cause of a dominantly inherited granulomatous disease named **Blau syndrome**, in which granulomas typically develop in the skin, eyes, and joints. Whereas

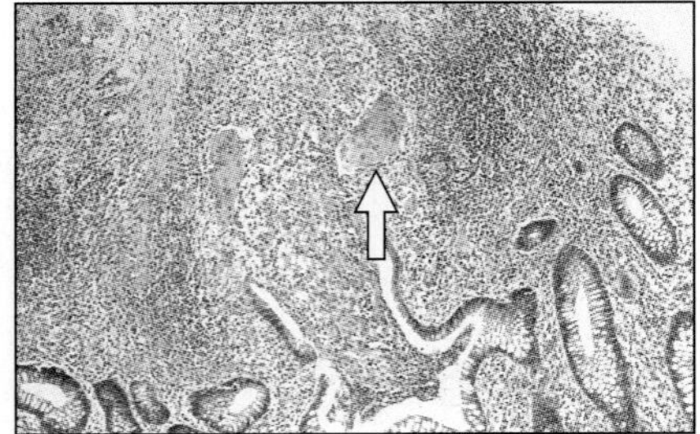

Fig. 13.34 Granulomatous inflammation in Crohn's disease. A section of bowel wall from a patient with Crohn's disease. The arrow marks a giant cell granuloma. There is a dense infiltrate of lymphocytes throughout the bowel submucosa. Photograph courtesy of H.T. Cook.

Crohn's disease represents a loss of function of NOD2, it is thought that Blau syndrome represents a gain of function.

NOD2 serves as an intracellular receptor for the muramyl dipeptide derived from bacterial peptidoglycan, and its stimulation leads to activation of the transcription factor NFκB and the induction of genes encoding pro-inflammatory cytokines (see Section 2-10). This pro-inflammatory response is believed to be important for the clearance of gut bacteria whose presence would otherwise lead to sustained chronic inflammation (see Section 11-11). The mutant forms of NOD2 have lost this function, and it is thought that this allows the development of chronic inflammation.

A further complication to the story is the identification of a deficiency in innate immunity in patients with Crohn's disease, in which a failure to clear pathogenic bacteria was found to be due to defective CXCL8 production and defective neutrophil accumulation. This may not lead to abnormal bowel pathology unless there is also a defect in NOD2, thereby promoting abnormal inflammation. Thus, it has been proposed that defects in innate immunity and in the regulation of inflammation act synergistically to promote the pathology of Crohn's disease.

Analysis of the autoinflammatory diseases has opened a new field of study in the medical sciences; it is likely that many other diseases will turn out to be caused by or modified by polymorphic genetic variants or mutants in the genes that regulate innate immune responses and the control of inflammation. A minor infection or physiological stress with no adverse consequences in most people might turn out to have devastating effects in a minority of genetically predisposed people. A second important message from these illnesses is that a more robust classification of diseases will be possible when we understand their underlying molecular basis.

Summary.

Hypersensitivity diseases reflect normal immune mechanisms that are inappropriately directed against innocuous antigens or inflammatory stimuli. They can be mediated by IgG antibodies bound to modified cell surfaces, or by complexes of antibodies bound to poorly catabolized antigens, as occurs in serum sickness. Hypersensitivity reactions mediated by T cells can be activated by modified self proteins or by injected proteins such as those in the mycobacterial extract tuberculin. These T cell-mediated responses require the induced synthesis of effector molecules and develop more slowly, which is why they are termed delayed-type hypersensitivity. A genetic failure to regulate inflammation gives rise to rare autoinflammatory syndromes, whereas Crohn's disease is associated with a failure to control commensal gut bacteria and prevent them from causing chronic inflammation.

Summary to Chapter 13.

In some people, immune responses to otherwise innocuous antigens produce allergic or hypersensitivity reactions upon reexposure to the same antigen. Most allergies involve the production of IgE antibody against common environmental allergens. Some people are intrinsically prone to making IgE antibodies against many allergens, and such people are said to be atopic. IgE production is driven by antigen-specific T_H2 cells; the response is polarized toward T_H2 by an array of chemokines and cytokines that engage specific signaling pathways. The IgE produced binds to the high-affinity IgE receptor FcεRI on mast cells and basophils. Specific effector T cells, mast cells, and eosinophils, in combination with T_H1 and T_H2 cytokines and chemokines, orchestrate chronic allergic inflammation, which is the major cause of the chronic morbidity of asthma. Failure to regulate these responses can occur at many levels of the immune system, including defects in regulatory T cells. Antibodies of other isotypes and antigen-specific effector T cells contribute to hypersensitivity to other antigens. The autoinflammatory syndromes are due to uncontrolled inflammation in the absence of disease, whereas Crohn's disease is thought to represent a failure to control the numbers of commensal gut bacteria.

Questions.

13.1 *List three hypersensitivities that involve IgE and three that involve other mechanisms.*

13.2 *Describe how a person becomes sensitized to an allergen.*

13.3 *Discuss the factors predisposing to the production of IgE.*

13.4 *What are the key features that differentiate acute and chronic allergic reactions?*

13.5 *How can the innate immune system contribute to allergy?*

13.6 *How do infectious agents modulate allergy?*

13.7 *Which types of white blood cells participate in allergic responses, and what do they do?*

13.8 *Describe how an ingested food allergen can give rise to the allergic skin reaction urticaria.*

13.9 *How does desensitization therapy work?*

13.10 *What are the main features of (a) type II hypersensitivity disease; (b) type III hypersensitivity disease; and (c) type IV hypersensitivity disease? Give an example of each type.*

13.11 *How does autoinflammatory disease differ from allergy?*

13.12 *How is the regulation of cell death and autoinflammatory disease linked?*

General references.

Johansson, S.G., Bieber, T., Dahl, R., Friedmann, P.S., Lanier, B.Q., Lockey, R.F., Motala, C., Ortega Martell, J.A., Platts-Mills, T.A., Ring, J., *et al.*: **Revised nomenclature for allergy for global use: Report of the Nomenclature Review Committee of the World Allergy Organization, October 2003.** *J. Allergy Clin. Immunol.* 2004, **113**:832–836.

Kay, A.B.: *Allergy and Allergic Diseases.* Oxford, Blackwell Science, 1997.

Kay, A.B.: **Allergy and allergic diseases. First of two parts.** *N. Engl. J. Med.* 2001, **344:**30–37.

Kay, A.B.: **Allergy and allergic diseases. Second of two parts.** *N. Engl. J. Med.* 2001, **344:**109–113

Kay, A.B.: **The role of T lymphocytes in asthma.** *Chem. Immunol. Allergy* 2006, **91:**59–75.

Maddox, L., and Schwartz, D.A.: **The pathophysiology of asthma.** *Annu. Rev. Med.* 2002, **53**:477–498.

Papageorgiou, F.S.: **Clinical aspects of food allergy.** *Biochem. Soc. Trans.* 2002, **30**:901–906.

Ring, J., Kramer, U., Schafer, T., and Behrendt, H.: **Why are allergies increasing?** *Curr. Opin. Immunol.* 2001, **13**:701–708.

Romagnani, S.: **The role of lymphocytes in allergic disease.** *J. Allergy Clin. Immunol.* 2000, **105**:399–408.

Rosen, F.S.: **Urticaria, angioedema, and anaphylaxis.** *Pediatr. Rev.* 1992, **13:**387–390.

Section references.

13-1 Allergens are often delivered transmucosally at low dose, a route that favors IgE production.

Holt, P.G.: **The role of airway dendritic cell populations in regulation of T-cell responses to inhaled antigens: atopic asthma as a paradigm.** *J. Aerosol Med.* 2002, **15**:161–168.

Lambrecht, B.N., De Veerman, M., Coyle, A.J., Gutierrez-Ramos, J.C., Thielemans, K., and Pauwels, R.A.: **Myeloid dendritic cells induce Th2 responses to inhaled antigen, leading to eosinophilic airway inflammation.** *J. Clin. Invest.* 2000, **106**:551–559.

O'Hehir, R.E., Garman, R.D., Greenstein, J.L., and Lamb, J.R.: **The specificity and regulation of T-cell responsiveness to allergens.** *Annu. Rev. Immunol.* 1991, **9**:67–95.

13-2 Enzymes are frequent triggers of allergy.

Grunstein, M.M., Veler, H., Shan, X., Larson, J., Grunstein, J.S., and Chuang, S.: **Proasthmatic effects and mechanisms of action of the dust mite allergen, Der p 1, in airway smooth muscle.** *J. Allergy Clin. Immunol.* 2005, **116**:94–101.

Kauffman, H.F., Tomee, J.F., van de Riet, M.A., Timmerman, A.J., and Borger, P.: **Protease-dependent activation of epithelial cells by fungal allergens leads to morphologic changes and cytokine production.** *J. Allergy Clin. Immunol.* 2000, **105**:1185–1193.

Nordlee, J.A., Taylor, S.L., Townsend, J.A., Thomas, L.A., and Bush, R.K.: **Identification of a Brazil-nut allergen in transgenic soybeans.** *N. Engl. J. Med.* 1996, **334**:688–692.

Sehgal, N., A. Custovic, and Woodcock, A.: **Potential roles in rhinitis for protease and other enzymatic activities of allergens.** *Curr. Allergy Asthma Rep.* 2005, **5**:221–226.

Sprecher, E., Tesfaye-Kedjela, A., Ratajczak, P., Bergman, R., and Richard, G.: **Deleterious mutations in SPINK5 in a patient with congenital ichthyosiform erythroderma: molecular testing as a helpful diagnostic tool for Netherton syndrome.** *Clin. Exp. Dermatol.* 2004, **29:**513–517.

Thomas, W.R., Smith, W., and Hales, B.J.: **House dust mite allergen characterisation: implications for T-cell responses and immunotherapy.** *Int. Arch. Allergy Immunol.* 1998, **115**:9–14.

Wan, H., Winton, H.L., Soeller, C., Tovey, E.R., Gruenert, D.C., Thompson, P.J., Stewart, G.A., Taylor, G.W., Garrod, D.R., Cannell, M.B., *et al.*: **Der p 1 facilitates transepithelial allergen delivery by disruption of tight junctions.** *J. Clin. Invest.* 1999, **104**:123–133.

13-3 Class switching to IgE in B lymphocytes is favored by specific signals.

Chen, Z., Lund, R., Aittokallio, T., Kosonen, M., Nevalainen, O., and Lahesmaa, R.: **Identification of novel IL-4/Stat6-regulated genes in T lymphocytes.** *J. Immunol.* 2003, **171**:3627–3635.

Gauchat, J.F., Henchoz, S., Mazzei, G., Aubry, J.P., Brunner, T., Blasey, H., Life, P., Talabot, D., Flores Romo, L., Thompson, J., *et al.*: **Induction of human IgE synthesis in B cells by mast cells and basophils.** *Nature* 1993, **365**:340–343.

Geha, R.S., Jabara, H.H., and Brodeur, S.R.: **The regulation of immunoglobulin E class-switch recombination.** *Nat. Rev. Immunol.* 2003, **3**:721–732.

Hoey, T., and Grusby, M.J.: **STATs as mediators of cytokine-induced responses.** *Adv. Immunol.* 1999, **71**:145–162.

Pease, J.E.: **Asthma, allergy and chemokines.** *Curr. Drug Targets* 2006, **7**:3–12.

Robinson, D.S.: **The Th1 and Th2 concept in atopic allergic disease.** *Chem Immunol.* 2000, **78**:50–61.

Romagnani, S.: **Cytokines and chemoattractants in allergic inflammation.** *Mol. Immunol.* 2002, **38**:881–885.

Shimoda, K., van Deursen, J., Sangster, M.Y., Sarawar, S.R., Carson, R.T., Tripp, R.A., Chu, C., Quelle, F.W., Nosaka, T., Vignali, D.A., *et al.*: **Lack of IL-4-induced Th2 response and IgE class switching in mice with disrupted *Stat6* gene.** *Nature* 1996, **380**:630–633.

Urban, J.F., Jr., Noben-Trauth, N., Donaldson, D.D., Madden, K.B., Morris, S.C., Collins, M., and Finkelman, F.D.: **IL-13, IL-4Rα, and Stat6 are required for the expulsion of the gastrointestinal nematode parasite *Nippostrongylus brasiliensis*.** *Immunity* 1998, **8**:255–264.

Zhu, J., Guo, L., Watson, C.J., Hu-Li, J., and Paul, W.E.: **Stat6 is necessary and sufficient for IL-4's role in Th2 differentiation and cell expansion.** *J. Immunol.* 2001, **166**:7276–7281.

13-4 Both genetic and environmental factors contribute to the development of IgE-mediated allergy.

Cookson, W.: **The immunogenetics of asthma and eczema: a new focus on the epithelium.** *Nat. Rev. Immunol.* 2004, **4**:978–988.

Culley, F.J., Pollott, J., and Openshaw, P.J.: **Age at first viral infection determines the pattern of T cell-mediated disease during reinfection in adulthood.** *J. Exp. Med.* 2002, **196**:1381–1386.

Dunne, D.W., and Cooke, A.: **Opinion: a worm's eye view of the immune system: consequences for evolution of human autoimmune disease.** *Nat. Rev. Immunol.* 2005, **5**:420-426.

Eder, W., and von Mutius, E.: **Hygiene hypothesis and endotoxin: what is the evidence?** *Curr. Opin. Allergy Clin. Immunol.* 2004, **4**:113–117.

Gilliland, F.D., Li, Y.F., Saxon, A., and Diaz-Sanchez, D.: **Effect of glutathione-S-transferase M1 and P1 genotypes on xenobiotic enhancement of allergic responses: randomised, placebo-controlled crossover study.** *Lancet* 2004, **363**:119–125.

Hershey, G.K., Friedrich, M.F., Esswein, L.A., Thomas, M.L., and Chatila, T.A.: **The association of atopy with a gain-of-function mutation in the α subunit of the interleukin-4 receptor.** *N. Engl. J. Med.* 1997, **337**:1720–1725.

Lynch, N.R., Hagel, I., Perez, M., Di Prisco, M.C., Lopez, R., and Alvarez, N.: **Effect of antihelminthic treatment on the allergic reactivity of children in a tropical slum.** *J. Allergy Clin. Immunol.* 1993, **92**:404–411.

Matricardi, P.M., Rosmini, F., Ferrigno, L., Nisini, R., Rapicetta, M., Chionne, P., Stroffolini, T., Pasquini, P., and D'Amelio, R.: **Cross sectional retrospective study of prevalence of atopy among Italian military students with antibodies against hepatitis A virus.** *BMJ* 1997, **314**:999–1003.

McIntire, J.J., Umetsu, S.E., Akbari, O., Potter, M., Kuchroo, V.K., Barsh, G.S., Freeman, G.J., Umetsu, D.T., and DeKruyff, R.H.: **Identification of Tapr (an airway hyperreactivity regulatory locus) and the linked Tim gene family.** *Nat. Immunol.* 2001, **2**:1109–1116.

Mitsuyasu, H., Yanagihara, Y., Mao, X.Q., Gao, P.S., Arinobu, Y., Ihara, K., Takabayashi, A., Hara, T., Enomoto, T., Sasaki, S., *et al.*: **Cutting edge: dominant effect of Ile50Val variant of the human IL-4 receptor α-chain in IgE synthesis.** *J. Immunol.* 1999, **162**:1227–1231.

Morahan, G., Huang, D., Wu, M., Holt, B.J., White, G.P., Kendall, G.E., Sly, P.D., and Holt, P.G.: **Association of IL12B promoter polymorphism with severity of atopic and non-atopic asthma in children.** *Lancet* 2002, **360**:455–459.

Palmer, L.J., Silverman, E.S., Weiss, S.T., and Drazen, J.M.: **Pharmacogenetics of asthma.** *Am. J. Respir. Crit. Care Med.* 2002, **165**:861–866.

Raitala, A., Karjalainen, J., Oja, S.S., Kosunen, T.U., and Hurme, M.: **Indoleamine 2,3-dioxygenase (IDO) activity is lower in atopic than in non-atopic individuals and is enhanced by environmental factors protecting from atopy.** *Mol. Immunol.* 2006, **43**:1054–1056.

Saxon, A., and Diaz-Sanchez, D. **Air pollution and allergy: you are what you breathe.** *Nat. Immunol.* 2005, **6**:223–226.

Shaheen, S.O., Aaby, P., Hall, A.J., Barker, D.J., Heyes, C.B., Shiell, A.W., and Goudiaby, A.: **Measles and atopy in Guinea-Bissau.** *Lancet* 1996, **347**:1792–1796.

Shapiro, S.D., and Owen, C.A.: **ADAM-33 surfaces as an asthma gene.** *N. Engl. J. Med.* 2002, **347**:936–938.

Strachan, D.P.: **Hay fever, hygiene, and household size.** *BMJ* 1989, **299**:1259–1260.

Summers, R.W., Elliott, D.E., Urban, J.F., Jr., Thompson, R.A., and Weinstock, J.V.: ***Trichuris suis*** **therapy for active ulcerative colitis: a randomized controlled trial.** *Gastroenterology* 2005, **128**:825–832.

Umetsu, D.T., McIntire, J.J., Akbari, O., Macaubas, C., and DeKruyff, R.H.: **Asthma: an epidemic of dysregulated immunity.** *Nat. Immunol.* 2002, **3**:715–720.

Van Eerdewegh, P., Little, R.D., Dupuis, J., Del Mastro, R.G., Falls, K., Simon, J., Torrey, D., Pandit, S., McKenny, J., Braunschweiger, K., *et al.*: **Association of the *ADAM33* gene with asthma and bronchial hyperresponsiveness.** *Nature* 2002, **418**:426–430.

von Mutius, E., Martinez, F.D., Fritzsch, C., Nicolai, T., Roell, G., and Thiemann, H.H.: **Prevalence of asthma and atopy in two areas of West and East Germany.** *Am. J. Respir. Crit Care Med.* 1994, **149**:358–364.

Wills-Karp, M.: **Asthma genetics: not for the TIMid?** *Nat. Immunol.* 2001, **2**:1095–1096.

Wills-Karp, M., Santeliz, J., and Karp, C.L.: **The germless theory of allergic disease: revisiting the hygiene hypothesis.** *Nat. Rev. Immunol.* 2001, **1**:69–75.

13-5 Regulatory T cells can control allergic responses.

Akdis, M., Blaser, K., and Akdis, C.A.: **T regulatory cells in allergy: novel concepts in the pathogenesis, prevention, and treatment of allergic diseases.** *J. Allergy Clin. Immunol.* 2005, **116**:961–968.

Haddeland, U., Karstensen, A.B., Farkas, L., Bo, K.O., Pirhonen, J., Karlsson, M., Kvavik, W., Brandtzaeg, P., and Nakstad, B.: **Putative regulatory T cells are impaired in cord blood from neonates with hereditary allergy risk.** *Pediatr. Allergy Immunol.* 2005, **16**:104–112.

Hawrylowicz, C.M.: **Regulatory T cells and IL-10 in allergic inflammation.** *J. Exp. Med.* 2005, **202**:1459–1463.

Hayashi, T., Beck, L., Rossetto, C., Gong, X., Takikawa, O., Takabayashi, K., Broide, D.H., Carson, D.A., and Raz, E.: **Inhibition of experimental asthma by indoleamine 2,3-dioxygenase.** *J. Clin. Invest.* 2004, **114**:270–279.

Kuipers, H., and Lambrecht, B.N.: **The interplay of dendritic cells, Th2 cells and regulatory T cells in asthma.** *Curr. Opin. Immunol.* 2004, **16**:702–708.

Lin, W., Truong, N., Grossman, W.J., Haribhai, D., Williams, C.B., Wang, J., Martin, M.G., and Chatila, T.A.: **Allergic dysregulation and hyperimmunoglobulinemia E in Foxp3 mutant mice.** *J. Allergy Clin. Immunol.* 2005, **116**:1106–1115.

Mellor, A.L., and Munn, D.H.: **IDO expression by dendritic cells: tolerance and tryptophan catabolism.** *Nat. Rev. Immunol.* 2004, **4**:762–774.

13-6 Most IgE is cell-bound and engages effector mechanisms of the immune system by different pathways from other antibody isotypes.

Conner, E.R., and Saini, S.S.: **The immunoglobulin E receptor: expression and regulation.** *Curr. Allergy Asthma Rep.* 2005, **5**:191–196.

Gilfillan, A.M., and Tkaczyk, C.: **Integrated signalling pathways for mast-cell activation.** *Nat. Rev. Immunol.* 2006, **6**:218–230.

Heyman, B.: **Regulation of antibody responses via antibodies, complement, and Fc receptors.** *Annu. Rev. Immunol.* 2000, **18**:709–737.

Kinet, J.P.: **The high-affinity IgE receptor (FcεRI): from physiology to pathology.** *Annu. Rev. Immunol.* 1999, **17**:931–972.

Payet, M., and Conrad, D.H.: **IgE regulation in CD23 knockout and transgenic mice.** *Allergy* 1999, **54**:1125–1129.

13-7 Mast cells reside in tissues and orchestrate allergic reactions.

Ali, K., Bilancio, A., Thomas, M., Pearce, W., Gilfillan, A.M., Tkaczyk, C., Kuehn, N., Gray, A., Giddings, J., Peskett, E., *et al.*: **Essential role for the p110δ phosphoinositide 3-kinase in the allergic response.** *Nature* 2004, **431**:1007–1011.

Austen, K.F.: **The Paul Kallos Memorial Lecture. From slow-reacting substance of anaphylaxis to leukotriene C4 synthase.** *Int. Arch. Allergy Immunol.* 1995, **107**:19–24.

Bingham, C.O., and Austen, K.F.: **Mast-cell responses in the development of asthma.** *J. Allergy Clin. Immunol.* 2000, **105**:S527–S534.

Galli, S.J., Nakae, S., and Tsai, M.: **Mast cells in the development of adaptive immune responses.** *Nat. Immunol.* 2005, **6**:135–142.

Gonzalez-Espinosa, C., Odom, S., Olivera, A., Hobson, J.P., Martinez, M.E., Oliveira-Dos-Santos, A., Barra, L., Spiegel, S., Penninger, J.M., and Rivera, J.: **Preferential signaling and induction of allergy-promoting lymphokines upon weak stimulation of the high affinity IgE receptor on mast cells.** *J. Exp. Med.* 2003 **197**:1453–1465.

Luster, A.D., and Tager, A.M.: **T-cell trafficking in asthma: lipid mediators grease the way.** *Nat Rev Immunol.* 2004, **4**:711–724.

Mekori, Y.A., and Metcalfe, D.D.: **Mast cell–T cell interactions.** *J. Allergy Clin. Immunol.* 1999, **104**:517–523.

Oguma, T., Palmer, L.J., Birben, E., Sonna, L.A., Asano, K., and Lilly, C.M.: **Role of prostanoid DP receptor variants in susceptibility to asthma.** *N. Engl. J. Med.* 2004, **351**:1752–1763.

Taube, C., Miyahara, N., Ott, V., Swanson, B., Takeda, K., Loader, J., Shultz, L.D., Tager, A.M., Luster, A.D., Dakhama, A., *et al.*: **The leukotriene B4 receptor (BLT1) is required for effector CD8+ T cell-mediated, mast cell-dependent airway hyperresponsiveness.** *J. Immunol.* 2006, **176**:3157–3164.

Williams, C.M., and Galli, S.J.: **The diverse potential effector and immunoregulatory roles of mast cells in allergic disease.** *J. Allergy Clin. Immunol.* 2000, **105**:847–859.

13-8 Eosinophils are normally under tight control to prevent inappropriate toxic responses.

Bisset, L.R., and Schmid-Grendelmeier, P.: **Chemokines and their receptors in the pathogenesis of allergic asthma: progress and perspective.** *Curr. Opin. Pulm. Med.* 2005, **11**:35–42.

Dombrowicz, D., and Capron, M.: **Eosinophils, allergy and parasites.** *Curr. Opin. Immunol.* 2001, **13**:716–720.

Lukacs, N.W.: **Role of chemokines in the pathogenesis of asthma.** *Nat. Rev. Immunol.* 2001, **1**:108–116.

Mattes, J., and Foster, P.S.: **Regulation of eosinophil migration and Th2 cell function by IL-5 and eotaxin.** *Curr. Drug Targets Inflamm. Allergy.* 2003, **2**:169–174.

Robinson, D.S., Kay, A.B., and Wardlaw, A.J.: **Eosinophils.** *Clin. Allergy Immunol.* 2002, **16**:43–75.

13-9 Eosinophils and basophils cause inflammation and tissue damage in allergic reactions.

Dvorak, A.M.: **Cell biology of the basophil.** *Int. Rev. Cytol.* 1998, **180**:87–236.

Kay, A.B., Phipps, S., and Robinson, D.S.: **A role for eosinophils in airway remodelling in asthma.** *Trends Immunol.* 2004, **25**:477–482.

MacGlashan, D., Jr., Gauvreau, G., and Schroeder, J.T.: **Basophils in airway disease.** *Curr. Allergy Asthma Rep.* 2002, **2**:126–132.

Odemuyiwa, S.O., Ghahary, A., Li, Y., Puttagunta, L., Lee, J.E., Musat-Marcu, S., and Moqbel, R.: **Cutting edge: human eosinophils regulate T cell subset selection through indoleamine 2,3-dioxygenase.** *J. Immunol.* 2004, **173**:5909–5913.

Plager, D.A., Stuart, S., and Gleich, G.J.: **Human eosinophil granule major basic protein and its novel homolog.** *Allergy* 1998, **53**:33–40.

Thomas, L.L.: **Basophil and eosinophil interactions in health and disease.** *Chem. Immunol.* 1995, **61**:186–207.

13-10 Allergic reactions can be divided into immediate and late-phase responses.

Bentley, A.M., Kay, A.B., and Durham, S.R.: **Human late asthmatic reactions.** *Clin. Exp. Allergy* 1997, **27 Suppl 1**:71–86.

Liu, M.C., Hubbard, W.C., Proud, D., Stealey, B.A., Galli, S.J., Kagey Sobotka, A., Bleecker, E.R., and Lichtenstein, L.M.: **Immediate and late inflammatory responses to ragweed antigen challenge of the peripheral airways in allergic asthmatics. Cellular, mediator, and permeability changes.** *Am. Rev. Respir. Dis.* 1991, **144**:51–58.

Macfarlane, A.J., Kon, O.M., Smith, S.J., Zeibecoglou, K., Khan, L.N., Barata, L.T., McEuen, A.R., Buckley, M.G., Walls, A.F., Meng, Q., *et al.*: **Basophils, eosinophils, and mast cells in atopic and nonatopic asthma and in late-phase allergic reactions in the lung and skin.** *J. Allergy Clin. Immunol.* 2000, **105**:99–107.

Pearlman, D.S.: **Pathophysiology of the inflammatory response.** *J. Allergy Clin. Immunol.* 1999, **104**:S132–S137.

Taube, C., Duez, C., Cui, Z.H., Takeda, K., Rha, Y.H., Park, J.W., Balhorn, A., Donaldson, D.D., Dakhama, A., and Gelfand, E.W.: **The role of IL-13 in established allergic airway disease.** *J. Immunol.* 2002, **169**:6482-6489.

13-11 The clinical effects of allergic reactions vary according to the site of mast-cell activation.

deShazo, R.D., and Kemp, S.F.: **Allergic reactions to drugs and biologic agents.** *JAMA* 1997, **278**:1895–1906.

Dombrowicz, D., Flamand, V., Brigman, K.K., Koller, B.H., and Kinet, J.P.: **Abolition of anaphylaxis by targeted disruption of the high affinity immunoglobulin E receptor α chain gene.** *Cell* 1993, **75**:969–976.

Fernandez, M., Warbrick, E.V., Blanca, M., and Coleman, J.W.: **Activation and hapten inhibition of mast cells sensitized with monoclonal IgE anti-penicillin antibodies: evidence for two-site recognition of the penicillin derived determinant.** *Eur. J. Immunol.* 1995, **25**:2486–2491.

Finkelman, F.D., Rothenberg, M.E., Brandt, E.B., Morris, S.C., and Strait, R.T.: **Molecular mechanisms of anaphylaxis: lessons from studies with murine models.** *J. Allergy Clin. Immunol.* 2005, **115**:449-457; quiz 458.

Kemp, S.F., Lockey, R.F., Wolf, B.L., and Lieberman, P.: **Anaphylaxis. A review of 266 cases.** *Arch. Intern. Med.* 1995, **155**:1749–1754.

Oettgen, H.C., Martin, T.R., Wynshaw Boris, A., Deng, C., Drazen, J.M., and Leder, P.: **Active anaphylaxis in IgE-deficient mice.** *Nature* 1994, **370**:367–370.

Padovan, E.: **T-cell response in penicillin allergy.** *Clin. Exp. Allergy* 1998, **28 Suppl 4**:33–36.

Reisman, R.E.: **Insect stings.** *N. Engl. J. Med.* 1994, **331**:523–527.

Schwartz, L.B.: **Effector cells of anaphylaxis: mast cells and basophils.** *Novartis Found Symp.* 2004, **257**:65–74; discussion 74–69, 98–100, 276–185.

Weltzien, H.U., and Padovan, E.: **Molecular features of penicillin allergy.** *J. Invest. Dermatol.* 1998, **110**:203–206.

13-12 Allergen inhalation is associated with the development of rhinitis and asthma.

Bousquet, J., Jeffery, P.K., Busse, W.W., Johnson, M., and Vignola, A.M.: **Asthma. From bronchoconstriction to airways inflammation and remodeling.** *Am. J. Respir. Crit. Care Med.* 2000, **161**:1720–1745.

Boxall, C., Holgate, S.T., and Davies, D.E.: **The contribution of transforming growth factor-β and epidermal growth factor signalling to airway remodelling in chronic asthma.** *Eur. Respir. J.* 2006, **27**:208–229.

Busse, W.W., and Lemanske, R.F., Jr.: **Asthma.** *N. Engl. J. Med.* 2001, **344**:350–362.

Dakhama, A., Park, J.W., Taube, C., Joetham, A., Balhorn, A., Miyahara, N., Takeda, K., and Gelfand, E.W.: **The enhancement or prevention of airway hyperresponsiveness during reinfection with respiratory syncytial virus is critically dependent on the age at first infection and IL-13 production.** *J. Immunol.* 2005, **175**:1876–1883.

Day, J.H., Ellis, A.K., Rafeiro, E., Ratz, J.D., and Briscoe, M.P.: **Experimental models for the evaluation of treatment of allergic rhinitis.** *Ann. Allergy Asthma Immunol.* 2006, **96**:263–277; quiz 277–268, 315.

Finotto, S., Neurath, M.F., Glickman, J.N., Qin, S., Lehr, H.A., Green, F.H., Ackerman, K., Haley, K., Galle, P.R., Szabo, S.J., *et al.*: **Development of spontaneous airway changes consistent with human asthma in mice lacking T-bet.** *Science* 2002, **295**:336–338.

Grunig, G., Warnock, M., Wakil, A.E., Venkayya, R., Brombacher, F., Rennick, D.M., Sheppard, D., Mohrs, M., Donaldson, D.D., Locksley, R.M., *et al.*: **Requirement for IL-13 independently of IL-4 in experimental asthma.** *Science* 1998, **282**:2261–2263.

Haselden, B.M., Kay, A.B., and Larche, M.: **Immunoglobulin E-independent major histocompatibility complex-restricted T cell peptide epitope-induced late asthmatic reactions.** *J. Exp. Med.* 1999, **189**:1885–1894.

Kuperman, D.A., Huang, X., Koth, L.L., Chang, G.H., Dolganov, G.M., Zhu, Z., Elias, J.A., Sheppard, D., and Erle, D.J.: **Direct effects of interleukin-13 on epithelial cells cause airway hyperreactivity and mucus overproduction in asthma.** *Nat. Med.* 2002, **8**:885–889.

Lee, N.A., Gelfand, E.W., and Lee, J.J.: **Pulmonary T cells and eosinophils: coconspirators or independent triggers of allergic respiratory pathology?** *J. Allergy Clin. Immunol.* 2001, **107**:945–957.

Louahed, J., Toda, M., Jen, J., Hamid, Q., Renauld, J.C., Levitt, R.C., and Nicolaides, N.C.: **Interleukin-9 upregulates mucus expression in the airways.** *Am. J. Respir. Cell Mol. Biol.* 2000, **22**:649–656.

Platts-Mills, T.A.: **The role of allergens in allergic airway disease.** *J. Allergy Clin. Immunol.* 1998, **101**:S364–S366.

Szabo, S.J., Sullivan, B.M., Stemmann, C., Satoskar, A.R., Sleckman, B.P., and Glimcher, L.H.: **Distinct effects of T-bet in TH1 lineage commitment and IFN-γ production in CD4 and CD8 T cells.** *Science* 2002, **295**:338–342.

Wills-Karp, M.: **Interleukin-13 in asthma pathogenesis.** *Immunol. Rev.* 2004, **202**:175–190.

Zureik, M., Neukirch, C., Leynaert, B., Liard, R., Bousquet, J., and Neukirch, F.: **Sensitisation to airborne moulds and severity of asthma: cross sectional study from European Community respiratory health survey.** *BMJ* 2002, **325**:411–414.

13-13 Skin allergy is manifested as urticaria or chronic eczema.

Grattan, C.E.: **Autoimmune urticaria.** *Immunol. Allergy Clin. North Am.* 2004, **24**:163–181.

Howell, M.D., Gallo, R.L., Boguniewicz, M., Jones, J.F., Wong, C., Streib, J.E., and Leung, D.Y.: **Cytokine milieu of atopic dermatitis skin subverts the innate immune response to vaccinia virus.** *Immunity* 2006, **24**:341–348.

Simpson, E.L., and Hanifin, J.M.: **Atopic dermatitis.** *Med. Clin. North Am.* 2006, **90**:149–167.

Tsutsui, H., Yoshimoto, T., Hayashi, N., Mizutani, H., and Nakanishi, K.: **Induction of allergic inflammation by interleukin-18 in experimental animal models.** *Immunol. Rev.* 2004, **202**:115–138.

Verhagen, J., Akdis, M., Traidl-Hoffmann, C., Schmid-Grendelmeier, P., Hijnen, D., Knol, E.F., Behrendt, H., Blaser, K., and Akdis, C.A.: **Absence of T-regulatory cell expression and function in atopic dermatitis skin.** *J. Allergy Clin. Immunol.* 2006, **117**:176–183.

13-14 Allergy to foods causes systemic reactions as well as symptoms limited to the gut.

Astwood, J.D., Leach, J.N., and Fuchs, R.L.: **Stability of food allergens to digestion *in vitro*.** *Nat. Biotechnol.* 1996, **14**:1269–1273.

Ewan, P.W.: **Clinical study of peanut and nut allergy in 62 consecutive patients: new features and associations.** *BMJ* 1996, **312**:1074–1078.

Lee, L.A., and Burks, A.W.: **Food allergies: prevalence, molecular characterization, and treatment/prevention strategies.** *Annu. Rev. Nutr.* 2006, **26:**539–565.

13-15 Celiac disease is a model of antigen-specific immunopathology.

Ciccocioppo, R., Di Sabatino, A., and Corazza, G.R.: **The immune recognition of gluten in celiac disease.** *Clin. Exp. Immunol.* 2005, **140**:408–416.

Koning, F.: **Celiac disease: caught between a rock and a hard place.** *Gastroenterology* 2005, **129:**1294–1301.

Shan, L., Molberg, O., Parrot, I., Hausch, F., Filiz, F., Gray, G.M., Sollid, L.M., and Khosla, C.: **Structural basis for gluten intolerance in celiac sprue.** *Science* 2002, **297:**2275–2279.

Sollid, L.M.: **Celiac disease: dissecting a complex inflammatory disorder.** *Nat. Rev. Immunol.* 2002, **2:**647–655.

13-16 Allergy can be treated by inhibiting either IgE production or the effector pathways activated by the cross-linking of cell-surface IgE.

Adkinson, N.F., Jr., Eggleston, P.A., Eney, D., Goldstein, E.O., Schuberth, K.C., Bacon, J.R., Hamilton, R.G., Weiss, M.E., Arshad, H., Meinert, C.L., *et al.*: **A controlled trial of immunotherapy for asthma in allergic children.** *N. Engl. J. Med.* 1997, **336**:324–331.

Ali, F.R., Kay, A.B., and Larche, M.: **The potential of peptide immunotherapy in allergy and asthma.** *Curr. Allergy Asthma Rep.* 2002, **2**:151–158.

Bertrand, C., and Geppetti, P.: **Tachykinin and kinin receptor antagonists: therapeutic perspectives in allergic airway disease.** *Trends Pharmacol. Sci.* 1996, **17**:255–259.

Bryan, S.A., O'Connor, B.J., Matti, S., Leckie, M.J., Kanabar, V., Khan, J., Warrington, S.J., Renzetti, L., Rames, A., Bock, J.A., *et al.*: **Effects of recombinant human interleukin-12 on eosinophils, airway hyper-responsiveness, and the late asthmatic response.** *Lancet* 2000, **356**:2149–2153.

Creticos, P.S., Reed, C.E., Norman, P.S., Khoury, J., Adkinson, N.F., Jr., Buncher, C.R., Busse, W.W., Bush, R.K., Gadde, J., Li, J.T., *et al.*: **Ragweed immunotherapy in adult asthma.** *N. Engl. J. Med.* 1996, **334**:501–506.

D'Amato, G.: **Role of anti-IgE monoclonal antibody (omalizumab) in the treatment of bronchial asthma and allergic respiratory diseases.** *Eur. J. Pharmacol.* 2006, **533**:302–307.

Kline, J.N.: **Effects of CpG DNA on Th1/Th2 balance in asthma.** *Curr. Top. Microbiol. Immunol.* 2000, **247**:211–225.

Leckie, M.J., ten Brinke, A., Khan, J., Diamant, Z., O'Connor, B.J., Walls, C.M., Mathur, A.K., Cowley, H.C., Chung, K.F., Djukanovic, R., *et al.*: **Effects of an interleukin-5 blocking monoclonal antibody on eosinophils, airway hyper-responsiveness, and the late asthmatic response.** *Lancet* 2000, **356**:2144–2148.

Oldfield, W.L., Larche, M., and Kay, A.B.: **Effect of T-cell peptides derived from Fel d 1 on allergic reactions and cytokine production in patients sensitive to cats: a randomised controlled trial.** *Lancet* 2002, **360**:47–53.

Peters-Golden, M., and Henderson, W.R., Jr.: **The role of leukotrienes in allergic rhinitis.** *Ann. Allergy Asthma Immunol.* 2005, **94**:609–618; quiz 618-620, 669.

Roberts, G., C. Hurley, V. Turcanu, and Lack, G.: **Grass pollen immunotherapy as an effective therapy for childhood seasonal allergic asthma.** *J. Allergy Clin. Immunol.* 2006, **117**:263-268.

Sabroe, I., Peck, M.J., Van Keulen, B.J., Jorritsma, A., Simmons, G., Clapham, P.R., Williams, T.J., and Pease, J.E.: **A small molecule antagonist of chemokine receptors CCR1 and CCR3. Potent inhibition of eosinophil function and CCR3-mediated HIV-1 entry.** *J. Biol. Chem.* 2000, **275**:25985–25992.

Verhagen, J., Taylor, A., Blaser, K., Akdis, M., and Akdis, C.A.: **T regulatory cells in allergen-specific immunotherapy.** *Int. Rev. Immunol.* 2005, **24**:533–548.

Verhoef, A., Alexander, C., Kay, A.B., and Larche, M.: **T cell epitope immunotherapy induces a CD4+ T cell population with regulatory activity.** 2005, *PLoS Med.* **2**:e78.

Youn, C.J., Miller, M., Baek, K.J., Han, J.W., Nayar, J., Lee, S.Y., McElwain, K., McElwain, S., Raz, E., and Broide, D.H.: **Immunostimulatory DNA reverses established allergen-induced airway remodeling.** *J Immunol.* 2004, **173**:7556–7564.

Zhu, D., Kepley, C.L., Zhang, K., Terada, T., Yamada, T., and Saxon, A.: **A chimeric human–cat fusion protein blocks cat-induced allergy.** *Nat. Med.* 2005, **11**:446–449.

13-17 Innocuous antigens can cause type II hypersensitivity reactions in susceptible individuals by binding to the surfaces of circulating blood cells.

Arndt, P.A., and Garratty, G.: **The changing spectrum of drug-induced immune hemolytic anemia.** *Semin Hematol.* 2005, **42:**137–144.

Greinacher, A., Potzsch, B., Amiral, J., Dummel, V., Eichner, A., and Mueller Eckhardt, C.: **Heparin-associated thrombocytopenia: isolation of the antibody and characterization of a multimolecular PF4–heparin complex as the major antigen.** *Thromb. Haemost.* 1994, **71**:247–251.

Semple, J.W., and Freedman, J.: **Autoimmune pathogenesis and autoimmune hemolytic anemia.** *Semin. Hematol.* 2005, **42**:122–130.

13-18 Systemic disease caused by immune-complex formation can follow the administration of large quantities of poorly catabolized antigens.

Bielory, L., Gascon, P., Lawley, T.J., Young, N.S., and Frank, M.M.: **Human serum sickness: a prospective analysis of 35 patients treated with equine anti-thymocyte globulin for bone marrow failure.** *Medicine (Baltimore)* 1988, **67**:40–57.

Davies, K.A., Mathieson, P., Winearls, C.G., Rees, A.J., and Walport, M.J.: **Serum sickness and acute renal failure after streptokinase therapy for myocardial infarction.** *Clin. Exp Immunol.* 1990, **80**:83–88.

Gamarra, R.M., McGraw, S.D. Drelichman, V.S., and Maas, L.C.: **Serum sickness-like reactions in patients receiving intravenous infliximab.** *J. Emerg. Med.* 2006, **30**:41–44.

Lawley, T.J., Bielory, L., Gascon, P., Yancey, K.B., Young, N.S., and Frank, M.M.: **A prospective clinical and immunologic analysis of patients with serum sickness.** *N. Engl. J. Med.* 1984, **311**:1407–1413.

Schifferli, J.A., Ng, Y.C., and Peters, D.K.: **The role of complement and its receptor in the elimination of immune complexes.** *N. Engl. J. Med.* 1986, **315**:488–495.

Schmidt, R.E., and Gessner, J.E.: **Fc receptors and their interaction with complement in autoimmunity.** *Immunol. Lett.* 2005, **100**:56-67.

Skokowa, J., Ali, S.R., Felda, O., Kumar, V., Konrad, S., Shushakova, N., Schmidt, R.E., Piekorz, R.P., Nurnberg, B., Spicher, K., *et al.*: **Macrophages induce the inflammatory response in the pulmonary Arthus reaction through $G\alpha_{i2}$ activation that controls C5aR and Fc receptor cooperation.** *J. Immunol.* 2005, **174**:3041–3050.

Theofilopoulos, A.N., and Dixon, F.J.: **Immune complexes in human diseases: a review.** *Am. J. Pathol.* 1980, **100**:529–594.

13-19 Delayed-type hypersensitivity reactions are mediated by T_H1 cells and CD8 cytotoxic T cells.

Bernhagen, J., Bacher, M., Calandra, T., Metz, C.N., Doty, S.B., Donnelly, T., and Bucala, R.: **An essential role for macrophage migration inhibitory factor in the tuberculin delayed-type hypersensitivity reaction.** *J. Exp. Med.* 1996, **183**:277–282.

Kalish, R.S., Wood, J.A., and LaPorte, A.: **Processing of urushiol (poison ivy) hapten by both endogenous and exogenous pathways for presentation to T cells *in vitro*.** *J. Clin. Invest.* 1994, **93**:2039–2047.

Kimber, I., and Dearman, R.J.: **Allergic contact dermatitis: the cellular effectors.** *Contact Dermatitis* 2002, **46**:1–5.

Larsen, C.G., Thomsen, M.K., Gesser, B., Thomsen, P.D., Deleuran, B.W., Nowak, J., Skodt, V., Thomsen, H.K., Deleuran, M., Thestrup Pedersen, K., *et al.*: **The delayed-type hypersensitivity reaction is dependent on IL-8. Inhibition of a tuberculin skin reaction by an anti-IL-8 monoclonal antibody.** *J. Immunol.* 1995, **155**:2151–2157.

Mark, B.J., and Slavin, R.G.: **Allergic contact dermatitis.** *Med. Clin. North Am.* 2006, **90**:169–185.

Muller, G., Saloga, J., Germann, T., Schuler, G., Knop, J., and Enk, A.H.: **IL-12 as mediator and adjuvant for the induction of contact sensitivity *in vivo*.** *J. Immunol.* 1995, **155**:4661–4668.

Tsuji, R.F., Szczepanik, M., Kawikova, I., Paliwal, V., Campos, R.A., Itakura, A., Akahira-Azuma, M., Baumgarth, N., Herzenberg, L.A., and Askenase, P.W.: **B cell-dependent T cell responses: IgM antibodies are required to elicit contact sensitivity.** *J. Exp. Med.* 2002, **196**:1277–1290.

Vollmer, J., Weltzien, H.U., and Moulon, C.: **TCR reactivity in human nickel allergy indicates contacts with complementarity-determining region 3 but excludes superantigen-like recognition.** *J. Immunol.* 1999, **163**:2723–2731.

13-20 Mutation or genetic variation in the molecular regulators of inflammation can cause hypersensitive inflammatory responses resulting in 'autoinflammatory disease.'

Chae, J.J., Komarow, H.D., Cheng, J., Wood, G., Raben, N., Liu, P.P., and Kastner, D.L.: **Targeted disruption of pyrin, the FMF protein, causes heightened sensitivity to endotoxin and a defect in macrophage apoptosis.** *Mol. Cell* 2003, **11**:591–604.

Delpech, M., and Grateau, G.: **Genetically determined recurrent fevers.** *Curr. Opin. Immunol.* 2001, **13**:539–542.

Drenth, J.P., and van der Meer, J.W.: **Hereditary periodic fever.** *N. Engl. J. Med.* 2001, **345**:1748–1757.

Hoffman, H.M., Mueller, J.L., Broide, D.H., Wanderer, A.A., and Kolodner, R.D.: **Mutation of a new gene encoding a putative pyrin-like protein causes familial cold autoinflammatory syndrome and Muckle–Wells syndrome.** *Nat. Genet.* 2001, **29**:301–305.

Houten, S.M., Frenkel, J., Rijkers, G.T., Wanders, R.J., Kuis, W., and Waterham, H.R.: **Temperature dependence of mutant mevalonate kinase activity as a pathogenic factor in hyper-IgD and periodic fever syndrome.** *Hum. Mol. Genet.* 2002, **11**:3115–3124.

INFEVERS [http://fmf.igh.cnrs.fr/infevers].

Inohara, N., Ogura, Y., and Nunez, G.: **Nods: a family of cytosolic proteins that regulate the host response to pathogens.** *Curr. Opin. Microbiol.* 2002, **5**:76–80.

Kastner, D.L., O'Shea, J.J.: **A fever gene comes in from the cold.** *Nat. Genet.* 2001, **29**:241–242.

McDermott, M.F., Aksentijevich, I., Galon, J., McDermott, E.M., Ogunkolade, B.W., Centola, M., Mansfield, E., Gadina, M., Karenko, L., Pettersson, T., *et al.*: **Germline mutations in the extracellular domains of the 55 kDa TNF receptor, TNFR1, define a family of dominantly inherited autoinflammatory syndromes.** *Cell* 1999, **97**:133–144.

Stehlik, C., and Reed, J.C.: **The PYRIN connection: novel players in innate immunity and inflammation.** *J Exp Med.* 2004, **200**:551–558.

Wise, C.A., Gillum, J.D., Seidman, C.E., Lindor, N.M., Veile, R., Bashiardes, S., and Lovett, M.: **Mutations in CD2BP1 disrupt binding to PTP PEST and are responsible for PAPA syndrome, an autoinflammatory disorder.** *Hum. Mol. Genet.* 2002, **11**:961–969.

13-21 Crohn's disease is a common inflammatory bowel disease.

Beutler, B.: **Autoimmunity and apoptosis: the Crohn's connection.** *Immunity* 2001, **15**:5–14.

Bonen, D.K., Ogura, Y., Nicolae, D.L., Inohara, N., Saab, L., Tanabe, T., Chen, F.F., Foster, S.J., Duerr, R.H., Brant, S.R., *et al.*.: **Crohn's disease-associated NOD2 variants share a signaling defect in response to lipopolysaccharide and peptidoglycan.** *Gastroenterology* 2003, **124**:140–146.

Hampe, J., Cuthbert, A., Croucher, P.J., Mirza, M.M., Mascheretti, S., Fisher, S., Frenzel, H., King, K., Hasselmeyer, A., Macpherson, A.J., *et al.*: **Association between insertion mutation in NOD2 gene and Crohn's disease in German and British populations.** *Lancet* 2001, **357**:1925–1928.

Hugot, J.P., Chamaillard, M., Zouali, H., Lesage, S., Cezard, J.P., Belaiche, J., Almer, S., Tysk, C., O'Morain, C.A., Gassull, M., *et al.*: **Association of NOD2 leucine-rich repeat variants with susceptibility to Crohn's disease.** *Nature* 2001, **411**:599–603.

Marks, D.J., Harbord, M.W., MacAllister, R., Rahman, F.Z., Young, J., Al-Lazikani, B., Lees, W., Novelli, M., Bloom, S., and Segal, A.W.: **Defective acute inflammation in Crohn's disease: a clinical investigation.** *Lancet* 2006, **367**:668-678.

Wang, X., Kuivaniemi, H., Bonavita, G., Mutkus, L., Mau, U., Blau, E., Inohara, N., Nunez, G., Tromp, G., and Williams, C.J.: **CARD15 mutations in familial granulomatosis syndromes: a study of the original Blau syndrome kindred and other families with large-vessel arteritis and cranial neuropathy.** *Arthritis Rheum.* 2002, **46**:3041–3045.

Autoimmunity and Transplantation

14

We have already seen in Chapter 13 how undesirable adaptive immune responses can be elicited by environmental antigens, and how this can cause serious disease in the form of allergic and hypersensitivity reactions. In this chapter we examine unwanted responses to two other medically important categories of antigens—those expressed on the body's cells and tissues. The first are responses to antigens on an individual's own cells and tissues. The response to self is called **autoimmunity** and can lead to **autoimmune diseases** that are characterized by tissue damage. The second is the response to nonself antigens on transplanted organs that leads to **graft rejection**.

The gene rearrangement that occurs during lymphocyte development in the central lymphoid organs is random, and thus inevitably results in the generation of some lymphocytes with affinity for self antigens. These are normally removed from the repertoire or held in check by a variety of mechanisms, many of which we have already encountered in Chapter 7. These generate a state of **self-tolerance** in which an individual's immune system does not attack the normal tissues of the body. Autoimmunity represents a breakdown or failure of the mechanisms of self-tolerance. We therefore first revisit the mechanisms that keep the lymphocyte repertoire self-tolerant and see how these may fail. We then discuss particular autoimmune diseases that illustrate the various pathogenetic mechanisms by which autoimmunity can damage the body. How genetic and environmental factors predispose to or trigger autoimmunity are then considered. In the remaining part of the chapter, we discuss the adaptive immune responses to nonself tissue antigens that cause transplant rejection.

The making and breaking of self-tolerance

To generate self-tolerance, the immune system must be able to distinguish self-reactive from nonself-reactive lymphocytes as they develop. As we learned in Chapter 7, the immune system takes advantage of surrogate markers of self and nonself to identify and delete potentially self-reactive lymphocytes. Despite this, some self-reactive lymphocytes escape elimination, leave the thymus, and can subsequently be activated to cause autoimmune disease. In part, autoreactivity occurs because the recognition of self-reactivity is indirect and therefore imperfect. In addition, many lymphocytes with some degree of self-reactivity can also make an immune response to foreign antigens; therefore, if all weakly self-reactive lymphocytes were eliminated, the function of the immune system would be impaired.

14-1 A critical function of the immune system is to discriminate self from nonself.

The immune system has very powerful effector mechanisms that can eliminate a wide variety of pathogens. Early in the study of immunity it was realized that these could, if turned against the host, cause severe tissue damage. The concept of autoimmunity was first presented at the beginning of the 20th century by **Paul Ehrlich**, who described it as '*horror autotoxicus*'. Autoimmune responses resemble normal immune responses to pathogens in that they are specifically activated by antigens, in this case self antigens or **autoantigens**, and give rise to autoreactive effector cells and to antibodies, called **autoantibodies**, against the self antigen. When reactions to self tissues do occur and are then improperly regulated, they cause a variety of chronic syndromes called autoimmune diseases. These syndromes are quite varied in their severity, in the tissues affected, and in the effector mechanisms that are most important in causing damage (Fig. 14.1).

With the exception of rheumatoid arthritis and thyroiditis, autoimmune diseases are individually uncommon, but collectively they affect approximately 5% of the populations of Western countries. Their relative rarity indicates that the immune system has evolved multiple mechanisms to prevent damage to self tissues. The most fundamental principle underlying these mechanisms is the discrimination of self from nonself, but this discrimination is not easy to achieve. B cells recognize the three-dimensional shape of an epitope on an antigen, and a pathogen epitope can be indistinguishable from one in humans. Similarly, the short peptides derived from the processing of pathogen antigens can be identical to self peptides. So how does a lymphocyte know what 'self' really is if there are no unique molecular signatures of self?

The first mechanism proposed for distinguishing between self and nonself was that recognition of antigen by an immature lymphocyte leads to a negative signal that causes lymphocyte death or inactivation. Thus, 'self' was thought to comprise those molecules recognized by a lymphocyte shortly after it has begun to express its antigen receptor. Indeed, this is an important mechanism of inducing self-tolerance in lymphocytes developing in the thymus and bone marrow (see Sections 7-20 and 7-21). The tolerance induced at this stage is known as **central tolerance**. Newly formed lymphocytes are especially sensitive to inactivation by strong signals through their antigen receptors, whereas the same signals would activate a mature lymphocyte.

Another antigenic quality that correlates with self is high and constant antigen concentration. Many self proteins are expressed by every cell in the body

Disease	Disease mechanism	Consequence
Graves' disease	Autoantibodies against the thyroid-stimulating-hormone receptor	Hyperthyroidism: overproduction of thyroid hormones
Rheumatoid arthritis	Autoreactive T cells against antigens of joint synovium	Joint inflammation and destruction causing arthritis
Hashimoto's thyroiditis	Autoantibodies and autoreactive T cells against thyroid antigens	Destruction of thyroid tissue leading to hypothyroidism: underproduction of thyroid hormones
Type 1 diabetes (insulin-dependent diabetes mellitus, IDDM)	Autoreactive T cells against pancreatic islet cell antigens	Destruction of pancreatic islet β cells leading to non-production of insulin
Multiple sclerosis	Autoreactive T cells against brain antigens	Formation of sclerotic plaques in brain with destruction of myelin sheaths surrounding nerve cell axons, leading to muscle weakness, ataxia, and other symptoms
Systemic lupus erythematosus	Autoantibodies and autoreactive T cells against DNA, chromatin proteins, and ubiquitous ribonucleoprotein antigens	Glomerulonephritis, vasculitis, rash
Sjögren's syndrome	Autoantibodies and autoreactive T cells against ribonucleoprotein antigens	Lymphocyte infiltration of exocrine glands, leading to dry eyes and/or dry mouth; other organs may be involved, leading to systemic disease

Fig. 14.1 Some common autoimmune diseases. The diseases listed are among the commonest autoimmune diseases and will be used as examples in this part of the chapter. A more comprehensive listing and discussion of autoimmune diseases is given later in the chapter.

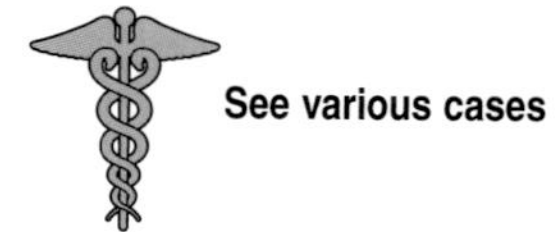

or abundantly on connective tissues. These can provide strong signals to lymphocytes, and even mature lymphocytes can be made tolerant to an antigen, or **tolerized**, by strong and constant signals through their antigen receptors. In contrast, pathogens and other foreign antigens are introduced to the immune system suddenly, and the concentrations of their antigens increase rapidly and exponentially as the pathogens replicate in the early stages of an infection. Naive mature lymphocytes are tuned to respond by activation to a sudden increase in antigen-receptor signals.

A third mechanism for discriminating between self and nonself relies on the innate immune system, which provides signals that are crucial in enabling the activation of an adaptive immune response to infection (see Chapter 2). In the absence of infection, these signals are not generated. In these circumstances, the encounter of a naive lymphocyte with a self antigen, especially when the antigen-presenting cell does not express co-stimulatory molecules, tends to lead to a negative, inactivating, signal, rather than no signal at all (see Section 7-26). This tolerance mechanism is particularly important for antigens that are encountered outside the thymus and bone marrow. Tolerance induced in the mature lymphocyte repertoire once cells have left the central lymphoid organs is known as **peripheral tolerance**.

Thus, several clues are used by lymphocytes to distinguish self ligands from nonself ligands: encounter with the ligand when the lymphocyte is still immature; a high and constant concentration of ligand; and binding the ligand in the absence of co-stimulatory signals. All these mechanisms are error-prone, however, because none of them particularly distinguishes a self ligand from a foreign one at the molecular level. The immune system therefore has several additional mechanisms for controlling autoimmune responses should they start.

14-2 Multiple tolerance mechanisms normally prevent autoimmunity.

The mechanisms that normally prevent autoimmunity may be considered as a succession of checkpoints. Each checkpoint is partly effective in preventing anti-self responses, and all of them together act synergistically to provide efficient protection against autoimmunity without inhibiting the ability of the immune system to mount effective responses to pathogens. Central tolerance mechanisms eliminate newly formed strongly autoreactive lymphocytes. On the other hand, mature self-reactive lymphocytes that do not sense self strongly in the central lymphoid organs—because their cognate self antigens are not expressed there, for example—may be killed or inactivated in the periphery. The principal mechanisms of peripheral tolerance are anergy (functional unresponsiveness), deletion (apoptotic cell death), and suppression by regulatory T cells (T_{reg}) (Fig. 14.2).

Each checkpoint strikes a balance between preventing autoimmunity but not impairing immunity too greatly, and in combination they provide an effective overall defense against autoimmune disease. It is relatively easy to find isolated breakdowns of one or even more layers of protection, even in healthy individuals. Thus, activation of autoreactive lymphocytes does not necessarily equal autoimmune disease. In fact, a low level of autoreactivity is physiological and crucial to normal immune function. Autoantigens help to form the repertoire of mature lymphocytes, and the survival of naive T cells and B cells in the periphery requires continuous exposure to autoantigens (see Chapter 7). Autoimmune disease develops only if enough of the safeguards are overcome to lead to a sustained reaction to self that includes the generation of effector cells and molecules that destroy tissues. Although the mechanisms by which this occurs are not completely known, autoimmunity is thought to result from a combination of genetic susceptibility, breakdown in natural tolerance mechanisms, and environmental triggers such as infections (Fig. 14.3).

Fig. 14.2 Self-tolerance depends on the concerted action of a variety of mechanisms that operate at different sites and stages of development. The different ways in which the immune system prevents activation of and damage caused by autoreactive lymphocytes are listed, along with the specific mechanism and where such tolerance predominantly occurs.

Layers of self-tolerance		
Type of tolerance	**Mechanism**	**Site of action**
Central tolerance	Deletion Editing	Thymus Bone marrow
Antigen segregation	Physical barrier to self-antigen access to lymphoid system	Peripheral organs (e.g. thyroid, pancreas)
Peripheral anergy	Cellular inactivation by weak signaling without co-stimulus	Secondary lymphoid tissue
Regulatory cells	Suppression by cytokines, intercellular signals	Secondary lymphoid tissue and sites of inflammation
Cytokine deviation	Differentiation to T_H2 cells, limiting inflammatory cytokine secretion	Secondary lymphoid tissue and sites of inflammation
Clonal deletion	Apoptosis post-activation	Secondary lymphoid tissue and sites of inflammation

14-3 Central deletion or inactivation of newly formed lymphocytes is the first checkpoint of self-tolerance.

Central tolerance mechanisms, which remove strongly autoreactive lymphocytes, are the first and most important checkpoints in self-tolerance and are covered in detail in Chapter 7. Without them, the immune system would be strongly self-reactive, and lethal autoimmunity would most certainly be present from birth. It is unlikely that the other, later-acting, mechanisms of tolerance would be sufficient to compensate for the failure to remove self-reactive lymphocytes during their primary development. Indeed, there are no known autoimmune diseases that are attributable to a complete failure of these basic mechanisms, although some are associated with a partial failure of central tolerance.

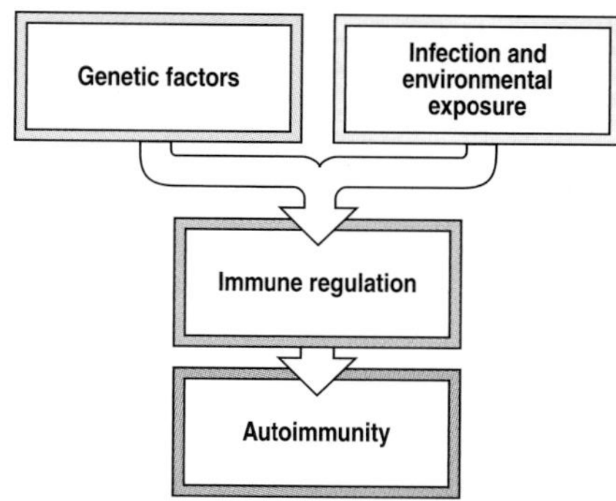

Fig. 14.3 Requirements for the development of autoimmune disease. In genetically predisposed individuals, autoimmunity may be triggered as a result of the failure of intrinsic tolerance mechanisms and/or environmental triggers such as infection.

Self-tolerance generated in the central lymphoid organs is effective, but for a long time it was thought that many self antigens were not expressed in the thymus or bone marrow, and that peripheral mechanisms must be the only way of generating tolerance to them. It is now clear, however, that many (but not all) tissue-specific antigens, such as insulin, are actually expressed in the thymus by a subset of dendritic-like cells, and thus self-tolerance against these antigens can be generated centrally. How these 'peripheral' genes are turned on ectopically in the thymus is not yet completely worked out, but an important clue has been found. A single transcription factor, AIRE (for *a*uto*i*mmune *re*gulator), is thought to be responsible for turning on many peripheral genes in the thymus (see Section 7-20). The *AIRE* gene is defective in patients with a rare inherited form of autoimmunity—**APECED** (**autoimmune polyendocrinopathy–candidiasis–ectodermal dystrophy**)—that leads to the destruction of multiple endocrine tissues, including insulin-producing pancreatic islets. This disease is also known as autoimmune polyglandular syndrome 1 (APS-1). Mice that have been engineered to lack the *AIRE* gene have a similar syndrome, although they do not seem to be susceptible to fungal infections such as candidiasis. Most importantly, these mice no longer express many of the peripheral genes in the thymus. This links the AIRE protein to the expression of these genes as well as suggesting that an inability to express these genes in the thymus leads to autoimmune disease (Fig. 14.4). The autoimmunity that accompanies AIRE deficiency takes time to develop and does not always affect all potential organ targets. So as well as emphasizing the importance of central tolerance, the disease shows that other layers of tolerance control have an important role.

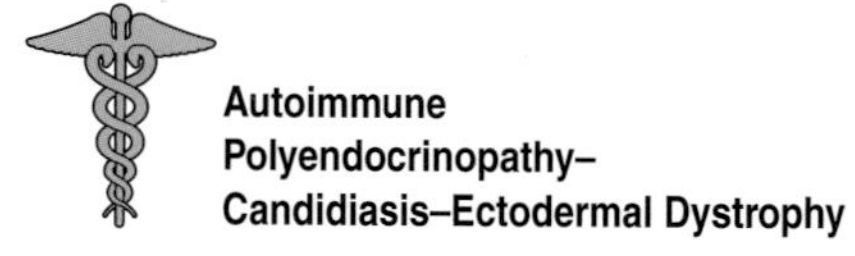

14-4 Lymphocytes that bind self antigens with relatively low affinity usually ignore them but in some circumstances become activated.

Some lymphocytes with a relatively low affinity for self antigens make no response to them, escape tolerance mechanisms altogether, and may be considered as 'ignorant' of self (see Section 7-6). Such ignorant but latently self-reactive cells can be recruited into autoimmune responses if the stimulus is strong enough. One such stimulus could be infection. Naive T cells with low affinity for a ubiquitous self antigen can become activated if they encounter an activated dendritic cell presenting that antigen and expressing high levels of co-stimulatory signals as a result of the presence of infection.

A particular situation in which ignorant lymphocytes may be activated is where their autoantigens are also the ligands for Toll-like receptors (TLRs). These receptors are usually considered to be pattern-recognition receptors specific for pathogen-associated molecular patterns (see Section 2-7), but these patterns are not exclusive to pathogens and can be found among self molecules. An example of this type of potential autoantigen is unmethylated CpG sequences in DNA that are recognized by TLR-9. Unmethylated CpG is

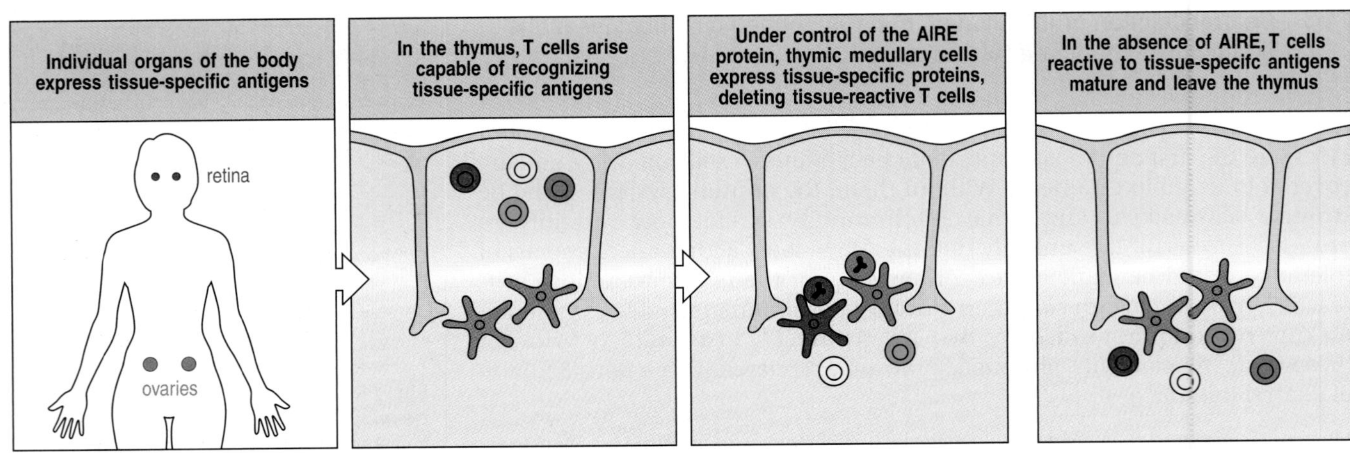

Individual organs of the body express tissue-specific antigens
retina
ovaries
In the thymus, T cells arise capable of recognizing tissue-specific antigens
Under control of the AIRE protein, thymic medullary cells express tissue-specific proteins, deleting tissue-reactive T cells
In the absence of AIRE, T cells reactive to tissue-specifc antigens mature and leave the thymus

Fig. 14.4 The 'autoimmune regulator' gene AIRE promotes the expression of some tissue-specific antigens in thymic medullary cells, causing the deletion of immature thymocytes that can react to these antigens. Although the thymus expresses many genes, and thus self proteins, common to all cells, it is not obvious how antigens that are specific to specialized tissues, such as retina or ovary (first panel), gain access to the thymus to promote the negative selection of immature autoreactive thymocytes. It is now known that a gene called *AIRE* promotes the expression of many tissue-specific proteins in thymic medullary cells. Some developing thymocytes will be able to recognize these tissue-specific antigens (second panel). Peptides from these proteins are presented to the developing thymocytes as they undergo negative selection in the thymus (third panel), causing deletion of these cells. In the absence of *AIRE*, such deletion does not occur; instead, the autoreactive thymocytes mature and are exported to the periphery (fourth panel), where they could cause autoimmune disease. Indeed, people and mice that lack expression of *AIRE* develop an autoimmune syndrome called APECED, or autoimmune polyendocrinopathy–candidiasis–ectodermal dystrophy.

normally much more common in bacterial DNA than in mammalian DNA but is enriched in mammalian cells undergoing apoptosis. In a scenario where there is extensive cell death coupled with inadequate clearance of apoptotic fragments (possibly as a result of infection), B cells specific for components of chromatin can internalize the CpG sequences via their B-cell receptors. These sequences encounter their receptor, TLR-9, intracellularly, leading to a co-stimulatory signal that, together with the signal from the B-cell receptor, activates the previously ignorant anti-chromatin B cell (Fig. 14.5). B cells activated in this way will proceed to produce anti-chromatin autoantibodies and also can act as antigen-presenting cells for autoreactive T cells. Ribonucleoprotein complexes containing uridine-rich RNA have similarly been shown to activate naive B cells through binding of the RNA by TLR-7 or TLR-8. Autoantibodies against DNA, chromatin proteins, and ribonucleoproteins are produced in the autoimmune disease systemic lupus erythematosus (SLE), and this may be one of the mechanisms by which self-reactive B cells are stimulated to produce them. These findings challenge the concept that Toll-like receptors are completely reliable at distinguishing self from nonself; their proposed role in autoimmunity has been called the 'Toll hypothesis'.

Another mechanism by which ignorant lymphocytes can be drawn into action is by changing the availability or form of a self antigen. Some antigens are normally intracellular and are not encountered by lymphocytes, but they may be released as a result of massive tissue death or inflammation. These antigens can then activate hitherto ignorant T and B cells, leading to autoimmunity. This can occur after myocardial infarction, when an autoimmune response is detectable some days after the release of cardiac antigens. Such reactions are typically transient and cease when the autoantigens have been removed; however, when clearance mechanisms are inadequate or genetically deficient, they can continue, causing clinical autoimmune disease.

Some autoantigens are present in great quantity but are usually in a nonimmunogenic form. IgG is a good example, because there are large quantities of

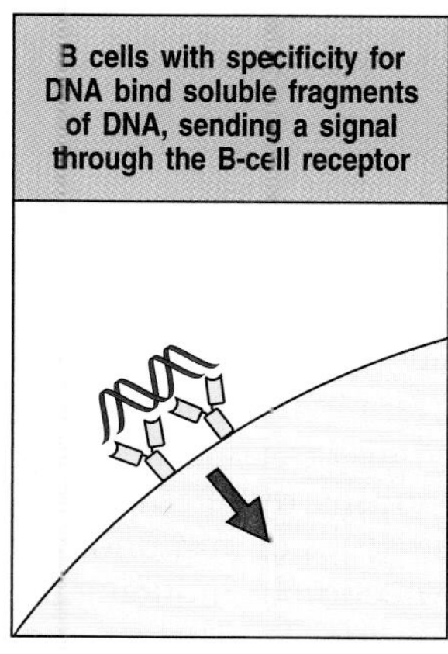

B cells with specificity for DNA bind soluble fragments of DNA, sending a signal through the B-cell receptor

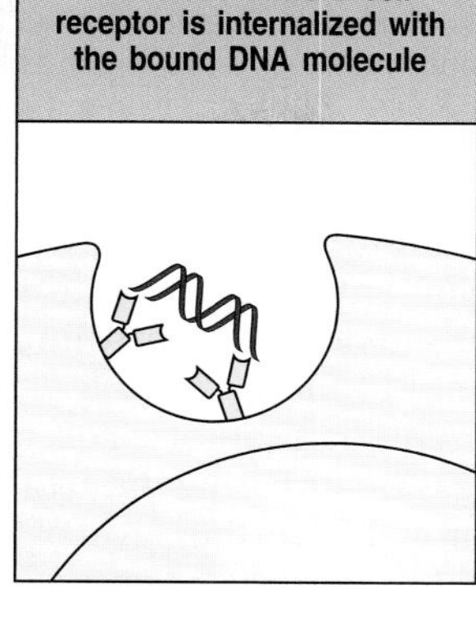

The cross-linked B-cell receptor is internalized with the bound DNA molecule

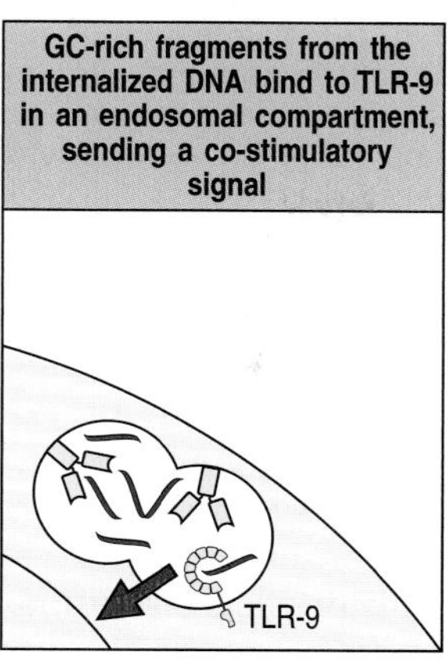

GC-rich fragments from the internalized DNA bind to TLR-9 in an endosomal compartment, sending a co-stimulatory signal
TLR-9

Fig. 14.5 Self antigens that are recognized by Toll-like receptors can activate autoreactive B cells by providing co-stimulation. The receptor TLR-9 promotes the activation of B cells specific for DNA, a common autoantibody in the autoimmune disease systemic lupus erythematosus (SLE) (see Fig. 14.1). Although B cells with strong affinity for DNA are eliminated in the bone marrow, some DNA-specific B cells with lower affinity escape and persist in the periphery but are not normally activated. Under some conditions and in genetically susceptible individuals, the concentration of DNA may increase, leading to the ligation of enough B-cell receptors to initiate activation of these B cells. B cells signal through their receptor (left panel) but also take up the DNA (center panel) and deliver it to an endosomal compartment (right panel). Here the DNA has access to TLR-9, which recognizes DNA that is enriched in unmethylated CpG DNA sequences. Such CpG-enriched sequences are much more common in microbial than eukaryotic DNA and normally this allows TLR-9 to distinguish pathogens from self. DNA in apoptotic mammalian cells is enriched in unmethylated CpG, however, and the DNA-specific B cell will also concentrate the self DNA in the endosomal compartment. This would provide sufficient ligand to activate TLR-9, potentiating the activation of the DNA-specific B cell and ultimately leading to the production of autoantibodies against DNA.

it in blood and in other extracellular fluid. B cells specific for the IgG constant region are not usually activated because the IgG is monomeric and cannot cross-link the B-cell receptor. However, when immune complexes form after a severe infection or an immunization, enough IgG is in multivalent form to evoke a response from these otherwise ignorant B cells. The anti-IgG autoantibody they produce is known as **rheumatoid factor** because it is commonly present in rheumatoid arthritis. Again, this response is normally short lived, as long as the immune complexes are cleared rapidly.

A unique situation occurs in peripheral lymphoid organs when activated B cells undergo somatic hypermutation in germinal centers (see Section 9-7). This can result in some already activated B cells becoming self-reactive or increasing their affinity for a self antigen (Fig. 14.6). Like the ignorant lymphocytes discussed above, such self-reactive B cells would have bypassed all the other tolerance mechanisms but would now be a source of potentially pathogenic autoantibodies. There seems, however, to be a mechanism to control germinal-center B cells that have acquired affinity for self. In this case, the self antigen is likely to be present within the germinal center, whereas a pathogen is less likely to be. If a hypermutated self-reactive B cell encounters strong cross-linking of its B-cell receptor in the germinal center, it undergoes apoptosis rather than further proliferation.

14-5 Antigens in immunologically privileged sites do not induce immune attack but can serve as targets.

Tissue grafts placed in some sites in the body do not elicit immune responses. For instance, the brain and the anterior chamber of the eye are sites in which tissues can be grafted without inducing rejection. Such locations are termed **immunologically privileged sites** (Fig. 14.7). It was originally believed that immunological privilege arose from the failure of antigens to leave privileged sites and induce immune responses. Subsequent studies have shown that antigens do leave these sites and that they do interact with T cells. Instead of eliciting a destructive immune response, however, they induce tolerance or a response that is not destructive to the tissue.

Immunologically privileged sites seem to be unusual in three ways. First, the communication between the privileged site and the body is atypical in that extracellular fluid in these sites does not pass through conventional lymphatics, although proteins placed in these sites do leave them and can have immunological effects. Privileged sites are generally surrounded by tissue barriers that exclude naive lymphocytes. The brain, for example, is guarded by the blood–brain barrier. Second, soluble factors, presumably cytokines,

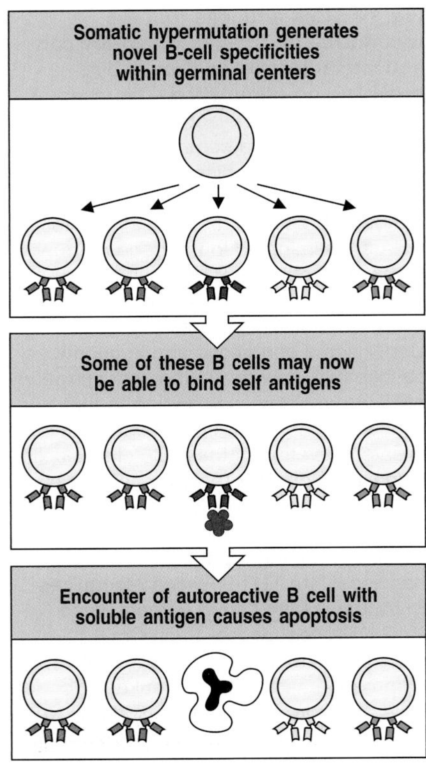

Somatic hypermutation generates novel B-cell specificities within germinal centers
Some of these B cells may now be able to bind self antigens
Encounter of autoreactive B cell with soluble antigen causes apoptosis

Fig. 14.6 Elimination of autoreactive B cells in germinal centers. During somatic hypermutation in germinal centers (top panel), B cells with autoreactive B-cell receptors can arise. Ligation of these receptors by soluble autoantigen (center panel) induces apoptosis of the autoreactive B cell by signaling through the B-cell antigen receptor in the absence of helper T cells (bottom panel).

that affect the course of an immune response are produced in privileged sites and leave them together with antigens. The anti-inflammatory transforming growth factor (TGF)-β seems to be particularly important in this regard: antigens mixed with TGF-β seem to induce mainly T-cell responses that do not damage tissues, such as noninflammatory T_H2 responses rather than T_H1 responses. Third, the expression of Fas ligand by the tissues of immunologically privileged sites may provide a further level of protection by inducing the apoptosis of Fas-bearing lymphocytes that enter these sites. This last mechanism of protection is not fully understood, because it seems that under some circumstances the expression of Fas ligand by tissues may trigger an inflammatory response by neutrophils.

Paradoxically, the antigens sequestered in immunologically privileged sites are often the targets of autoimmune attack; for example, brain autoantigens such as myelin basic protein are targeted in the autoimmune disease **multiple sclerosis**, a chronic inflammatory demyelinating disease of the central nervous system (see Fig. 14.1). It is therefore clear that the tolerance normally shown to this antigen cannot be due to clonal deletion of the self-reactive T cells. In the condition **experimental autoimmune encephalomyelitis** (**EAE**), a mouse model for multiple sclerosis, mice become diseased only when they are deliberately immunized with myelin basic protein, which causes substantial infiltration of the brain with antigen-specific T_H1 cells.

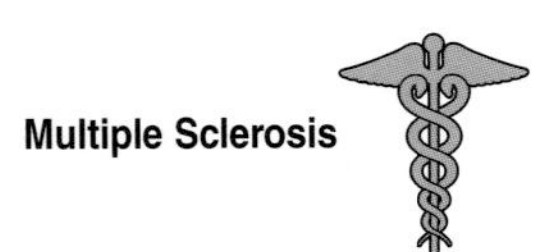

This shows that at least some antigens expressed in immunologically privileged sites induce neither tolerance nor lymphocyte activation in normal circumstances, but if autoreactive lymphocytes are activated elsewhere, these autoantigens can become targets for autoimmune attack. It seems plausible that T cells specific for antigens sequestered in immunologically privileged sites are most likely to be in a state of immunological ignorance. Further evidence comes from the eye disease **sympathetic ophthalmia** (Fig. 14.8). If one eye is ruptured by a blow or other trauma, an autoimmune response to eye proteins can occur, although this happens only rarely. Once the response is induced, it often attacks both eyes. Immunosuppression—and, rarely, removal of the damaged eye, the source of antigen—is required to preserve vision in the undamaged eye.

Immunologically privileged sites
Brain
Eye
Testis
Uterus (fetus)
Hamster cheek pouch

Fig. 14.7 Some sites in the body are immunologically privileged. Tissue grafts placed in these sites often last indefinitely, and antigens placed in these sites do not elicit destructive immune responses.

It is not surprising that effector T cells can enter immunologically privileged sites: such sites can become infected, and effector cells must be able to enter these sites during infection. As we learned in Chapter 10, effector T cells enter most or all tissues after activation, but accumulation of cells is seen only when antigen is recognized in the site, triggering the production of cytokines that alter tissue barriers.

14-6 Autoreactive T cells that express particular cytokines may be nonpathogenic or may suppress pathogenic lymphocytes.

We learned in Chapter 8 that, during the course of normal immune responses, CD4 T cells can differentiate into various types of effector cells, namely T_H1 and T_H2. T_H1 and T_H2 cells secrete different cytokines (interferon (IFN)-γ and tumor necrosis factor (TNF)-α for T_H1, and interleukin (IL)-4, IL-5, IL-10, and IL-13 for T_H2) and have different effects on antigen-presenting cells, on B cells, and on pathogen clearance. A similar paradigm holds true for autoimmunity. In particular, certain T cell-mediated autoimmune diseases such as

type 1 diabetes mellitus (also known as **insulin-dependent diabetes mellitus (IDDM)**) (see Fig. 14.1) and multiple sclerosis seem to depend on T_H1 cells to cause disease. In contrast, in SLE the generation of autoantibodies seems to require both T_H1 and T_H2 cells. In murine models of diabetes, when cytokines were infused to influence T-cell differentiation or when knockout mice predisposed to T_H2 differentiation were studied, the development of diabetes was inhibited. In some cases, potentially pathogenic T cells specific for pancreatic islet-cell components, and expressing T_H2 instead of T_H1 cytokines, are actually suppressive of disease caused by T_H1 cells of the same specificity. Attempts to control human autoimmune disease by switching cytokine profiles from T_H1 to T_H2, a procedure termed **immune modulation**, have not proved successful, however. Another important subset of CD4 T cells, the regulatory T cells, may prove to be more important in the natural prevention of autoimmune disease.

14-7 Autoimmune responses can be controlled at various stages by regulatory T cells.

Autoreactive cells that have escaped the tolerance-inducing mechanisms described previously can still be regulated so that they do not cause clinical disease. This regulation takes two forms: the first is extrinsic, coming from specific regulatory T cells that exert effects on activated T cells and on antigen-presenting cells. The second is intrinsic and has to do with limits on the size and duration of immune responses that are programmed into the lymphocytes themselves. We shall first discuss the role of regulatory T cells, which were introduced in Chapter 8.

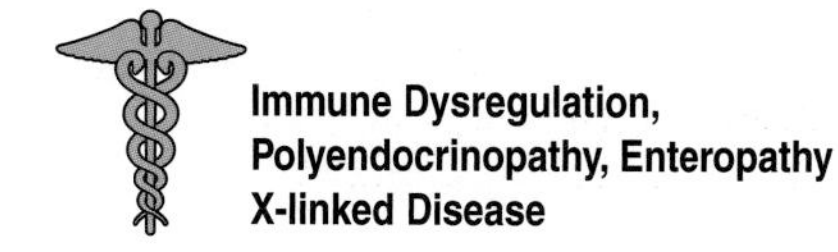
Immune Dysregulation,
Polyendocrinopathy, Enteropathy
X-linked Disease

Tolerance due to regulatory lymphocytes is distinguished from other forms of self-tolerance by the fact that regulatory T cells have the potential to suppress self-reactive lymphocytes that recognize antigens different from those recognized by the regulatory T cells. This type of tolerance is therefore sometimes known as **regulatory tolerance**, **dominant immune suppression**, or **infectious tolerance**. The key feature of dominant regulatory tolerance is that regulatory cells can suppress autoreactive lymphocytes that recognize a variety of different self antigens, as long as the antigens are all on the same tissue or are presented by the same antigen-presenting cell (Fig. 14.9). Regulatory T cells are thought to be moderately autoreactive T cells that escape deletion

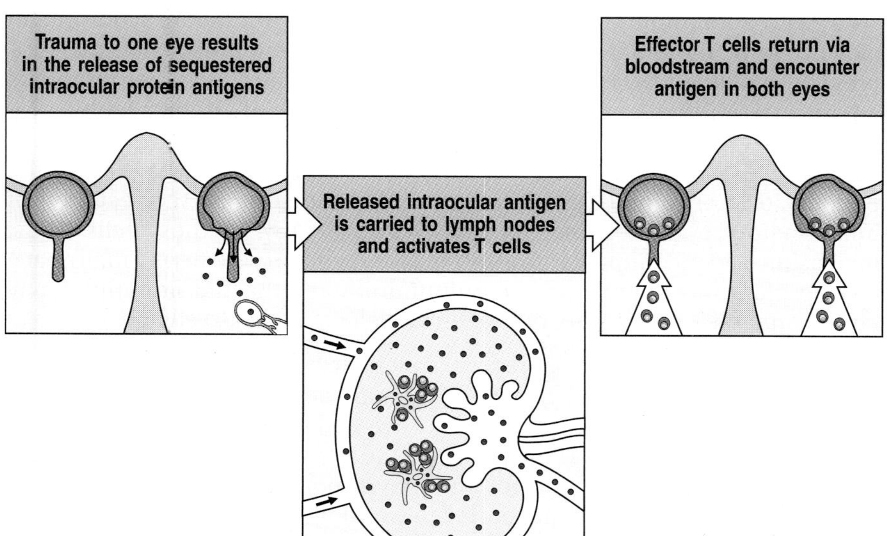
Trauma to one eye results in the release of sequestered intraocular protein antigens
Released intraocular antigen is carried to lymph nodes and activates T cells
Effector T cells return via bloodstream and encounter antigen in both eyes

Fig. 14.8 Damage to an immunologically privileged site can induce an autoimmune response. In the disease sympathetic ophthalmia, trauma to one eye releases the sequestered eye antigens into the surrounding tissues, making them accessible to T cells. The effector cells that are elicited attack the traumatized eye, and also infiltrate and attack the healthy eye. Thus, although the sequestered antigens do not induce a response by themselves, if a response is induced elsewhere they can serve as targets for attack.

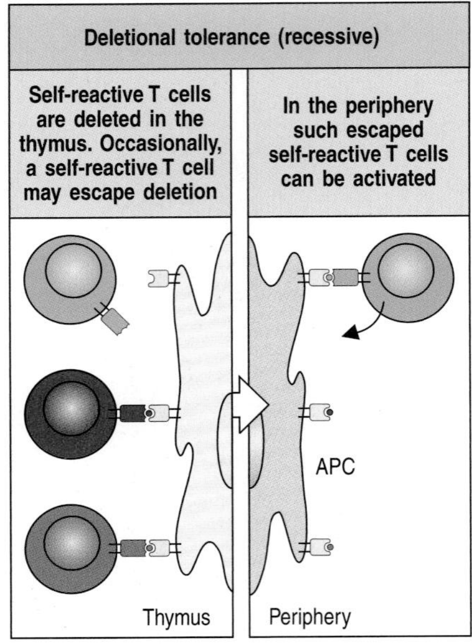

Deletional tolerance (recessive)
Self-reactive T cells are deleted in the thymus. Occasionally, a self-reactive T cell may escape deletion
In the periphery such escaped self-reactive T cells can be activated
APC
Thymus
Periphery

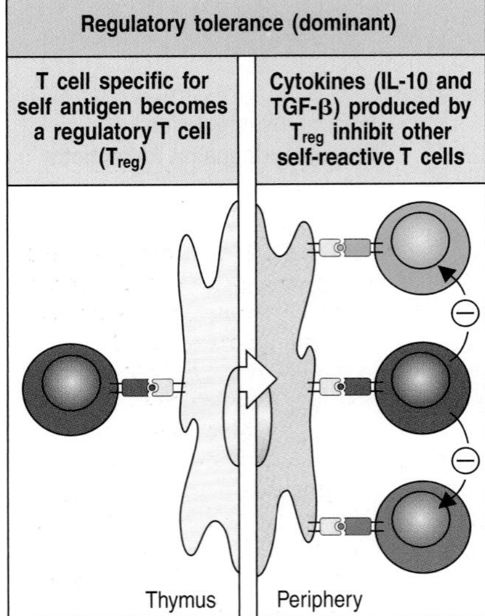

Fig. 14.9 Recessive tolerance occurs when self-reactive T cells are deleted, whereas a dominant form of tolerance mediated by regulatory T cells can inhibit multiple autoreactive T cells that all recognize the same tissue. As discussed previously, one of the main mechanisms of self-tolerance is the deletion of self-reactive T cells in the thymus by thymic dendritic cells that express self antigens (upper left). However, some autoreactive cells may not be deleted because their particular autoantigen is not available on the deleting cell (upper left, turquoise cell). Such T cells can cause damage in the periphery if they encounter their autoantigen on an antigen-presenting cell (APC) and become activated (upper right). One mechanism that suppresses this potentially harmful autoreactivity is known as regulatory tolerance (lower panels). This is mediated by specialized regulatory T cells (T_{reg}) that develop in the thymus in response to weak stimulation by self antigen that is not sufficient to cause deletion but is more than is required for simple positive selection (lower left). These cells migrate to the periphery where, if they encounter their self antigen (lower right) on an antigen-presenting cell, they secrete inhibitory cytokines such as IL-10 and TGF-β that inhibit all surrounding autoreactive T cells, regardless of their precise autoantigen specificity. This is a dominant form of tolerance in that a single cell can regulate many others.

in the thymus and when activated by self antigens do not differentiate into cells that can initiate an autoimmune response; instead, they differentiate into powerful suppressor cells that inhibit other self-reactive T cells that recognize antigens on the same tissue. Many investigators have therefore hypothesized that regulatory T cells could have therapeutic potential for the treatment of autoimmune disease if they could be isolated and infused into patients.

One of the best-characterized types of regulatory T cells bears CD4 and CD25 (the α chain of the IL-2 receptor) on its surface (see Section 8-20). They have been shown to have a protective role in several autoimmune syndromes in mice, including inflammation of the colon (colitis), diabetes, EAE, and SLE. A proposed model for the resolution of autoimmune colitis in mice by CD4 CD25 T cells is shown in Fig. 14.10. Experiments in mouse models of these diseases show that CD4 CD25 regulatory T cells suppress disease when transferred *in vivo* and that depletion of these cells exacerbates or causes disease. These regulatory T cells have also been shown to prevent or ameliorate other immunopathologic syndromes, such as graft-versus-host disease and graft rejection, which are described later in this chapter.

The importance of regulatory T cells has been demonstrated in several human autoimmune diseases. For example, in patients with multiple sclerosis or with autoimmune polyglandular syndrome type 2 (a rare syndrome in which two or more autoimmune diseases occur simultaneously), the suppressive activity of CD4 CD25 T_{reg} cells is defective, although their numbers are normal. A different picture emerges from studies of patients with active rheumatoid arthritis. Peripheral blood CD4 CD25 T_{reg} cells from these patients were found to be effective in suppressing the proliferation of the patients' own effector T cells *in vitro* but did not suppress the secretion of inflammatory cytokines, including TNF-α and IFN-γ, by these cells. Thus, increasing evidence supports the notion that regulatory T cells normally have an important role in preventing autoimmunity, and that autoimmunity may be accompanied by a variety of functional defects in these cells.

CD4 CD25 T cells are not the only type of regulatory lymphocyte that has been found. CD25-negative regulatory T cells include the T_H3 cells identified in the mucosal immune system (see Section 11-13) and the T_R1 cells characterized *in vitro* (see Section 8-20). The T_H3 cells of the mucosal immune system seem to function to suppress or control immune responses at the mucosa, which form barriers against the pathogen-laden world. T_R1 cells can be generated *in vitro* after stimulation with IL-10, and a similar type of IL-10-dependent regulatory T cell may be present in mucosa but has not yet been identified. Lack

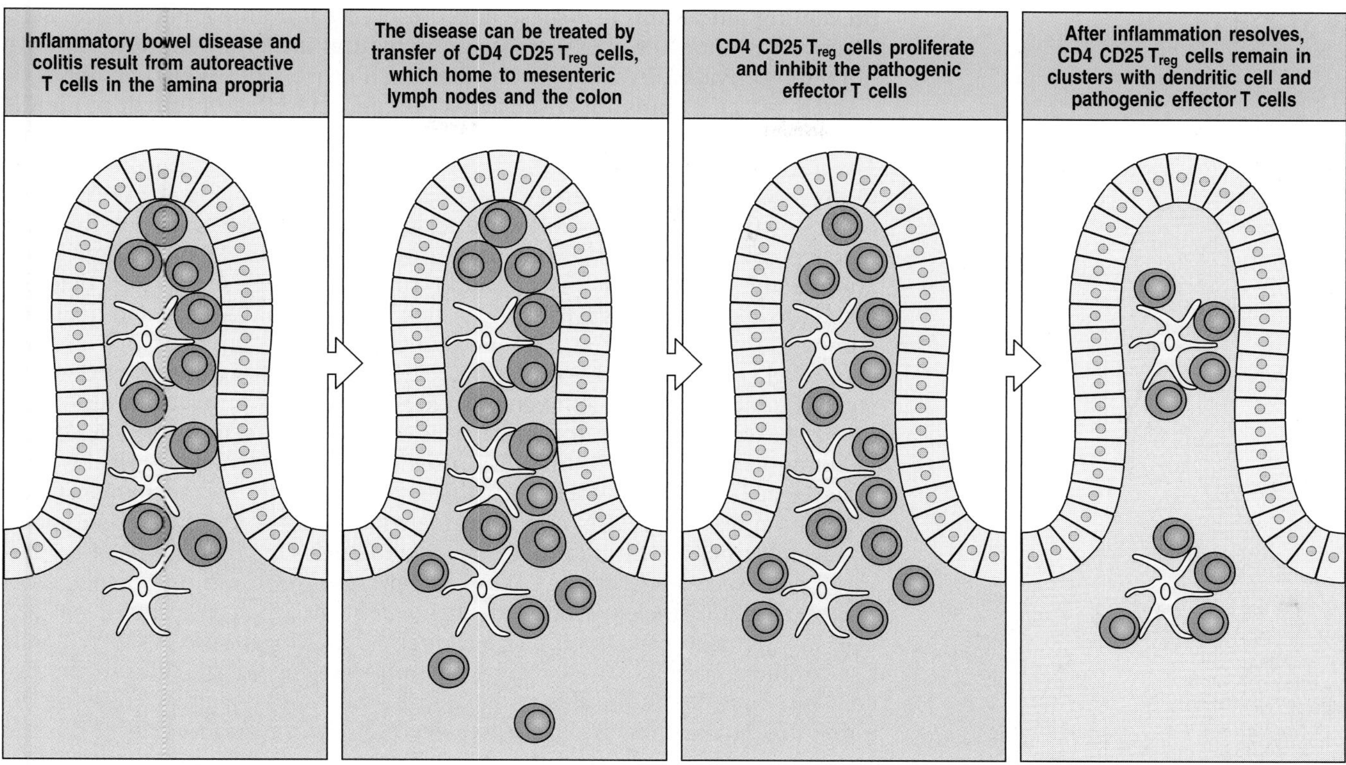

Inflammatory bowel disease and colitis result from autoreactive T cells in the lamina propria
The disease can be treated by transfer of CD4 CD25 T_{reg} cells, which home to mesenteric lymph nodes and the colon
CD4 CD25 T_{reg} cells proliferate and inhibit the pathogenic effector T cells
After inflammation resolves, CD4 CD25 T_{reg} cells remain in clusters with dendritic cell and pathogenic effector T cells

Fig. 14.10 CD4 CD25 regulatory T cells inhibit colitis by migrating to the colon and mesenteric lymph nodes, where they interact with dendritic and effector T cells. Naive T cells that contain some autoreactive clones (first panel, pink cells) cause colitis when transferred to T-cell-deficient mice. The naive population lacks CD4 CD25 T_{reg} cells, but if these are also transferred along with the naive T cells (second panel; blue cells are T_{reg} cells), colitis is blocked. The blocking mechanism includes migration of the T_{reg} cells to mesenteric lymph nodes (not shown) and later to the lamina propria of the colon. The T_{reg} cells proliferate and secrete regulatory cytokines (third panel), including IL-10, which is essential, and interact with both dendritic and autoreactive T cells, reducing activation (indicated by the smaller size of the pink cells) and ultimately reducing inflammation. Once inflammation has been quelled, regulatory T cells remain in the lamina propria (fourth panel). Based on a figure by F. Powrie.

of T_H3 cells is linked to autoimmune disease in the gut, and T_R1 cells have been shown experimentally to suppress inflammatory bowel disease in mice. Giving animals large amounts of self antigen orally, which induces so-called oral tolerance (see Section 11-13), can sometimes lead to unresponsiveness to these antigens when given by other routes, and can prevent autoimmune disease. This oral tolerance is accompanied by the generation or expansion of T_H3 cells, which could have a role in the mechanism.

Almost every type of lymphocyte has been shown to display regulatory activity in some circumstance. Even B cells can regulate experimentally induced autoimmune syndromes, including collagen-induced arthritis (CIA) and EAE in mice. This regulatory activity is probably mediated in a similar way to that of regulatory CD4 T cells, with the secretion of cytokines that inhibit T-cell proliferation and the differentiation of T_H1 cells being of major importance. Immature dendritic cells induce the differentiation of regulatory T cells, which contributes to the maintenance of tolerance in the absence of infection.

In addition to the extrinsic regulation of autoreactive T and B cells by regulatory cells, lymphocytes have intrinsic limits to proliferation and survival that can help to restrict autoimmune responses as well as normal immune responses (see Section 10-12). This is illustrated by the effects of mutations in the pathways that control apoptosis, such as the Bcl-2 pathway or the Fas pathway (see Section 6-25), that lead to spontaneous autoimmunity, as we shall see later in this chapter. This case of autoimmunity provides evidence that autoreactive cells are normally generated but are then controlled by apoptosis. This is evidently an important mechanism for both T- and B-cell tolerance.

Summary.

Discrimination between self and nonself is imperfect, partly because of its indirect nature and partly because a proper balance must be struck between preventing autoimmune disease and preserving immune competence. Self-reactive lymphocytes always exist in the natural immune repertoire but are not often activated. In autoimmune disease, however, these cells become activated by specific autoantigens. If activation persists, effector functions identical to those elicited in response to pathogens are generated and cause disease. The immune system has a remarkable set of mechanisms that work together to prevent autoimmune disease (see Fig. 14.2). This collective action means that each mechanism need not work perfectly nor apply to every possible self-reactive cell. Self-tolerance begins during lymphocyte development, when autoreactive T cells in the thymus and B cells in the bone marrow are deleted. Mechanisms of peripheral tolerance, such as peripheral anergy and deletion, complement these central tolerance mechanisms for antigens that are not expressed centrally. Weakly self-reactive lymphocytes are not removed at this stage; extending tolerance mechanisms such as deletion to weakly autoreactive cells would impose too great a limitation on the immune repertoire, resulting in impaired immune responses to pathogens. Instead, weakly self-reactive cells are suppressed only if they are activated, by mechanisms that include regulatory T cells and immune modulation—the differentiation of T cells to express noninflammatory T_H2 cytokines. One major type of regulatory T cell expresses CD4 and CD25, and fairly severe autoimmunity occurs in its absence. It is unclear what activates regulatory T cells, but the CD4 CD25 cells are themselves autoreactive, although not pathogenic. Regulatory T cells can inhibit a variety of self-reactive lymphocytes, as long as the regulatory cells are targeting autoantigens located in the same general vicinity of the autoantigens to which the self-reactive lymphocytes respond. This allows the regulatory cells to home to and suppress sites of autoimmune inflammation. A final mechanism that controls autoimmunity is the natural tendency of immune responses to be self-limited: intrinsic programs in activated lymphocytes make them prone to apoptosis. Activated lymphocytes also acquire sensitivity to external apoptosis-inducing signals, such as those mediated by Fas.

Autoimmune diseases and pathogenic mechanisms.

Here we describe some of the more common clinical autoimmune syndromes, and the ways in which loss of self-tolerance and expansion of self-reactive lymphocytes lead to tissue damage. These mechanisms of pathogenesis resemble in most ways those that target invading pathogens. Damage by autoantibodies, mediated through the complement and Fc receptor systems, has an important role in some diseases, such as SLE. Similarly, cytotoxic T cells directed at self tissues destroy them much as they would virus-infected cells, and this is one way in which pancreatic β cells are destroyed in diabetes. However, self proteins cannot normally be eliminated, with rare exceptions such as the islet cells in the pancreas, and so the response continues. Some pathogenic mechanisms are unique to autoimmunity, such as antibodies against receptors on cell surfaces that affect their function, as in the disease myasthenia gravis, as well as hypersensitivity-type reactions. In this part of the chapter we describe the pathogenic mechanisms of some clinical syndromes of autoimmune disease.

14-8 Specific adaptive immune responses to self antigens can cause autoimmune disease.

In certain genetically susceptible strains of experimental animals, autoimmune disease can be induced artificially by the injection of 'self' tissues from a genetically identical animal mixed with strong adjuvants containing bacteria (see Appendix I, Section A-4). This shows directly that autoimmunity can be provoked by inducing a specific adaptive immune response to self antigens. Such experimental systems highlight the importance of the activation of other components of the immune system, primarily dendritic cells, by the bacteria contained in the adjuvant. There are drawbacks to the use of such animal models for the study of autoimmunity. In humans and genetically autoimmune-prone animals, autoimmunity usually arises spontaneously: that is, we do not know what events initiate the immune response to self that leads to the autoimmune disease. By studying the patterns of autoantibodies and also the particular tissues affected, it has been possible to identify some of the self antigens that are targets of autoimmune disease, although it has still to be proved that the immune response was initiated in response to these same antigens. In animal models, and to a lesser degree in humans, it has sometimes been possible to identify self proteins that incite self-reactive T cells.

Some autoimmune disorders may be triggered by infectious agents that express epitopes resembling self antigens and that lead to sensitization of the patient against that tissue. There is, however, also evidence from animal models of autoimmunity that many autoimmune disorders are caused by internal dysregulation of the immune system without the apparent participation of infectious agents.

14-9 Autoimmune diseases can be classified into clusters that are typically either organ-specific or systemic.

The classification of disease is an uncertain science, especially in the absence of a precise understanding of causative mechanisms. This is well illustrated by the difficulty in classifying the autoimmune diseases. From a clinical perspective it is useful to distinguish between the following two major patterns of autoimmune disease: the diseases in which the expression of autoimmunity is restricted to specific organs of the body, known as 'organ-specific' autoimmune diseases; and those in which many tissues of the body are affected, the 'systemic' autoimmune diseases. Systemic autoimmune diseases affect multiple organs and have a tendency to become chronic, because the autoantigens can never be cleared from the body. Some autoimmune diseases seem to be dominated by the pathogenic effects of a particular immune effector pathway, either autoantibodies or activated autoreactive T cells. However, both of these pathways often contribute to the overall pathogenesis of autoimmune disease.

In organ-specific diseases, autoantigens from one or a few organs only are targeted, and disease is therefore limited to those organs. Examples of organ-specific autoimmune diseases are **Hashimoto's thyroiditis** and **Graves' disease**, both predominantly affecting the thyroid gland, and type 1 diabetes, which is caused by immune attack on insulin-producing pancreatic β cells. Examples of systemic autoimmune disease are SLE and primary Sjögren's syndrome, in which tissues as diverse as the skin, kidneys, and brain may all be affected (Fig. 14.11).

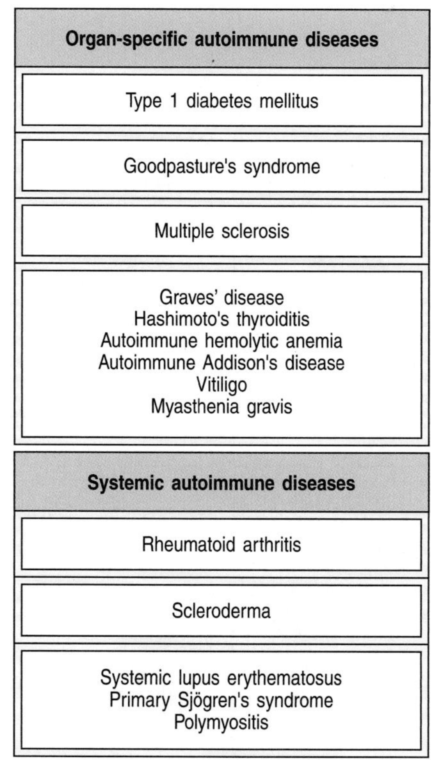

Fig. 14.11 Some common autoimmune diseases classified according to their 'organ-specific' or 'systemic' nature. Diseases that tend to occur in clusters are grouped in single boxes. Clustering is defined as more than one disease affecting a single patient or different members of a family. Not all autoimmune diseases can be classified according to this scheme. For example, autoimmune hemolytic anemia can occur in isolation or in association with systemic lupus erythematosus.

The autoantigens recognized in these two categories of disease are themselves organ-specific and systemic, respectively. Thus, Graves' disease is characterized by the production of antibodies against the thyroid-stimulating hormone (TSH) receptor, which is specific to the thyroid gland, Hashimoto's thyroiditis by antibodies against thyroid peroxidase, and type 1 diabetes by anti-insulin antibodies. By contrast, SLE is characterized by the presence of antibodies against antigens that are ubiquitous and abundant in every cell of the body, such as chromatin and the proteins of the pre-mRNA splicing machinery—the spliceosome complex.

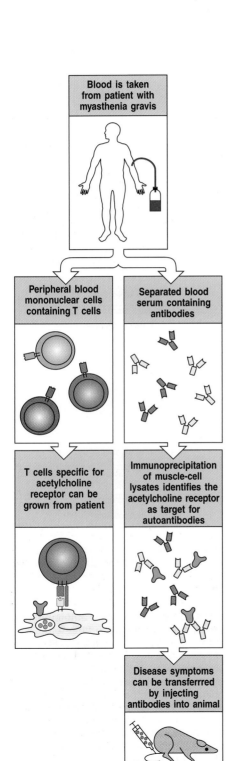

Systemic Lupus Erythematosus

It is likely that the organ-specific and systemic autoimmune diseases have somewhat different etiologies, which provides a biological basis for their division into two broad categories. Evidence for the validity of this classification also comes from observations that different autoimmune diseases cluster within individuals and within families. The organ-specific autoimmune diseases frequently occur together in many combinations; for example, autoimmune thyroid disease and the autoimmune depigmenting disease vitiligo are often found in the same person. Similarly, SLE and Sjögren's syndrome can coexist within a single individual or among different members of a family.

These clusters of autoimmune diseases provide the most useful classification into different subtypes, each of which may turn out to have a distinct mechanism. The classification of autoimmune diseases given in Fig. 14.11 is based on such clustering. A strict separation of diseases into organ-specific and systemic categories does, however, break down to some extent, as not all autoimmune diseases can be usefully classified in this manner. For example, autoimmune hemolytic anemia, in which red blood cells are destroyed, sometimes occurs as a solitary entity and could be classified as an organ-specific disease. In other circumstances it can occur in conjunction with SLE as part of a systemic autoimmune disease.

14-10 Multiple aspects of the immune system are typically recruited in autoimmune disease.

Immunologists have long been concerned with the issue of which parts of the immune system are important in different autoimmune syndromes, as this can be useful in understanding how a disease is caused and how it is maintained, with the ultimate goal of finding effective therapies. In myasthenia gravis, for example, autoantibodies seem to have the major role in causing the disease symptoms. Antibodies produced against the acetylcholine receptor cause blocking of receptor function at the neuromuscular junction, resulting in a syndrome of muscle weakness. In other autoimmune conditions, antibodies in the form of immune complexes are deposited in tissues and cause tissue damage as a result of complement activation and ligation of Fc receptors on inflammatory cells.

Relatively common autoimmune diseases in which effector T cells seem to be the main destructive agents include type 1 diabetes and multiple sclerosis. In these diseases, T cells specific for self-peptide:self-MHC complexes cause local inflammation by activating macrophages or by damaging tissue cells directly, and affected tissues are heavily infiltrated by T lymphocytes and activated macrophages.

When disease can be transferred from a diseased individual to a healthy one by transferring autoantibodies and/or self-reactive T cells, this both confirms that the disease is autoimmune in nature and also proves the involvement of

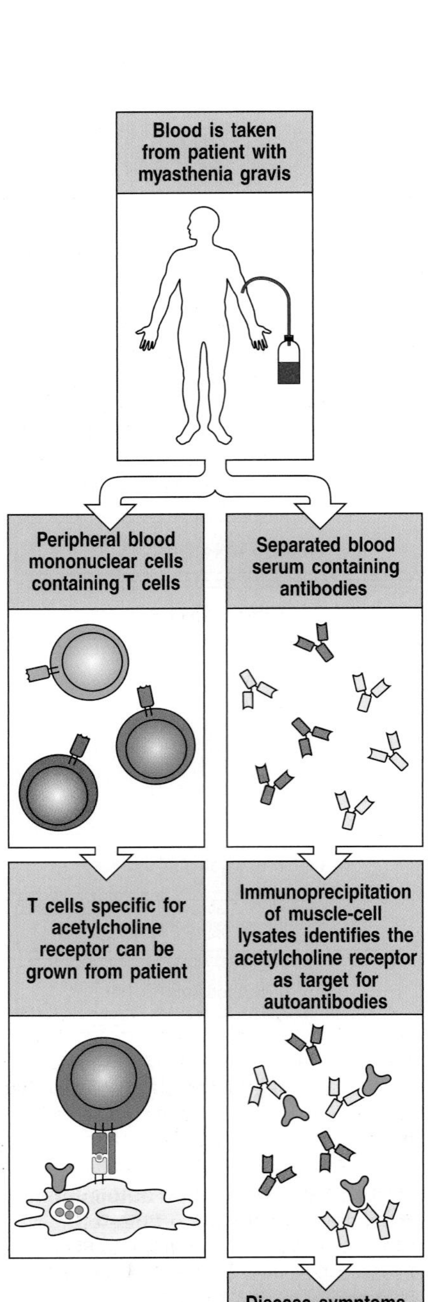
Blood is taken from patient with myasthenia gravis
Peripheral blood mononuclear cells containing T cells
Separated blood serum containing antibodies
T cells specific for acetylcholine receptor can be grown from patient
Immunoprecipitation of muscle-cell lysates identifies the acetylcholine receptor as target for autoantibodies
Disease symptoms can be transferrred by injecting antibodies into animal

Fig. 14.12 Identification of autoantibodies that can transfer disease in patients with myasthenia gravis. Autoantibodies from the serum of patients with myasthenia gravis immunoprecipitate the acetylcholine receptor from lysates of skeletal muscle cells (right-hand panels). Because they can bind to the murine as well as the human acetylcholine receptor, they can transfer disease when injected into mice (bottom panel). This experiment demonstrates that the antibodies are pathogenic. However, to be able to produce antibodies, the same patients should also have CD4 T cells that respond to a peptide derived from the acetylcholine receptor. To detect them, T cells from myasthenia gravis patients are isolated and grown in the presence of the acetylcholine receptor plus antigen-presenting cells of the correct MHC type (left-hand panels). T cells specific for epitopes of the acetylcholine receptor are stimulated to proliferate and can thus be detected.

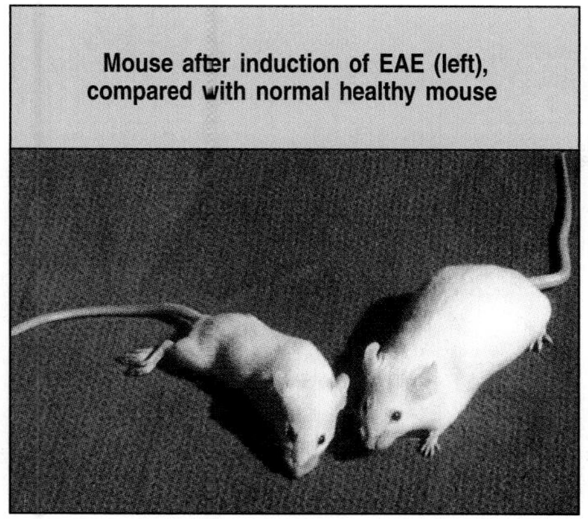

Mouse after induction of EAE (left), compared with normal healthy mouse

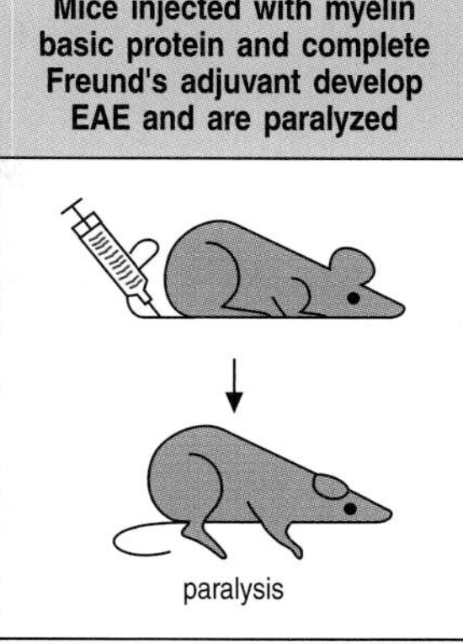

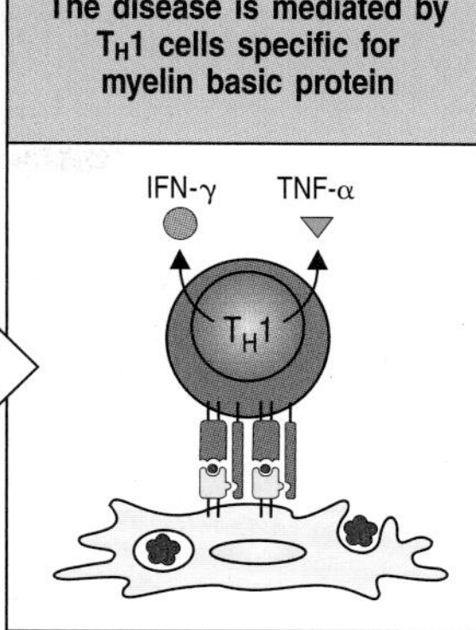

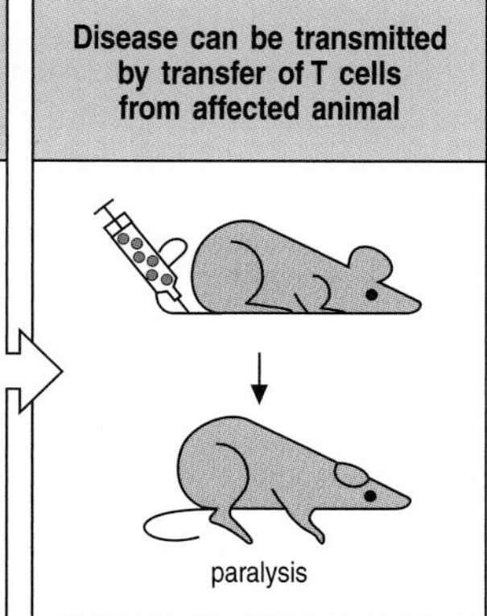

Disease can be transmitted by transfer of T cells from affected animal
paralysis

Fig. 14.13 T cells specific for myelin basic protein mediate inflammation of the brain in experimental autoimmune encephalomyelitis (EAE). This disease is produced in experimental animals by injecting them with isolated spinal cord homogenized in complete Freund's adjuvant. EAE is due to an inflammatory reaction in the brain that causes a progressive paralysis affecting first the tail and hind limbs (as shown in the mouse on the left of the photograph, compared with a healthy mouse on the right) before progressing to forelimb paralysis and eventual death. One of the autoantigens identified in the spinal cord homogenate is myelin basic protein (MBP). Immunization with MBP alone in complete Freund's adjuvant can also cause these disease symptoms. Inflammation of the brain and paralysis are mediated by T_H1 and T_H17 cells specific for MBP. Cloned MBP-specific T_H1 cells can transfer symptoms of EAE to naive recipients provided that the recipients carry the correct MHC allele. In this system it has therefore proved possible to identify the peptide:MHC complex recognized by the T_H1 clones that transfer disease. Other purified components of the myelin sheath can also induce the symptoms of EAE, so there is more than one autoantigen in this disease.

the transferred material in the pathological process. In myasthenia gravis, serum from affected patients can transfer similar disease symptoms to animal recipients, thus proving the pathogenic role of the anti-acetylcholine autoantibodies (Fig. 14.12). Similarly, in the animal model disease EAE, T cells from affected animals can transfer disease to normal animals (Fig. 14.13).

Pregnancy is an experiment of nature that can demonstrate a role for antibodies in the causation of disease. IgG antibodies, but not T cells, can cross the placenta (see Section 9-15). For some autoimmune diseases (Fig. 14.14), transmission of autoantibodies across the placenta leads to disease in the fetus or the neonate (Fig. 14.15). This provides proof in humans that such autoantibodies cause some of the symptoms of autoimmunity. The symptoms of disease in the newborn infant typically disappear rapidly as the maternal antibody is catabolized, but in some cases the antibodies cause

Autoimmune diseases transferred across the placenta to the fetus and newborn infant		
Disease	**Autoantibody**	**Symptom**
Myasthenia gravis	Anti-acetylcholine receptor	Muscle weakness
Graves' disease	Anti-thyroid-stimulating-hormone (TSH) receptor	Hyperthyroidism
Thrombocytopenic purpura	Anti-platelet antibodies	Bruising and hemorrhage
Neonatal lupus rash and/or congenital heart block	Anti-Ro antibodies Anti-La antibodies	Photosensitive rash and/or bradycardia
Pemphigus vulgaris	Anti-desmoglein-3	Blistering rash

Fig. 14.14 Some autoimmune diseases that can be transferred across the placenta by pathogenic IgG autoantibodies. These diseases are caused mostly by autoantibodies against cell-surface or tissue-matrix molecules. This suggests that an important factor determining whether an autoantibody that crosses the placenta causes disease in the fetus or newborn baby is the accessibility of the antigen to the autoantibody. Autoimmune congenital heart block is caused by fibrosis of the developing cardiac conducting tissue, which expresses abundant Ro antigen. Ro protein is a constituent of an intracellular small cytoplasmic ribonucleoprotein. It is not yet known whether it is expressed at the cell surface of cardiac conducting tissue to act as a target for autoimmune tissue injury. Nevertheless, autoantibody binding leads to tissue damage and results in slowing of the heart rate (bradycardia).

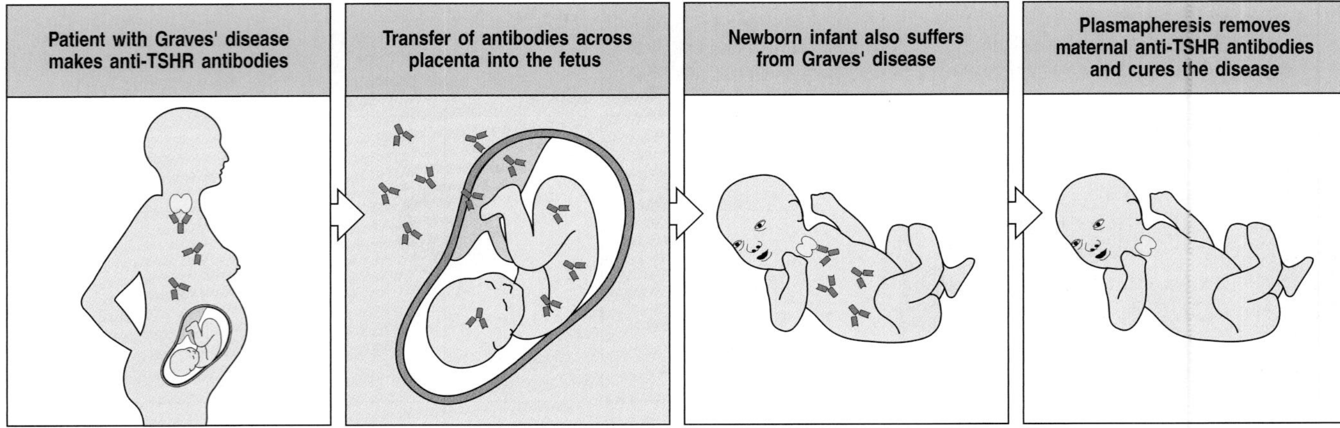

Patient with Graves' disease makes anti-TSHR antibodies
Transfer of antibodies across placenta into the fetus
Newborn infant also suffers from Graves' disease
Plasmapheresis removes maternal anti-TSHR antibodies and cures the disease

Fig. 14.15 Antibody-mediated autoimmune diseases can appear in the infants of affected mothers as a consequence of transplacental antibody transfer. In pregnant women, IgG antibodies cross the placenta and accumulate in the fetus before birth (see Fig. 9.22). Babies born to mothers with IgG-mediated autoimmune disease therefore frequently show symptoms similar to those of the mother in the first few weeks of life. Fortunately, there is little lasting damage because the symptoms disappear along with the maternal antibody. In Graves' disease, the symptoms are caused by antibodies against the thyroid-stimulating hormone receptor (TSHR). Children of mothers making thyroid-stimulating antibody are born with hyperthyroidism, but this can be corrected by replacing the plasma with normal plasma (plasmapheresis), thus removing the maternal antibody.

chronic organ injury before they are removed, such as damage to the conducting tissue of the heart in babies of mothers with SLE or Sjögren's syndrome. Antibody clearance can be speeded up by exchange of the infant's blood or plasma (plasmapheresis), although this is of no clinical use after permanent injury has occurred, as in congenital heart block.

Although the diseases noted above are clear examples that a particular effector function, once established, can cause disease, the idea that most autoimmune diseases are caused solely by a single effector pathway of the immune system is an oversimplification. It is more useful to consider autoimmune responses, like immune responses to pathogens, as engaging the integrated immune system and therefore typically involving T cells, B cells, and dendritic cells. In the nonobese diabetic (NOD) mouse model of type 1 diabetes, for example, a disease that is usually considered to be T-cell mediated, B cells are required for disease initiation. In this case, the B cells are probably functioning as essential antigen-presenting cells for T cells, although the exact details are not clear. A selection of autoimmune diseases showing which parts of the immune response contribute to pathogenesis is given in Fig. 14.16.

Fig. 14.16 Autoimmune diseases involve all aspects of the immune response. Although some autoimmune diseases have traditionally been thought to be mediated by B cells or T cells, it is useful to consider that, typically, all aspects of the immune system have a role. For four important autoimmune diseases, the figure lists the roles of T cells, B cells, and antibody. In some diseases, such as systemic lupus erythematosus, T cells can have multiple roles such as helping B cells to make autoantibody and directly promoting tissue damage, whereas B cells can have two roles as well—presenting autoantigens to stimulate T cells and secreting pathogenic autoantibodies.

Autoimmune diseases involve all aspects of the immune response			
Disease	**T cells**	**B cells**	**Antibody**
Systemic lupus erythematosus	Pathogenic Help for antibody	Present antigen to T cells	Pathogenic
Type 1 diabetes	Pathogenic	Present antigen to T cells	Present, but role unclear
Myasthenia gravis	Help for antibody	Antibody secretion	Pathogenic
Multiple sclerosis	Pathogenic	Present antigen to T cells	Present, but role unclear

14-11 Chronic autoimmune disease develops through positive feedback from inflammation, inability to clear the self antigen, and a broadening of the autoimmune response.

When normal immune responses are engaged to destroy a pathogen, the typical outcome is the elimination of the foreign invader, after which the immune response ceases, leaving only an expanded cohort of memory lymphocytes (see Chapter 10). In autoimmunity, however, the self antigen cannot easily be eliminated, because it is in vast excess or is ubiquitous, as with the SLE autoantigen, chromatin. Thus, a very important mechanism for limiting the extent of an immune response cannot apply to autoimmune diseases. Instead, autoimmune diseases tend to evolve into a chronic state (Fig. 14.17). There is no cure for such diseases once they are well established—short of a bone marrow transplant (see Section 14-35) that replaces much of the immune system from new cohorts of precursor cells. Even this may not be successful in curing disease.

In general, autoimmune diseases are characterized by an early activation phase with the involvement of only a few autoantigens, followed by a chronic stage. The constant presence of autoantigen leads to chronic inflammation. This in turn leads to the release of more autoantigens as a result of tissue damage, and this breaks an important barrier to autoimmunity known as 'sequestration', by which many self antigens are normally kept apart from the immune system. It also leads to the attraction of nonspecific effector cells such as macrophages and neutrophils that respond to the release of cytokines and chemokines from injured tissues (see Fig. 14.17). The result is a continuing and evolving self-destructive process.

The transition to the chronic stage is usually accompanied by an extension of the autoimmune response to new epitopes and initiating autoantigen, and to new autoantigens. This phenomenon is known as **epitope spreading** and plays an important role in perpetuating and amplifying the disease. As we saw in Chapter 9, activated B lymphocytes can efficiently take up antigens by receptor-mediated endocytosis, process them and present the derived peptides to T cells. An activated autoreactive B cell can thus take up and process the autoantigen to which it is specific, revealing a variety of novel, previously hidden, epitopes called **cryptic epitopes**, that it can then present to T cells. Autoreactive T cells responding to these epitopes can then provide help to any B cells presenting this peptide, recruiting additional B-cell clones to the autoimmune reaction, and resulting in the production of a greater variety of autoantibodies. B cells bind and neutralize their cognate antigen recognized

Fig. 14.17 Autoantibody-mediated inflammation can lead to release of autoantigens from damaged tissues, which in turn promotes further activation of autoreactive B cells. Autoantigens, particularly intracellular ones that are targets in SLE, stimulate B cells only when released from dying cells (first panel). The result is the activation of autoreactive T and B cells and the eventual secretion of autoantibodies (second and third panels). These autoantibodies can mediate tissue damage through a variety of effector functions (see Chapter 9) and this results in the further death of cells (fourth panel). A positive feedback loop is established because these additional autoantigens recruit and activate additional autoreactive B cells (fifth panel). These in turn can start the cycle over again, as shown in the first panel.

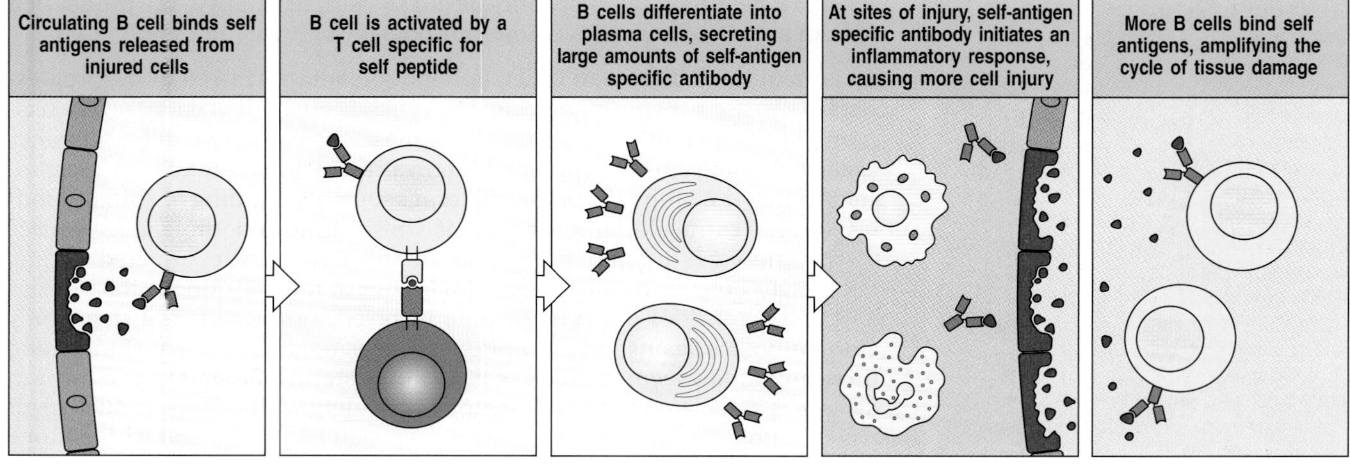
Circulating B cell binds self antigens released from injured cells
B cell is activated by a T cell specific for self peptide
B cells differentiate into plasma cells, secreting large amounts of self-antigen specific antibody
At sites of injury, self-antigen specific antibody initiates an inflammatory response, causing more cell injury
More B cells bind self antigens, amplifying the cycle of tissue damage

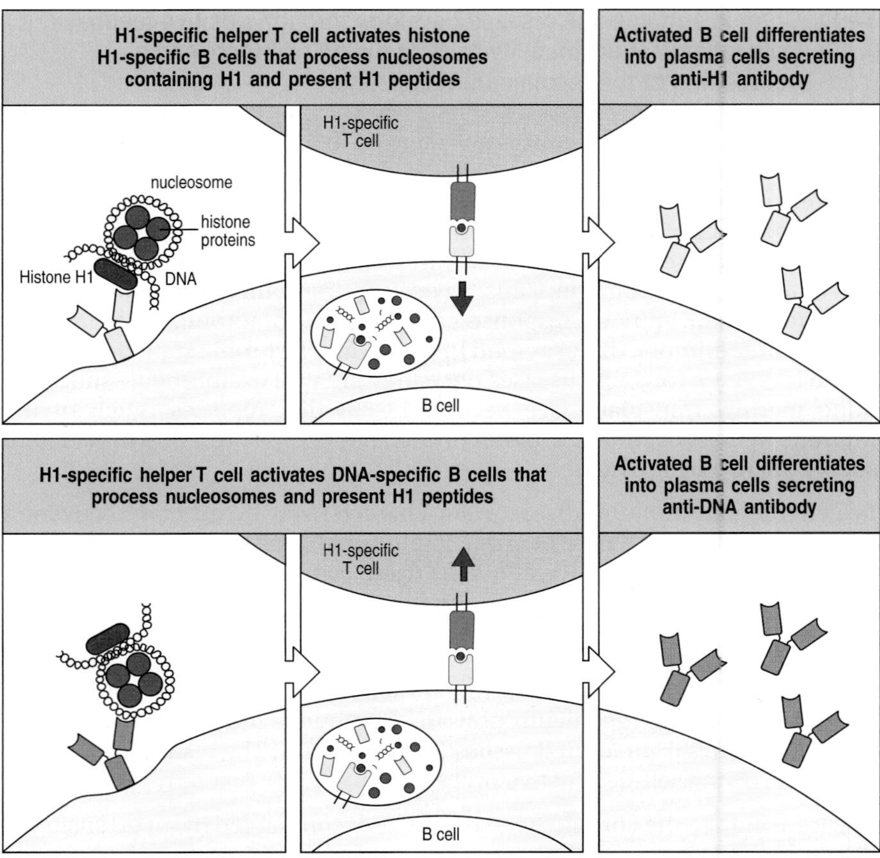
H1-specific helper T cell activates histone H1-specific B cells that process nucleosomes containing H1 and present H1 peptides
Activated B cell differentiates into plasma cells secreting anti-H1 antibody
H1-specific T cell
nucleosome
histone proteins
Histone H1
DNA
B cell
H1-specific helper T cell activates DNA-specific B cells that process nucleosomes and present H1 peptides
Activated B cell differentiates into plasma cells secreting anti-DNA antibody
H1-specific T cell
B cell

Fig. 14.18 Epitope spreading occurs when B cells specific for various components of a complex antigen are stimulated by an autoreactive helper T cell of a single specificity. In SLE, patients often produce autoantibodies against both the DNA and histone protein components of a nucleosome (a subunit of chromatin), or of some other complex antigen. The most likely explanation is that different autoreactive B cells have been activated by a single clone of autoreactive T cells specific for a peptide of one of the proteins in the complex. A B cell binding to any component of the complex through its surface immunoglobulin can internalize the whole complex, degrade it, and return peptides derived from the histone proteins to the cell surface bound to MHC class II molecules, where they stimulate helper T cells. These, in turn, activate the B cells. Thus, a T cell specific for the H1 histone protein of the nucleosome can activate both a B cell specific for the histone protein (top panels) and a B cell specific for double-stranded DNA (bottom panels). T cells of additional epitope specificities can also become recruited into the response in this way by antigen-presenting B cells bearing a variety of nucleosome-derived peptide:MHC complexes on their surface.

by their antibody receptor. But in doing so they may also internalize other molecules associated with the cognate antigen. The B cells may then act as antigen-presenting cells for peptides derived from proteins different from the original autoantigen that might have initiated the autoimmune reaction.

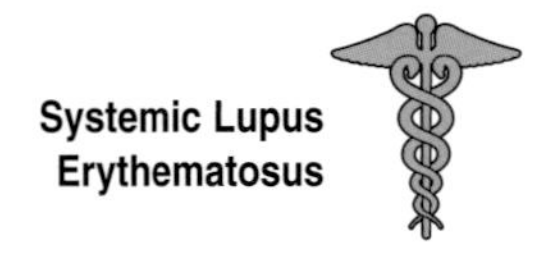
Systemic Lupus Erythematosus

The autoantibody response in SLE initiates this mechanism of epitopes and antigen spreading. In this disease, autoantibodies against both the protein and DNA components of chromatin are found. Fig. 14.18 shows how autoreactive B cells specific for DNA can recruit autoreactive T cells specific for histone proteins, another component of chromatin, into the autoimmune response. In turn, these T cells provide help not only to the original DNA-specific B cells but also to histone-specific B cells, resulting in the production of both anti-DNA and anti-histone antibodies.

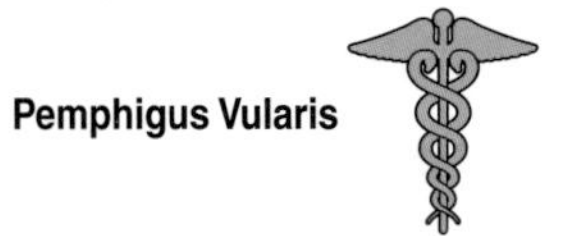
Pemphigus Vularis

An autoimmune disease in which epitope spreading is linked to the progression of disease is **pemphigus vulgaris**, which is characterized by severe blistering of the skin and mucosal membranes. It is caused by autoantibodies against desmogleins, a type of cadherin present in cell junctions (desmosomes) that hold the cells of the epidermis together. Binding of autoantibodies to the extracellular domains of these adhesion molecules causes dissociation of the junctions and dissolution of the affected tissue. Pemphigus vulgaris usually starts with lesions in the oral and genital mucosa and only later does the skin become involved. In the mucosal stage, only autoantibodies against certain epitopes on desmoglein Dsg-3 are found, and these antibodies seem unable to cause skin blistering. Progression to the skin disease is associated both with epitope spreading within Dsg-3, which gives rise to autoantibodies that can cause deep skin blistering, and with epitope spreading to another desmoglein, Dsg-1, which is more abundant in the epidermis. Dsg-1 is also the autoantigen in a less severe variant of the disease, pemphigus foliaceus. In that disease, the autoantibodies first produced against Dsg-1 cause no damage; disease appears only after autoantibodies are made against epitopes on parts of the protein involved in the adhesion of epidermal cells.

14-12 Both antibody and effector T cells can cause tissue damage in autoimmune disease.

The manifestations of autoimmune disease are caused by the effector mechanisms of the immune system being directed at the body's own tissues. As discussed previously, the response is usually amplified and maintained by the constant supply of new autoantigen. An important exception to this general rule is type 1 diabetes, in which the autoimmune response destroys the target organ completely. This leads to a failure to produce insulin—one of the major autoantigens in this disease—and it is the lack of insulin that is responsible for the disease symptoms.

The mechanisms of tissue injury in autoimmunity can be classified according to the scheme adopted for hypersensitivity reactions (Fig. 14.19; see also Fig. 13.1). It should be emphasized, however, that both B and T cells are involved in all autoimmune diseases, even in cases where a particular type of response predominates in causing tissue damage. The antigen, or group of antigens, against which the autoimmune response is directed, and the mechanism by which the antigen-bearing tissue is damaged, together determine the pathology and clinical expression of the disease.

Autoimmune diseases differ from hypersensitivity responses in that type I IgE-mediated responses do not seem to play a major role. By contrast, autoimmunity that damages tissues by mechanisms analogous to type II hypersensitivity reactions is quite common. In this form of autoimmunity, IgG or IgM responses to autoantigens located on cell surfaces or extracellular matrix cause the injury. In other cases of autoimmunity, tissue damage can be due to type III responses, which involve immune complexes containing autoantibodies against soluble autoantigens; these autoimmune diseases are systemic and are characterized by autoimmune vasculitis—inflammation of blood vessels. In SLE, autoantibodies cause damage by both type II and type III mechanisms. Finally, in several organ-specific autoimmune diseases, T_H1 cells and/or cytotoxic T cells are directly involved in causing tissue damage.

In most autoimmune diseases, several mechanisms of immunopathogenesis operate. Notably, helper T cells are almost always required for the production of pathogenic autoantibodies. Reciprocally, B cells often have an important role in the maximal activation of T cells that mediate tissue damage or help autoantibody production (see Section 14-10). In type 1 diabetes and rheumatoid arthritis, for example, which are classed as T cell-mediated diseases, both T-cell and antibody-mediated pathways cause tissue injury. SLE is an example of an autoimmune disease that was previously thought to be mediated solely by antibodies and immune complexes but is now known to have a component of T cell-mediated pathogenesis as well. We first examine how autoantibodies cause tissue damage, before considering self-reactive T-cell responses and their role in autoimmune disease.

14-13 Autoantibodies against blood cells promote their destruction.

IgG or IgM responses to antigens located on the surface of blood cells lead to the rapid destruction of these cells. An example of this is **autoimmune**

Fig. 14.19 Mechanisms of tissue damage in autoimmune diseases. Autoimmune diseases can be grouped in the same way as hypersensitivity reactions, according to the predominant type of immune response and the mechanism by which it damages tissues. The immunopathological mechanisms are as illustrated for the hypersensitivity reactions in Fig.13.1, with the exception of the type I IgE-mediated responses, which are not a known cause of autoimmune disease. Some additional autoimmune diseases in which the antigen is a cell-surface receptor, and in which the pathology is due to altered signaling, are listed later, in Fig. 14.23. In many autoimmune diseases, several immunopathogenic mechanisms operate in parallel. This is illustrated here for rheumatoid arthritis, which appears in more than one category of immunopathological mechanism.

Some common autoimmune diseases classified by immunopathogenic mechanism

Syndrome	Autoantigen	Consequence
Type II antibody against cell-surface or matrix antigens		
Autoimmune hemolytic anemia	Rh blood group antigens, I antigen	Destruction of red blood cells by complement and FcR^+ phagocytes, anemia
Autoimmune thrombocytopenic purpura	Platelet integrin GpIIb:IIIa	Abnormal bleeding
Goodpasture's syndrome	Noncollagenous domain of basement membrane collagen type IV	Glomerulonephritis, pulmonary hemorrhage
Pemphigus vulgaris	Epidermal cadherin	Blistering of skin
Acute rheumatic fever	Streptococcal cell-wall antigens. Antibodies cross-react with cardiac muscle	Arthritis, myocarditis, late scarring of heart valves
Type III immune-complex disease		
Mixed essential cryoglobulinemia	Rheumatoid factor IgG complexes (with or without hepatitis C antigens)	Systemic vasculitis
Rheumatoid arthritis	Rheumatoid factor IgG complexes	Arthritis
Type IV T-cell-mediated disease		
Type 1 diabetes	Pancreatic β-cell antigen	β-cell destruction
Rheumatoid arthritis	Unknown synovial joint antigen	Joint inflammation and destruction
Multiple sclerosis	Myelin basic protein, proteolipid protein, myelin oligodendrocyte glycoprotein	Brain invasion by CD4 T cells, muscle weakness, and other neurological symptoms

See various cases

Autoimmune Hemolytic Anemia

hemolytic anemia, in which antibodies against self antigens on red blood cells trigger destruction of the cells, leading to anemia. This can occur in two ways (Fig. 14.20). Red cells with bound IgG or IgM antibody are rapidly cleared from the circulation by interaction with Fc or complement receptors, respectively, on cells of the fixed mononuclear phagocytic system; this occurs particularly in the spleen. Alternatively, the autoantibody-sensitized red cells are lysed by formation of the membrane-attack complex of complement. In **autoimmune thrombocytopenic purpura**, autoantibodies against the GpIIb:IIIa fibrinogen receptor or other platelet-specific surface antigens can cause thrombocytopenia (a depletion of platelets), which can in turn cause hemorrhage.

Lysis of nucleated cells by complement is less common because these cells are better defended by complement regulatory proteins, which protect cells against immune attack by interfering with the activation of complement components and their assembly into a membrane-attack complex (see

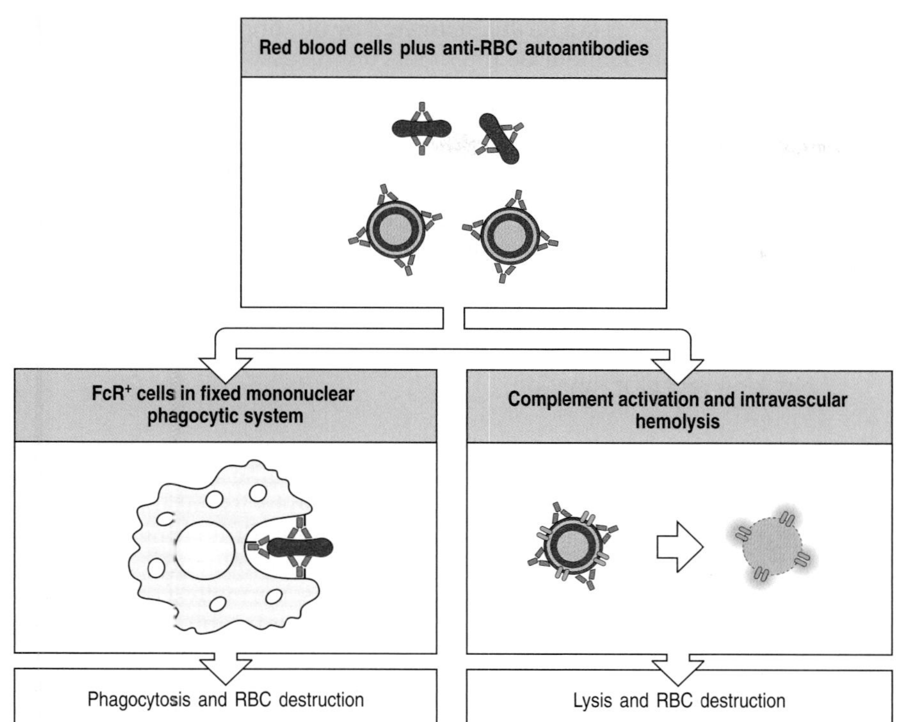

Fig. 14.20 Antibodies specific for cell-surface antigens can destroy cells. In autoimmune hemolytic anemias, red blood cells (RBC) coated with IgG autoantibodies against a cell-surface antigen are rapidly cleared from the circulation by uptake by Fc receptor-bearing macrophages in the fixed mononuclear phagocytic system (left panel). Red cells coated with IgM autoantibodies fix C3 and are cleared by CR1- and CR3-bearing macrophages in the fixed mononuclear phagocytic system (not shown). Uptake and clearance by these mechanisms occurs mainly in the spleen. The binding of certain rare autoantibodies that fix complement extremely efficiently causes the formation of the membrane-attack complex on the red cells, leading to intravascular hemolysis (right panel).

Section 2-21). Nevertheless, nucleated cells targeted by autoantibodies are still destroyed by cells of the mononuclear phagocytic system. Autoantibodies against neutrophils, for example, cause neutropenia, which increases susceptibility to infection with pyogenic bacteria. In all these cases, accelerated clearance of autoantibody-sensitized cells is the cause of their depletion in the blood. One therapeutic approach to this type of autoimmunity is removal of the spleen, the organ in which the main clearance of red cells, platelets, and leukocytes occurs. Another is the administration of large quantities of nonspecific IgG (termed IVIG, for intravenous immunoglobulin), which among other mechanisms inhibits the Fc receptor-mediated uptake of antibody-coated cells.

14-14 The fixation of sublytic doses of complement to cells in tissues stimulates a powerful inflammatory response.

The binding of IgG and IgM antibodies against cells in tissues (as in the case of blood cells) causes inflammatory injury by a variety of mechanisms. As in blood cells, one of these is the fixation of complement. Although nucleated cells are relatively resistant to lysis by complement, the assembly of sublytic amounts of the membrane-attack complex on their surface provides a powerful activating stimulus. Depending on the type of cell, the interaction of sublytic doses of the membrane-attack complex with the cell membrane can cause cytokine release, the generation of a respiratory burst, or the mobilization of membrane phospholipids to generate arachidonic acid—the precursor of prostaglandins and leukotrienes (lipid mediators of inflammation).

Most cells in tissues are fixed in place, and cells of the inflammatory system are attracted to them by chemoattractant molecules. One such molecule is the complement fragment C5a, which is released as a result of complement activation triggered by autoantibody binding. Other chemoattractants, such as leukotriene B4, can be released by the autoantibody-targeted cells.

Inflammatory leukocytes are further activated by binding to autoantibody Fc regions and fixed complement C3 fragments on the tissue cells. Tissue injury can then result from the products of the activated leukocytes and by antibody-dependent cellular cytotoxicity mediated by NK cells (see Section 9-23).

A probable example of this type of autoimmunity is Hashimoto's thyroiditis, in which autoantibodies against tissue-specific antigens such as thyroid peroxidase and thyroglobulin are found at extremely high levels for prolonged periods. Direct T cell-mediated cytotoxicity, which we discuss later, is probably also important in this disease.

14-15 Autoantibodies against receptors cause disease by stimulating or blocking receptor function.

A special class of type II hypersensitivity reaction occurs when the autoantibody binds to a cell-surface receptor. Antibody binding to a receptor can either stimulate the receptor or block its stimulation by its natural ligand. In Graves' disease, autoantibody against the thyroid-stimulating hormone receptor on thyroid cells stimulates the excessive production of thyroid hormone. The production of thyroid hormone is normally controlled by feedback regulation; high levels of thyroid hormone inhibit the release of thyroid-stimulating hormone (TSH) by the pituitary. In Graves' disease, feedback inhibition fails because the autoantibody continues to stimulate the TSH receptor in the absence of TSH, and the patient becomes hyperthyroid (Fig. 14.21).

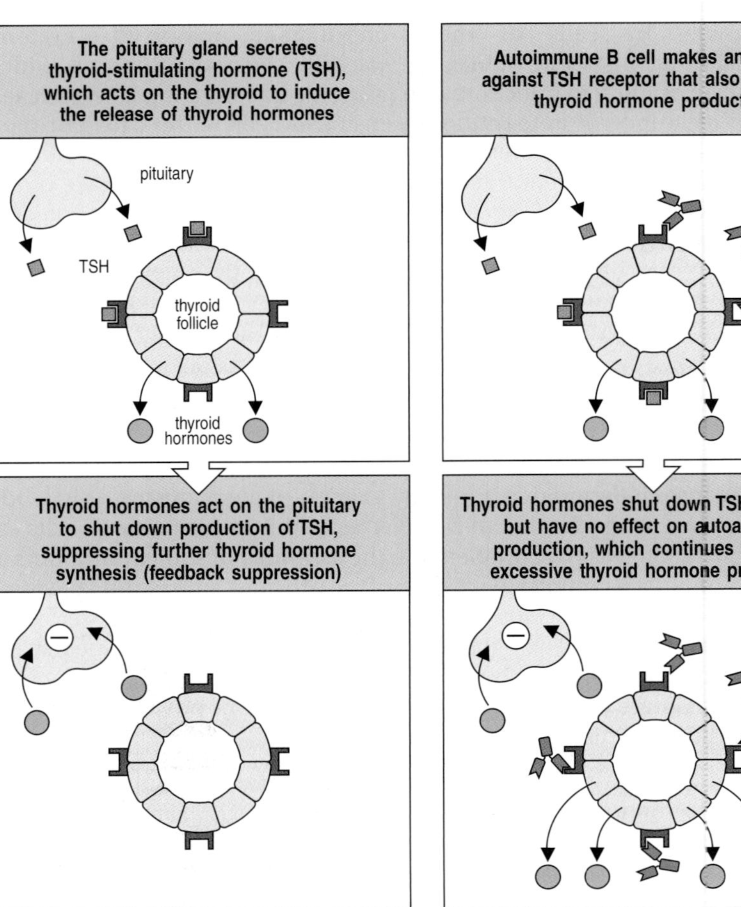

The pituitary gland secretes thyroid-stimulating hormone (TSH), which acts on the thyroid to induce the release of thyroid hormones
pituitary
TSH
thyroid follicle
thyroid hormones
Autoimmune B cell makes antibodies against TSH receptor that also stimulate thyroid hormone production
Thyroid hormones act on the pituitary to shut down production of TSH, suppressing further thyroid hormone synthesis (feedback suppression)
Thyroid hormones shut down TSH production but have no effect on autoantibody production, which continues to cause excessive thyroid hormone production

Fig. 14.21 Feedback regulation of thyroid hormone production is disrupted in Graves' disease. Graves' disease is caused by autoantibodies specific for the receptor for thyroid-stimulating hormone (TSH). Normally, thyroid hormones are produced in response to TSH and limit their own production by inhibiting the production of TSH by the pituitary (left panels). In Graves' disease, the autoantibodies are agonists for the TSH receptor and therefore stimulate the production of thyroid hormones (right panels). The thyroid hormones inhibit TSH production in the normal way but do not affect production of the autoantibody; the excessive thyroid hormone production induced in this way causes hyperthyroidism.

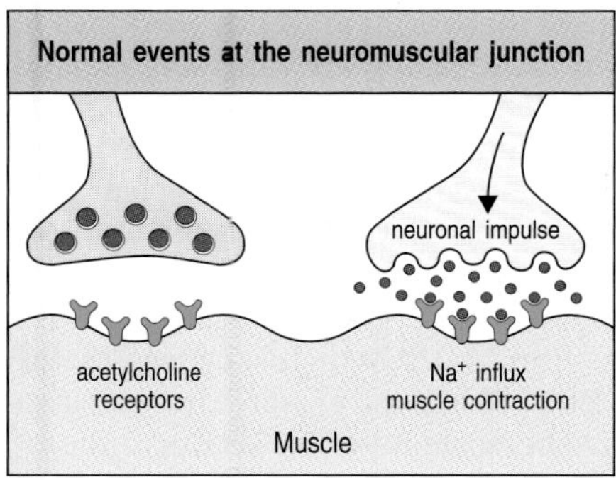

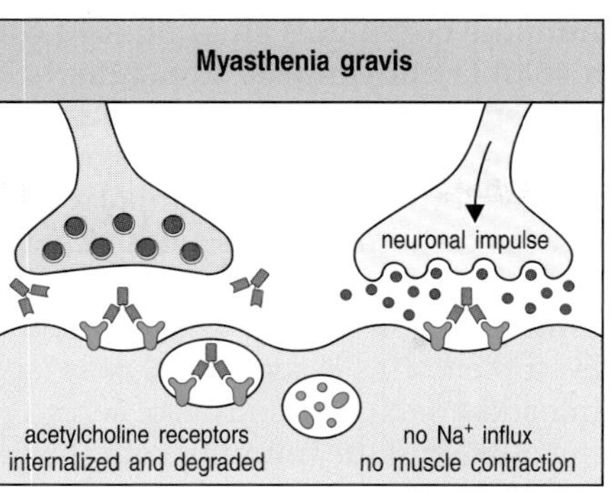

Fig. 14.22 Autoantibodies inhibit receptor function in myasthenia gravis. In normal circumstances, acetylcholine released from stimulated motor neurons at the neuromuscular junction binds to acetylcholine receptors on skeletal muscle cells, triggering muscle contraction (left panel). Myasthenia gravis is caused by autoantibodies against the α subunit of the receptor for acetylcholine. These autoantibodies bind to the receptor without activating it and also cause receptor internalization and degradation (right panel). As the number of receptors on the muscle is decreased, the muscle becomes less responsive to acetylcholine.

In myasthenia gravis, autoantibodies against the α chain of the nicotinic acetylcholine receptor, which is present on skeletal muscle cells at neuromuscular junctions, can block neuromuscular transmission. The antibodies are believed to drive the internalization and intracellular degradation of acetylcholine receptors (Fig. 14.22). Patients with myasthenia gravis develop potentially fatal progressive weakness as a result of their autoimmune disease. Diseases caused by autoantibodies that act as agonists or antagonists for cell-surface receptors are listed in Fig. 14.23.

14-16 Autoantibodies against extracellular antigens cause inflammatory injury by mechanisms akin to type II and type III hypersensitivity reactions.

Antibody responses to extracellular matrix molecules are infrequent, but they can be very damaging when they occur. In **Goodpasture's syndrome**, an example of a type II hypersensitivity reaction (see Fig. 13.1), antibodies are formed against the α_3 chain of basement membrane collagen (type IV collagen). These antibodies bind to the basement membranes of renal glomeruli (Fig. 14.24a) and, in some cases, to the basement membranes of pulmonary alveoli, causing a rapidly fatal disease if untreated. The autoantibodies bound to basement membrane ligate Fcγ receptors, leading to the activation of monocytes, neutrophils, and tissue basophils and mast cells. These release chemokines that attract a further influx of neutrophils into the glomeruli, causing severe tissue injury (Fig. 14.24b). The autoantibodies also cause a local activation of complement, which may amplify the tissue injury.

Diseases mediated by autoantibodies against cell-surface receptors

Syndrome	Antigen	Consequence
Graves' disease	Thyroid-stimulating hormone receptor	Hyperthyroidism
Myasthenia gravis	Acetylcholine receptor	Progressive weakness
Insulin-resistant diabetes (type 2 diabetes)	Insulin receptor (antagonist)	Hyperglycemia, ketoacidosis
Hypoglycemia	Insulin receptor (agonist)	Hypoglycemia
Chronic urticaria	Receptor-bound IgE or IgE receptor (agonist)	Persistent itchy rash

Fig. 14.23 Autoimmune diseases caused by autoantibodies against cell-surface receptors. These antibodies produce different effects depending on whether they are agonists (which stimulate the receptor) or antagonists (which inhibit it). Note that different autoantibodies against the insulin receptor can either stimulate or inhibit signaling.

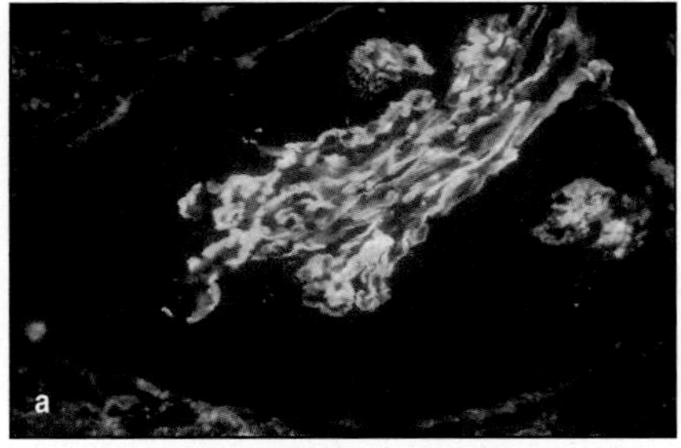

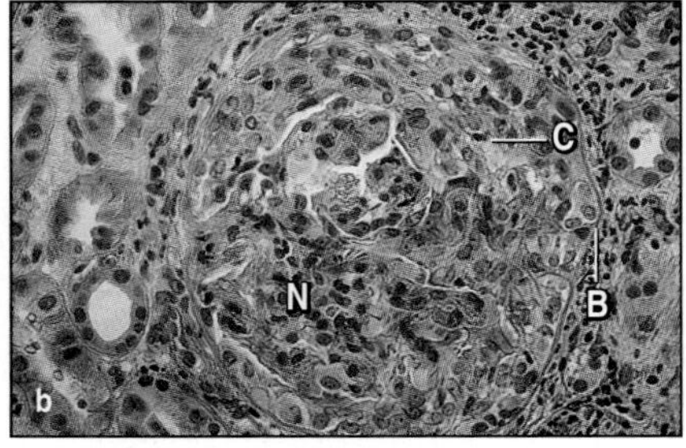

Fig. 14.24 Autoantibodies reacting with glomerular basement membrane cause the inflammatory glomerular disease known as Goodpasture's syndrome. The panels show sections of renal glomeruli in serial biopsies taken from patients with Goodpasture's syndrome. Panel a, glomerulus stained for IgG deposition by immunofluorescence. Anti-glomerular basement membrane antibody (stained green) is deposited in a linear fashion along the glomerular basement membrane. The autoantibody causes local activation of cells bearing Fc receptors, complement activation, and influx of neutrophils. Panel b, hematoxylin and eosin staining of a section through a renal glomerulus shows that the glomerulus is compressed by the formation of a crescent (C) of proliferating mononuclear cells within the Bowman's capsule (B) and there is an influx of neutrophils (N) into the glomerular tuft. Photographs courtesy of M. Thompson and D. Evans.

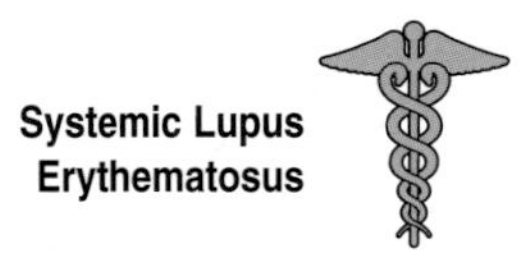

Immune complexes are produced whenever there is an antibody response to a soluble antigen (see Appendix I, Section A-8). They are normally cleared efficiently by red blood cells bearing complement receptors and by phagocytes of the mononuclear phagocytic system that have both complement and Fc receptors, and such complexes cause little tissue damage. This clearance system can, however, fail in three circumstances. The first follows the injection of large amounts of antigen, leading to the formation of large amounts of immune complexes that overwhelm the normal clearance mechanisms. An example of this is serum sickness (see Section 13-18), which is caused by the injection of large amounts of serum proteins. This is a transient disease, lasting only until the immune complexes have been cleared. The second circumstance is seen in chronic infections such as bacterial endocarditis, in which the immune response to bacteria lodged on a cardiac valve is incapable of clearing the infection. The persistent release of bacterial antigens from the valve infection in the presence of a strong antibacterial antibody response causes widespread immune-complex injury to small blood vessels in organs such as the kidney and the skin.

Third, part of the pathogenesis of SLE can also be attributed to the failure to clear immune complexes. In SLE there is chronic IgG antibody production directed at ubiquitous self antigens present in all nucleated cells, leading to a wide range of autoantibodies against common cellular constituents. The main antigens are three types of intracellular nucleoprotein particles—the nucleosome subunits of chromatin, the spliceosome, and a small cytoplasmic ribonucleoprotein complex containing two proteins known as Ro and La (named after the first two letters of the surnames of the two patients in whom autoantibodies against these proteins were discovered). For these autoantigens to participate in immune-complex formation, they must become extracellular. The autoantigens of SLE are exposed on dead and dying cells and are released from injured tissues. In SLE, large quantities of antigen are available, so large amounts of small immune complexes are produced continuously and are deposited in the walls of small blood vessels in the renal glomerulus, in glomerular basement membrane (Fig. 14.25), in joints, and in other organs. This leads to the activation of phagocytic cells through their Fc receptors. The consequent tissue damage releases more nucleoprotein complexes, which in turn form more immune complexes. During this process, autoreactive T cells also become activated, although much less is known about their specificity. The experimental animal models for SLE cannot be initiated without the help of T cells, and T cells can also be directly pathogenic, forming part of the cellular infiltrates in skin and the interstitial areas of the kidney. As we discuss in the next section, T cells contribute to autoimmune disease in two ways: by helping B cells to make antibodies, in an analogous manner to a normal T-dependent immune response, and by direct effector functions of T cells as they infiltrate and destroy target tissues such as skin, renal interstitium, and vessels. Eventually, the inflammation induced in these tissues can cause sufficient damage to kill the patient.

14-17 T cells specific for self antigens can cause direct tissue injury and sustain autoantibody responses.

It is much more difficult to demonstrate the existence of autoreactive T cells than it is to demonstrate the presence of autoantibodies. First, autoreactive human T cells cannot be used to transfer disease to experimental animals because T-cell recognition is MHC-restricted, and animals and humans have different MHC alleles. Second, it is difficult to identify the antigen recognized by a T cell; for example, autoantibodies can be used to stain self tissues to reveal the distribution of the autoantigen, whereas T cells cannot be used in the same way. Nevertheless, there is strong evidence for the involvement of

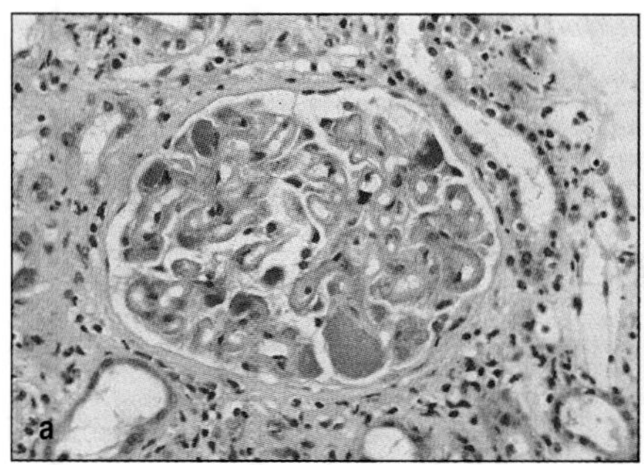

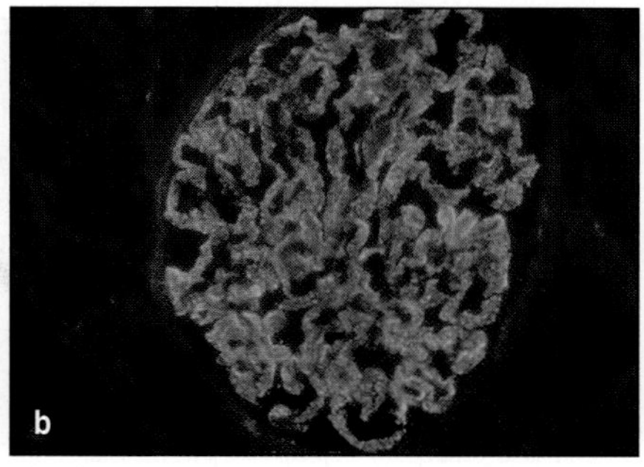

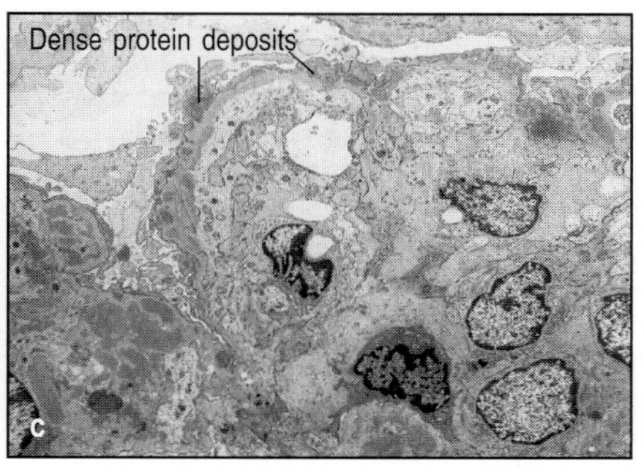

Fig. 14.25 Deposition of immune complexes in the renal glomerulus causes renal failure in systemic lupus erythematosus (SLE). Panel a, a section through a renal glomerulus from a patient with SLE, shows that the deposition of immune complexes has caused thickening of the glomerular basement membrane, seen as the clear 'canals' running through the glomerulus. Panel b, a similar section stained with fluorescent anti-immunoglobulin, reveals immunoglobulin deposits in the basement membrane. In panel c, the immune complexes are seen under the electron microscope as dense protein deposits between the glomerular basement membrane and the renal epithelial cells. Polymorphonuclear neutrophilic leukocytes are also present, attracted by the deposited immune complexes. Photographs courtesy of H.T. Cook and M. Kashgarian.

autoreactive T cells in several autoimmune diseases. In type 1 diabetes, for example, the insulin-producing β cells of the pancreatic islets of Langerhans are selectively destroyed by specific cytotoxic T cells. In rare cases in which diabetic patients were transplanted with half a pancreas from an identical twin donor, the β cells in the grafted tissue were rapidly and selectively destroyed by the recipient's T cells. Recurrence of disease can be prevented by the immunosuppressive drug cyclosporin A (see Chapter 15), which inhibits T-cell activation.

Autoantigens recognized by CD4 T cells can be identified by adding cells or tissues, containing autoantigens, to cultures of blood mononuclear cells and testing for recognition by CD4 cells derived from an autoimmune patient. If the autoantigen is present, it should be effectively presented, because phagocytes in the blood cultures can take up extracellular protein, degrade it in intracellular vesicles, and present the resulting peptides bound to MHC class II molecules. The identification of autoantigenic peptides is particularly difficult in autoimmune diseases in which CD8 T cells have a role, because autoantigens recognized by CD8 T cells are not effectively presented in such cultures. Peptides presented by MHC class I molecules must usually be made by the target cells themselves (see Chapter 5); intact cells of target tissue from the patient must therefore be used to study autoreactive CD8 T cells that cause tissue damage. Conversely, the pathogenesis of the disease can itself give clues to the identity of the antigen in some CD8 T cell-mediated diseases. For example, in type 1 diabetes, the insulin-producing β cells seem to be specifically targeted and destroyed by CD8 T cells (Fig. 14.26). This suggests that a protein unique to β cells is the source of the peptide recognized by the pathogenic CD8 T cells. Studies in the NOD mouse model of type 1 diabetes have shown that peptides from insulin itself are recognized by pathogenic CD8 cells, confirming insulin as one of the principal autoantigens in this diabetes model.

A Kidney Graft for Complications of Autoimmune Insulin-Dependent Diabetes Mellitus

Multiple sclerosis is an example of a T cell-mediated chronic neurological disease that is caused by a destructive immune response against several brain antigens, including myelin basic protein (MBP), proteolipid protein (PLP), and myelin oligodendrocyte glycoprotein (MOG). It takes its name from the hard (sclerotic) lesions, or plaques, that develop in the white matter of the central nervous system. These lesions show dissolution of the myelin that normally sheathes nerve cell axons, along with inflammatory infiltrates of lymphocytes and macrophages, particularly along the blood vessels. Patients with multiple sclerosis develop a variety of neurological symptoms, including muscle weakness, ataxia, blindness, and paralysis of the limbs. Lymphocytes and other blood cells do not normally cross the blood–brain barrier, but if the brain and its blood vessels become inflamed, the blood–brain barrier breaks down. When this happens, activated CD4 T cells autoreactive for brain antigens and expressing α_4:β_1 integrin can bind vascular cell adhesion molecules (VCAM) on the surface of the activated venule

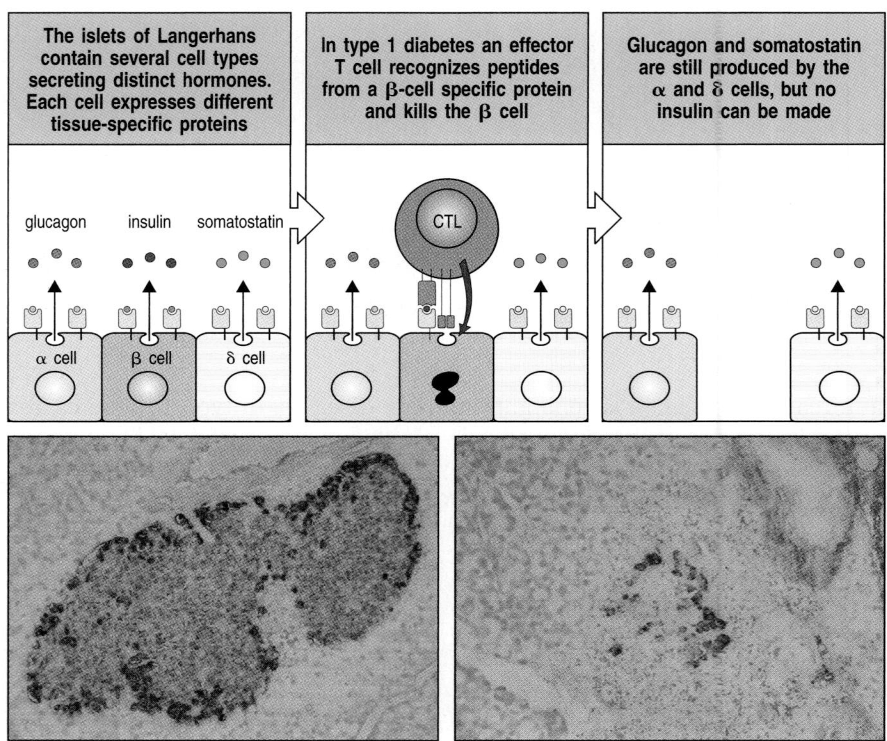

Fig. 14.26 Selective destruction of pancreatic β cells in type 1 diabetes indicates that the autoantigen is produced in β cells and recognized on their surface. In type 1 diabetes there is highly specific destruction of insulin-producing β cells in the pancreatic islets of Langerhans, sparing other islet cell types (α and δ). This is shown schematically in the upper panels. In the lower panels, islets from normal (left) and diabetic (right) mice are stained for insulin (brown), which shows the β cells, and for glucagon (black), which shows the α cells. Note the lymphocytes infiltrating the islet in the diabetic mouse (right) and the selective loss of the β cells (brown), whereas the α cells (black) are spared. The characteristic morphology of the islet is also disrupted with the loss of the β cells. Photographs courtesy of I. Visintin.

endothelium (see Section 10-6), enabling the T cells to migrate out of the blood vessel. There they reencounter their specific autoantigen presented by MHC class II molecules on microglial cells (Fig. 14.27). Microglia are phagocytic macrophage-like cells of the innate immune system resident in the central nervous system and, like macrophages, can act as antigen-presenting cells. Inflammation causes increased vascular permeability and the site becomes heavily infiltrated by T cells and activated macrophages, which produce T_H1 cytokines, such as IFN-γ, that exacerbate the inflammation, resulting in the further recruitment of T cells, B cells, macrophages, and dendritic cells to the site of the lesion. Autoreactive B cells produce autoantibodies against myelin antigens with help from T cells. Activated mast cells release

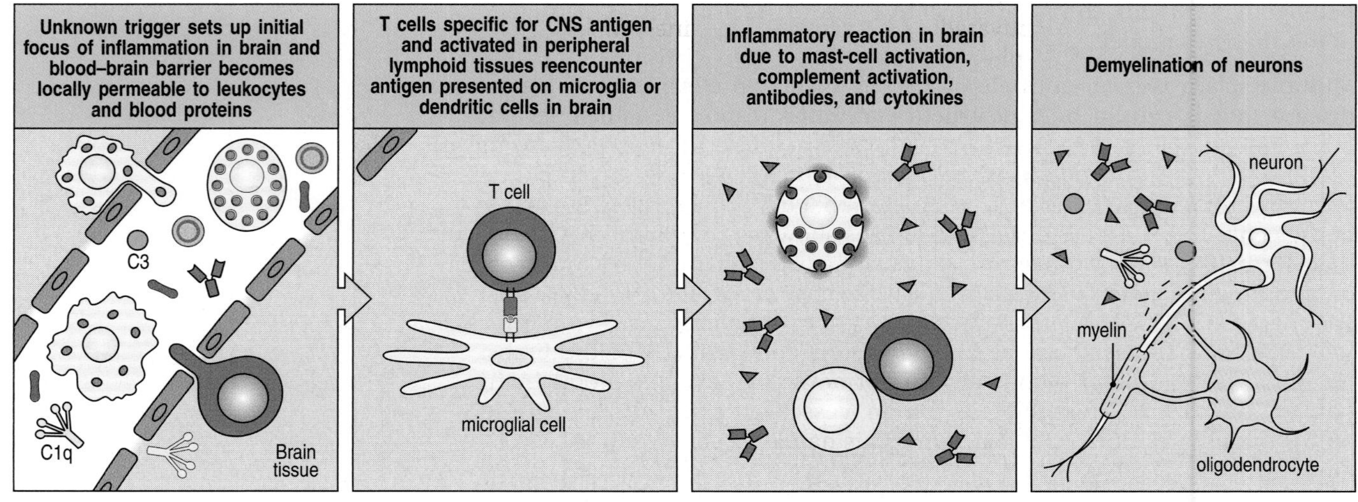

Fig. 14.27 The pathogenesis of multiple sclerosis. At sites of inflammation, activated T cells autoreactive for brain antigens can cross the blood–brain barrier and enter the brain, where they reencounter their antigens on microglial cells and secrete cytokines such as IFN-γ. The production of T-cell and macrophage cytokines exacerbates the inflammation and induces a further influx of blood cells (including macrophages, dendritic cells, and B cells) and blood proteins (such as complement) into the affected site. Mast cells also become activated. The individual roles of these components in demyelination and loss of neuronal function are still not well understood. CNS, central nervous system.

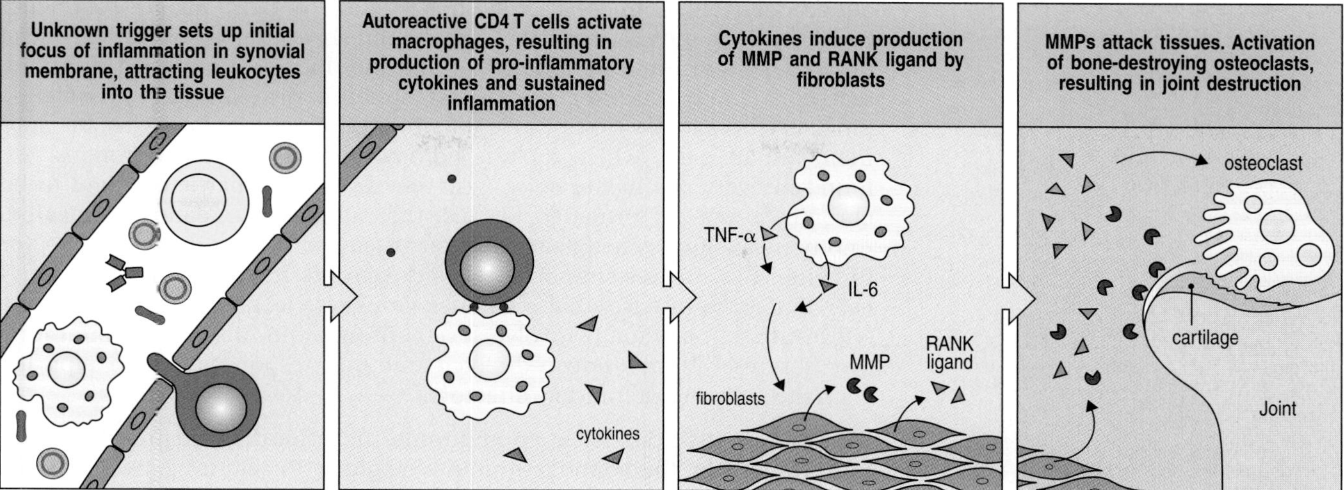
Unknown trigger sets up initial focus of inflammation in synovial membrane, attracting leukocytes into the tissue
Autoreactive CD4 T cells activate macrophages, resulting in production of pro-inflammatory cytokines and sustained inflammation
cytokines
Cytokines induce production of MMP and RANK ligand by fibroblasts
TNF-α
IL-6
RANK ligand
MMP
fibroblasts
MMPs attack tissues. Activation of bone-destroying osteoclasts, resulting in joint destruction
osteoclast
cartilage
Joint

Fig. 14.28 The pathogenesis of rheumatoid arthritis. Inflammation of the synovial membrane, initiated by some unknown trigger, attracts autoreactive lymphocytes and macrophages to the inflamed tissue. Autoreactive effector CD4 T cells activate macrophages, with the production of pro-inflammatory cytokines such as IL-1, IL-6, IL-17, and TNF-α. Fibroblasts activated by cytokines produce matrix metalloproteinases (MMPs), which contribute to tissue destruction. The TNF family cytokine RANK ligand, expressed by T cells and fibroblasts in the inflamed joint, is the primary activator of bone-destroying osteoclasts. Antibodies against several joint proteins are also produced (not shown), but their role in pathogenesis is uncertain.

histamine, contributing to the inflammation. These combined activities lead to demyelination and interference with neuronal function.

Rheumatoid arthritis (RA) is a chronic disease characterized by inflammation of the synovium (the thin lining of a joint). As the disease progresses, the inflamed synovium invades and damages the cartilage, followed by erosion of the bone (Fig 14.28). Patients with rheumatoid arthritis suffer chronic pain, loss of function, and disability. Rheumatoid arthritis was at first considered an autoimmune disease driven mainly by B cells producing anti-IgG autoantibodies called rheumatoid factor (see Section 14-4). However, the identification of rheumatoid factor in some healthy individuals, and its absence in some patients with rheumatoid arthritis, suggested that more complex mechanisms orchestrate this pathology. The discovery that rheumatoid arthritis has an association with particular class II HLA-DR genes of the major histocompatibility complex (MHC) suggested that T cells were involved in the pathogenesis of this disease. In rheumatoid arthritis, as in multiple sclerosis, autoreactive CD4 T cells become activated by dendritic cells and by inflammatory cytokines produced by macrophages. Once activated, the autoreactive T cells provide help to B cells to differentiate into plasma cells producing arthritogenic antibodies. Autoantigens such as type II collagen, proteoglycans, aggrecan, cartilage link protein and heat-shock proteins have been proposed as potential antigens because of their ability to induce arthritis in mice. Their pathogenic role in humans remains to be ascertained, however. The activated T cells produce cytokines, which in turn stimulate monocytes/macrophages, endothelial cells, and fibroblasts to produce more pro-inflammatory cytokines such as TNF-α, IL-1 and IFN-γ, or chemokines (CXCL8, CCL2), and finally matrix metalloproteinases, which are responsible for tissue destruction. However, it needs to be realized that in rheumatoid arthritis, as in many other autoimmune diseases, we do not yet know how disease starts. Mouse models of rheumatoid arthritis teach us that both T cells and B cells are needed to initiate the disease, because mice lacking $CD3^+$ T cells or B cells are resistant to developing it.

Rheumatoid Arthritis

Summary.

Autoimmune diseases can be broadly classified into those that affect a specific organ and those that affect tissues throughout the body. Organ-specific autoimmune diseases include diabetes, multiple sclerosis, myasthenia gravis, and Graves' disease. In each case the effector functions target autoantigens that are restricted to particular organs: insulin-producing β cells of the pancreas (diabetes), the myelin sheathing axons in the central nervous system (multiple sclerosis), and the thyroid-stimulating hormone receptor (Graves' disease). In contrast, systemic diseases such as systemic lupus erythematosus (SLE) cause inflammation in multiple tissues because their autoantigens, which include chromatin and ribonucleoproteins, are found in every cell of the body. Systemic diseases in particular tend to be chronically active if untreated, because their autoantigens cannot be cleared. Another way of classifying autoimmune diseases is according to the effector functions that are most important in pathogenesis. It is becoming clear, however, that many diseases that were once thought to be mediated solely by one effector function actually involve several of them. In this way, autoimmune diseases resemble pathogen-directed immune responses, which typically elicit the activities of multiple effectors.

For a disease to be defined as autoimmune, the tissue damage must be shown to be caused by the adaptive immune response to self antigens. The most convincing proof that the immune response is causal in autoimmunity is the transfer of disease by transferring the active component of the immune response to an appropriate recipient. Autoimmune diseases are mediated by autoreactive lymphocytes and/or their soluble products, pro-inflammatory cytokines, and autoantibodies responsible for inflammation and tissue injury. A few autoimmune diseases are caused by antibodies that bind to cell-surface receptors, causing either excess activity or inhibition of receptor function. In these diseases, transplacental passage of natural IgG autoantibodies can cause disease in the fetus and in the neonate. T cells can be involved directly in inflammation or cellular destruction, and they are also required in sustaining autoantibody responses. Similarly, B cells are important antigen-presenting cells for sustaining autoantigen-specific T-cell responses and for causing epitope spreading. In spite of our knowledge of the mechanisms of tissue damage and the therapeutic approaches that this information has engendered, the deeper, more important question is how the autoimmune response is induced.

The genetic and environmental basis of autoimmunity.

Given the complex and varied mechanisms that exist to prevent autoimmunity, it is not surprising that autoimmune diseases are the result of multiple factors, both genetic and environmental. We first discuss the genetic basis of autoimmunity, attempting to understand how genetic defects perturb the various tolerance mechanisms. Genetic defects alone are not, however, always sufficient to cause autoimmune disease. Environmental factors such as toxins, drugs and infections also play a part, although these factors are poorly understood. As we shall see, genetic and environmental factors together can overcome tolerance mechanisms and result in autoimmune disease.

14-18 Autoimmune diseases have a strong genetic component.

Although the causes of autoimmunity are still being worked out, it is clear that some individuals are genetically predisposed to autoimmunity. Perhaps the clearest demonstration of this is found in the several inbred mouse strains that are prone to various types of autoimmune diseases. For example, mice of the NOD strain are very likely to get diabetes. The female mice become diabetic more quickly than the males (Fig. 14.29). Many autoimmune diseases are more common in females than in males (see Fig. 14.33), although occasionally the opposite is true. Autoimmune diseases in humans also have a genetic component. Some autoimmune diseases, including type 1 diabetes, run in families, suggesting a role for genetic susceptibility. Most convincingly, if one of two identical (monozygotic) twins is affected, then the other twin is quite likely to be as well, whereas concordance of disease is much less in nonidentical (dizygotic) twins.

Environmental influences are also clearly involved. For example, although most of a colony of NOD mice are destined to get diabetes, they will do so at different ages (see Fig. 14.29). Moreover, the timing of disease onset often differs from one investigator's animal colony to the next, even though all the mice are genetically identical. Thus, environmental variables must be, in part, determining the rate of development of diabetes; a few even escape the disease entirely. Identical twins tell a similar story. With SLE, although disease occurs in both twins in about 25% of monozygotic twins, the overall rate is much higher than the normal chance of getting SLE. Nevertheless, the concordance rate is far from 100%. The explanation for the incomplete concordance could lie in environmental variables or it could simply be random.

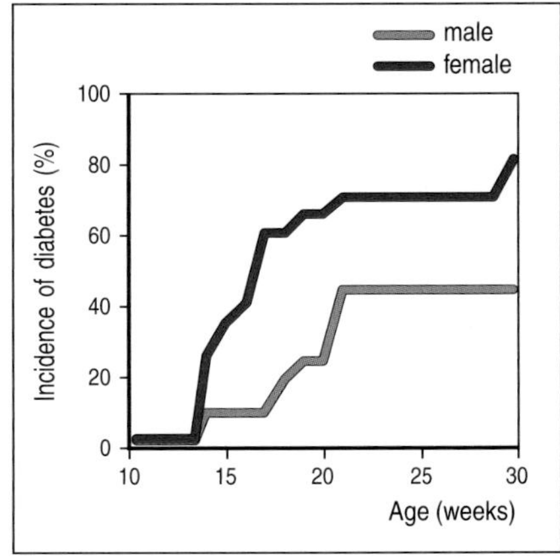

Fig. 14.29 Sex differences in the incidence of autoimmune disease. Many autoimmune diseases are more common in females than males, as illustrated here by the cumulative incidence of diabetes in a population of diabetes-prone NOD mice. Females (red line) get diabetes at a much younger age than do males, indicating their greater predisposition. Data kindly provided by S. Wong.

14-19 A defect in a single gene can cause autoimmune disease.

Predisposition to most of the common autoimmune diseases is due to the combined effects of multiple genes, but there are a very small number of known monogenic autoimmune diseases. In these, possession of the predisposing allele confers a very high risk of disease on the individual, but the overall impact on the population is minimal because these variants are rare (Fig. 14.30). The existence of monogenic autoimmune disease was first observed in mutant mice, in which the inheritance of an autoimmune syndrome followed a pattern consistent with a single-gene defect. Autoimmune disease alleles are usually recessive or X-linked. For example, the disease APECED, discussed in Section 14-3, is a recessive autoimmune disease caused by a defect in the gene *AIRE*.

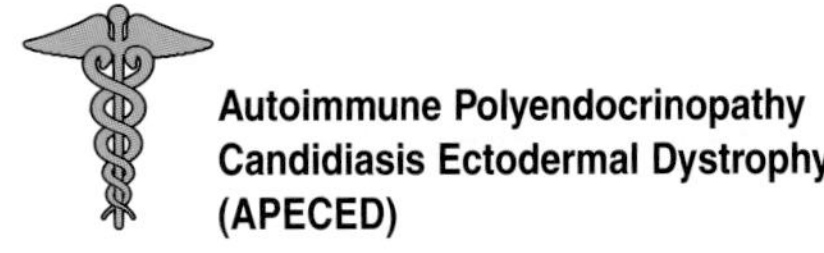

Two monogenic autoimmune syndromes have been linked to defects in regulatory T cells. The X-linked recessive autoimmune syndrome **IPEX** (immune dysregulation, polyendocrinopathy, enteropathy, X-linked syndrome) is caused by a mutation in the gene for the transcription factor FoxP3, which is a key factor in the differentiation of some types of regulatory T cells (see Section 8-20). Also known as XLAAD (X-linked autoimmunity-allergic dysregulation syndrome), this disease is characterized by severe allergic inflammation, autoimmune polyendocrinopathy, secretory diarrhea, hemolytic anemia, and thrombocytopenia, and usually leads to early death. Despite the mutation, in this group of patients the numbers of CD4 CD25 T_{reg} cells, the cells normally involved in maintaining peripheral tolerance (see Section 14-7), were comparable to those measured in the blood of healthy individuals; however, their suppressive function was reduced. A spontaneous mutation in the mouse *Foxp3* gene (the *scurfy* mutation) leads to an analogous systemic autoimmune disease, in this case associated with the absence of CD4 CD25 T_{reg} cells.

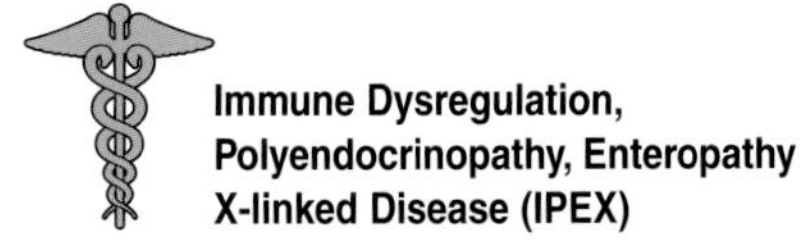

A second instance of autoimmunity resulting from a genetic defect in regulatory T-cell function has been identified in a single patient with a deficiency of CD25 as a result of a deletion in *CD25* and impaired peripheral tolerance. This patient suffered from multiple immunological deficiencies and autoimmune diseases and was highly susceptible to infections. These findings further confirm the important roles of CD25 CD4 T_{reg} cells in the regulation of the immune system.

Single-gene traits associated with autoimmunity			
Gene	Human disease	Mouse mutant or knockout	Mechanism of autoimmunity
AIRE	APECED (APS-1)	Knockout	Decreased expression of self antigens in the thymus, resulting in defective negative selection of self-reactive T cells
CTLA4	Association with Graves' disease, type 1 diabetes and others	Knockout	Failure of T-cell anergy and reduced activation threshold of self-reactive T cells
FOXP3	IPEX	Knockout and mutation (*scurfy*)	Decreased function of CD4 CD25 regulatory T cells
FAS	ALPS	*lpr/lpr;gld/gld* mutants	Failure of apoptotic death of self-reactive B and T cells
C1q	SLE	Knockout	Defective clearance of immune complexes and apoptotic cells

Fig. 14.30 Single-gene traits associated with autoimmunity. APECED, autoimmune polyendocrinopathy–candidiasis–ectodermal dystrophy; APS-1, autoimmune polyglandular syndrome 1; IPEX, immune dysregulation, polyendocrinopathy, enteropathy, X-linked syndrome; ALPS, autoimmune lymphoproliferative syndrome. The *lpr* mutation in mice affects the gene for Fas, whereas the *gld* mutation affects the gene for FasL. Reprinted with permission from Macmillan Publishers Ltd: *Nature*. J.D. Rioux, A.K. Abbas, **435**: 584–589, © 2005.

Autoimmine Lymphoproliferative Syndrome (ALPS)

An interesting case of a monogenic autoimmune disease is the systemic autoimmune syndrome caused by mutations in the gene for Fas. Fas is normally present on the surface of activated T and B cells, and when ligated by FasL it signals the Fas-bearing cell to undergo apoptosis (see Section 8-27). In this way it functions to limit the extent of immune responses. Mutations that eliminate or inactivate Fas lead to a massive accumulation of lymphocytes, especially T cells, and in mice the production of large quantities of pathogenic autoantibodies. The resulting disease resembles SLE, although typical SLE in humans has not been associated with mutations in Fas. A mutation that led to this autoimmune syndrome was first observed in the mouse strain MRL and dubbed *lpr*, for lymphoproliferation; it was subsequently identified as a mutation in *Fas*. Researchers studying a group of human patients with the rare **autoimmune lymphoproliferative syndrome (ALPS)**, a syndrome similar to that in the MRL/*lpr* mice, identified and cloned the mutant gene responsible for most of these cases, which also turned out to be *Fas* (see Fig. 14.30).

Autoimmune diseases caused by single genes are not common. They are nonetheless of great interest, as the mutations that cause them identify some of the important pathways that normally prevent the development of autoimmune responses.

14-20 Several approaches have given us insight into the genetic basis of autoimmunity.

Since the advent of gene knockout technology in mice (see Section A-47, Appendix I), many genes that encode proteins of the immune system have been experimentally disrupted. Several of these mutant mouse strains have signs of autoimmune disease, including autoantibodies and, in some cases, infiltration of organs by T cells. The study of these mice has greatly expanded our knowledge of the genetic pathways that can contribute to autoimmunity and that therefore might be candidates for naturally occurring mutations. At least 20 genes whose deletion or overexpression can contribute to the pathogenesis of autoimmunity have been identified. These encode cytokines, co-receptors, members of cytokine- or antigen-signaling cascades, co-stimulatory molecules, proteins involved in pathways that promote apoptosis and those that inhibit it, and proteins that clear antigen or antigen:antibody complexes. Some of the cytokines and signaling proteins implicated in

Defects in cytokine production or signaling that can lead to autoimmunity		
Defect	**Cytokine or intracellular signal**	**Result**
Overexpression	TNF-α	Inflammatory bowel disease, arthritis, vasculitis
	IL-2, IL-7, IL-10, IL-2R, IL-10R	Inflammatory bowel disease
	IL-3	Demyelinating syndrome
	IFN-γ	Overexpression in skin leads to SLE
	STAT4	Inflammatory bowel disease
Underexpression	TNF-α	SLE
	IL-1 receptor agonist	Arthritis
	STAT3	Inflammatory bowel disease
	TGF-β	Ubiquitous underexpression leads to inflammatory bowel disease. Underexpression specifically in T cells leads to SLE

Fig. 14.31 Defects in cytokine production or signaling that can lead to autoimmunity. Some of the signaling pathways involved in autoimmunity have been identified by genetic analysis, mainly in animal models. The effects of overexpression or underexpression of some of the cytokines and intracellular signaling molecules involved are listed here (see the text for further discussion).

autoimmune disease are listed in Fig. 14.31, and Fig. 14.32 lists some of the known associations for other categories of proteins.

In humans, the association of autoimmunity with a particular gene or genetic region can be assessed by large-scale family studies, or by association studies in the general population, that look for a correlation between disease frequency and variant alleles, genetic markers, duplications or deletions, and most recently with **single-nucleotide polymorphisms (SNPs)**, positions in the genome that differ by a single base between individuals. These studies have supported the concept that genetic susceptibility to autoimmune disease in humans is typically due to a combination of susceptibility alleles at multiple loci. For example, in large association studies looking at candidate susceptibility genes in humans, several of the commonest autoimmune diseases, including type 1 diabetes, Graves' disease, Hashimoto's thyroiditis, Addison's disease, rheumatoid arthritis, and multiple sclerosis, show genetic association with the *CTLA4* locus on chromosome 2. The cell-surface protein CTLA-4 is produced by activated T cells and is an inhibitory receptor for B7 co-stimulatory molecules (see Section 8-14). The effects of genetic variation in *CTLA4* on susceptibility to type 1 diabetes have been studied in mice. *CTLA4* is located on mouse chromosome 1 in a cluster with the genes for the other co-stimulatory receptors, CD28 and ICOS. When this genetic region in the diabetes-susceptible NOD mouse strain was replaced with the same region from the autoimmune-resistant B10 strain, it conferred diabetes resistance on the NOD mice. It seems that genetic variation in the splicing of the *CTLA4* mRNA may contribute to the difference in susceptibility. Splice variants of CTLA-4 that lack a portion essential for binding to its ligands B7.1 and B7.2 were still resistant to activation, and there was increased expression of this variant in the memory and regulatory T cells of diabetes-resistant mice.

A second locus, *PTPN22*, has been implicated in susceptibility to type 1 diabetes and rheumatoid arthritis. This gene encodes a lymphoid-associated protein tyrosine phosphatase that, like CTLA-4, is normally involved in suppressing T-cell activation.

Fig. 14.32 Categories of genetic defects that lead to autoimmune syndromes. Many genes have been identified in which mutations predispose to autoimmunity in humans and animal models. These are best understood by the type of process affected by the genetic defect. A list of such genes is given here, organized by process (see the text for further discussion). In some cases, the same gene has been identified in mice and humans. In other cases, different genes affecting the same mechanism are implicated in mice and humans. The smaller number of human genes identified so far undoubtedly reflects the difficulty of identifying the genes responsible in outbred human populations.

Proposed mechanism	Murine models	Disease phenotype	Human gene affected	Disease phenotype
Antigen clearance and presentation	C1q knockout	Lupus-like	C1q	Lupus-like
	C4 knockout		C2 C4	
			Mannose-binding lectin	
	AIRE knockout	Multiorgan autoimmunity resembling APECED	AIRE	APECED
	Mer knockout	Lupus-like		
Signaling	SHP-1 knockout	Lupus-like		
	Lyn knockout			
	CD22 knockout			
	CD45 E613R point mutation			
	B cells deficient in all Src-family kinases (triple knockout)			
	FcγRIIB knockout (inhibitory signaling molecule)		FcγRII	Lupus
Co-stimulatory molecules	CTLA-4 knockout (blocks inhibitory signal)	Lymphocyte infiltration into organs		
	PD-1 knockout (blocks inhibitory signal)	Lupus-like		
	BAFF overexpression (transgenic mouse)			
Apoptosis	Fas knockout (*lpr*)	Lupus-like with lymphocyte infiltrates	*Fas* and *FasL* mutations (ALPS)	Lupus-like with lymphocyte infiltrates
	FasL knockout (*gld*)			
	Bcl-2 overexpression (transgenic mouse)	Lupus-like		
	Pten heterozygous deficiency			

14-21 Genes that predispose to autoimmunity fall into categories that affect one or more of the mechanisms of tolerance.

The genes identified as predisposing to autoimmunity can be classified as follows: genes that affect autoantigen availability and clearance; those that affect apoptosis; those that affect signaling thresholds; those involved in cytokine gene expression; and those affecting the expression of co-stimulatory molecules (see Figs 14.31 and 14.32).

Genes that control antigen availability and clearance are important either centrally in the thymus in making self proteins available for inducing tolerance in developing lymphocytes or peripherally in controlling how self molecules are made available in an immunogenic form to peripheral lymphocytes. In peripheral tolerance, a hereditary deficiency of some complement proteins is strongly associated with the development of SLE in humans. Specifically C1q, C3, and C4 are important in clearing apoptotic cells and immune complexes. If apoptotic cells and immune complexes are not cleared, this increases the immunogenicity for low-affinity self-reactive lymphocytes in the periphery. Genes that control apoptosis, such as *Fas*, are important in regulating the duration and vigor of immune responses. Failure to regulate immune responses properly could cause excessive destruction of self tissues, releasing autoantigens. In addition, because clonal deletion and anergy are not absolute, immune responses can include some self-reactive cells. As long as their numbers are limited by apoptotic mechanisms, they may not be sufficient to cause autoimmune disease, but they could cause a problem if apoptosis is not properly regulated.

Perhaps the largest category of mutations associated with autoimmunity comprises those associated with signals that control lymphocyte activation. One subset contains mutations that inactivate negative regulators of lymphocyte activation and thus lead to the hyperproliferation of lymphocytes and exaggerated immune responses. These include mutations in CTLA-4 (as discussed in Section 14-20), in inhibitory Fc receptors, and in receptors containing ITIMs (see Section 6-20), such as CD22 on B cells. Another subset contains mutations in proteins involved in signal transduction through the antigen receptor itself. Adjusting thresholds in either direction, to make signaling more or less sensitive, can result in autoimmunity, depending on the situation. A decrease in sensitivity in the thymus, for example, can lead to a failure of negative selection and thereby to autoreactivity in the periphery. In contrast, increasing receptor sensitivity in the periphery can lead to greater and prolonged activation, again resulting in an exaggerated immune response with the side effect of autoimmunity. A final subset of mutations comprises those that affect the expression of genes for cytokines and co-stimulatory molecules.

14-22 MHC genes have an important role in controlling susceptibility to autoimmune disease.

Among all the genetic loci that could contribute to autoimmunity, susceptibility to autoimmune disease has so far been most consistently associated with MHC genotype. Human autoimmune diseases that show associations with HLA (MHC) type are shown in Fig. 14.33. For most of these diseases, susceptibility is linked most strongly with MHC class II alleles, but in some cases there are strong associations with particular MHC class I alleles. In some cases, class III alleles such as those for TNF-α or complement protein have been associated with disease. The development of experimental diabetes or arthritis in

Fig. 14.33 Associations of HLA serotype and sex with susceptibility to autoimmune disease. The 'relative risk' for an HLA allele in an autoimmune disease is calculated by comparing the observed number of patients carrying the HLA allele with the number that would be expected, given the prevalence of the HLA allele in the general population. For type 1 insulin-dependent diabetes mellitus, the association is in fact with the HLA-DQ gene, which is tightly linked to the DR genes but is not detectable by serotyping. Some diseases show a significant bias in the sex ratio; this is taken to imply that sex hormones are involved in pathogenesis. Consistent with this, the difference in the sex ratio in these diseases is greatest between the menarche and the menopause, when levels of such hormones are highest.

Associations of HLA serotype with susceptibility to autoimmune disease

Disease	HLA allele	Relative risk	Sex ratio (♀:♂)
Ankylosing spondylitis	B27	87.4	0.3
Acute anterior uveitis	B27	10	<0.5
Goodpasture's syndrome	DR2	15.9	~1
Multiple sclerosis	DR2	4.8	10
Graves' disease	DR3	3.7	4–5
Myasthenia gravis	DR3	2.5	~1
Systemic lupus erythematosus	DR3	5.8	10–20
Type 1 (insulin-dependent) diabetes mellitus	DR3/DR4 heterozygote	~25	~1
Rheumatoid arthritis	DR4	4.2	3
Pemphigus vulgaris	DR4	14.4	~1
Hashimoto's thyroiditis	DR5	3.2	4–5

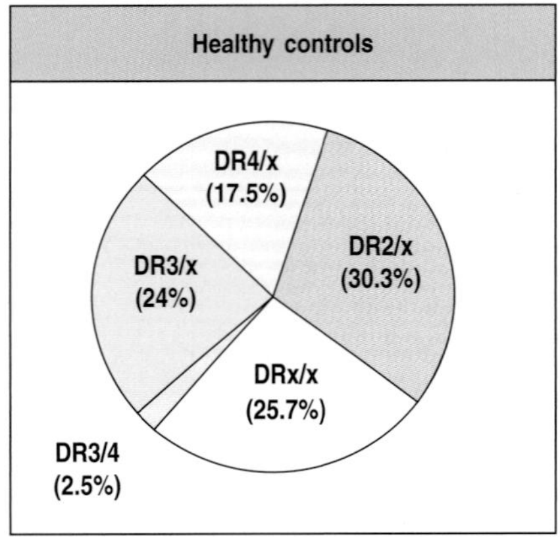

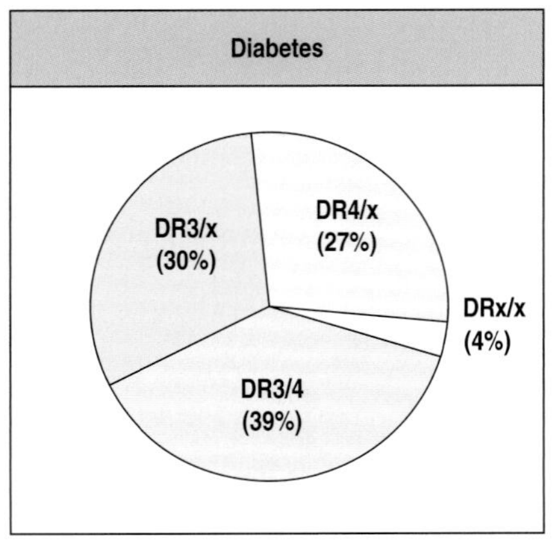

transgenic mice expressing specific human HLA antigens strongly suggests that particular MHC alleles can confer susceptibility to disease.

The association of MHC genotype with disease is assessed initially by comparing the frequency of different alleles in patients with their frequency in the normal population. For type 1 diabetes, this approach originally demonstrated an association with the HLA-DR3 and HLA-DR4 alleles identified by serotyping (Fig. 14.34). Such studies also showed that the MHC class II allele HLA-DR2 has a dominant protective effect: individuals carrying HLA-DR2, even in association with one of the susceptibility alleles, rarely develop diabetes. Another way of determining whether MHC genes are important in autoimmune disease is to study the families of affected patients; it has been shown that two siblings affected with the same autoimmune disease are far more likely than expected to share the same MHC haplotypes (Fig. 14.35). As HLA genotyping has become more exact through the DNA sequencing of HLA alleles, disease associations that were originally discovered through HLA serotyping have been defined more precisely. For example, the association

Fig. 14.34 Population studies show association of susceptibility to type 1 diabetes with HLA genotype. The HLA genotypes (determined by serotyping) of diabetic patients (bottom panel) are not representative of those found in the general population (top panel). Almost all diabetic patients express HLA-DR3 and/or HLA-DR4, and HLA-DR3/DR4 heterozygosity is greatly over-represented in diabetics compared with controls. These alleles are linked tightly to HLA-DQ alleles that confer susceptibility to type 1 diabetes. By contrast, HLA-DR2 protects against the development of diabetes and is found only extremely rarely in diabetic patients. The small letter x represents any allele other than DR2, DR3, or DR4.

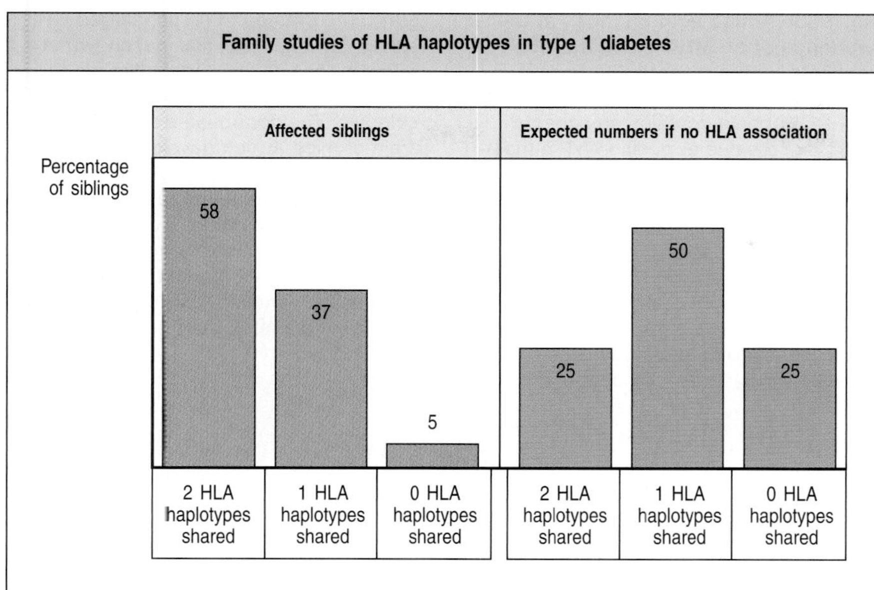

Fig. 14.35 Family studies show strong linkage of susceptibility to type 1 diabetes with HLA genotype. In families in which two or more siblings have type 1 diabetes, it is possible to compare the HLA genotypes of affected siblings. Affected siblings share two HLA haplotypes much more frequently than would be expected if the HLA genotype did not influence disease susceptibility.

between type 1 diabetes and the DR3 and DR4 alleles is now known to be due to their tight genetic linkage to DQβ alleles that actually confer susceptibility to the disease. Indeed, susceptibility is most closely associated with polymorphisms at a particular position in the DQβ amino-acid sequence. The most common DQβ amino-acid sequence has an aspartic acid residue at position 57 that is able to form a salt bridge across the end of the peptide-binding cleft of the DQ molecule. In contrast, diabetic patients in Caucasoid populations mostly have valine, serine, or alanine at that position, and thus make DQ molecules that lack this salt bridge (Fig. 14.36). The NOD strain of mice, which develops spontaneous diabetes, also has a serine at that position in the homologous mouse MHC class II molecule, known as I-A^{g7}.

The association of MHC genotype with autoimmune disease is not surprising, because autoimmune responses involve T cells, and the ability of T cells to respond to a particular antigen depends on MHC genotype. Thus, the associations can be explained by a simple model in which susceptibility to an autoimmune disease is determined by differences in the ability of different allelic variants of MHC molecules to present autoantigenic peptides to autoreactive T cells. This would be consistent with what we know of T-cell involvement in particular diseases. In diabetes, for example, there are associations with both MHC class I and MHC class II alleles, and this is consistent with the finding that both CD8 and CD4 T cells, which respond to antigens presented by MHC class I and MHC class II molecules, respectively, mediate the autoimmune response.

An alternative hypothesis for the association between MHC genotype and susceptibility to autoimmune diseases emphasizes the role of MHC alleles in shaping the T-cell receptor repertoire (see Chapter 7). This hypothesis proposes that self peptides associated with certain MHC molecules may drive the positive selection of developing thymocytes that are specific for particular autoantigens. Such autoantigenic peptides might be expressed at too low a level or bind too poorly to self MHC molecules to drive negative selection in the thymus, but might be present at a sufficient level or bind strongly enough to drive positive selection. This hypothesis is supported by observations that I-A^{g7}, the disease-associated MHC class II molecule in NOD mice, binds many peptides very poorly and may therefore be less effective in driving intrathymic negative selection of T cells that bind self peptides.

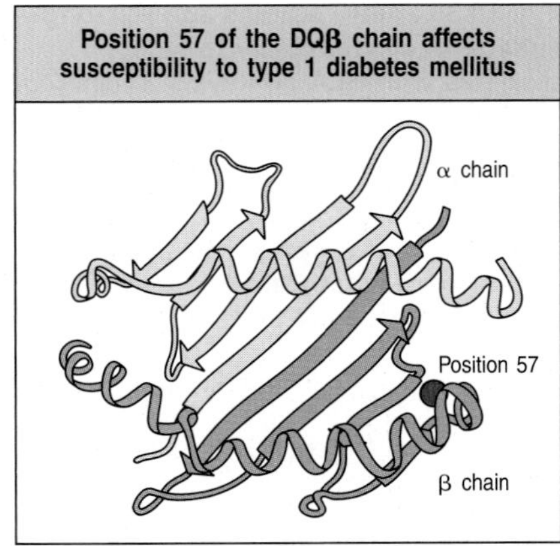

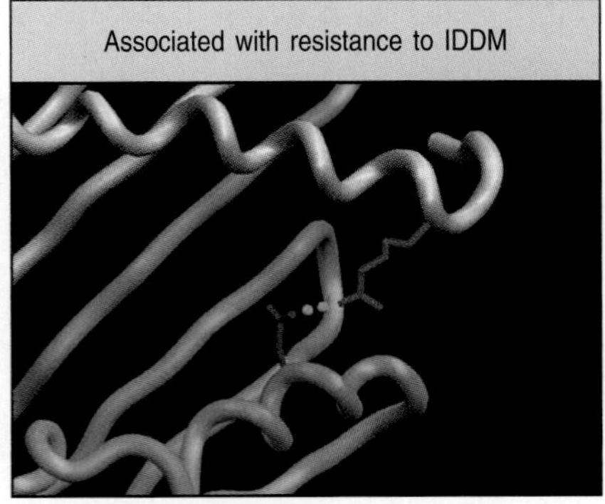

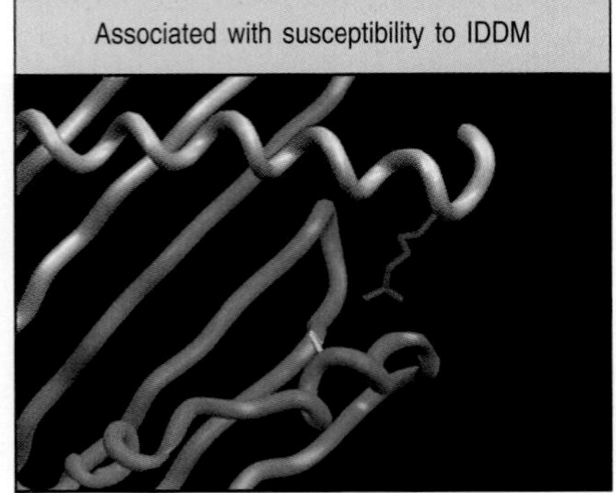

Fig. 14.36 Amino acid changes in the sequence of an MHC class II protein correlate with susceptibility to and protection from diabetes. The HLA-DQβ_1 chain contains an aspartic acid (Asp) residue at position 57 in most people; in Caucasoid populations, patients with type 1 diabetes more often have valine, serine, or alanine at this position instead, as well as other differences. Asp 57, shown in red on the backbone structure of the DQβ chain, forms a salt bridge (shown in green in the center panel) to an arginine residue (shown in pink) in the adjacent α chain (gray). The change to an uncharged residue (for example, alanine, shown in yellow in the bottom panel) disrupts this salt bridge, altering the stability of the DQ molecule. The nonobese diabetic (NOD) strain of mice, which develops spontaneous diabetes, shows a similar substitution of serine for aspartic acid at position 57 of the homologous I-Aβ chain, and NOD mice transgenic for β chains with Asp 57 have a marked reduction in diabetes incidence. IDDM, insulin dependent diabetes mellitus. Courtesy of C. Thorpe.

14-23 External events can initiate autoimmunity.

The geographic distribution of autoimmune diseases reveals a heterogeneous distribution between continents, countries, and ethnic groups. For example, the incidence of disease seems to decrease from north to south in the northern hemisphere. This gradient is particularly prominent in diseases such as multiple sclerosis and type 1 diabetes in Europe, which have a higher incidence in the northern countries than in the Mediterranean regions. Several studies have also shown a reduced incidence of autoimmunity in developing countries compared with the more developed world.

There are numerous contributing factors to these geographic variations besides genetic susceptibility—socioeconomic status and diet seem to play a part. An example of how factors beside genetic background influence the onset of disease is the fact that even genetically identical mice develop autoimmunity at different rates and severity (see Fig. 14.29). In humans, exposure to infections and environmental toxins may be factors that help trigger autoimmunity. However, it should be noted that epidemiological and clinical studies over the past century have also shown a negative correlation between exposure to some types of infection in early life and the development of allergy and autoimmune diseases. This 'hygiene hypothesis' is discussed in detail in Section 13-4; it proposes that a lack of infection during childhood may affect the regulation of the immune system in later life, leading to a greater likelihood of allergic and autoimmune responses.

14-24 Infection can lead to autoimmune disease by providing an environment that promotes lymphocyte activation.

How might pathogens initiate or modulate autoimmunity? During an infection and the consequent immune response, the combination of the inflammatory mediators released from activated antigen-presenting cells and lymphocytes and the increased expression of co-stimulatory molecules can have effects on so-called bystander cells—lymphocytes that are not themselves specific for the antigens of the infectious agent. Self-reactive lymphocytes can become activated in these circumstances, particularly if tissue destruction by the infection leads to an increase in the availability of the self antigen (Fig. 14.37, first panel).

In general, any infection will lead to an inflammatory response and recruitment of inflammatory cells to the site of the infection. The perpetuation, and even exacerbation, of autoimmune disease by viral or bacterial infections has been shown in experimental animal models. For example, the severity of type 1 diabetes in NOD mice is exacerbated by Coxsackie virus B4 infection, which leads to inflammation, tissue damage and the release of sequestered islet antigens, and the generation of autoreactive T cells.

We discussed earlier the ability of self ligands such as unmethylated CpG DNA sequences and RNA to directly activate ignorant autoreactive B cells via their Toll-like receptors and thus break tolerance to self (see Section 14-4). Microbial ligands for Toll-like receptors may also promote autoimmunity by stimulating dendritic cells and macrophages to produce large quantities of cytokines that cause local inflammation and help stimulate and maintain already activated autoreactive T cells and B cells. This mechanism might be relevant to the flares that follow infection in patients with autoimmune vasculitis associated with anti-neutrophil cytoplasmic antibodies.

One example of how exposure to Toll-like receptor ligands can induce local inflammation derives from an animal model of arthritis in which injection of bacterial CpG DNA into the joints of healthy mice induces an aseptic arthritis characterized by macrophage infiltration. These macrophages express chemokine receptors on their surface and produce large amounts of CC chemokines, which promote leukocyte recruitment to the site of injection.

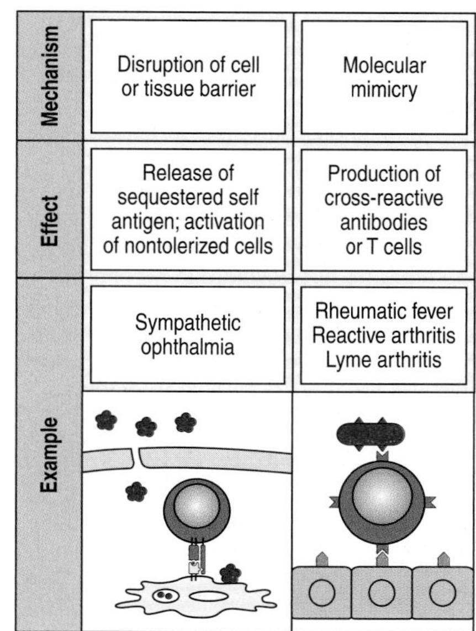

Mechanism
Disruption of cell or tissue barrier
Molecular mimicry
Effect
Release of sequestered self antigen; activation of nontolerized cells
Production of cross-reactive antibodies or T cells
Example
Sympathetic ophthalmia
Rheumatic fever
Reactive arthritis
Lyme arthritis

Fig. 14.37 Infectious agents could break self-tolerance in several different ways. Left panel: because some antigens are sequestered from the circulation, either behind a tissue barrier or within the cell, an infection that breaks cell and tissue barriers might expose hidden antigens. Right panel: molecular mimicry might result in infectious agents inducing either T- or B-cell responses that can cross-react with self antigens.

14-25 Cross-reactivity between foreign molecules on pathogens and self molecules can lead to anti-self responses and autoimmune disease.

Infection with certain pathogens is particularly associated with autoimmune sequelae. Some pathogens express protein or carbohydrate antigens that resemble host molecules, a phenomenon dubbed **molecular mimicry**. In such cases, antibodies produced against a pathogen epitope may cross-react with a self protein (see Fig. 14.37, second panel). Such structures do not necessarily have to be identical: it is sufficient if they are similar enough to be recognized by the same antibody. Molecular mimicry may also activate autoreactive naive or effector T cells if a processed peptide from a pathogen antigen is identical or similar to a host peptide, resulting in attack on self tissues. A model system to demonstrate molecular mimicry has been generated using transgenic mice that express a viral antigen in the pancreas. Normally, there is no response to this virus-derived 'self' antigen. But if the mice are infected with the virus that was the source of the transgenic antigen, they develop diabetes, because the virus activates T cells that are cross-reactive with the 'self' viral antigen and attack the pancreas (Fig. 14.38).

One might wonder why these self-reactive lymphocytes have not been deleted or inactivated by the usual mechanisms of self-tolerance. One reason, as discussed earlier in the chapter, is that lower-affinity self-reactive B and T cells are not removed efficiently and are present in the naive lymphocyte repertoire as ignorant lymphocytes (see Section 14-4). Second, the strong pro-inflammatory stimulus that accompanies an infection could be sufficient to activate even anergic T and B cells in the periphery, thus drawing into the response cells that would usually be silent. Third, pathogens may provide substantially higher local doses of the eliciting antigen in an immunogenic form, whereas normally it would be relatively unavailable to lymphocytes. Some examples of autoimmune syndromes thought to involve molecular mimicry are the rheumatic fever that sometimes follows streptococcal infection, and the reactive arthritis that can occur after enteric infection.

Once self-reactive lymphocytes are activated by such a mechanism, their effector functions can destroy self tissues. Autoimmunity of this type is sometimes transient and remits when the inciting pathogen is eliminated. This is the case in the autoimmune hemolytic anemia that follows mycoplasma infection, in which antibodies against the pathogen cross-react

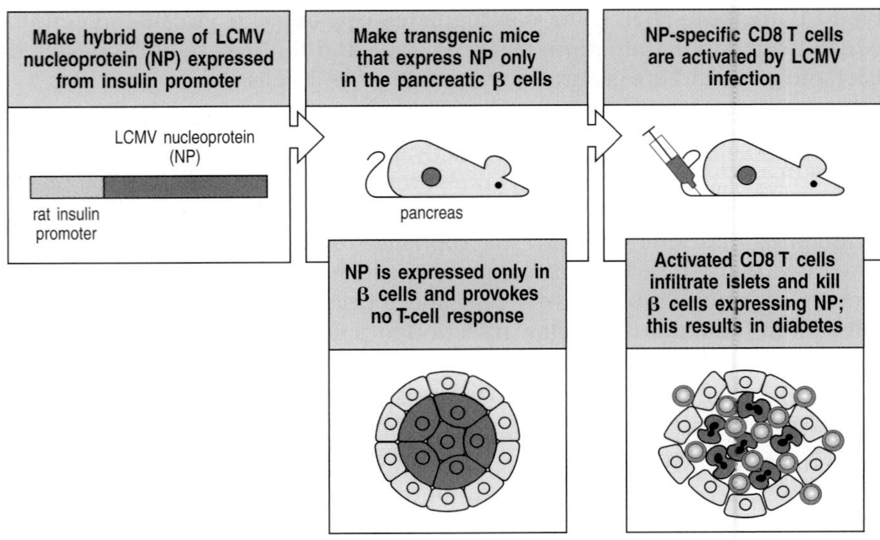

Fig. 14.38 Virus infection can break tolerance to a transgenic viral protein expressed in pancreatic β cells. Mice made transgenic for the lymphocytic choriomeningitis virus (LCMV) nucleoprotein under the control of the rat insulin promoter express the nucleoprotein in their pancreatic β cells but do not respond to this protein and therefore do not develop an autoimmune diabetes. However, if the transgenic mice are infected with LCMV, a potent antiviral cytotoxic T-cell response is elicited, and this kills the β cells, leading to diabetes. It is thought that infectious agents can sometimes elicit T-cell responses that cross-react with self peptides (a process known as molecular mimicry) and that this could cause autoimmune disease in a similar way.

with an antigen on red blood cells, leading to hemolysis (see Section 14-13). The autoantibodies disappear when the patient recovers from the infection. Sometimes, however, the autoimmunity persists well beyond the initial infection. This is true in some cases of **rheumatic fever**, which occasionally follows sore throat or scarlet fever caused by *Streptococcus pyogenes*. The similarity of epitopes on streptococcal antigens to epitopes on some tissues leads to antibody-mediated, and possibly T cell-mediated, damage to a variety of tissues, including heart valves. Although rheumatic fever is often transient, especially with antibiotic treatment, it can sometimes become chronic. Similarly, Lyme disease, an infection with the spirochete *Borrelia burgdorferi*, is followed by later-developing autoimmunity, causing so-called Lyme arthritis. In this case, the mechanism is not entirely clear, but it is likely to involve cross-reactivity of pathogen and host components, leading to a self-perpetuating autoimmune reaction.

14-26 Drugs and toxins can cause autoimmune syndromes.

Perhaps some of the clearest evidence of external causative agents in human autoimmunity comes from the effects of certain drugs, which elicit autoimmune reactions as side effects in a small proportion of patients. Procainamide, a drug used to treat heart arrhythmias, is particularly notable for inducing autoantibodies similar to those in SLE, although these are rarely pathogenic. Several drugs are associated with the development of autoimmune hemolytic anemia, in which autoantibodies against surface components of red blood cells attack and destroy these cells (see Section 14-13). Toxins in the environment can also cause autoimmunity. When heavy metals, such as gold or mercury, are administered to genetically susceptible strains of mice, a predictable autoimmune syndrome, including the production of autoantibodies, ensues. The extent to which heavy metals promote autoimmunity in humans is debatable, but the animal models clearly show that environmental factors such as toxins could have key roles in certain syndromes.

The mechanisms by which drugs and toxins cause autoimmunity are uncertain. For some drugs it is thought that they react chemically with self proteins and form derivatives that the immune system recognizes as foreign. The immune response to these haptenated self proteins can lead to inflammation, complement deposition, destruction of tissue, and finally immune responses to the original underivatized self proteins.

14-27 Random events may be required for the initiation of autoimmunity.

Although scientists and physicians would like to attribute the onset of 'spontaneous' diseases to some specific cause, this may not always be possible. There may not be one virus or bacterium, or even any understandable pattern of events that precedes the onset of autoimmune disease. The chance encounter in the peripheral lymphoid tissues of a few autoreactive B and T cells that can interact with each other, at just the moment when an infection is providing pro-inflammatory signals, may be all that is needed. This could be a rare event, and in a genetically resistant individual it could even be brought under control. But in a susceptible individual such events could be more frequent and/or more difficult to control.

Thus, the onset or incidence of autoimmunity can seem to be random. Genetic predisposition represents, in part, an increased chance of occurrence of this random event. This view, in turn, may explain why many autoimmune diseases appear in early adulthood or later, after enough time has elapsed to permit low-frequency random events to occur. It may also explain why after certain kinds of experimental aggressive therapies for these diseases, such as bone marrow transplantation or B-cell depletion, the disease eventually recurs after a long interval of remission.

Summary.

The specific causes of most autoimmune diseases are in most cases not known. Genetic risk factors including particular alleles of MHC class II molecules and other genes have been identified, but many individuals with genetic variants that predispose to a particular autoimmune disease do not get the disease. Epidemiological studies of genetically identical populations of animals have highlighted the role of environmental factors for the initiation of autoimmunity, but although environmental factors have at least as strong an influence on the outcome as genetics, they are even less well understood. Some toxins and drugs are known to cause autoimmune syndromes, but their role in the common varieties of autoimmune disease is unclear. Similarly, some autoimmune syndromes can follow viral or bacterial infections. Pathogens can promote autoimmunity by causing nonspecific inflammation and tissue damage. They can also sometimes elicit responses to self proteins if they express molecules that resemble self, a phenomenon known as molecular mimicry. Much more progress needs to be made to define environmental factors. It may be that there will not be a single, or even an identifiable, environmental factor that contributes to most diseases, and chance may have an important role in determining disease onset.

Responses to alloantigens and transplant rejection.

The transplantation of tissues to replace diseased organs is now an important medical therapy. In most cases, adaptive immune responses to the grafted tissues are the major impediment to successful transplantation. Rejection is caused by immune responses to alloantigens on the graft, which are proteins that vary from individual to individual within a species and are therefore perceived as foreign by the recipient. When tissues containing nucleated cells are transplanted, T-cell responses to the highly polymorphic MHC molecules almost always trigger a response against the grafted organ. Matching the MHC type of the donor and the recipient increases the success rate of grafts, but perfect matching is possible only when donor and recipient are related and, in these cases, genetic differences at other loci can still trigger rejection, although less severely. Nevertheless, advances in immunosuppression and transplantation medicine now mean that the precise matching of tissues for transplantation is no longer considered the major restricting factor in graft survival. In blood transfusion, which was the earliest tissue transplant and is still the most common, MHC matching is not necessary as red blood cells and platelets express only small amounts of MHC class I molecules and do not express MHC class II at all; thus, they are not targets for the T cells of the recipient. Blood must, however, be matched for ABO and Rh blood group antigens to avoid the rapid destruction of mismatched red blood cells by antibodies in the recipient (see Appendix I, Section A-11). Because there are only four major ABO types and two Rh types, this is relatively easy. In this part of the chapter we examine the immune response to tissue grafts and also ask why such responses do not reject the one foreign tissue graft that is tolerated routinely—the mammalian fetus.

14-28 Graft rejection is an immunological response mediated primarily by T cells.

The basic rules of tissue grafting were first elucidated by skin transplantation between inbred strains of mice. Skin can be grafted with 100% success between different sites on the same animal or person (an **autograft**), or between genetically identical animals or people (a **syngeneic graft**). However, when skin is grafted between unrelated or **allogeneic** individuals (an **allograft**), the graft initially survives but is then rejected about 10–13 days after grafting (Fig. 14.39). This response is called an **acute rejection** and is quite consistent. It depends on a T-cell response in the recipient, because skin grafted onto *nude* mice, which lack T cells, is not rejected. The ability to reject skin can be restored to *nude* mice by the adoptive transfer of normal T cells.

When a recipient that has previously rejected a graft is regrafted with skin from the same donor, the second graft is rejected more rapidly (6–8 days) in

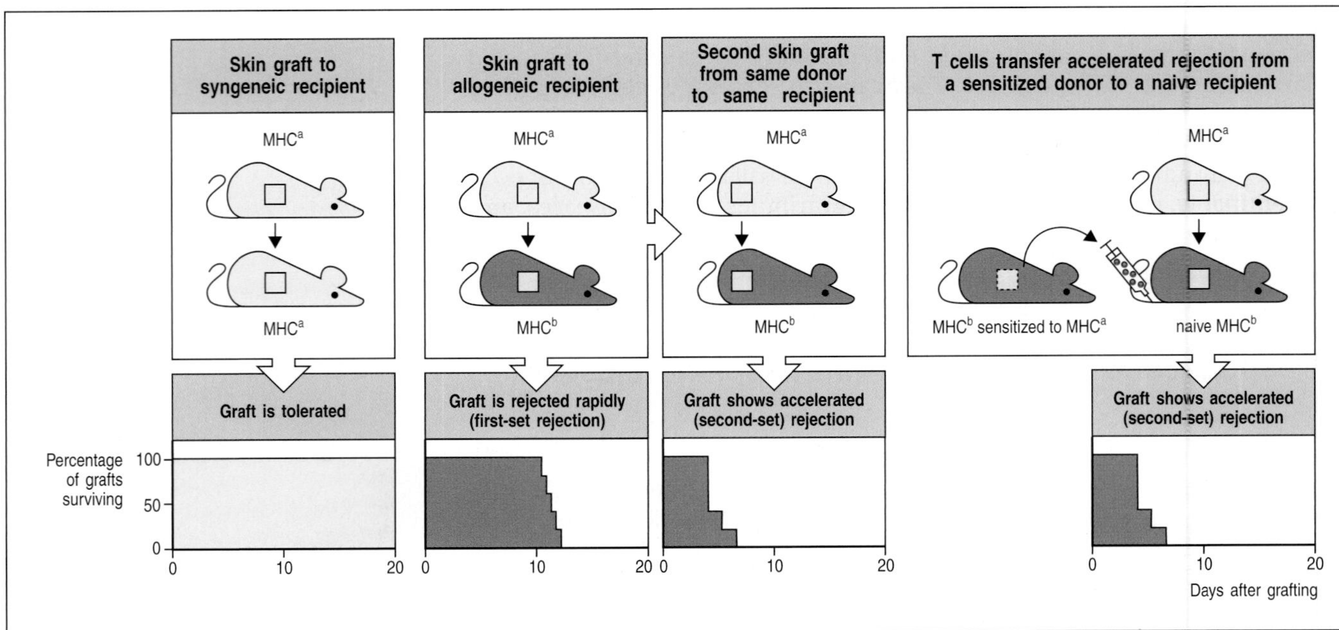

Fig. 14.39 Skin graft rejection is the result of a T cell-mediated anti-graft response. Grafts that are syngeneic are permanently accepted (first panels), but grafts differing at the MHC are rejected about 10–13 days after grafting (first-set rejection, second panels). When a mouse is grafted for a second time with skin from the same donor, it rejects the second graft faster (third panels). This is called a second-set rejection and the accelerated response is MHC-specific; skin from a second donor of the same MHC type is rejected equally fast, whereas skin from an MHC-different donor is rejected in a first-set pattern (not shown). Naive mice that are given T cells from a sensitized donor behave as if they had already been grafted (final panels).

an **accelerated rejection** (see Fig. 14.39). Skin from a third-party donor grafted onto the same recipient at the same time does not show this faster response but follows a first-set rejection course. The rapid course of second-set rejection can also be transferred to normal or irradiated recipients by T cells from the initial recipient, showing that second-set rejection is caused by a memory-type immune response (see Chapter 10) from clonally expanded and primed T cells specific for the donor skin.

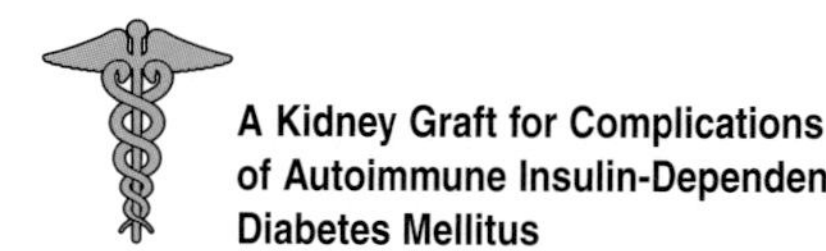

A Kidney Graft for Complications of Autoimmune Insulin-Dependent Diabetes Mellitus

Immune responses are a major barrier to effective tissue transplantation, destroying grafted tissue by an adaptive immune response to its foreign proteins. These responses can be mediated by CD8 T cells, by CD4 T cells, or by both. Antibodies can also contribute to second-set rejection of tissue grafts.

14-29 Matching donor and recipient at the MHC improves the outcome of transplantation.

Antigens that differ between members of the same species are known as **alloantigens**, and an immune response against such antigens is known as an **alloreactive** response. When donor and recipient differ at the MHC, an alloreactive immune response is directed at the nonself allogeneic MHC molecule or molecules present on the graft. In most tissues these will be predominantly MHC class I antigens. Once a recipient has rejected a graft of a particular MHC type, any further graft bearing the same nonself MHC molecule will be rapidly rejected in a second-set response. As we learned in Chapter 5, the frequency of T cells specific for any nonself MHC molecule is relatively high, making differences at MHC loci the most potent trigger of the rejection of initial grafts; indeed, the major histocompatibility complex was originally so named because of its central role in graft rejection.

Once it became clear that recognition of nonself MHC molecules is a major determinant of graft rejection, a considerable amount of effort was put into MHC matching between recipient and donor. Although matching at the MHC locus, known as the HLA locus in humans, significantly improves the success rate of clinical organ transplantation, it does not in itself prevent rejection reactions. There are two main reasons for this. First, clinically applicable methods for HLA typing are imprecise, because of the polymorphism and complexity of the human MHC; unrelated individuals who type as HLA-identical with antibodies against MHC proteins rarely have identical MHC genotypes. This should not be a problem with HLA-identical siblings: because siblings inherit their MHC genes as a haplotype, one sibling in four should be truly HLA-identical. Nevertheless, grafts between HLA-identical siblings invariably incite a rejection reaction, albeit more slowly, unless donor and recipient are identical twins. This reaction is the result of differences between minor histocompatibility antigens, which is the second reason for the failure of HLA matching to prevent rejection reactions. These minor histocompatibility antigens, which are peptides from non-MHC proteins that also vary between individuals, are discussed in the next section.

Thus, unless donor and recipient are identical twins, all graft recipients must be given immunosuppressive drugs to prevent rejection. Indeed, the current success of clinical transplantation of solid organs is more the result of advances in immunosuppressive therapy, discussed in Chapter 15, than of improved tissue matching. The limited supply of cadaveric organs, coupled with the urgency of identifying a recipient once a donor organ becomes available, means that accurate matching of tissue types is achieved only rarely, with the notable exception of matched sibling donation of kidneys.

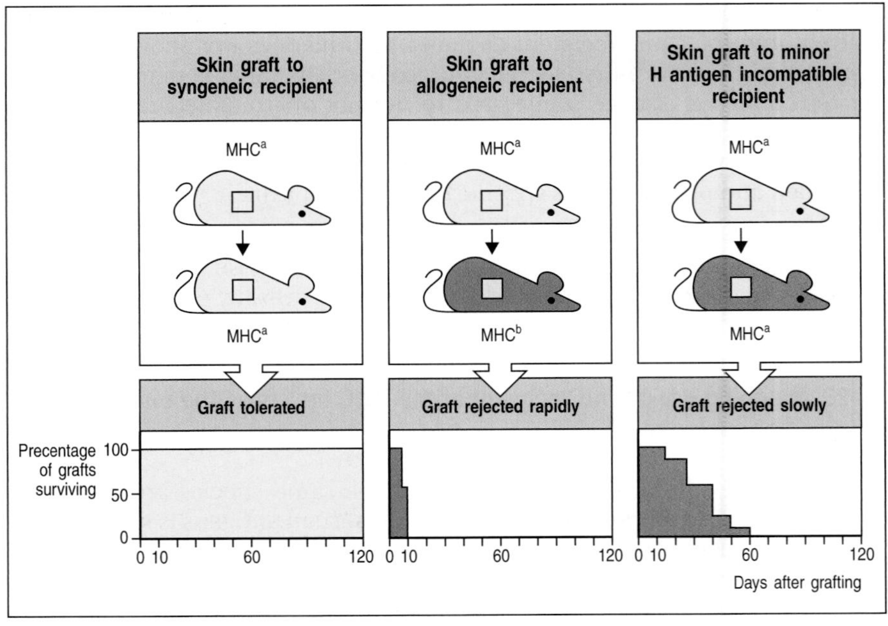

Fig. 14.40 Even complete matching at the MHC does not ensure graft survival. Although syngeneic grafts are not rejected (left panels), MHC-identical grafts from donors that differ at other loci (minor H antigen loci) are rejected (right panels), albeit more slowly than MHC-disparate grafts (center panels).

14-30 In MHC-identical grafts, rejection is caused by peptides from other alloantigens bound to graft MHC molecules.

When donor and recipient are identical at the MHC but differ at other genetic loci, graft rejection is not so rapid (Fig. 14.40). MHC class I and class II molecules bind and present a selection of peptides derived from proteins made in the cell, and if polymorphisms in these proteins mean that different peptides are produced in different members of a species, these can be recognized as **minor histocompatibility antigens** (Fig. 14.41). One set of proteins that induce minor histocompatibility responses are encoded on the male-specific Y chromosome. Responses induced by these proteins are known collectively as H-Y. As these Y-chromosome-specific genes are not expressed in females, female anti-male minor histocompatibility responses occur; however, male

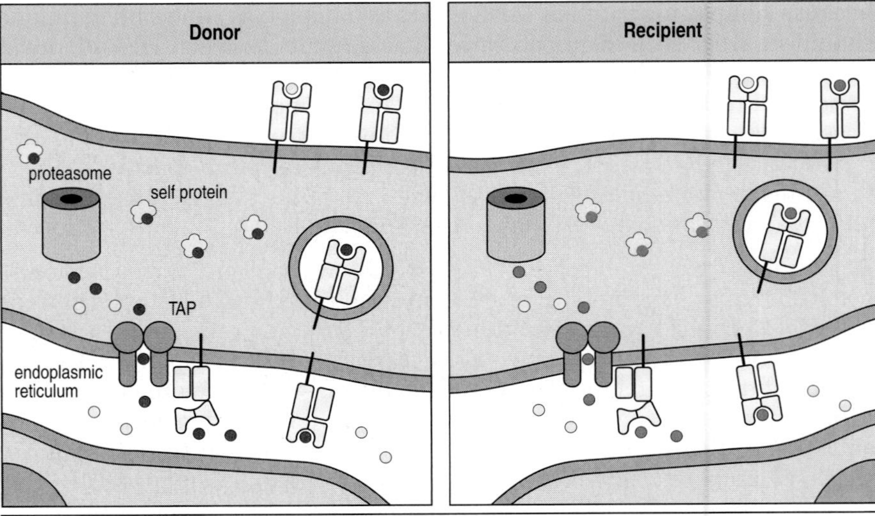

Fig. 14.41 Minor H antigens are peptides derived from polymorphic cellular proteins bound to MHC class I molecules. Self proteins are routinely digested by proteasomes within the cell's cytosol, and peptides derived from them are delivered to the endoplasmic reticulum, where they can bind to MHC class I molecules and be delivered to the cell surface. If a polymorphic protein differs between the graft donor (shown in red on the left) and the recipient (shown in blue on the right), it can give rise to an antigenic peptide (red on the donor cell) that can be recognized by the recipient's T cells as nonself and elicit an immune response. Such antigens are the minor H antigens.

anti-female responses are not seen, because both males and females express X-chromosome genes. One H-Y antigen has been identified in mice and humans as peptides from a protein encoded by the Y-chromosome gene *Smcy*. An X-chromosome homolog of *Smcy*, called *Smcx*, does not contain these peptide sequences, which are therefore expressed uniquely in males. Most minor histocompatibility antigens are encoded by autosomal genes and their identity is largely unknown, although 10 have now been identified at the genetic level.

The response to minor histocompatibility antigens is in most ways analogous to the response to viral infection. However, an antiviral response eliminates only infected cells, whereas all cells in the graft express these antigens, and thus the entire graft is destroyed in the response against the antigens. Given the virtual certainty of mismatches in minor histocompatibility antigens between any two individuals, and the potency of the reactions they incite, it is no wonder that successful transplantation requires the use of powerful immunosuppressive drugs.

14-31 There are two ways of presenting alloantigens on the transplant to the recipient's T lymphocytes.

Before naive alloreactive T cells can cause rejection, they must be activated by antigen-presenting cells that both bear the allogeneic MHC molecules and have co-stimulatory activity. Organ grafts carry with them antigen-presenting cells of donor origin, sometimes called passenger leukocytes, and these are an important stimulus to alloreactivity. This route for sensitization of the recipient to a graft seems to involve donor antigen-presenting cells leaving the graft and migrating via the lymph to regional lymph nodes. Here they can activate those host T cells that bear the corresponding T-cell receptors. The activated alloreactive effector T cells are then carried back to the graft, which they attack directly (Fig. 14.42). This recognition pathway is known as **direct allorecognition** (Fig. 14.43, left panel). Indeed, if the grafted tissue is depleted of antigen-presenting cells by treatment with antibodies or by prolonged incubation, rejection occurs only after a much longer time. Also, if the site of grafting lacks lymphatic drainage, no response to the graft results.

A second mechanism of allograft recognition leading to graft rejection is the uptake of allogeneic proteins by the recipient's own antigen-presenting cells and their presentation to T cells, including T_{reg}, by self-MHC molecules. The recognition of allogeneic proteins presented in this way is known as **indirect allorecognition** (Fig. 14.43, right panel). Among the graft-derived peptides presented by the recipient's antigen-presenting cells are the minor histocompatibility antigens and also peptides from the foreign MHC molecules

Fig. 14.42 The initiation of graft rejection normally involves the migration of donor antigen-presenting cells from the graft to local lymph nodes. The example of a skin graft is illustrated here, in which Langerhans cells are the antigen-presenting cells. They display peptides from the graft on their surface. After travelling to a lymph node, these antigen-presenting cells encounter recirculating naive T cells specific for graft antigens, and stimulate these T cells to divide. The resulting activated effector T cells migrate via the thoracic duct to the blood and home to the grafted tissue, which they rapidly destroy. Destruction is highly specific for donor-derived cells, suggesting that it is mediated by direct cytotoxicity and not by nonspecific inflammatory processes.

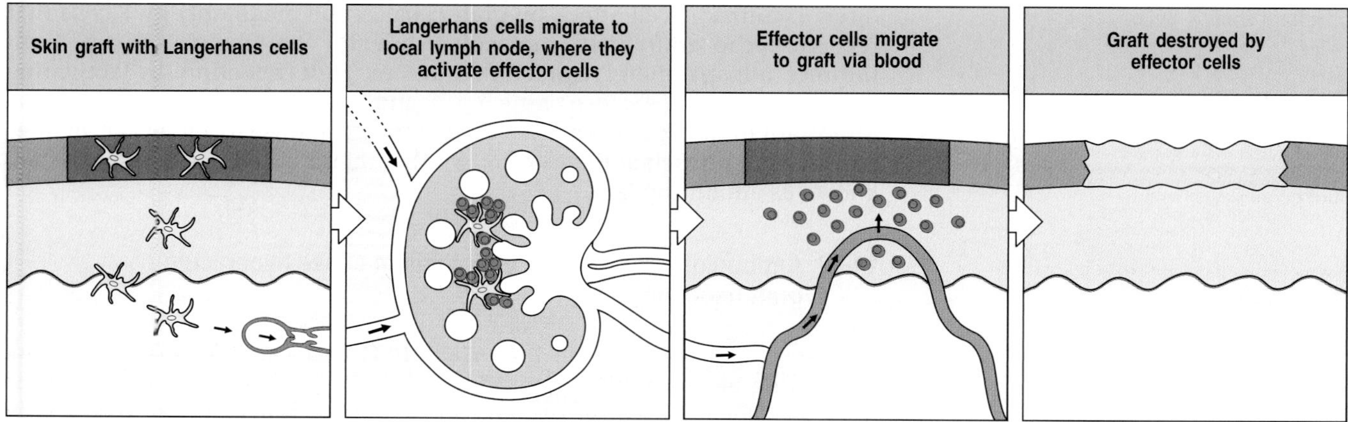
Skin graft with Langerhans cells
Langerhans cells migrate to local lymph node, where they activate effector cells
Effector cells migrate to graft via blood
Graft destroyed by effector cells

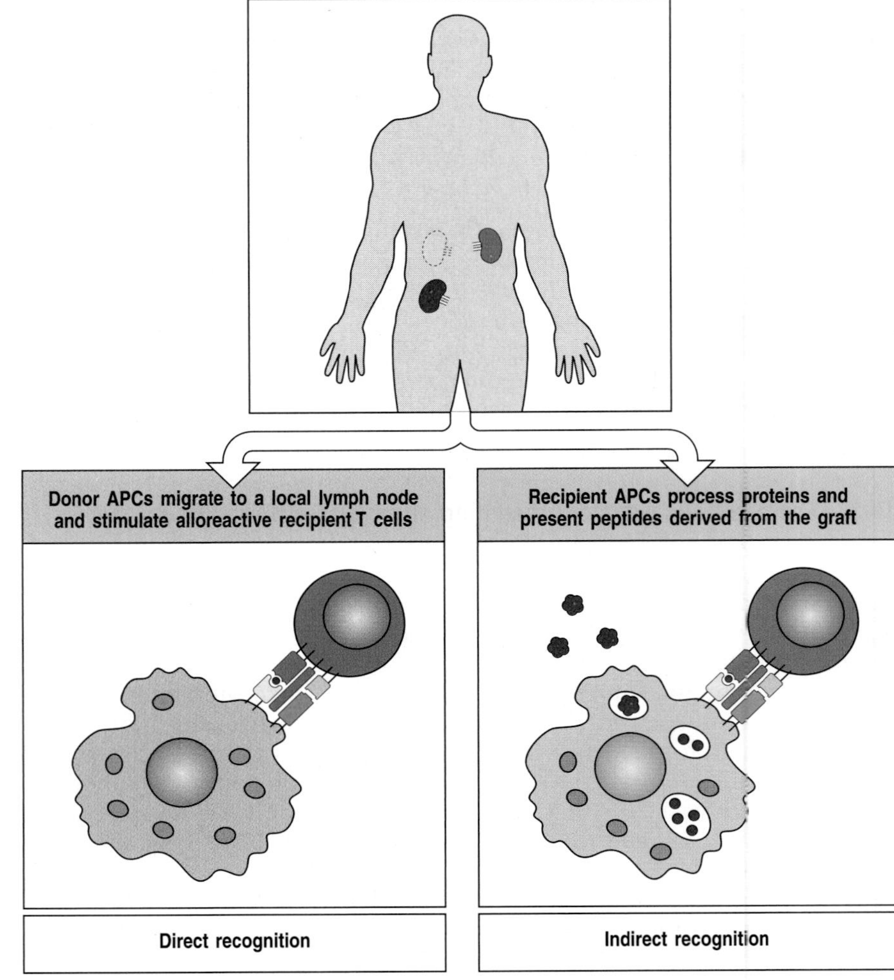

Donor APCs migrate to a local lymph node and stimulate alloreactive recipient T cells
Recipient APCs process proteins and present peptides derived from the graft
Direct recognition
Indirect recognition

Fig. 14.43 Alloantigens in grafted organs are recognized in two different ways. Direct recognition of a grafted organ (red in upper panel) is by T cells whose receptors have specificity for the allogeneic MHC class I or class II molecule in combination with peptide. These alloreactive T cells are stimulated by donor antigen-presenting cells (APCs), which express both the allogeneic MHC molecule and co-stimulatory activity (bottom left panel). Indirect recognition of the graft (bottom right panel) involves T cells whose receptors are specific for allogeneic peptides derived from the grafted organ. Proteins from the graft (red) are taken up and processed by the recipient's antigen-presenting cells and are therefore presented by self (recipient) MHC class I or class II molecules.

themselves, which are a major source of the polymorphic peptides recognized by the recipient's T cells, although in this case only if the graft is MHC-mismatched with the recipient.

The relative contributions of direct and indirect allorecognition in graft rejection are not known. Direct allorecognition is thought to be largely responsible for acute rejection, especially when MHC mismatches mean that the frequency of directly alloreactive recipient T cells is high. Furthermore, a direct cytotoxic T-cell attack on graft cells can be made only by T cells that recognize the graft MHC molecules directly. Nonetheless, T cells with indirect allospecificity can contribute to graft rejection by activating macrophages, which cause tissue injury and fibrosis, and are also likely to be important in the development of an antibody response to a graft. Antibodies produced to nonself antigens from another member of the same species are known as **alloantibodies**.

14-32 Antibodies reacting with endothelium cause hyperacute graft rejection.

Antibody responses are an important potential cause of graft rejection. Preexisting alloantibodies against blood group antigens and polymorphic MHC antigens can cause the rapid rejection of transplanted organs in a complement-dependent reaction that can occur within minutes of transplantation. This type of reaction is known as **hyperacute graft rejection**. Most grafts that are transplanted routinely in clinical medicine are vascularized organ grafts linked directly to the recipient's circulation. In some cases the recipient might already have circulating antibodies against donor graft antigens. Antibodies of the ABO type can bind to all tissues, not just red blood cells. They are preformed and are relevant in all ABO-mismatched individuals. In addition, antibodies against other antigens can be produced in response to a previous transplant or a blood transfusion. All such preexisting antibodies can cause the very rapid rejection of vascularized grafts because they react with antigens on the vascular endothelial cells of the graft and initiate the complement and blood clotting cascades. The vessels of the graft become blocked, causing its immediate death. Such grafts become engorged and purple-colored from hemorrhaged blood, which becomes deoxygenated (Fig. 14.44). This problem can be avoided by ABO-matching as well as **cross-matching** donor and recipient. Cross-matching involves determining whether the recipient has antibodies that react with the white blood cells of the donor. If antibodies of this type are found, they have hitherto been considered a serious contraindication to transplantation, as in the absence of any treatment they lead to near-certain hyperacute rejection.

These dogmas are changing, however. The presence of donor-specific MHC alloantibodies and a positive cross-match is no longer considered a major restricting factor in graft survival. The desensitization of patients by treatment with intravenous immunoglobulin has been successful in a proportion of patients in whom antibodies to the donor tissue were already present. Thus a positive cross-match is now not an absolute contraindication for transplantation.

A very similar problem prevents the routine use of animal organs—**xenografts**—in transplantation. If xenogeneic grafts could be used, it would circumvent the major limitation in organ replacement therapy, namely the severe shortage of donor organs. Pigs have been suggested as a potential source of organs for xenografting, because they are of a similar size to humans and are easily farmed. Most humans and other primates have natural antibodies that react with a ubiquitous cell-surface carbohydrate antigen (α-Gal) of other mammalian species, including pigs. When pig xenografts are placed in humans, these antibodies trigger hyperacute rejection by binding to the endothelial cells of the graft and initiating the complement and clotting cascades. The problem of hyperacute rejection is exacerbated in xenografts because complement-regulatory proteins such as CD59, DAF (CD55), and MCP (CD46) (see Section 2-22) work less efficiently across a species barrier and so pig regulatory proteins, for example, cannot protect the graft from attack by human complement.

A recent step toward xenotransplantation has been the development of transgenic pigs expressing human DAF as well as pigs that lack α-Gal. These approaches might one day reduce or eliminate hyperacute rejection in xenotransplantation. However, hyperacute rejection is only the first barrier faced by a xenotransplanted organ. The T lymphocyte-mediated graft rejection mechanisms might be extremely difficult to overcome with the immunosuppressive regimes currently available.

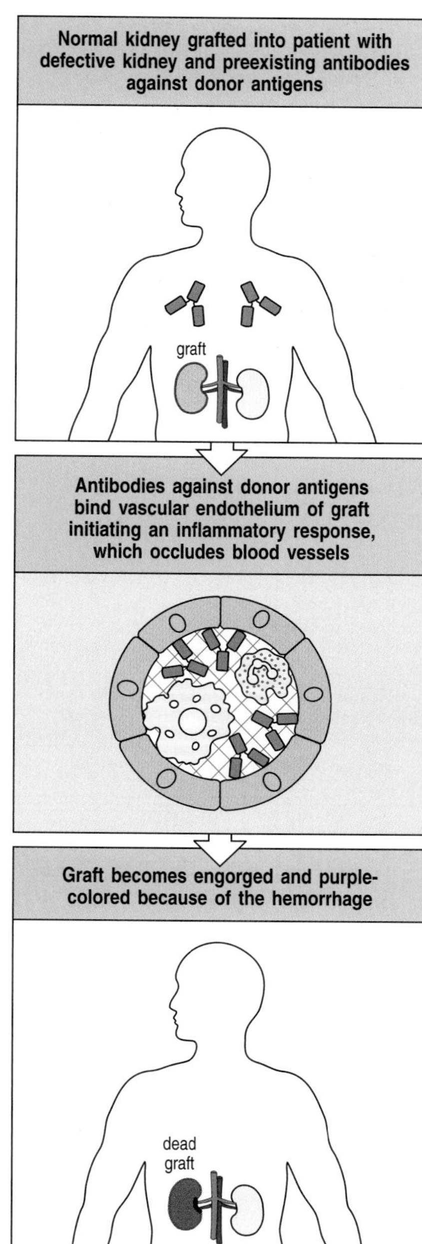

Fig. 14.44 Preexisting antibody against donor graft antigens can cause hyperacute graft rejection. In some cases, recipients already have antibodies against donor antigens, which are often blood group antigens. When the donor organ is grafted into such recipients, these antibodies bind to vascular endothelium in the graft, initiating the complement and clotting cascades. Blood vessels in the graft become obstructed by clots and leak, causing hemorrhage of blood into the graft. This becomes engorged and turns purple from the presence of deoxygenated blood.

14-33 Chronic organ rejection is caused by inflammatory vascular injury to the graft.

The success of modern immunosuppression means that about 90% of cadaveric kidney grafts are still functioning a year after transplantation. There has, however, been no improvement in rates of long-term graft survival: the half-life for functional survival of renal allografts remains about 8 years. The major cause of late organ failure is chronic rejection, characterized by concentric arteriosclerosis of graft blood vessels, accompanied by glomerular and tubular fibrosis and atrophy.

Mechanisms that contribute to chronic rejection can be divided into those due to alloreactivity and those due to other pathways, and into early and late events after transplantation. Alloreactivity can occur days and weeks after transplantation and can cause acute graft rejection. Alloreactive responses can also occur months to years after transplantation, and may be associated with a clinically hard-to-detect gradual loss of graft function. Other important causes of chronic graft rejection include ischemia–reperfusion injury, which occurs at the time of grafting but may have late adverse effects on the grafted organ, and later-developing adverse factors such as chronic cyclosporin toxicity or cytomegalovirus infection.

Infiltration of the graft vessels and tissues by macrophages, followed by scarring, are prominent histological features of late graft rejection. A model of injury has been developed in which alloreactive T cells infiltrating the graft secrete cytokines that stimulate the expression of endothelial adhesion molecules and also secrete chemokines such as CCL5 (see Fig. 2.46), which attracts monocytes that mature into macrophages in the graft. A second phase of chronic inflammation then supervenes, dominated by macrophage products including IL-1, TNF-α, and the chemokine CCL2, which leads to further recruitment of macrophages. These mediators conspire to cause chronic inflammation and scarring, which eventually leads to irreversible organ failure. Animal models of chronic rejection also show that alloreactive IgG antibodies may induce accelerated atherosclerosis in transplanted solid organs.

Tissue transplanted	No. of grafts in USA (2006)*	5-year graft survival
Kidney	18,017	71.9%
Liver	6650	67.4%
Heart	2192	71.5%
Pancreas	1387	53.2%#
Lung	1405	46.3%
Cornea	~40,000†	~70%
Bone marrow	15,000‡	40%/60%‡

Fig. 14.45 Tissues commonly transplanted in clinical medicine. The numbers of organ grafts performed in the United States in 2006 are shown. *Number of grafts includes multiple organ grafts. Data from the United Network for Organ Sharing. #Pancreas survival is 53.2% when transplanted alone, or 76.3% when transplanted with a kidney. †Data for 2000 courtesy of National Eye Institute. ‡Data for 2005 from the International Bone Marrow Transplant Registry for allogeneic transplants only; survival depends on disease and is 40% for patients with acute and 60% for patients with chronic forms of myelogenous leukemia. All grafts except corneas require long-term immunosuppression.

14-34 A variety of organs are transplanted routinely in clinical medicine.

Although the immune response makes organ transplantation difficult, there are few alternative therapies for organ failure. Three major advances have made it possible to use organ transplantation routinely in the clinic. First, the technical skill to perform organ replacement surgery has been mastered by many people. Second, networks of transplantation centers have been organized to ensure that the few healthy organs that are available are HLA-typed and so are matched with the most suitable recipient. Third, the use of powerful immunosuppressive drugs, especially cyclosporin A and FK-506 (tacrolimus) to inhibit T-cell activation (see Chapter 15), or blockade of IL-2 receptor signal with rapamycin, which provokes apoptosis of allospecifically activated CD4 lymphocytes, has markedly increased graft-survival rates. The different organs that are transplanted in the clinic are listed in Fig. 14.45. Some of these operations are performed routinely with a very high success rate. By far the most frequently transplanted solid organ is the kidney, the organ first successfully transplanted between identical twins in the 1950s. Transplantation of the cornea is even more frequent; this tissue is a special case because it is not vascularized, and corneal grafts between unrelated people are usually successful even without immunosuppression.

Many problems other than graft rejection are associated with organ transplantation. First, donor organs are difficult to obtain; this is especially a problem when the organ involved is a vital one, such as the heart or liver. Second, the disease that destroyed the patient's organ might also destroy the graft, as in pancreatic β-cell destruction in diabetes. Third, the immunosuppression required to prevent graft rejection increases the risk of cancer and infection. All of these problems need to be addressed before clinical transplantation can become commonplace. The problems most amenable to scientific solution are the development of more effective means of immunosuppression, the induction of graft-specific tolerance, and the development of xenografts as a practical solution to organ availability.

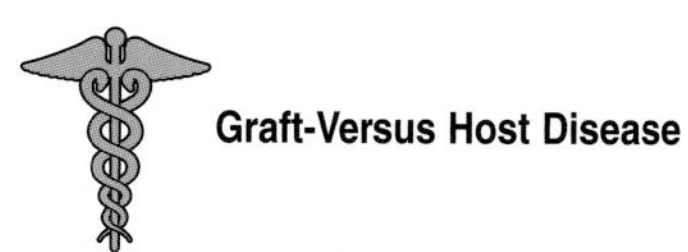

14-35 The converse of graft rejection is graft-versus-host disease.

Transplantation of hematopoietic stem cells by means of bone marrow transplantation is a successful therapy for some tumors derived from bone marrow precursor cells, such as certain leukemias and lymphomas. It can also be used to cure some primary immunodeficiency diseases (see Chapter 12) and inherited hematopoietic stem-cell diseases, such as the severe forms of thalassemia, by replacing genetically defective stem cells with normal donor ones. In leukemia therapy, the recipient's bone marrow, the source of the leukemia, must first be destroyed by a combination of irradiation and aggressive cytotoxic chemotherapy. One of the major complications of allogeneic bone marrow transplantation is **graft-versus-host disease** (**GVHD**), in which mature donor T cells that contaminate the allogeneic bone marrow recognize the tissues of the recipient as foreign, causing a severe inflammatory disease characterized by rashes, diarrhea, and liver disease. GVHD is particularly virulent when there is a mismatch of a major MHC class I or class II antigens. Most transplants are therefore undertaken only when the donor and recipient are HLA-matched siblings or, less frequently, when there is an HLA-matched unrelated donor. As in organ transplantation, GVHD also occurs in the context of disparities between minor histocompatibility antigens, so immunosuppression must also be used in every stem-cell transplant.

The presence of alloreactive donor T cells can easily be demonstrated experimentally by the **mixed lymphocyte reaction** (MLR), in which lymphocytes from a potential donor are mixed with irradiated lymphocytes from the potential recipient. If the donor lymphocytes contain naive T cells that recognize alloantigens on the recipient lymphocytes, they will respond by proliferating (Fig. 14.46). The MLR is sometimes used in the selection of donors for bone marrow transplants, when the lowest possible alloreactive response is essential. However, the limitation of the MLR in the selection of bone marrow

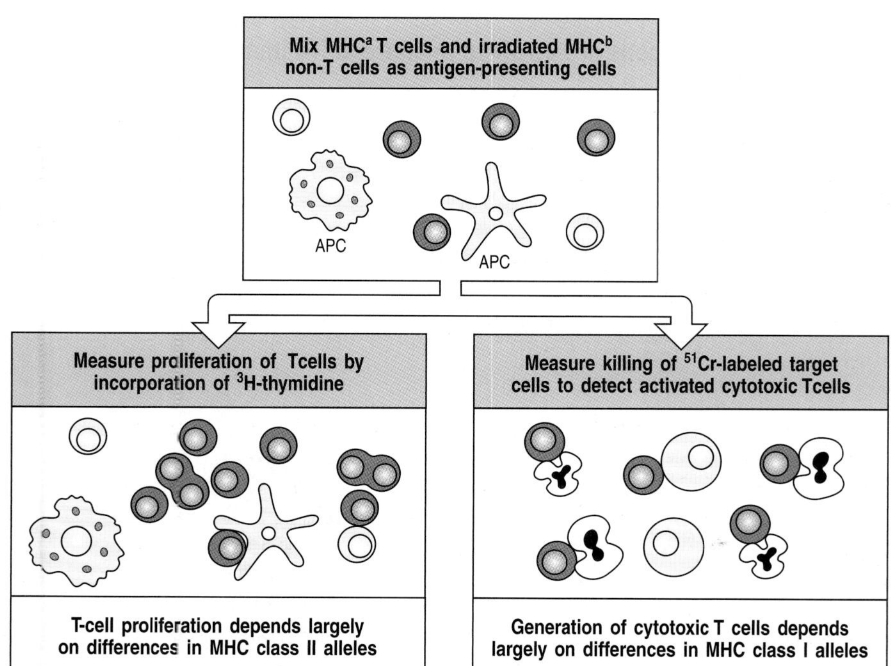

Fig. 14.46 The mixed lymphocyte reaction (MLR) can be used to detect histoincompatibility. Lymphocytes from the two individuals who are to be tested for compatibility are isolated from peripheral blood. The cells from one person (yellow), which will also contain antigen-presenting cells, are either irradiated or treated with mitomycin C; they will act as stimulator cells but cannot now respond by DNA synthesis and cell division to antigenic stimulation by the other person's cells. The cells from the two individuals are then mixed (top panel). If the unirradiated lymphocytes (the responders, blue) contain alloreactive T cells, these will be stimulated to proliferate and differentiate to effector cells. Between 3 and 7 days after mixing, the culture is assessed for T-cell proliferation (bottom left panel), which is mainly the result of CD4 T cells recognizing differences in MHC class II molecules, and for the generation of activated cytotoxic T cells (bottom right panel), which respond to differences in MHC class I molecules.

donors is that the test does not accurately quantify alloreactive T cells. A more accurate test is a version of the limiting-dilution assay (see Appendix I, Section A-25), which precisely counts the frequency of alloreactive T cells.

Although GVHD is harmful to the recipient of a bone marrow transplant, it can have some beneficial effects that are crucial to the success of the therapy. Much of the therapeutic effect of bone marrow transplantation for leukemia can be due to a **graft-versus-leukemia effect**, in which the allogeneic bone marrow recognizes minor histocompatibility antigens or tumor-specific antigens expressed by the leukemic cells, leading the donor cells to kill the leukemic cells. One of the treatment options for suppressing the development of GVHD is the elimination of mature T cells from the donor bone marrow *in vitro* before transplantation, thereby removing alloreactive T cells. Those T cells that subsequently mature from the donor marrow *in vivo* in the recipient are tolerant to the recipient's antigens. Although the elimination of GVHD has benefits for the patient, there is an increased risk of leukemic relapse, which provides strong evidence in support of the graft-versus-leukemia effect.

Immunodeficiency is another complication of donor T-cell depletion. Because most of the recipient's T cells are destroyed by the combination of high-dose chemotherapy and irradiation used to treat the recipient before the transplant, donor T cells are the major source for reconstituting a mature T-cell repertoire after the transplant. This is particularly true in adults, who have poor residual thymic function. If too many T cells are depleted from the graft, therefore, transplant recipients experience, and often die from, numerous opportunistic infections. The need to balance the beneficial effects of the graft-versus-leukemia effect and immunocompetence with the adverse effects of GVHD caused by donor T cells has spawned much research. One particularly promising approach is to prevent donor T cells from reacting with recipient antigens that they could meet shortly after the transplant. This is accomplished by depleting the recipient's antigen-presenting cells, chiefly dendritic cells (Fig. 14.47). Evidently, in this situation, the donor T cells are not activated during the initial inflammation that accompanies the transplant, and thereafter they do not promote GVHD. However, it is unclear whether there would be a graft-versus-leukemia effect in this context.

14-36 Regulatory T cells are involved in alloreactive immune responses.

As in all immune responses, regulatory CD25 CD4 T cells are now thought to play an important immunoregulatory role in the alloreactive immune responses involved in graft rejection. Experiments on the transplantation of

Fig. 14.47 Recipient type antigen-presenting cells are required for the efficient initiation of graft-versus-host disease (GVHD). T cells that accompany the hematopoietic stem cells from the donor (left panel) can recognize minor histocompatibility antigens of the recipient and start an immune response against the recipient's tissues. In stem-cell transplantation, minor antigens could be presented by either recipient- or donor-derived antigen-presenting cells, the latter deriving from the stem-cell graft and from precursor cells that differentiate after the transplant. Antigen-presenting cells are shown here as dendritic cells in a lymph node (middle panel). In mice, it has been possible to inactivate the host antigen-presenting cells by using gene knockouts. Such recipients are entirely resistant to GVHD mediated by donor CD8 T cells (right panel). Thus, cross-presentation of the recipient's minor histocompatibility antigens on donor dendritic cells is not sufficient to stimulate GVHD; those antigens endogenously synthesized and presented by the recipient's antigen-presenting cells are required to stimulate donor T cells. For this strategy to be useful for preventing GVHD in human patients, ways of depleting the recipient's antigen-presenting cells will be needed. These are the focus of research in several laboratories.

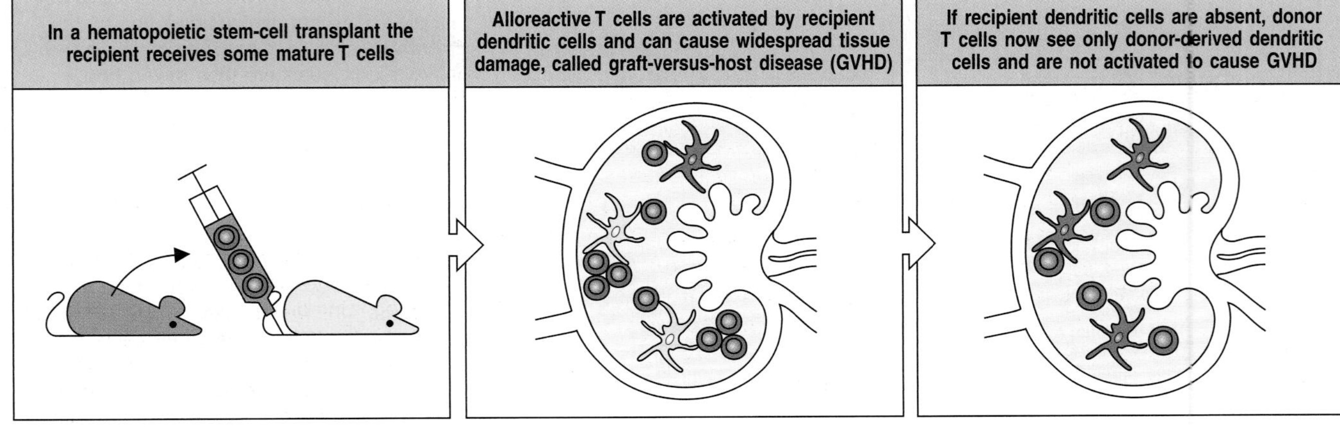
In a hematopoietic stem-cell transplant the recipient receives some mature T cells
Alloreactive T cells are activated by recipient dendritic cells and can cause widespread tissue damage, called graft-versus-host disease (GVHD)
If recipient dendritic cells are absent, donor T cells now see only donor-derived dendritic cells and are not activated to cause GVHD

allogeneic hematopoietic stem cells in mice have thrown some light on this question. Here, depletion of either CD4 CD25 T_{reg} cells in the recipient or of CD4 CD25 T_{reg} cells in the bone marrow graft before transplantation accelerated the onset of GVHD and subsequent death. In contrast, supplementing the graft with fresh CD4 CD25 T_{reg} cells or CD4 CD25 T_{reg} cells activated and expanded *ex vivo* delayed or even prevented death from GVHD. These experiments suggest that enriching or generating T_{reg} cells in bone-marrow grafts might provide a possible therapy for GVHD in the future.

Another class of regulatory T cells, $CD8^+$ $CD28^-$ T_{reg} cells, have an anergic phenotype and are thought to maintain T-cell tolerance indirectly by inhibiting the capacity of antigen-presenting cells to activate helper T cells. Cells of this type have been isolated from transplant patients. They can be distinguished from alloreactive CD8 cytotoxic T cells because they do not display cytotoxic activity against donor cells and express high levels of the inhibitory killer receptor CD94 (see Section 2-31). This finding suggests the possibility that $CD8^+$ $CD28^-$ T_{reg} cells interfere with the activation of antigen-presenting cells and have a role in the maintenance of transplant tolerance.

14-37 The fetus is an allograft that is tolerated repeatedly.

All of the transplants discussed so far are artifacts of modern medical technology. However, one 'foreign' tissue that is repeatedly grafted and repeatedly tolerated is the mammalian fetus. The fetus carries paternal MHC and minor histocompatibility antigens that differ from those of the mother (Fig. 14.48), and yet a mother can successfully bear many children expressing the same nonself MHC proteins derived from the father. The mysterious lack of rejection of the fetus has puzzled generations of immunologists, and no comprehensive explanation has yet emerged. One problem is that acceptance of the fetal allograft is so much the norm that it is difficult to study the mechanism that prevents rejection; if the mechanism for rejecting the fetus is rarely activated, how can one analyze the mechanisms that control it?

Various hypotheses have been advanced to account for the tolerance shown to the fetus. It has been proposed that the fetus is simply not recognized as foreign. There is evidence against this hypothesis, because women who have borne several children usually make antibodies directed against the father's MHC proteins and red blood cell antigens; indeed, this is the best source of antibodies for human MHC typing. However, the placenta, which is a fetus-derived tissue, seems to sequester the fetus from the mother's T cells. The outer layer of the placenta, the interface between fetal and maternal tissues, is the trophoblast. This does not express classical MHC class I and class II proteins, making it resistant to recognition and attack by maternal T cells. Tissues lacking MHC class I expression are, however, vulnerable to attack by NK cells (see Section 2-31). The trophoblast might be protected from attack by NK cells by the expression of a nonclassical and minimally polymorphic HLA class I molecule—HLA-G. This protein has been shown to bind to the two major inhibitory NK receptors, KIR1 and KIR2, and to inhibit NK killing.

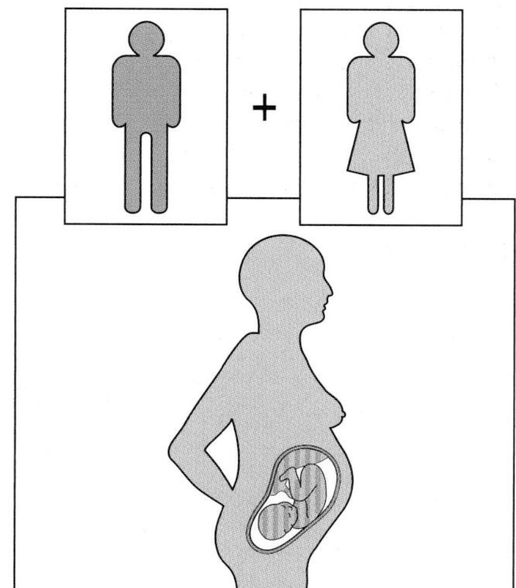

Fig. 14.48 The fetus is an allograft that is not rejected. Although the fetus carries MHC molecules derived from the father, and other foreign antigens, it is not rejected. Even when the mother bears several children to the same father, no sign of immunological rejection is seen.

The placenta might also sequester the fetus from the mother's T cells by an active mechanism of nutrient depletion. The enzyme indoleamine 2,3-dioxygenase (IDO) is expressed at a high level by cells at the maternal–fetal interface. This enzyme catabolizes, and thereby depletes, the essential amino acid tryptophan at this site, and T cells starved of tryptophan show reduced responsiveness. Inhibition of IDO in pregnant mice, using the inhibitor 1-methyltryptophan, causes rapid rejection of allogeneic but not syngeneic fetuses. This supports the hypothesis that maternal T cells, alloreactive to paternal MHC proteins, might be held in check in the placenta by tryptophan depletion.

It is likely that fetal tolerance is a multifactorial process. The trophoblast does not act as an absolute barrier between mother and fetus, and fetal blood cells can cross the placenta and be detected in the maternal circulation, albeit in very low numbers. There is direct evidence from experiments in mice for specific T-cell tolerance against paternal MHC alloantigens. Pregnant female mice whose T cells bear a transgenic receptor specific for a paternal alloantigen showed reduced expression of this T-cell receptor during pregnancy. These same mice lost the ability to control the growth of an experimental tumor bearing the same paternal MHC alloantigen. After pregnancy, tumor growth was controlled and the level of the T-cell receptor increased. This experiment demonstrates that the maternal immune system must have been exposed to paternal MHC alloantigens and that the immune response to these antigens was temporarily suppressed.

Yet another factor that might contribute to maternal tolerance of the fetus is the secretion of cytokines at the maternal–fetal interface. Both the uterine epithelium and the trophoblast secrete cytokines, including TGF-β, IL-4, and IL-10. This cytokine pattern tends to suppress T_H1 responses (see Section 10-5). The induction or injection of cytokines such as IFN-γ and IL-12, which promote T_H1 responses in experimental animals, promotes fetal resorption, the equivalent of spontaneous abortion in humans. Finally, it is possible that regulatory T cells could have a role in suppressing responses to the fetus.

The fetus is thus tolerated for two main reasons: it occupies a site protected by a nonimmunogenic tissue barrier, and it promotes a local immunosuppressive response in the mother. Several sites in the body, such as the eye, have these characteristics and allow the prolonged acceptance of foreign tissue grafts. They are usually called immunologically privileged sites (see Section 14-5).

Summary.

Clinical transplantation is now an everyday reality, its success built on MHC matching, immunosuppressive drugs, and technical skill. However, even accurate MHC matching does not prevent graft rejection; other genetic differences between host and donor can result in allogeneic proteins whose peptides are presented as minor histocompatibility antigens by MHC molecules on the grafted tissue, and responses to these can lead to rejection. Because we lack the ability to specifically suppress the response to the graft without compromising host defense, most transplants require generalized immunosuppression of the recipient. This can be toxic and increases the risk of cancer and infection. The fetus is a natural allograft that must be accepted—it almost always is—or the species will not survive. Tolerance to the fetus might hold the key to inducing specific tolerance to grafted tissues, or it might be a special case not applicable to organ replacement therapy.

Summary to Chapter 14.

Ideally, the effector functions of the immune system would be targeted only to foreign pathogens and never to self tissues. In practice, because foreign and self proteins are chemically similar, strict discrimination between self and nonself is impossible. Yet the immune system maintains tolerance to self tissues. This is accomplished by layers of regulation, all of which use surrogate markers to distinguish self from nonself, thus properly directing the immune response. When these regulatory mechanisms break down, autoimmune disease can result. Minor breaches of single regulatory barriers probably occur every day but are quelled by the effects of other regulatory layers: thus, tolerance operates at the level of the overall immune system. For disease to occur, multiple layers of tolerance have to be overcome and the effect needs to be chronic. These layers begin with central tolerance in the bone marrow and thymus, and include peripheral mechanisms such as anergy, cytokine deviation, and regulatory T cells. Sometimes immune responses do not occur simply because the antigens are not available, as in immune sequestration.

Perhaps because of selective pressure to mount effective immune responses to pathogens, the damping of immune responses to promote self-tolerance is limited and prone to failure. Genetic predisposition has an important role in determining which individuals will get an autoimmune disease. The MHC region has an important effect in many diseases. There are many other genes that contribute to immune regulation and thus, when defective, can cause or predispose to autoimmune disease. Environmental forces also have a significant role, because even identical twins are not always both affected by the same autoimmune disease. Influences from the environment could include infections, toxins, and chance events.

When self-tolerance is broken and autoimmune disease ensues, the effector mechanisms are quite similar to those used in pathogen responses. Although the details vary from disease to disease, both antibody and T cells can be involved. Much has been learned about immune responses made to tissue antigens by examining the response to nonself transplanted organs and tissues; lessons learned in the study of graft rejection apply to autoimmunity and vice versa. Transplantation of solid organs and bone marrow has brought on syndromes of rejection that are in many ways similar to autoimmune disease, but the targets are either major or minor histocompatibility antigens. The latter come from polymorphic genes. T cells are the main effectors in graft rejection and graft-versus-host disease.

For each of the undesirable categories of response discussed here (along with allergy, discussed in Chapter 13), the question is how to control the response without adversely affecting protective immunity to infection. The answer might lie in a more complete understanding of the regulation of the immune response, especially the suppressive mechanisms that seem to be important in tolerance. The deliberate control of the immune response is examined further in Chapter 15.

Questions.

14.1 *(a) Discuss the multiple layers of self-tolerance. (b) Name at least four of these layers and describe the mechanism of each in a few sentences.*

14.2 *What is the difference between 'dominant' and 'recessive' tolerance? Explain, and give an example of each.*

14.3 *What is the evidence that genetic predisposition has an important role in autoimmune disease? Give two examples, and for each explain why the example implicates genetics.*

14.4 *(a) Discuss one compelling piece of evidence that environment also has a role in the development of autoimmunity. (b) Name two potential environmental factors, and for one of them describe in more detail how it might work to incite autoimmunity.*

14.5 There are several different pathogenesis mechanisms in autoimmunity. Provide an example of four, and briefly describe an example of each. Include both antibody-dependent and Tcell dependent mechanisms.

14.6 A person with leukemia receives a bone marrow transplant from his HLA-identical brother. Two weeks later he develops a skin rash and nausea, although his leukemia is in remission. (a) What is this syndrome called? (b) What type of lymphocyte causes it? (c) What antigens are being recognized?

14.7 Why is systemic lupus erythematosus (SLE) thought to be an autoimmune disease?

14.8 How is the interaction of T cells with B cells relevant to the pathogenesis of SLE?

14.9 What is the cause of autoimmune diabetes?

14.10 What is the role of TNF-α in the rheumatoid arthritis? Which cells does it come from?

14.11 Distinguish the immunological and thyroid functional aspects of destructive (Hashimoto's) thyroiditis and hyperthyroid Graves' disease.

14.12 Give three ways in which autoimmunity and allergy differ and three ways in which they are similar.

Section references.

14-1 A critical function of the immune system is to discriminate self from nonself.

Ehrlich, P., and Morgenroth, J.: **On haemolysins**, in Himmelweit, F. (ed): *The Collected Papers of Paul Ehrlich.* London, Pergamon, 1957: 246–255.

Janeway, C.A., Jr.: **The immune system evolved to discriminate infectious nonself from noninfectious self.** *Immunol. Today* 1992, **13**:11–16.

14-2 Multiple tolerance mechanisms normally prevent autoimmunity.

Goodnow, C.C.: **Balancing immunity and tolerance: deleting and tuning lymphocyte repertoires.** *Proc. Natl Acad. Sci. USA* 1996, **93**:2264–2271.

14-3 Central deletion or inactivation of newly formed lymphocytes is the first checkpoint of self-tolerance.

Goodnow, C.C., Adelstein, S., and Basten, A.: **The need for central and peripheral tolerance in the B cell repertoire.** *Science* 1990, **248**:1373–1379.

Kisielow, P., Bluthmann, H., Staerz, U.D., Steinmetz, M., and von Boehmer, H.: **Tolerance in T-cell-receptor transgenic mice involves deletion of nonmature $CD4^+8^+$ thymocytes.** *Nature* 1988, **333**:742–746.

Nemazee, D.A., and Burki, K.: **Clonal deletion of B lymphocytes in a transgenic mouse bearing anti-MHC class-I antibody genes.** *Nature* 1989, **337**:562–566.

Nossal, G.J.V., and Pike, B.L.: **Clonal anergy: persistence in tolerant mice of antigen-binding B lymphocytes incapable of responding to antigen or mitogen.** *Proc. Natl Acad. Sci. USA* 1980, **77**:1602–1606.

Nossal, G.J.V., and Pike, B.L.: **Mechanisms of clonal abortion tolerogenesis: I. Response of immature hapten-specific B lymphocytes.** *J. Exp. Med.* 1978, **148**:1161–1170.

14-4 Lymphocytes that bind self antigens with relatively low affinity usually ignore them but in some circumstances become activated.

Billingham, R.E., Brent, L., and Medawar, P.B.: **Actively acquired tolerance of foreign cells.** *Nature* 1953, **172**:603–606.

Goverman, J., Woods, A., Larson, L., Weiner, L.P., Hood, L., and Zaller, D.M.: **Transgenic mice that express a myelin basic protein-specific T cell receptor develop spontaneous autoimmunity.** *Cell* 1993, **72**:551–560.

Hannum, L.G., Ni, D., Haberman, A.M., Weigert, M.G., and Shlomchik, M.J.: **A disease-related RF autoantibody is not tolerized in a normal mouse: implications for the origins of autoantibodies in autoimmune disease.** *J. Exp. Med.* 1996, **184**:1269–1278.

Katz, J.D., Wang, B., Haskins, K., Benoist, C., and Mathis, D.: **Following a diabetogenic T cell from genesis through pathogenesis.** *Cell* 1993, **74**:1089–1100.

Kurts, C., Sutherland, R.M., Davey, G., Li, M., Lew, A.M., Blanas, E., Carbone, F.R., Miller, J.F., and Heath, W.R.: **CD8 T cell ignorance or tolerance to islet antigens depends on antigen dose.** *Proc. Natl Acad. Sci. USA* 1999, **96**:12703–12707.

Marshak-Rothstein, A.: **Toll-like receptors in systemic autoimmune disease.** *Nat. Rev. Immunol.* 2006, **6**:823–835.

Martin, D.A., and Elkon, K.B.: **Autoantibodies make a U-turn: the toll hypothesis for autoantibody specificity.** *J. Exp. Med.* 2005, **202**:1465–1469.

Miller, J.F., and Heath, W.R.: **Self-ignorance in the peripheral T-cell pool.** *Immunol. Rev.* 1993, **133**:131–150.

14-5 Antigens in immunologically privileged sites do not induce immune attack but can serve as targets.

Alison, J., Georgiou, H.M., Strasser, A., and Vaux, D.L.: **Transgenic expression of CD95 ligand on islet β cells induces a granulocytic infiltration but does not confer immune privilege upon islet allografts.** *Proc. Natl Acad. Sci. USA* 1997, 94: 3943–3947.

Ferguson, T.A., and Griffith, T.S.: **A vision of cell death: insights into immune privilege.** *Immunol. Rev.* 1997, **156**:167–184.

Green, D.R., and Ware, C.F.: **Fas-ligand: privilege and peril.** *Proc. Natl Acad. Sci. USA* 1997, **94**:5986–5990.

Streilein, J.W., Ksander, B.R., and Taylor, A.W.: **Immune deviation in relation to ocular immune privilege.** *J. Immunol.* 1997, **158**:3557–3560.

14-6 Autoreactive T cells that express particular cytokines may be nonpathogenic or may suppress pathogenic lymphocytes.

von Herrath, M.G., and Harrison, L.C.: **Antigen-induced regulatory T cells in autoimmunity.** *Nat. Rev. Immunol.* 2003, **3**:223–232.

14-7 Autoimmune responses can be controlled at various stages by regulatory T cells.

Asano, M., Toda, M., Sakaguchi, N., and Sakaguchi, S.: **Autoimmune disease as a consequence of developmental abnormality of a T cell subpopulation.** *J. Exp. Med.* 1996, **184**:387–396.

Faria, A.M., and Weiner, H.L.: **Oral tolerance: mechanisms and therapeutic applications.** *Adv. Immunol.* 1999, **73**:153–264.

Fillatreau, S., Sweenie, C.H., McGeachy, M.J., Gray, D., and Anderton, S.M.: **B cells regulate autoimmunity by provision of IL-10.** *Nat. Immunol.* 2002, **3**:944–950.

Fontenot, J.D., Gavin, M.A., and Rudensky, A.Y.: **Foxp3 programs the development and function of CD4⁺CD25⁺ regulatory T cells.** *Nat. Immunol.* 2003, **4**:330–336.

Hara, M., Kingsley, C.I., Niimi, M., Read, S., Turvey, S.E., Bushell, A.R., Morris, P.J., Powrie, F., and Wood, K.J.: **IL-10 is required for regulatory T cells to mediate tolerance to alloantigens *in vivo*.** *J. Immunol.* 2001, **166**:3789–3796.

Johnson, B.D., Becker, E.E., LaBelle, J.L., and Truitt, R.L.: **Role of immunoregulatory donor T cells in suppression of graft-versus-host disease following donor leukocyte infusion therapy.** *J. Immunol.* 1999, **163**:6479–6487.

Jordan, M.S., Boesteanu, A., Reed, A.J., Petrone, A.L., Holenbeck, A.E., Lerman, M.A., Naji, A., and Caton, A.J.: **Thymic selection of $CD4^+CD25^+$ regulatory T cells induced by an agonist self-peptide.** *Nat. Immunol.* 2001, **2**:301–306.

Khattri, R., Cox, T., Yasayko, S.A., and Ramsdell, F.: **An essential role for Scurfin in $CD4^+CD25^+$ T regulatory cells.** *Nat. Immunol.* 2003, **4**:337–342.

Ku, C.C., Murakami, M., Sakamoto, A., Kappler, J., and Marrack, P.: **Control of homeostasis of $CD8^+$ memory T cells by opposing cytokines.** *Science* 2000, **288**:675–678.

Mauri, C., Gray, D., Mushtaq, N., and Londei, M.: **Prevention of arthritis by interleukin 10-producing B cells.** *J. Exp. Med.* 2003, **197**:489–501.

Maloy, K.J., and Powrie, F.: **Regulatory T cells in the control of immune pathology.** *Nat. Immunol.* 2001, **2**:816–822.

Mottet, C., Uhlig, H.H., and Powrie, F.: **Cutting edge: cure of colitis by $CD4^+CD25^+$ regulatory T cells.** *J. Immunol.* 2003, **170**:3939–3943.

Plas, D.R., Rathmell, J.C., and Thompson, C.B.: **Homeostatic control of lymphocyte survival: potential origins and implications.** *Nat. Immunol.* 2002, **3**:515–521.

Qin, S., Cobbold, S.P., Pope, H., Elliott, J., Kioussis, D., Davies, J., and Waldmann, H.: **'Infectious' transplantation tolerance.** *Science* 1993, **259**:974–977.

Rioux, J.D., and Abbas, A.K.: **Paths to understanding the genetic basis of autoimmune disease.** *Nature* 2005, 435: 584–589.

Roncarolo, M.G., and Levings, M.K.: **The role of different subsets of T regulatory cells in controlling autoimmunity.** *Curr. Opin. Immunol.* 2000, **12**:676–683.

Sakaguchi, S.: **Regulatory T cells: key controllers of immunologic self-tolerance.** *Cell* 2000, **101**:455–458.

Seo, S.J., Fields, M.L., Buckler, J.L., Reed, A.J., Mandik-Nayak, L., Nish, S.A., Noelle, R.J., Turka, L.A., Finkelman, F.D., Caton, A.J., et al.: **The impact of T helper and T regulatory cells on the regulation of anti-double-stranded DNA B cells.** *Immunity* 2002, **16**:535–546.

Shevach, E.M.: **$CD4^+$ $CD25^+$ suppressor T cells: more questions than answers.** *Nat. Rev. Immunol.* 2002, **2**:389–400.

Singer, G.G., and Abbas, A.K.: **The fas antigen is involved in peripheral but not thymic deletion of T lymphocytes in T cell receptor transgenic mice.** *Immunity* 1994, **1**:365–371.

Ueda, H., Howson, J.M., Esposito, L., Heward, J., Snook, H., Chamberlain, G., Rainbow, D.B., Hunter, K.M., Smith, A.N., DiGenova, G., *et al.*, **Association of the T-cell regulatory gene CTLA4 with susceptibility to autoimmune disease.** *Nature* 2003, **423**, 506–511.

Wang, B., Geng, Y.B., and Wang, C.R.: **CD1-restricted NK T cells protect nonobese diabetic mice from developing diabetes.** *J. Exp. Med.* 2001, **194**:313–320.

Weiner, H.L.: **Oral tolerance: immune mechanisms and the generation of Th3-type TGF-β-secreting regulatory cells.** *Microbes Infect.* 2001, **3**:947–954.

Wildin, R.S., Ramsdell, F., Peake, J., Faravelli, F., Casanova, J.L., Buist, N., Levy-Lahad, E., Mazzella, M., Goulet, O., Perroni, L., *et al.*: **X-linked neonatal diabetes mellitus, enteropathy and endocrinopathy syndrome is the human equivalent of mouse scurfy.** *Nat. Genet.* 2001, **27**:18–20.

Yamanouchi, J., Rainbow, D., Serra, P., Howlett, S., Hunter, K., Garner, V.E.S., Gonzalez-Munoz, A., Clark, J., Veijola, R., Cubbon, R., *et al.*: **Interleukin-2 gene variation impairs regulatory T cell function and causes autoimmunity.** *Nat. Genet.* 2007, **39**:329–337.

14-8 Specific adaptive immune responses to self antigens can cause autoimmune disease.

Hardin, J.A., and Craft, J.E.: **Patterns of autoimmunity to nucleoproteins in patients with systemic lupus erythematosus.** *Rheum. Dis. Clinics N. Am.* 1987, **13**:37–46.

Lotz, P.H.: **The autoantibody repertoire: searching for order.** *Nat. Rev. Immunol.* 2003, **3**:73–78.

Shlomchik, M.J., Marshak-Rothstein, A., Wolfowicz, C.B., Rothstein, T.L., and Weigert, M.G.: **The role of clonal selection and somatic mutation in autoimmunity.** *Nature* 1987, **328**:805–811.

Steinman, L.: **Multiple sclerosis: a coordinated immunological attack against myelin in the central nervous system.** *Cell* 1996, **85**:299–302.

14-9 Autoimmune diseases can be classified into clusters that are typically either organ-specific or systemic.

Bach, J.F.: **Organ-specific autoimmunity.** *Immunol. Today* 1995, **16**:353–355.

King, C., and Sarvetnick, N.: **Organ-specific autoimmunity.** *Curr. Opin. Immunol.* 1997, **9**:863–871.

14-10 Multiple aspects of the immune system are typically recruited in autoimmune disease.

Christensen, S.R., Shupe, J., Nickerson, K., Kashgarian, M., Flavell, R.A., and Shlomchik, M.J.: **Toll-like receptor 7 and TLR9 dictate autoantibody specificity and have opposing inflammatory and regulatory roles in a murine model of lupus.** *Immunity* 2006, **25**:417–428.

Couser, W.G.: **Pathogenesis of glomerulonephritis.** *Kidney Int. Suppl.* 1993, **42**:S19–S26.

Green, E.A., and Flavell, R.A.: **The initiation of autoimmune diabetes.** *Curr. Opin. Immunol.* 1999, **11**:663–669.

Huang, X.R., Tipping, P.G., Apostolopoulos, C., Oettinger, C., D'Souza, M., Milton, G., and Holdsworth, S.R.: **Mechanisms of T cell-induced glomerular injury in anti-glomerular basement membrane (GBM) glomerulonephritis in rats.** *Clin. Exp. Immunol.* 1997, **157**:134–142.

Shlomchik, M.J., and Madaio, M.P.: **The role of antibodies and B cells in the pathogenesis of lupus nephritis.** *Springer Semin. Immunopathol.* 2003, **24**:363–375.

14-11 Chronic autoimmune disease develops through positive feedback from inflammation, inability to clear the self antigen, and a broadening of the autoimmune response.

Salato, V.K., Hacker-Foegen, M.K., Lazarova, Z., Fairley, J.A., and Lin, M.S.: **Role of intramolecular epitope spreading in pemphigus vulgaris.** *Clin. Immunol.* 2005, **116**:54–64.

Shlomchik, M.J., Craft, J., and Mamula, M.J.: **From T to B and back again: positive feedback in systemic autoimmune disease.** *Nat. Rev. Immunol.* 2001, **1**:147–153.

Steinman, L.: **A few autoreactive cells in an autoimmune infiltrate control a vast population of nonspecific cells: a tale of smart bombs and the infantry.** *Proc. Natl Acad. Sci. USA* 1996, **93**:2253–2256.

14-12 Both antibody and effector T cells can cause tissue damage in autoimmune disease.

Chan, O.T.M., Madaio, M.P., and Shlomchik, M.J.: **The central and multiple roles of B cells in lupus pathogenesis.** *Immunol. Rev.* 1999, **169**:107–121.

Naparstek, Y., and Plotz, P.H.: **The role of autoantibodies in autoimmune disease.** *Annu. Rev. Immunol.* 1993, **11**:79–104.

Vlahakos, D., Foster, M.H., Ucci, A.A., Barrett, K.J., Datta, S.K., and Madaio, M.P.: **Murine monoclonal anti-DNA antibodies penetrate cells, bind to nuclei, and induce glomerular proliferation and proteinuria *in vivo*.** *J. Am. Soc. Nephrol.* 1992, **2**:1345–1354.

14-13 Autoantibodies against blood cells promote their destruction.

Beardsley, D.S., and Ertem, M.: **Platelet autoantibodies in immune thrombocytopenic purpura.** *Transfus. Sci.* 1998, **19**:237–244.

Clynes, R., and Ravetch, J.V.: **Cytotoxic antibodies trigger inflammation through Fc receptors.** *Immunity* 1995, **3**:21–26.

Domen, R.E.: **An overview of immune hemolytic anemias.** *Cleveland Clin. J. Med.* 1998, **65**:89–99.

Silberstein, L.E.: **Natural and pathologic human autoimmune responses to carbohydrate antigens on red blood cells.** *Springer Semin. Immunopathol.* 1993, **15**:139–153.

14-14 The fixation of sublytic doses of complement to cells in tissues stimulates a powerful inflammatory response.

Brandt, J., Pippin, J., Schulze, M., Hansch, G.M., Alpers, C.E., Johnson, R.J., Gordon, K., and Couser, W.G.: **Role of the complement membrane attack complex (C5b-9) in mediating experimental mesangioproliferative glomerulonephritis.** *Kidney Int.* 1996, **49**:335–343.

Hansch, G.M.: **The complement attack phase: control of lysis and non-lethal effects of C5b-9.** *Immunopharmacol.* 1992, **24**:107–117.

Shin, M.L., and Carney, D.F.: **Cytotoxic action and other metabolic consequences of terminal complement proteins.** *Prog. Allergy* 1988, **40**:44–81.

14-15 Autoantibodies against receptors cause disease by stimulating or blocking receptor function.

Bahn, R.S., and Heufelder, A.E.: **Pathogenesis of Graves' ophthalmopathy.** *N. Engl. J. Med.* 1993, **329**:1468–1475.

Feldmann, M., Dayan, C., Grubeck Loebenstein, B., Rapoport, B., and Londei, M.: **Mechanism of Graves thyroiditis: implications for concepts and therapy of autoimmunity.** *Int. Rev. Immunol.* 1992, **9**:91–106.

Vincent, A., Lily, O., and Palace, J.: **Pathogenic autoantibodies to neuronal proteins in neurological disorders.** *J. Neuroimmunol.* 1999, **100**:169–180.

14-16 Autoantibodies against extracellular antigens cause inflammatory injury by mechanisms akin to type II and type III hypersensitivity reactions.

Casciola-Rosen, L.A., Anhalt, G., and Rosen, A.: **Autoantigens targeted in systemic lupus erythematosus are clustered in two populations of surface structures on apoptotic keratinocytes.** *J. Exp. Med.* 1994, **179**:1317–1330.

Clynes, R., Dumitru, C., and Ravetch, J.V.: **Uncoupling of immune complex formation and kidney damage in autoimmune glomerulonephritis.** *Science* 1998, **279**:1052–1054.

Kotzin, B.L.: **Systemic lupus erythematosus.** *Cell* 1996, **85**:303–306.

Lawley, T.J., Bielory, L., Gascon, P., Yancey, K.B., Young, N.S., and Frank, M.M.: **A prospective clinical and immunologic analysis of patients with serum sickness.** *N. Engl. J. Med.* 1984, **311**:1407–1413.

Tan, E.M.: **Antinuclear antibodies: diagnostic markers for autoimmune diseases and probes for cell biology.** *Adv. Immunol.* 1989, **44**:93–151.

Mackay, M., Stanevsky, A., Wang, T., Aranow, C., Li, M., Koenig, S., Ravetch, J.V., and Diamond, B.: **Selective dysregulation of the FcγIIB receptor on memory B cells in SLE.** *J. Exp. Med.* 2006 **203**:2157–2164.

Xiang, Z., Cutler, A.J., Brownlie, R.J., Fairfax, K., Lawlor, K.E., Severinson, E., Walker, E.U., Manz, R.A., Tarlinton, D.M., and Smith, K.G.: **FcγRIIb controls bone marrow plasma cell persistence and apoptosis.** *Nat. Immunol.* 2007, **8**:419–429.

14-17 T cells specific for self antigens can cause direct tissue injury and sustain autoantibody responses.

Feldmann, M., and Steinman, L.: **Design of effective immunotherapy for human autoimmunity.** *Nature* 2005, **435**:612–619.

Firestein, G.S.: **Evolving concepts of rheumatoid arthritis.** *Nature* 2003, **423**:356–361.

Haskins, K., and Wegmann, D.: **Diabetogenic T-cell clones.** *Diabetes* 1996, **45**:1299–1305.

Peng, S.L., Madaio, M.P., Hughes, D.P., Crispe, I.N., Owen, M.J., Wen, L., Hayday, A.C., and Craft, J.: **Murine lupus in the absence of αβ T cells.** *J. Immunol.* 1996, **156**:4041–4049.

Zamvil, S., Nelson, P., Trotter, J., Mitchell, D., Knobler, R., Fritz, R., and Steinman, L.: **T-cell clones specific for myelin basic protein induce chronic relapsing paralysis and demyelination.** *Nature* 1985, **317**:355–358.

Zekzer, D., Wong, F.S., Ayalon, O., Altieri, M., Shintani, S., Solimena, M., and Sherwin, R.S.: **GAD-reactive CD4$^+$ Th1 cells induce diabetes in NOD/SCID mice.** *J. Clin. Invest.* 1998, **101**:68–73.

14-18 Autoimmune diseases have a strong genetic component.

Gonzalez, A., Katz, J.D., Mattei, M.G., Kikutani, H., Benoist, C., and Mathis, D.: **Genetic control of diabetes progression.** *Immunity* 1997, **7**:873–883.

Morel, L., Rudofsky, U.H., Longmate, J.A., Schiffenbauer, J., and Wakeland, E.K.: **Polygenic control of susceptibility to murine systemic lupus erythematosus.** *Immunity* 1994, **1**:219–229.

14-19 A defect in a single gene can cause autoimmune disease.

Anderson, M.S., Venanzi, E.S., Chen, Z., Berzins, S.P., Benoist, C., and Mathis, D.: **The cellular mechanism of Aire control of T cell tolerance.** *Immunity* 2005, **23**:227–239.

Bacchetta, R., Passerini, L., Gambineri, E., Dai, M., Allan, S.E., Perroni, L., Dagna-Bricarelli, F., Sartirana, C., Matthes-Martin, S., Lawitschka, A., *et al.*: **Defective regulatory and effector T cell functions in patients with FOXP3 mutations.** *J. Clin. Invest.* 2006, **116**:1713–1722.

Rieux-Laucat, F., Le Deist, F., and Fischer, A.: **Autoimmune lymphoproliferative syndromes: genetic defects of apoptosis pathways.** *Cell Death Differ.* 2003, **10**:124–133.

Rizzi, M., Ferrera, F., Filaci, G., and Indiveri, F.: **Disruption of immunological tolerance: role of AIRE gene in autoimmunity.** *Autoimmun Rev* 2006, **5**:145–147.

Santiago-Raber, M.L., Laporte, C., Reininger, L., and Izui, S.: **Genetic basis of murine lupus.** *Autoimmun. Rev.* 2004, **3**:33–39.

Singer, G.G., Carrera, A.C., Marshak-Rothstein, A., Martinez, C., and Abbas, A.K.: **Apoptosis, Fas and systemic autoimmunity: the MRL-lpr/lpr model.** *Curr. Opin. Immunol.* 1994, **6**:913–920.

14-20 Several approaches have given us insight into the genetic basis of autoimmunity.

Gregersen, P.K. **Pathways to gene identification in rheumatoid arthritis: PTPN22 and beyond.** *Immunol Rev.* 2005, **204**:74–86.

Kumar, K.R., Li L., Yan, M., Bhaskarabhatla, M., Mobley, A.B., Nguyen, C., Mooney, J.M., Schatzle, J.D., Wakeland, E.K., and Mohan, C.: **Regulation of B cell tolerance by the lupus susceptibility gene *Ly108*.** *Science* 2006, **312**:1665–1669.

Nishimura, H., Nose, M., Hiai, H., Minato, N., and Honjo, T.: **Development of lupus-like autoimmune diseases by disruption of the PD-1 gene encoding an ITIM motif-carrying immunoreceptor.** *Immunity* 1999, **11**:141–151.

Okazaki, T., and Wang, J.: **PD-1/PD-L pathway and autoimmunity.** *Autoimmunity* 2005, **38**:353–357.

Vyse, T.J., and Todd, J.A.: **Genetic analysis of autoimmune disease.** *Cell* 1996, **85**:311–318.

Wakeland, E.K., Wandstrat, A.E., Liu, K., and Morel, L.: **Genetic dissection of systemic lupus erythematosus.** *Curr. Opin. Immunol.* 1999, **11**:701–707.

14-21 Genes that predispose to autoimmunity fall into categories that affect one or more of the mechanisms of tolerance.

Goodnow, C.C.: **Polygenic autoimmune traits: Lyn, CD22, and SHP-1 are limiting elements of a biochemical pathway regulating BCR signaling and selection.** *Immunity* 1998 **8**:497–508.

Tivol, E.A., Borriello, F., Schweitzer, A.N., Lynch, W.P., Bluestone, J.A., and Sharpe, A.H.: **Loss of CTLA-4 leads to massive lymphoproliferation and fatal multiorgan tissue destruction, revealing a critical negative regulatory role of CTLA-4.** *Immunity* 1995, **3**:541–547.

Wakeland, E.K., Liu, K., Graham, R.R., and Behrens, T.W.: **Delineating the genetic basis of systemic lupus erythematosus.** *Immunity* 2001, **15**:397–408.

Walport, M.J.: **Lupus, DNase and defective disposal of cellular debris.** *Nat. Genet.* 2000, **25**:135–136.

Whitacre, C.C., Reingold, S.C., and O'Looney, P.A.: **A gender gap in autoimmunity.** *Science* 1999, **283**:1277–1278.

14-22 MHC genes have an important role in controlling susceptibility to autoimmune disease.

Haines, J.L., Ter Minassian, M., Bazyk, A., Gusella, J.F., Kim, D.J., Terwedow, H., Pericak-Vance, M.A., Rimmler, J.B., Haynes, C.S., Roses, A.D., *et al.*: **A complete genomic screen for multiple sclerosis underscores a role for the major histocompatibility complex. The Multiple Sclerosis Genetics Group.** *Nat. Genet.* 1996, **13**:469–471.

McDevitt, H.O.: **Discovering the role of the major histocompatibility complex in the immune response.** *Annu. Rev. Immunol.* 2000, **18**:1–17.

14-23 External events can initiate autoimmunity.

Klareskog, L., Padyukov, L., Ronnelid, J., and Alfredsson, L.: **Genes, environment and immunity in the development of rheumatoid arthritis.** *Curr. Opin. Immunol.* 2006 **18**:650–655.

14-24 Infection can lead to autoimmune disease by providing an environment that promotes lymphocyte activation.

Aichele, P., Bachmann, M.F., Hengarter, H., and Zinkernagel, R.M.: **Immunopathology or organ-specific autoimmunity as a consequence of virus infection.** *Immunol. Rev.* 1996, **152**:21–45.

Bach, J.F.: **Infections and autoimmune diseases.** *J. Autoimmunity* 2005, **25**:74–80.

Moens, U., Seternes, O.M., Hey, A.W., Silsand, Y., Traavik, T., Johansen, B., and Rekvig, O.P.: ***In vivo* expression of a single viral DNA-binding protein generates systemic lupus erythematosus-related autoimmunity to double-stranded DNA and histones.** *Proc. Natl Acad. Sci. USA* 1995, **92**:12393–12397.

Steinman, L., and Conlon, P.: **Viral damage and the breakdown of self-tolerance.** *Nat. Med.* 1997, **3**:1085–1087.

von Herrath, M.G., Evans, C.F., Horwitz, M.S., and Oldstone, M.B.: **Using transgenic mouse models to dissect the pathogenesis of virus-induced autoimmune disorders of the islets of Langerhans and the central nervous system.** *Immunol. Rev.* 1996, **152**:111–143.

von Herrath, M.G., Holz, A., Homann, D., and Oldstone, M.B.: **Role of viruses in type I diabetes.** *Semin. Immunol.* 1998, **10**:87–100.

14-25 Cross-reactivity between foreign molecules on pathogens and self molecules can lead to anti-self responses and autoimmune disease.

Barnaba, V., and Sinigaglia, F.: **Molecular mimicry and T cell-mediated autoimmune disease.** *J. Exp. Med.* 1997, **185**:1529–1531.

Rose, N.R.: **Infection, mimics, and autoimmune disease.** *J. Clin. Invest.* 2001, **107**:943–944.

Rose, N.R., Herskowitz, A., Neumann, D.A., and Neu, N.: **Autoimmune myocarditis: a paradigm of post-infection autoimmune disease.** *Immunol. Today* 1988, **9**:117–120.

Steinman, L., and Oldstone, M.B.: **More mayhem from molecular mimics.** *Nat. Med.* 1997, **3**:1321–1322.

14-26 Drugs and toxins can cause autoimmune syndromes.

Bagenstose, L.M., Salgame, P., and Monestier, M.: **Murine mercury-induced autoimmunity: a model of chemically related autoimmunity in humans.** *Immunol. Res.* 1999, **20**:67–78.

Yoshida, S., and Gershwin, M.E.: **Autoimmunity and selected environmental factors of disease induction.** *Semin. Arthritis Rheum.* 1993, **22**:399–419.

14-27 Random events may be required for the initiation of autoimmunity.

Eisenberg, R.A., Craven, S.Y., Warren, R.W., and Cohen, P.L.: **Stochastic control of anti-Sm autoantibodies in MRL/Mp-lpr/lpr mice.** *J. Clin. Invest.* 1987, **80**:691–697.

Todd, J.A., and Steinman, L.: **The environment strikes back.** *Curr. Opin. Immunol.* 1993, **5**:863–865.

14-28 Graft rejection is an immunological response mediated primarily by T cells.

Arakelov, A., and Lakkis, F.G.: **The alloimmune response and effector mechanisms of allograft rejection.** *Semin. Nephrol.* 2000, **20**:95–102.

Rosenberg, A.S., and Singer, A.: **Cellular basis of skin allograft rejection: an in vivo model of immune-mediated tissue destruction.** *Annu. Rev. Immunol.* 1992, **10**:333–358.

Strom, T.B., Roy-Chaudhury, P., Manfro, R., Zheng, X.X., Nickerson, P.W., Wood, K., and Bushell, A.: **The Th1/Th2 paradigm and the allograft response.** *Curr. Opin. Immunol.* 1996, **8**:688–693.

Zelenika, D., Adams, E., Humm, S., Lin, C.Y., Waldmann, H., and Cobbold, S.P.: **The role of CD4$^+$ T-cell subsets in determining transplantation rejection or tolerance.** *Immunol. Rev.* 2001, **182**:164–179.

14-29 Matching donor and recipient at the MHC improves the outcome of transplantation.

Opelz, G.: **Factors influencing long-term graft loss. The Collaborative Transplant Study.** *Transplant. Proc.* 2000, **32**:647–649.

Opelz, G., and Wujciak, T.: **The influence of HLA compatibility on graft survival after heart transplantation. The Collaborative Transplant Study.** *N. Engl. J. Med.* 1994, **330**:816–819.

14-30 In MHC-identical grafts, rejection is caused by peptides from other alloantigens bound to graft MHC molecules.

den Haan, J.M., Meadows, L.M., Wang, W., Pool, J., Blokland, E., Bishop, T.L., Reinhardus, C., Shabanowitz, J., Offringa, R., Hunt, D.F., *et al.*: **The minor histocompatibility antigen HA-1: a diallelic gene with a single amino acid polymorphism.** *Science* 1998, **279**:1054–1057.

Mutis, T., Gillespie, G., Schrama, E., Falkenburg, J.H., Moss, P., and Goulmy, E.: **Tetrameric HLA class I-minor histocompatibility antigen peptide complexes demonstrate minor histocompatibility antigen-specific cytotoxic T lymphocytes in patients with graft-versus-host disease.** *Nat. Med.* 1999, **5**:839–842.

14-31 There are two ways of presenting alloantigens on the transplant to the recipient's T lymphocytes.

Benichou, G., Takizawa, P.A., Olson, C.A., McMillan, M., and Sercarz, E.E.: **Donor major histocompatibility complex (MHC) peptides are presented by recipient MHC molecules during graft rejection.** *J. Exp. Med.* 1992, **175**:305–308.

Carbone, F.R., Kurts, C., Bennett, S.R., Miller, J.F., and Heath, W.R.: **Cross-presentation: a general mechanism for CTL immunity and tolerance.** *Immunol. Today* 1998, **19**:368–373.

14-32 Antibodies reacting with endothelium cause hyperacute graft rejection.

Kissmeyer Nielsen, F., Olsen, S., Petersen, V.P., and Fjeldborg, O.: **Hyperacute rejection of kidney allografts, associated with pre-existing humoral antibodies against donor cells.** *Lancet* 1966, **ii**:662–665.

Robson, S.C., Schulte am Esche, J., and Bach, F.H.: **Factors in xenograft rejection.** *Ann. N.Y. Acad. Sci.* 1999, **875**:261–276.

Sharma, A., Okabe, J., Birch, P., McClellan, S.B., Martin, M.J., Platt, J.L., and Logan, J.S.: **Reduction in the level of Gal(α1,3)Gal in transgenic mice and pigs by the expression of an α(1,2)fucosyltransferase.** *Proc. Natl Acad. Sci. USA* 1996, **93**:7190–7195.

Williams, G.M., Hume, D.M., Hudson, R.P., Jr., Morris, P.J., Kano, K., and Milgrom, F.: **'Hyperacute' renal-homograft rejection in man.** *N. Engl. J. Med.* 1968, **279**:611–618.

14-33 Chronic organ rejection is caused by inflammatory vascular injury to the graft.

Orosz, C.G., and Peletier, R.P.: **Chronic remodeling pathology in grafts.** *Curr. Opin. Immunol.* 1997, **9**:676–680.

Paul, L.C.: **Current knowledge of the pathogenesis of chronic allograft dysfunction.** *Transplant. Proc.* 1999, **31**:1793–1795.

Womer, K.L., Vella, J.P., and Sayegh, M.H.: **Chronic allograft dysfunction: mechanisms and new approaches to therapy.** *Semin. Nephrol.* 2000, **20**:126–147.

14-34 A variety of organs are transplanted routinely in clinical medicine.

Murray, J.E.: **Human organ transplantation: background and consequences.** *Science* 1992, **256**:1411–1416.

14-35 The converse of graft rejection is graft-versus-host disease.

Dazzi, F., and Goldman, J.: **Donor lymphocyte infusions.** *Curr. Opin. Hematol.* 1999, **6**:394–399.

Goulmy, E., Schipper, R., Pool, J., Blokland, E., Flakenburg, J.H., Vossen, J., Grathwohl, A., Vogelsang, G.B., van Houwelingen, H.C., and van Rood, J.J.: **Mismatches of minor histocompatibility antigens between HLA-identical donors and recipients and the development of graft-versus-host disease after bone marrow transplantation.** *N. Engl. J. Med.* 1996, **334**:281–285.

Murphy, W.J., and Blazar, B.R.: **New strategies for preventing graft-versus-host disease.** *Curr. Opin. Immunol.* 1999, **11**:509–515.

Porter, D.L., and Antin, J.H.: **The graft-versus-leukemia effects of allogeneic cell therapy.** *Annu. Rev. Med.* 1999, **50**:369–386.

Ruggeri, L., Capanni, M., Urbani, E., Perruccio, K., Shlomchik, W.D., Tosti, A., Posati, S., Rogaia, D., Frassoni, F., Aversa, F., *et al.*: **Effectiveness of donor natural killer cell alloreactivity in mismatched hematopoietic transplants.** *Science* 2002, **295**:2097–2100.

Shlomchik, W.D., Couzens, M.S., Tang, C.B., McNiff, J., Robert, M.E., Liu, J., Shlomchik, M.J., and Emerson, S.G.: **Prevention of graft versus host disease by inactivation of host antigen-presenting cells.** *Science* 1999, **285**:412–415.

14-36 Regulatory T cells are involved in alloreactive immune responses.

Joffre, O., and van Meerwijk, J.P.: **$CD4^+CD25^+$ regulatory T lymphocytes in bone marrow transplantation.** *Semin. Immunol.* 2006, **18**:128–135.

Li, J,, Liu, Z., Jiang, S., Coresini, R., Lederman, S., and Suciu-Foca, N.: **T suppressor lymphocytes inhibit NF-κB-mediated transcription of CD86 gene in APC.** *J. Immunol.* 1999, **163**:6386–6392.

Lu, L.F., Lind, E.F., Gondek, D.C., Bennett, K.A., Gleeson, M.W., Pino-Lagos, K., Scott, Z.A., Coyle, A.J., Reed, J.L., Van Snick, J., *et al.*: **Mast cells are essential intermediaries in regulatory T-cell tolerance.** *Nature* 2006, **31**:997–1002.

14-37 The fetus is an allograft that is tolerated repeatedly.

Carosella, E.D., Rouas-Freiss, N., Paul, P., and Dausset, J.: **HLA-G: a tolerance molecule from the major histocompatibility complex.** *Immunol. Today* 1999, **20**:60–62.

Mellor, A.L., and Munn, D.H.: **Immunology at the maternal–fetal interface: lessons for T cell tolerance and suppression.** *Annu. Rev. Immunol.* 2000, **18**:367–391.

Munn, D.H., Zhou, M., Attwood, J.T., Bondarev, I., Conway, S.J., Marshall, B., Brown, C., and Mellor, A.L.: **Prevention of allogeneic fetal rejection by tryptophan catabolism.** *Science* 1998, **281**:1191–1193.

Parham, P.: **Immunology: keeping mother at bay.** *Curr. Biol.* 1996, **6**:638–641.

Schust, D.J., Tortorella, D., and Ploegh, H.L.: **HLA-G and HLA-C at the feto-maternal interface: lessons learned from pathogenic viruses.** *Semin. Cancer Biol.* 1999, **9**:37–46.

Manipulation of the Immune Response

15

Most of this book has been concerned with the mechanisms by which the immune system successfully protects us from disease. In the preceding three chapters, however, we have seen examples of the failure of immunity to some important infections, and how inappropriate immune responses can cause allergy and autoimmune disease. We have also discussed the problems arising from immune responses to grafted tissues.

In this chapter we consider the ways in which the immune system can be manipulated or controlled, both to suppress unwanted immune responses in autoimmunity, allergy, and graft rejection, and to stimulate protective immune responses. It has long been felt that it should be possible to deploy the powerful and specific mechanisms of adaptive immunity to destroy tumors, and we discuss the present state of progress toward that goal. In the final section of the chapter we discuss current vaccination strategies and how a more rational approach to the design and development of vaccines promises to increase their efficacy and widen their usefulness and application.

Extrinsic regulation of unwanted immune responses.

Although the unwanted immune responses that occur in autoimmune disease, transplant rejection, and allergy each present somewhat different problems, the therapeutic goal in all cases is to downregulate the harmful immune response and thus avoid damage to tissues or disruption of their function. From the point of view of management, the single most important difference between allograft rejection on the one hand and allergy and autoimmunity on the other is that allografts are a deliberate surgical intervention and the immune response to them can be foreseen, whereas autoimmune and allergic responses are not detected until they are already established. Effective treatment of an established immune response is much harder to achieve than prevention of a response before it has had a chance to develop, and so autoimmune diseases are generally harder to control than a *de novo* immune response to an allograft. For allergic responses, the best treatment is avoidance of the allergen, but this is not always possible. The relative difficulty of suppressing established immune responses is seen in animal models of autoimmunity, in which treatments that are able to prevent the induction of the disease generally fail to halt established disease.

Conventional treatments for immunological disorders are nearly all empirical in origin, using immunosuppressive drugs identified by screening large numbers of natural and synthetic compounds. The conventional drugs currently used to suppress the immune system can be divided into three categories: first, powerful anti-inflammatory drugs of the corticosteroid family such as prednisone; second, cytotoxic drugs such as azathioprine and cyclophosphamide; and third, fungal and bacterial derivatives such as cyclosporin A, tacrolimus (FK506 or fujimycin), and rapamycin (sirolimus), which inhibit intracellular signaling pathways within T lymphocytes. These drugs are all very broad in their actions and inhibit the protective actions of the immune system as well as the harmful ones. Opportunistic infection is therefore a common complication of immunosuppressive drug therapy.

In recent years, newer treatments have been introduced that are designed to target specific components of the harmful immune response. A much-researched approach to avoiding wholesale immunosuppression is to target those aspects of the immune response that cause the tissue damage. Even this therapeutic intervention does not come free of side-effects, however, because the cells or molecules targeted often have important functions in normal immune responses against infectious disease. Antibodies themselves, by virtue of their exquisite specificity, offer the most immediate possibility of inhibiting a particular part of an immune response. Approaches that were noted as experimental in previous editions of this book, such as treatment with anti-cytokine monoclonal antibodies, are now part of established medical practice, and new 'experimental' treatments are being tested all the time. These experimental treatments include targeting specific cells, neutralizing local excesses of cytokines and chemokines, and manipulating the immune response to enhance natural regulatory mechanisms, such as those involving regulatory T cells (T_{reg}).

15-1 Corticosteroids are powerful anti-inflammatory drugs that alter the transcription of many genes.

Corticosteroid drugs are powerful anti-inflammatory and immunosuppressive agents that are used widely to attenuate the harmful effects of immune responses of autoimmune or allergic origin, as well as those induced by organ transplants. **Corticosteroids** are pharmacological derivatives of members of the glucocorticoid family of steroid hormones; one of the most widely used is **prednisone**, which is a synthetic analog of cortisol. Cortisol acts both through intracellular receptors of the steroid receptor superfamily and through poorly characterized membrane-bound receptors that are expressed in almost every cell of the body. After they have bound ligand, the intracellular receptors either bind directly to specific sites in the DNA, thereby altering transcription, or interact with other transcription factors, such as NFκB, to modulate their function (Fig. 15.1). Cortisol can also act directly on cellular processes, leading to a much more rapid production of anti-inflammatory proteins than could be explained by new gene transcription.

As many as 20% of the genes expressed in leukocytes may be regulated by glucocorticoids, which either induce or suppress the transcription of responsive genes. The therapeutic effects of corticosteroid drugs are due to exposure of the glucocorticoid receptors to levels of ligand much higher than they would normally encounter in the body. This causes exaggerated responses, which have both beneficial and toxic effects.

Given the large number of genes regulated by corticosteroids and the fact that different genes are regulated in different tissues, it is hardly surprising that the effects of steroid therapy are very complex. A further layer of complexity is provided by the variability of different tissues to these agents over time,

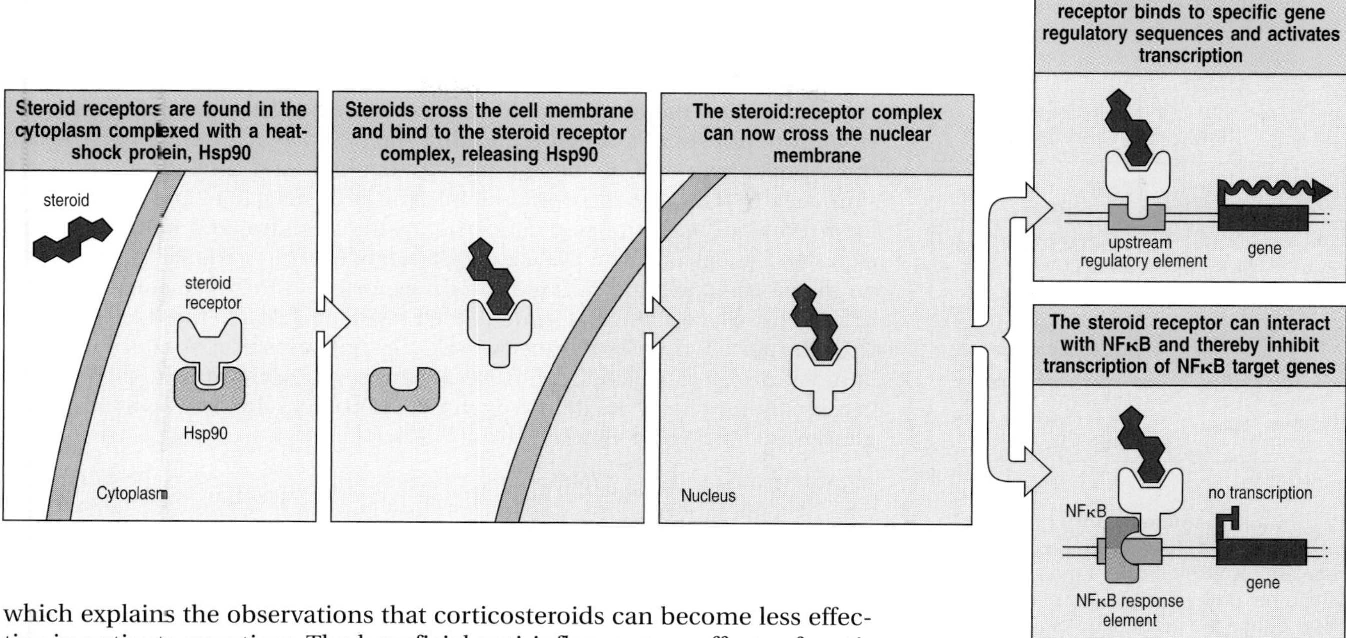

Fig. 15.1 Mechanism of steroid action. Corticosteroids are lipid-soluble molecules that enter cells by diffusing across the plasma membrane and bind to receptors in the cytosol. Binding of corticosteroid to their receptors displaces molecular chaperones, including heat-shock proteins, exposing the DNA-binding region of the receptor. The steroid:receptor complex can act by entering the nucleus and either binding to specific DNA sequences in the promoter regions of steroid-responsive genes or interacting with other transciption factors such as NFκB. Many of the actions of corticosteroids occur rapidly and thus are mediated by non-genetic mechanisms such as through putative membrane receptors.

which explains the observations that corticosteroids can become less effective in patients over time. The beneficial anti-inflammatory effects of corticosteroids are briefly summarized in Fig. 15.2: these drugs target the functions of monocytes and macrophages and reduce the number of CD4 T cells. They have many adverse effects, however, including fluid retention, weight gain, diabetes, bone mineral loss, and thinning of the skin. Tremendous efforts are currently being made to identify compounds with the anti-inflammatory effects of steroids but without their side-effects. The therapeutic use of corticosteroids entails keeping a careful balance between helping the patient by reducing the inflammatory manifestations of disease and causing harm from the toxic side-effects. For this reason, corticosteroids used in transplant recipients and for treating inflammatory autoimmune and allergic disease are often administered in combination with other drugs in an effort to keep the dose and toxic effects to a minimum. In the treatment of autoimmunity and allograft rejection, corticosteroids are commonly combined with cytotoxic immunosuppressive drugs; however, these present their own problems.

15-2 Cytotoxic drugs cause immunosuppression by killing dividing cells and have serious side-effects.

The three cytotoxic drugs most commonly used as immunosuppressants are **azathioprine**, **cyclophosphamide**, and **mycophenolate**. These drugs interfere with DNA synthesis, and their major pharmacological action is on tissues in which cells are continually dividing. They were developed originally to treat cancer and, after observations that they were cytotoxic to dividing lymphocytes, were found to be immunosuppressive as well. Azathioprine also interferes with CD28 co-stimulation, leading to the generation of an apoptotic signal through the blockade of a critical signaling molecule, the small GTPase Rac1. The use of these compounds is limited by a range of toxic effects on tissues in which cells are dividing, such as the skin, gut lining, and bone marrow. Effects include decreased immune function, as well as anemia, leukopenia, thrombocytopenia, damage to intestinal epithelium, hair loss, and fetal death or injury. As a result of their toxicity, these drugs are used at high doses only when the aim is to eliminate all dividing lymphocytes, and in these cases treated patients require subsequent bone marrow transplantation

Corticosteroid therapy	
Effect on	**Physiological effects**
↓ IL-1, TNF-α, GM-CSF IL-3, IL-4, IL-5, CXCL8	↓ Inflammation caused by cytokines
↓ NOS	↓ NO
↓ Phospholipase A_2 ↓ Cyclooxygenase type 2 ↑ Annexin-1	↓ Prostaglandins ↓ Leukotrienes
↓ Adhesion molecules	Reduced emigration of leukocytes from vessels
↑ Endonucleases	Induction of apoptosis in lymphocytes and eosinophils

Fig. 15.2 Anti-inflammatory effects of corticosteroid therapy. Corticosteroids regulate the expression of many genes, with a net anti-inflammatory effect. First, they reduce the production of inflammatory mediators, including cytokines, prostaglandins, and nitric oxide. Second, they inhibit inflammatory cell migration to sites of inflammation by inhibiting the expression of adhesion molecules. Third, corticosteroids promote the death by apoptosis of leukocytes and lymphocytes. The layers of complexity are illustrated by the actions of annexin-1 (originally identified as a factor induced by corticosteroids and named lipocortin),which has now been shown to participate in all of the effects of corticosteroids listed on the right.

to restore their hematopoietic function. They are used at lower doses and in combination with other drugs such as corticosteroids to treat unwanted immune responses.

Azathioprine is converted *in vivo* to the purine analog 6-thioguanine (6-TG), which is in turn metabolized to 6-thioinosinic acid. This competes with inosine monophosphate, thereby blocking the *de novo* synthesis of adenosine monophosphate and guanosine monophosphate and thus inhibiting DNA synthesis. 6-TG is also incorporated into the DNA instead of guanine, and the long-term side-effect of skin cancer in azathioprine-treated patients can be explained by the fact that the accumulation of 6-TG in patients' DNA confers an increased likelihood of mutation on exposure to the ultraviolet radiation in sunlight. Mycophenolate mofetil, the newest addition to the family of cytotoxic immunosuppressive drugs, works in a similar fashion to azathioprine. It is metabolized to mycophenolic acid, which is an inhibitor of the enzyme inosine monophosphate dehydrogenase. This blocks the *de novo* synthesis of guanosine monophosphate.

Azathioprine and mycophenolate are less toxic than cyclophosphamide, which is metabolized to phosphoramide mustard, which alkylates DNA. Cyclophosphamide is a member of the nitrogen mustard family of compounds, which were originally developed as chemical weapons. With this pedigree goes a range of highly toxic effects including inflammation of and hemorrhage from the bladder, known as hemorrhagic cystitis, and induction of bladder neoplasia.

15-3 Cyclosporin A, tacrolimus (FK506), and rapamycin (sirolimus) are powerful immunosuppressive agents that interfere with T-cell signaling.

Relatively nontoxic alternatives to the cytotoxic drugs are available for use as immunosuppressants. The systematic screening of bacterial and fungal products has given us many important medicines, including three immunosuppressive drugs **cyclosporin A**, **tacrolimus** (previously known as **FK506**), and **rapamycin** (also known as **sirolimus**), which are now widely used to treat transplant recipients. Cyclosporin A is a cyclic decapeptide derived from a soil fungus, *Tolypocladium inflatum*, from Norway. Tacrolimus is a macrolide compound from the filamentous bacterium *Streptomyces tsukabaensis* found in Japan; macrolides are compounds that contain a many-membered lactone ring to which is attached one or more deoxysugars. Rapamycin, another *Streptomyces* macrolide, has become important in the prevention of transplant rejection; rapamycin is derived from *Streptomyces hygroscopicus*, found on Easter Island ('Rapa Nui' in Polynesian—hence the name of the drug). All three compounds exert their pharmacological effects by binding to members of a family of intracellular proteins known as the **immunophilins**, forming complexes that interfere with signaling pathways important for the clonal expansion of lymphocytes.

Cyclosporin A and tacrolimus block T-cell proliferation by inhibiting the phosphatase activity of a Ca^{2+}-activated enzyme called **calcineurin**, and are effective at nanomolar concentrations. Their mechanism of action, which we discuss further in the next section, revealed a role for calcineurin in transmitting signals from the T-cell receptor to the nucleus (see Section 6-15). Both drugs reduce the expression of several cytokine genes that are normally induced on T-cell activation (Fig. 15.3). These include interleukin (IL)-2, which is an important growth factor for T cells (see Section 8-13). Cyclosporin A and tacrolimus inhibit T-cell proliferation in response to either specific antigens or allogeneic cells and are used extensively in medical practice to prevent the rejection of allogeneic organ grafts. Although the major

Immunological effects of cyclosporin A and tacrolimus	
Cell type	**Effects**
T lymphocyte	Reduced expression of IL-2, IL-3, IL-4, GM-CSF, TNF-α Reduced proliferation following decreased IL-2 production Reduced Ca^{2+}-dependent exocytosis of granule-associated serine esterases Inhibition of antigen-driven apoptosis
B lymphocyte	Inhibition of proliferation secondary to reduced cytokine production by T lymphocytes Inhibition of proliferation following ligation of surface immunoglobulin Induction of apoptosis following B-cell activation
Granulocyte	Reduced Ca^{2+}-dependent exocytosis of granule-associated serine esterases

Fig. 15.3 Cyclosporin A and tacrolimus inhibit lymphocyte and some granulocyte responses.

immunosuppressive effects of both drugs are probably the result of inhibition of T-cell proliferation, they also act on other cells and have a large variety of other immunological effects (see Fig. 15.3), some of which might turn out to be important pharmacologically.

Cyclosporin A and tacrolimus are effective, but they are not problem-free. First, as with the cytotoxic agents, they affect all immune responses indiscriminately. The only way of controlling their immunosuppressive action is by varying the dose; at the time of grafting, high doses are required but, once a graft is established, the dose can be decreased to allow useful protective immune responses while maintaining adequate suppression of the residual response to the grafted tissue. This is a difficult balance that is not always achieved. Furthermore, although T cells are particularly sensitive to the actions of these drugs, their molecular targets are found in other cell types and therefore these drugs have effects on many other tissues. Both cyclosporin A and tacrolimus are toxic to kidneys and other organs, for example. Finally, treatment with these drugs is expensive, because they are complex natural products that must be taken for long periods. Thus, there is room for improvement in these compounds, and better and less expensive analogs are being sought. Nevertheless, at present they are the drugs of choice in clinical transplantation, and they are also being tested in a variety of autoimmune diseases, especially those that, like graft rejection, are mediated by T cells.

15-4 Immunosuppressive drugs are valuable probes of intracellular signaling pathways in lymphocytes.

The mechanism of action of cyclosporin A and tacrolimus is now fairly well understood. Each binds to a different group of immunophilins: cyclosporin A to the cyclophilins, and tacrolimus to the FK-binding proteins (FKBP). These immunophilins are peptidyl-prolyl *cis–trans* isomerases but their isomerase activity does not seem to be relevant to the immunosuppressive activity of the drugs that bind them. Rather, the immunophilin:drug complexes bind and inhibit the Ca^{2+}-activated serine/threonine phosphatase calcineurin. Calcineurin is activated in T cells when intracellular calcium ion levels rise after T-cell receptor binding; on activation it dephosphorylates the NFAT family of transcription factors in the cytoplasm, allowing them to migrate to the nucleus, where they form complexes with nuclear partners, including the transcription factor AP-1, and induce the transcription of genes that include those for IL-2, CD40 ligand, and Fas ligand (Fig. 15.4), all of which are necessary for proper immune function. This pathway is inhibited by cyclosporin A and tacrolimus, which thus inhibit the clonal expansion of activated T cells. Calcineurin is found in other cells besides T cells but at higher levels; T cells are therefore particularly susceptible to the inhibitory effects of these drugs.

Rapamycin has a different mode of action from either cyclosporin A or tacrolimus. Like tacrolimus it binds to the FKBP family of immunophilins. However, the rapamycin:immunophilin complex has no effect on calcineurin activity, but instead inhibits a serine/threonine kinase known as mTOR (mammalian target of rapamycin), which is involved in the phosphatidylinositol

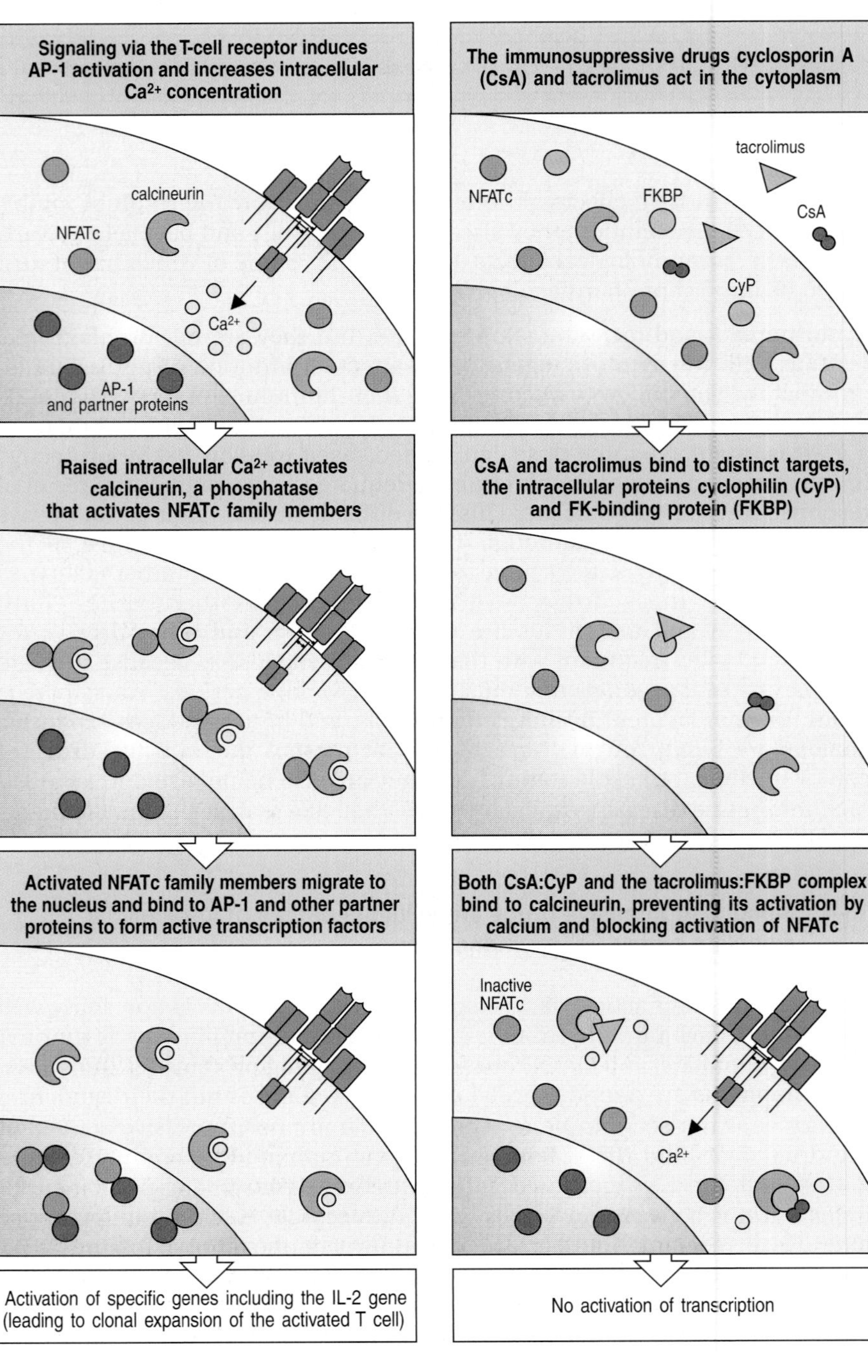

Fig. 15.4 Cyclosporin A and tacrolimus inhibit T-cell activation by interfering with the serine/threonine-specific phosphatase calcineurin. Signaling via T-cell receptor-associated tyrosine kinases leads to the activation and increased synthesis of the transcription factor AP-1 and other partner proteins, as well as increasing the concentration of Ca^{2+} in the cytoplasm (left panels). The Ca^{2+} binds to calcineurin and thereby activates it to dephosphorylate the cytoplasmic form of members of the family of nuclear factors of activated T cells (NFATc). Once dephosphorylated, the active NFATc family members migrate to the nucleus to form a complex with AP-1 and other partner proteins; the NFATc:AP-1 complexes can then induce the transcription of genes required for T-cell activation, including the gene encoding IL-2. When cyclosporin A (CsA) or tacrolimus are present, they form complexes with their immunophilin targets, cyclophilin (CyP) and FK-binding protein (FKBP), respectively (right panels). The complex of cyclophilin with cyclosporin A can bind to calcineurin and block its ability to activate NFATc family members. The complex of tacrolimus with FKBP binds to calcineurin at the same site, also blocking its activity.

3-kinase (PI 3-kinase)/Akt (protein kinase B) signaling pathway (see Section 6-19). Blockade of this pathway has a dramatic effect on T-cell proliferation. It causes arrest of cells in the G_1 phase of the cell cycle and the cells die by apoptosis. The drug similarly inhibits lymphocyte proliferation driven by growth factors such as IL-2, IL-4, and IL-6. Intriguingly, rapamycin increases the number of regulatory T cells, perhaps because these cells use different signaling pathways from those of effector T cells.

15-5 Antibodies against cell-surface molecules have been used to remove specific lymphocyte subsets or to inhibit cell function.

Cytotoxic drugs indiscriminately affect all types of activated lymphocyte and any other dividing cells. Cyclosporin A, tacrolimus, and rapamycin are more selective, but still inhibit most adaptive immune responses. In contrast, antibodies can interfere with immune responses in a nontoxic and much more specific manner. The potential of antibodies for the removal of unwanted lymphocytes is demonstrated by **anti-lymphocyte globulin**, a preparation of immunoglobulin from horses immunized with human lymphocytes, which has been used for many years to treat episodes of acute graft rejection. Anti-lymphocyte globulin does not, however, discriminate between useful lymphocytes and those responsible for the unwanted responses. Horse immunoglobulin is also highly antigenic in humans, and the large doses used in therapy are often followed by the development of serum sickness, caused by the formation of immune complexes of the horse immunoglobulin and human anti-horse immunoglobulin antibodies (see Section 13-18).

Anti-lymphocyte globulins are nevertheless still in use to treat acute rejection and have stimulated the quest for monoclonal antibodies (see Appendix I, Section A-12) that would achieve more specifically targeted effects. One such antibody is Campath-1H (also known as alemtuzumab), which is directed at the cell-surface protein CD52 expressed by most lymphocytes. It has similar actions to anti-lymphocyte globulin, causing long-standing lymphopenia, and is used as an alternative in some clinical situations.

Immunosuppressive monoclonal antibodies act by one of two general mechanisms. Some, such as Campath-1H, trigger the destruction of lymphocytes *in vivo* and are referred to as **depleting antibodies**, whereas others are **non-depleting** and act by blocking the function of their target protein without killing the cell that bears it. Monoclonal IgG antibodies that cause lymphocyte depletion target these cells to macrophages and NK cells, which bear Fc receptors and kill the lymphocytes by phagocytosis and antibody-dependent cell-mediated cytotoxicity (ADCC), respectively. Complement-mediated lysis may also play a part in lymphocyte destruction. Many antibodies are being tested for their ability to inhibit allograft rejection and to modify the expression of autoimmune disease. We discuss some of these examples after looking at the measures being taken to produce monoclonal antibodies for therapy in humans.

15-6 Antibodies can be engineered to reduce their immunogenicity in humans.

The major impediment to therapy with monoclonal antibodies in humans is that these antibodies are most readily made with the use of mouse cells (see Appendix I, Section A-12), and humans rapidly develop antibody responses to mouse antibodies. This not only blocks the actions of the mouse antibodies but also leads to allergic reactions, and if treatment is continued it can result in anaphylaxis (see Section 13-11). Once this has happened, future treatment with any mouse monoclonal antibody is ruled out. This problem

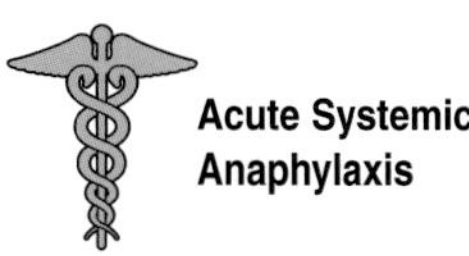

can, in principle, be avoided by making antibodies that are not recognized as foreign by the human immune system, and three strategies for their construction are being explored. One approach is to clone human V regions into a phage display library and select for binding to human cells, as described in Appendix I (see Section A-13). In this way, monoclonal antibodies that are entirely human in origin can be obtained. Second, mice lacking endogenous immunoglobulin genes can be made transgenic (see Appendix I, Section A-46) for human immunoglobulin heavy- and light-chain loci by using yeast artificial chromosomes. B cells in these mice, sometimes known as humanized mice, have receptors encoded by human immunoglobulin genes but are not tolerant to most human proteins. In these mice, it is possible to induce the production of human monoclonal antibodies against epitopes on human cells or proteins.

Finally, one can graft the complementarity-determining regions (CDRs) of a mouse monoclonal antibody, which form the antigen-binding loops, onto the framework of a human immunoglobulin molecule, a process known as **humanization**. Because antigen-binding specificity is determined by the structure of the CDRs (see Chapter 3), and because the overall frameworks of mouse and human antibodies are so similar, this approach produces a monoclonal antibody that is antigenically identical to human immunoglobulin but binds the same antigen as the mouse monoclonal antibody from which the CDR sequences were derived. Although these recombinant antibodies are far less immunogenic in humans than the parent mouse monoclonal antibodies, it is becoming apparent that these 'chimeric' antibodies can still cause hypersensitivity reactions. Fully human antibodies against many target antigens are therefore being developed to overcome this problem, often after the chimeric equivalent has been shown to be therapeutically effective.

15-7 Monoclonal antibodies can be used to prevent allograft rejection.

Antibodies specific for various physiological targets are being used, or are under investigation, to prevent the rejection of transplanted organs by inhibiting the development of harmful inflammatory and cytotoxic responses. For example, Campath-1H has been used successfully in both solid organ and bone marrow transplantation.

Elimination of mature T lymphocytes from donor bone marrow before infusion into a recipient is very effective at reducing the incidence of graft-versus-host disease (see Section 14-35). In this disease, the T lymphocytes in the donor bone marrow recognize the recipient as foreign and mount a damaging alloreaction against them, causing rashes, diarrhea, and hepatitis, which is often fatal. It had been thought that the elimination of mature donor T cells might not be so advantageous when the bone marrow graft is being given as a treatment for leukemia, because the anti-leukemic action of the donor T cells could be lost, but this has been shown not to be the case when Campath-1H is used. This antibody is also licensed for the treatment of certain leukemias and may be used as a treatment in its own right before bone marrow transplantation is considered.

More specific antibodies have been used to treat episodes of graft rejection that occur after transplantation. The antibody OKT3 targets the CD3 complex and leads to T-cell immunosuppression by inhibiting signaling through the T-cell receptor. It has been used clinically in solid organ transplantation but is often associated with an unwanted stimulation of cytokine release, and its use is declining. The cytokine release is related to an intact Fc region, which when mutated (as in the antibody called OKT3γ1(Ala-Ala)) no longer produces this potentially dangerous side-effect. The latter antibody retains the antigen-binding region of OKT3, but amino acids 234 and 235 of the human

IgG1 Fc region have been changed to alanines, preventing the interactions that lead to cytokine release (see Section 15-11).

Monoclonal antibodies against other targets have also had some success in preventing graft rejection in animals. Certain nondepleting anti-CD4 antibodies, when given for a short time during the first exposure to grafted tissue, induce a state of tolerance to graft antigens in the recipient (Fig. 15.5). This tolerant state is an example of the immune regulation by regulatory T cells discussed in Section 14-7. The tolerance-inducing cells are CD4 CD25 T_{reg} cells, although other regulatory T-cell subpopulations may have similar effects. The tolerance is specific: thus, animals of strain A that are tolerant to strain B still reject grafts of strain C. This tolerance is also 'infectious'—a naive T-cell population exposed to allografts in the presence of regulatory T cells specific for that allograft acquires tolerance to the allograft antigens. We do not yet know precisely how the anti-CD4 antibodies induce the regulatory T cells.

A different approach to inhibiting allograft rejection is to blockade the co-stimulatory signals needed to activate T cells that recognize donor antigens. The co-stimulatory molecules B7.1 and B7.2 are present on the surface of specialized antigen-presenting cells, such as dendritic cells, and both bind to the receptor CD28 and to its homolog CTLA-4 on CD4 T cells and some CD8 T cells (see Section 8-14). In animal studies of graft rejection, the soluble recombinant protein CTLA-4–Ig, which binds tightly to B7 molecules and thus prevents their interaction with the co-stimulatory receptors on T cells,

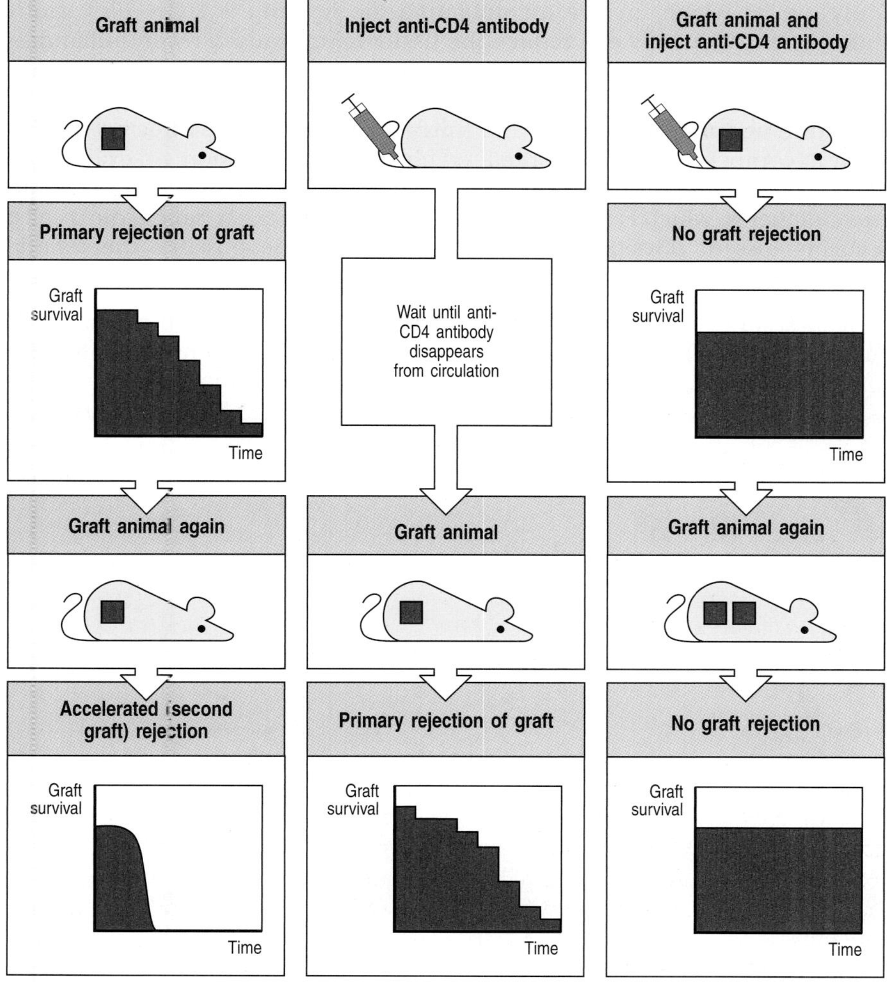

Fig. 15.5 A tissue graft given together with anti-CD4 antibody can induce specific tolerance. Mice grafted with tissue from a genetically different mouse reject that graft. Having been primed to respond to the antigens in the graft, they then reject a subsequent graft of identical tissue more rapidly (left panels). Mice injected with anti-CD4 antibody alone can recover immune competence when the antibody disappears from the circulation, as shown by a normal primary rejection of graft tissue (center panels). However, when tissue is grafted and anti-CD4 antibody is administered at the same time, the primary rejection response is markedly inhibited (right panels). An identical graft made later in the absence of anti-CD4 antibody is not rejected, showing that the animal has become tolerant to the graft antigen. This tolerance can be transferred with T cells to naive recipients (not shown).

has allowed the long-term survival of certain grafted tissues, presumably by suppressing T-cell activation. CTLA-4–Ig is composed of CTLA-4 fused to the Fc portion of human immunoglobulin.

Even more effective in a primate model of kidney transplant rejection was a humanized monoclonal antibody against the co-stimulatory molecule CD40 ligand, which is present on the surface of T cells (see Section 8-14). CD40 ligand binds to CD40, expressed on dendritic and endothelial cells, stimulating them to secrete cytokines such as IL-6, IL-8, and IL-12. The mechanism of the immunosuppressive effect of anti-CD40 ligand antibody is not known, but it is most likely to be a consequence of blocking the activation of dendritic cells by helper T cells that recognize donor antigens. There have only been preliminary studies of the use of anti-CD40 ligand antibodies in humans. One antibody was associated with thromboembolic complications and was withdrawn; a different anti-CD40 ligand antibody was administered to patients with the autoimmune disease systemic lupus erythematosus (SLE) without significant complications but also with little evidence of effectiveness.

15-8 Biological agents can be used to alleviate and suppress autoimmune disease.

We shall next look at some approaches to treating another unwanted immune response—autoimmunity. Autoimmune disease is only detected once the autoimmune response has caused tissue damage or has disturbed specific physiological functions. There are three main approaches to its treatment, only two of which involve manipulating the immune system. First, anti-inflammatory therapy can reduce the tissue injury caused by an inflammatory autoimmune response; second, therapy can be aimed at modifying and reducing the autoimmune response—this comes under the broad remit of **immunomodulatory therapy**; and third, treatment can be directed specifically at compensating for the impaired physiological function. An example of this third, non-immunological, approach is the use of injected insulin to treat diabetes, which is induced by autoimmune attack on pancreatic β cells, causing loss of physiological insulin secretion. The possible therapeutic targets in an autoimmune response are illustrated in Fig. 15.6.

The first line of anti-inflammatory therapy for autoimmune disease is normally drug therapy. The typical sequence is to use anti-inflammatory

Fig. 15.6 The potential targets of immune intervention strategies.

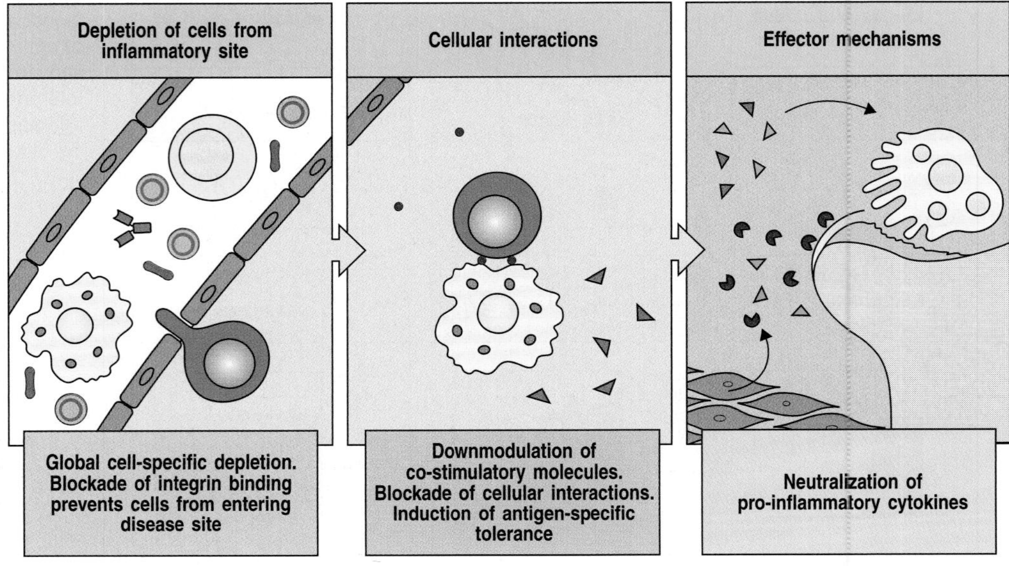
Depletion of cells from inflammatory site
Cellular interactions
Effector mechanisms
Global cell-specific depletion. Blockade of integrin binding prevents cells from entering disease site
Downmodulation of co-stimulatory molecules. Blockade of cellular interactions. Induction of antigen-specific tolerance
Neutralization of pro-inflammatory cytokines

agents such as aspirin, other nonsteroidal anti-inflammatory drugs, and sometimes low-dose corticosteroids for mild disease. For more severe disease, immunosuppressive and anti-inflammatory therapy are combined, with corticosteroids at higher doses often combined with one of the cytotoxic drugs described in Section 15-2. In addition, there is a new class of therapy that has been named **biological therapy**. This term designates treatments comprising natural proteins such as antibodies and cytokines or fragments of proteins or synthetic peptides. It also encompasses the use of anti-lymphocyte globulin and antibodies to inhibit autoreactive lymphocytes, and the use of whole cells, such as adoptive T-cell transfer in cancer immunotherapy. Biological therapy has become established as part of the anti-inflammatory treatment of certain autoimmune diseases—in particular, treatments directed toward neutralizing the effects of the pro-inflammatory cytokine tumor necrosis factor (TNF)-α—and we discuss this aspect first.

Anti-TNF-α antibodies have been found to induce striking remissions in rheumatoid arthritis (Fig. 15.7) and to reduce tissue inflammation in Crohn's disease, an inflammatory bowel disease (see Section 13-21). There are two established means of antagonizing TNF-α in clinical practice. The first is the use of humanized or fully human monoclonal antibodies, such as infliximab and adalumimab, respectively, that bind to TNF-α and block its activity. The second is the use of a recombinant human TNF receptor (TNFR) subunit p75–Fc fusion protein called etanercept, which binds TNF-α, thereby neutralizing its activity. These biological agents are extremely potent anti-inflammatories, and the number of diseases in which they have been shown to be effective is growing as further clinical trials are performed. In addition to rheumatoid arthritis, the rheumatic diseases ankylosing spondylitis, psoriatic arthropathy, and juvenile chronic arthritis each respond well to blockade of TNF-α, and this treatment is now routine for many of these diseases. Indeed, more than a million people have been treated with anti-TNF-α worldwide. In medicine, however, most treatments that have powerful effects also carry the risk of major side-effects. With TNF-α blockade there is a small but increased risk to patients of developing serious infections, including tuberculosis. This is an excellent illustration of the participation of TNF-α in host defense against tuberculosis, as noted in Section 12-17. Anti-TNF-α therapy has not been successful in all diseases. TNF-α blockade in experimental autoimmune encephalomyelitis (EAE, the mouse model of multiple sclerosis), led to amelioration of the disease, but in patients with multiple sclerosis treated with anti-TNF-α relapses became more frequent, possibly because of an increase in T-cell activation. This illustrates the potential pitfalls of using animal models for designing therapies for human disease (see also Section 15-13).

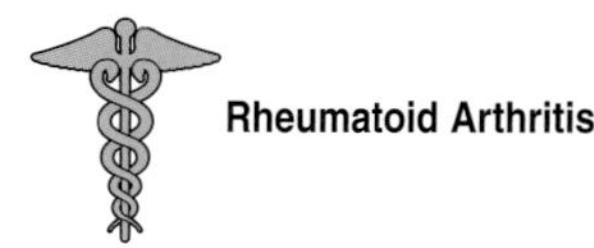

Fig. 15.7 Anti-inflammatory effects of anti-TNF-α therapy in rheumatoid arthritis. The clinical course of 24 patients was followed for 4 weeks after treatment with either a placebo or a monoclonal antibody against TNF-α at a dose of 10 mg kg^{-1}. The antibody therapy was associated with a reduction in both subjective and objective parameters of disease activity (as measured by pain score and swollen-joint count, respectively) and in the systemic inflammatory acute-phase response, measured as a fall in the concentration of the acute-phase C-reactive protein. Data courtesy of R.N. Maini.

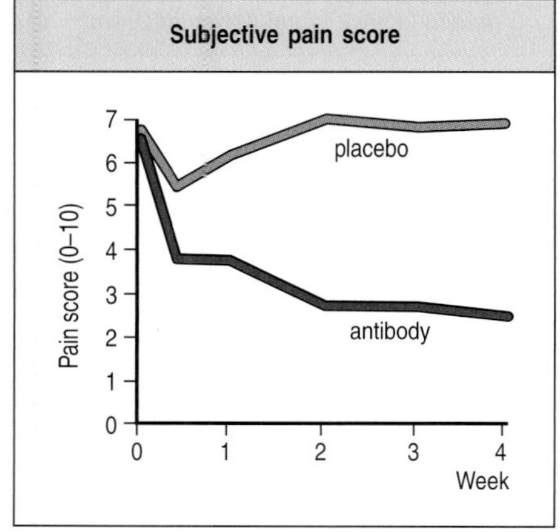

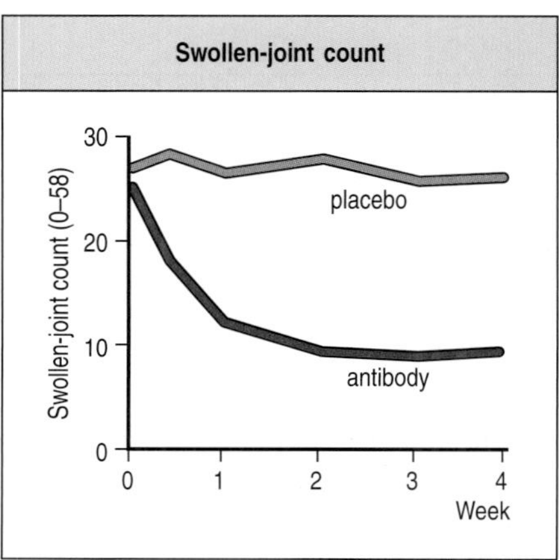

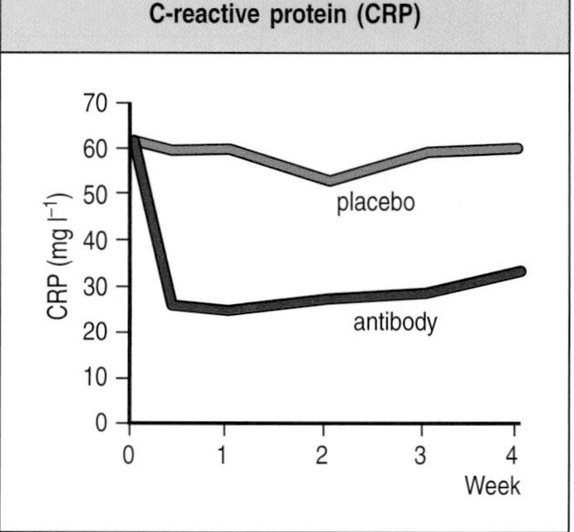

Anti-TNF therapy was the first specific biological therapeutic to enter the clinical armamentarium. Close behind as a licensed treatment was anti-IL-1 therapy, but this has not proved as effective in humans as TNF-α blockade, despite being equally powerful in animal models of arthritis. Other antagonists to cytokines are in clinical trials: one is a humanized antibody against the IL-6 receptor and blocks the effects of IL-6, an important pro-inflammatory cytokine. This seems to be as effective as anti-TNF-α in patients with rheumatoid arthritis.

Antibodies can also block cell migration to sites of inflammation. Effector lymphocytes expressing the integrin α_4:β_1 (VLA-4) bind to VCAM-1 on endothelium in the central nervous system, while those expressing α_4:β_7 (lamina propria-associated molecule 1) bind to MAdCAM-1 on endothelium in the gut. The humanized monoclonal antibody natalizumab is specific for the α_4 integrin subunit and binds both VLA-4 and α_4:β_7, preventing their interaction with their ligands (Fig. 15.8). This antibody has shown therapeutic benefit in placebo-controlled trials in patients with Crohn's disease or with multiple sclerosis. The early signs that this treatment could be successful illustrate the fact that disease depends on the continuing emigration of lymphocytes, monocytes, and macrophages from the circulation into the tissues of the brain in multiple sclerosis, and into the gut wall in Crohn's disease. However, blockade of α_4:β_1 integrin is not specific and, like anti-TNF therapy, could lead to reduced defense against infection. Three patients being treated with natalizumab developed a rare fatal multifocal leukoencephalopathy caused by the JC virus, leading to the withdrawal of this drug from the market in 2005, but in June 2006 it was allowed to be re-prescribed to a limited set of patients with multiple sclerosis. Chemokines and their receptors are also excellent potential targets for drugs intended to prevent the migration of immune effector cells to sites of autoimmune disease. FTY720, a sphingosine 1-phosphate analog, is a new drug that causes the retention of lymphocytes in peripheral lymphoid organs and inhibits dendritic cell migration (see Section 8-3). It is showing promise in the treatment of rejection in kidney transplantation and of autoimmune diseases such as multiple sclerosis and asthma.

15-9 Depletion or inhibition of autoreactive lymphocytes can treat autoimmune disease.

Ways of suppressing the autoimmune response by the direct targeting of autoreactive lymphocytes are also being explored, and in some cases these

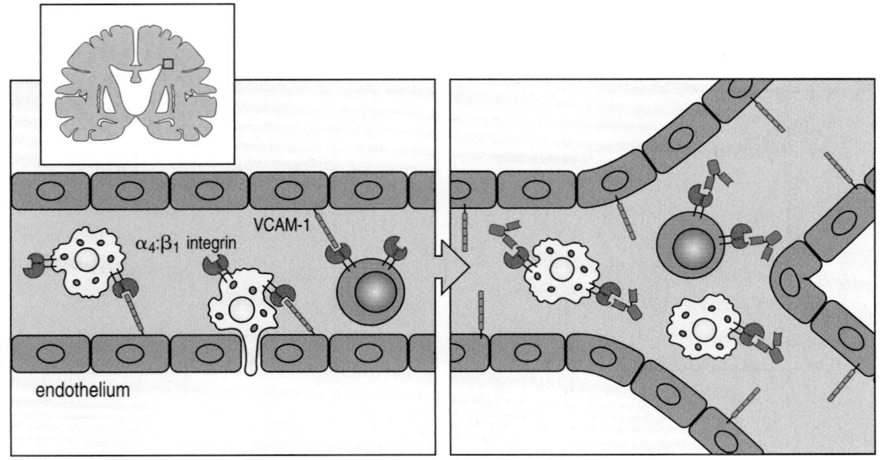

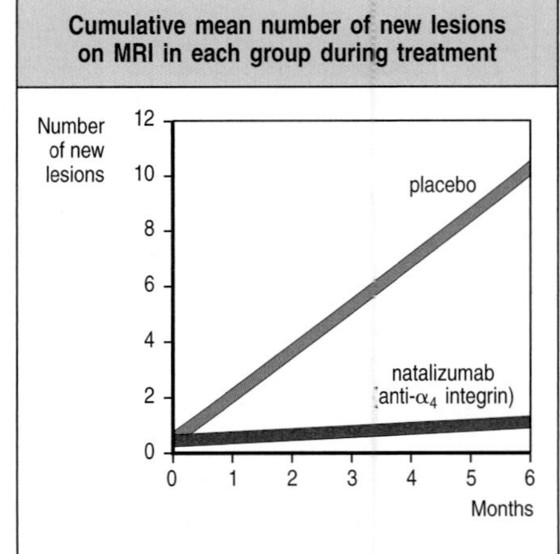

Fig. 15.8 Treatment with anti-α_4 integrin humanized monoclonal antibody reduces relapses of multiple sclerosis. Left panel: interaction between α_4:β_1 integrin (VLA-4) on lymphocytes and macrophages and VCAM-1 expressed on endothelial cells allows the adhesion of these cells to brain endothelium. This facilitates the migration of these cells into the plaques of inflammation in multiple sclerosis. Center panel: the monoclonal antibody natalizumab binds to the α_4 chain of the integrin and blocks adhesive interactions between lymphocytes and monocytes and VCAM-1 on endothelial cells, thus preventing the cells from entering the tissue and exacerbating the inflammation. The future of this treatment is unclear because of the development of a rare infection as a side effect (see the text). Right panel: the number of new lesions detected on magnetic resonance imaging (MRI) of the brain is greatly reduced in patients treated with natalizumab compared with a placebo. Data courtesy of D. Miller.

have had some therapeutic success. Pathogenic lymphocytes can be removed in a crude way by depleting whole populations of lymphocytes—of which only a small subfraction are actually pathogenic. Polyclonal anti-lymphocyte globulin is one means of doing this; we considered the effects and side-effects of this treatment in Section 15-5. Here we look at antibodies that are more selective in their massacre of lymphocytes. For example, if the clonally restricted T-cell receptors or immunoglobulins on the lymphocytes that are causing disease can be identified, they can be targeted by antibodies against idiotypic determinants on the receptor (see Appendix I, Section A-10).

Monoclonal antibodies that react with lymphocytes have various effects on the target cells. Some cause cell depletion, as described in Section 11-5. Nondepleting antibodies, in contrast, are not associated with any change in cell numbers. It is an apparent paradox that some nondepleting antibodies seem to be more effective treatments for autoimmunity than depleting antibodies that bind to the identical target proteins on lymphocytes. The most likely explanation is that nondepleting antibodies exert their effects by modifying the function of the cells that they have bound in some beneficial way. The effects of this latter type of antibody are considered in the next section.

Treatment with anti-CD4 antibodies that cause depletion of helper T cells (see Fig. 15.5) has been tried experimentally in rheumatoid arthritis and multiple sclerosis, with disappointing results. In controlled studies the antibodies showed only small therapeutic effects but caused depletion of T lymphocytes from peripheral blood for more than 6 years after treatment. Subsequent studies showed that the likely explanation for the failure was that these antibodies failed to delete primed CD4 T_H1 cells secreting the pro-inflammatory cytokine interferon (IFN)-γ, and may thus have missed their target. This cautionary tale shows that it is possible to deplete large numbers of lymphocytes and yet completely fail to kill the cells that matter.

The monoclonal antibody Campath-1H has a similar killing profile to that of anti-lymphocyte globulin (see Section 15-5) and showed some beneficial effect in studies of small numbers of patients with multiple sclerosis, but immediately after its infusion most patients suffered a frightening, although fortunately brief, flare-up of their illness. This flare illustrates another potential complication of antibody therapy. While Campath-1H was binding and killing cells by complement- and Fc-dependent mechanisms, cytokines were released, including TNF-α, IFN-γ, and IL-6. One of the effects of this was a transient blockade of nerve conduction in nerve fibers previously affected by demyelination, which caused the dramatic exacerbation of symptoms. Nevertheless, Campath-1H could be useful at early stages of disease when the inflammatory response is maximal, but this has yet to be determined.

It has also been possible to explore the effects of depleting B cells by the use of a chimeric mouse/human monoclonal anti-CD20 antibody, named rituximab, that was originally developed for treating B-cell lymphomas. Ligation and clustering of CD20 by the antibody transduces a signal that causes lymphocyte apoptosis. Infusions of rituximab cause B-cell depletion for several months, and the drug has been used in trials in autoimmune diseases in which autoantibody-mediated pathogenesis is believed to be important. There is evidence for the efficacy of this antibody in some patients with autoimmune hemolytic anemia, SLE, rheumatoid arthritis, or type II mixed cryoglobulinemia (see Fig. 14.16). Although CD20 is not expressed on antibody-producing plasma cells, their B-cell precursors are targeted by anti-CD20, resulting in a substantial reduction in the short-lived, but not the long-lived, plasma-cell population. Alternative strategies to remove these antibody-producing cells include targeting other cell-surface molecules, including the B-cell co-receptor component CD19, which is expressed by all B cells.

15-10 Interference with co-stimulatory pathways for the activation of lymphocytes could be a treatment for autoimmune disease.

We learned in Section 15-7 that interference with co-stimulatory pathways that lead to T-cell activation could be a useful therapy to prevent allograft rejection. This pathway is also an obvious target for autoimmune therapy, and various biological agents are undergoing trials. For example, the B7 blocker CTLA-4–Ig (see Section 15-7) has been shown to be effective in randomized, double-blind clinical trials in patients with rheumatoid arthritis or psoriasis. Psoriasis is an inflammatory skin disease driven primarily by T cells, leading to the production of pro-inflammatory cytokines. When CTLA-4–Ig was given to patients with psoriasis, there was an improvement in the psoriatic rash and histological evidence of loss of activation of keratinocytes, T cells, and dendritic cells within the damaged skin.

Another co-stimulatory pathway that has been targeted in psoriasis is the interaction between the adhesion molecules CD2 on T cells and CD58 (LFA-3) on antigen-presenting cells. Patients were treated with a recombinant CD58–IgG1 fusion protein, called alefacept, which inhibits the interaction between CD2 and CD58, or with placebo. A marked improvement in symptoms could be attributed to the alefacept treatment, and there was a reduction in both CD4 and CD8 memory effector lymphocytes in peripheral blood. Alafacept is now in routine clinical use for psoriasis and has a good safety record, and although memory T cells are targeted by this therapy, responses to vaccination such as anti-tetanus remain intact. Another new treatment for psoriasis is the monoclonal antibody efalizumab, which is directed at the integrin α_L (CD11a, a subunit of the integrin LFA-1). Efalizumab blocks the interaction between LFA-1 on T cells and the adhesion molecule ICAM-1 on antigen-presenting cells (see Section 8-11). The number of T cells and inflammatory dendritic cells in psoriatic skin lesions is substantially reduced, and this is associated with a marked improvement in disease (Fig. 15.9). These dendritic cells, which express HLA-DR, CD40, and B7.2, are not only important effector cells in psoriasis, through their production of TNF-α and nitric oxide, but also act to prime T cells.

As well as giving very encouraging therapeutic results, the inhibition of co-stimulation tells us something important about psoriasis, as it demonstrates the importance of T cells in the induction of the skin lesions. This fits with the fact that cyclosporin A is also an established treatment for the disease.

15-11 Induction of regulatory T cells by antibody therapy can inhibit autoimmune disease.

The ultimate goal of immunotherapy for autoimmune disease is specific intervention to restore tolerance to the relevant autoantigens. The aim is to try to turn a pathological autoimmune response into an innocuous one. Currently, the main focus of experimental immunotherapy in this context is the expansion or restoration of function of regulatory T cells. This approach is being pursued because tolerance to tissue antigens does not always depend on the absence of a T-cell response; instead, it can be actively maintained by regulatory T cells that suppress the development of a harmful, inflammatory T-cell response.

One partial success in this field has been the use of anti-CD3 antibodies (see section 15-7), which have shown promise for the treatment of type 1 diabetes mellitus both in animal models of autoimmunity and in clinical trials. The anti-CD3 antibody now in use lacks the Fc portion, unlike the first generation of anti-CD3 antibodies, and does not lead to a massive cytokine release and consequent fever and illness. In contrast to many immunomodulatory

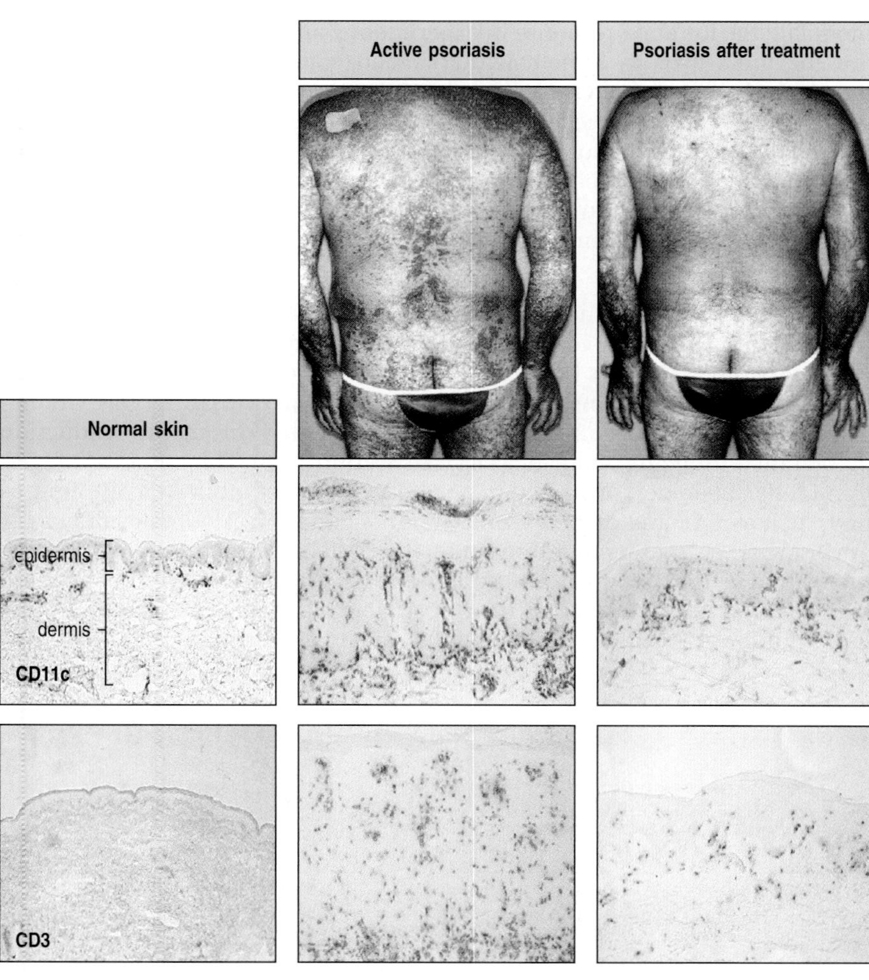

Fig. 15.9 The anti-CD11a antibody (efalizumab) inhibits the migration of dendritic cells and T cells into psoriatic skin lesions. The top two panels illustrate the excellent clinical response seen in a patient with psoriasis given eight weekly infusions of the monoclonal antibody efalizumab. The lower panels show skin biopsies from a healthy individual (left panels) and a patient before (center panels) and after treatment with efalizumab (day 56, right panels). The skin samples were stained for $CD11c^+$ dendritic cells (top row) or for $CD3^+$ T cells (bottom row) with peroxidase-conjugated antibody (brown). There was a 41% reduction in $CD11c^+$ cells and a 47% reduction in $CD3^+$ cells in a cohort of patients treated with efalizumab. Top panel: Papp, K., *et al.*, *J. Am. Acad. Dermatol.* 2001, **45**:665–674. Bottom panel: Lowes, M., *et al.*: *Proc. Natl. Acad. Sci.* 2005, **102**: 19057-19062.

agents, this anti-CD3 antibody restored tolerance to pancreatic β cells in the NOD murine model of diabetes but was ineffective in preventing the onset of disease. This intriguing finding could suggest that tolerizing autoantigens can only be suitably presented in the context of ongoing inflammation. Other forms of immune intervention, for example anti-cytokine antibodies, are usually able to suppress disease onset but do not lead to long-term tolerance when the treatment is stopped. Treatment with the anti-CD3 antibody was associated with the induction and expansion of regulatory T cells, and its effects could be partly blocked through the inhibition of TGF-β, which is thought to be important in both the generation and function of these cells. These findings have been successfully translated to the clinic, and in a controlled trial in patients with type 1 diabetes mellitus, anti-CD3 considerably reduced the requirement for insulin for 18 months after the treatment. Regulatory T cells are also induced after anti-TNF therapy in patients with rheumatoid arthritis, but only in those patients who respond positively to the therapy, raising the possibility that induction of regulatory T cells is an additional mechanism by which anti-TNF exerts its effects.

15-12 A number of commonly used drugs have immunomodulatory properties.

A number of existing medications, such as the statins and angiotensin blockers widely used in the prevention and treatment of cardiovascular disease, can also modulate the immune response in experimental animals. Statins, which block the enzyme 3-hydroxy-3-methylglutaryl-coenzyme A (HMG-CoA) reductase, thereby reducing cholesterol levels, also reduce the increased level of expression of MHC class II molecules in some autoimmune diseases. These effects may be due to an alteration in the cholesterol content of membranes, thereby disrupting lipid rafts and lymphocyte signaling (see Section 6-6). These drugs also result in a switch from a more pathogenic T_H1 response to a more protective T_H2 response in animal models, although whether this occurs in human patients is not clear.

Another potential immunomodulatory agent is vitamin D_3, which is familiar as an essential hormone for bone and mineral homeostasis. As shown in Fig. 15.10, vitamin D_3 targets both dendritic cells and effector T cells, leading to inhibition of T_H1 cytokines and an increase in T_H2 cytokines. This vitamin also leads to an expansion in regulatory T cells, partly through the induction of tolerance-inducing dendritic cells (see Section 10-3). The potential of vitamin D_3 has been demonstrated in a variety of animal models of autoimmunity, such as EAE and diabetes, and in transplantation. The major drawback of vitamin D_3 is that its immunomodulatory effects are seen only at dosages that would lead to hypercalcemia and bone resorption in humans. There is a major search under way for structural analogs of vitamin D_3 that retain the immunomodulatory effects but do not cause hypercalcemia.

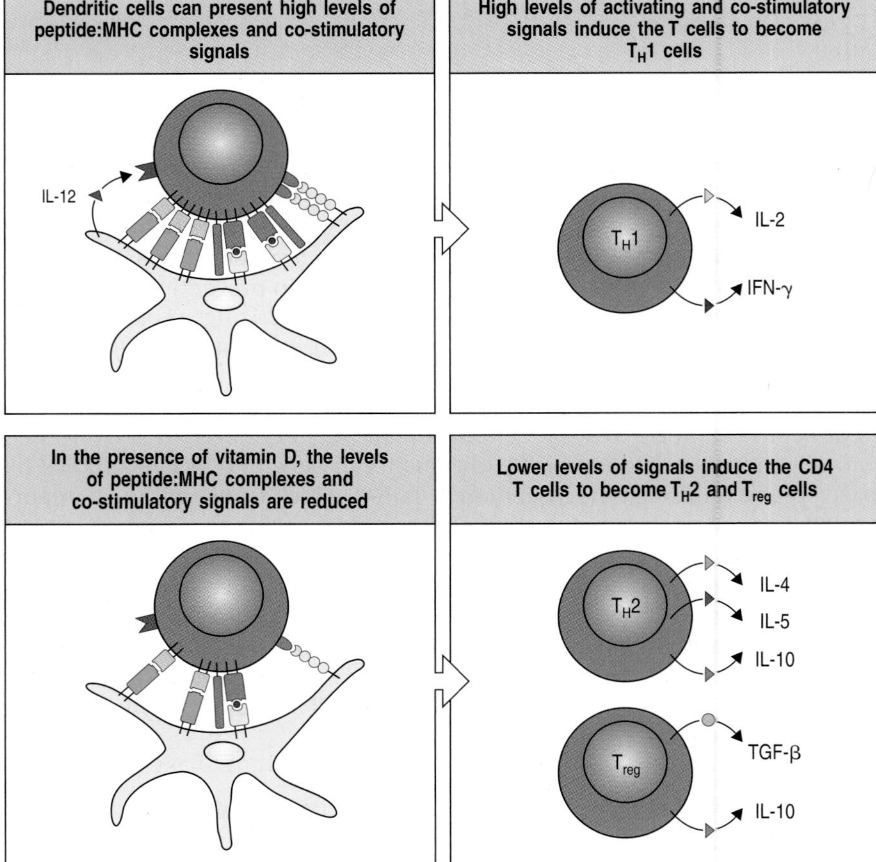

Fig. 15.10 The immunomodulatory effects of vitamin D_3. Vitamin D_3 inhibits the expression of peptide:MHC II complexes and co-stimulatory molecules on the surface of antigen-presenting cells such as dendritic cells, thus reducing the efficiency of antigen presentation. It also inhibits production of the cytokine IL-12 by dendritic cells. This results in a shift in T lymphocyte differentiation from a T_H1 towards a T_H2 phenotype. Vitamin D_3 also exerts its immunomodulatory effects directly on T lymphocytes by inhibiting the production of the T_H1 cytokines IL-2 and IFN-γ, and stimulating the production of T_H2 cytokines. Vitamin D_3 also favors the induction of regulatory T lymphocytes (T_{reg}).

15-13 Controlled administration of antigen can be used to manipulate the nature of an antigen-specific response.

When the target antigen of an unwanted response is identified, it is sometimes possible to manipulate the response by using antigen directly rather than using antibodies or relying on the bystander effects discussed in the previous section. This is because the way in which antigen is presented to the immune system affects the nature of the response, and the induction of one type of response to an antigen can inhibit a pathogenic response to the same antigen. As discussed in Chapter 13, this principle has been applied with some success to the treatment of allergies caused by an IgE response to very low doses of antigen. Repeated treatment of allergic individuals with increasing doses of allergen seems to divert the allergic response to one dominated by T cells that favor the production of IgG and IgA antibodies. These antibodies are thought to desensitize the patient by binding the small amounts of allergen normally encountered and preventing it from binding to IgE.

There has been considerable interest in using peptide antigens to suppress pathogenic responses in T-cell-mediated autoimmune disease. The type of CD4 T-cell response induced by a peptide depends on the way in which it is presented to the immune system. For instance, peptides given orally tend to prime regulatory T cells that make predominantly TGF-β, without activating T_H1 cells or inducing a great deal of systemic antibody. Indeed, experiments in animals indicate that oral antigens can protect against induced autoimmune disease. EAE in mice is induced by the injection of myelin basic protein (MBP) in complete Freund's adjuvant and resembles multiple sclerosis, whereas collagen arthritis is similarly induced in mice by injection of collagen type II and has features in common with rheumatoid arthritis. Oral administration of MBP or type II collagen, respectively, inhibits the development of these diseases in animals, and has some beneficial effects in reducing the activity of already established disease. In general, however, the oral administration of the whole antigen in people with multiple sclerosis or rheumatoid arthritis has proved of only marginal therapeutic effect. Similarly, a large study to examine whether giving low-dose parenteral insulin to those at high risk of developing diabetes could delay the onset of the disease showed absolutely no protective effect.

Other approaches to shifting the autoimmune T-cell response to a less damaging T_H2 response have been more effective in humans. The peptide drug glatiramer acetate is an approved drug for multiple sclerosis, reducing relapse rates by up to 30%. It mimics the amino acid composition of MBP and induces a T_H2-type protective response.

A still experimental approach to manipulating antigen-specific responses in animals is through the intramuscular injection of DNA constructs that encode the relevant self antigen, leading to its presentation by dendritic cells without upregulation of co-stimulatory molecules. Another strategy uses altered peptide ligands (APLs), in which amino acid substitutions have been made in T-cell receptor contact positions in the antigenic peptide. APLs can be designed to act as partial agonists or antagonists, or even to induce the differentiation of regulatory T cells. But despite their success in ameliorating EAE in mice, the trial of these peptides for multiple sclerosis led to exacerbated disease in some patients, highlighting again the potential problems in moving from animal models of autoimmunity to the human disease (see Section 15-8). Allergic reactions, associated with a vigorous T_H2 response, occurred in some patients with multiple sclerosis who were given APL, which has led to the development of a rodent model of allergy to enable future drugs to be tested for this side-effect. Whether such approaches can be effective in manipulating the established immune responses that drive human autoimmune diseases remains to be seen.

Summary.

Existing treatments for unwanted immune responses, such as allergic reactions, autoimmunity, and graft rejection, depend largely on three types of drugs, namely anti-inflammatory, cytotoxic and immunosuppressive. Anti-inflammatory drugs, of which the most potent are the corticosteroids, are used for all three types of responses. However, these have a broad spectrum of actions and a correspondingly wide range of toxic side-effects; their dose must be controlled carefully. They are therefore normally used in combination with either cytotoxic or immunosuppressive drugs. The cytotoxic drugs kill all dividing cells and thereby prevent lymphocyte proliferation, but they suppress all immune responses indiscriminately and also kill other types of dividing cells. The immunosuppressive drugs act by intervening in the intracellular signaling pathways of T cells. They are less generally toxic than the cytotoxic drugs, but they also suppress all immune responses indiscriminately. They are also much more expensive than cytotoxic drugs.

Immunosuppressive drugs are now the drugs of choice in the treatment of transplant patients: they can be used to suppress the immune response to the graft before it has become established. Autoimmune responses are already well established at the time of diagnosis and are consequently much more difficult to suppress. They are therefore less responsive to the immunosuppressive drugs, and for that reason they are usually controlled with a combination of corticosteroids and cytotoxic drugs. In animal experiments, attempts have been made to target immunosuppression more specifically, by blocking the response to autoantigen with the use of antibodies or antigenic peptides, or by diverting the immune response into a nonpathogenic pathway by manipulating the cytokine environment, or by administering antigen through the oral route, where a nonpathogenic immune response is likely to be invoked. Many of these approaches to treatment are now being tried in humans, in some cases with great success. The development and introduction of antagonists to TNF-α has been one of the triumphs of immunotherapy. Many biological agents are under development and some will enter clinical practice (Fig. 15.11). All have the disadvantage that they are expensive to produce and cumbersome to administer. An important goal of the pharmaceutical industry is to produce small-molecule drugs that have targets and effects similar to those of current biological therapies.

Using the immune response to attack tumors.

Cancer is one of the three leading causes of death in industrialized nations, the others being infectious disease and cardiovascular disease. As treatments for infectious diseases and the prevention of cardiovascular disease continue to improve, and the average life expectancy increases, cancer is likely to become the most common fatal disease in these countries. Cancers are caused by the progressive growth of the progeny of a single transformed cell. Curing cancer therefore requires that all the malignant cells be removed or destroyed without killing the patient. An attractive way of achieving this would be to induce an immune response against the tumor that would discriminate between the cells of the tumor and their normal cell counterparts, in the same way that vaccination against a viral or bacterial pathogen induces a specific

Therapeutic agents used to treat human autoimmune diseases				
Target	**Therapeutic agent**	**Disease**	**Disease outcome**	**Disadvantages**
Integrins	α_4:β_1 integrin-specific monoclonal antibody (mAb)	Relapsing/remitting multiple sclerosis (MS) Rheumatoid arthritis (RA) Inflammatory bowel disease	Reduction in relapse rate; delay in disease progression	Increased risk of infection; progressive multifocal encephalopathy
B cells	CD20-specific mAb	RA Systemic lupus erythematosus (SLE) MS	Improvement in arthritis, possibly in SLE	Increased risk of infection
HMG-coenzyme A reductase	Statins	MS	Reduction in disease activity	Hepatotoxicity; rhabdomyolysis
T cells	CD3-specific mAb	Type 1 diabetes mellitus	Reduced insulin use	Increased risk of infection
	CTLA4-immuno-globulin fusion protein	RA Psoriasis MS	Improvement in arthritis	
Cytokines	TNF-specific mAb and soluble TNFR fusion protein	RA Crohn's disease Psoriatic arthritis Ankylosing spondylitis	Improvement in disability; joint repair in arthritis	Increased risk of tuberculosis and other infections; slight increase in risk of lymphoma
	IL-1 receptor antagonist	RA	Improves disability	Low efficacy
	IL-15-specific mAb	RA	May improve disability	Increased risk of opportunistic infection
	IL-6 receptor-specific mAb	RA	Decreased disease activity	Increased risk of opportunistic infection
	Type I interferons	Relapsing/remitting MS	Reduction in relapse rate	Liver toxicity; influenza-like syndrome is common

Fig. 15.11 New therapeutic agents for human autoimmunity. The category of therapeutic agent has been color-coded according to the target pathways identified in Fig. 15.6.

immune response that provides protection only against that pathogen. Immunological approaches to the treatment of cancer have been attempted for over a century, but it is only in the past decade that immunotherapy of cancer has shown real promise. An important conceptual advance has been the integration of conventional approaches such as surgery or chemotherapy, which substantially reduce tumor load, with immunotherapy.

15-14 The development of transplantable tumors in mice led to the discovery of protective immune responses to tumors.

The finding that tumors could be induced in mice after treatment with chemical carcinogens or irradiation, coupled with the development of inbred

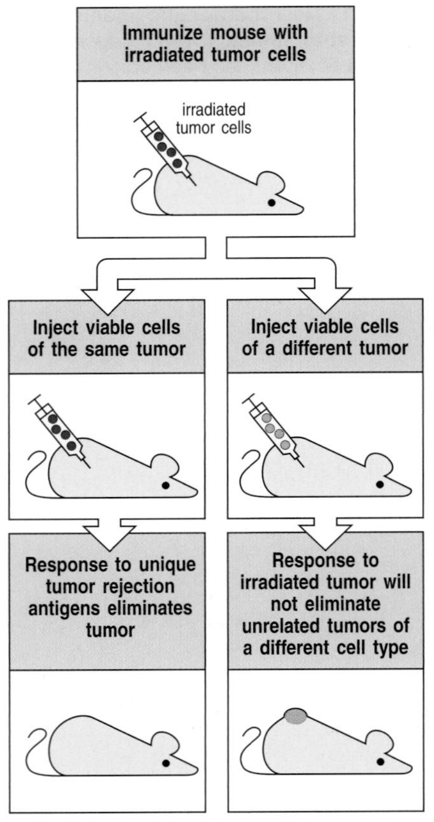

Fig. 15.12 Tumor rejection antigens are specific to individual tumors. Mice immunized with irradiated tumor cells and challenged with viable cells of the same tumor can, in some cases, reject a lethal dose of that tumor (left panels). This is the result of an immune response to tumor rejection antigens. If the immunized mice are challenged with viable cells of a different tumor, there is no protection and the mice die (right panels).

strains of mice, made it possible to undertake the key experiments that led to the discovery of immune responses to tumors. These tumors could be transplanted between mice, and the experimental study of tumor rejection has generally been based on the use of such tumors. If these bear MHC molecules foreign to the mice into which they are transplanted, the tumor cells are readily recognized and destroyed by the immune system, a fact that was exploited to develop the first MHC-congenic strains of mice. Specific immunity to tumors must therefore be studied within inbred strains, so that host and tumor can be matched for their MHC type.

Transplantable tumors in mice exhibit a variable pattern of growth when injected into syngeneic recipients. Most tumors grow progressively and eventually kill the host. However, if mice are injected with irradiated tumor cells that cannot grow, they are frequently protected against subsequent injection with a normally lethal dose of viable cells of the same tumor. There seems to be a spectrum of immunogenicity among transplantable tumors: injections of irradiated tumor cells seem to induce varying degrees of protective immunity against a challenge injection of viable tumor cells at a distant site. These protective effects are not seen in T-cell-deficient mice but can be conferred by adoptive transfer of T cells from immune mice, showing the need for T cells to mediate all these effects.

These observations indicate that the tumors express antigenic peptides that can become targets of a tumor-specific T-cell response. The antigens expressed by experimentally induced murine tumors, often termed **tumor rejection antigens** (**TRAs**), are usually specific for an individual tumor. Thus, immunization with irradiated tumor cells from tumor X protects a syngeneic mouse from challenge with live cells from tumor X but not from challenge with a different syngeneic tumor Y, and vice versa (Fig. 15.12).

15-15 Tumors can escape rejection in many ways.

F.M. Burnet called the ability of the immune system to detect tumor cells and destroy them '**immune surveillance**.' It has become clear, however, that the relationship between the immune system and cancer is considerably more complex. The concept of immune surveillance has been modified and is now considered in three phases. The first is the 'elimination phase,' which is what was previously called immune surveillance and in which the immune system recognizes and destroys potential tumor cells (Fig. 15.13). Then follows an 'equilibrium phase,' which occurs if elimination is not completely successful and in which tumor cells undergo changes or mutations that aid their survival as a result of the selection pressure imposed by the immune system. This process is known as **immunoediting** because it shapes the properties of the tumor cells that survive. The final phase is the 'escape phase,' which occurs when some tumor cells have accumulated sufficient mutations to elude the attentions of the immune system; the tumor is now able to grow unimpeded and become clinically detectable.

Mice with targeted gene deletions that remove specific components of innate and adaptive immunity have provided the best evidence that immune surveillance does influence the development of certain types of tumor. For example, mice lacking perforin, part of the killing mechanism of NK cells and CD8 cytotoxic T cells (see Section 8-28), show an increased frequency of lymphomas—tumors of the lymphoid system. Strains of mice lacking the RAG and STAT1 proteins, and thus deficient in both adaptive and certain innate immune mechanisms, develop gut epithelial and breast tumors. Mice lacking T lymphocytes expressing γ:δ receptors show markedly increased susceptibility to skin tumors induced by the topical application of carcinogens, illustrating a role for intraepithelial γ:δ T cells (see Section 11-10) in surveying

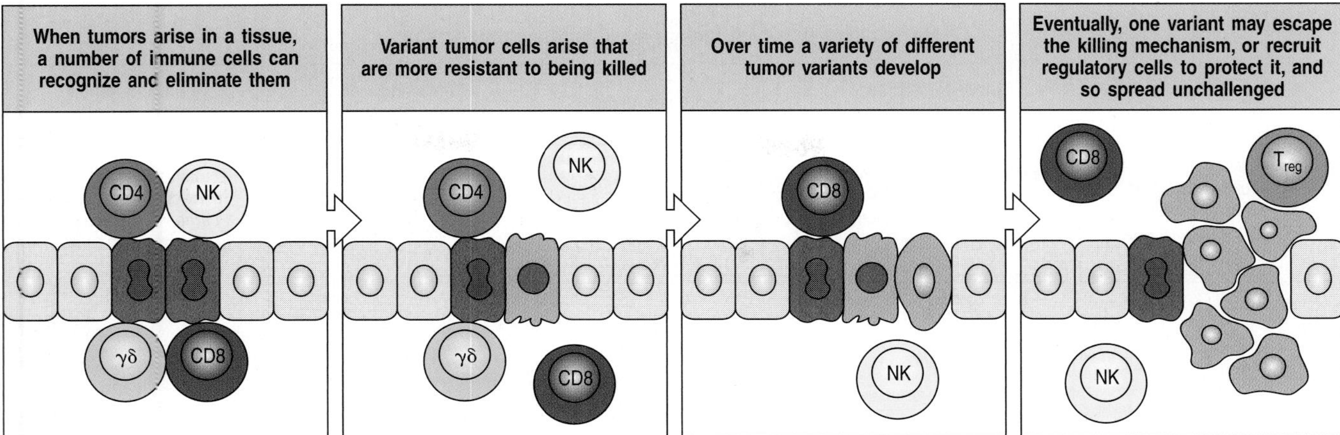

When tumors arise in a tissue, a number of immune cells can recognize and eliminate them
Variant tumor cells arise that are more resistant to being killed
Over time a variety of different tumor variants develop
Eventually, one variant may escape the killing mechanism, or recruit regulatory cells to protect it, and so spread unchallenged
CD4
NK
γδ
CD8
T_{reg}

Fig. 15.13 Malignant cells can be controlled by immune surveillance. Some types of tumor cells are recognized by a variety of immune-system cells, which can eliminate them. If the tumor cells are not completely eliminated, variants occur that eventually escape the immune system and proliferate to form a tumor.

and killing abnormal epithelial cells. Studies of the various effector cells of the immune system have identified both IFN-γ and IFN-α as important in the elimination of tumor cells, either directly, or indirectly through their actions on other cells. γ:δ T cells are a major source of IFN-γ, which may explain their importance in the removal of cancer cells noted above.

According to the immunoediting hypothesis, those tumor cells that survive the equilibrium phase have acquired many mutations that prevent their elimination by the immune system. In an immunocompetent individual, non-mutant cells are continually being removed by the immune response, thus delaying tumor growth, but when the immune system is compromised, the equilibrium phase quickly turns into escape as no tumor cells at all are removed. An excellent clinical example to support the presence of the equilibrium phase is the occurrence of cancer in recipients of organ transplants. One study reported the development of melanoma between 1 and 2 years after transplant in two patients who had received kidneys from the same donor, a patient who had had malignant melanoma, successfully treated at the time, 16 years before her death. One would presume that melanoma cells, which are known to spread easily to other organs, were present in the kidney of this patient but were in an equilibrium phase with the immune system. Thus, the melanoma cells are not killed off completely by the immune system; an immunocompetent immune system does, however, hold the number of cells in check. Because the recipients' immune systems were immunosuppressed, that allowed the melanoma cells to divide rapidly and spread to other parts of the body.

Most of the common spontaneous tumors, however, are not more common in immunodeficient individuals, and thus do not seem to be subject to immune surveillance. The major tumor types that occur with increased frequency in immunodeficient mice or humans are virus-associated tumors; immune surveillance therefore seems to be critical for the control of virus-associated tumors, and indeed tumor immunotherapy is generally more effective for virus-induced tumors.

It is not surprising that tumors arising spontaneously are rarely rejected by T cells, because in general they probably lack either distinctive antigenic peptides or the adhesion or co-stimulatory molecules needed to elicit a primary T-cell response (Fig. 15.14, first panel). Even tumors that express tumor-specific antigens may be treated as 'self' if they cause no inflammation. If the antigens are taken up by antigen-presenting cells such as immature dendritic cells and are presented to T lymphocytes in the absence of co-stimulatory signals, this will result in T-cell anergy or deletion (see Section 7-26).

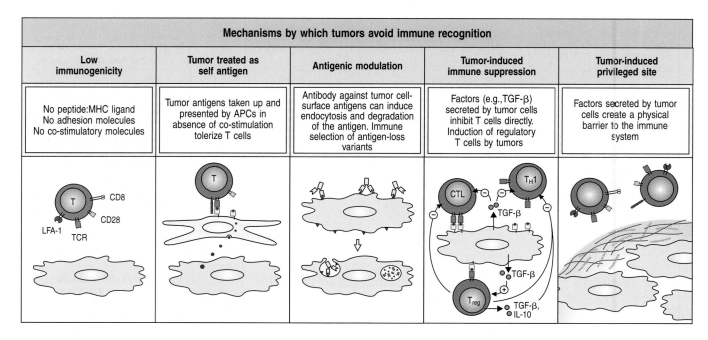

Mechanisms by which tumors avoid immune recognition

Low immunogenicity	Tumor treated as self antigen	Antigenic modulation	Tumor-induced immune suppression	Tumor-induced privileged site
No peptide:MHC ligand No adhesion molecules No co-stimulatory molecules	Tumor antigens taken up and presented by APCs in absence of co-stimulation tolerize T cells	Antibody against tumor cell-surface antigens can induce endocytosis and degradation of the antigen. Immune selection of antigen-loss variants	Factors (e.g.,TGF-β) secreted by tumor cells inhibit T cells directly. Induction of regulatory T cells by tumors	Factors secreted by tumor cells create a physical barrier to the immune system

Fig. 15.14 Tumors can avoid immune recognition in a variety of ways. First panel: tumors can have low immunogenicity. Some tumors do not have peptides of novel proteins that can be presented by MHC molecules, and therefore appear normal to the immune system. Others have lost one or more MHC molecules, and most do not express co-stimulatory proteins, which are required to activate naive T cells. Second panel: tumor antigens presented in the absence of co-stimulatory signals will make the responding T cells tolerant to that antigen. Third panel: tumors can initially express antigens to which the immune system responds but lose them by antibody-induced internalization or antigenic variation. The process of genetic instability leading to antigenic change is now considered to be part of an equilibrium phase, which can lead to outgrowth of the tumor when the immune system loses the race and is no longer able to adapt. When a tumor is attacked by cells responding to a particular antigen, any tumor cell that does not express that antigen will have a selective advantage. Fourth panel: tumors often secrete molecules, such as TGF-β, that suppress immune responses directly or can recruit regulatory T cells that can themselves secrete immunosuppressive cytokines. Fifth panel: tumor cells can secrete molecules such as collagen that form a physical barrier around the tumor, preventing lymphocyte access. APC, antigen-presenting cell.

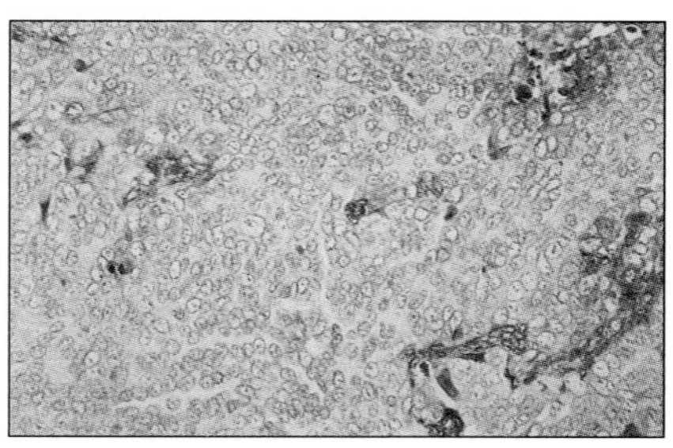

Fig. 15.15 Loss of MHC class I expression in a prostatic carcinoma. Some tumors can evade immune surveillance by a loss of expression of MHC class I molecules, preventing their recognition by CD8 T cells. A section of a human prostate cancer that has been stained with a peroxidase-conjugated antibody against HLA class I molecules is shown. The brown stain correlating with HLA class I expression is restricted to infiltrating lymphocytes and tissue stromal cells. The tumor cells that occupy most of the section show no staining. Photograph courtesy of G. Stamp.

During the equilibrium phase there are numerous mechanisms by which tumors can either avoid stimulating an immune response or evade it when it occurs (see Fig. 15.14). Tumors tend to be genetically unstable and can lose their antigens by mutation; in the presence of an immune response, mutants that have lost antigens and thus escape the immune response would be selected for. Some tumors, such as colon and cervical cancers, lose the expression of a particular MHC class I molecule (Fig. 15.15), perhaps through immunoselection by T cells specific for a peptide presented by that MHC class I molecule. In experimental studies, when a tumor loses expression of all MHC class I molecules it can no longer be recognized by cytotoxic T cells, although it might become susceptible to NK cells (Fig. 15.16). However, tumors that lose only one MHC class I molecule might be able to avoid recognition by specific CD8 cytotoxic T cells while remaining resistant to NK cells, conferring a selective advantage *in vivo*.

Yet another way in which tumors might evade immune attack is by recruiting the suppressor effects of regulatory T cells. CD4 CD25 T_{reg} cells have been found in a variety of cancers and may well be expanded specifically in response to tumor antigens. In mouse models of cancer, removal of regulatory T cells increases resistance to cancer, whereas their transfer into a T_{reg}-negative recipient allows cancers to develop. The expansion of CD4 CD25 T_{reg} cells may also be the reason for the relatively low effectiveness of IL-2 treatment in melanoma. Although approved for clinical use, IL-2 leads to a long-term beneficial response in relatively few patients. Therefore, a

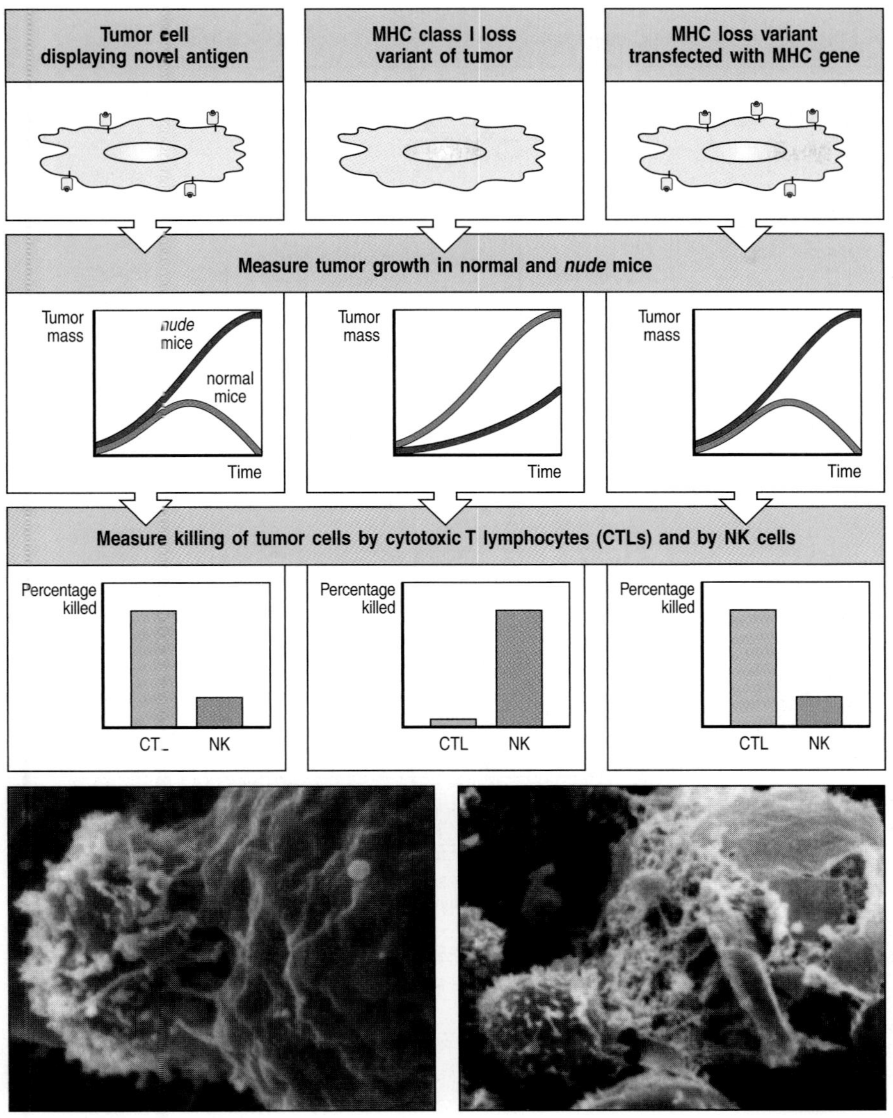

Fig. 15.16 Tumors that lose expression of all MHC class I molecules as a mechanism of escape from immune surveillance are more susceptible to being killed by NK cells. Regression of transplanted tumors is largely due to the actions of cytotoxic T lymphocytes (CTLs), which recognize novel peptides bound to MHC class I antigens on the surface of the cell (left panels). NK cells have inhibitory receptors that bind MHC class I molecules, so variants of the tumor that have low levels of MHC class I, although less sensitive to CD8 cytotoxic T cells, become susceptible to NK cells (center panels). *nude* mice lack T cells but have higher than normal levels of NK cells, and so tumors that are sensitive to NK cells grow less well in *nude* mice than in normal mice. Transfection with MHC class I genes can restore both resistance to NK cells and susceptibility to CD8 cytotoxic T cells (right panels). The bottom panels show scanning electron micrographs of NK cells attacking leukemia cells. Left panel: shortly after binding to the target cell, the NK cell has put out numerous microvillous extensions and established a broad zone of contact with the leukemia cell. The NK cell is the smaller cell on the left in both photographs. Right panel: 60 minutes after mixing, long microvillous processes can be seen extending from the NK cell (bottom left) to the leukemia cell and there is extensive damage to the leukemia cell; the plasma membrane has rolled up and fragmented. Photographs courtesy of J.C. Hiserodt.

possible additional therapy would be to deplete or inactivate regulatory T cells together with the IL-2 administration.

Many tumors evade an immune response by making immunosuppressive cytokines. TGF-β was first identified in the culture supernatant of a tumor (hence its name, transforming growth factor-β) and, as we have seen, it tends to suppress inflammatory T-cell responses and cell-mediated immunity, which are needed to control tumor growth. Interestingly, TGF-β has also been shown to induce the development of regulatory T cells. Several tumors of different tissue origins, such as melanoma, ovarian carcinoma, and B-cell lymphoma, have also been shown to produce the immunosuppressive cytokine IL-10, which can reduce dendritic cell development and activity as well as directly inhibit T-cell activation.

Some tumors avoid the immune system by creating their own immunologically privileged sites (see Section 14-5). They grow in nodules surrounded by physical barriers such as collagen and fibrin. These tumors may be invisible to the immune system, which is thus ignorant of their existence, and they can grow in this way until the tumor mass is too great to be controlled, even if the

Fig. 15.17 Proteins selectively expressed in human tumors are candidate tumor rejection antigens. The molecules listed here have all been shown to be recognized by cytotoxic T lymphocytes raised from patients with the tumor type listed.

Potential tumor rejection antigens have a variety of origins			
Class of antigen	**Antigen**	**Nature of antigen**	**Tumor type**
Tumor-specific mutated oncogene or tumor suppressor	Cyclin-dependent kinase 4	Cell-cycle regulator	Melanoma
	β-Catenin	Relay in signal transduction pathway	Melanoma
	Caspase-8	Regulator of apoptosis	Squamous cell carcinoma
	Surface Ig/ Idiotype	Specific antibody after gene rearrangements in B-cell clone	Lymphoma
Germ cell	MAGE-1 MAGE-3	Normal testicular proteins	Melanoma Breast Glioma
Differentiation	Tyrosinase	Enzyme in pathway of melanin synthesis	Melanoma
Abnormal gene expression	HER-2/neu	Receptor tyrosine kinase	Breast Ovary
	Wilms' tumor	Transcription factor	Leukemia
Abnormal post-translational modification	MUC-1	Underglycosylated mucin	Breast Pancreas
Abnormal post-transcriptional modification	GP100 TRP2	Retention of introns in the mRNA	Melanoma
Oncoviral protein	HPV type 16, E6 and E7 proteins	Viral transforming gene products	Cervical carcinoma

physical barrier is destroyed or inflammation ensues. Thus, there are many different ways in which tumors avoid recognition and destruction by the immune system.

15-16 T lymphocytes can recognize specific antigens on human tumors, and adoptive T-cell transfer is being tested in cancer patients.

The tumor rejection antigens recognized by the immune system are peptides of tumor-cell proteins that are presented to T cells by MHC molecules (see Section 15-14). These peptides become the targets of a tumor-specific T-cell response even though they can also be present on normal tissues. For instance, strategies to induce immunity to the relevant antigens in melanoma patients can induce vitiligo, an autoimmune destruction of pigmented cells in healthy skin. Several categories of tumor rejection antigens can be distinguished, and examples of each are given in Fig. 15.17. The first category consists of antigens that are strictly tumor specific. These antigens are the result of point mutations or gene rearrangements, which often arise as part of

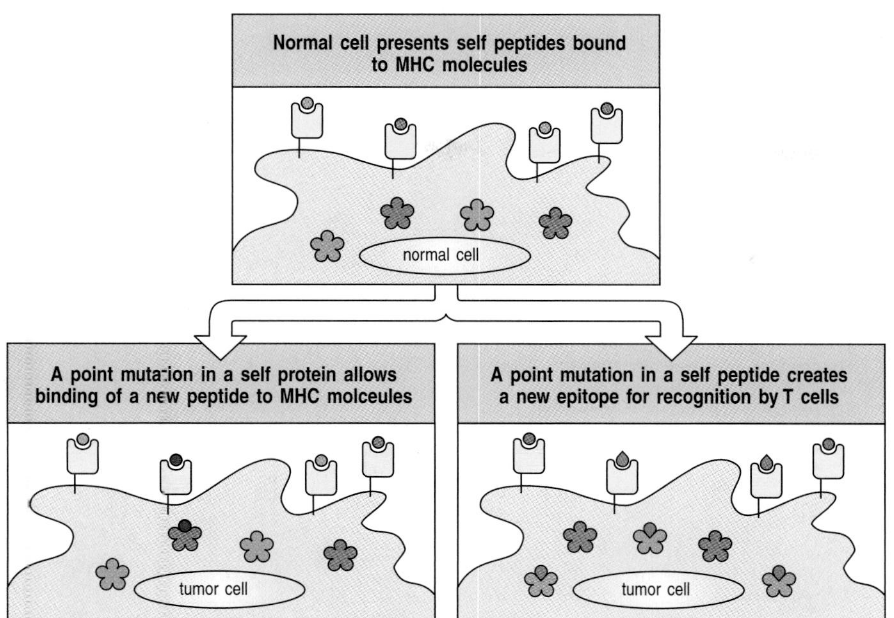

Fig. 15.18 Tumor rejection antigens may arise by point mutations in self proteins, which occur during the process of oncogenesis. In some cases a point mutation in a self protein may allow a new peptide to associate with MHC class I molecules (lower left panel). In other cases, a point mutation occurring within a self peptide that can bind self MHC proteins causes the expression of a new epitope for T-cell binding (lower right panel). In both cases, these mutated peptides will not have induced tolerance by the clonal deletion of developing T cells and can be recognized by mature T cells.

the process of oncogenesis. Point mutations may evoke a T-cell response either by allowing *de novo* binding of a peptide to MHC class I molecules or by creating a new epitope for T cells by modification of a peptide that already binds class I molecules (Fig. 15.18). These mutant peptides may, however, associate only poorly with MHC molecules or not be properly processed, and are thus less able to stimulate an effective response. In B- and T-cell tumors, which are derived from single clones of lymphocytes, a special class of tumor-specific antigen comprises the idiotypes (see Appendix I, Section A-10) unique to the antigen receptor expressed by the clone.

The second category comprises proteins encoded by genes that are normally expressed only in male germ cells, which do not express MHC molecules and therefore cannot present peptides from these molecules to T lymphocytes. Tumor cells show widespread abnormalities of gene expression, including the activation of genes encoding germ-cell proteins, such as the MAGE antigens on melanomas (see Fig. 15.17). Peptides derived from these can be presented to T cells by tumor-cell MHC class I molecules; these germ-cell proteins are therefore effectively tumor specific in their expression as antigens (Fig. 15.19).

The third category of tumor rejection antigens comprises differentiation antigens encoded by genes that are expressed only in particular types of tissues. The best examples of these are the differentiation antigens expressed in melanocytes and melanoma cells; several of these antigens are proteins in the pathways that produce the black pigment melanin. The fourth category consists of antigens that are strongly overexpressed in tumor cells compared with their normal counterparts (see Fig. 15.19). An example is HER-2/neu (also known as c-Erb-2), which is a receptor tyrosine kinase homologous to the epidermal growth factor receptor (EGFR). This receptor is overexpressed in many adenocarcinomas, including breast and ovarian cancers, where it is linked with a poor prognosis. MHC class I-restricted, CD8-positive cytotoxic T lymphocytes have been found infiltrating solid tumors overexpressing HER-2/neu but are not capable of destroying such tumors *in vivo*. The fifth category of tumor rejection antigens comprises molecules that display abnormal posttranslational modifications. An example is underglycosylated mucin, MUC-1, which is expressed by several tumors, including breast and pancreatic cancers. The sixth category comprises novel proteins that are generated

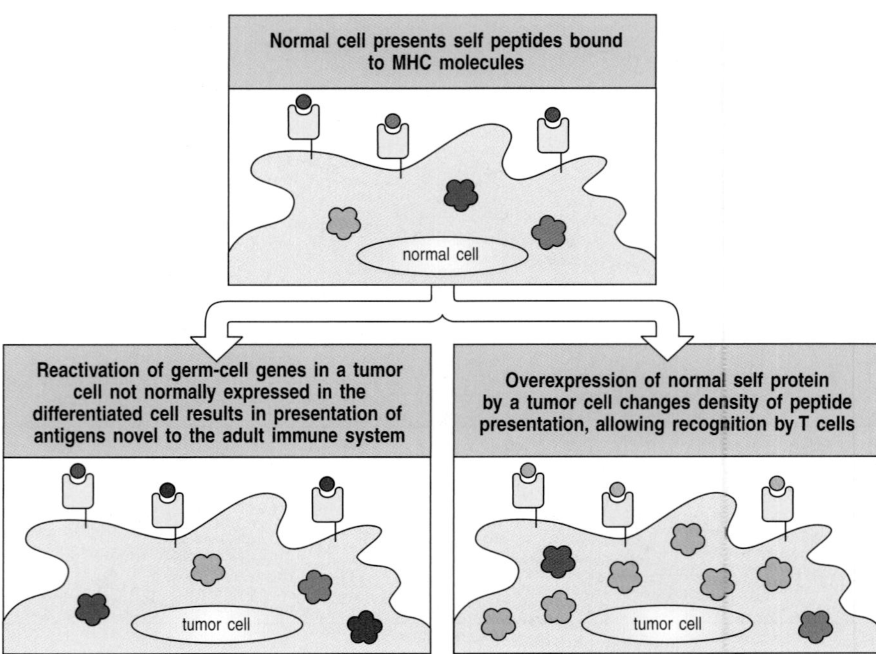

Fig. 15.19 Tumor rejection antigens are peptides of cell proteins presented by self MHC class I molecules. This figure shows two ways in which tumor rejection antigens may arise from unmutated proteins. In some cases, proteins that are normally expressed only in male germ cell tissues are reexpressed by the tumor cells (lower left panel). As these proteins are normally expressed only during germ cell development and in cells lacking MHC antigens, T cells are not tolerant of these self antigens and can respond to them as though they were foreign proteins. In other tumors, overexpression of a self protein increases the density of presentation of a normal self peptide on tumor cells (lower right panel). Such peptides are then presented at high enough levels to be recognized by low-avidity T cells. It is often the case that the same germ cell or self proteins are overexpressed in many tumors of a given tissue origin, giving rise to shared tumor rejection antigens.

when one or more introns are retained in the mRNA, which occurs in melanoma. Proteins encoded by viral oncogenes comprise the seventh category of tumor rejection antigen. These oncoviral proteins may have a critical role in the oncogenic process and, because they are foreign, they can evoke a T-cell response. Examples are the human papilloma virus type 16 proteins, E6 and E7, which are expressed in cervical carcinoma (see Section 15-18).

Although each of these categories of tumor rejection antigens may evoke an antitumor response *in vitro* and *in vivo*, it is exceptional for such a response to be able to eliminate an established tumor spontaneously. It is the goal of tumor immunotherapy to harness and augment such responses to treat cancer more effectively. The spontaneous remission occasionally observed in malignant melanoma and renal-cell carcinoma, even when disease is quite advanced, offers hope that this goal is achievable.

In melanoma, tumor-specific antigens were discovered by culturing irradiated tumor cells with autologous lymphocytes, a reaction known as the mixed lymphocyte–tumor cell culture. From such cultures, cytotoxic T lymphocytes could be identified that would kill, in an MHC-restricted fashion, tumor cells bearing the relevant tumor-specific antigen. Melanomas have been studied in detail with this approach. Cytotoxic T cells reactive against melanoma peptides have been cloned and used to characterize melanomas by the array of tumor-specific antigens displayed. These studies have yielded three important findings. First, melanomas carry at least five different antigens that can be recognized by cytotoxic T lymphocytes. Second, cytotoxic T lymphocytes reactive against melanoma antigens are not expanded *in vivo*, suggesting that these antigens are not immunogenic *in vivo*. Third, the expression of these antigens can be selected against *in vitro* and possibly also *in vivo* by the presence of specific cytotoxic T cells. These discoveries offer hope for tumor immunotherapy, an indication that these antigens are not naturally strongly immunogenic, and also a caution about the possibility of selecting, *in vivo*, tumor cells that can escape recognition and killing by cytotoxic T cells.

Consistent with these findings is the observation that melanoma-specific T cells can be propagated from peripheral blood lymphocytes, from tumor-infiltrating lymphocytes, or by draining the lymph nodes of patients in whom the melanoma is growing. Interestingly, none of the peptides recognized by these T cells derives from proteins encoded by the mutant proto-oncogenes or tumor suppressor genes that are likely to be responsible for the initial transformation of the cell into a cancer cell, although a few are the products of other mutant genes. The rest derive from normal proteins but are now displayed on tumor cells at levels detectable by T cells for the first time. For example, antigens of the MAGE family are not expressed in any normal adult tissues, with the exception of the testis, which is an immunologically privileged site (see Fig. 15.17). They probably represent early developmental antigens reexpressed in the process of tumorigenesis. Only a minority of melanoma patients have T cells reactive to the MAGE antigens, indicating that these antigens either are not expressed or are not immunogenic in most cases.

The most common melanoma antigens are peptides from the enzyme tyrosinase or from three other proteins—gp100, MART1, and gp75. These are differentiation antigens specific to the melanocyte lineage from which melanomas arise. It is likely that overexpression of these antigens in tumor cells leads to an abnormally high density of specific peptide:MHC complexes and this makes them immunogenic. Although tumor rejection antigens are usually presented as peptides complexed with MHC class I molecules, tyrosinase has been shown to stimulate CD4 T-cell responses in some melanoma patients by being ingested and presented by cells expressing MHC class II molecules. It is important to note that both CD4 and CD8 T cells are likely to be important in achieving immunological control of tumors. CD8 cells can kill the tumor cells directly, while CD4 T cells have a role in the activation of CD8 cytotoxic T cells and the establishment of memory. CD4 T cells may also kill tumor cells by means of the cytokines, such as TNF-α, that they secrete.

In addition to the human tumor antigens that have been shown to induce cytotoxic T-cell responses (see Fig. 15.17), many other candidate tumor rejection antigens have been identified by studies of the molecular basis of cancer development. These include the products of mutated cellular oncogenes or tumor suppressors, such as Ras and p53, and also fusion proteins, such as the Bcr–Abl tyrosine kinase that results from the chromosomal translocation (t9;22) found in chronic myeloid leukemia (CML). It is intriguing that, in each case, no specific cytotoxic T-cell response has been identified when the patient's lymphocytes are cultured with tumor cells bearing these mutated antigens.

When present on CML cells, the HLA class I molecule, HLA-A*0301 can display a peptide derived from the fusion site between Bcr and Abl. This peptide was detected by a powerful technique known as 'reverse' immunogenetics, in which peptides eluted from the grooves of polymorphic variants of MHC molecules are recovered and sequenced using highly sensitive mass spectrometry, enabling the peptide sequences spontaneously bound by MHC molecules to be identified. The technique has been used to detect HLA-bound peptides from other tumor antigens, for example peptides derived from the MART1 and gp100 tumor antigens of melanomas. It has also been used to identify candidate peptide sequences for vaccination against infectious diseases.

T cells specific for the Bcr–Abl fusion peptide can be identified in peripheral blood from patients with CML by using as specific ligands tetramers of HLA-A*0301 carrying the peptide (see Appendix I, Section A-28). Cytotoxic T lymphocytes specific for this and other tumor antigens can be selected *in vitro* by using peptides derived from the mutated or fused portions of these oncogenic proteins; these cytotoxic T cells are able to recognize and kill tumor cells.

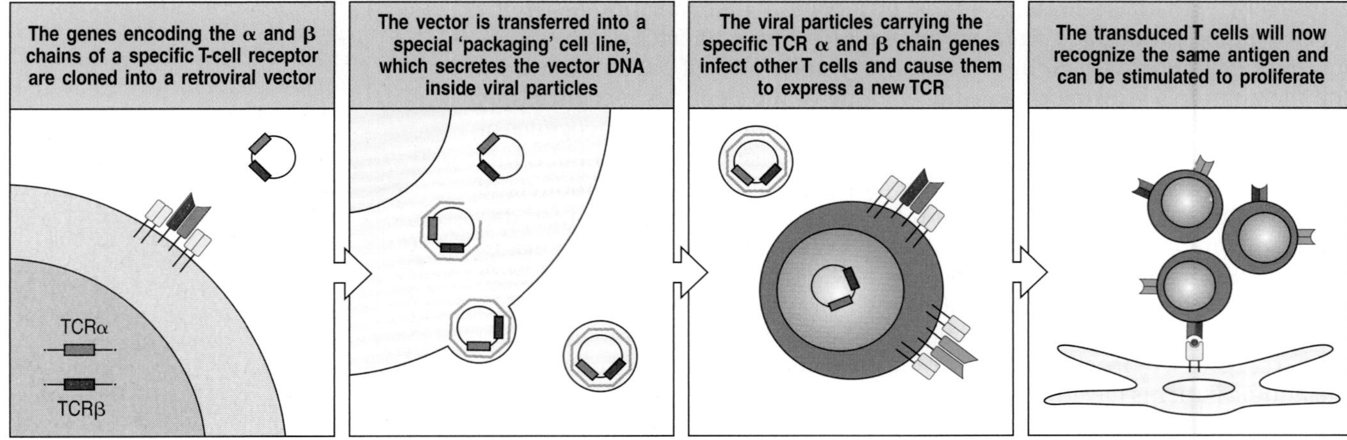

The genes encoding the α and β chains of a specific T-cell receptor are cloned into a retroviral vector
The vector is transferred into a special 'packaging' cell line, which secretes the vector DNA inside viral particles
The viral particles carrying the specific TCR α and β chain genes infect other T cells and cause them to express a new TCR
The transduced T cells will now recognize the same antigen and can be stimulated to proliferate
TCRα
TCRβ

Fig. 15.20 Retroviral transfer of T-cell receptor genes. Retroviral DNA constructs are transfected into packaging cells to produce viral particles. Peripheral blood lymphocytes are polyclonally activated, using anti-CD3 antibodies or beads coated with anti-CD3/CD28 antibodies. Two days after activation, lymphocytes are exposed to viral particles, and 5 days after activation T-cell receptor expression can be demonstrated by FACS analysis. Antigen stimulation *in vitro* or *in vivo* leads to the expansion of cells expressing the introduced T-cell receptor.

After a bone marrow transplant to treat CML, mature lymphocytes from the bone marrow donor infused into the patient can help to eliminate any residual tumor. This technique is known as donor lymphocyte infusion (DLI). At present, it is not clear to what extent the clinical response is due to a graft-versus-host effect, in which the donor lymphocytes are responding to alloantigens expressed on the leukemia cells, or whether a specific antileukemic response is important (see Section 14-35). It is encouraging that it has been possible to separate T lymphocytes *in vitro* that mediate either a graft-versus-host effect or a graft-versus-leukemia effect. The ability to prime the donor cells against leukemia-specific peptides offers the prospect of enhancing the antileukemic effect while minimizing the risk of graft-versus-host disease.

There is now good reason to believe that T-cell immunotherapy against tumor antigens is a feasible clinical approach. Adoptive T-cell therapy involves the *ex vivo* expansion of tumor-specific T cells to large numbers and the infusion of those T cells into patients. Cells are expanded *in vitro* by culture with IL-2, anti-CD3 antibodies, and allogeneic antigen-presenting cells, which provide a co-stimulatory signal. Adoptive T-cell therapy is more effective if the patient is immunosuppressed before treatment, and its effects are enhanced by the systemic administration of IL-2. T cells directed at malignancies expressing Epstein–Barr virus (EBV) antigens can also be expanded in an antigen-specific manner by using EBV-transformed B-lymphoblastoid cell lines from the patient. Another approach that has excited much interest is to transfer tumor-specific T-cell receptor genes using retroviral vectors into patients' T cells before reinfusion. This can have long-lasting effects as a result of the ability of T cells to become memory cells, and there is no requirement for histocompatibility as the transfused cells are derived from the patient (Fig. 15.20).

15-17 Monoclonal antibodies against tumor antigens, alone or linked to toxins, can control tumor growth.

The advent of monoclonal antibodies suggested the possibility of targeting and destroying tumors with antibodies against tumor-specific antigens (Fig. 15.21). This approach depends on finding a tumor-specific antigen that is a cell-surface molecule. Some of the cell-surface molecules targeted in clinical trials are shown in Fig. 15.22, and some of these treatments have now been licensed for therapy. Some striking results have been reported in the treatment of breast cancer with a humanized monoclonal antibody called trastuzumab (Herceptin), which targets the receptor HER-2/neu, which is

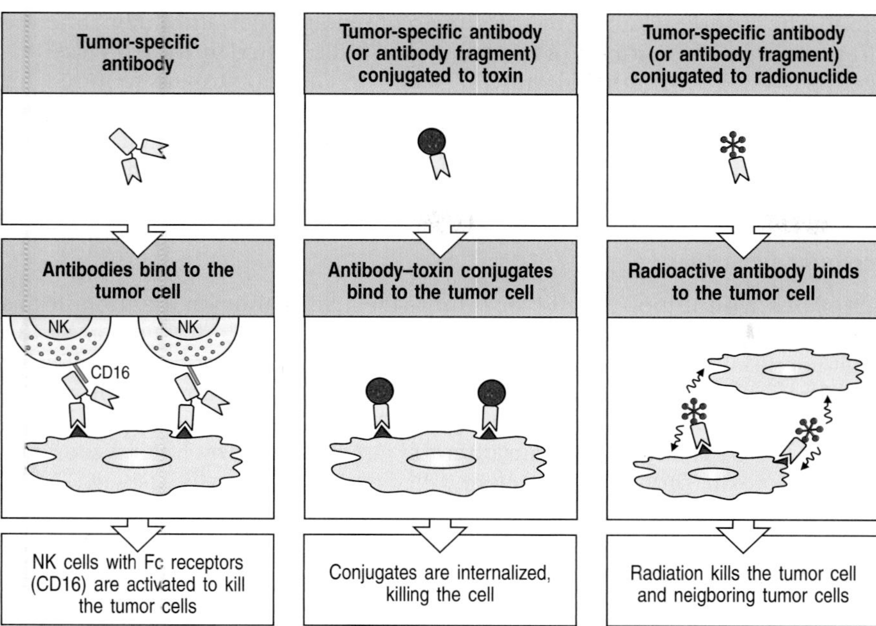

Fig.15.21 Monoclonal antibodies that recognize tumor-specific antigens have been used to help eliminate tumors. Tumor-specific antibodies of the correct isotypes can lyse tumor cells by recruiting effector cells such as NK cells, activating the NK cells via their Fc receptors (left panels). Another strategy has been to couple the antibody to a powerful toxin (center panels). When the antibody binds to the tumor cell and is endocytosed, the toxin is released from the antibody and can kill the tumor cell. If the antibody is coupled to a radioisotope (right panels), binding of the antibody to a tumor cell will deliver a dose of radiation sufficient to kill the tumor cell. In addition, nearby tumor cells could also receive a lethal radiation dose, even though they do not bind the antibody. Antibody fragments have started to replace whole antibodies for coupling to toxins or radioisotopes.

overexpressed in about one-quarter of breast cancer patients. As discussed in Section 15-16, the overexpression accounts for HER-2/neu evoking an antitumor T-cell response, although HER-2/neu is also associated with a poorer prognosis. Herceptin is thought to act by blocking the interaction between the receptor and its natural ligand and by downregulating the level of expression of the receptor. The effects of this antibody can be enhanced when it is combined with conventional chemotherapy. A second monoclonal antibody that has yielded excellent results in the treatment of non-Hodgkin's B-cell lymphoma is the anti-CD20 antibody rituximab, which triggers apoptosis on binding to CD20 on B cells (see Section 15-9).

Tumor tissue origin	Type of antigen	Antigen	Tumor type
Lymphoma/ leukemia	Differentiation antigen	CD5 Idiotype CD52 (CAMPATH1)	T-cell lymphoma B-cell lymphoma T- and B-cell lymphoma/ leukemia
	B-cell signaling receptor	CD20	Non-Hodgkin's B-cell lymphoma
Solid tumors	Cell-surface antigens Glycoprotein Carbohydrate	 CEA, mucin-1 Lewis^y CA-125	 Epithelial tumors (breast, colon, lung) Epithelial tumors Ovarian carcinoma
	Growth factor receptors	Epidermal growth factor receptor HER-2/neu IL-2 receptor Vascular endothelial growth factor (VEGF)	Lung, breast, head, and neck tumors Breast, ovarian tumors T- and B-cell tumors Colon cancer Lung, prostate, breast
	Stromal extracellular antigen	FAP-α Tenascin Metalloproteinases	Epithelial tumors Glioblastoma multiforme Epithelial tumors

Fig. 15.22 Examples of tumor antigens that have been targeted by monoclonal antibodies in therapeutic trials. CEA, carcinoembryonic antigen.

Solid tumors are sustained by the growth of blood vessels into them, and the importance of this process in tumor survival is illustrated by the effects of targeting vascular endothelial growth factor (VEGF), a cytokine required for blood-vessel growth. Significant improvements in the survival of patients with advanced colorectal cancer were seen when they were treated with a humanized anti-VEGF antibody, bevacizumab, in combination with conventional chemotherapy. This antibody, together with another antibody, cetuximab, that targets the EGF receptor, are now licensed for the treatment of colorectal cancer.

Problems with tumor-specific or tumor-selective monoclonal antibodies as therapeutic agents include antigenic variation of the tumor (see Fig. 15.13), inefficient killing of cells after binding the monoclonal antibody, inefficient penetration of the antibody into the tumor mass (which can be improved by using small antibody fragments), and soluble target antigens mopping up the antibody. The first problem can often be circumvented by linking the antibody to a toxin, producing a reagent called an **immunotoxin** (see Fig. 15.21): two favored toxins are ricin A chain and *Pseudomonas* toxin. The antibody must be internalized to allow the cleavage of the toxin from the antibody in the endocytic compartment, allowing the toxin chain to penetrate and kill the cell. Toxins coupled to native antibodies have had limited success in cancer therapy, but fragments of antibodies such as single-chain Fv molecules (see Section 3-3) show more promise. An example of a successful immunotoxin is a recombinant Fv anti-CD22 antibody fused to a fragment of *Pseudomonas* toxin. This induced complete remissions in two-thirds of a group of patients with a type of B-cell leukemia known as hairy-cell leukemia, in which the disease was resistant to conventional chemotherapy.

Two other approaches using monoclonal antibody conjugates involve linking the antibody molecule to chemotherapeutic drugs such as adriamycin or to radioisotopes. In the case of a drug-linked antibody, the specificity of the monoclonal antibody for a cell-surface antigen on the tumor concentrates the drug to the site of the tumor. After internalization, the drug is released in the endosomes and exerts its cytostatic or cytotoxic effect. A variation on this approach is to link an antibody to an enzyme that metabolizes a nontoxic pro-drug to the active cytotoxic drug, a technique known as antibody-directed enzyme/pro-drug therapy (ADEPT). This technique has the potential advantage that a small amount of enzyme, localized by the antibody to the tumor, can generate much larger amounts of active cytotoxic drug in the immediate vicinity of tumor cells than could be coupled directly to the targeting antibody. Monoclonal antibodies linked to radioisotopes (see Fig. 15.21) concentrate the radioactive source in the tumor site. This strategy has been successfully used to treat refractory B-cell lymphoma with anti-CD20 antibodies linked to yttrium-90 (ibritumomab tiuxetan). Monoclonal antibodies coupled to γ-emitting radioisotopes have also been used successfully to image tumors for the purpose of diagnosis and monitoring tumor spread (Fig. 15.23).

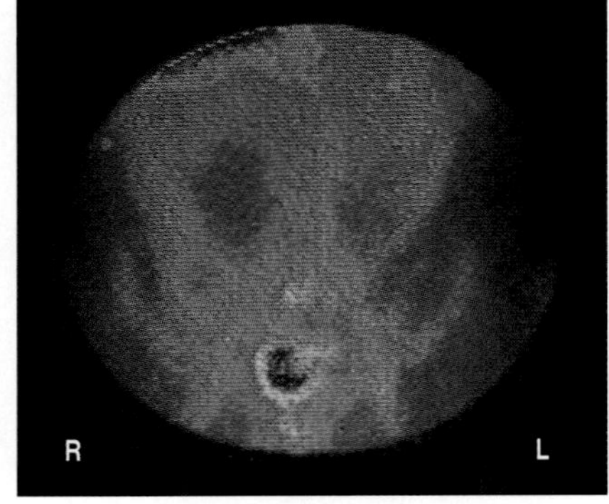

Fig. 15.23 Recurrent colorectal cancer can be detected with a radiolabeled monoclonal antibody against carcinoembryonic antigen. A patient with a possible recurrence of a colorectal cancer was injected intravenously with an indium-111-labeled monoclonal antibody against carcinoembryonic antigen. The recurrent tumor is seen as two red spots located in the pelvic region. The blood vessels are faintly outlined by circulating antibody that has not bound to the tumor. Courtesy of A.M. Peters.

These approaches have the advantage of also killing neighboring tumor cells, because the released drug or radioactive emissions can affect cells adjacent to those that bind the antibody. Ultimately, combinations of toxin-, drug-, or radioisotope-linked monoclonal antibodies, together with vaccination strategies aimed at inducing T-cell-mediated immunity, might provide the most effective cancer immunotherapy.

15-18 Enhancing the immune response to tumors by vaccination holds promise for cancer prevention and therapy.

The major breakthrough in cancer vaccines since the last edition of this book has been in the prevention of a virus-induced cancer. Toward the end of 2005, a large randomized trial involving 12,167 women showed that a recombinant vaccine against human papilloma virus (HPV) was 100% effective in preventing cervical cancer caused by two key strains of HPV-16 and HPV-18, which are associated with 70% of cervical cancers.

In contrast, attempts to use vaccines to treat tumors have been consistently disappointing. Vaccines based on tumor antigens are, in principle, the ideal approach to T-cell-mediated cancer immunotherapy. Such vaccines are difficult to develop, however; it is not clear how widely the relevant epitopes will be shared between tumors, and peptides of tumor rejection antigens will be presented only by particular MHC alleles. To be effective, a tumor vaccine may therefore need to include a range of tumor antigens. MAGE-1 antigens, for example, are recognized only by T cells in melanoma patients expressing the HLA-A1 haplotype, but the range of MAGE-type proteins has now been characterized that encompasses peptide epitopes presented by many HLA class I and II molecules. It is clear that cancer vaccines for therapy should be used only where the tumor burden is low, such as after adequate surgery and chemotherapy.

Until recently, most cancer vaccines have used the individual patient's tumor removed at surgery as a source of vaccine antigens. These cell-based vaccines are prepared by mixing either irradiated tumor cells or tumor extracts with bacterial adjuvants such as Bacille Calmette–Guérin (BCG) or *Corynebacterium parvum*, which enhance their immunogenicity (see Appendix I, Section A-4). Although vaccination using BCG adjuvants has had variable results in the past, there is renewed interest as a result of the better understanding of Toll-like receptors. Stimulation of TLR-4 by BCG and other ligands has been tested in melanoma and other solid tumors. CpG DNA, which binds to TLR-9, has also been used to increase the immunogenicity of cancer vaccines.

Where candidate tumor rejection antigens have been identified, for example in melanoma, experimental vaccination strategies include the use of whole proteins, peptide vaccines based on sequences recognized by cytotoxic T lymphocytes and helper T lymphocytes (either administered alone or presented by the patient's own dendritic cells), and recombinant viruses encoding these peptide epitopes. Tumor antigens expressed by B-cell lymphomas are seen as unique and suitable for vaccine-based immunotherapy, but this approach has not yet been successful clinically. A novel experimental approach to tumor vaccination is the use of heat-shock proteins isolated from tumor cells. The underlying principle is that heat-shock proteins act as intracellular chaperones for antigenic peptides, and there is evidence for receptors on the surface of dendritic cells that take up certain heat-shock proteins together with any bound peptides. This delivers the peptide into the antigen-processing pathways leading to peptide presentation by MHC class I molecules. This experimental technique for anti-tumor vaccination has the advantage that it does not depend on knowing the nature of the relevant tumor rejection antigens, but the disadvantage that the heat-shock proteins purified from the tumor cell include very many peptides, so that a tumor rejection antigen might constitute only a tiny fraction of these peptides.

A further experimental approach to tumor vaccination in mice is to increase the immunogenicity of tumor cells by introducing genes that encode co-stimulatory molecules or cytokines. This is intended to make the tumor itself more immunogenic. The basic scheme of such experiments is shown in Fig. 15.24. A tumor cell transfected with the gene encoding the co-stimulatory molecule B7 is implanted in a syngeneic animal. These B7-positive cells can activate tumor-specific naive T cells to become effector T cells able to reject the tumor cells. They are also able to stimulate further proliferation of the effector cells that reach the site of implantation. These T cells can then target the tumor cells whether they express B7 or not; this can be shown by reimplanting nontransfected tumor cells, which are also rejected. However, B7 can also activate CTLA-4, thereby inhibiting T-cell responses. Blockade of CTLA-4, using anti-CTLA-4 antibodies, has shown some promise in treating melanoma, by causing enhanced activation of both helper T cells and cytotoxic T cells, although autoimmune phenomena have also developed in these patients. An alternative to B7 is to use CD40 ligand; when the gene for CD40 ligand was transfected into tumor cells this could promote the maturation of dendritic cells, thereby priming the immune system.

The second strategy, that of introducing cytokine genes into tumors so that they secrete the relevant cytokine, is aimed at attracting antigen-presenting

Fig. 15.24 Transfection of tumors with the gene for B7 or for GM-CSF enhances tumor immunogenicity. A tumor that does not express co-stimulatory molecules will not induce an immune response, even though it might express tumor rejection antigens (TRAs), because naive CD8 T cells specific for the TRA cannot be activated by the tumor. The tumor therefore grows progressively in normal mice and eventually kills the host (top panels). If such tumor cells are transfected with a co-stimulatory molecule, such as B7, TRA-specific CD8 T cells now receive both signal 1 and signal 2 from the same cell and can therefore be activated (center panels). The same effect can be obtained by transfecting the tumor with the gene encoding GM-CSF, which attracts and stimulates the differentiation of dendritic cell precursors (bottom panels). Both these strategies have been tested in mice and shown to elicit memory T cells, although results with GM-CSF are more impressive. Because TRA-specific CD8 cells have now been activated, even the original B7-negative or GM-CSF-negative tumor cells can be rejected.

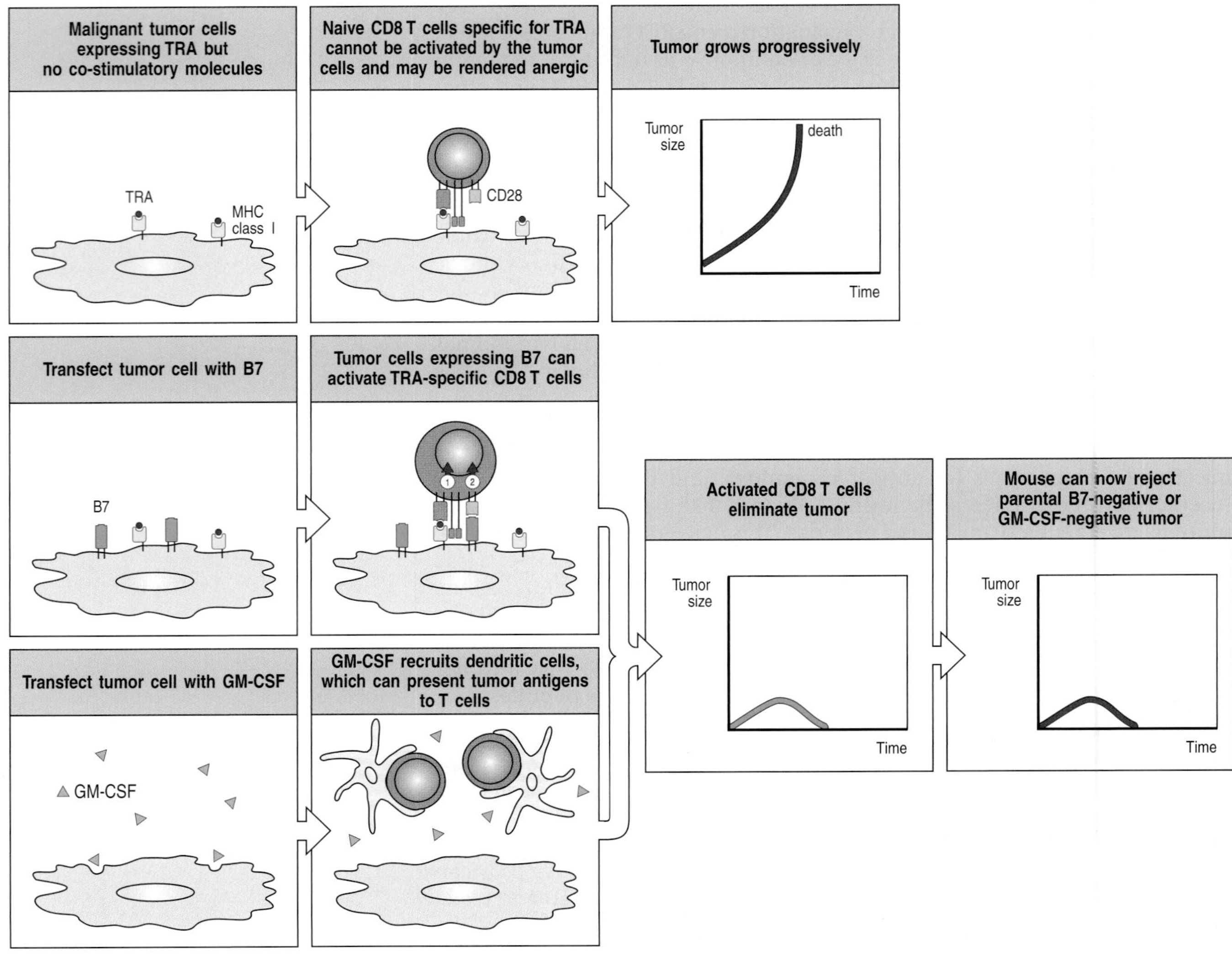

cells to the tumor and takes advantage of the paracrine actions of cytokines. In mice, the most effective tumor vaccines so far are tumor cells that secrete granulocyte-macrophage colony-stimulating factor (GM-CSF), which induces the differentiation of hematopoietic precursors into dendritic cells and attracts these to the site. GM-CSF is also thought to function as an adjuvant, activating the dendritic cells. It is believed that these cells process the tumor antigens and migrate to the local lymph nodes, where they induce potent antitumor responses. The B7-transfected cells seem less potent in inducing antitumor responses, perhaps because the bone marrow derived dendritic cells express more of the molecules required to activate naive T cells than do B7-transfected tumor cells. In addition, the tumor cells do not share the dendritic cells' special ability to migrate into the T-cell areas of the lymph nodes, where they are optimally placed to interact with passing naive T cells (see Section 8-4). GM-CSF has had limited success in patients because of the transient nature of the immune response it stimulates.

The potency of dendritic cells in activating T-cell responses provides the rationale for yet another anti-tumor vaccination strategy. The use of antigen-loaded dendritic cells to stimulate therapeutically useful cytotoxic T-cell responses to tumors has been developed in animals, and there have been initial trials in humans with cancer. Other methods under trial include loading dendritic cells *ex vivo* with DNA encoding the tumor antigen or with mRNA derived from tumor cells, and the use of apoptotic or necrotic tumor cells as sources of antigens. Dendritic cell vaccination against tumors is a very active research field, and many variables are being explored in early-phase studies in patients.

Summary.

Some tumors elicit specific immune responses that suppress or modify their growth. A partially functioning immune system can lead to the outgrowth of tumors, suggesting that the immune system does play an important role in suppressing tumor development. Tumors evade or suppress the immune system in a number of ways, and regulatory T cells have received much attention in this area. Monoclonal antibodies have been successfully developed for tumor immunotherapy in several cases, including anti-CD20 for B-cell lymphoma, and anti-VEGF antibodies in colorectal cancer. Attempts are also being made to develop vaccines incorporating peptides designed to generate effective cytotoxic and helper T-cell responses. The efficiency of dendritic cells in presenting tumor antigens has been improved by pulsing the individual's dendritic cells *in vitro* with modified tumor cells or tumor antigens and then replacing them in the body. This approach has been extended in animal experiments to the transfection of tumor cells with genes encoding co-stimulatory molecules or cytokines that attract and activate dendritic cells. The possibility of the near eradication of cervical cancer has been brought a step closer by the development of an effective vaccine against specific strains of the cancer-causing human papilloma virus.

Manipulating the immune response to fight infection.

Infectious diseases are the leading cause of death worldwide (see, for example, Fig. 11.2). The two most important contributions to public health in the past 100 years have been sanitation and vaccination, which together have dramatically reduced deaths from infectious disease. Modern immunology grew from the success of Jenner's and Pasteur's vaccines against smallpox and chicken cholera, respectively, and its greatest triumph has been the global eradication of smallpox, announced by the World Health Organization in 1980. Unfortunately, we may not have seen the last of this lethal disease, if laboratory stocks of the organism have been illegally held and were to fall into the hands of terrorists. As part of a state of cautious preparedness, smallpox vaccine doses are once more being stockpiled around the world. A global campaign to eradicate polio is now under way.

Adaptive immunity to a specific infectious agent can be achieved in several ways. One early strategy was to deliberately cause a mild infection with the unmodified pathogen. This was the principle of variolation, in which the inoculation of a small amount of dried material from a smallpox pustule would cause a mild infection followed by long-lasting protection against reinfection. However, infection after variolation was not always mild: fatal smallpox ensued in about 3% of cases, which would not meet modern criteria of safety. Jenner's achievement was the realization that infection with a bovine analog of smallpox, vaccinia (from *vacca*, Latin for cow), which caused cowpox, would provide protective immunity against smallpox in humans without the risk of significant disease. He named the process **vaccination**, and Pasteur, in his honour, extended the term to the stimulation of protection to other infectious agents. Humans are not a natural host of vaccinia, which establishes only a brief and limited subcutaneous infection. But it contains antigens that stimulate an immune response that is cross-reactive with smallpox antigens and thereby confers protection against the human disease.

This established the general principles of safe and effective vaccination, and vaccine development in the early part of the 20th century followed two empirical pathways. The first was the search for **attenuated** organisms with reduced pathogenicity, which would stimulate protective immunity but not cause disease; the second was the development of vaccines based on killed organisms and, subsequently, on purified components of organisms that would be as effective as live whole organisms. Killed vaccines were desirable because any live vaccine, including vaccinia, can cause lethal systemic infection in immunosuppressed people.

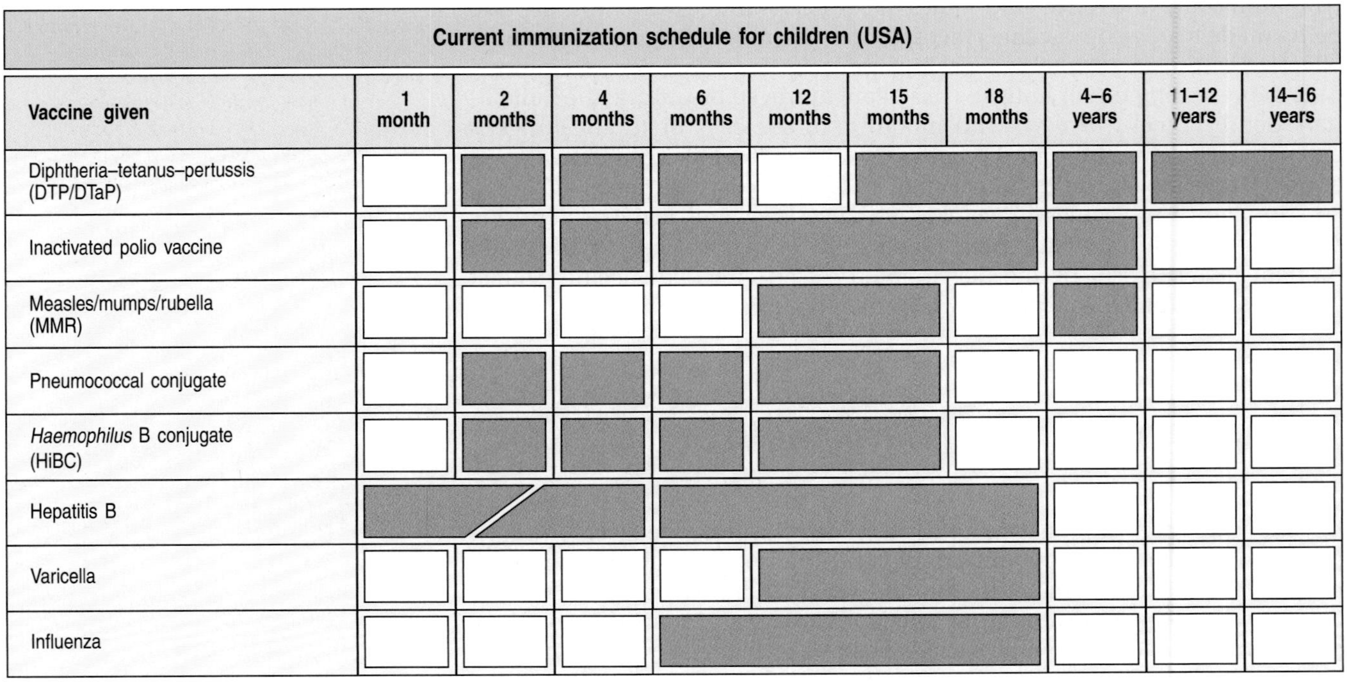

Fig. 15.25 Recommended childhood vaccination schedules (in red) in the United States. Each red bar denotes a time range during which a vaccine dose should be given. Bars spanning multiple months indicate a range of times during which the vaccine may be given.

Current immunization schedule for children (USA)

Vaccine given	1 month	2 months	4 months	6 months	12 months	15 months	18 months	4–6 years	11–12 years	14–16 years
Diphtheria–tetanus–pertussis (DTP/DTaP)										
Inactivated polio vaccine										
Measles/mumps/rubella (MMR)										
Pneumococcal conjugate										
Haemophilus B conjugate (HiBC)										
Hepatitis B										
Varicella										
Influenza										

Immunization is now considered so safe and so important that most states in the United States require all children to be immunized against measles, mumps, and polio viruses with live-attenuated vaccines as well as against tetanus (caused by *Clostridium tetani*), diphtheria (caused by *Corynebacterium diphtheriae*), and whooping cough (caused by *Bordetella pertussis*), with inactivated toxins or toxoids prepared from these bacteria (see Fig. 1.33). More recently, a vaccine has become available against *Haemophilus influenzae* type b, one of the causative agents of meningitis. Current vaccination schedules for children in the United States are shown in Fig. 15.25. Impressive as these accomplishments are, there are still many diseases for which we lack effective vaccines, as shown in Fig. 15.26. Even when a vaccine such as measles can be used effectively in developed countries, technical and economic problems can prevent its widespread use in developing countries, where mortality from these diseases is still high. The development of vaccines therefore remains an important goal of immunology, and the latter half of the 20th century saw a shift to a more rational approach based on a detailed molecular understanding of microbial pathogenicity, analysis of the protective host response to pathogenic organisms, and the understanding of the regulation of the immune system to generate effective T- and B-lymphocyte responses.

15-19 There are several requirements for an effective vaccine.

The particular requirements for successful vaccination vary depending on the nature of the infecting organism. For extracellular organisms, antibody provides the most important adaptive mechanism of host defense, whereas for the control of intracellular organisms an effective CD8 T-lymphocyte response is also essential. The ideal vaccination provides host defense at the point of entry of the infectious agent; stimulation of mucosal immunity is therefore an important goal of vaccination against those many organisms that enter through mucosal surfaces.

Effective protective immunity against some microorganisms requires the presence of preexisting antibody at the time of exposure to the infection. For

Some infections for which effective vaccines are not yet available	
Disease	**Estimated annual mortality**
Malaria	1,272,000
Schistosomiasis	15,000
Intestinal worm infestation	12,000
Tuberculosis	1,566,000
Diarrheal disease	1,798,000
Respiratory infections	3,963,000
HIV/AIDS	2,777,000
Measles†	611,000

Fig. 15.26 Diseases for which effective vaccines are still needed. †Current measles vaccines are effective but heat-sensitive, which makes their use difficult in tropical countries. Estimated mortality data for 2002 from World Health Report 2004 (World Health Organization).

example, the clinical manifestations of tetanus and diphtheria are due entirely to the effects of extremely powerful exotoxins (see Fig. 9.23), and preexisting antibody against the exotoxin is necessary to provide a defense against these diseases. Indeed, the tetanus exotoxin is so powerful that the tiny amount that can cause disease may be insufficient to lead to a protective immune response. This means that even survivors of tetanus require vaccination to be protected against the risk of subsequent attack. Preexisting antibodies are also required to protect against some intracellular pathogens, such as the polio virus, which infect critical host cells within a short period after entering the body and are not easily controlled by T lymphocytes once intracellular infection has been established.

Immune responses to infectious agents usually involve antibodies directed at multiple epitopes, and only some of these antibodies confer protection. The particular T-cell epitopes recognized can also affect the nature of the response. For example, as we saw in Chapter 12, the predominant epitope recognized by T cells after vaccination with respiratory syncytial virus induces a vigorous inflammatory response but fails to elicit neutralizing antibodies and thus causes pathology without protection. Thus, an effective vaccine must lead to the generation of antibodies and T cells directed at the correct epitopes. For some of the modern vaccine techniques, in which only one or a few epitopes are used, this consideration is particularly important.

Several very important additional requirements need to be satisfied by a successful vaccine (Fig. 15.27). First, it must be safe. Vaccines must be given to huge numbers of people, relatively few of whom are likely to die of, or sometimes even catch, the disease that the vaccine is designed to prevent. This means that even a low level of toxicity is unacceptable. Second, the vaccine must be able to produce protective immunity in a very high proportion of the people to whom it is given. Third, because it is impracticable to give large or dispersed rural populations regular 'booster' vaccinations, a successful vaccine must generate long-lived immunological memory. This means that the vaccine must prime both B and T lymphocytes. Fourth, vaccines must be very cheap if they are to be administered to large populations. Vaccines are one of the most cost-effective measures in health care, but this benefit is eroded as the cost per dose rises.

Fig. 15.27 There are several criteria for an effective vaccine.

Features of effective vaccines	
Safe	Vaccine must not itself cause illness or death
Protective	Vaccine must protect against illness resulting from exposure to live pathogen
Gives sustained protection	Protection against illness must last for several years
Induces neutralizing antibody	Some pathogens (such as polio virus) infect cells that cannot be replaced (e.g., neurons). Neutralizing antibody is essential to prevent infection of such cells
Induces protective T cells	Some pathogens, particularly intracellular, are more effectively dealt with by cell-mediated responses
Practical considerations	Low cost per dose Biological stability Ease of administration Few side-effects

An effective vaccination program provides herd immunity—by lowering the number of susceptible members of a population, the natural reservoir of infected individuals in that population falls, reducing the probability of transmission of infection. Thus, even unvaccinated members of a population can be protected from infection if the majority are vaccinated. The herd immunity effect is only seen at relatively high levels of vaccine uptake; in the case of mumps it is estimated to be around 80% and below this level sporadic epidemics can occur. This is illustrated by a dramatic increase in mumps in the United Kingdom in 2004–2005 in young adults as a result of the variable use in the mid-1990s of a measles/rubella vaccine, rather than the combined measles/mumps/rubella vaccine (MMR), which was in short supply at that time.

15-20 The history of vaccination against *Bordetella pertussis* illustrates the importance of developing an effective vaccine that is perceived to be safe.

The history of vaccination against the bacterium that causes whooping cough, *Bordetella pertussis*, provides a good example of the challenges of developing and disseminating an effective vaccine. At the beginning of the 20th century, whooping cough killed approximately 0.5% of American children under the age of 5 years. In the early 1930s, a trial of a killed, whole bacterial cell vaccine on the Faroe Islands provided evidence of a protective effect. In the United States, systematic use of a whole-cell vaccine in combination with diphtheria and tetanus toxoids (the DTP vaccine) from the 1940s resulted in a decline in the annual infection rate from 200 to fewer than 2 cases per 100,000 of the population. First vaccination with DTP was typically given at the age of 3 months.

Whole-cell pertussis vaccine causes side-effects, typically redness, pain, and swelling at the site of the injection; less commonly, vaccination is followed by high temperature and persistent crying. Very rarely, fits and a short-lived sleepiness or a floppy unresponsive state ensue. During the 1970s, widespread concern developed after several anecdotal observations that encephalitis leading to irreversible brain damage might very rarely follow pertussis vaccination. In Japan, in 1972, approximately 85% of children were given the pertussis vaccine, and fewer than 300 cases of whooping cough and no deaths were reported. As a result of two deaths after vaccination in Japan in 1975, DTP was temporarily suspended and then reintroduced with the first vaccination at 2 years of age rather than 3 months. In 1979 there were approximately 13,000 cases of whooping cough and 41 deaths. The possibility that pertussis vaccine very rarely causes severe brain damage has been studied extensively and expert consensus is that pertussis vaccine is not a primary cause of brain injury. There is no doubt that there is greater morbidity from whooping cough than from the vaccine.

The public and medical perception that whole-cell pertussis vaccination might be unsafe provided a powerful incentive to develop safer pertussis vaccines. Study of the natural immune response to *B. pertussis* showed that infection induced antibodies against four components of the bacterium—pertussis toxin, filamentous hemagglutinin, pertactin, and fimbrial antigens. Immunization of mice with these antigens in purified form protected them against challenge with pertussis. This has led to the development of acellular pertussis vaccines, all of which contain purified pertussis toxoid, that is, toxin inactivated by chemical treatment, for example with hydrogen peroxide or formaldehyde, or more recently by genetic engineering of the toxin. Some also contain one or more of the filamentous hemagglutinin, pertactin, and fimbrial antigens. Current evidence shows that these are probably as effective as whole-cell pertussis vaccine and are free of the common minor side-effects of the whole-cell vaccine. The acellular vaccine is more expensive, however, thus restricting its use in poorer countries.

The main messages of the history of pertussis vaccination are, first, that vaccines must be extremely safe and free of side-effects; second, that the public and the medical profession must perceive the vaccine to be safe; and third, that careful study of the nature of the protective immune response can lead to acellular vaccines that are safer than whole-cell vaccines but still as effective.

Public concerns about vaccination remain high. Unwarranted fears of a link between the combined live-attenuated MMR vaccine and autism saw the uptake of MMR vaccine in England fall from a peak of 92% of children in 1995–1996 to 84% in 2001–2002. Small clustered outbreaks of measles during 2002 in London illustrate the importance of maintaining high vaccine uptake to maintain herd immunity.

15-21 Conjugate vaccines have been developed as a result of understanding how T and B cells collaborate in an immune response.

Although acellular vaccines are inevitably safer than vaccines based on whole organisms, a fully effective vaccine cannot normally be made from a single isolated constituent of a microorganism, and it is now clear that this is because of the need to activate more than one cell type to initiate an immune response. One consequence of this insight has been the development of conjugate vaccines. We have already briefly described one of the most important of these in Section 9-3.

Many bacteria, including *Neisseria meningitidis* (meningococcus), *Streptococcus pneumoniae* (pneumococcus), and *Haemophilus influenzae*, have an outer capsule composed of polysaccharides that are species- and type-specific for particular strains of the bacterium. The most effective defense against these microorganisms is opsonization of the polysaccharide coat with antibody. The aim of vaccination is therefore to elicit antibodies against the polysaccharide capsules of the bacteria.

Capsular polysaccharides can be harvested from bacterial growth medium and, because they are T-cell independent antigens, they can be used on their own as vaccines. However, young children under the age of 2 years cannot make good T-cell independent antibody responses and cannot be vaccinated effectively with polysaccharide vaccines. An efficient way of overcoming this problem (see Fig. 9.5) is to conjugate bacterial polysaccharides chemically to protein carriers, which provide peptides that can be recognized by antigen-specific T cells, thus converting a T-cell independent response into a T-cell dependent anti-polysaccharide antibody response. By using this approach, various conjugate vaccines have been developed against *Haemophilus influenzae* type b, an important cause of serious childhood chest infections and meningitis, and against *N. meningitidis* serogroup C, an important cause of meningitis, and these are now widely applied. The success of the latter vaccine in the United Kingdom is illustrated in Fig. 15.28, which illustrates that the incidence of meningitis C has been dramatically reduced in comparison with meningitis B, against which there is currently no vaccine.

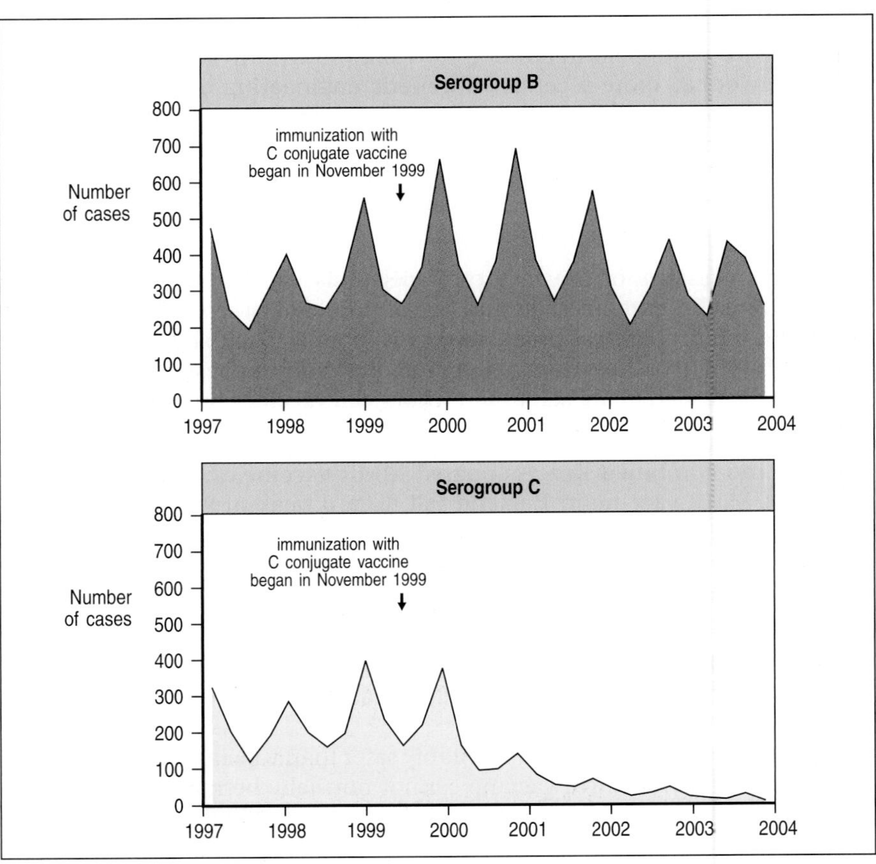

Fig. 15.28 The effect of vaccination against group C *Neisseria meningitidis* (meningococcus) on the number of cases of group B and group C meningococcal disease in England and Wales. Meningococcal infection affects approximately 5 in 100,000 people a year in the UK, with groups B and C meningococci accounting for almost all the cases. Before the introduction of the meningitis C vaccine, group C disease was the second most common cause of meningococcal disease, accounting for around 40% of cases. Group C disease now accounts for less than 10% of cases, with group B disease accounting for over 80% of cases. After the introduction of the vaccine, there was a significant fall in the number of laboratory-confirmed cases of group C disease in all age groups. The impact was greatest in the immunized groups, with reductions of over 90% in all age groups. An impact has also been seen in the unimmunized age groups, with a reduction of around 70%, suggesting that this vaccine has had a herd immunity effect.

15-22 The use of adjuvants is another important approach to enhancing the immunogenicity of vaccines.

Purified antigens are not usually strongly immunogenic on their own, and most acellular vaccines require the addition of **adjuvants**, which are defined as substances that enhance the immunogenicity of antigens (see Appendix I, Section A-4). For example, tetanus toxoid is not immunogenic in the absence of adjuvants, and tetanus toxoid vaccines often contain aluminum salts, which bind polyvalently to the toxoid by ionic interactions and selectively stimulate antibody responses. Pertussis toxin has adjuvant properties in its own right and when given mixed as a toxoid with tetanus and diphtheria toxoids, not only vaccinates against whooping cough but also acts as an adjuvant for the other two toxoids. This mixture makes up the DTP triple vaccine given to infants in the first year of life.

Many important adjuvants are sterile constituents of bacteria, particularly of their cell walls. For example, Freund's complete adjuvant, widely used in experimental animals to augment antibody responses, is an oil–water emulsion containing killed mycobacteria. A complex glycolipid, muramyl dipeptide, which can be extracted from mycobacterial cell walls or synthesized, contains much of the adjuvant activity of whole killed mycobacteria. Other bacterial adjuvants include killed *B. pertussis*, bacterial polysaccharides, bacterial heat-shock proteins, and bacterial DNA. Many of these adjuvants cause quite marked inflammation and are not suitable for use in vaccines for humans.

It is thought that most, if not all, adjuvants act on antigen-presenting cells, especially on dendritic cells, and reflect the importance of these cells in initiating immune responses. Dendritic cells are widely distributed throughout the body, acting as sentinels to detect potential pathogens at their portals of entry. These tissue dendritic cells take up antigens from their environment by phagocytosis and macropinocytosis, and they are tuned to respond to the presence of infection by migrating into lymphoid tissue and presenting these antigens to T cells. They seem to detect the presence of pathogens in two main ways. The first of these is direct, and follows the ligation and activation of receptors for invading microorganisms. These include complement receptors, Toll-like receptors, and other pattern recognition receptors of the innate immune system (see Chapter 2).

The discovery that the effects of many adjuvants are mediated by the activation of Toll-like receptors on dendritic cells opens the door to the rational development of novel adjuvants for vaccine therapy. Lipopolysaccharide (LPS) is a component of the cell wall of Gram-negative bacteria. It has adjuvant effects but these are limited by its toxicity. Small amounts of injected LPS can induce a state of shock and systemic inflammation that mimics Gram-negative sepsis. A key question is: can the adjuvant effects of LPS be separated from the toxic effects? A derivative of LPS, monophosphoryl lipid A, partly achieves this, retaining adjuvant effects but being associated with much lower toxicity than LPS. Both LPS and monophosphoryl lipid A are ligands for TLR-4, which seems to be the most important receptor mediating the adjuvant effect of LPS and its derivatives. Other adjuvants use other Toll-like receptors: unmethylated CpG DNA binds to TLR-9, and lipoprotein components of many Gram-positive bacteria bind to TLR-2. Muramyl dipeptide binds to NOD2, which is implicated in the intracellular recognition of bacteria (see Section 13-21).

These discoveries have transformed our understanding of the mechanisms of action of adjuvants. When dendritic cells are activated through the ligation of Toll-like receptors, they respond by secreting cytokines and expressing co-stimulatory molecules, which in turn stimulate the activation and differentiation of antigen-specific T cells. However, even with this greater understanding, it is likely that the therapeutic window between the efficacy and the toxicity of adjuvants will remain narrow. This is because both the adjuvant and the toxic effects of these compounds are due to the same mechanism. The physiological function of Toll-like receptors is to stimulate an inflammatory and immune response to infection. Pharmacological ligation of these receptors by adjuvants in the context of vaccination therefore steers a fine course between stimulating beneficial immunity and damaging inflammation.

The second mechanism of stimulation of dendritic cells by invading organisms is indirect and involves their activation by cytokine signals derived from the inflammatory response (see Chapter 2). Cytokines such as GM-CSF are particularly effective in activating dendritic cells to express co-stimulatory signals, and, in the context of viral infection, dendritic cells also express IFN-α and IL-12.

Adjuvants trick the immune system into responding as though there were an active infection, and just as different classes of infectious agent stimulate different types of immune responses (see Chapter 11), different adjuvants may promote different types of responses, for example an inflammatory T_H1 response or an antibody-dominated response. Some proteins, for example pertussis toxin, cholera toxin, and *E. coli* heat-labile enterotoxin, act as adjuvants and stimulate mucosal immune responses, which are a particularly important defense against organisms entering through the digestive or respiratory tracts. The use of these proteins as adjuvants is discussed further in Section 15-26.

As a result of our better understanding of the mechanisms of action of adjuvants, rational approaches to improving the activity of vaccines in clinical settings are being implemented. One approach is to co-administer cytokines. The cytokine IL-12, for example, is produced by macrophages, dendritic cells, and B cells, stimulating T lymphocytes and NK cells to release IFN-γ and promoting a T_H1 response. It has been used as an adjuvant to promote protective immunity against the protozoan parasite *Leishmania major*. Certain strains of mice are susceptible to severe cutaneous and systemic infection by *L. major*; these mice mount an immune response that is predominantly T_H2 in type and is ineffective in eliminating the organism (see Section 10-5). The co-administration of IL-12 with a vaccine containing leishmania antigens generated a T_H1 response and protected the mice against challenge with *L. major*. The use of IL-12 to promote a T_H1 response has also proved valuable in reducing the pathogenic consequences of experimental infection with the helminth parasite *Schistosoma mansoni*. These are important examples of how an understanding of the regulation of immune responses can enable rational intervention to enhance the effectiveness of vaccines.

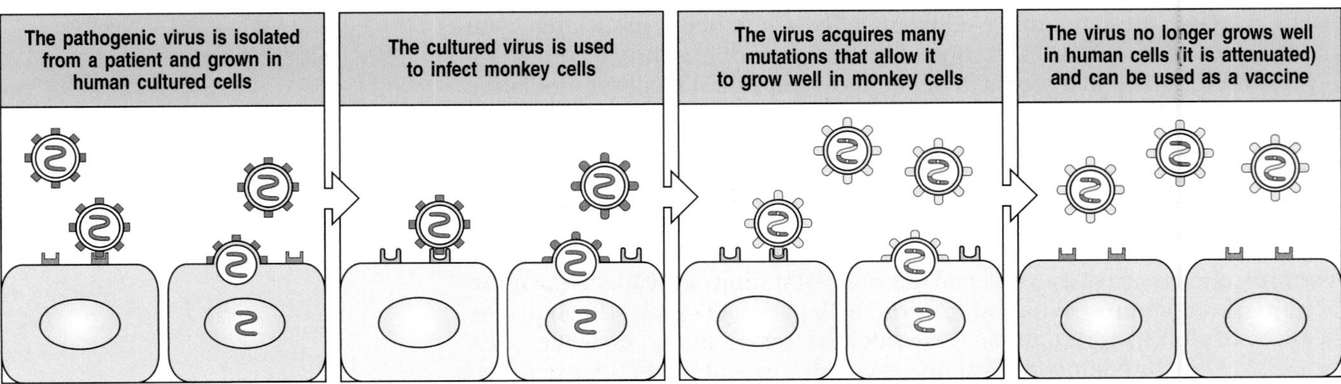

The pathogenic virus is isolated from a patient and grown in human cultured cells
The cultured virus is used to infect monkey cells
The virus acquires many mutations that allow it to grow well in monkey cells
The virus no longer grows well in human cells (it is attenuated) and can be used as a vaccine

Fig. 15.29 Viruses are traditionally attenuated by selecting for growth in nonhuman cells. To produce an attenuated virus, the virus must first be isolated by growing it in cultured human cells. The adaptation to growth in cultured human cells can cause some attenuation in itself; the rubella vaccine, for example, was made in this way. In general, however, the virus is then adapted to growth in cells of a different species, until it grows only poorly in human cells. The adaptation is a result of mutation, usually a combination of several point mutations. It is usually hard to tell which of the mutations in the genome of an attenuated viral stock are critical to attenuation. An attenuated virus will grow poorly in the human host and will therefore produce immunity but not disease.

15-23 Live-attenuated viral vaccines are usually more potent than 'killed' vaccines and can be made safer by the use of recombinant DNA technology.

Most antiviral vaccines currently in use consist of inactivated or live attenuated viruses. Inactivated, or 'killed,' viral vaccines consist of viruses treated so that they are unable to replicate. Live-attenuated viral vaccines are generally far more potent, perhaps because they elicit a greater number of relevant effector mechanisms, including cytotoxic CD8 T cells: inactivated viruses cannot produce proteins in the cytosol, so peptides from the viral antigens cannot be presented by MHC class I molecules and thus cytotoxic CD8 T cells are not generated by these vaccines. Attenuated viral vaccines are now in use for polio, measles, mumps, rubella, and varicella.

Traditionally, attenuation is achieved by growing the virus in cultured cells. Viruses are usually selected for preferential growth in nonhuman cells and, in the course of selection, become less able to grow in human cells (Fig. 15.29). Because these attenuated strains replicate poorly in human hosts, they induce immunity but not disease when given to people. Although attenuated virus strains contain multiple mutations in genes encoding several of their proteins, it might be possible for a pathogenic virus strain to re-emerge by a further series of mutations. For example, the type 3 Sabin polio vaccine strain differs from a wild-type progenitor strain at only 10 of 7429 nucleotides. On extremely rare occasions, reversion of the vaccine to a neurovirulent strain can occur, causing paralytic disease in the unfortunate recipient.

Attenuated viral vaccines can also pose particular risks to immunodeficient recipients, in whom they often behave as virulent opportunistic infections. Immunodeficient infants who are vaccinated with live attenuated polio before their inherited immunoglobulin deficiencies have been diagnosed are at risk because they cannot clear the virus from their gut, and there is therefore an increased chance that mutation of the virus associated with its continuing uncontrolled replication in the gut will lead to fatal paralytic disease.

An empirical approach to attenuation is still in use but might be superseded by two new approaches that use recombinant DNA technology. One is the isolation and *in vitro* mutagenesis of specific viral genes. The mutated genes are used to replace the wild-type gene in a reconstituted virus genome, and this deliberately attenuated virus can then be used as a vaccine (Fig. 15.30). The advantage of this approach is that mutations can be engineered so that reversion to wild type is virtually impossible.

Such an approach might be useful in developing live influenza vaccines. As described in Chapter 12, the influenza virus can reinfect the same host several times, because it undergoes antigenic shift and thus predominantly escapes the original immune response. A weak protection conferred by previous infections with a different subtype of influenza is observed in adults, but not in children, and is called heterosubtypic immunity. The current approach to vaccination against influenza is to use a killed virus vaccine that is reformulated annually on the basis of the prevalent strains of virus. The vaccine is moderately effective reducing mortality in elderly people and illness in healthy adults. The ideal influenza vaccine would be an attenuated live organism that matched the prevalent virus strain. This could be created by first introducing a series of attenuating mutations into the gene encoding a viral polymerase protein, PB2. The mutated gene segment from the attenuated virus could then be substituted for the wild-type gene in a virus carrying the relevant hemagglutinin and neuraminidase antigen variants of the current epidemic or pandemic strain. This last procedure could be repeated as necessary to keep pace with the antigenic shift of the virus. Public attention has recently been directed toward the possibility of a flu pandemic caused by the H5N1 avian flu

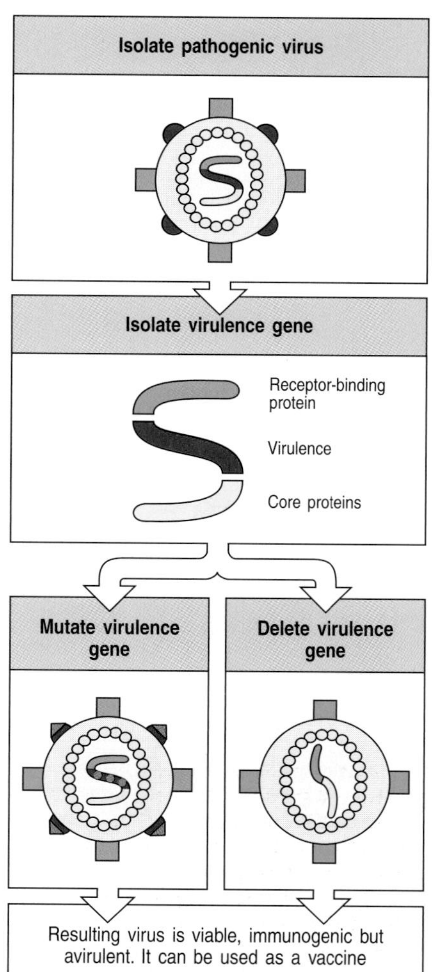

Fig. 15.30 Attenuation can be achieved more rapidly and reliably with recombinant DNA techniques. If a gene in the virus that is required for virulence but not for growth or immunogenicity can be identified, this gene can be either multiply mutated (left lower panel) or deleted from the genome (right lower panel) by using recombinant DNA techniques. This procedure creates an avirulent (nonpathogenic) virus that can be used as a vaccine. The mutations in the virulence gene are usually large, so that it is very difficult for the virus to revert to the wild type.

strain. This strain can be passed between bird and human with a high mortality rate, but a pandemic would occur only if human-to-human transmission could occur. A live-attenuated vaccine would be used only if a pandemic occurred, because to give it beforehand would introduce new influenza virus genes that might recombine with existing influenza viruses.

15-24 Live-attenuated bacterial vaccines can be developed by selecting nonpathogenic or disabled mutants.

Similar approaches have been used for bacterial vaccine development. The most important example of an attenuated vaccine is that of BCG, which is quite effective at protecting against tuberculosis in children but is less so in adults. The current BCG vaccine, which remains the most widely used vaccine in the world, was obtained from a pathogenic isolate of *Mycobacterium bovis* and passaged in the laboratory at the beginning of the 20th century. Since then, several genetically diverse strains of BCG have evolved. The level of protection afforded by BCG is extremely variable, ranging from none in some countries, such as Malawi, to 50–80% in the UK. Considering that tuberculosis remains one of the biggest killers worldwide, there is an urgent need for a new vaccine, but there are substantial hurdles to overcome. One approach is to randomly mutate or delete various virulence genes, for instance, to produce a variant, known as an auxotroph, that requires an external supply of an essential nutrient that wild-type bacteria can make themselves.

15-25 Synthetic peptides of protective antigens can elicit protective immunity.

A new route to vaccine development that does not depend on giving the whole organism, whether killed or attenuated, in a vaccine, is the identification of the T-cell peptide epitopes that stimulate protective immunity. This can be approached in two ways. One possibility is to synthesize systematically overlapping peptides from immunogenic proteins and to test each in turn for its ability to stimulate protective immunity. An alternative, but no less arduous, approach—'reverse' immunogenetics—has been used in developing a vaccine against malaria (Fig. 15.31). We met this approach in Section 15-16 in the context of the characterization of tumor antigens. The entire genome of *Plasmodium falciparum*, the major cause of fatal malaria, has now been sequenced, and has aided the effort to find an effective vaccine. Partly on the basis of this information, peptides that enhance both protective T-cell and antibody responses are being identified.

The immunogenicity of T-cell peptide epitopes depends on their specific associations with particular polymorphic variants of MHC molecules. The starting point for the studies on malaria was an association between the

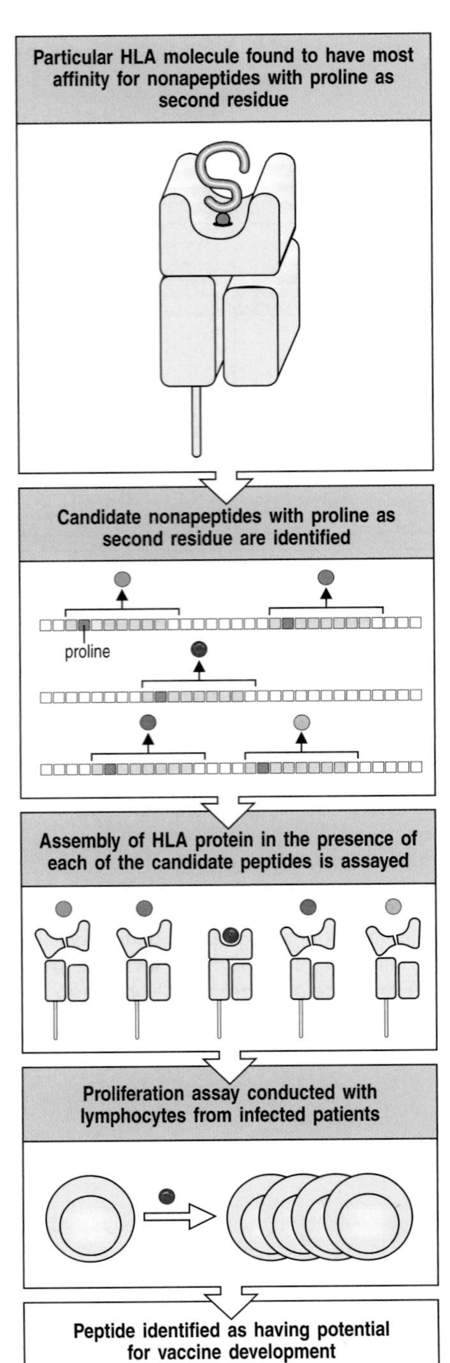

Particular HLA molecule found to have most affinity for nonapeptides with proline as second residue
Candidate nonapeptides with proline as second residue are identified
proline
Assembly of HLA protein in the presence of each of the candidate peptides is assayed
Proliferation assay conducted with lymphocytes from infected patients
Peptide identified as having potential for vaccine development

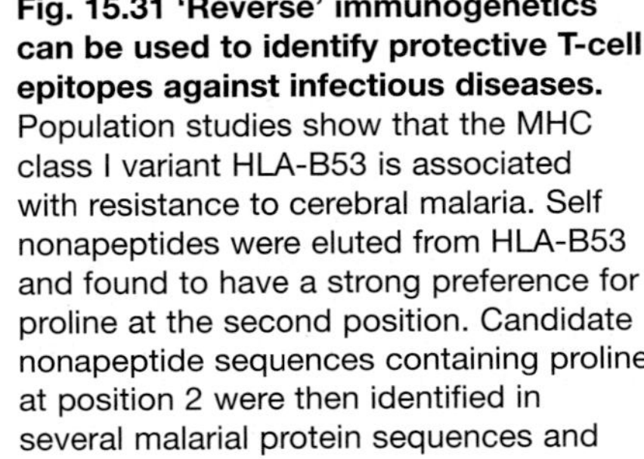

Fig. 15.31 'Reverse' immunogenetics can be used to identify protective T-cell epitopes against infectious diseases. Population studies show that the MHC class I variant HLA-B53 is associated with resistance to cerebral malaria. Self nonapeptides were eluted from HLA-B53 and found to have a strong preference for proline at the second position. Candidate nonapeptide sequences containing proline at position 2 were then identified in several malarial protein sequences and synthesized. These synthetic nonapeptides were then tested to see whether they fitted well into the peptide groove of HLA-B53 by assaying whether HLA-B53 would assemble to form a stable cell-surface heterodimer in the presence of peptide. Peptide sequences identified by this approach were then tested to see whether they would induce the proliferation of T cells from patients infected by malaria. Such sequences are good candidates for incorporation into vaccines.

human MHC class I molecule HLA-B53 and resistance to cerebral malaria—a relatively infrequent, but usually fatal, complication of infection. The hypothesis is that these MHC molecules are protective because they present peptides that are particularly good at activating naive cytotoxic T lymphocytes. A direct route to identifying the relevant peptides is to elute them from the MHC molecules of cells infected with the pathogen. A high proportion of the peptides eluted from HLA-B53 had proline as the second of their nine amino acids; this information was used to identify candidate protective peptides from four proteins of *P. falciparum* expressed in the early phase of hepatocyte infection, an important phase of infection to target in an effective immune response. One of the candidate peptides, from liver stage antigen-1, is recognized by cytotoxic T cells when bound to HLA-B53.

This approach is being extended to other MHC class I and class II molecules associated with protective immune responses against infection. Recently, a protective peptide epitope was eluted from MHC class II molecules in *Leishmania*-infected macrophages and used as a guide to isolate the gene from *Leishmania*. The gene was then used to make a protein-based vaccine that primed mice from susceptible strains to make responses to a *Leishmania* infection.

These results show considerable promise, but they also illustrate one of the major drawbacks of this approach. A malaria peptide that is restricted by HLA-B53 might not be immunogenic in an individual lacking HLA-B53: indeed, this presumably accounts for the higher susceptibility of these individuals to natural infections. Because of the very high polymorphism of the MHC in humans it will be necessary to identify panels of protective T-cell epitopes and construct vaccines containing arrays of these to develop vaccines that will protect the majority of a susceptible population.

There are other problems with peptide vaccines. Peptides are not strongly immunogenic and it is particularly difficult to generate MHC class I-specific responses by *in vivo* immunization with peptides. One approach to this problem is to integrate peptides by genetic engineering into carrier proteins within a viral vector, such as hepatitis B core antigen, which are then processed *in vivo* through natural antigen-processing pathways. A second possible technique is the use of **ISCOMs** (immune stimulatory complexes). These are lipid carriers that act as adjuvants but have minimal toxicity. They induce powerful antibody and cell-mediated responses both in animal models of infection and in humans, although their precise mechanism of action is not clear. Another approach to delivering protective peptides is the genetic engineering of infectious microorganisms to create vaccines that stimulate immunity without causing disease. Plant viruses, which are nonpathogenic to humans, are one source of novel vaccine vectors, because they can be engineered to incorporate foreign peptides into the viral coat protein. The success of this approach relies on the successful identification of protective peptide antigens as well as the natural immunogenicity of the vaccine. Mice have been protected against a lethal challenge with rabies virus by feeding them with spinach leaves infected by recombinant alfalfa mosaic virus incorporating a rabies virus peptide.

15-26 The route of vaccination is an important determinant of success.

Most vaccines are given by injection. This route has two disadvantages: the first is practical, the second immunological. Injections are painful and expensive, requiring needles, syringes, and a trained injector. They are unpopular with the recipient, which reduces vaccine uptake, and mass vaccination by this approach is laborious. The immunological drawback is that injection may not be the most effective way of stimulating an appropriate immune response because it does not mimic the usual route of entry of the majority of pathogens against which vaccination is directed.

Many important pathogens infect mucosal surfaces or enter the body through mucosal surfaces. Examples include respiratory microorganisms such as *B. pertussis*, rhinoviruses, and influenza viruses, and enteric microorganisms such as *Vibrio cholerae*, *Salmonella typhi*, enteropathogenic *E. coli*, and *Shigella*. Intranasally administered live-attenuated vaccine against influenza virus induces mucosal antibodies, which are more effective than systemic antibodies in the control of upper respiratory tract infection. However, the systemic antibodies induced by injection are effective in controlling lower respiratory tract disease, which is responsible for the severe morbidity and mortality in this disease. Thus, a realistic goal of any pandemic influenza vaccine is to prevent the lower respiratory tract disease but accept the fact that mild illness will not be prevented.

The power of the mucosal approach is illustrated by the effectiveness of live-attenuated polio vaccines. The Sabin oral polio vaccine consists of three attenuated polio virus strains and is highly immunogenic. Moreover, just as polio itself can be transmitted by fecal contamination of public swimming pools and other failures of hygiene, the vaccine can be transmitted from one individual to another by the orofecal route. Infection with *Salmonella* likewise stimulates a powerful mucosal and systemic immune response.

The rules of mucosal immunity are poorly understood. On the one hand, presentation of soluble protein antigens by the oral route often results in tolerance, which is important given the enormous load of foodborne and airborne antigens presented to the gut and respiratory tract (see Chapter 11). The ability to induce tolerance by oral administration of antigens is being explored as a therapeutic mechanism for reducing unwanted immune responses (see Section 15-13). On the other hand, the mucosal immune system can respond to and eliminate mucosal infections, such as pertussis, cholera, and polio, that enter by the oral route. The proteins from these microorganisms that stimulate immune responses are therefore of special interest. One group of powerfully immunogenic proteins at mucosal surfaces is a group of bacterial toxins that have the property of binding to eukaryotic cells and are resistant to protease. A recent finding of potential practical importance is that certain of these proteins, such as the *E. coli* heat-labile toxin and pertussis toxin, have adjuvant properties that are retained even when the parent molecule has been engineered to eliminate its toxic properties. These molecules can be used as adjuvants for oral or nasal vaccines. In mice, nasal insufflation of either of these mutant toxins together with tetanus toxoid resulted in the development of protection against lethal challenge with tetanus toxin.

15-27 Protective immunity can be induced by injecting DNA encoding microbial antigens and human cytokines into muscle.

The latest development in vaccination has come as a surprise even to the scientists who first discovered the method. The story begins with attempts to use nonreplicating bacterial plasmids encoding proteins for gene therapy: proteins expressed *in vivo* from these plasmids were found to stimulate an immune response. When DNA encoding a viral immunogen is injected intramuscularly, it leads to the development of antibody responses and cytotoxic T cells that allow the mice to reject a later challenge with whole virus (Fig. 15.32). This response does not seem to damage the muscle tissue, is safe and effective, and, because it uses only a single microbial gene or DNA encoding sets of antigenic peptides, does not carry the risk of active infection. This procedure has been termed **DNA vaccination**. DNA coated onto minute

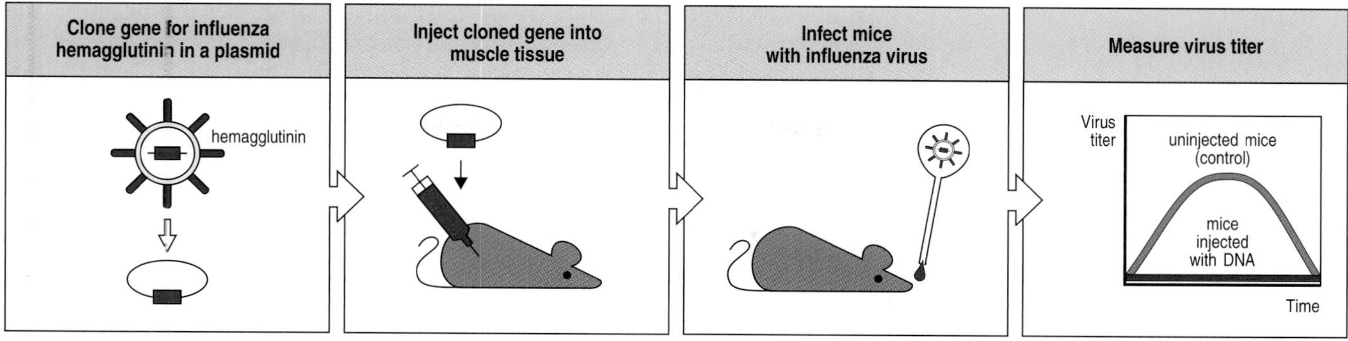
Clone gene for influenza hemagglutinin in a plasmid
hemagglutinin
Inject cloned gene into muscle tissue
Infect mice with influenza virus
Measure virus titer
Virus titer
uninjected mice (control)
mice injected with DNA
Time

Fig. 15.32 DNA vaccination by injection of DNA encoding a protective antigen and cytokines directly into muscle. Influenza hemagglutinin contains both B- and T-cell epitopes. When a DNA plasmid containing the gene for hemagglutinin is injected directly into muscle, an influenza-specific immune response consisting of both antibody and cytotoxic CD8 T cells results. The response can be enhanced by including a plasmid encoding GM-CSF in the injection. The plasmid DNA coated onto metal beads is taken up by dendritic cells in the muscle tissue into which the plasmids are injected, provoking an immune response that involves both antibody and cytotoxic T cells.

metal projectiles can be administered by a gene gun, so that several particles penetrate the skin and enter the muscle beneath. This technique has been shown to be effective in animals and might be suitable for mass immunization. Mixing in plasmids that encode cytokines such as IL-12, IL-23, or GM-CSF makes immunization with genes encoding protective antigens much more effective (see Section 15-22). Unmethylated CpG DNA is a ligand for TLR-9, and the targets for DNA vaccines are probably dendritic cells, and other antigen-presenting cells, that take up and express the DNA, undergoing activation through TLR-9 in the process. DNA vaccines for the prevention of malaria, influenza and HIV are being tested in human trials.

15-28 The effectiveness of a vaccine can be enhanced by targeting it to sites of antigen presentation.

One way of enhancing the effectiveness of a vaccine is to target it efficiently to antigen-presenting cells. This is an important mechanism of action of vaccine adjuvants. There are three complementary approaches. The first is to prevent proteolysis of the antigen on its way to antigen-presenting cells. Preserving antigen structure is an important reason why so many vaccines are given by injection rather than by the oral route, which exposes the vaccine to digestion in the gut. The second and third approaches are to target the vaccine selectively, once in the body, to antigen-presenting cells and to devise methods of engineering the selective uptake of the vaccine into antigen-processing pathways within the cell.

Techniques to enhance the uptake of antigens by antigen-presenting cells include coating the antigen with mannose to enhance uptake by mannose receptors on the cells, and presenting the antigen as an immune complex to take advantage of antibody and complement binding by Fc and complement receptors. The effects of DNA vaccination have been enhanced experimentally by injecting DNA encoding a fusion protein composed of the antigen coupled to CTLA-4. The expressed protein will then selectively bind to antigen-presenting cells carrying B7, the receptor for CTLA-4.

A more complicated strategy is to target vaccine antigens selectively into the antigen-presenting pathways within the cell. For example, HPV E7 antigen has been coupled to the signal peptide that targets lysosomal-associated membrane proteins to lysosomes and endosomes. This directs the E7 antigen directly to the intracellular compartments in which antigens are cleaved to peptides before binding to MHC class II molecules (see Section 5-7). A vaccinia virus incorporating this chimeric antigen induced a greater response in mice to E7 antigen than did vaccinia incorporating wild-type E7 antigen alone. Antigen coupled to antibodies directed at dendritic-cell receptors has been shown to lead to long-lasting immunity, providing a further approach to improving vaccination intended to activate T cells.

An improved understanding of the mechanisms of mucosal immunity (see Chapter 11) has led to the development of techniques to target antigens to M cells overlying the Peyer's patches (see Fig. 1.20). These specialized epithelial cells lack the mucin barrier and digestive properties of other mucosal epithelial cells. Instead, they can bind and endocytose macromolecules and microorganisms, which are transcytosed intact and delivered to the underlying lymphoid tissue, and some pathogens target M cells to gain entry to the body. The counterattack by immunologists is to gain a detailed molecular understanding of this mechanism of bacterial pathogenesis and subvert it as a delivery system for vaccines. For example, the outer membrane fimbrial proteins of *Salmonella typhimurium* have a key role in the binding of these bacteria to M cells. It might be possible to use these fimbrial proteins or, ultimately, just their binding motifs, as targeting agents for vaccines. A related strategy to encourage the uptake of mucosal vaccines by M cells is to encapsulate antigens in particulate carriers that are taken up selectively by M cells.

15-29 An important question is whether vaccination can be used therapeutically to control existing chronic infections.

There are many chronic diseases in which infection persists because of a failure of the immune system to eliminate disease. These can be divided into two groups: those infections in which there is an obvious immune response that fails to eliminate the organism, and those in which the infection seems to be invisible to the immune system and evokes a barely detectable immune response.

In the first category, the immune response is often partly responsible for the pathogenic effects. Infection by the helminth *Schistosoma mansoni* is associated with a powerful T_H2-type response, characterized by high IgE levels, circulating and tissue eosinophilia, and a harmful fibrotic response to schistosome ova, leading to hepatic fibrosis. Other common parasites, such as *Plasmodium* and *Leishmania* species, cause damage because they are not eliminated effectively by the immune response in many patients. Mycobacteria causing tuberculosis and leprosy cause persistent intracellular infection; a T_H1 response helps to contain these infections but also causes granuloma formation and tissue necrosis (see Fig. 8.44).

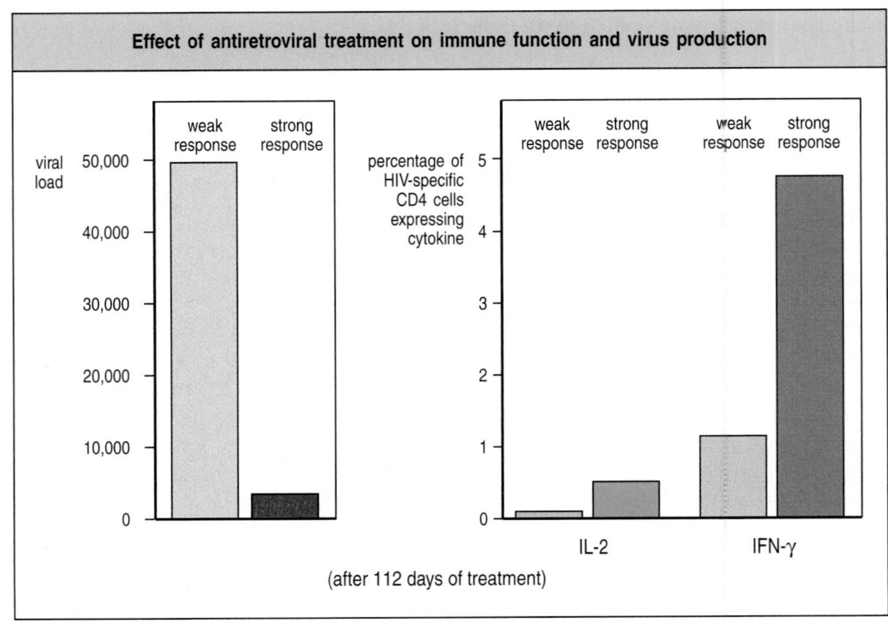

Fig. 15.33 Vaccination with dendritic cells loaded with HIV substantially reduces viral load and generates T-cell immunity. Left panel: viral load is shown for a weak and transient response to treatment (pink); the red bar represents individuals who made a strong and durable response. Right panel: CD4 T-cell IL-2 and interferon-γ production for individuals who made a weak or strong response. The production of both these cytokines, indicating T-cell activity, correlates with the response to treatment.

Among viruses, hepatitis B and hepatitis C infections are commonly followed by persistent viral carriage and hepatic injury, resulting in death from hepatitis or from hepatoma. HIV infection, as we have seen in Chapter 12, persists despite an ongoing immune response. In a preliminary trial involving patients with HIV infection, therapeutic dendritic cell vaccination reduced viral load by 80%, and in almost half of the patients this suppression of viremia lasted for more than a year. Dendritic cells derived from the patients' own bone marrow were loaded with chemically inactivated HIV. After immunization with the loaded cells, a robust T-cell response to HIV was observed that was associated with the production of IL-2 and IFN-γ (Fig. 15.33).

There is a second category of chronic infection, predominantly viral, in which the immune response fails to clear infection because of the relative invisibility of the infectious agent to the immune system. A good example is herpes simplex type 2, which is transmitted venereally, becomes latent in nerve tissue, and causes genital herpes, which is frequently recurrent. This invisibility seems to be caused by a viral protein, ICP-47, that binds to the TAP complex (see Section 5-2) and inhibits peptide transport into the endoplasmic reticulum in infected cells. Thus, viral peptides are not presented to the immune system by MHC class I molecules. Another example in this category of chronic infection is genital warts, caused by certain papilloma viruses that evoke very little immune response. In circumstances in which immunity is reduced, such as after bone marrow transplantation, T cells specific for viral antigens have been used to treat or prevent cytomegalovirus or EBV infections, viruses that remain dormant in immunocompetent hosts but are fatal when immunity is compromised. There is substantial pharmaceutical investment in therapeutic vaccination, but it is too early to know whether it will be successful.

15-30 Modulation of the immune system might be used to inhibit immunopathological responses to infectious agents.

The other approach to immunotherapy for chronic infections is to try to boost or change the host's immune response by using cytokines or anti-cytokine antibodies. The experimental treatment of leprosy gives some hope that this avenue might be successful: one can clear certain leprosy lesions by the injection of cytokines directly into the lesion, which causes reversal of the type of leprosy seen. Cytokine therapy has also been effective experimentally in treating established leishmania infection when combined with an antiparasitic drug. In mice infected with *Leishmania* and subsequently treated with a combination of drug therapy and IL-12, the immune response deviated from a T_H2 to a T_H1 pattern in a number of individuals and the infection was cleared. In most animal studies, however, it seems that the cytokine or the anti-cytokine antibody needs to be present at the first encounter with the antigen to modulate the response effectively. For example, in experimental leishmaniasis in mice, susceptible BALB/c mice injected with anti-IL-4 antibody (to suppress the T_H2-inducing effect of IL-4) at the time of infection clear their infection (Fig. 15.34). If administration of anti-IL-4 antibody is delayed by just 1 week, there is progressive growth of the parasite and a dominant T_H2 response, which does not clear the infection (see Section 10-5).

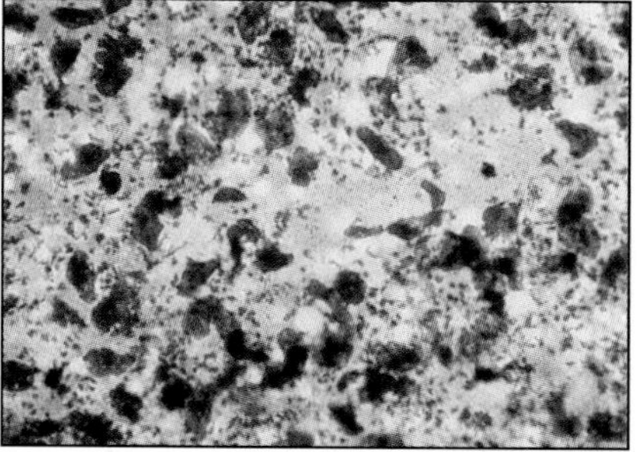

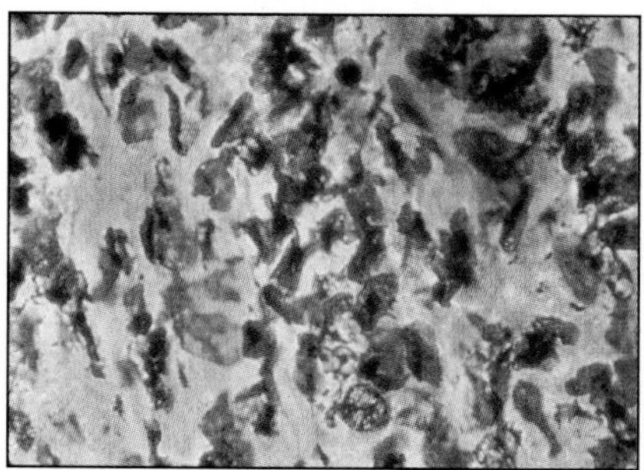

Fig. 15.34 Treatment with anti-IL-4 antibody at the time of infection with *Leishmania major* allows normally susceptible mice to clear the infection. The top panel shows a hematoxylin–eosin-stained section through the footpad of a mouse of the BALB/c strain infected with *Leishmania major* (small red dots). Large numbers of parasites are present in tissue macrophages. The bottom panel shows a similar preparation from a mouse infected in the same experiment but simultaneously treated with a single injection of anti-IL-4 monoclonal antibody. Very few parasites are present. Photographs courtesy of R.M. Locksley.

The cytokine approach is being explored as a means of inhibiting harmful immune responses to a number of important infections. As we saw in Section 15-29, liver fibrosis in schistosomiasis results from the powerful T_H2-type response. The co-administration of *S. mansoni* ova together with IL-12 does not protect mice against subsequent infection with *S. mansoni* cercariae but has a striking effect in reducing hepatic granuloma formation and fibrosis in response to schistosome ova. IgE levels are reduced, with reduced tissue eosinophilia, and the cytokine response indicates the activation of T_H1 rather than T_H2 cells. Although these results suggest that it might be possible to use a combination of antigen and cytokines to prevent the pathological consequences of diseases for which a fully protective vaccine is unavailable, they do not solve the problem of whether this approach can be effectively applied in patients whose infection is already established.

Summary.

The greatest triumphs of modern immunology have come from vaccination, which has eradicated or virtually eliminated several human diseases. It is the single most successful manipulation of the immune system so far, because it takes advantage of the immune system's natural specificity and inducibility. Nevertheless, there are many important infectious diseases for which there is still no effective vaccine. The most effective vaccines are based on attenuated live microorganisms, but these carry some risk and are potentially lethal to immunosuppressed or immunodeficient individuals. Better techniques for developing live-attenuated vaccines, or vaccines that incorporate both immunogenic components and protective antigens of pathogens, are therefore being sought. Most current viral vaccines are based on live attenuated virus, but many bacterial vaccines are based on components of the microorganism, including components of the toxins that it produces. Protective responses to carbohydrate antigens can be enhanced by conjugation to a protein. Vaccines based on peptide epitopes are still at an experimental stage and have the problem that the peptide is likely to be specific for particular variants of the MHC molecules to which they must bind, as well as being only very weakly immunogenic. A vaccine's immunogenicity often depends on adjuvants that can help, directly or indirectly, to activate antigen-presenting cells that are necessary for the initiation of immune responses. Adjuvants activate these cells by engaging the innate immune system and providing ligands for Toll-like and other receptors on antigen-presenting cells. The development of oral vaccines is particularly important for stimulating immunity to the many pathogens that enter through the mucosa. Cytokines have been used experimentally as adjuvants to boost the immunogenicity of vaccines or to bias the immune response along a specific path.

Summary to Chapter 15.

One of the great future challenges in immunology is to be able to control the immune response so that unwanted immune responses can be suppressed and desirable responses elicited. Current methods of suppressing unwanted responses rely, to a great extent, on drugs that suppress adaptive immunity indiscriminately and are thus inherently flawed. We have seen in this book that the immune system can suppress its own responses in an antigen-specific manner and, by studying these endogenous regulatory events, it might be possible to devise strategies to manipulate specific responses while sparing general immune competence. Using this approach, new treatments are starting to be developed that selectively suppress the responses that lead to allergy, autoimmunity, or the rejection of grafted organs. Similarly, as we understand more about tumors and infectious agents, better strategies to mobilize the immune system against cancer and infection are becoming possible. To achieve all this, we need to learn more about the induction of immunity and the biology of the immune system, and to apply what we have learned to human disease.

Questions.

15.1 How can regulatory T cells be induced to treat autoimmune disease and transplantation?

15.2 What are the roles of antibodies in the treatment of disease?

15.3 What approaches are there to induce tolerance in autoimmunity?

15.4 Discuss the question of whether immunotherapy is a realistic approach to the treatment of tumors.

15.5 How do tumors evade the immune response?

15.6 What is an adjuvant and how does it work?

15.7 Discuss the importance of herd immunity.

15.8 How might vaccines treat established infections?

15.9 Discuss the different uses of CTLA-4 Ig and anti-CTLA antibodies.

General references.

Ada, G.: **Vaccines and vaccination.** *N. Engl. J. Med.* 2001, **345**:1042–1053.

Curtiss, R., III: **Bacterial infectious disease control by vaccine development.** *J. Clin. Invest.* 2002, **110**:1061–1066.

Feldmann, M., and Steinman, L.: **Design of effective immunotherapy for human autoimmunity.** *Nature* 2005, **435**:612–619.

Goodnow, C.C.: **Pathways for self-tolerance and the treatment of autoimmune diseases.** *Lancet* 2001, **357**:2115–2121.

Steinman, L., and Zamvil, S.S.: **Virtues and pitfalls of EAE for the development of therapies for multiple sclerosis.** *Trends Immunol.* 2005, **26**:565–571.

Ulmer, J.B., and Liu, M.A.: **Ethical issues for vaccines and immunization.** *Nat. Rev. Immunol.* 2002, **2**:291–296.

Yu, X., Carpenter, P., and Anasetti, C.: **Advances in transplantation tolerance.** *Lancet* 2001, **357**:1959–1963.

Section references.

15-1 Corticosteroids are powerful anti-inflammatory drugs that alter the transcription of many genes.

Boumpas, D.T., Chrousos, G.P., Wilder, R.L., Cupps, T.R., and Balow, J.E.: **Glucocorticoid therapy for immune-mediated diseases: basic and clinical correlates.** *Ann. Intern. Med.* 1993, **119**:1198–1208.

Galon, J., Franchimont, D., Hiroi, N., Frey, G., Boettner, A., Ehrhart-Bornstein, M., O'Shea, J.J., Chrousos, G.P., Bornstein, S.R.: **Gene profiling reveals unknown enhancing and suppressive actions of glucocorticoids on immune cells.** *FASEB J.* 2002, **16**:61–71.

Kampa, M., and Castanas, E.: **Membrane steroid receptor signaling in normal and neoplastic cells.** *Mol. Cell. Endocrinol.* 2006, **246**:76–82.

Rhen, T., and Cidlowski J.A.: **Antiinflammatory action of glucocorticoids—new mechanisms for old drugs.** *N. Engl. J. Med.* 2005, **353**:1711–1723.

15-2 Cytotoxic drugs cause immunosuppression by killing dividing cells and have serious side-effects.

Aarbakke, J., Janka-Schaub, G., and Elion, G.B.: **Thiopurine biology and pharmacology.** *Trends Pharmacol. Sci.* 1997, **18**:3–7.

Allison, A.C., and Eugui, E.M.: **Mechanisms of action of mycophenolate mofetil in preventing acute and chronic allograft rejection.** *Transplantation* 2005, **80** (Suppl):S181–S190.

O'Donovan, P., Perrett, C.M., Zhang, X., Montaner, B., Xu, Y.Z., Harwood, C.A., McGregor, J.M., Walker, S.L., Hanaoka, F., Karran, P.: **Azathioprine and UVA light generate mutagenic oxidative DNA damage.** *Science* 2005, **309**:1871–1874.

Taylor, A.L., Watson, C.J., and Bradley, J.A.: **Immunosuppressive agents in solid organ transplantation: mechanisms of action and therapeutic efficacy.** *Crit. Rev. Oncol. Hematol.* 2005, **56**:23–46.

Zhu, L.P., Cupps, T.R., Whalen, G., and Fauci, A.S.: **Selective effects of cyclophosphamide therapy on activation, proliferation, and differentiation of human B cells.** *J. Clin. Invest.* 1987, **79**:1082–1090.

15-3 Cyclosporin A, tacrolimus (FK506), and rapamycin (sirolimus) are powerful immunosuppressive agents that interfere with T-cell signaling.

Brazelton, T.R., and Morris, R.E.: **Molecular mechanisms of action of new xenobiotic immunosuppressive drugs: tacrolimus (FK506), sirolimus (rapamycin), mycophenolate mofetil and leflunomide.** *Curr. Opin. Immunol.* 1996, **8**:710–720.

Crabtree, G.R.: **Generic signals and specific outcomes: signaling through Ca^{2+}, calcineurin, and NF-AT.** *Cell* 1999, **96**:611–614.

15-4 Immunosuppressive drugs are valuable probes of intracellular signaling pathways in lymphocytes.

Battaglia, M., Stabilini, A., and Roncarolo, M.G.: **Rapamycin selectively expands $CD4^+CD25^+FoxP3^+$ regulatory T cells.** *Blood* 2005, **105**:4743–4748.

Bierer, B.E., Mattila, P.S., Standaert, R.F., Herzenberg, L.A., Burakoff, S.J., Crabtree, G., and Schreiber, S.L.: **Two distinct signal transmission pathways in T lymphocytes are inhibited by complexes formed between an immunophilin and either FK506 or rapamycin.** *Proc. Natl Acad. Sci. USA* 1990, **87**:9231–9235.

Brown, E.J., and Schreiber, S.L.: **A signaling pathway to translational control.** *Cell* 1996, **86**:517–520.

Crespo, J.L., and Hall, M.N.: **Elucidating TOR signaling and rapamycin action: lessons from *Saccharomyces cerevisiae*.** *Microbiol. Mol. Biol. Rev.* 2002, **66**:579–591.

Gingras, A.C., Raught, B., and Sonenberg, N.: **Regulation of translation initiation by FRAP/mTOR.** *Genes Dev.* 2001, **15**:807–826.

15-5 Antibodies against cell-surface molecules have been used to remove specific lymphocyte subsets or to inhibit cell function.

Graca, L., Le Moine, A., Cobbold, S.P., and Waldmann, H.: **Antibody-induced transplantation tolerance: the role of dominant regulation.** *Immunol. Res.* 2003, **28**:181–191.

Waldmann, H., and Hale, G.: **CAMPATH: from concept to clinic.** *Phil. Trans. R. Soc. Lond. B* 2005, **360**:1707–1711.

15-6 Antibodies can be engineered to reduce their immunogenicity in humans.

Kim, S.J., Park, Y., and Hong, H.J.: **Antibody engineering for the development of therapeutic antibodies.** *Mol. Cells* 2005, **20**:17–29.

Little, M., Kipriyanov, S.M., Le Gall, F., and Moldenhauer, G.: **Of mice and men: hybridoma and recombinant antibodies.** *Immunol. Today* 2000, **21**:364–370.

Winter, G., Griffiths, A.D., Hawkins, R.E., and Hoogenboom, H.R.: **Making antibodies by phage display technology.** *Annu. Rev. Immunol.* 1994, **12**:433–455.

15-7 Monoclonal antibodies can be used to prevent allograft rejection.

Graca, L., Cobbold, S.P., and Waldmann, H.: **Identification of regulatory T cells in tolerated allografts.** *J. Exp. Med.* 2002, **195**:1641–1646.

Graca, L., Thompson, S., Lin, C.Y., Adams, E., Cobbold, S.P., and Waldmann, H.: **Both $CD4^+CD25^+$ and $CD4^+CD25^-$ regulatory cells mediate dominant transplantation tolerance.** *J. Immunol.* 2002, **168**:5558–5565.

Kingsley, C.I., Karim, M., Bushell, A.R., and Wood, K.J.: **$CD25^+CD4^+$ regulatory T cells prevent graft rejection: CTLA-4- and IL-10-dependent immunoregulation of alloresponses.** *J. Immunol.* 2002, **168**:1080–1086.

Kirk, A.D., Burkly, L.C., Batty, D.S., Baumgartner, R.E., Berning, J.D., Buchanan, K., Fechner, J.H., Jr., Germond, R.L., Kampen, R.L., Patterson, N.B., *et al.*: **Treatment with humanized monoclonal antibody against CD154 prevents acute renal allograft rejection in nonhuman primates.** *Nat. Med.* 1999, **5**:686–693.

Li, X.C., Strom, T.B., Turka, L.A., and Wells, A.D.: **T-cell death and transplantation tolerance.** *Immunity* 2001, **14**:407–416.

Li, Y., Li, X.C., Zheng, X.X., Wells, A.D., Turka, L.A., and Strom, T.B.: **Blocking both signal 1 and signal 2 of T-cell activation prevents apoptosis of alloreactive T cells and induction of peripheral allograft tolerance.** *Nat. Med.* 1999, **5**:1298–1302.

Lin, C.Y., Graca, L., Cobbold, S.P., and Waldmann, H.: **Dominant transplantation tolerance impairs $CD8^+$ T cell function but not expansion.** *Nat. Immunol.* 2002, **3**:1208–1213.

Waldmann, H.: **Reprogramming the immune system.** *Immunol. Rev.* 2002, **185**:227–235.

Waldmann, H.: **Therapeutic approaches for transplantation.** *Curr. Opin. Immunol.* 2001, **13**:606–610.

15-8 Biological agents can be used to alleviate and suppress autoimmune disease.

Cyster, J.G.: **Chemokines, sphingosine-1-phosphate, and cell migration in secondary lymphoid organs.** *Annu. Rev. Immunol.* 2005, **23**:127–159.

Feldmann, M., and Maini, R.N.: **Lasker Clinical Medical Research Award. TNF defined as a therapeutic target for rheumatoid arthritis and other autoimmune diseases.** *Nat. Med.* 2003, **9**:1245–1250.

Hallegua, D.S., and Weisman, M.H.: **Potential therapeutic uses of interleukin 1 receptor antagonists in human diseases.** *Ann. Rheum. Dis.* 2002, **61**:960–967.

Idzko, M., Hammad, H., van Nimwegen, M., Kool, M., Muller, T., Soullie, T., Willart, M.A., Hijdra, D., Hoogsteden, H.C., and Lambrecht, B.N.: **Local application of FTY720 to the lung abrogates experimental asthma by altering dendritic cell function.** *J. Clin. Invest.* 2006, **116**:2935–2944.

Mackay, C.R.: **New avenues for anti-inflammatory therapy.** *Nat. Med.* 2002, **8**:117–118.

Miller, D.H., Khan, O.A., Sheremata, W.A., Blumhardt, L.D., Rice, G.P., Libonati, M.A., Willmer-Hulme, A.J., Dalton, C.M., Miszkiel, K.A., and O'Connor, P.W.: **A controlled trial of natalizumab for relapsing multiple sclerosis.** *N. Engl. J. Med.* 2003, **348**:15–23.

Podolsky, D.K.: **Selective adhesion-molecule therapy and inflammatory bowel disease—a tale of Janus?** *N. Engl. J. Med.* 2005, **353**:1965–1968.

Sandborn, W.J., and Targan, S.R.: **Biologic therapy of inflammatory bowel disease.** *Gastroenterology* 2002, **122**:1592–1608.

15-9 Depletion or inhibition of autoreactive lymphocytes can treat autoimmune disease.

Coles, A., Deans, J., and Compston, A.: **Campath-1H treatment of multiple sclerosis: lessons from the bedside for the bench.** *Clin. Neurol. Neurosurg.* 2004, **106**:270–274.

Edwards, J.C., Leandro, M.J., and Cambridge, G.: **B lymphocyte depletion in rheumatoid arthritis: targeting of CD20.** *Curr. Dir. Autoimmun.* 2005, **8**:175–192.

Rep, M.H., van Oosten, B.W., Roos, M.T., Ader, H.J., Polman, C.H., and van Lier, R.A.: **Treatment with depleting CD4 monoclonal antibody results in a preferential loss of circulating naive T cells but does not affect IFN-γ secreting TH1 cells in humans.** *J. Clin. Invest.* 1997, **99**:2225–2231.

Singh, R., Robinson, D.B., and El-Gabalawy, H.S.: **Emerging biologic therapies in rheumatoid arthritis: cell targets and cytokines.** *Curr. Opin. Rheumatol.* 2005, **17**:274–279.

Willis, F., Marsh, J.C., Bevan, D.H., Killick, S.B., Lucas, G., Griffiths, R., Ouwehand, W., Hale, G., Waldmann, H., and Gordon-Smith, E.C. **The effect of treatment with Campath-1H in patients with autoimmune cytopenias.** *Br. J. Haematol.* 2001, **114**:891–898.

Yazawa, N., Hamaguchi, Y., Poe, J.C., and Tedder, T.F.: **Immunotherapy using unconjugated CD19 monoclonal antibodies in animal models for B lymphocyte malignancies and autoimmune disease.** *Proc. Natl Acad. Sci. USA* 2005, **102**:15178–15783.

Zaja, F., De Vita, S., Mazzaro, C., Sacco, S., Damiani, D., De Marchi, G., Michelutti, A., Baccarani, M., Fanin, R., and Ferraccioli, G.: **Efficacy and safety of rituximab in type II mixed cryoglobulinemia.** *Blood* 2003, **101**:3827–3834.

15-10 Interference with co-stimulatory pathways for the activation of lymphocytes could be a treatment for autoimmune disease.

Abrams, J.R., Kelley, S.L., Hayes, E., Kikuchi, T., Brown, M.J., Kang, S., Lebwohl, M.G., Guzzo, C.A., Jegasothy, B.V., Linsley, P.S., and Krueger, J.G.: **Blockade of**

T lymphocyte costimulation with cytotoxic T lymphocyte-associated antigen 4-immunoglobulin (CTLA4Ig) reverses the cellular pathology of psoriatic plaques, including the activation of keratinocytes, dendritic cells, and endothelial cells. *J. Exp. Med.* 2000, **192**:681–694.

Aruffo, A., and Hollenbaugh, D.: **Therapeutic intervention with inhibitors of co-stimulatory pathways in autoimmune disease.** *Curr. Opin. Immunol.* 2001, **13**:683–686.

Ellis, C.N., and Krueger, G.G.: **Treatment of chronic plaque psoriasis by selective targeting of memory effector T lymphocytes.** *N. Engl. J. Med.* 2001, **345**:248–255.

Kraan, M.C., van Kuijk, A.W., Dinant, H.J., Goedkoop, A.Y., Smeets, T.J., de Rie, M.A., Dijkmans, B.A., Vaishnaw, A.K., Bos, J.D., and Tak, P.P.: **Alefacept treatment in psoriatic arthritis: reduction of the effector T cell population in peripheral blood and synovial tissue is associated with improvement of clinical signs of arthritis.** *Arthritis Rheum.* 2002, **46**:2776–2784.

Lowes, M.A., Chamian, F., Abello, M.V., Fuentes-Duculan, J., Lin, S.L., Nussbaum, R., Novitskaya, I., Carbonaro, H., Cardinale, I., Kikuchi, T., *et al.*: **Increase in TNF-α and inducible nitric oxide synthase-expressing dendritic cells in psoriasis and reduction with efalizumab (anti-CD11a).** *Proc. Natl Acad. Sci. USA* 2005, **102**:19057–19062.

15-11 Induction of regulatory T cells by antibody therapy can inhibit autoimmune disease.

Chatenoud, L.: **CD3-specific antibodies restore self-tolerance: mechanisms and clinical applications.** *Curr. Opin. Immunol.* 2005, **17**:632–637.

Ehrenstein, M.R., Evans, J.G., Singh, A., Moore, S., Warnes, G., Isenberg, D.A., and Mauri, C.: **Compromised function of regulatory T cells in rheumatoid arthritis and reversal by anti-TNFα therapy.** *J. Exp. Med.* 2004, **200**:277–285.

Hafler, D.A., Kent S.C., Pietrusewicz, M.J., Khoury, S.J., Weiner, H.L., and Fukaura, H.: **Oral administration of myelin induces antigen-specific TGF-β 1 secreting T cells in patients with multiple sclerosis.** *Ann. N.Y. Acad. Sci.* 1997, **835**:120–131.

Herold, K.C., Burton, J.B., Francois, F., Poumian-Ruiz, E., Glandt, M., and Bluestone, J.A.: **Activation of human T cells by FcR nonbinding anti-CD3 mAb, hOKT3γ1(Ala-Ala).** *J. Clin. Invest.* 2003, **111**:409–418.

Herold, K.C., Hagopian, W., Auger, J.A., Poumian-Ruiz, E., Taylor, L., Donaldson, D., Gitelman, S.E., Harlan, D.M., Xu, D., Zivin, R.A., *et al.*: **Anti-CD3 monoclonal antibody in new-onset type 1 diabetes mellitus.** *N. Engl. J. Med.* 2002, **346**:1692–1698.

Masteller, E.L., and Bluestone, J.A.: **Immunotherapy of insulin-dependent diabetes mellitus.** *Curr. Opin. Immunol.* 2002, **14**:652–659.

Roncarolo, M.G., Bacchetta, R., Bordignon, C., Narula, S., and Levings, M.K.: **Type 1 T regulatory cells.** *Immunol. Rev.* 2001, **182**:68–79.

15-12 A number of commonly used drugs have immunomodulatory properties.

van Etten, E., and Mathieu, C.: **Immunoregulation by 1,25-dihydroxyvitamin D3: basic concepts.** *J. Steroid Biochem. Mol. Biol.* 2005, **97**:93–101.

Youssef, S., Stuve, O., Patarroyo, J.C., Ruiz, P.J., Radosevich, J.L., Hur, E.M., Bravo, M., Mitchell, D.J., Sobel, R.A., Steinman, L., *et al.*: **The HMG-CoA reductase inhibitor, atorvastatin, promotes a Th2 bias and reverses paralysis in central nervous system autoimmune disease.** *Nature* 2002, **420**:78–84.

15-13 Controlled administration of antigen can be used to manipulate the nature of an antigen-specific response.

Diabetes Prevention Trial: Type 1 Diabetes Study Group: **Effects of insulin in relatives of patients with type 1 diabetes mellitus.** *N. Engl. J. Med.* 2002, **346**:1685–1691.

Liblau, R., Tisch, R., Bercovici, N., and McDevitt, H.O.: **Systemic antigen in the treatment of T-cell-mediated autoimmune diseases.** *Immunol. Today* 1997, **18**:599–604.

Magee, C.C., and Sayegh, M.H.: **Peptide-mediated immunosuppression.** *Curr. Opin. Immunol.* 1997, **9**:669–675.

Steinman, L., Utz, P.J., and Robinson, W.H.: **Suppression of autoimmunity via microbial mimics of altered peptide ligands.** *Curr. Top. Microbiol. Immunol.* 2005, **296**:55–63.

Weiner, H.L.: **Oral tolerance for the treatment of autoimmune diseases.** *Annu. Rev. Med.* 1997, **48**:341–351.

15-14 The development of transplantable tumors in mice led to the discovery of protective immune responses to tumors.

Jaffee, E.M., and Pardoll, D.M.: **Murine tumor antigens: is it worth the search?** *Curr. Opin. Immunol.* 1996, **8**:622–627.

15-15 Tumors can escape rejection in many ways.

Ahmadzadeh, M., and Rosenberg, S.A.: **IL-2 administration increases CD4$^+$CD25^{hi}Foxp3$^+$ regulatory T cells in cancer patients.** *Blood* 2006, **107**:2409–2414.

Bodmer, W.F., Browning, M.J., Krausa, P., Rowan, A., Bicknell, D.C., and Bodmer, J.G.: **Tumor escape from immune response by variation in HLA expression and other mechanisms.** *Ann N.Y. Acad. Sci.* 1993, **690**:42–49.

Dunn, G.P., Old, L.J., and Schreiber, R.D.: **The immunobiology of cancer immunosurveillance and immunoediting.** *Immunity* 2004, **21**:137–148.

Girardi, M., Oppenheim, D.E., Steele, C.R., Lewis, J.M., Glusac, E., Filler, R., Hobby, P., Sutton, B., Tigelaar, R.E., and Hayday, A.C.: **Regulation of cutaneous malignancy by γδ T cells.** *Science* 2001, **294**:605–609.

Ikeda, H., Lethe, B., Lehmann, F., van Baren, N., Baurain, J.F., de Smet, C., Chambost, H., Vitale, M., Moretta, A., Boon, T., *et al.*: **Characterization of an antigen that is recognized on a melanoma showing partial HLA loss by CTL expressing an NK inhibitory receptor.** *Immunity* 1997, **6**:199–208.

Koopman, L.A., Corver, W.E., van der Slik, A.R., Giphart, M.J., and Fleuren, G.J.: **Multiple genetic alterations cause frequent and heterogeneous human histocompatibility leukocyte antigen class I loss in cervical cancer.** *J. Exp. Med.* 2000, **191**:961–976.

Ochsenbein, A.F., Klenerman, P., Karrer, U., Ludewig, B., Pericin, M., Hengartner, H., and Zinkernagel, R.M.: **Immune surveillance against a solid tumor fails because of immunological ignorance.** *Proc. Natl Acad. Sci. USA* 1999, **96**:2233–2238.

Ochsenbein, A.F., Sierro, S., Odermatt, B., Pericin, M., Karrer, U., Hermans, J., Hemmi, S., Hengartner, H., and Zinkernagel, R.M.: **Roles of tumour localization, second signals and cross priming in cytotoxic T-cell induction.** *Nature* 2001, **411**:1058–1064.

Pardoll, D.: **T cells and tumours.** *Nature* 2001, **411**:1010–1012.

Tada, T., Ohzeki, S., Utsumi, K., Takiuchi, H., Muramatsu, M., Li, X.F., Shimizu, J., Fujiwara, H., and Hamaoka, T.: **Transforming growth factor-β-induced inhibition of T cell function. Susceptibility difference in T cells of various phenotypes and functions and its relevance to immunosuppression in the tumor-bearing state.** *J. Immunol.* 1991, **146**:1077–1082.

Torre Amione, G., Beauchamp, R.D., Koeppen, H., Park, B.H., Schreiber, H., Moses, H.L., and Rowley, D.A.: **A highly immunogenic tumor transfected with a murine transforming growth factor type β1 cDNA escapes immune surveillance.** *Proc. Natl Acad. Sci. USA* 1990, **87**:1486–1490.

Wang, H.Y., Lee, D.A., Peng, G., Guo, Z., Li, Y., Kiniwa, Y., Shevach, E.M., and Wang, R.F.: **Tumor-specific human CD4$^+$ regulatory T cells and their ligands: implications for immunotherapy.** *Immunity* 2004, **20**:107–118.

15-16 T lymphocytes can recognize specific antigens on human tumors, and adoptive T-cell transfer is being tested in cancer patients.

Boon, T., Coulie, P.G., and Van den Eynde, B.: **Tumor antigens recognized by T cells.** *Immunol. Today* 1997, **18**:267–268.

Chaux, P., Vantomme, V., Stroobant, V., Thielemans, K., Corthals, J., Luiten, R., Eggermont, A.M., Boon, T., and van der Bruggen, P.: **Identification of MAGE-3 epitopes presented by HLA-DR molecules to CD4$^+$ T lymphocytes.** *J. Exp. Med.* 1999, **189**:767–778.

Clark, R.E., Dodi, I.A., Hill, S.C., Lill, J.R., Aubert, G., Macintyre, A.R., Rojas, J., Bourdon, A., Bonner, P.L., Wang, L., *et al.*: **Direct evidence that leukemic cells present HLA-associated immunogenic peptides derived from the BCR-ABL b3a2 fusion protein.** *Blood* 2001, **98**:2887–2893.

Comoli, P., Pedrazzoli, P., Maccario, R., Basso, S., Carminati, O., Labirio, M., Schiavo, R., Secondino, S., Frasson, C., Perotti, C., *et al.*: **Cell therapy of Stage IV nasopharyngeal carcinoma with autologous Epstein–Barr virus-targeted cytotoxic t lymphocytes.** *J. Clin. Oncol.* 2005, **23**:8942–8949.

de Smet, C., Lurquin, C., Lethe, B., Martelange, V., and Boon, T.: **DNA methylation is the primary silencing mechanism for a set of germ line- and tumor-specific genes with a CpG-rich promoter.** *Mol. Cell. Biol.* 1999, **19**:7327–7335.

Disis, M.L., and Cheever, M.A.: **HER-2/neu oncogenic protein: issues in vaccine development.** *Crit. Rev. Immunol.* 1998, **18**:37–45.

Disis, M.L., and Cheever, M.A.: **Oncogenic proteins as tumor antigens.** *Curr. Opin. Immunol.* 1996, **8**:637–642.

Dudley, M.E., Wunderlich, J.R., Yang, J.C., Sherry, R.M., Topalian, S.L., Restifo, N.P., Royal, R.E., Kammula, U., White, D.E., Mavroukakis, S.A., *et al.*: **Adoptive cell transfer therapy following non-myeloablative but lymphodepleting chemotherapy for the treatment of patients with refractory metastatic melanoma.** *J. Clin. Oncol.* 2005, **23**:2346–2357.

Michalek, J., Collins, R.H., Durrani, H.P., Vaclavkova, P., Ruff, L.E., Douek, D.C., and Vitetta, E.S.: **Definitive separation of graft-versus-leukemia- and graft-versus-host-specific CD4⁺ T cells by virtue of their receptor β loci sequences.** *Proc. Natl Acad. Sci. USA* 2003, **100**:1180–1184.

Morris, E.C., Tsallios, A., Bendle, G.M., Xue, S.A., and Stauss, H.J.: **A critical role of T cell antigen receptor-transduced MHC class I-restricted helper T cells in tumor protection.** *Proc. Natl Acad. Sci. USA* 2005, **102**:7934–7939.

Robbins, P.F., and Kawakami, Y.: **Human tumor antigens recognized by T cells.** *Curr. Opin. Immunol.* 1996, **8**:628–636.

15-17 Monoclonal antibodies against tumor antigens, alone or linked to toxins, can control tumor growth.

Alekshun, T., and Garrett, C.: **Targeted therapies in the treatment of colorectal cancers.** *Cancer Control* 2005, **12**:105–110.

Bagshawe, K.D., Sharma, S.K., Burke, P.J., Melton, R.G., and Knox, R.J.: **Developments with targeted enzymes in cancer therapy.** *Curr. Opin. Immunol.* 1999, **11**:579–583.

Cragg, M.S., French, R.R., and Glennie, M.J.: **Signaling antibodies in cancer therapy.** *Curr. Opin. Immunol.* 1999, **11**:541–547.

Fan, Z., and Mendelsohn, J.: **Therapeutic application of anti-growth factor receptor antibodies.** *Curr. Opin. Oncol.* 1998, **10**:67–73.

Hortobagyi, G.N.: **Trastuzumab in the treatment of breast cancer.** *N. Engl. J. Med.* 2005, **353**:1734–1736.

Houghton, A.N., and Scheinberg, D.A.: **Monoclonal antibody therapies—a 'constant' threat to cancer.** *Nat. Med.* 2000, **6**:373–374.

Kreitman, R.J., Wilson, W.H., Bergeron, K., Raggio, M., Stetler-Stevenson, M., FitzGerald, D.J., and Pastan, I.: **Efficacy of the anti-CD22 recombinant immunotoxin BL22 in chemotherapy-resistant hairy-cell leukemia.** *N. Engl. J. Med.* 2001, **345**:241–247.

White, C.A., Weaver, R.L., and Grillo-Lopez, A.J.: **Antibody-targeted immunotherapy for treatment of malignancy.** *Annu. Rev. Med.* 2001, **52**:125–145.

15-18 Enhancing the immune response to tumors by vaccination holds promise for cancer prevention and therapy.

Bendandi, M., Gocke, C.D., Kobrin, C.B., Benko, F.A., Sternas, L.A., Pennington, R., Watson, T.M., Reynolds, C.W., Gause, B.L., Duffey, P.L., *et al.*: **Complete molecular remissions induced by patient-specific vaccination plus granulocyte-monocyte colony-stimulating factor against lymphoma.** *Nat. Med.* 1999, **5**:1171–1177.

Hellstrom, K.E., Gladstone, P., and Hellstrom, I.: **Cancer vaccines: challenges and potential solutions.** *Mol. Med. Today* 1997, **3**:286–290.

Kugler, A., Stuhler, G., Walden, P., Zoller, G., Zobywalski, A., Brossart, P., Trefzer, U., Ullrich, S., Muller, C.A., Becker, V., *et al.*: **Regression of human metastatic renal cell carcinoma after vaccination with tumor cell-dendritic cell hybrids.** *Nat. Med.* 2000, **6**:332–336.

Li, Y., Hellstrom, K.E., Newby, S.A., and Chen, L.: **Costimulation by CD48 and B7–1 induces immunity against poorly immunogenic tumors.** *J. Exp. Med.* 1996, **183**:639–644.

Melief, C.J., Offringa, R., Toes, R.E., and Kast, W.M.: **Peptide-based cancer vaccines.** *Curr. Opin. Immunol.* 1996, **8**:651–657.

Morse, M.A., Chui, S., Hobeika, A., Lyerly, H.K., and Clay, T.: **Recent developments in therapeutic cancer vaccines.** *Nat. Clin. Pract. Oncol.* 2005, **2**:108–113.

Murphy, A., Westwood, J.A., Teng, M.W., Moeller, M., Darcy, P.K., and Kershaw, M.H.: **Gene modification strategies to induce tumor immunity.** *Immunity* 2005, **22**:403–414.

Nestle, F.O., Banchereau, J., and Hart, D.: **Dendritic cells: on the move from bench to bedside.** *Nat. Med.* 2001, **7**:761–765.

Pardoll, D.M.: **Cancer vaccines.** *Nat. Med.* 1998, **4**:525–531.

Pardoll, D.M.: **Paracrine cytokine adjuvants in cancer immunotherapy.** *Annu. Rev. Immunol.* 1995, **13**:399–415.

Phan, G.Q., Yang, J.C., Sherry, R.M., Hwu, P., Topalian, S.L., Schwartzentruber, D.J., Restifo, N.P., Haworth, L.R., Seipp, C.A., Freezer, L.J., *et al.*: **Cancer regression and autoimmunity induced by cytotoxic T lymphocyte-associated antigen 4 blockade in patients with metastatic melanoma.** *Proc. Natl Acad. Sci. USA* 2003, **100**:8372–8377.

Przepiorka, D., and Srivastava, P.K.: **Heat shock protein–peptide complexes as immunotherapy for human cancer.** *Mol. Med. Today* 1998, **4**:478–484.

Ragnhammar, P.: **Anti-tumoral effect of GM-CSF with or without cytokines and monoclonal antibodies in solid tumors.** *Med. Oncol.* 1996, **13**:167–176.

Stanley, M.: **Prophylactic HPV vaccines: prospects for eliminating ano-genital cancer.** *Br. J. Cancer* 2007, **96**:1320–1323.

Steinman, R.M., and Pope, M.: **Exploiting dendritic cells to improve vaccine efficacy.** *J. Clin. Invest.* 2002, **109**:1519–1526.

15-19 There are several requirements for an effective vaccine.

Ada, G.L.: **The immunological principles of vaccination.** *Lancet* 1990, **335**:523–526.

Anderson, R.M., Donnelly, C.A., and Gupta, S.: **Vaccine design, evaluation, and community-based use for antigenically variable infectious agents.** *Lancet* 1997, **350**:1466–1470.

Gupta, R.K., Best, J., and MacMahon, E.: **Mumps and the UK epidemic 2005.** *BMJ* 2005, **330**:1132–1135.

Levine, M.M., and Levine, O.S.: **Influence of disease burden, public perception, and other factors on new vaccine development, implementation, and continued use.** *Lancet* 1997, **350**:1386–1392.

Nichol, K.L., Lind, A., Margolis, K.L., Murdoch, M., McFadden, R., Hauge, M., Magnan, S., and Drake, M.: **The effectiveness of vaccination against influenza in healthy, working adults.** *N. Engl. J. Med.* 1995, **333**:889–893.

Palese, P., and Garcia-Sastre, A.: **Influenza vaccines: present and future.** *J. Clin. Invest.* 2002, **110**:9–13.

Rabinovich, N.R., McInnes, P., Klein, D.L., and Hall, B.F.: **Vaccine technologies: view to the future.** *Science* 1994, **265**:1401–1404.

15-20 The history of vaccination against *Bordetella pertussis* illustrates the importance of developing an effective vaccine that is perceived to be safe.

Decker, M.D., and Edwards, K.M.: **Acellular pertussis vaccines.** *Pediatr. Clin. North Am.* 2000, **47**:309–335.

Madsen, K.M., Hviid, A., Vestergaard, M., Schendel, D., Wohlfahrt, J., Thorsen, P., Olsen, J., and Melbye, M.: **A population-based study of measles, mumps, and rubella vaccination and autism.** *N. Engl. J. Med.* 2002, **347**:1477–1482.

Mortimer, E.A.: **Pertussis vaccines**, in Plotkin, S.A., and Mortimer, E.A.: *Vaccines*, 2nd edn. Philadelphia, W.B. Saunders Co., 1994.

Poland, G.A.: **Acellular pertussis vaccines: new vaccines for an old disease.** *Lancet* 1996, **347**:209–210.

15-21 Conjugate vaccines have been developed as a result of understanding how T and B cells collaborate in an immune response.

Kroll, J.S., and Booy, R.: ***Haemophilus influenzae*: capsule vaccine and capsulation genetics.** *Mol. Med. Today* 1996, **2**:160–165.

Peltola, H., Kilpi, T., and Anttila, M.: **Rapid disappearance of *Haemophilus influenzae* type b meningitis after routine childhood immunisation with conjugate vaccines.** *Lancet* 1992, **340**:592–594.

Rosenstein, N.E., and Perkins, B.A.: **Update on *Haemophilus influenzae* serotype b and meningococcal vaccines.** *Pediatr. Clin. N. Am.* 2000, **47**:337–352.

van den Dobbelsteen, G.P., and van Rees, E.P.: **Mucosal immune responses to pneumococcal polysaccharides: implications for vaccination.** *Trends Microbiol.* 1995, **3**:155–159.

15-22 The use of adjuvants is another important approach to enhancing the immunogenicity of vaccines.

Alving, C.R., Koulchin, V., Glenn, G.M., and Rao, M.: **Liposomes as carriers of peptide antigens: induction of antibodies and cytotoxic T lymphocytes to conjugated and unconjugated peptides.** *Immunol. Rev.* 1995, **145**:5–31.

Gupta, R.K., and Siber, G.R.: **Adjuvants for human vaccines—current status, problems and future prospects.** *Vaccine* 1995, **13**:1263–1276.

Hartmann, G., Weiner, G.J., and Krieg, A.M.: **CpG DNA: a potent signal for growth, activation, and maturation of human dendritic cells.** *Proc. Natl Acad. Sci. USA* 1999, **96**:9305–9310.

Kersten, G.F., and Crommelin, D.J.: **Liposomes and ISCOMs.** *Vaccine* 2003, **21**:915–920.

Persing, D.H., Coler, R.N., Lacy, M.J., Johnson, D.A., Baldridge, J.R., Hershberg, R.M., and Reed, S.G.: **Taking toll: lipid A mimetics as adjuvants and immunomodulators.** *Trends Microbiol.* 2002, **10**:S32–S37.

Scott, P., and Trinchieri, G.: **IL-12 as an adjuvant for cell-mediated immunity.** *Semin. Immunol.* 1997, **9**:285–291.

Takeda, K., Kaisho, T., and Akira, S.: **Toll-like receptors.** *Annu. Rev. Immunol.* 2003, **21**:335–376.

van Duin, D., Medzhitov, R., and Shaw, A.C.: **Triggering TLR signaling in vaccination.** *Trends Immunol.* 2005, **27**:49–55.

Vogel, F.R.: **Immunologic adjuvants for modern vaccine formulations.** *Ann. NY Acad. Sci.* 1995, **754**:153–160.

15-23 Live-attenuated viral vaccines are usually more potent than 'killed' vaccines and can be made safer by the use of recombinant DNA technology.

Brochier, B., Kieny, M.P., Costy, F., Coppens, P., Bauduin, B., Lecocq, J.P., Languet, B., Chappuis, G., Desmettre, P., Afiademanyo, K., *et al.*: **Large-scale eradication of rabies using recombinant vaccinia–rabies vaccine.** *Nature* 1991, **354**:520–522.

Murphy, B.R., and Collins, P.L.: **Live-attenuated virus vaccines for respiratory syncytial and parainfluenza viruses: applications of reverse genetics.** *J. Clin. Invest.* 2002, **110**:21–27.

Parkin, N.T., Chiu, P., and Coelingh, K.: **Genetically engineered live attenuated influenza A virus vaccine candidates.** *J. Virol.* 1997, **71**:2772–2778.

Subbarao, K., Murphy, B.R., and Fauci, A.S.: **Development of effective vaccines against pandemic influenza.** *Immunity* 2006, **24**:5–9.

15-24 Live-attenuated bacterial vaccines can be developed by selecting nonpathogenic or disabled mutants.

Guleria, I., Teitelbaum, R., McAdam, R.A., Kalpana, G., Jacobs, W.R., Jr., and Bloom, B.R.: **Auxotrophic vaccines for tuberculosis.** *Nat. Med.* 1996, **2**:334–337.

Martin, C.: **The dream of a vaccine against tuberculosis; new vaccines improving or replacing BCG?** *Eur. Respir. J.* 2005, **26**:162–167.

15-25 Synthetic peptides of protective antigens can elicit protective immunity.

Alonso, P.L., Sacarlal, J., Aponte, J.J., Leach, A., Macete, E., Aide, P., Sigauque, B., Milman, J., Mandomando, I., Bassat, Q., *et al.*: **Duration of protection with RTS,S/AS02A malaria vaccine in prevention of *Plasmodium falciparum* disease in Mozambican children: single-blind extended follow-up of a randomised controlled trial.** *Lancet* 2005, **366**:2012–2018.

Berzofsky, J.A.: **Epitope selection and design of synthetic vaccines. Molecular approaches to enhancing immunogenicity and cross-reactivity of engineered vaccines.** *Ann. N.Y. Acad. Sci.* 1993, **690**:256–264.

Berzofsky, J.A.: **Mechanisms of T cell recognition with application to vaccine design.** *Mol. Immunol.* 1991, **28**:217–223.

Canizares, M., Nicholson, L., and Lomonossoff, G.P.: **Use of viral vectors for vaccine production in plants.** *Immunol. Cell Biol.* 2005, **83**:263–270.

Davenport, M.P., and Hill, A.V.: **Reverse immunogenetics: from HLA-disease associations to vaccine candidates.** *Mol. Med. Today* 1996, **2**:38–45.

Hill, A.V.: **Pre-erythrocytic malaria vaccines: towards greater efficacy.** *Nat. Rev. Immunol.* 2006, **6**:21–32.

Hoffman, S.L., Rogers, W.O., Carucci, D.J., and Venter, J.C.: **From genomics to vaccines: malaria as a model system.** *Nat. Med.* 1998, **4**:1351–1353.

Modelska, A., Dietzschold, B., Sleysh, N., Fu, Z.F., Steplewski, K., Hooper, D.C., Koprowski, H., and Yusibov, V.: **Immunization against rabies with plant-derived antigen.** *Proc. Natl Acad. Sci. USA* 1998, **95**:2481–2485.

Sanders, M.T., Brown, L.E., Deliyannis, G., and Pearse, M.J.: **ISCOM-based vaccines: the second decade.** *Immunol. Cell Biol.* 2005, **83**:119–128.

15-26 The route of vaccination is an important determinant of success.

Burnette, W.N.: **Bacterial ADP-ribosylating toxins: form, function, and recombinant vaccine development.** *Behring Inst. Mitt.* 1997, **98**:434–441.

Douce, G., Fontana, M., Pizza, M., Rappuoli, R., and Dougan, G.: **Intranasal immunogenicity and adjuvanticity of site-directed mutant derivatives of cholera toxin.** *Infect. Immun.* 1997, **65**:2821–2828.

Dougan, G.: **The molecular basis for the virulence of bacterial pathogens: implications for oral vaccine development.** *Microbiology* 1994, **140**:215–224.

Dougan, G., Ghaem-Maghami, M., Pickard, D., Frankel, G., Douce, G., Clare, S., Dunstan, S., and Simmons, C.: **The immune responses to bacterial antigens encountered in vivo at mucosal surfaces.** *Phil. Trans. R. Soc. Lond. B* 2000, **355**:705–712.

Eriksson, K., and Holmgren, J.: **Recent advances in mucosal vaccines and adjuvants.** *Curr. Opin. Immunol.* 2002, **14**:666–672.

Ivanoff, B., Levine, M.M., and Lambert, P.H.: **Vaccination against typhoid fever: present status.** *Bull. World Health Org.* 1994, **72**:957–971.

Levine, M.M.: **Modern vaccines. Enteric infections.** *Lancet* 1990, **335**:958–961.

15-27 Protective immunity can be induced by injecting DNA encoding microbial antigens and human cytokines into muscle.

Donnelly, J.J., Ulmer, J.B., Shiver, J.W., and Liu, M.A.: **DNA vaccines.** *Annu. Rev. Immunol.* 1997, **15**:617–648.

Gurunathan, S., Klinman, D.M., and Seder, R.A.: **DNA vaccines: immunology, application, and optimization.** *Annu. Rev. Immunol.* 2000, **18**:927–974.

Wolff, J.A., and Budker, V.: **The mechanism of naked DNA uptake and expression.** *Adv. Genet.* 2005, **54**:3–20.

15-28 The effectiveness of a vaccine can be enhanced by targeting it to sites of antigen presentation.

Bonifaz, L.C., Bonnyay, D.P., Charalambous, A., Darguste, D.I., Fujii, S., Soares, H., Brimnes, M.K., Moltedo, B., Moran, T.M., and Steinman, R.M.: **In vivo targeting of antigens to maturing dendritic cells via the DEC-205 receptor improves T cell vaccination.** *J. Exp. Med.* 2004, **199**:815–824.

Deliyannis, G., Boyle, J.S., Brady, J.L., Brown, L.E., and Lew, A.M.: **A fusion DNA vaccine that targets antigen-presenting cells increases protection from viral challenge.** *Proc. Natl Acad. Sci. USA* 2000, **97**:6676–6680.

Hahn, H., Lane-Bell, P.M., Glasier, L.M., Nomellini, J.F., Bingle, W.H., Paranchych, W., and Smit, J.: **Pilin-based anti-*Pseudomonas* vaccines: latest developments and perspectives.** *Behring Inst. Mitt.* 1997, **98**:315–325.

Neutra, M.R.: **Current concepts in mucosal immunity. V. Role of M cells in transepithelial transport of antigens and pathogens to the mucosal immune system.** *Am. J. Physiol.* 1998, **274**:G785–G791.

Shen, Z., Reznikoff, G., Dranoff, G., and Rock, K.L.: **Cloned dendritic cells can present exogenous antigens on both MHC class I and class II molecules.** *J. Immunol.* 1997, **158**:2723–2730.

Tan, M.C., Mommaas, A.M., Drijfhout, J.W., Jordens, R., Onderwater, J.J., Verwoerd, D., Mulder, A.A., van der Heiden, A.N., Scheidegger, D., Oomen, L.C., *et al.*: **Mannose receptor-mediated uptake of antigens strongly enhances HLA class II-restricted antigen presentation by cultured dendritic cells.** *Eur. J. Immunol.* 1997, **27**:2426–2435.

Thomson, S.A., Burrows, S.R., Misko, I.S., Moss, D.J., Coupar, B.E., and Khanna, R.: **Targeting a polyepitope protein incorporating multiple class II-restricted viral epitopes to the secretory/endocytic pathway facilitates immune recognition by CD4+ cytotoxic T lymphocytes: a novel approach to vaccine design.** *J. Virol.* 1998, **72**:2246–2252.

15-29 An important question is whether vaccination can be used therapeutically to control existing chronic infections.

Burke, R.L.: **Contemporary approaches to vaccination against herpes simplex virus.** *Curr. Top. Microbiol. Immunol.* 1992, **179**:137–158.

Grange, J.M., and Stanford, J.L.: **Therapeutic vaccines.** *J. Med. Microbiol.* 1996, **45**:81–83.

Hill, A., Jugovic, P., York, I., Russ, G., Bennink, J., Yewdell, J., Ploegh, H., and Johnson, D.: **Herpes simplex virus turns off the TAP to evade host immunity.** *Nature* 1995, **375**:411–415.

Lu, W., Arraes, L.C., Ferreira, W.T., and Andrieu, J.M.: **Therapeutic dendritic-cell vaccine for chronic HIV-1 infection.** *Nat. Med.* 2004, **10**:1359–1365.

Modlin, R.L.: **Th1–Th2 paradigm: insights from leprosy.** *J. Invest. Dermatol.* 1994, **102**:828–832.

Plebanski, M., Proudfoot, O., Pouniotis, D., Coppel, R.L., Apostolopoulos, V., and Flannery, G.: **Immunogenetics and the design of *Plasmodium falciparum* vaccines for use in malaria-endemic populations.** *J. Clin. Invest.* 2002, **110**:295–301.

Reiner, S.L., and Locksley, R.M.: **The regulation of immunity to *Leishmania major*.** *Annu. Rev. Immunol.* 1995, **13**:151–177.

Stanford, J.L.: **The history and future of vaccination and immunotherapy for leprosy.** *Trop. Geogr. Med.* 1994, **46**:93–107.

15-30 Modulation of the immune system might be used to inhibit immunopathological responses to infectious agents.

Biron, C.A., and Gazzinelli, R.T.: **Effects of IL-12 on immune responses to microbial infections: a key mediator in regulating disease outcome.** *Curr. Opin. Immunol.* 1995, **7**:485–496.

Grau, G.E., and Modlin, R.L.: **Immune mechanisms in bacterial and parasitic diseases: protective immunity versus pathology.** *Curr. Opin. Immunol.* 1991, **3**:480–485.

Kaplan, G.: **Recent advances in cytokine therapy in leprosy.** *J. Infect. Dis.* 1993, **167** (Suppl 1):S18–S22.

Locksley, R.M.: **Interleukin 12 in host defense against microbial pathogens.** *Proc. Natl Acad. Sci. USA* 1993, **90**:5879–5880.

Murray, H.W.: **Interferon-γ and host antimicrobial defense: current and future clinical applications.** *Am. J. Med.* 1994, **97**:459–467.

Sher, A., Gazzinelli, R.T., Oswald, I.P., Clerici, M., Kullberg, M., Pearce, E.J., Berzofsky, J.A., Mosmann, T.R., James, S.L., and Morse, H.C.: **Role of T-cell derived cytokines in the downregulation of immune responses in parasitic and retroviral infection.** *Immunol. Rev.* 1992, **127**:183–204.

Sher, A., Jankovic, D., Cheever, A., and Wynn, T.: **An IL-12-based vaccine as an approach for preventing immunopathology in schistosomiasis.** *Ann. NY Acad. Sci.* 1996, **795**:202–207.

PART VI THE ORIGINS OF IMMUNE RESPONSES

Evolution of the Immune System

16

We began this book with an overview of immunology and its fascination for scientists throughout the 20th century. In this chapter we reexamine how the basic machinery of immunology has evolved over eons. We start, as we started this book, with the innate immune system, some aspects of which are as old as the first multicellular organisms. We then look at the fascinating question of how immune systems have evolved the ability to recognize and react to increasing numbers of different antigens from the universe of possible pathogens.

We have distinguished between innate and adaptive immunity by the way in which the organism encodes the molecules that recognize pathogens. Innate immunity uses receptors encoded directly in the genome, and in the species we have considered so far—humans and mice—the number of these receptors is limited. The Toll-like receptors and NOD proteins described in Chapter 2 are examples of this limited repertoire of pathogen-recognizing receptors. Adaptive immunity overcomes this limitation by generating a much larger repertoire of clonally diverse receptors—in the form of antibodies and T-cell receptors—produced through the somatic gene rearrangements described in Chapter 5. Because this results in a very significant increase in the diversity of antigen recognition, we call this kind of repertoire 'anticipatory,' in the sense that it is large enough to anticipate encounter with a seemingly infinite number of antigens.

Until very recently, it was thought that anticipatory, or adaptive, immunity arose uniquely in the jawed vertebrates, through the actions of the *RAG1* and *RAG2* genes that are unique to this group. New discoveries have forced us to change this viewpoint. We now recognize that very large repertoires of molecules acting in immune responses can be generated by different kinds of genetic mechanisms in organisms as diverse as insects, echinoderms, molluscs, and the jawless vertebrates (agnathans). As we shall see in this chapter, some organisms increase the diversity of pathogen recognition simply through an enormous increase in the numbers of receptors encoded by somatic cells—a highly elaborated innate immune system. However, other species, including the fruitfly *Drosophila melanogaster*, increase diversity in their response even further, but use genetic mechanisms other than somatic cell gene rearrangement. Finally, a system of somatic cell gene rearrangement that produces soluble 'antibody'-like proteins, but that is distinct from the RAG-dependent V(D)J rearrangement found in jawed vertebrates, has been discovered in the surviving species of jawless vertebrates—the lampreys and hagfish.

Thus, we must now view our own adaptive immune system, that of the jawed vertebrates, as only one solution to the problem of generating enormously diverse systems for pathogen recognition. The essence of adaptive immunity can no longer be defined as the possession of V(D)J rearrangement in lymphocytes. Rather, it is the generation of a significantly diverse anticipatory repertoire of effector molecules, through whatever mechanism, and the clonal selection from this repertoire for effector responses that can change throughout the lifetime of the organism.

Evolution of the innate immune system.

16-1 The evolution of the immune system can be studied by comparing the genes expressed by different species.

It seems likely that the concept of an immune system—the defense of the individual against infectious agents—is ubiquitous, because all organisms are attacked by pathogens, and natural selection will always provide a pressure to develop protection against them. Even bacteria have some protection against the parasites (plasmids) and pathogens (bacteriophages) that infect them: these are the restriction enzymes that cleave incoming DNA and the modification systems that modify the bacterium's own DNA so that the enzymes cannot cleave it. It is unlikely that such systems have any direct counterparts in higher organisms, but bacteria also produce antimicrobial peptides that are active against competing bacteria, and this type of host defense does have a counterpart in multicellular organisms. Nevertheless, when evaluating similarities between organisms, one always has to keep in mind the possibility that they may represent convergent evolution—the independent evolution of similar solutions to the same problem. The antimicrobial peptides of higher organisms might therefore represent independent evolution of the same function, rather than direct evolution from a common ancestral peptide in bacteria.

Our focus will be the evolution of an immune system in multicellular organisms with a defined body that can be invaded and colonized by pathogens. The problem with evolutionary studies is that the direct ancestors of the animal species that exist today are no longer available for study, and so we cannot say exactly what molecules or what immunological functions were present in these organisms. This does not mean that we can know nothing of the evolutionary past, however; we can make use of the fact that the presence or absence of individual components of the immune system in different species sheds light on its evolutionary history.

When studying the evolution of any biological system, such as the immune system, the underlying assumption we make is that if a gene is found in the same (or similar) form in two different species, then that gene was also present in the common ancestor of those species. The more widely diverged the species, the more distant is the common ancestor. An evolutionary 'tree' showing the organisms discussed in this chapter, and the order in which the different lineages diverged, is shown in Fig. 16.1. Thus, the divergence of plants from their common ancestor with animals occurred earlier than the divergence of the insect lineage from that leading to the deuterostomes (echinoderms and chordates). Advances in DNA sequencing have led to knowledge of the complete sequence of a number of organisms. This information has revealed a tremendous similarity in strategies of innate immunity across phyla, and has revealed unexpected diversity in responses. Complete or near-complete genome assemblies have been produced for many of the organisms shown in Fig. 16.1, including plants (*Arabidopsis thaliana*), insects (*Drosophila melanogaster*), echinoderms (*Strongylocentrotus purpuratus*), urochordates (*Ciona intestinalis* and *C. savignyi*), as well as several fish species, amphibians (*Xenopus tropicalis*), birds (*Gallus gallus*), and mammals (*Homo sapiens, Mus musculus*).

Using this information we are able to trace the evolution of host-defense mechanisms from our more ancient ancestors, such as those we have in common with insects, through our common ancestor with echinoderms and eventually to our common ancestor with the urochordate ascidians (the tunicates or sea squirts), a sister group of the lineage that leads to the vertebrates. Within the vertebrates, we can trace the development of immune functions from the agnathans (jawless fishes, such as the lampreys and hagfishes), through the cartilaginous fishes (sharks, skates, and rays) to the bony fishes, then to the amphibians, to reptiles and birds, and finally to mammals. In some cases we do not yet know whether a particular immune-system gene is present in all the 'intermediate' groups. However, if it is present in, for example, mammals and invertebrates, then we would assume that it is also present (or was at one time) in all vertebrate lineages. It is not quite as straightforward to draw the opposite conclusion—that a gene absent from one species must therefore not be present in the common ancestor—because it is possible for genes and functions to be lost in individual lineages.

An innate immune system is well developed in *Drosophila,* a favorite model organism for many aspects of biological research, and in many other invertebrates. What *Drosophila* shares with vertebrates are invariant receptors—the pattern-recognition receptors—that recognize common molecular patterns of pathogens, and intracellular signaling pathways leading from these receptors to the activation of the transcription factor NFκB (see Chapters 2 and 6). Many multicellular animal species have a cassette of genes that encode the proteins of this pathway. This suggests that the activation of NFκB is the original and central signaling pathway in innate immunity, leading in turn to the activation of a set of genes that depend on NFκB for their transcription. This is an almost universal pathway that leads to the activation of many different host-defense systems.

16-2 Antimicrobial peptides are likely to be the most ancient immune defenses.

One form of host defense found in both plants and animals, and therefore pre-dating the separation of the plant and animal lineages, is the production of antimicrobial peptides. There are many different antimicrobial peptides, with a great variety of physical and chemical characteristics and effects on different microbial pathogens. One widely distributed class comprises the small peptides known as defensins (see Section 2-3). Defensins from mammals, insects, and plants differ in structural detail (Fig. 16.2), but it is clear that they are all related and derive from the same ancestral system of host defense.

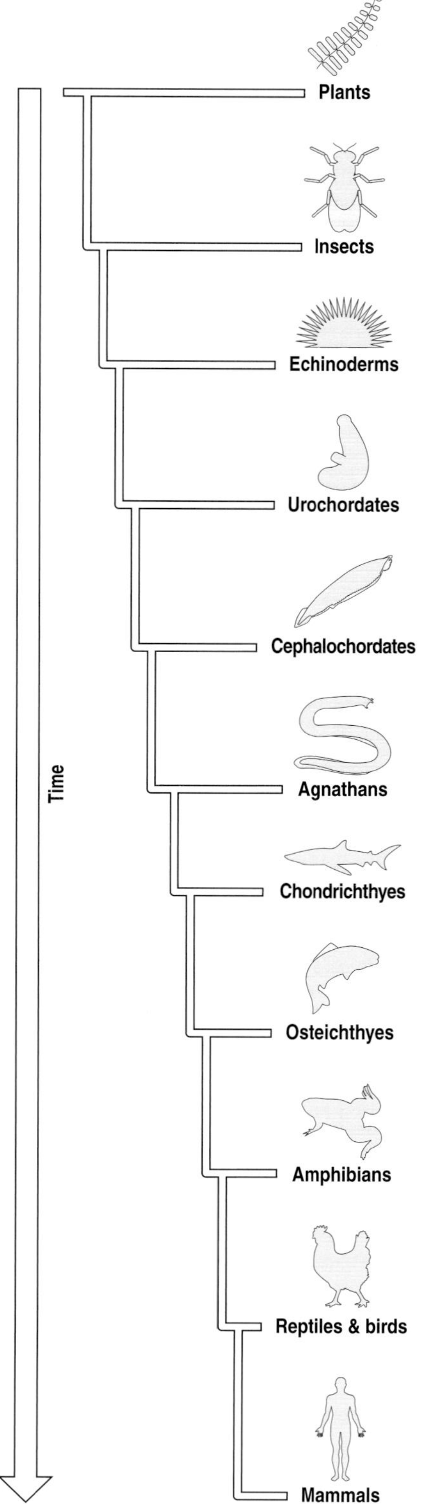

Fig.16.1 The evolutionary history of the organisms mentioned in this chapter. The branching of the highly schematic evolutionary 'tree' shown here represents the order of divergence of the different lineages-that is, the plant lineage diverged from its last common ancestor with animals before the separation of the insect lineage, and so on. Note that this tree does not depict the relative times involved. The chordates (the phylum that includes the vertebrates) comprise the invertebrate urochordates (for example, sea squirts) and cephalochordates (for example, lancelets) and the vertebrate agnathans (jawless fishes), cartilaginous fishes (Chondrichthyes), bony fishes (Osteichthyes), amphibians, reptiles, birds and mammals.

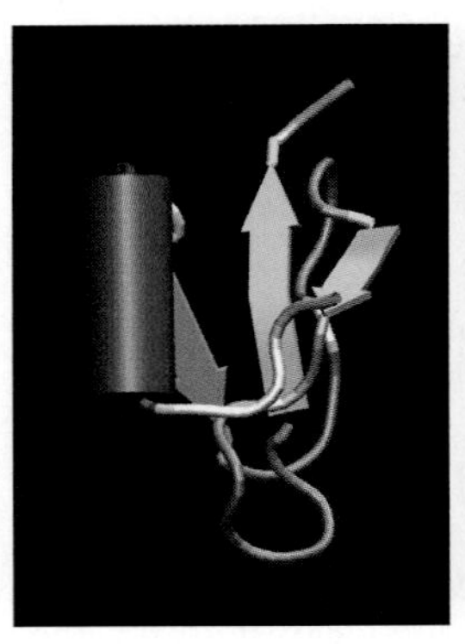

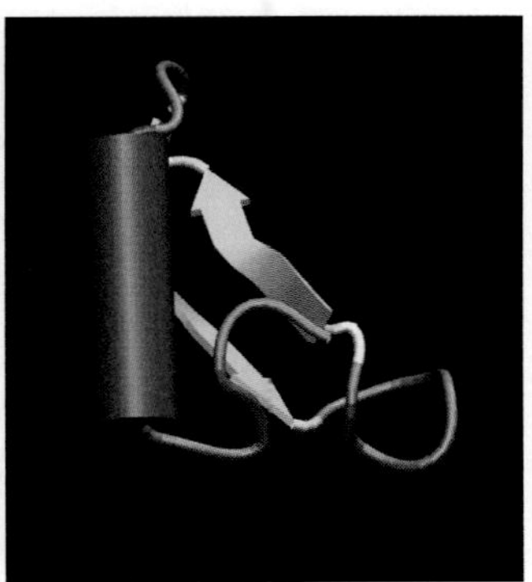

Fig. 16.2 The antimicrobial defensins of plants, insects, and mammals are structurally related. The structures of the plant defensin AFP-1 from the radish *Raphanus sativus* (left panel), the insect antimicrobial peptide drosomycin (center panel) from *Drosophila*, and the human β_2-defensin (right panel) are shown. The colors indicate the nature of the peptide structure, with the α helices shown in purple and the β strands in yellow. Unstructured regions are shown in cyan; turns are shown in white. All three defensins have a similar structure, with a short segment of α helix resting against two or three strands of antiparallel β sheet, which has been conserved since before the divergence of the plant and animal kingdoms.

The defensins are thought to act by disrupting the cell membranes of bacteria and fungi and also the membrane envelope of some viruses, although some may be able to cross microbial cell membranes and enter the cell.

Most multicellular organisms make many different defensins—the plant *Arabidopsis thaliana* produces 13, *Drosophila* at least 15, and in humans a single gut cell can make as many as 21. The various defensins have distinct activities, some being active against Gram-positive bacteria and some against Gram-negative bacteria, whereas others are specific for fungal pathogens. Multicellular organisms also make other types of antimicrobial peptides.

The production of antimicrobial peptides by both plants and animals suggests that this means of defense must have evolved before these two lineages diverged. The common precursor of plants and animals is likely to have been a single-celled organism. Many of the other lineages that diverged at about that time are single-celled eukaryotic organisms known generally as protists (some of which, such as the parasitic protozoa, cause human disease today). Whether antimicrobial peptides exist in extant protists is not known, nor is it clear whether such peptides would necessarily have a protective function in these organisms. Many free-living protists regard bacteria as a source of food rather than a threat to their well-being.

However, when we consider the behavior of phagocytic cells in multicellular organisms, such as macrophages in vertebrates, it is not unreasonable to speculate that at least some aspects of innate immunity evolved from the phagocytic feeding mechanisms of single-celled eukaryotes. All vertebrates, and many invertebrates, have phagocytic cells that patrol their blood vessels and tissues, as described in Chapter 2, and that have much in common with protists such as amoebae. It is possible that phagocytic cells in animals could have derived from a population of cells that retained the morphology and behavior of the unicellular ancestors.

16-3 Toll-like receptors may represent the most ancient pathogen-recognition system.

If antimicrobial peptides are thought to be the earliest form of defense against infection, then receptors that recognize pathogens and induce the production of antimicrobial peptides have a good claim to be among the first receptors dedicated to host defense. Such receptors have been discovered, and seem also to have been conserved over a long evolutionary period. The Toll receptor, first discovered in *Drosophila*, induces the expression of several host-defense mechanisms, including antimicrobial peptides primarily acting against Gram-positive bacteria and fungal pathogens.

The Toll receptor gene was first identified as a gene that acted during *Drosophila* embryonic development to specify correct dorso-ventral patterning. Only later was the function of the Toll receptor in host defense in the adult insect recognized through the work of other scientists, with the observation that mutations in Toll or in the signaling proteins activated by Toll affected the production of the antimicrobial peptide drosomycin, and the consequent susceptibility of *Drosophila* to fungal infections. Later, homologs of Toll were found in other species, ranging from plants to mammals, in which they are associated with resistance to viral, bacterial, and fungal infection. In plants, as in insects, Toll-like proteins are involved in the production of antimicrobial peptides, indicating their ancient association with this means of host defense. They have acquired additional functions in vertebrates (see Chapter 2), but because of their apparently ancient origin they are good candidates for at least one type of primordial pathogen receptor.

Humans and mice have around 10 functional Toll-like receptors, which recognize components of pathogens such as bacterial, yeast, and fungal cell walls, bacterial flagella, viral RNA, and bacterial DNA (see Fig. 2.16). The first Toll-like receptor identified, now known as Toll-like receptor 4 (TLR-4), is required for an innate immune response to bacterial lipopolysaccharide (LPS), a component of the cell surface of Gram-negative bacteria (see Fig. 2.14).

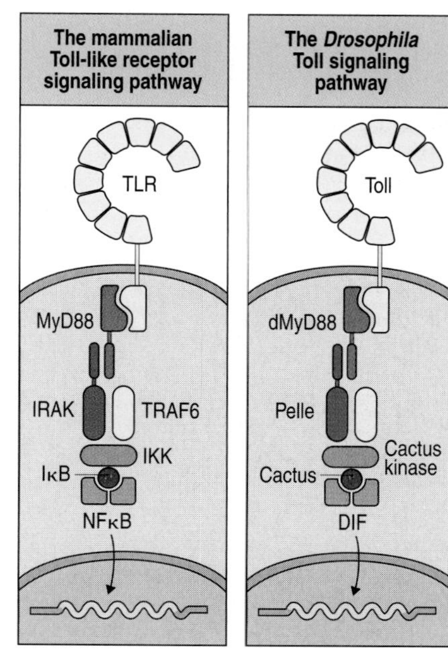

Fig. 16.3 A comparison of the *Drosophila* and mammalian Toll signaling pathways. The components of the mammalian Toll-like receptor signaling pathway that culminates in the activation of NFκB have direct parallels in the components of the signaling pathway from the Toll receptor of *Drosophila*. The intracellular domain of the Toll-like receptors interacts with a homologous domain in the adaptor protein MyD88. A similar interaction occurs between the intracellular domain of Toll and dMyD88. The next step in both signaling pathways occurs via the interaction of death domains, between MyD88 and IRAK in the mammalian cells and between dMyD88 and Pelle in *Drosophila*. Both IRAK and Pelle are serine kinases. At this point the mammalian signaling pathway passes through an adaptor, TRAF6, which is activated by IRAK and in turn activates IKK. IKK phosphorylates the inhibitor of NFκB, IκB, targeting it for degradation and releasing the active dimeric transcription factor, NFκB. In *Drosophila*, homologs of MyD88, TRAF6, and the kinase IKK, which phosphorylates the *Drosophila* IκB homolog Cactus, are found. Moreover, the terminal parts of the pathway are also homologous between *Drosophila* and mammals; phosphorylation of Cactus initiates its degradation and the release of the DIF dimer, which is a member of the NFκB family of transcription factors.

The signaling pathway from *Drosophila* Toll also seems to be conserved between widely different species, and there is a good correspondence between its components in *Drosophila* and mammals, as shown in Fig. 16.3. The end of the pathway in *Drosophila* is the activation of transcription factors of the Rel family (homologs of the mammalian NFκB transcription factors), which then translocate to the nucleus to induce new gene transcription. In *Drosophila*, the Rel transcription factors involved in the induction of antimicrobial peptides in response to Toll stimulation are the Dorsal-related immunity factor (DIF) and, to a smaller extent, the transcription factor Dorsal. A third member of the Rel family, Relish, also induces the production of antimicrobial peptides but in response to a different signaling pathway, which we describe later.

Drosophila Toll does not directly recognize pathogen products, but instead binds a cleaved version of a self protein, Spätzle. The exact sequence of events leading to the cleavage of Spätzle during *Drosophila* immune responses is not known, because the best-described pathway for Spätzle cleavage operates during embryogenesis but has no role in host defense. The immune-response pathway seems to involve specific pathogen-recognition molecules that interact with serine proteases to initiate the cleavage of Spätzle. One of these molecules has been identified, a protein encoded by a gene called *semmelweis* (in honor of **Ignaz Semmelweis**, a pioneer of the prevention of infection in hospitals). This protein is a member of a family of peptidoglycan-recognition proteins (PGRPs), which bind the peptidoglycan components of bacterial cell walls (see Fig. 2.14); *Drosophila* has some 13 PGRP genes. The protein encoded by *semmelweis*, PGRP-SA, is involved in the recognition of Gram-positive bacteria. Another family of *Drosophila* proteins, the so-called Gram-negative binding proteins (GNBPs), which recognize β-1–3-linked glucans, is involved in the recognition of fungi and, rather unexpectedly, also of Gram-positive bacteria. The protein GNBP1 cooperates with PGRP-SA in the recognition of peptidoglycan from Gram-positive bacteria. The Spätzle-activating serine protease of the Gram-positive bacterial recognition pathway has not yet been identified, but a serine protease involved in the detection of fungal infections has. This protease, called Persephone, has similarities to the proteases of the clotting pathway of both insect hemolymph (a fluid in some ways analogous to vertebrate blood serum) and mammalian blood, and seems to be activated directly by a fungal virulence factor. A fungus-specific recognition protein, GNBP3, can also activate Toll in a manner analogous to that of PGRP-SA, but it is not yet clear whether another serine protease is involved in this pathway.

Of the mammalian Toll-like receptors, the one whose mode of recognition most closely resembles that of *Drosophila* Toll may be TLR-4, which rather than binding the bacterial ligand LPS directly, instead binds it indirectly through a soluble LPS-binding protein, which in turn binds TLR-4. A more relevant functional parallel, however, might be with the complement system, in which the proteolytic activation of a series of proteases generates ligands for cell-surface receptors; in the case of complement, these receptors are involved in stimulating phagocytosis (see Chapter 2). Although the recognition specificities of the mammalian Toll-like receptors seem now to be fairly completely determined (see Chapter 2), it is still not established whether they recognize pathogen-derived components directly, as is often presumed, or whether this requires additional components, as in *Drosophila* Toll and TLR-4. In particular, there is no structural determination of a direct Toll-like receptor recognition of ligand, although a direct interaction between TLR-5 and its flagellin ligand has been shown by other means.

16-4 Toll-like receptor genes have undergone extensive diversification in some invertebrate species.

Although the mammalian Toll-like receptor system is somewhat more extensive than that in *Drosophila*, at least one case of a much greater diversification of these receptors has been discovered. The genome sequence of the sea urchin *S. purpuratus* reveals an unprecedented complexity of innate immune recognition. Altogether, the sea urchin genome contains 222 different *TLR* genes; the specificities of the encoded proteins remain to be determined. There is also an increase in the number of proteins that are likely to be involved in signaling from these receptors, with four genes similar to mammalian *MyD88*, which encodes an adaptor molecule (see Section 6-27). It is interesting to note that despite the much greater number of TLR genes, there is no apparent increase in the number of downstream targets, such as the family of NFκB transcription factors, suggesting that the ultimate outcome of Toll-like receptor signaling in the sea urchin will be very similar to that in other organisms.

The external portions of Toll-like receptors are composed of a series of protein domains called **leucine-rich repeats** (**LRRs**). These multiple LRRs are thought to form a scaffold that is generally adaptable for binding and recognition. In the sea urchin genome, the 222 *TLR* genes fall into two broad categories: a small set of 11 divergent genes and a large family of 211 genes, in which there is evidence of hypervariability located within particular LRR regions. This fact, and the large number of pseudogenes in this family, has been taken as evidence of rapid evolutionary turnover, which could reflect rapidly changing receptor specificities. This would contrast with the limited and stable Toll-like receptor repertoire in vertebrates, which recognizes a relatively small number of invariant pathogen-associated molecular patterns (PAMPs). Although we do not yet know the pathogen specificity of the sea urchin Toll-like receptors, it seems that in this group of organisms, hypervariability in the LRR domain has been used to generate a highly diversified pathogen-recognition system based on Toll-like receptors. As we will see later, the same strategy has also arisen independently in a vertebrate lineage.

One might ask whether this huge diversification of TLR-based recognition implies a primitive form of adaptive immunity in the sea urchin. We do not yet know, however, whether all these TLR genes are expressed together in one type of immune cell, or whether they are expressed in a clonally restricted manner. In the mammalian adaptive immune system, antigen receptors of distinct specificities are expressed in individual clones of lymphocytes. This clonal expression allows the character of the immune response to change over the lifetime of the organism by clonal selection of lymphocytes with particular specificities. We cannot yet say whether the diversification of Toll-like receptors in the sea urchin has simply led to an increased capacity for pathogen recognition, or whether there is selection and clonal expansion of cells expressing particular Toll-like receptor specificities, which would be the beginnings of a truly adaptive immunity.

16-5 A second recognition system in Drosophila homologous to the mammalian TNF receptor pathway provides protection from Gram-negative bacteria.

In mammals, the Toll-like receptors recognize a variety of pathogens, including Gram-positive and Gram-negative bacteria and fungi. In *Drosophila*, the Toll receptor does not seem to be involved in the recognition of Gram-negative bacteria. Instead, a second pathway, the **Imd (immunodeficiency) pathway**, is used. Two receptors that recognize Gram-negative bacteria have been identified in *Drosophila*, both of which are members of the PGRP family. One is PGRP-LC, which is associated with the cell membrane; the other, PGRP-LE, is secreted. Some of the steps in the signaling pathway from these receptors have been determined through the analysis of *Drosophila* mutants that are susceptible to infections with Gram-negative, but not Gram-positive, bacteria. The Imd pathway is strikingly similar to the mammalian tumor necrosis factor (TNF) receptor pathway that initiates new gene transcription (Fig. 16.4). The Imd protein itself is homologous to the TNF-receptor-binding protein RIP. The end result of the Imd pathway is the activation of the transcription factor Relish, which activates the expression of several immune-response genes, including those encoding the antimicrobial peptides diptericin, attacin, and cecropin; these are distinct from the peptides induced by Toll signaling. Thus, the Toll and Imd pathways activate equivalent effector mechanisms to eliminate infections. It is likely that these two distinct signaling pathways have arisen by the duplication of a more ancient common pathway of host defense, but it is impossible to say whether that pathway resembled the Toll pathway or the Imd pathway. In mammals, however, it seems that the host-defense functions of the Imd pathway have been taken over by equivalent Toll-like receptor pathways.

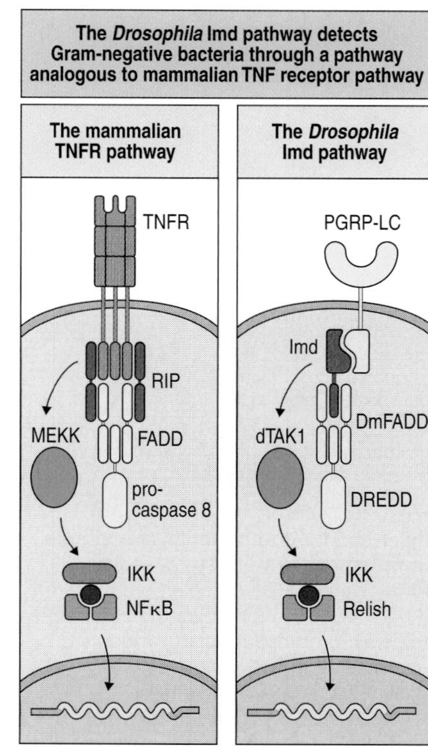

Fig. 16.4 *Drosophila* detects Gram-negative bacteria through the Imd signaling receptor pathway, which is analogous to the mammalian TNF receptor pathway. The TNF receptor transduces signals that lead either to new gene expression or to cell death. In the mammalian TNFR-I pathway, the binding of ligand to the receptor leads to the recruitment of the adaptor TRADD (TNF receptor-associated death-domain protein, not shown here), which in turn can either recruit FADD (Fas-associated death-domain protein), which leads to apoptosis, or RIP (receptor-interacting protein), which is a serine/threonine kinase. Each of these initiates a different signaling pathway. FADD activates caspase-8, initiating a protease cascade that leads to apoptosis, whereas RIP acts in a pathway via another kinase, MEKK3, which activates the Iκ kinase IKK, leading ultimately to the activation of NFκB and the induction of new gene expression. The Imd pathway seems to be a *Drosophila* homolog of the TNFR pathway and leads to the same outcomes. Imd itself is a homolog of RIP, whereas DmFADD is a homolog of FADD, and DREDD is a homolog of caspase-8. In this pathway, dTAK1 may be the homolog of MEKK3, activating the Iκ kinase (IKK) and resulting in the activation of the transcription factor Relish and the induction of several immune-related genes, including those for defensins.

16-6 An ancestral complement system opsonizes pathogens for uptake by phagocytic cells.

The complement system is another ancient means of host defense (see Chapter 2). The most primitive function of complement is likely to have been opsonization, a means of increasing the efficiency of pathogen uptake by the scavenger phagocytes that patrol animal body spaces. Even before complement components were discovered in invertebrates, it had been suggested that a primitive complement system would contain a minimum of three components. The central component would be C3, which would be activated spontaneously, as it is in the alternative pathway of complement activation in mammals today (see Section 2-16). Activated C3 would bind the equivalent of factor B, forming a C3 convertase that amplified the original signal by cleaving and activating many more molecules of C3. The third component of this system would be a C3 receptor expressed by phagocytes and capable of activating phagocytosis of the C3-coated pathogens.

This prediction has been borne out by the discovery of complement components in invertebrates, as shown in Fig. 16.5. A homolog of C3 has been found

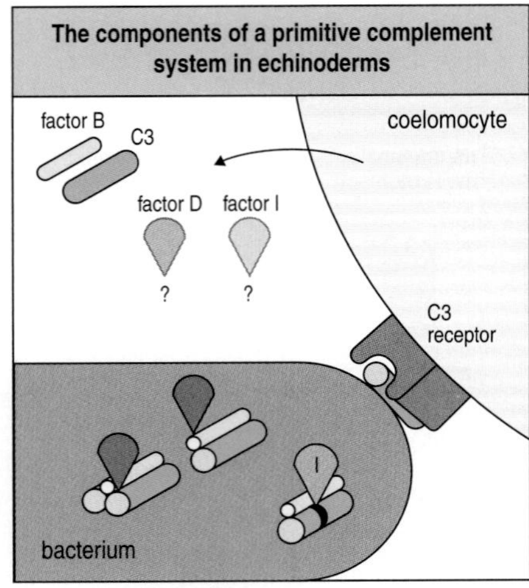

Fig. 16.5 The components of a simple complement system are present in echinoderms. The complement system of echinoderms resembles the alternative pathway of complement activation in mammals. Echinoderms possess equivalents of the complement components C3 and factor B, which are produced by coelomocytes, and are also thought to possess equivalents of factor D and the complement regulatory protein factor I. In this system, spontaneously activated C3 would bind to pathogen surfaces, where it would in turn bind factor B. Cleavage of factor B by a protease in the coelomic fluid, an as yet unidentified equivalent of factor D, would create the C3 convertase C3bBb, which can cleave and activate many more molecules of C3. Because echinoderm coelomocytes are phagocytes that will readily take up cells coated with C3, it is thought that they must express a C3 receptor. Finally, the C3 convertase is inactivated by another unidentified humoral protease, thought to be the equivalent of factor I.

in echinoderms: it is produced by amoeboid coelomocytes, phagocytic cells in the echinoderm coelomic fluid, and its expression is increased when bacteria are present. A homolog of factor B has also been identified in echinoderms. In mammals, factor B is activated by another protease—factor D—and, although no equivalent of factor D has yet been identified in echinoderms, the site at which factor D cleaves is conserved in echinoderm factor B. Thus, the echinoderms seem to possess the components of the amplification loop of the alternative pathway of complement activation, in which spontaneously activated C3 binds factor B, which is then cleaved by factor D to create an active C3 convertase, which cleaves more C3. As far as the function of the cleaved C3 is concerned, although no echinoderm C3 receptor has yet been identified, it is known that cells coated with C3 are taken up more efficiently than uncoated cells by echinoderm phagocytes, and so it seems that a functional, opsonizing complement system equivalent to the predicted ancestral system does exist in these invertebrates.

The spontaneous activation of C3 and its amplification by factor B poses the same problem for echinoderms as it does for mammals: how is such a system regulated to prevent tissue damage (see Sections 2-17 and 2-22)? How this is achieved in echinoderms is not known, although there is indirect evidence for the presence of a 'factor I' that can inactivate C3, and it is possible that the relevant genes and their complement regulatory products are present but have not yet been identified. The factor I cleavage site is conserved in echinoderm C3, and fragments of C3 consistent with cleavage at that site can be found in the coelomic fluid. In echinoderms, however, the C3 and factor B proteins are produced by the phagocytic cells themselves (see Fig. 16.5), and it is possible that they are secreted directly onto the surface of microbes, much as mammalian T cells secrete their effector molecules directly into the interface between the T cell and its target. In such a case there is less need for regulatory proteins that prevent complement from attacking the organism's own cells.

With the emergence of the chordates, the main components of the complement system seem to have been well established. In the urochordate *Ciona*, for which the complete genome sequence has been determined, homologs of C3 and factor B have been identified, as well as several genes homologous to integrins, which could encode complement receptors. In another urochordate, *Halocynthia*, an integrin family CR3-like receptor is known to have a function in C3-mediated phagocytosis. The characteristic marker of many mammalian complement regulatory proteins is a small domain called a short consensus repeat (SCR) or complement control protein (CCP) repeat (see Section 2-22). Several genes encoding proteins bearing such SCR domains have been identified in the *Ciona* genome, and it is expected that some of these will be found to have complement regulatory functions.

It is not known how ancient this opsonizing complement system is. C3 homologs have been found in invertebrates more distantly related to vertebrates than are echinoderms or urochordates, notably horseshoe crabs and *Drosophila*, but their function has not been defined. C3, which is cleaved and activated by serine proteases, is clearly evolutionarily related to, and seems to have been duplicated from, the serine protease inhibitor α_2-macroglobulin. In *Drosophila* there seem to be at least four C3 homologs containing the thioester linkage characteristic of this family of proteins (see Section 2-15); this linkage allows the activated protein to bind covalently to pathogen surfaces. These proteins are known as **thioester-containing proteins (TEPs)**.

The TEPs are thought to have some immune function in *Drosophila*, because the expression of at least three of them increases when *Drosophila* is infected with bacteria. *Drosophila* has phagocytic cells (hemocytes) in the hemolymph, but there is no evidence so far of opsonizing activity in the hemolymph.

Moreover, the TEPs are synthesized by the *Drosophila* fat body, the equivalent of the mammalian liver, rather than by the phagocytic cells themselves, as is the case for the echinoderm C3 homolog. Thus, although the *Drosophila* TEPs are clearly evolutionarily related to C3, they might have some quite different role. A clearer picture is revealed in another insect, the mosquito *Anopheles gambiae*, in which the protein TEP1 is produced by hemocytes and is induced in response to infection. In *Anopheles*, there is also direct evidence of binding of TEP1 to bacterial surfaces, and of the involvement of TEPs in the phagocytosis of Gram-negative bacteria. The origin of the complement system might therefore pre-date the ancient separation of the Bilateria (multicellular animals other than the sponges and the coelenterates) into the protostomes, which include the insects, and the deuterostomes, which include the echinoderms and chordates (and thus the vertebrates).

16-7 The lectin pathway of complement activation evolved in invertebrates.

After its initial appearance, the complement system seems to have evolved by the acquisition of new activation pathways, allowing microbial surfaces to be specifically targeted. The first of these new complement-activating systems to appear is likely to have been the ficolin pathway, which is present both in vertebrates and in some closely related invertebrates, such as the urochordates. Ficolins are related to the collectins (see Section 2-14), the family to which the vertebrate mannose-binding lectin (MBL) belongs. Like the collectins, the ficolins have a collagen-like domain and a carbohydrate-binding domain and form a similar 'bunch of tulips' multimeric structure. However, the carbohydrate-binding domain of ficolins is not related to C-type lectins, as in MBL, but is similar to fibrinogen. The carbohydrate-binding domain of ficolin is able to bind *N*-acetylglucosamine, as does MBL, although the latter is also able to bind mannose-containing carbohydrates, which the ficolins do not recognize. Evolutionarily, the ficolins may pre-date the collectins, which are also first seen in the urochordates.

Homologs of both MBL and the classical pathway complement component C1q, another collectin, have been identified in the *Ciona* genome. This suggests that in the evolution of the antibody-mediated classical pathway of complement activation (see Section 2-13) the ancestral immunoglobulin molecule, which did not appear until much later in evolution, took advantage of an already diversified family of collectins rather than driving the diversification of C1q from an MBL-like ancestor.

The activation of complement by ficolins and collectins is mediated through serine proteases called MASPs (MBL-associated serine proteases) which are able to cleave and activate C2, C4, and C3. In vertebrates, two different MASPs—MASP1 and MASP2—are associated with the ficolins and collectins, and this seems also to be true of the invertebrate ficolins. Two distinct invertebrate homologs of mammalian MASPs have been identified in the same ascidian species from which the ficolins were identified. The specificity of the invertebrate MASPs has not been determined, but it seems likely that they are able to cleave and activate C3. This invertebrate ficolin complement system, illustrated in Fig. 16.6, is functionally identical to the ficolin and MBL-mediated pathways found in mammals. Thus, the minimal complement system of the echinoderms has been supplemented in the urochordates by the recruitment of a specific activation system that can target the deposition of C3 onto microbial surfaces. The complement activation system evolved further by diversification of a C1q-like collectin and its associated MASPs to become the initiating components (C1q, C1r, and C1s) of the classical complement pathway. This could occur only after the evolution of the

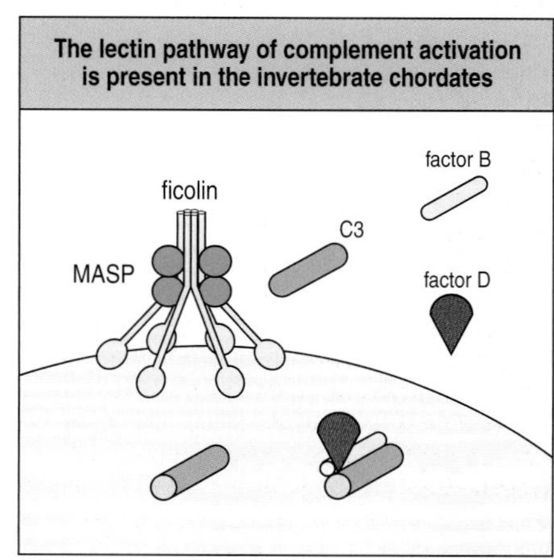
The lectin pathway of complement activation is present in the invertebrate chordates
ficolin
factor B
C3
MASP
factor D

Fig. 16.6 A lectin pathway of complement activation is present in invertebrate chordates.
A lectin-mediated pathway of complement activation has been defined in an ascidian, a urochordate. Ficolin, which uses a fibrinogen-like domain rather than a C-type lectin domain to bind carbohydrate ligands on a pathogen surface, is associated with serine proteases homologous to the MBL-associated serine proteases MASP1 and MASP2. Binding of ficolin to a cell surface allows the MASPs to cleave and activate C3. Activated C3b binds to the pathogen and initiates an amplification loop, in which the bound C3b binds factor B, allowing it to be cleaved by factor D to create a C3 convertase, C3bBb, which cleaves C3 to produce more C3b.

specific antigen-recognition molecules of the adaptive immune system, and it is to those that we turn next.

Summary.

The innate immune system provides an early defense against pathogen attack and alerts the adaptive immune system to the fact that a pathogen invasion has begun. This dual function seems to operate through a very ancient signaling pathway, the Toll pathway, that long pre-dates the adaptive immune system and is present in invertebrates as well as in vertebrates. In some organisms, the innate immune system has undergone extensive diversification, such as the expansion of the Toll-like receptor family in the sea urchin. The first defense molecules to arise in multicellular organisms were probably antimicrobial peptides, which are produced by both plants and animals. Another component of innate immunity in animals, the phagocytic cells that scavenge incoming pathogens, could have had their origins in unicellular amoeba-like eukaryotes. The complement system also pre-dates the vertebrates, being found in echinoderms and urochordates.

Evolution of the adaptive immune response.

The evolution of adaptive immunity is a fascinating problem. For a long time its origin was obscure, partly because it seemed to have emerged suddenly as a complete biological system at roughly the same time as the jawed vertebrates. The picture is now becoming clearer, as molecular methods are applied to a wider range of species. As we will see in this part of the chapter, the evolution of adaptive immunity in jawed vertebrates seems to have been made possible by the invasion of a putative immunoglobulin-like gene by a transposable genetic element. This conferred on the ancestral immunoglobulin gene the ability to undergo somatic gene rearrangement, and thus to generate diversity. When a mobile DNA element excises itself from a piece of DNA, alterations in the original sequence are introduced in the host DNA when the cut ends are rejoined; this is the origin of the diversity of antigen receptors in the adaptive immune system of higher vertebrates.

However, recent discoveries have revealed that species outside the jawed vertebrates have also developed ways to diversify pathogen-recognizing receptors that, in at least one case, seem to generate a truly adaptive immune system. We shall see that diversification can also be based on alternative splicing of an extensive array of alternative exons within an immunoglobulin-like gene, or on the introduction of somatic mutations, or on somatic gene rearrangement of genes similar in structure to the Toll-like receptors.

There are also many unanswered questions regarding vertebrate adaptive immunity. What was the nature of the gene that was invaded? It must have resembled a member of the immunoglobulin gene superfamily and may have already been functioning as some type of antigen receptor, which would have enabled it to operate appropriately in its changed form. This should help to narrow the search considerably. What was the function of the cell type in which this immunoglobulin ancestor was expressed and which could make good use of this new ability to generate a diversity of antigen-recognition molecules? It may have already been something like a lymphocyte, but it could equally have been a phagocytic cell like a macrophage or a polymorphonuclear leukocyte that then lost the capacity to phagocytose as it evolved new functions made possible by the expression of a variable antigen receptor.

The answer is that we really do not know. It might even have resembled a primitive NK cell, in which an invariant immunoglobulin superfamily receptor was already involved in the recognition of a primordial MHC-like molecule. Or perhaps it was some quite different cell type that no longer exists in vertebrates.

16-8 Some invertebrates generate extensive diversity in a repertoire of immunoglobulin-like genes.

Until very recently, it was thought that invertebrate immunity was limited to an innate system that had a very restricted diversity in recognizing pathogens. This idea was based on the knowledge that innate immunity in vertebrates relied on around 10 distinct Toll-like receptors and a similar number of other receptors that also recognize PAMPs, and also on the assumption that there were no greater numbers in invertebrates. Recent studies have, however, uncovered at least two invertebrate examples of extensive diversification of an immunoglobulin superfamily member, which could potentially provide an extended range of recognition of pathogens.

In *Drosophila*, fat-body cells and hemocytes act as part of the immune system. Fat-body cells secrete proteins, such as the antimicrobial defensins, into the hemolymph. Another protein found in hemolymph is the **Down syndrome cell adhesion molecule** (**Dscam**), a member of the immunoglobulin superfamily. Dscam was originally discovered in the fly as a protein involved in the specification of neuronal wiring. It is also made in fat-body cells and hemocytes, which can secrete it into the hemolymph. Here it is thought to opsonize invading bacteria and aid their engulfment by phagocytes.

The Dscam protein contains multiple, usually 10, immunoglobulin-like domains. The gene that encodes Dscam has, however, evolved to contain a large number of alternative exons for several of these domains (Fig. 16.7). Exon 4 of the Dscam protein can be encoded by any one of 12 different exons, each specifying an immunoglobulin domain of differing sequence. Exon cluster 6 has 48 alternative exons, cluster 9 another 33, and cluster 17 contains 2: it is estimated that the Dscam gene could encode around 38,000 protein isoforms. A role for Dscam in immunity was proposed when it was found that *in vitro* phagocytosis of *E. coli* by isolated hemocytes lacking Dscam was less efficient than normal. These observations suggest that at least some of this extensive repertoire of alternative exons may have evolved to diversify insects' ability to recognize pathogens. This role for Dscam has also been

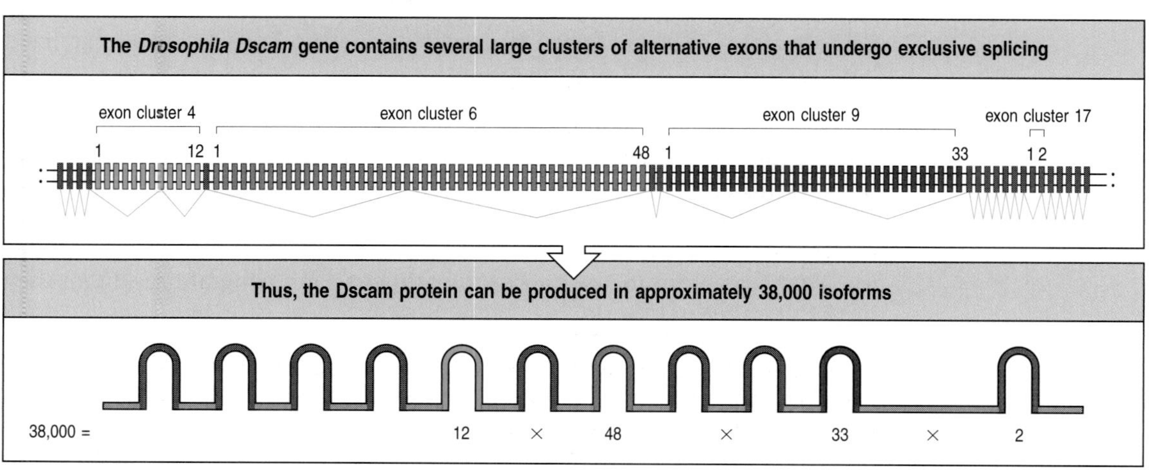

Fig. 16.7 The Dscam protein of *Drosophila* innate immunity contains multiple immunoglobulin domains and is highly diversified through alternative splicing. The gene encoding Dscam in *Drosophila* contains several large clusters of alternative exons. The clusters encoding exon 4 (green), exon 6 (light blue), exon 9 (red) and exon 17 (dark blue) contain 12, 48, 33, and 2 alternative exons respectively. For each of these clusters, only one alternative exon is used in the complete *Dscam* mRNA. There is some differential usage of exons in neurons, fat-body cells, and hemocytes. All three cell types use the entire range of alternative exons for exons 4 and 6. For exon 9, there is a restricted use of alternative exons in hemocytes and fat-body cells. The combinatorial use of alternative exons in the *Dscam* gene makes it possible to generate more than 38,000 protein isoforms. Adapted from Anastassiou, D.: *Genome Biol.* 2006, 7:R2.

confirmed for *Anopheles gambiae*, in which silencing of the Dscam homolog AgDscam has been shown to weaken the mosquito's normal resistance to bacteria and the malaria parasite *Plasmodium*.

Another invertebrate, this time a mollusc, uses a different strategy to diversify an immunoglobulin superfamily protein for use in immunity. The freshwater snail *Biomphalaria glabrata* expresses a small family of **fibrinogen-related proteins** (**FREPs**) thought to have a role in innate immunity. FREPs are produced by hemocytes and secreted into the hemolymph. Their concentration increases when the snail is infected by parasites—it is the intermediate host for the parasitic schistosomes that cause human schistosomiasis. FREPs have one or two immunoglobulin domains at their amino-terminal end and a fibrinogen domain at their carboxy-terminal end. The immunoglobulin domains may interact with pathogens, while the fibrinogen domain may confer on the FREP lectin-like properties that help precipitate the complex.

The *B. glabrata* genome contains many copies of FREP genes that can be divided into approximately 13 subfamilies. A study of the sequences of expressed FREP3 subfamily members has revealed that the FREPs expressed in an individual organism are extensively diversified compared with the germline genes. There are fewer than five genes in the FREP3 subfamily, but an individual snail was found to generate more than 45 distinct FREP3 proteins, all with slightly different sequences. An analysis of the protein sequences suggested that this diversification was due to the accumulation of point mutations in one of the germline FREP3 genes. Although the precise mechanism of this diversification, and the cell type in which it occurs, is not yet known, it does suggest some similarity to somatic hypermutation, which occurs during humoral immune responses in vertebrates (see Section 4-18). The *Biomphalaria* mechanism seems to represent a way of diversifying molecules involved in immune defense, again resembling in some ways the strategy of an adaptive immune response.

In neither of the above examples do we know whether receptor diversification is accompanied by a clonally distributed expression of receptors of differing specificity. Thus, we cannot yet say whether these mechanisms could provide what we generally require in a definition of adaptive immunity—the capacity for selection of particular variants and the capacity for immunological memory. The next section, however, provides such an example.

16-9 Agnathans possess an adaptive immune system that uses somatic gene rearrangement to diversify receptors built from LRR domains.

It has been known for at least 50 years that all jawed fish (the gnathostomes) can mount an adaptive immune response. Even the cartilaginous fish, the earliest group of jawed fishes to survive to the present day, have organized lymphoid tissue, T-cell receptors, and immunoglobulins, and the ability to mount adaptive immune responses. Adaptive immunity in all these organisms is based on the assembly of antigen receptors by the mechanism of RAG-based somatic recombination. Until very recently, it was thought that all invertebrates and all agnathans lacked all signs of an adaptive immune system. This notion has now been completely overturned. A closer examination of the surviving agnathan species reveals that they do in fact possess the ability to generate immune responses to pathogens and allografts, and that they exhibit features of immunological memory.

For some time it has been known that the hagfish and the lamprey could mount a form of accelerated rejection of transplanted skin grafts and exhibit a kind of delayed-type hypersensitivity. They also seemed to have in their serum an activity that behaved as a specific agglutinin, increasing in titer after secondary immunizations, in a similar fashion to antibodies in higher vertebrates. These animals also had cells that seemed to undergo rapid activation—blast transformation—after stimulation with mitogens, similarly to lymphocytes. However, there was no evidence of a thymus or of immunoglobulins.

With advances in molecular techniques, recent work has focused on characterizing the genes expressed by the lymphocyte-like cells of the lamprey *Petromyzon marinus*. No genes related to those for T-cell receptors or immunoglobulins have been found. These cells do, however, express large amounts of mRNAs for genes encoding multiple LRR domains, the same protein domain from which the pathogen-recognizing Toll-like receptors are built.

This might simply mean that these cells are specialized for recognizing and reacting to pathogens, but the LRR proteins expressed had some surprises in store. Instead of being present in just a few forms, as in most organisms, they had highly variable amino acid sequences, and showed a type of LRR rearrangement, with a large number of variable LRR units placed between less variable amino-terminal and carboxy-terminal LRR units. These LRR-containing proteins, called **variable lymphocyte receptors** (**VLRs**), have an invariant stalk region connecting them to the plasma membrane by a glycosylphosphatidylinositol linkage, and can either be tethered to the cell or, at other times, be secreted into the serum like antibodies.

Analysis of the organization of the expressed lamprey VLR genes has indicated that they are assembled by a process of somatic gene rearrangement (Fig. 16.8). In the germline configuration there is a single, but incomplete, VLR gene. This encodes a signal peptide, a partial amino-terminal LRR unit and a partial carboxy-terminal LRR unit, but these three blocks of coding sequence are separated by non-coding DNA that does not contain typical signals for RNA splicing, nor the recombination signal sequences (RSSs) present

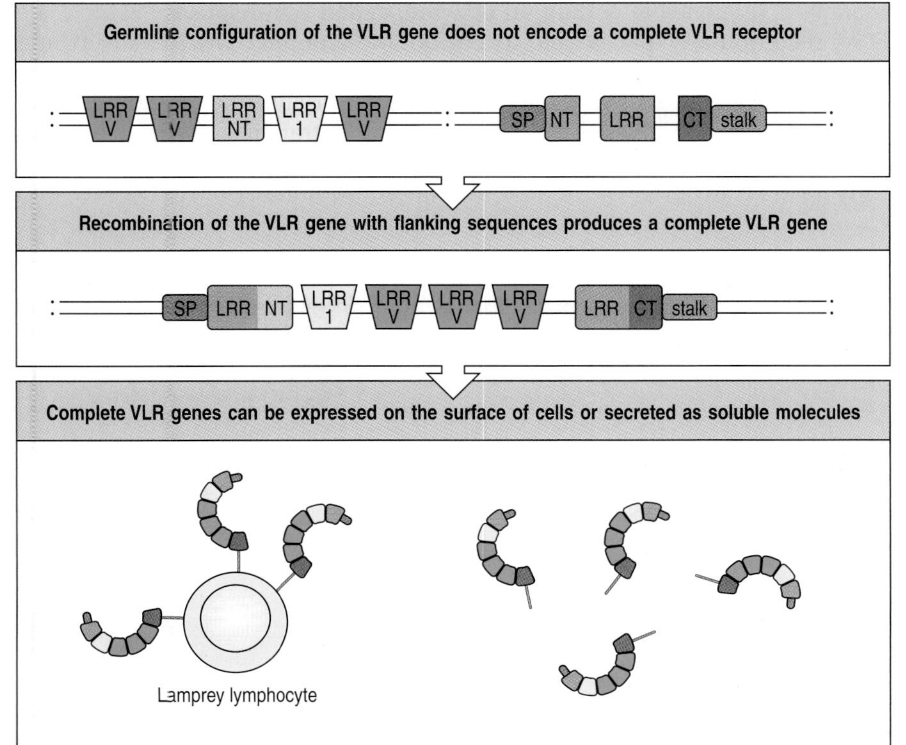
Germline configuration of the VLR gene does not encode a complete VLR receptor
LRR V
LRR V
LRR NT
LRR 1
LRR V
SP
NT
LRR
CT
stalk
Recombination of the VLR gene with flanking sequences produces a complete VLR gene
SP
LRR
NT
LRR 1
LRR V
LRR V
LRR V
LRR
CT
stalk
Complete VLR genes can be expressed on the surface of cells or secreted as soluble molecules
Lamprey lymphocyte

Fig. 16.8 Somatic recombination of an incomplete germline VLR gene generates a diverse repertoire of complete VLR genes in the lamprey. Top panel: the single incomplete copy of the lamprey VLR gene contains a framework for the complete gene: the signal peptide (SP), part of an amino-terminal LRR unit (NT, dark blue), and a carboxy-terminal LRR unit (red) that is split into two parts (LRR and CT) by intervening non-coding DNA sequences. In nearby flanking regions of the chromosome are multiple copies of other parts of the VLR gene-'cassettes' containing single or double copies of variable LRR domains (green) and cassettes that encode part of the amino-terminal LRR domains (light blue and yellow). Middle panel: by some process of somatic cell recombination, alternative LRR units are used to form a complete VLR gene. The complete VLR gene contains the assembled amino-terminal LRR cassette (LRR NT) and first LRR (yellow) followed by several variable LRR units (green) and the completed carboxy-terminal LRR unit. The receptor is attached to the cell membrane by glycosylphosphatidylinositol (GPI) linkage of the stalk region (purple). Bottom panel: an individual lymphocyte undergoes somatic gene rearrangement to produce a unique VLR receptor. These receptors can be tethered to the surface of the lymphocyte via the GPI linkage or can be secreted into the serum. Unique somatic rearrangement events in each developing lymphocyte generate a repertoire of VLR receptors of differing specificities. Adapted from Pancer, Z. and Cooper, M.D.: *Annu. Rev. Immunol.* 2006, **24**:497-518.

in immunoglobulin genes (see Section 4-4). However, in the regions flanking the incomplete VLR gene, there are a large number of DNA 'cassettes' that contain LRR units—one, two, or three LRR domains at a time.

Each lamprey lymphocyte expresses a complete and unique VLR gene that has undergone recombination of these flanking regions with the germline VLR gene. Flanking LRR units seem to be randomly incorporated into the VLR gene in steps that lead to the completion of the amino-terminal capping LRR subunit, followed by the addition of internal LRR domains, and finally, removal of the internal non-coding regions to complete the formation of the carboxy-terminal LRR domain. Researchers are currently searching for the molecular mechanism leading to this rearrangement: gene conversion is an attractive candidate. It is estimated that this somatic rearrangement mechanism can generate as much diversity in the VLR proteins as is possible for immunoglobulins. Thus, the diversity of the anticipatory repertoire of agnathans may be limited not by the numbers of possible receptors they can generate but by the number of lymphocytes present in any individual, as in the adaptive immune system of their evolutionary cousins the gnathostomes.

16-10 Adaptive immunity based on a diversified repertoire of immunoglobulin-like genes appeared abruptly in the cartilaginous fish.

In jawed fish and all higher vertebrates, adaptive immunity is possible because of a specific event that occurred in some ancestor of the jawed fish, when a mobile DNA carrying the ancestral *RAG* recombinases inserted itself into a stretch of DNA, presumably into a gene that was similar to an immunoglobulin gene or a T-cell receptor gene V region. Both prokaryotic and eukaryotic genomes contain a variety of mobile DNA elements, known as transposable elements or transposons, which can move themselves, or copies of themselves, to different positions on the chromosomes in the process known as transposition. Transposons contain two essential elements—a sequence encoding an enzyme, called a transposase, which is a DNA recombinase that is able to cut double-stranded DNA to insert and excise the element, and terminal repeat sequences that are recognized by the transposase and are required for the element to be able to undergo excision and insertion (Fig. 16.9). A key feature of transposition is that both insertion and excision cause changes to the 'host' target DNA. Insertion of a transposon leads to the formation of short additional sequences at each end

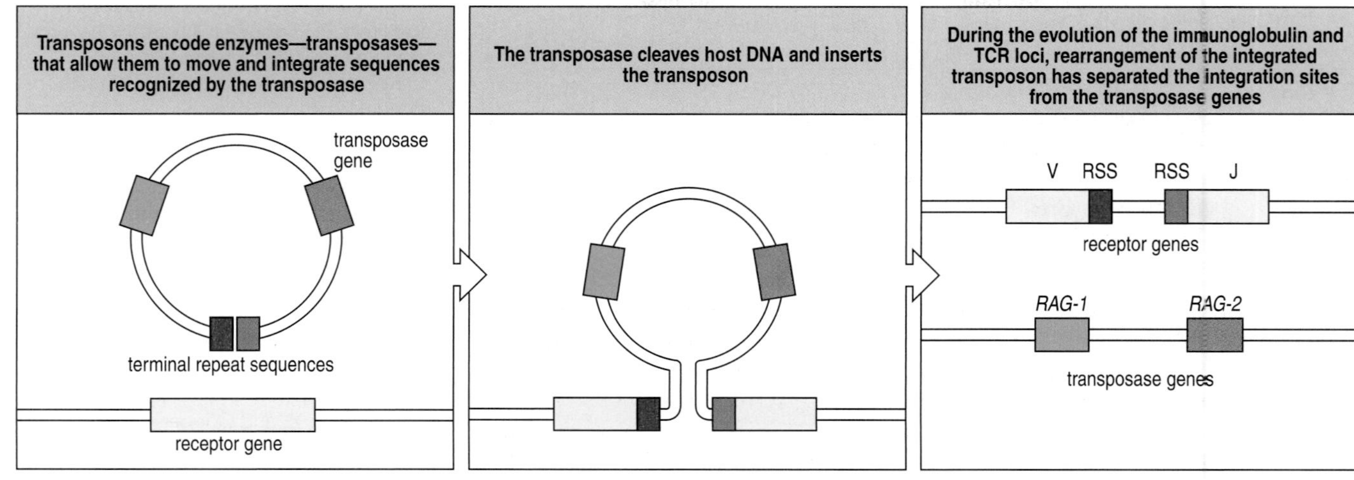

Fig. 16.9 The integration of a transposon into a receptor gene ultimately gave rise to the immunoglobulin and T-cell receptor genes and their capacity for somatic recombination. Transposons are DNA sequences that can move around the genome by excising themselves from one site and inserting into another. Left panel: a transposon must contain two functional elements-a sequence encoding a transposase, the enzyme that mediates the excision and integration of the transposon, and specific recognition sequences for the transposase, which are present at each end of the transposon and are required for the transposon to excise from or integrate into DNA. Center panel: after excision from DNA (not shown) the transposon reinserts elsewhere. The transposase cleaves genomic DNA at a random site, and then joins the free ends of the transposon to the cut ends of the genomic DNA. A transposon excises by the reverse of this process, the transposase bringing the terminal sequences together and then cutting the transposon out of the genomic DNA. Right panel: in the evolution of the immunoglobulin and T-cell receptor (TCR) genes, an initial integration event into the middle of a cell-surface receptor has been followed by DNA rearrangements that separated the transposase genes, which we now know as the *RAG-1* and *RAG-2* genes, from the transposon terminal sequences, which we now term the recombination signal sequences (RSSs).

of the integrated element; the excision process leaves these sequences in the DNA and also makes a gap in the host DNA that is repaired by the cell's error-prone DNA repair mechanisms.

In the case of the transposon thought to have kicked off the evolution of the vertebrate adaptive immune system, the transposase would have been the ancestral RAG recombinase. After the original integration event, it seems that the transposon sequences encoding the recombinase became separated from their recognition sequences. This could have occurred most simply by the deletion of the transposase genes from the transposon integrated into the primordial immune-receptor gene, while a copy of the transposon elsewhere in the genome retained the transposase genes but lost the associated recognition sequences (the terminal repeats). The terminal repeats left in the immune-receptor gene became the recombination signal sequences (RSSs) that flank gene segments in immunoglobulin and T-cell receptor genes, whereas the sequences encoding the transposases became the *RAG-1* and *RAG-2* genes, which now encode the recombinase that is essential for the rearrangement of antigen-receptor genes (see Section 4-5). A transposon had been proposed for many years as the origin of the *RAG* genes, because, unusually for mammals, these genes lack introns, which is a feature of transposons. That the action of the RAG proteins on the RSSs was similar to the mechanism of transposon excision was well known, but it has only quite recently been shown that the present-day RAG proteins can catalyze the insertion into DNA of a DNA fragment containing the RSSs, a process identical to transposition.

The origin of somatic gene rearrangement in the excision of a transposable element makes sense of an apparent paradox in the rearrangement of immune-system genes. This is that the RSSs are joined precisely in the excised DNA, which has no further function and whose fate is irrelevant to the cell, whereas the cut ends in the genomic DNA, which form part of the immunoglobulin or T-cell receptor gene, are joined by an error-prone process, which could be viewed as a disadvantage. However, when looked at from the transposon's point of view this makes sense, because the transposon preserves its integrity by this excision mechanism, whereas the fate of the DNA it leaves behind is of no significance to it. As it turned out, the error-prone joining in the primitive immunoglobulin gene generated useful diversity in molecules used for antigen recognition and was clearly strongly selected for. Subsequent duplication, reduplication, and recombination of the immune-receptor gene and its inserted RSSs eventually led to the multi-segmented immunoglobulin and T-cell receptor loci of present-day vertebrates.

16-11 The target of the transposon is likely to have been a gene encoding a cell-surface receptor containing an immunoglobulin-like V domain.

Proteins containing immunoglobulin-like domains are ubiquitous throughout the plant, animal, and bacterial kingdoms, making this one of the most abundant protein superfamilies; in species whose genomes have been fully sequenced, the immunoglobulin superfamily is found to be one of the largest families of protein domains in the genome. The functions of the members of this superfamily are very disparate, and they are a striking example of natural selection taking a useful structure—the basic immunoglobulin-domain fold—and adapting it to different purposes.

The immunoglobulin superfamily domains can be divided into four families—V (resembling an immunoglobulin variable domain), C1 and C2 (resembling constant-region domains), and the more diverse I domains—on the basis of differences in structure and sequence. V, C1, and C2 domains are found in many immune-system molecules. For example, immunoglobulins and T-cell receptors have V and C1 domains, CD4 and CD8 molecules have V and, in the case of CD4, C2 domains, and MHC class I and class II molecules have C1 domains. The adhesion molecules VCAM and ICAM contain C2 and I domains. The I domains seem to be the most disparate. As well as being found in adhesion molecules of the immune system, they are also present in proteins outside the immune system, such as the muscle proteins titin and twitchin.

The ancestral domain into which the transposon inserted to create the capacity for rearrangement was certainly a V domain. This domain was most probably linked to a C1 domain to form a transmembrane receptor, as this is the basic organization of both T-cell receptors and immunoglobulins. It is possible that the original receptor could have been a V domain coupled to a C2 domain, a combination found, for example, in the KIR receptors of NK cells and in other members of the extended leukocyte receptor family (see Section 2-31), with a subsequent evolution of the C2 domain into a C1 domain within the lineage that led to the immunoglobulins and T-cell receptors, although this is less likely. Genes with both of these types of organization have been found in the urochordate *Ciona*: two genes contain V domains in association with C2 domains, and two others contain V domains linked to C1-like and C2 domains. The latter two are the most likely candidates for the ancestor of the vertebrate antigen receptors.

Two additional invertebrate protein families containing V domains have been identified in the cephalochordate *Branchiostoma* (the amphioxus). One family comprises proteins containing immunoglobulin V domains associated with chitin-binding domains, rather than with immunoglobulin C domains. The second family is represented by a protein containing a V domain linked to a multispan transmembrane domain. In both cases, there is as yet no evidence for any immune function associated with this protein.

16-12 Different species generate immunoglobulin diversity in different ways.

Most of the animals we are familiar with generate a large part of their antigen receptor diversity in the same way as humans, by putting together gene segments in different combinations, as described in Chapters 3 and 4. We noted a few exceptions, however (see Section 4-19), and we return to these now. Some animals use gene rearrangement to always join together the same V and J gene segment initially, and then diversify this recombined V region. In chickens and rabbits, it is diversified by gene conversion in the bursa of Fabricius (in chickens) or another intestinal lymphoid organ (in rabbits) (Fig. 16.10). Other animals may generate their diverse repertoire mainly by somatic hypermutation of a fairly invariant recombined V region. The generation of some immunoglobulin diversity within the sheep ileal Peyer's patch may depend on this mechanism, although gene conversion also operates in this species.

The immunoglobulin loci in bony fish and higher vertebrates are organized such that separate blocks containing repeated V regions lie upstream of blocks of D regions (in the V_H locus) and blocks of J regions. In contrast, the cartilaginous fish have multiple copies of discrete V_L–J_L–C_L and V_H–D_H–J_H–C_H cassettes, and activate rearrangement within individual cassettes (see Fig. 16.10). Although these mechanisms differ from the canonical process described in Chapter 4, in which diversity is generated through combinatorial gene rearrangement, in most cases there is still a requirement for a somatic rearrangement event.

In the rays and carcharine sharks, some of the immunoglobulin genes are not generated by rearrangement. Instead these organisms have multiple

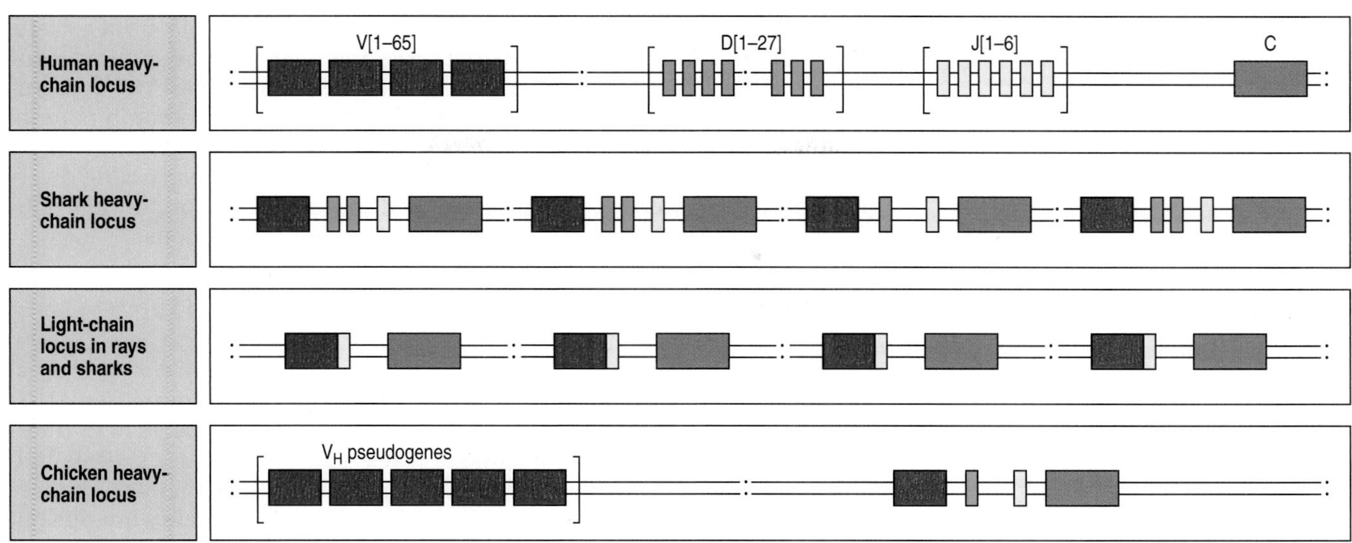

Fig. 16.10 The organization of immunoglobulin genes is different in different species, but all can generate a diverse repertoire of receptors. The organization of the immunoglobulin heavy-chain genes in mammals, in which there are separated clusters of repeated V, D, and J gene segments, is not the only solution to the problem of generating a diverse repertoire of receptors. Other vertebrates have found alternative solutions. In 'primitive' groups, such as the sharks, the locus consists of multiple repeats of a basic unit composed of a V gene segment, one or two D gene segments, a J gene segment, and a C gene segment. A more extreme version of this organization is found in the λ-like light-chain locus of some cartilaginous fishes such as the rays and the carcharine sharks, in which the repeated unit consists of already rearranged VJ-C genes, from which a random choice is made for expression. In chickens, there is a single rearranging set of gene segments at the heavy-chain locus but multiple copies of V-segment pseudogenes. Diversity in this system is created by gene conversion, in which sequences from the V_H pseudogenes are copied onto the single rearranged V_H gene.

'rearranged' V_L regions (and sometimes rearranged V_H regions) in the germline genome (see Fig. 16.10) and apparently generate diversity by activating the transcription of different copies. These are examples of non-combinatorial immunoglobulin systems, although in the strict sense there is still combinatorial diversity, which is generated by the subsequent pairing of heavy and light chains. This organization of the light-chain loci is unlikely to represent an intermediate evolutionary stage, because in that case the heavy-chain and light-chain genes would have had to independently acquire the capacity for rearrangement by a process of convergent rather than divergent evolution. It is much more likely that, after the divergence of the cartilaginous fishes, some of the immunoglobulin loci in a common ancestor of this group became rearranged in the germline through activation of the *RAG* genes in germ cells, with the consequent inheritance of rearranged loci by the offspring. In these species, the rearranged germline loci may confer some advantages, perhaps in early development before a complex repertoire becomes established, or in ensuring rapid responses to common pathogens by using a preformed set of immunoglobulin chains.

The predominant form of immunoglobulin in cartilaginous fish is IgM, and this is also true of bony fish. The cartilaginous fishes also have at least two types of immunoglobulin heavy chains not found in more recently evolved species. One, IgW, has six constant-region domains, whereas the second, IgNAR (for new antigen receptor) seems to be related to IgW but has lost the first constant domain and does not pair with light chains, instead forming a homodimer in which each V domain forms a separate antigen-binding site. The IgW molecule is thought to be present only as a cell-surface receptor on B cells, and this function may have been taken over by IgD, which is first found in bony fish. This variability does suggest that in the early cartilaginous fish the immunoglobulins had only recently evolved and were throwing up variants to be tested by natural selection.

16-13 Both α:β and γ:δ T-cell receptors are present in cartilaginous fish.

Neither the T-cell receptors nor the immunoglobulins have been found in any species evolutionarily earlier than the cartilaginous fishes. What is surprising is that by the time we first observe them, they have essentially the same form that we see in mammals. The identification of TCRβ-chain and δ-chain homologs from sharks, and of distinct TCRα, β, γ, and δ chains from a skate, show that even at the earliest time that these adaptive immune system receptors can be identified, they had already diversified into at least two recognition systems. Moreover, each shows diversity resulting from combinatorial somatic rearrangement. Although we still do not fully understand the role of γ:δ T cells in the mammalian adaptive immune system, the very early divergence of the two sets of T-cell receptors and their conservation through subsequent evolution suggests an early separation of their functions.

16-14 MHC class I and class II molecules are also first found in the cartilaginous fishes.

One would expect to see the specific ligands of T-cell receptors, the MHC molecules, emerge at around the same time in evolution. Indeed, MHC molecules are present in the cartilaginous fishes and in all higher species but, like the T-cell receptors, they have not been found in agnathans or invertebrates. Both MHC class I and class II α-chain and β-chain genes are present in sharks, and their products seem to function in an identical way to mammalian MHC molecules. The key residues of the peptide-binding cleft that interact with the ends of the peptide, in MHC class I molecules, or with the central region of the peptide, in MHC class II molecules, are conserved in shark MHC molecules.

Moreover, the MHC genes are also polymorphic in sharks, with multiple alleles of class I and class II loci. In some species, more than 20 MHC class I alleles have been identified so far. For the shark MHC class II molecules, both the class II α and the class II β chains are polymorphic. Thus, not only has the function of the MHC molecules in selecting peptides for presentation evolved during the divergence of the agnathans and the cartilaginous fishes, but the continuous selection imposed by pathogens has resulted in the polymorphism that is a characteristic feature of the MHC.

The MHC class I genes can be classified into the classical MHC class I genes (sometimes called class Ia) and the non-classical MHC class Ib genes (see Section 5-17). This is also true in cartilaginous fishes, because the class I genes of sharks include some that resemble mammalian class Ib molecules. However, it is thought that the shark class Ib genes are not the direct ancestors of the mammalian class Ib genes. Instead, some class Ib genes, notably CD1 and some that have functions distinct from antigen presentation, such as zinc-α_2-glycoprotein and the MHC-like neonatal Fc receptor, FcRn (see Section 9-15), seem to have evolved early, before the divergence of the cartilaginous fishes from the vertebrate line, and are likely to have homologs in all vertebrates. For the other class I genes, it seems that within each of the five major vertebrate lineages studied (cartilaginous fishes, lobe-finned fishes, ray-finned fishes, amphibians, and mammals) these genes have independently separated into classical and non-classical loci.

Thus, the characteristic features of the MHC molecules are all present when these molecules are first encountered, and there are no intermediate forms to guide our understanding of their evolution. Thus, although we can trace the evolution of the components of the innate immune system, the mystery of the origin of the adaptive immune system still largely persists.

What was the selective force that drove the evolution of adaptive immunity in higher vertebrates? There has been intriguing speculation that it might have been a side effect of acquiring jaws, which led to the ability to eat a wider variety of foodstuffs. The consequent exposure of the gut tissue to hard shells or chitinous exoskeletons might have led to increased infections. However, the acquisition of a jaw was only one of several changes taking place during the transition from agnathans to jawed vertebrates, both in the organization of the vertebrate body and in the development and lifestyle of the organisms. Some molluscs, notably the beaked cephalopods such as octopuses and squid, also eat shelled or bony prey, and so this feature by itself does not seem to be a sufficient selective force for the development of adaptive immunity.

Indeed, we now recognize that agnathans have their own form of adaptive immunity, although it is built upon a different set of building blocks. And so, although we may not have a sure answer to the question of what forces led to RAG-dependent elaboration of adaptive immunity, it has never been clearer that, as Charles Darwin remarked about evolution in general, "from so simple a beginning endless forms most beautiful and most wonderful have been, and are being evolved."

Summary.

Once considered a wholly inexplicable 'immunological Big Bang,' the evolution of an adaptive immune response in the jawed vertebrates is now thought to be related to the chance insertion of a transposon into a member of the immunoglobulin gene superfamily. This event must have taken place in a germline cell in an ancestor of the vertebrates. By chance, the transposon terminal sequences were placed in an appropriate location within the primordial antigen-receptor gene to enable intramolecular somatic recombination, thus paving the way for the full-blown somatic gene rearrangement seen in present-day immunoglobulin and T-cell receptor genes. The transposase genes (the *RAG* genes), presumably from the same transposon, became separated from the transposon terminal sequences and are now carried on a different chromosome.

Many animals other than the jawed vertebrates have the potential to generate a previously unsuspected amount of diversity in the receptor repertoire that recognizes and defends against pathogens. The extensive genomic content of Toll-like receptors in the sea urchin is paralleled by extensive alternative splicing of an array of immunoglobulin-domain-encoding exons in *Drosophila*, and by a mechanism of somatic mutation in the mollusc *Biomphalaria*. Most notably, our close vertebrate cousins, the jawless fishes, have evolved an adaptive immune system built on a completely different basis—the diversification of LRR domains rather than immunoglobulin domains—but which otherwise seems to have the essential features of clonal selection and immunological memory of a true adaptive immune system.

Summary to Chapter 16.

The evolution of the immune system, summarized in Fig. 16.11, has mostly been a gradual process of increasing diversification from a small number of very ancient recognition and effector pathways, this gradual process being broken into by the dramatic acquisition of adaptive immunity. After that, the gradual pace of steady development and diversification returned. From the time of the common ancestors of animals and plants, antimicrobial peptides have been a basic defense mechanism, supplemented later by the retention of mobile phagocytic cells capable of disposing of invading microbes. Systems of innate immunity evolved that targeted pathogens to phagocytes more efficiently; the first of these was a simple version of the alternative pathway of complement activation, which was followed by the evolution of a lectin-mediated pathway. We now recognize that a form of adaptive immunity exists in our close relatives, the agnathans, based on a system of rearranging LRR-containing genes rather than on immunoglobulins and T-cell receptors. Adaptive immunity in jawed fishes emerged from an as-yet unknown ancestral immune system, with the rapid evolution of a full complement of T-cell and immunoglobulin receptors, together with the MHC class I and class II antigen-presenting molecules. Subsequent evolution has served to refine the adaptive immune system, but its essential nature has remained unchanged.

Fig. 16.11 Summary of the evolutionary emergence of innate and adaptive immune characteristics.

	Drosophila (insect)	Sea urchin (echinoderm)	Sea squirt (ascidian)	Lamprey (agnathan)	Shark (elasmo-branch)	Carp (teleost)	Frog (amphibian)	Snake (reptile)	Chicken (bird)	Human (mammal)
Adaptive immunity	No	No	No	Yes	Yes	Yes	Yes	Yes	Yes	Yes
Immunoglobulin rearrangement	No	No	No	No	Yes	Yes	Yes	Yes	Yes	Yes
VLR gene rearrangement	No	No	No	Yes	No	No	No	No	No	No
Combinatorial T-cell receptor rearrangement	No	No	No	No	Yes	Yes	Yes	Yes	Yes	Yes
Polymorphic MHC molecules	No	No	No	No	Yes	Yes	Yes	Yes	Yes	Yes
Classical complement pathway	No	No	No	No	Yes	Yes	Yes	Yes	Yes	Yes
C3 and factor B	No	Yes	Yes	Yes	Yes	Yes	Yes	Yes	Yes	Yes
Mannose-binding lectin	No	?	Yes	Inferred	Inferred	Yes	Inferred	Inferred	Yes	Yes
Ficolins	No	?	Yes	Inferred	Inferred	Inferred	Inferred	Inferred	Inferred	Yes
MASPs	No	?	Yes	Yes	Yes	Yes	Yes	Yes	Yes	Yes
Toll-like receptors	Yes	Yes	Yes	Yes	Yes	Yes	Yes	Yes	Yes	Yes
Antibacterial peptides	Yes	Inferred	Inferred	Inferred	Inferred	Inferred	Yes	Yes	Yes	Yes

Questions.

16.1 *Discuss the features that distinguish innate immunity from adaptive immunity.*

16.2 *(a) Could an adaptive immune system be based on a repertoire of receptors that do not undergo somatic gene rearrangement? (b) The sea urchin genome contains a gene thought to be related to the ancestral RAG transposon. What does this imply about the alternative evolution of adaptive immunity in agnathans and the jawed vertebrates?*

16.3 *Drosophila melanogaster can express a diverse repertoire of Dscam isoforms. Does this imply that it has adaptive immunity? Explain your answer.*

Section references.

16-1 The evolution of the immune system can be studied by comparing the genes expressed by various organisms.

Adams, M.D., Celniker, S.E., Holt, R.A., Evans, C.A., Gocayne, J.D., Amanatides, P.G., Scherer, S.E., Li, P.W., Hoskins, R.A., Galle, R.F., *et al.*: **The genome sequence of *Drosophila melanogaster*.** *Science* 2000, **287**:2185–2195.

Gregory, S.G., Sekhon, M., Schein, J., Zhao, S., Osoegawa, K., Scott, C.E., Evans, R.S., Burridge, P.W., Cox, T.V., Fox, C.A., *et al.*: **A physical map of the mouse genome.** *Nature* 2002, **418**:743–750.

Mural, R.J., *et al.*: **A comparison of whole-genome shotgun-derived mouse chromosome 16 and the human genome.** *Science* 2002, **296**:1617–1618.

16-2 Antimicrobial peptides are likely to be the most ancient immune defenses.

Ganz, T.: **Defensins and host defense.** *Science* 1999, **286**:420–421.

Ganz, T.: **Defensins: antimicrobial peptides of innate immunity.** *Nat. Rev. Immunol.* 2003, **3**:710–720.

Gura, T.: **Innate immunity. Ancient system gets new respect.** *Science* 2001, **291**:2068–2071.

Hoffmann, J.A.: **Innate immunity of insects.** *Curr. Opin. Immunol.* 1995, **7**:4–10.

Thomma, B.P., Cammue, B.P., and Thevissen, K.: **Plant defensins.** *Planta* 2002, **216**:193–202.

16-3 Toll-like receptors may represent the most ancient pathogen-recognition system.

Hetru, C., Troxler, L., and Hoffmann, J.A.: ***Drosophila melanogaster* antimicrobial defense.** *J. Infect. Dis.* 2003, **187 Suppl 2**:S327–S334.

Hoffmann, J.A., Kafatos, F.C., Janeway, C.A., and Ezekowitz, R.A.: **Phylogenetic perspectives in innate immunity.** *Science* 1999, **284**:1313–1318.

Imler, J.L., and Hoffmann, J.A.: **Toll and Toll-like proteins: an ancient family of receptors signaling infection.** *Rev. Immunogenet.* 2000, **2**:294–304.

Imler, J.L., and Hoffmann, J.A.: **Toll receptors in innate immunity.** *Trends Cell Biol.* 2001, **11**:304–311.

Imler, J.L., and Hoffmann, J.A.: **Toll signaling: the TIReless quest for specificity.** *Nat. Immunol.* 2003, **4**:105–106.

Gottar, M., Gobert, V., Matskevich, A.A., Reichhart, J.M., Wang, C., Butt, T.M., Belvin, M., Hoffmann, J.A., and Ferrandon, D.: **Dual detection of fungal infections in *Drosophila* via recognition of glucans and sensing of virulence factors.** *Cell* 2006, **127**:1425–1437.

Pili-Floury, S., Leulier, F., Takahashi, K., Saigo, K., Samain, E., Ueda, R., and Lemaitre, B.: ***In vivo* RNA interference analysis reveals an unexpected role for GNBP1 in the defense against Gram-positive bacterial infection in *Drosophila* adults.** *J. Biol. Chem.* 2004, **279**:12848–12853.

Royet, J., Reichhart, J.M., and Hoffmann, J.A.: **Sensing and signaling during infection in *Drosophila*.** *Curr. Opin. Immunol.* 2005, **17**:11–17.

16-4 Toll-like receptor genes have undergone extensive diversification in some invertebrate species.

Rast, J.P., Smith, L.C., Loza-Coll, M., Hibino, T., Litman, G.W.: **Genomic insights into the immune system of the sea urchin.** *Science* 2006, **314**:952–956.

Samanta, M.P., Tongprasit, W., Istrail, S., Cameron, R.A., Tu, Q., Davidson, E.H., Stolc, V.: **The transcriptome of the sea urchin embryo.** *Science* 2006, **314**:960–962.

16-5 A second recognition system in *Drosophila* homologous to the mammalian TNF receptor pathway provides protection from Gram-negative bacteria.

Ferrandon, D., Jung, A.C., Criqui, M., Lemaitre, B., Uttenweiler-Joseph, S., Michaut, L., Reichhart, J., and Hoffmann, J.A.: **A drosomycin-GFP reporter transgene reveals a local immune response in *Drosophila* that is not dependent on the Toll pathway.** *EMBO J.* 1998, **17**:1217–1227.

Georgel, P., Naitza, S., Kappler, C., Ferrandon, D., Zachary, D., Swimmer, C., Kopczynski, C., Duyk, G., Reichhart, J.M., and Hoffmann, J.A.: ***Drosophila* immune deficiency (IMD) is a death domain protein that activates antibacterial defense and can promote apoptosis.** *Dev. Cell* 2001, **1**:503–514.

Hoffmann, J.A., and Reichhart, J.M.: ***Drosophila* innate immunity: an evolutionary perspective.** *Nat. Immunol.* 2002, **3**:121–126.

Rutschmann, S., Jung, A.C., Zhou, R., Silverman, N., Hoffmann, J.A., and Ferrandon, D.: **Role of *Drosophila* IKKγ in a *toll*-independent antibacterial immune response.** *Nat. Immunol.* 2000, **1**:342–347.

16-6 An ancestral complement system opsonizes pathogens for uptake by phagocytic cells.

Gross, P.S., Al-Sharif, W.Z., Clow, L.A., and Smith, L.C.: **Echinoderm immunity and the evolution of the complement system.** *Dev. Comp. Immunol.* 1999, **23**:429–442.

Smith, L.C.: **The complement system in sea urchins.** *Adv. Exp. Med. Biol.* 2001, **484**:363–372.

Smith, L.C., Clow, L.A., and Terwilliger, D.P.: **The ancestral complement system in sea urchins.** *Immunol. Rev.* 2001, **180**:16–34.

Smith, L.C., Shih, C.S., and Dachenhausen, S.G.: **Coelomocytes express SpBf, a homologue of factor B, the second component in the sea urchin complement system.** *J. Immunol.* 1998, **161**:6784–6793.

16-7 The lectin pathway of complement activation evolved in invertebrates.

Fujita, T.: **Evolution of the lectin-complement pathway and its role in innate immunity.** *Nat. Rev. Immunol.* 2002, **2**:346–353.

Holmskov, U., Thiel, S., and Jensenius, J.C.: **Collections and ficolins: humoral lectins of the innate immune defense.** *Annu. Rev. Immunol.* 2003, **21**:547–578.

Matsushita, M., and Fujita, T.: **Ficolins and the lectin complement pathway.** *Immunol. Rev.* 2001, **180**:78–85.

Matsushita, M., and Fujita, T.: **The role of ficolins in innate immunity.** *Immunobiology* 2002, **205**:490–497.

Nonaka, M., Azumi, K., Ji, X., Namikawa-Yamada, C., Sasaki, M., Saiga, H., Dodds, A.W., Sekine, H., Homma, M.K., Matsushita, M., *et al.*: **Opsonic complement component C3 in the solitary ascidian, *Halocynthia roretzi*.** *J. Immunol.* 1999, **162**:387–391.

Raftos, D., Green, P., Mahajan, D., Newton, R., Pearce, S., Peters, R., Robbins, J., and Nair, S.: **Collagenous lectins in tunicates and the proteolytic activation of complement.** *Adv. Exp. Med. Biol.* 2001, **484**:229–236.

Smith, L.C., Azumi, K., and Nonaka, M.: **Complement systems in invertebrates. The ancient alternative and lectin pathways.** *Immunopharmacology* 1999, **42**:107–120.

16-8 Some invertebrate species generate extensive diversity in a repertoire of immunoglobulin-like genes.

Dong, Y., Taylor, H.E., and Dimopoulos, G.: **AgDscam, a hypervariable immunoglobulin domain-containing receptor of the *Anopheles gambiae* innate immune system.** *PLoS Biol* 2006, **4**:e229.

Loker, E.S., Adema, C.M., Zhang, S.M., Kepler, T.B.: **Invertebrate immune systems—not homogeneous, not simple, not well understood.** *Immunol. Rev.* 2004, **198**:10–24.

Watson, F.L., Puttmann-Holgado, R., Thomas, F., Lamar, D.L., Hughes, M., Kondo, M., Rebel, V.I., and Schmucker, D.: **Extensive diversity of Ig-superfamily proteins in the immune system of insects.** *Science* 2005, **309**:1826–1827.

Zhang, S.M., Adema, C.M., Kepler, T.B., and Loker, E.S.: **Diversification of Ig superfamily genes in an invertebrate.** *Science* 2004, **305**:251–254.

16-9 Agnathans possess an adaptive immune system that uses somatic gene rearrangement to diversify receptors built from LRR domains.

Cooper, M.D., Alder, M.N.: **The evolution of adaptive immune systems.** *Cell* 2006, **124**:815–822.

Finstad, J. and Good, R.A. **The evolution of the humoral immune response. 3. Immunologic responses in the lamprey.** *J. Exp. Med.* 1964, **120**:1151–1168.

Litman, G.W., Finstad, F.J., Howell, J., Pollara, B.W., and Good, R.A.: **The evolution of the immune response. 3. Structural studies of the lamprey immuoglobulin.** *J. Immunol.* 1970, **105**:1278-85.

Pancer, Z., Amemiya, C.T., Ehrhardt, G.R., Ceitlin, J., Gartland, G.L., and Cooper, M.D.: **Somatic diversification of variable lymphocyte receptors in the agnathan sea lamprey.** *Nature* 2004, **430**:174–180.

16-10 Adaptive immunity based on a diversified repertoire of immunoglobulin-like genes appeared abruptly in the cartilaginous fish.

Agrawal, A.: **Amersham Pharmacia Biotech & Science Prize. Transposition and evolution of antigen-specific immunity.** *Science* 2000, **290**:1715–1716.

Agrawal, A., Eastman, Q.M., and Schatz, D.G.: **Transposition mediated by RAG1 and RAG2 and its implications for the evolution of the immune system.** *Nature* 1998, **394**:744–751.

Hansen, J.D., and McBlane, J.F.: **Recombination-activating genes, transposition, and the lymphoid-specific combinatorial immune system: a common evolutionary connection.** *Curr. Top. Microbiol. Immunol.* 2000, **248**:111–135.

van Gent, D.C., Mizuuchi, K., and Gellert, M.: **Similarities between initiation of V(D)J recombination and retroviral integration.** *Science* 1996, **271**:1592–1594.

Schatz, D.G.: **Transposition mediated by RAG1 and RAG2 and the evolution of the adaptive immune system.** *Immunol. Res.* 1999, **19**:169–182.

16-11 The target of the transposon is likely to have been a gene encoding a cell-surface receptor containing an immunoglobulin-like V domain.

Cannon, J.P., Haire, R.N., and Litman, G.W.: **Identification of diversified genes that contain immunoglobulin-like variable regions in a protochordate.** *Nat. Immunol.* 2002, **3**:1200–1207.

Rast, J.P., and Litman, G.W.: **Towards understanding the evolutionary origins and early diversification of rearranging antigen receptors.** *Immunol. Rev.* 1998, **166**:79–86.

16-12 Different species generate immunoglobulin diversity in different ways.

Anderson, M.K., Shamblott, M.J., Litman, R.T., and Litman, G.W.: **Generation of immunoglobulin light chain gene diversity in *Raja erinacea* is not associated with somatic rearrangement, an exception to a central paradigm of B cell immunity.** *J. Exp. Med.* 1995, **182**:109–119.

Anderson, M.K., Strong, S.J., Litman, R.T., Luer, C.A., Amemiya, C.T., Rast, J.P., and Litman, G.W.: **A long form of the skate IgX gene exhibits a striking resemblance to the new shark IgW and IgNARC genes.** *Immunogenetics* 1999, **49**:56–67.

Rast, J.P., Amemiya, C.T., Litman, R.T., Strong, S.J., and Litman, G.W.: **Distinct patterns of IgH structure and organization in a divergent lineage of chrondrichthyan fishes.** *Immunogenetics* 1998, **47**:234–245.

Yoder, J.A., and Litman, G.W.: **Immune-type diversity in the absence of somatic rearrangement.** *Curr. Top. Microbiol. Immunol.* 2000, **248**:271–282.

16-13 Both α:β and γ:δ T-cell receptors are present in cartilaginous fish.

Rast, J.P., and Litman, G.W.: **T-cell receptor gene homologs are present in the most primitive jawed vertebrates.** *Proc. Natl. Acad. Sci. USA* 1994, **91**:9248–9252.

Rast, J.P., Anderson, M.K., Strong, S.J., Luer, C., Litman, R.T., and Litman, G.W. **α, β, γ, and δ T-cell antigen receptor genes arose early in vertebrate phylogeny.** *Immunity* 1997, **6**:1–11.

16-14 MHC class I and class II molecules are also first found in the cartilaginous fishes.

Hashimoto, K., Okamura, K., Yamaguchi, H., Ototake, M., Nakanishi, T., and Kurosawa, Y.: **Conservation and diversification of MHC class I and its related molecules in vertebrates.** *Immunol. Rev.* 1999, **167**:81–100.

Kurosawa, Y., and Hashimoto, K.: **How did the primordial T cell receptor and MHC molecules function initially?** *Immunol. Cell Biol.* 1997, **75**:193–196.

Ohta, Y., Okamura, K., McKinney, E.C., Bartl, S., Hashimoto, K., and Flajnik, M.F.: **Primitive synteny of vertebrate major histocompatibility complex class I and class II genes.** *Proc. Natl. Acad. Sci. USA* 2000, **97**:4712–4717.

Okamura, K., Ototake, M., Nakanishi, T., Kurosawa, Y., and Hashimoto, K.: **The most primitive vertebrates with jaws possess highly polymorphic MHC class I genes comparable to those of humans.** *Immunity* 1997, **7**:777–790.

Immunologists' Toolbox

APPENDIX I

Immunization.

Natural adaptive immune responses are normally directed at antigens borne by pathogenic microorganisms. The immune system can also be induced to respond to simple nonliving antigens, and experimental immunologists have focused on the responses to these simple antigens in developing our understanding of the immune response. The deliberate induction of an immune response is known as **immunization**. Experimental immunizations are routinely carried out by injecting the test antigen into the animal or human subject. The route, dose, and form in which antigen is administered can profoundly affect whether a response occurs and the type of response that is produced, and are considered in Sections A-1–A-4. The induction of protective immune responses against common microbial pathogens in humans is often called vaccination, although this term is correctly only applied to the induction of immune responses against smallpox by immunizing with the cross-reactive cowpox virus, vaccinia.

To determine whether an immune response has occurred and to follow its course, the immunized individual is monitored for the appearance of immune reactants directed at the specific antigen. Immune responses to most antigens elicit the production of both specific antibodies and specific effector T cells. Monitoring the antibody response usually involves the analysis of relatively crude preparations of **antiserum** (plural: **antisera**). The **serum** is the fluid phase of clotted blood, which, if taken from an immunized individual, is called antiserum because it contains specific antibodies against the immunizing antigen as well as other soluble serum proteins. To study immune responses mediated by T cells, blood lymphocytes or cells from lymphoid organs such as the spleen are tested; T-cell responses are more commonly studied in experimental animals than in humans.

Any substance that can elicit an immune response is said to be **immunogenic** and is called an **immunogen**. There is a clear operational distinction between an immunogen and an antigen. An antigen is defined as any substance that can bind to a specific antibody. All antigens therefore have the potential to elicit specific antibodies, but some need to be attached to an immunogen in order to do so. This means that although all immunogens are antigens, not all antigens are immunogenic. The antigens used most frequently in experimental immunology are proteins, and antibodies to proteins are of enormous utility in experimental biology and medicine. Purified proteins are, however, not always highly immunogenic and to provoke an immune response have to be administered with an adjuvant (see Section A-4). Carbohydrates, nucleic acids, and other types of molecules are all potential antigens, but will often only induce an immune response if attached to a protein carrier. Thus, the immunogenicity of protein antigens determines the outcome of virtually every immune response.

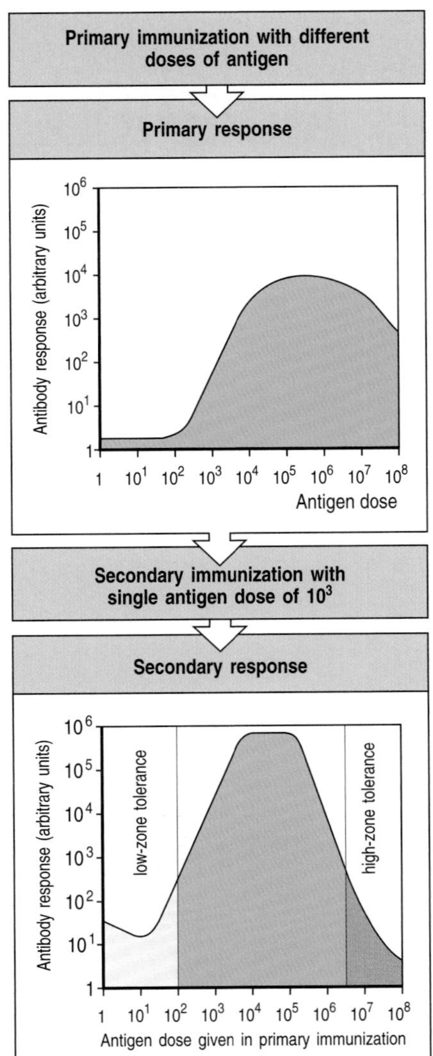

Fig. A.1 The dose of antigen used in an initial immunization affects the primary and secondary antibody response. The typical antigen dose–response curve shown here illustrates the influence of dose on both a primary antibody response (amounts of antibody produced expressed in arbitrary units) and the effect of the dose used for priming on a secondary antibody response elicited by a dose of antigen of 10^3 arbitrary mass units. Very low doses of antigen do not cause an immune response at all. Slightly higher doses appear to inhibit specific antibody production, an effect known as low-zone tolerance. Above these doses there is a steady increase in the response with antigen dose to reach a broad optimum. Very high doses of antigen also inhibit immune responsiveness to a subsequent challenge, a phenomenon known as high-zone tolerance.

Antisera generated by immunization with even the simplest antigen will contain many different antibody molecules that bind to the immunogen in slightly different ways. Some of the antibodies in an antiserum are cross-reactive. A **cross-reaction** is defined as the binding of an antibody to an antigen other than the immunogen; most antibodies cross-react with closely related antigens but, on occasion, some bind antigens having no clear relationship to the immunogen. These cross-reacting antibodies can create problems when the antiserum is used to detect a specific antigen. They can be removed from an antiserum by **absorption** with the cross-reactive antigen, leaving behind the antibodies that bind only to the immunogen. Absorption can be performed by affinity chromatography using immobilized antigen, a technique that is also used for purification of antibodies or antigens (see Section A-5). Most problems of cross-reactivity can be avoided, however, by making monoclonal antibodies (see Section A-12).

Although almost any structure can be recognized by antibody as an antigen, usually only proteins elicit fully developed adaptive immune responses. This is because proteins have the ability to engage T cells, which contribute to inducing most antibody responses and are required for immunological memory. Proteins engage T cells because the T cells recognize antigens as peptide fragments of proteins bound to major histocompatibility complex (MHC) molecules. An adaptive immune response that includes immunological memory can be induced by nonpeptide antigens only when they are attached to a protein carrier that can engage the necessary T cells (see Section 9-3 and Fig. 9.4).

Immunological memory is produced as a result of the initial or **primary immunization**, which evokes the **primary immune response**. This is also known as **priming**, as the animal or person is now 'primed' like a pump to mount a more potent response to subsequent challenges with the same antigen. The response to each immunization is increasingly intense, so that **secondary**, **tertiary**, and subsequent responses are of increasing magnitude (Fig. A.1). Repetitive challenge with antigen to achieve a heightened state of immunity is known as **hyperimmunization**.

Certain properties of a protein that favor the priming of an adaptive immune response have been defined by studying antibody responses to simple natural proteins like hen egg-white lysozyme and to synthetic polypeptide antigens (Fig. A.2). The larger and more complex a protein, and the more distant its relationship to self proteins, the more likely it is to elicit a response. This is because such responses depend on the proteins being degraded into peptides that can bind to MHC molecules, and on the subsequent recognition of these peptide:MHC complexes by T cells. The larger and more distinct the protein antigen, the more likely it is to contain such peptides. Particulate or aggregated antigens are more immunogenic because they are taken up more efficiently by the specialized antigen-presenting cells responsible for initiating a response. Indeed small soluble proteins are unable to induce a response unless they are made to aggregate in some way. Many vaccines, for example, use aggregated protein antigens to potentiate the immune response.

A-1 Haptens.

Small organic molecules of simple structure, such as phenyl arsonates and nitrophenyls, do not provoke antibodies when injected by themselves. However, antibodies can be raised against them if the molecule is attached covalently, by simple chemical reactions, to a protein carrier. Such small molecules were termed **haptens** (from the Greek *haptein*, to fasten) by the immunologist Karl Landsteiner, who first studied them in the early 1900s. He found that animals immunized with a hapten–carrier conjugate produced three distinct sets of antibodies (Fig. A.3). One set comprised hapten-specific

Factors that influence the immunogenicity of proteins		
Parameter	**Increased immunogenicity**	**Decreased immunogenicity**
Size	Large	Small (MW<2500)
Dose	Intermediate	High or low
Route	Subcutaneous > intraperitoneal > intravenous or intragastric	
Composition	Complex	Simple
Form	Particulate	Soluble
	Denatured	Native
Similarity to self protein	Multiple differences	Few differences
Adjuvants	Slow release	Rapid release
	Bacteria	No bacteria
Interaction with host MHC	Effective	Ineffective

Fig. A.2 Intrinsic properties and extrinsic factors that affect the immunogenicity of proteins.

antibodies that reacted with the same hapten on any carrier, as well as with free hapten. The second set of antibodies was specific for the carrier protein, as shown by their ability to bind both the hapten-modified and unmodified carrier protein. Finally, some antibodies reacted only with the specific conjugate of hapten and carrier used for immunization. Landsteiner studied mainly the antibody response to the hapten, as these small molecules could be synthesized in many closely related forms. He observed that antibodies raised against a particular hapten bind that hapten but, in general, fail to bind even very closely related chemical structures. The binding of haptens by anti-hapten antibodies has played an important part in defining the precision of antigen binding by antibody molecules. Anti-hapten antibodies are also important medically as they mediate allergic reactions to penicillin and other compounds that elicit antibody responses when they attach to self proteins (see Section 13-11).

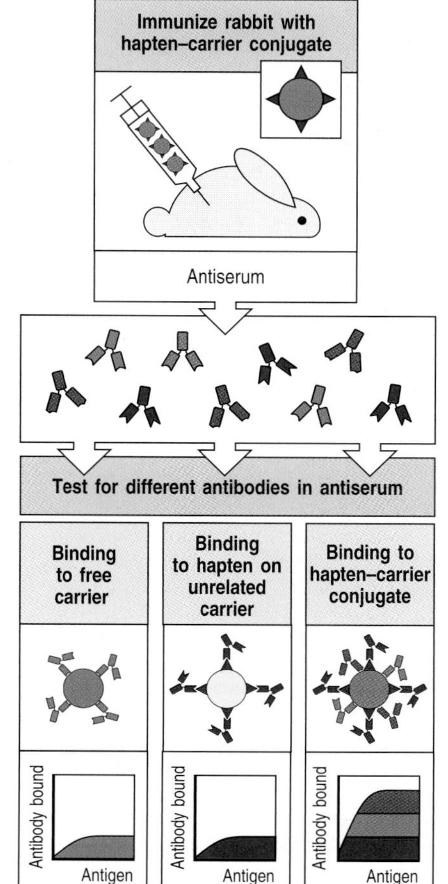

Fig. A.3 Antibodies can be elicited by small chemical groups called haptens only when the hapten is linked to an immunogenic protein carrier. Three types of antibodies are produced. One set (blue) binds the carrier protein alone and is called carrier-specific. One set (red) binds to the hapten on any carrier or to free hapten in solution and is called hapten-specific. One set (purple) only binds the specific conjugate of hapten and carrier used for immunization, apparently binding to sites at which the hapten joins the carrier, and is called conjugate-specific. The amount of antibody of each type in this serum is shown schematically in the graphs at the bottom; note that the original antigen binds more antibody than the sum of anti-hapten and anti-carrier antibodies owing to the additional binding of conjugate-specific antibody.

A-2 Routes of immunization.

The route by which antigen is administered affects both the magnitude and the type of response obtained. The most common routes by which antigen is introduced experimentally or as a vaccine into the body are injection into tissue by **subcutaneous (s.c.)** injection between the epidermis and dermal layers, or by **intradermal (i.d.)** injection, or **intramuscular (i.m.)** injection; by **intravenous (i.v.)** injection or transfusion directly into the bloodstream; into the gastrointestinal tract by oral administration; into the respiratory tract by **intranasal (i.n.)** administration or inhalation.

Antigens injected subcutaneously generally elicit the strongest responses, most probably because the antigen is taken up by Langerhans' cells and efficiently presented in local lymph nodes, and so this is the method most commonly used when the object of the experiment is to elicit specific antibodies or T cells against a given antigen. Antigens injected or transfused directly into the bloodstream tend to induce immune unresponsiveness or tolerance unless they bind to host cells or are in the form of aggregates that are readily taken up by antigen-presenting cells.

Antigen administration via the gastrointestinal tract is used mostly in the study of allergy. It has distinctive effects, frequently eliciting a local antibody response in the intestinal lamina propria, while producing a systemic state of tolerance that manifests as a diminished response to the same antigen if subsequently administered in immunogenic form elsewhere in the body. This 'split tolerance' may be important in avoiding allergy to antigens in food, as the local response prevents food antigens from entering the body, while the inhibition of systemic immunity helps to prevent the formation of IgE antibodies, which are the cause of such allergies (see Chapter 13).

Introduction of antigen into the respiratory tract is also used mainly in the study of allergy. Protein antigens that enter the body through the respiratory epithelium tend to elicit allergic responses, for reasons that are not clear.

A-3 Effects of antigen dose.

The magnitude of the immune response depends on the dose of immunogen administered. Below a certain threshold dose, most proteins do not elicit any immune response. Above the threshold dose, there is a gradual increase in the response as the dose of antigen is increased, until a broad plateau level is reached, followed by a decline at very high antigen doses (see Fig. A.1). As most infectious agents enter the body in small numbers, immune responses are generally elicited only by pathogens that multiply to a level sufficient to exceed the antigen dose threshold. The broad response optimum allows the system to respond to infectious agents across a wide range of doses. At very high antigen doses the immune response is inhibited, which may be important in maintaining tolerance to abundant self proteins such as plasma proteins. In general, secondary and subsequent immune responses occur at lower antigen doses and achieve higher plateau values, which is a sign of immunological memory. However, under some conditions, very low or very high doses of antigen may induce specific unresponsive states, known respectively as acquired **low-zone** or **high-zone tolerance**.

A-4 Adjuvants.

Most proteins are poorly immunogenic or nonimmunogenic when administered by themselves. Strong adaptive immune responses to protein antigens almost always require that the antigen be injected in a mixture known as an

adjuvant. An adjuvant is any substance that enhances the immunogenicity of substances mixed with it. Adjuvants differ from protein carriers in that they do not form stable linkages with the immunogen. Furthermore, adjuvants are needed primarily for initial immunizations, whereas carriers are required to elicit not only primary but also subsequent responses to haptens. Commonly used adjuvants are listed in Fig. A.4.

Adjuvants can enhance immunogenicity in two different ways. First, adjuvants convert soluble protein antigens into particulate material, which is more readily ingested by antigen-presenting cells such as macrophages. For example, the antigen can be adsorbed on particles of the adjuvant (such as alum), made particulate by emulsification in mineral oils, or incorporated into the colloidal particles of ISCOMs. This enhances immunogenicity somewhat, but such adjuvants are relatively weak unless they also contain bacteria or bacterial products. Such microbial constituents are the second means by which adjuvants enhance immunogenicity, and although their exact contribution to enhancing immunogenicity is unknown, they are clearly the more important component of an adjuvant. Microbial products may signal macrophages or dendritic cells to become more effective antigen-presenting cells (see Chapter 2). One of their effects is to induce the production of inflammatory cytokines and potent local inflammatory responses; this effect is probably intrinsic to their activity in enhancing responses, but precludes their use in humans.

Nevertheless, some human vaccines contain microbial antigens that can also act as effective adjuvants. For example, purified constituents of the bacterium *Bordetella pertussis*, which is the causal agent of whooping cough, are used as both antigen and adjuvant in the triplex DPT (diphtheria, pertussis, tetanus) vaccine against these diseases.

Adjuvants that enhance immune responses

Adjuvant name	Composition	Mechanism of action
Incomplete Freund's adjuvant	Oil-in-water emulsion	Delayed release of antigen; enhanced uptake by macrophages
Complete Freund's adjuvant	Oil-in-water emulsion with dead mycobacteria	Delayed release of antigen; enhanced uptake by macrophages; induction of co-stimulators in macrophages
Freund's adjuvant with MDP	Oil-in-water emulsion with muramyldipeptide (MDP), a constituent of mycobacteria	Similar to complete Freund's adjuvant
Alum (aluminum hydroxide)	Aluminum hydroxide gel	Delayed release of antigen; enhanced macrophage uptake
Alum plus *Bordetella pertussis*	Aluminum hydroxide gel with killed *B. pertussis*	Delayed release of antigen; enhanced uptake by macrophages; induction of co-stimulators
Immune stimulatory complexes (ISCOMs)	Matrix of Quil A containing viral proteins	Delivers antigen to cytosol; allows induction of cytotoxic T cells

Fig. A.4 Common adjuvants and their use. Adjuvants are mixed with the antigen and usually render it particulate, which helps to retain the antigen in the body and promotes uptake by macrophages. Most adjuvants include bacteria or bacterial components that stimulate macrophages, aiding in the induction of the immune response. ISCOMs (immune stimulatory complexes) are small micelles of the detergent Quil A; when viral proteins are placed in these micelles, they apparently fuse with the antigen-presenting cell, allowing the antigen to enter the cytosol. Thus, the antigen-presenting cell can stimulate a response to the viral protein, much as a virus infecting these cells would stimulate an anti-viral response.

The detection, measurement, and characterization of antibodies and their use as research and diagnostic tools.

B cells contribute to adaptive immunity by secreting antibodies, and the response of B cells to an injected immunogen is usually measured by analyzing the specific antibody produced in a **humoral immune response**. This is most conveniently achieved by assaying the antibody that accumulates in the fluid phase of the blood or **plasma**; such antibodies are known as circulating antibodies. Circulating antibody is usually measured by collecting blood, allowing it to clot, and then isolating the serum from the clotted blood. The amount and characteristics of the antibody in the resulting antiserum are then determined using the assays we will describe in Sections A-5–A-11.

The most important characteristics of an antibody response are the specificity, amount, isotype or class, and affinity of the antibodies produced. The **specificity** determines the ability of the antibody to distinguish the immunogen from other antigens. The amount of antibody can be determined in many different ways and is a function of the number of responding B cells, their rate of antibody synthesis, and the persistence of the antibody after production. The persistence of an antibody in the plasma and extracellular fluid bathing the tissues is determined mainly by its isotype or class (see Sections 4-12 and 9-14); each isotype has a different half-life *in vivo*. The isotypic composition of an antibody response also determines the biological functions these antibodies can perform and the sites in which antibody will be found. Finally, the strength of binding of the antibody to its antigen in terms of a single antigen-binding site binding to a monovalent antigen is termed its **affinity** (the total binding strength of a molecule with more than one binding site is called its **avidity**). Binding strength is important, since the higher the affinity of the antibody for its antigen, the less antibody is required to eliminate the antigen, as antibodies with higher affinity will bind at lower antigen concentrations. All these parameters of the humoral immune response help to determine the capacity of that response to protect the host from infection.

Antibody molecules are highly specific for their corresponding antigen, being able to detect one molecule of a protein antigen out of more than 10^8 similar molecules. This makes antibodies both easy to isolate and study, and invaluable as probes of biological processes. Whereas standard chemistry would have great difficulty in distinguishing between two such closely related proteins as human and pig insulin, or two such closely related structures as *ortho*- and *para*-nitrophenyl, antibodies can be made that discriminate between these two structures absolutely. The value of antibodies as molecular probes has stimulated the development of many sensitive and highly specific techniques to measure their presence, to determine their specificity and affinity for a range of antigens, and to ascertain their functional capabilities. Many standard techniques used throughout biology exploit the specificity and stability of antigen binding by antibodies. Comprehensive guides to the conduct of these antibody assays are available in many books on immunological methodology; we will illustrate here only the most important techniques, especially those used in studying the immune response itself.

Some assays for antibody measure the direct binding of the antibody to its antigen. Such assays are based on **primary interactions**. Others determine the amount of antibody present by the changes it induces in the physical state of the antigen, such as the precipitation of soluble antigen or the clumping of antigenic particles; these are called secondary interactions. Both types of assay can be used to measure the amount and specificity of the antibodies produced after immunization, and both can be applied to a wide range of other biological questions.

As assays for antibody were originally conducted with antisera from immune individuals, they are commonly referred to as **serological assays**, and the use of antibodies is often called **serology**. The amount of antibody is usually determined by antigen-binding assays after titration of the antiserum by serial dilution, and the point at which binding falls to 50% of the maximum is usually referred to as the **titer** of an antiserum.

A-5 Affinity chromatography.

Specific antibody can be isolated from an antiserum by **affinity chromatography**, which exploits the specific binding of antibody to antigen held on a solid matrix (Fig. A.5). Antigen is bound covalently to small, chemically reactive beads, which are loaded into a column, and the antiserum is allowed to pass over the beads. The specific antibodies bind, while all the other proteins in the serum, including antibodies to other substances, can be washed away. The specific antibodies are then eluted, typically by lowering the pH to 2.5 or raising it to greater than 11. Antibodies bind stably under physiological conditions of salt concentration, temperature, and pH, but the binding is reversible as the bonds are noncovalent. Affinity chromatography can also be used to purify antigens from complex mixtures by using beads coated with specific antibody. The technique is known as affinity chromatography because it separates molecules on the basis of their affinity for one another.

A-6 Radioimmunoassay (RIA), enzyme-linked immunosorbent assay (ELISA), and competitive inhibition assay.

Radioimmunoassay (RIA) and **enzyme-linked immunosorbent assay (ELISA)** are direct binding assays for antibody (or antigen) and both work on the same principle, but the means of detecting specific binding is different. Radioimmunoassays are commonly used to measure the levels of hormones in blood and tissue fluids, while ELISA assays are frequently used in viral

Fig. A.5 Affinity chromatography uses antigen–antibody binding to purify antigens or antibodies. To purify a specific antigen from a complex mixture of molecules, a monoclonal antibody is attached to an insoluble matrix, such as chromatography beads, and the mixture of molecules is passed over the matrix. The specific antibody binds the antigen of interest; other molecules are washed away. Specific antigen is then eluted by altering the pH, which can usually disrupt antibody–antigen bonds. Antibodies can be purified in the same way on beads coupled to antigen (not shown).

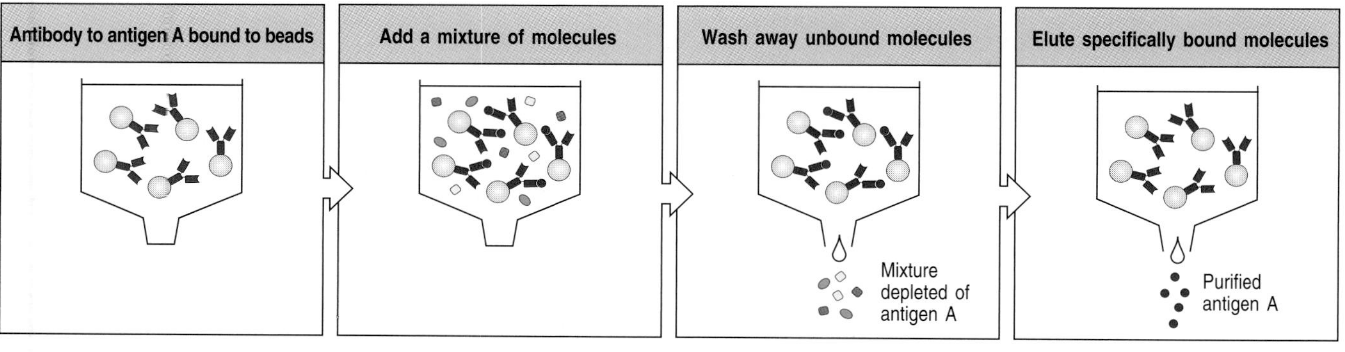
Antibody to antigen A bound to beads
Add a mixture of molecules
Wash away unbound molecules
Elute specifically bound molecules
Mixture depleted of antigen A
Purified antigen A

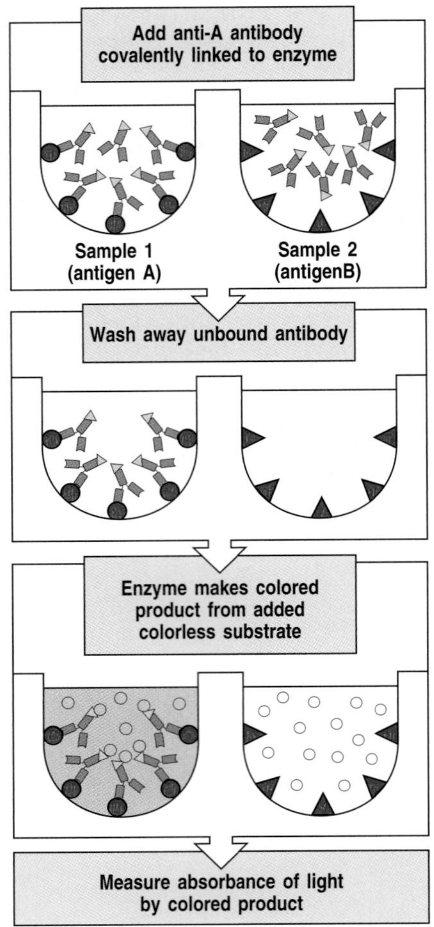

Fig. A.6 The principle of the enzyme-linked immunosorbent assay (ELISA). To detect antigen A, purified antibody specific for antigen A is linked chemically to an enzyme. The samples to be tested are coated onto the surface of plastic wells to which they bind nonspecifically; residual sticky sites on the plastic are blocked by adding irrelevant proteins (not shown). The labeled antibody is then added to the wells under conditions where nonspecific binding is prevented, so that only binding to antigen A causes the labeled antibody to be retained on the surface. Unbound labeled antibody is removed from all wells by washing, and bound antibody is detected by an enzyme-dependent color-change reaction. This assay allows arrays of wells known as microtiter plates to be read in fiberoptic multichannel spectrometers, greatly speeding the assay. Modifications of this basic assay allow antibody or antigen in unknown samples to be measured as shown in Figs A.7 and A.30 (see also Section A-10).

diagnostics, for example in detecting cases of HIV infection. For both these methods one needs a pure preparation of a known antigen or antibody, or both, in order to standardize the assay. We will describe the assay with a sample of pure antibody, which is the more usual case, but the principle is similar if pure antigen is used instead. In RIA for an antigen, pure antibody against that antigen is radioactively labeled, usually with ^{125}I; for the ELISA, an enzyme is linked chemically to the antibody. The unlabeled component, which in this case would be antigen, is attached to a solid support, such as the wells of a plastic multiwell plate, which will adsorb a certain amount of any protein.

The labeled antibody is allowed to bind to the unlabeled antigen, under conditions where nonspecific adsorption is blocked, and any unbound antibody and other proteins are washed away. Antibody binding in RIA is measured directly in terms of the amount of radioactivity retained by the coated wells, whereas in ELISA, binding is detected by a reaction that converts a colorless substrate into a colored reaction product (Fig. A.6). The color change can be read directly in the reaction tray, making data collection very easy, and ELISA also avoids the hazards of radioactivity. This makes ELISA the preferred method for most direct-binding assays. Labeled anti-immunoglobulin antibodies (see Section A-10) can also be used in RIA or ELISA to detect binding of unlabeled antibody to unlabeled antigen-coated plates. In this case, the labeled anti-immunoglobulin antibody is used in what is termed a 'second layer.' The use of such a second layer also amplifies the signal, as at least two molecules of the labeled anti-immunoglobulin antibody are able to bind to each unlabeled antibody. RIA and ELISA can also be carried out with unlabeled antibody stuck to the plates and labeled antigen added.

A modification of ELISA known as a **capture** or **sandwich ELISA** (or more generally as an **antigen-capture assay**) can be used to detect secreted products such as cytokines. Rather than the antigen being directly attached to a plastic plate, antigen-specific antibodies are bound to the plate. These are able to bind antigen with high affinity, and thus concentrate it on the surface of the plate, even with antigens that are present in very low concentrations in the initial mixture. A separate labeled antibody that recognizes a different epitope to the immobilized first antibody is then used to detect the bound antigen.

These assays illustrate two crucial aspects of all serological assays. First, at least one of the reagents must be available in a pure, detectable form in order to obtain quantitative information. Second, there must be a means of separating the bound fraction of the labeled reagent from the unbound, free fraction so that the percentage of specific binding can be determined. Normally, this separation is achieved by having the unlabeled partner trapped on a solid support. Labeled molecules that do not bind can then be washed away, leaving just the labeled partner that has bound. In Fig. A.6, the unlabeled antigen is attached to the well and the labeled antibody is trapped by binding to it. The separation of bound from free is an essential step in every assay that uses antibodies.

RIA and ELISA do not allow one to measure directly the amount of antigen or antibody in a sample of unknown composition, as both depend on the binding of a pure labeled antigen or antibody. There are various ways around this problem, one of which is to use a **competitive inhibition assay**, as shown in Fig. A.7. In this type of assay, the presence and amount of a particular antigen in an unknown sample is determined by its ability to compete with a labeled reference antigen for binding to an antibody attached to a plastic well. A standard curve is first constructed by adding varying amounts of a known, unlabeled standard preparation; the assay can then measure the amount of antigen in unknown samples by comparison with the standard.

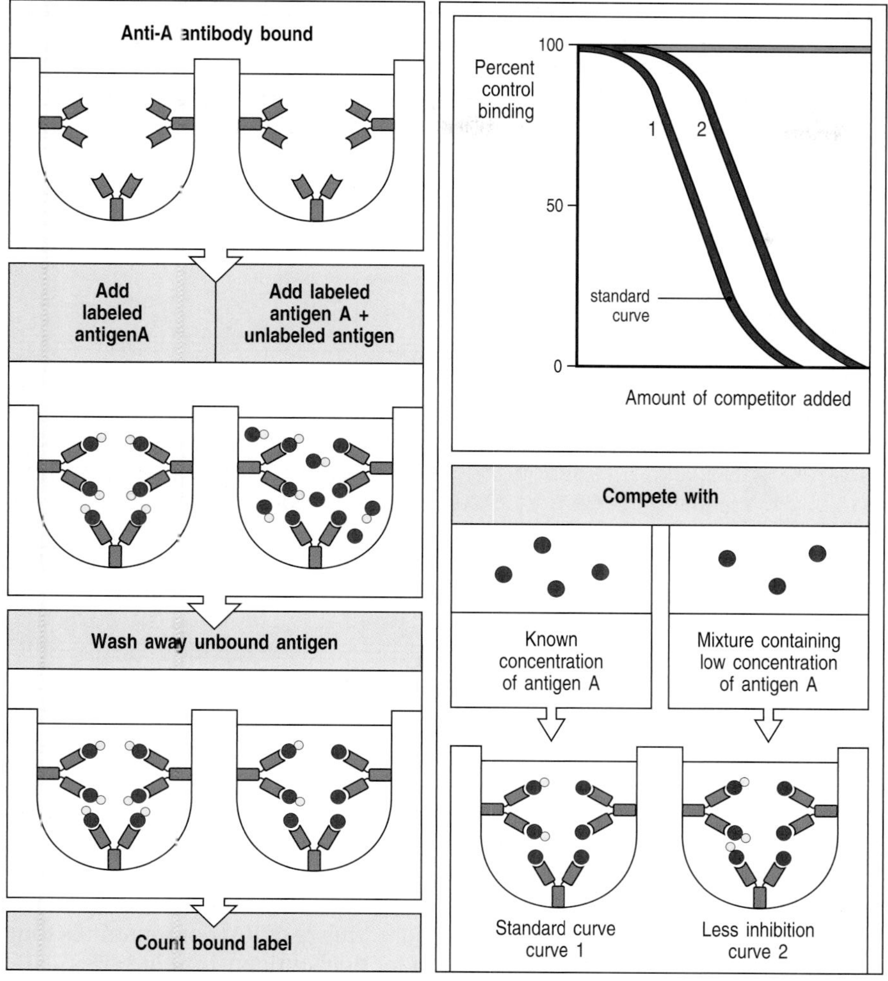

Fig. A.7 Competitive inhibition assay for antigen in unknown samples. A fixed amount of unlabeled antibody is attached to a set of wells, and a standard reference preparation of a labeled antigen is bound to it. Unlabeled standard or test samples are then added in various amounts and the displacement of labeled antigen is measured, generating characteristic inhibition curves. A standard curve is obtained by using known amounts of unlabeled antigen identical to that used as the labeled species, and comparison with this curve allows the amount of antigen in unknown samples to be calculated. The green line on the graph represents a sample lacking any substance that reacts with anti-A antibodies.

The competitive binding assay can also be used for measuring antibody in a sample of unknown composition by attaching the appropriate antigen to the plate and measuring the ability of the test sample to inhibit the binding of a labeled specific antibody.

A-7 Hemagglutination and blood typing.

The direct measurement of antibody binding to antigen is used in most quantitative serological assays. However, some important assays are based on the ability of antibody binding to alter the physical state of the antigen it binds to. These secondary interactions can be detected in a variety of ways. For instance, when the antigen is displayed on the surface of a large particle such as a bacterium, antibodies can cause the bacteria to clump or **agglutinate**. The same principle applies to the reactions used in blood typing, only here the target antigens are on the surface of red blood cells and the clumping reaction caused by antibodies against them is called **hemagglutination** (from the Greek *haima*, blood).

Hemagglutination is used to determine the **ABO blood group** of blood donors and transfusion recipients. Clumping or agglutination is induced by antibodies or agglutinins called anti-A or anti-B that bind to the A or B blood-group substances, respectively (Fig. A.8). These blood-group antigens are arrayed in

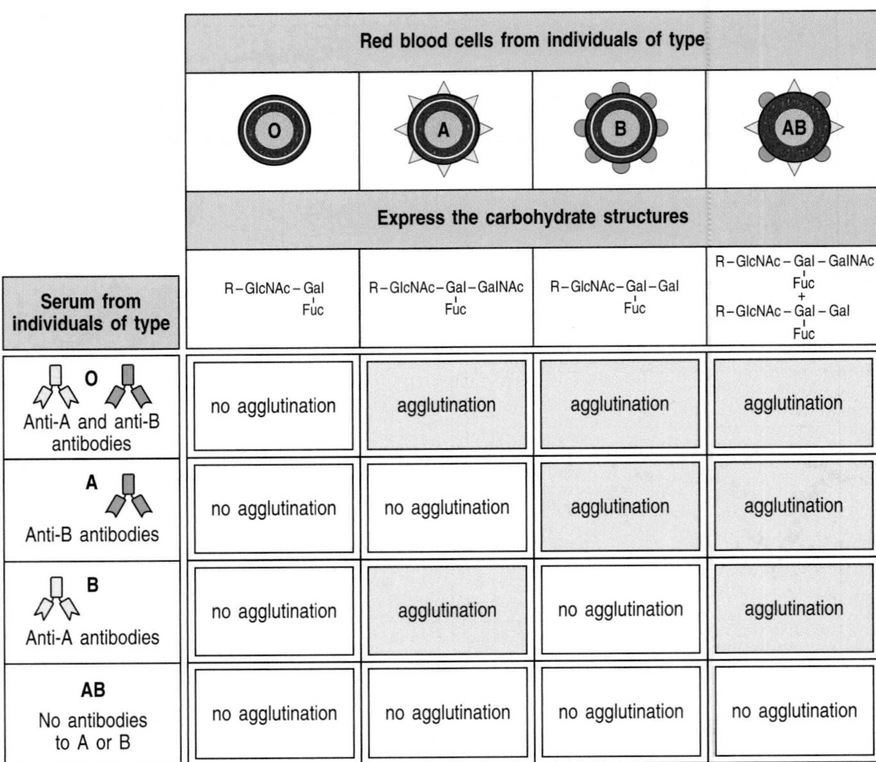

Serum from individuals of type	Red blood cells from individuals of type O	A	B	AB
Express the carbohydrate structures	R–GlcNAc–Gal, Fuc	R–GlcNAc–Gal–GalNAc, Fuc	R–GlcNAc–Gal–Gal, Fuc	R–GlcNAc–Gal–GalNAc, Fuc + R–GlcNAc–Gal–Gal, Fuc
O: Anti-A and anti-B antibodies	no agglutination	agglutination	agglutination	agglutination
A: Anti-B antibodies	no agglutination	no agglutination	agglutination	agglutination
B: Anti-A antibodies	no agglutination	agglutination	no agglutination	agglutination
AB: No antibodies to A or B	no agglutination	no agglutination	no agglutination	no agglutination

Fig. A.8 Hemagglutination is used to type blood groups and match compatible donors and recipients for blood transfusion. Common gut bacteria bear antigens that are similar or identical to blood-group antigens, and these stimulate the formation of antibodies to these antigens in individuals who do not bear the corresponding antigen on their own red blood cells (left column); thus, type O individuals, who lack A and B, have both anti-A and anti-B antibodies, while type AB individuals have neither. The pattern of agglutination of the red blood cells of a transfusion donor or recipient with anti-A and anti-B antibodies reveals the individual's ABO blood group. Before transfusion, the serum of the recipient is also tested for antibodies that agglutinate the red blood cells of the donor, and vice versa, a procedure called a cross-match, which may detect potentially harmful antibodies to other blood groups that are not part of the ABO system.

many copies on the surface of the red blood cell, allowing the cells to agglutinate when cross-linked by antibodies. Because hemagglutination involves the cross-linking of blood cells by the simultaneous binding of antibody molecules to identical antigens on different cells, this reaction demonstrates that each antibody molecule has at least two identical antigen-binding sites.

A-8 Precipitin reaction.

When sufficient quantities of antibody are mixed with soluble macromolecular antigens, a visible precipitate consisting of large aggregates of antigen cross-linked by antibody molecules can form. The amount of precipitate depends on the amounts of antigen and antibody, and on the ratio between them (Fig. A.9). This **precipitin reaction** provided the first quantitative assay for antibody, but is now seldom used in immunology. However, it is important to understand the interaction of antigen with antibody that leads to this reaction, as the production of **antigen:antibody complexes**, also known as **immune complexes**, *in vivo* occurs in almost all immune responses and occasionally can cause significant pathology (see Chapters 13 and 14).

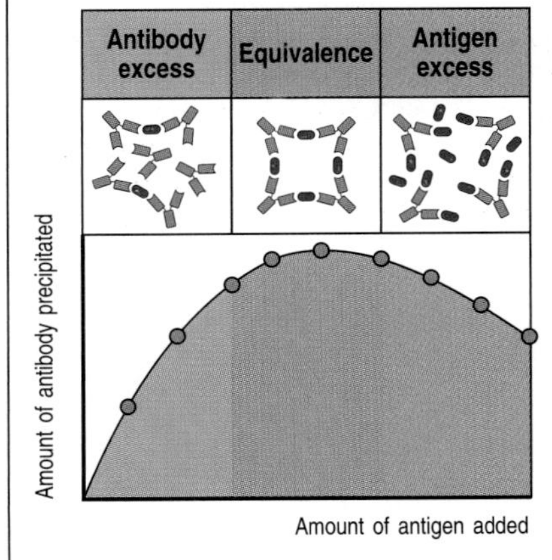

Fig. A.9 Antibody can precipitate soluble antigen. Analysis of the precipitate can generate a precipitin curve. Different amounts of antigen are added to a fixed amount of antibody, and precipitates form by antibody cross-linking of antigen molecules. The precipitate is recovered and the amount of precipitated antibody measured; the supernatant is tested for residual antigen or antibody. This defines zones of antibody excess, equivalence, and antigen excess. At equivalence, the largest antigen:antibody complexes form. In the zone of antigen excess, some of the immune complexes are too small to precipitate. These soluble immune complexes can cause pathological damage to small blood vessels when they form *in vivo* (see Chapter 14).

In the precipitin reaction, various amounts of soluble antigen are added to a fixed amount of serum containing antibody. As the amount of antigen added increases, the amount of precipitate generated also increases up to a maximum and then declines (see Fig. A.9). When small amounts of antigen are added, antigen:antibody complexes are formed under conditions of antibody excess so that each molecule of antigen is bound by antibody and cross-linked to other molecules of antigen. When large amounts of antigen are added, only small antigen:antibody complexes can form and these are often soluble in this zone of antigen excess. Between these two zones, all of the antigen and antibody is found in the precipitate, generating a zone of equivalence. At equivalence, very large lattices of antigen and antibody are formed by cross-linking. While all antigen:antibody complexes can potentially produce disease, the small, soluble immune complexes formed in the zone of antigen excess may persist and cause pathology *in vivo*.

The precipitin reaction is affected by the number of binding sites that each antibody has for antigen, and by the maximum number of antibodies that can be bound by an antigen molecule or particle at any one time. These quantities are defined as the **valence** of the antibody and the valence of the antigen: the valence of both the antibodies and the antigen must be two or greater before any precipitation can occur. The valence of an antibody depends on its structural class (see Section 4-16).

Antigen will be precipitated only if it has several antibody-binding sites. This condition is usually satisfied in macromolecular antigens, which have a complex surface with binding sites for several different antibodies. The site on an antigen to which each distinct antibody molecule binds is called an **antigenic determinant** or an **epitope**. Steric considerations limit the number of distinct antibody molecules that can bind to a single antigen molecule at any one time however, because antibody molecules binding to epitopes that partially overlap will compete for binding. For this reason, the valence of an antigen is almost always less than the number of epitopes on the antigen (Fig. A.10).

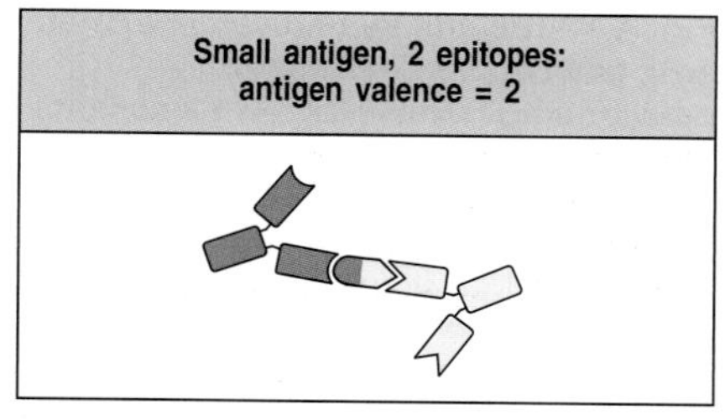

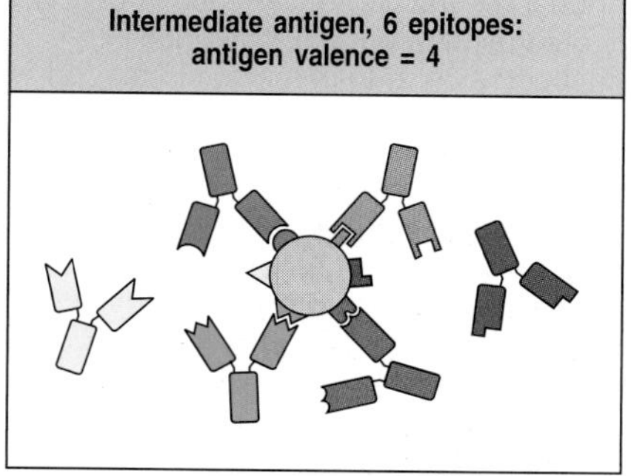

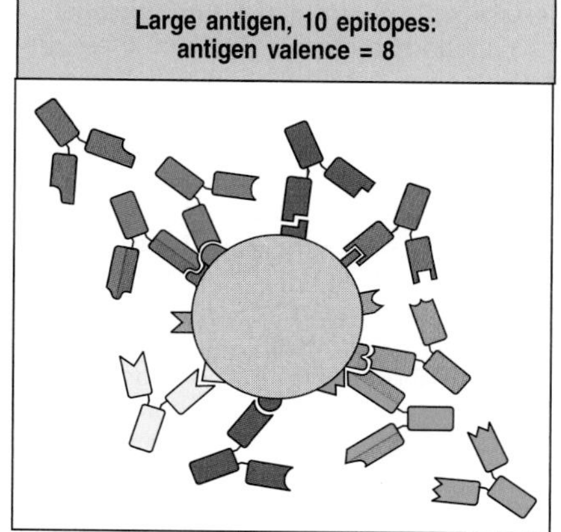

Fig. A.10 Different antibodies bind to distinct epitopes on an antigen molecule. The surface of an antigen possesses many potential antigenic determinants or epitopes, distinct sites to which an antibody can bind. The number of antibody molecules that can bind to a molecule of antigen at one time defines the antigen's valence. Steric considerations can limit the number of different antibodies that bind to the surface of an antigen at any one time (center and bottom panels) so that the number of epitopes on an antigen is always greater than or equal to its valence.

A-9 Equilibrium dialysis: measurement of antibody affinity and avidity.

The affinity of an antibody is the strength of binding of a monovalent ligand to a single antigen-binding site on the antibody. The affinity of an antibody that binds small antigens, such as haptens, that can diffuse freely across a dialysis membrane can be determined directly by the technique of **equilibrium dialysis**. A known amount of antibody, whose molecules are too large to cross a dialysis membrane, is placed in a dialysis bag and offered various amounts of antigen. Molecules of antigen that bind to the antibody are no longer free to diffuse across the dialysis membrane, so only the unbound molecules of antigen equilibrate across it. By measuring the concentration of antigen inside the bag and in the surrounding fluid, one can determine the amount of the antigen that is bound as well as the amount that is free when equilibrium has been achieved. Given that the amount of antibody present is known, the affinity of the antibody and the number of specific binding sites for the antigen per molecule of antibody can be determined from this information. The data are usually analyzed using **Scatchard analysis** (Fig. A.11); such analyses were used to demonstrate that a molecule of IgG antibody has two identical antigen-binding sites.

Whereas affinity measures the strength of binding of an antigenic determinant to a single antigen-binding site, an antibody reacting with an antigen that has multiple identical epitopes or with the surface of a pathogen will often bind the same molecule or particle with both of its antigen-binding sites. This increases the apparent strength of binding, since both binding sites must release at the same time in order for the two molecules to dissociate.

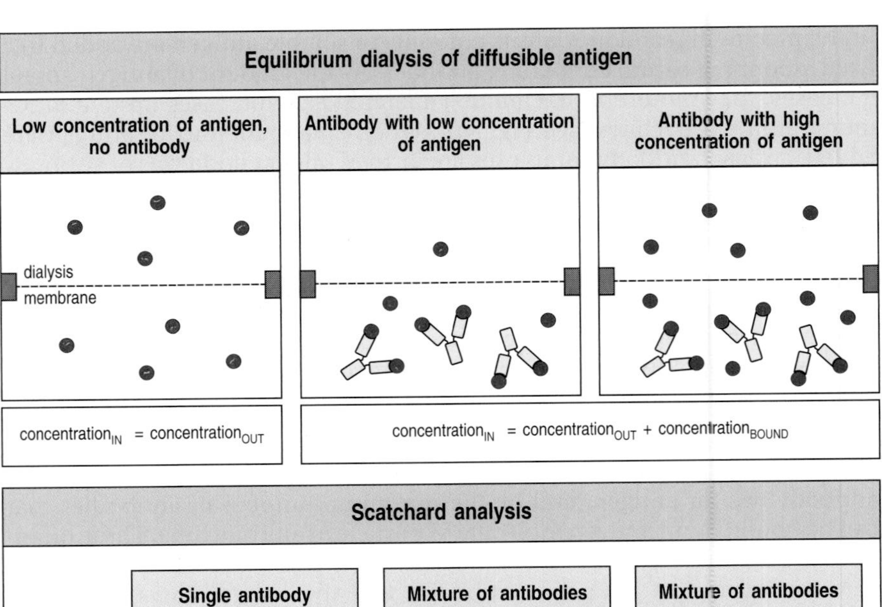

Fig. A.11 The affinity and valence of an antibody can be determined by equilibrium dialysis. A known amount of antibody is placed in the bottom half of a dialysis chamber and exposed to different amounts of a diffusible monovalent antigen, such as a hapten. At equilibrium, the concentration of free antigen will be the same on each side of the membrane, so that at each concentration of antigen added, the fraction of the antigen bound is determined from the difference in concentration of total antigen in the top and bottom chambers. This information can be transformed into a Scatchard plot as shown here. In Scatchard analysis, the ratio r/c (where r = moles of antigen bound per mole of antibody and c = molar concentration of free antigen) is plotted against r. The number of binding sites per antibody molecule can be determined from the value of r at infinite free-antigen concentration, where r/free = 0, in other words at the x-axis intercept. The analysis of a monoclonal IgG antibody molecule, in which there are two identical antigen-binding sites per molecule, is shown in the left panel. The slope of the line is determined by the affinity of the antibody molecule for its antigen; if all the antibody molecules in a preparation are identical, as for this monoclonal antibody, then a straight line is obtained whose slope is equal to $-K_a$, where K_a is the association (or affinity) constant and the dissociation constant $K_d = 1/K_a$. However, antiserum raised even against a simple antigenic determinant such as a hapten contains a heterogeneous population of antibody molecules (see Section A-1). Each antibody molecule would, if isolated, make up part of the total and give a straight line whose x-axis intercept is less than two, as this antibody molecule contains only a fraction of the total binding sites in the population (middle panel). As a mixture, they give curved lines with an x-axis intercept of two for which an average affinity ($\overline{K}_a$) can be determined from the slope of this line at a concentration of antigen where 50% of the sites are bound, or at x = 1 (right panel). The association constant determines the equilibrium state of the reaction Ag + Ab = Ag:Ab, where antigen = Ag and antibody = Ab, and K_a = [Ag:Ab]/[Ag][Ab]. This constant reflects the 'on' and 'off' rates for antigen binding to the antibody; with small antigens such as haptens, binding is usually as rapid as diffusion allows, whereas differences in 'off' rates determine the affinity constant. However, with larger antigens the 'on' rate may also vary as the interaction becomes more complex.

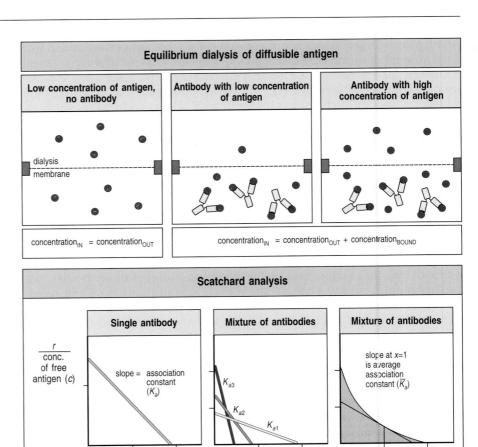

This is often referred to as **cooperativity** in binding, but it should not be confused with the cooperative binding found in a protein such as hemoglobin, in which binding of ligand at one site enhances the affinity of a second binding site for its ligand. The overall strength of binding of an antibody molecule to an antigen or particle is called its avidity (Fig. A.12). For IgG antibodies, bivalent binding can significantly increase avidity; in IgM antibodies, which have ten identical antigen-binding sites, the affinity of each site for a monovalent antigen is usually quite low, but the avidity of binding of the whole antibody to a surface such as a bacterium that displays multiple identical epitopes can be very high.

A-10 Anti-immunoglobulin antibodies.

A general approach to the detection of bound antibody that avoids the need to label each preparation of antibody molecules is to detect bound, unlabeled antibody with a labeled antibody specific for immunoglobulins themselves. Immunoglobulins, like other proteins, are immunogenic when used to immunize individuals of another species. The majority of **anti-immunoglobulin antibodies** raised in this way recognize conserved features shared by all immunoglobulin molecules of the immunizing species. Others can be specific for immunoglobulin chains, heavy or light chains, for example, or for individual isotypes. Antibodies raised by immunization of goats with mouse IgG are commonly used in experimental immunology. Such goat anti-mouse IgG antibodies can be purified using affinity chromatography, then labeled and used as a general probe for bound IgG antibodies. Anti-immunoglobulin antisera have found many uses in clinical medicine and biological research since their introduction. Fluorescently labeled anti-immunoglobulin antibodies are now widely used both in immunology and other areas of biology as secondary reagents for detecting specific antibodies bound, for example, to cell structures (see Sections A-14 and A-16). Labeled anti-immunoglobulin antibodies can also be used in radioimmunoassay or ELISA (see Section A-6) to detect binding of unlabeled antibody to antigen-coated plates.

When an immunoglobulin is used as an antigen to immunize a different species of animal, it will be treated like any other foreign protein and will elicit an antibody response. Anti-immunoglobulin antibodies can be made that recognize the amino acids that characterize the isotype of the injected antibody. Such **anti-isotypic antibodies** recognize all immunoglobulins of the same isotype in all members of the species from which the injected antibody came.

It is also possible to raise antibodies that recognize differences in immunoglobulins from members of the same species that are due to the presence of multiple alleles of the individual C genes in the population (genetic polymorphism). Such allelic variants are called **allotypes**. In contrast to anti-isotypic antibodies, anti-allotypic antibodies will recognize immunoglobulin of a particular isotype only in some members of a species. Finally, as individual antibodies differ in their variable regions, one can raise antibodies against unique features of the antigen-binding site, which are called **idiotypes**.

A schematic picture of the differences between idiotypes, allotypes, and isotypes is given in Fig. A.13. Historically, the main features of immunoglobulins were defined by using isotypic and allotypic genetic markers identified by antisera raised in different species or in genetically distinct members of the same species (see Section A-10). The independent segregation of allotypic and isotypic markers revealed the existence of separate heavy-chain, κ, and λ genes. Such anti-idiotypic, allotypic, and isotypic antibodies are still enormously useful in detecting antibodies and B cells in scientific experimentation and medical diagnostics.

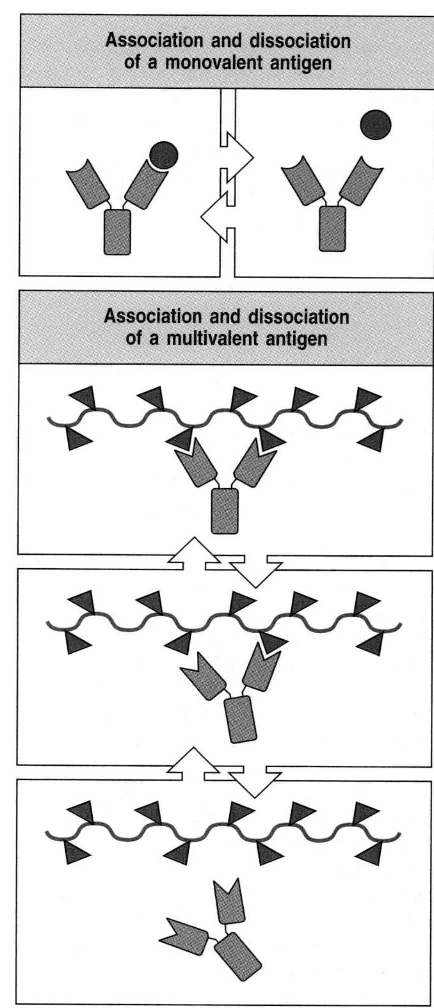

Fig. A.12 The avidity of an antibody is its strength of binding to intact antigen. When an IgG antibody binds a ligand with multiple identical epitopes, both binding sites can bind the same molecule or particle. The overall strength of binding, called avidity, is greater than the affinity, the strength of binding of a single site, since both binding sites must dissociate at the same time for the antibody to release the antigen. This property is very important in the binding of antibody to bacteria, which usually have multiple identical epitopes on their surfaces.

Antibodies specific for individual immunoglobulin isotypes can be produced by immunizing an animal of a different species with a pure preparation of one isotype and then removing those antibodies that cross-react with immunoglobulins of other isotypes by using affinity chromatography (see Section A-5). Anti-isotype antibodies can be used to measure how much antibody of a particular isotype in an antiserum reacts with a given antigen. This reaction is particularly important for detecting small amounts of specific antibodies of the IgE isotype, which are responsible for most allergies. The presence in an individual's serum of IgE binding to an antigen correlates with allergic reactions to that antigen.

An alternative approach to detecting bound antibodies exploits bacterial proteins that bind to immunoglobulins with high affinity and specificity. One of these, **Protein A** from the bacterium *Staphylococcus aureus*, has been exploited widely in immunology for the affinity purification of immuno-globulin and for the detection of bound antibody. The use of standard second reagents such as labeled anti-immunoglobulin antibodies or Protein A to detect antibody bound specifically to its antigen allows great savings in reagent labeling costs, and also provides a standard detection system so that results in different assays can be compared directly.

A-11 Coombs tests and the detection of Rhesus incompatibility.

These tests use anti-immunoglobulin antibodies (see Section A-10) to detect the antibodies that cause **hemolytic disease of the newborn**, or **erythroblastosis**

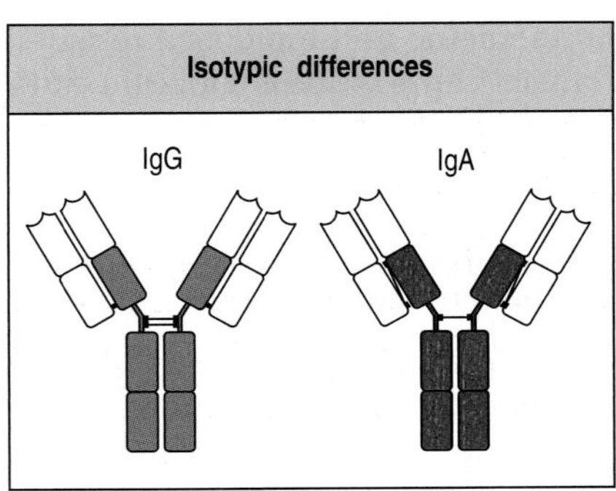

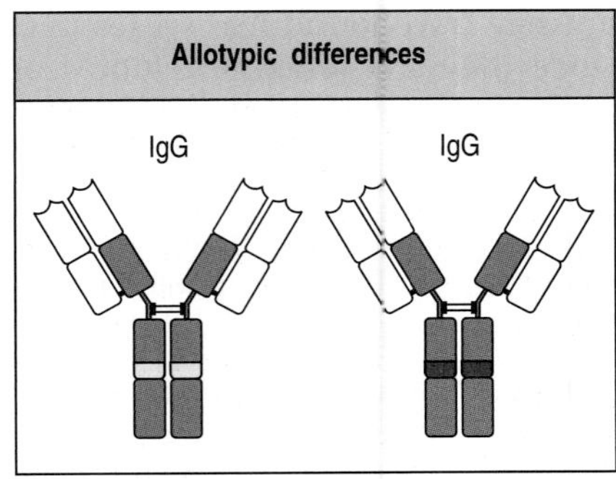

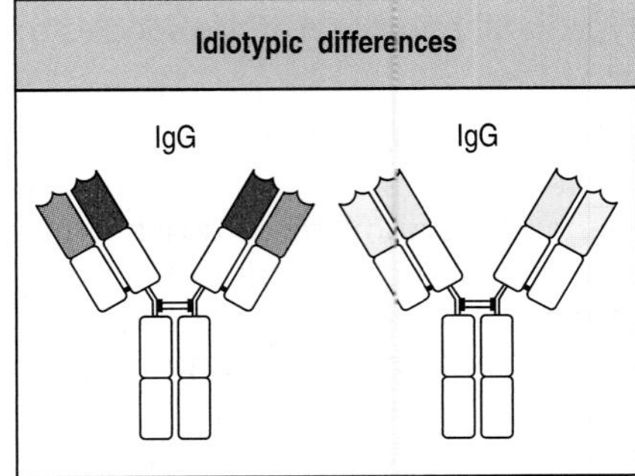

Fig. A.13 Different types of variation between immunoglobulins. Differences between constant regions due to usage of different C-region genes are called isotypes; differences due to different alleles of the same C gene are called allotypes; differences due to particular rearranged V_H and V_L genes are called idiotypes.

fetalis. Anti-immunoglobulin antibodies were first developed by Robin Coombs and the test for this disease is still called the Coombs test. Hemolytic disease of the newborn occurs when a mother makes IgG antibodies specific for the **Rhesus** or **Rh blood group antigen** expressed on the red blood cells of her fetus. Rh-negative mothers make these antibodies when they are exposed to Rh-positive fetal red blood cells bearing the paternally inherited Rh antigen. Maternal IgG antibodies are normally transported across the placenta to the fetus, where they protect the newborn infant against infection. However, IgG anti-Rh antibodies coat the fetal red blood cells, which are then destroyed by phagocytic cells in the liver, causing a hemolytic anemia in the fetus and newborn infant.

Since the Rh antigens are widely spaced on the red blood cell surface, the IgG anti-Rh antibodies cannot fix complement and cause lysis of red blood cells *in vitro*. Furthermore, for reasons that are not fully understood, antibodies to Rh blood group antigens do not agglutinate red blood cells as do antibodies to the ABO blood group antigens. Thus, detecting these antibodies was difficult until anti-human immunoglobulin antibodies were developed. With these, maternal IgG antibodies bound to the fetal red blood cells can be detected after washing the cells to remove unbound immunoglobulin that is present in the fetal serum. Adding anti-human immunoglobulin antibodies to the washed fetal red blood cells agglutinates any cells to which maternal antibodies are bound. This is the **direct Coombs test** (Fig. A.14), so called because it directly detects antibody bound to the surface of the fetal red blood cells. An **indirect Coombs test** is used to detect nonagglutinating anti-Rh antibody in maternal serum; the serum is first incubated with Rh-positive red blood cells, which bind the anti-Rh antibody, after which the antibody-coated cells are washed to remove unbound immunoglobulin and are then agglutinated with anti-immunoglobulin antibody (see Fig. A.14). The indirect Coombs test allows Rh incompatibilities that might lead to hemolytic disease of the newborn to be detected, and this knowledge allows the disease to be prevented (see Section 10-19). The Coombs test is also commonly used to detect antibodies to drugs that bind to red blood cells and cause hemolytic anemia.

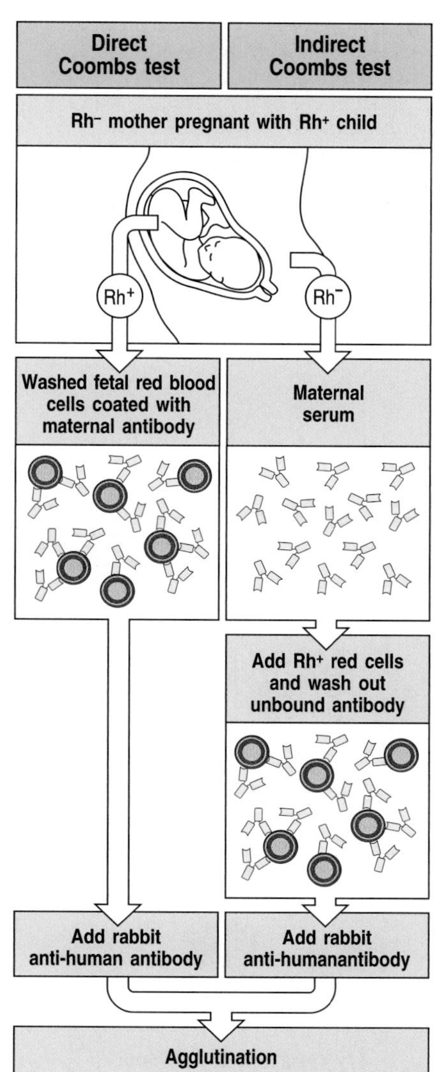

Fig. A.14 The Coombs direct and indirect anti-globulin tests for antibody to red blood cell antigens. A Rh⁻ mother of a Rh⁺ fetus can become immunized to fetal red blood cells that enter the maternal circulation at the time of delivery. In a subsequent pregnancy with a Rh⁺ fetus, IgG anti-Rh antibodies can cross the placenta and damage the fetal red blood cells. In contrast to anti-Rh antibodies, maternal anti-ABO antibodies are of the IgM isotype and cannot cross the placenta, and so do not cause harm. Anti-Rh antibodies do not agglutinate red blood cells but their presence on the fetal red blood cell surface can be shown by washing away unbound immunoglobulin and then adding antibody to human immunoglobulin, which agglutinates the antibody-coated cells. Anti-Rh antibodies can be detected in the mother's serum in an indirect Coombs test; the serum is incubated with Rh⁺ red blood cells, and once the antibody binds, the red blood cells are treated as in the direct Coombs test.

A-12 Monoclonal antibodies.

The antibodies generated in a natural immune response or after immunization in the laboratory are a mixture of molecules of different specificities and affinities. Some of this heterogeneity results from the production of antibodies that bind to different epitopes on the immunizing antigen, but even antibodies directed at a single antigenic determinant such as a hapten can be markedly heterogeneous, as shown by **isoelectric focusing.** In this technique, proteins are separated on the basis of their isoelectric point, the pH at which their net charge is zero. By electrophoresing proteins in a pH gradient for long enough, each molecule migrates along the pH gradient until it reaches the pH at which it is neutral and is thus concentrated (focused) at that point. When antiserum containing anti-hapten antibodies is treated in this way and then transferred to a solid support such as nitrocellulose paper, the anti-hapten antibodies can be detected by their ability to bind labeled hapten. The binding of antibodies of various isoelectric points to the hapten shows that even antibodies that bind the same antigenic determinant can be heterogeneous.

Antisera are valuable for many biological purposes but they have certain inherent disadvantages that relate to the heterogeneity of the antibodies they contain. First, each antiserum is different from all other antisera, even if raised in a genetically identical animal by using the identical preparation of antigen and the same immunization protocol. Second, antisera can be produced in only limited volumes, and thus it is impossible to use the identical serological reagent in a long or complex series of experiments or clinical tests. Finally, even antibodies purified by affinity chromatography (see Section A-5) may include minor populations of antibodies that give unexpected cross-reactions, which confound the analysis of experiments. To avoid these problems, and to harness the full potential of antibodies, it was necessary to develop a way of making an unlimited supply of antibody molecules of homogeneous structure and known specificity. This has been achieved through the production of monoclonal antibodies from cultures of hybrid antibody-forming cells or, more recently, by genetic engineering.

Biochemists in search of a homogeneous preparation of antibody that they could subject to detailed chemical analysis turned early to proteins produced by patients with multiple myeloma, a common tumor of plasma cells. It was known that antibodies are normally produced by plasma cells and since this disease is associated with the presence of large amounts of a homogeneous gamma globulin called a **myeloma protein** in the patient's serum, it seemed likely that myeloma proteins would serve as models for normal antibody molecules. Thus, much of the early knowledge of antibody structure came from studies on myeloma proteins. These studies showed that monoclonal antibodies could be obtained from immortalized plasma cells. However, the antigen specificity of most myeloma proteins was unknown, which limited their usefulness as objects of study or as immunological tools.

This problem was solved by Georges Köhler and Cesar Milstein, who devised a technique for producing a homogeneous population of antibodies of known antigenic specificity. They did this by fusing spleen cells from an immunized mouse to cells of a mouse myeloma to produce hybrid cells that both proliferated indefinitely and secreted antibody specific for the antigen used to immunize the spleen cell donor. The spleen cell provides the ability to make specific antibody, while the myeloma cell provides the ability to grow indefinitely in culture and secrete immunoglobulin continuously. By using a myeloma cell partner that produces no antibody proteins itself, the antibody produced by the hybrid cells comes only from the immune spleen cell partner. After fusion, the hybrid cells are selected using drugs that kill the myeloma parental cell, while the unfused parental spleen cells have a limited life-span and soon die, so that only hybrid myeloma cell lines or **hybridomas** survive. Those hybridomas producing antibody of the desired specificity are then identified and cloned by regrowing the cultures from single cells (Fig. A.15). Since each hybridoma is a **clone** derived from fusion with a single B cell, all the antibody molecules it produces are identical in structure, including their antigen-binding site and isotype. Such antibodies are called **monoclonal antibodies**. This technology has revolutionized the use of antibodies by providing a limitless supply of antibody of a single and known specificity. Monoclonal antibodies are now used in most serological assays, as diagnostic probes, and as therapeutic agents. So far, however, only mouse monoclonals are routinely produced and efforts to use this same approach to make human monoclonal antibodies have met with very limited success.

A-13 Phage display libraries for antibody V-region production.

This is a technique for producing antibody-like molecules. Gene segments encoding the antigen-binding variable or V domains of antibodies are fused to genes encoding the coat protein of a bacteriophage. Bacteriophage containing such gene fusions are used to infect bacteria, and the resulting phage particles have coats that express the antibody-like fusion protein, with the antigen-binding domain displayed on the outside of the bacteriophage. A collection of recombinant phage, each displaying a different antigen-binding domain on its surface, is known as a **phage display library**. In much the same

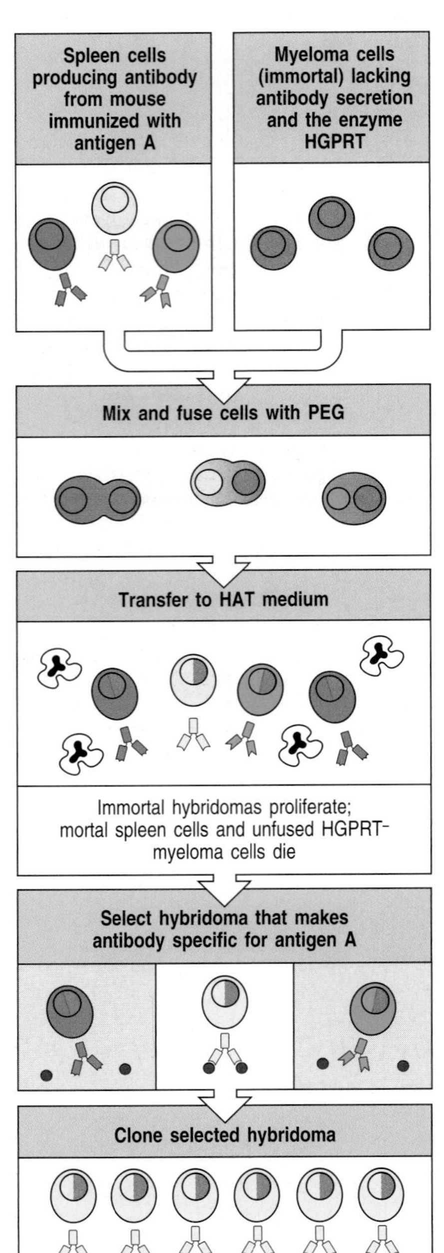
Spleen cells producing antibody from mouse immunized with antigen A
Myeloma cells (immortal) lacking antibody secretion and the enzyme HGPRT
Mix and fuse cells with PEG
Transfer to HAT medium
Immortal hybridomas proliferate; mortal spleen cells and unfused HGPRT⁻ myeloma cells die
Select hybridoma that makes antibody specific for antigen A
Clone selected hybridoma

Fig. A.15 The production of monoclonal antibodies. Mice are immunized with antigen A and given an intravenous booster immunization three days before they are killed, in order to produce a large population of spleen cells secreting specific antibody. Spleen cells die after a few days in culture. In order to produce a continuous source of antibody they are fused with immortal myeloma cells by using polyethylene glycol (PEG) to produce a hybrid cell line called a hybridoma. The myeloma cells are selected beforehand to ensure that they are not secreting antibody themselves and that they are sensitive to the hypoxanthine-aminopterin-thymidine (HAT) medium that is used to select hybrid cells because they lack the enzyme hypoxanthine:guanine phosphoribosyl transferase (HGPRT). The HGPRT gene contributed by the spleen cell allows hybrid cells to survive in the HAT medium, and only hybrid cells can grow continuously in culture because of the malignant potential contributed by the myeloma cells. Therefore, unfused myeloma cells and unfused spleen cells die in the HAT medium, as shown here by cells with dark, irregular nuclei. Individual hybridomas are then screened for antibody production, and cells that make antibody of the desired specificity are cloned by growing them up from a single antibody-producing cell. The cloned hybridoma cells are grown in bulk culture to produce large amounts of antibody. As each hybridoma is descended from a single cell, all the cells of a hybridoma cell line make the same antibody molecule, which is thus called a monoclonal antibody.

way that antibodies specific for a particular antigen can be isolated from a complex mixture by affinity chromatography (see Section A-5), phage expressing antigen-binding domains specific for a particular antigen can be isolated by selecting the phage in the library for binding to that antigen. The phage particles that bind are recovered and used to infect fresh bacteria. Each phage isolated in this way will produce a monoclonal antigen-binding particle analogous to a monoclonal antibody (Fig. A.16). The genes encoding the antigen-binding site, which are unique to each phage, can then be recovered from the phage DNA and used to construct genes for a complete antibody molecule by joining them to parts of immunoglobulin genes that encode the invariant parts of an antibody. When these reconstructed antibody genes are introduced into a suitable host cell line, such as the nonantibody-producing myeloma cells used for hybridomas, the transfected cells can secrete antibodies with all the desirable characteristics of monoclonal antibodies produced from hybridomas.

In much the same way that a collection of phage can display a wide variety of potential antigen-binding sites, the phage can also be engineered to display a wide variety of antigens; such a library is known as an **antigen display library**. In such cases, the antigens displayed are often short peptides encoded by chemically synthesized DNA sequences that have mixtures of all four nucleotides in some positions, so that all possible amino acids are incorporated. It is not usual for every position in a peptide to be allowed to vary in this way, since the number of different peptide sequences increases dramatically with the number of variable positions; there are over 2×10^{10} possible sequences of eight amino acids.

A-14 Immunofluorescence microscopy.

Since antibodies bind stably and specifically to their corresponding antigen, they are invaluable as probes for identifying a particular molecule in cells, tissues, or biological fluids. Antibody molecules can be used to locate their target molecules accurately in single cells or in tissue sections by a variety of different labeling techniques. When the antibody itself, or the anti-immunoglobulin antibody used to detect it, is labeled with a fluorescent dye the technique is known as **immunofluorescence microscopy**. As in all serological techniques, the antibody binds stably to its antigen, allowing unbound antibody to be removed by thorough washing. As antibodies to proteins recognize the surface features of the native, folded protein, the native structure of the protein being sought usually needs to be preserved, either by using only the most gentle chemical fixation techniques or by using frozen tissue sections that are fixed only after the antibody reaction has been performed. Some

Fig. A.16 The production of antibodies by genetic engineering. Short primers to consensus sequences in heavy- and light-chain variable (V) regions of immunoglobulin genes are used to generate a library of heavy- and light-chain V-region DNAs by the polymerase chain reaction, with spleen DNA as the starting material. These heavy- and light+chain V-region genes are cloned randomly into a filamentous phage such that each phage expresses one heavy-chain and one light-chain V region as a surface fusion protein with antibody-like properties. The resulting phage display library is multiplied in bacteria, and the phage are then bound to a surface coated with antigen. The unbound phage are washed away; the bound phage are recovered, multiplied in bacteria, and again bound to antigen. After a few cycles, only specific high-affinity antigen binding phage are left. These can be used like antibody molecules, or their V genes can be recovered and engineered into anti-body genes to produce genetically engineered antibody molecules (not shown). This technology may replace the hybridoma technology for producing monoclonal antibodies and has the advantage that humans can be used as the source of DNA.

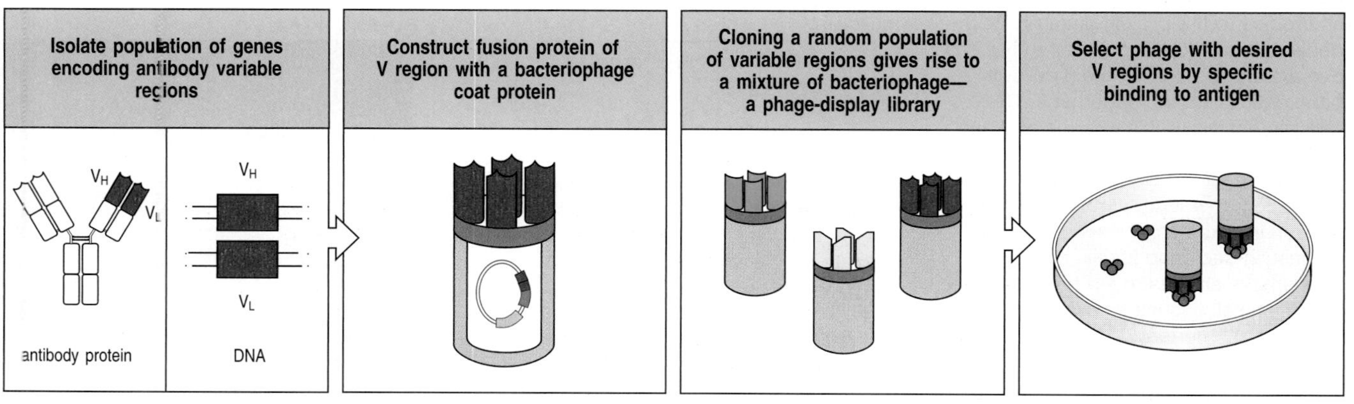

Isolate population of genes encoding antibody variable regions
V_H
V_L
antibody protein
DNA
Construct fusion protein of V region with a bacteriophage coat protein
Cloning a random population of variable regions gives rise to a mixture of bacteriophage— a phage-display library
Select phage with desired V regions by specific binding to antigen

Fig. A.17 Excitation and emission wavelengths for common fluorochromes.

Excitation and emission wavelengths of some commonly used fluorochromes		
Probe	Excitation (nm)	Emission (nm)
R-phycoerythrin (PE)	480; 565	578
Fluorescein	495	519
PerCP	490	675
Texas Red	589	615
Rhodamine	550	573

antibodies, however, bind proteins even if they are denatured, and such antibodies will bind specifically even to protein in fixed tissue sections.

The fluorescent dye can be covalently attached directly to the specific antibody, but more commonly, the bound antibody is detected by fluorescent anti-immunoglobulin, a technique known as **indirect immunofluorescence**. The dyes chosen for immunofluorescence are excited by light of one wavelength, usually blue or green, and emit light of a different wavelength in the visible spectrum. The most common fluorescent dyes are fluorescein, which emits green light, Texas Red and Peridinin chlorophyll protein (PerCP), which emit red light, and rhodamine and phycoerythrin (PE) which emit orange/red light (Fig. A.17). By using selective filters, only the light coming from the dye or fluorochrome used is detected in the fluorescence microscope (Fig. A.18). Although Albert Coons first devised this technique to identify the plasma cell as the source of antibody, it can be used to detect the distribution of any protein. By attaching different dyes to different antibodies, the distribution of two or more molecules can be determined in the same cell or tissue section (see Fig. A.18).

The recent development of the **confocal fluorescent microscope**, which uses computer-aided techniques to produce an ultrathin optical section of a cell or tissue, gives very high resolution immunofluorescence microscopy without the need for elaborate sample preparation. The resolution of the confocal microscope can be further increased using low-intensity illumination so that two photons are required to excite the fluorochrome. A pulsed laser beam is used, and only when it is focused into the focal plane of the microscope is the intensity sufficient to excite fluorescence. In this way the fluorescence emission itself can be restricted to the optical section.

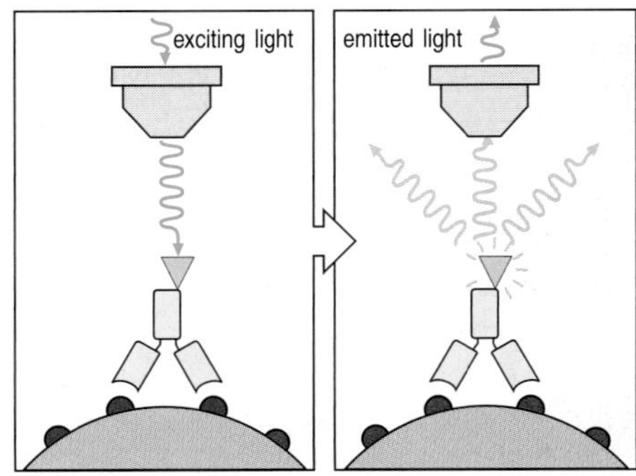

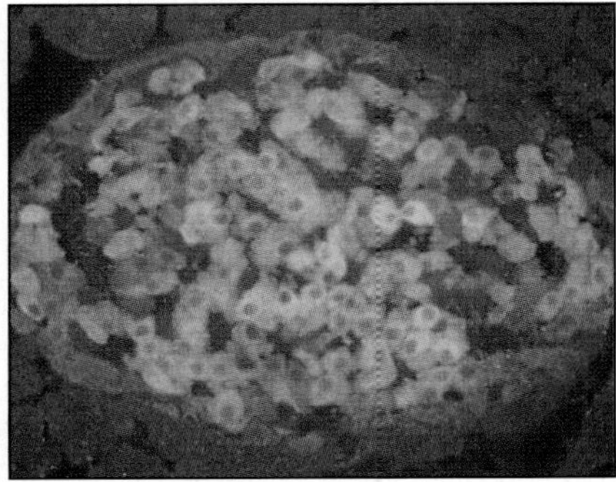

Fig. A.18 Immunofluorescence microscopy. Antibodies labeled with a fluorescent dye such as fluorescein (green triangle) are used to reveal the presence of their corresponding antigens in cells or tissues. The stained cells are examined in a microscope that exposes them to blue or green light to excite the fluorescent dye. The excited dye emits light at a characteristic wavelength, which is captured by viewing the sample through a selective filter. This technique is applied widely in biology to determine the location of molecules in cells and tissues. Different antigens can be detected in tissue sections by labeling antibodies with dyes of distinctive color. Here, antibodies to the protein glutamic acid decarboxylase (GAD) coupled to a green dye are shown to stain the β cells of pancreatic islets of Langerhans. The α cells do not make this enzyme and are labeled with antibodies to the hormone glucagon coupled with an orange fluorescent dye. GAD is an important antigen in diabetes, a disease in which the insulin-secreting β cells of the islets of Langerhans are destroyed by an immune attack on self tissues (see Chapter 14). Photograph courtesy of M. Solimena and P. De Camilli.

One important development in the area of microscopy has been the use of **time-lapse video microscopy**, in which sensitive digital video cameras record the movement of fluorescently labeled molecules in cell membranes and their redistribution when cells come into contact with each other. Cell-surface molecules can be fluorescently labeled in two main ways. One is by the binding of fluorochrome-labeled Fab fragments of antibodies specific for the protein of interest; the other is by generating a fusion protein, in which the protein of interest has been attached to one of a family of fluorescent proteins obtained from jellyfish. The first of these fluorescent proteins to come into wide use was green fluorescent protein (GFP), isolated from the jellyfish, *Aequorea victoria*. Other variants of this protein with different properties are now available and the list of available fluorescent labels now includes red, blue, cyan, or yellow fluorescent proteins. Using cells transfected with the genes encoding such fusion proteins, it is possible to show the redistribution of T-cell receptors, co-receptors, adhesion molecules, and other signaling molecules, such as CD45, that takes place when a T cell makes contact with a target cell. It is clear from these observations that the interface between the T cell and its target does not represent a simple apposition of two cell membranes, but is an actively organized and dynamic structure, often now referred to as the 'immunological synapse.'

A-15 Immunoelectron microscopy.

Antibodies can be used to detect the intracellular location of structures or particular proteins at high resolution by electron microscopy, a technique known as **immunoelectron microscopy**. Antibodies against the required antigen are labeled with gold particles and then applied to ultrathin sections, which are then examined in the transmission electron microscope. Antibodies labeled with gold particles of different diameters enable two or more proteins to be studied simultaneously (see Fig. 5.10). The difficulty with this technique is in staining the ultrathin section adequately, as few molecules of antigen are present in each section.

A-16 Immunohistochemistry.

An alternative to immunofluorescence (see Section A-14) for detecting a protein in tissue sections is **immunohistochemistry**, in which the specific antibody is chemically coupled to an enzyme that converts a colorless substrate into a colored reaction product *in situ*. The localized deposition of the colored product where antibody has bound can be directly observed under a light microscope. The antibody binds stably to its antigen, allowing unbound antibody to be removed by thorough washing. This method of detecting bound antibody is analogous to ELISA (see Section A-6) and frequently uses the same coupled enzymes, the difference in detection being primarily that in immunohistochemistry the colored products are insoluble and precipitate at the site where they are formed. Horseradish peroxidase and alkaline phosphatase are the two enzymes most commonly used in these applications. Horseradish peroxidase oxidises the substrate diaminobenzidine to produce a brown precipitate, while alkaline phosphatase can produce red or blue dyes depending on the substrates used; a common substrate is 5-bromo-4-chloro-3-indolyl phosphate plus nitroblue tetrazolium (BCIP/NBT), which gives rise to a dark blue or purple stain. As with immunofluorescence, the native structure of the protein being sought usually needs to be preserved, so that it will be recognized by the antibody. Tissues are fixed by the most gentle chemical fixation techniques or frozen tissue sections are used that are fixed only after the antibody reaction has been performed.

A-17 Immunoprecipitation and co-immunoprecipitation.

In order to raise antibodies to membrane proteins and other cellular structures that are difficult to purify, mice are often immunized with whole cells or crude cell extracts. Antibodies to the individual molecules are then obtained by using these immunized mice to produce hybridomas making monoclonal antibodies (see Section A-12) that bind to the cell type used for immunization. To characterize the molecules identified by the antibodies, cells of the same type are labeled with radioisotopes and dissolved in nonionic detergents that disrupt cell membranes but do not interfere with antigen–antibody interactions. This allows the labeled protein to be isolated by binding to the antibody in a reaction known as **immunoprecipitation**. The antibody is usually attached to a solid support, such as the beads that are used in affinity chromatography (see Section A-5), or to Protein A. Cells can be labeled in two main ways for immunoprecipitation analysis. All the proteins in a cell can be labeled metabolically by growing the cell in radioactive amino acids that are incorporated into cellular protein (Fig. A.19). Alternatively, one can label only the cell-surface proteins by radioiodination under conditions that prevent iodine from crossing the plasma membrane and labeling proteins inside the cell, or by a reaction that labels only membrane proteins with biotin, a small molecule that is detected readily by labeled avidin, a protein found in egg whites that binds biotin with very high affinity.

Once the labeled proteins have been isolated by the antibody, they can be characterized in several ways. The most common is polyacrylamide gel electrophoresis (PAGE) of the proteins after dissociating them from antibody in the strong ionic detergent sodium dodecyl sulfate (SDS), a technique generally

Fig. A.19 Cellular proteins reacting with an antibody can be characterized by immunoprecipitation of labeled cell lysates. All actively synthesized cellular proteins can be labeled metabolically by incubating cells with radioactive amino acids (shown here for methionine) or one can label just the cell-surface proteins by using radioactive iodine in a form that cannot cross the cell membrane or by a reaction with the small molecule biotin, detected by its reaction with labeled avidin (not shown). Cells are lysed with detergent and individual labeled cell-associated proteins can be precipitated with a monoclonal antibody attached to beads. After unbound proteins have been washed away, the bound protein is eluted in the detergent sodium dodecyl sulfate (SDS), which dissociates it from the antibody and also coats the protein with a strong negative charge, allowing it to migrate according to its size in polyacrylamide gel electrophoresis (PAGE). The positions of the labeled proteins are determined by autoradiography using X-ray film. This technique of SDS-PAGE can be used to determine the molecular weight and subunit composition of a protein. Patterns of protein bands observed with metabolic labeling are usually more complex than those revealed by radioiodination, owing to the presence of precursor forms of the protein (right panel). The mature form of a surface protein can be identified as being the same size as that detected by surface iodination or biotinylation (not shown).

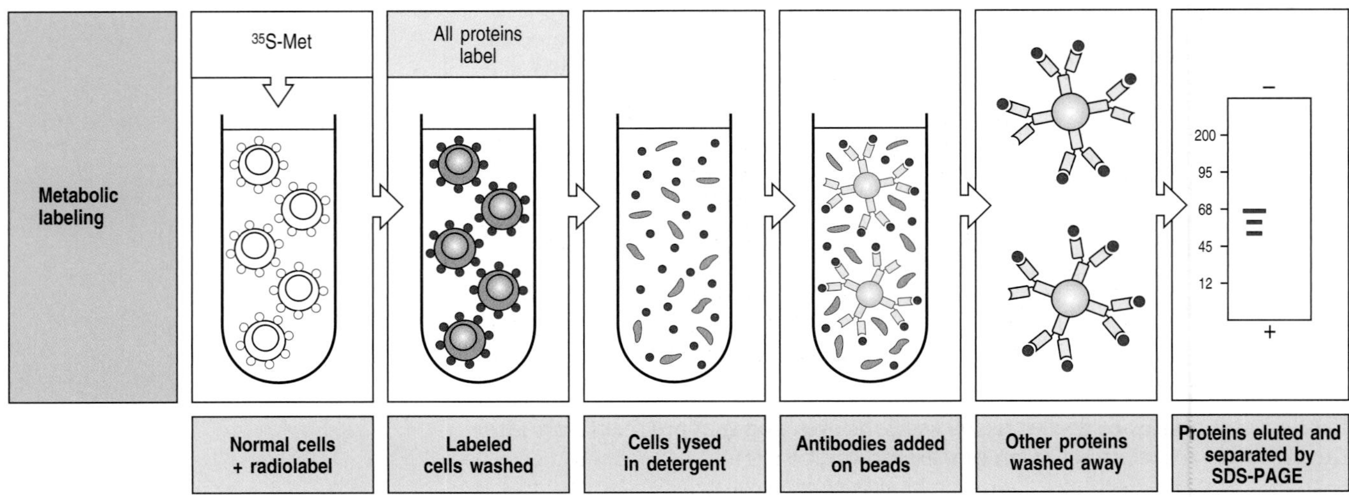
Metabolic labeling
35S-Met
All proteins label
–
200
95
68
45
12
+
Normal cells + radiolabel
Labeled cells washed
Cells lysed in detergent
Antibodies added on beads
Other proteins washed away
Proteins eluted and separated by SDS-PAGE

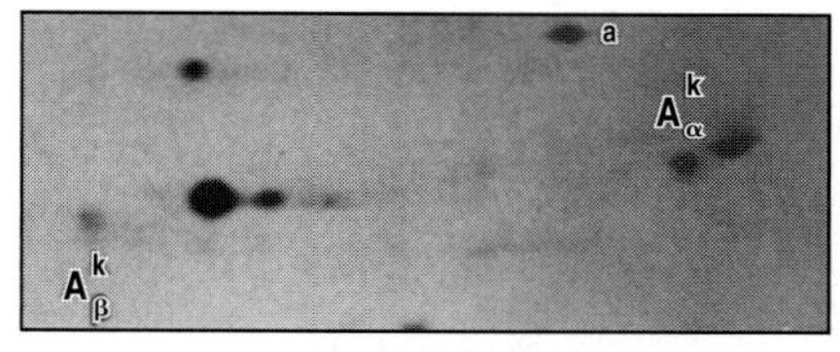

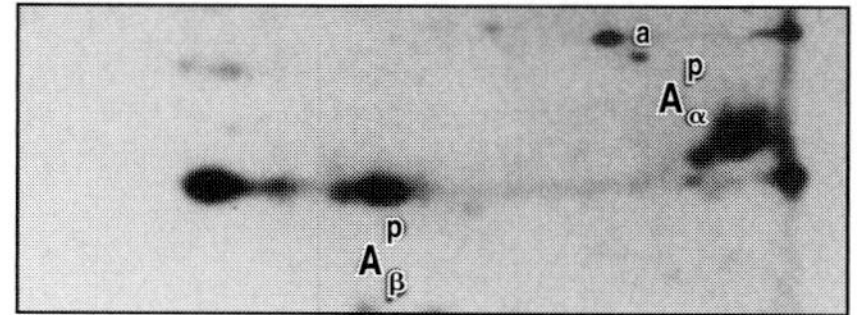

Fig. A.20 Two-dimensional gel electrophoresis of MHC class II molecules. Proteins in mouse spleen cells have been labeled metabolically (see Fig. A.19), precipitated with a monoclonal antibody against the mouse MHC class II molecule H2-A, and separated by isoelectric focusing in one direction and SDS-PAGE in a second direction at right angles to the first; hence the term two-dimensional gel electrophoresis. This allows one to distinguish molecules of the same molecular weight on the basis of their charge. The separated proteins are detected by autoradiography. The MHC class II molecules are composed of two chains, α and β, and in the different MHC class II molecules these have different isoelectric points (compare upper and lower panels). The MHC genotype of mice is indicated by lower case superscripts (k,p). Actin, a common contaminant, is marked a. Photographs courtesy of J.F. Babich.

abbreviated as **SDS-PAGE**. SDS binds relatively homogeneously to proteins, conferring a charge that allows the electrophoretic field to drive protein migration through the gel. The rate of migration is controlled mainly by protein size (see Fig. A.19). Proteins of differing charges can be separated using isoelectric focusing (see Section A-12). This technique can be combined with SDS-PAGE in a procedure known as **two-dimensional gel electrophoresis**. For this, the immunoprecipitated protein is eluted in urea, a nonionic solubilizing agent, and run on an isoelectric focusing gel in a narrow tube of polyacrylamide. This first-dimensional isoelectric focusing gel is then placed across the top of an SDS-PAGE slab gel, which is then run vertically to separate the proteins by molecular weight (Fig. A.20). Two-dimensional gel electro-phoresis is a powerful technique that allows many hundreds of proteins in a complex mixture to be distinguished from one another.

Immunoprecipitation and the related technique of immunoblotting (see Section A-18) are useful for determining the molecular weight and isoelectric point of a protein as well as its abundance, distribution, and whether, for example, it undergoes changes in molecular weight and isoelectric point as a result of processing within the cell.

A-18 Immunoblotting (Western blotting).

Like immunoprecipitation (see Section A-17), **immunoblotting** is used for identifying the presence of a given protein in a cell lysate, but it avoids the problem of having to label large quantities of cells with radioisotopes. Unlabeled cells are placed in detergent to solubilize all cell proteins and the lysate is run on SDS-PAGE to separate the proteins (see Section A-17). The size-separated proteins are then transferred from the gel to a stable support such as a nitrocellulose membrane. Specific proteins are detected by treatment with antibodies able to react with SDS-solubilized proteins (mainly those that react with denatured sequences); the bound antibodies are revealed by anti-immunoglobulin antibodies labeled with radioisotopes or an enzyme. The term **Western blotting** as a synonym for immunoblotting arose because the comparable technique for detecting specific DNA sequences is known as Southern blotting, after Ed Southern who devised it, which in turn provoked the name Northern for blots of size-separated RNA, and Western for blots of size-separated proteins. Western blots have many applications in basic research and clinical diagnosis. They are often used to test sera for the presence of antibodies to specific proteins, for example to detect antibodies to different constituents of the human immunodeficiency virus, HIV (Fig. A.21).

Co-immunoprecipitation is an extension of the immunoprecipitation technique which is used to determine whether a given protein interacts physically with another given protein. Cell extracts containing the presumed interaction

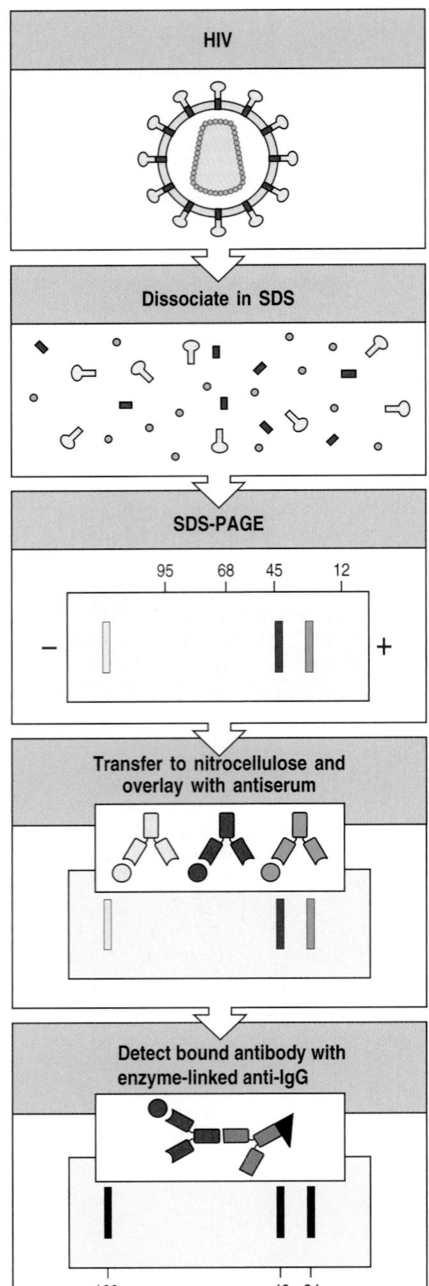

Fig. A.21 Western blotting is used to identify antibodies to the human immunodeficiency virus (HIV) in serum from infected individuals. The virus is dissociated into its constituent proteins by treatment with the detergent SDS, and its proteins are separated using SDS-PAGE. The separated proteins are transferred to a nitro-cellulose sheet and reacted with the test serum. Anti-HIV antibodies in the serum bind to the various HIV proteins and are detected with enzyme-linked anti-human immunoglobulin, which deposits colored material from a colorless substrate. This general methodology will detect any combination of antibody and antigen and is used widely, although the denaturing effect of SDS means that the technique works most reliably with antibodies that recognize the antigen when it is denatured.

complex are first immunoprecipitated with antibody against one of the proteins. The material identified by this means is then tested for the presence of the other protein by immunoblotting with a specific antibody.

A-19 Use of antibodies in the isolation and identification of genes and their products.

As a first step in isolating the gene that codes for a particular protein, antibodies specific for the protein can be used to isolate the purified protein from cells using affinity chromatography (see Section A-5). Small amounts of amino acid sequence can then be obtained from the protein's amino-terminal end or from peptide fragments generated by proteolysis. The information in these amino acid sequences is used to construct a set of synthetic oligonucleotides corresponding to the possible DNA sequences, which are then used as probes to isolate the gene encoding the protein from either a library of DNA sequences complementary to mRNA (a cDNA library) or a genomic DNA library (a library of chromosomal DNA fragments).

An alternative approach to gene identification uses antibodies to identify the protein product of a gene that has been introduced into a cell that does not normally express it. This technique is most commonly applied to the identification of genes encoding cell-surface proteins. A set of cDNA-containing expression vectors is first made from a cDNA library prepared from the total mRNA from a cell type that does express the protein of interest. The vectors are used to transfect a cell type that does not normally express the protein of interest, and the vector drives expression of the cDNA it contains without integrating into the host cell DNA. Cells expressing the required protein after transfection are then isolated by binding to specific antibodies that detect the presence of the protein on the cell surface. The vector containing the required gene can then be recovered from these cells (Fig. A.22).

The recovered vectors are then introduced into bacterial cells where they replicate rapidly, and these amplified vectors are used in a second round of transfection in mammalian cells. After several cycles of transfection, isolation, and amplification in bacteria, single colonies of bacteria are picked and the vectors prepared from cultures of each colony are used in a final transfection to identify a cloned vector carrying the cDNA of interest, which is then isolated and characterized. This methodology has been used to isolate many genes encoding cell-surface molecules.

The full amino acid sequence of the protein can be deduced from the nucleotide sequence of its cDNA, and this often gives clues to the nature of the protein and its biological properties. The nucleotide sequence of the gene and its regulatory regions can be determined from genomic DNA clones. The gene can be manipulated and introduced into cells by transfection for larger-scale production and functional studies. This approach has been used to

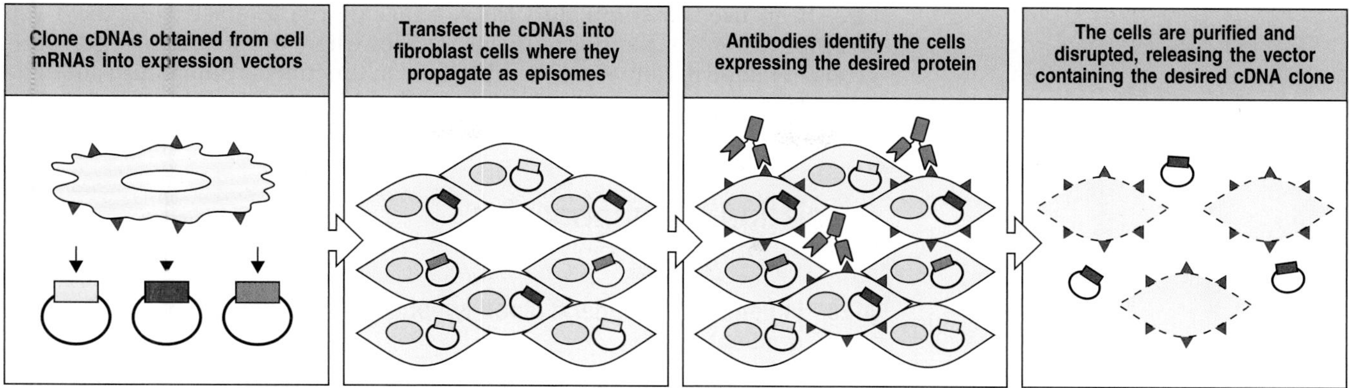

Clone cDNAs obtained from cell mRNAs into expression vectors
Transfect the cDNAs into fibroblast cells where they propagate as episomes
Antibodies identify the cells expressing the desired protein
The cells are purified and disrupted, releasing the vector containing the desired cDNA clone

Fig. A.22 The gene encoding a cell-surface molecule can be isolated by expressing it in fibroblasts and detecting its protein product with monoclonal antibodies. Total mRNA from a cell line or tissue expressing the protein is isolated, converted into cDNA, and cloned as cDNAs in a vector designed to direct expression of the cDNA in fibroblasts. The entire cDNA library is used to transfect cultured fibroblasts. Fibroblasts that have taken up cDNA encoding a cell-surface protein express the protein on their surface; they can be isolated by binding a monoclonal antibody against that protein. The vector containing the gene is isolated from the cells that express the antigen and used for more rounds of transfection and reisolation until uniform positive expression is obtained, ensuring that the correct gene has been isolated. The cDNA insert can then be sequenced to determine the sequence of the protein it encodes and can also be used as the source of material for large-scale expression of the protein for analysis of its structure and function. The method illustrated is limited to cloning genes for single-chain proteins (that is, those encoded by only one gene) that can be expressed in fibroblasts. It has been used to clone many genes of immunological interest such as that for CD4.

characterize many immunologically important proteins, such as the MHC glycoproteins.

The converse approach is taken to identify the unknown protein product of a cloned gene. The gene sequence is used to construct synthetic peptides of 10–20 amino acids that are identical to part of the deduced protein sequence, and antibodies are then raised against these peptides by coupling them to carrier proteins; the peptides behave as haptens. These anti-peptide antibodies often bind the native protein and so can be used to identify its distribution in cells and tissues and to try to ascertain its function (Fig. A.23). This approach to identifying the function of a gene is often called 'reverse genetics' as it works from gene to phenotype rather than from phenotype to gene, which is the classical genetic approach. The great advantage of reverse genetics over the classical approach is that it does not require a detectable phenotypic genetic trait in order to identify a gene.

Antibodies can also be used in the determination of the function of gene products. Some antibodies are able to act as agonists, when the binding of the antibody to the molecule mimics the binding of the natural ligand and activates the function of the gene product. For example, antibodies to the CD3 molecule

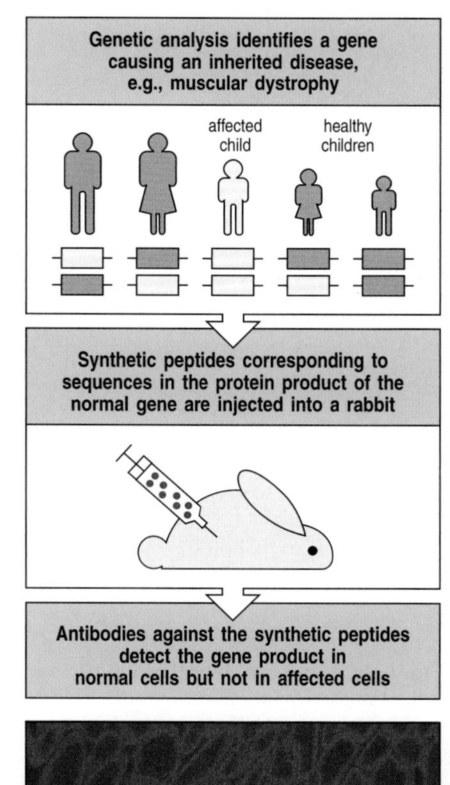

Fig. A.23 The use of antibodies to detect the unknown protein product of a known gene is called reverse genetics. When the gene responsible for a genetic disorder such as Duchenne muscular dystrophy is isolated, the amino acid sequence of the unknown protein product of the gene can be deduced from the nucleotide sequence of the gene, and synthetic peptides representing parts of this sequence can be made. Antibodies are raised against these peptides and purified from the antiserum by affinity chromatography on a peptide column (see Fig. A.5). Labeled antibody is used to stain tissue from individuals with the disease and from unaffected individuals to determine differences in the presence, amount, and distribution of the normal gene product. The product of the dystrophin gene is present in normal mouse skeletal muscle cells, as shown in the bottom panel (red fluorescence), but is missing from the cells of mice bearing the mutation mdx, the mouse equivalent of Duchenne muscular dystrophy (not shown). Photograph (× 15) courtesy of H.G.W. Lidov and L. Kunkel.

have been used to stimulate T cells, replacing the interaction of the T-cell receptor with MHC:peptide antigens in cases where the specific peptide antigen is not known. Conversely, antibodies can function as antagonists, inhibiting the binding of the natural ligand and thus blocking the function.

Isolation of lymphocytes

A-20 Isolation of peripheral blood lymphocytes by Ficoll-Hypaque™ gradient.

The first step in studying lymphocytes is to isolate them so that their behavior can be analyzed *in vitro*. Human lymphocytes can be isolated most readily from peripheral blood by density centrifugation over a step gradient consisting of a mixture of the carbohydrate polymer Ficoll™ and the dense iodine-containing compound metrizamide. This yields a population of mononuclear cells at the interface that has been depleted of red blood cells and most polymorphonuclear leukocytes or granulocytes (Fig. A.24). The resulting population, called **peripheral blood mononuclear cells (PBMCs)**, consists mainly of lymphocytes and monocytes. Although this population is readily accessible, it is not necessarily representative of the lymphoid system, as only recirculating lymphocytes can be isolated from blood.

A particular cell population can be isolated from a sample or culture by binding to antibody-coated plastic surfaces, a technique known as **panning**, or by removing unwanted cells by treatment with specific antibody and complement to kill them. Cells can also be passed over columns of antibody-coated, nylon-coated steel wool and different populations differentially eluted. This technique extends affinity chromatography to cells, and is now a very popular way to separate cells. All these techniques can also be used as a pre-purification step prior to sorting out highly purified populations by FACS (see Section A-22).

A-21 Isolation of lymphocytes from tissues other than blood.

In experimental animals, and occasionally in humans, lymphocytes are isolated from lymphoid organs, such as spleen, thymus, bone marrow, lymph nodes, or mucosal-associated lymphoid tissues, most commonly the palatine tonsils in humans (see Fig. 11.5). A specialized population of lymphocytes resides in surface epithelia; these cells are isolated by fractionating the epithelial layer after its detachment from the basement membrane. Finally, in

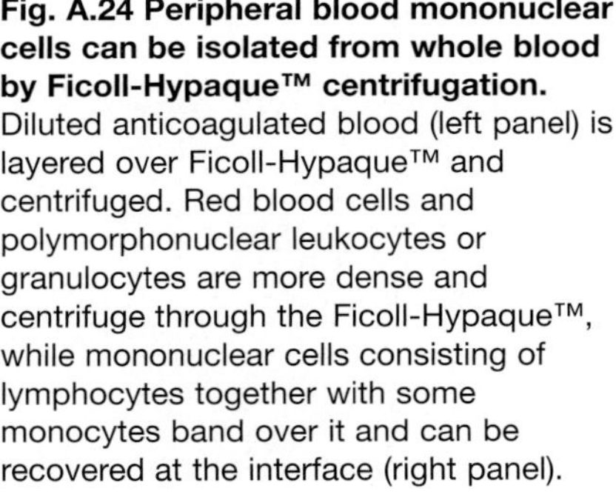

Fig. A.24 Peripheral blood mononuclear cells can be isolated from whole blood by Ficoll-Hypaque™ centrifugation. Diluted anticoagulated blood (left panel) is layered over Ficoll-Hypaque™ and centrifuged. Red blood cells and polymorphonuclear leukocytes or granulocytes are more dense and centrifuge through the Ficoll-Hypaque™, while mononuclear cells consisting of lymphocytes together with some monocytes band over it and can be recovered at the interface (right panel).

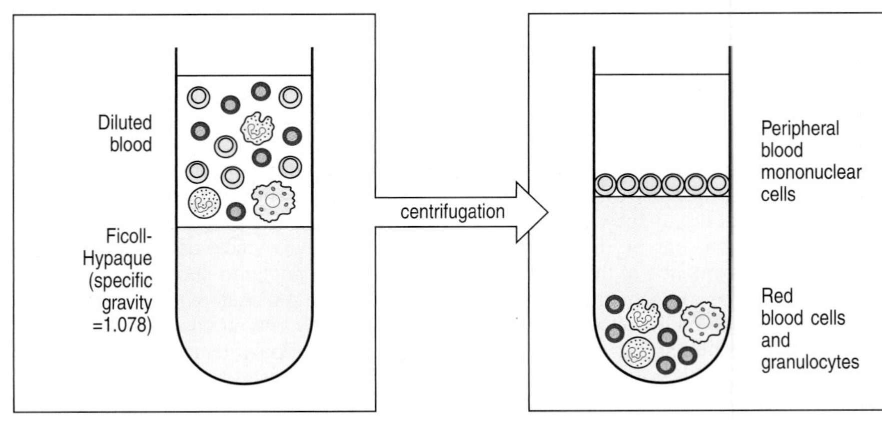

situations where local immune responses are prominent, lymphocytes can be isolated from the site of the response itself. For example, in order to study the autoimmune reaction that is thought to be responsible for rheumatoid arthritis, an inflammatory response in joints, lymphocytes are isolated from the fluid aspirated from the inflamed joint space.

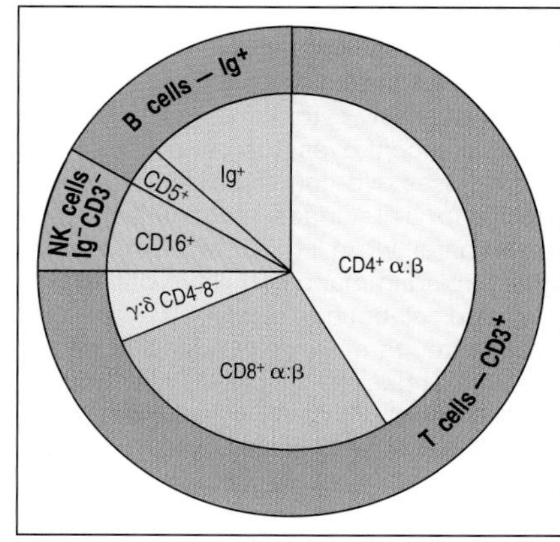

Fig. A.25 The distribution of lymphocyte subpopulations in human peripheral blood. As shown on the outside of the circle, lymphocytes can be divided into T cells bearing T-cell receptors (detected with anti-CD3 antibodies), B cells bearing immuno-globulin receptors (detected with anti-immunoglobulin antibodies), and null cells including natural killer (NK) cells, that label with neither. Further divisions of the T-cell and B-cell populations are shown inside. Using anti-CD4 and anti-CD8 antibodies, α:β T cells can be subdivided into two populations, whereas γ:δ T cells are identified with antibodies against the γ:δ T-cell receptor and mainly lack CD4 and CD8. A minority population of B cells express CD5 on their surface (see Section 7-28).

A-22 Flow cytometry and FACS analysis.

Resting lymphocytes present a deceptively uniform appearance, all being small round cells with a dense nucleus and little cytoplasm (see Fig. 1.6). However, these cells comprise many functional subpopulations, which are usually identified and distinguished from each other on the basis of their differential expression of cell-surface proteins, which can be detected using specific antibodies (Fig. A.25). B and T lymphocytes, for example, are identified unambiguously and separated from each other by antibodies to the constant regions of the B- and T-cell antigen receptors. T cells are further subdivided on the basis of expression of the co-receptor proteins CD4 and CD8.

An immensely powerful tool for defining and enumerating lymphocytes is the flow cytometer, which detects and counts individual cells passing in a stream through a laser beam. A flow cytometer equipped to separate the identified cells is called a **fluorescence-activated cell sorter** (**FACS**). These instruments are used to study the properties of cell subsets identified using monoclonal antibodies to cell-surface proteins. Individual cells within a mixed population are first tagged by treatment with specific monoclonal antibodies labeled with fluorescent dyes, or by specific antibodies followed by labeled anti-immunoglobulin antibodies. The mixture of labeled cells is then forced with a much larger volume of saline through a nozzle, creating a fine stream of liquid containing cells spaced singly at intervals. As each cell passes through a laser beam it scatters the laser light, and any dye molecules bound to the cell will be excited and will fluoresce. Sensitive photomultiplier tubes detect both the scattered light, which gives information on the size and granularity of the cell, and the fluorescence emissions, which give information on the binding of the labeled monoclonal antibodies and hence on the expression of cell-surface proteins by each cell (Fig. A.26).

In the cell sorter, the signals passed back to the computer are used to generate an electric charge, which is passed from the nozzle through the liquid stream at the precise time that the stream breaks up into droplets, each containing no more than a single cell; droplets containing a charge can then be deflected from the main stream of droplets as they pass between plates of opposite charge, so that positively charged droplets are attracted to a negatively charged plate, and vice versa. In this way, specific subpopulations of cells, distinguished by the binding of the labeled antibody, can be purified from a mixed population of cells. Alternatively, to deplete a population of cells, the same fluorochrome can be used to label different antibodies directed at marker proteins expressed by the various undesired cell types. The cell sorter can be used to direct labeled cells to a waste channel, retaining only the unlabeled cells.

When cells are labeled with a single fluorescent antibody, the data from a flow cytometer are usually displayed in the form of a histogram of fluorescence intensity versus cell numbers. If two or more antibodies are used, each coupled to different fluorescent dyes, then the data are more usually displayed in the form of a two-dimensional scatter diagram or as a contour diagram, where the fluorescence of one dye-labeled antibody is plotted against that of a second, with the result that a population of cells labeling with one antibody can be further subdivided by its labeling with the second antibody (see Fig. A.26). By examining large numbers of cells, flow cytometry can give quantitative data on the percentage of cells bearing different molecules, such as surface immunoglobulin, which characterizes B cells, the T-cell receptor-associated molecules known as CD3, and the CD4 and CD8 co-receptor proteins that distinguish the major T-cell subsets. Likewise, FACS analysis has been instrumental in defining stages in the early development of B and T cells. As the power of

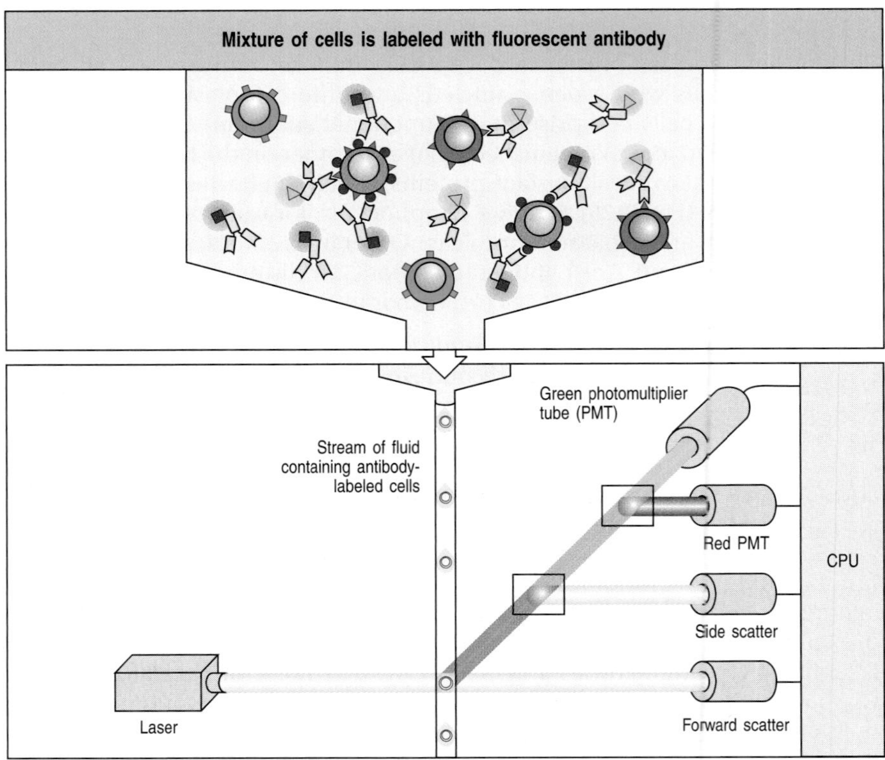

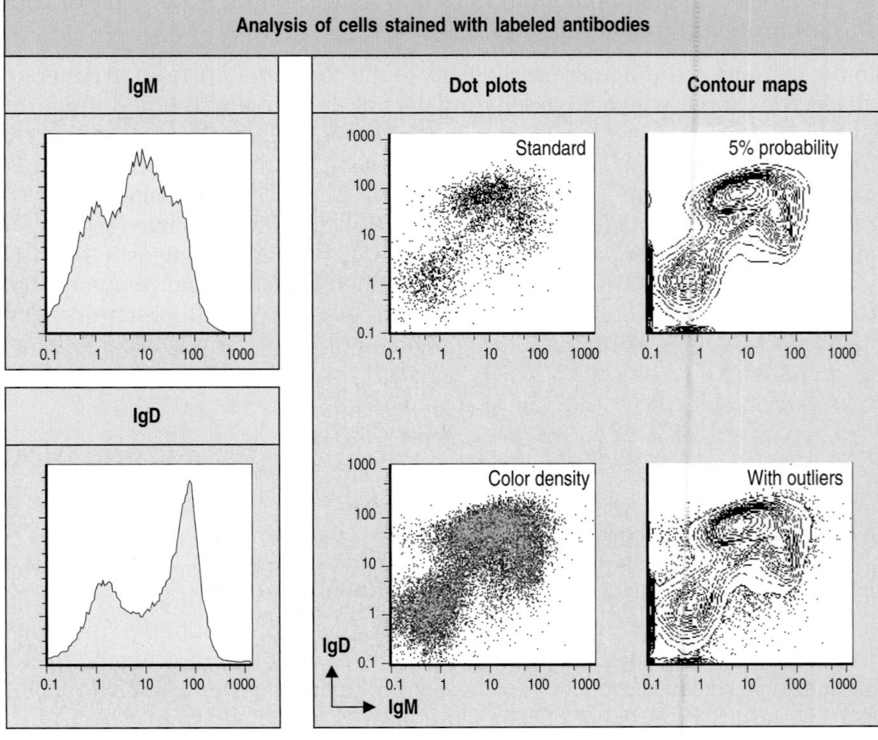

Fig. A.26 The FACS™ allows individual cells to be identified by their cell-surface antigens and to be sorted. Cells to be analyzed by flow cytometry are first labeled with fluorescent dyes (top panel). Direct labeling uses dye-coupled antibodies specific for cell-surface antigens (as shown here), while indirect labeling uses a dye-coupled immunoglobulin to detect unlabeled cell-bound antibody. The cells are forced through a nozzle in a single-cell stream that passes through a laser beam (second panel). Photomultiplier tubes (PMTs) detect the scattering of light, which is a sign of cell size and granularity, and emissions from the different fluorescent dyes. This information is analyzed by computer (CPU). By examining many cells in this way, the number of cells with a specific set of characteristics can be counted and levels of expression of various molecules on these cells can be measured. The lower part of the figure shows how this data can be represented, using the expression of two surface immuno-globulins, IgM and IgD, on a sample of B cells from a mouse spleen. The two immunoglobulins have been labeled with different colored dyes. When the expression of just one type of molecule is to be analyzed (IgM or IgD), the data is usually displayed as a histogram, as in the left-hand panels. Histograms display the distribution of cells expressing a single measured parameter (for example, size, granularity, fluorescence intensity). When two or more parameters are measured for each cell (IgM and IgD), various types of two-color plots can be used to display the data, as shown in the right-hand panel. All four plots represent the same data. The horizontal axis represents intensity of IgM fluorescence and the vertical axis the intensity of IgD fluorescence. Two-color plots provide more information than histo-grams; they allow recognition, for example, of cells that are 'bright' for both colors, 'dull' for one and bright for the other, dull for both, negative for both, and so on. For example, the cluster of dots in the extreme lower left portions of the plots represents cells that do not express either immunoglobulin, and are mostly T cells. The standard dot plot (upper left) places a single dot for each cell whose fluorescence is measured. It is good for picking up cells that lie outside the main groups but tends to saturate in areas containing a large number of cells of the same type. A second means of presenting these data is the color dot plot (lower left), which uses color density to indicate high-density areas. A contour plot (upper right) draws 5% 'probability' contours, with 5% of the cells lying between each contour providing the best monochrome visualization of regions of high and low density. The lower right plot is a 5% probability contour map which also shows outlying cells as dots.

the FACS technology has grown, progressively more antibodies labeled with distinct fluorescent dyes can be used at the same time. Three-, four-, and even five-color analyses can now be handled by very powerful machines. FACS analysis has been applied to a broad range of problems in immunology; indeed, it played a vital role in the early identification of AIDS as a disease in which T cells bearing CD4 are depleted selectively (see Chapter 12).

A-23 Lymphocyte isolation using antibody-coated magnetic beads.

Although the FACS is superb for isolating small numbers of cells in pure form, when large numbers of lymphocytes must be prepared quickly, mechanical means of separating cells are preferable. A powerful and efficient way of isolating lymphocyte populations is to couple paramagnetic beads to monoclonal antibodies that recognize distinguishing cell-surface molecules. These antibody-coated beads are mixed with the cells to be separated, and run through a column containing material that attracts the paramagnetic beads when the column is placed in a strong magnetic field. Cells binding the magnetically labeled antibodies are retained; cells lacking the appropriate surface molecule can be washed away (Fig. A.27). The bound cells are positively selected for expression of the particular cell-surface molecule, while the unbound cells are negatively selected for its absence.

A-24 Isolation of homogeneous T-cell lines.

The analysis of specificity and effector function in T cells depends heavily on the study of monoclonal populations of T lymphocytes. These can be obtained in four distinct ways. First, as for B-cell hybridomas (see Section A-12), normal T cells proliferating in response to specific antigen can be fused to malignant T-cell lymphoma lines to generate **T-cell hybrids**. The hybrids express the receptor of the normal T cell, but proliferate indefinitely owing to the cancerous state of the lymphoma parent. T-cell hybrids can be cloned to yield a population of cells all having the same T-cell receptor. When stimulated by their specific antigen these cells release cytokines such as the T-cell growth factor interleukin-2 (IL-2), and the production of cytokines is used as an assay to assess the antigen specificity of the T-cell hybrid.

T-cell hybrids are excellent tools for the analysis of T-cell specificity, as they grow readily in suspension culture. However, they cannot be used to analyze the regulation of specific T-cell proliferation in response to antigen because they are continually dividing. T-cell hybrids also cannot be transferred into an animal to test for function *in vivo* because they would give rise to tumors. Functional analysis of T-cell hybrids is also confounded by the fact that the malignant partner cell affects their behavior in functional assays. Therefore, the regulation of T-cell growth and the effector functions of T cells must be studied using **T-cell clones**. These are clonal cell lines of a single T-cell type and antigen specificity, which are derived from cultures of heterogeneous T cells,

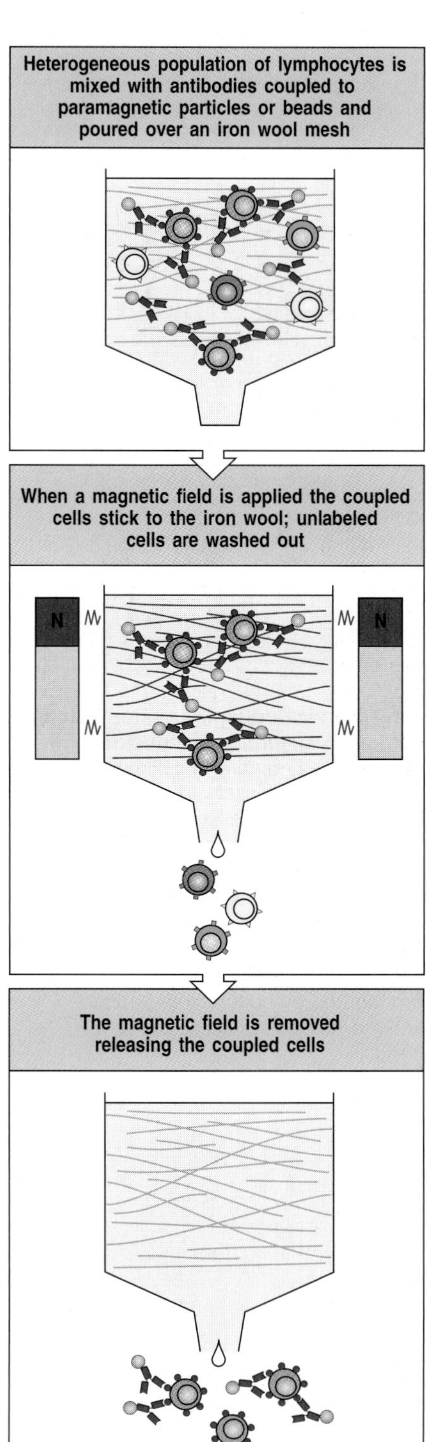

Heterogeneous population of lymphocytes is mixed with antibodies coupled to paramagnetic particles or beads and poured over an iron wool mesh
When a magnetic field is applied the coupled cells stick to the iron wool; unlabeled cells are washed out
N
N
The magnetic field is removed releasing the coupled cells

Fig. A.27 Lymphocyte subpopulations can be separated physically by using antibodies coupled to paramagnetic particles or beads. A mouse monoclonal antibody specific for a particular cell-surface molecule is coupled to paramagnetic particles or beads. It is mixed with a heterogeneous population of lymphocytes and poured over an iron wool mesh in a column. A magnetic field is applied so that the antibody-bound cells stick to the iron wool while cells which have not bound antibody are washed out; these cells are said to be negatively selected for lack of the molecule in question. The bound cells are released by removing the magnetic field; they are said to be positively selected for presence of the antigen recognized by the antibody.

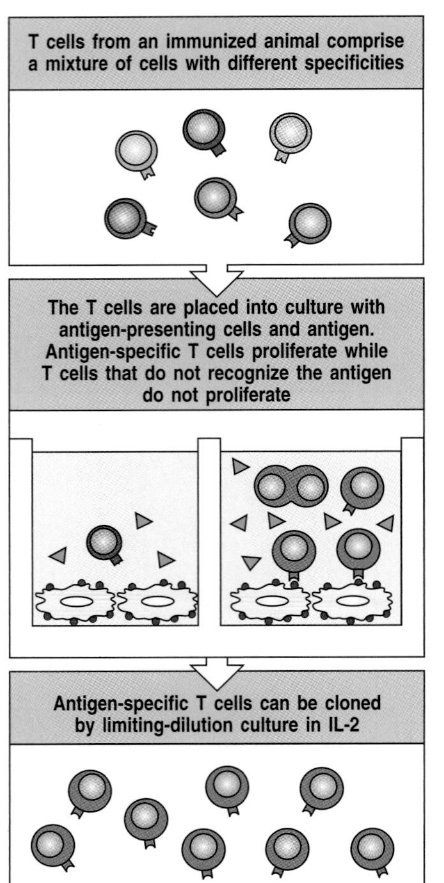

T cells from an immunized animal comprise a mixture of cells with different specificities
The T cells are placed into culture with antigen-presenting cells and antigen. Antigen-specific T cells proliferate while T cells that do not recognize the antigen do not proliferate
Antigen-specific T cells can be cloned by limiting-dilution culture in IL-2

Fig. A.28 Production of cloned T-cell lines. T cells from an immune donor, comprising a mixture of cells with different specificities, are activated with antigen and antigen-presenting cells. Single responding cells are cultured by limiting dilution in the T-cell growth factor IL-2, which selectively stimulates the responding cells to proliferate. From these single cells, cloned lines specific for antigen are identified and can be propagated by culture with antigen, antigen-presenting cells, and IL-2.

called **T-cell lines**, whose growth is dependent on periodic restimulation with specific antigen and, frequently, on the addition of T-cell growth factors (Fig. A.28). T-cell clones also require periodic restimulation with antigen and are more tedious to grow than T-cell hybrids but, because their growth depends on specific antigen recognition, they maintain antigen specificity, which is often lost in T-cell hybrids. Cloned T-cell lines can be used for studies of effector function both *in vitro* and *in vivo*. In addition, the proliferation of T cells, a critical aspect of clonal selection, can be characterized only in cloned T-cell lines, where such growth is dependent on antigen recognition. Thus, both types of monoclonal T-cell line have valuable applications in experimental studies.

Studies of human T cells have relied largely on T-cell clones because a suitable fusion partner for making T-cell hybrids has not been identified. However, a human T-cell lymphoma line, called Jurkat, has been characterized extensively because it secretes IL-2 when its antigen receptor is cross-linked with anti-receptor monoclonal antibodies. This simple assay system has yielded much information about signal transduction in T cells. One of the Jurkat cell line's most interesting features, shared with T-cell hybrids, is that it stops growing when its antigen receptor is cross-linked. This has allowed mutants lacking the receptor or having defects in signal transduction pathways to be selected simply by culturing the cells with anti-receptor antibody and selecting those that continue to grow. Thus, T-cell tumors, T-cell hybrids, and cloned T-cell lines all have valuable applications in experimental immunology.

Finally, primary T cells from any source can be isolated as single, antigen-specific cells by limiting dilution rather than by first establishing a mixed population of T cells in culture as a T-cell line and then deriving clonal subpopulations. During the growth of T-cell lines, particular T-cell clones can come to dominate the cultures and give a false picture of the number and specificities in the original sample. Direct cloning of primary T cells avoids this artifact.

Characterization of lymphocyte specificity, frequency, and function.

B cells are relatively easy to characterize as they have only one function—antibody production. T cells are more difficult to characterize as there are several different classes with different functions. It is also technically more difficult to study the membrane-bound T-cell receptors than the antibodies secreted in large amounts by B cells. All the methods in this part of the appendix can be used for T cells. Some are also used to detect and count B cells.

On many occasions it is important to know the frequency of antigen-specific lymphocytes, especially T cells, in order to measure the efficiency with which an individual responds to a particular antigen, for example, or the degree to which specific immunological memory has been established. There are a number of methods for doing this, either by detecting the cells directly by the specificity of their receptor, or by detecting activation of the cells to provide some particular function, such as cytokine secretion or cytotoxicity.

The first technique of this type to be established was the limiting-dilution culture (see Section A-25), in which the frequency of specific T or B cells responding to a particular antigen could be estimated by plating the cells into 96-well plates at increasing dilutions and measuring the number of wells in which there was no response. However, in this type of assay it became laborious to ask detailed questions about the phenotype of the responding cells, and to compare responses from different cell subpopulations.

A simpler assay for measuring the responses of T-cell populations has been developed from a variant of the antigen-capture ELISA method (see Section A-6), called the ELISPOT assay (see Section A-26). It assays T cells on the basis of cytokine production. In the ELISPOT assay, cytokine secreted by individual activated T cells is immobilized as discrete spots on a plastic plate, which are counted to give the number of activated T cells. The ELISPOT assay suffers from many of the same problems as the limiting-dilution assay in giving information about the nature of the activated cells, and it can be difficult to determine whether individual cells are capable of secreting mixtures of cytokines. It was therefore important to develop assays that could make these measurements on single cells. Measurements based on flow cytometry (see Section A-22) proved the answer, with the development of methods for detecting fluorescently labeled cytokines within activated T cells. The drawback of intracellular cytokine staining (see Section A-27) was that the T cells have to be killed and permeabilized by detergents to enable the cytokines to be detected. This led to the more sophisticated technique of capturing secreted labeled cytokines on the surfaces of the living T cells (see Section A-27).

Finally, methods for directly detecting T cells on the basis of the specificity of their receptor, using fluorochrome-tagged tetramers of specific MHC:peptide complexes (see Section A-28), have revolutionized the study of T-cell responses in a similar way to the use of monoclonal antibodies.

A-25 Limiting-dilution culture.

The response of a lymphocyte population is a measure of the overall response, but the frequency of lymphocytes able to respond to a given antigen can be determined only by **limiting-dilution culture**. This assay makes use of the Poisson distribution, a statistical function that describes how objects are distributed at random. For instance, when a sample of heterogeneous T cells is distributed equally into a series of culture wells, some wells will receive no T cells specific for a given antigen, some will receive one specific T cell, some two, and so on. The T cells in the wells are activated with specific antigen, antigen-presenting cells, and growth factors. After allowing several days for their growth and differentiation, the cells in each well are tested for a response to antigen, such as cytokine release or the ability to kill specific target cells (Fig. A.29). The assay is replicated with different numbers of T cells in the samples. The logarithm of the proportion of wells in which there is no response is plotted against the number of cells initially added to each well. If cells of one type, typically antigen-specific T cells because of their rarity, are the only limiting factor for obtaining a response, then a straight line is obtained. From the Poisson distribution, it is known that there is, on average, one antigen-specific cell per well when the proportion of negative wells is 37%. Thus, the frequency of antigen-specific cells in the population equals the reciprocal of the number of cells added to each well when 37% of the wells are negative. After priming, the frequency of specific cells goes up substantially, reflecting the antigen-driven proliferation of antigen-specific cells. The limiting-dilution assay can also be used to measure the frequency of B cells that can make antibody to a given antigen.

A-26 ELISPOT assays.

A modification of the ELISA antigen-capture assay (see Section A-6), called the **ELISPOT assay**, has provided a powerful tool for measuring the frequency of T-cell responses. Populations of T cells are stimulated with the antigen of interest, and are then allowed to settle onto a plastic plate coated with antibodies to the cytokine that is to be assayed (Fig. A.30). If an activated T cell is

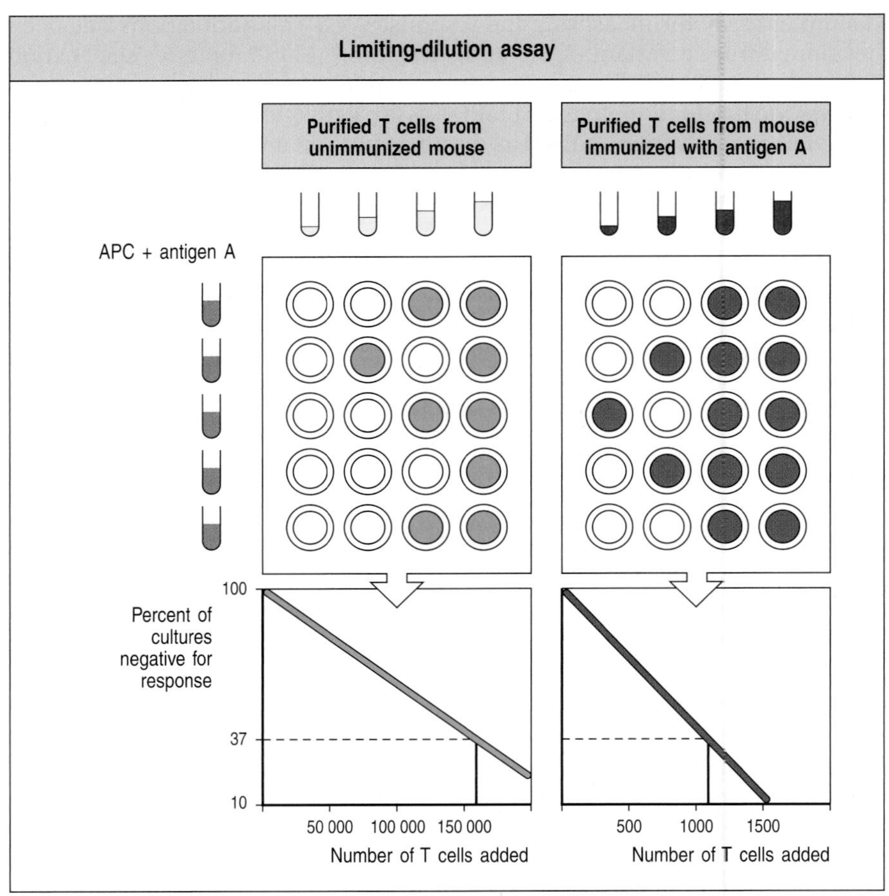

Fig. A.29 The frequency of specific lymphocytes can be determined using limiting-dilution assay. Varying numbers of lymphoid cells from normal or immunized mice are added to individual culture wells and stimulated with antigen and antigen-presenting cells (APCs) or polyclonal mitogen and added growth factors. After several days, the wells are tested for a specific response to antigen, such as cytotoxic killing of target cells. Each well that initially contained a specific T cell will make a response to its target, and from the Poisson distribution one can determine that when 37% of the wells are negative, each well contained, on average, one specific T cell at the beginning of the culture. In the example shown, for the unimmunized mouse 37% of the wells are negative when 160,000 T cells have been added to each well; thus the frequency of antigen-specific T cells is 1 in 160,000. When the mouse is immunized, 37% of the wells are negative when only 1100 T cells have been added; hence the frequency of specific T cells after immunization is 1 in 1100, an increase in responsive cells of 150-fold.

secreting that cytokine, it is captured by the antibody on the plastic plate. After a period the cells are removed, and a second antibody to the cytokine is added to the plate to reveal a circle of bound cytokine surrounding the position of each activated T cell; counting each spot, and knowing the number of T cells originally added to the plate allows a simple calculation of the frequency of T cells secreting that particular cytokine, giving the ELISPOT assay its name. ELISPOT can also be used to detect specific antibody secretion by B cells, in this case by using antigen-coated surfaces to trap specific antibody and labeled anti-immunoglobulin to detect the bound antibody.

A-27 Identification of functional subsets of T cells by staining for cytokines.

One problem with the detection of cytokine production on a single-cell level is that the cytokines are secreted by the T cells into the surrounding medium, and any association with the originating cell is lost. Two methods have been devised that allow the cytokine profile produced by individual cells to be determined. The first, that of **intracellular cytokine staining** (Fig. A.31), relies on the use of metabolic poisons that inhibit protein export from the cell. The cytokine thus accumulates within the endoplasmic reticulum and vesicular network of the cell. If the cells are subsequently fixed and rendered permeable by the use of mild detergents, antibodies can gain access to these intracellular compartments and detect the cytokine. The T cells can be stained for other markers simultaneously, and thus the frequency, for example, of IL-10-producing $CD25^+$ CD4 T cells can be easily obtained.

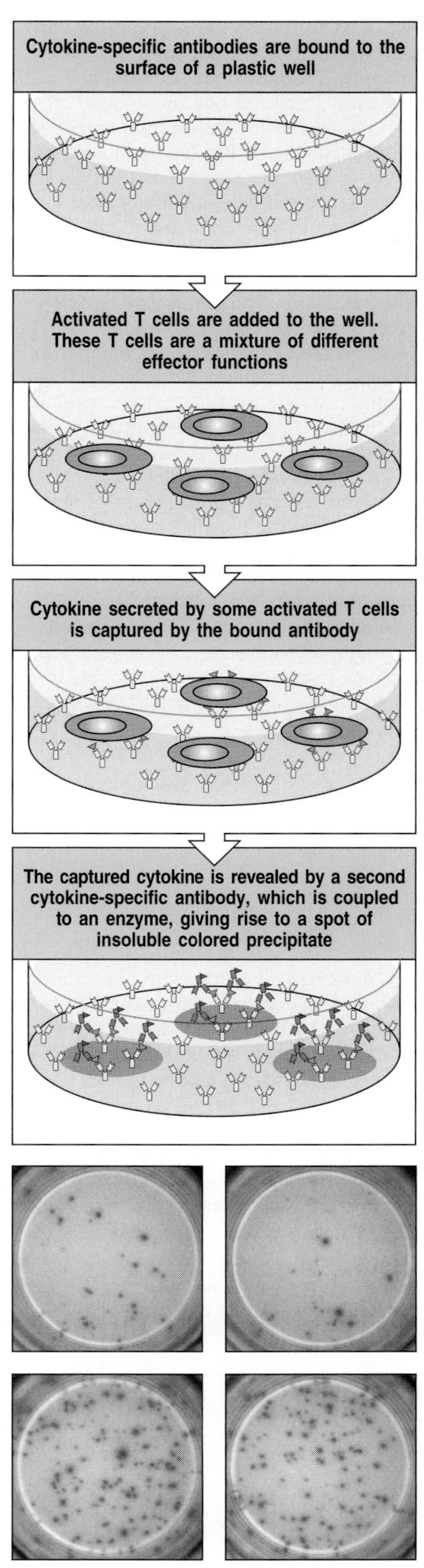

Cytokine-specific antibodies are bound to the surface of a plastic well
Activated T cells are added to the well. These T cells are a mixture of different effector functions
Cytokine secreted by some activated T cells is captured by the bound antibody
The captured cytokine is revealed by a second cytokine-specific antibody, which is coupled to an enzyme, giving rise to a spot of insoluble colored precipitate

Fig. A.30 The frequency of cytokine-secreting T cells can be determined by the ELISPOT assay. The ELISPOT assay is a variant of the ELISA assay in which antibodies bound to a plastic surface are used to capture cytokines secreted by individual T cells. Usually, cytokine specific antibodies are bound to the surface of a plastic tissue-culture well and the unbound antibodies are removed (top panel). Activated T cells are then added to the well and settle onto the antibody-coated surface (second panel). If a T cell is secreting the appropriate cytokine, this will then be captured by the antibody molecules on the plate surrounding the T cell (third panel). After a period of time the T cells are removed, and the presence of the specific cytokine is detected using an enzyme-labeled second antibody specific for the same cytokine. Where this binds, a colored reaction product can be formed (fourth panel). Each T cell that originally secreted cytokine gives rise to a single spot of color, hence the name of the assay. The results of such an ELISPOT assay for T cells secreting IFN-γ in response to different stimuli are shown in the last panel. In this example, T cells from a stem cell transplant recipient were treated with a control peptide (top two panels) or a peptide from cytomegalovirus (bottom two panels). You can see a greater number of spots in the bottom two panels, indicating clearly that the patient's T cells are able to respond to the viral peptide and produce IFNγ. Photographs courtesy of S. Nowack.

A second method, which has the advantage that the cells being analyzed are not killed in the process, is called **cytokine capture**. This technique uses hybrid antibodies, in which the two separate heavy- and light-chain pairs from different antibodies are combined to give a mixed antibody molecule in which the two antigen-binding sites recognize different ligands (Fig. A.32). In the bispecific antibodies used to detect cytokine production, one of the antigen-binding sites is specific for a T-cell surface marker, while the other is specific for the cytokine in question. The bispecific antibody binds to the T cells through the binding site for the cell-surface marker, leaving the cytokine-binding site free. If that T cell is secreting the particular cytokine, it is captured by the bound antibody before it diffuses away from the surface of the cell. It can then be detected by adding a fluorochrome-labeled second antibody specific for the cytokine to the cells.

A-28 Identification of T-cell receptor specificity using MHC:peptide tetramers.

For many years, the ability to identify antigen-specific T cells directly through their receptor specificity eluded immunologists. Foreign antigen could not be used directly to identify T cells, since, unlike B cells, they do not recognize antigen alone but rather the complexes of peptide fragments of antigen bound to self MHC molecules. Moreover, the affinity of interaction between the T-cell receptor and the MHC:peptide complex was in practice so low that attempts to label T cells with their specific MHC:peptide complexes routinely failed. The breakthrough in labeling antigen-specific T cells came with the idea of making multimers of the MHC:peptide complex, so as to increase the avidity of the interaction.

Peptides can be biotinylated using the bacterial enzyme BirA, which recognizes a specific amino acid sequence. Recombinant MHC molecules containing this target sequence are used to make MHC:peptide complexes which are then biotinylated. Avidin, or the bacterial counterpart streptavidin, contains four sites that bind biotin with extremely high affinity. Mixing the biotinylated MHC:peptide complex with avidin or streptavidin results in the formation of an **MHC:peptide tetramer**—four specific MHC:peptide complexes bound to a single molecule of streptavidin (Fig. A.33). Routinely, the streptavidin moiety is labeled with a fluorochrome to allow detection of those T cells capable of binding the MHC:peptide tetramer.

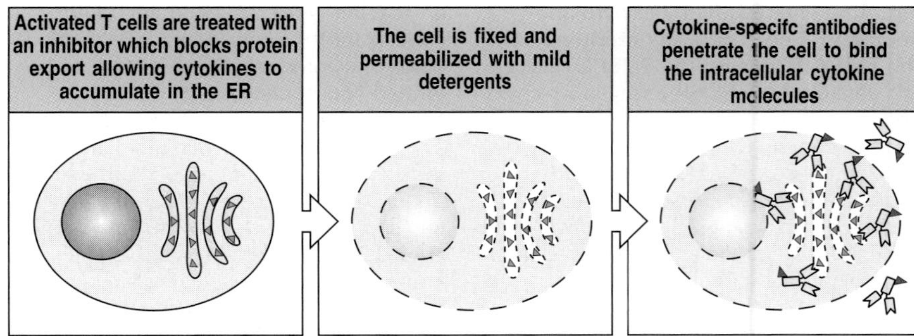

Activated T cells are treated with an inhibitor which blocks protein export allowing cytokines to accumulate in the ER
The cell is fixed and permeabilized with mild detergents
Cytokine-specifc antibodies penetrate the cell to bind the intracellular cytokine mclecules

Fig. A.31 Cytokine-secreting cells can be identified by intracellular cytokine staining. The cytokines secreted by activated T cells can be determined by using fluorochrome-labeled antibodies to detect cytokine molecules that have been allowed to accumulate inside the cell. The accumulation of cytokine molecules, to allow them to reach a high enough concentration for efficient detection, is achieved by treating the activated T cells with inhibitors of protein export. In such treated cells, proteins destined to be secreted are instead retained within the endoplasmic reticulum (first panel). These treated cells are then fixed, to cross-link the proteins inside the cell and in the cell membranes, so that they are not lost when the cell is permeabilized by dissolving the cell membrane in a mild detergent (center panel). Fluorochrome-labeled antibodies can now enter the permeabilized cell and bind to the cytokines inside the cell (last panel). Cells labeled in this way can also be labeled with antibodies that bind to cell-surface proteins to determine which subsets of T cells are secreting particular cytokines.

MHC:peptide tetramers have been used to identify populations of antigen-specific T cells in, for example, patients with acute Epstein–Barr virus infections (infectious mononucleosis), showing that up to 80% of the peripheral T cells in infected individuals can be specific for a single MHC:peptide complex. They have also been used to follow responses over longer timescales in individuals with HIV or, in the example we show, cytomegalovirus infections. These reagents have also been important in identifying the cells responding, for example, to nonclassical class I molecules such as HLA-E or HLA-G, in both cases showing that these nonclassical molecules are recognized by subsets of NK receptors.

A-29 Assessing the diversity of the T-cell repertoire by 'spectratyping.'

The extent of the diversity of the T-cell repertoire, either generally or during specific immune responses, is often of interest. In particular, as T cells do not undergo somatic hypermutation and affinity maturation in the same way that B cells do, the relationship between the repertoire of T cells making a primary response to antigen and the repertoire of T cells involved in secondary

Fig. A.32 Hybrid antibodies containing cell-specific and cytokine-specific binding sites can be used to assay cytokine secretion by living cells and to purify cells secreting particular cytokines. Hybrid antibodies can be made by mixing together heavy- and light-chain pairs from antibodies of different specificities, for example, an antibody to an MHC class I molecule and an antibody specific for a cytokine such as IL-4 (first panel). The hybrid antibodies are then added to a population of activated T cells, and bind to each cell via the MHC class I binding arm (second panel). If some of the cells in the population are secreting the appropriate cytokine, IL-4, this is captured by the cytokine specific arm of the hybrid antibody (third panel). The presence of the cytokine can then be revealed, for example using a fluorochrome-labeled second antibody specific for the same cytokine, but binding to a different site to the one used for the hybrid antibody (last panel). Such labeled cells may be analyzed by flow cytometry, or can be isolated using a fluorescence activated cell sorter. Alternatively, the second cytokine specific antibody may be coupled to magnetic beads, and the cytokine producing cells isolated magnetically.

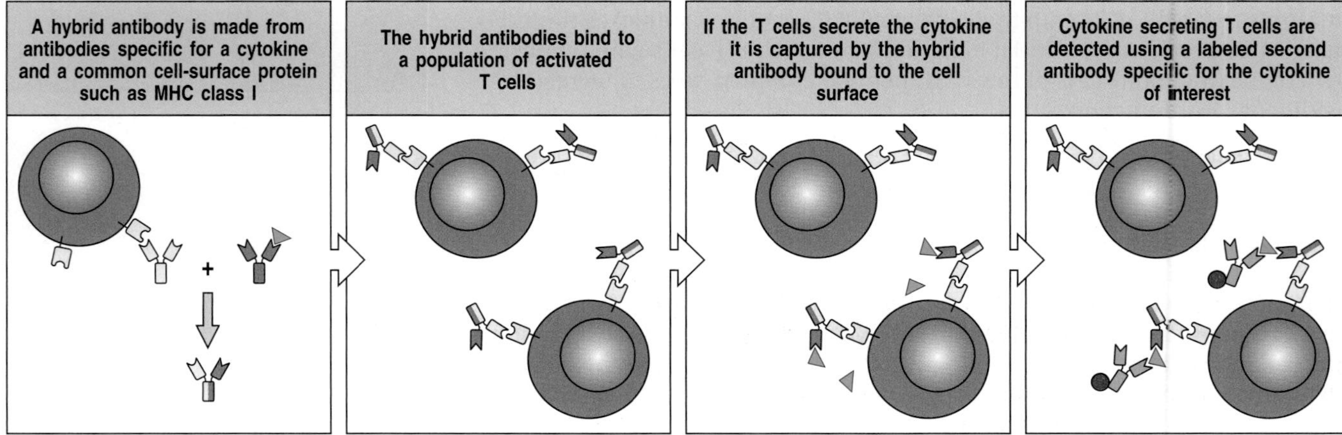

A hybrid antibody is made from antibodies specific for a cytokine and a common cell-surface protein such as MHC class I
The hybrid antibodies bind to a population of activated T cells
If the T cells secrete the cytokine it is captured by the hybrid antibody bound to the cell surface
Cytokine secreting T cells are detected using a labeled second antibody specific for the cytokine of interest

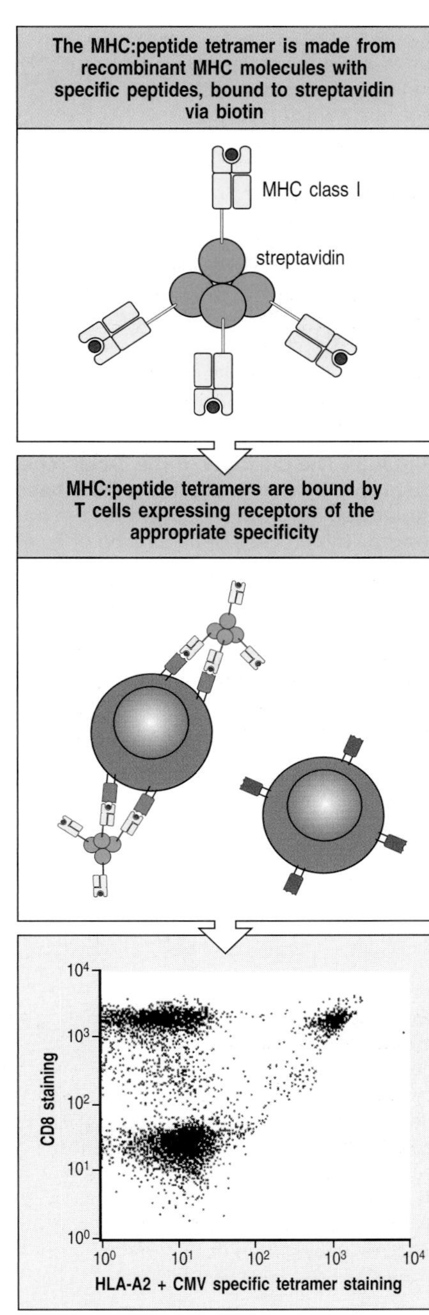

The MHC:peptide tetramer is made from recombinant MHC molecules with specific peptides, bound to streptavidin via biotin
MHC class I
streptavidin
MHC:peptide tetramers are bound by T cells expressing receptors of the appropriate specificity
CD8 staining
HLA-A2 + CMV specific tetramer staining
10^0
10^1
10^2
10^3
10^4

Fig. A.33 MHC:peptide complexes coupled to streptavidin to form tetramers are able to stain antigen-specific T cells. MHC:peptide tetramers are formed from recombinant refolded MHC:peptide complexes containing a single defined peptide epitope. The MHC molecules can be chemically derivatized to contain biotin, but more usually the recombinant MHC heavy chain is linked to a bacterial biotinylation sequence, a target for the E. coli enzyme BirA, which is used to add a single biotin group to the MHC molecule. Streptavidin is a tetramer, each subunit having a single binding site for biotin, hence the streptavidin/MHC:peptide complex creates a tetramer of MHC:peptide complexes (top panel). While the affinity between the T-cell receptor and its MHC:peptide ligand is too low for a single complex to bind stably to a T cell, the tetramer, by being able to make a more avid interaction with multiple MHC:peptide complexes binding simultaneously, is able to bind to T cells whose receptors are specific for the particular MHC:peptide complex (center panel). Routinely, the streptavidin molecules are coupled to a fluoro-chrome, so that the binding to T cells can be monitored by flow cytometry. In the example shown in the bottom panel, T cells have been stained simultaneously with antibodies specific for CD3 and CD8, and with a tetramer of HLA-A2 molecules containing a cytomegalovirus peptide. Only the $CD3^+$ cells are shown, with the staining of CD8 displayed on the vertical axis and the tetramer staining displayed along the horizontal axis. The $CD8^-$ cells (mostly $CD4^+$) on the bottom left of the figure show no specific tetramer staining, while the bulk of the $CD8^+$ cells, on the top left, likewise show no tetramer staining. However, a discrete population of tetramer positive $CD8^+$ cells, at the top right of the panel, comprising some 5% of the total $CD8^+$ cells, can clearly be demonstrated. Data courtesy of G. Aubert.

and subsequent responses to antigen has been difficult to determine. This information has usually been obtained through the laborious process of cloning the T cells involved in specific responses (see Section A-24), and the cloning and sequencing of their T-cell receptors.

It is possible, however, to estimate the diversity of T-cell responses by making use of the junctional diversity generated when T-cell receptors are created by somatic recombination, a technique known as **spectratyping**. Variability in the length of the CDR3 segments is created during the recombination process, both by variation in the exact positions at which the junctions between gene segments occur, and by variation in the number of N-nucleotides added. Both these processes result in the length of the V_β CDR3 varying by up to nine amino acids. The problem in detecting this variability is that there are 24 families of V_β gene segments in humans and it is not possible to design a single oligonucleotide primer that will anneal to all of these families. Specific oligonucleotide primers can, however, be designed for each family, and these can be used in the polymerase chain reaction (PCR), together with a primer specific for the C_β region, to amplify, for each individual family, a segment of the mRNA for the T-cell receptor β chain that spans the CDR3 region. A population of TCR V_β genes will therefore show a distribution, or 'spectrum,' of CD3 lengths, and will give rise to PCR products of different lengths that can be resolved, usually by polyacrylamide gel electrophoresis (Fig. A.34). The deletion and addition of nucleotides during the generation of T-cell receptors by rearrangement is random, and so in a normal individual the CDR3 lengths follow a Gaussian distribution. Deviations from this Gaussian distribution, such as an excess of one particular CDR3 length, indicate the presence of clonal expansions of T cells, such as occurs during a T-cell response.

A-30 Biosensor assays for measuring the rates of association and disassociation of antigen receptors for their ligands.

Two of the important questions that are always asked of any receptor–ligand interaction is: what is the strength of binding, or affinity, of the interaction,

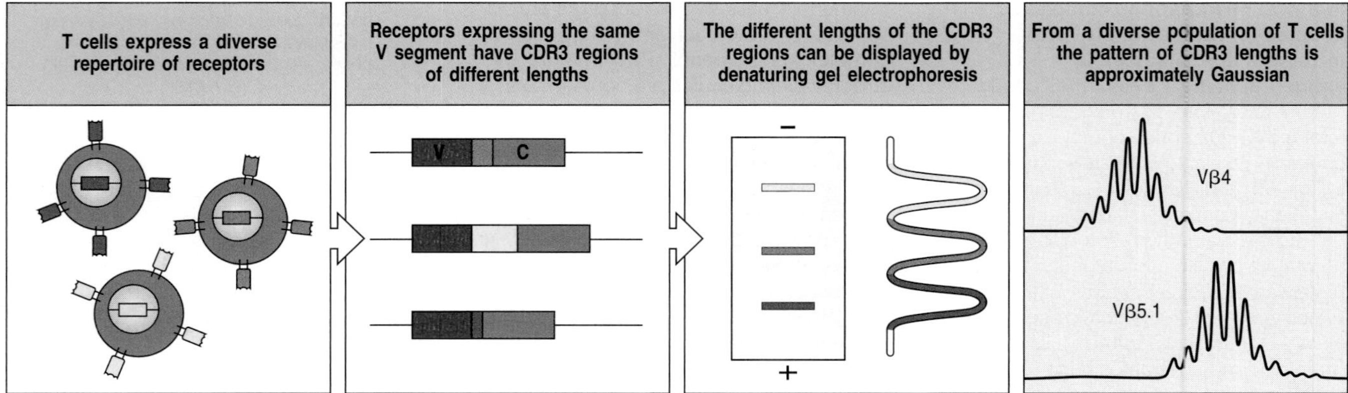
T cells express a diverse repertoire of receptors
Receptors expressing the same V segment have CDR3 regions of different lengths
V
C
The different lengths of the CDR3 regions can be displayed by denaturing gel electrophoresis
−
+
From a diverse population of T cells the pattern of CDR3 lengths is approximately Gaussian
Vβ4
Vβ5.1

Fig. A.34 The diversity of the T-cell receptor repertoire can be displayed by spectratyping, a PCR-based technique that separates different receptors on the basis of their CDR3 length. The process of generation of T-cell receptors is stochastic, giving rise to a population of mature T cells whose receptors are clonally distributed (first panel). In each of the cells expressing a particular V_{β} gene segment, all of the differences between the unique receptors are restricted to the CDR3 region, where there will be differences in length as well as sequence as a consequence of the imprecision of the rearrangement process (second panel). Using sets of primers for the PCR reaction that are specific for individual V_{β} gene segments at one end and for a conserved part of the C region at the other, it is possible to generate a set of DNA fragments that span the CDR3 region. If these are separated by denaturing acrylamide gel electrophoresis then a series of bands are formed or, since these fragments can be labeled with fluorochromes and analyzed by automated gel readers, a series of peaks corresponding to the different length fragments (third panel). The pattern of peaks obtained in this way is known as a spectratype. From a diverse population of cells, the distribution of fragment lengths is Gaussian, as shown in the last panel, where the spectratypes of two different V_{β} regions from the same individual are shown. In this case, both of the patterns are approximately Gaussian; deviations from a Gaussian distribution may indicate expansion of particular clones of T cells, perhaps in response to antigenic challenge. Data courtesy of L. McGreavey.

and what are the rates of association and disassociation? Traditionally, measurements of affinity have been made by equilibrium binding measurements (see Section A-9), and measurements of rates of binding have been difficult to obtain. Equilibrium binding assays also cannot be performed on T-cell receptors, which have large macromolecular ligands and which cannot be isolated and purified in large quantity.

It is now possible to measure binding rates directly, by following the binding of ligands to receptors immobilized on gold-plated glass slides, using a phenomenon known as **surface plasmon resonance (SPR)** to detect the binding (Fig. A.35). A full explanation of surface plasmon resonance is beyond the scope of this textbook, as it is based on advanced physical and quantum mechanical principles. Briefly, it relies on the total internal reflection of a beam of light from the surface of a gold-coated glass slide. As the light is reflected, some of its energy excites electrons in the gold coating and these excited electrons are in turn affected by the electric field of any molecules binding to the surface of the glass coating. The more molecules that bind to the surface, the greater the effect on the excited electrons, and this in turn affects the reflected light beam. The reflected light thus becomes a sensitive measure of the number of atoms bound to the gold surface of the slide.

If a purified receptor is immobilized on the surface of the gold-coated glass slide, to make a biosensor 'chip,' and a solution containing the ligand is flowed over that surface, the binding of ligand to the receptor can be followed until it reaches equilibrium (see Fig. A.35). If the ligand is then washed out, dissociation of ligand from the receptor can easily be followed and the dissociation rate calculated. A new solution of the ligand at a different concentration can then be flowed over the chip and the binding once again measured. The affinity of binding can be calculated in a number of ways in this type of assay. Most simply, the ratio of the rates of association and dissociation will give an estimate of

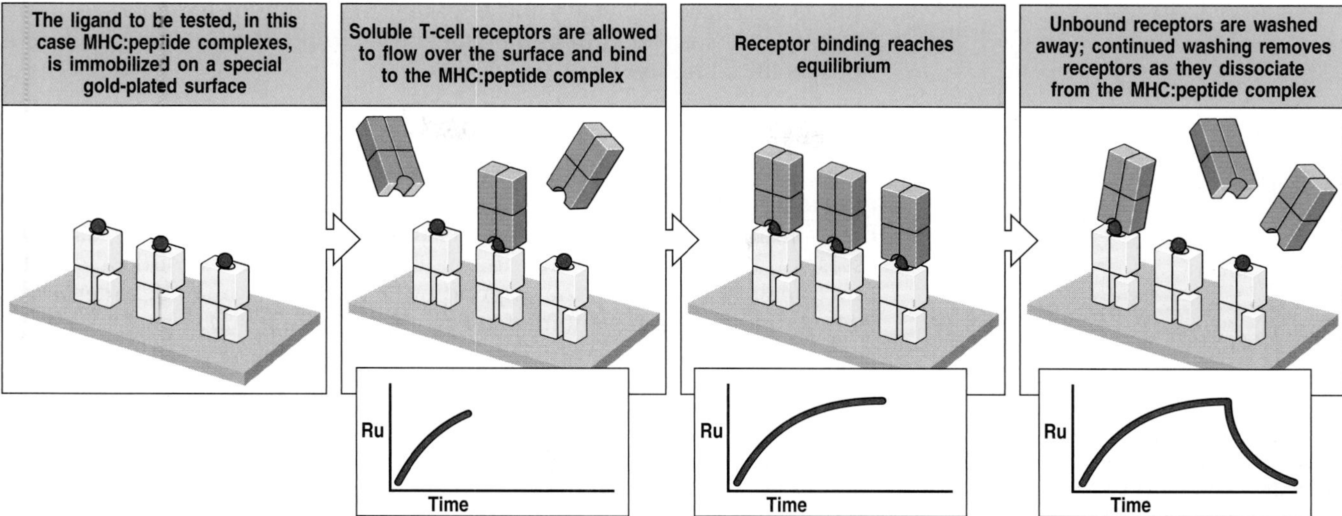

Fig. A.35 Measurement of receptor–ligand interactions can be made in real time using a biosensor. Biosensors are able to measure the binding of molecules on the surface of gold-plated glass chips through the indirect effects of the binding on the total internal reflection of a beam of polarized light at the surface of the chip. Changes in the angle and intensity of the reflected beam are measured in 'resonance units' (Ru) and plotted against time in what is termed a 'sensorgram.' Depending on the exact nature of the receptor–ligand pair to be analyzed, either the receptor or the ligand can be immobilized on the surface of the chip. In the example shown, MHC:peptide complexes are immobilized on such a surface (first panel). T-cell receptors in solution are now allowed to flow over the surface, and to bind to the immobilized MHC:peptide complexes (second panel). As the T-cell receptors bind, the sensorgram (inset panel below the main panel) reflects the increasing amount of protein bound. As the binding reaches either saturation or equilibrium (third panel), the sensorgram shows a plateau, as no more protein binds. At this point, unbound receptors can be washed away. With continued washing, bound receptors now start to dissociate and are removed in the flow of the washing solution (last panel). The sensorgram now shows a declining curve, reflecting the rate at which the receptor and ligand dissociation occurs.

the affinity, but more accurate estimates can be obtained from the measurements of the binding at different concentrations of ligand. From measurements of binding at equilibrium, a Scatchard plot (see Fig. A.11) will give a measurement of the affinity of the receptor–ligand interaction.

A-31 Stimulation of lymphocyte proliferation by treatment with polyclonal mitogens or specific antigen.

To function in adaptive immunity, rare antigen-specific lymphocytes must proliferate extensively before they differentiate into functional effector cells in order to generate sufficient numbers of effector cells of a particular specificity. Thus, the analysis of induced lymphocyte proliferation is a central issue in their study. It is, however, difficult to detect the proliferation of normal lymphocytes in response to specific antigen because only a minute proportion of cells will be stimulated to divide. Great impetus was given to the field of lymphocyte culture by the finding that certain substances induce many or all lymphocytes of a given type to proliferate. These substances are referred to collectively as **polyclonal mitogens** because they induce mitosis in lymphocytes of many different specificities or clonal origins. T and B lymphocytes are stimulated by different polyclonal mitogens (Fig. A.36). Polyclonal mitogens seem to trigger essentially the same growth response mechanisms as antigen. Lymphocytes normally exist as resting cells in the G_0 phase of the cell cycle. When stimulated with polyclonal mitogens, they rapidly enter the G_1 phase and progress through the cell cycle. In most studies, lymphocyte proliferation is most simply measured by the incorporation of ^{3}H-thymidine into DNA.

Mitogen	Responding cells
Phytohemagglutinin (PHA) (red kidney bean)	T cells
Concanavalin (ConA) (Jack bean)	T cells
Pokeweed mitogen (PWM) (Pokeweed)	T and B cells
Lipopolysaccharide (LPS) (*Escherichia coli*)	B cells (mouse)

Fig. A.36 Polyclonal mitogens, many of plant origin, stimulate lymphocyte proliferation in tissue culture. Many of these mitogens are used to test the ability of lymphocytes in human peripheral blood to proliferate.

This assay is used clinically for assessing the ability of lymphocytes from patients with suspected immunodeficiencies to proliferate in response to a nonspecific stimulus.

Once lymphocyte culture had been optimized using the proliferative response to polyclonal mitogens as an assay, it became possible to detect antigen-specific T-cell proliferation in culture by measuring ^{3}H-thymidine uptake in response to an antigen to which the T-cell donor had been previously immunized (Fig. A.37). This is the assay most commonly used for assessing T-cell responses after immunization, but it reveals little about the functional capabilities of the responding T cells. These must be ascertained by functional assays, as outlined in Sections A-33 and A-34.

A-32 Measurements of apoptosis by the TUNEL assay.

Apoptotic cells can be detected by a procedure known as **TUNEL staining**. In this technique, the 3′ ends of the DNA fragments generated in apoptotic cells are labeled with biotin-coupled uridine by using the enzyme terminal deoxynucleotidyl transferase (TdT). The biotin label is then detected with enzyme tagged streptavidin, which binds to biotin. When the colorless substrate of the enzyme is added to a tissue section or cell culture, it is reacted upon to produce a colored precipitate only in cells that have undergone apoptosis (Fig. A.38). This technique has revolutionized the detection of apoptotic cells.

A-33 Assays for cytotoxic T cells.

Activated CD8 T cells generally kill any cells that display the specific peptide:MHC class I complex they recognize. Thus CD8 T-cell function can be determined using the simplest and most rapid T-cell bioassay—the killing of a target cell by a cytotoxic T cell. This is usually detected in a ^{51}Cr-release assay. Live cells will take up, but do not spontaneously release, radioactively labeled sodium chromate, $Na_2{}^{51}CrO_4$. When these labeled cells are killed, the radioactive chromate is released and its presence in the supernatant of mixtures of target cells and cytotoxic T cells can be measured (Fig. A.39). In a similar assay, proliferating target cells such as tumor cells can be labeled with ^{3}H-thymidine, which is incorporated into the replicating DNA. On attack by a cytotoxic T cell, the DNA of the target cells is rapidly fragmented and retained in the filtrate, while large, unfragmented DNA is collected on a filter, and one can measure either the release of these fragments or the retention of ^{3}H-thymidine in chromosomal DNA. These assays provide a rapid, sensitive, and specific measure of the activity of cytotoxic T cells.

A-34 Assays for CD4 T cells.

CD4 T-cell functions usually involve the activation rather than the killing of cells bearing specific antigen. The activating effects of CD4 T cells on B cells or macrophages are mediated in large part by nonspecific mediator proteins

Immunize with antigen A
After 5–10 days remove cells from lymph node
Culture cells with antigens A or B
no antigen
antigen A
antigen B
Measure T-cell proliferation
Proliferation
antigen A
antigen B
no antigen
3
5
Days of culture

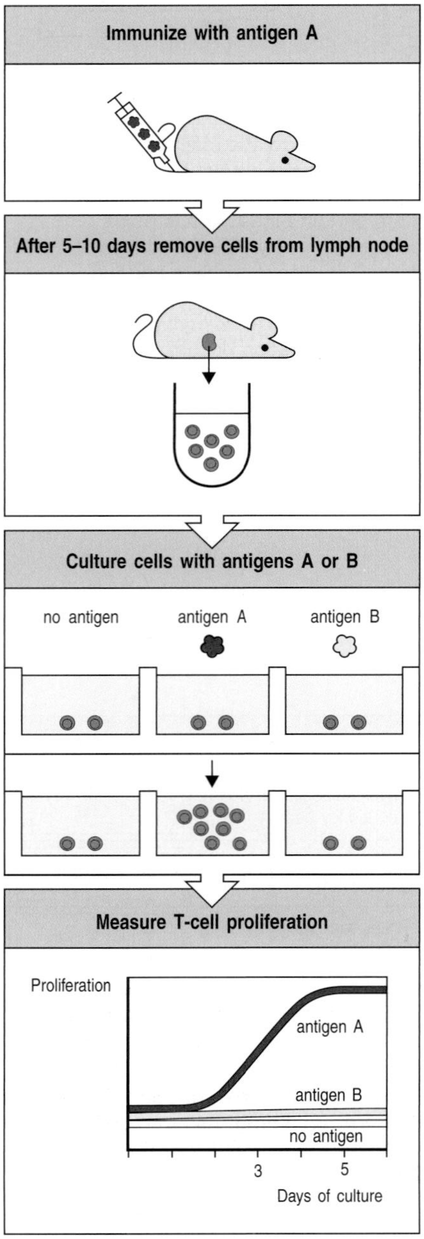

Fig. A.37 Antigen-specific T-cell proliferation is used frequently as an assay for T-cell responses. T cells from mice or humans that have been immunized with an antigen (A) proliferate when they are exposed to antigen A and antigen-presenting cells but not when cultured with unrelated antigens to which they have not been immunized (antigen B). Proliferation can be measured by incorporation of ^{3}H-thymidine into the DNA of actively dividing cells. Antigen-specific proliferation is a hallmark of specific CD4 T-cell immunity.

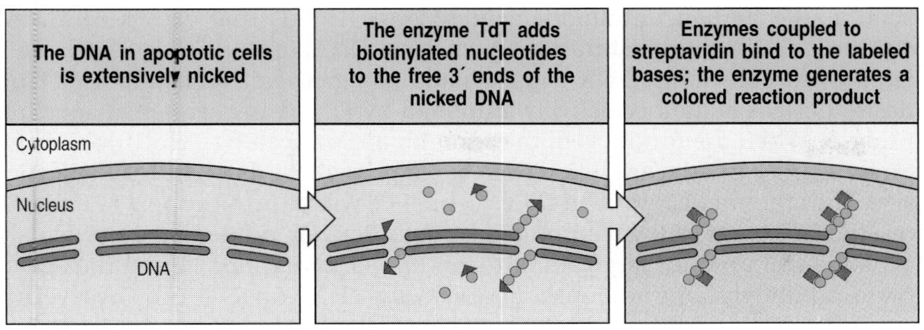

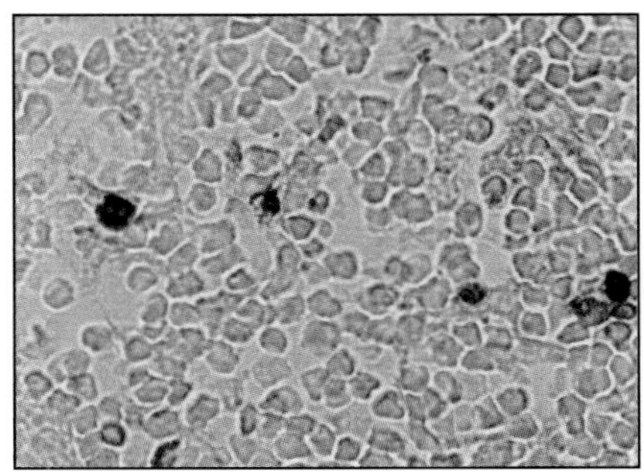

Fig. A.38 Fragmented DNA can be labeled by terminal deoxynucleotidyl transferase (TdT) to reveal apoptotic cells. When cells undergo programmed cell death, or apoptosis, their DNA becomes fragmented (left panel). The enzyme TdT is able to add nucleotides to the ends of DNA fragments; most commonly in this assay, biotin-labeled nucleotides (usually dUTP) are added (second panel). The biotinylated DNA can be detected by using streptavidin, which binds to biotin, coupled to enzymes that convert a colorless substrate into a colored insoluble product (third panel). Cells stained in this way can be detected by light microscopy, as shown in the photograph of apoptotic cells (stained red) in the thymic cortex. Photograph courtesy of R. Budd and J. Russell.

called cytokines, which are released by the T cell when it recognizes antigen. Thus, CD4 T-cell function is usually studied by measuring the type and amount of these released proteins. As different effector T cells release different amounts and types of cytokines, one can learn about the effector potential of that T cell by measuring the proteins it produces.

Cytokines can be detected by their activity in biological assays of cell growth, where they serve either as growth factors or growth inhibitors. A more specific assay is a modification of ELISA known as a capture or sandwich ELISA (see Section A-6). In this assay, the cytokine is characterized by its ability to bridge between two monoclonal antibodies reacting with different epitopes on the cytokine molecule. Cytokine-secreting cells can also be detected by ELISPOT (see Section A-26).

Sandwich ELISA and ELISPOT avoid a major problem of cytokine bioassays, the ability of different cytokines to stimulate the same response in a bioassay. Bioassays must always be confirmed by inhibition of the response with neutralizing monoclonal antibodies specific for the cytokine. Another way of identifying cells actively producing a given cytokine is to stain them with a fluorescently tagged anti-cytokine monoclonal antibody and identify and count them by FACS (see Section A-22).

A quite different approach to detecting cytokine production is to determine the presence and amount of the relevant cytokine mRNA in stimulated T cells. This can be done for single cells by *in situ* hybridization and for cell populations by **reverse transcriptase–polymerase chain reaction (RT–PCR)**. Reverse transcriptase is an enzyme used by certain RNA viruses,

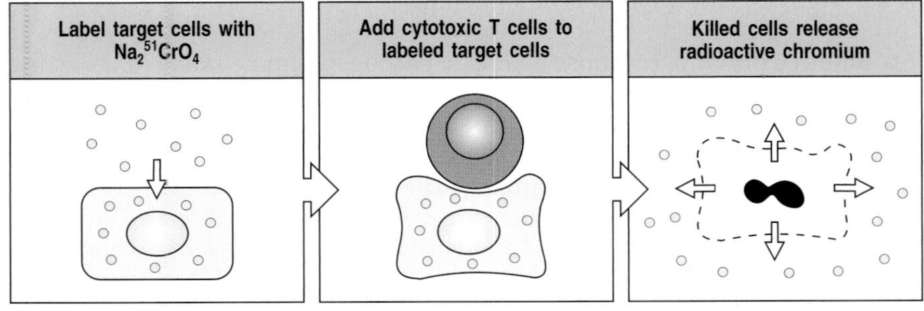

Fig. A.39 Cytotoxic T-cell activity is often assessed by chromium release from labeled target cells. Target cells are labeled with radioactive chromium as $Na_2{}^{51}CrO_4$, washed to remove excess radioactivity and exposed to cytotoxic T cells. Cell destruction is measured by the release of radioactive chromium into the medium, detectable within 4 hours of mixing target cells with T cells.

such as the human immunodeficiency virus (HIV-1) that causes AIDS, to convert an RNA genome into a DNA copy, or cDNA. In RT–PCR, mRNA is isolated from cells and cDNA copies made using reverse transcriptase. The desired cDNA is then selectively amplified by PCR using sequence-specific primers. When the products of the reaction are subjected to electrophoresis on an agarose gel, the amplified DNA can be visualized as a band of a specific size. The amount of amplified cDNA sequence will be proportional to its representation in the mRNA; stimulated T cells actively producing a particular cytokine will produce large amounts of that particular mRNA and thus give correspondingly large amounts of the selected cDNA on RT–PCR. The level of cytokine mRNA in the original tissue is usually determined by comparison with the outcome of RT–PCR on the mRNA produced by a so-called 'housekeeping gene' expressed by all cells.

A-35 DNA microarrays.

Any cell expresses, at any one time, many hundreds or even thousands of genes. Some of the products are expressed at high levels, the actin that forms the cytoskeleton of the cell is one example, while others may only be expressed in a few copies per cell. Different cell types, or cells at different stages of maturation, or even tumor cells compared to their normal counterparts, will express different sets of genes, and trying to identify these differences is an important field of research, in immunology as well as in other areas of biology. One important new technique in analyzing these differences makes use of arrays of hundreds of DNA sequences attached to a glass surface—a so-called **DNA microarray** or 'DNA chip.' The array contains a range of DNA sequences from known genes, arranged in a fixed pattern, and the differential expression of those genes in a particular cell type or tissue is tested by exposing the array to labeled mRNA (or cDNA made from it) from the tissue. Hybridization of labeled mRNAs to their corresponding DNA sequences in the array is detected by standard techniques and the whole technique is readily automated. Many different samples can be examined in parallel, which makes this a powerful analytical technique, as can be seen from the example we illustrate in Fig. A.40. Here the DNA microarray has been constructed with almost 18,000 cDNA clones known to be expressed in either B or T cells and B-cell tumors. This array was then probed with fluorochrome-labeled cDNAs from 96 normal and malignant cells, and the level of expression of approximately 18,000 genes in each of the cell lines was measured simultaneously. In this particular case the patterns of expression of the different genes revealed that the malignant B cells formed discrete subtypes, which were then found to have different clinical prognoses.

Detection of immunity *in vivo*.

A-36 Assessment of protective immunity.

An adaptive immune response against a pathogen often confers long-lasting immunity against infection with that pathogen; successful vaccination achieves the same end. The very first experiment in immunology, Jenner's successful vaccination against smallpox, is still the model for assessing the presence of such protective immunity. The assessment of protective immunity conferred by vaccination has three essential steps. First, an immune response is elicited by immunization with a candidate vaccine. Second, the immunized individuals, along with unimmunized controls, are challenged

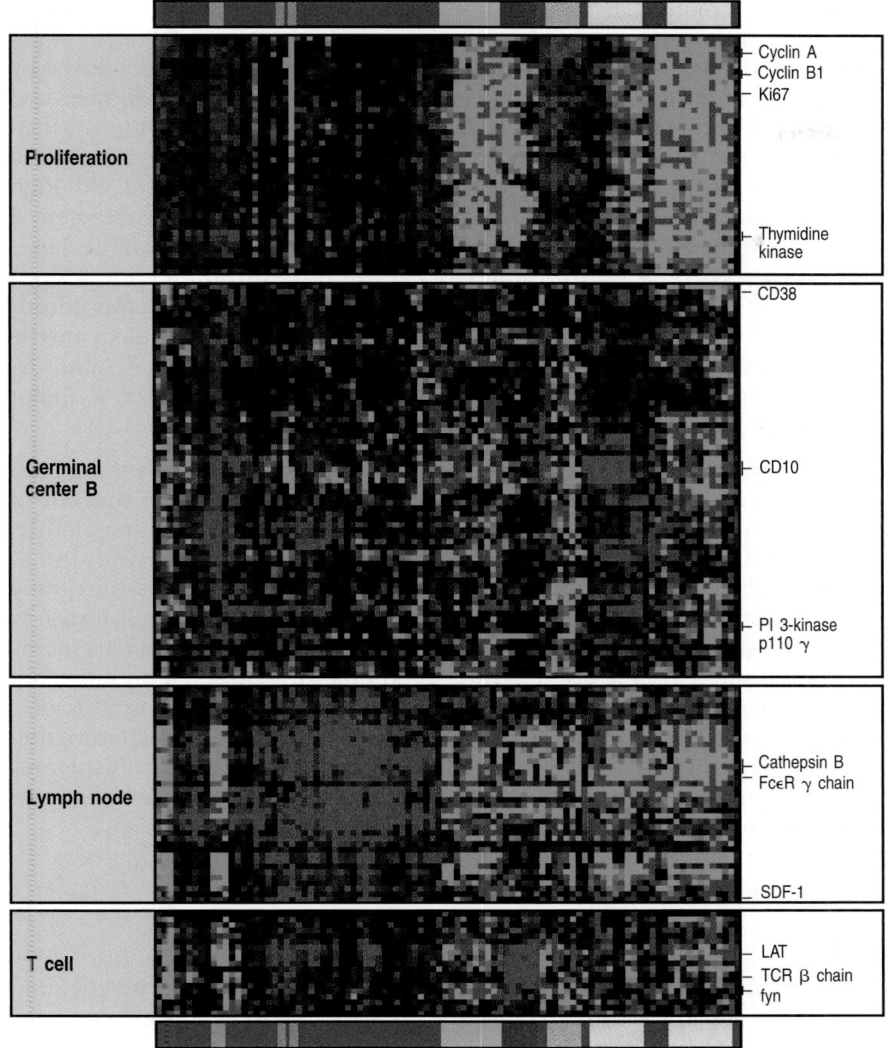

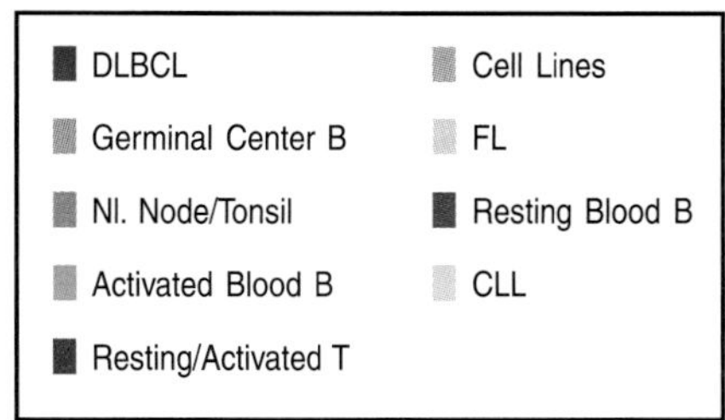

Fig. A.40 DNA microarrays allow a rapid, simultaneous screening of a great many genes for changes in expression between different cells. In the experiment whose results are shown here, nearly 18,000 cDNA clones made from lymphoid cells and lymphoid tumors were arrayed horizontally across the chip. Some of these cDNAs represent the products of known genes, and examples of these are indicated on the right hand edge of the diagram. To these cDNAs mRNA was hybridized from 96 normal cells, cell lines and lymphoid tumor cells, with the mRNA from each individual cell arrayed vertically. The types of cells used are indicated by the colored bars at the top, with the colors explained in the key. The malignant cells used were from diffuse large B-cell lymphoma (DLBCL), follicular lymphoma (FL) or mantle cell lymphoma, and chronic lymphocytic leukemia (CLL). A number of established lymphoid cell lines were included. Normal cells are represented by resting B cells from peripheral blood (Resting Blood B), B cells activated by cross-linking cell-surface IgM with or without additional cytokines and co-stimulation (Activated blood B) germinal center B cells from tonsil (Germinal Center B), and normal, noninflamed tonsil and lymph node (Nl. Node/Tonsil) were used as representatives of different stages in B-cell maturation. Normal T cells, CD4 T cells, either resting or stimulated with PMA and ionomycin (Resting/Activated T) were also used. Each point on the array represents, therefore, the hybridization of the mRNA from one of these cell lines to one of the cDNAs, and is displayed in color to represent the level of expression of the mRNA in question, green being those expressed at lower levels than in a control cell, while red represents those expressed at a higher level. The data shown has been clustered by patterns of expression of the various genes, to give clusters of genes upregulated in proliferating cells, germinal center B cells, lymph node B cells, and in T cells. Courtesy of L. M. Staudt.

with the infectious agent (Fig. A.41). Finally, the prevalence and severity of infection in the immunized individual is compared with the course of the disease in the unimmunized controls. For obvious reasons, such experiments are usually carried out first in animals, if a suitable animal model for the infection exists. However, eventually a trial must be carried out in humans. In this case, the infectious challenge is usually provided naturally by carrying out the trial in a region where the disease is prevalent. The efficacy of the vaccine is determined by assessing the prevalence and severity of new infections in the immunized and control populations. Such studies necessarily give less precise results than a direct experiment but, for most diseases, they are the only way of assessing a vaccine's ability to induce protective immunity in humans.

A-37 Transfer of protective immunity.

The tests described in Section A-36 show that protective immunity has been established, but cannot show whether it involves humoral immunity, cell-mediated immunity, or both. When these studies are carried out in inbred mice, the nature of protective immunity can be determined by transferring

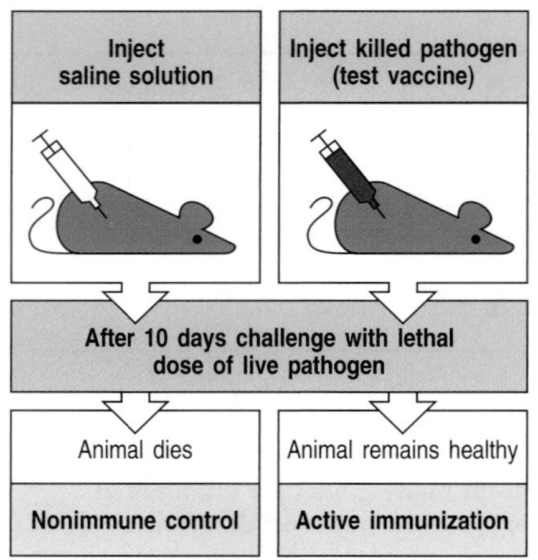
Inject
saline solution
Inject killed pathogen
(test vaccine)
After 10 days challenge with lethal
dose of live pathogen
Animal dies
Animal remains healthy
Nonimmune control
Active immunization

Fig. A.41 *In vivo* assay for the presence of protective immunity after vaccination in animals. Mice are injected with the test vaccine or a control such as saline solution. Different groups are then challenged with lethal or pathogenic doses of the test pathogen or with an unrelated pathogen as a specificity control (not shown). Unimmunized animals die or become severely infected. Successful vaccination is seen as specific protection of immunized mice against infection with the test pathogen. This is called active immunity and the process is called active immunization.

serum or lymphoid cells from an immunized donor animal to an unimmunized syngeneic recipient (that is, a genetically identical animal of the same inbred strain) (Fig. A.42). If protection against infection can be conferred by the transfer of serum, the immunity is provided by circulating antibodies and is called **humoral immunity**. Transfer of immunity by antiserum or purified antibodies provides immediate protection against many pathogens and against toxins such as those of tetanus and snake venom. However, although protection is immediate, it is temporary, lasting only so long as the transferred antibodies remain active in the recipient's body. This type of transfer is therefore called **passive immunization**. Only **active immunization** with antigen can provide lasting immunity. Moreover, the recipient may become immunized to the antiserum used to transfer immunity. Horse or sheep sera are the usual sources of anti-snake venoms used in humans, and repeated administration can lead either to serum sickness (see Section 13-18) or, if the recipient becomes allergic to the foreign serum, to anaphylaxis (see Section 13-11).

Protection against many diseases cannot be transferred with serum but can be transferred by lymphoid cells from immunized donors. The transfer of lymphoid cells from an immune donor to a normal syngeneic recipient is called **adoptive transfer** or **adoptive immunization**, and the immunity transferred is called **adoptive immunity**. Immunity that can be transferred only with lymphoid cells is called **cell-mediated immunity**. Such cell transfers must be between genetically identical donors and recipients, such as members of the same inbred strain of mouse, so that the donor lymphocytes are not rejected by the recipient and do not attack the recipient's tissues. Adoptive transfer of immunity is used clinically in humans in experimental approaches to cancer therapy or as an adjunct to bone marrow transplantation; in these cases, the patient's own T cells, or the T cells of the bone marrow donor, are given.

A-38 The tuberculin test.

Local responses to antigen can indicate the presence of active immunity. Active immunity is often studied *in vivo*, especially in humans, by injecting antigens locally in the skin. If a reaction appears, this indicates the presence of antibodies or immune lymphocytes that are specific for that antigen; the **tuberculin test** is an example of this. When people have had tuberculosis they develop cell-mediated immunity that can be detected as a local response when their skin is injected with a small amount of tuberculin, an extract of *Mycobacterium tuberculosis*, the pathogen that causes tuberculosis. The response typically appears a day or two after the injection and consists of a raised, red, and hard (or indurated) area in the skin, which then disappears as the antigen is degraded.

A-39 Testing for allergic responses.

Local **intracutaneous** injections of minute doses of the antigens that cause allergies are used to determine which antigen triggers a patient's allergic reactions. Local responses that happen in the first few minutes after antigen injection in immune recipients are called **immediate hypersensitivity reactions**, and they can be of several forms, one of which is the wheal-and-flare response (see Fig. 13.14). Immediate hypersensitivity reactions are mediated by specific antibodies of the IgE class formed as a result of earlier exposures to the antigen. Responses that take hours to days to develop, such as the tuberculin test, are referred to as **delayed-type hypersensitivity responses** and are caused by preexisting immune T cells. This latter type of response was observed by Jenner when he tested vaccinated individuals with a local injection of vaccinia virus.

These tests work because the local deposit of antigen remains concentrated in the initial site of injection, eliciting responses in local tissues. They do not cause generalized reactions if sufficiently small doses of antigen are used. However, local tests carry a risk of systemic allergic reactions, and they should be used with caution in people with a history of hypersensitivity.

A-40 Assessment of immune responses and immunological competence in humans.

The methods used for testing immune function in humans are necessarily more limited than those used in experimental animals, but many different tests are available. They fall into several groups depending on the reason the patient is being studied.

Assessment of protective immunity in humans generally relies on tests conducted *in vitro*. To assess humoral immunity, specific antibody levels in the patient's serum are assayed by RIA or, more commonly, ELISA (see Section A-6), using the test microorganism or a purified microbial product as antigen. To test for humoral immunity against viruses, antibody production is often measured by the ability of serum to neutralize the infectivity of live virus for tissue culture cells. In addition to providing information about protective immunity, the presence of antibody to a particular pathogen indicates that the patient has been exposed to it, making such tests of crucial importance in epidemiology. At present, testing for antibody to HIV is the main screening test for infection with this virus, critical both for the patient and in blood banking, where blood from infected donors must be excluded from the supply. Essentially similar tests are used in investigating allergy, where allergens are used as the antigens in tests for specific IgE antibody by ELISA or RIA (see Section A-6), which may be used to confirm the results of skin tests.

Cell-mediated immunity, that is immunity mediated by T cells, is technically more difficult to measure than humoral immunity. This is principally because T cells do not make a secreted antigen-binding product, so there is no simple binding assay for their antigen-specific responses. T-cell activity can be divided into an induction phase, in which T cells are activated to divide and differentiate, and an effector phase, in which their function is expressed. Both phases require that the T cell interacts with another cell and that it recognizes specific antigen displayed in the form of peptide:MHC complexes on the surface of this interacting cell. In the induction phase, the interaction must be with an antigen-presenting cell able to deliver co-stimulatory signals, whereas, in the effector phase, the nature of the target cell depends on the type of effector T cell that has been activated. Most commonly, the presence of T cells that have responded to a specific antigen is detected by their subsequent *in vitro* proliferation when reexposed to the same antigen (see Section A-31).

T-cell proliferation indicates only that cells able to recognize that antigen have been activated previously; it does not reveal what effector function they mediate. The effector function of a T cell is assayed by its effect on an appropriate target cell. Assays for cytotoxic CD8 T cells (see Section A-33) and for cytokine production by CD4 T cells (see Sections A-26, A-27, and A-34) are used to characterize the immune response. Cell-mediated immunity to infectious agents can also be tested by skin test with extracts of the pathogen, as in the tuberculin test (see Section A-36). These tests provide information about the exposure of the patient to the disease and also about their ability to mount an adaptive immune response to it.

Patients with immune deficiency (see Chapter 12) are usually detected clinically by a history of recurrent infection. To determine the competence of the

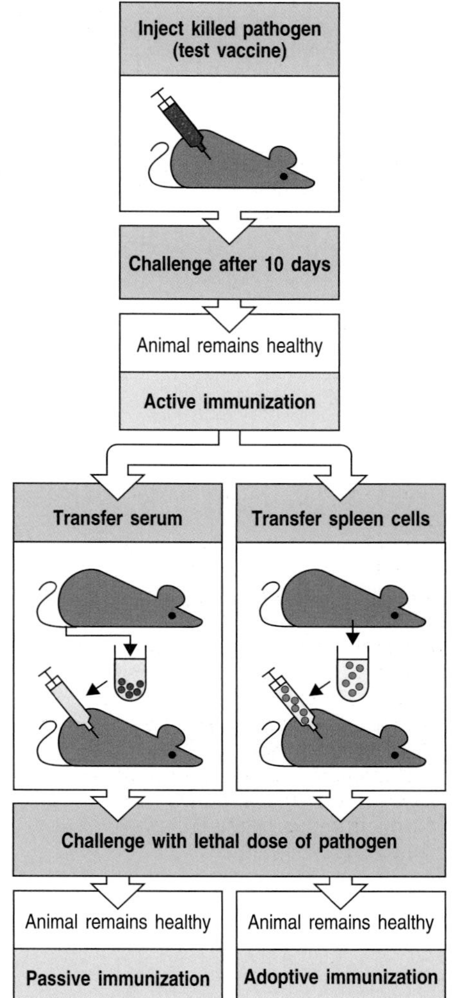

Fig. A.42 Immunity can be transferred by antibodies or by lymphocytes. Successful vaccination leads to a long-lived state of protection against the specific immunizing pathogen. If this immune protection can be transferred to a normal syngeneic recipient with serum from an immune donor, then immunity is mediated by antibodies; such immunity is called humoral immunity and the process is called passive immunization. If immunity can only be transferred by infusing lymphoid cells from the immune donor into a normal syngeneic recipient, then the immunity is called cell-mediated immunity and the transfer process is called adoptive transfer or adoptive immunization. Passive immunity is short-lived, as antibody is eventually catabolized, but adoptively transferred immunity is mediated by immune cells, which can survive and provide longer-lasting immunity.

immune system in such patients, a battery of tests are usually conducted (see Appendix V); these focus with increasing precision as the nature of the defect is narrowed down to a single element. The presence of the various cell types in blood is determined by routine hematology, often followed by FACS analysis of lymphocyte subsets (see Section A-22), and the measurement of serum immunoglobulins. The phagocytic competence of freshly isolated polymorphonuclear leukocytes and monocytes is tested, and the efficiency of the complement system (see Chapters 2 and 9) is determined by testing the dilution of serum required for lysis of 50% of antibody-coated red blood cells (this is denoted the CH_{50}).

In general, if such tests reveal a defect in one of the broad compartments of immune function, more specialized testing is then needed to determine the precise nature of the defect. Tests of lymphocyte function are often valuable, starting with the ability of polyclonal mitogens to induce T-cell proliferation and B-cell secretion of immunoglobulin in tissue culture (see Section A-31). These tests can eventually pinpoint the cellular defect in immunodeficiency.

In patients with autoimmune diseases (see Chapter 14), the same parameters are usually analyzed to determine whether there is a gross abnormality in the immune system. However, most patients with such diseases show few abnormalities in general immune function. To determine whether a patient is producing antibody against their own cellular antigens, the most informative test is to react their serum with tissue sections, which are then examined for bound antibody by indirect immunofluorescence using anti-human immunoglobulin labeled with fluorescent dye (see Section A-14). Most autoimmune diseases are associated with the production of broadly characteristic patterns of autoantibodies directed at self tissues. These patterns aid in the diagnosis of the disease and help to distinguish autoimmunity from tissue inflammation due to infectious causes.

It is also possible to investigate allergies by administration of possible allergens by routes other than intracutaneous administration. Allergen may be given by inhalation to test for asthmatic allergic responses (see Fig. 13.14); this is mainly done for experimental purposes in studies of the mechanisms and treatment of asthma. Similarly, food allergens may be given by mouth. The administration of allergens is potentially very dangerous because of the risk of causing anaphylaxis, and must only be carried out by trained and experienced investigators in an environment in which full resuscitation facilities are available.

A-41 The Arthus reaction.

This is an experimental method using only animal models for studying the formation of immune complexes in tissues and how immune complexes cause inflammation (see Section 13-18). The original reaction described by Maurice Arthus was induced by the repeated injection of horse serum into rabbits. Initial injections of horse serum into the skin induced no reaction, but later injections, following the production of antibodies to the proteins in horse serum, induced an inflammatory reaction at the site of injection after several hours, characterized by the presence of edema, hemorrhage, and neutrophil infiltration, which frequently progressed to tissue necrosis. Most investigators now use passive models of the Arthus reaction in which either antibody is infused systemically and antigen given locally (passive Arthus reaction) or antigen is infused systemically and antibody injected locally (reverse passive Arthus reaction).

Manipulation of the immune system.

A-42 Adoptive transfer of lymphocytes.

Ionizing radiation from X-ray or γ-ray sources kills lymphoid cells at doses that spare the other tissues of the body. This makes it possible to eliminate immune function in a recipient animal before attempting to restore immune function by adoptive transfer, and allows the effect of the adoptively transferred cells to be studied in the absence of other lymphoid cells. James Gowans originally used this technique to prove the role of the lymphocyte in immune responses. He showed that all active immune responses could be transferred to irradiated recipients by small lymphocytes from immunized donors. This technique can be refined by transferring only certain lymphocyte subpopulations, such as B cells, CD4 T cells, and so on. Even cloned T-cell lines have been tested for their ability to transfer immune function, and have been shown to confer adoptive immunity to their specific antigen. Such adoptive transfer studies are a cornerstone in the study of the intact immune system, as they can be carried out rapidly, simply, and in any strain of mouse.

A-43 Hematopoietic stem-cell transfers.

All cells of hematopoietic origin can be eliminated by treatment with high doses of X rays, allowing replacement of the entire hematopoietic system, including lymphocytes, by transfusion of donor bone marrow or purified hematopoietic stem cells from another animal. The resulting animals are called **radiation bone marrow chimeras** from the Greek word *chimera*, a mythical animal that had the head of a lion, the tail of a serpent, and the body of a goat. This technique is used experimentally to examine the development of lymphocytes, as opposed to their effector functions, and has been particularly important in studying T-cell development. Essentially the same technique is used in humans to replace the hematopoietic system when it fails, as in aplastic anemia or after nuclear accidents, or to eradicate the bone marrow and replace it with normal marrow in the treatment of certain cancers. In man, bone marrow is the main source of hematopoietic stem cells, but increasingly they are being obtained from peripheral blood after the donor has been treated with hematopoietic growth factors such as GM-CSF, or from umbilical cord blood, which is rich in such stem cells.

A-44 *In vivo* depletion of T cells.

The importance of T-cell function *in vivo* can be ascertained in mice with no T cells of their own. Under these conditions, the effect of a lack of T cells can be studied, and T-cell subpopulations can be restored selectively to analyze their specialized functions. T lymphocytes originate in the thymus, and neonatal **thymectomy**, the surgical removal of the thymus of a mouse at birth, prevents T-cell development from occurring because the export of most functionally mature T cells only occurs after birth in the mouse. Alternatively, adult mice can be thymectomized and then irradiated and reconstituted with bone marrow; such mice will develop all hematopoietic cell types except mature T cells.

The recessive *nude* mutation in mice is caused by a mutation in the gene for the transcription factor Wnt and in homozygous form causes hairlessness and absence of the thymus. Consequently, these animals fail to develop T cells from bone marrow progenitors. Grafting thymectomized or *nude/nude* mice with thymic epithelial elements depleted of lymphocytes allows the graft recipients to develop normal mature T cells. This procedure allows the role of the nonlymphoid thymic stroma to be examined; it has been crucial in determining the role of thymic stromal cells in T-cell development (see Chapter 7).

A-45 *In vivo* depletion of B cells.

There is no single site of B-cell development in mice, so techniques such as thymectomy cannot be applied to the study of B-cell function and development in rodents. However, **bursectomy**, the surgical removal of the Bursa of Fabricius in birds, can inhibit the development of B cells in these species. In fact, it was the effect of thymectomy versus bursectomy that led to the naming of T cells for thymus-derived lymphocytes and B cells for bursal-derived lymphocytes. There are no known spontaneous mutations (analogous to the *nude* mutation) in mice that produce animals with T cells but no B cells. However, such mutations exist in humans, leading to a failure to mount humoral immune responses or make antibody. The diseases produced by such mutations are called agammaglobulinemias because they were originally detected as the absence of gamma globulins. The genetic basis for one form of this disease in humans has now been established (see Chapter 12), and some features of the disease can be reproduced in mice by targeted disruption of the corresponding gene (see Section A-47). Several different mutations in crucial regions of immunoglobulin genes have already been produced by gene targeting and have provided mice lacking B cells.

A-46 Transgenic mice.

The function of genes has traditionally been studied by observing the effects of spontaneous mutations in whole organisms and, more recently, by analyzing the effects of targeted mutations in cultured cells. The advent of gene cloning and *in vitro* mutagenesis now make it possible to produce specific mutations in whole animals. Mice with extra copies or altered copies of a gene in their genome can be generated by **transgenesis**, which is now a well established procedure. To produce **transgenic mice**, a cloned gene is introduced into the mouse genome by microinjection into the male pronucleus of a fertilized egg, which is then implanted into the uterus of a pseudopregnant female mouse. In some of the eggs, the injected DNA becomes integrated randomly into the genome, giving rise to a mouse that has an extra genetic element of known structure, the transgene (Fig. A.43).

The transgene, to be studied in detail, needs to be introduced onto a stable, well-characterized genetic background. However, it is difficult to prepare transgenic embryos successfully in inbred strains of mice, and transgenic

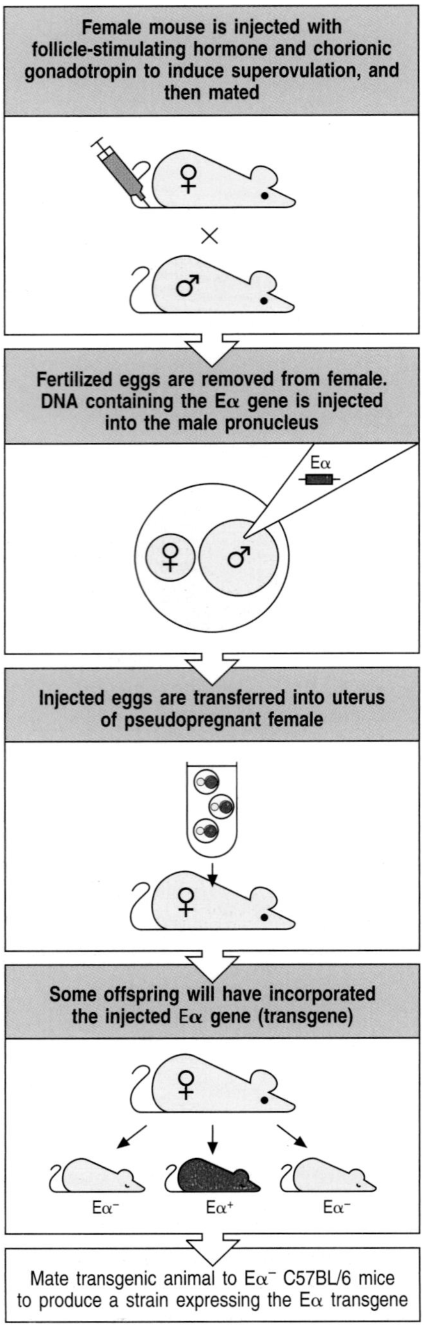

Female mouse is injected with follicle-stimulating hormone and chorionic gonadotropin to induce superovulation, and then mated

♀ × ♂

Fertilized eggs are removed from female. DNA containing the $E\alpha$ gene is injected into the male pronucleus

$E\alpha$

Injected eggs are transferred into uterus of pseudopregnant female

Some offspring will have incorporated the injected $E\alpha$ gene (transgene)

$E\alpha^-$ $E\alpha^+$ $E\alpha^-$

Mate transgenic animal to $E\alpha^-$ C57BL/6 mice to produce a strain expressing the $E\alpha$ transgene

Fig. A.43 The function and expression of genes can be studied in vivo by using transgenic mice. DNA encoding a protein of interest, here the mouse MHC class II protein $E\alpha$, is purified and microinjected into the male pronuclei of fertilized eggs. The eggs are then implanted into pseudopregnant female mice. The resulting offspring are screened for the presence of the transgene in their cells, and positive mice are used as founders that transmit the transgene to their offspring, establishing a line of transgenic mice that carry one or more extra genes. The function of the $E\alpha$ gene used here is tested by breeding the transgene into C57BL/6 mice that carry an inactivating mutation in their endogenous $E\alpha$ gene.

mice are routinely prepared in F_2 embryos (that is, the embryo formed after the mating of two F_1 animals). The transgene must then be bred onto a well-characterized genetic background; this requires 10 generations of back-crossing with an inbred strain to assure that the integrated transgene is largely (>99%) free of heterogeneous genes from the founder mouse of the transgenic mouse line (Fig. A.44).

This technique allows one to study the impact of a newly discovered gene on development, to identify the regulatory regions of a gene required for its normal tissue-specific expression, to determine the effects of its over-expression or expression in inappropriate tissues, and to find out the impact of mutations on gene function. Transgenic mice have been particularly useful in studying the role of T-cell and B-cell receptors in lymphocyte development, as described in Chapter 7.

A-47 Gene knockout by targeted disruption.

In many cases, the functions of a particular gene can be fully understood only if a mutant animal that does not express the gene can be obtained. Whereas genes used to be discovered through identification of mutant phenotypes, it is now far more common to discover and isolate the normal gene and then determine its function by replacing it *in vivo* with a defective copy. This procedure is known as **gene knockout**, and it has been made possible by two fairly recent developments: a powerful strategy to select for targeted mutation by homologous recombination, and the development of continuously growing lines of **embryonic stem cells (ES cells)**. These are embryonic cells which, on implantation into a blastocyst, can give rise to all cell lineages in a chimeric mouse.

The technique of **gene targeting** takes advantage of the phenomenon known as **homologous recombination** (Fig. A.45). Cloned copies of the target gene are altered to make them nonfunctional and are then introduced into the ES cell where they recombine with the homologous gene in the cell's genome, replacing the normal gene with a nonfunctional copy. Homologous recombination is a rare event in mammalian cells, and thus a powerful selection strategy is required to detect those cells in which it has occurred. Most commonly, the introduced gene construct has its sequence disrupted by an inserted antibiotic-resistance gene such as that for neomycin resistance. If this construct undergoes homologous recombination with the endogenous copy of the gene, the endogenous gene is disrupted but the antibiotic-resistance gene remains functional, allowing cells that have incorporated the gene to be selected in culture for resistance to the neomycin-like drug G418. However, antibiotic resistance on its own shows only that the cells have taken up and integrated the neomycin-resistance gene. To be able to select for those cells in which homologous recombination has occurred, the ends of the construct usually carry the thymidine kinase gene from the herpes simplex virus

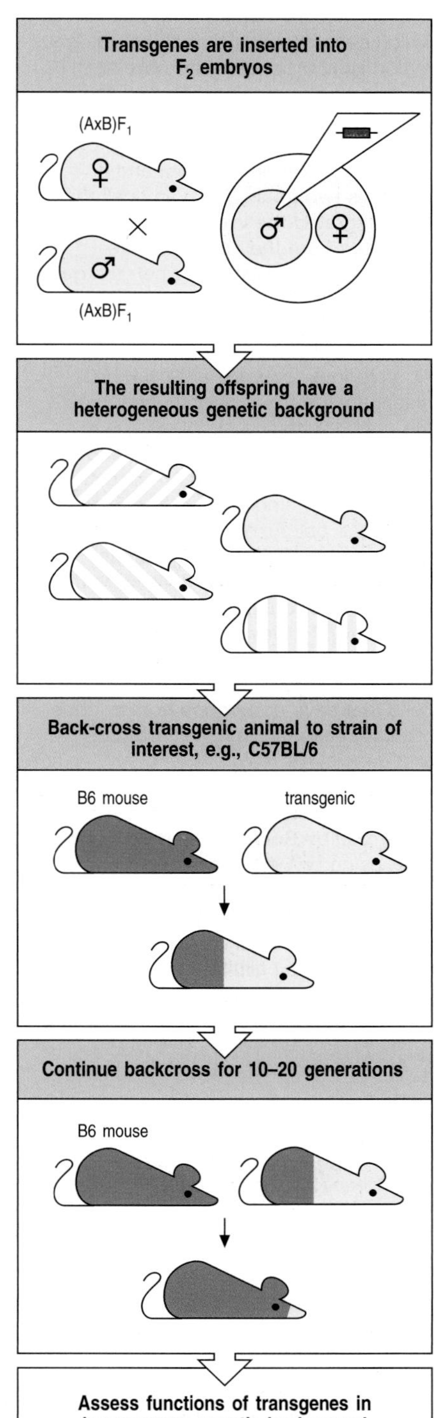
Transgenes are inserted into F_2 embryos
(AxB)F_1
♀
×
♂
(AxB)F_1
♂
♀
The resulting offspring have a heterogeneous genetic background
Back-cross transgenic animal to strain of interest, e.g., C57BL/6
B6 mouse
transgenic
Continue backcross for 10–20 generations
B6 mouse
Assess functions of transgenes in homogenous genetic background

Fig. A.44 The breeding of transgenic co-isogeneic or congenic mouse strains. Transgenic mouse strains are routinely made in F2 mice. To produce mice on an inbred background, the transgene is introgressively back-crossed onto a standard strain, usually C57BL/6 (B6). The presence of the transgene is tracked by carrying out PCR on genomic DNA extracted from the tail of young mice. After 10 generations of back-crossing, mice are >99% genetically identical, so that any differences observed between the mice are likely to be due to the transgene itself. The same technique can be used to breed a gene knockout into a standard strain of mice, as most gene knockouts are made in the 129 strain of mice (see Fig. A.46). The mice are then intercrossed and homozygous knockout mice detected by an absence of an intact copy of the gene of interest (as determined by PCR).

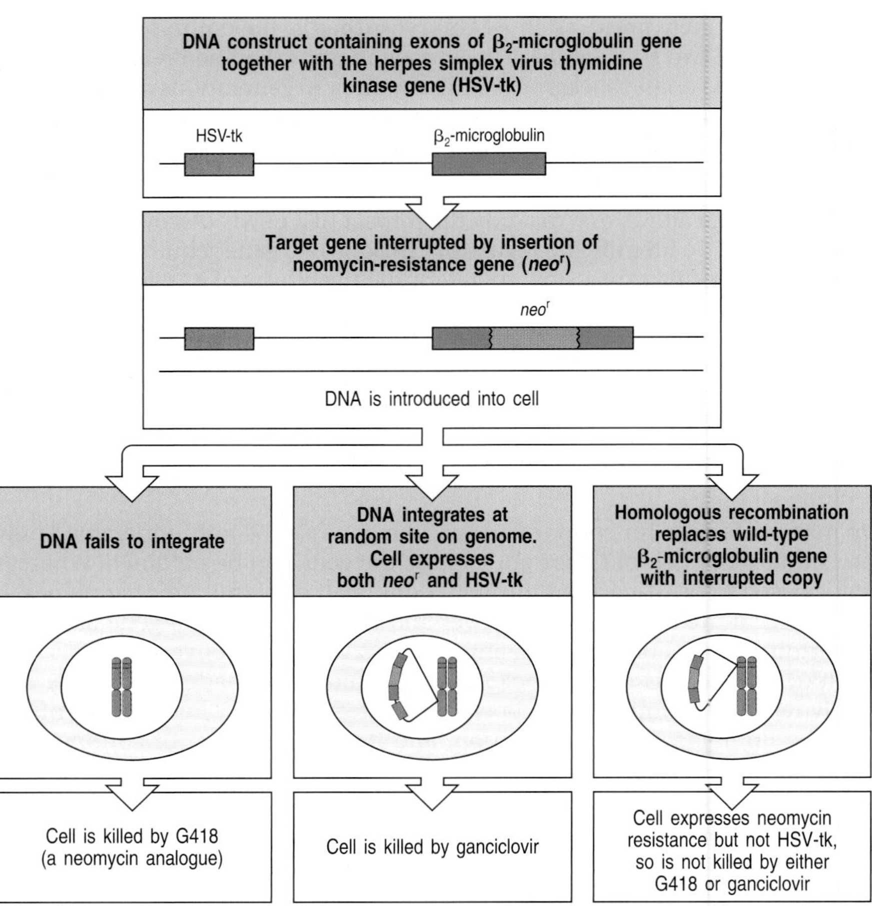

Fig. A.45 The deletion of specific genes can be accomplished by homologous recombination. When pieces of DNA are introduced into cells, they can integrate into cellular DNA in two different ways. If they randomly insert into sites of DNA breaks, the whole piece is usually integrated, often in several copies. However, extrachromosomal DNA can also undergo homologous recombination with the cellular copy of the gene, in which case only the central, homologous region is incorporated into cellular DNA. Inserting a selectable marker gene such as resistance to neomycin (neo^r) into the coding region of a gene does not prevent homologous recombination, and it achieves two goals. First, any cell that has integrated the injected DNA is protected from the neomycin-like antibiotic G418. Second, when the gene recombines with homologous cellular DNA, the neo^r gene disrupts the coding sequence of the modified cellular gene. Homologous recombinants can be discriminated from random insertions if the gene for herpes simplex virus thymidine kinase (HSV-tk) is placed at one or both ends of the DNA construct, which is often known as a 'targeting construct' because it targets the cellular gene. In random DNA integrations, HSV-tk is retained. HSV-tk renders the cell sensitive to the antiviral agent ganciclovir. However, as HSV-tk is not homologous to the target DNA, it is lost from homologous recombinants. Thus, cells that have undergone homologous recombination are uniquely both G418 and ganciclovir resistant, and survive in a mixture of the two antibiotics. The presence of the disrupted gene has to be confirmed by Southern blotting or by PCR using primers in the neo^r gene and in cellular DNA lying outside the region used in the targeting construct. By using two different resistance genes one can disrupt the two cellular copies of a gene, making a deletion mutant (not shown).

(HSV-tk). Cells that incorporate DNA randomly usually retain the entire DNA construct including HSV-tk, whereas homologous recombination between the construct and cellular DNA, the desired result, involves the exchange of homologous DNA sequences so that the nonhomologous HSV-tk genes at the ends of the construct are eliminated. Cells carrying HSV-tk are killed by the antiviral drug ganciclovir, and so cells with homologous recombinations have the unique feature of being resistant to both neomycin and ganciclovir, allowing them to be selected efficiently when these drugs are added to the cultures (see Fig. A.45).

This technique can be used to produce homozygous mutant cells in which the effects of knocking-out a specific gene can be analyzed. Diploid cells in which both copies of a gene have been mutated by homologous recombination can be selected after transfection with a mixture of constructs in which the gene to be targeted has been disrupted by one or other of two different antibiotic-resistance genes. Having obtained a mutant cell with a functional defect, the defect can be ascribed definitively to the mutated gene if the mutant phenotype can be reverted with a copy of the normal gene transfected into the mutant cell. Restoration of function means that the defect in the mutant gene has been complemented by the normal gene's function. This technique is very powerful as it allows the gene that is being transferred to be mutated in precise ways to determine which parts of the protein are required for function.

To knock out a gene *in vivo*, it is only necessary to disrupt one copy of the cellular gene in an ES cell. ES cells carrying the mutant gene are produced by targeted mutation (see Fig. A.45), and injected into a blastocyst which is reimplanted into the uterus. The cells carrying the disrupted gene become incorporated into the developing embryo and contribute to all tissues of the resulting chimeric offspring, including those of the germline. The mutated gene can therefore be transmitted to some of the offspring of the original chimera, and further breeding of the mutant gene to homozygosity produces mice that completely lack the expression of that particular gene product (Fig. A.46). The effects of the absence of the gene's function can then be studied. In addition, the parts of the gene that are essential for its function can be identified by determining whether function can be restored by introducing different mutated copies of the gene back into the genome by transgenesis. The manipulation of the mouse genome by gene knockout and transgenesis is revolutionizing our understanding of the role of individual genes in lymphocyte development and function.

Because the most commonly used ES cells are derived from a poorly characterized strain of mice known as strain 129, the analysis of the function of a gene knockout often requires extensive back-crossing to another strain, just as in transgenic mice (see Fig. A.44). One can track the presence of the mutant copy of the gene by the presence of the *neo*^{r} gene. After sufficient back-crossing, the mice are intercrossed to produce mutants on a stable genetic background.

A problem with gene knockouts arises when the function of the gene is essential for the survival of the animal; in such cases the gene is termed a **recessive lethal gene** and homozygous animals cannot be produced. However, by making chimeras with mice that are deficient in B and T cells, it is possible to analyze the function of recessive lethal genes in lymphoid cells. To do this, ES cells with homozygous lethal loss-of-function mutations are injected into blastocysts of mice lacking the ability to rearrange their antigen-receptor genes because of a mutation in their recombinase-activating genes (*RAG* knockout mice). As these chimeric embryos develop, the *RAG*-deficient cells can compensate for any developmental failure resulting from the gene knockout in the ES cells in all except the lymphoid lineage. So long as the mutated ES cells can develop into hematopoietic progenitors in the bone marrow, the embryos will survive and all of the lymphocytes in the resulting chimeric mouse will be derived from the mutant ES cells (Fig. A.47).

A second powerful technique achieves tissue-specific or developmentally regulated gene deletion by employing the DNA sequences and enzymes used by bacteriophage P1 to excise itself from a host cell's genome. Integrated

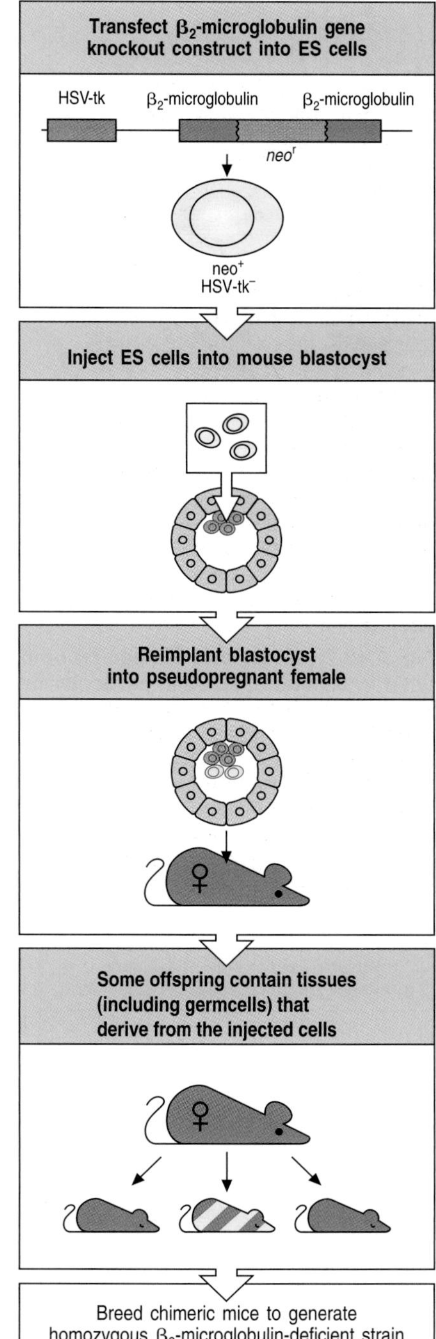

Fig. A.46 Gene knockout in embryonic stem cells enables mutant mice to be produced. Specific genes can be inactivated by homologous recombination in cultures of embryonic stem cells (ES cells). Homologous recombination is carried out as described in Fig. A.45. In this example, the gene for β_2-microglobulin in ES cells is disrupted by homologous recombination with a targeting construct. Only a single copy of the gene needs to be disrupted. ES cells in which homologous recombination has taken place are injected into mouse blastocysts. If the mutant ES cells give rise to germ cells in the resulting chimeric mice (striped in the figure), then the mutant gene can be transferred to their offspring. By breeding the mutant gene to homozygosity, a mutant phenotype is generated. These mutant mice are usually of the 129 strain as gene knockout is generally conducted in ES cells derived from the 129 strain of mice. In this case, the homozygous mutant mice lack MHC class I molecules on their cells, as MHC class I molecules have to pair with β_2-microglobulin for surface expression. The β_2-microglobulin-deficient mice can then be bred with mice transgenic for subtler mutants of the deleted gene, allowing the effect of such mutants to be tested *in vivo*.

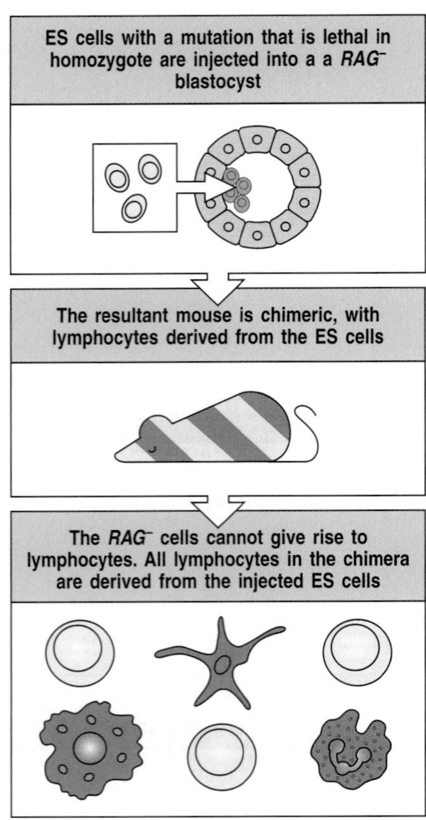

ES cells with a mutation that is lethal in homozygote are injected into a a *RAG⁻* blastocyst
The resultant mouse is chimeric, with lymphocytes derived from the ES cells
The *RAG⁻* cells cannot give rise to lymphocytes. All lymphocytes in the chimera are derived from the injected ES cells

Fig. A.47 The role of recessive lethal genes in lymphocyte function can be studied using RAG-deficient chimeric mice. ES cells homozygous for the lethal mutation are injected into a RAG-deficient blastocyst (top panel). The RAG-deficient cells can give rise to all the tissues of a normal mouse except lymphocytes, and so can compensate for any deficiency in the developmental potential of the mutant ES cells (middle panel). If the mutant ES cells are capable of differentiating into hematopoietic stem cells, that is, if the gene function that has been deleted is not essential for this developmental pathway, then all the lymphocytes in the chimeric mouse will be derived from the ES cells (bottom panel), as RAG-deficient mice cannot make lymphocytes of their own.

bacteriophage P1 DNA is flanked by recombination signal sequences called *loxP* sites. A recombinase, Cre, recognizes these sites, cuts the DNA and joins the two ends, thus excising the intervening DNA in the form of a circle. This mechanism can be adapted to allow the deletion of specific genes in a transgenic animal only in certain tissues or at certain times in development. First, *loxP* sites flanking a gene, or perhaps just a single exon, are introduced by homologous recombination (Fig. A.48). Usually, the introduction of these sequences into flanking or intronic DNA does not disrupt the normal function of the gene. Mice containing such *loxP* mutant genes are then mated with mice made transgenic for the Cre recombinase, under the control of a tissue-specific or inducible promoter. When the Cre recombinase is active, either in the appropriate tissue or when induced, it excises the DNA between the inserted *loxP* sites, thus inactivating the gene or exon. Thus, for example, using a T-cell specific promoter to drive expression of the Cre recombinase, a gene can be deleted only in T cells, while remaining functional in all other cells of the animal. This is an extremely powerful genetic technique that while still in its infancy, was used to demonstrate the importance of B-cell receptors in B-cell survival. It is certain to yield exciting results in the future.

Fig. A.48 The P1 bacteriophage recombination system can be used to eliminate genes in particular cell lineages. The P1 bacteriophage protein Cre excises DNA that is bounded by recombination signal sequences called loxP sequences. These sequences can be introduced at either end of a gene by homologous recombination (left panel). Animals carrying genes flanked by loxP can also be made transgenic for the gene for the Cre protein, which is placed under the control of a tissue-specific promoter so that it is expressed only in certain cells or only at certain times during development (middle panel). In the cells in which the Cre protein is expressed, it recognizes the loxP sequences and excises the DNA lying between them (right panel). Thus, individual genes can be deleted only in certain cell types or only at certain times. In this way, genes that are essential for the normal development of a mouse can be analyzed for their function in the developed animal and/or in specific cell types. Genes are shown as boxes, RNA as squiggles, and proteins as colored balls.

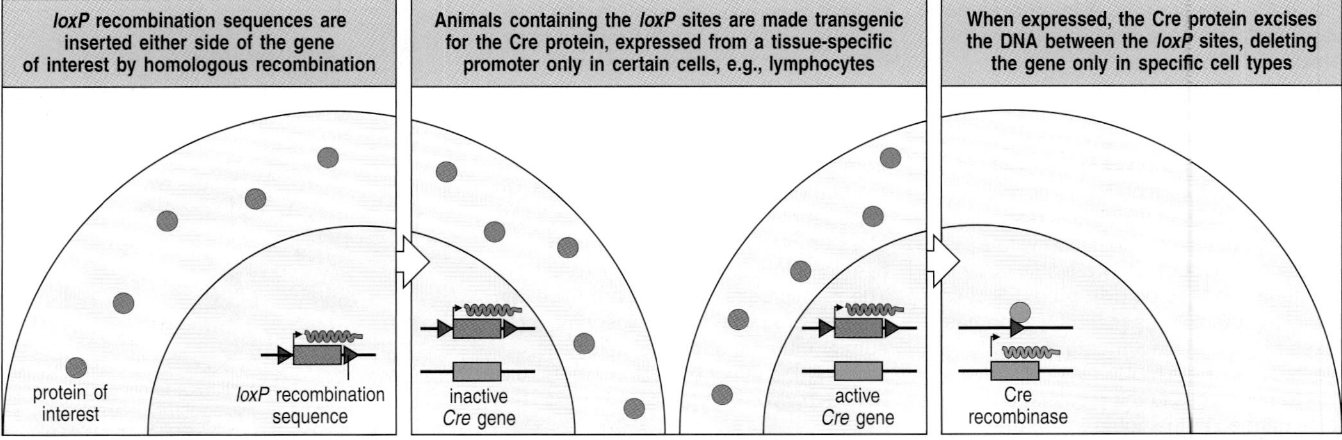

APPENDICES II – V

Appendix II. CD antigens.					
CD antigen	**Cellular expression**	**Molecular weight (kDa)**	**Functions**	**Other names**	**Family relationships**
CD1a,b,c,d	Cortical thymocytes, Langerhans cells, dendritic cells, B cells (CD1c), intestinal epithelium, smooth muscle, blood vessels (CD1d)	43–49	MHC class I-like molecule, associated with β_2-microglobulin. Has specialized role in presentation of lipid antigens		Immunoglobulin
CD2	T cells, thymocytes, NK cells	45–58	Adhesion molecule, binding CD58 (LFA-3). Binds Lck intracellularly and activates T cells	T11, LFA-2	Immunoglobulin
CD3	Thymocytes, T cells	γ: 25–28 δ: 20 ε: 20	Associated with the T-cell antigen receptor (TCR). Required for cell-surface expression of and signal transduction by the TCR	T3	Immunoglobulin
CD4	Thymocyte subsets, T_H1 and T_H2 T cells (about two thirds of peripheral T cells), monocytes, macrophages	55	Co-receptor for MHC class II molecules. Binds Lck on cytoplasmic face of membrane. Receptor for HIV-1 and HIV-2 gp120	T4, L3T4	Immunoglobulin
CD5	Thymocytes, T cells, subset of B cells	67		T1, Ly1	Scavenger receptor
CD6	Thymocytes, T cells, B cells in chronic lymphatic leukemia	100–130	Binds CD166	T12	Scavenger receptor
CD7	Pluripotential hematopoietic cells, thymocytes, T cells	40	Unknown, cytoplasmic domain binds PI 3-kinase on cross-linking. Marker for T cell acute lymphatic leukemia and pluripotential stem cell leukemias		Immunoglobulin
CD8	Thymocytes subsets, cytotoxic T cells (about one third of peripheral T cells)	α: 32–34 β: 32–34	Co-receptor for MHC class I molecules. Binds Lck on cytoplasmic face of membrane	T8, Lyt2,3	Immunoglobulin
CD9	Pre-B cells, monocytes, eosinophils, basophils, platelets, activated T cells, brain and peripheral nerves, vascular smooth muscle	24	Mediates platelet aggregation and activation via FcγRIIa, may play a role in cell migration		Tetraspanning membrane protein, also called transmembrane 4 (TM4)
CD10	B- and T-cell precursors, bone marrow stromal cells	100	Zinc metalloproteinase, marker for pre-B acute lymphatic leukemia (ALL)	Neutral endopeptidase, common acute lymphocytic leukemia antigen (CALLA)	
CD11a	Lymphocytes, granulocytes, monocytes and macrophages	180	αL subunit of integrin LFA-1 (associated with CD18); binds to CD54 (ICAM-1), CD102 (ICAM-2), and CD50 (ICAM-3)	LFA-1	Integrin α

CD antigen	Cellular expression	Molecular weight (kDa)	Functions	Other names	Family relationships
CD11b	Myeloid and NK cells	170	αM subunit of integrin CR3 (associated with CD18): binds CD54, complement component iC3b, and extracellular matrix proteins	Mac-1	Integrin α
CD11c	Myeloid cells	150	αX subunit of integrin CR4 (associated with CD18); binds fibrinogen	CR4, p150, 95	Integrin α
CD11d	Leukocytes	125	αD subunits of integrin; associated with CD18; binds to CD50		Integrin α
CDw12	Monocytes, granulocytes, platelets	90–120	Unknown		
CD13	Myelomonocytic cells	150–170	Zinc metalloproteinase	Aminopeptidase N	
CD14	Myelomonocytic cells	53–55	Receptor for complex of lipopolysaccharide and lipopolysaccharide binding protein (LBP)		
CD15	Neutrophils, eosinophils, monocytes		Terminal trisaccharide expressed on glycolipids and many cell-surface glycoproteins	Lewis^x (Le^x)	
CD15s	Leukocytes, endothelium		Ligand for CD62E, P	Sialyl-Lewis^x (sLe^x)	poly-N-acetyl-lactosamine
CD15u			Sulphated CD15		Carbohydrate structures
CD16	Neutrophils, NK cells, macrophages	50–80	Component of low affinity Fc receptor, FcγRIII, mediates phagocytosis and antibody-dependent cell-mediated cytotoxicity	FcγRIII	Immunoglobulin
CDw17	Neutrophils, monocytes, platelets		Lactosyl ceramide, a cell-surface glycosphingolipid		
CD18	Leukocytes	95	Intergrin β2 subunit, associates with CD11a, b, c, and d		Integrin β
CD19	B cells	95	Forms complex with CD21 (CR2) and CD81 (TAPA-1); co-receptor for B cells—cytoplasmic domain binds cytoplasmic tyrosine kinases and PI 3-kinase		Immunoglobulin
CD20	B cells	33–37	Oligomers of CD20 may form a Ca^{2+} channel; possible role in regulating B-cell activation		Contains 4 transmembrane segments
CD21	Mature B cells, follicular dendritic cells	145	Receptor for complement component C3d, Epstein–Barr virus. With CD19 and CD81, CD21 forms co-receptor for B cells	CR2	Complement control protein (CCP)
CD22	Mature B cells	α: 130 β: 140	Binds sialoconjugates	BL-CAM	Immunoglobulin
CD23	Mature B cells, activated macrophages, eosinophils, follicular dendritic cells, platelets	45	Low-affinity receptor for IgE, regulates IgE synthesis; ligand for CD19:CD21:CD81 co-receptor	FcεRII	C-type lectin
CD24	B cells, granulocytes	35–45	Unknown	Possible human homologue of mouse heat stable antigen (HSA)	
CD25	Activated T cells, B cells, and monocytes	55	IL-2 receptor α chain	Tac	CCP
CD26	Activated B and T cells, macrophages	110	Exopeptidase, cleaves N terminal X-Pro or X-Ala dipeptides from polypeptides	Dipeptidyl peptidase IV	Type II transmembrane glycoprotein

CD antigen	Cellular expression	Molecular weight (kDa)	Functions	Other names	Family relationships
CD27	Medullary thymocytes, T cells, NK cells, some B cells	55	Binds CD70; can function as a co-stimulator for T and B cells		TNF receptor
CD28	T-cell subsets, activated B cells	44	Activation of naive T cells, receptor for co-stimulatory signal (signal 2) binds CD80 (B7.1) and CD86 (B7.2)	Tp44	Immunoglobulin and CD86 (B7.2)
CD29	Leukocytes	130	Integrin β1 subunit, associates with CD49a in VLA-1 integrin		Integrin β
CD30	Activated T, B, and NK cells, monocytes	120	Binds CD30L (CD153); cross-linking CD30 enhances proliferation of B and T cells	Ki-1	TNF receptor
CD31	Monocytes, platelets, granulocytes, T-cell subsets, endothelial cells	130–140	Adhesion molecule, mediating both leukocyte–endothelial and endothelial–endothelial interactions	PECAM-1	Immunoglobulin
CD32	Monocytes, granulocytes, B cells, eosinophils	40	Low affinity Fc receptor for aggregated immunoglobulin:immune complexes	FcγRII	Immunoglobulin
CD33	Myeloid progenitor cells, monocytes	67	Binds sialoconjugates		Immunoglobulin
CD34	Hematopoietic precursors, capillary endothelium	105–120	Ligand for CD62L (L-selectin)		Mucin
CD35	Erythrocytes, B cells, monocytes, neutrophils, eosinophils, follicular dendritic cells	250	Complement receptor 1, binds C3b and C4b, mediates phagocytosis	CR1	CCP
CD36	Platelets, monocytes, endothelial cells	88	Platelet adhesion molecule; involved in recognition and phagocytosis of apoptosed cells	Platelet GPIV, GPIIIb	
CD37	Mature B cells, mature T cells, myeloid cells	40–52	Unknown, may be involved in signal transduction; forms complexes with CD53, CD81, CD82, and MHC class II		Transmembrane 4
CD38	Early B and T cells, activated T cells, germinal center B cells, plasma cells	45	NAD glycohydrolase, augments B cell proliferation	T10	
CD39	Activated B cells, activated NK cells, macrophages, dendritic cells	78	Unknown, may mediate adhesion of B cells		
CD40	B cells, macrophages, dendritic cells, basal epithelial cells	48	Binds CD154 (CD40L); receptor for co-stimulatory signal for B cells, promotes growth, differentiation, and isotype switching of B cells, and cytokine production by macrophages and dendritic cells		TNF receptor
CD41	Platelets, megakaryocytes	Dimer: GPIIba: 125 GPIIbb: 22	αIIb integrin, associates with CD61 to form GPIIb, binds fibrinogen, fibronectin, von Willebrand factor, and thrombospondin	GPIIb	Integrin α
CD42a,b,c,d	Platelets, megakaryocytes	a: 23 b: 135, 23 c: 22 d: 85	Binds von Willebrand factor, thrombin; essential for platelet adhesion at sites of injury	a: GPIX b: GPIbα c: GPIbβ d: GPV	Leucine-rich repeat
CD43	Leukocytes, except resting B cells	115–135 (neutrophils) 95–115 (T cells)	Has extended structure, approx. 45 nm long and may be anti-adhesive	Leukosialin, sialophorin	Mucin
CD44	Leukocytes, erythrocytes	80–95	Binds hyaluronic acid, mediates adhesion of leukocytes	Hermes antigen, Pgp-1	Link protein

CD antigen	Cellular expression	Molecular weight (kDa)	Functions	Other names	Family relationships
CD45	All hematopoietic cells	180–240 (multiple isoforms)	Tyrosine phosphatase, augments signaling through antigen receptor of B and T cells, multiple isoforms result from alternative splicing (see below)	Leukocyte common antigen (LCA), T200, B220	Fibronectin type III
CD45RO	T-cell subsets, B-cell subsets, monocytes, macrophages	180	Isoform of CD45 containing none of the A, B, and C exons		Fibronectin type II
CD45RA	B cells, T-cell subsets (naive T cells), monocytes	205–220	Isoforms of CD45 containing the A exon		Fibronectin type II
CD45RB	T-cell subsets, B cells, monocytes, macrophages, granulocytes	190–220	Isoforms of CD45 containing the B exon	T200	Fibronectin type II
CD46	Hematopoietic and non-hematopoietic nucleated cells	56/66 (splice variants)	Membrane co-factor protein, binds to C3b and C4b to permit their degradation by Factor I	MCP	CCP
CD47	All cells	47–52	Adhesion molecule, thrombospondin receptor	IAP, MER6, OA3	Immunoglobulin superfamily
CD48	Leukocytes	40–47	Putative ligand for CD244	Blast-1	Immunoglobulin
CD49a	Activated T cells, monocytes, neuronal cells, smooth muscle	200	α1 integrin, associates with CD29, binds collagen, laminin-1	VLA-1	Integrin α
CD49b	B cells, monocytes, platelets, megakaryocytes, neuronal, epithelial and endothelial cells, osteoclasts	160	α2 integrin, associates with CD29, binds collagen, laminin	VLA-2, platelet GPIa	Integrin α
CD49c	B cells, many adherent cells	125, 30	α3 integrin, associates with CD29, binds laminin-5, fibronectin, collagen, entactin, invasin	VLA-3	Integrin α
CD49d	Broad distribution includes B cells, thymocytes, monocytes, granulocytes, dendritic cells	150	α4 integrin, associates with CD29, binds fibronectin, MAdCAM-1, VCAM-1	VLA-4	Integrin α
CD49e	Broad distribution includes memory T cells, monocytes, platelets	135, 25	α5 integrin, associates with CD29, binds fibronectin, invasin	VLA-5	Integrin α
CD49f	T lymphocytes, monocytes, platelets, megakaryocytes, trophoblasts	125, 25	α6 integrin, associates with CD29, binds laminin, invasin, merosine	VLA-6	Integrin α
CD50	Thymocytes, T cells, B cells, monocytes, granulocytes	130	Binds integrin CD11a/CD18	ICAM-3	Immunoglobulin
CD51	Platelets, megakaryocytes	125, 24	αV integrin, associates with CD61, binds vitronectin, von Willebrand factor, fibrinogen, and thrombospondin; may be receptor for apoptotic cells	Vitronectin receptor	Integrin α
CD52	Thymocytes, T cells, B cells (not plasma cells), monocytes, granulocytes, spermatozoa	25	Unknown, target for antibodies used therapeutically to deplete T cells from bone marrow	CAMPATH-1, HE5	
CD53	Leukocytes	35–42	Unknown	MRC OX44	Transmembrane 4
CD54	Hematopoietic and non-hematopoietic cells	75–115	Intercellular adhesion molecule (ICAM)-1 binds CD11a/CD18 integrin (LFA-1) and CD11b/CD18 integrin (Mac-1), receptor for rhinovirus	ICAM-1	Immunoglobulin
CD55	Hematopoietic and non-hematopoietic cells	60–70	Decay accelerating factor (DAF), binds C3b, disassembles C3/C5 convertase	DAF	CCP
CD56	NK cells	135–220	Isoform of neural cell-adhesion molecule (NCAM), adhesion molecule	NKH-1	Immunoglobulin
CD57	NK cells, subsets of T cells, B cells, and monocytes		Oligosaccharide, found on many cell-surface glycoproteins	HNK-1, Leu-7	

CD antigen	Cellular expression	Molecular weight (kDa)	Functions	Other names	Family relationships
CD58	Hematopoietic and non-hematopoietic cells	55–70	Leukocyte function-associated antigen-3 (LFA-3), binds CD2, adhesion molecule	LFA-3	Immunoglobulin
CD59	Hematopoietic and non-hematopoietic cells	19	Binds complement components C8 and C9, blocks assembly of membrane-attack complex	Protectin, Mac inhibitor	Ly-6
CD60a			Disialyl ganglioside D3 (GD3)		Carbohydrate structures
CD60b			9-*O*-acetyl-GD3		Carbohydrate structures
CD60c			7-*O*-acetyl-GD3		Carbohydrate structures
CD61	Platelets, megakaryocytes, macrophages	110	Intergrin β3 subunit, associates with CD41 (GPIIb/IIIa) or CD51 (vitronectin receptor)		Integrin β
CD62E	Endothelium	140	Endothelium leukocyte adhesion molecule (ELAM), binds sialyl-Lewis^x, mediates rolling interaction of neutrophils on endothelium	ELAM-1, E-selectin	C-type lectin, EGF, and CCP
CD62L	B cells, T cells, monocytes, NK cells	150	Leukocyte adhesion molecule (LAM), binds CD34, GlyCAM, mediates rolling interactions with endothelium	LAM-1, L-selectin, LECAM-1	C-type lectin, EGF, and CCP
CD62P	Platelets, megakaryocytes, endothelium	140	Adhesion molecule, binds CD162 (PSGL-1), mediates interaction of platelets with endothelial cells, monocytes and rolling leukocytes on endothelium	P-selectin, PADGEM	C-type lectin, EGF, and CCP
CD63	Activated platelets, monocytes, macrophages	53	Unknown, is lysosomal membrane protein translocated to cell surface after activation	Platelet activation antigen	Transmembrane 4
CD64	Monocytes, macrophages	72	High-affinity receptor for IgG, binds IgG3>IgG1>IgG4>>>IgG2, mediates phagocytosis, antigen capture, ADCC	FcγRI	Immunoglobulin
CD65	Myeloid cells		Oligosaccharide component of a ceramide dodecasaccharide		
CD66a	Neutrophils	160–180	Unknown, member of carcino-embryonic antigen (CEA) family (see below)	Biliary glyco-protein-1 (BGP-1)	Immunoglobulin
CD66b	Granulocytes	95–100	Unknown, member of carcino-embryonic antigen (CEA) family	Previously CD67	Immunoglobulin
CD66c	Neutrophils, colon carcinoma	90	Unknown, member of carcino-embryonic antigen (CEA) family	Nonspecific cross-reacting antigen (NCA)	Immunoglobulin
CD66d	Neutrophils	30	Unknown, member of carcino-embryonic antigen (CEA) family		Immunoglobulin
CD66e	Adult colon epithelium, colon carcinoma	180–200	Unknown, member of carcino-embryonic antigen (CEA) family	Carcino-embryonic antigen (CEA)	Immunoglobulin
CD66f	Unknown		Unknown, member of carcino-embryonic antigen (CEA) family	Pregnancy specific glycoprotein	Immunoglobulin
CD68	Monocytes, macrophages, neutrophils, basophils, large lymphocytes	110	Unknown	Macrosialin	Mucin

CD antigen	Cellular expression	Molecular weight (kDa)	Functions	Other names	Family relationships
CD69	Activated T and B cells, activated macrophages and NK cells	28, 32 homodimer	Unknown, early activation antigen	Activation inducer molecule (AIM)	C-type lectin
CD70	Activated T and B cells, and macrophages	75, 95, 170	Ligand for CD27, may function in co-stimulation of B and T cells	Ki-24	TNF
CD71	All proliferating cells, hence activated leukocytes	95 homodimer	Transferrin receptor	T9	
CD72	B cells (not plasma cells)	42 homodimer	Unknown	Lyb-2	C-type lectin
CD73	B-cell subsets, T-cell subsets	69	Ecto-5′-nucleotidase, dephosphorylates nucleotides to allow nucleoside uptake		
CD74	B cells, macrophages, monocytes, MHC class II positive cells	33, 35, 41, 43 (alternative initiation and splicing)	MHC class II-associated invariant chain	Ii, Iγ	
CD75	Mature B cells, T-cells subsets		Lactosamines, ligand for CD22, mediates B-cell–B-cell adhesion		
CD75s			α-2,6-sialylated lactosamines		Carbohydrate structures
CD77	Germinal center B cells		Neutral glycosphingolipid (Galα1→4Galβ1→4Galcβ1 →ceramide), binds Shiga toxin, cross-linking induces apoptosis	Globotriaocyl-ceramide (Gb3) Pk blood group	
CD79α,β	B cells	α: 40–45 β: 37	Components of B-cell antigen receptor analogous to CD3, required for cell-surface expression and signal transduction	Igα, Igβ	Immunoglobulin
CD80	B-cell subset	60	Co-stimulator, ligand for CD28 and CTLA-4	B7 (now B7.1), BB1	Immunoglobulin
CD81	Lymphocytes	26	Associates with CD19, CD21 to form B cell co-receptor	Target of anti-proliferative anti-body (TAPA-1)	Transmembrane 4
CD82	Leukocytes	50–53	Unknown	R2	Transmembrane 4
CD83	Dendritic cells, B cells, Langerhans cells	43	Unknown	HB15	Immunoglobulin
CDw84	Monocytes, platelets, circulating B cells	73	Unknown	GR6	Immunoglobulin
CD85	Dendritic cells		ILT/LIR family	GR4	Immunoglobulin superfamily
CD86	Monocytes, activated B cells, dendritic cells	80	Ligand for CD28 and CTLA4	B7.2	Immunoglobulin
CD87	Granulocytes, monocytes, macrophages, T cells, NK cells, wide variety of nonhematopoietic cell types	35–59	Receptor for urokinase plasminogen activator	uPAR	Ly-6
CD88	Polymorphonuclear leukocytes, macrophages, mast cells	43	Receptor for complement component C5a	C5aR	G protein-coupled receptor
CD89	Monocytes, macrophages, granulocytes, neutrophils, B-cell subsets, T-cell subsets	50–70	IgA receptor	FcαR	Immunoglobulin
CD90	CD34^{+} prothymocytes (human), thymocytes, T cells (mouse)	18	Unknown	Thy-1	Immunoglobulin
CD91	Monocytes, many non-hematopoietic cells	515, 85	α2-macroglobulin receptor		EGF, LDL receptor

CD antigen	Cellular expression	Molecular weight (kDa)	Functions	Other names	Family relationships
CD92	Neutrophils, monocytes, platelets, endothelium	70	Unknown	GR9	
CD93	Neutrophils, monocytes, endothelium	120	Unknown	GR11	
CD94	T-cell subsets, NK cells	43	Unknown	KP43	C-type lectin
CD95	Wide variety of cell lines, *in vivo* distribution uncertain	45	Binds TNF-like Fas ligand, induces apoptosis	Apo-1, Fas	TNF receptor
CD96	Activated T cells, NK cells	160	Unknown	T-cell activation increased late expression (TACTILE)	Immunoglobulin
CD97	Activated B and T cells, monocytes, granulocytes	75–85	Binds CD55	GR1	EGF, G protein-coupled receptor
CD98	T cells, B cells, natural killer cells, granulocytes, all human cell lines	80, 45 heterodimer	May be amino acid transporter	4F2, FRP-1	
CD99	Peripheral blood lymphocytes, thymocytes	32	Unknown	MIC2, E2	
CD100	Hematopoietic cells	150 homodimer	Unknown	GR3	Semaphorin
CD101	Monocytes, granulocytes, dendritic cells, activated T cells	120 homodimer	Unknown	BPC#4	Immunoglobulin
CD102	Resting lymphocytes, monocytes, vascular endothelium cells (strongest)	55–65	Binds CD11a/CD18 (LFA-1) but not CD11b/CD18 (Mac-1)	ICAM-2	Immunoglobulin
CD103	Intraepithelial lymphocytes, 2–6% peripheral blood lymphocytes	150, 25	αE integrin	HML-1, α6, αE integrin	Integrin α
CD104	$CD4^-$ $CD8^-$ thymocytes, neuronal, epithelial, and some endothelial cells, Schwann cells, trophoblasts	220	Integrin β4 associates with CD49f, binds laminins	β4 integrin	Integrin β
CD105	Endothelial cells, activated monocytes and macrophages, bone marrow cell subsets	90 homodimer	Binds TGF-β	Endoglin	
CD106	Endothelial cells	100–110	Adhesion molecule, ligand for VLA-4 ($\alpha_4\beta_1$ integrin)	VCAM-1	Immunoglobulin
CD107a	Activated platelets, activated T cells, activated neutrophils, activated endothelium	110	Unknown, is lysosomal membrane protein translocated to the cell surface after activation	Lysosomal associated membrane protein-1 (LAMP-1)	
CD107b	Activated platelets, activated T cells, activated neutrophils, activated endothelium	120	Unknown, is lysosomal membrane protein translocated to the cell surface after activation	LAMP-2	
CD108	Erythrocytes, circulating lymphocytes, lymphoblasts	80	Unknown	GR2, John Milton-Hagen blood group antigen	
CD109	Activated T cells, activated platelets, vascular endothelium	170	Unknown	Platelet activation factor, GR56	
CD110	Platelets		MPL, TPO R		
CD111	Myeloid cells		PPR1/Nectin1		
CD112	Myeloid cells		PRR2		
CD114	Granulocytes, monocytes	150	Granulocytes colony-stimulating factor (G-CSF) receptor		Immunoglobulin, fibronectin type III

CD antigen	Cellular expression	Molecular weight (kDa)	Functions	Other names	Family relationships
CD115	Monocytes, macrophages	150	Macrophage colony-stimulating factor (M-CSF) receptor	M-CSFR, c-fms	Immunoglobulin, tyrosine kinase
CD116	Monocytes, neutrophils, eosinophils, endothelium	70–85	Granulocyte-macrophage colony-stimulating factor (GM-CSF) receptor α chain	GM-CSFRα	Cytokine receptor, fibronectin type III
CD117	Hematopoietic progenitors	145	Stem-cell factor (SCF) receptor	c-Kit	Immunoglobulin, tyrosine kinase
CD118	Broad cellular expression		Interferon-α, β receptor	IFN-α, βR	
CD119	Macrophages, monocytes, B cells, endothelium	90–100	Interferon-γ receptor	IFN-γR	Fibronectin type III
CD120a	Hematopoietic and nonhematopoietic cells, highest on epithelial cells	50–60	TNF receptor, binds both TNF-α and TNF-β	TNFR-I	TNF receptor
CD120b	Hematopoietic and nonhematopoietic cells, highest on myeloid cells	75–85	TNF receptor, binds both TNF-α and TNF-β	TNFR-II	TNF receptor
CD121a	Thymocytes, T cells	80	Type I interleukin-1 receptor, binds IL-1α and IL-1β	IL-1R type I	Immunoglobulin
CDw121b	B cells, macrophages, monocytes	60–70	Type II interleukin-1 receptor, binds IL-1α and IL-1β	IL-1R type II	Immunoglobulin
CD122	NK cells, resting T-cell subsets, some B-cell lines	75	IL-2 receptor β chain	IL-2Rβ	Cytokine receptor, fibronectin type III
CD123	Bone marrow stem cells, granulocytes, monocytes, megakaryocytes	70	IL-3 receptor α chain	IL-3Rα	Cytokine receptor, fibronectin type III
CD124	Mature B and T cells, hematopoietic precursor cells	130–150	IL-4 receptor	IL-4R	Cytokine receptor, fibronectin type III
CD125	Eosinophils, basophils, activated B cells	55–60	IL-5 receptor	IL-5R	Cytokine receptor, fibronectin type III
CD126	Activated B cells and plasma cells (strong), most leukocytes (weak)	80	IL-6 receptor α subunit	IL-6Rα	Immunoglobulin, cytokine receptor, fibronectin type III
CD127	Bone marrow lymphoid precursors, pro-B cells, mature T cells, monocytes	68–79, possibly forms homodimers	IL-7 receptor	IL-7R	Fibronectin type III
CDw128	Neutrophils, basophils, T-cell subsets	58–67	IL-8 receptor	IL-8R	G protein-coupled receptor
CD129	Not yet assigned				
CD130	Most cell types, strong on activated B cells and plasma cells	130	Common subunit of IL-6, IL-11, oncostatin-M (OSM) and leukemia inhibitory factor (LIF) receptors	IL-6Rβ, IL-11Rβ, OSMRβ, LIFRβ, IFRβ	Immunoglobulin, cytokine receptor, fibronectin type III
CDw131	Myeloid progenitors, granulocytes	140	Common β subunit of IL-3, IL-5, and GM-CSF receptors	IL-3Rβ, IL-5Rβ, GM-CSFRβ	Cytokine receptor, fibronectin type III
CD132	B cells, T cells, NK cells, mast cells, neutrophils	64	IL-2 receptor γ chain, common subunit of IL-2, IL-4, IL-7, IL-9, and IL-15 receptors		Cytokine receptor
CD133	Stem/progenitor cells		AC133		
CD134	Activated T cells	50	May act as adhesion molecule costimulator	OX40	TNF receptor
CD135	Multipotential precursors, myelomonocytic and B-cell progenitors	130, 155	Growth factor receptor	FLK2, STK-1	Immunoglobulin, tyrosine kinase
CDw136	Monocytes, epithelial cells, central and peripheral nervous system	180	Chemotaxis, phagocytosis, cell growth, and differentiation	MSP-R, RON	Tyrosine kinase

CD antigen	Cellular expression	Molecular weight (kDa)	Functions	Other names	Family relationships
CDw137	T and B lymphocytes, monocytes, some epithelial cells		Co-stimulator of T-cell proliferation	ILA (induced by lymphocyte activation), 4-1BB	TNF receptor
CD138	B cells		Heparan sulphate proteoglycan binds collagen type I	Syndecan-1	
CD139	B cells	209, 228	Unknown		
CD140a,b	Stromal cells, some endothelial cells	a: 180 b: 180	Platelet derived growth factor (PDGF) receptor α and β chains		
CD141	Vascular endothelial cells	105	Anticoagulant, binds thrombin, the complex then activates protein C	Thrombomodulin fetomodulin	C-type lectin, EGF
CD142	Epidermal keratinocytes, various epithelial cells, astrocytes, Schwann cells. Absent from cells in direct contact with plasma unless induced by inflammatory mediators	45–47	Major initiating factor of clotting. Binds Factor VIIa; this complex activates Factors VII, IX, and X	Tissue factor, thromboplastin	Fibronectin type III
CD143	Endothelial cells, except large blood vessels and kidney, epithelial cells of brush borders of kidney and small intestine, neuronal cells, activated macrophages and some T cells. Soluble form in plasma	170–180	Zn^{2+} metallopeptidase dipeptidyl peptidase, cleaves angiotensin I and bradykinin from precursor forms	Angiotensin converting enzyme (ACE)	
CD144	Endothelial cells	130	Organizes adherens junction in endothelial cells	Cadherin-5, VE-cadherin	Cadherin
CD145	Endothelial cells, some stromal cells	25, 90, 110	Unknown		
CD146	Endothelium	130	Potential adhesion molecule, localized at cell–cell junctions	MCAM, MUC18, S-ENDO	Immunoglobulin
CD147	Leukocytes, red blood cells, platelets, endothelial cells	55–65	Potential adhesion molecule	M6, neurothelin, EMMPRIN, basigin, OX-47	Immunoglobulin
CD148	Granulocytes, monocytes, dendritic cells, T cells, fibroblasts, nerve cells	240–260	Contact inhibition of cell growth	HPTPη	Fibronectin type III, protein tyrosine phosphatase
CD150	Thymocytes, activated lymphocytes	75–95	Unknown	SLAM	Immunoglobulin
CD151	Platelets, megakaryocytes, epithelial cells, endothelial cells	32	Associates with β1 integrins	PETA-3, SFA-1	Transmembrane 4
CD152	Activated T cells	33	Receptor for B7.1 (CD80), B7.2 (CD86); negative regulator of T-cell activation	CTLA-4	Immunoglobulin
CD153	Activated T cells, activated macrophages, neutrophils, B cells	38–40	Ligand for CD30, may co-stimulate T cells	CD30L	TNF
CD154	Activated CD4 T cells	30 trimer	Ligand for CD40, inducer of B cell proliferation and activation	CD40L, TRAP, T-BAM, gp39	TNF receptor
CD155	Monocytes, macrophages, thymocytes, CNS neurons	80–90	Normal function unknown; receptor for poliovirus	Poliovirus receptor	Immunoglobulin
CD156a	Neutrophils, monocytes	69	Unknown, may be involved in integrin leukocyte extravasation	MS2, ADAM 8 (A disintegrin and metallo-protease)	
CD156b			TACE/ADAM17. Adhesion structures		
CD157	Granulocytes, monocytes, bone marrow stromal cells, vascular endothelial cells, follicular dendritic cells	42–45 (50 on monocytes)	ADP-ribosyl cyclase, cyclic ADP-ribose hydrolase	BST-1	
CD158	NK cells		KIR family		

CD antigen	Cellular expression	Molecular weight (kDa)	Functions	Other names	Family relationships
CD158a	NK-cell subsets	50 or 58	Inhibits NK cell cytotoxicity on binding MHC class I molecules	p50.1, p58.1	Immunoglobulin
CD158b	NK-cell subsets	50 or 58	Inhibits NK cell cytotoxicity on binding HLA-Cw3 and related alleles	p50.2, p58.2	Immunoglobulin
CD159a	NK cells		Binds CD94 to form NK receptor; inhibits NK cell cytotoxicity on binding MHC class I molecules	NKG2A	
CD160	T cells			BY55	
CD161	NK cells, T cells	44	Regulates NK cytotoxicity	NKRP1	C-type lectin
CD162	Neutrophils, lymphocytes, monocytes	120 homodimer	Ligand for CD62P	PSGL-1	Mucin
CD162R	NK cells			PEN5	
CD163	Monocytes, macrophages	130	Unknown	M130	
CD164	Epithelial cells, monocytes, bone marrow stromal cells	80	Unknown	MUC-24 (multi-glycosylated protein 24)	Mucin
CD165	Thymocytes, thymic epithelial cells, CNS neurons, pancreatic islets, Bowman's capsule	37	Adhesion between thymocytes and thymic epithelium	Gp37, AD2	
CD166	Activated T cells, thymic epithelium, fibroblasts, neurons	100–105	Ligand for CD6, involved integrin neurite extension	ALCAM, BEN, DM-GRASP, SC-1	Immunoglobulin
CD167a	Normal and transformed epithelial cells	63, 64 dimer	Binds collagen	DDR1, trkE, cak, eddr1	Receptor tyrosine kinase, discoidin-related
CD168	Breast cancer cells	Five isoforms: 58, 60, 64, 70, 84	Adhesion molecule. Receptor for hyaluronic acid-mediated motility—mediated cell migration	RHAMM	
CD169	Subsets of macrophages	185	Adhesion molecule. Binds sialylated carbohydrates. May mediate macrophage binding to granulocytes and lymphocytes	Sialoadhesin	Immunoglobulin superfamily, sialoadhesin family
CD170	Neutrophils	67 homodimer	Adhesion molecule. **S**ialic acid-binding **I**g-like **lec**tin (Siglec). Cytoplasmic tail contains ITIM motifs	Siglec-5, OBBP2, CD33L2	Immunoglobulin superfamily, sialoadhesin family
CD171	Neurons, Schwann cells, lymphoid and myelomonocytic cells, B cells, CD4 T cells (not CD8 T cells)	200–220, exact MW varies with cell type	Adhesion molecule, binds CD9, CD24, CD56, also homophilic binding	L1, NCAM-L1	Immunoglobulin superfamily
CD172a		115–120	Adhesion molecule; the transmembrane protein is a substrate of activated receptor tyrosine kinases (RTKs) and binds to SH2 domains	SIRP, SHPS1, MYD-1, SIRP-α-1, protein tyrosine phosphatase, non-receptor type substrate 1 (PTPNS1)	Immunoglobulin superfamily
CD173	All cells		Blood group H type 2. Carbohydrate moiety		
CD174	All cells		Lewis y Blood group. Carbohydrate moiety		
CD175	All cells		Tn Blood group. Carbohydrate moiety		
CD175s	All cells		Sialyl-Tn Blood group. Carbohydrate moiety		
CD176	All cells		TF Blood group. Carbohydrate moiety		

CD antigen	Cellular expression	Molecular weight (kDa)	Functions	Other names	Family relationships
CD177	Myeloid cells	56–64	NB1 is a GPI-linked neutrophil-specific antigen, found on only a subpopulation of neutrophils present in NB1-positive adults (97% of healthy donors) NB1 is first expressed at the myelocyte stage of myeloid differentiation	NB1	
CD178	Activated T cells	38–42	Fas ligand; binds to Fas to induce apoptosis	FasL	TNF superfamily
CD179a	Early B cells	16–18	Immunoglobulin iota chain associates non-covalently with CD179b to form a surrogate light chain which is a component of the pre-B-cell receptor that plays a critical role in early B-cell differentiation	VpreB, IGVPB, IGι	Immunoglobulin superfamily
CD179b	B cells	22	Immunoglobulin λ-like polypeptide 1 associates noncovalently with CD179a to form a surrogate light chain that is selectively expressed at the early stages of B-cell development. Mutations in the CD179b gene have been shown to result in impairment of B-cell development and agammaglobulinemia in humans	IGLL1, λ5 (IGL5), IGVPB, 14.	Immunoglobulin superfamily
CD180	B cells	95–105	Type 1 membrane protein consisting of extracellular leucine-rich repeats (LRR). Is associated with a molecule called MD-1 and forms the cell-surface receptor complex, RP105/MD-1, which by working in concert with TLR4, controls B-cell recognition and signaling of lipopolysaccharide (LPS)	LY64, RP105	Toll-like receptors (TLR)
CD183	Particularly on malignant B cells from chronic lymphoproliferative disorders	46–52	CXC chemokine receptor involved in chemotaxis of malignant B lymphocytes. Binds INP10 and MIG[3]	CXCR3, G protein-coupled receptor 9 (GPR 9)	Chemokine receptors, G protein-coupled receptor superfamily
CD184	Preferentially expressed on the more immature CD34$^+$ haematopoietic stem cells	46–52	Binding to SDF-1 (LESTR/fusin); acts as a cofactor for fusion and entry of T-cell line; trophic strains of HIV-1	CXCR4, NPY3R, LESTR, fusin, HM89	Chemokine receptors, G protein-coupled receptor superfamily
CD195	Promyelocytic cells	40	Receptor for a CC type chemokine. Binds to MIP-1α, MIP-1β and RANTES. May play a role in the control of granulocytic lineage proliferation or differentiation. Acts as co-receptor with CD4 for primary macrophage-tropic isolates of HIV-1	CMKBR5, CCR5, CKR-5, CC-CKR-5, CKR5	Chemokine receptors, G protein-coupled receptor superfamily
CDw197	Activated B and T lymphocytes, strongly upregulated in B cells infected with EBV and T cells infected with HHV6 or 7	46–52	Receptor for the MIP-3β chemokine; probable mediator of EBV effects on B lymphocytes or of normal lymphocyte functions	CCR7. EBI1 (Epstein–Barr virus induced gene 1), CMKBR7, BLR2	Chemokine receptors, G protein-coupled receptor superfamily
CD200	Normal brain and B-cell lines	41 (rat thymocytes) 47 (rat brain)	Antigen identified by MoAb MRC OX-2. Nonlineage molecules. Function unknown	MOX-2, MOX-1	Immunoglobulin superfamily
CD201	Endothelial cells	49	Endothelial cell-surface receptor (EPCR) that is capable of high-affinity binding of protein C and activated protein C. It is downregulated by exposure of endothelium to tumor necrosis factor	EPCR	CD1major histocompatibility complex family
CD202b	Endothelial cells	140	Receptor tyrosine kinase, binds angiopoietin-1; important in angiogenesis, particularly for vascular network formation in endothelial cells. Defects in TEK are associated with inherited venous malformations; the TEK signaling pathway appears to be critical for endothelial cell–smooth muscle cell communication in venous morphogenesis	VMCM, TEK (tyrosine kinase, endothelial), TIE2 (tyrosine kinase with Ig and EGF homology domains), VMCM1	Immunoglobulin superfamily, tyrosine kinase

CD antigen	Cellular expression	Molecular weight (kDa)	Functions	Other names	Family relationships
CD203c	Myeloid cells (uterus, basophils, and mast cells)	101	Belongs to a series of ectoenzymes that are involved in hydrolysis of extracellular nucleotides. They catalyze the cleavage of phosphodiester and phosphosulfate bonds of a variety of molecules, including deoxynucleotides, NAD, and nucleotide sugars	NPP3, B10, PDNP3, PD-Iβ, gp130RB13-6	Type II transmembrane proteins, Ecto-nucleotide pyrophosphatase/phosphodiesterase (E-NPP) family
CD204	Myeloid cells	220	Mediate the binding, internalization, and processing of a wide range of negatively charged macromolecules. Implicated in the pathologic deposition of cholesterol in arterial walls during atherogenesis	Macrophage scavenger R (MSR1)	Scavenger receptor family, collagen-like
CD205	Dendritic cells	205	Lymphocyte antigen 75; putative antigen-uptake receptor on dendritic cells	LY75, DEC-205, GP200-MR6	Type I transmembrane protein
CD206	Macrophages, endothelial cells	175–190	Type I membrane glycoprotein; only known example of a C-type lectin that contains multiple C-type CRDs (carbohydrate-recognition domains); it binds high-mannose structures on the surface of potentially pathogenic viruses, bacteria, and fungi	Macrophage mannose receptor (MMR), MRC1	C-type lectin superfamily
CD207	Langerhans cells	40	Type II transmembrane protein; Langerhans cell specific C-type lectin; potent inducer of membrane superimposition and zippering leading to BG (Birbeck granules) formation	Langerin	C-type lectin superfamily
CD208	Interdigitating dendritic cells in lymphoid organs	70–90	Homologous to CD68, DC-LAMP is a lysosomal protein involved in remodeling of specialized antigen-processing compartments and in MHC class II-restricted antigen presentation. Upregulated in mature DCs induced by CD40L, TNF-α and LPS.	D lysosome-associated membrane protein, DC-LAMP	Major histocompatibility complex family
CD209	Dendritic cells	44	C-type lectin; binds ICAM3 and HIV-1 envelope glycoprotein gp120 enables T-cell receptor engagement by stabilization of the DC/T-cell contact zone, promotes efficient infection in *trans* cells that express CD4 and chemokine receptors; type II transmembrane protein	DC-SIGN (dendritic cell-specific ICAM3-grabbing non-integrin)	C-type lectin superfamily
CDw210	B cells, T helper cells, and cells of the monocyte/macrophage lineage	90–110	Interleukin 10 receptor α and β	IL-10Rα, IL-10RA, HIL-10R, IL-10Rβ, IL-10RB, CRF2-4, CRFB4	Class II cytokine receptor family
CD212	Activated CD4, CD8, and NK cells	130	IL-12 receptor β chain; a type I transmembrane protein involved in IL-12 signal transduction.	IL-12R IL-12RB	Haemopoietin cytokine receptor superfamily
CD213a1	B cells, monocytes, fibroblasts, endothelial cells	60–70	Receptor which binds Il-13 with a low affinity; together with IL-4Rα can form a functional receptor for IL-13, also serves as an alternate accessory protein to the common cytokine receptor gamma chain for IL-4 signaling	IL-13Rα 1, NR4, IL-13Ra	Haemopoietic cytokine receptor superfamily
CD213a2	B cells, monocytes, fibroblasts, endothelial cells		IL-13 receptor which binds as a monomer with high affinity to interleukin-13 (IL-13), but not to IL-4; human cells expressing IL-13RA2 show specific IL-13 binding with high affinity	IL-13Rα 2, IL-13BP	Haemopoietic cytokine receptor superfamily
CDw217	Activated memory T cells	120	Interleukin 17 receptor homodimer	IL-17R, CTLA-8	Chemokine/ cytokine receptors

CD antigen	Cellular expression	Molecular weight (kDa)	Functions	Other names	Family relationships
CD220	Nonlineage molecules	α: 130 β: 95	Insulin receptor; integral transmembrane glycoprotein comprised of two α and two β subunits; this receptor binds insulin and has a tyrosine-protein kinase activity—autophosphorylation activates the kinase activity	Insulin receptor	Insulin receptor family of tyrosine-protein kinases, EGFR family
CD221	Nonlineage molecules	α: 135 β: 90	Insulin-like growth factor I receptor binds insulin-like growth factor with a high affinity. It has tyrosine kinase activity and plays a critical role in transformation events. Cleavage of the precursor generates α and β subunits.	IGF1R, JTK13	Insulin receptor family of tyrosine-protein kinases, EGFR family
CD222	Nonlineage molecules	250	Ubiquitously expressed multifunctional type I transmembrane protein. Its main functions include internalization of IGF-II, internalization or sorting of lysosomal enzymes and other M6P-containing proteins	IGF2R, CIMPR, CI-MPR, IGF2R, M6P-R (Mannose-6-phosphate receptor)	Mammalian lectins
CD223	Activated T and NK cells	70	Involved in lymphocyte activation; binds to HLA class-II antigens; role in down-regulating antigen specific response; close relationship of LAG3 to CD4	Lymphocyte-activation gene 3 LAG-3	Immunoglobulin superfamily
CD224	Nonlineage molecules	62 (unprocessed precursor)	Predominantly a membrane-bound enzyme; plays a key role in the γ-glutamyl cycle, a pathway for the synthesis and degradation of glutathione. This enzyme consists of two polypeptide chains, which are synthesized in precursor form from a single polypeptide	γ-glutamyl transferase, GGT1, D22S672 D22S732	γ-glutamyl transferase protein family
CD225	Leukocytes and endothelial cells	16–17	Interferon-induced transmembrane protein 1 is implicated in the control of cell growth. It is a component of a multimeric complex involved in the transduction of antiproliferative and homotypic adhesion signals	Leu 13, IFITM1, IFI17	IFN-induced transmembrane proteins
CD226	NK cells, platelets, monocytes, and a subset of T cells	65	Adhesion glycoprotein; mediates cellular adhesion to other cells bearing an unidentified ligand and cross-linking CD226 with antibodies causes cellular activation	DNAM-1 (PTA1), DNAX, TLiSA1	Immunoglobulin superfamily
CD227	Human epithelial tumors, such as breast cancer	122 (non-glycosylated)	Epithelial mucin containing a variable number of repeats with a length of twenty amino acids, resulting in many different alleles. Direct or indirect interaction with actin cytoskeleton.	PUM (peanut-reactive urinary mucin), MUC.1, mucin 1	Mucin
CD228	Predominantly in human melanomas	97	Tumor-associated antigen (melanoma) identified by monoclonal antibodies 133.2 and 96.5,involved in cellular iron uptake.	Melanotransferrin, P97	Transferrin superfamily
CD229	Lymphocytes	90–120	May participate in adhesion reactions between T lymphocytes and accessory cells by homophilic interaction	Ly9	Immunoglobulin superfamily (CD2 subfamily)
CD230	Expressed both in normal and infected cells	27–30	The function of PRP is not known. It is encoded in the host genome found in high quantity in the brain of humans and animals infected with neurodegenerative diseases known as transmissible spongiform encephalopathies or prion diseases (Creutzfeld–Jakob disease, Gerstmann–Strausler–Scheinker syndrome, fatal familial insomnia)	CJD, PRIP, prion protein (p27-30)	Prion family

CD antigen	Cellular expression	Molecular weight (kDa)	Functions	Other names	Family relationships
CD231	T-cell acute lymphoblastic leukemia, neuroblastoma cells and normal brain neuron	150	The function of CD231 is currently unknown. It is cell-surface glycoprotein which is a specific marker for T-cell acute lymphoblastic leukemia. Also found on neuroblastomas	TALLA-1, TM4SF2, A15, MXS1, CCG-B7	Transmembrane 4 superfamily (TM4SF also known as tetraspanins)
CD232	Nonlineage molecules	200	Receptor for an immunologically active semaphorin (virus-encoded semaphorin protein receptor)	VESPR, PLXN, PLXN-C1	Plexin family
CD233	Erythroid cells	93	Band 3 is the major integral glycoprotein of the erythrocyte membrane. It has two functional domains. Its integral domain mediates a 1:1 exchange of inorganic anions across the membrane, whereas its cytoplasmic domain provides binding sites for cytoskeletal proteins, glycolytic enzymes, and hemoglobin. Multifunctional transport protein	SLC4A1, Diego blood group, D1, AE1, EPB3	Anion exchanger family
CD234	Erythroid cells and nonerythroid cells	35	Fy-glycoprotein; Duffy blood group antigen; nonspecific receptor for many chemokines such as IL-8, GRO, RANTES, MCP-1 and TARC. It is also the receptor for the human malaria parasites *Plasmodium vivax* and *Plasmodium knowlesi* plays a role in inflammation and in malaria infection	GPD, CCBP1, DARC (duffy antigen/receptor for chemokines)	Family 1 of G protein-coupled receptors, chemokine receptors superfamily
CD235a	Erythroid cells	31	Major carbohydrate rich sialoglycoprotein of human erythrocyte membrane which bears the antigenic determinants for the MN and Ss blood groups. The N-terminal glycosylated segment, which lies outside the erythrocyte membrane, has MN blood group receptors and also binds influenza virus	Glycophorin A, GPA, MNS	Glycophorin A family
CD235b	Erythroid cells	GYPD is smaller than GYPC (24 kD vs 32 kD)	This protein is a minor sialoglycoprotein in human erythrocyte membranes. Along with GYPA, GYPB is responsible for the MNS blood group system. The Ss blood group antigens are located on glycophorin B	Glycophorin B, MNS, GPB	Glycophorin A family
CD236	Erythroid cells	24	Glycophorin C (GPC) and glycophorin D (GPD) are closely related sialoglycoproteins in the human red blood cell (RBC) membrane. GPD is a ubiquitous shortened isoform of GPC, produced by alternative splicing of the same gene. The Webb and Duch antigens, also known as glycophorin D, result from single point mutations of the glycophorin C gene	Glycophorin D, GPD, GYPD	Type III membrane proteins
CD236R	Erythroid cells	32	Glycophorin C (GPC) is associated with the Gerbich (Ge) blood group deficiency. It is a minor red cell-membrane component, representing about 4% of the membrane sialoglycoproteins, but shows very little homology with the major red cell-membrane glycophorins A and B. It plays an important role in regulating the mechanical stability of red cells and is a putative receptor for the merozoites of *Plasmodium falciparum*	Glycophorin C, GYPC, GPC	Type III membrane proteins
CD238	Erythroid cells	93	KELL blood group antigen; homology to a family of zinc metalloglycoproteins with neutral endopeptidase activity, type II transmembrane glycoprotein	KELL	Belongs to peptidase family m13 (zinc metalloprotease); also known as the neprilysin subfamily

CD antigen	Cellular expression	Molecular weight (kDa)	Functions	Other names	Family relationships
CD239	Erythroid cells	78	A type I membrane protein.The human F8/G253 antigen, B-CAM, is a cell surface glycoprotein that is expressed with restricted distribution pattern in normal fetal and adult tissues, and is upregulated following malignant transformation in some cell types. Its overall structure is similar to that of the human tumor marker MUC 18 and the chicken neural adhesion molecule SC1	B-CAM (B-cell adhesion molecule), LU, Lutheran blood group	Immunoglobulin superfamily
CD240CE	Erythroid cells	45.5	Rhesus blood group, CcEe antigens. May be part of an oligomeric complex which is likely to have a transport or channel function in the erythrocyte membrane. It is highly hydrophobic and deeply buried within the phospholipid bilayer	RHCE, RH30A, RHPI, Rh4	Rh family
CD240D	Erythroid cells	45.5 (product—30)	Rhesus blood group, D antigen. May be part of an oligomeric complex which is likely to have a transport or channel function in the erythrocyte membrane. Absent in the Caucasian RHD-negative phenotype	RhD, Rh4, RhPI, RhII, Rh30D	Rh family
CD241	Erythroid cells	50	Rhesus blood group-associated glycoprotein RH50, component of the RH antigen multisubunit complex; required for transport and assembly of the Rh membrane complex to the red blood cell surface. Highly homologous to RH, 30kD components. Defects in RhAg are a cause of a form of chronic hemolytic anemia associated with stomatocytosis, and spherocytosis, reduced osmotic fragility, and increased cation permeability	RhAg, RH50A	Rh family
CD242	Erythroid cells	42	Intercellular adhesion molecule 4, Landsteiner-Wiener blood group. LW molecules may contribute to the vaso-occlusive events associated with episodes of acute pain in sickle cell disease	ICAM-4, LW	Immunoglobulin superfamily, intercellular adhesion molecules (ICAMs)
CD243	Stem/progenitor cells	170	Multidrug resistance protein 1 (P-glycoprotein). P-gp has been shown to utilise ATP to pump hydrophobic drugs out of cells, thus increasing their intracellular concentration and hence their toxicity. The MDR 1 gene is amplified in multidrug resistant cell lines	MDR-1, p-170	ABC superfamily of ATP-binding transport proteins
CD244	NK cells	66	2B4 is a cell-surface glycoprotein related to CD2 and implicated in the regulation of natural killer and T-lymphocyte function. It appears that the primary function of 2B4 is to modulate other receptor–ligand interactions to enhance leukocyte activation	2B4, NK cell activation inducing ligand (NAIL)	Immunoglobulin superfamily
CD245	T cells	220–240	Cyclin E/Cdk2 interacting protein p220. NPAT is involved in a key S phase event and links cyclical cyclin E/Cdk2 kinase activity to replication-dependent histone gene transcription. NPAT gene may be essential for cell maintenance and may be a member of the housekeeping genes	NPAT	

CD antigen	Cellular expression	Molecular weight (kDa)	Functions	Other names	Family relationships
CD246	Expressed in the small intestine, testis, and brain but not in normal lymphoid cells	177 kDa; after glycosylation, produces a 200 kDa mature glyco-protein	Anaplastic ($CD30^+$ large cell) lymphoma kinase; plays an important role in brain development, involved in anaplastic nodal non Hodgkin lymphoma or Hodgkin's disease with translocation t(2;5)(p23;q35) or inv2(23;q35). Oncogenesis via the kinase function is activated by oligomerization of NPM1-ALK mediated by the NPM1 part	ALK	Insulin receptor family of tyrosine-protein kinases
CD247	T cells, NK cells	16	T-cell receptor ζ; has a probable role in assembly and expression of the TCR complex as well as signal transduction upon antigen triggering. TCRζ together with TCRα:β and γ:δ heterodimers and CD3-γ, -δ, and -ε, forms the TCR-CD3 complex. The ζ chain plays an important role in coupling antigen recognition to several intracellular signal-transduction pathways. Low expression of the antigen results in impaired immune response	ζ chain, CD3Z	Immunoglobulin superfamily

Compiled by Laura Herbert, Royal Free Hospital, London. Data based on CD designations made at the 7th Workshop on Human Leucocyte Differentiation Antigens, provided by Protein Reviews on the Web (www.ncbi.nlm.nih.gov/prow/).

Appendix III. Cytokines and their receptors.

Family	Cytokine (alternative names)	Size (no. of amino acids and form)	Receptors (c denotes common subunit)	Producer cells	Actions	Effect of cytokine or receptor knock-out (where known)
Colony-stimulating factors	G-CSF	174, monomer*	G-CSFR	Fibroblasts and monocytes	Stimulates neutrophil development and differentiation	G-CSF, G-CSFR: defective neutrophil production and mobilization
	GM-CSF (granulocyte macrophage colony stimulating factor)	127, monomer*	CD116, βc	Macrophages, T cells	Stimulates growth and differentiation of myelomonocytic lineage cells, particularly dendritic cells	GM-CSF, GM-CSFR: pulmonary alveolar proteinosis
	M-CSF (CSF-1)	α: 224 β: 492 γ: 406 active forms are homo- or heterodimeric	CSF-1R (c-fms)	T cells, bone marrow stromal cells, osteoblasts	Stimulates growth of cells of monocytic lineage	Osteopetrosis
Interferons	IFN-α (at least 12 distinct proteins)	166, monomer	CD118, IFNAR2	Leukocytes, dendritic cells	Antiviral, increased MHC class I expression	CD118: impaired antiviral activity
	IFN-β	166, monomer	CD118, IFNAR2	Fibroblasts	Antiviral, increased MHC class I expression	IFN-β: increased susceptibility to certain viruses
	IFN-γ	143, homodimer	CD119, IFNGR2	T cells, natural killer cells	Macrophage activation, increased expression of MHC molecules and antigen processing compone+F22nts, Ig class switching, supresses T_H2	IFN-γ, CD119: decreased resistance to bacterial infection and tumors
Interleukins	IL-1$\cdot\alpha$	159, monomer	CD121a (IL-1RI) and CD121b (IL-1RII)	Macrophages, epithelial cells	Fever, T-cell activation, macrophage activation	IL-1RI: decreased IL-6 production
	IL-1β	153, monomer	CD121a (IL-1RI) and CD121b (IL-1RII)	Macrophages, epithelial cells	Fever, T-cell activation, macrophage activation	IL-1β: impaired acute-A12phase response
	IL-1 RA	152, monomer	CD121a	Monocytes, macrophages, neutrophils, hepatocytes	Binds to but doesn't trigger IL-1 receptor, acts as a natural antagonist of IL-1 function	IL-1RA: reduced body mass, increased sensitivity to endotoxins (septic shock)
	IL-2 (T-cell growth factor)	133, monomer	CD25α, CD122β, CD132 (γc)	T cells	T-cell proliferation	IL-2: deregulated T-cell proliferation, colitis IL-2Rα: incomplete T-cell development autoimmunity IL-2Rβ: increased T-cell autoimmunity IL-2Rγc: severe combined immunodeficiency
	IL-3 (multicolony CSF)	133, monomer	CD123, βc	T cells, thymic epithelial cells	Synergistic action in early hematopoiesis	IL-3: impaired eosinophil development. Bone marrow unresponsive to IL-5, GM-CSF
	IL-4 (BCGF-1, BSF-1)	129, monomer	CD124, CD132 (γc)	T cells, mast cells	B-cell activation, IgE switch, induces differentiation into T_H2 cells	IL-4: decreased IgE synthesis
	IL-5 (BCGF-2)	115, homodimer	CD125, βc	T cells, mast cells	Eosinophil growth, differentiation	IL-5: decreased IgE, IgG1 synthesis (in mice); decreased levels of IL-9, IL-10, and eosinophils
	IL-6 (IFN-β2, BSF-2, BCDF)	184, monomer	CD126, CD130	T cells, macrophages, endothelial cells	T- and B-cell growth and differentiation, acute phase protein production, fever	IL-6: decreased acute phase reaction, reduced IgA production
	IL-7	152, monomer*	CD127, CD132 (γc)	Non-T cells	Growth of pre-B cells and pre-T cells	IL-7: early thymic and lymphocyte expansion severely impaired

*May function as dimers

Family	Cytokine (alternative names)	Size (no. of amino acids) and form	Receptors (c denotes common subunit)	Producer cells	Actions	Effect of cytokine or receptor knock-out (where known)
	IL-9	125, monomer	IL-9R, CD132 (γc)	T cells	Mast-cell enhancing activity, stimulates T_H2	Defects in mast-cell expansion
	IL-10 (cytokine synthesis inhibitory factor)	160, homodimer	IL-10Rα, IL-10Rβc (CRF2-4, IL-10R2)	Monocytes	Potent suppressant of macrophage functions	IL-10 and IL20Rβc-: reduced growth, anemia, chronic enterocolitis
	IL-11	178, monomer	IL-11R, CD130	Stromal fibroblasts	Synergistic action with IL-3 and IL-4 in hematopoiesis	IL-11R: defective decidualization
	IL-12 (NK-cell stimulatory factor)	197 (p35) and 306 (p40c), heterodimer	IL-12Rβ1c +IL-12Rβ2	Macrophages, dendritic cells	Activates NK cells, induces CD4 T-cell differentiation into T_H1-like cells	IL-12: impaired IFN-γ production and T_H1 responses
	IL-13 (p600)	132, monomer	IL-13R, CD132 (γc) (may also include CD24)	T cells	B-cell growth and differentiation, inhibits macrophage inflammatory cytokine production and T_H1 cells, induces allergy/asthma	IL-13: defective regulation of isotype specific responses
	IL-15 (T-cell growth factor)	114, monomer	IL-15Rα, CD122 (IL-2Rβ) CD132 (γc)	Many non-T cells	IL-2-like, stimulates growth of intestinal epithelium, T cells, and NK cells, enhances CD8 memory T cell survival	IL-15: reduced numbers of NK cells and memory phenotype $CD8^+$ T cells IL-15Ra: lymphopenia
	IL-16	130, homotetramer	CD4	T cells, mast cells, eosinophils	Chemoattractant for CD4 T cells, monocytes, and eosinophils, anti-apoptotic for IL-2-stimulated T cells	
	IL-17A (mCTLA-8)	150, homodimer	IL-17AR (CD217)	T_H17, CD8 T cells, NK cells, γδ T cells, neutrophils	Induces cytokine production by epithelia, endothelia, and fibroblasts, proinflamatory	IL-17R: reduced neutrophil migration into infected sites
	IL-17F (ML-1)	134, homodimer	IL-17AR (CD217)	T_H17, CD8 T cells, NK cells, γδ T cells, neutrophils	Induces cytokine production by epithelia, endothelia, and fibroblasts, proinflamatory	
	IL-18 (IGIF, interferon-α inducing factor)	157, monomer	IL-1Rrp (IL-1R related protein)	Activated macrophages and Kupffer cells	Induces IFN-γ production by T cells and NK cells, promotes T_H1 induction	Defective NK activity and T_H1 responses
	IL-19	153, monomer	IL-20Rα +IL-10Rβc	Monocytes	Induces IL-6 and TNF-α expression by monocytes	
	IL-20	152	IL-20Rα +ILa10Rβc; IL-22Rαc +IL-10Rβc	T_H1 cells	Stimulates keratinocyte proliferation and TNF-α production	
	IL-21	133	IL-21R, +CD132(γc)	T_H2 cells, developing T_H17 cells	Induces proliferation of B, T and NK cells	Increased IgE production
	IL-22 (IL-TIF)	146	IL-22Rαc+ IL-10Rβc	NK cells	Induces liver acute-phase proteins, pro-inflammatory agents	
	IL-23	170 (p19) and 306 (p40c), heterodimer	IL-12Rβ1 +IL-23R	Dendritic cells	Induces proliferation of memory T cells, increased IFN-γ production	Defective inflammation
	IL-24 (MDA-7)	157	IL-22Rαc +IL-10Rβc; IL-20Rα+ IL-10Rβc	Monocytes, T cells	Inhibits tumor growth	
	IL-25 (IL-17E)	145	IL-17BR (IL-17Rh1)	T_H2 cells, mast cells	Promotes T_H2 cytokine production	Defective+G35 T_H2 response
	IL-26(AK155)	150	IL-20Rα +IL-10Rβc	T cells (type 1), NK cells		
	IL-27	142 (p28) and 209 (EBI3), heterodimer	WSX-1 +CD130c	Monocytes, macrophages, dendritic cells	Induces IL-12R on T cells via T-bet induction	EBI3: reduced NKT cells. WSX-1: overreaction to *Toxoplasma gondii* infection and death from inflammation

Family	Cytokine (alternative names)	Size (no. of amino acids) and form	Receptors (c denotes common subunit)	Producer cells	Actions	Effect of cytokine or receptor knock-out (where known)
	IL-28A,B (IFN-λ2,3)	175	IL-28Rαc +IL-10Rβc		Antiviral	
	IL-29 (IFN-λ1)	181	IL-28Rαc +IL-10Rβc		Antiviral	
	LIF (leukemia inhibitory factor)	179, monomer	LIFR, CD130	Bone marrow stroma, fibroblasts	Maintains embryonic stem cells, like IL-6, IL-11, OSM	LIFR: die at or soon after birth; decreased hematopoietic stem cells
	OSM (OM, oncostatin M)	196, monomer	OSMR or LIFR, CD130	T cells, macrophages	Stimulates Kaposi's sarcoma cells, inhibits melanoma growth	OSMR: defective liver regeneration
TNF family	TNFα· (cachectin)	157, trimers	p55 (CD120a), p75 (CD120b)	Macrophages, NK cells, T cells	Promotes inflammation, endothelial activation	p55: resistance to septic shock, susceptibility to *Listeria*, STNFαR: periodic febrile attacks
	LT-α (lymphotoxin-α)	171, trimers	p55 (CD120a), p75 (CD120b)	T cells, B cells	Killing, endothelial activation	LT-α: absent lymph nodes, decreased antibody, increased IgM
	LT-β	Transmembrane, trimerizes with LT-α	LTβR or HVEM	T cells, B cells	Lymph node development	Defective development of peripheral lymph nodes, Peyer's patches, and spleen
	CD40 ligand (CD40L)	Trimers	CD40	T cells, mast cells	B-cell activation, class switching	CD40L: poor antibody response, no class switching, diminished T+cell priming (hyper IgM syndrome)
	Fas ligand (FasL)	Trimers	CD95 (Fas)	T cells, stroma(?)	Apoptosis, Ca^{2+}-independent cytotoxicity	Fas, FasL: mutant forms lead to lymphoproliferation, and autoimmunity
	CD27 ligand (CD27L)	Trimers (?)	CD27	T cells	Stimulates T-cell proliferation	
	CD30 ligand (CD30L)	Trimers (?)	CD30	T cells	Stimulates T- and B-cell proliferation	CD30: increased thymic size, alloreactivity
	4-1BBL	Trimers (?)	4-1BB	T cells	Co-stimulates T and B cells	
	Trail (AP0-2L)	281, trimers	DR4, DR5 DCR1, DCR2 and OPG	T cells, monocytes	Apoptosis of activated T cells and tumor cells	tumor-prone phenotype
	OPG-L (RANK-L)	316, trimers	RANK/OPG	Osteoblasts, T cells	Stimulates osteoclasts and bone resorption	OPG-L: osteopetrotic, runted, toothless OPG: osteoporosis
	APRIL	86	TAC1 or BCMA	Activated T cells	B-cell proliferation	Impaired IgA-class switching
	LIGHT	240	HVEM, LT,R	T cells,	Dendritic cell activation	Defective $CD8^+$ T-cell expansion
	TWEAK	102	TWEAKR (Fn14)	macrophages, EBV transformed cells	Angiogenesis	
	BAFF (CD257, BlyS)	153	TAC1 or BCMA or BR3	B cells	B-cell proliferation	BAFF: B-cell dysfunction
Unassigned	TGFβ1	112, homo- and heterotrimers	TGFβR	Chondrocytes, monocytes, T cells	Inhibits cell growth, anti-inflammatory, induces switch to IgA production	TGF-β: lethal inflammation
	MIF	115, monomer	MIF-R	T cells, pituitary cells	Inhibits macrophage migration, stimulates macrophage activation, induces steroid resistance	MIF: resistance to septic shock, hyporesponsive to Gram-negative bacteria

Appendix IV. Chemokines and their receptors.				
Chemokine systematic name	**Common names**	**Chromosome**	**Target cell**	**Specific receptor**
CXCL (†ELR⁺)				
1	GROα	4	Neutrophil, fibroblast, melanoma cell	CXCR2
2	GROβ	4	Neutrophil, fibroblast, melanoma cell	CXCR2
3	GROγ	4	Neutrophil, fibroblast, melanoma cell	CXCR2
5	ENA-78	4	Neutrophil, endothelial cell	CXCR2>>1
6	GCP-2	4	Neutrophil, endothelial cell	CXCR2>1
7	NAP-2 (PBP/CTAP-IIIβ-B44TG)	4	Fibroblast, neutrophil, endothelial cell	CXCR2
8	IL-8	4	Neutrophil, basophil, CD8, T cell subset, endothelial cell	CXCR1, 2
14	BRAK/bolekine	5	T cell, monocyte, B cell	Unknown
15	Lungkine/WECHE	5	Neutrophil, epithelial cell, endothelial cell	Unknown
(†ELR⁻)				
4	PF4	4	Fibroblast, endothelial cell	CXCR3B (alternative splice)
9	Mig	4	Activated T cell ($T_H1 > T_H2$), natural killer cell, B cell, endothelial cell, plasmacytoid dendritic cell	CXCR3A and B
10	IP-10	4	Activated T cell ($T_H1 > T_H2$), natural killer cell, B cell, endothelial cell	CXCR3A and B
11	I-TAC	4	Activated T cell ($T_H1 > T_H2$), natural killer cell, B cell, endothelial cell	CXCR3A and B, CXCR7
12	SDF-1α/β	10	CD34+ bone marrow cell, thymocytes, monocytes/macrophages, naive activated T cell, B cell, plasma cell, neutrophil immature dendritic cells, mature dendritic cells, plasmacytoid dendritic cells	CXCR4, CXCR7
13	BLC/BCA-1	4	Naive B cells, activated CD4 T cells, immature dendritic cells, mature dendritic cells	CXCR5>>CXCR3
16	(None)	17	Activated T cell, natural killer T cell, endothelial cells	CXCR6
CCL				
1	I-309	17	Neutrophil (TCA-3 only), T cell, monocyte	CCR8
2	MCP-1	17	T cell, monocyte, basophil, immature dendritic cells, natural killer cells	CCR2
3	MIP-1α	17	Monocyte/macrophage, T cell ($T_H1 > T_H2$), natural killer cell, basophil, immature dendritic cell, eosinophil, neutrophil, astrocyte, fibroblast, osteoclast	CCR1, 5
4	MIP-1β	17	Monocyte/macrophage, T cell ($T_H1 > T_H2$), natural killer cell, basophil, immature dendritic cell, eosinophil, B cell	CCR5>>1
5	RANTES	17	Monocyte/macrophage, T cell (memory T cell > T cell; $T_H1 > T_H2$), natural killer cell, basophil, eosinophil, immature dendritic cell	CCR1, 3, 5
6	C10/MRP-1	11 (mouse only)	Monocyte, B cell, CD_4^+ T cell, natural killer cell	CCR 1
7	MCP-3	17	T cell, monocyte, eosinophil, basophil, immature dendritic cell, natural killer cell	CCR1, 2, 3, 5, 10
8	MCP-2	17	T cell, monocyte, eosinophil, basophil, immature dendritic cell, natural killer cell	CCR2, 3, 5>1
9	MRP-2/MIP-1γ	11 (mouse only)	T cell, monocyte, adipocyte	CCR1
11	Eotaxin	17	Eosinophil, basophil, T_H2 cell	CCR3>>CCR5
12	MCP-5	11 (mouse only)	Eosinophil, monocyte, T cell, B cell	CCR2
13	MCP-4	17	T cell, monocyte, eosinophil, basophil, dendritic cell	CCR1, 2, 3>5
14a	HCC-1	17	Monocyte	CCR1, 5
14b	HCC-3	17	Monocyte	Unknown

Chemokine systematic name	Common names	Chromosome	Target cell	Specific receptor
15	MIP-5/HCC-2	17	T cell, monocyte, eosinophil, dendritic cell	CCR1, 3
16	HCC-4/LEC	17	Monocyte, T cell, natural killer cell, immature dendritic cell	CCR1, 2, 5
17	TARC	16	T cell ($T_H2 > T_H1$), immature dendritic cells, thymocyte, regulatory T cell	CCR4>>8
18	DC-CK1/PARC	17	Naive T cell > activated T cell, immature dendritic cells, mantle zone B cells	Unknown
19	MIP-3β/ELC	9	Naive T cell, mature dendritic cell, B cell	CCR7
20	MIP-3α/LARC	2	T cell (memory T cell > T cell), peripheral blood mononuclear cell, immature dendritic cell, activated B cells, natural killer T cells	CCR6
21	6Ckine/SLC	9	Naive T cell, B cell, thymocytes, natural killer cell, mature dendritic cells	CCR7
22	MDC	16	Immature dendritic cell, natural killer cell, T cell ($T_H2 > T_H1$), thymocyte, endothelial cells, monocyte, regulatory T cell	CCR4
23	MPIF-1/CK-β\8	17	Monocyte, T cell, resting neutrophil	CCR1, 5
24	Eotaxin-2/MPIF-2	7	Eosinophil, basophil, T cell	CCR3
25	TECK	19	Macrophage, thymocytes, dendritic cell, intraepithelial lymphocyte, IgA+ plasma cell D118	CCR9
26	Eotaxin-3	7	Eosinophil, basophil, fibroblast	CCR3
27	CTACK	9	Skin homing memory T cell, B cell	CCR10
28	MEC	5	T cell, eosinophil, IgA+B cell	CCR10>3
C and CX3C				
XCL 1	Lymphotactin	1 (1)	T cell, natural killer cell	XCR1
XCL 2	SCM-1β	1	T cell, natural killer cell	XCR1
CX3CL 1	Fractalkine	16	Activated T cell, monocyte, neutrophil, natural killer cell, immature dendtritic cells, mast cells, astrocytes, neurons	CX3CR1

Chromosome locations are for humans. Chemokines for which there is no human homolog are listed with the mouse chromosome.

† ELR refers to the three amino acids that precede the first cysteine residue of the CXC motif. If these amino acids are Glu-Leu-Arg (ie ELR+), then the chemokine is chemotactic for neutrophils; if they are not (ELR–) then the chemokine is chemotactic for lymphocytes.

Appendix V. Immunological Constants.

	Evaluation of the cellular components of the human immune system		
	B cells	**T cells**	**Phagocytes**
Normal numbers ($\times 10^9$ per liter of blood)	Approximately 0.3	Total 1.0–2.5 CD4 0.5–1.6 CD8 0.3–0.9	Monocytes 0.15–0.6 Polymorphonuclear leukocytes Neutrophils 3.00–5.5 Eosinophils 0.05–0.25 Basophils 0.02
Measurement of function *in vivo*	Serum Ig levels Specific antibody levels	Skin test	—
Measurement of function *in vitro*	Induced antibody production in response to pokeweed mitogen	T-cell proliferation in response to phytohemagglutinin or to tetanus toxoid	Phagocytosis Nitro blue tetrazolium uptake Intracellular killing of bacteria
Specific defects	See Fig. 11.8	See Fig. 11.8	See Fig. 11.8

	Evaluation of the humoral components of the human immune system				
	Immunoglobulins				**Complement**
Component	IgG	IgM	IgA	IgE	
Normal levels	600–1400 $mg\,dl^{-1}$	40–345 $mg\,dl^{-1}$	60–380 $mg\,dl^{-1}$	0–200 $IU\,ml^{-1}$	CH_{50} of 125–300 $IU\,ml^{-1}$

BIOGRAPHIES

Emil von Behring (1854–1917) discovered antitoxin antibodies with Shibasaburo Kitasato.

Baruj Benacerraf (1920–) discovered immune response genes and collaborated in the first demonstration of MHC restriction.

Jules Bordet (1870–1961) discovered complement as a heat-labile component in normal serum that would enhance the antimicrobial potency of specific antibodies.

Frank MacFarlane Burnet (1899–1985) proposed the first generally accepted clonal selection hypothesis of adaptive immunity.

Jean Dausset (1916–) was an early pioneer in the study of the human major histocompatibility complex or HLA.

Peter Doherty (1940–) and **Rolf Zinkernagel** (1944–) showed that antigen recognition by T cells is MHC-restricted, thereby establishing the biological role of the proteins encoded by the major histocompatibility complex and leading to an understanding of antigen processing and its importance in the recognition of antigen by T cells.

Gerald Edelman (1929–) made crucial discoveries about the structure of immunoglobulins, including the first complete sequence of an antibody molecule.

Paul Ehrlich (1854–1915) was an early champion of humoral theories of immunity, and proposed a famous side-chain theory of antibody formation that bears a striking resemblance to current thinking about surface receptors.

James Gowans (1924–) discovered that adaptive immunity is mediated by lymphocytes, focusing the attention of immunologists on these small cells.

Michael Heidelberger (1888–1991) developed the quantitative precipitin assay, ushering in the era of quantitative immunochemistry.

Charles A. Janeway, Jr. (1945–2003) recognized the importance of co-stimulation for initiating adaptive immune responses. He predicted the existence of receptors of the innate immune system that would recognize pathogen-associated molecular patterns and would signal activation of the adaptive immune system. His laboratory discovered the first mammalian Toll-like receptor that had this function. He was also the principal original author of this textbook.

Edward Jenner (1749–1823) described the successful protection of humans against smallpox infection by vaccination with cowpox or vaccinia virus. This founded the field of immunology.

Niels Jerne (1911–1994) developed the hemolytic plaque assay and several important immunological theories, including an early version of clonal selection, a prediction that lymphocyte receptors would be inherently biased to MHC recognition, and the idiotype network.

Shibasaburo Kitasato (1852–1931) discovered antibodies in collaboration with Emil von Behring.

Robert Koch (1843–1910) defined the criteria needed to characterize an infectious disease, known as Koch's postulates.

Georges Köhler (1946–1995) pioneered monoclonal antibody production from hybrid antibody-forming cells with César Milstein.

Karl Landsteiner (1868–1943) discovered the ABO blood group antigens. He also carried out detailed studies of the specificity of antibody binding using haptens as model antigens.

Peter Medawar (1915–1987) used skin grafts to show that tolerance is an acquired characteristic of lymphoid cells, a key feature of clonal selection theory.

Elie Metchnikoff (1845–1916) was the first champion of cellular immunology, focusing his studies on the central role of phagocytes in host defense.

CÎsar Milstein (1927–2002) pioneered monoclonal antibody production with Georges Köhler.

Louis Pasteur (1822–1895) was a French microbiologist and immunologist who validated the concept of immunization first studied by Jenner. He prepared vaccines against chicken cholera and rabies.

Rodney Porter (1917–1985) worked out the polypeptide structure of the antibody molecule, laying the groundwork for its analysis by protein sequencing.

Ignác Semmelweis (1818–1865) German-Hungarian physician who first determined a connection between hospital hygiene and an infectious disease, puerperal fever, and consquently introduced antisepsis into medical practice.

George Snell (1903–1996) worked out the genetics of the murine major histocompatibility complex and generated the congenic strains needed for its biological analysis, laying the groundwork for our current understanding of the role of the MHC in T-cell biology.

Susumu Tonegawa (1939–) discovered the somatic recombination of immunological receptor genes that underlies the generation of diversity in human and murine antibodies and T-cell receptors.

Don C. Wiley (1944–2001) solved the first crystal structure of an MHC I protein, providing a startling insight into how T cells recognize their antigen in the in the context of MHC molecules.

GLOSSARY

The **12/23 rule** states that gene segments of immunoglobulin or T-cell receptors can be joined only if one has a recognition signal sequence with a 12 base pair spacer, and the other has a 23 base pair spacer.

In the context of immunoglobulins, **α** is the type of heavy chain in IgA. Chains called α are also found in a many other proteins, for example MHC molecules and T-cell receptors.

α:β T cell: see **T cell**.

α:β T-cell receptor: see **T-cell receptor**.

The **ABO blood group system** antigens are expressed on red blood cells. They are used for typing human blood for transfusion. Matching is necessary because individuals who do not express A or B antigens on their red blood cells naturally form anti-A and anti-B antibodies that interact with and destroy red blood cells bearing A or B antigens if they are transfused into the bloodstream.

The removal of antibodies specific for one antigen from an antiserum to render it specific for another antigen or antigens is called **absorption**.

Accelerated rejection refers to the fact that when a recipient that has previously rejected a graft is regrafted with skin from the same donor, the second graft is rejected more rapidly. It was one of the pieces of evidence that showed that graft rejection was due to an adaptive immune response.

Accessory effector cells in adaptive immunity are cells that aid in the response but do not directly mediate specific antigen recognition. They include phagocytes, mast cells, and NK cells.

Acquired immune deficiency syndrome (**AIDS**) is a disease caused by infection with the human immunodeficiency virus (HIV-1). AIDS occurs when an infected patient has lost most of his or her CD4 T cells, so that infections with opportunistic pathogens occur.

Acquired immune response: see **adaptive immune response**.

Activation-induced cell death is the normal process by which all immune responses end in the death of most of the responding cells, leaving only a small number of resting memory cells.

The enzyme **activation-induced cytidine deaminase** (**AID**) contributes to somatic hypermutation of immunoglobulin gene variable regions by deaminating DNA directly at cytosine. Depending on how this DNA lesion is repaired, it can lead to a permanent base change at the deaminated site. The enzyme is also involved in isotype switching and gene conversion. An inherited deficiency of the enzyme—**AID deficiency**—blocks both somatic hypermutation and isotype switching, leading to a type of hyper IgM immunodeficiency syndrome.

Immunization with antigen to provoke adaptive immunity is called **active immunization** to distinguish it from the transfer of antibody to an unimmunized individual, which is called passive immunization.

Acute lymphoblastic leukemia is a highly aggressive, undifferentiated form of lymphoid malignancy derived from a progenitor cell that is thought to be able to give rise to both T and B lineages of lymphoid cells. Most of these leukemias show partial differentiation toward the B-cell lineage (so called B-ALL) whereas a minority show features of T cells (T-ALL).

The **acute phase** of HIV infection occurs soon after a person becomes infected and is characterized by an influenza-like illness, abundant virus in the blood, and a decrease in the number of circulating CD4 T cells.

Acute-phase proteins are found in the blood shortly after the onset of an infection. These proteins participate in early phases of host defense against infection. An example is the mannose-binding lectin.

The **acute-phase response** is a change in the blood that occurs during early phases of an infection. It includes the production of acute-phase proteins.

Acute rejection of a tissue or organ graft from a genetically unrelated donor occurs within 10–13 days of transplantation.

The **adaptive immune response,** or **adaptive immunity,** is the response of antigen-specific lymphocytes to antigen, including the development of immunological memory. Adaptive immune responses are distinct from the innate and nonadaptive phases of immunity, which are not mediated by clonal selection of antigen-specific lymphocytes. Adaptive immune responses are also known as **acquired immune responses**.

Adaptive regulatory T cells are regulatory CD4 T cells that are thought to differentiate from naive CD4 T cells in the periphery under the influence of particular environmental conditions. *cf.* **natural regulatory T cells**.

Adaptor proteins are non-enzymatic proteins that form physical links between members of a signaling pathway, particularly between a receptor and other signaling proteins. They serve to recruit members of the signaling pathway into functional protein complexes.

ADCC: see **antibody-dependent cell-mediated cytotoxicity**.

The **adenoids** are mucosal-associated lymphoid tissues located in the nasal cavity.

The enzyme defect **adenosine deaminase deficiency** (**ADA deficiency**) leads to the accumulation of toxic purine nucleosides and nucleotides, resulting in the death of most developing lymphocytes within the thymus. It is a cause of severe combined immunodeficiency.

Adhesion molecules: see **cell-adhesion molecules**.

An **adjuvant** is any substance that enhances the immune response to an antigen with which it is mixed.

Adoptive immunity is immunity conferred on a naive or irradiated recipient by transfer of lymphoid cells from an actively immunized donor. This is called **adoptive transfer** or **adoptive immunization**.

Afferent lymphatic vessels drain fluid from the tissues and carry macrophages and dendritic cells from sites of infection to the lymph nodes.

Affinity is the strength of binding of one molecule to another at a single site, such as the binding of a monovalent Fab fragment of antibody to a monovalent antigen. See also **avidity**.

Affinity chromatography is the purification of a substance by means of its affinity for another substance immobilized on a solid support. For example, an antigen can be purified by affinity chromatography on a column of beads to which specific antibody molecules are covalently linked.

Affinity maturation refers to the increase in the affinity for the specific antigen of the antibodies produced during the course of a humoral immune response. It is particularly prominent in secondary and subsequent immunizations.

Agammaglobulinemia: see **X-linked agammaglobulinemia** (**XLA**).

Agglutination is the clumping together of particles, usually by antibody molecules binding to antigens on the surfaces of adjacent particles. Such particles are said to **agglutinate**.

Agonist selection is a process by which T cells are positively selected in the thymus by relatively high-affinity ligands.

AID: see **activation-induced cytidine deaminase**.

AID deficiency: see **activation-induce cytidine deaminase deficiency**

AIDS: see **acquired immune deficiency syndrome**.

Airway remodeling is a thickening of the airway walls due to hyperplasia and hypertrophy of the smooth-muscle layer and mucous glands, with the eventual development of fibrosis, that occurs in chronic asthma.

Alleles are variants of a single genetic locus.

Allelic exclusion refers to the fact that in a heterozygous individual, only one of the alternative C-region alleles of the heavy or light chain is expressed in a single B cell and in an immunoglobulin molecule. The term has come to be used more generally to describe the expression of a single receptor specificity in cells with the potential to express two or more receptors.

Allergens are antigens that elicit hypersensitivity or allergic reactions.

Allergic asthma is an allergic reaction to inhaled antigen, which causes constriction of the bronchi and difficulty in breathing.

The lining of the eye, called the conjunctiva, manifests **allergic conjunctivitis** in sensitized individuals exposed to allergens.

An **allergic reaction** is a response to innocuous environmental antigens, or allergens, due to preexisting antibody or primed T cells. There are various mechanisms underlying allergic reactions, but the most common is the binding of allergen to IgE bound to mast cells, which causes the release of histamine and other biologically active molecules from the cell that cause the symptoms of asthma, hay fever, and other common allergic reactions.

Allergic rhinitis is an allergic reaction in the nasal mucosa, also known as hay fever, that causes runny nose, sneezing, and tears.

Allergy is a symptomatic reaction to a normally innocuous environmental antigen. It results from the interaction between the antigen and antibody or primed T cells produced by earlier exposure to the same antigen.

Antibodies produced against antigens from another member of the same species (**alloantigens**) are known as **alloantibodies**.

Two individuals or two mouse strains that differ at the MHC are said to be **allogeneic**. The term can also be used for allelic differences at other loci. Rejection of grafted tissues from unrelated donors usually results from T-cell responses to allogeneic MHC molecules (alloantigens) expressed by the grafted tissues. See also **syngeneic**, **xenogeneic**.

An **allograft** is a graft of tissue from an allogeneic or nonself donor of the same species; such grafts are invariably rejected unless the recipient is immunosuppressed.

Alloreactivity describes the stimulation of T cells by MHC molecules other than self; it marks the recognition of allogeneic MHC molecules. Such responses are also called **alloreactions** or **alloreactive** responses.

Allotypes are allelic polymorphisms that can be detected by antibodies specific for the polymorphic gene products. In immunology, **allotypic** differences in the constant regions of immunoglobulin molecules were important in deciphering the genetics of antibodies.

ALPS: see **autoimmune lymphoproliferative syndrome**.

The **alternative pathway** of complement activation is triggered by the presence of a pathogen in the absence of specific antibodies and is thus part of the innate immune system. It leads to the production of complement protein C3b and its binding to the surface of the pathogen, after which the pathway is the same as the classical and lectin pathways of complement activation.

Anaphylactic shock or systemic anaphylaxis is an allergic reaction to systemically administered antigen that causes circulatory collapse and suffocation due to tracheal swelling. It results from binding of antigen to IgE antibody on connective tissue mast cells throughout the body, leading to the disseminated release of inflammatory mediators.

Anaphylatoxins are small fragments of complement proteins, released by cleavage during complement activation. These small fragments are recognized by specific receptors, and they recruit fluid and inflammatory cells to sites of their release. The fragments C5a, C3a, and C4a are all anaphylatoxins, listed in order of decreasing potency *in vivo*.

Peptide fragments of antigens are bound to specific MHC class I molecules by **anchor residues**. These are residues of the peptide that have amino acid side chains that bind into pockets lining the peptide-binding groove of the MHC class I molecule. Each MHC class I molecule binds different patterns of anchor residues, called anchor motifs, giving some specificity to peptide binding. Anchor residues exist but are less obvious for peptides that bind to MHC class II molecules.

Anergy is a state of nonresponsiveness to antigen. People are said to be **anergic** when they cannot mount delayed-type hypersensitivity reactions to challenge antigens, whereas T cells and B cells are said to be **anergic** when they cannot respond to their specific antigen under optimal conditions of stimulation.

An **antibody** is a protein that binds specifically to a particular substance—its antigen. Each antibody molecule has a unique structure that enables it to bind specifically to its corresponding antigen, but all antibodies have the same overall structure and are known collectively as immunoglobulins or Igs. Antibodies are produced by plasma cells in response to infection or immunization, and bind to and neutralize pathogens or prepare them for uptake and destruction by phagocytes.

Antibody combining site: see **antigen-binding site**.

Antibody-dependent cell-mediated cytotoxicity (**ADCC**) is the killing of antibody-coated target cells by cells with Fc receptors that recognize the constant region of the bound antibody. Most ADCC is mediated by NK cells that have the Fc receptor FcγRIII or CD16 on their surface.

The **antibody repertoire** or immunoglobulin repertoire describes the total variety of antibodies in the body of an individual.

An **antigen** is any molecule that can bind specifically to an antibody. Their name arises from their ability to **gen**erate **anti**bodies. However, some antigens do not, by themselves, elicit antibody production; those antigens that can induce antibody production are called immunogens.

Antigen:antibody complexes are noncovalently associated groups of antigen and antibody molecules that can vary in size from small soluble complexes to large insoluble complexes that precipitate out of solution; they are also known as immune complexes.

The **antigen-binding site** of an antibody, or **antibody combining site**, is found at the surface of the antibody molecule that makes physical contact with the antigen. Antigen-binding sites are made up of six hypervariable loops, three from the light-chain varaible region and three from the heavy-chain variable region.

In an **antigen-capture assay**, the antigen binds to a specific antibody, and its presence is detected by a second antibody that must be labeled and directed at a different epitope.

Antigen display libraries are libraries of cDNA clones in expression vectors or bacteriophage libraries encoding random peptide sequences that can be expressed as part of the phage coat. They are used to identify the targets of specific antibodies and, in some cases, of T cells.

An **antigenic determinant** is that portion of an antigenic molecule that is bound by the antigen-binding site of a given antibody or antigen receptor; it is also known as an epitope.

Influenza virus varies genetically from year to year by a process of **antigenic drift**, in which point mutations in viral genes cause small differences in the structure of the viral surface antigens. Periodically, influenza viruses undergo an **antigenic shift** through reassortment of their segmented genome with another influenza virus, changing their surface antigens radically. Such antigenic shift variants are not recognized by individuals immune to influenza, so when an antigenic shift occurs, there is widespread and severe disease.

Some pathogens evade the immune system by altering their surface antigens, a phenomenon known as **antigenic variation**.

Antigen presentation describes the display of antigen as peptide fragments bound to MHC molecules on the surface of a cell. T cells recognize antigen when it is presented in this way.

The term **antigen-presenting cells** (**APCs**) usually refers to highly specialized cells that can process antigens and display their peptide fragments on the cell surface together with other, co-stimulatory, proteins required for activating naive T cells. The main antigen-presenting cells for naive T cells are dendritic cells, macrophages, and B cells.

Antigen processing is the degradation of proteins into peptides that can bind to MHC molecules for presentation to T cells. All protein antigens must be processed into peptides before they can be presented by MHC molecules.

T and B lymphocytes collectively bear on their surface highly diverse **antigen receptors** capable of recognizing a wide diversity of antigens. Each individual lymphocyte bears receptors of a single antigen specificity.

Anti-idiotype antibodies are antibodies raised against antigenic determinants unique to the variable region of a single antibody.

Anti-isotypic antibodies are antibodies against universal features of a given constant-region isotype (such as γ or μ) of one species that are made by immunizing a member of another species with that isotype. Such antibodies will bind any antibody of that isotype, and are thus useful for detecting bound antibody molecules in immunoassays and other applications.

Anti-immunoglobulin antibodies: see **anti-isotypic antibodies**.

Anti-lymphocyte globulin is antibody raised in another species against human T cells.

An **antiserum** (plural: **antisera**) is the fluid component of clotted blood from an immune individual that contains antibodies against the molecule used for immunization. Antisera contain heterogeneous collections of antibodies, which bind the antigen used for immunization, but each has its own structure, its own epitope on the antigen, and its own set of cross-reactions. This heterogeneity makes each antiserum unique.

Bites by poisonous snakes can be treated by identification of the snake and injection of an **antivenin** specific for that snake's venom.

AP-1 is a transcription factor that is produced as a result of intracellular signaling from the antigen receptors of lymphocytes.

APECED: see **autoimmune polyendocrinopathy-candidiasis-ectodermal dystrophy**.

Aplastic anemia is a failure of bone marrow stem cells so that formation of all cellular elements of the blood ceases; it can be treated by bone marrow transplantation.

Apoptosis, or programmed cell death, is a form of cell death in which the cell activates an internal death program. It is characterized by nuclear DNA degradation, nuclear degeneration and condensation, and the phagocytosis of cell remains. Proliferating cells frequently undergo apoptosis, which is a natural process in development, and proliferating lymphocytes undergo high rates of apoptosis in development and during immune responses. Apoptosis contrasts with necrosis, death caused by external factors, which occurs in situations such as poisoning and oxygen starvation.

The **appendix** is a gut-associated lymphoid tissue located at the beginning of the colon.

Artemis is an endonuclease involved in the gene rearrangements that generate functional immunoglobulin and T-cell receptor genes.

The **Arthus reaction** is a skin reaction in which antigen is injected into the dermis and reacts with IgG antibodies in the extracellular spaces, activating complement and phagocytic cells to produce a local inflammatory response.

Ascertainment artifact refers to data that seem to demonstrate some finding, but fail to do so because they are collected from a population that is selected in a biased fashion.

The **asymptomatic period** of HIV infection is the phase, which may last for many years, when the infection is being partly held in check and no symptoms occur.

Ataxia telangiectasia (**AT**) is a disease characterized by staggering, multiple disorganized blood vessels, and an immunodeficiency in a protein called ATM, which contains a kinase thought to be important in signaling of double-stranded DNA breaks.

Atopic dermatitis: see **eczema**.

Atopy is the increased tendency seen in some people to produce immediate hypersensitivity reactions (usually mediated by IgE antibodies) against innocuous substances.

Pathogens are said to be **attenuated** when they can grow in their host and induce immunity without producing serious clinical disease.

Antibodies specific for self antigens are called **autoantibodies**.

Self antigens to which the immune system makes a response are called **autoantigens**.

A graft of tissue from one site to another on the same individual is called an **autograft**.

Diseases in which the pathology is caused by adaptive immune responses to self antigens are called **autoimmune diseases**.

Autoimmune hemolytic anemia is a pathological condition with low levels of red blood cells (anemia), which is caused by autoantibodies that bind red blood cell surface antigens and target the red blood cell for destruction.

Autoimmune lymphoproliferative syndrome (**ALPS**) is an inherited syndrome in which a defect in the Fas gene leads to a failure in normal apoptosis, causing unregulated immune responses, including autoimmune responses.

In the disease **autoimmune polyendocrinopathy-candidiasis-ectodermal dystrophy** (**APECED**), there is a loss of tolerance to self antigens due to a breakdown of negative selection in the thymus. This is due to defects in the gene *AIRE*, which encodes a transcriptional regulatory protein that enables many self antigens to be expressed by thymic epithelial cells. This disease is also called autoimmune polyglandular syndrome type I.

An adaptive immune response directed at self antigens is called an **autoimmune response**; likewise, adaptive immunity specific for self antigens is called **autoimmunity**.

In the disease **autoimmune thrombocytopenic purpura**, antibodies against a patient's platelets are made. Antibody binding to platelets causes them to be taken up by cells with Fc receptors and complement receptors, causing a fall in platelet counts that leads to purpura (bleeding).

Autoinflammatory disease is characterized by unregulated inflammation in the absence of infection, and has a variety of causes.

Autophagy is the digestion and breakdown by a cell of its own organelles and proteins in lysosomes. It may be one route by which cytosolic proteins can be processed for presentation on MHC class II molecules.

Autoreactivity describes immune responses directed at self antigens.

Avidity is the sum total of the strength of binding of two molecules or cells to one another at multiple sites. It is distinct from affinity, which is the strength of binding of one site on a molecule to its ligand.

Azathioprine is a potent immunosuppressive drug that is converted to its active form *in vivo* and then kills rapidly proliferating cells, including lymphocytes responding to grafted tissues.

4-1BB is a member of the TNF receptor family that specifically binds to the TNF family protein **4-1BB ligand**.

B-1 cell, B-2 cell: see **B cell**.

The major T-cell co-stimulatory molecules are the **B7 molecules**, **B7.1** (CD80) and **B7.2** (CD86). They are closely related members of the immunoglobulin gene superfamily and both bind to the CD28 molecule on T cells. They are expressed differentially on various antigen-presenting cell types. We use the term **B7 molecules** to refer to both B7.1 and B7.2.

B7-RP is a ligand for B7 molecules.

β barrel: see **β sheet**.

β-defensins are antimicrobial peptides made by virtually all multicellular organisms. In mammals they are produced by the epithelia of the respiratory and urogenital tracts, skin, and tongue.

A **β sheet** is one of the fundamental structural building blocks of proteins, consisting of adjacent, extended strands of amino acids (**β strands**) that are bonded together by interactions between backbone amide and carbonyl groups. β sheets can be parallel, in which case the adjacent strands run in the same direction, or antiparallel, where adjacent strands run in opposite directions. All immunoglobulin domains are made up of antiparallel β-sheet structures. A **β barrel** or a **β sandwich** is another way of describing the structure of the immunoglobulin domain.

The light chain of the MHC class I proteins is called **β_2-microglobulin.** It binds noncovalently to the heavy or α chain.

The **B and T lymphocyte attenuator (BTLA)** is an inhibitory CD28-related receptor expressed by B and T lymphocytes.

Many infectious diseases are caused by **bacteria**, which are prokaryotic microorganisms that exist as many different species and strains. Bacteria can live on body surfaces, in extracellular spaces, in cellular vesicles, or in the cytosol, and different bacterial species cause distinctive infectious diseases.

Bacterial lipolysaccharide: see **LPS**.

BALT: see **bronchial-associated lymphoid tissues**.

Bare lymphocyte syndrome: see **MHC class I deficiency**; **MHC class II deficiency**.

Basophils are white blood cells containing granules that stain with basic dyes, and which are thought to have a function similar to mast cells.

Bb is the large active fragment of complement component factor B. It is produced when factor B is captured by bound C3b and cleaved by factor D. Bb remains associated with C3b and is the serine protease component of the alternative pathway C3 convertase.

A **B cell**, or **B lymphocyte**, is one of the two major types of lymphocyte. The antigen receptor on B lymphocytes, usually called the B-cell receptor, is a cell-surface immunoglobulin. On activation by antigen, B cells differentiate into cells producing antibody molecules of the same antigen specificity as this receptor. B cells are divided into two classes. **B-1 cells**, also known as CD5 B cells, are a class of atypical, self-renewing B cells found mainly in the peritoneal and pleural cavities in adults. They have a far less diverse repertoire of receptors than do **B-2 cells** (also known as conventional B cells). The latter are generated in the bone marrow throughout life, emerging to populate the blood and lymphoid tissues.

The **B-cell antigen receptor**, or **B-cell receptor** (**BCR**), is the cell-surface receptor on B cells for specific antigen. It is composed of a transmembrane immunoglobulin molecule associated with the invariant Igα and Igβ chains in a noncovalent complex.

The proteins CD19, CD81, and CR2 make up the **B-cell co-receptor complex**; co-ligation of this complex with the B-cell antigen receptor increases responsiveness to antigen by about a 100-fold.

The **B-cell corona** in the spleen is the zone of the white pulp primarily made up of B cells.

B-cell mitogens are substances that cause B cells to proliferate.

The protein known as **Bcl-2** protects cells from apoptosis by binding to the mitochondrial membrane. It is encoded by the *bcl-2* gene, which was discovered at the breakpoint of an oncogenic chromosomal translocation in B-cell leukemia.

Biological therapy is the name given to treatments comprising natural proteins such as antibodies and cytokines, and antisera or whole cells.

Blau syndrome is an inherited granulomatous disease caused by gain-of-function mutations in the *NOD2* gene.

BLIMP-1 (B-lymphocyte-induced maturation protein 1) is a transcriptional repressor that acts in plasmablasts to direct their differentiation into plasma cells.

Blk: see **tyrosine kinase**.

BLNK (B-cell linker protein) is a scaffold protein in B cells that recruits proteins involved in the intracellular signaling pathway from the antigen receptor.

Blood group antigens are surface molecules on red blood cells that are detectable with antibodies from other individuals. The major blood group antigens are called ABO and Rh (Rhesus), and are used routinely to type blood. There are many other blood group antigens.

Blood typing is used to determine whether donor and recipient have compatible ABO and Rh blood group antigens before blood is transfused. A cross-match, in which serum from the donor is tested on the cells of the recipient, and vice versa, is used to rule out other incompatibilities. Transfusion of incompatible blood causes a transfusion reaction, in which red blood cells are destroyed and the released hemoglobin causes toxicity.

Bloom's syndrome is a disease characterized by low T-cell numbers, reduced antibody levels, and an increased susceptibility to respiratory infections, cancer, and radiation damage. It is caused by mutations in a DNA helicase.

B lymphocyte: see **B cell**.

B-lymphocyte chemokine (**BLC**): see **CXCL13**.

The **bone marrow** is the site where all the cellular elements of blood, including red blood cells, monocytes, polymorphonuclear leukocytes, and platelets, are generated. The bone marrow is also the site of B-cell development in mammals and the source of stem cells that give rise to T cells upon migration to the thymus. Thus, bone-marrow transplantation can restore all the cellular elements of the blood, including the cells required for adaptive immunity.

A **bone marrow chimera** is formed by transferring bone marrow from one mouse to an irradiated recipient mouse, so that all of the lymphocytes and blood cells are of donor genetic origin. Bone marrow chimeras have been crucial in elucidating the development of lymphocytes and other blood cells.

A **booster immunization** is commonly given after a primary immunization, to increase the amount, or titer, of antibodies.

Bradykinin is a vasoactive peptide that is produced as a result of tissue damage and acts as an inflammatory mediator.

The lymphoid cells and organized lymphoid tissues in the respiratory tract are termed the **bronchus-associated lymphoid tissues (BALT)**. These tissues are very important in the induction of immune responses to inhaled antigens and to respiratory infections.

Bruton's X-linked agammaglobulinemia: see **X-linked agammaglobulinemia**.

Burkitt's lymphoma is a tumor of germinal center B cells and is caused by Epstein–Barr virus (EBV); it occurs mainly in sub-Saharan Africa.

The **bursa of Fabricius** in chickens is the site of B-cell development.

The **C1 complex** of complement comprises one molecule of **C1q** bound to two molecules each of the zymogens **C1r** and **C1s**. C1q initiates the classical pathway of complement activation by binding to a pathogen surface or to bound antibody. This binding activates the associated C1r, which in turn cleaves and activates C1s. The active form of C1s then cleaves the next two components in the pathway, **C4** and **C2**.

C1 inhibitor (C1INH) is a protein that inhibits the activity of activated complement component C1 by binding to and inactivating its C1r:C1s enzymatic activity. It also inhibits other serine proteases including kallikrein. Deficiency in C1INH is the cause of the disease hereditary angioneurotic edema, in which the production of vasoactive peptides, kinins, leads to subcutaneous and laryngeal swelling.

The complement fragment **C3b** is the major product of the enzyme C3 convertase, and the principal effector molecule of the complement system. It has a highly reactive thioester bond which allows it to bind covalently to the surface on which it is generated. Once bound, it acts as an opsonin to promote the destruction of pathogens by phagocytes and removal of immune complexes; C3b is bound by the complement receptor CR1, while its proteolytic derivative, iC3b, is bound by the complement receptors CR1, CR2, and CR3.

The generation of the enzyme **C3 convertase** on the surface of a pathogen or cell is a crucial step in complement activation. The classical and lectin pathway C3 convertase is formed from membrane-bound C4b complexed with the protease C2b. The alternative pathway of complement activation uses a homologous C3 convertase formed from membrane-bound C3b complexed with the protease Bb. These C3 convertases have the same activity, catalyzing the deposition of large numbers of C3b molecules that bind covalently to the pathogen surface.

C3dg is a breakdown product of C3b that remains attached to the microbial surface, where it can bind to CD21, the complement receptor CR2.

C4b-binding protein (C4BP) can inactivate the classical pathway C3 convertase if it forms on host cells, by displacing C2b from the C4bC2b complex. It binds to C4b attached to host cells, but cannot bind C4b attached to pathogens. This is because it has a second binding site specific for sialic acid, a terminal sugar on vertebrate cell surfaces, but not on pathogens.

C5 is an inactive complement component that is cleaved by the **C5 convertase** to release the potent inflammatory peptide **C5a** and a larger fragment, **C5b**, that initiates the formation of a membrane-attack complex from the terminal components of complement.

The receptor for the C5a fragment of complement, the **C5a receptor**, is a seven-transmembrane spanning receptor that couples to a heterotrimeric G protein. Similar receptors bind to **C3a** and **C4a**.

The complement components **C6**, **C7**, and **C8** form a complex with the active complement fragment **C5b** in the late events of complement activation. This complex inserts into the membrane and induces polymerization of **C9** to form a pore known as the membrane-attack complex.

The cytosolic serine/threonine phosphatase **calcineurin** has a crucial role in signaling via the T-cell receptor. The immunosuppressive drugs cyclosporin A and tacrolimus (also known as FK506) form complexes with cellular proteins called immunophilins that bind and inactivate calcineurin, suppressing T-cell responses.

Ca^{2+} acts as an intracellular signal mainly by binding to the protein **calmodulin**, which is then able to bind to and regulate the activity of a wide variety of enzymes, including calcineurin.

The protein **calnexin** binds to partly folded members of the immunoglobulin superfamily of proteins and retains them in the endoplasmic reticulum until folding is completed.

Calreticulin is the molecular chaperone that binds initially to MHC class I, MHC class II, and other proteins that contain immunoglobulin-like domains, such as the T-cell and B-cell antigen receptors.

Antibodies or antigens can be measured in various capture assays, such as a **capture ELISA**. Antigens are captured by antibodies bound to plastic (or vice versa). Antibody binding to a plate-bound antigen can be measured using labeled antigen or anti-immunoglobulin. Antigen binding to plate-bound antibody can be measured by using an antibody that binds to a different epitope on the antigen.

Carriers are foreign proteins to which small non-immunogenic antigens, or haptens, can be coupled to render the hapten immunogenic. *In vivo*, self proteins can also serve as carriers if they are correctly modified by the hapten; this is important in allergy to drugs.

Caseation necrosis is a form of necrosis seen in the center of large granulomas, such as the granulomas in tuberculosis. The term comes from the white cheesy appearance of the central necrotic area.

Caspases are a family of closely related cysteine proteases that cleave proteins at aspartic acid residues. They have important roles in apoptosis.

Cbl is a ubiquitin ligase that recognizes tyrosine-phosphorylated proteins and targets them for destruction.

The chemokines **CCL18** (DC-CK), **CCL19** and **CCL21** (SLC) are produced by cells in peripheral lymphoid organs and attract T cells.

CCR7: the receptor on T cells for the chemokine CCL21.

CD: see **clusters of differentiation** and Appendix II.

CD2 is the immunoglobulin superfamily cell-adhesion molecule also known as LFA-2.

The **CD3 complex** is the complex of α:β or γ:δ T-cell receptor chains with the invariant subunits CD3γ, δ, and ε, and the dimeric ζ chains.

The cell-surface protein **CD4** is important for recognition by the T-cell receptor of antigenic peptides bound to MHC class II molecules. It acts as a co-receptor by binding to the lateral face of the MHC class II molecules.

CD4 T cells are T cells that carry the co-receptor protein CD4. They recognize peptides derived from intravesicular sources, which are bound to MHC class II molecules, and differentiate into CD4 T_H1 and CD4 T_H2 effector cells that activate macrophages and B-cell responses to antigen.

CD5⁺ B cells are a class of atypical, self-renewing B cells found mainly in the peritoneal and pleural cavities in adults. They have a far less diverse receptor repertoire than conventional B cells, and since they are the first B cells to be produced they are also known as B-1 cells.

The cell-surface protein **CD8** is important for recognition by the T-cell receptor of antigenic peptides bound to MHC class I molecules. It acts as a co-receptor by binding to the lateral face of MHC class I molecules.

CD8 T cells are T cells that carry the co-receptor CD8. They recognize antigens, for example viral antigens, that are synthesized in the cytoplasm of a cell. Peptides derived from these antigens are transported by TAP, assembled with MHC class I molecules in the endoplasmic reticulum, and displayed as peptide:MHC class I complexes on the cell surface. CD8 T cells differentiate into cytotoxic CD8 T cells.

CD11b:CD18: see **CR3**.

CD11c:CD18: see **CR4**.

CD19:CR2:TAPA-1 complex: see **B-cell co-receptor complex**.

CD21: see **CR2**.

CD27 is a TNF receptor family protein constitutively expressed on naive T cells that binds CD70 on dendritic cells and delivers a potent co-stimulatory signal to T cells early in the activation process.

CD28 on T cells is the receptor for the B7 co-stimulatory molecules on specialized antigen-presenting cells such as dendritic cells. It is a member of the immunoglobulin superfamily.

CD30 on B cells and **CD30 ligand** on helper T cells are co-stimulatory molecules involved in stimulating the proliferation of antigen-activated naive B cells.

CD31: see **PECAM**.

CD34 is a cell-surface protein present on hematopoietic stem cells. It is a ligand for L-selectin.

CD35: see **CR1**.

B-cell growth is triggered in part by the binding of **CD40 ligand** (**CD154**), expressed on activated helper T cells, to **CD40** on the B-cell surface.

CD45, or the leukocyte common antigen, is a transmembrane tyrosine phosphatase found on all leukocytes. It is expressed in different isoforms on different cell types, including the different subtypes of T cells. These isoforms are commonly denoted by the designation of CD45R followed by the exon whose presence gives rise to distinctive antibody-binding patterns.

CD46: see **membrane cofactor of proteolysis** (**MCP**).

CD55: see **decay-accelerating factor** (**DAF**).

CD58 is the immunoglobulin family cell adhesion molecule also known as LFA-3. See **leukocyte function antigens**.

CD59: see **protectin**.

CD70 is the ligand for CD27.

CD80 and **CD86**: see **B7.1** and **B7.2**.

C domain: see **constant domain**.

CDRs: see **complementarity-determining regions**.

Cell-adhesion molecules mediate the binding of one cell to other cells or to extracellular matrix proteins. Integrins, selectins, members of the immunoglobulin gene superfamily (e.g. ICAM-1), and CD44 and related proteins are all cell-adhesion molecules important in the operation of the immune system.

Celiac disease is a chronic condition of the upper small intestine caused by an immune response directed at gluten, a complex of proteins present in wheat, oats, and barley. The gut wall becomes chronically inflamed, the villi are destroyed and the gut's ability to absorb nutrients is compromised.

Cell-mediated immunity or **cell-mediated immune response** describes any adaptive immune response in which antigen-specific T cells have the main role. It is defined operationally as all adaptive immunity that cannot be transferred to a naive recipient by serum antibody. A primary cell-mediated immune response is the T-cell response that occurs the first time a particular antigen is encountered. *Cf.* **humoral immunity**.

Cell-surface immunoglobulin is the B-cell receptor for antigen. See also **B-cell antigen receptor**.

Cellular immunology is the study of the cellular basis of immunity.

The **central lymphoid organs** are sites of lymphocyte development and in humans are the bone marrow and thymus. B lymphocytes develop in bone marrow, whereas T lymphocytes develop within the thymus from bone marrow-derived progenitors. They are also sometimes known as the primary lymphoid organs.

Central memory cells are a class of memory cells with characteristic activation properties that are thought to reside in the T-cell areas of peripheral lymphoid tissues.

Central tolerance is tolerance to self antigens that is established in lymphocytes developing in central lymphoid organs. *Cf.* **peripheral tolerance**.

Centroblasts are large, rapidly dividing cells found in germinal centers, and are the cells in which somatic hypermutation is believed to occur. Antibody-secreting and memory B cells derive from these cells.

Centrocytes are the small B cells in germinal centers that derive from centroblasts. They may mature into antibody-secreting plasma cells or memory B cells, or may undergo apoptosis, depending on their receptor's interaction with antigen.

Chediak–Higashi syndrome is caused by a defect in a protein involved in intracellular vesicle fusion. Phagocytic cell function is affected as lysosomes fail to fuse properly with phagosomes and there is impaired killing of ingested bacteria.

Chemokines are small chemoattractant proteins that stimulate the migration and activation of cells, especially phagocytic cells and lymphocytes. They have a central role in inflammatory responses. Chemokines and their receptors are listed in Appendix IV.

Most lymphoid tumors, and many other tumors, bear **chromosomal translocations** that mark points of breakage and rejoining of different chromosomes. These chromosomal breaks are particularly frequent in lymphomas and leukemias.

Chronic allergic inflammation of the airways is seen in chronic asthma, and is a consequence of the cell-mediated late-phase allergic response.

Chronic granulomatous disease is an immunodeficiency disease in which multiple granulomas form as a result of defective elimination of bacteria by phagocytic cells. It is caused by defects in the NADPH oxidase system of enzymes that generate the superoxide radical involved in bacterial killing.

Chronic lymphocytic leukemias (**CLLs**) are B-cell tumors that are found in the blood. The great majority express CD5 and unmutated variable regions and are therefore thought to arise from B-1 cells.

The **class** of an antibody is defined by the type of heavy chain it contains. There are five main classes of antibodies: IgA, IgD, IgM, IgG, and IgE, containing heavy chains α, δ, μ, γ, and ε, respectively. The IgG class has several subclasses. See also **isotype**.

The **class II-associated invariant chain peptide** (**CLIP**) is a peptide of variable length cleaved from the invariant chain (Ii) by proteases. It remains associated with the MHC class II molecule in an unstable form until it is removed by the HLA-DM protein.

Class II transactivator (**CIITA**): see **MHC class II transactivator**.

The **classical pathway** of complement activation is the pathway activated by C1 binding either directly to bacterial surfaces or to antibody, that serves as a means of flagging the bacteria as foreign. See also **alternative pathway**; **lectin pathway**.

Activated B cells first produce IgM but subsequently undergo **class switching** to secrete antibodies of different classes: IgG, IgA, and IgE. This does not affect antibody specificity, but alters the effector functions that an antibody can engage by replacing one heavy-chain constant region with another. Class switching takes place only in rearranged immunoglobulin heavy-chain genes in antigen-activated B cells.

Class switching involves the process of **class switch recombination**, in which recombination occurs between a rearranged variable region and a selected constant-region sequence at so-called switch (S) regions to produce a functional immunoglobulin gene with a different constant region.

CLIP: see **class II-associated invariant-chain peptide**.

Clonal deletion is the elimination of immature lymphocytes when they bind to self antigens, which produces tolerance to self as required by the clonal selection theory. Clonal deletion is the main mechanism of central tolerance and can also occur in peripheral tolerance.

Clonal expansion is the proliferation of antigen-specific lymphocytes in response to antigenic stimulation and precedes their differentiation into effector cells. It is an essential step in adaptive immunity, allowing rare antigen-specific cells to increase in number so that they can effectively combat the pathogen that elicited the response.

The **clonal selection theory** is a central paradigm of adaptive immunity. It states that adaptive immune responses derive from individual antigen-specific lymphocytes that are self-tolerant. These specific lymphocytes proliferate in response to antigen and differentiate into antigen-specific effector cells that eliminate the eliciting pathogen, and into memory cells to sustain immunity. The theory was formulated by Sir Macfarlane Burnet and in earlier forms by Niels Jerne and David Talmage.

A **clone** is a population of cells all derived from a single progenitor cell.

A feature unique to individual cells or members of a clone is said to be **clonotypic**. Thus, a monoclonal antibody that reacts with the receptor on a cloned T-cell line is said to be a clonotypic antibody and to recognize its clonotype, or the clonotypic receptor of that cell.

CLP: **see common lymphoid progenitor**.

Clusters of differentiation (**CD**) are groups of monoclonal antibodies that identify the same cell-surface molecule. The cell-surface molecule is designated CD followed by a number (e.g., CD1, CD2, etc.). For a current listing of CDs see Appendix II.

The **coagulation system** is a proteolytic cascade of plasma enzymes that triggers blood clotting when blood vessels are damaged.

The expression of a gene is said to be **codominant** when both alleles at one locus are expressed in roughly equal amounts in heterozygotes. Most genes show this property, including the highly polymorphic MHC genes.

A **coding joint** is formed by the imprecise joining of a V gene segment to a (D)J gene segment in immunoglobulin or T-cell receptor genes.

A B cell is given help by a **cognate** T cell, that is, a helper T cell primed by the same antigen.

The technique of **co-immunoprecipitation** is used to isolate a particular protein together with other proteins that bind to it, by using a labeled antibody against the first protein to precipitate the protein complex from a cell extract.

Collectins are a structurally related family of calcium-dependent sugar-binding proteins or lectins containing collagen-like sequences. An example is mannose-binding lectin.

Antigen receptors manifest two distinct types of **combinatorial diversity** generated by the combination of separate units of genetic information. Receptor gene segments are joined in many different combinations to generate diverse receptor chains, and then two different receptor chains (heavy and light in immunoglobulins; α and β, or γ and δ, in T-cell receptors) are combined to make the antigen-recognition site.

Microorganisms that normally live harmlessly in symbiosis with their host are known as **commensal microorganisms**. Many commensals confer a positive benefit on their host in some way.

The **common γ chain** (γ_c) is a transmembrane polypeptide chain (CD132) that is common to a subgroup of cytokine receptors.

Common lymphoid progenitors (**CLPs**) are stem cells that give rise to all lymphocytes. They are derived from pluripotent hematopoietic stem cells.

The term **common mucosal immune system** for the mucosal immune system describes the fact that lymphocytes that have been primed in one part of the mucosal system can recirculate as effector cells to other parts of the mucosal system.

The **common myeloid progenitor** is the precursor of the macrophages, granulocytes, mast cells and dendritic cells of the innate immune system, and also of megakaryocytes and red blood cells.

Common variable immunodeficiency (**CVID**) is a relatively common deficiency in antibody production whose pathogenesis is not yet fully understood. There is a strong association with genes mapping within the MHC.

Competitive binding assays are serological assays in which unknowns are detected and quantitated by their ability to inhibit the binding of a labeled known ligand to its specific antibody. When known sources of antibody or antigen are used as competitive inhibitors of antigen–antibody interactions, this assay is referred to as a **competitive inhibition assay**.

Complement or the **complement system** is a set of plasma proteins that act together to attack extracellular forms of pathogens. **Complement activation** can occur spontaneously on certain pathogens or by antibody binding to the pathogen. The pathogen becomes coated with complement proteins that facilitate pathogen removal by phagocytes and can also kill certain pathogens directly.

Complement receptors (**CRs**) are cell-surface proteins on various cells that recognize and bind complement proteins that have bound an antigen such as a pathogen. Complement receptors on phagocytes allow them to identify pathogens coated with complement proteins and take them up and destroy them. Complement receptors include CR1, CR2, CR3, CR4, and the receptor for C1q.

The **complementarity-determining regions** (**CDRs**) of immunoglobulins and T-cell receptors are the parts of these molecules that determine their specificity and make contact with specific ligand. The CDRs are the most variable part of the molecule, and contribute to the diversity of these molecules. There are three such regions (**CDR1**, **CDR2**, and **CDR3**) in each V domain.

Confocal fluorescent microscopy produces optical images at very high resolution by having two origins of fluorescent light that come together only at one plane of a thicker section.

When a protein binds a ligand, the protein often undergoes a change in its tertiary structure, or a **conformational change**, that has an effect on its function, either activating it or inhibiting it.

Conformational epitopes, or discontinuous epitopes, on a protein antigen are formed from several separate regions in the primary sequence of a protein brought together by protein folding. Antibodies that bind conformational epitopes bind only native folded proteins.

Conjugate vaccines are vaccines made from capsular polysaccharides bound to proteins of known immunogenicity, such as tetanus toxoid.

The **constant region** (**C region**) of an immunoglobulin or T-cell receptor is that part of the molecule that is relatively constant in amino acid sequence between different molecules. In an antibody molecule, the constant regions of each chain are composed of one or more **constant domains** (**C domains**). The constant region of an antibody determines its particular effector function. *Cf.* **variable region.**

A **contact hypersensitivity reaction** is a form of delayed-type hypersensitivity in which T cells respond to antigens that are introduced by contact with the skin.

Continuous epitopes, or linear epitopes, are antigenic determinants on proteins that are contiguous in the amino acid sequence and therefore do not require the protein to be folded into its native conformation for antibody to bind. The epitopes detected by T cells are continuous.

Conventional dendritic cells are the dendritic-cell lineage that mainly participates in antigen presentation to and activation of naive T cells. *Cf.* **plasmacytoid dendritic cells**.

A **convertase** is an enzyme that converts a complement protein into its reactive form by cleaving it.

The **Coombs test** is a test to detect antibody bound to red blood cells. Red blood cells that are coated with antibody are agglutinated if they are exposed to an anti-immunoglobulin antibody. The Coombs test is important in detecting the nonagglutinating antibodies against red blood cells produced by Rh incompatibility in pregnancy.

Two binding sites are said to demonstrate **cooperativity** in binding to their ligand when the binding of ligand to one site enhances the binding of ligand to the second site.

A **co-receptor** is a cell-surface protein that increases the sensitivity of the antigen receptor to antigen by binding to associated ligands and participating in signaling for activation. CD4 and CD8 are MHC-binding co-receptors on T cells, and CD19 is part of a complex that makes up the co-receptor on B cells.

Corona: see **B-cell corona**.

Corticosteroids are a family of drugs related to steroids such as cortisone that are naturally produced in the adrenal cortex. Corticosteroids can kill lymphocytes, especially developing thymocytes, inducing apoptotic cell death. They are useful anti-inflammatory, anti-lymphoid tumor, and immunosuppressive agents.

The activation and proliferation of lymphocytes after they first encounter antigen also requires the receipt of a separate **co-stimulatory signal**. Such signals are usually delivered to T cells by **co-stimulatory molecules** on the surface of the antigen-presenting cell. The most important of such molecules in the activation of naive T cells are B7.1 and B7.2, which are bound by CD28 on the T-cell surface. B cells may receive co-stimulatory signals from common pathogen components such as LPS, from complement fragments, or from CD40 ligand expressed on the surface of an activated antigen-specific helper T cell.

The **counter-regulation hypothesis** proposes that all types of infection early in childhood might protect against the development of atopy by driving the production of cytokines such as IL-10 and transforming growth factor-β, which downregulate both T_H1 and T_H2 responses.

Cowpox is the common name of the disease produced by vaccinia virus, used by Edward Jenner in the successful vaccination against smallpox, which is caused by the related variola virus.

CR: see **complement receptor**.

CR1 (**CD35**) is one of several receptors on cells for various components of complement. It is used to remove immune complexes from the plasma.

CR2 (**CD21**) is part of the B-cell co-receptor complex along with CD19 and CD81. It binds to antigens that have various breakdown products of C3, especially C3dg, bound to them and, by cross-linking to the B-cell receptor, enhances sensitivity to antigen by at least a 100-fold. It is also used by the Epstein–Barr virus to invade B cells.

CR3 (**CD11b:CD18**) is a β_2 integrin that functions both as an adhesion molecule and as a complement receptor. It binds iC3b, and stimulates phagocytosis.

CR4 (**CD11c:CD18**) is a β_2 integrin that binds iC3b and stimulates phagocytosis.

CRAC channels are calcium release-activated calcium channels in the plasma membrane that open to let calcium flow into the cell during the response of a lymphocyte to antigen.

C region: see **constant region**.

C-reactive protein is an acute-phase protein that binds to phosphocholine, which is a constituent of the C-polysaccharide of the bacterium *Streptococcus pneumoniae*. Many other bacteria also have surface phosphocholine, so C-reactive protein can bind many different bacteria and opsonize them for uptake by phagocytes. C-reactive protein does not bind to mammalian tissues.

Crohn's disease is a chronic inflammatory bowel disease thought to result from an abnormal overresponsiveness to the commensal gut flora.

When the antigen receptors on a lymphocyte are linked together by a multivalent antigen, they are said to be **cross-linked**.

Cross-matching is used in blood typing and histocompatibility testing to determine whether donor and recipient have antibodies against each other's cells that might interfere with successful transfusion or grafting.

Extracellular proteins taken up by dendritic cells can give rise to peptides presented by MHC class I molecules by the phenomenon of **cross-presentation**. This enables antigens from extracellular sources to be presented by MHC class I molecules and activate CD8 T cells.

A **cross-reaction** is the binding of antibody or a T-cell to an antigen not used to elicit that antibody.

The **cryptdins** are α-defensins (antimicrobial polypeptides) made by the Paneth cells of the small intestine.

A **cryptic epitope** is any epitope that cannot be recognized by a lymphocyte receptor until the antigen has been broken down and processed.

Cryptopatches are aggregates of lymphoid tissue in the wall of the intestine.

c-SMAC: see **supramolecular adhesion complex**.

The protein **C-terminal Src kinase** (**Csk**) is constitutively active in lymphocytes and has the function of phosphorylating the C-terminal tyrosine of Src-family kinases, thus inactivating them.

CTLA-4 is the high-affinity receptor for B7 molecules on T cells.

CXCL13 is a chemokine that attracts B cells and activated T cells into the follicles of peripheral lymphoid tissues by binding to the CXCR5 receptor on these cells.

Cutaneous lymphocyte antigen (**CLA**) is a cell-surface molecule that is involved in lymphocyte homing to the skin in humans.

Cutaneous T-cell lymphoma is a malignant growth of T cells that home to the skin.

CVID: see **common variable immunodeficiency**.

Cyclic neutropenia is a dominantly inherited disease in which neutrophil numbers fluctuate from near normal to very low or none, with an approximate cycle time of 21 days.

Cyclophosphamide is a DNA alkylating agent that is used as an immunosuppressive drug. It acts by killing rapidly dividing cells, including lymphocytes proliferating in response to antigen.

Cyclosporin A is a powerful immunosuppressive drug that inhibits signaling from the T-cell receptor, preventing T-cell activation and effector function. It binds to cyclophilin, and this complex binds to and inactivates the serine/threonine phosphatase calcineurin.

Cystic fibrosis is an inherited disease in which a defect in a membrane transport protein results in, among other symptoms, the secretion of thick sticky mucus in the airways, which clogs up the lungs and increases the risk of respiratory failure and lung infections.

Cytokine capture: see **antigen-capture assay**.

A cytokine is any small protein made by a cell that affects the behavior of other cells. Cytokines made by lymphocytes are often called lymphokines or interleukins (abbreviated IL), but the generic term cytokine is used in this book and in most of the literature. Cytokines act via specific cytokine receptors on the cells that they affect. Cytokines and their receptors are listed in Appendix III. See also **chemokines**.

Cytokine receptors are cellular receptors for cytokines. Binding of the cytokine to the cytokine receptor induces new activities in the cell, such as growth, differentiation, or death. Cytokine receptors are listed in Appendix III.

Cytotoxic granules containing the cytotoxic proteins perforin, granzymes, and granulysin are a defining characteristic of effector CD8 cytotoxic T cells and NK cells.

T cells that can kill other cells are called **cytotoxic T cells**. Most cytotoxic T cells are MHC class I-restricted CD8 T cells, but CD4 T cells can also kill in some cases. Cytotoxic T cells are important in host defense against cytosolic pathogens.

Cytotoxins are proteins made by cytotoxic T cells and NK cells that participate in the destruction of target cells. Perforins, granzymes, and granulysins are the major defined cytotoxins.

In the context of immunoglobulins, **δ** is the type of heavy chain in IgD. δ is also the name of one of the chains of the antigen receptor of a subset of T cells called γ:δ T cells.

Dark zone: see **germinal centers**.

ICAM-3 binds with high affinity to a lectin called **DC-SIGN**, which is found only on dendritic cells.

Death domains are protein-interaction domains originally discovered in proteins involved in programmed cell death or apoptosis.

Death receptors are cell-surface receptors whose engagement by extracellular ligands stimulates apoptosis in the receptor-bearing cell.

The **decay-accelerating factor** (**DAF** or **CD55**) is a cell-surface molecule that protects cells from lysis by complement. Its absence causes the disease paroxysmal nocturnal hemoglobinuria.

Defective ribosomal products (DRiPs) are peptides translated from introns in improperly spliced mRNAs, translations of frameshifts, and improperly folded proteins, which are recognized and tagged by ubiquitin for rapid degradation by the proteasome.

Defensins: see **β-defensins**; **cryptdins**.

Delayed-type hypersensitivity, or type IV hypersensitivity, is a form of cell-mediated immunity elicited by antigen in the skin and is mediated by CD4 T_H1 cells. It is called delayed-type hypersensitivity because the reaction appears hours to days after antigen is injected. *Cf.* **immediate hypersensitivity**.

Dendritic cells are bone-marrow-derived cells found in most tissues, including lymphoid tissues. Two main functional subsets are distinguished. Conventional dendritic cells take up antigen in peripheral tissues, are activated by contact with pathogens, and travel to the peripheral lymphoid organs, where they are the most potent stimulators of T-cell responses. Plasmacytoid dendritic cells also take up and present antigen, but their main function in an infection is to produce large amounts of the antiviral interferons. Both these types of dendritic cells are distinct from the follicular dendritic cell that presents antigen to B cells in lymphoid follicles.

Dendritic epidermal T cells (**dETCs**) are a specialized class of γ:δ T cells found in the skin of mice and some other species, but not humans. All dETCs have the same γ:δ T-cell receptor; their function is unknown.

Immunosuppressive monoclonal antibodies that trigger the destruction of lymphocytes *in vivo* are known as **depleting antibodies.** They are used for treating episodes of acute graft rejection.

Desensitization is a procedure in which an allergic individual is exposed to increasing doses of allergen in hope of inhibiting their allergic reactions. It probably involves shifting the balance between CD4 T_H1 and T_H2 cells and thus changing the antibody produced from IgE to IgG.

D gene segments, or **diversity gene segments**, are short DNA sequences that join the V and J gene segments in rearranged immunoglobulin heavy-chain genes and in T-cell receptor β- and δ-chain genes. See **gene segments**.

Diabetes: see **type 1 diabetes mellitus**.

Diacylglycerol (**DAG**) is most commonly formed from membrane inositol phospholipids by the action of phospholipase C-γ as the result of the activation of many different receptors. The diacylglycerol stays in the membrane, where it acts as an intracellular signaling molecule, activating protein kinase C, which further propagates the signal.

Diapedesis is the movement of blood cells, particularly leukocytes, from the blood across blood vessel walls into tissues.

DiGeorge's syndrome is a recessive genetic immunodeficiency disease in which there is a failure to develop thymic epithelium. Parathyroid glands are also absent and there are anomalies in the large blood vessels. It seems to be due to a developmental defect in neural crest cells.

Direct allorecognition of a grafted tissue involves donor antigen-presenting cells leaving the graft, migrating via the lymph to regional lymph nodes and activating host T cells bearing the corresponding T-cell receptors.

The **direct Coombs test** uses anti-immunoglobulin to agglutinate red blood cells as a way of detecting whether they are coated with antibody *in vivo* due to autoimmunity or maternal anti-fetal immune responses (see **Coombs test**; **indirect Coombs test**).

Discontinuous epitopes: see **conformational epitopes**.

Diversity gene segment: see **D gene segment**.

DN1, DN2, DN3, and **DN4** are substages in the development of double-positive T cells in the thymus. Rearrangement of the TCRβ-chain locus starts at DN2 and is completed by DN4.

DNA ligase IV is the enzyme that joins together the ends of double-stranded DNA broken during the gene rearrangements that generate functional genes for immunoglobulins or T-cell receptors.

DNA microarrays are created by placing a different DNA on a small part of a microchip, and using them to assess RNA expression in normal or malignant cells.

DNA vaccination is a novel means of raising an adaptive immune response. For unknown reasons, when DNA is injected into muscle, it is expressed and elicits antibody and T-cell responses to the protein encoded by the DNA.

Dominant immune suppression: see **regulatory tolerance**.

Double-negative thymocytes are immature T cells in the thymus that lack expression of the two co-receptors, CD4 and CD8. In a normal thymus, these represent about 5% of thymocytes.

Double-positive thymocytes are immature T cells in the thymus that are characterized by expression of both the CD4 and the CD8 co-receptor proteins. They represent the majority (~80%) of thymocytes.

Down syndrome cell adhesion molecule (**Dscam**) is a member of the immunoglobulin superfamily which in insects is thought to opsonize invading bacteria and aid their engulfment by phagocytes. It can be made in a multiplicity of different forms as a result of alternative splicing.

A **draining lymph node** is a lymph node downstream of a site of infection that receives antigens and microbes from the site via the lymphatic system. Draining lymph nodes often enlarge enormously during an immune response and can be palpated; they were originally called swollen glands.

Dscam: see **Down syndrome cell adhesion molecule**.

In the context of immunoglobulins, ε (epsilon) is the heavy chain of IgE.

EAE: see **experimental allergic encephalomyelitis.**

The **early induced responses** or early nonadaptive responses are a series of host defense responses that are triggered by infectious agents early in infection. They are distinct from innate immunity because there is an inductive phase, and from adaptive immunity in that they do not operate by clonal selection of rare antigen-specific lymphocytes.

The **early lymphoid progenitor** (**ELP**) is a bone marrrow cell that can give rise both to the common lymphoid progenitor and to T-cell precursors that migrate from the bone marrow to the thymus.

Early pro-B cell: see **pro-B cells**.

Eczema or **atopic dermatitis** is an allergic skin condition seen mainly in children; its etiology is poorly understood.

In immunology, **edema** is the swelling caused by the entry of fluid and cells from the blood into the tissues, which is one of the cardinal features of the process of inflammation.

Effector caspases are intracellular proteases that are activated as a result of an apoptotic signal and initiate the cellular changes associated with apoptosis.

Effector lymphocytes develop from naive lymphocytes after initial activation by antigen and can mediate the removal of pathogens from the body without the need for further differentiation. In this they are distinct from naive lymphocytes, and from memory lymphocytes, which must differentiate and often proliferate before they become effector lymphocytes.

Effector mechanisms are those processes by which pathogens are destroyed and cleared from the body. Innate and adaptive immune responses use most of the same effector mechanisms to eliminate pathogens.

Effector memory cells are memory cells that are thought to be specialized for quickly entering inflamed tissues after restimulation with antigen.

Effector T cells are the T cells that carry out the functions of an immune response, such as cell killing and cell activation, that directly result in the clearance of the infectious agent from the body. There are several different subsets, each with a specific role in immune responses.

Lymphocytes leave a lymph node through the **efferent lymphatic vessel.**

Electrophoresis is the movement of molecules in a charged field. In immunology, techniques based on electrophoresis are used to separate molecules, especially protein molecules, and to determine their charge, size, and subunit composition.

ELISA: see **enzyme-linked immunosorbent assay**.

ELISPOT assay is an adaptation of ELISA in which cells are placed over antibodies or antigens attached to a plastic surface. The antigen or antibody traps the cells' secreted products, which can then be detected using an enzyme-coupled antibody that cleaves a colorless substrate to make a localized colored spot.

ELP: **early lymphoid progenitor**.

Embryonic stem (**ES**) **cells** are early embryonic cells that will grow continuously in culture and that retain the ability to contribute to all cell lineages. Mouse ES cells can be genetically manipulated in tissue culture and then inserted into mouse blastocysts to generate mutant lines of mice.

Encapsulated bacteria have thick carbohydrate coats that protect them from phagocytosis. They cause extracellular infections and are effectively engulfed and destroyed by phagocytes only if the bacteria are first coated with antibody and complement.

Cytokines that can induce a rise in body temperature are called **endogenous pyrogens**, as distinct from exogenous pyrogens such as endotoxin from Gram-negative bacteria that induce fever by triggering the synthesis and release of endogenous pyrogens.

The **endoplasmic reticulum-associated aminopeptidase associated with antigen processing** (**ERAAP**) is an enzyme in the endoplasmic reticulum that trims longer polypeptides to a size at which they can bind to MHC class I molecules.

Endosomes are membrane-bounded intracellular vesicles. Antigen taken up by phagocytosis generally enters the endosomal system.

The **endosteum** in bone marrow is the region adjacent to the inner surface of the bone and is the location of the earliest hematopoietic stem cells.

The changes that occur in the endothelial walls of small blood vessels as a result of inflammation, such as increased permeability and the increased production of cell adhesion molecules and cytokines, are known generally as **endothelial activation**.

Endotoxins are bacterial toxins that are released only when the bacterial cell is damaged, as opposed to exotoxins, which are secreted. The most important endotoxin medically is the lipopolysaccharide (LPS) of Gram-negative bacteria, which is a potent inducer of cytokine synthesis; when present in large amounts in the blood it can cause a systemic shock reaction called endotoxic shock.

The **enzyme-linked immunosorbent assay** (**ELISA**) is a serological assay in which bound antigen or antibody is detected by a linked enzyme that converts a colorless substrate into a colored product. The ELISA assay is widely used in biology and medicine as well as in immunology.

Eosinophils are white blood cells thought to be important chiefly in defense against parasitic infections. The level of eosinophils in the blood is normally quite low. It can increase markedly in several situations, such as atopy, resulting in **eosinophilia**, an abnormally large number of eosinophils in the blood.

Eotaxin-1 (**CCL11**), **eotaxin-2** (**CCL24**), and **eotaxin-3** (**CCL26**) are CC chemokines that act predominantly on eosinophils.

An **epitope** is a site on an antigen recognized by an antibody or an antigen receptor; epitopes are also called antigenic determinants. A T-cell epitope is a short peptide derived from a protein antigen. It binds to an MHC molecule and is recognized by a particular T cell. B-cell epitopes are antigenic determinants recognized by B cells and are typically structural motifs on the surface of the antigen.

Epitope spreading describes the fact that responses to autoantigens tend to become more diverse as the response persists, due to responses being made to epitopes other than the original one.

The **Epstein–Barr virus** (**EBV**) is a herpesvirus that selectively infects human B cells by binding to complement receptor 2 (CD21). It causes infectious mononucleosis and establishes a lifelong latent infection in B cells that is controlled by T cells. Some B cells latently infected with EBV will proliferate *in vitro* to form lymphoblastoid cell lines.

The affinity of an antibody for its antigen can be determined by **equilibrium dialysis**, a technique in which antibody in a dialysis bag is exposed to varying amounts of a small antigen able to diffuse across the dialysis membrane. The amount of antigen inside and outside the bag at the equilibrium diffusion state is determined by the amount and affinity of the antibody in the bag.

ERAAP: see **endoplasmic reticulum-associated aminopeptidase associated with antigen processing**.

Erp57 is a chaperone molecule involved in loading peptide onto MHC class I molecules in the endoplasmic reticulum.

Erythroblastosis fetalis is a severe form of Rh hemolytic disease in which maternal anti-Rh antibody enters the fetus and produces a hemolytic anemia so severe that the fetus has mainly immature erythroblasts in the peripheral blood.

E-selectin: see **selectins**.

An **exogenous pyrogen** is any substance originating outside the body that can induce fever, such as the bacterial lipopolysaccharide LPS. *Cf.* **endogenous pyrogen**.

Experimental allergic encephalomyelitis (**EAE**) is an inflammatory disease of the central nervous system that develops after mice are immunized with neural antigens in a strong adjuvant.

The movement of cells or fluid from within blood vessels to the surrounding tissues is called **extravasation**.

The **extrinsic pathway of apoptosis** is triggered by extracellular ligands binding to specific cell-surface receptors (death receptors), that then signal the cell to undergo programmed cell death.

A **Fab fragment** of an antibody molecule cleaved by papain consists of a single arm of the antibody composed of a light chain and the amino-terminal half of a heavy chain held together by an interchain disulfide bond. The enzyme pepsin cleaves an antibody molecule to produce the **F(ab′)$_2$ fragment**, in which the two arms of the antibody molecule remain linked. See also **Fc fragment**.

FACS®: see **fluorescence-activated cell sorter**.

Factor B, **factor D**, **factor H**, **factor I**, and **factor P** are all components of the alternative pathway of complement activation. **Factor B** plays a role very similar to that of C2b in the classical pathway. **Factor D** is a serine protease that cleaves factor B. **Factor H** is an inhibitory protein with a role similar to decay-accelerating factor. **Factor I** is a protease that breaks down various components of the alternative pathway. **Factor P**, or **properdin**, is a positive regulatory component of the alternative pathway. It stabilizes the C3 convertase of the alternative pathway on the surface of bacterial cells.

Factor I deficiency is a genetically determined lack of the complement-regulatory protein factor I. This results in uncontrolled complement activation, so that complement proteins rapidly become depleted and people suffer repeated bacterial infections, especially with ubiquitous pyogenic bacteria.

Familial cold autoinflammatory syndrome (FCAS) is an episodic autoinflammatory disease caused by mutations in the gene *CSA1*, encoding cryopyrin. It is induced by exposure to cold.

Familial hemophagocytic lymphohistiocytosis (**FHL**) is a progressive and potentially lethal inflammatory disease caused by an inherited deficiency of perforin. Large numbers of polyclonal CD8-positive T cells accumulate in lymphoid and other organs, and this is associated with activated macrophages that phagocytose blood cells, including erythrocytes and leukocytes.

Familial Mediterranean fever (FMF) is a severe autoinflammatory disease, inherited as an autosomal recessive disorder. It is caused by mutation in the gene encoding the protein pyrin, but how this results in the disease is not known.

Farmer's lung is a hypersensitivity disease caused by the interaction of IgG antibodies with large amounts of an inhaled allergen in the alveolar wall of the lung, causing alveolar wall inflammation and compromising gas exchange.

Fas is a member of the TNF receptor family; it is expressed on certain cells and makes them susceptible to killing by cells expressing **Fas ligand** (**FasL**), a cell-surface member of the TNF family of proteins. Binding of Fas ligand to Fas triggers apoptosis in the Fas-bearing cell.

The **Fc fragment** of an antibody cleavaed by papain consists of the carboxy-terminal halves of the two heavy chains disulfide-bonded to each other by the residual hinge region. See also **Fab fragments**.

FCAS: see **familial cold autoinflammatory syndrome**.

Fc receptors bind the Fc portions of immunoglobulins. There are different Fc receptors for different isotypes: FcγR binds IgG, for example, and FcεR binds IgE.

The high-affinity **Fcε receptor** (**FcεRI**) on the surface of mast cells and basophils binds the Fc region of free IgE. When antigen binds this IgE and cross-links FcεRI, it causes mast-cell activation.

Fcγ receptors, including **FcγRI**, **RII**, and **RIII**, are cell-surface receptors that bind the Fc portion of IgG molecules. Most Fcγ receptors bind only aggregated IgG, allowing them to discriminate bound antibody from free IgG. They are expressed on phagocytes, B lymphocytes, NK cells, and follicular dendritic cells. They have a key role in humoral immunity, linking antibody binding to effector cell functions.

FDC: see **follicular dendritic cell**.

FHL: see **familial hemophagocytic lymphohistiocytosis.**

Fibrinogen-related proteins (FREPs) are members of the immunoglobulin superfamily that are thought to have a role in innate immunity in the freshwater snail *Biomphalaria glabrata*.

Ficolins are carbohydrate-binding proteins that initiate the lectin pathway of complement activation. They are members of the collectin family and bind to the *N*-acetylglucosamine present on the surface of some pathogens.

When tissue or organ grafts are placed in an unmatched recipient, they are rejected by a **first-set rejection**, which is an immune response by the host against foreign antigens in the graft. *Cf.* **second-set rejection**.

FK506: see **tacrolimus**.

Individual cells can be characterized and separated in a machine called a **fluorescence-activated cell sorter** (**FACS®**) that measures cell size, granularity, and fluorescence due to bound fluorescent antibodies as single cells pass in a stream past photodetectors. The analysis of single cells in this way is called flow cytometry and the instruments that carry out the measurements and/or sort cells are called flow cytometers or cell sorters.

FMF: see **familial Mediterranean fever**.

Peripheral lymphoid tissues, such as lymph nodes, spleen, and Peyer's patches, contain large areas of B cells called **follicles**, which are organized around follicular dendritic cells.

A **follicular center cell lymphoma** is a type of B-cell lymphoma that tends to grow in the follicles of lymphoid tissues.

The **follicular dendritic cells (FDCs)** of lymphoid follicles are cells of uncertain origin. They are characterized by long branching processes that make intimate contact with many different B cells. They have Fc receptors that are not internalized by receptor-mediated endocytosis and thus hold antigen:antibody complexes on the surface for long periods. These cells are crucial in selecting antigen-binding B cells during antibody responses.

Follicular helper cells are a subset of CXCR5-positive central memory T cells that produce IL-2 and provide help for B cells.

The V domains of immunoglobulins and T-cell receptors contain relatively invariant **framework regions** that provide a protein scaffold for the hypervariable regions that make contact with antigen.

Fungi are single-celled and multicellular eukaryotic organisms, including the yeasts and molds, that can cause a variety of diseases. Immunity to fungi is complex and involves both humoral and cell-mediated responses.

Fv: see **single-chain Fv**.

Fyn: see **tyrosine kinase**.

In the context of immunoglobulins, **γ** is the heavy chain of IgG.

A subset of T lymphocytes bears a distinct **γ:δ T-cell receptor** composed of different antigen-recognition chains, γ and δ, assembled in a γ:δ heterodimer. Cells bearing these receptors are called **γ:δ T cells** and the antigens they recognize and their function are not yet clear.

GALT: see **gut-associated lymphoid tissues**.

Plasma proteins can be separated on the basis of electrophoretic mobility into albumin and the α, β, and γ globulins. Most antibodies migrate in electrophoresis as **γ globulins** (or **gamma globulins**), and patients who lack antibodies are said to have agammaglobulinemia.

GAP: see **GTPase-activating protein**.

GEFs: see **guanine nucleotide exchange factors**.

In birds and rabbits, immunoglobulin receptor diversity is generated mainly by **gene conversion**, in which homologous inactive V gene segments exchange short sequences with an active, rearranged variable-region sequence.

Gene knockout or gene targeting is a way of disabling a specific gene by homologous recombination with an introduced DNA construct designed for that purpose. Mice carrying such gene knockouts in their genomes can be produced.

Gene rearrangement is the recombination of gene segments in the immunoglobulin and T-cell receptor loci to produce a functional variable-region sequence.

The variable domains of the polypeptide chains of antigen receptors are encoded in sets of **gene segments** that must undergo somatic recombination to form a complete variable-domain exon. There are three types of gene segment: V gene segments encode the first 95 amino acids, D gene segments (in heavy-chain and TCRα chain loci only) encode about 5 amino acids, and J gene segments form the last 10–15 amino acids of the variable domain. There are multiple copies of each type of gene segment in the germline DNA, but only one of each type are joined together to form the variable domain.

A **genetic locus** (plural **loci**) is the site of a gene on a chromosome. In the case of the genes for the immunoglobulin and T-cell receptor chains, the term locus refers to the complete collection of gene segments and C regions for the given chain.

Gene targeting: see **gene knockout**.

Mice that are raised in the complete absence of intestinal and other flora are called **germ-free** or **gnotobiotic** mice. Such mice have very depleted immune systems, but they can respond virtually normally to any specific antigen, provided it is mixed with a strong adjuvant.

Germinal centers in lymphoid follicles in peripheral lymphoid tissues are sites of intense B-cell proliferation, differentiation, somatic hypermutation, and class switching during antibody responses.

Immunoglobulin and T-cell receptor genes are said to be in the **germline configuration** in the DNA of germ cells and in all somatic cells in which somatic recombination has not occurred.

The **germline diversity** of antigen receptors is due to the inheritance of multiple gene segments that encode variable domains. This type of diversity is distinct from the diversity that is generated during gene rearrangement or after the expression of an antigen-receptor gene, which is somatically generated.

The **germline theory** of antibody diversity proposed that each antibody was encoded in a separate germline gene. This is now known not to be true for humans, mice, and most other vertebrates, but appears to be true in elasmobranch fishes, which have rearranged genes in the germline.

GlyCAM-1 is a mucinlike molecule found on the high endothelial venules of peripheral lymphoid tissues. It is a ligand for the cell-adhesion protein L-selectin on naive lymphocytes, directing these cells to leave the blood and enter the lymphoid tissues.

Gnotobiotic: see **germ-free**.

Goodpasture's syndrome is an autoimmune disease in which autoantibodies against type IV collagen (found in basement membranes) are produced, causing extensive inflammation in kidneys and lungs.

G proteins are intracellular proteins that bind GTP and convert it to GDP in the process of cell signal transduction. There are two kinds: the heterotrimeric (α, β, γ subunits) receptor-associated G proteins, and the small G proteins, such as Ras and Raf, that act downstream of many transmembrane signaling events.

Tissue and organ grafts between genetically distinct individuals almost always elicit an adaptive immune response that causes **graft rejection**, the destruction of the grafted tissue by attacking lymphocytes.

In bone marrow transplantation between gentically non-identical people, mature T cells in the transplanted bone marrow attack the recipient's tissues, causing **graft-versus-host disease** (**GVHD**).

Some of the therapeutic effect of bone marrow transplantation for leukemia can be due to a **graft-versus-leukemia effect**, in which T cells in the donor bone marrow recognize minor histocompatibility antigens or tumor-specific antigens on the recipient's leukemic cells and attack them.

Granulocyte: see **polymorphonuclear leukocytes**.

Granulocyte-macrophage colony-stimulating factor (**GM-CSF**) is a cytokine involved in the growth and differentiation of cells of the myeloid lineage, including dendritic cells, monocytes and tissue macrophages, and granulocytes.

A **granuloma** is a site of chronic inflammation usually triggered by persistent infectious agents such as mycobacteria or by a non-degradable foreign body. Granulomas have a central area of macrophages, often fused into multinucleate giant cells, surrounded by T lymphocytes.

Granulysin is a cytotoxic protein present in the cytotoxic granules of cytotoxic CD8 T cells and NK cells.

Granzymes are serine proteases present in cytotoxic CD8 T cells and NK cells and are involved in inducing apoptosis in the target cell.

Graves' disease is an autoimmune disease in which antibodies against the thyroid-stimulating hormone receptor cause overproduction of thyroid hormone and thus hyperthyroidism.

The inherited immunodeficiency disease **Griscelli syndrome** (type 2) affects the pathway for secretion of lysosomes. It is caused by mutations in a small GTPase Rab27a, which controls the movement of vesicles within cells.

GTPase-activating proteins (GAPs) are regulatory proteins that accelerate the intrinsic GTPase activity of G proteins and thus facilitate the conversion of G proteins from the active (GTP-bound) state to the inactive (GDP-bound) state.

The **guanine nucleotide exchange factors** (**GEFs**) are proteins that can remove the bound GDP from G proteins; this allows GTP to bind and activate the G protein.

The **gut-associated lymphoid tissues** (**GALT**) are peripheral lymphoid tissues closely associated with the gastrointestinal tract, including the palatine tonsils, Peyer's patches, isolated lymphoid follicles and intraepithelial lymphocytes. GALT have a distinctive biology related to their exposure to antigens from food and normal intestinal microbial flora.

GVHD: see **graft-versus-host disease**.

The major histocompatibility complex of the mouse is called **H-2** (for **histocompatibility-2**). Haplotypes are designated by a lower-case superscript, as in H-2^b.

HAART: see **highly active antiretroviral therapy**.

H antigens or **histocompatibility antigens**, are known as major histocompatibility antigens when they encode proteins (the MHC molecules) that present foreign peptides to T cells and as minor H antigens when they present polymorphic self peptides to T cells. See also **histocompatibility**.

A **haplotype** is a linked set of genes associated with one haploid genome. The term is used mainly in connection with the linked genes of the major histocompatibility complex (MHC), which are usually inherited as one haplotype from each parent.

Haptens are small molecules that can bind antibody but cannot by themselves elicit an adaptive immune response. Haptens must be chemically linked to protein carriers to elicit antibody and T-cell responses.

Hashimoto's thyroiditis is an autoimmune disease characterized by persistent high levels of antibody against thyroid-specific antigens. These antibodies recruit NK cells to the thyroid, leading to damage and inflammation.

All immunoglobulin molecules have two types of chain, a **heavy chain** (**H chain**) and a light chain. The basic immunoglobulin unit consists of two identical heavy chains and two identical light chains. Heavy chains come in a variety of **heavy-chain classes** (isotypes), each of which confers a distinctive functional activity on the antibody molecule.

Helper CD4 T cells are CD4 T cells that stimulate or 'help' B cells to make antibody in response to antigenic challenge. Both T_H2 and T_H1 subsets of effector CD4 T cells can carry out this function.

A **hemagglutinin** is any substance that causes red blood cells to agglutinate, a process known as **hemagglutination**. The hemagglutinins in human blood are antibodies that recognize the ABO blood group antigens. Influenza and some other viruses have hemagglutinin proteins that bind to glycoproteins on host cells to initiate the infectious process.

Hematopoiesis is the generation of all the cellular elements of blood, and in humans occurs in the bone marrow. All blood cells originate from pluripotent **hematopoietic stem cells** in the marrow and subsequently differentiate into the different blood cell types.

The **hematopoietin family** is a large family of structurally related cytokines that includes growth factors and many interleukins with roles in both adaptive and innate immunity

Hemolytic disease of the newborn: see **erythroblastosis fetalis**.

In individuals with inherited deficiencies in complement regulatory proteins, uncontrolled complement activation typically leads to **hemolytic uremic syndrome**, characterized by damage to platelets and red blood cells and inflammation of the kidneys.

In **hemophagocytic syndrome** there is a dysregulated expansion of CD8-positive lymphocytes which is associated with macrophage activation. The activated macrophages phagocytose blood cells, including erythrocytes and leukocytes.

The **hepatobiliary route** is one of the routes by which mucosally produced dimeric IgA reaches the intestine. The antibodies are taken up into the portal veins in the lamina propria, transported to the liver, and from there reach the bile duct by transcytosis. This pathway is not of great significance in humans.

In immunoglobulin and T-cell receptor loci, the conserved seven-nucleotide DNA sequence in the recombination signal sequences (RSSs) flanking gene segments is known as the **heptamer**.

Hereditary angioneurotic edema is the clinical name for a genetic deficiency of the C1 inhibitor of the complement system. In the absence of C1 inhibitor, spontaneous activation of the complement system can cause diffuse fluid leakage from blood vessels, the most serious consequence of which is swelling of the epiglottis (the throat) leading to suffocation.

Individuals **heterozygous** for a particular gene have two different alleles of that gene.

High endothelial venules (**HEVs**) are specialized venules found in lymphoid tissues. Lymphocytes migrate from the blood into lymphoid tissues by attaching to the **high endothelial cells** that make up the walls of these blood vessels and migrating between them.

The **highly active antiretroviral therapy (HAART)** used to control HIV infection is a combination of nucleoside analogs, which prevent reverse transcription, and drugs that inhibit the viral protease.

Tolerance to injected protein anitigens occurs at low or high doses of antigen. Tolerance induced by the injection of high doses of antigen is called **high-zone tolerance**, whereas tolerance produced with low doses of antigen is called low-zone tolerance.

The **hinge region** of antibody molecules is a flexible domain that joins the Fab arms to the Fc piece. The flexibility of the hinge region in IgG and IgA molecules allows the Fab arms to adopt a wide range of angles, permitting binding to epitopes spaced variable distances apart.

Histamine is a vasoactive amine stored in mast-cell granules. Histamine released by antigen binding to IgE antibodies bound to mast cells causes dilation of local blood vessels and smooth-muscle contraction, producing some of the symptoms of immediate hypersensitivity reactions. Antihistamines are drugs that counter histamine action.

Histocompatibility refers to the ability of tissues from one individual to be accepted, or rejected, if transplanted to another individual, and to the biological mechanisms that determine acceptance or rejection.

Histocompatibility-2: see **H-2**.

HIV: see **human immunodeficiency virus**.

HLA, the acronym for **H**uman **L**eukocyte **A**ntigen, is the genetic designation for the human MHC. Individual loci are designated by uppercase letters, as in HLA-A, and alleles are designated by numbers, as in HLA-A*0201.

The invariant **HLA-DM** protein in humans is involved in loading peptides onto MHC class II molecules. It is encoded in the MHC within a set of genes resembling MHC class II genes. A homologous protein in mice is called H-2M.

The atypical MHC class II molecule **HLA-DO** acts as a negative regulator of HLA-DM, binding to it and inhibiting the release of CLIP from MHC class II molecules in intracellular vesicles.

Hodgkin's disease is an immune system tumor characterized by large cells called Reed-Sternberg cells, which derive from mutated B-lineage cells.

Homeostasis is the status of physiological normality. In the case of the immune system, homeostasis refers to its state (e.g. numbers of lymphocytes) in an uninfected individual.

Homing receptors on lymphocytes are receptors for chemokines, cytokines and adhesion molecules specific to particular tissues, and which enable the lymphocyte to enter that tissue. The direction of a lymphocyte into a particular tissue is known as **homing**.

Genes can be disrupted by **homologous recombination** with copies of the gene into which erroneous sequences have been inserted. When these exogenous DNA fragments are introduced into cells, they recombine selectively with the cellular gene through the remaining regions of sequence homology, replacing the functional gene with a nonfunctional copy.

Host-versus-graft disease (**HVGD**) is another name for the allograft rejection reaction. The term is used mainly in relation to bone marrow transplantation.

The **human immunodeficiency virus** (**HIV**) is the causative agent of the acquired immune deficiency syndrome (AIDS). HIV is a retrovirus of the lentivirus family that selectively infects macrophages and CD4 T cells, leading to their slow depletion, which eventually results in immunodeficiency. There are two major strains of the virus, HIV-1 and HIV-2, of which HIV-1 causes most disease worldwide. HIV-2 is endemic to West Africa but is spreading.

Humanization is the genetic engineering of mouse hypervariable loops of a desired specificity into otherwise human antibodies for use as therapeutic agents. Such antibodies are less likely to cause an immune response in people treated with them than are wholly mouse antibodies.

Human leukocyte antigen: see **HLA**.

Humoral immunity is immunity due to antibodies and is produced as a result of a **humoral immune response**. Humoral immunity can be transferred to unimmunized recipients by the transfer of serum containing specific antibody.

Hybridomas are hybrid cell lines formed by fusing a specific antibody-producing B lymphocyte with a myeloma cell that is selected for its ability to grow in tissue culture and for an absence of immunoglobulin chain synthesis. The antibodies produced are all of a single specificity and are called monoclonal antibodies.

hygiene hypothesis: see **counter-regulation hypothesis**.

Hyperacute graft rejection is an immediate rejection reaction caused by preformed natural antibodies that react against antigens on the transplanted organ. The antibodies bind to endothelium and trigger the blood-clotting cascade, leading to an engorged, ischemic graft and rapid death of the organ.

Hyper IgD syndrome (HIDS) is an autoinflammatory disease due to mutations that lead to a partial deficiency of mevalonate kinase.

Hyper IgM immunodeficiency: see **hyperIgM type II syndrome**; **hypohydrotic ectodermal dysplasia with immunodeficiency**; **X-linked hyper IgM syndrome**.

Hyper IgM type 2 syndrome is an inherited immunodeficiency characterized by an abundance of relatively low-affinity IgM antibodies and a lack of antibodies of any other isotype. It is due to defects in the gene for AID (activation-induced cytidine deaminase), an enzyme required for both somatic hypermutation and isotype switching in immunoglobulin genes. *Cf.* X-linked hyper IgM syndrome.

Repeated immunization to achieve a heightened state of immunity is called **hyperimmunization**.

Hyperreactivity is the general hyperresponsiveness of the airways to non-immunological stimuli, such as cold or smoke, that develops in chronic asthma.

Immune responses to innocuous antigens that lead to symptomatic reactions upon reexposure to the antigen are called **hypersensitivity reactions**. These can cause **hypersensitivity diseases** if they occur repeatedly. This state of heightened reactivity to an antigen is called **hypersensitivity**. Hypersensitivity reactions are classified by mechanism: type I hypersensitivity reactions involve IgE antibody triggering of mast cells; type II hypersensitivity reactions involve IgG antibodies against cell-surface or matrix antigens; type III hypersensitivity reactions involve antigen:antibody complexes; and type IV hypersensitivity reactions are T cell-mediated.

The **hypervariable** (**HV**) **regions** of immunoglobulin and T-cell receptor V domains are small regions that make contact with the antigen and differ extensively from one receptor to the next. They are more often known as the complementarity-determining regions. *Cf.* **framework regions**.

Hypohydrotic ectodermal dysplasia with immunodeficiency, also known as NEMO deficiency, is an inherited syndrome with some features resembling hyper IgM syndrome. It is caused by defects in the protein NEMO a component of the NFκB signaling pathway.

The inactive complement fragment **iC3b** is produced by cleavage of C3b and is the first step in C3b inactivation.

The **ICAMs** (intercellular adhesion molecules) are cell-surface ligands for the leukocyte integrins and are crucial in the binding of lymphocytes and other leukocytes to certain cells, including antigen-presenting cells and endothelial cells. They are members of the immunoglobulin superfamily. **ICAM-1** is the most prominent ligand for the integrin CD11a:CD18 (LFA-1). It is rapidly inducible on endothelial cells by infection, and plays a major role in local inflammatory responses. **ICAM-2** is constitutively expressed at relatively low levels by endothelium. **ICAM-3** is expressed only on leukocytes and is thought to play an important part in adhesion between T cells and antigen-presenting cells, particularly dendritic cells.

Iccosomes are small fragments of membrane coated with immune complexes that fragment off the processes of follicular dendritic cells in lymphoid follicles early in a secondary or subsequent antibody response.

ICOS (inducible co-stimulatory protein) is a CD28-related protein that is induced on activated T cells and can enhance T-cell responses. It binds a ligand known as **LICOS** (ligand of ICOS), which is distinct from the B7 molecules.

Each immunoglobulin molecule has a set of unique features which is known as its **idiotype**.

IEL: see **intraepithelial lymphocyte**.

IFN-α, IFN-β, IFN-γ: see **interferon-α and -β**; **interferon-γ**.

Ig is the standard abbreviation for immunoglobulin.

Igα, **Igβ**: see **B-cell antigen receptor**.

IgA is the class of immunoglobulin characterized by α heavy chains. IgA is the main antibody class secreted by mucosal lymphoid tissues.

IgD is the class of immunoglobulin characterized by δ heavy chains. It appears as surface immunoglobulin on mature naive B cells but its function is unknown.

IgE is the class of immunoglobulin characterized by ε heavy chains. It is involved in the defense against parasite infections and in allergic reactions.

IgG is the class of immunoglobulin characterized by γ heavy chains. It is the most abundant class of immunoglobulin found in the plasma.

IgM is the class of immunoglobulin characterized by μ heavy chains. It is the first immunoglobulin to appear on the surface of B cells and the first to be secreted.

Ii: see **invariant chain**.

IL: see **interleukin**.

ILLs: see **innate-like lymphocytes.**

The **Imd pathway** (immunodeficiency pathway) is a defense against Gram-negative bacteria in insects that results in the production of antimicrobial peptides such as diptericin, attacin, and cecropin.

Immature B cells are B cells that have rearranged a heavy- and a light-chain V-region gene and express surface IgM, but have not yet matured sufficiently to express surface IgD as well.

Tissues throughout the body contain phagocytic **immature dendritic cells**, which leave the tissues and mature in response to inflammation or an infection. See also **dendritic cells**.

In the context of allergy, the **immediate reaction** or **immediate hypersensitivity reaction** is the reaction that occurs within seconds of encounter with antigen. *Cf.* **late-phase reaction**; **delayed-type hypersensitivity**.

The binding of antibody to a soluble antigen forms an **immune complex**. Large immune complexes form when sufficient antibody is available to cross-link the antigen; these are readily cleared by the reticuloendothelial system of cells bearing Fc receptors and complement receptors. Small, soluble immune complexes that form when antigen is in excess can be deposited in small blood vessels and damage them.

Immune effector functions are all those components and functions of the immune system that contain an infection and eliminate it, for example, complement, macrophages, neutrophils and other leukocytes, antibodies, and effector T cells.

Immune modulation is the deliberate attempt to change the course of an immune response, for example by altering the bias towards T_H1 or T_H2 dominance.

Immune regulation is the capacity of the immune system in normal circumstances to regulate itself so that an immune response does not go out of control and cause tissue damage, autoimmune reactions or allergic reactions.

The **immune response** is any response made by an organism to defend itself against a pathogen.

Immune response (Ir) gene is a term used in the past to describe a genetic polymorphism that controls the intensity of the immune response to a particular antigen. Virtually all Ir phenotypes are now known to be due to differences between alleles of the genes for MHC molecules, especially MHC class II molecules, leading to differences in the ability of MHC molecules to bind particular peptide antigens.

Immune surveillance is the recognition, and in some cases the elimination, of tumor cells by the immune system before they become clinically detectable.

The **immune system** is the tissues, cells, and molecules involved in innate immunity and adaptive immunity.

Immunity is the ability to resist infection with a particular pathogen. See also **protective immunity**.

Immunization is the deliberate provocation of an adaptive immune response by introducing antigen into the body. See also **active immunization**; **passive immunization**.

Immunobiology is the study of the biological basis for host defense against infection.

Immunoblotting is a common technique in which proteins separated by gel electrophoresis are blotted onto a nitrocellulose membrane and revealed by the binding of specific labeled antibodies.

Immunodeficiency diseases are a group of inherited or acquired disorders in which some aspect or aspects of host defense are absent or functionally defective.

Specific antibodies can be used to reveal ultramicroscopic structures in cells by the technique of **immunoelectron microscopy**. Different-sized gold particles are linked to antibodies against proteins of the structure and detected as bound gold particles by electron microscopy.

Immunodominant epitopes are those epitopes in an antigen that are preferentially recognized by T cells, such that T-cells specific for those epitopes come to dominate the immune response.

An **immunoediting** phase of immune surveillance is thought to occur if tumor cells are not completely eliminated as a result of their initial recognition by the immune system. During this phase, further mutation of the tumor cell occurs and cells that escape elimination by the immune system are selected for survival.

Immunoevasins are proteins produced by some viruses that prevent the appearance of peptide:MHC class I complexes on the infected cell, thus preventing the recognition of virus-infected cells by cytotoxic T cells.

Immunofluorescence is a technique for detecting molecules using antibodies labeled with fluorescent dyes. The bound fluorescent antibody can be detected by microscopy (**immunofluorescence microscopy**), or by flow cytometry, depending on the application being used. **Indirect immunofluorescence** uses anti-immunoglobulin antibodies labeled with fluorescent dyes to detect the binding of a specific unlabeled antibody.

An **immunogen** is any molecule that can elicit an adaptive immune response on injection into a person or animal.

The **immunoglobulins (Ig)** are the protein family to which antibodies and B-cell receptors belong. **Immunoglobulin A**: see **IgA**.

Immunoglobulin D: see **IgD**.

Many proteins are partly or entirely composed of protein domains known as **immunoglobulin domains** because they were first described in antibody molecules. The immunoglobulin domain consists of a sandwich of two β sheets held together by a disulfide bond and is called the **immunoglobulin fold**. There are two main types of immunoglobulin domain: C domains and V domains. Domains less closely related to the canonical Ig domains are sometimes also called **immunoglobulin-like domains**.

Immunoglobulin E: see **IgE**.

Immunoglobulin G: see **IgG**.

Immunoglobulin M: see **IgM**.

The **immunoglobulin repertoire**, also known as the antibody repertoire, describes the variety of antigen-specific immunoglobulins (antibodies and B-cell receptors) present in an individual.

Many proteins involved in antigen recognition and cell–cell interaction in the immune system and other biological systems are members of a protein family called the **immunoglobulin superfamily**, or **Ig superfamily**, because their shared structural features were first defined in immunoglobulin molecules. All members of the immunoglobulin superfamily have at least one immunoglobulin domain or immunoglobulin-like domain.

Immunohistochemistry is the detection of antigens in tissues by means of visible products produced by the degradation of a colorless substrate by antibody-linked enzymes. This technique has the advantage that it can be combined with other stains to be viewed in the light microscope, whereas immunofluorescence microscopy requires a special dark-field or UV microscope.

Immunological ignorance describes a form of self tolerance in which reactive lymphocytes and their target antigen are both detectable within an individual, yet no autoimmune attack occurs. Most autoimmune diseases probably reflect the loss of other lymphocytes known as regulatory or suppressor T cells.

Allogeneic tissue placed in certain sites in the body, such as the brain, does not elicit graft rejection. Such sites are called **immunologically privileged sites**. Immunological privilege can be due both to physical barriers to cell and antigen migration and to the presence of immunosuppressive cytokines.

Immunological memory is the ability of the immune system to respond more rapidly and more effectively on a second encounter with an antigen. Immunological memory is specific for a particular antigen and is long-lived.

Immunological recognition is the general term for the ability of the cells of the innate and adaptive immune system to recognize the presence of an infection.

The contact between a T cell and an antigen-presenting cell is a highly organized interface known as the **immunological synapse** or supramolecular adhesion complex, in which the spatial organization of the signaling molecules and their temporal organization contribute to the overall signal that is generated on antigen recognition.

Immunological tolerance: see **tolerance**.

Immunology is the study of all aspects of host defense against infection and of adverse consequences of immune responses.

Treatments that aim to modify an immune respose in a beneficial way, for example reduce or prevent an autoimmune or allergic response, are known as **immunomodulatory therapy**.

Immunopathology is the damage caused to tissues as the result of an immune response.

Immunophilins are proteins in T cells that are bound by the immunosuppressive drugs cyclosporin A, tacrolimus, and rapamycin. The complexes thus formed interfere with intracellular signaling pathways and prevent the clonal expansion of lymphocytes following antigen activation.

The presence of a particular protein in a cell can be determined by its **immunoprecipitation** from a cell extract using labeled antibodies specific for that protein.

The **immunoproteasome** is a form of proteasome found in cells exposed to interferons and contains three different subunits compared with the normal proteasome.

The T-cell and B-cell antigen receptors are associated with transmembrane signaling molecules that have **immunoreceptor tyrosine-based activation motifs (ITAMs)** in their cytoplasmic domains. These motifs are sites of tyrosine phosphorylation and of association with tyrosine kinases and other phosphotyrosine-binding proteins involved in intracellular signaling. **Immunoreceptor tyrosine-based inhibitory motifs (ITIMs)** are related motifs found in other receptors that inhibit cell activation; these motifs recruit phosphatases to the signaling pathway, which remove phosphate groups added by tyrosine kinases.

Compounds that inhibit adaptive immune responses are called **immunosuppressive drugs**. They are used mainly in the treatment of graft rejection and severe autoimmune disease.

Immunotoxins are antibodies that are chemically coupled to toxic proteins usually derived from plants or microbes. The antibody targets the toxin moiety to the required cells. Immunotoxins are being tested as anticancer agents and as immunosuppressive drugs.

Indirect allorecognition of a grafted tissue involves uptake of allogeneic proteins by the recipient's antigen-presenting cells and their presentation to T cells by self-MHC molecules.

The **indirect Coombs test** is a variation of the direct Coombs test in which an unknown serum is tested for antibodies against normal red blood cells by first mixing the two and then washing out the serum from the red blood cells and reacting them with anti-immunoglobulin antibody. If antibody in the unknown serum binds to the red blood cells, the red blood cells will be agglutinated by the anti-immunoglobulin.

Indirect immunofluorescence: see **immunofluorescence**.

Inducible co-stimulatory protein: see **ICOS**.

Infectious mononucleosis, or glandular fever, is the common form of infection with the Epstein–Barr virus. It consists of fever, malaise, and swollen lymph nodes.

Infectious tolerance: see **regulatory tolerance**.

Inflammation is a general term for the local accumulation of fluid, plasma proteins, and white blood cells that is initiated by physical injury, infection, or a local immune response. This is also known as an **inflammatory response**. Acute inflammation is the term used to describe early and often transient episodes, whereas chronic inflammation occurs when the infection persists or during autoimmune diseases. Many different forms of inflammation are seen in different diseases. The cells that invade tissues undergoing inflammatory responses are often called **inflammatory cells** or an inflammatory infiltrate.

In the signaling pathway that leads to apoptotis, **initiator caspases** promote apoptosis by cleaving and activating other caspases.

The early phases of the host response to infection depend on **innate immunity** in which a variety of innate resistance mechanisms recognize and respond to the presence of a pathogen in an **innate immune response**. Innate immunity is present in all individuals at all times, does not increase with repeated exposure to a given pathogen, and discriminates between groups of similar pathogens.

The production or exacerbation of a hypersensitivity-type response to an antigen because of innate immune responses due to activation of Toll-like receptors is known as **innate-type allergy**.

Innate-like lymphocytes (ILLs) are a type of lymphocyte that contribute to rapid responses to infection by acting early but which use a limited set of antigen-receptor gene segments to make immunoglobulins and T-cell receptors.

When inositol phospholipid is cleaved by phospholipase C-γ, it yields **inositol 1,4,5-trisphosphate (IP$_3$)** and diacylglycerol. Inositol trisphosphate acts as a mobile second messenger that releases calcium ions from intracellular stores in the endoplasmic reticulum.

Insulin-dependent diabetes mellitus (IDDM): see **type 1 diabetes mellitus**.

Integrase is the enzyme in the human immunodeficiency virus (HIV) and other retroviruses that mediates the integration of the DNA copy of the viral genome into the host-cell genome.

Integrins are heterodimeric cell-surface proteins involved in cell–cell and cell–matrix interactions. They are important in adhesive interactions between lymphocytes and antigen-presenting cells and in lymphocyte and leukocyte adherence to blood vessel walls and migration into tissues.

Intercellular adhesion molecules: see **ICAMs**.

Interdigitating dendritic cells: see **dendritic cells**.

Interferon-α (IFN-α) and **interferon-β (IFN-β)** are antiviral cytokines produced by a wide variety of cells in response to infection by a virus, and which also help healthy cells resist viral infection. They act through a common **interferon receptor** that signals through a Janus-family tyrosine kinase.

Interferon-γ (IFN-γ) is a cytokine produced by effector CD4 T_H1 cells, CD8 T cells, and NK cells, and its primary function is the activation of macrophages.

Interferon-producing cells (IPCs) are a subset of dendritic cells, also called plasmacytoid dendritic cells, that are specialized to produce large amounts of interferon in response to viral infections.

Interleukin, abbreviated **IL**, is a generic term for cytokines produced by leukocytes. We use the more general term cytokine in this book, but the term interleukin is used in the naming of specific cytokines such as IL-2. The interleukins are listed in Appendix III.

Interleukin-2 (IL-2) is a cytokine produced by activated naive T cells and is essential for their further proliferation and differentiation. It is one of the key cytokines in the development of an adaptive immune response.

Staining for cytokines in cells that produce them can be achieved by permeabilizing the cell and reacting it with a labeled fluorescent anti-cytokine antibody. This procedure is called **intracellular cytokine staining**.

An **intracellular signaling pathway** is the collection of proteins that interact with each other to carry a signal from an activated receptor to that part of the cell that will respond to the signal.

An **intradermal (i.d.)** injection delivers antigen into the dermis of the skin.

Intraepithelial lymphocytes (IELs) are lymphocytes present in the epithelium of mucosal surfaces such as the gut. They are predominantly T cells, and in the gut are predominantly CD8 T cells.

An **intramuscular (i.m.)** injection delivers antigen into muscle tissue.

Intranasal (i.n.) administration of antigen delivers it directly into the nose, usually in the form of an aerosol.

An **intravenous (i.v.)** injection delivers antigen directly into a vein.

Intrathymic dendritic cells: see **dendritic cells**.

The **intrinsic pathway** of apoptosis mediates apoptosis in response to noxious stimuli including UV irradiation, chemotherapeutic drugs, starvation, or lack of the growth factors required for survival. It is initiated by mitochondrial damage. Also known as the mitochondrial pathway of apoptosis.

The **invariant chain (Ii)** assembles as part of a major histocompatibility complex (MHC) class II protein in the endoplasmic reticulum and shields the MHC molecule from binding peptides there. When the MHC molecule reaches an endosome, Ii is degraded, leaving the MHC class II molecule able to bind antigenic peptides.

IPC: see **interferon-producing cell**.

IPEX (immune dysregulation, polyendocrinopathy, enteropathy, X-linked syndrome) is a very rare inherited condition in which regulatory CD4 CD25 T cells are lacking, due to a mutation in the gene for the transcription factor FoxP3, leading to the development of autoimmunity.

Ir genes: see **immune response genes**.

ISCOMs are immune stimulatory complexes of antigen held within a lipid matrix that acts as an adjuvant and enables the antigen to be taken up into the cytoplasm after fusion of the lipid with the plasma membrane.

Isolated lymphoid follicles are a type of organized gut mucosal tissue composed mainly of B cells.

Isoelectric focusing is an electrophoretic technique in which proteins migrate in a pH gradient until they reach the place in the gradient at which their net charge is neutral—their isoelectric point. Uncharged proteins no longer migrate; thus each protein is focused at its isoelectric point.

The **isotype** of an immunoglobulin chain is determined by the type of constant (C) region it has. Light chains can have either a κ or a λ C region. Heavy chains can be of μ, δ, γ, α, or ε isotypes. The different heavy-chain C regions are encoded by C-region exons 3′ to the V(D)J rearrangement site in the heavy-chain locus. In activated B cells, a rearranged heavy-chain variable region can be linked to different heavy-chain C-region exons as a result of a DNA recombination process known as class switching or **isotype switching**. The different heavy-chain isotypes have different effector functions and determine the class and functional properties of antibodies (IgM, IgD, IgG, IgA, IgE, respectively).

Isotypic exclusion describes the use of one or other of the light-chain isotypes, κ or λ, by a given B cell or antibody.

The **J gene segments**, or **joining gene segments**, are found some distance 5′ to the constant-region genes in immunoglobulin and T-cell receptor loci. In the light-chain locus, the TCRα locus, and the TCRγ locus, a V gene segment joins directly with a J gene segment to form a complete V-region exon. In the case of the heavy-chain locus, the TCRβ locus, and the TCRδ locus, a D gene segment first combines with a J gene segment and the DJ segment then recombines with a V gene segment.

Many cytokine receptors signal via **Janus-family tyrosine kinases (JAKs)**—tyrosine kinases that are activated by the aggregation of cytokine receptors. These kinases phosphorylate proteins known as STATs, for **S**ignal **T**ransducers and **A**ctivators of **T**ranscription. STATs are normally found in the cytosol, but move to the nucleus on phosphorylation and activate a variety of genes.

Junctional diversity is the diversity present in antigen-specific receptors that is created during the process of joining V, D, and J gene segments.

In the context of immunoglobulins, $\boldsymbol{\kappa}$ is one of the two classes or isotypes of light chain.

Killer cell immunoglobulin-like receptors (KIRs) and **killer lectin-like receptors (KLRs)** are two large families of receptors present on NK cells that are involved in activating and inhibiting the NK cell's killing activity. Both families contain activating and inhibitory receptors.

The **kinin system** is an enzymatic cascade of plasma proteins that is triggered by tissue damage to produce several inflammatory mediators, including the vasoactive peptide bradykinin.

Ku is a DNA repair protein required for immunoglobulin and T-cell receptor gene rearrangement.

Kupffer cells are phagocytes lining the hepatic sinusoids; they remove debris and dying cells from the blood, but are not known to elicit immune responses.

In the context of immunoglobulins, $\boldsymbol{\lambda}$ is one of the two classes or isotypes of light chain.

λ5: see **surrogate light chain**.

The **lamina propria** is a layer of connective tissue underlying a mucosal epithelium. It contains lymphocytes and other immune-system cells.

Langerhans cells are phagocytic immature dendritic cells found in the epidermis. They can migrate from the epidermis to regional lymph nodes via the afferent lymphatics. In the lymph node they differentiate into mature dendritic cells.

In B-cell development, the **large pre-B cell** stage has a rearranged heavy-chain gene and expresses a cell-surface pre-B-cell receptor.

LAT: see **linker of activation in T cells**.

Some viruses can enter a cell but not replicate, a state known as **latency**. Latency can be established in various ways; when the virus is reactivated and replicates, it can produce disease.

In type I immediate hypersensitivity reactions, the **late-phase reaction** occurs some hours after initial encounter with the antigen. It is resistant to treatment with antihistamine.

The **late pro-B cell** is the stage in B-cell development in which V_H to DJ_H joining occurs.

L chain: see **light chain**.

The tyrosine kinase **Lck** associates strongly with the cytoplasmic tails of CD4 and CD8 and is involved in helping to activate signaling from the T-cell receptor complex once antigen has bound.

The **lectin pathway** of complement activation is initiated by opsonins such as mannose-binding lectin (MBLs) and ficolins bound to bacteria.

Lentiviruses are a group of retroviruses that include the human immunodeficiency virus, HIV-1. They cause disease after a long incubation period.

Leprosy is caused by *Mycobacterium leprae* and occurs in a variety of forms. There are two polar forms: **lepromatous leprosy**, which is characterized by abundant replication of leprosy bacilli and abundant antibody production without cell-mediated immunity; and **tuberculoid leprosy**, in which few organisms are seen in the tissues, there is little or no antibody, but cell-mediated immunity is very active. The other forms of leprosy are intermediate between the polar forms.

The extracellular portions of Toll-like receptors are formed of multiple protein motifs called **leucine-rich repeats (LRRs)**.

Leukemia is the unrestrained proliferation of a malignant white blood cell characterized by very high numbers of the malignant cells in the blood. Leukemias can be lymphocytic, myelocytic, or monocytic, depending on the type of white blood cell involved.

Leukocyte is a general term for a white blood cell. Leukocytes include lymphocytes, polymorphonuclear leukocytes, and monocytes.

Leukocyte adhesion deficiency is an immunodeficiency disease in which the common β chain of the leukocyte integrins is not produced. This mainly affects the ability of leukocytes to enter sites of infection with extracellular pathogens, so that this type of infection cannot be effectively eradicated.

The **leukocyte functional antigens** or **LFAs**, are cell adhesion molecules initially defined with monoclonal antibodies: **LFA-1** is a β_2 integrin; **LFA-2** (now often called CD2) is a member of the immunoglobulin superfamily, as is **LFA-3** (now called CD58). LFA-1 is particularly important in T-cell adhesion to endothelial cells and antigen-presenting cells.

The **leukocyte integrins** are those integrins typically found on leukacytes. They have a common β_2 chain with distinct α chains and include LFA-1 and the very late activation antigens (VLAs).

Leukocytosis is the presence of increased numbers of leukocytes in the blood. It is commonly seen in acute infection.

Leukotrienes are lipid mediators of inflammation that are derived from arachidonic acid. They are produced by macrophages and other cells.

LFA-1, **LFA-2**, **LFA-3**: see **leukocyte functional antigens**.

The activation of a dendritic cell so that it is able to present antigen to and activate naive T cells is sometimes called **licensing**.

LICOS is the ligand for ICOS, a CD28-related protein that is induced on activated T cells and can enhance T-cell responses. LICOS is expressed on the surface of activated dendritic cells, monocytes, and B cells.

The immunoglobulin **light chain** (**L chain**) is the smaller of the two types of polypeptide chain that make up an immunoglobulin molecule. It consists of one V and one C domain, and is disulfide-bonded to the heavy chain. There are two classes, or isotypes, of light chain, known as κ and λ, which are produced from separate genetic loci.

Light zone: see **germinal centers**.

Linear epitopes: see **continuous epitopes**.

The **lingual tonsils** are mucosal peripheral lymphoid tissue situated at the base of the tongue.

Alleles at linked loci within the major histocompatibility complex are said to be in **linkage disequilibrium** if they are inherited together more frequently than predicted from their individual frequencies.

Epitopes recognized by B cells and helper T cells must be physically linked for the helper T cell to activate the B cell. This is called **linked recognition**.

The adaptor protein known as **linker of activation in T cells** (**LAT**) is a cytoplasmic protein with several tyrosines that become phosphorylated by the tyrosine kinase ZAP-70. It becomes associated with membrane lipid rafts and coordinates downstream signaling events in T-cell activation.

Lipid rafts are small, cholesterol-rich areas in the cell membrane that are relatively resistant to solubilization by mild detergents.

low-zone tolerance: see **high-zone tolerance**.

LPS is the abbreviation for the surface lipopolysaccharide of Gram-negative bacteria, which stimulates a Toll-like receptor on macrophages and dendritic cells. See also **endotoxin**.

A molecule of bacterial lipopolysaccharide (LPS) has first to be bound by the **LPS-binding protein** (**LBP**) before it can interact with CD14, an LPS:LBP-binding protein on cells such as macrophages.

LRR: see **leucine-rich repeats**.

L-selectin is an adhesion molecule of the selectin family found on lymphocytes. L-selectin binds to CD34 and GlyCAM-1 on high endothelial venules to initiate the migration of naive lymphocytes into lymphoid tissue. Also called CD62L.

Lyme disease is a chronic infection with *Borrelia burgdorferi*, a spirochete that can evade the immune response.

Lymph is the extracellular fluid that accumulates in tissues and is carried by lymphatic vessels back through the lymphatic system to the thoracic duct and into the blood.

The **lymphatic system** is the system of lymphoid channels and tissues that drains extracellular fluid from tissues via the thoracic duct to the blood. It includes the lymph nodes, Peyer's patches, and other organized lymphoid elements apart from the spleen, which communicates directly with the blood.

Lymphatic vessels, or **lymphatics**, are thin-walled vessels that carry lymph through the lymphatic system.

Lymph nodes are a type of peripheral lymphoid organ. They are found in many locations thoughout the body where lymphatic vessels converge, and are sites where adaptive immune responses are initiated. Antigen-presenting cells and antigen delivered by the lymphatic vessels from a site of infection are displayed to the naive T and B lymphocytes that are continually recirculating through the lymph nodes. Some of these lymphocytes will recognize the antigen and respond to it, triggering an adaptive immune response.

A **lymphoblast** is a lymphocyte that has enlarged and increased its rate of RNA and protein synthesis.

The **lymphocyte receptor repertoire** is the totality of the highly variable antigen receptors carried by B and T lymphocytes.

All adaptive immune responses are mediated by **lymphocytes**. Lymphocytes are a class of white blood cells that bear variable cell-surface receptors for antigen. These receptors are encoded in rearranging gene segments. There are two main classes of lymphocyte—B lymphocytes (B cells) and T lymphocytes (T cells)—which mediate humoral and cell-mediated immunity, respectively. Small lymphocytes have little cytoplasm and condensed nuclear chromatin; on antigen recognition, the cell enlarges to form a lymphoblast and then proliferates and differentiates into an antigen-specific effector cell.

Tissues composed of lymphocytes are known as **lymphoid** tissues.

Peripheral lymphoid tissues such as lymph nodes contain **lymphoid follicles** made up of follicular dendritic cells and B lymphocytes. The **primary lymphoid follicles** contain resting B lymphocytes. When a primary follicle is entered by activated B cells, a germinal center forms at this site, and the follicle is called a **secondary lymphoid follicle**.

Lymphoid organs are organized tissues characterized by very large numbers of lymphocytes interacting with a nonlymphoid stroma. The central or primary lymphoid organs, where lymphocytes are generated, are the thymus and bone marrow. The main peripheral lymphoid organs, in which adaptive immune responses are initiated, are the lymph nodes, spleen, and mucosa-associated lymphoid tissues such as tonsils and Peyer's patches.

Lymphokines are cytokines produced by lymphocytes.

Lymphomas are tumors of lymphocytes that grow in lymphoid and other tissues but do not enter the blood in large numbers. There are many types of lymphoma, which represent the transformation of various developmental stages of B or T lymphocytes.

Lymphopoiesis is the differentiation of lymphoid cells from a common lymphoid progenitor.

Lymphotoxin (**LT**) is a cytokine of the tumor necrosis factor (TNF) family and was formerly known as TNF-β. It is secreted by inflammatory CD4 T cells and is directly cytotoxic for some cells.

Lysosomes are acidified organelles that contain many degradative hydrolytic enzymes. Material taken up into endosomes by phagocytosis or receptor-mediated endocytosis is eventually delivered to lysosomes.

In the context of immunoglobulins, μ is the heavy chain of IgM.

Antigens and pathogens enter the body from the intestine through cells called **M cells** (**microfold cells**), which are specialized for this function. They are found over the gut-associated lymphoid tissue, such as Peyer's patches. They may provide a route of infection for HIV.

Macroautophagy is the engulfment by a cell of large quantities of its own cytoplasm and its delivery to lysosomes for degradation; it is induced in cells by starvation.

Macrophage activation is the enhancement of the capacity of the macrophage to kill engulfed pathogens and to produce cytokines that follows its antigen-specific interaction with an effector T cell.

The **macrophage mannose receptor** is a receptor on macrophages that is highly specific for certain mannose-containing carbohydrates that occur on the surface of some pathogens but not on host cells.

Macrophages are large mononuclear phagocytic cells important as scavenger cells, as pathogen-recognition cells and a source of pro-inflammatory cytokines in innate immunity, as antigen-presenting cells, and as effector phagocytic cells in humoral and cell-mediated immunity. They are migratory cells deriving from bone marrow precursors and are found in most tissues of the body. They have a crucial role in host defense.

Dendritic cells are unique in being able to carry out **macropinocytosis**, a process in which large amounts of extracellular fluid are taken up into an intracellular vesicle. This is one way in which these cells can take up a variety of antigens from their surroundings.

MAdCAM-1 is the mucosal cell adhesion molecule-1 or mucosal addressin that is recognized by the lymphocyte surface proteins L-selectin and VLA-4, allowing specific homing of lymphocytes to mucosal tissues.

Eosinophils can be triggered to release their **major basic protein**, which can then act on mast cells to cause their degranulation, with the release of histamine and other inflammatory mediators.

The **major histocompatibility complex** (**MHC**) is a cluster of genes on human chromosome 6 or mouse chromosome 17. It encodes a set of membrane glycoproteins called the MHC molecules. The **MHC class I molecules** present antigenic peptides generated in the cytosol to CD8 T cells, and the **MHC class II molecules** present antigenic peptides generated in intracellular vesicles to CD4 T cells. The MHC also encodes proteins involved in antigen processing and other aspects of host defense. The MHC is the most polymorphic gene cluster in the human genome, having large numbers of alleles at several different loci. Because this polymorphism is usually detected by using antibodies or specific T cells, the MHC molecules are often called major histocompatibility antigens.

MALT: see **mucosa-associated lymphoid tissue**.

The **mannose-binding lectin** (**MBL**) is an acute-phase protein present in the blood that binds to mannose residues. It can opsonize pathogens bearing mannose on their surfaces and can activate the complement system via the lectin pathway, an important part of innate immunity.

The follicular **mantle zone** is a rim of B lymphocytes that surrounds lymphoid follicles. The precise nature and role of mantle zone lymphocytes have not yet been determined.

MAP kinases, MAP kinase cascade: see **mitogen-activated protein kinases**.

Each area of white pulp in the spleen is is demarcated by a **marginal sinus**, a blood-filled vascular network that branches from the central arteriole.

The **marginal zone** of the lymphoid tissue of the spleen lies at the border of the white pulp. It contains a unique population of B cells, the **marginal zone B cells**, which do not circulate and are distinguished by a distinct set of surface proteins.

MASP-1 and **MASP-2** are serine proteases of the lectin pathway of complement activation; they bind to mannose-binding lectin and cleave C4.

Mast cells are large cells found in connective tissues throughout the body, most abundantly in the submucosal tissues and the dermis. They contain large granules that store a variety of mediator molecules including the vasoactive amine histamine. Mast cells have high-affinity Fcε receptors (FcεRI) that allow them to bind IgE monomers. Antigen-binding to this IgE triggers mast-cell degranulation and mast-cell activation, producing a local or systemic immediate hypersensitivity reaction. Mast cells have a crucial role in allergic reactions.

Mastocytosis indicates an overproduction of mast cells.

Mature B cells are B cells that have acquired surface IgM and IgD and have become able to respond to antigen.

MBL: see **mannose-binding lectin**.

The **medulla** is generally the central or collecting point of an organ. The thymic medulla is the central area of each thymic lobe, rich in bone marrow-derived antigen-presenting cells and the cells of a distinctive medullary epithelium. The medulla of the lymph node is a site of macrophage and plasma cell concentration through which the lymph flows on its way to the efferent lymphatics.

The **membrane-attack complex** is made up of the terminal complement components, which assemble to generate a membrane-spanning hydrophilic pore, damaging the membrane.

Membrane cofactor of proteolysis (**MCP** or **CD46**) is a host-cell membrane protein that acts in conjunction with factor I to cleave C3b to its inactive derivative iC3b and thus prevent convertase formation.

B cells carry on their surfaces many molecules of **membrane immunoglobulin** (**mIg**) of a single specificity, which acts as the receptor for antigen.

Memory cells are the lymphocytes that mediate immunological memory. They are more sensitive to antigen than are naive lymphocytes and respond rapidly on reexposure to the antigen that originally induced them. Both **memory B cells** and **memory T cells** have been defined.

The **mesenteric lymph nodes** are located in the connective tissue tethering the intestine to the rear wall of the abdomen. They drain the Peyers patches and isolated lymphoid follicles of the gut.

MHC; **MHC class I molecules**; **MHC class II molecules**: see **major histocompatibility complex**.

The **MHC class Ib** molecules encoded within the MHC are not highly polymorphic like the MHC class I and MHC class II molecules, and present a restricted set of antigens.

In **MHC class I deficiency**, MHC class I molecules are not present on the cell surface, usually due to inherited deficiency of either TAP-1 or TAP-2.

The **MHC class II compartment** (**MIIC**) is a site in the cell where MHC class II molecules accumulate, encounter HLA-DM, and bind antigenic peptides, before migrating to the surface of the cell.

In **MHC class II deficiency**, MHC class II molecules are not present on cells as a result of one of various inherited defects in regulatory genes. Patients are severely immunodeficient and have few CD4 T cells.

The protein that activates the transcription of MHC class II genes, the **MHC class II transactivator** (**CIITA**), is one of several defective genes in the disease bare lymphocyte syndrome, in which MHC class II molecules are lacking on all cells.

MHC genes are inherited in most cases as an **MHC haplotype**, the set of genes in a haploid genome inherited from one parent. Thus, if the parents are designated as ab and cd, then the offspring are most likely to be ac, ad, bc, or bd.

MHC molecules is the general name given to the highly polymorphic glycoproteins encoded by MHC class I and MHC class II genes, which are involved in the presentation of peptide antigens to T cells. They are also known as histocompatibility antigens.

MHC:peptide tetramers are fluorescently labeled complexes of specific peptides with their MHC molecules that are used to detect and stain the corresponding specific T cells.

MHC restriction refers to the fact that a given T cell will recognize a peptide antigen only when it is bound to a particular MHC molecule.

Normally, as T cells are stimulated only in the presence of self-MHC molecules, antigen is recognized only as peptides bound to self-MHC molecules.

MIC molecules are MHC class I-like molecules that are expressed in the gut under conditions of stress and are encoded within the class I region of the human MHC. They are not found in mice.

Microautophagy is the continuous internalization of the cytosol into the vesicular system.

Microfold cells: see **M cells**.

Microorganisms are microscopic organisms, unicellular except for some fungi, that include bacteria, yeasts and other fungi, and protozoa. All these groups contain some microorganisms that cause human disease.

mIg: see **membrane immunoglobulin**.

MIIC: see **MHC class II compartment**.

Minor histocompatibility antigens (**minor H antigens**) are peptides of polymorphic cellular proteins bound to MHC molecules that can lead to graft rejection when they are recognized by T cells.

Mitogen-activated protein kinases (**MAP kinases**) are kinases that become phosphorylated and activated on cellular stimulation by a variety of ligands, and lead to new gene expression by phosphorylating key transcription factors. They act as a series of three kinases, called the **MAP kinase cascade**, with each kinase phosphorylating and activating the next. The MAP kinases are part of many signaling pathways, especially those leading to cell proliferation, and have different names in different organisms.

When lymphocytes from two unrelated individuals are cultured together, the T cells proliferate in response to the allogeneic MHC molecules on the cells of the other donor. This **mixed lymphocyte reaction** (**MLR**) is used in testing for histocompatibility.

It has been proposed that infectious agents could provoke autoimmunity by **molecular mimicry**, the induction of antibodies and T cells that react against the pathogen but also cross-react with self antigens.

Monoclonal antibodies are produced by a single clone of B lymphocytes. Monoclonal antibodies are usually produced by making hybrid antibody-forming cells from a fusion of nonsecreting myeloma cells with immune spleen cells.

Monocytes are white blood cells with a bean-shaped nucleus; they are precursors of macrophages.

Some antibodies recognize all allelic forms of a polymorphic molecule such as an MHC class I protein; these antibodies are thus said to recognize a **monomorphic** epitope.

Mucins are highly glycosylated cell-surface proteins. Mucinlike molecules are bound by L-selectin in lymphocyte homing.

Muckle–Wells syndrome is an inherited episodic autoinflammatory disease caused by mutations in the gene for cryopyrin (*CIAS1*).

The **mucosa-associated lymphoid tissue** (**MALT**) or mucosal immune system comprises all lymphoid cells in epithelia and in the lamina propria lying below the body's mucosal surfaces. The main sites of mucosa-associated lymphoid tissues are the gut-associated lymphoid tissues (GALT), and the bronchus-associated lymphoid tissues (BALT).

The body's internal cavities that connect with the outside (e.g. gut, airways, vaginal tract) are lined with epithelium that is coated with mucus, and are called **mucosal epithelia**.

The **mucosal immune system** protects internal mucosal surfaces (e.g. the linings of the gut, respiratory tract, and urinogenital tracts), which are the site of entry for virtually all pathogens and other antigens. It comprises organized peripheral lymphoid tissues located within the mucosa as well as lymphocytes and other immune-system cells scattered more diffusely throughout the mucosa. See also **mucosa-associated lymphoid tissue**.

Mucosal mast cells are specialized mast cells present in mucosa. They produce little histamine but large amounts of other inflammatory mediators such as prostaglandins and leukotrienes.

Mucosal tolerance is the suppression of subsequent immune responses observed after administration of nonliving antigens into the respiratory tract.

Multiple myeloma is a tumor of plasma cells, almost always first detected as multiple foci in bone marrow. Myeloma cells produce a monoclonal immunoglobulin, called a myeloma protein, that is detectable in the patient's plasma.

Multiple sclerosis is a neurological disease characterized by focal demyelination in the central nervous system, lymphocytic infiltration in the brain, and a chronic progressive course. It is caused by an autoimmune response to various antigens found in the myelin sheath.

Multipotent progenitor cells are bone marrow cells that can give rise to both lymphoid and myeloid cells but are no longer self-renewing stem cells.

Mx is an interferon-inducible protein required for cellular resistance to influenza virus replication.

Myasthenia gravis is an autoimmune disease in which autoantibodies against the acetylcholine receptor on skeletal muscle cells cause a block in neuromuscular junctions, leading to progressive weakness and eventually death.

Mycophenolate is an inhibitor of the synthesis of guanosine monophosphate and acts as a cytotoxic immunosuppressive drug. It acts by killing rapidly dividing cells, including lymphocytes proliferating in response to antigen.

The **myeloid** lineage of blood cells includes all leukocytes except lymphocytes.

Myeloma proteins are immunoglobulins secreted by myeloma tumors and are found in the patient's plasma.

Naive lymphocytes are lymphocytes that have never encountered their specific antigen and thus have never responded to it, as distinct from memory or effector lymphocytes. All lymphocytes leaving the central lymphoid organs are naive lymphocytes, those from the thymus being **naive T cells** and those from bone marrow being **naive B cells**.

Nasal-associated lymphoid tissue (NALT) is the lymphoid tissue found in the mucosa lining the nsal passages.

Natural antibodies are antibodies produced by the immune system in the apparent absence of any infection. They have a broad specificity for self and microbial antigens, can react with many pathogens and can activate complement.

Natural cytotoxicity receptors (**NCRs**) are activating receptors on NK cells that recognize infected cells and stimulate cell killing by the NK cell.

Natural killer cells (**NK cells**) are large granular, non-T, non-B lymphocytes, which kill virus-infected cells and some tumor cells. They bear a wide variety of invariant activating and inhibitory receptors, but do not rearrange immunoglobulin or T-cell receptor genes. NK cells are important in innate immunity to viruses and other intracellular pathogens, and in antibody-dependent cell-mediated cytotoxicity (ADCC).

Natural regulatory T cells (**T_{reg}**) are those regulatory CD4 T cells that are thought to be specified in the thymus. They express Fox3P and carry CD25 and CD4 on their surface.

Necrosis is the death of cells or tissues due to chemical or physical injury, as opposed to apoptosis, which is a biologically programmed form of cell death. Necrosis leaves extensive cellular debris that needs to be removed by phagocytes, whereas apoptosis does not.

During intrathymic development, thymocytes that recognize self are deleted from the repertoire, a process known as **negative selection**. Autoreactive B cells undergo a similar process in bone marrow.

NEMO deficiency: see **hypohydrotic ectodermal dysplasia with immunodeficiency**.

Antibodies that can inhibit the infectivity of a virus or the toxicity of a toxin molecule are said to **neutralize** them. Such antibodies are known as **neutralizing antibodies** and the process of inactivation as **neutralization**.

Neutropenia describes the situation in which there are fewer neutrophils in the blood than normal.

Neutrophils, also known as **polymorphonuclear neutrophilic leukocytes**, are the major class of white blood cell in human peripheral blood. They have a multilobed nucleus and neutrophilic granules. Neutrophils are phagocytes and have an important role in entering infected tissues and engulfing and killing extracellular pathogens.

NFAT: see **nuclear factor of activated T cells**.

The transcription factor called **NFκB** is made up of two chains of 50 kDa and 65 kDa. In the absence of cell stimulation is found in the cytosol, where it is bound to a third chain called IκB, which is an inhibitor of NFκB transcription. It is one of the transcription factors activated by stimulation of Toll-like receptors.

NK cells: see **natural killer cells**.

NK T cells are a class of T cells that express the cell-surface marker NK1.1 normally associated with NK cells, but also have an α:β T-cell receptor, which is, however virtually invariant.

N-nucleotides are inserted into the junctions between gene segments of T-cell receptor and immunoglobulin heavy-chain V-regions during gene segment joining. These N-regions are not encoded in either gene segment, but are inserted by the enzyme terminal deoxynucleotidyl transferase (TdT). They markedly increase the diversity of these receptors.

NOD1 and **NOD2** are intracellular proteins which bind microbial components and activate the NFκB pathway, initiating inflammatory responses.

Recombination signal sequences (RSS) flanking gene segments consist of a seven-nucleotide heptamer and a nine-nucleotide **nonamer** of conserved sequence, separated by 12 or 23 nucleotides. RSSs form the target for the site-specific recombinase that joins the gene segments in antigen receptor gene rearrangement.

Nondepleting antibodies being developed for immunosuppression block the function of target proteins on cells without causing the cells to be destroyed.

When T- and B-cell receptor gene segments rearrange, they often form **nonproductive rearrangements** that cannot encode a protein because the coding sequences are in the wrong translational reading frame.

The transcription factor **nuclear factor of activated T cells (NFAT)** is a complex of a protein called NFATc (which is held in the cytosol by serine/threonine phosphorylation), and the Fos/Jun dimer known as AP-1. When activated in response to signaling from the antigen receptor in lymphocytes, it moves from the cytosol to the nucleus on cleavage of the phosphate residues by calcineurin, a serine/threonine protein phosphatase.

N-regions: see **N-nucleotides**.

The ***nude*** mutation of mice produces hairlessness and defective formation of the thymic stroma, so that nude mice, which are homozygous for this mutation, have no mature T cells.

When a person's work induces allergy, this is called **occupational allergy**.

Oligoadenylate synthetase is an enzyme produced in response to interferon stimulation of cells. It synthesizes unusual nucleoide polymers which in turn activate a ribonuclease that degrades viral RNA.

Patients with the inherited disease **Omenn's syndrome** have defects in either of the *RAG* genes but make small amounts of functional RAG protein, allowing a small amount of V(D)J recombination. They suffer from a severe immunodeficiency with increased susceptibility to multiple opportunistic infections.

Oncogenes are genes involved in regulating cell growth. When these genes are defective in structure or expression, they can cause cells to grow continuously to form a tumor.

An **opportunistic pathogen** is a microorganism that does not normally cause disease in healthy individuals but can cause disease in individuals with compromised host defenses, as occurs e.g. in AIDS.

Opsonization is the alteration of the surface of a pathogen or other particle so that it can be ingested by phagocytes. Antibody and complement opsonize extracellular bacteria for destruction by neutrophils and macrophages.

Taking in an antigen as food leads typically to a state of specific and active unresponsiveness of the rest of the immune system to that antigen, a phenomenon known as **oral tolerance**.

Original antigenic sin describes the tendency of humans to make antibody responses to those epitopes shared between the first strain of a virus they encountered and subsequent related viruses, while ignoring other highly immunogenic epitopes on the second and subsequent viruses.

The **palatine tonsils** are mucosal peripheral lymphoid tissues located on either side of the throat.

PAMPs: see **pathogen-associated molecular patterns**.

Lymphocyte subpopulations can be isolated by **panning** on petri dishes coated with monoclonal antibodies against cell-surface markers, to which the lymphocytes bind.

PAPA: see **pyogenic arthritis, pyoderma gangrenosum, and acne**.

The **paracortical area**, or **paracortex**, is the T-cell area of lymph nodes, lying just below the follicular cortex, which is primarily composed of B cells.

Parasites are organisms that obtain sustenance from a live host. In medical practice, the term is restricted to worms and protozoa, the subject matter of parasitology.

Paroxysmal nocturnal hemoglobinuria (PNH) is a disease in which complement regulatory proteins are defective, so that complement activation leads to episodes of spontaneous hemolysis.

Passive hemagglutination is a technique for detecting antibody in which red blood cells are coated with antigen and the antibody is detected by agglutination of the coated red blood cells.

The injection of antibody or immune serum into a naive recipient is called **passive immunization**. *Cf.* **active immunization**.

Pathogen-associated molecular patterns (PAMPs) describe the molecules associated with groups of pathogens, and which are recognized by cells of the innate immune system.

Pathogenesis refers to the origin or cause of the pathology of a disease.

Pathogenic microorganisms, or **pathogens**, are microorganisms that can cause disease when they infect a host.

Pathology is the scientific study of disease. The term pathology is also used to describe detectable damage to tissues.

Pattern-recognition receptors (PRRs) are receptors of the innate immune system that recognize common molecular patterns on pathogen surfaces.

PD-1 is a receptor on T cells that when bound by its ligands **PD-L1** and **PD-L2** inhibits signaling from the antigen receptor.

The cell-adhesion molecule **PECAM (CD31)** is found both on lymphocytes and at endothelial cell junctions. It is believed that

CD31–CD31 interactions enable leukocytes to leave blood vessels and enter tissues.

Pemphigus vulgaris is an autoimmune disease characterized by severe blistering of the skin and mucosal membranes.

Pentraxins are a family of acute-phase proteins formed of five identical subunits, to which C-reactive protein and serum amyloid protein belong.

The **peptide-binding cleft** or **peptide-binding groove** is the longitudinal cleft in the top surface of an MHC molecule into which the antigenic peptide is bound.

In the context of antigen processing and presentation, **peptide editing** is the removal of unstably bound peptides from MHC class II molecules by HLA-DM.

Perforin is a protein that can polymerize to form the membrane pores that are an important part of the killing mechanism in cell-mediated cytotoxicity. Perforin is produced by cytotoxic T cells and NK cells and is stored in granules that are released by the cell when it contacts a specific target cell.

The **periarteriolar lymphoid sheath** (**PALS**) is part of the inner region of the white pulp of the spleen, and contains mainly T cells.

Peripheral lymphoid organs and **tissues** are the lymph nodes, spleen, and mucosal-associated lymphoid tissues, in which immune responses are induced, as opposed to the central lymphoid organs, in which lymphocytes develop. They are also called **secondary lymphoid organs** and **tissues**.

Peripheral blood mononuclear cells (**PBMCs**) are lymphocytes and monocytes isolated from peripheral blood, usually by Ficoll-Hypaque™ density centrifugation.

Peripheral tolerance is tolerance acquired by mature lymphocytes in the peripheral tissues, as opposed to central tolerance, which is acquired by immature lymphocytes during their development.

Peyer's patches are organized lymphoid organs along the small intestine, especially the ileum. The contain lymphoid follicles and T-cell areas.

Antibody-like phage can be produced by cloning immunoglobulin V-region genes in filamentous phage, which thus express antigen-binding domains on their surfaces, forming a **phage display library**. Antigen-binding phage can be replicated in bacteria. This technique can be used to develop novel antibodies of any specificity.

Phagocytosis is the internalization of particulate matter by cells. Usually, the **phagocytic cells** or **phagocytes** are macrophages or neutrophils, and the particles are bacteria or virus particles that are taken up and destroyed. The ingested material is contained in a vesicle called a **phagosome**, which then fuses with one or more lysosomes to form a **phagolysosome**. The lysosomal enzymes and other molecules are important in the killing of the pathogen and its degradation.

Phosphatidylinositol 3,4-bisphosphate (**PIP_2**) is a membrane-associated phospholipid that is cleaved by phospholipase C-γ to give the signaling molecules diacylglycerol and inositol trisphosphate.

Phosphatidylinositol 3-kinase (**PI 3-kinase**) is an enzyme that phosphorylates the membrane phospholipid PIP_2 to give PIP_3 (phosphatidylinositol-3,4,5-trisphosphate). It is part of many different intracellular signaling pathways.

Phospholipase C-γ (**PLC-γ**) is a key enzyme in signal transduction. It is activated by protein tyrosine kinases that are themselves activated by receptor ligation, and activated phospholipase C-γ cleaves inositol phospholipids, such as PIP_2, into inositol trisphosphate and diacylglycerol.

The normal healthy intestine is in a state of **physiological inflammation**, containing large numbers of lymphocytes and other cells that in other orrgans are associated with chronic inflammation and disease. This is thought to be the result of continual stimuation by commensal organisms and food antigens.

IFN-α and IFN-β activate a serine/threonine kinase called **PKR kinase**, which phosphorylates the eukaryotic protein synthesis initiation factor eIF-2, inhibiting translation and thus contributing to the inhibition of viral replication.

Plasma is the fluid component of blood containing water, electrolytes, and the plasma proteins.

A **plasmablast** is a B cell in a lymph node that already shows some features of a plasma cell.

Plasma cells are terminally differentiated B lymphocytes and are the main antibody-secreting cells of the body. They are found in the medulla of the lymph nodes, in splenic red pulp, and in bone marrow.

Plasmacytoid dendritic cells are a distinct lineage of dendritic cells that secrete large amounts of interferon on activation by pathogens and their products via receptors such as Toll-like receptors. *Cf.* **conventional dendritic cells**.

Platelet-activating factor (**PAF**) is a lipid mediator that activates the blood clotting cascade and several other components of the innate immune system.

Platelets are small cell fragments found in the blood that are crucial for blood clotting. They are formed from megakaryocytes.

PMNs: see **polymorphonuclear leukocytes**.

P-nucleotides are nucleotides found in junctions between gene segments of the rearranged V-region genes of antigen receptors. They are an inverse repeat of the sequence at the end of the adjacent gene segment, being generated from a hairpin intermediate during recombination, and hence are called palindromic or P-nucleotides.

Antigen activates specific lymphocytes, whereas all mitogens, by definition, activate most or all lymphocytes, a process known as **polyclonal activation** because it involves multiple clones of diverse specificity. Such mitogens are known as **polyclonal mitogens**.

The major histocompatibility complex is both **polygenic**, containing several loci encoding proteins of identical function, and polymorphic, having multiple alleles at each locus.

The **poly-Ig receptor** (the **polymeric immunoglobulin receptor**) binds polymeric immunoglobulins, especially IgA, at the basolateral membrane of epithelia and transports them across the cell, where they are released from the apical surface. This transcytosis transfers IgA from its site of synthesis to its site of action at epithelial surfaces.

The **polymerase chain reaction** (**PCR**) uses high temperature and thermostable DNA polymerases to replicate specific sequences in DNA, producing thousands of copies of the replicated sequence.

Polymorphism literally means existing in a variety of different shapes. Genetic polymorphism is variability at a gene locus in which all variants occur at a frequency greater than 1%. The major histocompatibility complex is the most **polymorphic** gene cluster known in humans.

Polymorphonuclear leukocytes (**PMNs**) are white blood cells with multilobed nuclei and cytoplasmic granules. There are three types of polymorpho-nuclear leukocyte: the neutrophils with granules that stain with neutral dyes, the eosinophils with granules that stain with eosin, and the basophils with granules that stain with basic dyes.

Some antibodies show **polyspecificity**, the ability to bind to many different antigens. This is also known as **polyreactivity**.

Only those developing T cells whose receptors can recognize antigens presented by self-MHC molecules can mature in the thymus, a process known as **positive selection**. All other developing T cells die before reaching maturity.

Expression of the **pre-B-cell receptor**, or **pre-B-cell receptor complex**, is a critical event in B-cell development. Expression of this receptor, which is a complex of at least five proteins, including an immunoglobulin heavy chain, causes the pre-B cell to enter the cell cycle, to turn off the *RAG* genes, to degrade the RAG proteins, and to expand by several cell divisions. Then the signal ceases, and the pre-B cell is ready to rearrange its light chains.

During B-cell development, **pre-B cells** are cells that have rearranged their heavy-chain genes but not their light-chain genes.

The **precipitin reaction** was the first quantitative technique for measuring antibody production. The amount of antibody is determined from the amount of precipitate obtained with a fixed amount of antigen. The precipitin reaction also can be used to define antigen valence and zones of antibody or antigen excess in mixtures of antigen and antibody.

Prednisone is a synthetic steroid with potent anti-inflammatory and immunosuppressive activity used in treating acute graft rejection, autoimmune disease, and lymphoid tumors.

In T-cell development, TCRβ chains expressed by CD44^{low}CD25^{+} thymocytes pair with a surrogate α chain called pTα (pre-T-cell α) to form a **pre-T-cell receptor** that exits the endoplasmic reticulum via the Golgi as a complex with the CD3 molecules.

Primary cell-mediated immune reponse: see **cell-mediated immune response**.

During T-cell dependent antibody responses, a **primary focus** of B-cell activation forms in the vicinity of the margin between T and B cell areas of lymphoid tissue. Here, the T and B cells interact and B cells can differentiate directly into antibody-forming cells or migrate to lymphoid follicles for further proliferation and differentiation.

The **primary immune response** is the adaptive immune response to an initial exposure to antigen. **Primary immunization**, also known as priming, generates both the primary immune response and immunological memory.

The binding of antibody molecules to antigen is called a **primary interaction**, as distinct from secondary interactions in which binding is detected by some associated change such as the precipitation of soluble antigen or agglutination of particulate antigen.

The **primary lymphoid follicles** of peripheral lymphoid organs are aggregates of resting B lymphocytes. *Cf.* **secondary lymphoid follicle**.

Primary lymphoid organs: see **central lymphoid organs**.

Priming of antigen-specific naive lymphocytes occurs when antigen is presented to them in an immunogenic form; the primed cells will differentiate either into armed effector cells or into memory cells that can respond in second and subsequent immune responses.

During B-cell development, **pro-B cells** are cells that have displayed B-cell surface marker proteins but have not yet completed heavy-chain gene rearrangement. They are divided into early pro-B cells and late pro-B cells.

Any lymphocyte receptor-chain gene can be rearranged in either of two ways, productive and nonproductive. **Productive rearrangements** are in the correct reading frame for the receptor chain in question.

Pro-enzymes are inactive forms of enzymes, often proteases, that must be modified in some way, for example by selective cleavage of the protein chain, before they can become active.

Programmed cell death: see **apoptosis**.

Properdin: see **factor P**.

Prostaglandins, like leukotrienes, are lipid products of the metabolism of arachidonic acid that have a variety of effects on a variety of tissues, including activities as inflammatory mediators.

Cytosolic proteins are degraded by a large catalytic multisubunit protease called a **proteasome**. It is thought that peptides that are presented by MHC class I molecules are generated by the action of proteasomes, and two interferon-inducible subunits of some proteasomes are encoded in the MHC.

Protectin (**CD59**) is a cell-surface protein that protects host cells from being damaged by complement. It inhibits the formation of the membrane-attack complex by preventing the binding of C8 and C9 to the C5b67 complex.

Protective immunity is the resistance to a specific pathogen that results from infection or vaccination. It is due to the adaptive immune response, which set up immunological memory of that pathogen.

Protein A is a membrane component of *Staphylococcus aureus* that binds to the Fc region of IgG and is thought to protect the bacteria from IgG antibodies by inhibiting their interactions with complement and Fc receptors. It is useful for purifying IgG antibodies.

Protein interaction domains are protein domains, usually with no enzymatic activity themselves, that specifically interact with particular sites (e.g. phosphorylated tyrosines, proline-rich regions, membrane phospholipids) on other proteins or cellular structures.

Protein kinase C (**PKC**) is a family of serine/threonine kinases that are activated by diacylglycerol and calcium as a result of signaling from many different receptors.

Protein kinases add phosphate groups to proteins, and **protein phosphatases** remove these phosphate groups. Enzymes that add phosphate groups to tyrosine residues are called **protein tyrosine kinases**. These enzymes have crucial roles in signal transduction and regulation of cell growth. Their activity is regulated by a second set of molecules called **protein tyrosine phosphatases** that remove the phosphate from the tyrosine residues. Protein kinases that add phosphate groups to serines or threonine residues are known as **protein serine/threonine kinases**.

Protein phosphorylation is the covalent addition of a phosphate group to a specific site on a protein. Phosphorylation can alter the activity of a protein and also provides new binding sites for other proteins to interact with it.

Proto-oncogenes are cellular genes that regulate growth control. When mutated or aberrantly expressed, they can contribute to the malignant transformation of cells, leading to cancer. *Cf.* **oncogenes**.

A **provirus** is the DNA form of a retrovirus when it is integrated into the host cell genome, where it can remain transcriptionally inactive for long periods of time.

P-selectin: see **selectins**.

P-selectin glycoprotein ligand-1 (**PSGL-1**) expressed by activated effector T cells is a ligand for P-selectin on endothelial cells, and may enable activated T cells to enter all tissues in small numbers.

p-SMAC: see **supramolecular adhesion complex**.

pTα: see **pre-T-cell receptor**.

PTB (phosphotyrosine-binding domain) is a protein domain that binds phosphorylated tyrosine residues. It is found in many proteins that participate in intracellular signaling pathways.

Purine nucleotide phosphorylase (**PNP**) **deficiency** is an enzyme defect that results in severe combined immunodeficiency. This enzyme is important in purine metabolism, and its deficiency causes the accumulation of purine nucleosides, which are toxic for developing T cells, causing the immune deficiency.

Pus is the mixture of cell debris and dead neutrophils that is present in wounds and abscesses infected with extracellular encapsulated bacteria.

Pyogenic arthritis, pyoderma gangrenosum, and acne (**PAPA**) is an inherited autoinflammatory syndrome due to mutations in a protein that interacts with pyrin.

Bacteria with large capsules are difficult for phagocytes to ingest. Such encapsulated bacteria often produce pus at the site of infection, and are thus called **pus-forming bacteria** or **pyogenic bacteria**. Pyogenic organisms used to kill many young people. Now, pyogenic infections are largely limited to the elderly.

The RNA genome of the human immunodeficiency virus mutates rapidly, leading to the formation of large numbers of different genetic forms, or **quasi-species**, of the virus throughout the course of an infection.

Radiation bone marrow chimeras are mice that have been heavily irradiated and then reconstituted with bone marrow cells of a different strain of mouse, so that the lymphocytes differ genetically from the environment in which they develop. Such chimeric mice have been important in studying lymphocyte development.

Antigen–antibody interaction can be studied by **radioimmunoassay (RIA)** in which antigen or antibody is labeled with radioactivity. An unlabeled antigen or antibody is attached to a solid support such as a plastic surface, and the fraction of the labeled antibody or antigen retained on the surface is determined in order to measure binding.

The recombination-activating genes ***RAG-1*** and ***RAG-2*** encode the recombinase proteins **RAG-1** and **RAG-2,** which are essential for immunoglobulin and T-cell receptor gene rearrangement. Mice lacking either of these genes cannot form receptors and thus have no lymphocytes.

Rapamycin, or sirolimus, is an immunosuppressive drug that blocks cytokine action.

The **Ras** proteins are a family of small G proteins with important roles in intracellular signaling pathways, including those from lymphocyte antigen receptors.

The antigen receptors of lymphocytes are associated with **receptor-associated tyrosine kinases**, mainly of the Src family, which bind to receptor tails via their SH2 domains.

Receptor-mediated endocytosis is the internalization into endosomes of molecules bound to cell-surface receptors. Antigens bound to B-cell receptors are internalized by this process.

The replacement of a light chain of a self-reactive antigen receptor on immature B cells with a light chain that does not confer autoreactivity is known as **receptor editing**. This has also been shown with heavy chains.

A **recessive lethal gene** encodes a protein that is needed for the human or animal to develop to adulthood; when both copies are defective, the human or animal dies *in utero* or early after birth.

In any situation in which cells or tissues are transplanted, they come from a donor and are placed in a **recipient** or host.

Recombination-activating genes: see ***RAG-1*** and ***RAG-2***.

Recombination signal sequences (RSSs) are short stretches of DNA that flank the gene segments that are rearranged to generate a V-region exon. They always consist of a conserved heptamer and nonamer separated by 12 or 23 base pairs. Gene segments are only joined if one is flanked by an RSS containing a 12 base pair spacer and the other is flanked by an RSS containing a 23 base pair spacer—the 12/23 rule of gene segment joining.

The nonlymphoid area of the spleen in which red blood cells are broken down is called the **red pulp**.

Reed-Sternberg cells are large malignant B cells that are found in Hodgkin's disease.

Regulatory CD4 T cells are effector CD4 T cells that inhibit T-cell responses. They are also called **regulatory T cells**. Several different subsets have been distinguished.

Tolerance due to the actions of regulatory T cells is called **regulatory tolerance**.

When neutrophils and macrophages take up opsonized particles this triggers an oxygen-requiring metabolic change in the cell called the **respiratory burst**. It leads to the production of a number of mediators that are involved in killing engulfed microorganisms.

The virus known as **respiratory syncytial virus (RSV)** is a human pathogen that is a common cause of severe chest infection in young children, often associated with wheezing, as well as in immunocompromised patients and patients with AIDS.

Retrograde translocation or **retrotranslocation** returns endoplasmic reticulum proteins to the cytosol.

The enzyme **reverse transcriptase** is an essential component of retroviruses, as it translates the RNA genome into DNA before integration into host cell DNA. Reverse transcriptases also enable RNA sequences to be converted into complementary DNA (cDNA), and so to be cloned, and thus is an essential reagent in molecular biology.

The **reverse transcriptase–polymerase chain reaction (RT-PCR)** is used to amplify RNA sequences. The enzyme reverse transcriptase is used to convert an RNA sequence into a cDNA sequence, which is then amplified by PCR.

The **Rhesus (Rh) blood group antigen** is a red cell membrane antigen that is also detectable on the red blood cells of rhesus monkeys. Anti-Rh antibodies do not agglutinate human red blood cells, so antibody to Rh antigen must be detected by using a Coombs test.

Rheumatic fever is caused by antibodies elicited by infection with some *Streptococcus* species. These antibodies cross-react with kidney, joint, and heart antigens.

Rheumatoid arthritis is a common inflammatory joint disease that is probably due to an autoimmune response. The disease is accompanied by the production of **rheumatoid factor**, an IgM anti-IgG antibody that can also be produced in normal immune responses.

RIG-1 is an intracellular protein that detects the presence of viral RNA, leading to interferon production.

An **R-loop** is a bubble-like structure formed when transcribed RNA displaces the non-template strand of the DNA double helix at switch regions in the immunoglobulin constant-region gene cluster. R-loops are thought to promote class switch recombination.

RSSs: see **recombination signal sequences**.

RSV: see **respiratory syncytial virus**.

The technique of **sandwich ELISA** uses antibody bound to a surface to trap a protein by binding to one of its epitopes. The trapped protein is then detected by an enzyme-linked antibody specific for a different epitope on the protein's surface. This gives the assay a high degree of specificity.

Scaffold proteins are adaptor-type proteins with multiple binding sites for proteins, which bring together specific proteins into a functional signaling complex.

Scatchard analysis is a mathematical analysis of equilibrium binding that allows the affinity and valence of a receptor–ligand interaction to be determined.

Scavenger receptors on macrophages and other cells bind to numerous ligands and remove them from the blood. The Kupffer cells in the liver are particularly rich in scavenger receptors.

SCID, *scid*: see **severe combined immunodeficiency**.

SDS-PAGE is the common abbreviation for polyacrylamide gel electro-phoresis (PAGE) of proteins dissolved in the detergent sodium dodecyl sulfate (SDS). This technique is widely used to characterize proteins, especially after labeling and immunoprecipitation.

SE: see **staphylococcal enterotoxin**.

A **secondary antibody response** is the antibody response induced by a second or booster injection of antigen—a **secondary immunization**. The secondary response starts sooner after antigen injection, reaches higher levels, is of higher affinity than the primary response, and is dominated by IgG antibodies. Therefore, the response to each immunization is increasingly intense, so the secondary, tertiary, and subsequent responses are of increasing magnitude.

Secondary interactions: see **primary interaction**.

Secondary lymphoid organ: see **peripheral lymphoid organ**.

A lymphoid follicle develops into a **secondary lymphoid follicle** after it is entered by activated B cells, which proliferate and mature there, forming a germinal center.

Second messengers are small molecules or ions (such as Ca^{2+}) that are produced in response to a signal and which act to amplify the signal and carry it to the next stage within the cell.

When the recipient of a first tissue or organ graft has rejected that graft, a second graft from the same donor is rejected more rapidly and vigorously in what is called a **second-set rejection**.

The co-stimulatory signal required for lymphocyte activation is often called a **second signal**, with the first signal coming from the binding of antigen by the antigen receptor. Both signals are required for the activation of most lymphocytes.

The **secretory component** attached to IgA antibodies in body secretions is a fragment of the poly-Ig receptor left attached to the IgA after transport across epithelial cells.

Secretory IgA is the dimeric IgA antibody secreted across mucosal surfaces.

A cell is said to be **selected** by antigen when its receptors bind that antigen. If the cell starts to proliferate as a result, this is called clonal selection, and the cell founds a clone; if the cell is killed as a result of binding antigen, this is called negative selection or clonal deletion.

Selectins are a family of cell-surface adhesion molecules of leukocytes and endothelial cells that bind to sugar moieties on specific glycoproteins with mucinlike features.

Selective IgA deficiency is the most common inherited form of immunoglobulin deficiency in populations of European origin. No obvious disease susceptibility is associated with the defect.

Antigens in the body of an individual are by convention called **self antigens**. Lymphocytes are screened during their immature stages for reactivity with self antigens, and those that do respond undergo apoptosis.

Self-tolerance is the failure to make an immune response against the body's own antigens.

Allergic reactions require prior immunization, called **sensitization**, by the allergen that elicits the acute response. Allergic reactions occur only in **sensitized** individuals.

Sepsis is infection of the bloodstream. This is a very serious and frequently fatal condition. Infection of the blood with Gram-negative bacteria triggers **septic shock** through the release of the cytokine TNF-α.

A **sequence motif** is a pattern of nucleotides or amino acids shared by different genes or proteins that often have related functions. Sequence motifs observed in peptides that bind a particular MHC glycoprotein are based on the requirements for particular amino acids to achieve binding to that MHC molecule.

Serine/threonine protein kinases are enzymes that phosphorylate proteins on either serine or threonine residues.

Seroconversion is the phase of an infection when antibodies against the infecting agent are first detectable in the blood.

Serology is the use of antibodies to detect and measure antigens using **serological assays**, so called because these assays were originally carried out with serum, the fluid component of clotted blood, from immunized individuals.

Different strains of bacteria and other pathogens can sometimes be distinguished by their **serotype**, the ability of an immune serum to agglutinate or lyse some strains of bacteria and not others.

Serpins are a large family of protease inhibitors.

Serum is the fluid component of clotted blood.

Serum sickness occurs when foreign serum or serum proteins are injected into a person. It is caused by the formation of immune complexes between the injected protein and the antibodies formed against it. It is characterized by fever, arthralgias, and nephritis.

Severe combined immunodeficiency (**SCID**) is an immune deficiency disease, fatal if untreated, in which neither antibody nor T-cell responses are made. It is usually the result of T-cell deficiencies. The *scid* mutation causes severe combined immune deficiency in mice.

In **severe congenital neutropenia**, which can be inherited as a dominant or recessive trait, the neutrophil count is persistently extremely low.

SH2 domain: see **Src-family tyrosine kinases**.

Shingles is the disease caused when herpes zoster virus (the virus that causes chickenpox) is reactivated later in life in a person who has had chickenpox.

SHIP is an SH2-containing inositol phosphatase which removes the phosphate from PIP_3 to give PIP_2.

Shock is the name given to the potentially fatal circulatory collapse caused by the systemic actions of cytokines such as TNF-α.

SHP is an SH2-containing protein phosphatase.

Shwachman–Diamond syndrome is a rare genetic conditions in which some patients have a deficiency of neutrophils.

A **signal joint** is formed by the precise joining of recognition signal sequences in the process of somatic recombination that generates T-cell receptor and immunoglobulin genes. The piece of chromosome containing the signal joint is excised from the chromosome as a small circle of DNA.

Signal transducers and activators of transcription (**STATs**): see **Janus-family tyrosine kinases**.

Signal transduction describes the general process by which cells perceive changes in their environment. More specifically, it refers to the processes by which a cell transforms one type of signal, for example, antigen binding to a lymphocyte antigen receptor, into the intracellular events that signal the cell to make a particular type of response.

A **single-chain Fv** fragment, comprising a V region of a heavy chain linked by a stretch of synthetic peptide to a V region of a light chain, can be made by genetic engineering.

Single-nucleotide polymorphisms (**SNP**) are positions in the genome that differ by a single base between individuals.

During T-cell maturation in the thymus, mature T cells are detected by the expression of either the CD4 or the CD8 co-receptor and are therefore called **single-positive thymocytes**.

Sirolimus is the drug name that has been adopted for the chemical rapamycin; the two terms are used interchangeably in the literature.

SLP-76 is a scaffold protein involved in the antigen-receptor signaling pathway in lymphocytes.

Small G proteins are monomeric G proteins such as Ras, that act as intracellular signaling molecules downstream of many transmembrane signaling events. In their active form they bind GTP, which they hydrolyze to GDP, becoming inactive.

Small pre-B cells: see **large pre-B cells**.

When immunologists discovered that antibodies were variable, they entertained various theories, including the **somatic diversification theory** that postulated that the genes for immunoglobulin were constant, and that they diversified in somatic cells. This turned out to be partly true, as somatic hypermutation is now well established. However, other theories were needed to explain other features of antibody diversity, including somatic gene rearrangement and isotype switching.

Somatic gene therapy is the introduction of functional genes into somatic cells and their reintroduction into the body to treat disease.

During B-cell responses to antigen, the V-region DNA sequence undergoes **somatic hypermutation,** resulting in the generation of variant immunoglobulins, some of which bind antigen with a higher affinity. This allows the affinity of the antibody response to increase. These mutations affect only somatic cells, and are not inherited through germline transmission.

During lymphocyte development, receptor gene segments undergo **somatic recombination** to generate intact V-region exons that encode the V region of each immunoglobulin and T-cell receptor chain. These events occur only in somatic cells; the changes are not inherited.

Spacer: see **12/23 rule**.

Specific allergen immunotherapy: see **desensitization**.

The **specificity** of an antibody determines its ability to distinguish the immunogen from other antigens.

One uses **spectratyping** to define certain types of DNA gene segments that give a repetitive spacing of three nucleotides, or one codon.

Sphingosine 1-phosphate (**S1P**) is a lipid with chemotactic activity that controls the egress of T cells from lymph nodes. There appears to be a S1P concentration gradient between the lymphoid tissues and lymph or blood, such that naive T cells expressing a S1P receptor are drawn away from the lymphoid tissues and back into circulation.

The **spleen** is an organ in the upper left side of the peritoneal cavity containing a red pulp, involved in removing senescent blood cells, and a white pulp of lymphoid cells that respond to antigens delivered to the spleen by the blood.

The **Src-family tyrosine kinases** are receptor-associated protein tyrosine kinases. They have several domains, called Src-homology 1, 2, and 3. The SH1 domain contains the active site of the kinase, the SH2 domain can bind to phosphotyrosine residues, and the SH3 domain is involved in interactions with proline-rich regions in other proteins.

Staphylococcal enterotoxins (**SEs**) cause food poisoning and also stimulate many T cells by binding to MHC class II molecules and the V_β domain of certain T-cell receptors; the staphylococcal enterotoxins are thus superantigens.

STATs: see **Janus-family tyrosine kinases**.

The development of B lymphocytes and T lymphocytes occurs in association with **stromal cells**, which provide various soluble and cell-bound signals to the developing lymphocyte.

Antigens can be injected into the **subcutaneous** (s.c.) layer (ie, below the skin or dermis) to induce an adaptive immune response.

Superantigens are molecules that stimulate a subset of T cells by binding to MHC class II molecules and V_β domains of T-cell receptors, stimulating the activation of T cells expressing particular V_β gene segments. The staphylococcal enterotoxins are one of the sources of superantigens.

Suppressor T cells: see **regulatory T cells**.

Clustering of T-cell or B-cell receptors after binding their ligands leads to the formation of an organized structure, called the **supramolecular adhesion complex (SMAC)**, in which the antigen receptors are co-localized with other cell-surface signaling and adhesion molecules. It has a central zone known as c-SMAC comprising the T-cell receptors and co-receptors, and a peripheral zone, called p-SMAC, comprising cell-adhesion molecules.

The membrane-bound immunoglobulin that acts as the antigen receptor on B cells is often known as **surface immunoglobulin** (**sIg**).

surfactant proteins A and D (**SP-A** and **SP-D**) are acute-phase proteins that help protect the epithelial surfaces of the lung against infection.

The **surrogate light chain** is made up of two molecules called VpreB and λ5. Together, this chain can pair with an in-frame heavy chain, move to the cell surface, and signal for pre-B-cell growth.

When isotype switching occurs, the active heavy-chain V-region exon undergoes somatic recombination with a 3′ constant-region gene at a **switch region** of DNA. These DNA joints do not need to occur at precise sites, because they occur in intronic DNA. Thus, all **switch recombinations** are productive.

When one eye is damaged, there is often an autoimmune response that damages the other eye, a syndrome known as **sympathetic ophthalmia**.

A **syngeneic graft** is a graft between two genetically identical individuals. It is accepted as self.

Syphilis is a chronic disease caused by the spirochete *Treponema pallidum*, a spirochete that can evade the immune response.

Systemic anaphylaxis is the most dangerous form of immediate hypersensitivity reaction. It involves antigen in the bloodstream triggering mast cells all over the body. The activation of these mast cells causes widespread vasodilation, tissue fluid accumulation, epiglottal swelling, and often death.

The lymph nodes and spleen are sometimes called the **systemic immune system** to distinguish them from the mucosal immune system.

Systemic lupus erythematosus (**SLE**) is an autoimmune disease in which autoantibodies against DNA, RNA, and proteins associated with nucleic acids form immune complexes that damage small blood vessels, especially of the kidney.

Tacrolimus, or FK506, is an immunosuppressive polypeptide drug that inactivates T cells by inhibiting signal transduction from the T-cell receptor. Tacrolimus and cyclosporin A are the most commonly used immunosuppressive drugs in organ transplantation.

TAP-1 and **TAP-2** (transporters associated with antigen processing) are ATP-binding cassette proteins involved in transporting short peptides from the cytosol into the lumen of the endoplasmic reticulum, where they associate with MHC class I molecules.

Tapasin, or the **TAP-associated protein**, is a key molecule in the assembly of MHC class I molecules; a cell deficient in this protein has only unstable MHC class I molecules on the cell surface.

The functions of effector T cells are always assayed by the changes that they produce in antigen-bearing **target cells**. These cells can be B cells, which are activated to produce antibody; macrophages, which are activated to kill bacteria or tumor cells; or labeled cells that are killed by cytotoxic T cells.

The **Tat** protein is a product of the *tat* gene of HIV. It is produced when latently infected cells are activated, and it binds to a transcriptional enhancer in the long terminal repeat of the provirus, increasing the transcription of the proviral genome.

T-cell antigen receptor: see **T-cell receptor**.

The **T-cell areas** in peripheral lymphoid organs are enriched in naive T cells and are distinct from the B-cell zones and the stromal elements. They are the sites at which adaptive immune responses are initiated.

A **T-cell clone** is derived from a single progenitor T cell.

T-cell hybrids are formed by fusing an antigen-specific, activated T cell with a T-cell lymphoma. The hybrid cells bear the receptor of the specific T-cell parent and grow in culture like the lymphoma.

T-cell lines are cultures of T cells grown by repeated cycles of stimulation, usually with antigen and antigen-presenting cells. When single T cells from these lines are propagated, they can give rise to T-cell clones.

The **T-cell receptor** (**TCR**) consists of a disulfide-linked heterodimer of the highly variable α and β chains expressed at the cell membrane as a complex with the invariant CD3 chains. T cells carrying this type of receptor are often called α:β T cells. An alternative receptor made up of variable γ and δ chains is expressed with CD3 on a subset of

T cells. Both of these receptors are expressed with a disulfide-linked homodimer of ζ chains, which carries out the intracellular signaling function of the receptor.

T cells, or **T lymphocytes**, are a subset of lymphocytes defined by their development in the thymus and by heterodimeric receptors associated with the proteins of the CD3 complex. Most T cells have α:β heterodimeric receptors but γ:δ T cells have a γ:δ heterodimeric receptor. Effector T cells carry out a variety of functions in an immune response, acting always by interacting with another cell in an antigen-specific manner. Some T cells activate macrophages, some help B cells produce antibody and some T cells kill cells infected with viruses and other intracellular pathogens.

T-cell zones: see T-cell areas.

T lymphocytes: see **T cells**.

TCRα and **TCRβ** are the two chains of the α:β T-cell receptor.

TdT: see **terminal deoxynucleotidyl transferase**.

Activation of the lymphocyte antigen receptors is linked to activation of PLC-γ through members of the **Tec kinase** family of Src-like tyrosine kinases. Other Tec kinases are Btk in B cells, which is mutated in the human immunodeficiency disease X-linked agammaglobulinemia (XLA), and Itk in T lymphocytes.

The complement system can be activated directly or by antibody, but both pathways converge with the activation of the **terminal complement components,** which may assemble to form the membrane-attack complex.

The enzyme **terminal deoxynucleotidyl transferase** (**TdT**) inserts nontemplated or N-nucleotides into the junctions between gene segments in T-cell receptor and immunoglobulin V-region genes. The N-nucleotides contribute greatly to junctional diversity in V regions.

When the same antigen is injected a third time, the response elicited is called a **tertiary response** and the injection a **tertiary immunization**.

T_H1 cells are a subset of CD4 T cells that are characterized by the cytokines they produce. They are mainly involved in activating macrophages, and are sometimes called inflammatory CD4 T cells.

T_H2 cells are a subset of CD4 T cells that are characterized by the cytokines they produce. They are mainly involved in stimulating B cells to produce antibody, and are often called helper CD4 T cells.

T_H3 cells are a subset of regulatory CD4 T cells produced in the mucosal immune response to antigens that are presented orally. They produce mainly transforming growth factor-β.

T_H17 cells are a subset of CD4 T cells that are characterized by production of the cytokine IL-17. They are thought to help recruit neutrophils to sites of infection.

The **thioester-containing proteins** (**TEPs**) are homologs of complement component C3 that are found in insects and are thought to have some function in innate immunity.

The lymph from most of the body, except the head, neck, and right arm, is gathered in a large lymphatic vessel, the **thoracic duct**, which runs parallel to the aorta through the thorax and drains into the left subclavian vein. The thoracic duct thus returns the lymphatic fluid and lymphocytes back into the peripheral blood circulation.

Surgical removal of the thymus is called **thymectomy**.

The **thymic anlage** is the tissue from which the thymic stroma develops during embryogenesis.

The **thymic cortex** is the outer region of each thymic lobule in which thymic progenitor cells proliferate, rearrange their T-cell receptor genes, and undergo thymic selection, especially positive selection on **thymic cortical epithelial cells**.

The **thymic stroma** consists of epithelial cells and connective tissue that form the essential microenvironment for T-cell development.

Thymocytes are lymphoid cells found in the thymus. They consist mainly of developing T cells, although a few thymocytes have achieved functional maturity.

Thymoma is a tumor of thymic stroma.

The **thymus**, the site of T-cell development, is a lymphoepithelial organ in the upper part of the middle of the chest, just behind the breastbone.

Some antigens elicit responses only in individuals that have T cells; they are called **thymus-dependent antigens** or **TD antigens**. Other antigens can elicit antibody production in the absence of T cells and are called **thymus-independent antigens** or **TI antigens**. There are two types of TI antigen: the **TI-1 antigens,** which have intrinsic B-cell activating activity, and the **TI-2 antigens,** which seem to activate B cells by having multiple identical epitopes that cross-link the B-cell receptor.

T cell and T lymphocyte are short designations for **thymus-dependent lymphocyte**, the lymphocyte population that fails to develop in the absence of a functioning thymus.

Time-lapse video microscopy can be used to examine all sorts of processes in biology, from cell migration (fast) to a flower blossoming (slow).

During the process of germinal center formation, cells called **tingible body macrophages** appear. These are phagocytic cells engulfing apoptotic B cells, which are produced in large numbers during the height of the germinal center response.

Transplantation of organs or **tissue grafts** such as skin grafts is used medically to repair organ or tissue deficits.

Some autoimmune diseases attack particular tissues, such as the β cell in the islets of Langerhans in autoimmune diabetes mellitus; such diseases are called **tissue-specific autoimmune disease**.

The **titer** of an antiserum is a measure of its concentration of specific antibodies based on serial dilution to an end point, such as a certain level of color change in an ELISA assay.

TLR-2 is a mammalian Toll-like receptor that recognizes the lipoteichoic acid of Gram-positive bacteria and lipoproteins of Gram-negative bacteria.

TLR-3 is a mammalian Toll-like receptor that recognizes viral double-stranded RNA.

TLR-4 is a mammalian Toll-like receptor that, in conjunction with the macrophage LPS receptor, recognizes bacterial lipopolysaccharide (LPS).

TLR-5 is a mammalian Toll-like receptor that recognizes the flagellin protein of bacterial flagella.

The **TNF family** of cytokines includes both secreted (e.g. tumor necrosis factor-α (TNF) and lymphotoxin) and membrane-bound (e.g. CD40 ligand) members.

The **TNF receptor** (**TNFR**) family of cytokine receptors includes some that lead to apoptosis of the cell on which they are expressed (e.g. **TNFR-I**, TNF-RII, Fas), while others lead to activation (e.g. CD40, 4-1BB). All of them signal as trimeric proteins.

TNF-receptor associated periodic syndrome (TRAPS): see **familial Mediterranean fever**.

Tolerance is the failure to respond to an antigen; the immune system is said to be **tolerant** to self antigens. Tolerance to self antigens is an essential feature of the immune system; when tolerance is lost, the immune system can destroy self tissues, as happens in autoimmune disease. The immune system mainly becomes tolerant to self antigens—is **tolerized**—during lymphocyte development.

An antigen that induces tolerance is said to be **tolerogenic**.

The **Toll-like receptors** (**TLR**) are innate immune system receptors on macrophages and dendritic cells, and some other cells, that recognize pathogens and their products, e.g. bacterial lipopolysaccharide. Recognition stimulates the receptor-bearing cells to produce cytokines and initiate immune responses.

The **Toll pathway** is an ancient signaling pathway used by the Toll-like receptors that activates the transcription factor NFκB by degrading its inhibitor IκB.

Tonsils: see lingual tonsils; palatine tonsils.

Toxic shock syndrome is a systemic toxic reaction caused by the massive production of cytokines by CD4 T cells activated by the bacterial superantigen **toxic shock syndrome toxin-1 (TSST-1)**, which is secreted by *Staphylococcus aureus*.

Inactivated toxins called **toxoids** are no longer toxic but retain their immunogenicity so that they can be used for immunization.

T_R1: see **regulatory T cells**.

The family of proteins known as TNF receptor-associated factors, or **TRAFs**, consists of at least six members that bind to various TNF-family receptors. They share a domain known as a TRAF domain, and have a crucial role as signal transducers between upstream members of the TNFR family and downstream transcription factors.

The active transport of molecules across epithelial cells is called **transcytosis**. Transcytosis of IgA molecules involves transport across intestinal epithelial cells in vesicles that originate on the basolateral surface and fuse with the apical surface in contact with the intestinal lumen.

The insertion of small pieces of DNA into cells is called **transfection**. If the DNA is expressed without integrating into host cell DNA, this is called a transient transfection; if the DNA integrates into host cell DNA, then it replicates whenever host cell DNA is replicated, producing a stable transfection.

Foreign genes can be placed in the mouse genome by **transgenesis**. This generates **transgenic mice** that are used to study the function of the inserted gene, or transgene, and the regulation of its expression.

Some cancers have chromosomal **translocations**, in which a piece of one chromosome is abnormally linked to another chromosome.

The grafting of organs or tissues from one individual to another is called **transplantation**. The **transplanted organs** or grafts can be rejected by the immune system unless the host is tolerant to the graft antigens or immunosuppressive drugs are used to prevent rejection.

Transporters associated with antigen processing-1 and -2: see **TAP-1** and **TAP-2**.

T_{reg} cells: see **natural regulatory T cells**.

The **tropism** of a pathogen describes the cell types it will infect.

TSLP (thymic stroma-derived lymphopoietin) is a cytokine thought to be involved in promoting B-cell development in the embryonic liver.

The **tuberculin test** is a clinical test in which a purified protein derivative (PPD) of *Mycobacterium tuberculosis*, the causative agent of tuberculosis, is injected subcutaneously. PPD elicits a delayed-type hypersensitivity reaction in individuals who have had tuberculosis or have been immunized against it.

Tuberculoid leprosy: see **leprosy**.

Tumor immunology is the study of host defenses against tumors, usually studied by tumor transplantation.

Tumor necrosis factor-α (TNF-α) is a cytokine produced by macrophages and T cells that has multiple functions in the immune response. It is the defining member of the TNF family of cytokines. These cytokines function as cell-associated or secreted proteins that interact with receptors of the **tumor necrosis factor receptor (TNFR)** family, which in turn communicates with the interior of the cell via components known as TRAFs (tumor necrosis factor receptor-associated factors).

Tumor necrosis factor-β (TNF-β): see **lymphotoxin**.

Tumors transplanted into syngeneic recipients can grow progressively or can be rejected through T-cell recognition of **tumor rejection antigens (TRAs)**. TRAs are peptides of mutant or overexpressed cellular proteins bound to MHC class I molecules on the tumor cell surface.

The **TUNEL assay** (**T**dT-dependent d**U**TP–biotin **n**ick **e**nd **l**abeling) identifies apoptotic cells *in situ* by the characteristic fragmentation of their DNA. Biotin-tagged dUTP added to the free 3′ ends of the DNA fragments by the enzyme TdT can be detected by immunohistochemical staining with enzyme-linked streptavidin.

In **two-dimensional gel electrophoresis,** proteins are separated by isoelectric focusing in one dimension, followed by SDS-PAGE on a slab gel at right-angles to the first dimension. This can separate and identify large numbers of distinct proteins.

Type 1 diabetes mellitus is a disease in which the β cells of the pancreatic islets of Langerhans are destroyed so that no insulin is produced. The disease is believed to result from an autoimmune attack on the β cells. It is also known as insulin-dependent diabetes mellitus (IDDM), as the symptoms can be ameliorated by injections of insulin.

Hypersensitivity reactions are classified by mechanism: **type I hypersensitivity reactions** involve IgE antibody triggering of mast cells; **type II hypersensitivity reactions** involve IgG antibodies against cell surface or matrix antigens; **type III hypersensitivity reactions** involve antigen:antibody complexes; and **type IV hypersensitivity reactions** are T-cell mediated.

A **tyrosine protein kinase** is an enzyme that specifically phosphorylates tyrosine residues in proteins. They are critical in T- and B-cell activation. The kinases that are critical for B-cell activation are Blk, Fyn, Lyn, and Syk. The tyrosine kinases that are critical for T-cell activation are called Lck, Fyn, and ZAP-70.

Ubiquitin is a small protein that can be attached to other proteins to target them for degradation in the proteasome.

Urticaria is the technical term for hives, which are red, itchy skin welts usually brought on by an allergic reaction.

Vaccination is the deliberate induction of adaptive immunity to a pathogen by injecting a **vaccine**, a dead or attenuated (nonpathogenic) form of the pathogen.

The **valence** of an antibody or antigen is the number of different molecules that it can combine with at one time.

The **variability plot** of a protein is a measure of the difference between the amino acid sequences of different variants of that protein. The most variable proteins known are antibodies and T-cell receptors.

Variable gene segments: see **V gene segments**.

Variable lymphocyte receptors (VLRs) are non-immunoglobulin LRR-containing variable rceptors and secreted proteins expressed by the lymphocyte-like cells of the lamprey. They are generated by a process of somatic gene rearrangement, and may be a means of generating an adaptive immune response.

The **variable region (V region)** of an immunoglobulin or T-cell receptor is formed of the amino-terminal domains of its component polypeptide chains. These are called the **variable domains (V domains)** and are the most variable parts of the molecule. They contain the antigen-binding sites.

Vascular addressins are molecules on endothelial cells to which leukocyte adhesion molecules bind. They have a key role in selective homing of leukocytes to particular sites in the body.

The adhesion molecule **VCAM-1** is expressed by vascular endothelium at sites of inflammation; it binds the integrin VLA-4, which allows armed effector T cells to enter sites of infection.

The enzyme that joins the gene segments of B-cell and T-cell receptor genes is called the **V(D)J recombinase**. It is made up of several enzymes, but most important are the products of the recombination-activating genes *RAG-1* and *RAG-2* whose protein products RAG-1 and RAG-2 are expressed in developing lymphocytes and make up the only known lymphoid-specific components of the V(D)J recombinase.

The process of **V(D)J recombination** is found exclusively in lymphocytes in vertebrates, and allows the recombination of different gene segments into sequences encoding complete protein chains of immunoglobulins and T-cell receptors.

The variable region of the polypeptide chains of an immunoglobulin or T-cell receptor is composed of a single amino-terminal **V domain**. Paired V domains form the antigen-binding sites of immunoglobulins and T-cell receptors.

The first 95 amino acids or so of immunoglobulin or T-cell receptor V domains are encoded in inherited **V gene segments**. There are multiple different V gene segments in the germline genome. To produce a complete exon encoding a V domain, one V gene segment must be rearranged to join up with a J or a rearranged DJ gene segment.

The **very late activation antigens** (**VLA**) are members of the β_1 family of integrins involved in cell–cell and cell–matrix interactions. Some VLAs are important in leukocyte and lymphocyte migration.

Vesicles are small membrane-bounded compartments within the cytosol.

Many viruses that are produced by mammalian cells are enclosed in a **viral envelope** of host cell membrane lipid and proteins bound to the viral core by viral envelope proteins.

The **viral protease** of the human immunodeficiency virus cleaves the long polyprotein products of the viral genes into individual proteins.

Virions are complete virus particles, the form in which viruses spread from cell to cell or from one individual to another.

Viruses are pathogens composed of a nucleic acid genome enclosed in a protein coat. They can replicate only in a living cell, as they do not possess the metabolic machinery for independent life.

VpreB: see **surrogate light chain**.

WAS: see **Wiskott-Aldrich syndrome**.

Weibel–Palade bodies are granules within endothelial cells that contain P-selectin. Activation of the endothelial cell by mediators such as histamine and C5a leads to rapid translocation of P-selectin to the cell surface.

In **Western blotting**, a mixture of proteins is separated, usually by gel electrophoresis and transferred by blotting to a nitrocellulose membrane; labeled antibodies are then used as probes to detect specific proteins.

When small amounts of allergen are injected into the dermis of an allergic individual, a **wheal-and-flare reaction** is observed. This consists of a raised area of skin containing fluid and a spreading, red, itchy circular reaction.

The discrete areas of lymphoid tissue in the spleen are known as the **white pulp**.

WHIM (warts, hypogammaglobulinemia, infections and myelokathexis syndrome) is

The **Wiskott–Aldrich syndrome** (**WAS**) is characterized by defects in the cytoskeleton of cells due to a mutation in the protein **WASP**, which is involved in interactions with the actin cytoskeleton. Patients with this disease are highly susceptible to infections with pyogenic bacteria.

Animals of different species are **xenogeneic**.

The use of **xenografts**, organs from a different species, is being explored as a solution to the severe shortage of human organs for transplantation. The main problem with xenografting is the presence of natural antibodies against xenograft antigens; attempts are being made to modify these reactions by creating transgenic animals.

Mice with mutations in the *btk* gene have a deficiency in making antibodies, especially in primary responses. These mice are called **xid**, for X-linked immunodeficiency, the mouse equivalent of X-linked agammaglobulinemia, the human form of this disease.

X-linked agammaglobulinemia (**XLA**) is a genetic disorder in which B-cell development is arrested at the pre-B-cell stage and no mature B cells or antibodies are formed. The disease is due to a defect in the gene encoding the protein tyrosine kinase Btk.

X-linked hyper IgM syndrome is a disease in which little or no IgG, IgE, or IgA antibody is produced and even IgM responses are deficient, but serum IgM levels are normal to high. It is due to a defect in the gene encoding the CD40 ligand or CD154.

X-linked lymphoproliferative syndrome is a rare immunodeficiency that results from mutations in a gene named SH2-domain containing gene 1A (*SH2D1A*). Boys with this deficiency typically develop overwhelming EBV infection during childhood, and sometimes lymphomas.

X-linked severe combined immunodeficiency (**X-linked SCID**) is a disease in which T-cell development fails at an early intrathymic stage and no production of mature T cells or T-cell dependent antibody occurs. It is due to a defect in a gene that encodes the γ_c chain that is a component of the receptors for several different cytokines.

The protein kinase **ZAP-70** is found in T cells and is a relative of Syk in B cells. It contains two SH2 domains that, when bound to the phosphorylated ζ chain, leads to activation of kinase activity. The main cellular substrate of ZAP-70 is a large adaptor protein called LAT.

Zymogen: see **pro-enzymes**.

INDEX

Figures in the Appendix are labeled in the form **Fig. A.1** and those in Chapters 1 to 16 are labeled in the form **Fig. 1.1**, **Fig. 2.1** etc.

To save space in the index the following abbreviations have been used:

APC – antigen-presenting cells
BCR – B-cell receptor
Btk – Bruton's tyrosine kinase
CDRs – complementarity determining regions
ICAM – intracellular adhesion molecule
IDDM – insulin dependent diabetes mellitus (type I)
IL– interleukin
ITAMs – immunoreceptor tyrosine-based activating motifs
ITIMs – immunoreceptor tyrosine-based inhibitory motifs
LFA – lymphocyte function-associated antigen
TCR – T-cell receptor
TNF – tumor necrosis factor
vs. indicates a differential diagnosis or comparison.